AF746635

THE SADTLER HANDBOOK OF PROTON NMR SPECTRA

THE SADTLER HANDBOOK OF PROTON NMR SPECTRA

Editor:
William W. Simons
Spectroscopist
Sadtler Research Laboratories, Inc.

UNIV. COLL. N.W.
BANGOR
LIBRARY

SADTLER | HEYDEN

Science Reference
QC 793.5.P72752

Published by
Sadtler Research Laboratories, Inc.
Subsidiary of Block Engineering, Inc.
3316 Spring Garden Street
Philadelphia, Pennsylvania 19104

ISBN 0-8456-0035-4

Library of Congress Catalog Card Number: 78-54281

792927

Co-published for exclusive distribution in Europe by
Heyden & Son Ltd.
Spectrum House, Hillview Gardens
London N.W. 4, England 2JQ

ISBN 0-8550-1442-3

Copyright © 1978, by Sadtler Research Laboratories, Inc. All rights reserved. No part of this book may be reproduced or reformatted in any form, by photostat, microfilm, retrieval system, or any other means, without written permission from Sadtler Research Laboratories, Inc.

Printed in the United States of America

PREFACE

The Sadtler Handbook of NMR Spectra is the second in the series of three sets of reference spectra. The proton NMR spectra in this volume possess the same spectrum numbers as the compounds published in The Sadtler Handbook of Infrared Spectra and represent the identical molecular structures.

The ordering system by functional group is somewhat untraditional from an NMR spectroscopists point of view but is found to offer special advantages to the educator, student and chemist who wishes to utilize PMR as a primary or secondary analytical technique.

The emphasis of the text and tables is on the chemical shifts exhibited by the groups of spectra as determined by the shielding or deshielding effect of the functional groups under discussion. For those desiring a more detailed treatment of the spin-spin coupling patterns which arise from the various hydrocarbon groups I refer you to texts which emphasize this approach [(1)].

The ppm delta scale (TMS = 0.0 ppm) is used throughout the text and tables, and in the listings of proton assignments for the individual spectra. These shifts were read from the original calibrated NMR charts and are considered accurate to within ± 0.05 ppm.

The majority of the spectra were obtained using one of six common NMR solvents, and were selected in the following order of preference: Carbon tetrachloride (for liquid samples), Deuterochloroform, Polysol (50% $CDCl_3$/50% DMSO-d_6), Dimethylsulfoxide-d_6, D_2O and Trifluoroacetic acid. A limited number of samples were scanned as solutions in undeuterated Acetone or undeuterated Dimethylsulfoxide. Because of the importance of solvent effects on proton chemical shifts, all tabulated data in the chapter headings list the solvent that was employed to record the spectrum.

In order to define simple molecular structures in a linear format suitable for typing, several shorthand notations are used throughout the text. Standard chemical notations are employed with the three arbitrary additions described below.

> C_n— is used to describe an alicyclic ring containing "n" carbon atoms, where C_3 represents cyclopropane, C_4 represents cyclobutane, etc.
>
> R_n— is used to define normal (straight chain) alkyl groups consisting of "n" number of carbons, so that R_4— represents a butyl group and R_{12}— describes a dodecyl group. The letter R is occasionally used without a numerical subscript, in which case it represents a branched or unspecified aliphatic group in which branching or other substituents are not considered to have a pronounced effect upon the chemical shifts of the proton group under examination.
>
> Ø— is used to represent a phenyl group or a substituted benzene ring in which the other substituents do not have a pronounced effect upon the shifts being described.

(1) a. Abraham, "Analysis of High Resolution NMR Spectra", Elsevier (1971).
b. Becker, J. Chem. Ed., 42, 591 (1965).
c. Roberts, "An Introduction to the Analysis of Spin-Spin Splitting in High Resolution Nuclear Magnetic Resonance Spectra", Benjamin (1962).
d. Simons and Zanger, "Guide to NMR Spectra", Sadtler (1972).
e. Wilberg and Nist, "Interpretation of NMR Spectra", Benjamin (1962).

Examples of notation system:

R_9—Ø would represent nonyl benzene or a ring substituted derivative of nonyl benzene,

R_2—NH—C(=S)—NH_2 describes n-ethyl thiourea,

CH_3—C(=O)—O—CH_2CH(R_2)—R_4 would indicate acetic acid, 2-ethyl hexyl ester, and

C_5—Cl would represent chlorocyclopentane.

The first chemical shift table presented in each chapter heading deals primarily with the type of compounds that are to follow. When the spectra of such compounds are present in the chapter, their spectrum number in the book will be given. In many cases, supplemental chemical shift information will be supplied for compounds which are not in this volume but are contained in the Sadtler collection of Standard NMR spectra. These chemical shift listings will show a series of dashes (- - - -) in the spectrum column.

To determine whether a specific derivative of the parent compound, which may contain a different functional group, is present in the text but in a different section, please consult the alphabetical index in volume 2.

NOTE

The spectrum numbers listed in this volume apply only to the spectra in the Sadtler Handbooks of Reference Spectra (Infrared, Proton NMR, Ultraviolet) and do not correspond in any way to the spectrum numbers in the Sadtler Standard Spectra volumes.

TABLE OF CONTENTS

TABLE OF CONTENTS

TABLE OF CONTENTS

TABLE OF CONTENTS

THE NORMAL ALKANES

The normal alkanes are an easily recognized group of compounds consisting of two higher order bands resonating in a narrow chemical shift range at high field.

Chemical Shifts

CH_3	near 0.9 ppm – a distorted triplet
$(CH_2)_n$	near 1.3 ppm -- a complex multiplet in the shorter (C_4, C_5, C_6) alkanes gradually becoming a broad, single peak as the number of carbons in the chain increases.

Coupling and Coupling Constants

Due to the higher order patterns which result from the very narrow chemical shift range, it is not possible to measure accurately the vicinal coupling constants (H—C—C—H). However, because the distorted triplet at 0.9 ppm is nearly identical to those observed for the substituted alkanes, it would appear that the coupling constants are similar to those of the substituted alkanes, i.e. $J_{H—C—C—H}$ = 6-8 Hz.

Solubility and Solvent Effects

The normal alkanes, as indeed all of the hydrocarbons, are most readily soluble in the halogenated solvents, CCl_4 and $CDCl_3$. Their solubility in even these liquids decreases markedly as the molecular weight (chain length) increases beyond molecular weight 200 (C_{12} to C_{15}). Allowing the sample-solvent slurry to stand overnight, agitation of the mixture and warming, are helpful in obtaining higher sample solution concentrations.

Impurities

Because most commercially available alkanes are obtained from the fractional distillation of petroleum, impurities such as the cyclic alkanes, and simple aromatic hydrocarbons such as toluene, ethyl benzene and the xylenes, may be observed in their NMR spectra.

Characterization

Differentiating between the various alkanes is usually accomplished by careful measurement of the methyl and methylene integration values. Determinations accurate to within one carbon are routinely obtained with chain lengths up to about Triacontane (C_{30}).

Note

Other nuclei which possess a weak deshielding effect similar to that of hydrocarbon groups may produce spectra almost identical in appearance and chemical shift to those of the normal alkanes. Included in this group would be nuclei such as Phosphorus, Tin, Lead, Mercury, Boron and Silicon. Since many of these nuclei possess a spin greater than zero, their spectra may display isotope side-bands slightly above and/or below the primary chemical shift range of the sample. These isotope sidebands are helpful in determining that a high field pattern is not that of a normal alkane.

PENTANE

C_5H_{12}

Mol. Wt. 72.15

B.P. 36.1°C

Solvent: CCl_4

Assignments

a = 0.91

b = ca 1.28

1

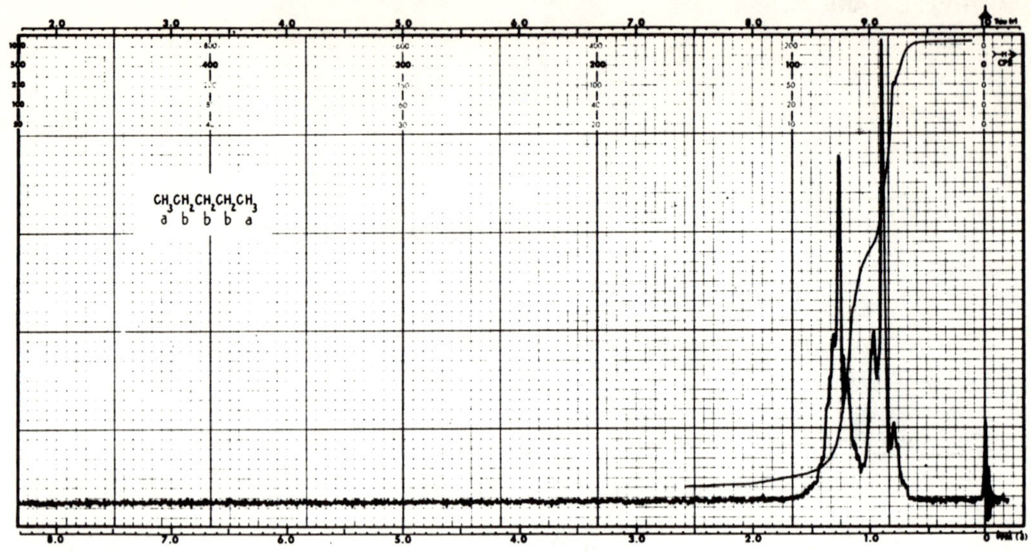

© 1978 Sadtler Research Laboratories, Inc.

HEXANE

C_6H_{14}

Mol. Wt. 86.18

B.P. 68.8°C

Solvent: CCl_4

Assignments

a = 0.89

b = ca 1.25

2

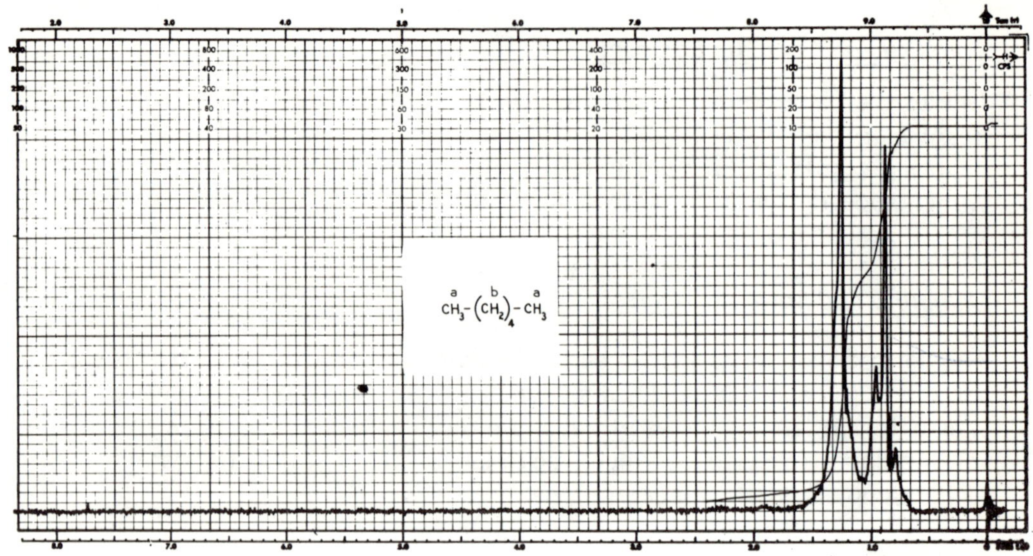

© 1978 Sadtler Research Laboratories, Inc.

HEPTANE

C_7H_{16}

Mol. Wt. 100.21

B.P. 98-99°C

Solvent: CCl_4

Assignments

a = 0.99

b = ca 1.29

3

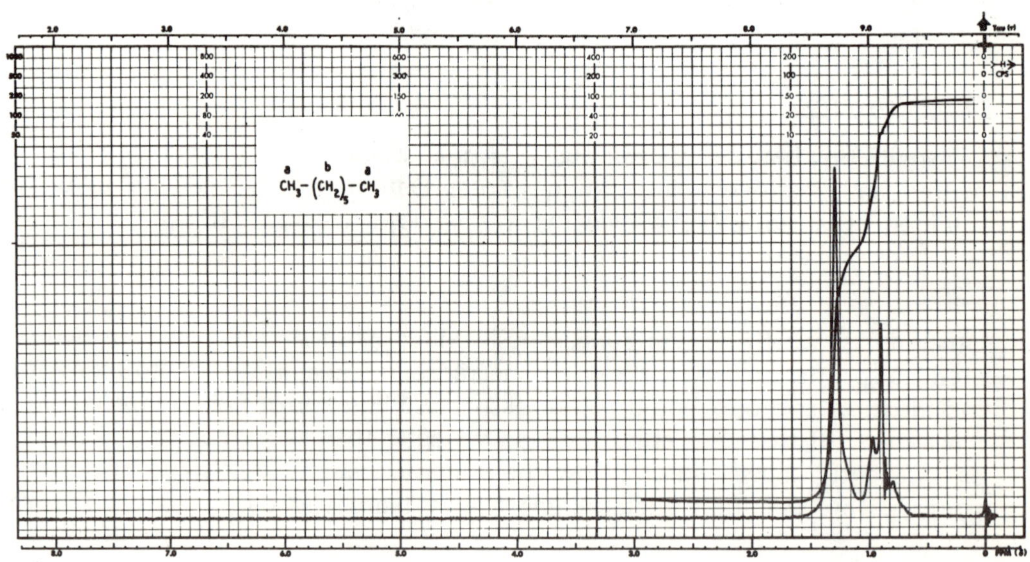

© 1978 Sadtler Research Laboratories, Inc.

OCTANE

C_8H_{16}

Mol. Wt. 114.23

B.P. 124-126°C

Solvent: CCl_4

Assignments
a = 0.90
b = ca 1.28

4

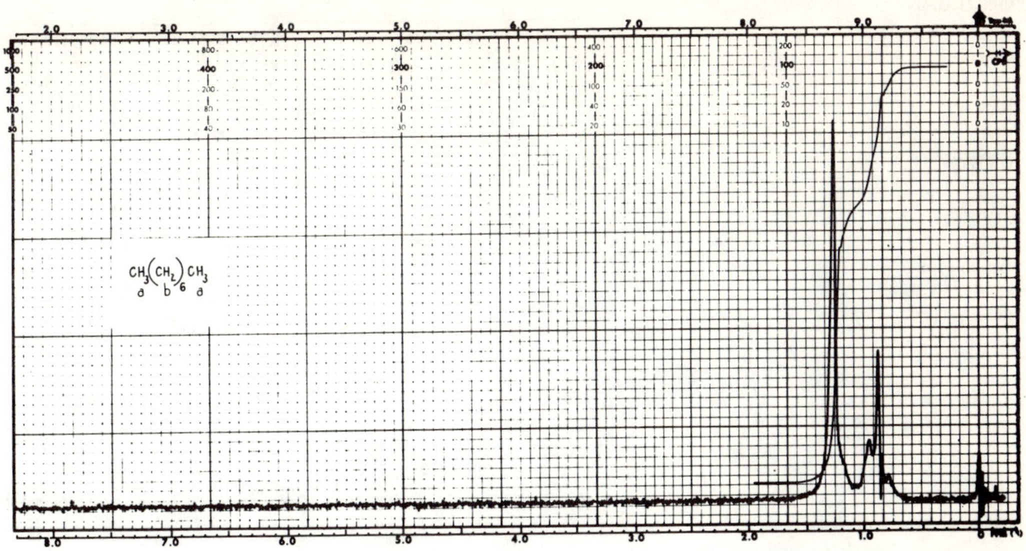

© 1978 Sadtler Research Laboratories, Inc.

NONANE

C_9H_{20}

Mol. Wt. 128.25

B.P. 150-152°C

Solvent: $CDCl_3$

Assignments
a = 0.88
b = ca 1.28

5

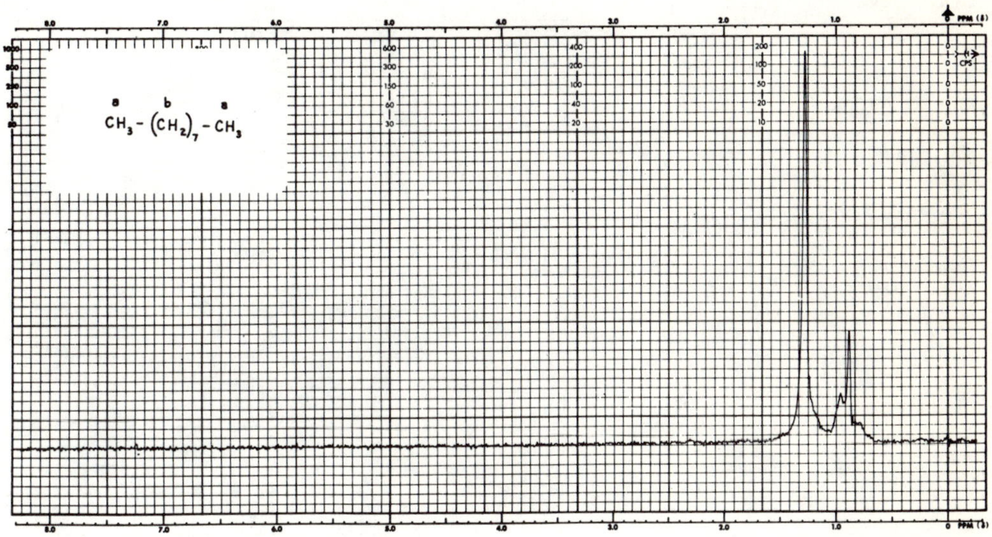

© 1978 Sadtler Research Laboratories, Inc.

DECANE

$C_{10}H_{22}$

Mol. Wt. 142.28

B.P. 174°C (lit.)

Solvent: CCl_4

Assignments
a = 0.90
b = ca 1.28

6

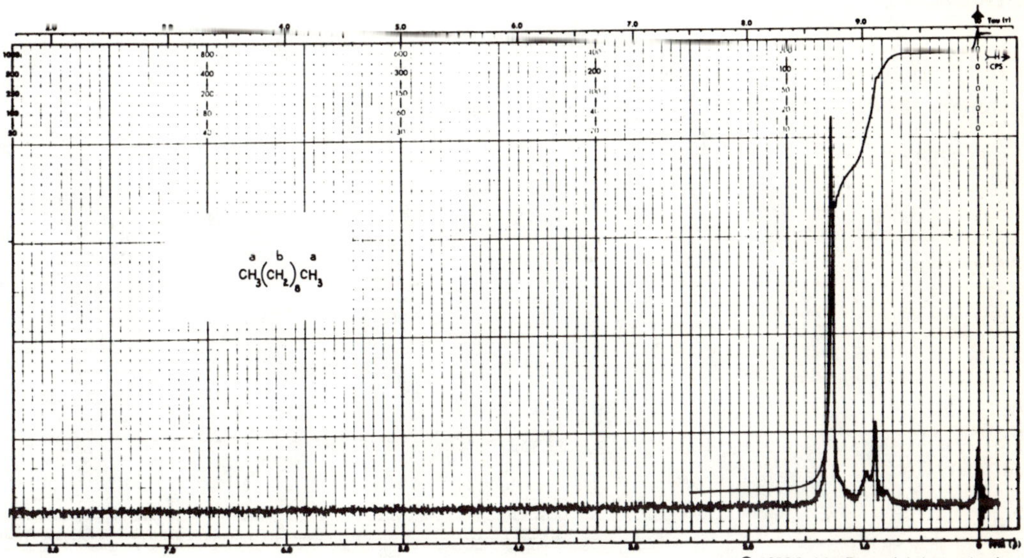

© 1978 Sadtler Research Laboratories, Inc.

UNDECANE

$C_{11}H_{24}$

Mol. Wt. 150.30

B.P. 384°F

Solvent: CCl_4

Assignments
a = 0.88
b = ca 1.25

7

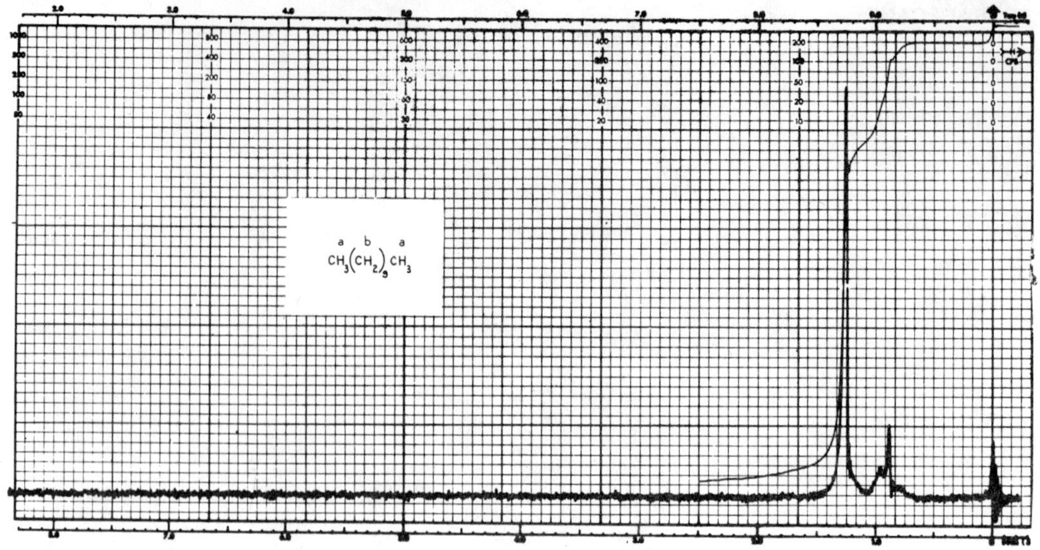

© 1978 Sadtler Research Laboratories, Inc.

DODECANE

$C_{12}H_{26}$

Mol. Wt. 170.34

B.P. 212.8°C

Solvent: CCl_4

Assignments
a = 0.89
b = ca 1.27

8

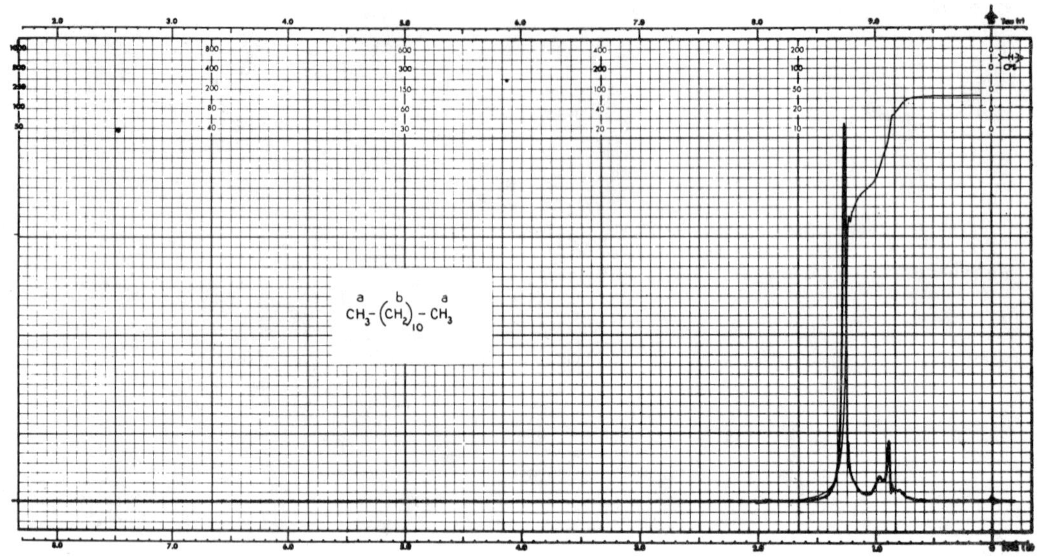

© 1978 Sadtler Research Laboratories, Inc.

TRIDECANE

$C_{13}H_{28}$

Mol. Wt. 184.37

B.P. 225.5°C

Solvent: CCl_4

Assignments
a = 0.89
b = ca 1.26

9

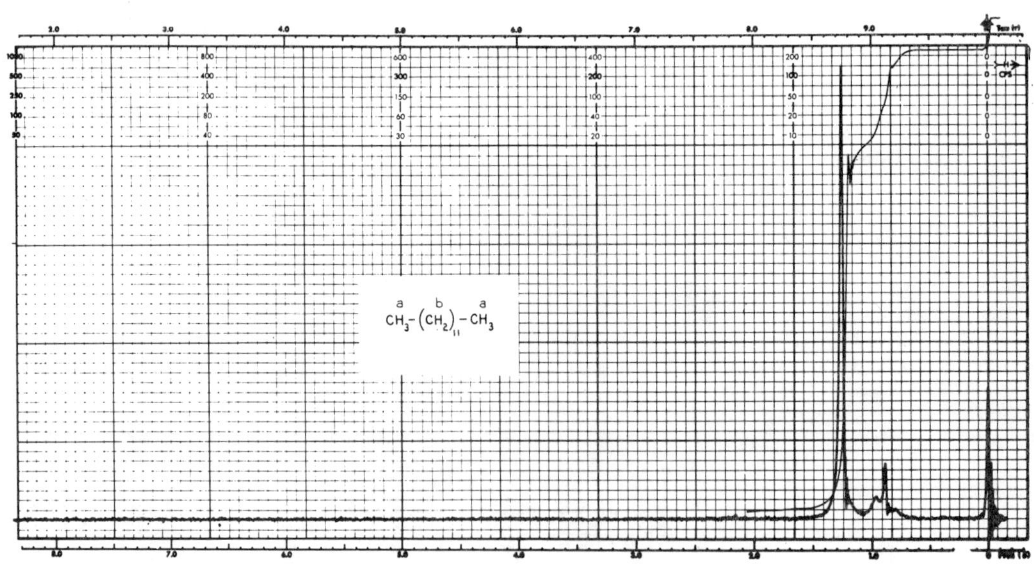

© 1978 Sadtler Research Laboratories, Inc.

TETRADECANE

$C_{14}H_{30}$

Mol. Wt. 198.40

B.P. 250°C

Solvent: CCl_4

Assignments
a = 0.89
b = ca 1.25

10

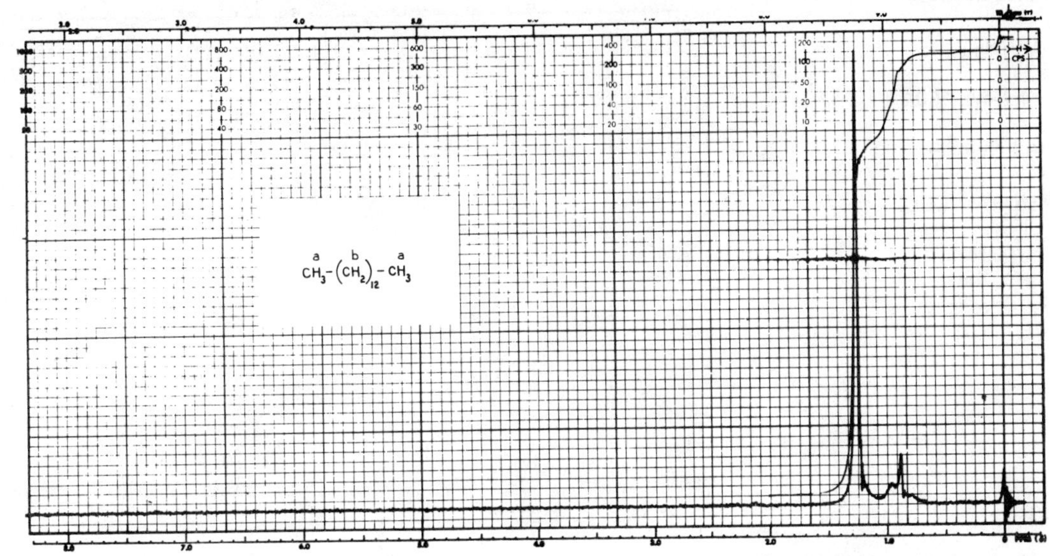

© 1978 Sadtler Research Laboratories, Inc.

PENTADECANE

$C_{15}H_{32}$

Mol. Wt. 212.42

B.P. 270.5°C

Solvent: CCl_4

Assignments
a = 0.88
b = ca 1.26

11

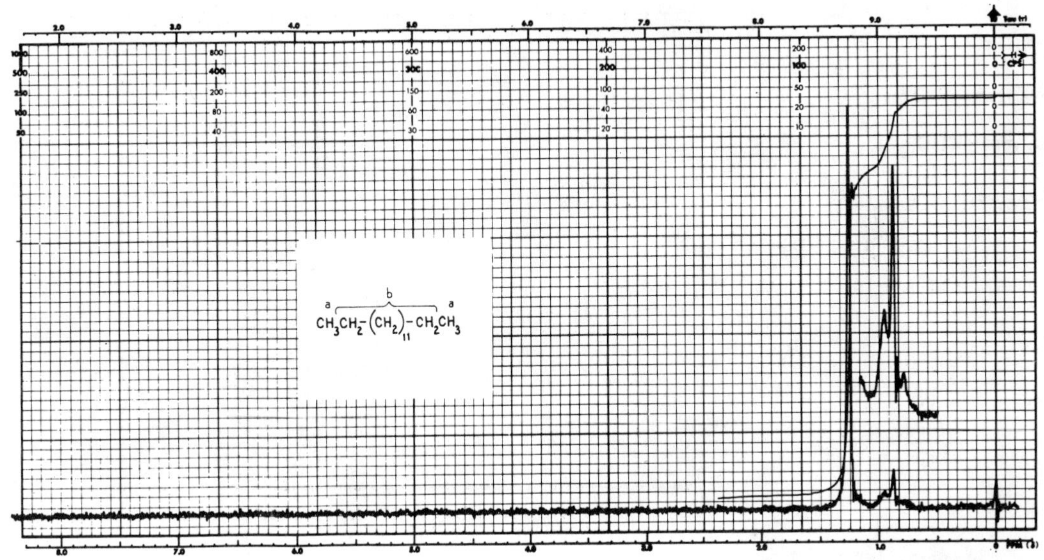

© 1978 Sadtler Research Laboratories, Inc.

OCTADECANE

$C_{18}H_{38}$

Mol. Wt. 254.50

Solvent: $CDCl_3$

Assignments
a = 0.90
b = ca 1.28

12

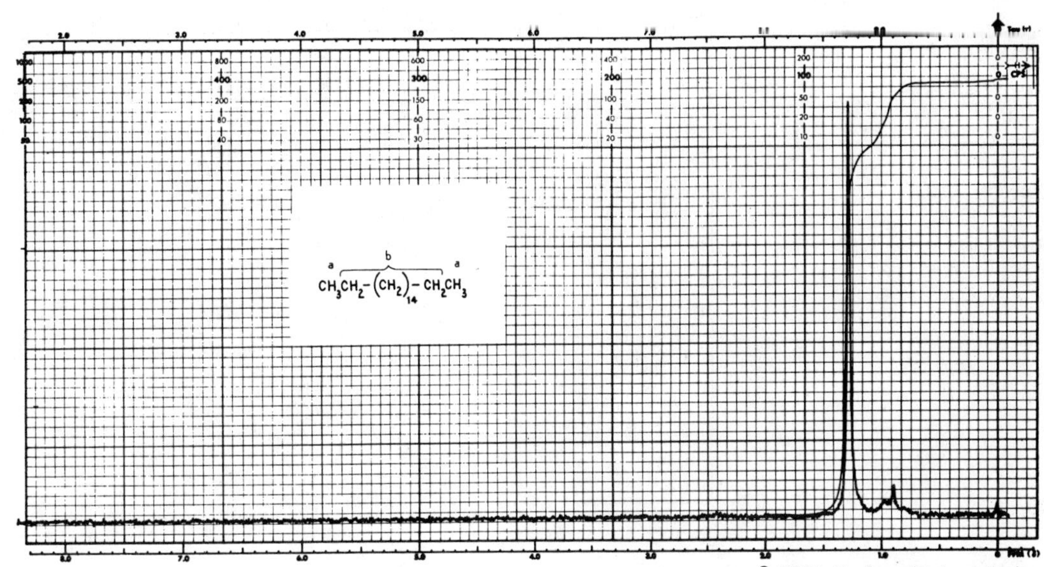

© 1978 Sadtler Research Laboratories, Inc.

EICOSANE

$C_{20}H_{42}$

Mol. Wt. 282.56

M.P. 36-37°C

Solvent: $CDCl_3$

Assignments
a = 0.87
b = ca 1.25

13

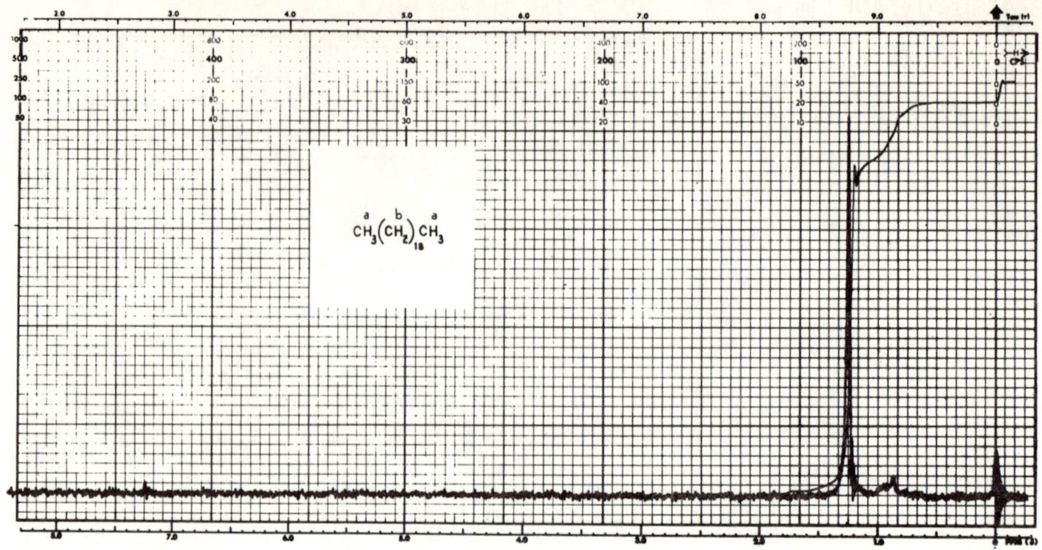

© 1978 Sadtler Research Laboratories, Inc.

DOCOSANE

$C_{22}H_{46}$

Mol. Wt. 310.61

Solvent: $CDCl_3$

Assignments
a = 0.89
b = ca 1.26

14

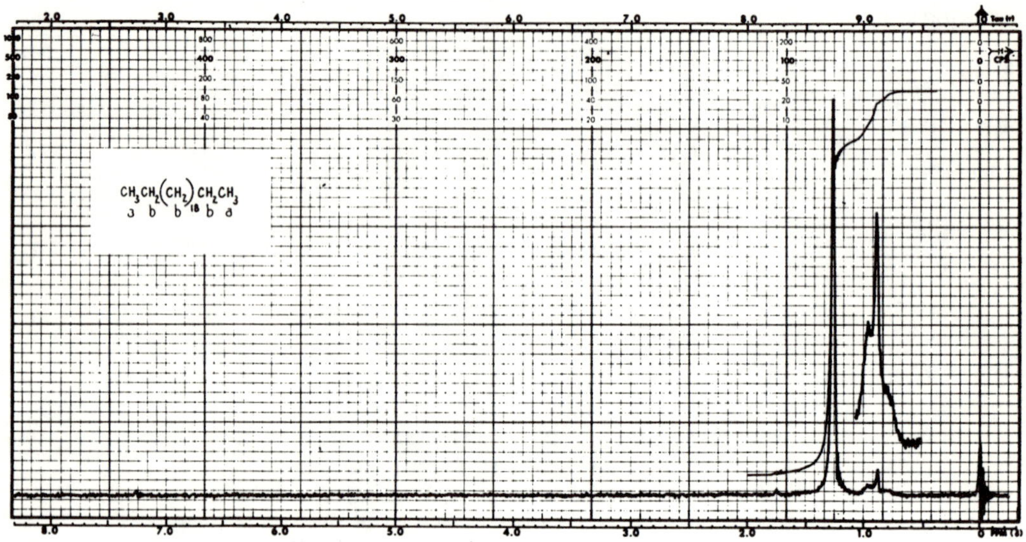

© 1978 Sadtler Research Laboratories, Inc.

TRICOSANE

$C_{23}H_{48}$

Mol. Wt. 324.64

M.P. 47-48°C (lit.)

Solvent: CCl_4

Assignments
a = 0.88
b = ca 1.24

15

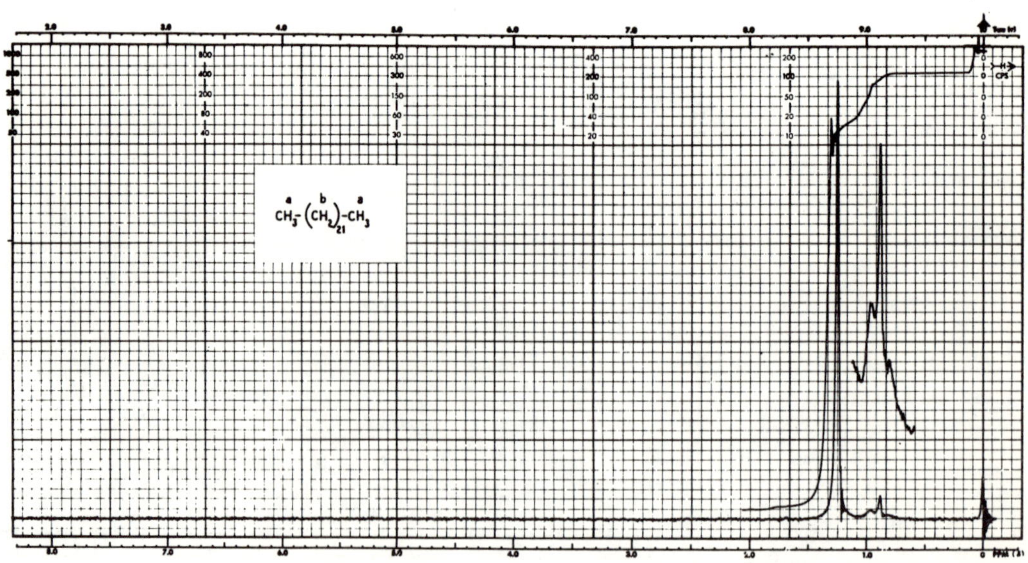

© 1978 Sadtler Research Laboratories, Inc.

HEXATRIACONTANE

$C_{36}H_{74}$

Mol. Wt. 506.99

M.P. 74-76°C B.P. 497°C

Solvent: Polysol-d

Assignments
a = 0.89
b = ca 1.29
c = 2.52 DMSO-d_5
d = 2.70 H_2O

16

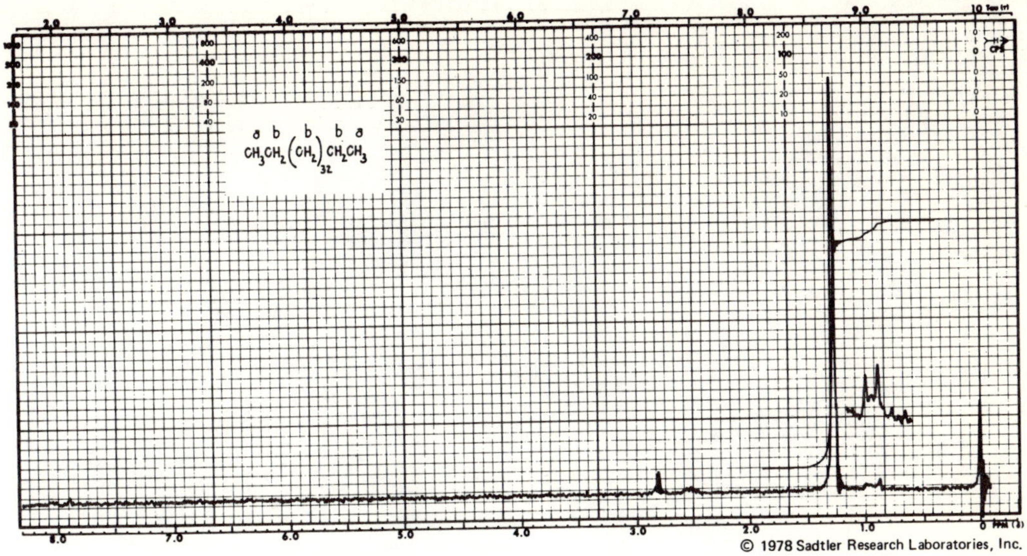

© 1978 Sadtler Research Laboratories, Inc.

NOTES

THE BRANCHED ALKANES

The presence of one or more branching groups along the hydrocarbon chain increases the complexity and chemical shift ranges of the branched alkanes in comparison to those of the normal alkanes. As a consequence of the larger number of methyl groups that are present, the intensity of the bands at higher field is increased.

Chemical Shifts

The overall range of chemical shifts for the branched alkanes is 0.6-2.0 ppm, with the methyl (CH_3) resonances on the higher field side, the methylene (CH_2) resonances in the intermediate portion of the range and the methine (CH) groups resonating in the lower field area (1.3-2.0 ppm).

Methyl Groups   0.8-0.95 ppm

$R-CH_2-\underline{CH_3}$	--	distorted triplet
$R-CH-\underline{(CH_3)_2}$	--	broadened doublet
$R-C-\underline{(CH_3)_3}$	--	sharp singlet

Methylene and Methine Groups

The methylene and methine groups are almost always complex, higher order, overlapping multiplets that cannot be easily characterized by first order approximations.

Coupling and Coupling Constants

As with the normal alkanes, higher order effects prevent the direct measurement of coupling constants, however, the separation of the peaks of methyl doublets indicates that the $J_{H-C-C-H}$ (vicinal) coupling constants are of the order of 6-9 Hz.

Solubility and Solvent Effects

The branched alkanes are readily soluble in the halogenated hydrocarbons normally utilized as NMR solvents. The presence of branching groups makes the branched hydrocarbons more soluble than a normal hydrocarbon of comparable molecular weight.

Impurities

Branched hydrocarbons obtained from petroleum sources may display impurities arising from the cyclic alkanes and low molecular weight aromatic hydrocarbons.

Characterization

The branched alkanes are probably the most difficult compounds to identify without the aid of known reference spectra. These spectra do however produce unique "fingerprint" patterns which are well represented in the various collections of NMR reference spectra which are currently available.

If the methyl, methylene and methine resonance bands are sufficiently well separated, a comparison of the integration values can be used to advantage in determining the relative number of the various types of carbon atoms present (methyl, methylene, methine).

2-METHYLBUTANE

C_5H_{12}

Mol. Wt. 72.15

B.P. 27.8°C

Solvent: CCl_4

Assignments
a = 0.86
b = 0.87
c = 0.90-1.60

17

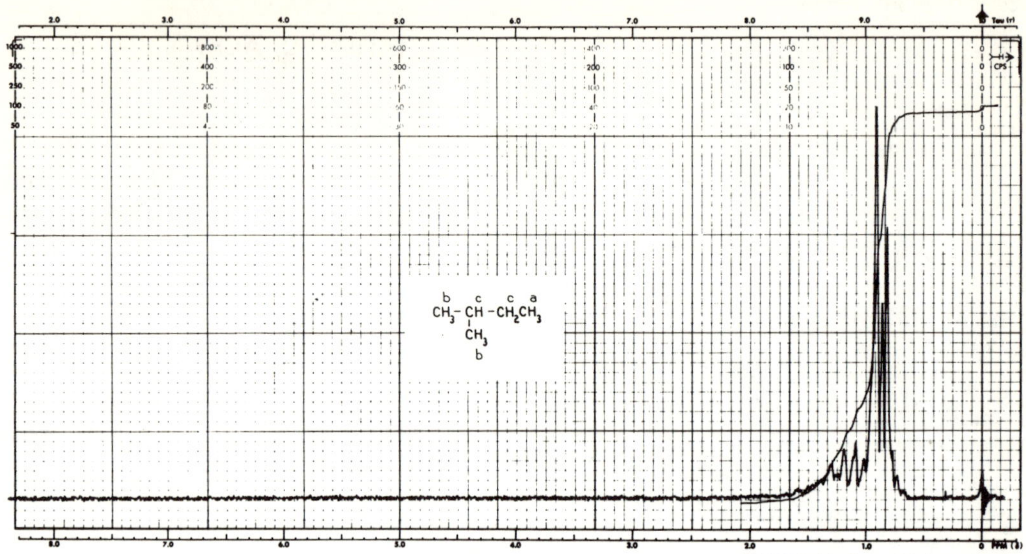

© 1978 Sadtler Research Laboratories, Inc.

2,3-DIMETHYLBUTANE

C_6H_{14}

Mol. Wt. 86.18

B.P. 58°C

Solvent: CCl_4

Assignments
a = 0.88
b = 1.40

18

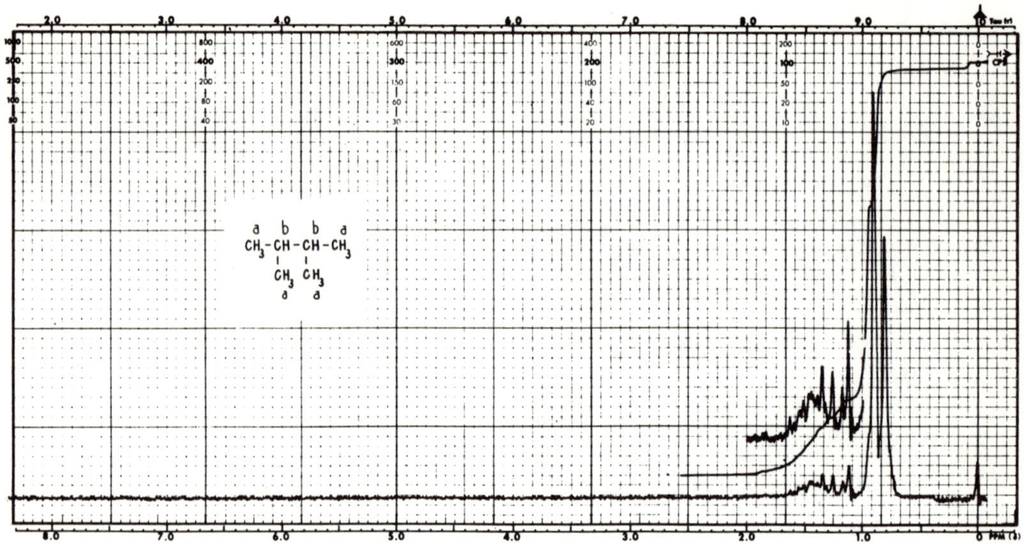

© 1978 Sadtler Research Laboratories, Inc.

2-METHYLPENTANE

C_6H_{14}

Mol. Wt. 86.18

B.P. 60.1°C

Solvent: CCl_4

Assignments
a = 0.85
b = 0.89
c = 1.05-1.90

19

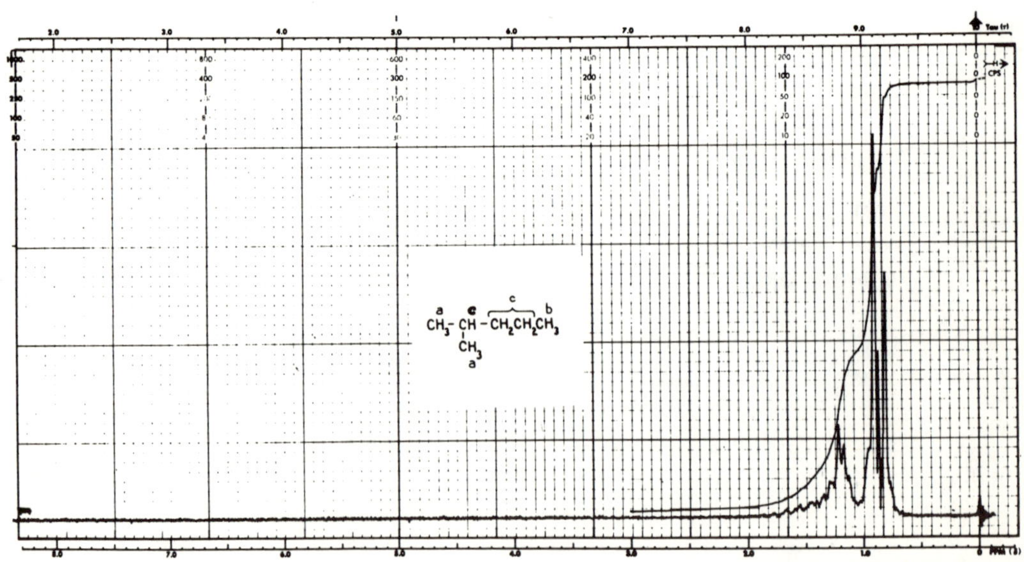

© 1978 Sadtler Research Laboratories, Inc.

3-METHYLPENTANE

C_6H_{14}

Mol. Wt. 86.18

B.P. 60.1°C

Solvent: CCl_4

Assignments
a = 0.62-1.00
b = 1.00-1.50

20

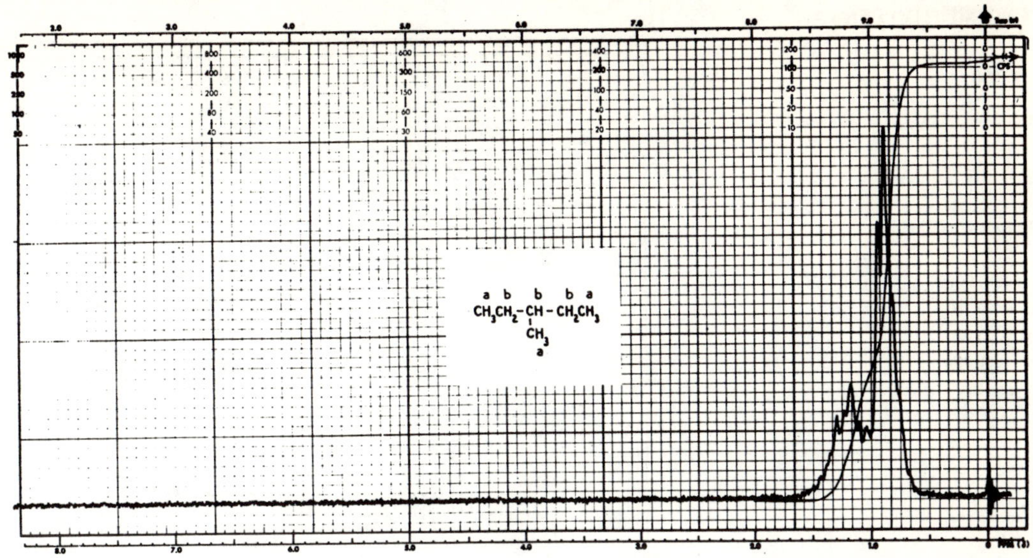

© 1978 Sadtler Research Laboratories, Inc.

2,2-DIMETHYLPENTANE

C_7H_{16}

Mol. Wt. 100.21

B.P. 78°C/43mm

Solvent: $CDCl_3$

Assignments
a = 0.89
b = 0.90
c = 1.03-1.45

21

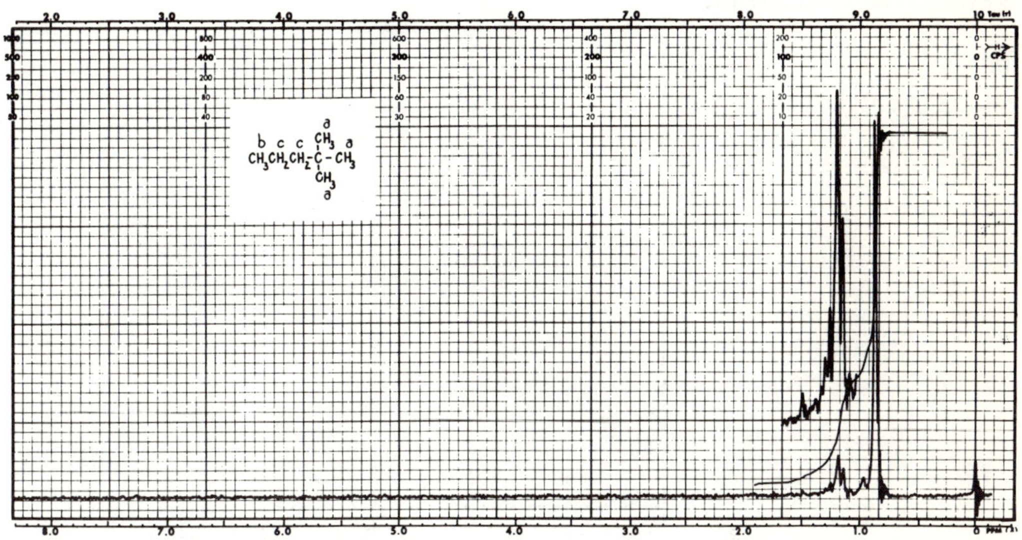

© 1978 Sadtler Research Laboratories, Inc.

2,4-DIMETHYLPENTANE

C_7H_{16}

Mol. Wt. 100.21

B.P. 80.3°C

Solvent: CCl_4

Assignments
a = 0.86
b = 1.09
c = 1.60

22

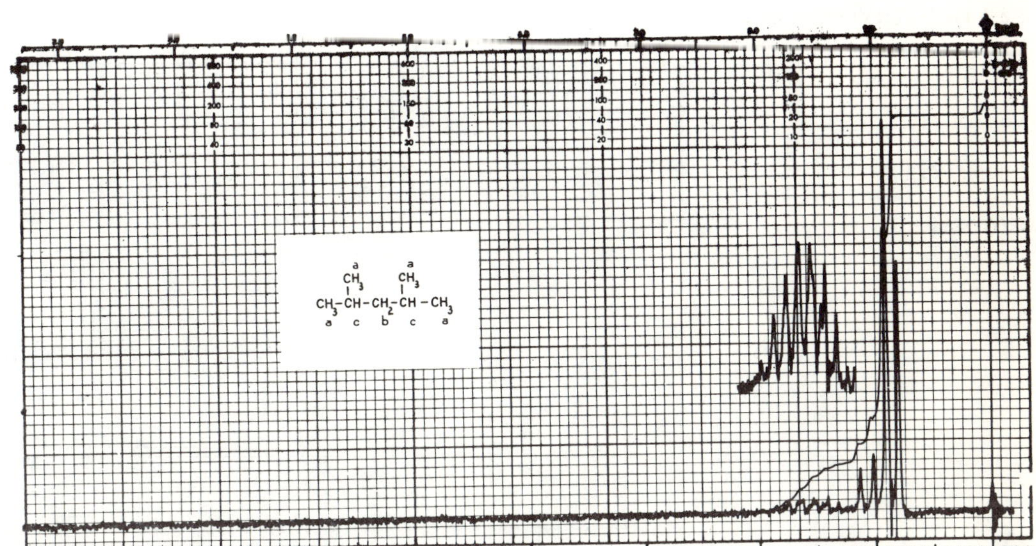

© 1978 Sadtler Research Laboratories, Inc.

3,3-DIMETHYLPENTANE

C_7H_{16}

Mol. Wt. 100.21

B.P. 86°C

Solvent: $CDCl_3$

Assignments
a = 0.79
b = 0.84
c = 1.19

23

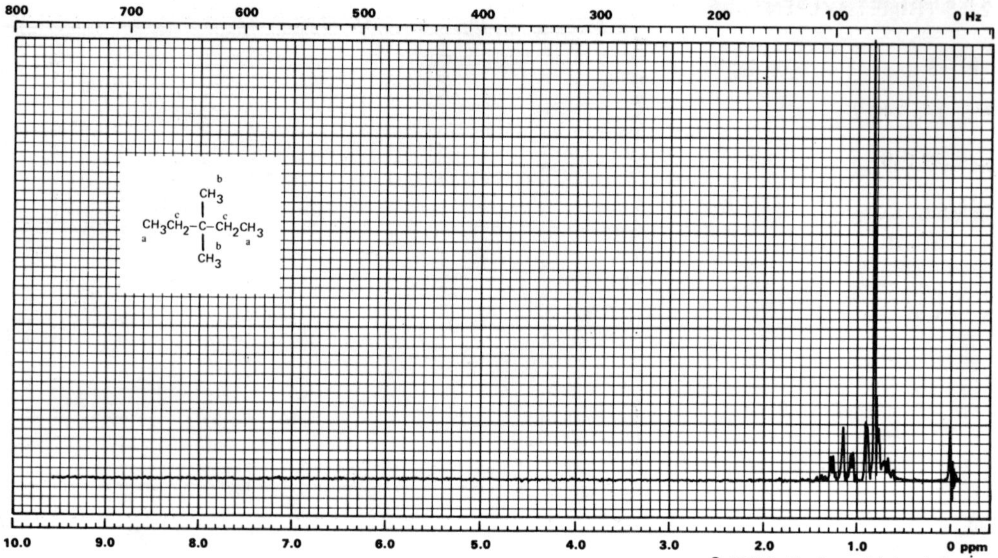

© 1978 Sadtler Research Laboratories, Inc.

3-ETHYL-2-METHYLPENTANE

C_8H_{18}

Mol. Wt. 114.23

Solvent: CCl_4

Assignments
a = 0.86
b = 0.91
c = 1.00-2.00

24

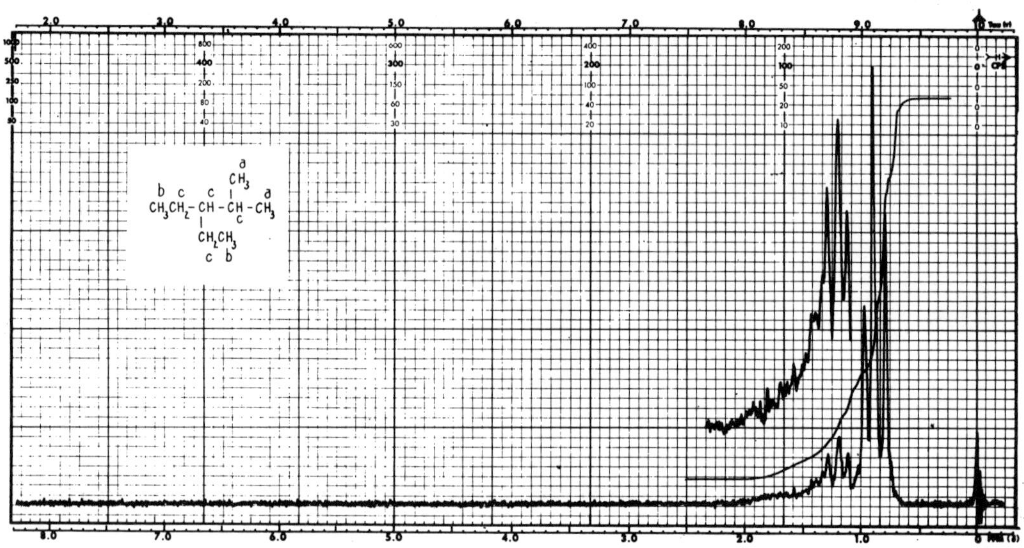

© 1978 Sadtler Research Laboratories, Inc.

2,2,3-TRIMETHYLPENTANE

C_8H_{18}

Mol. Wt. 114.23

Solvent: CCl_4

Assignments
a = ca 0.88
b = 1.10-1.90

25

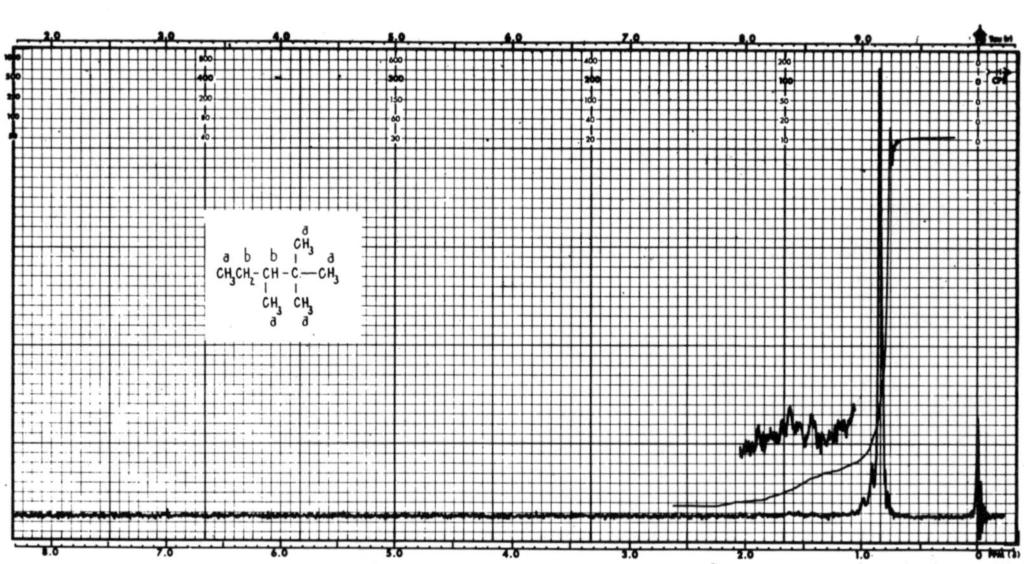

© 1978 Sadtler Research Laboratories, Inc.

2,3,3-TRIMETHYLPENTANE

C_8H_{18}

Mol. Wt. 114.23

Solvent: CCl_4

Assignments
a = ca 0.79
b = 0.82
c = 1.00-1.82

26

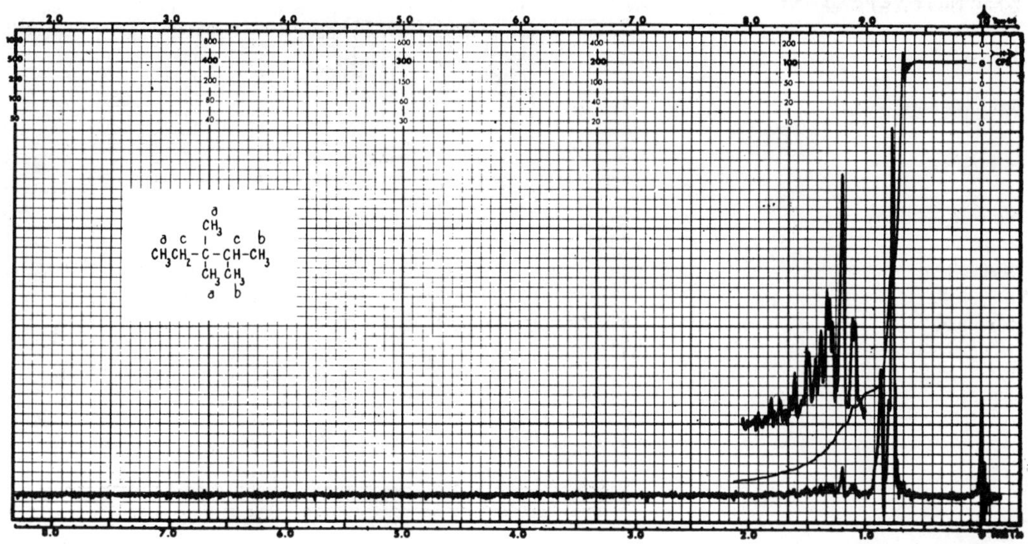

© 1978 Sadtler Research Laboratories, Inc.

2,2,4-TRIMETHYLPENTANE

C_8H_{18}

Mol. Wt. 114.23

Solvent: CCl_4

Assignments
a = 0.90
b = 0.92
c = 1.12
d = 1.68

27

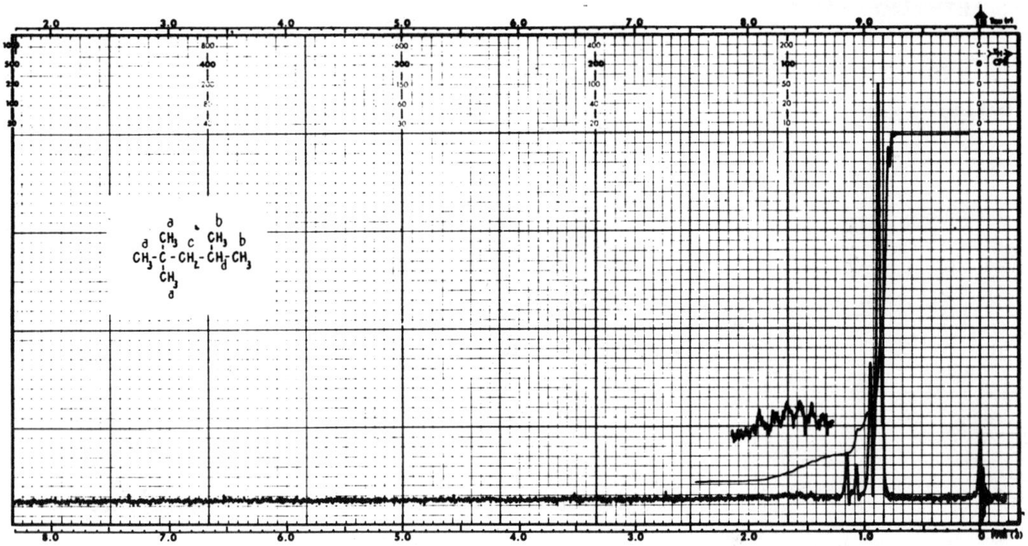

© 1978 Sadtler Research Laboratories, Inc.

2,3,4-TRIMETHYLPENTANE

C_8H_{18}

Mol. Wt. 114.23

B.P. 113.3°C

Solvent: CCl_4

Assignments
a = 0.58-1.07
b = 1.38-1.98

28

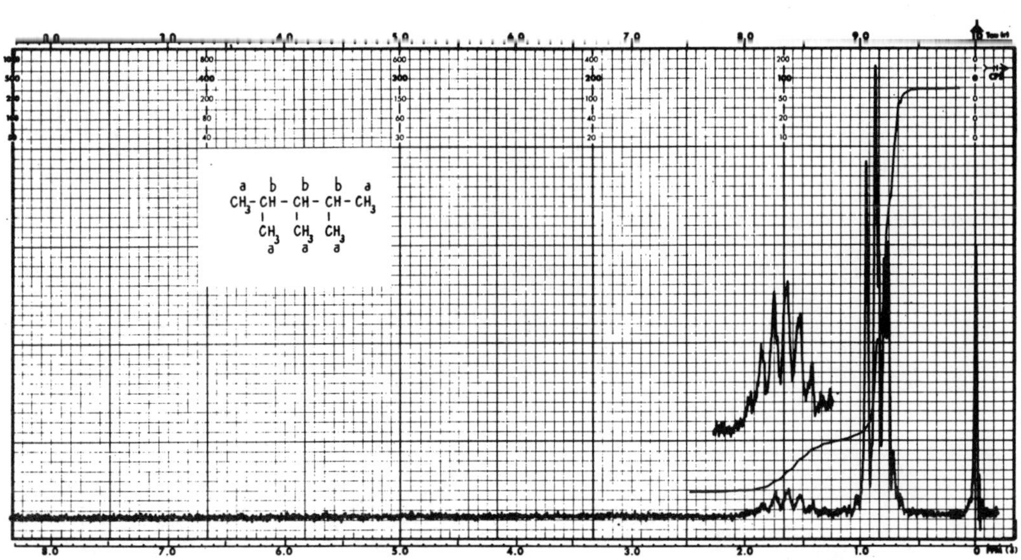

© 1978 Sadtler Research Laboratories, Inc.

2-METHYLHEXANE

C_7H_{16}

Mol. Wt. 100.21

B.P. 90°C

Solvent: $CDCl_3$

Assignments
a = 0.88
b = 0.90
c = 1.00-1.80

29

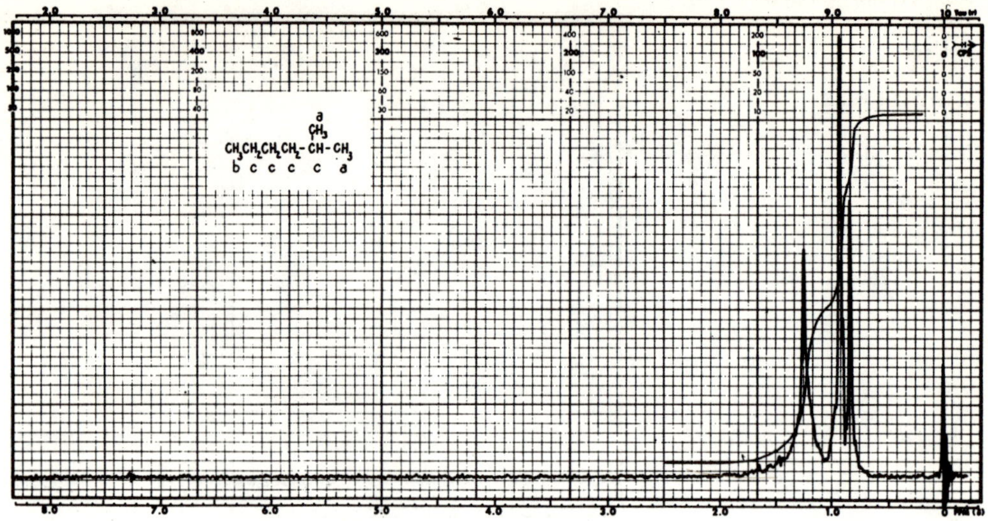

© 1978 Sadtler Research Laboratories, Inc.

3-METHYLHEXANE

C_7H_{16}

Mol. Wt. 100.21

Solvent: CCl_4

Assignments
a = ca 0.89
b = 1.00-1.60

30

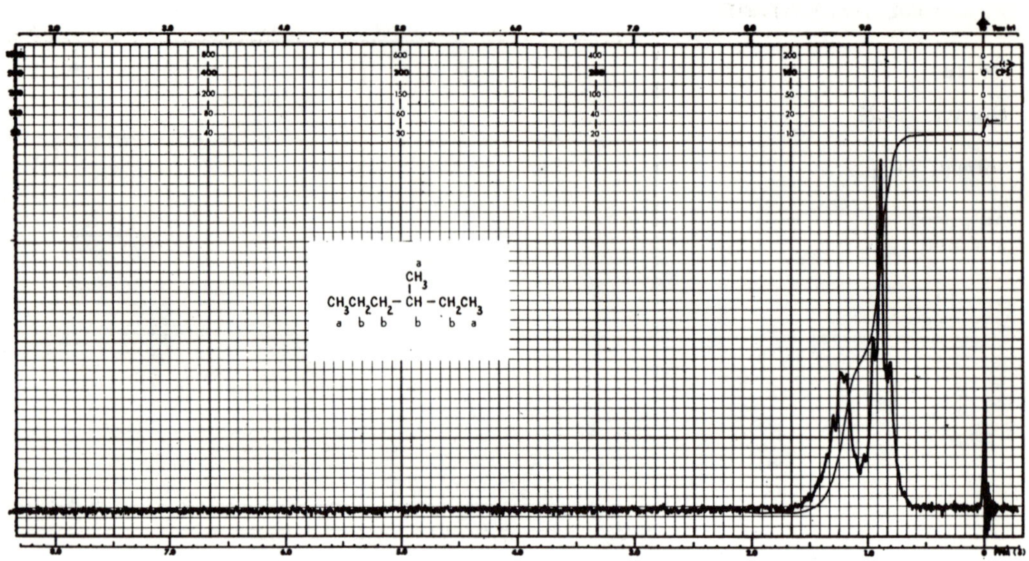

© 1978 Sadtler Research Laboratories, Inc.

2,5-DIMETHYLHEXANE

C_8H_{18}

Mol. Wt. 114.23

Solvent: CCl_4

Assignments
a = 0.88
b = 1.08-1.83

31

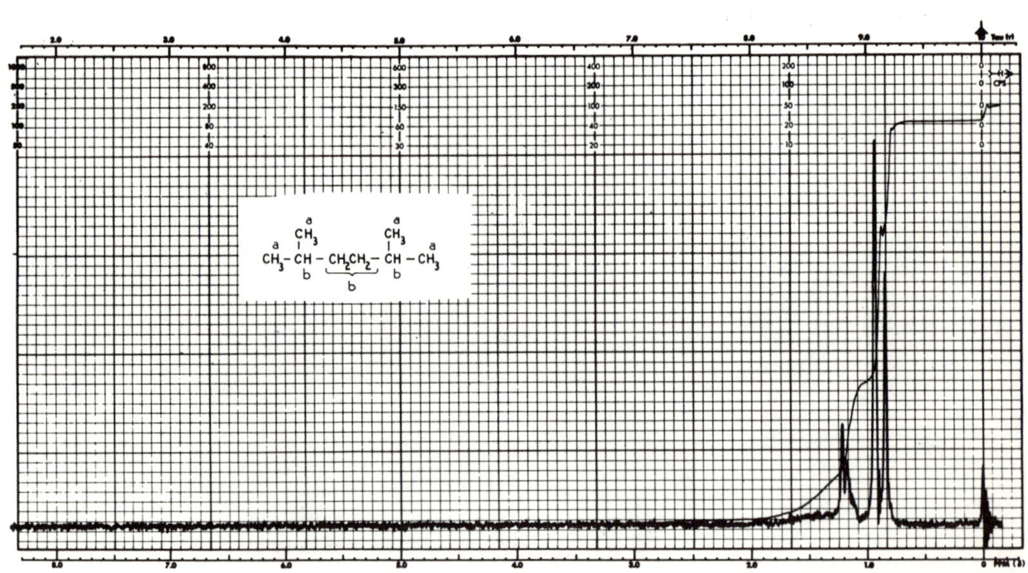

© 1978 Sadtler Research Laboratories, Inc.

2-METHYLHEPTANE

C_8H_{18}

Mol. Wt. 114.23

Solvent: CCl_4

Assignments
a = 0.89
b = 0.91
c = 1.26

32

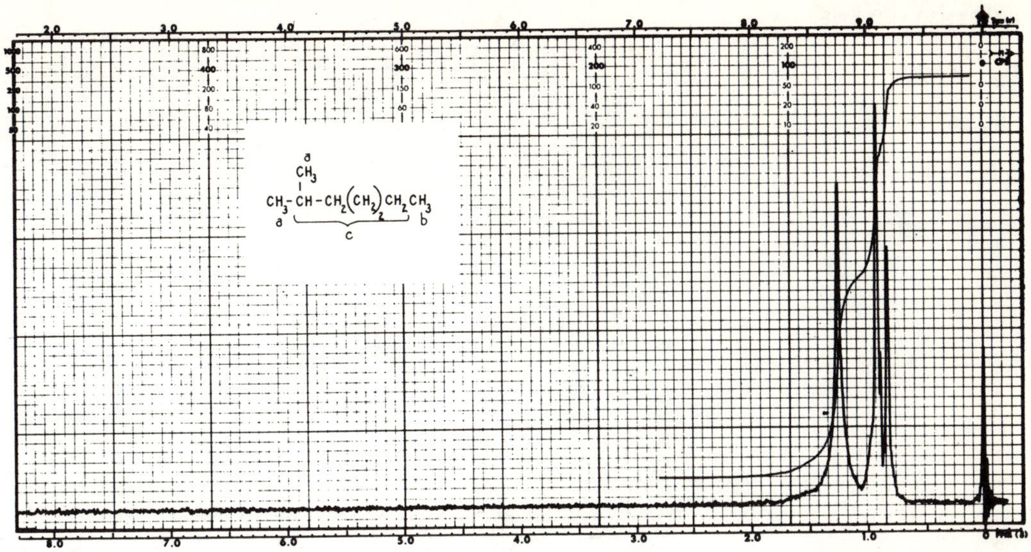

© 1978 Sadtler Research Laboratories, Inc.

3,3-DIMETHYLHEPTANE

C_9H_{20}

Mol. Wt. 128.26

B.P. 137°C

Solvent: CCl_4

Assignments
a = ca 0.81
b = ca 1.20

33

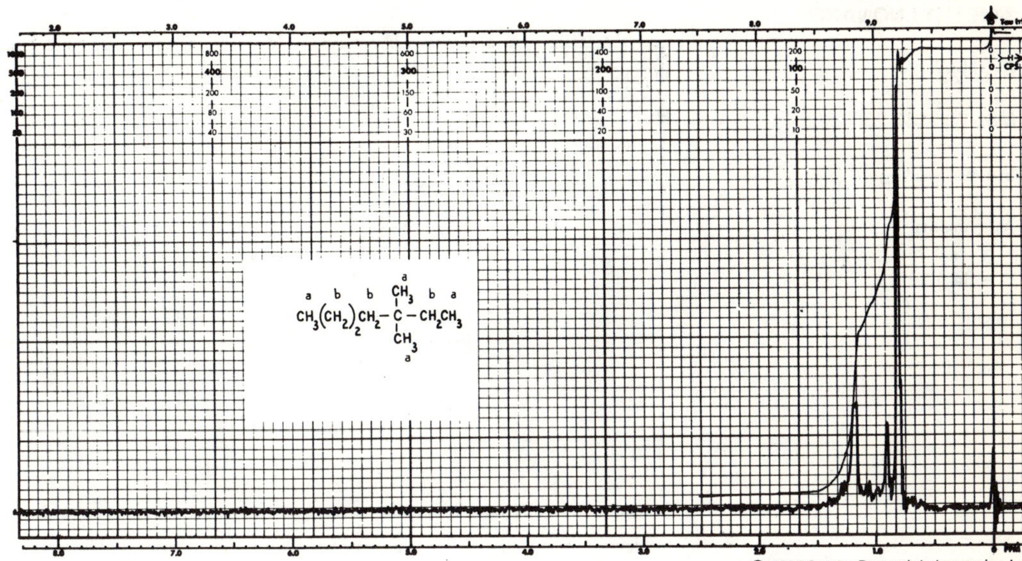

© 1978 Sadtler Research Laboratories, Inc.

2-METHYLOCTANE

C_9H_{20}

Mol. Wt. 128.26

B.P. 142°C

Solvent: CCl_4

Assignments
a = 0.89
b = 0.91
c = ca 1.29

34

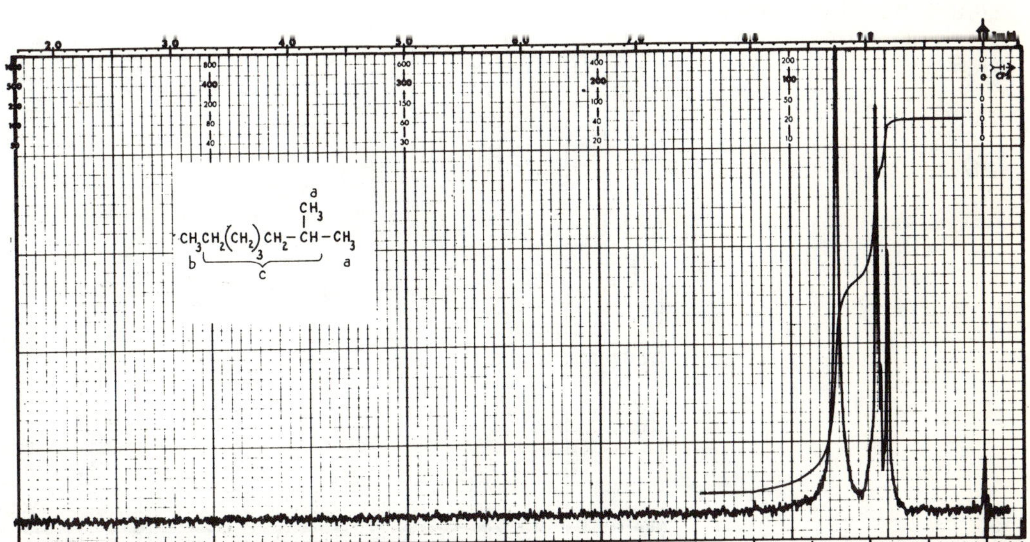

© 1978 Sadtler Research Laboratories, Inc.

4-METHYLOCTANE

C_9H_{20}

Mol. Wt. 128.26

B.P. 142°C

Solvent: $CDCl_3$

Assignments
a = 0.84
b = 0.89
c = ca 1.25

35

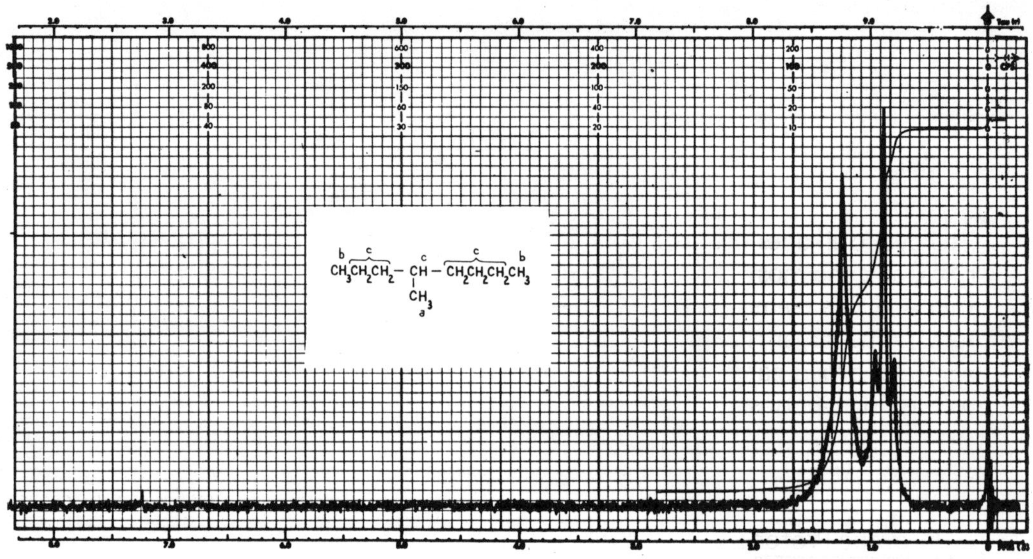

© 1978 Sadtler Research Laboratories, Inc.

3-METHYLNONANE

$C_{10}H_{22}$

Mol. Wt. 142.29

Solvent: CCl_4

Assignments
a = ca 0.90
b = ca 1.29

36

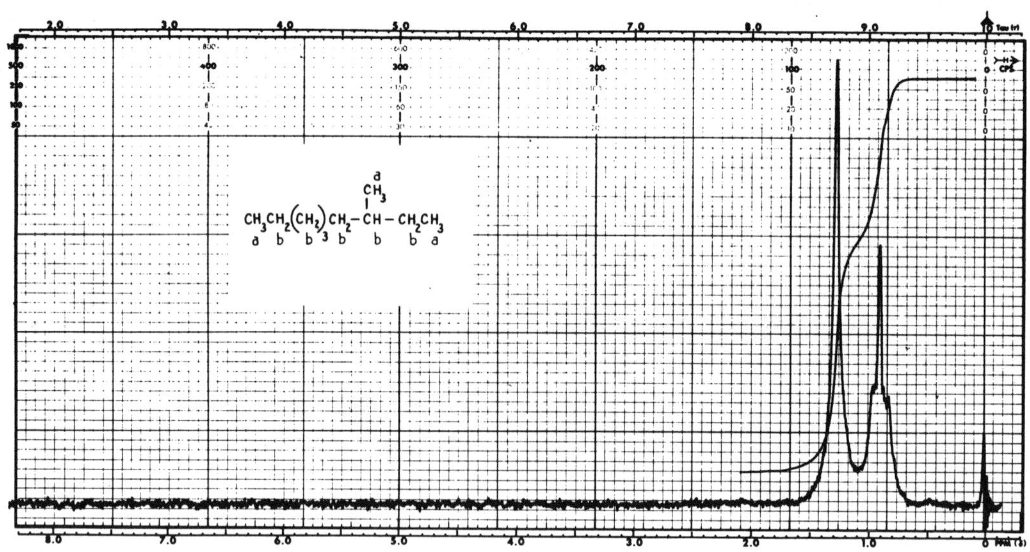

© 1978 Sadtler Research Laboratories, Inc.

2,6,10,14-TETRAMETHYLPENTADECANE

$C_{19}H_{40}$

Mol. Wt. 268.53

Solvent: CCl_4

Assignments
a = 0.86
b = ca 1.22

37

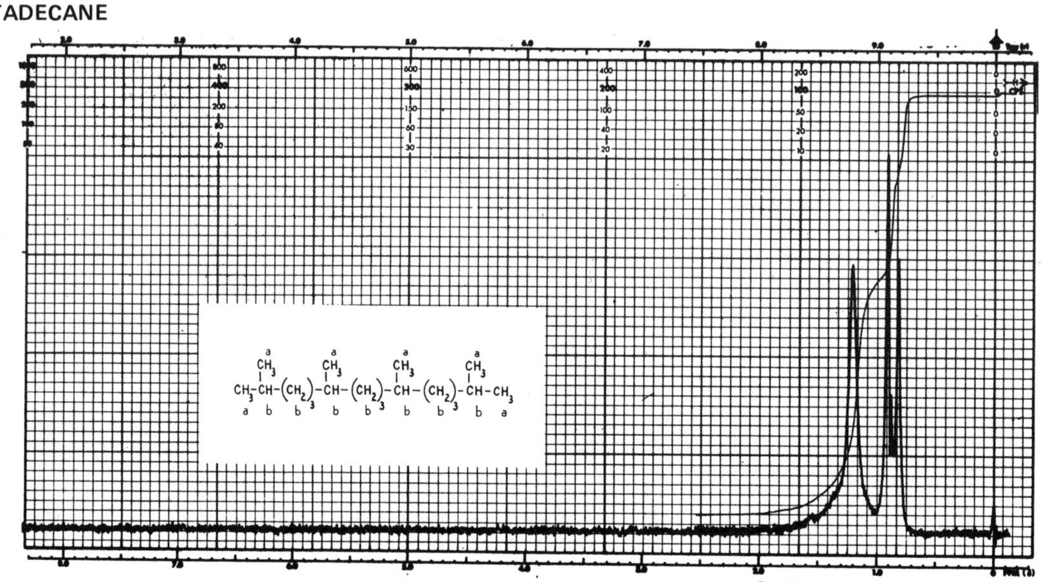

© 1978 Sadtler Research Laboratories, Inc.

THE CYCLIC ALKANES

The cyclic alkanes produce both the simplest patterns (one single peak) for the unsubstituted parent rings, and the most complex, poorly resolved patterns, for the substituted derivatives. The three, four and five membered rings usually produce complex but relatively well resolved absorption patterns. The larger rings (C_6 and higher) due to the slow interchange in ring shape produce poorly resolved, broad bands often covering more than a full one ppm, arising from the hydrogens attached to the ring carbons.

Chemical Shifts

Alicyclic Protons

		Solvent	Spectrum
0.22	Cyclopropane	(lit.)	----
1.96	Cyclobutane	(lit.)	----
1.50	Cyclopentane	CCl_4	42
1.42	Cyclohexane	CCl_4	45
1.53	Cycloheptane	$CDCl_3$	50
1.52	Cyclooctane	CCl_4	51

Coupling and Coupling Constants

The spectral patterns of the cyclic alkanes are usually too complex or too poorly resolved to provide any useful measurements of the coupling constants.

Solubility and Solvent Effects

Most representatives of the cyclic alkanes are readily soluble in the halogenated hydrocarbon solvents.

Characterization

As with the branched alkanes, if the resonance bands are sufficiently well separated, then a comparison of the integration values may be useful. Generally, though, a comparison with known reference spectra will be found to be the most reliable method of identifying an unknown of this group.

Alicyclic Derivatives

Substitution of the cyclic alkanes by a deshielding substituent leads to a characteristic chemical shift for the hydrogen attached to the alpha carbon depending upon the size of the ring and the deshielding effect of the substituting group as listed below.

Cyclopropane Derivatives (C_3–X)

$(CH_2)_2$–	CH	–X	Solvent
0.1-0.7 ppm	0.98	$-CH_2-Ø$	$CDCl_3$
0.1-0.6	1.05	$-C(CH_3,OH,Ø)$	CCl_4
0.2-0.9	1.10	$-CH_2-NH_2 \cdot HCl$	D_2O
0.3-0.8	1.29	$-C(CH_3)=CH_2$	CCl_4
0.5-1.1	1.36	$-C(=O)-NH-NH_2$	$CDCl_3$
0.5-1.2	1.40	$-C(=O)-NH_2$	$CDCl_3$
0.6-1.1	1.50	$-C(=O)-O-CH_3$	CCl_4
0.7-1.3	1.53	$-C(=O)-OH$	CCl_4
0.6-1.2	1.79	$-C(=O)-NH-Ø$	Polysol
0.7-1.1	1.97	$-C(=O)-CH_3$	$CDCl_3$
0.4-1.2	2.00	$-C(=O)-C_3$	CCl_4
0.9-1.5	2.07	$-C(=O)-Cl$	CCl_4
0.2-0.7	2.31	$-NH_2$	$CDCl_3$
0.5-1.2	2.49	$-C(C_3)=N-OH$	$CDCl_3$
0.7-1.4	2.61	$-C(=O)-Ø$ (4–F)	$CDCl_3$
0.63, 1.80	3.35	$P^+(Ø)_3\ Br^-$	Polysol

With the exception of the last substituent on the list, the hydrogens bonded to the beta carbons produce a complex higher order pattern at high field.

Cyclobutane Derivatives (C_4–X)

$(CH_2)_3$–	CH	–X	
1.4-2.3	2.50	$-CH_2-OH$	$CDCl_3$
1.7-2.5	3.08	$-C(=O)-O-R_2$	CCl_4
1.6-2.7	3.19	$-C(=O)-OH$	$CDCl_3$
1.7-2.6	3.45	$-C(=O)-NH-C(=O)-R$	$CDCl_3$
1.6-2.7	3.60	$-C(=O)-Cl$	CCl_4
1.5-2.6	3.85	$-NH_2 \cdot HCl$	D_2O
1.6-2.7	3.95	$-C(=O)-Ø$	CCl_4
1.1-2.5	4.16	$-OH$	$CDCl_3$

Cyclopentane Derivatives $(C_5{-}X)$

$(CH_2)_4-$	CH	–X	Solvent
1.2-2.3	1.81	$-CH_2$	CCl_4
1.3-2.2	2.69	–C(=O)–OH	CCl_4
1.6-2.2	2.70	$-C(=O)-O-CH_3$	CCl_4
1.4-2.3	2.76	–C(=O)–NH–Ø	$CDCl_3$
1.3-2.2	2.90	–C(=S)–NH–R	$CDCl_3$
1.2-2.2	3.00	$-NH-CH_3$	$CDCl_3$
1.4-2.3	3.19	–C(=O)–Cl	CCl_4
1.0-2.1	3.31	$-NH_2$	$CDCl_3$
1.3-2.2	3.65	–C(=O)–Ø	CCl_4
1.1-2.2	4.19	–NH–C(=O)–R	$CDCl_3$
1.3-2.2	4.21	–OH	CCl_4
1.3-2.4	4.32	–I	CCl_4
1.4-2.3	4.35	–Cl	CCl_4
1.4-2.4	4.38	–Br	CCl_4
1.1-2.1	4.49	$-N(CH_3)-C(=O)-NH_2$	Polysol

Cyclohexane Derivatives $(C_6{-}X)$

$(CH_2)_5-$	CH	–X	
0.9-2.2	2.26	$-C(=O)-O-CH_2CH_3$	CCl_4
0.7-2.1	2.32	$-NH-R_2$	CCl_4
0.9-2.2	2.34	–C(=O)–NH–Ø	Polysol
0.6-2.1	2.35	$-NH-CH_3$	$CDCl_3$
0.8-2.1	2.39	$-N(R_2)_2$	CCl_4
0.8-2.2	2.40	$-NH-CH_2-CH_2-SH$	$CDCl_3$
0.9-2.1	2.40	–Ø	CCl_4
0.8-2.1	2.42	$-C(=O)-C_6$	CCl_4
0.5-2.2	2.42	$-NH-CH_2-CH{=}CH_2$	CCl_4
0.7-2.1	2.49	$-NH-C_6$	CCl_4
0.6-2.1	2.64	$-NH_2$	CCl_4
1.2-2.1	2.64	–C≡N	CCl_4
1.0-2.4	2.71	–C(=O)–Cl	CCl_4
1.0-2.2	2.74	–SH	CCl_4
1.0-2.5	2.92	$-SO_2-R$	$CDCl_3$

Cyclohexane Derivatives continued $(C_6{-}X)$

$(CH_2)_5-$	CH	$-X$	Solvent
0.7-2.2	3.00	$-NH-SO_2-NH-C_6$	Polysol
0.8-2.1	3.10	$-NH-SO_2-Ø$	$CDCl_3$
0.7-2.2	3.17	$-N{=}C{=}N-C_6$	$CDCl_3$
0.6-2.3	3.21	$-NH-Ø$	$CDCl_3$
0.8-2.3	3.25	$-C(=O)-Ø$	$CDCl_3$
0.7-2.2	3.27	$-O-R$	$CDCl_3$
0.6-2.2	3.36	$-NH-C(=O)-NH_2$	Polysol
0.9-2.2	3.46	$-N(Ø)-CH_3$	$CDCl_3$
1.0-2.2	3.48	$-N{=}C{=}O$	$CDCl_3$
0.8-2.1	3.49	$-OH$	CCl_4
1.0-2.2	3.60	$-N{\equiv}C$	$CDCl_3$
0.7-2.3	3.70	$-NH-C(=O)-S-CH_3$	$CDCl_3$
0.8-2.2	3.79	$-NH-C(=O)-H$	$CDCl_3$
0.9-2.8	3.80	$-NH-SO_2-OH$	TFA
0.8-2.4	3.95	$-Cl$	CCl_4
0.7-2.3	4.03	$-NH-C(=S)-R$	DMSO-d_6
0.9-2.2	4.04	$-O-B(-O-C_6)_2$	$CDCl_3$
1.1-2.5	4.13	$-Br$	CCl_4
1.0-2.5	4.29	$-NO_2$	CCl_4
0.9-2.5	4.36	$-I$	$CDCl_3$
0.9-2.2	4.71	$-O-C(=O)-NH-Ø$	CCl_4
1.0-2.1	4.76	$-O-C(=O)-CH_3$	$CDCl_3$
0.9-2.3	3.75, 4.83	$-N(N{=}O)-C_6$	$CDCl_3$
0.9-2.2	4.89	$-O-C(=O)-H$	$CDCl_3$
1.0-2.2	4.95	$-O-C(=O)-CF_3$	CCl_4
1.1-2.4	5.16	$-O-C(=O)-Ø$	$CDCl_3$

Cycloheptane Derivatives

$-(CH_2)_4-$	$(CH_2)_2$	CH	–X	Solvent
1.2-1.9	1.80	2.53	–C(=O)–OH	$CDCl_3$
1.3-2.3	1.3-2.3	2.78	–C≡N	CCl_4
1.0-2.2	1.0-2.2	2.92	$-NH_2$	$CDCl_3$
1.2-2.2	1.2-2.2	3.80	–OH	$CDCl_3$
ca 1.66	2.10	3.85	$-NH-C(=O)-NH_2$	TFA
1.2-1.8	1.90	3.86	–NH–C(=O)–NH–X	$CDCl_3$
ca 1.61	2.00	4.10	–Cl	CCl_4
ca 1.53	2.17	4.33	–Br	$CDCl_3$

Cyclooctane Derivatives

$-(CH_2)_5-$	$(CH_2)_2$	CH	–X	
1.2-2.0	1.2-2.0	2.53	$-NH-CH_3$	$CDCl_3$
1.1-2.1	1.1-2.1	2.90	$-NH_2$	CCl_4
1.2-2.1	1.2-2.1	3.72	–OH	CCl_4
1.4-2.2	1.4-2.2	3.88	$-NH-C(=O)-NH_2$	TFA
ca 1.59	2.19	4.31	–Br	CCl_4
1.3-2.1	1.3-2.1	4.83	$-O-C(=O)-CH_3$	CCl_4
ca 1.59	1.85	5.09	–O–C(=O)–H	CCl_4
ca 1.59	1.85	5.15	$-O-C(=O)-CF_3$	CCl_4

ISOPROPYLCYCLOPROPANE

C_6H_{12}

Mol. Wt. 84.16

Solvent: CCl_4

Assignments
a = -0.02
b = 0.31
c = ca 0.90
$CHCl_3$ used as internal reference

38

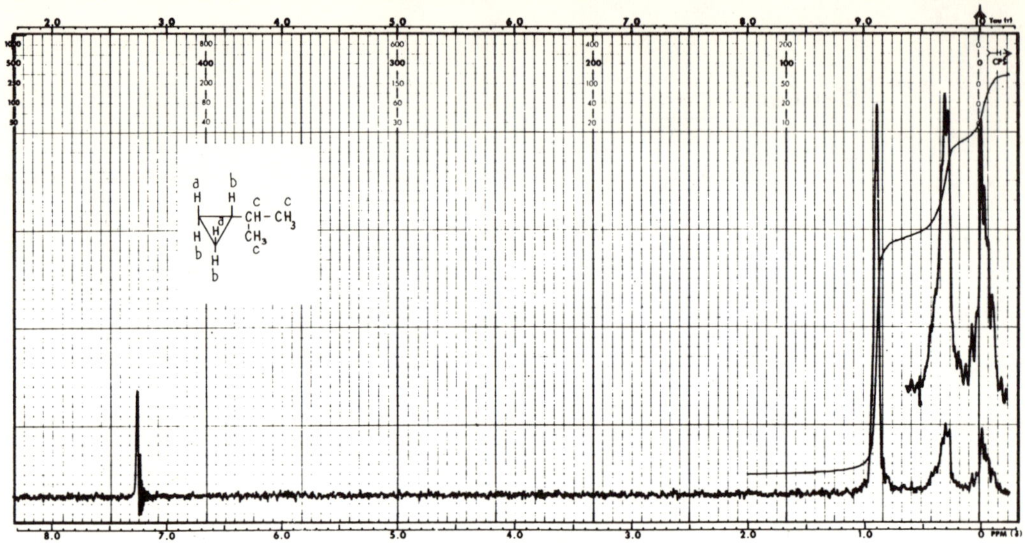

© 1978 Sadtler Research Laboratories, Inc.

cis-1,2-DIMETHYLCYCLOPROPANE

C_5H_{10}

Mol. Wt. 70.14

Solvent: CCl_4

Assignments
a = -0.11
b = 0.75
c = ca 0.90
d = 1.15
$CHCl_3$ used as internal reference

39

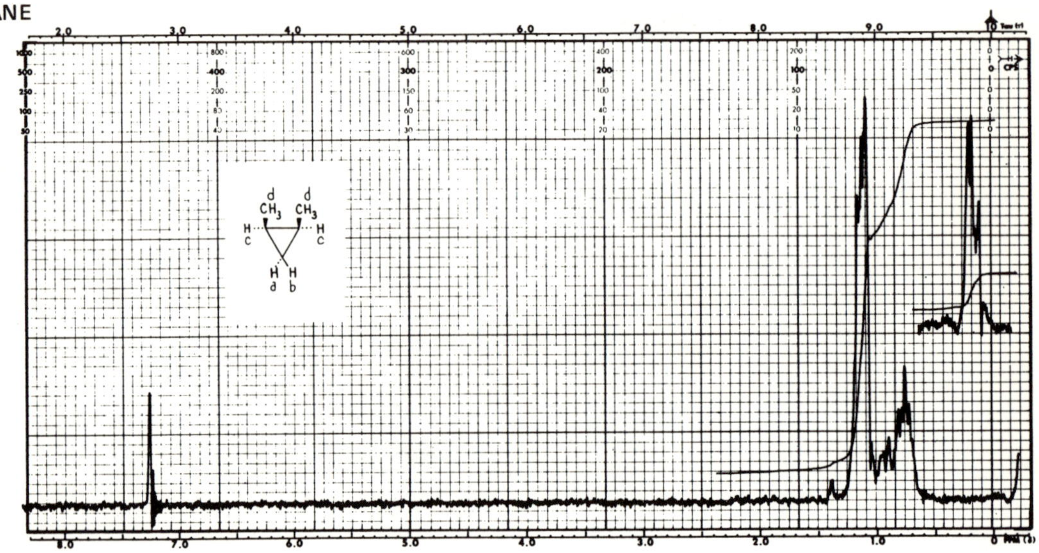

© 1978 Sadtler Research Laboratories, Inc.

1,1,2-TRIMETHYLCYCLOPROPANE

C_6H_{12}

Mol. Wt. 84.16

Solvent: CCl_4

Assignments
a = -0.05
b = 0.33-0.97
c = 1.14
$CHCl_3$ used as internal reference

40

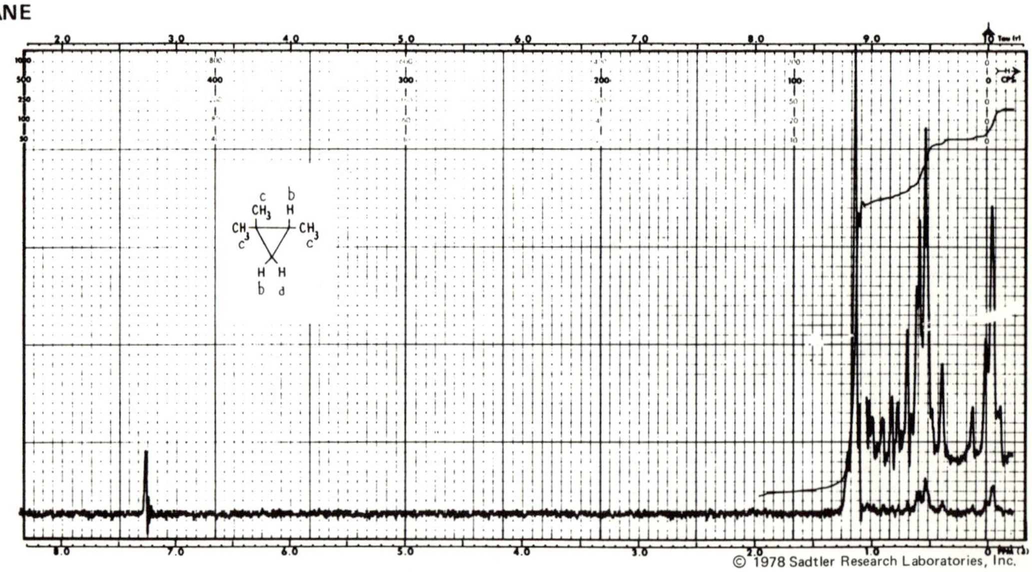

© 1978 Sadtler Research Laboratories, Inc.

DICYCLOPROPYLMETHANE

C_7H_{12}

Mol. Wt. 96.17

Solvent: CCl_4

Assignments
a = 0.03
b = 0.41
c = ca 0.75
d = 1.19
$CHCl_3$ used as the internal reference

41

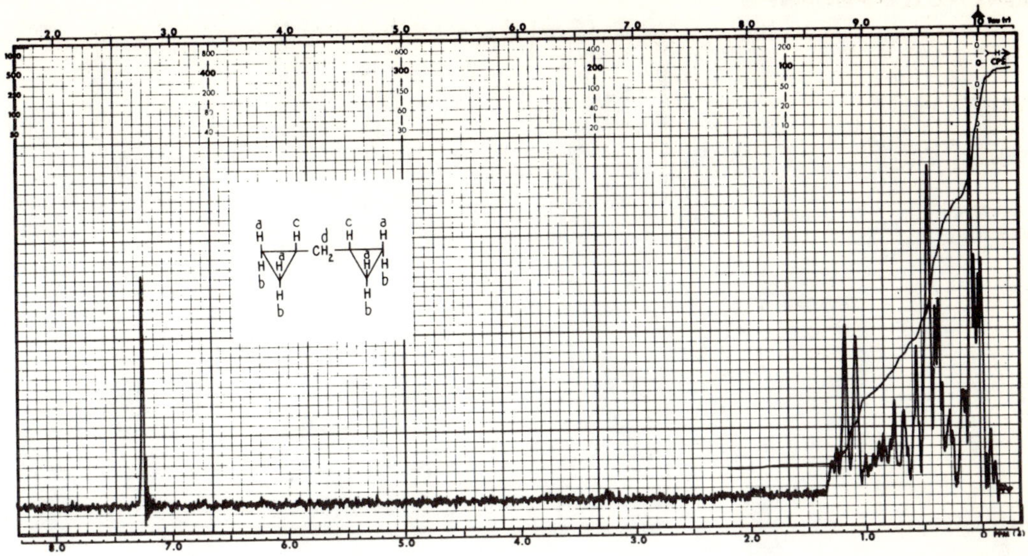

© 1978 Sadtler Research Laboratories, Inc.

CYCLOPENTANE

C_5H_{10}

Mol. Wt. 70.14

B.P. 49.3°C

Solvent: CCl_4

Assignments
a = 1.50

42

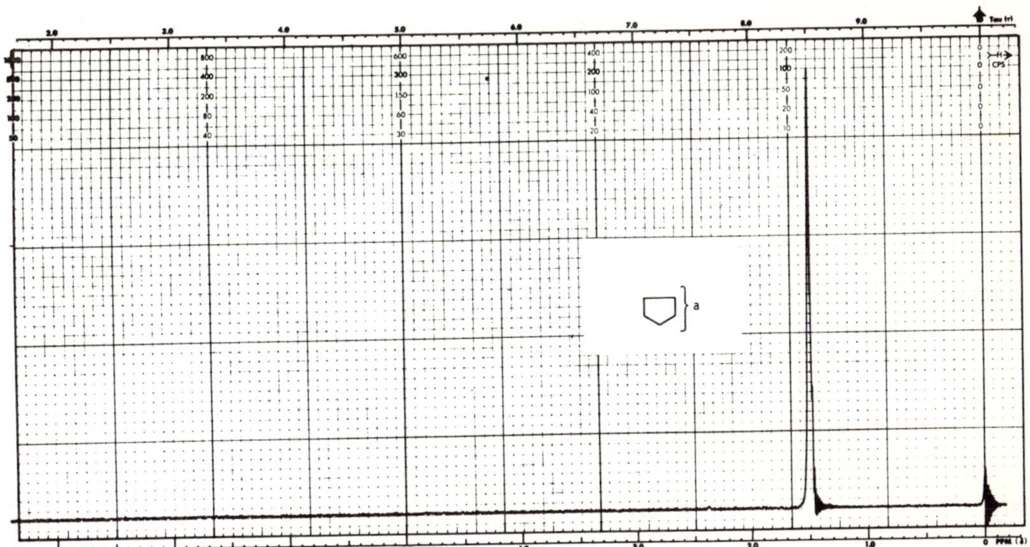

© 1978 Sadtler Research Laboratories, Inc.

METHYLCYCLOPENTANE

C_6H_{12}

Mol. Wt. 84.16

B.P. 71.8°C

Solvent: CCl_4

Assignments
a = 0.97
b = ca 1.60
c = ca 1.81

43

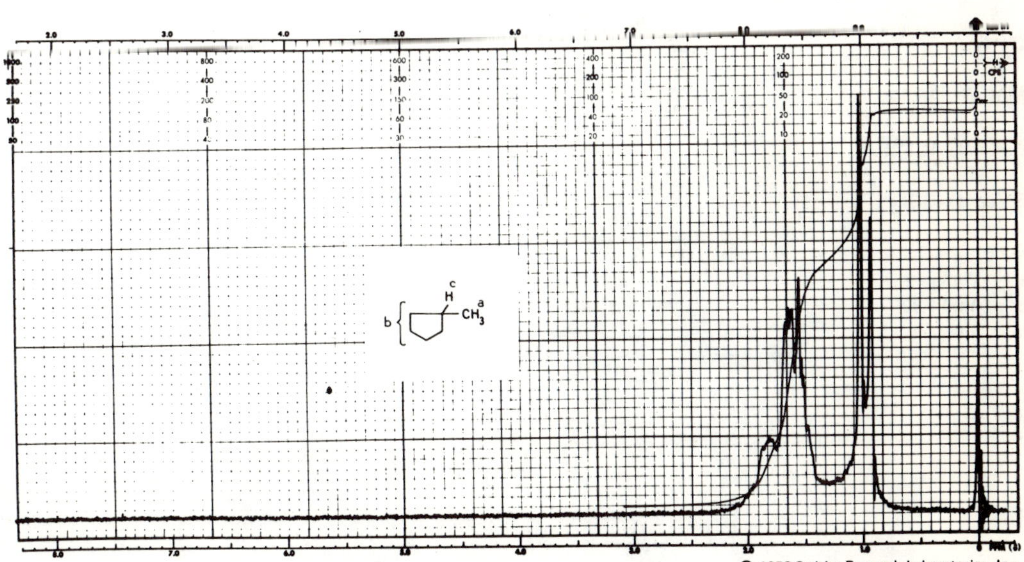

© 1978 Sadtler Research Laboratories, Inc.

PENTYLCYCLOPENTANE

$C_{10}H_{20}$

Mol. Wt. 140.27

Solvent: CCl_4

Assignments
a = 0.90
b = 1.30
c = 1.40-2.00

44

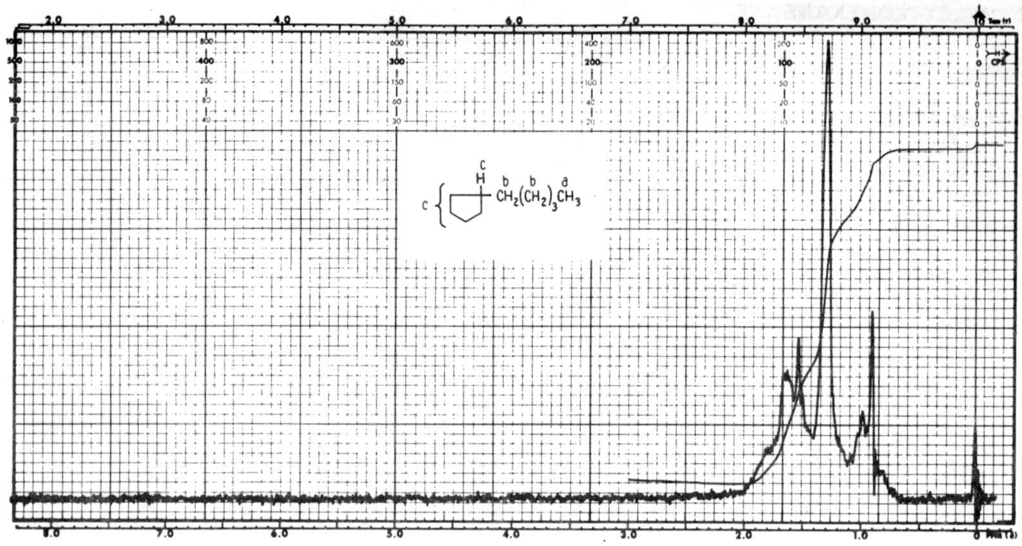

© 1978 Sadtler Research Laboratories, Inc.

CYCLOHEXANE

C_6H_{12}

Mol. Wt. 84.16

M.P. 5-6°C

Solvent: CCl_4

Assignments
a = 1.42

45

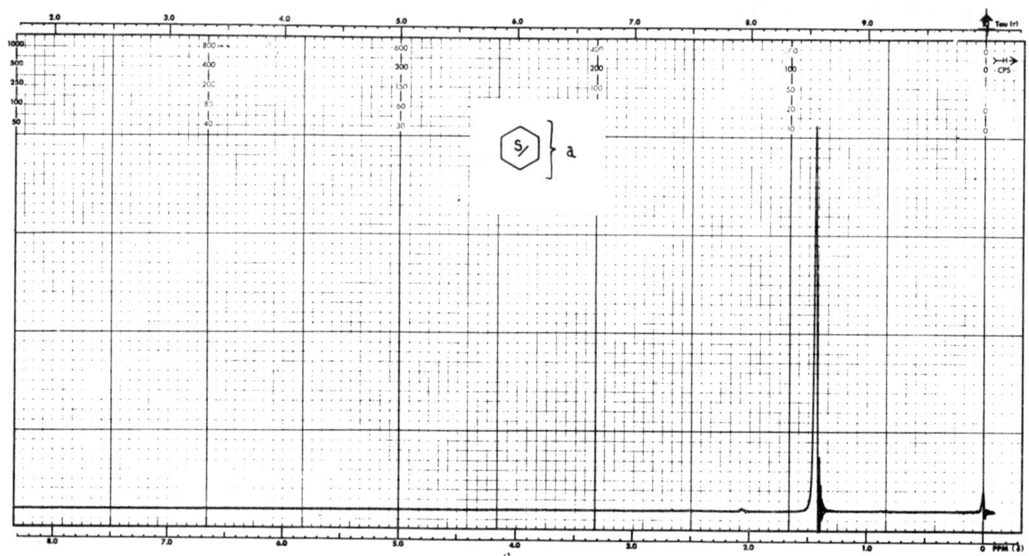

© 1978 Sadtler Research Laboratories, Inc.

METHYLCYCLOHEXANE

C_7H_{14}

Mol. Wt. 98.19

B.P. 100-101°C

Solvent: CCl_4

Assignments
a = 0.90
b = 1.00-1.80
c = 2.06 imp.

46

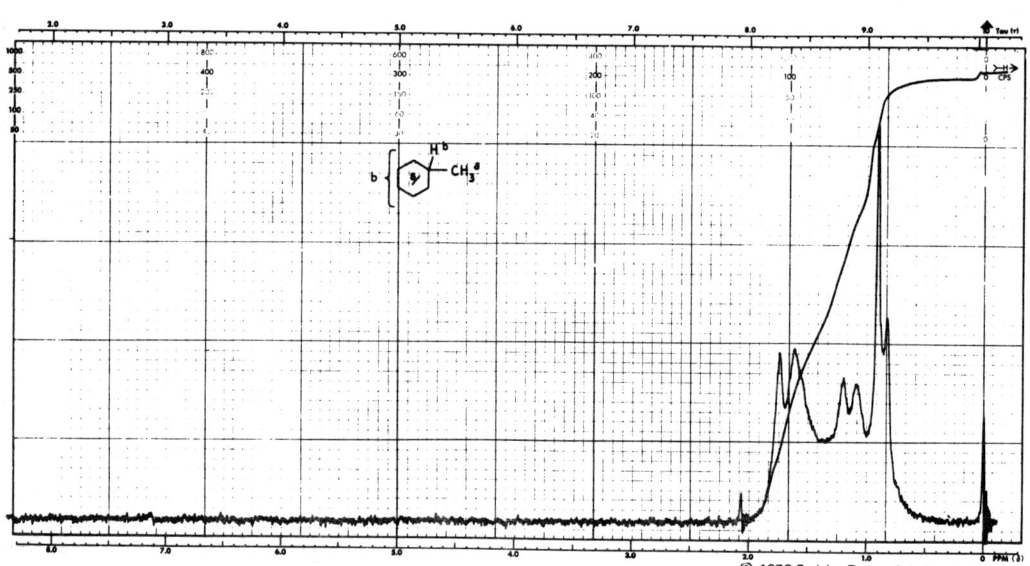

© 1978 Sadtler Research Laboratories, Inc.

ETHYLCYCLOHEXANE

C_8H_{16}

Mol. Wt. 112.21

B.P. 130-132°C

Solvent: CCl_4

Assignments
a = 0.87
b = 0.90-1.90
c = 2.03 imp.

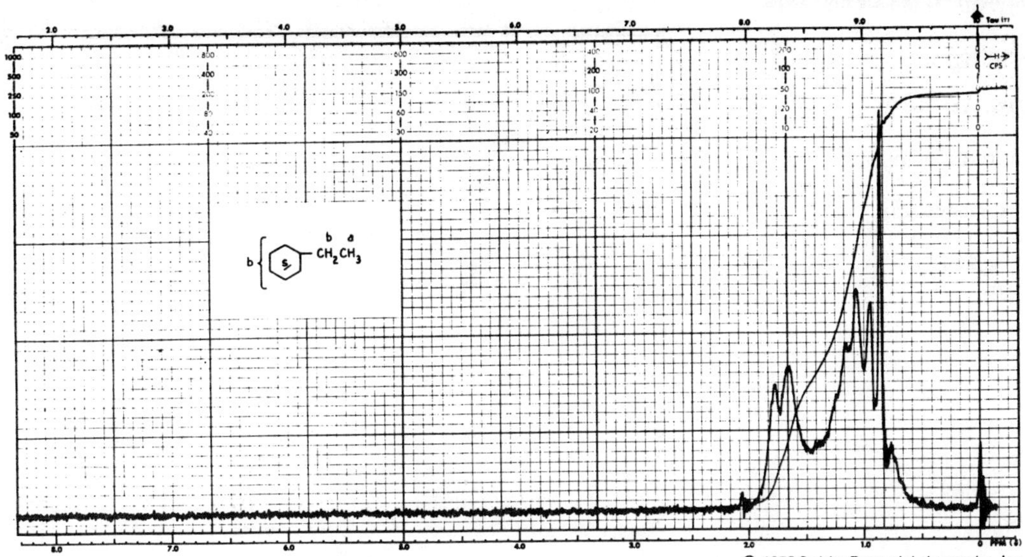

47

1-CYCLOHEXYL-2-METHYLPROPANE

$C_{10}H_{20}$

Mol. Wt. 140.27

Solvent: CCl_4

Assignments
a = 0.89
b = 0.70-2.00

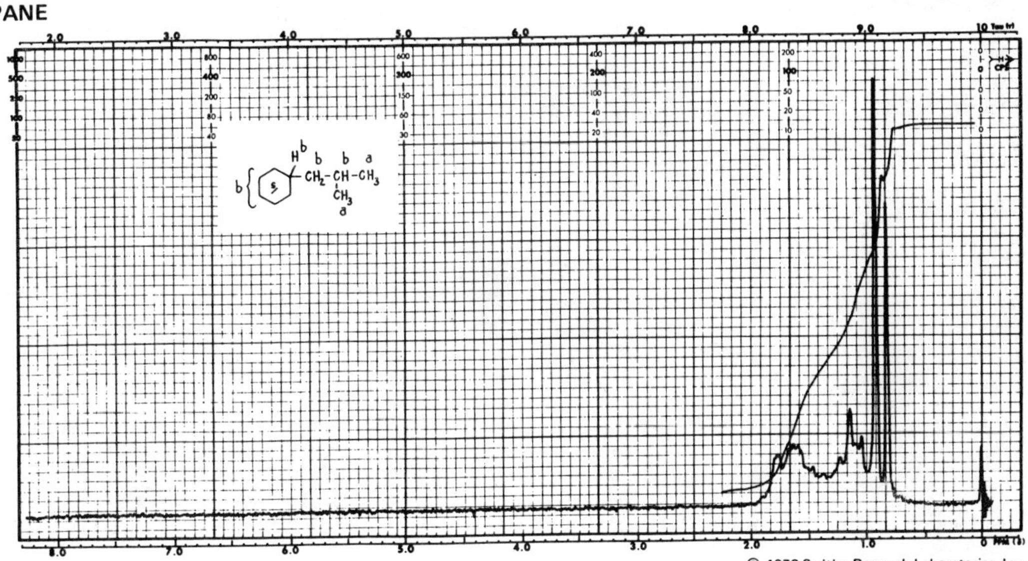

48

BICYCLOHEXYL

$C_{12}H_{22}$

Mol. Wt. 166.31

Solvent: CCl_4

Assignments
a = 0.71-2.00

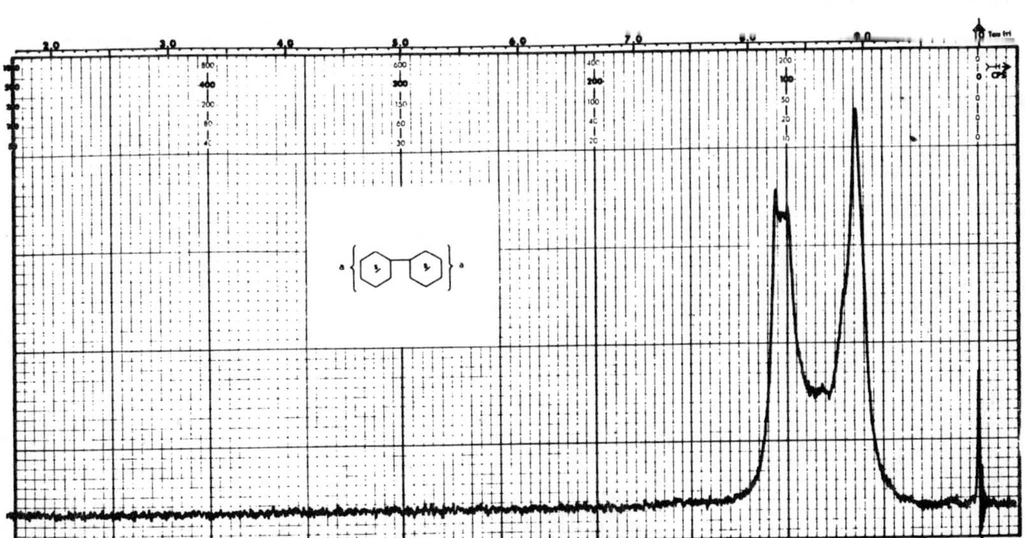

49

CYCLOHEPTANE

C_7H_{14}

Mol. Wt. 98.19

B.P. 118.5°C

Solvent: $CDCl_3$

Assignments
a = 1.53

50

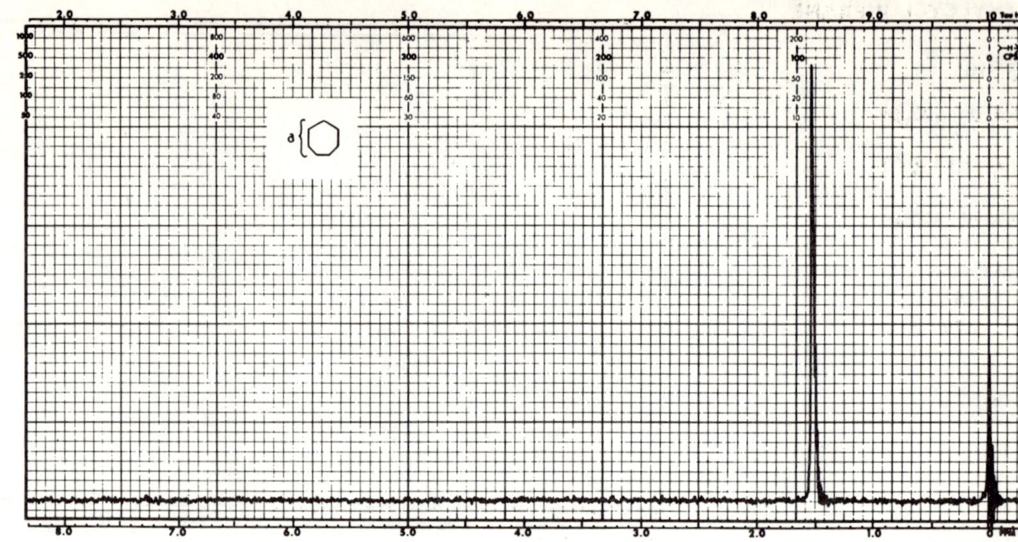

© 1978 Sadtler Research Laboratories, Inc.

CYCLOOCTANE

C_8H_{16}

Mol. Wt. 112.22

Solvent: CCl_4

Assignments
a = 1.52

51

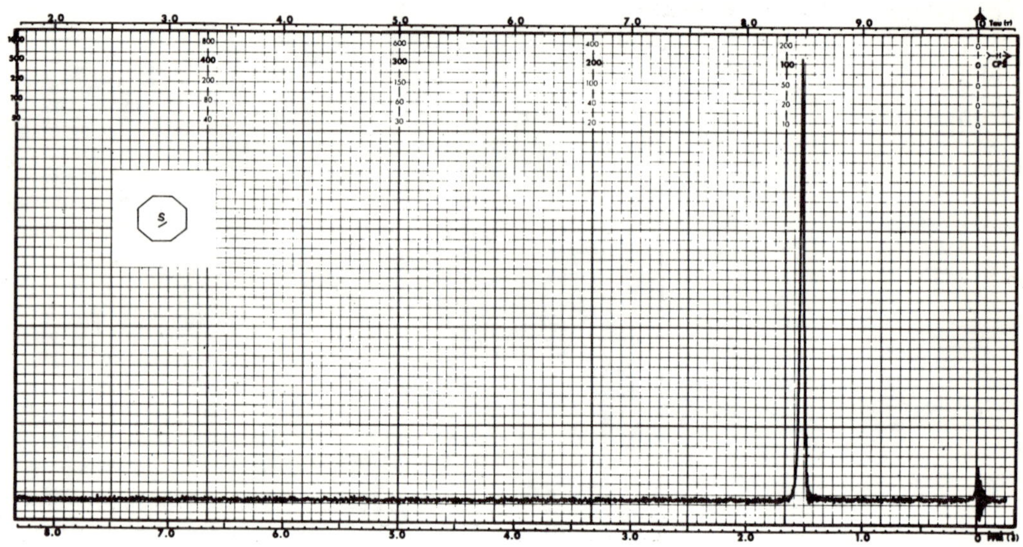

© 1978 Sadtler Research Laboratories, Inc.

trans-DECAHYDRONAPHTHALENE

$C_{10}H_{18}$

Mol. Wt. 138.25

B.P. 185°C

Solvent: CCl_4

Assignments
a = 0.80-1.90

52

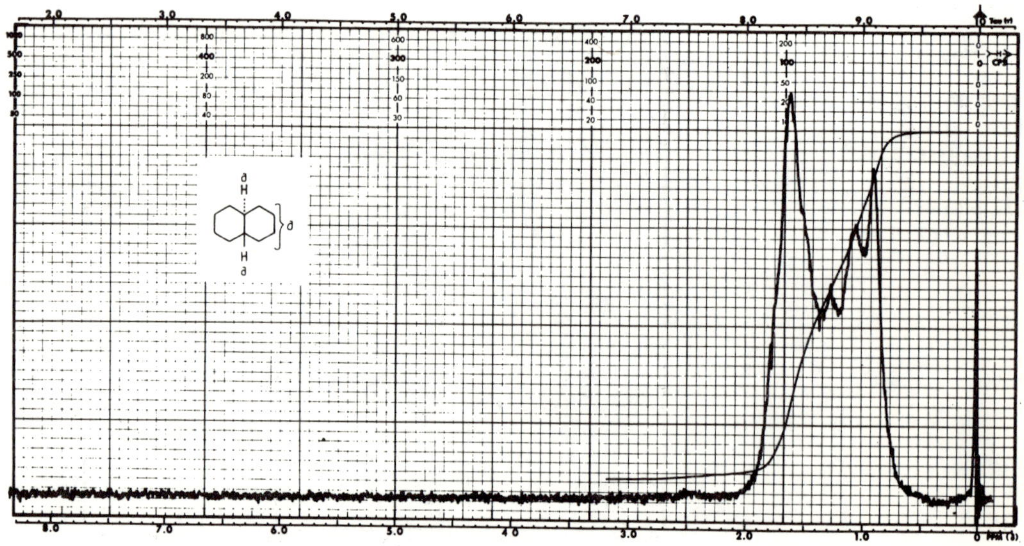

© 1978 Sadtler Research Laboratories, Inc.

DODECAHYDROFLUORENE

$C_{13}H_{22}$

Mol. Wt. 178.32

Solvent: $CDCl_3$

Assignments

a = 0.93-2.17

53

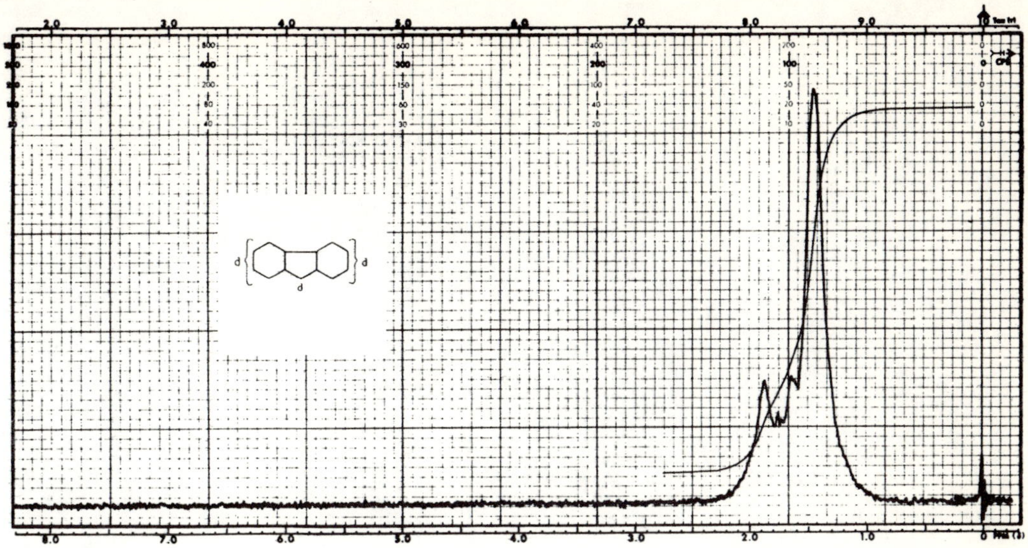

© 1978 Sadtler Research Laboratories, Inc.

NOTES

THE ACYCLIC ALKENES

No other type of organic compound produces such a wide variety of multiplet types over such a large chemical shift range as the acyclic alkenes. Proton-proton coupling through four bonds is common. Many samples are found to contain both cis and trans isomers producing a spectrum more complex than might be expected from a proposed structure.

The olefinic double bond is a weak deshielder of both aliphatic and aromatic hydrogens often resulting in higher order, overlapping multiplets.

The aliphatic chemical shift ranges below were abstracted from a large number of representative compounds.

Chemical Shifts

Aliphatic Protons (General ranges)

HC–3	HC–2	HC–1		Solvent
		1.6–2.0	CH_3–C=C	CCl_4, $CDCl_3$
		(1.6–1.8)	$(CH_3)_2$–C=C	CCl_4, $CDCl_3$
	0.9–1.3	1.9–2.1	CH_3–CH_2–C=C	CCl_4, $CDCl_3$
0.9–1.0	1.3–1.7	1.9–2.2	CH_3–CH_2–CH_2–C=C	CCl_4, $CDCl_3$
	(0.9–1.2)	1.9–2.7	$(CH_3)_2$–CH–C=C	CCl_4, $CDCl_3$
	(0.9–1.2)	- - - -	$(CH_3)_3$–C–C=C	CCl_4, $CDCl_3$

Coupling and Coupling Constants

The wide variety of coupling constants observed in the spectra of the alkenes is quite helpful in determining the molecular arrangement of such structures and aid the analyst in differentiating the spectra of cis and trans isomers. The general coupling constant ranges provided below are the values observed for a large number of alkene compounds.

CH=CH	trans	J	=	13–17 Hz
CH=CH	cis	J	=	6–14 Hz
C=CH–CH		J	=	4–8 Hz
H_2C=C	geminal	J	=	0–3 Hz
HC–C=CH		J	=	0–2 Hz

Vinyl Compounds

CH_2=CH–X

trans	cis	HC–2	–X	Solvent
5.10	5.01	5.79	–CH(R_2)–OH	CCl_4
4.82	4.78	5.80	–C(CH_3)$_3$	CCl_4
5.53	6.20	6.00	–C(=O)–NH–R	$CDCl_3$
5.02	5.10	6.34	–C(CH_3)=C$\dot{H}_2$	CCl_4
5.12	5.53	6.60	–Ø (4–F)	$CDCl_3$
5.04	5.50	6.62	–Ø (4-O–CH_3)	CCl_4
4.43	4.74	7.18	–O–C(=O)–CH_3	CCl_4
4.49	4.80	7.23	–O–C(=O)–R_{11}	$CDCl_3$

CH_2=C(CH_3)–X

trans or	cis	(CH_3)	–X	
4.60	4.60	1.62	–C_3	CCl_4
4.60	4.60	1.68	–CH(CH_3)$_2$	CCl_4
4.70	4.89	1.70	–CH(CH_3)–OH	CCl_4
4.93	4.93	1.82	–C$\dot{H}$=CH_2	CCl_4
5.39	5.71	1.89	–C(=O)–O–Na	D_2O
4.59	4.77	1.90	–C(CH_3)$_3$	CCl_4
5.47	6.02	1.90	–C(=O)–O–R_{18}	CCl_4
5.47	6.00	1.91	–C(=O)–O–CH_3	CCl_4
5.60	6.20	1.95	–C(=O)–OH	CCl_4
5.40	5.79	1.97	–C(=O)–NH_2	$CDCl_3$
ca 5.79		1.98	–C≡N	CCl_4
5.68	6.30	2.03	–C(=O)–O–Ø	$CDCl_3$
5.01	5.31	2.10	–Ø	CCl_4

1-Propenes

CH_2=CH	$-CH_2$	$-X$	Solvent
4.6–6.0	1.91	$-C(CH_3)_3$	CCl_4
	1.93	$-CH(CH_3)_3$	CCl_4
	2.07	$-R_6$	$CDCl_3$
	2.17	$-C(CH_3)_2-OH$	CCl_4
	2.79	$-CH=CH_2$	$CDCl_3$
	3.10	$-C\equiv N$	CCl_4
	3.12	$-C(=O)-OH$	$CDCl_3$
	3.29	–S–S–R	CCl_4
	3.30	$-NH_2$	$CDCl_3$
	3.33	–Ø	CCl_4
	3.63	–1(Naphthyl)	CCl_4
	3.80	–I	CCl_4
	3.88	–Br	CCl_4
	3.99	–Cl	CCl_4
	4.02	$-O-CH_2CH_2-O-R$	$CDCl_3$
	4.05	–OH	CCl_4
	4.11	–NH–C(=S)–NH–R	$CDCl_3$
	4.48	–O–C(=O)–R	CCl_4
	4.60	–O–C(=O)–CH=CH–Ø	CCl_4
	4.62	$-O-$Ø$(2-C(=O)-NH_2)$	$CDCl_3$
	4.82	–O–2(Pyridyl)	CCl_4

2-Methyl Propenes

CH_2=	$C(CH_3)$	$-CH_2$	$-X$	
	1.68	1.85	$-CH(CH_3)_2$	CCl_4
	1.75	1.90	$-C(CH_3)_3$	$CDCl_3$
	1.69	1.98	$-R_2$	CCl_4
	1.69	2.01	$-CH_3$	CCl_4
	1.70	2.00	$-R_5$	CCl_4
	1.62	2.11	$-CH_2-CH=CH_2$	$CDCl_3$
	1.71	2.13	$-CH_2-C(CH_3)=CH_2$	CCl_4
	1.69	2.34	$-CH_2-OH$	$CDCl_3$
	1.67	2.37	$-CH_2-$Ø	CCl_4

Chemical Shifts continued

2-Methyl Propenes continued

$CH_2=C(CH_3)$	$-CH_2$	$-X$	Solvent
1.71	2.78	$-N(-CH_2-C(CH_3)=CH_2)_2$	CCl_4
1.73	3.12	$-NH_2$	CCl_4
1.87	3.98	$-Cl$	CCl_4
1.79	4.36	$-O-Ø$	CCl_4

1-Butenes

CH_2 trans	CH_2 cis	$=CH$	$-CH_2$	$-CH_2$	$-X$	
4.88	4.92	5.68	1.99	1.32	$-CH_3$	CCl_4
4.89	4.86	5.69	2.00	1.45	$-R_{11}$	CCl_4
4.87	4.91	5.68	2.05	1.45	$-R_4$	CCl_4
4.91	4.96	5.70	2.11	2.11	$-CH=CH_2$	CCl_4
4.97	5.01	5.81	2.37	2.67	$-Ø$	$CDCl_3$
5.10	5.14	5.80	2.37	3.65	$-OH$	$CDCl_3$
5.02	5.08	5.80	2.40	2.40	$-C(=O)-OH$	$CDCl_3$

2-Butenes

	CH_3	$-CH$	$=CH$	$-CH_2$	$-X$	
	1.59	5.33	5.33	1.96	$-R_2$	CCl_4
	1.59	5.31	5.31	1.97	$-R_3$	CCl_4
	1.60	5.34	5.34	1.98	$-CH_3$	CCl_4
	1.68	5.47	5.47	3.27	$-Ø$	CCl_4

X–	CH_2	$-CH$	$=CH$	$-CH_2$	$-X$	
$(R_2)_2N-$	3.07	5.63	5.63	3.07	$-N(R_2)_2$	$CDCl_3$
Br–	3.99	6.01	6.01	3.07	$-Br$	$CDCl_3$
HO–	4.18	5.72	5.72	4.18	$-OH$	D_2O

CH_3	$-CH$	$=CH$	$-X$	
1.53	4.27	5.82	$-O-R_2$	CCl_4
1.74	5.35	5.35	$-C(CH_3)_3$	$CDCl_3$
2.03	6.62	5.38	$-C\equiv N$	$CDCl_3$

Olefinic Protons (CH_3–CH=CH–X)

CH_3	−CH	=CH	−X (all trans)	Solvent
1.53	4.68	6.12	−O−R_2	CCl_4
1.61	5.32	5.32	−CH$(CH_3)_2$	CCl_4
1.63	5.40	5.40	−R_4	$CDCl_3$
1.63	5.58	5.58	−C$(CH_3)_3$	$CDCl_3$
1.78	5.91	6.21	−Ø (4−O−CH_3)	CCl_4
1.80	6.00	6.34	−Ø	CCl_4
1.79	6.08	6.37	−Ø	$CDCl_3$
1.81	6.59	5.84	−C(=O)−O−K	D_2O
1.92	6.62	5.38	−C≡N	$CDCl_3$
1.89	6.70	6.00	−C(=O)−CH_3	CCl_4
2.00	6.80	6.04	−C(=O)−H	CCl_4
1.90	7.01	5.84	−C(=O)−O−O−R	$CDCl_3$
1.93	7.10	5.91	−C(=O)−O−R	CCl_4

Ø −CH	=CH	−X	Cis/trans	
7.35	5.53	−C≡N	cis	$CDCl_3$
7.71	5.86	−C≡N	trans	$CDCl_3$
6.21	5.91	−CH_3	trans	CCl_4
6.48	6.11	−Cl	cis	CCl_4
6.73	6.42	−Cl	trans	CCl_4
6.57	6.31	−CH_2−OH	trans	$CDCl_3$
7.56	6.31	−C(=O)−O−R	trans	CCl_4
7.60	6.38	−C(=O)−OH	trans	DMSO-d6
7.80	6.43	−C(=O)−OH	trans	$CDCl_3$
7.71	6.45	−C(=O)−N=C=S	trans	$CDCl_3$
7.02	6.61	−Br	trans	CCl_4
7.44	6.67	−C(=O)−CH_3	trans	CCl_4
7.52	6.67	−C(=O)−R	trans	$CDCl_3$
7.60	6.82	−SO_2−CH=CH−Ø	trans	$CDCl_3$
7.79	6.91	−C(=O)−H	trans	DMSO-d6
7.70	7.02	−C(=O)−CH=CH−Ø	trans	$CDCl_3$
7.79	7.29	−SO_2−Cl	trans	$CDCl_3$
7.97	7.45	−NO_2	trans	$CDCl_3$

X–	CH	=CH	–X	Solvent
CH_3–	5.32	5.32	–$C(CH_3)_3$	$CDCl_3$ cis
R_7–	5.34	5.34	–R_7	CCl_4
R_2–	5.37	5.37	–R_2	CCl_4 trans
CH_3–	5.45	5.45	–$C(CH_3)_3$	$CDCl_3$ trans
CH_3–O–C(=O)–	5.79	6.98	–R_4	$CDCl_3$
Ø–C(=O)–	5.89	7.46	–NH–Ø	$CDCl_3$ cis
2-(Naphthyl)–C(=O)–	6.00	7.64	–$N(R)_2$	$CDCl_3$ trans
R–O–C(=O)–	6.24	6.48	–C(=O)–NH–C(=O)–NH_2	Polysol cis
R–O–C(=O)–	6.25	6.25	–C(=O)–O–R	CCl_4 cis
HO–C(=O)–	6.32	6.82	–C(=O)–NH–Ø	$CDCl_3$
CH_3–O–C(=O)–	6.59 or	7.03	–C(=O)–CH_3	$CDCl_3$
Br–	6.64	6.64	–Br	CCl_4 cis
R_4–O–C(=O)–	6.83	6.83	–C(=O)–O–R_4	$CDCl_3$ trans
Br–	7.04	7.04	–Br	CCl_4 trans

Aromatic Protons

The aromatic patterns produced by the phenyl protons of alkene substituted benzenes are represented by a wide variety of chemical shifts and patterns depending on the site and type of substituents on the C=C moiety. A relatively sharp single peak may be observed as high as 7.13 ppm (Cl–CH=CH–, trans) or as low as 7.54 ppm (Cl–SO_2–CH=CH–). Generally, a complex band of overlapping multiplets is observed in the range from 7.1--7.6 ppm (CH_2=CH–, CH_2=C(CH_3)–, R–C(=O)–CH=CH–). In at least one case, the ortho aromatic hydrogens are strongly deshielded in relation to the meta and para hydrogens producing two sets of bands, one at 7.55 and a range from 7.1-7.4 ppm (Cl–CH=CH–, cis).

A series of para substituted styrene derivatives is presented below.

X–	ortho	meta	para	
CH_3–O–	6.65	7.08	–CH=CH–CH_3	CCl_4
F–	6.90	7.23	–CH=CH_2	$CDCl_3$
CH_3–O–	6.76	7.22	–CH=CH_2	CCl_4
Cl–	7.21	7.21	–CH=CH_2	CCl_4
Cl–	7.23	7.23	–C(CH_3)=CH_2	$CDCl_3$
Br–	7.34	7.12	–CH=CH_2	CCl_4
Cl–	7.32	7.47	–CH=CH–C(=O)–R	$CDCl_3$
$(CH_3)_2$CH–	7.28	7.68	–CH=CH–C(=O)–Ø	Polysol
NO_2–	8.07	7.49	–CH=CH–CH_2–OH	Polysol

3,3-DIMETHYL-1-BUTENE

C_6H_{12}

Mol. Wt. 84.16

Solvent: CCl_4

Assignments
a = 1.00
b = 4.77
c = 4.82
d = 5.80

54

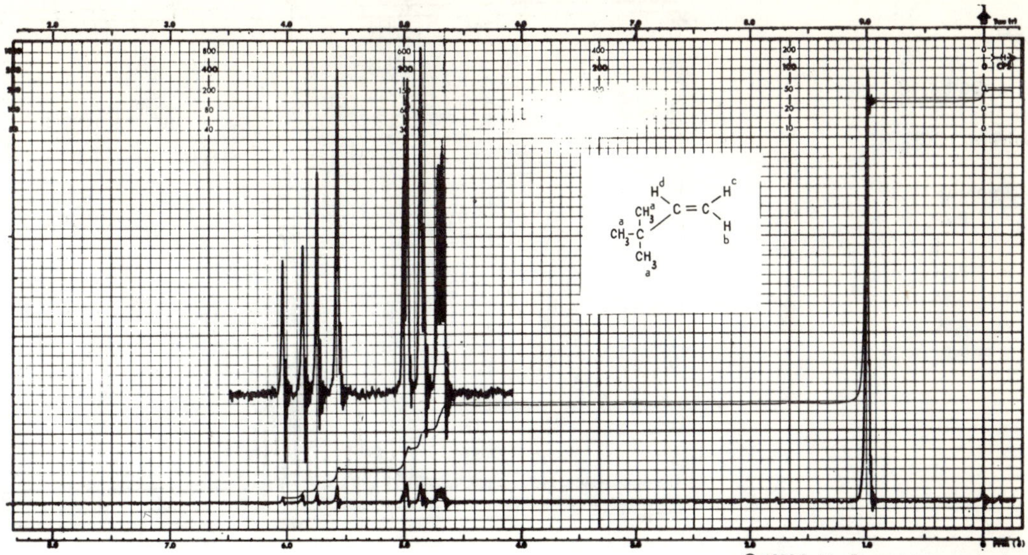

© 1978 Sadtler Research Laboratories, Inc.

3-METHYL-1-PENTENE

C_6H_{12}

Mol. Wt. 84.16

Solvent: CCl_4

Assignments
a = 0.88
b = 0.97
c = 1.31
d = 2.06
e = 4.88
f = 4.91
g = 5.70

55

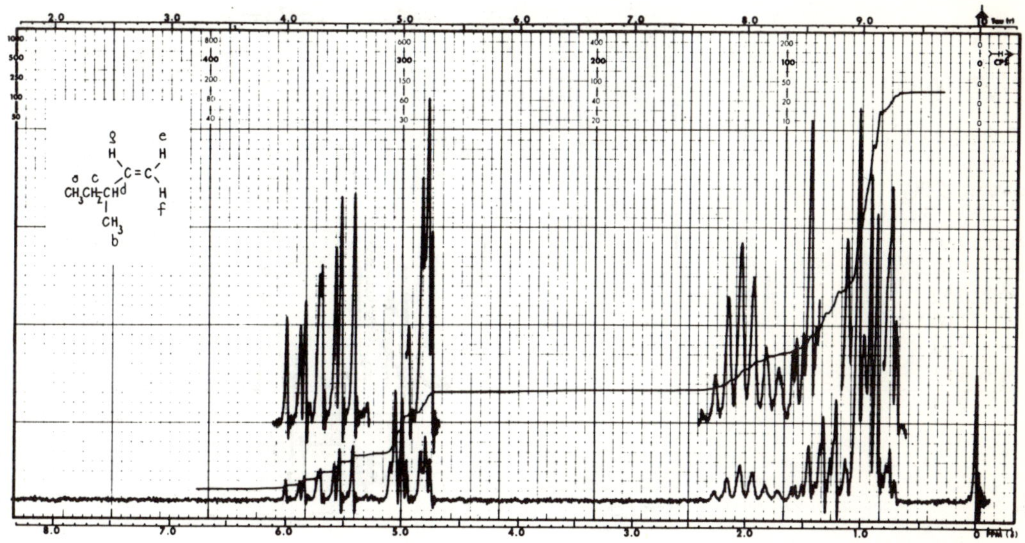

© 1978 Sadtler Research Laboratories, Inc.

3-ETHYL-1-PENTENE

C_7H_{14}

Mol. Wt. 98.19

Solvent: $CDCl_3$

Assignments
a = 0.86
b = 1.10-2.10
c = 4.90
d = 4.90
e = 5.60

56

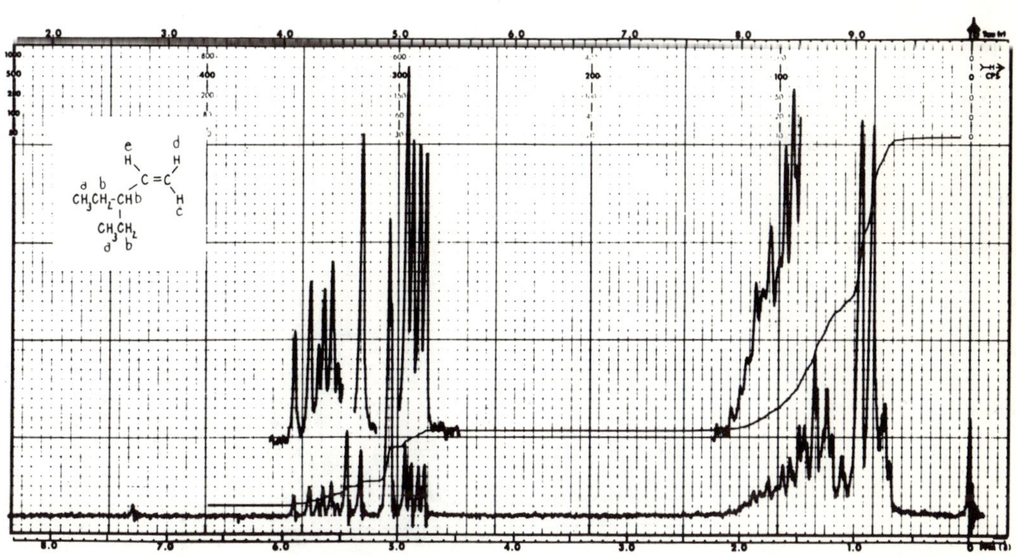

© 1978 Sadtler Research Laboratories, Inc.

1-HEXENE

C_6H_{12}

Mol. Wt. 84.16

Solvent: CCl_4

Assignments

a = 0.89
b = 1.35
c = 2.04
d = 4.86
e = 4.90
f = ca 5.67

57

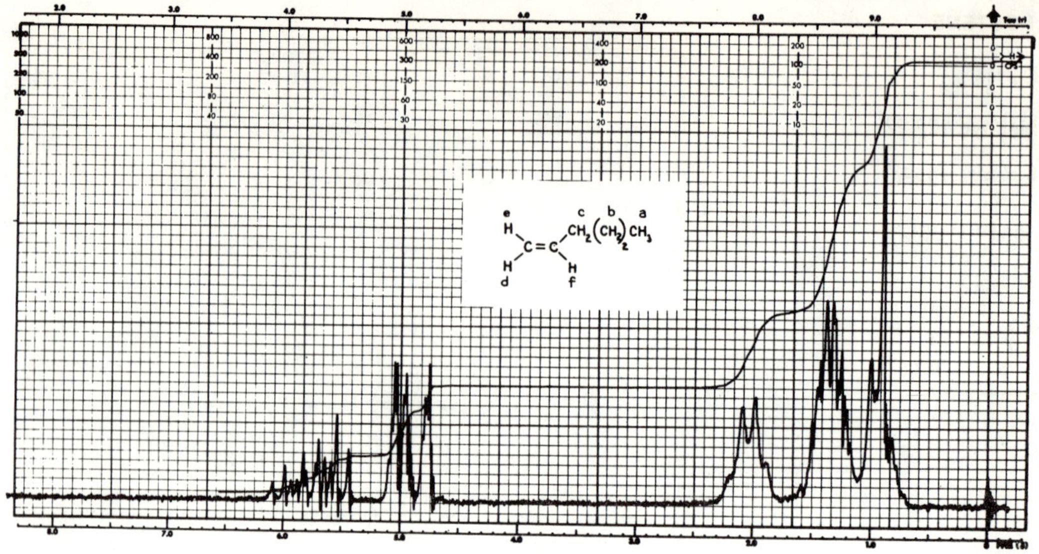

© 1978 Sadtler Research Laboratories, Inc.

4-METHYL-1-HEXENE

C_7H_{14}

Mol. Wt. 98.19

B.P. 87.2-87.5°C

Solvent: CCl_4

Assignments

a = 0.89
b = 0.92
c = 1.00-1.80
d = 1.98
e = ca 4.91
f = 5.72

58

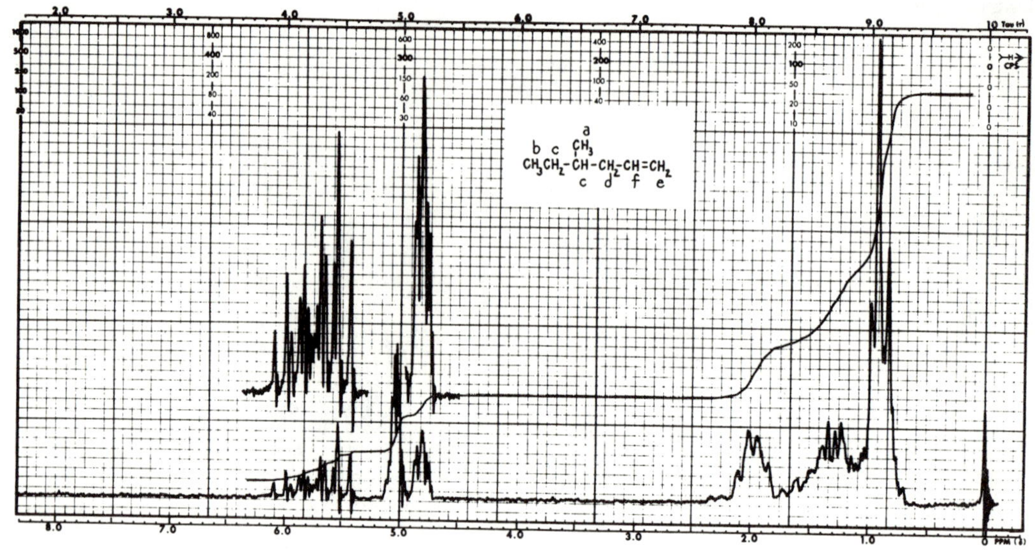

© 1978 Sadtler Research Laboratories, Inc.

1-HEPTENE

C_7H_{14}

Mol. Wt. 98.19

M.P. −10°C B.P. 94.9°C

Solvent: CCl_4

Assignments

a = 0.90
b = 1.10-1.70
c = 2.03
d = 4.87
e = 4.91
f = ca 5.66

59

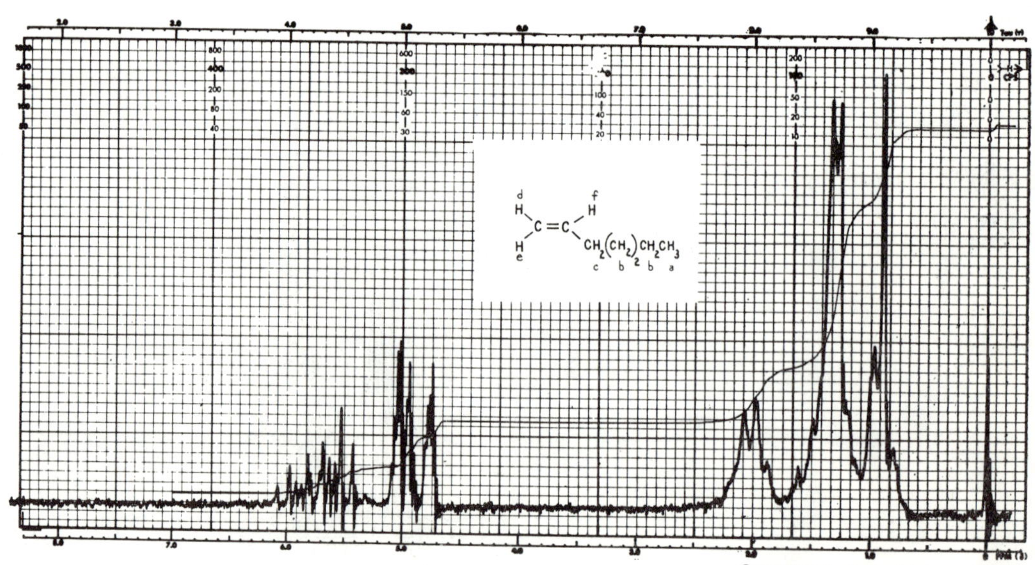

© 1978 Sadtler Research Laboratories, Inc.

1-OCTENE

C_8H_{16}

Mol. Wt. 122.22

B.P. 121-122°C

Solvent: CCl_4

Assignments

a = 0.88

b = ca 1.32

c = 1.78-2.47

d = 4.87

e = 4.91

f = ca 5.68

60

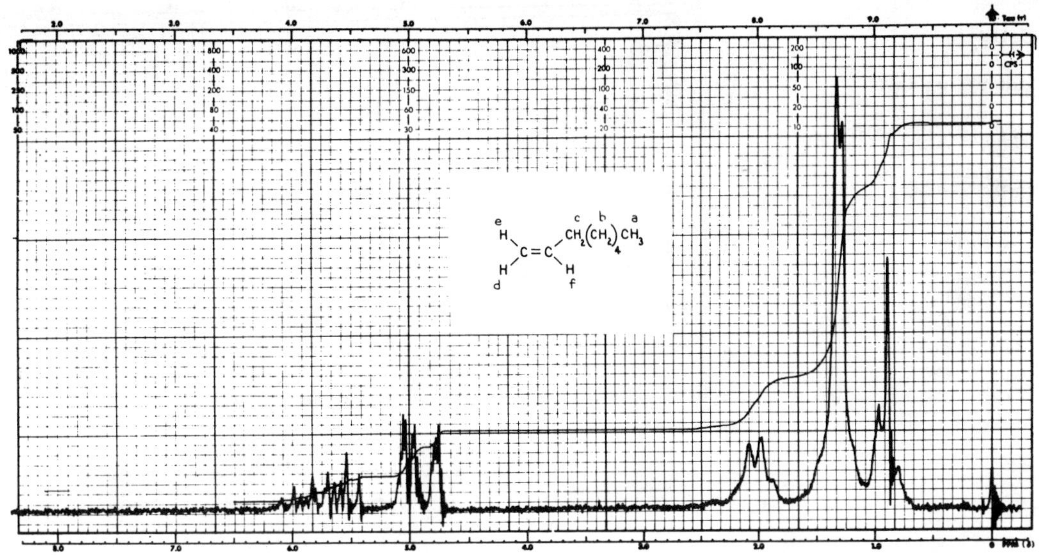

© 1978 Sadtler Research Laboratories, Inc.

1-DECENE

$C_{10}H_{20}$

Mol. Wt. 140.27

B.P. 171-172°C

Solvent: CCl_4

Assignments

a = 0.91

b = 1.30

c = 2.05

d = 4.89

e = 4.94

f = 5.72

61

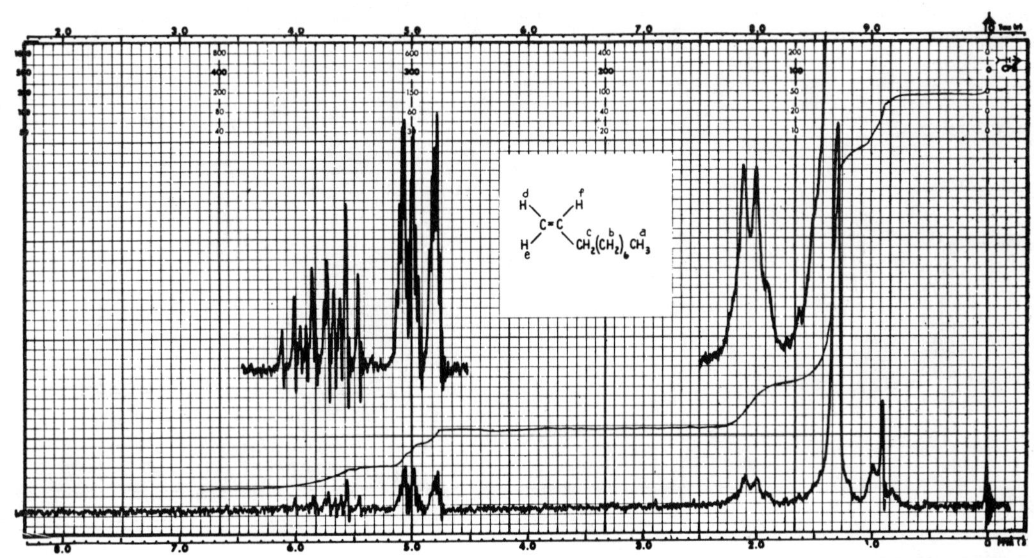

© 1978 Sadtler Research Laboratories, Inc.

1-DODECENE

$C_{12}H_{24}$

Mol. Wt. 168.31

B.P. 95-96°C/15mm

Solvent: CCl_4

Assignments

a = 0.88

b = 1.28

c = 2.04

d = 4.77-6.06

62

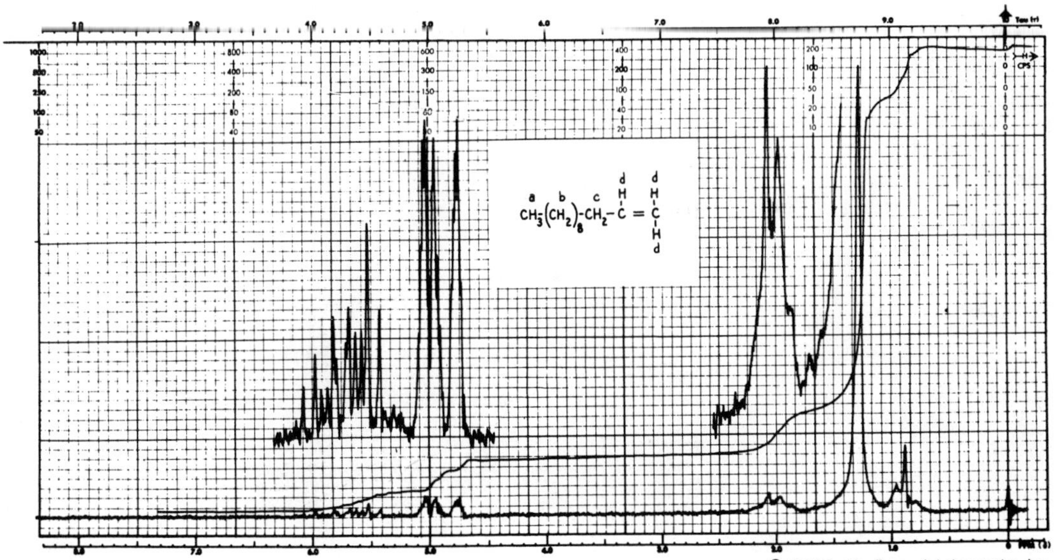

© 1978 Sadtler Research Laboratories, Inc.

1-TETRADECENE

$C_{14}H_{28}$

Mol. Wt. 196.38

B.P. 126-127°C/15mm

Solvent: CCl_4

Assignments

a = 0.90
b = ca 1.30
c = 2.00
d = 4.88
e = 4.92
f = 5.70

63

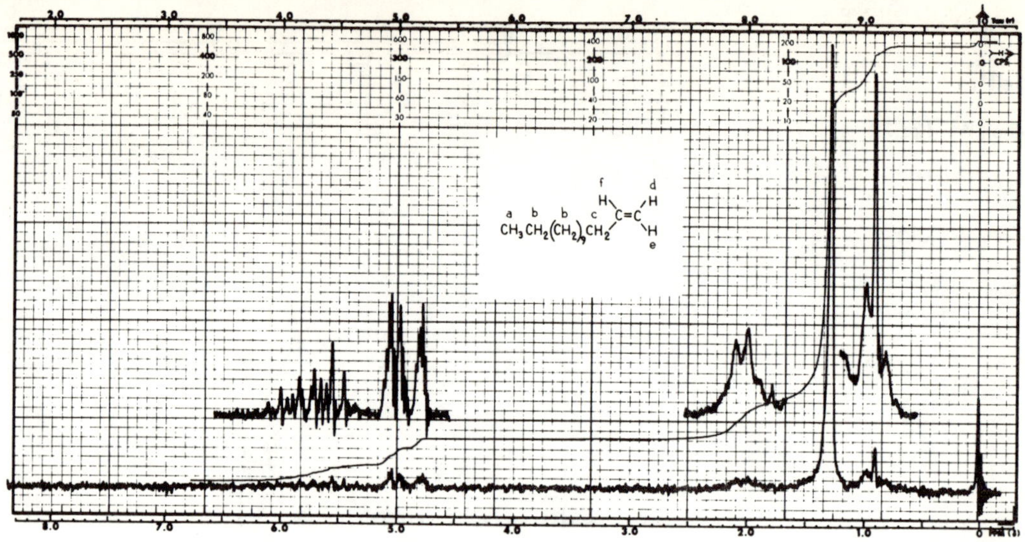

© 1978 Sadtler Research Laboratories, Inc.

1-HEXADECENE

$C_{16}H_{32}$

Mol. Wt. 224.43

B.P. 274°C

Solvent: CCl_4

Assignments

a = 0.88
b = ca 1.27
c = 2.08
d = 4.86
e = 4.90
f = ca 5.66

64

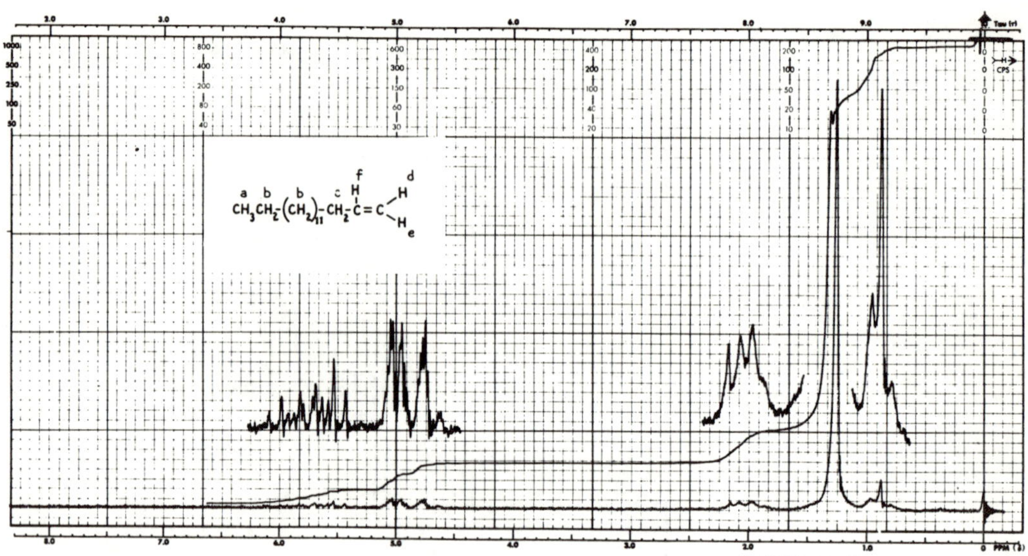

© 1978 Sadtler Research Laboratories, Inc.

1-OCTADECENE

$C_{18}H_{36}$

Mol. Wt. 252.49

B.P. 135-136°C/2mm

Solvent: CCl_4

Assignments

a = 0.90
b = 1.28
c = 2.00
d = 4.88
e = 4.90
f = ca 5.66

65

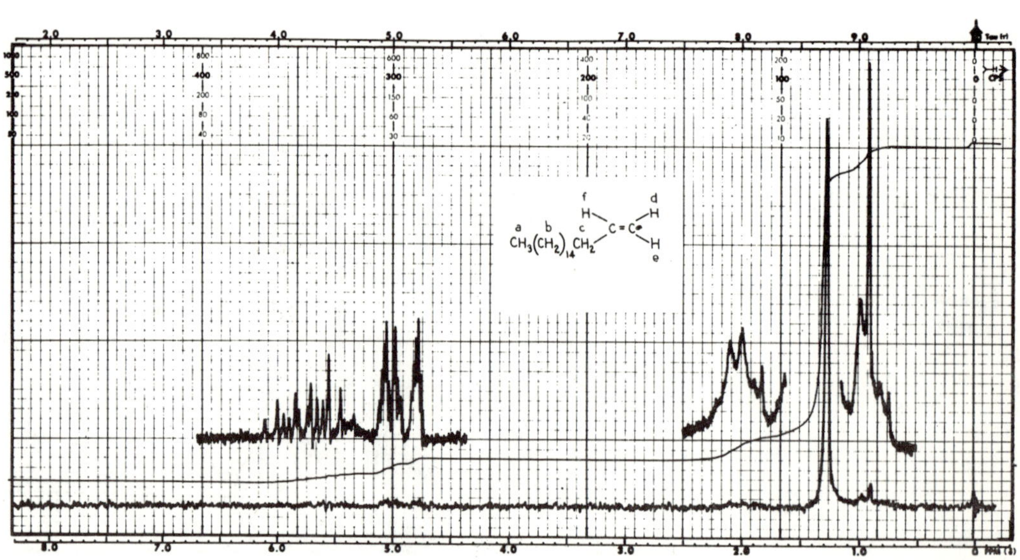

© 1978 Sadtler Research Laboratories, Inc.

1,5-HEXADIENE

C_6H_{10}

Mol. Wt. 82.15

B.P. 58-60°C

Solvent: CCl_4

Assignments
a = 2.11
b = 4.91
c = 4.96
d = 5.70

66

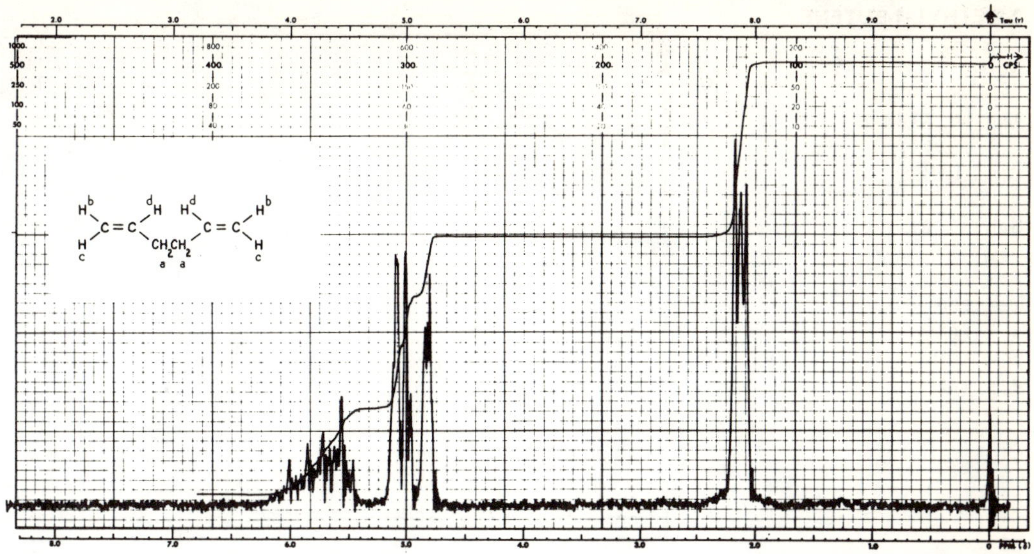

© 1978 Sadtler Research Laboratories, Inc.

3-METHYL-1,5-HEXADIENE

C_7H_{12}

Mol. Wt. 96.17

Solvent: CCl_4

Assignments
a = 1.00
b = 1.90-2.50
c = 4.75-5.20
d = 5.45-6.15

67

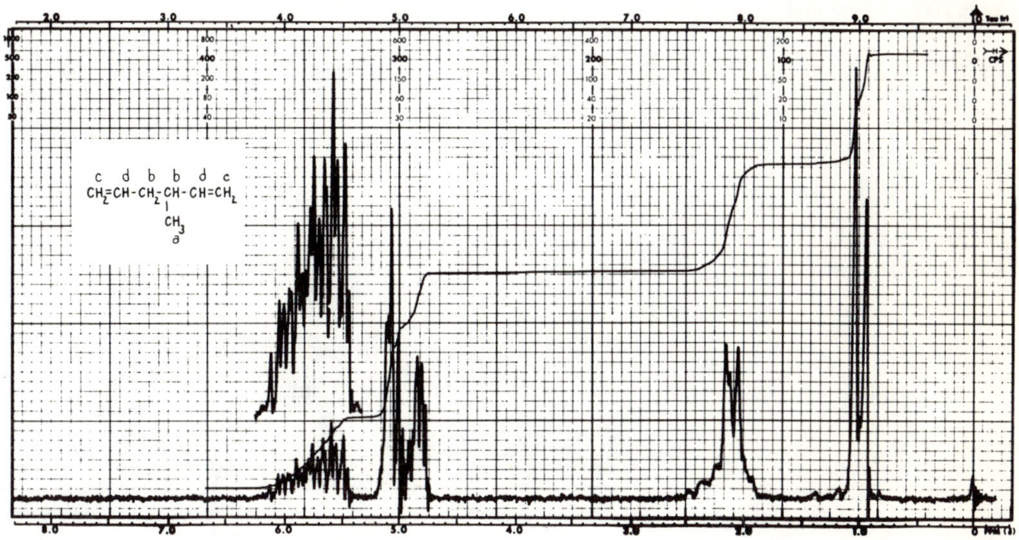

© 1978 Sadtler Research Laboratories, Inc.

2-METHYLPROPENE

C_4H_8

Mol. Wt. 56.11

Solvent: $CDCl_3$

Assignments
a = 1.73
b = 4.60

68

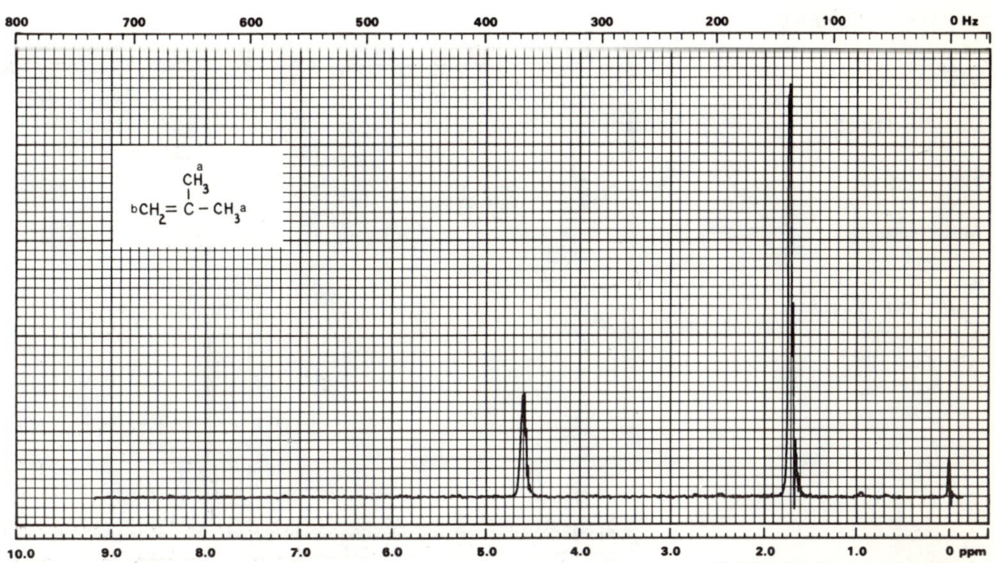

© 1978 Sadtler Research Laboratories, Inc.

2-METHYL-1-BUTENE

C_5H_{10}

Mol. Wt. 70.14

B.P. 31.1°C

Solvent: CCl_4

Assignments
a = 1.00
b = 1.69
c = 2.01
d = 4.62

69

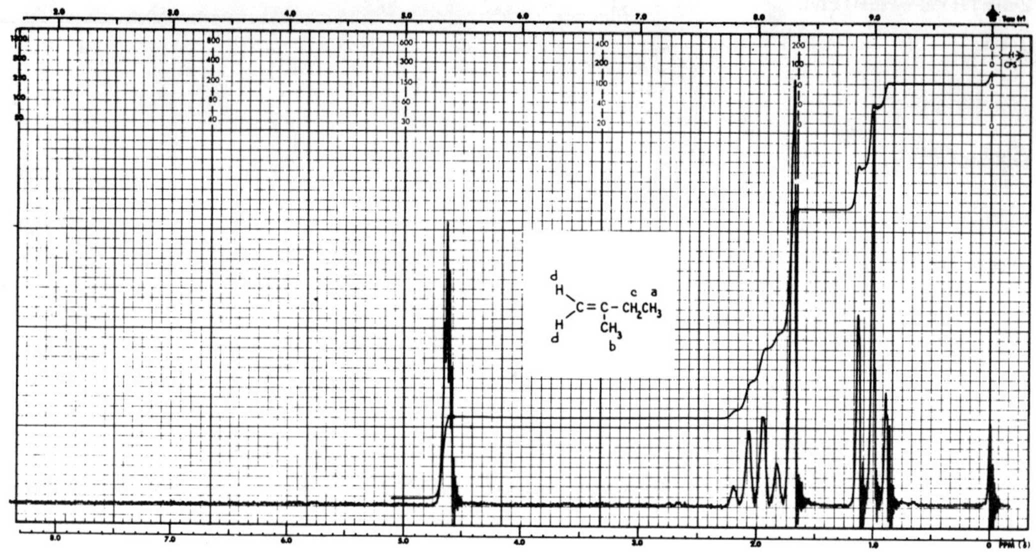

© 1978 Sadtler Research Laboratories, Inc.

2-METHYL-1-HEXENE

C_7H_{14}

Mol. Wt. 98.19

Solvent: CCl_4

Assignments
a = 0.92
b = 1.20-1.60
c = 1.70
d = 2.00
e = 4.61

70

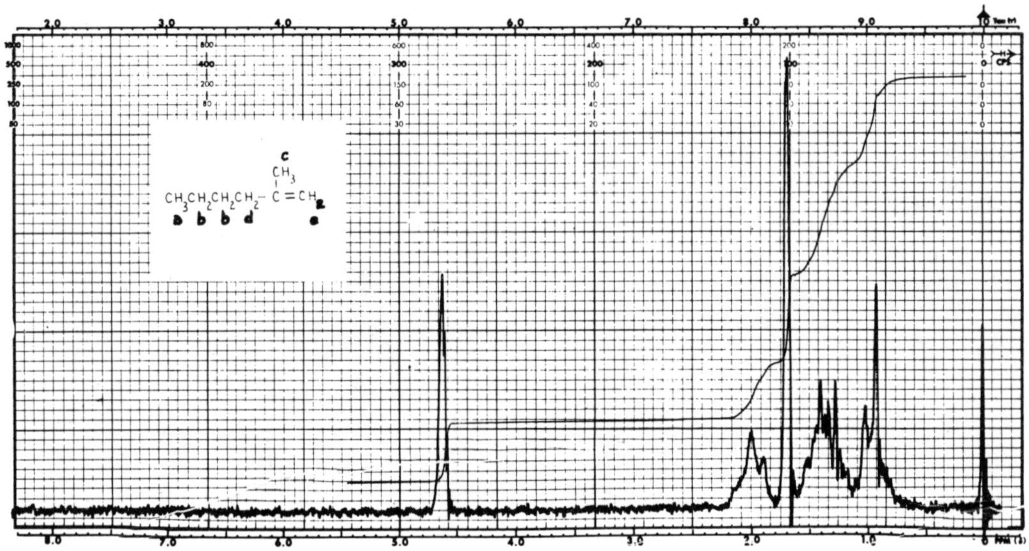

© 1978 Sadtler Research Laboratories, Inc.

2-ETHYL-1-HEXENE

C_8H_{16}

Mol. Wt. 112.22

Solvent: CCl_4

Assignments
a = 0.92
b = 1.01
c = 1.10-1.70
d = 1.80-2.25
e = 4.66
f = 5.08 imp.

71

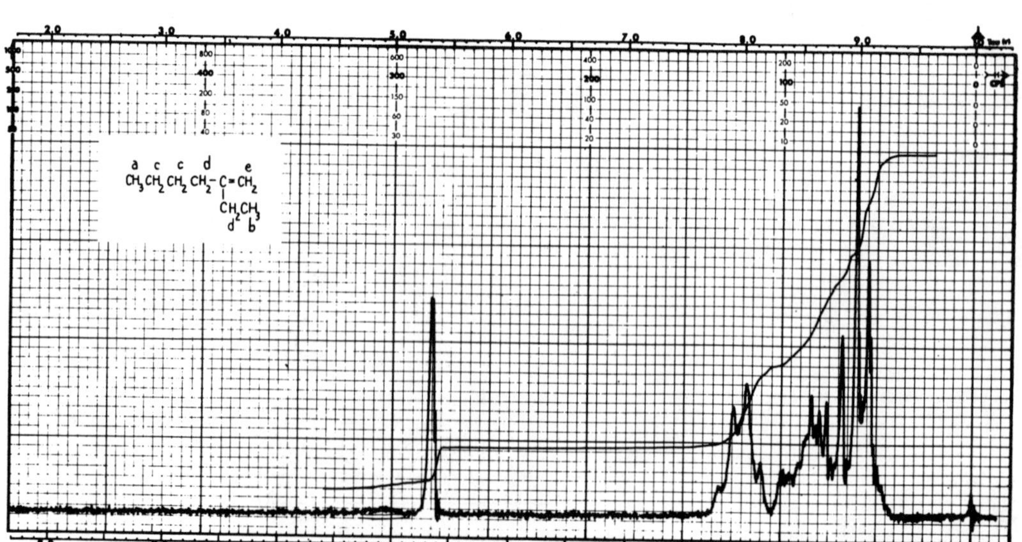

© 1978 Sadtler Research Laboratories, Inc.

2-METHYL-1-HEPTENE

C_8H_{16}

Mol. Wt. 112.22

Solvent: CCl_4

Assignments
a = 0.91
b = ca 1.31
c = 1.70
d = 1.99
e = 4.62

72

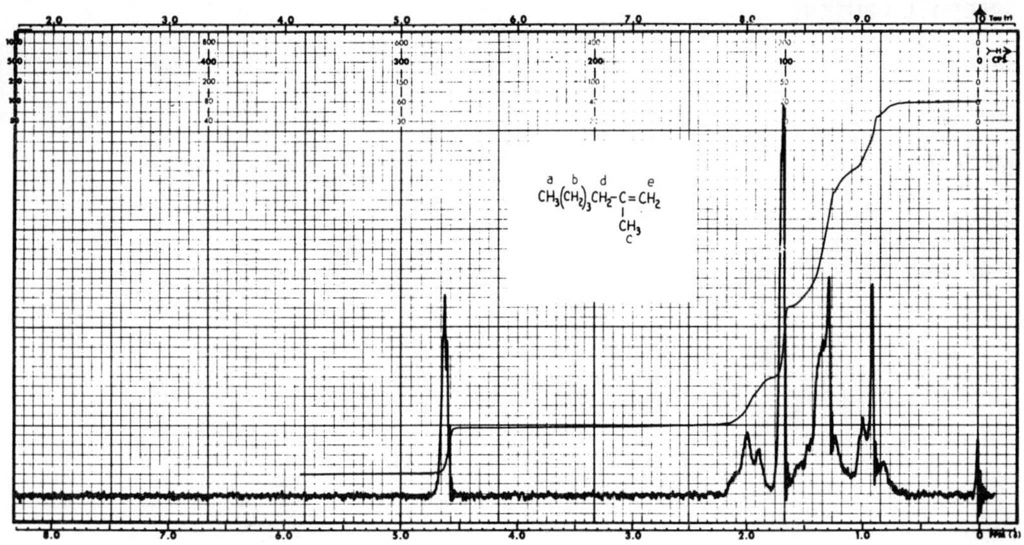

© 1978 Sadtler Research Laboratories, Inc.

ISOPRENE

C_5H_8

Mol. Wt. 68.12

B.P. 34°C

Solvent: CCl_4

Assignments
a = 1.82
b = 4.93
c = 5.02
d = 5.10
e = ca 6.34

73

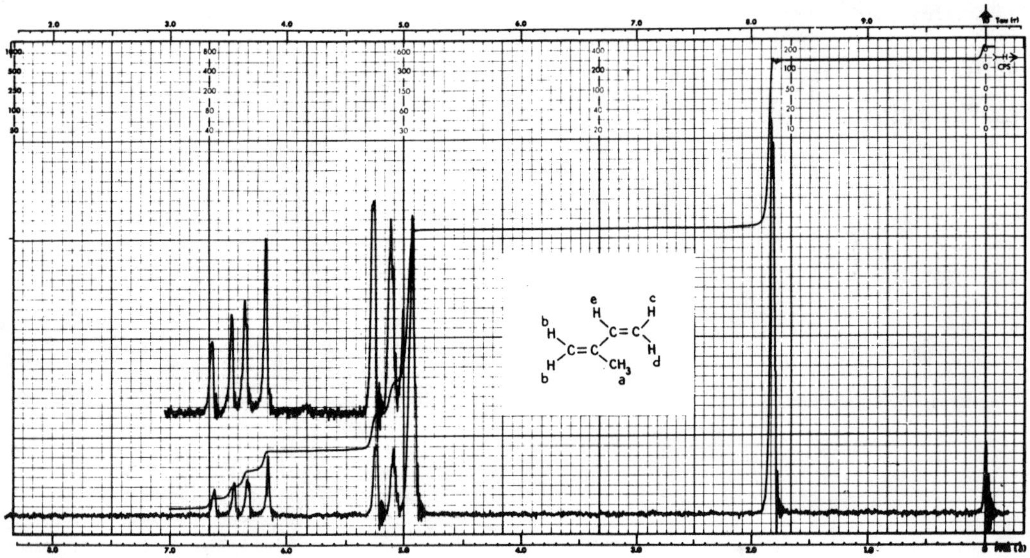

© 1978 Sadtler Research Laboratories, Inc.

2,3-DIMETHYL-1,3-BUTADIENE

C_6H_{10}

Mol. Wt. 82.15

Solvent: CCl_4

Assignments
a = 1.87
b = 4.90
c = 5.00

74

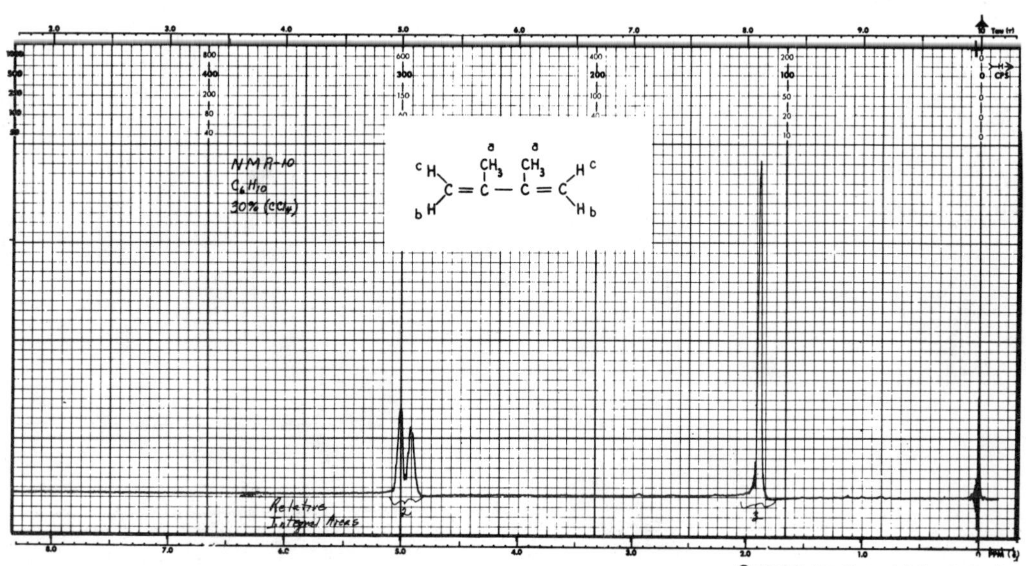

© 1978 Sadtler Research Laboratories, Inc.

2-PENTENE

C_5H_{10}

Mol. Wt. 70.14

B.P. 36.6°C

Solvent: CCl_4

Assignments
a = 0.95
b = 1.60
c = 1.98
d = ca 5.34

75

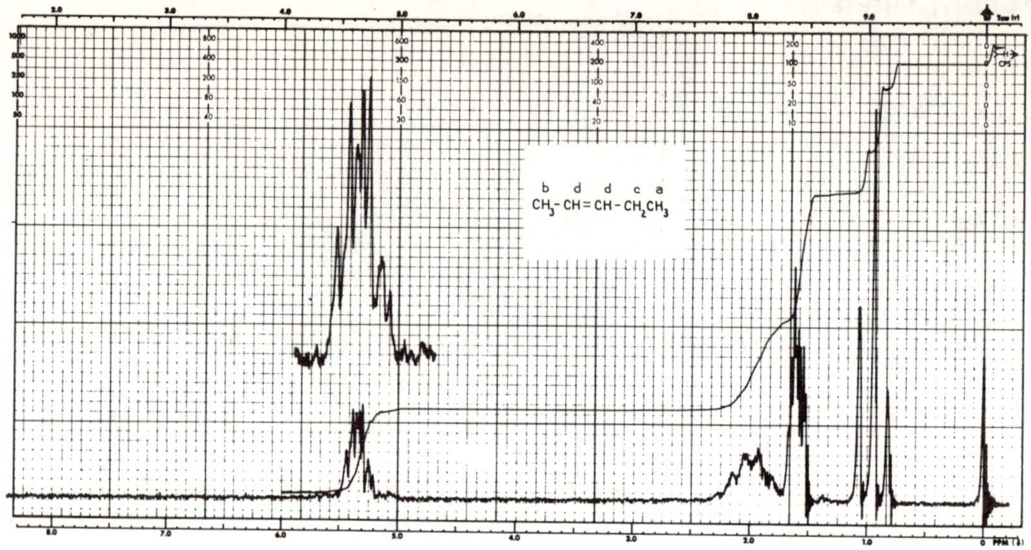

© 1978 Sadtler Research Laboratories, Inc.

4-METHYL-2-PENTENE

C_6H_{12}

Mol. Wt. 84.16

B.P. 57°C

Solvent: CCl_4

Assignments
a = 0.97
b = 1.59
c = 2.59
d = 4.90-5.41
Mixture of cis and trans forms

76

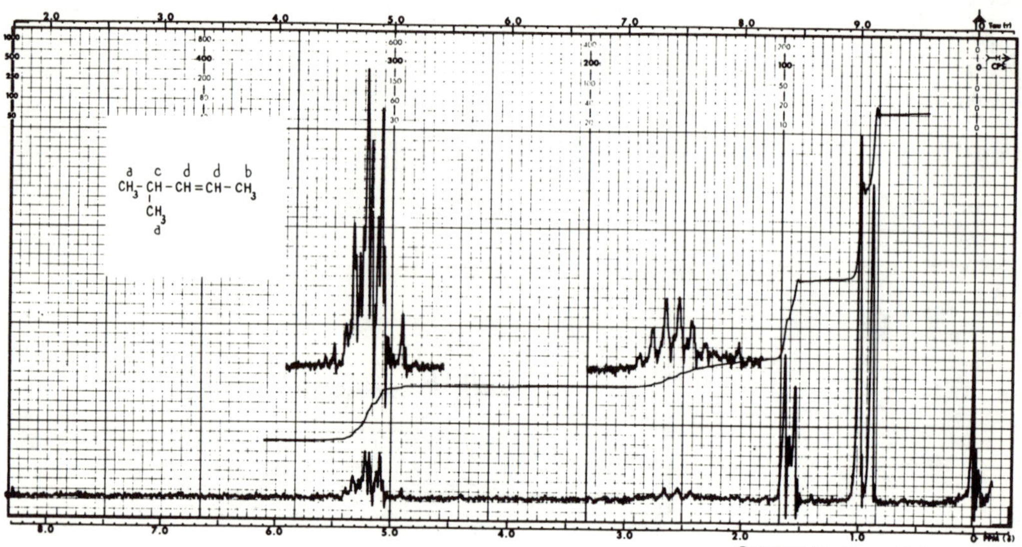

© 1978 Sadtler Research Laboratories, Inc.

trans-4-METHYL-2-PENTENE

C_6H_{12}

Mol. Wt. 84.16

B.P. 57-58°C

Solvent: CCl_4

Assignments
a = 0.96
b = 1.62
c = 2.19
d = 5.31

77

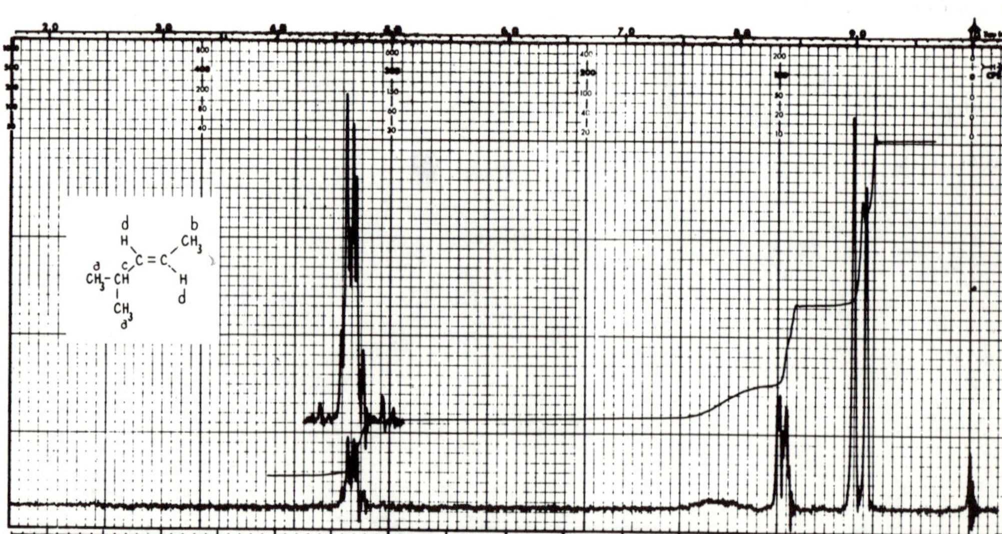

© 1978 Sadtler Research Laboratories, Inc.

cis-3-HEXENE

C_6H_{12}

Mol. Wt. 84.16

B.P. 67°C

Solvent: $CDCl_3$

Assignments

a = 0.95

b = 2.03

c = 5.30

78

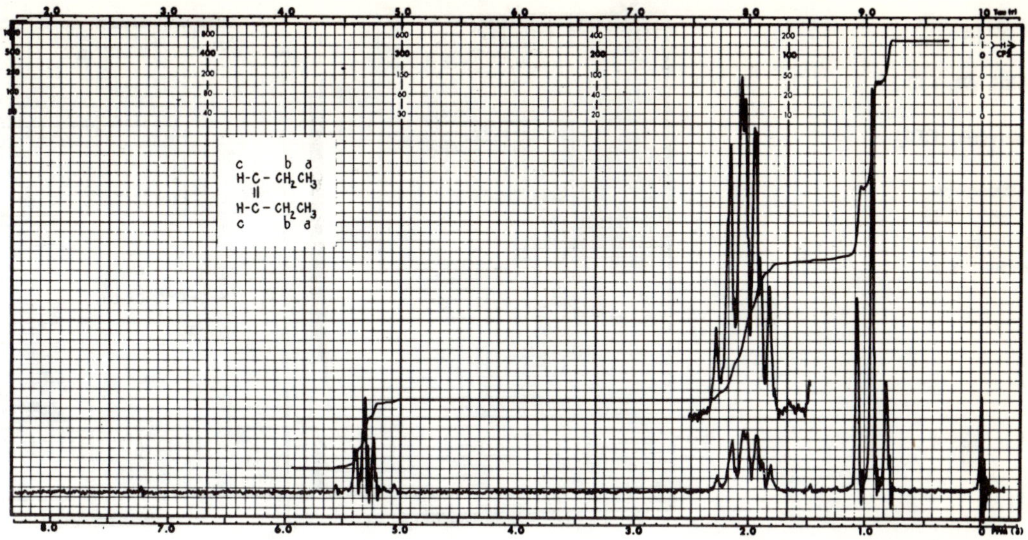

© 1978 Sadtler Research Laboratories, Inc.

trans-3-HEXENE

C_6H_{12}

Mol. Wt. 84.16

Solvent: CCl_4

Assignments

a = 0.92

b = 1.70-2.24

c = 5.37

79

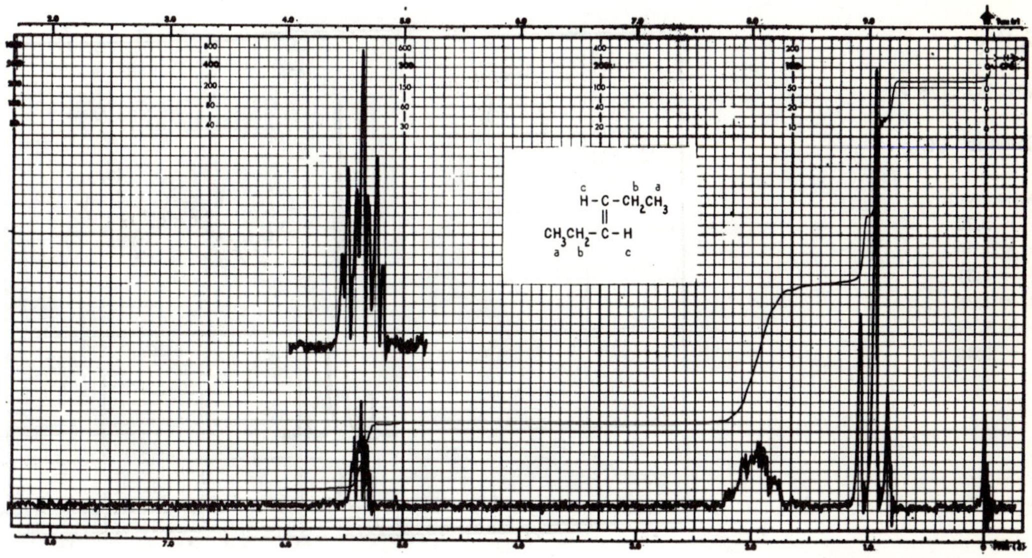

© 1978 Sadtler Research Laboratories, Inc.

cis-2,5-DIMETHYL-3-HEXENE

C_8H_{16}

Mol. Wt. 112.22

Solvent; CCl_4

Assignments

a = 0.95

b = 2.60

c = 5.00

80

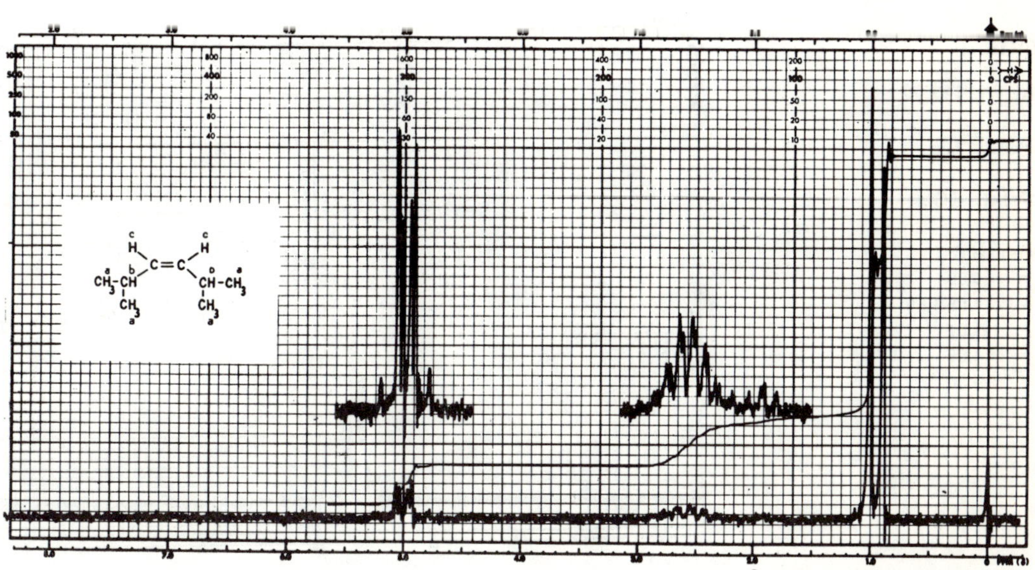

© 1978 Sadtler Research Laboratories, Inc.

trans-2,5-DIMETHYL-3-HEXENE

C_8H_{16}

Mol. Wt. 112.22

Solvent: CCl_4

Assignments
a = 0.98
b = ca 2.20
c = 5.27

81

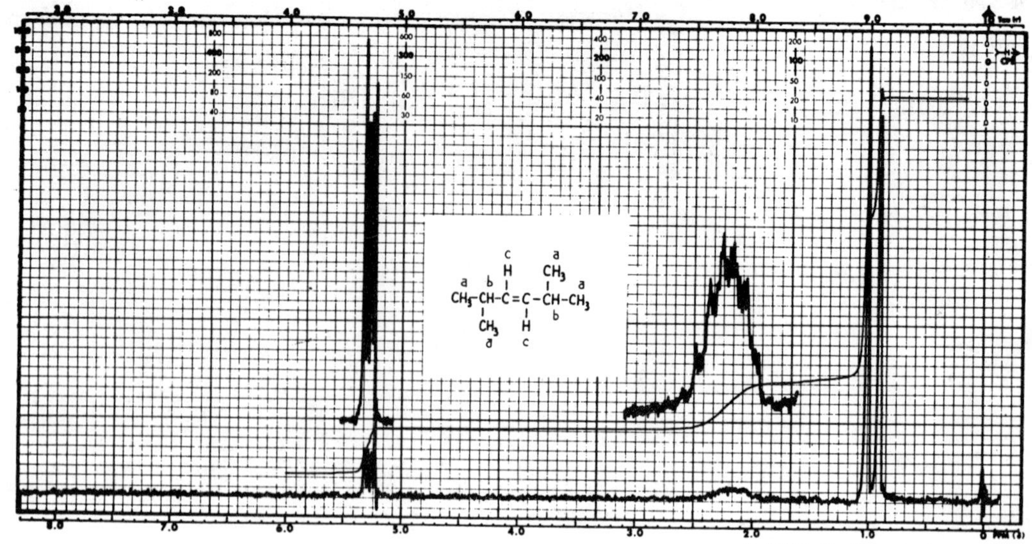

© 1978 Sadtler Research Laboratories, Inc.

trans-2,2-DIMETHYL-3-HEPTENE

C_9H_{18}

Mol. Wt. 126.24

Solvent: CCl_4

Assignments
a = 0.90
b = 1.00
c = 1.31
d = 1.96
e = 5.24
f = 5.40

82

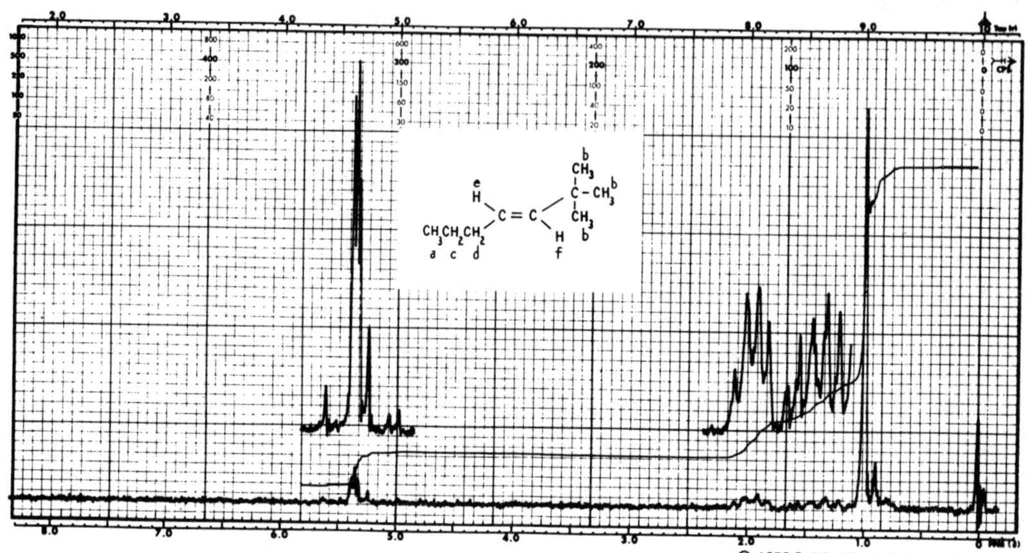

© 1978 Sadtler Research Laboratories, Inc.

2-OCTENE

C_8H_{16}

Mol. Wt. 112.22

B.P. 125°C

Solvent: CCl_4

Assignments
a = 0.89
b = 1.05-1.50
c = 1.57
d = ca 1.95
e = ca 5.30

83

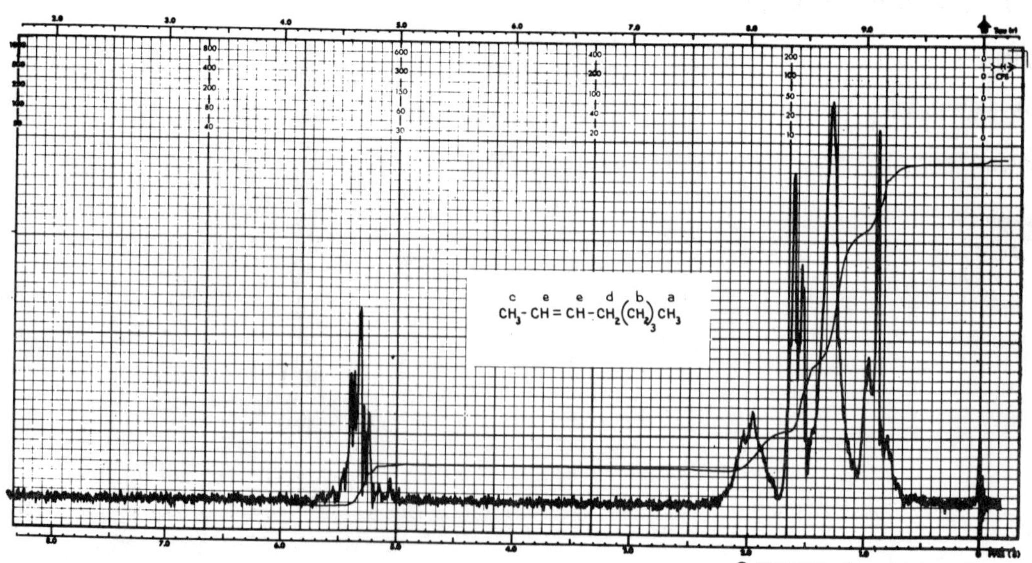

© 1978 Sadtler Research Laboratories, Inc.

trans-4-NONENE

C_9H_{18}

Mol. Wt. 126.24

Solvent: CCl_4

Assignments
a = ca 0.91
b = 1.10-1.70
c = ca 1.97
d = ca 5.32

84

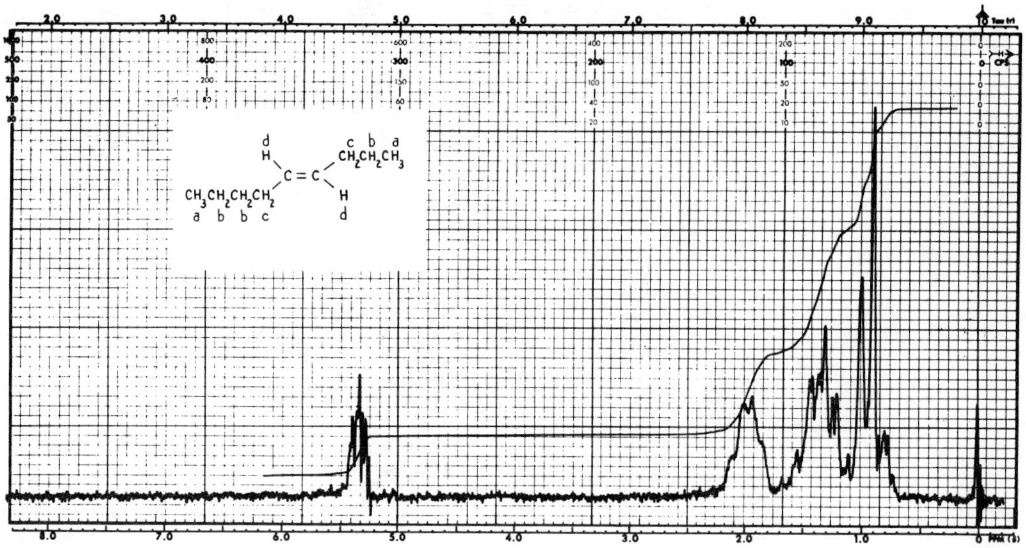

© 1978 Sadtler Research Laboratories, Inc.

7-TETRADECENE

$C_{14}H_{28}$

Mol. Wt. 196.38

Solvent: CCl_4

Assignments
a = 0.91
b = 1.30
c = 1.95
d = 5.34

85

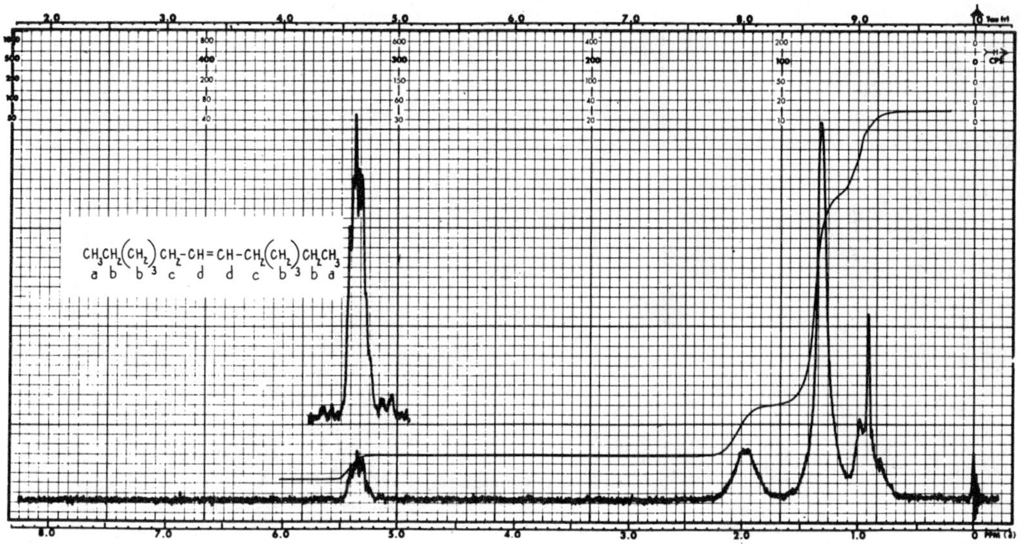

© 1978 Sadtler Research Laboratories, Inc.

17-PENTATRIACONTENE

$C_{35}H_{70}$

Mol. Wt. 490.95

B.P. 260-265°C/0.5mm

Solvent: $CDCl_3$

Assignments
a = 0.88
b = 1.27
c = 1.96
d = 5.33

86

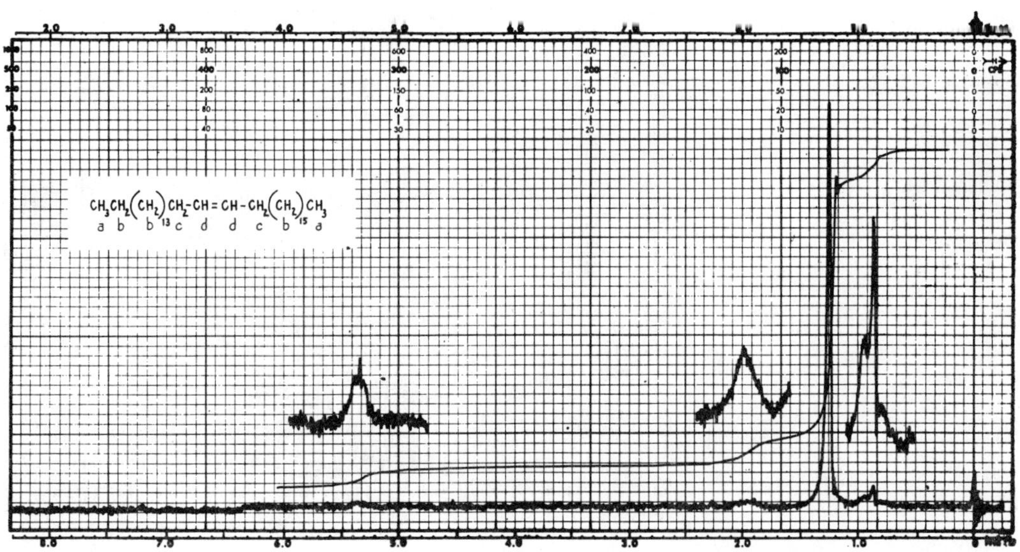

© 1978 Sadtler Research Laboratories, Inc.

2,4-HEXADIENE

C_6H_{10}

Mol. Wt. 82.15

B.P. 82-84°C

Solvent: CCl_4

Assignments
a = 1.73
b = 5.20-5.80
c = 5.93

87

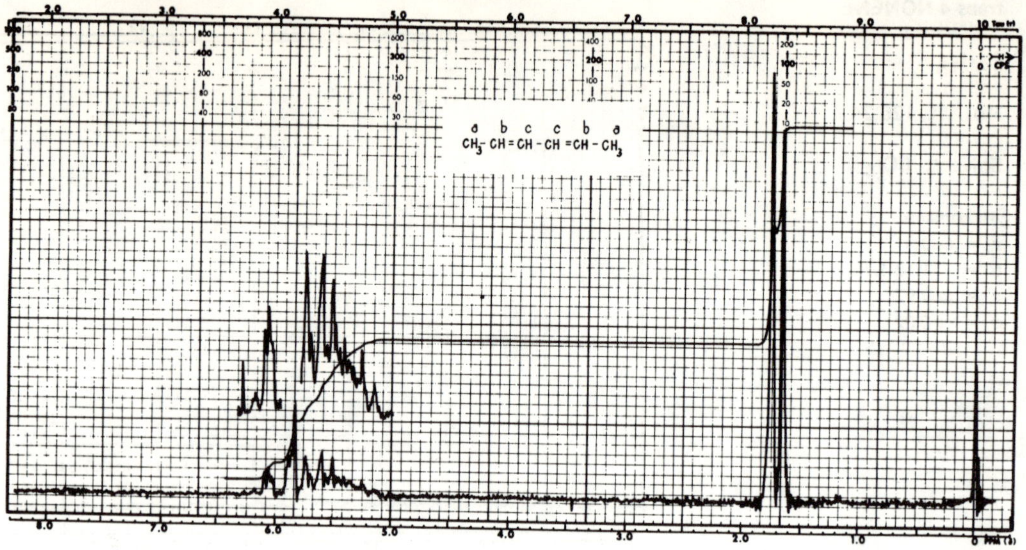

© 1978 Sadtler Research Laboratories, Inc.

3-METHYL-1,5-HEPTADIENE

C_8H_{14}

Mol. Wt. 110.20

B.P. 111-112°C/733mm (lit.)

Solvent: CCl_4

Assignments
a = 0.97
b = 1.63
c = 1.80-2.40
d = 4.83
e = 4.95
f = ca 5.35
g = ca 5.63

88

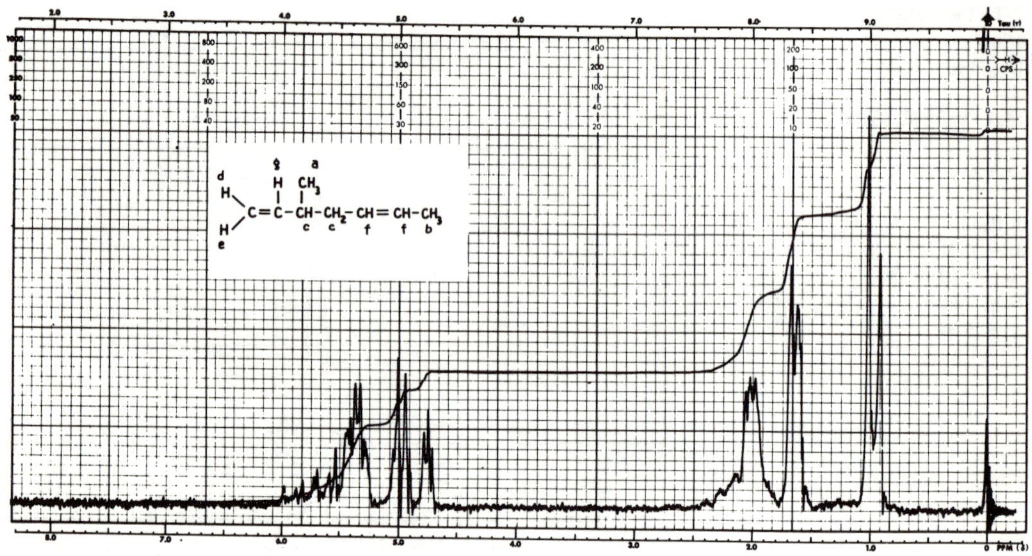

© 1978 Sadtler Research Laboratories, Inc.

5-METHYL-1,3,6-HEPTATRIENE

C_8H_{12}

Mol. Wt. 108.18

B.P. 117°C

Solvent: CCl_4

Assignments
a = 1.10
b = 2.88
c = ca 5.02
d = 5.30-6.60

89

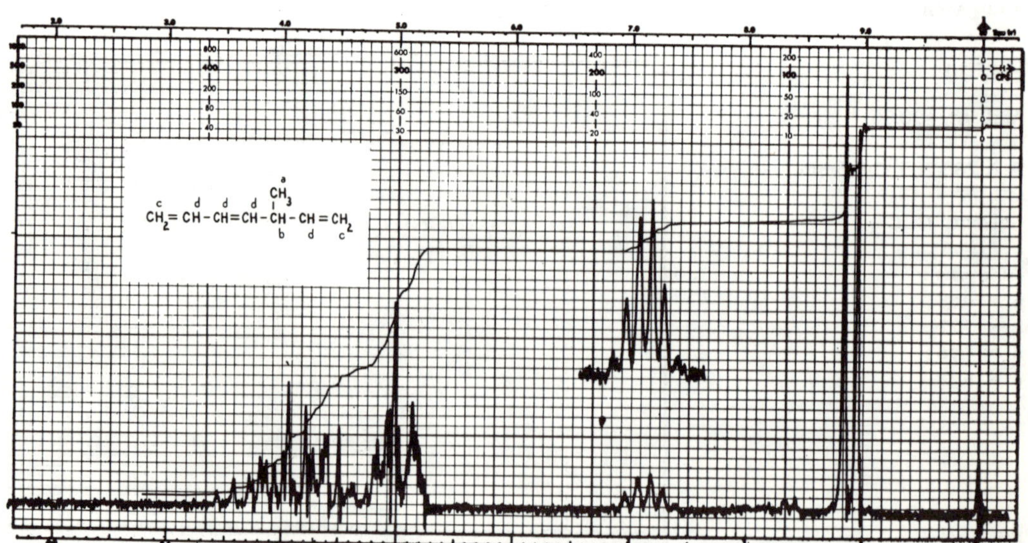

© 1978 Sadtler Research Laboratories, Inc.

2-METHYL-2-BUTENE

C_5H_{10}

Mol. Wt. 71.14

B.P. 38.5°C

Solvent: CCl_4

Assignments
a = 1.48
b = ca 1.58
c = 5.12

90

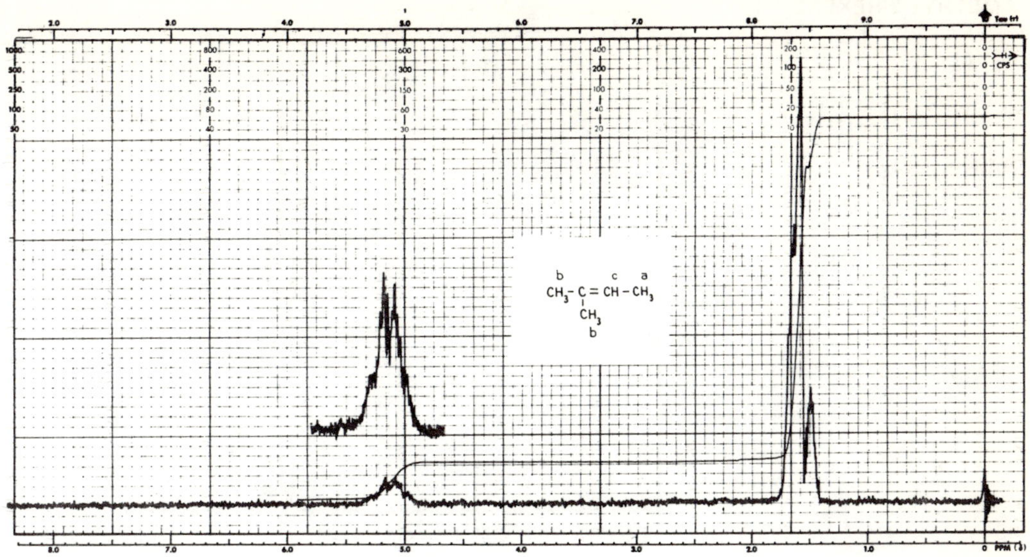

© 1978 Sadtler Research Laboratories, Inc.

2,4,4-TRIMETHYL-2-PENTENE

C_8H_{16}

Mol. Wt. 112.22

B.P. 104°C

Solvent: CCl_4

Assignments
a = 1.09
b = 1.65 or c
c = 1.71 or b
d = 5.11
e = 0.94 imp.

91

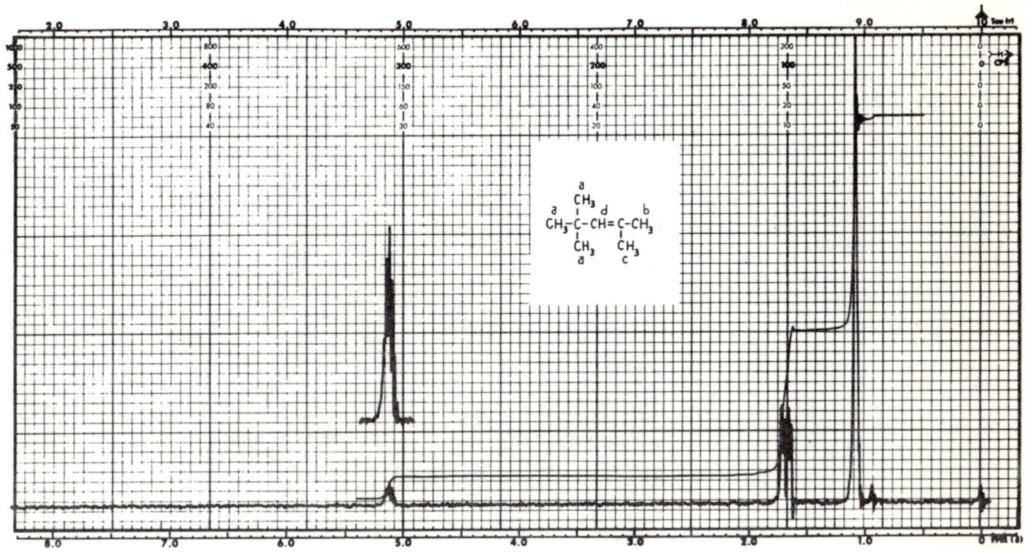

© 1978 Sadtler Research Laboratories, Inc.

cis-3-METHYL-2-HEXENE

C_7H_{14}

Mol. Wt. 98.19

Solvent: CCl_4

Assignments
a = 0.89
b = 1.33
c = 1.55
d = 1.59
e = 1.97
f = 5.16

92

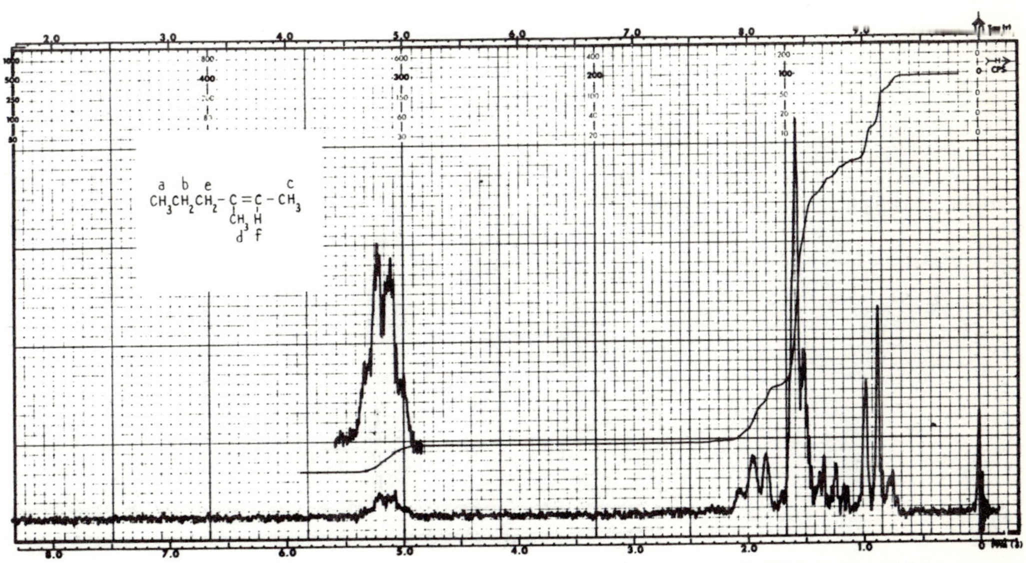

© 1978 Sadtler Research Laboratories, Inc.

2-METHYL-2-HEXENE

C_7H_{14}

Mol. Wt. 98.19

Solvent: CCl_4

Assignments
a = 0.90
b = 1.29
c = 1.60 or d
d = 1.69 or c
e = 1.95
f = 5.09

93

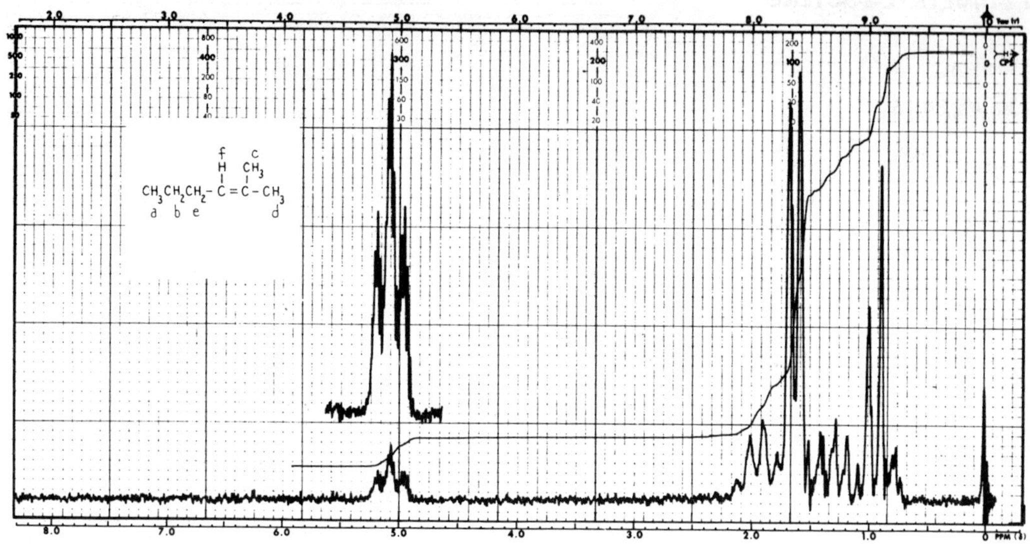

© 1978 Sadtler Research Laboratories, Inc.

2,5-DIMETHYL-2-HEXENE

C_8H_{16}

Mol. Wt. 112.22

Solvent: CCl_4

Assignments
a = 0.90
b = 1.60 or c
c = 1.70 or b
d = 1.26-2.08
e = 5.10

94

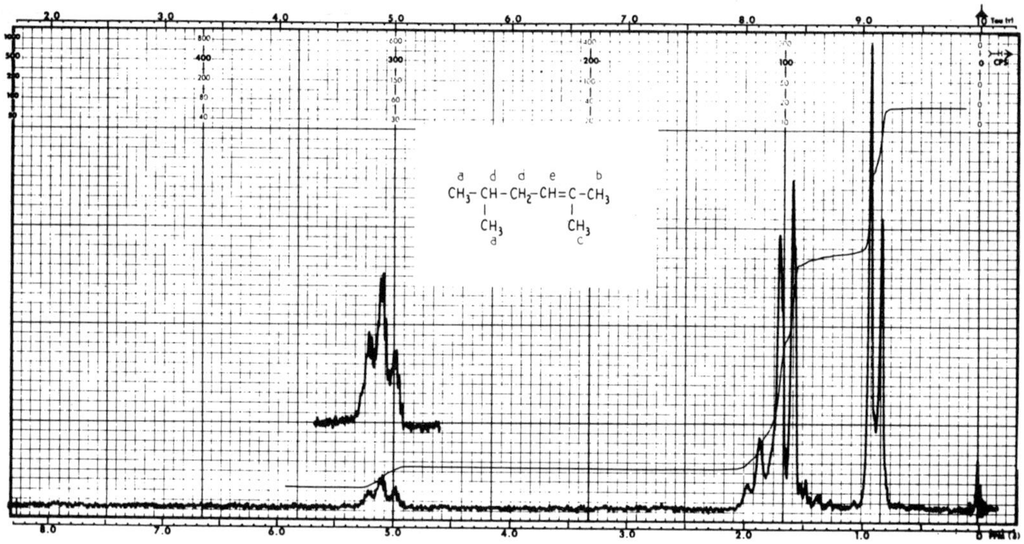

© 1978 Sadtler Research Laboratories, Inc.

2,5-DIMETHYL-2,4-HEXADIENE

C_8H_{14}

Mol. Wt. 110.20

M.P. 12-14°C

Solvent: CCl_4

Assignments
a = 1.71
b = 1.78
c = 5.89
d = 1.28 imp.

95

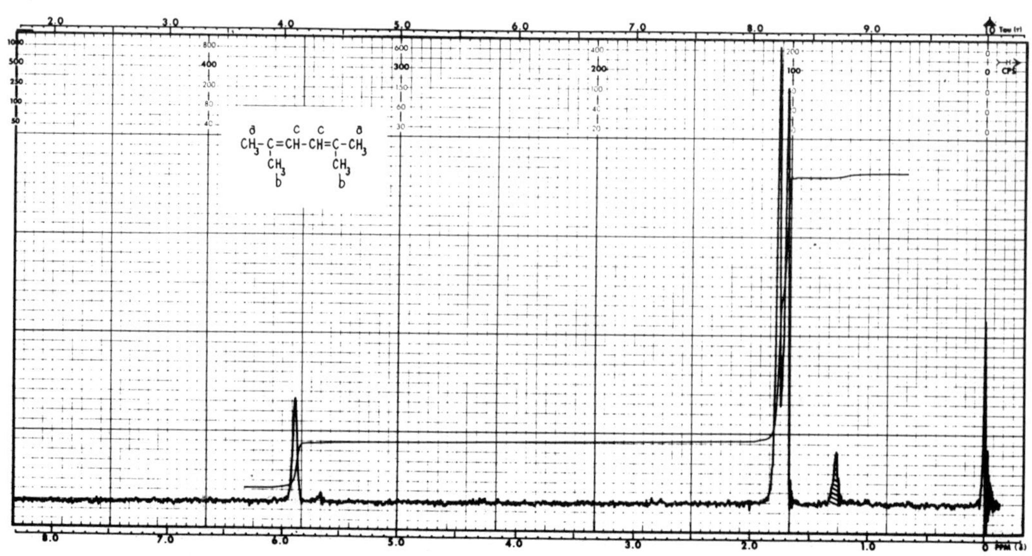

© 1978 Sadtler Research Laboratories, Inc.

2,3-DIMETHYL-2-BUTENE

C_6H_{12}

Mol. Wt. 84.16

Solvent: $CDCl_3$

Assignments

a = 1.62

96

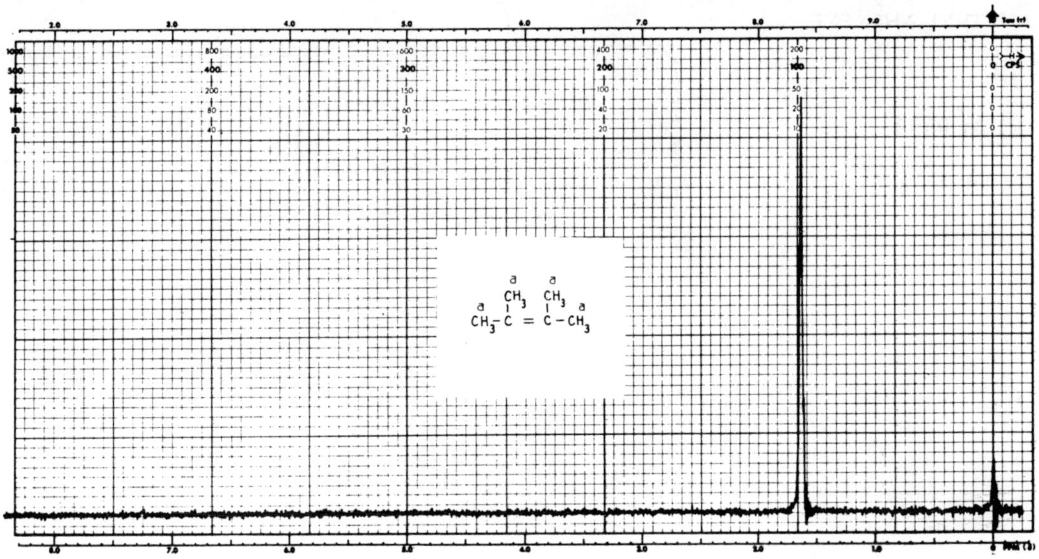

© 1978 Sadtler Research Laboratories, Inc.

NOTES

The cyclic alkenes are usually a relatively simple group to identify from their PMR spectra. The spectra display three sets of resonance bands with the olefinic protons resonating in the range from 5-6 ppm, the methylene groups adjacent to the double bonds deshielded to about 2.1 ppm and the remaining methylene groups forming a separate band at slightly higher field. In the spectra of the smaller rings, cyclopentene and cyclohexene, the coupling constant between the aliphatic and olefinic hydrogens (CH_2–CH=C) is quite small resulting in a single, slightly broadened peak for the olefinic hydrogens. As the ring sizes increase, this coupling constant increases in magnitude, to about 4 Hz for cycloheptene and about 5 Hz for cyclooctene.

Chemical Shifts

$-CH_2$	$-CH{=}CH$	$-CH_2$	$-(CH_2)_n-$	Compound	Solvent	Spectrum
2.29	5.66	2.29	(1.85)	Cyclopentene	CCl_4	97
1.97	5.58	1.97	$(1.62)_2$	Cyclohexene	CCl_4	99
2.11	5.71	2.11	$(1.64)_3$	Cycloheptene	CCl_4	----
2.12	5.59	2.12	$(1.53)_4$	Cyclooctene (cis)	CCl_4	106

The deshielding effect of the olefinic bond on aliphatic groups bonded to it is similar to that observed for the acyclic alkenes, i.e. a weakly deshielding effect.

HC–2	HC–1	Group	Ring System		
		CH_3-	Cyclopentene	CCl_4	----
1.03	2.19	CH_3-CH_2-	Cyclopentene	CCl_4	----
	1.60	CH_3-	Cyclohexene	CCl_4	102
0.96	1.95	CH_3-CH_2-	Cyclohexene	CCl_4	103

The chemical shift changes brought about by substituents bonded to the various positions of the cyclic alkenes are often dramatic, particularly on the position-2 olefinic hydrogen. A series of monosubstituted compounds is listed below.

Because of their smaller ring size, the cyclopentenes usually display two or three distinct bands at high field for the ring methylene groups depending on the position and deshielding effect of the substituent.

Cyclopentenes

HC–5	HC–5	HC–3	HC–2	HC–1		–X	Solvent
2.29	1.85	2.29	5.66	5.66		----	CCl_4
	(1.5–2.5)		5.19	----	-	$-1-CH_2-CH_3$	CCl_4
2.30	1.40	3.30	5.63	5.63	-	$-3-CH_2-C(=O-CH_3$	CCl_4
	(1.0–2.5)		5.61	5.61	-	$-4-CH{=}CH_2$	CCl_4
2.00	2.98	2.00	5.61	5.61	-	$-4-C(=O)-Cl$	CCl_4

Chemical Shifts continued

Cyclohexenes

HC–6	HC–5	HC–4	HC–3	HC–2	HC–1	–X	Solvent
1.89	1.57	1.57	1.89	5.23	----	$-1-CH_2-CH_3$	CCl_4
2.09	1.72	1.72	2.09	5.27	----	$-1-O-C(=O)-CH_3$	CCl_4
1.88	1.60	1.60	1.88	5.30	----	$-1-CH_3$	CCl_4
1.98	1.65	1.65	1.98	5.46	----	$-1-CH_2-CH_2-OH$	CCl_4
2.15	1.69	1.69	2.15	5.49	----	–1–Ø	CCl_4
1.97	1.62	1.62	1.97	5.58	5.58	----	CCl_4
2.00	1.60	1.60	2.00	5.71	----	$-1-Cl$	CCl_4
2.05	1.65	1.65	2.05	5.78	----	$-1-CH_2-C{\equiv}N$	$CDCl_3$
2.22	1.68	1.68	2.22	6.58	----	$-1-C{\equiv}N$	CCl_4
2.28	1.68	1.68	2.28	7.11	----	$-1-C(=O)-OH$	CCl_4
2.32	1.69	1.69	2.32	7.39	----	$-1-C(=O)-Cl$	CCl_4

3-Substituted Cyclohexenes

HC–6	HC–5	HC–4	HC–3	HC–2	HC–1	–X	
	(1.1–2.5)			5.56	5.56	$-3-CH_3$	$CDCl_3$
1.95	(1.3–1.9)		3.61	5.74	5.74	$-3-O-CH_3$	CCl_4
2.11	(1.3–2.1)		4.78	5.80	5.80	$-3-Br$	CCl_4

4-Substituted Cyclohexenes

HC–6	HC–5	HC–4	HC–3	HC–2	HC–1	–X	
	(1.0–2.4)			5.63	5.63	$-4-CH_2-OH$	$CDCl_3$
(1.6–2.5)		2.71		5.65	5.65	$-4-C{\equiv}N$	CCl_4
	(1.6–2.8)			5.66	5.66	$-4-C(=O)-O-R$	$CDCl_3$
(1.6–2.6)		3.00		5.79	5.79	$-4-C(=O)-Cl$	CCl_4

CYCLOPENTENE

C_5H_8

Mol. Wt. 68.12

B.P. 44.2°C

Solvent: CCl_4

Assignments
a = 1.85
b = 2.29
c = 5.66

97

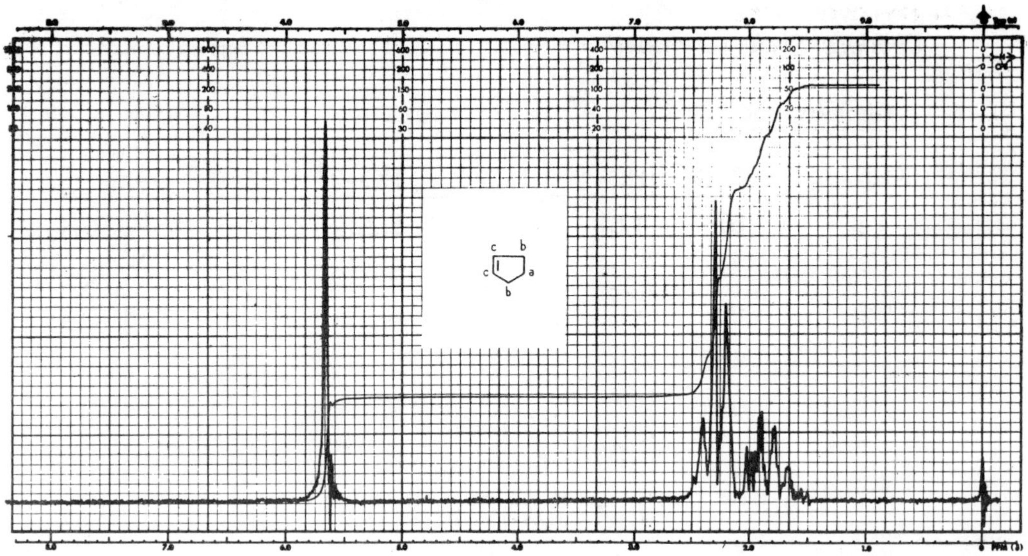

© 1978 Sadtler Research Laboratories, Inc.

1,2-DIMETHYLCYCLOPENTENE

C_7H_{12}

Mol. Wt. 96.17

Solvent: CCl_4

Assignments
a = 1.57
b = 1.48-1.97
c = 2.22

98

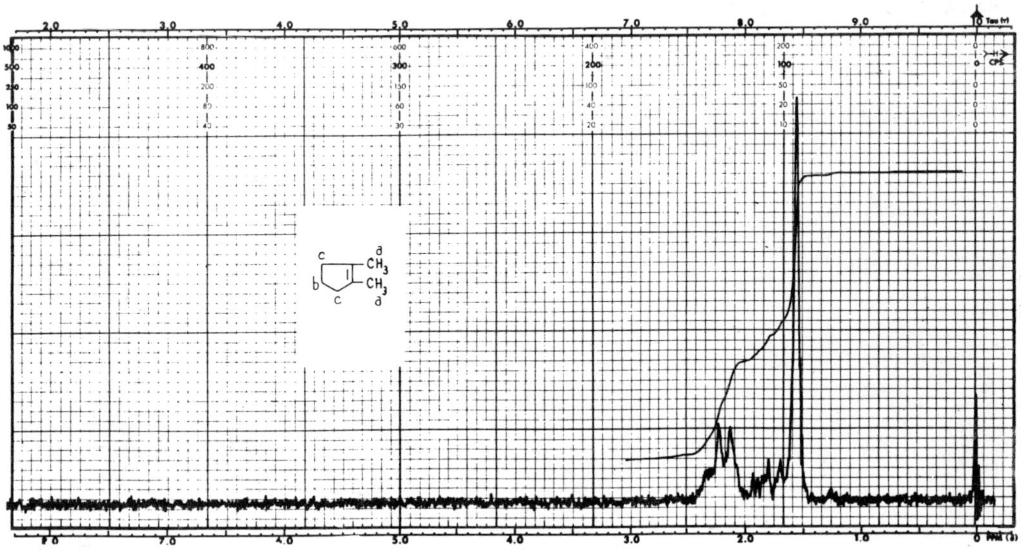

© 1978 Sadtler Research Laboratories, Inc.

CYCLOHEXENE

C_6H_{10}

Mol. Wt. 82.15

B.P. 83.2°C

Solvent: CCl_4

Assignments
a = ca 1.62
b = 1.97
c = 5.58

99

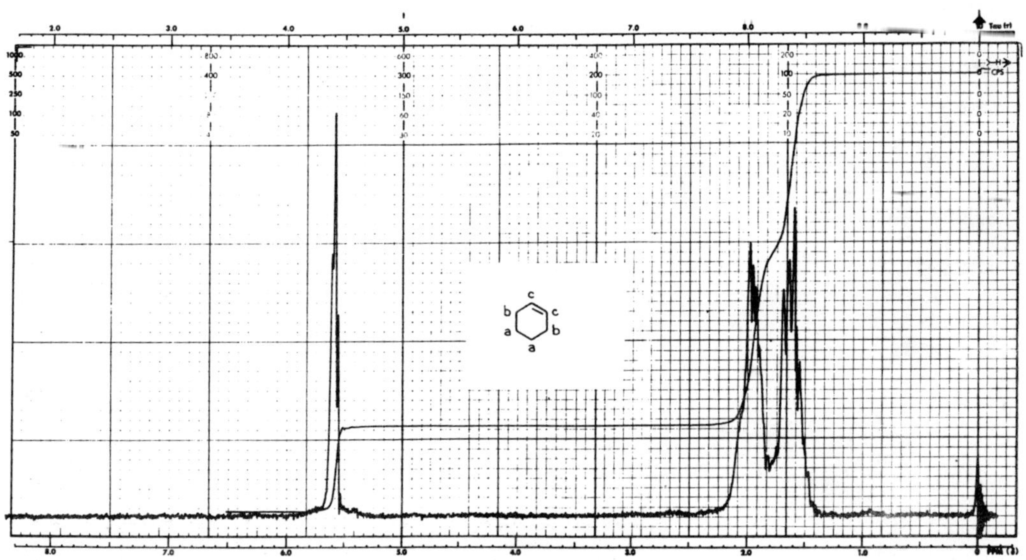

© 1978 Sadtler Research Laboratories, Inc.

4-tert-PENTYLCYCLOHEXENE

$C_{11}H_{20}$

Mol. Wt. 152.28

Solvent: CCl_4

Assignments

a = 0.79
b = 1.26
c = 1.00-1.60
d = ca 1.90
e = 5.48

100

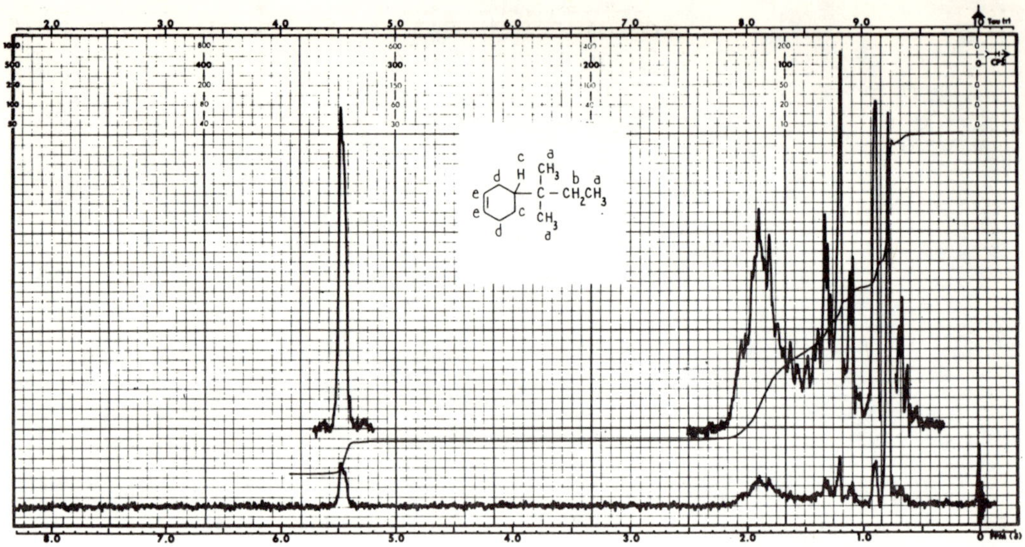

© 1978 Sadtler Research Laboratories, Inc.

4-VINYL-1-CYCLOHEXENE

C_8H_{12}

Mol. Wt. 108.18

B.P. 125-130°C

Solvent: CCl_4

Assignments

a = 0.98-2.51
b = 4.90
c = 4.98
d = ca 5.61
e = 5.88

101

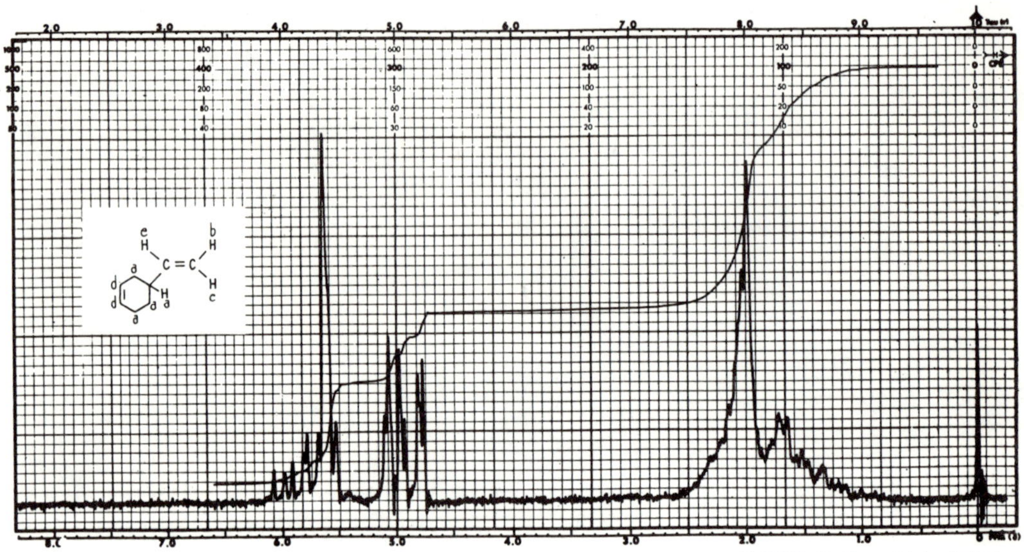

© 1978 Sadtler Research Laboratories, Inc.

1-METHYL-1-CYCLOHEXENE

C_7H_{12}

Mol. Wt. 96.17

B.P. 109.9°C

Solvent: CCl_4

Assignments

a = ca 1.60
b = 1.88
c = ca 5.30

102

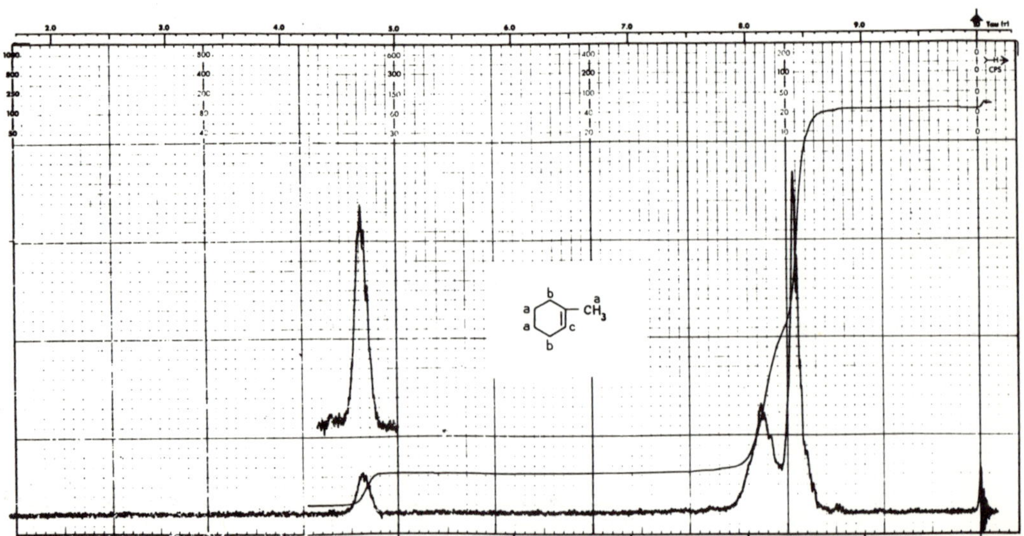

© 1978 Sadtler Research Laboratories, Inc.

1-ETHYLCYCLOHEXENE

C_8H_{14}

Mol. Wt. 110.20

Solvent: CCl_4

Assignments
a = 0.96
b = 1.57
c = 1.89
d = 5.23

103

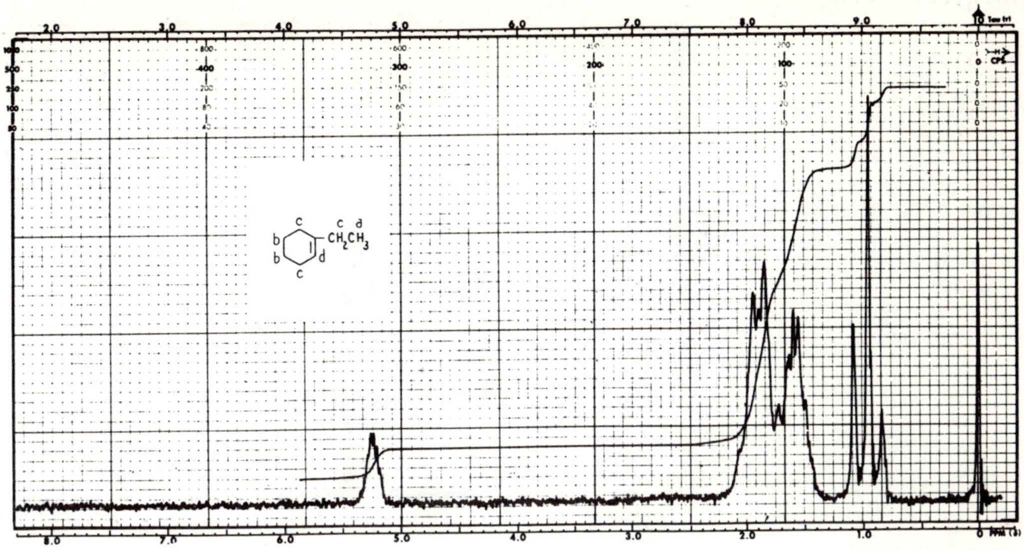

© 1978 Sadtler Research Laboratories, Inc.

1,2-DIMETHYLCYCLOHEXENE

C_8H_{14}

Mol. Wt. 110.20

Solvent: CCl_4

Assignments
a = 1.40-1.69
b = 1.56
c = 1.84

104

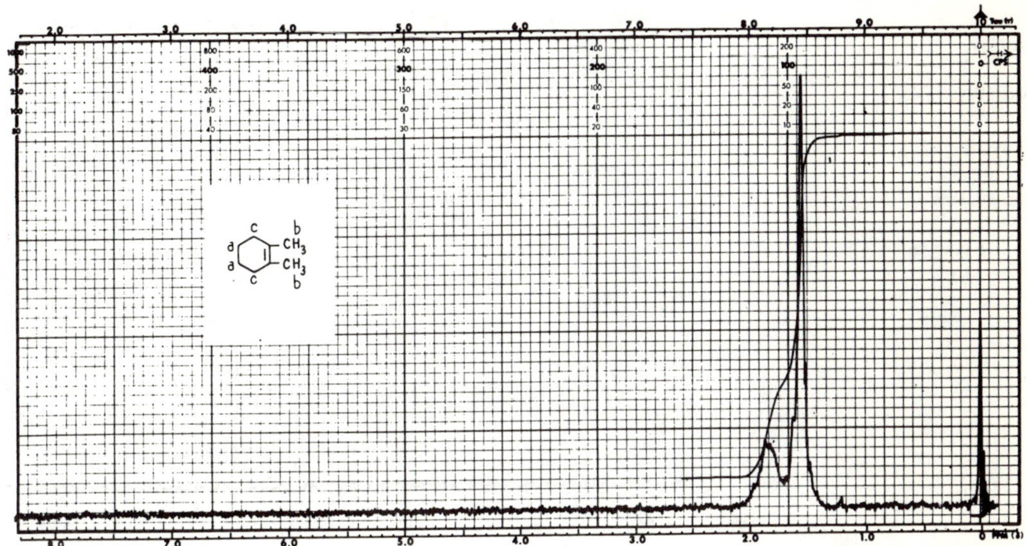

© 1978 Sadtler Research Laboratories, Inc.

BI-1-CYCLOHEXEN-1-YL

$C_{12}H_{18}$

Mol. Wt. 162.28

Solvent: CCl_4

Assignments
a = 1.00-1.90
b = 1.90-2.50
c = ca 5.70

105

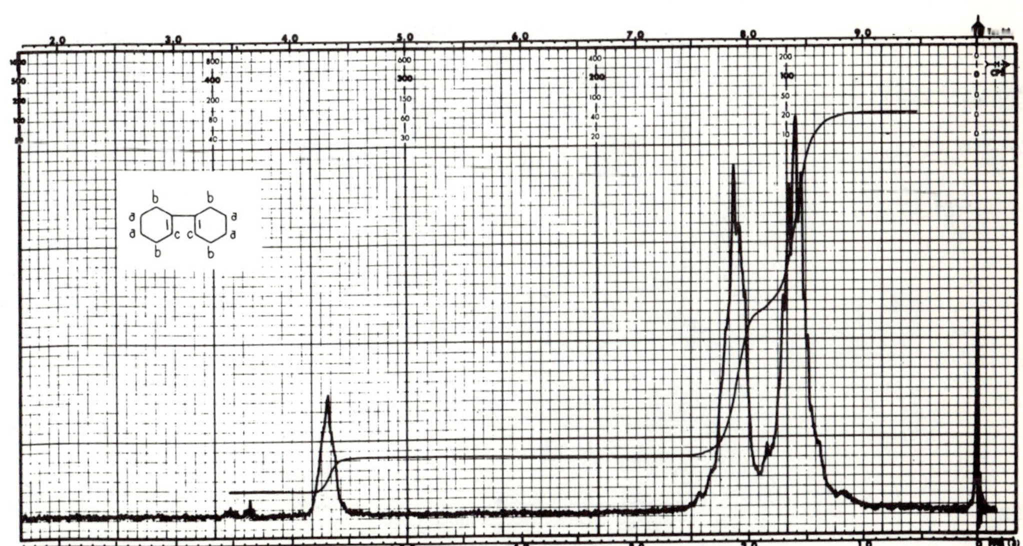

© 1978 Sadtler Research Laboratories, Inc.

trans-CYCLOOCTENE

C_8H_{14}

Mol. Wt. 110.20

B.P. 135°C

Solvent: CCl_4

Assignments
a = ca 1.50
b = 2.12
c = 5.30-5.90

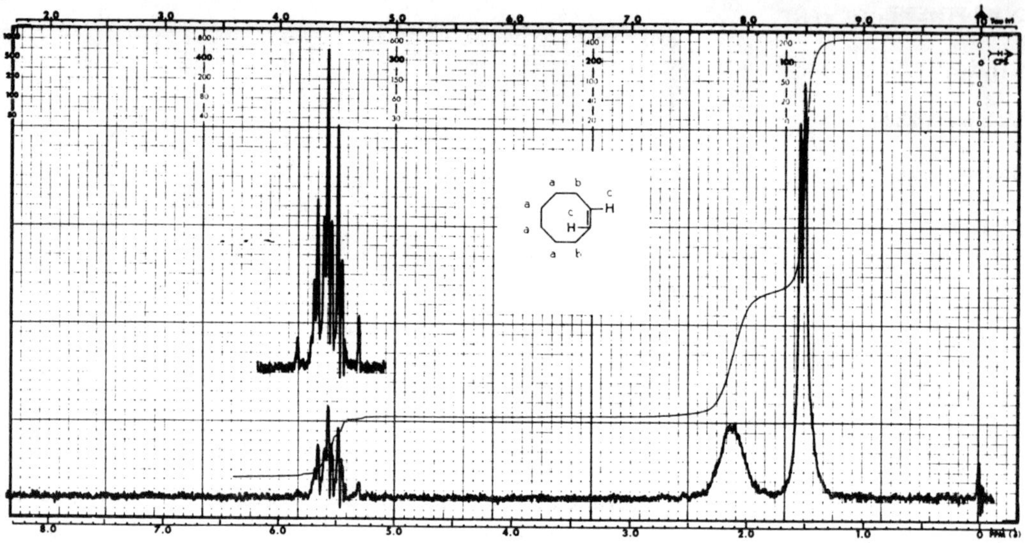

106

© 1978 Sadtler Research Laboratories, Inc.

---

cis-1,3-DIMETHYL-2-METHYLENECYCLOHEXANE

C_9H_{16}

Mol. Wt. 124.23

Solvent: CCl_4

Assignments
a = 1.03
b = 0.80-2.30
c = 4.51

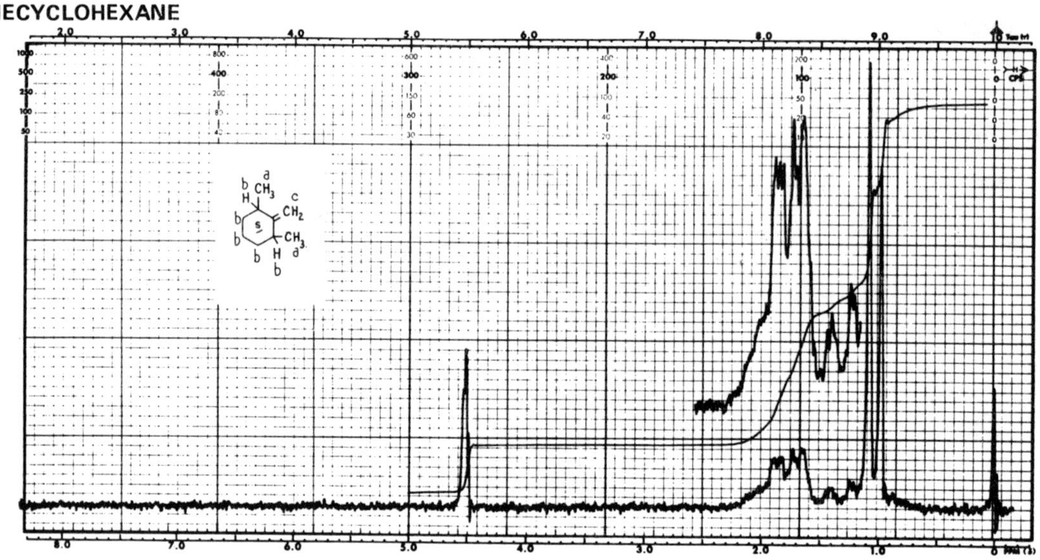

107

© 1978 Sadtler Research Laboratories, Inc.

3,3-DIMETHYL-2-METHYLENENORBORNANE

$C_{10}H_{16}$

Mol. Wt. 136.24

Solvent: $CDCl_3$

Assignments
a = 0.82
b = 1.02
c = 1.07
d = 1.10-1.85
e = 1.89
f = 2.65
g = 4.47 or h
h = 4.69 or g

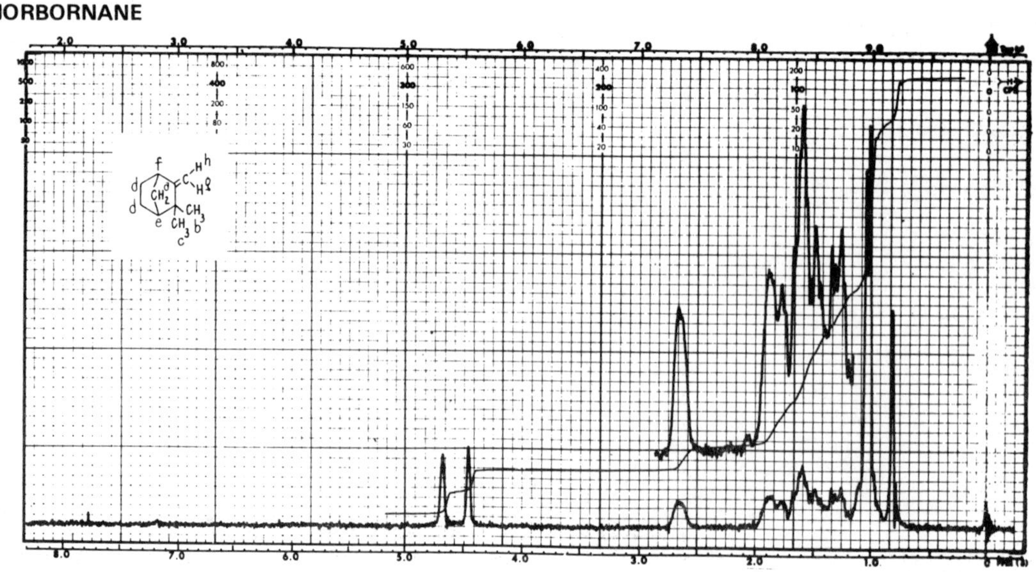

108

© 1978 Sadtler Research Laboratories, Inc.

2(10)-PINENE

$C_{10}H_{16}$

Mol. Wt. 136.24

Solvent: CCl_4

Assignments
a = 0.72
b = 1.25
c = 1.31-2.60
d = ca 4.57

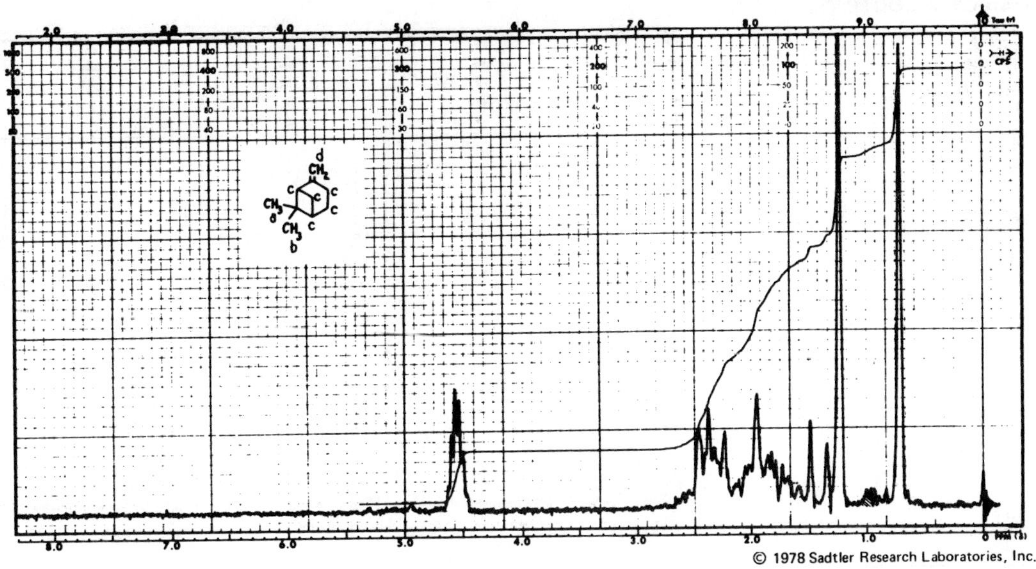

© 1978 Sadtler Research Laboratories, Inc.

109

2-PINENE

$C_{10}H_{16}$

Mol. Wt. 136.24

B.P. 152°C

Solvent: CCl_4

Assignments
a = 0.84
b = 1.27
c = 1.65
d = 1.80-2.60
e = 5.17

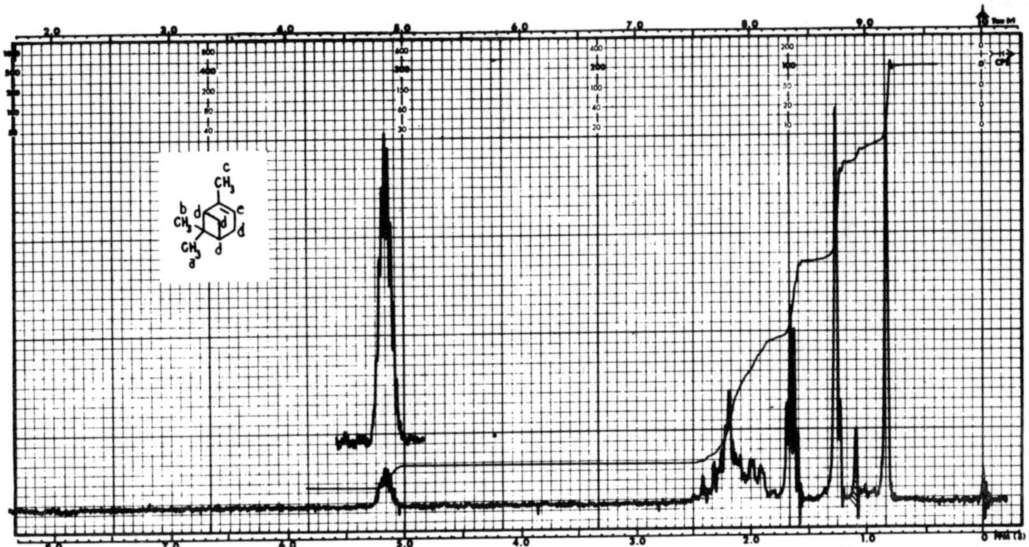

© 1978 Sadtler Research Laboratories, Inc.

110

---

p-MENTHA-1,8-DIENE

$C_{10}H_{16}$

Mol. Wt. 136.24

Solvent: CCl_4

Assignments
a = 1.63
v = 1.71
c = 1.60-2.40
d = 4.69
e = 5.39

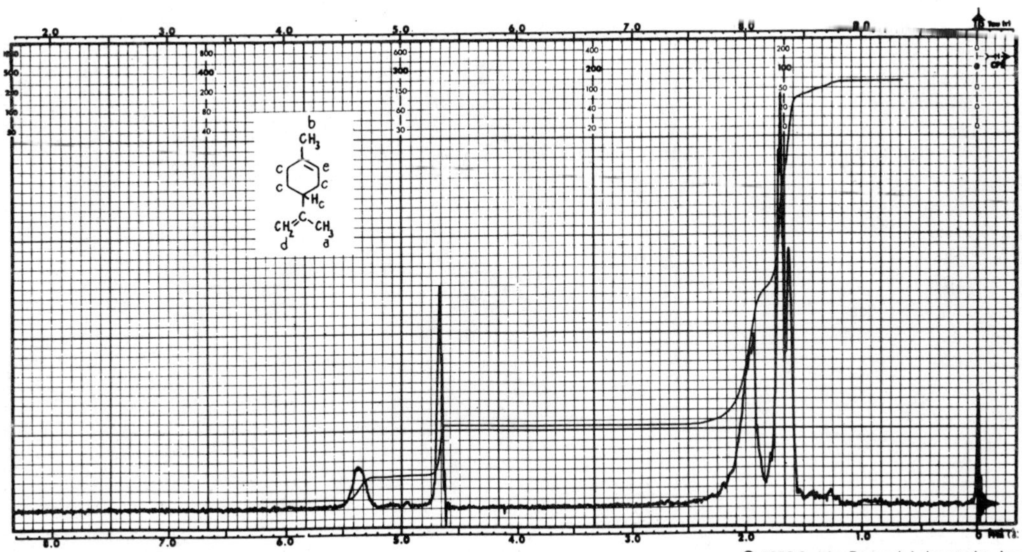

© 1978 Sadtler Research Laboratories, Inc.

111

1,3-CYCLOHEXADIENE

C_6H_8

Mol. Wt. 80.13

Solvent: $CDCl_3$

Assignments
a = 2.14
b = 5.83

112

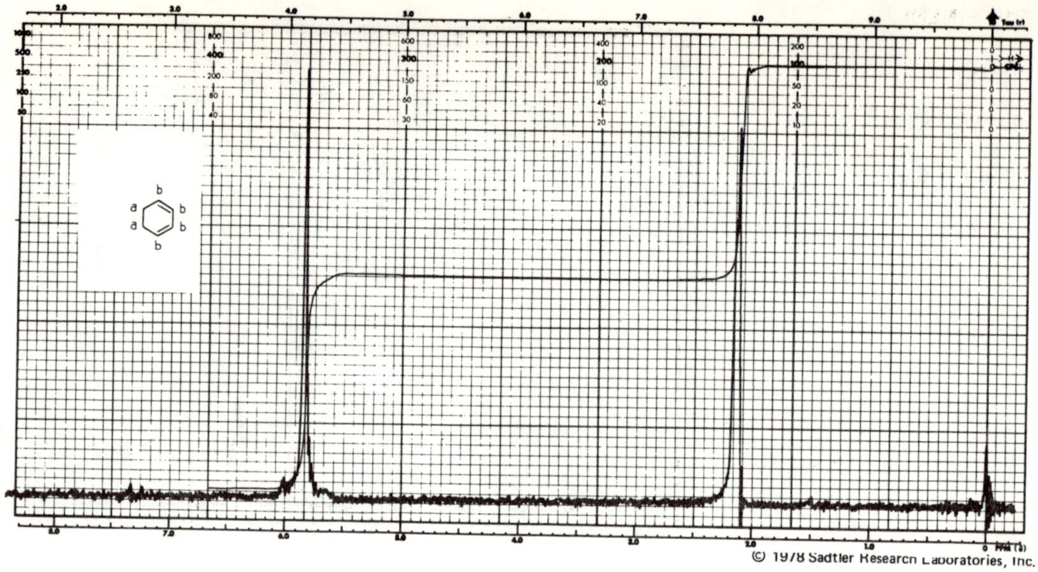

© 1978 Sadtler Research Laboratories, Inc.

1,4-CYCLOHEXADIENE

C_6H_8

Mol. Wt. 80.13

Solvent: $CDCl_3$

Assignments
a = 2.65
b = 5.68

113

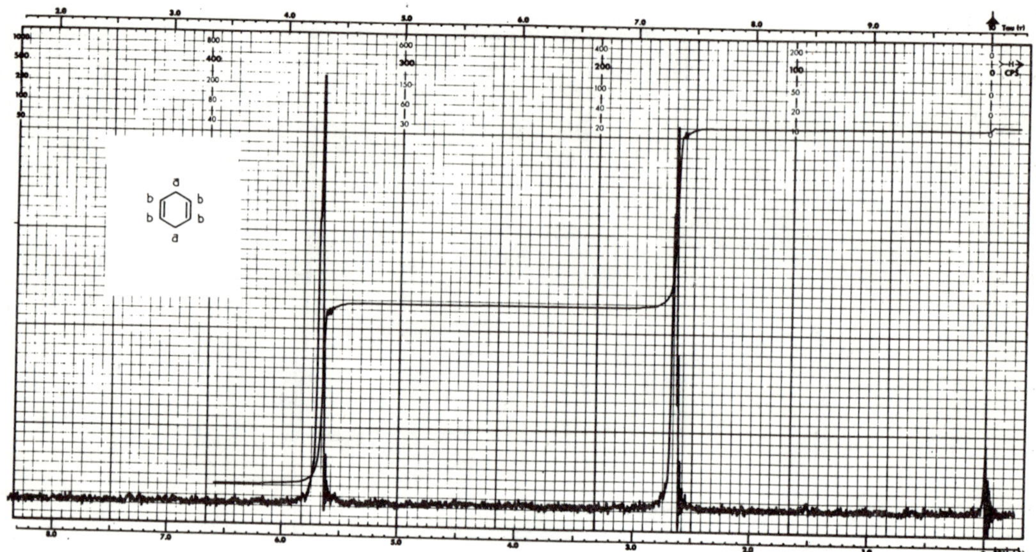

© 1978 Sadtler Research Laboratories, Inc.

1,5-CYCLOOCTADIENE

C_8H_{12}

Mol. Wt. 108.18

B.P. 144-153°C

Solvent: CCl_4

Assignments
a = 2.33
b = 5.48

114

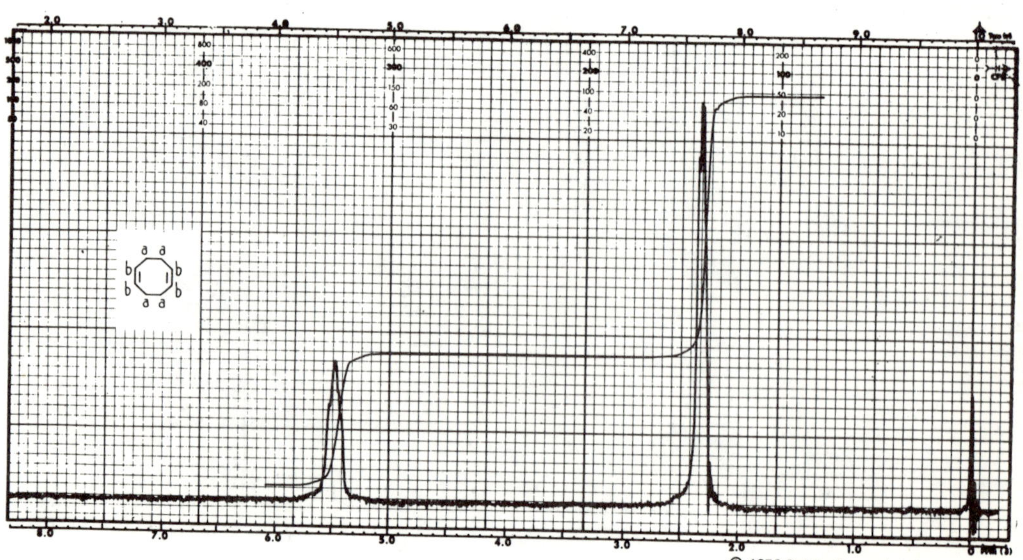

© 1978 Sadtler Research Laboratories, Inc.

3a,4,7,7a-TETRAHYDRO-4,7-METHANOINDENE

$C_{10}H_{12}$

Mol. Wt. 132.21

M.P. 33°C

Solvent: CCl_4

Assignments

a = 1.27 or b

b = 1.50 or a

c = 1.50-3.40

d = ca 5.41

e = ca 5.90

115

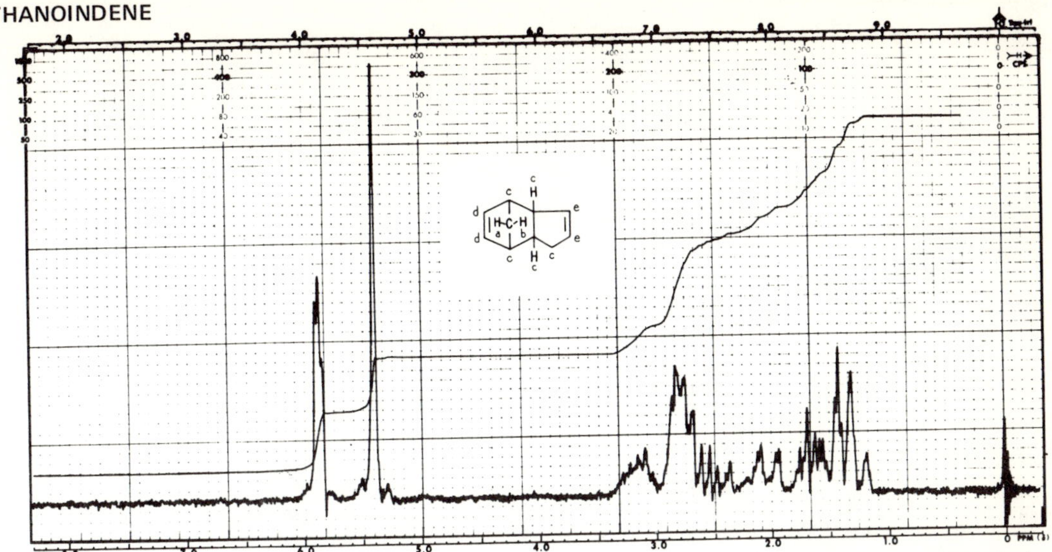

© 1978 Sadtler Research Laboratories, Inc.

NOTES

The monosubstituted acetylenes (H–C≡C–X) are usually easily characterized because the acetylenic hydrogen appears over a relatively limited chemical shift range (2.0–3.0 ppm in CCl_4) and displays coupling between non-equivalent proton groups on opposite sides of the triple bond linkage (J = 2.0–3.2 Hz). The absence of this hydrogen in the disubstituted acetylenes makes the identification of these compounds somewhat more difficult although the coupling across the triple bond may still be observed.

The deshielding effect of the C≡C linkage is similar to that of the C=C group, i.e., it is a weak deshielder of both aliphatic and aromatic protons.

Chemical Shifts

Aliphatic Protons

HC–4	HC–3	HC–2	HC–1		Solvent	Spectrum
			1.80	CH_3–C≡C–H	$CDCl_3$	- - - -
	1.01	1.55	2.12	CH_3–CH_2–CH_2–C≡C–H	CCl_4	117
0.93	1.40	1.53	2.11	CH_3–CH_2–CH_2–CH_2–C≡C–H	CCl_4	118
			1.70	CH_3–C≡C–R_1	CCl_4	125
		(1.18)	- - - -	$(CH_3)_3$–C–C≡C–R_1	CCl_4	126
			1.71	CH_3–C≡C–Ø	$CDCl_3$	167
	1.01	1.59	2.32	CH_3–CH_2–CH_2–C≡C–Ø	CCl_4	- - - -

Monosubstituted ethynes

H–C≡C	–X	Solvent
1.79	–R_3	CCl_4
1.00	–CH_3	$CDCl_3$
2.25	–C(–CH_3, –CH_3, –R_2)	CCl_4
2.29	–cyclohexyl	$CDCl_3$
2.30	–C(–CH_3, –R_2, –OH)	CCl_4
2.33	–CH(–R_2)–OH	CCl_4
2.33	–CH(–R_3)–OH	CCl_4
2.37	–cyclopentyl	CCl_4
2.71	–C$(CH_3)_2$–NH–SO_2–Ø	Polysol
2.86	–C(=O)–O–CH$(CH_3)_2$	$CDCl_3$
3.04	–Ø	$CDCl_3$
3.19	–C(=O)–OH	$CDCl_3$
3.40	–C(=O–CH_3	$CDCl_3$
4.06	–Ø	DMSO-d_6

Chemical Shifts continued

Aliphatic Protons continued

1-Propynes

H–C≡C	–CH_2	–X	Solvent
2.23	3.29	–$N(CH_3)_2$	$CDCl_3$
2.31	3.99	–O–CH_3	CCl_4
2.39	3.97	–SO_2–Ø	$CDCl_3$
2.41	3.83	–Br	CCl_4
2.42	4.06	–Cl	CCl_4
2.49	4.25	–SO_2–Ø	TFA
2.54	4.23	–OH	$CDCl_3$

1-Butynes

H–C≡C	–CH_2	–CH_2	–X	
1.89	2.20	1.40	–$CH(CH_3)_2$	CCl_4
1.90	2.39	2.39	–C≡C–H	CCl_4
1.96	2.38	3.67	–OH	CCl_4
2.53	2.55	3.43	–SO_2–Ø	Polysol

2-Butynes (1,4-disubstituted)

X	–CH_2–C≡C		–CH_2	–Y	
R_5–	2.20		4.19	–OH	CCl_4
R_4–	2.22		4.89	–O–C(=O)–CF_3	$CDCl_3$
CH_3–	2.23		4.17	–OH	CCl_4
R–	2.24		3.27	–$N(R_2)_2$	CCl_4
CH_3–	2.26		4.90	–O–C(=O)–CF_3	$CDCl_3$
Ø–S–	3.60		3.60	–S–Ø	Polysol
Ø–S–	3.68		4.61	–O–Ø	$CDCl_3$
Br–	3.98		3.98	–Br	$CDCl_3$
Cl–	4.12		4.12	–Cl	CCl_4
Cl–	4.12		4.68	–O–Ø (p–Br)	$CDCl_3$
Cl–	4.16		4.82	–O–C(=O)–NH–Ø	$CDCl_3$
Cl–	4.32	or	4.41	–SO_2–Ø	Polysol
Ø–SO_2–	4.32		4.79	–O–Ø	Polysol
(p–CH_3) Ø–O–	4.65		4.65	–O–Ø (p–CH_3)	$CDCl_3$

Chemical Shifts continued

Aromatic Protons

The placement of various substituents on the triple bond opposite a phenyl group has only a slight effect upon the chemical shift ranges observed for the resulting complex, higher order bands. The fact that the lower field limit of these ranges is only about 0.2 ppm downfield from the chemical shift of unsubstituted benzene (7.37 ppm) indicates that the C≡C group has only a slight deshielding effect on the ortho aromatic hydrogens.

Phenyl acetylenes

Ø–C≡C	–X	Solvent
7.00–7.65	–H	$CDCl_3$
7.05–7.50	$-R_7$	CCl_4
7.10–7.50	$-CH_3$	$CDCl_3$
7.15–7.60	$-CH(-O-R_2)_2$	$CDCl_3$
7.15–7.65	–C≡C–Ø	$CDCl_3$
7.15–7.65	–Ø	CCl_4
7.20–7.65	$-C(=O)-O-R_2$	CCl_4
7.20–7.70	$-C(=O)-CH_3$	$CDCl_3$
7.25–7.70	–C(=O)–OH	Polysol

Para substituted compounds

	ortho	meta	para	
Ø–C≡C–	7.42	6.85	$-O-CH_3$	$CDCl_3$
H–C≡C–	7.43 or	7.30	–Br	$CDCl_3$

Solubility and Solvent Effects

The low molecular weight alkynes are readily soluble in the halogenated solvents normally used for the preparation of NMR solutions. It has been noted that the hydrogen bonded directly to the triple bond carbon is strongly deshielded in the presence of DMSO-d_6 in comparison to CCl_4 and $CDCl_3$. For example, the acetylenic hydrogen of phenyl acetylene appears at about 3.0 ppm in $CDCl_3$ solution but at 4.1 ppm in DMSO-d_6.

3,3-DIMETHYL-1-BUTYNE

C_6H_{10}

Mol. Wt. 82.15

Solvent: CCl_4

Assignments
a = 1.21
b = 1.89

116

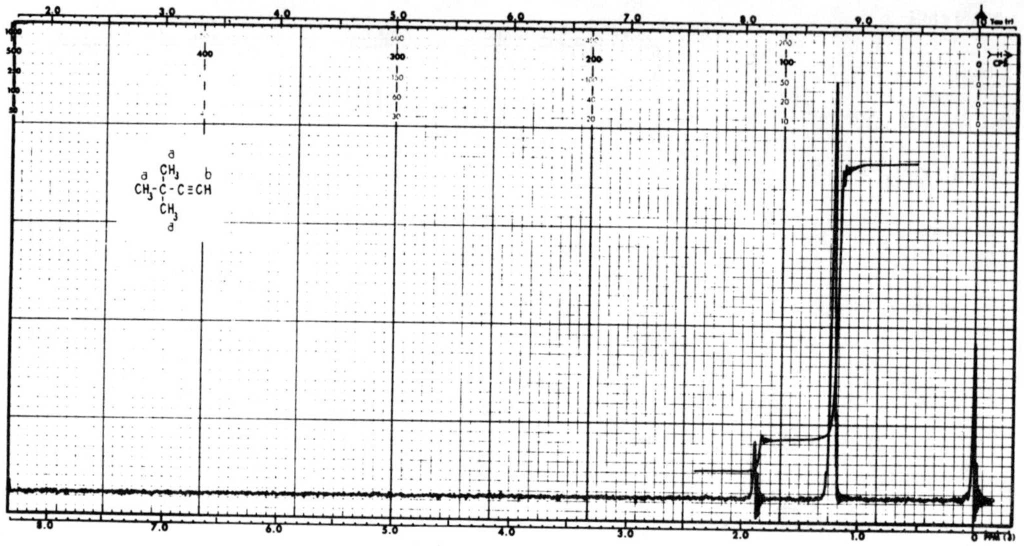

© 1978 Sadtler Research Laboratories, Inc.

1-PENTYNE

C_5H_8

Mol. Wt. 68.12

Solvent: CCl_4

Assignments
a = 1.01
b = 1.55
c = 1.79
d = 2.12

117

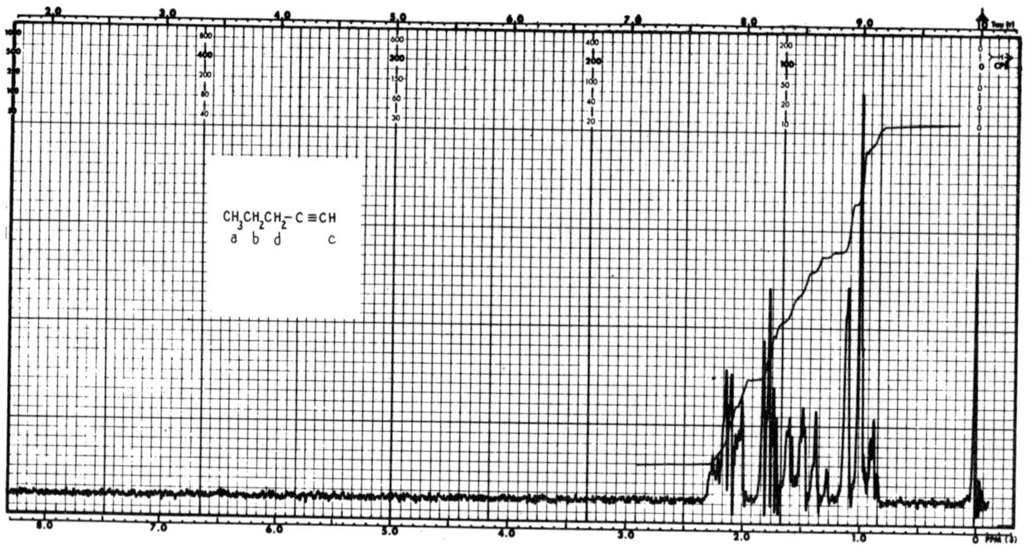

© 1978 Sadtler Research Laboratories, Inc.

1-HEXYNE

C_6H_{10}

Mol. Wt. 82.15

Solvent: CCl_4

Assignments
a = 0.93
b = 1.18-1.65
c = 1.73
d = 2.11

118

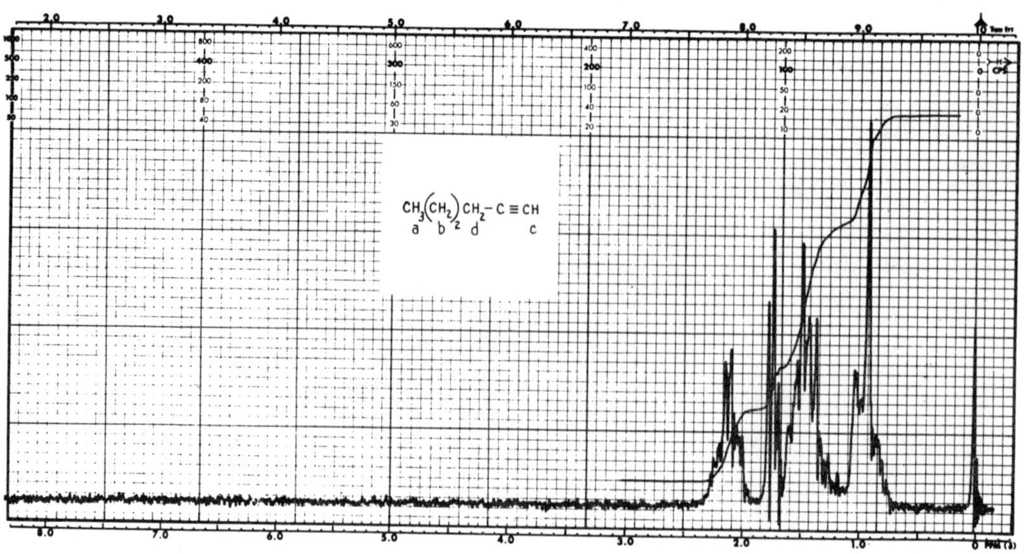

© 1978 Sadtler Research Laboratories, Inc.

1-NONYNE

C_9H_{16}

Mol. Wt. 124.23

Solvent: CCl_4

Assignments

a = 0.89
b = 1.32
c = 1.72
d = ca 2.09

119

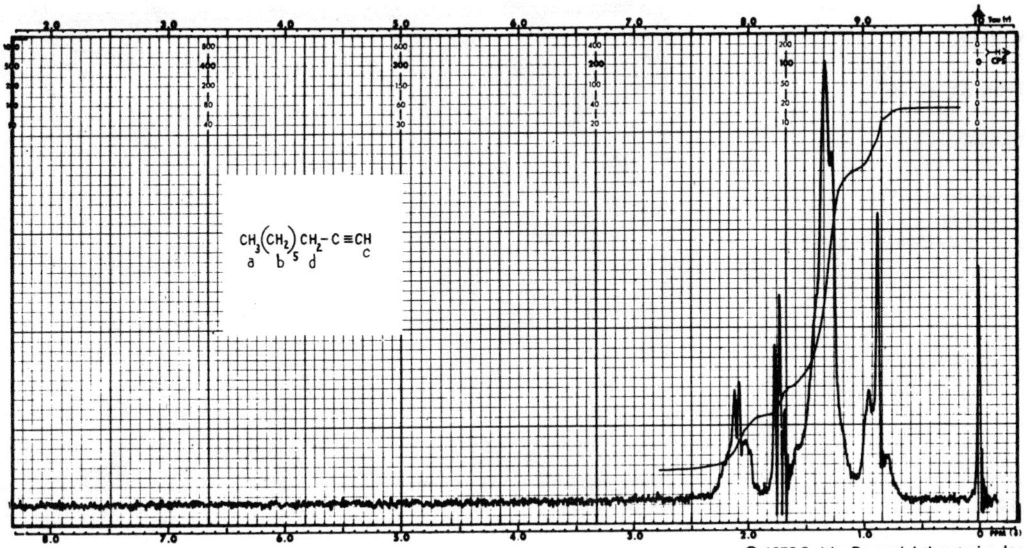

© 1978 Sadtler Research Laboratories, Inc.

1-DECYNE

$C_{10}H_{18}$

Mol. Wt. 132.25

M.P. 44°C B.P. 162-165°C

Solvent: CCl_4

Assignments

a = 0.88
b = 1.30
c = 1.74
d = 2.10

120

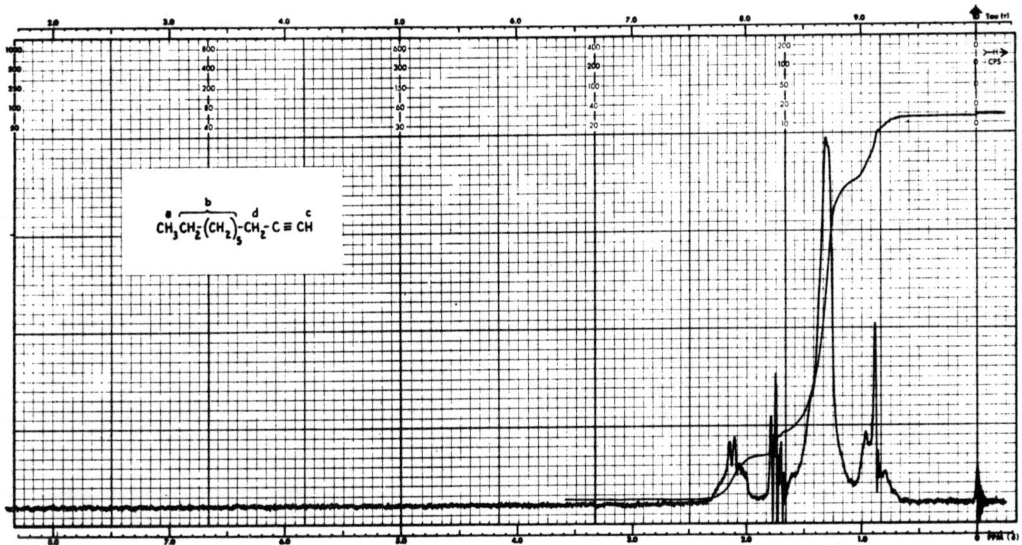

© 1978 Sadtler Research Laboratories, Inc.

1-UNDECYNE

$C_{11}H_{20}$

Mol. Wt. 152.28

Solvent: CCl_4

Assignments

a = 0.90
b = 1.10-1.85
c = 1.79
d = 2.15

121

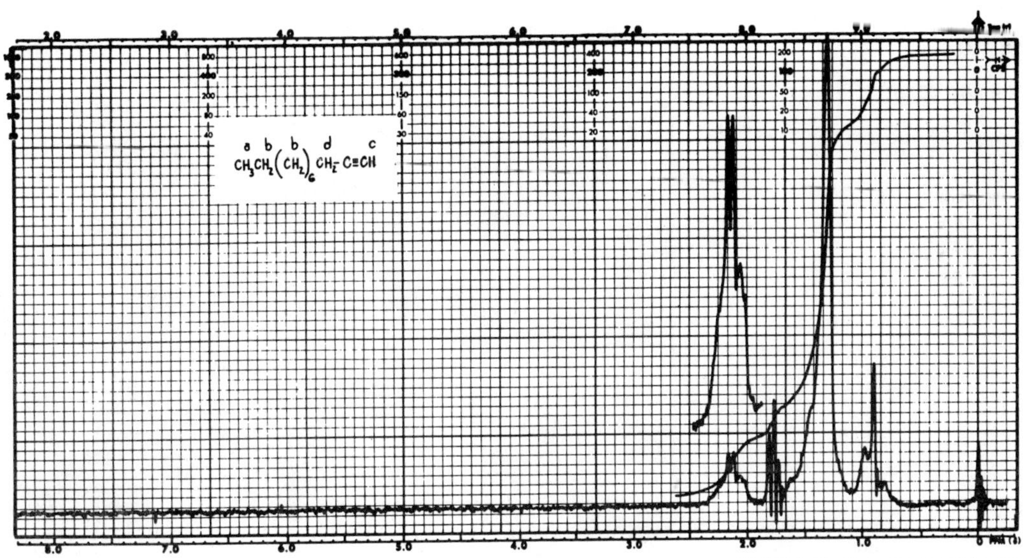

© 1978 Sadtler Research Laboratories, Inc.

1-OCTADECYNE

$C_{18}H_{34}$

Mol. Wt. 250.47

Solvent: CCl_4

Assignments
a = 0.90
b = 1.10-1.70
c = 1.75
d = 2.11

122

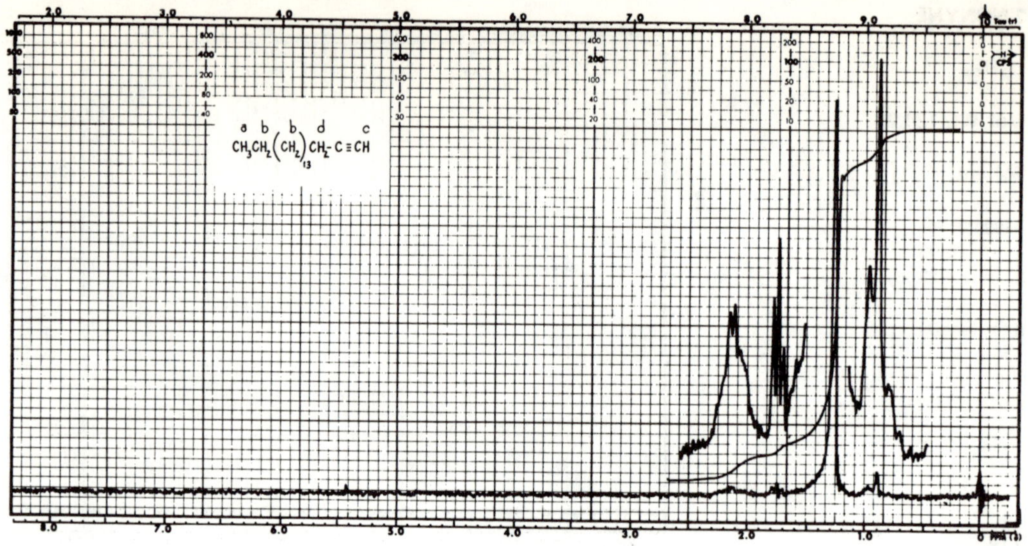

© 1978 Sadtler Research Laboratories, Inc.

1,7-OCTADIYNE

C_8H_{10}

Mol. Wt. 106.17

Solvent: $CDCl_3$

Assignments
a = 1.68
b = 1.98
c = 2.22

123

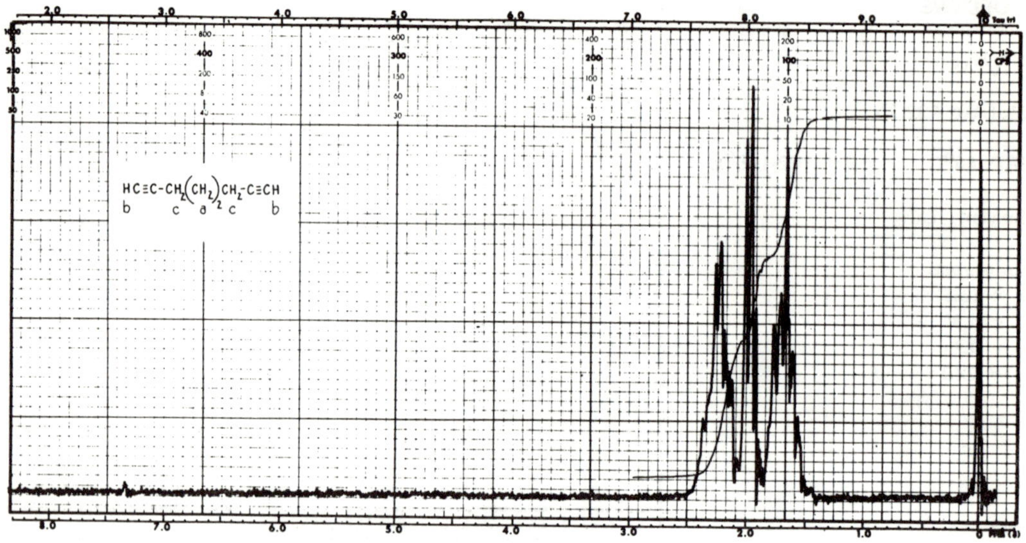

© 1978 Sadtler Research Laboratories, Inc.

1,8-NONADIYNE

C_9H_{12}

Mol. Wt. 120.20

Solvent: $CDCl_3$

Assignments
a = ca 1.52
b = 1.93
c = 2.20

124

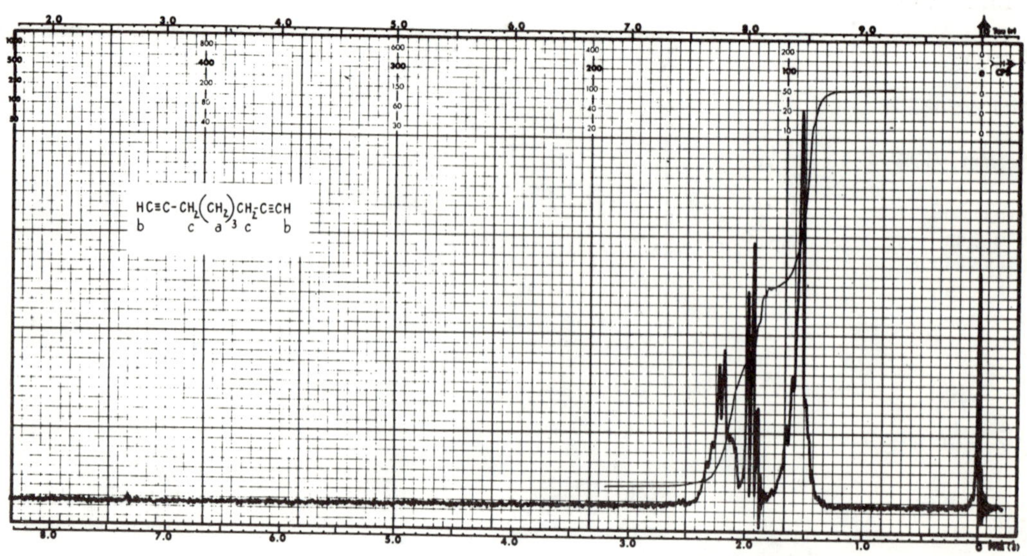

© 1978 Sadtler Research Laboratories, Inc.

2-BUTYNE

C_4H_6

Mol. Wt. 54.09

Solvent: CCl_4

Assignments
a = 1.70

125

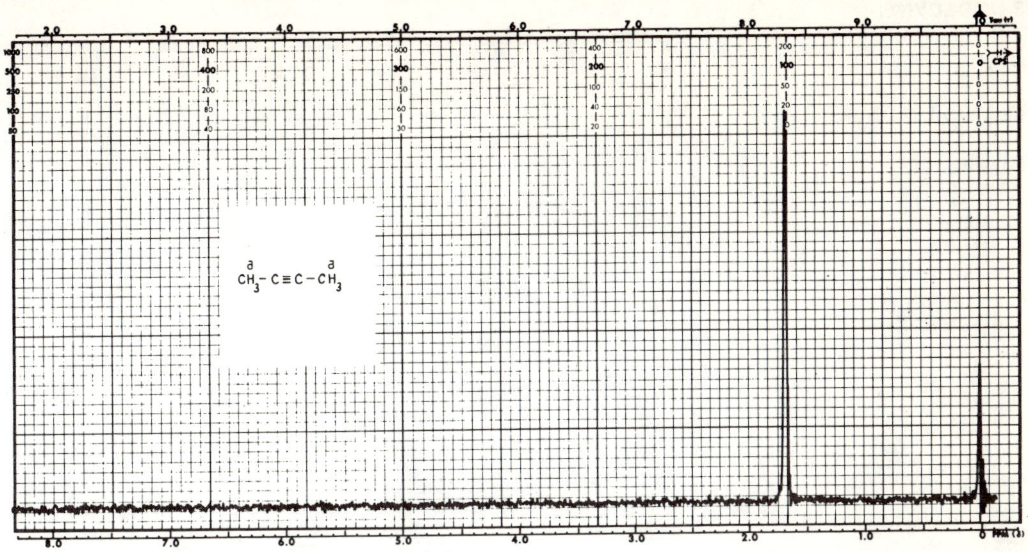

© 1978 Sadtler Research Laboratories, Inc.

4,4-DIMETHYL-2-PENTYNE

C_7H_{12}

Mol. Wt. 96.17

Solvent: CCl_4

Assignments
a = 1.18
b = 1.71

126

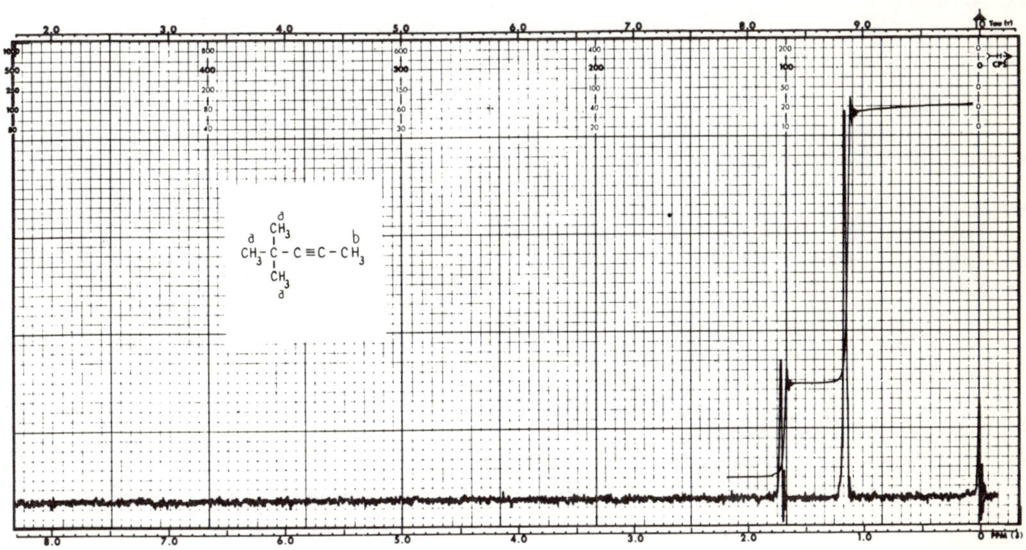

© 1978 Sadtler Research Laboratories, Inc.

4-DECYNE

$C_{10}H_{18}$

Mol. Wt. 138.25

Solvent: CCl_4

Assignments
a = 0.91
b = 0.98
c = 1.10-1.80
d = 2.10

127

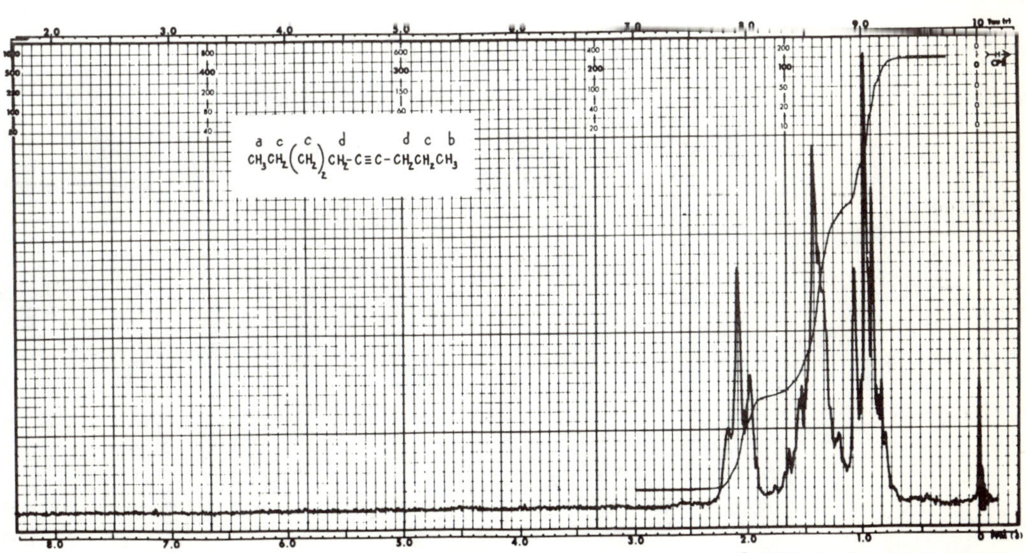

© 1978 Sadtler Research Laboratories, Inc.

3-UNDECYNE

$C_{11}H_{20}$

Mol. Wt. 152.28

Solvent: CCl_4

Assignments
a = 0.91
b = 1.11
c = 1.10-1.75
d = ca 2.10

128

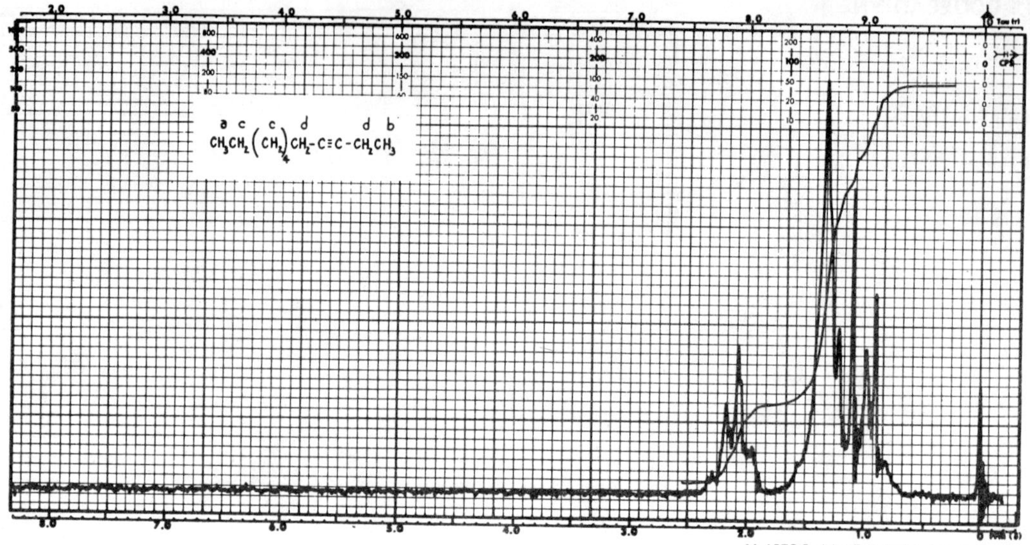

© 1978 Sadtler Research Laboratories, Inc.

6-DODECYNE

$C_{12}H_{22}$

Mol. Wt. 166.31

Solvent: CCl_4

Assignments
a = 0.91
b = 1.10-1.80
c = 2.09

129

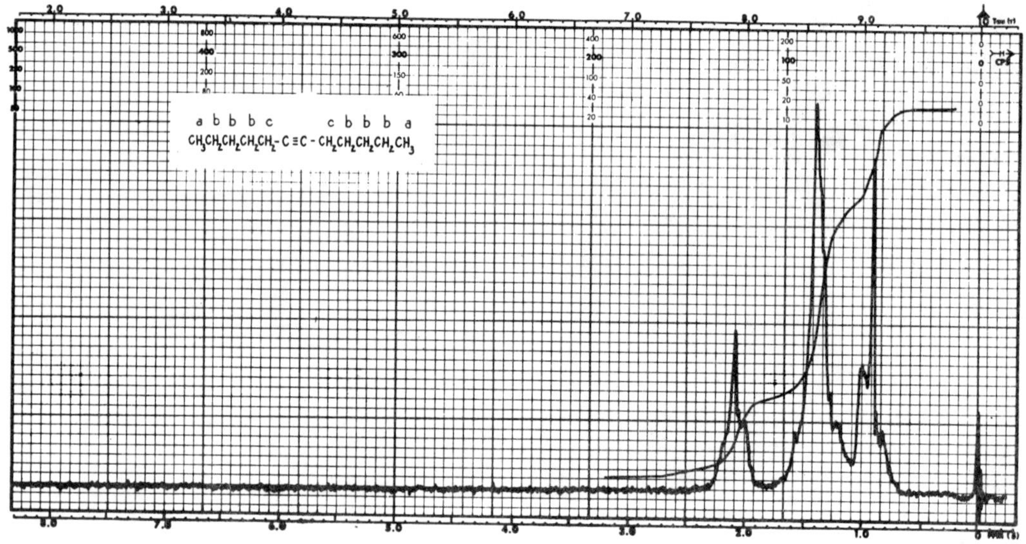

© 1978 Sadtler Research Laboratories, Inc.

2,9-DIMETHYL-5-DECYNE

$C_{12}H_{22}$

Mol. Wt. 166.31

Solvent: CCl_4

Assignments
a = 0.90
b = 1.36
c = 1.68
d = 2.10

130

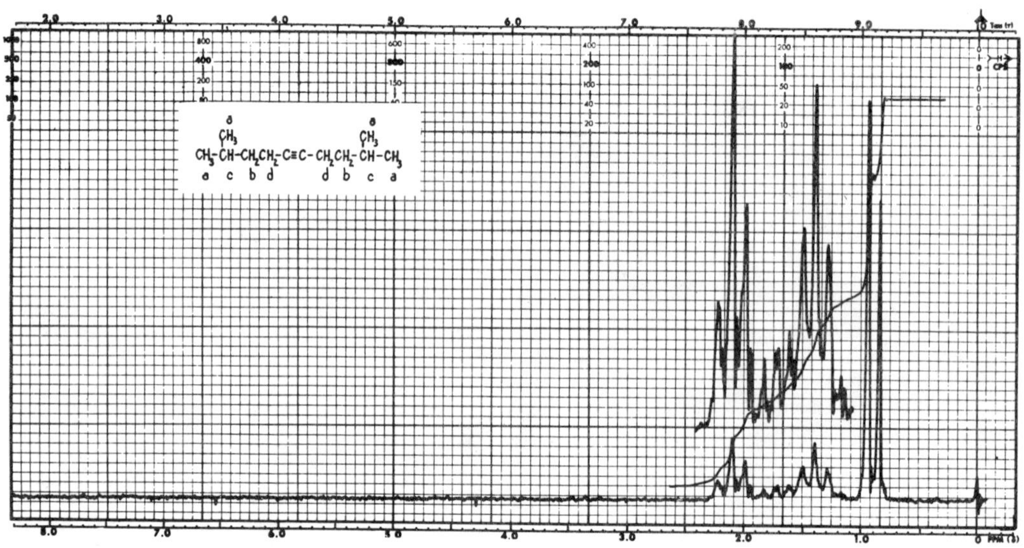

© 1978 Sadtler Research Laboratories, Inc.

3,9-DODECADIYNE

$C_{12}H_{18}$

Mol. Wt. 162.28

Solvent: CCl_4

Assignments

a = 1.11

b = 1.52

c = ca 2.11

131

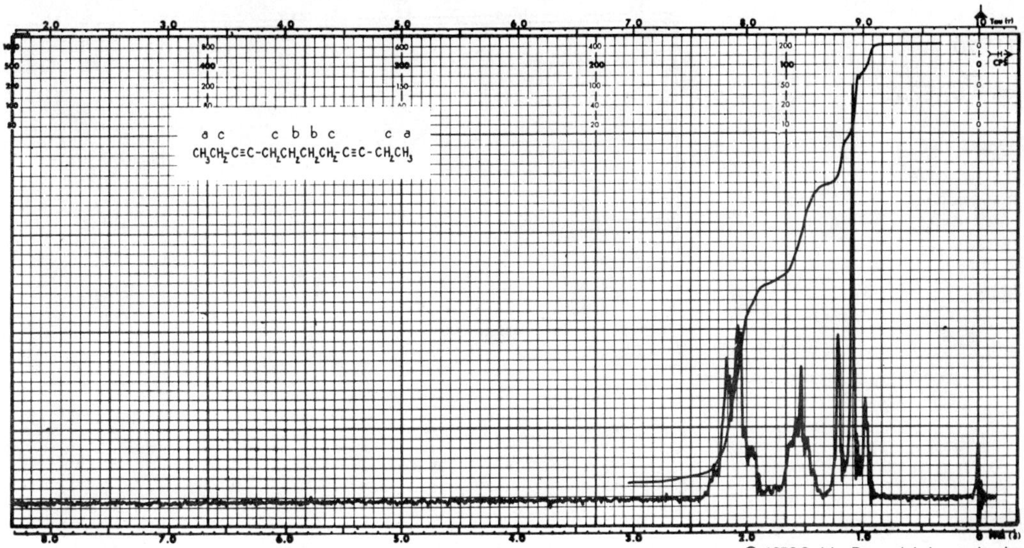

© 1978 Sadtler Research Laboratories, Inc.

NOTES

The aromatic protons of alkyl substituted benzene rings usually appear as a single broad peak near 7.1 ppm or a complex multiplet in the range from 6.9–7.5 ppm for highly branched chains such as the tert-butyl group. Aliphatic groups shield the ortho aromatic hydrogens by a factor of about 0.34 ppm, as evidenced by the aromatic resonance of mesitylene (1,3,5-trimethyl benzene) which appears at 6.69 ppm.

		Solvent	Spectrum
Benzene	7.37 ppm	CCl_4	132
Toluene	7.04 ppm	CCl_4	133
p-Xylene	7.07 ppm	CCl_4	181
Mesitylene	6.69 ppm	CCl_4	189

Chemical Shifts

Aliphatic Protons

HC–3	HC–2	HC–1			
		2.29	CH_3–Ø	CCl_4	133
	1.19	2.53	CH_3–CH_2–Ø	CCl_4	134
0.93	1.63	2.58	CH_3–CH_2–CH_2–Ø	$CDCl_3$	135
	(1.22)	2.83	$(CH_3)_2$–CH–Ø	CCl_4	136
	(1.32)	- - - -	$(CH_3)_3$–C–Ø	CCl_4	140
		2.46	CH_3–1(naphthalene)	$CDCl_3$	- - - -
		2.49	CH_3–2(naphthalene)	$CDCl_3$	202
	1.37	3.03	CH_3–CH_2–1(naphthalene)	CCl_4	201
	1.30	2.77	CH_3–CH_2–2(naphthalene)	$CDCl_3$	- - - -

As shown above, the alkyl protons of aliphatic groups bonded to the naphthalene ring system resonate at lower field than those bonded to benzene, in addition, the groups bonded to carbons 1,4,5,8 resonate at lower field than similar groups situated at positions 2,3,6,7. The aromatic resonances of naphthalene ring systems generally appear as a complex, higher order series of multiplets in the range from 7.0–8.0 ppm.

Chemical Shifts continued

Aliphatic Protons continued

Alpha-substituted toluenes

Phenyl–	CH_2	–X	Solvent
ca 7.07	2.53	$-CH_3$	CCl_4
ca 7.24	2.53	$-C_3$	$CDCl_3$
ca 6.94	2.91	$-Si(\emptyset)_3$	$CDCl_3$
ca 7.14	3.33	$-CH{=}CH_2$	CCl_4
ca 7.27	3.61	–C(=O)–OH	$CDCl_3$
ca 7.27	3.62	–C≡N	CCl_4
ca 7.22	3.64	–SH	CCl_4
7.0-7.4	3.67	–C(=O)–R	$CDCl_3$
ca 7.29	3.85	–S(=O)–R	$CDCl_3$
ca 7.29	3.85	$-NH_2$	$CDCl_3$
ca 7.59	4.27	$-NH_2$ (salt)	D_2O
ca 7.22	4.34	–Br	CCl_4
7.07-7.5	4.38	–I	$CDCl_3$
ca 7.19	4.41	–OH	CCl_4
ca 7.28	4.48	–Cl	CCl_4
ca 7.24	4.55	–S–C(=S)–S–R	$CDCl_3$
ca 7.33	4.59	–N≡C	$CDCl_3$
ca 7.29	4.59	–N=C=S	$CDCl_3$
ca 7.44	4.83	$-SO_2-Cl$	$CDCl_3$
ca 7.42	5.07	$-SO_2-F$	Polysol
ca 7.27	5.24	–F	CCl_4

Aromatic Protons

Para substituted toluenes (in increasing "meta" shift)

CH_3–	Benzene ring ortho	meta	para	Solvent
2.18	6.79	6.33	$-NH_2$	CCl_4
2.21	6.92	6.40	$-NH-CH_3$	$CDCl_3$
2.18	6.89	6.49	$-N(CH_3)_2$	CCl_4
2.21	7.01	6.59	$-N(R_2)_2$	$CDCl_3$
2.22	6.93	6.62	$-O-R_{12}$	CCl_4
2.20	6.92	6.69	$-OH$	$CDCl_3$
2.28	7.09	6.79	$-O-CH_3$	$CDCl_3$
2.23	7.00	6.81	$-F$	CCl_4
2.28	7.01	6.82	$-O-C(=O)-CH_3$	CCl_4
2.27	6.90	6.90	$-CH_3$	CCl_4
2.27	7.14	6.92	$-O-C(=O)-NH_2$	DMSO-d_6
2.28	6.99	6.99	$-CH(CH_3)_2$	CCl_4
2.28	7.01	7.01	$-CH_2-OH$	CCl_4
2.30	7.04	7.04	$-CH_2-CH_3$	CCl_4
2.30	7.04	7.04	$-CH_2-NH_2$	CCl_4
2.28	7.05	7.05	$-S-CH_3$	CCl_4
2.29	7.05	7.05	$-CH_2-C{\equiv}N$	CCl_4
2.34	7.07	7.07	$-O-C(=O)-\emptyset$	CCl_4
2.20	6.91	7.09	$-SH$	$CDCl_3$
2.30	7.09	7.09	$-CH_2-CH_2-NH_2$	$CDCl_3$
2.31	7.10	7.10	$-CH_2-CH_2-OH$	$CDCl_3$
2.29	6.98	7.12	$-Cl$	CCl_4
2.29	7.00	7.12	$-C(CH_3)_3$	CCl_4
2.30	7.16	7.16	$-NH-SO_2-R$	$CDCl_3$
2.29	6.99	7.18	$-CH_2-Br$	CCl_4
2.28	7.02	7.18	$-CH_2-Cl$	CCl_4
2.33	7.19	7.19	$-NH-C(=S)-R$	$CDCl_3$
2.30	7.07	7.21	$-S-\emptyset$	$CDCl_3$
2.25	6.95	7.30	$-Br$	CCl_4
2.26	7.08	7.30	$-Hg-Cl$	DMSO-d_6
2.30	7.05	7.34	$-S-S-\emptyset$	$CDCl_3$

Para substituted toluenes (in increasing "meta" shift) continued

	Benzene ring			
CH_3–	ortho	meta	para	Solvent
2.41	7.23	7.47	–C≡N	CCl_4
2.26	6.89	7.55	–I	$CDCl_3$
2.35	7.20	7.55	–S(=O)–OH	$CDCl_3$
2.36	7.22	7.55	–P(=O)–(Ø)$_2$	$CDCl_3$
2.32	7.24	7.58	–P(=O)–(OH)$_2$	DMSO-d_6
2.36	7.35	7.58	–S(=O)–O–Na	D_2O
2.33	7.21	7.59	–CH=CH–C(=O)–OH	DMSO-d_6
2.30	7.21	7.61	–SO_2–OH (salt)	DMSO-d_6
2.39	7.22	7.65	–C(=O)–Ø	$CDCl_3$
2.32	7.09	7.69	–C(=O)–CH_3	CCl_4
2.33	7.10	7.69	–N=S=O	CCl_4
2.33	7.12	7.70	–SO_2–OH (salt)	$CDCl_3$
2.43	7.32	7.71	–SO_2–O–CH_3	CCl_4
2.42	7.29	7.70	–SO_2–O–R_{10}	CCl_4
2.41	7.31	7.76	–C(=O)–H	$CDCl_3$
2.43	7.37	7.79	–SO_2–O–CH_2–Ø	$CDCl_3$
2.39	7.22	7.83	–N=N–Ø	$CDCl_3$
2.46	7.37	7.84	–SO_2–NH_2	TFA
2.39	7.25	7.84	–N=N–Ø	$CDCl_3$
2.29	7.26	7.88	–SO_2–O–Na	D_2O
2.30	7.15	7.89	–C(=O)–O–CH_3	$CDCl_3$
2.50	7.41	7.90	–SO_2–F	$CDCl_3$
2.49	7.41	7.90	–SO_2–Cl	$CDCl_3$
2.49	7.43	7.92	–C(=O)–NH_2	TFA
2.44	7.24	7.96	–C(=O)–Cl	CCl_4
2.47	7.30	8.02	–C(=O)–OH	TFA
2.45	7.30	8.10	–NO_2	$CDCl_3$

BENZENE

C_6H_6

Mol. Wt. 78.11

M.P. 5.2°C B.P. 80.1°C

Solvent: CCl_4

Assignments
a = 7.23

132

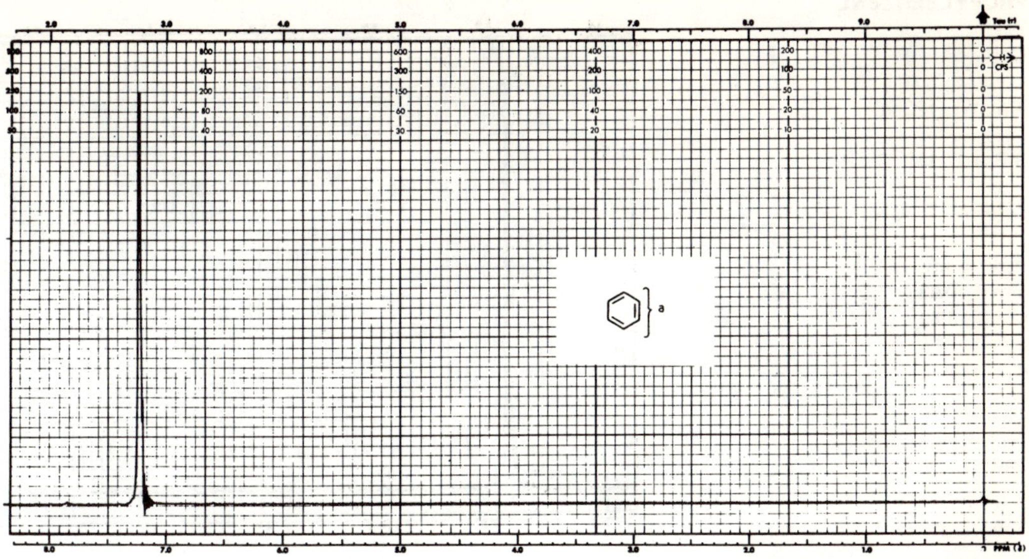

© 1978 Sadtler Research Laboratories, Inc.

TOLUENE

C_7H_8

Mol. Wt. 92.14

B.P. 110.6°C

Solvent: CCl_4

Assignments
a = 2.29
b = 7.04

133

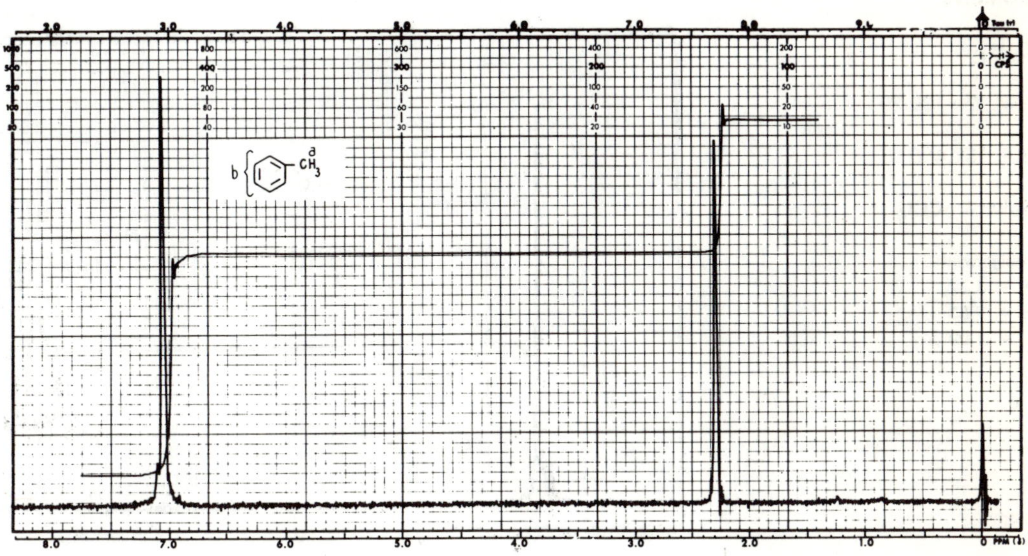

© 1978 Sadtler Research Laboratories, Inc.

ETHYLBENZENE

C_8H_{10}

Mol. Wt. 106.17

B.P. 134-136°C

Solvent: CCl_4

Assignments
a = 1.19
b = 2.53
c = 7.07

134

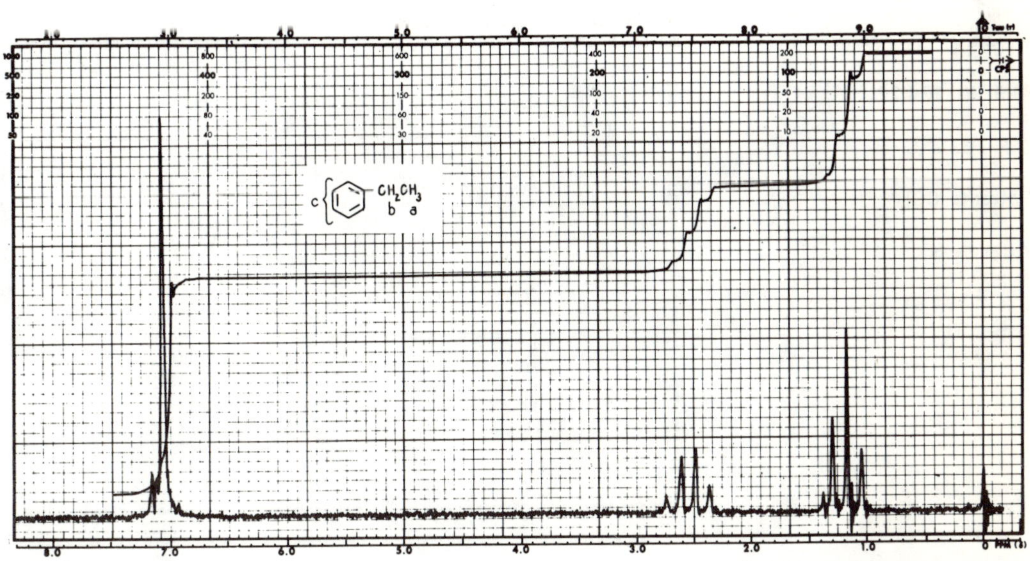

© 1978 Sadtler Research Laboratories, Inc.

PROPYLBENZENE

C_9H_{12}

Mol. Wt. 120.20

B.P. 157-159°C

Solvent: $CDCl_3$

Assignments

a = 0.93
b = 1.63
c = 2.58
d = ca 7.19

135

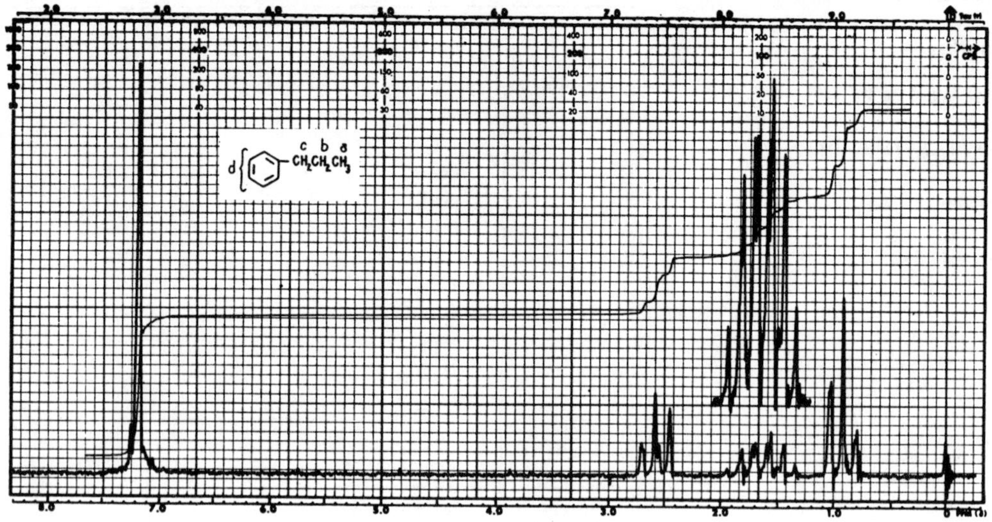

© 1978 Sadtler Research Laboratories, Inc.

CUMENE

C_9H_{12}

Mol. Wt. 120.20

B.P. 152.4°C

Solvent: CCl_4

Assignments

a = 1.22
b = 2.83
c = 7.09

136

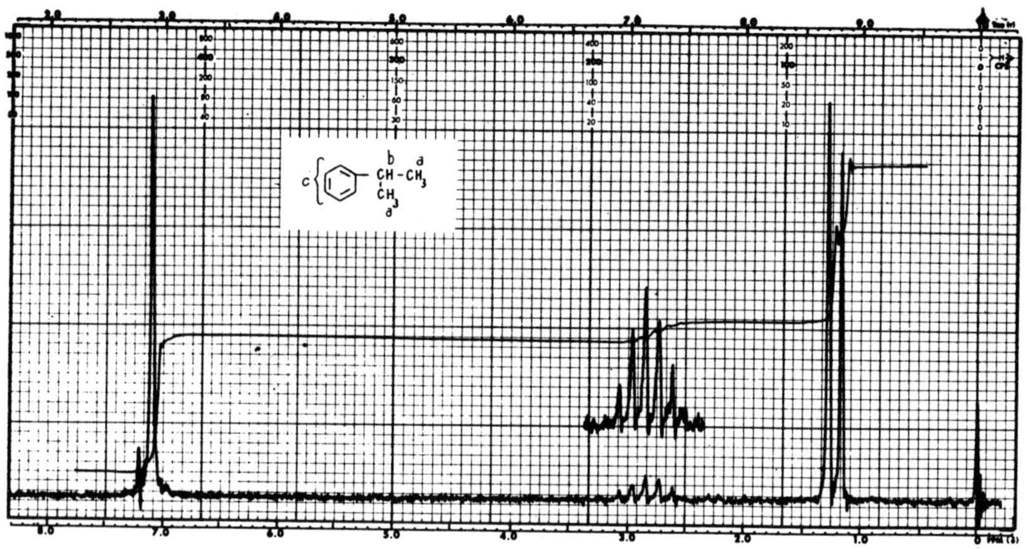

© 1978 Sadtler Research Laboratories, Inc.

BUTYLBENZENE

$C_{10}H_{14}$

Mol. Wt. 134.22

B.P. 182.2°C

Solvent: CCl_4

Assignments

a = 0.92
b = 1.10-1.80
c = 2.58
d = 7.09

137

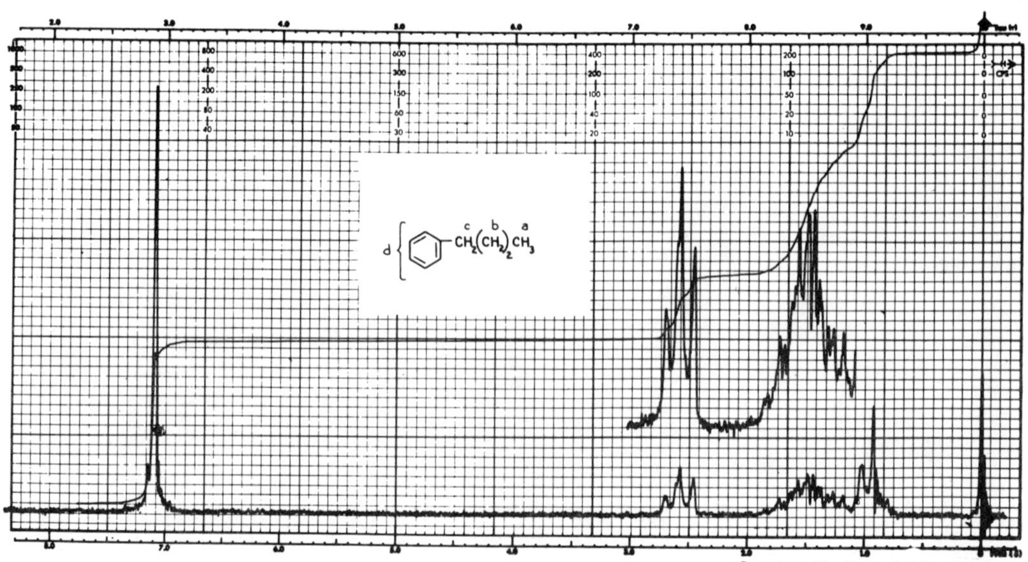

© 1978 Sadtler Research Laboratories, Inc.

sec-BUTYLBENZENE

$C_{10}H_{14}$

Mol. Wt. 134.21

B.P. 173-175°C (lit.)

Solvent: CCl_4

Assignments
a = 0.80
b = 1.20
c = ca 1.59
d = ca 2.52
e = 7.10

138

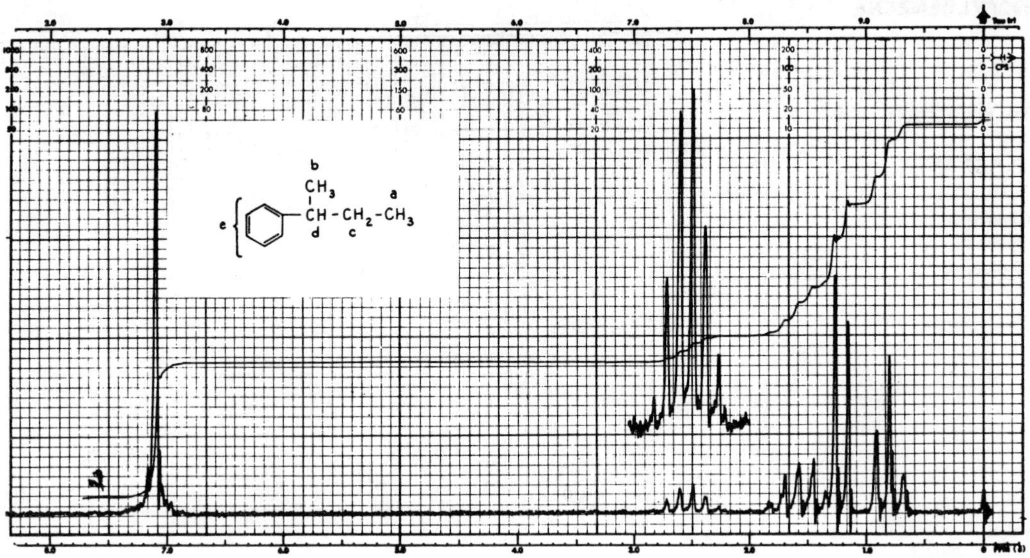

© 1978 Sadtler Research Laboratories, Inc.

ISOBUTYLBENZENE

$C_{10}H_{14}$

Mol. Wt. 134.22

B.P. 172°C

Solvent: CCl_4

Assignments
a = 0.90
b = 1.88
c = 2.42
d = 7.05

139

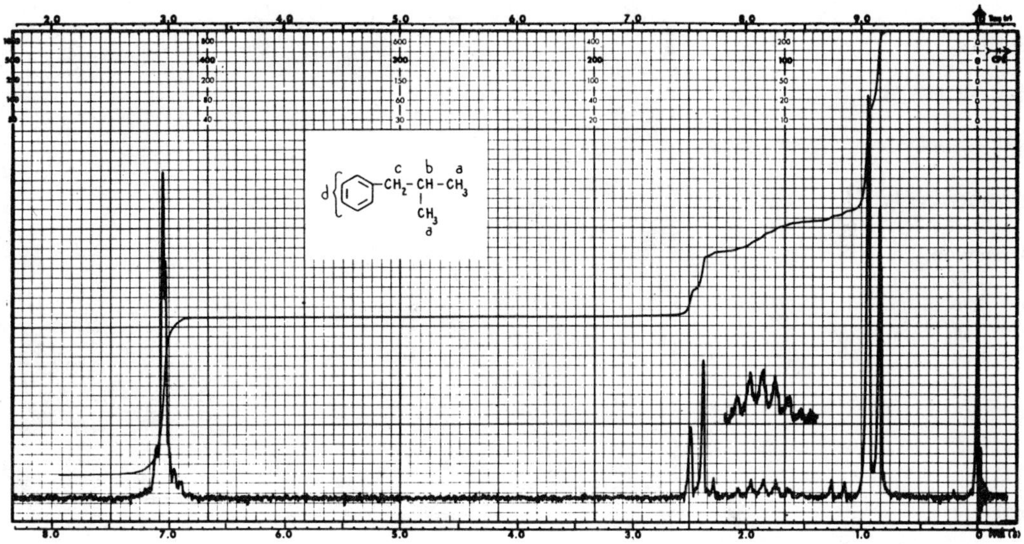

© 1978 Sadtler Research Laboratories, Inc.

tert-BUTYLBENZENE

$C_{10}H_{14}$

Mol. Wt. 134.11

B.P. 168-170°C

Solvent: CCl_4

Assignments
a = 1.32
b = 7.00-7.35

140

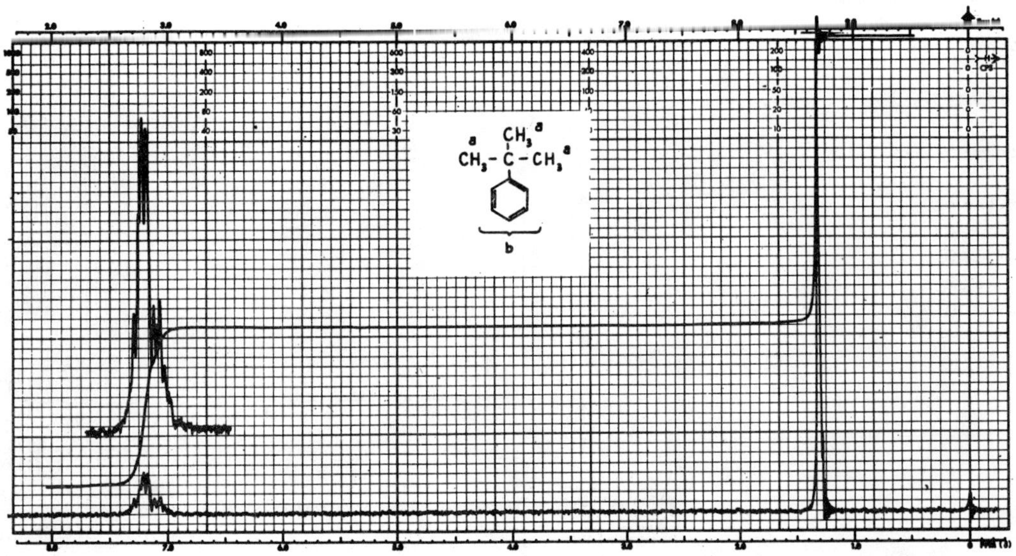

© 1978 Sadtler Research Laboratories, Inc.

UNIV. COLL. N.W. BANGOR LIBRARY

PENTYLBENZENE

$C_{11}H_{16}$

Mol. Wt. 148.25

Solvent: CCl_4

Assignments
a = 0.90
b = 1.08-1.91
c = 2.58
d = 7.13

141

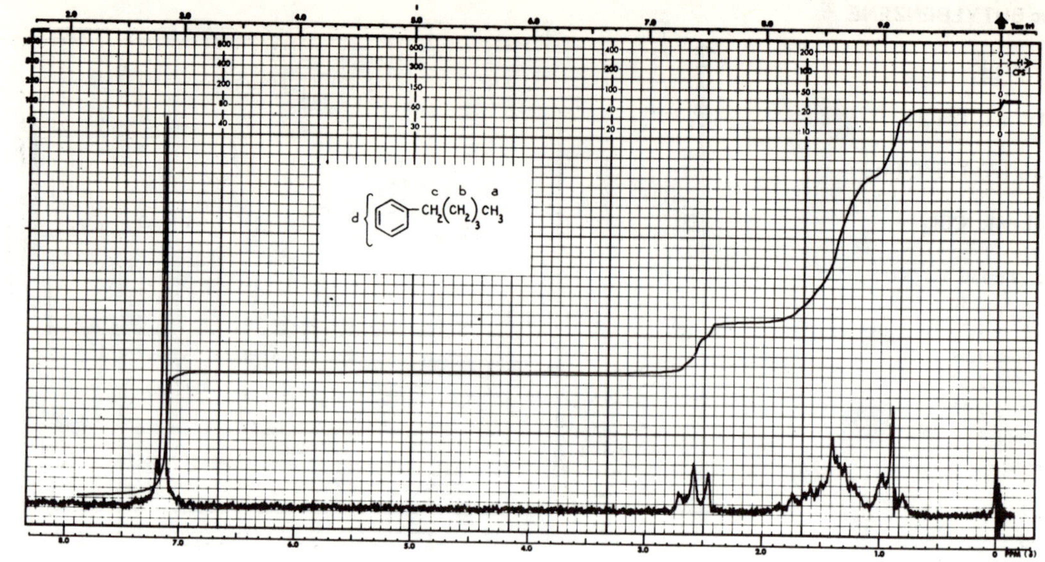

© 1978 Sadtler Research Laboratories, Inc.

1-PHENYLHEXANE

$C_{12}H_{18}$

Mol. Wt. 162.28

B.P. 224-226°C

Solvent: $CDCl_3$

Assignments
a = 0.90
b = 1.10-1.50
c = 1.59
d = 2.59
e = 7.15

142

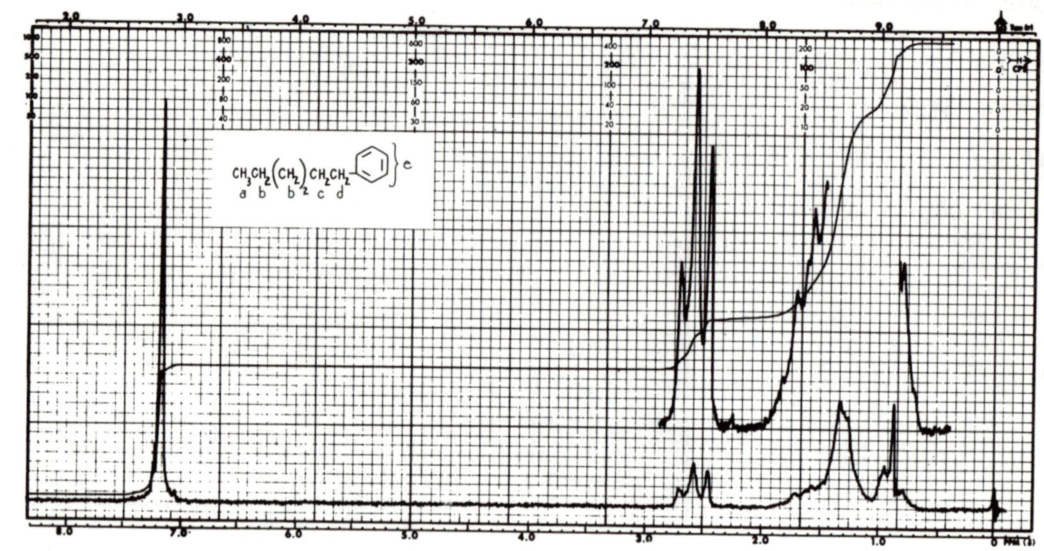

© 1978 Sadtler Research Laboratories, Inc.

1-PHENYLOCTANE

$C_{14}H_{22}$

Mol. Wt. 190.33

Solvent: CCl_4

Assignments
a = 0.88
b = 1.28
c = 2.57
d = 7.10

143

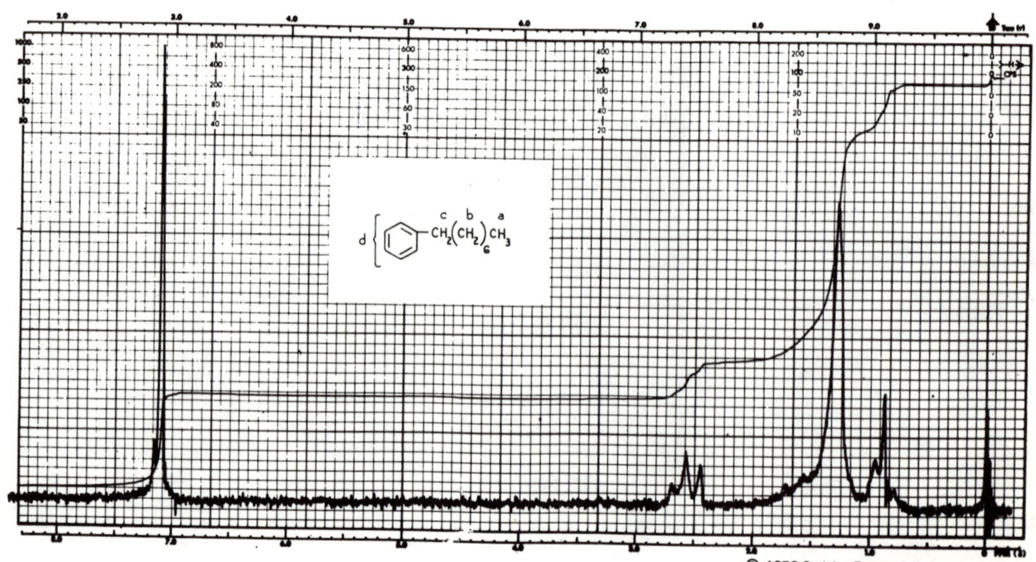

© 1978 Sadtler Research Laboratories, Inc.

BENZYLCYCLOPROPANE

$C_{10}H_{12}$

Mol. Wt. 132.21

Solvent: $CDCl_3$

Assignments
a = 0.08-0.74
b = ca 0.98
c = 2.53
d = 7.24

144

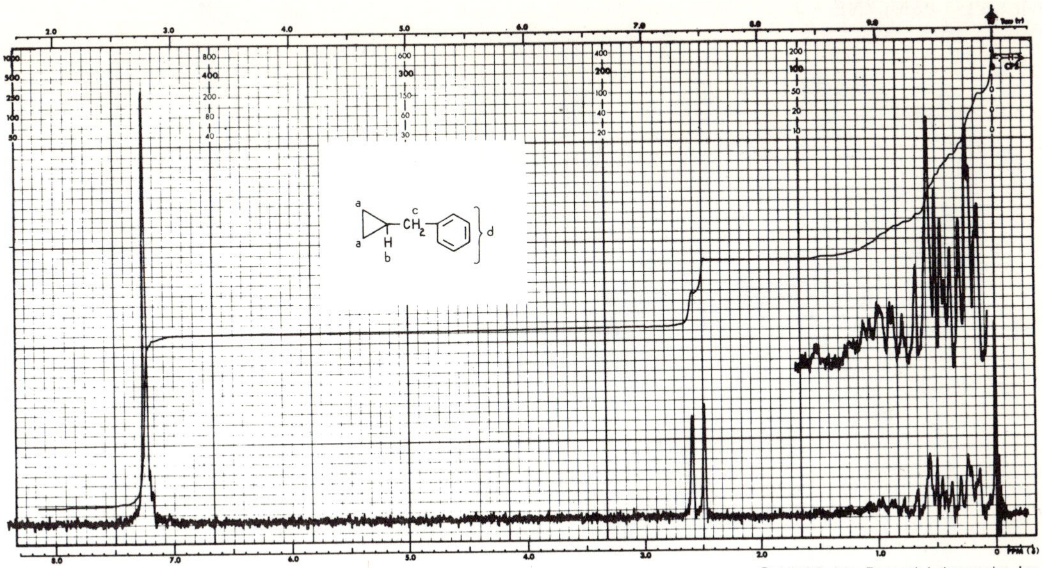

© 1978 Sadtler Research Laboratories, Inc.

CYCLOHEXYLBENZENE

$C_{12}H_{16}$

Mol. Wt. 160.25

B.P. 238°C

Solvent: CCl_4

Assignments
a = 0.80-2.10
b = ca 2.40
c = 7.09

145

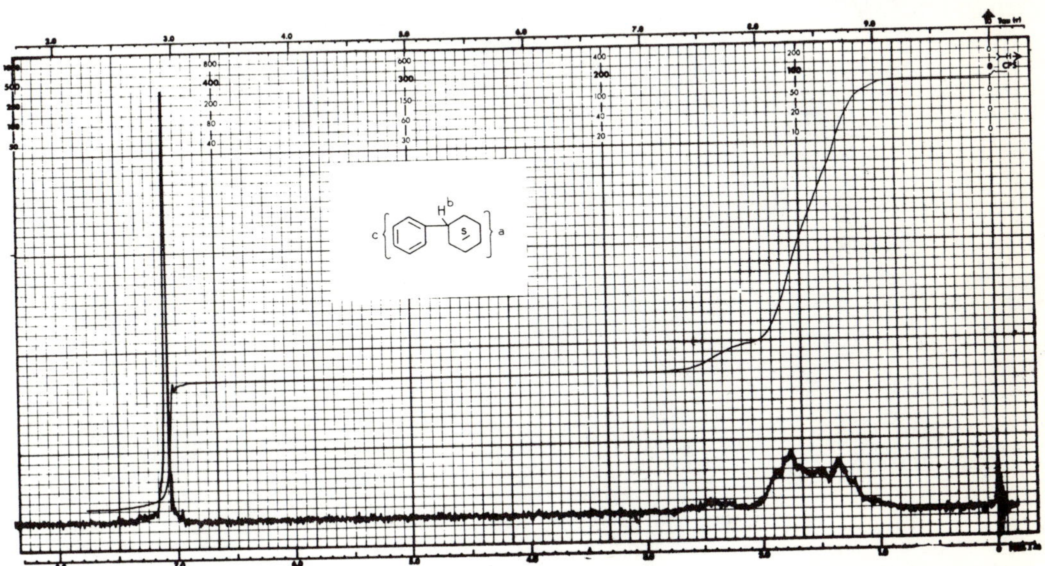

© 1978 Sadtler Research Laboratories, Inc.

(1-CYCLOHEXEN-1-YL)BENZENE

$C_{12}H_{14}$

Mol. Wt. 158.25

M.P. −11°C B.P. 251-253°C

Solvent: CCl_4

Assignments
a = 1.30-2.00
b = 2.00-2.50
c = 6.01
d = 7.00-7.45

146

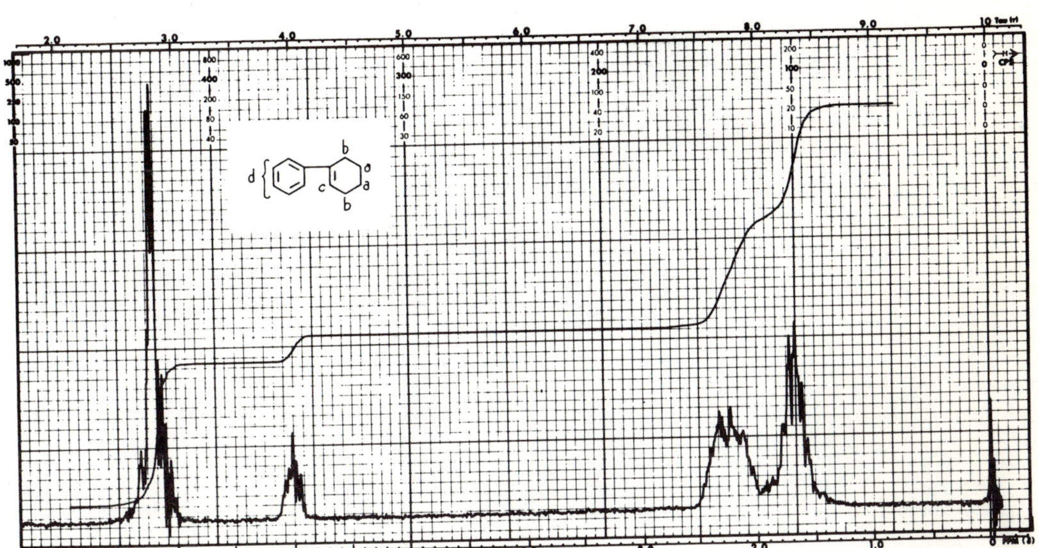

© 1978 Sadtler Research Laboratories, Inc.

5-PHENYL-1-PENTYNE

$C_{11}H_{12}$

Mol. Wt. 144.22

Solvent: $CDCl_3$

Assignments
a = 1.60-2.33
b = 2.70
c = 7.14

147

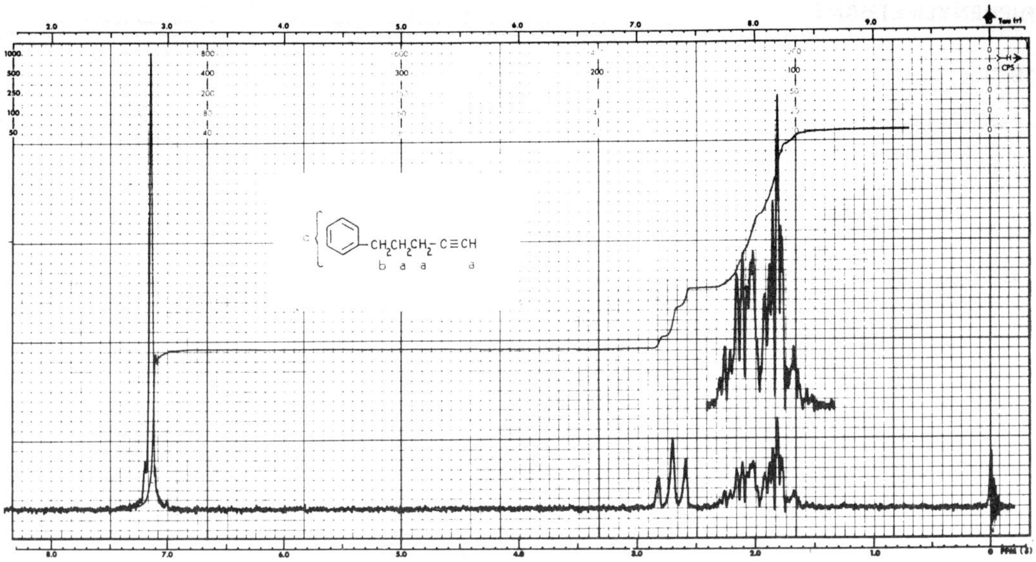

© 1978 Sadtler Research Laboratories, Inc.

DIPHENYLMETHANE

$C_{13}H_{12}$

Mol. Wt. 168.24

M.P. 22-24°C

Solvent: CCl_4

Assignments
a = 3.82
b = ca 7.07

148

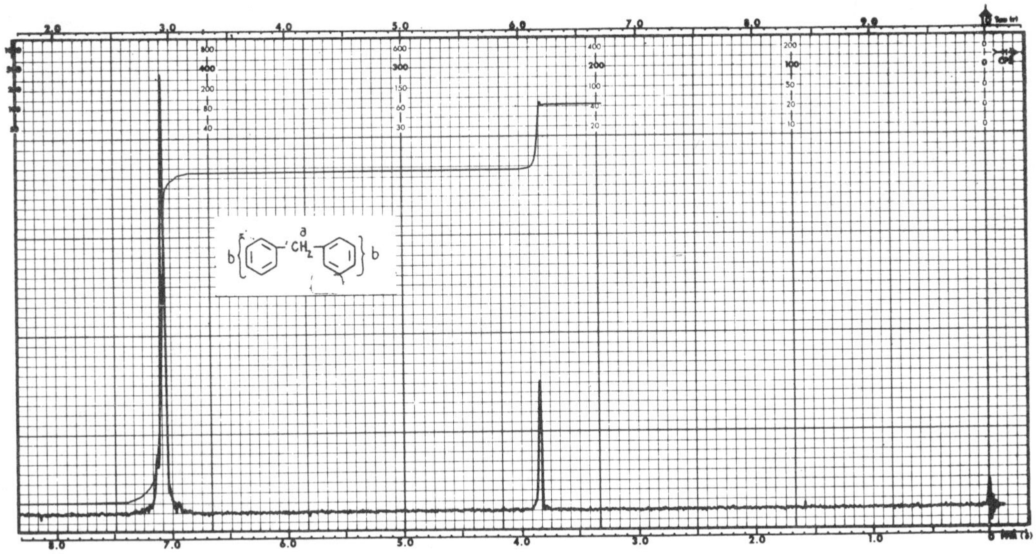

© 1978 Sadtler Research Laboratories, Inc.

BIBENZYL

$C_{14}H_{14}$

Mol. Wt. 182.25

M.P. 50-51°C

Solvent: $CDCl_3$

Assignments
a = 2.89
b = 7.19

149

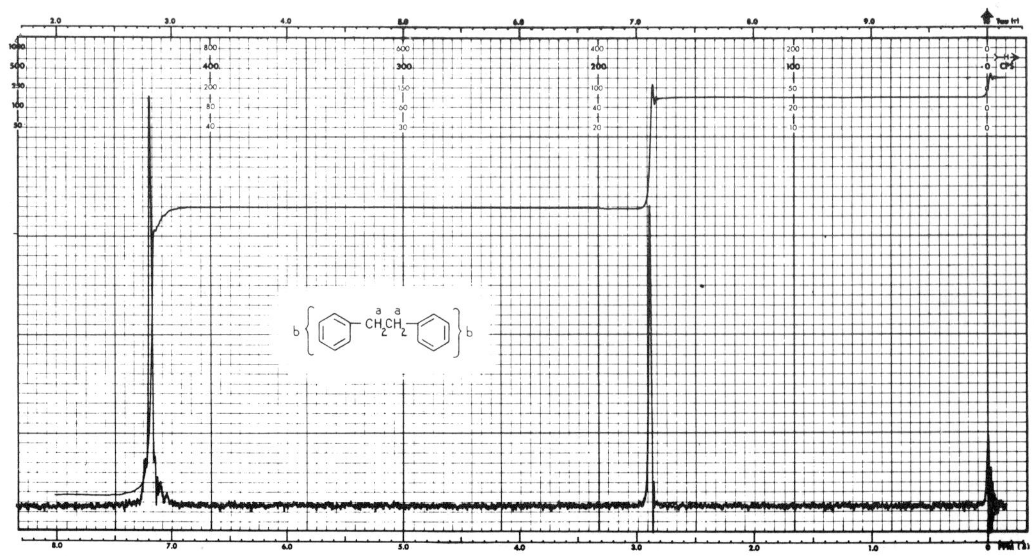

© 1978 Sadtler Research Laboratories, Inc.

TRIPHENYLMETHANE

$C_{19}H_{16}$

Mol. Wt. 244.32

M.P. 93-94°C

Solvent: $CDCl_3$

Assignments
a = 5.50
b = ca 7.15

150

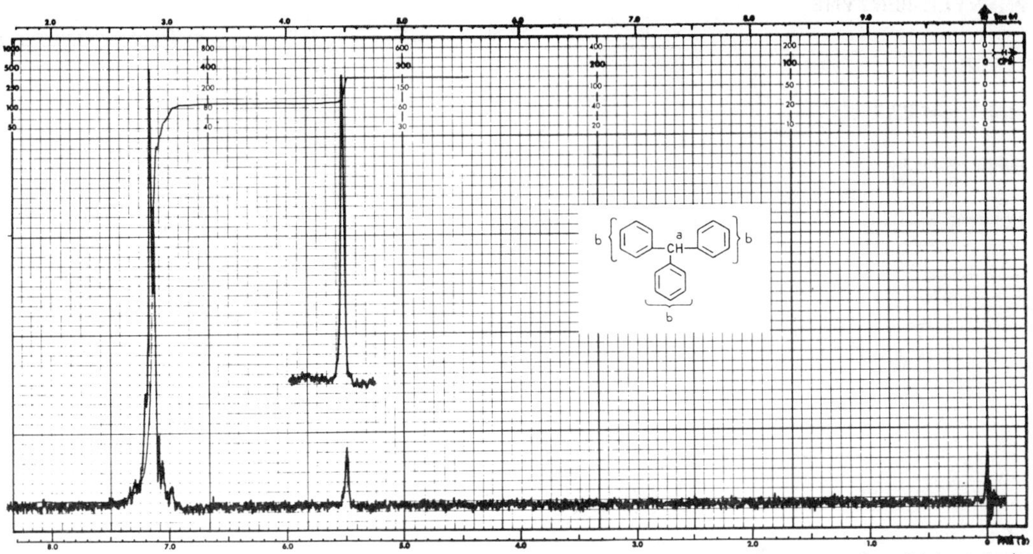

© 1978 Sadtler Research Laboratories, Inc.

1,1,2-TRIPHENYLETHANE

$C_{20}H_{18}$

Mol. Wt. 258.37

M.P. 53-55°C

Solvent: $CDCl_3$

Assignments
a = 3.35
b = 4.21
c = 6.80-7.30

151

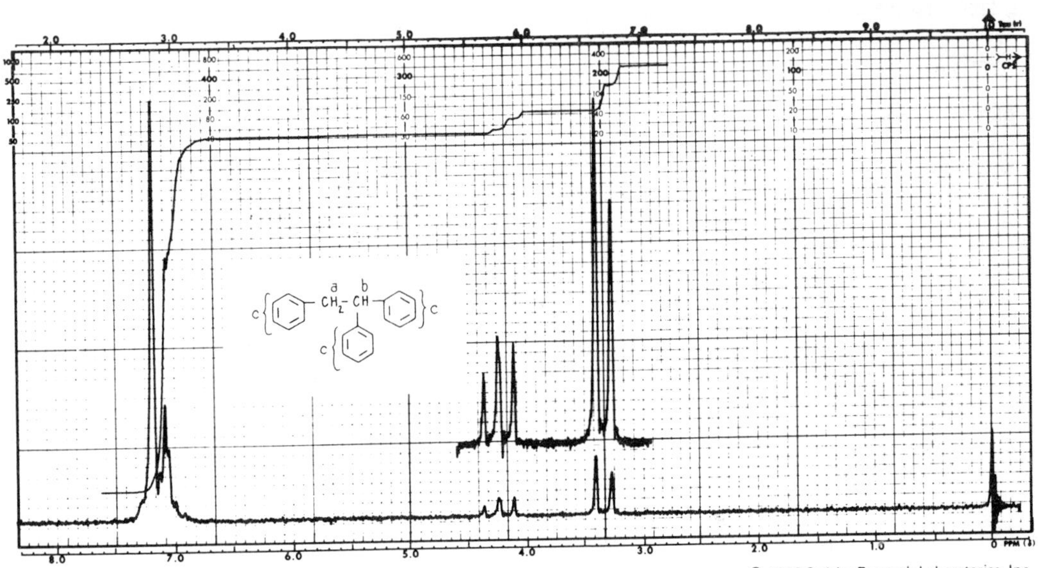

© 1978 Sadtler Research Laboratories, Inc.

meso-1,2,3,4-TETRAPHENYLBUTANE

$C_{28}H_{26}$

Mol. Wt. 362.52

M.P. 188°C

Solvent: $CDCl_3$

Assignments
a = 2.50-3.30
b = 6.50-7.30
c = ca 7.18

152

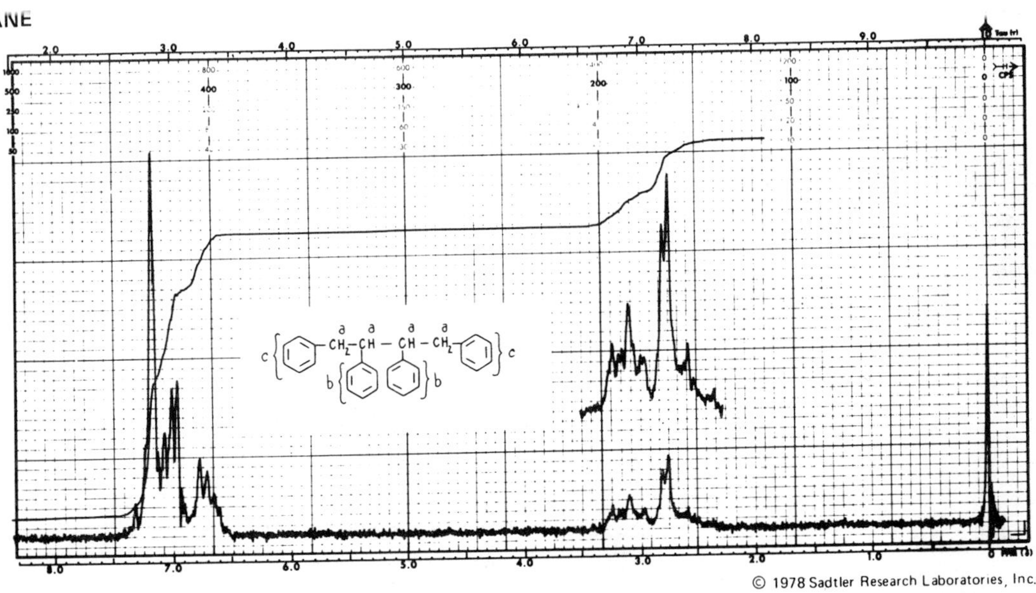

© 1978 Sadtler Research Laboratories, Inc.

ALLYLBENZENE

C_9H_{10}

Mol. Wt. 118.18

B.P. 155-157°C

Solvent: CCl_4

Assignments

a = 3.33
b = 4.99
c = 5.02
d = ca 5.85
e = 7.14

153

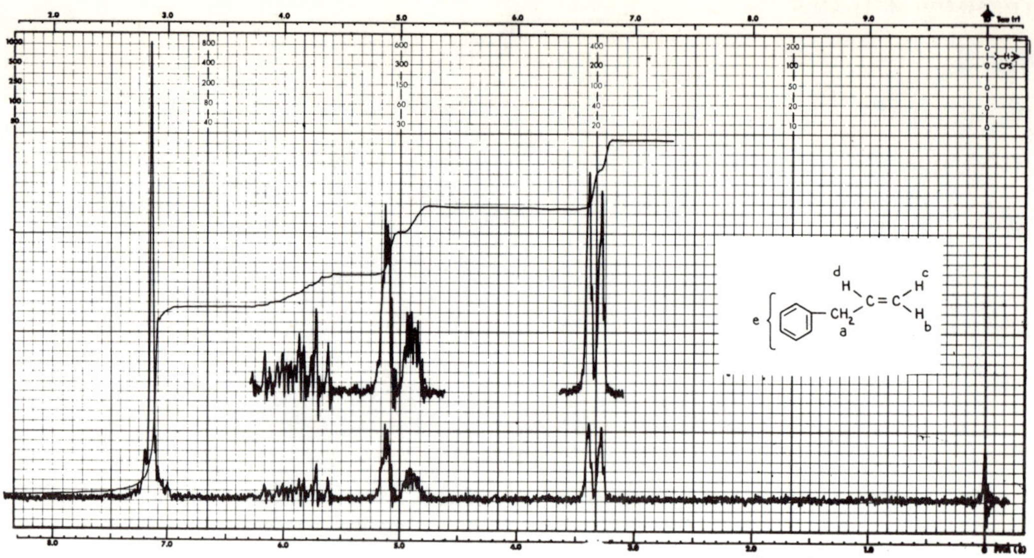

© 1978 Sadtler Research Laboratories, Inc.

1-PHENYL-2-BUTENE

$C_{10}H_{12}$

Mol. Wt. 132.21

Solvent: CCl_4

Assignments

a = 1.68
b = 3.27
c = ca 5.47
d = 7.11

154

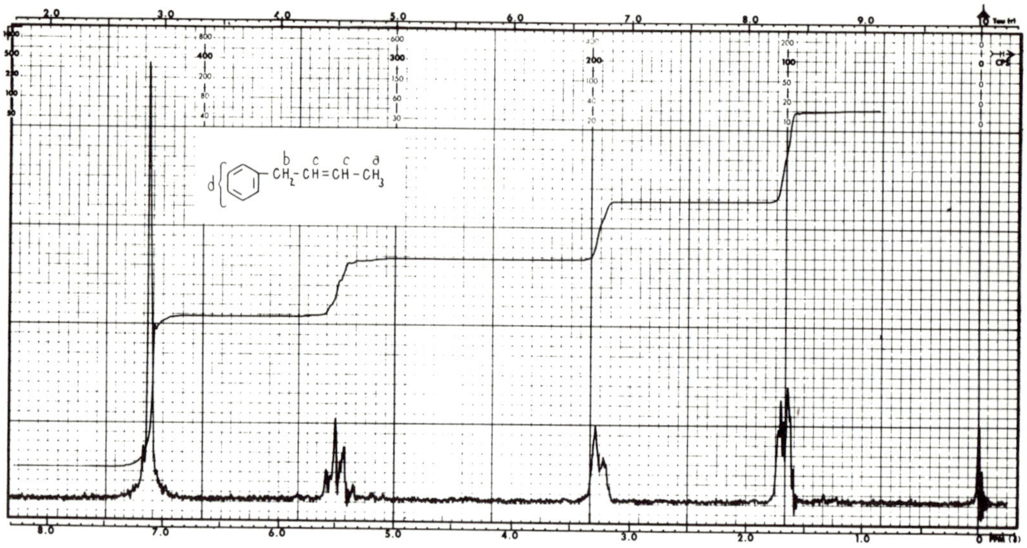

© 1978 Sadtler Research Laboratories, Inc.

4-PHENYL-1-BUTENE

$C_{10}H_{12}$

Mol. Wt. 132.21

B.P. 180-182°C

Solvent: $CDCl_3$

Assignments

a = 2.39
b = 2.67
c = 4.97
d = 5.01
e = 5.81
f = 7.19

155

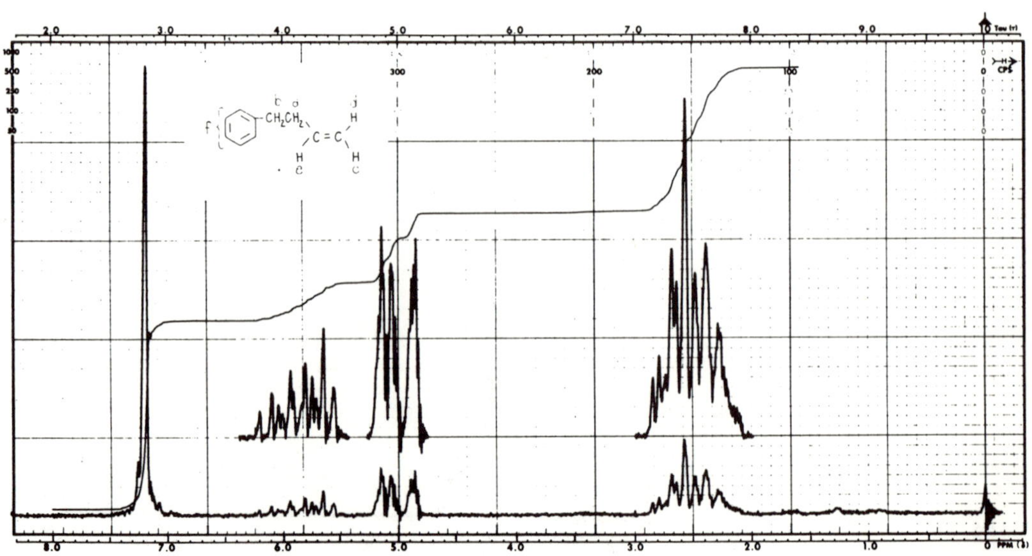

© 1978 Sadtler Research Laboratories, Inc.

STYRENE

C_8H_8

Mol. Wt. 104.15

B.P. 146°C

Solvent: CCl_4

Assignments
a = 5.11
b = 5.59
c = 6.59
d = 7.00-7.40

156

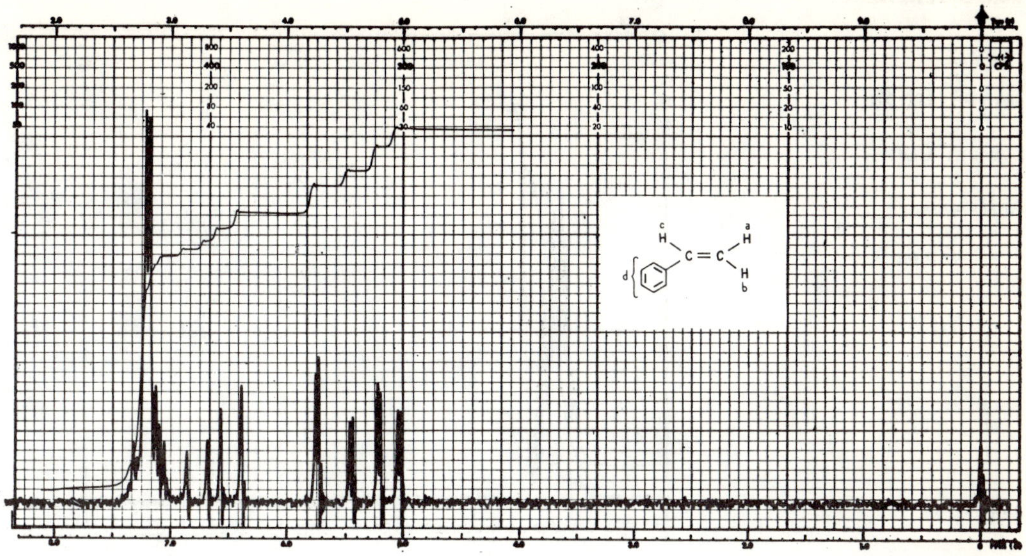

© 1978 Sadtler Research Laboratories, Inc.

α-METHYLSTYRENE

C_9H_{10}

Mol. Wt. 118.18

Solvent: CCl_4

Assignments
a = 2.10
b = 5.01
c = 5.31
d = 7.10-7.52

157

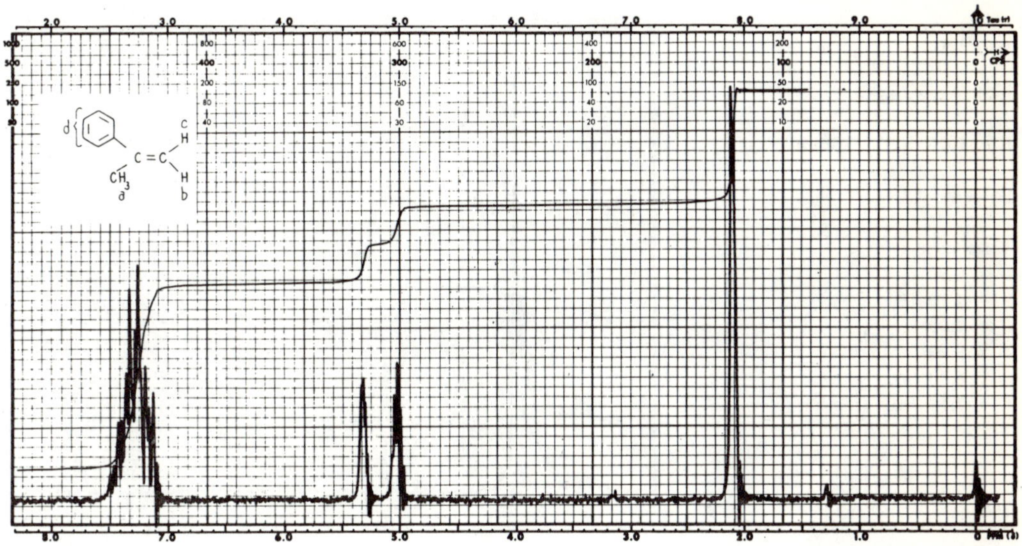

© 1978 Sadtler Research Laboratories, Inc.

---

cis-STILBENE

$C_{14}H_{12}$

Mol. Wt. 180.25

B.P. 144-145°C/13mm

Solvent: $CDCl_3$

Assignments
a = 6.58
b = 7.20

158

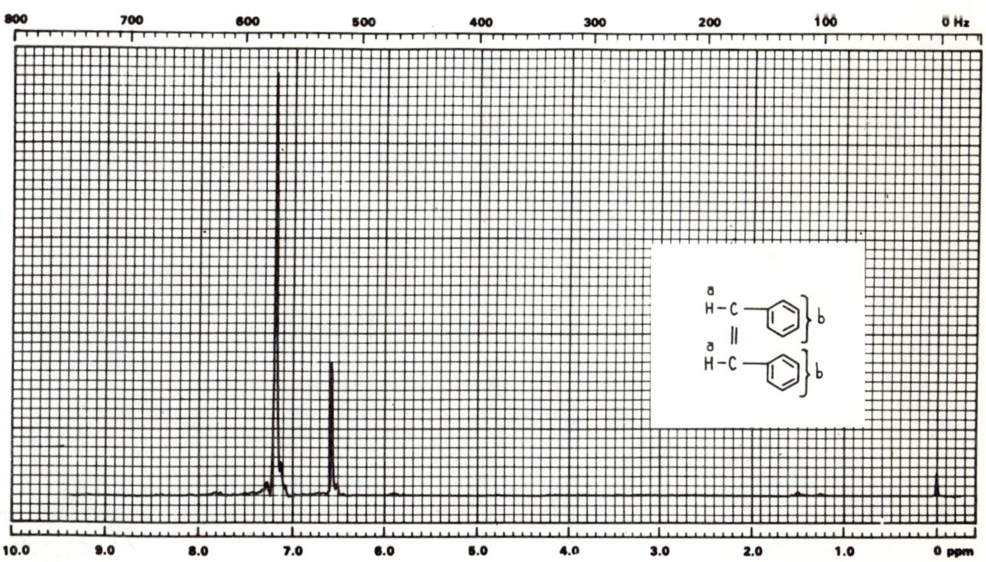

© 1978 Sadtler Research Laboratories, Inc.

trans-STILBENE

$C_{14}H_{12}$

Mol. Wt. 180.25

Solvent: $CDCl_3$

Assignments
a = 7.08
b = 7.15-7.65

159

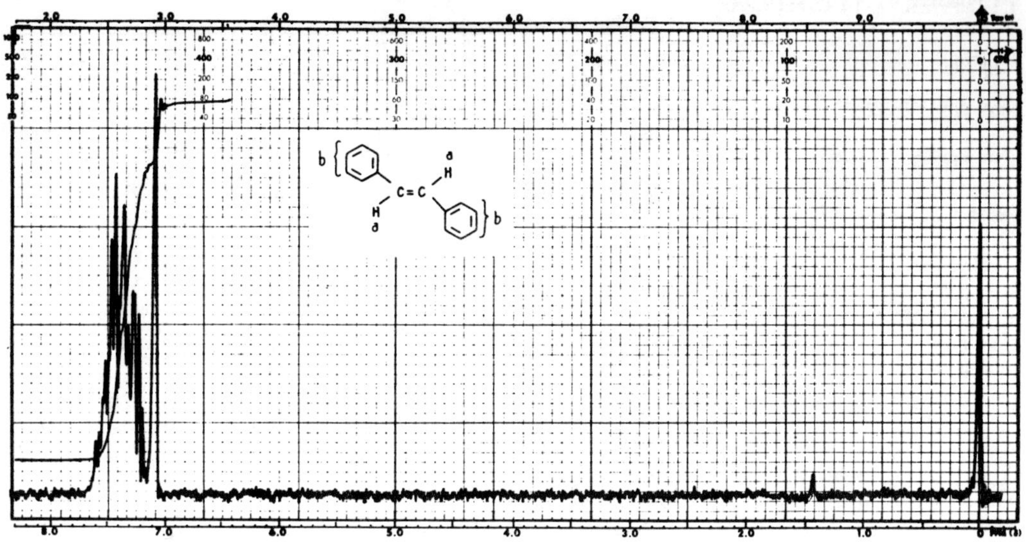

© 1978 Sadtler Research Laboratories, Inc.

1,1-DIPHENYL-1-HEXENE

$C_{18}H_{20}$

Mol. Wt. 236.36

B.P. 188-192°C/15mm

Solvent: CCl_4

Assignments
a = 0.88
b = 1.10-1.80
c = 2.09
d = 5.99
ė = 6.90-7.60
f = 7.70 imp.

160

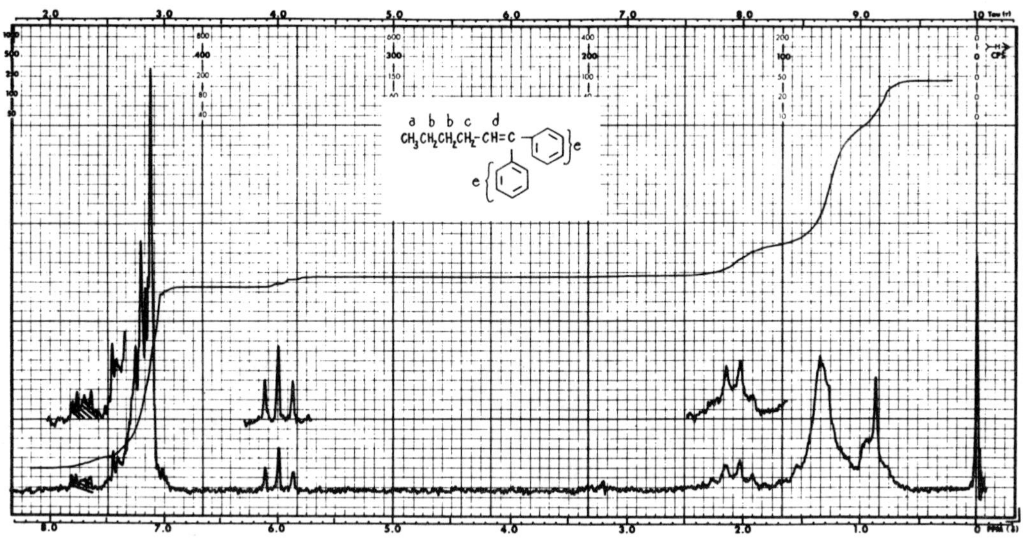

© 1978 Sadtler Research Laboratories, Inc.

1,1-DIPHENYL-1-NONENE

$C_{21}H_{26}$

Mol. Wt. 278.44

B.P. 176°C/1mm

Solvent: CCl_4

Assignments
a = 0.89
b = 1.00-1.70
c = 2.08
d = 5.98
e = 6.90-7.50

161

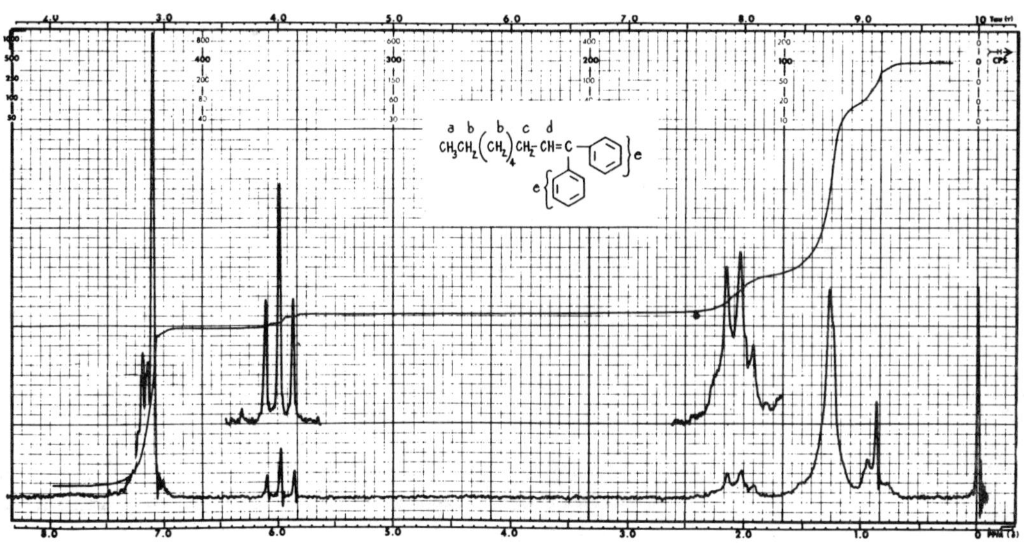

© 1978 Sadtler Research Laboratories, Inc.

1,1-DIPHENYL-1-DODECENE

$C_{24}H_{32}$

Mol. Wt. 320.52

B.P. 210°C/0.6mm

Solvent: $CDCl_3$

Assignments
a = 0.89
b = 1.10-1.70
c = 2.10
d = 6.07
e = 7.00-7.50

162

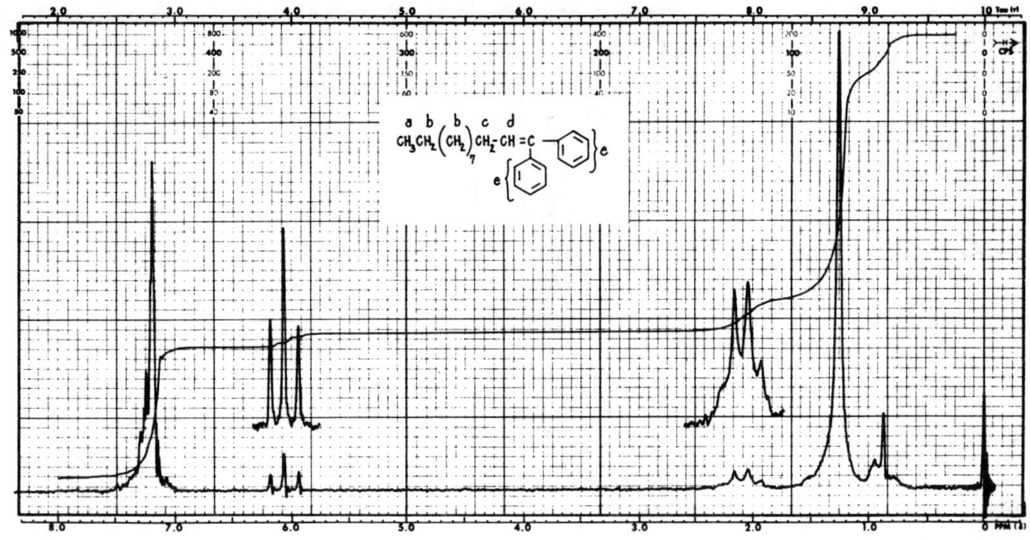

© 1978 Sadtler Research Laboratories, Inc.

TRIPHENYLETHYLENE

$C_{20}H_{16}$

Mol. Wt. 256.35

M.P. 72-73°C (lit.)

B.P. 220-221°C/14mm

Solvent: $CDCl_3$

Assignments
a = 6.99
b = 7.08
c = 7.29

163

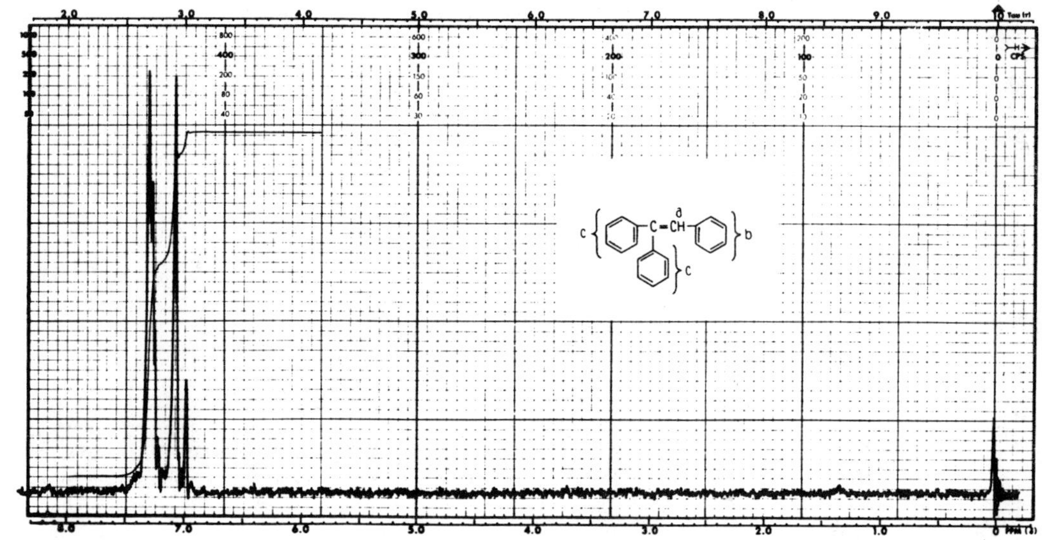

© 1978 Sadtler Research Laboratories, Inc.

1,1,4,4-TETRAPHENYL-1,3-BUTADIENE

$C_{28}H_{22}$

Mol. Wt. 358.49

M.P. 203-204°C

Solvent: $CDCl_3$

Assignments
a = 6.80
b = 7.20 or c
c = 7.38 or b

164

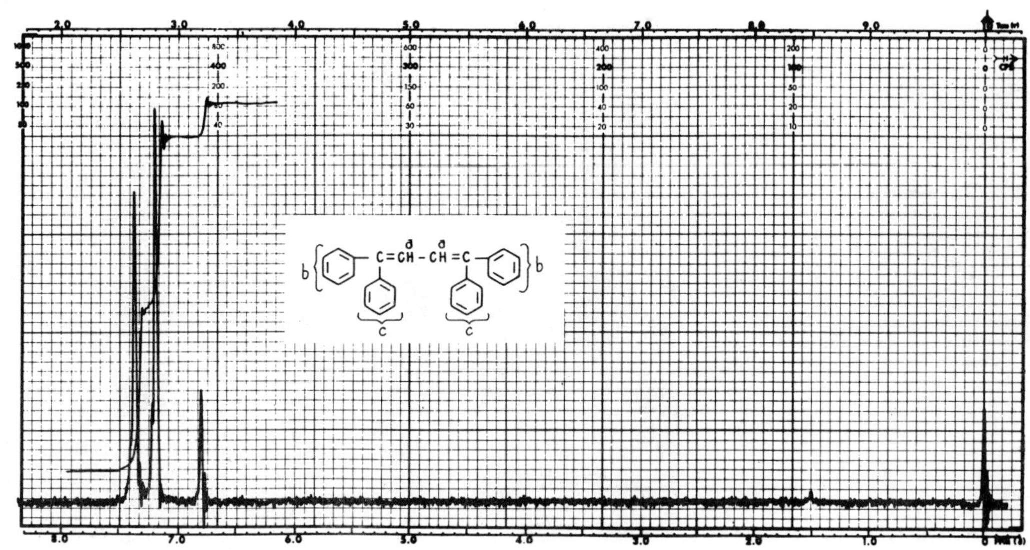

© 1978 Sadtler Research Laboratories, Inc.

TETRAPHENYLPROPADIENE

$C_{27}H_{20}$

Mol. Wt. 344.46

M.P. 164°C

Solvent: $CDCl_3$

Assignments
a = 7.10-7.50

165

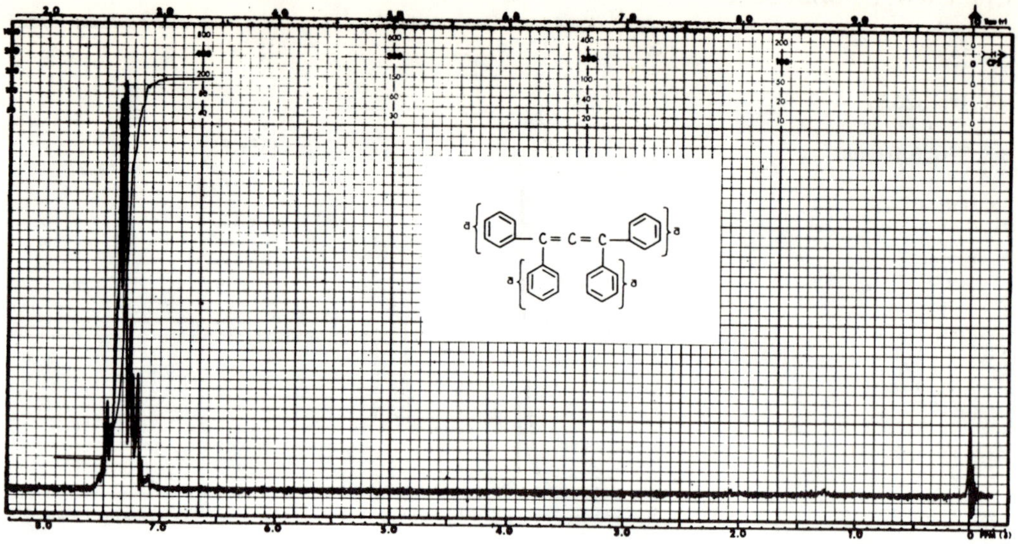

© 1978 Sadtler Research Laboratories, Inc.

ETHYNYLBENZENE

C_8H_6

Mol. Wt. 102.13

B.P. 142-144°C

Solvent: $CDCl_3$

Assignments
a = 3.08

166

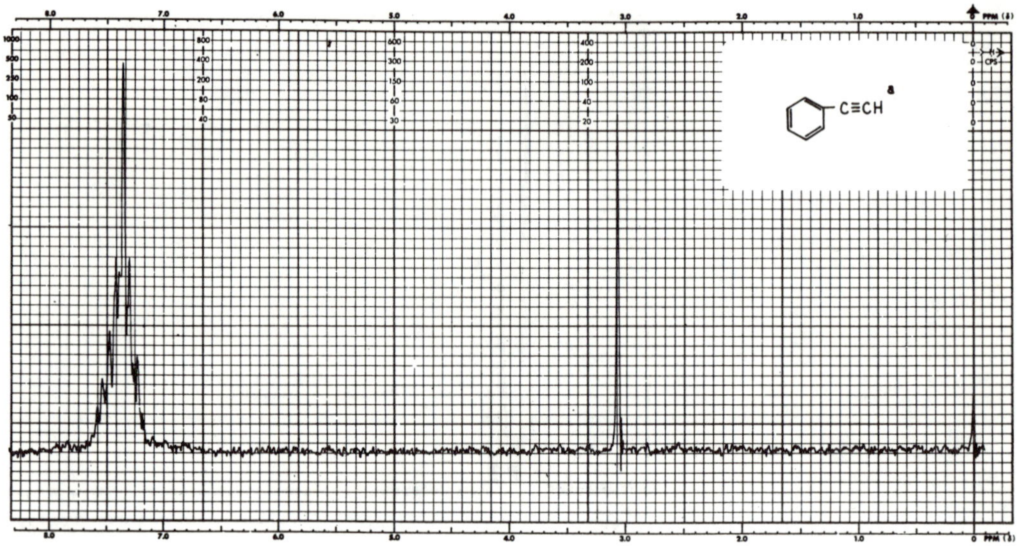

© 1978 Sadtler Research Laboratories, Inc.

(1-PROPYNYL)BENZENE

C_9H_8

Mol. Wt. 116.16

B.P. 185°C

Solvent: $CDCl_3$

Assignments
a = 1.99
b = 7.10-7.50

167

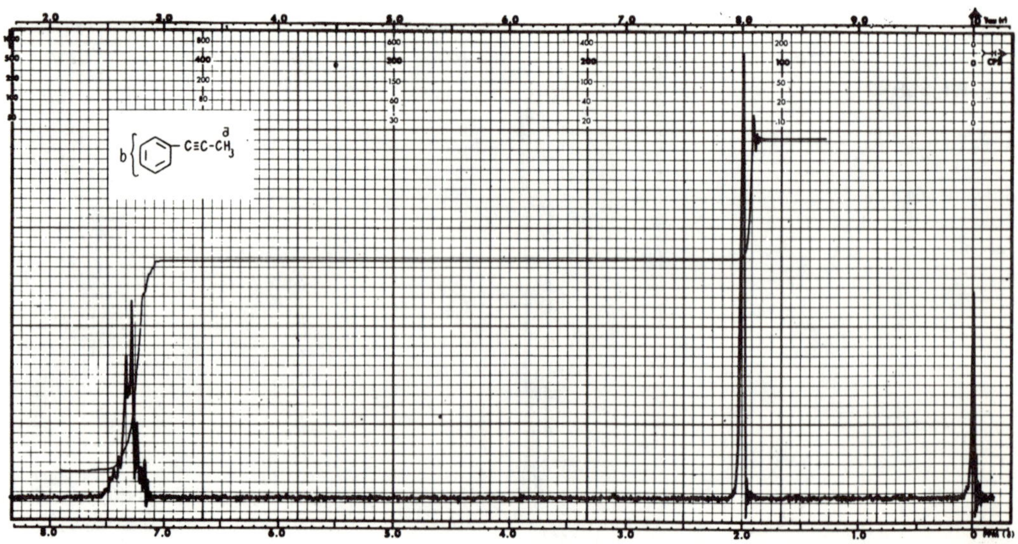

© 1978 Sadtler Research Laboratories, Inc.

1-PHENYL-1-NONYNE

$C_{15}H_{20}$

Mol. Wt. 200.33

Solvent: CCl_4

Assignments
a = 0.90
b = 1.10-2.00
c = 2.37
d = 7.05-7.50

168

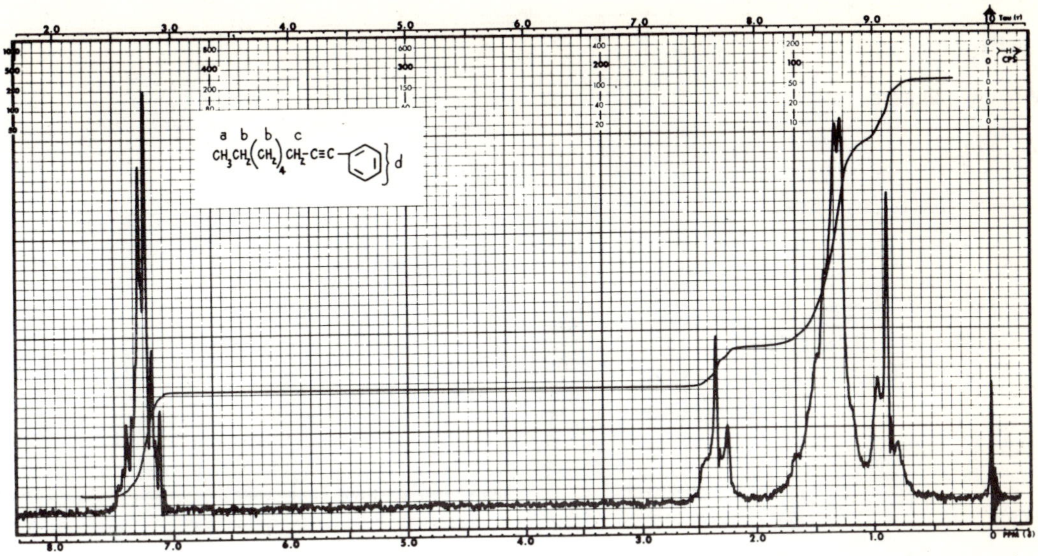

© 1978 Sadtler Research Laboratories, Inc.

DIPHENYLACETYLENE

$C_{14}H_{10}$

Mol. Wt. 178.24

M.P. 59-60°C (lit.)

Solvent: CCl_4

169

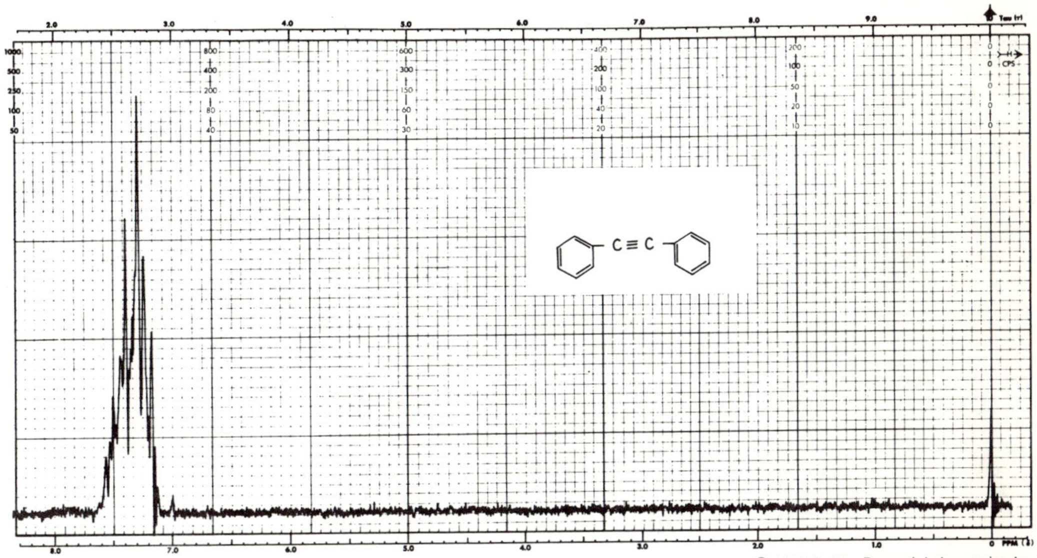

© 1978 Sadtler Research Laboratories, Inc.

1,4-DIPHENYLBUTADIYNE

$C_{16}H_{10}$

Mol. Wt. 202.26

Solvent: $CDCl_3$

Assignments
a = 7.10-7.70

170

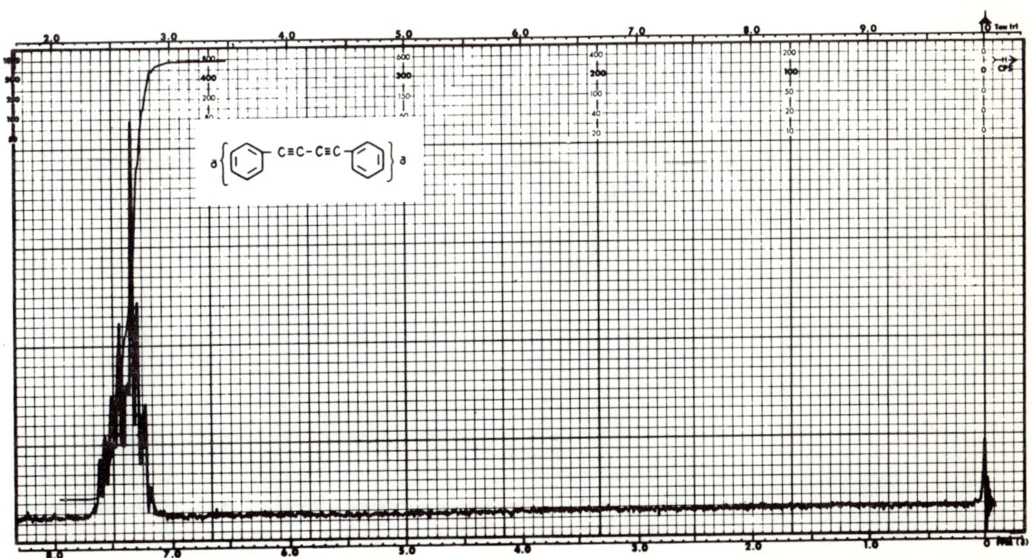

© 1978 Sadtler Research Laboratories, Inc.

BIPHENYL

$C_{12}H_{10}$

Mol. Wt. 154.21

B.P. 254.9°C

Solvent: $CDCl_3$

Assignments
a = 7.12-7.63

171

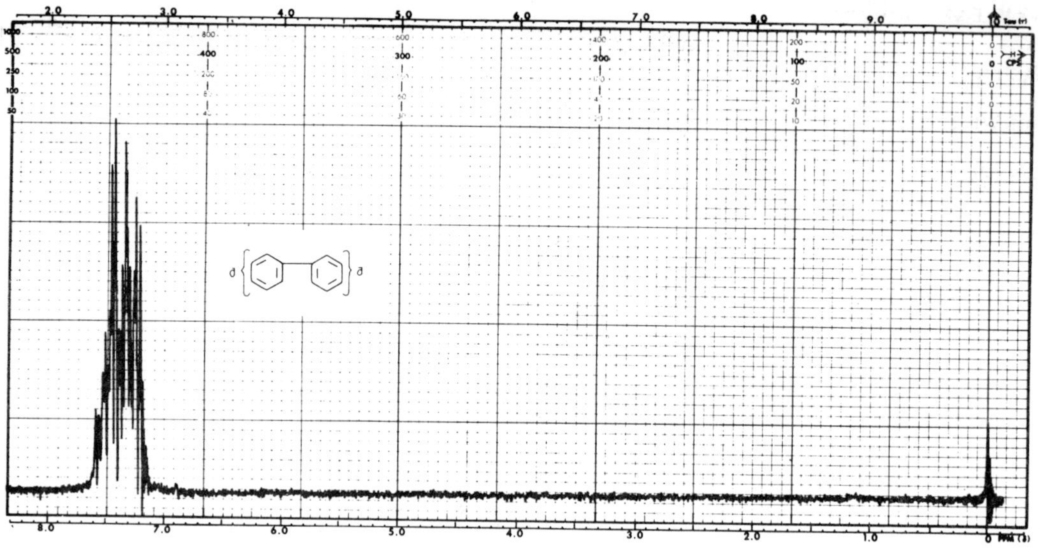

© 1978 Sadtler Research Laboratories, Inc.

2-ETHYL-4'-METHYLBIPHENYL

$C_{15}H_{16}$

Mol. Wt. 196.29

Solvent: CCl_4

Assignments
a = 1.09
b = 2.38
c = 2.59
d = 7.08-7.28

172

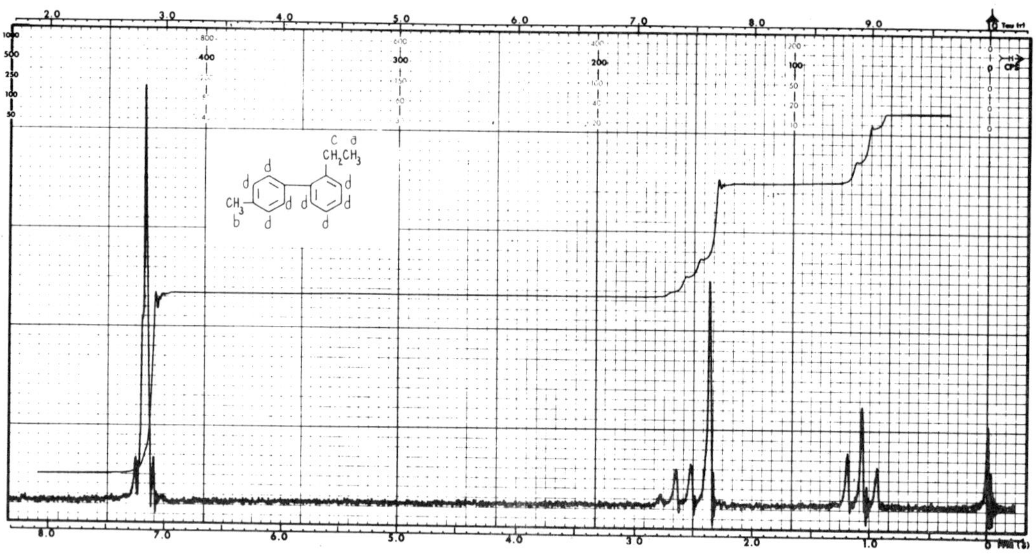

© 1978 Sadtler Research Laboratories, Inc.

4-PHENYLBIBENZYL

$C_{20}H_{18}$

Mol. Wt. 258.37

M.P. 106-108°C

Solvent: $CDCl_3$

Assignments
a = 2.97
b = 7.10-7.70

173

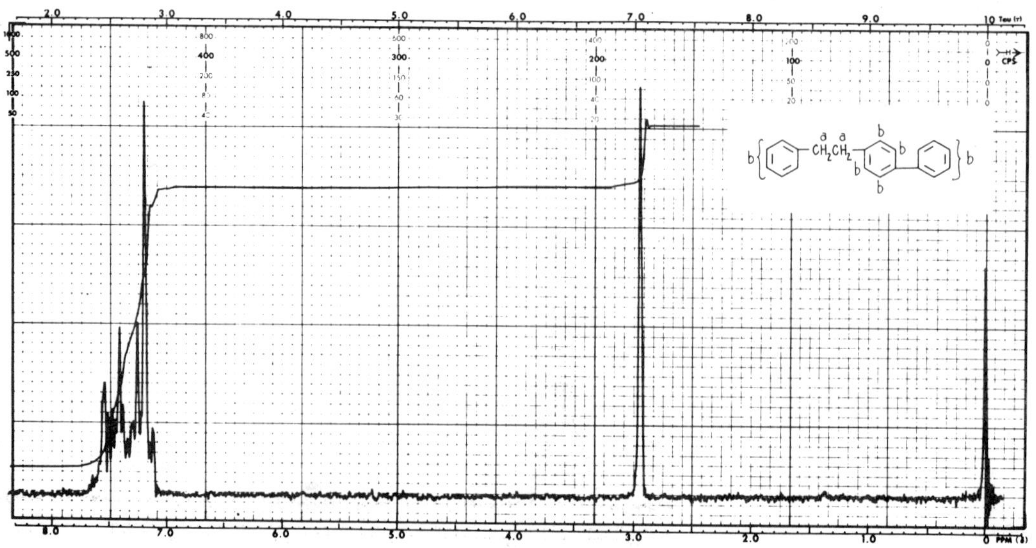

© 1978 Sadtler Research Laboratories, Inc.

o-XYLENE

C_8H_{10}

Mol. Wt. 106.17

B.P. 144°C

Solvent: CCl_4

Assignments
a = 2.20
b = 6.95

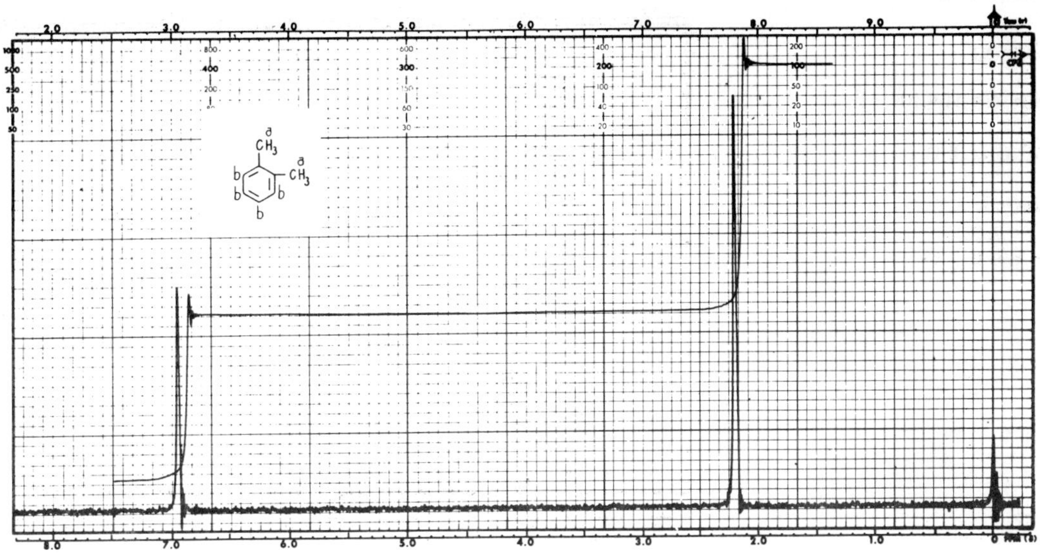

174

© 1978 Sadtler Research Laboratories, Inc.

o-(1-CYCLOHEXEN-1-YL)TOLUENE

$C_{13}H_{16}$

Mol. Wt. 172.27

Solvent: CCl_4

Assignments
a = ca 1.69
b = ca 2.15
c = 2.24
d = 5.49
e = ca 7.01

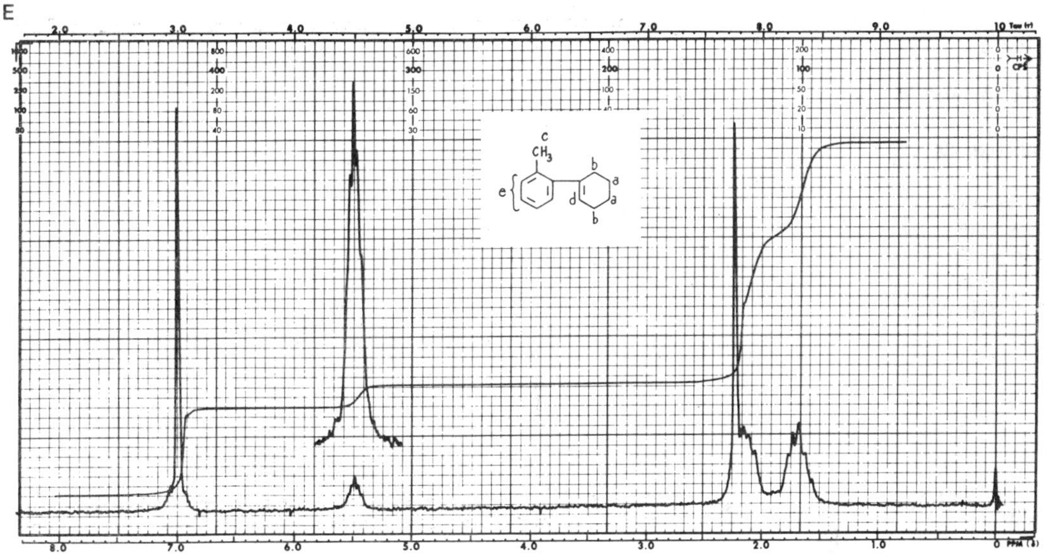

175

© 1978 Sadtler Research Laboratories, Inc.

o-TERPHENYL

$C_{18}H_{14}$

Mol. Wt. 230.31

M.P. 56-57°C

Solvent: $CDCl_3$

Assignments
a = 7.11
b = 7.37

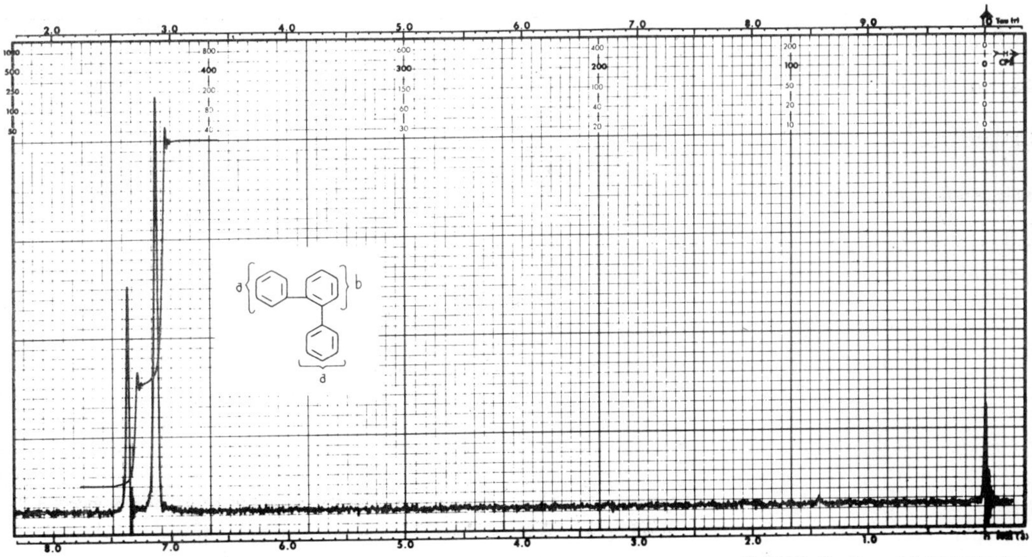

176

© 1978 Sadtler Research Laboratories, Inc.

m-XYLENE

C_8H_{10}

Mol. Wt. 106.17

Solvent: CCl_4

Assignments
a = 2.82
b = 6.62-7.08

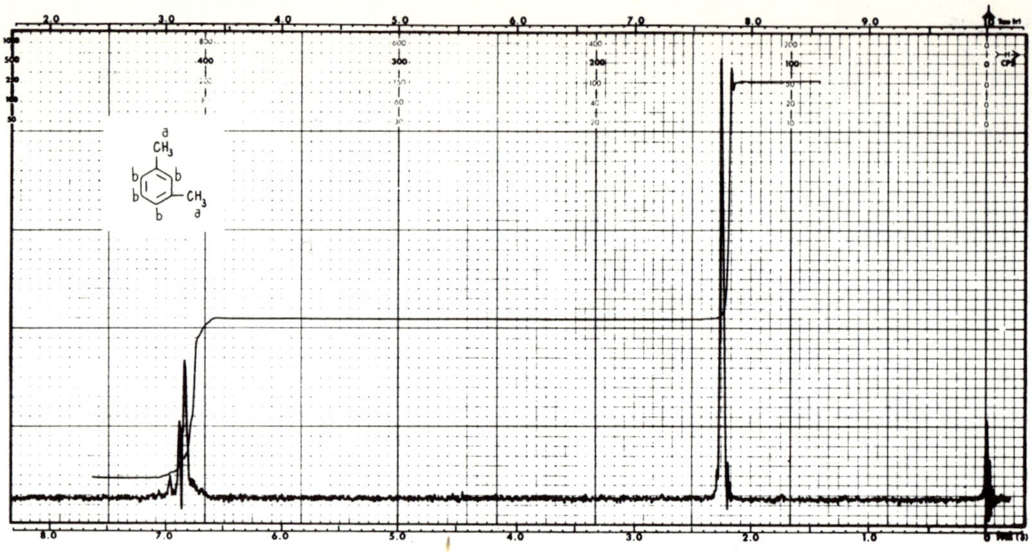

177

© 1978 Sadtler Research Laboratories, Inc.

---

m-ETHYLTOLUENE

C_9H_{12}

Mol. Wt. 120.20

Solvent: $CDCl_3$

Assignments
a = 1.21
b = 2.31
c = 2.60
d = 6.87-7.39

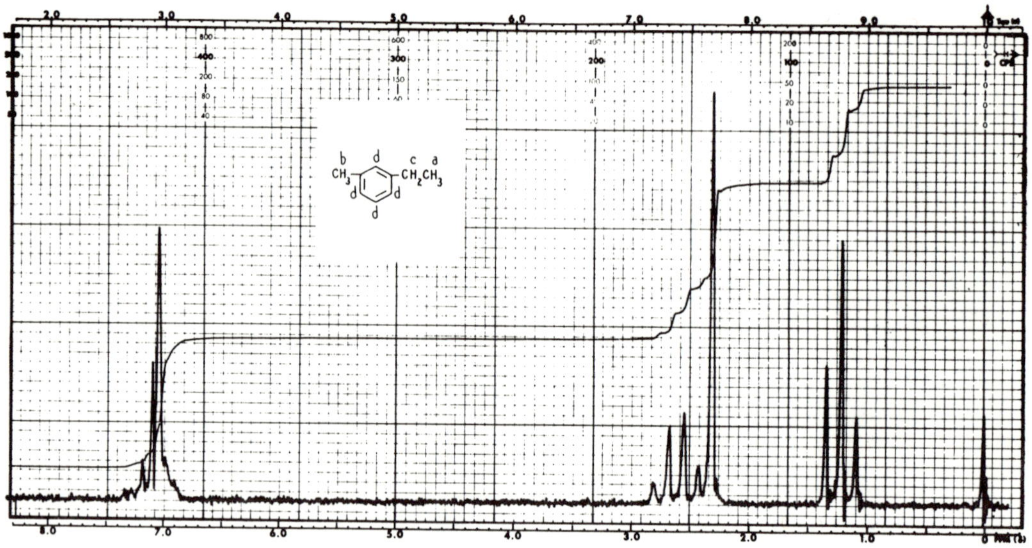

178

© 1978 Sadtler Research Laboratories, Inc.

---

m-DIETHYLBENZENE

$C_{10}H_{14}$

Mol. Wt. 134.22

Solvent: CCl_4

Assignments
a = 1.20
b = 2.58
c = ca 6.92

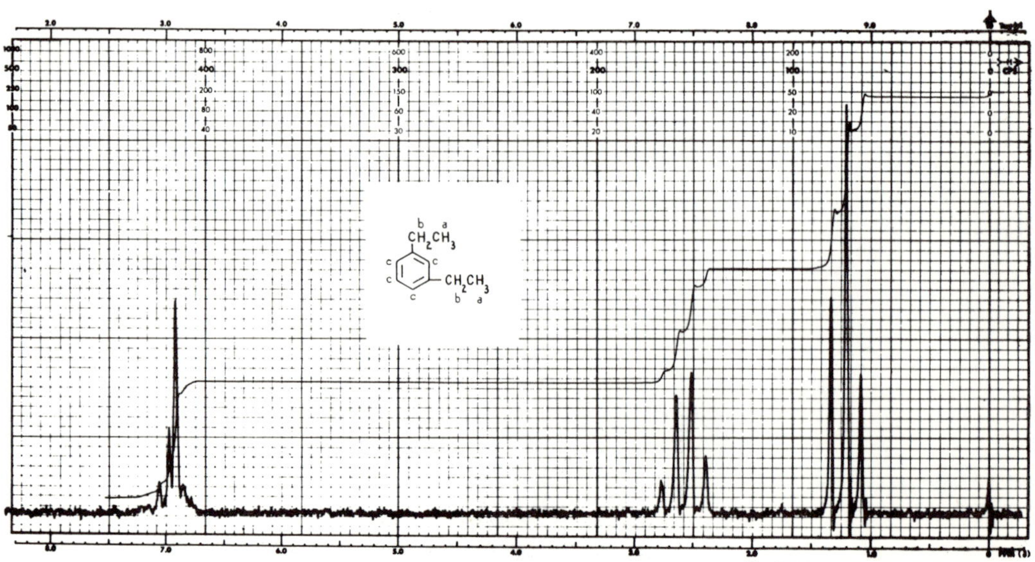

179

© 1978 Sadtler Research Laboratories, Inc.

m-TERPHENYL

$C_{18}H_{14}$

Mol. Wt. 230.30

M.P. 86-87°C (lit.)

Solvent: $CDCl_3$

Assignments
a = 7.00-7.70
b = 7.78

180

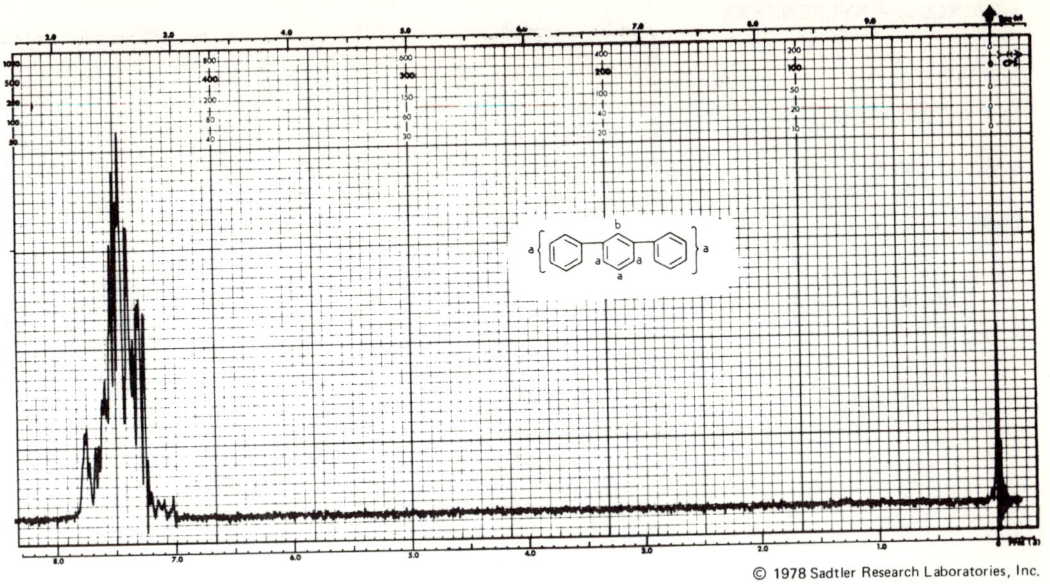

© 1978 Sadtler Research Laboratories, Inc.

---

p-XYLENE

C_8H_{10}

Mol. Wt. 106.17

M.P. 12-13°C

Solvent: CCl_4

Assignments
a = 2.27
b = 6.90

181

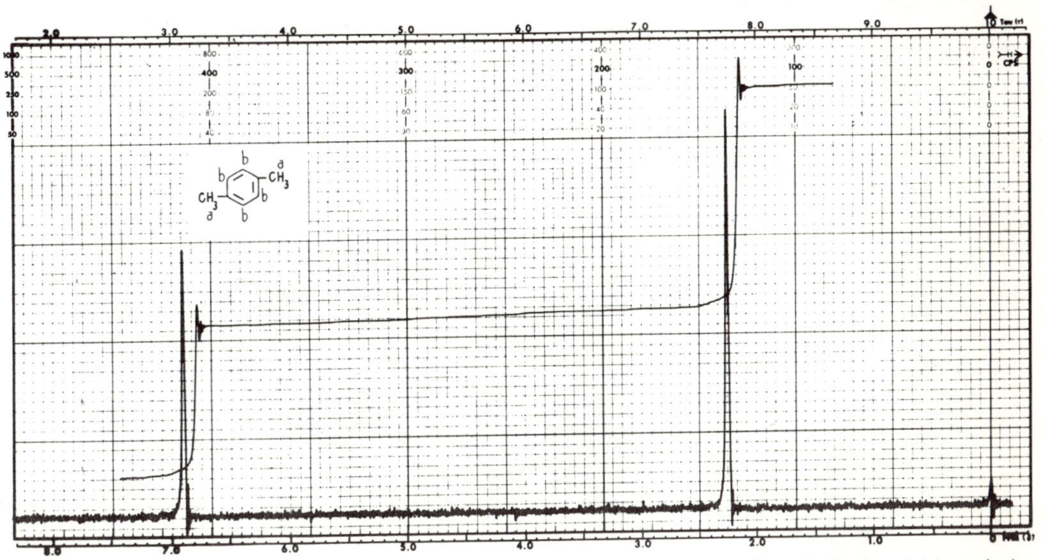

© 1978 Sadtler Research Laboratories, Inc.

P-CYMENE

$C_{10}H_{14}$

Mol. Wt. 134.22

Solvent: CCl_4

Assignments
a = 1.21
b = 2.28
c = 2.80
d = 6.99

182

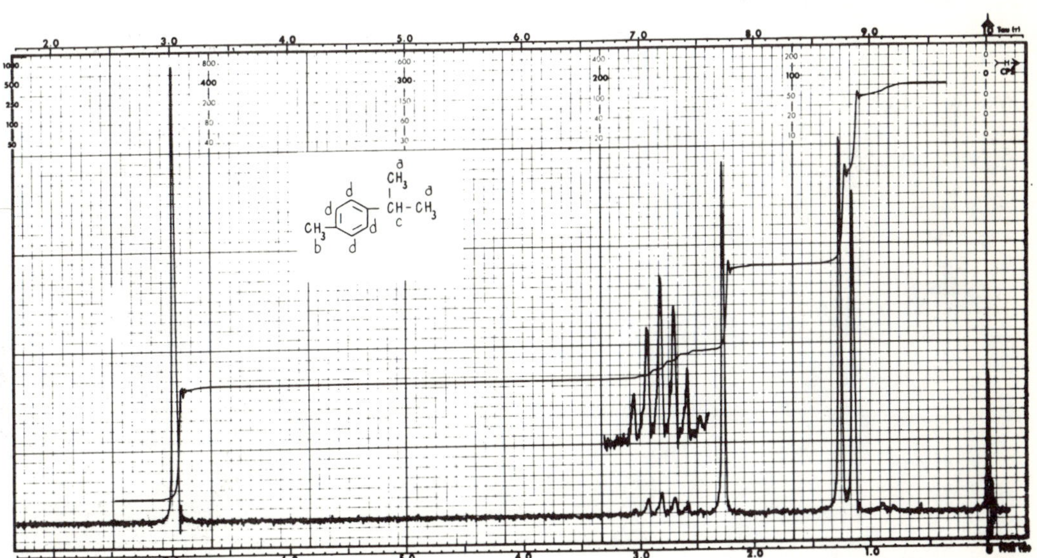

© 1978 Sadtler Research Laboratories, Inc.

p-DICYCLOHEXYLBENZENE

$C_{18}H_{26}$

Mol. Wt. 242.41

M.P. 102-103.5°C

Solvent: $CDCl_3$

Assignments
a = 0.90-2.10
b = 2.45
c = 7.11

183

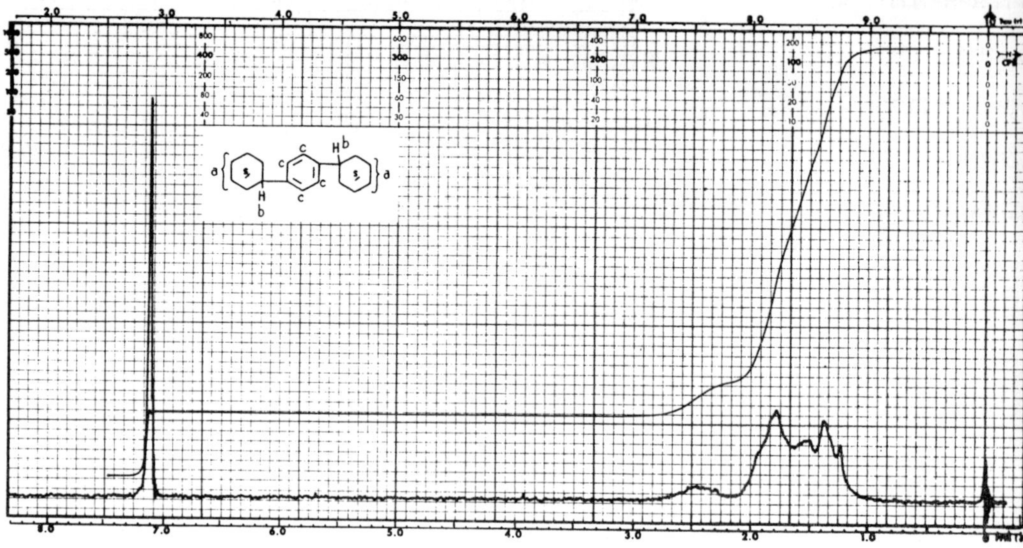

© 1978 Sadtler Research Laboratories, Inc.

1-(p-TOLYL)-1-CYCLOHEXENE

$C_{13}H_{16}$

Mol. Wt. 172.27

Solvent: CCl_4

Assignments
a = 1.30-2.00
b = 2.00-2.50
c = 2.30
d = 5.98
e = 6.99
f = 7.16

184

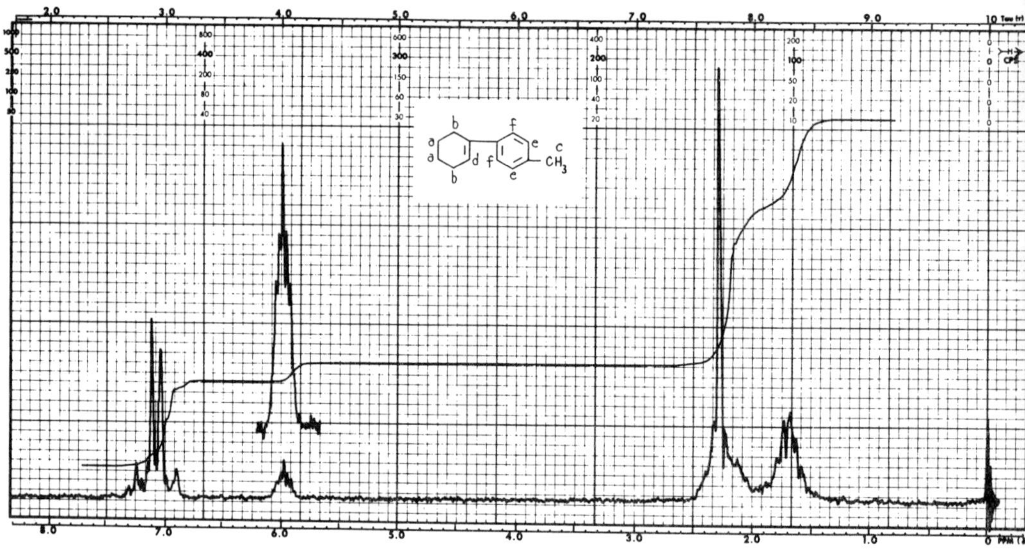

© 1978 Sadtler Research Laboratories, Inc.

4,4'-DIMETHYLBIBENZYL

$C_{16}H_{18}$

Mol. Wt. 210.32

Solvent: $CDCl_3$

Assignments
a = 2.30
b = 2.86
c = 7.08

185

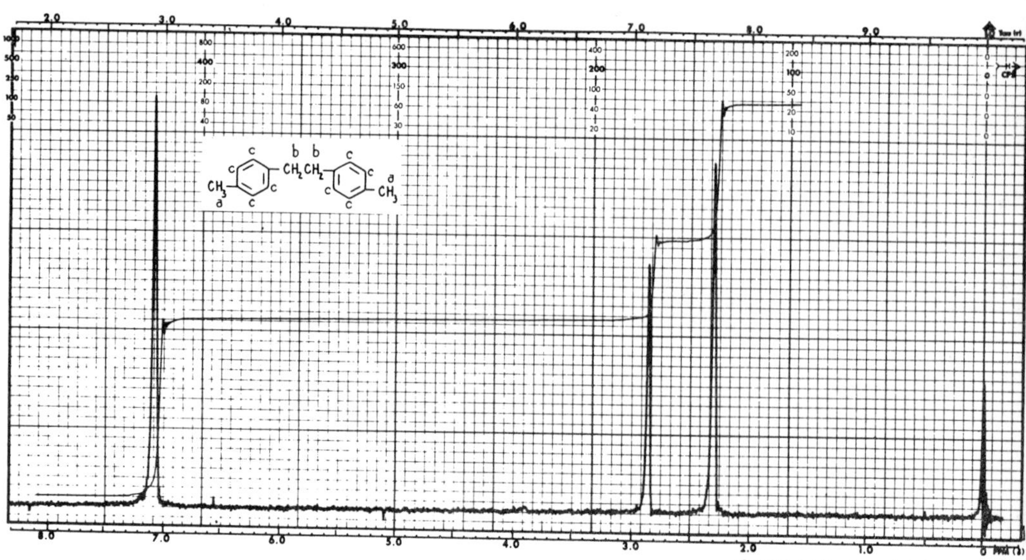

© 1978 Sadtler Research Laboratories, Inc.

-TERPHENYL

$_{18}H_{14}$

Aol. Wt. 230.31

A.P. 213°C (lit.)

olvent: $CDCl_3$

ssignments
= 7.30-7.75
= 7.66
= 7.20 $CHCl_3$

186

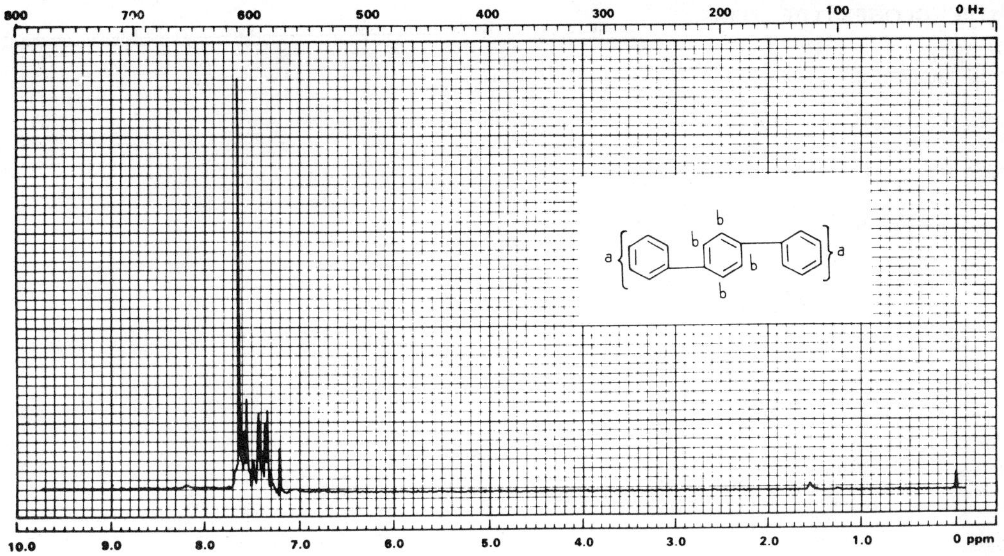

© 1978 Sadtler Research Laboratories, Inc.

1,2,3-TRIMETHYLBENZENE

C_9H_{12}

Mol. Wt. 120.20

B.P. 175-175.5°C

Solvent: $CDCl_3$

Assignments
a = 2.19
b = 2.29
c = 7.00

187

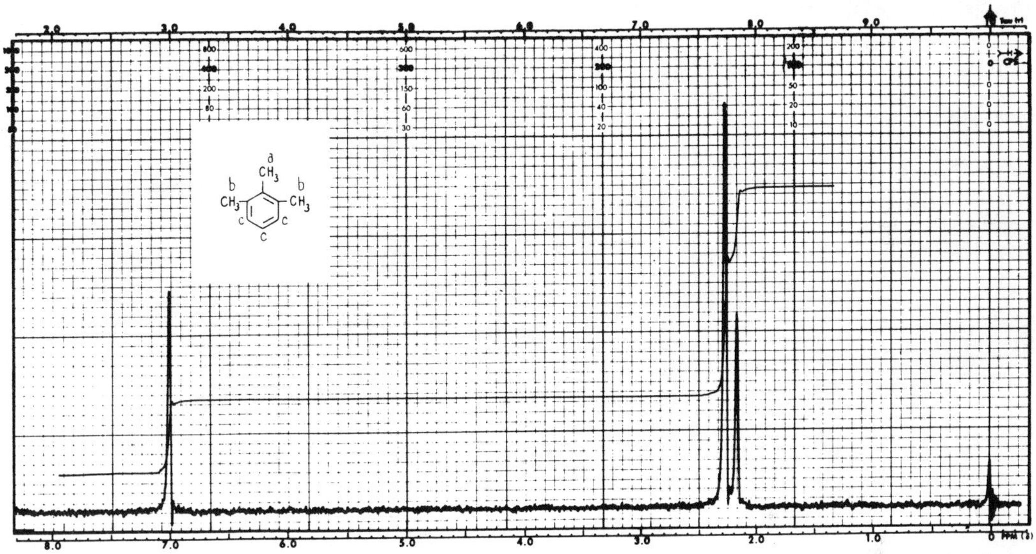

© 1978 Sadtler Research Laboratories, Inc.

1,2,4-TRIMETHYLBENZENE

C_9H_{12}

Mol. Wt. 120.20

B.P. 168.9°C

Solvent: CCl_4

Assignments
a = ca 2.19
b = ca 6.79

188

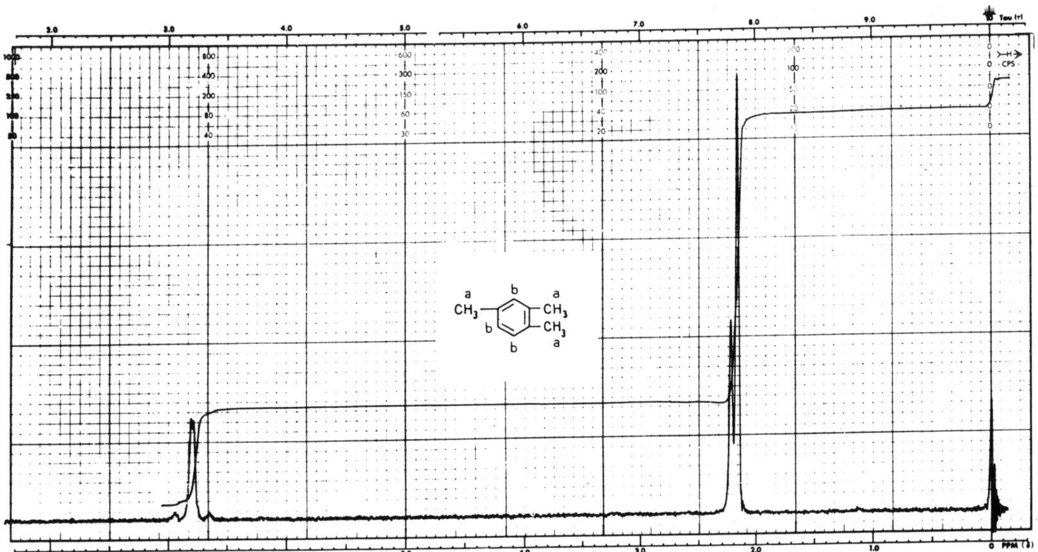

© 1978 Sadtler Research Laboratories, Inc.

MESITYLENE

C_9H_{12}

Mol. Wt. 120.20

Solvent: CCl_4

Assignments
a = 2.23
b = 6.69

189

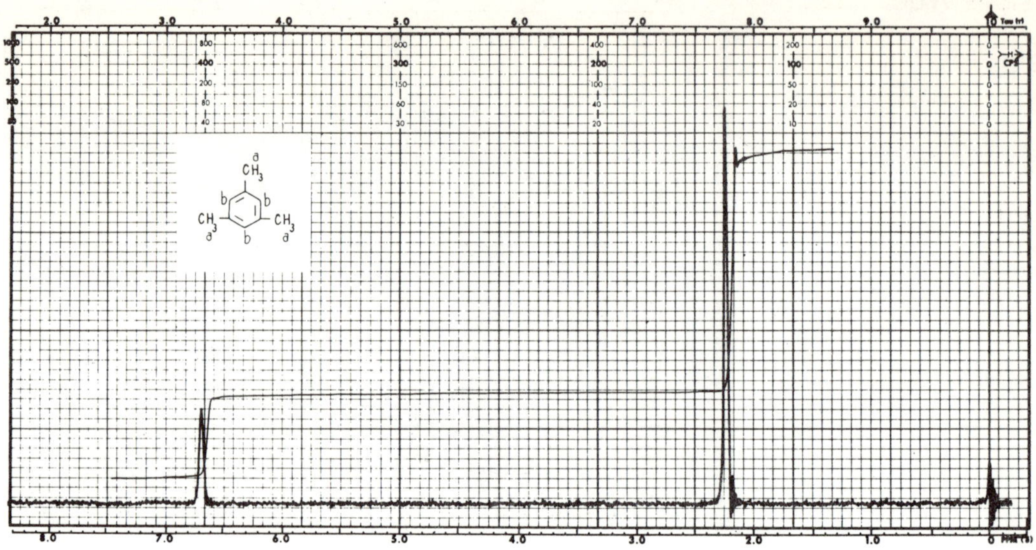

© 1978 Sadtler Research Laboratories, Inc.

1,3,5-TRIETHYLBENZENE

$C_{12}H_{18}$

Mol. Wt. 162.28

Solvent: CCl_4

Assignments
a = 1.23
b = 2.55
c = 6.73

190

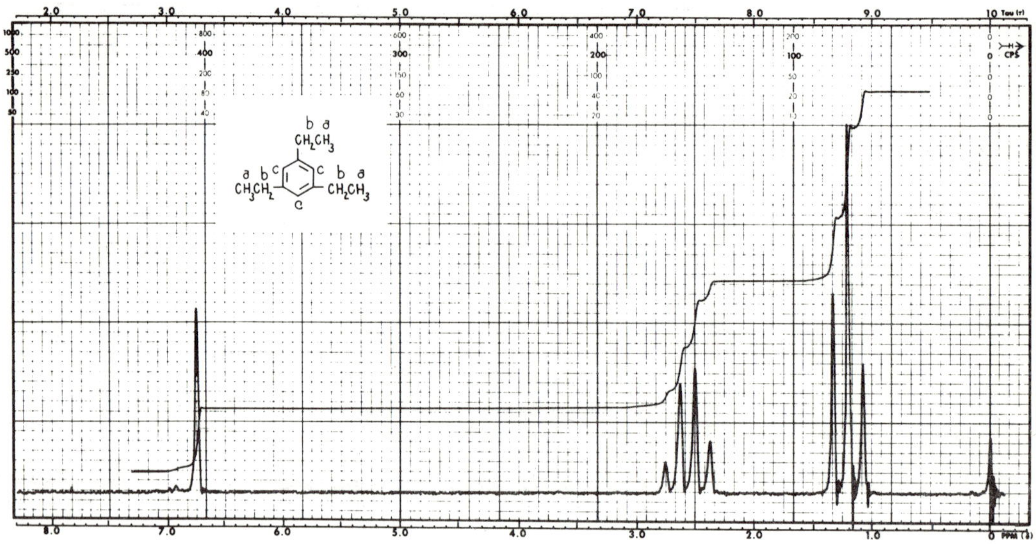

© 1978 Sadtler Research Laboratories, Inc.

1,2,3,4-TETRAMETHYLBENZENE

$C_{10}H_{14}$

Mol. Wt. 134.22

B.P. 204°C M.P. −4°C

Solvent: CCl_4

Assignments
a = 2.13
b = 2.19
c = 6.72

191

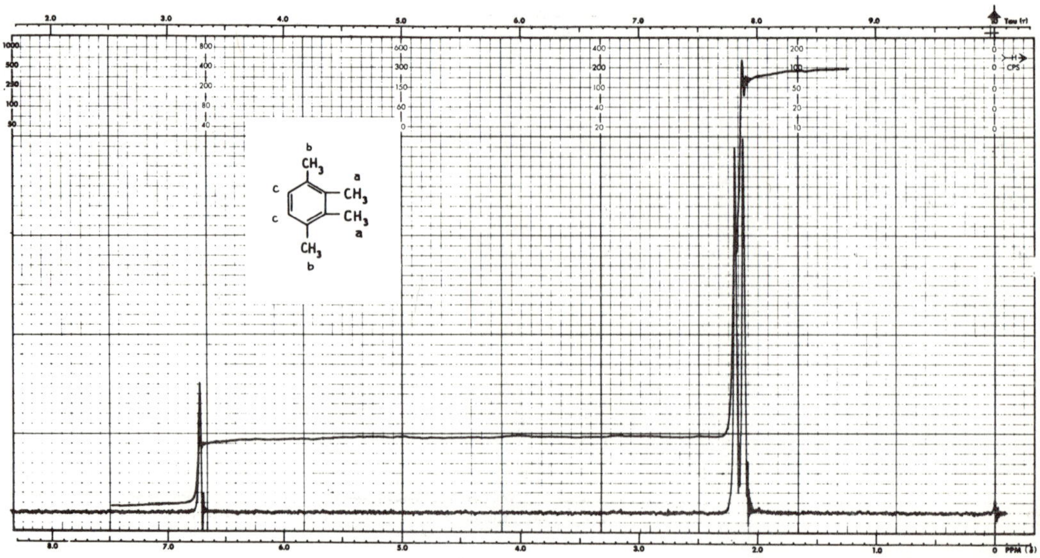

© 1978 Sadtler Research Laboratories, Inc.

1,2,3,5-TETRAMETHYLBENZENE

$C_{10}H_{14}$

Mol. Wt. 134.22

Solvent: CCl_4

Assignments
a = 2.04 or b
b = 2.09 or a
c = 2.15
d = 6.54

192

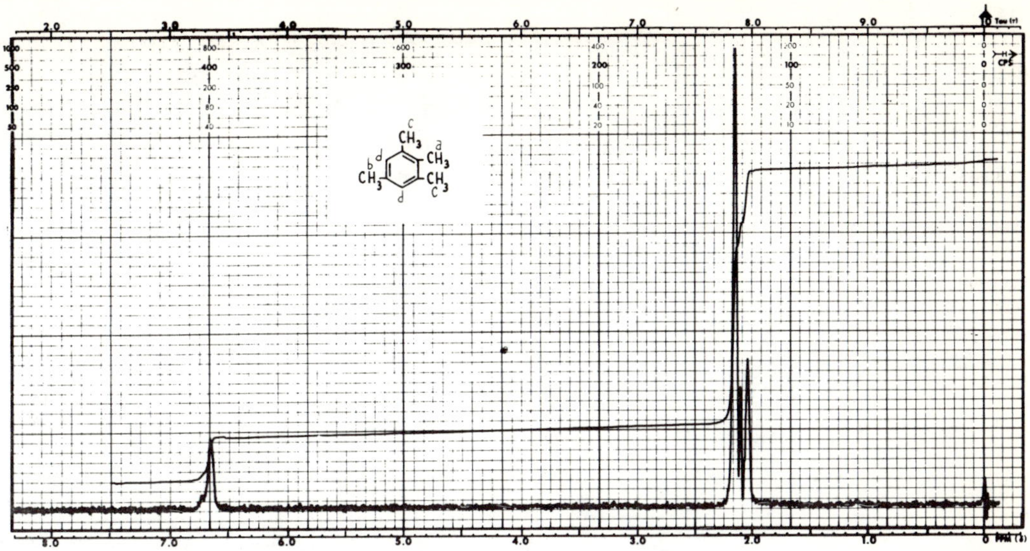

© 1978 Sadtler Research Laboratories, Inc.

1,2,4,5-TETRAMETHYLBENZENE

$C_{10}H_{14}$

Mol. Wt. 134.22

M.P. 78-80°C

Solvent: $CDCl_3$

Assignments
a = 2.20
b = 6.91

193

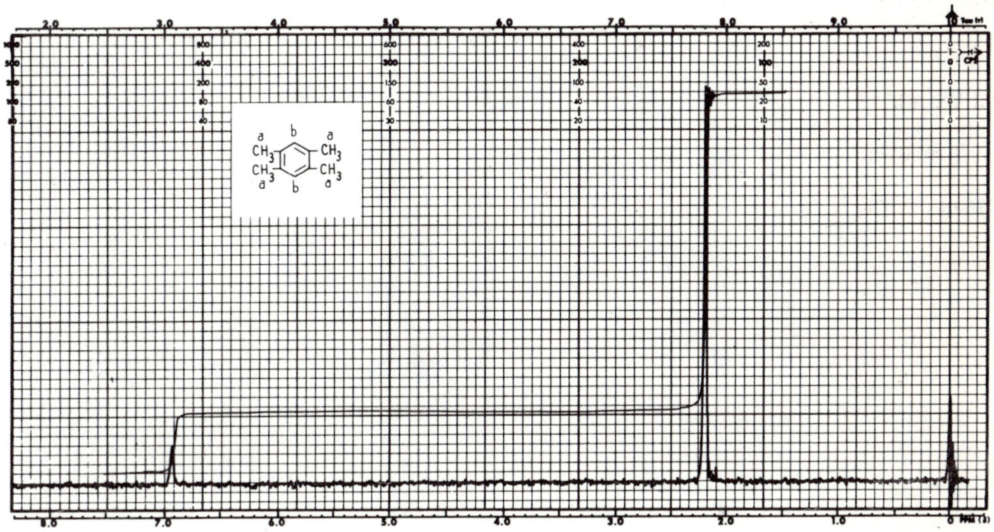

© 1978 Sadtler Research Laboratories, Inc.

PENTAMETHYLBENZENE

$C_{11}H_{16}$

Mol. Wt. 148.25

Solvent: $CDCl_3$

Assignments
a = 2.20
b = 6.81

194

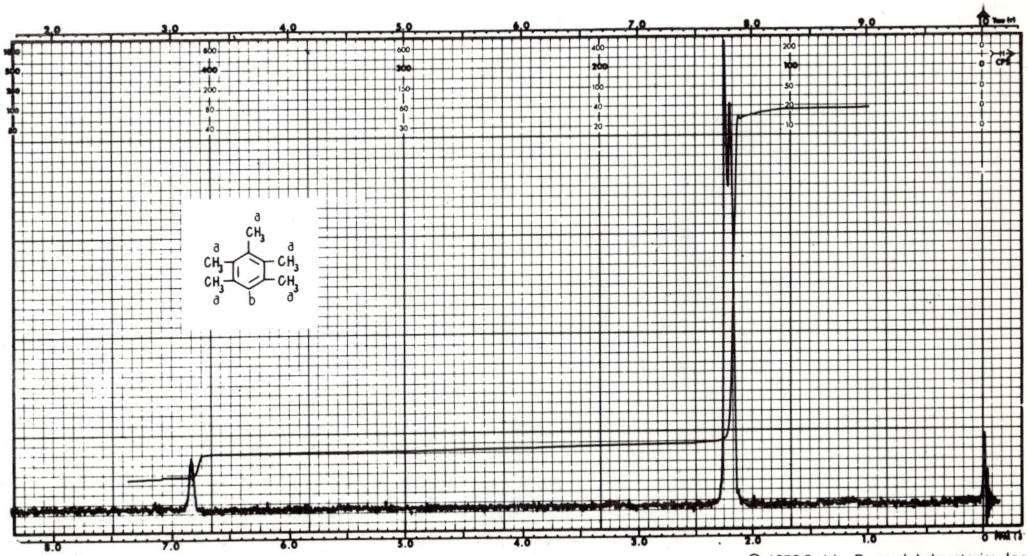

© 1978 Sadtler Research Laboratories, Inc.

INDAN

C_9H_{10}

Mol. Wt. 118.18

B.P. 173-177°C

Solvent: $CDCl_3$

Assignments
a = 2.02
b = 2.90
c = ca 7.25

195

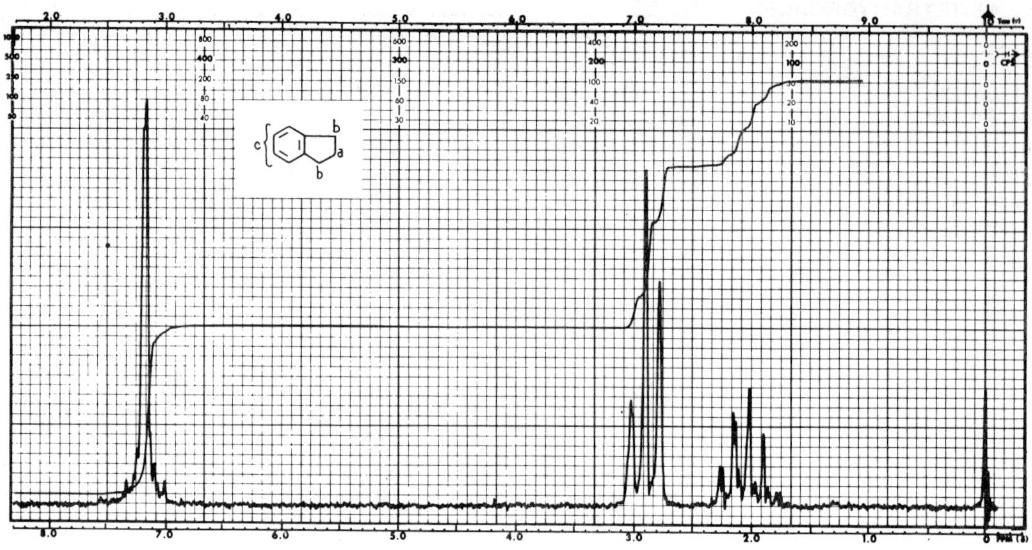

© 1978 Sadtler Research Laboratories, Inc.

1,2,3,4-TETRAHYDRONAPHTHALENE

$C_{10}H_{12}$

Mol. Wt. 132.21

B.P. 204-207°C

Solvent: $CDCl_3$

Assignments
a = 1.79
b = 2.75
c = ca 7.02

196

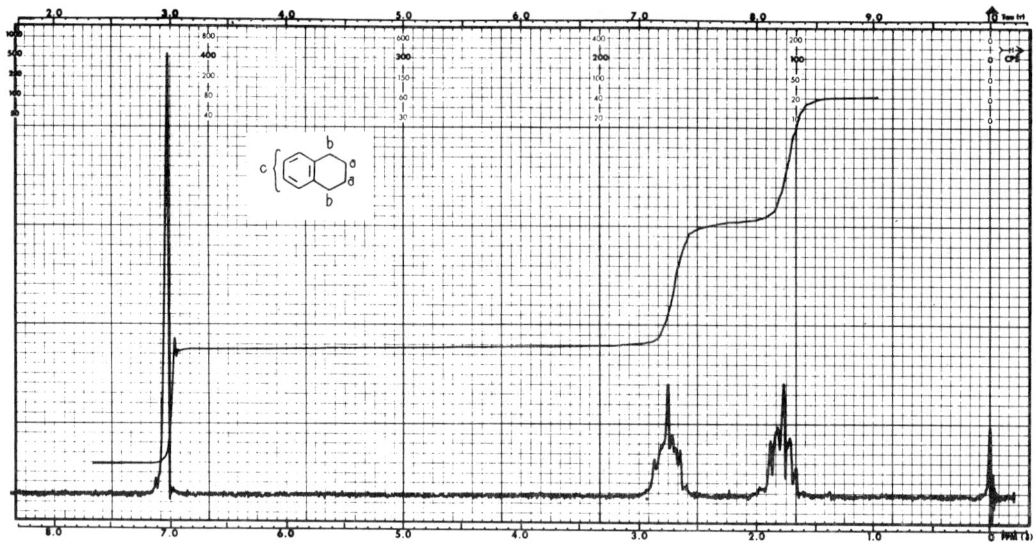

© 1978 Sadtler Research Laboratories, Inc.

1-ETHYL-1,2,3,4-TETRAHYDRONAPHTHALENE

$C_{12}H_{16}$

Mol. Wt. 160.26

B.P. 239-246°C

Solvent: CCl_4

Assignments
a = 0.95
b = 1.40-2.10
c = 2.70
d = 6.99

197

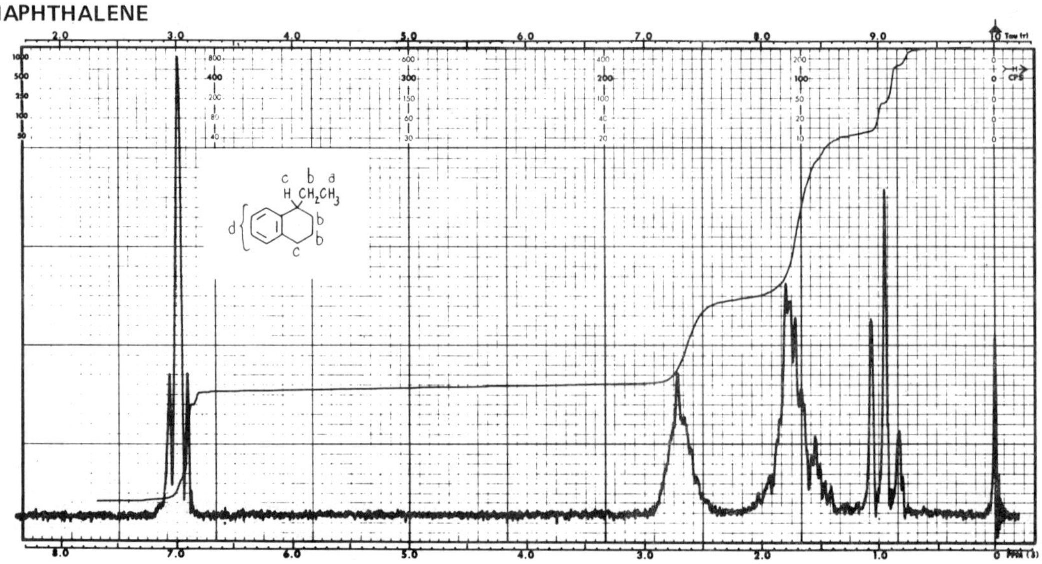

© 1978 Sadtler Research Laboratories, Inc.

2a,3,4,5-TETRAHYDROACENAPHTHENE

$C_{12}H_{14}$

Mol. Wt. 158.25

M.P. 12°C B.P. 246°C

Solvent: CCl_4

Assignments
a = 0.80-2.60
b = 2.50-3.00
c = 6.60-7.10

198

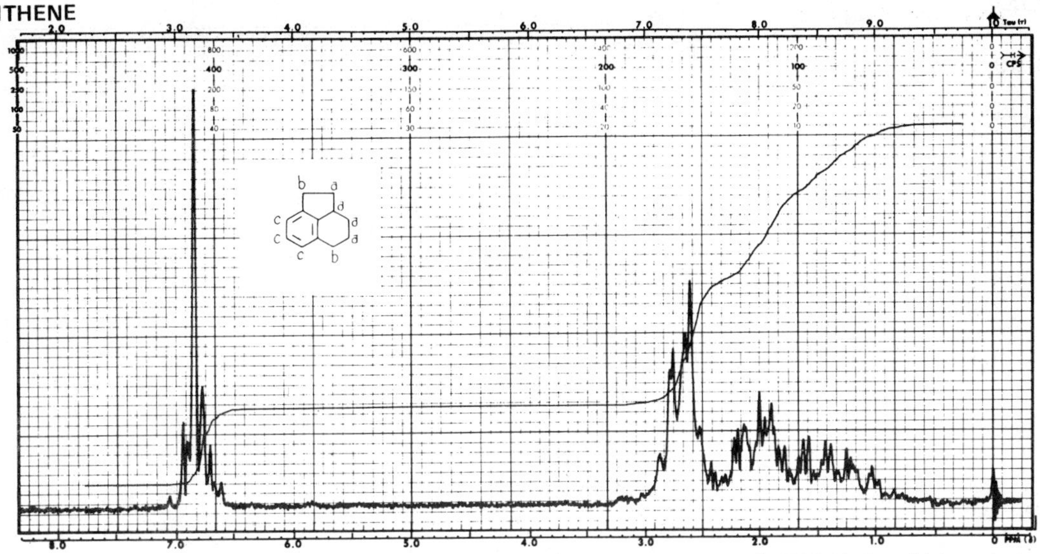

© 1978 Sadtler Research Laboratories, Inc.

INDENE

C_9H_8

Mol. Wt. 116.16

Solvent: CCl_4

Assignments
a = 3.12
b = 6.28
c = 6.68
d = 6.90-7.40

199

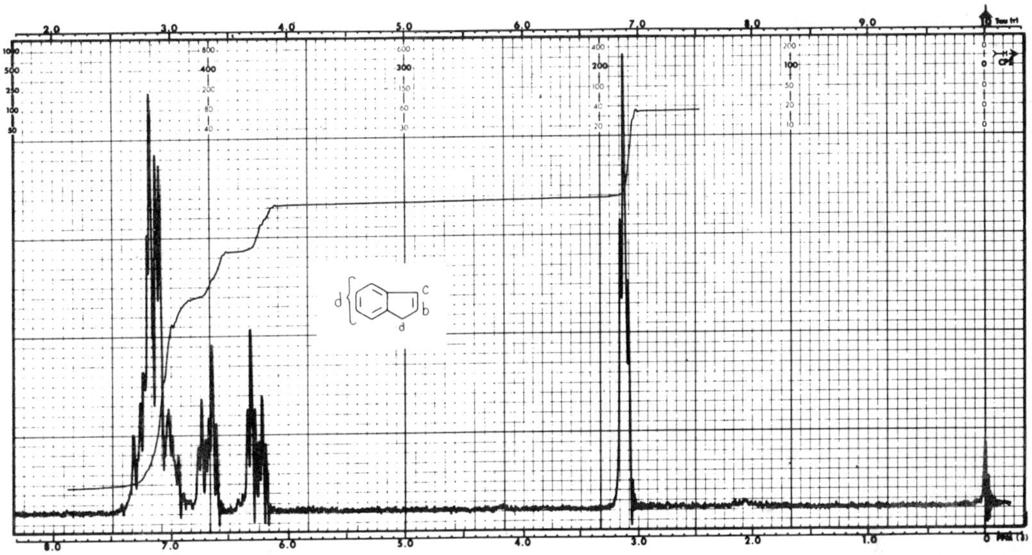

© 1978 Sadtler Research Laboratories, Inc.

NAPHTHALENE

$C_{10}H_8$

Mol. Wt. 128.18

M.P. 80.2°C

Solvent: $CDCl_3$

Assignments
a = 7.41
b = 7.79

200

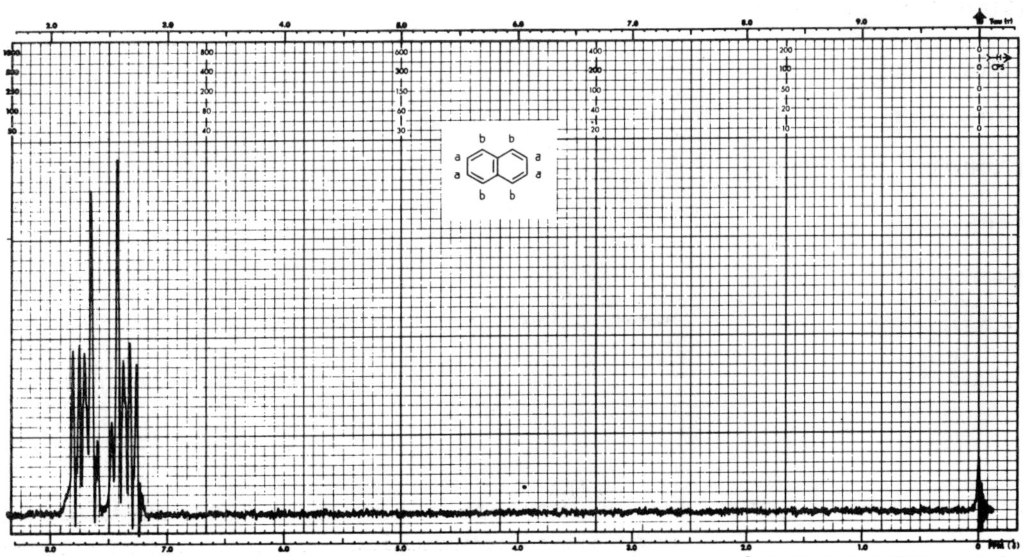

© 1978 Sadtler Research Laboratories, Inc.

1-ETHYLNAPHTHALENE

$C_{12}H_{12}$

Mol. Wt. 156.23

Solvent: CCl_4

Assignments
a = 1.37
b = 3.03
c = 7.10-7.55
d = 7.62
e = 7.85

201

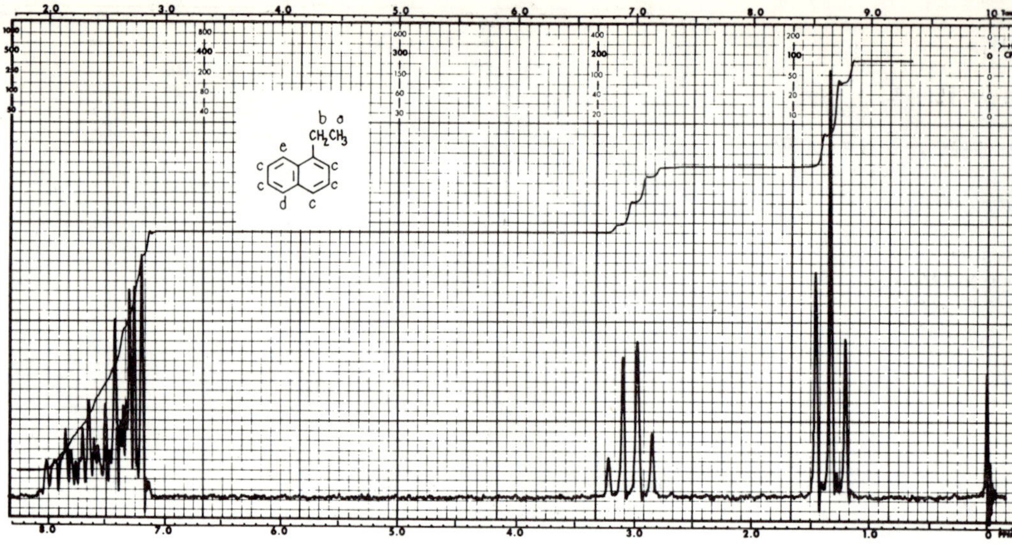

© 1978 Sadtler Research Laboratories, Inc.

2-METHYLNAPHTHALENE

$C_{11}H_{10}$

Mol. Wt. 142.20

M.P. 35-36°C

Solvent: $CDCl_3$

Assignments
a = 2.49
b = 7.12-7.91

202

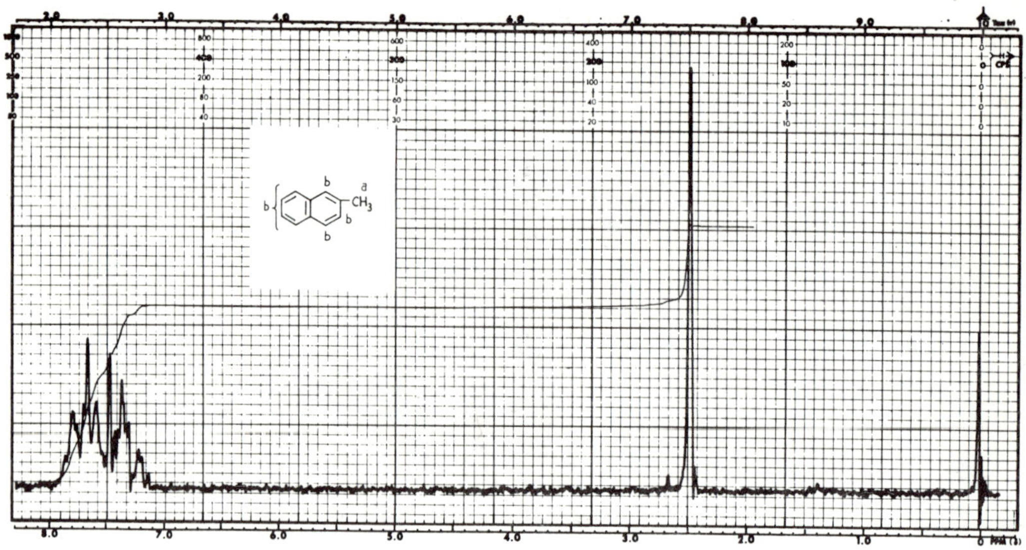

© 1978 Sadtler Research Laboratories, Inc.

2,3-DIMETHYLNAPHTHALENE

$C_{12}H_{12}$

Mol. Wt. 156.23

M.P. 104°C B.P. 265°C

Solvent: $CDCl_3$

Assignments
a = 2.31
b = 7.31
c = 7.48
d = 7.62

203

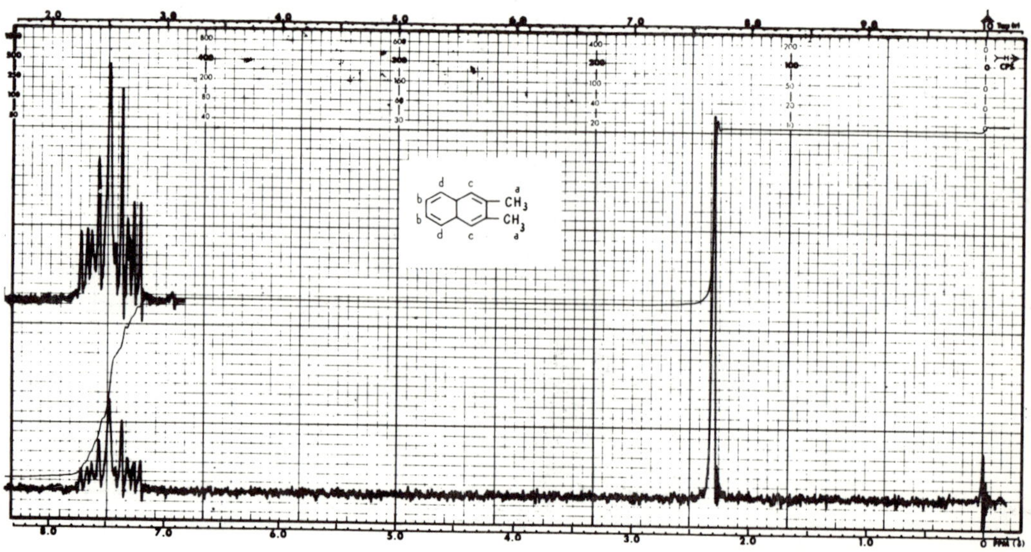

© 1978 Sadtler Research Laboratories, Inc.

2,6-DIMETHYLNAPHTHALENE

$C_{12}H_{12}$

Mol. Wt. 156.23

M.P. 109-111°C

Solvent: CCl_4

Assignments
a = 2.42
b = 7.12
c = 7.35-7.65

204

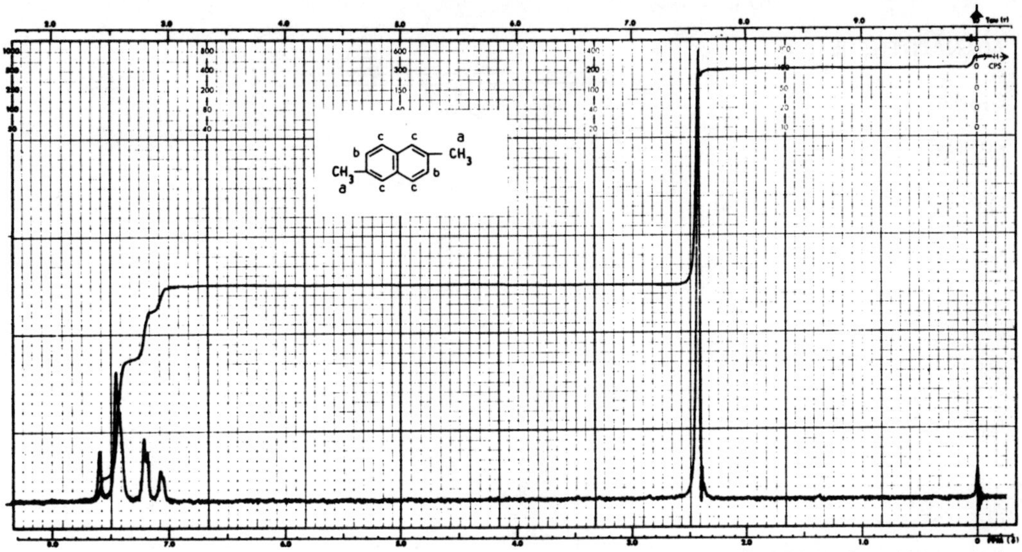

© 1978 Sadtler Research Laboratories, Inc.

2,3,6-TRIMETHYLNAPHTHALENE

$C_{13}H_{14}$

Mol. Wt. 170.26

M.P. 102°C B.P. 286°C/762mm

Solvent: $CDCl_3$

Assignments
a = 2.31
b = 2.40
c = 7.12
d = ca 7.38
e = 7.53

205

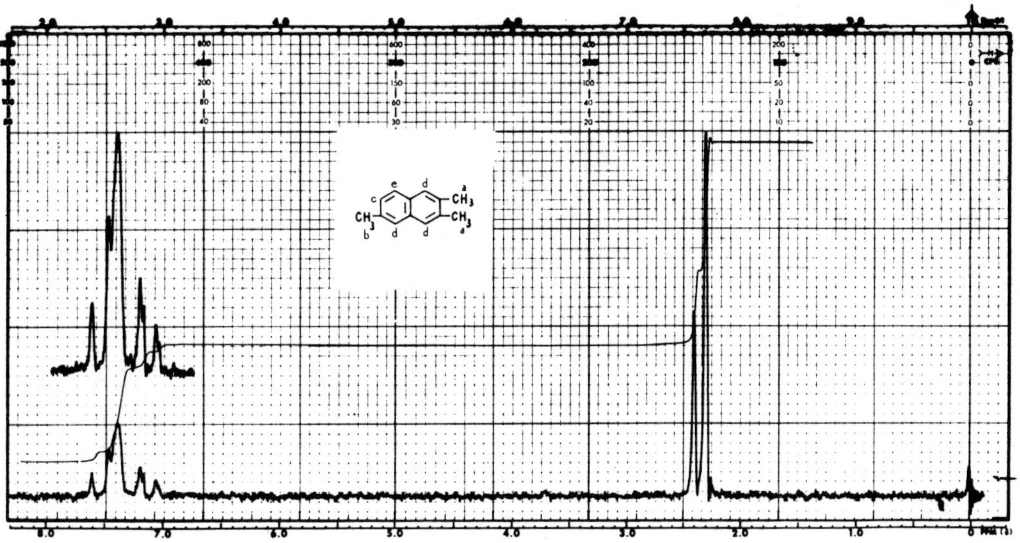

© 1978 Sadtler Research Laboratories, Inc.

2,2'-BINAPHTHYL

$C_{20}H_{14}$

Mol. Wt. 254.33

Solvent: $CDCl_3$

Assignments
a = ca 7.46
b = 7.65-8.05
c = 8.11
d = 7.20 $CHCl_3$

206

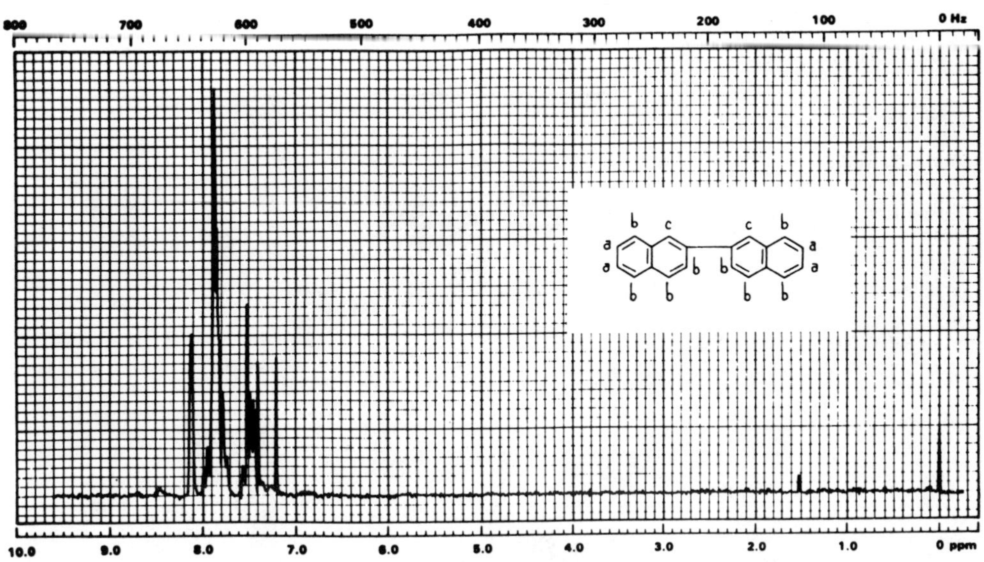

© 1978 Sadtler Research Laboratories, Inc.

ACENAPHTHENE

$C_{12}H_{10}$

Mol. Wt. 154.20

M.P. 95°C

Solvent: $CDCl_3$

Assignments
a = 3.30
b = 7.16
c = 7.38
d = 7.53

207

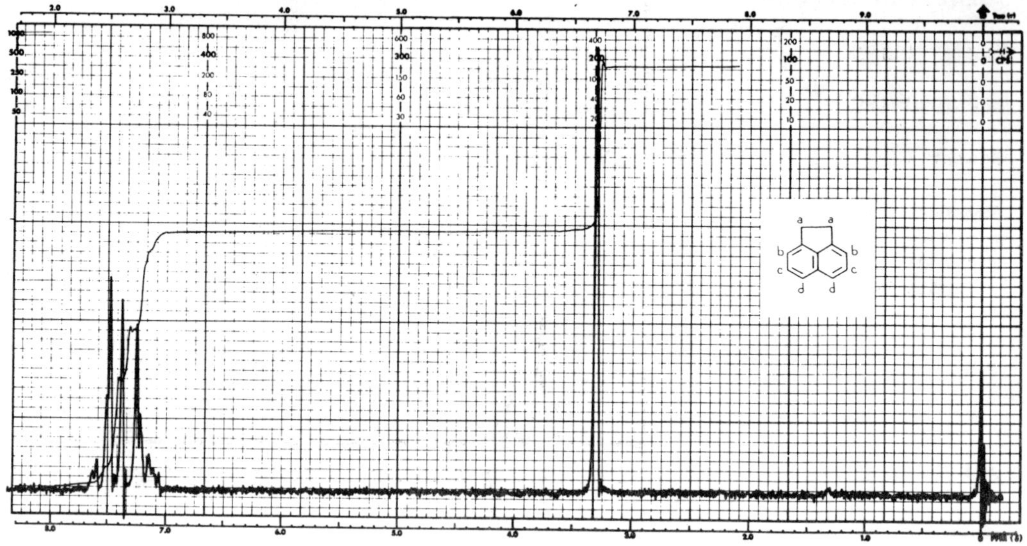

© 1978 Sadtler Research Laboratories, Inc.

FLUORENE

$C_{13}H_{10}$

Mol. Wt. 166.21

M.P. 116°C

Solvent: $CDCl_3$

Assignments
a = 3.76

208

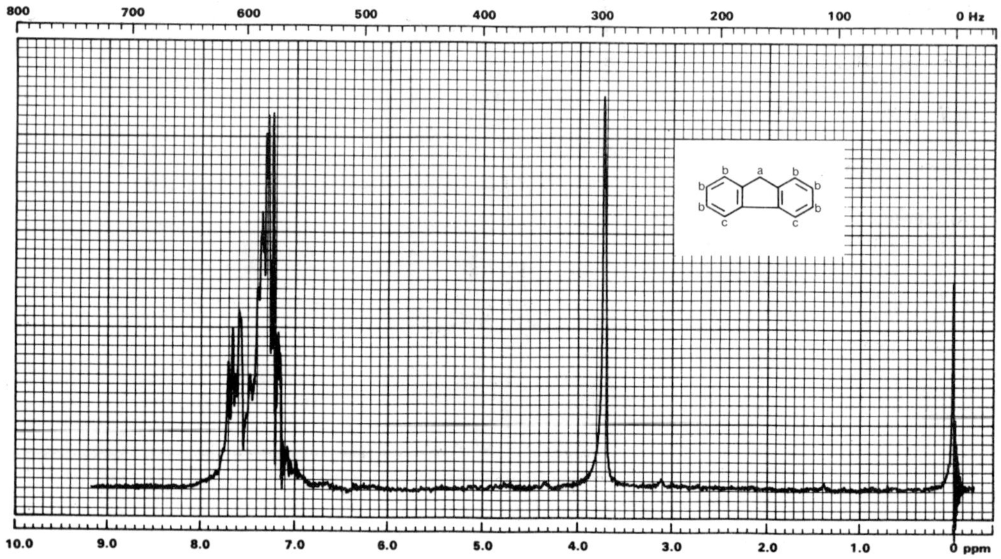

© 1978 Sadtler Research Laboratories, Inc.

2-METHYLFLUORENE

$C_{14}H_{12}$

Mol. Wt. 180.25

Solvent: $CDCl_3$

Assignments
a = 2.41
b = 3.82
c = 6.91-7.98
d = 1.43 imp.

209

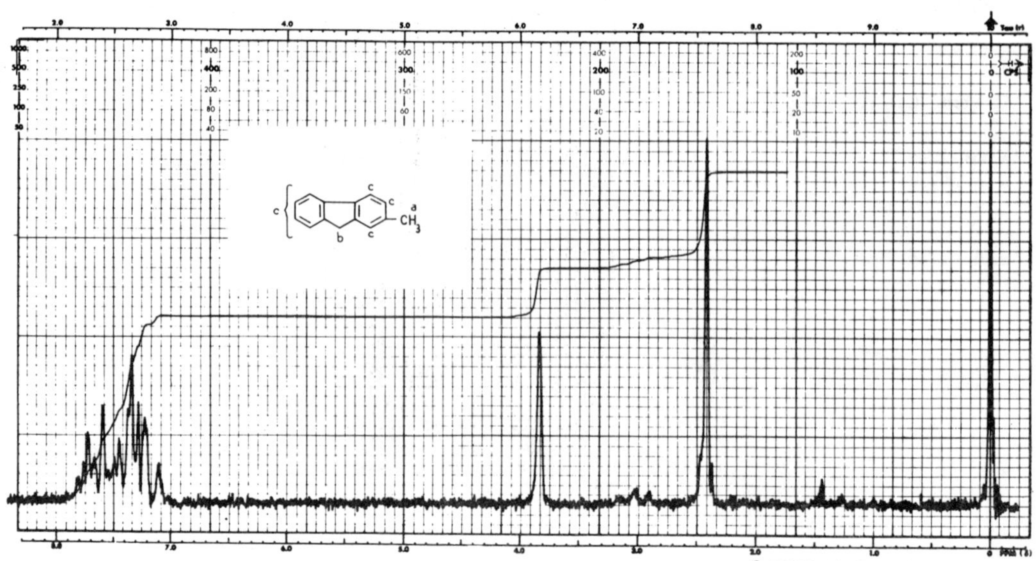

© 1978 Sadtler Research Laboratories, Inc.

9-BENZYLIDENEFLUORENE

$C_{20}H_{14}$

Mol. Wt. 254.33

M.P. 76-77°C

Solvent: $CDCl_3$

Assignments
a = 6.80-7.80

210

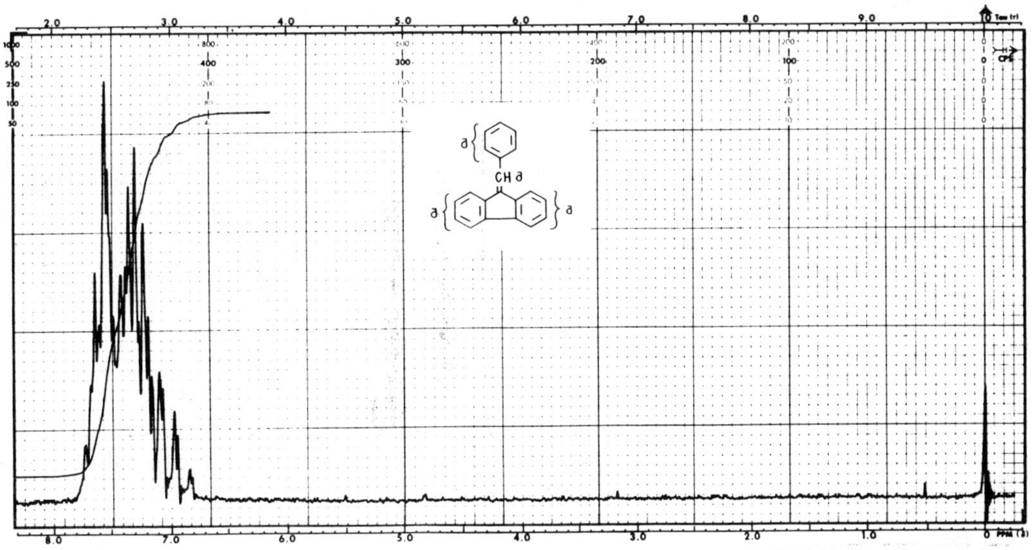

© 1978 Sadtler Research Laboratories, Inc.

9,10-DIHYDROPHENANTHRENE

$C_{14}H_{12}$

Mol. Wt. 180.25

Solvent: $CDCl_3$

Assignments
a = 2.86
b = 7.10-7.50
c = 7.71

211

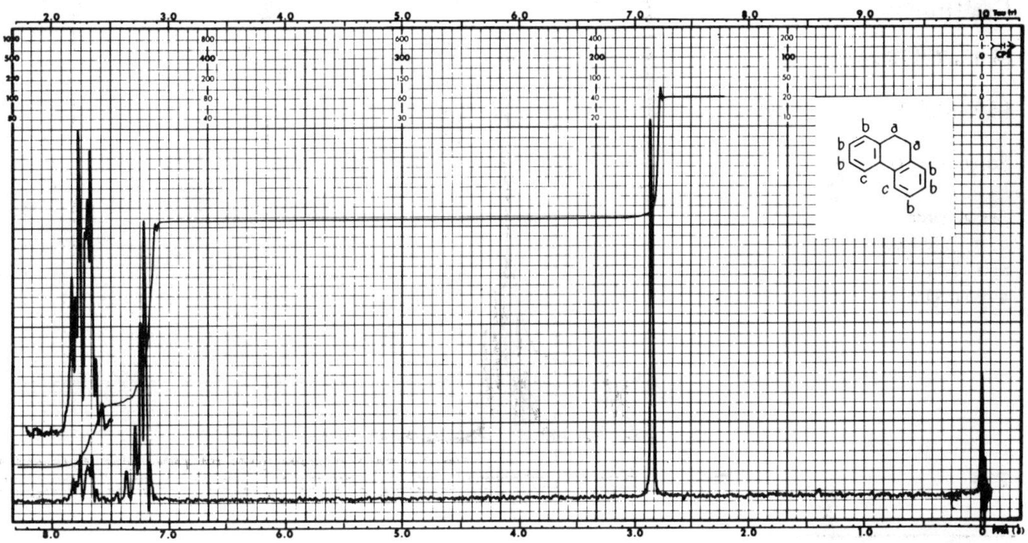

© 1978 Sadtler Research Laboratories, Inc.

FLUORANTHENE

$C_{16}H_{10}$

Mol. Wt. 202.26

M.P. 110°C B.P. 250°C/60mm

Solvent: $CDCl_3$

Assignments
a = 7.15-7.90

212

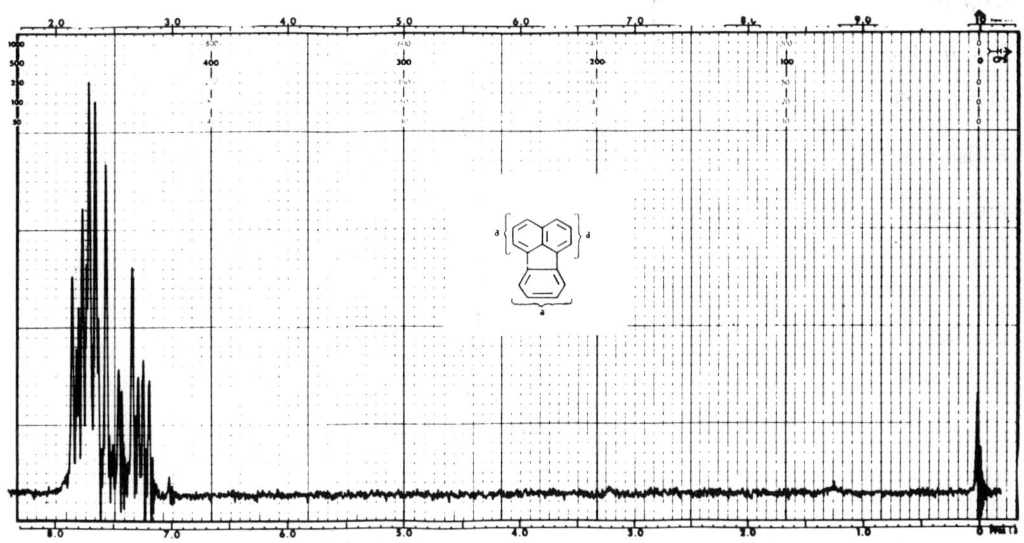

© 1978 Sadtler Research Laboratories, Inc.

ANTHRACENE

$C_{14}H_{10}$

Mol. Wt. 178.24

M.P. 217-218°C   B.P. 342°C

Solvent: $CDCl_3$

Assignments
a = 7.41
b = 7.95
c = 8.40
d = 7.21 $CHCl_3$

213

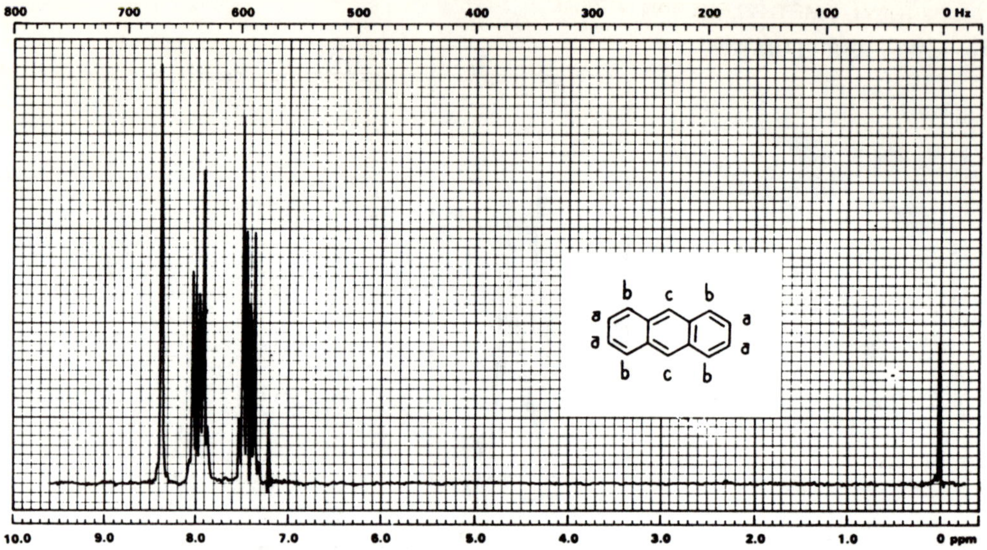

© 1978 Sadtler Research Laboratories, Inc.

PHENANTHRENE

$C_{14}H_{10}$

Mol. Wt. 178.24

M.P. 100°C

Solvent: $CDCl_3$

Assignments
a = 7.20-7.80
b = 7.50
c = 8.41

214

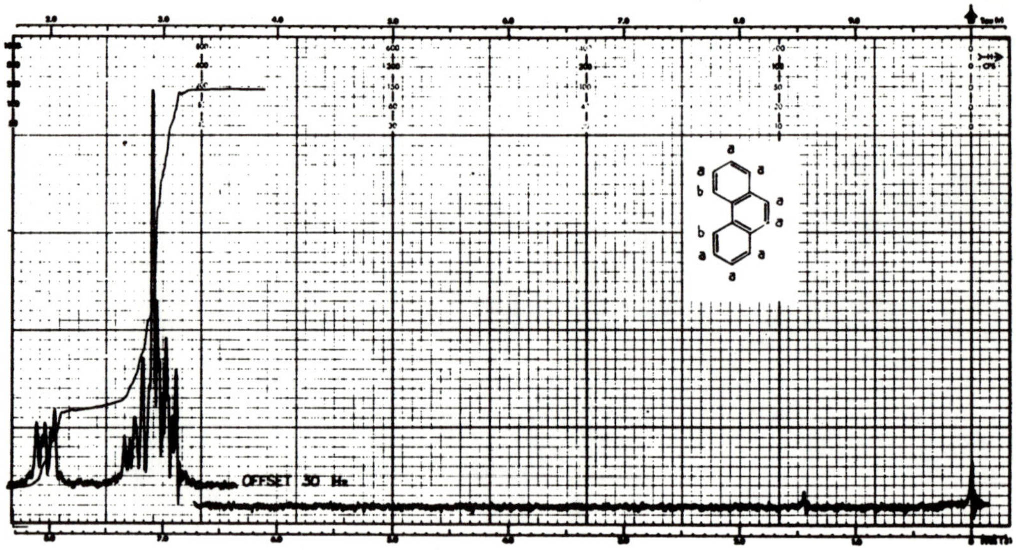

© 1978 Sadtler Research Laboratories, Inc.

2-METHYLPHENANTHRENE

$C_{15}H_{12}$

Mol. Wt. 192.26

M.P. 56-57°C

Solvent: $CDCl_3$

Assignments
a = 2.44
b = 7.38
c = 7.46-7.87
d = 7.62
e = ca 8.54

215

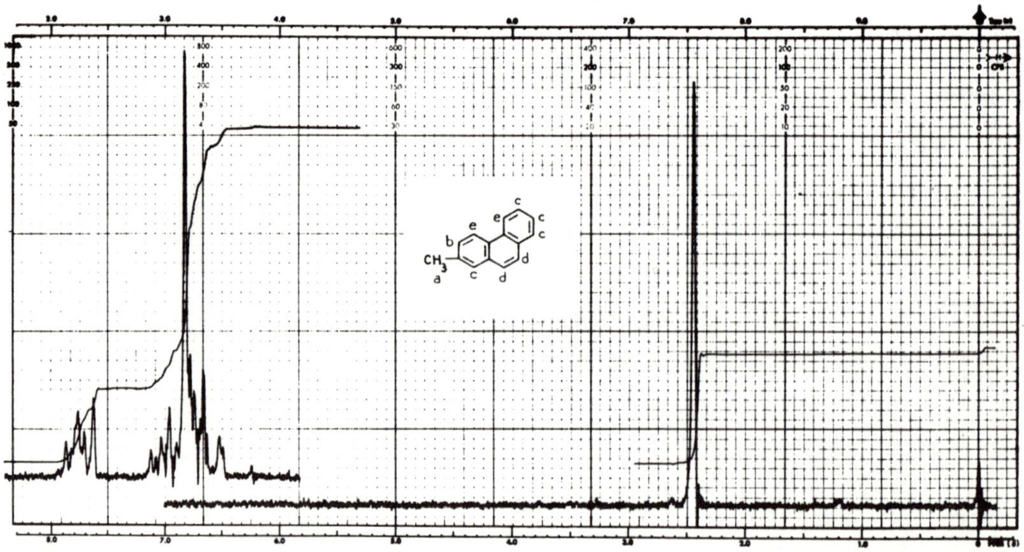

© 1978 Sadtler Research Laboratories, Inc.

7-ISOPROPYL-1-METHYLPHENANTHRENE

$C_{18}H_{18}$

Mol. Wt. 234.34

Solvent: $CDCl_3$

Assignments
a = 1.23
b = 2.28
c = 2.80
d = 7.15-7.85
e = ca 8.40

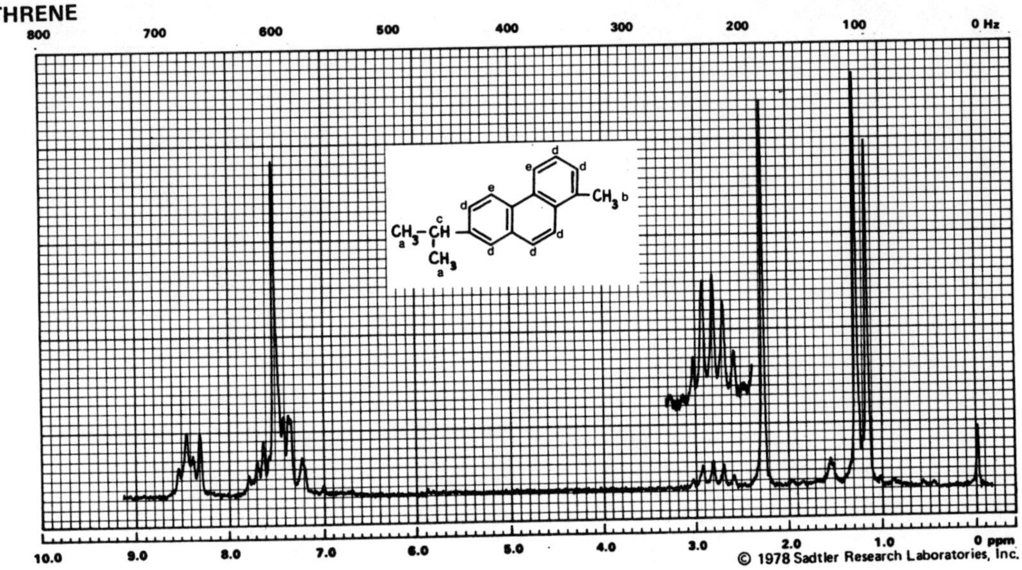

216

PYRENE

$C_{16}H_{10}$

Mol. Wt. 202.24

M.P. 150°C

Solvent: CCl_4

Assignments
a = 7.60-8.15

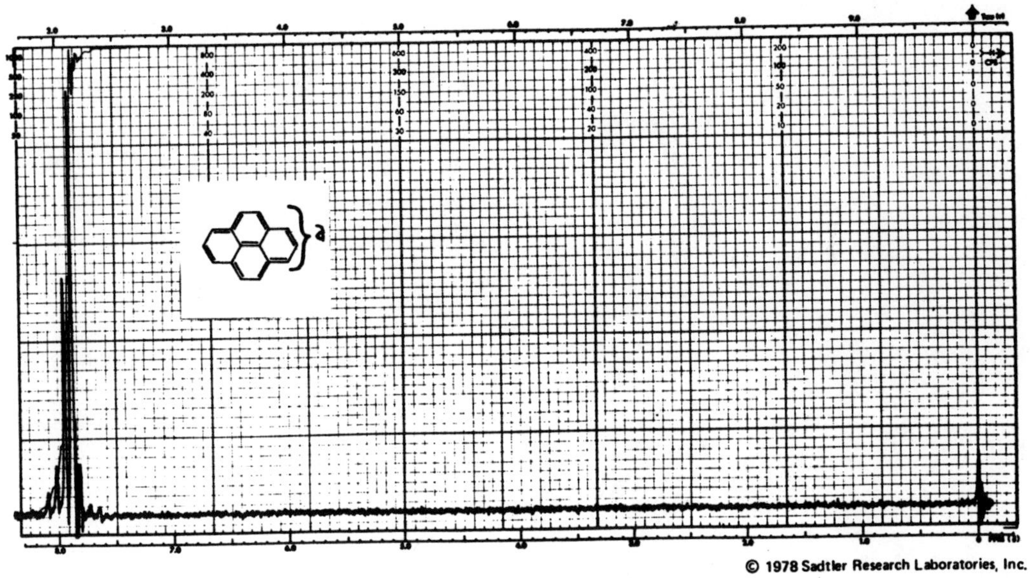

217

CHRYSENE

$C_{18}H_{12}$

Mol. Wt. 228.30

M.P. 251°C (lit.)

Solvent: $CDCl_3$

Assignments
a = ca 7.63
b = ca 7.90
c = 7.97
d = 8.67
e = 8.75

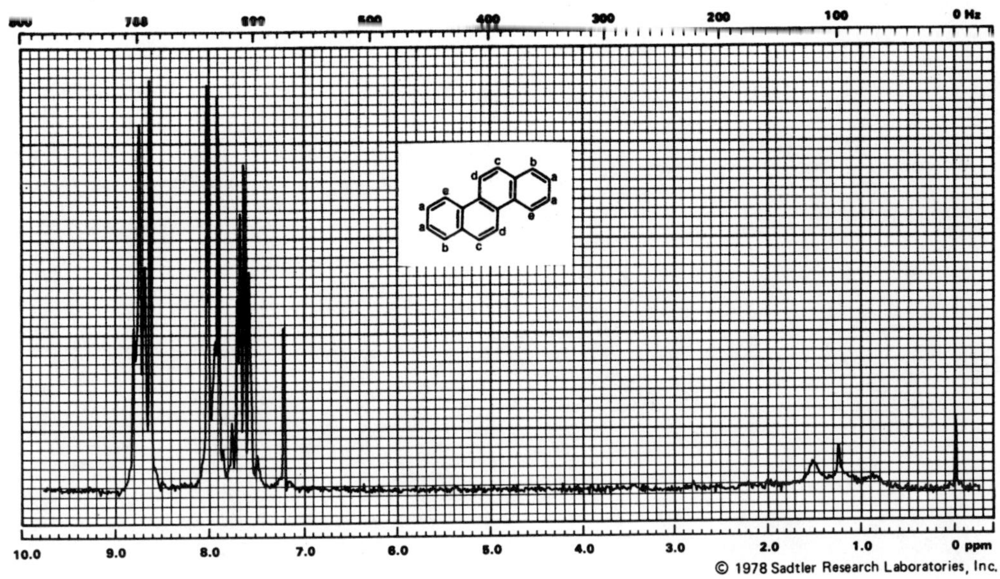

218

NOTES

FLUORINATED HYDROCARBONS

Fluorine containing compounds characteristically display coupling between the fluorine nuclei and nearby hydrogens. The multiplets which are produced are usually well resolved and the magnitude of the Fluorine-Hydrogen coupling constants is quite large in comparison to those produced by Hydrogen-Hydrogen and Phosphorus-Hydrogen coupling.

Fluorine is a strong deshielder of aliphatic groups but has a moderately strong shielding effect upon the aromatic hydrogens that are ortho and para to it.

Chemical Shifts

Aliphatic Protons

HC–2	HC–1		Solvent	Spectrum
3.72	4.37	HO–CH_2–CH_2–F	$CDCl_3$	- - - -
1.8–1.9	4.3–4.4	R–CH_2–CH_2–F	$CDCl_3$	219, 220

Substituted Fluoromethanes

F–	CH_2	–X		
	4.75	–C(=O)–O–Na	D_2O	2653
	4.88	–C(=O)–O–R	$CDCl_3$	- - - -
	5.02	–C(=O)–NH_2	TFA	- - - -
	5.07	–C(=O)–CH_3	D_2O	- - - -
	5.11	–C(=O)–R	$CDCl_3$	1990
	5.24	Ø	CCl_4	221

Coupling and Coupling Constants

Aliphatic Protons

J_{F-C-H} = 46.7–51.9 Hz

$J_{F-C-C-H}$ = 20.0–30.0 Hz

$J_{F_2-C-C-H}$ = 3.9–13.0 Hz

Aromatic Protons

J_{F-H} (ortho) = 8.0–9.0 Hz

J_{F-H} (meta) = 5.0–6.0 Hz

J_{F-H} (para) = 2.0–3.0 Hz

Chemical Shifts continued

Aromatic Protons

Fluorine substituents on aromatic rings have a moderately strong shielding effect upon the ortho and para hydrogens and display coupling to the ortho, meta and para hydrogens.

The chemical shifts of several para substituted fluorobenzenes are listed below.

F–	ortho	meta	para	Solvent
	6.81	7.00	$-CH_3$	CCl_4
	6.89	7.09	–R	CCl_4
	6.90	7.12	$-CH_2-OH$	$CDCl_3$
	6.90	7.15	–N=C=S	CCl_4
	6.90	7.23	$-CH{=}CH_2$	$CDCl_3$
	6.91	7.19	–Cl	$CDCl_3$
	6.92	7.29	$-CH_2-Br$	CCl_4
	6.93	7.40	–Ø	CCl_4
	6.99	7.19	$-CH_2-C{\equiv}N$	CCl_4
	7.00	6.81	$-O-CH_3$	$CDCl_3$
	7.11	7.99	$-C(=O)-CH_3$	$CDCl_3$
	7.15	8.01	–C(=O)–R	$CDCl_3$
	7.17	7.66	$-C{\equiv}N$	$CDCl_3$
	7.20	8.20	$-NO_2$	CCl_4

Several para substituted ***a,a,a***-trifluorotoluenes are presented below.

CF_3	ortho	meta	para	
CF_3–	7.46	6.91	–OH	$CDCl_3$
	7.49	7.60	–Br	CCl_4
	7.51	7.37	–CH(OH)–Ø	$CDCl_3$
	7.51	8.08	$-C(=O)-CH_3$	$CDCl_3$
	7.60	7.60	–NH–C(=O)–R	$CDCl_3$
	7.60	7.60	$-CH(-O-R_2)_2$	$CDCl_3$
	7.61	7.31	$-CH_2-$Ø	DMSO-d_6
	7.64	7.13	–F	CCl_4
	7.69	8.03	$-C(=O)-R_2$	CCl_4
	7.71	8.18	–C(=O)–OH	Polysol
	7.73	8.20	–C(=O)–Cl	$CDCl_3$
	7.80	7.80	$-C{\equiv}N$	$CDCl_3$

1-FLUOROHEXANE

$C_6H_{13}F$

Mol. Wt. 104.17

B.P. 92-93°C

Solvent: CCl_4

Assignments
a = 0.92
b = 1.10-1.65
c = 1.86
d = 4.35

219

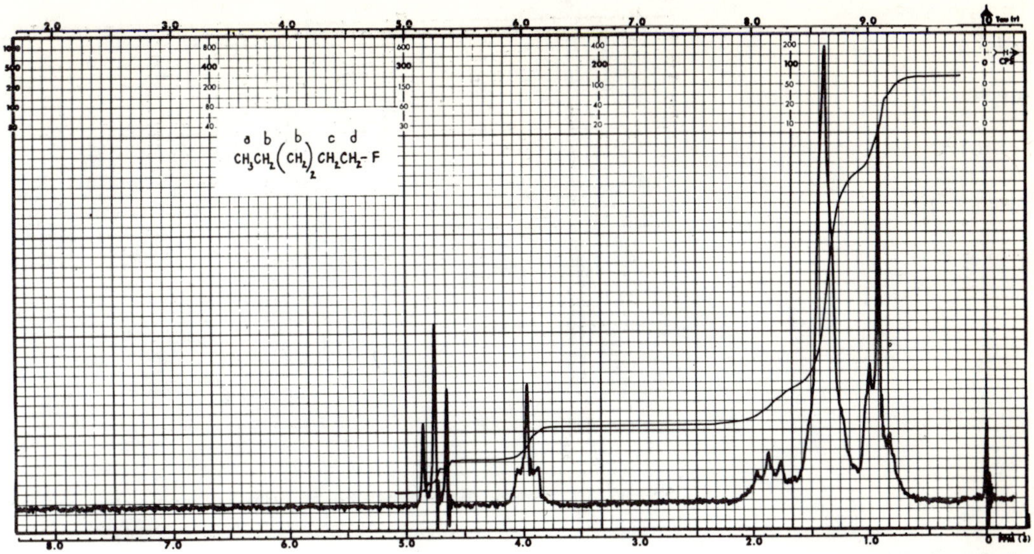

© 1978 Sadtler Research Laboratories, Inc.

1-FLUORODECANE

$C_{10}H_{21}F$

Mol. Wt. 160.28

B.P. 185-187°C

Solvent: CCl_4

Assignments
a = 0.90
b = 1.10-1.60
c = 1.88
d = 4.31

220

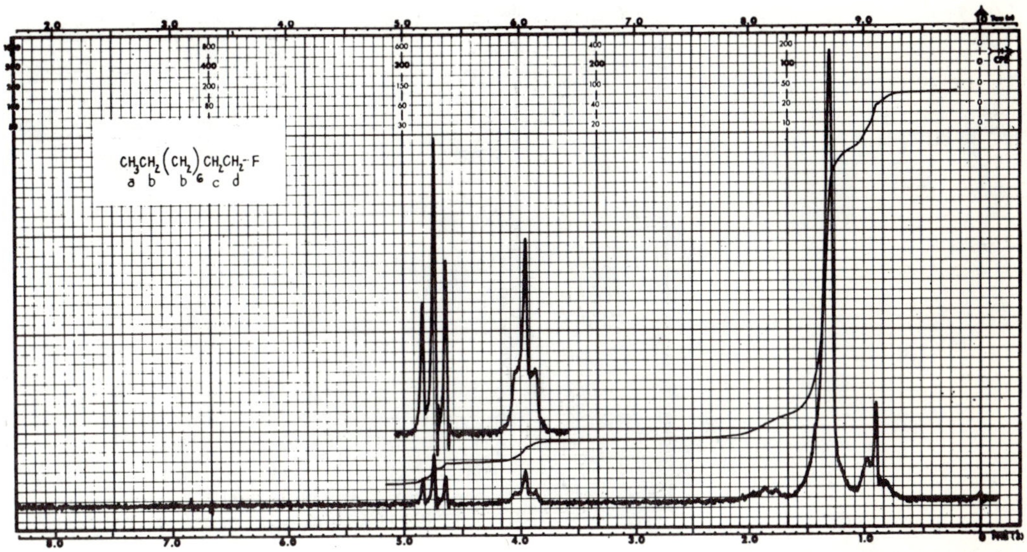

© 1978 Sadtler Research Laboratories, Inc.

α-FLUOROTOLUENE

C_7H_7F

Mol. Wt. 110.13

Solvent: CCl_4

Assignments
a = 5.27
b = ca 7.27

221

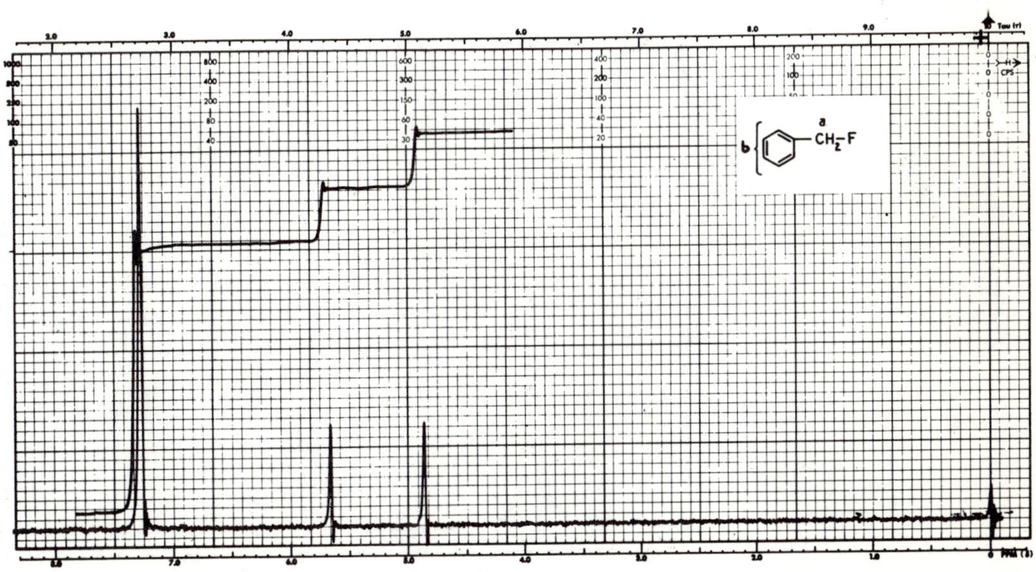

© 1978 Sadtler Research Laboratories, Inc.

a,a,a-**TRIFLUOROTOLUENE**

$C_7H_5F_3$

Mol. Wt. 146.11

B.P. 101°C

Solvent: CCl_4

Assignments
a = 7.30-7.70

222

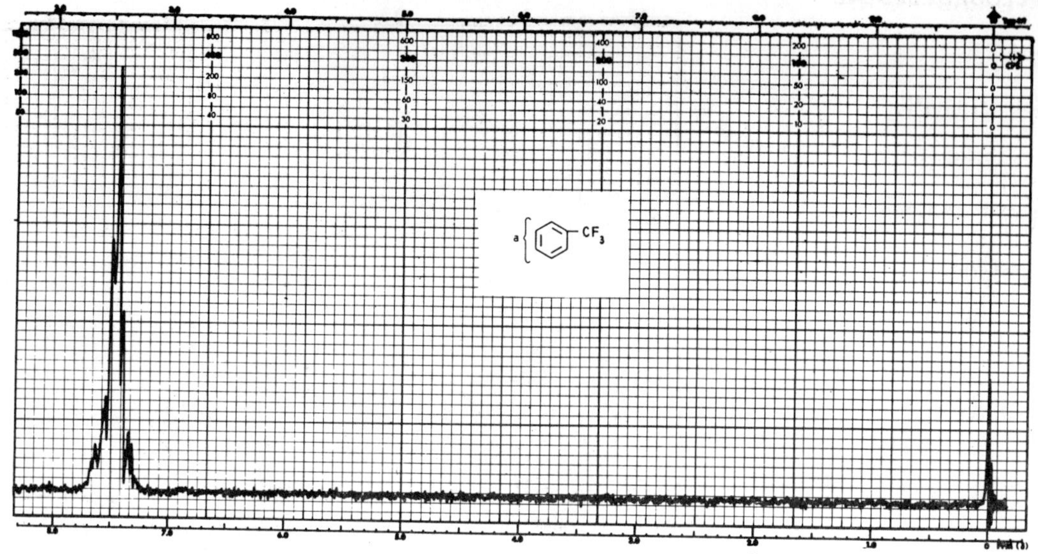

© 1978 Sadtler Research Laboratories, Inc.

FLUOROBENZENE

C_6H_5F

Mol. Wt. 96.11

B.P. 85-86°C (lit.)

Solvent: CCl_4

Assignments
a = 6.71-7.43

223

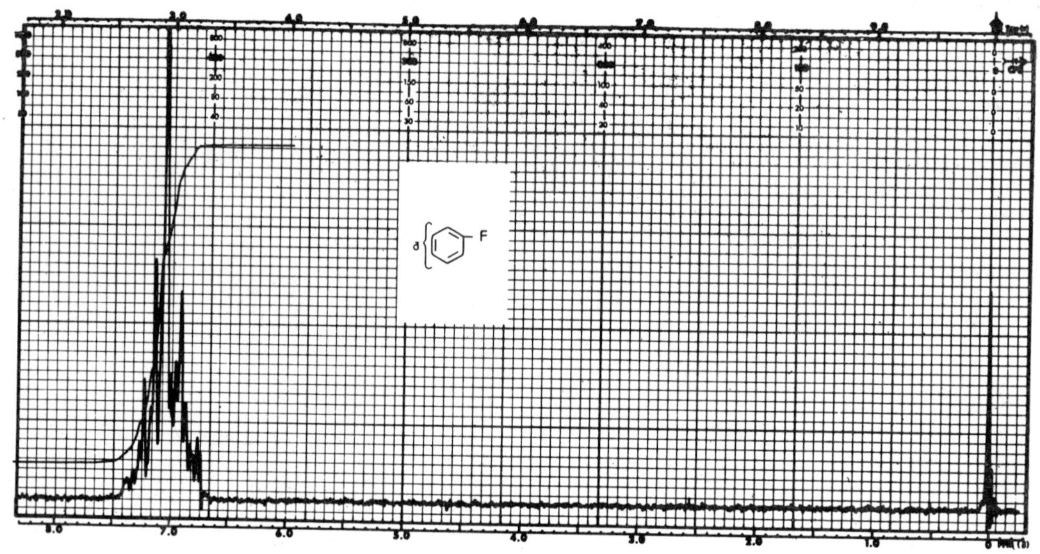

© 1978 Sadtler Research Laboratories, Inc.

---

o-FLUOROTOLUENE

C_7H_7F

Mol. Wt. 110.13

B.P. 113-114°C

Solvent: CCl_4

Assignments
a = 2.22
b = 6.65-7.27

224

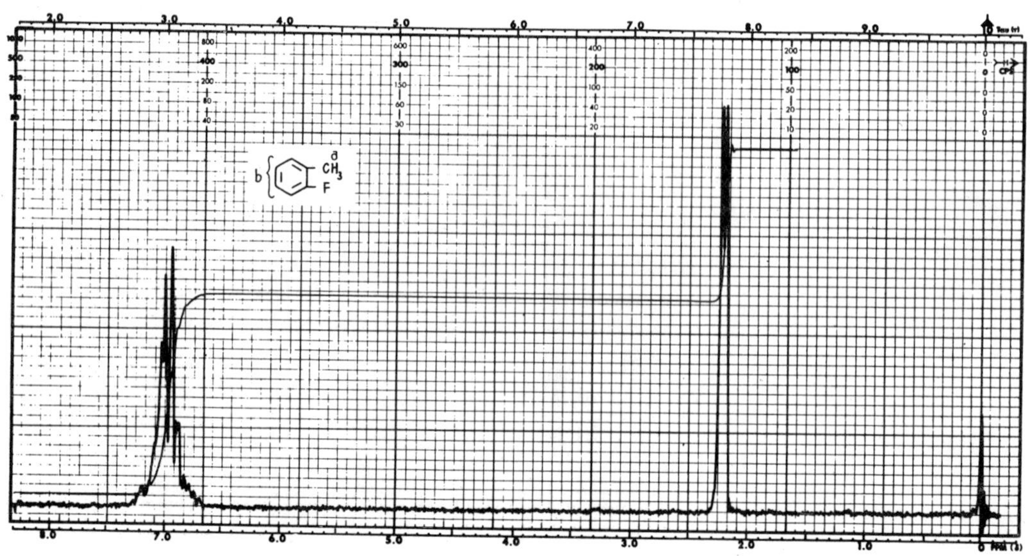

© 1978 Sadtler Research Laboratories, Inc.

m-FLUOROTOLUENE

C_7H_7F

Mol. Wt. 110.13

B.P. 115°C

Solvent: CCl_4

Assignments
a = 2.32
b = 6.50-7.40

225

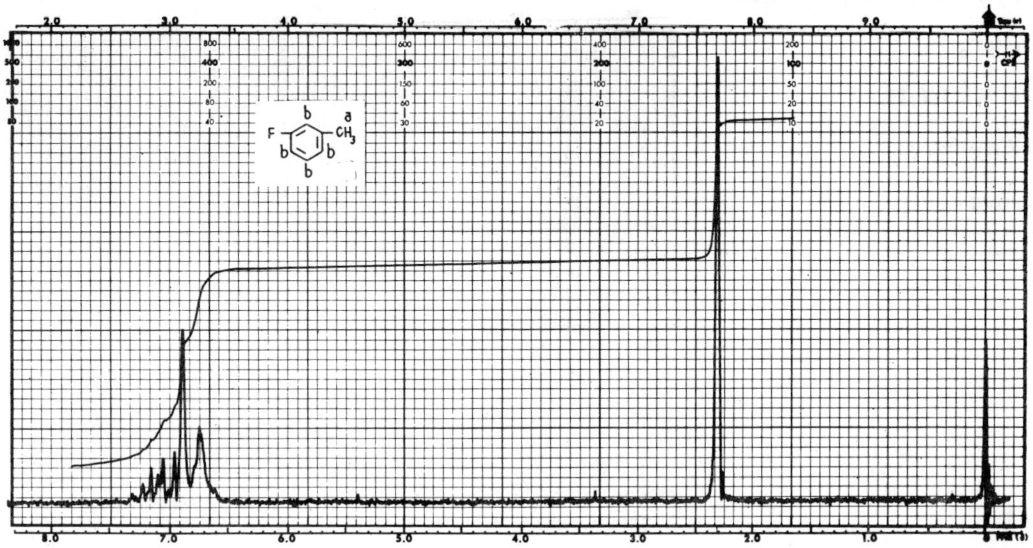

© 1978 Sadtler Research Laboratories, Inc.

p-FLUOROTOLUENE

C_7H_7F

Mol. Wt. 110.13

B.P. 116-117°C

Solvent: CCl_4

Assignments
a = 2.23
b = 6.81
c = 7.00

226

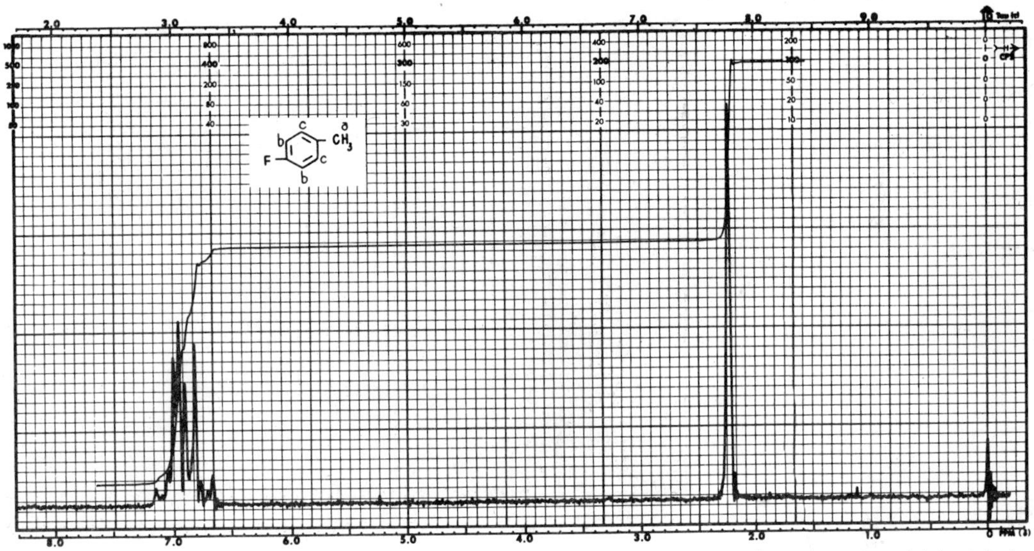

© 1978 Sadtler Research Laboratories, Inc.

o,a,a,a-TETRAFLUOROTOLUENE

$C_7H_4F_4$

Mol. Wt. 164.10

Solvent: CCl_4

Assignments
a = 6.80-7.70
b = 2.01 imp.

227

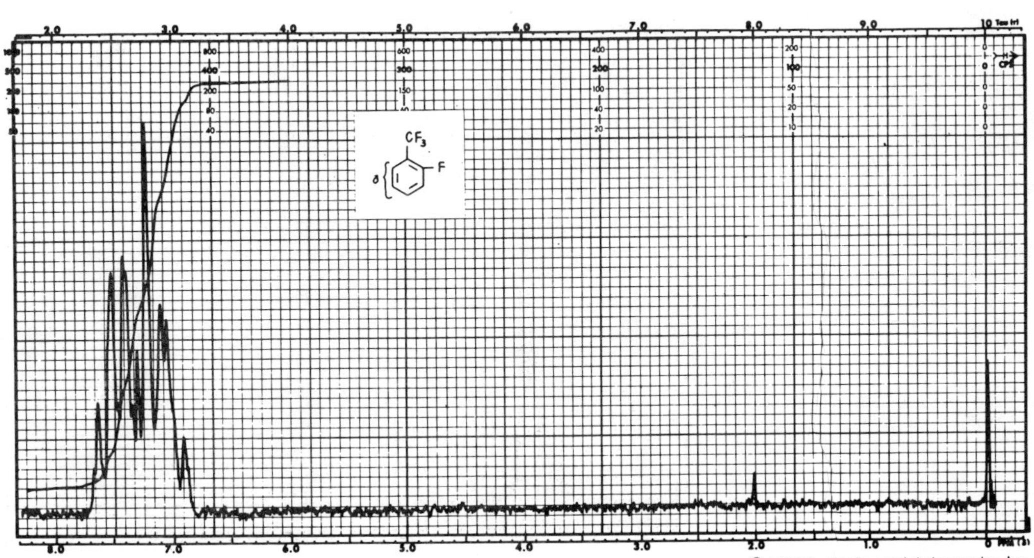

© 1978 Sadtler Research Laboratories, Inc.

m,a,a,a-TETRAFLUOROTOLUENE

$C_7H_4F_4$

Mol. Wt. 164.10

Solvent: $CDCl_3$

Assignments
a = 6.90-7.65

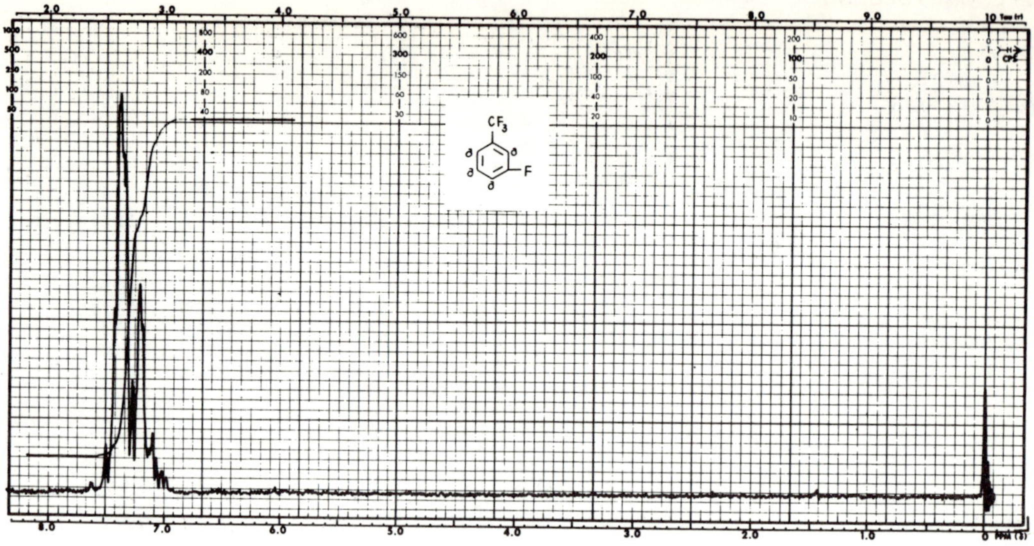

228

© 1978 Sadtler Research Laboratories, Inc.

---

a,a,a-p-TETRAFLUOROTOLUENE

$C_7H_4F_4$

Mol. Wt. 164.10

B.P. 102-105°C/760mm (lit.)

Solvent: CCl_4

Assignments
a = 7.13
b = 7.64
c = 1.29 imp.

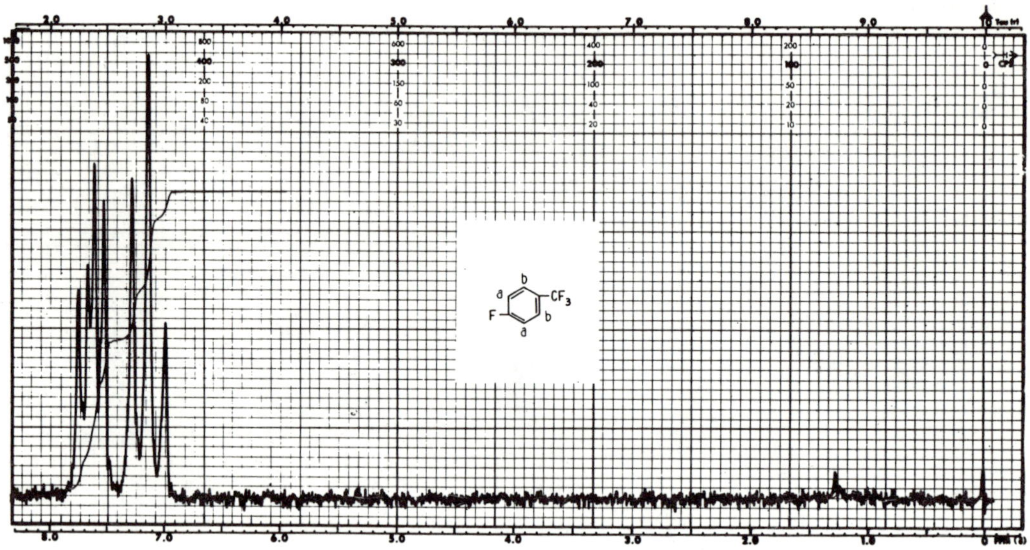

229

© 1978 Sadtler Research Laboratories, Inc.

---

o-DIFLUOROBENZENE

$C_6H_4F_2$

Mol. Wt. 114.09

M.P. –34°C

Solvent: $CDCl_3$

Assignments
a = 6.90-7.40

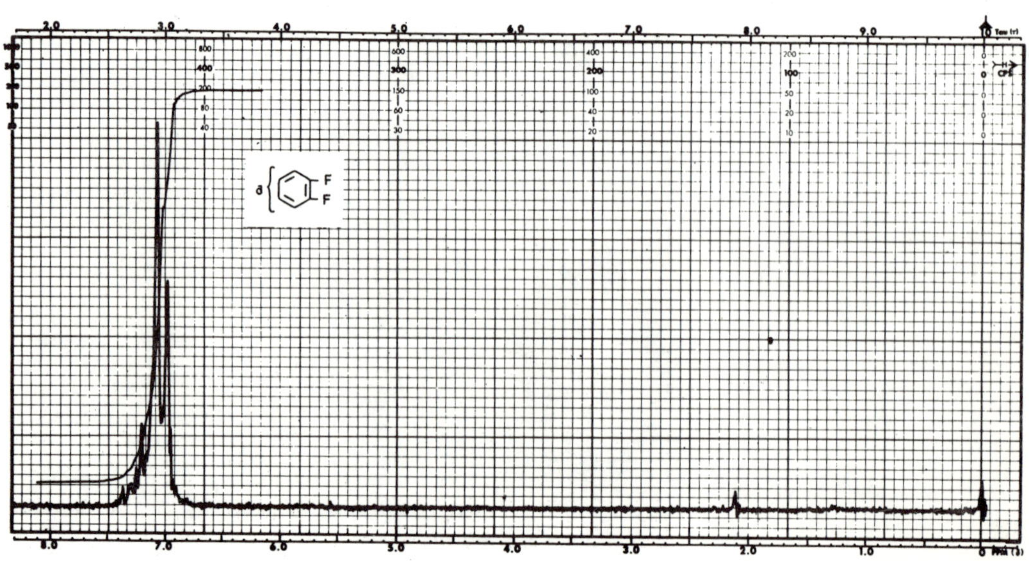

230

© 1978 Sadtler Research Laboratories, Inc.

m-DIFLUOROBENZENE

$C_6H_4F_2$

Mol. Wt. 114.09

B.P. 81-83°C

Solvent: $CDCl_3$

Assignments
a = 6.55-7.10
b = 7.27
c = 1.29, 1.48 imps
d = 2.11 imp.

231

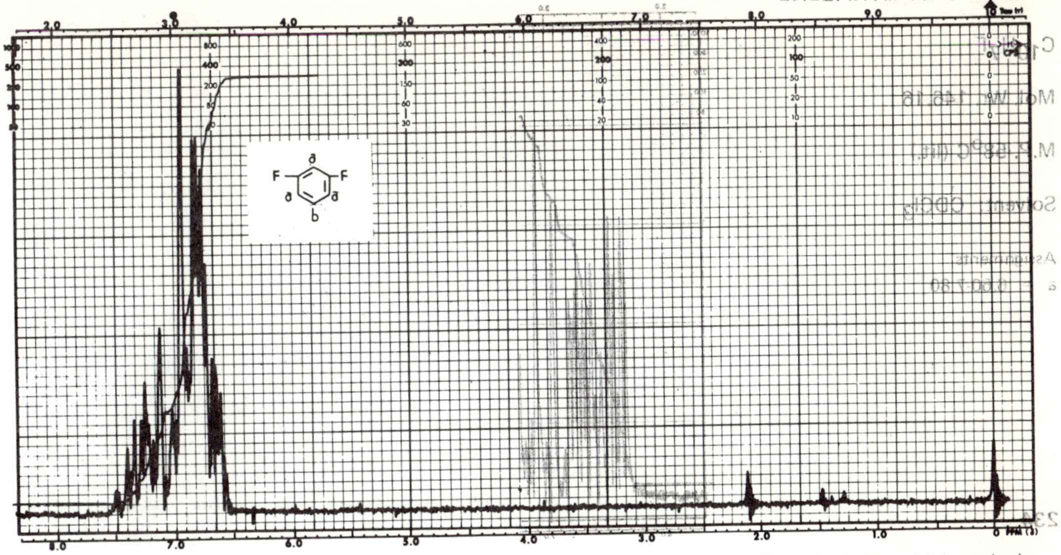

© 1978 Sadtler Research Laboratories, Inc.

p-DIFLUOROBENZENE

$C_6H_4F_2$

Mol. Wt. 114.10

Solvent: CCl_4

Assignments
a = 6.95

232

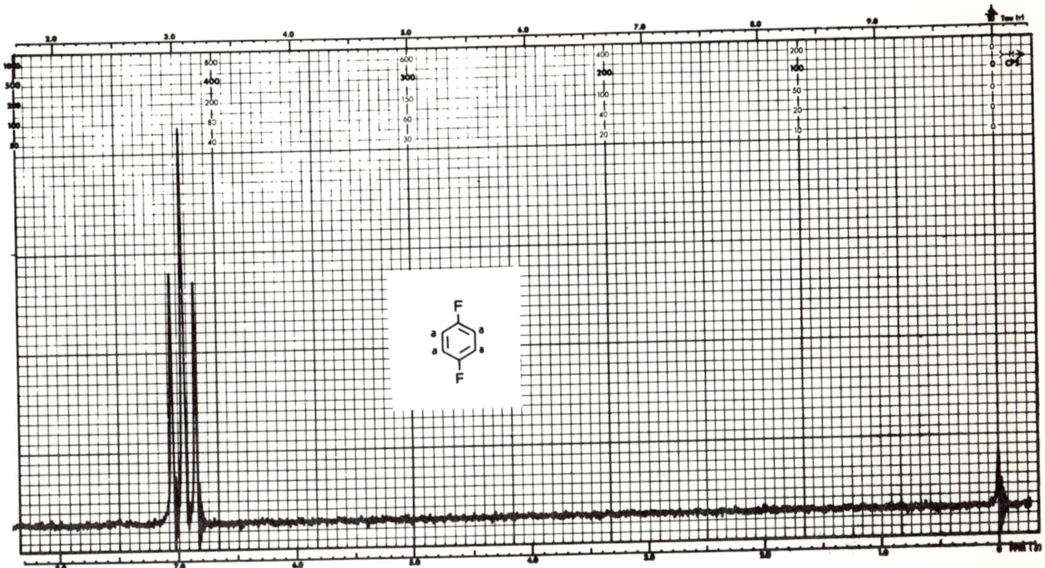

© 1978 Sadtler Research Laboratories, Inc.

1-FLUORONAPHTHALENE

$C_{10}H_7F$

Mol. Wt. 146.17

B.P. 93-94°C/14mm

Solvent: $CDCl_3$

Assignments
a = 6.80-7.81
b = 8.01

233

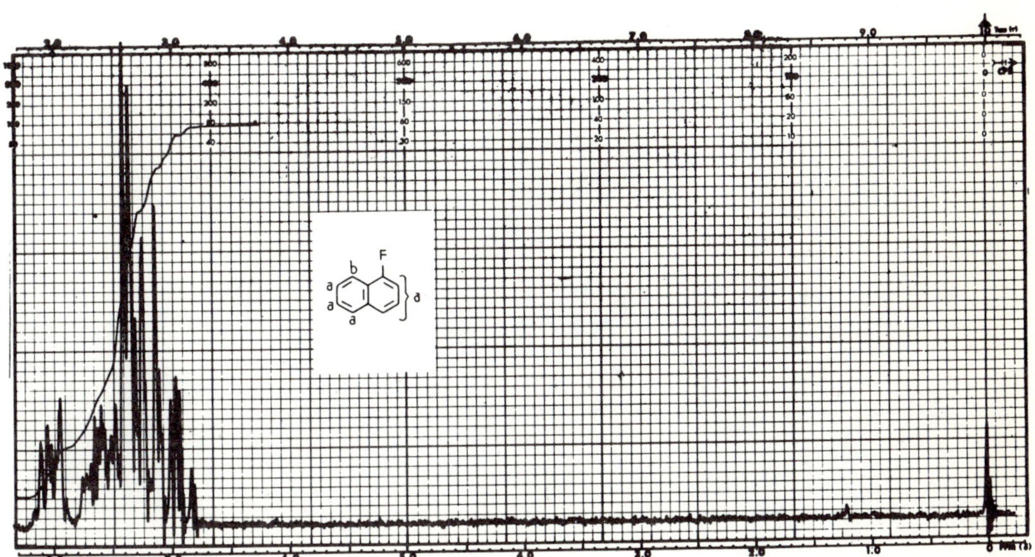

© 1978 Sadtler Research Laboratories, Inc.

2-FLUORONAPHTHALENE

$C_{10}H_7F$

Mol. Wt. 146.16

M.P. 58°C (lit.)

Solvent: $CDCl_3$

Assignments
a = 6.60-7.80

234

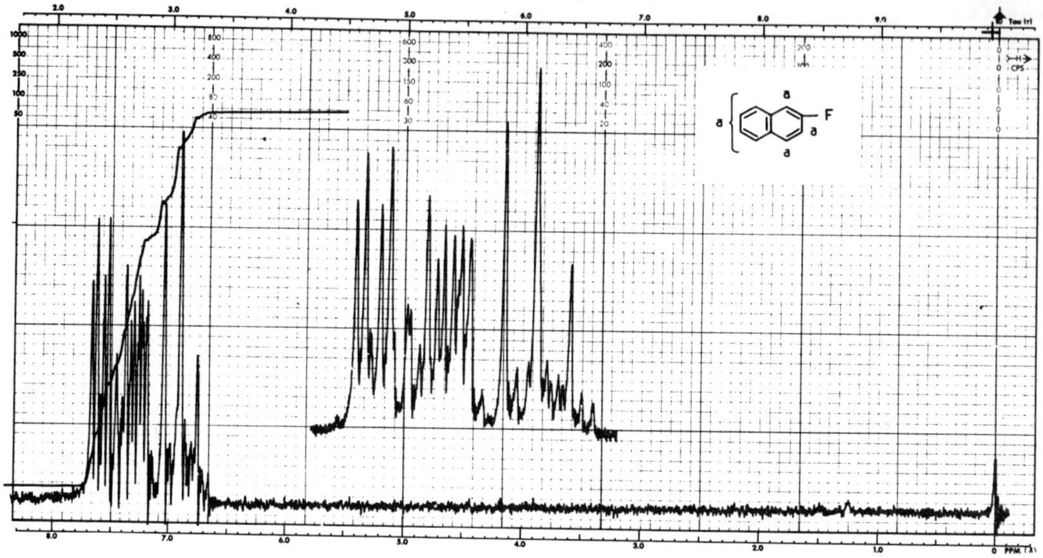

© 1978 Sadtler Research Laboratories, Inc.

CHLORINATED HYDROCARBONS

Because Chlorine does not possess a spin as Fluorine does, nor does it possess any exchangeable hydrogens, the identification of chlorine substituents via NMR must be based solely upon the observed chemical shifts.

Fortunately, chlorine has a strong deshielding effect upon aliphatic hydrogens and thus the analysis of such materials is relatively straightforward. However, since it has only a very weak shielding/deshielding effect upon aromatic hydrogens, it becomes quite difficult to identify chlorine groups bonded to an aromatic ring without the use of alternate techniques.

Chemical Shifts

Aliphatic Protons

HC–4	HC–3	HC–2	HC–1		Solvent	Spectrum
			3.05	CH_3–Cl	(lit.)	- - - -
		1.33	3.47	CH_3–CH_2–Cl	(lit.)	- - - -
	1.05	1.77	3.45	CH_3–CH_2–CH_2–Cl	CCl_4	235
0.95	1.45	1.75	3.49	CH_3–CH_2–CH_2–CH_2–Cl	CCl_4	236
		(1.51)	4.11	$(CH_3)_2$–CH–Cl	CCl_4	238
		(1.61)	- - - -	$(CH_3)_3$–C–Cl	CCl_4	239

Substituted Chloromethanes

Cl–	CH_2–	X	
	4.05	–C(=O)–OH	CCl_4
	4.06	–C≡C–R	CCl_4
	4.11	–C≡N	CCl_4
	4.12	–C(=O)–CH_3	$CDCl_3$
	4.18	–C(=O)–NH_2	D_2O
	4.20	–C(=O)–O–Ø	$CDCl_3$
	4.33	–C(=O)–R	$CDCl_3$
	4.48	–Ø	CCl_4
	4.50	–C(=O)–Cl	CCl_4
	4.66	–C(=O)–Ø	$CDCl_3$
	5.73	–O–Ø	CCl_4

2-Substituted Chloroethanes

Cl— CH$_2$—	CH$_2$—	X	Solvent
3.61	3.00	—Ø	CCl_4
3.61	3.61	—N=C=O	$CDCl_3$
3.63	3.79	—OH	$CDCl_3$
3.64	4.29	—O—C(=O)—CH_3	CCl_4
3.67	3.67	—Cl	CCl_4
3.68	2.72	—C(=O)—O—CH_3	CCl_4
3.70	2.80	—C≡N	CCl_4
3.70	3.56	—Br	CCl_4
3.76	4.47	—O—(O_2)S—CH_3	$CDCl_3$
4.01	4.01	—SO_2—Cl	CCl_4

Aromatic Protons

As noted previously, chlorine does not significantly alter the chemical shifts of aromatic hydrogens in comparison with the effects noted for other substituents. The NMR spectrum of chlorobenzene displays only a broad, single band at about 7.2 ppm. A series of para substituted chlorobenzenes is listed below.

Para substituted chlorobenzenes

Cl— ortho	meta	para	Solvent
7.00	6.39	—NH—R	$CDCl_3$
7.09	7.61	—I	$CDCl_3$
7.12	6.91	—CH_2—Ø	$CDCl_3$
7.12	6.98	—CH_3	CCl_4
7.13	7.37	—Br	CCl_4
7.19	6.91	—F	$CDCl_3$
7.21	7.21	—CH=CH_2	CCl_4
7.22	6.85	—O—R	Polysol
7.23	7.23	—Cl	CCl_4
7.27	7.02	—N=S=O	$CDCl_3$
7.30	7.30	—CH_2—Cl	CCl_4
7.31	7.80	—C(=O)—CH_3	CCl_4
7.32	7.11	—O—C(=O)—Ø	$CDCl_3$
7.36	7.84	—CCl_3	CCl_4

Chemical Shifts continued

Cl–	ortho	meta	para	Solvent
	7.40	7.81	–N=N–Ø	$CDCl_3$
	7.45	7.45	–NH–Ø	Polysol
	7.46	8.02	–C(=O)–Cl	$CDCl_3$
	7.55	7.97	–C(=O)–NH_2	DMSO

1-CHLOROPROPANE

C_3H_7Cl

Mol. Wt. 78.54

B.P. 46-48°C

Solvent: CCl_4

Assignments
a = 1.05
b = 1.77
c = 3.45

235

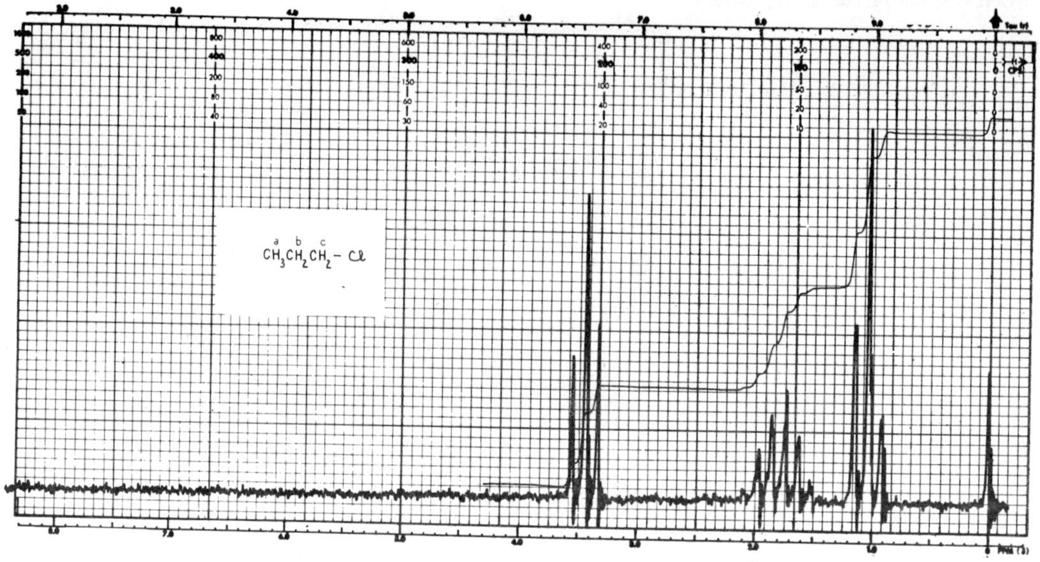

© 1978 Sadtler Research Laboratories, Inc.

1-CHLOROBUTANE

C_4H_9Cl

Mol. Wt. 92.57

B.P. 77-78°C

Solvent: CCl_4

Assignments
a = 0.95
b = 1.10-2.00
c = 3.49

236

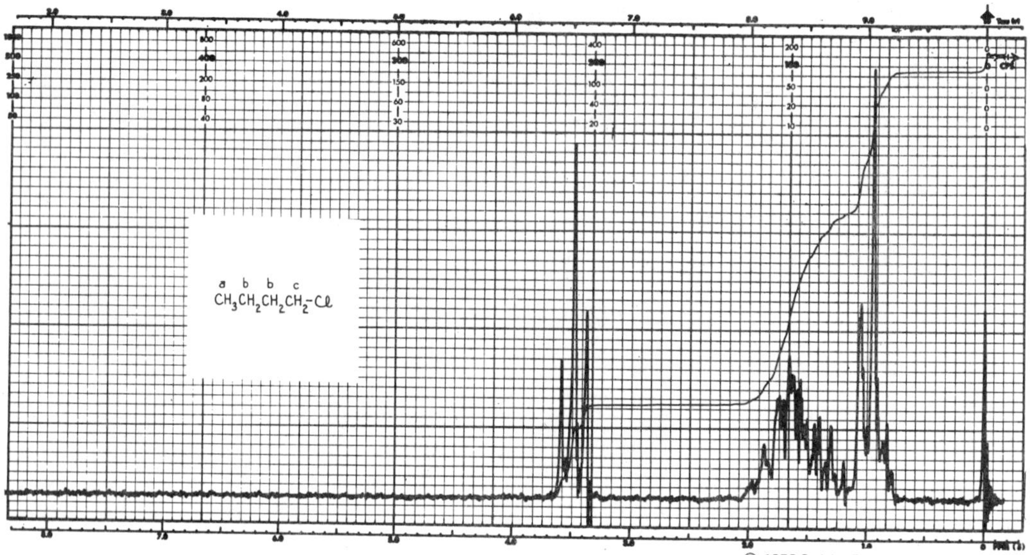

© 1978 Sadtler Research Laboratories, Inc.

2-CHLOROBUTANE

C_4H_9Cl

Mol. Wt. 92.57

Solvent: CCl_4

Assignments
a = 1.01
b = 1.49
c = 1.72
d = 3.88

237

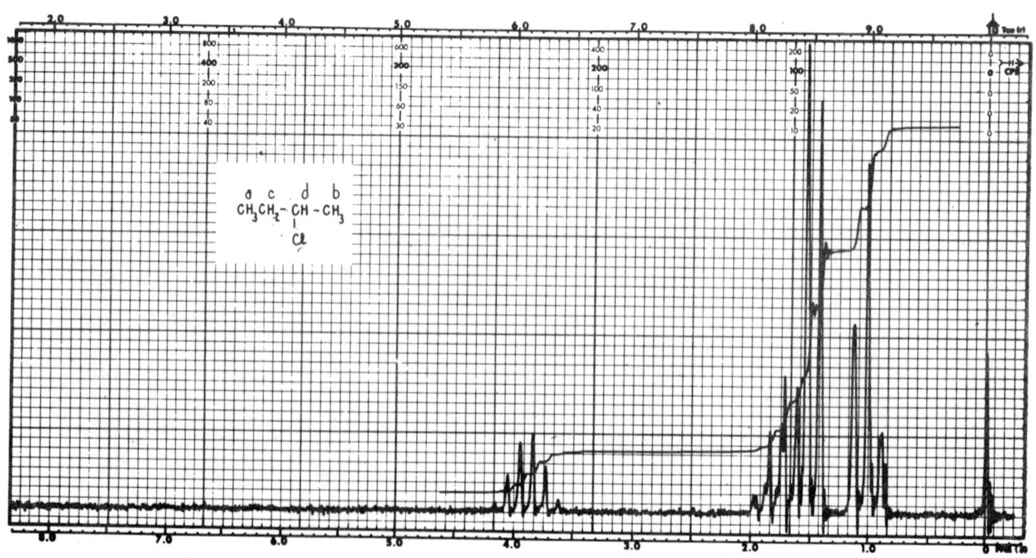

© 1978 Sadtler Research Laboratories, Inc.

1-CHLORO-2-METHYLPROPANE

C_4H_9Cl

Mol. Wt. 92.57

M.P. 68-70°C

Solvent: CCl_4

Assignments
a = 1.03
b = 1.91
c = 3.32

238

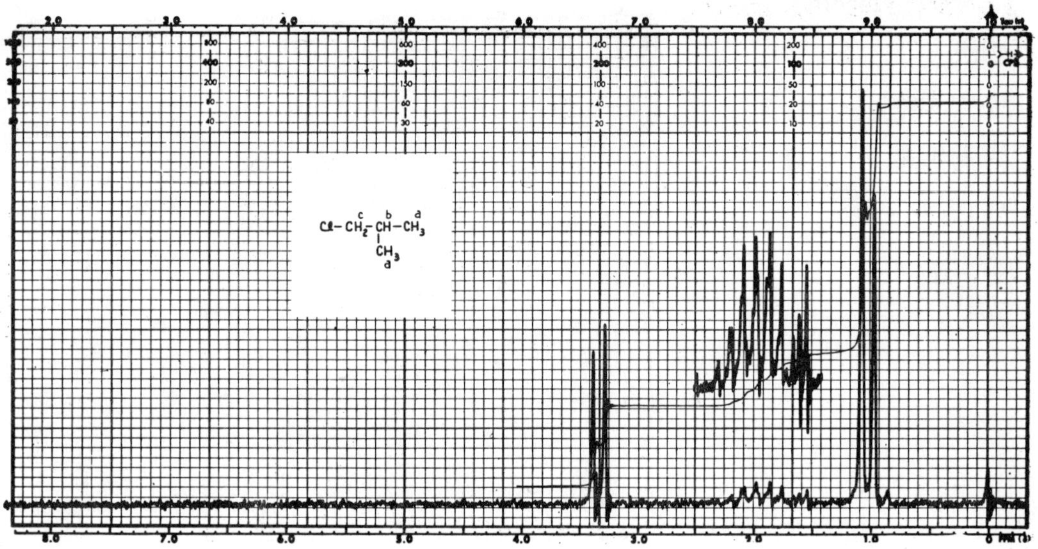

© 1978 Sadtler Research Laboratories, Inc.

2-CHLORO-2-METHYLPROPANE

C_4H_9Cl

Mol. Wt. 92.57

B.P. 51-52°C

Solvent: CCl_4

Assignments
a = 1.61

239

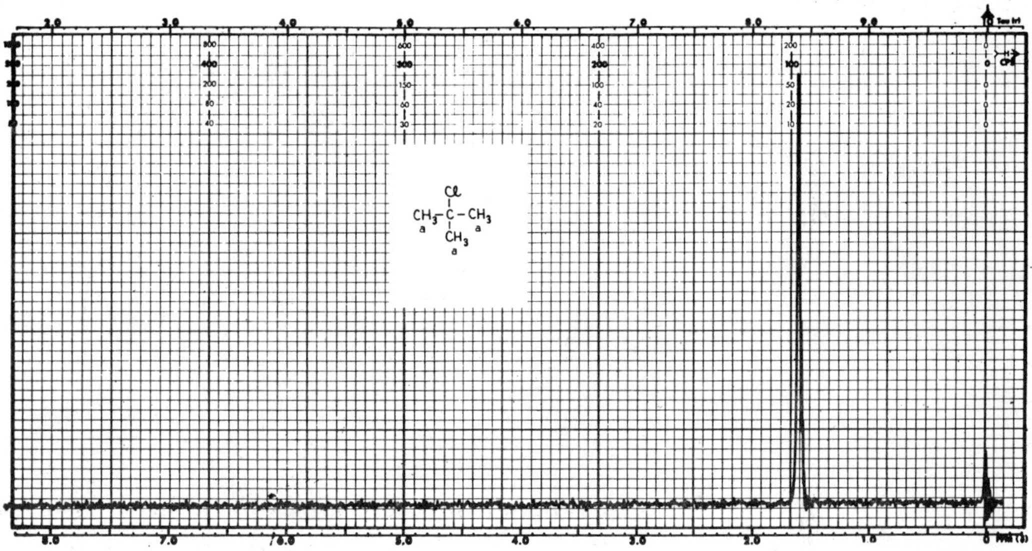

© 1978 Sadtler Research Laboratories, Inc.

1-CHLOROPENTANE

$C_5H_{11}Cl$

Mol. Wt. 106.60

B.P. 108.2-108.8°C (lit.)

Solvent: CCl_4

Assignments
a = 0.93
b = 1.18-2.01
c = 1.77
d = 3.40

240

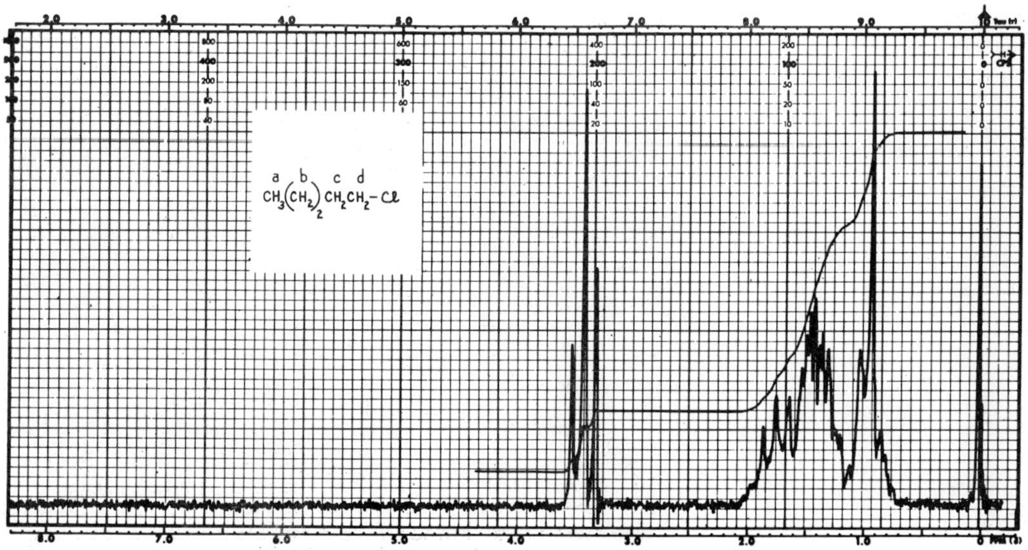

© 1978 Sadtler Research Laboratories, Inc.

2-CHLORO-2-METHYLBUTANE

$C_5H_{11}Cl$

Mol. Wt. 106.60

B.P. 84-86°C

Solvent: CCl_4

Assignments
a = 1.01
b = 1.51
c = 1.73

241

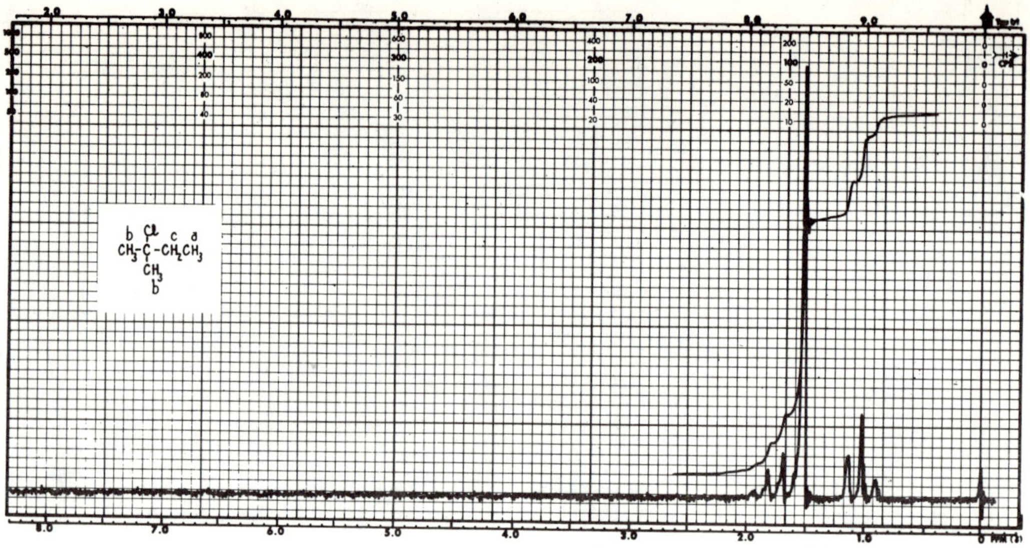

© 1978 Sadtler Research Laboratories, Inc.

1-CHLOROHEPTANE

$C_7H_{15}Cl$

Mol. Wt. 134.65

B.P. 159-161°C

Solvent: CCl_4

Assignments
a = 0.90
b = 1.06-2.03
c = 3.44

242

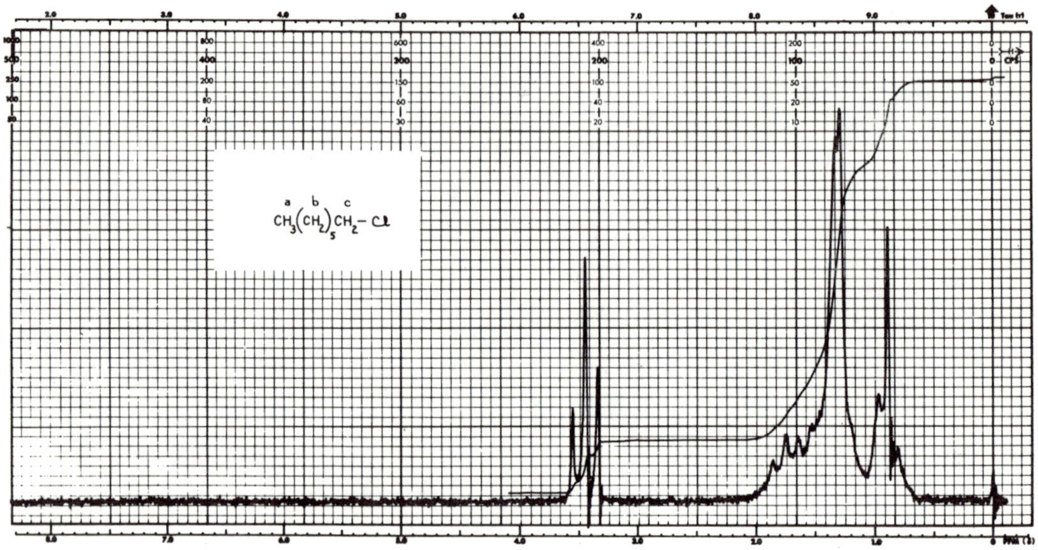

© 1978 Sadtler Research Laboratories, Inc.

1-CHLORODECANE

$C_{10}H_{21}Cl$

Mol. Wt. 176.73

B.P. 103-106°C/15mm

Solvent: CCl_4

Assignments
a = 0.90
b = 1.30
c = 1.76
d = 3.43

243

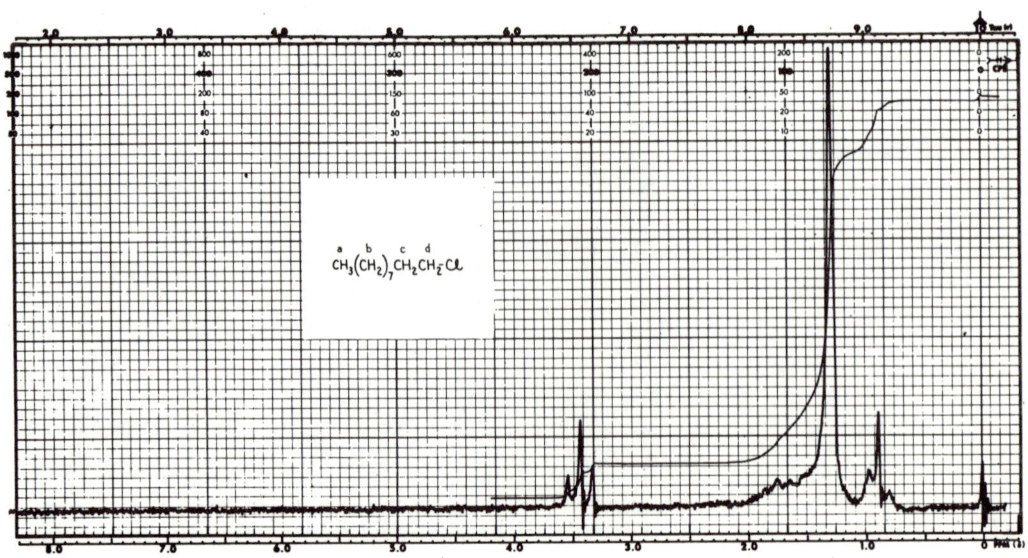

© 1978 Sadtler Research Laboratories, Inc.

1-CHLOROUNDECANE

$C_{11}H_{23}Cl$

Mol. Wt. 190.76

Solvent: CCl_4

Assignments
a = 0.88
b = 1.28
c = 1.74
d = 3.44

244

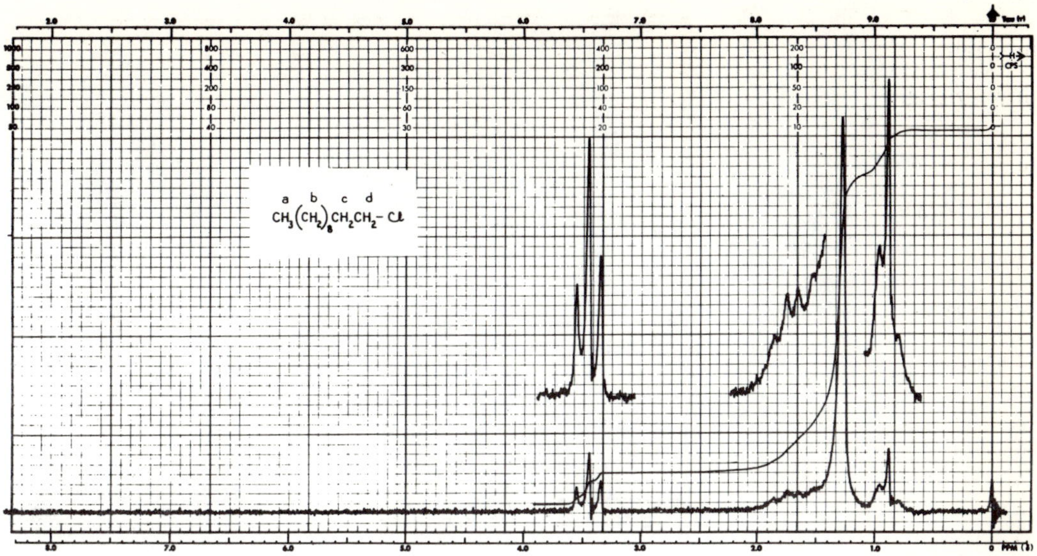

© 1978 Sadtler Research Laboratories, Inc.

1-CHLORODODECANE

$C_{12}H_{25}Cl$

Mol. Wt. 204.79

B.P. 113-115°C/5mm

Solvent: CCl_4

Assignments
a = 0.91
b = ca 1.30
c = 1.77
d = 3.46

245

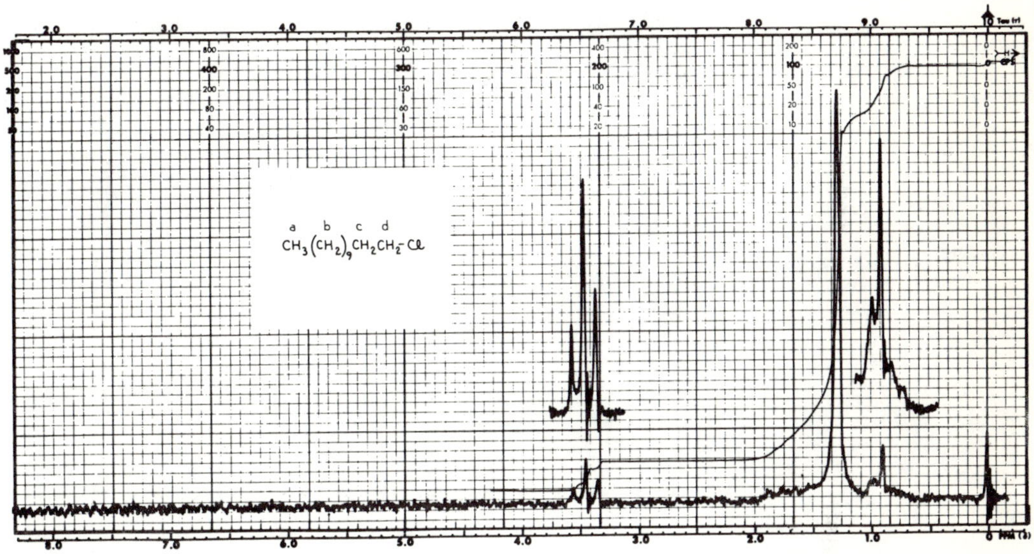

© 1978 Sadtler Research Laboratories, Inc.

1-CHLOROTETRADECANE

$C_{14}H_{29}Cl$

Mol. Wt. 232.84

B.P. 154-155°C/15mm

Solvent: CCl_4

Assignments
a = 0.91
b = ca 1.29
c = 3.46

246

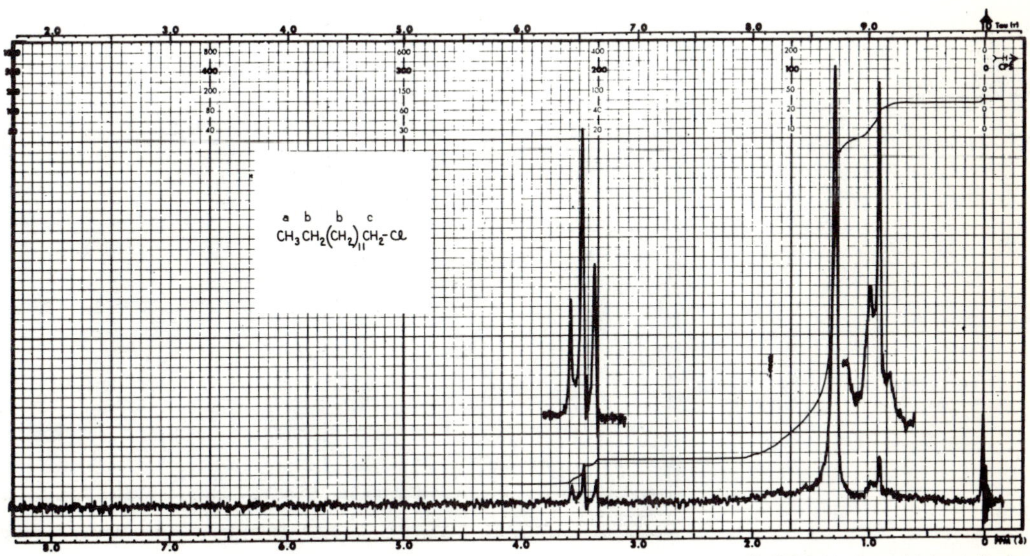

© 1978 Sadtler Research Laboratories, Inc.

1-CHLOROHEXADECANE

$C_{16}H_{33}Cl$

Mol. Wt. 260.89

M.P. 12°C

Solvent: CCl_4

Assignments
a = 0.89
b = ca 1.28
c = 1.74
d = 3.48

247

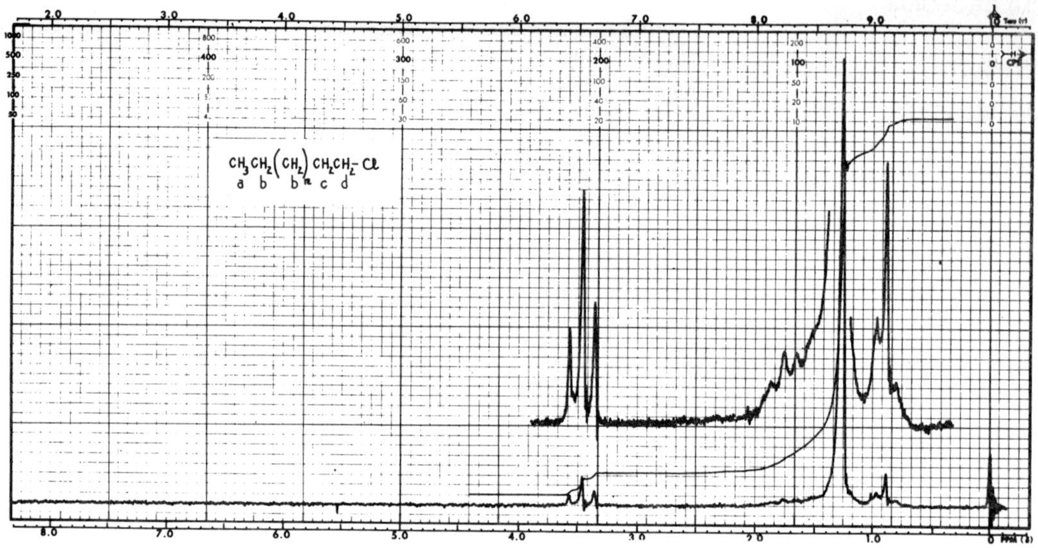

© 1978 Sadtler Research Laboratories, Inc.

1-CHLOROOCTADECANE

$C_{18}H_{37}Cl$

Mol. Wt. 288.95

M.P. 18-19°C

Solvent: CCl_4

Assignments
a = 0.90
b = 1.28
c = ca 1.70
d = 3.46

248

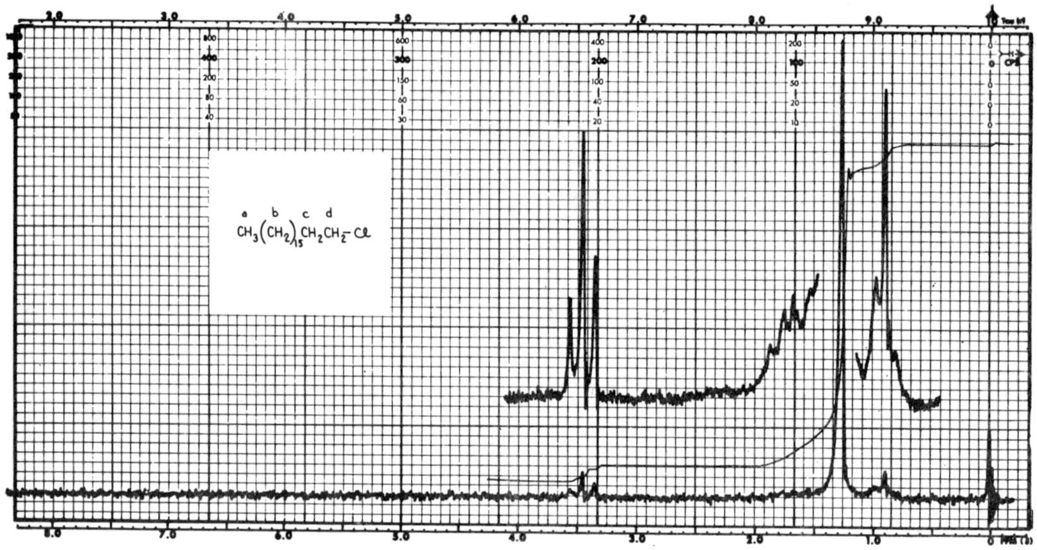

© 1978 Sadtler Research Laboratories, Inc.

METHYLENE DICHLORIDE

CH_2Cl_2

Mol. Wt. 84.93

B.P. 39.5-40.3°C/760mm

Solvent: CCl_4

Assignments
a = 5.28

249

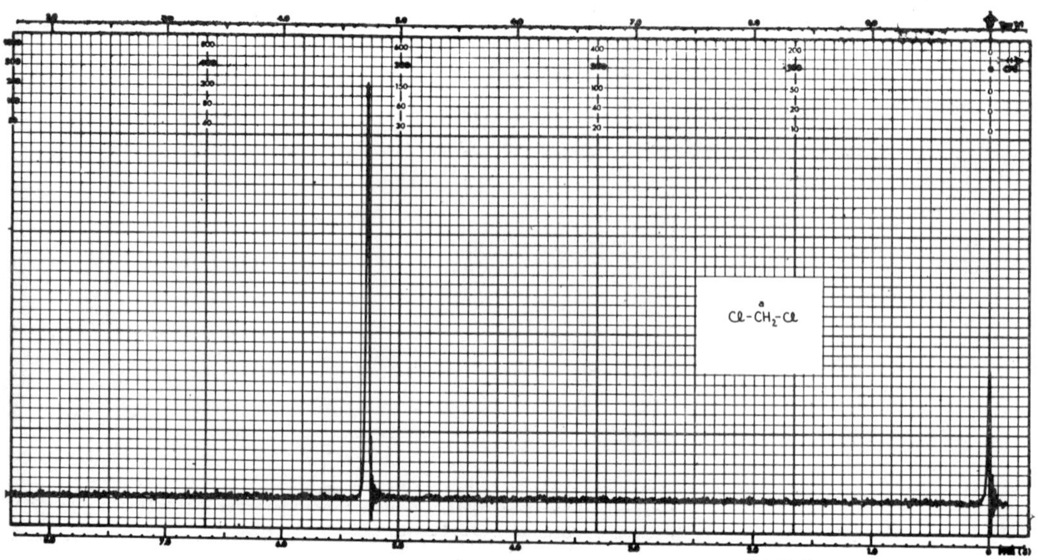

© 1978 Sadtler Research Laboratories, Inc.

CHLOROFORM

$CHCl_3$

Mol. Wt. 119.38

Solvent: CCl_4

Assignments
a = 7.21
b = 1.22 imp.

250

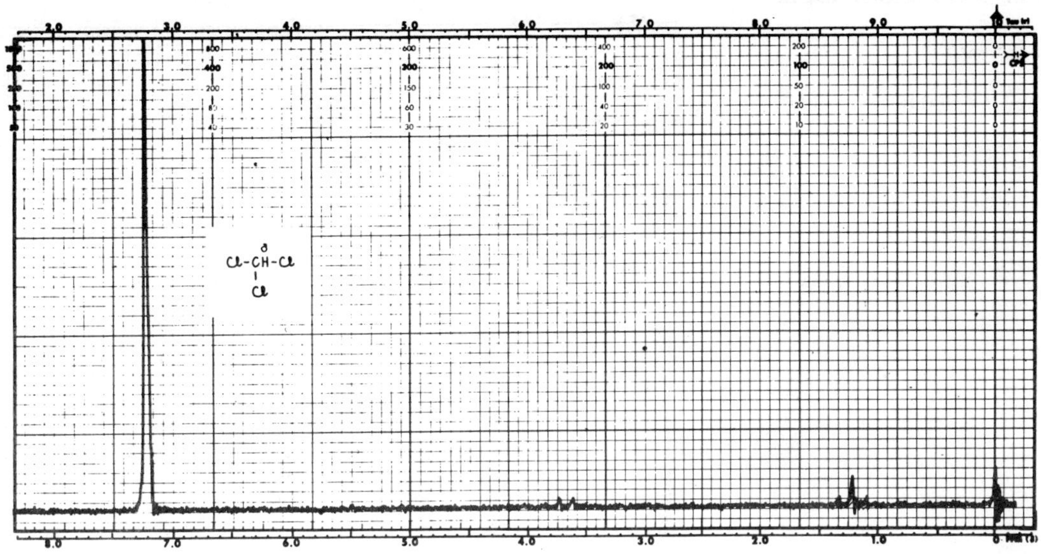

© 1978 Sadtler Research Laboratories, Inc.

1,1-DICHLOROETHANE

$C_2H_4Cl_2$

Mol. Wt. 98.96

B.P. 57-58°C

Solvent: CCl_4

Assignments
a = 2.07
b = 5.89

251

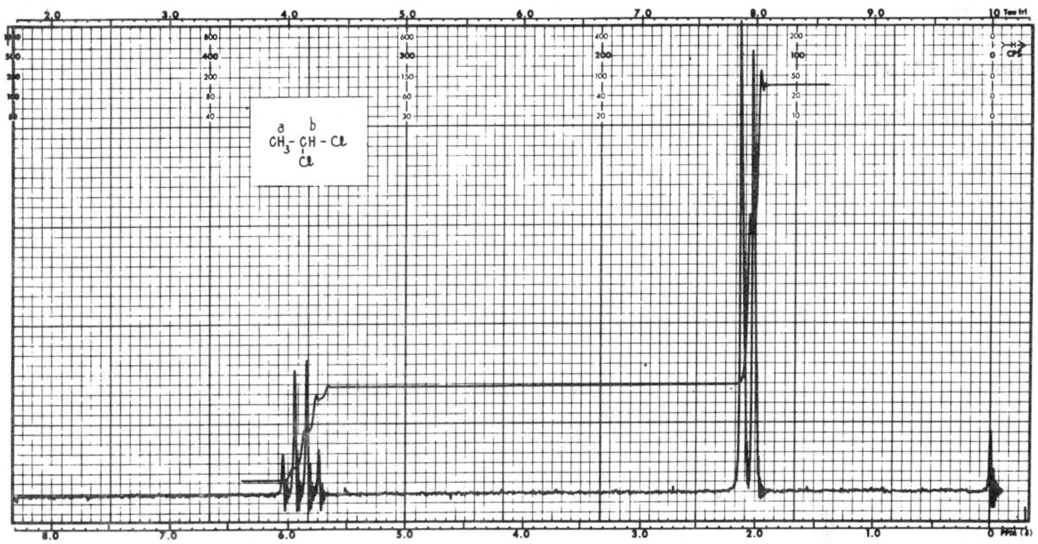

© 1978 Sadtler Research Laboratories, Inc.

1,2-DICHLOROETHANE

$C_2H_4Cl_2$

Mol. Wt. 98.96

B.P. 83.7°C

Solvent: CCl_4

Assignments
a = 3.67

252

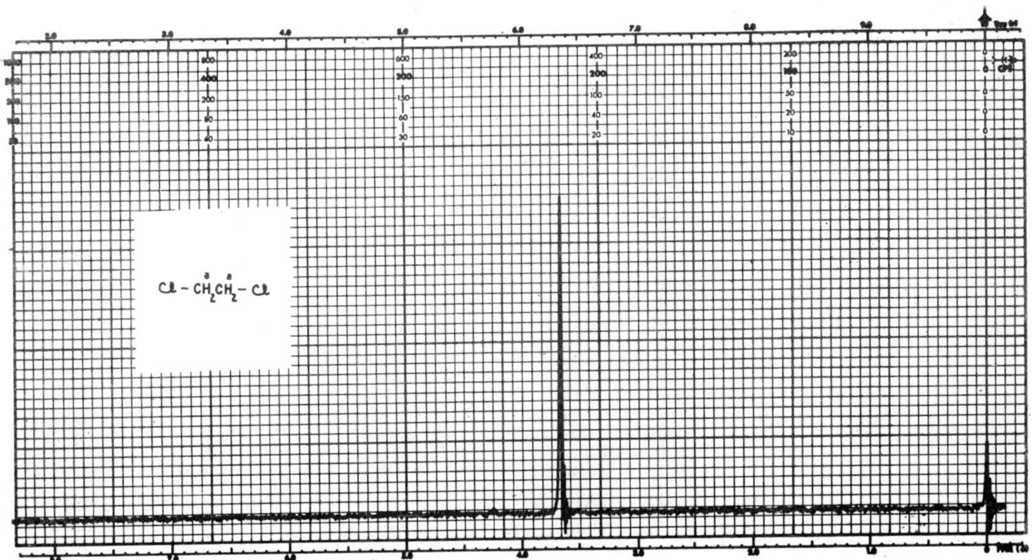

© 1978 Sadtler Research Laboratories, Inc.

1,1,1-TRICHLOROETHANE

$C_2H_3Cl_3$

Mol. Wt. 133.41

Solvent: CCl_4

Assignments
a = 2.70
b = 3.56 imp.

253

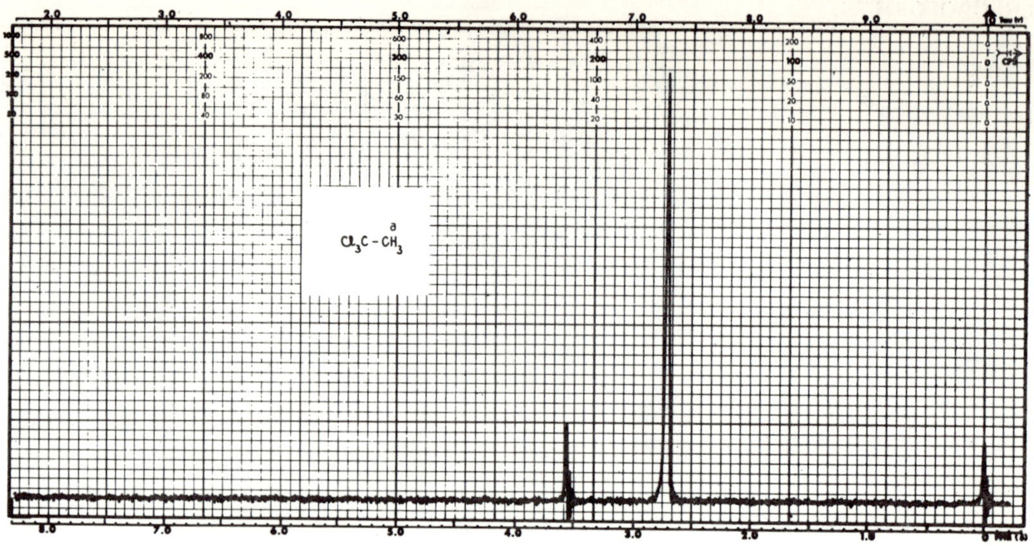

© 1978 Sadtler Research Laboratories, Inc.

1,1,2-TRICHLOROETHANE

$C_2H_3Cl_3$

Mol. Wt. 133.41

B.P. 110-117°C

Solvent: $CDCl_3$

Assignments
a = 3.97
b = 5.79

254

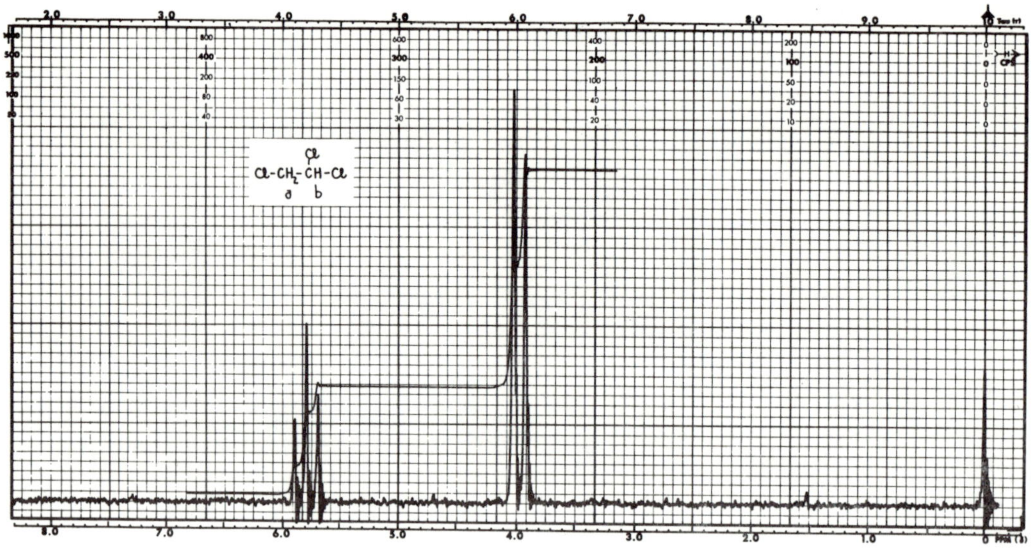

© 1978 Sadtler Research Laboratories, Inc.

1,1,2,2-TETRACHLOROETHANE

$C_2H_2Cl_4$

Mol. Wt. 167.85

B.P. 144-146°C

Solvent: $CDCl_3$

Assignments
a = 5.91

255

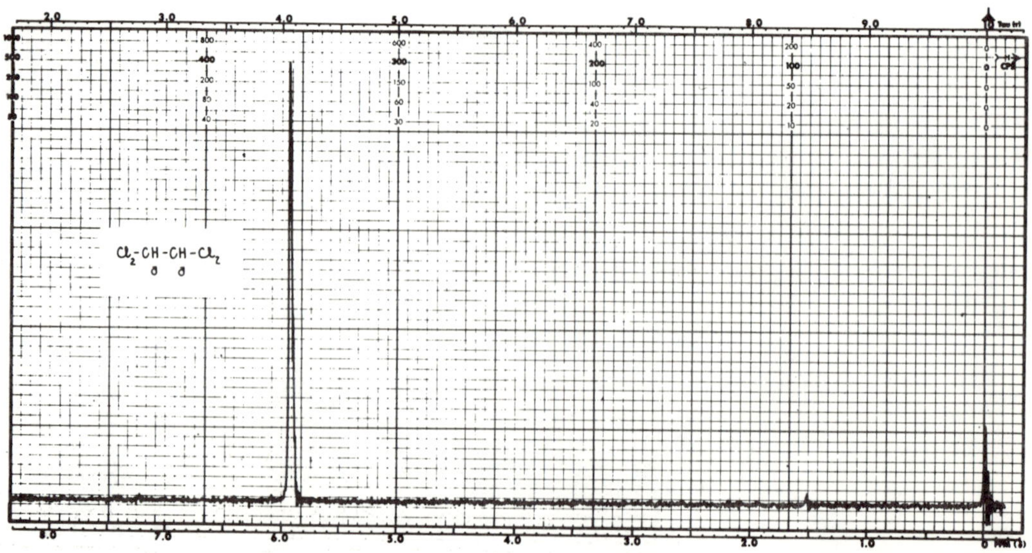

© 1978 Sadtler Research Laboratories, Inc.

PENTACHLOROETHANE

C_2HCl_5

Mol. Wt. 202.30

B.P. 161-162°C

Solvent: CCl_4

Assignments
a = 6.10

256

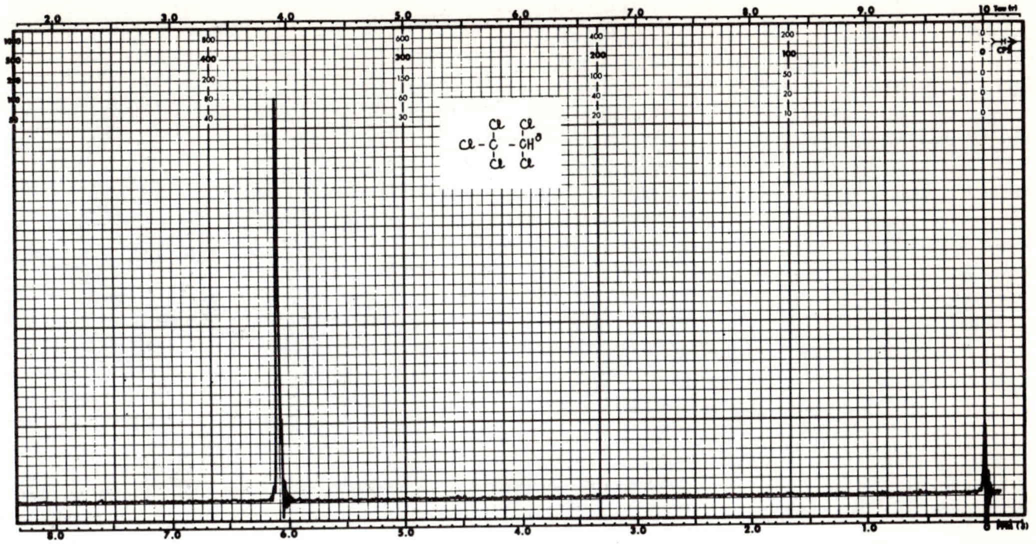

© 1978 Sadtler Research Laboratories, Inc.

1,2-DICHLOROPROPANE

$C_3H_6Cl_2$

Mol. Wt. 112.99

B.P. 95-96°C

Solvent: CCl_4

Assignments
a = 1.60
b = 3.28-3.78
c = 4.03

257

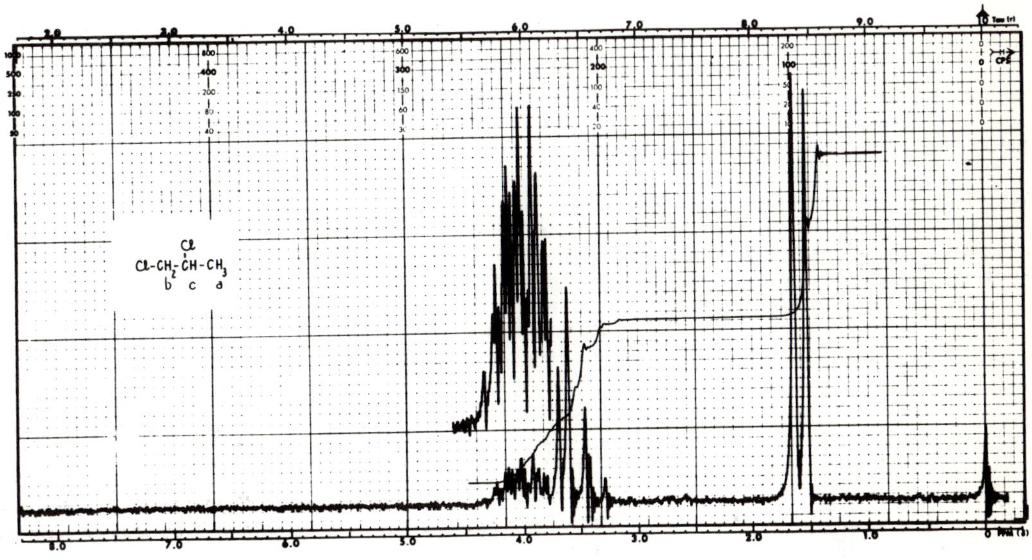

© 1978 Sadtler Research Laboratories, Inc.

2,2-DICHLOROPROPANE

$C_3H_6Cl_2$

Mol. Wt. 112.99

Solvent: CCl_4

Assignments
a = 2.20

258

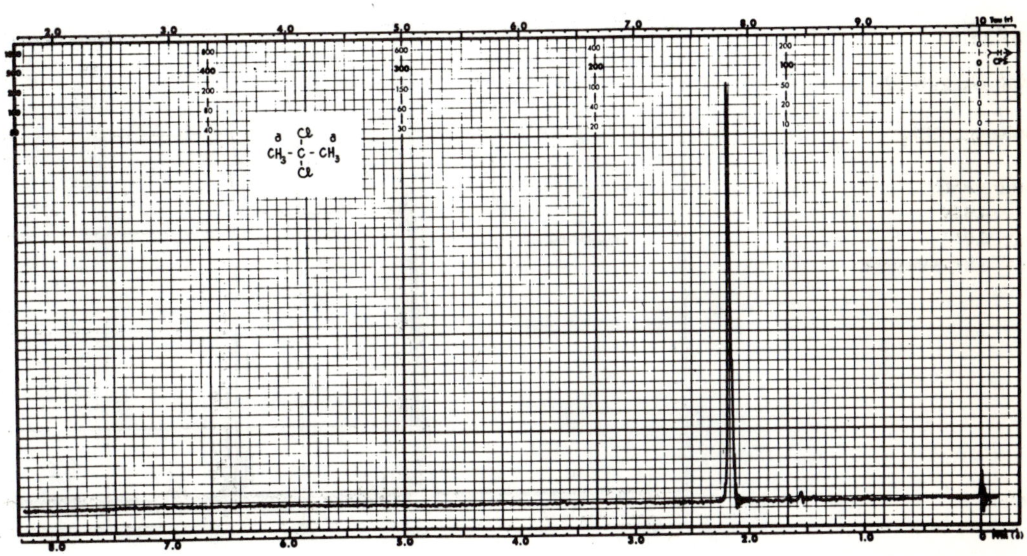

© 1978 Sadtler Research Laboratories, Inc.

1,2,3-TRICHLOROPROPANE

$C_3H_5Cl_3$

Mol. Wt. 147.63

B.P. 155-157°C

Solvent: CCl_4

Assignments
a = 3.88
b = 4.19

259

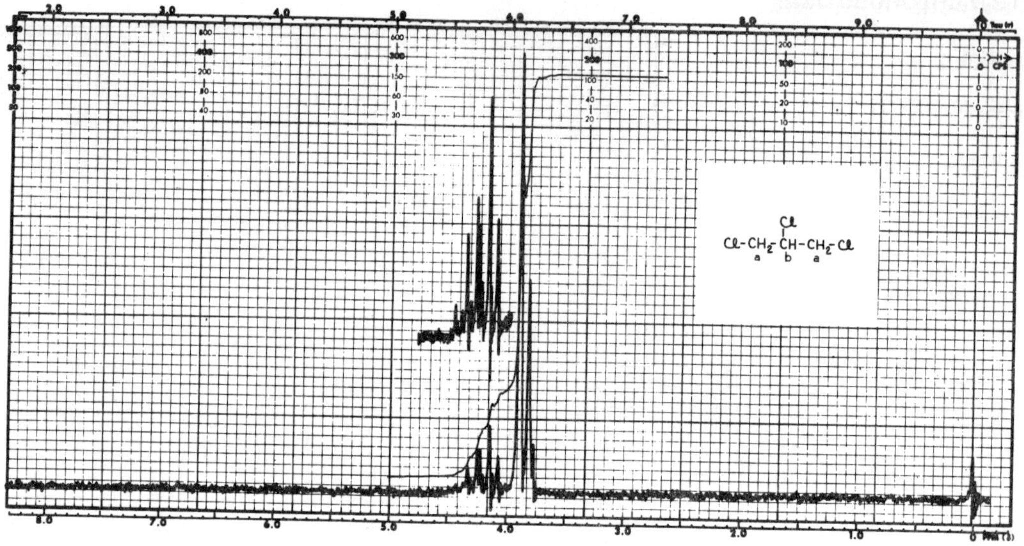

© 1978 Sadtler Research Laboratories, Inc.

1,1,1,2-TETRACHLOROPROPANE

$C_3H_4Cl_4$

Mol. Wt. 181.88

Solvent: CCl_4

Assignments
a = 1.88
b = 4.55
c = 2.50, 4.10 imps.

260

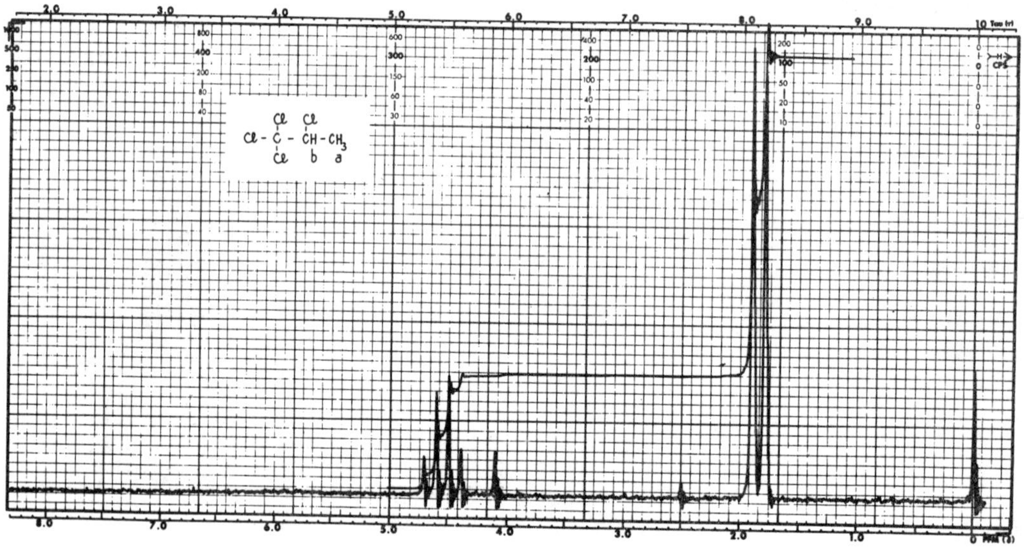

© 1978 Sadtler Research Laboratories, Inc.

1,1,2,3,3-PENTACHLOROPROPANE

$C_3H_3Cl_5$

Mol. Wt. 216.32

Solvent: CCl_4

Assignments
a = 4.44
b = 6.01
c = 3.87, 3.99 imps.

261

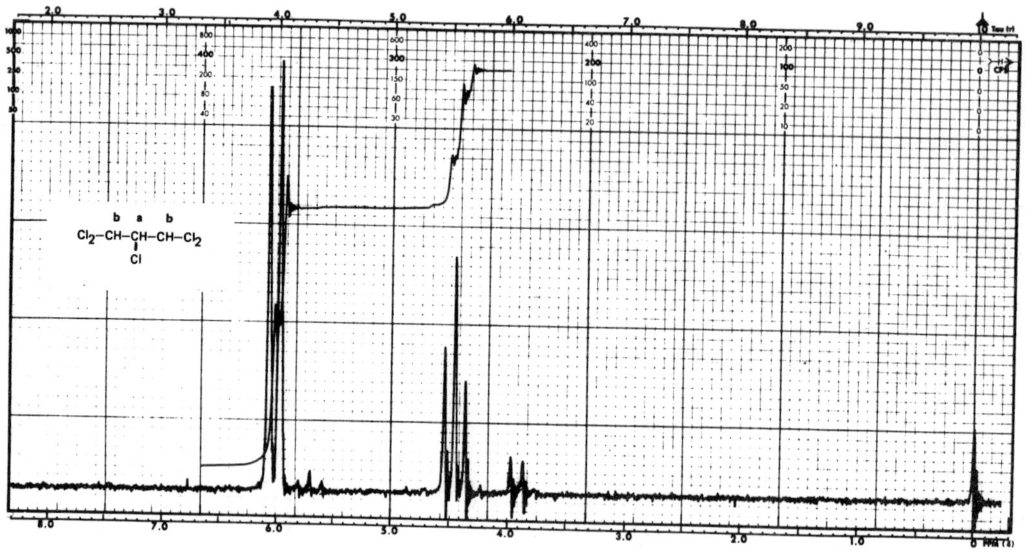

© 1978 Sadtler Research Laboratories, Inc.

1,2-DICHLOROBUTANE

$C_4H_8Cl_2$

Mol. Wt. 127.01

Solvent: CCl_4

Assignments
a = 1.05
b = 1.88
c = 3.31-4.16

262

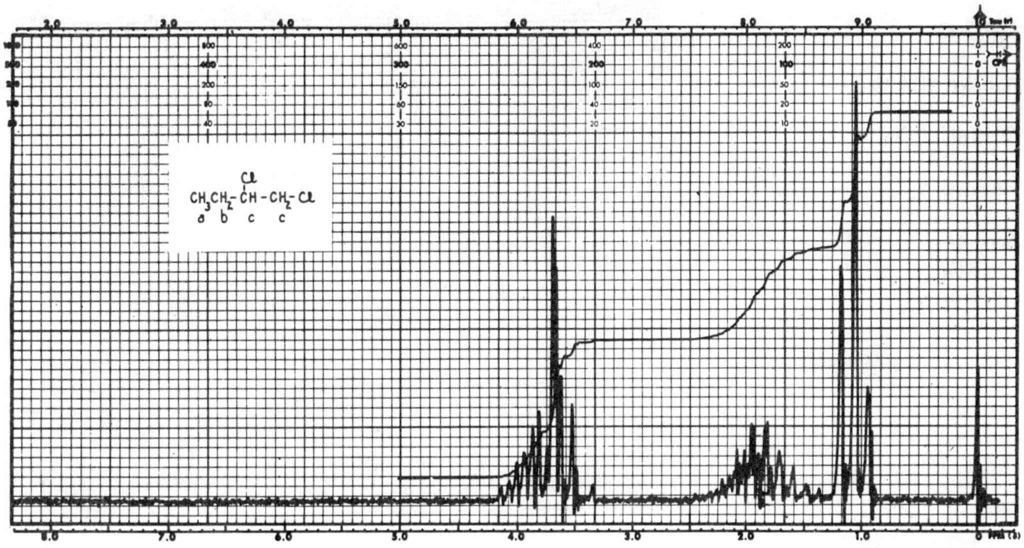

© 1978 Sadtler Research Laboratories, Inc.

1,4-DICHLOROBUTANE

$C_4H_8Cl_2$

Mol. Wt. 127.01

B.P. 153-155°C

Solvent: CCl_4

Assignments
a = 1.92
b = 3.52

263

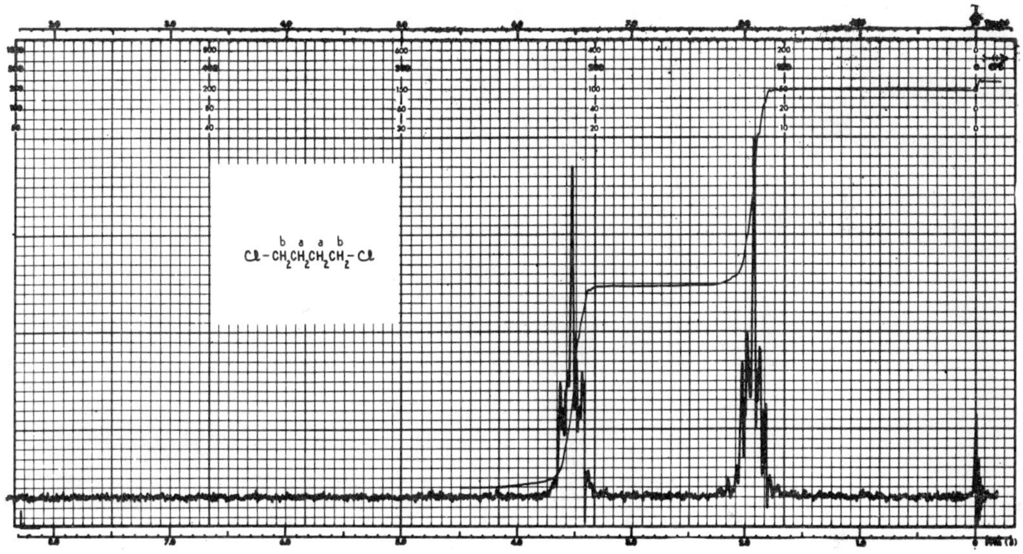

© 1978 Sadtler Research Laboratories, Inc.

1,5-DICHLOROPENTANE

$C_5H_{10}Cl_2$

Mol. Wt. 141.04

B.P. 52-54°C/5mm

Solvent: CCl_4

Assignments
a = 1.35-2.12
b = 3.51

264

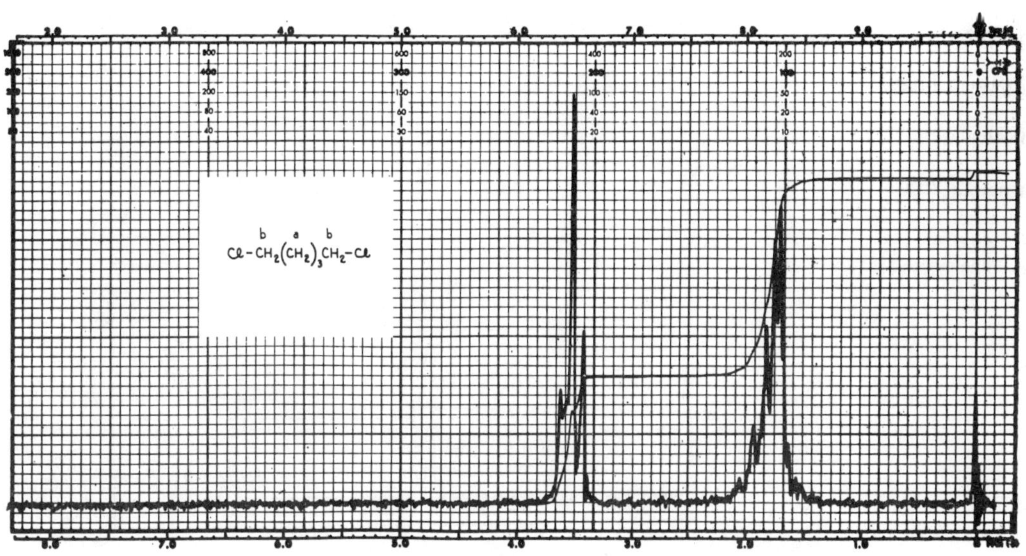

© 1978 Sadtler Research Laboratories, Inc.

1,1-DICHLORO-3,3-DIMETHYLBUTANE

$C_6H_{12}Cl_2$

Mol. Wt. 155.07

Solvent: CCl_4

Assignments
a = 1.07
b = 2.28
c = 5.77

265

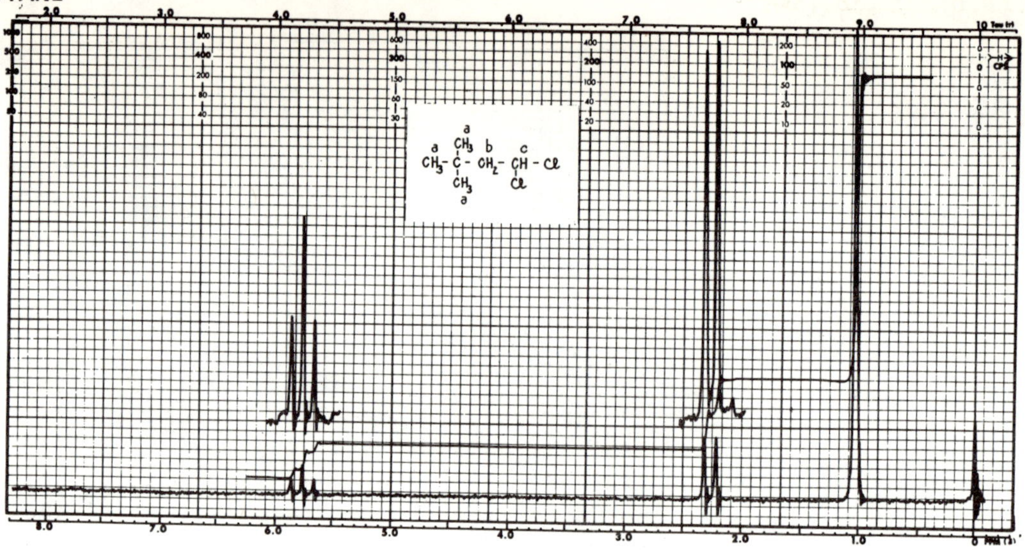

© 1978 Sadtler Research Laboratories, Inc.

CHLOROCYCLOPENTANE

C_5H_9Cl

Mol. Wt. 104.58

B.P. 106-107°C/630mm

Solvent: CCl_4

Assignments
a = 1.38-2.30
b = 4.35

266

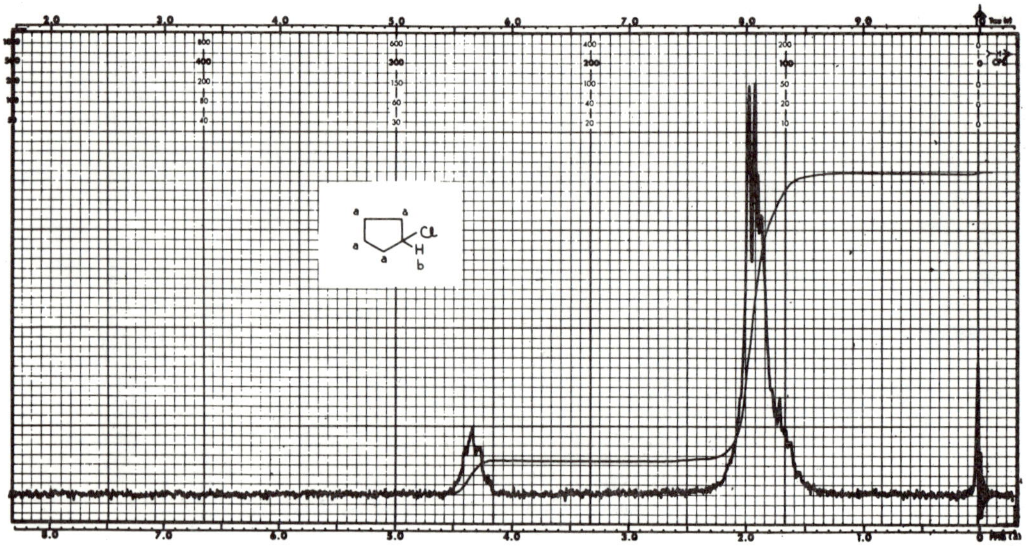

© 1978 Sadtler Research Laboratories, Inc.

CHLOROCYCLOHEXANE

$C_6H_{11}Cl$

Mol. Wt. 118.61

B.P. 141-143°C

Solvent: CCl_4

Assignments
a = 1.00-2.40
b = 3.98

267

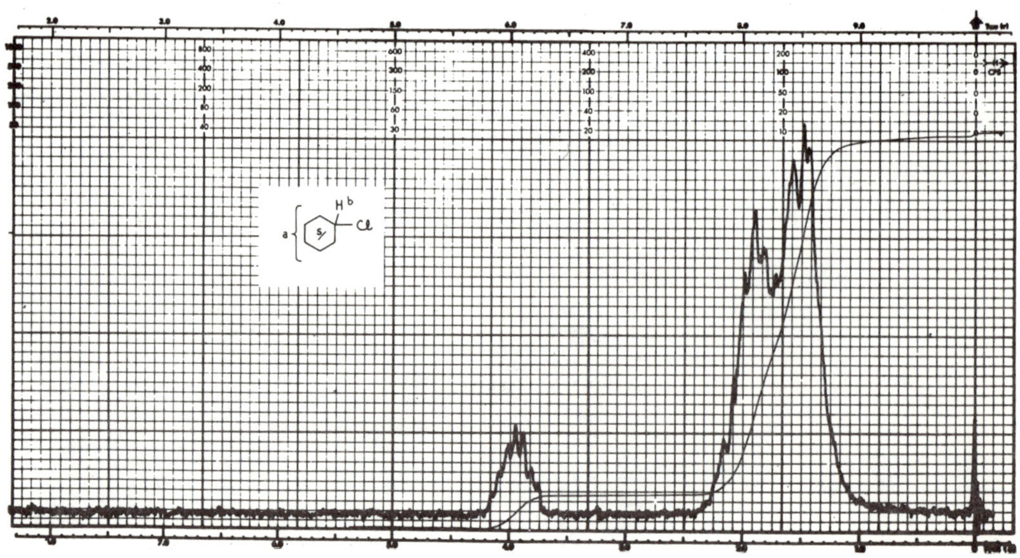

© 1978 Sadtler Research Laboratories, Inc.

1-CHLOROADAMANTANE

$C_{10}H_{15}Cl$

Mol. Wt. 170.68

M.P. 163°C

Solvent: $CDCl_3$

Assignments
a = 1.71
b = ca 2.16

268

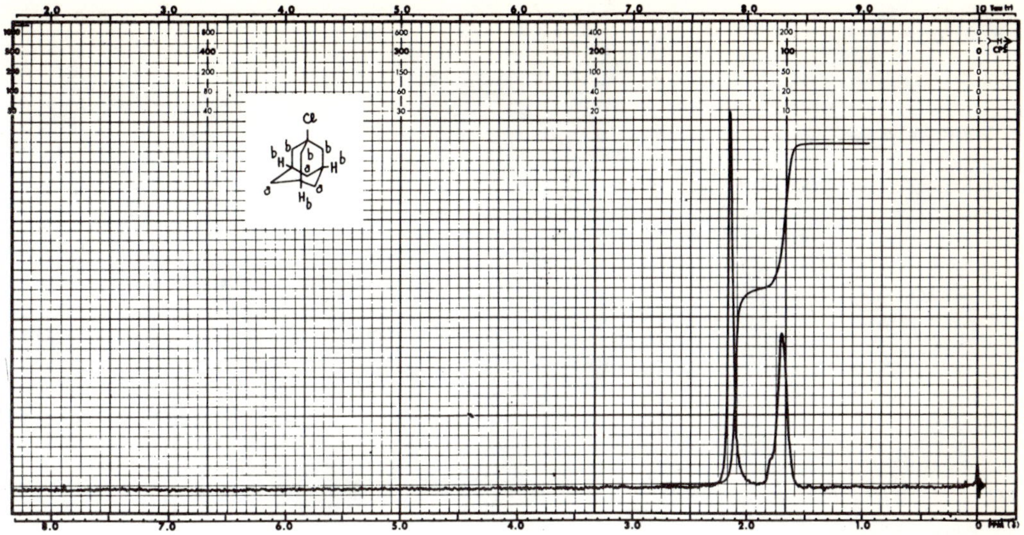

© 1978 Sadtler Research Laboratories, Inc.

2-CHLOROADAMANTANE (exo & endo)

$C_{10}H_{15}Cl$

Mol. Wt. 170.68

M.P. 193°C

Solvent: $CDCl_3$

Assignments
a = 1.30-2.50
b = 4.25-4.75

269

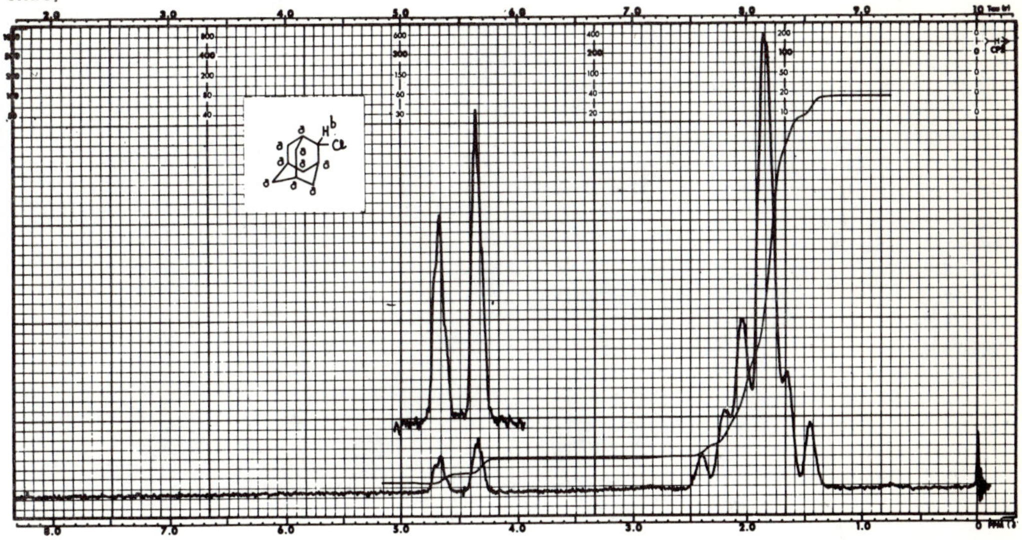

© 1978 Sadtler Research Laboratories, Inc.

2,2-DICHLORONORCARANE

$C_7H_{10}Cl_2$

Mol. Wt. 165.06

Solvent: CCl_4

Assignments
a = 0.90-2.50

270

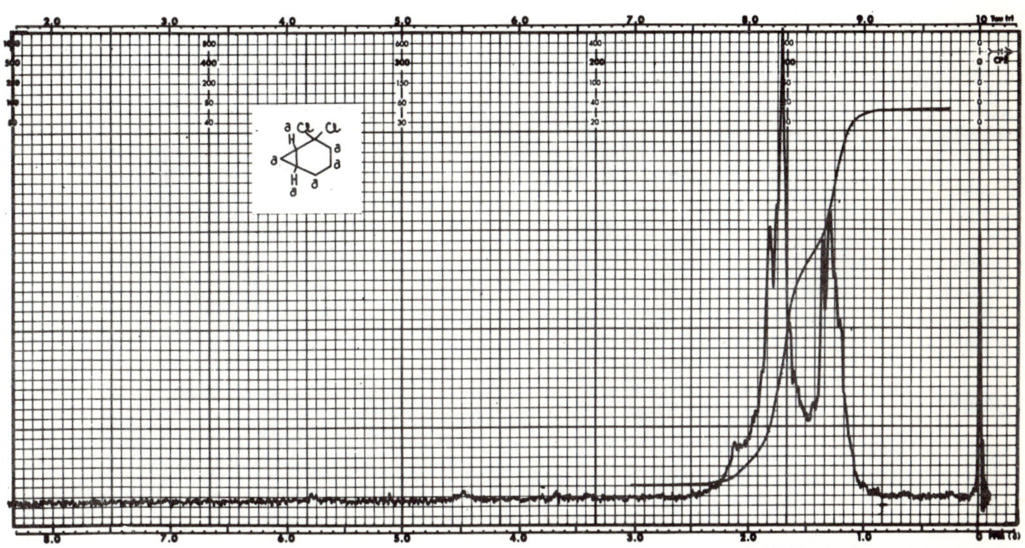

© 1978 Sadtler Research Laboratories, Inc.

1,2,3,4,5,6-HEXACHLOROCYCLOHEXANE δ-ISOMER

$C_6H_6Cl_6$

Mol. Wt. 290.83

Solvent: Polysol-d

Assignments
a = 4.10-4.90

271

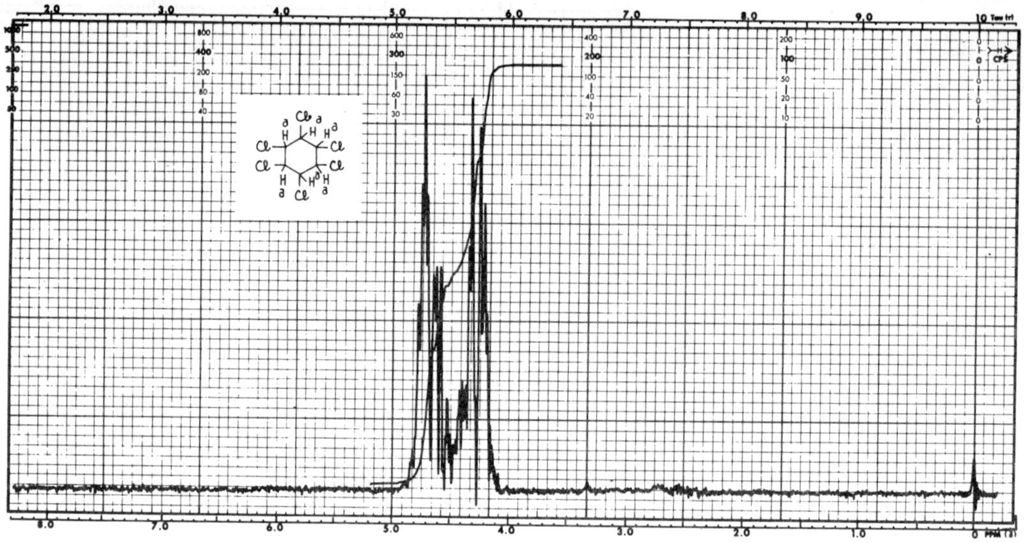

© 1978 Sadtler Research Laboratories, Inc.

3-CHLOROPROPENE

C_3H_5Cl

Mol. Wt. 76.53

B.P. 45°C

Solvent: CCl_4

Assignments
a = 3.99
b = 5.15
c = 5.29
d = ca 5.90

272

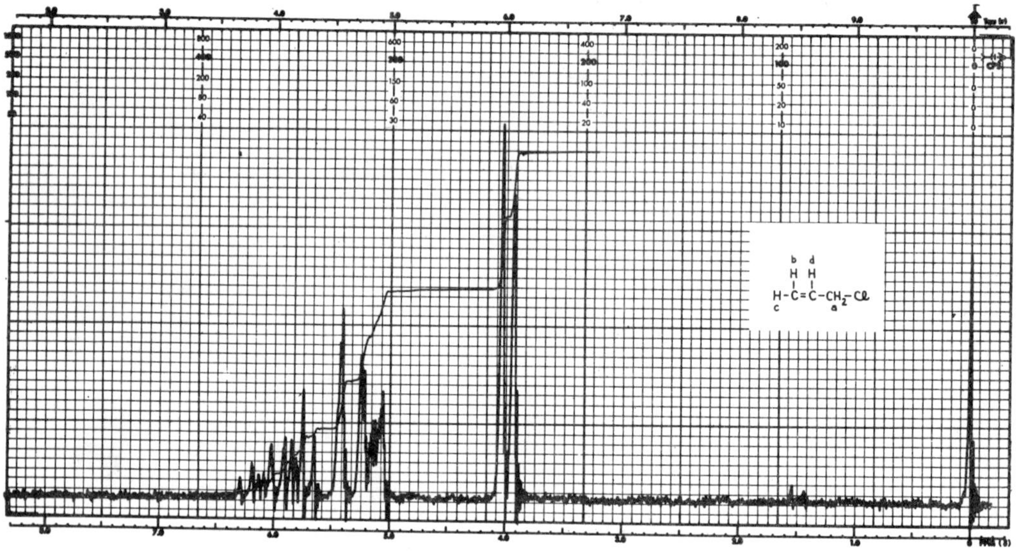

© 1978 Sadtler Research Laboratories, Inc.

2-METHYLALLYL CHLORIDE

C_4H_7Cl

Mol. Wt. 905.53

B.P. 72.2°C F.P. −80°C

Solvent: CCl_4

Assignments
a = 1.87
b = 3.98
c = 4.90 or d
d = 5.03 or c
e = 1.07, 2.17, imps.
f = 3.48 imp.

273

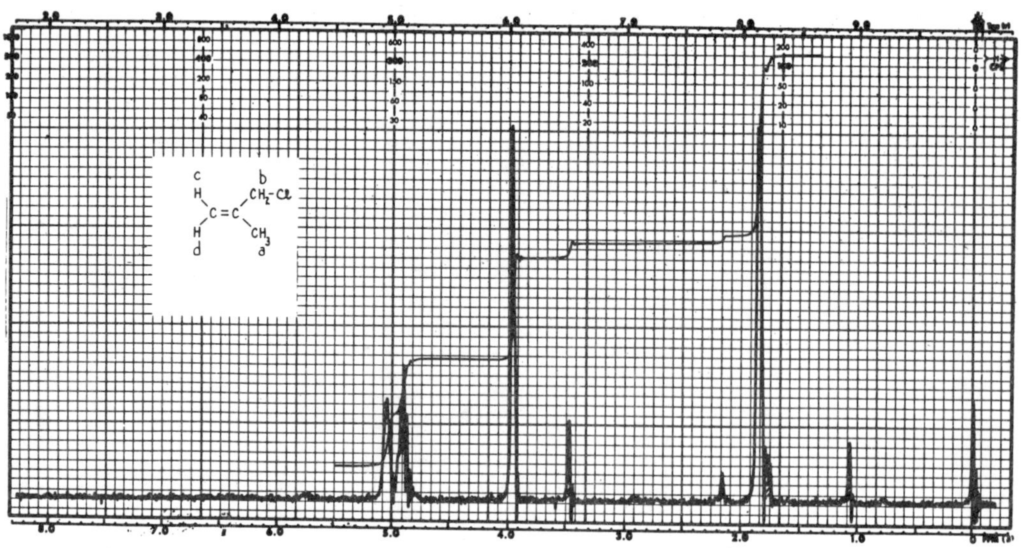

© 1978 Sadtler Research Laboratories, Inc.

1-CHLORO-5,5,7,7-TETRAMETHYL-2-OCTENE

$C_{12}H_{23}Cl$

Mol. Wt. 202.77

B.P. 127-137°C/25mm

Solvent: $CDCl_3$

Assignments
a = 1.00
b = 1.28
c = 2.05
d = 4.07
e = 5.30-6.15

274

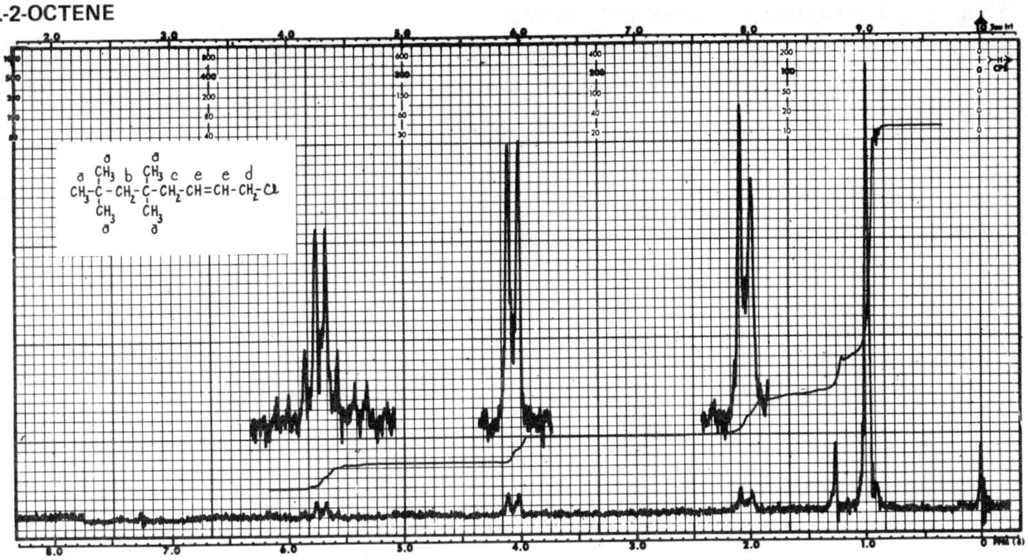

© 1978 Sadtler Research Laboratories, Inc.

3,4-DICHLORO-1-BUTENE

$C_4H_6Cl_2$

Mol. Wt. 125.00

Solvent: CCl_4

Assignments
a = 3.61 or b
b = 3.78 or a
c = 4.48
d = ca 5.46
e = ca 5.58
f = ca 5.95

275

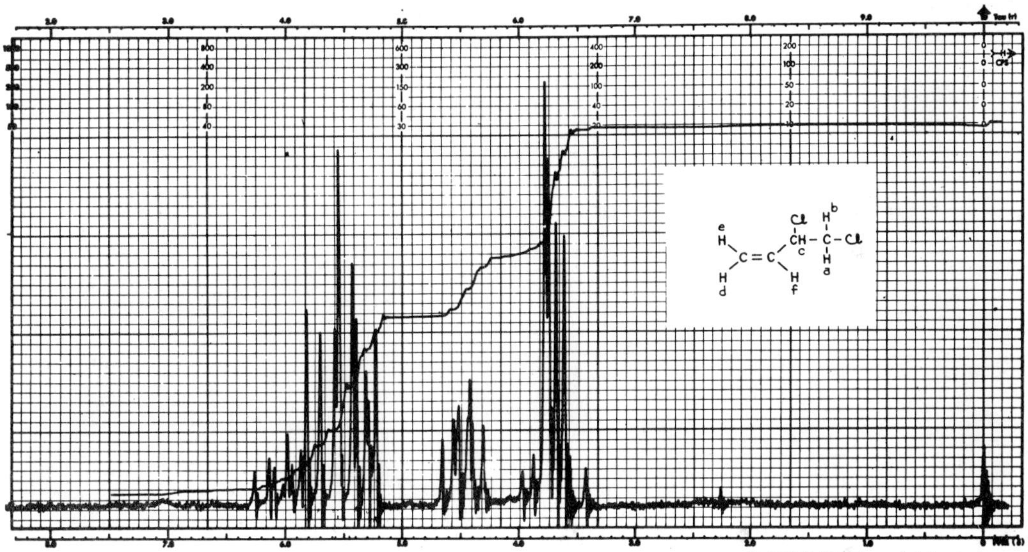

© 1978 Sadtler Research Laboratories, Inc.

1,4-DICHLORO-2-BUTENE

$C_4H_6Cl_2$

Mol. Wt. 125.00

B.P. 72-75°C/40mm

Solvent: CCl_4

Assignments
a = 4.05
b = 5.93

276

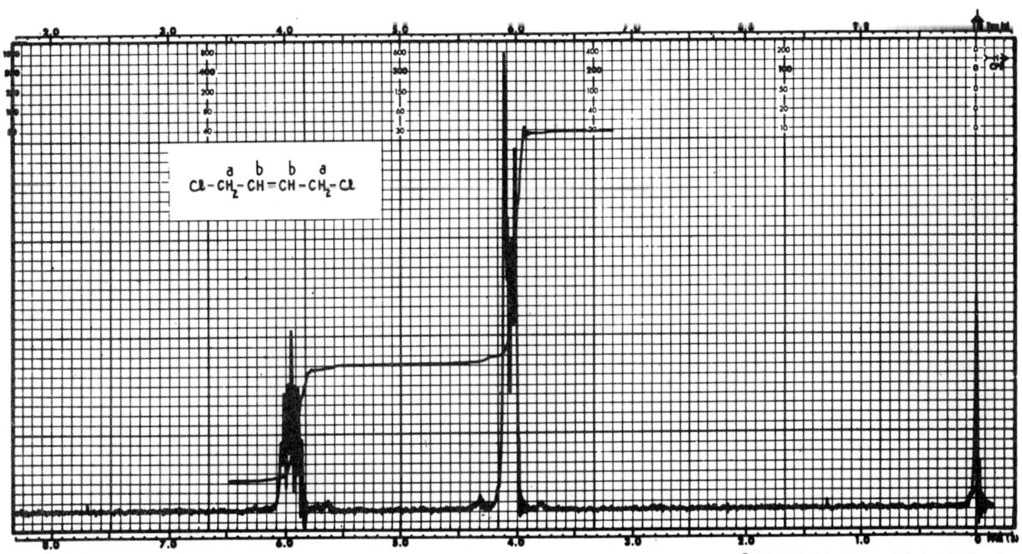

© 1978 Sadtler Research Laboratories, Inc.

1,3-DICHLOROPROPENE

$C_3H_4Cl_2$

Mol. Wt. 110.97

Solvent: $CDCl_3$

Assignments
a = 4.03 or b
b = 4.20 or a
c = 5.80-6.50

277

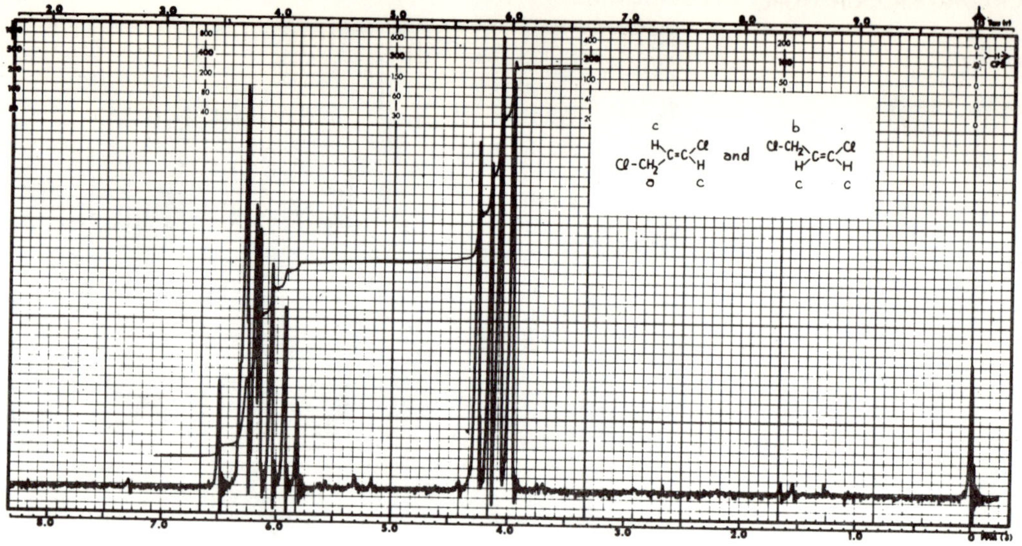

© 1978 Sadtler Research Laboratories, Inc.

1,3-DICHLORO-2-BUTENE

$C_4H_6Cl_2$

Mol. Wt. 125.00

Solvent: CCl_4

Assignments
a = 2.19
b = 4.12
c = 5.71
d = 3.92 imp.

278

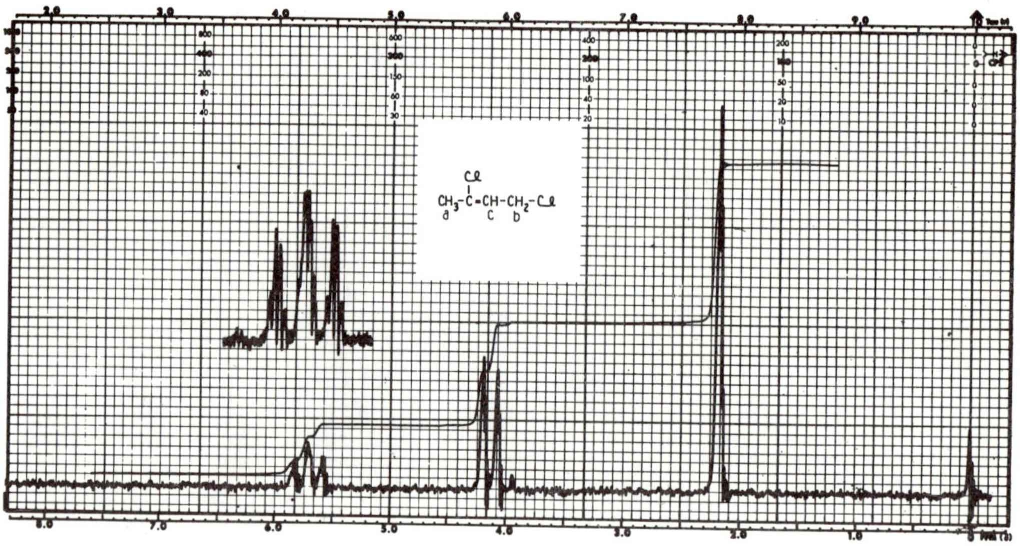

© 1978 Sadtler Research Laboratories, Inc.

1,1-DICHLOROETHYLENE

$C_2H_2Cl_2$

Mol. Wt. 96.94

Solvent: CCl_4

Assignments
a = 5.50

279

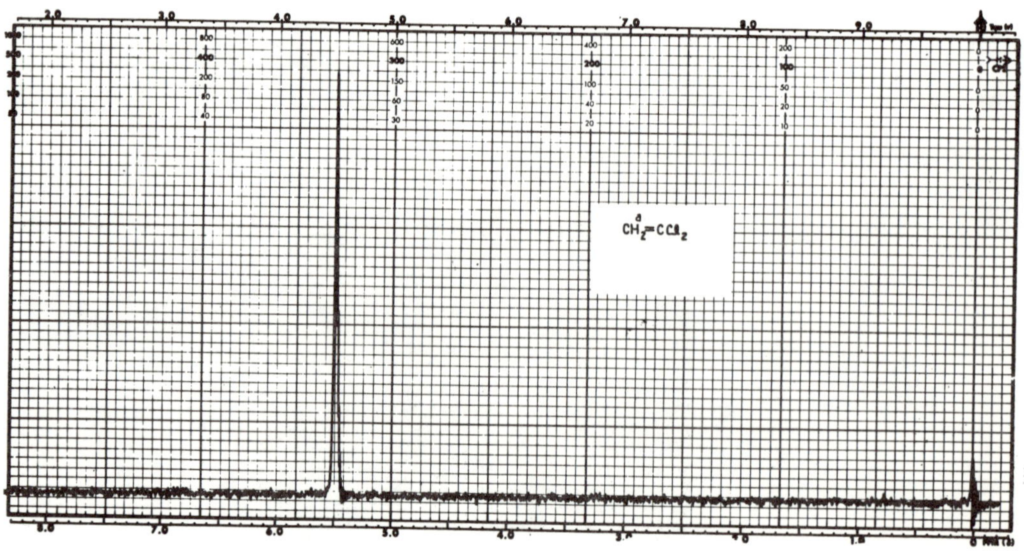

© 1978 Sadtler Research Laboratories, Inc.

1,1-DICHLORO-1-PROPENE

$C_3H_4Cl_2$

Mol. Wt. 110.97

B.P. 77°C

Solvent: $CDCl_3$

Assignments
a = 1.73
b = 5.86

280

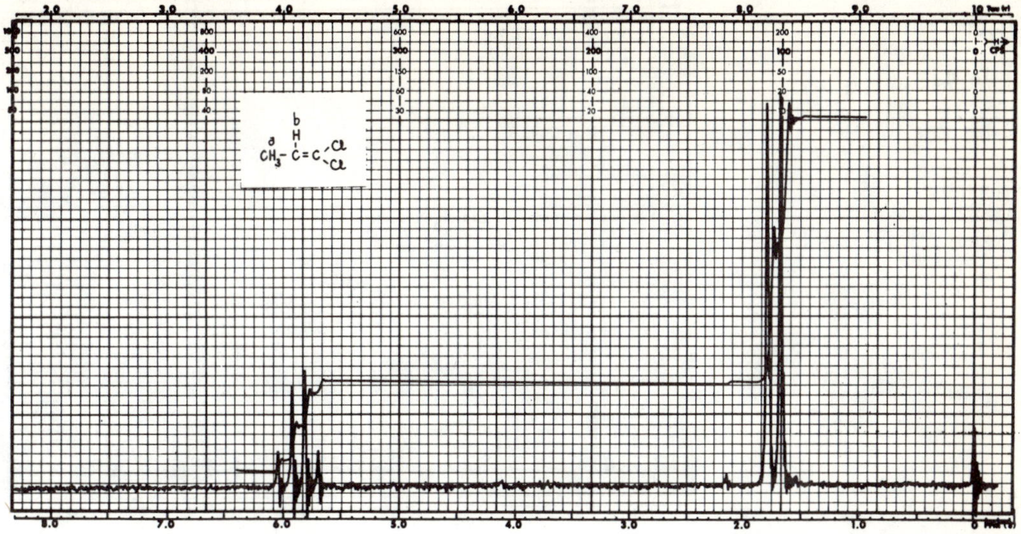

© 1978 Sadtler Research Laboratories, Inc.

trans-1,2-DICHLOROETHYLENE

$C_2H_2Cl_2$

Mol. Wt. 96.94

B.P. 48-48.5°C

Solvent: CCl_4

Assignments
a = 6.36

281

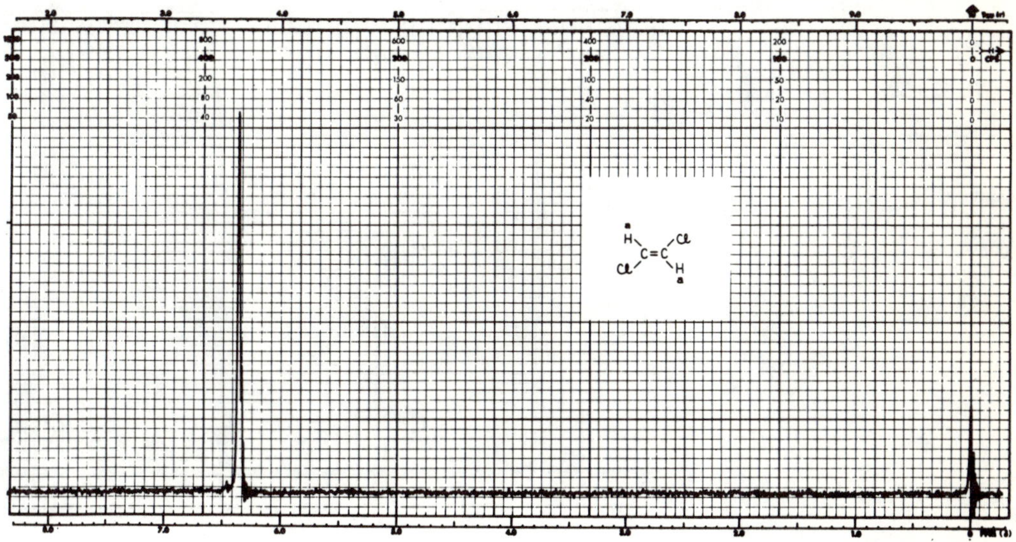

© 1978 Sadtler Research Laboratories, Inc.

TRICHLOROETHYLENE

C_2HCl_3

Mol. Wt. 131.39

M.P. 86.7°C(lit.)

Solvent: CCl_4

Assignments
a = 6.49

282

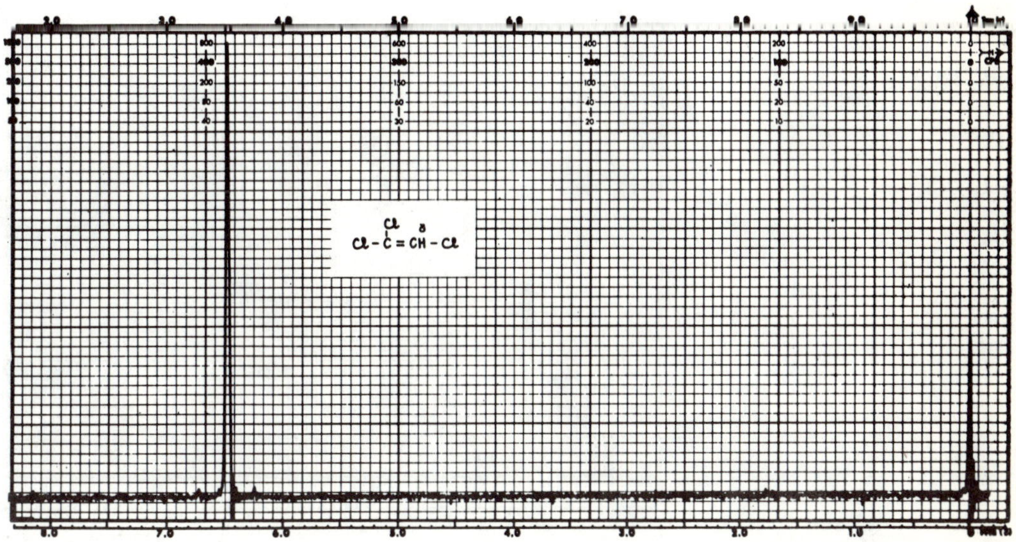

© 1978 Sadtler Research Laboratories, Inc.

3-CHLOROPROPYNE

C_3H_3Cl

Mol. Wt. 74.51

Solvent: CCl_4

Assignments
a = 2.42
b = 4.06

283

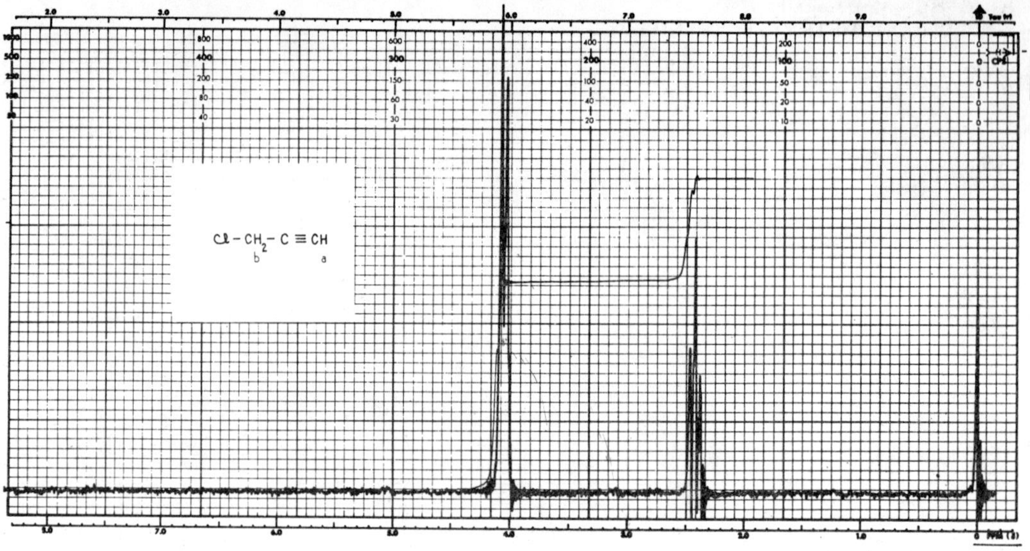

© 1978 Sadtler Research Laboratories, Inc.

α-CHLOROTOLUENE

C_7H_7Cl

Mol. Wt. 126.59

B.P. 177-178°C

Solvent: CCl_4

Assignments
a = 4.48
b = 7.28

284

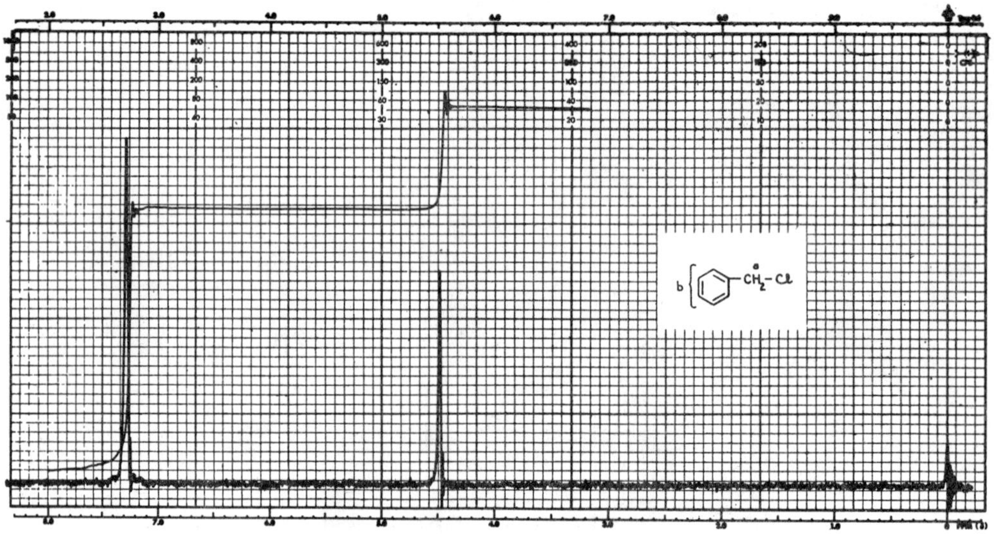

© 1978 Sadtler Research Laboratories, Inc.

α-CHLORO-p-XYLENE

C_8H_9Cl

Mol. Wt. 140.61

B.P. 192°C

Solvent: CCl_4

Assignments
a = 2.30
b = 4.43
c = 7.05
d = 7.19

285

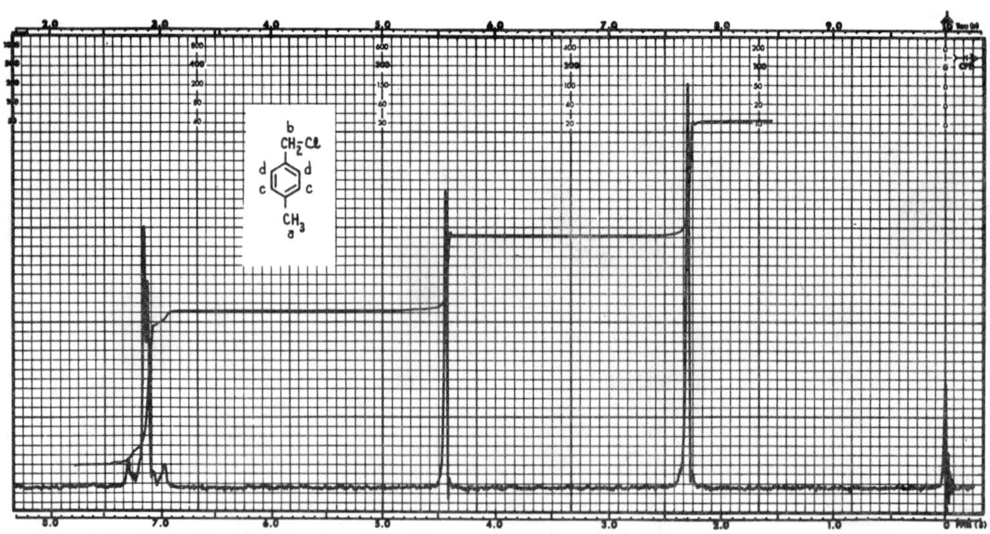

© 1978 Sadtler Research Laboratories, Inc.

4,6-BIS(CHLOROMETHYL)-m-XYLENE

$C_{10}H_{12}Cl_2$

Mol. Wt. 203.11

Solvent: $CDCl_3$

Assignments
a = 2.34
b = 4.50
c = 6.98
d = 7.20

286

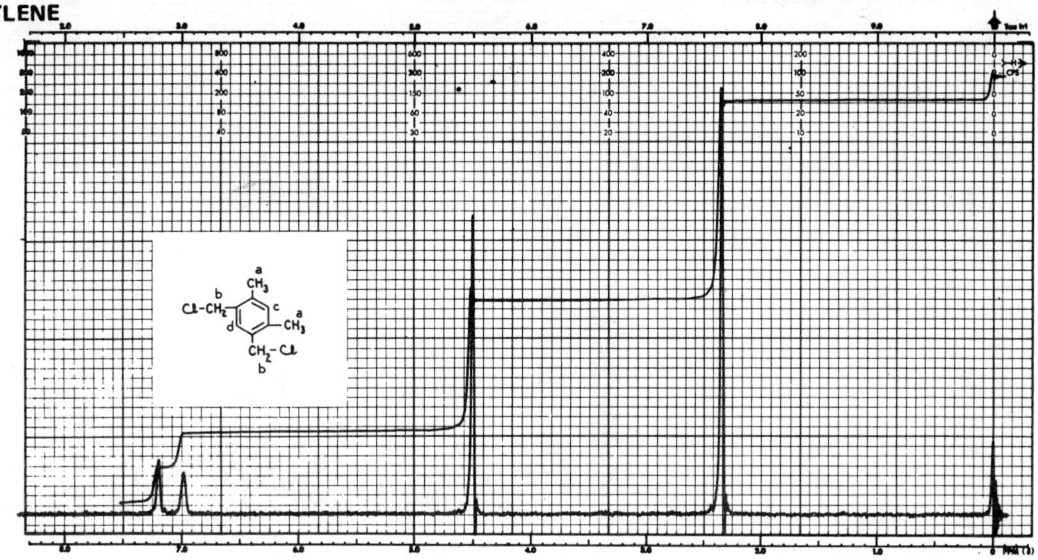

© 1978 Sadtler Research Laboratories, Inc.

***a,a*-DICHLOROTOLUENE**

$C_7H_6Cl_2$

Mol. Wt. 161.03

M.P. –17 to –15°C

B.P. 98-100°C/11mm

Solvent: CCl_4

Assignments
a = 6.60
b = ca 7.37

287

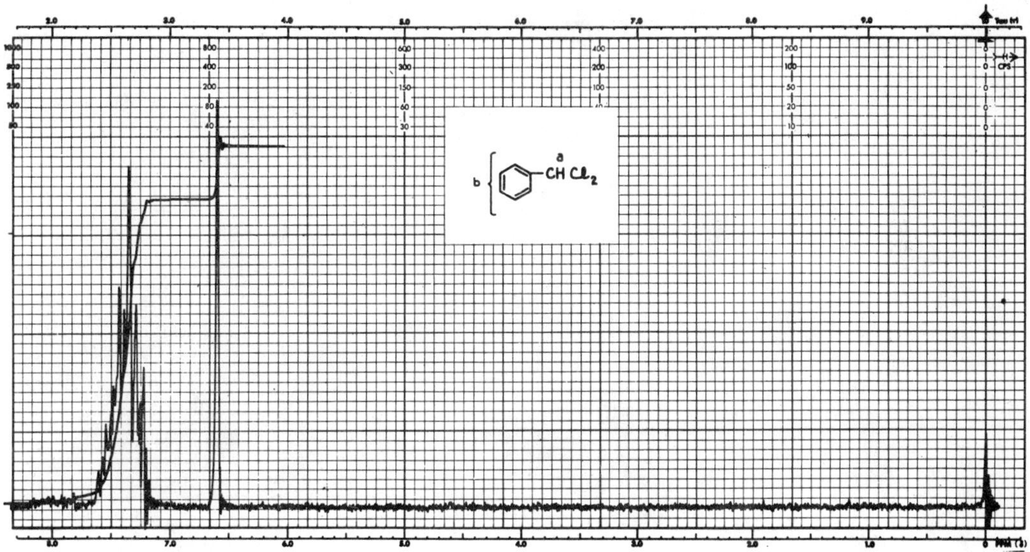

© 1978 Sadtler Research Laboratories, Inc.

DICHLORODIPHENYLMETHANE

$C_{13}H_{10}Cl_2$

Mol. Wt. 237.13

Solvent: CCl_4

288

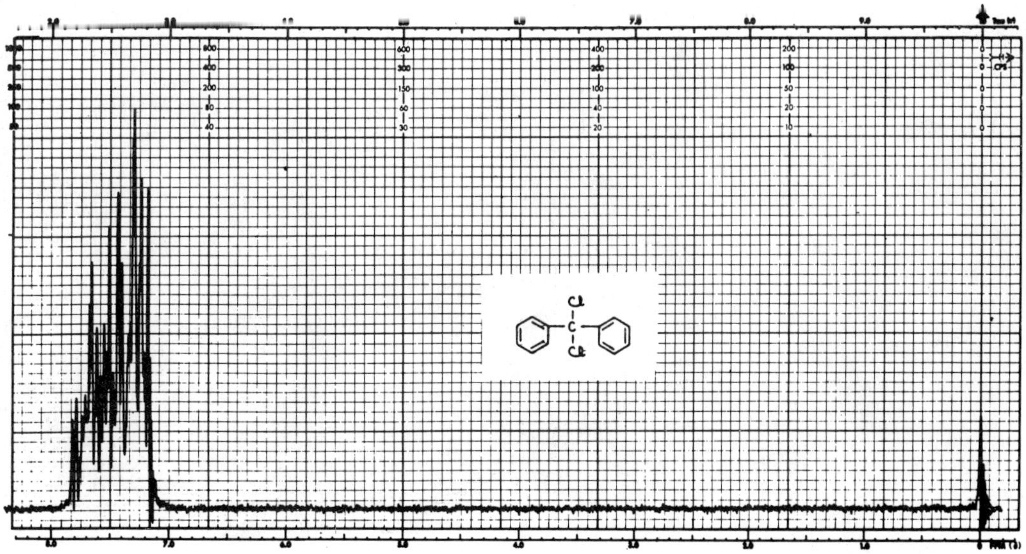

© 1978 Sadtler Research Laboratories, Inc.

a,a,a-TRICHLOROTOLUENE

$C_7H_5Cl_3$

Mol. Wt. 195.48

B.P. 220.8°C

Solvent: CCl_4

Assignments
a = 7.34
b = 7.83

289

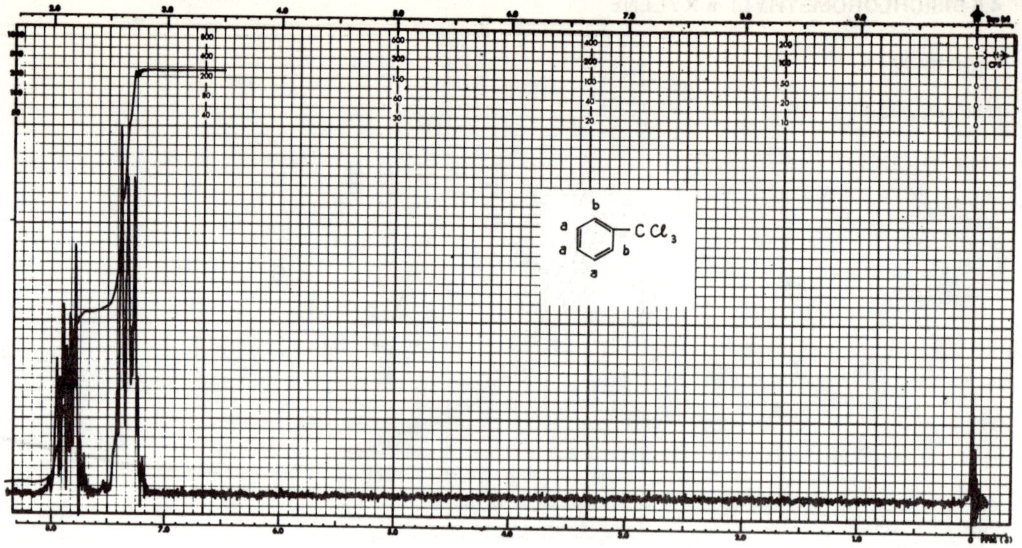

© 1978 Sadtler Research Laboratories, Inc.

---

p-FLUORO-*a,a,a*-TRICHLOROTOLUENE

$C_7H_4Cl_3F$

Mol. Wt. 213.47

B.P. 98°C/16mm

Solvent: CCl_4

Assignments
a = 7.05
b = 7.88

290

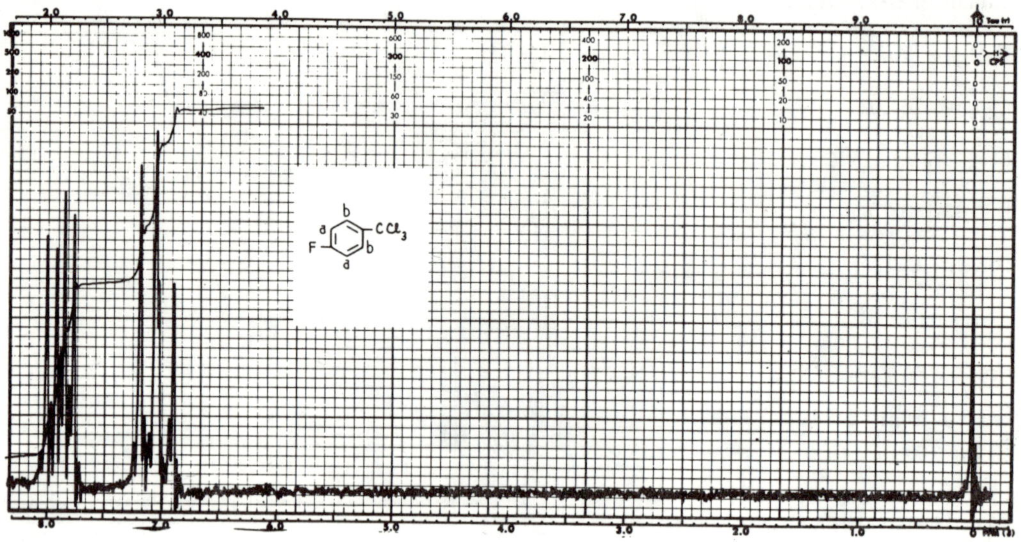

© 1978 Sadtler Research Laboratories, Inc.

CHLOROBENZENE

C_6H_5Cl

Mol. Wt. 112.56

B.P. 130°C (lit.)

Solvent: CCl_4

Assignments
a = ca 7.22

291

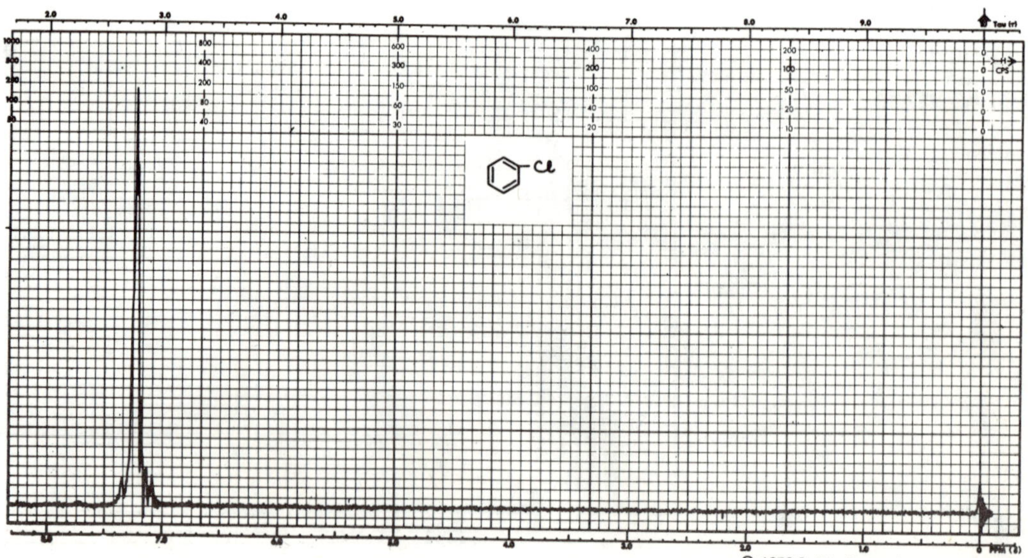

© 1978 Sadtler Research Laboratories, Inc.

o-CHLOROTOLUENE

C_7H_7Cl

Mol. Wt. 126.58

B.P. 159.5°C

Solvent: CCl_4

Assignments
a = 2.35

292

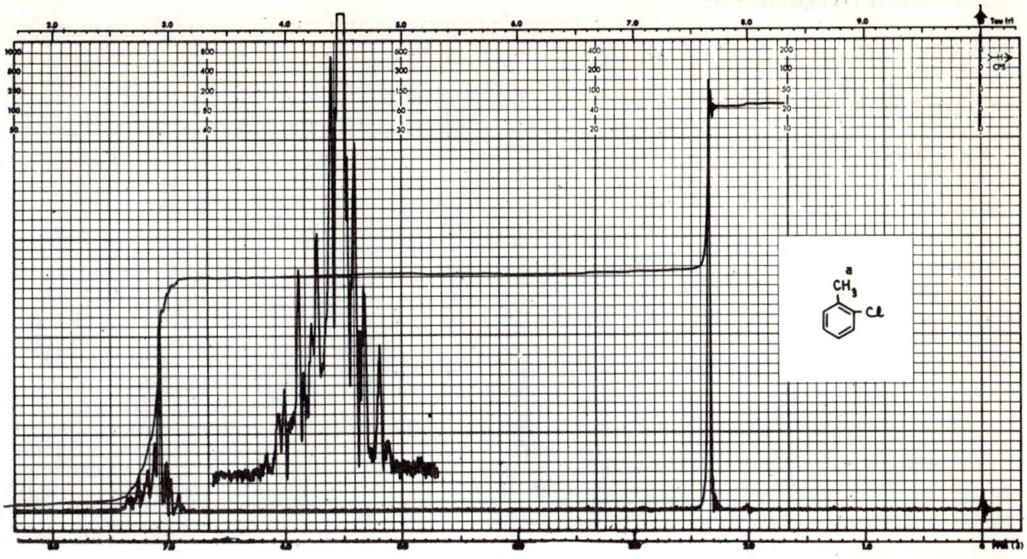

© 1978 Sadtler Research Laboratories, Inc.

---

m-CHLOROTOLUENE

C_7H_7Cl

Mol. Wt. 126.59

B.P. 159-161°C

Solvent: CCl_4

Assignments
a = 2.30
b = ca 7.07

293

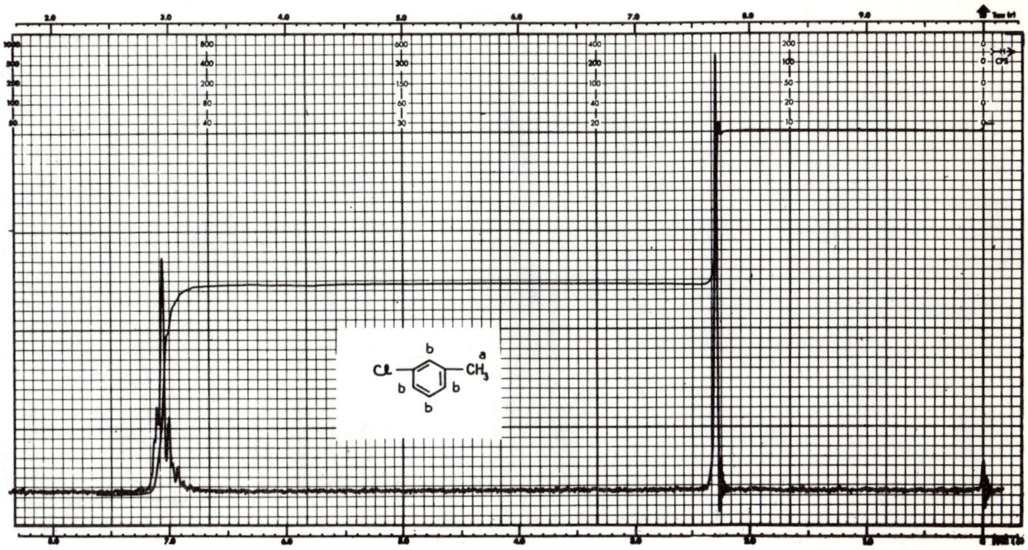

© 1978 Sadtler Research Laboratories, Inc.

---

p-CHLOROTOLUENE

C_7H_7Cl

Mol. Wt. 125.57

M.P. 6-8°C

B.P. 159-162°C

Solvent: CCl_4

Assignments
a = 2.29
b = 6.98 or c
c = 7.12 or b

294

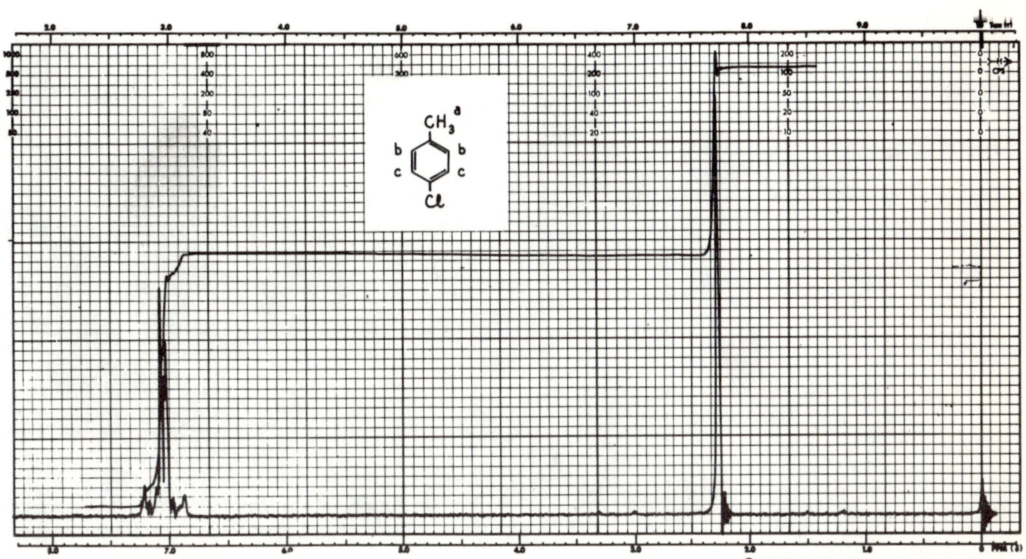

© 1978 Sadtler Research Laboratories, Inc.

1-CHLORO-2-ETHYLBENZENE

C_8H_9Cl

MOl. Wt. 140.61

B.P. 178.4°C (lit.)

Solvent: CCl_4

Assignments
a = 1.22
b = 2.73
c = ca 7.15

295

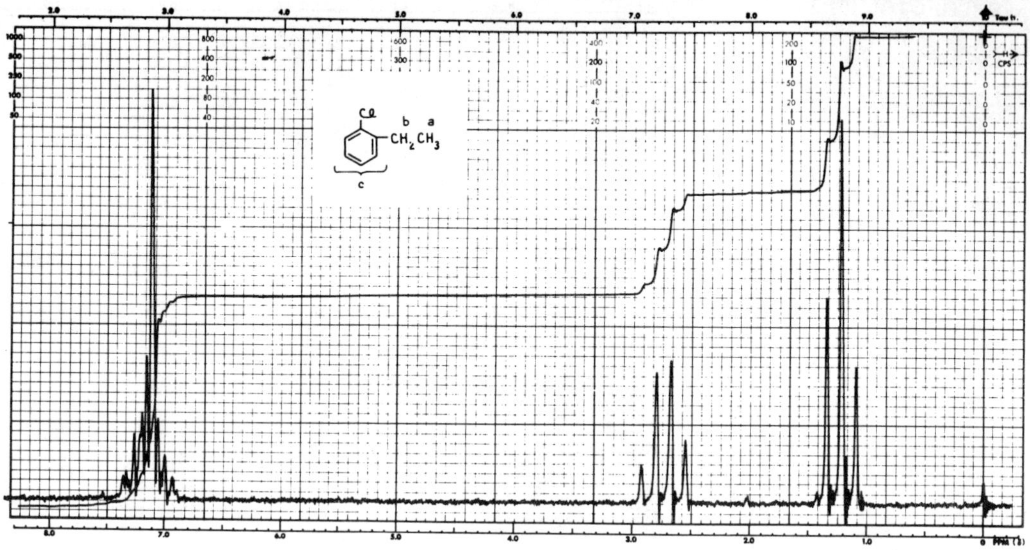

© 1978 Sadtler Research Laboratories, Inc.

---

o-CHLOROSTYRENE

C_8H_7Cl

Mol. Wt. 138,60

B.P. 58-60°C/7mm

Solvent: CCl_4

Assignments
a = 5.29
b = 5.60
c = 7.08
d = 7.00-7.30
e = 7.40

296

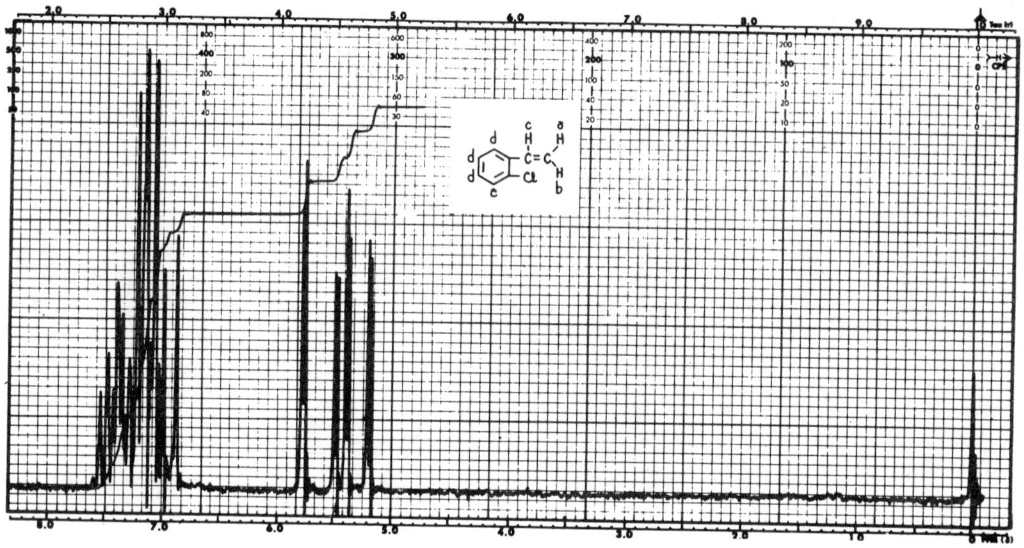

© 1978 Sadtler Research Laboratories, Inc.

2-CHLOROBIPHENYL

$C_{12}H_9Cl$

Mol. Wt. 188.66

Solvent: CCl_4

Assignments
a = 6.85-7.55

297

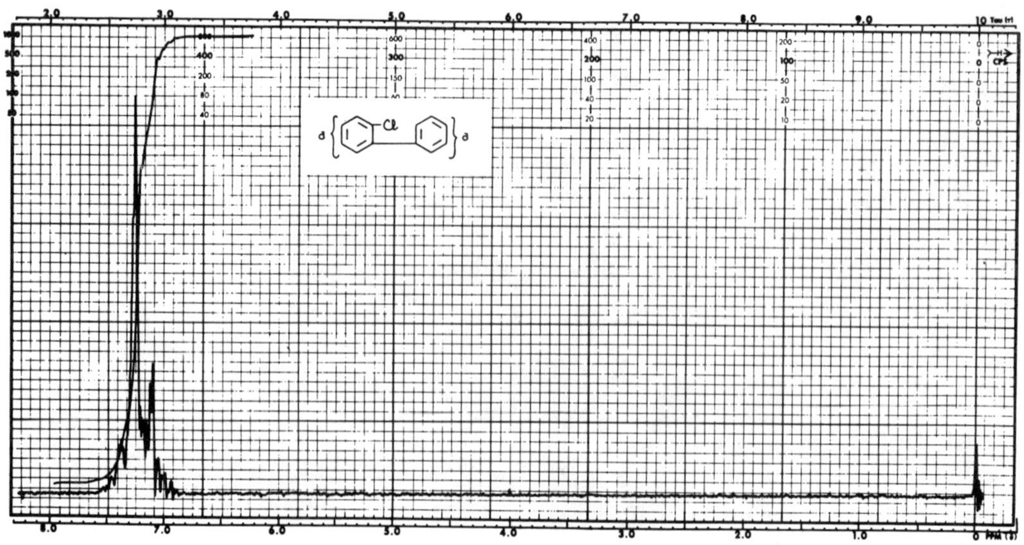

© 1978 Sadtler Research Laboratories, Inc.

4-CHLOROBIPHENYL

$C_{12}H_9Cl$

Mol. Wt. 188.66

M.P. 77°C

Solvent: $CDCl_3$

Assignments
a = 7.15-7.60

298

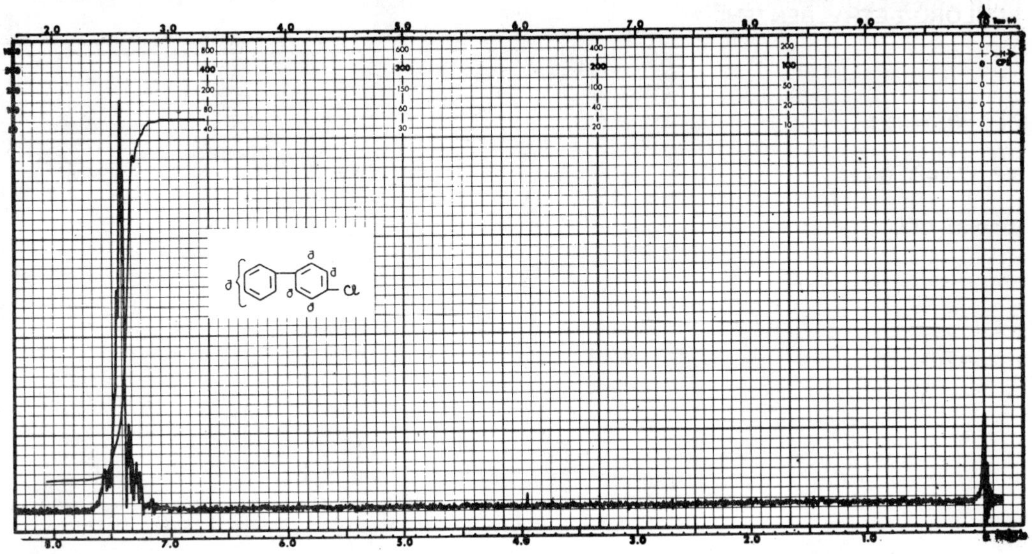

© 1978 Sadtler Research Laboratories, Inc.

2-CHLORO-p-XYLENE

C_8H_9Cl

Mol. Wt. 140.61

M.P. 2°C B.P. 186°C

Solvent: CCl_4

Assignments
a = 2.31
b = 6.87
c = 7.02
d = 7.11

299

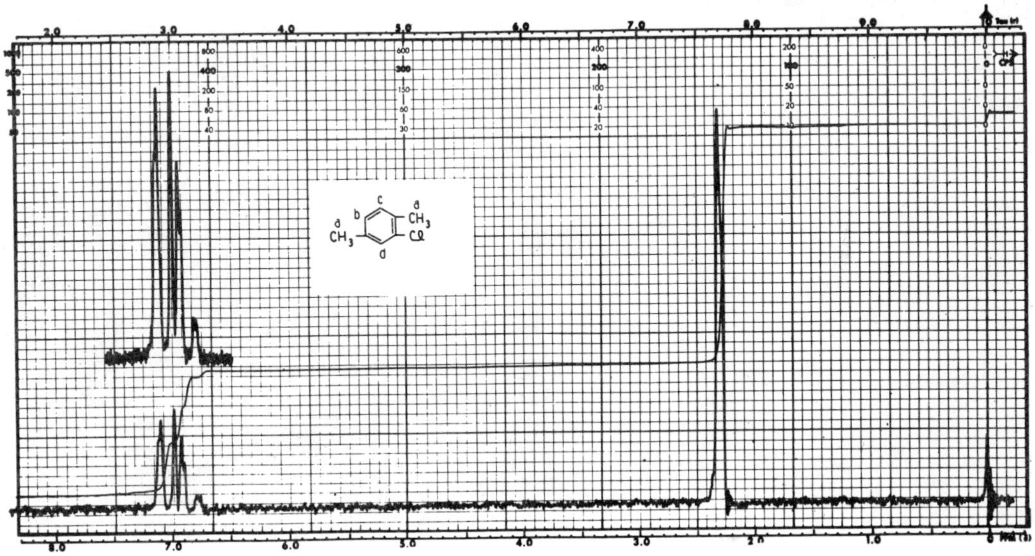

© 1978 Sadtler Research Laboratories, Inc.

---

p-CHLORO-α,α,α-TRIFLUOROTOLUENE

$C_7H_4ClF_3$

Mol. Wt. 180.56

B.P. 139°C

Solvent: CCl_4

Assignments
a = 7.42
b = 7.53

300

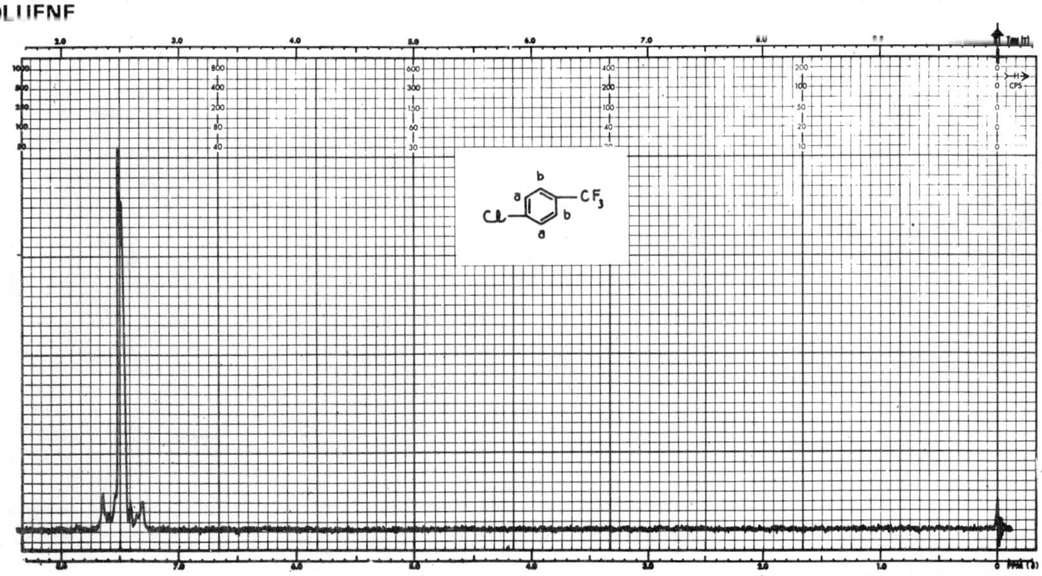

© 1978 Sadtler Research Laboratories, Inc.

1-CHLORO-2-FLUOROBENZENE

C_6H_4ClF

Mol. Wt. 130.55

B.P. 137-138°C

Solvent: CCl_4

Assignments
a = 6.80-7.25
b = 7.36

301

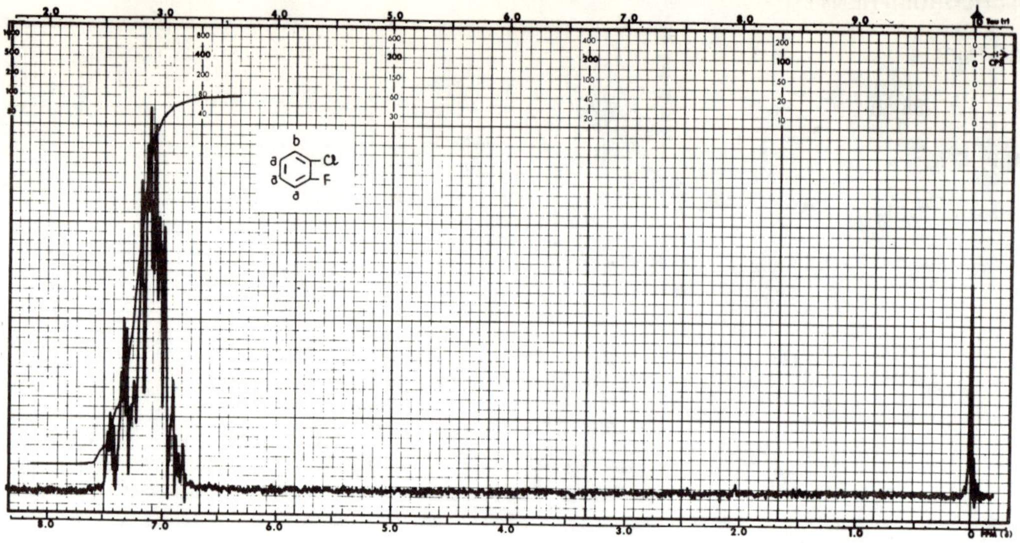

© 1978 Sadtler Research Laboratories, Inc.

1-CHLORO-3-FLUOROBENZENE

C_6H_4ClF

Mol. Wt. 130.55

B.P. 126-128°C

Solvent: CCl_4

Assignments
a = 6.67-7.40

302

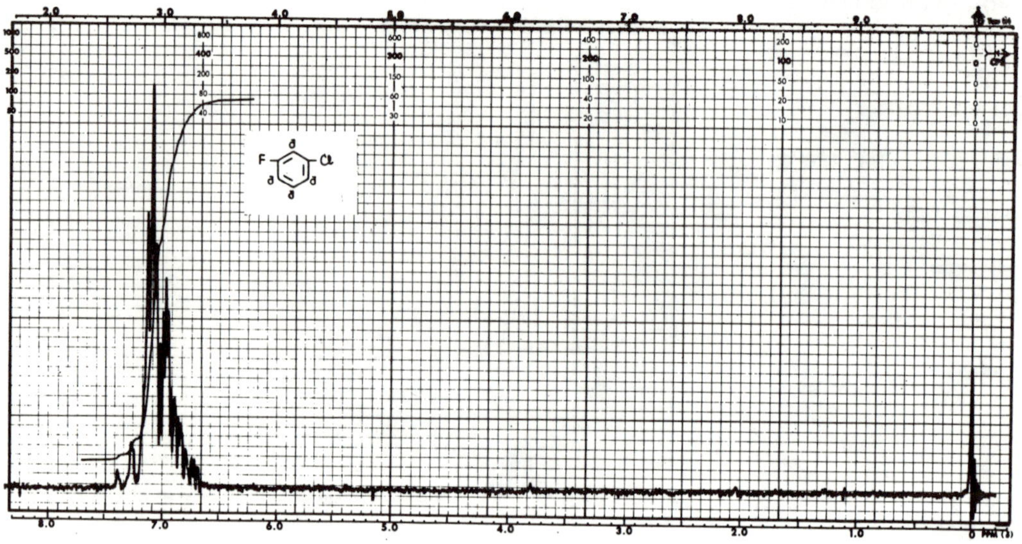

© 1978 Sadtler Research Laboratories, Inc.

1-CHLORO-4-FLUOROBENZENE

C_6H_4ClF

Mol. Wt. 130.55

M.P. −21.5°C

Solvent: $CDCl_3$

Assignments
a = 6.91
b = 7.19

303

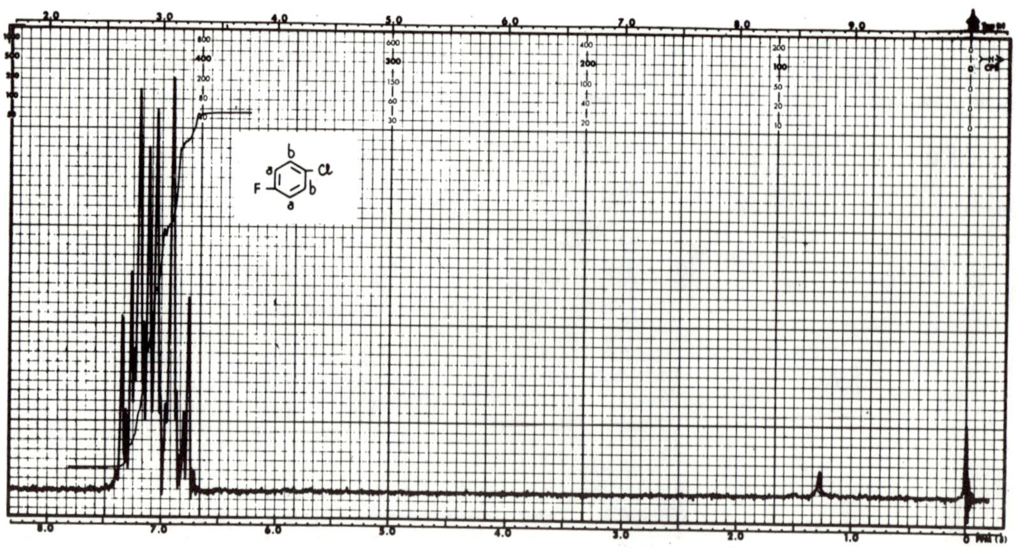

© 1978 Sadtler Research Laboratories, Inc.

2-CHLORO-4-FLUOROTOLUENE

C_7H_6ClF

Mol. Wt. 144.58

Solvent: CCl_4

Assignments
a = 2.29
b = 6.73
c = 6.99
d = 7.04

304

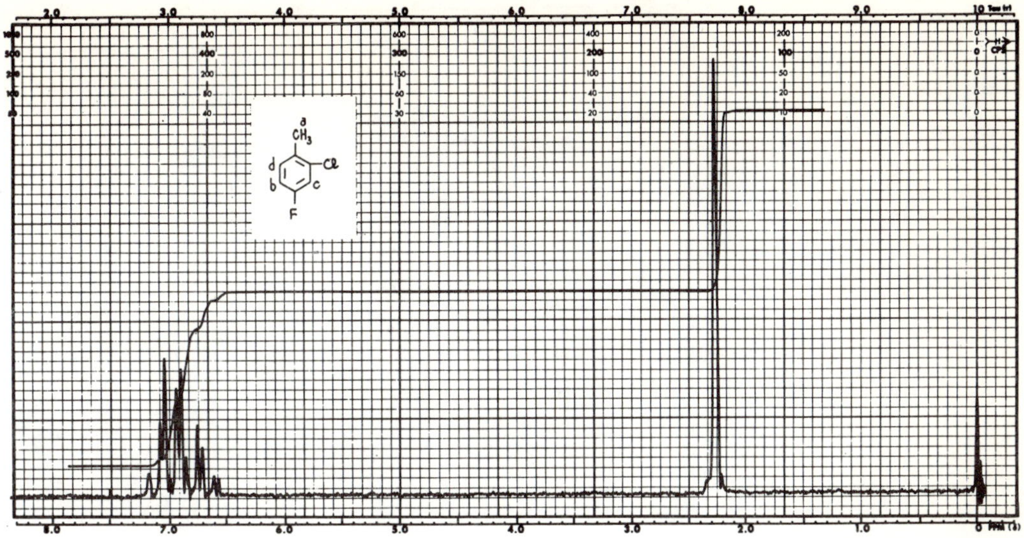

© 1978 Sadtler Research Laboratories, Inc.

o-DICHLOROBENZENE

$C_6H_4Cl_2$

Mol. Wt. 147.01

Solvent: CCl_4

Assignments
a = 6.90-7.55

305

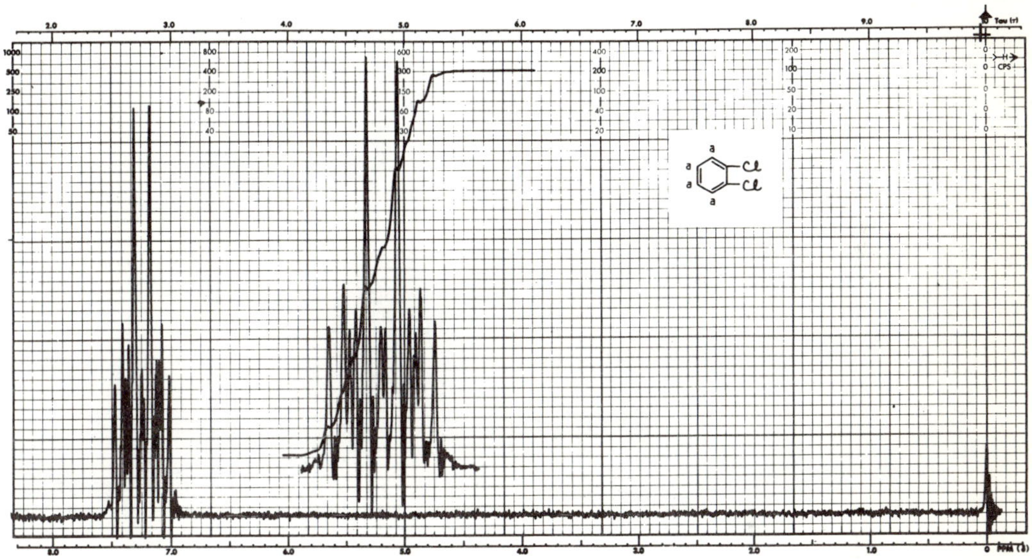

© 1978 Sadtler Research Laboratories, Inc.

m-DICHLOROBENZENE

$C_6H_4Cl_2$

Mol. Wt. 147.00

M.P. –24.8°C

B.P. 172°C

Solvent: CCl_4

Assignments
a = 7.22
b = 7.39

306

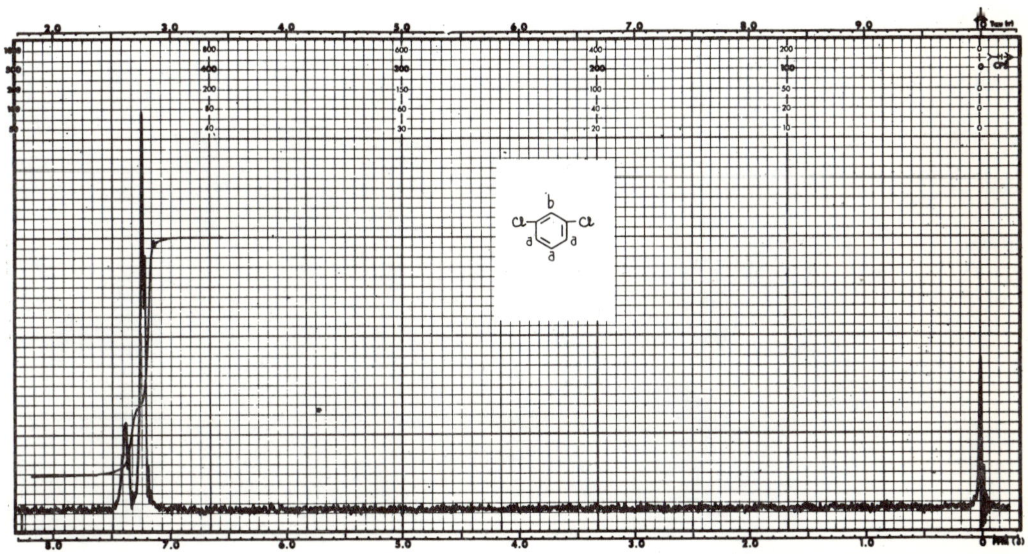

© 1978 Sadtler Research Laboratories, Inc.

p-DICHLOROBENZENE

$C_6H_4Cl_2$

Mol. Wt. 147.00

M.P. 50.6°C (lit.)

Solvent: CCl_4

Assignments
a = 7.23

307

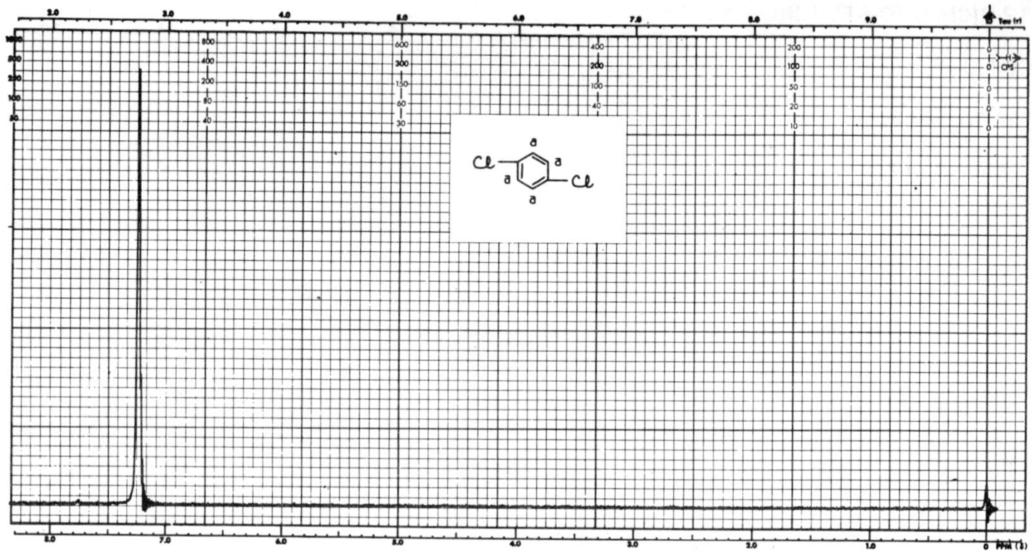

© 1978 Sadtler Research Laboratories, Inc.

2,4-DICHLOROTOLUENE

$C_7H_6Cl_2$

Mol. Wt. 161.03

B.P. 200°C

Solvent: CCl_4

Assignments
a = 2.34
b = ca 7.04
c = 7.30

308

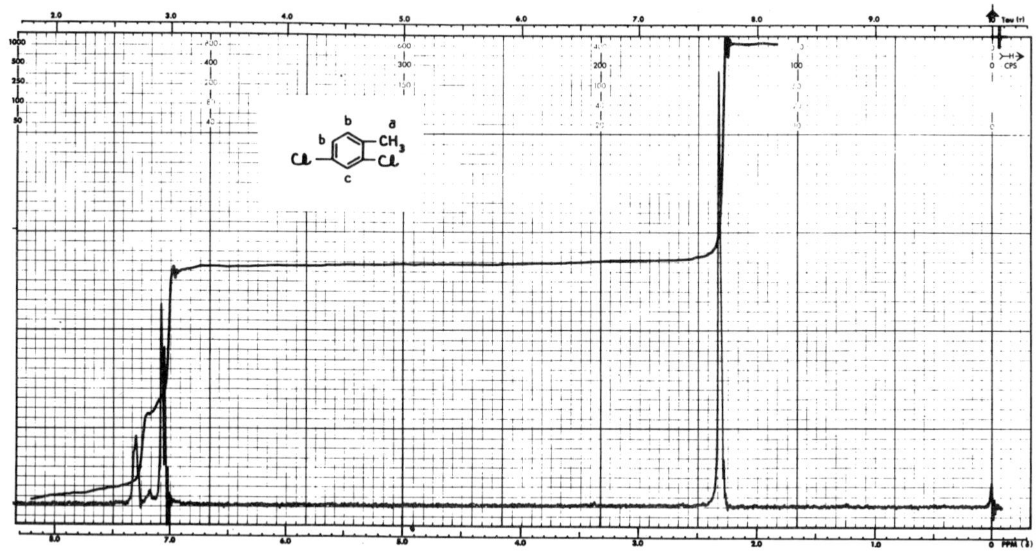

© 1978 Sadtler Research Laboratories, Inc.

2,6-DICHLOROTOLUENE

$C_7H_6Cl_2$

Mol. Wt. 161.03

B.P. 73-76°C/11mm

Solvent: CCl_4

Assignments
a = 2.41
b = 6.79-7.28

309

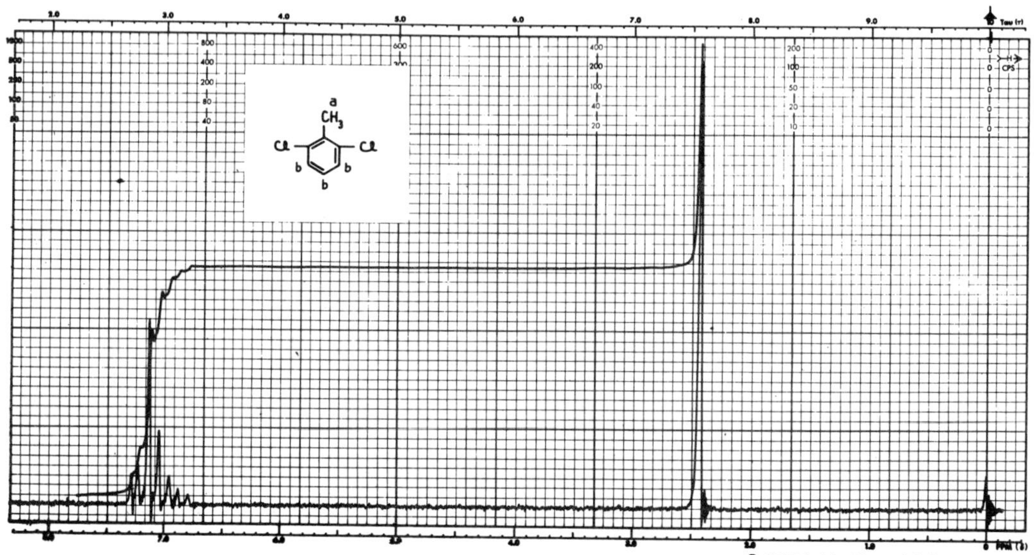

© 1978 Sadtler Research Laboratories, Inc.

1,2-DICHLORO-4-FLUOROBENZENE

$C_6H_3Cl_2F$

Mol. Wt. 164.99

Solvent: $CDCl_3$

Assignments
a = 6.91
b = 7.15
c = 7.32

310

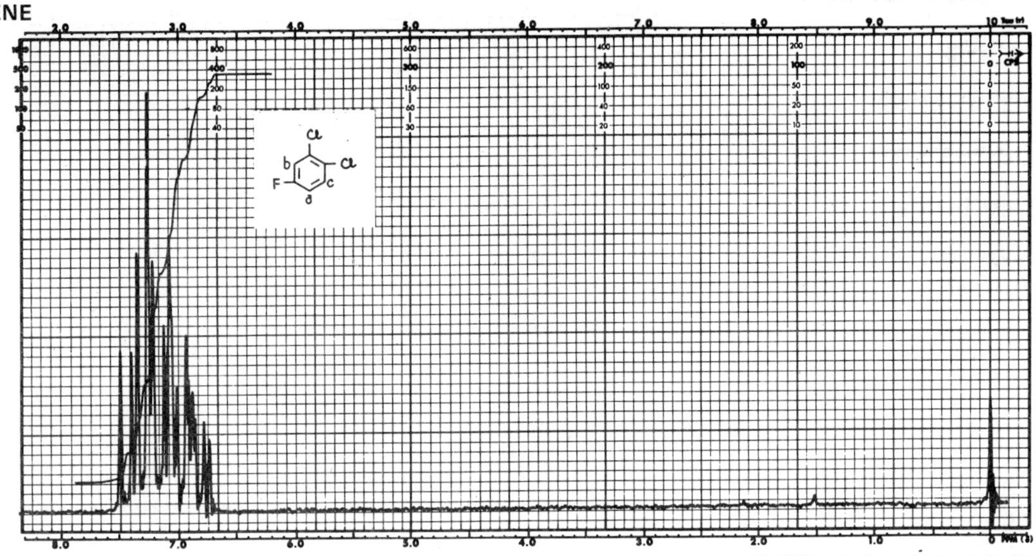

© 1978 Sadtler Research Laboratories, Inc.

α,3,4-TRICHLOROTOLUENE

$C_7H_5Cl_3$

Mol. Wt. 195.48

Solvent: CCl_4

Assignments
a = 4.42
b = 7.11
c = 7.36
d = 7.39

311

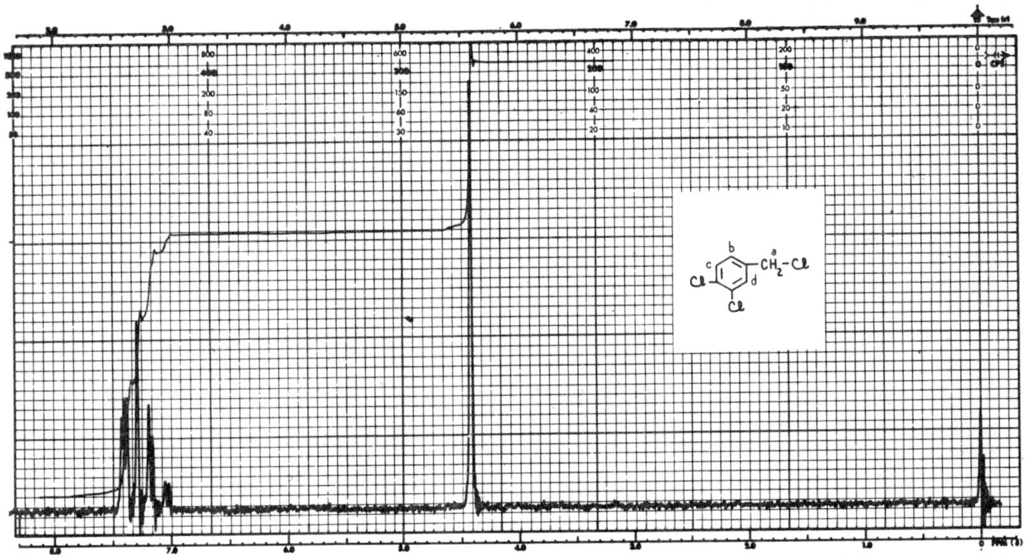

© 1978 Sadtler Research Laboratories, Inc.

1,2,3-TRICHLOROBENZENE

$C_6H_3Cl_3$

Mol. Wt. 181.46

M.P. 52°C

B.P. 219°C/760mm (lit.)

Solvent: $CDCl_3$

Assignments
a = 6.89-7.46

312

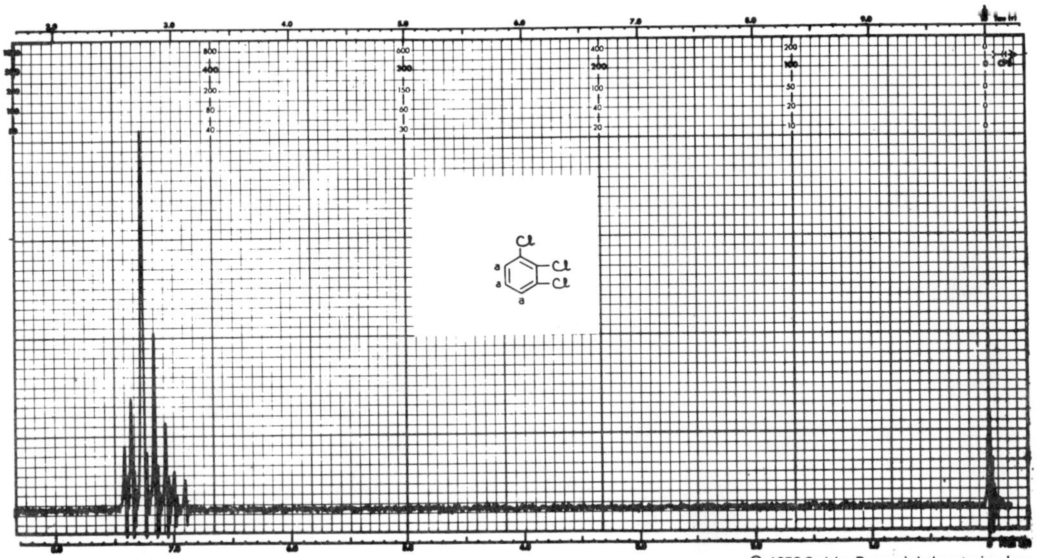

© 1978 Sadtler Research Laboratories, Inc.

1,2,4-TRICHLOROBENZENE

$C_6H_3Cl_3$

Mol. Wt. 181.45

M.P. 15-17°C

Solvent: CCl_4

Assignments
a = 7.12
b = 7.35
c = 7.43

313

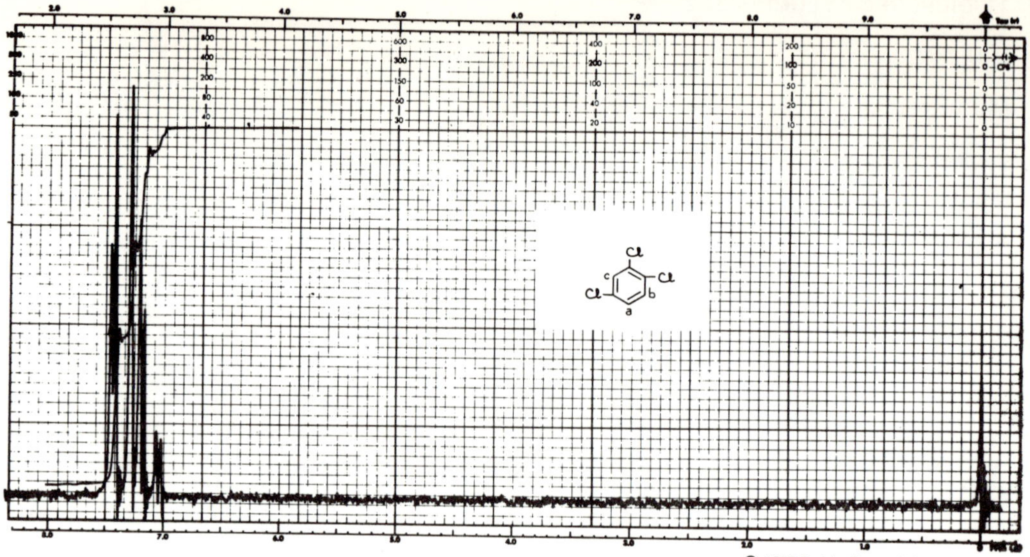

© 1978 Sadtler Research Laboratories, Inc.

1,3,5-TRICHLOROBENZENE

$C_6H_3Cl_3$

Mol. Wt. 181.45

M.P. 63.5°C (lit.)

Solvent: $CDCl_3$

Assignments
a = 7.30

314

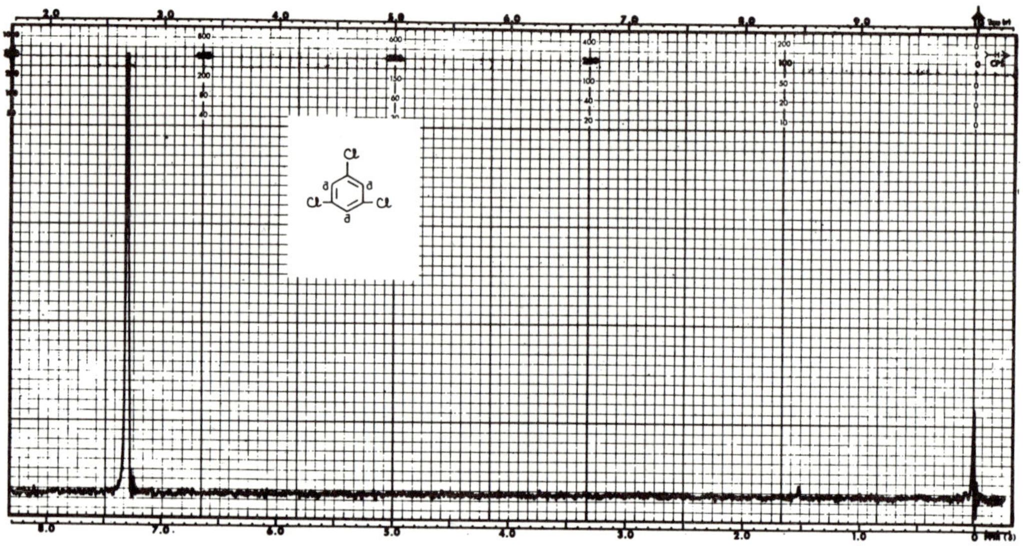

© 1978 Sadtler Research Laboratories, Inc.

1,2,4,5-TETRACHLOROBENZENE

$C_6H_2Cl_4$

Mol. Wt. 215.89

B.P. 240-246°C

Solvent: $CDCl_3$

Assignments
a = 7.52
b = 1.12 imp.

315

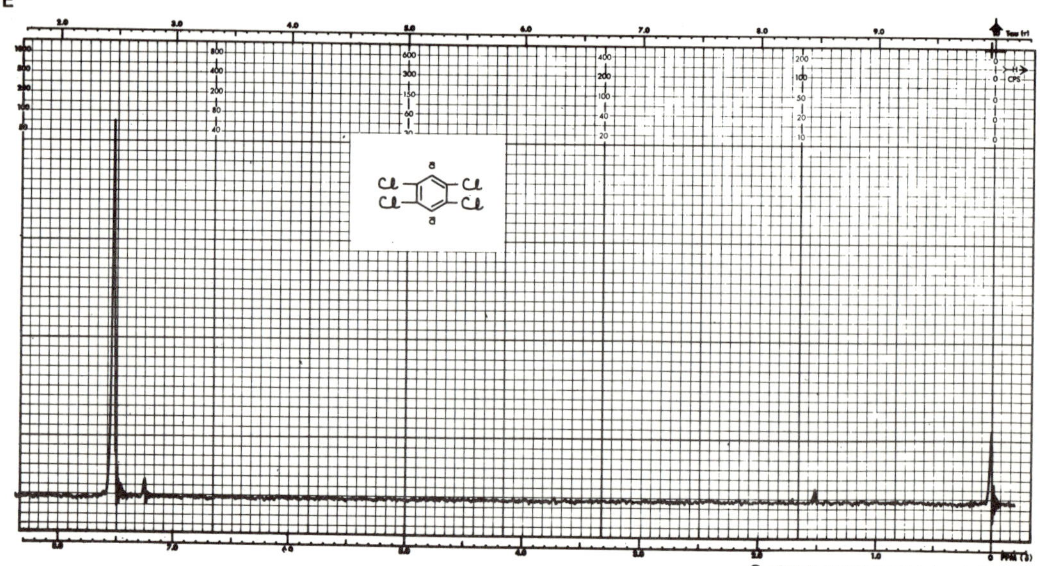

© 1978 Sadtler Research Laboratories, Inc.

1-(CHLOROMETHYL)NAPHTHALENE

$C_{11}H_9Cl$

Mol. Wt. 176.65

B.P. 155-158°C/13mm

Solvent: CCl_4

Assignments
a = 4.83
b = 7.18-8.12
c = 4.55 imp.

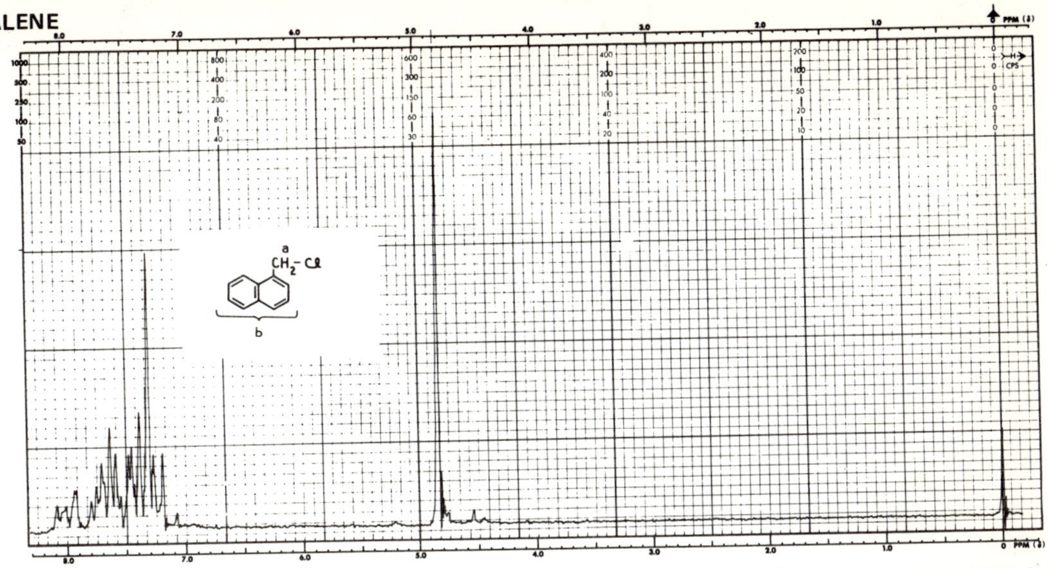

316

© 1978 Sadtler Research Laboratories, Inc.

1-CHLORONAPHTHALENE

$C_{10}H_7Cl$

Mol. Wt. 162.62

B.P. 111-113°C/5mm

Solvent: CCl_4

Assignments
a = 7.11
b = 7.18-7.77
c = 8.15

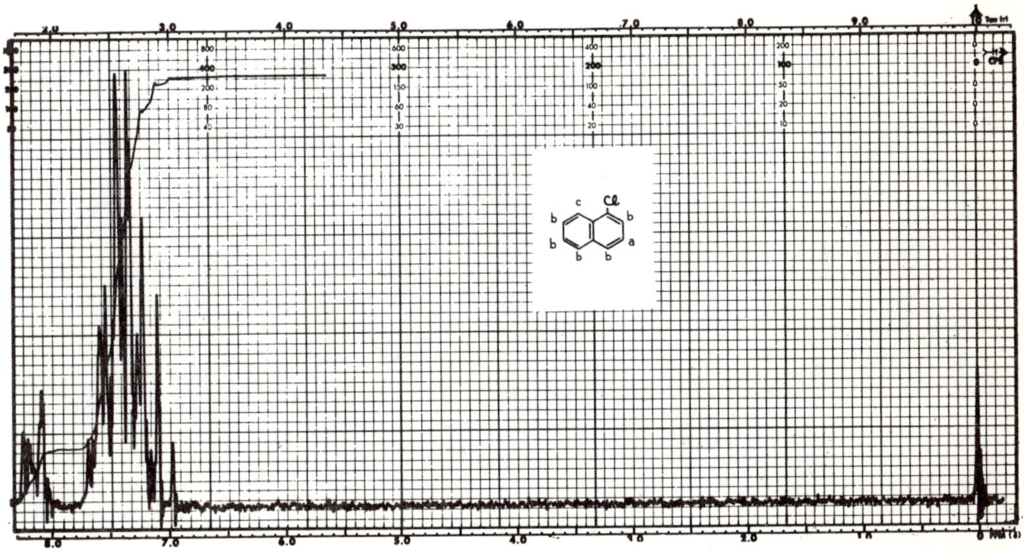

317

© 1978 Sadtler Research Laboratories, Inc.

2-CHLORONAPHTHALENE

$C_{10}H_7Cl$

Mol. Wt. 162.62

M.P. 58°C

Solvent: $CDCl_3$

Assignments
a = 7.20-7.90

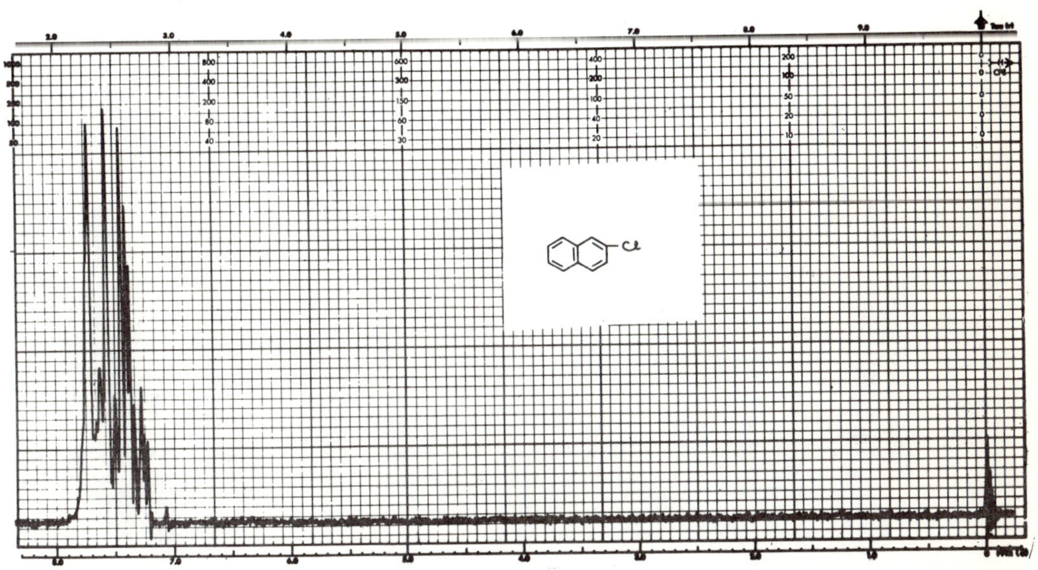

318

© 1978 Sadtler Research Laboratories, Inc.

NOTES

BROMINATED HYDROCARBONS

The Bromine nucleus is an intermediate deshielding group in relation to both aliphatic and aromatic hydrogens. It does not couple to nearby hydrogens and thus its presence in a molecule must be inferred from the observed chemical shifts.

Chemical Shifts

Aliphatic Protons

HC–4	HC–3	HC–2	HC–1		Solvent	Spectrum
		1.66	3.34	CH_3–CH_2–Br	CCl_4	319
	1.02	1.89	3.36	CH_3–CH_2–CH_2–Br	CCl_4	320
0.99	1.43	1.82	3.39	CH_3–CH_2–CH_2–CH_2–Br	CCl_4	322
	(1.08)	1.93	3.24	$(CH_3)_2$–CH–CH_2–Br	CCl_4	324
	(1.02)	----	3.15	$(CH_3)_3$–C–CH_2–Br	CCl_4	329
		(1.70)	4.21	$(CH_3)_2$–CH–Br	CCl_4	321
		(1.77)	----	$(CH_3)_3$–C–Br	CCl_4	325

Substituted Bromoethanes

Br–	CH_2	–X	
Br–	2.38	–Si$(CH_3)_3$	CCl_4
	3.21	–Cyclohexane	Polysol
	3.77	–C(=O)–O–CH_3	CCl_4
	3.83	–C≡C–H	CCl_4
	3.88	–CH=CH_2	CCl_4
	3.91	–C(=O)–NH_2	D_2O
	3.92	–C(=O)–OH	$CDCl_3$
	3.98	–C≡C–R	$CDCl_3$
	4.38	–C(=O)–Br	CCl_4
	4.34	–Ø	CCl_4
	4.40	–CCl_3	CCl_4
	4.94	–Br	CCl_4
	5.18	–Cl	CCl_4

2-Substituted Bromoethanes

Br–	CH_2–	CH_2–	X	Solvent
	3.40	3.65	–O–CH_3	CCl_4
	3.41	2.04	–$(CH_2)_2$–Br	CCl_4
	3.45	3.85	–OH	CCl_4
	3.51	2.36	–CH_2–Br	CCl_4
	3.53	2.97	–C≡N	CCl_4
	3.56	2.97	–C(=O)–OH	$CDCl_3$
	3.56	3.70	–Cl	CCl_4
	3.68	3.68	–Br	$CDCl_3$

Aromatic Protons

Bromine has a weakly deshielding effect upon the ortho aromatic hydrogens. As spectrum 381 shows, the ortho hydrogens are deshielded to 7.41 ppm while the meta and para hydrogens appear as a complex higher order pattern centered t about 7.2 ppm.

A series of para substituted bromobenzenes is provided below.

Br–	ortho	meta	para	
	7.18	7.50	–I	$CDCl_3$
	7.20	6.49	–NH–R	$CDCl_3$
	7.29	6.69	–O–CH_3	CCl_4
	7.29	7.29	–Br	$CDCl_3$
	7.30	6.92	–CH_3	CCl_4
	7.30	7.07	–SH	$CDCl_3$
	7.31	6.98	–R_2	$CDCl_3$
	7.34	7.12	–CH=CH_2	CCl_4
	7.37	7.13	–Cl	CCl_4
	7.39	7.11	–S–CH_3	$CDCl_3$
	7.40	6.90	–F	CCl_4
	7.45	7.62	–C≡N	$CDCl_3$
	7.49	7.60	–CF_3	CCl_4
	7.61	7.87	–C(=O)–NH_2	DMSO-d_6
	7.68	7.68	–C(=O)–H	$CDCl_3$
	7.68	7.79	–SO_2–O–Na	D_2O
	7.71	7.90	–C(=O)–OH	DMSO-d_6

BROMOETHANE

C_2H_5Br

Mol. Wt. 108.98

B.P. 38-40°C

Solvent: CCl_4

Assignments
a = 1.66
b = 3.34

319

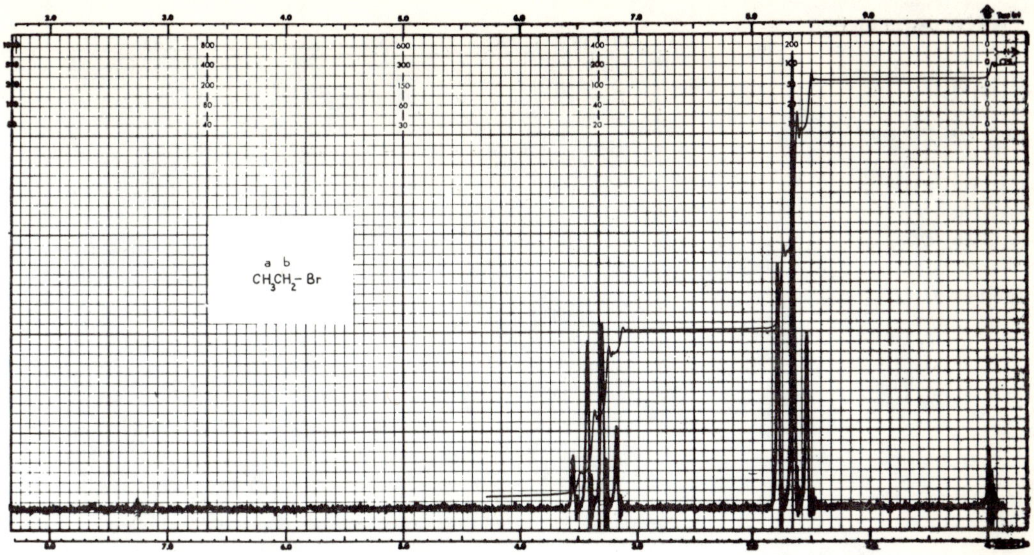

© 1978 Sadtler Research Laboratories, Inc.

1-BROMOPROPANE

C_3H_7Br

Mol. Wt. 123.00

B.P. 73-73.3°C

Solvent: CCl_4

Assignments
a = 1.02
b = 1.89
c = 3.36

320

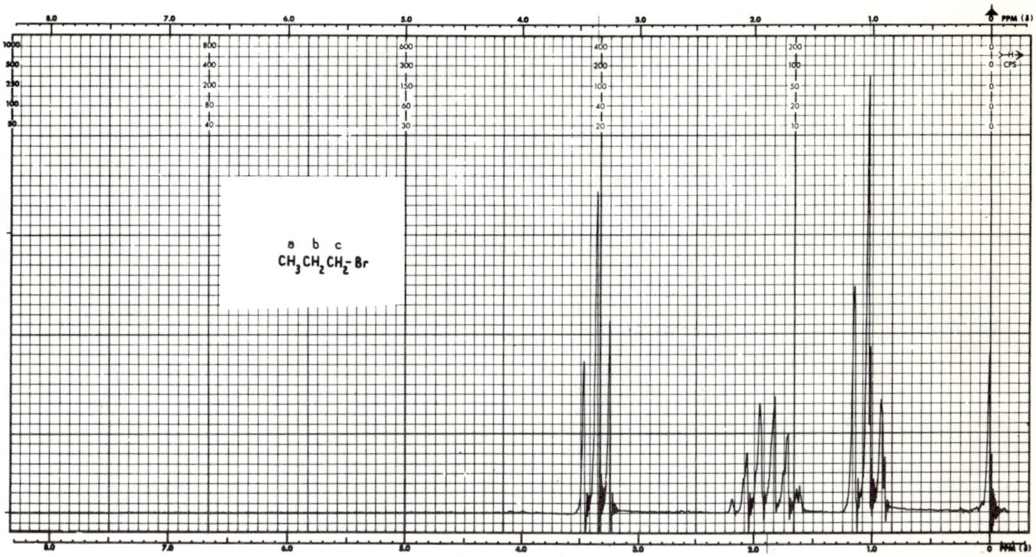

© 1978 Sadtler Research Laboratories, Inc.

2-BROMOPROPANE

C_3H_7Br

Mol. Wt. 123.00

Solvent: CCl_4

Assignments
a = 1.70
b = 4.21

321

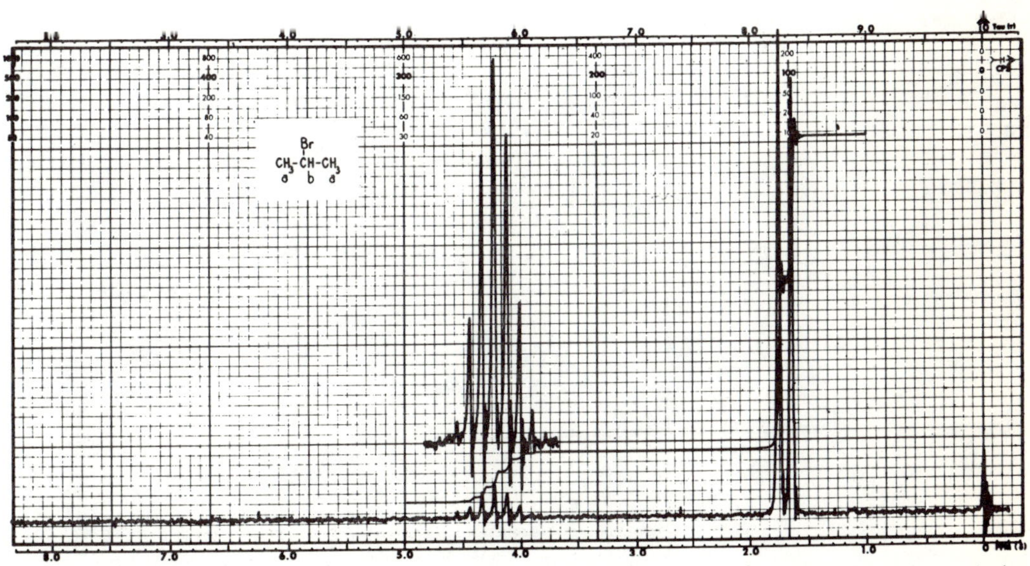

© 1978 Sadtler Research Laboratories, Inc.

1-BROMOBUTANE

C_4H_9Br

Mol. Wt. 137.02

B.P. 100-101°C

Solvent: CCl_4

Assignments
a = 0.99
b = ca 1.43
c = 1.82
d = 3.39

322

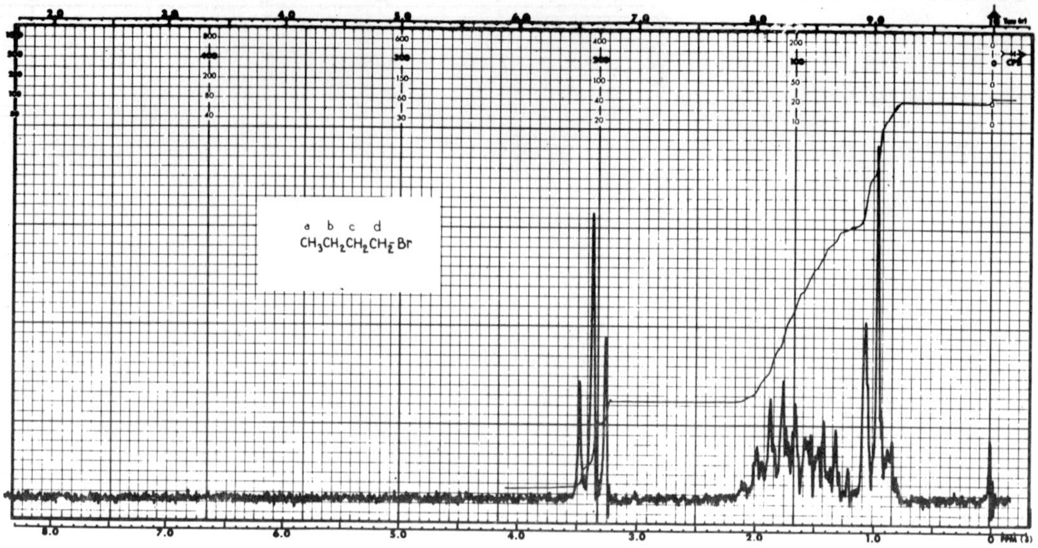

© 1978 Sadtler Research Laboratories, Inc.

2-BROMOBUTANE

C_4H_9Br

Mol. Wt. 137.03

B.P. 89-91°C

Solvent: $CDCl_3$

Assignments
a = 1.00
b = 1.70
c = 1.82
d = 4.10

323

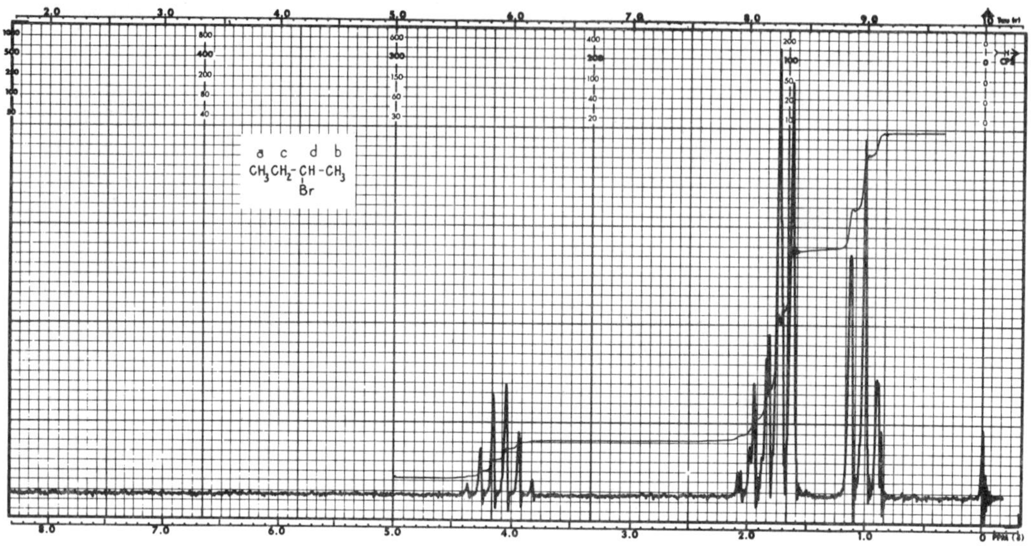

© 1978 Sadtler Research Laboratories, Inc.

1-BROMO-2-METHYLPROPANE

C_4H_9Br

Mol. Wt. 137.02

B.P. 90-92°C

Solvent: CCl_4

Assignments
a = 1.08
b = 1.93
c = 3.24

324

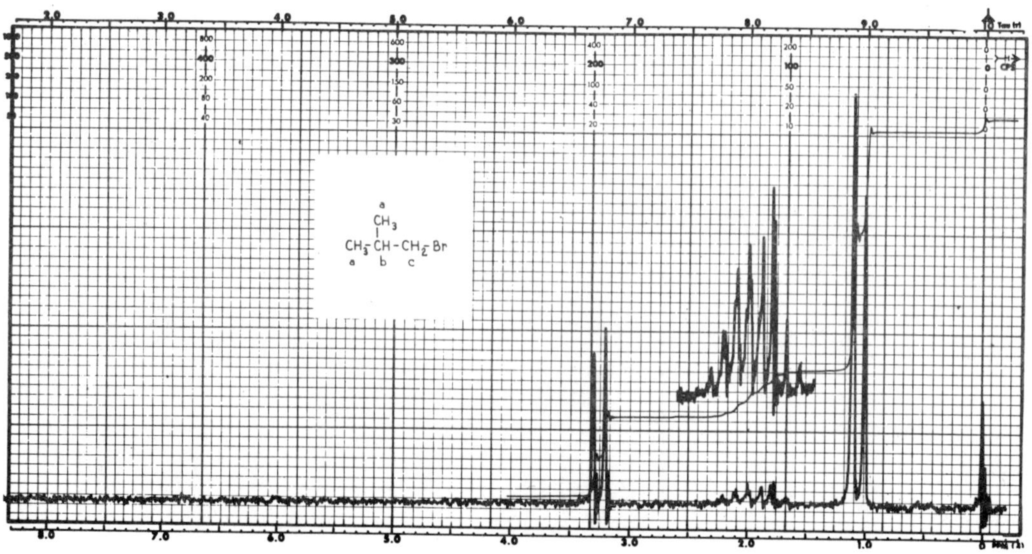

© 1978 Sadtler Research Laboratories, Inc.

2-BROMO-2-METHYLPROPANE

C_4H_9Br

Mol. Wt. 137.03

B.P. 72-74°C

Solvent: CCl_4

Assignments
a = 1.77

325

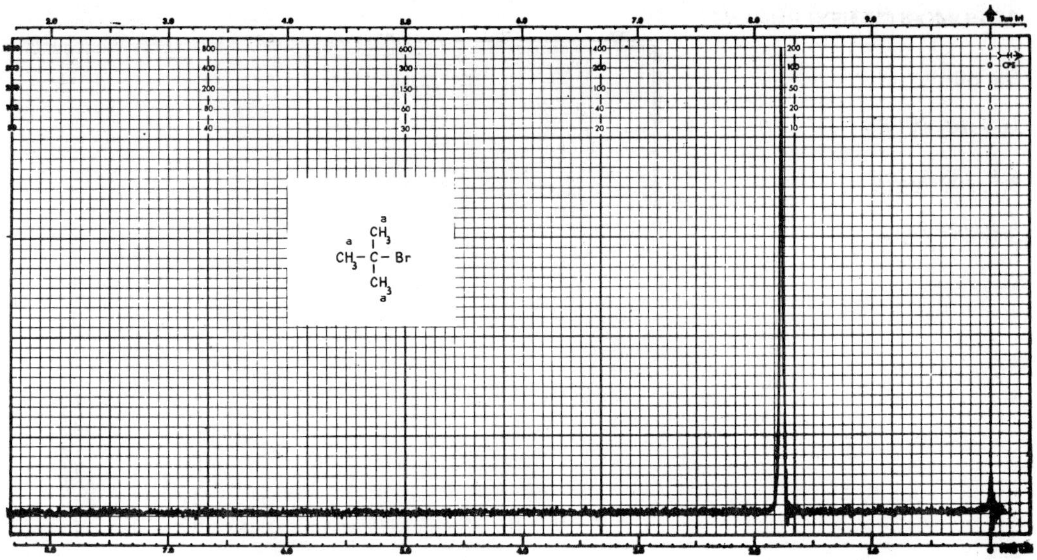

© 1978 Sadtler Research Laboratories, Inc.

1-BROMOPENTANE

$C_5H_{11}Br$

Mol. Wt. 151.05

B.P. 127.5-130°C

Solvent: CCl_4

Assignments
a = 0.92
b = 1.16-1.70
c = 1.85
d = 3.33

326

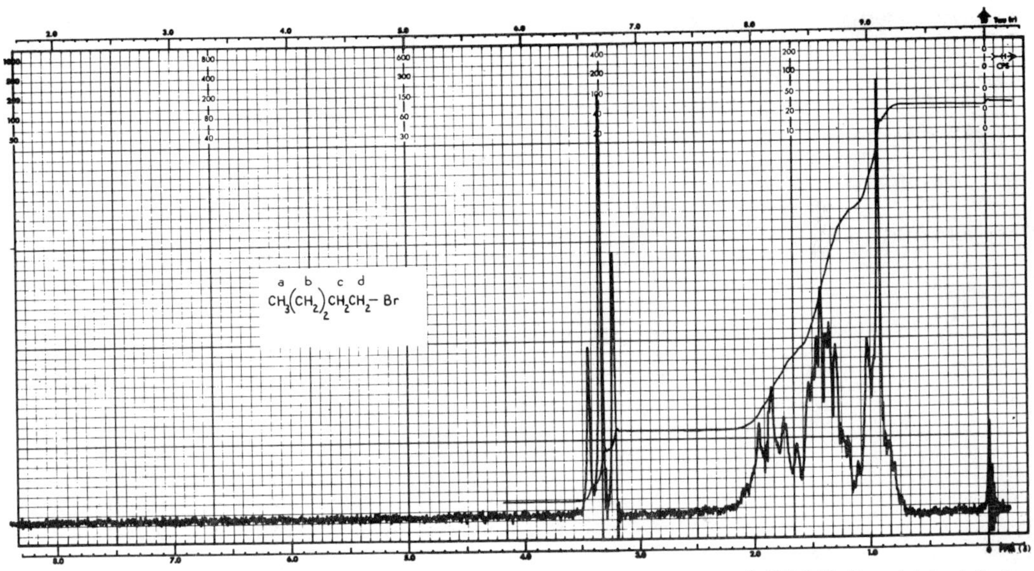

© 1978 Sadtler Research Laboratories, Inc.

2-BROMOPENTANE

$C_5H_{11}Br$

Mol. Wt. 151.05

B.P. 113°C/760mm

Solvent: CCl_4

Assignments
a = 0.95
b = 1.10-2.10
c = 1.71
d = 4.04

327

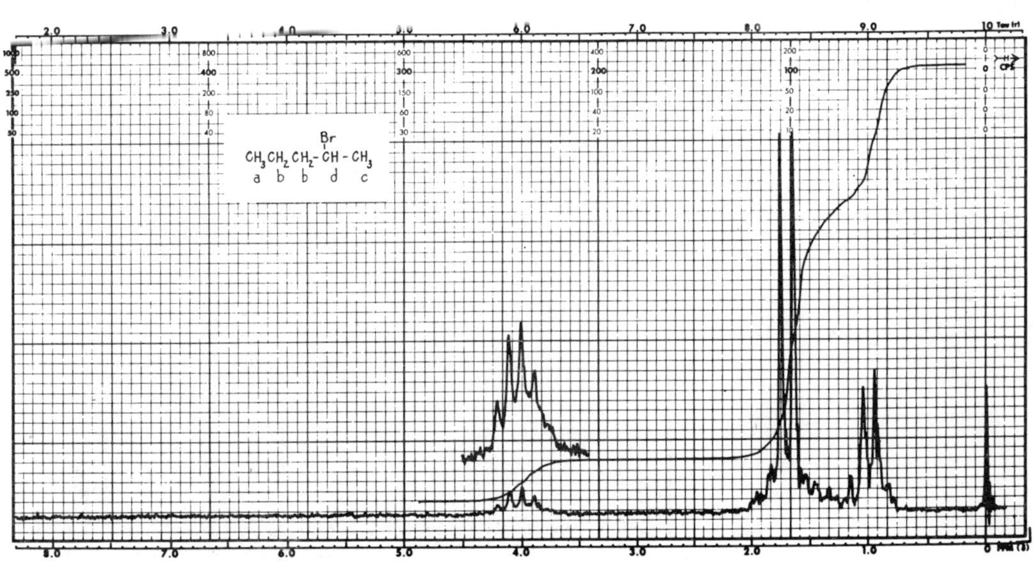

© 1978 Sadtler Research Laboratories, Inc.

1-BROMO-3-METHYLBUTANE

$C_5H_{11}Br$

Mol. Wt. 151.05

B.P. 118-120°C

Solvent: CCl_4

Assignments
a = 0.94
b = 1.76
c = 3.33

328

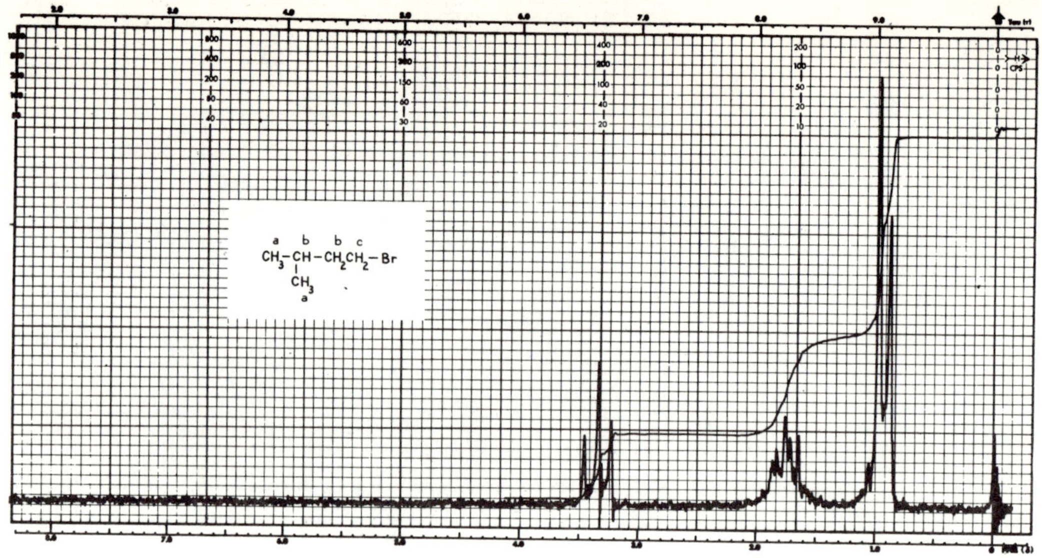

© 1978 Sadtler Research Laboratories, Inc.

1-BROMO-2,2-DIMETHYLPROPANE

$C_5H_{11}Br$

Mol. Wt. 151.05

B.P. 104-106°C (lit.)

Solvent: CCl_4

Assignments
a = 1.02
b = 3.15

329

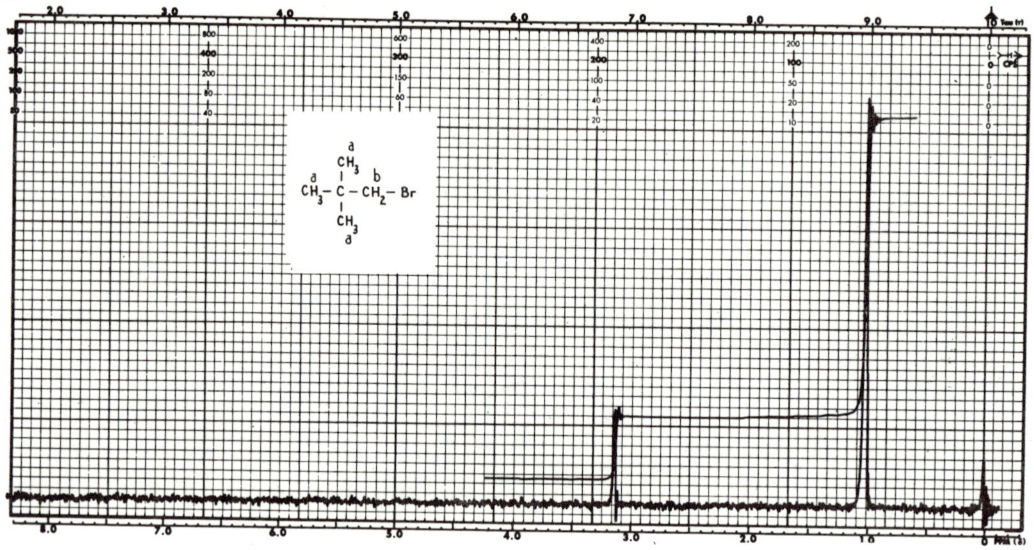

© 1978 Sadtler Research Laboratories, Inc.

1-BROMOHEXANE

$C_6H_{13}Br$

Mol. Wt. 165.07

B.P. 75-78°C/65mm

Solvent: CCl_4

Assignments
a = 0.93
b = 1.10-1.80
c = 1.88
d = 3.38

330

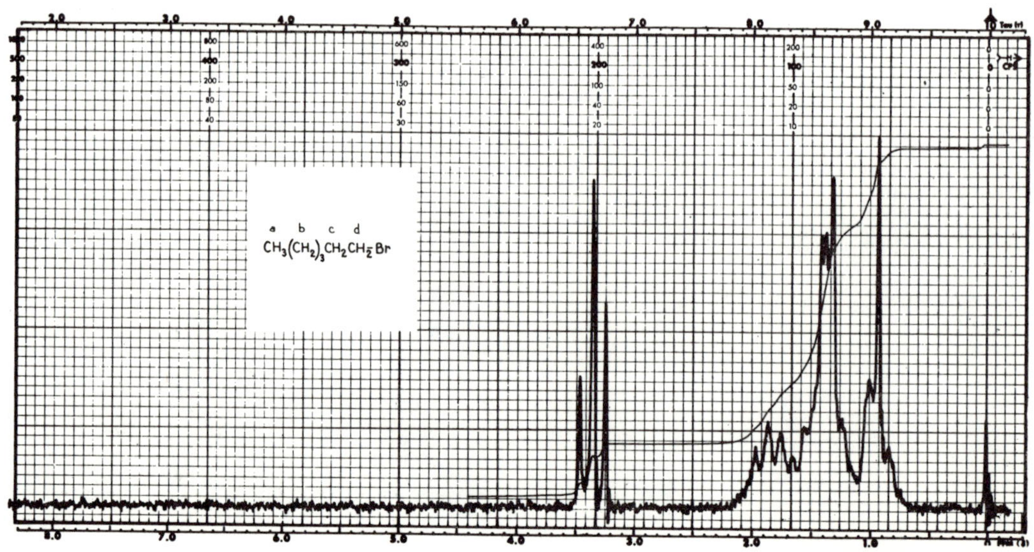

© 1978 Sadtler Research Laboratories, Inc.

3-BROMOHEXANE

$C_6H_{13}Br$

Mol. Wt. 165.08

Solvent: CCl_4

Assignments
a = 0.94
b = 1.07
c = ca 1.60
d = ca 1.80
e = 3.91

331

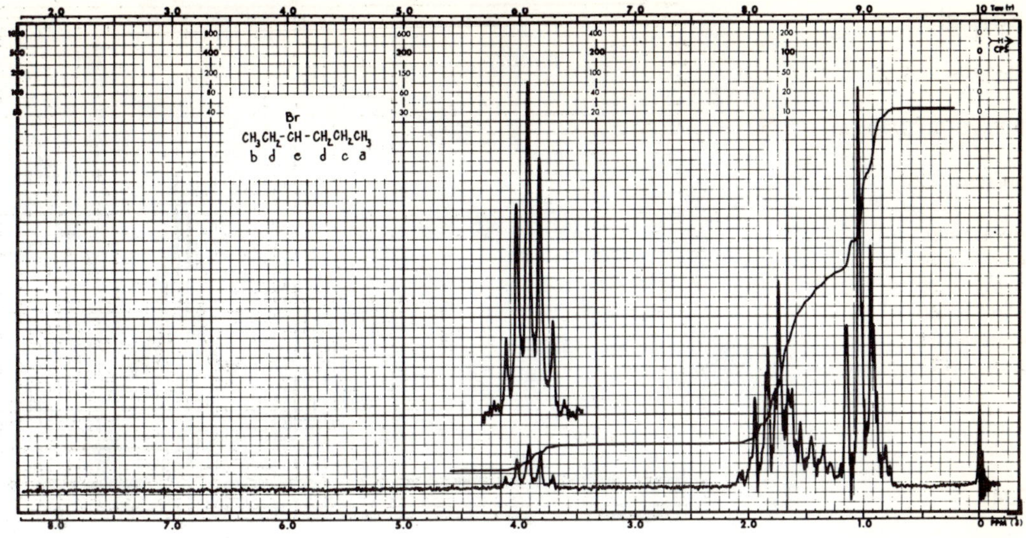

© 1978 Sadtler Research Laboratories, Inc.

1-BROMOHEPTANE

$C_7H_{15}Br$

Mol. Wt. 179.11

B.P. 176-178°C

Solvent: CCl_4

Assignments
a = 0.90
b = 1.08-1.60
c = 1.85
d = 3.33

332

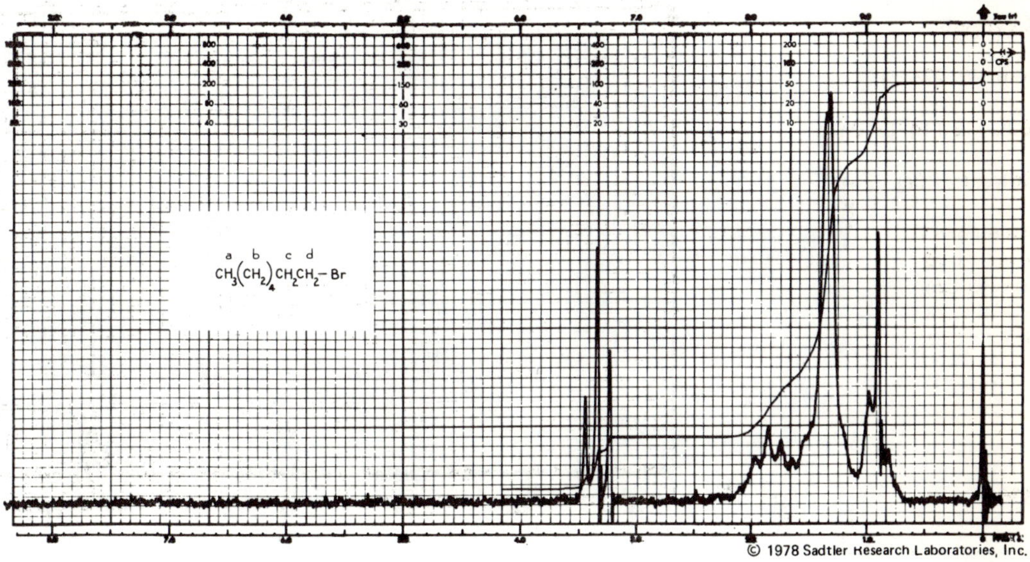

© 1978 Sadtler Research Laboratories, Inc.

1-BROMOOCTANE

$C_8H_{17}Br$

Mol. Wt. 193.13

B.P. 90-92°C/20mm

Solvent: CCl_4

Assignments
a = 0.91
b = 1.10-1.60
c = 1.87
d = 3.35

333

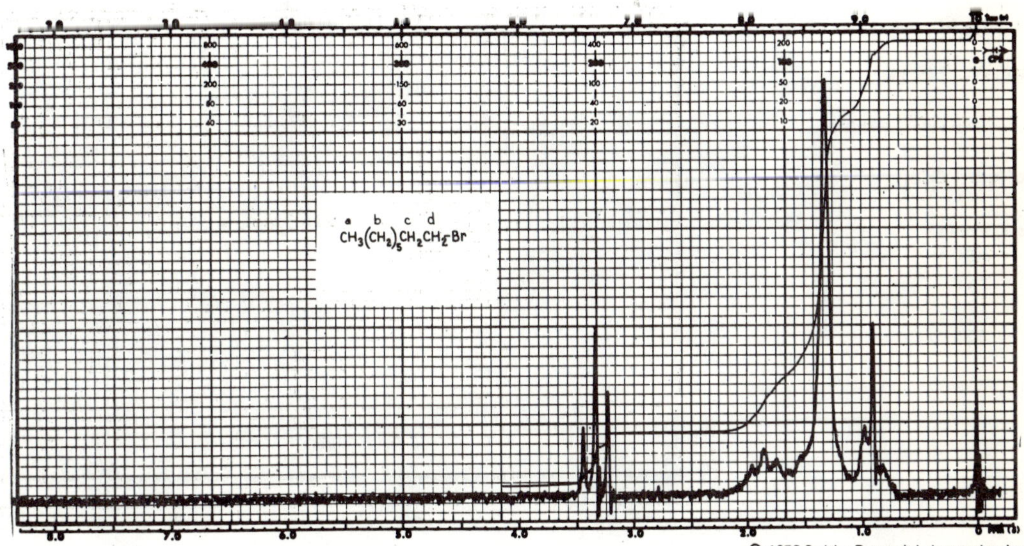

© 1978 Sadtler Research Laboratories, Inc.

2-BROMOOCTANE

$C_8H_{17}Br$

Mol. Wt. 193.13

B.P. 76-78°C/10mm

Solvent: CCl_4

Assignments
a = 0.90
b = 1.30
c = 1.69
d = ca 1.81
e = 3.99

334

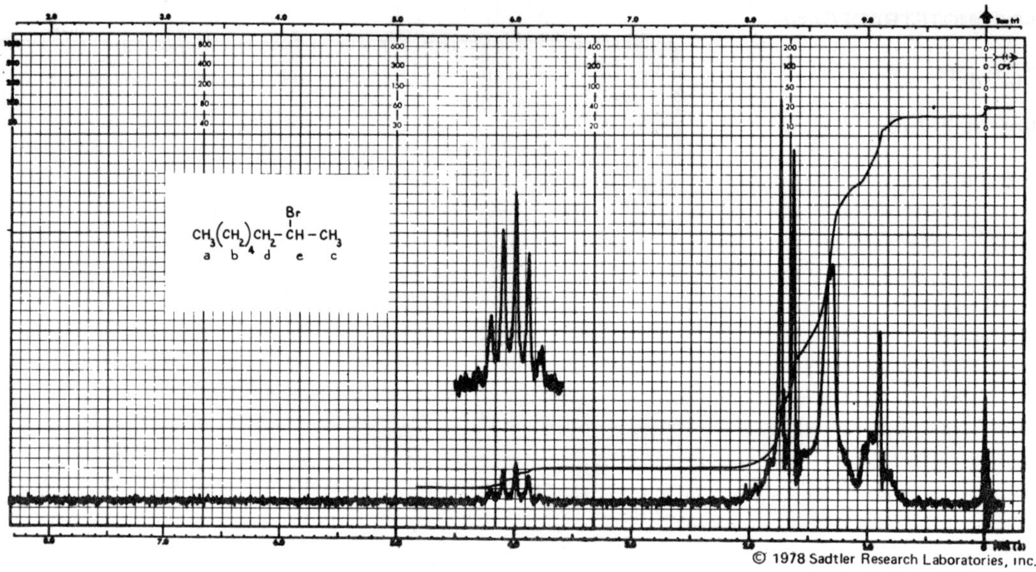

© 1978 Sadtler Research Laboratories, Inc.

1-BROMOUNDECANE

$C_{11}H_{23}Br$

Mol. Wt. 235.21

B.P. 112-114°C/10mm

Solvent: CCl_4

Assignments
a = 0.90
b = ca 1.29
c = 1.86
d = 3.32

335

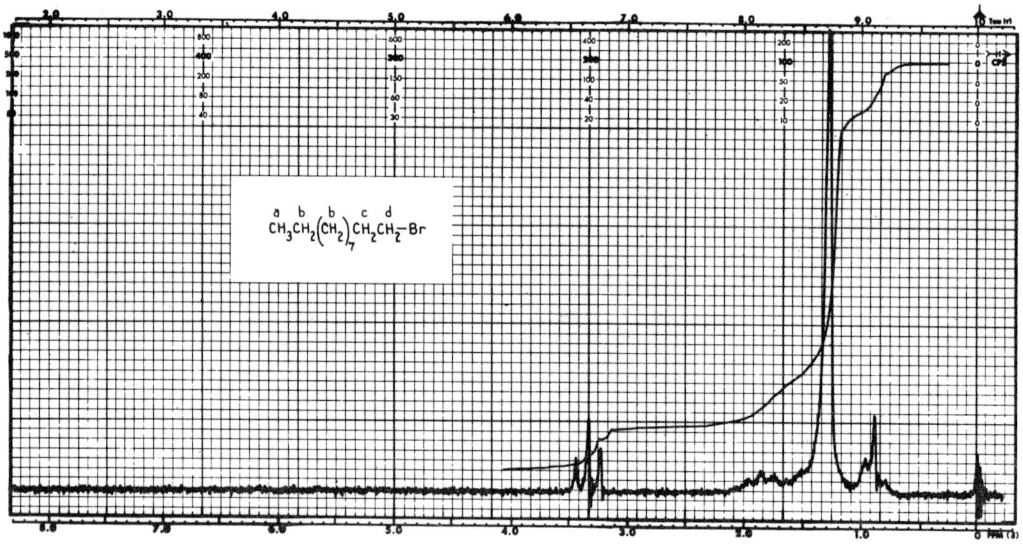

© 1978 Sadtler Research Laboratories, Inc.

1-BROMODODECANE

$C_{12}H_{25}Br$

Mol. Wt. 249.24

B.P. 175-180°C/45mm

Solvent: CCl_4

Assignments
a = 0.90
b = 1.29
c = 1.87
d = 3.34

336

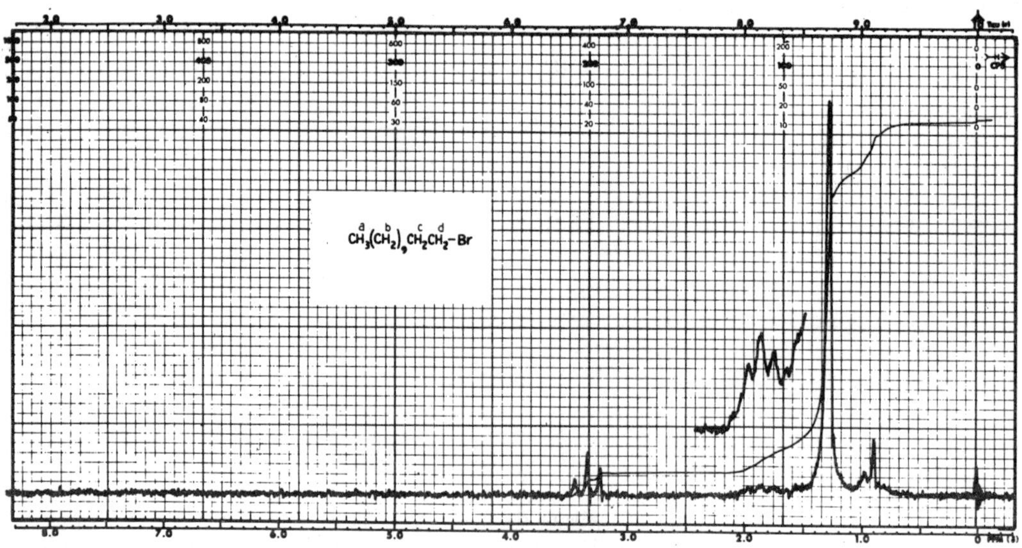

© 1978 Sadtler Research Laboratories, Inc.

1-BROMOTETRADECANE

$C_{14}H_{29}Br$

Mol. Wt. 277.29

M.P. 2-4°C

Solvent: CCl_4

Assignments
a = 0.91
b = ca 1.29
c = 1.87
d = 3.34

337

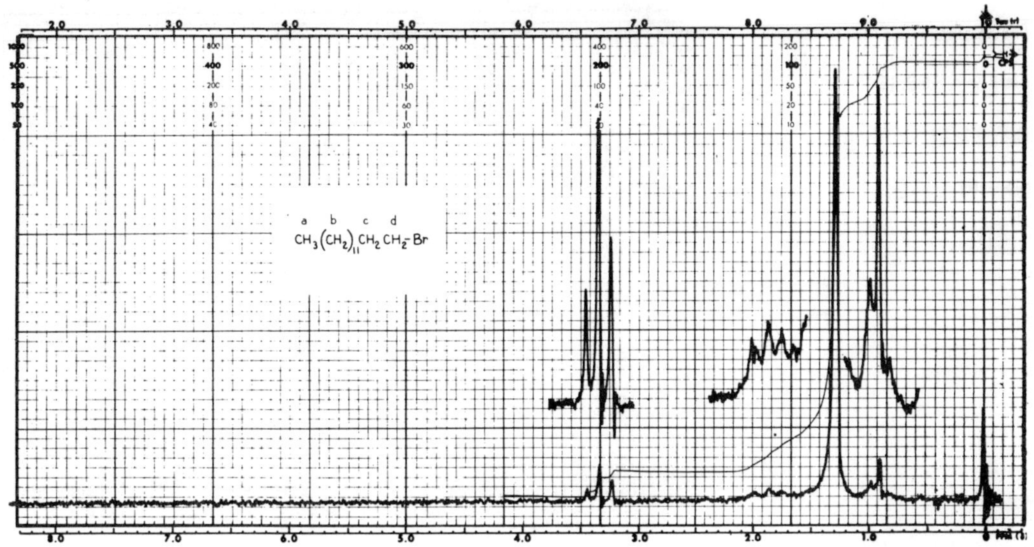

© 1978 Sadtler Research Laboratories, Inc.

1-BROMOPENTADECANE

$C_{15}H_{31}Br$

Mol. Wt. 291.32

M.P. 16.5-18.5°C

Solvent: CCl_4

Assignments
a = 0.89
b = ca 1.25
d = 1.84
e = 3.31

338

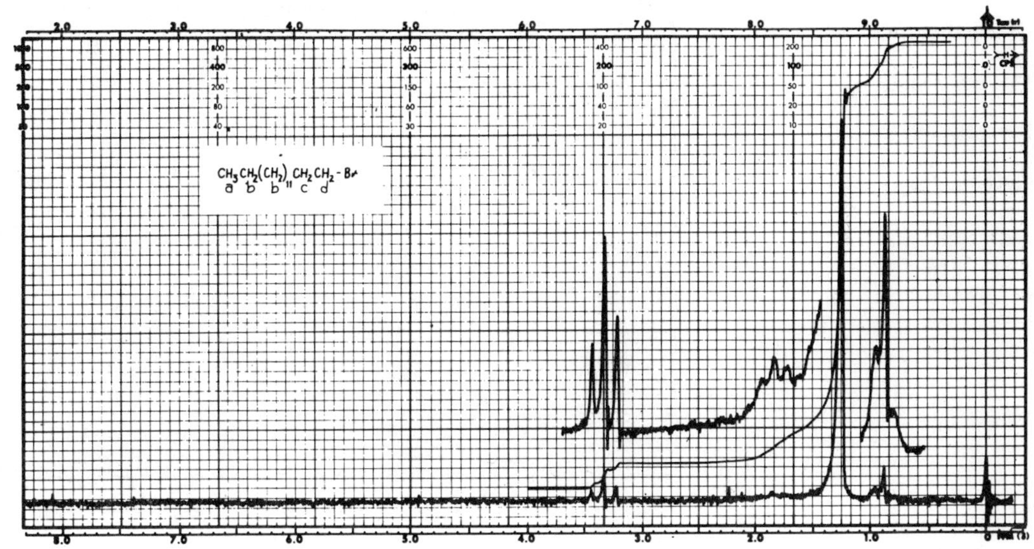

© 1978 Sadtler Research Laboratories, Inc.

1-BROMOHEXADECANE

$C_{16}H_{33}Br$

Mol. Wt. 305.35

M.P. 15°C

Solvent: CCl_4

Assignments
a = 0.89
b = ca 1.29
c = 1.86
d = 3.35

339

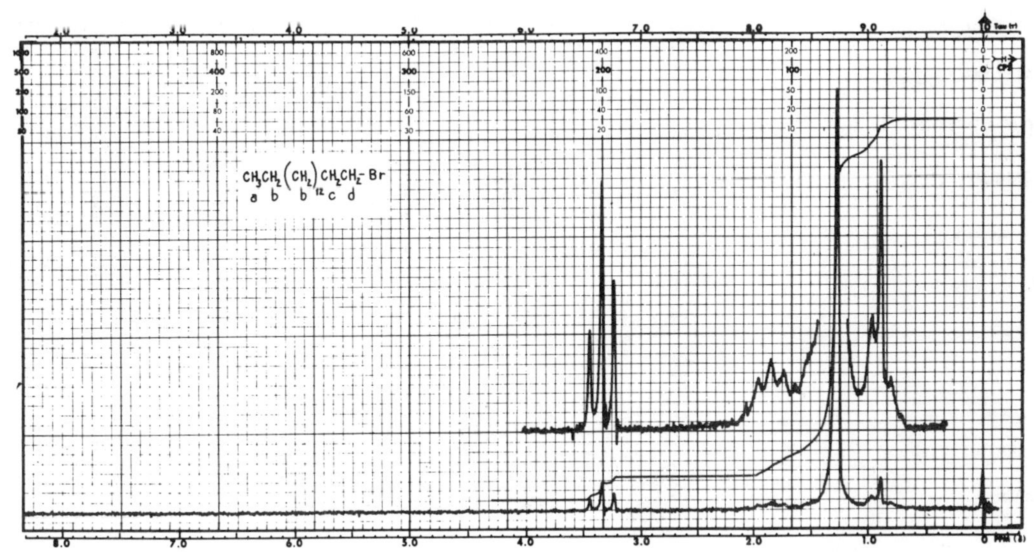

© 1978 Sadtler Research Laboratories, Inc.

1-BROMOHEPTADECANE

$C_{17}H_{35}Br$

Mol. Wt. 319.38

M.P. 32°C

Solvent: $CDCl_3$

Assignments

a = 0.90

b = ca 1.29

c = 1.88

d = 3.41

340

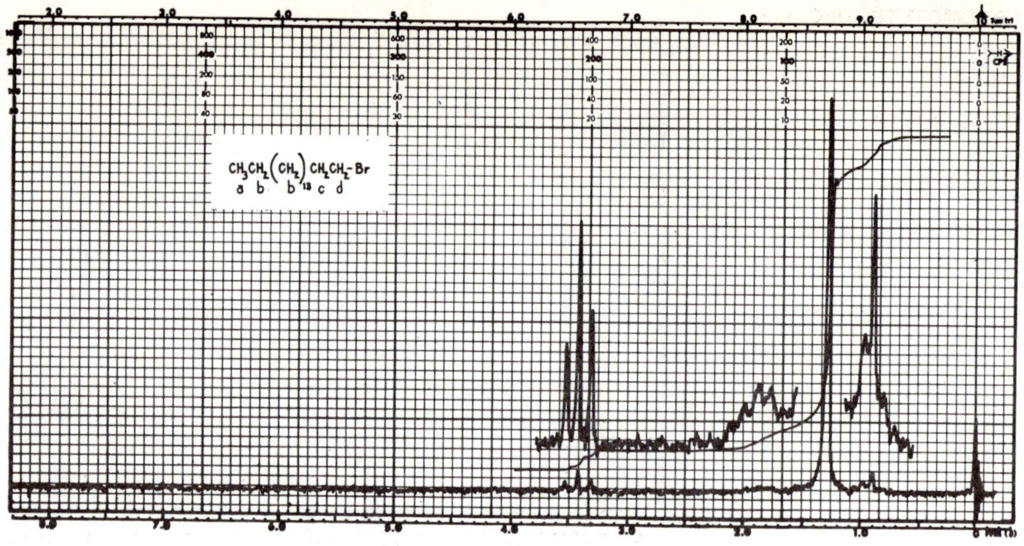

© 1978 Sadtler Research Laboratories, Inc.

1-BROMONONADECANE

$C_{19}H_{39}Br$

Mol. Wt. 347.43

M.P. 36-37°C

Solvent: $CDCl_3$

Assignments

a = 0.89

b = ca 1.29

c = 1.80

d = 3.39

341

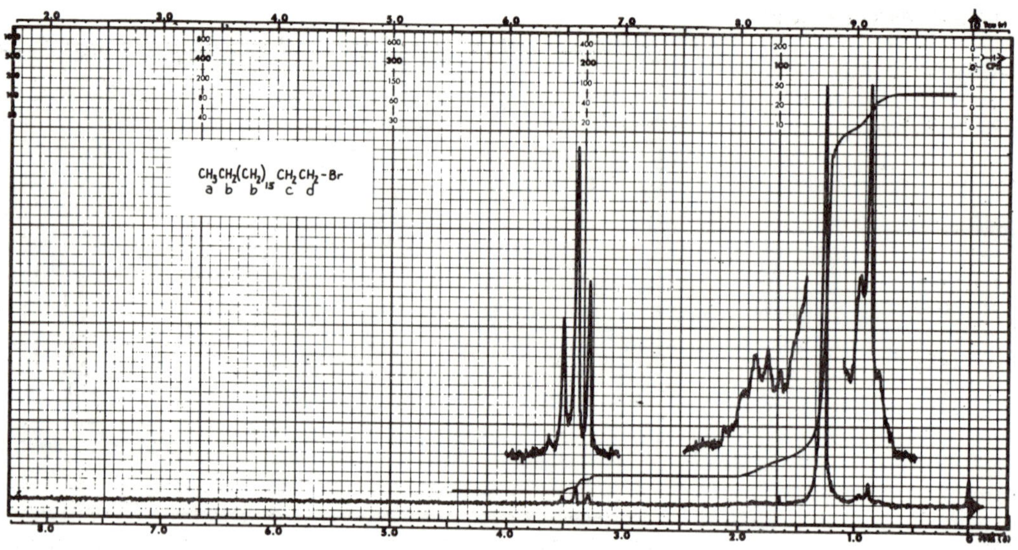

© 1978 Sadtler Research Laboratories, Inc.

BROMODICHLOROMETHANE

$CHBrCl_2$

Mol. Wt. 163.83

B.P. 91°C

Solvent: CCl_4

Assignments

a = 7.20

b = 4.92 imp.

342

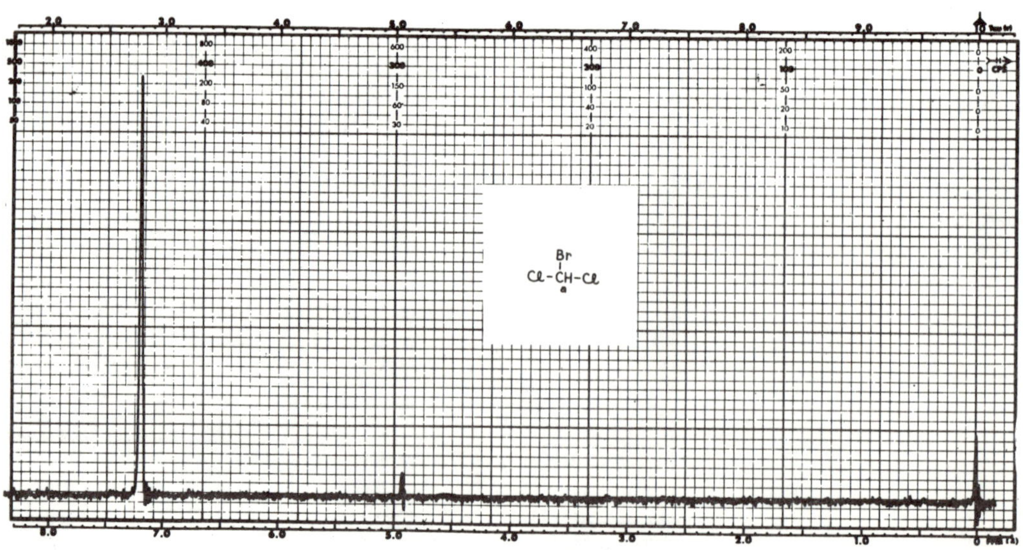

© 1978 Sadtler Research Laboratories, Inc.

1-BROMO-2-CHLOROETHANE

C_2H_4BrCl

Mol. Wt. 143.42

Solvent: CCl_4

Assignments
a = 3.56
b = 3.70

343

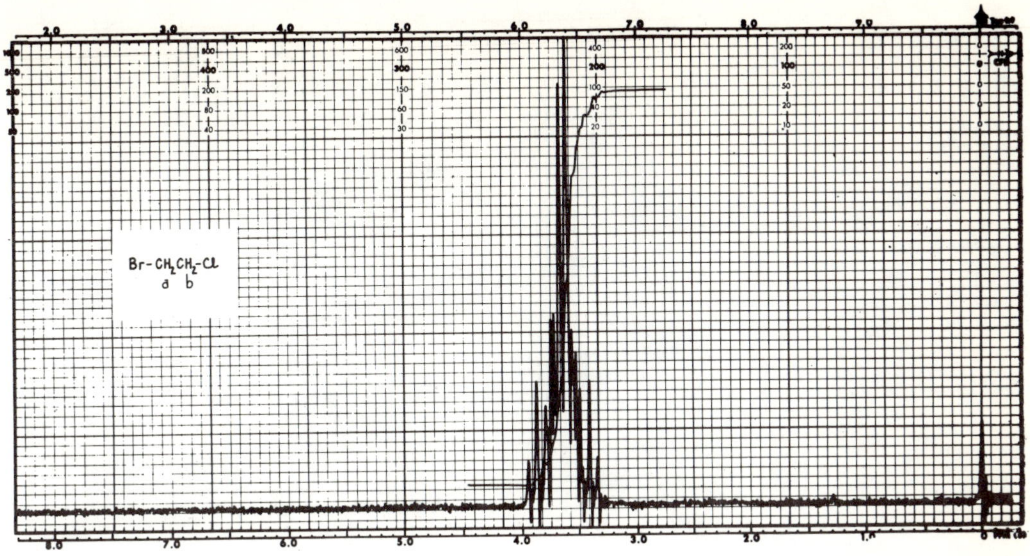

© 1978 Sadtler Research Laboratories, Inc.

1-BROMO-3-CHLORO-2-METHYLPROPANE

C_4H_8BrCl

Mol. Wt. 171.47

B.P. 154-156°C

Solvent: $CDCl_3$

Assignments
a = 1.17
b = 2.21
c = 3.55

344

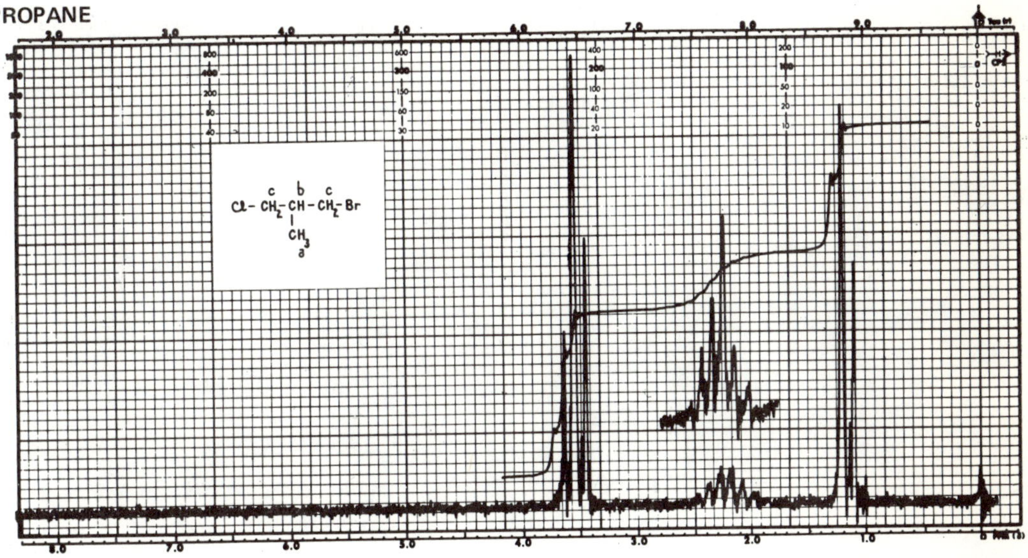

© 1978 Sadtler Research Laboratories, Inc.

1-BROMO-2-CHLORO-1,1,2-TRIFLUOROETHANE

$C_2HBrClF_3$

Mol. Wt. 197.39

Solvent: CCl_4

Assignments
a = 6.10

345

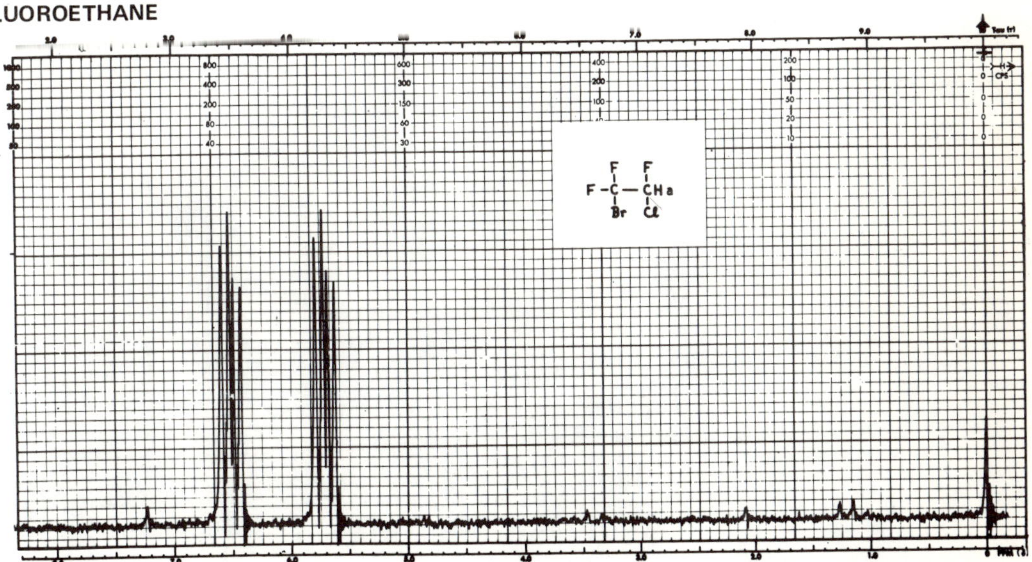

© 1978 Sadtler Research Laboratories, Inc.

DIBROMOMETHANE

CH_2Br_2

Mol. Wt. 173.84

B.P. 96-98°C

Solvent: CCl_4

Assignments
a = 4.94

346

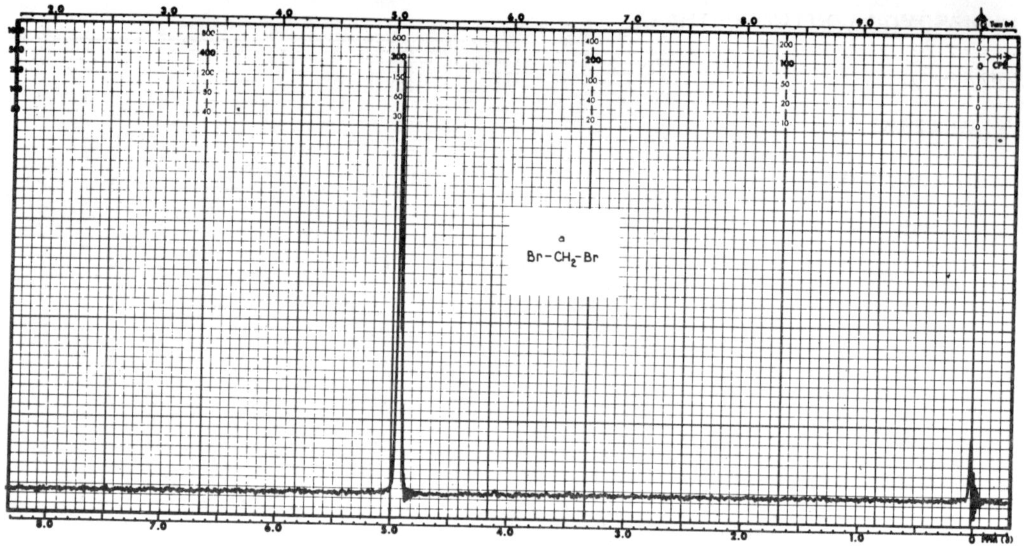

© 1978 Sadtler Research Laboratories, Inc.

TRIBROMOMETHANE

$CHBr_3$

Mol. Wt. 252.73

M.P. 6-7°C

B.P. 149.5°C/760mm

Solvent: CCl_4

Assignments
a = 6.80

347

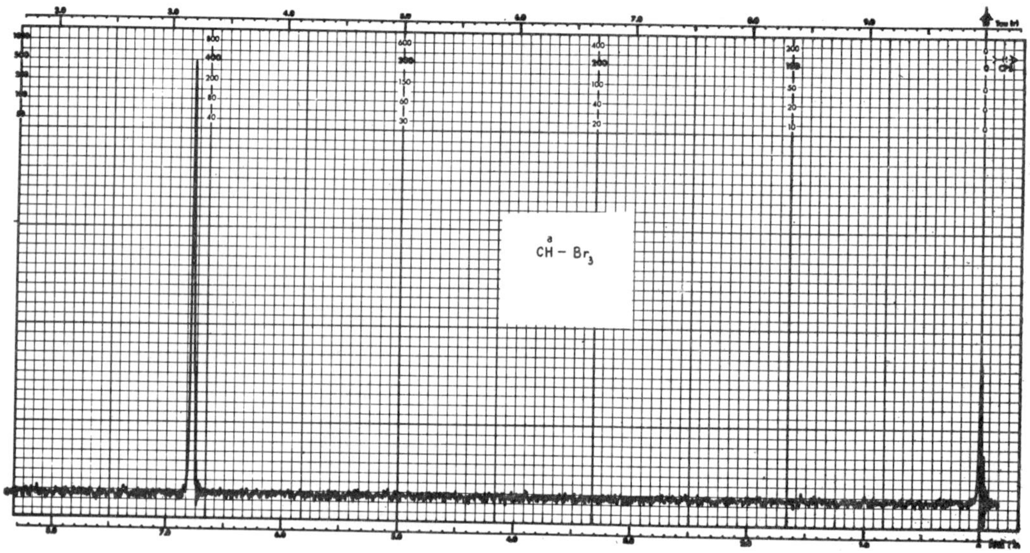

© 1978 Sadtler Research Laboratories, Inc.

1,2-DIBROMOETHANE

$C_2H_4Br_2$

Mol. Wt. 187.86

B.P. 130.5°C

Solvent: $CDCl_3$

Assignments
a = 3.68

348

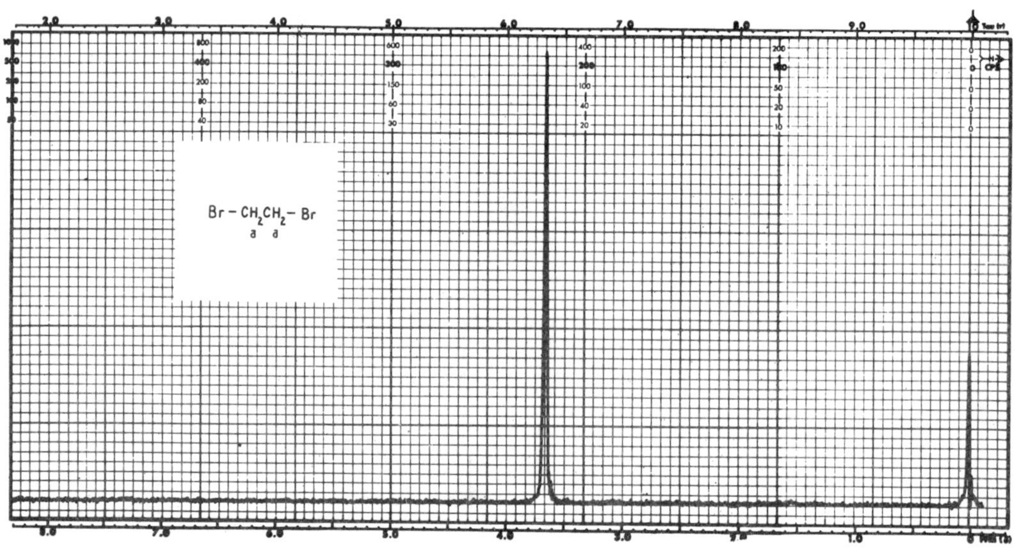

© 1978 Sadtler Research Laboratories, Inc.

1,2-DIBROMO-1,1-DIFLUOROETHANE

$C_2H_2Br_2F_2$

Mol. Wt. 223.85

B.P. 92.7-93.2°C (lit.)

Solvent: CCl_4

Assignments
a = 4.02

349

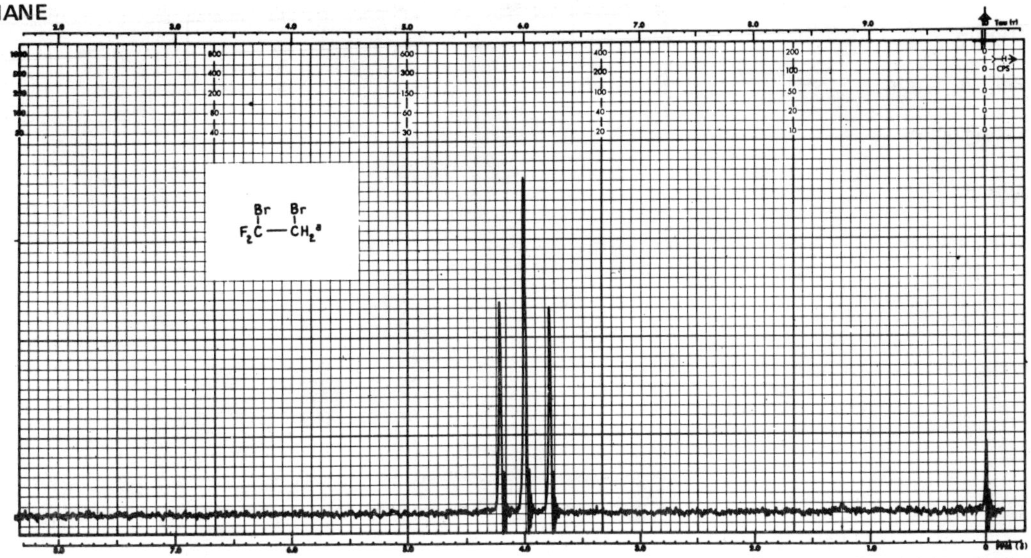

© 1978 Sadtler Research Laboratories, Inc.

1,2-DIBROMO-1,1-DICHLOROETHANE

$C_2H_2Br_2Cl_2$

Mol. Wt. 256.75

B.P. 173-176°C/760mm

Solvent: CCl_4

Assignments
a = 4.40

350

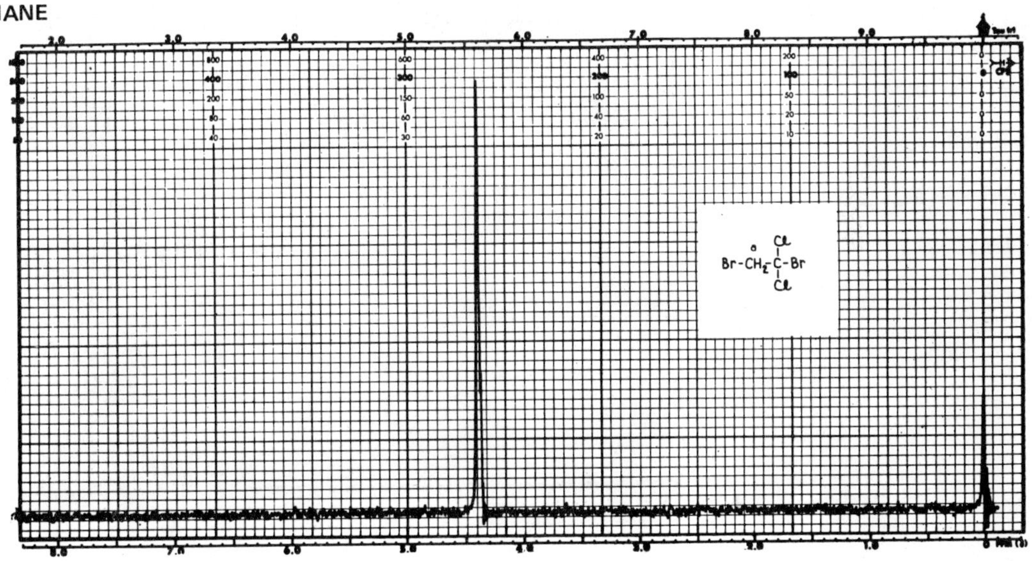

© 1978 Sadtler Research Laboratories, Inc.

1,1,2-TRIBROMOETHANE

$C_2H_3Br_3$

Mol. Wt. 266.77

B.P. 66-68°C/10mm

Solvent: CCl_4

Assignments
a = 4.09
b = 5.69

351

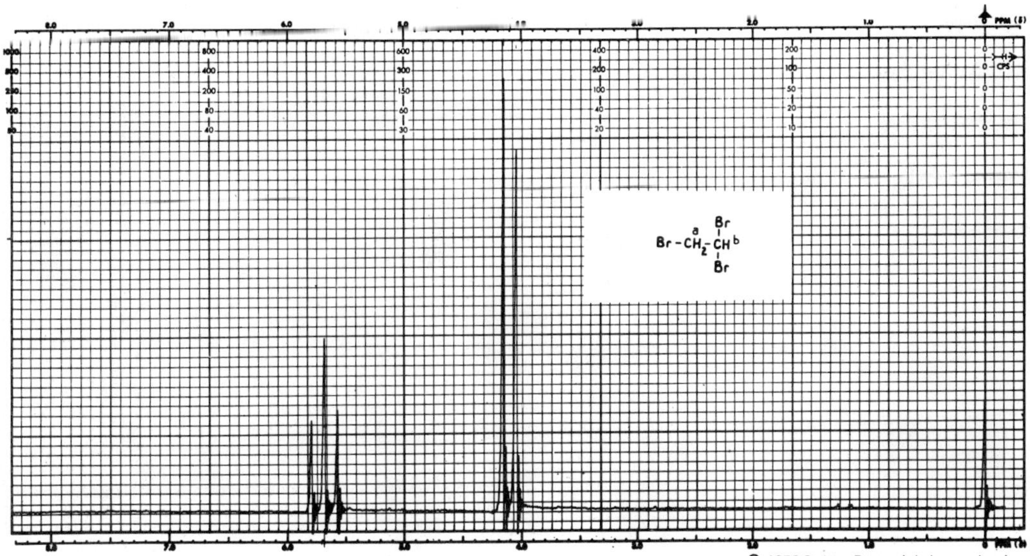

© 1978 Sadtler Research Laboratories, Inc.

1,1,2,2-TETRABROMOETHANE

$C_2H_2Br_4$

Mol. Wt. 345.65

M.P. 0-2°C

Solvent: CCl_4

Assignments
a = 6.04

352

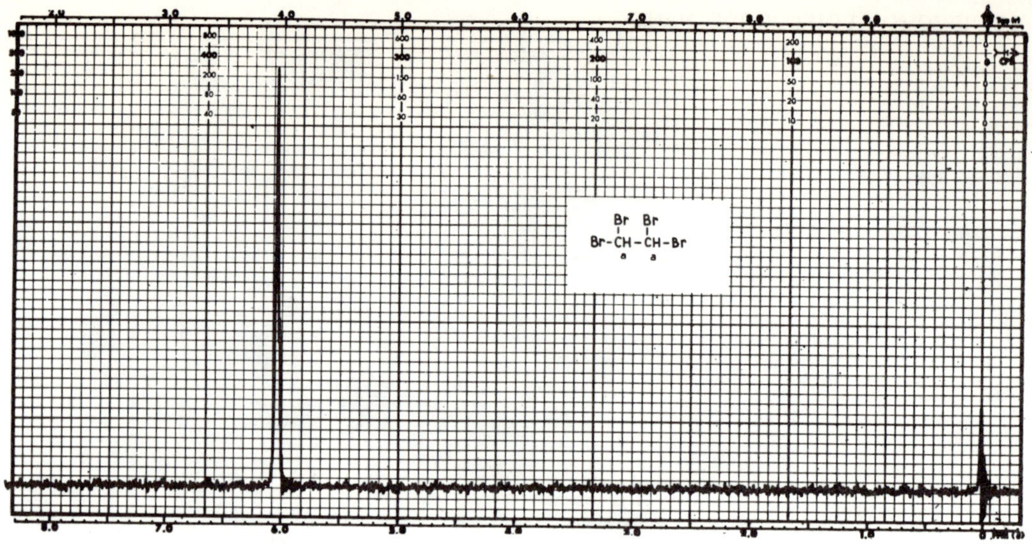

© 1978 Sadtler Research Laboratories, Inc.

1,2-DIBROMOPROPANE

$C_3H_6Br_2$

Mol. Wt. 201.91

B.P. 138-140°C

Solvent: CCl_4

Assignments
a = 1.84
b = 3.48 or c
c = 3.80 or b
d = ca 4.15

353

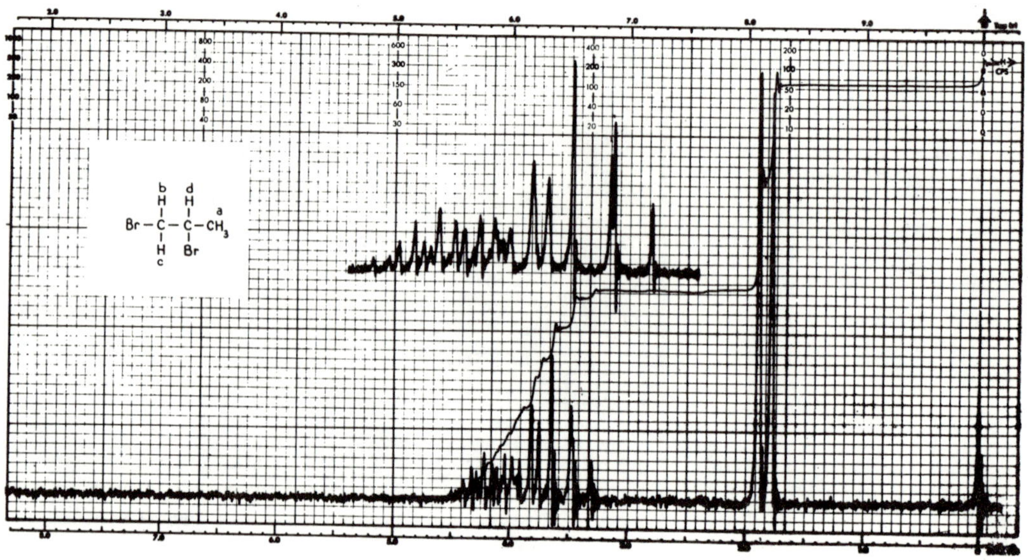

© 1978 Sadtler Research Laboratories, Inc.

1,3-DIBROMOPROPANE

$C_3H_6Br_2$

Mol. Wt. 201.91

B.P. 165-167°C

Solvent: CCl_4

Assignments
a = 2.36
b = 3.51

354

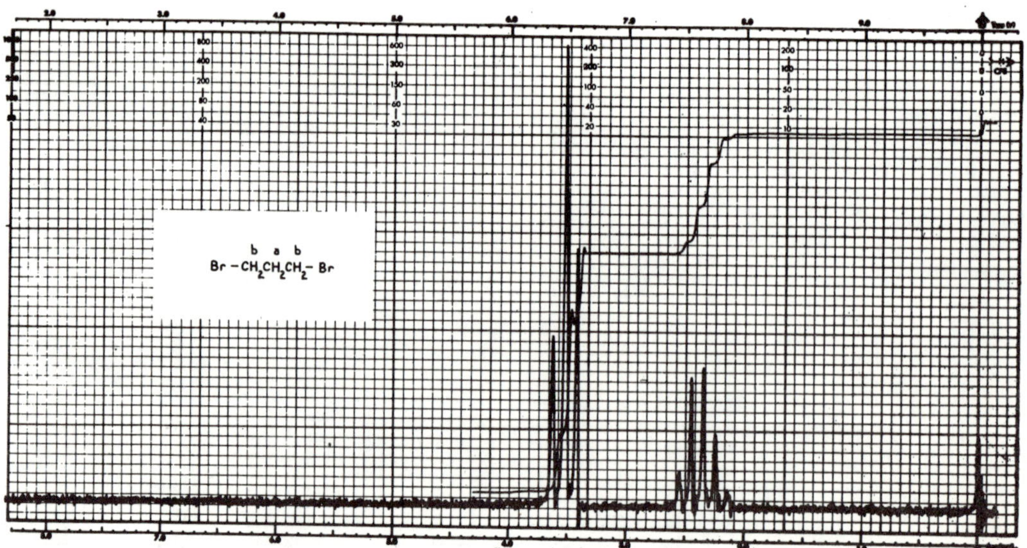

© 1978 Sadtler Research Laboratories, Inc.

1,2-DIBROMO-3-CHLOROPROPANE

$C_3H_5Br_2Cl$

Mol. Wt. 236.33

Solvent: CCl_4

Assignments
a = 3.82
b = 3.96
c = 4.32

355

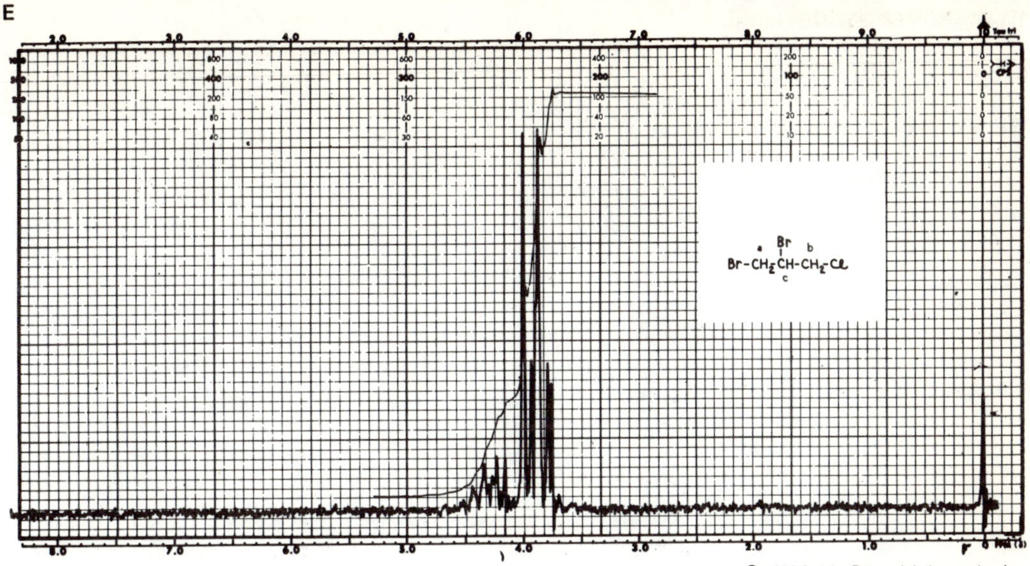

© 1978 Sadtler Research Laboratories, Inc.

1,2,3-TRIBROMOPROPANE

$C_3H_5Br_3$

Mol. Wt. 280.82

B.P. 219-221°C

Solvent: CCl_4

Assignments
a = 3.80 or b
b = 3.90 or a
c = 4.30

356

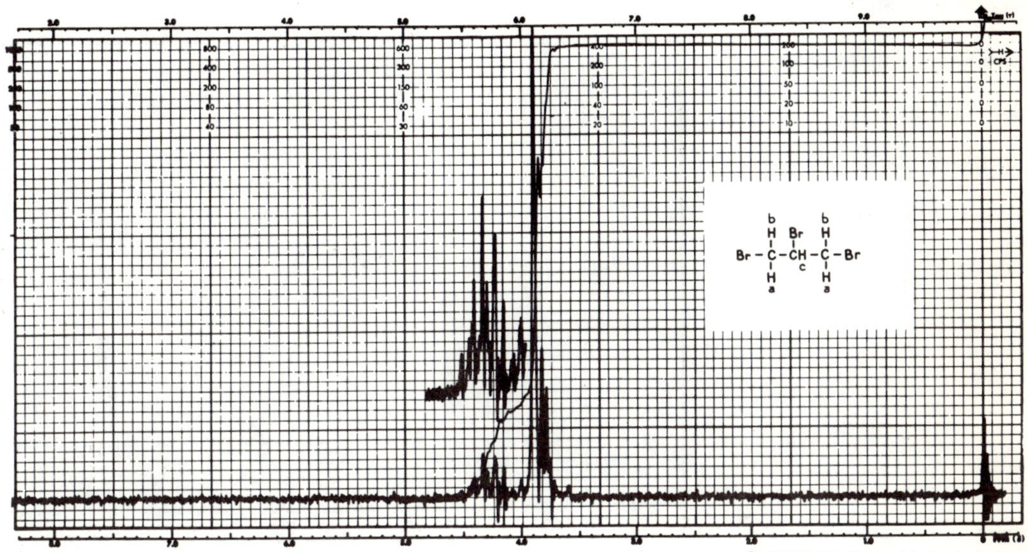

© 1978 Sadtler Research Laboratories, Inc.

1,2-DIBROMOBUTANE

$C_4H_8Br_2$

Mol. Wt. 215.93

B.P. 166.3°C/760mm

Solvent: $CDCl_3$

Assignments
a = 1.08
b = 1.40-2.50
c = 3.60
d = 3.81
e = 4.10

357

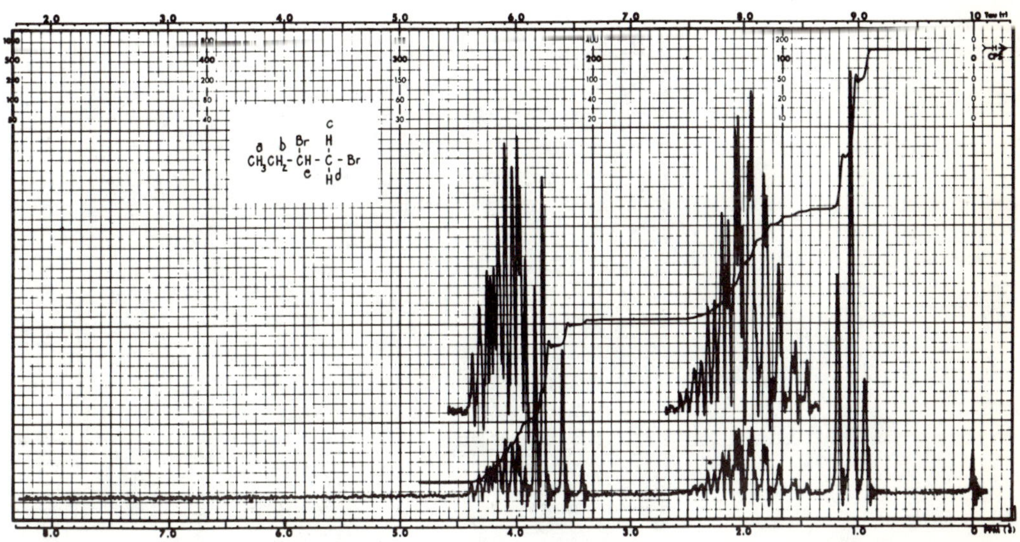

© 1978 Sadtler Research Laboratories, Inc.

1,3-DIBROMOBUTANE

$C_4H_8Br_2$

Mol. Wt. 215.92

B.P. 70-71°C/20mm

Solvent: CCl_4

Assignments
a = 1.78
b = 2.28
c = 3.52
d = 4.20

358

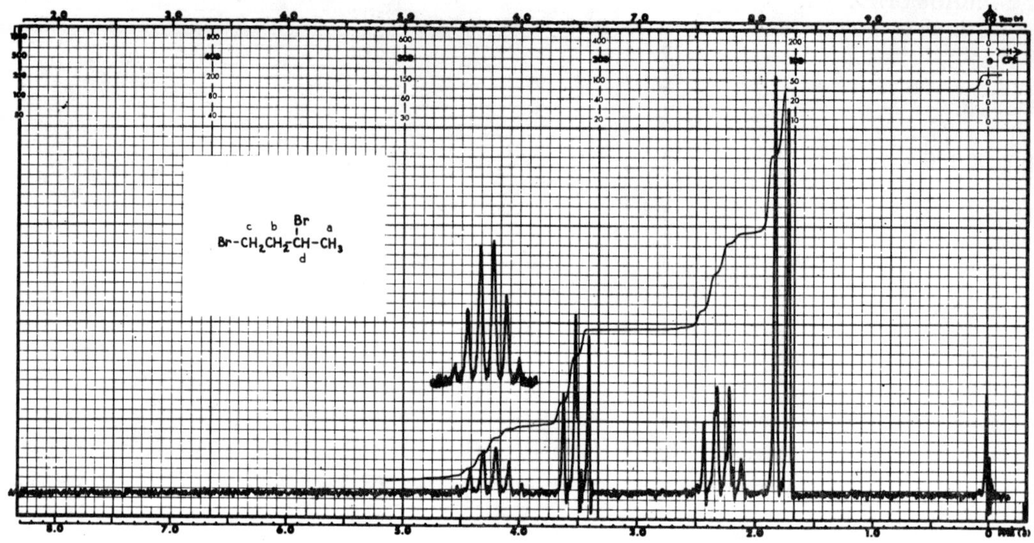

© 1978 Sadtler Research Laboratories, Inc.

1,4-DIBROMOBUTANE

$C_4H_8Br_2$

Mol. Wt. 215.92

B.P. 63-65°C/6mm

Solvent: CCl_4

Assignments
a = 2.04
b = 3.41

359

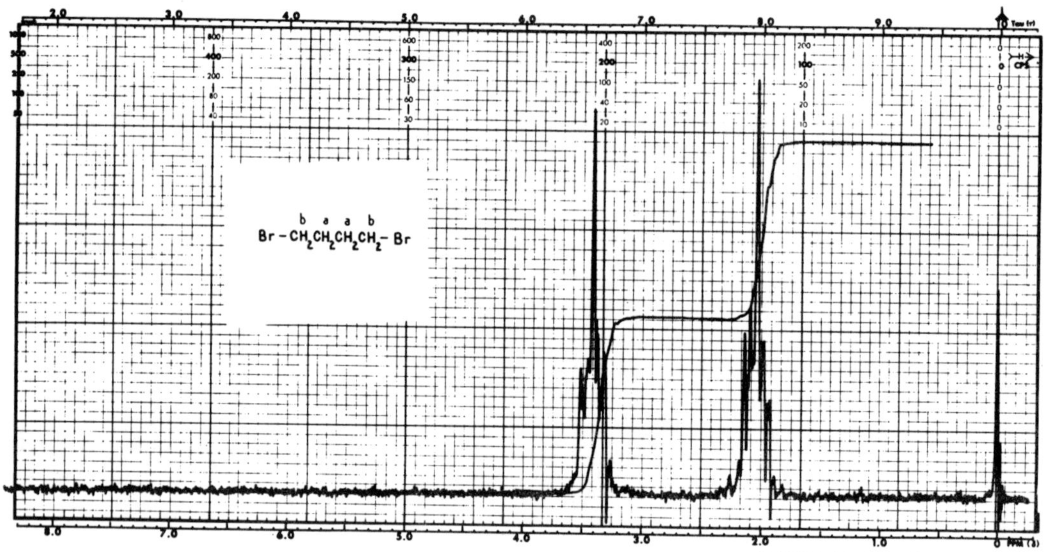

© 1978 Sadtler Research Laboratories, Inc.

1,5-DIBROMOPENTANE

$C_5H_{10}Br_2$

Mol. Wt. 229.94

B.P. 79-80°C/5mm

Solvent: CCl_4

Assignments
a = 1.32-2.20
b = 3.38

360

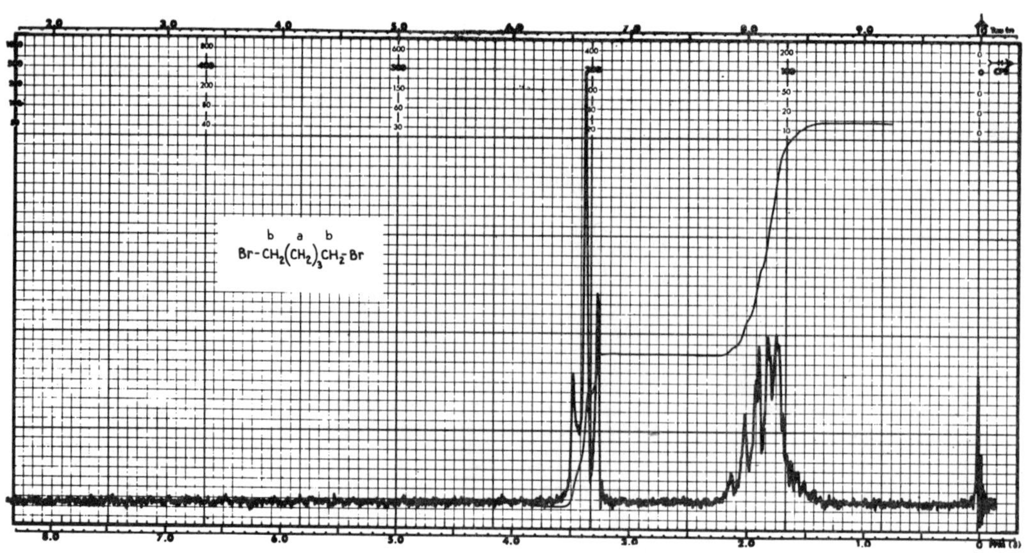

© 1978 Sadtler Research Laboratories, Inc.

2,5-DIBROMOHEXANE

$C_6H_{12}Br_2$

Mol. Wt. 243.99

B.P. 98-100°C/20mm

Solvent: CCl_4

Assignments
a = 1.76
b = 1.98
c = 4.08

361

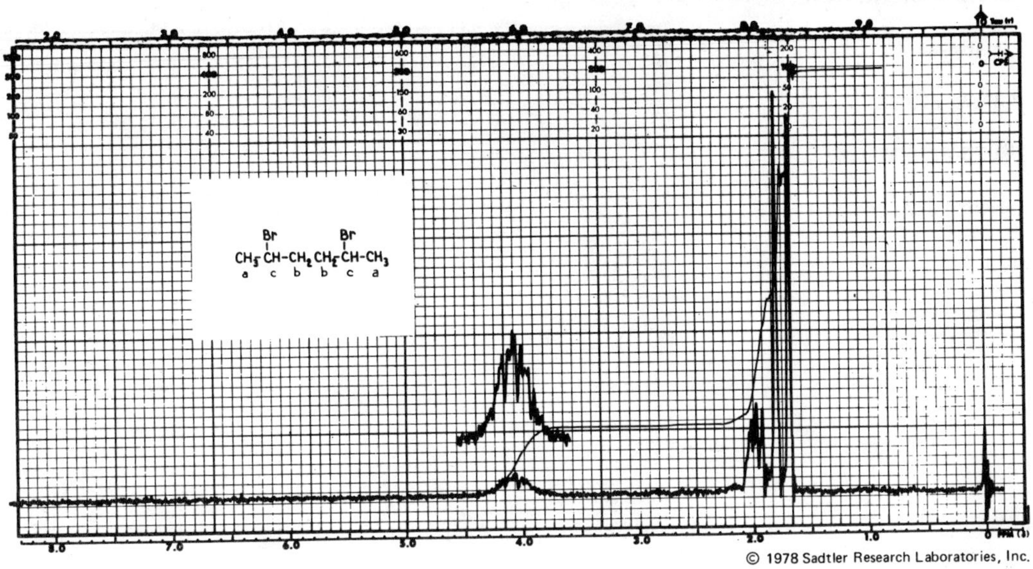

© 1978 Sadtler Research Laboratories, Inc.

1,6-DIBROMOHEXANE

$C_6H_{12}Br_2$

Mol. Wt. 243.98

M.P. 112-114°C

Solvent: CCl_4

Assignments
a = ca 1.52
b = ca 1.89
c = 3.37

362

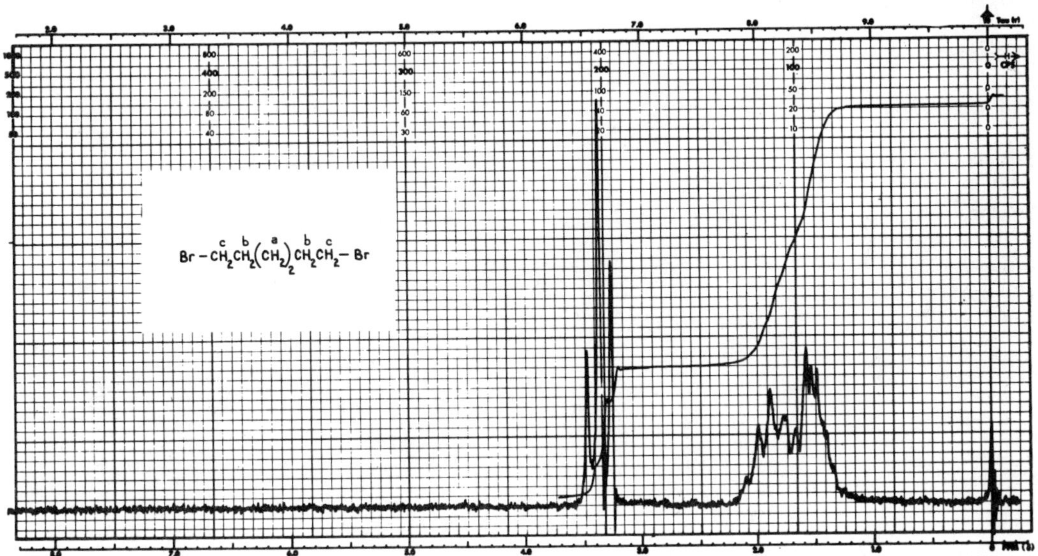

© 1978 Sadtler Research Laboratories, Inc.

BROMOCYCLOPENTANE

C_5H_9Br

Mol. Wt. 149.03

Solvent: CCl_4

Assignments
a = 1.40-2.32
b = 4.38

363

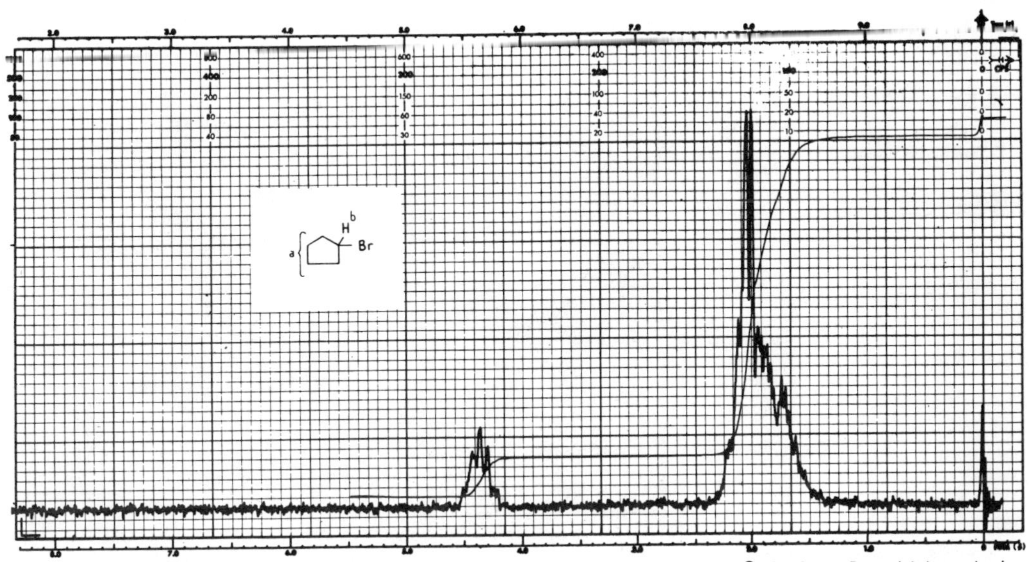

© 1978 Sadtler Research Laboratories, Inc.

(3-BROMOPROPYL)CYCLOHEXANE

$C_9H_{17}Br$

Mol. Wt. 205.14

Solvent: CCl_4

Assignments
a = 0.50-2.15
b = 3.30

364

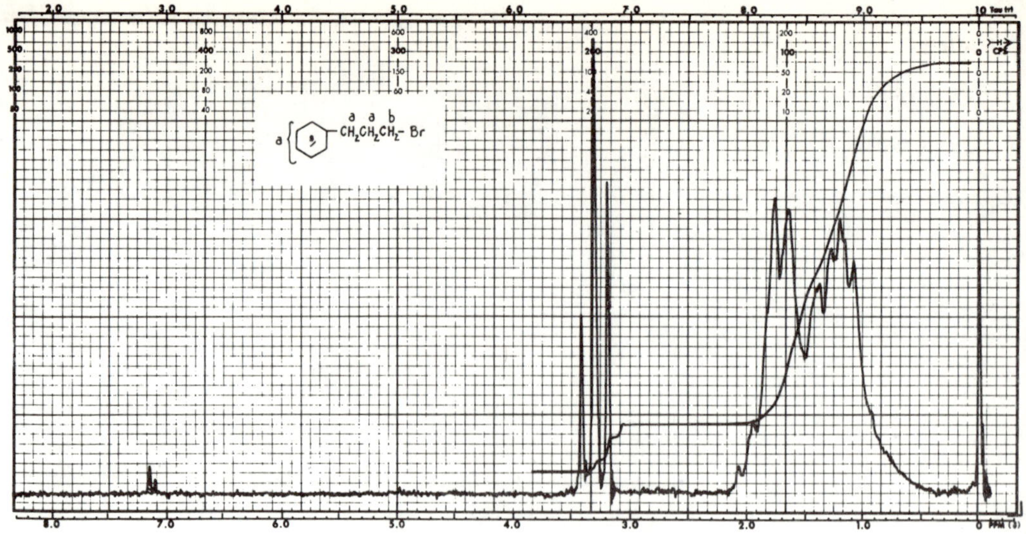

© 1978 Sadtler Research Laboratories, Inc.

BROMOCYCLOHEXANE

$C_6H_{11}Br$

Mol. Wt. 163.06

B.P. 58-59°C/12mm

Solvent: CCl_4

Assignments
a = 1.10-2.50
b = 4.13

365

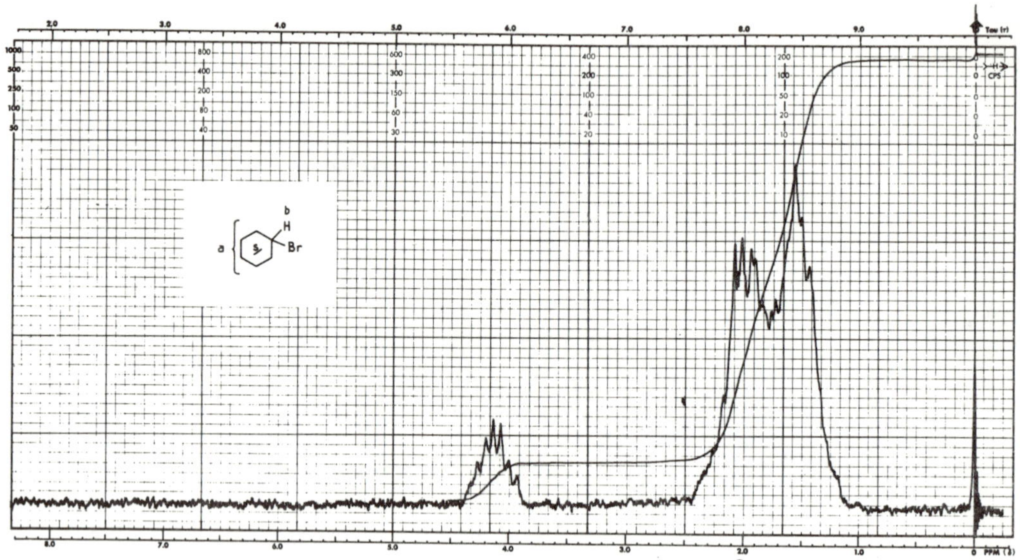

© 1978 Sadtler Research Laboratories, Inc.

1-BROMO-3-METHYLCYCLOHEXANE

$C_7H_{13}Br$

Mol. Wt. 177.09

B.P. 58-60°C/10mm

Solvent: CCl_4

Assignments
a = 0.91
b = 1.09-2.44
c = 4.52

366

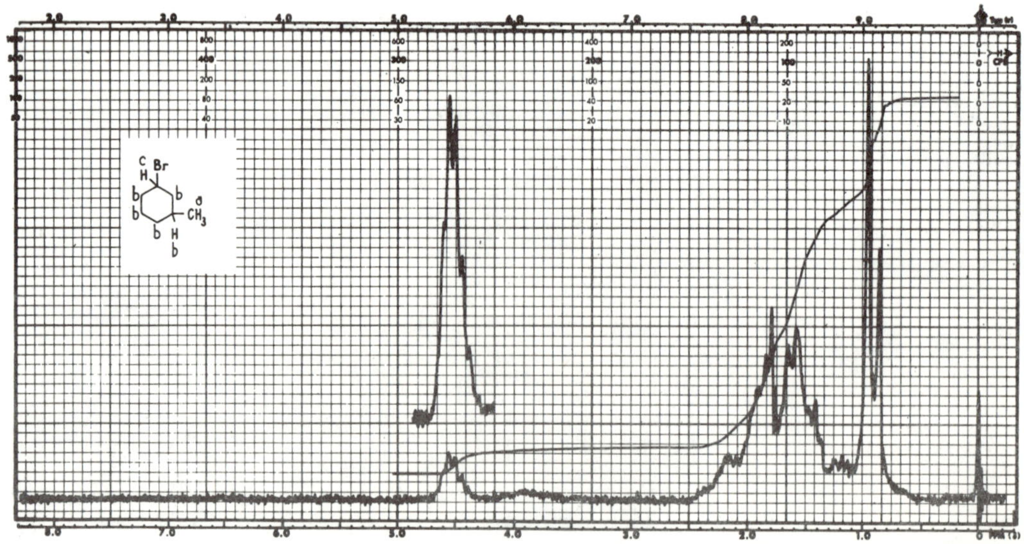

© 1978 Sadtler Research Laboratories, Inc.

1,4-DIBROMOCYCLOHEXANE

$C_6H_{10}Br_2$

Mol. Wt. 241.96

Solvent: $CDCl_3$

Assignments
a = 1.10-2.60
b = 4.32
c = 7.20 imp.

367

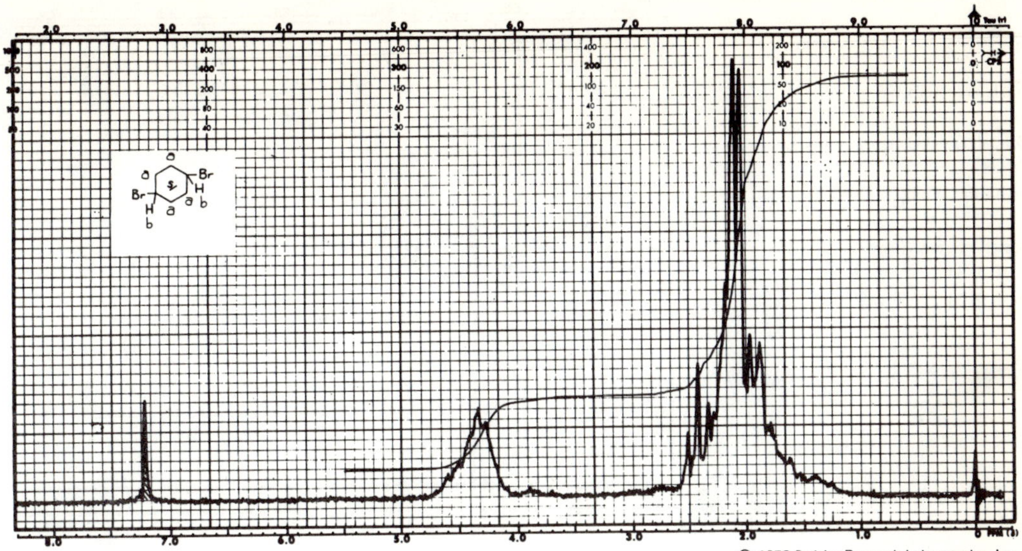

© 1978 Sadtler Research Laboratories, Inc.

3-BROMO-1-PROPENE

C_3H_5Br

Mol. Wt. 120.98

B.P. 70-71°C

Solvent: CCl_4

Assignments
a = 3.88
b = 5.09
c = 5.26
d = 5.91

368

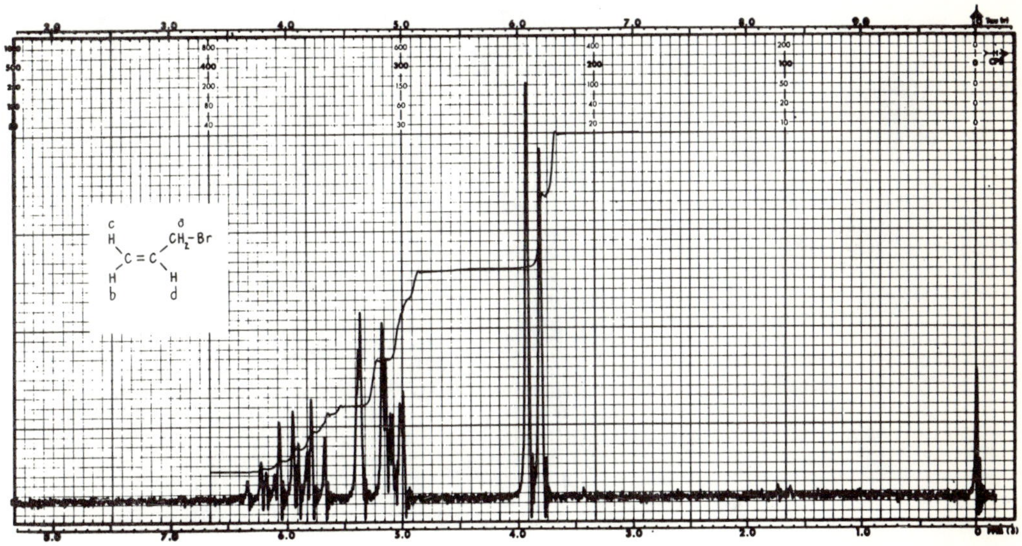

© 1978 Sadtler Research Laboratories, Inc.

1,4-DIBROMO-2-BUTENE

$C_4H_6Br_2$

Mol. Wt. 213.91

M.P. 47-51°C

Solvent: $CDCl_3$

Assignments
a = 3.99
b = 6.01

369

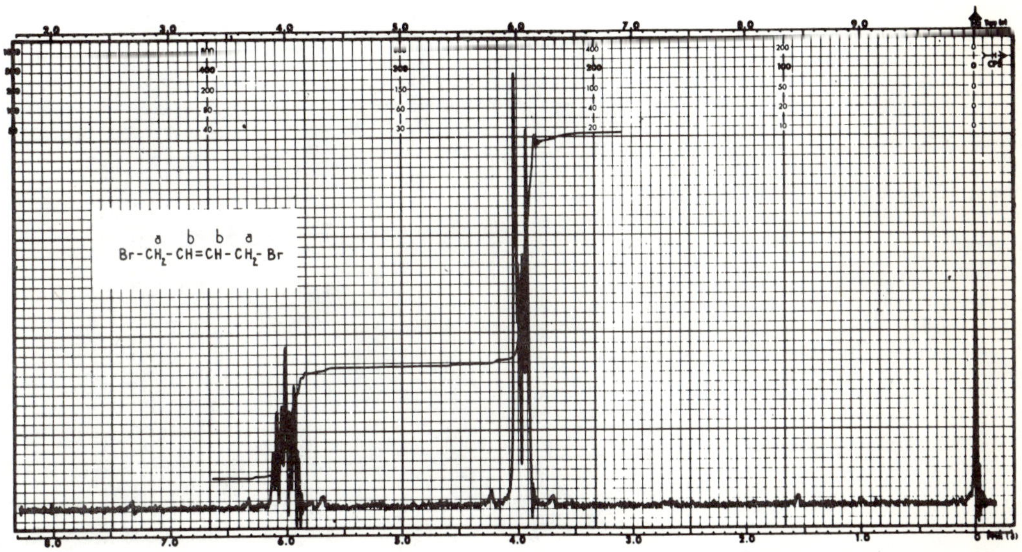

© 1978 Sadtler Research Laboratories, Inc.

1,2-DIBROMOETHYLENE

$C_2H_2Br_2$

Mol. Wt. 185.86

B.P. 107-110°C

Solvent: CCl_4

Assignments
a = 6.64
b = 7.04

370

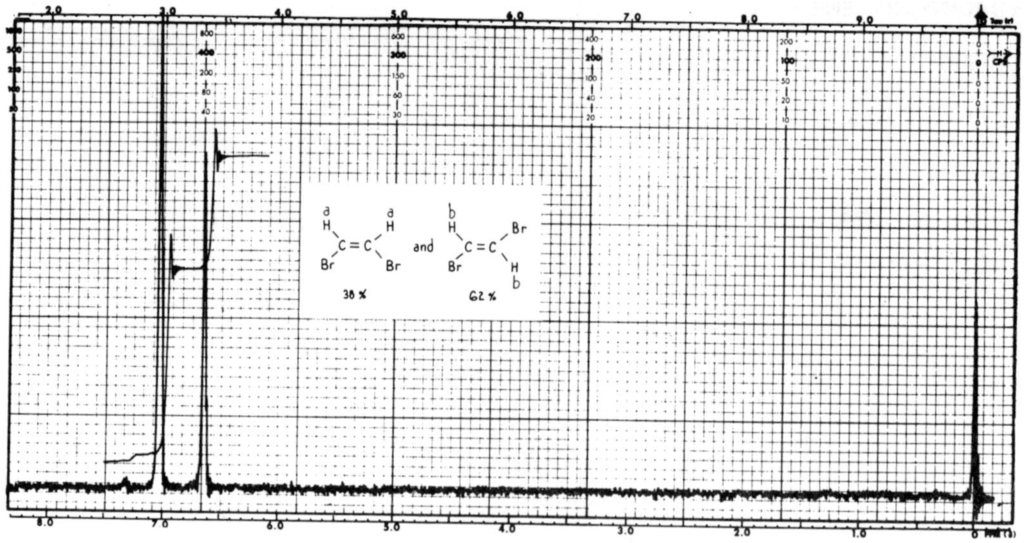

© 1978 Sadtler Research Laboratories, Inc.

α-BROMOTOLUENE

C_7H_7Br

Mol. Wt. 171.04

B.P. 197-199°C

Solvent: CCl_4

Assignments
a = 4.34
b = 7.22

371

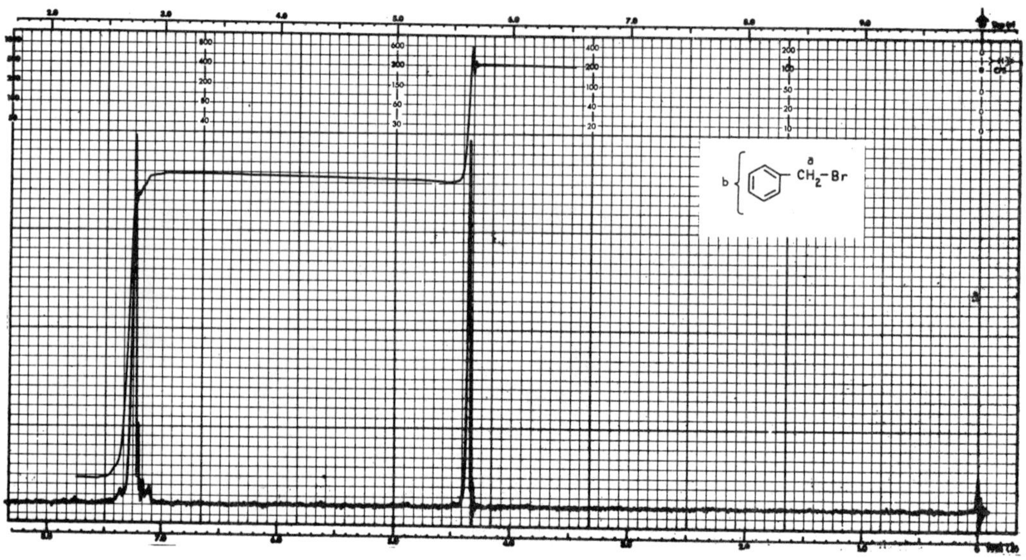

© 1978 Sadtler Research Laboratories, Inc.

α-BROMO-o-XYLENE

C_8H_9Br

Mol. Wt. 185.07

B.P. 80-82°C/5mm

Solvent: CCl_4

Assignments
a = 2.36
b = 4.39
c = 7.00-7.30

372

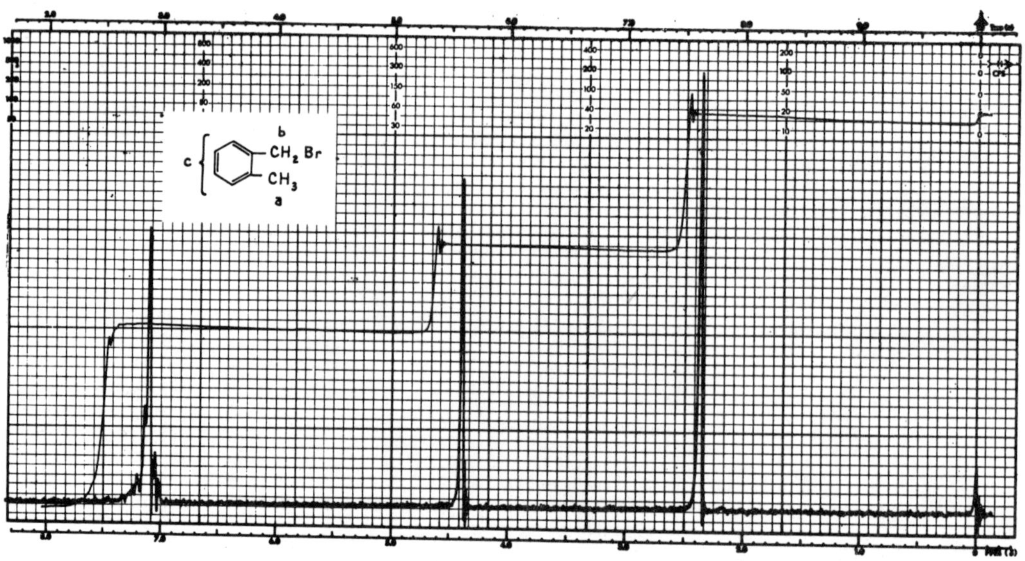

© 1978 Sadtler Research Laboratories, Inc.

α-BROMO-p-XYLENE

C_8H_9Br

Mol. Wt. 185.07

M.P. 35-37°C

B.P. 218-220°C/740mm (lit.)

Solvent: CCl_4

Assignments
a = 2.29
b = 4.32
c = 6.99
d = 7.18

373

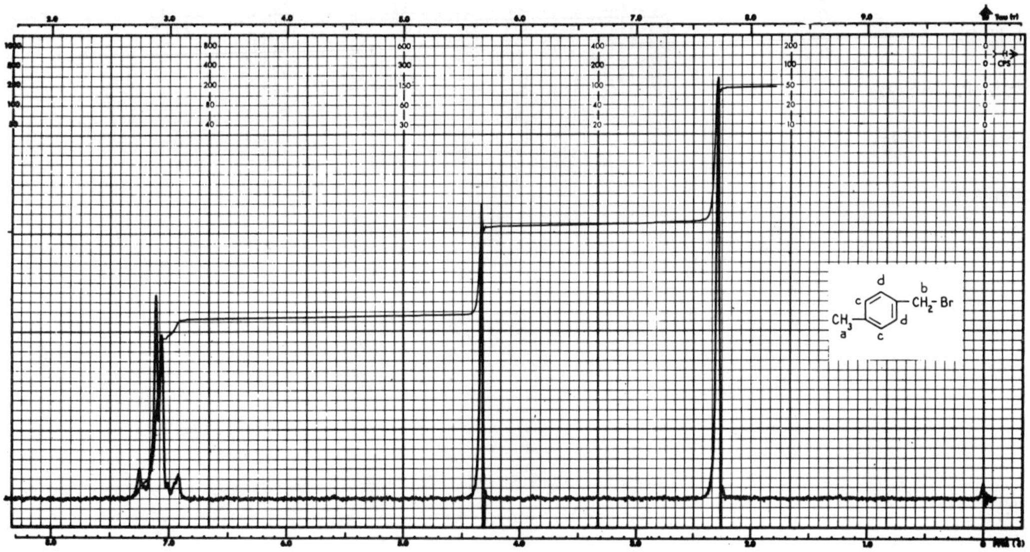

© 1978 Sadtler Research Laboratories, Inc.

BROMODIPHENYLMETHANE

$C_{13}H_{11}Br$

Mol. Wt. 247.14

Solvent: $CDCl_3$

Assignments
a = 6.19
b = 7.02-7.51
c = 5.38 imp.

374

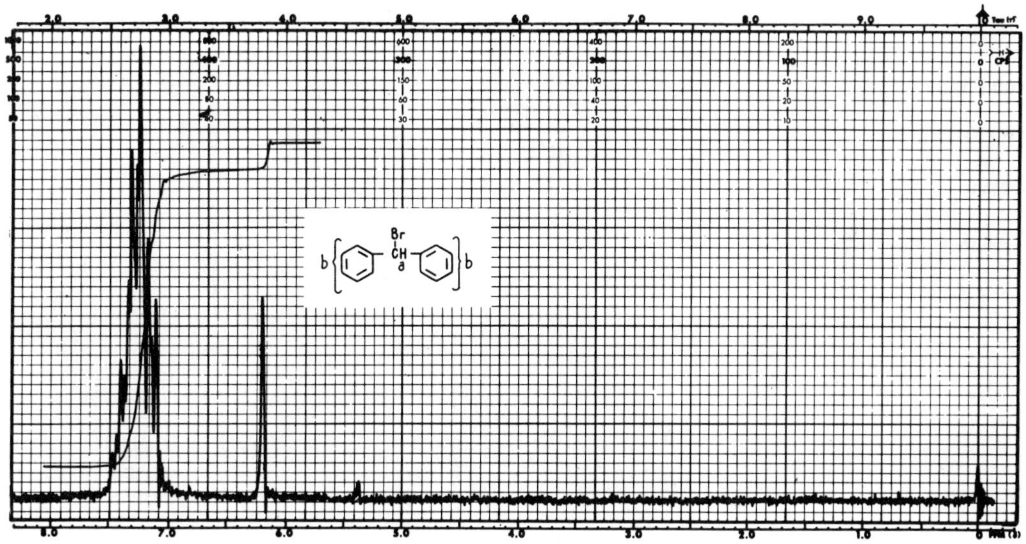

© 1978 Sadtler Research Laboratories, Inc.

1-BROMO-2-ETHYLBENZENE

C_8H_9Br

Mol. Wt. 185.06

B.P. 155-158°C/5mm

Solvent: CCl_4

Assignments
a = 1.22
b = 2.74
c = 6.79-7.23
d = 7.46

375

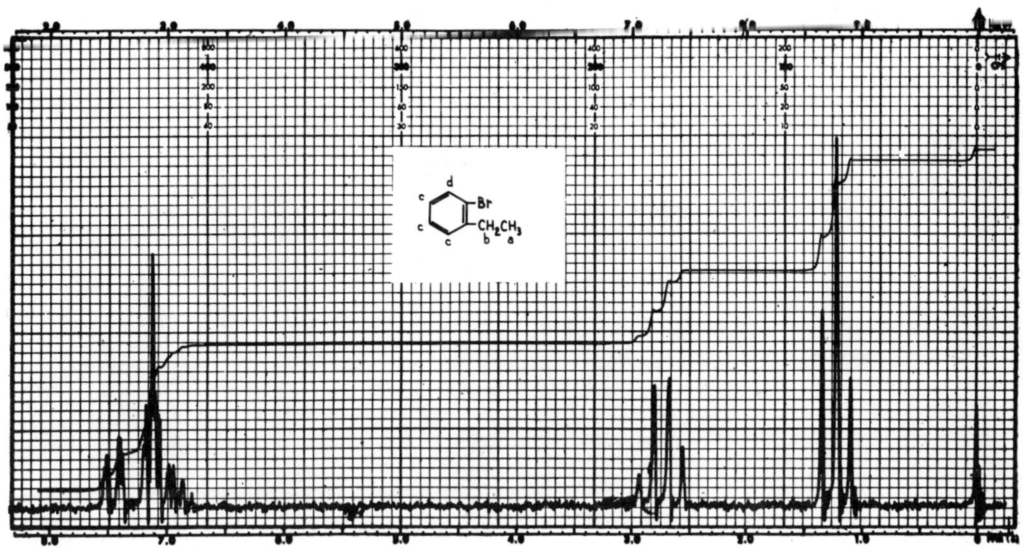

© 1978 Sadtler Research Laboratories, Inc.

a,a-**DIBROMOTOLUENE**

$C_7H_6Br_2$

Mol. Wt. 249.94

B.P. 156°C/23mm

Solvent: $CDCl_3$

Assignments
a = 6.61
b = 7.15-7.68

376

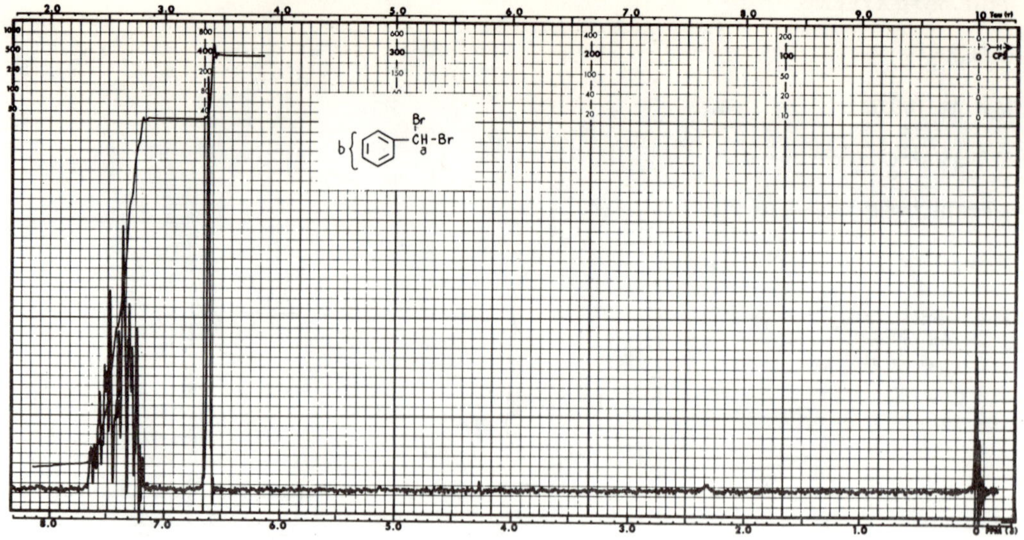

© 1978 Sadtler Research Laboratories, Inc.

a,a-**DIBROMO-m-XYLENE**

$C_8H_8Br_2$

Mol. Wt. 263.97

B.P. 112-114°C/6mm

Solvent: CCl_4

Assignments
a = 2.29
b = 6.49
c = 6.80-7.35

377

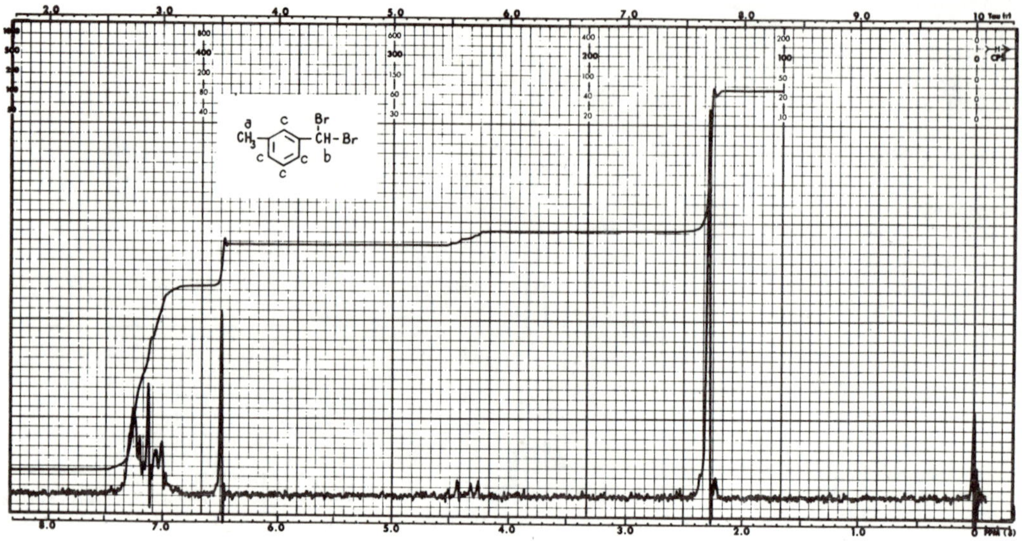

© 1978 Sadtler Research Laboratories, Inc.

(1,2-DIBROMOETHYL)BENZENE

$C_8H_8Br_2$

Mol. Wt. 263.97

M.P. 72-74°C

Solvent: $CDCl_3$

Assignments
a = ca 3.89
b = ca 4.09
c = 5.10
d = 7.31

378

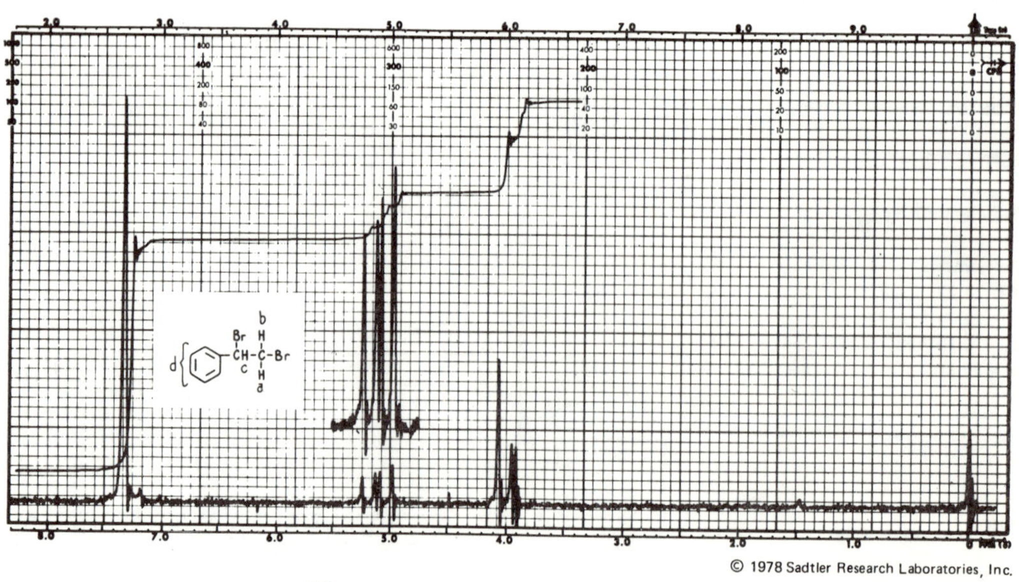

© 1978 Sadtler Research Laboratories, Inc.

1,5-DIBROMO-3-PHENYLPENTANE

$C_{11}H_{14}Br_2$

Mol. Wt. 306.04

B.P. 117-119°C/0.5mm

Solvent: CCl_4

Assignments
a = 2.11
b = 2.75-3.40
c = ca 7.19

379

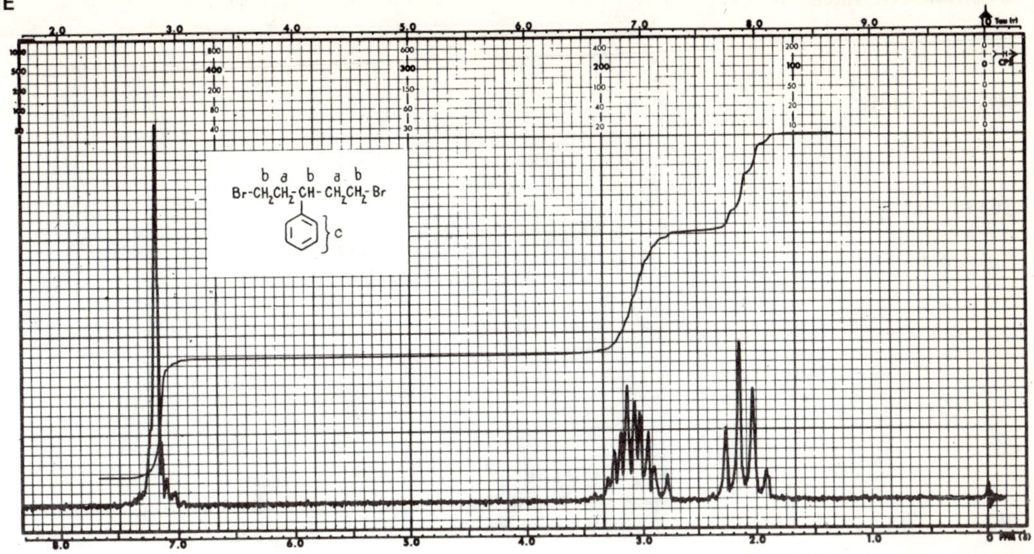

© 1978 Sadtler Research Laboratories, Inc.

β-BROMOSTYRENE

C_8H_7Br

Mol. Wt. 183.05

Solvent: CCl_4

Assignments
a = 6.61
b = 7.02
c = ca 7.19

380

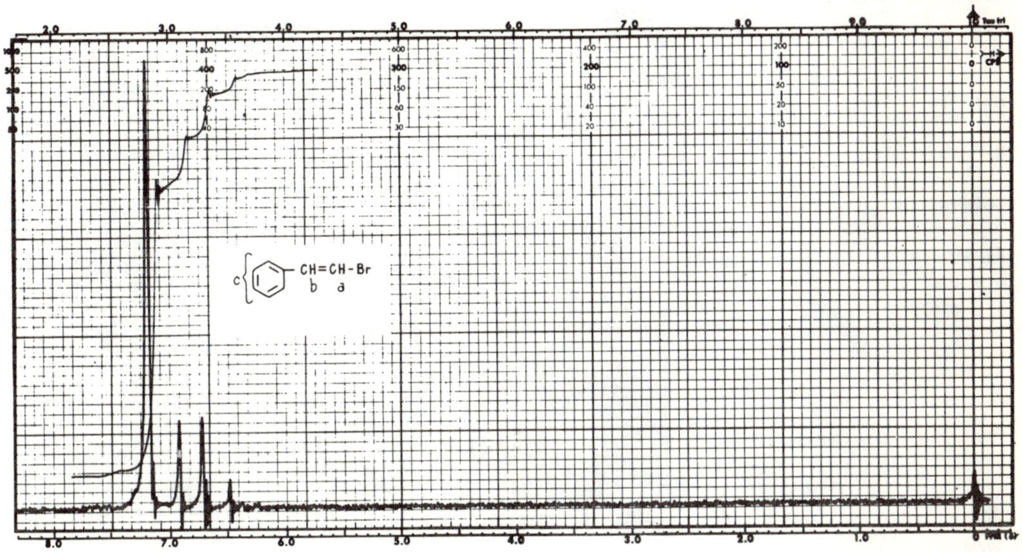

© 1978 Sadtler Research Laboratories, Inc.

BROMOBENZENE

C_6H_5Br

Mol. Wt. 157.02

B.P. 154-155°C

Solvent: CCl_4

Assignments
a = 7.00-7.35
b = 7.41

381

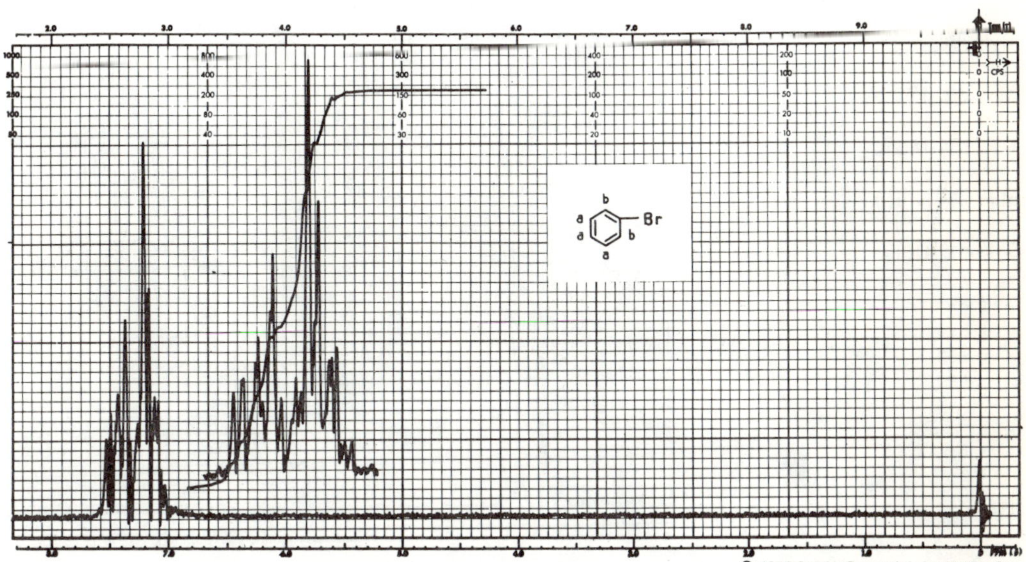

© 1978 Sadtler Research Laboratories, Inc.

o-BROMOTOLUENE

C_7H_7Br

Mol. Wt. 171.04

B.P. 181.8°C (lit.)

Solvent: CCl_4

Assignments
a = 2.36
b = 7.10
c = 7.46

382

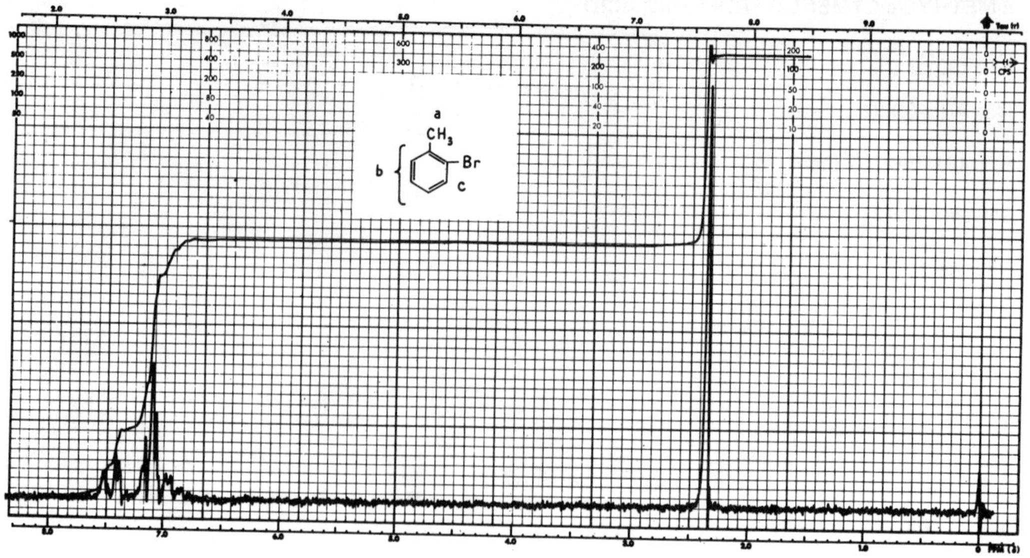

© 1978 Sadtler Research Laboratories, Inc.

m-BROMOTOLUENE

C_7H_7Br

Mol. Wt. 171.04

B.P. 182-184°C

Solvent: CCl_4

Assignments
a = 2.30
b = 6.95-7.30

383

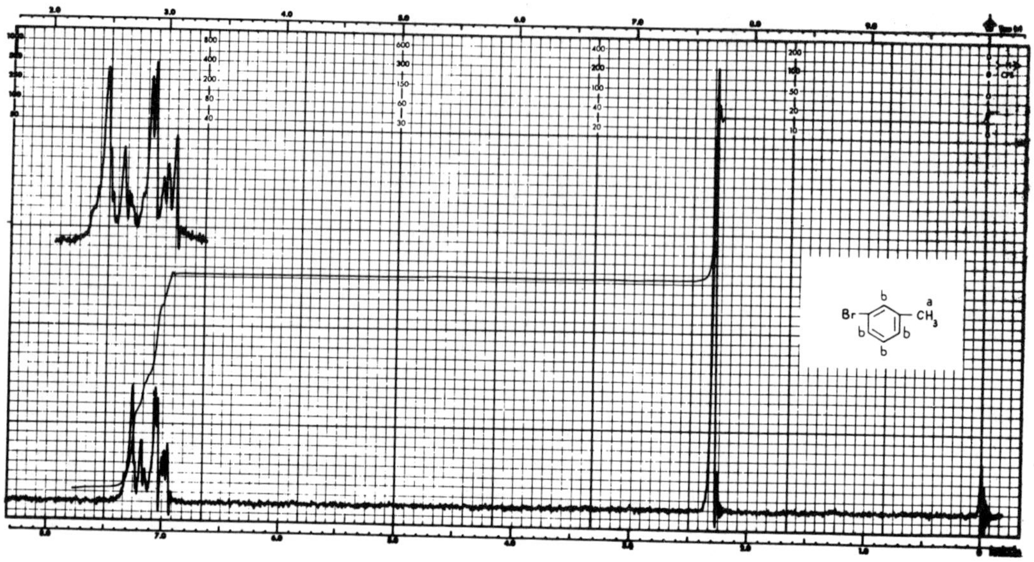

© 1978 Sadtler Research Laboratories, Inc.

p-BROMOTOLUENE

C_7H_7Br

Mol. Wt. 171.04

B.P. 184-185°C

M.P. 28°C (lit.)

Solvent: CCl_4

Assignments
a = 2.25
b = 6.92
c = 7.30

384

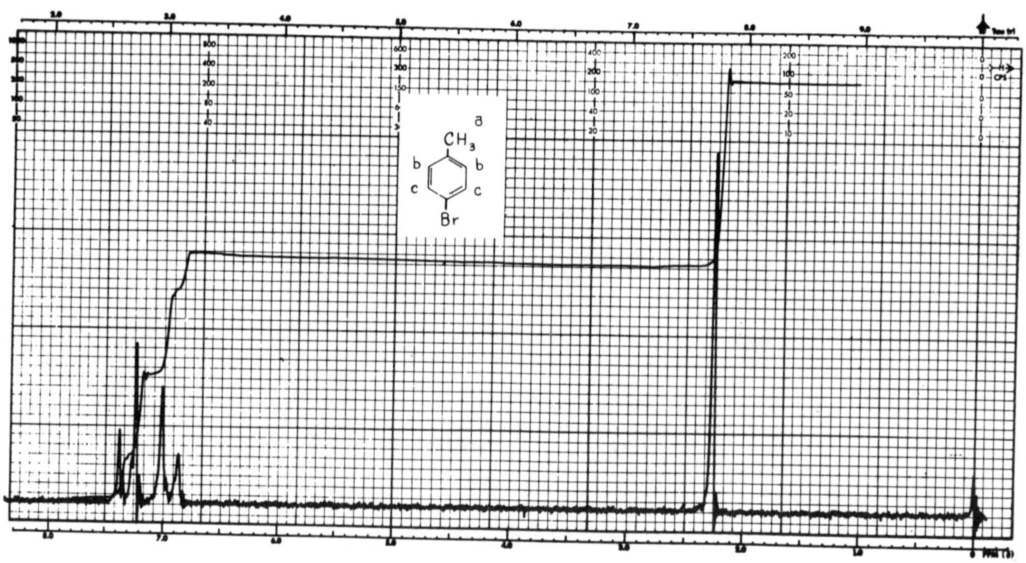

© 1978 Sadtler Research Laboratories, Inc.

α-METHYL-p-CYMENE-3-PROPIONIC ACID

$C_{14}H_{20}O_2$

Mol. Wt. 220.31

B.P. 197-198°C/23mm

Solvent: $CDCl_3$

Assignments

a = ca 1.21
b = 2.29
c = 2.45-3.40
d = 6.90-7.25
e = 11.60

385

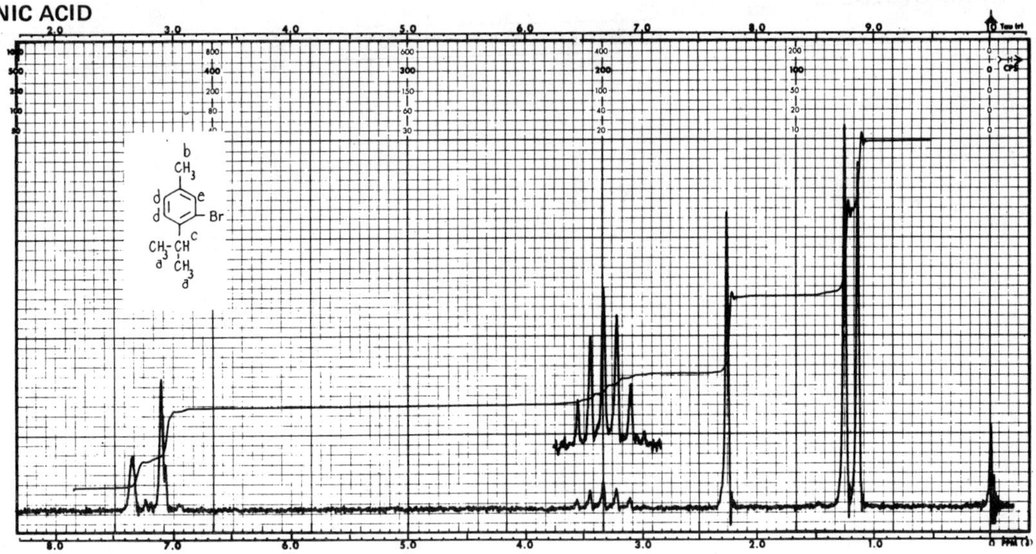

© 1978 Sadtler Research Laboratories, Inc.

1-BROMO-2,4,5-TRIMETHYLBENZENE

$C_9H_{11}Br$

Mol. Wt. 199.10

Solvent: $CDCl_3$

Assignments

a = 2.16
b = 2.31
c = 6.98
d = 7.27

386

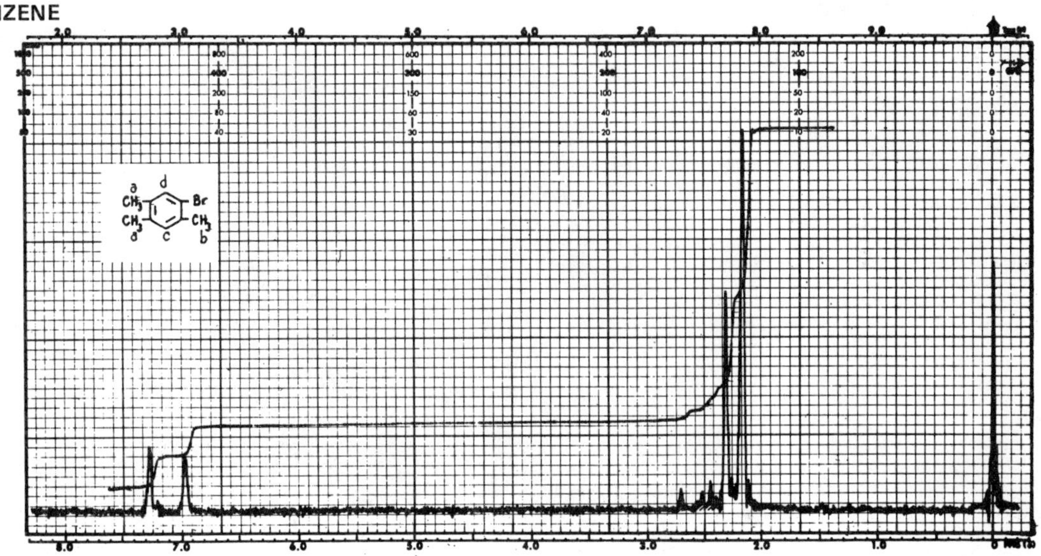

© 1978 Sadtler Research Laboratories, Inc.

2-BROMOMESITYLENE

$C_9H_{11}Br$

Mol. Wt. 199.10

Solvent: CCl_4

Assignments

a = 2.21
b = 2.32
c = 6.78

387

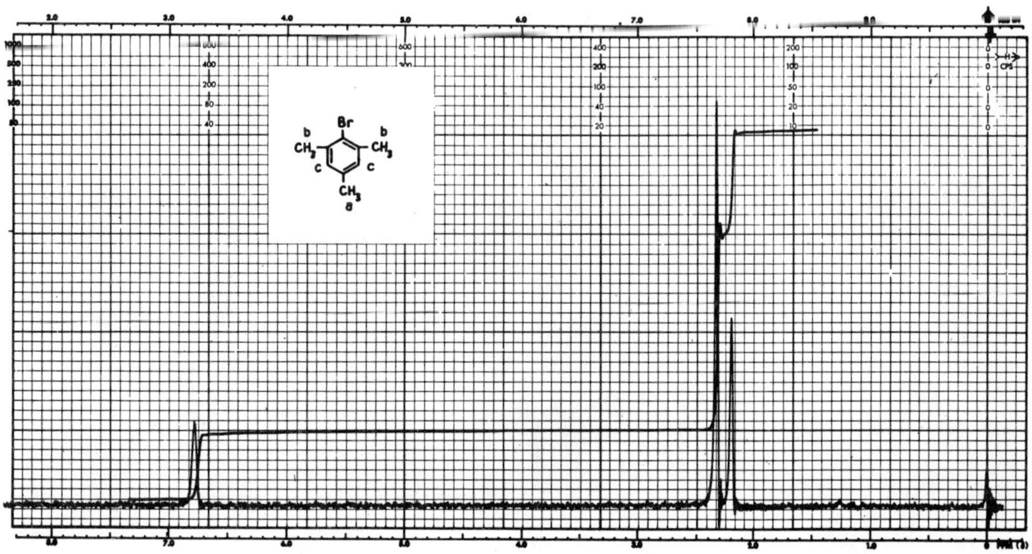

© 1978 Sadtler Research Laboratories, Inc.

p-BROMO-α,α,α-TRIFLUOROTOLUENE

$C_7H_4BrF_3$

Mol. Wt. 225.01

B.P. 154-155°C

Solvent: CCl_4

Assignments
a = 7.49
b = 7.60

388

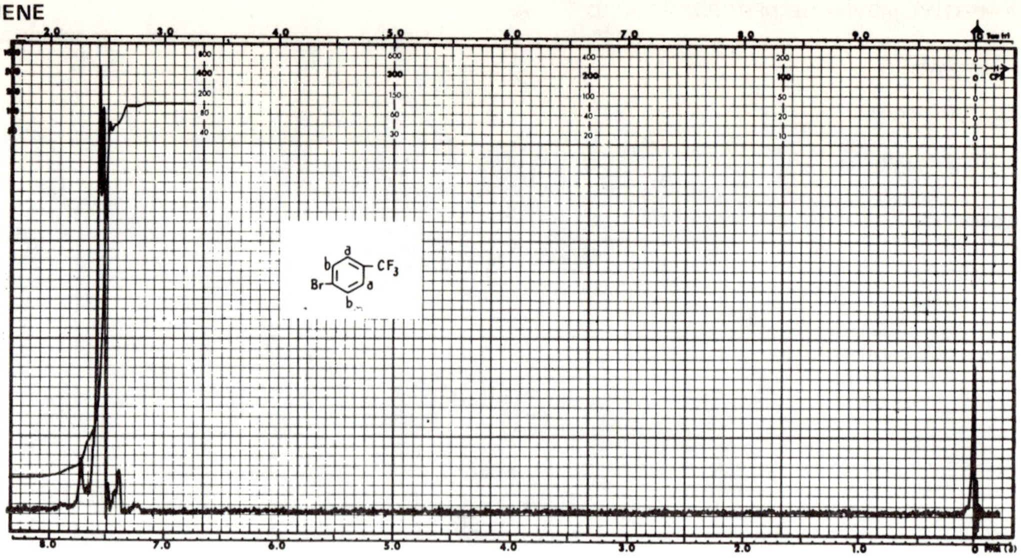

© 1978 Sadtler Research Laboratories, Inc.

1-BROMO-2-FLUOROBENZENE

C_6H_4BrF

Mol. Wt. 175.10

Solvent: CCl_4

Assignments
a = 6.70-7.60

389

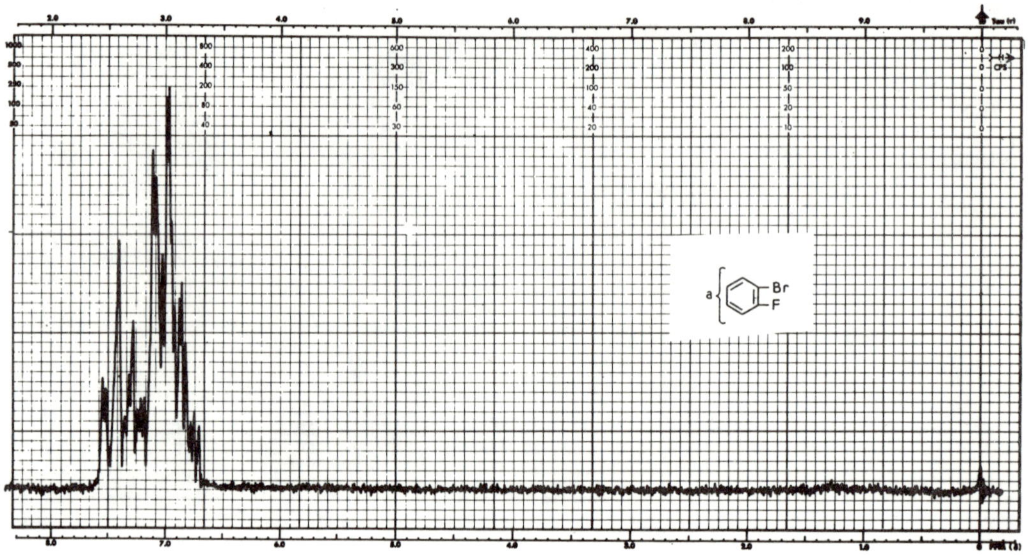

© 1978 Sadtler Research Laboratories, Inc.

1-BROMO-3-FLUOROBENZENE

C_6H_4BrF

Mol. Wt. 175.01

Solvent: CCl_4

Assignments
a = 6.97
b = 7.09-7.43

390

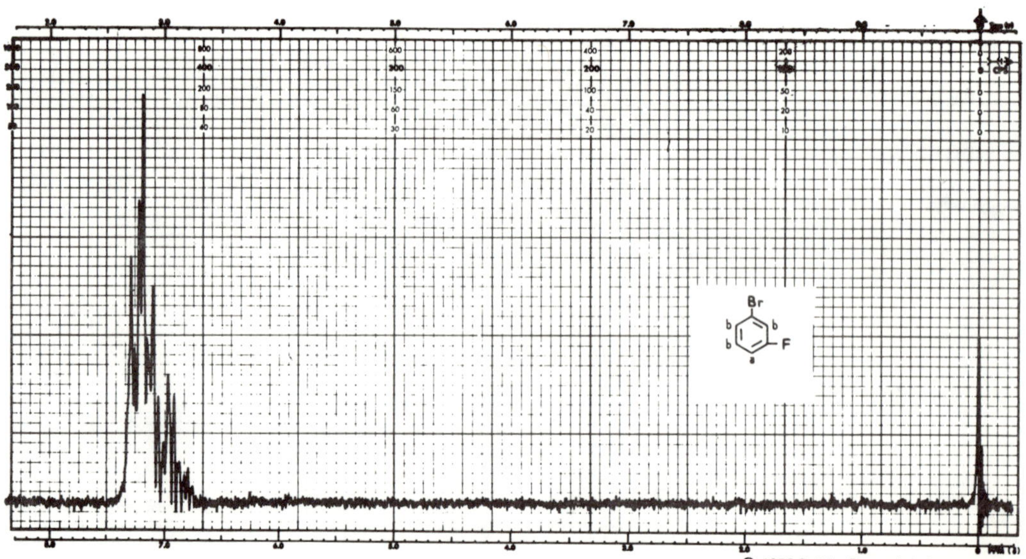

© 1978 Sadtler Research Laboratories, Inc.

1-BROMO-4-FLUOROBENZENE

C_6H_4BrF

Mol. Wt. 175.00

Solvent: CCl_4

Assignments
a = 6.90
b = 7.40

391

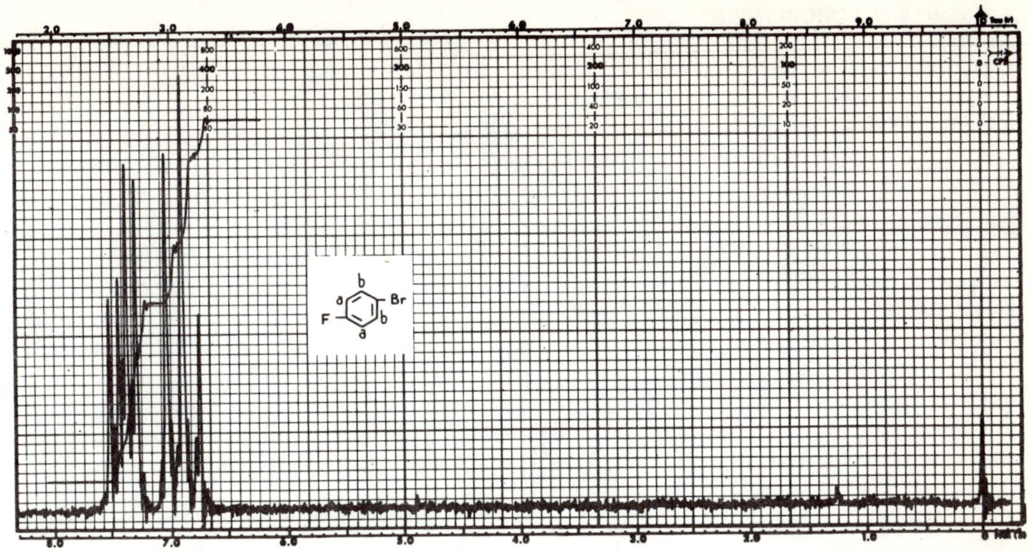

© 1978 Sadtler Research Laboratories, Inc.

1-BROMO-2-CHLOROBENZENE

C_6H_4BrCl

Mol. Wt. 191.46

Solvent: CCl_4

Assignments
a = 6.78-7.60

392

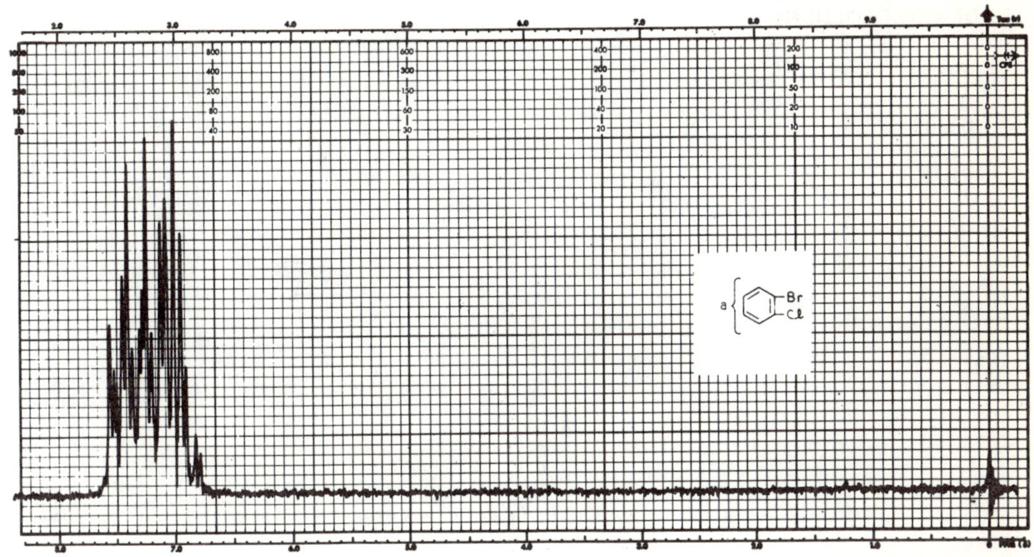

© 1978 Sadtler Research Laboratories, Inc.

1-BROMO-3-CHLOROBENZENE

C_6H_4BrCl

Mol. Wt. 191.46

B.P. 81-82°C/20mm

Solvent: CCl_4

Assignments
a = 7.03
b = ca 7.20
c = 7.46

393

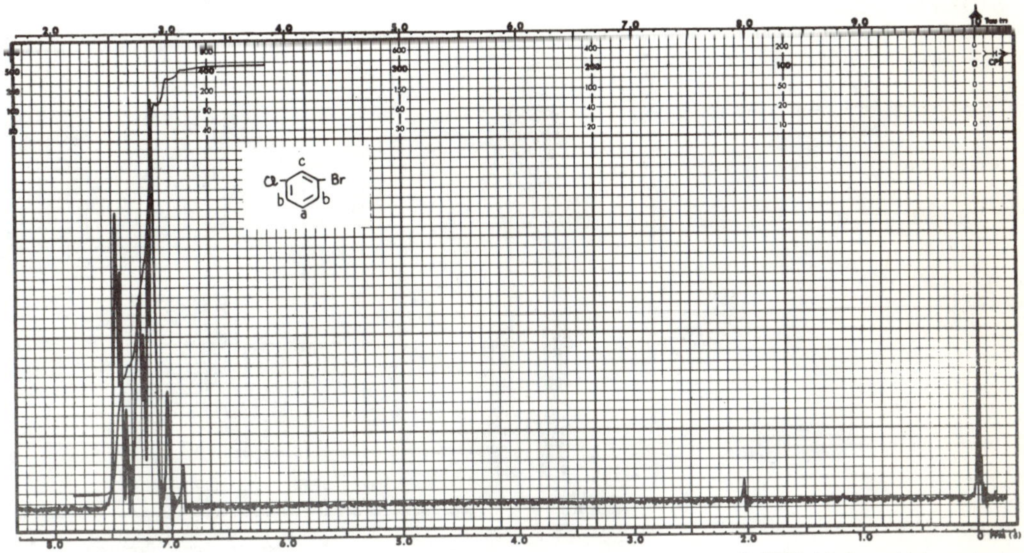

© 1978 Sadtler Research Laboratories, Inc.

1-BROMO-4-CHLOROBENZENE

C_6H_4BrCl

Mol. Wt. 191.46

M.P. 64-66°C

Solvent: CCl_4

Assignments
a = 7.13
b = 7.37

394

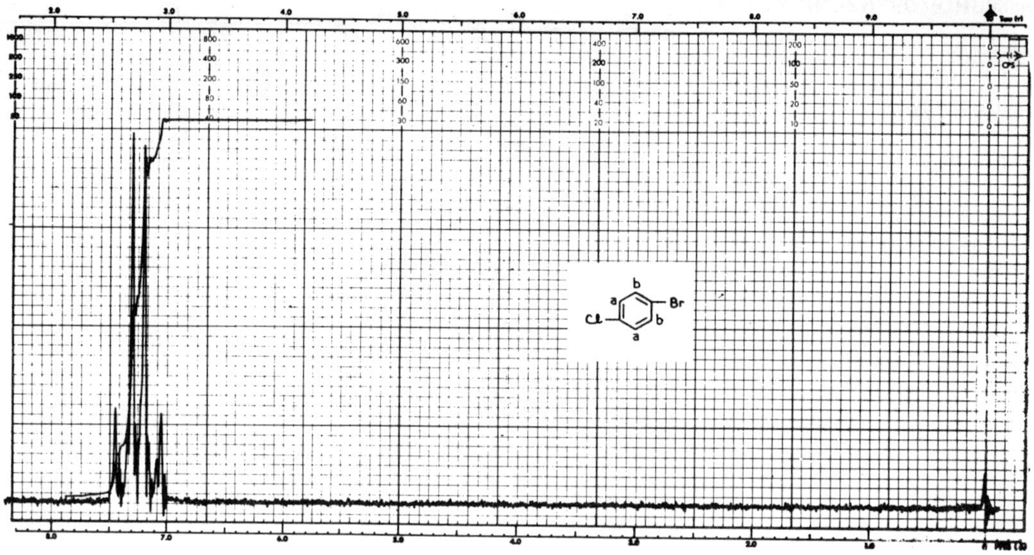

© 1978 Sadtler Research Laboratories, Inc.

m,α-DIBROMOTOLUENE

$C_7H_6Br_2$

Mol. Wt. 249.94

M.P. 41-43°C

Solvent: $CDCl_3$

Assignments
a = 4.33
b = 7.10
c = 7.22
d = 7.36
e = 7.47

395

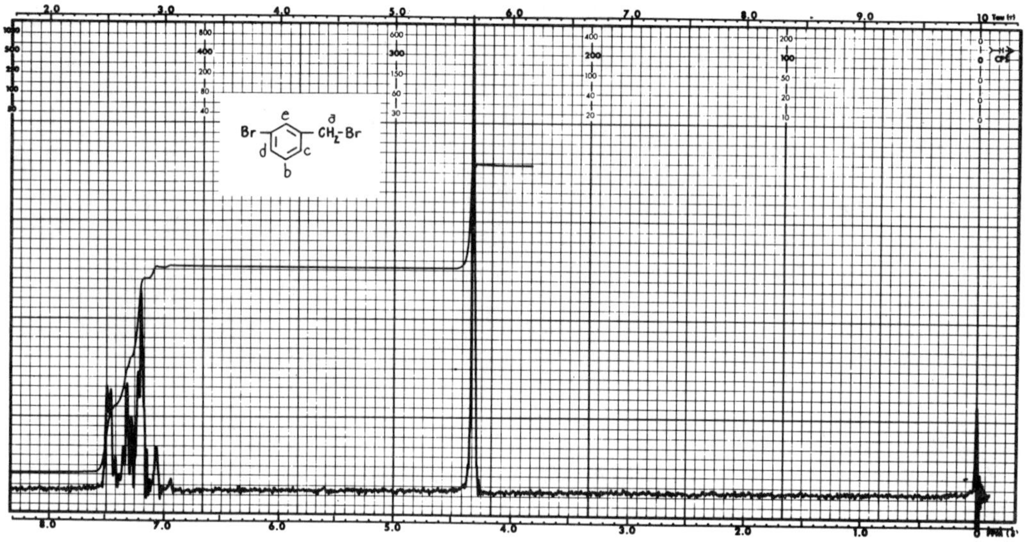

© 1978 Sadtler Research Laboratories, Inc.

o-DIBROMOBENZENE

$C_6H_4Br_2$

Mol. Wt. 235.92

Solvent: CCl_4

Assignments
a = 7.06
b = 7.49

396

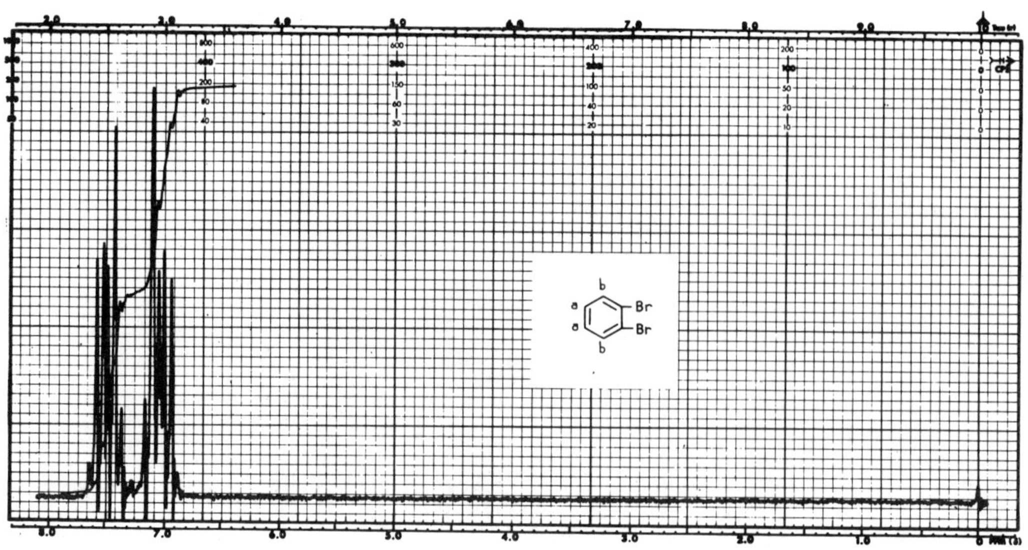

© 1978 Sadtler Research Laboratories, Inc.

m-DIBROMOBENZENE

$C_6H_4Br_2$

Mol. Wt. 235.92

M.P. −7° to 5°C

Solvent: CCl_4

Assignments
a = 6.92
b = 7.29
c = 7.55

397

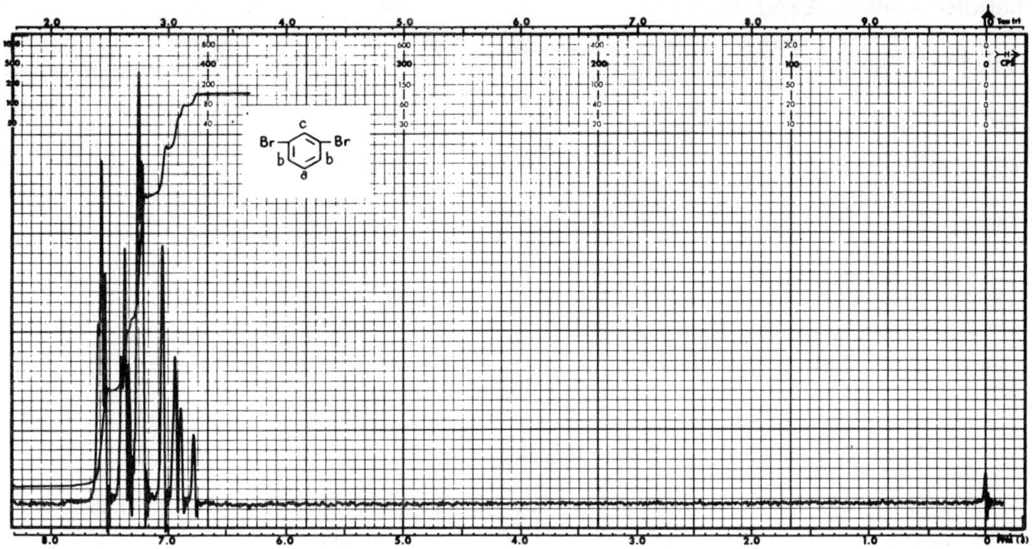

© 1978 Sadtler Research Laboratories, Inc.

p-DIBROMOBENZENE

$C_6H_4Br_2$

Mol. Wt. 235.91

M.P. 86.9°C

B.P. 218-219°C/760mm

Solvent: $CDCl_3$

Assignments
a = 7.29

398

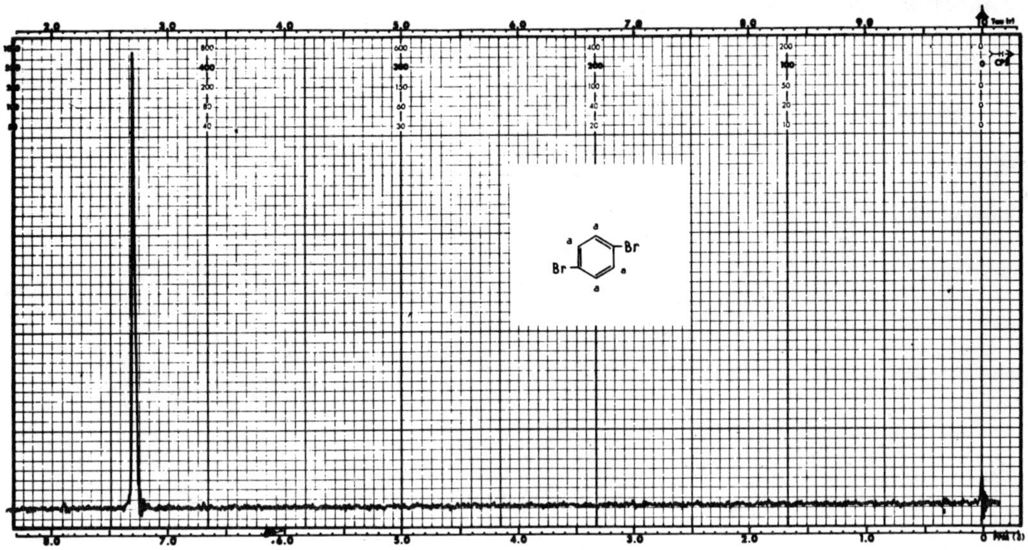

© 1978 Sadtler Research Laboratories, Inc.

4-BROMOBIPHENYL

$C_{12}H_9Br$

Mol. Wt. 233.11

M.P. 90-91°C

Solvent: $CDCl_3$

Assignments
a = 7.11-7.63

399

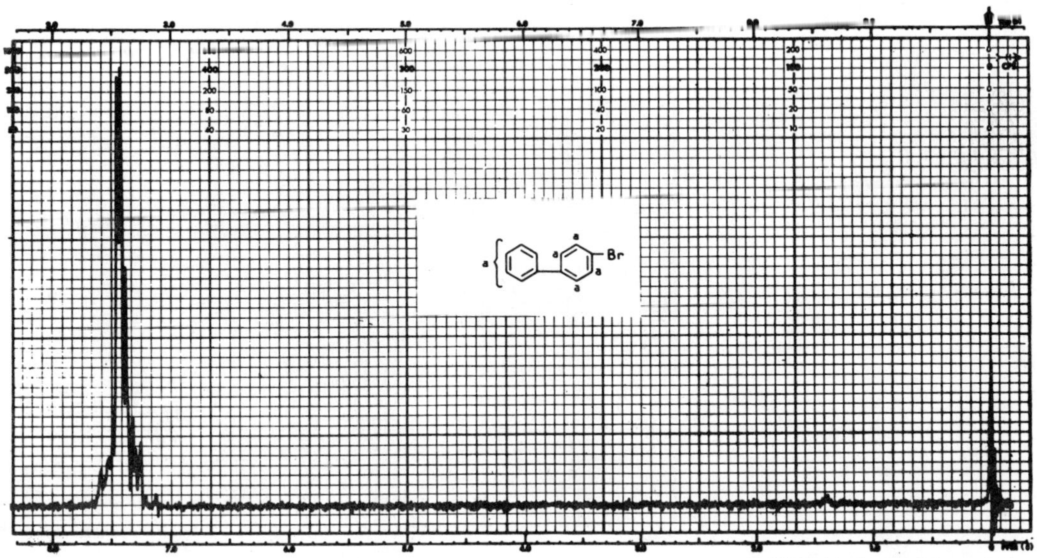

© 1978 Sadtler Research Laboratories, Inc.

1-BROMONAPHTHALENE

$C_{10}H_7Br$

Mol. Wt. 207.07

M.P. 2-4°C (lit.)

Solvent: CCl_4

Assignments
a = 7.00-7.80
b = 8.15

400

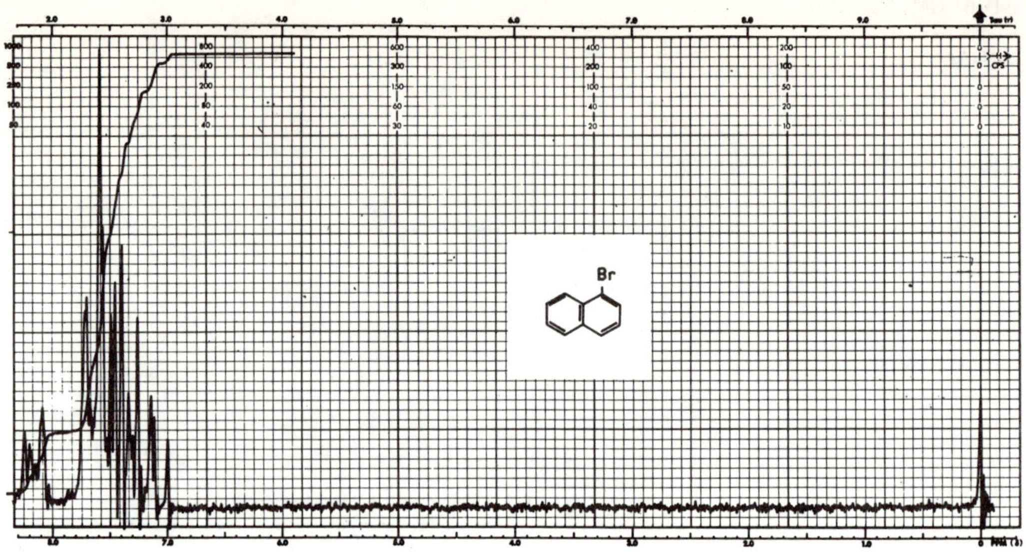

© 1978 Sadtler Research Laboratories, Inc.

2-BROMONAPHTHALENE

$C_{10}H_7Br$

Mol. Wt. 207.08

M.P. 55-56°C

Solvent: CCl_4

Assignments
a = 7.23-7.80
b = 7.91

401

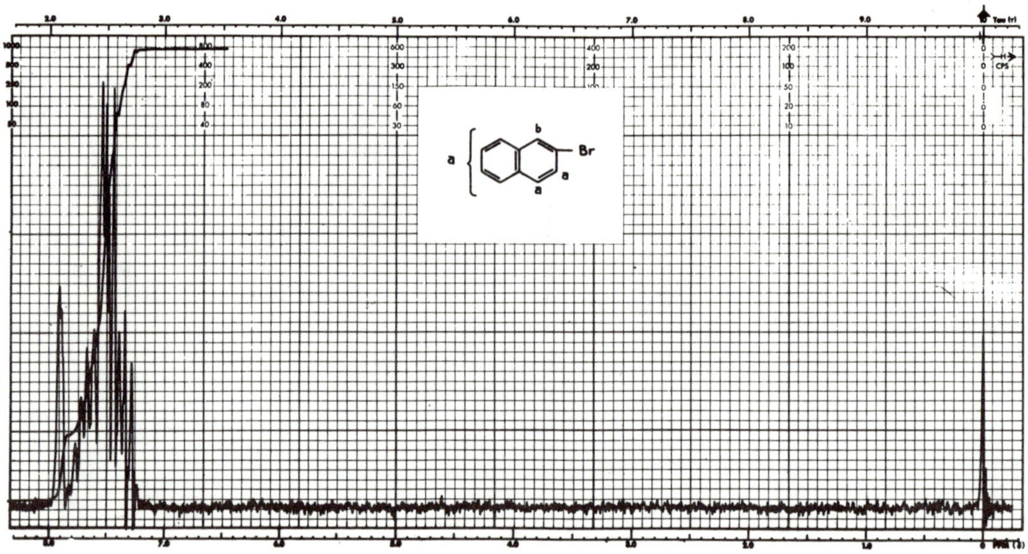

© 1978 Sadtler Research Laboratories, Inc.

1,4-DIBROMONAPHTHALENE

$C_{10}H_6Br_2$

Mol. Wt. 285.98

M.P. 81-83°C

Solvent: $CDCl_3$

Assignments
a = 7.40
b = 7.43
c = 8.06

402

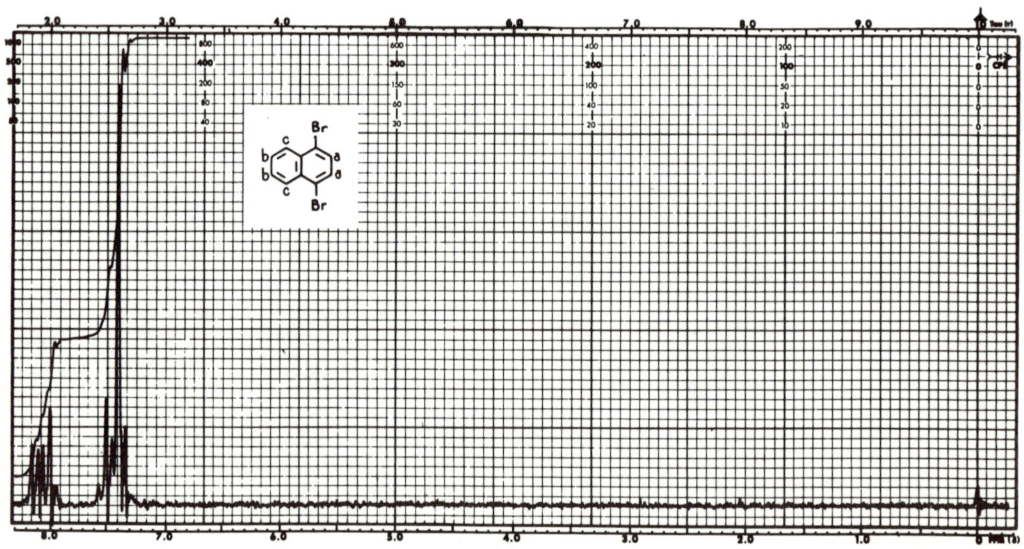

© 1978 Sadtler Research Laboratories, Inc.

Of the halogens, Iodine has the weakest deshielding effect upon aliphatic hydrogens, but the strongest deshielding effect on the ortho aromatic hydrogens. It is observed to have an unusually strong deshielding effect upon the hydrogens bonded to beta aliphatic carbon atoms.

Chemical Shifts

Aliphatic Protons

HC–4	HC–3	HC–2	HC–1		Solvent	Spectrum
			2.20	CH_3–I	$CDCl_3$	403
		1.84	3.13	CH_3–CH_2–I	CCl_4	404
	1.00	1.85	3.15	CH_3–CH_2–CH_2–I	CCl_4	405
0.99	1.40	1.80	3.19	CH_3–CH_2–CH_2–CH_2–I	CCl_4	407
	(1.01)	1.73	3.10	$(CH_3)_2$–CH–CH_2–I	CCl_4	409
		(1.91)	4.29	$(CH_3)_2$–CH–I	CCl_4	406

Substituted Iodomethanes

CH_2	–X	
3.55	–CF_2–CF_3	$CDCl_3$
3.63	–C(=O)–O–Na	D_2O
3.69	–C(=O)–O–R	CCl_4
3.71	–C(=O)–OH	$CDCl_3$
3.80	–CH=CH_2	CCl_4
3.88	–I	CCl_4
3.98	–C(=O)–NH_2	TFA
4.38	–Ø	$CDCl_3$

2-Substituted Iodoethanes

CH_2	–CH_2	–X	
3.17	1.94	–$(CH_2)_2$–I	CCl_4
3.19	3.08	–Ø	$CDCl_3$
3.28	3.81	–OH	$CDCl_3$
3.30	3.09	–C(=O)–OH	$CDCl_3$
3.29	2.29	–CH_2–I	$CDCl_3$
3.70	3.70	–I	CCl_4

Aromatic Protons

Of the four halogens, iodine has the strongest deshielding effect on aromatic hydrogens, producing a shift of the ortho hydrogens to 7.65 ppm (spectrum 426).

A series of para substituted Iodobenzenes is listed below.

I–	ortho	meta	para	Solvent
	7.40	7.40	–I	$CDCl_3$
	7.50	7.18	–Br	$CDCl_3$
	7.55	6.89	$-CH_3$	$CDCl_3$
	7.55	7.95	–C≡N	Polysol
	7.58	6.78	–F	$CDCl_3$
	7.61	7.09	–Cl	$CDCl_3$
	7.61	7.88	–C(=O)–R	$CDCl_3$
	7.69	7.99	$-SO_2-Cl$	$CDCl_3$
	7.71	7.90	–C(=O)–OH	DMSO-d_6
	7.78	7.78	$-C(=O)-O-CH_3$	Polysol
	7.70	7.09	$-NH-C(=O)-CH_3$	TFA
	7.79	7.30	$-CF_3$	$CDCl_3$

METHYL IODIDE

CH_3I

Mol. Wt. 141.94

Solvent: $CDCl_3$

Assignments
a = 2.20

403

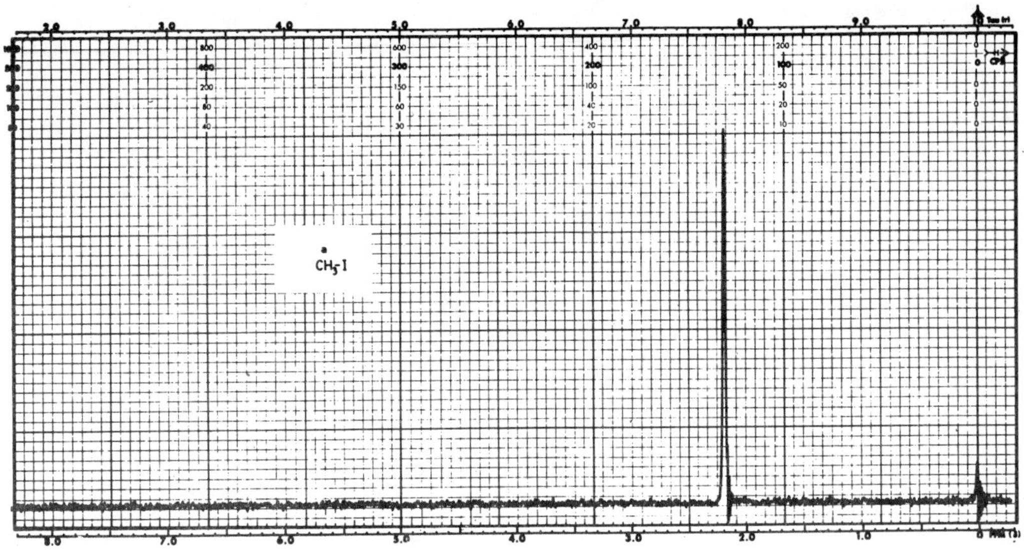

© 1978 Sadtler Research Laboratories, Inc.

IODOETHANE

C_2H_5I

Mol. Wt. 155.97

B.P. 68-74°C

Solvent: CCl_4

Assignments
a = 1.84
b = 3.13

404

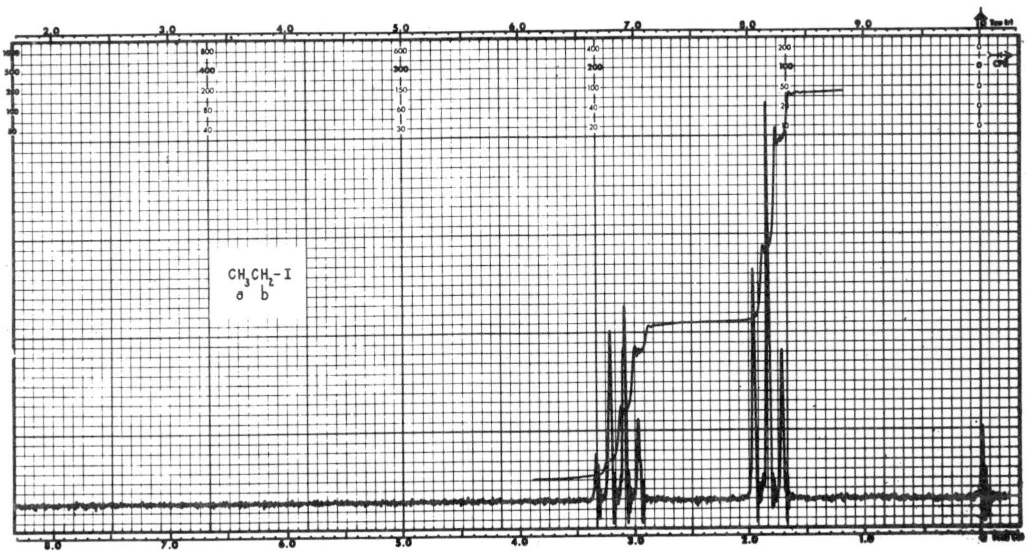

© 1978 Sadtler Research Laboratories, Inc.

1-IODOPROPANE

C_3H_7I

Mol. Wt. 170.00

B.P. 100-102°C

Solvent: CCl_4

Assignments
a = 1.00
b = 1.85
c = 3.15

405

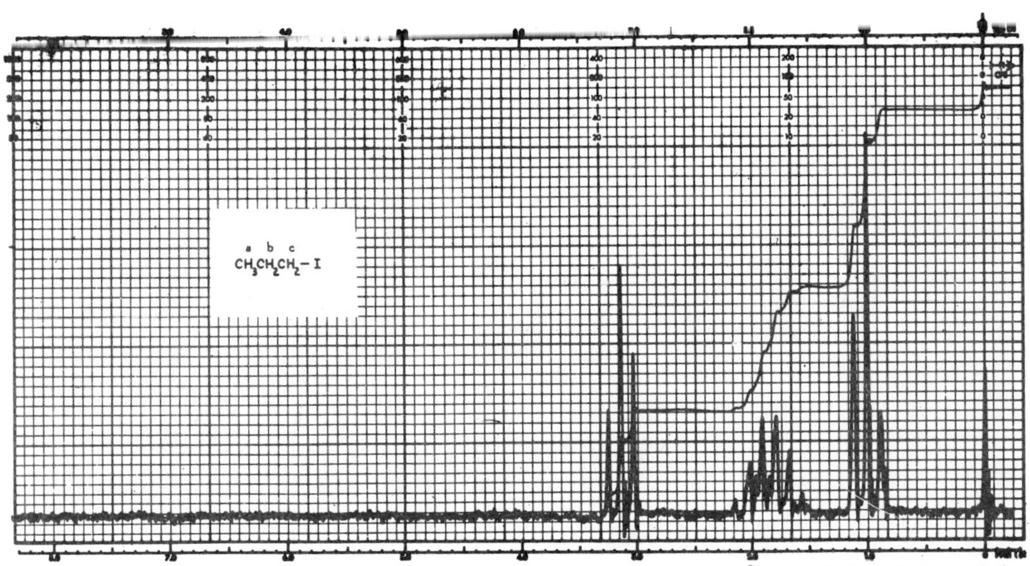

© 1978 Sadtler Research Laboratories, Inc.

2-IODOPROPANE

C_3H_7I

Mol. Wt. 169.99

B.P. 87-89°C

Solvent: CCl_4

Assignments
a = 1.91
b = 4.29

406

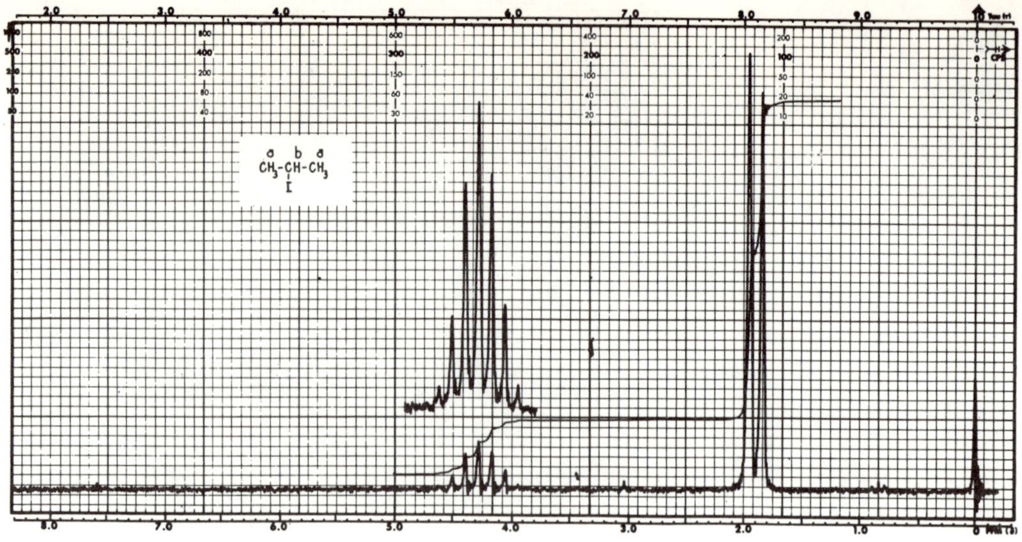

© 1978 Sadtler Research Laboratories, Inc.

1-IODOBUTANE

C_4H_9I

Mol. Wt. 184.02

B.P. 129-131°C

Solvent: CCl_4

Assignments
a = 0.99
b = ca 1.40
c = ca 1.80
d = 3.19

407

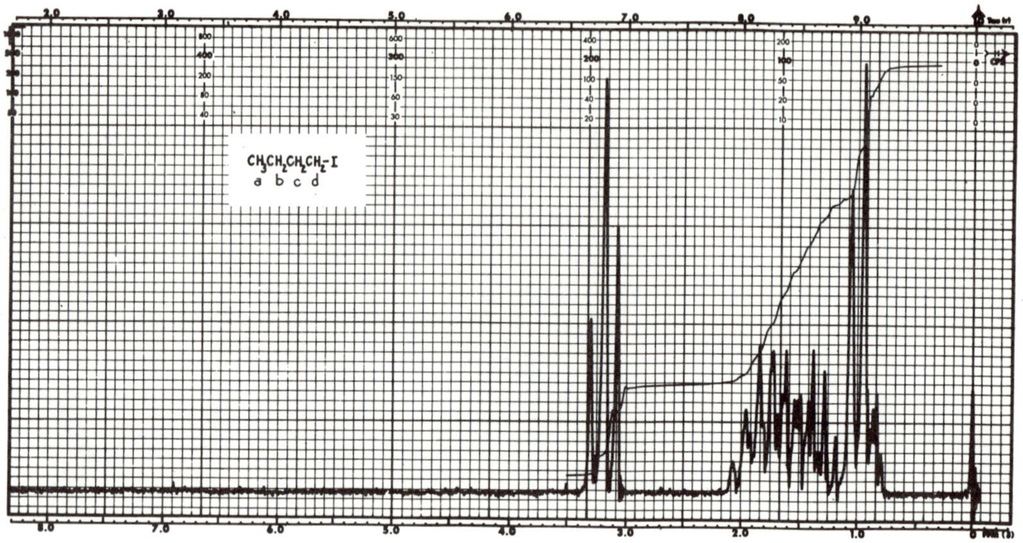

© 1978 Sadtler Research Laboratories, Inc.

2-IODOBUTANE

C_4H_9I

Mol. Wt. 184.02

B.P. 119-122°C

Solvent: $CDCl_3$

Assignments
a = 0.99
b = 1.75
c = 1.90
d = 4.07

408

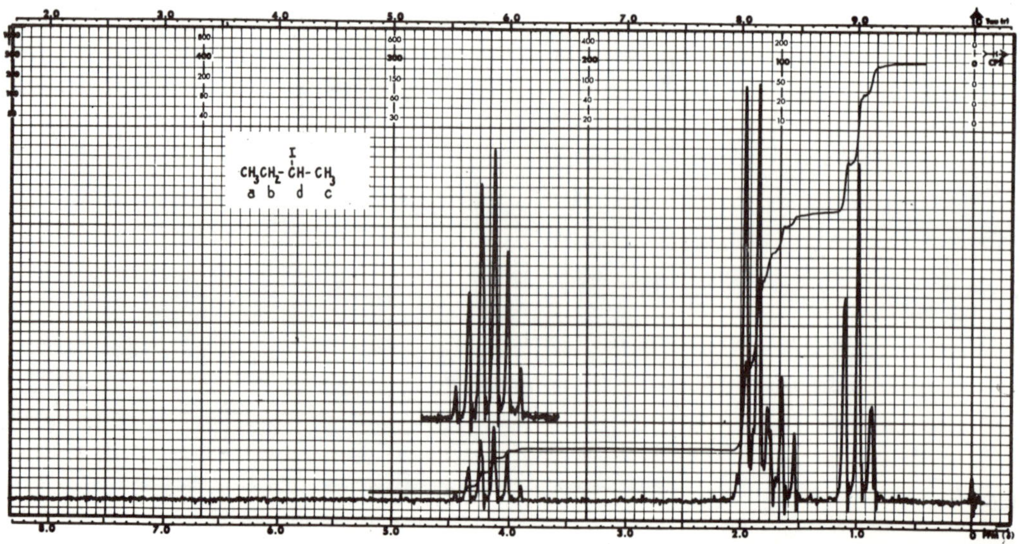

© 1978 Sadtler Research Laboratories, Inc.

1-IODO-2-METHYLPROPANE

C_4H_9I

Mol. Wt. 184.02

B.P. 119-121^oC

Solvent: CCl_4

Assignments
a = 1.01
b = 1.73
c = 3.10

409

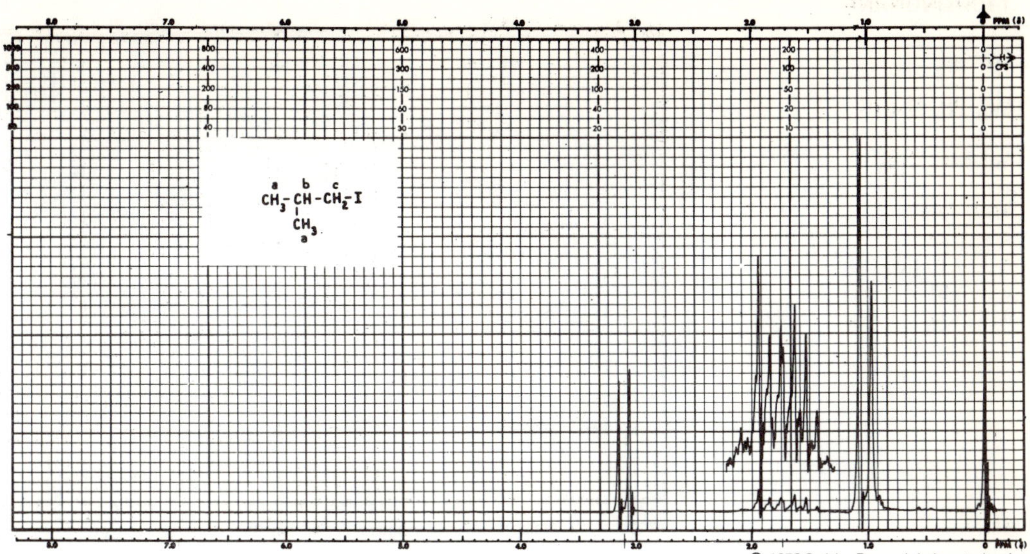

© 1978 Sadtler Research Laboratories, Inc.

1-IODO-3-METHYLBUTANE

$C_5H_{11}I$

Mol. Wt. 98.05

B.P. 145-147^oC

Solvent: CCl_4

Assignments
a = 0.96
b = ca 1.73
c = 3.17

410

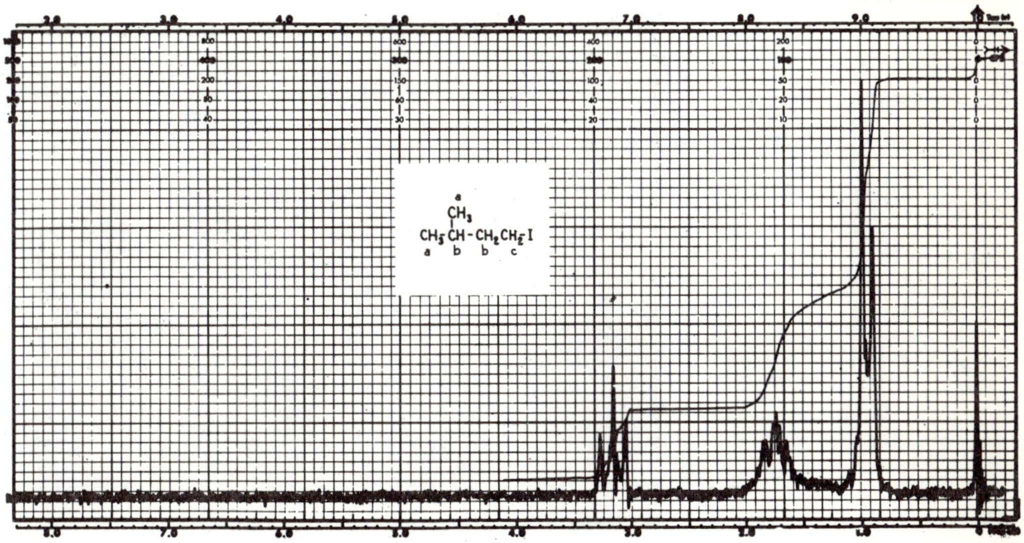

© 1978 Sadtler Research Laboratories, Inc.

1-IODOHEXANE

$C_6H_{13}I$

Mol. Wt. 212.07

B.P. 70-73^oC/20mm

Solvent: CCl_4

Assignments
a = 0.90
b = 1.10-1.65
c = 1.82
d = 3.14

411

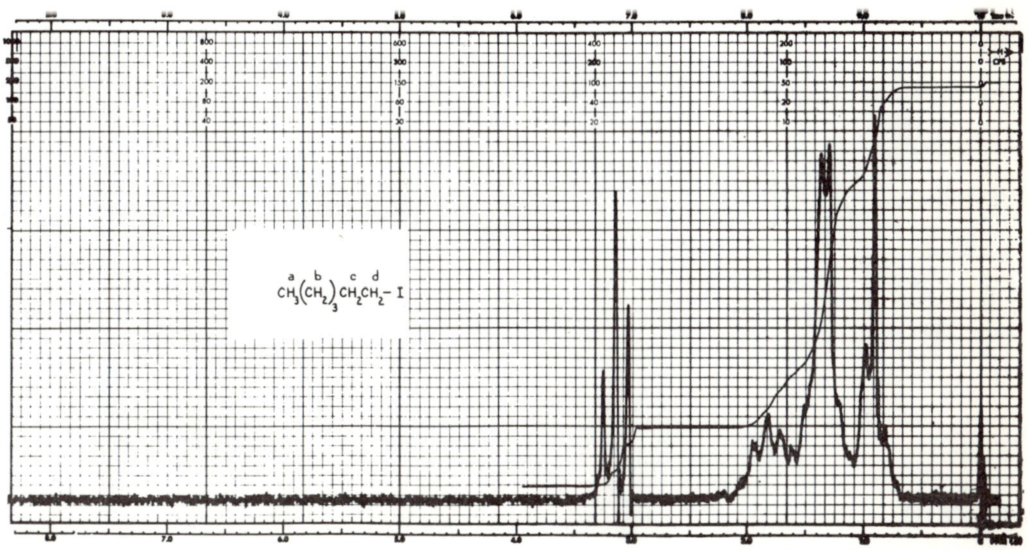

© 1978 Sadtler Research Laboratories, Inc.

1-IODONONANE

$C_9H_{19}I$

Mol. Wt. 254.16

Solvent: CCl_4

Assignments
a = 0.88
b = 1.10-1.60
c = 1.81
d = 3.12

412

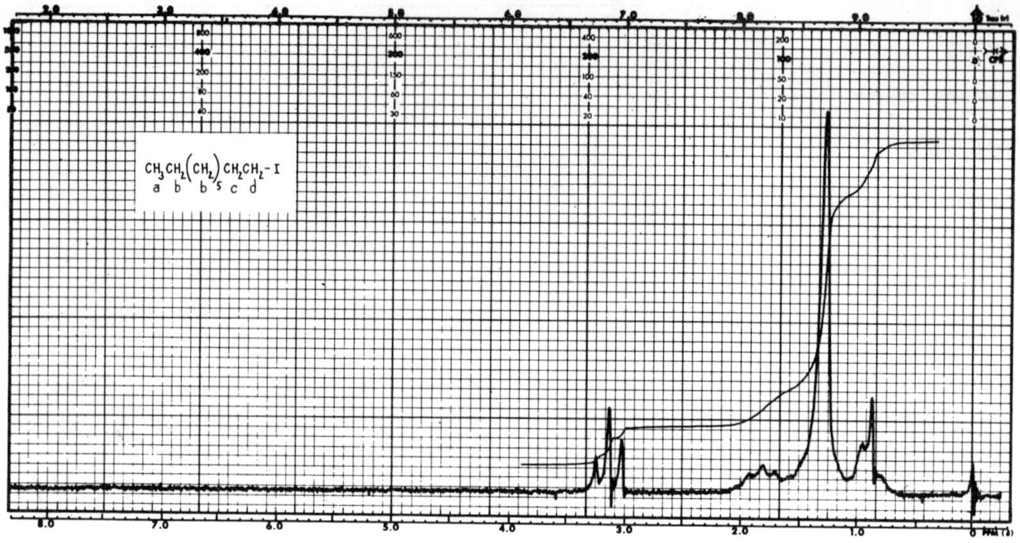

© 1978 Sadtler Research Laboratories, Inc.

1-IODODODECANE

$C_{12}H_{25}I$

Mol. Wt. 296.24

M.P. –5 to –2°C

Solvent: CCl_4

Assignments
a = 0.89
b = ca 1.28
c = 1.83
d = 3.14

413

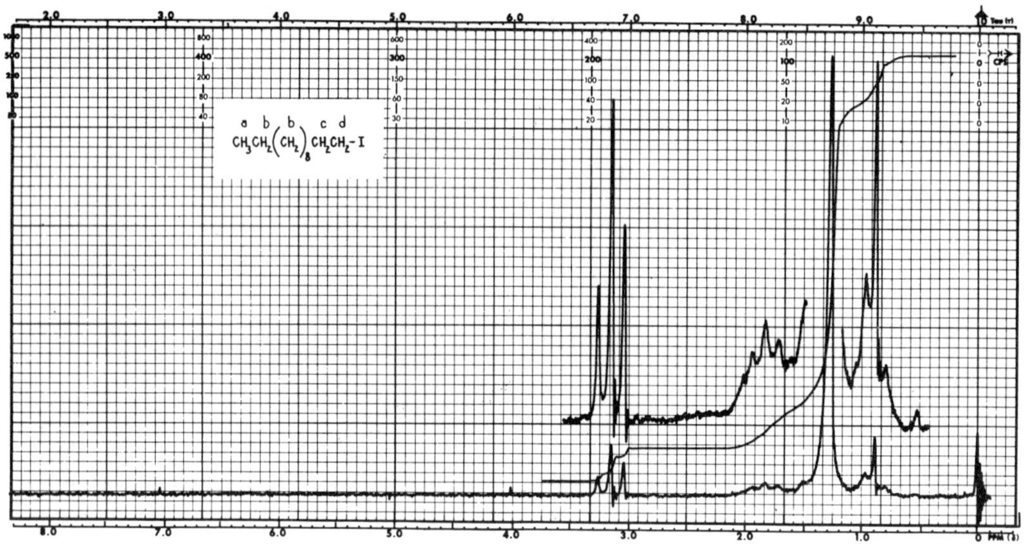

© 1978 Sadtler Research Laboratories, Inc.

1-IODOHEXADECANE

$C_{16}H_{33}I$

Mol. Wt. 352.35

M.P. 21-22°C

Solvent: CCl_4

Assignments
a = 0.89
b = ca 1.29
c = 1.82
d = 3.14

414

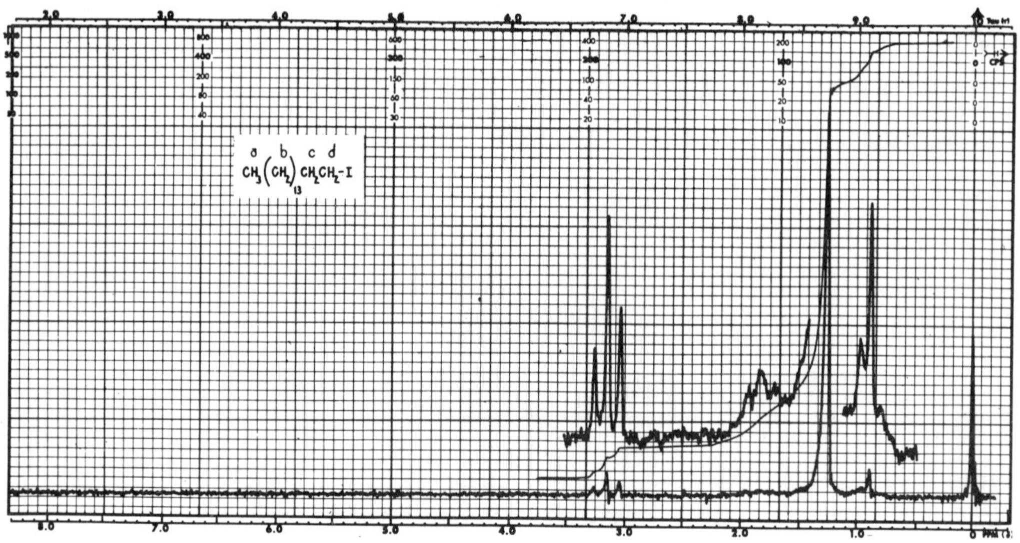

© 1978 Sadtler Research Laboratories, Inc.

DIIODOMETHANE

CH_2I_2

Mol. Wt. 267.85

M.P. 5-6°C

Solvent: CCl_4

Assignments
a = 3.88

415

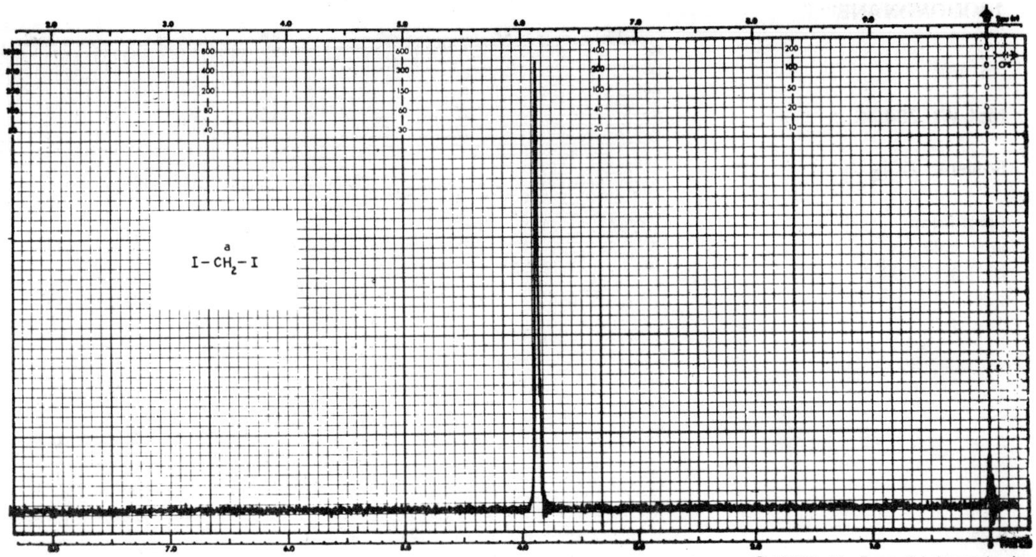

© 1978 Sadtler Research Laboratories, Inc.

1,2-DIIODOETHANE

$C_2H_4I_2$

Mol. Wt. 281.86

M.P. 81-82°C (lit.)

Solvent: $CDCl_3$

Assignments
a = 3.70

416

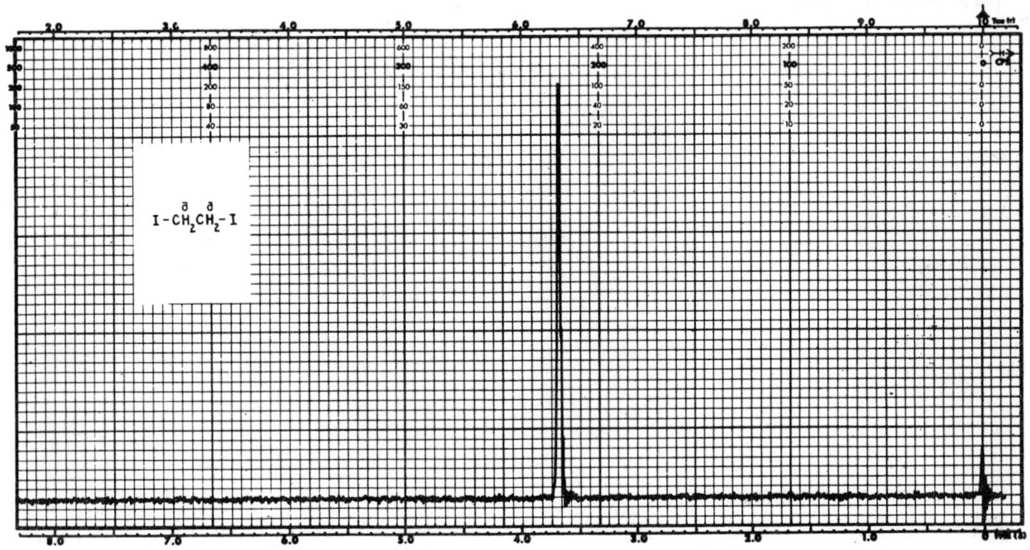

© 1978 Sadtler Research Laboratories, Inc.

1,3-DIIODOPROPANE

$C_3H_6I_2$

Mol. Wt. 295.89

Solvent: CCl_4

Assignments
a = 2.29
b = 3.29

417

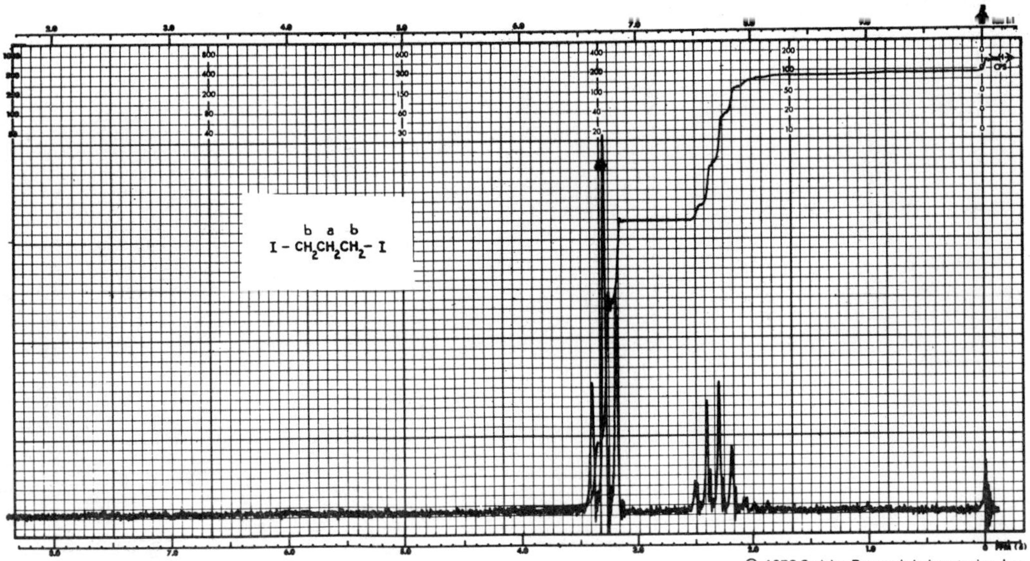

© 1978 Sadtler Research Laboratories, Inc.

1,4-DIIODOBUTANE

$C_4H_8I_2$

Mol. Wt. 309.95

Solvent: CCl_4

Assignments
a = 1.94
b = 3.17

418

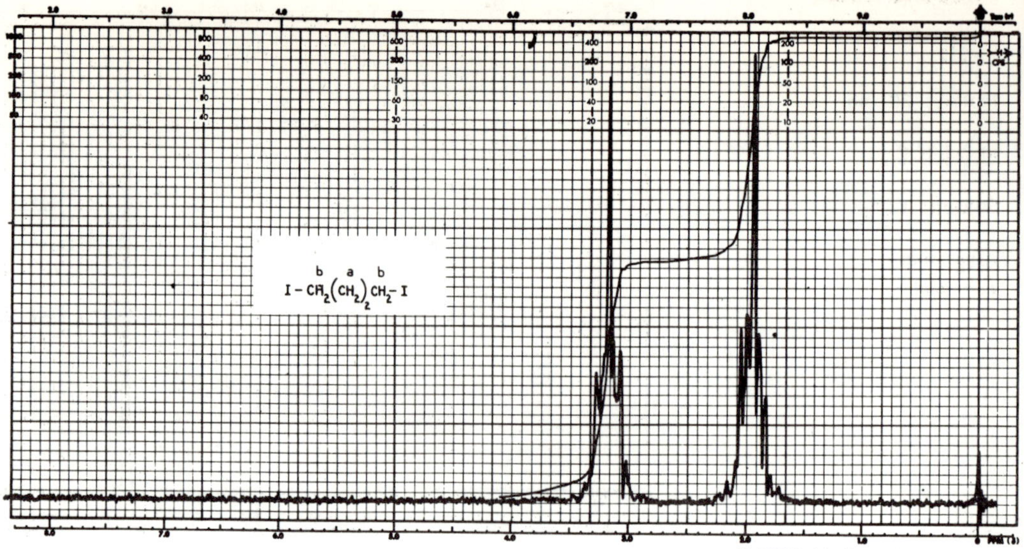

© 1978 Sadtler Research Laboratories, Inc.

1,5-DIIODOPENTANE

$C_5H_{10}I_2$

Mol. Wt. 323.94

M.P. 8-10°C

Solvent: $CDCl_3$

Assignments
a = 1.20-2.20
b = 3.20

419

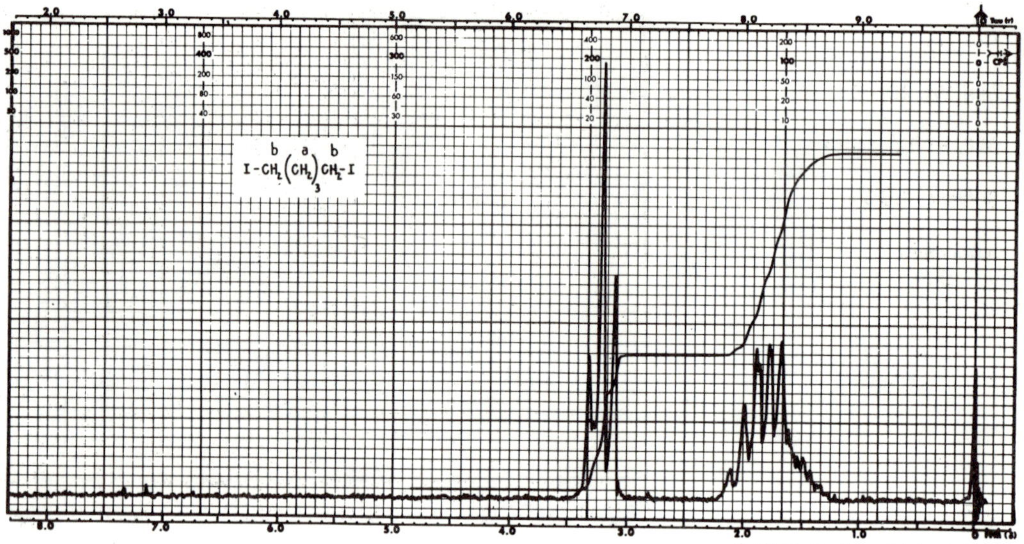

© 1978 Sadtler Research Laboratories, Inc.

1,6-DIIODOHEXANE

$C_6H_{12}I_2$

Mol. Wt. 333.97

B.P. 135-138°C/6mm

Solvent: CCl_4

Assignments
a = 1.50
b = 1.88
c = 3.20

420

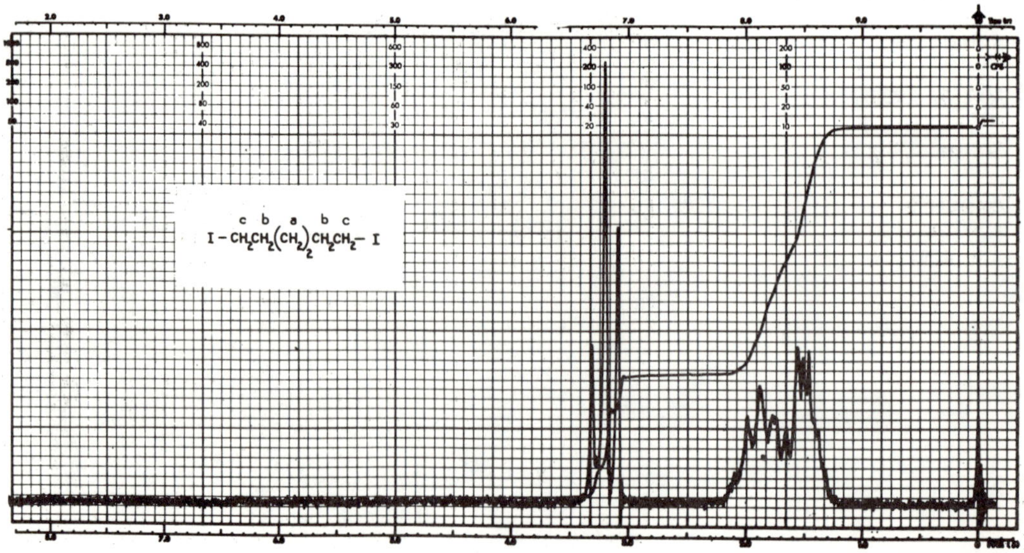

© 1978 Sadtler Research Laboratories, Inc.

IODOCYCLOPENTANE

C_5H_9I

Mol. Wt. 196.03

Solvent: CCl_4

Assignments
a = 1.78
b = 2.03
c = 4.32

421

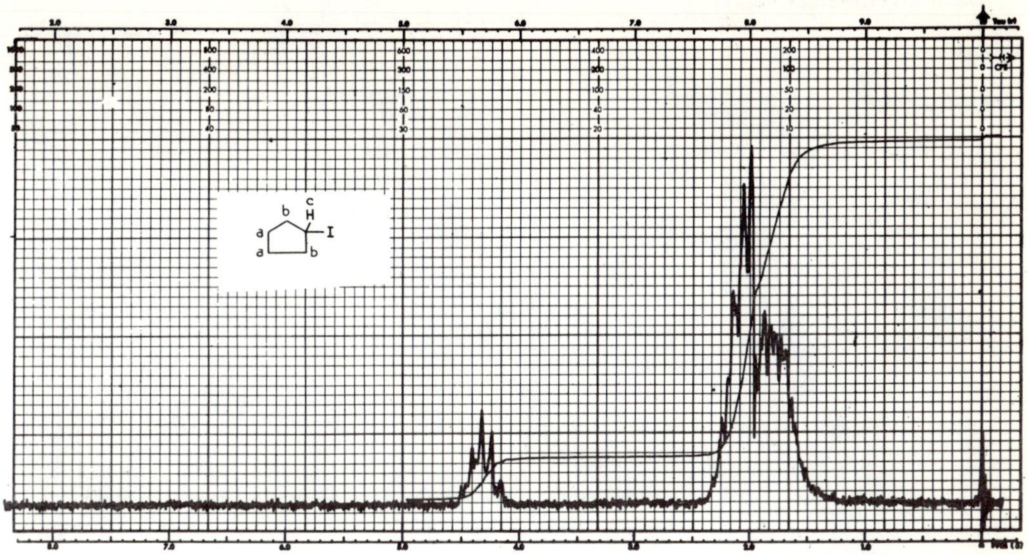

© 1978 Sadtler Research Laboratories, Inc.

IODOCYCLOHEXANE

$C_6H_{11}I$

Mol. Wt. 210.06

B.P. 81.5°C/20mm

Solvent: $CDCl_3$

Assignments
a = 1.00-2.00
b = ca 2.08
c = 4.36

422

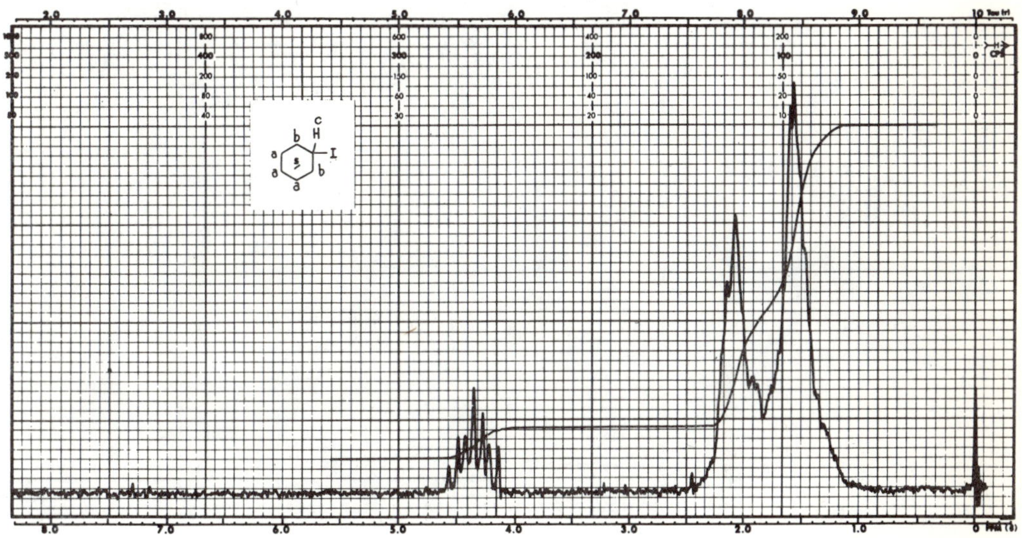

© 1978 Sadtler Research Laboratories, Inc.

3-IODOPROPENE

C_3H_5I

Mol. Wt. 167.98

Solvent: CCl_4

Assignments
a = 3.80
b = 4.91
c = 5.16
d = 5.98

423

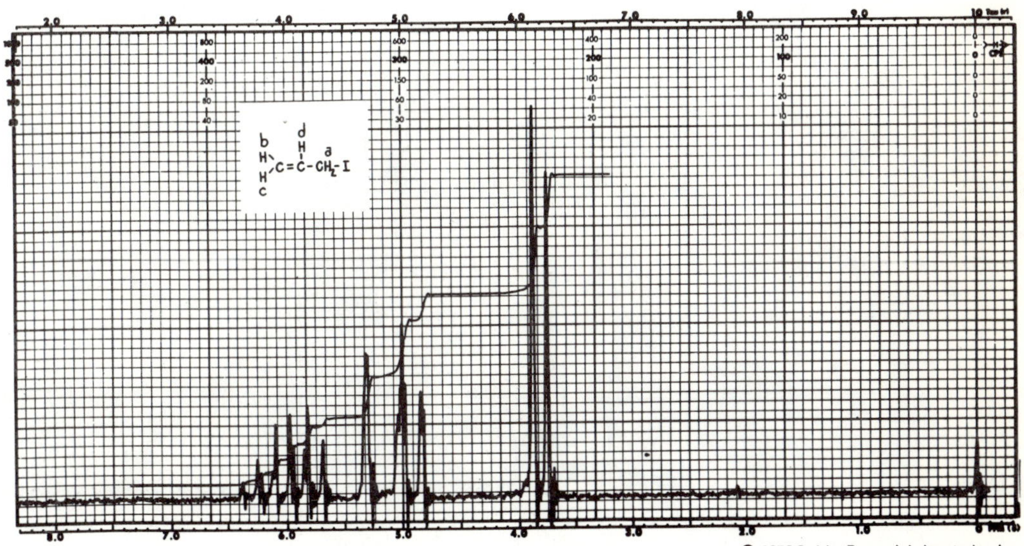

© 1978 Sadtler Research Laboratories, Inc.

(2-IODOETHYL)BENZENE

C_8H_9I

Mol. Wt. 232.07

B.P. 114-116°C/14mm

Solvent: $CDCl_3$

Assignments
a = 3.08
b = 3.19
c = 6.90-7.35

424

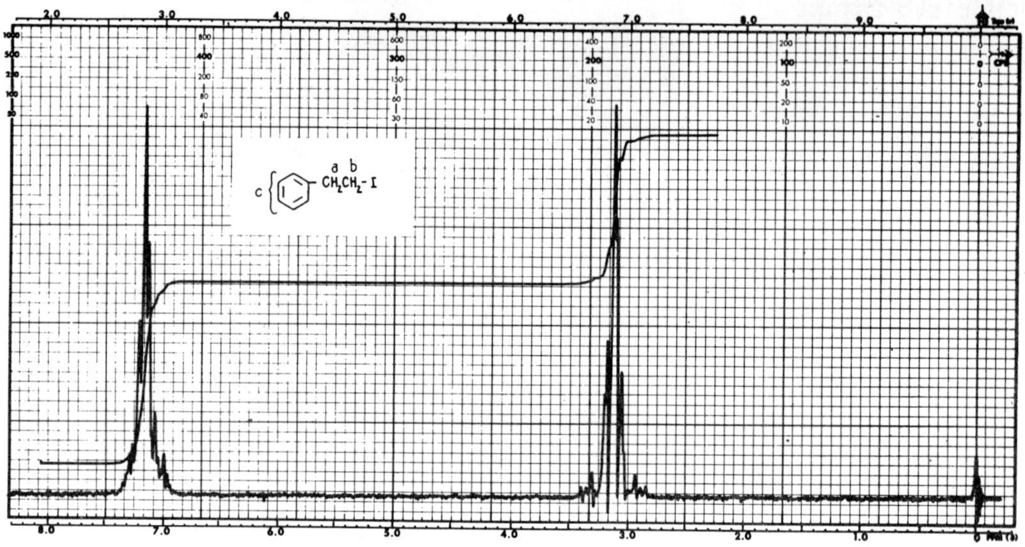

© 1978 Sadtler Research Laboratories, Inc.

α-IODOTOLUENE

C_7H_7I

Mol. Wt. 218.04

M.P. 23-24°C

Solvent: $CDCl_3$

Assignments
a = 4.38
b = 7.05-7.45

425

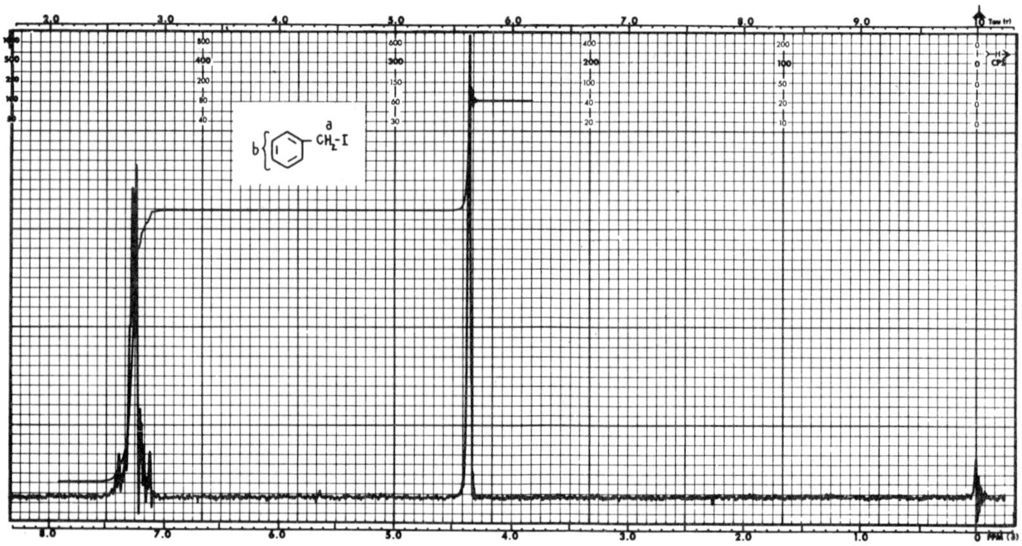

© 1978 Sadtler Research Laboratories, Inc.

IODOBENZENE

C_6H_5I

Mol. Wt. 204.02

B.P. 70-71°C/12mm

Solvent: CCl_4

Assignments
a = 6.82-7.45
b = 7.65

426

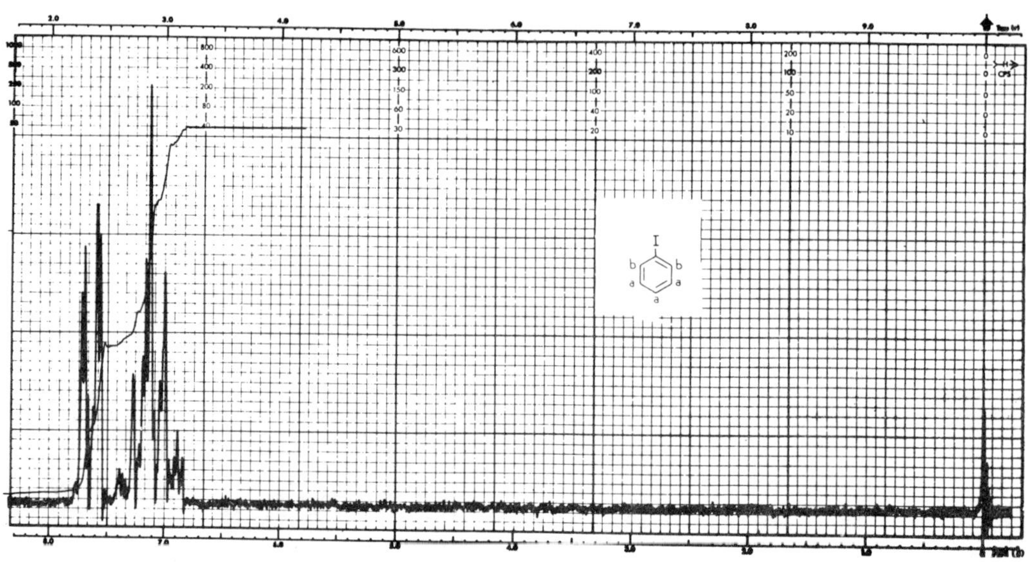

© 1978 Sadtler Research Laboratories, Inc.

o-IODOTOLUENE

C_7H_7I

Mol. Wt. 218.04

B.P. 90-92°C/18mm

Solvent: $CDCl_3$

Assignments
a = 2.36
b = 6.76
c = ca 7.11
d = 7.71

427

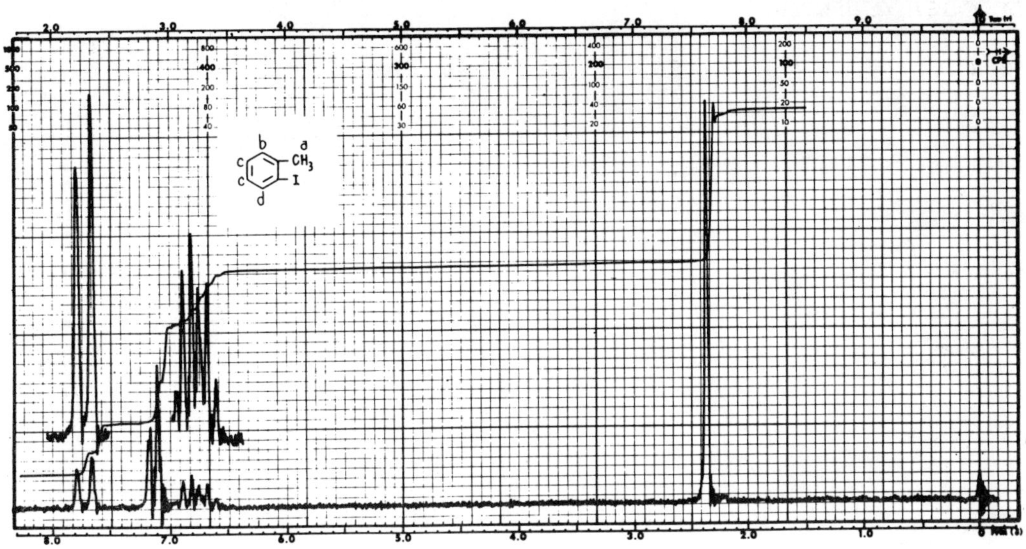

© 1978 Sadtler Research Laboratories, Inc.

m-IODOTOLUENE

C_7H_7I

Mol. Wt. 218.04

B.P. 92-94°C/15mm

Solvent: CCl_4

Assignments
a = 2.24
b = 6.90
c = 7.07
d = 7.48
e = 7.51

428

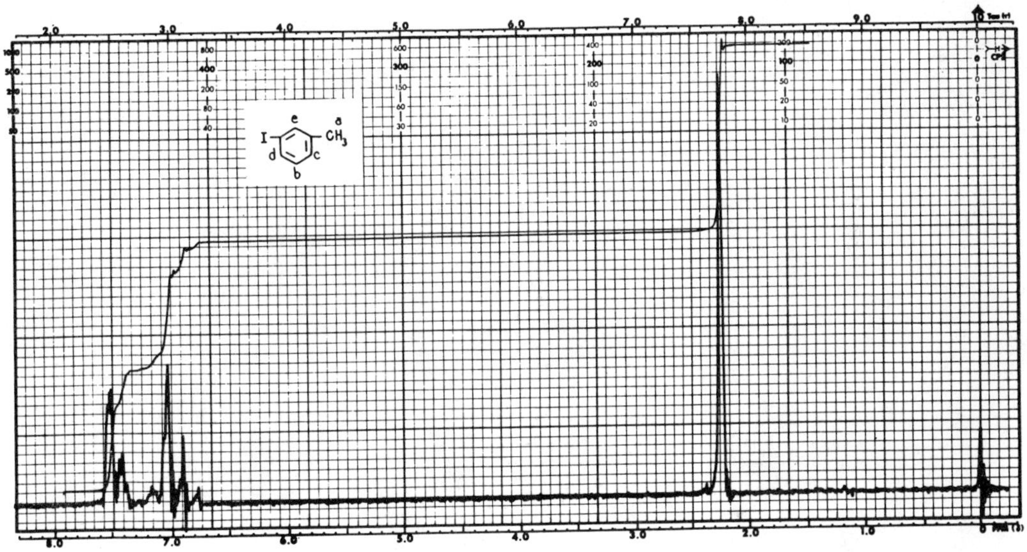

© 1978 Sadtler Research Laboratories, Inc.

p-IODOTOLUENE

C_7H_7I

Mol. Wt. 218.05

M.P. 34-35°C

Solvent: $CDCl_3$

Assignments
a = 2.26
b = 6.89
c = 7.55

429

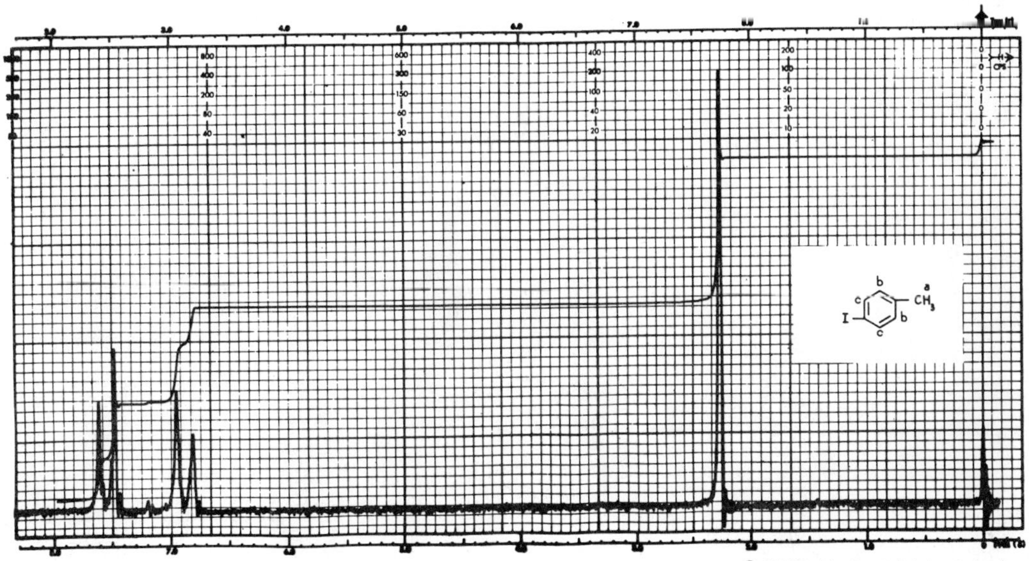

© 1978 Sadtler Research Laboratories, Inc.

2-IODO-p-XYLENE

C_8H_9I

Mol. Wt. 232.07

B.P. 113-114°C/20mm

Solvent: CCl_4

Assignments
a = 2.18
b = 2.30
c = ca 6.97
d = 7.55

430

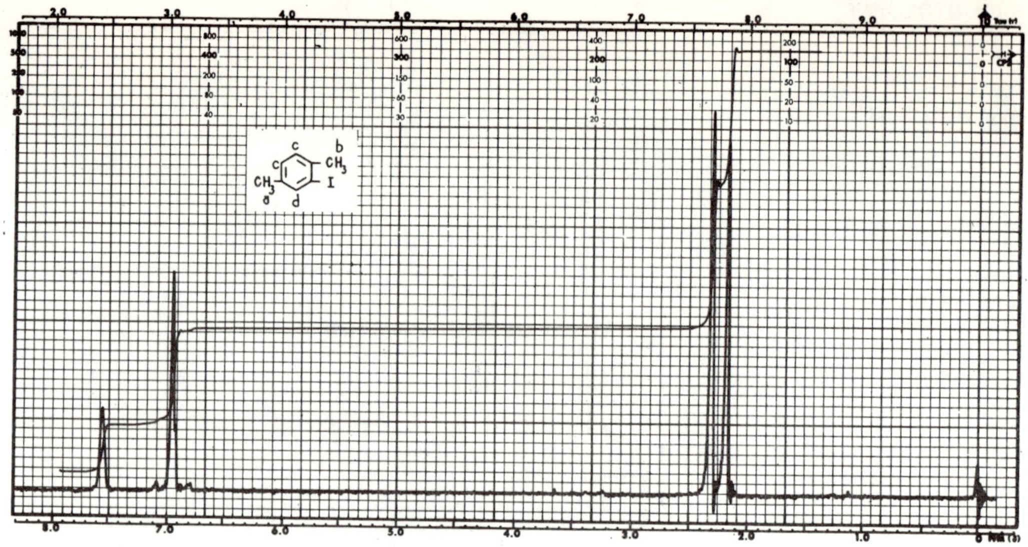

© 1978 Sadtler Research Laboratories, Inc.

---

m-IODO-*a,a,a*-TRIFLUOROTOLUENE

$C_7H_4F_3I$

Mol. Wt. 272.01

Solvent: CCl_4

Assignments
a = 7.20
b = 7.59
c = 7.90
d = 7.97
e = 1.29 imp.

431

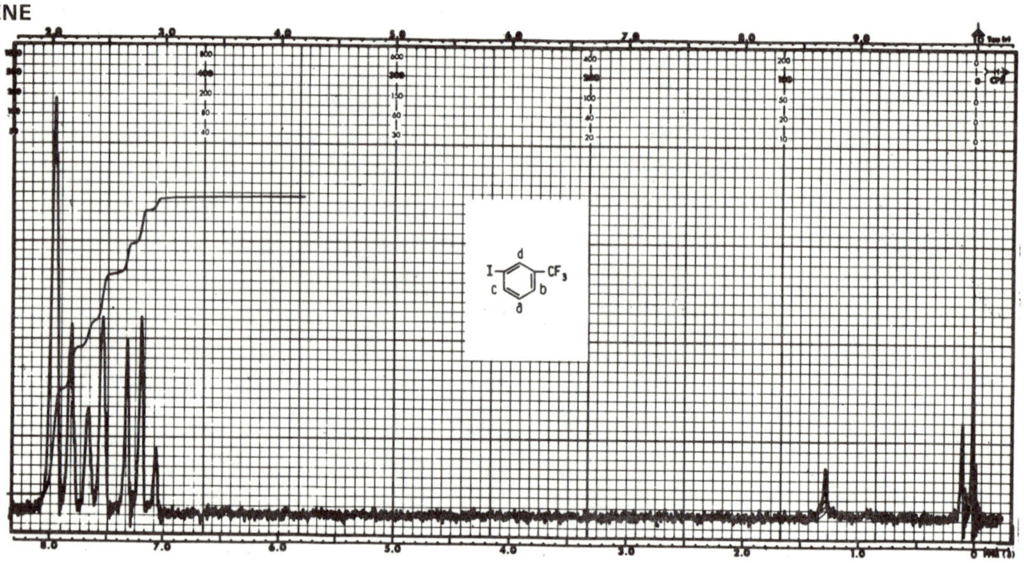

© 1978 Sadtler Research Laboratories, Inc.

1-FLUORO-4-IODOBENZENE

C_6H_4FI

Mol. Wt. 222.00

Solvent: $CDCl_3$

Assignments
a = 6.78
b = 7.58

432

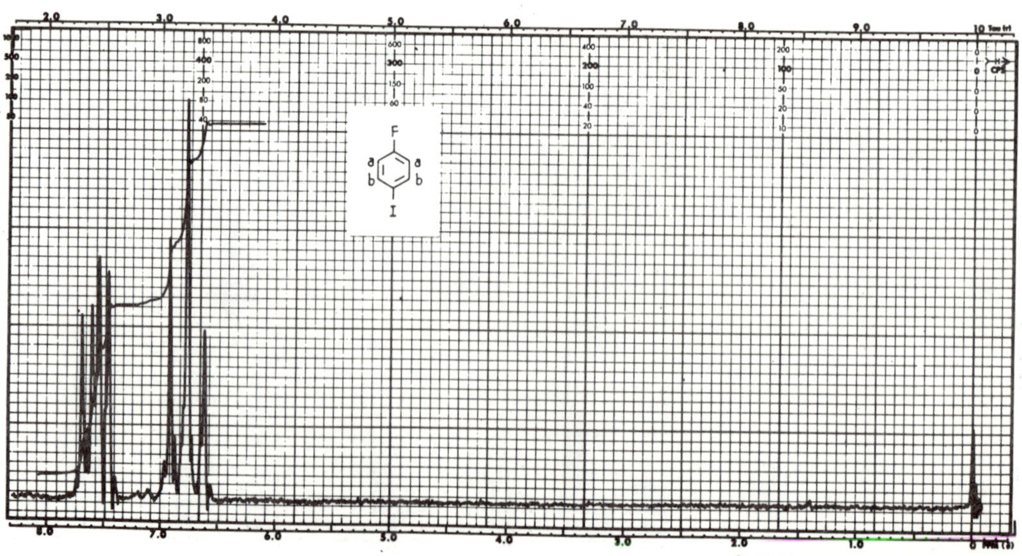

© 1978 Sadtler Research Laboratories, Inc.

1-CHLORO-2-IODOBENZENE

C_6H_4ClI

Mol. Wt. 238.46

B.P. 119-120°C/27mm

Solvent: CCl_4

Assignments
a = 6.88
b = ca 7.38
c = ca 7.77

433

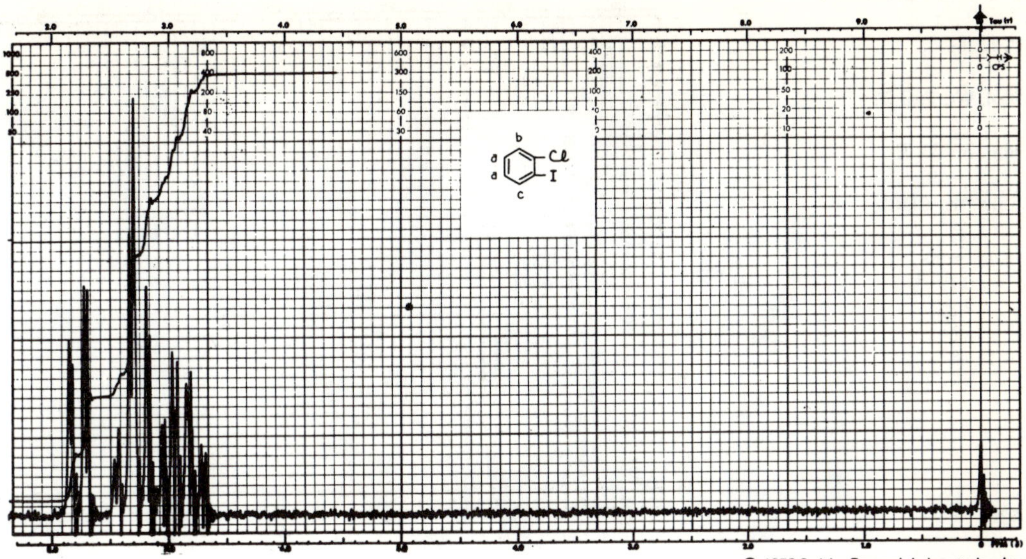

© 1978 Sadtler Research Laboratories, Inc.

1-BROMO-2-IODOBENZENE

C_6H_4BrI

Mol. Wt. 282.91

B.P. 124-127°C/17mm

Solvent: CCl_4

Assignments
a = 6.90
b = 7.11
c = 7.65 or d
d = 7.79 or c

434

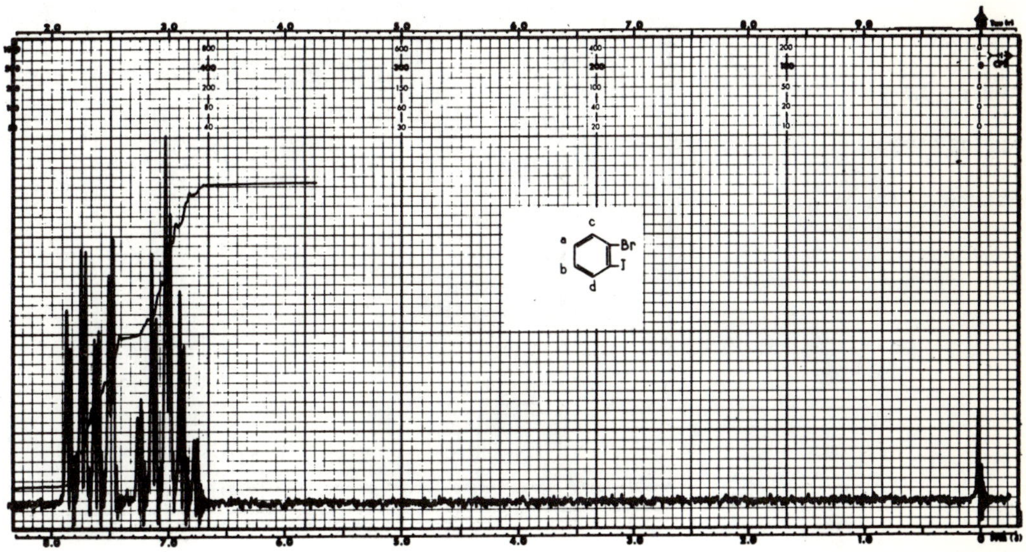

© 1978 Sadtler Research Laboratories, Inc.

1-BROMO-4-IODOBENZENE

C_6H_4BrI

Mol. Wt. 282.91

M.P. 90-91°C

Solvent: $CDCl_3$

Assignments
a = 7.18
b = 7.50

435

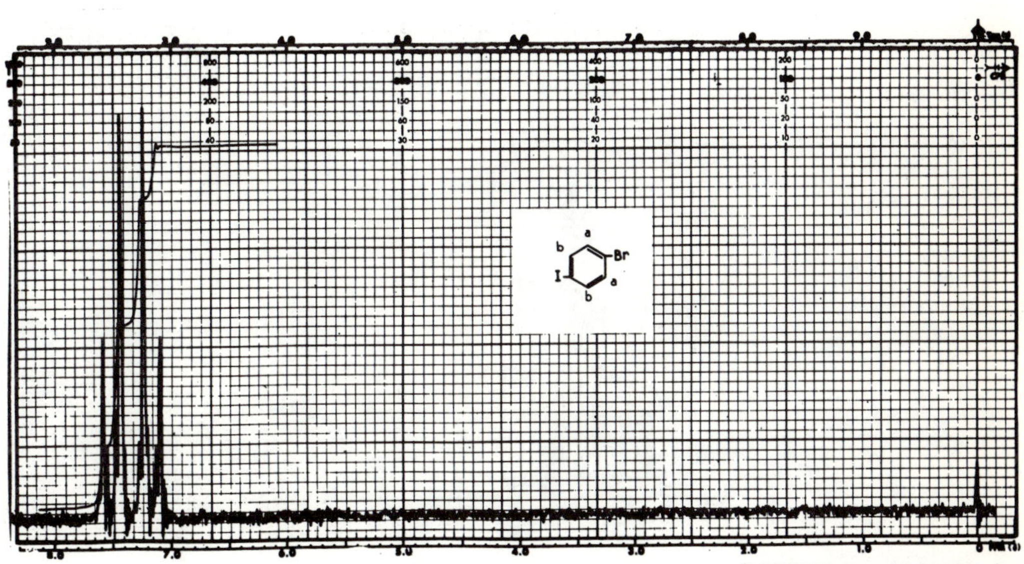

© 1978 Sadtler Research Laboratories, Inc.

m-DIIODOBENZENE

$C_6H_4I_2$

Mol. Wt. 329.91

M.P. 35-37°C

Solvent: $CDCl_3$

Assignments
a = 6.79
b = 7.60
c = 8.04

436

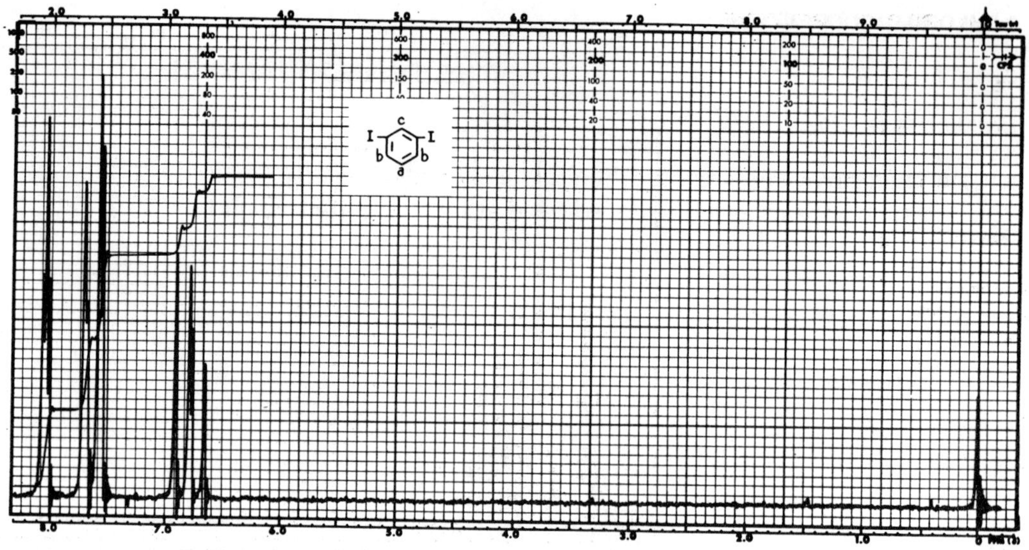

© 1978 Sadtler Research Laboratories, Inc.

p-DIIODOBENZENE

$C_6H_4I_2$

Mol. Wt. 329.91

M.P. 127-129°C

Solvent: $CDCl_3$

Assignments
a = 7.40

437

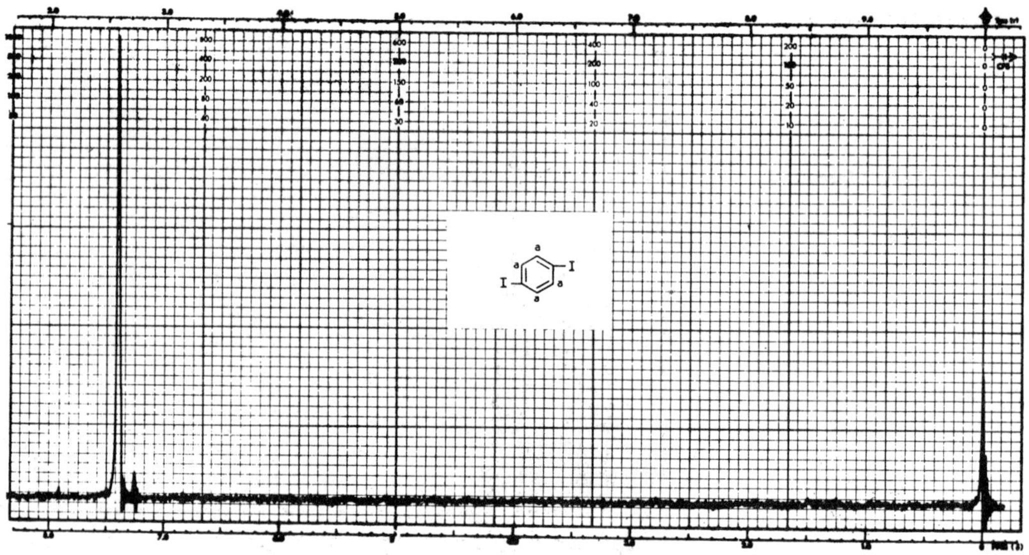

© 1978 Sadtler Research Laboratories, Inc.

1-IODONAPHTHALENE

$C_{10}H_7I$

Mol. Wt. 254.07

M.P. 10-12°C

Solvent: $CDCl_3$

Assignments
a = 7.00
b = 7.20-7.60
c = 7.65
d = 7.99
e = 8.01

438

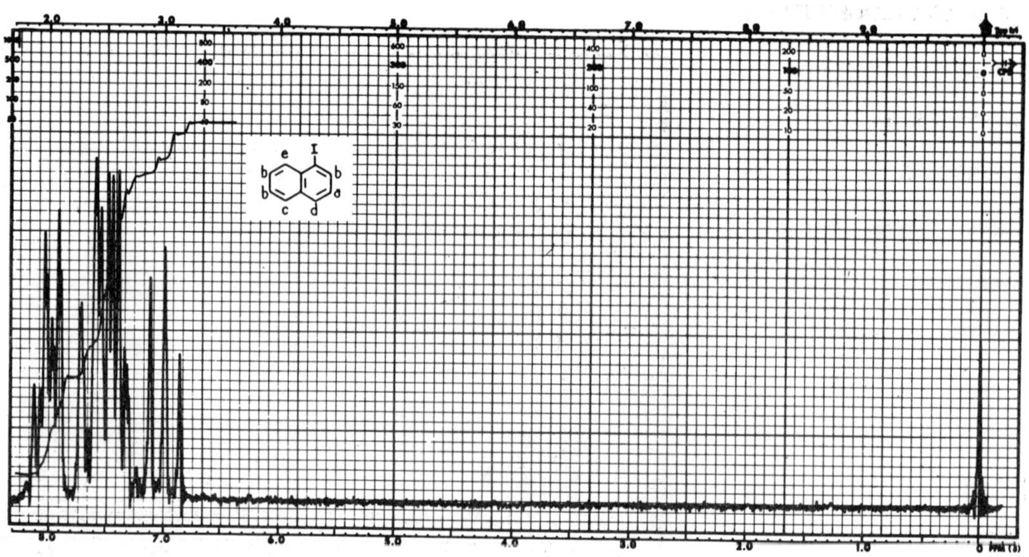

© 1978 Sadtler Research Laboratories, Inc.

THE PRIMARY AMINES

The primary amines are often relatively easy to characterize due to the presence of the $-NH_2$ group which appears as a broadened band at intermediate to high field (6.7–0.6 ppm).

The primary amine group acts as a weak deshielding substituent on methyl, methylene and methine groups, but has a strong shielding effect upon the ortho and para aromatic hydrogens.

Chemical Shifts

Aliphatic Protons

HC–3	HC–2	HC–1		Solvent	Spectrum
	1.05	2.61	$CH_3-CH_2-NH_2$	D_2O	- - - -
0.91	1.48	2.65	$CH_3-CH_2-CH_2-NH_2$	CCl_4	439
	(1.01)	3.04	$(CH_3)_2-CH-NH_2$	D_2O	440
	(1.30)	- - - -	$(CH_3)_3-C-NH_2$	$CDCl_3$	- - - -

Substituted methylamines

H_2N	$-CH_2$	$-X$	
	2.52	–cyclohexyl	$CDCl_3$
	3.12	$-P(=O)(OH)_2$	D_2O
	3.30	$-CH=CH_2$	$CDCl_3$
	3.58	–C(=O)–OH	D_2O
	3.76	–2–Furyl	D_2O
	3.85	–Ø	$CDCl_3$
	3.86	–4–Pyridyl	$CDCl_3$

2 Substituted ethylamines

H_2N	$-CH_2$	$-CH_2$	$-X$	
- - - -	2.65	1.48	$-CH_3$	CCl_4
1.55	2.99	2.49	–CN	D_2O
- - - -	3.18	2.53	–C(=O)–OH	D_2O
- - - -	2.86	2.59	–SH	D_2O
1.19	2.60	2.60	$-NH_2$	CCl_4
0.93	2.82	2.68	–Ø	CCl_4

Aromatic Protons

The strong shielding effect of the primary amine group on the hydrogens of benzene is evident from the chemical shifts of the parent compound, aniline.

Aniline	H_2N-	ortho	meta	para	
		6.40	7.00	6.59	CCl_4

Several examples of para substituted anilines are presented below.

H_2N-	ortho	meta	para	Solvent
	6.34	6.34	$-O-\emptyset$	$CDCl_3$
	6.32	6.50	$-S-CH_3$	CCl_4
	6.41	6.55	$-R_2$	$CDCl_3$
	6.38	6.57	$-SH$	$CDCl_3$
	6.59	6.59	$-N(R)_2$	$CDCl_3$
	6.62	6.62	$-N(CH_3)_2$	$CDCl_3$
	6.49	6.65	$-O-R_2$	$CDCl_3$
	6.68	6.72	$-O-CH_3$	$CDCl_3$
	6.49	6.81	$-CH_2-\emptyset$	DMSO-d_6
	6.75	6.92	$-CF_3$	$CDCl_3$
	6.60	7.02	$-Cl$	DMSO
	6.62	7.09	$-C(CH_3)_2-\emptyset$	$CDCl_3$
	6.57	7.21	$-Br$	$CDCl_3$
	6.55	7.26	$-NH-C(=O)-R$	DMSO
	6.77	7.36	$-C(=O)-O-CH_3$	$CDCl_3$
	6.37	7.37	$-I$	$CDCl_3$
	6.70	7.39	$-As(=O)(OH)_2$	DMSO
	6.67	7.60	$-SO_2-NH-R$	DMSO-d_6
	6.79	7.77	$-C(=O)-CH_3$	$CDCl_3$
	6.60	7.79	$-N=N-\emptyset$	$CDCl_3$

Exchangeable Protons

The NH_2 Group

Range (ppm)	Type	
0.66–1.52	Aliphatic–NH_2	CCl_4, $CDCl_3$
1.72–1.78	Alicyclic–NH_2	CCl_4, $CDCl_3$
3.30–6.70	Aromatic–NH_2	CCl_4, $CDCl_3$

ROPYLAMINE

$_3H_9N$

lol. Wt. 59.11

.P. –83°C

olvent: CCl_4

ssignments
= 0.91
= 1.48
= 1.80-2.40
= 2.65

39

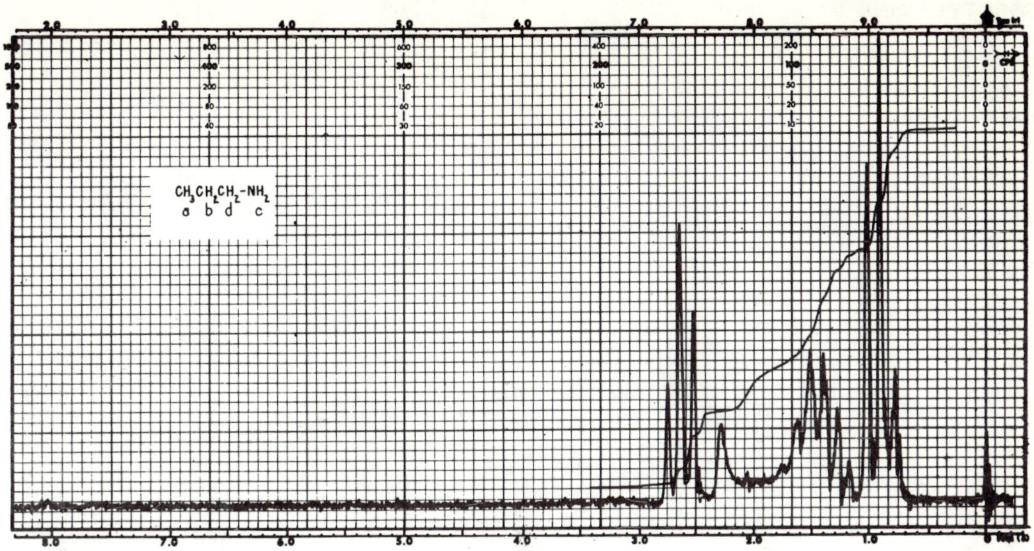

© 1978 Sadtler Research Laboratories, Inc.

SOPROPYLAMINE

C_3H_9N

Mol. Wt. 59.11

M.P. 101°C B.P. 34°C

Solvent: D_2O

Assignments
a = 1.01
b = 3.04
c = 4.60 HDO

440

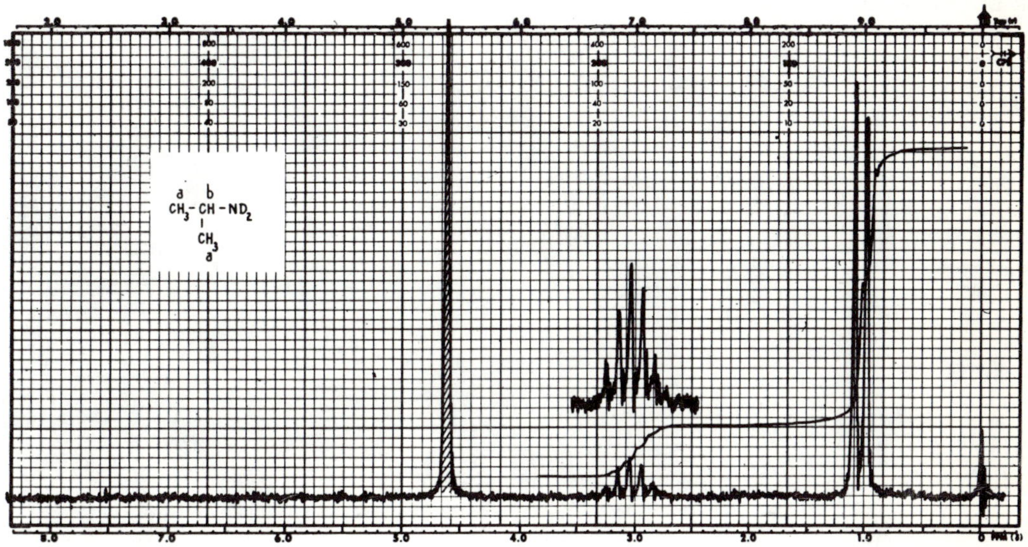

© 1978 Sadtler Research Laboratories, Inc.

BUTYLAMINE

$C_4H_{11}N$

Mol. Wt. 73.14

B.P. 76-78°C

Solvent: $CDCl_3$

Assignments
a = 0.91
b = 1.17-1.57
c = 1.60
d = 2.61
e = 1.81 imp.

441

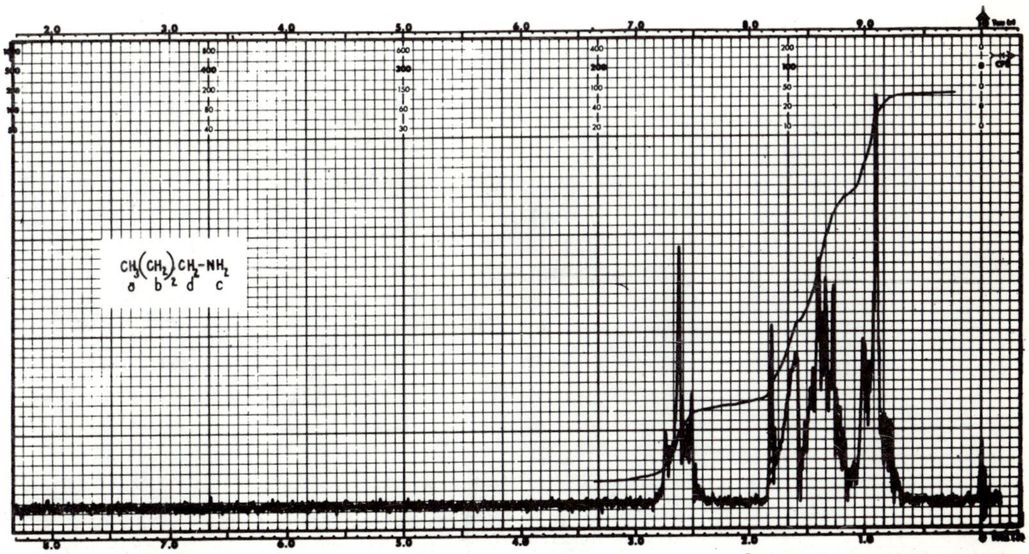

© 1978 Sadtler Research Laboratories, Inc.

ISOBUTYLAMINE

$C_4H_{11}N$

Mol. Wt. 73.14

B.P. 66-70°C

Solvent: $CDCl_3$

Assignments
a = 0.90
b = 1.59
c = 1.82
d = 2.49
e = 7.48 imp.

442

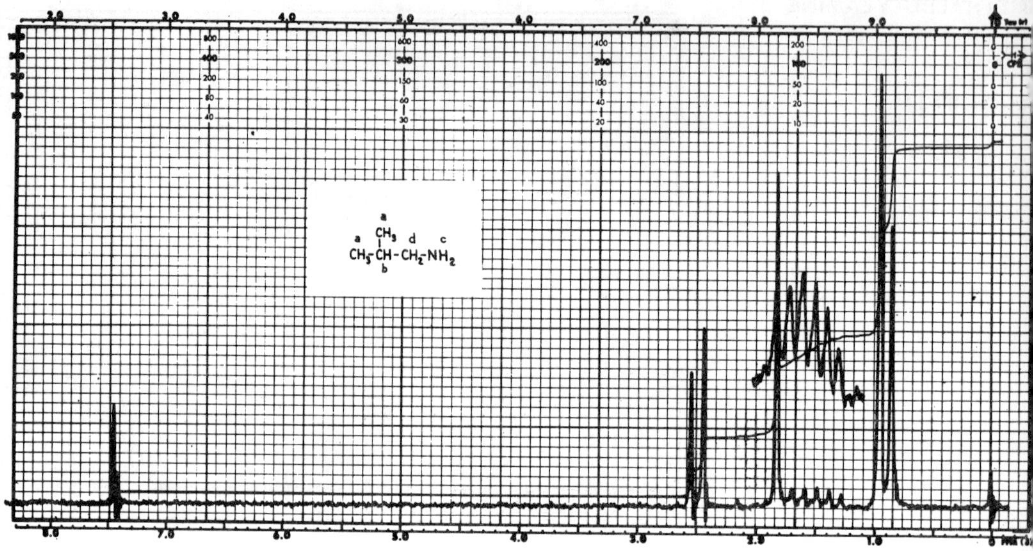

© 1978 Sadtler Research Laboratories, Inc.

sec-BUTYLAMINE

$C_4H_{11}N$

Mol. Wt. 73.14

B.P. 62-64°C

Solvent: $CDCl_3$

Assignments
a = 0.92
b = 1.05
c = 1.17
d = 1.31
e = 2.81

443

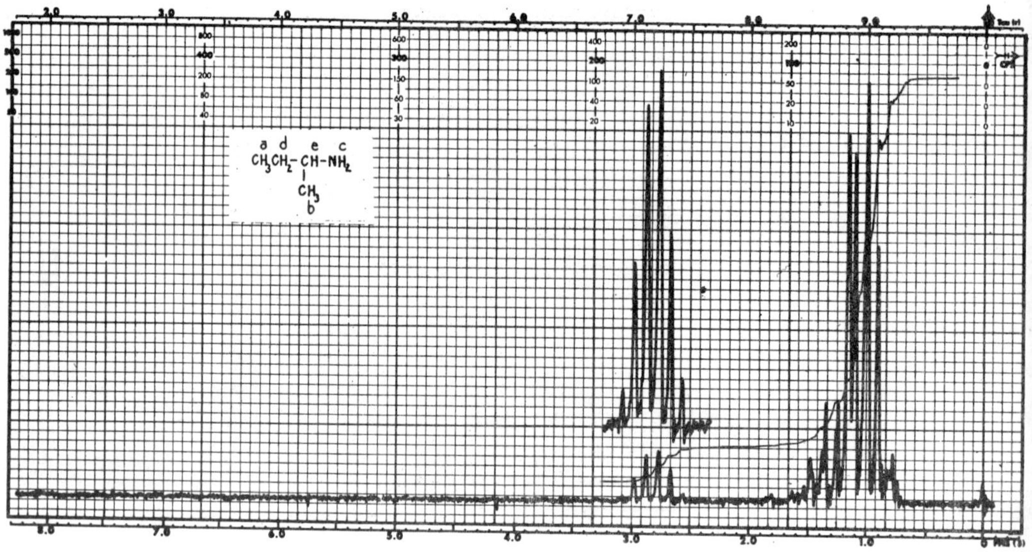

© 1978 Sadtler Research Laboratories, Inc.

PENTYLAMINE

$C_5H_{13}N$

Mol. Wt. 87.17

B.P. 103-106°C

Solvent: $CDCl_3$

Assignments
a = 0.90
b = 1.09-1.66
c = 1.51
d = 2.69

444

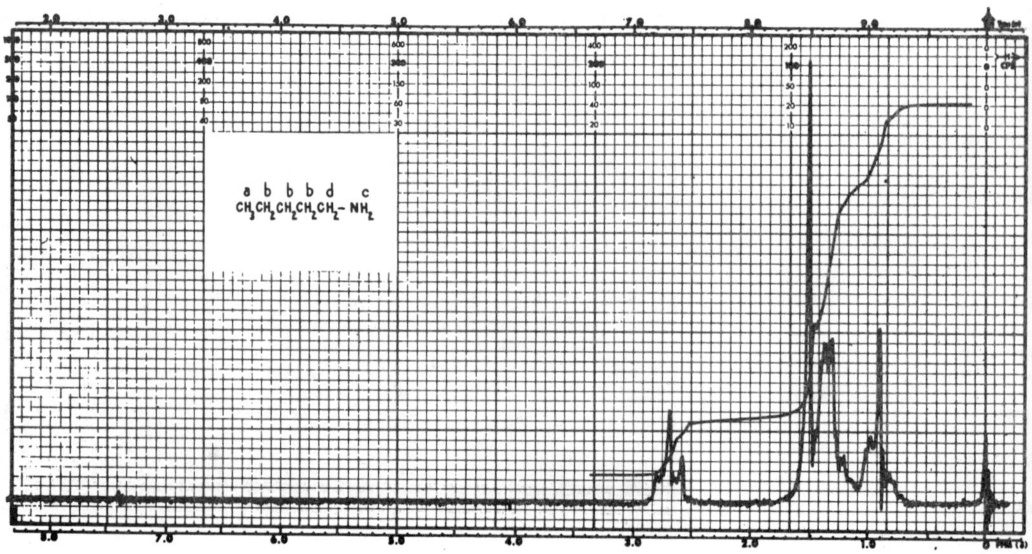

© 1978 Sadtler Research Laboratories, Inc.

-METHYLBUTYLAMINE

$_5H_{13}N$

lol. Wt. 87.17

.P. 96-100°C

olvent: $CDCl_3$

ssignments
= 0.92
= 0.94
= 1.09-1.68
= 1.21
= 2.57

45

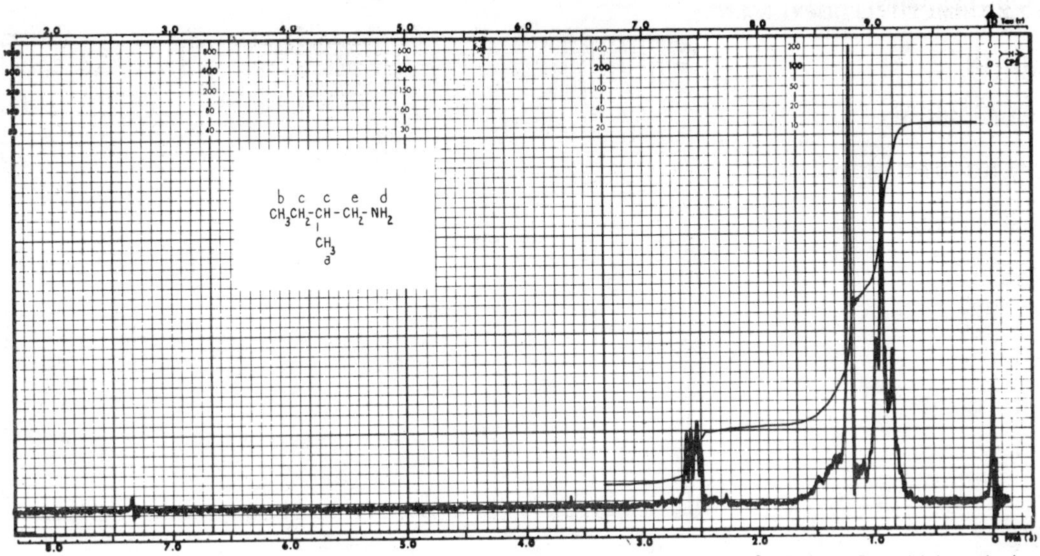

© 1978 Sadtler Research Laboratories, Inc.

SOPENTYLAMINE

$C_5H_{13}N$

Mol. Wt. 87.17

3.P. 95-96°C/760mm (lit.)

Solvent: $CDCl_3$

Assignments
= 0.91
= 1.32
= 1.49
= 1.70
= 2.71
= 7.70 $CHCl_3$

446

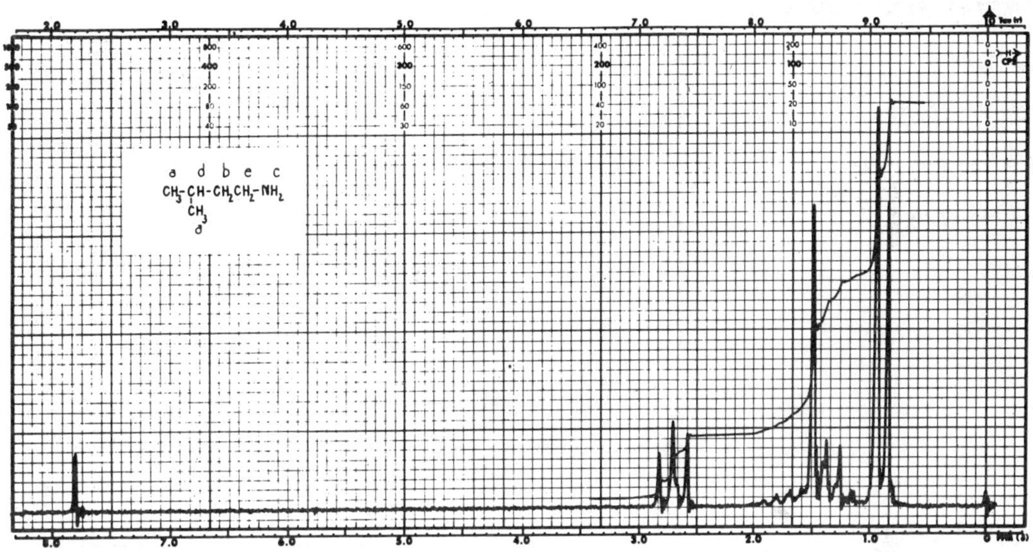

© 1978 Sadtler Research Laboratories, Inc.

tert-PENTYLAMINE

$C_5H_{13}N$

Mol. Wt. 87.16

Solvent: $CDCl_3$

Assignments
a = 0.89
b = 1.10
c = 1.22
d = 1.35

447

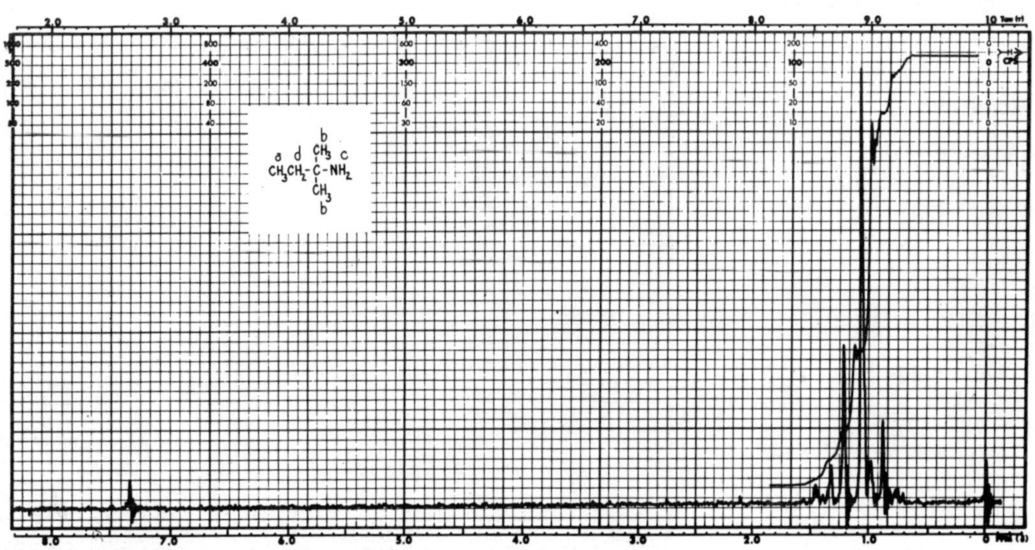

© 1978 Sadtler Research Laboratories, Inc.

1,2,2-TRIMETHYLPROPYLAMINE

$C_6H_{15}N$

Mol. Wt. 101.19

B.P. 103-105°C/760mm (lit.)

Solvent: CCl_4

Assignments
a = 0.85
b = 0.96
c = 1.37
d = 2.59
e = 1.08-1.11 imps.

448

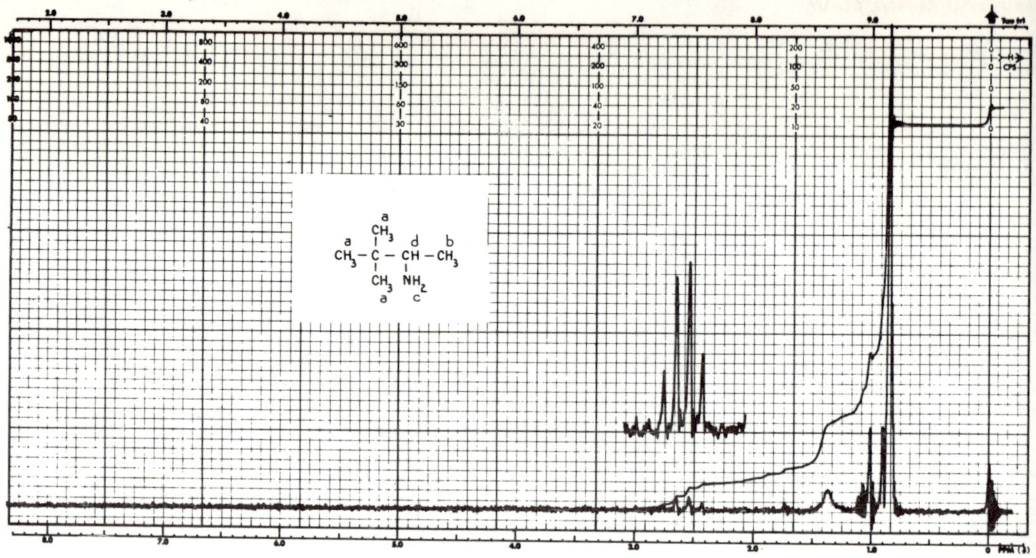

© 1978 Sadtler Research Laboratories, Inc.

1-METHYLPENTYLAMINE

$C_6H_{16}N$

Mol. Wt. 101.19

B.P. 141.4°C/755mm

Solvent: $CDCl_3$

Assignments
a = 0.87
b = 1.05
c = 1.25
d = 3.20

449

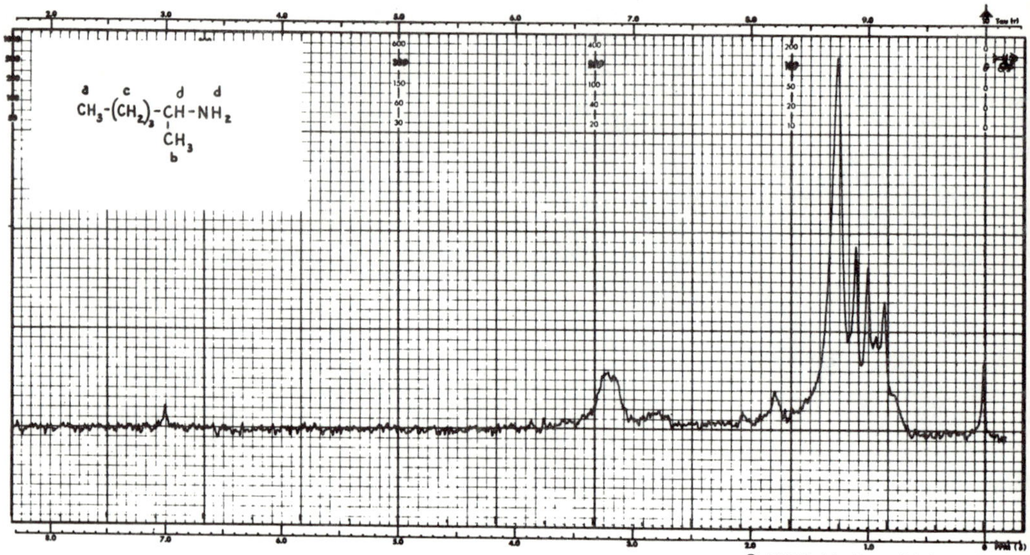

© 1978 Sadtler Research Laboratories, Inc.

1,3-DIMETHYLBUTYLAMINE

$C_6H_{15}N$

Mol. Wt. 101.19

B.P. 105-110°C

Solvent: CCl_4

Assignments
a = 0.91
b = 1.08
c = 1.20
d = ca 1.38-2.14
e = ca 2.19
f = 2.98
D_2O added

450

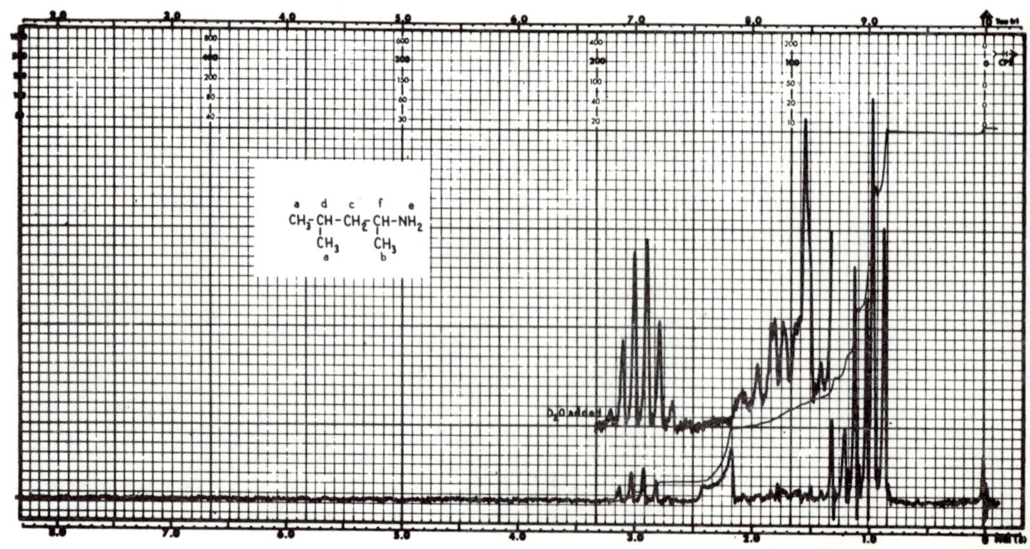

© 1978 Sadtler Research Laboratories, Inc.

HEPTYLAMINE

$C_7H_{17}N$

Mol. Wt. 152.20

B.P. 157-158°C

Solvent: $CDCl_3$

Assignments
a = 0.90
b = 1.20
c = 1.35
d = 2.70

451

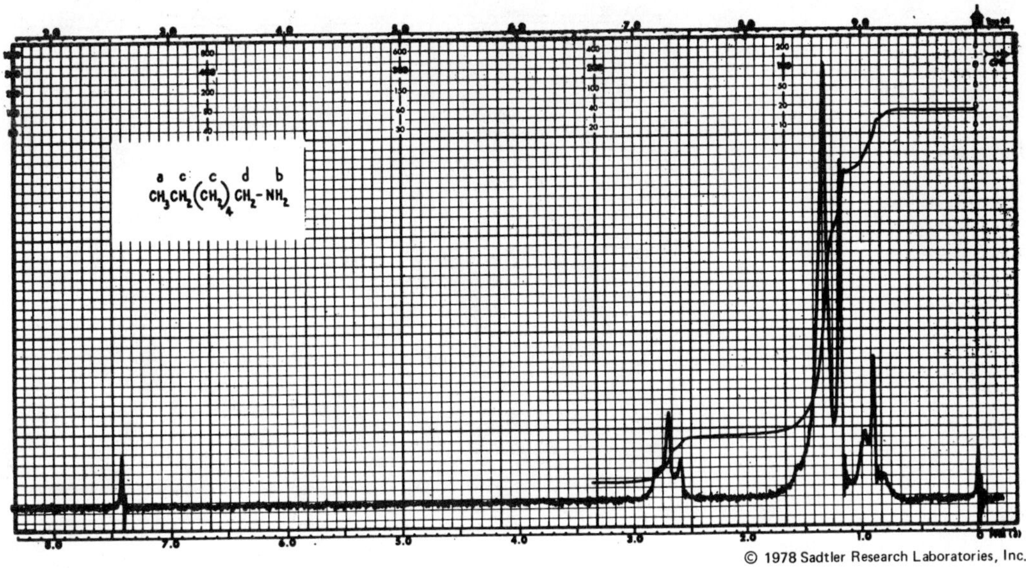

© 1978 Sadtler Research Laboratories, Inc.

1,1-DIMETHYLPENTYLAMINE

$C_7H_{17}N$

Mol. Wt. 115.22

B.P. 125-126°C

Solvent: $CDCl_3$

Assignments
a = 0.91
b = 1.11
c = ca 1.29

452

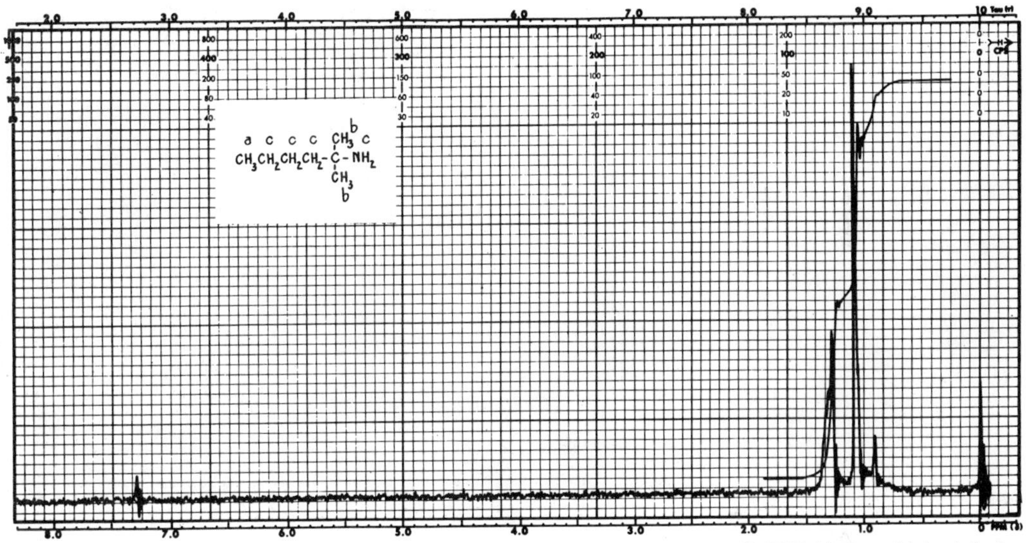

© 1978 Sadtler Research Laboratories, Inc.

1,4-DIMETHYLPENTYLAMINE

$C_7H_{17}N$

Mol. Wt. 115.22

Solvent: $CDCl_3 + D_2O$

Assignments
a = 0.89
b = 1.06
c = 1.00-1.90
d = 2.81
e = 4.58 HDO

453

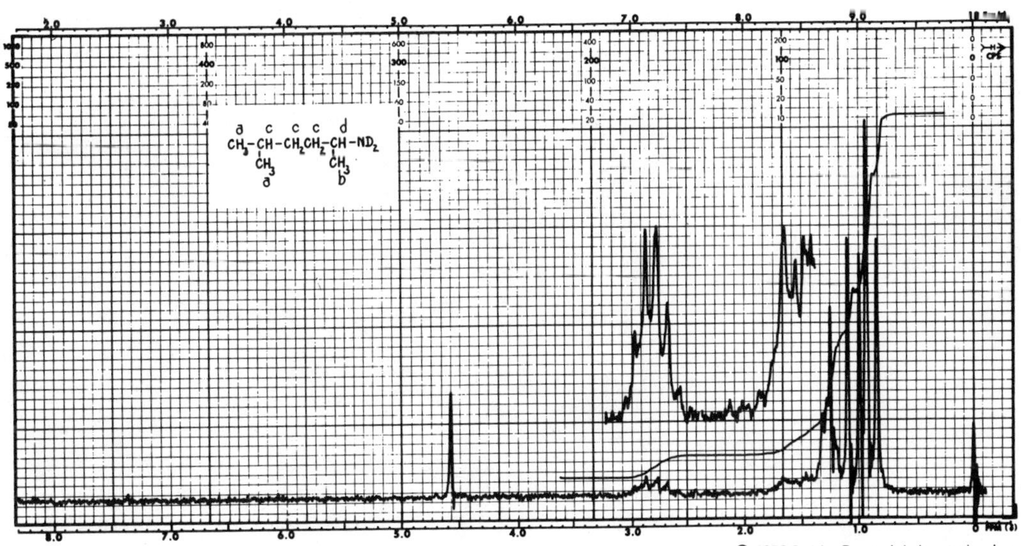

© 1978 Sadtler Research Laboratories, Inc.

2-ETHYLHEXYLAMINE

$C_8H_{19}N$

Mol. Wt. 129.25

B.P. 168.5-169.5°C

Solvent: $CDCl_3$

Assignments
a = 0.90
b = 1.29
c = 2.61

454

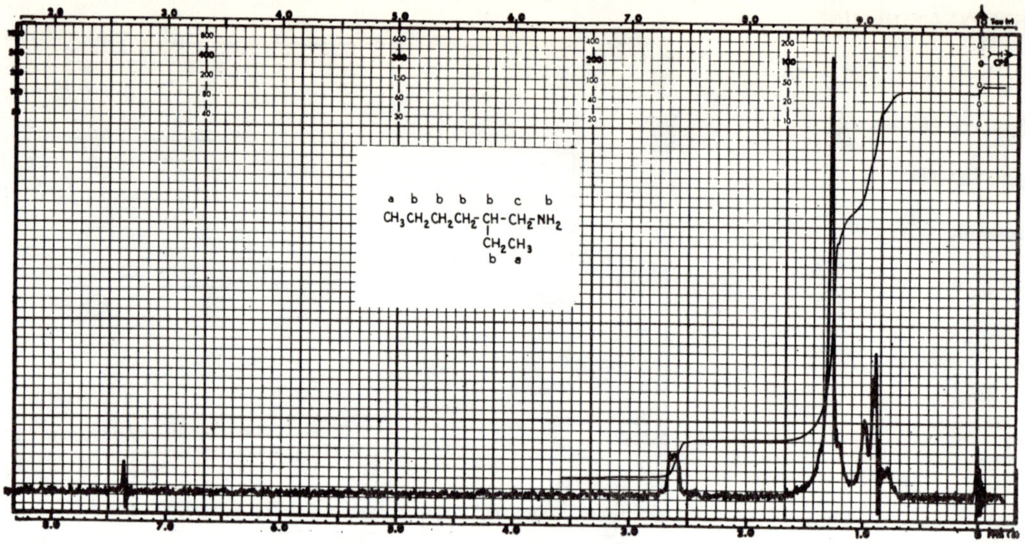

© 1978 Sadtler Research Laboratories, Inc.

1-BUTYLPENTYLAMINE

$C_9H_{21}N$

Mol. Wt. 143.27

B.P. 70-72°C/11mm

Solvent: $CDCl_3$

Assignments
a = 0.92
b = 1.29
c = 1.00-1.60
d = 2.70

455

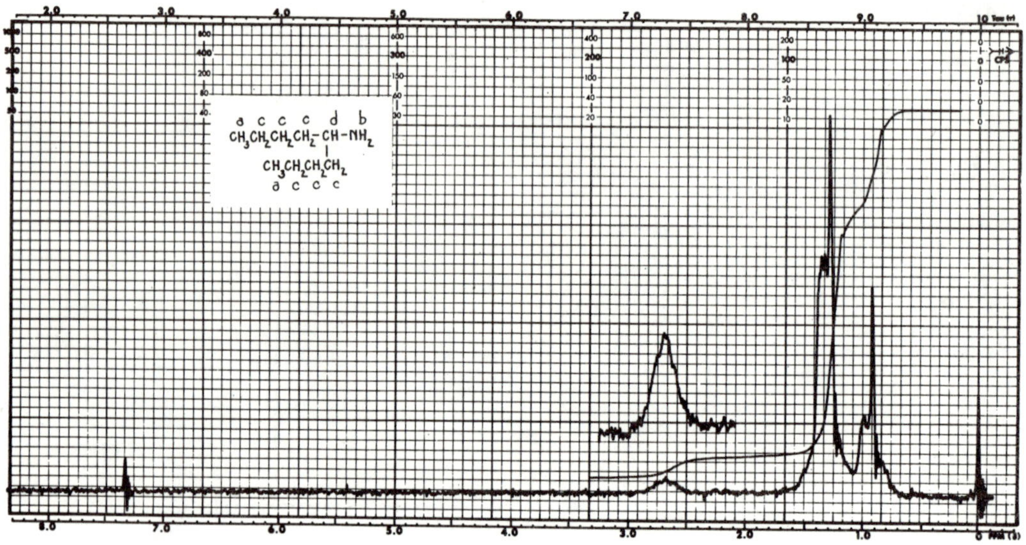

© 1978 Sadtler Research Laboratories, Inc.

1,1-DIBUTYLPENTYLAMINE

$C_{13}H_{29}N$

Mol. Wt. 199.38

B.P. 85°C/0.65mm

Solvent: $CDCl_3$

Assignments
a = 0.93
b = 1.19
c = ca 1.30

456

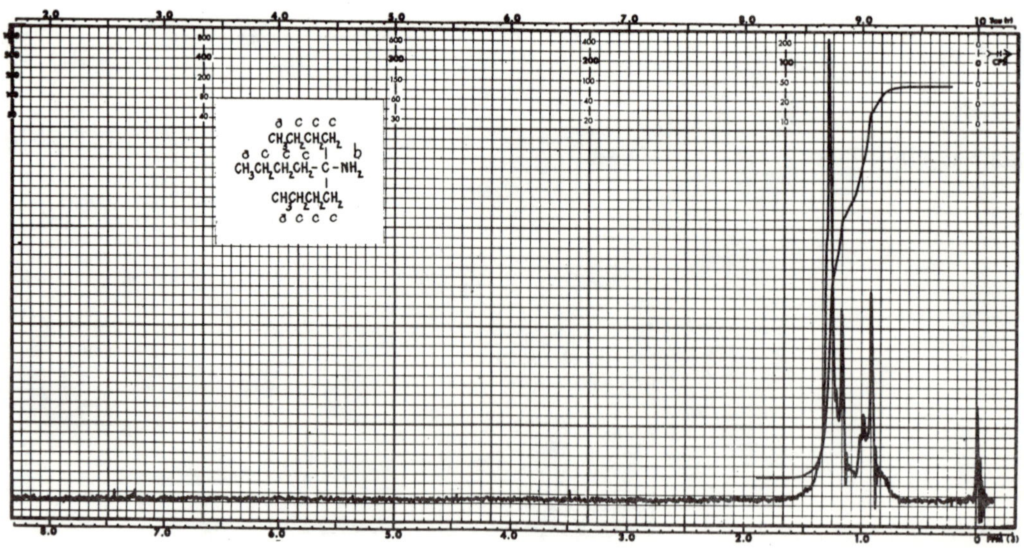

© 1978 Sadtler Research Laboratories, Inc.

ETHYLENEDIAMINE

$C_2H_8N_2$

Mol. Wt. 60.10

B.P. 117.2°C

Solvent: CCl_4

Assignments
a = 1.19
b = 2.60

457

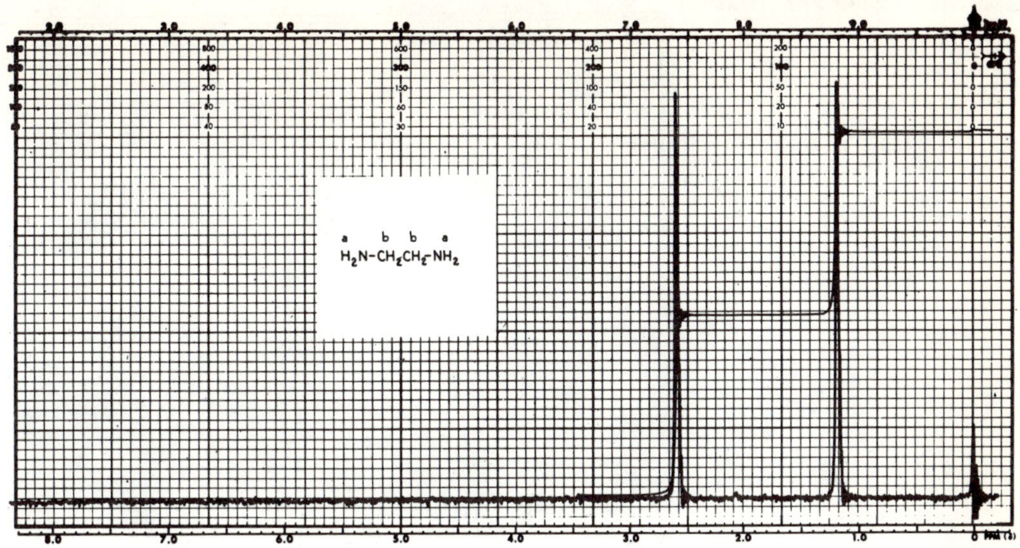

© 1978 Sadtler Research Laboratories, Inc.

1,3-PROPANEDIAMINE

$C_3H_{10}N_2$

Mol. Wt. 74.13

B.P. 120-122°C

Solvent: D_2O

Assignments
a = 1.53
b = 2.62
c = 4.54 HDO

458

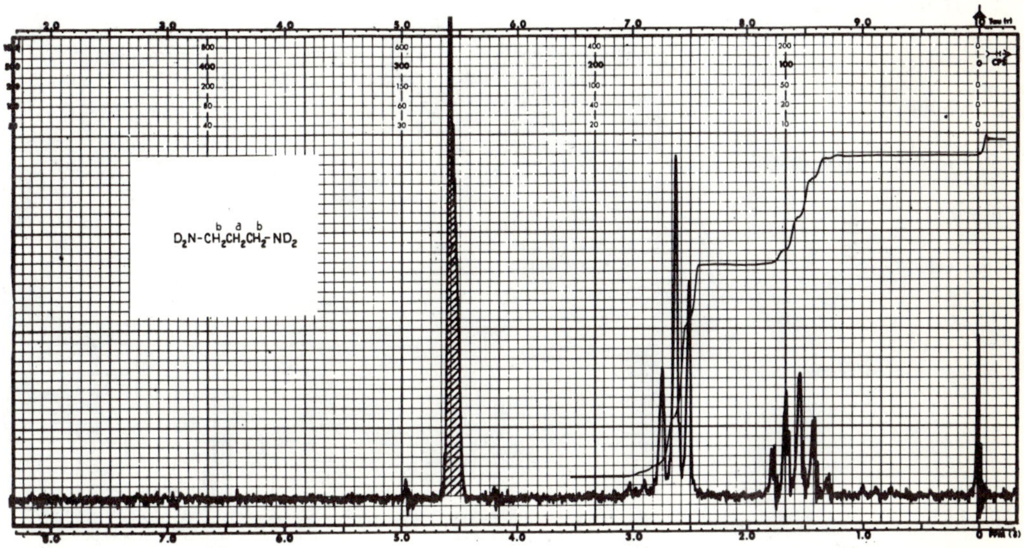

© 1978 Sadtler Research Laboratories, Inc.

1,2-PROPANEDIAMINE

$C_3H_{10}N_2$

Mol. Wt. 74.13

B.P. 119-120°C

Solvent: $CDCl_3$

Assignments
a = 1.07
b = 1.20
c = 2.20-2.65
d = 2.84

459

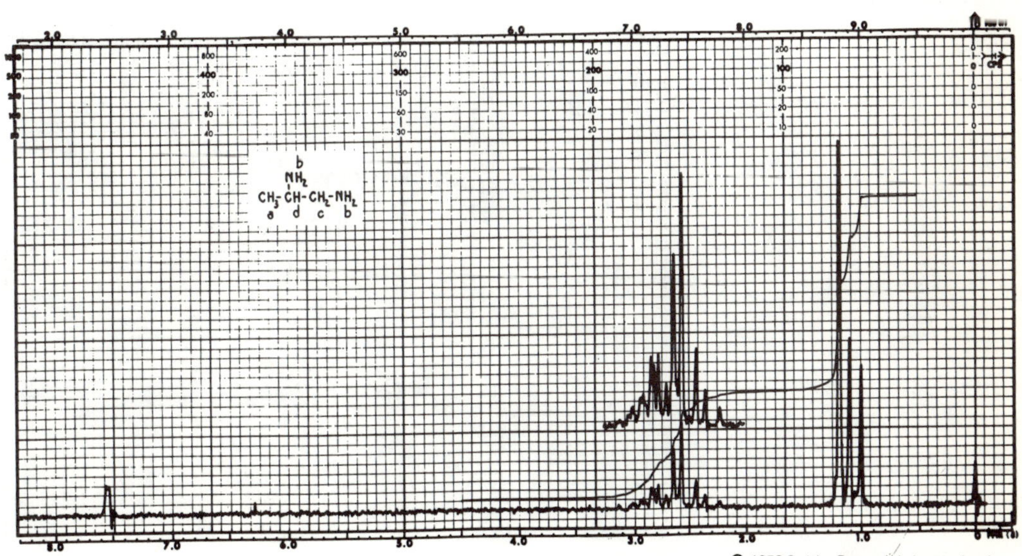

© 1978 Sadtler Research Laboratories, Inc.

2-METHYL-1,2-PROPANEDIAMINE

$C_4H_{12}N_2$

Mol. Wt. 88.15

Solvent: CCl_4

Assignments
a = 1.00
b = 1.14
c = 2.40

460

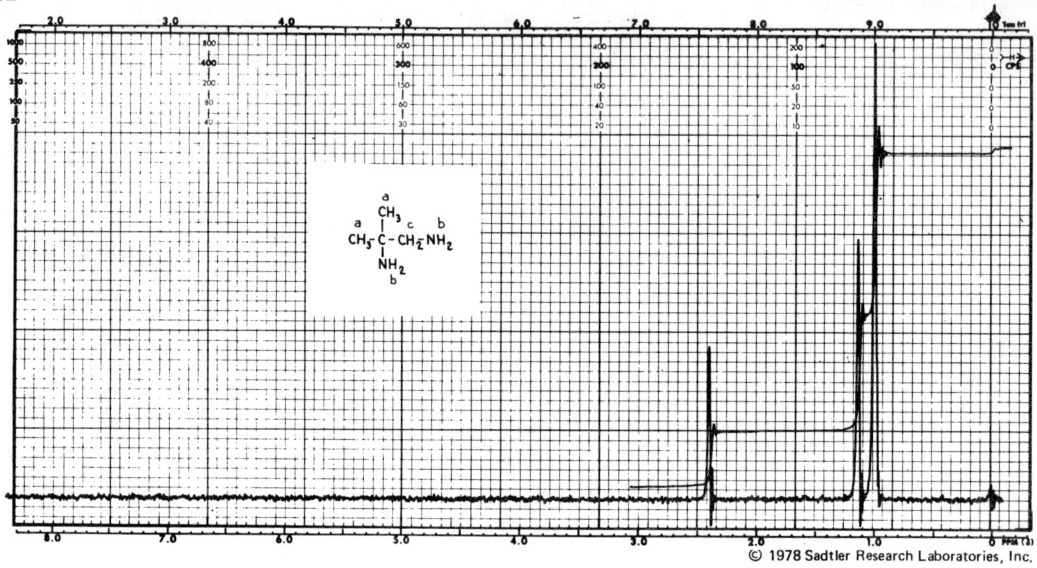

1,4-BUTANEDIAMINE

$C_4H_{12}N_2$

Mol. Wt. 88.15

M.P. 25-27°C B.P. 60-61°C/11mm

Solvent: D_2O

Assignments
a = 1.47
b = 2.60
c = 4.55 HDO
d = 3.76, 5.40 HDO side bands

461

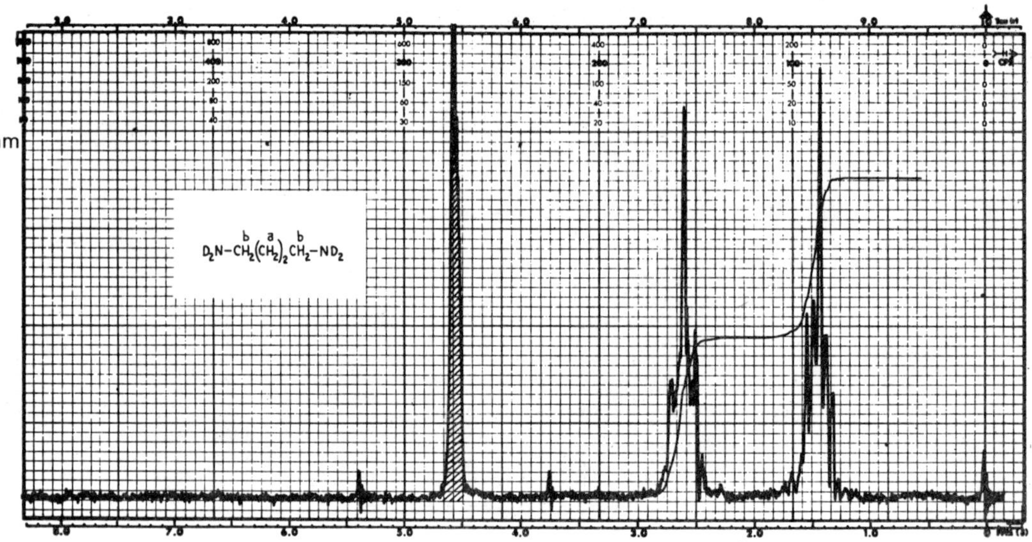

1,12-DODECANEDIAMINE

$C_{12}H_{28}N_2$

Mol. Wt. 200.37

M.P. 66-67°C

Solvent: $CDCl_3$

Assignments
a = 1.30
b = 1.59
c = 2.68

462

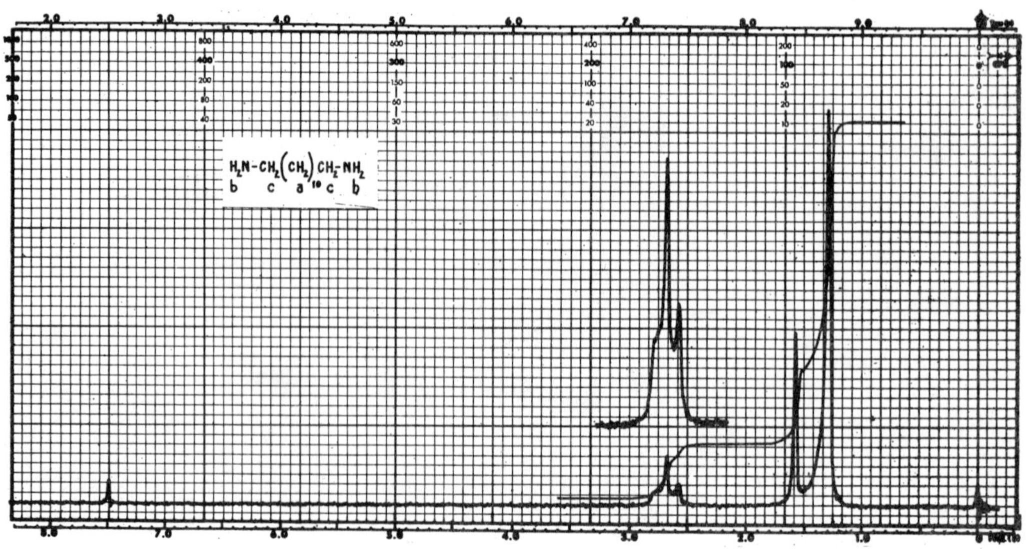

N-ETHYLETHYLENEDIAMINE

$C_4H_{12}N_2$

Mol. Wt. 88.15

B.P. 126-129°C/760mm (lit.)

Solvent: D_2O

Assignments
a = 1.07
b = 2.60
c = ca 2.65
d = 4.69 HDO

463

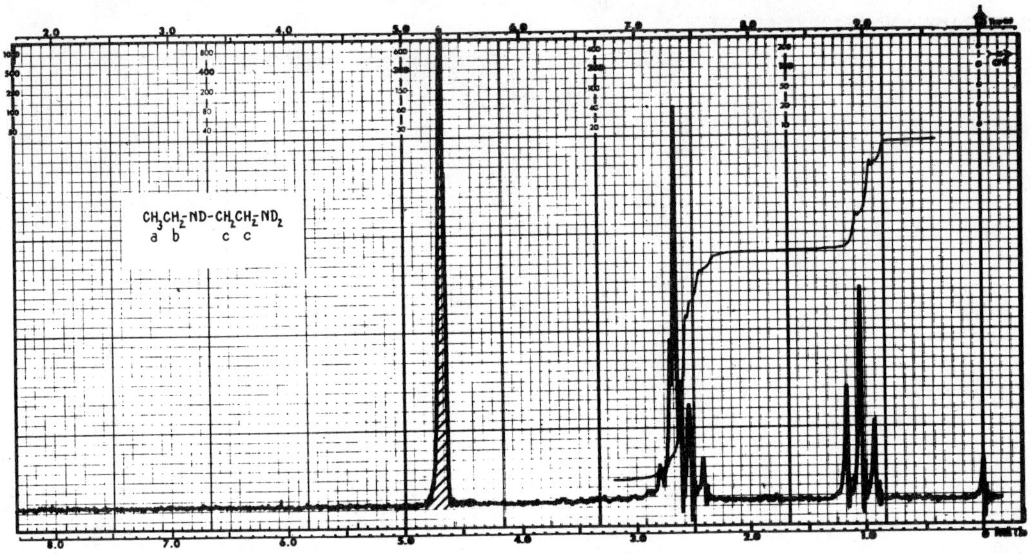

© 1978 Sadtler Research Laboratories, Inc.

N-METHYL-1,3-PROPANEDIAMINE

$C_4H_{12}N_2$

Mol. Wt. 88.15

Solvent: D_2O

Assignments
a = 1.58
b = 2.29
c = 2.53
d = 2.61
e = 4.65 HDO

464

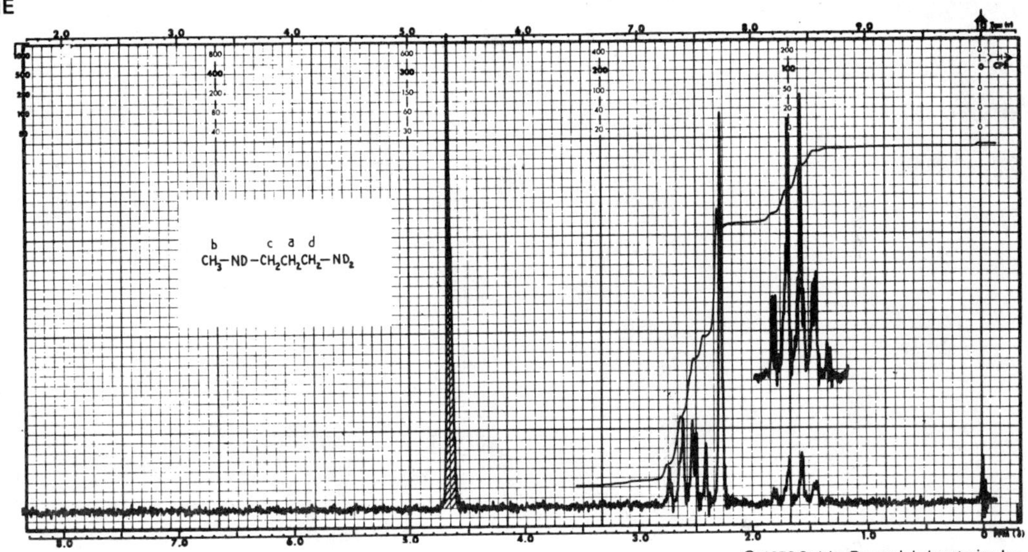

© 1978 Sadtler Research Laboratories, Inc.

N,N-DIBUTYL-1,3-PROPANEDIAMINE

$C_{11}H_{26}N_2$

Mol. Wt. 186.34

Solvent: CCl_4

Assignments
a = 0.90
b = 1.06-1.50
c = 1.61
d = 2.10-2.55
e = 2.68

465

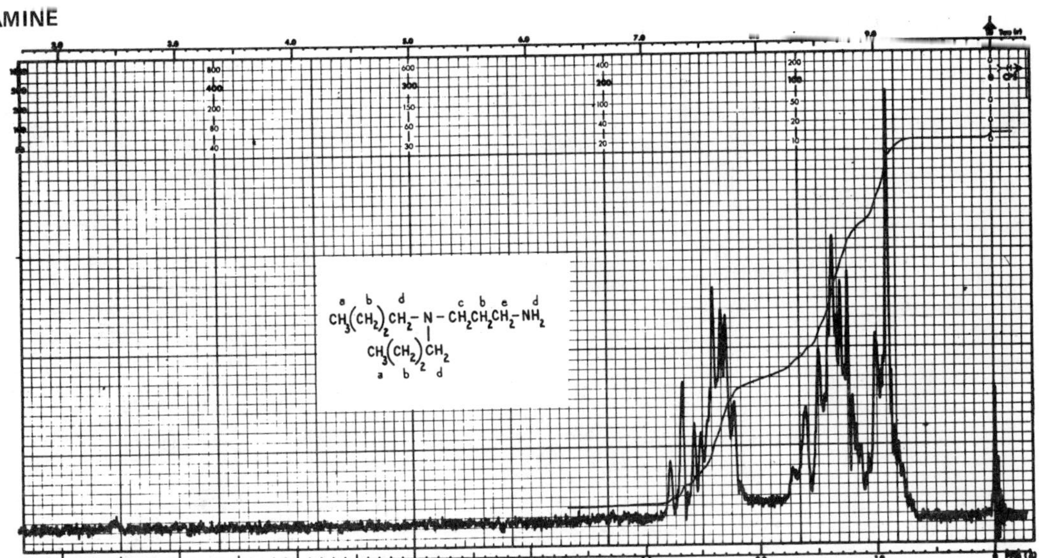

© 1978 Sadtler Research Laboratories, Inc.

3,3′-DIAMINODIPROPYLAMINE

$C_6H_{17}N_3$

Mol. Wt. 131.22

B.P. 116-118°C/10mm

Solvent: $CDCl_3$

Assignments
a = 1.18
b = 1.63
c = 2.70
Band at 7.7 ppm due to exchange of NH & NH_2 with $CDCl_3$

466

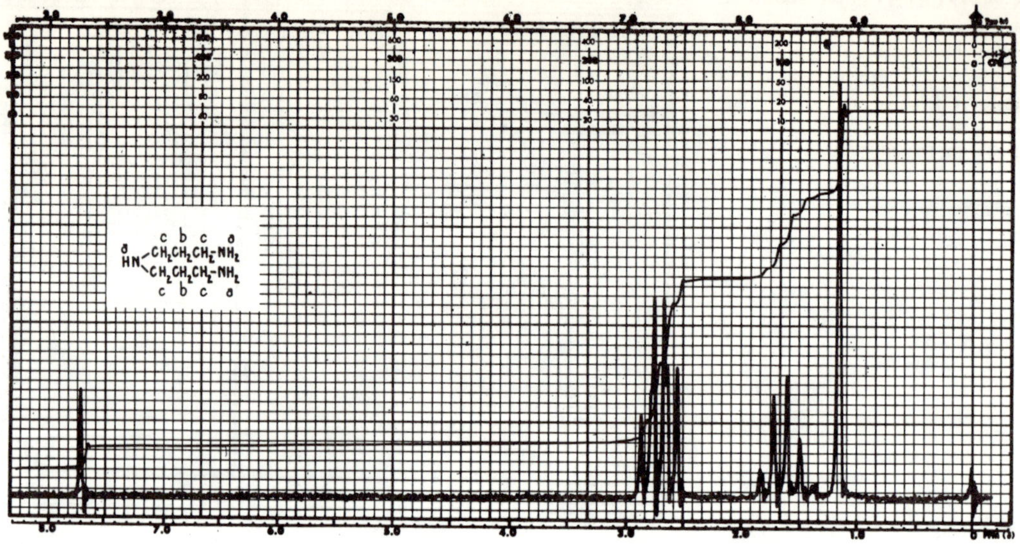

© 1978 Sadtler Research Laboratories, Inc.

N,N-BIS(2-AMINOETHYL)DODECYLAMINE

$C_{16}H_{37}N_3$

Mol. Wt. 271.49

B.P. 143-145°C/2mm

Solvent: CCl_4

Assignments
a = 0.88
b = ca 1.27
c = 2.32-2.88

467

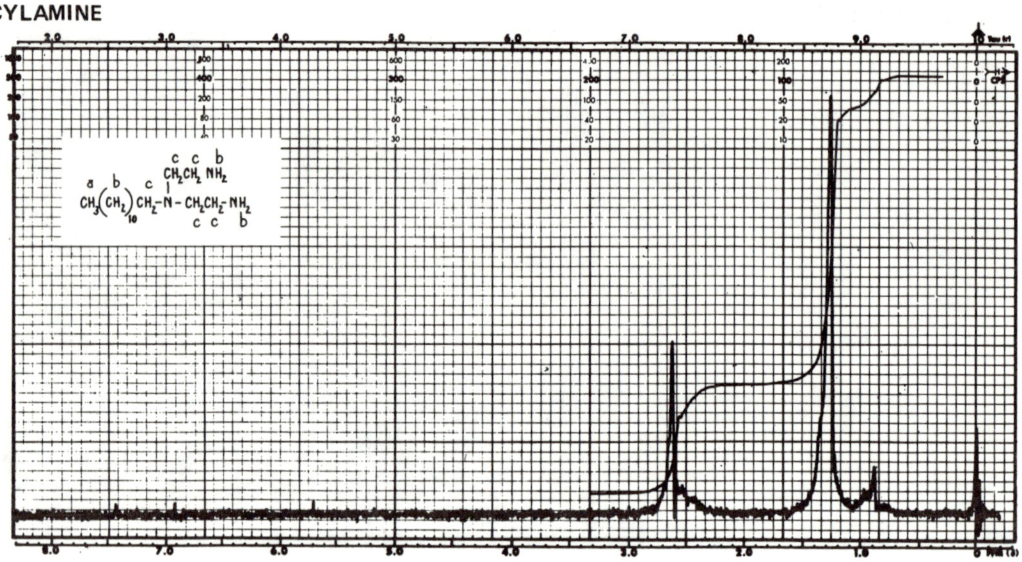

© 1978 Sadtler Research Laboratories, Inc.

CYCLOPENTANEAMINE

$C_5H_{11}N$

Mol. Wt. 85.15

B.P. 106-108°C

Solvent: $CDCl_3$

Assignments
a = 1.00-2.10
b = 1.47
c = 3.31

468

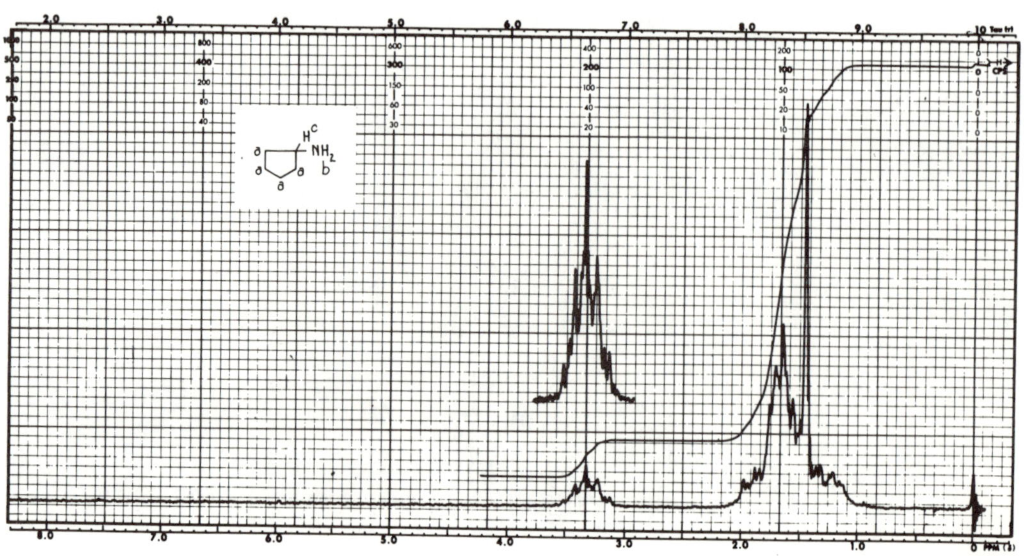

© 1978 Sadtler Research Laboratories, Inc.

CYCLOHEXANEMETHYLAMINE

$C_7H_{15}N$

Mol. Wt. 113.20

Solvent: $CDCl_3$

Assignments
a = 0.80-2.00
b = 1.49
c = 2.52

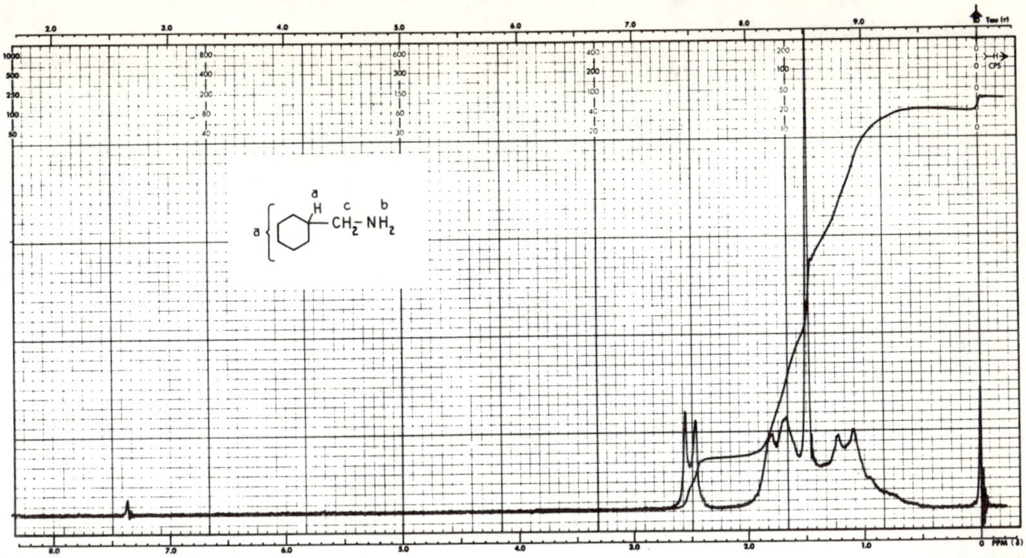

469

© 1978 Sadtler Research Laboratories, Inc.

CYCLOHEXYLAMINE

$C_6H_{13}N$

Mol. Wt. 99.18

B.P. 133°C

Solvent: CCl_4

Assignments
a = 0.80-2.02
b = 1.51
c = 2.64

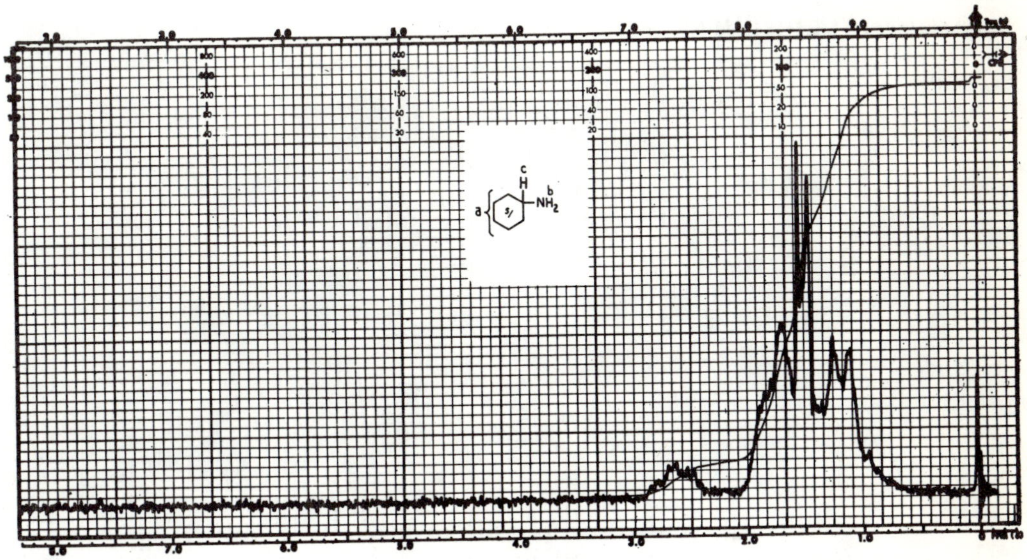

470

© 1978 Sadtler Research Laboratories, Inc.

4-METHYLCYCLOHEXYLAMINE

$C_7H_{15}N$

Mol. Wt. 113.20

Solvent: $CDCl_3$

Assignments
a = ca 0.92
b = 1.31
c = 0.80-2.20
d = 2.20-3.10
e = 7.32 imp. ($CHCl_3$)

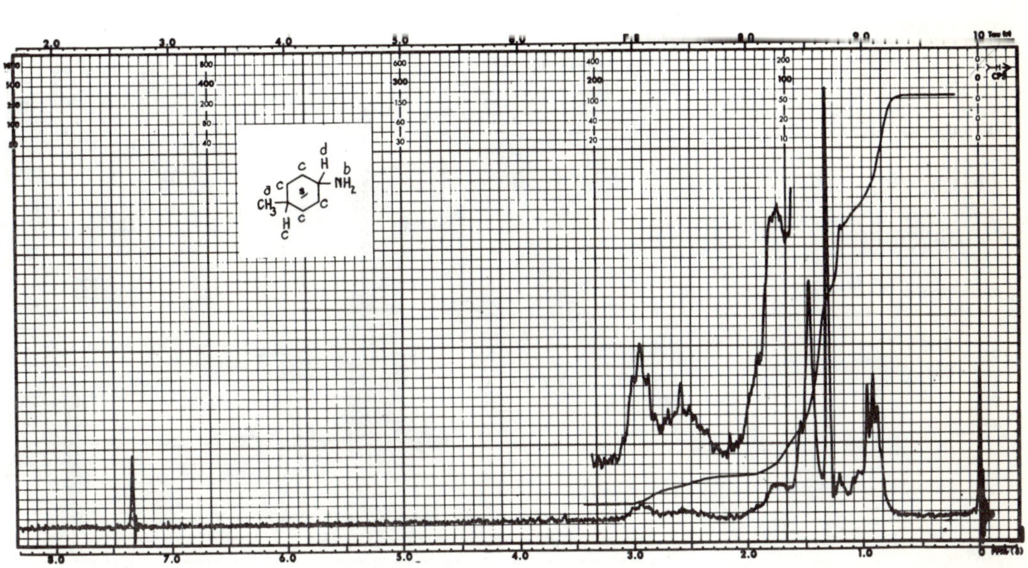

471

© 1978 Sadtler Research Laboratories, Inc.

CYCLOHEPTYLAMINE

$C_7H_{15}N$

Mol. Wt. 113.20

B.P. 54°C/11mm

Solvent: $CDCl_3$

Assignments
a = 1.36
b = 1.00-2.20
c = 2.92

472

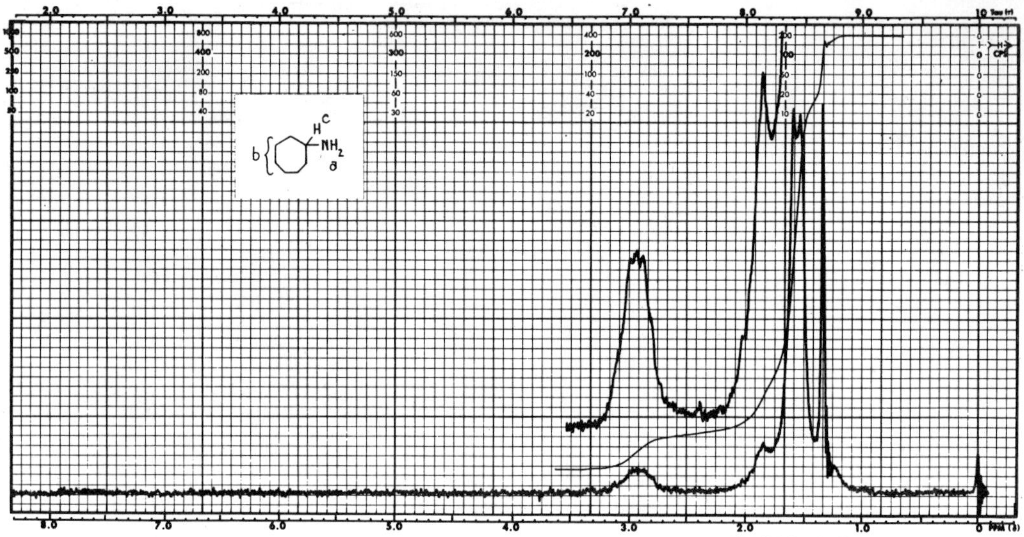

© 1978 Sadtler Research Laboratories, Inc.

1-ADAMANTANEMETHYLAMINE

$C_{11}H_{19}N$

Mol. Wt. 165.28

Solvent: $CDCl_3$

Assignments
a = 1.21
b = 1.48
c = 1.71
d = 1.97
e = 2.31

473

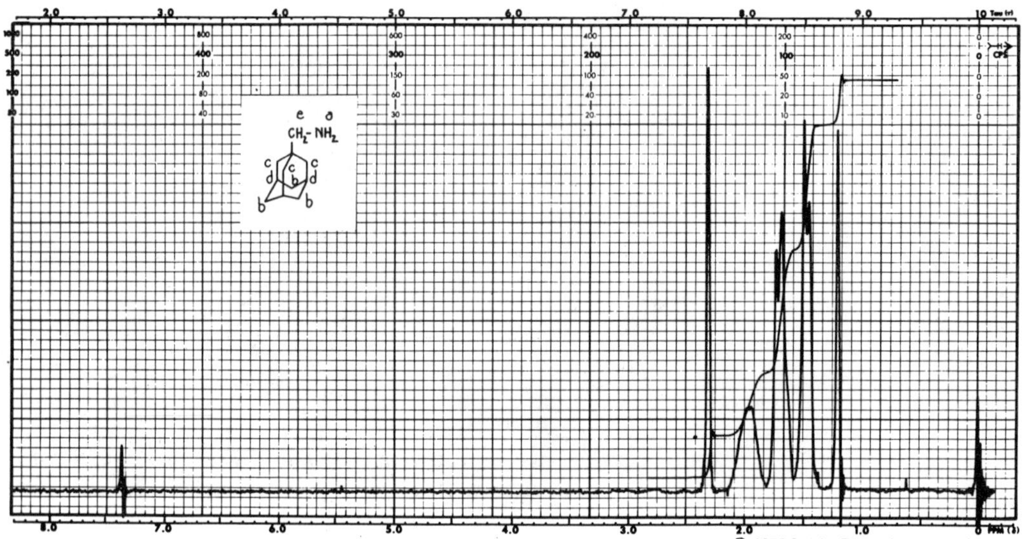

© 1978 Sadtler Research Laboratories, Inc.

ALLYLAMINE

C_3H_7N

Mol. Wt. 57.10

B.P. 53-54°C

Solvent: $CDCl_3$

Assignments
a = 1.21
b = 3.30
c = 5.08
d = 5.12
e = 6.00

474

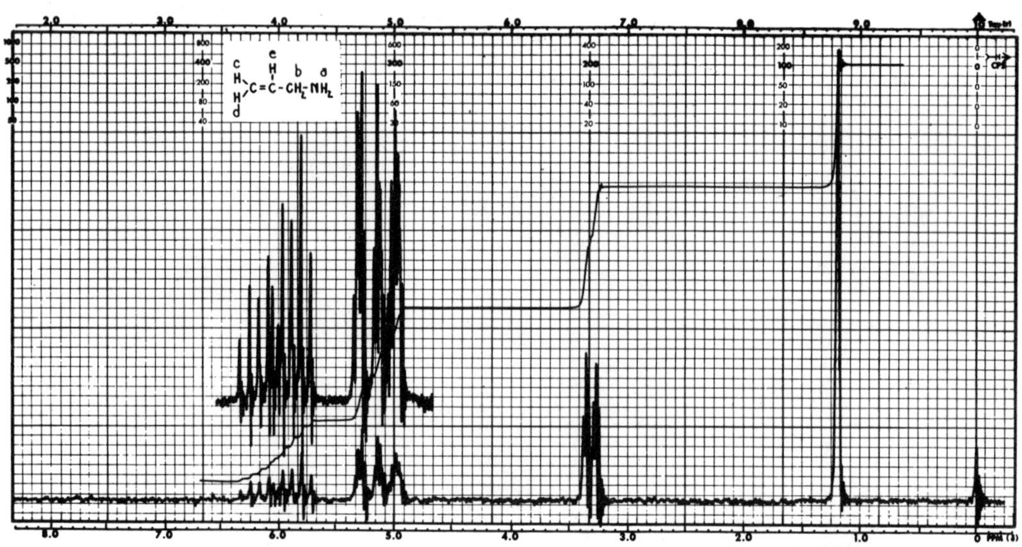

© 1978 Sadtler Research Laboratories, Inc.

2-METHYLALLYLAMINE

C_4H_9N

Mol. Wt. 71.12

B.P. 79°C

Solvent: CCl_4 CC

Assignments
a = 1.10
b = 1.73
c = 3.12
d = 4.70
e = 4.80

475

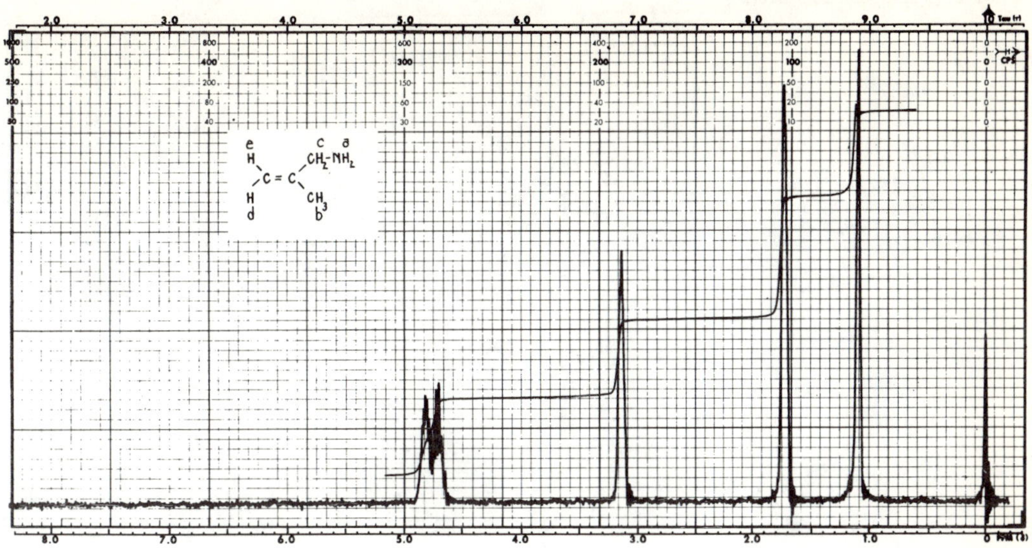

© 1978 Sadtler Research Laboratories, Inc.

BENZYLAMINE

C_7H_9N

Mol. Wt. 107.15

B.P. 185°C

Solvent: $CDCl_3$

Assignments
a = 1.52
b = 3.85
c = 7.29

476

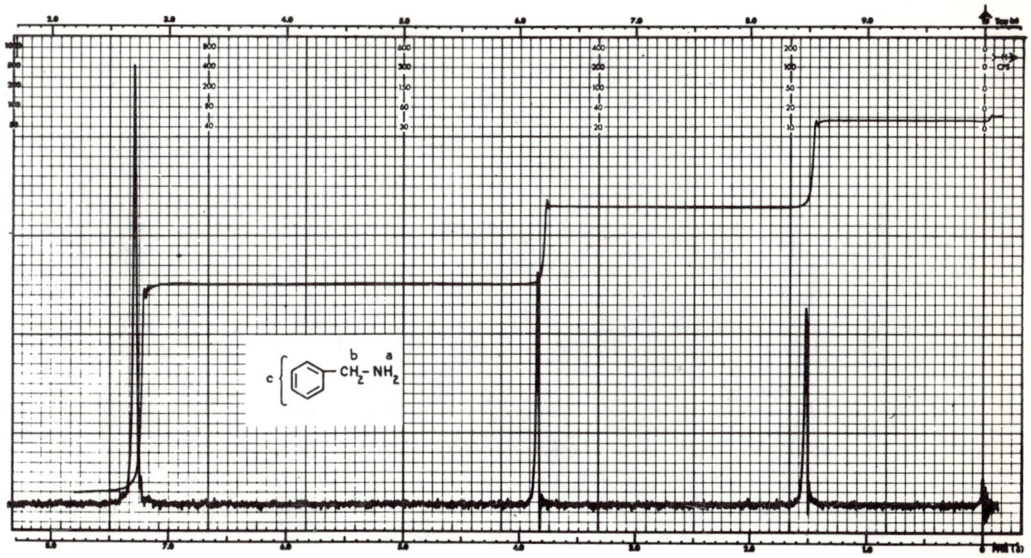

© 1978 Sadtler Research Laboratories, Inc.

PHENETHYLAMINE

$C_8H_{11}N$

Mol. Wt. 121.18

Solvent: CCl_4

Assignments
a = ca 0.93
b = 2.64
c = 2.82
d = ca 7.14

477

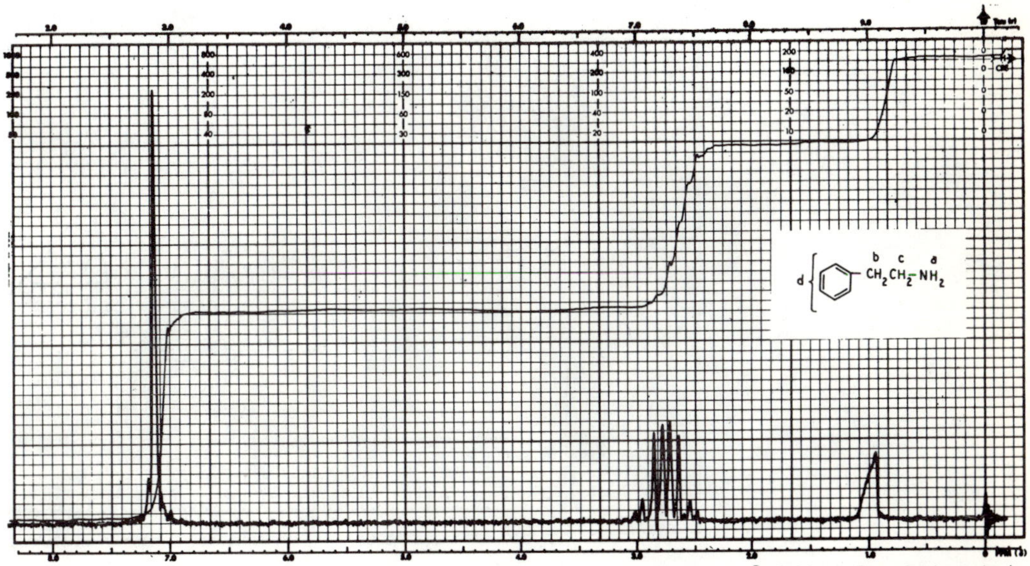

© 1978 Sadtler Research Laboratories, Inc.

(+)-α-METHYLBENZYLAMINE

$C_8H_{11}N$

Mol. Wt. 121.18

Solvent: CCl_4

Assignments
a = 1.29
b = 1.47
c = 4.01
d = 7.21

478

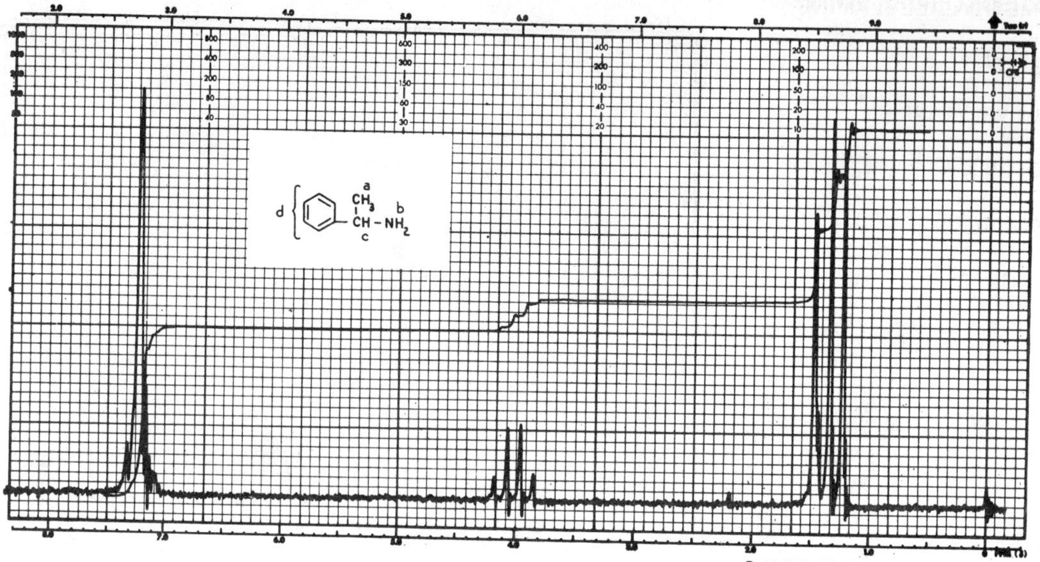

© 1978 Sadtler Research Laboratories, Inc.

β-METHYLPHENETHYLAMINE

$C_9H_{13}N$

Mol. Wt. 135.21

Solvent: $CDCl_3$

Assignments
a = 0.99
b = 1.21
c = ca 2.71
d = ca 7.20

479

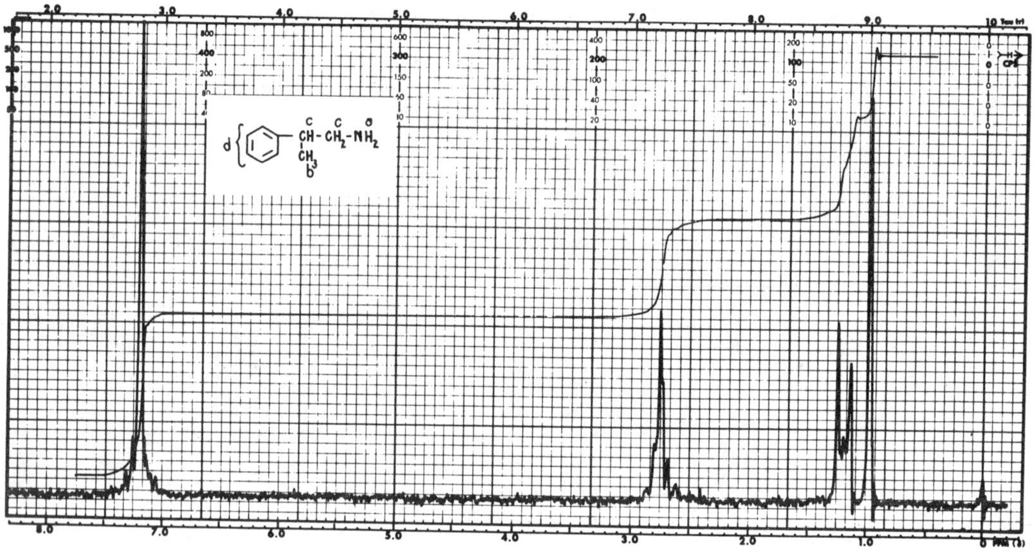

© 1978 Sadtler Research Laboratories, Inc.

α,α-DIMETHYLBENZYLAMINE

$C_9H_{13}N$

Mol. Wt. 135.21

Solvent: CCl_4

Assignments
a = 2.19
b = 3.35
c = 7.19

480

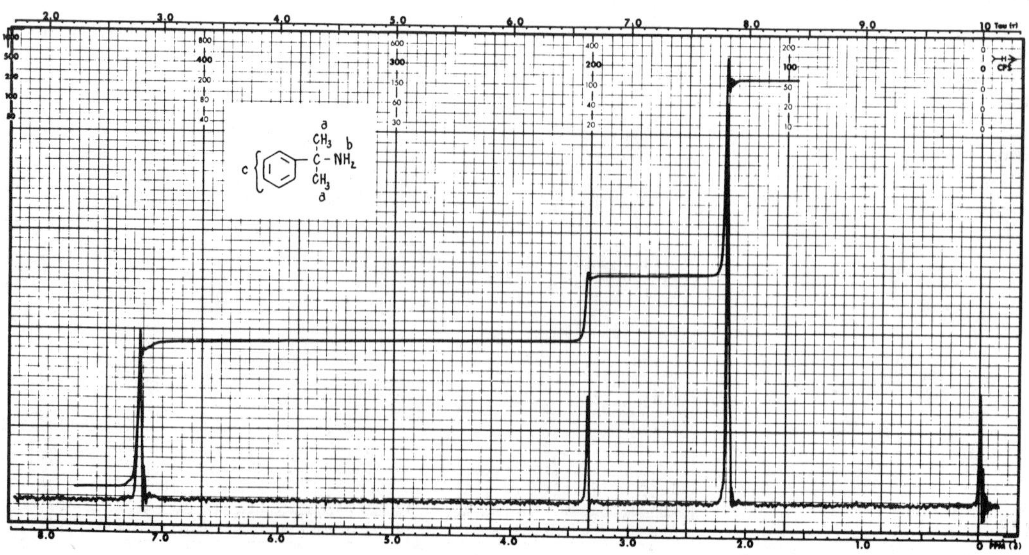

© 1978 Sadtler Research Laboratories, Inc.

3-PHENYLPROPYLAMINE

$C_9H_{13}N$

Mol. Wt. 135.21

Solvent: $CDCl_3$

Assignments

a = 1.01
b = 1.80
c = 2.64
d = 2.70
e = ca 7.20

481

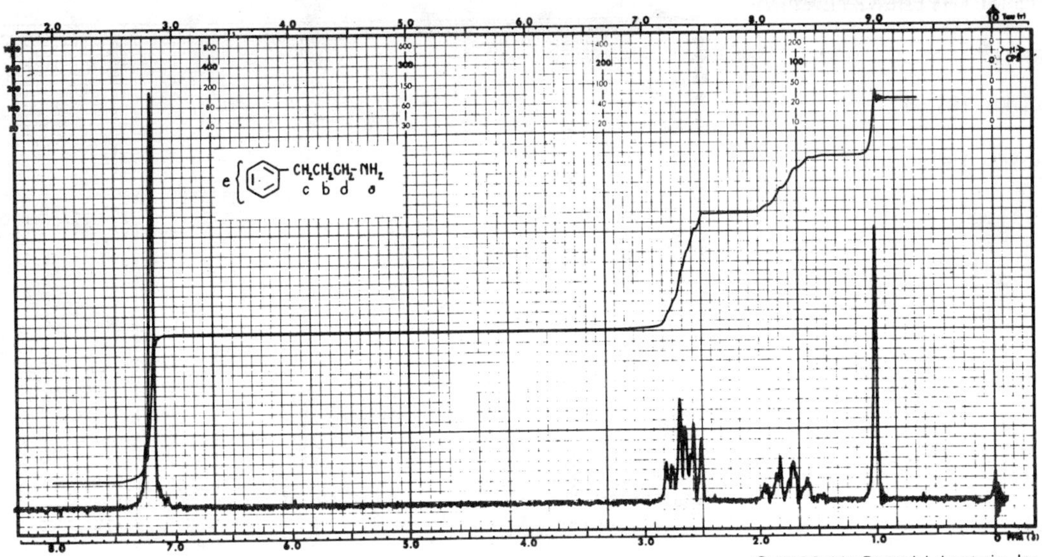

© 1978 Sadtler Research Laboratories, Inc.

4-PHENYLBUTYLAMINE

$C_{10}H_{15}N$

Mol. Wt. 149.24

Solvent: CCl_4

Assignments

a = 0.99
b = ca 1.49
c = 2.54
d = 2.60
e = 7.12

482

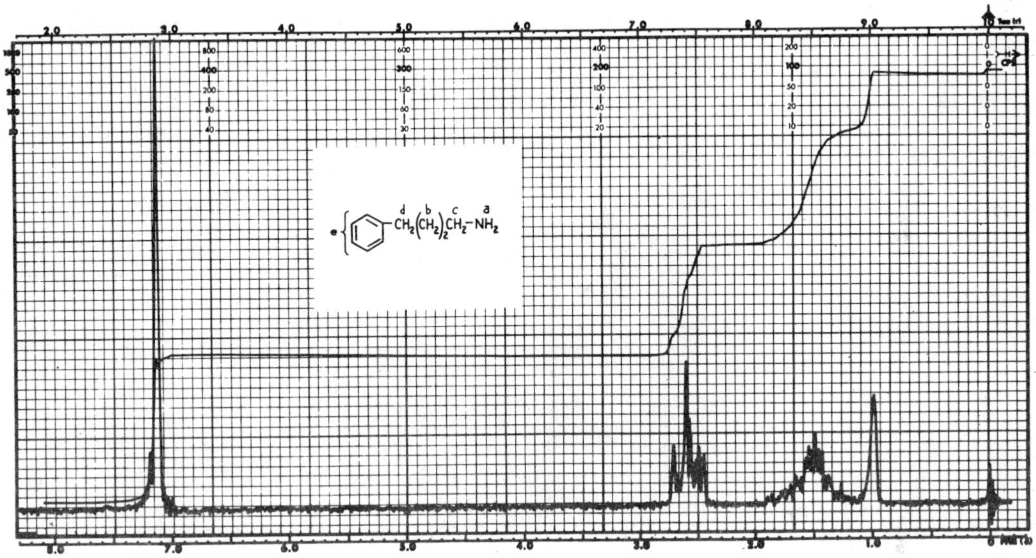

© 1978 Sadtler Research Laboratories, Inc.

o-METHYLBENZYLAMINE

$C_8H_{11}N$

Mol. Wt. 121.18

Solvent: $CDCl_3$

Assignments

a = 1.31
b = 2.30
c = 3.79
d = 7.03-7.32

483

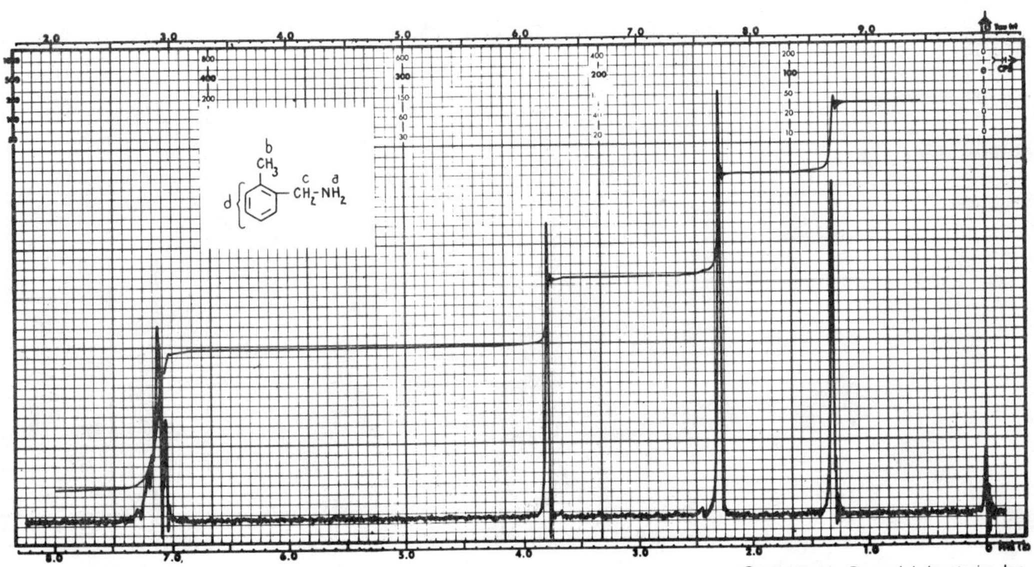

© 1978 Sadtler Research Laboratories, Inc.

m-METHYLBENZYLAMINE

$C_8H_{11}N$

Mol. Wt. 121.18

Solvent: $CDCl_3$

Assignments
a = 1.80
b = 2.33
c = 3.80
d = 7.12
e = 4.76 imp.

484

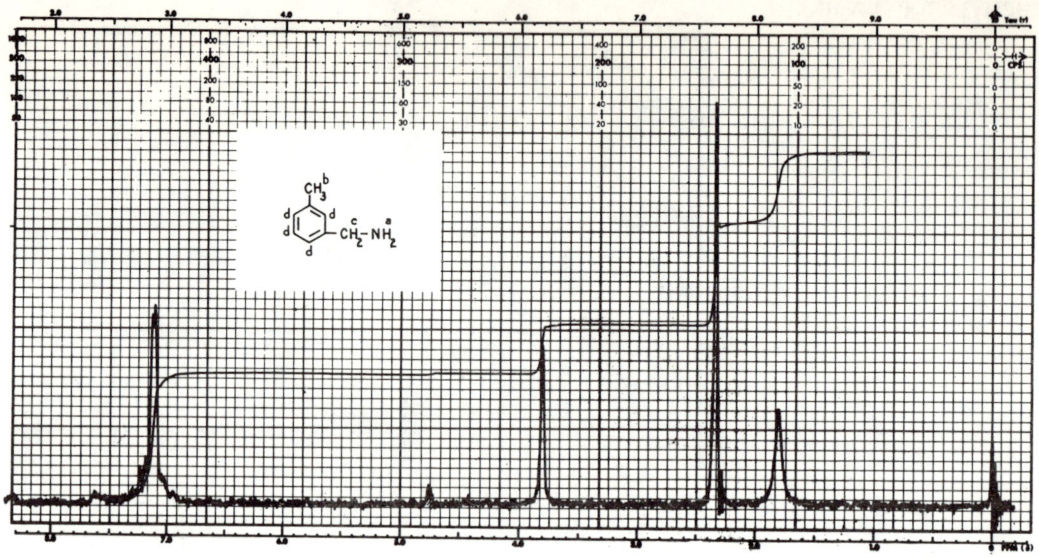

© 1978 Sadtler Research Laboratories, Inc.

p-METHYLBENZYLAMINE

$C_8H_{11}N$

Mol. Wt. 121.18

Solvent: CCl_4

Assignments
a = 1.18
b = 2.30
c = 3.72
d = 7.04

485

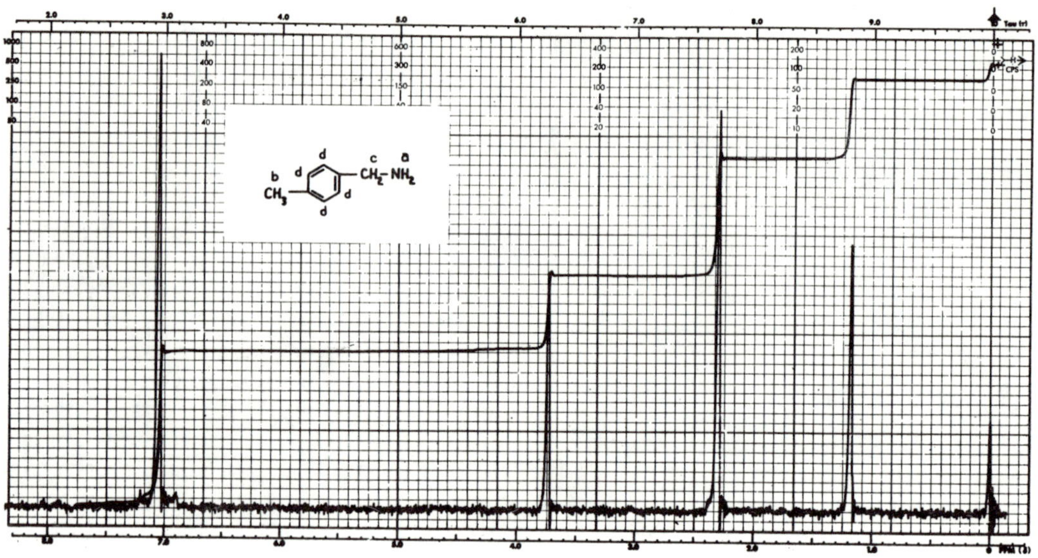

© 1978 Sadtler Research Laboratories, Inc.

p-ISOPROPYLBENZYLAMINE

$C_{10}H_{15}N$

Mol. Wt. 149.24

Solvent: CCl_4

Assignments
a = 1.21
b = 1.51
c = 2.85
d = 3.71
e = ca 7.11

486

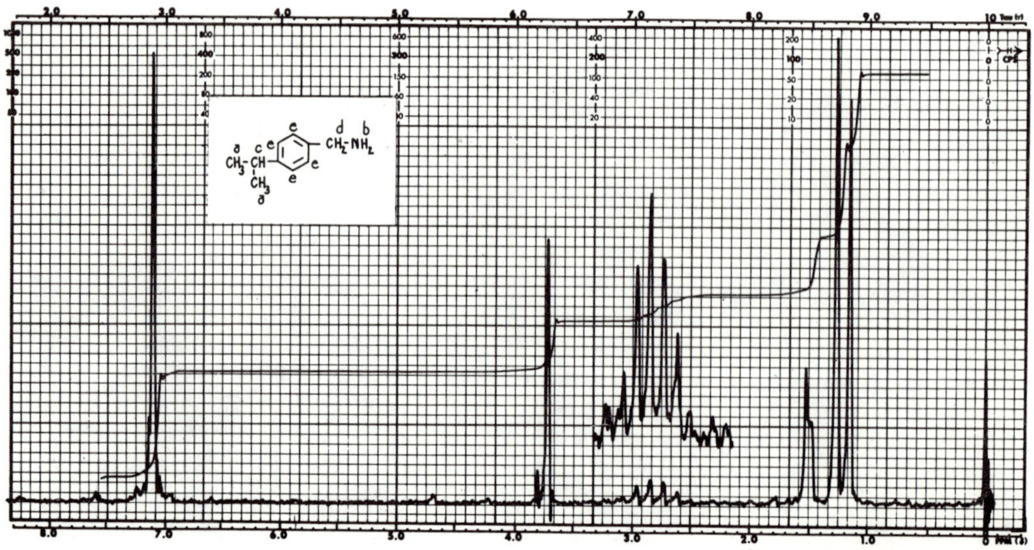

© 1978 Sadtler Research Laboratories, Inc.

ANILINE

C_6H_7N

Mol. Wt. 93.12

B.P. 184.4°C (lit.)

Solvent: CCl_4

Assignments
a = 3.32
b = 6.44
c = ca 6.59
d = 6.99

487

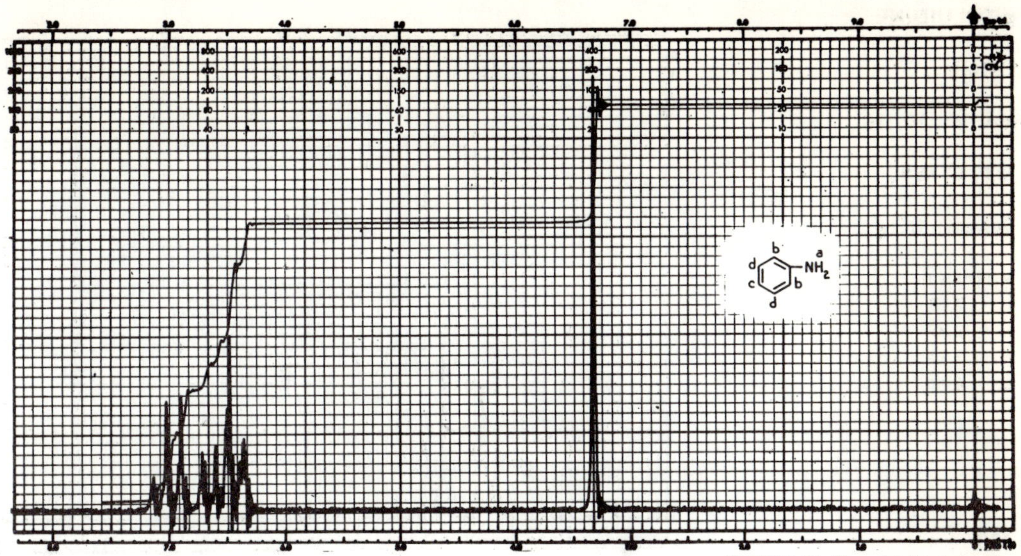

© 1978 Sadtler Research Laboratories, Inc.

---

o-TOLUIDINE

C_7H_9N

Mol. Wt. 107.15

B.P. 83-85°C/15mm (lit.)

Assignments
a = 2.04
b = 3.23
c = 6.30-7.10

488

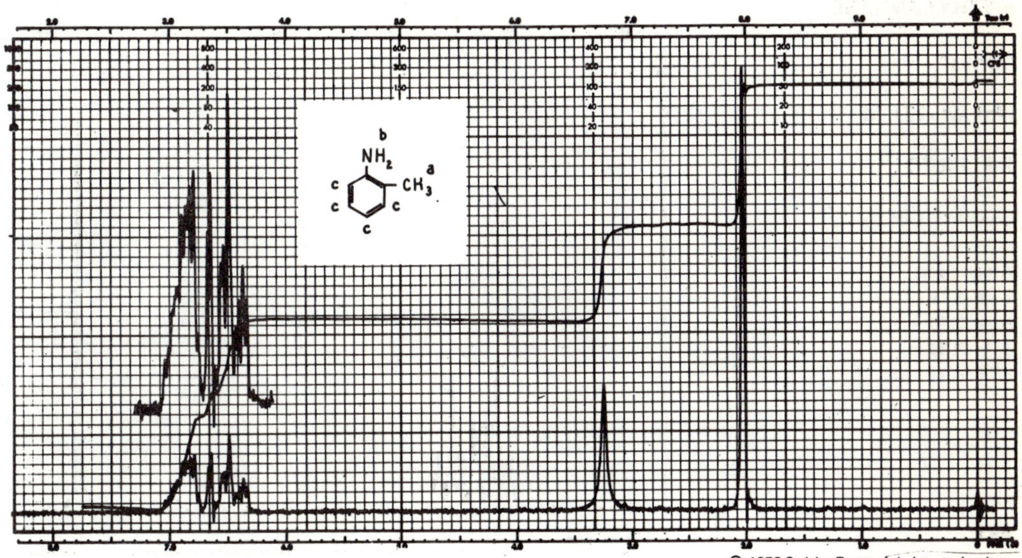

© 1978 Sadtler Research Laboratories, Inc.

---

m-TOLUIDINE

C_7H_9N

Mol. Wt. 107.15

B.P. 92-93°C/15mm

Assignments
a = 2.20
b = 3.32
c = 6.15-6.55
d = 6.90

489

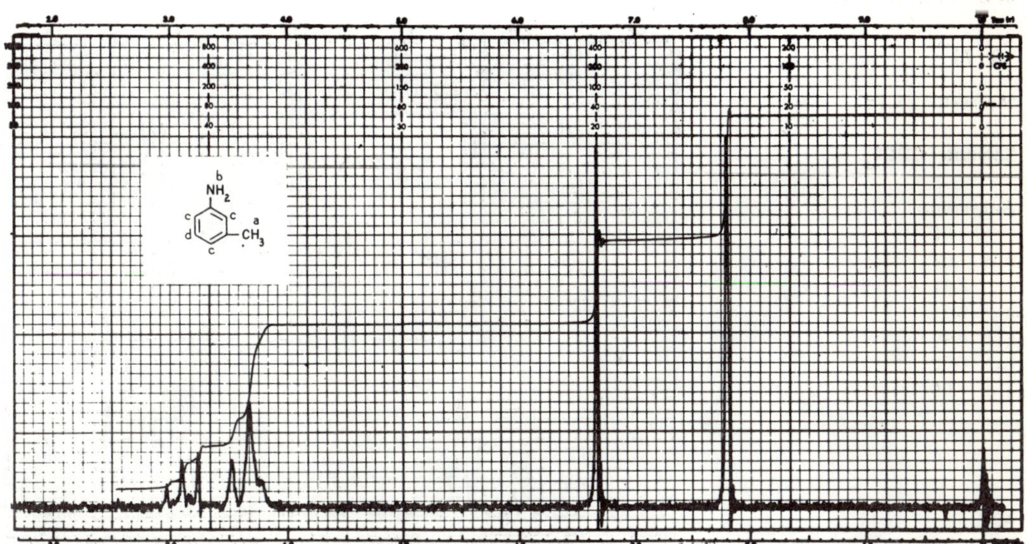

© 1978 Sadtler Research Laboratories, Inc.

p-TOLUIDINE

C_7H_9N

Mol. Wt. 107.16

M.P. 44-46°C

Solvent: CCl_4

Assignments
a = 2.18
b = 3.25
c = 6.33
d = 6.79

490

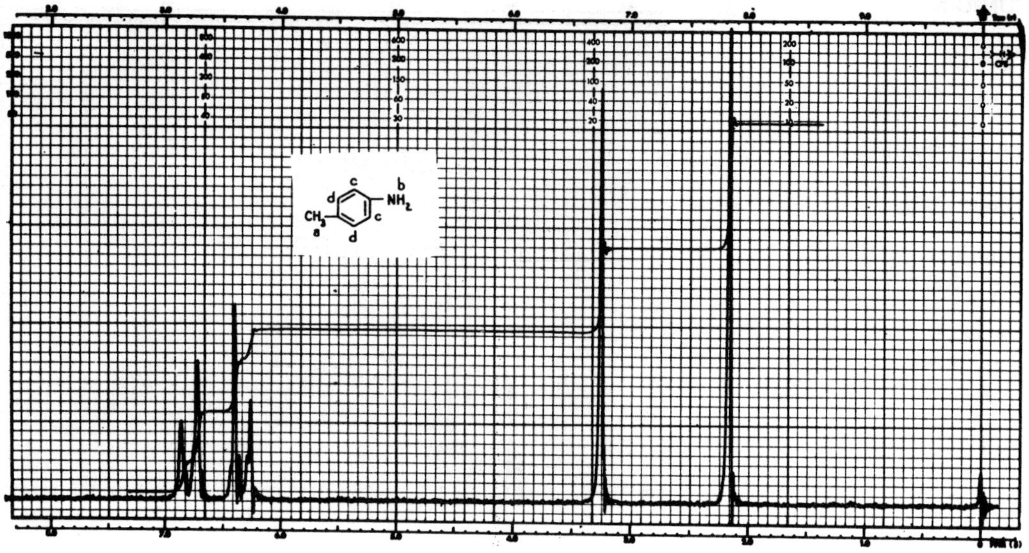

© 1978 Sadtler Research Laboratories, Inc.

---

o-ETHYLANILINE

$C_8H_{11}N$

Mol. Wt. 121.18

B.P. 215°C

Solvent: $CDCl_3$

Assignments
a = 1.18
b = 2.40
c = 3.30
d = 6.30-7.10

491

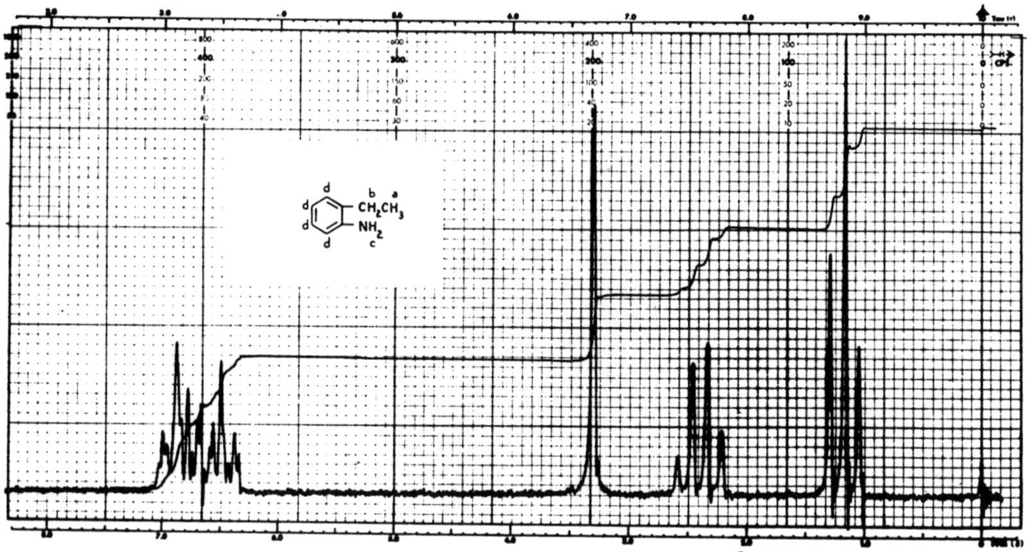

© 1978 Sadtler Research Laboratories, Inc.

---

m-ETHYLANILINE

$C_8H_{11}N$

Mol. Wt. 121.18

Solvent: $CDCl_3$

Assignments
a = 1.21
b = 2.52
c = 3.47
d = 6.41
e = 6.49
f = 6.55
g = 7.05

492

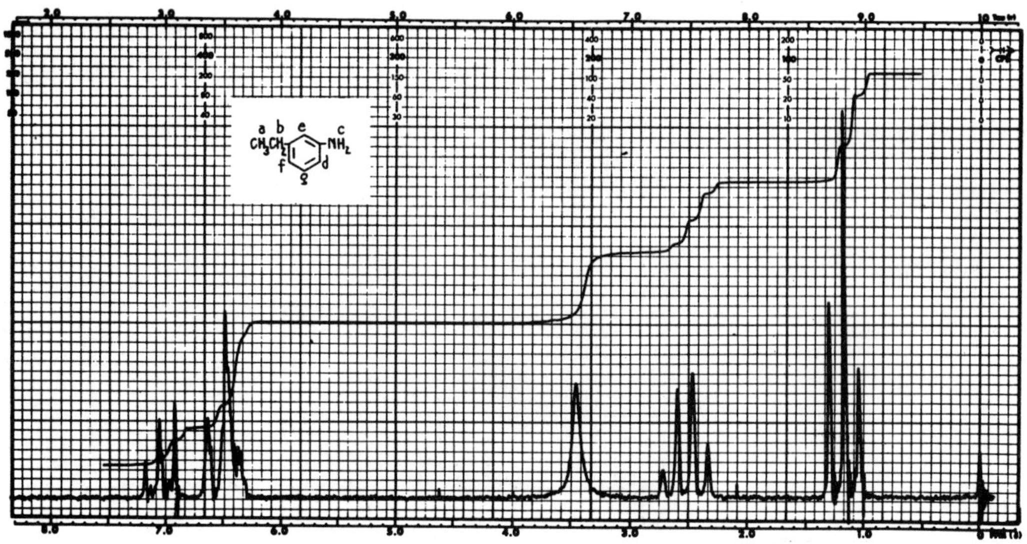

© 1978 Sadtler Research Laboratories, Inc.

CUMIDINE

$C_9H_{13}N$

Mol. Wt. 135.21

Solvent: $CDCl_3$

Assignments

a = 1.22
b = 2.80
c = 3.43
d = 6.60
e = 7.03

493

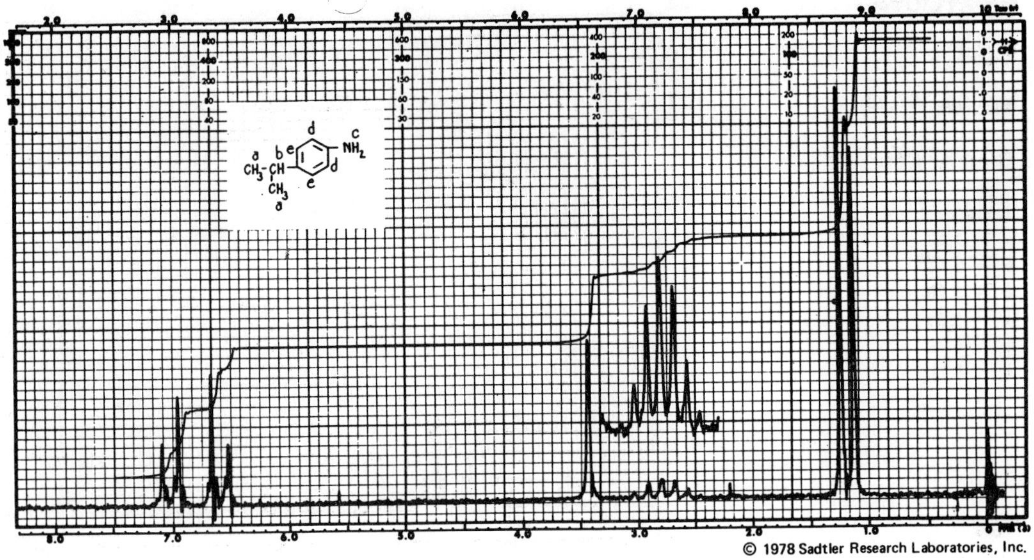

© 1978 Sadtler Research Laboratories, Inc.

2-sec-BUTYLANILINE

$C_{10}H_{15}N$

Mol. Wt. 149.24

B.P. 107°C

Solvent: CCl_4

Assignments

a = 0.88
b = 1.20
c = 1.53
d = 2.51
e = 3.34
f = 6.48
g = 6.60-7.11

494

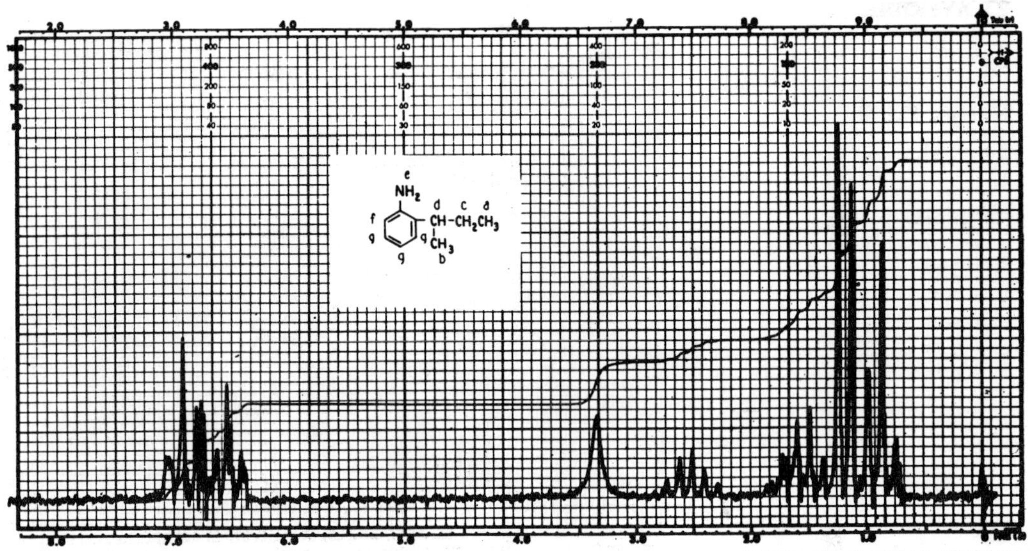

© 1978 Sadtler Research Laboratories, Inc.

2,3-XYLIDINE

$C_8H_{11}N$

Mol. Wt. 121.18

B.P. 94-96°C/11mm

Solvent: CCl_4

Assignments

a = 2.00 or b
b = 2.21 or a
c = 3.35
d = 6.46
e = 6.80

495

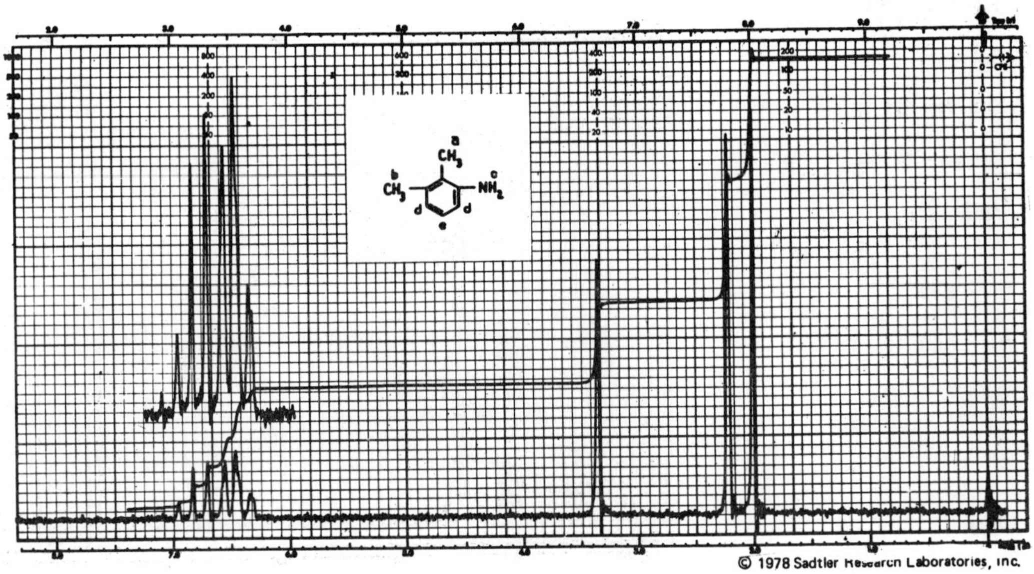

© 1978 Sadtler Research Laboratories, Inc.

2,4-XYLIDINE

$C_8H_{11}N$

Mol. Wt. 121.18

Solvent: CCl_4

Assignments
a = 2.02
b = 2.16
c = 3.16
d = 6.36
e = 6.66
f = 6.72

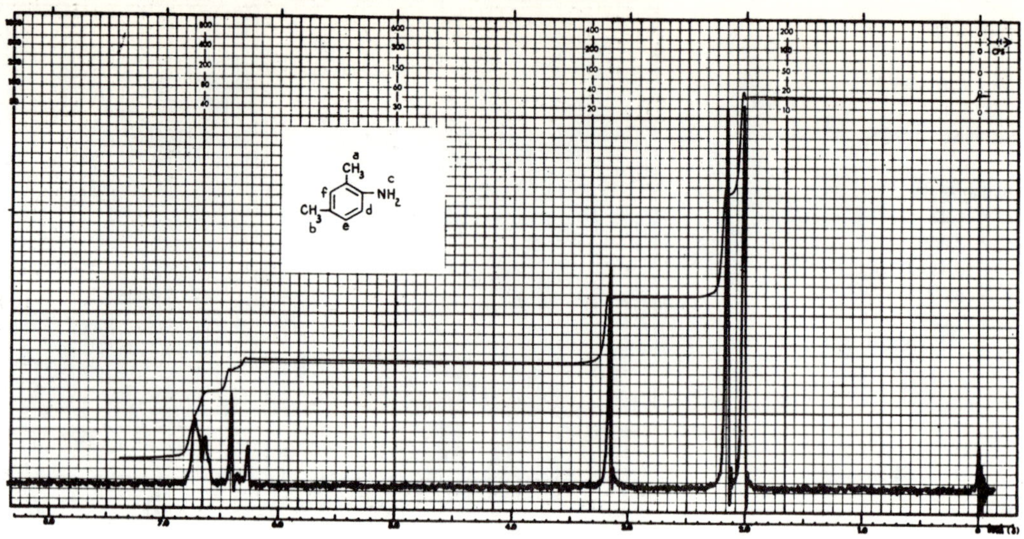

496

© 1978 Sadtler Research Laboratories, Inc.

2,5-XYLIDINE

$C_8H_{11}N$

Mol. Wt. 121.18

Solvent: CCl_4

Assignments
a = 2.00
b = 2.17
c = 3.21
d = 6.25
e = 6.34
f = 6.76

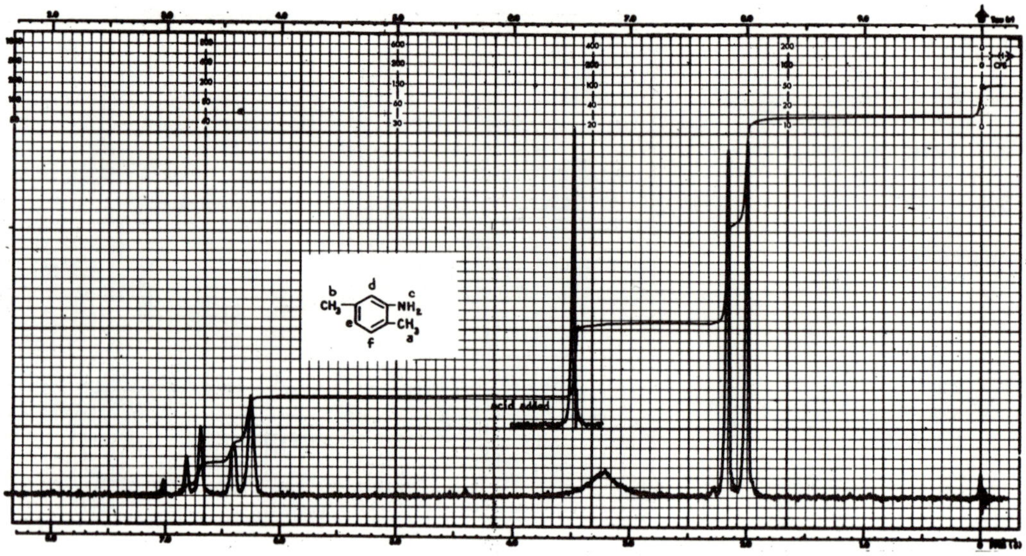

497

© 1978 Sadtler Research Laboratories, Inc.

3,4-XYLIDINE

$C_8H_{11}N$

Mol. Wt. 121.18

Solvent: $CDCl_3$

Assignments
a = 2.19
b = 3.43
c = 6.49
d = 6.53
f = 6.93

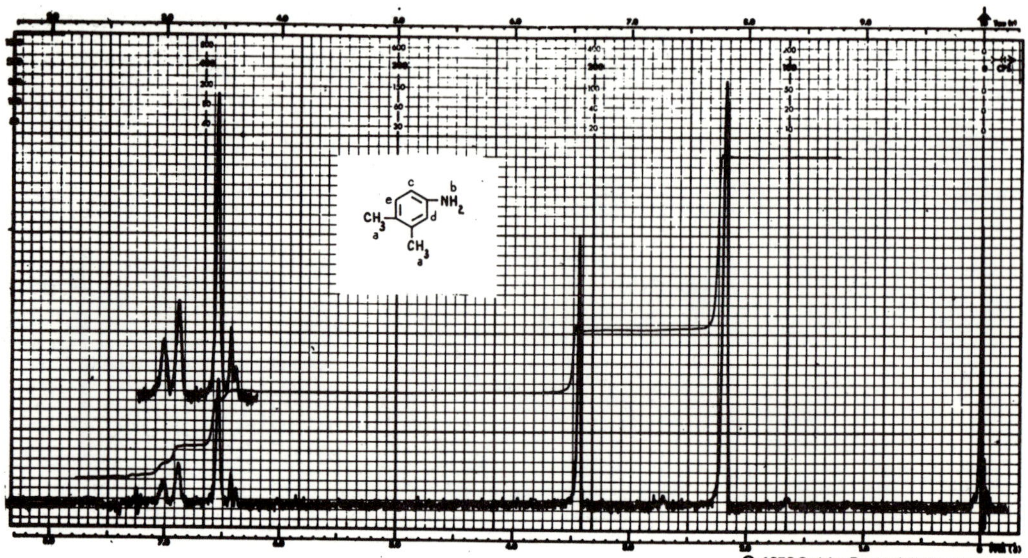

498

© 1978 Sadtler Research Laboratories, Inc.

3,5-XYLIDINE

$C_8H_{11}N$

Mol. Wt. 121.18

Solvent: CCl_4

Assignments
a = 2.15
b = 3.29
c = 6.09
d = 6.29

499

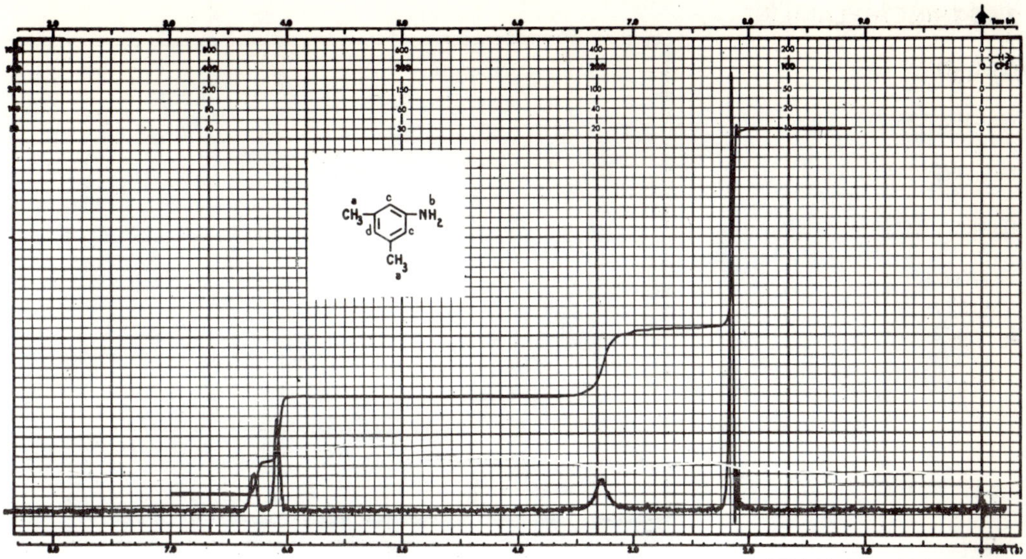

© 1978 Sadtler Research Laboratories, Inc.

6-ETHYL-o-TOLUIDINE

$C_9H_{13}N$

Mol. Wt. 135.21

M.P. −33°C B.P. 231°C

Solvent: $CDCl_3$

Assignments
a = 1.19
b = 2.08
c = 2.42
d = 3.38
e = 6.64
f = 6.90

500

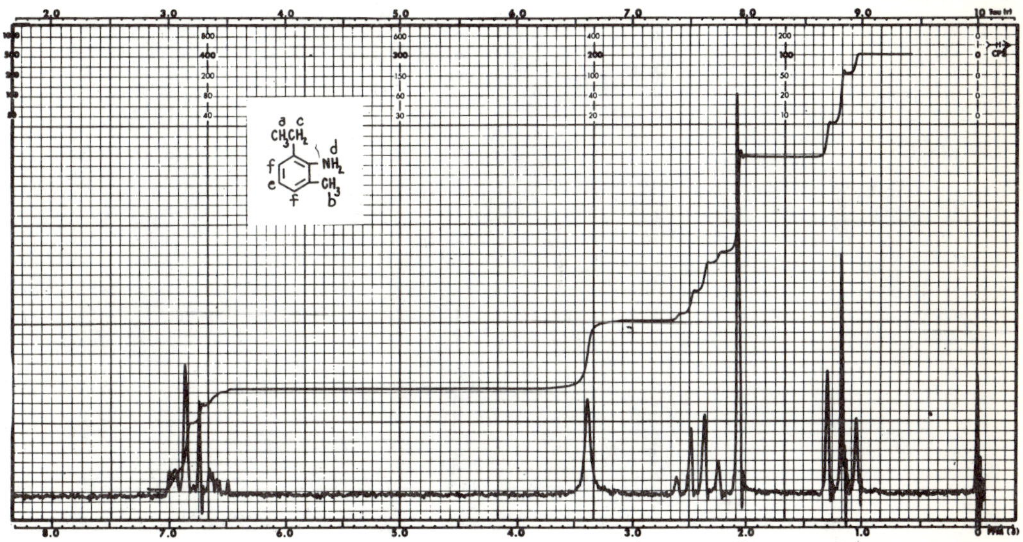

© 1978 Sadtler Research Laboratories, Inc.

2,6-DIISOPROPYLANILINE

$C_{12}H_{19}N$

Mol. Wt. 177.29

M.P. −45°C B.P. 257°C

Solvent: $CDCl_3$

Assignments
a = 1.25
b = 2.89
c = 3.58
d = 6.80
e = 7.03

501

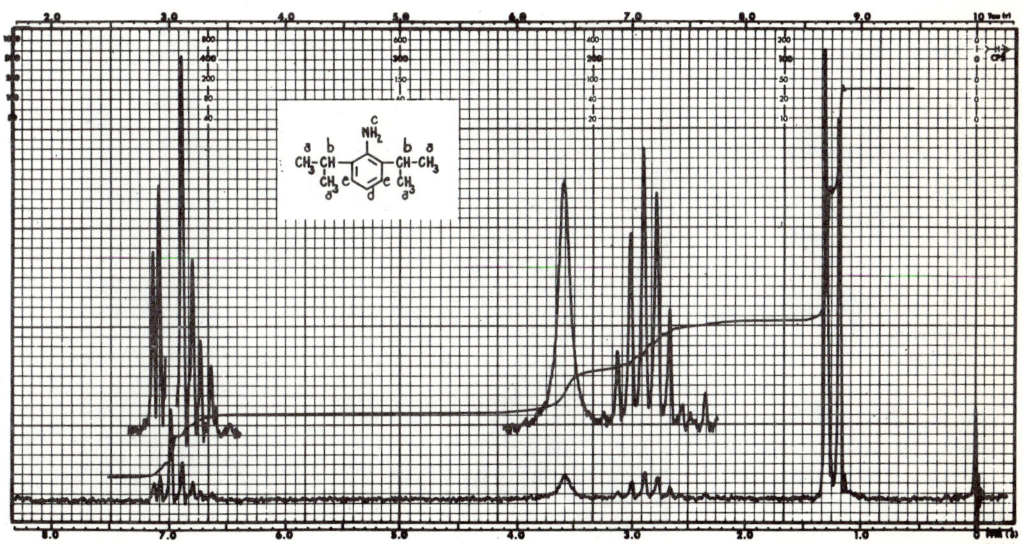

© 1978 Sadtler Research Laboratories, Inc.

2,4,6-TRIMETHYLANILINE

$C_9H_{13}N$

Mol. Wt. 135.21

M.P. 233°C (lit.)

Solvent: CCl_4

Assignments
a = 2.07
b = 2.16
c = 3.19
d = 6.63

502

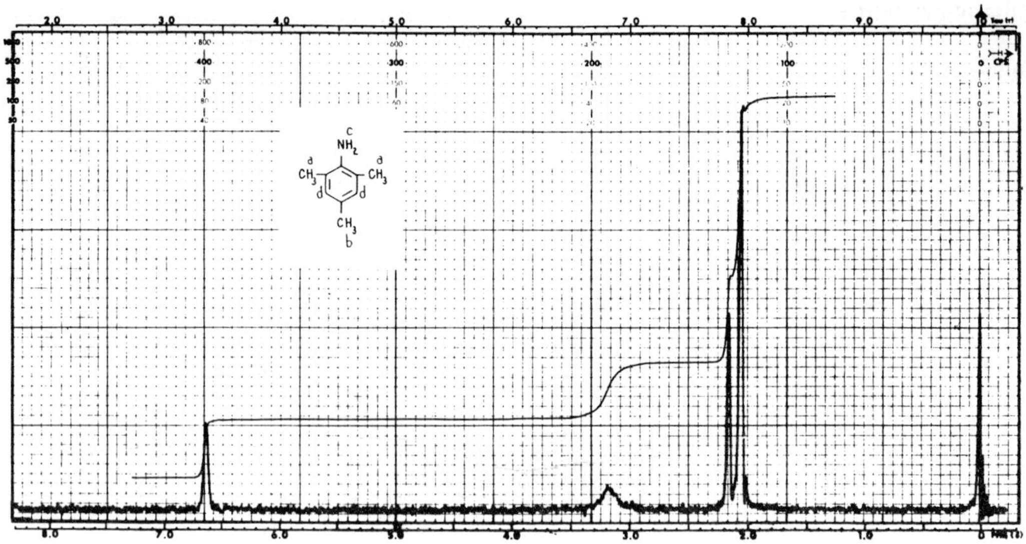

© 1978 Sadtler Research Laboratories, Inc.

2,6-DIETHYL-p-TOLUIDINE

$C_{11}H_{17}N$

Mol. Wt. 163.26

B.P. 122°C

Solvent: CCl_4

Assignments
a = 1.19
b = 2.18
c = 2.40
d = 3.21
e = 6.61

503

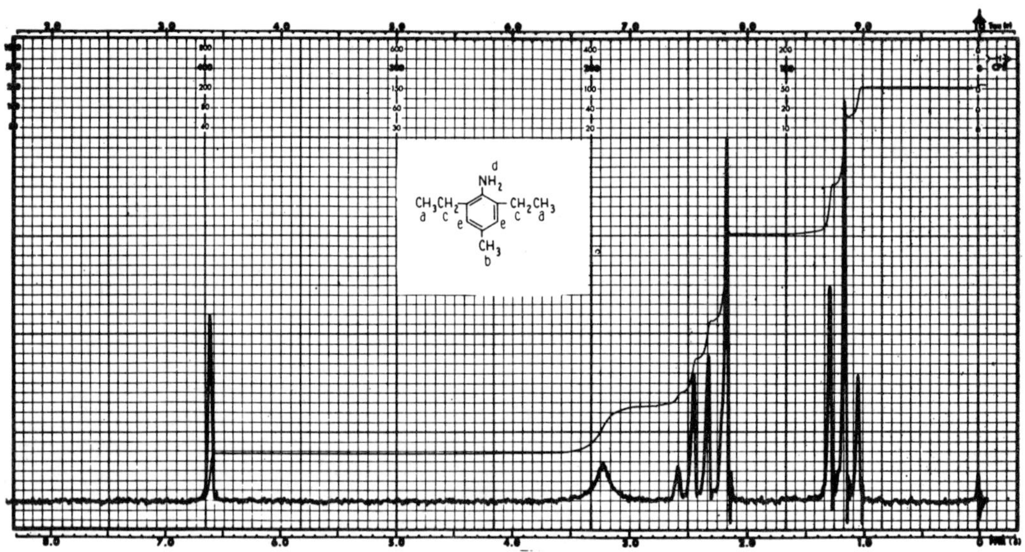

© 1978 Sadtler Research Laboratories, Inc.

2-AMINOBIPHENYL

$C_{12}H_{11}N$

Mol. Wt. 169.23

Solvent: Dimethyl sulfoxide-d_6

Assignments
a = 4.58
b = 6.58-7.20
c = 7.43
d = 2.34 DMSO-d_5
e = 3.61 H_2O

504

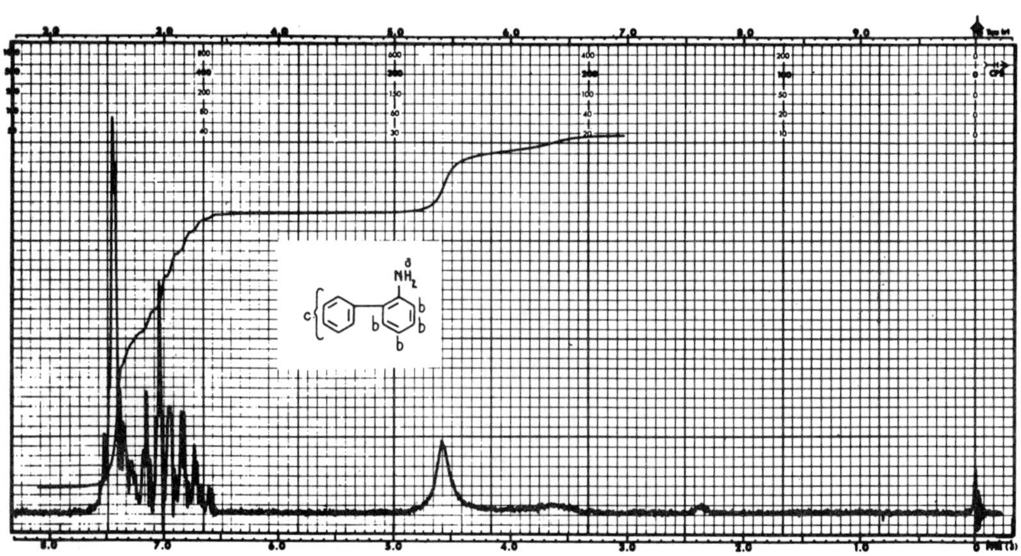

© 1978 Sadtler Research Laboratories, Inc.

4-BIPHENYLAMINE

$C_{12}H_{11}N$

Mol. Wt. 169.23

M.P. 54-55°C (lit.)

Solvent: Acetone

Assignments
a = 3.93
b = 6.73
c = 7.05-7.62

505

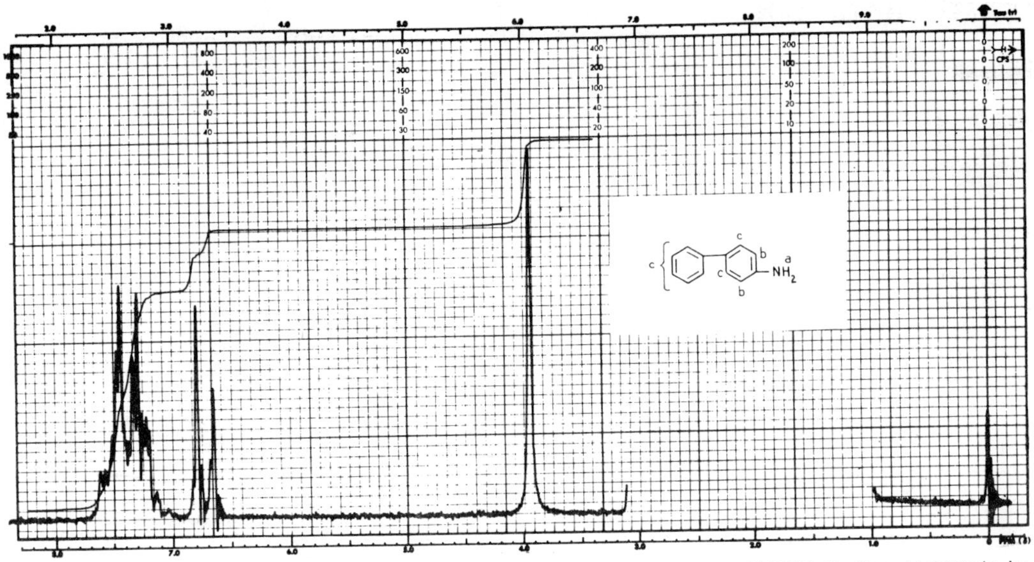

© 1978 Sadtler Research Laboratories, Inc.

---

a,a,a-**TRIFLUORO-m-TOLUIDINE**

$C_7H_6F_3N$

Mol. Wt. 161.13

Solvent: $CDCl_3$

Assignments
a = 3.70
b = 6.75
c = 6.81
d = 6.92
e = 7.20

506

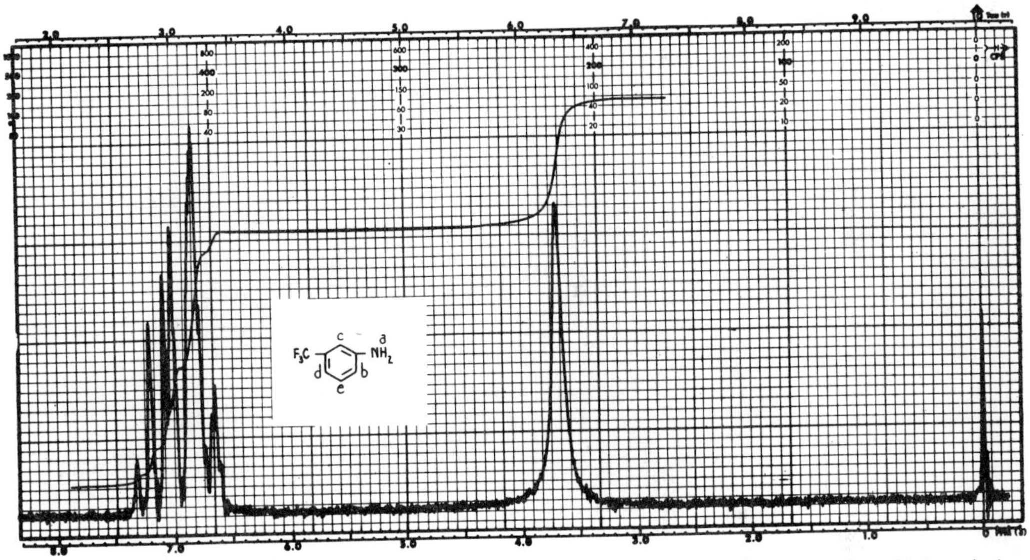

© 1978 Sadtler Research Laboratories, Inc.

---

a,a,a-**TRIFLUORO-p-TOLUIDINE**

$C_7H_6F_3N$

Mol. Wt. 161.13

Solvent: $CDCl_3$

Assignments
a = 3.77
b = 6.56
c = 7.35

507

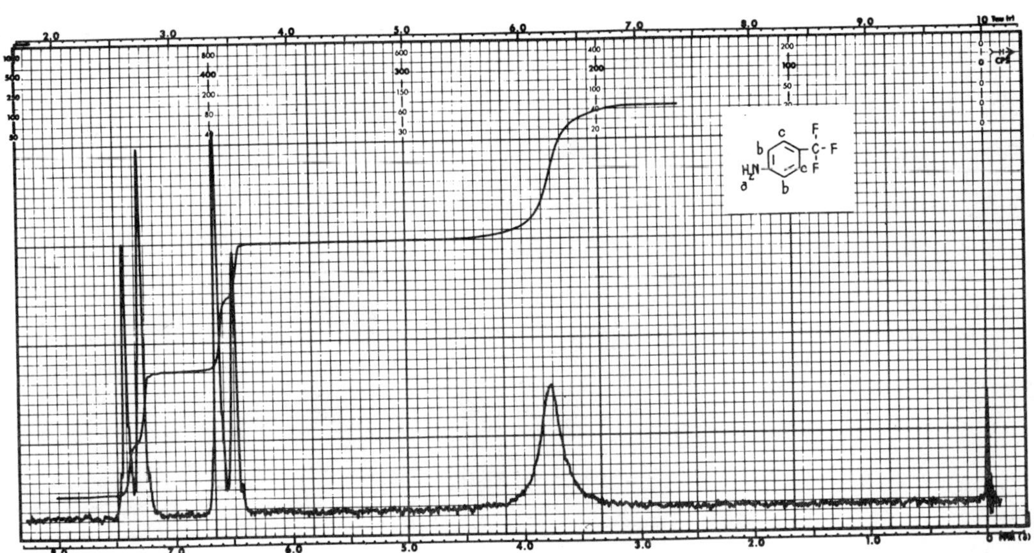

© 1978 Sadtler Research Laboratories, Inc.

o-FLUOROANILINE

C_6H_6FN

Mol. Wt. 111.12

Solvent: CCl_4

Assignments
a = 3.50
b = 6.29-7.10

508

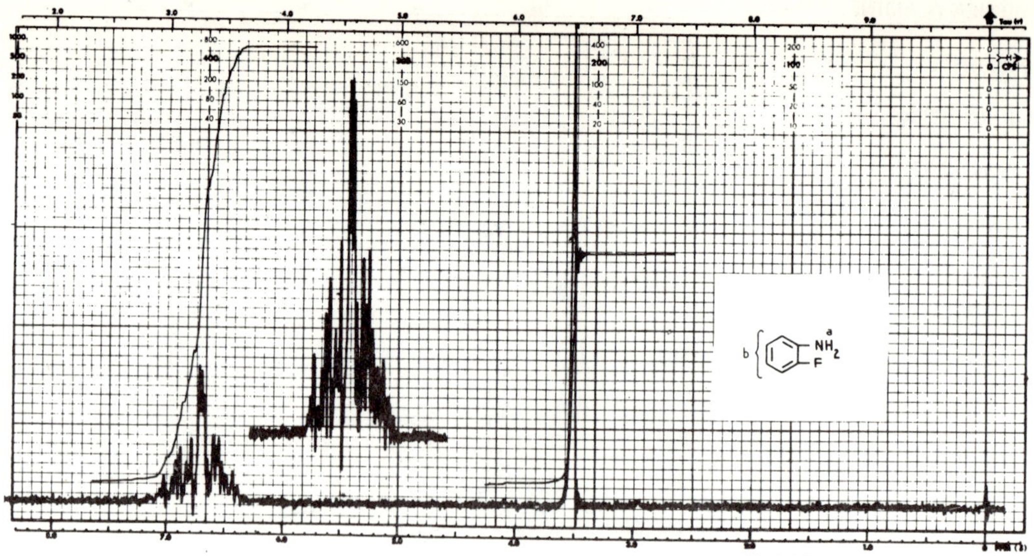

© 1978 Sadtler Research Laboratories, Inc.

m-FLUOROANILINE

C_6H_6FN

Mol. Wt. 111.12

Solvent: $CDCl_3$

Assignments
a = 3.60
b = 6.15-6.60
c = 6.97

509

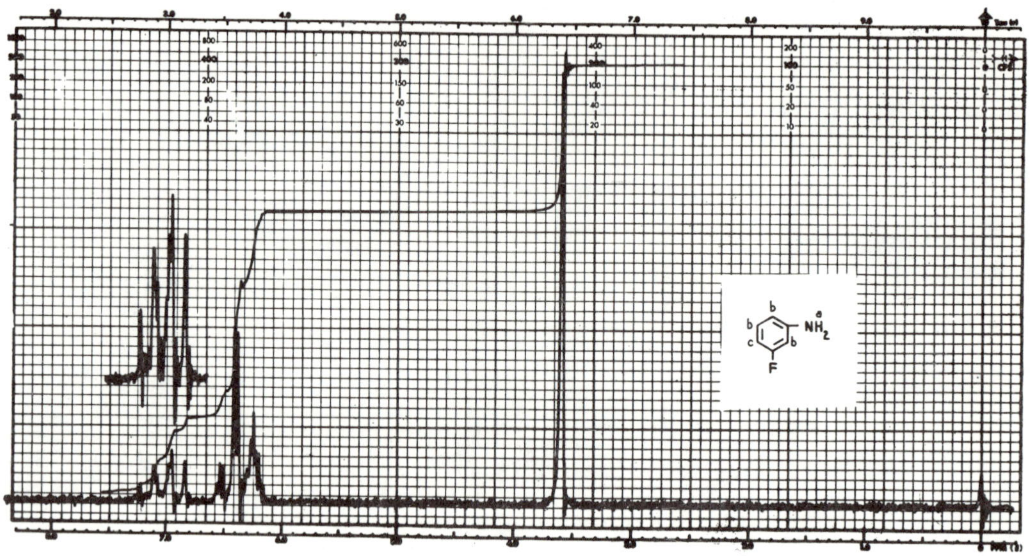

© 1978 Sadtler Research Laboratories, Inc.

p-FLUOROANILINE

C_6H_6FN

Mol. Wt. 111.12

Solvent: CCl_4

Assignments
a = 3.33
b = 6.42
c = 6.75

510

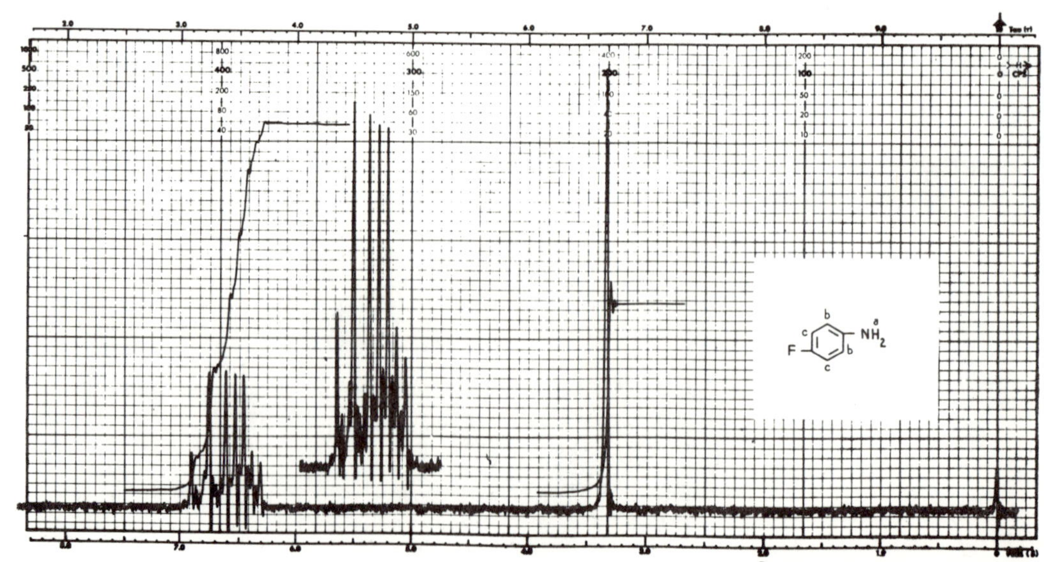

© 1978 Sadtler Research Laboratories, Inc.

4-FLUORO-m-TOLUIDINE

C_7H_8FN

Mol. Wt. 125.15

Solvent: CCl_4

Assignments
a = 2.11
b = 3.32
c = 6.05-6.37
d = 6.69
$J_{F^{19}\text{-}a}$ = 1.8 Hz

511

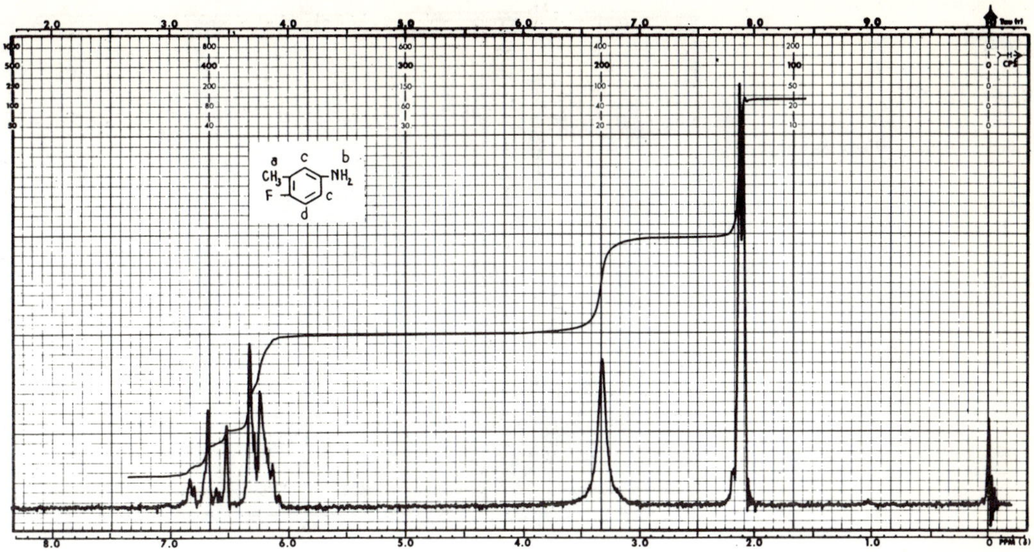

© 1978 Sadtler Research Laboratories, Inc.

---

o-CHLOROANILINE

C_6H_6ClN

Mol. Wt. 127.57

B.P. 208-210°C

Solvent: CCl_4

Assignments
a = 3.81
b = 6.40-7.30

512

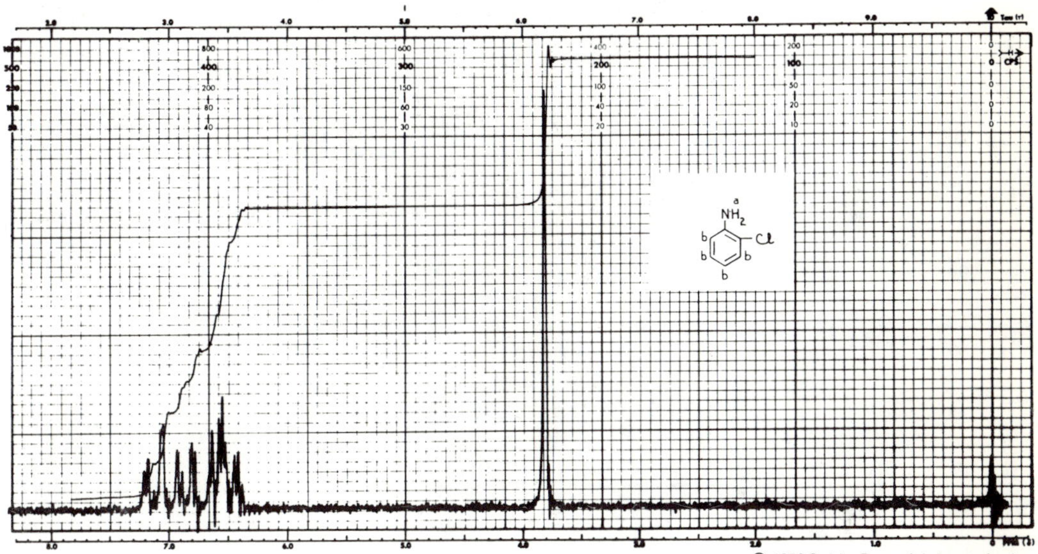

© 1978 Sadtler Research Laboratories, Inc.

---

m-CHLOROANILINE

C_6H_6ClN

Mol. Wt. 127.57

B.P. 228-231°C

Solvent: CCl_4

Assignments
a = 3.50
b = 6.37
c = 6.49
d = 6.59
e = 6.96

513

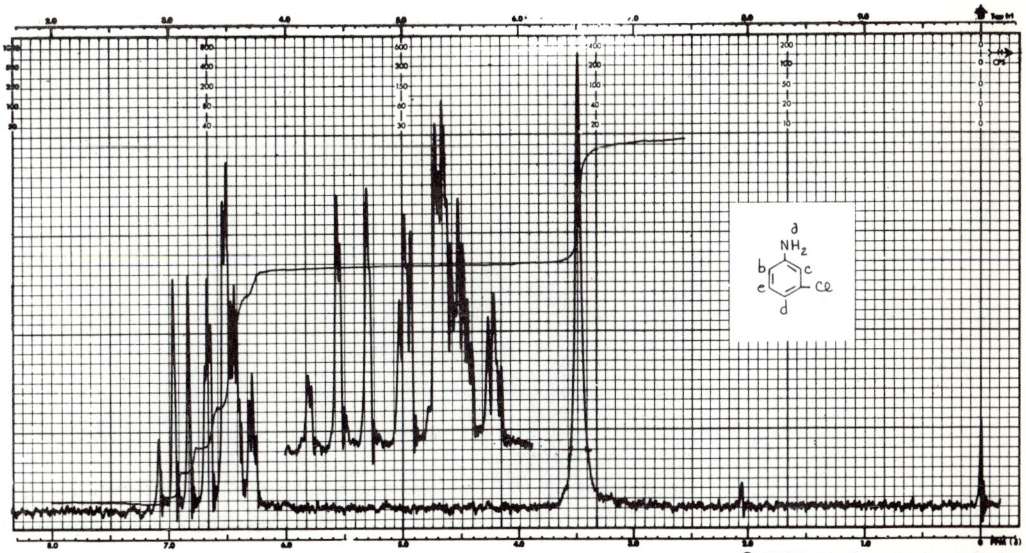

© 1978 Sadtler Research Laboratories, Inc.

p-CHLOROANILINE

C_6H_6ClN

Mol. Wt. 127.57

Solvent: Dimethyl sulfoxide

Assignments
a = 5.03
b = 6.60
c = 7.02

514

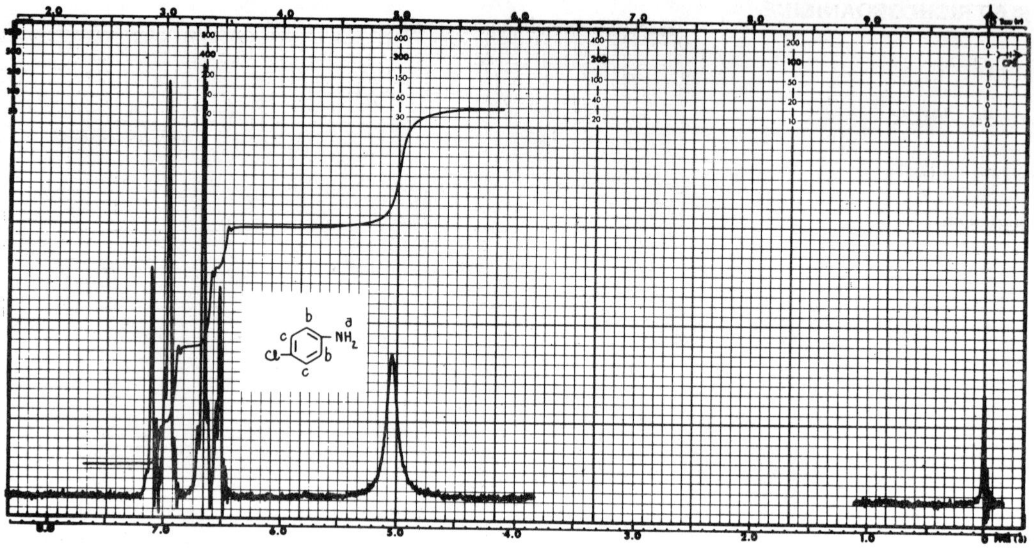

© 1978 Sadtler Research Laboratories, Inc.

4-CHLORO-o-TOLUIDINE

C_7H_8ClN

Mol. Wt. 141.61

M.P. 26-29°C

Solvent: CCl_4

Assignments
a = 2.08
b = 3.49
c = 6.50
d = ca 6.95

515

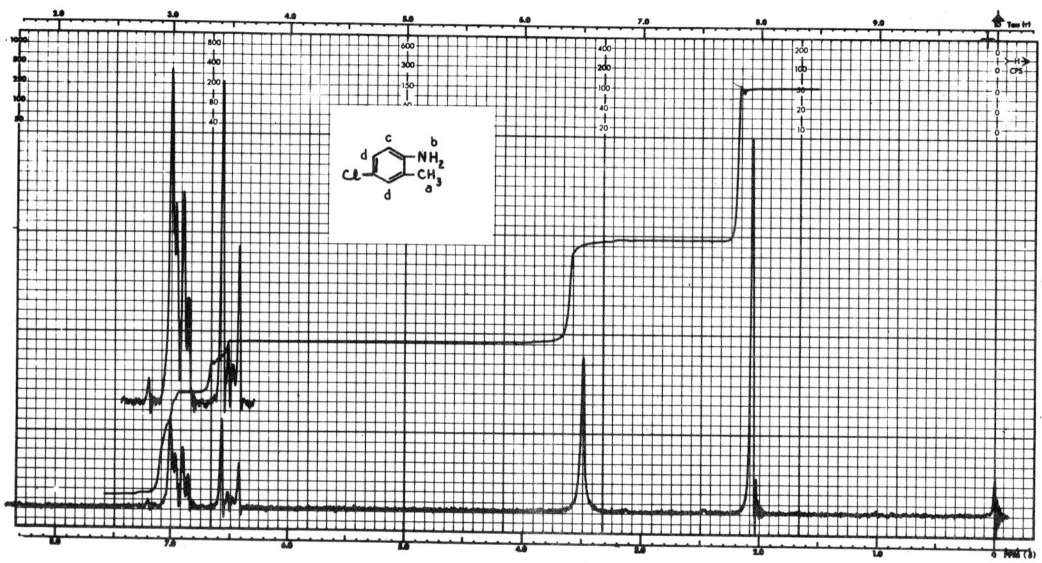

© 1978 Sadtler Research Laboratories, Inc.

2,4-DICHLOROANILINE

$C_6H_5Cl_2N$

Mol. Wt. 162.02

Solvent: $CDCl_3$

Assignments
a = 3.99
b = 6.60
c = 7.00
d = 7.22

516

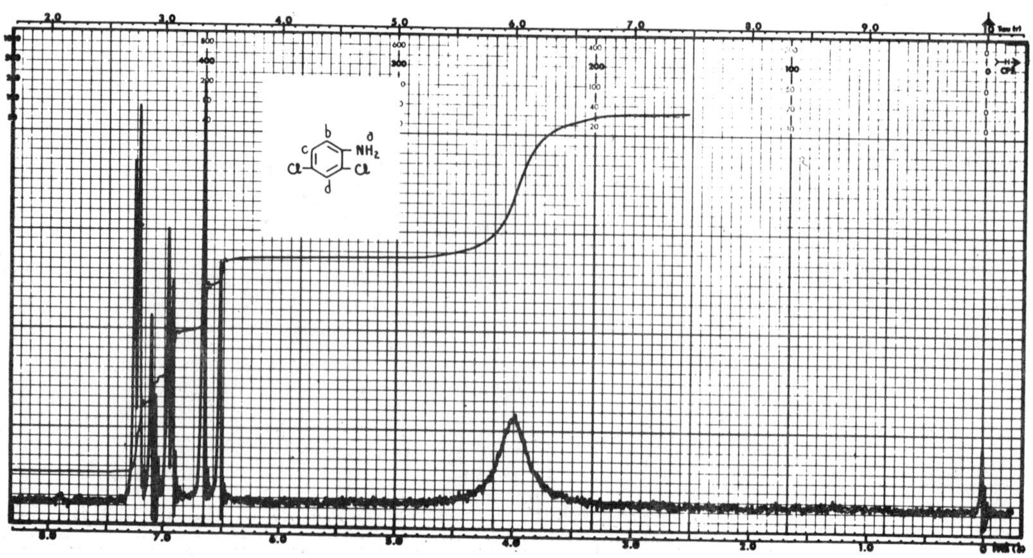

© 1978 Sadtler Research Laboratories, Inc.

2,4,6-TRICHLOROANILINE

$C_6H_4Cl_3N$

Mol. Wt. 196.46

Solvent: $CDCl_3$

Assignments
a = 4.39
b = 7.18

517

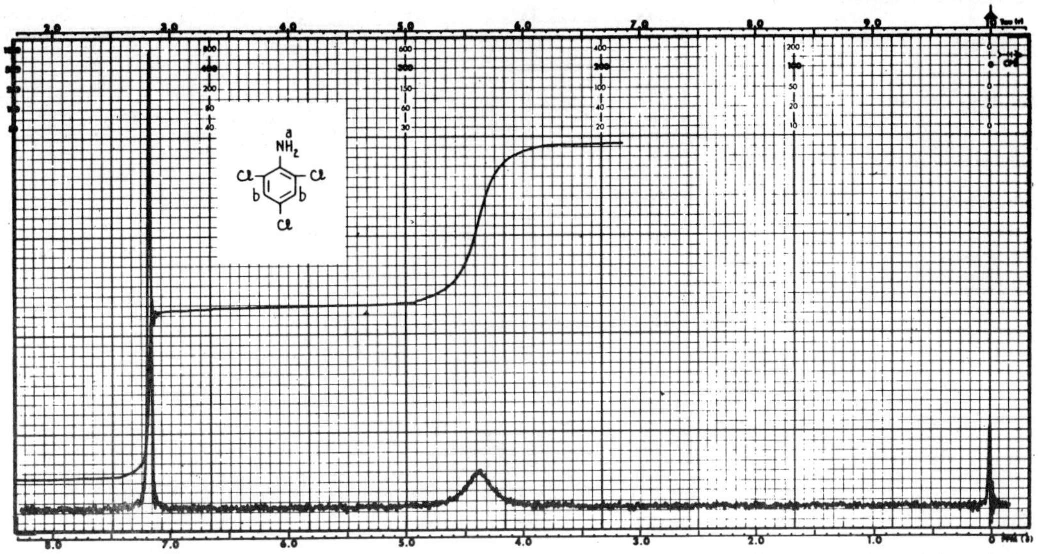

© 1978 Sadtler Research Laboratories, Inc.

---

o-BROMOANILINE

C_6H_6BrN

Mol. Wt. 172.03

M.P. 28-29°C

Solvent: $CDCl_3$

Assignments
a = 3.95
b = 6.54
c = 6.60
d = 7.03
e = 7.34

518

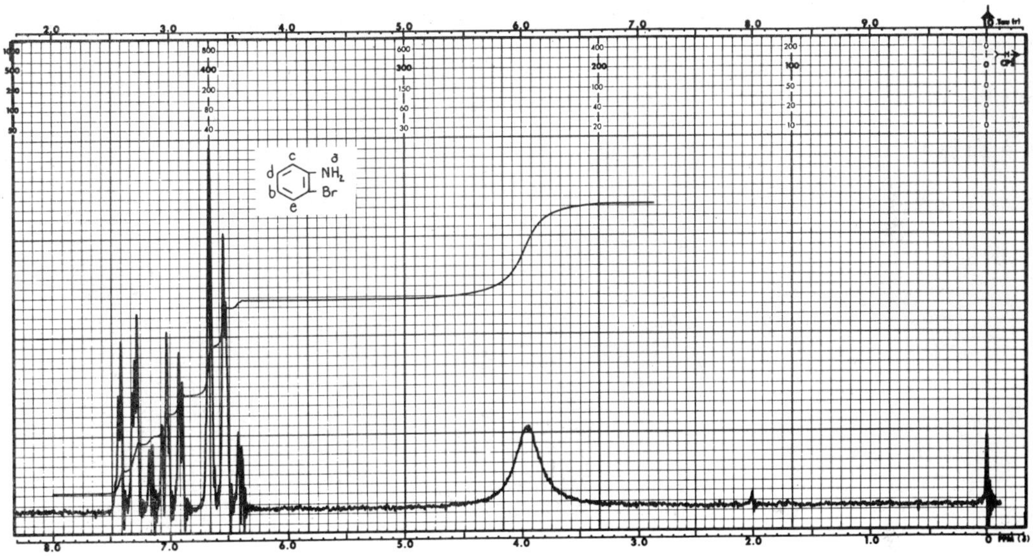

© 1978 Sadtler Research Laboratories, Inc.

---

m-BROMOANILINE

C_6H_6BrN

Mol. Wt. 172.03

M.P. 15-17°C

Solvent: $CDCl_3$

Assignments
a = 3.58
b = 6.41
c = 6.69
d = 6.75-7.06

519

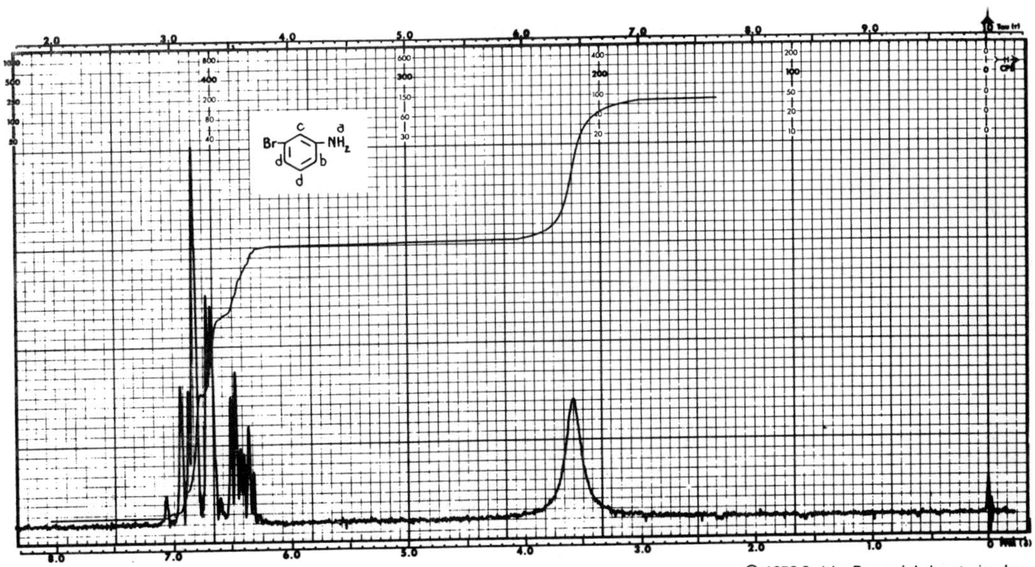

© 1978 Sadtler Research Laboratories, Inc.

p-BROMOANILINE

C_6H_6BrN

Mol. Wt. 172.03

B.P. 62-64°C

Solvent: $CDCl_3$

Assignments
a = 3.53
b = 6.57
c = 7.21

520

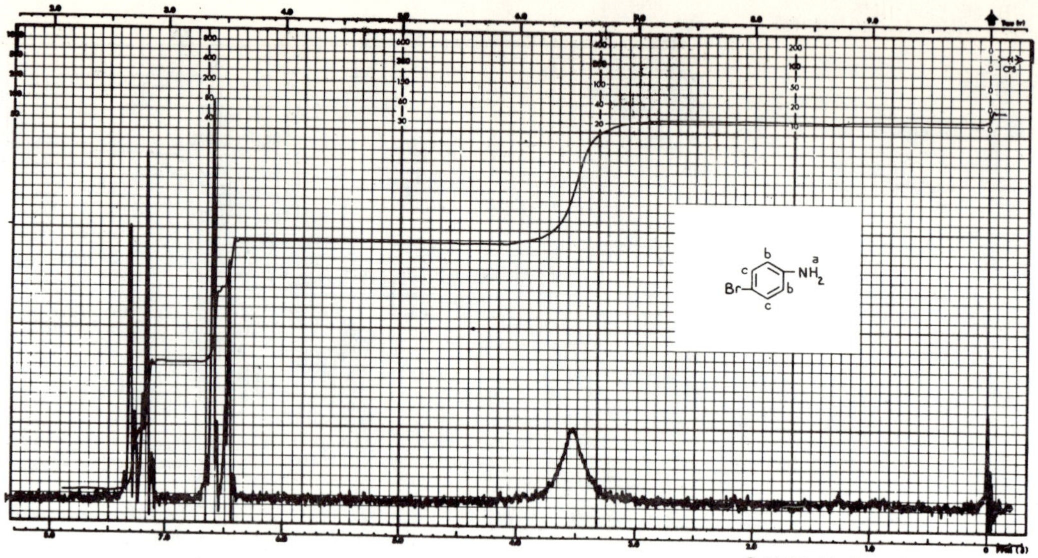

© 1978 Sadtler Research Laboratories, Inc.

2-BROMO-p-TOLUIDINE

C_7H_8BrN

Mol. Wt. 186.06

M.P. 14-16°C

Solvent: $CDCl_3$

Assignments
a = 2.17
b = 3.82
c = 6.55
d = 6.85
e = 7.19

521

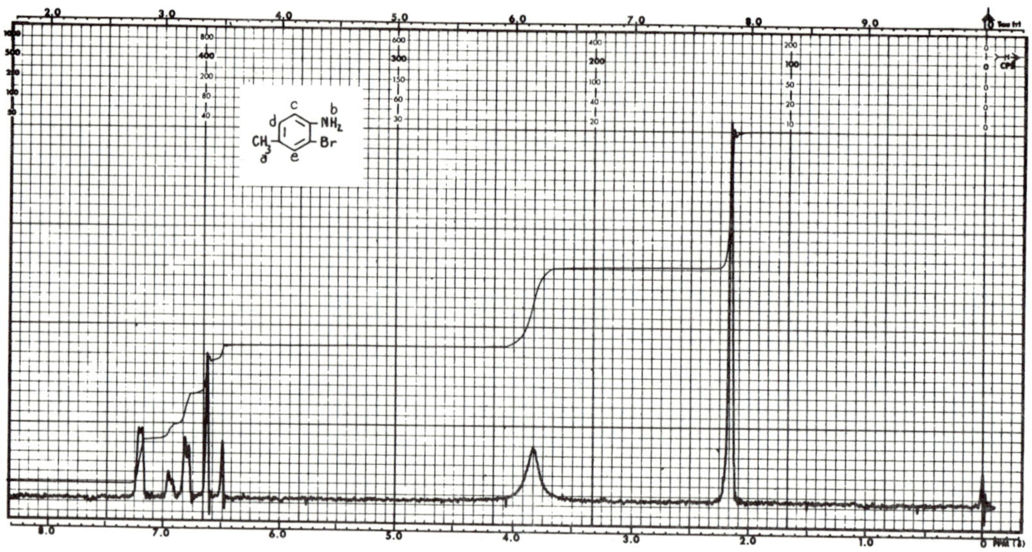

© 1978 Sadtler Research Laboratories, Inc.

4-BROMO-3-CHLOROANILINE

C_6H_5BrClN

Mol. Wt. 206.48

M.P. 60°C

Solvent: $CDCl_3$

Assignments
a = 3.65
b = 6.34
c = 6.69
d = 7.25

522

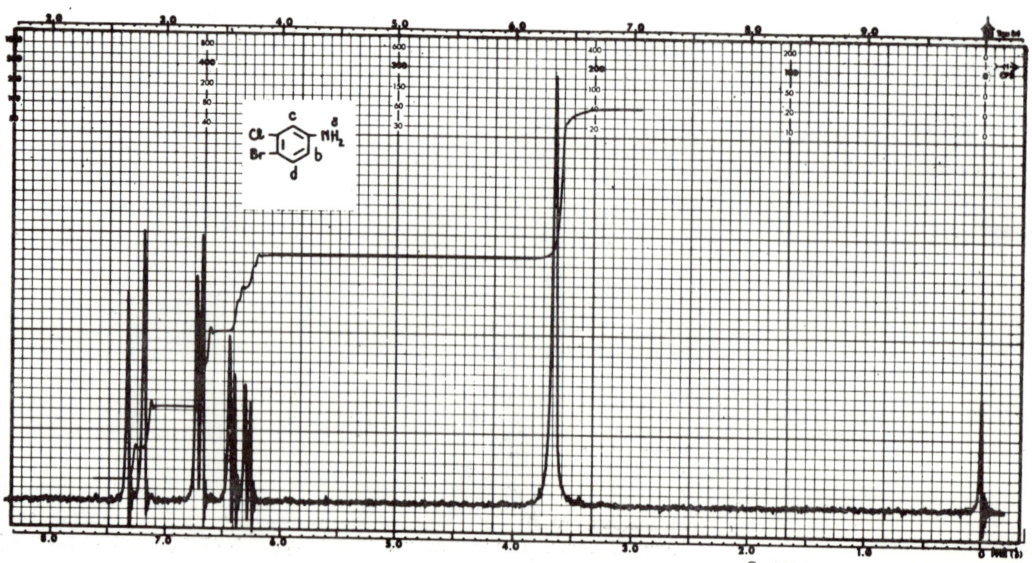

© 1978 Sadtler Research Laboratories, Inc.

2,4-DIBROMOANILINE

$C_6H_5Br_2N$

Mol. Wt. 250.95

M.P. 78-79°C

Solvent: $CDCl_3$

Assignments
a = 3.99
b = 6.58
c = 7.18
d = 7.51

523

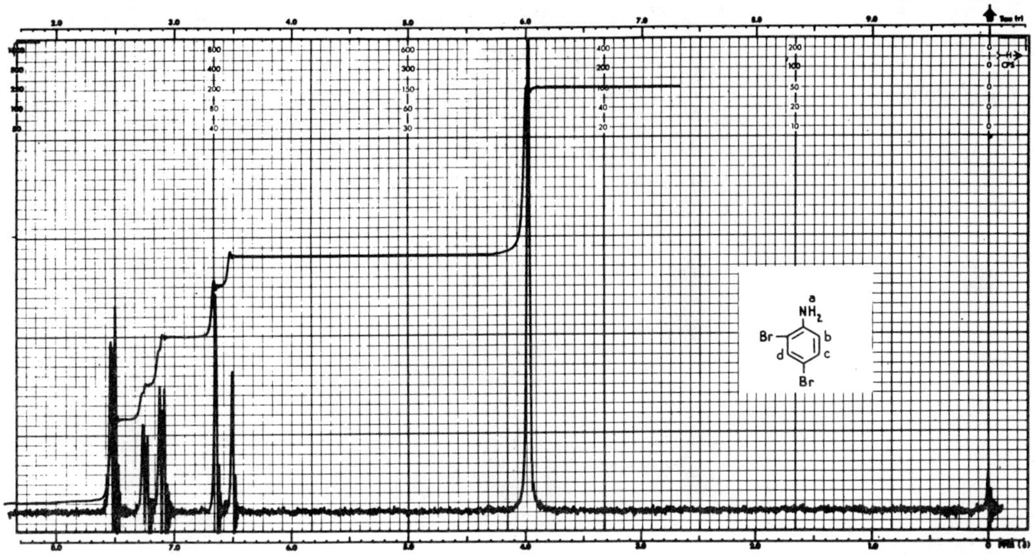

© 1978 Sadtler Research Laboratories, Inc.

2,4,6-TRIBROMOANILINE

$C_6H_4Br_3N$

Mol. Wt. 329.82

M.P. 121-123°C

Solvent: $CDCl_3$

Assignments
a = 4.59
b = 7.50

524

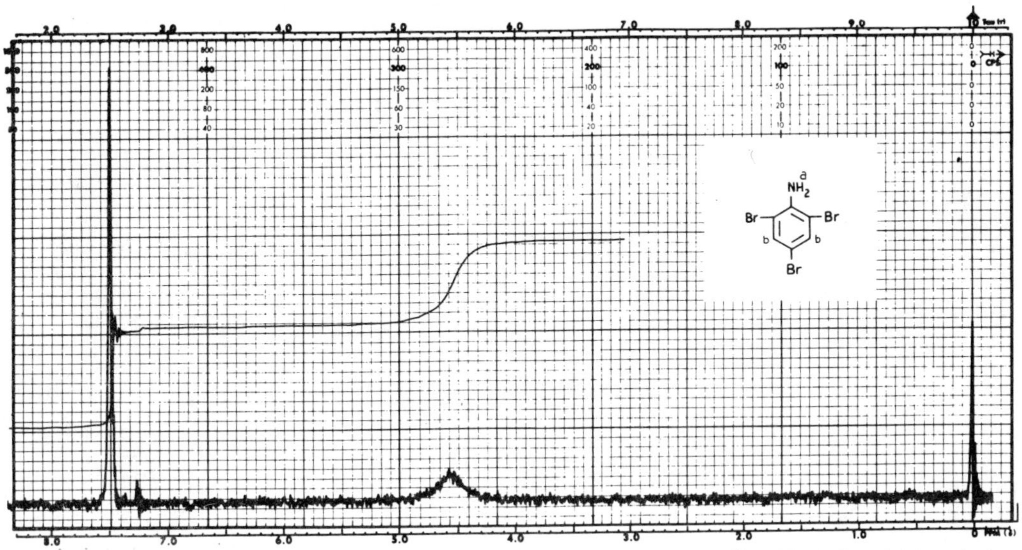

© 1978 Sadtler Research Laboratories, Inc.

o-IODOANILINE

C_6H_6IN

Mol. Wt. 219.03

M.P. 56-58°C

Solvent: $CDCl_3$

Assignments
a = 4.02
b = 6.47
c = 6.71
d = 7.12
e = 7.63

525

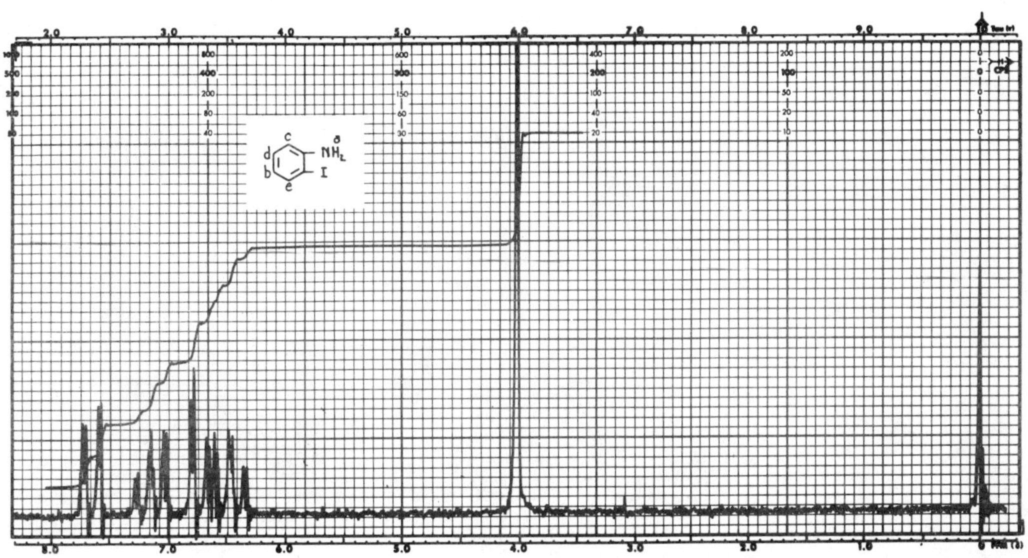

© 1978 Sadtler Research Laboratories, Inc.

m-IODOANILINE

C_6H_6IN

Mol. Wt. 219.03

M.P. 27-29°C

Solvent: $CDCl_3$

Assignments
a = 3.54
b = 6.49
c = 6.79
d = 6.97
e = 7.03

526

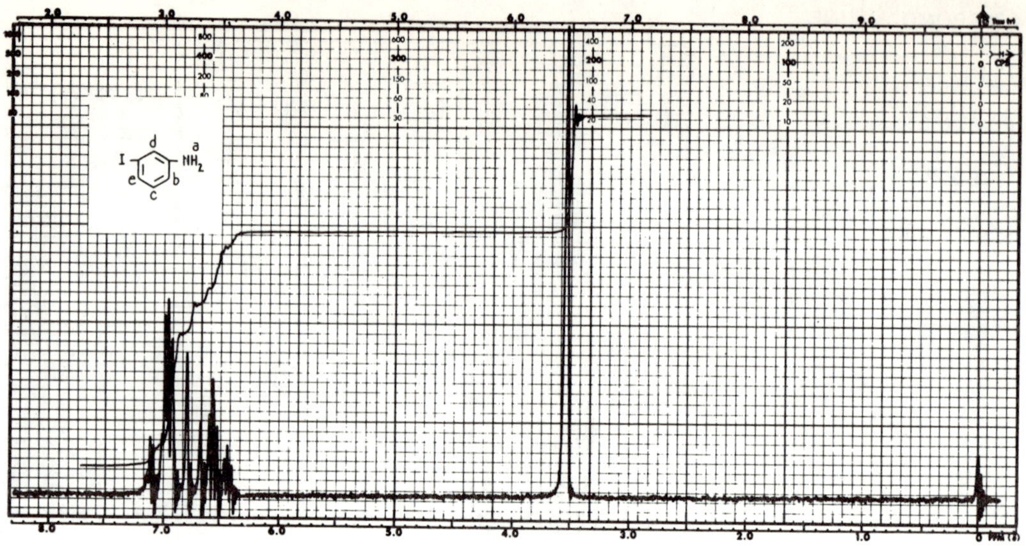

© 1978 Sadtler Research Laboratories, Inc.

p-IODOANILINE

C_6H_6IN

Mol. Wt. 219.02

M.P. 62-63°C

Solvent: $CDCl_3$

Assignments
a = 3.59
b = 6.37
c = 7.37

527

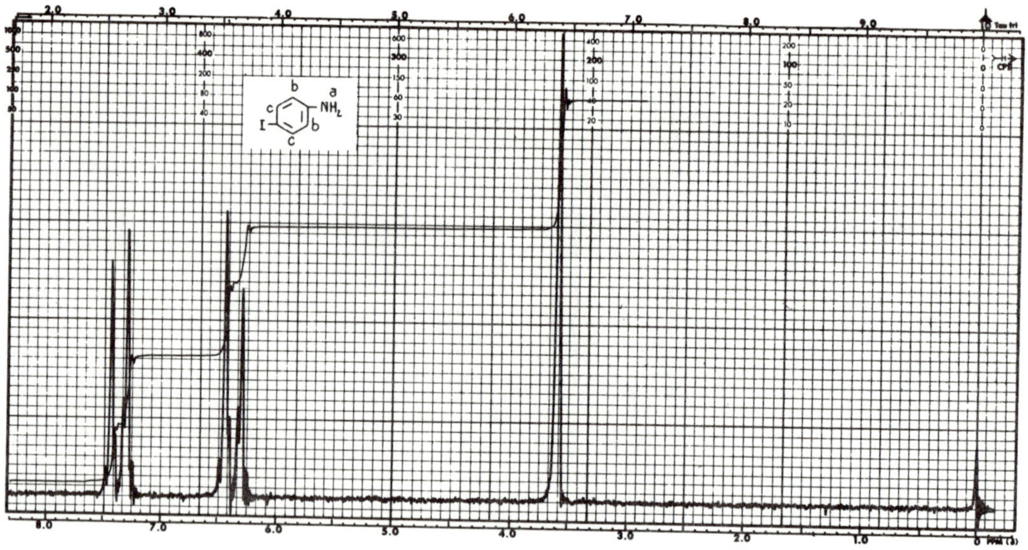

© 1978 Sadtler Research Laboratories, Inc.

4-IODO-o-TOLUIDINE

C_7H_8IN

Mol. Wt. 233.05

Solvent: Polysol-d

Assignments
a = 2.04
b = 4.56
c = 6.42
d = 7.12
e = 7.20

528

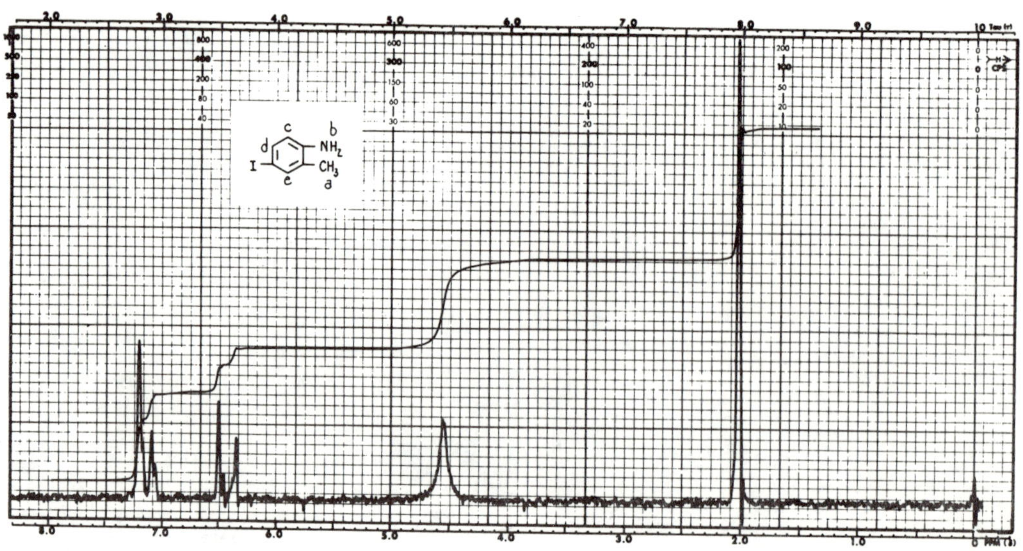

© 1978 Sadtler Research Laboratories, Inc.

p-AMINOPHENETHYLAMINE

$C_8H_{12}N_2$

Mol. Wt. 136.20

Solvent: $CDCl_3$

Assignments
a = 2.61
b = 2.86
c = 6.61
d = 6.98
e = 4.53 HDO

529

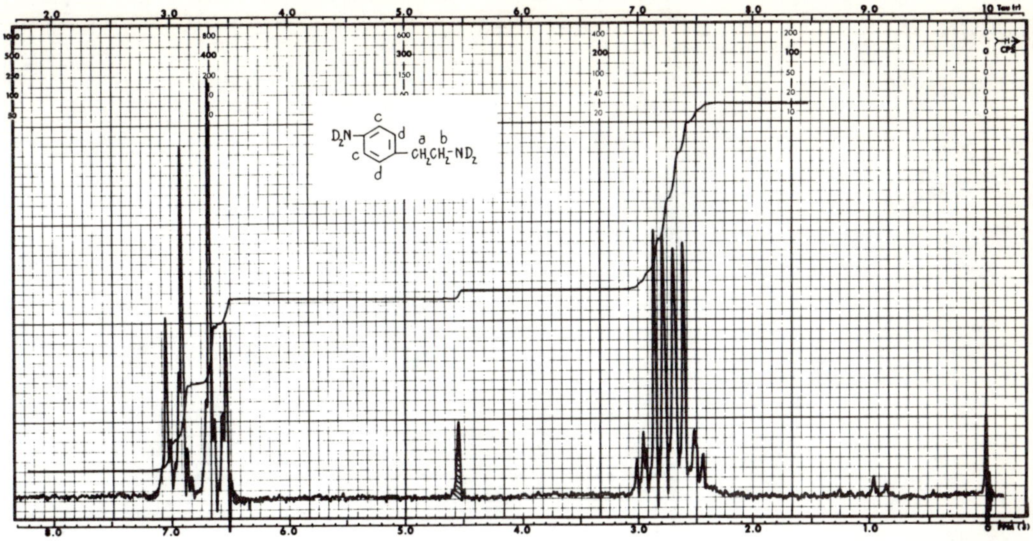

© 1978 Sadtler Research Laboratories, Inc.

BIS(p-AMINOPHENYL)METHANE

$C_{13}H_{14}N_2$

Mol. Wt. 198.27

M.P. 70°C

Solvent: Dimethyl sulfoxide-d_6

Assignments
a = 3.57
b = 4.63
c = 6.49
d = 6.81
e = 2.46 DMSO-d_5
f = 3.30 H_2O

530

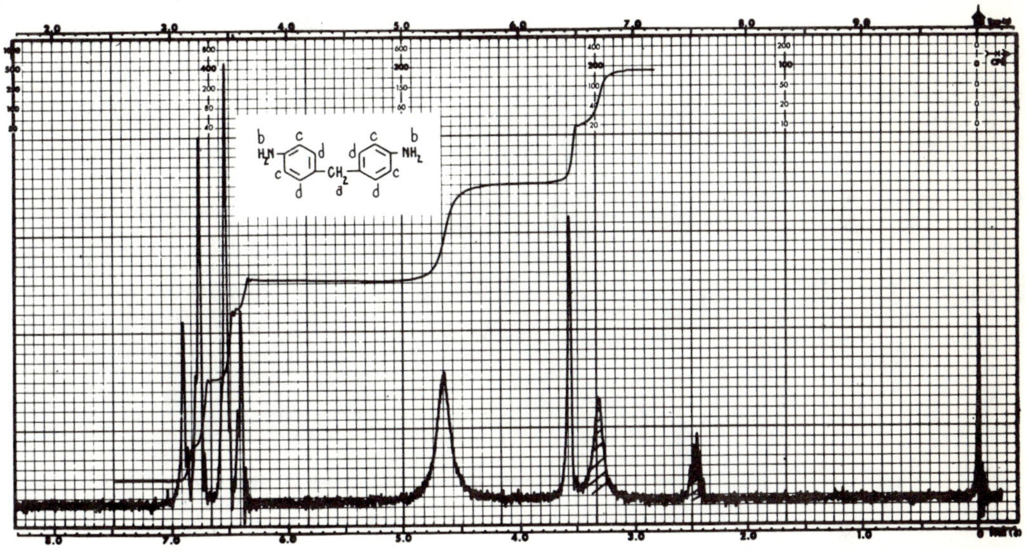

© 1978 Sadtler Research Laboratories, Inc.

BENZIDINE

$C_{12}H_{12}N_2$

Mol. Wt. 184.24

M.P. 127-129°C

Solvent: Dimethyl sulfoxide-D_6

Assignments
a = 4.41
b = 6.62
c = 7.24
d = 2.50 DMSO-D_5

531

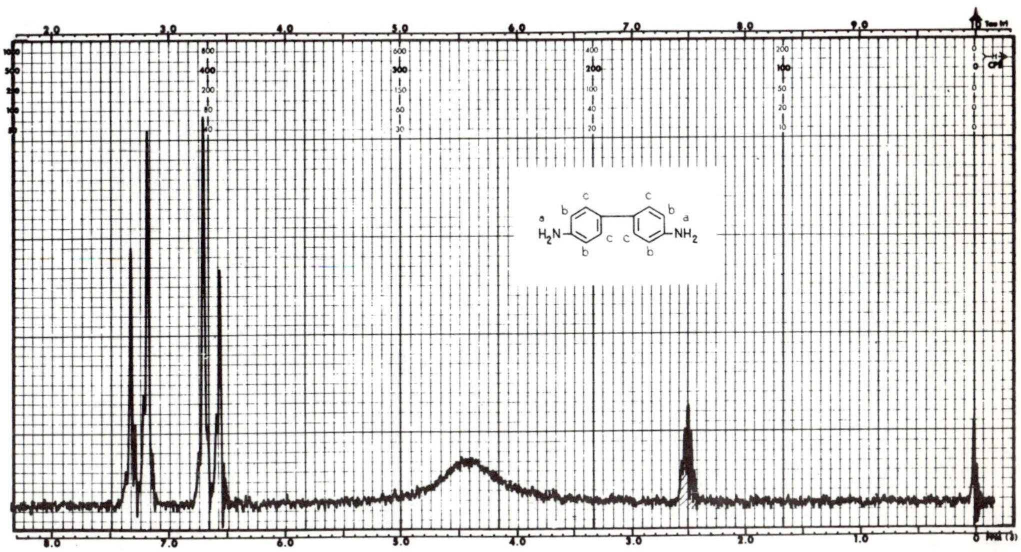

© 1978 Sadtler Research Laboratories, Inc.

o-TOLIDINE

$C_{14}H_{16}N_2$

Mol. Wt. 212.28

M.P. 129-131°C

Solvent: $CDCl_3$

Assignments
a = 2.19
b = 3.45
c = 6.68
d = 7.20
e = 7.23

532

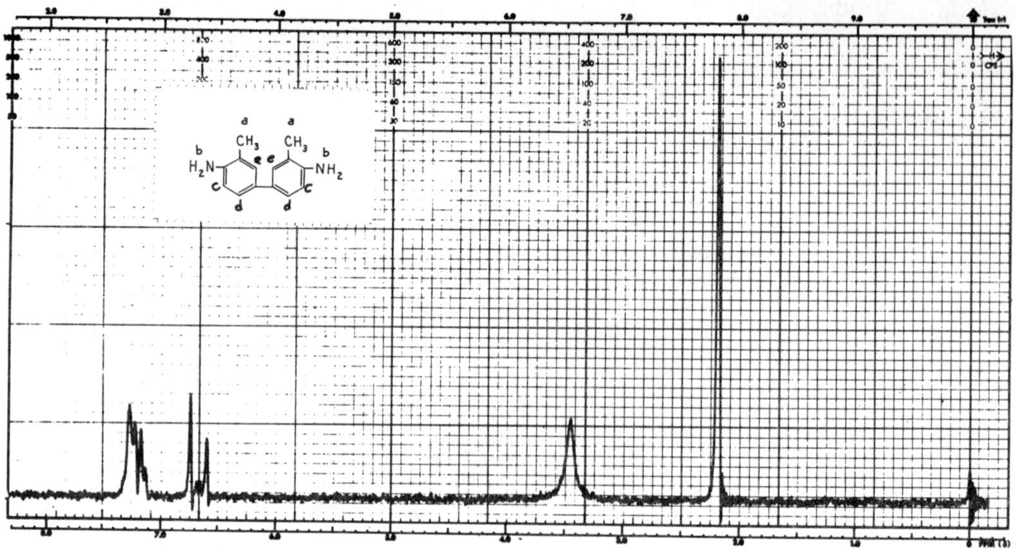

© 1978 Sadtler Research Laboratories, Inc.

o-PHENYLENEDIAMINE

$C_6H_8N_2$

Mol. Wt. 108.41

M.P. 101-102°C

Solvent: $CDCl_3$

Assignments
a = 3.28
b = 6.67

533

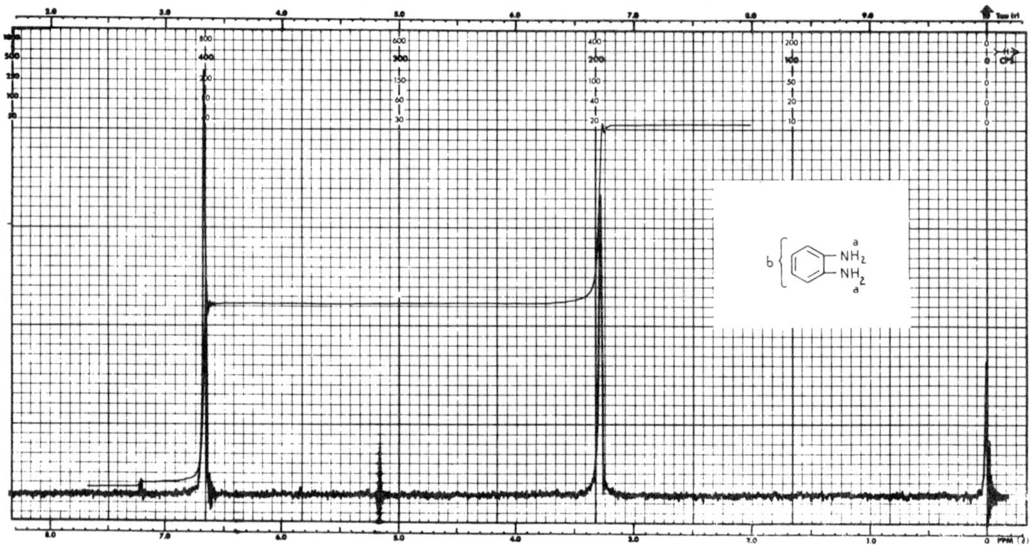

© 1978 Sadtler Research Laboratories, Inc.

m-PHENYLENEDIAMINE

$C_6H_8N_2$

Mol. Wt. 108.14

Solvent: Polysol-d

Assignments
a = 4.08
b = 5.84
c = 5.91
d = 6.74

534

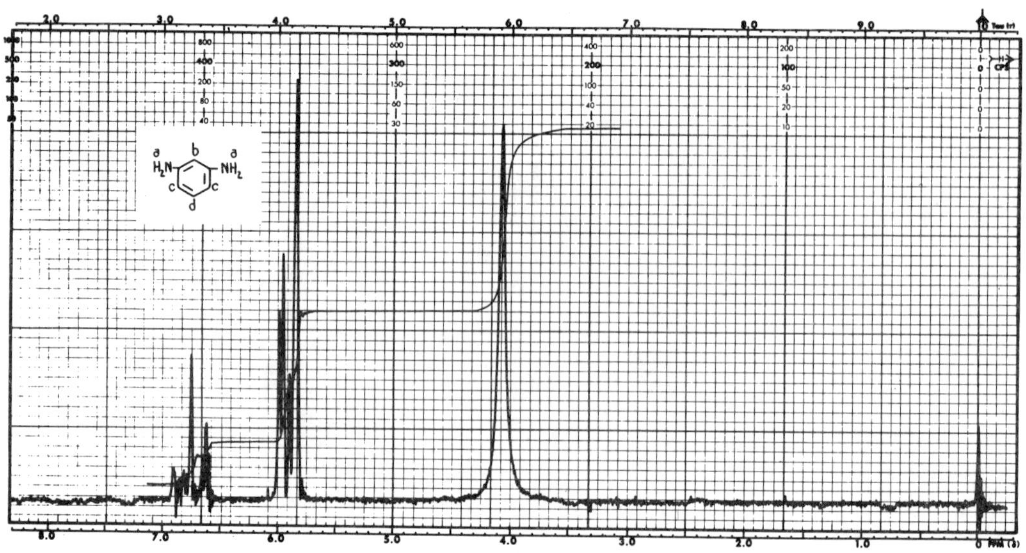

© 1978 Sadtler Research Laboratories, Inc.

p-PHENYLENEDIAMINE

$C_6H_8N_2$

Mol. Wt. 108.14

M.P. 139°C (lit.)

Solvent: Dimethyl sulfoxide-D_6

Assignments
a = 4.00
b = 6.39
c = 2.49 DMSO-D_5

535

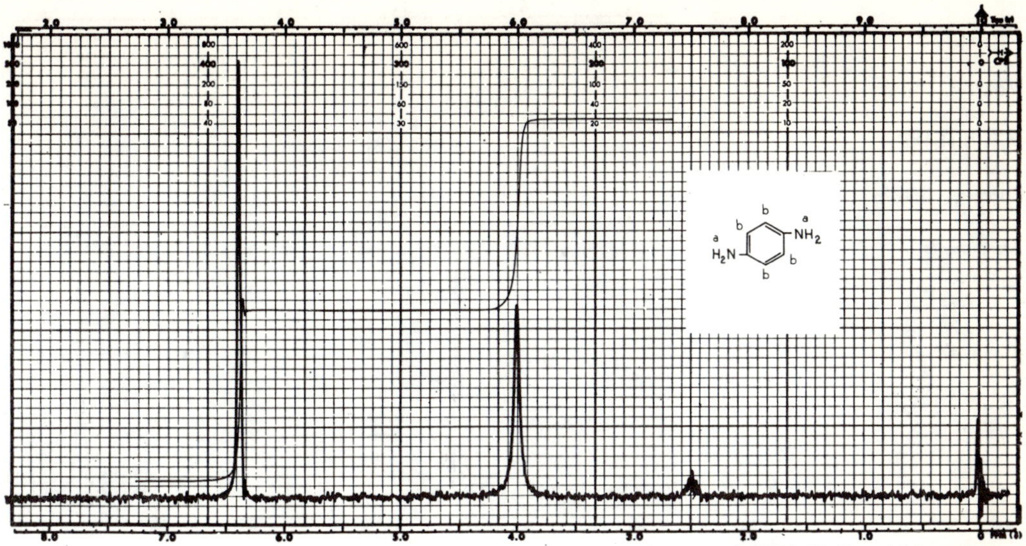

© 1978 Sadtler Research Laboratories, Inc.

TOLUENE-2,4-DIAMINE

$C_7H_{10}N_2$

Mol. Wt. 122.17

M.P. 99°C (lit.)

Solvent: $CDCl_3$

Assignments
a = 2.03
b = 3.39
c = 5.97
d = 6.03
e = 6.79

536

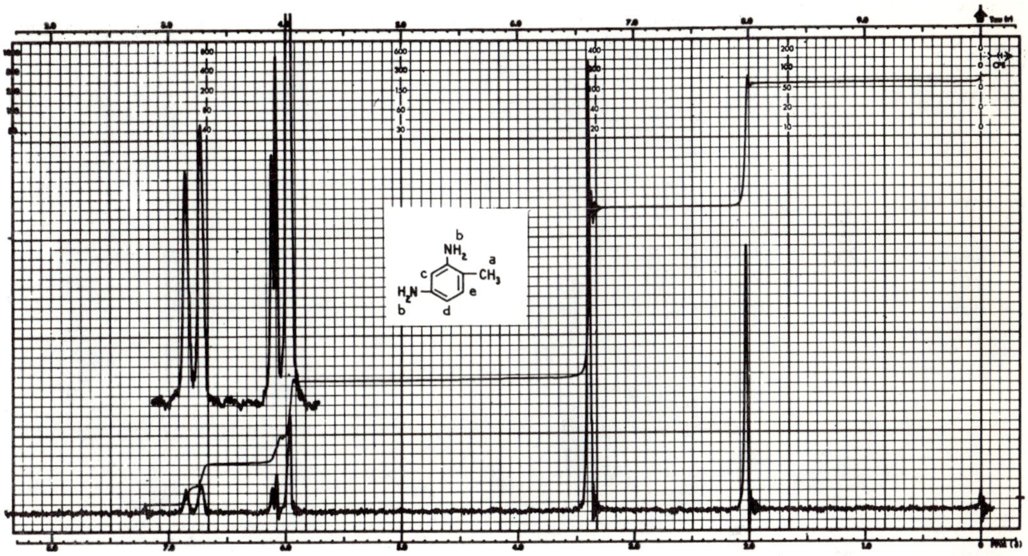

© 1978 Sadtler Research Laboratories, Inc.

2,6-DIMETHYL-p-PHENYLENEDIAMINE

$C_8H_{12}N_2$

Mol. Wt. 136.20

M.P. 104°C

Solvent: $CDCl_3$

Assignments
a = 2.10
b = 3.17
c = 6.31

537

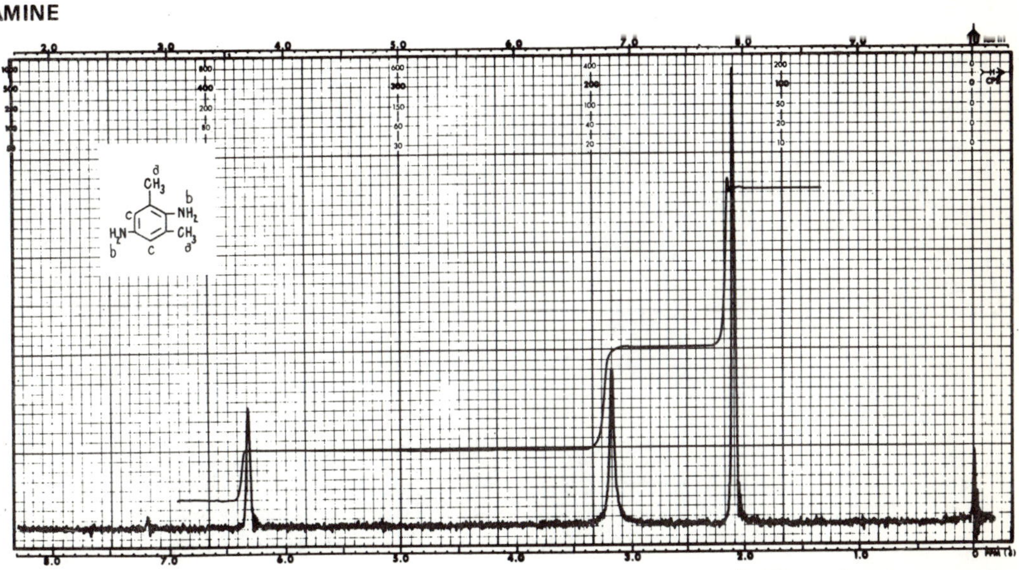

© 1978 Sadtler Research Laboratories, Inc.

2,5-DIETHYL-p-PHENYLENEDIAMINE

$C_{10}H_{16}N_2$

Mol. Wt. 164.25

M.P. 80-81°C

Solvent: $CDCl_3$

Assignments
a = 1.20
b = 2.43
c = 3.23
d = 6.47

538

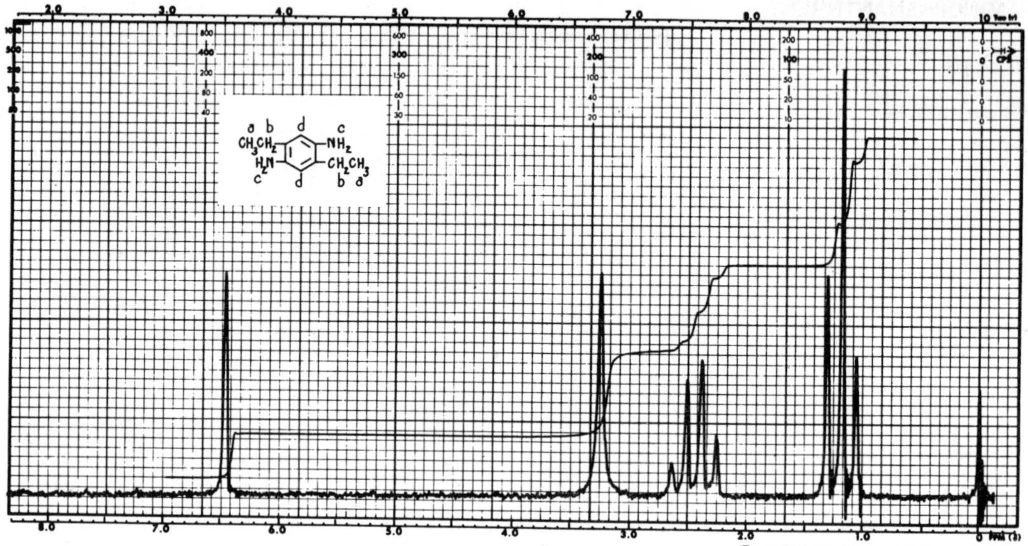

© 1978 Sadtler Research Laboratories, Inc.

1-NAPHTHYLAMINE

$C_{10}H_9N$

Mol. Wt. 143.19

M.P. 48-50°C

Solvent: $CDCl_3$

Assignments
a = 3.81
b = 6.59
c = 7.00-7.85

539

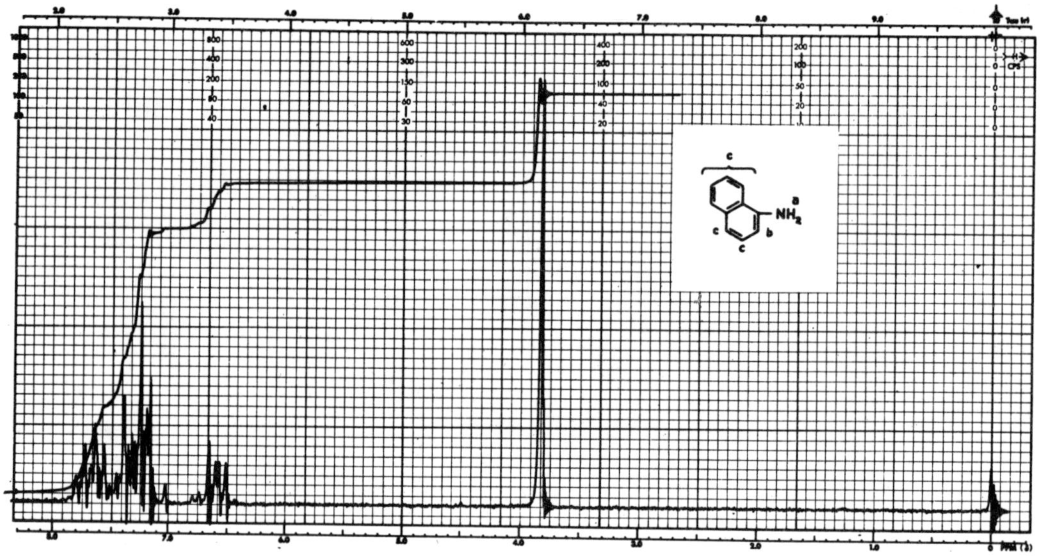

© 1978 Sadtler Research Laboratories, Inc.

2,7-NAPHTHALENEDIAMINE

$C_{10}H_{10}N_2$

Mol. Wt. 158.20

M.P. 157-160°C

Solvent: Polysol-d

Assignments
a = 4.42
b = 6.52
c = 6.60
d = 7.34

540

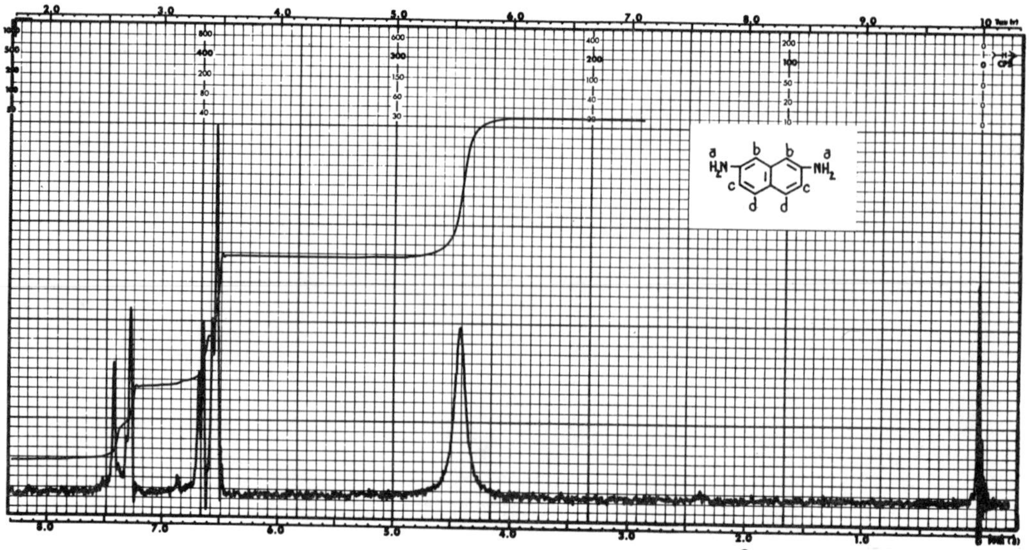

© 1978 Sadtler Research Laboratories, Inc.

9-AMINOPHENANTHRENE

$C_{14}H_{11}N$

Mol. Wt. 193.25

M.P. 137-139°C

Solvent: Dimethyl sulfoxide

Assignments
a = 5.76
b = 6.98
c = 7.10-7.80
d = 8.26
e = 8.57
f = 8.75

541

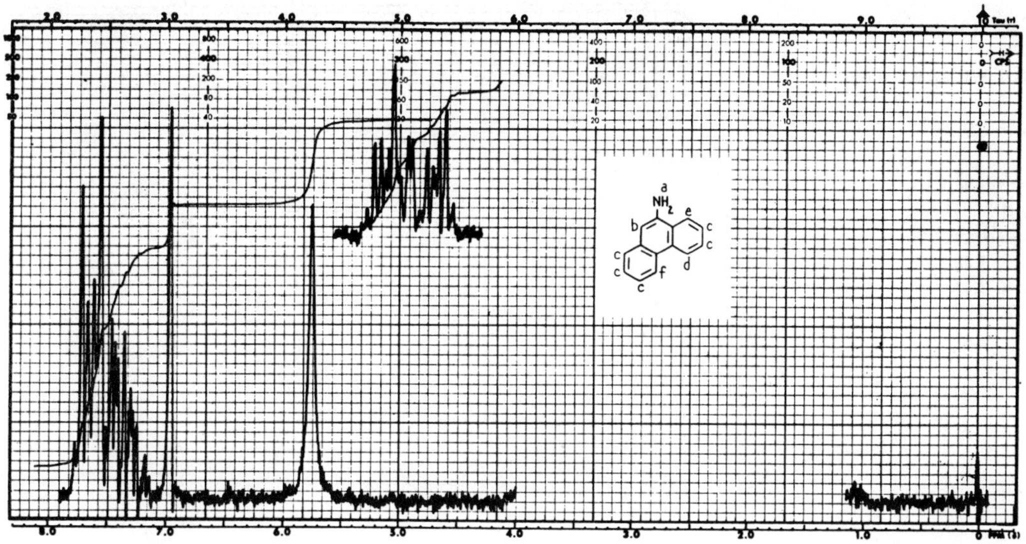

© 1978 Sadtler Research Laboratories, Inc.

NOTES

SECONDARY AMINES

The secondary amines are similar in many respects to the primary amines except for the presence of one instead of two exchangeable hydrogens. A wider range of chemical shifts is observed because of the different types of groups bonded to the nitrogen nucleus. As illustrated below, phenyl substituted secondary amines display a stronger deshielding capability than their aliphatic counterparts.

Separate listings for the aromatic and aliphatic secondary amines are presented.

Like the primary and tertiary amines, the secondary type is capable of forming amine-acid salts upon the addition of acid to the sample solution resulting in shifts to lower field of about 0.8 ppm.

Chemical Shifts

Aliphatic Protons

Alkyl secondary amines

HC–3	HC–2	HC–1	NH		Solvent	Spectrum
		2.32	----	CH_3–NH–R_2	D_2O	----
		2.32	----	CH_3–NH–R_4	D_2O	544
		2.40	1.49	CH_3–NH–R_{18}	$CDCl_3$	552
	1.09	2.60	----	CH_3–CH_2–NH–R_1	D_2O	----
	1.11	2.65	1.23	CH_3–CH_2–NH–R	$CDCl_3$	555
0.90	1.49	2.58	0.85	CH_3–CH_2–CH_2–NH–R_1	$CDCl_3$	----
(0.88)	1.60	2.31	0.79	$(CH_3)_2$CH–CH_2–NH–R	CCl_4	546
	(1.02)	2.76	----	$(CH_3)_2$–CH–NH–R_2	D_2O	----
	(1.00)	2.88	0.67	$(CH_3)_2$–CH–NH–R	CCl_4	543
	(1.01)	2.98	1.20	$(CH_3)_2$–CH–NH–C_6	$CDCl_3$	559
	(1.00)	----	0.74	$(CH_3)_3$–C–NH–R_4	$CDCl_3$	547
		2.67	3.34	CH_3–NH–Ø	CCl_4	582
	1.12	3.03	3.23	CH_3–CH_2–NH–Ø	CCl_4	583
0.92	1.54	2.97	3.31	CH_3–CH_2–CH_2–NH–Ø	CCl_4	584
	(1.12)	3.54	3.19	$(CH_3)_2$–CH–NH–Ø	CCl_4	----

Aliphatic Protons continued

R–NH–	CH_2	–X	Solvent
	2.07	$-Si(CH_3)_3$	$CDCl_3$
	2.48	$-C(CH_3)_2-NH_2$	$CDCl_3$
	2.60	$-C_4$	$CDCl_3$
	2.60	$-CH_3$	D_2O
	2.65	$-CH_3$	$CDCl_3$
	3.13	–C(=O)–O–Na	D_2O
	3.22	$-CH=CH_2$	$CDCl_3$
	3.38	–C(=O)–O–R	CCl_4
	3.41	$-C\equiv C-CH_2-NH-R_2$	$CDCl_3$
	3.70	–∅	CCl_4
	3.68	–C(=O)–OH	D_2O

R–NH–	CH_2	$-CH_2$	–X	Solvent
	2.58	1.49	$-CH_3$	$CDCl_3$
	2.61	2.61	$-NH_2$	D_2O
	2.62	0.80	$-Si(CH_3)_3$	$CDCl_3$
	2.62	3.51	–OH	CCl_4
	2.70	3.41	$-O-CH_3$	CCl_4
	2.76	2.90	–SH	$CDCl_3$
	2.76	2.76	$-NH_2$	$CDCl_3$
	2.79	2.79	–∅	$CDCl_3$
	2.96	2.55	$-C\equiv N$	$CDCl_3$

∅–NH–	CH_2	–X	Solvent
	2.83	$-C(CH_3)_2-NH_2$	CCl_4
	3.60	$-CH=CH_2$	$CDCl_3$
	3.71	$-C(=O)-NH-C_6$	$CDCl_3$
	3.77	$-C\equiv N$	$CDCl_3$
	3.78	$-C(=O)-O-R_2$	$CDCl_3$
	3.81	–C(=O)–OH	DMSO-d_6
	3.97	$-C\equiv CH$	$CDCl_3$
	4.22	–∅	$CDCl_3$
	4.40	$-SO_2-Na$	DMSO
	4.49	$-SO_2-R_8$	$CDCl_3$
	4.75	–S–∅	$CDCl_3$

Aliphatic Protons continued

Ø–NH	–CH_2	–CH_2	–X	Solvent
	2.97	1.54	–CH_3	CCl_4
	3.00	2.75	–NH_2	CCl_4
	3.22	3.78	–OH	$CDCl_3$
	3.28	2.50	–C(=O)–OH	DMSO-d_6
	3.46	2.55	–C≡N	$CDCl_3$
	3.46	2.92	–Ø	$CDCl_3$

Aromatic Protons

	ortho	meta	para	
CH_3–NH–	6.40	7.08	6.57	CCl_4
Ø–NH–	6.96	7.20	6.83	$CDCl_3$

Para substituted secondary amines

R–NH–	ortho	meta	para	
	6.52	6.52	–NH–R	$CDCl_3$
	6.52	6.52	–OH	$CDCl_3$
	6.45	6.69	–O–CH_3	Polysol
	6.80	6.93	–R_4	Polysol
	6.45	6.96	–CH_3	$CDCl_3$
	6.47	7.01	–$CH(CH_3)_2$	$CDCl_3$
	6.55	7.07	–Cl	DMSO-d_6
	6.83	7.10	–S–CH_3	Polysol
	6.50	7.11	–CH_2–S–Ø	$CDCl_3$
	6.49	7.20	–Br	$CDCl_3$
	6.62	7.55	–SO_2–NH_2	Polysol
	6.47	7.78	–N=N–Ø	$CDCl_3$
	6.54	7.80	–C(=O)–CH_3	$CDCl_3$
	6.72	7.84	–N=O	Polysol

Ø–NH–	ortho	meta	para	
	6.81	6.47	–NH_2	$CDCl_3$
	7.20	6.96	–O–CH_3	$CDCl_3$
	7.45	7.45	–Cl	Polysol
	7.08	7.77	–N=O	Polysol
	7.09	8.05	–NO_2	Acetone

DIPROPYLAMINE

$C_6H_{15}N$

Mol. Wt. 101.19

B.P. 108-110°C

Solvent: $CDCl_3$

Assignments
a = ca 0.90
b = 1.49
c = 2.58

542

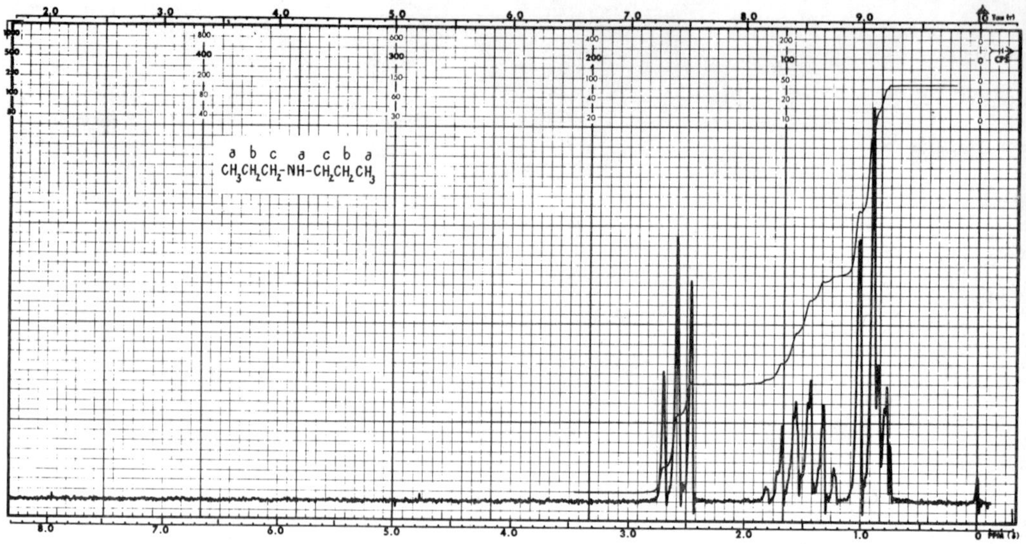

© 1978 Sadtler Research Laboratories, Inc.

DIISOPROPYLAMINE

$C_6H_{15}N$

Mol. Wt. 101.19

B.P. 83-84°C

Solvent: CCl_4

Assignments
a = 0.67
b = 1.00
c = 2.88

543

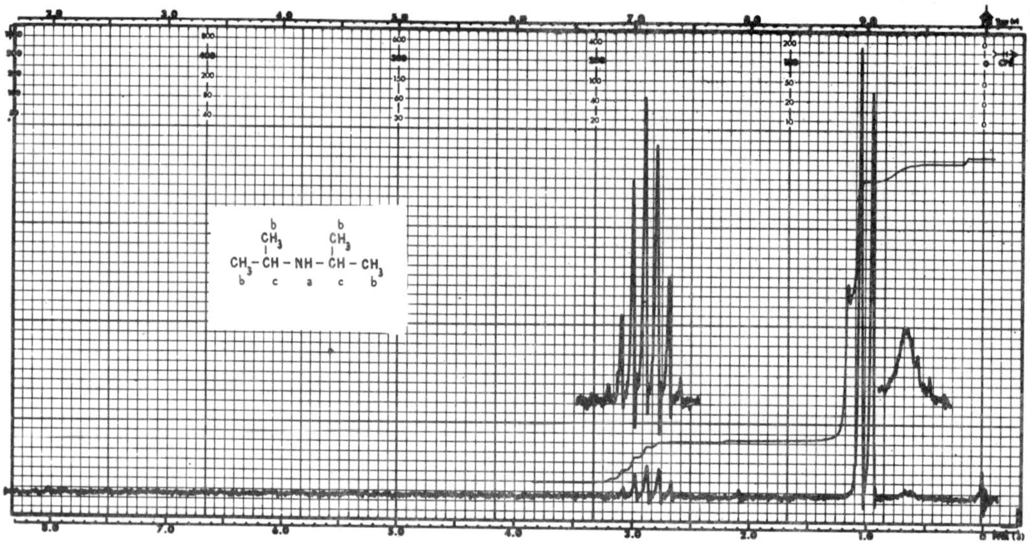

© 1978 Sadtler Research Laboratories, Inc.

N-METHYLBUTYLAMINE

$C_5H_{13}N$

Mol. Wt. 87.17

Solvent: D_2O

Assignments
a = 0.92
b = 1.17-1.63
c = 2.32
d = 2.52
e = 4.63 HDO

544

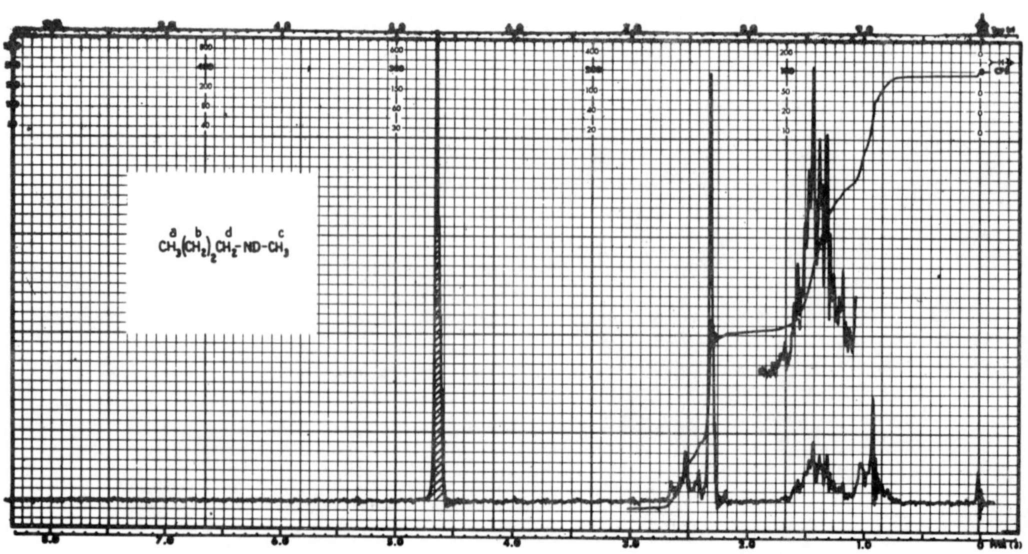

© 1978 Sadtler Research Laboratories, Inc.

DI-SEC-BUTYLAMINE

$C_8H_{19}N$

Mol. Wt. 129.25

B.P. 135-138°C

Solvent: CCl_4

Assignments

a = 0.90
b = ca 1.00
c = 1.00-1.70
d = 2.63

545

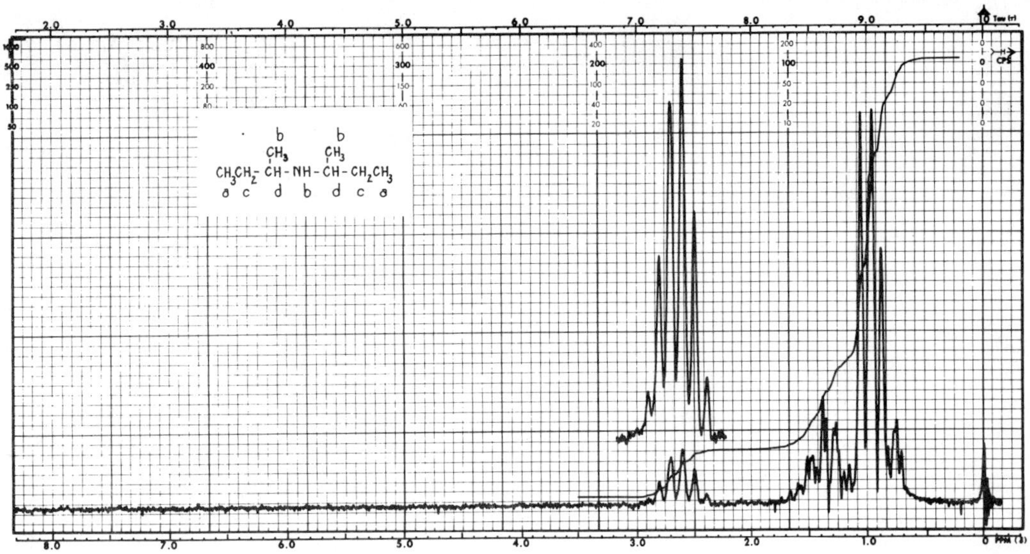

© 1978 Sadtler Research Laboratories, Inc.

DIISOBUTYLAMINE

$C_8H_{19}N$

Mol. Wt. 129.25

B.P. 136-141°C

Solvent: CCl_4

Assignments

a = 0.79
b = 0.88
c = 1.60
d = 2.31

546

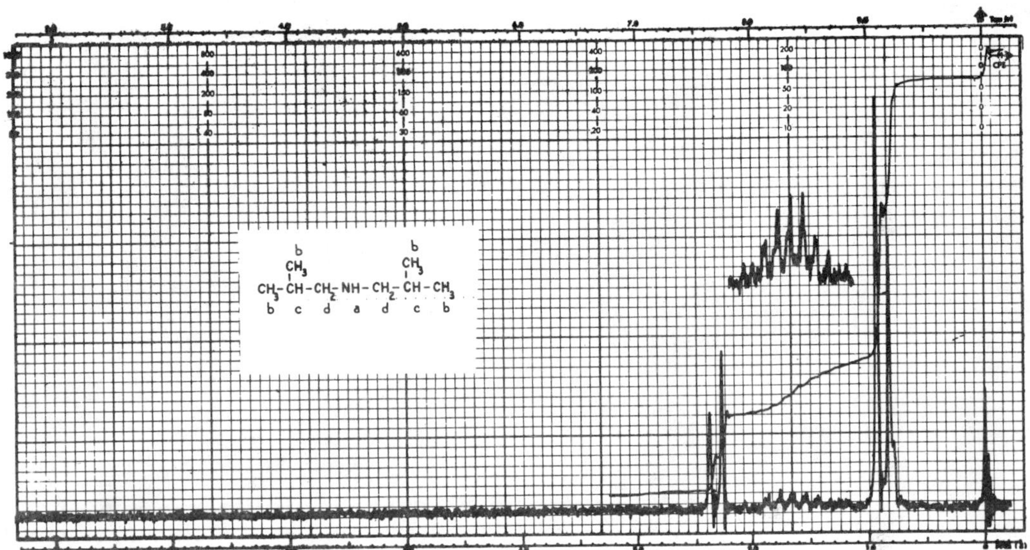

© 1978 Sadtler Research Laboratories, Inc.

N-tert-BUTYLBUTYLAMINE

$C_8H_{19}N$

Mol. Wt. 129.25

B.P. 135-137°C

Solvent: $CDCl_3$

Assignments

a = 0.74
b = 0.93
c = 1.00
d = 1.17-1.65
e = 2.56

547

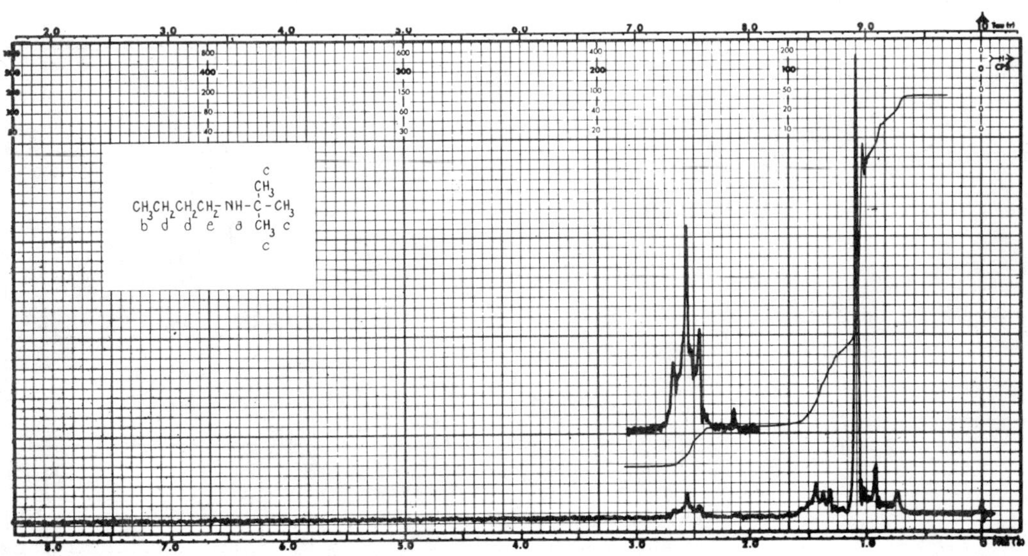

© 1978 Sadtler Research Laboratories, Inc.

DIPENTYLAMINE

$C_{10}H_{23}N$

Mol. Wt. 157.30

B.P. 93-95°C/15mm

Solvent: $CDCl_3$

Assignments
a = 0.90
b = 1.00-1.90
c = 2.60

548

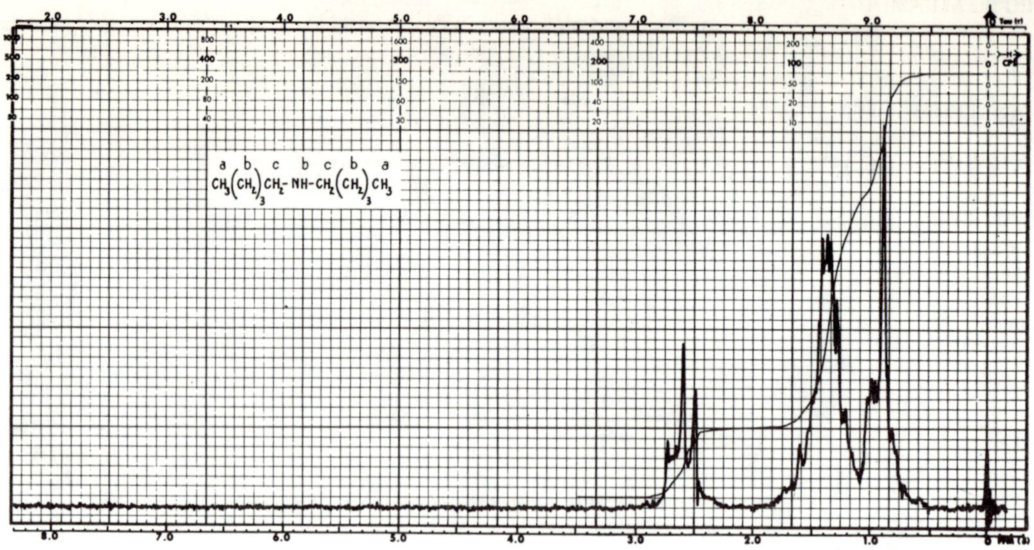

© 1978 Sadtler Research Laboratories, Inc.

BIS(3-METHYLBUTYL)AMINE

$C_{10}H_{23}N$

Mol. Wt. 157.30

B.P. 185-188°C

Solvent: CCl_4

Assignments
a = 0.90
b = 1.07-1.81
c = 1.68
d = 2.52

549

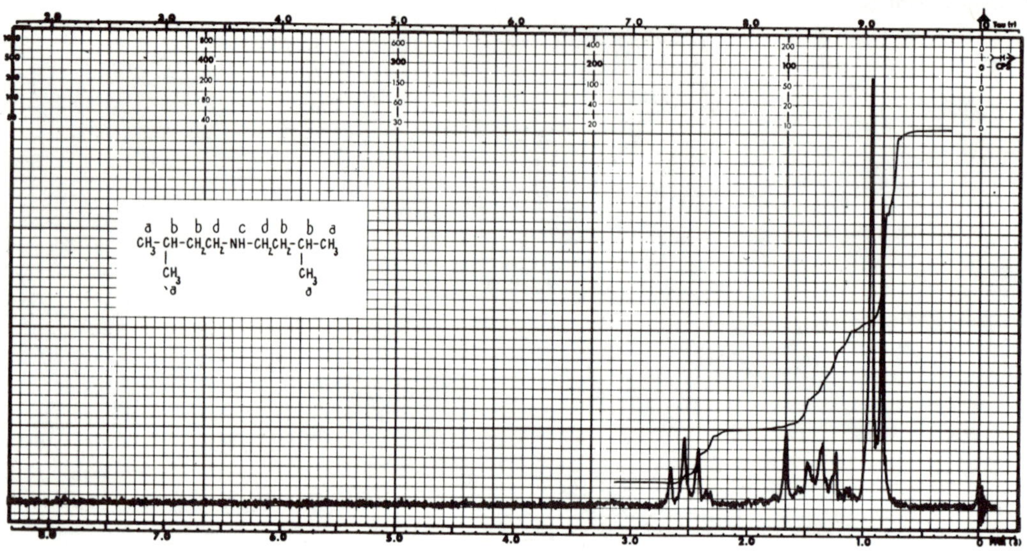

© 1978 Sadtler Research Laboratories, Inc.

DIHEXYLAMINE

$C_{12}H_{27}N$

Mol. Wt. 185.36

B.P. 122°C/15mm (lit.)

Solvent: $CDCl_3$

Assignments
a = 0.90
b = 1.10-1.80
c = 1.79
d = 2.40

550

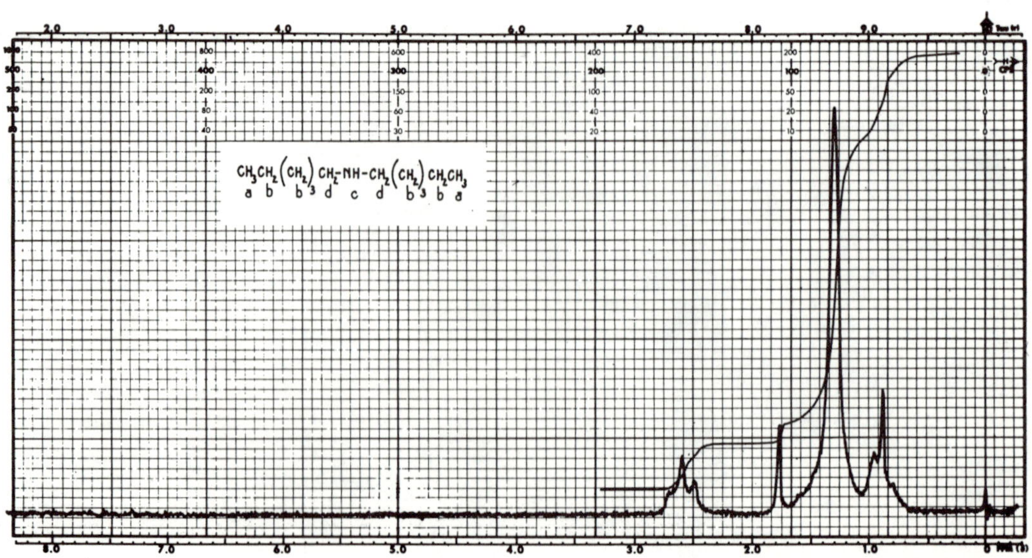

© 1978 Sadtler Research Laboratories, Inc.

DIHEXYLAMINE

$C_{12}H_{27}N$

Mol. Wt. 185.36

B.P. 122°C/15mm (lit.)

Solvent: $CDCl_3$

Assignments
a = 0.90
b = 1.10-1.80
c = 1.79
d = 2.40

551

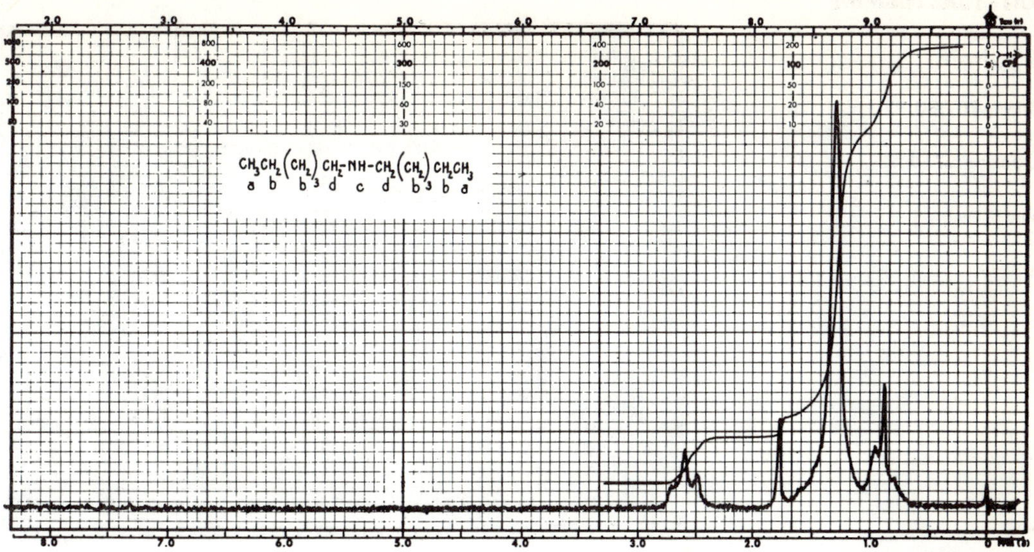

© 1978 Sadtler Research Laboratories, Inc.

N-METHYLOCTADECYLAMINE

$C_{19}H_{41}N$

Mol. Wt. 283.55

B.P. 155°C/0.5mm (lit.)

Solvent: $CDCl_3$

Assignments
a = 0.86
b = ca 1.23
c = 1.49
d = 2.40
e = 2.54

552

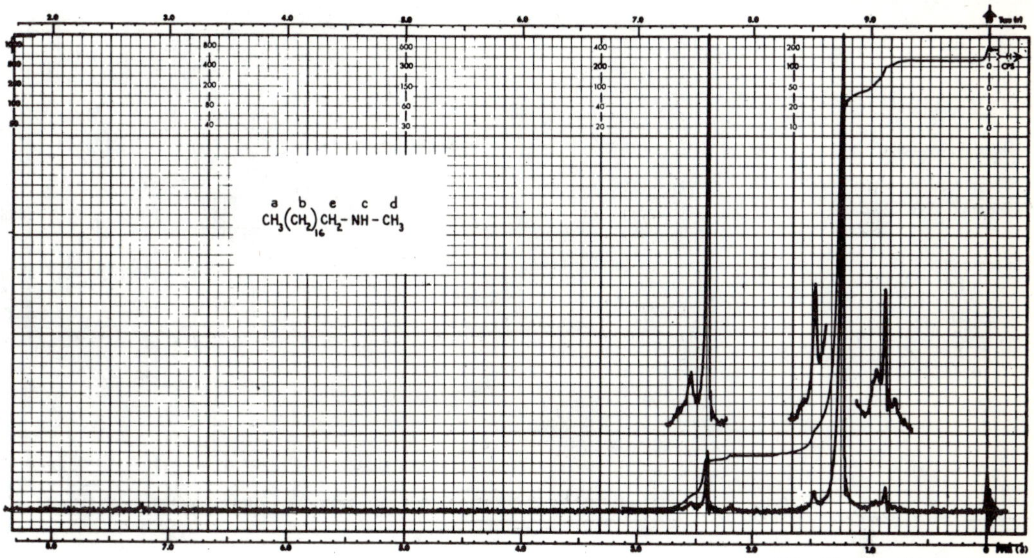

© 1978 Sadtler Research Laboratories, Inc.

DIDECYLAMINE

$C_{20}H_{43}N$

Mol. Wt. 297.57

B.P. 164°C/1mm (lit.)

Solvent: CCl_4

Assignments
a = 0.66
b = 0.89
c = 1.27
d = 2.51

553

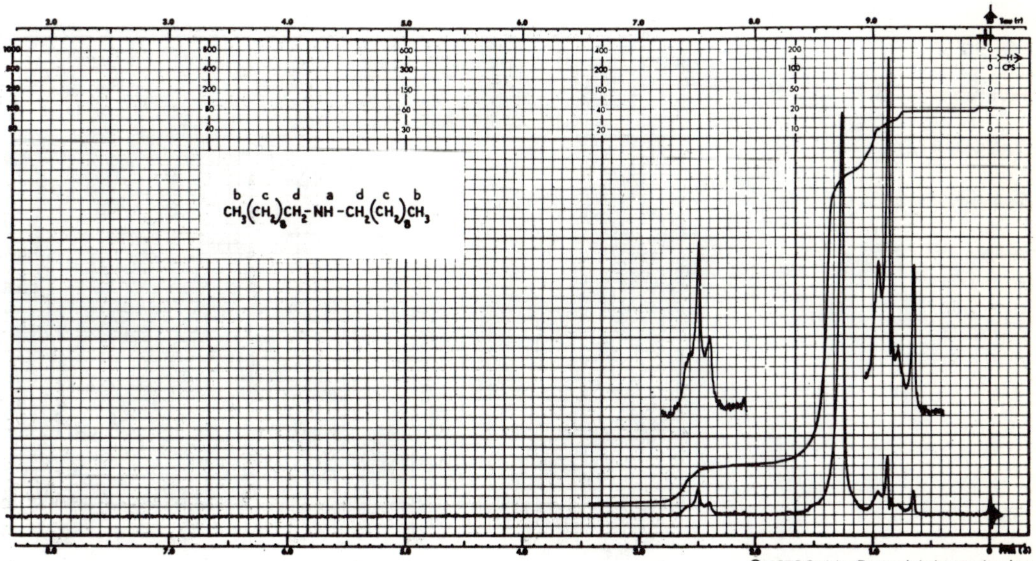

© 1978 Sadtler Research Laboratories, Inc.

DITRIDECYLAMINE

$C_{26}H_{55}N$

Mol. Wt. 381.73

Solvent: $CDCl_3$

Assignments
a = 0.88
b = 1.10-1.80
c = 1.59
d = 2.59

554

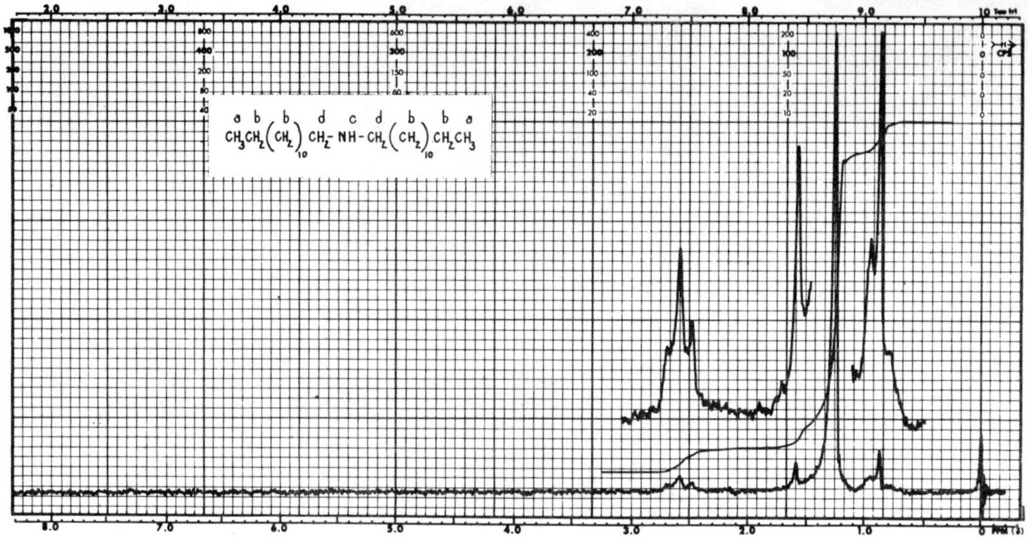

© 1978 Sadtler Research Laboratories, Inc.

N,N'-DIETHYLETHYLENEDIAMINE

$C_6H_{16}N_2$

Mol. Wt. 116.21

B.P. 152-154°C

Solvent: $CDCl_3$

Assignments
a = 1.11
b = 1.23
c = 2.65
d = 2.72

555

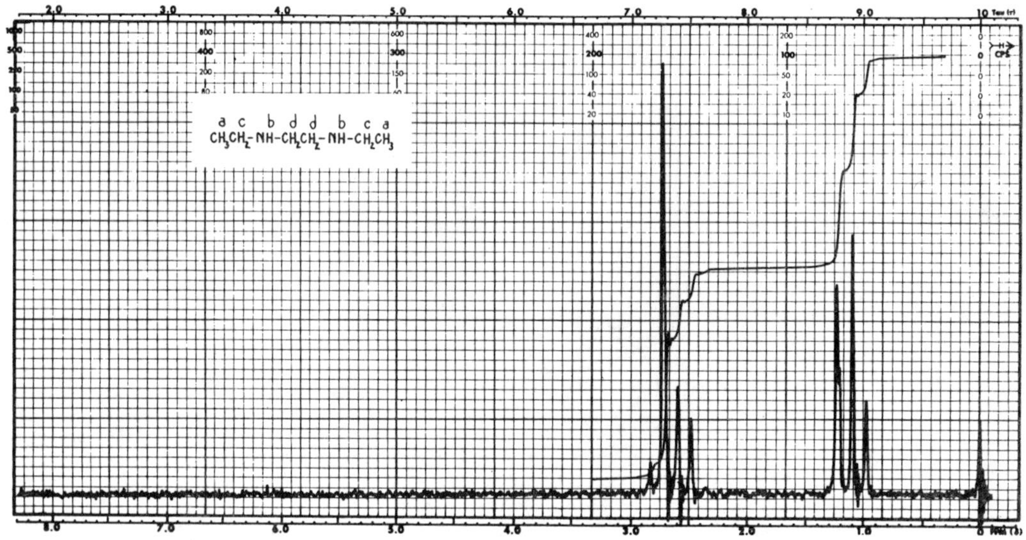

© 1978 Sadtler Research Laboratories, Inc.

N,N'-DIISOPROPYLETHYLENEDIAMINE

$C_8H_{20}N_2$

Mol. Wt. 144.26

B.P. 169-172°C

Solvent: D_2O

Assignments
a = 1.03
b = 2.62
c = 2.78
d = 4.60 HDO

556

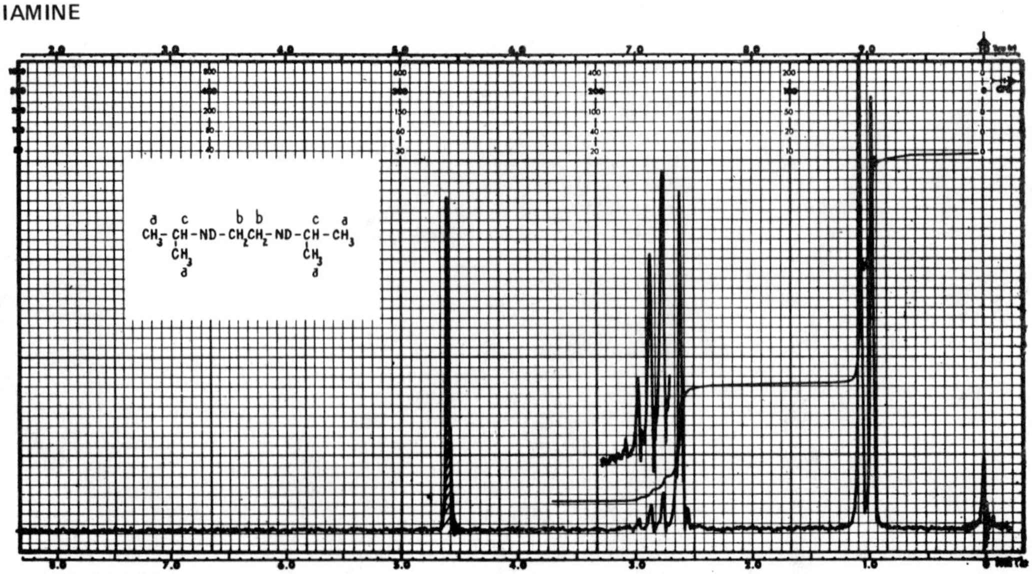

© 1978 Sadtler Research Laboratories, Inc.

N,N'-BIS(1-METHYLHEPTYL)ETHYLENEDIAMINE

$C_{18}H_{40}N_2$

Mol. Wt. 284.53

Solvent: $CDCl_3$

Assignments
a = 0.99
b = 1.03
c = 1.31
d = 2.58
e = 2.70

557

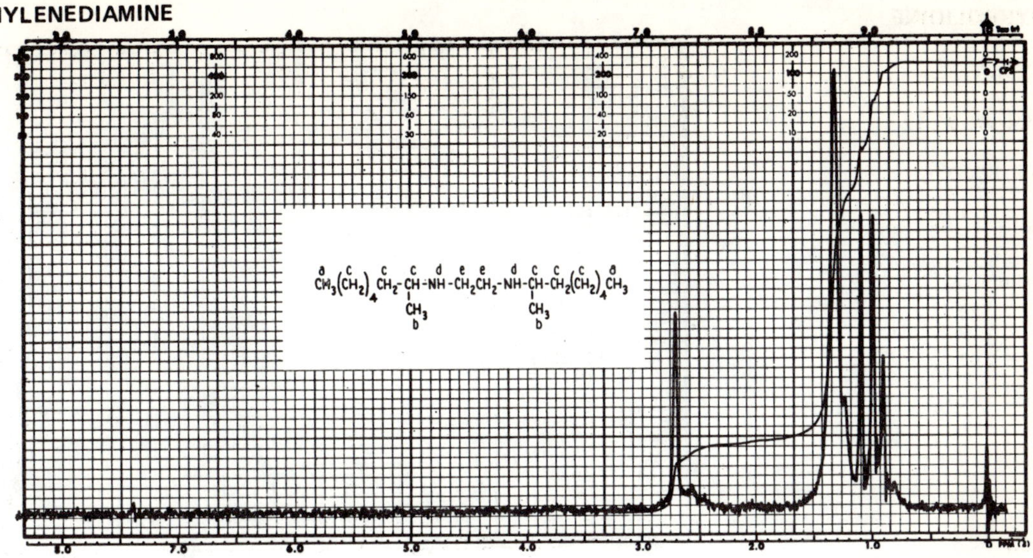

© 1978 Sadtler Research Laboratories, Inc.

N-METHYLCYCLOHEXYLAMINE

$C_7H_{15}N$

Mol. Wt. 113.20

B.P. 145-147°C (lit.)

Solvent: $CDCl_3$

Assignments
a = 0.82-2.07
b = 2.39

558

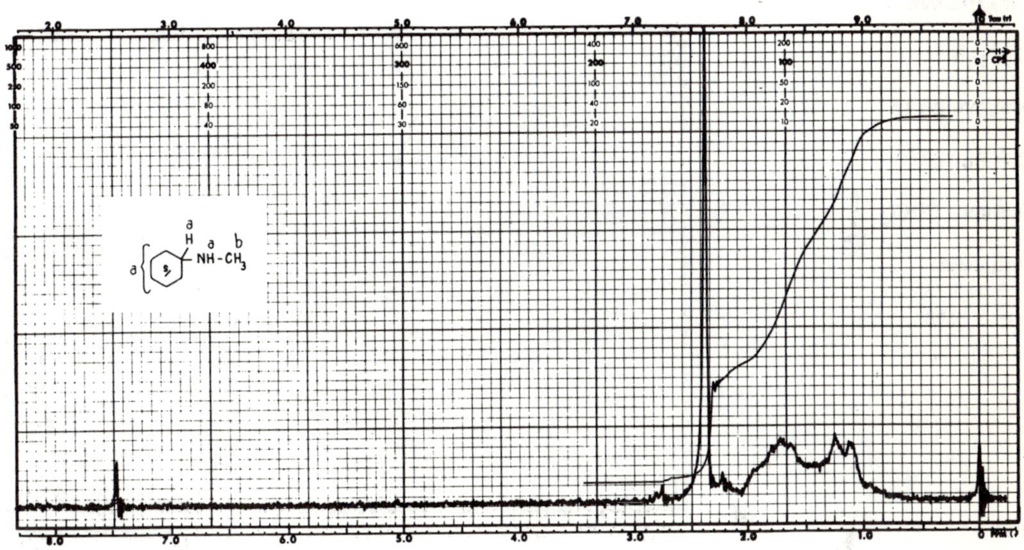

© 1978 Sadtler Research Laboratories, Inc.

N-ISOPROPYLCYCLOHEXYLAMINE

$C_9H_{19}N$

Mol. Wt. 141.26

B.P. 60-65°C/12mm

Solvent: $CDCl_3$

Assignments
a = 1.01
b = 0.80-2.10
c = 2.49
d = 2.98

559

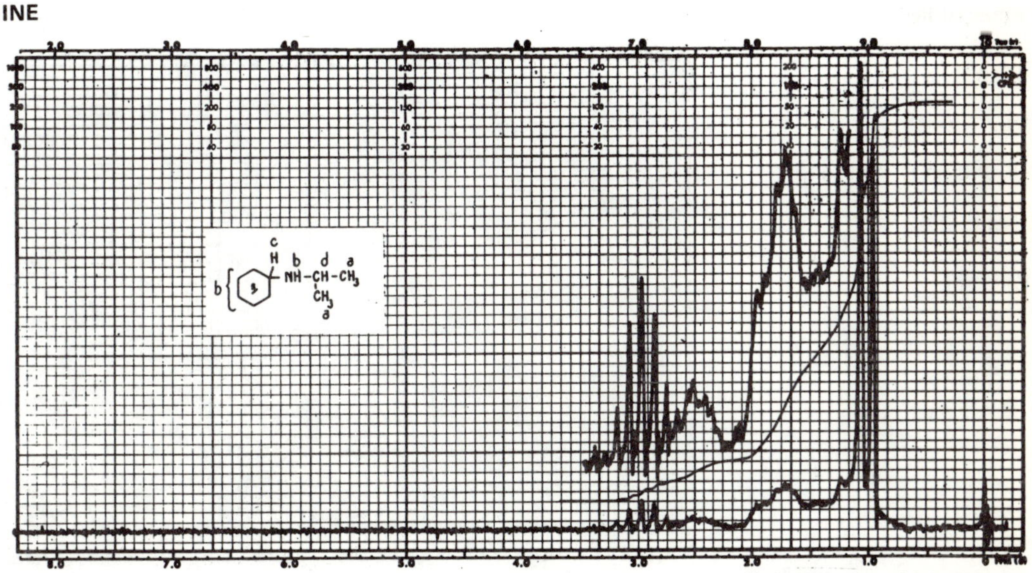

© 1978 Sadtler Research Laboratories, Inc.

PYRROLIDINE

C_4H_9N

Mol. Wt. 71.12

B.P. 86-87°C

Solvent: Benzene

Assignments
a = 1.45
b = 2.39
c = 2.69

560

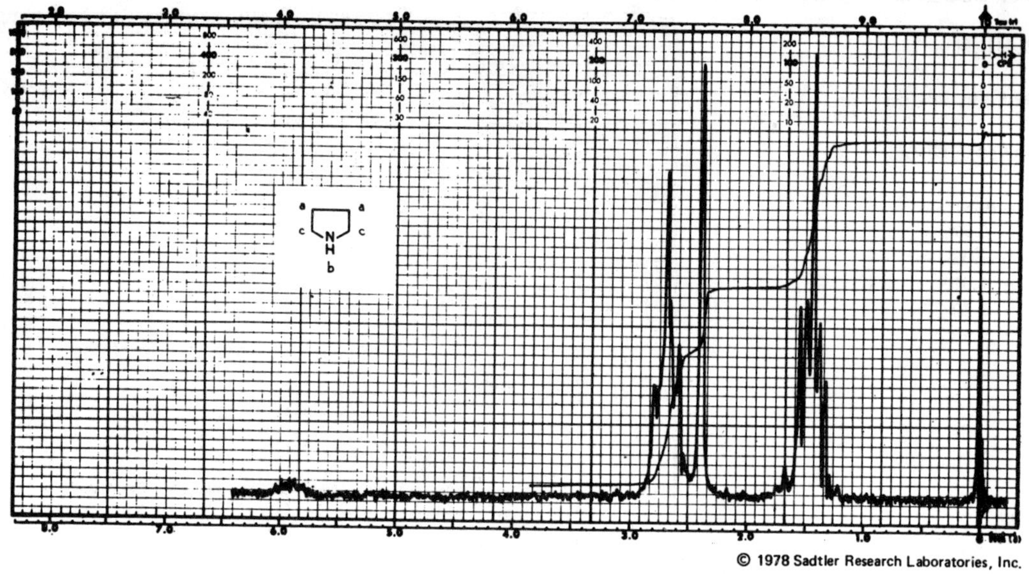

© 1978 Sadtler Research Laboratories, Inc.

PIPERIDINE

$C_5H_{11}N$

Mol. Wt. 85.15

B.P. 98-108°C

Solvent: $CDCl_3$

Assignments
a = ca 1.53
b = 2.80

561

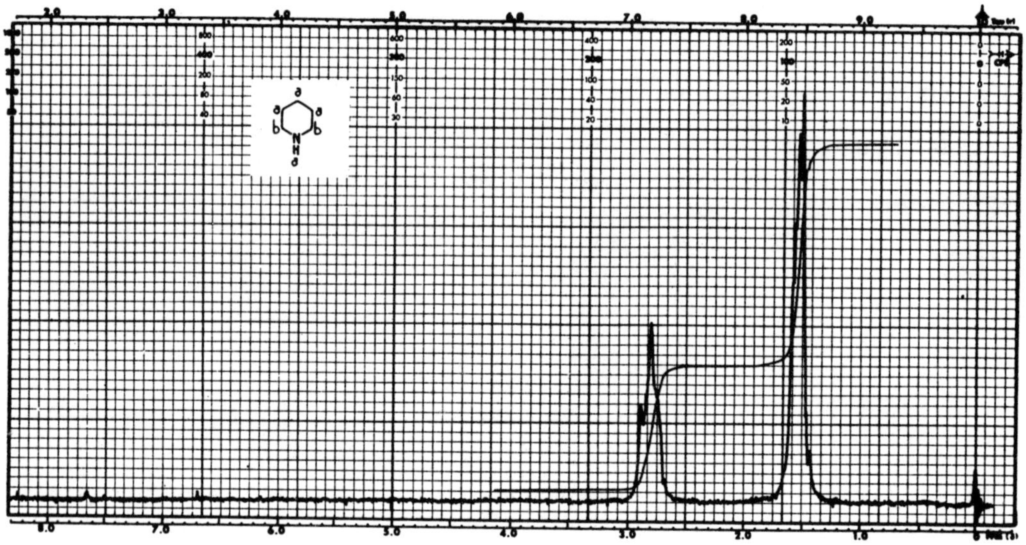

© 1978 Sadtler Research Laboratories, Inc.

2-PIPECOLINE

$C_6H_{13}N$

Mol. Wt. 99.18

M.P. 9°C (lit.) B.P. 119°C

Solvent: CCl_4

Assignments
a = 1.00
b = 1.00-1.90
c = 2.31
d = 2.30-3.20
e = 5.15 imp.

562

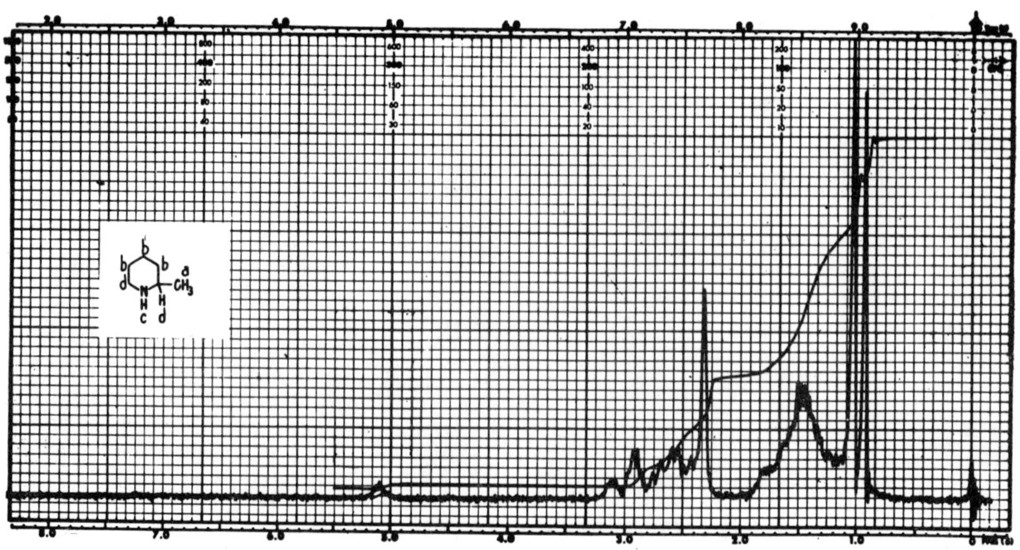

© 1978 Sadtler Research Laboratories, Inc.

4-PIPECOLINE

$C_6H_{13}N$

Mol. Wt. 99.18

Solvent: CCl_4

Assignments
a = 0.91
b = 0.90-1.80
c = 1.55
d = 2.49
e = 2.96

563

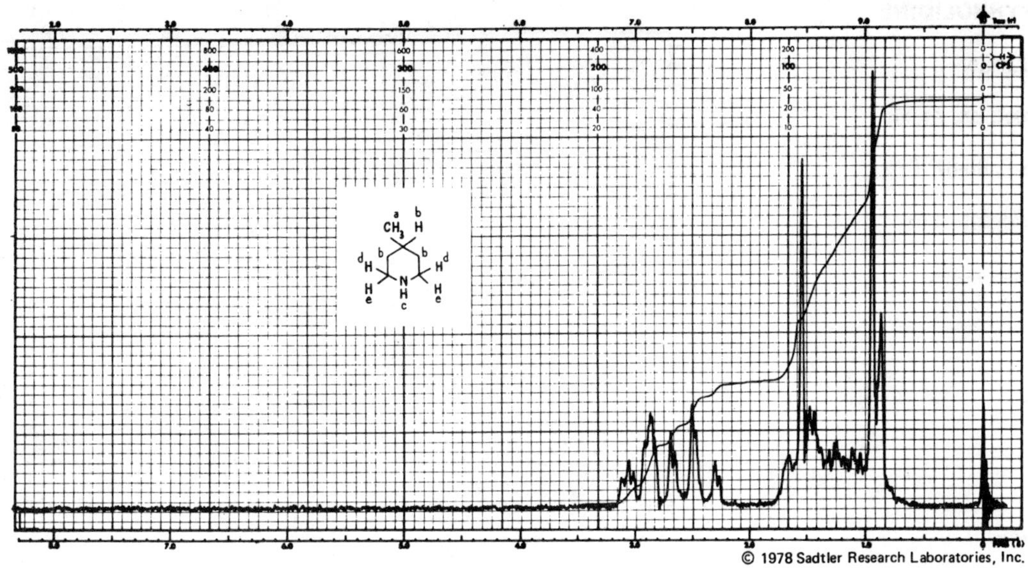

© 1978 Sadtler Research Laboratories, Inc.

2,6-DIMETHYLPIPERIDINE

$C_7H_{14}N$

Mol. Wt. 112.20

Solvent: CCl_4

Assignments
a = 1.00
b = 0.71-1.89
c = 2.60

564

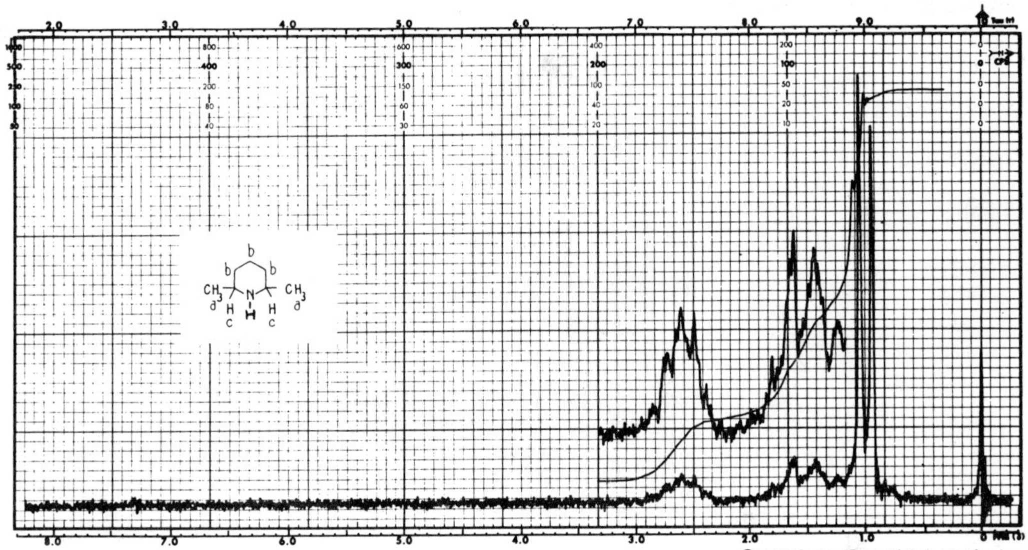

© 1978 Sadtler Research Laboratories, Inc.

AZACYCLOTRIDECANE

$C_{12}H_{25}N$

Mol. Wt. 183.34

Solvent: $CDCl_3$

Assignments
a = 0.95
b = ca 1.39
c = 2.64

565

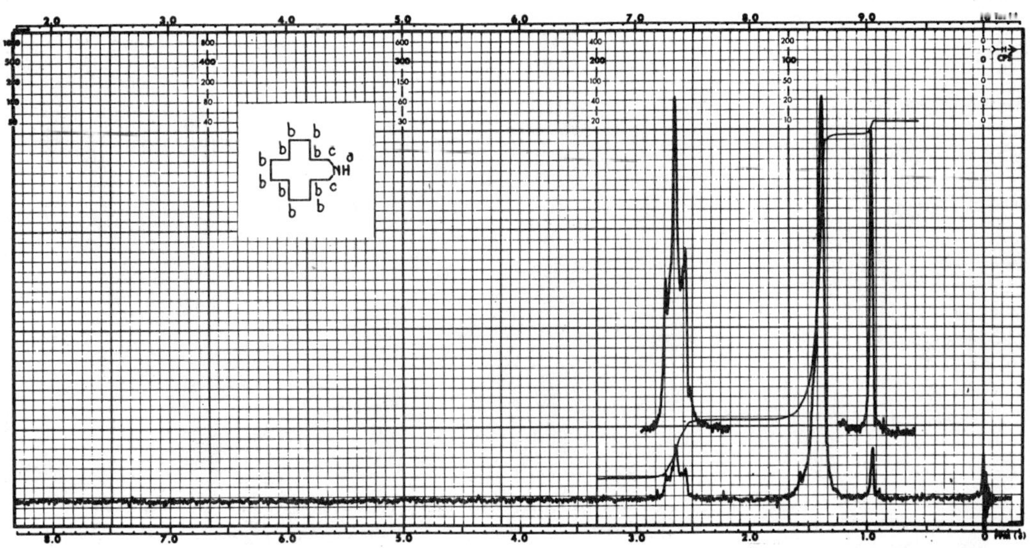

© 1978 Sadtler Research Laboratories, Inc.

PIPERAZINE

$C_4H_{10}N_2$

Mol. Wt. 86.14

M.P. 109-112°C

Solvent: $CDCl_3$

Assignments
a = 2.03
b = 2.83

566

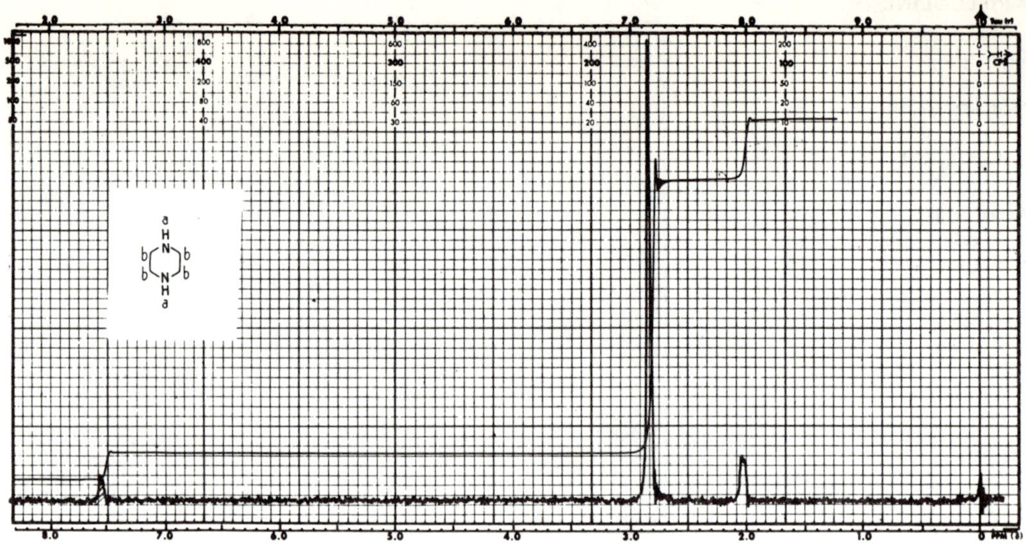

© 1978 Sadtler Research Laboratories, Inc.

cis-2,5-DIMETHYLPIPERAZINE

$C_6H_{14}N_2$

Mol. Wt. 114.19

B.P. 165°C

Solvent: $CDCl_3$

Assignments
a = 1.13
b = 1.49
c = 2.30-3.10

567

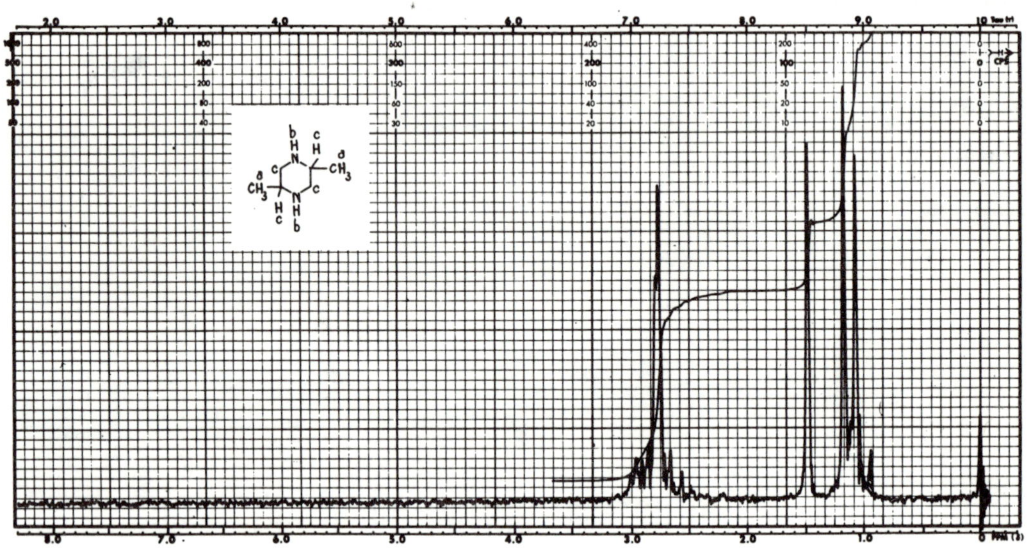

© 1978 Sadtler Research Laboratories, Inc.

TRANS-2,5-DIMETHYLPIPERAZINE

$C_6H_{14}N_2$

Mol. Wt. 114.19

M.P. 117-118°C

Solvent: $CDCl_3$

Assignments
a = 0.99
b = 1.43
c = 2.08-3.09

568

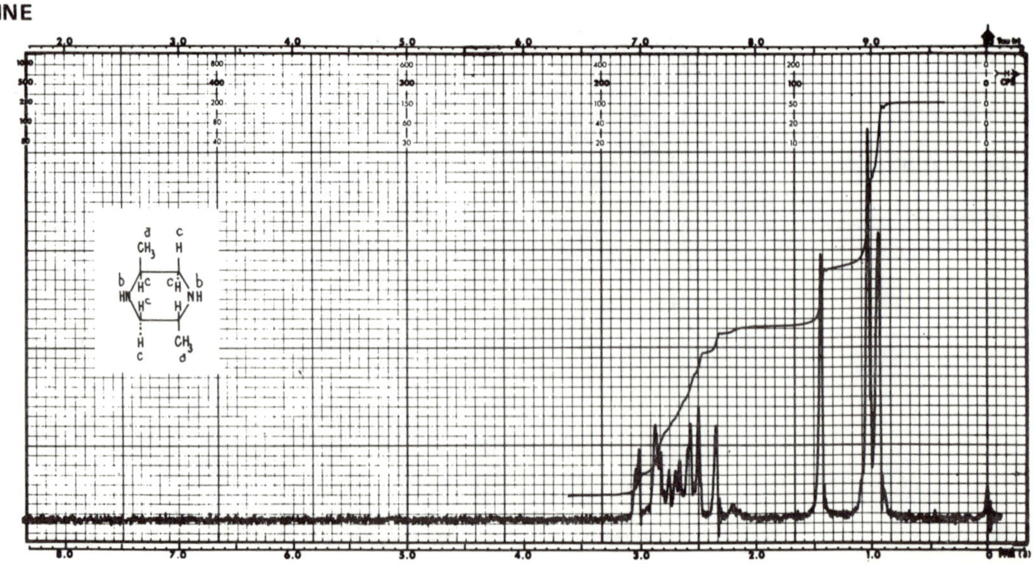

© 1978 Sadtler Research Laboratories, Inc.

DECAHYDROQUINOLINE

$C_9H_{17}N$

Mol. Wt. 139.24

Solvent: $CDCl_3$

Assignments
a = 0.80-1.90
b = 1.34
c = 2.09
d = 2.40-3.27

569

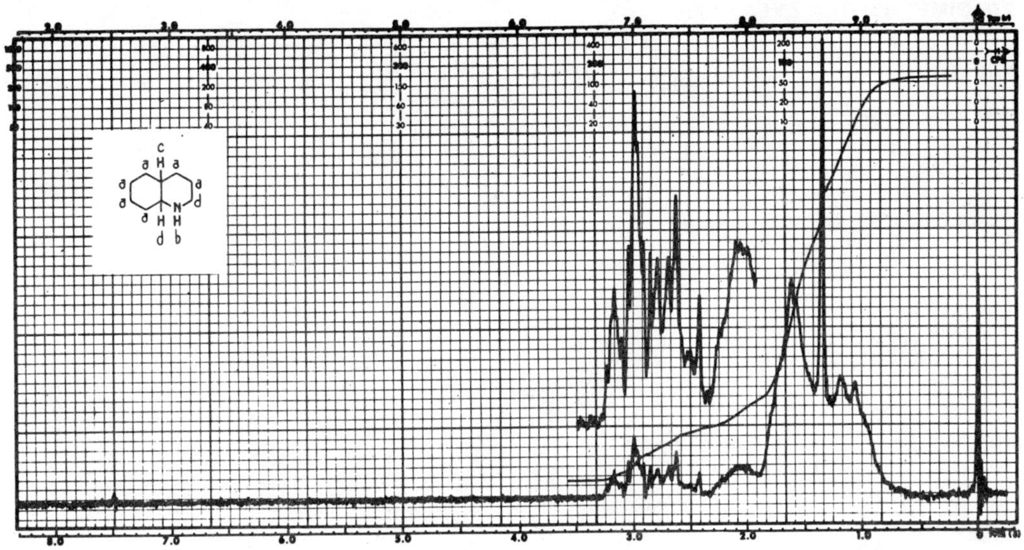

© 1978 Sadtler Research Laboratories, Inc.

N-ALLYLCYCLOHEXYLAMINE

$C_9H_{17}N$

Mol. Wt. 139.24

Solvent: CCl_4

Assignments
a = 0.90-2.20
b = ca 2.42
c = 3.20
d = ca 5.00
e = ca 5.05
f = ca 5.75

570

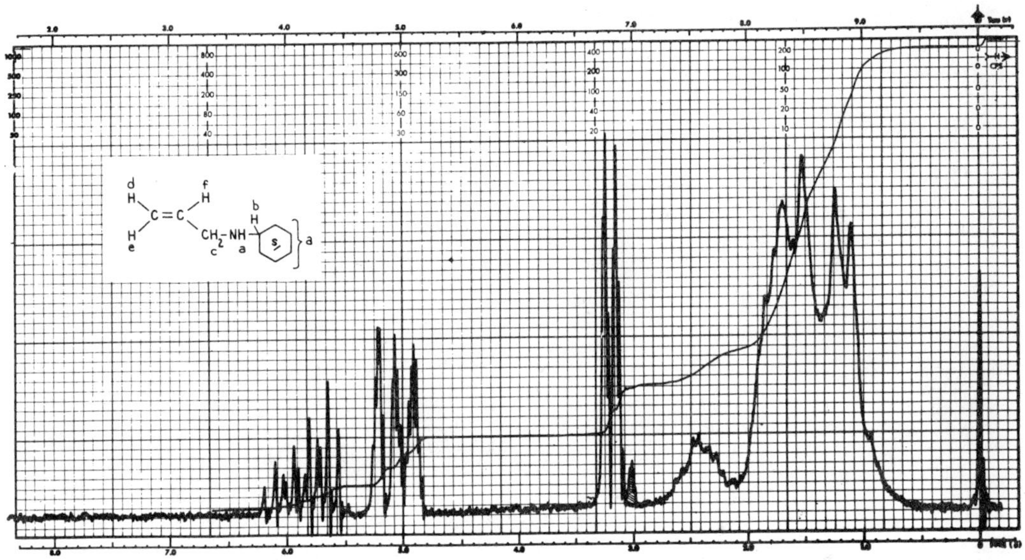

© 1978 Sadtler Research Laboratories, Inc.

DIALLYLAMINE

$C_6H_{11}N$

Mol. Wt. 97.16

B.P. 110-112°C

Solvent: $CDCl_3$

Assignments
a = 1.56
b = 3.22
c = 5.03
d = 5.11
e = 5.83

571

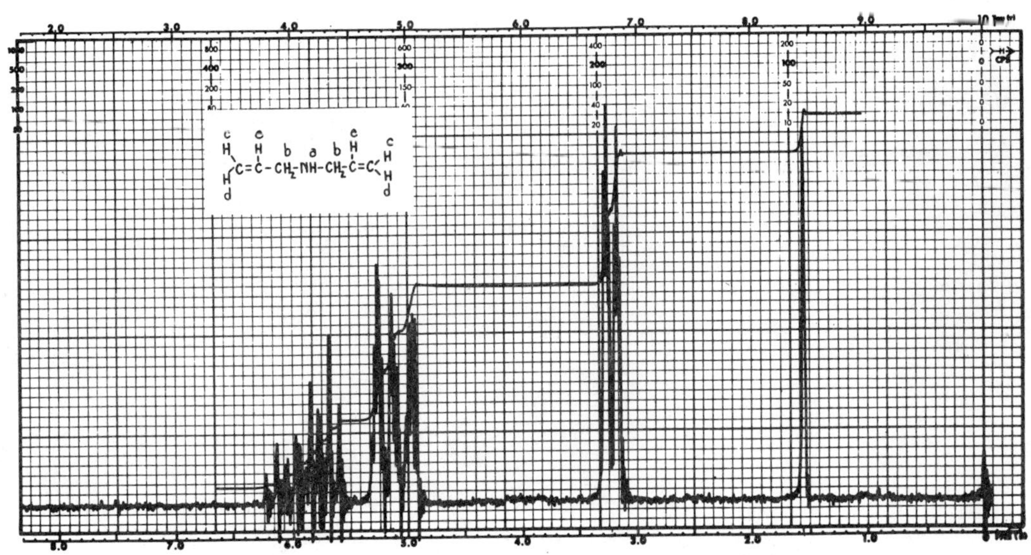

© 1978 Sadtler Research Laboratories, Inc.

2,2′-DIMETHYLDIALLYLAMINE

$C_8H_{15}N$

Mol. Wt. 125.22

B.P. 149°C

Solvent: $CDCl_3$

Assignments
a = 1.39
b = 1.78
c = 3.13
d = 4.87

572

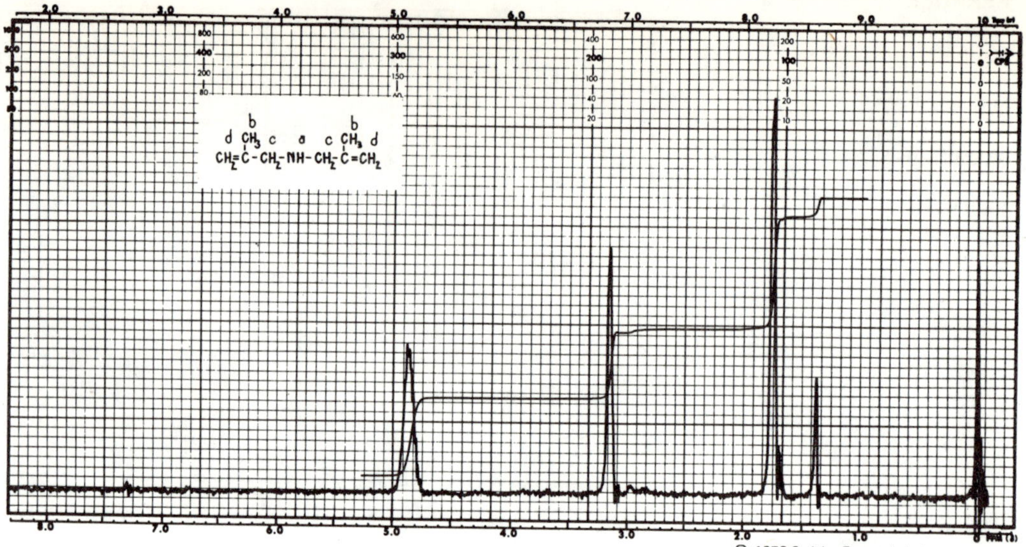

© 1978 Sadtler Research Laboratories, Inc.

3,3′-DIMETHYL-DI-2-BUTENYLAMINE

$C_{10}H_{19}N$

Mol. Wt. 153.27

B.P. 92-94°C/15mm

Solvent: $CDCl_3$

Assignments
a = 1.34
b = 1.68
c = 1.72
d = 3.19
e = 5.26

573

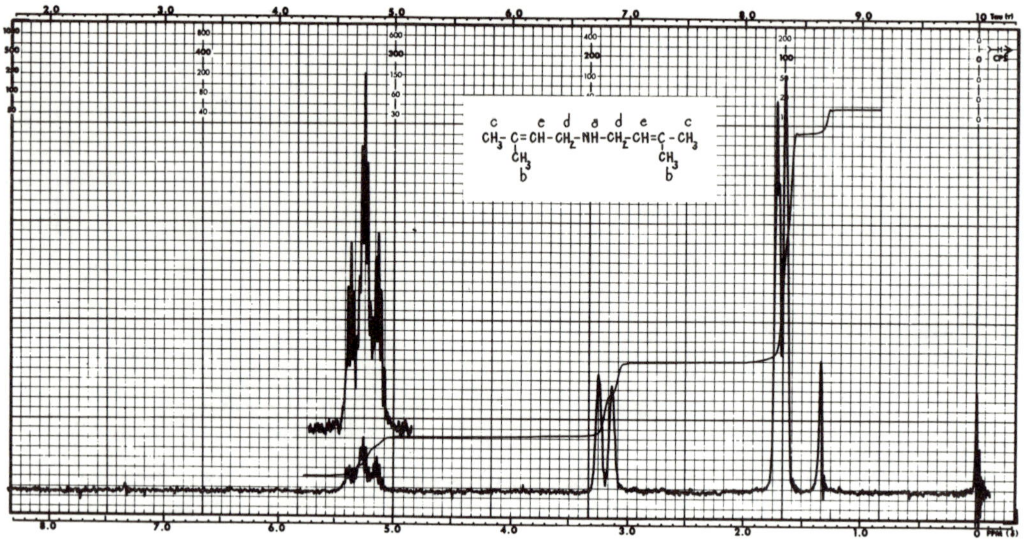

© 1978 Sadtler Research Laboratories, Inc.

N,N′-DIETHYL-2-BUTENE-1,4-DIAMINE

$C_8H_{18}N_2$

Mol. Wt. 142.25

B.P. 80-83°C/8mm

Solvent: CCl_4

Assignments
a = 0.86
b = 1.04
c = 2.56
d = 3.12
e = 5.55

574

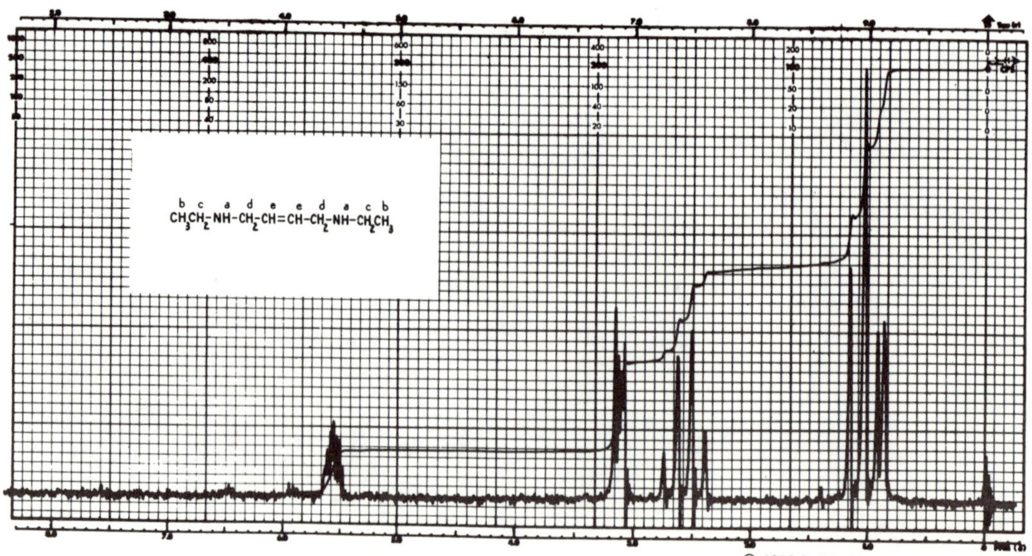

© 1978 Sadtler Research Laboratories, Inc.

PYRROLE

C_4H_5N

Mol. Wt. 67.09

Solvent: CCl_4

Assignments
a = 6.07
b = 6.41
c = 7.48

575

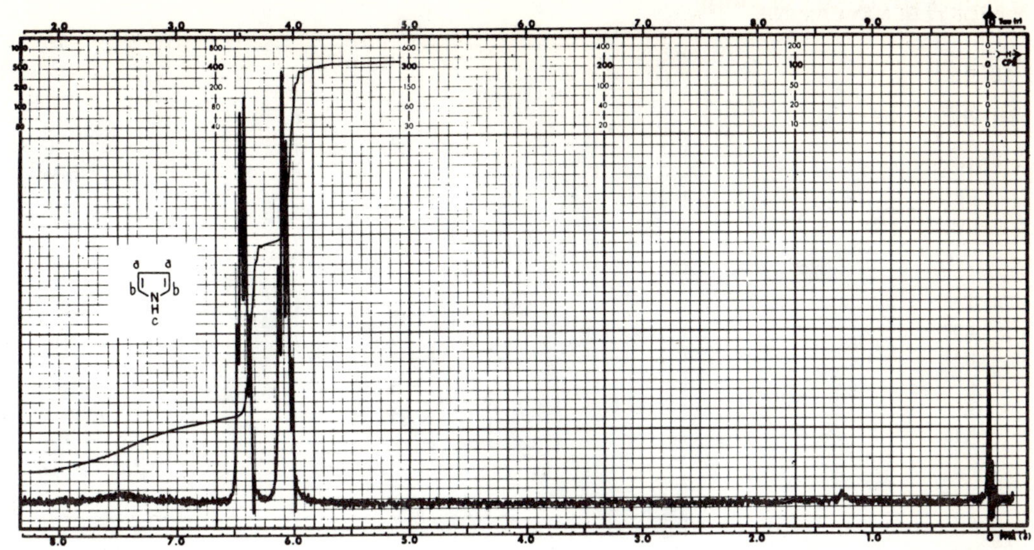

© 1978 Sadtler Research Laboratories, Inc.

1,2,3,6-TETRAHYDROPYRIDINE

C_5H_9N

Mol. Wt. 83.13

M.P. 61°F B.P. 108°C

Solvent: $CDCl_3$

Assignments
a = 1.94
b = ca 2.08
c = 2.97
d = 3.30
e = ca 5.76

576

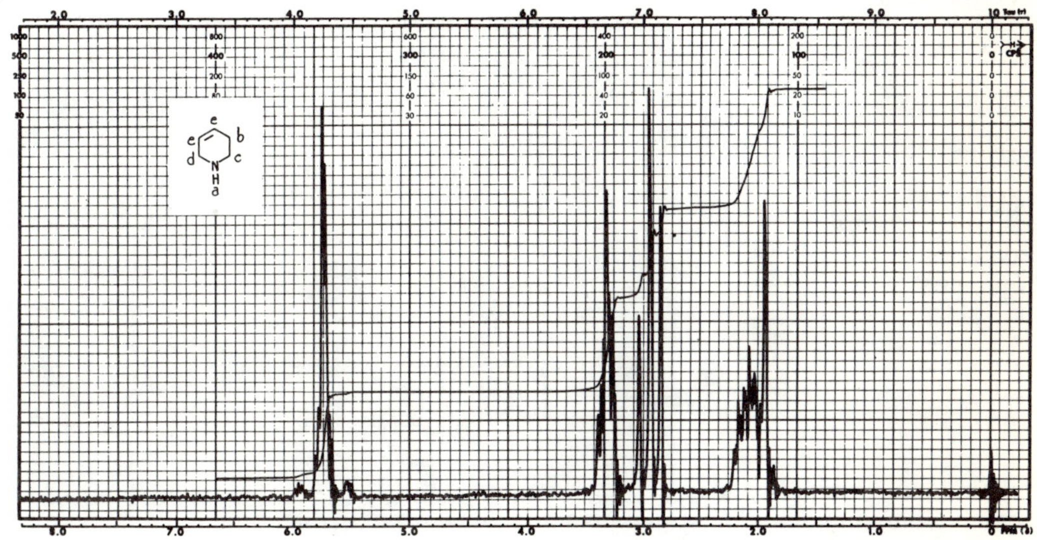

© 1978 Sadtler Research Laboratories, Inc.

N-ETHYLBENZYLAMINE

$C_9H_{13}N$

Mol. Wt. 135.21

B.P. 79-83°C/12mm (lit.)

Solvent: CCl_4

Assignments
a = 0.91
b = 1.07
c = 2.6C
d = 3.70
e = 7.18

577

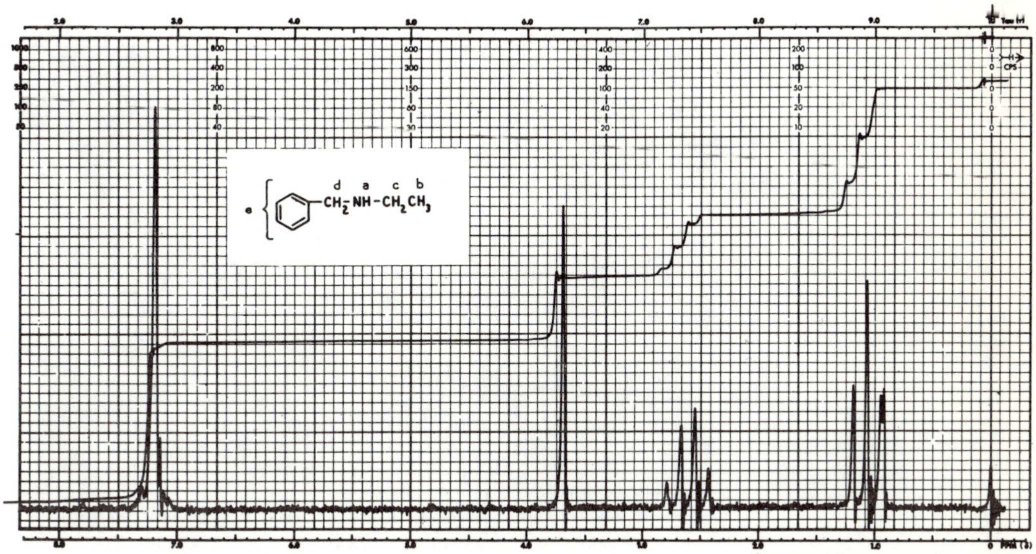

© 1978 Sadtler Research Laboratories, Inc.

N-PROPYLBENZYLAMINE

$C_{10}H_{15}N$

Mol. Wt. 149.24

B.P. 102-108°C/12mm (lit.)

Solvent: CCl_4

Assignments
a = 0.90
b = 1.00
c = 1.47
d = 2.53
e = 3.68
f = 7.19

578

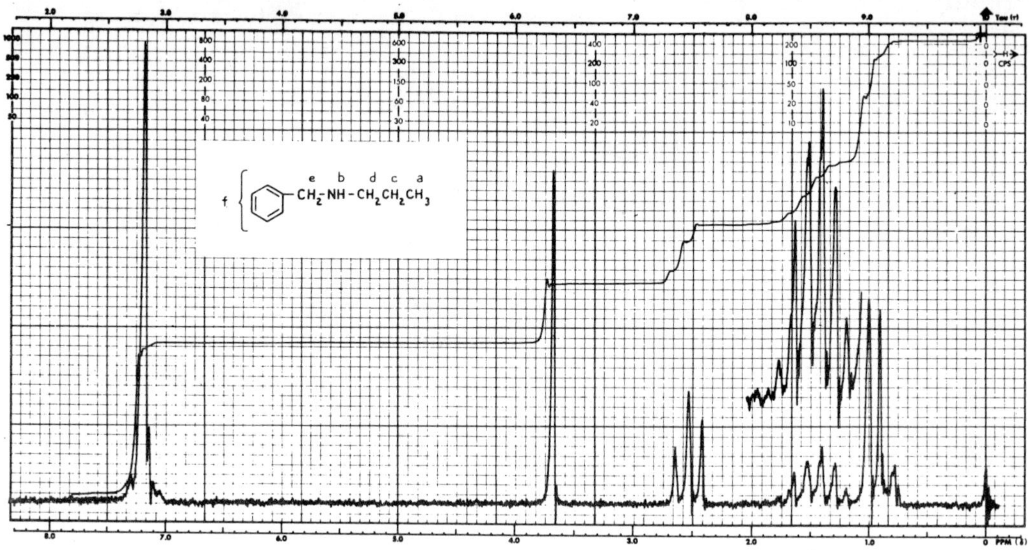

© 1978 Sadtler Research Laboratories, Inc.

N-BUTYLBENZYLAMINE

$C_{11}H_{17}N$

Mol. Wt. 163.26

B.P. 120-127°C/16mm (lit.)

Solvent: CCl_4

Assignments
a = 0.89
b = ca 1.41
c = 2.57
d = 3.69
e = 7.19

579

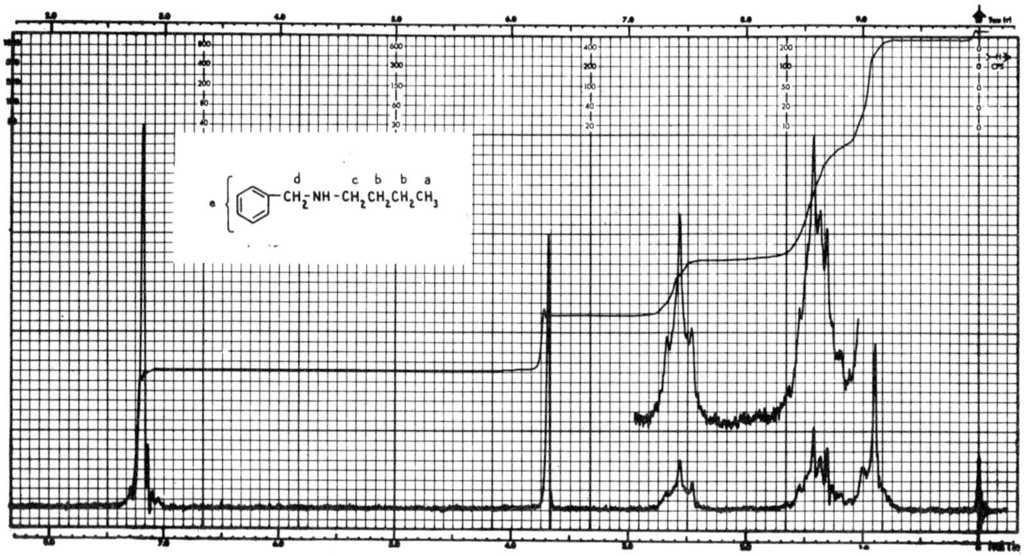

© 1978 Sadtler Research Laboratories, Inc.

N-BENZYLETHYLENEDIAMINE

$C_9H_{14}N_2$

Mol. Wt. 150.23

B.P. 129-130°C/11mm

Solvent: D_2O

Assignments
a = ca 2.60
b = 3.64
c = ca 7.29
d = 4.70 HDO

580

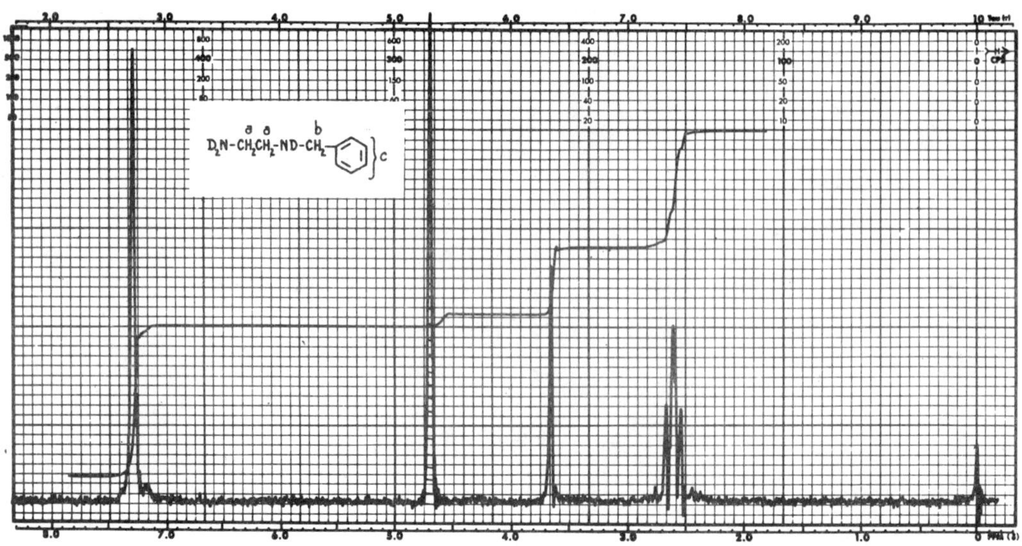

© 1978 Sadtler Research Laboratories, Inc.

DIBENZYLAMINE

$C_{14}H_{15}N$

Mol. Wt. 197.28

B.P. 151-153°C/5mm

Solvent: CCl_4

Assignments
a = 1.32
b = 3.70
c = 7.20

581

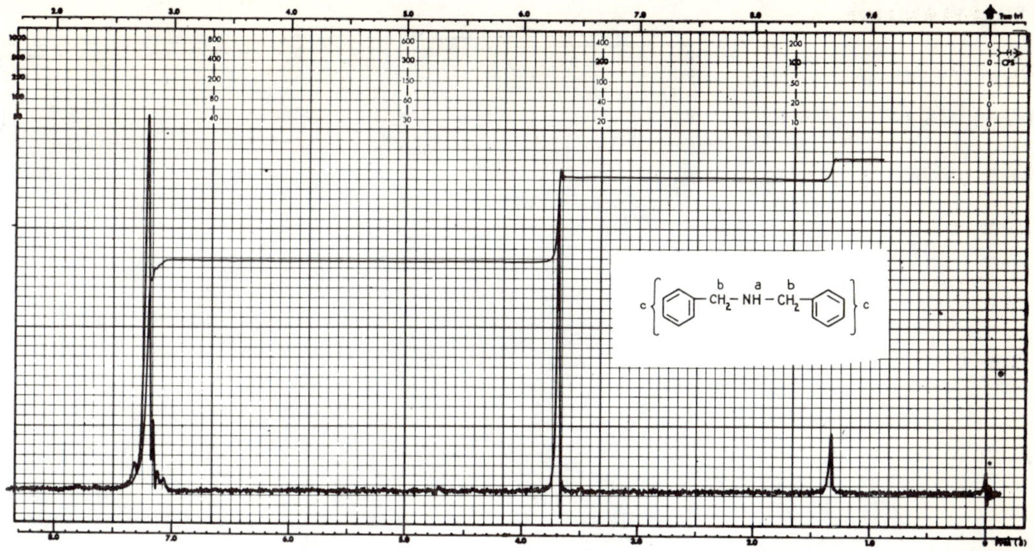

© 1978 Sadtler Research Laboratories, Inc.

N-METHYLANILINE

C_7H_9N

Mol. Wt. 107.16

B.P. 76-77°C/10mm

Solvent: CCl_4

Assignments
a = 2.67
b = 3.34
c = 6.40
d = 6.57
e = 7.08

582

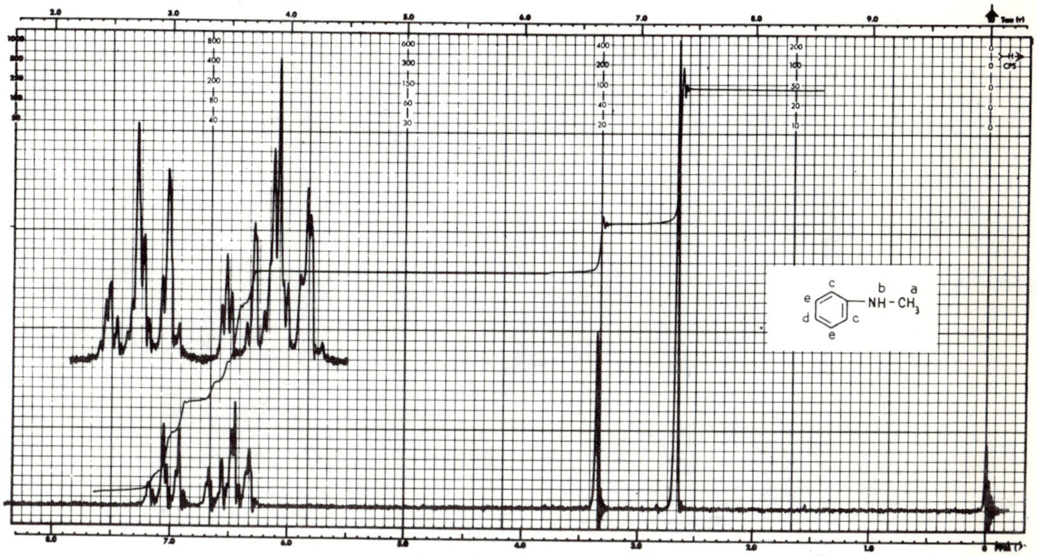

© 1978 Sadtler Research Laboratories, Inc.

N-ETHYLANILINE

$C_8H_{11}N$

Mol. Wt. 121.18

Solvent: CCl_4

Assignments
a = 1.12
b = 3.03
c = 3.23
d = 6.32-7.23

583

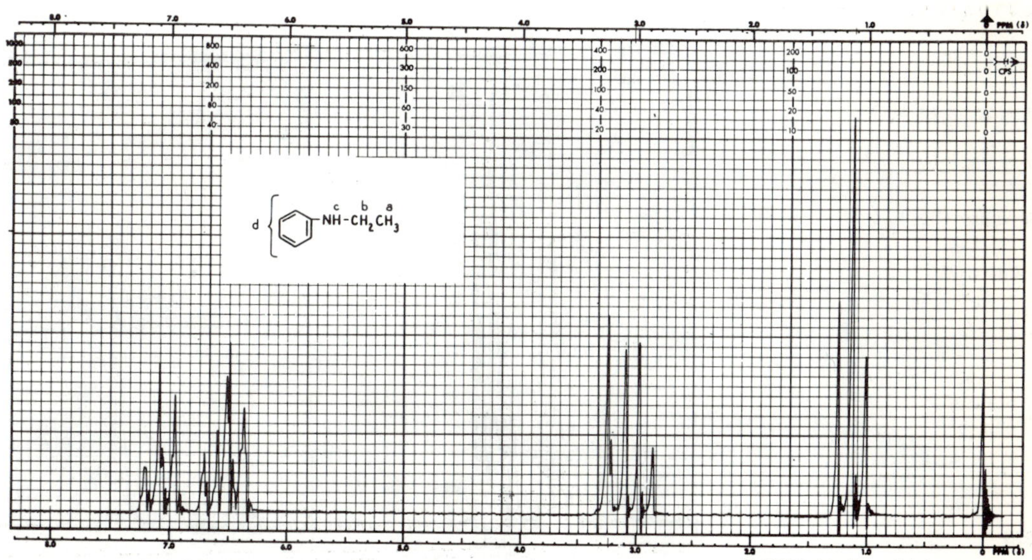

© 1978 Sadtler Research Laboratories, Inc.

N-PROPYLANILINE

$C_9H_{13}N$

Mol. Wt. 135.27

Solvent: CCl_4

Assignments
a = 0.92
b = 1.54
c = 2.97
d = 3.31
e = 6.32-7.25

584

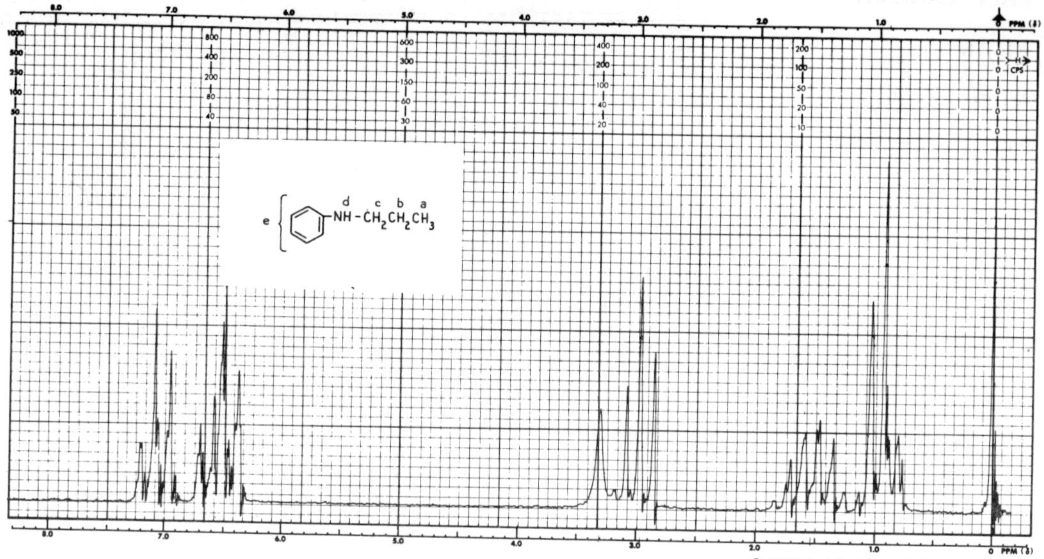

© 1978 Sadtler Research Laboratories, Inc.

N-ISOPENTYLANILINE

$C_{11}H_{17}N$

Mol. Wt. 163.26

B.P. 117-119°C/10mm

Solvent: $CDCl_3$

Assignments
a = 0.91
b = 1.20-2.10
c = 3.04
d = 3.35
e = 6.51
f = 6.61
g = 7.13

585

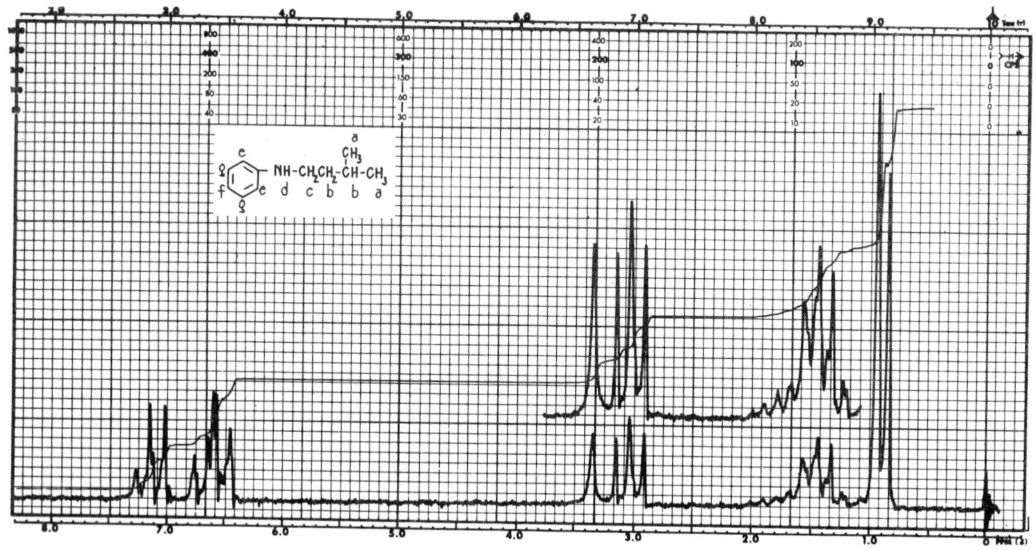

© 1978 Sadtler Research Laboratories, Inc.

N-CYCLOHEXYLANILINE

$C_{12}H_{17}N$

Mol. Wt. 175.28

M.P. 11-13°C

Solvent: $CDCl_3$

Assignments
a = 0.70-2.20
b = 3.21
c = 3.38
d = 6.52
e = 6.59
f = 7.12

586

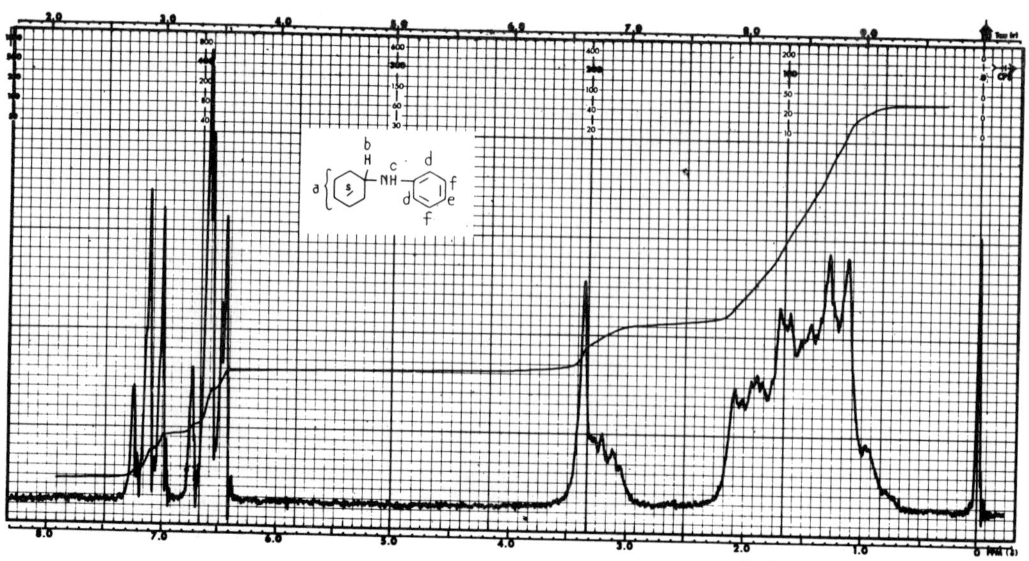

© 1978 Sadtler Research Laboratories, Inc.

N-ALLYLANILINE

$C_9H_{11}N$

Mol. Wt. 133.19

B.P. 218-220°C

Solvent: $CDCl_3$

Assignments
a = 3.59
b = 3.60
c = 5.08
d = 5.13
e = 5.80
f = 6.50
g = 6.64
h = 7.12

587

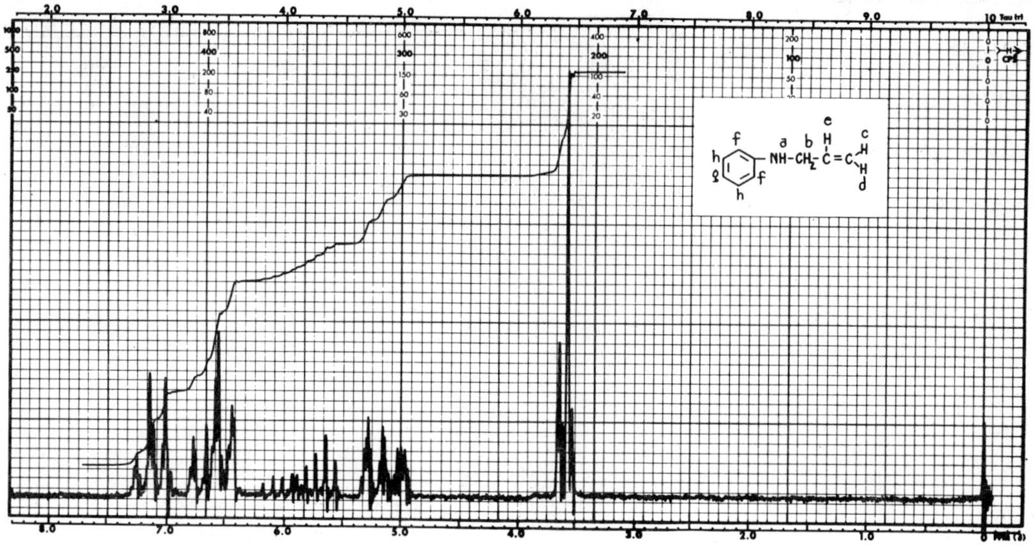

© 1978 Sadtler Research Laboratories, Inc.

N-(1,1-DIMETHYL-2-PROPYNYL)ANILINE

$C_{11}H_{13}N$

Mol. Wt. 159.23

M.P. 49.5-51°C

Solvent: $CDCl_3$

Assignments
a = 1.61
b = 2.33
c = 3.41
d = 6.60-7.40

588

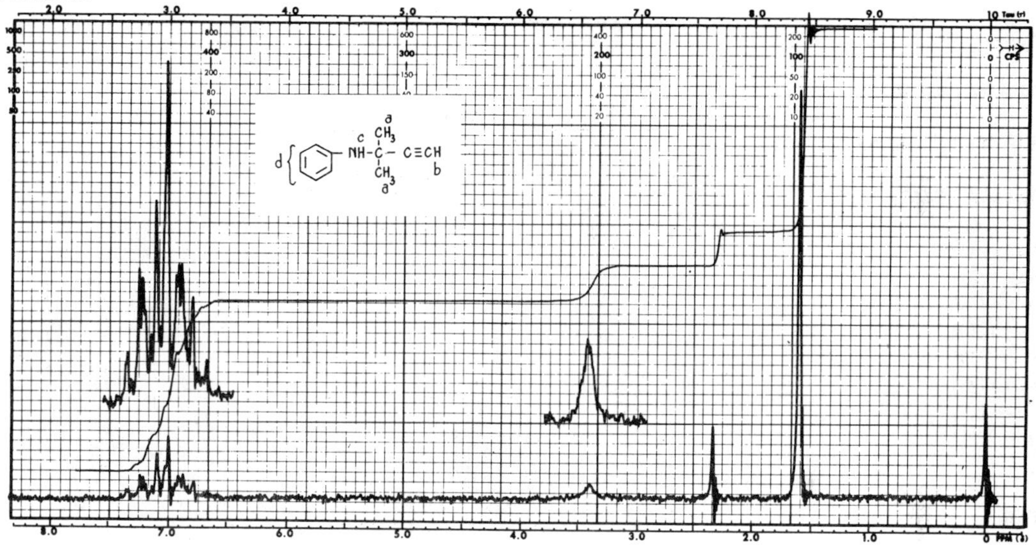

© 1978 Sadtler Research Laboratories, Inc.

N-METHYL-o-TOLUIDINE

$C_8H_{11}N$

Mol. Wt. 121.18

B.P. 98-100°C/12mm

Solvent: $CDCl_3$

Assignments
a = 2.10
b = 2.85
c = 3.39
d = 6.60
e = 6.65
f = 7.08
g = 7.13

589

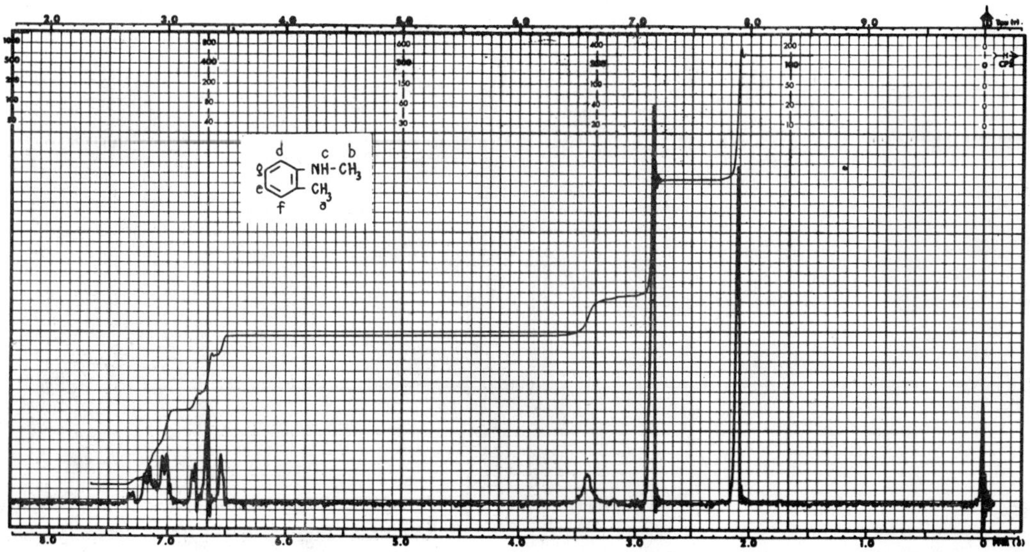

© 1978 Sadtler Research Laboratories, Inc.

N-METHYL-m-TOLUIDINE

$C_8H_{11}N$

Mol. Wt. 121.18

B.P. 86-87°C/7mm

Solvent: $CDCl_3$

Assignments

a = 2.24
b = 2.69
c = 3.39
d = ca 6.39
e = 6.50
f = 7.06

590

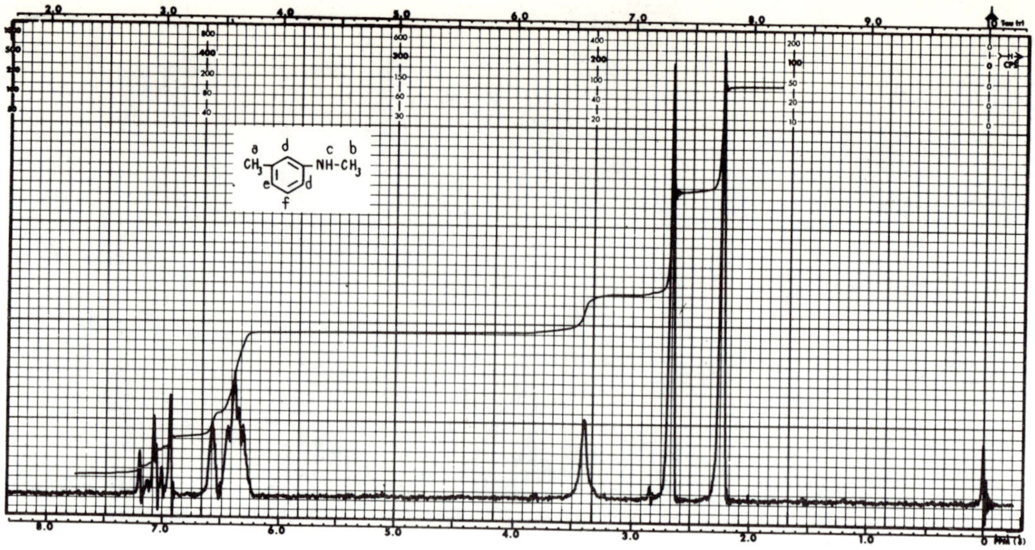

© 1978 Sadtler Research Laboratories, Inc.

N-METHYL-p-TOLUIDINE

$C_8H_{11}N$

Mol. Wt. 121.18

B.P. 88-90°C/10mm

Solvent: $CDCl_3$

Assignments

a = 2.21
b = 2.62
c = 3.29
d = 6.40
e = 6.92

591

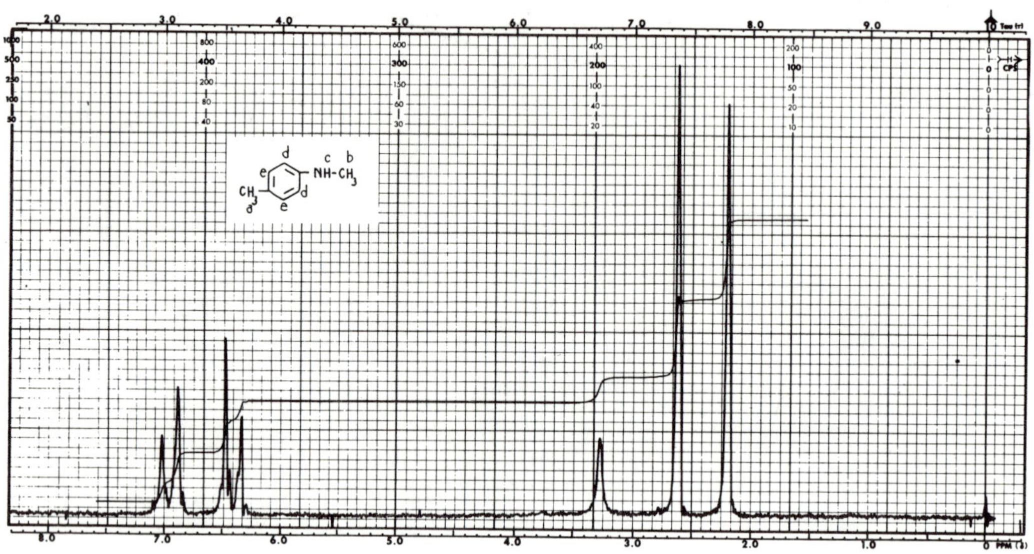

© 1978 Sadtler Research Laboratories, Inc.

N-METHYLCUMIDINE

$C_{10}H_{15}N$

Mol. Wt. 149.24

Solvent: $CDCl_3$

Assignments

a = 1.22
b = 2.70
c = 2.76
d = 3.36
e = 6.47
f = 7.01

592

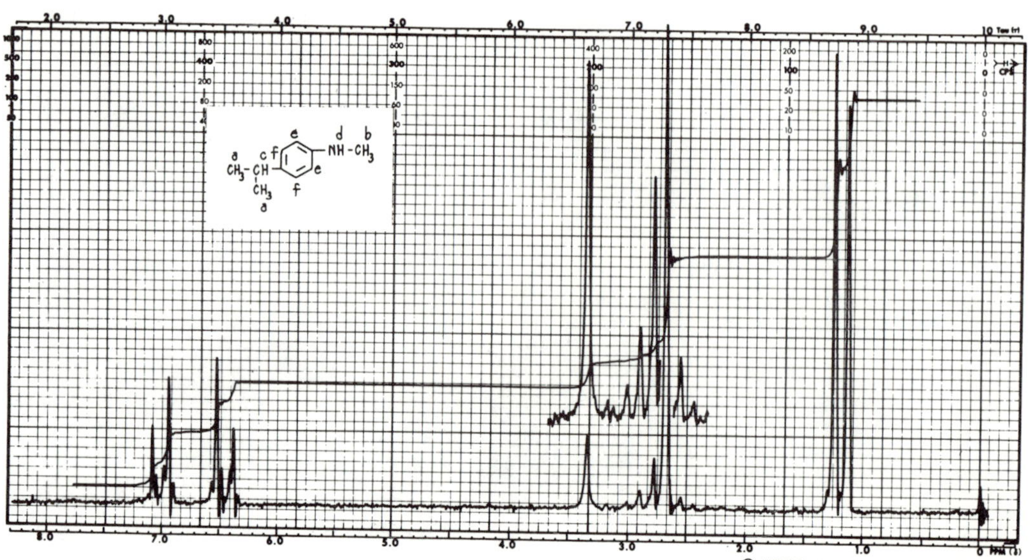

© 1978 Sadtler Research Laboratories, Inc.

N-ETHYL-p-TOLUIDINE

$C_9H_{13}N$

Mol. Wt. 135.21

B.P. 101-102°C/10mm

Solvent: $CDCl_3$

Assignments
a = 1.10
b = 2.20
c = 3.01
d = 3.20
e = 6.43
f = 6.96

593

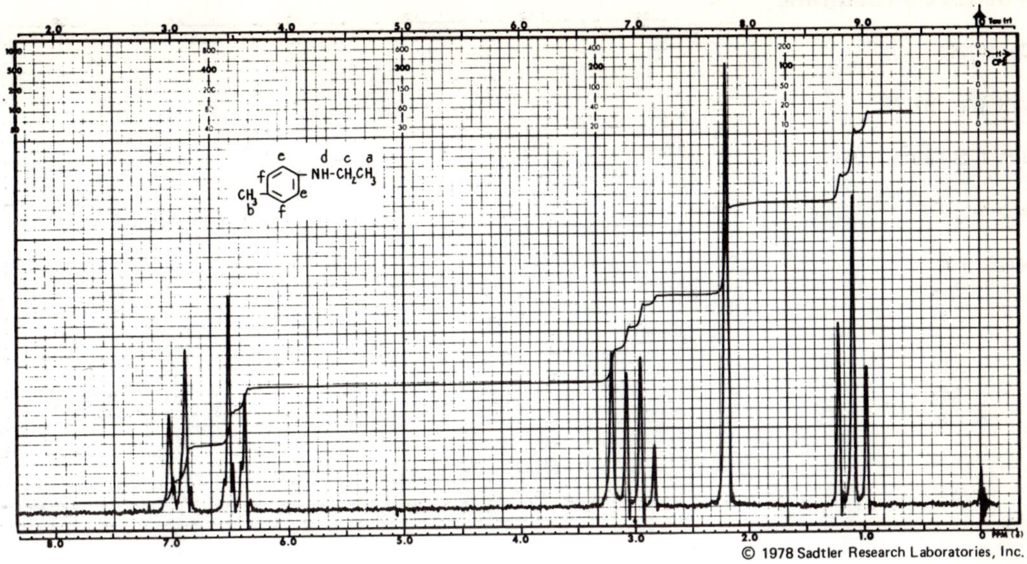

© 1978 Sadtler Research Laboratories, Inc.

N-PHENYLBENZYLAMINE

$C_{13}H_{13}N$

Mol. Wt. 183.24

M.P. 34-36°C

Solvent: $CDCl_3$

Assignments
a = 3.78
b = 4.22
c = 6.58
d = 6.65
e = 7.12
f = 7.27

594

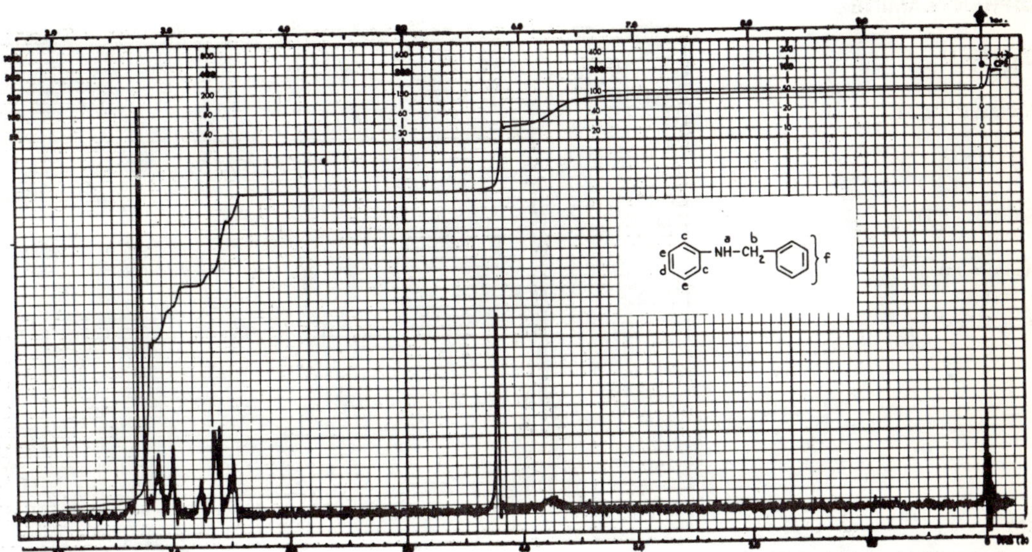

© 1978 Sadtler Research Laboratories, Inc.

N-BENZYL-o-TOLUIDINE

$C_{14}H_{15}N$

Mol. Wt. 197.28

M.P. 58-60°C

Solvent: $CDCl_3$

Assignments
a = 2.00
b = 3.62
c = 4.18
d = 6.52
e = 6.59
f = 6.90-7.35
g = ca 7.21

595

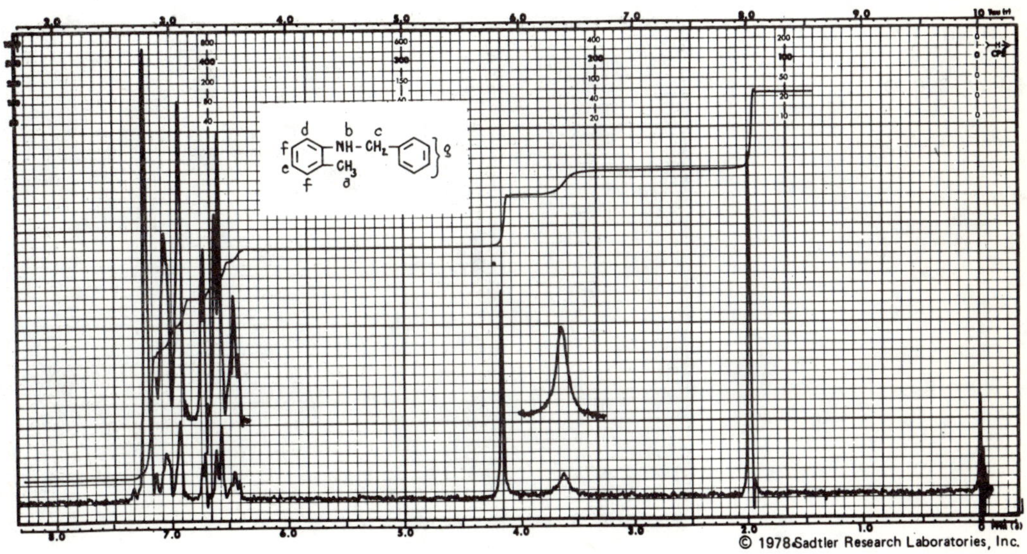

© 1978 Sadtler Research Laboratories, Inc.

N-BENZYL-p-TOLUIDINE

$C_{14}H_{15}N$

Mol. Wt. 197.28

B.P. 178-180°C/10mm

Solvent: CCl_4

Assignments
a = 2.14
b = 3.48
c = 4.07
d = 6.32
e = 6.83
f = ca 7.19

596

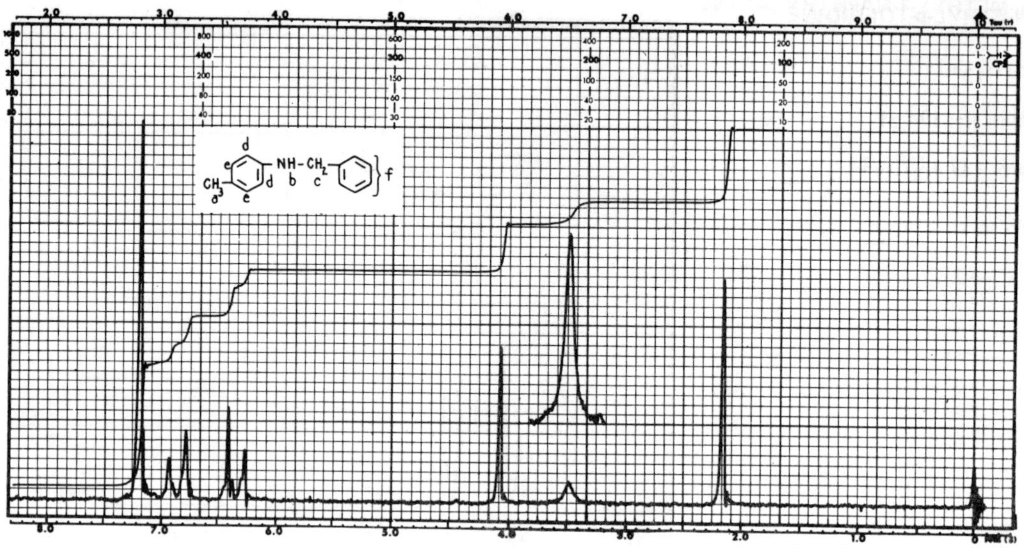

© 1978 Sadtler Research Laboratories, Inc.

DIPHENYLAMINE

$C_{12}H_{11}N$

Mol. Wt. 69.22

M.P. 53°C

Solvent: $CDCl_3$

Assignments
a = 5.42
b = 6.68-7.38

597

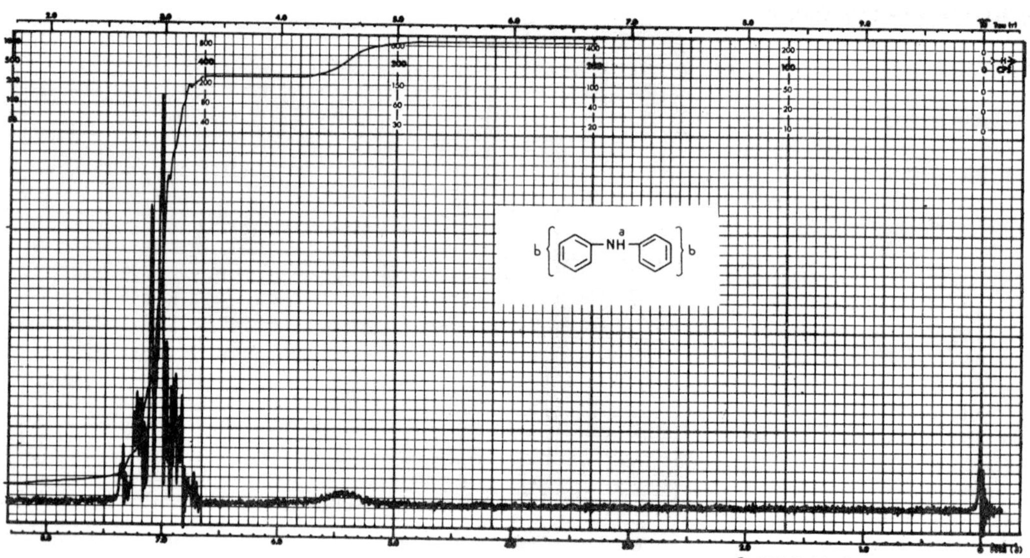

© 1978 Sadtler Research Laboratories, Inc.

N,N'-DI-sec-BUTYL-p-PHENYLENEDIAMINE

$C_{14}H_{14}N_2$

Mol. Wt. 220.36

Solvent: $CDCl_3$

Assignments
a = 0.92
b = 1.12
c = 1.10-1.80
d = 2.63
e = 3.27
f = 6.52

598

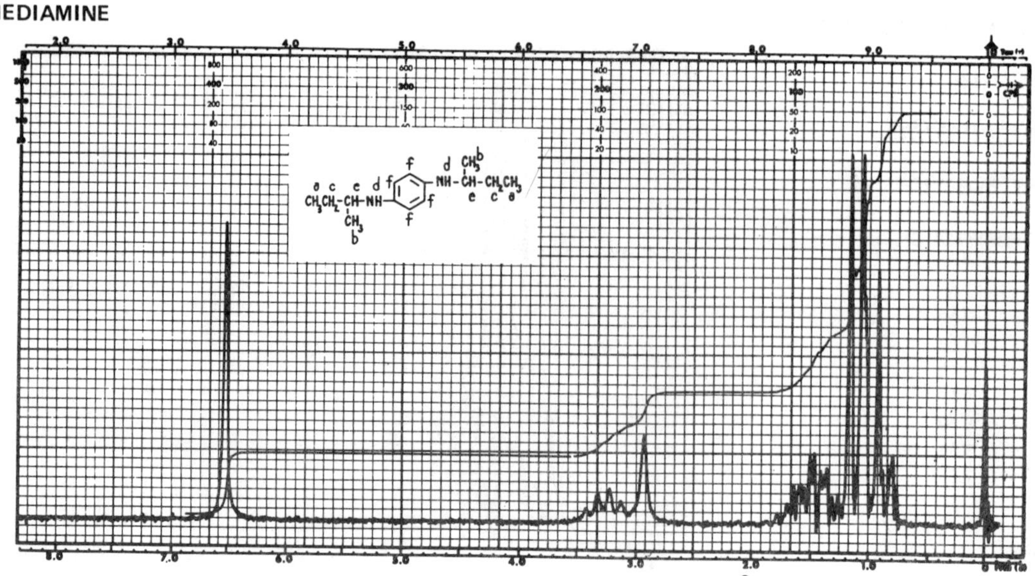

© 1978 Sadtler Research Laboratories, Inc.

N-METHYL-1-NAPHTHYLAMINE

$C_{11}H_{11}N$

Mol. Wt. 157.22

B.P. 171-173°C/17mm

Solvent: $CDCl_3$

Assignments
a = 2.79
b = 4.03
c = 6.49
d = 7.10-7.85

599

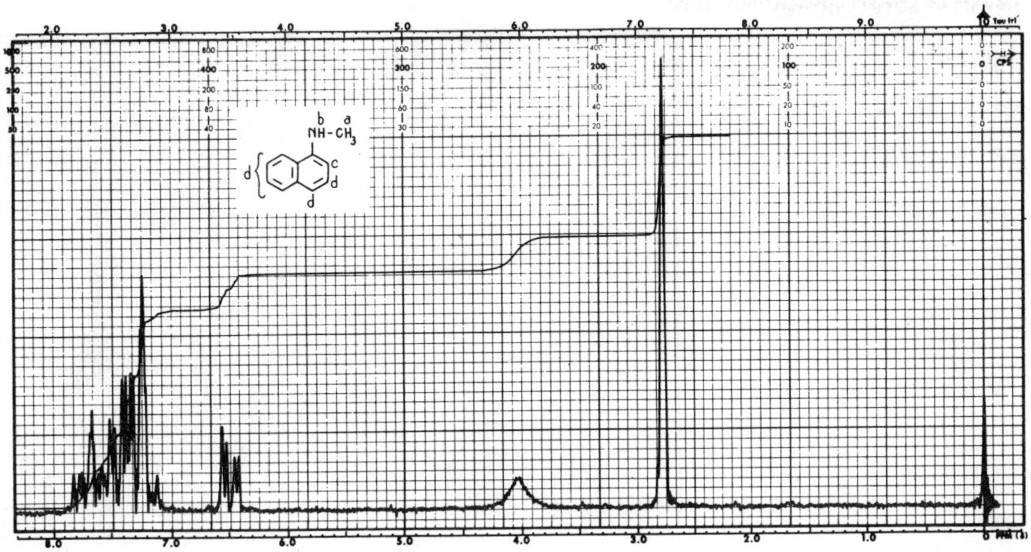

© 1978 Sadtler Research Laboratories, Inc.

N-ETHYL-1-NAPHTHYLAMINE

$C_{12}H_{13}N$

Mol. Wt. 171.24

Solvent: CCl_4

Assignments
a = 1.32
b = 3.20
c = 3.95
d = 6.42
e = 7.00-7.50
f = ca 7.64

600

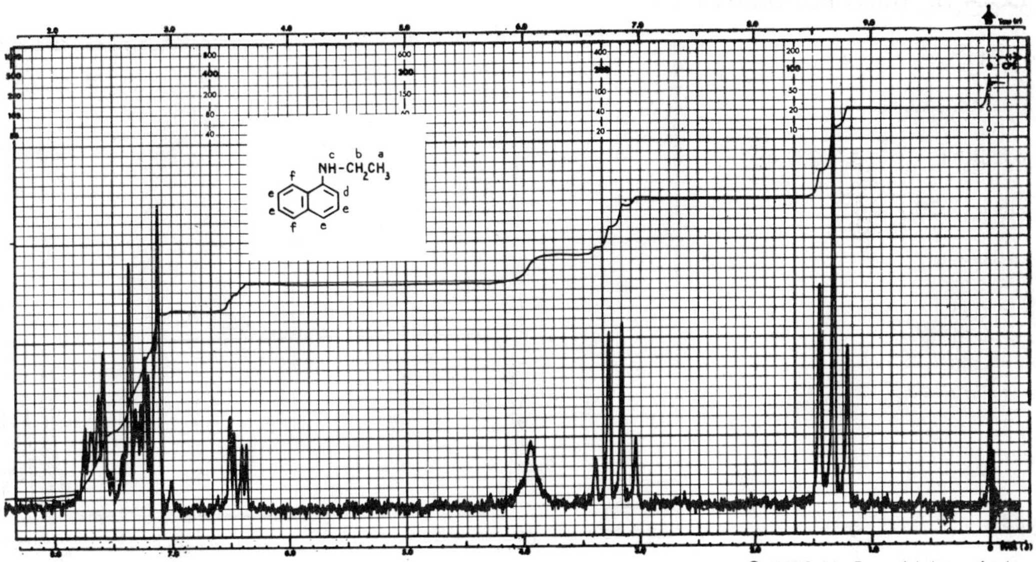

© 1978 Sadtler Research Laboratories, Inc.

1,2,3,4-TETRAHYDROQUINOLINE

$C_9H_{11}N$

Mol. Wt. 133.19

Solvent: $CDCl_3$

Assignments
a = 1.83
b = 2.09
c = 3.14
d = 3.62
e = 6.34
f = 6.57
g = 6.89
h = 6.92

601

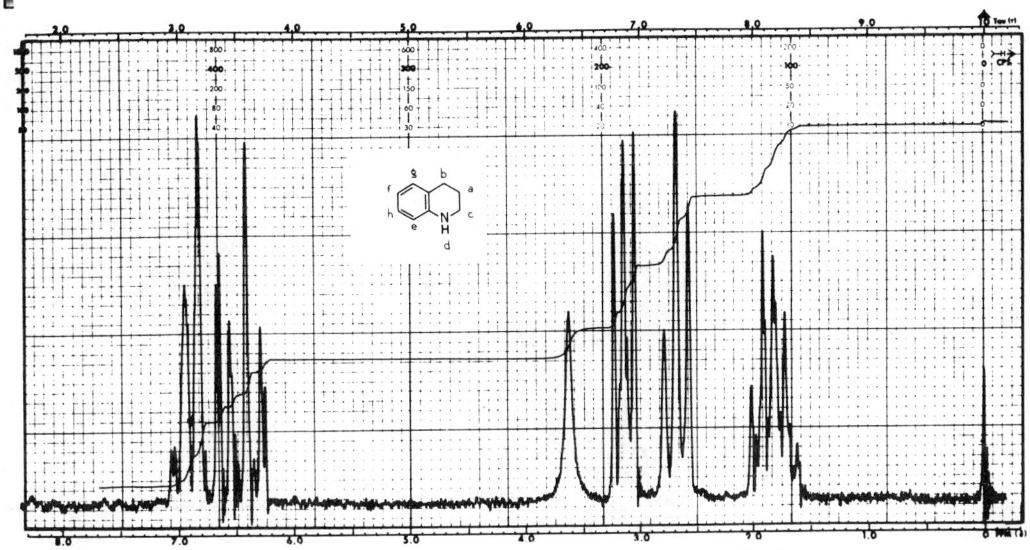

© 1978 Sadtler Research Laboratories, Inc.

1,2,3,4-TETRAHYDROQUINALDINE

$C_{10}H_{13}N$

Mol. Wt. 147.22

B.P. 250-252°C

Solvent: $CDCl_3$

Assignments

a = 1.16
b = 1.20-2.15
c = 2.40-2.90
d = 3.30
e = 3.48
f = 6.39
g = 6.54
h = 6.75-7.20

602

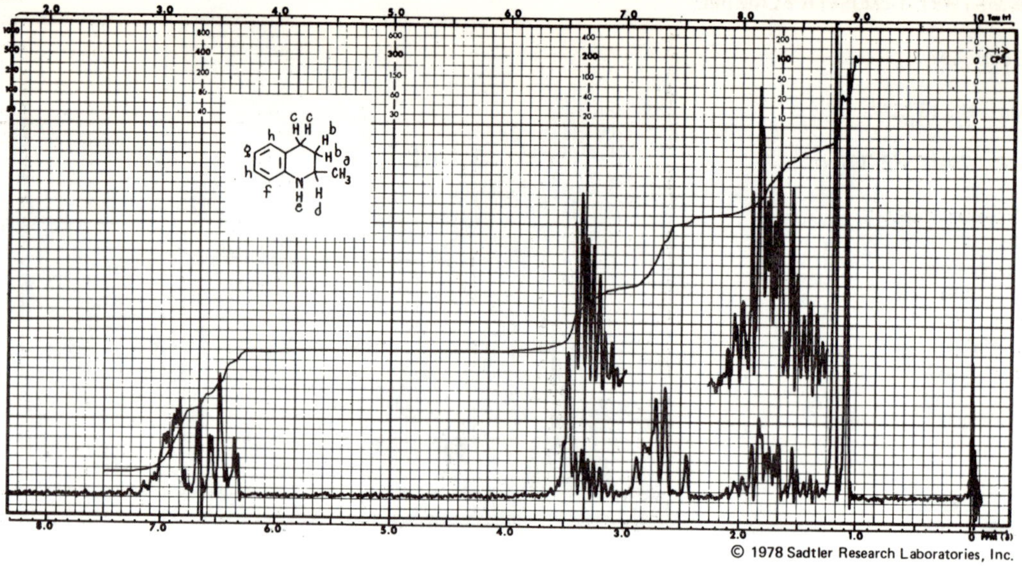

© 1978 Sadtler Research Laboratories, Inc.

1,2,3,4-TETRAHYDROISOQUINOLINE

$C_9H_{11}N$

Mol. Wt. 133.19

M.P. 30°C B.P. 237.4°C

Solvent: CCl_4

Assignments

a = 2.08
b = 2.61
c = 2.91
d = 3.80
e = 6.70-7.20

603

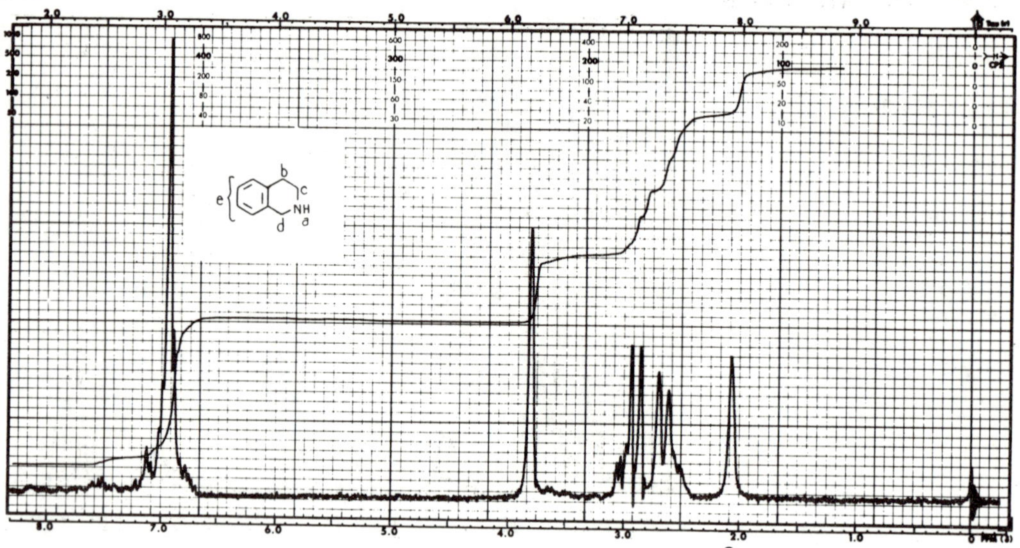

© 1978 Sadtler Research Laboratories, Inc.

N-PHENYL-1-NAPHTHYLAMINE

$C_{16}H_{13}N$

Mol. Wt. 219.27

M.P. 59.5-60.5°C (lit.)

Solvent: CCl_4

Assignments

a = 5.61
b = 6.60-7.55
c = 7.80

604

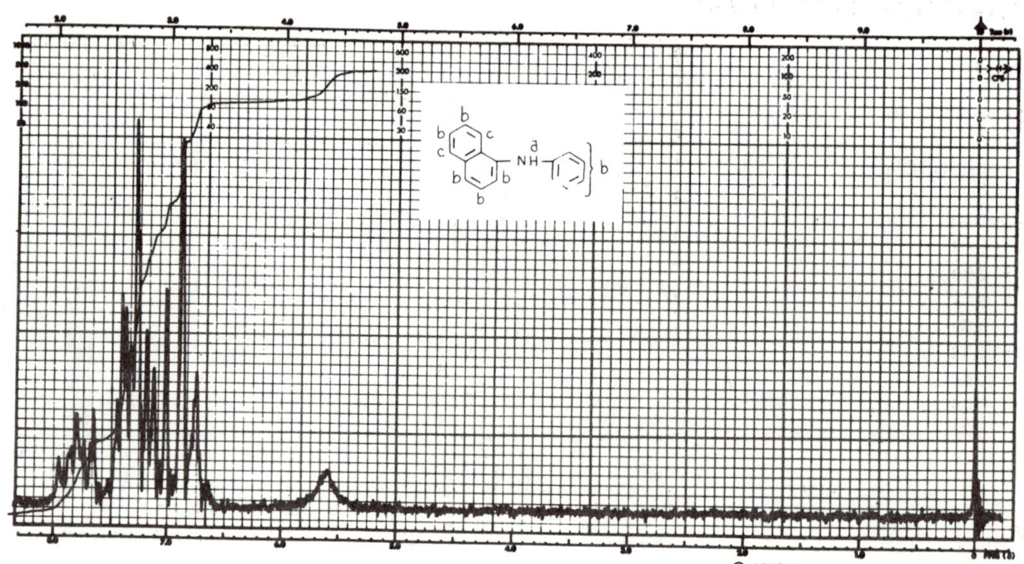

© 1978 Sadtler Research Laboratories, Inc.

INDOLE

C_8H_7N

Mol. Wt. 117.15

M.P. 52-53°C

Solvent: CCl_4

Assignments

a = 6.38

b = 6.76

c = ca 7.03

d = ca 7.04

e = 7.50

605

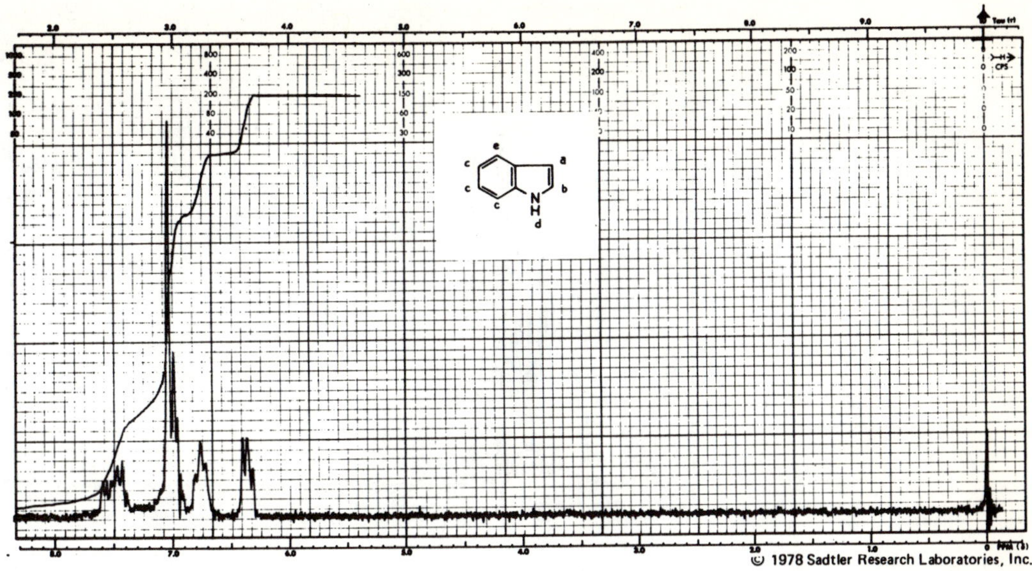

© 1978 Sadtler Research Laboratories, Inc.

2-METHYLINDOLE

C_9H_9N

Mol. Wt. 131.18

M.P. 58-60°C

Solvent: $CDCl_3$

Assignments

a = 2.30

b = 6.18

c = 7.04

d = 7.42

e = 7.56

606

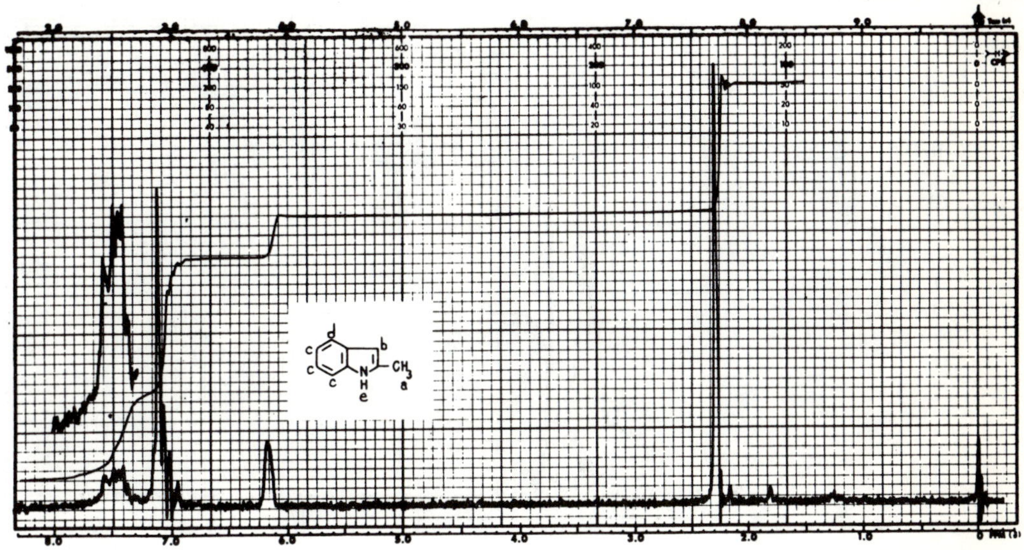

© 1978 Sadtler Research Laboratories, Inc.

3-METHYLINDOLE

C_9H_9N

Mol. Wt. 131.18

M.P. 92-94°C

Solvent: $CDCl_3$

Assignments

a = 2.29

b = 6.75

c = 7.17

d = ca 7.49

e = 7.59

607

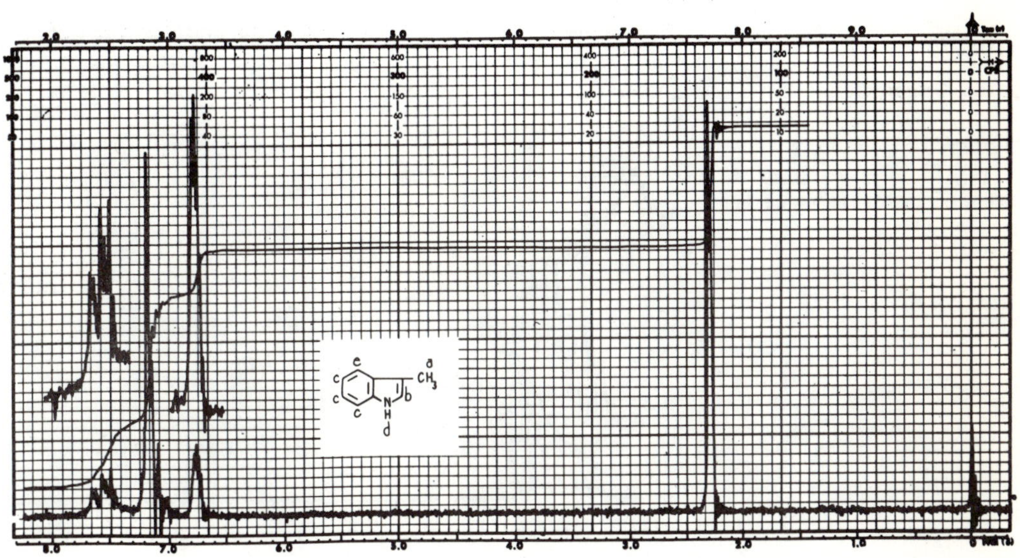

© 1978 Sadtler Research Laboratories, Inc.

4-METHYLINDOLE

C_9H_9N

Mol. Wt. 131.18

Solvent: $CDCl_3$

Assignments
a = 2.48
b = 6.40
c = 6.69
d = 6.80-7.20
e = ca 7.19

608

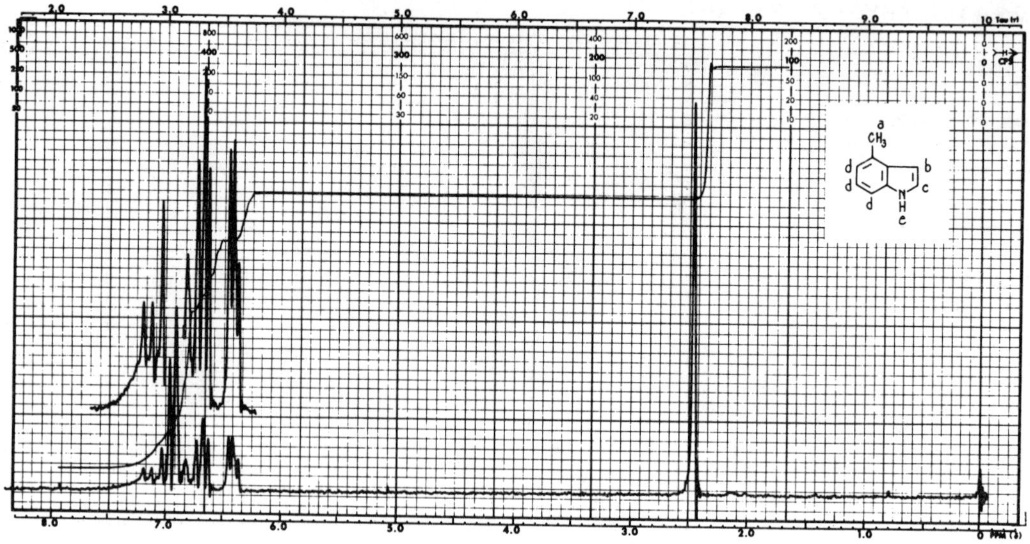

© 1978 Sadtler Research Laboratories, Inc.

2,3-DIMETHYLINDOLE

$C_{10}H_{11}N$

Mol. Wt. 145.21

M.P. 105-107°C B.P. 285°C

Solvent: $CDCl_3$

Assignments
a = 2.21
b = 2.23
c = 6.90-7.25
d = ca 7.41

609

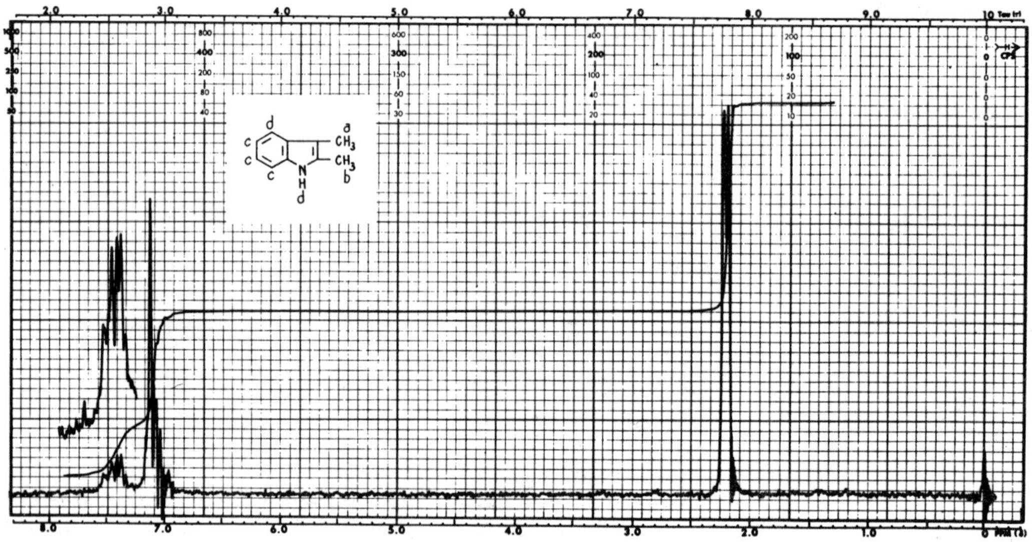

© 1978 Sadtler Research Laboratories, Inc.

3-[(DIMETHYLAMINO)METHYL] INDOLE

$C_{11}H_{14}N_2$

Mol. Wt. 174.25

M.P. 134°C

Solvent: Dimethyl sulfoxide-d_6

Assignments
a = 2.14
b = 3.52
c = 6.76-7.71
d = 10.83
e = 2.47 DMSO-d_5

610

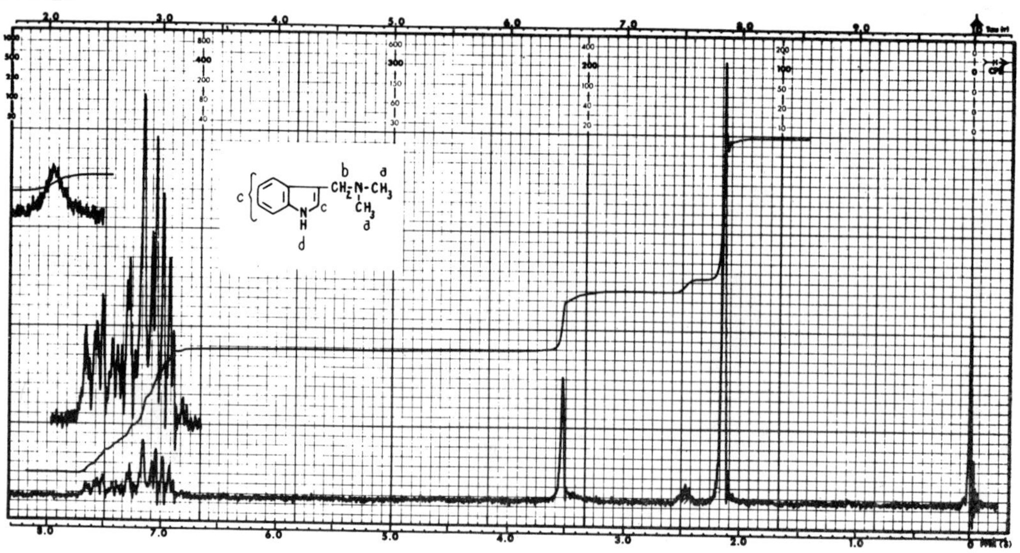

© 1978 Sadtler Research Laboratories, Inc.

1,2,3,4-TETRAHYDROCARBAZOLE

$C_{12}H_{13}N$

Mol. Wt. 171.24

M.P. 113-116°C

Solvent: $CDCl_3$

Assignments
a = 1.83
b = 2.60
c = 7.09
d = 7.39

611

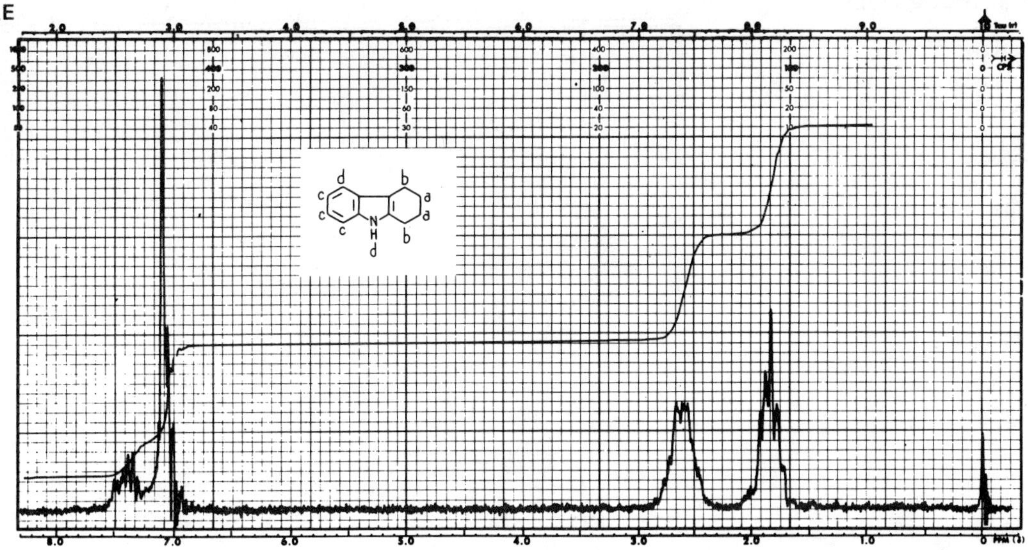

© 1978 Sadtler Research Laboratories, Inc.

CARBAZOLE

$C_{12}H_9N$

Mol. Wt. 167.21

Solvent: Acetone

Assignments
a = 7.17
b = ca 7.45
c = 7.49
d = 8.12
NH is unobserved

612

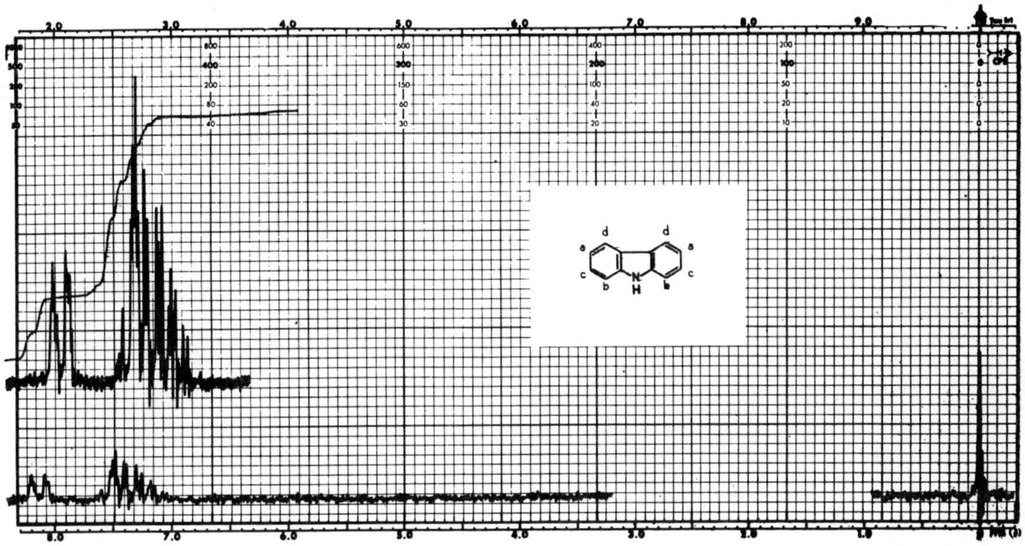

© 1978 Sadtler Research Laboratories, Inc.

NOTES

The absence of an exchangeable hydrogen attached to the tertiary amine group makes this amine more difficult to characterize than either the primary or secondary amines. The large number of aliphatic and aromatic beta shift effects produce wider chemical shift ranges than nearly any other common functional group. A simple test for the presence of any amine including the tertiary variety is the addition of a few drops of weak mineral acid. The formation of the amine salt produces significant shifts to lower field.

Chemical Shifts

Aliphatic Protons

HC–4	HC–3	HC–2	HC–1		Solvent	Spectrum
			(2.14)	$(CH_3)_2$–N–CH_2–Ø	CCl_4	644
			(2.21)	$(CH_3)_2$–N–R_8	$CDCl_3$	----
			(2.27)	$(CH_3)_2$–N–R_2	$CDCl_3$	613
			(2.39)	$(CH_3)_2$–N–CH=C$(R)_2$	$CDCl_3$	643
			(2.85)	$(CH_3)_2$–N–Ø	CCl_4	648
		1.09	2.33	CH_3–CH_2–N$(CH_3)_2$	$CDCl_3$	613
		0.98	2.42	CH_3–CH_2–N$(R_2)_2$	CCl_4	614
		1.10	3.27	CH_3–CH_2–N(R_2)–Ø	CCl_4	649
	0.88	1.36	2.31	CH_3–CH_2–CH_2–N$(R_3)_2$	CCl_4	615
0.90	1.25	1.35	2.31	CH_3–CH_2–CH_2–CH_2–N$(R_4)_2$	CCl_4	617
		(0.97)	3.02	$(CH_3)_2$–CH–N$(R_2)_2$	CCl_4	616

Substituted Trimethylamines

$(CH_3)_2$N –CH_2	–X	Solvent	Spectrum
2.33	–CH_3	$CDCl_3$	613
2.72	–N$(CH_3)_2$	$CDCl_3$	630
2.93	–CH=CH_2	$CDCl_3$	642
3.07	–C(=O)–NH–Ø	$CDCl_3$	----
3.07	–C(=O)–O–R	CCl_4	----
3.20	–P(=O)–$Ø_2$	$CDCl_3$	----
3.29	–C≡C–H	$CDCl_3$	----
3.33	–Ø	CCl_4	644
3.42	–2–Pyrrole	$CDCl_3$	677
3.49	–C≡N	$CDCl_3$	----
3.69	–2–Pyridine	$CDCl_3$	----

Aliphatic Protons continued

$CH_3(\emptyset)-N$ $-CH_2$	$-X$	Solvent
3.30	$-CH_3$	CCl_4
3.84	$-CH=CH_2$	CCl_4
3.95	$-C\equiv C-R$	$CDCl_3$
3.99	$-C(=O)-NH_2$	DMSO-d_6
4.41	$-\emptyset$	CCl_4
4.70	$-SO_2-\emptyset$	$CDCl_3$
4.89	$-S-\emptyset$	$CDCl_3$

$(R)_2N$ $-CH_2$	$-CH_2$	$-X$	
2.41	2.41	$-N(CH_3)_2$	CCl_4
2.81	2.45	$-C(=O)-NH_2$	D_2O
2.63	2.63	$-SH$	$CDCl_3$
2.41	2.69	$-NH_2$	D_2O
2.70	2.70	$-C(=O)-CH_3$	$CDCl_3$
2.70	2.70	$-\emptyset$	$CDCl_3$
2.70	4.15	$-O-C(=O)-R$	CCl_4

$\emptyset(R)N$ $-CH_2$	$-CH_2$	$-X$	
3.13	1.59	$-CH_3$	CCl_4
3.21	3.50	$-OH$	CCl_4
3.60	2.49	$-C(=O)-OH$	DMSO-d_6
3.61	3.61	$-Cl$	Polysol
3.65	2.50	$-C\equiv N$	$CDCl_3$

Aromatic Protons

Phenyl Amines

	ortho	meta	para	
$(CH_3)_2N-$	6.57	7.10	6.57	CCl_4
$(R_2)_2N-$	6.56	7.07	6.48	CCl_4
$(R_3)_2N-$	6.52	7.02	6.48	CCl_4
(Piperazine)N–	6.86	7.23	6.82	$CDCl_3$
(Pyrrole)N–	7.05–7.60			$CDCl_3$
$(\emptyset)_2N-$	6.80–7.40			$CDCl_3$

Aromatic Protons continued

Para substituted aromatics

$(CH_3)_2N-$	ortho	meta	para	Solvent
	6.62	6.62	$-NH_2$	$CDCl_3$
	6.79	6.79	$-O-CH_3$	$CDCl_3$
	6.77	7.00	$-CH(\emptyset)_2$	$CDCl_3$
	6.62	7.01	$-CH_2-\emptyset$	$CDCl_3$
	6.53	7.25	$-Br$	$CDCl_3$
	6.63	7.33	$-S-\emptyset$	$CDCl_3$
	6.63	7.63	$-C(=O)-OH$	$CDCl_3$
	6.60	7.70	$-N=O$	$CDCl_3$
	6.65	7.73	$-C(=O)-\emptyset$	$CDCl_3$
	6.61	7.73	$-C(=S)-\emptyset$	$CDCl_3$
	6.60	7.80	$-N=N-\emptyset$	$CDCl_3$
	6.71	7.95	$-C(=O)-CH_3$	$CDCl_3$
	6.69	7.99	$-C(=O)-O-R$	$CDCl_3$
	6.58	8.08	$-NO_2$	$CDCl_3$

$(R)_2N-$	ortho	meta	para	
	6.59	6.59	$-NH_2$	$CDCl_3$
	6.76	6.76	$-O-CH_3$	$CDCl_3$
	6.59	7.01	$-CH_3$	$CDCl_3$
	6.70	7.15	$-Cl$	DMSO-d_6
	6.48	7.20	$-Br$	$CDCl_3$
	6.58	7.35	$-C\equiv N$	$CDCl_3$
	6.57	7.56	$-C(=O)-OH$	$CDCl_3$
	6.62	7.75	$-C(=O)-\emptyset$	$CDCl_3$
	6.78	7.77	$-C(=O)-H$	$CDCl_3$
	6.65	7.79	$-N=O$	$CDCl_3$
	6.71	7.84	$-N=N-\emptyset$	$CDCl_3$
	6.70	8.02	$-C(=O)-O-R$	$CDCl_3$
	6.59	8.08	$-NO_2$	$CDCl_3$

N,N-DIMETHYLETHYLAMINE

$C_4H_{11}N$

Mol. Wt. 73.14

B.P. 36-37°C

Solvent: $CDCl_3$

Assignments
a = 1.09
b = 2.27
c = 2.33

613

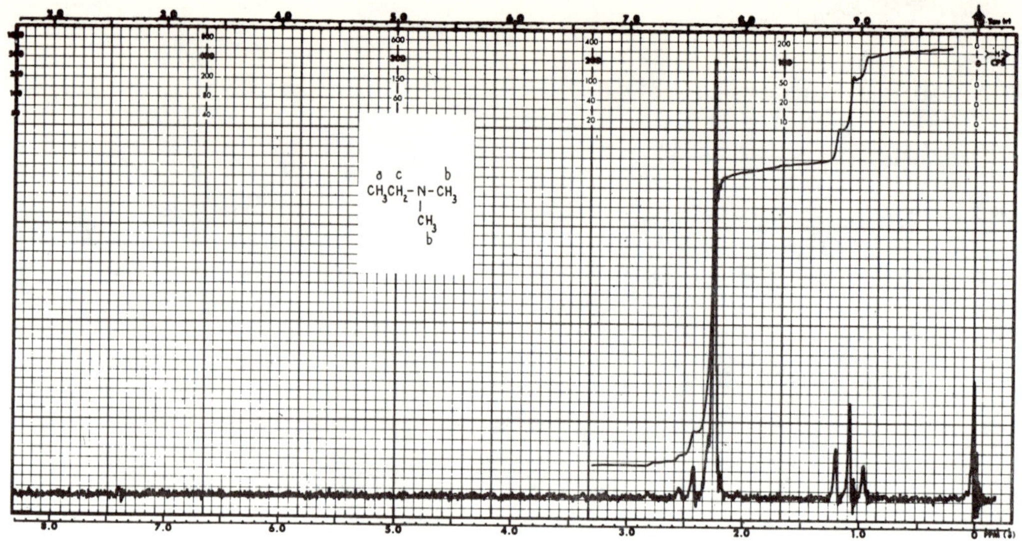

© 1978 Sadtler Research Laboratories, Inc.

TRIETHYLAMINE

$C_6H_{15}N$

Mol. Wt. 101.19

B.P. 89.6°C/755mm

Solvent: CCl_4

Assignments
a = 0.98
b = 2.42
c = 2.07 imp.

614

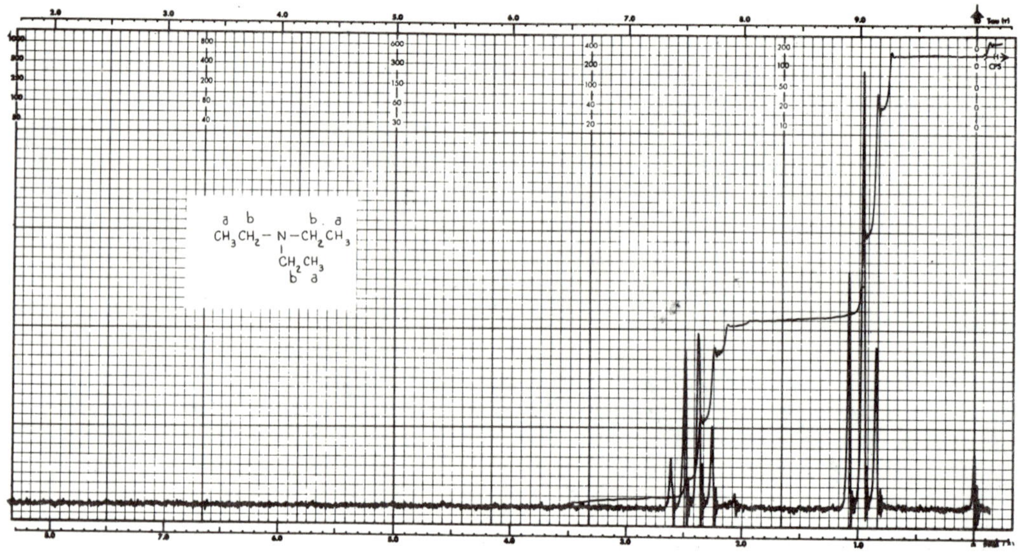

© 1978 Sadtler Research Laboratories, Inc.

TRIPROPYLAMINE

$C_9H_{21}N$

Mol. Wt. 143.27

B.P. 155-158°C

Solvent: CCl_4

Assignments
a = 0.88
b = 1.11-1.75
c = 2.31

615

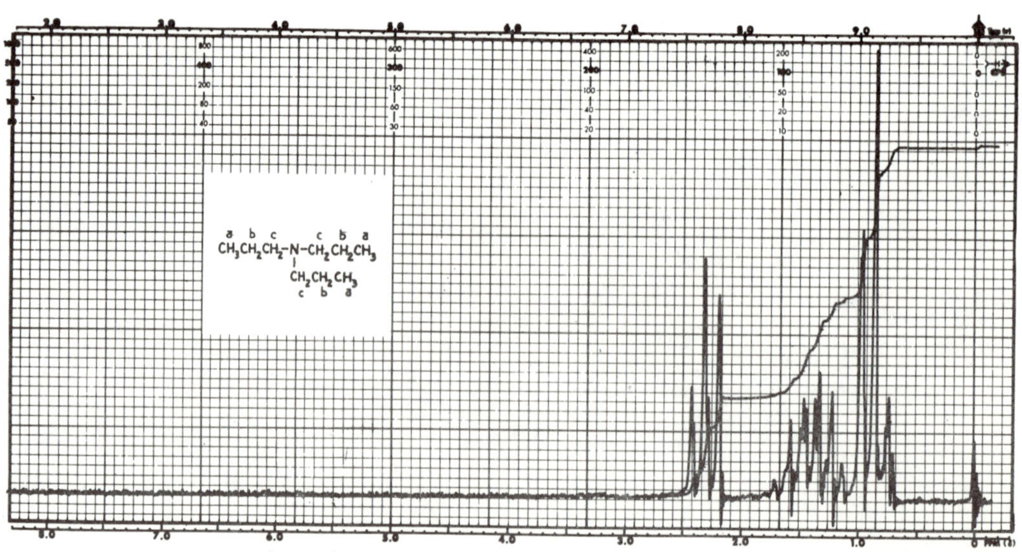

© 1978 Sadtler Research Laboratories, Inc.

1,1'-DIMETHYLTRIETHYLAMINE

$C_8H_{19}N$

Mol. Wt. 129.25

B.P. 126.5°C

Solvent: CCl_4

Assignments
a = 0.97
b = 2.48
c = 3.02

616

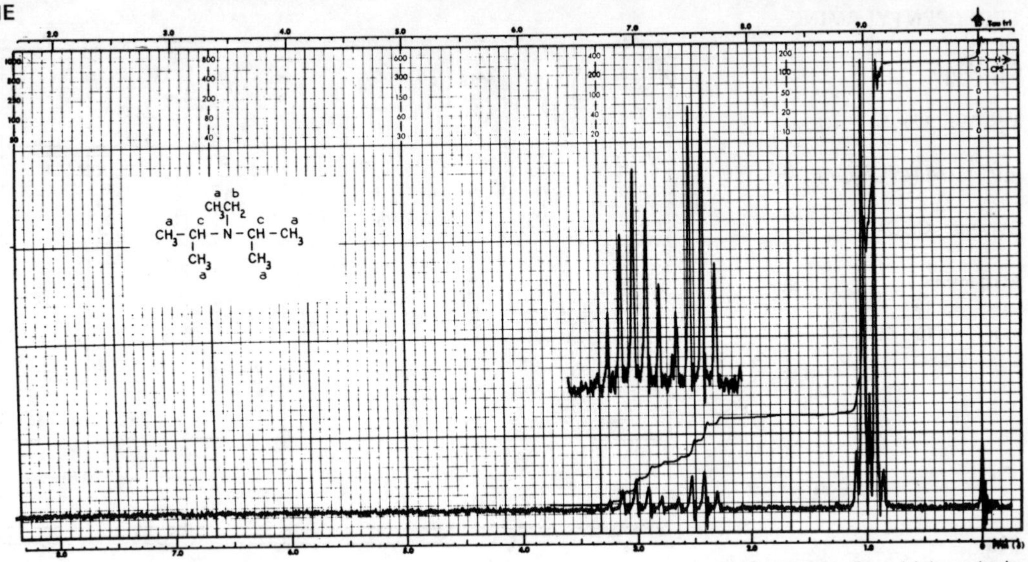

© 1978 Sadtler Research Laboratories, Inc.

TRIBUTYLAMINE

$C_{12}H_{27}N$

Mol. Wt. 185.34

B.P. 216.5°C

Solvent: CCl_4

Assignments
a = 0.90
b = 1.10-1.60
c = 2.31

617

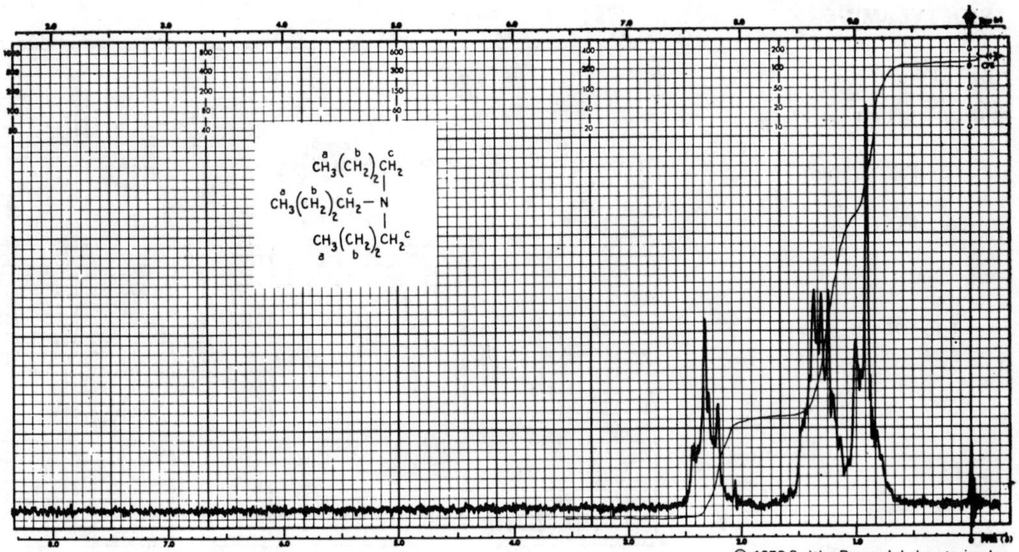

© 1978 Sadtler Research Laboratories, Inc.

TRIPENTYLAMINE

$C_{15}H_{33}N$

Mol. Wt. 227.44

B.P. 125-128°C/12mm

Solvent: CCl_4

Assignments
a = 0.90
b = 1.10-1.80
c = 2.40

618

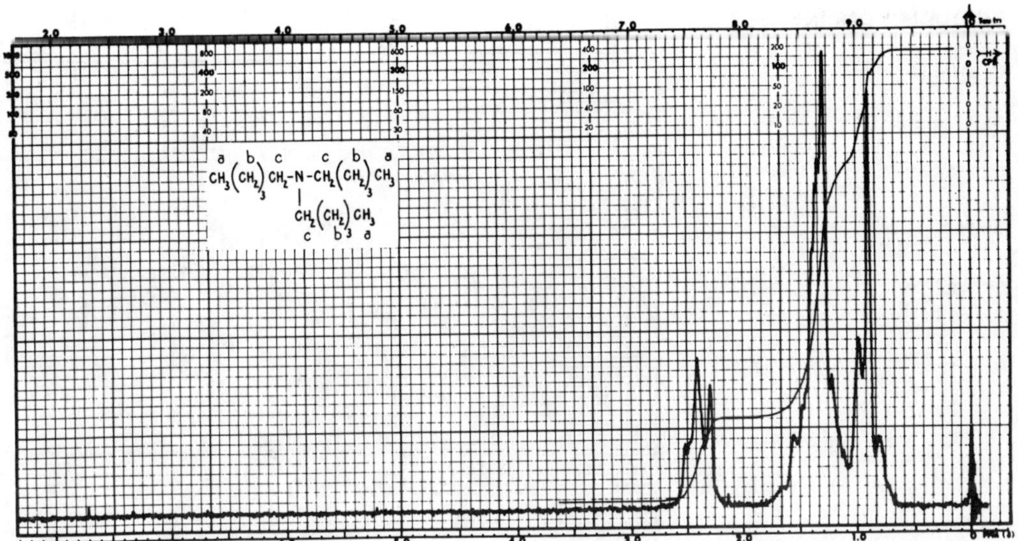

© 1978 Sadtler Research Laboratories, Inc.

TRIISOPENTYLAMINE

$C_{15}H_{33}N$

Mol. Wt. 227.44

B.P. 110-114°C/10mm

Solvent: $CDCl_3$

Assignments

a = 0.89

b = 1.10-1.88

c = ca 2.38

619

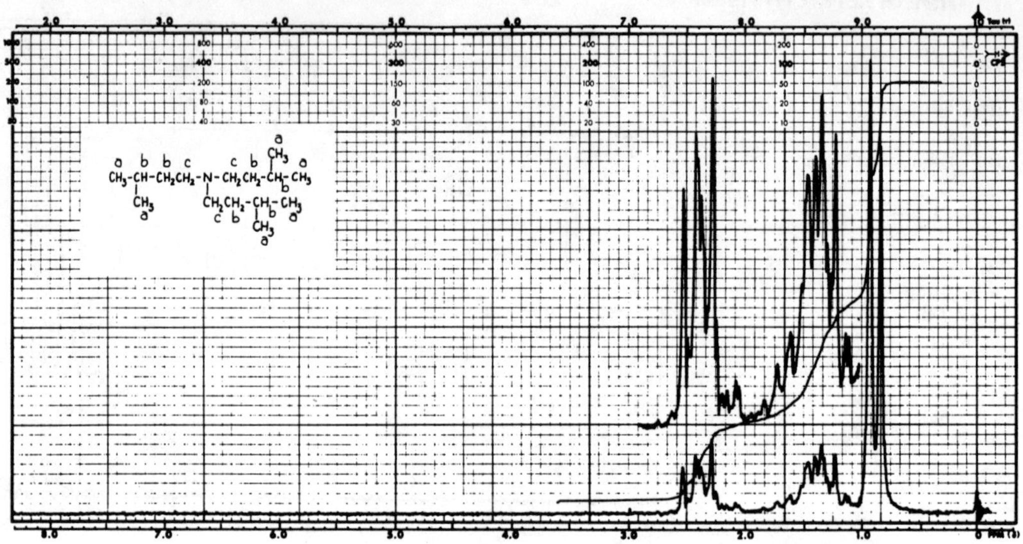

© 1978 Sadtler Research Laboratories, Inc.

TRIOCTYLAMINE

$C_{24}H_{51}N$

Mol. Wt. 353.69

B.P. 167-169°C/1mm

Solvent: CCl_4

Assignments

a = 0.90

b = 1.10-1.70

c = 2.29

620

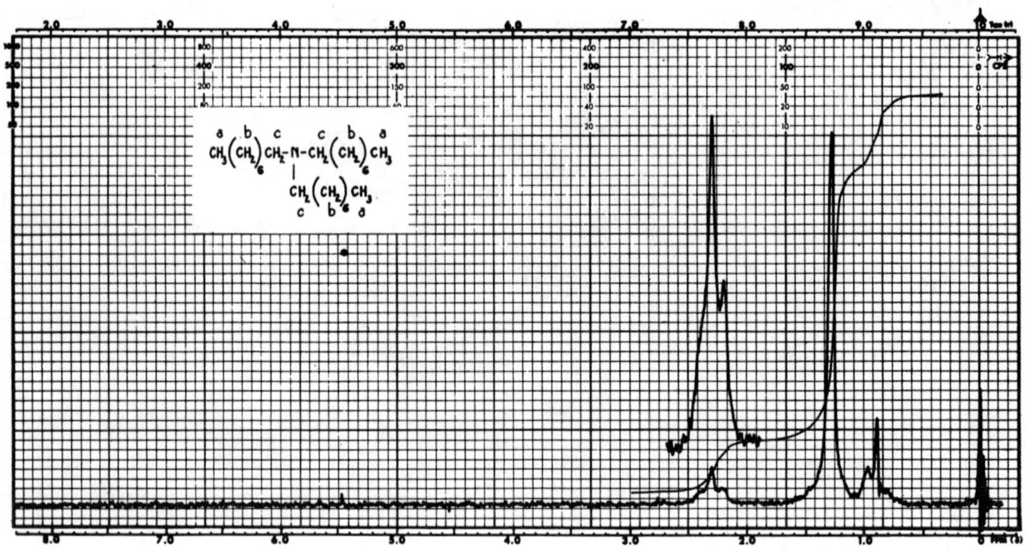

© 1978 Sadtler Research Laboratories, Inc.

TRIDODECYLAMINE

$C_{36}H_{75}N$

Mol. Wt. 522.01

M.P. 15-17°C

Solvent: $CDCl_3$

Assignments

a = 0.90

b = ca 1.29

c = 2.38

621

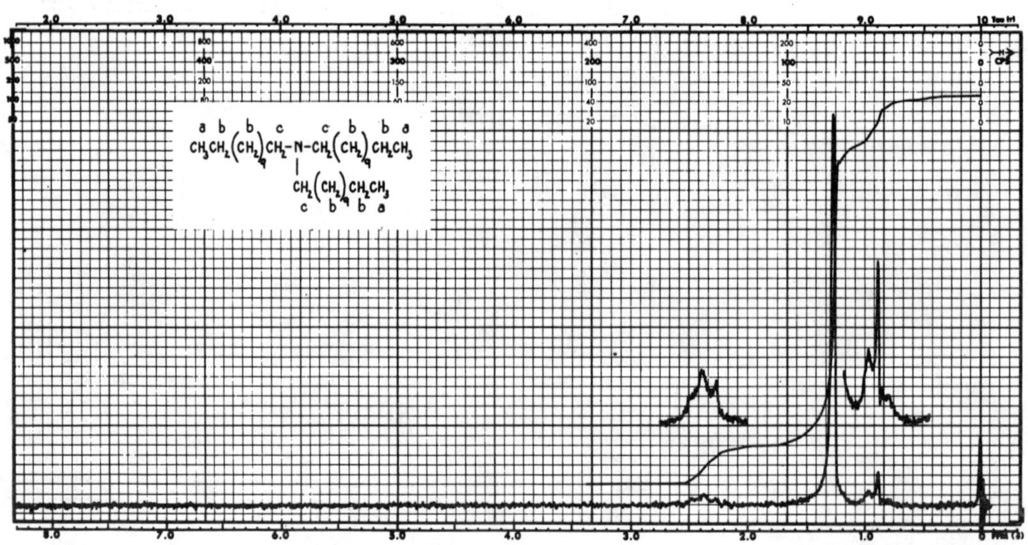

© 1978 Sadtler Research Laboratories, Inc.

N,N-DIMETHYLDECYLAMINE

$C_{12}H_{27}N$

Mol. Wt. 185.36

B.P. 103°C/10mm

Solvent: $CDCl_3$

Assignments
a = 0.89
b = ca 1.29
c = 2.21
d = 2.23

622

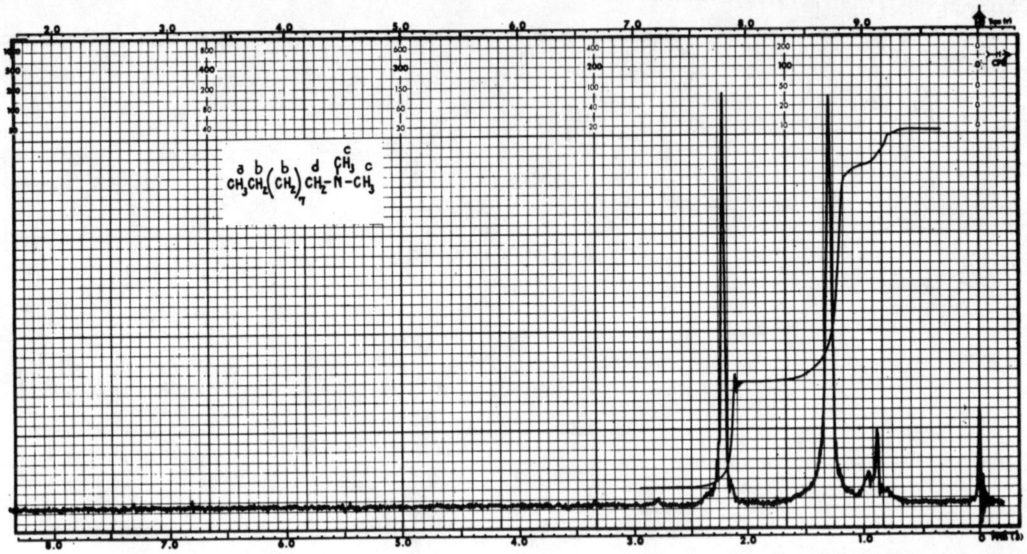

© 1978 Sadtler Research Laboratories, Inc.

N-METHYLDIOCTYLAMINE

$C_{17}H_{37}N$

Mol. Wt. 255.49

Solvent: $CDCl_3$

Assignments
a = 0.90
b = 1.10-1.80
c = 2.21
d = 2.31
e = 4.61 imp.

623

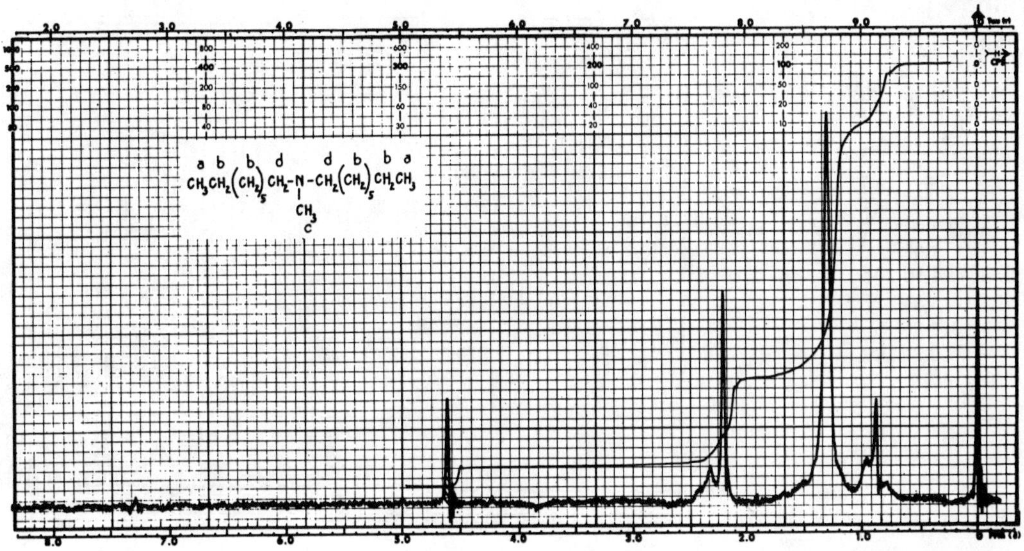

© 1978 Sadtler Research Laboratories, Inc.

N,N-DIMETHYLDOCOSYLAMINE

$C_{24}H_{51}N$

Mol. Wt. 353.68

Solvent: $CDCl_3$

Assignments
a = 0.89
b = ca 1.26
c = ca 2.21

624

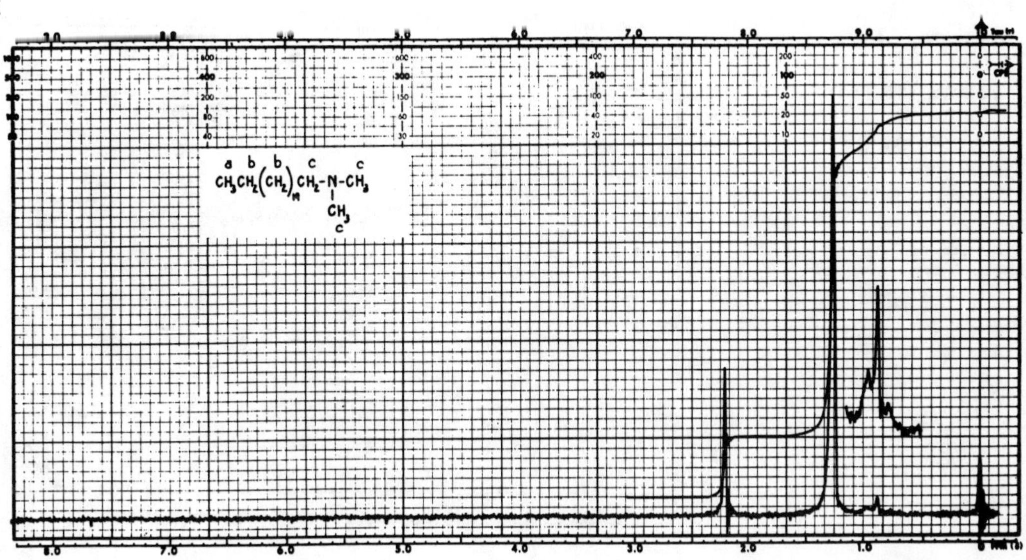

© 1978 Sadtler Research Laboratories, Inc.

N^2,N^2-DIMETHYL-1,2-PROPANEDIAMINE

$C_5H_{14}N_2$

Mol. Wt. 102.18

Solvent: D_2O

Assignments
a = 1.00
b = 2.19
c = 2.29-2.95
d = 4.69 HDO

625

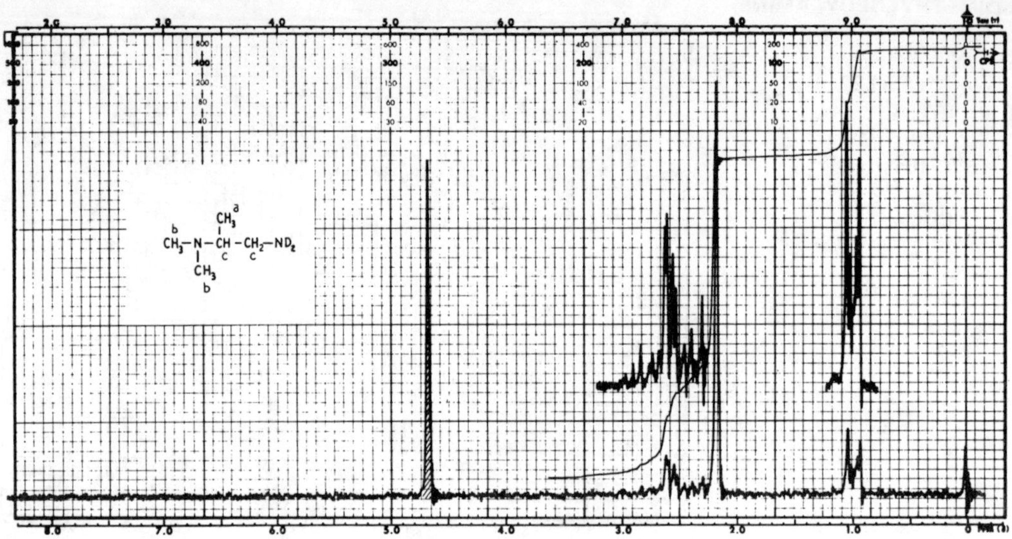

© 1978 Sadtler Research Laboratories, Inc.

N^4,N^4-DIETHYL-1-METHYL-1,4-BUTANEDIAMINE

$C_9H_{22}N_2$

Mol. Wt. 158.29

Solvent: $CDCl_3$

Assignments
a = 1.00
b = 1.05
c = 1.10-1.70
d = ca 1.39
e = ca 2.50
f = 2.88

626

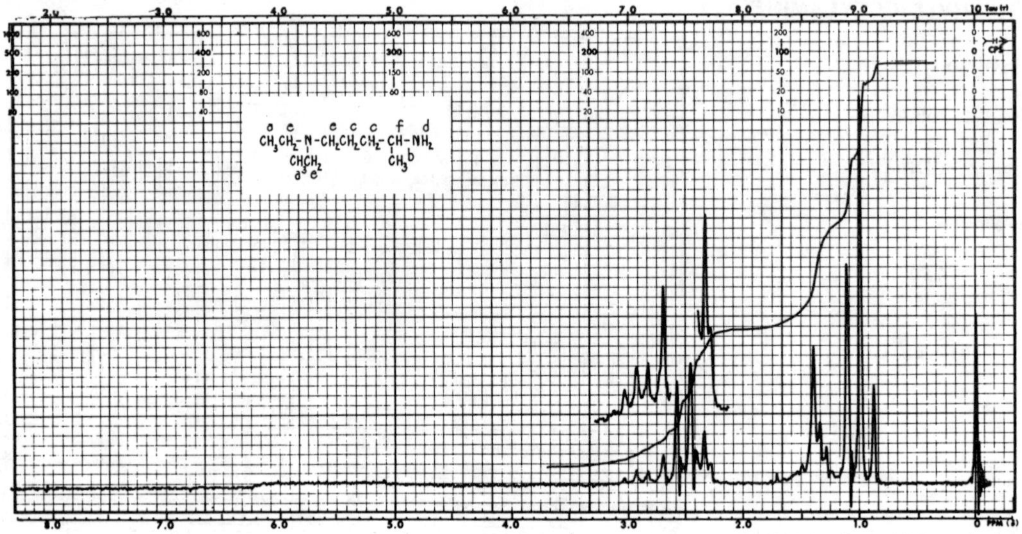

© 1978 Sadtler Research Laboratories, Inc.

3,3′-(METHYLIMINO)DIPROPYLAMINE

$C_7H_{19}N_3$

Mol. Wt. 145.25

B.P. 190-194°C

Solvent: $CDCl_3$

Assignments
a = 1.60
b = 1.69
c = 2.20
d = 2.40
e = 2.71

627

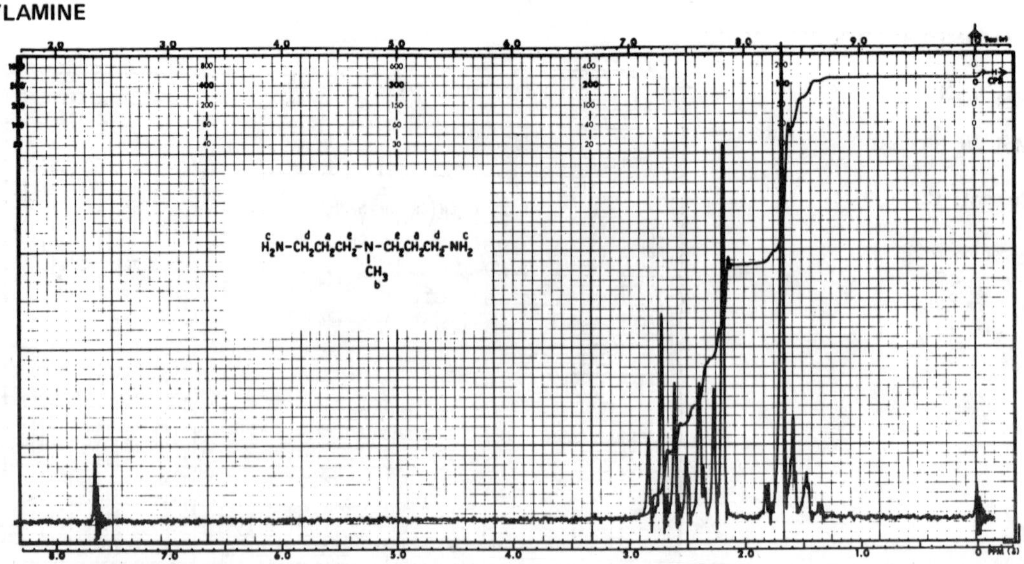

© 1978 Sadtler Research Laboratories, Inc.

1,1-DIMETHYLDIETHYLENETRIAMINE

$C_6H_{17}N_3$

Mol. Wt. 131.22

B.P. 199-201°C

Solvent: D_2O

Assignments
a = 2.19
b = 2.30-2.81
c = 4.61 HDO

628

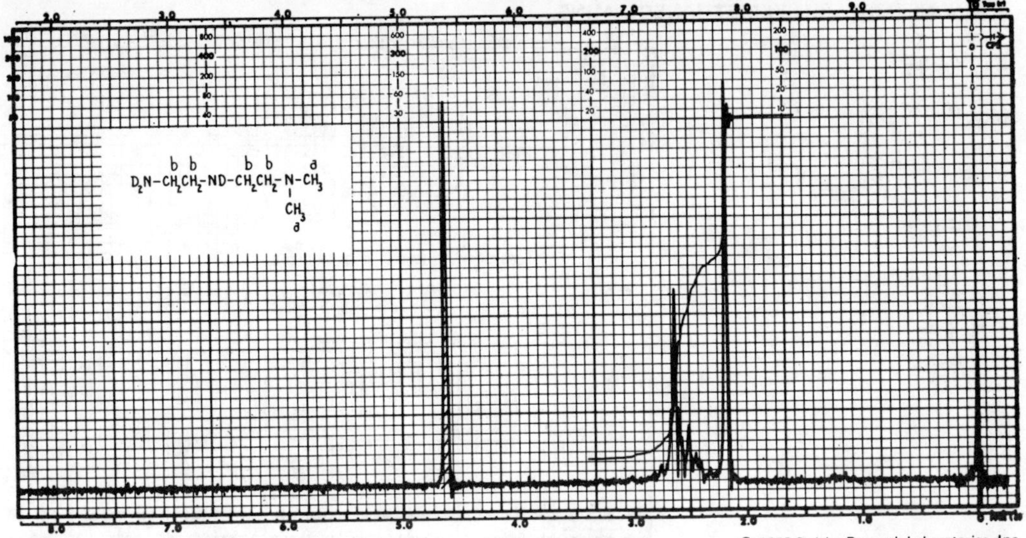

© 1978 Sadtler Research Laboratories, Inc.

N,N-DIETHYL-N'-(1-METHYLDODECYL)-1,3-PROPANEDIAMINE

$C_{20}H_{44}N_2$

Mol. Wt. 312.59

B.P. 165-170°C/1mm

Solvent: $CDCl_3$

Assignments
a = 0.90
b = 1.00
c = 1.10-1.90
d = ca 2.50

629

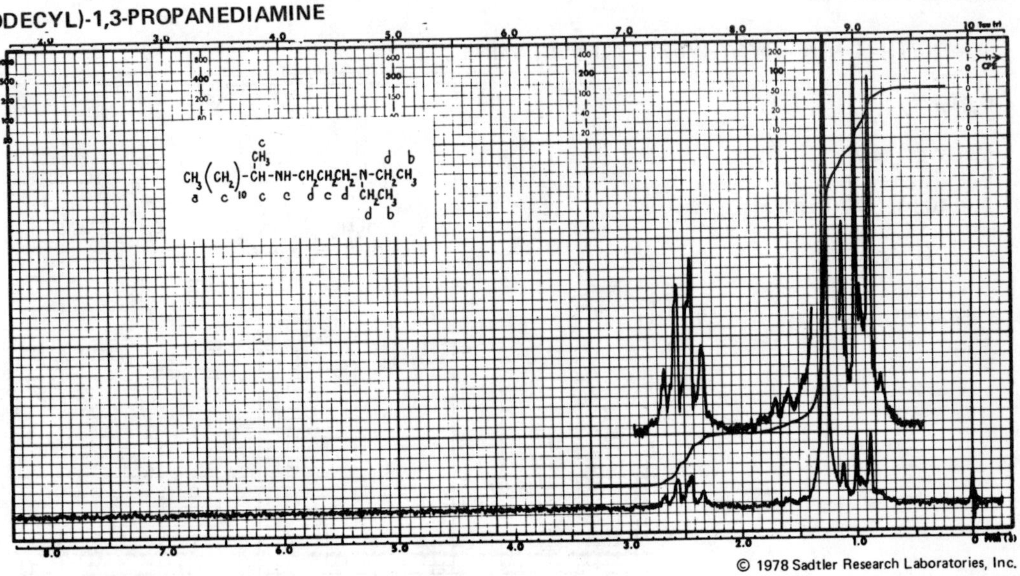

© 1978 Sadtler Research Laboratories, Inc.

N,N'-METHYLENEBIS(DIMETHYLAMINE)

$C_5H_{14}N_2$

Mol. Wt. 102.18

Solvent: $CDCl_3$

Assignments
a = 2.22
b = 2.72

630

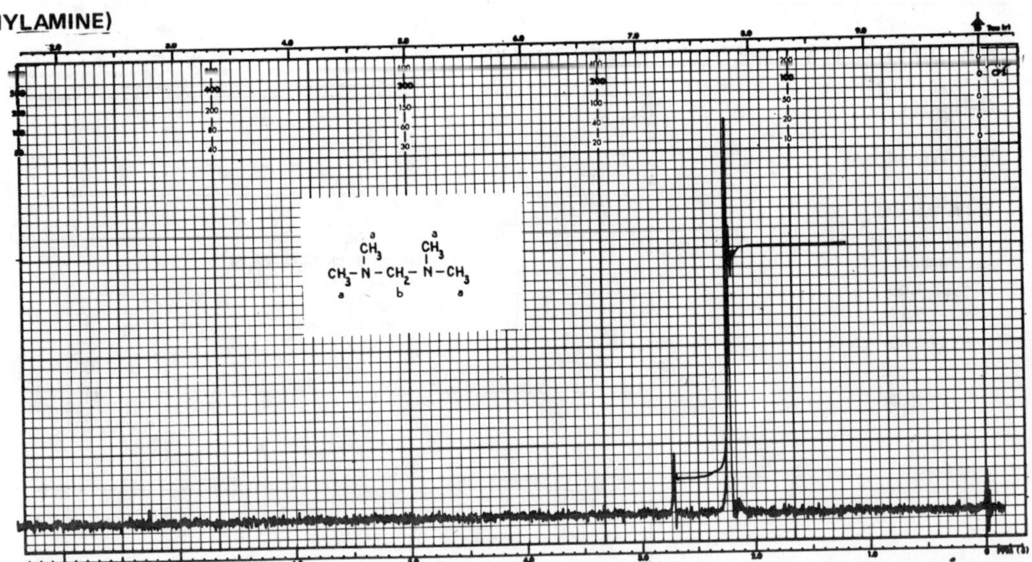

© 1978 Sadtler Research Laboratories, Inc.

N,N,N',N'-TETRABUTYLMETHANEDIAMINE

$C_{17}H_{38}N_2$

Mol. Wt. 270.51

Solvent: CCl_4

Assignments
a = 0.93
b = 1.12-1.70
c = 2.42
d = 2.96

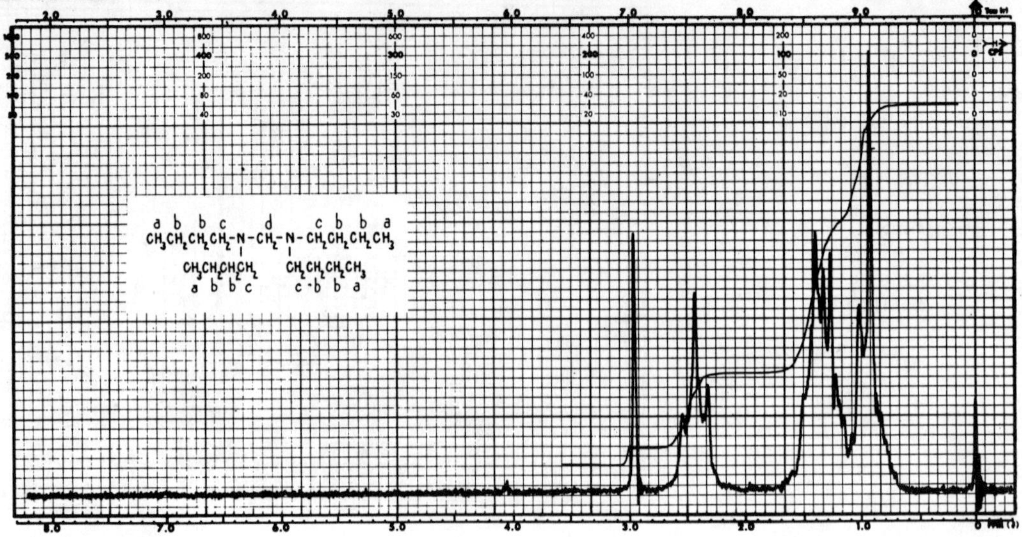

631

© 1978 Sadtler Research Laboratories, Inc.

N,N-DIETHYL-N',N'-DIMETHYLETHYLENEDIAMINE

$C_8H_{20}N_2$

Mol. Wt. 144.26

B.P. 163-165°C (lit.)

Solvent: CCl_4

Assignments
a = 0.97
b = 2.12
c = ca 2.40

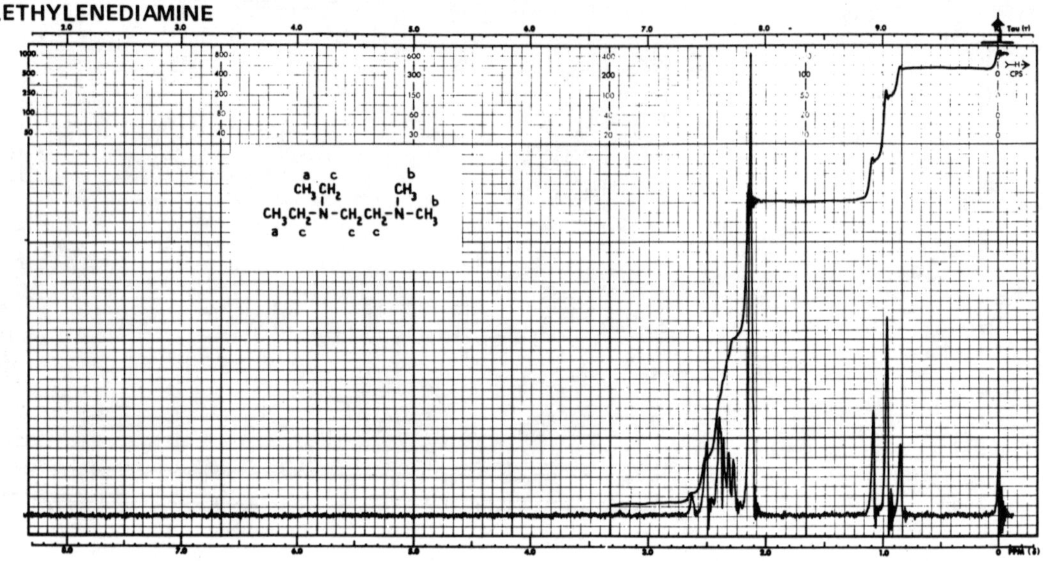

632

© 1978 Sadtler Research Laboratories, Inc.

N,N'-DIETHYL-N,N'-DIMETHYLETHYLENEDIAMINE

$C_8H_{20}N_2$

Mol. Wt. 144.26

Solvent: CCl_4

Assignments
a = 1.00
b = 2.12
c = ca 2.35

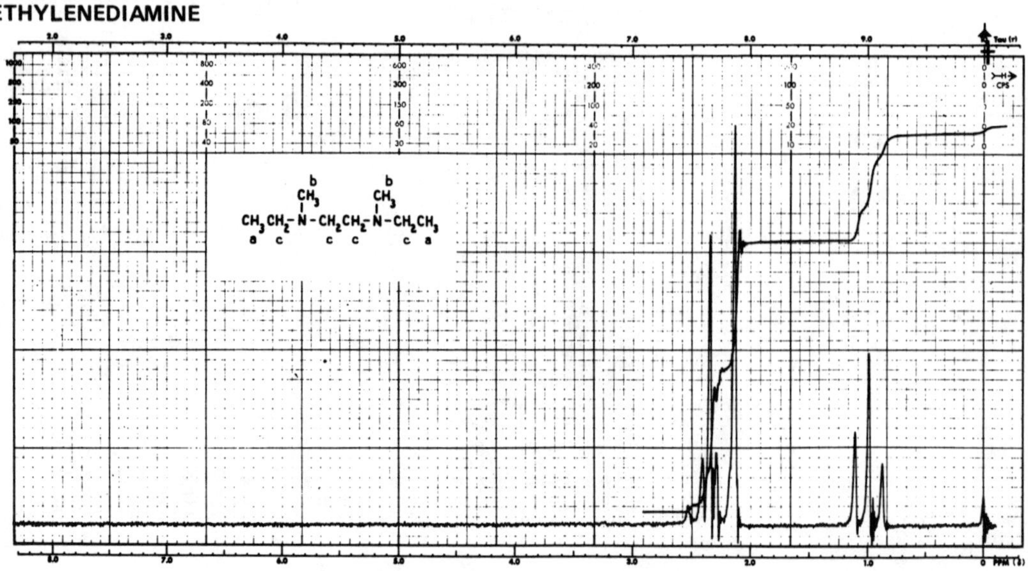

633

© 1978 Sadtler Research Laboratories, Inc.

N,N,N',N'-TETRAMETHYL-1,3-PROPANEDIAMINE

$C_7H_{18}N_2$

Mol. Wt. 130.24

Solvent: $CDCl_3$

Assignments
a = 1.67
b = 2.20
c = ca 2.29

634

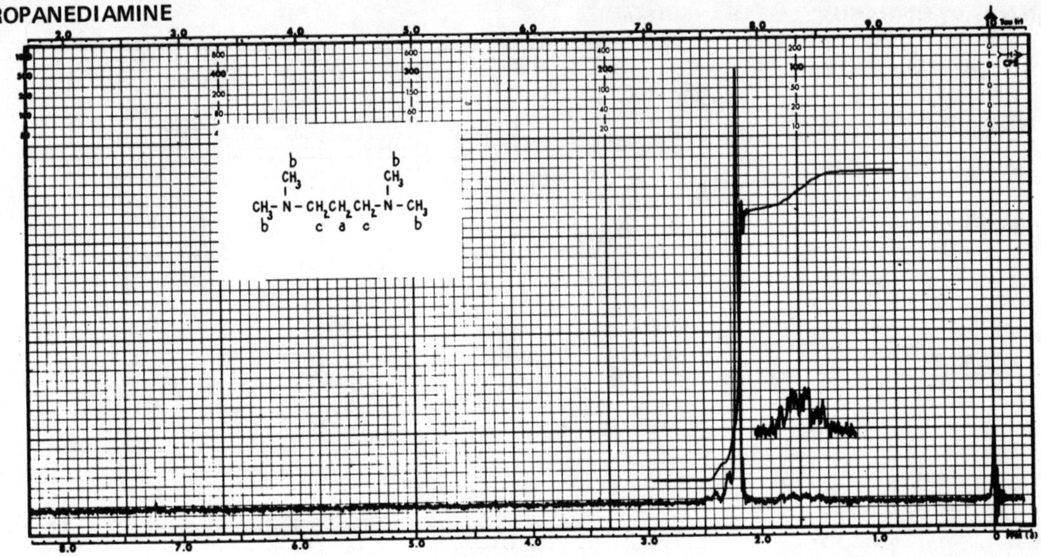

© 1978 Sadtler Research Laboratories, Inc.

N,N-DIETHYLCYCLOHEXYLAMINE

$C_{10}H_{21}N$

Mol. Wt. 155.29

Solvent: CCl_4

Assignments
a = 1.00
b = 1.00-2.00
c = ca 2.39
d = 2.48

635

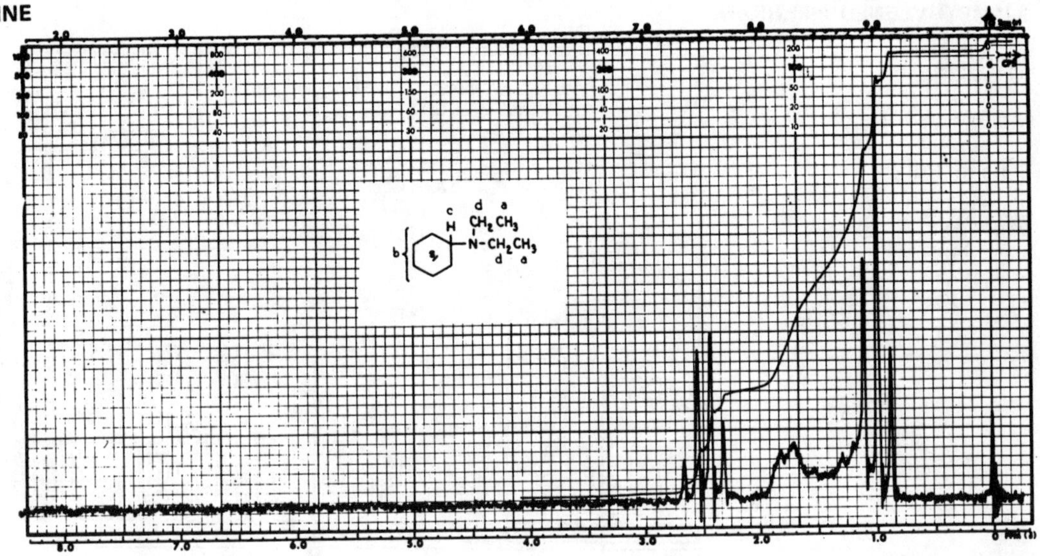

© 1978 Sadtler Research Laboratories, Inc.

1-METHYLPYRROLIDINE

$C_5H_{11}N$

Mol. Wt. 85.15

Solvent: $CDCl_3$

Assignments
a = 1.78
b = 2.34
c = 2.48
d = 3.03 imp.

636

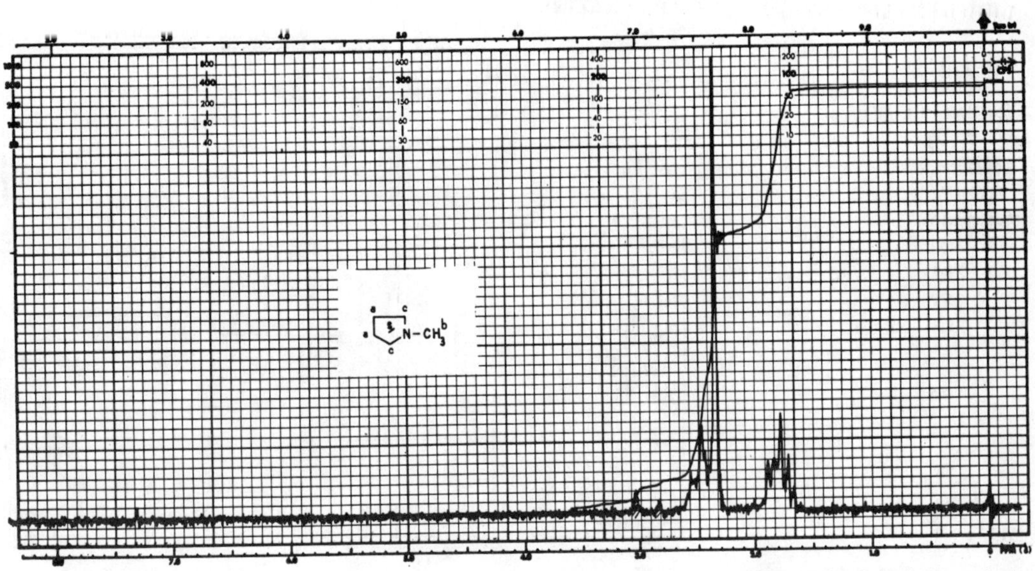

© 1978 Sadtler Research Laboratories, Inc.

1-METHYLPIPERIDINE

$C_6H_{13}N$

Mol. Wt. 99.18

B.P. 100-106°C

Solvent: CCl_4

Assignments
a = ca 1.42
b = 2.10
c = ca 2.23

637

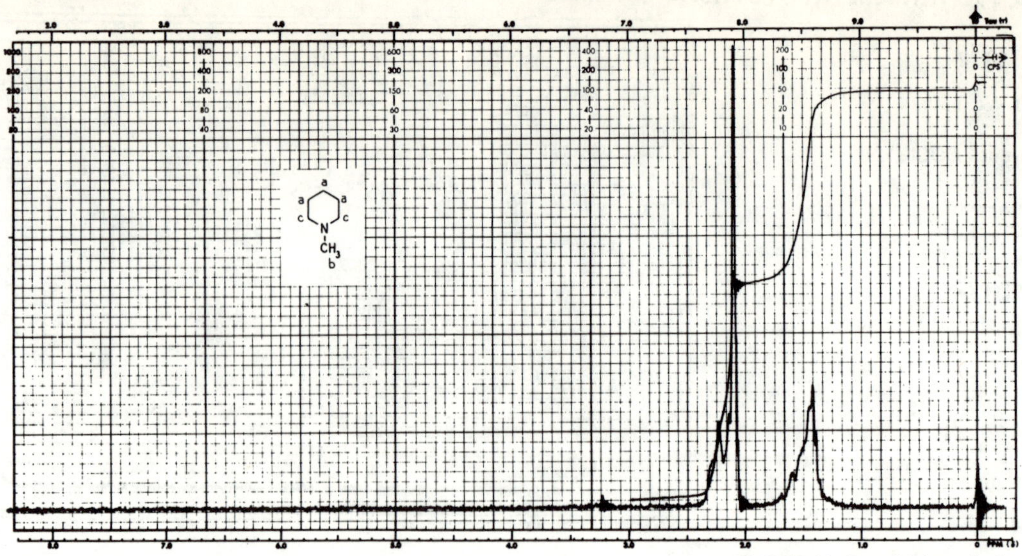

© 1978 Sadtler Research Laboratories, Inc.

1,1'-METHYLENEDIPIPERIDINE

$C_{11}H_{22}N_2$

Mol. Wt. 182.31

Solvent: CCl_4

Assignments
a = 1.20-1.80
b = ca 2.39
c = 2.71

638

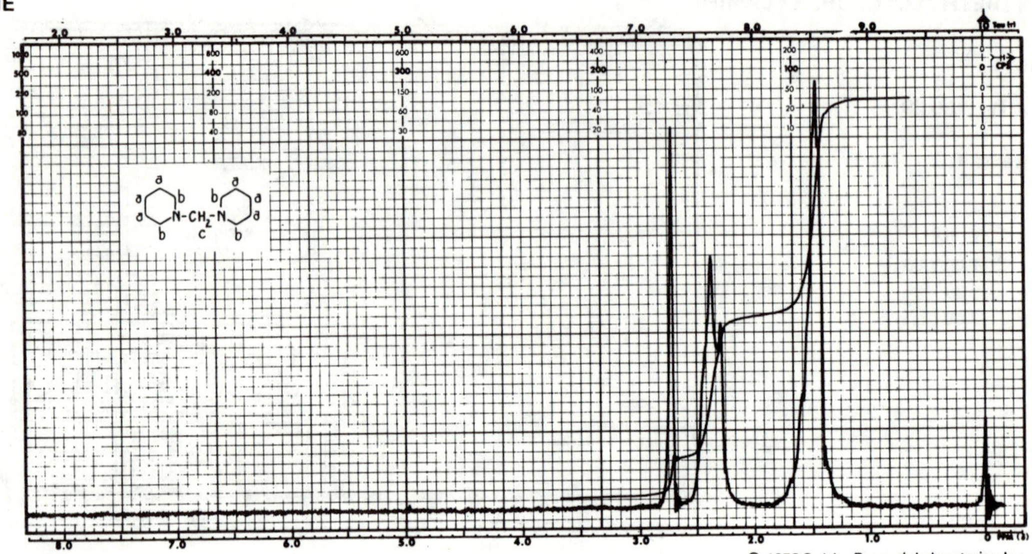

© 1978 Sadtler Research Laboratories, Inc.

1-(DIPHENYLMETHYL)-4-METHYLPIPERAZINE

$C_{18}H_{22}N_2$

Mol. Wt. 266.39

Solvent: $CDCl_3$

Assignments
a = 2.26
b = 2.42
c = 4.22
d = 7.03-7.53

639

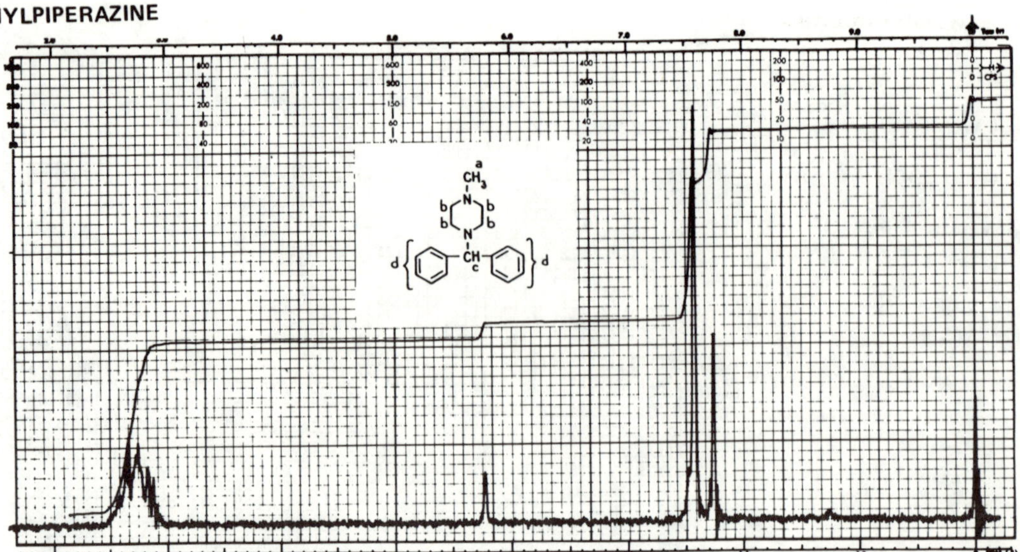

© 1978 Sadtler Research Laboratories, Inc.

1-PHENYLPIPERAZINE

$C_{10}H_{14}N_2$

Mol. Wt. 162.23

B.P. 156-157°C/10mm

Solvent: $CDCl_3$

Assignments
a = 1.78
b = 3.08
c = 6.88
d = 7.05
e = 7.18

640

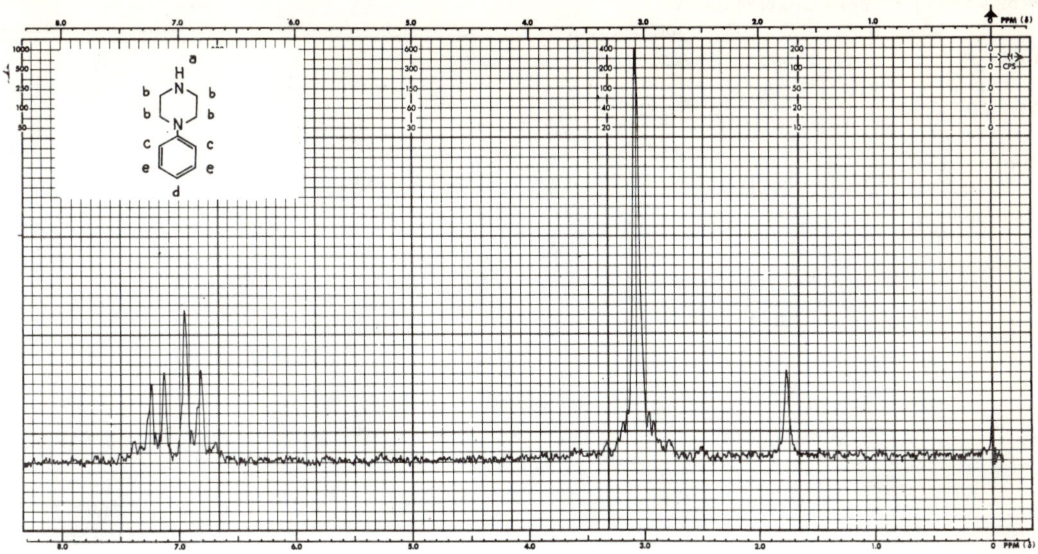

© 1978 Sadtler Research Laboratories, Inc.

2-METHYL-1-PHENYLPIPERAZINE

$C_{11}H_{16}N_2$

Mol. Wt. 176.26

Solvent: $CDCl_3$

Assignments
a = 1.01
b = 1.90
c = 2.60-3.26
d = ca 3.01
e = 3.67
f = 6.82
g = 6.89
h = 7.21
i = 2.07, 4.58 imps.

641

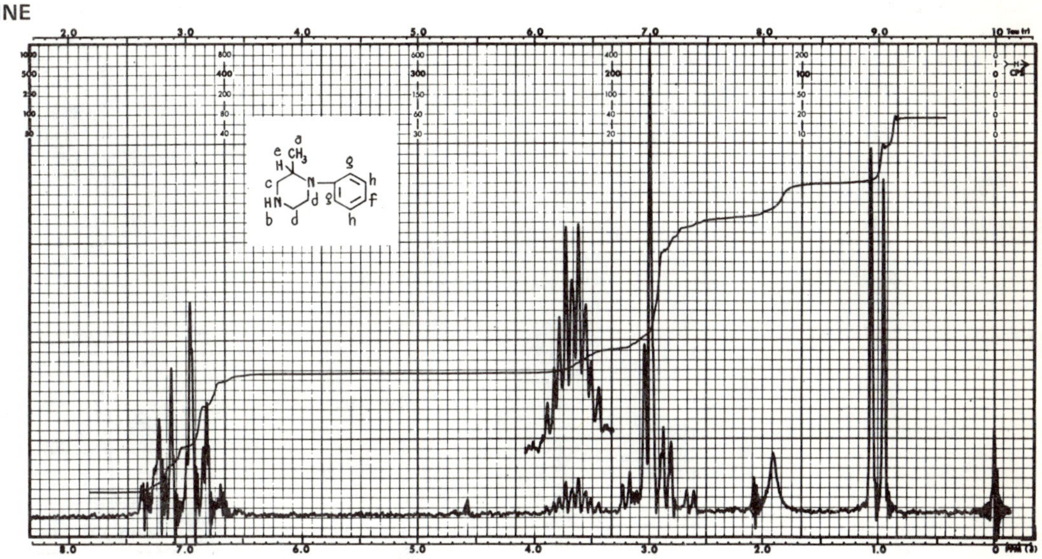

© 1978 Sadtler Research Laboratories, Inc.

N,N-DIMETHYLALLYLAMINE

$C_5H_{11}N$

Mol. Wt. 85.15

B.P. 63-64°C

Solvent: $CDCl_3$

Assignments
a = 2.21
b = 2.93
c = 5.11
d = 5.16
e = 5.82

642

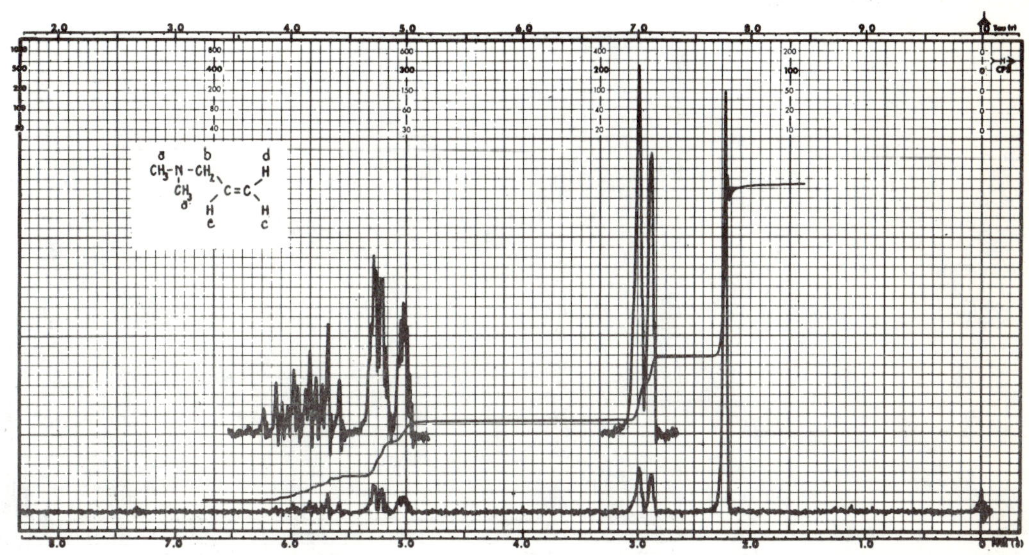

© 1978 Sadtler Research Laboratories, Inc.

N,N,2-TRIMETHYLPROPENYLAMINE

$C_6H_{13}N$

Mol. Wt. 99.18

Solvent: $CDCl_3$

Assignments
a = 1.65
b = 2.39
c = 5.31
d = 1.12, 2.43 imps.

643

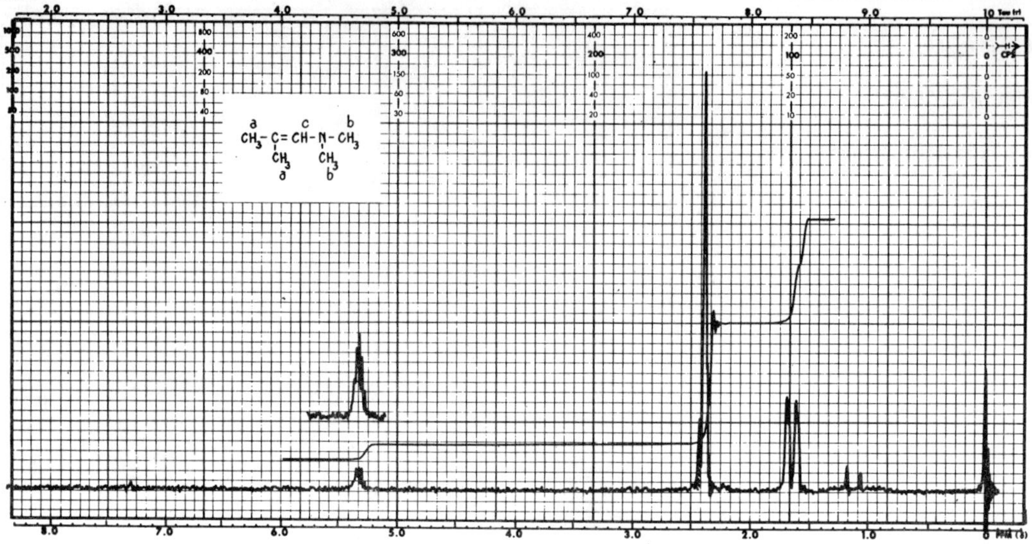

© 1978 Sadtler Research Laboratories, Inc.

N,N-DIMETHYLBENZYLAMINE

$C_9H_{13}N$

Mol. Wt. 135.21

B.P. 73-74°C/15mm (lit.)

Solvent: CCl_4

Assignments
a = 2.14
b = 3.33
c = 7.19

644

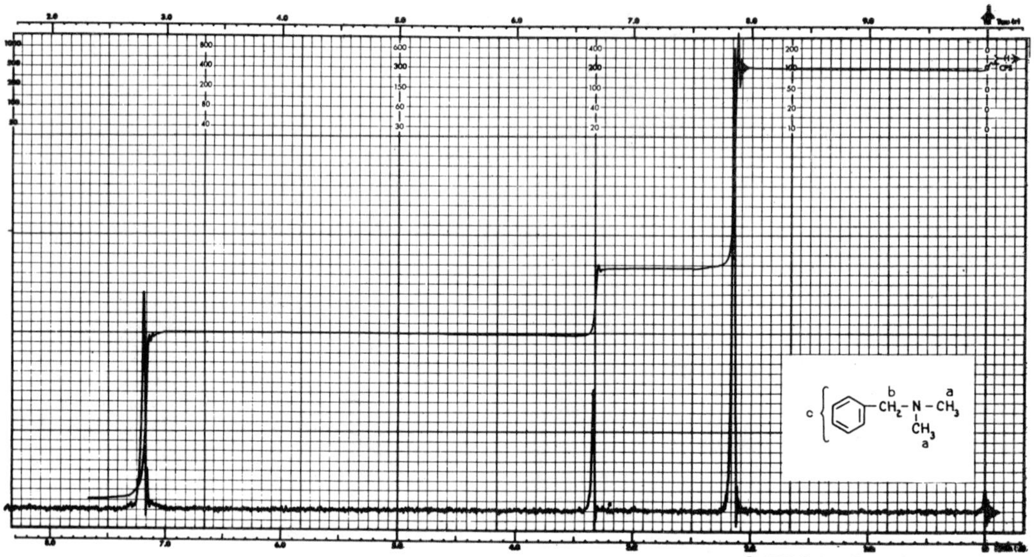

© 1978 Sadtler Research Laboratories, Inc.

N,N-DIETHYLBENZYLAMINE

$C_{11}H_{17}N$

Mol. Wt. 163.25

B.P. 210-212°C

Solvent: $CDCl_3$

Assignments
a = 1.02
b = 2.52
c = 3.56
d = 7.30

645

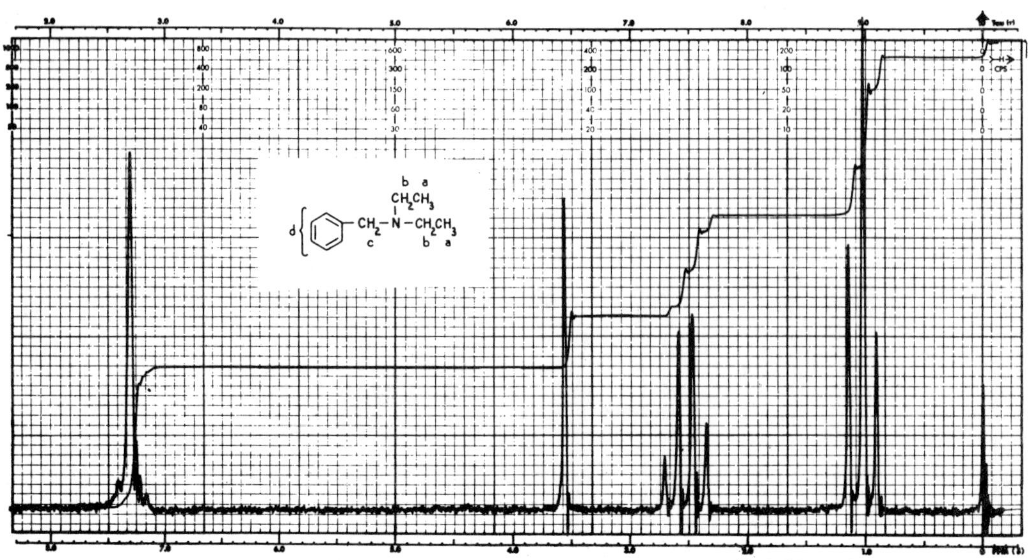

© 1978 Sadtler Research Laboratories, Inc.

N-ETHYLDIBENZYLAMINE

$C_{16}H_{19}N$

Mol. Wt. 225.34

B.P. 155-156°C/9mm

Solvent: $CDCl_3$

Assignments
a = 1.02
b = 2.49
c = 3.56
d = 7.10-7.50

646

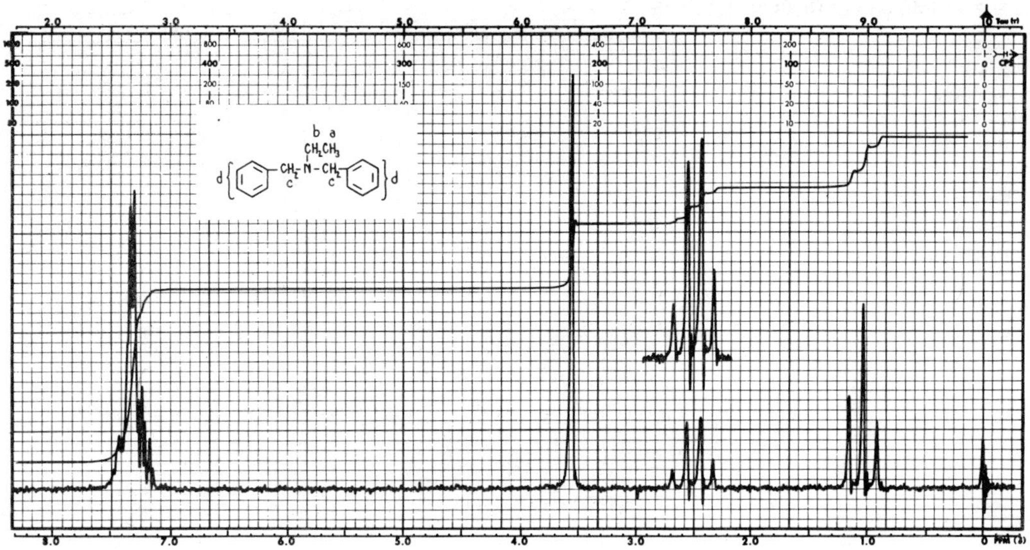

© 1978 Sadtler Research Laboratories, Inc.

TRIBENZYLAMINE

$C_{21}H_{21}N$

Mol. Wt. 287.41

Solvent: CCl_4

Assignments
a = 3.51
b = 7.27

647

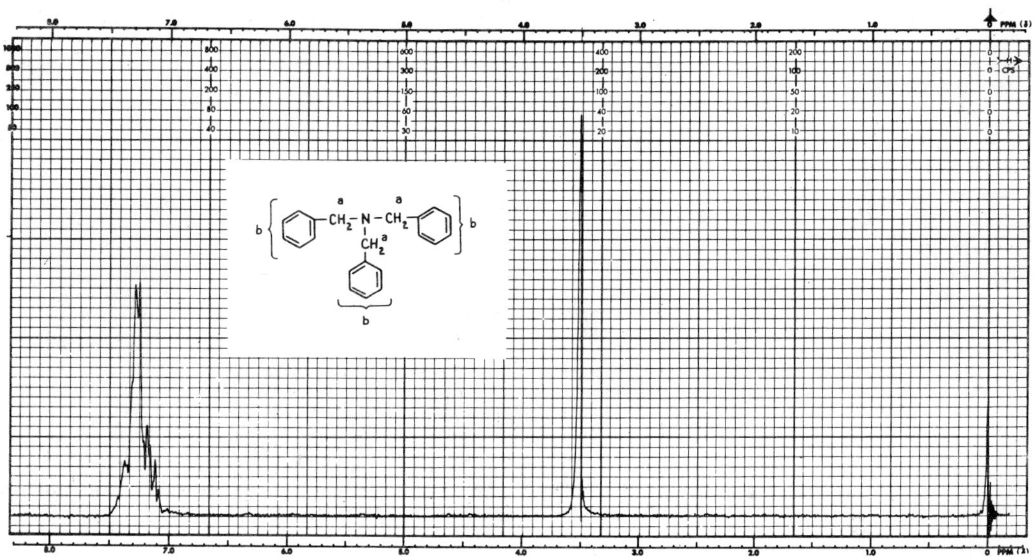

© 1978 Sadtler Research Laboratories, Inc.

N,N-DIMETHYLANILINE

$C_8H_{11}N$

Mol. Wt. 121.18

M.P. 2.5°C B.P. 192.5°C

Solvent: CCl_4

Assignments
a = 2.85
b = 6.42-6.72
c = 7.10

648

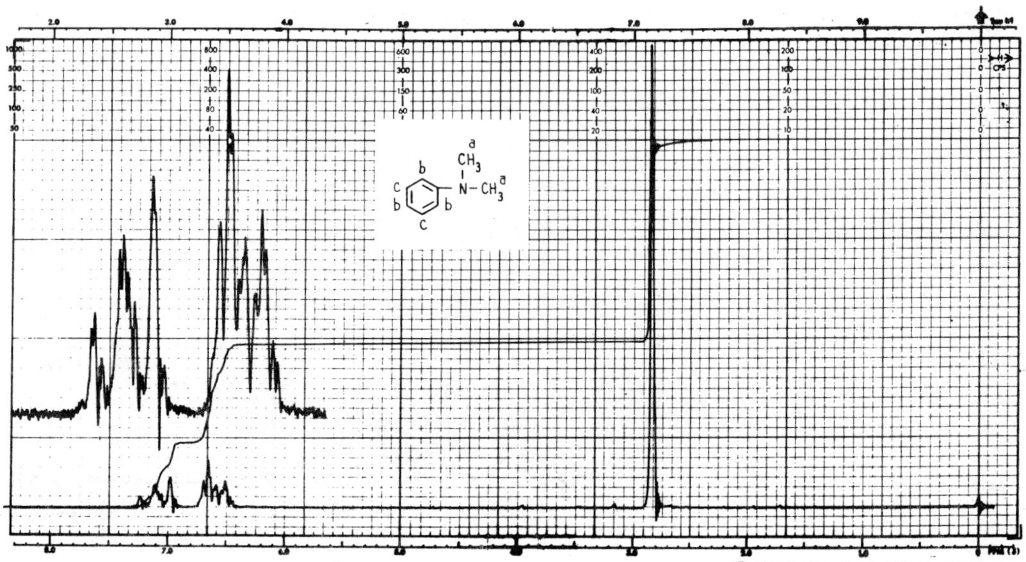

© 1978 Sadtler Research Laboratories, Inc.

N,N-DIETHYLANILINE

$C_{10}H_{15}N$

Mol. Wt. 149.24

B.P. 93-94°C/10mm (lit.)

Solvent: CCl_4

Assignments
a = 1.10
b = 3.27
c = ca 6.55
d = ca 7.05

649

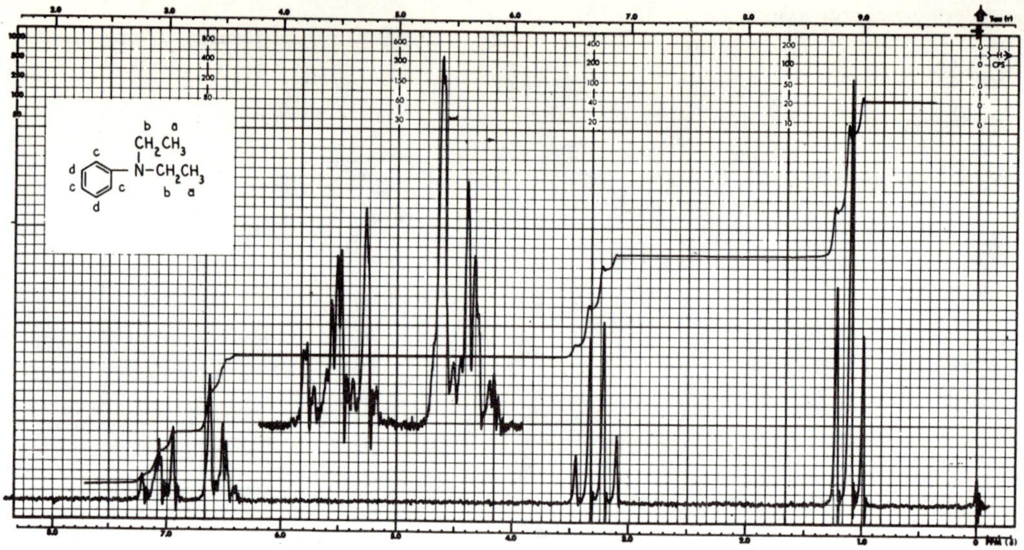

© 1978 Sadtler Research Laboratories, Inc.

N,N-DIPROPYLANILINE

$C_{12}H_{19}N$

Mol. Wt. 177.29

B.P. 116-118°C/12mm

Solvent: CCl_4

Assignments
a = 0.99
b = 1.58
c = 3.17
d = 6.36-7.20

650

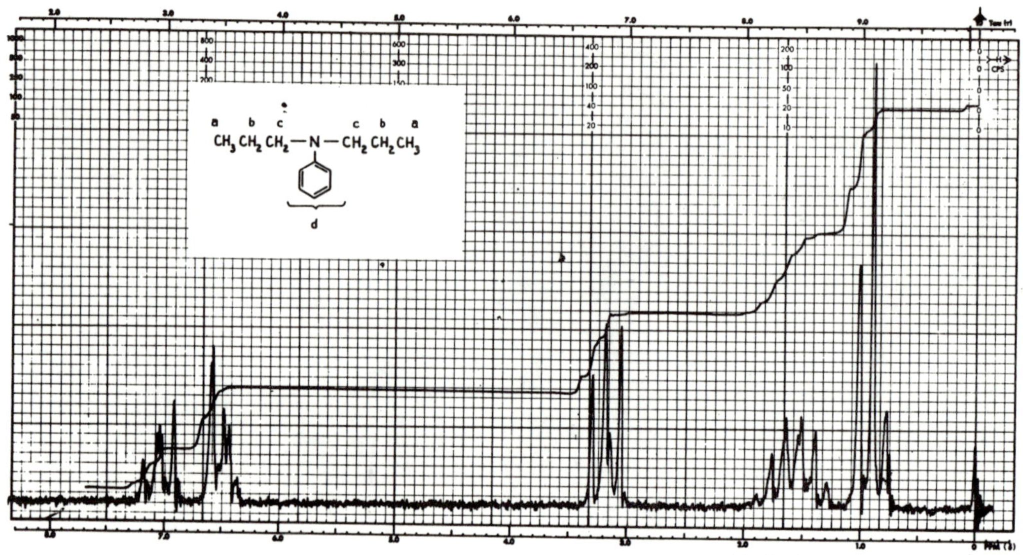

© 1978 Sadtler Research Laboratories, Inc.

N,N-DIBUTYLANILINE

$C_{14}H_{24}N$

Mol. Wt. 205.35

B.P. 268-275°C (lit.)

Solvent: CCl_4

Assignments
a = 0.95
b = 1.10-1.85
c = 3.22
d = 6.30-7.25

651

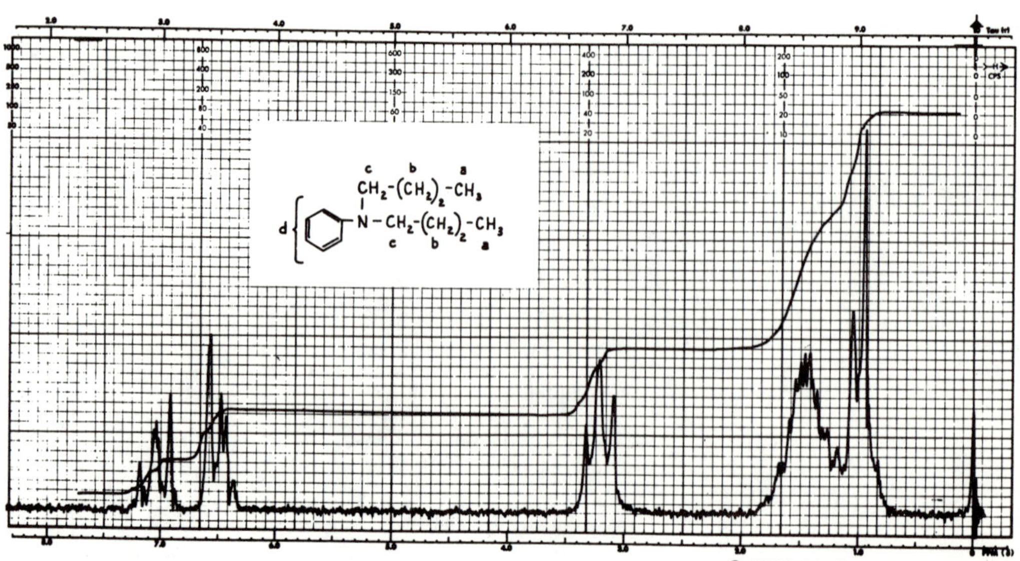

© 1978 Sadtler Research Laboratories, Inc.

N-METHYL-N-PHENYLBENZYLAMINE

$C_{14}H_{15}N$

Mol. Wt. 197.28

Solvent: CCl_4

Assignments
a = 2.90
b = 4.41
c = 6.45-6.75
d = 7.03
e = 7.12

652

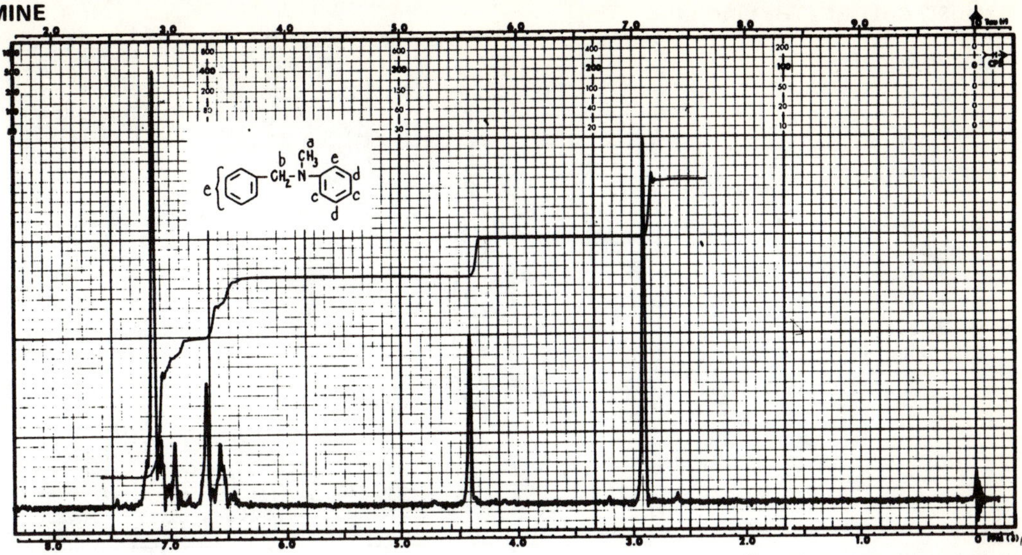

© 1978 Sadtler Research Laboratories, Inc.

N-ETHYL-N-PHENYLBENZYLAMINE

$C_{15}H_{17}N$

Mol. Wt. 211.31

Solvent: $CDCl_3$

Assignments
a = 1.17
b = 3.43
c = 4.48
d = 6.69
e = 7.13
f = 7.22

653

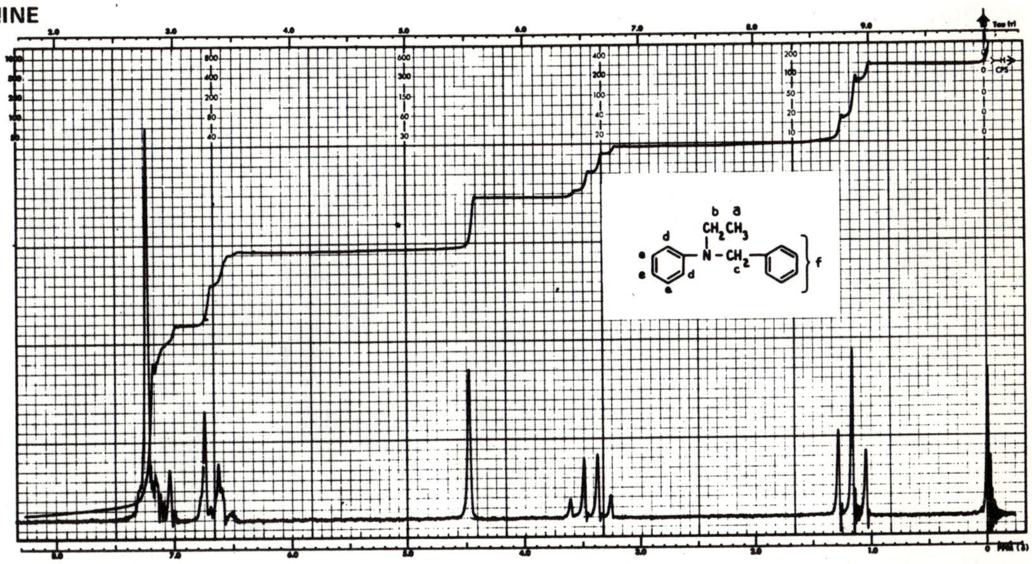

© 1978 Sadtler Research Laboratories, Inc.

N-PHENYLDIBENZYLAMINE

$C_{20}N_{19}N$

Mol. Wt. 273.38

M.P. 68-70°C

Solvent: $CDCl_3$

Assignments
a = 4.59
b = 6.50-6.80
c = 7.11
d = ca 7.21

654

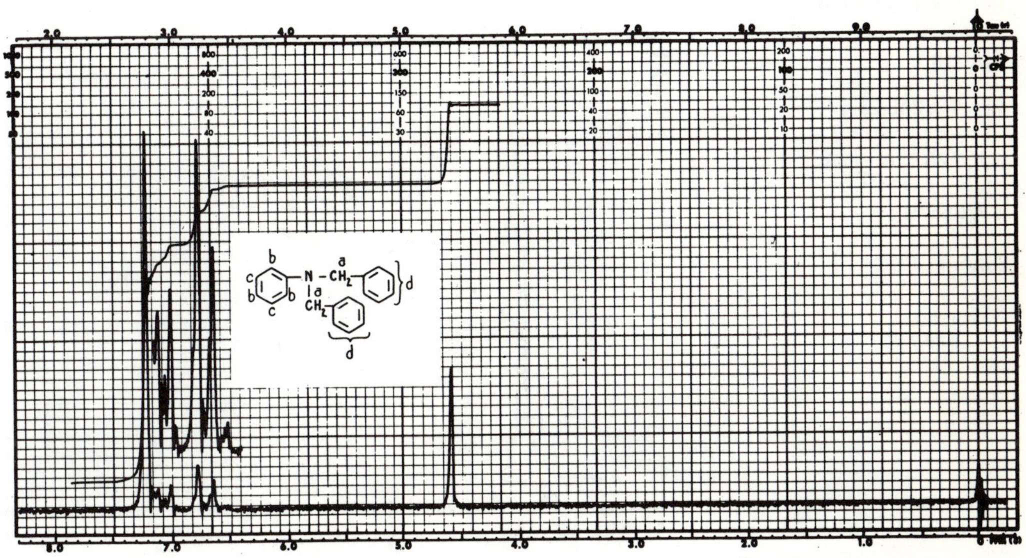

© 1978 Sadtler Research Laboratories, Inc.

N,N-BIS(2-CHLOROETHYL)ANILINE

$C_{10}H_{13}Cl_2N$

Mol. Wt. 218.13

Solvent: Polysol-d

Assignments
a = ca 3.61
b = 6.63
c = 6.70
d = 7.19

655

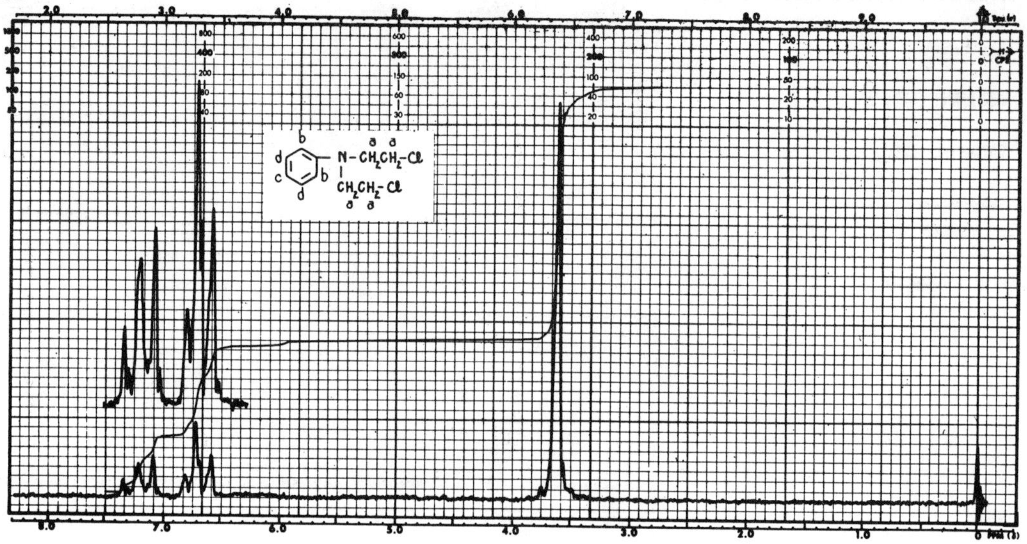

© 1978 Sadtler Research Laboratories, Inc.

N,N-DIALLYLANILINE

$C_{12}H_{15}N$

Mol. Wt. 173.26

Solvent: CCl_4

Assignments
a = 3.85
b = ca 5.10
c = 5.77
d = 6.57
e = 6.61
f = 7.07

656

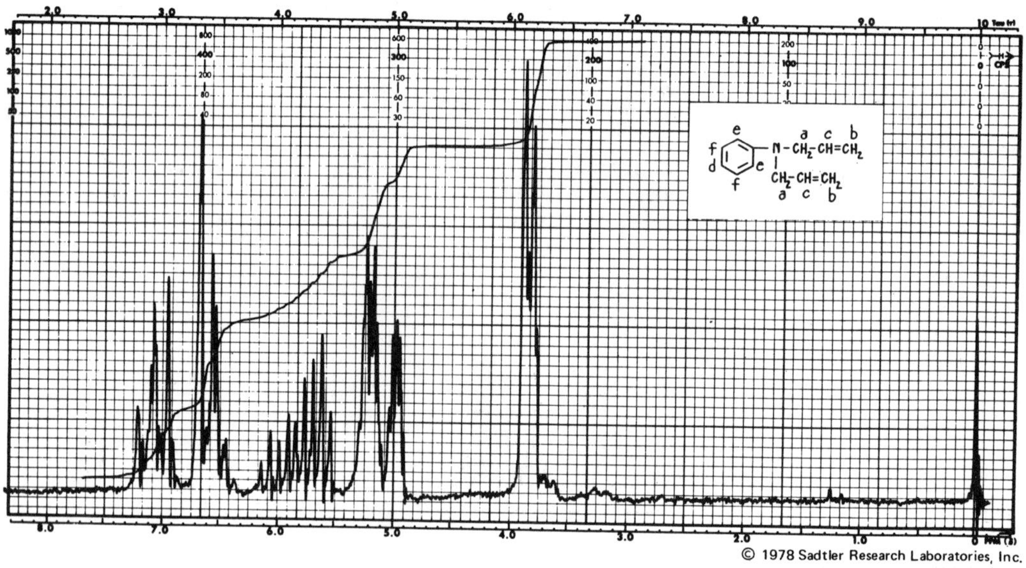

© 1978 Sadtler Research Laboratories, Inc.

TRIPHENYLAMINE

$C_{18}H_{15}N$

Mol. Wt. 245.33

M.P. 126.5°C

B.P. 365°C

Solvent: $CDCl_3$

Assignments
a = 6.80-7.40

657

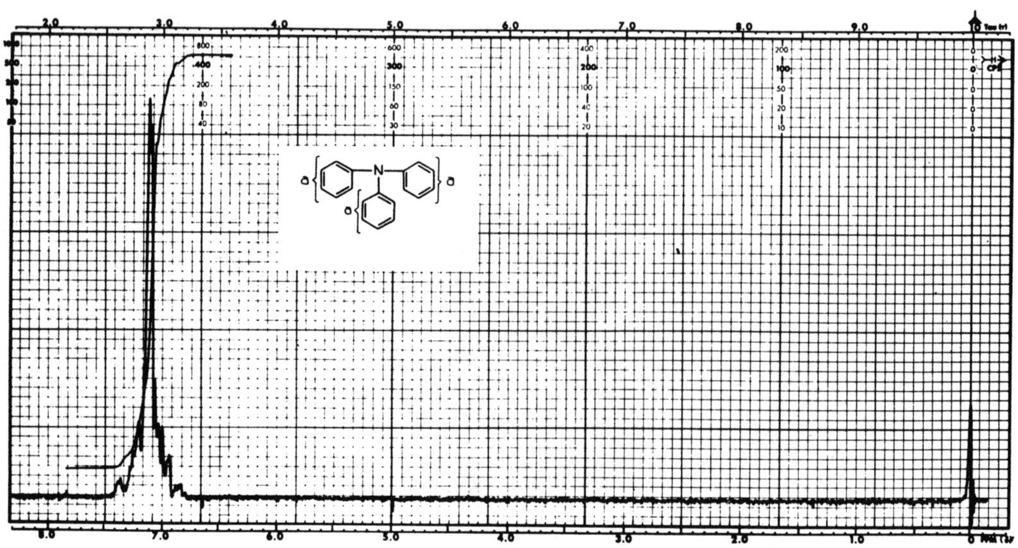

© 1978 Sadtler Research Laboratories, Inc.

N,N-DIMETHYL-o-TOLUIDINE

$C_9H_{13}N$

Mol. Wt. 135.21

Solvent: CCl_4

Assignments
a = 2.28
b = 2.62
c = 6.70-7.18

658

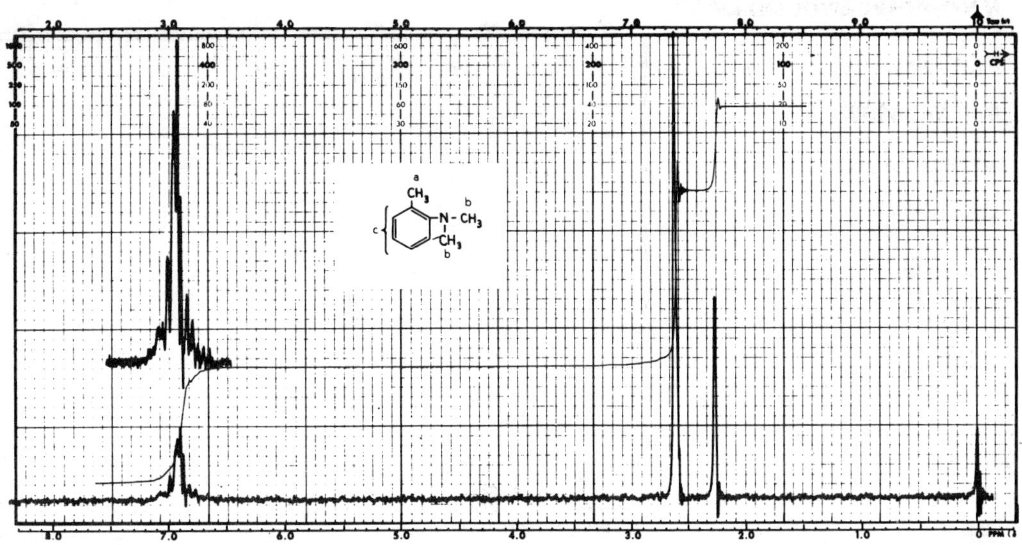

© 1978 Sadtler Research Laboratories, Inc.

N,N-DIMETHYL-m-TOLUIDINE

$C_9H_{13}N$

Mol. Wt. 135.21

Solvent: CCl_4

Assignments
a = 2.25
b = 2.79
c = ca 6.41
d = 6.99

659

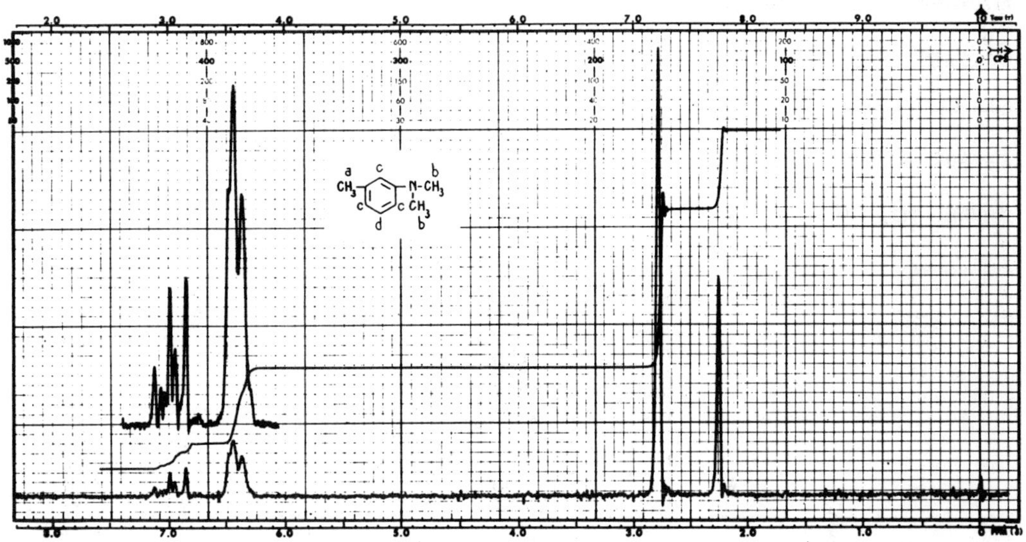

© 1978 Sadtler Research Laboratories, Inc.

N,N-DIMETHYL-p-TOLUIDINE

$C_9H_{13}N$

Mol. Wt. 135.21

B.P. 71-73°C/5mm

Solvent: CCl_4

Assignments
a = 2.19
b = 2.79
c = 6.49
d = 6.89
e = 2.57 imp.

660

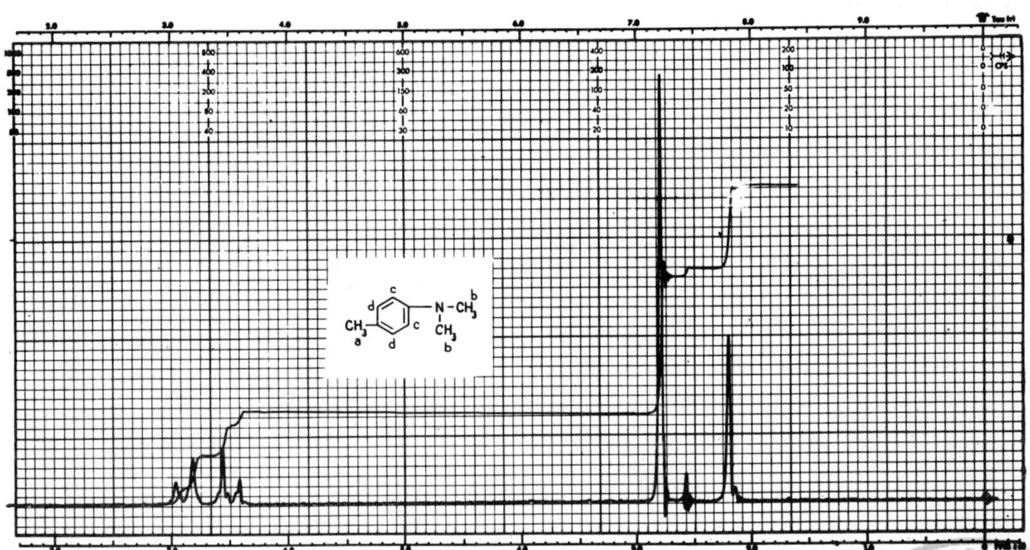

© 1978 Sadtler Research Laboratories, Inc.

BANGOR LIBRARY

N,N-DIETHYL-o-TOLUIDINE

$C_{11}H_{17}N$

Mol. Wt. 163.26

B.P. 85-87°C/10mm

Solvent: $CDCl_3$

Assignments
a = 0.97
b = 2.29
c = 2.98
d = 6.80-7.30

661

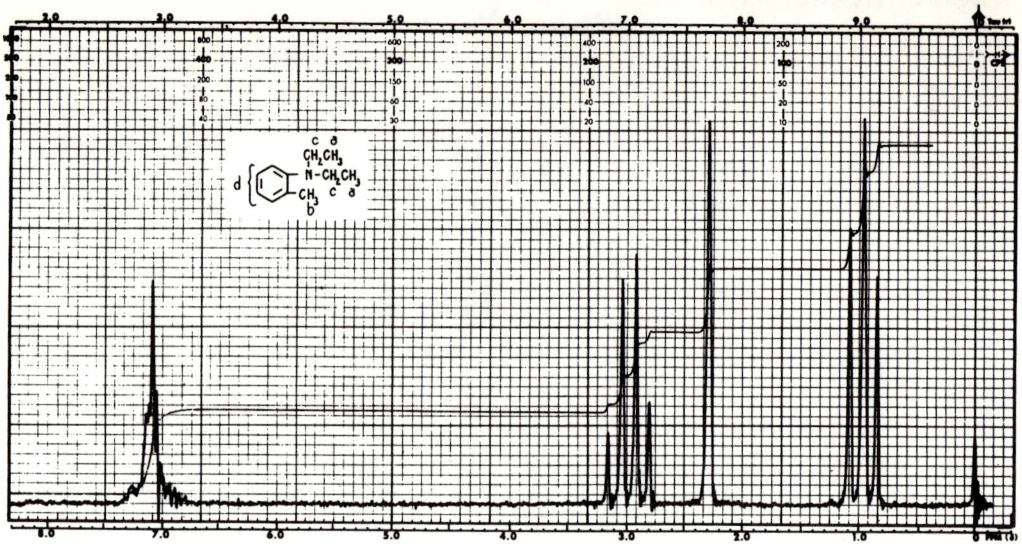

© 1978 Sadtler Research Laboratories, Inc.

N,N-DIETHYL-m-TOLUIDINE

$C_{11}H_{17}N$

Mol. Wt. 163.26

B.P. 228-230°C

Solvent: CCl_4

Assignments
a = 1.07
b = 2.23
c = 3.22
d = ca 6.35
e = 6.91

662

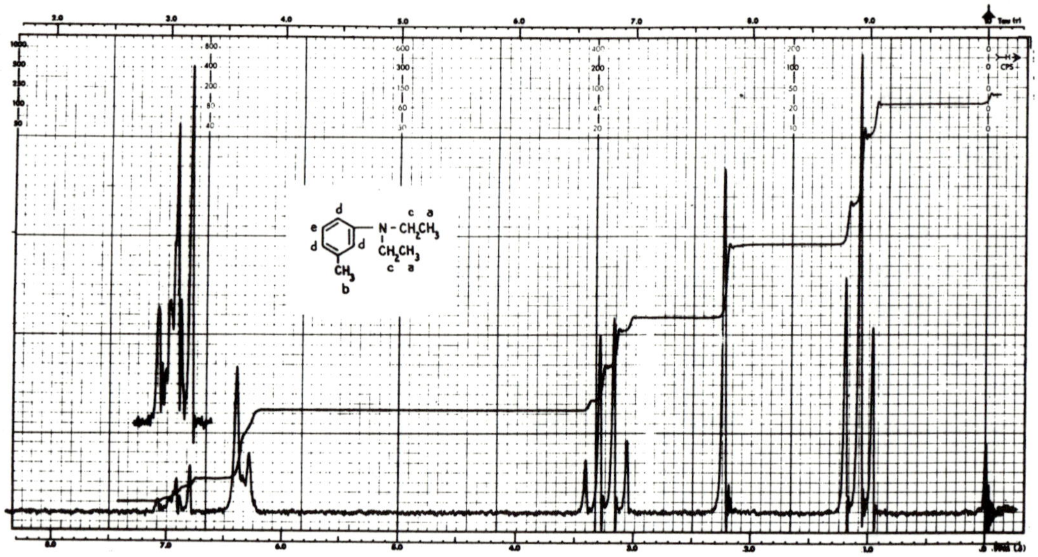

© 1978 Sadtler Research Laboratories, Inc.

N,N-DIETHYL-p-TOLUIDINE

$C_{11}H_{17}N$

Mol. Wt. 163.26

B.P. 102-104°C/10mm

Solvent: $CDCl_3$

Assignments
a = 1.08
b = 2.21
c = 3.23
d = 6.59
e = 7.01

663

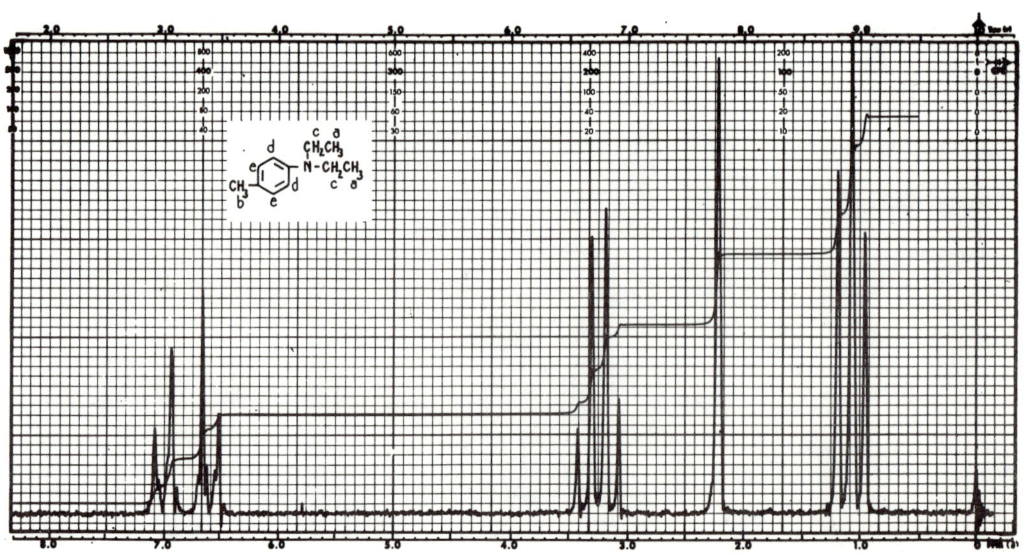

© 1978 Sadtler Research Laboratories, Inc.

N,N-DIETHYL-2,5-DIMETHYLANILINE

$C_{12}H_{19}N$

Mol. Wt. 177.29

B.P. 95-96°C/11mm

Solvent: CCl_4

Assignments
a = 0.95
b = 2.22
c = 2.28
d = 2.93
e = 6.71
f = 6.80
g = 6.99

664

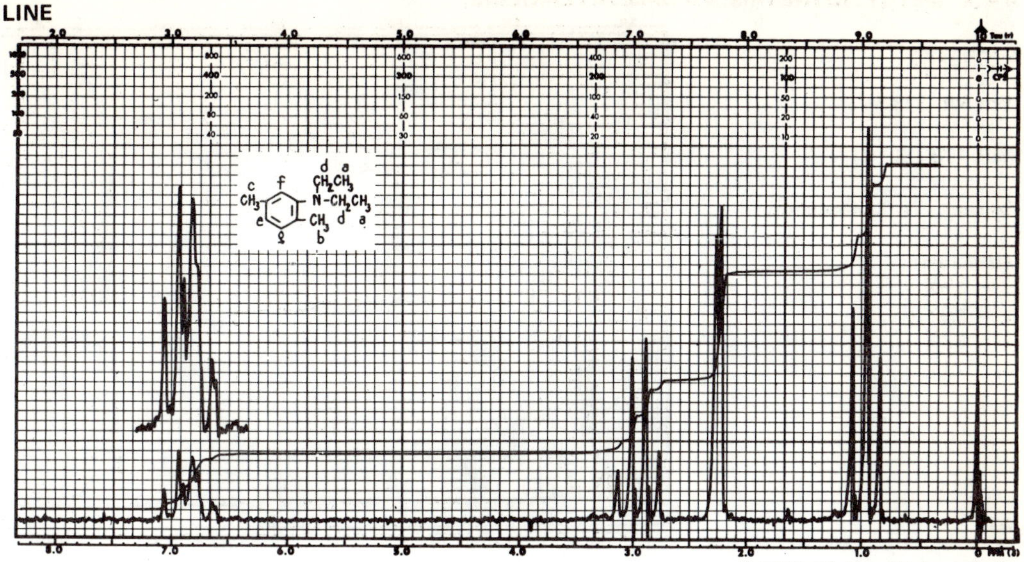

© 1978 Sadtler Research Laboratories, Inc.

N,N-DIMETHYL-2,6-XYLIDINE

$C_{10}H_{15}N$

Mol. Wt. 149.24

B.P. 82-83°C/17mm

Solvent: $CDCl_3$

Assignments
a = 2.26
b = 2.79
c = 6.91

665

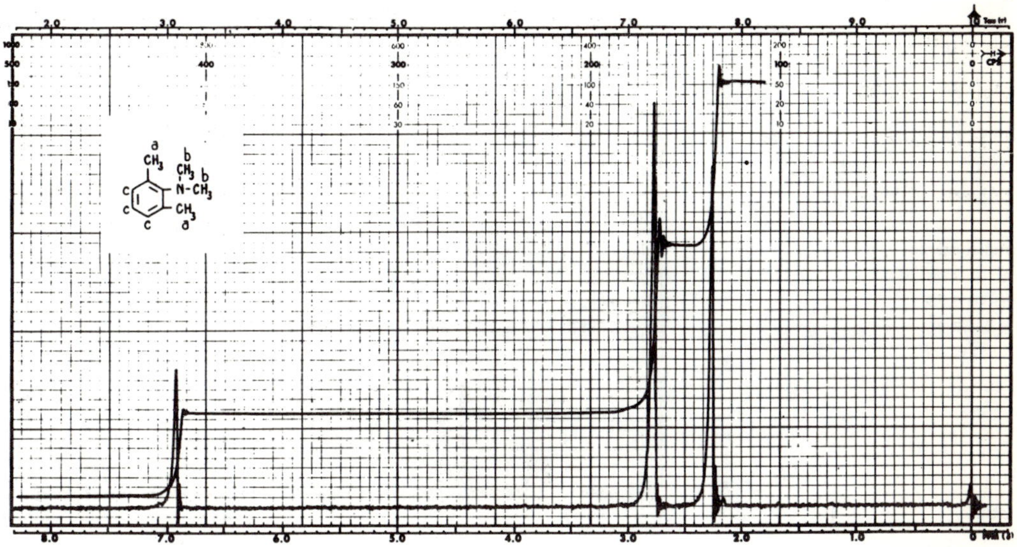

© 1978 Sadtler Research Laboratories, Inc.

4,4'-METHYLENEBIS(N,N-DIMETHYLANILINE)

$C_{17}H_{22}N_2$

Mol. Wt. 354.38

M.P. 89-90°C

Solvent: $CDCl_3$

Assignments
a = 2.85
b = 3.78
c = 6.62
d = 7.01

666

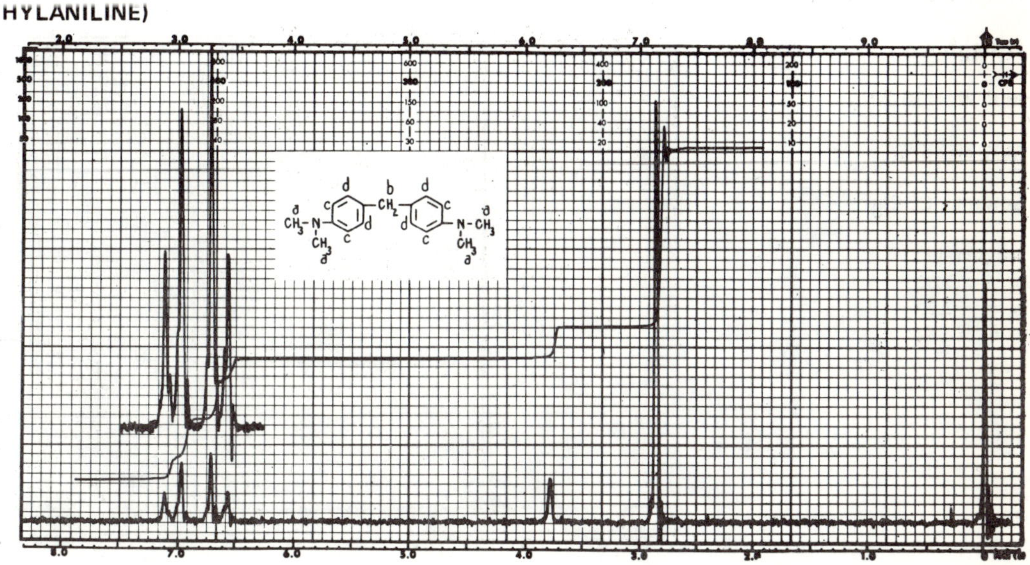

© 1978 Sadtler Research Laboratories, Inc.

4,4',4''-METHYLIDYNETRIS[N,N-DIMETHYLANILINE]

$C_{25}H_{31}N_3$

Mol. Wt. 373.54

M.P. 178-180°C

Solvent: Trifluoroacetic acid

Assignments
a = 3.52
b = 5.97
c = 7.49
d = 7.71

667

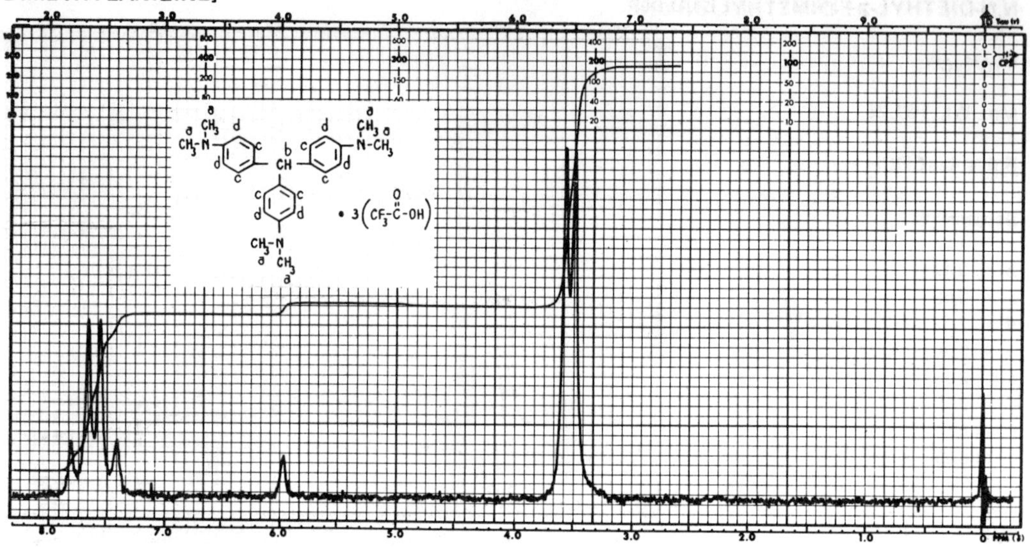

© 1978 Sadtler Research Laboratories, Inc.

p-BROMO-N,N-DIMETHYLANILINE

$C_8H_{10}BrN$

Mol. Wt. 200.08

M.P. 51-54°C

Solvent: $CDCl_3$

Assignments
a = 2.86
b = 6.53
c = 7.25

668

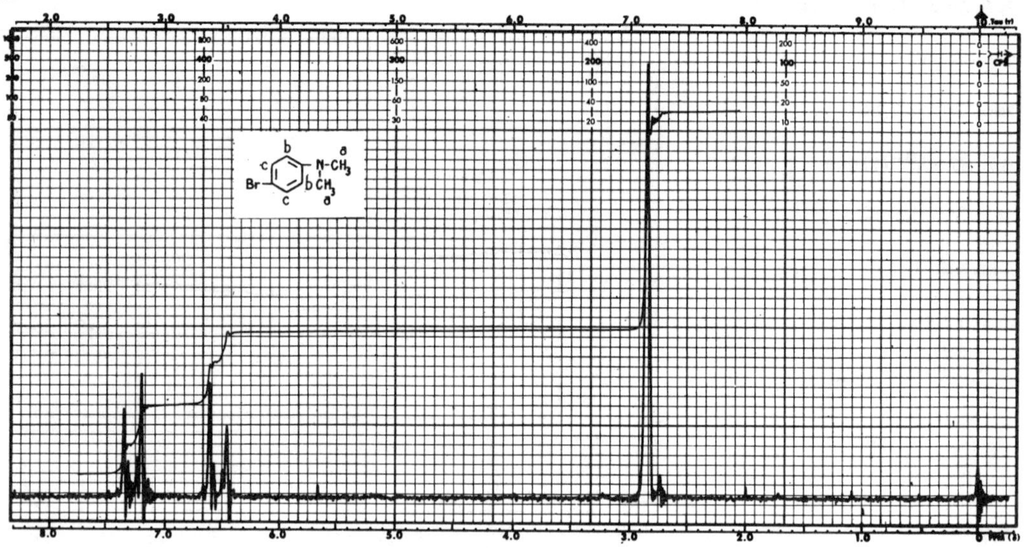

© 1978 Sadtler Research Laboratories, Inc.

N,N-DIMETHYL-p-PHENYLENEDIAMINE

$C_8H_{12}N_2$

Mol. Wt. 136.20

M.P. 36-38°C

Solvent: $CDCl_3$

Assignments
a = 2.80
b = 3.29
c = ca 6.62

669

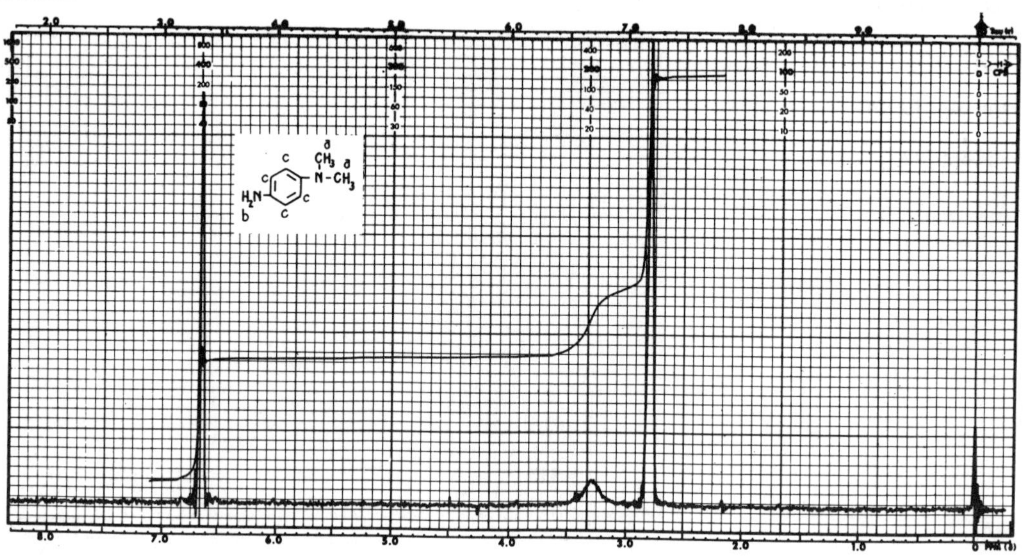

© 1978 Sadtler Research Laboratories, Inc.

N,N-DIETHYL-p-PHENYLENEDIAMINE

$C_{10}H_{16}N_2$

Mol. Wt. 164.25

Solvent: $CDCl_3$

Assignments
a = 1.08
b = 3.17
c = 6.59
d = 7.19 imp.
NH_2 unobserved

670

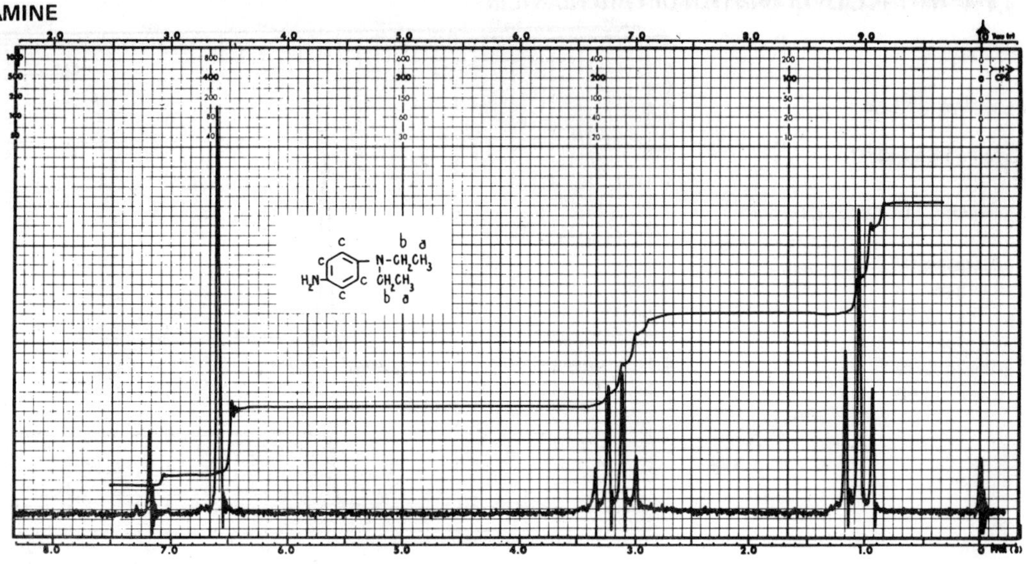

© 1978 Sadtler Research Laboratories, Inc.

N,N-DIMETHYL-1-NAPHTHYLAMINE

$C_{12}H_{13}N$

Mol. Wt. 171.24

M.P. 52-53°C

B.P. 305°C/760mm (lit.)

Solvent: CCl_4

Assignments
a = 2.81
b = 6.80-8.30

671

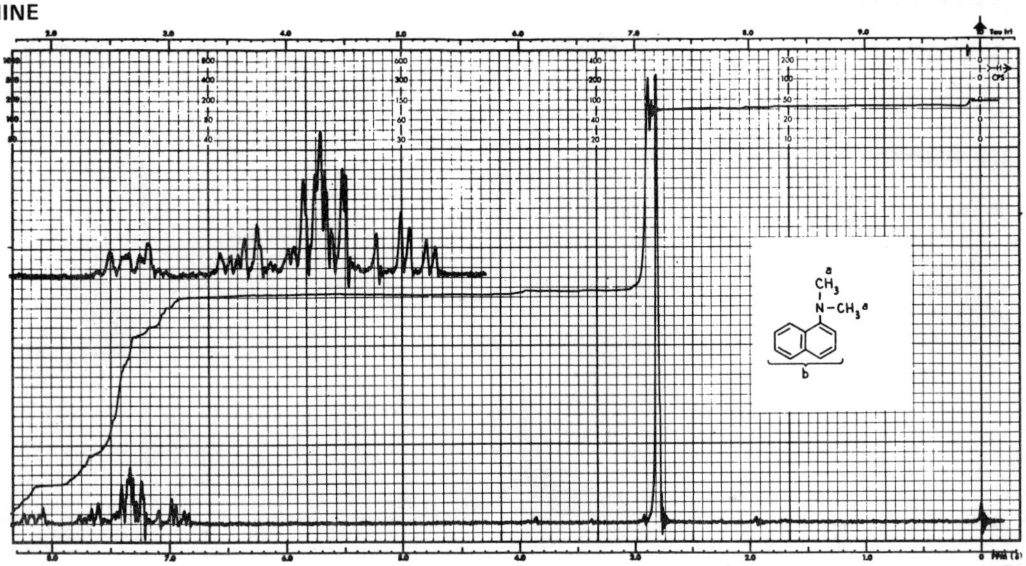

© 1978 Sadtler Research Laboratories, Inc.

N,N-DIETHYL-1-NAPHTHYLAMINE

$C_{14}H_{17}N$

Mol. Wt. 199.30

Solvent: CCl_4

Assignments
a = 0.99
b = 3.12
c = 6.93-8.42
d = 1.29 imp.

672

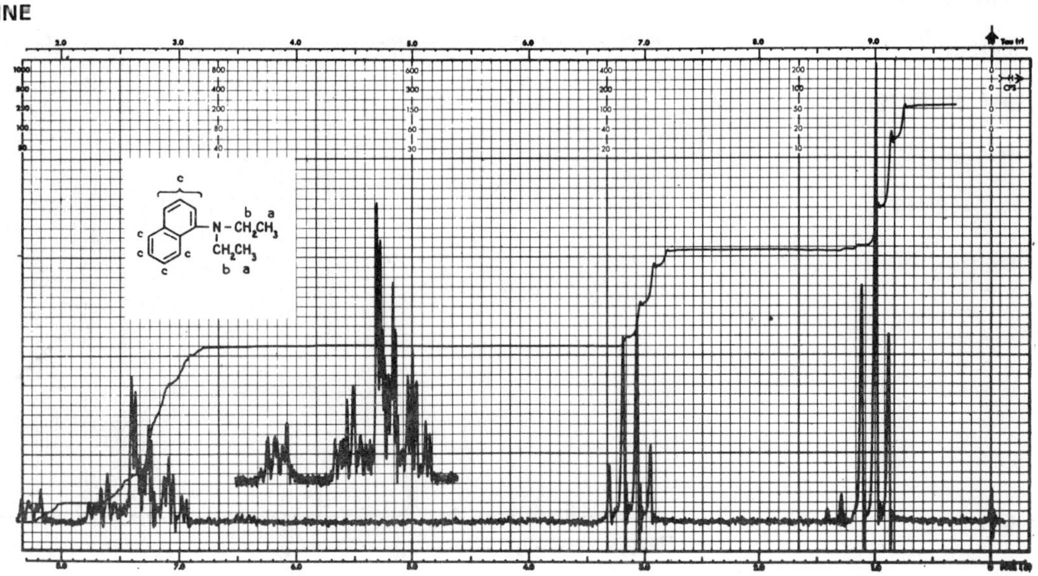

© 1978 Sadtler Research Laboratories, Inc.

1-METHYL-1,2,3,6-TETRAHYDROPYRIDINE

$C_6H_{11}N$

Mol. Wt. 97.16

M.P. 113-114°C

Solvent: CCl_4

Assignments
a = ca 2.14
b = ~2.21
c = ca 2.37
d = 2.79
e = 5.58

673

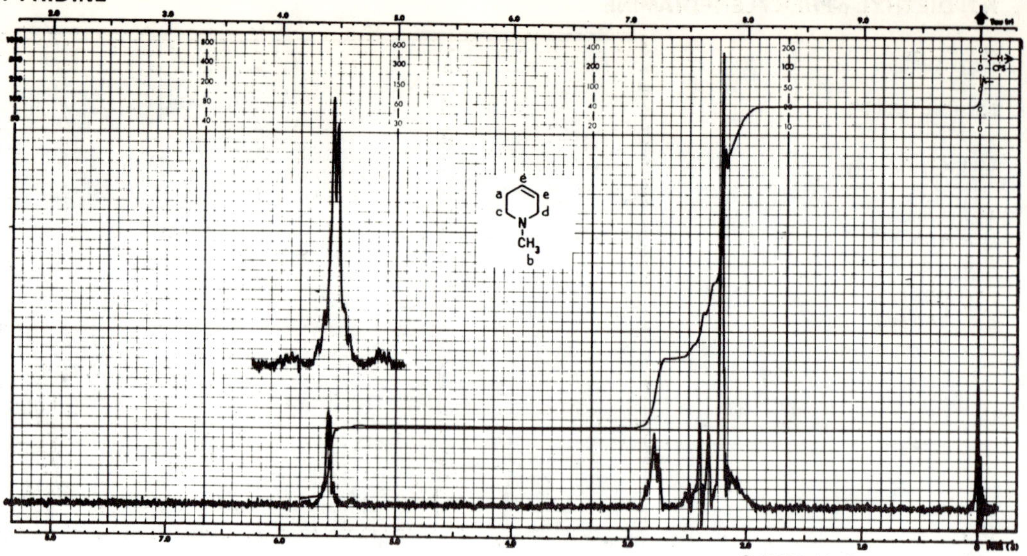

© 1978 Sadtler Research Laboratories, Inc.

1,2-DIPHENYL-2-IMIDAZOLINE

$C_{15}H_{14}N_2$

Mol. Wt. 222.29

Solvent: $CDCl_3$

Assignments
a = 4.00
b = 6.72
c = 6.80-7.20
d = 7.10-7.35
e = 7.42

674

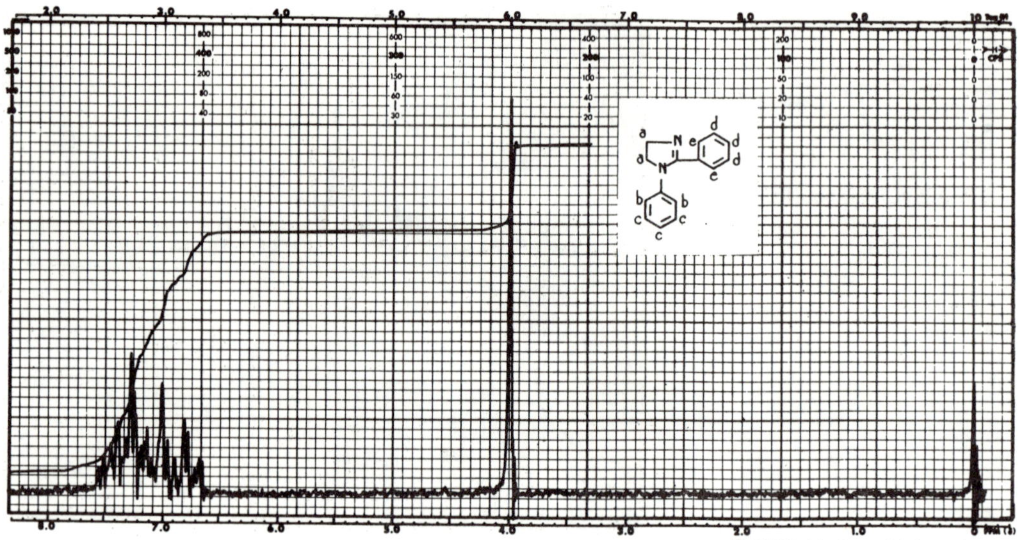

© 1978 Sadtler Research Laboratories, Inc.

1-METHYLPYRROLE

C_5H_7N

Mol. Wt. 81.12

M.P. –57°C

M.P. 112-113°C

Solvent: $CDCl_3$

Assignments
a = 3.62
b = 6.11
c = 6.57

675

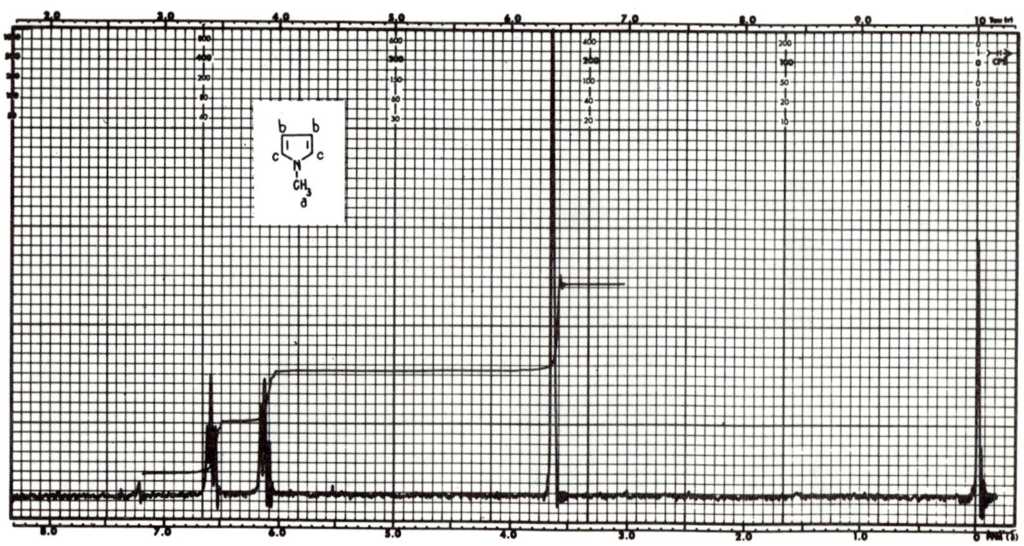

© 1978 Sadtler Research Laboratories, Inc.

2,5-DIMETHYL-1-PHENYLPYRROLE

$C_{12}H_{13}N$

Mol. Wt. 171.24

M.P. 51-53°C

Solvent: $CDCl_3$

Assignments
a = 2.02
b = 5.91
c = 7.05-7.60

676

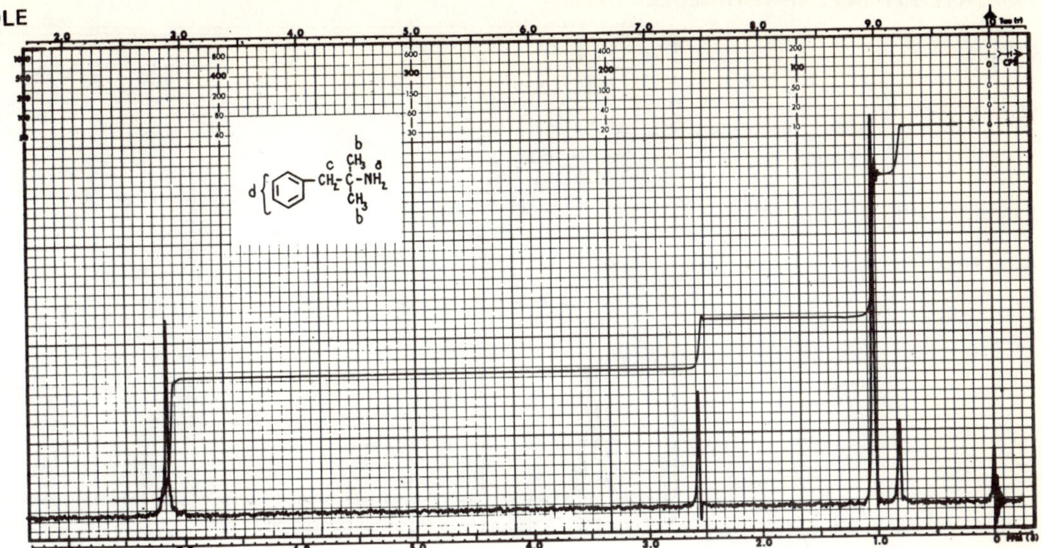

© 1978 Sadtler Research Laboratories, Inc.

2-[(DIMETHYLAMINO)METHYL] PYRROLE

$C_7H_{12}N$

Mol. Wt. 124.19

M.P. 60-63°C

Solvent: $CDCl_3$

Assignments
a = 2.22
b = 3.43
c = ca 6.09
d = 6.69
e = unobserved

677

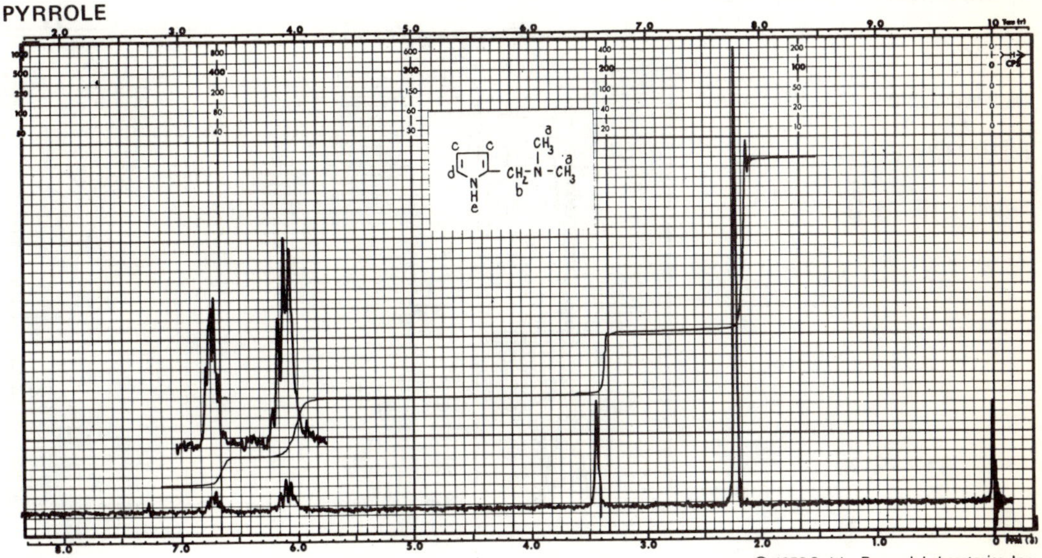

© 1978 Sadtler Research Laboratories, Inc.

1-PHENYLPYRAZOLE

$C_9H_8N_2$

Mol. Wt. 144.18

Solvent: $CDCl_3$

Assignments
a = 6.36
b = 7.00-7.55
c = 7.50-7.90
d = 1.71, 1.94 imps.
e = 3.43 imp.

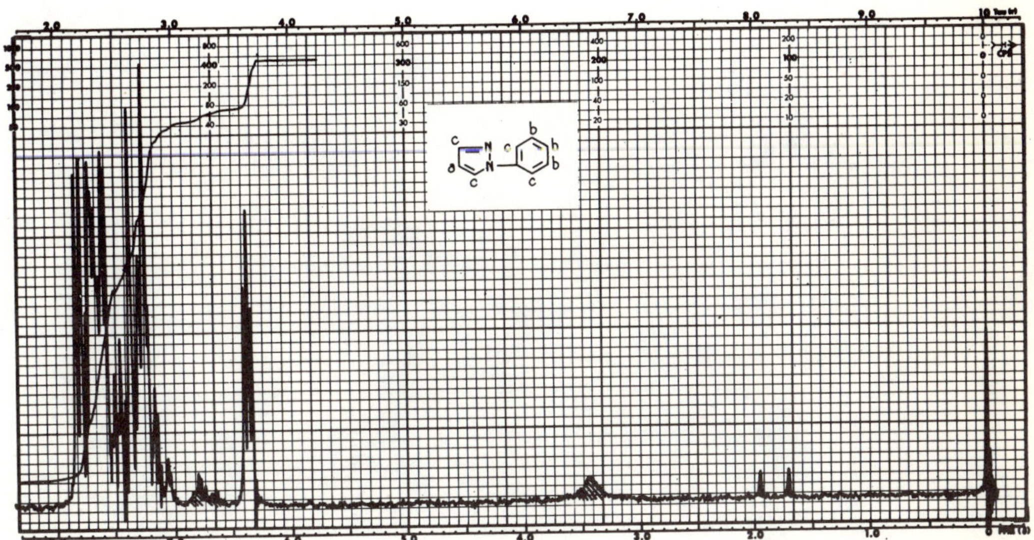

© 1978 Sadtler Research Laboratories, Inc.

1-METHYL-5-PHENYL-1H-TETRAZOLE

$C_8H_8N_4$

Mol. Wt. 160.18

M.P. 103-104°C

Solvent: $CDCl_3$

Assignments
a = 4.16
b = 7.42-7.84

679

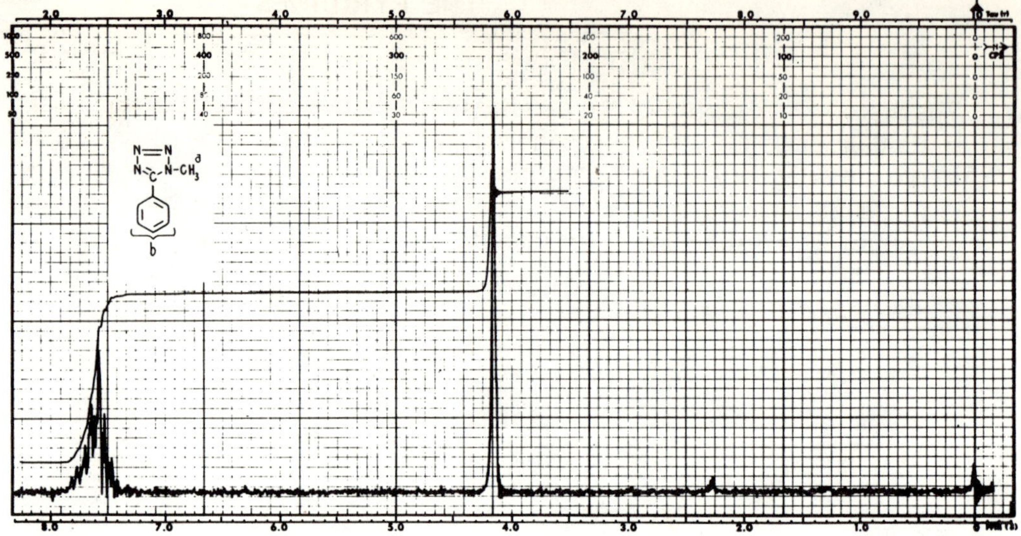

© 1978 Sadtler Research Laboratories, Inc.

1,5-DIPHENYL-1H-TETRAZOLE

$C_{13}H_{10}N_4$

Mol. Wt. 222.25

M.P. 144.7-145.3°C

Solvent: $CDCl_3$

Assignments
a = 7.19-7.69

680

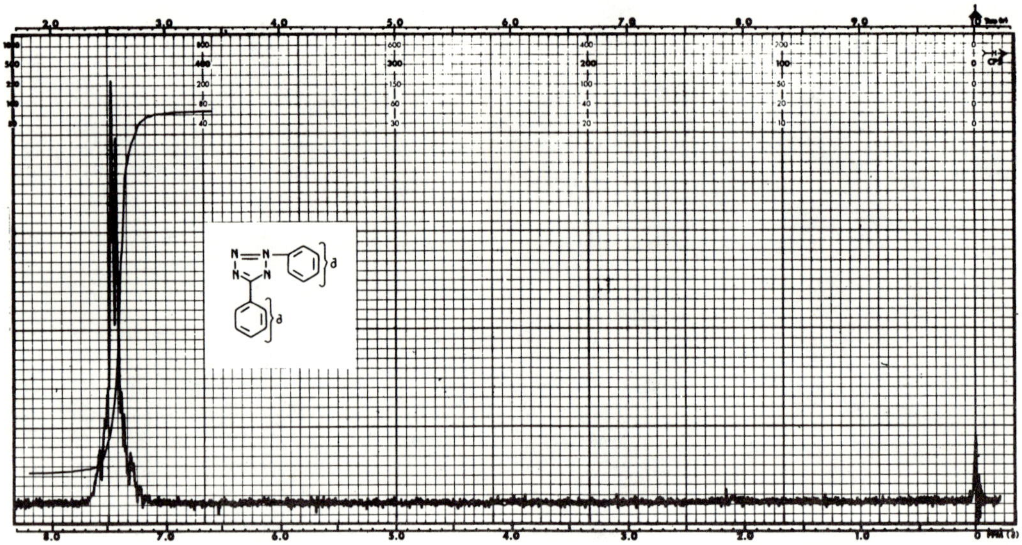

© 1978 Sadtler Research Laboratories, Inc.

The proton NMR spectra of the pyridines produce characteristic patterns over a wide range of chemical shifts. The chemical shifts of the individual ring protons and the coupling constants with other protons on the ring vary with their position relative to the pyridine nitrogen atom. The characteristic low field chemical shifts observed for the hydrogens at positions 2 and 6 (adjacent to the ring nitrogen atom) are a distinct aid in the identification of the NMR spectra of this class of compounds.

Chemical Shifts

Aromatic Protons

Pyridine

HC–4	HC–3,5	HC–2,6		Solvent	Spectrum
7.55	7.14	8.51	Pyridine	CCl_4	681

2-Substituted Pyridines

HC–6	HC–5	HC–4	HC–3	HC–2	
7.93	6.48	7.22	6.36	$-NH_2$	$CDCl_3$
8.07	6.45	7.36	6.45	–NH–R	$CDCl_3$
8.14	6.83	7.52	6.68	$-O-CH_3$	CCl_4
8.18	7.14	7.77	6.89	–F	CCl_4
8.46	6.98	7.48	7.05	$-CH_2-CH_3$	CCl_4
8.37	7.20	7.68	7.29	–Cl	CCl_4
8.27	7.20	7.48	7.38	–Br	CCl_4
8 76	7.60	7.75	7.89	–C≡N	CCl_4
8.71	7.39	7.81	7.90	$-C(=O)-CH_3$	CCl_4
8.78	7.58	7.91	7.91	–C(=O)–H	CCl_4
8.70	7.46	7.81	8.03	$-C(=O)-O-CH_3$	CCl_4
8.57	7.41	7.83	8.21	$-C(=O)-NH_2$	$CDCl_3$
8.27	7.03	7.70	8.27	$-NH-C(=O)-CH_3$	$CDCl_3$
8.67	7.25	7.77	8.42	–2(Pyridine)	$CDCl_3$

Aromatic Protons continued

3-Substituted Pyridines

HC–6	HC–5	HC–4	HC–3	HC–2	–X	Solvent
7.92	6.95	6.95	–X	8.04	$-NH_2$	$CDCl_3$
8.32	7.01	7.35	–X	8.36	$-CH_3$	CCl_4
8.33	7.10	7.42	–X	8.38	$-CH_2-CH_3$	CCl_4
8.41	7.15	7.59	–X	8.52	–Cl	CCl_4
8.50	7.21	7.62	–X	8.53	$-CH_2-C(=O)-O-R$	$CDCl_3$
8.48	7.10	7.74	–X	8.66	–Br	CCl_4

4-Substituted Pyridines

HC–2,6	HC–3,5	C–4	
8.03	6.50	$-NH_2$	Polysol
8.25	6.50	$-N(CH_3)_2$	$CDCl_3$
8.37	6.98	$-CH_3$	CCl_4
8.41	7.02	$-CH_2-CH_3$	CCl_4
8.51	7.26	$-CH_2-NH-CH_3$	$CDCl_3$
8.63	7.40	–Ø	$CDCl_3$
8.71	7.51	–4(Pyridine)	$CDCl_3$
8.82	7.64	–C(=O)–H	CCl_4

Coupling and Coupling Constants

The spin-spin couplings of the pyridines is limited to those between hydrogens on the ring. The protons at positions 2 and 6 often display an observable degree of broadening due to the adjacent nitrogen nucleus.

The coupling constants observed for the pyridines are unusual in that long range "para" couplings are observed through five bonds and that the two "ortho" couplings J_{2-3} and J_{3-4} are different in magnitude.

J_{2-3} = 4–7 Hz
J_{3-4} = 7–9 Hz
J_{2-4} = 1–3 Hz

J_{2-5} = 0.1–1.1 Hz
J_{3-5} = 1.1–2.5 Hz

PYRIDINE

C_5H_5N

Mol. Wt. 79.10

B.P. 115.0-115.5°C

Solvent: CCl_4

Assignments
a = 7.14
b = 7.55
c = 8.51
d = 3.93 H_2O

681

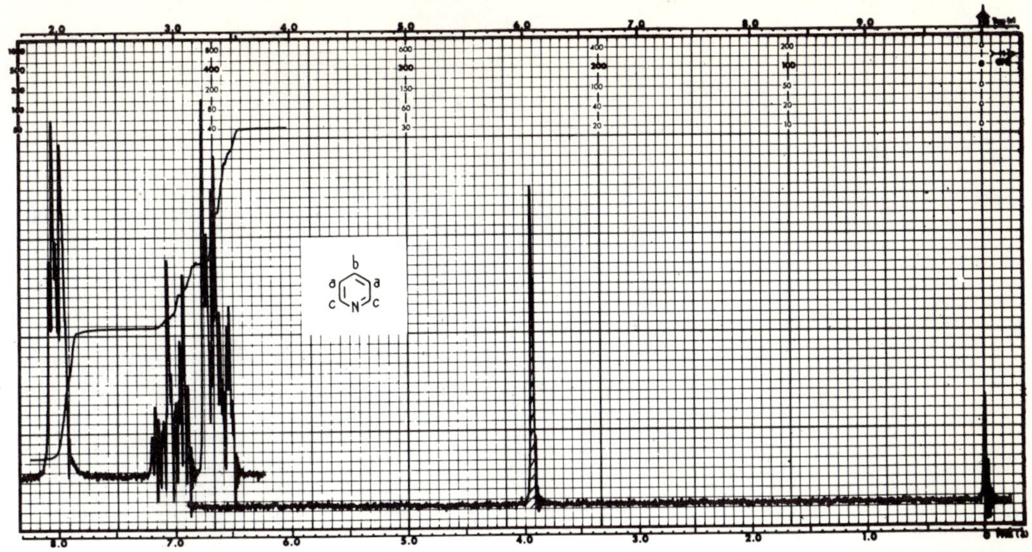

© 1978 Sadtler Research Laboratories, Inc.

2-PICOLINE

C_6H_7N

Mol. Wt. 93.13

B.P. 127-129°C

Solvent: CCl_4

Assignments
a = 2.47
b = ca 6.91
c = 7.40
d = 8.37

682

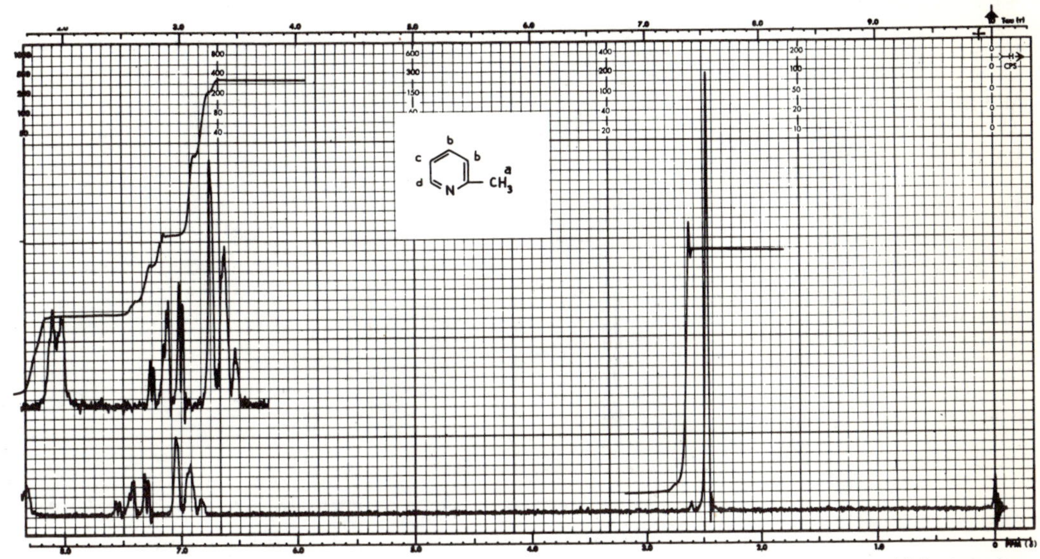

© 1978 Sadtler Research Laboratories, Inc.

3-PICOLINE

C_6H_7N

Mol. Wt. 93.13

B.P. 141-144°C

Solvent: CCl_4

Assignments
a = 2.27
b = 7.01 or c
c = 7.35 or b
d = 8.35
e = 2.42 imp.

683

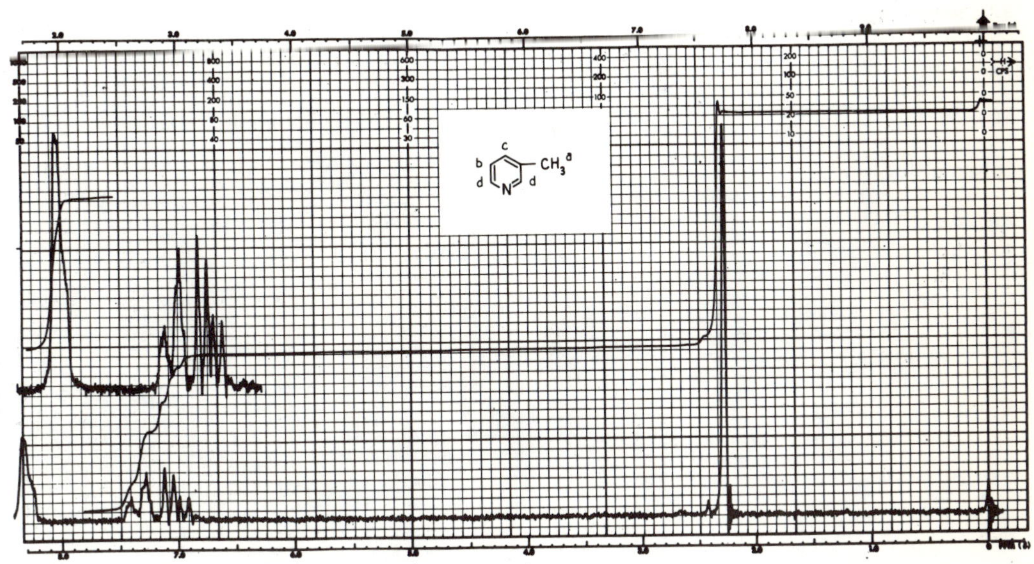

© 1978 Sadtler Research Laboratories, Inc.

4-PICOLINE

C_6H_7N

Mol. Wt. 93.13

B.P. 142-145°C

Solvent: CCl_4

Assignments

a = 2.29
b = 6.98
c = 8.37

684

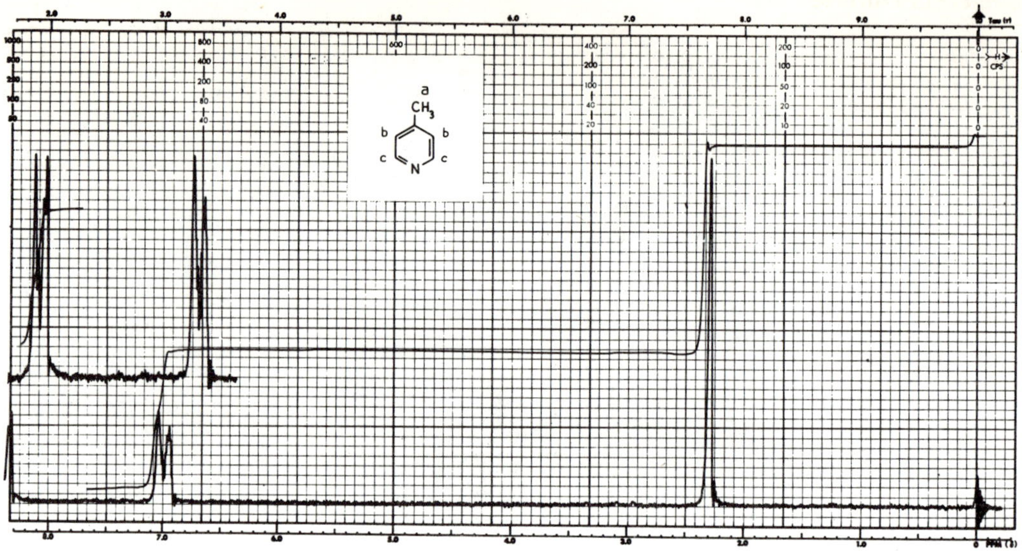

© 1978 Sadtler Research Laboratories, Inc.

4-ETHYLPYRIDINE

C_7H_9N

Mol. Wt. 107.16

B.P. 167.7°C/751mm

F.P. –90.7 to 91.5°C

Solvent: CCl_4

Assignments

a = 1.21
b = 2.59
c = 7.02
d = 8.41

685

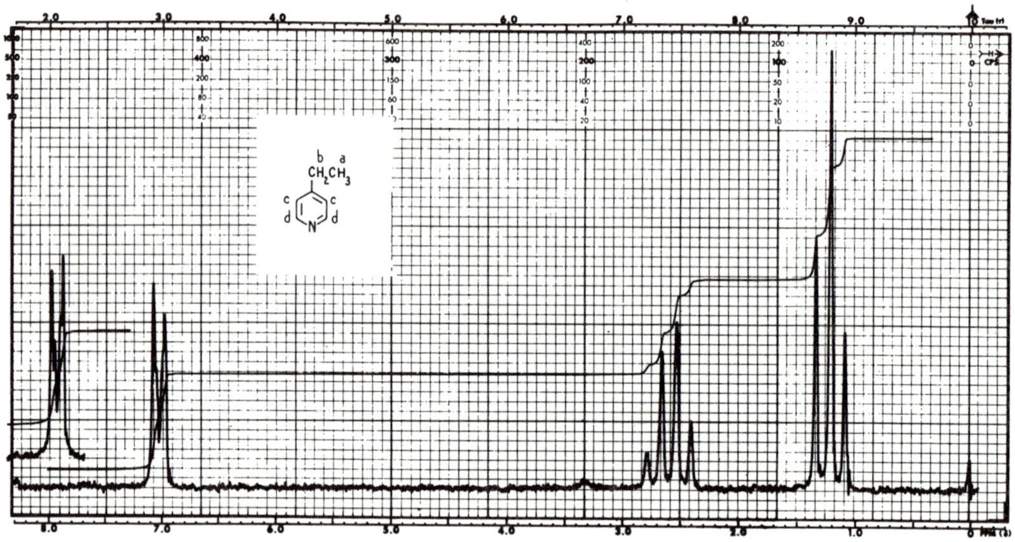

© 1978 Sadtler Research Laboratories, Inc.

4-ISOPROPYLPYRIDINE

$C_8H_{11}N$

Mol. Wt. 121.18

Solvent: $CDCl_3$

Assignments

a = 1.21
b = 2.83
c = 7.09
d = 8.43
e = 1.59 imp.

686

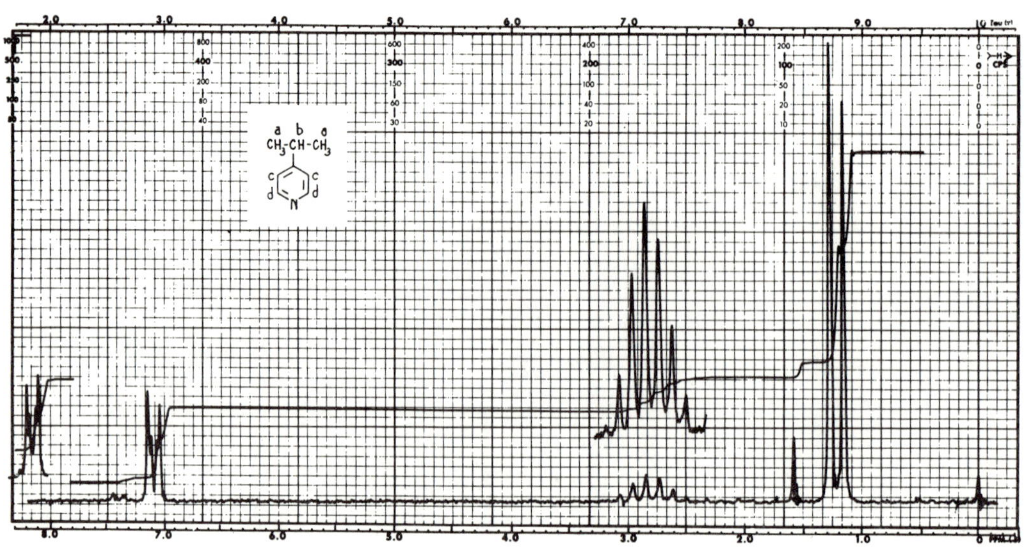

© 1978 Sadtler Research Laboratories, Inc.

2-PENTYLPYRIDINE

$C_{10}H_{15}N$

Mol. Wt. 149.24

B.P. 81-84°C/10mm

Solvent: CCl_4

Assignments
a = 0.91
b = ca 1.37
c = ca 1.72
d = 2.73
e = 6.95
f = 7.02
g = 7.45
h = 8.43

687

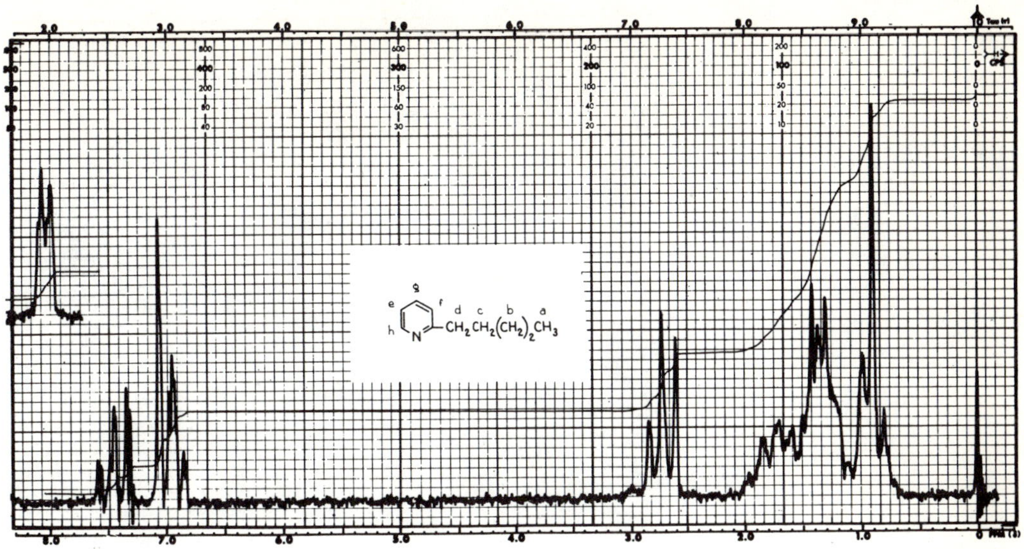

© 1978 Sadtler Research Laboratories, Inc.

4-PENTYLPYRIDINE

$C_{10}H_{15}N$

Mol. Wt. 149.24

B.P. 103-106°C/11mm

Solvent: CCl_4

Assignments
a = 0.88
b = ca 1.33
c = 2.57
d = 6.96
e = 8.36

688

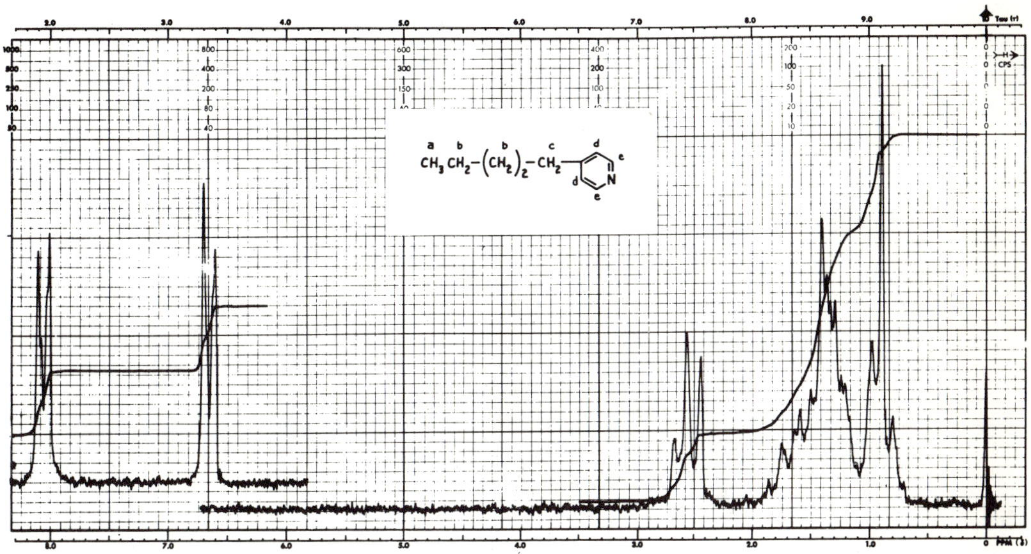

© 1978 Sadtler Research Laboratories, Inc.

2-(1-BUTYLPENTYL)PYRIDINE

$C_{14}H_{23}N$

Mol. Wt. 205.35

Solvent: CCl_4

Assignments
a = 0.88
b = 0.95-1.50
c = ca 1.68
d = 2.67
e = ca 6.97
f = 7.47
g = 8.49

689

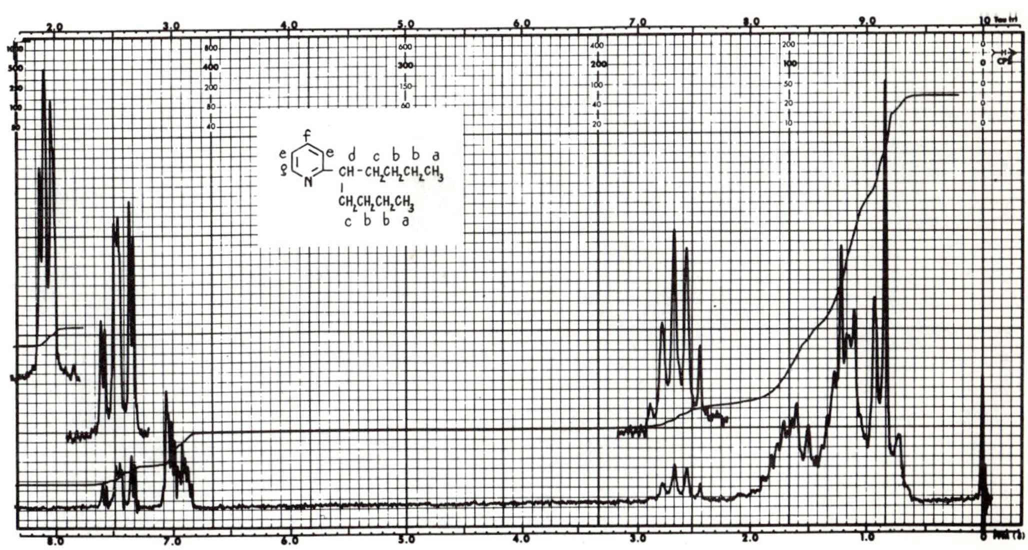

© 1978 Sadtler Research Laboratories, Inc.

2,4-LUTIDINE

C_7H_9N

Mol. Wt. 107.16

B.P. 157°C

Solvent: $CDCl_3$

Assignments
a = 2.29
b = 2.50
c = 6.97
d = 6.99
e = 8.35

690

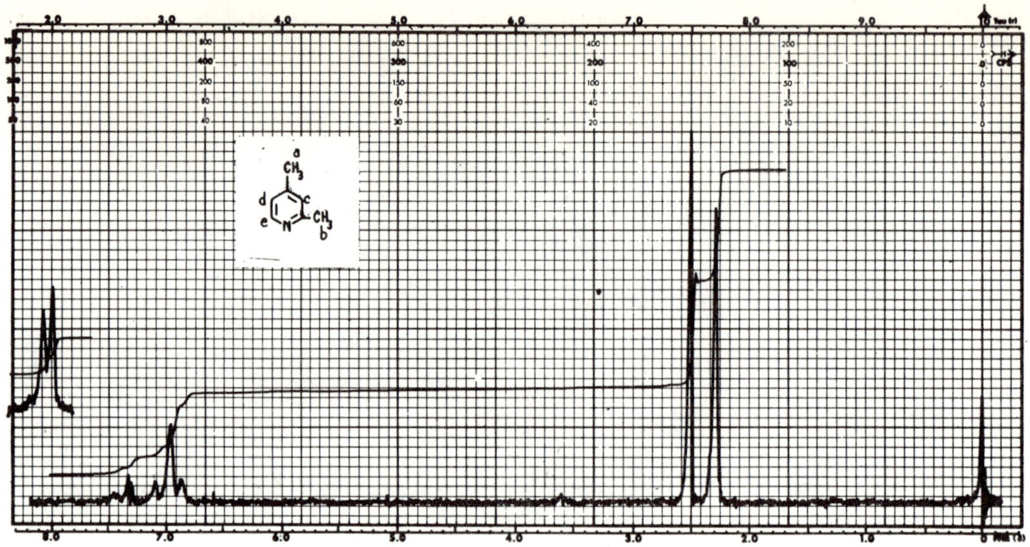

© 1978 Sadtler Research Laboratories, Inc.

2,5-LUTIDINE

C_7H_9N

Mol. Wt. 107.16

B.P. 156.5°C (lit.)

Solvent: $CDCl_3$

Assignments
a = 2.20
b = 2.46
c = 6.95
d = 7.20
e = 8.23
f = 4.00 imp.

691

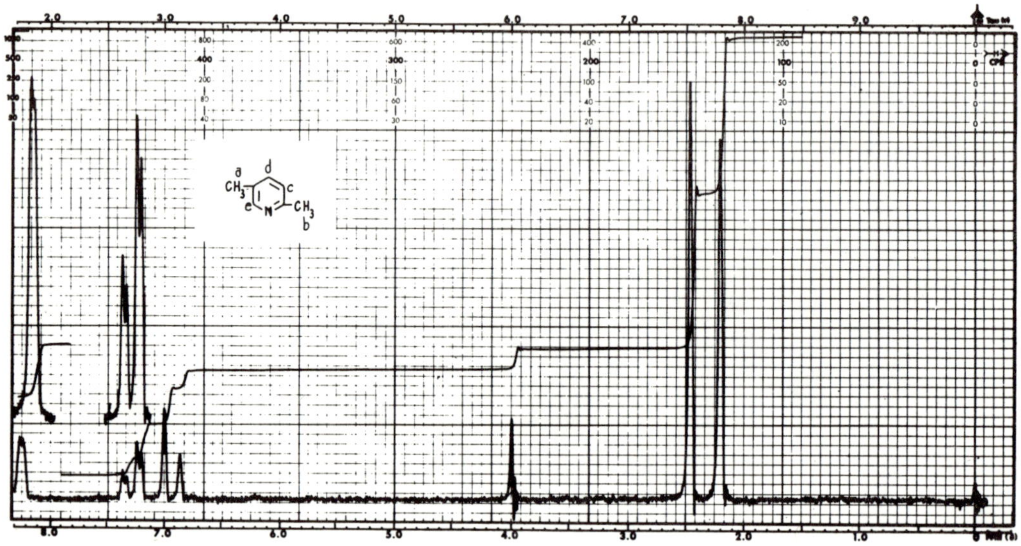

© 1978 Sadtler Research Laboratories, Inc.

3,5-LUTIDINE

C_7H_9N

Mol. Wt. 107.16

M.P. −65°C

B.P. 171°C

Solvent: CCl_4

Assignments
a = 2.24
b = 7.17
c = 8.12

692

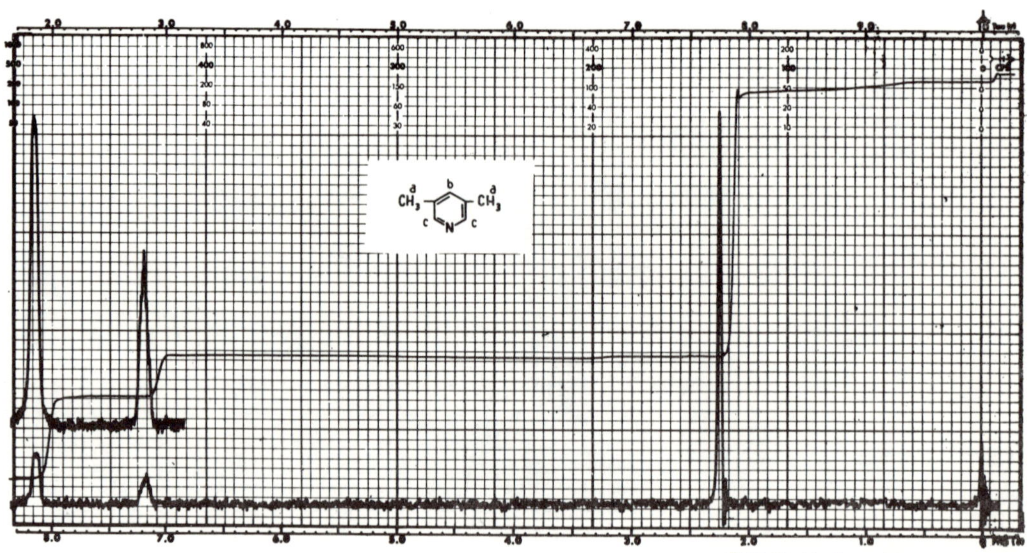

© 1978 Sadtler Research Laboratories, Inc.

3,5-DIBUTYLPYRIDINE

$C_{13}H_{21}N$

Mol. Wt. 191.32

Solvent: $CDCl_3$

Assignments
a = 0.93
b = 1.10-1.90
c = 2.60
d = 7.31
e = 8.30

693

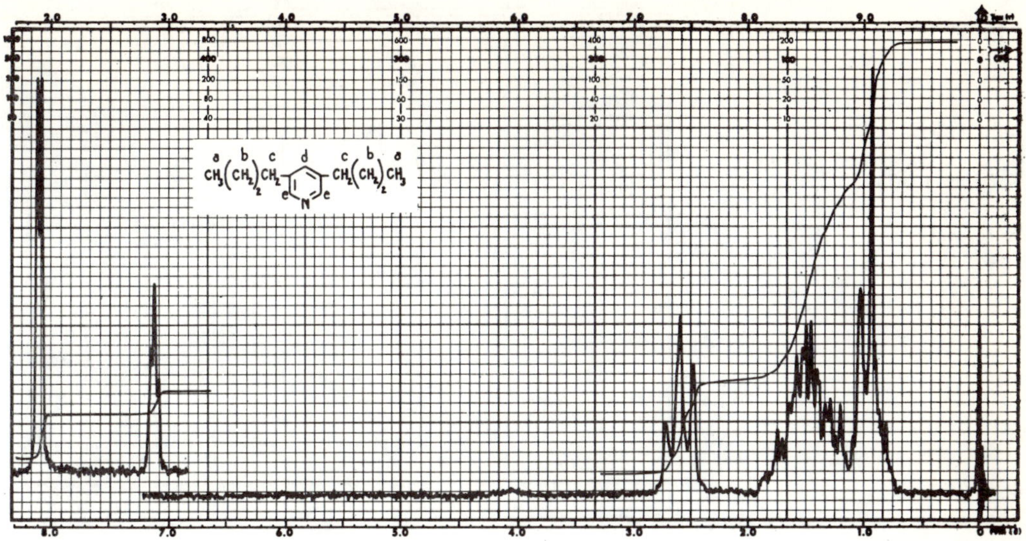

© 1978 Sadtler Research Laboratories, Inc.

2,6-DI-tert-BUTYLPYRIDINE

$C_{13}H_{21}N$

Mol. Wt. 191.32

B.P. ∿205°C; 120-121°C/20mm

Solvent: CCl_4

Assignments
a = 1.34
b = 7.04
c = 7.43

694

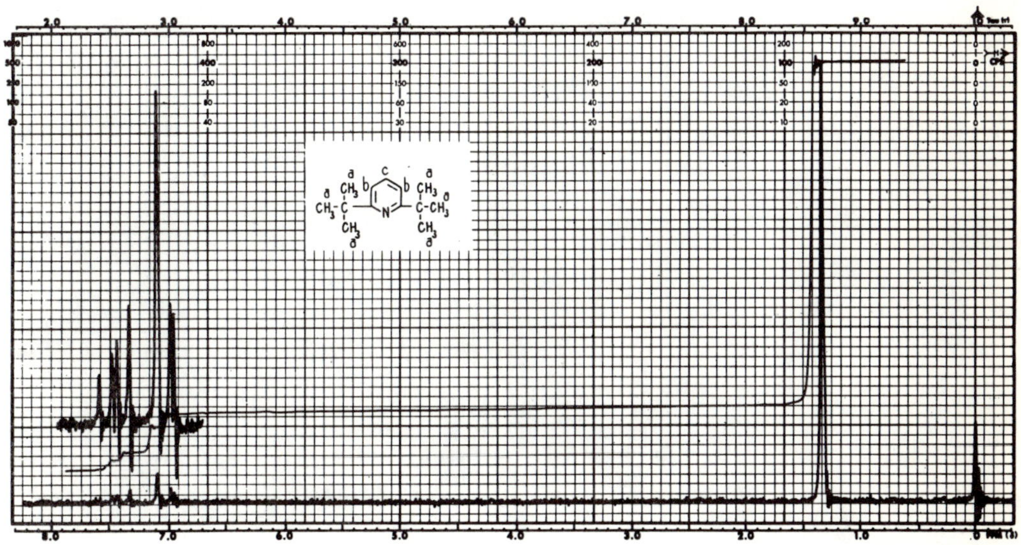

© 1978 Sadtler Research Laboratories, Inc.

2,4,6-TRIMETHYLPYRIDINE

$C_8H_{11}N$

Mol. Wt. 121.18

B.P. 171-172°C

Solvent: CCl_4

Assignments
a = 2.18
b = 2.37
c = 6.59

695

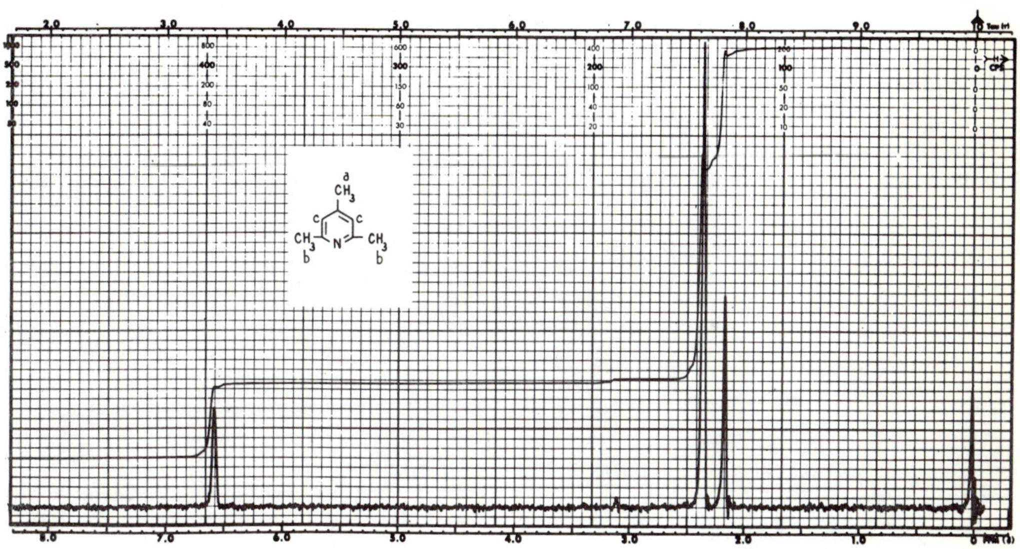

© 1978 Sadtler Research Laboratories, Inc.

2,6-DI-tert-BUTYL-4-PICOLINE

$C_{14}H_{23}N$

Mol. Wt. 205.35

M.P. 33.5-33.8°C

Solvent: $CDCl_3$

Assignments
a = 1.32
b = 2.31
c = 6.90

696

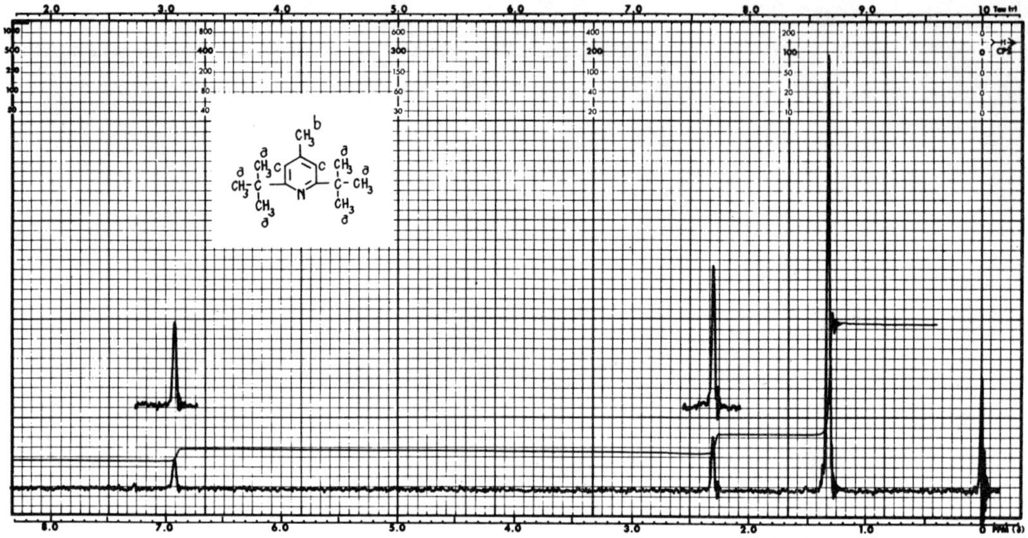

© 1978 Sadtler Research Laboratories, Inc.

2-PHENYLPYRIDINE

$C_{11}H_9N$

Mol. Wt. 155.20

B.P. 270°C/760mm (lit.)

Solvent: $CDCl_3$

Assignments
a = 6.99
b = 7.10-7.65
c = 7.93
d = 8.59

697

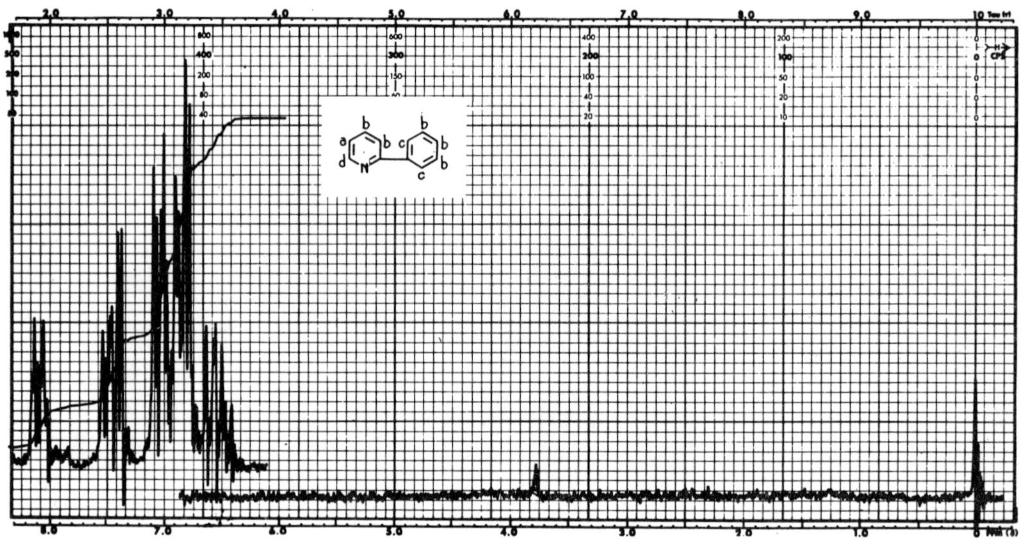

© 1978 Sadtler Research Laboratories, Inc.

4-PHENYLPYRIDINE

$C_{11}H_9N$

Mol. Wt. 155.20

M.P. 76-78°C

Solvent: $CDCl_3$

Assignments
a - 7.32-7.73
b = 8.63

698

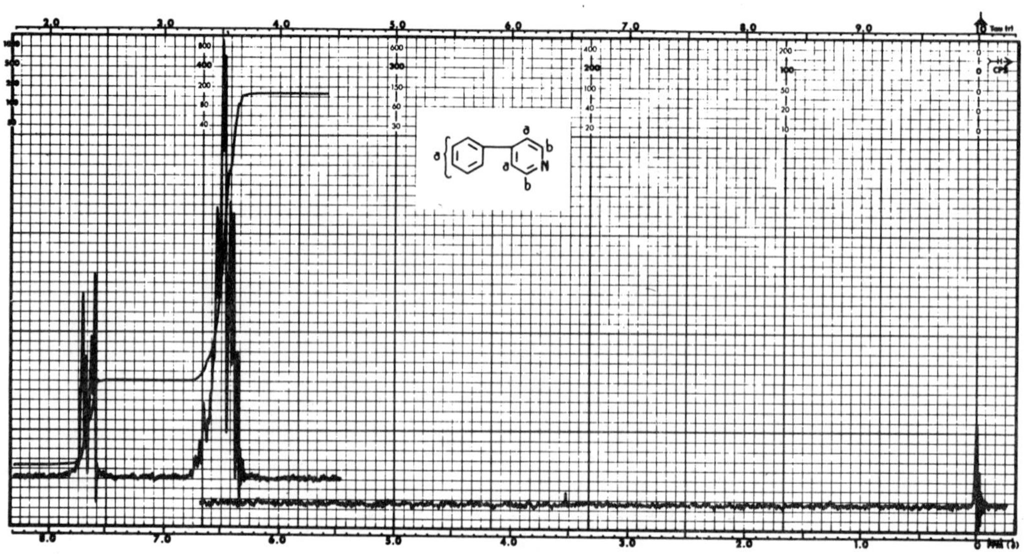

© 1978 Sadtler Research Laboratories, Inc.

2-FLUOROPYRIDINE

C_5H_4FN

Mol. Wt. 97.09

M.P. 125°C/758mm

Solvent: CCl_4

Assignments
a = 6.89
b = 7.14
c = 7.77
d = 8.18

699

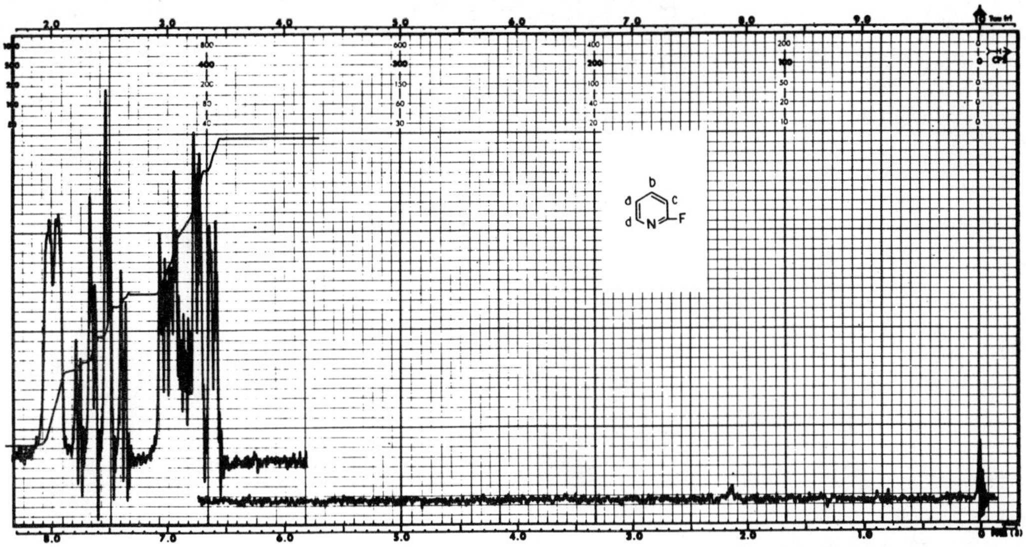

© 1978 Sadtler Research Laboratories, Inc.

2-CHLOROPYRIDINE

C_5H_4ClN

Mol. Wt. 113.55

B.P. 62-63.5°C/10mm

Solvent: CCl_4

Assignments
a = 7.20
b = 7.29
c = 7.68
d = 8.37

700

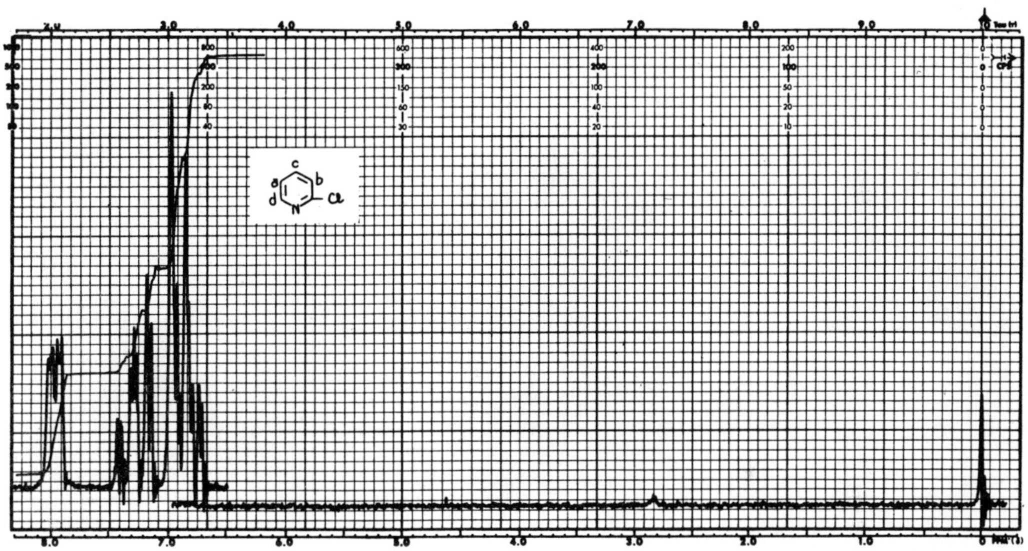

© 1978 Sadtler Research Laboratories, Inc.

3-CHLOROPYRIDINE

C_5H_4ClN

Mol. Wt. 113.55

Solvent: CCl_4

Assignments
a = 7.15
b = 7.59
c = 8.41
d = 8.52

701

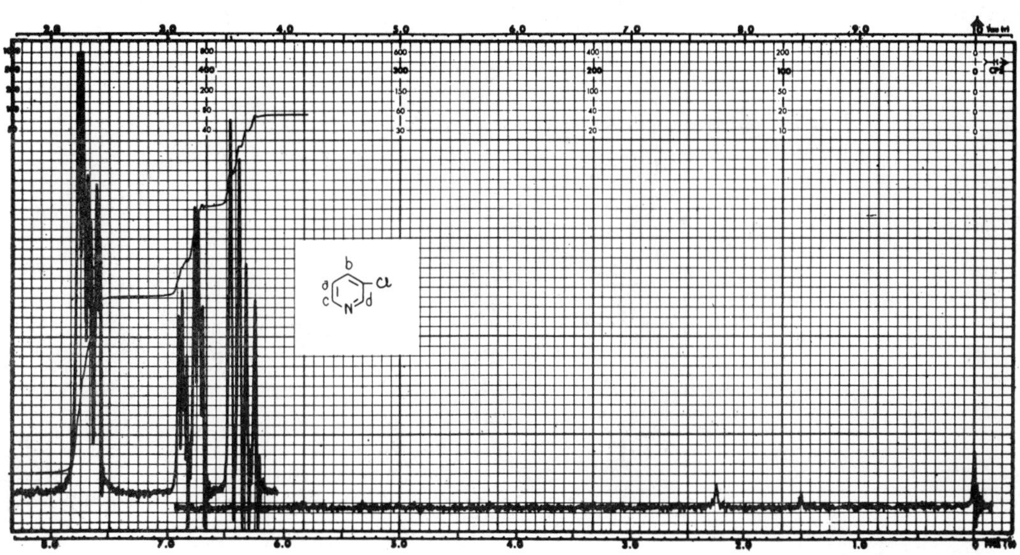

© 1978 Sadtler Research Laboratories, Inc.

2,3-DICHLOROPYRIDINE

$C_5H_3Cl_2N$

Mol. Wt. 147.99

M.P. 65-67°C

Solvent: $CDCl_3$

Assignments
a = 7.22
b = 7.78
c = 8.30

702

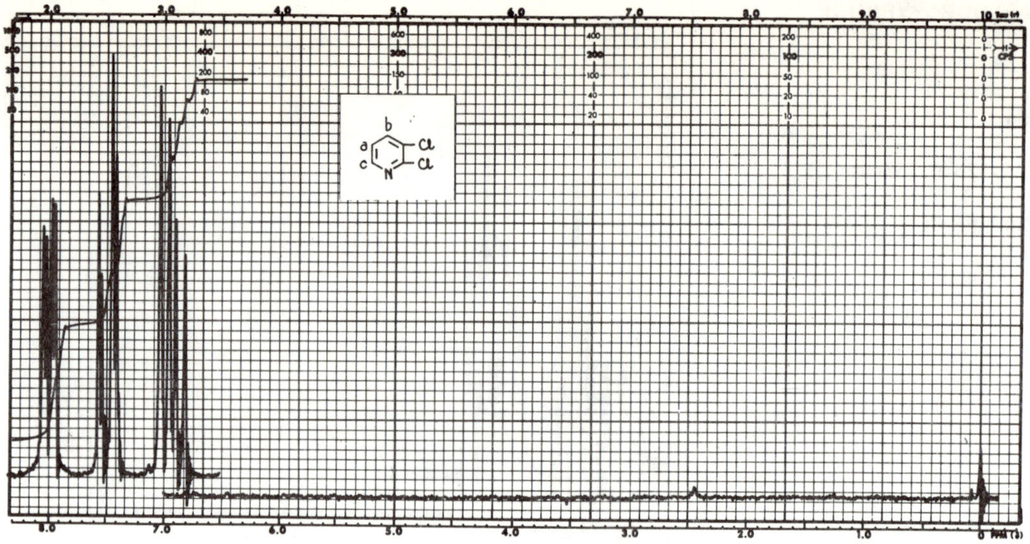

© 1978 Sadtler Research Laboratories, Inc.

2-BROMOPYRIDINE

C_5H_4BrN

Mol. Wt. 158.00

B.P. 192-194°C

Solvent: CCl_4

Assignments
a = 7.20
b = ca 7.40
c = 8.27

703

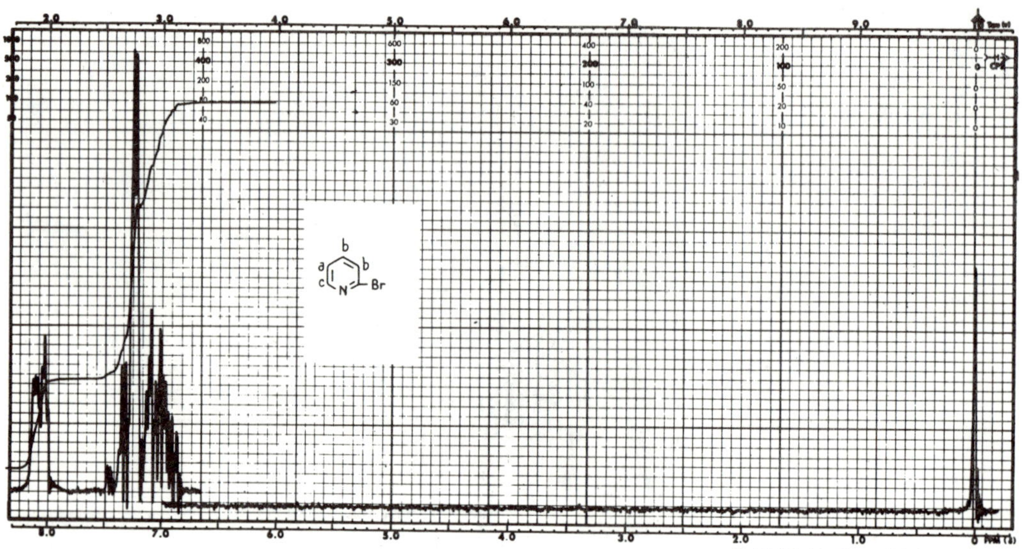

© 1978 Sadtler Research Laboratories, Inc.

3-BROMOPYRIDINE

C_5H_4BrN

Mol. Wt. 158.00

Solvent: CCl_4

Assignments
a = 7.10
b = 7.74
c = 8.48
d = 8.66

704

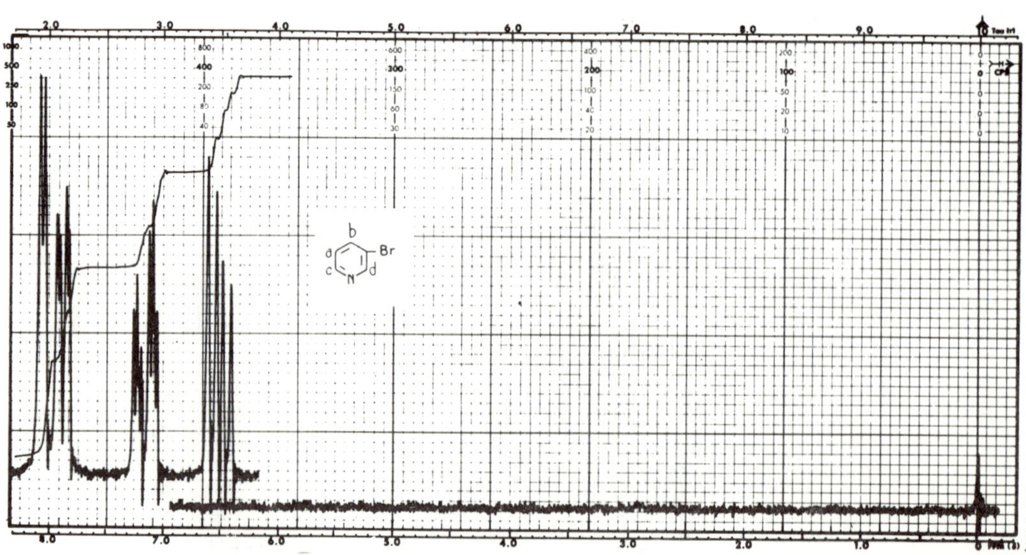

© 1978 Sadtler Research Laboratories, Inc.

2-[(DIETHYLAMINO)METHYL] PYRIDINE

$C_{10}H_{16}N_2$

Mol. Wt. 164.25

Solvent: $CDCl_3$

Assignments
a = 1.07
b = 2.59
c = 3.75
d = 7.06
e = 7.35-7.75
f = 8.52

705

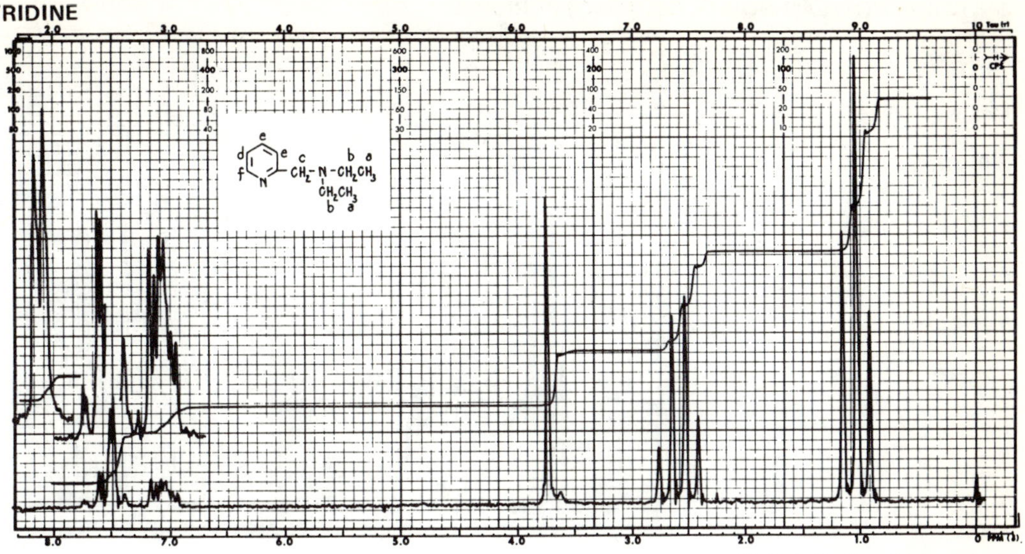

© 1978 Sadtler Research Laboratories, Inc.

2-AMINOPYRIDINE

$C_5H_6N_2$

Mol. Wt. 94.12

Solvent: $CDCl_3$

Assignments
a = 5.10
b = 6.36
c = 6.48
d = 7.22
e = 7.93

706

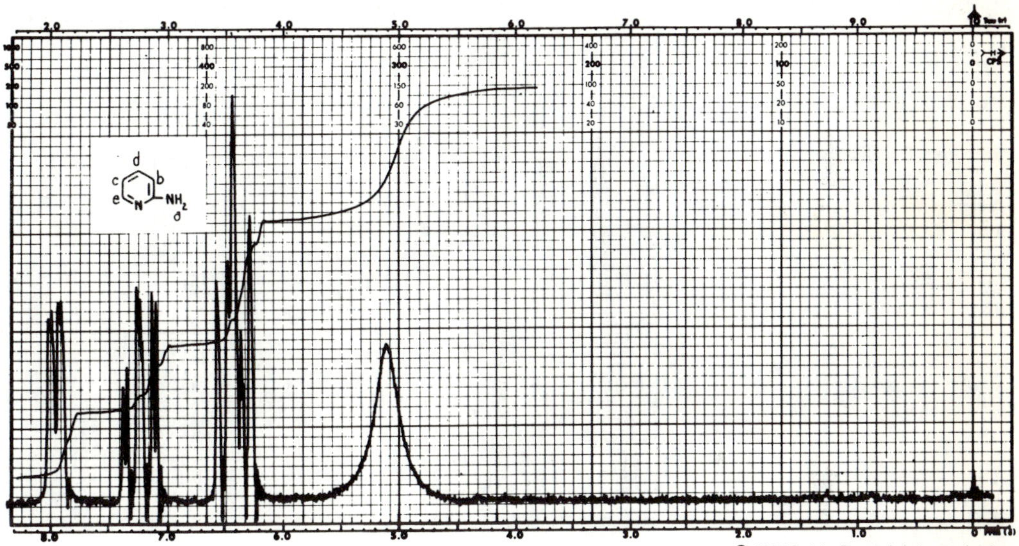

© 1978 Sadtler Research Laboratories, Inc.

2-AMINO-5-CHLOROPYRIDINE

$C_5H_5ClN_2$

Mol. Wt. 128.56

Solvent: Dimethyl sulfoxide-D_6

Assignments
a = 6.11
b = 6.58
c = 7.42
d = 7.94
e = 2.57 DMSO-D_5

707

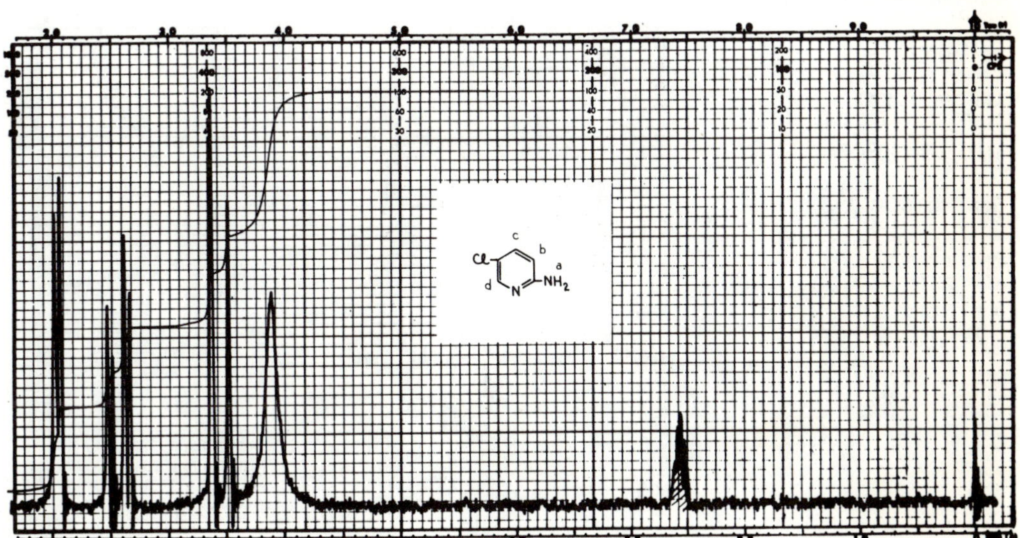

© 1978 Sadtler Research Laboratories, Inc.

2- {[2-(DIMETHYLAMINO)ETHYL] AMINO} PYRIDINE

$C_9H_{15}N_3$

Mol. Wt. 165.24

B.P. 265°C

Solvent: $CDCl_3$

Assignments
a = 2.24
b = 2.51
c = 3.36
d = 3.69
e = 6.45
f = 7.36
g = 8.07

708

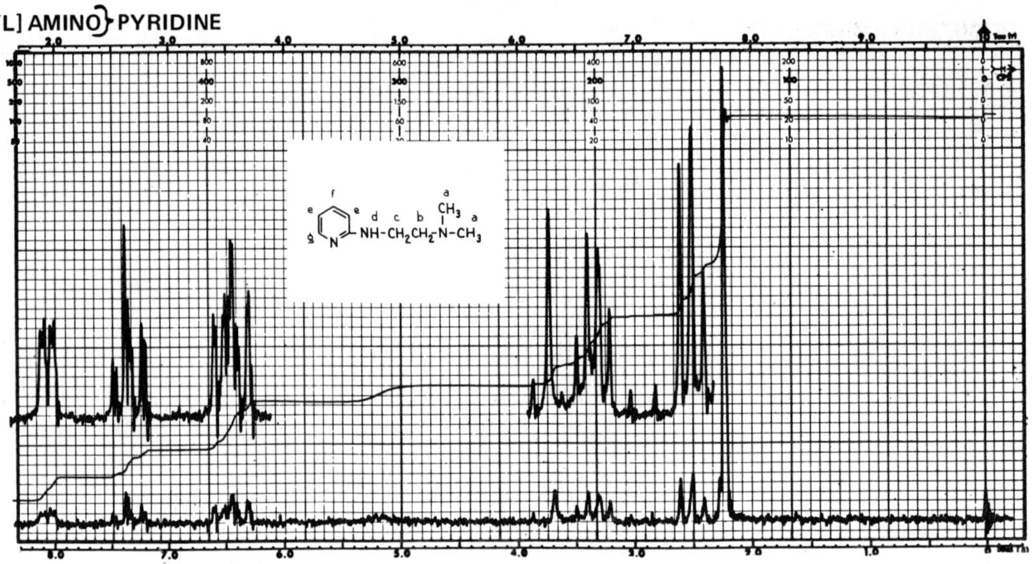

© 1978 Sadtler Research Laboratories, Inc.

3-(BENZYLAMINO)PYRIDINE

$C_{12}H_{12}N_2$

Mol. Wt. 184.24

Solvent: DMSO-d_6

Assignments
a = 4.29
b = 6.23
c = ca 6.91
d = 7.10-7.50
e = 7.79
f = 8.10
g = 2.43 DMSO-d_5
h = 3.39 H_2O

709

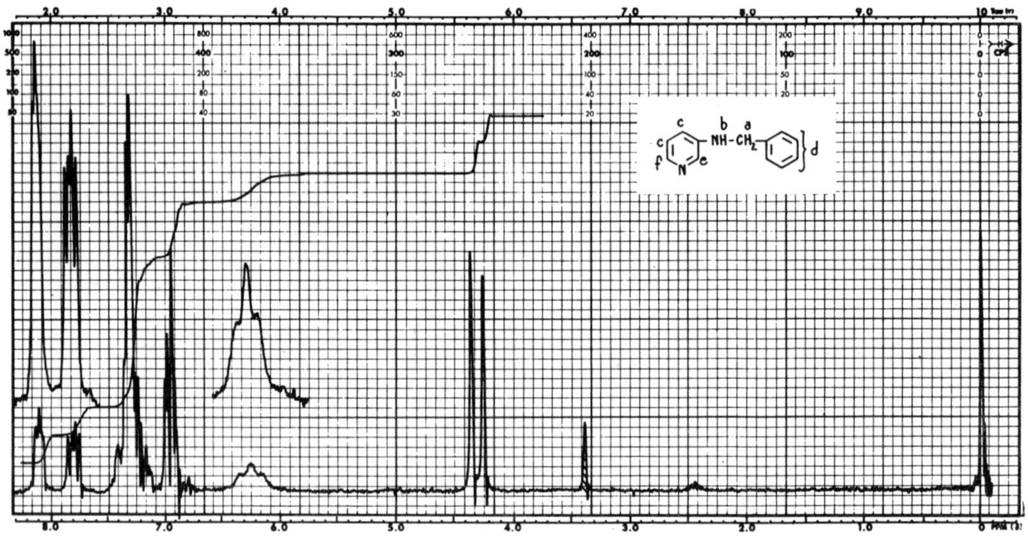

© 1978 Sadtler Research Laboratories, Inc.

4-(DIMETHYLAMINO)PYRIDINE

$C_7H_{10}N_2$

Mol. Wt. 122.17

M.P. 108-110°C

Solvent: $CDCl_3$

Assignments
a = 2.97
b = 6.50
c = 8.25

710

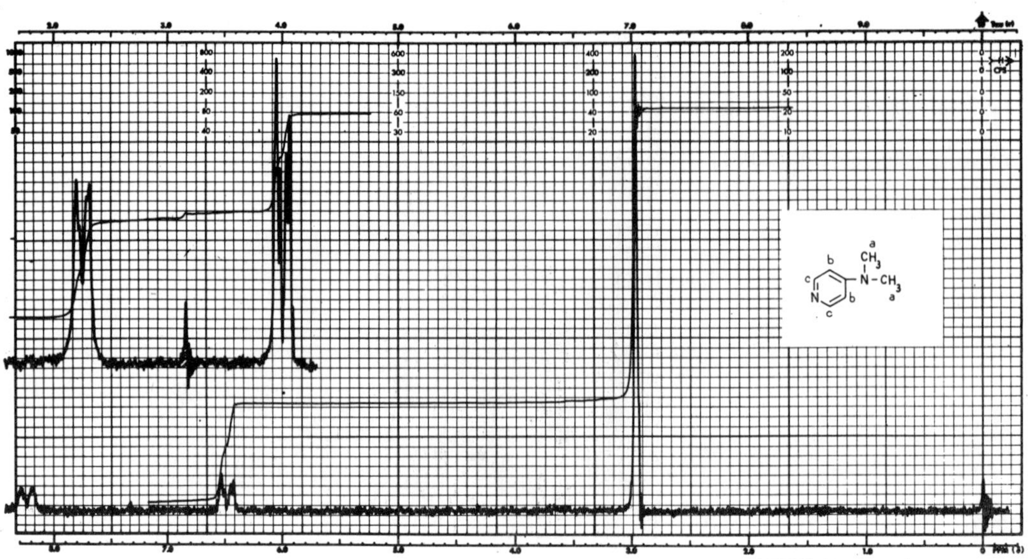

© 1978 Sadtler Research Laboratories, Inc.

NICOTINE

$C_{10}H_{14}N_2$

Mol. Wt. 162.23

Solvent: CCl_4

Assignments
a = 1.46-2.55
b = 2.15
c = 3.12
d = 7.13
e = 7.62
f = 8.52
g = 8.58

711

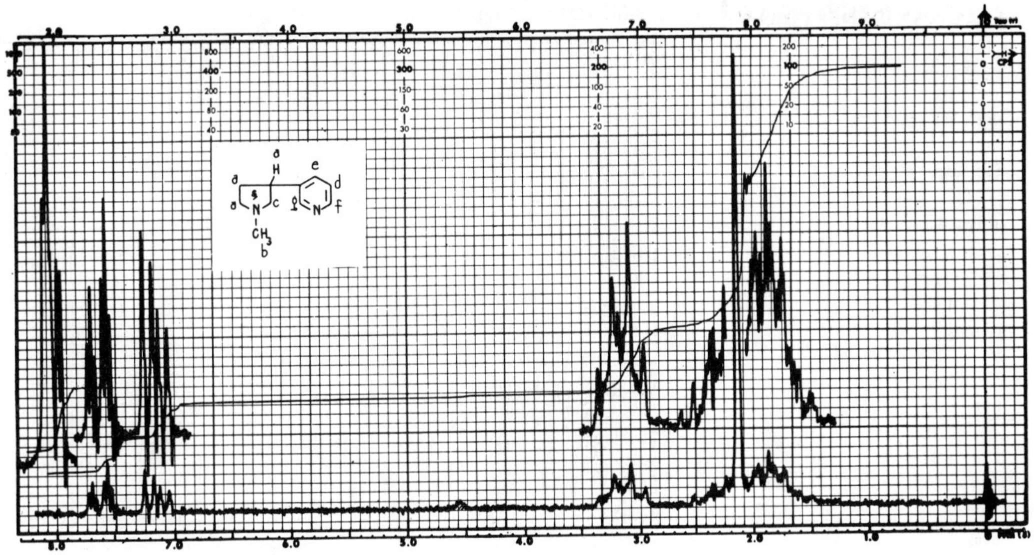

© 1978 Sadtler Research Laboratories, Inc.

2-(PIPERIDINOMETHYL)PYRIDINE

$C_{11}H_{16}N_2$

Mol. Wt. 176.26

Solvent: $CDCl_3$

Assignments
a = 1.10-1.80
b = 2.43
c = 3.62
d = 7.06
e = 7.40
f = 7.59
g = 8.52
h = 3.71 imp.

712

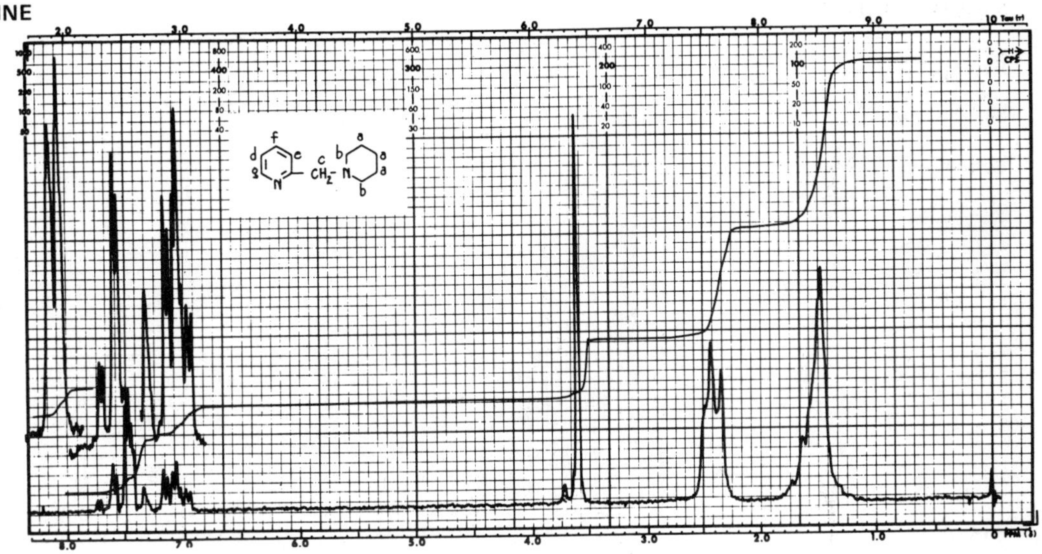

© 1978 Sadtler Research Laboratories, Inc.

3,4'-ETHYLENEDIPYRIDINE

$C_{12}H_{12}N_2$

Mol. Wt. 184.24

Solvent: $CDCl_3$

Assignments
a = 2.93
b = 7.06
c = 7.17
d = 7.43
e = ca 8.49

713

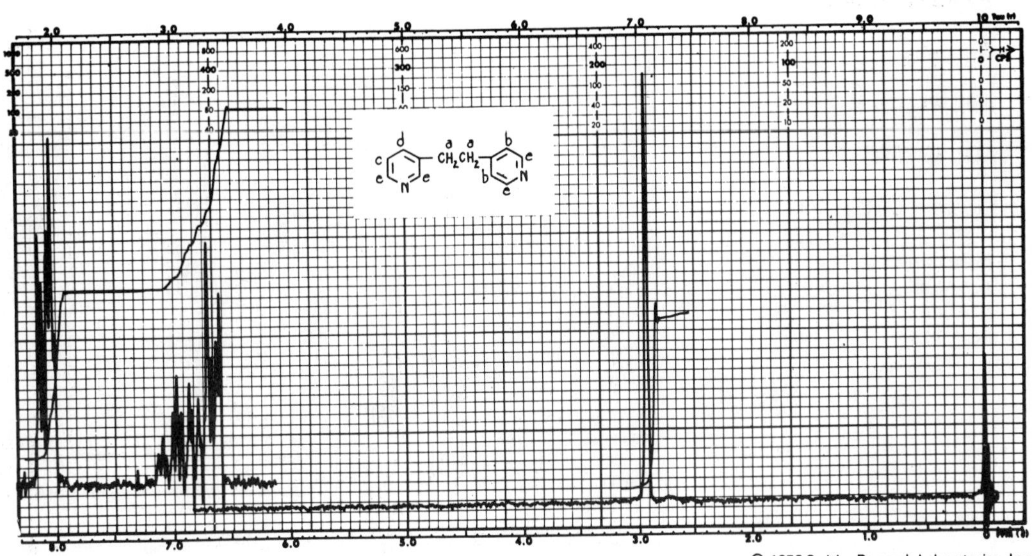

© 1978 Sadtler Research Laboratories, Inc.

2,3'-VINYLIDENEPYRIDINE

$C_{12}H_{10}N_2$

Mol. Wt. 182.23

Solvent: $CDCl_3$

Assignments
a = 6.90-7.90
b = 8.49
c = 8.59
d = 8.79

714

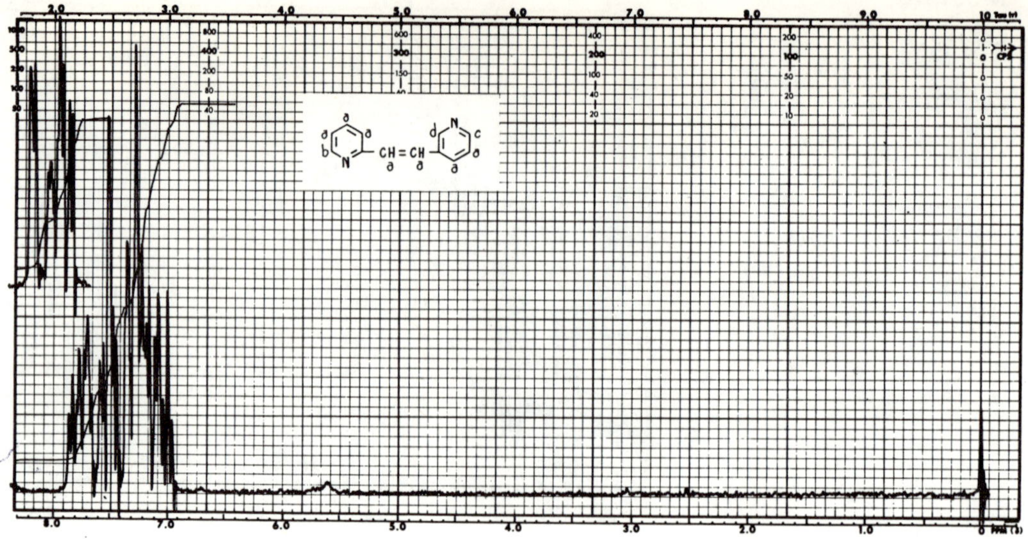

© 1978 Sadtler Research Laboratories, Inc.

2,4'-VINYLENEDIPYRIDINE

$C_{12}H_{10}N_2$

Mol. Wt. 182.23

Solvent: $CDCl_3$

Assignments
a = 6.90-7.70
b = ca 8.52

715

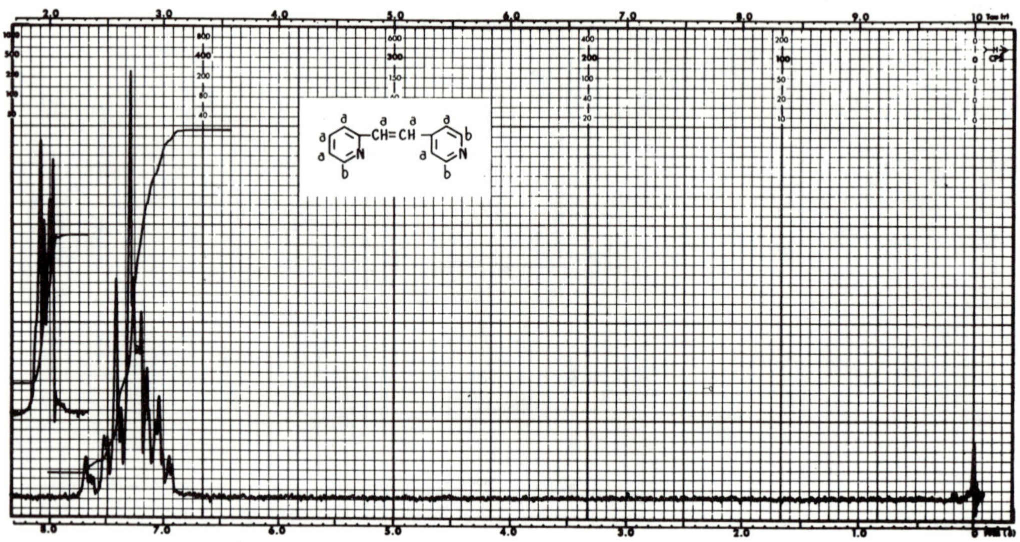

© 1978 Sadtler Research Laboratories, Inc.

2,2'-BIPYRIDINE

$C_{10}H_8N_2$

Mol. Wt. 156.19

M.P. 69-70°C

Solvent: $CDCl_3$

Assignments
a = 7.25
b = 7.77
c = 8.42
d = 8.67

716

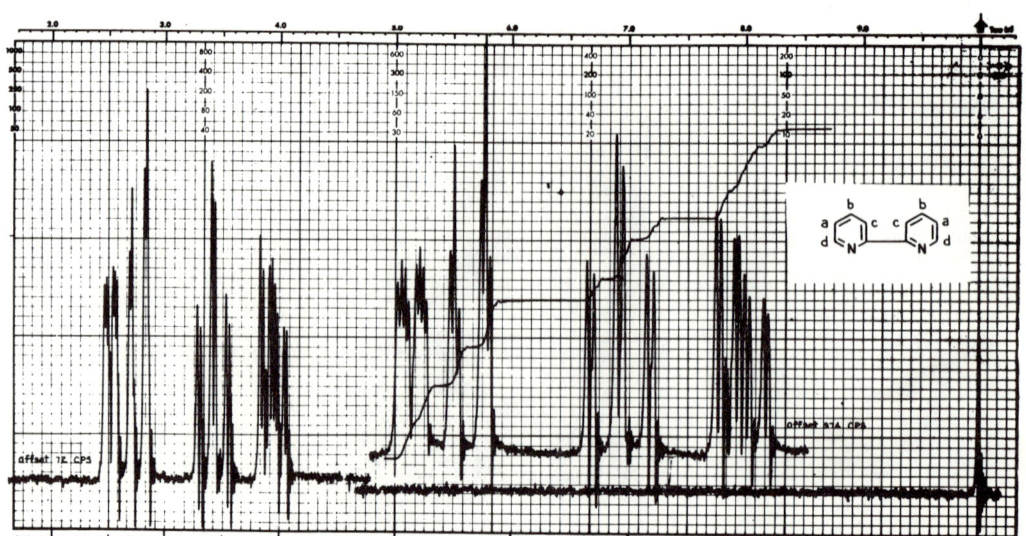

© 1978 Sadtler Research Laboratories, Inc.

4,4'-BIPYRIDINE

$C_{10}H_8N_2$

Mol. Wt. 156.19

M.P. 114°C (lit.)

Solvent: $CDCl_3$

Assignments
a = 7.51
b = 8.71
c = 4.01 imp.

717

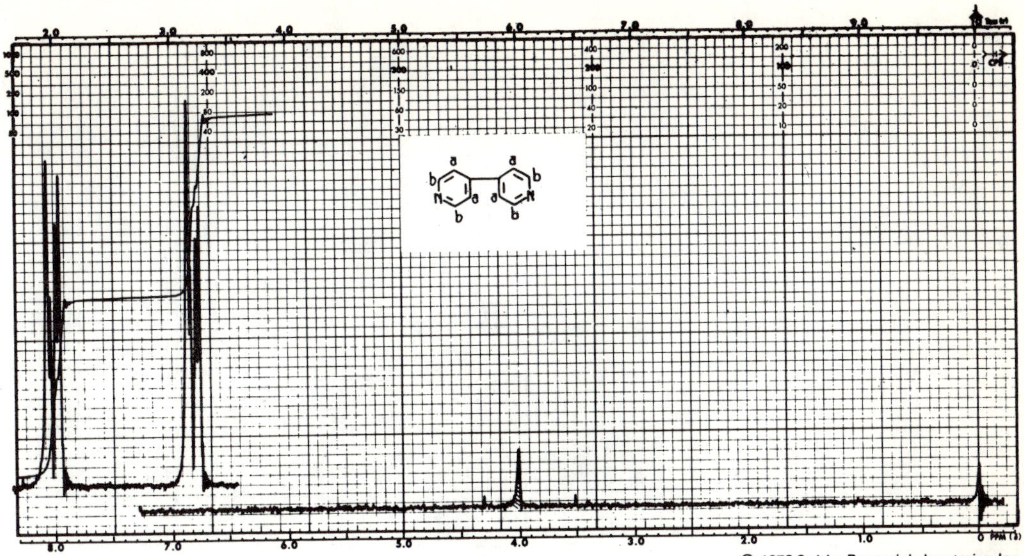

© 1978 Sadtler Research Laboratories, Inc.

QUINOLINE

C_9H_7N

Mol. Wt. 129.16

B.P. 235-237°C

Solvent: CCl_4

Assignments
a = 7.13
b = 7.25-8.30
c = 8.84

718

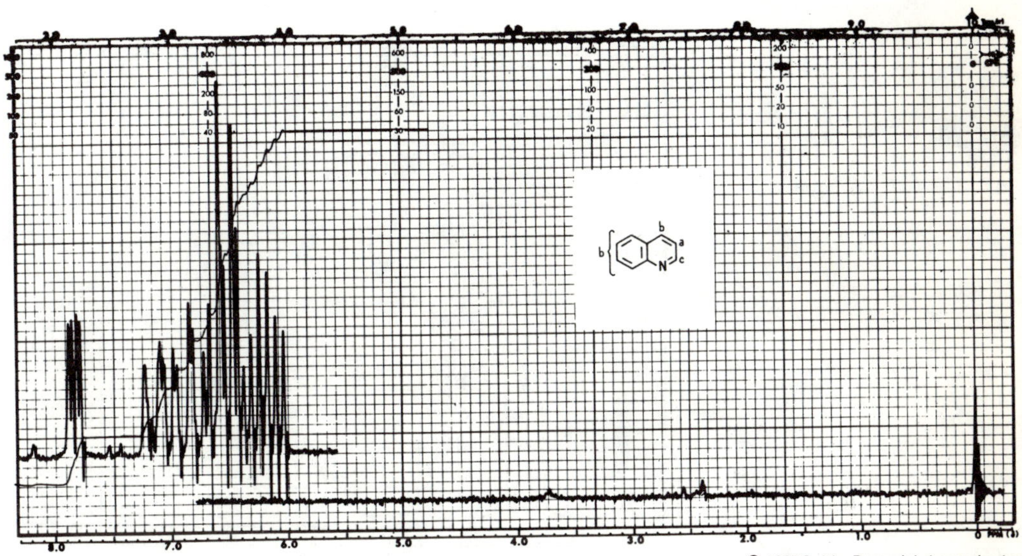

© 1978 Sadtler Research Laboratories, Inc.

QUINALDINE

$C_{10}H_9N$

Mol. Wt. 143.19

M.P. −2°C

Solvent: CCl_4

Assignments
a = 2.64
b = 7.05
c = 7.26-7.70
d = 7.82
e = 8.01

719

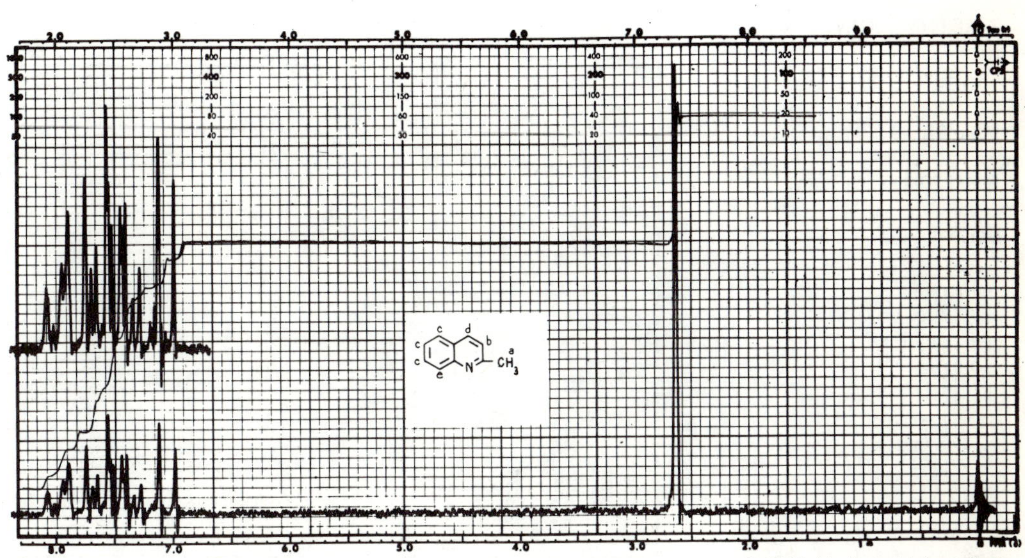

© 1978 Sadtler Research Laboratories, Inc.

LEPIDINE

$C_{10}H_9N$

Mol. Wt. 143.18

M.P. 9-10°C

B.P. 261-263°C

Solvent: $CDCl_3$

Assignments
a = 2.67
b = 7.18
c = ca 7.58
d = 8.04
e = 8.73

720

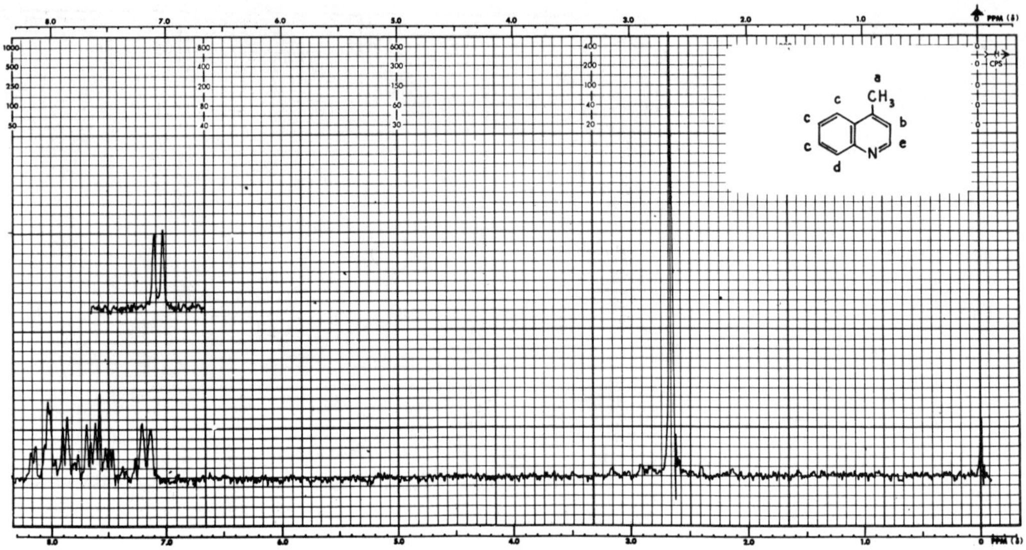

© 1978 Sadtler Research Laboratories, Inc.

6-METHYLQUINOLINE

$C_{10}H_9N$

Mol. Wt. 143.19

B.P. 129-131°C/15mm

Solvent: $CDCl_3$

Assignments
a = 2.36
b = 7.15
c = 7.29
d = 7.35
e = 7.82
f = 7.99
g = 8.78

721

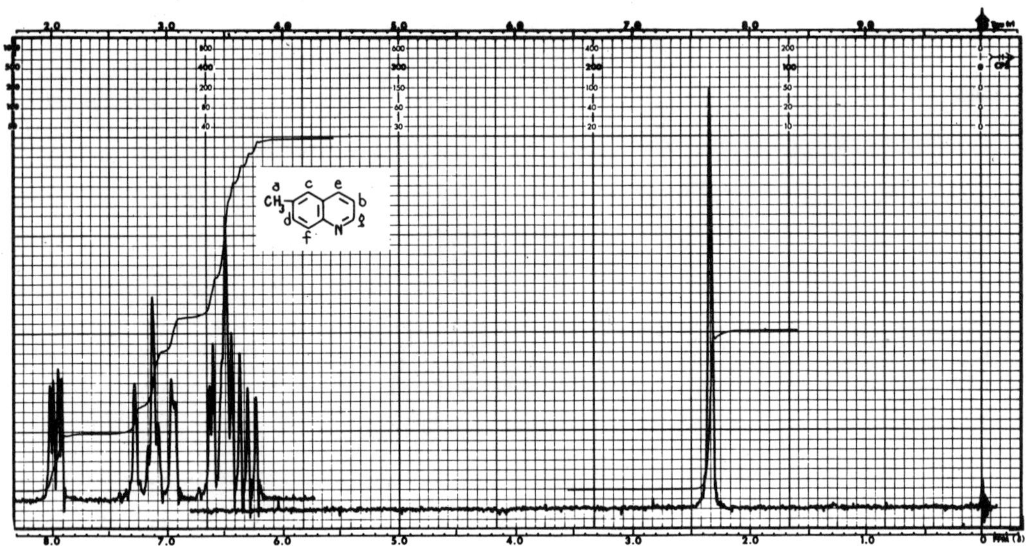

© 1978 Sadtler Research Laboratories, Inc.

7-METHYLQUINOLINE

$C_{10}H_9N$

Mol. Wt. 143.19

B.P. 120-122°C/7mm

Solvent: CCl_4

Assignments
a = 2.40
b = 6.90-7.30
c = 7.48
d = 7.82
e = 7.90
f = 8.79

722

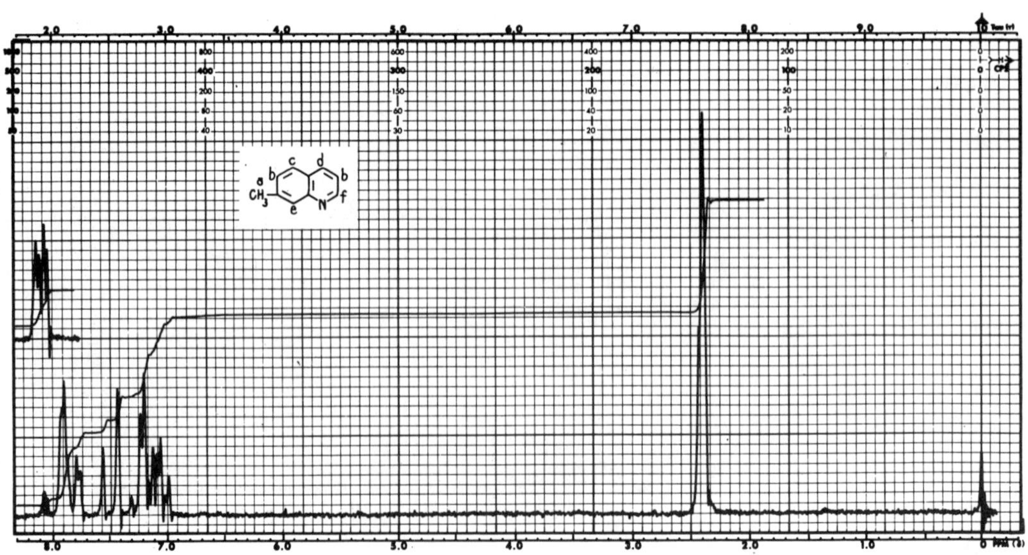

© 1978 Sadtler Research Laboratories, Inc.

8-METHYLQUINOLINE

$C_{10}H_9N$

Mol. Wt. 143.19

B.P. 112-114°C/8mm

Solvent: CCl_4

Assignments
a = 2.80
b = 7.12
c = 7.20-7.60
d = 7.84
e = 8.79

723

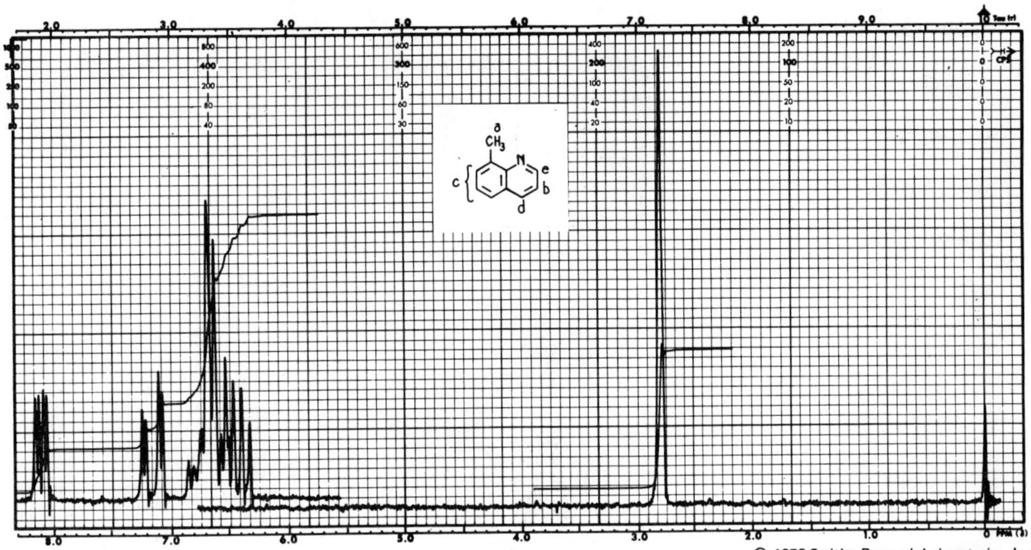

© 1978 Sadtler Research Laboratories, Inc.

2-CHLOROQUINOLINE

C_9H_6ClN

Mol. Wt. 163.61

M.P. 35.2-36.7°C

Solvent: $CDCl_3$

Assignments
a = 7.29
b = 7.40-7.82
c = 7.95

724

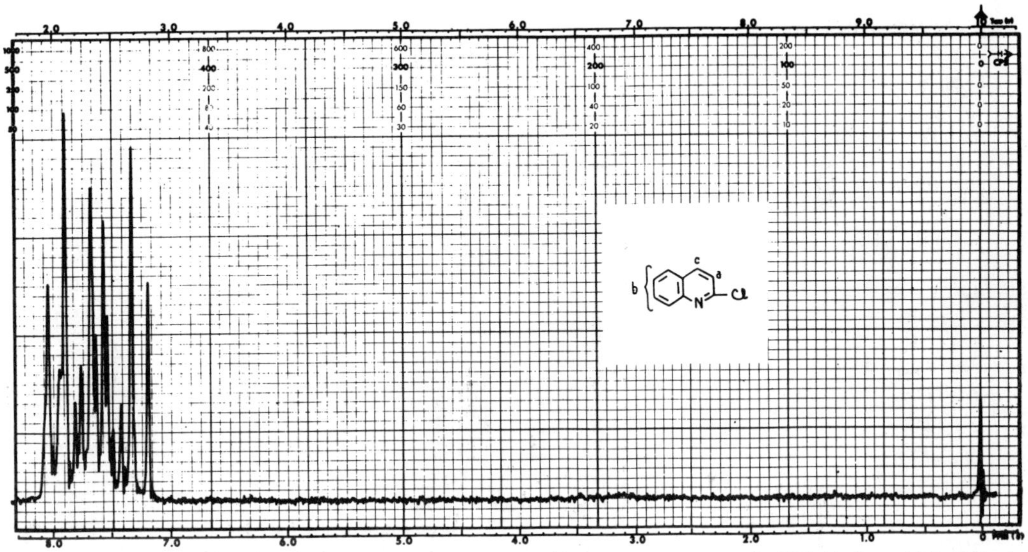

© 1978 Sadtler Research Laboratories, Inc.

6-BROMOQUINOLINE

C_9H_6BrN

Mol. Wt. 208.06

M.P. 17-19°C

Solvent: CCl_4

Assignments
a = 7.22
b = 7.50-8.05
c = 8.82

725

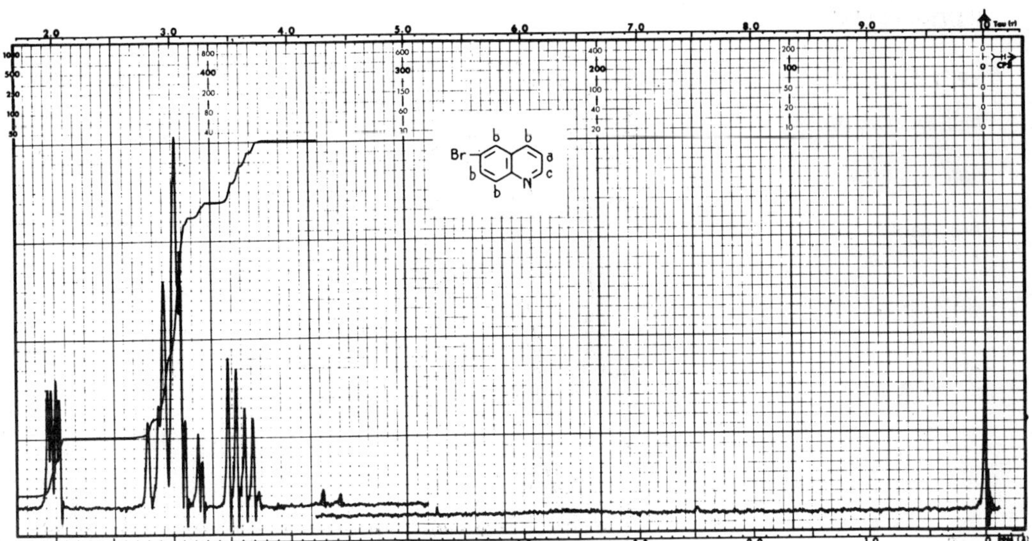

© 1978 Sadtler Research Laboratories, Inc.

2-IODOQUINOLINE

C_9H_6IN

Mol. Wt. 255.06

M.P. 54-56°C

Solvent: $CDCl_3$

Assignments
a = 7.20-7.80
b = 7.82

726

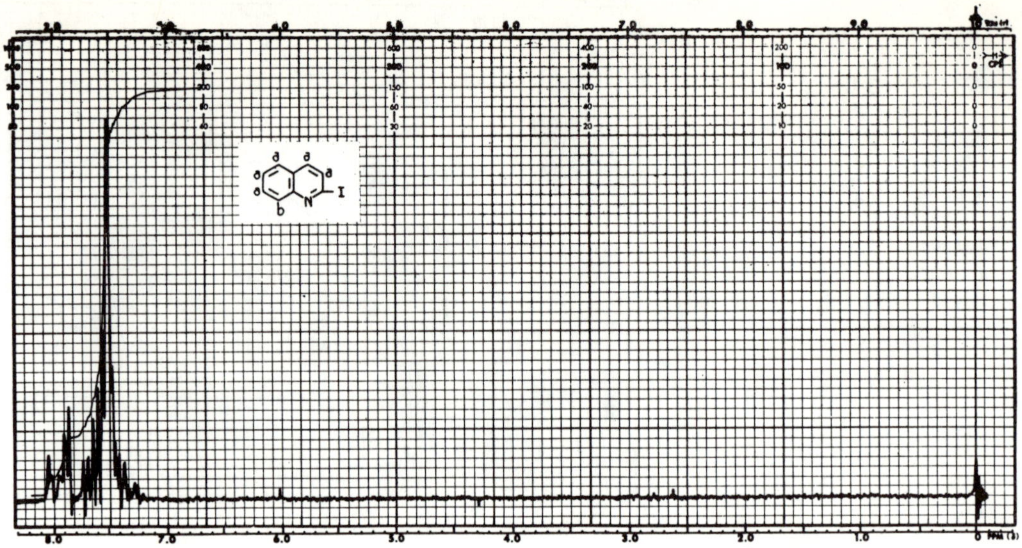

© 1978 Sadtler Research Laboratories, Inc.

3-AMINOQUINOLINE

$C_9H_8N_2$

Mol. Wt. 144.18

M.P. 92-94°C

Solvent: $CDCl_3$

Assignments
a = 3.99
b = 7.10
c = 7.22-7.65
d = 7.88
e = 8.48

727

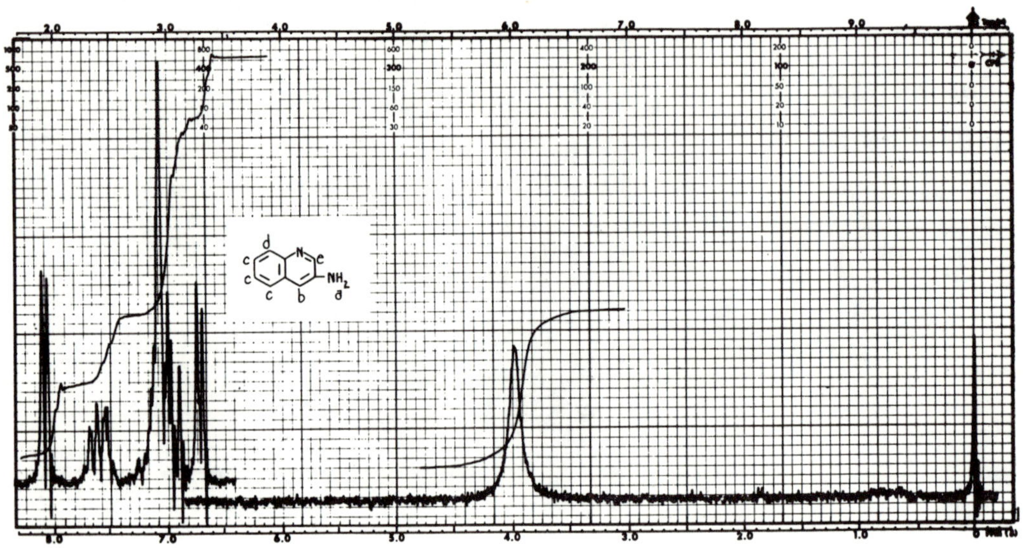

© 1978 Sadtler Research Laboratories, Inc.

ISOQUINOLINE

C_9H_7N

Mol. Wt. 129.16

Solvent: CCl_4

Assignments
a = 7.18-7.89
b = 8.43
c = 3.56 imp.

728

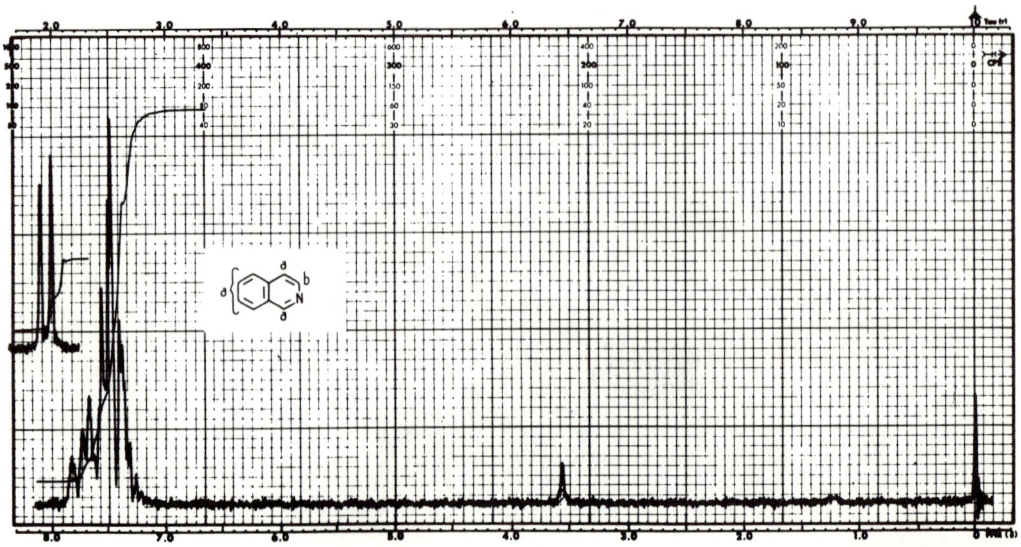

© 1978 Sadtler Research Laboratories, Inc.

1-CHLOROISOQUINOLINE

C_9H_6ClN

Mol. Wt. 163.61

M.P. 35-37°C

Solvent: $CDCl_3$

Assignments
a = 7.35
b = 7.40-7.65
c = 8.00
d = 8.16

729

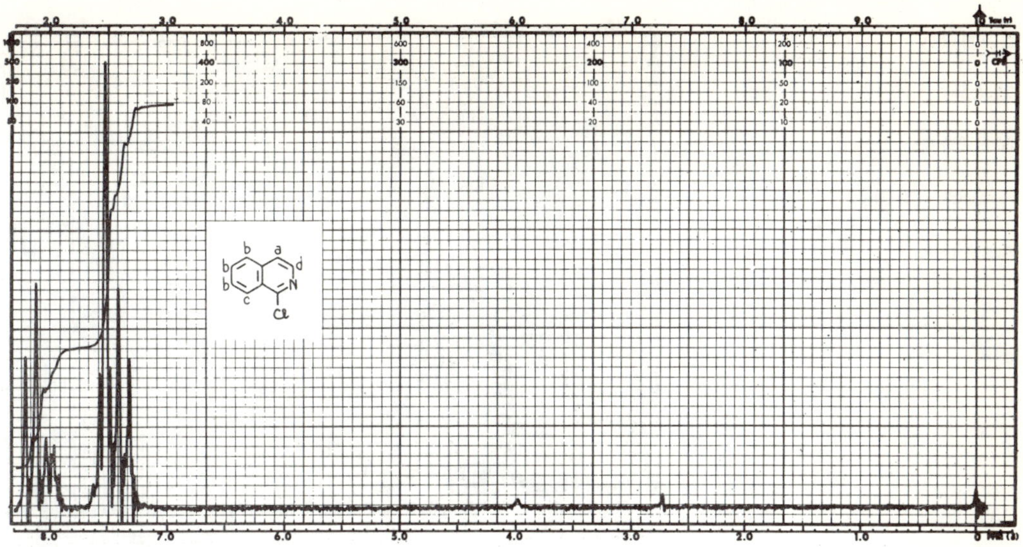

© 1978 Sadtler Research Laboratories, Inc.

1-CHLORO-3-METHYLISOQUINOLINE

$C_{10}H_8ClN$

Mol. Wt. 177.64

Solvent: $CDCl_3$

Assignments
a = 2.50
b = 7.11
c = 7.20-7.50
d = 8.02

730

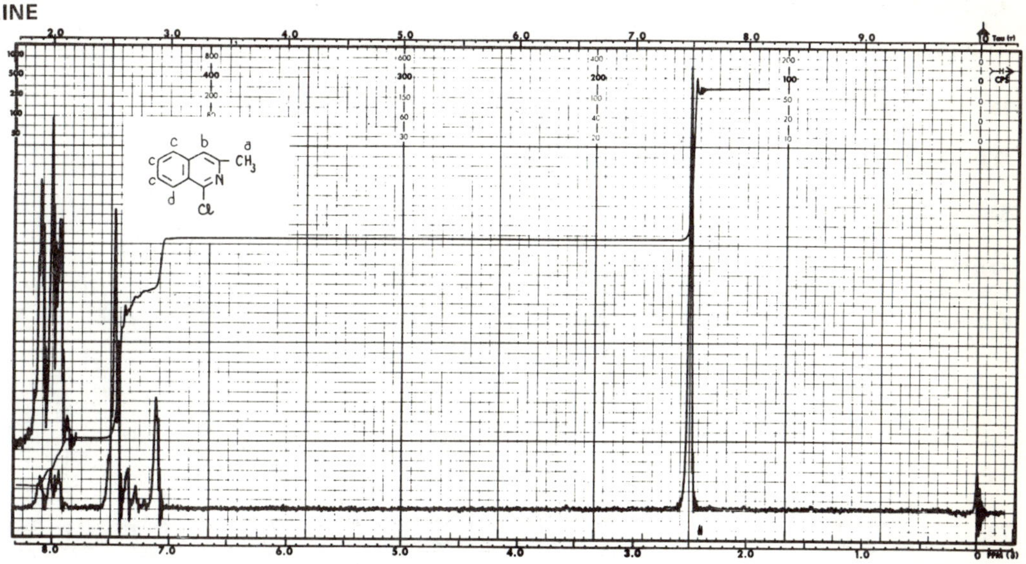

© 1978 Sadtler Research Laboratories, Inc.

4-BROMOISOQUINOLINE

C_9H_6BrN

Mol. Wt. 208.06

Solvent: $CDCl_3$

Assignments
a = 7.30-7.85
b = 7.95
c = 8.62
d = 9.02

731

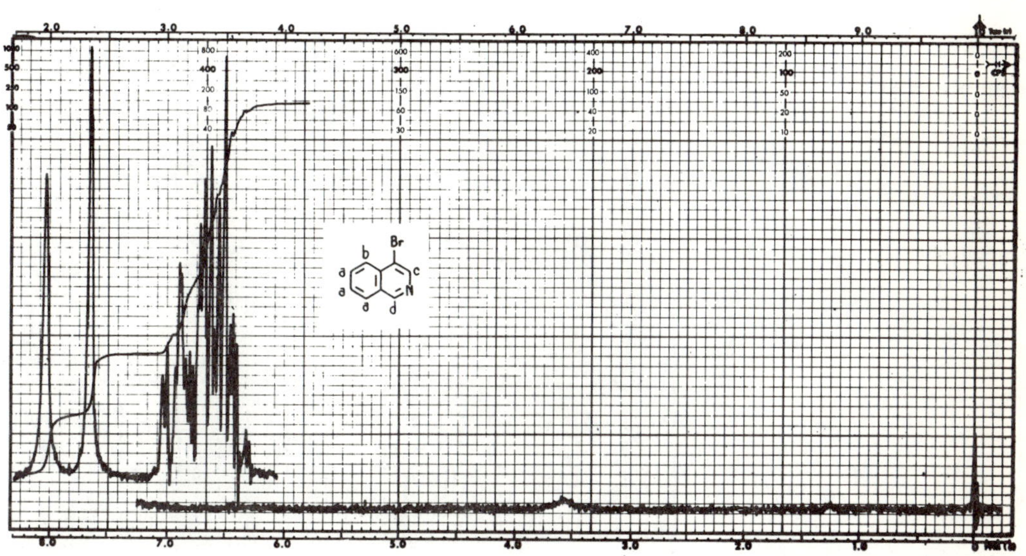

© 1978 Sadtler Research Laboratories, Inc.

4-PHENYLPYRIMIDINE

$C_{10}H_8N_2$

Mol. Wt. 156.19

M.P. 60-63°C

Solvent: $CDCl_3$

Assignments
a = 7.30-7.55
b = 7.62
c = ca 8.08
d = 8.70
e = 9.28

732

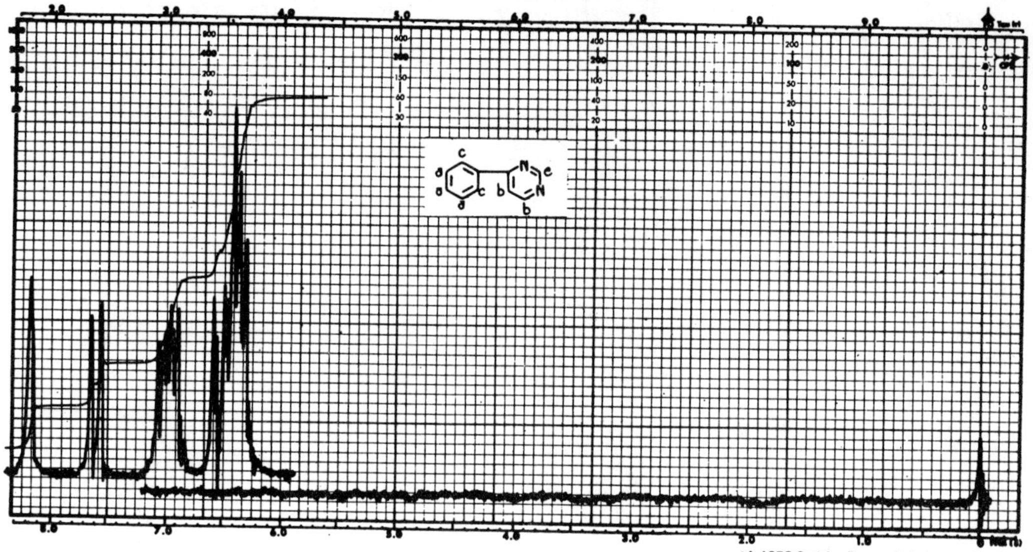

© 1978 Sadtler Research Laboratories, Inc.

4,6-DICHLOROPYRIMIDINE

$C_4H_2Cl_2N_2$

Mol. Wt. 148.98

M.P. 64-66°C

Solvent: $CDCl_3$

Assignments
a = 7.52
b = 8.89

733

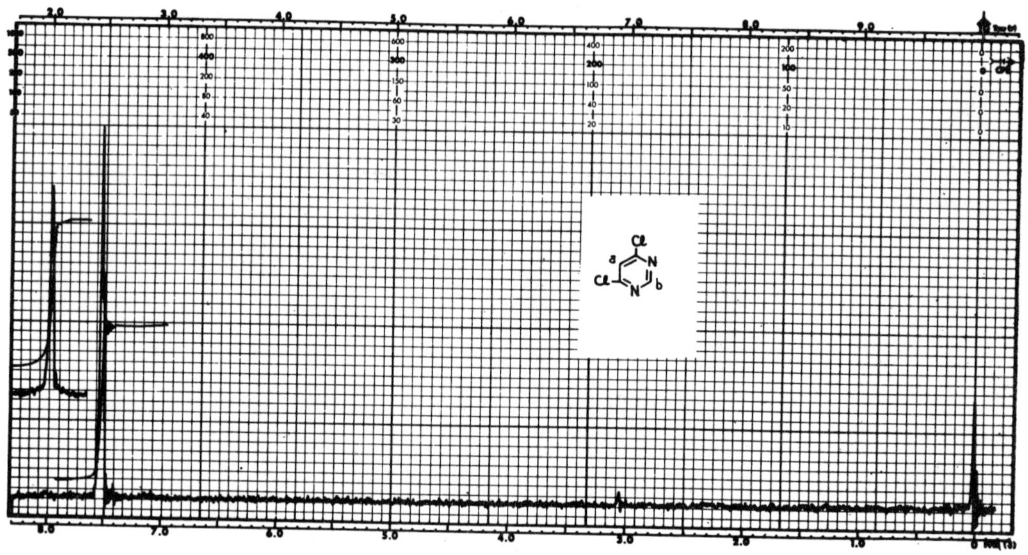

© 1978 Sadtler Research Laboratories, Inc.

2-AMINOPYRIMIDINE

$C_4H_5N_3$

Mol. Wt. 95.11

M.P. 125-127°C

Solvent: D_2O

Assignments
a = 6.79
b = 8.39
c = 4.53 HDO

734

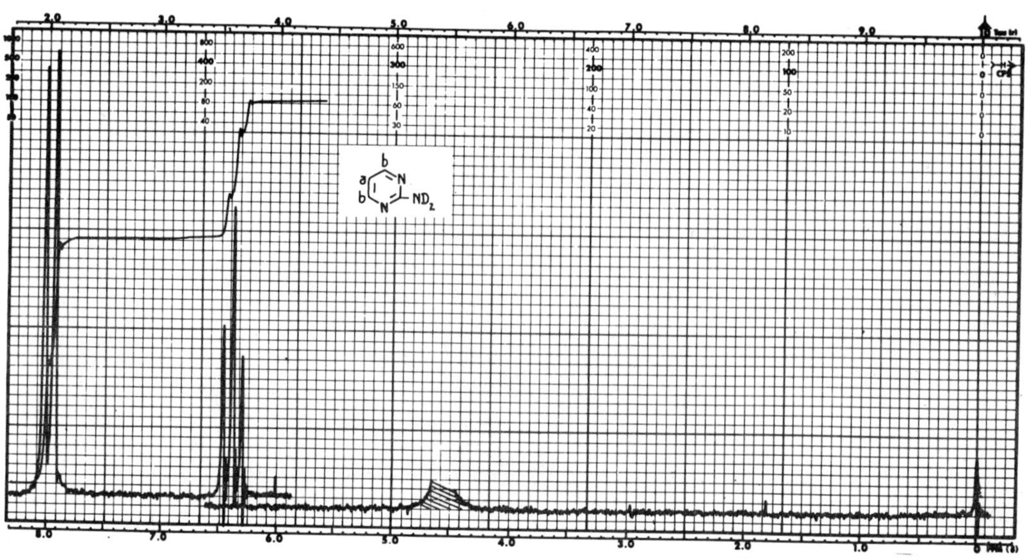

© 1978 Sadtler Research Laboratories, Inc.

TETRAMETHYLPYRAZINE

$C_8H_{12}N_2$

Mol. Wt. 136.20

Solvent: $CDCl_3$

Assignments
a = 2.43

735

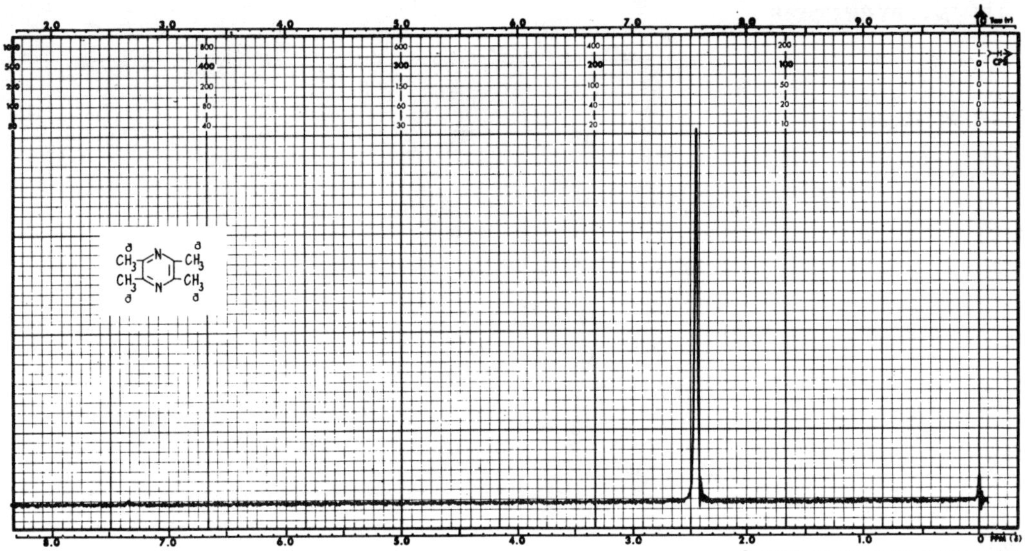

© 1978 Sadtler Research Laboratories, Inc.

2,4,6-TRIPHENYL-s-TRIAZINE

$C_{21}H_{15}N_3$

Mol. Wt. 309.37

M.P. 237-239°C

Solvent: Trifluoroacetic acid

Assignments
a = 7.40-7.90
b = 8.48
c = unobserved *
* Under solvent band

736

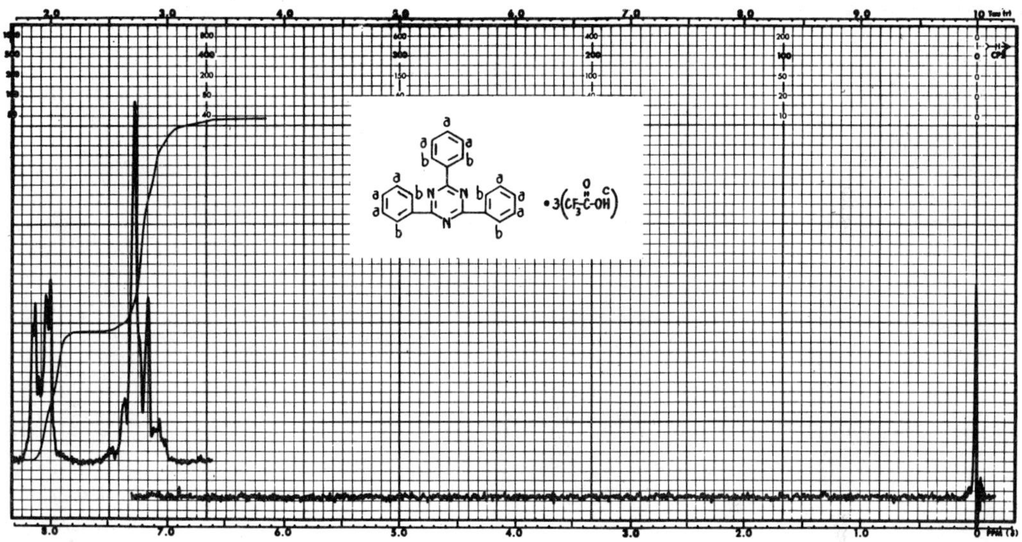

© 1978 Sadtler Research Laboratories, Inc.

2,4,6-TRIS[DI(2-ETHYLHEXYL)AMINO]-s-TRIAZINE

$C_{51}H_{102}N_6$

Mol. Wt. 799.42

M.P. 245-247°C

Solvent: CCl_4

Assignments
a = 0.90
b = ca 1.27
c = 1.79
d = 3.43

737

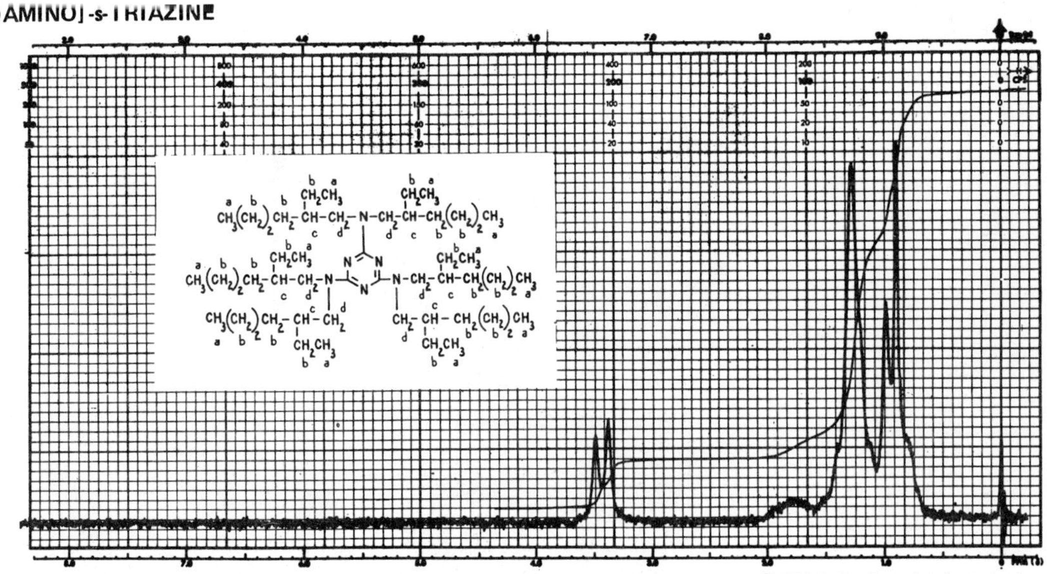

© 1978 Sadtler Research Laboratories, Inc.

BENZIMIDAZOLE

$C_7H_6N_2$

Mol. Wt. 118.14

M.P. 170-173°C

Solvent: Dimethyl sulfoxide-D_6

Assignments
a = 7.22
b = 7.66
c = 8.30
d = 2.52 DMSO-D_5
NH is unobserved

738

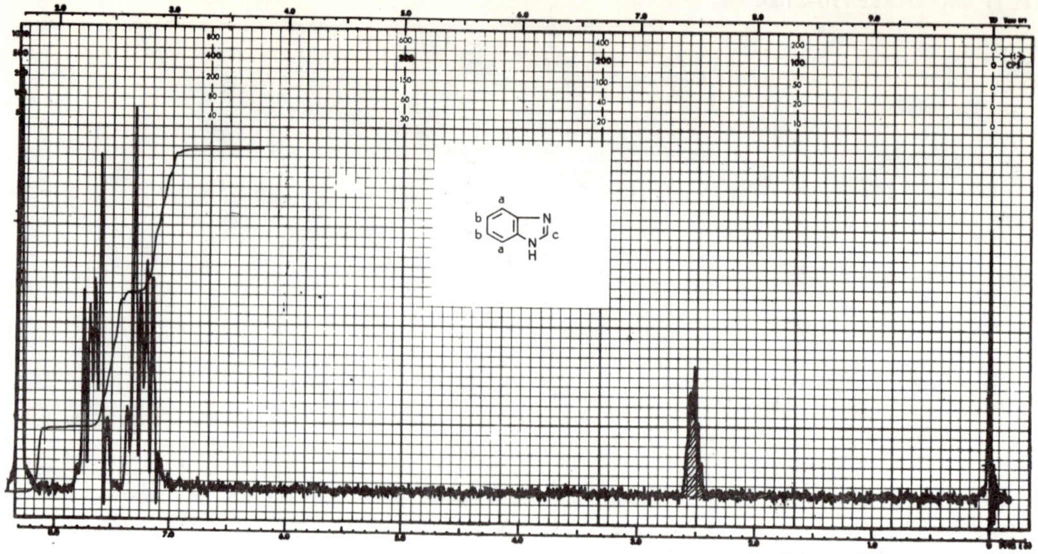

© 1978 Sadtler Research Laboratories, Inc.

2-PROPYLBENZIMIDAZOLE

$C_{10}H_{12}N_2$

Mol. Wt. 160.22

Solvent: Polysol-d

Assignments
a = 1.00
b = 1.90
c = 2.90
d = 7.12
e = 7.51
f = 11.58

739

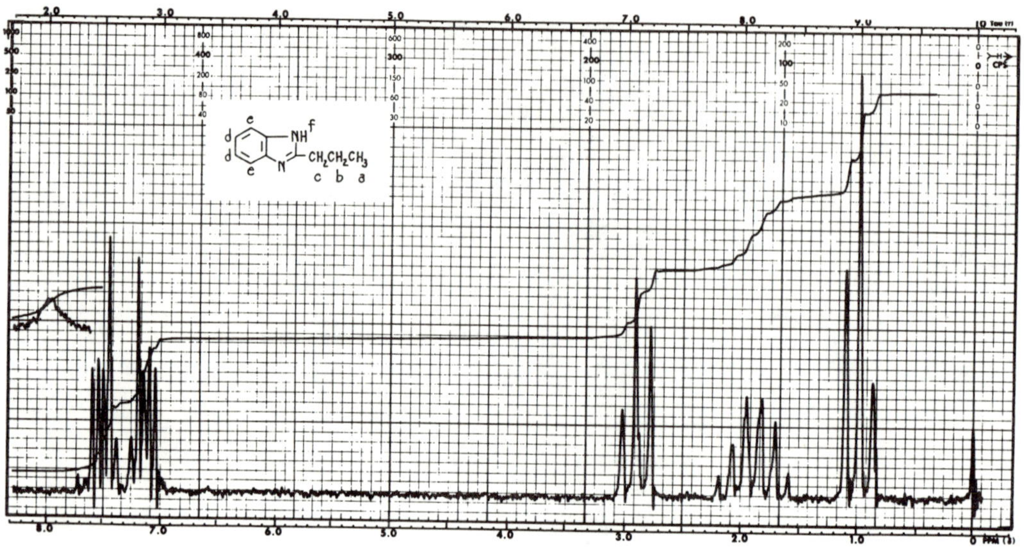

© 1978 Sadtler Research Laboratories, Inc.

5-METHYLBENZIMIDAZOLE

$C_8H_8N_2$

Mol. Wt. 132.17

M.P. 116-118°C

Solvent: $CDCl_3$

Assignments
a = 2.47
b = 7.11
c = 7.49
d = 7.58
e = 8.12
f = 12.49

740

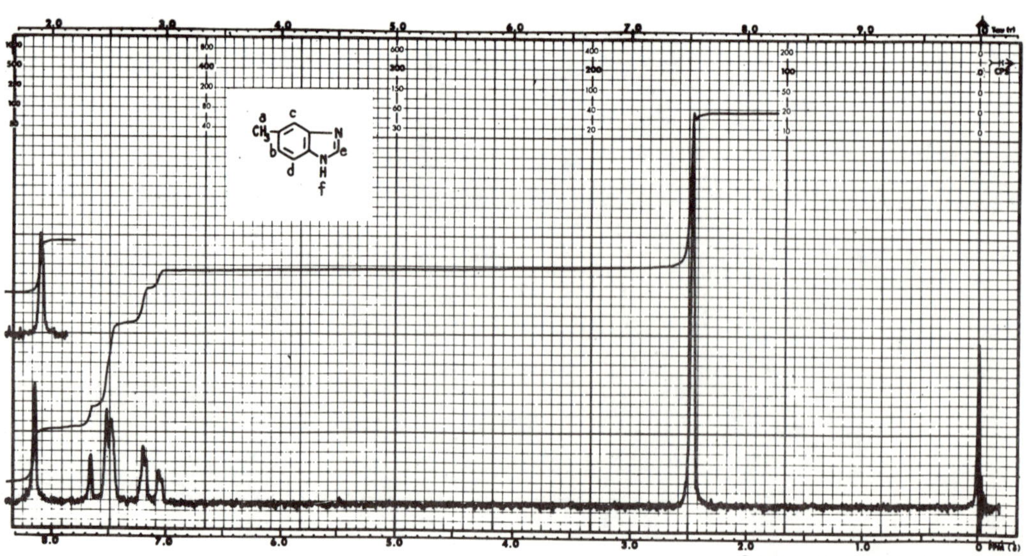

© 1978 Sadtler Research Laboratories, Inc.

1H-BENZOTRIAZOLE

$C_6H_5N_3$

Mol. Wt. 119.13

M.P. 96.8-98.8°C

Solvent: $CDCl_3$

Assignments
a = 7.44
b = 8.01
c = 14.49

741

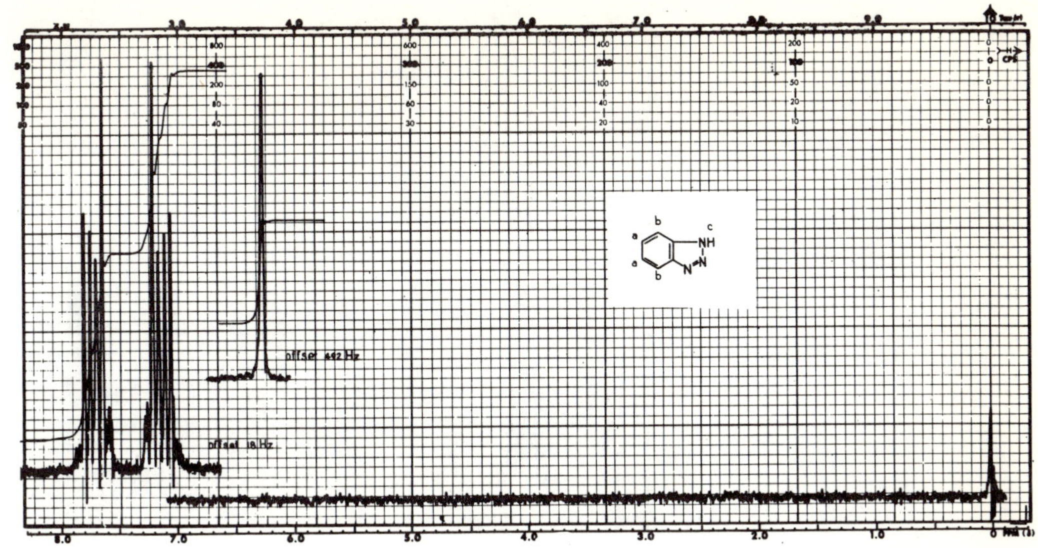

© 1978 Sadtler Research Laboratories, Inc.

4-CHLOROQUINAZOLINE

$C_8H_5ClN_2$

Mol. Wt. 164.60

Solvent: Trifluoroacetic acid

Assignments
a = 7.70-8.30
b = 8.56
c = 9.55

742

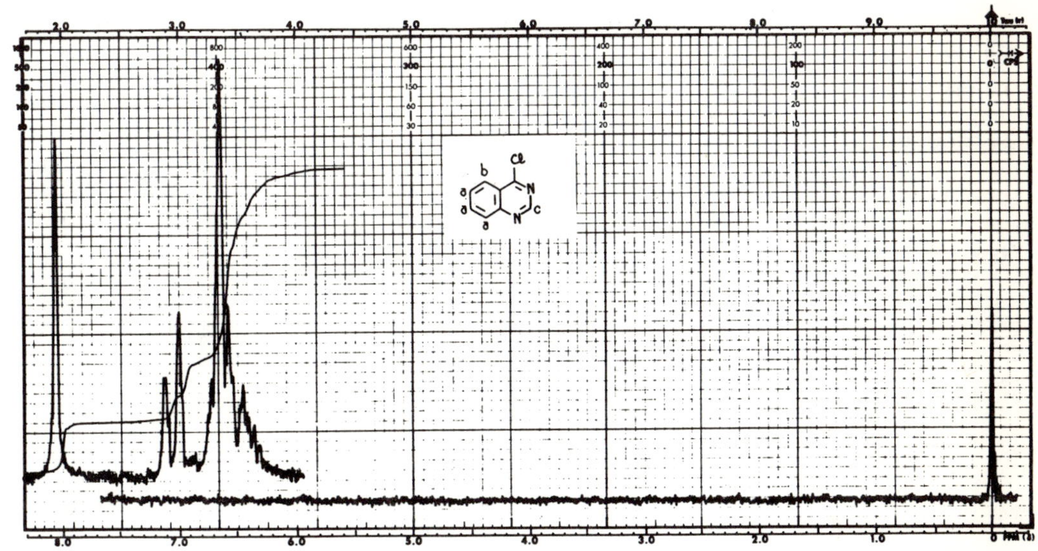

© 1978 Sadtler Research Laboratories, Inc.

PURINE

$C_5H_4N_4$

Mol. Wt. 120.11

M.P. 216-217°C

Solvent: Dimethyl sulfoxide-D_6

Assignments
a = 8.60-9.12
b = 2.52 DMSO-D_5
-NH is unobserved

743

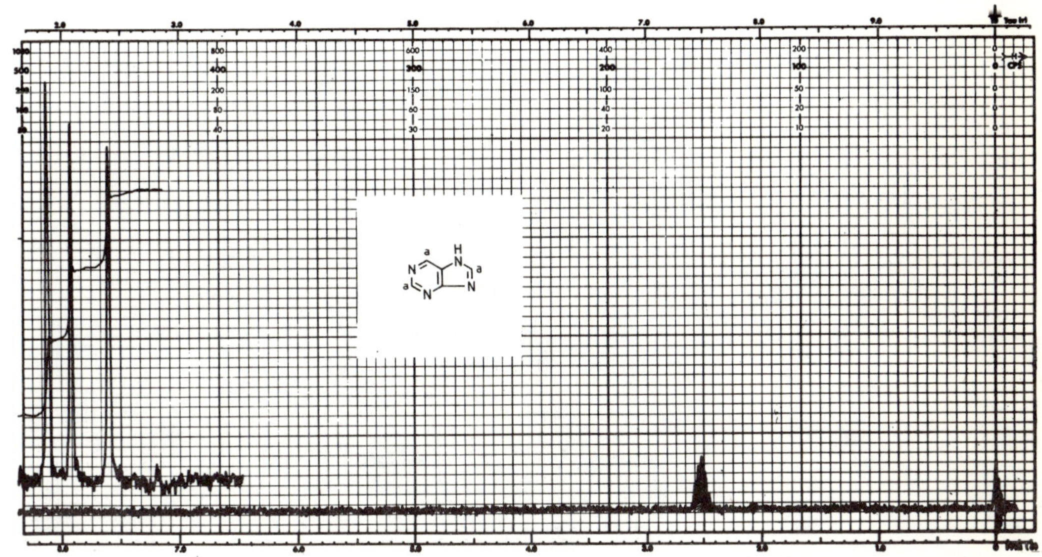

© 1978 Sadtler Research Laboratories, Inc.

9-ETHYLCARBAZOLE

$C_{14}H_{13}N$

Mol. Wt. 182.66

Solvent: $CDCl_3$

Assignments
a = 1.21
b = 4.10
c = 7.00-7.60
d = 8.04

744

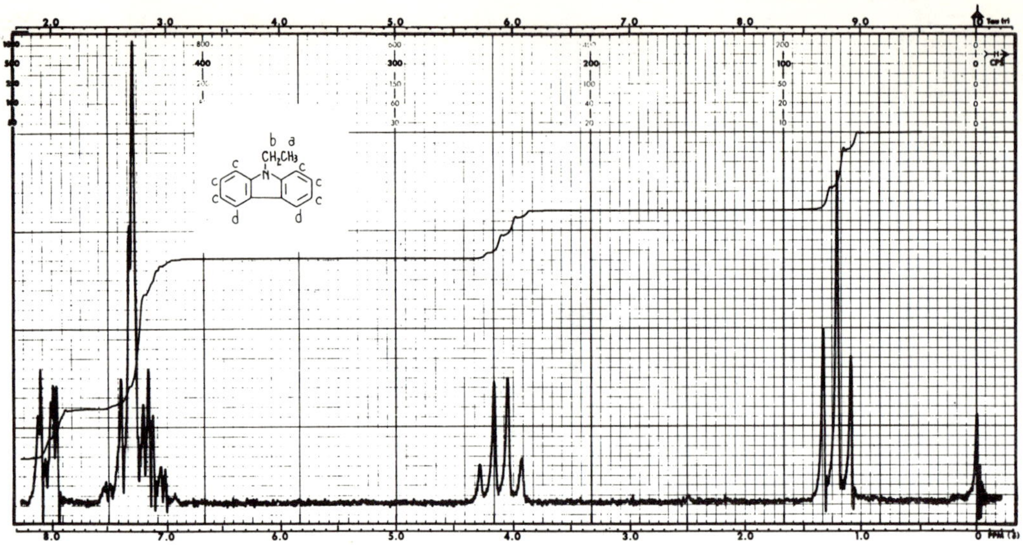

© 1978 Sadtler Research Laboratories, Inc.

9-PHENYLCARBAZOLE

$C_{18}H_{13}N$

Mol. Wt. 243.31

M.P. 91-93°C

Solvent: $CDCl_3$

Assignments
a = 7.10-7.70
b = 8.13
c = 1.43 H_2O

745

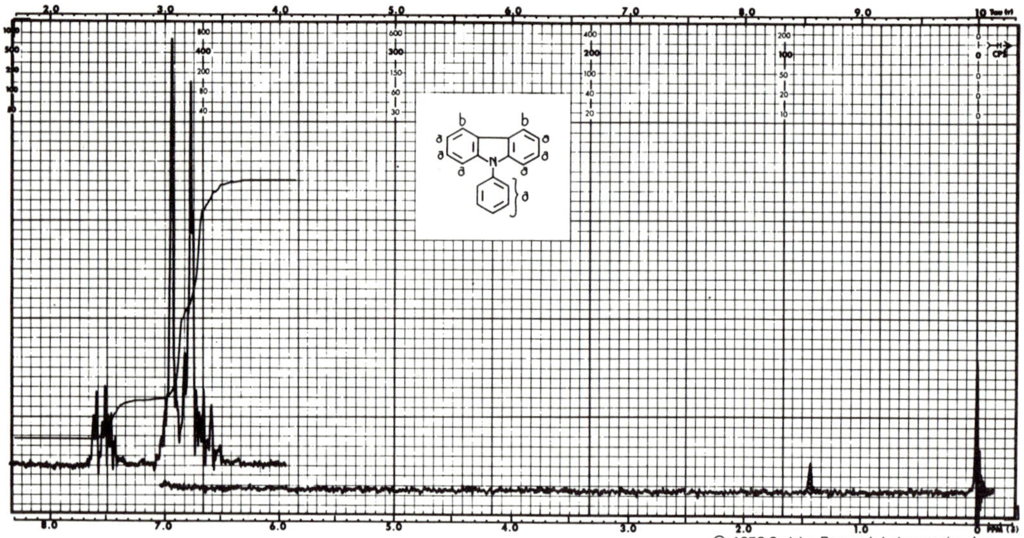

© 1978 Sadtler Research Laboratories, Inc.

ACRIDINE

$C_{13}H_9N$

Mol. Wt. 179.21

M.P. 110°C

B.P. 346°C

Solvent: CCl_4

Assignments
a = 1.97 imp.

746

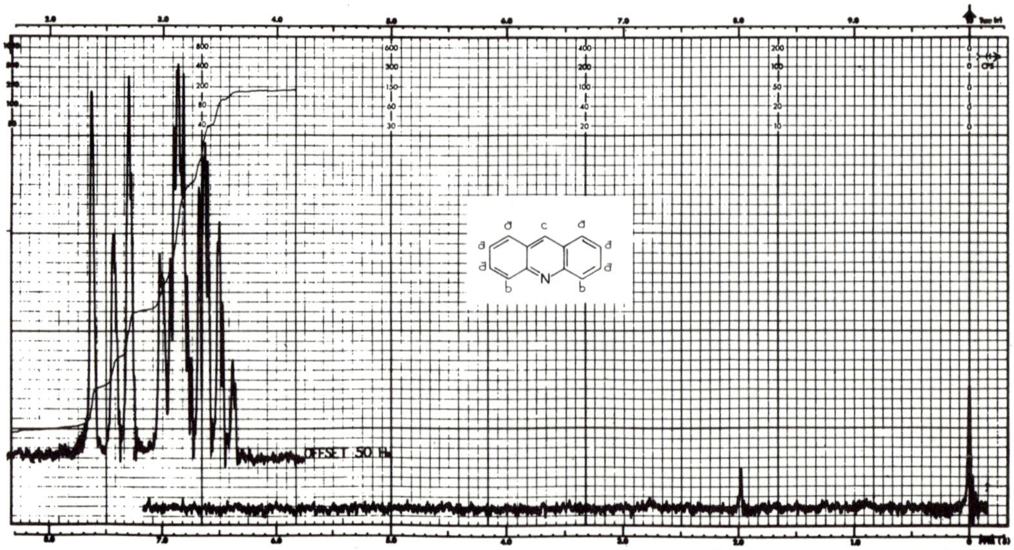

© 1978 Sadtler Research Laboratories, Inc.

9-AMINOACRIDINE

$C_{13}H_{10}N_2$

Mol. Wt. 194.24

M.P. 234.5-235°C (lit.)

Solvent: Dimethyl sulfoxide-d_6

Assignments
a = 7.18-7.83
b = 7.98
c = 8.48
d = 2.55 DMSO-d_5
e = 2.11 imp.
NH_2 unobserved

747

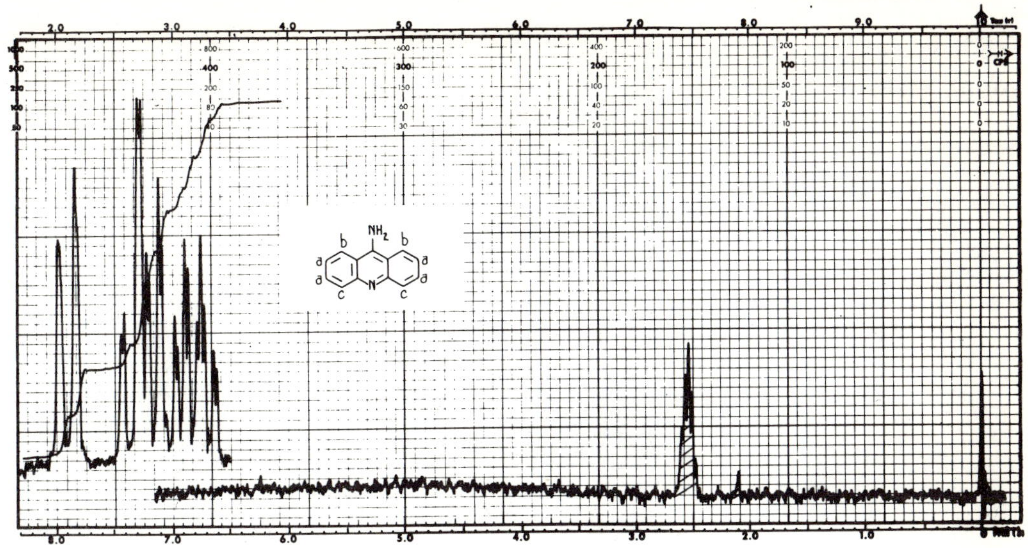

© 1978 Sadtler Research Laboratories, Inc.

PHENANTHRIDINE

$C_{13}H_9N$

Mol. Wt. 179.22

M.P. 105-107°C

B.P. 349°C/769mm

Solvent: $CDCl_3$

Assignments
a = 7.30-7.72
b = 7.81
c = 8.17
d = ca 8.35
e = 9.18

748

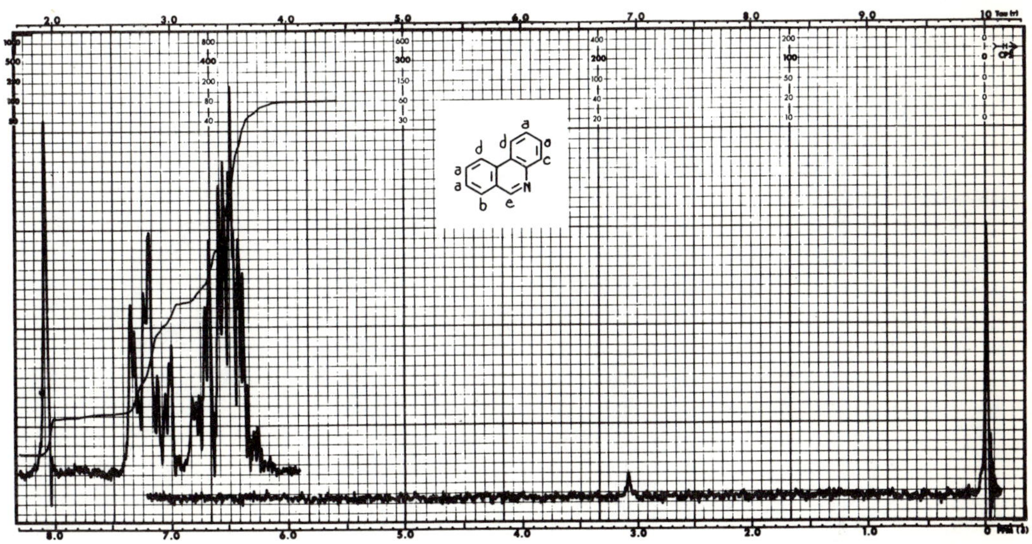

© 1978 Sadtler Research Laboratories, Inc.

BENZO[f] QUINOLINE

$C_{13}H_9N$

Mol. Wt. 179.22

M.P. 90-93°C

Solvent: $CDCl_3$

Assignments
a = 7.23
b = 7.30-7.80
c = 7.78
d = 7.91
e = 8.26
f = 8.55
g = 8.83

749

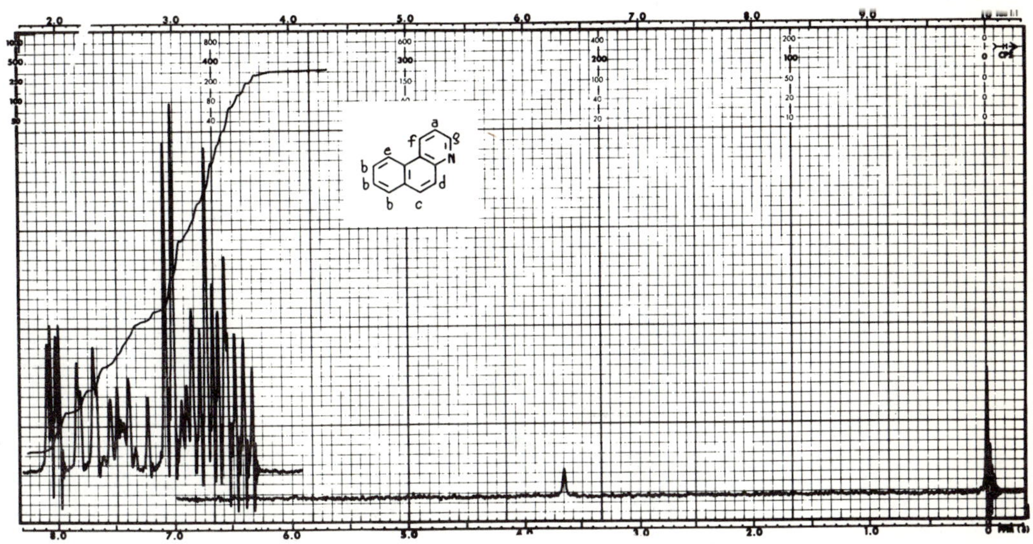

© 1978 Sadtler Research Laboratories, Inc.

1,10-PHENANTHROLINE, HYDRATE

$C_{12}H_8N_2 \cdot H_2O$

Mol. Wt. 198.23

M.P. 99-100°C

Solvent: Acetone

Assignments

a = 3.34
b = 7.72
c = 7.91
d = 8.41
e = 9.19

750

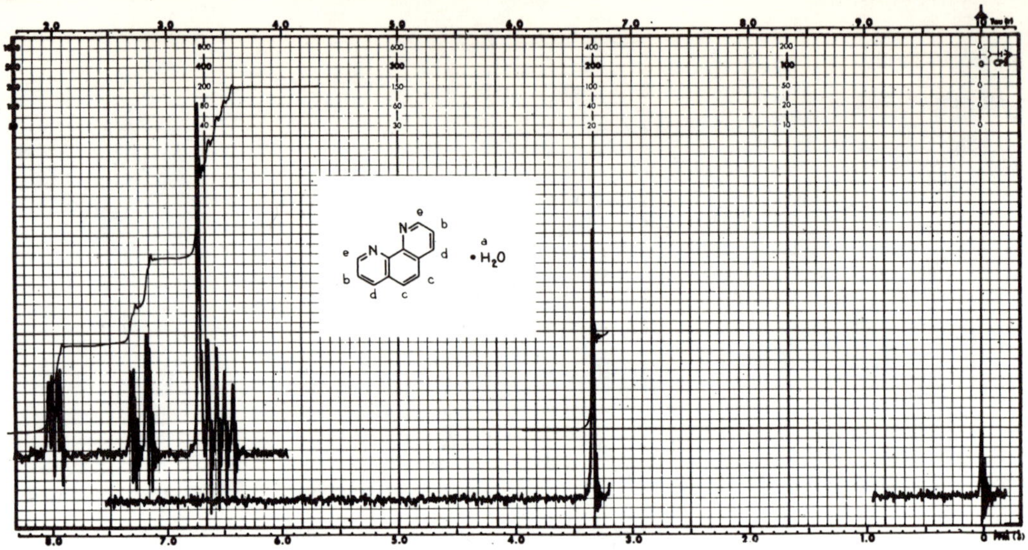

© 1978 Sadtler Research Laboratories, Inc.

2,9-DIMETHYL-1,10-PHENANTHROLINE

$C_{14}H_{12}N_2$

Mol. Wt. 208.27

Solvent: Polysol-d

Assignments

a = 2.89
b = 7.47
c = 7.69
d = 8.13
e = 3.24 imp.

751

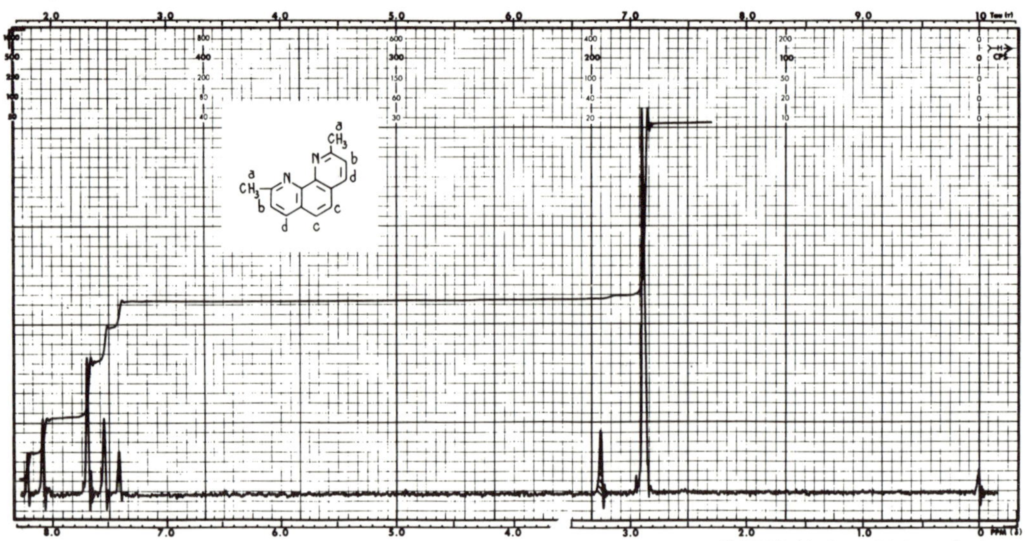

© 1978 Sadtler Research Laboratories, Inc.

NAPHTHO(1,8-de)TRIAZINE

$C_{10}H_7N_3$

Mol. Wt. 169.19

Solvent: Dimethyl sulfoxide

Assignments

a = 6.49
b = 6.98-7.37
* H_2O present

752

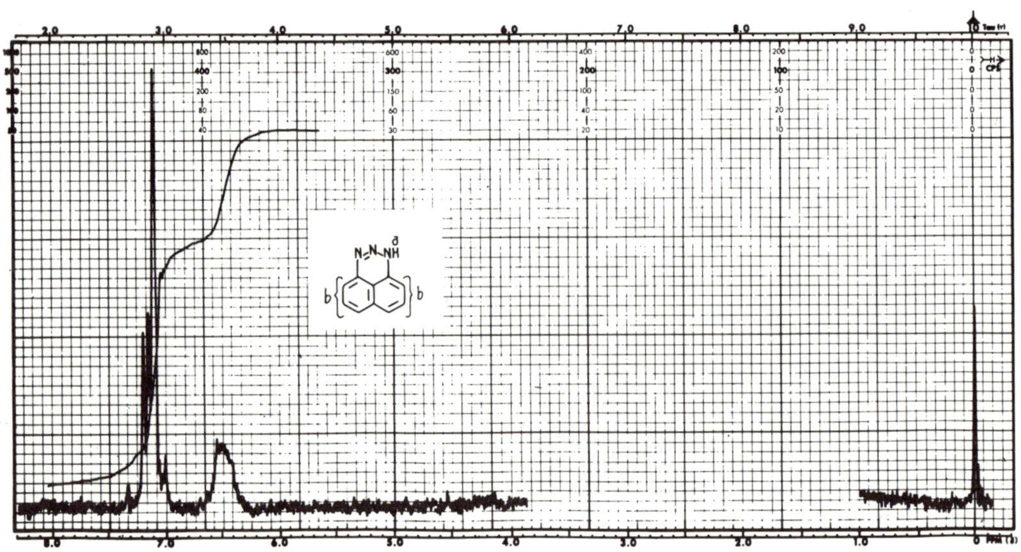

© 1978 Sadtler Research Laboratories, Inc.

The chemical shifts produced by Hydrazine substituents are similar to those observed for the corresponding primary, secondary or tertiary amines.

The hydrazine protons, on the average, resonate midway between the high field resonance of the aliphatic amines and the low field resonance of the aromatic amine protons.

The presence of a Hydrazine linkage in a molecule could be detected most easily if the integration ratio indicated either more hydrocarbon groups or more exchangeable hydrogens than a simple amine group could accommodate.

As with the amines, the Hydrazine group undergoes salt formation upon the addition of acid to the sample solution producing a shift to lower field for the Hydrazine protons and the protons of aliphatic or aromatic groups bonded to it.

Chemical Shifts

Aliphatic Protons

HC–1	NH	NH_2		Solvent	Spectrum
(2.35)	----	3.00	$(CH_3)_2$–N–NH_2	CCl_4	754
2.61	3.36	3.36	CH_3–NH–NH_2	CCl_4	753

The series of substituted ethanols presented below illustrates the comparative deshielding effect of the amines and hydrazines.

HO–	CH_2–	CH_2	–X		
	3.47	2.38	–N$(CH_3)_2$	CCl_4	----
	3.69	2.69	–NH–CH_3	D_2O	----
	3.54	2.73	–NH_2	$CDCl_3$	1642
	3.74	2.93	–NH–NH_2	D_2O	1654

Aromatic Protons

para	meta	ortho	–X		
6.52	7.00	6.52	–NH–NH_2	CCl_4	756
6.60	7.07	6.76	–NH–NH–Ø	DMSO-d_6	760
	6.70–7.30		–N(Ø)–NH_2	$CDCl_3$	759

METHYLHYDRAZINE

CH_6N_2

Mol. Wt. 46.07

B.P. 87-89°C

M.P. –52.4°C (lit.)

Solvent: CCl_4

Assignments
a = 2.61
b = 3.36
c = 7.60 imp.

753

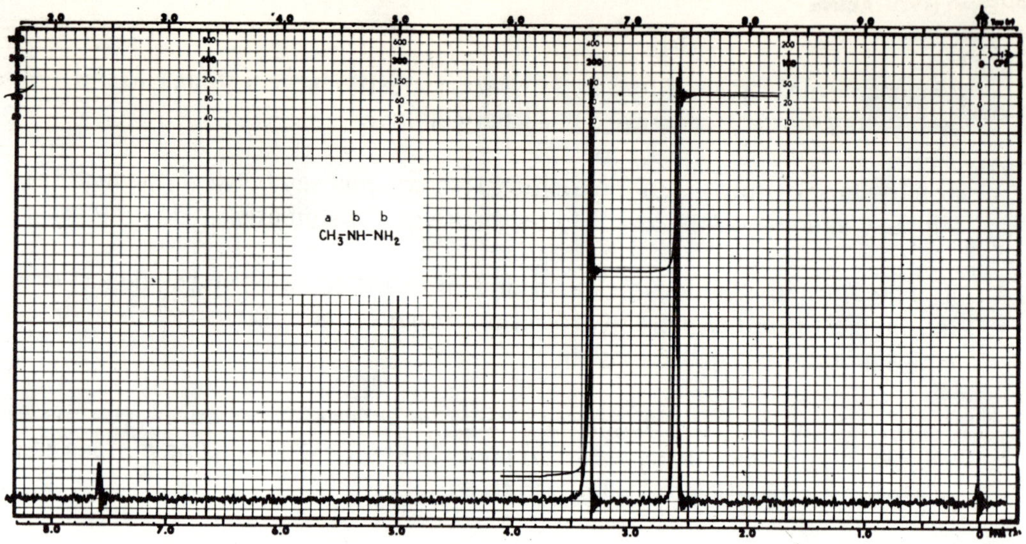

© 1978 Sadtler Research Laboratories, Inc.

1,1-DIMETHYLHYDRAZINE

$C_2H_8N_2$

Mol. Wt. 60.10

B.P. 61-63°C

Solvent: CCl_4

Assignments
a = 2.35
b = 3.00

754

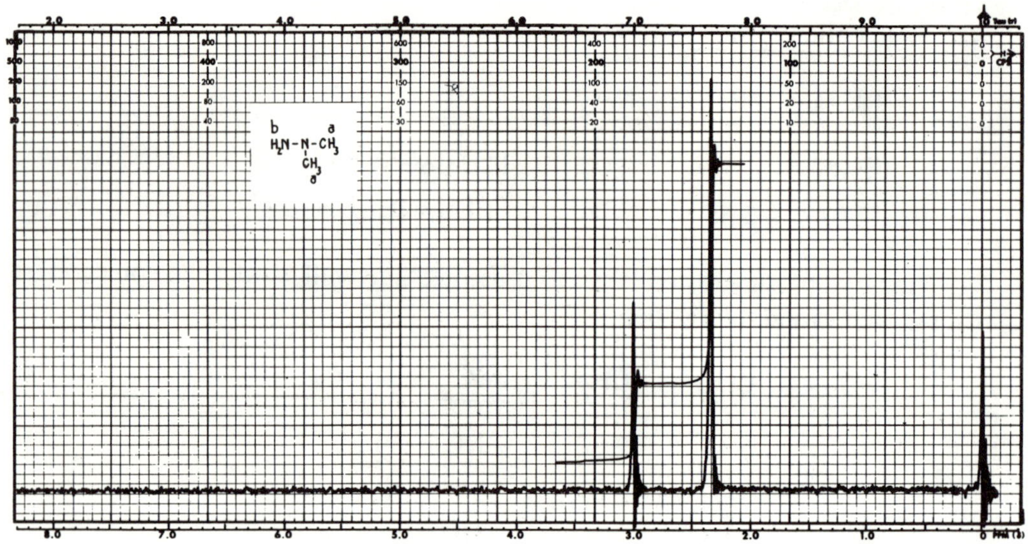

© 1978 Sadtler Research Laboratories, Inc.

1-AMINO-4-METHYLPIPERAZINE

$C_5H_{13}N_3$

Mol. Wt. 115.18

Solvent: $CDCl_3$

Assignments
a = 2.29
b = 2.20-3.00
c = 3.21

755

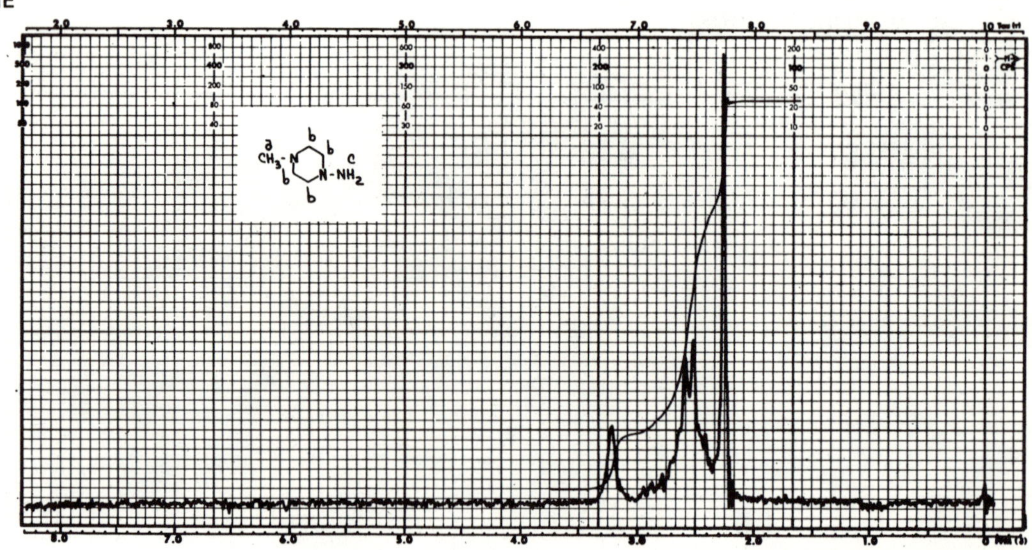

© 1978 Sadtler Research Laboratories, Inc.

PHENYLHYDRAZINE

$C_6H_8N_2$

Mol. Wt. 108.14

Solvent: $CDCl_3$

Assignments
a = 4.04
b = 6.70
c = 6.76
d = 7.19

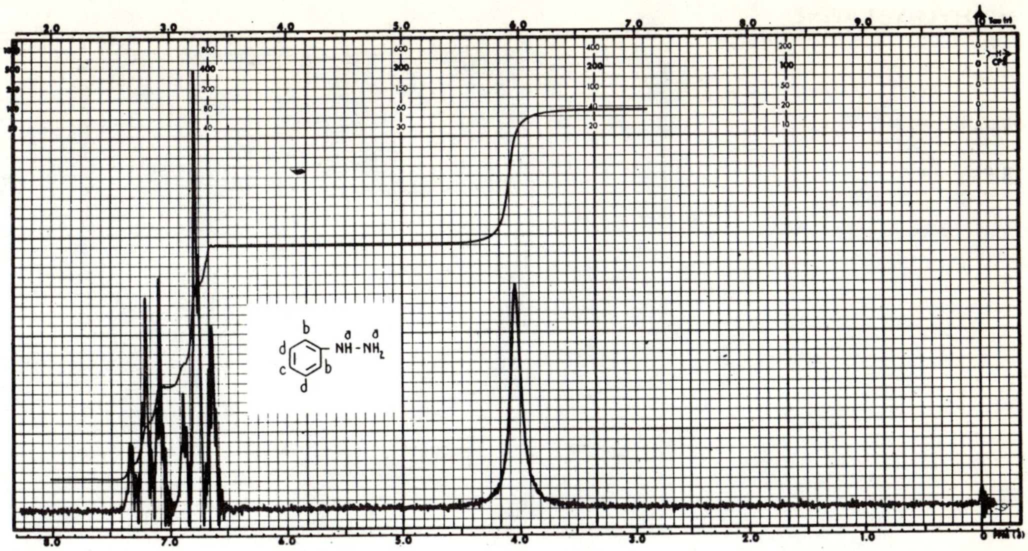

756

© 1978 Sadtler Research Laboratories, Inc.

(2,5-DICHLOROPHENYL)HYDRAZINE

$C_6H_6Cl_2N_2$

Mol. Wt. 177.03

M.P. 102-103°C

Solvent: $CDCl_3$

Assignments
a = ca 3.87 *
b = 6.65
c = ca 7.12
* Plus H_2O

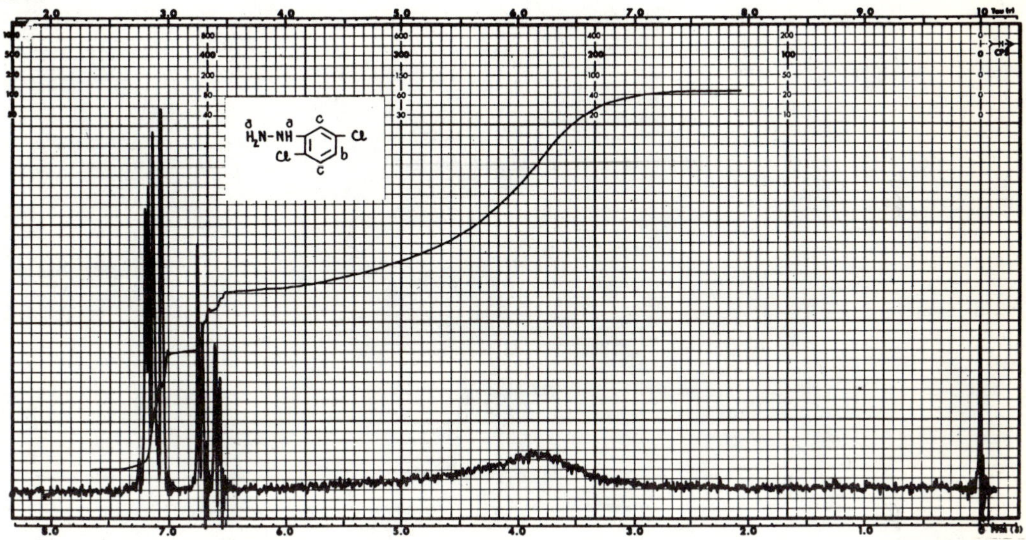

757

© 1978 Sadtler Research Laboratories, Inc.

(2,4,6-TRICHLOROPHENYL)HYDRAZINE

$C_6H_5Cl_3N_2$

Mol. Wt. 211.48

M.P. 142-143°C

Solvent: $CDCl_3$

Assignments
a = ca 4.18
b = 7.28

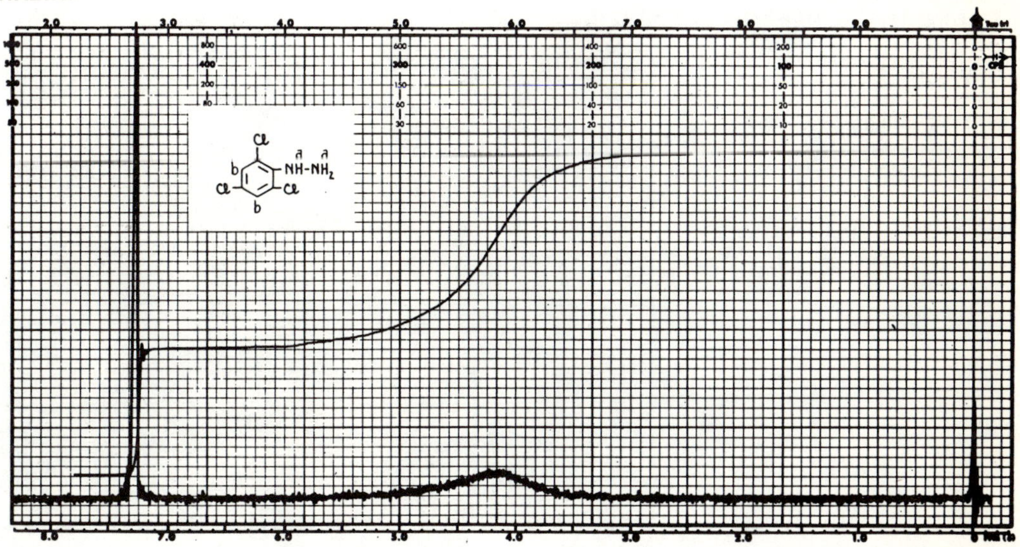

758

© 1978 Sadtler Research Laboratories, Inc.

1,1-DIPHENYLHYDRAZINE

$C_{12}H_{12}N_2$

Mol. Wt. 184.24

B.P. 22°C/40-50mm

Solvent: $CDCl_3$

Assignments
a = 3.97
b = 6.70-7.30

759

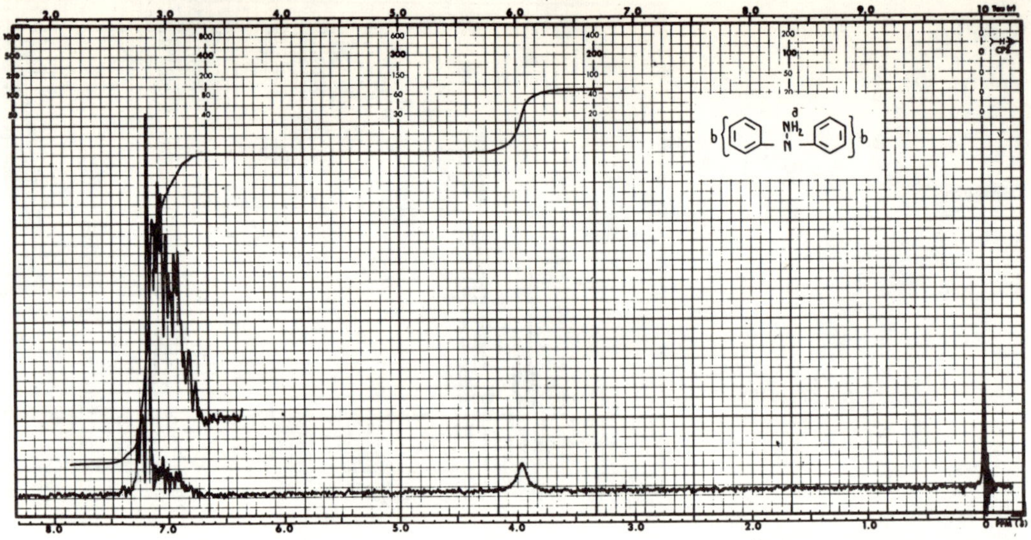

© 1978 Sadtler Research Laboratories, Inc.

HYDRAZOBENZENE

$C_{12}H_{12}N_2$

Mol. Wt. 184.24

Solvent: Dimethyl sulfoxide-D_6

Assignments
a = 6.60
b = 6.76
b = 7.07
d = 7.49
e = 2.43 DMSO-D_5

760

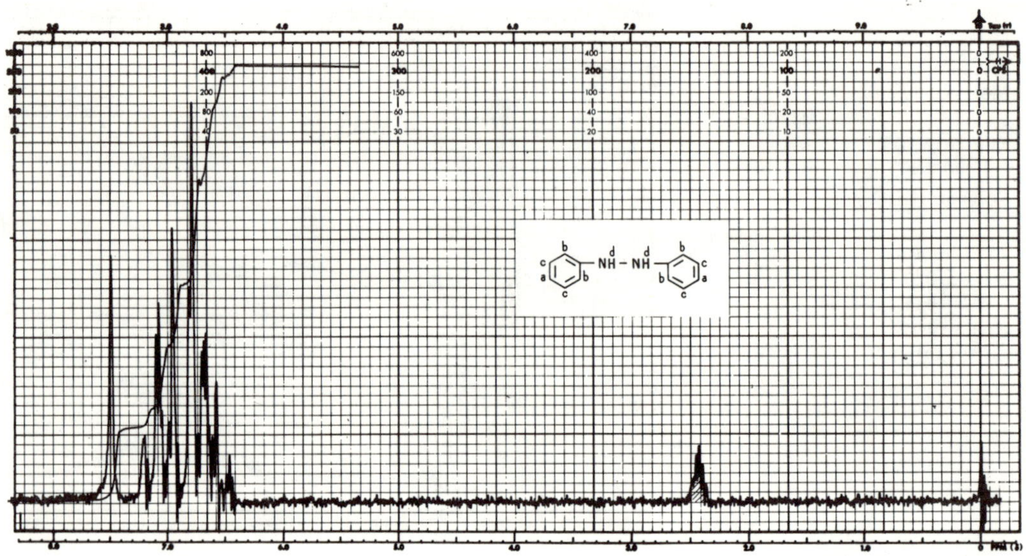

© 1978 Sadtler Research Laboratories, Inc.

1,2-DIMETHYL-,2-DI-p-TOLYLHYDRAZINE

$C_{16}H_{20}N_2$

Mol. Wt. 240.35

M.P. 66-67°C

Solvent: $CDCl_3$

Assignments
a = 2.25
b = 2.89
c = 6.72
d = 7.05

761

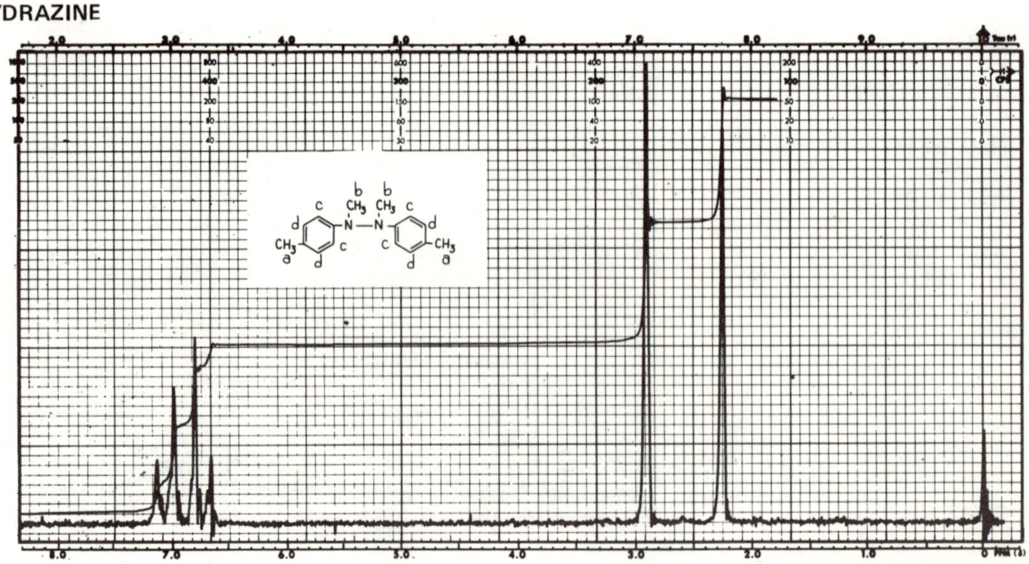

© 1978 Sadtler Research Laboratories, Inc.

1,2-BIS(4-BIPHENYLYL)-1,2-DIMETHYLHYDRAZINE

$C_{26}H_{24}N_2$

Mol. Wt. 364.49

M.P. 173-174°C

Solvent: $CDCl_3$

Assignments
a = 3.02
b = 6.90
c = 7.10-7.68

762

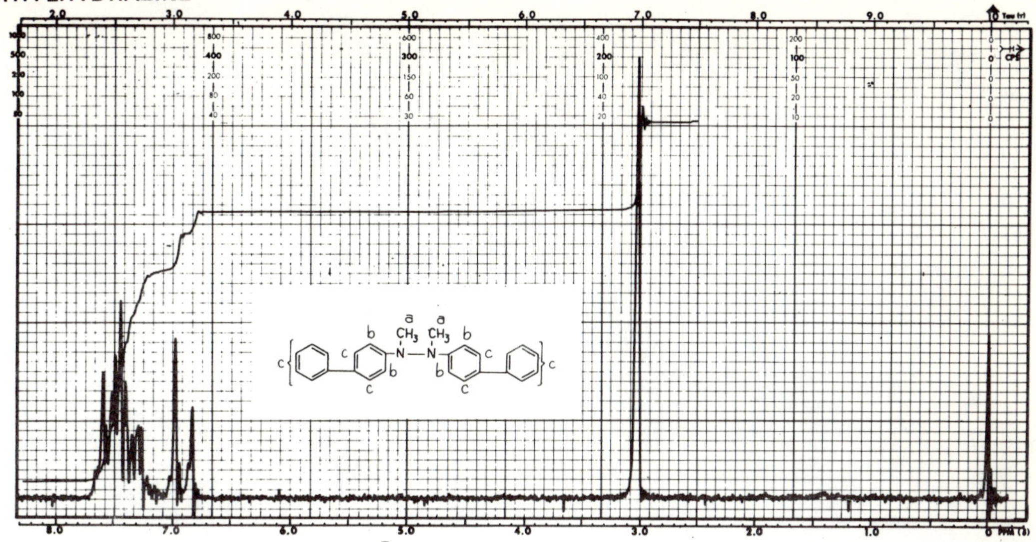

© 1978 Sadtler Research Laboratories, Inc.

1,2-DIMESITYL-1,2-DIMETHYLHYDRAZINE

$C_{20}H_{28}N_2$

Mol. Wt. 296.46

M.P. 118-119°C

Solvent: $CDCl_3$

Assignments
a = 2.21
b = 2.30
c = 2.89
d = 6.79

763

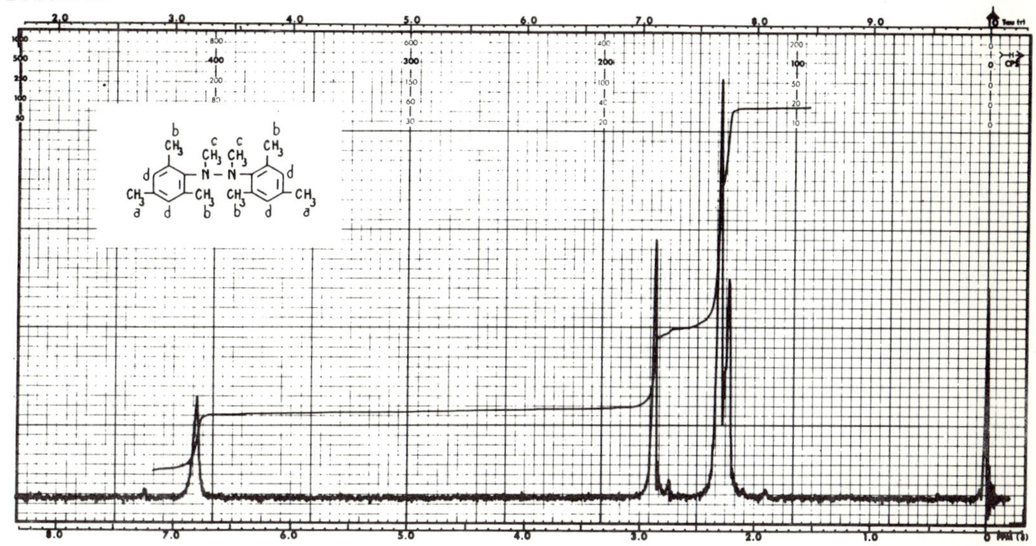

© 1978 Sadtler Research Laboratories, Inc.

2-HYDRAZINOQUINOLINE

$C_9H_9N_3$

Mol. Wt. 159.19

M.P. 141.5-142.5°C

Solvent: Polysol-d

Assignments
a = 4.87 *
b = 6.85
c = 7.00-7.85
* Plus H_2O

764

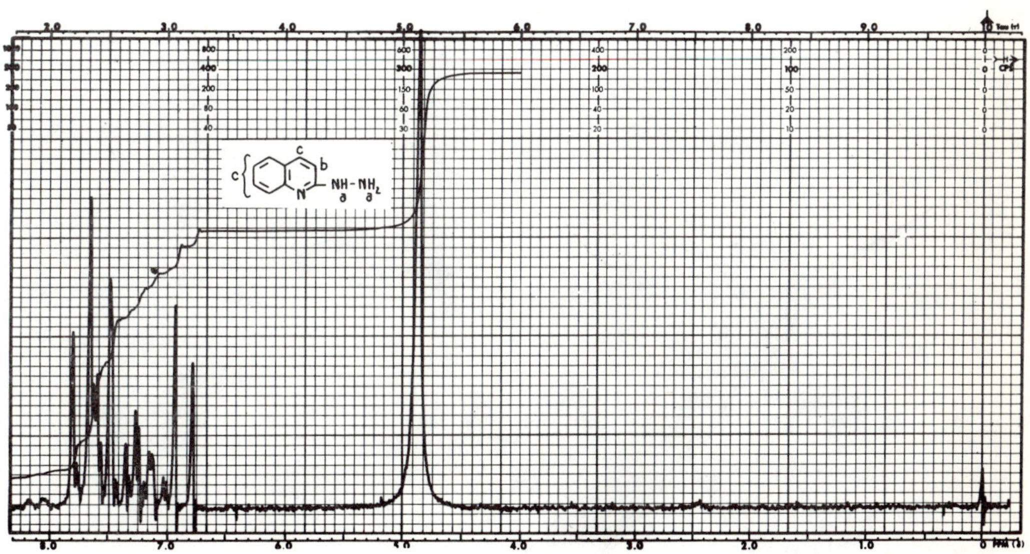

© 1978 Sadtler Research Laboratories, Inc.

NOTES

The reaction of mineral acids with primary, secondary and tertiary amines to form amine salts not only makes the compounds more soluble in polar solvents such as D_2O, but increases the deshielding effect of the amine group on both aliphatic and aromatic hydrogens.

The exchangeable hydrogens attached to the nitrogen nucleus normally resonate at lower field than the corresponding hydrogens of the free amine.

The amine salts can be neutralized by the addition of a few drops of a sodium bicarbonate solution to the sample. The amine salt is thus converted to the free base form with an attendant shift to higher field for proton containing groups bonded to the nitrogen atom.

Chemical Shifts

Aliphatic Protons

HC–3	HC–2	HC–1		Solvent	Spectrum
		2.62	CH_3–NH_2(HCl)	D_2O	765
		(2.76)	$(CH_3)_2$–NH(HCl)	D_2O	----
		(2.94)	$(CH_3)_3$–N (HCl)	D_2O	808
	(1.45	3.03)	$(CH_3$–$CH_2)_2$–NH(HCl)	$CDCl_3$	797
	1.30	3.08	CH_3–CH_2–NH_2 (HCl)	D_2O	766
	1.22	3.10	CH_3–CH_2–NH_2 (HBr)	D_2O	----
	(1.30	3.10)	$(CH_3$–$CH_2)_2$–NH (H_3PO_4)	D_2O	----
	1.41	3.31	CH_3–CH_2–$N(R)_2$ (HBr)	D_2O	----
	1.50	3.37	CH_3–CH_2–$N(R)_2$ (HBr)	$CDCl_3$	----
	(1.29)	3.33	$(CH_3)_2$–CH–NH_2 (HCl)	DMSO-d_6	----
0.93	1.76	2.91	CH_3–CH_2–CH_2–NH–R (HCl)	$CDCl_3$	----
1.11	1.99	3.39	CH_3–CH_2–CH_2–NH–R (HCl)	TFA	----

2-Substituted ethylamine salts

(XH)H_2N–	CH_2	–CH_2	–X	
	2.75	1.64	–CH_3	DMSO-d_6
	3.19	2.22	–CH_2–NH_2(HX)	D_2O
	3.22	2.85	–SH	D_2O
	3.36	2.85	–C(=O)–O–R	D_2O
	3.27	3.40	–SO_3H	D_2O
	3.42	3.42	–NH_2(HX)	D_2O
	3.75	3.55	–Br	D_2O
	3.55	3.55	–$N(CH_3)_2$(HX)	D_2O
	3.18	3.85	–OH	D_2O
	3.48	4.01	–Cl	D_2O

Chemical Shifts continued

Aliphatic Protons continued

Substituted methylamine salts

(HX) H_2N–	CH_2	–X	Solvent
	2.90	–Cyclopropyl	D_2O
	3.08	$-CH_3$	D_2O
	3.09	$-R_3$	$CDCl_3$
	3.60	$-CH{=}CH_2$	Polysol
	3.88	$-C(=O)-NH_2$	D_2O
	3.99	$-C(=O)-O-CH_3$	D_2O
	4.19	$-C{\equiv}N$	D_2O
	4.27	–Ø	D_2O
	4.29	–C(-O)–R	D_2O
	5.05	–C(=O)–Ø	TFA

Aromatic Protons

As a substituent on an aromatic ring, the amine salt group has an unusually uniform effect on the chemical shifts of the ortho, meta and para hydrogens. The series of aniline salts presented below indicate the minimal effect that the type of acid involved, the solvent employed and the degree of amine substitution have on the chemical shift of the phenyl protons. The aromatic resonance for all of these aniline salts appears as a relatively sharp peak in the narrow chemical shift range from 7.50–7.55 ppm.

ppm	Compound	
7.51	Ø–NH_2 (HCl)	D_2O
7.51	Ø–$NH_2(H_2SO_4)$	TFA
7.54	Ø–NH_2 (HBr)	D_2O
7.54	Ø–$N(CH_3)_2$(Picrate)	DMSO-d_6
7.55	Ø–NH–R_2 (HCl)	D_2O

Para substituted aniline salts

(XH) H_2N–	ortho	meta	para	
	7.19	6.67	$-O-R_2$	D_2O
	7.22	6.88	–OH	DMSO-d_6
	7.44	7.09	$-O-CH_3$	D_2O
	7.33	7.33	$-CH_3$	TFA
	7.67	7.39	–Br	D_2O

Aromatic Protons continued

Para substituted aniline salts continued

(XH) H_2N–	ortho	meta	para	Solvent
	7.46	7.46	$-CH_2-\emptyset$	D_2O
	7.56	7.56	–Cl	D_2O
	7.78	7.78	$-NH_2(HX)$	D_2O
	7.78	7.78	$-NH-CH_3(HX)$	D_2O
	7.79	7.79	$-N(CH_3)_2(HX)$	D_2O
	7.79	7.79	$-N(R_2)_2(HX)$	D_2O
	7.20	7.90	$-C(=O)-CH_3$	DMSO-d_6

Exchangeable Protons

The chemical shifts of the exchangeable protons of the amine salts are especially unreliable. Because they are sensitive to sample concentration, the presence of H_2O and structural differences involving substitution of the amine nitrogen nucleus, they are observed to resonate over a wide range of chemical shifts from 5.0 to 12.0 ppm.

The corresponding chemical shift range for primary and secondary amines (not salts) is about 1.5 ppm (1.0–2.5 ppm).

There does not appear to be any relationship between the type of acid used to form the salt and the chemical shift of the resultant exchangeable hydrogens.

METHYLAMINE, HYDROCHLORIDE

$CH_5N \cdot HCl$

Mol. Wt. 67.52

M.P. 226-228°C

Solvent: D_2O

Assignments
a = 2.62
b = 4.72 HDO

765

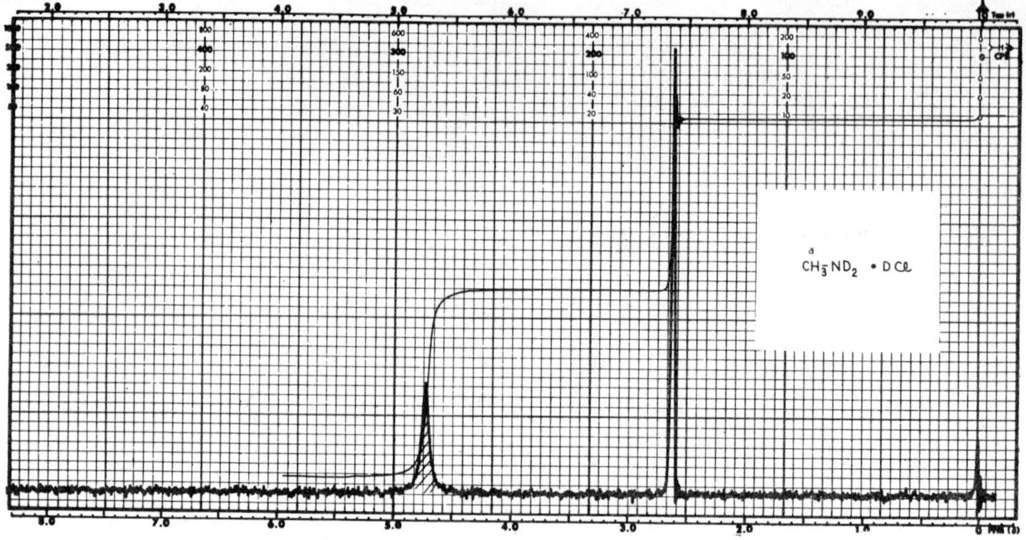

© 1978 Sadtler Research Laboratories, Inc.

ETHYLAMINE, HYDROCHLORIDE

$C_2H_7 \cdot HCl$

Mol. Wt. 81.55

M.P. 108°C

Solvent: D_2O

Assignments
a = 1.30
b = 3.08
c = 4.50 D_2O

766

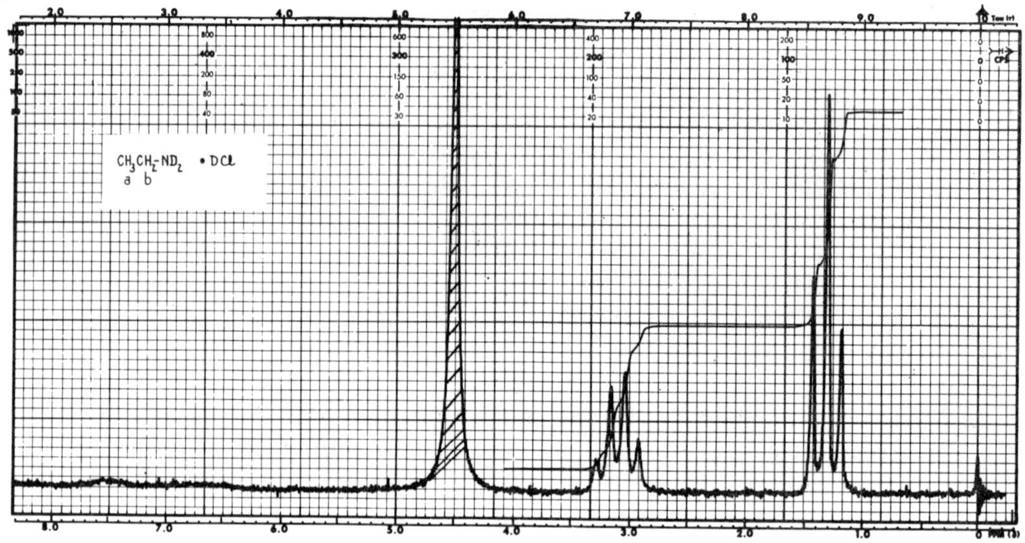

© 1978 Sadtler Research Laboratories, Inc.

PROPYLAMINE, HYDROCHLORIDE

$C_3H_9N \cdot HCl$

Mol. Wt. 95.57

M.P. 163-164°C

Solvent: Dimethyl sulfoxide-D_6

Assignments
a = 0.90
b = 1.64
c = 2.75
d = 8.32
e = 2.58 DMSO-D_5

767

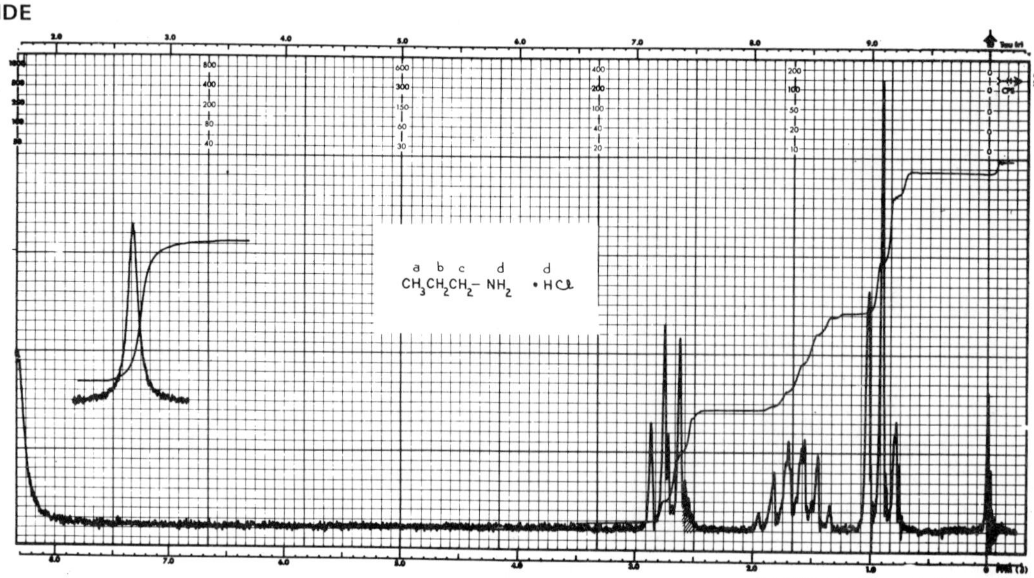

© 1978 Sadtler Research Laboratories, Inc.

BUTYLAMINE, HYDROCHLORIDE

$C_4H_{11}N \cdot HCl$

Mol. Wt. 109.60

Solvent: $CDCl_3$

Assignments
a = 0.98
b = 1.10-2.20
c = 3.09
d = 8.07

768

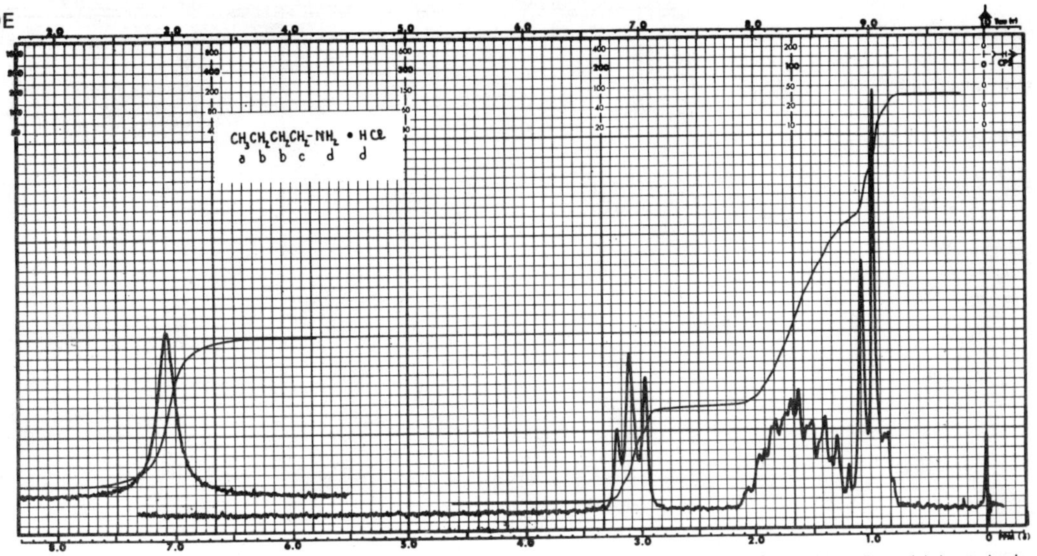

© 1978 Sadtler Research Laboratories, Inc.

tert-BUTYLAMINE, COMPOUND WITH BORANE

$C_4H_{11}N \cdot BH_3$

Mol. Wt. 86.97

Solvent: $CDCl_3$

Assignments
a = 1.30
b = 3.79

769

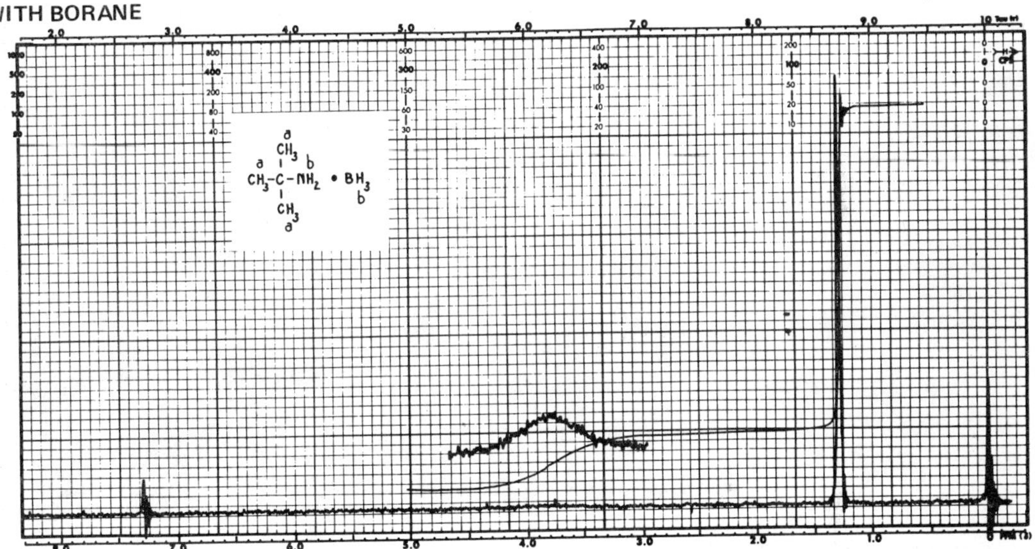

© 1978 Sadtler Research Laboratories, Inc.

OCTYLAMINE, HYDROCHLORIDE

$C_8H_{19}N \cdot HCl$

Mol. Wt. 165.71

M.P. 195°C

Solvent: D_2O

Assignments
a = 0.89
b = 1.34
c = 1.70
d = 3.03
e = 4.57 HDO

770

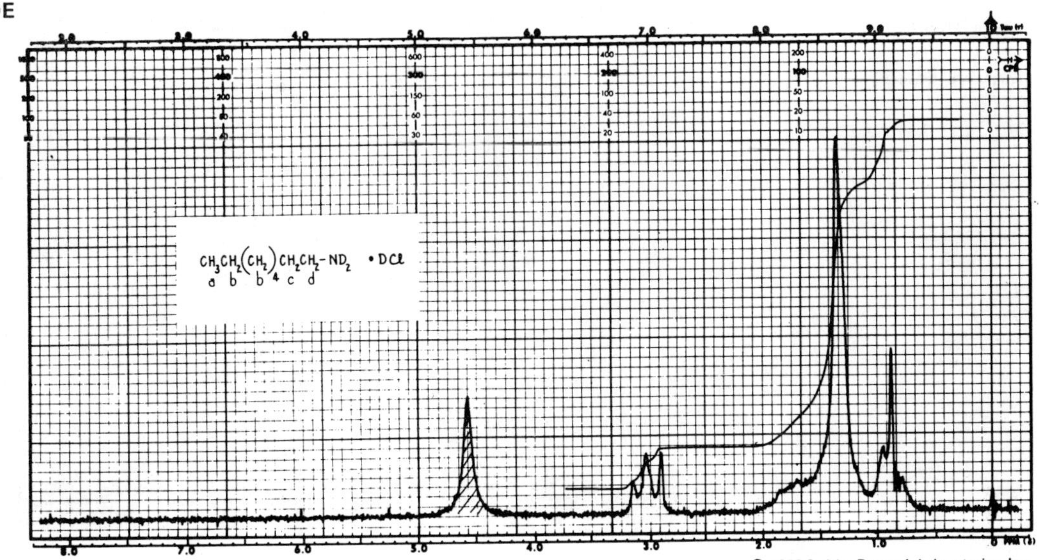

© 1978 Sadtler Research Laboratories, Inc.

2,2,2-TRIFLUOROETHYLAMINE, HYDROCHLORIDE

$C_2H_4F_3N \cdot HCl$

Mol. Wt. 135.52

M.P. 195°C (sublimes)

Solvent: D_2O

Assignments
a = 3.89
b = 4.79 HDO

771

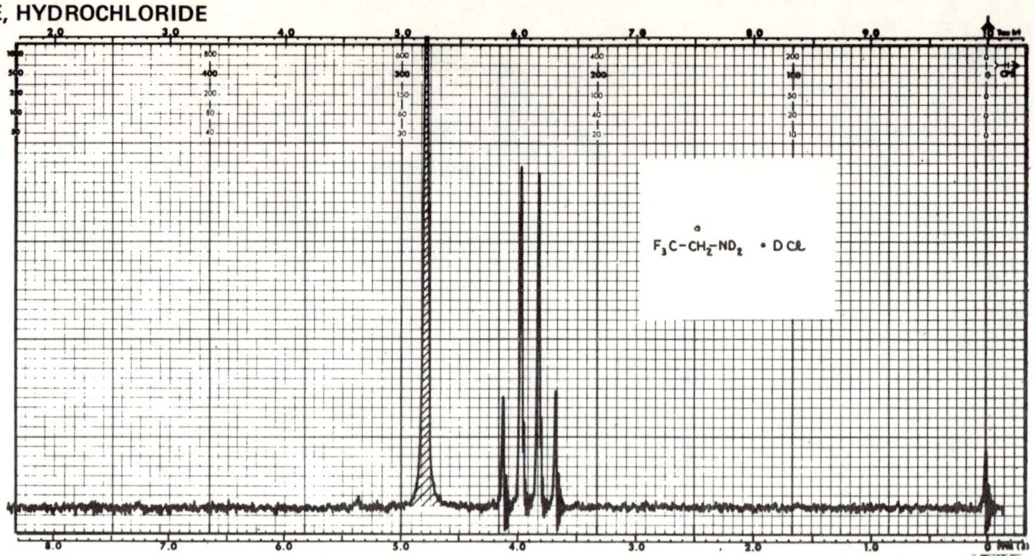

© 1978 Sadtler Research Laboratories, Inc.

CYCLOBUTYLAMINE, HYDROCHLORIDE

$C_4H_9N \cdot HCl$

Mol. Wt. 107.58

M.P. 182-184°C

Solvent: D_2O

Assignments
a = 1.50-2.60
b = 3.85
c = 4.53 HDO

772

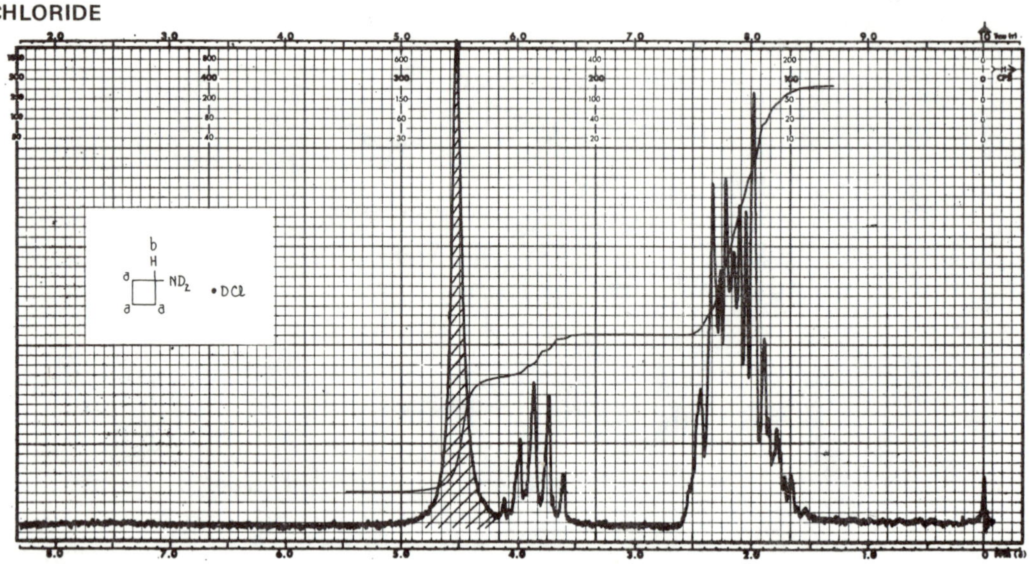

© 1978 Sadtler Research Laboratories, Inc.

2-NORBORNANAMINE, HYDROCHLORIDE

$C_7H_{13}N \cdot HCl$

Mol. Wt. 147.65

M.P. 301.5-302.5°C (dec.)

Solvent: D_2O

Assignments
a = 0.80-2.20
b = 2.29
c = 2.56
d = 3.60
e = 4.69 HDO

773

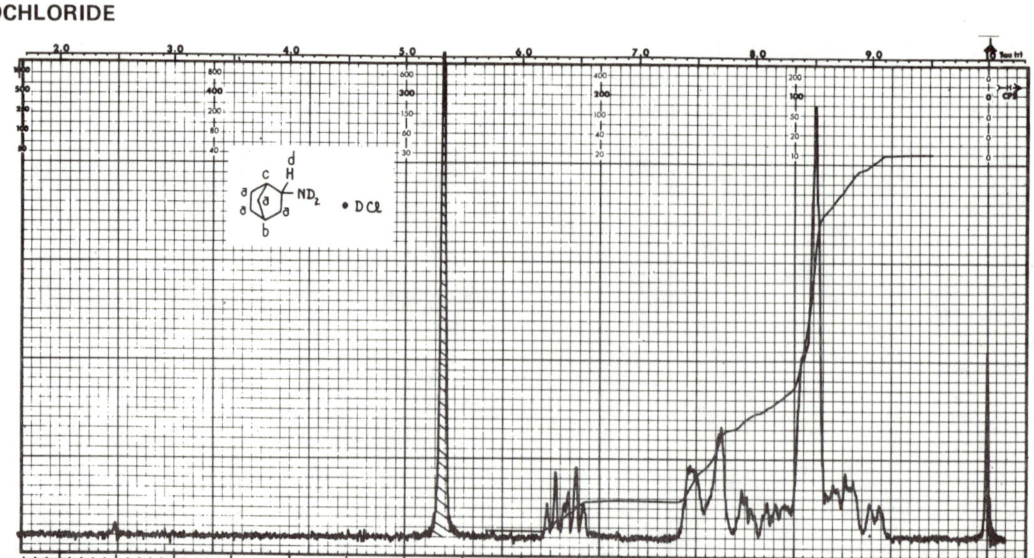

© 1978 Sadtler Research Laboratories, Inc.

ETHYLENEDIAMINE, DIHYDROCHLORIDE

$C_2H_8N_2 \cdot 2HCl$

Mol. Wt. 133.02

M.P. 167-169°C

Solvent: D_2O

Assignments
a = 3.43
b = 4.80 HDO

774

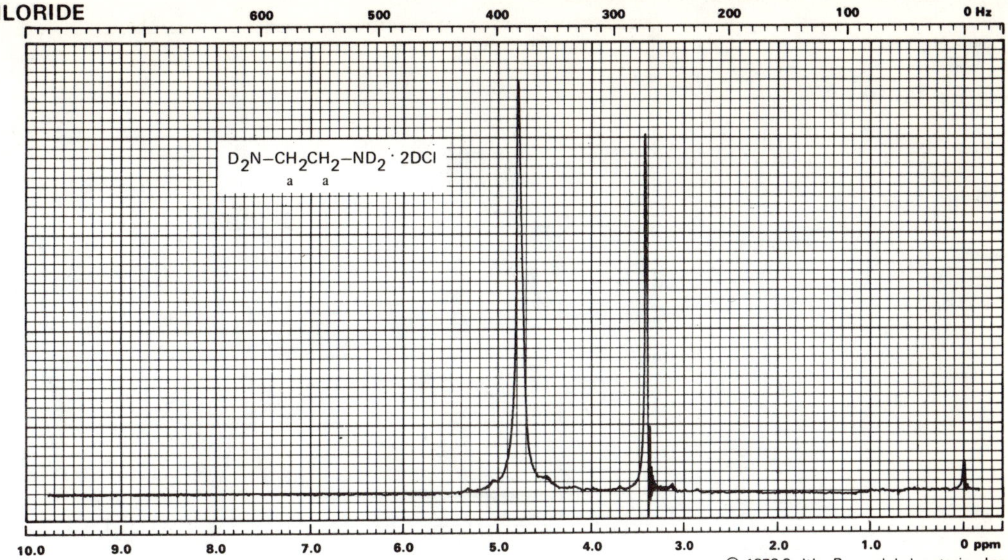

© 1978 Sadtler Research Laboratories, Inc.

ETHYLENEDIAMINE, DIHYDROBROMIDE

$C_2H_8N_2 \cdot 2HBr$

Mol. Wt. 221.92

Solvent: D_2O

Assignments
a = 3.40
b = 4.63 HDO

775

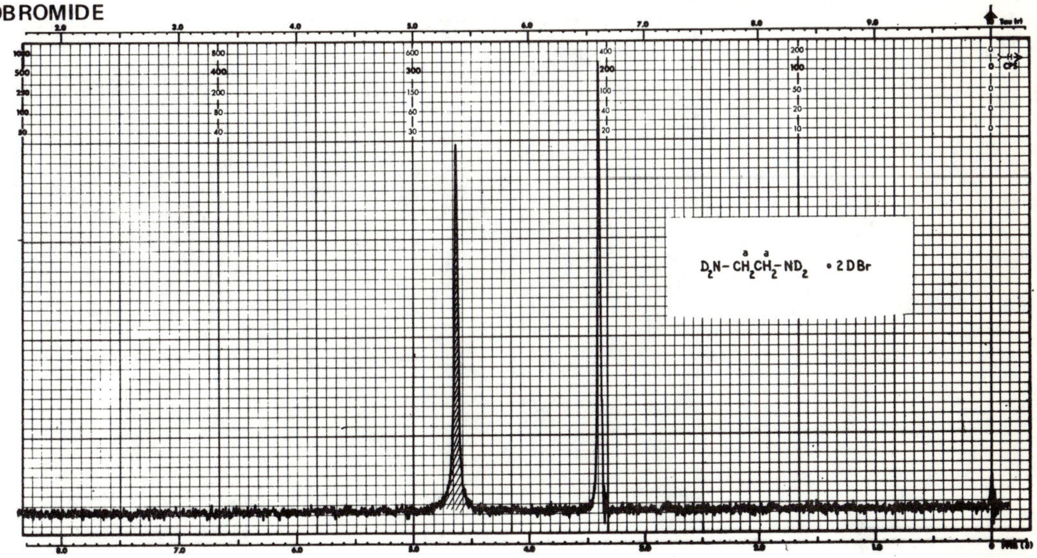

© 1978 Sadtler Research Laboratories, Inc.

ETHYLENEDIAMINE, SULFATE

$C_2H_8N_2 \cdot H_2SO_4$

Mol. Wt. 158.18

M.P. ∼ 300°C (dec.)

Solvent: D_2O

Assignments
a = 3.42
b = 4.80 HDO

776

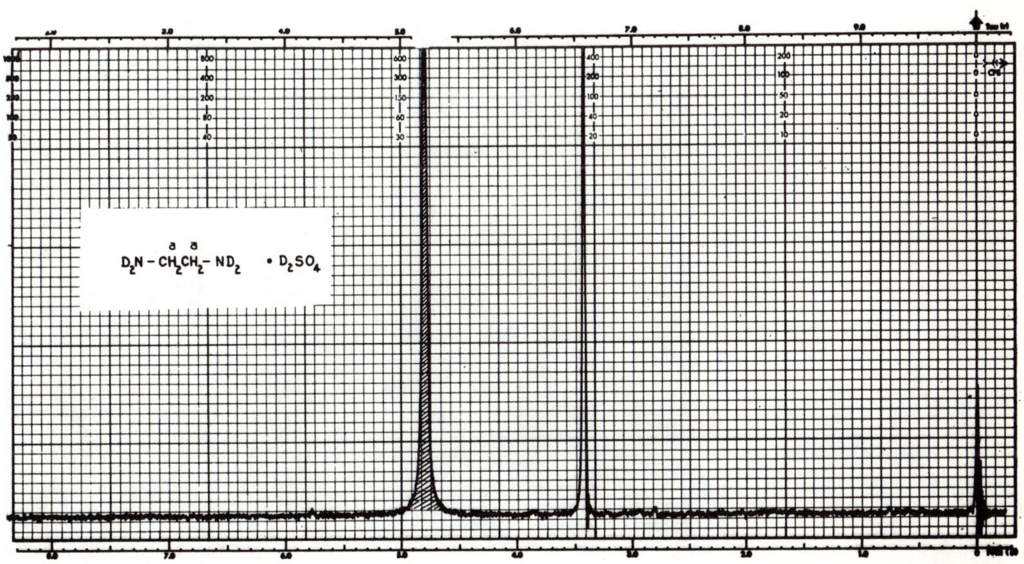

© 1978 Sadtler Research Laboratories, Inc.

1,4-BUTANEDIAMINE, DIHYDROCHLORIDE

$C_4H_{12}N_2 \cdot 2HCl$

Mol. Wt. 161.08

M.P. > 300°C

Solvent: D_2O

Assignments
a = 1.78
b = 3.07
c = 4.77 HDO

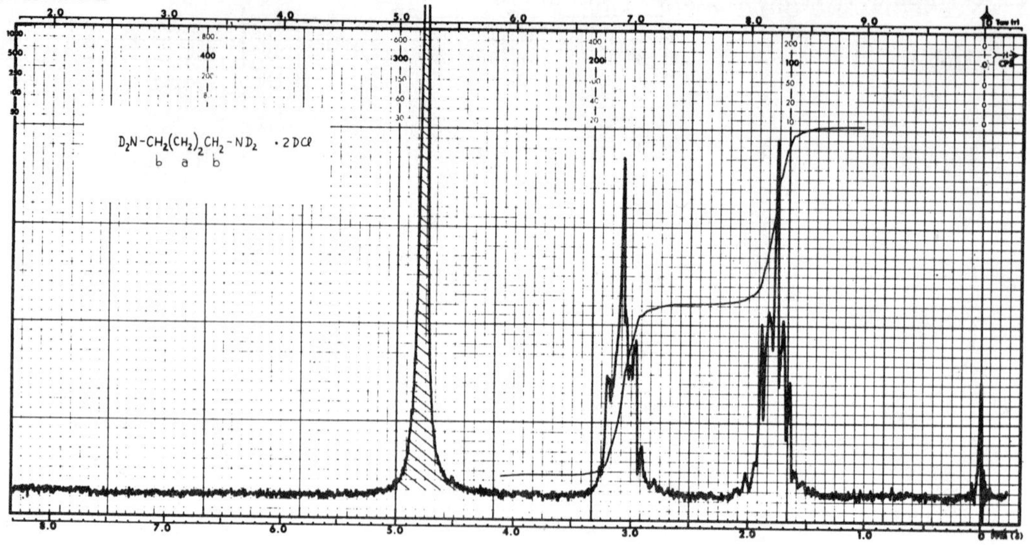

777

© 1978 Sadtler Research Laboratories, Inc.

1,5-PENTANEDIAMINE, DIHYDROCHLORIDE

$C_5H_{14}N_2 \cdot 2HCl$

Mol. Wt. 175.10

M.P. 256-258°C

Solvent: Trifluoroacetic acid

Assignments
a = 1.30-2.30
b = 3.00-3.70
c = ca 6.95

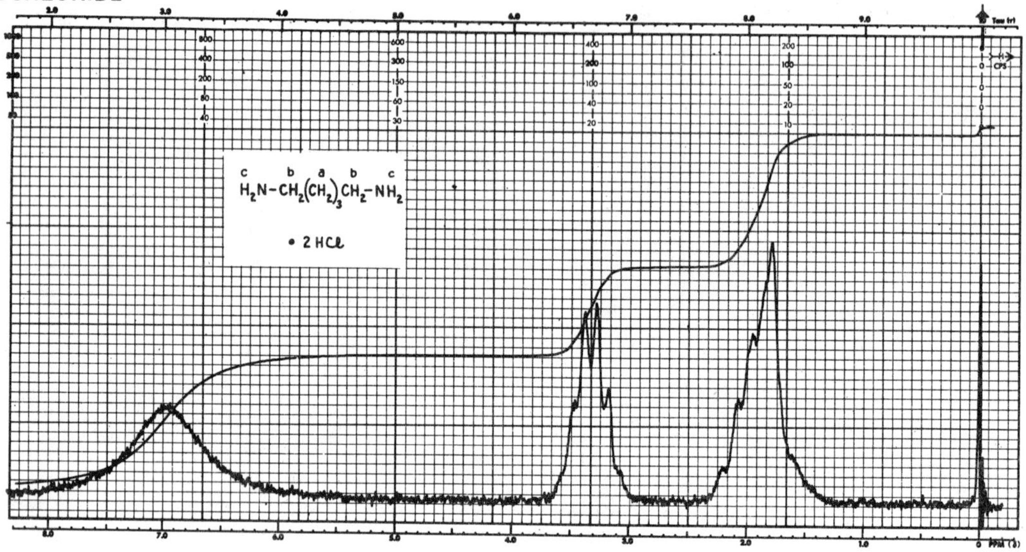

778

© 1978 Sadtler Research Laboratories, Inc.

1,6-HEXANEDIAMINE, DIHYDROCHLORIDE

$C_6H_{16}N_2 \cdot 2HCl$

Mol. Wt. 189.13

Solvent: D_2O

Assignments
a = ca 1.50
b = 1.70
c = 3.05
d = 4.71 HDO

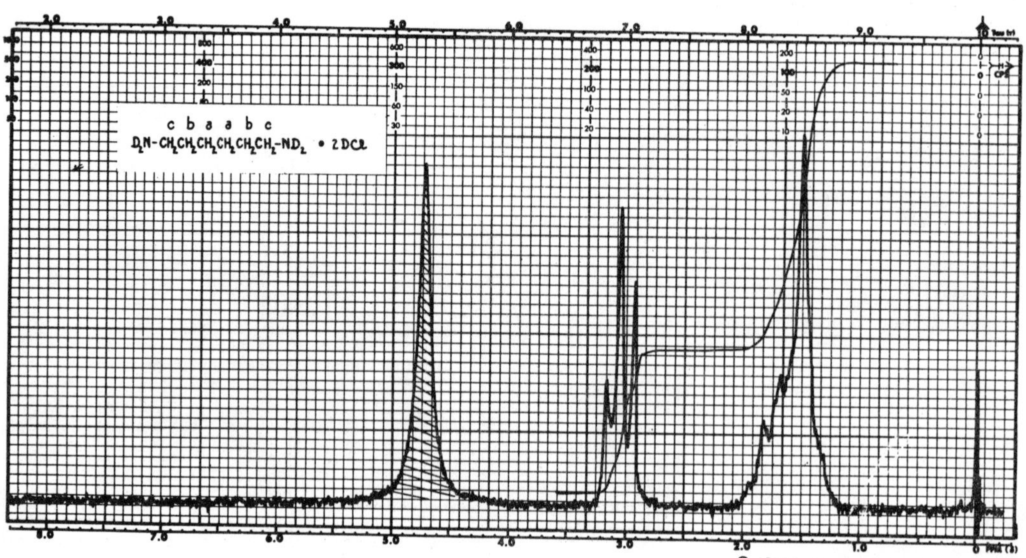

779

© 1978 Sadtler Research Laboratories, Inc.

2,2'-DITHIOBISETHYLAMINE, DIHYDROCHLORIDE

$C_4H_{12}N_2S_2 \cdot 2HCl$

Mol. Wt.225.20

M.P. 203-214°C (dec.)

Solvent: D_2O

Assignments
a = 3.07 or b
b = 3.39 or a
c = 4.68 HDO

780

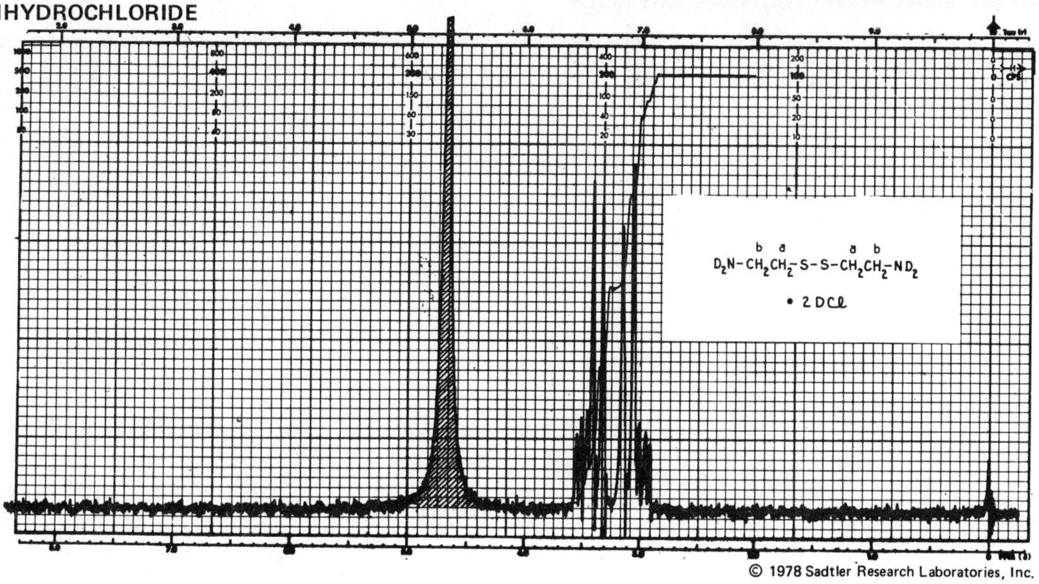

© 1978 Sadtler Research Laboratories, Inc.

ALLYLAMINE, HYDROCHLORIDE

$C_3H_7N \cdot HCl$

Mol. Wt. 93.56

M.P. 118-120°C

Solvent: Polysol-d

Assignments
a = 3.60
b = 5.32
c = 5.50
d = 6.00
e = 8.56

781

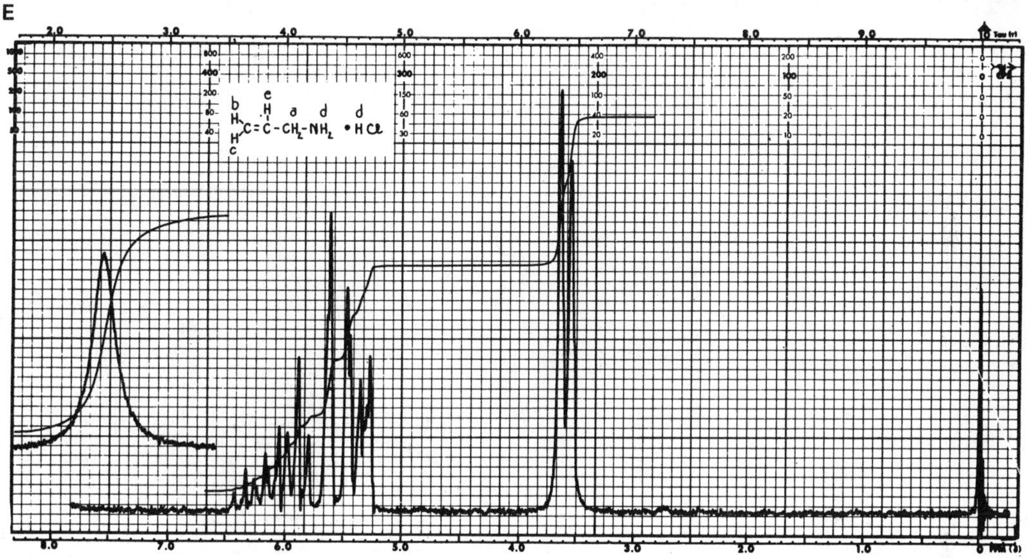

© 1978 Sadtler Research Laboratories, Inc.

BENZYLAMINE, HYDROCHLORIDE

$C_7H_9N \cdot HCl$

Mol. Wt. 143.68

M.P. 253-255°C

Solvent: D_2O

Assignments
a = 4.27
b = 7.59
c = 4.92 HDO

782

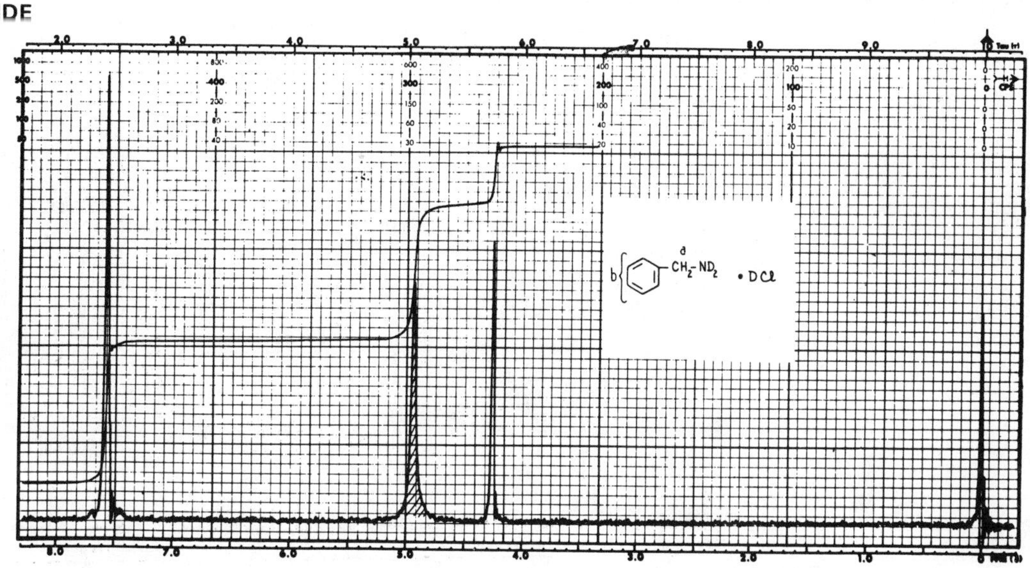

© 1978 Sadtler Research Laboratories, Inc.

m-BROMOBENZYLAMINE, HYDROCHLORIDE

$C_7H_8BrN \cdot HCl$

Mol. Wt. 222.52

M.P. 226-227°C

Solvent: Polysol-d

Assignments
a = 4.06
b = 7.20-7.75
c = 7.81
d = 8.29
e = 2.57 DMSO-d_5

783

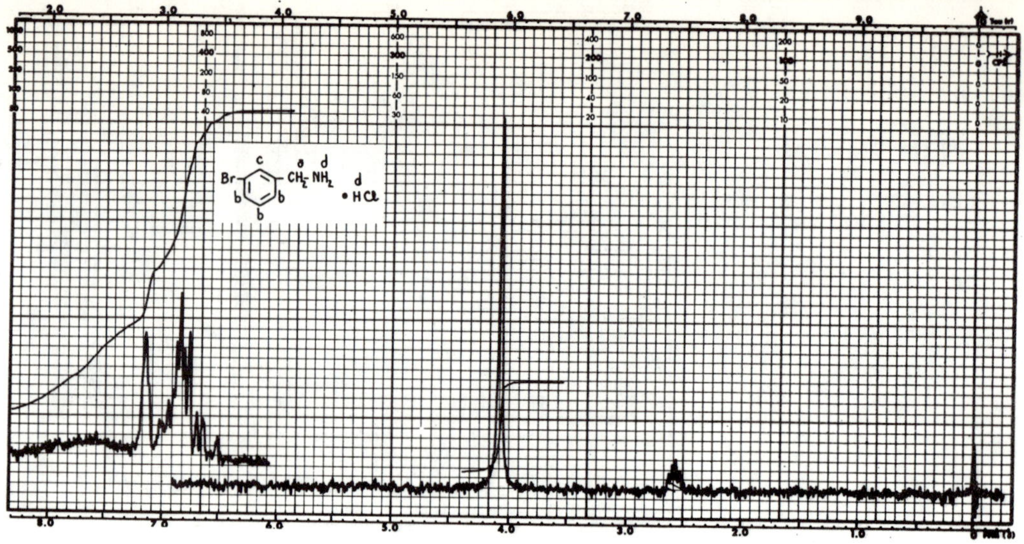

© 1978 Sadtler Research Laboratories, Inc.

α-METHYLBENZYLAMINE, HYDROCHLORIDE

$C_8H_{11}N \cdot HCl$

Mol. Wt. 157.64

M.P. 158°C

Solvent: Trifluoroacetic acid

Assignments
a = 1.87
b = 4.77
c = ca 7.18
d = 7.51

784

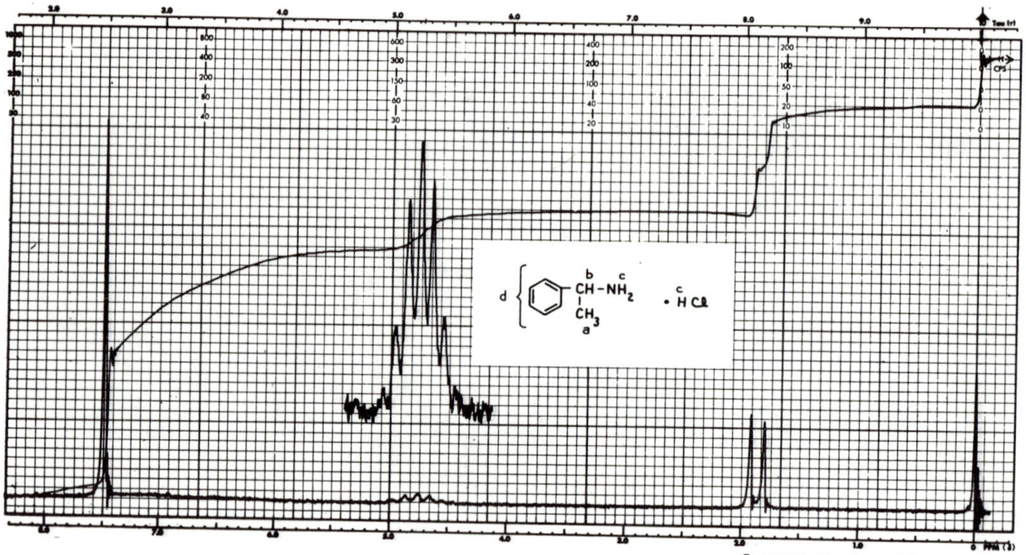

© 1978 Sadtler Research Laboratories, Inc.

1-METHYL-3-PHENYLPROPYLAMINE, HYDROCHLORIDE

$C_{10}H_{15}N \cdot HCl$

Mol. Wt. 185.70

M.P. 146-149°C

Solvent: Trifluoroacetic acid

Assignments
a = 1.52
b = 2.19
c = 2.83
d = 3.62
e = 5.40-7.00
f = 7.31

785

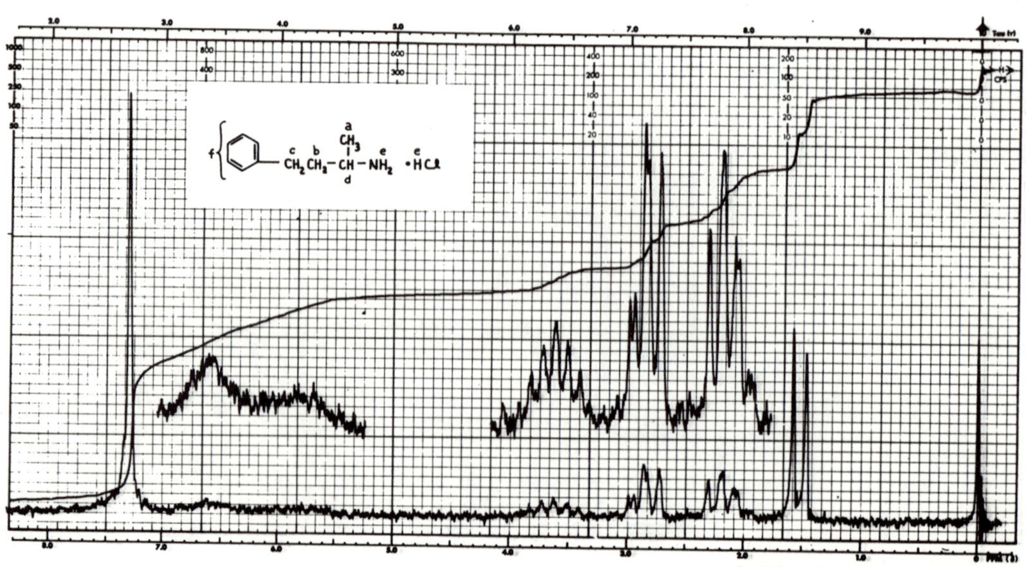

© 1978 Sadtler Research Laboratories, Inc.

p-CHLORO-α,α-DIMETHYLPHENETHYLAMINE, HYDROCHLORIDE

$C_{10}H_{14}ClN \cdot HCl$

Mol. Wt. 220.15

M.P. 231-233°C

Solvent: D_2O

Assignments
a = 1.40
b = 3.02
c = 7.31
d = 7.46
e = 4.06 HDO

786

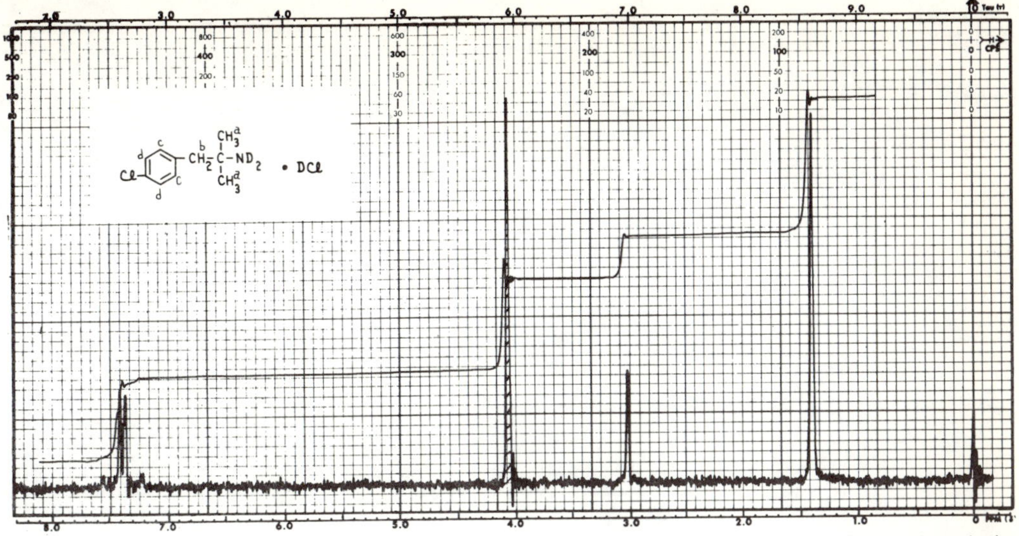

© 1978 Sadtler Research Laboratories, Inc.

ANILINE, HYDROCHLORIDE

C_6H_7N

Mol. Wt. 93.13

Solvent: Dimethyl sulfoxide-d_6

Assignments
a = 7.51
b = 9.78
c = 2.56 DMSO-d_5

787

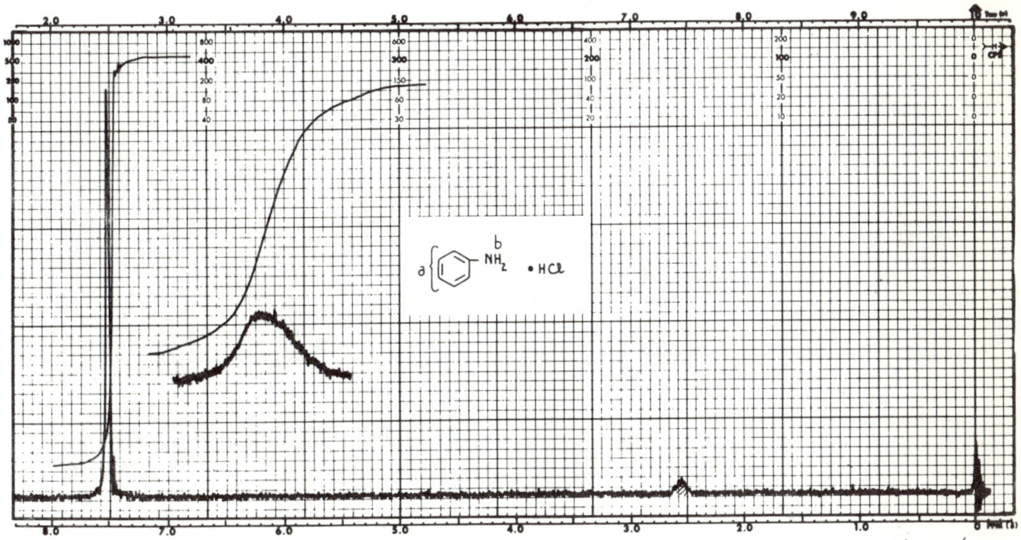

© 1978 Sadtler Research Laboratories, Inc.

ANILINE, SULFATE

$C_{12}H_{14}N_2 \cdot H_2SO_4$

Mol. Wt. 284.34

Solvent: Trifluoroacetic acid

Assignments
a = 7.51
NH$_2$ unobserved

788

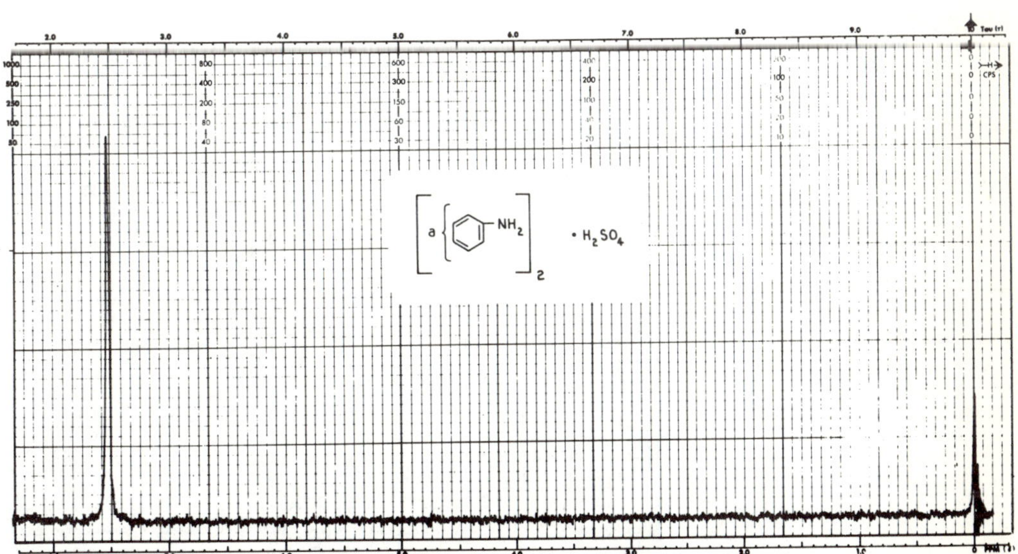

© 1978 Sadtler Research Laboratories, Inc.

o-TOLUIDINE, HYDROCHLORIDE

$C_7H_9N \cdot HCl$

Mol. Wt. 143.62

M.P. 218-220°C

Solvent: D_2O

Assignments
a = 2.49
b = 7.38
c = 5.18 HDO

789

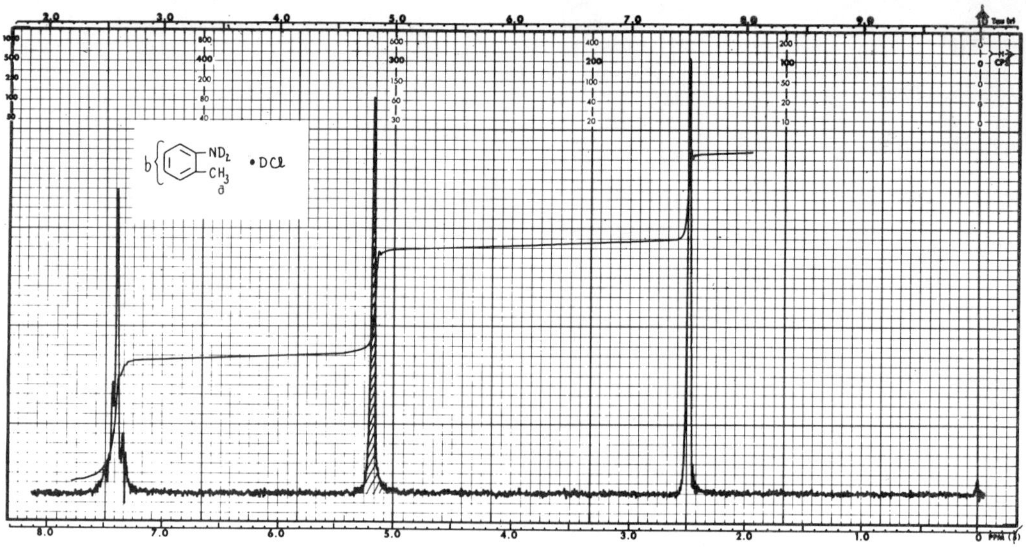

© 1978 Sadtler Research Laboratories, Inc.

m-TOLUIDINE, HYDROCHLORIDE

$C_7H_9N \cdot HCl$

Mol. Wt. 143.62

Solvent: Polysol-d

Assignments
a = 2.29
b = 7.05-7.40
c = 10.31

790

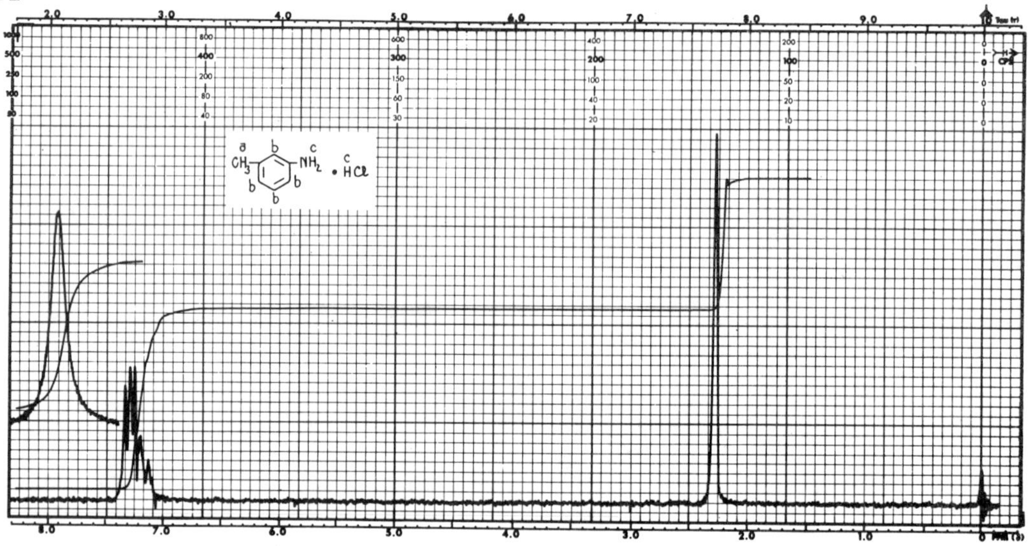

© 1978 Sadtler Research Laboratories, Inc.

p-ANISIDINE, HYDROCHLORIDE

$C_7H_9NO \cdot HCl$

Mol. Wt. 159.62

M.P. 216-218°C

Solvent: D_2O

Assignments
a = 3.84
b = 7.09
c = 7.44
d = 4.87 HDO

791

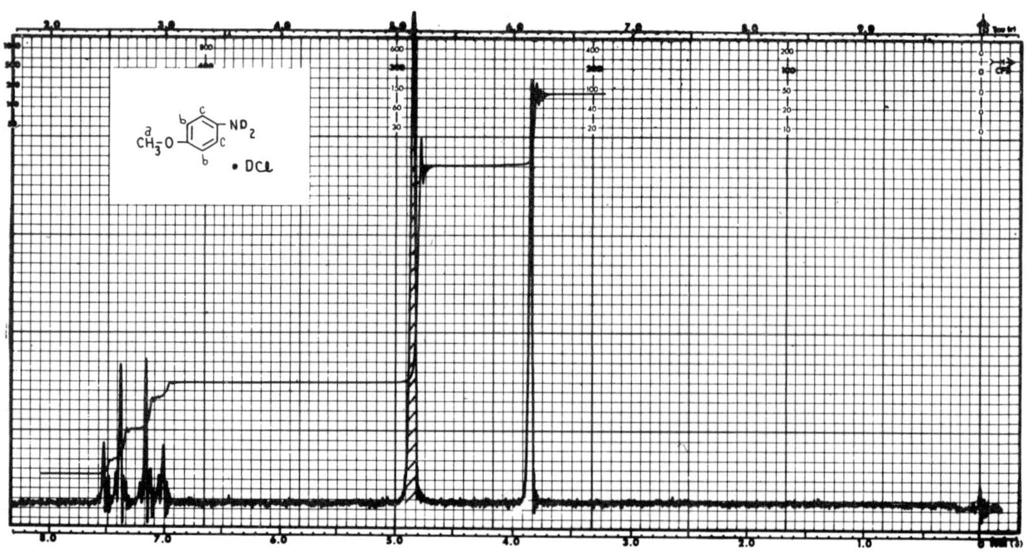

© 1978 Sadtler Research Laboratories, Inc.

2,5-XYLIDINE, HYDROCHLORIDE

$C_8H_{11}N \cdot HCl$

Mol. Wt. 157.64

M.P. 227-229°C

Solvent: D_2O

Assignments
a = 2.32
b = 2.43
c = 7.00-7.50
d = 4.51 HDO

792

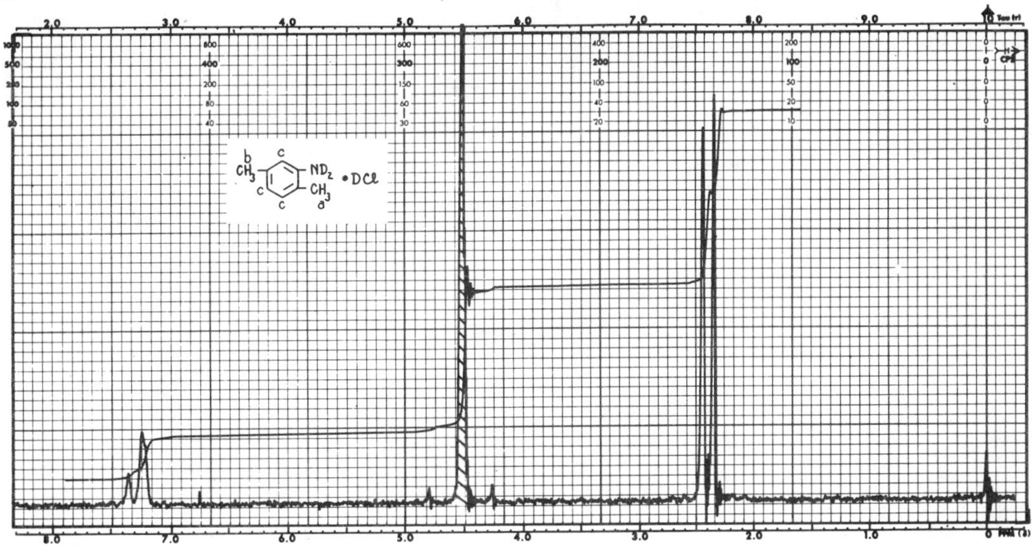

© 1978 Sadtler Research Laboratories, Inc.

o-PHENYLENEDIAMINE, DIHYDROCHLORIDE

$C_6H_8N_2 \cdot 2HCl$

Mol. Wt. 181.07

Solvent: D_2O

Assignments
a = 7.66
b = 5.57 HDO

793

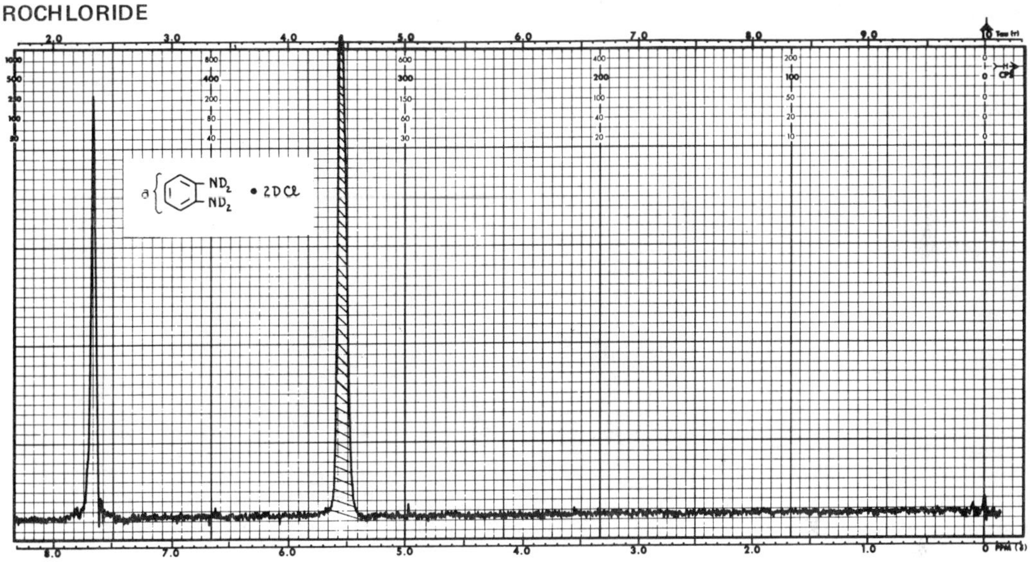

© 1978 Sadtler Research Laboratories, Inc.

m-PHENYLENEDIAMINE, DIHYDROCHLORIDE

$C_6H_8N_2 \cdot 2HCl$

Mol. Wt. 181.07

Solvent: D_2O

Assignments
a = ca 7.67
b = 5.00 HDO

794

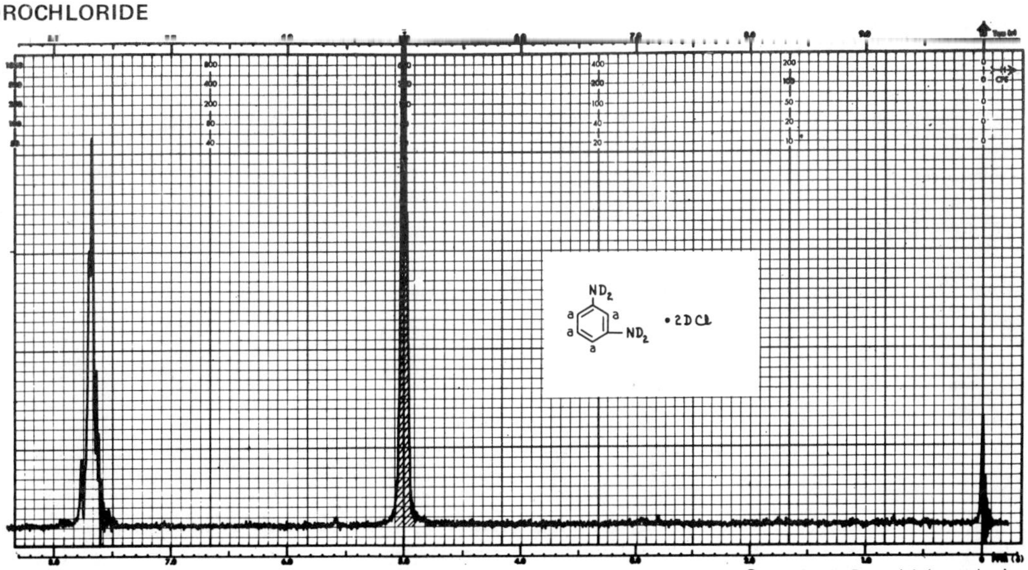

© 1978 Sadtler Research Laboratories, Inc.

p-PHENYLENEDIAMINE, DIHYDROCHLORIDE

$C_6H_8N_2 \cdot 2HCl$

Mol. Wt. 181.07

Solvent: D_2O

Assignments
a = 7.78
b = 5.00 HDO

795

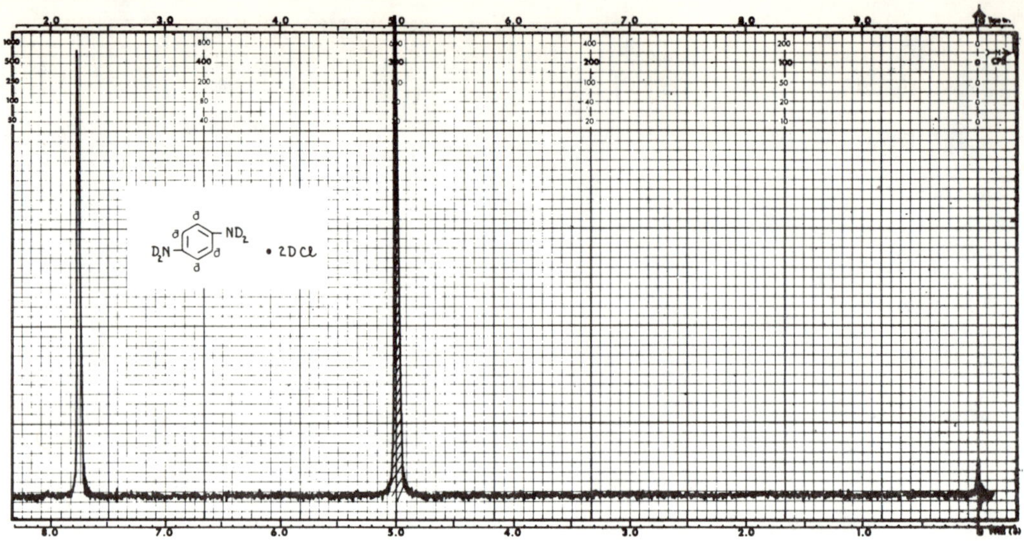

© 1978 Sadtler Research Laboratories, Inc.

2-CHLORO-p-PHENYLENEDIAMINE, DIHYDROCHLORIDE

$C_6H_7ClN_2 \cdot 2HCl$

Mol. Wt. 215.51

Solvent: D_2O

Assignments
a = ca 7.59
b = 7.85
c = 5.19 HDO

796

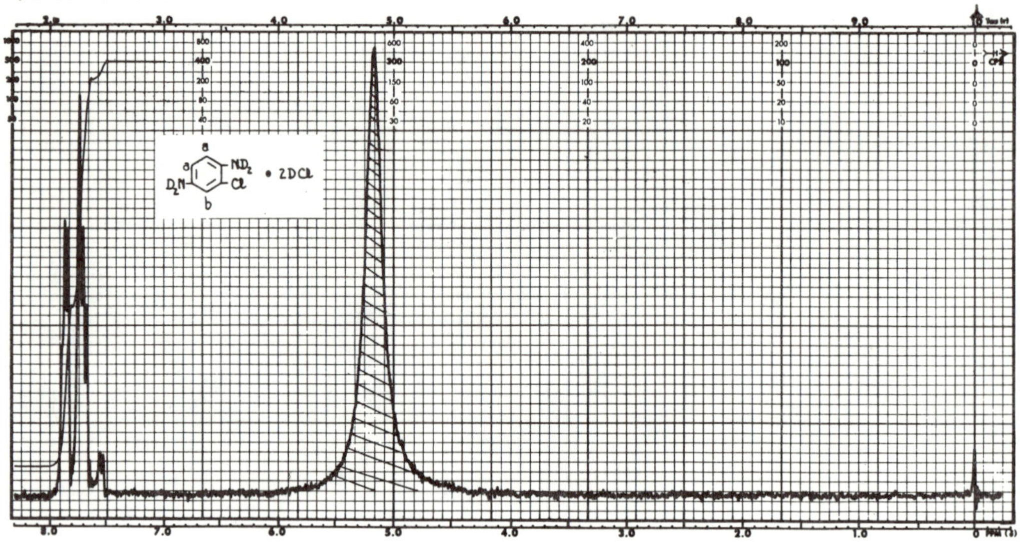

© 1978 Sadtler Research Laboratories, Inc.

DIETHYLAMINE, HYDROCHLORIDE

$C_4H_{11}N \cdot HCl$

Mol. Wt. 109.60

M.P. 228-229°C

Solvent: $CDCl_3$

Assignments
a = 1.45
b = 3.03
c = 9.35
Trace of D_2O added

797

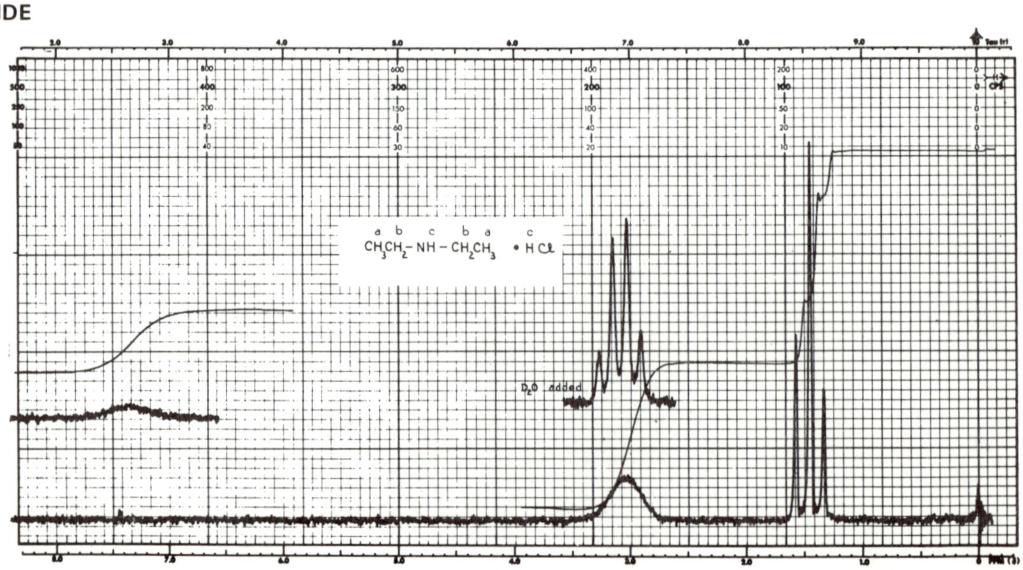

© 1978 Sadtler Research Laboratories, Inc.

DIISOPROPYLAMINE, HYDROBROMIDE

$C_6H_{15}N \cdot HBr$

Mol. Wt. 182.11

Solvent: D_2O

Assignments
a = 1.33
b = 3.51
c = 4.13 HDO

798

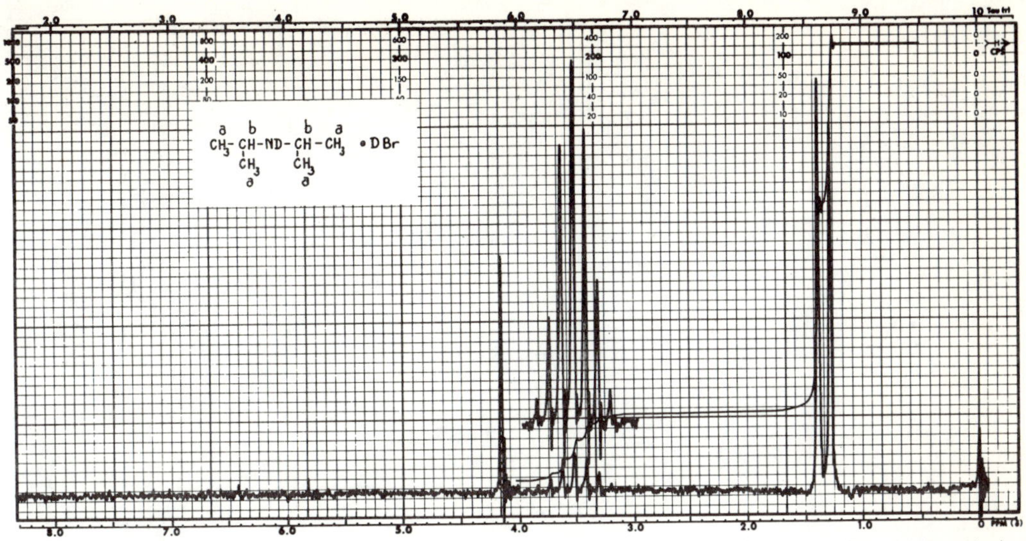

© 1978 Sadtler Research Laboratories, Inc.

DIBUTYLAMINE, HYDROCHLORIDE

$C_8H_{19}N \cdot HCl$

Mol. Wt. 165.71

Solvent: D_2O

Assignments
a = 0.93
b = 1.00-2.00
c = 3.07
d = 4.62 HDO

799

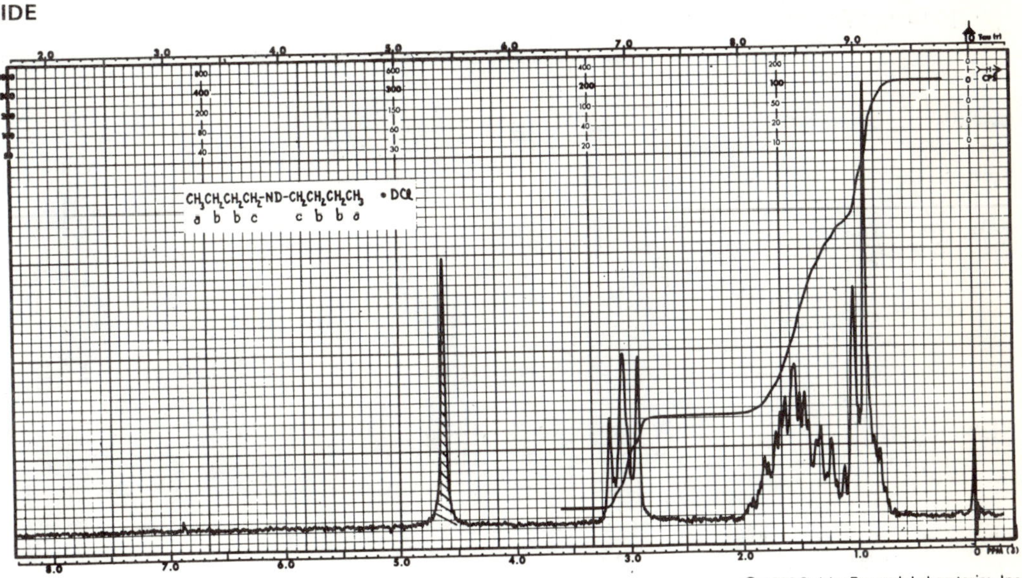

© 1978 Sadtler Research Laboratories, Inc.

DIBUTYLAMINE, HEXAFLUOROSILICATE

$(C_8H_{19}N)_2 \cdot H_2F_6Si$

Mol. Wt. 402.59

Solvent: D_2O

Assignments
a = 0.91
b = 1.53
c = 3.03
d = 4.55 HDO

800

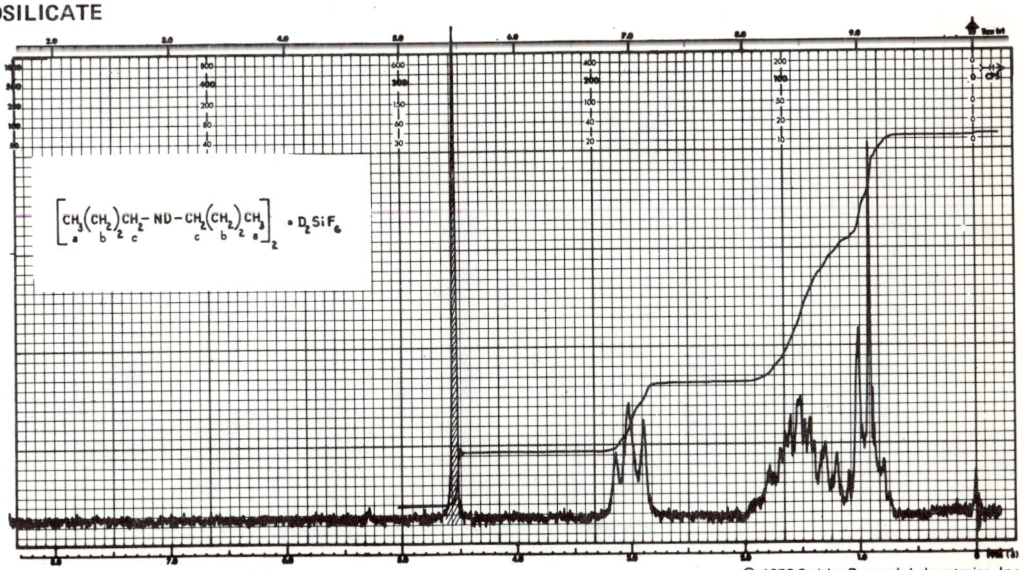

© 1978 Sadtler Research Laboratories, Inc.

PIPERIDINE, HYDROCHLORIDE

$C_5H_{11}N \cdot HCl$

Mol. Wt. 121.61

M.P. 142°C (dec.)

Solvent: $CDCl_3$

Assignments
a = 1.49-2.18
b = 3.20
c = 8.15

801

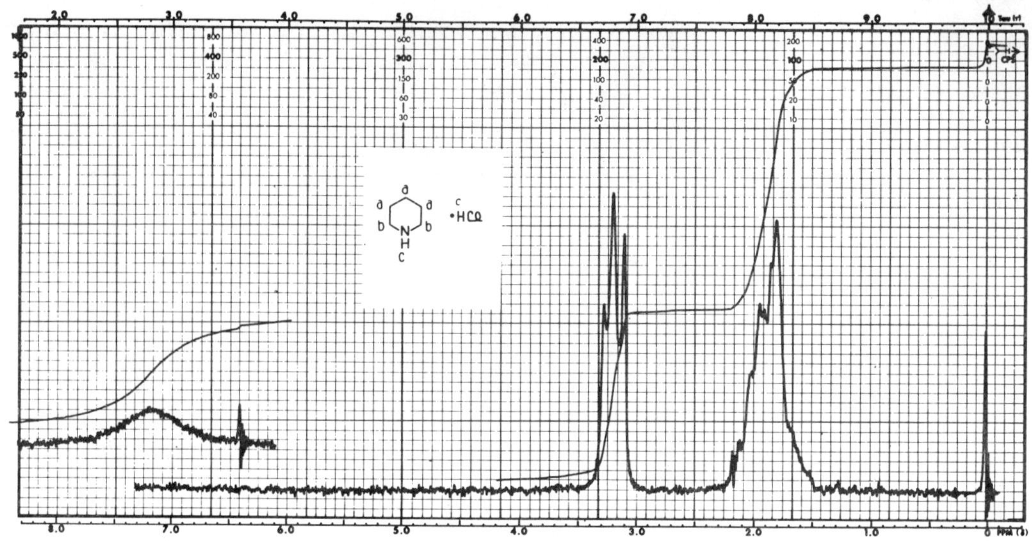

© 1978 Sadtler Research Laboratories, Inc.

PIPERAZINE, MONOHYDROCHLORIDE

$C_4H_{10}N_2 \cdot HCl$

Mol. Wt. 122.61

Solvent: D_2O

Assignments
a = 3.14
b = 4.65 HDO

802

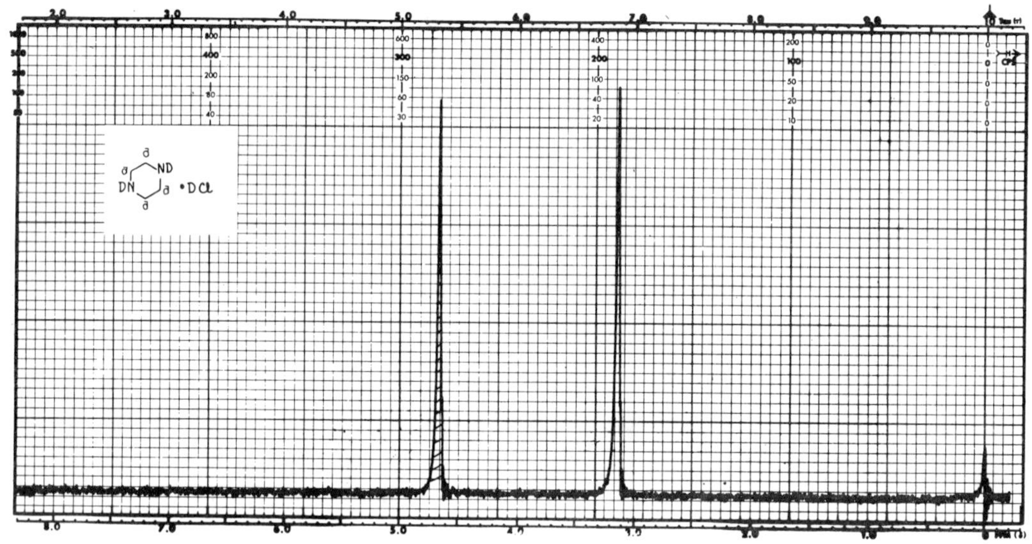

© 1978 Sadtler Research Laboratories, Inc.

PIPERAZINE, PHOSPHATE (3:2)

$3C_4H_{10}N_2 \cdot 2H_3PO_4$

Mol. Wt. 396.23

Solvent: D_2O

Assignments
a = 3.25
b = 4.56 HDO

803

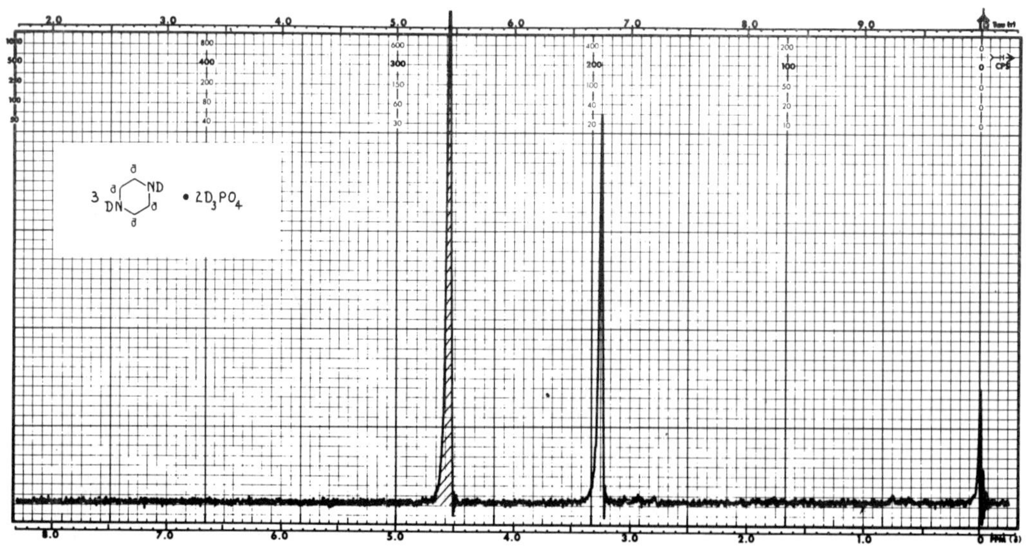

© 1978 Sadtler Research Laboratories, Inc.

DL-N,α-DIMETHYLPHENETHYLAMINE, HYDROCHLORIDE

$C_{10}H_{15}N \cdot HCl$

Mol. Wt. 185.70

M.P. 135-138°C

Solvent: $CDCl_3$

Assignments
a = 1.38
b = 2.78
c = 2.60-3.12
d = 3.51
e = 7.25
f = 9.21

804

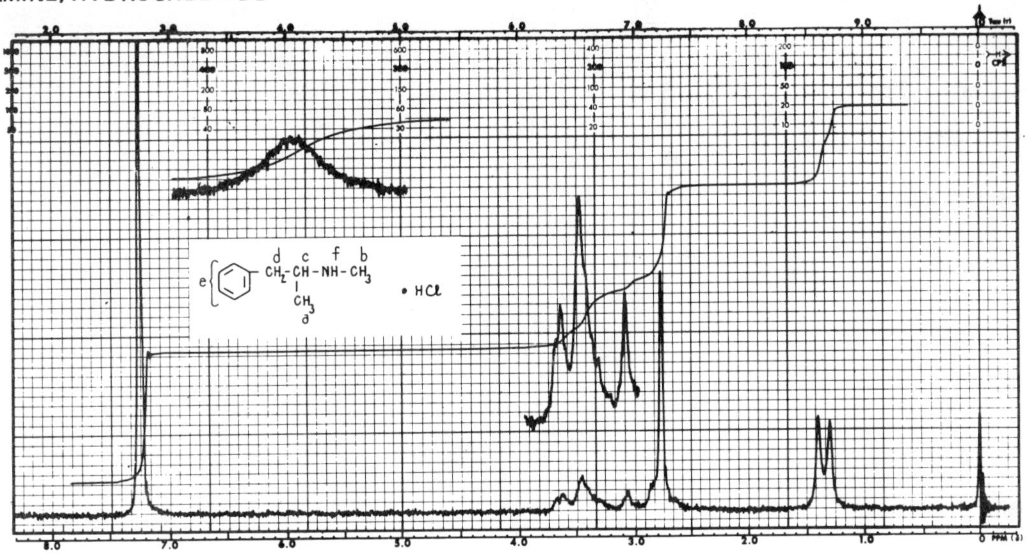

© 1978 Sadtler Research Laboratories, Inc.

DIBENZYLAMINE, HYDROCHLORIDE

$C_{14}H_{15}N \cdot HCl$

Mol. Wt. 233.74

M.P. 260-262°C

Solvent: Trifluoroacetic acid

Assignments
a = 4.37
b = 7.44
c = ca 7.60

805

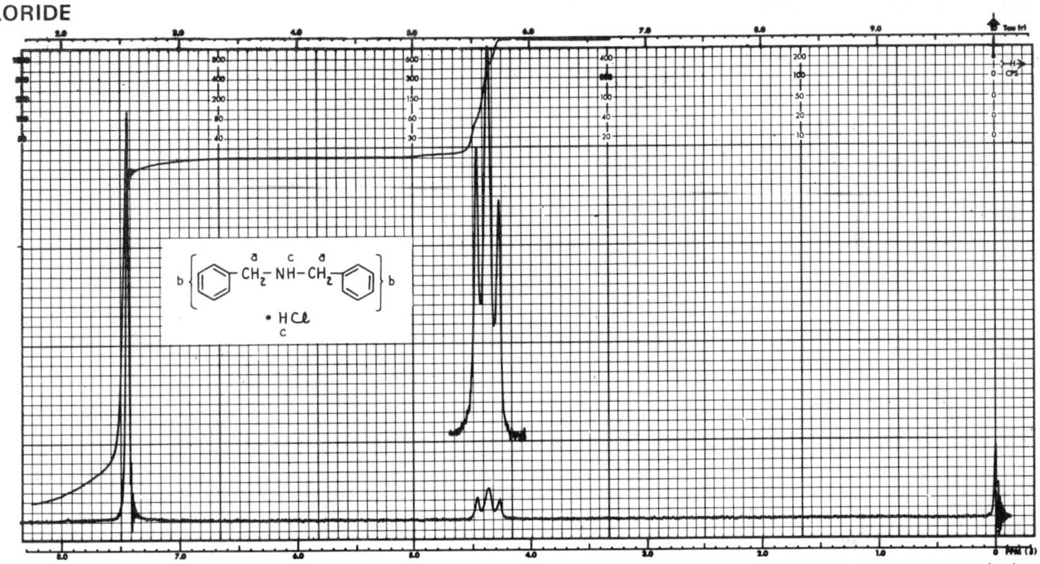

© 1978 Sadtler Research Laboratories, Inc.

N-ETHYLANILINE, HYDROCHLORIDE

$C_8H_{11}N \cdot HCl$

Mol. Wt. 157.64

M.P. 177-180°C

Solvent: D_2O

Assignments
a = 1.39
b = 3.51
c = ca 7.55
d = 4.81 HDO

806

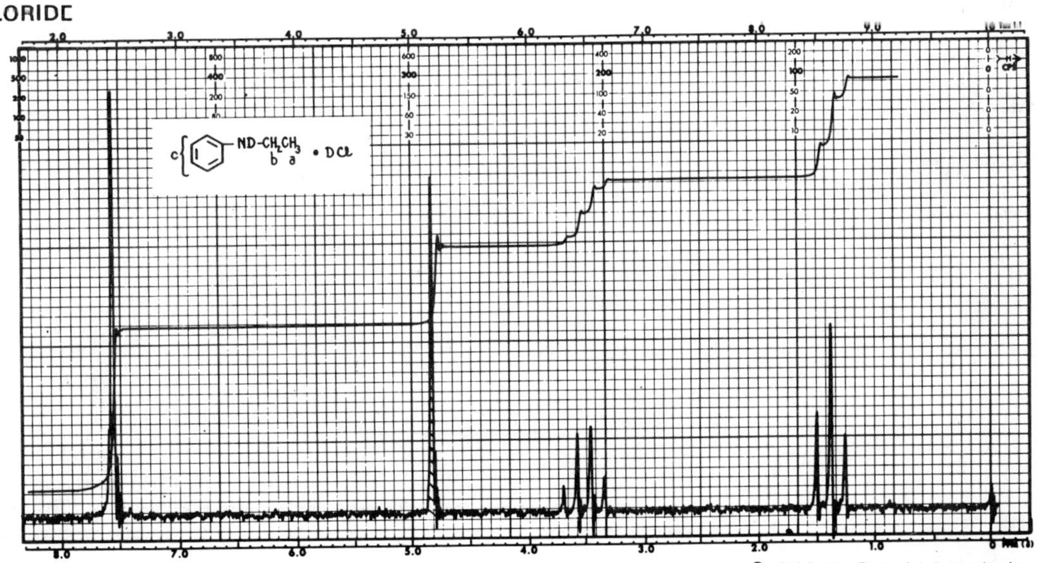

© 1978 Sadtler Research Laboratories, Inc.

DIPHENYLAMINE, HYDROCHLORIDE

$C_{12}H_{11}N \cdot HCl$

Mol. Wt. 205.69

Solvent: Dimethy l sulfoxide

Assignments
a = 6.88
b = 7.10-7.40
c = 10.26

807

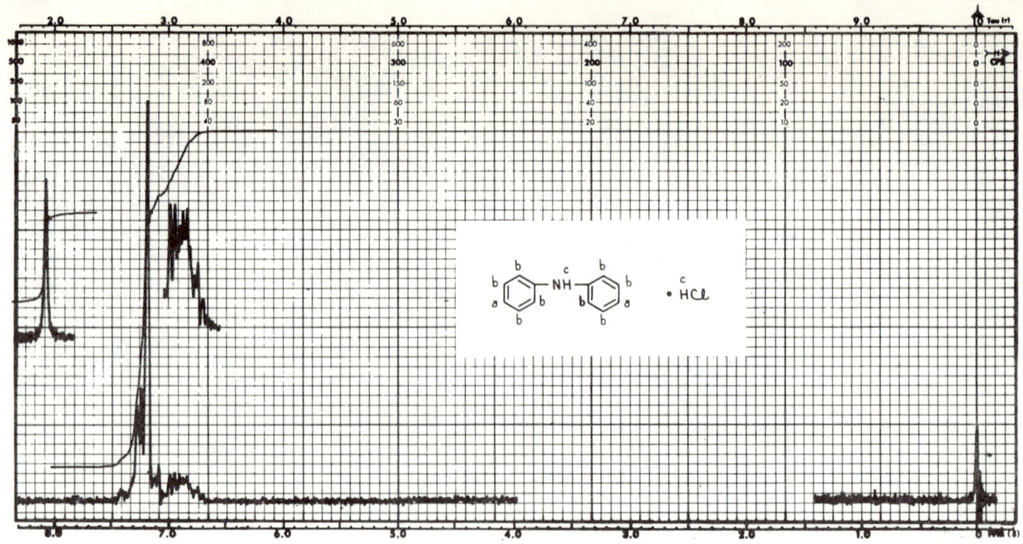

© 1978 Sadtler Research Laboratories, Inc.

TRIMETHYLAMINE, HYDROCHLORIDE

$C_3H_9N \cdot HCl$

Mol. Wt. 95.57

M.P. 271-273°C (lit.)

Solvent: D_2O

Assignments
a = 2.94
b = 4.68 HDO

808

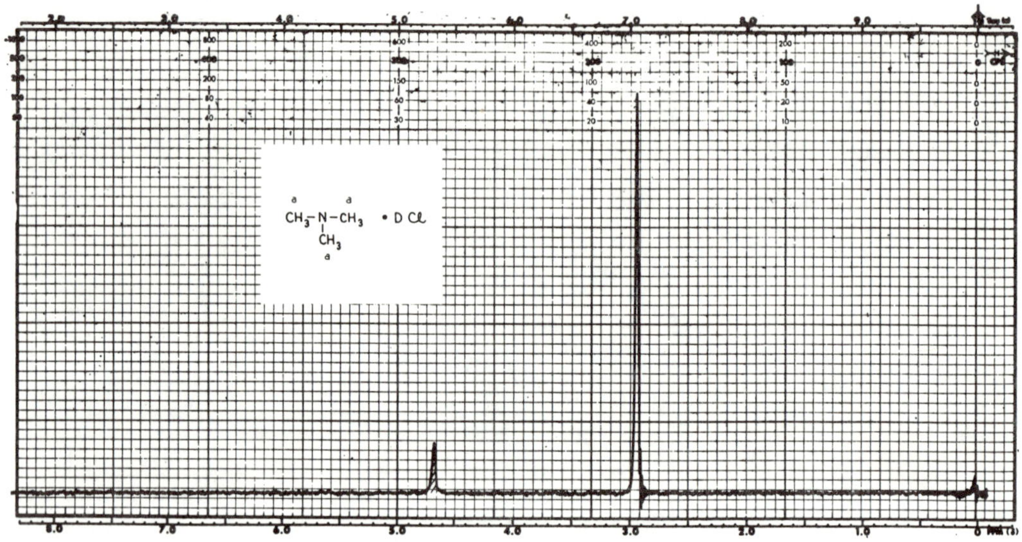

© 1978 Sadtler Research Laboratories, Inc.

TRIETHYLAMINE, HYDROCHLORIDE

$C_6H_{15}N \cdot HCl$

Mol. Wt. 137.65

Solvent: $CDCl_3$

Assignments
a = 1.45
b = 3.21

809

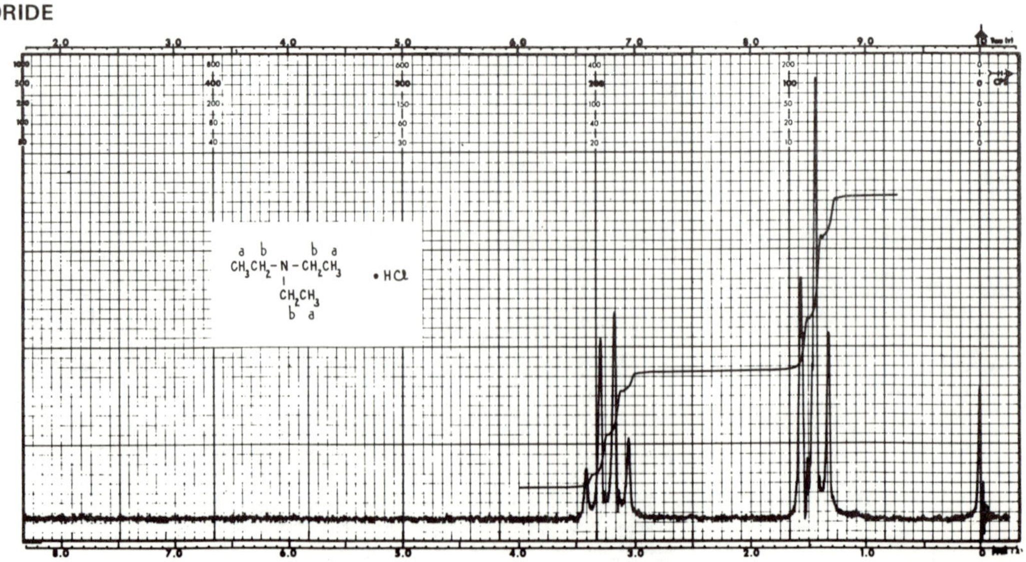

© 1978 Sadtler Research Laboratories, Inc.

TRIOCTYLAMINE, HYDROCHLORIDE

$C_{24}H_{51}N \cdot HCl$

Mol. Wt. 390.14

M.P. 72-74°C

Solvent: $CDCl_3$

Assignments
a = 0.89
b = 1.10-1.60
c = 1.84
d = 3.01
e = 11.51

810

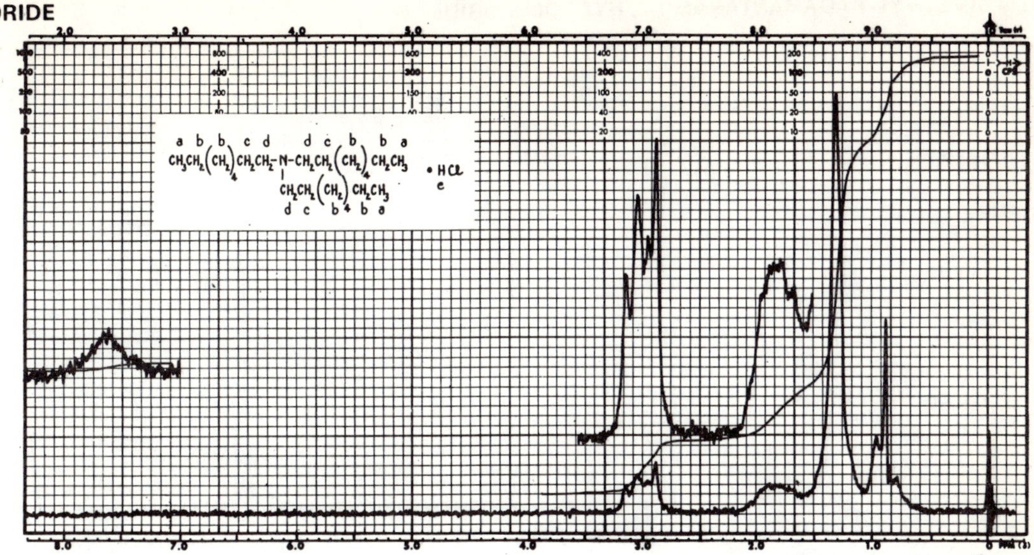

© 1978 Sadtler Research Laboratories, Inc.

2-CHLORO-N,N-DIMETHYLETHYLAMINE, HYDROCHLORIDE

$C_4H_{10}ClN \cdot HCl$

Mol. Wt. 144.04

M.P. 199°C

Solvent: D_2O

Assignments
a = 2.92
b = 3.56
c = 3.91
d = 4.53 HDO

811

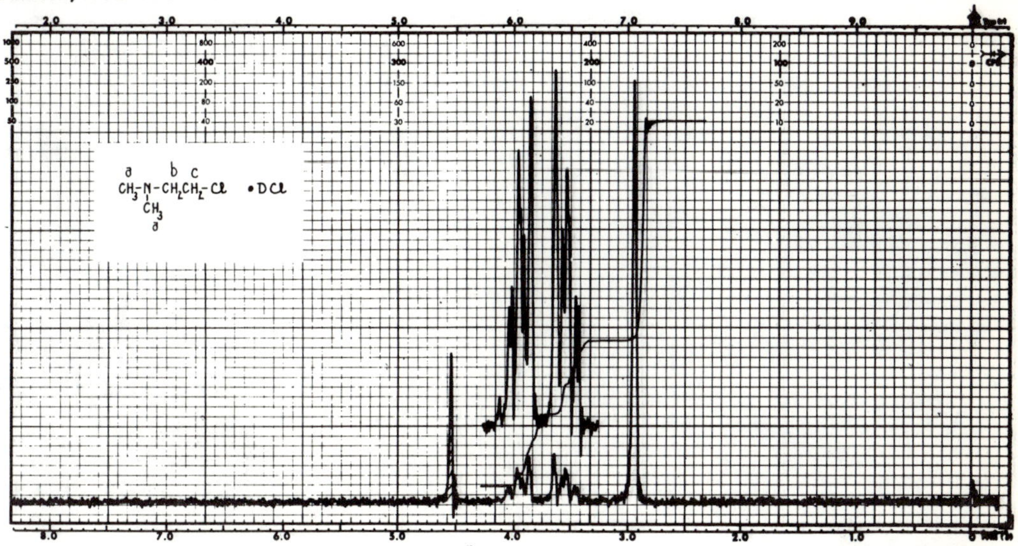

© 1978 Sadtler Research Laboratories, Inc.

2-CHLOROTRIETHYLAMINE, HYDROCHLORIDE

$C_6H_{14}ClN \cdot HCl$

Mol. Wt. 172.10

M.P. 200°C

Solvent: D_2O

Assignments
a = 1.38
b = 3.32
c = 3.57
d = 3.99
e = 4.17 HDO

812

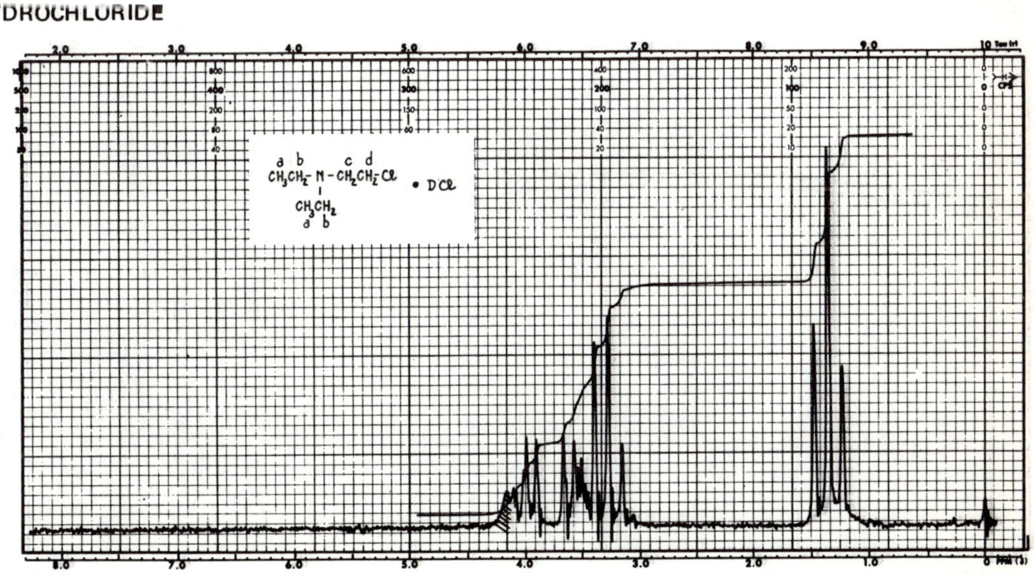

© 1978 Sadtler Research Laboratories, Inc.

N,N-DIMETHYL-1-ADAMANTANAMINE, HYDROCHLORIDE

$C_{12}H_{21}N \cdot HCl$

Mol. Wt. 215.77

M.P. 280°C

Solvent: D_2O

Assignments
a = 1.70
b = 1.94
c = 2.22
d = 2.78
e = 4.60 HDO

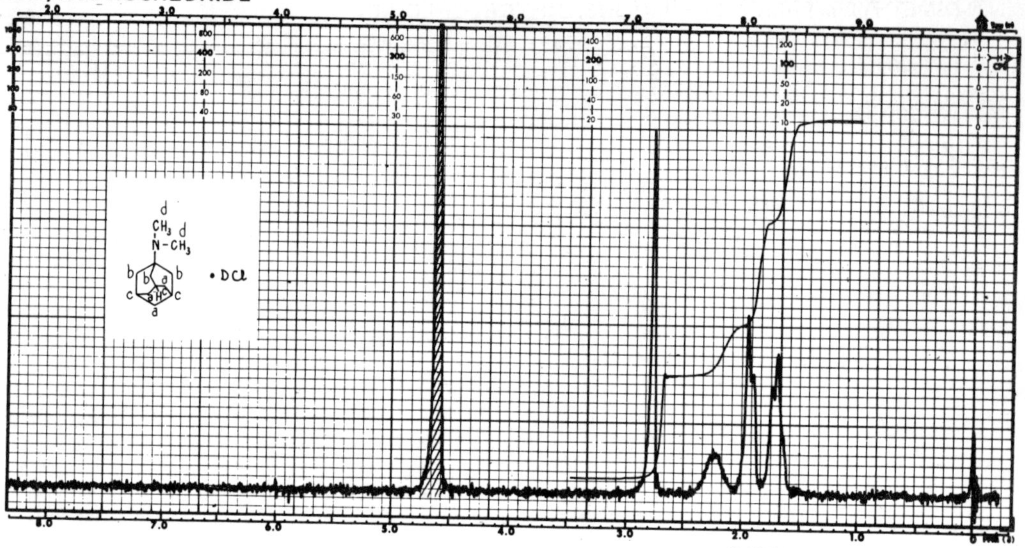

813

© 1978 Sadtler Research Laboratories, Inc.

4-(DIMETHYLAMINO)PIPERIDINE, DIHYDROCHLORIDE

$C_7H_{16}N_2 \cdot 2HCl$

Mol. Wt. 201.14

Solvent: D_2O

Assignments
a = 1.60-2.70
b = 2.97
c = 3.10-3.95
d = 4.30 HDO
e = 2.70 imp.

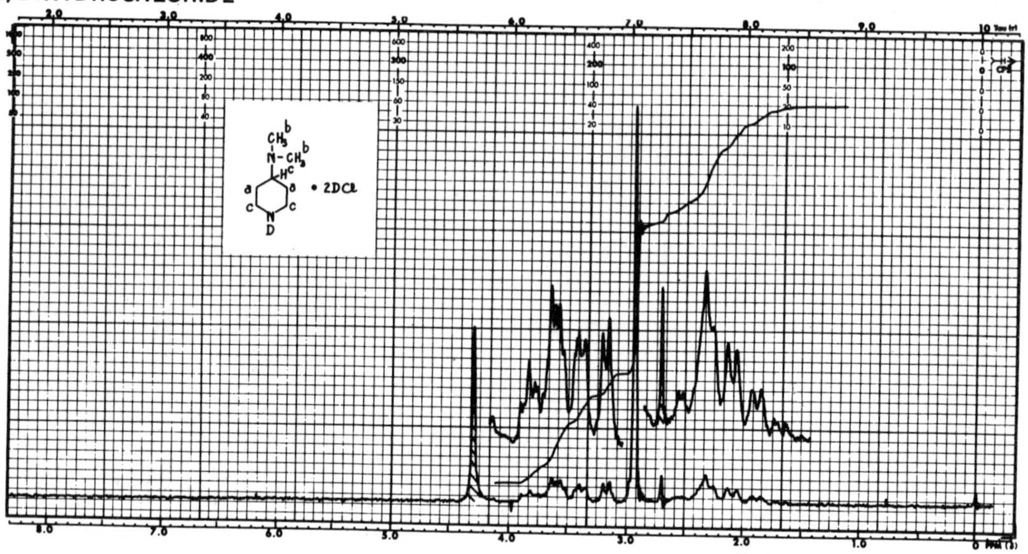

814

© 1978 Sadtler Research Laboratories, Inc.

1-(2-CHLOROETHYL)PIPERIDINE, HYDROCHLORIDE

$C_7H_{14}ClN \cdot HCl$

Mol. Wt. 184.11

Solvent: D_2O

Assignments
a = 1.30-2.20
b = 2.70-3.80
c = 3.53
d = 3.98
e = 4.51 HDO

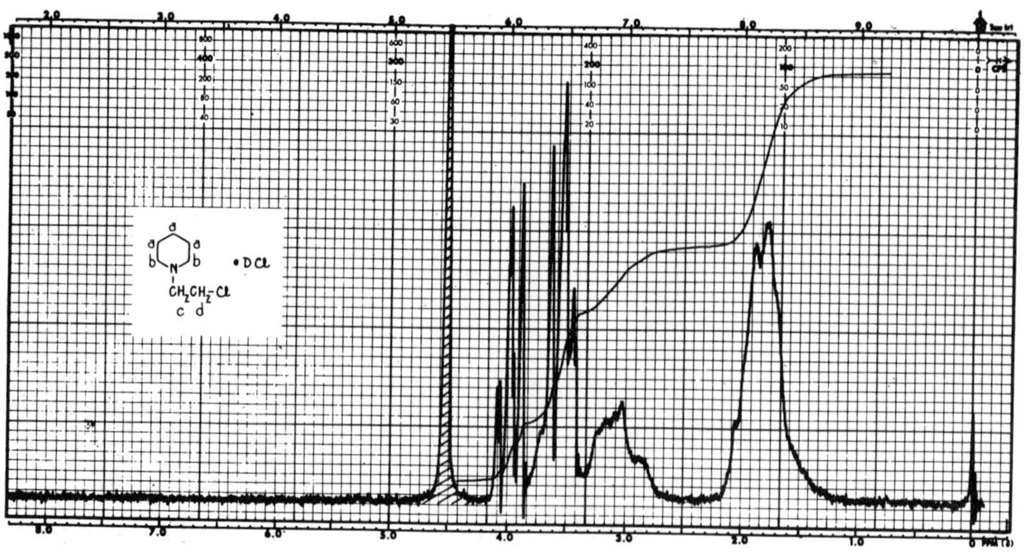

815

© 1978 Sadtler Research Laboratories, Inc.

N,N-DIMETHYL-m-PHENYLENEDIAMINE, DIHYDROCHLORIDE

$C_8H_{12}N_2 \cdot 2HCl$

Mol. Wt. 209.12

M.P. 217°C (dec.)

Solvent: D_2O

Assignments

a = 3.43

b = 7.65-8.00

c = 4.82 HDO

816

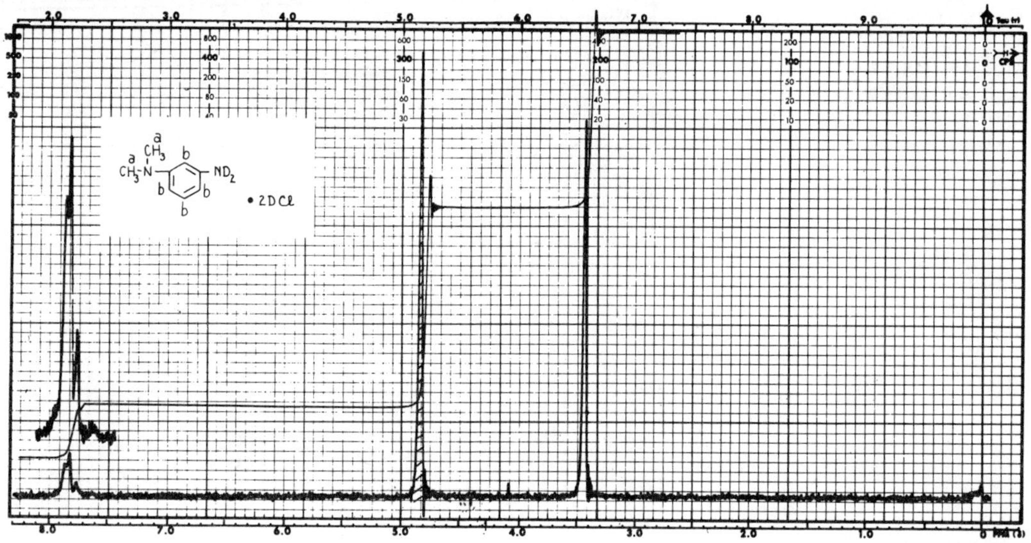

© 1978 Sadtler Research Laboratories, Inc.

N,N-DIETHYL-p-PHENYLENEDIAMINE, MONOHYDROCHLORIDE

$C_{10}H_{16}N_2 \cdot HCl$

Mol. Wt. 200.71

Solvent: D_2O

Assignments

a = 1.14

b = 3.63

c = 7.06

d = 7.40

e = 4.88 HDO

817

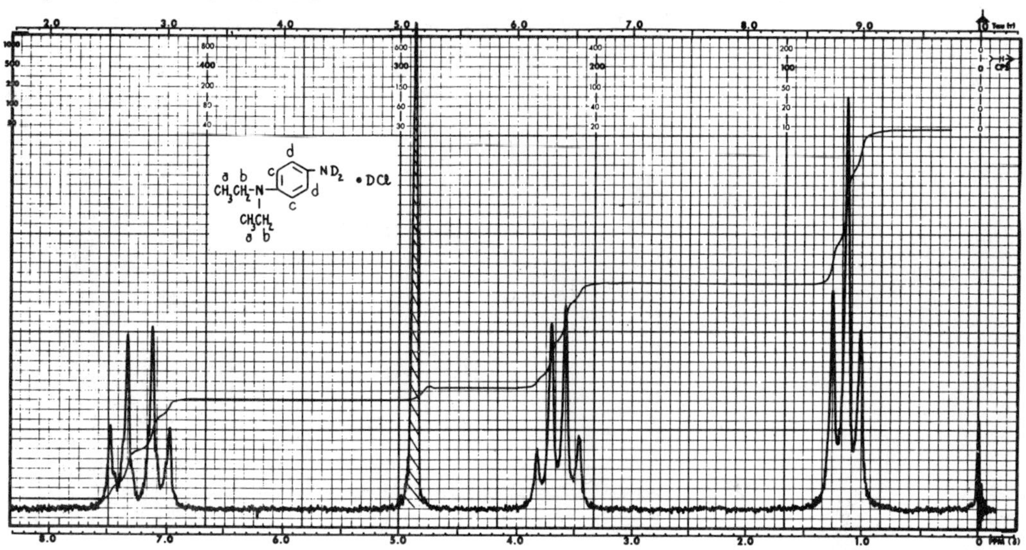

© 1978 Sadtler Research Laboratories, Inc.

N,N-DIMETHYL-p-PHENYLENEDIAMINE, DIHYDROCHLORIDE

$C_8H_{12}N_2 \cdot 2HCl$

Mol. Wt. 209.12

Solvent: D_2O

Assignments

a = 3.44

b = 7.79 or c

c = 7.99 or b

818

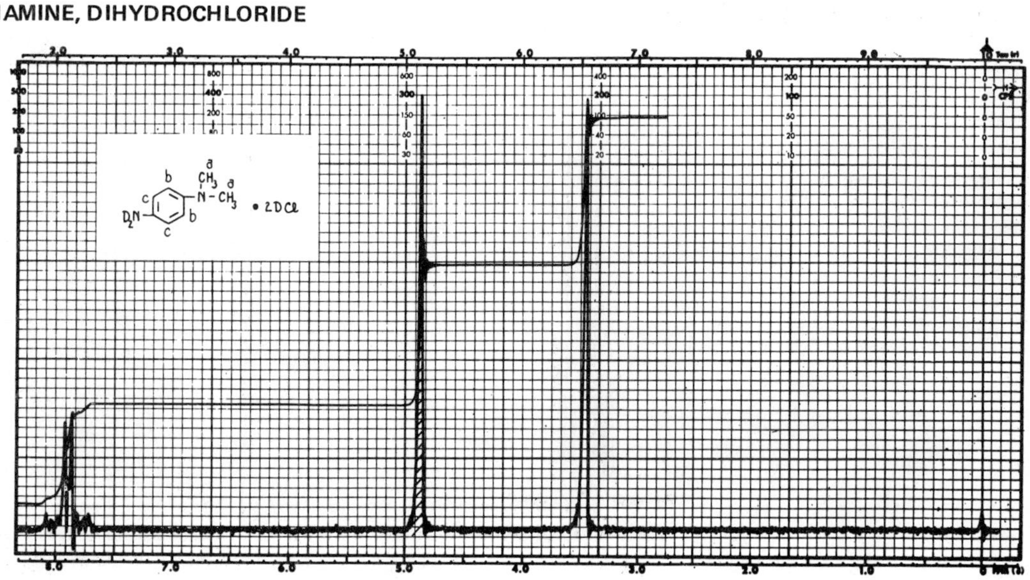

© 1978 Sadtler Research Laboratories, Inc.

N,N,N',N'-TETRAMETHYL-p-PHENYLENEDIAMINE, DIHYDROCHLORIDE

$C_{10}H_{16}N_2 \cdot 2HCl$

Mol. Wt. 237.17

Solvent: D_2O

Assignments
a = 3.36
b = 7.84
c = 4.71 HDO

819

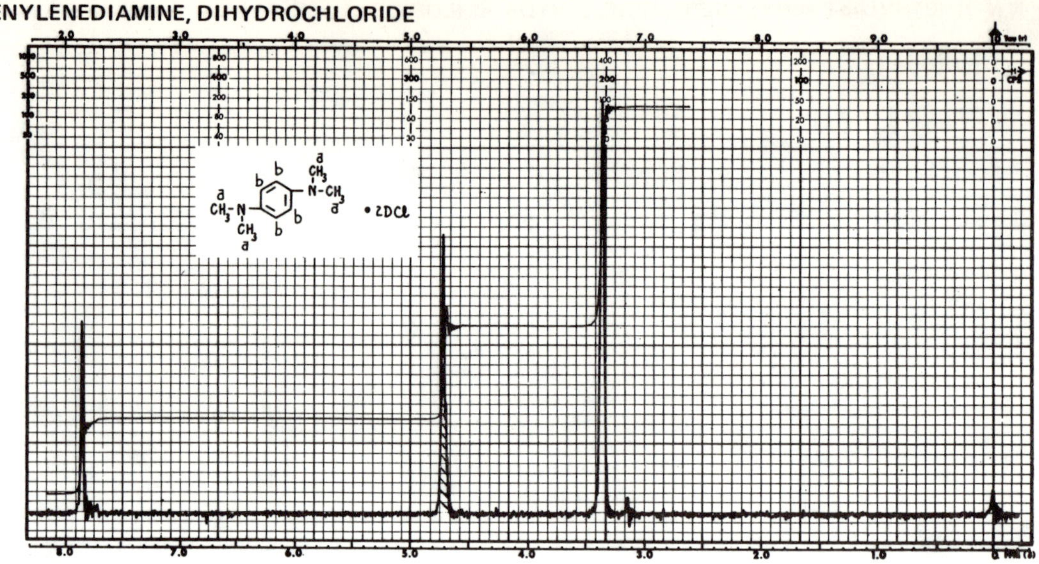

© 1978 Sadtler Research Laboratories, Inc.

HISTAMINE, MONOHYDROCHLORIDE

$C_5H_9N_3 \cdot HCl$

Mol. Wt. 147.61

Solvent: D_2O

Assignments
a = 3.32
b = 3.49
c = 7.58
d = 8.84
e = 5.15 HDO

820

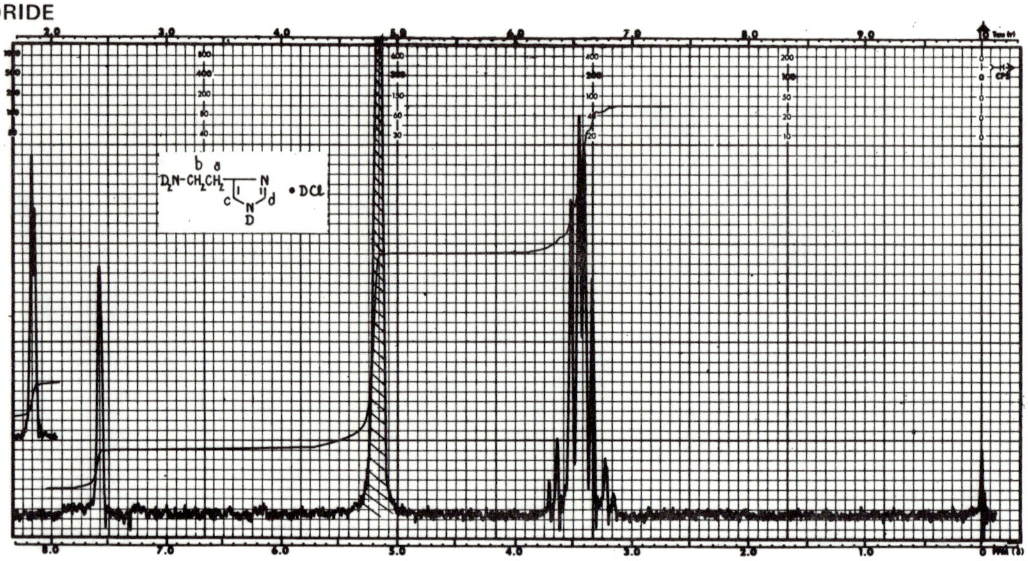

© 1978 Sadtler Research Laboratories, Inc.

4-BROMOPYRIDINE, HYDROBROMIDE

$C_5H_4BrN \cdot HBr$

Mol. Wt. 238.92

Solvent: D_2O

Assignments
a = 8.41
b = 8.77
c = 4.73 HDO

821

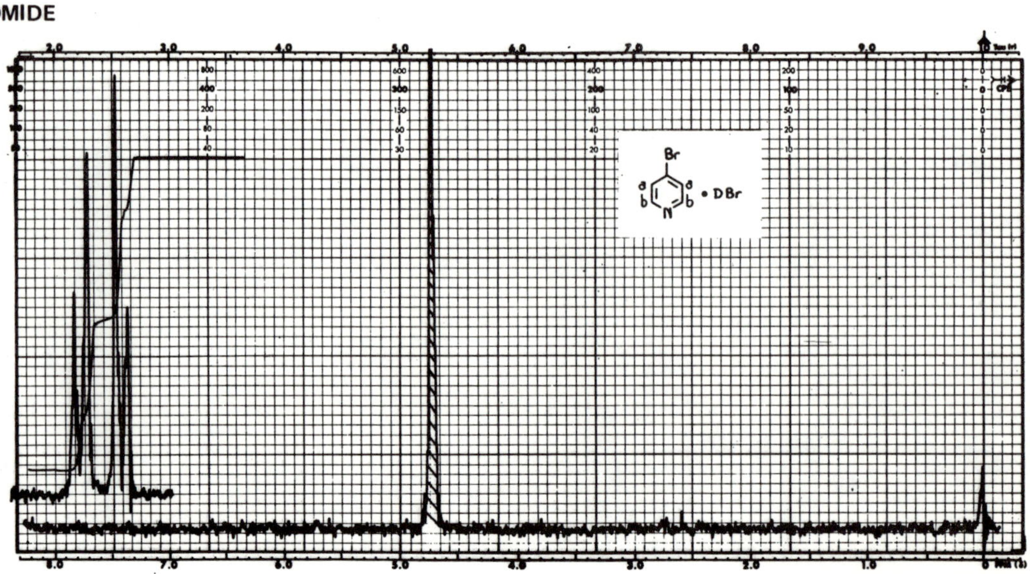

© 1978 Sadtler Research Laboratories, Inc.

METHYLHYDRAZINE, SULFATE

$CH_6N_2 \cdot H_2SO_4$

Mol. Wt. 144.15

M.P. 141-142°C

Solvent: D_2O

Assignments
a = 2.91
b = 4.92 HDO

822

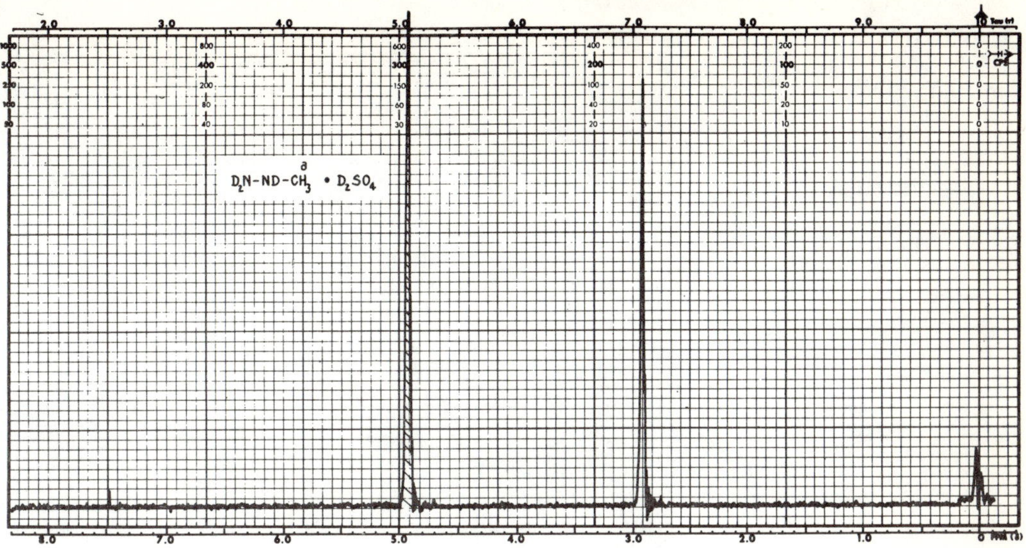

© 1978 Sadtler Research Laboratories, Inc.

(3-CHLORO-p-TOLYL)HYDRAZINE, MONOHYDROCHLORIDE

$C_7H_9ClN_2 \cdot HCl$

Mol. Wt. 193.08

M.P. 247°C (dec.)

Solvent: Dimethyl sulfoxide-d_6

Assignments
a = 2.25
b = 6.91
c = 7.14
d = 7.20
e = 7.50-9.50
f = 2.52 DMSO-d_5

823

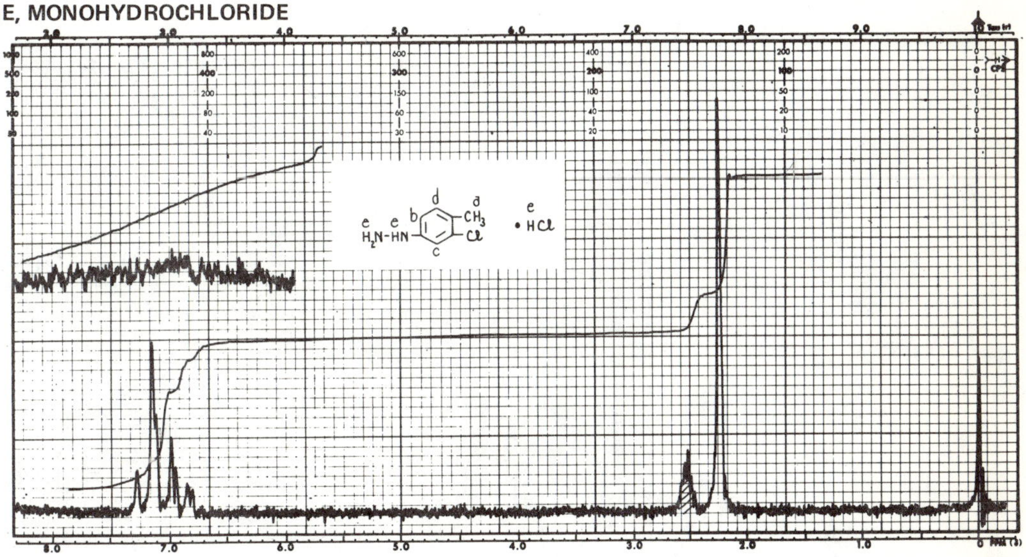

© 1978 Sadtler Research Laboratories, Inc.

(p-METHOXYPHENYL)HYDRAZINE, MONOHYDROCHLORIDE

$C_7H_{10}N_2O \cdot HCl$

Mol. Wt. 174.63

M.P. 153-153.5°C (dec.)

Solvent: D_2O

Assignments
a = 3.85
b = 7.08
c = 7.21
d = 3.40 imp.
e = 4.71 HDO

824

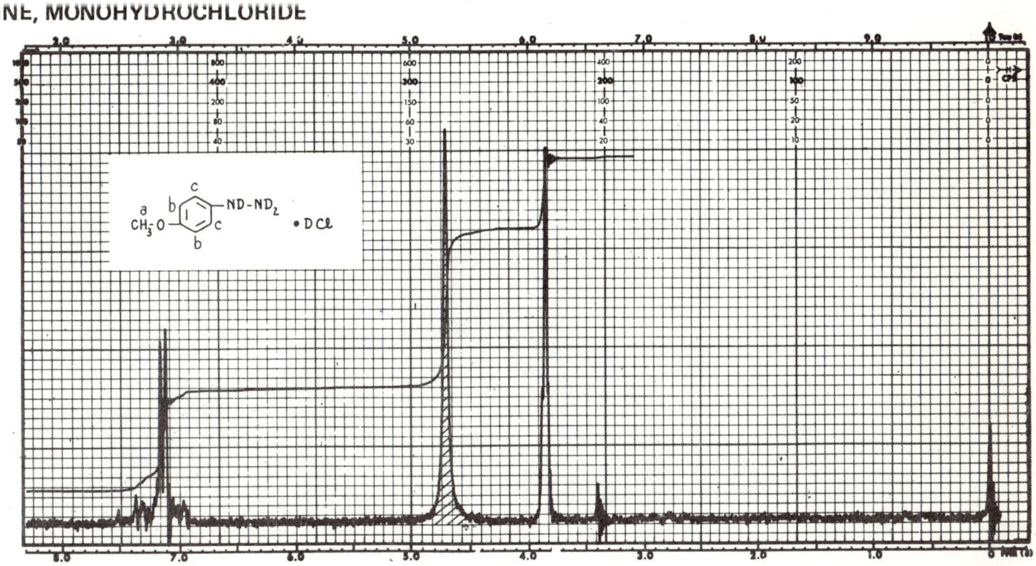

© 1978 Sadtler Research Laboratories, Inc.

(2-NAPHTHYL)HYDRAZINE, MONOHYDROCHLORIDE

$C_{10}H_{10}N_2 \cdot HCl$

Mol. Wt. 194.67

M.P. 236-238°C (dec.)

Solvent: DMSO-d_6

Assignments

a = 7.15-7.60

b = 7.71

c = 7.80

d = 9.63 *

e = 2.53 DMSO-d_5

* Plus H_2O

825

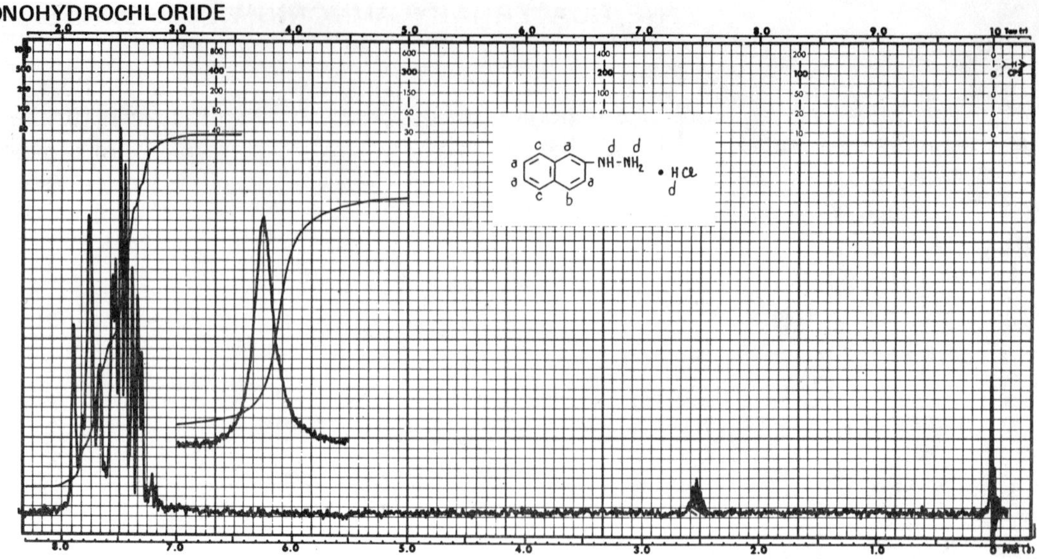

© 1978 Sadtler Research Laboratories, Inc.

3-HYDRAZINOQUINOLINE, DIHYDROCHLORIDE

$C_9H_9N_3 \cdot 2HCl$

Mol. Wt. 232.11

Solvent: D_2O

Assignments

a = 7.50-8.15

b = 8.46

c = 8.95

d = 4.93 HDO

826

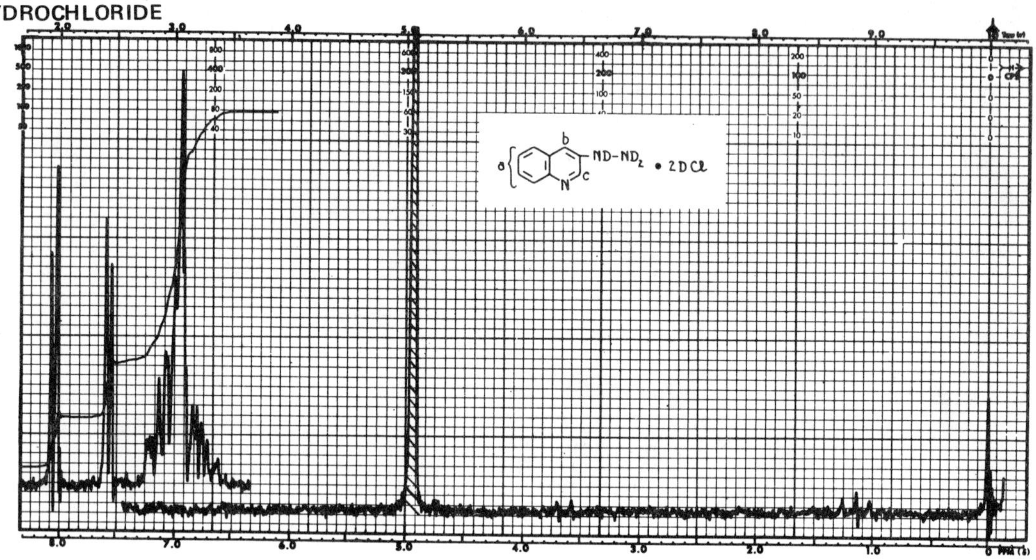

© 1978 Sadtler Research Laboratories, Inc.

The quaternary ammonium compounds display low field chemical shifts for the aliphatic groups bonded to the nitrogen atom. The groups often show a certain degree of broadening, possibly due to unresolved coupling to the nitrogen nucleus. The compounds are more soluble in D_2O and DMSO-d_6 than the corresponding tertiary amines.

As indicated below, relatively large ranges of chemical shifts are observed for similar groups on different environments. No consistant correlation with solvent, concentration or anion has been observed to explain these variations.

Chemical Shifts

Aliphatic Protons

HC–4	HC–3	HC–2	HC–1		Solvent	Spectrum
			(2.99)	$(CH_3)_3$–N^+(adamantane) (I^-)	D_2O	834
			(3.15)	$(CH_3)_4$–N^+(OH^-)	D_2O	----
			(3.21)	$(CH_3)_4$–N^+(Br^-)	D_2O	----
			(3.30)	$(CH_3)_3$–N^+–CH_2–Ø (I^-)	Polysol	837
			(3.40)	$(CH_3)_3$–N^+–CH_2–Ø (Cl^-)	$CDCl_3$	835
			(3.48)	$(CH_3)_3$–N^+–R_{12} (Br^-)	$CDCl_3$	830
		(1.26	3.24)	$(CH_3$–$CH_2)_4$–N^+ (I^-)	D_2O	----
		(1.30	3.40)	$(CH_3$–$CH_2)_3$–N^+–CH_2–C≡C–H (Br^-)	Polysol	834
		(1.28	3.95)	$(CH_3$–$CH_2)_3$–N^+–Ø (I^-)	D_2O	841
	(0.92	1.70	3.16)	$(CH_3$–CH_2–$CH_2)_4$–N^+ (Br^-)	D_2O	----
	(1.08	1.92	3.35)	CH_3–CH_2–CH_2–$N^+(R_2)_3$ (I^-)	$CDCl_3$	----
	(1.08	1.81	3.39)	$(CH_3$–CH_2–$CH_2)_4$–N^+ (Cl^-)	$CDCl_3$	----
(1.02	1.45	1.70	3.22)	$(CH_3$–CH_2–CH_2–$CH_2)_4$–N^+ (ClO_4^-)	TFA	----
(1.03	1.45	1.60	3.41)	$(CH_3$–CH_2–CH_2–$CH_2)_4$–N^+ (I^-)	$CDCl_3$	----
(1.01	1.43	1.65	3.42)	$(CH_3$–CH_2–CH_2–$CH_2)_4$–N^+ (Br^-)	$CDCl_3$	832

Substituted Methyl Ammonium Compounds

$(CH_3)_3$–N^+	–CH_2	–X		
(3.29)	3.48	–C_3	(I^-)	Polysol
(3.12)	3.95	–CH=CH_2	(Br^-)	D_2O
(3.31)	4.09	–C(=O)–NH–R	(Cl^-)	D_2O
(3.30)	4.22	–C(=O)–OH	(Cl^-)	D_2O
(3.15)	4.53	–Ø	(Br^-)	D_2O

Aliphatic Protons continued

2-Substituted Ethyl Ammonium Compounds

$(CH_3)_3-N^+$	$-CH_2$	$-CH_2$	$-X$		Solvent
3.29	3.55	3.31	$-S-C(=O)-CH_3$	(I^-)	Polysol
3.19	3.80	3.80	$-Br$	(Br^-)	D_2O
3.18	3.49	4.02	$-OH$	(tartrate)	D_2O
3.26	3.81	4.03	$-Cl$	(Cl^-)	D_2O
3.26	3.79	4.55	$-O-C(=O)-CH_3$	(Br^-)	D_2O
3.30	3.80	4.57	$-O-C(=O)-R_2$	(I^-)	D_2O

Olefinic Protons

The trimethylammonium group has an unusually strong and long range deshielding effect on the two terminal olefinic protons of allyl groups. In the spectrum of Allyltrimethylammonium Bromide, all three olefinic protons resonate in the chemical shift range from 5.4 to 6.2 ppm as a higher order ABC pattern. Normally, the terminal olefinic protons of allyl groups resonate at about 5.1 ppm as the AB portion of an ABX system.

Aromatic Protons

The chemical shifts of the quaternary ammonium aromatic compounds are dependent to a significant degree upon the solvent employed and/or the amount of water present in the sample solution. As an example, when benzyl trimethyl ammonium chloride (spectrum 836) was examined in $CDCl_3$ the ortho protons are strongly deshielded in relation to the meta and para hydrogens. In Polysol and DMSO-d_6 solution, all five protons resonate as a single complex band similar to that of spectrum 837. In D_2O solution, the five protons appear as a single sharp peak at about 7.5 ppm identical to the aromatic resonance band observed in spectrum 838. A somewhat similar case of solvent deshielding is noted in spectra 840 and 841, in which the aromatic protons of the compound examined in $CDCl_3$ solution resonate at lower field than those of a similar compound in D_2O.

TETRAETHYLAMMONIUM IODIDE

$C_8H_{20}IN$

Mol. Wt. 257.16

M.P. > 200°C

Solvent: D_2O

Assignments
a = 1.26
b = 3.24
c = 4.51 HDO

827

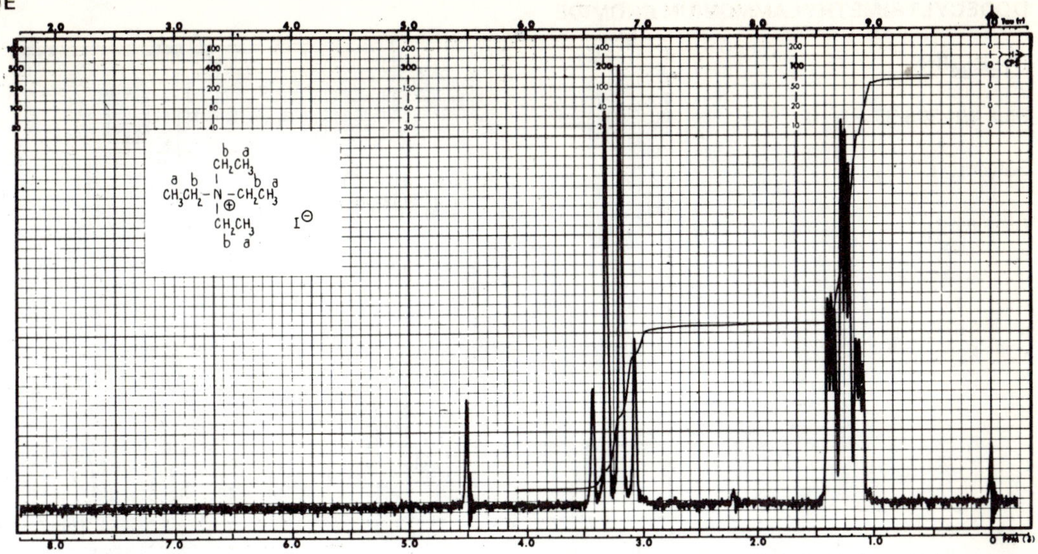

© 1978 Sadtler Research Laboratories, Inc.

TETRABUTYLAMMONIUM TETRAFLUOROBORATE(1–)

$C_{16}H_{36}BF_4N$

Mol. Wt. 329.28

M.P. 160-162°C

Solvent: $CDCl_3$

Assignments
a = 1.00
b = 1.10-2.00
c = 3.20

828

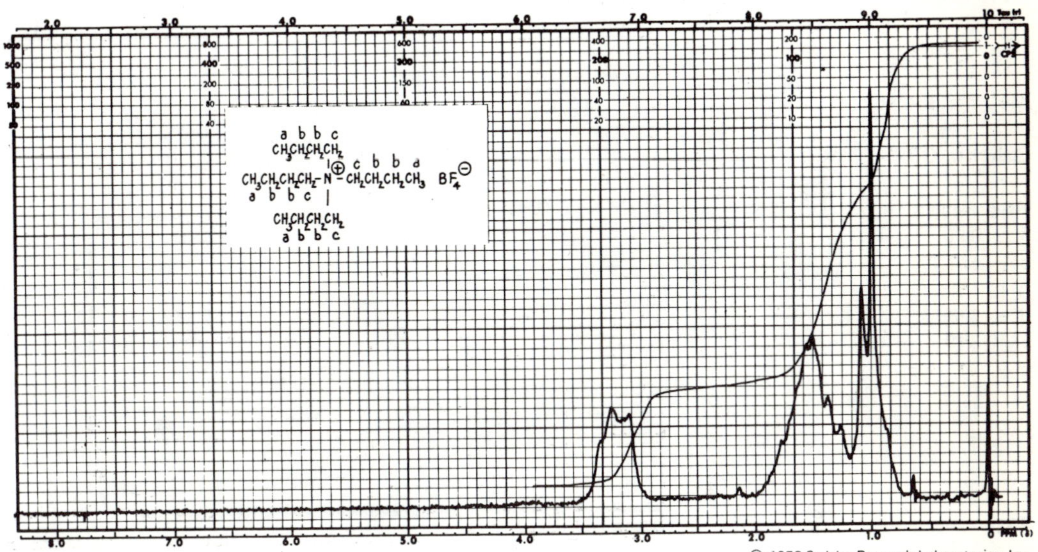

© 1978 Sadtler Research Laboratories, Inc.

DODECYLTRIMETHYLAMMONIUM CHLORIDE, MONOHYDRATE

$C_{15}H_{34}ClN \cdot H_2O$

Mol. Wt. 281.91

Solvent: $CDCl_3$

Assignments
a = 0.89
b = 1.10-2.10
c = 3.41
d = 3.52
e = 3.57

829

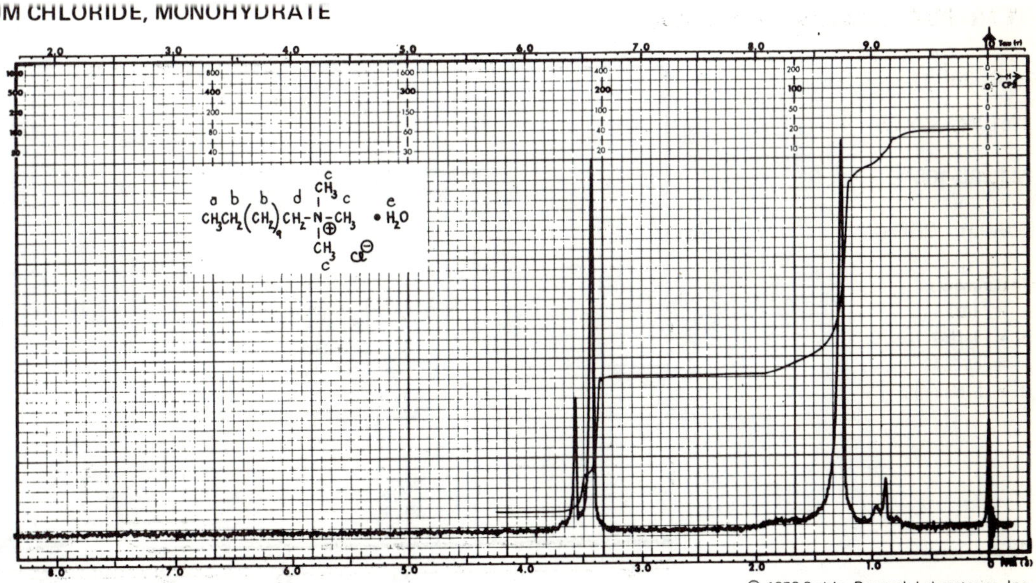

© 1978 Sadtler Research Laboratories, Inc.

DODECYLTRIMETHYLAMMONIUM BROMIDE

$C_{15}H_{34}BrN$

Mol. Wt. 308.35

Solvent: $CDCl_3$

Assignments
a = 0.89
b = 1.10-2.00
c = 3.48
d = 3.67
e = 2.93 imp.

830

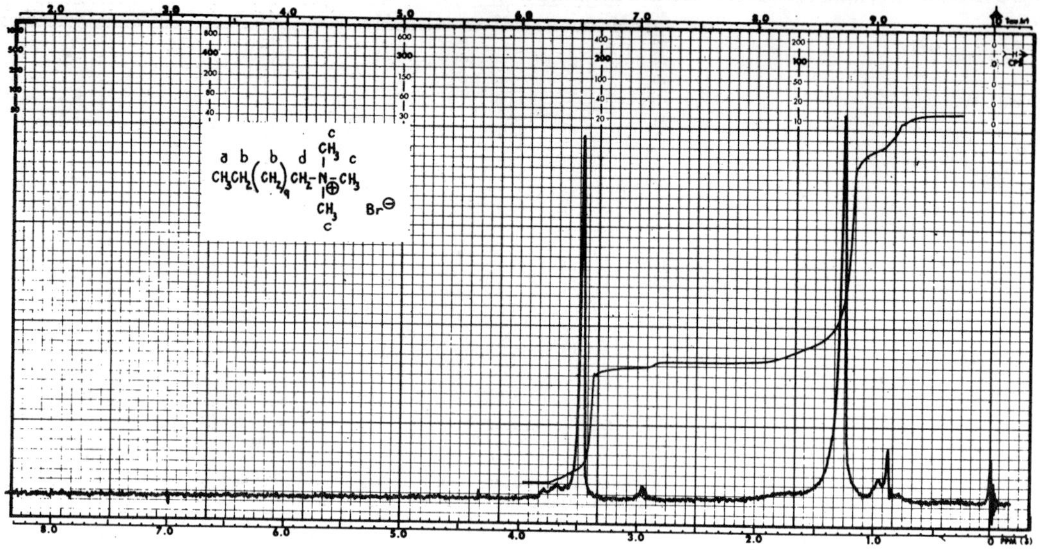

© 1978 Sadtler Research Laboratories, Inc.

HEXADECYLTRIMETHYLAMMONIUM BROMIDE

$C_{19}H_{42}BrN$

Mol. Wt. 364.46

Solvent: $CDCl_3$

Assignments
a = 0.90
b = ca 1.29
c = 1.81
d = 3.49
e = 3.69
f = 3.11 imp.

831

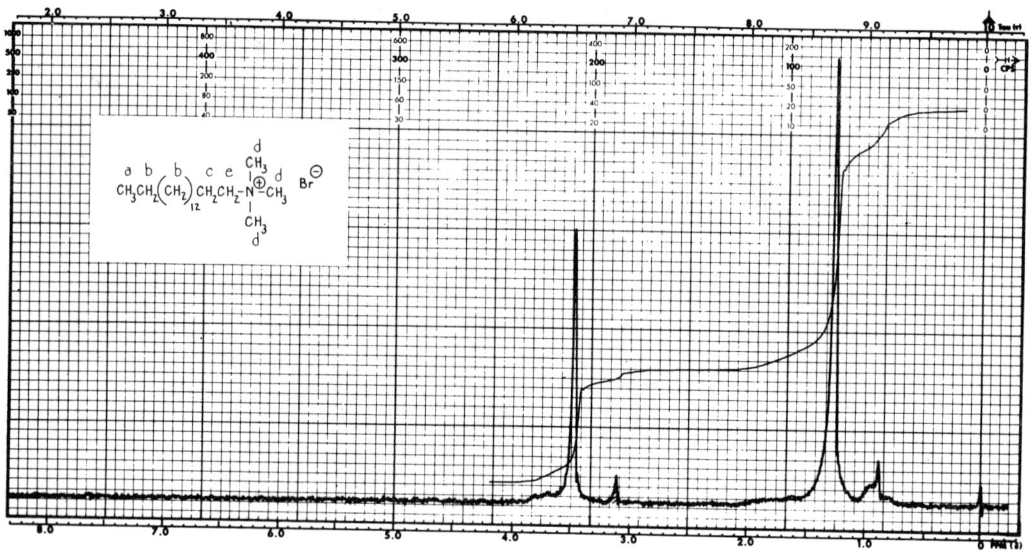

© 1978 Sadtler Research Laboratories, Inc.

TETRABUTYLAMMONIUM BROMIDE

$C_{16}H_{36}BrN$

Mol. Wt. 322.38

M.P. 98-100°C

Solvent: $CDCl_3$

Assignments
a = 1.01
b = 1.19-2.10
c = ca 3.42
d = 2.69 imp.

832

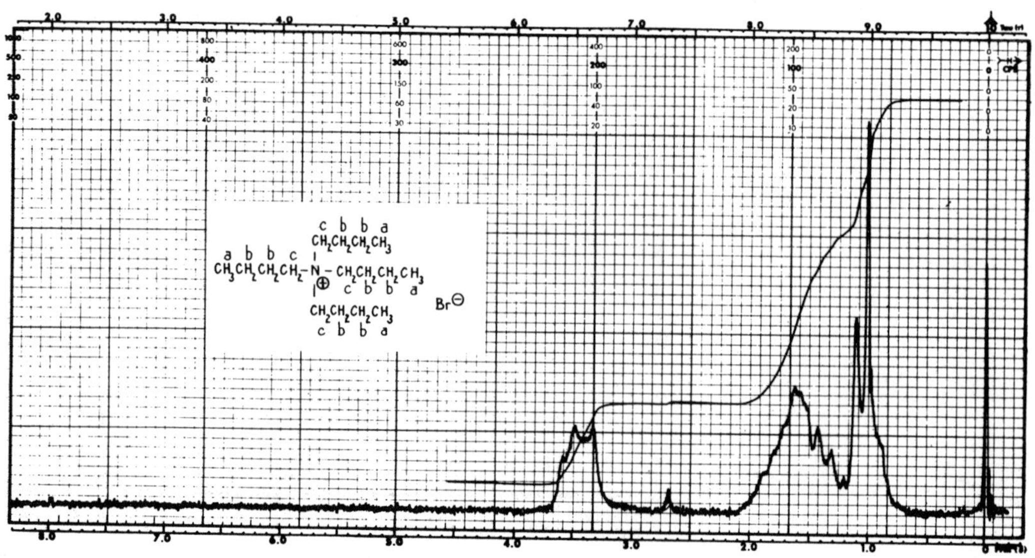

© 1978 Sadtler Research Laboratories, Inc.

OCTADECYLTRIMETHYLAMMONIUM BROMIDE

$C_{21}H_{46}BrN$

Mol. Wt. 392.52

Solvent: $CDCl_3$

Assignments
a = 0.89
b = 1.10-2.10
c = 3.50
d = 3.58

833

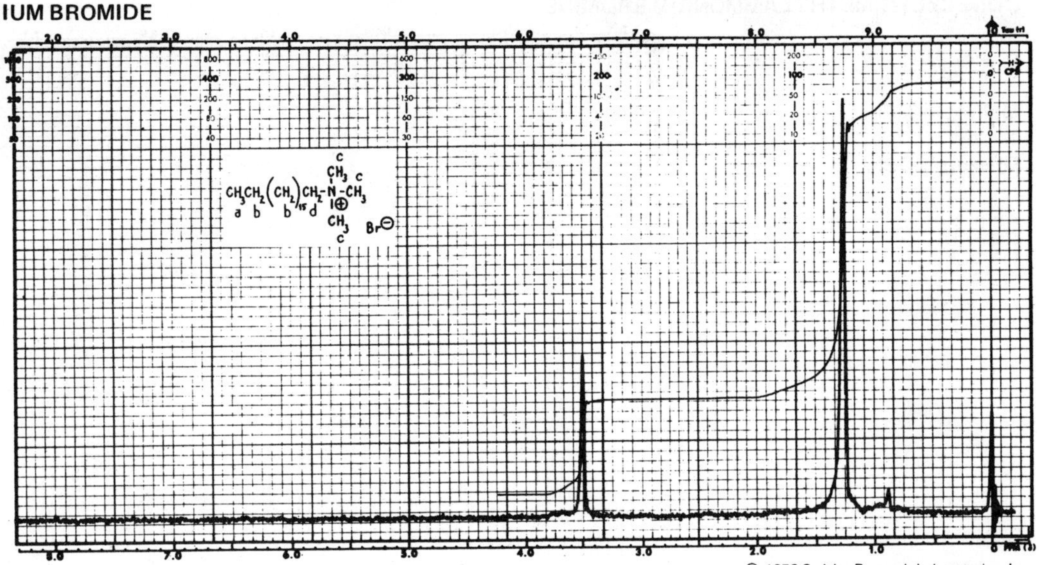

© 1978 Sadtler Research Laboratories, Inc.

(2-PROPYNYL)TRIETHYLAMMONIUM BROMIDE

$C_9H_{18}BrN$

Mol. Wt. 220.16

M.P. 198-200°C

Solvent: Polysol-d

Assignments
a = 1.30
b = 3.40
c = 3.88
d = 4.39
e = 2.52 DMSO-d_5

834

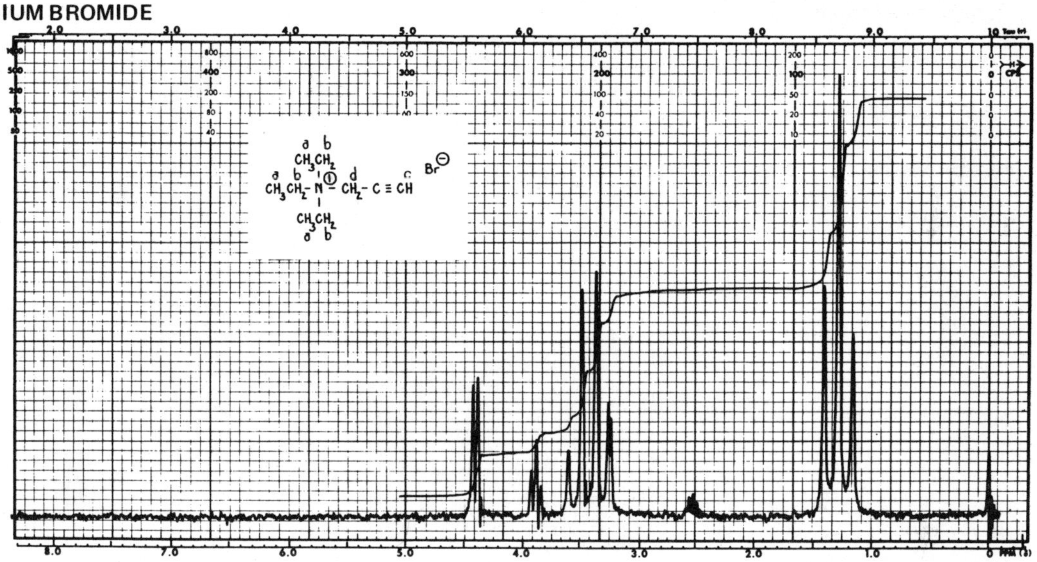

© 1978 Sadtler Research Laboratories, Inc.

1-ADAMANTYLTRIMETHYLAMMONIUM IODIDE

$C_{13}H_{24}IN$

Mol. Wt. 321.25

M.P. 302°C (dec.)

Solvent: D_2O

Assignments
a = 1.69
b = 2.07
c = 2.29
d = 2.99
e = 4.52 HDO

835

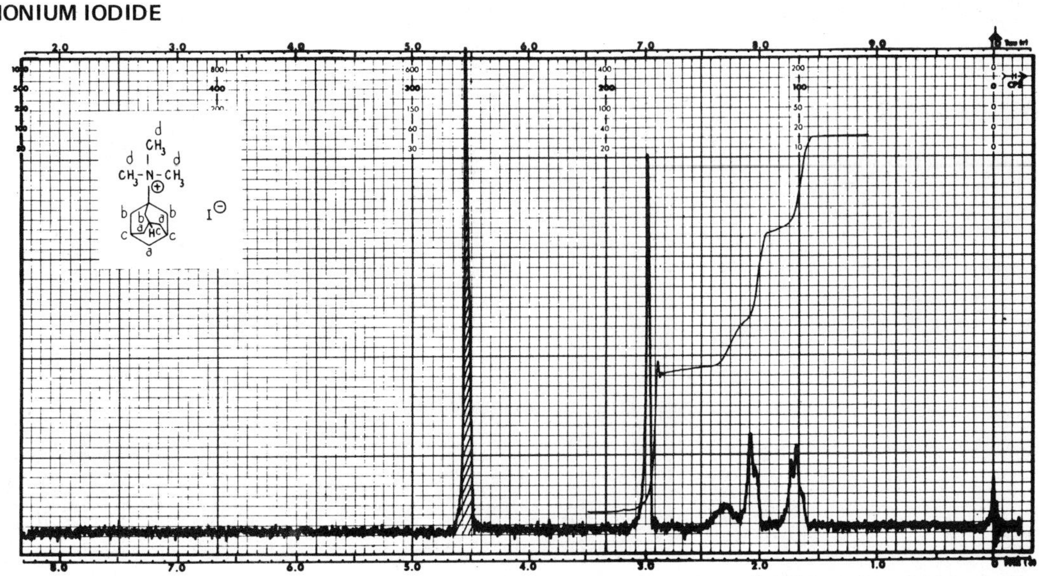

© 1978 Sadtler Research Laboratories, Inc.

BENZYLTRIMETHYLAMMONIUM CHLORIDE

$C_{10}H_{16}ClN$

Mol. Wt. 185.70

Solvent: $CDCl_3$

Assignments
a = 3.40
b = 5.06
c = 7.30-7.60
d = 7.69
e = 3.96 imp.

836

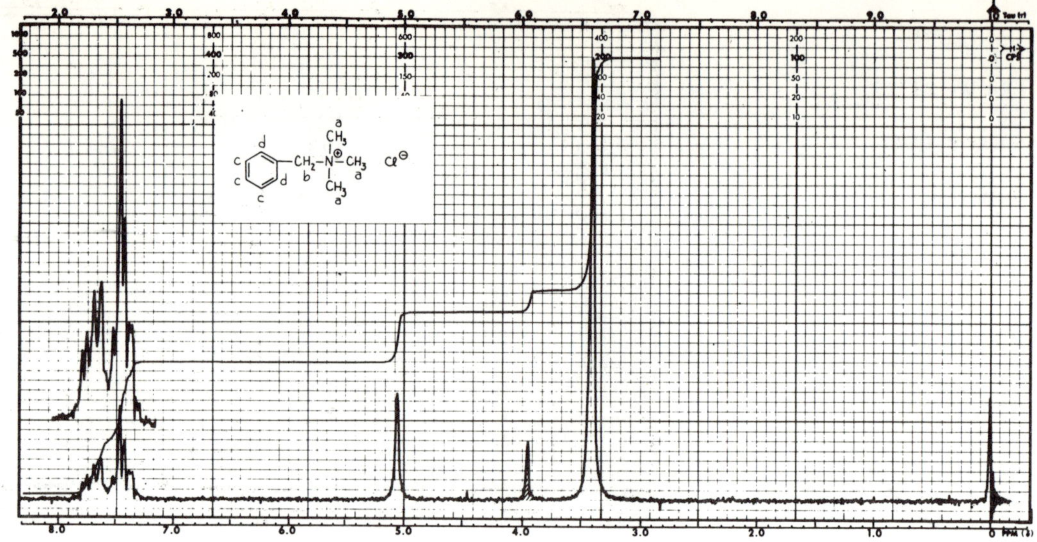

© 1978 Sadtler Research Laboratories, Inc.

BENZYLTRIMETHYLAMMONIUM IODIDE

$C_{10}H_{16}IN$

Mol. Wt. 277.15

M.P. 181-182°C

Solvent: Polysol-d

Assignments
a = 3.30
b = 4.93
c = 7.40-7.90

837

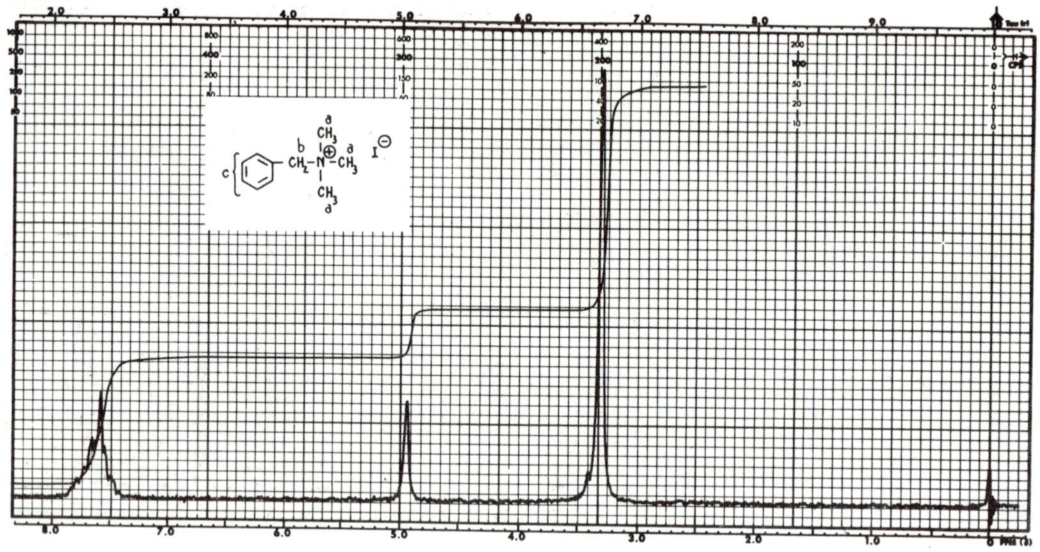

© 1978 Sadtler Research Laboratories, Inc.

BENZYLTRIMETHYLAMMONIUM HEXAFLUOROANTIMONATE (1–)

$C_{10}H_{16}F_6NSb$

Mol. Wt. 385.99

Solvent: Polysol-d

Assignments
a = 3.10
b = 4.48
c = ca 7.50

838

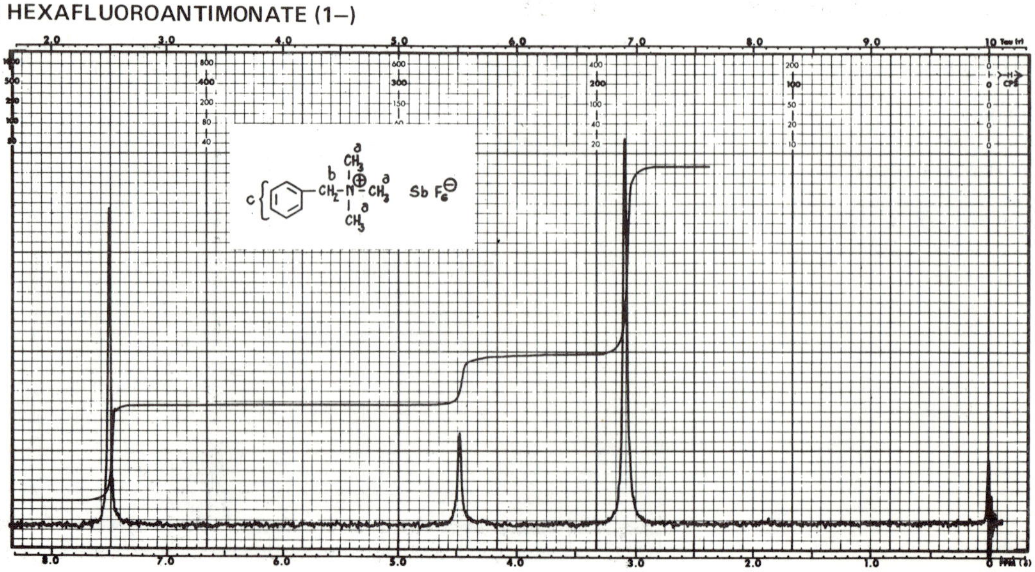

© 1978 Sadtler Research Laboratories, Inc.

BENZYLDIMETHYLHEXADECYLAMMONIUM CHLORIDE, HYDRATE

$C_{25}H_{46}ClN \cdot H_2O$

Mol. Wt. 414.12

M.P. 55-56°C (anhyd.) (lit.)

Solvent: $CDCl_3$

Assignments
- a = 0.90
- b = ca 1.29
- c = 1.83
- d = 3.05
- e = 3.29
- f = ca 3.50
- g = 4.99
- h = 7.30-7.80

839

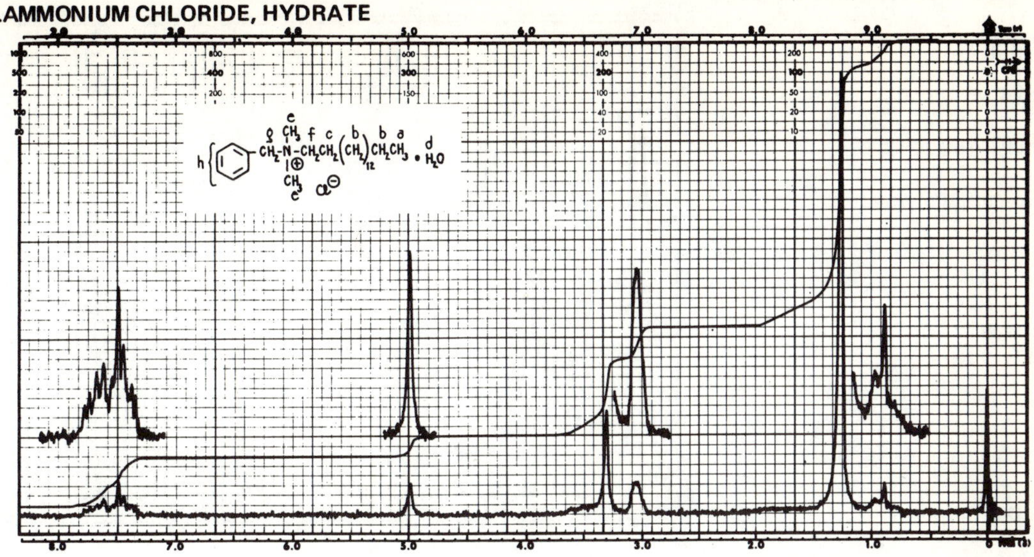

© 1978 Sadtler Research Laboratories, Inc.

PHENYLTRIMETHYLAMMONIUM CHLORIDE

$C_9H_{14}ClN$

Mol. Wt. 171.67

Solvent: D_2O

Assignments
- a = 3.69
- b = 7.55-8.00
- c = 4.69 HDO

840

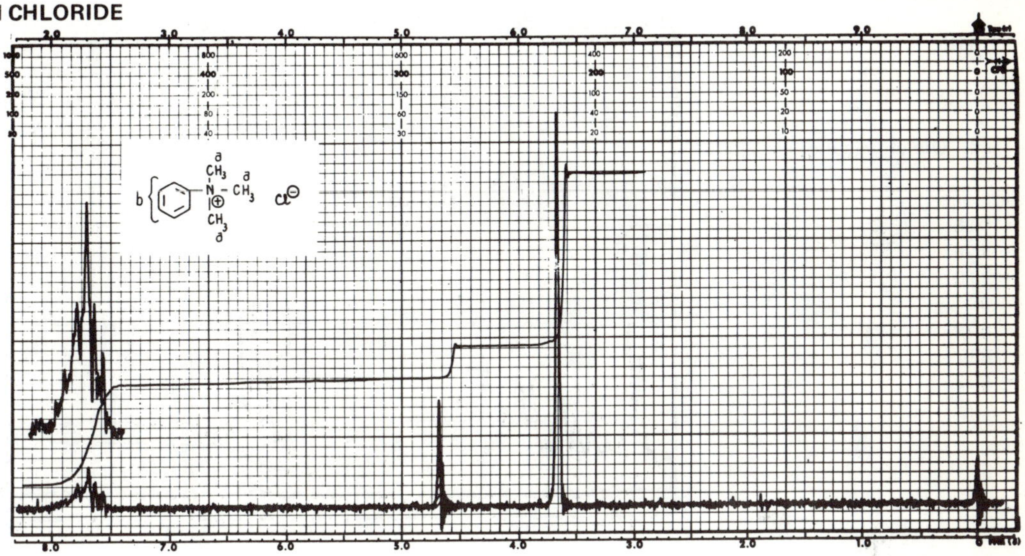

© 1978 Sadtler Research Laboratories, Inc.

PHENYLTRIETHYLAMMONIUM IODIDE

$C_{12}H_{20}IN$

Mol. Wt. 305.20

Solvent: $CDCl_3$

Assignments
- a = 1.30
- b = 4.14
- c = 7.40-7.85
- d = 8.11

841

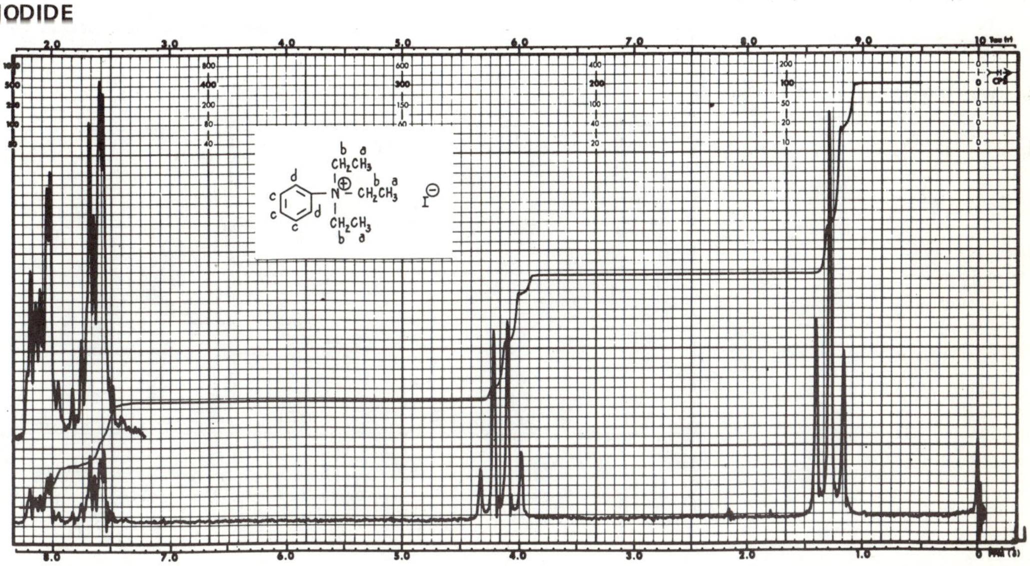

© 1978 Sadtler Research Laboratories, Inc.

1-METHYLPYRIDINIUM IODIDE

C_6H_8IN

Mol. Wt. 221.04

M.P. 118-120°C

Solvent: Polysol-d

Assignments
a = 4.54
b = 8.26
c = 8.72
d = 9.18

842

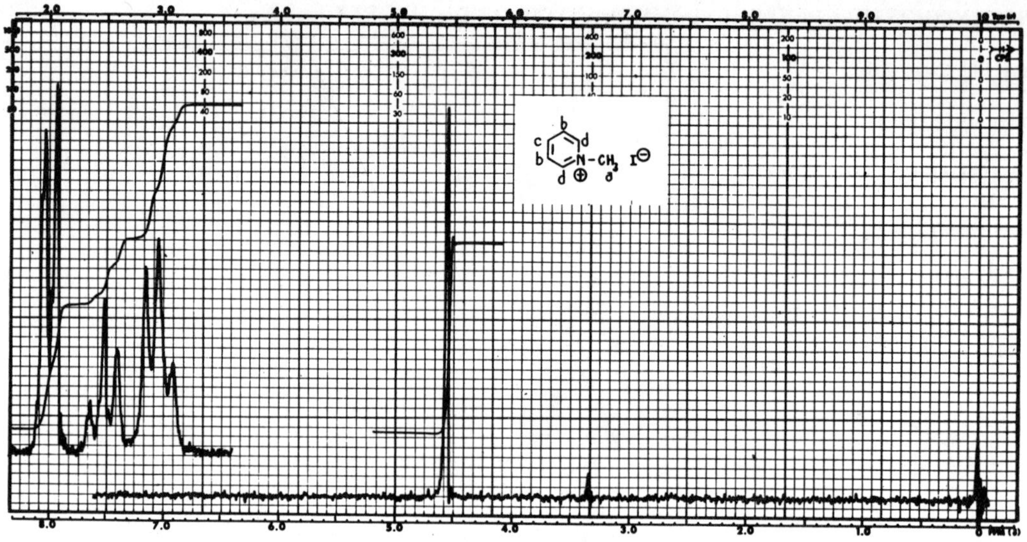

© 1978 Sadtler Research Laboratories, Inc.

1-HEXADECYLPYRIDINIUM IODIDE

$C_{21}H_{38}IN$

Mol. Wt. 431.45

Solvent: Dimethyl sulfoxide-d_6

Assignments
a = 0.88
b = 1.28
c = 1.98
d = 4.71
e = 8.19
f = 8.67
g = 9.18

843

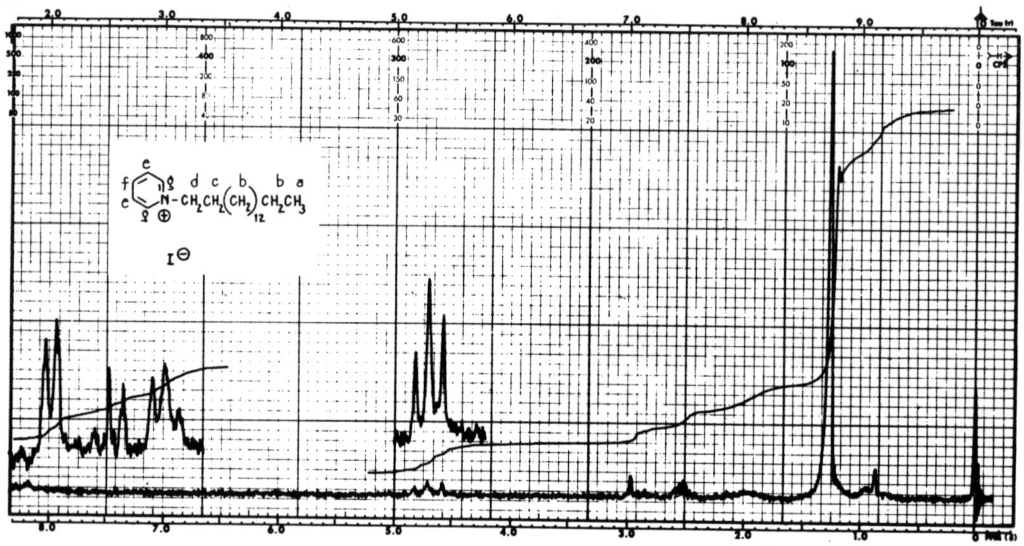

© 1978 Sadtler Research Laboratories, Inc.

3-AMINO-1-METHYLPYRIDINIUM BROMIDE

$C_6H_9BrN_2$

Mol. Wt. 189.06

M.P. 178-180°C

Solvent: D_2O

Assignments
a = 4.34
b = ca 7.79
c = ca 8.13
d = 4.20 HDO

844

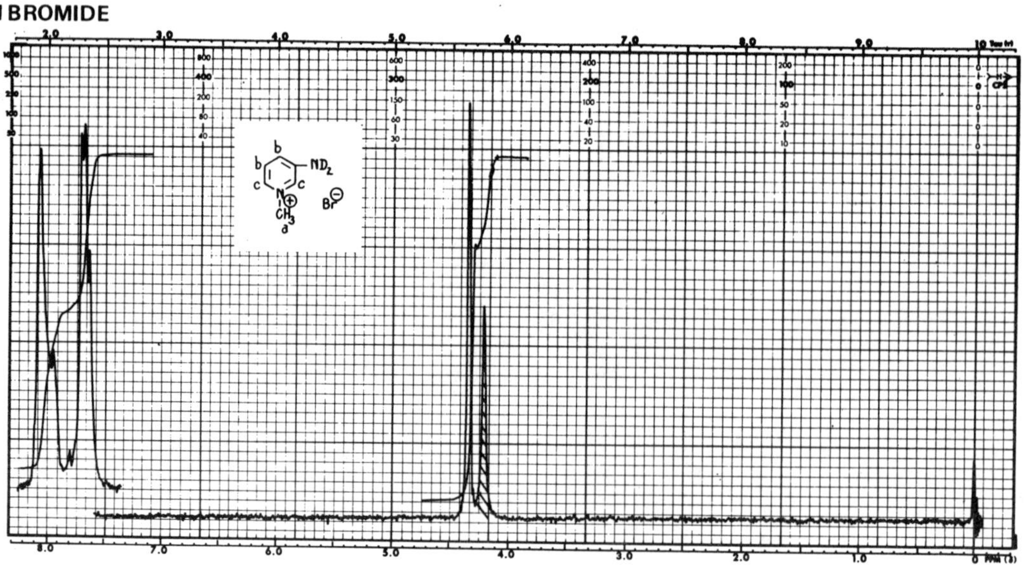

© 1978 Sadtler Research Laboratories, Inc.

1,1′-ETHYLENEBIS[PYRIDINIUM BROMIDE]

$C_{12}H_{14}Br_2N_2$

Mol. Wt. 346.07

M.P. 290°C (dec.)

Solvent: D_2O

Assignments
a = 5.43
b = 8.29
c = 8.81
d = 9.01
e = 4.92 HDO

845

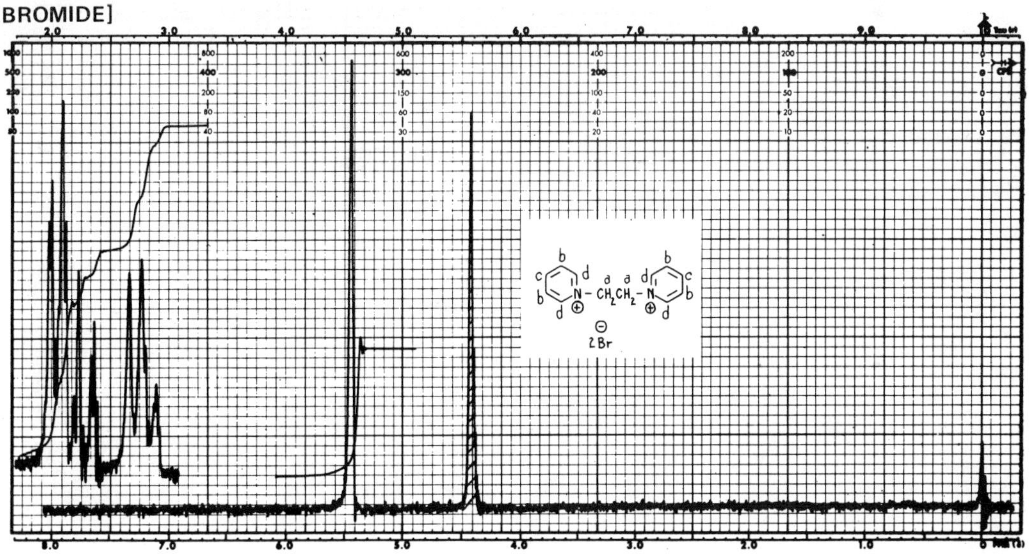

© 1978 Sadtler Research Laboratories, Inc.

o-METHYLHYDROXYLAMINE, HYDROCHLORIDE

$CH_5NO{\cdot}HCl$

Mol. Wt. 83.52

Solvent: D_2O

Assignments
a = 3.89
b = 4.71 HDO

846

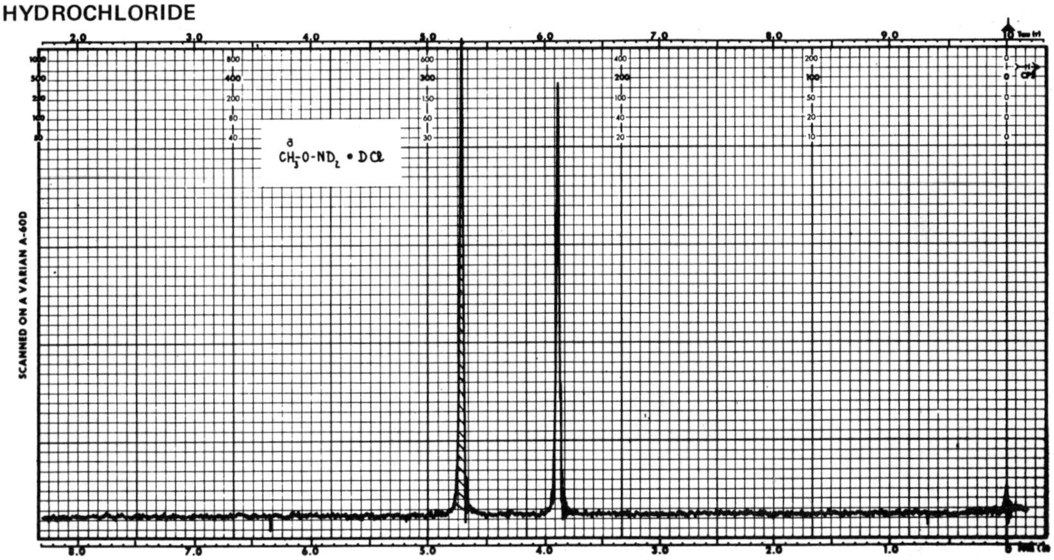

© 1978 Sadtler Research Laboratories, Inc.

PYRIDINE 1-OXIDE, HYDROCHLORIDE

$C_5H_5NO{\cdot}HCl$

Mol. Wt. 131.57

M.P. 175.8-179.1°C

Solvent: D_2O

Assignments
a = 8.10
b = 8.41
c = 8.68
d = 5.14 HDO

847

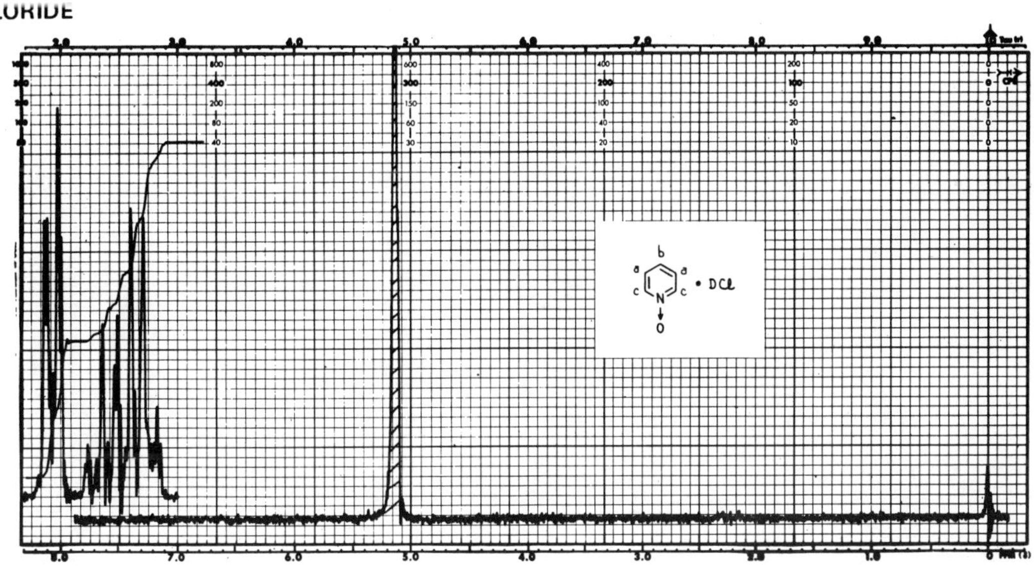

© 1978 Sadtler Research Laboratories, Inc.

NOTES

THE OXIMES

The Oximes of aldehydes which contain an HO—N=CH— group are easily identified by the presence of two clear n+1 multiplets in the range from 6.4 to 7.5 ppm arising from the syn and (anti) forms of the CH=N proton. The Oximes of ketones (HO—N=C(R)—R) do not possess such a proton and are thus more difficult to characterize. Both forms possess an N—OH hydroxyl group which usually appears as a rather broad resonance band in the chemical shift range from 7.8 to 9.6 ppm.

The deshielding effect of the oxime group is similar to that of the C=C group of the alkenes.

Chemical Shifts

Aliphatic Protons

CH_3	—C(=N—OH)	—X	Solvent	Spectrum
1.88	7.98	—H	$CDCl_3$	- - -
1.88	9.08	$-R_3$	$CDCl_3$	854
1.89	9.55	$-R_2$	$CDCl_3$	853
2.25		—Ø	$CDCl_3$	859

HC—3	HC—2	HC—1		Solvent	Spectrum
	1.09	2.22	CH_3—CH_2—C(=N—OH)—CH_3	$CDCl_3$	853
0.91	1.56	2.25	CH_3—CH_2—CH_2—C(=N—OH)—CH_3	$CDCl_3$	854
	(1.10)	2.44	$(CH_3)_2$—CH—CH=N—OH (anti)	CCl_4	848
	(1.08)	3.19	$(CH_2)_2$—CH—CH=N—OH (syn)	CCl_4	848

The HO—N=CH— group

HO	—N	=CH	—X	Solvent
9.00		6.44	$-R_6$	$CDCl_3$
8.45		6.44	—CH$(CH_3)_2$	CCl_4
9.80		6.71	$-R_6$	$CDCl_3$
7.98		6.83	$-CH_3$	$CDCl_3$
8.45		7.23	—CH$(CH_3)_2$	CCl_4
7.98		7.45	$-CH_3$	$CDCl_3$
- - - -		7.82	—CH=N—OH	DMSO
9.91		8.21	—Ø	$CDCl_3$
7.52		9.42	—8(Quinoline)	$CDCl_3$

Chemical Shifts continued

Aromatic Protons

meta/para	ortho	−X	Solvent
7.1–7.4	7.52	−CH=N−OH	$CDCl_3$
7.2–7.6	7.78	−C(Ø)=N−OH	$CDCl_3$
7.2–7.6	7.88	$-C(CH_3)=N-OH$	CCl_4

Coupling and Coupling Constants

Coupling between the CH=N proton and adjacent aliphatic groups is similar in magnitude to that of the corresponding coupling of the alkenes, $J_{CH-CH=N}$ = 6.0–7.0 Hz.

ISOBUTYRALDEHYDE, OXIME

C_4H_9NO

Mol. Wt. 87.12

Solvent: CCl_4

Assignments
a = 1.08
b = 1.10
c = 2.44
d = 3.19
e = 6.44
f = 7.23
g = 8.45

848

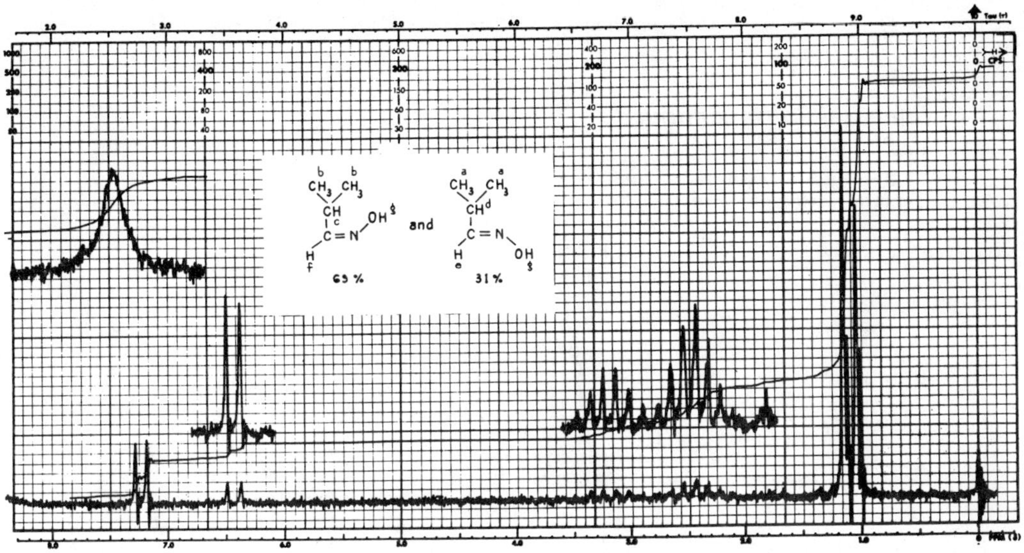

© 1978 Sadtler Research Laboratories, Inc.

HEPTANAL, OXIME

$C_7H_{15}NO$

Mol. Wt. 129.20

M.P. 54-56°C

Solvent: $CDCl_3$

Assignments
a = 0.89
b = 1.10-1.80
c = 2.00-2.60
d = 6.71
e = 6.44
f = 9.80

849

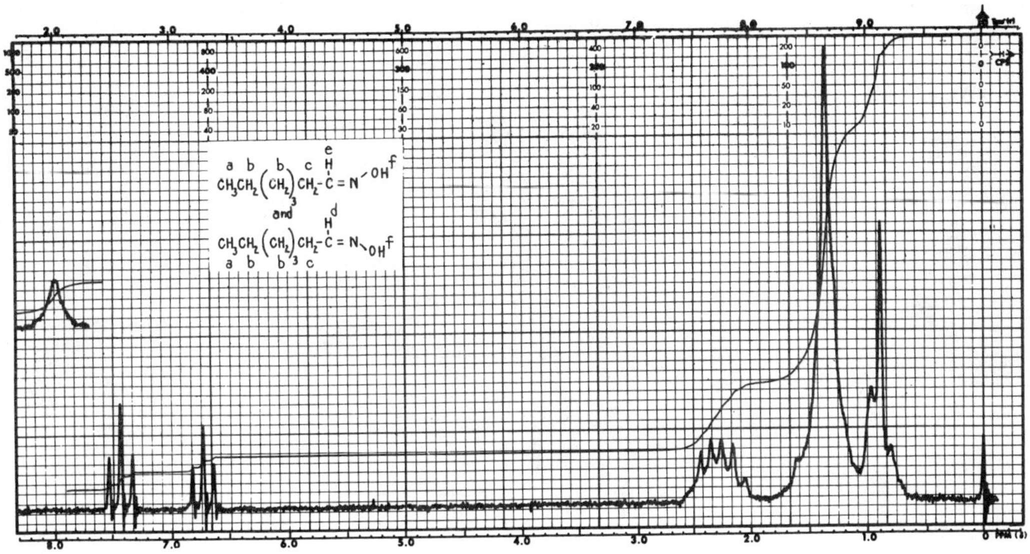

© 1978 Sadtler Research Laboratories, Inc.

GLYOXIME

$C_2H_4N_2O_2$

Mol. Wt. 88.07

M.P. 178°C (lit.)

Solvent: Dimethyl sulfoxide

Assignments
a = 7.82
OH is unobserved

850

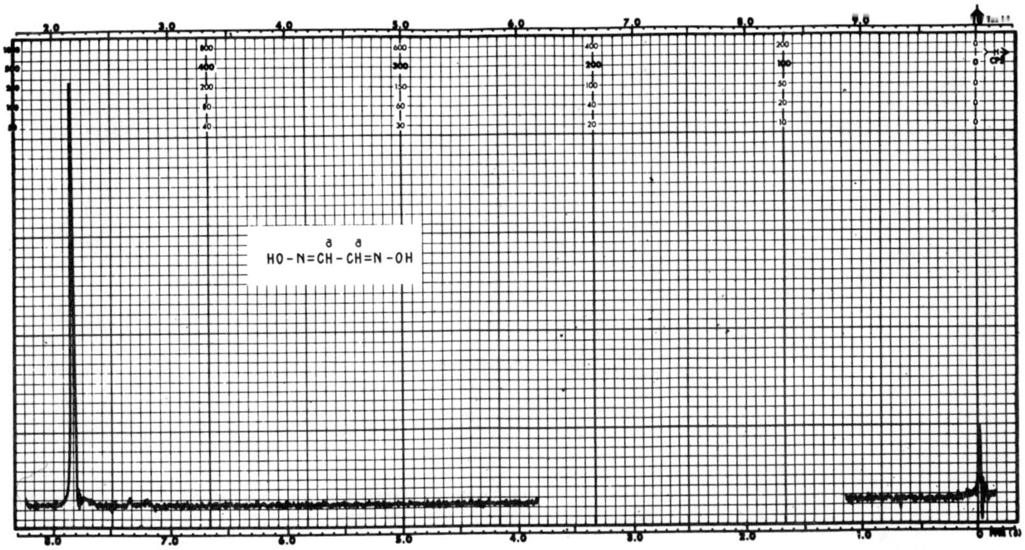

© 1978 Sadtler Research Laboratories, Inc.

CINNAMALDEHYDE, OXIME

C_9H_9NO

Mol. Wt. 147.18

M.P. 132-136°C

Solvent: Dimethyl sulfoxide

Assignments
a = 7.04
b = 7.22-7.82

851

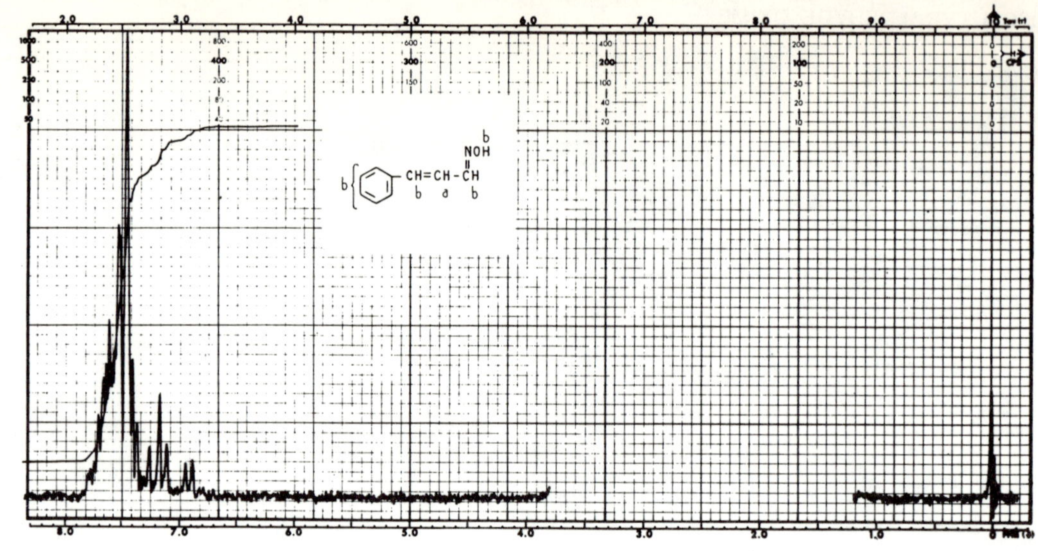

© 1978 Sadtler Research Laboratories, Inc.

BENZALDEHYDE, OXIME

C_7H_7NO

Mol. Wt. 121.14

M.P. 33-35°C

Solvent: $CDCl_3$

Assignments
a = ca 7.28
b = 7.52
c = 8.21
d = 9.91

852

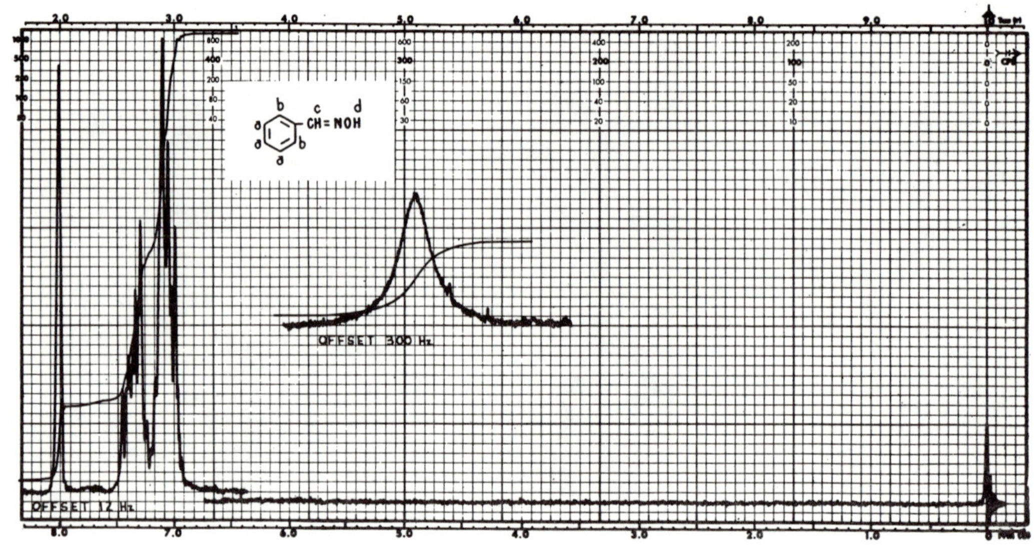

© 1978 Sadtler Research Laboratories, Inc.

2-BUTANONE, OXIME

C_4H_9NO

Mol. Wt. 87.12

Solvent: $CDCl_3$

Assignments
a = 1.09
b = 1.89
c = 2.22
d = 2.40
e = 9.55

853

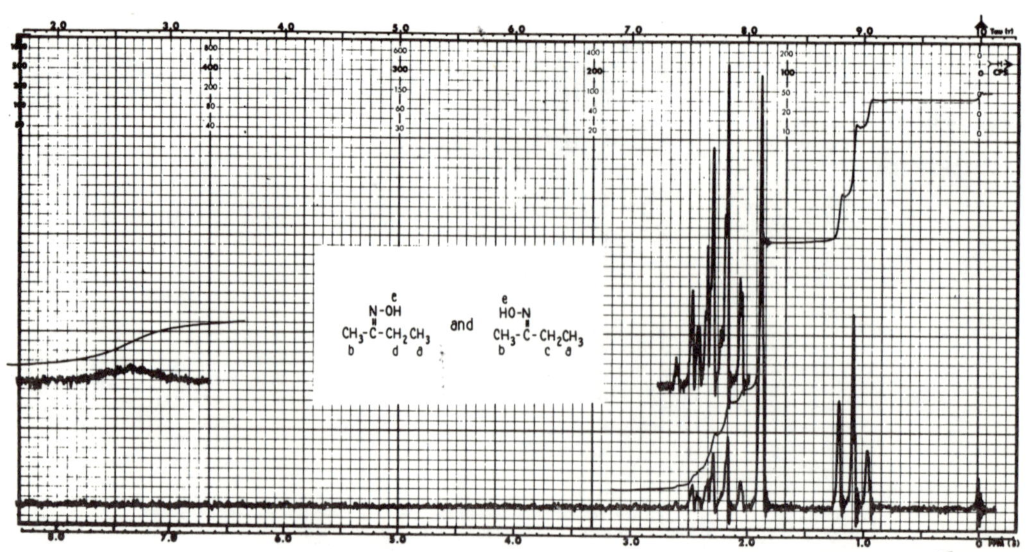

© 1978 Sadtler Research Laboratories, Inc.

2-PENTANONE, OXIME

$C_5H_{11}NO$

Mol. Wt. 101.15

Solvent: $CDCl_3$

Assignments
a = 0.91
b = 1.56
c = 1.88
d = 2.25
e = 9.08

854

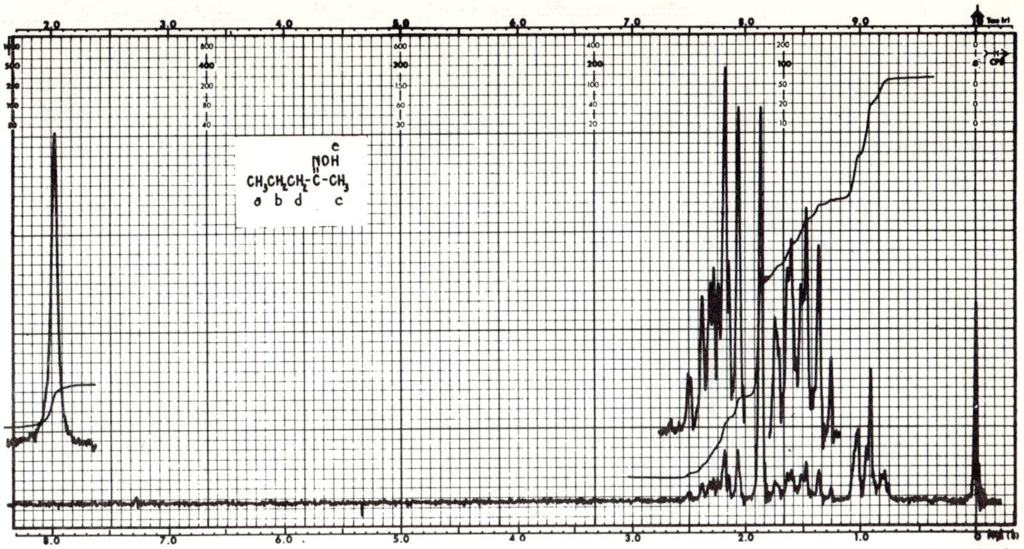

© 1978 Sadtler Research Laboratories, Inc.

4-METHYL-2-PENTANONE, OXIME

$C_6H_{13}NO$

Mol. Wt. 115.18

B.P. 77-78°C/5mm

Solvent: $CDCl_3$

Assignments
a - 0.92
b = 1.89
c = 1.98-2.40
d = 9.08

855

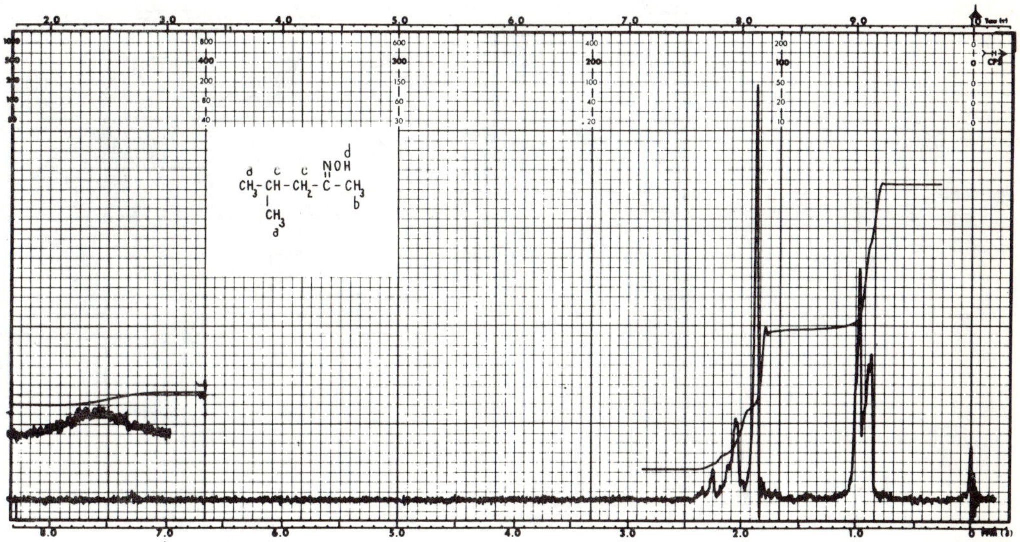

© 1978 Sadtler Research Laboratories, Inc.

5-METHYL-2-HEXANONE, OXIME

$C_7H_{15}NO$

Mol. Wt. 129.20

B.P. 68-72°C/1-2mm

Solvent: $CDCl_3$

Assignments
a = 0.91
b = 1.44
c = 1.90
d = ca 2.28
e = 9.83

856

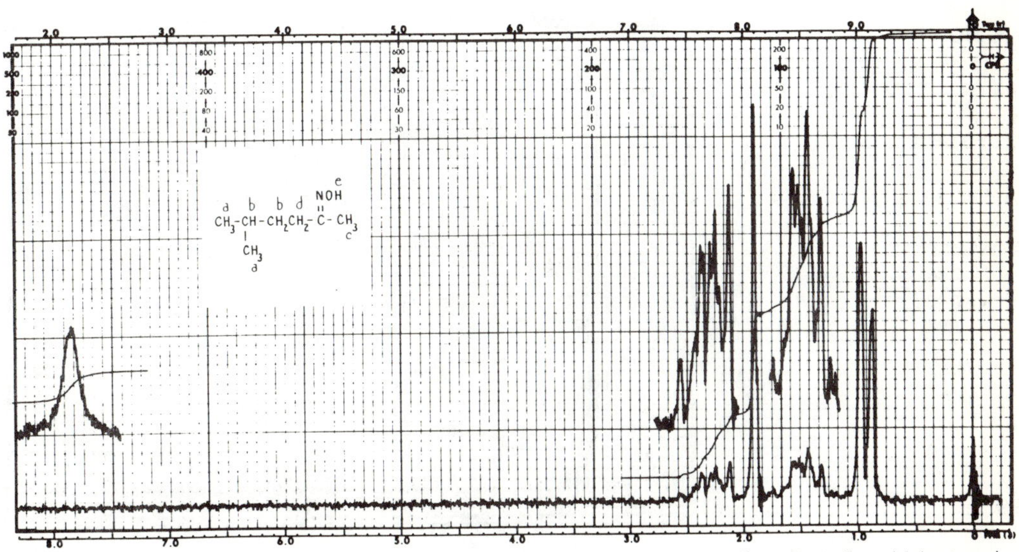

© 1978 Sadtler Research Laboratories, Inc.

2-OCTANONE, OXIME

$C_8H_{17}NO$

Mol. Wt. 143.23

B.P. 98-99°C/6mm

Solvent: $CDCl_3$

Assignments
a = 0.90
b = 1.10-1.70
c = 1.88
d = 2.21
e = 9.39

857

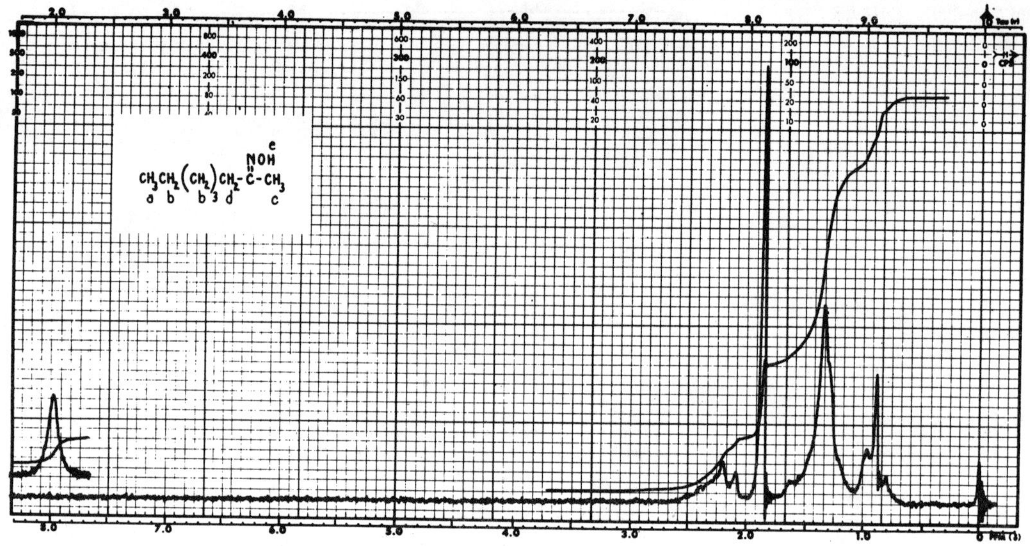

© 1978 Sadtler Research Laboratories, Inc.

CYCLOOCTANONE, OXIME

$C_8H_{15}NO$

Mol. Wt. 141.21

Solvent: $CDCl_3$

Assignments
a = 1.20-2.10
b = 2.10-2.65

858

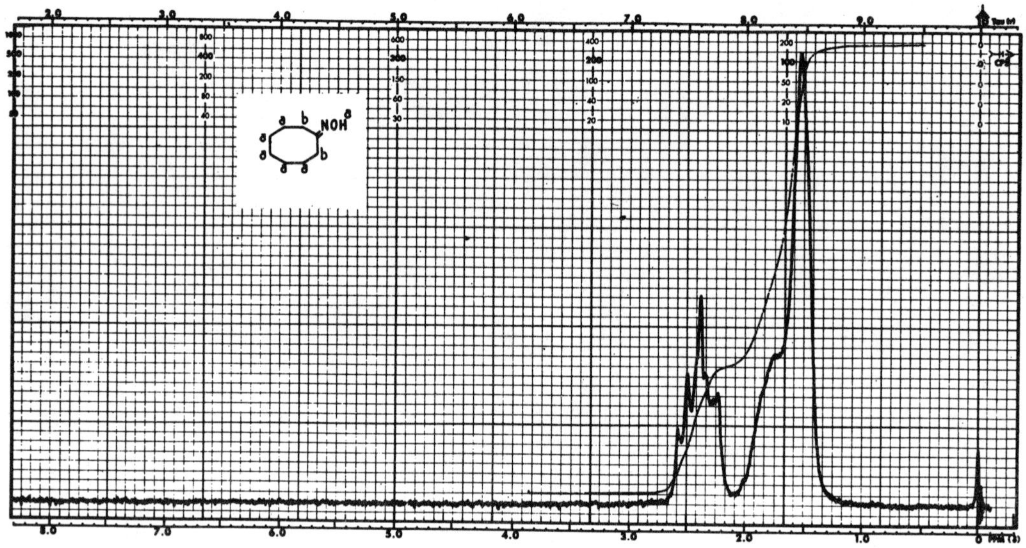

© 1978 Sadtler Research Laboratories, Inc.

ACETOPHENONE, OXIME

C_8H_9NO

Mol. Wt. 135.17

M.P. 56-58°C

Solvent: CCl_4

Assignments
a = 2.48
b = ca 7.40
c = 7.88

859

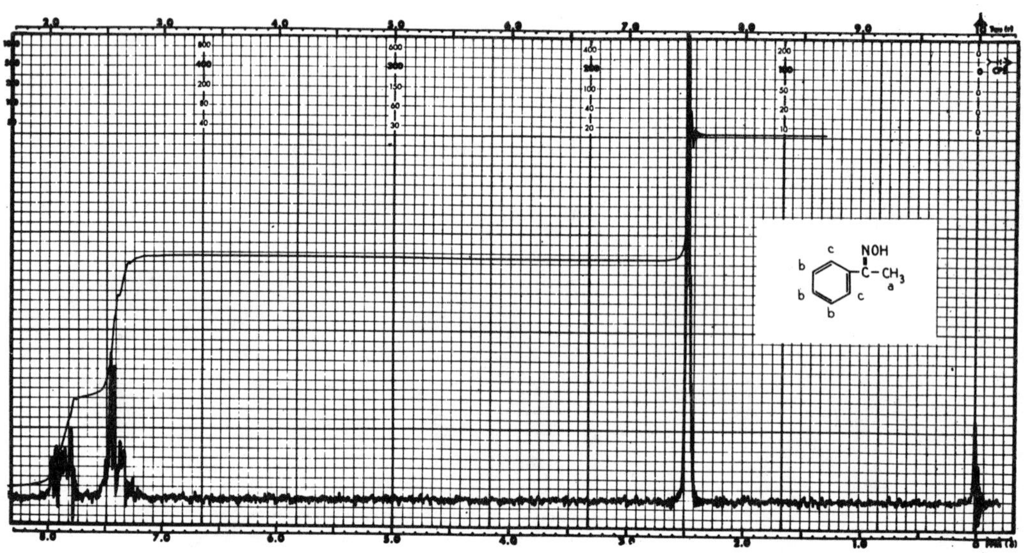

© 1978 Sadtler Research Laboratories, Inc

4′-CHLOROACETOPHENONE, OXIME

C_8H_8ClNO

Mol. Wt. 169.61

M.P. 95°C

Solvent: $CDCl_3$

Assignments
a = 2.25
b = 7.29
c = 7.50
d = 9.33

860

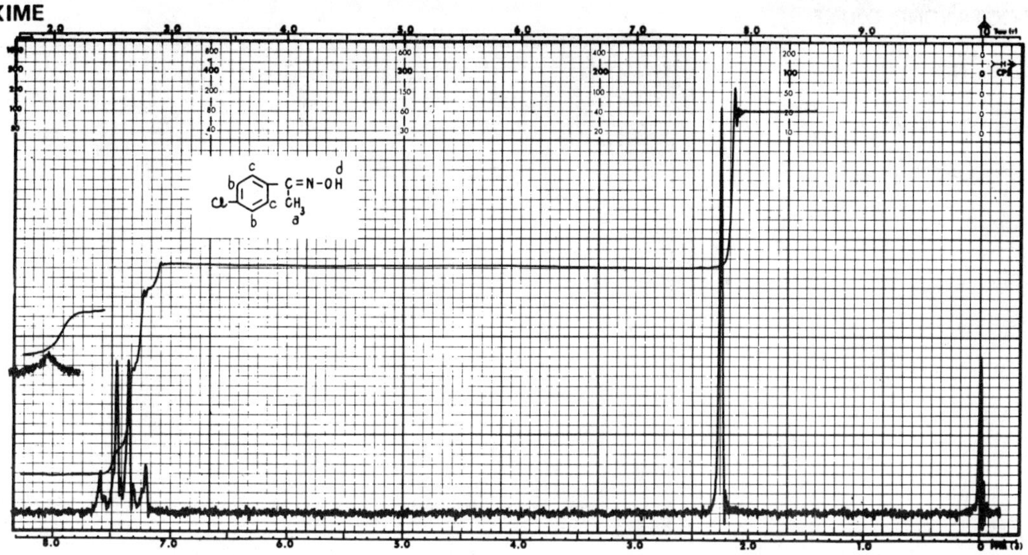

© 1978 Sadtler Research Laboratories, Inc.

BENZOPHENONE, OXIME

$C_{13}H_{11}NO$

Mol. Wt. 197.24

M.P. 143-145°C

Solvent: $CDCl_3$

Assignments
a = 7.46
b = 7.78
OH is unobserved

861

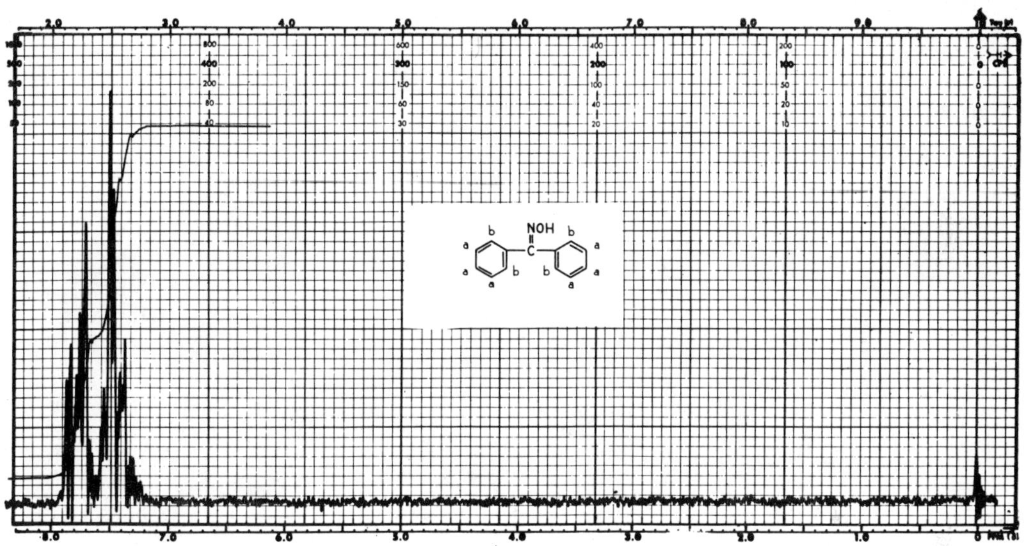

© 1978 Sadtler Research Laboratories, Inc.

2-AMINOBENZOPHENONE, OXIME

$C_{13}H_{12}N_2O$

Mol. Wt. 212.25

M.P. 151-153°C

Solvent: Polysol-d

Assignments
a = 4.62
b = 6.61
c = 6.79
d = 7.11
e = 7.10-7.60
f = 10.78 imp.
g = 2.46 DMSO-d_5

862

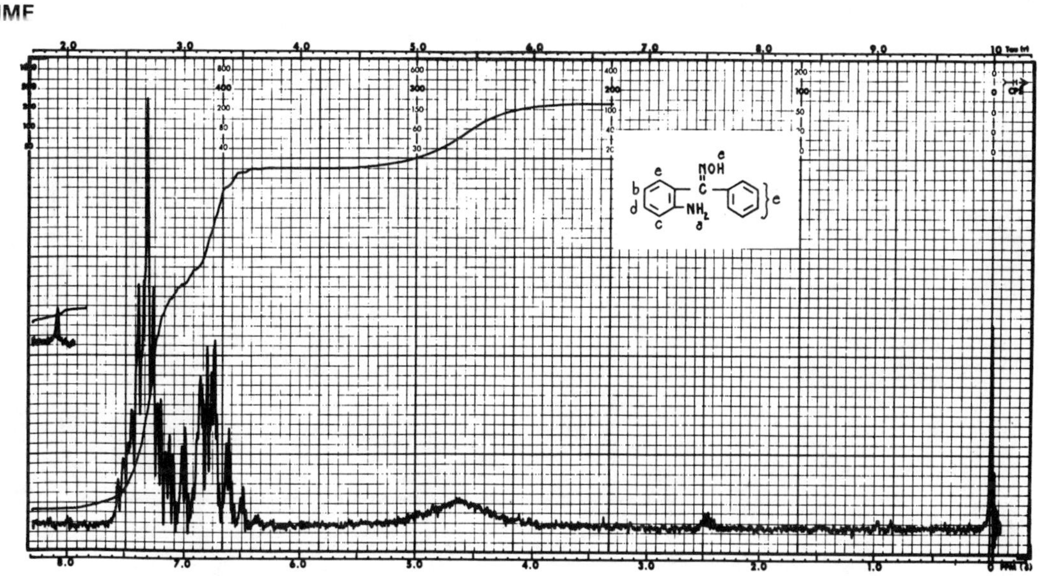

© 1978 Sadtler Research Laboratories, Inc.

2-INDANONE, OXIME

C_9H_9NO

Mol. Wt. 147.18

M.P. 152-153°C

B.P. (dec.)

Solvent: $CDCl_3$

Assignments
a = ca 3.83
b = ca 7.24
c = unobserved

863

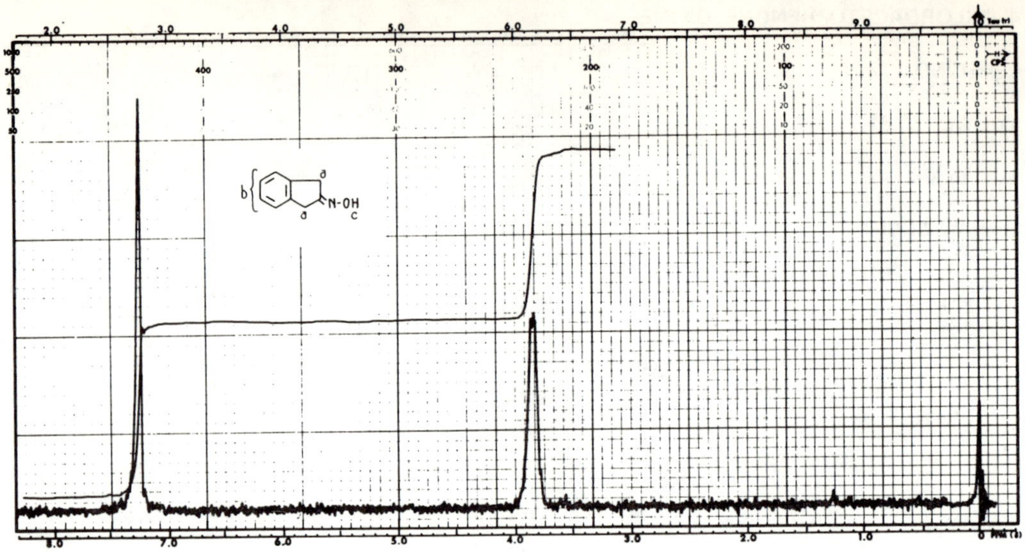

© 1978 Sadtler Research Laboratories, Inc.

2,3-BUTANEDIONE, DIOXIME

$C_4H_8N_2O_2$

Mol. Wt. 116.12

M.P. 240-241°C

Solvent: Polysol-d

Assignments
a = 2.01
b = 10.48

864

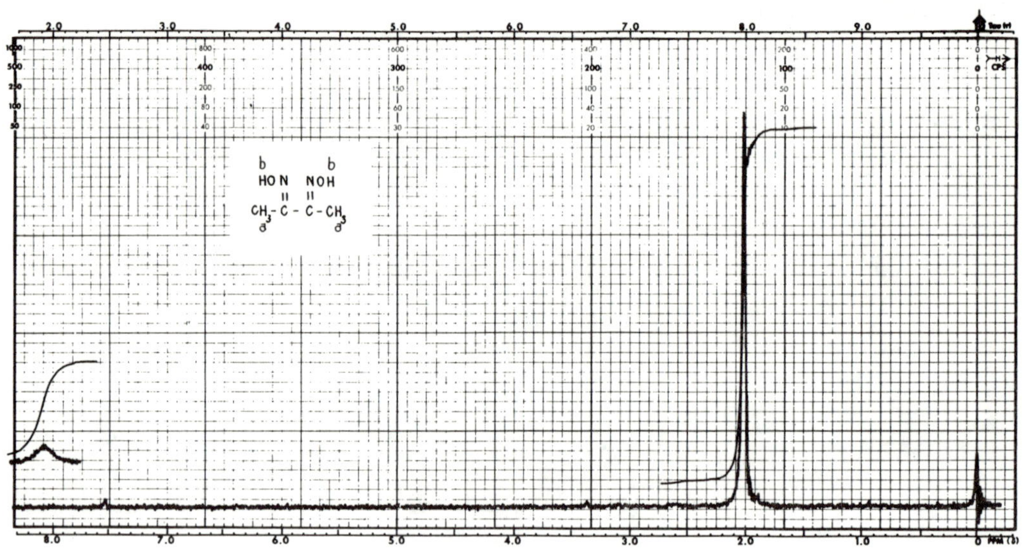

© 1978 Sadtler Research Laboratories, Inc.

1,2-CYCLOHEXANEDIONE, DIOXIME

$C_6H_{10}N_2O_2$

Mol. Wt. 142.16

B.P. 189-191°C

Solvent: $CDCl_3$ & Trifluoroacetic acid

Assignments
a = ca 1.87
b = ca 2.89
OH unobserved

865

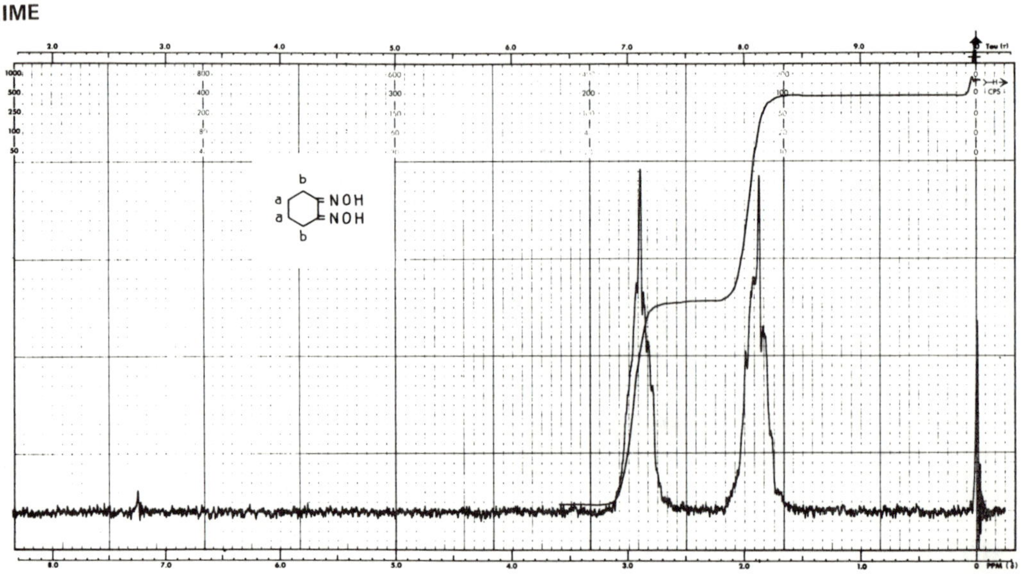

© 1978 Sadtler Research Laboratories, Inc.

THE –YLIDENE COMPOUNDS

The Benzylidene and Cinnamylidene compounds correspond to Oxime-like structures in which the –OH has been replaced by a hydrocarbon group. Because both sides of the CH=N group are substituted by bulky hydrocarbon groups, syn-/anti- isomerism is not ordinarily observed in the NMR spectra. It is assumed that the compounds exist primarily in the anti- form (substituents on opposite sides of the CH=N bond).

The spectra characteristically display a single band at relatively low field (7.9–8.4 ppm). For the Benzylidene compounds the band is a sharp singlet, for the cinnamylide compounds the band appears as a sharp triplet or doublet of doublets.

Chemical Shifts

Aliphatic Protons

HC–3	HC–2	HC–1		Solvent	Spectrum
0.95	1.69	3.49	CH_3–CH_2–CH_2–N=CH–Ø	$CDCl_3$	----
	(1.23)	3.48	$(CH_3)_2$–CH–N=CH–Ø	CCl_4	----
		3.41	CH_3–N=CH–Ø	$CDCl_3$	866

The –CH=N– proton

X–	CH=N	–Y	Solvent	Spectrum
Ø–CH=CH–	7.95	–R	$CDCl_3$	----
Ø–	8.07	–CH_2CH_2–Ø	$CDCl_3$	----
Ø–	8.11	–CH_3	$CDCl_3$	----
Ø–	8.13	–R_3	$CDCl_3$	----
Ø–CH=CH–	8.15	–Ø	$CDCl_3$	869
Ø–	8.18	–$CH(CH_3)_2$	CCl_4	----
Ø–	8.40	–Ø	$CDCl_3$	871

Coupling and Coupling Constants

Aliphatic groups are observed to couple weakly across the –CH=N– bond. The couplings vary from a slight broadening effect (J less than 0.8 Hz) to clear 5 Hz multiplets in the case of the Cinnamylidene compounds.

$J_{CH_3\text{–}N=CH\text{–}}$ = 2 Hz

$J_{CH_2\text{–}N=CH\text{–}}$ = 0 – 1 Hz

$J_{C=CH\text{–}CH=N\text{–}}$ = 4 – 5 Hz

N-BENZYLIDENEMETHYLAMINE

C_8H_9N

Mol. Wt. 119.17

B.P. 50-51°C/3mm

Solvent: $CDCl_3$

Assignments

a = 3.41
b = 7.29
c = 7.62
d = 8.11

866

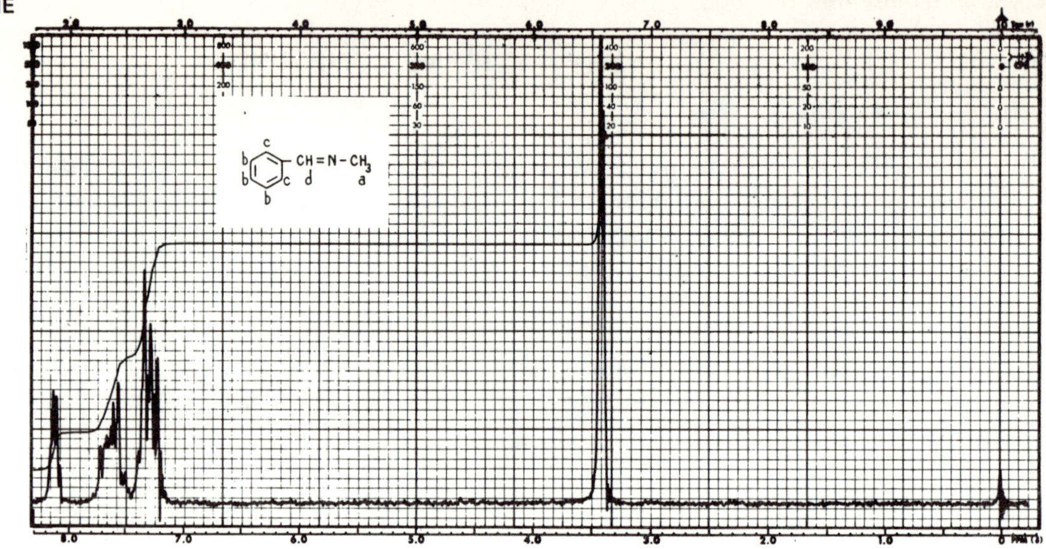

© 1978 Sadtler Research Laboratories, Inc.

N,N'-DICINNAMYLIDENEETHYLENEDIAMINE

$C_{20}H_{20}N_2$

Mol. Wt. 288.40

M.P. 105-110°C

Solvent: $CDCl_3$

Assignments

a = 3.80
b = ca 6.81
c = 7.05-7.50
d = 7.95
e = 2.93 imp.

867

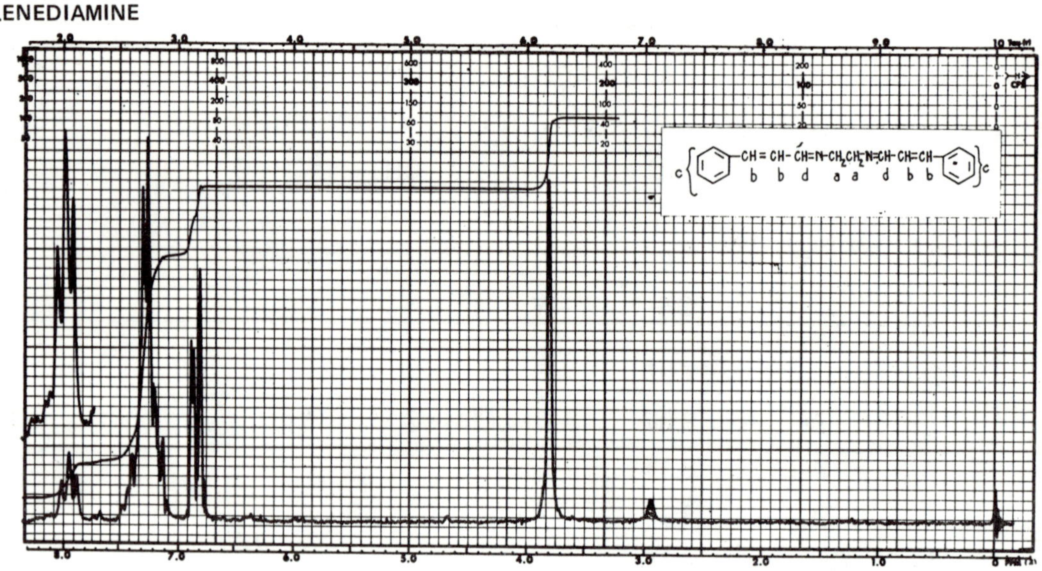

© 1978 Sadtler Research Laboratories, Inc.

N,N'-DIBENZYLIDENEETHYLENEDIAMINE

$C_{16}H_{16}N_2$

Mol. Wt. 236.32

Solvent: $CDCl_3$

Assignments

a = 3.96
b = 7.35
c = 7.67
d = 8.28

868

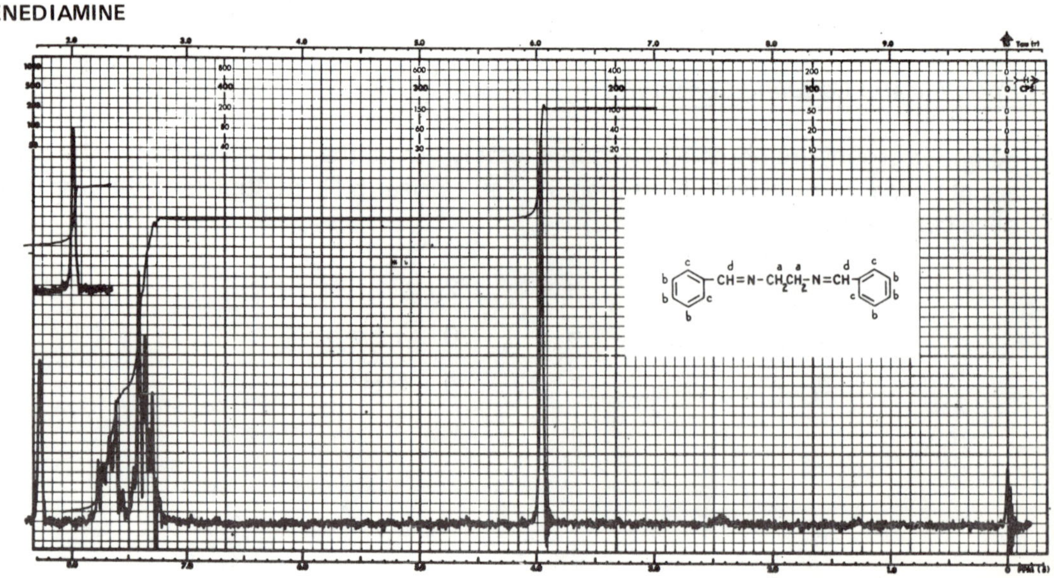

© 1978 Sadtler Research Laboratories, Inc.

N-CINNAMYLIDENEANILINE

$C_{15}H_{13}N$

Mol. Wt. 207.28

Solvent: $CDCl_3$

Assignments
a = 7.00-7.60
b = 8.15

869

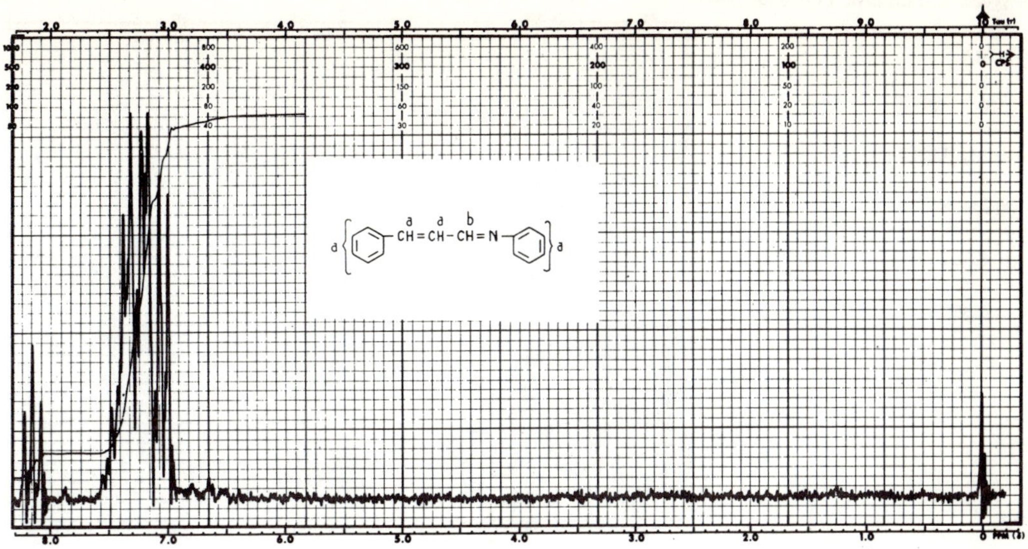

© 1978 Sadtler Research Laboratories, Inc.

N-CINNAMYLIDENE-p-TOLUIDINE

$C_{16}H_{15}N$

Mol. Wt. 221.30

Solvent: $CDCl_3$

Assignments
a = 2.30
b = 7.08
c = 6.95-7.56
d = 8.14

870

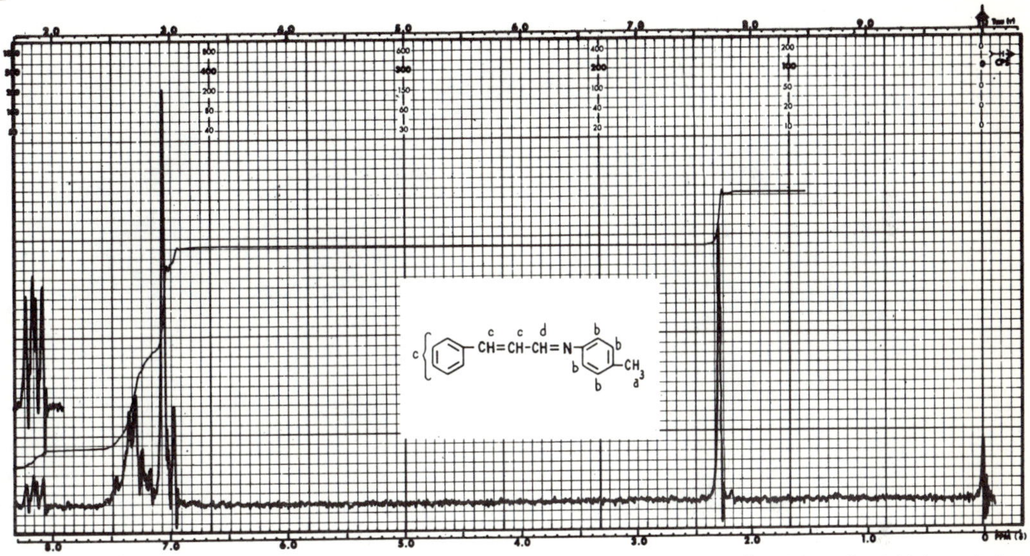

© 1978 Sadtler Research Laboratories, Inc.

N-BENZYLIDENEANILINE

$C_{13}H_{11}N$

Mol. Wt. 181.24

M.P. 56°C

Solvent: $CDCl_3$

Assignments
a = 7.34
b = 7.87
c = 8.40

871

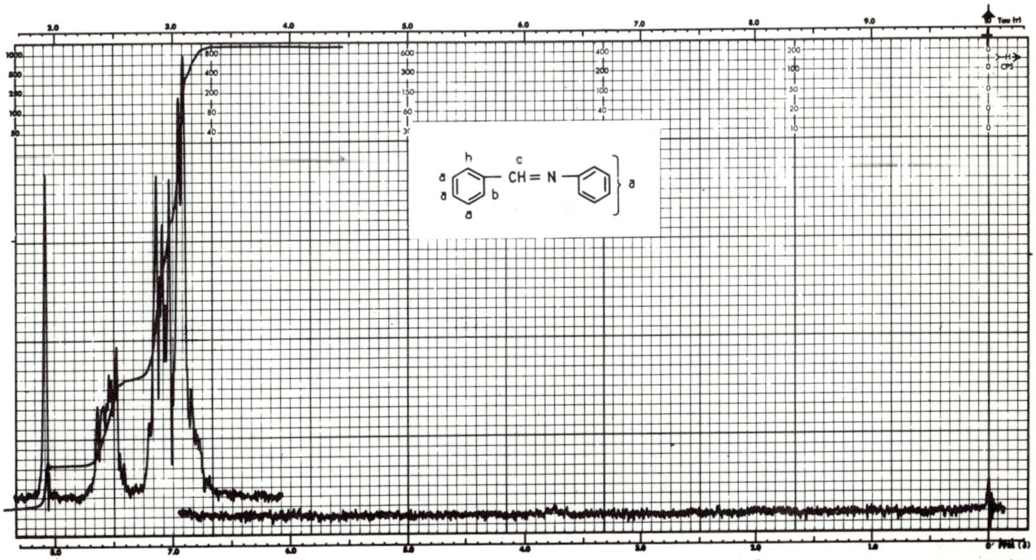

© 1978 Sadtler Research Laboratories, Inc.

N-(p-METHYLBENZYLIDENE)ANILINE

$C_{14}H_{13}N$

Mol. Wt. 195.27

Solvent: $CDCl_3$

Assignments
a = 2.35
b = 6.85-7.55
c = 7.76
d = 8.39

872

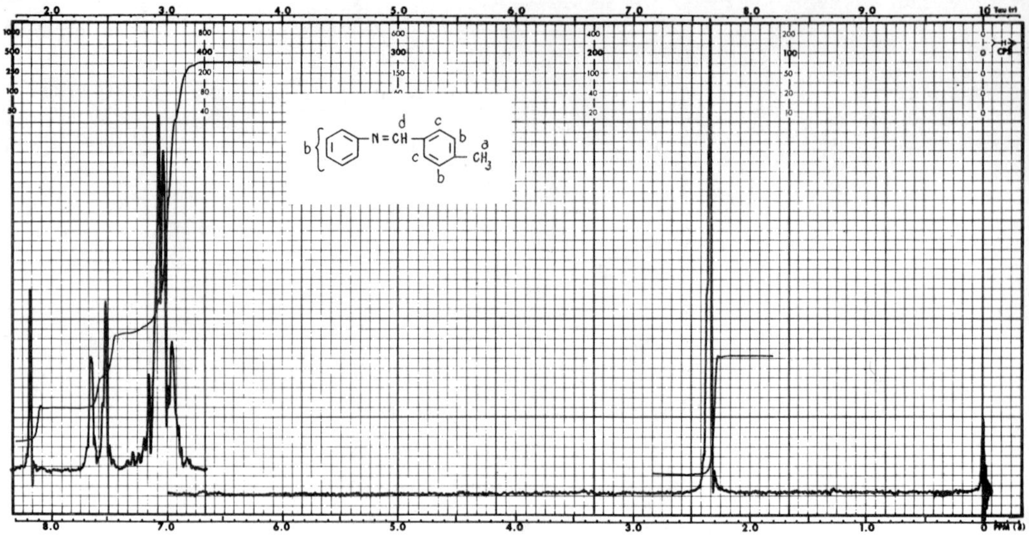

© 1978 Sadtler Research Laboratories, Inc.

THE HYDRAZONES

The Hydrazones which are primarily used as derivatives for the characterization of ketones and aldehydes produce rather wide ranges of chemical shifts for the —CH=N— proton and for the various types of NH hydrogens. Both groups are quite sensitive to the substituent and its position on adjacent aromatic rings, various nitro-phenyl hydrazones being a common variety.

Chemical Shifts

Aliphatic Protons

HC–3	HC–2	HC–1		Solvent	Spectrum
0.96	1.51	2.12	CH_3—CH_2—CH_2—CH=N—NH—Ø	CCl_4	874

Range (ppm)	Group	Structure
3.35–3.42	CH_3—	CH_3—N(Ø)—N=CH—Ø
5.37–6.53	NH_2—	NH_2—N=CH—Ø
6.90–10.98	—NH—	Ø—NH—N=CH—Ø
7.27–8.06	—N=CH—	Ø—NH—N=CH—Ø

The aromatic hydrogens of phenyl groups bonded to the hydrazone group cover a relatively wide range of chemical shifts. The carbon side of the group is a moderately strong deshielding group in relation to the ortho protons while the amine side of the group shields the para hydrogen rather strongely. Because two different rings are often present the aromatic patterns can become quite complex.

Benzaldehyde, phenylhydrazone

para	meta	ortho		
6.73	6.9–7.4	6.9–7.4	—NH—N=CH—Ø	Acetone
6.9–7.4	6.9–7.4	7.60	—CH=N=NH—Ø	Acetone

1-NAPHTHALDEHYDE, HYDRAZONE

$C_{11}H_{10}N_2$

Mol. Wt. 170.21

M.P. 93°C

Solvent: Polysol-d

Assignments
a = 5.37
b = 7.10-7.55
c = 7.55-8.00
d = 8.14
e = 8.54

873

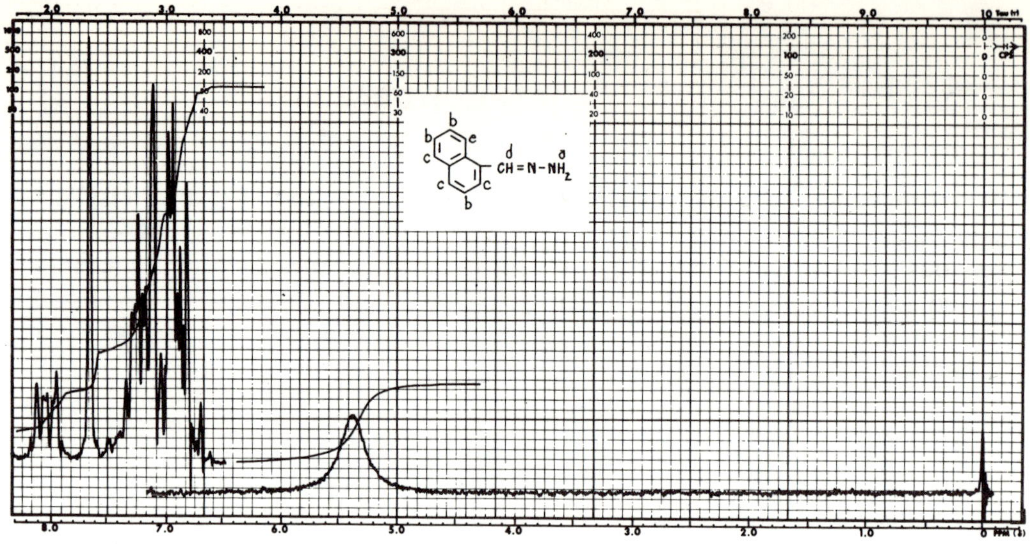

© 1978 Sadtler Research Laboratories, Inc.

BUTYRALDEHYDE, PHENYLHYDRAZONE

$C_{10}H_{14}N_2$

Mol. Wt. 162.24

B.P. 152-155°C/18mm

Solvent: CCl_4

Assignments
a = 0.96
b = 1.51
c = 2.12
d = 6.26-7.40

874

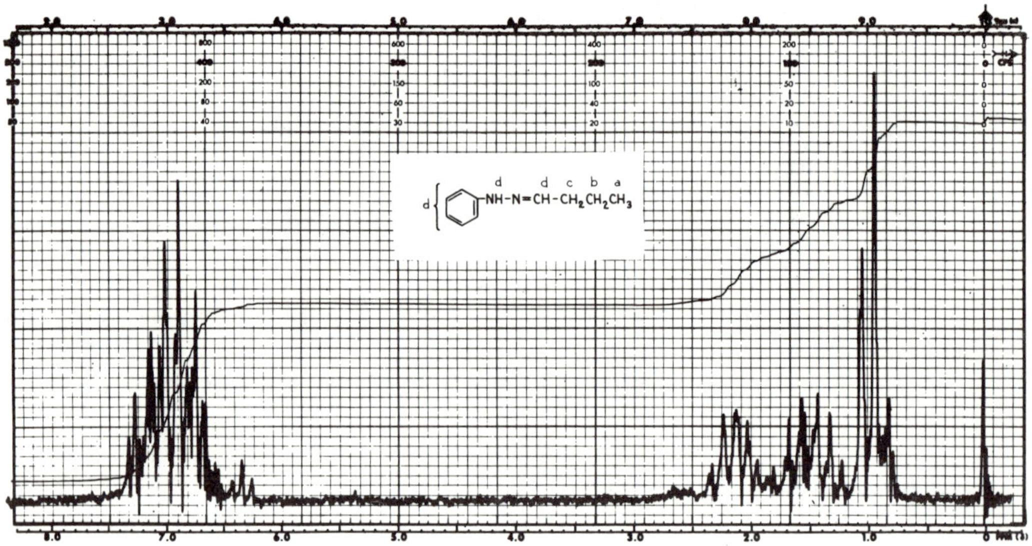

© 1978 Sadtler Research Laboratories, Inc.

CINNAMALDEHYDE, PHENYL HYDRAZONE

$C_{15}H_{14}N_2$

Mol. Wt. 222.29

M.P. 170-171°C

Solvent: Dimethyl sulfoxide

Assignments
a = 6.54-7.66
b = 7.80
c = 8.04

875

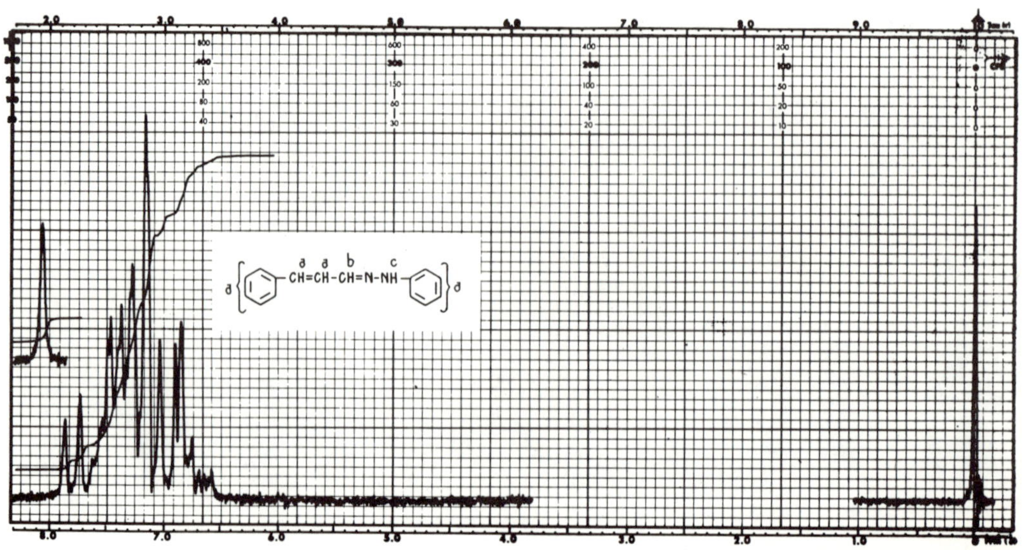

© 1978 Sadtler Research Laboratories, Inc.

BENZALDEHYDE, PHENYLHYDRAZONE

$C_{13}H_{12}N_2$

Mol. Wt. 196.25

M.P. 157°C

Solvent: Acetone

Assignments
a = 6.50-7.70
b = 7.73
c = 9.16

876

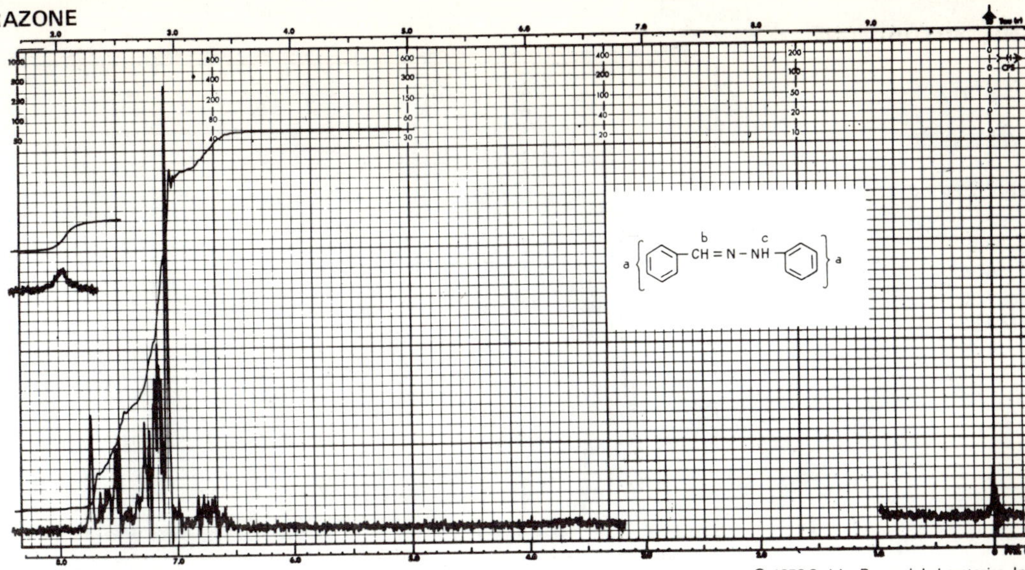

© 1978 Sadtler Research Laboratories, Inc.

---

o-CHLOROBENZALDEHYDE, PHENYLHYDRAZONE

$C_{13}H_{11}ClN_2$

Mol. Wt. 230.70

M.P. 84°C

Solvent: $CDCl_3$

Assignments
a = 6.64-7.39
b = 7.59
c = 7.90
d = 7.94

877

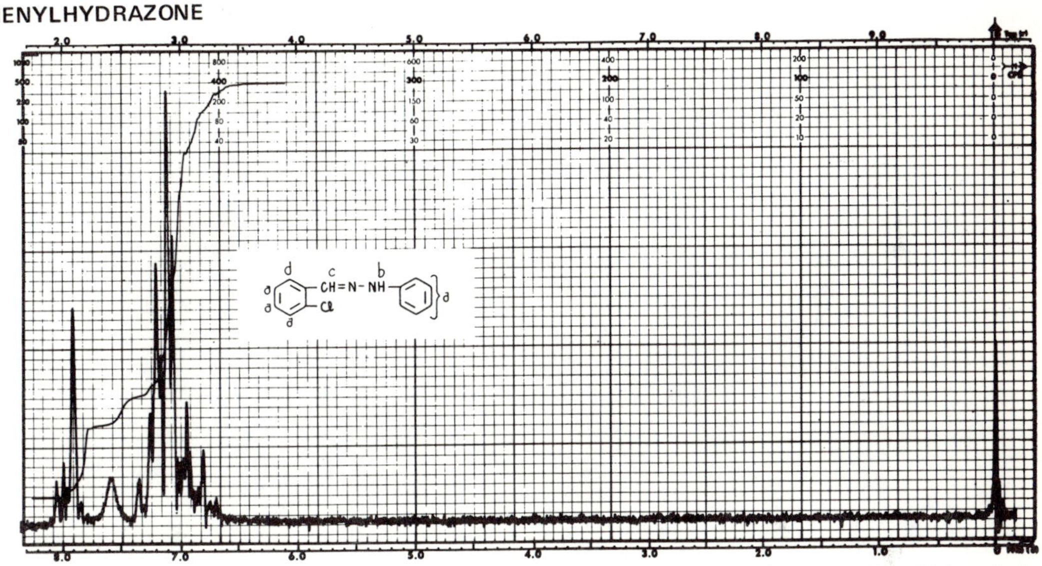

© 1978 Sadtler Research Laboratories, Inc.

METHYL 2-PYRIDYL KETONE, PHENYLHYDRAZONE

$C_{13}H_{13}N_3$

Mol. Wt. 211.27

M.P. 155°C

Solvent: Dimethyl sulfoxide-d_6

Assignments
a = 2.39
b = 6.80
c = 7.06-7.46
d = 7.71
e = 8.10
f = 8.51
g = 9.43
h = 2.50 DMSO-d_5
i = 3.49 H_2O

878

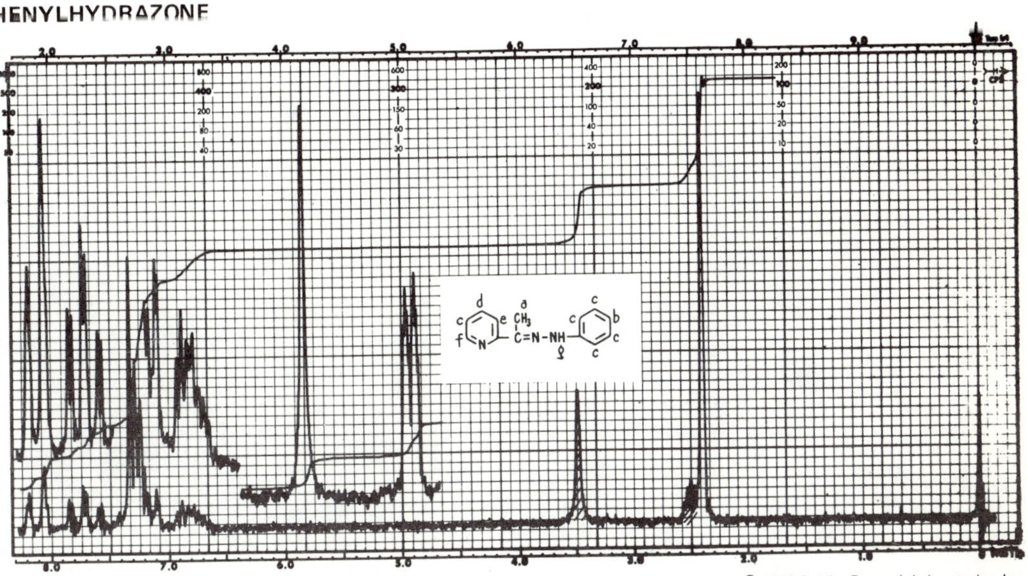

© 1978 Sadtler Research Laboratories, Inc.

BENZIL, DIHYDRAZONE

$C_{14}H_{14}N_4$

Mol. Wt. 238.29

Solvent: $CDCl_3$

Assignments

a = 5.35

b = ca 7.28

c = ca 7.55

879

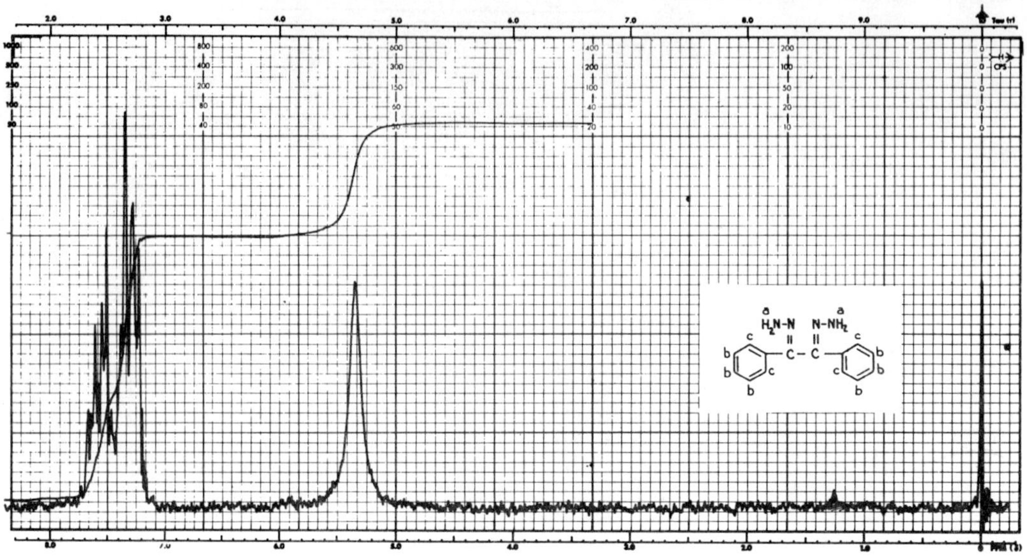

© 1978 Sadtler Research Laboratories, Inc.

The symmetrical structure of the Azine group generally produces less complex spectra than the preceeding groups such as the Hydrazones. Azine derivatives are usually produced from only one type of aromatic aldehyde so that the two protons of the Azine group are equivalent. In compounds synthesized utilizing two different aromatic aldehydes the Azine protons will appear as separate resonances at low field.

The overall range of chemical shifts for about 15 Azines was found to be as follows:

X–<u>CH=N–N=CH</u>–X	8.51-9.01 ppm	$CDCl_3$, Polysol

As observed for the Hydrazones, the –CH=N– group deshields the ortho aromatic hydrogens which resonate at about 7.78 ppm for a phenyl group.

Chemical Shifts

Aromatic Protons

meta/para	ortho		Solvent	Spectrum
7.1–7.5	7.78 ppm	–CH=N–N=CH–Ø	CCl_4	- - - -

para substituted Azines

Ø–CH=N–N=CH–	ortho	meta	para	Solvent	Spectrum
	7.78	6.95	$-O-CH_3$	$CDCl_3$	- - - -
	7.72	7.22	$-CH_3$	$CDCl_3$	881

Coupling and Coupling Constants

The band arising from the azine protons usually appears slightly broadened indicating the possibility that a very small coupling may exist between it and the ortho aromatic hydrogens. If the coupling in fact does exist it is quite small, less than 0.8 Hz. Derivatives of Cinnamaldehyde which result in an olefinic bond adjacent to the Azine linkage display clear coupling between the Azine protons and the olefinic hydrogens. Although this proton appears as a triplet in spectrum 880, equal coupling across the C–C double bond is unlikely and the triplet probably represents a higher order multiplet of the ABX type. A coupling constant of about 5 Hz. for the protons of C=<u>CH–CH</u>=N– appears reasonable.

Solvent Effects

The simple aromatic Azines are soluble in the halogenated hydrocarbons such as CCl_4 and $CDCl_3$. More polar solvents such as Polysol or DMSO-d6 may be required depending upon the type of substituents on the aromatic rings. There does not appear to be any special chemical shift relationship between the Azine protons and the solvent employed.

CINNAMALDEHYDE, AZINE

$C_{18}H_{16}N_2$

Mol. Wt. 260.34

M.P. 162°C

Solvent: $CDCl_3$

Assignments
a = 7.11
b = 7.25-7.70
c = 8.41

880

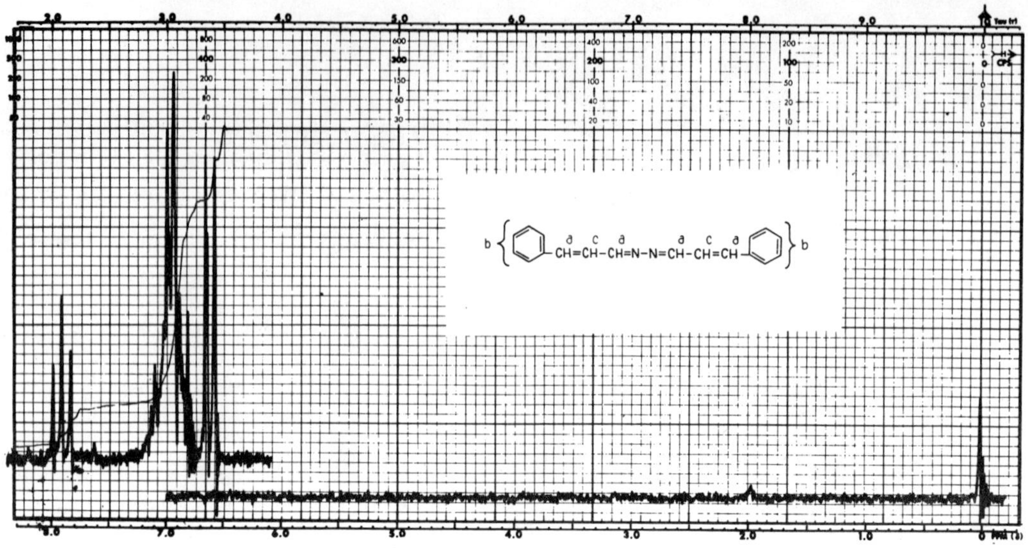

© 1978 Sadtler Research Laboratories, Inc

p-TOLUALDEHYDE, AZINE

$C_{16}H_{16}N_2$

Mol. Wt. 236.32

M.P. 158°C

Solvent: $CDCl_3$

Assignments
a = 2.38
b = 7.22
c = 7.72
d = 8.62

881

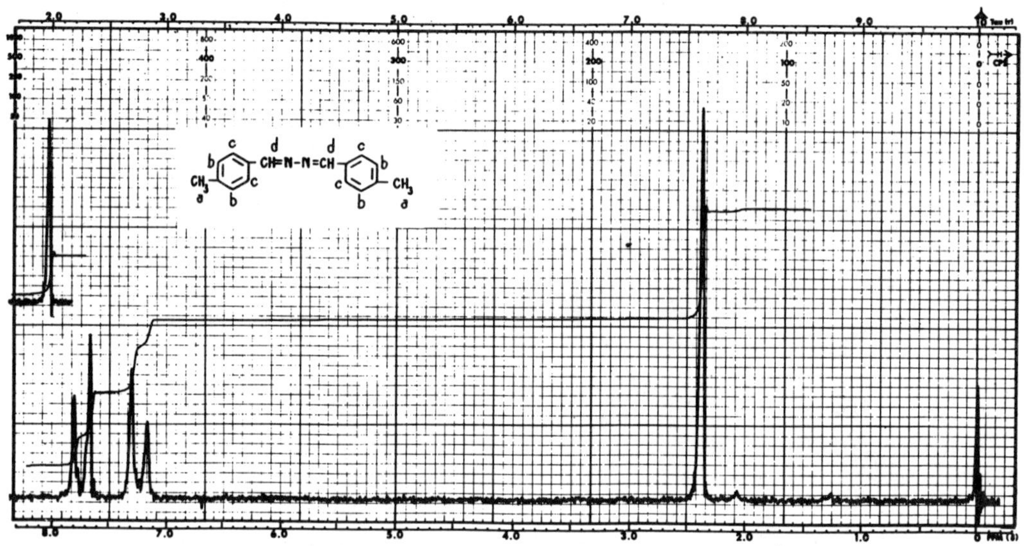

© 1978 Sadtler Research Laboratories, Inc.

PICOLINALDEHYDE, AZINE

$C_{12}H_{10}N_4$

Mol. Wt. 210.24

M.P. 152°C

Solvent: $CDCl_3$

Assignments
a = 7.31
b = 7.75
c = 8.13
d = 8.67
e = 8.70

882

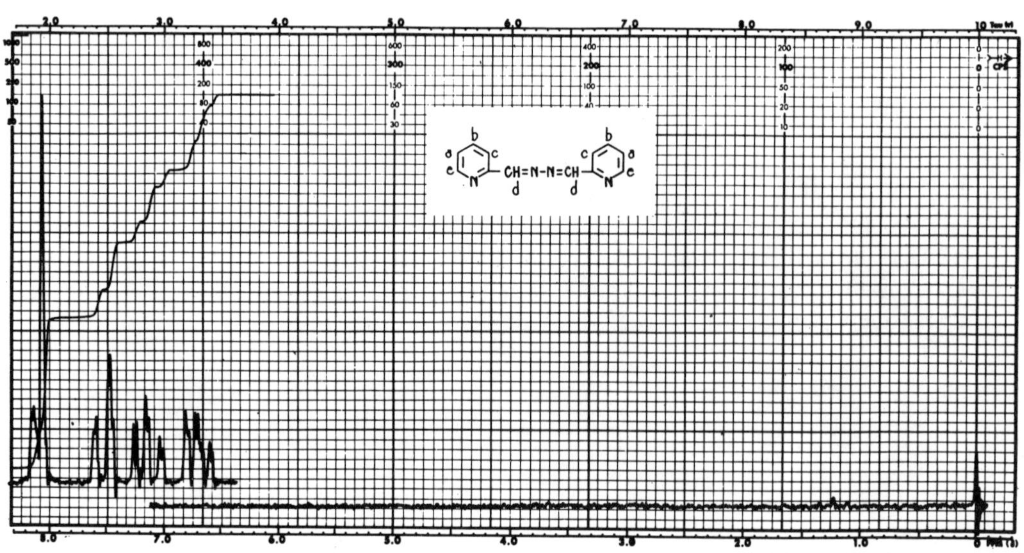

© 1978 Sadtler Research Laboratories, Inc.

THE AMIDINES

The most characteristic feature of the Amidines is the appearance of the —CH=N— proton as a relatively sharp singlet near 7.3 ppm. The chemical shifts and multiplets observed in the spectra can appear in a large number of possible combinations of aliphatic and aromatic groups.

The methyl resonance of the dimethyl formamidines is fairly constant in chemical shift resonating in the range from 2.8 to 3.0 ppm (spectra 883, 884, 885).

As a substituent, the nitrogen nucleus on the N=CH side of the linkage is observed to be a relatively strong shielding group in its effect on the ortho aromatic hydrogens (spectra 884, 885, 886).

Exchangeable —NH— protons, when present can appear as very broad bands at low field (spectrum 887) or as relatively sharp bands at much higher field (spectrum 890).

Because the CH=N proton is isolated from other proton groups by the two nitrogen atoms, it does not display any clear couplings.

The chemical shifts of several phenyl amidines are presented below.

Chemical Shifts

Aromatic Protons

meta/para	ortho		Solvent	Spectrum
6.8–7.3	6.75	—N=CH—N (Piperidine)	CCl_4	886
6.9–7.3	6.80	—N=CH—N$(CH_3)_2$	CCl_4	884
6.9–7.5	6.85	—NH—C(=NH)—Ø	DMSO-d_6	887
7.2–7.5	7.91	—C(=NH)—NH—Ø	DMSO-d_6	888

The three spectra, 887, 888 and 890, display extremely similar aromatic patterns for these symmetrically substituted amidine groups. It can be inferred that one is observing an "averaged" structure via resonance in which the double bond is shared by the nitrogen atoms endowing them with identical shielding/deshielding effects. This phenomenon is not observed in the unsymmetrical structures of this type.

N'-CYCLOHEXYL-N,N-DIMETHYLFORMAMIDINE

$C_9H_{18}N_2$

Mol. Wt. 154.26

B.P. 51-52°C/0.2mm

M.P. 41-43°C

Solvent: $CDCl_3$

Assignments
a = 0.82-1.98
b = ca 2.80
c = 7.30

883

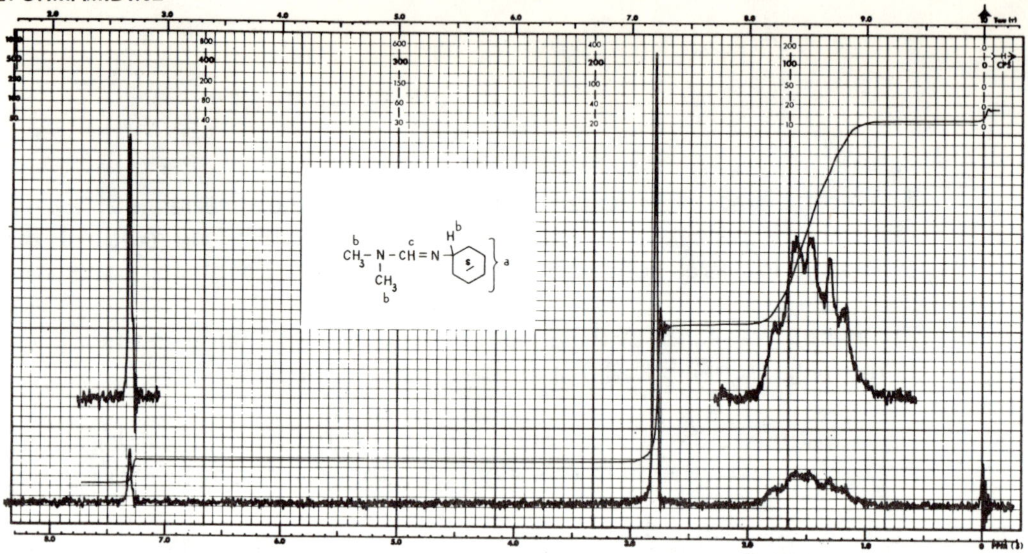

© 1978 Sadtler Research Laboratories, Inc.

N,N-DIMETHYL-N'-PHENYLFORMAMIDINE

$C_9H_{12}N_2$

Mol. Wt. 148.21

B.P. 127-128.5°C/11mm

Solvent: CCl_4

Assignments
a = 2.90
b = 6.70-7.30
c = 7.35

884

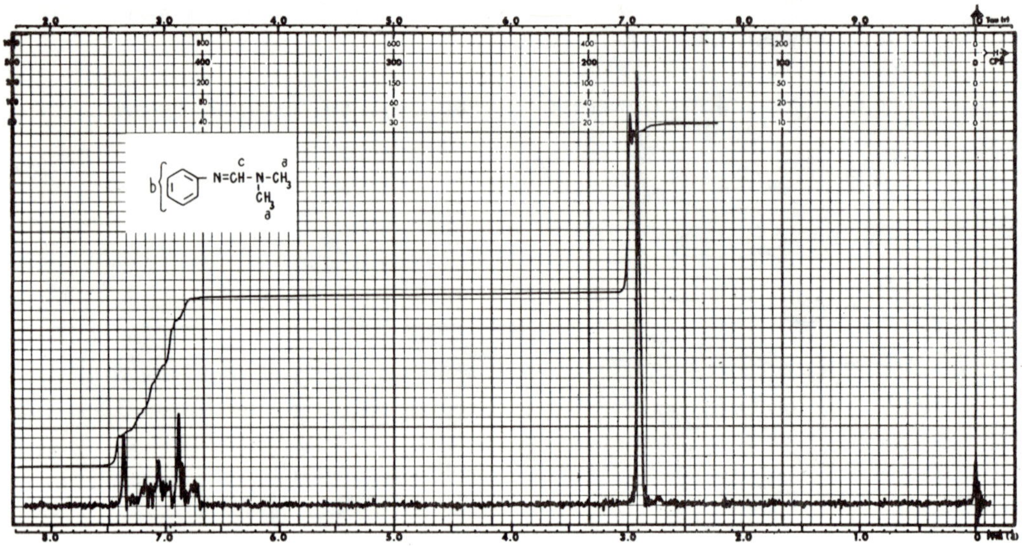

© 1978 Sadtler Research Laboratories, Inc.

N,N-DIMETHYL-N'-p-TOLYLFORMAMIDINE

$C_{10}H_{14}N_2$

Mol. Wt. 162.24

B.P. 103-105°C/0.5mm

Solvent: CCl_4

Assignments
a = 2.24
b = 2.87
c = 6.63
d = 6.88
e = 7.29

885

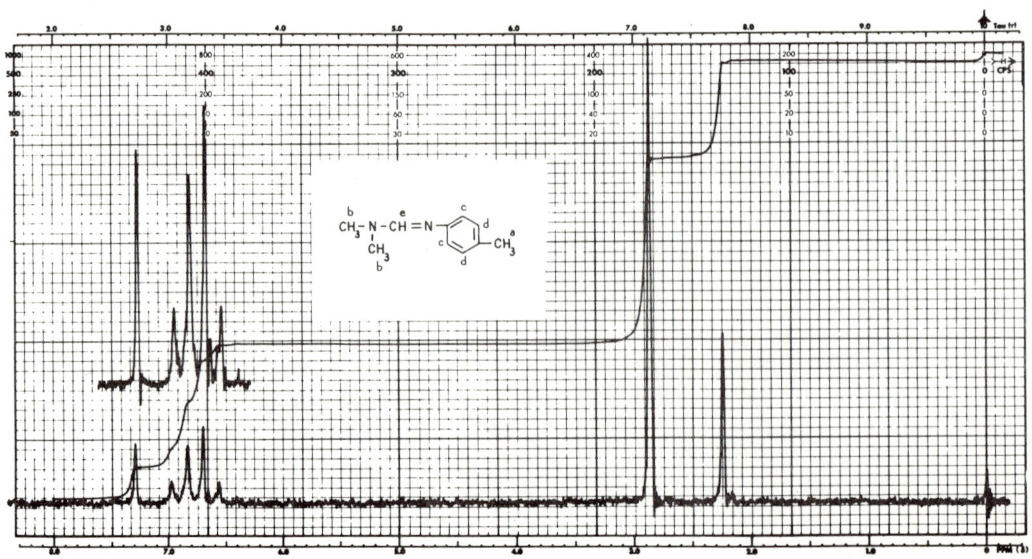

© 1978 Sadtler Research Laboratories, Inc.

1-(N-PHENYLFORMIMIDOYL)PIPERIDINE

$C_{12}H_{16}N_2$

Mol. Wt. 188.27

B.P. 135-137°C/0.5mm

Solvent: CCl_4

Assignments
a = 1.54
b = 3.31
c = 6.58-7.25
d = 7.28

886

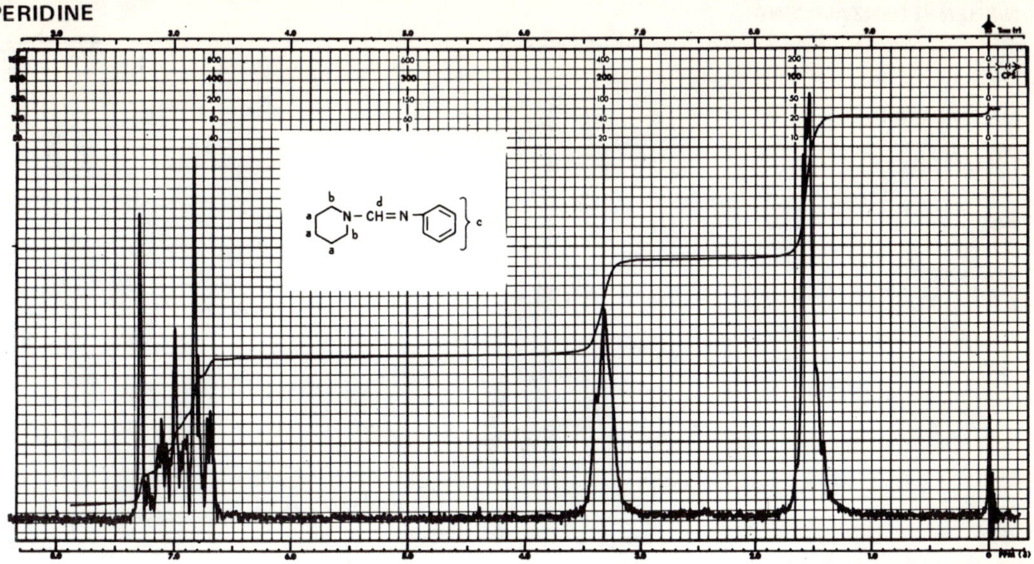

© 1978 Sadtler Research Laboratories, Inc.

N,N'-DIPHENYLFORMAMIDINE

$C_{13}H_{12}N_2$

Mol. Wt. 196.25

M.P. 130-132°C

Solvent: Polysol-d

Assignments
a = 6.90-7.45
b = ca 7.21
c = 8.11
d = 8.30-9.30

887

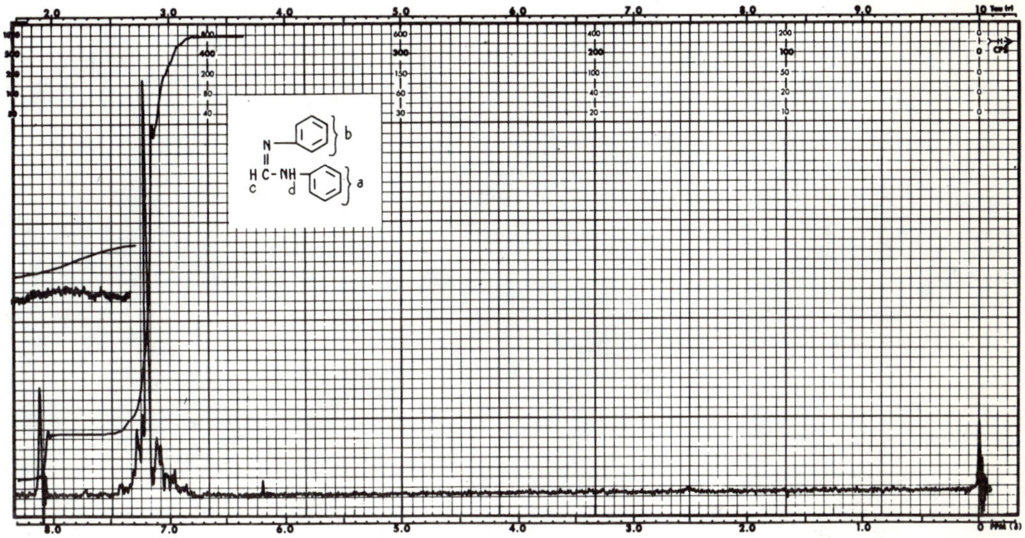

© 1978 Sadtler Research Laboratories, Inc.

N,N'-DIPHENYLACETAMIDINE

$C_{14}H_{14}N_2$

Mol. Wt. 210.28

M.P. 132-133°C

Solvent: $CDCl_3$

Assignments
a = 1.90
b = 6.21
c = 6.80-7.50

888

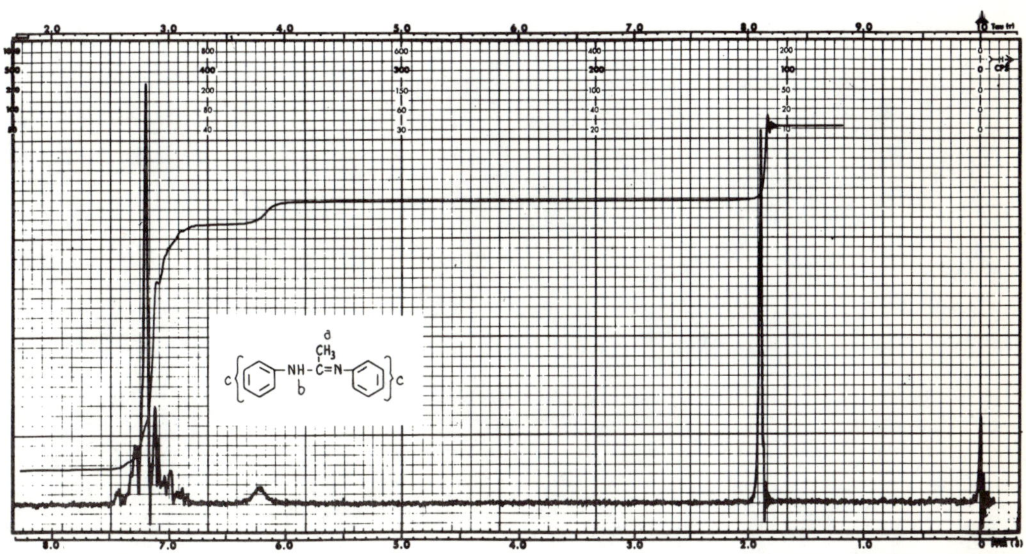

© 1978 Sadtler Research Laboratories, Inc.

N-PHENYLBENZAMIDINE

$C_{13}H_{12}N_2$

Mol. Wt. 196.25

Solvent: DMSO-d_6

Assignments
a = 5.71
b = 6.75-7.10
c = 7.10-7.50
d = 7.91

889

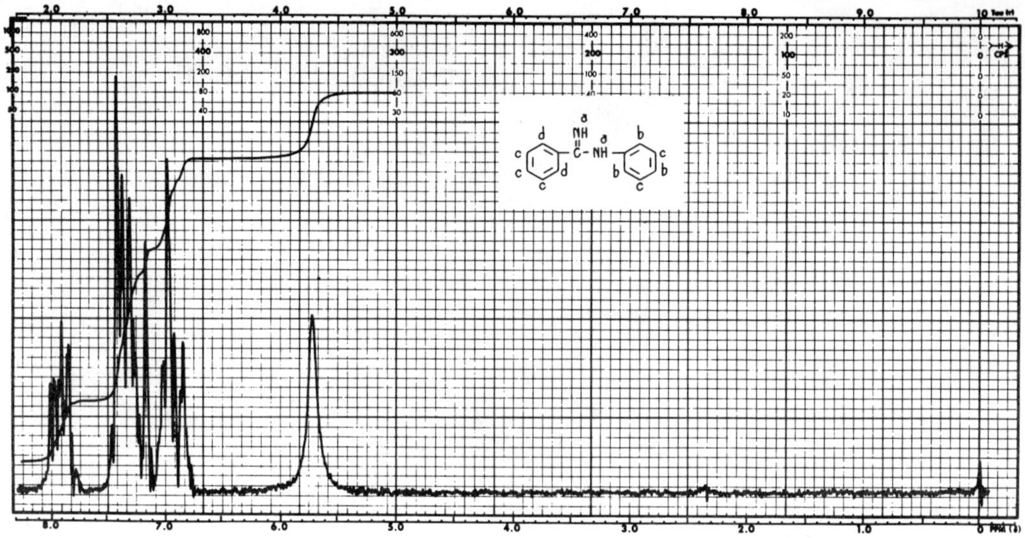

© 1978 Sadtler Research Laboratories, Inc.

1,2,3-TRIPHENYLGUANIDINE

$C_{19}H_{17}N_3$

Mol. Wt. 287.37

M.P. 145-147°C

Solvent: $CDCl_3$

Assignments
a = 5.89
b = 6.80-7.40

890

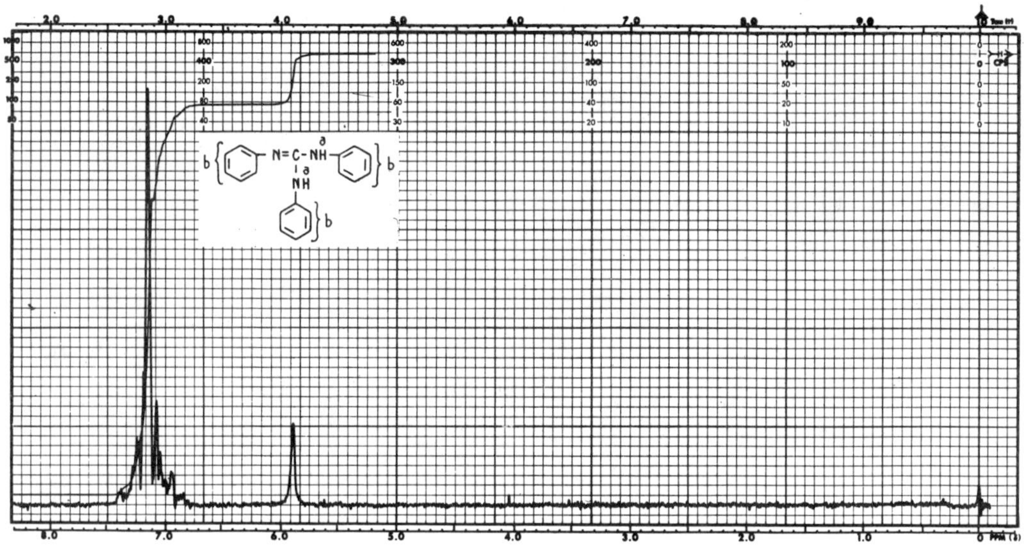

© 1978 Sadtler Research Laboratories, Inc.

THE HYDROXAMIC ACIDS

Although traditionally structured as N-Hydroxyamides, infrared spectral evidence indicates that the Hydroxamic Acids possess a Hydroxyoxime-like structure of the type R—C(OH)=N—OH.

The group has a weakly deshielding effect on adjacent hydrocarbon group protons but a strongly deshielding effect on the ortho aromatic hydrogens.

The two OH protons are usually in exchange and resonate as a single, rather broad band at low field (8—11 ppm).

The Hydroxamic acids are generally more soluble in DMSO-d_6, Polysol and Trifluoroacetic acid than in Deuterochloroform or Carbon Tetrachloride.

Chemical Shifts

Aliphatic Protons

X—	CH_2	—C(OH)=N—OH	Solvent	Spectrum
R_4—	2.20	ca. 8.60	$CDCl_3$	891
R_{15}—	2.51	- - - -	TFA	892
$(R)_2$—C=CH—	2.81	8.1—9.6	Polysol	- - - -
C_6—O—	4.04	ca. 8.54	Polysol	- - - -

Aromatic Protons

Benzohydroxamic Acid

para/meta	ortho	$(OH)_2$			
7.2—7.6	7.80	ca. 10.2	Ø—C(OH)=N—OH	Polysol	- - - -

4-Nitrobenzohydroxamic Acid

para	meta	ortho			
NO_2—	8.29	8.07	—C(OH)=N—OH	Polysol	- - - -

HEXANOHYDROXAMIC ACID

$C_6H_{13}NO_2$

Mol. Wt. 131.17

M.P. 65°C

Solvent: $CDCl_3$

Assignments
a = 0.91
b = 1.15-1.94
c = 2.20
d = 8.60

891

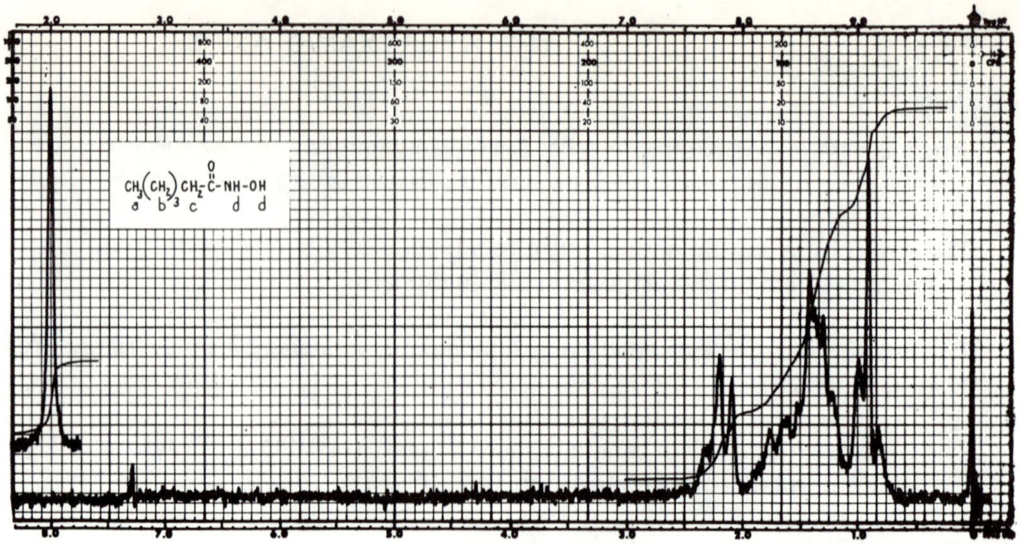

© 1978 Sadtler Research Laboratories, Inc.

STEAROHYDROXAMIC ACID

$C_{18}H_{37}NO_2$

Mol. Wt. 299.50

M.P. 102-104°C

Solvent: Trifluoroacetic acid

Assignments
a = 0.91
b = 1.10-1.60
c = 1.76
d = 2.51
e = unobserved *
* Under solvent band

892

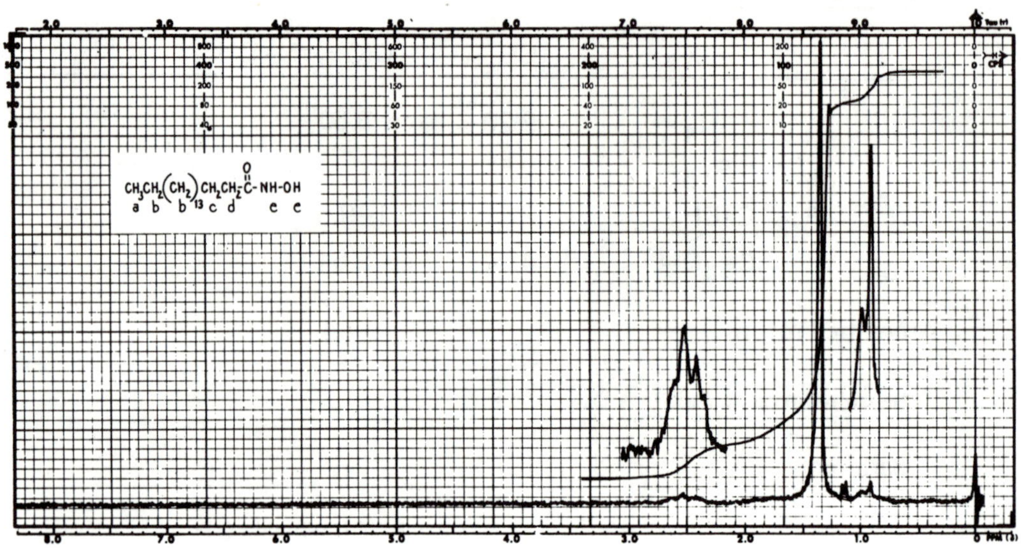

© 1978 Sadtler Research Laboratories, Inc.

3,7-DIMETHYL-6-OCTENOHYDROXAMIC ACID

$C_{10}H_{19}NO_2$

Mol. Wt. 185.27

M.P. 74-75°C

Solvent: $CDCl_3$

Assignments
a = 0.93
b = 1.59 or c
c = 1.68 or b
d = 1.11-2.33
e = 5.05
f = 8.69

893

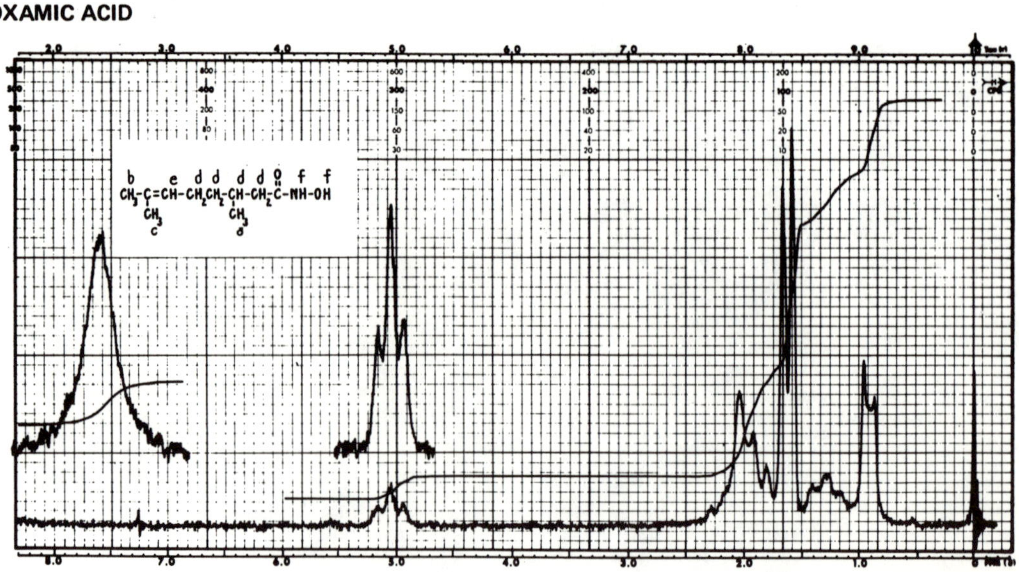

© 1978 Sadtler Research Laboratories, Inc.

SORBOHYDROXAMIC ACID

$C_6H_9NO_2$

Mol. Wt. 127.14

M.P. 131°C

Solvent: Polysol-d

Assignments
a = 1.81
b = 5.91
c = ca 6.06
d = 6.26
e = 7.11
f = 8.24 *
g = 2.59 DMSO-d_5
h = 1.20 imp.
* Plus H_2O

894

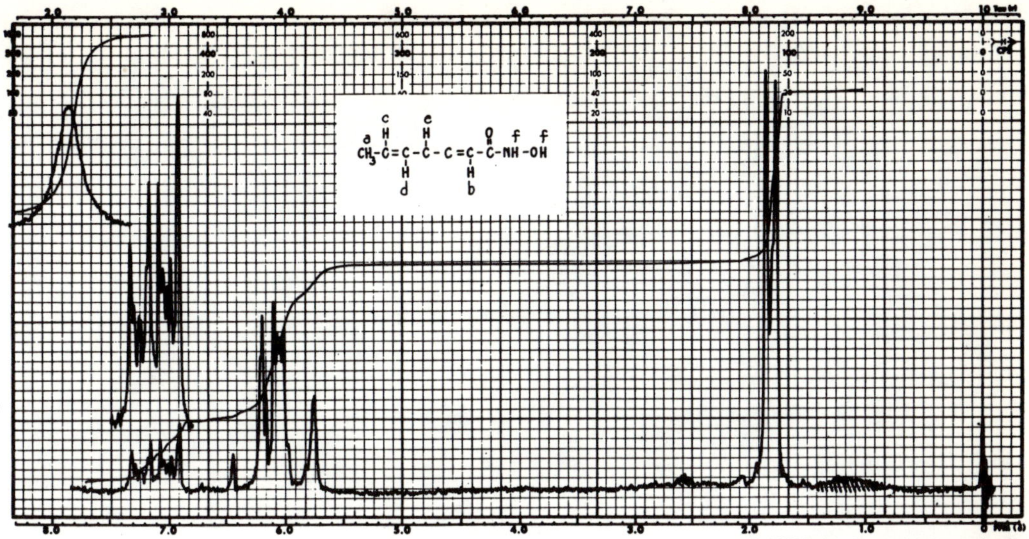

© 1978 Sadtler Research Laboratories, Inc.

ISONICOTINOHYDROXAMIC ACID

$C_6H_6N_2O_2$

Mol. Wt. 138.12

M.P. 161°C

Solvent: Dimethyl sulfoxide

Assignments
a = 7.73
b = 8.76
NH & OH unobserved

895

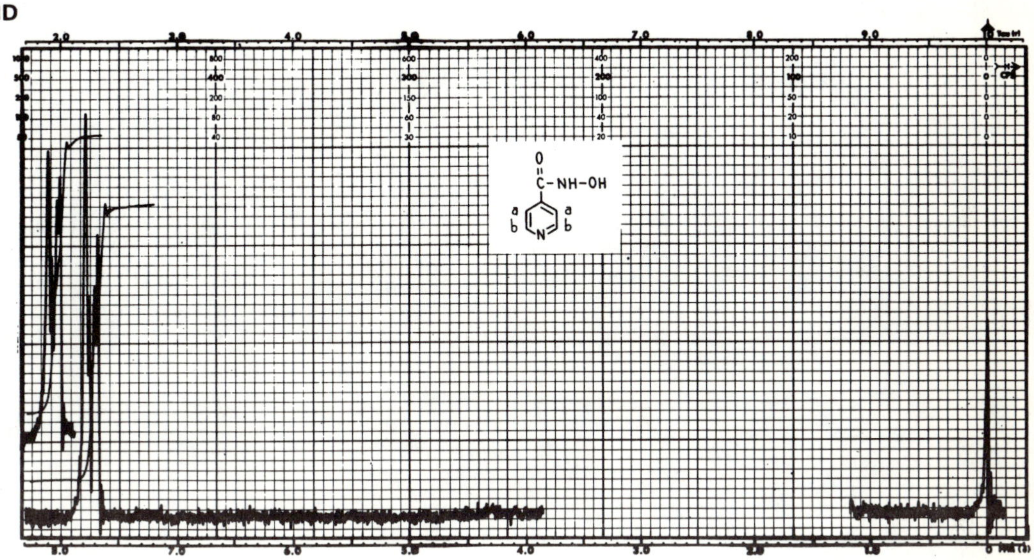

© 1978 Sadtler Research Laboratories, Inc.

NOTES

The Azo linkage between aromatic rings acts as a strong deshielding group on the ortho hydrogens. Due to the bulk of the aromatic rings, these compounds most likely exist only in the anti- form.

The only major solvent effect upon the deshielding ability of the Azo linkage is observed when these compounds are scanned as solutions in Trifluoroacetic acid. For these solutions, the hydrogens ortho to the Azo linkage are deshielded by an additional 0.3–0.4 ppm. The Azo compounds are readily soluble in $CDCl_3$, depending upon the nature of the other substituents present in the compound.

The chemical shifts for the parent compound, Azobenzene and a number of para-substituted derivatives are given below.

Chemical Shifts

Aromatic Protons

Azobenzene

meta/para	ortho		Solvent	Spectrum
7.4–7.8	7.96	–N=N–Ø	$CDCl_3$	896

Para substituted Azobenzenes

Ø–N=N–	ortho	meta	para	
	7.78	6.47	$-NH-CH_3$	$CDCl_3$
	7.83	6.63	$-N(CH_3)_2$	$CDCl_3$
	7.80	7.00	–OH	DMSO-d_6
	7.90	7.00	$-O-CH_3$	$CDCl_3$
	7.82	7.11	–F	$CDCl_3$
	7.87	7.20	–O–C(=O)–R	$CDCl_3$
	7.80	7.30	$-CH_3$	$CDCl_3$
	7.92	7.30	$-R_2$	$CDCl_3$
	7.73	7.31	$-S-CH_3$	$CDCl_3$
	7.81	7.40	–Cl	$CDCl_3$
	7.70	7.70	$-NH-C(=O)-CH_3$	Polysol
	7.88	7.88	$-SO_3Na$	DMSO
	7.79	7.98	$-C(=O)-CH_3$	$CDCl_3$
	8.01	8.01	$-As(=O)(OH)_2$	DMSO
	8.08	8.17	$-SO_2-Cl$	$CDCl_3$
	7.90	8.21	–C(=O)–Cl	$CDCl_3$
	7.98	8.24	–C(=O)–OH	DMSO
	7.92	8.32	$-NO_2$	$CDCl_3$

AZOBENZENE

$C_{12}H_{10}N_2$

Mol. Wt. 182.22

M.P. 67-68°C

Solvent: $CDCl_3$

Assignments
a = 7.37
b = 7.88

896

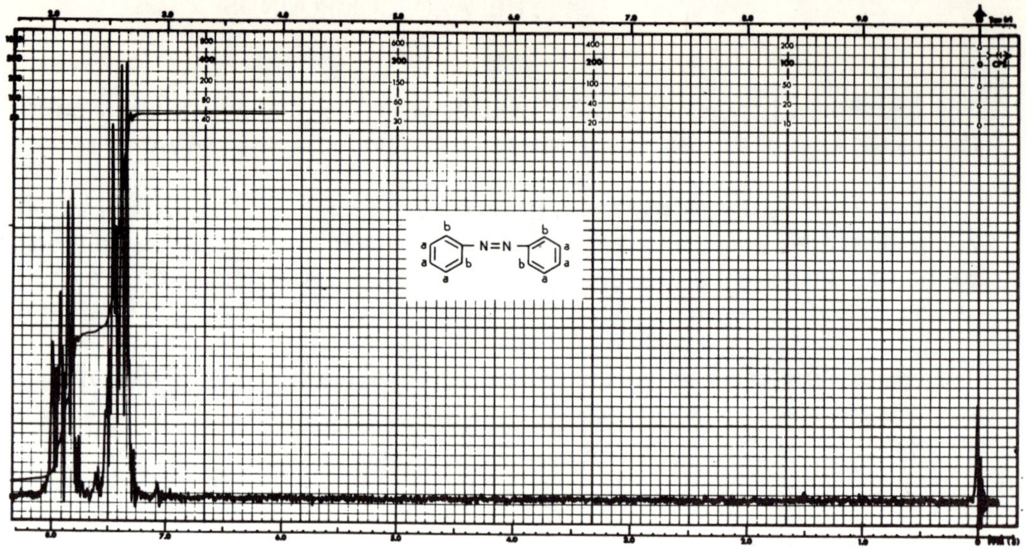

© 1978 Sadtler Research Laboratories, Inc.

---

m,m'-AZOTOLUENE

$C_{14}H_{14}N_2$

Mol. Wt. 210.28

Solvent: $CDCl_3$

Assignments
a = 2.40
b = ca 7.28
c = 7.37
d = 7.52-7.85

897

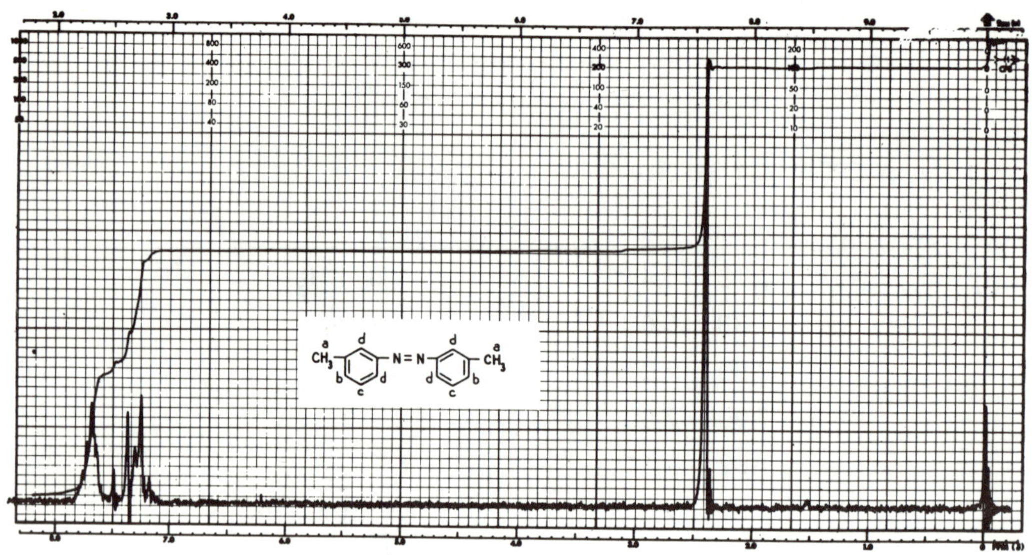

© 1978 Sadtler Research Laboratories, Inc.

3,3'-DICHLOROAZOBENZENE

$C_{12}H_8Cl_2N_2$

Mol. Wt. 251.12

M.P. 103-104°C

Solvent: $CDCl_3$

Assignments
a = ca 7.40
b = 7.60-7.90

898

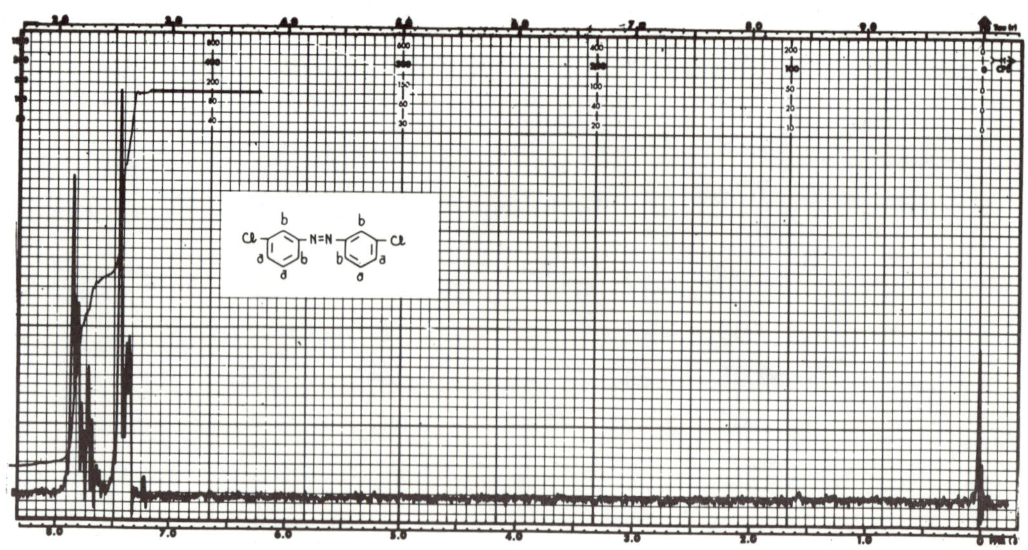

© 1978 Sadtler Research Laboratories, Inc

p-(PHENYLAZO)ANILINE

$C_{12}H_{11}N_3$

Mol. Wt. 197.24

M.P. 124-125°C

Solvent: $CDCl_3$

Assignments
a = 3.84
b = 6.60
c = 7.15-7.60
d = ca 7.79

899

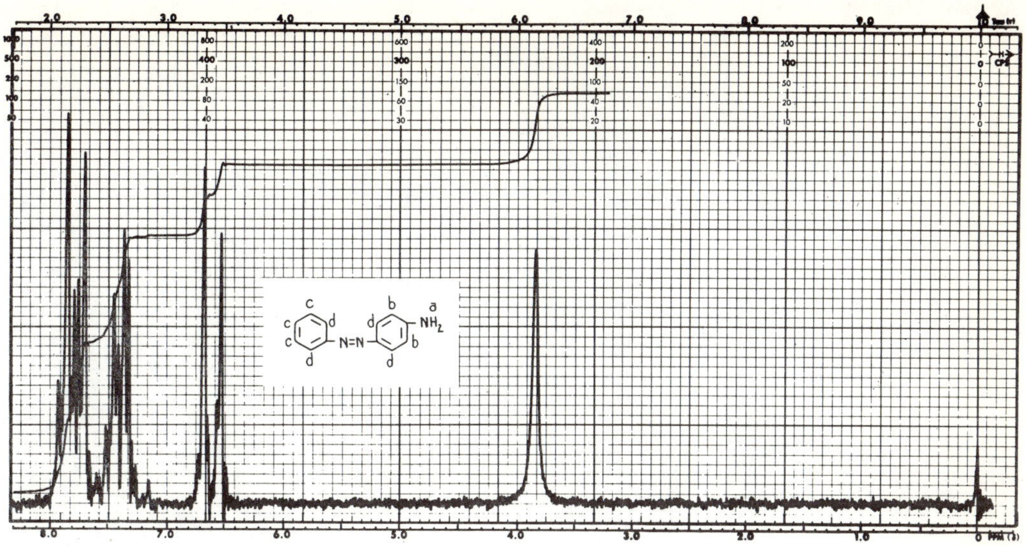

© 1978 Sadtler Research Laboratories, Inc.

N,N-DIMETHYL-p-PHENYLAZOANILINE

$C_{14}H_{15}N_3$

Mol. Wt. 225.30

M.P. 117.5-118°C

Solvent: $CDCl_3$

Assignments
a = 2.98
b = 6.67
c = 7.20-7.60
d = 7.82

900

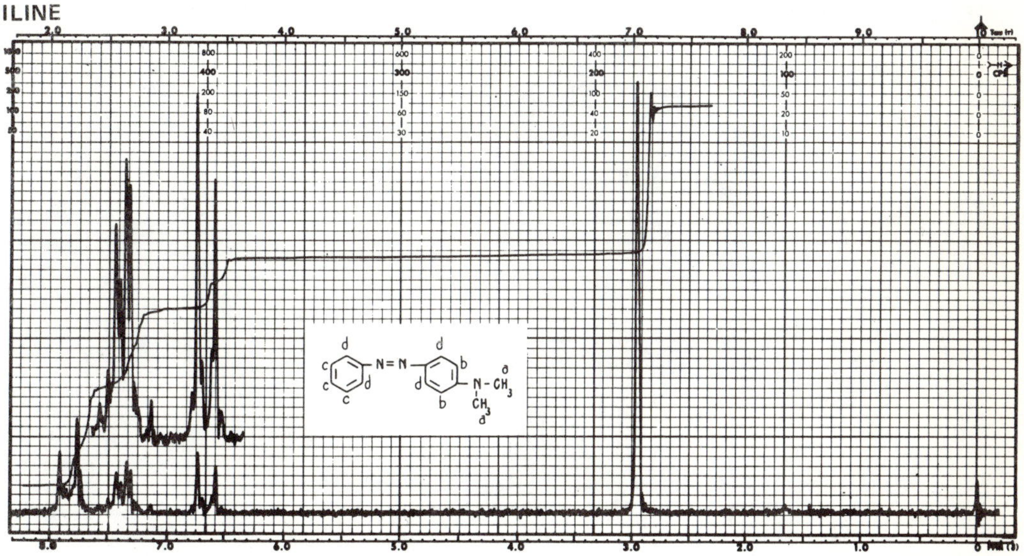

© 1978 Sadtler Research Laboratories, Inc.

N,N-DIMETHYL-p[(p-AMINOPHENYL)AZO] ANILINE

$C_{14}H_{16}N_4$

Mol. Wt. 240.31

M.P. 185-187°C

Solvent: Dimethyl sulfoxide-d_6

Assignments
a = 2.98
b = 5.70
c = 6.70 or d
d = 6.76 or c
e = 7.63 or f
f = 7.71 or e
g = 2.50 DMSO-d_5
h = 3.39 imp.

901

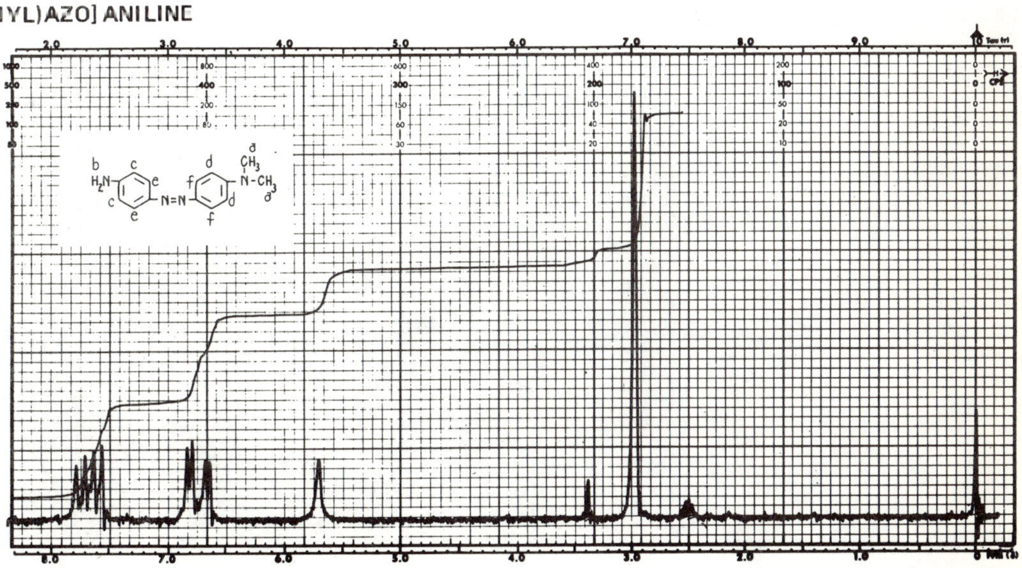

© 1978 Sadtler Research Laboratories, Inc.

AZOXYBENZENE

$C_{12}H_{10}N_2O$

Mol. Wt. 198.23

Solvent: $CDCl_3$

Assignments
a = 7.10-7.59
b = ca 8.20

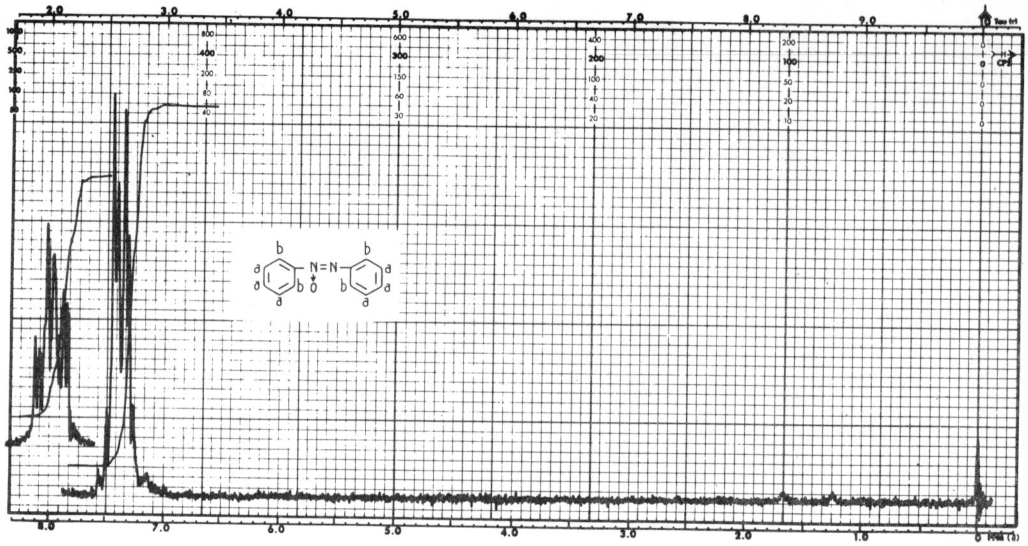

902

© 1978 Sadtler Research Laboratories, Inc.

4,4'-DICHLOROAZOXYBENZENE

$C_{12}H_8Cl_2N_2O$

Mol. Wt. 267.12

M.P. 154-156°C

Solvent: $CDCl_3$

Assignments
a = ca 7.39
b = 8.09
c = 8.18

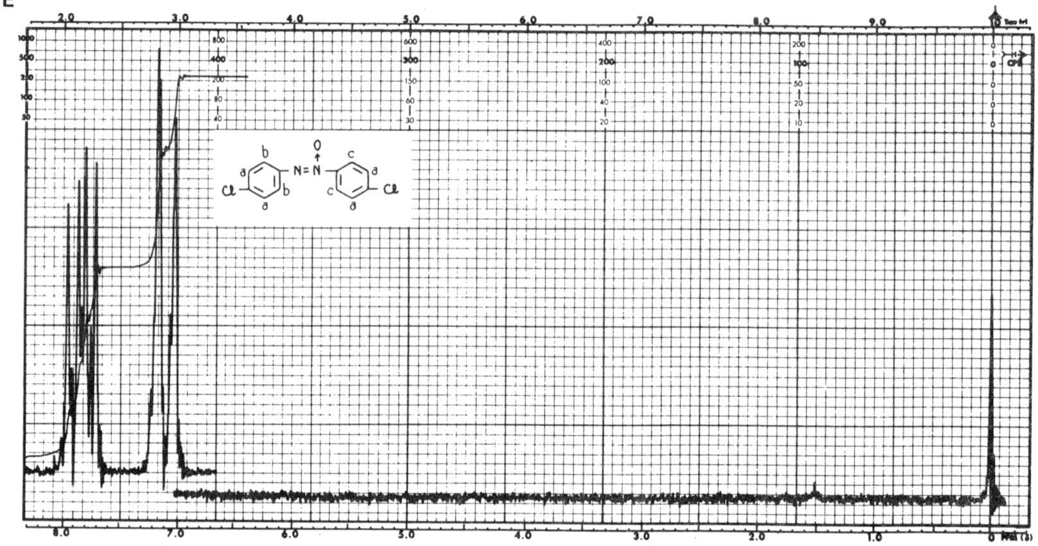

903

© 1978 Sadtler Research Laboratories, Inc.

3-METHYL-1-p-TOLYLTRIAZENE

$C_8H_{11}N_3$

Mol. Wt. 149.20

M.P. 70-80°C (lit.)

Solvent: $CDCl_3$

Assignments
a = 2.30
b = 3.21
c = 7.10
d = 7.22
e = 8.01

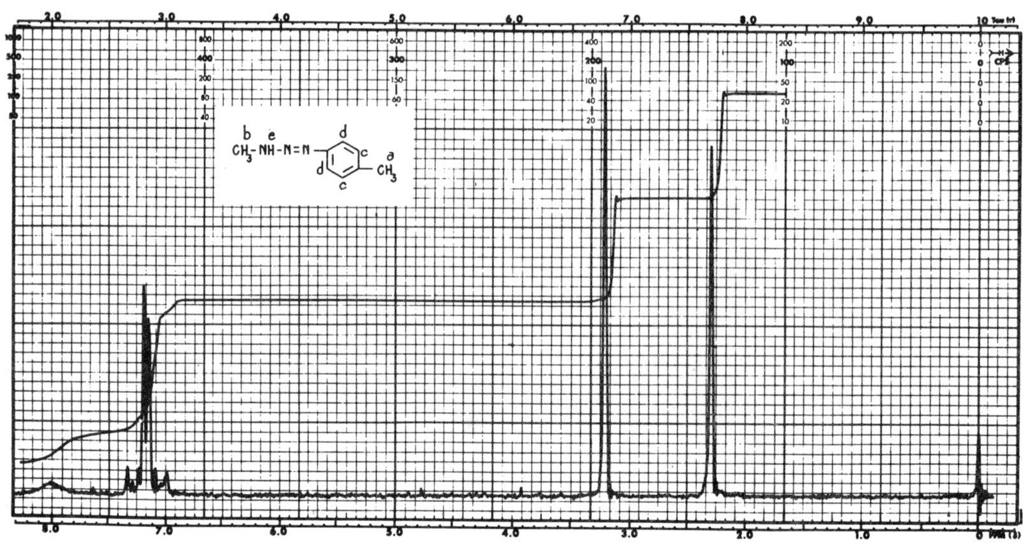

904

© 1978 Sadtler Research Laboratories, Inc.

1,3-DIPHENYLTRIAZENE

$C_{12}H_{11}N_3$

Mol. Wt. 197.24

M.P. 96-98°C

Solvent: $CDCl_3$

Assignments
a = 6.50-8.05

905

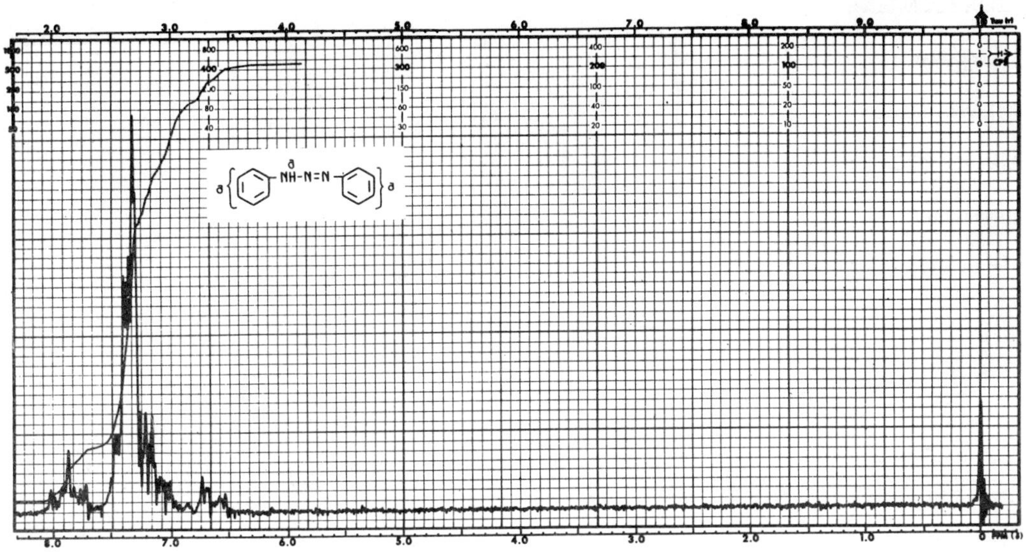

© 1978 Sadtler Research Laboratories, Inc.

3,3-DIMETHYL-1-PHENYLTRIAZENE

$C_8H_{11}N_3$

Mol. Wt. 149.20

Solvent: CCl_4

Assignments
a = 3.23
b = 6.88-7.50

906

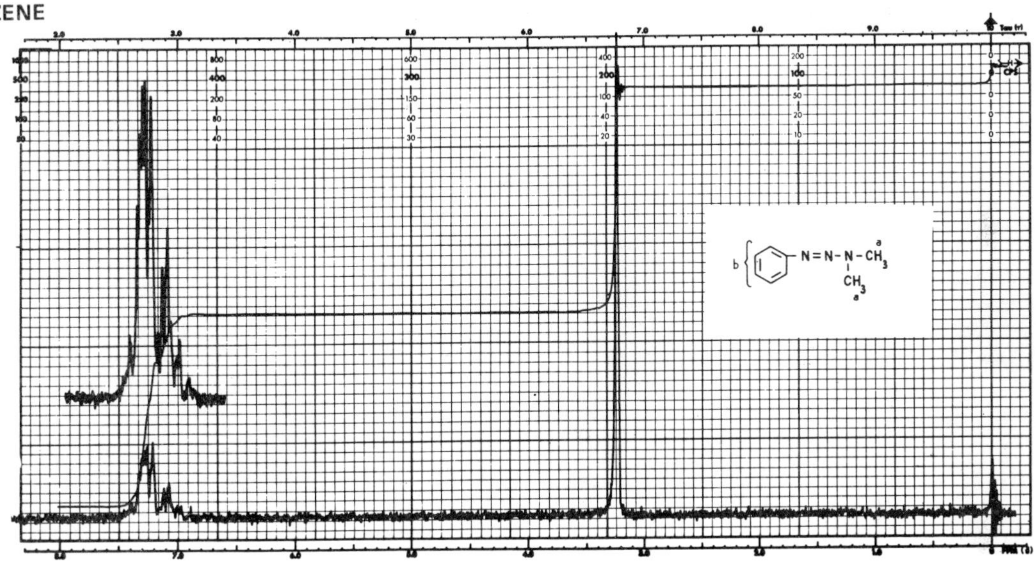

© 1978 Sadtler Research Laboratories, Inc.

NOTES

THE ISOCYANATES

The Isocyanate group has a moderate deshielding effect on aliphatic protons but a slight shielding effect on the ortho aromatic hydrogens. The chemical shifts of several representative aliphatic and aromatic compounds are listed below.

Chemical Shifts

Aliphatic Protons

HC–3	HC–2	HC–1		Solvent	Spectrum
		3.01	$CH_3–N=C=O$	$CDCl_3$	907
0.98	1.61	3.29	$CH_3–CH_2–CH_2–N=C=O$	$CDCl_3$	908
(1.1 – 1.9)		3.24	$CH_3–(CH_2)_4–CH_2–N=C=O$	CCl_4	909
(1.1 – 1.9)		329	$CH_3–(CH_2)_{10}–CH_2–N=C=O$	CCl_4	910

Aromatic Protons

Phenyl Isocyanates

O=C=N– ortho	meta	para	Solvent	Spectrum
	6.85–7.55		$CDCl_3$	912
7.02	6.81	$–O–CH_3$	$CDCl_3$	----
6.98	7.07	$–CH_2–Ø$	Polysol	918
7.10	7.10	$–O–CF_3$	CCl_4	----

Solubility and Solvent Effects

The Isocyanates are readily soluble in the halogenated solvents normally used to prepare NMR solutions. The aromatic Isocyanates may require more polar solvents such as D_2O or DMSO-d_6 depending on the character of the other substituents on the phenylisocyanate ring.

Characterization

Because the Isocyanate group is neither strongly shielding nor deshielding, possesses no exchangeable protons and displays no coupling to nearby protons, it is a very difficult functional group to identify based only on NMR data. Fortunately, this group is readily identified via its infrared absorption bands.

ISOCYANIC ACID, METHYL ESTER

C_2H_3NO

Mol. Wt. 57.05

B.P. 37-39°C

Solvent: $CDCl_3$

Assignments
a = 3.01

907

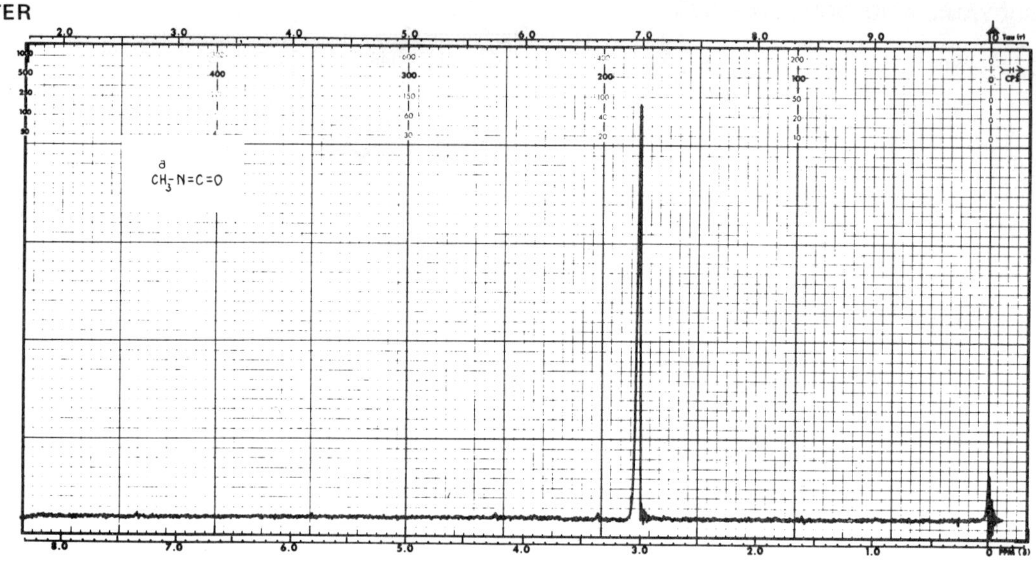

© 1978 Sadtler Research Laboratories, Inc.

ISOCYANIC ACID, PROPYL ESTER

C_4H_7NO

Mol. Wt. 85.11

B.P. 87-88°C

Solvent: $CDCl_3$

Assignments
a = 0.98
b = 1.61
c = 3.29

908

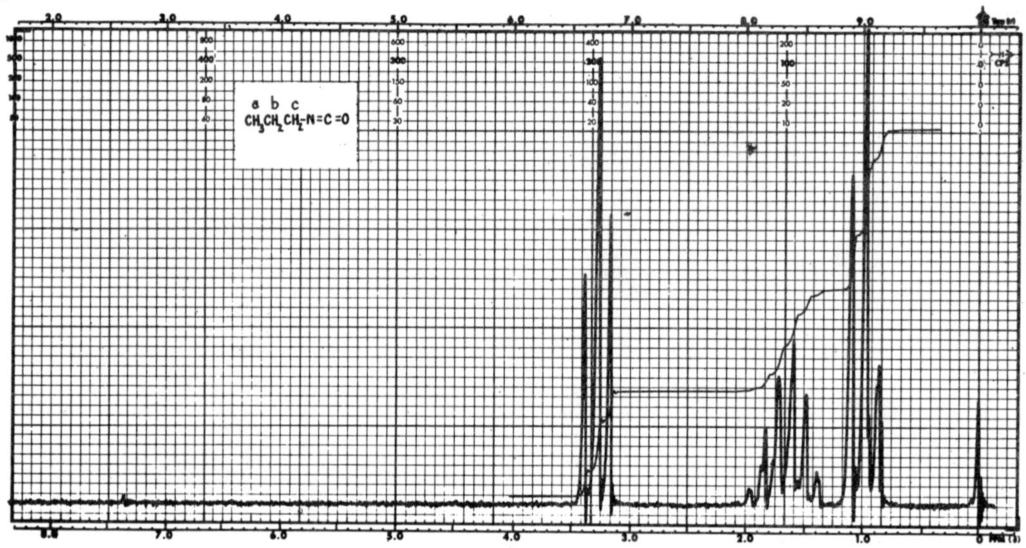

© 1978 Sadtler Research Laboratories, Inc.

ISOCYANIC ACID, HEXYL ESTER

$C_7H_{13}NO$

Mol. Wt. 127.19

B.P. 43-45°C/7mm

Solvent: CCl_4

Assignments
a = 0.90
b = 1.10-1.90
c = 3.24
d = 2.31, 4.51 imps.
e = 7.09, 7.28 imps.

909

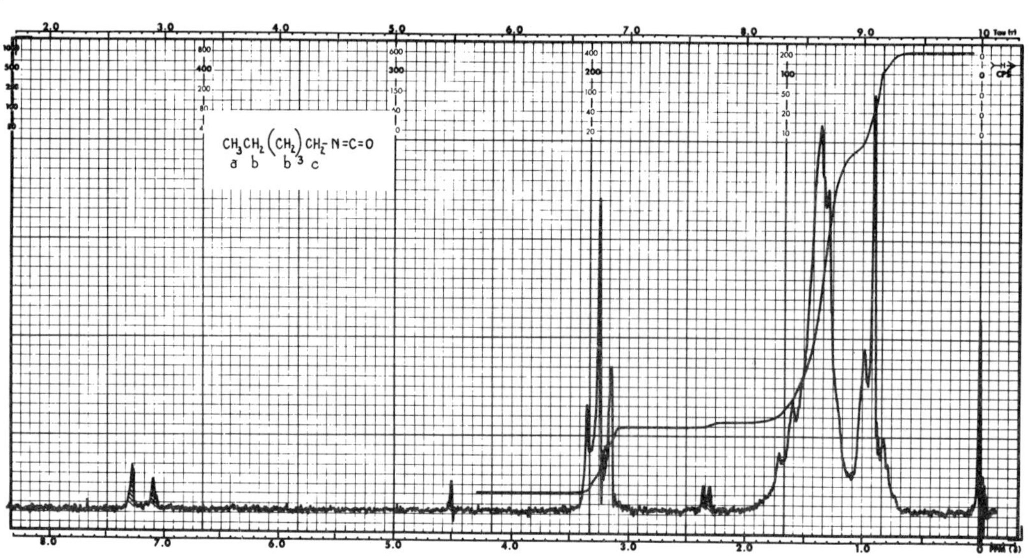

© 1978 Sadtler Research Laboratories, Inc.

ISOCYANIC ACID, DODECYL ESTER

$C_{13}H_{25}NO$

Mol. Wt. 211.35

B.P. 130-133°C/5mm

Solvent: CCl_4

Assignments
a = 0.90
b = 1.10-1.90
c = 3.29

910

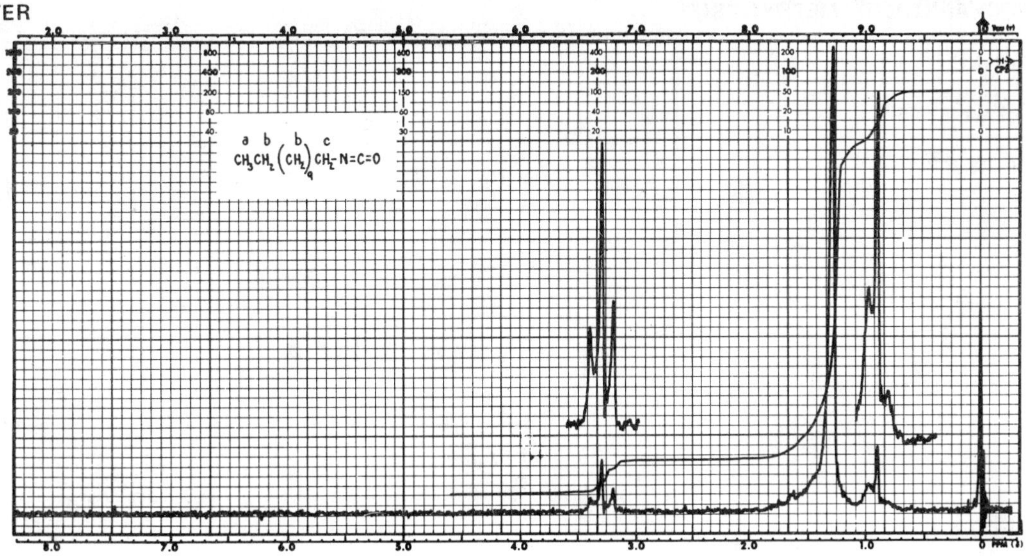

© 1978 Sadtler Research Laboratories, Inc.

ISOCYANIC ACID, BICYCLO[2.2.2] OCT-1,4-YLENE ESTER

$C_{10}H_{12}N_2O_2$

Mol. Wt. 192.32

M.P. 111.0-112.5°C

Solvent: $CDCl_3$

Assignments
a = 1.90

911

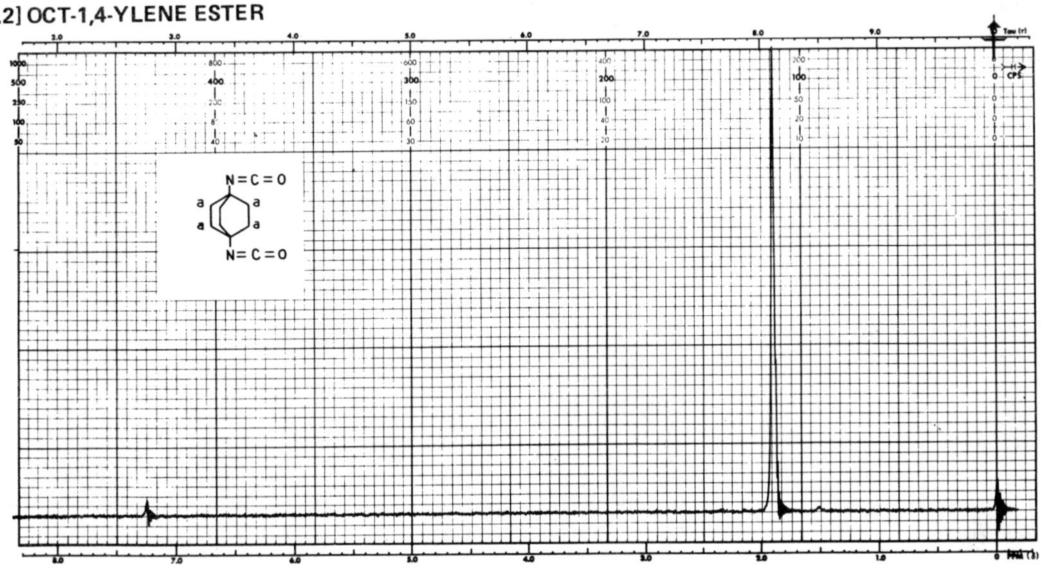

© 1978 Sadtler Research Laboratories, Inc.

ISOCYANIC ACID, PHENYL ESTER

C_7H_5NO

Mol. Wt. 119.12

B.P. 163.7°C

Solvent: $CDCl_3$

Assignments
a = 7.13
b = 7.23

912

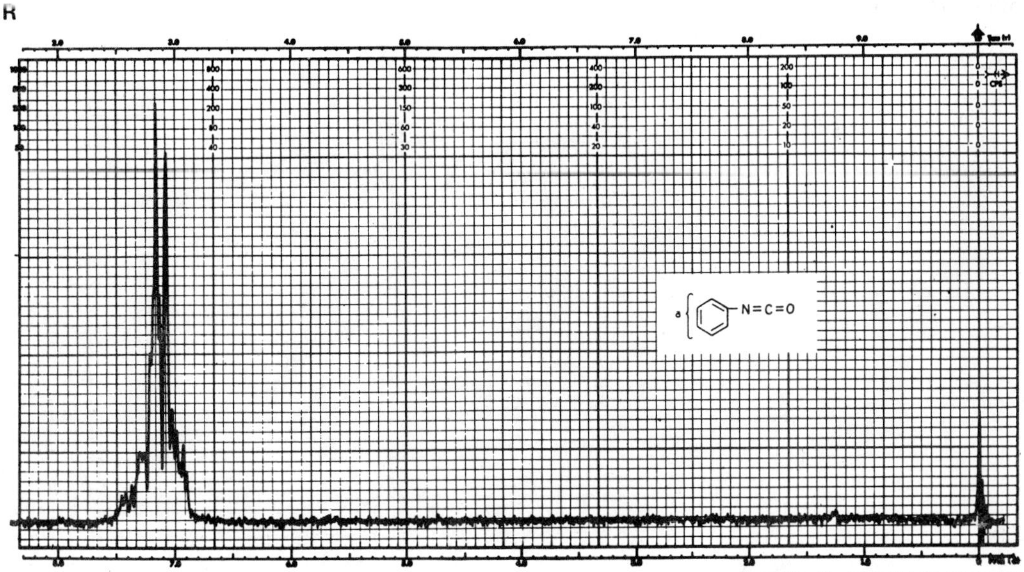

© 1978 Sadtler Research Laboratories, Inc.

ISOCYANIC ACID, o-TOLYL ESTER

C_8H_7NO

Mol. Wt. 133.15

B.P. 77-78°C/12mm

Solvent: CCl_4

Assignments
a = 2.29
b = 6.90-7.20

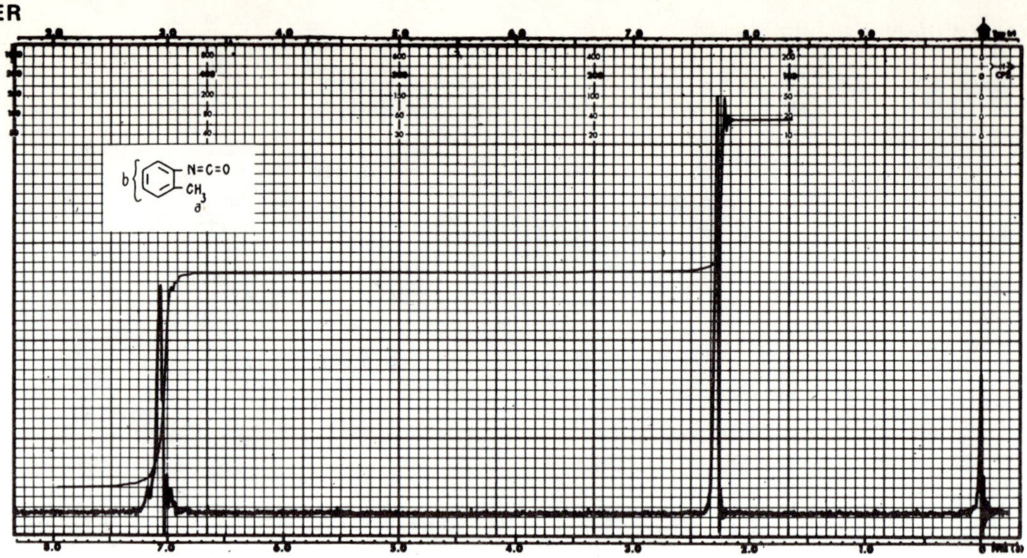

913

© 1978 Sadtler Research Laboratories, Inc.

ISOCYANIC ACID, *a,a,a*-TRIFLUORO-o-TOLYL ESTER

$C_8H_4F_3NO$

Mol. Wt. 187.12

Solvent: $CDCl_3$

Assignments
a = 7.10-7.80

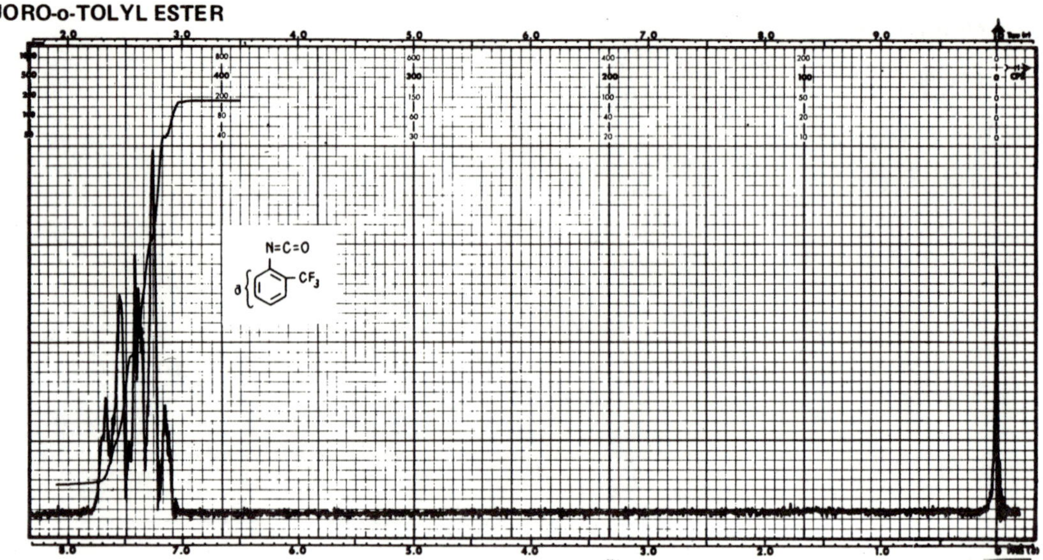

914

© 1978 Sadtler Research Laboratories, Inc.

ISOCYANIC ACID, *a,a,a*-TRIFLUORO-m-TOLYL ESTER

$C_8H_4F_3NO$

Mol. Wt. 187.12

B.P. 54°C/11mm

Solvent: $CDCl_3$

Assignments
a = 7.15-7.55

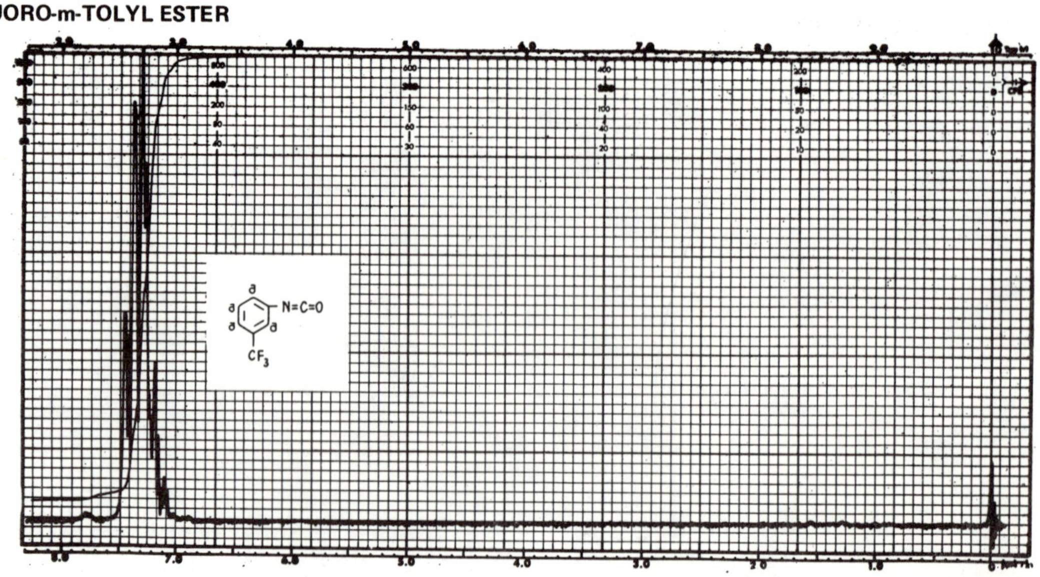

915

© 1978 Sadtler Research Laboratories, Inc.

ISOCYANIC ACID, o-FLUOROPHENYL ESTER

C_7H_4FNO

Mol. Wt. 137.11

Solvent: CCl_4

Assignments
a = 6.75-7.20

916

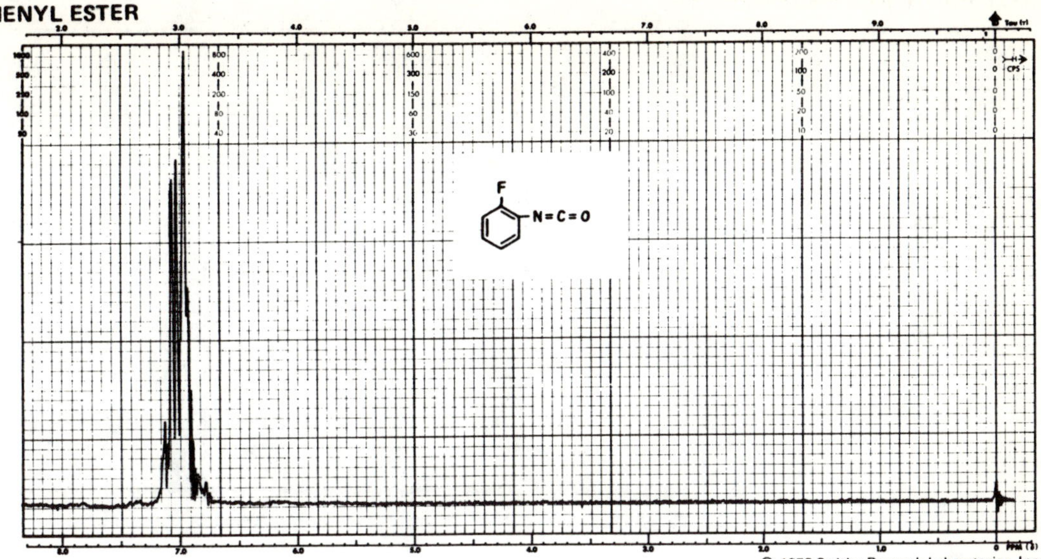

© 1978 Sadtler Research Laboratories, Inc.

ISOCYANIC ACID, m-FLUOROPHENYL ESTER

C_7H_4FNO

Mol. Wt. 137.11

Solvent: $CDCl_3$

Assignments
a = 6.60-7.42

917

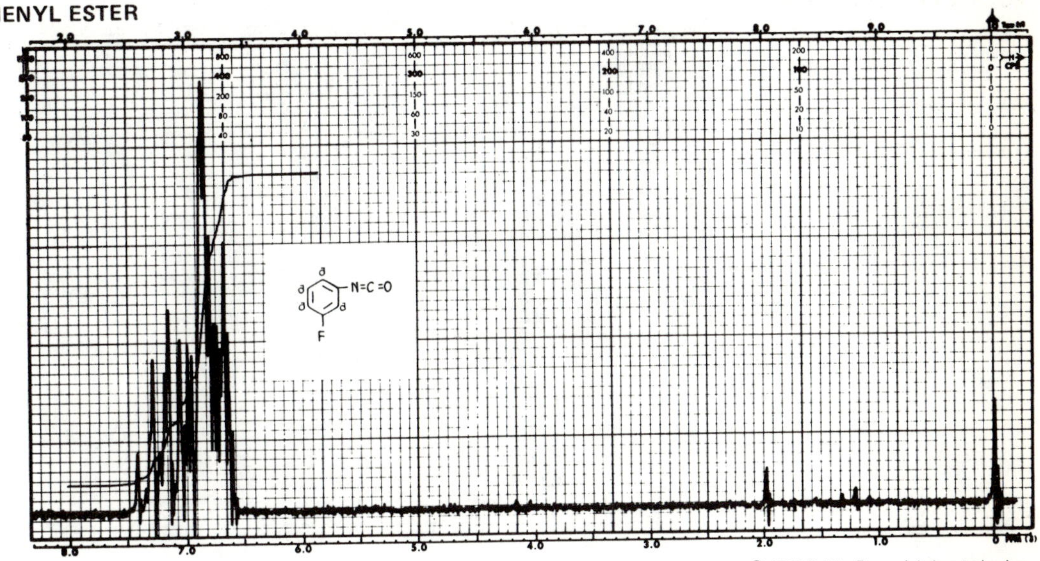

© 1978 Sadtler Research Laboratories, Inc.

ISOCYANIC ACID, METHYLENEDI-p-PHENYLENE ESTER

$C_{15}H_{10}N_2O_2$

Mol. Wt. 250.26

M.P. 40-45°C

Solvent: Polysol-d

Assignments
a = 3.87
b = 6.98
c = 7.07

918

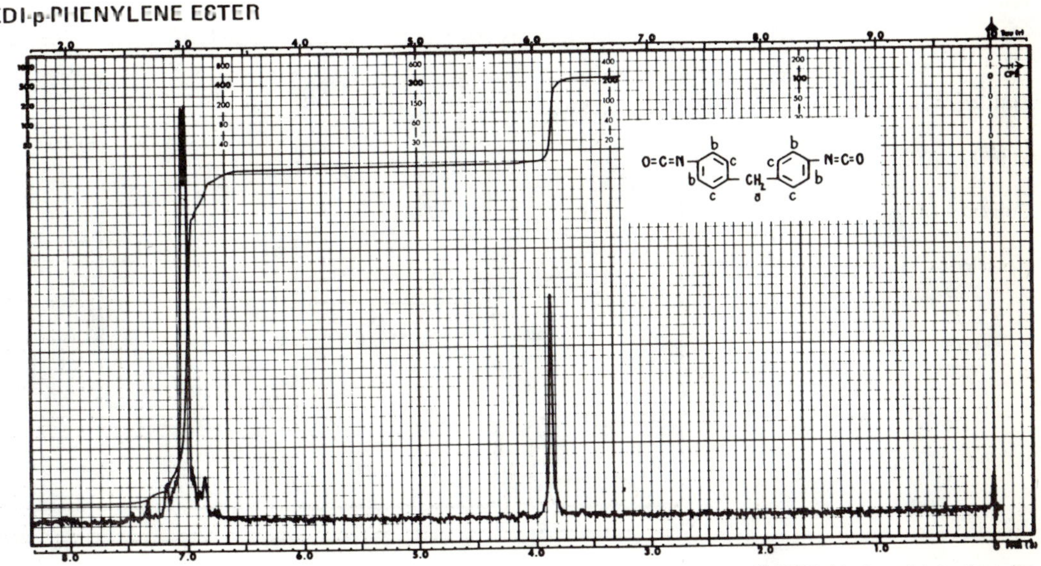

© 1978 Sadtler Research Laboratories, Inc.

ISOCYANIC ACID, 2,5-XYLYL ESTER

C_9H_9NO

Mol. Wt. 147.18

M.P. 7-9°C

Solvent: CCl_4

Assignments
a = 2.29
b = 6.74-6.90
c = 7.02

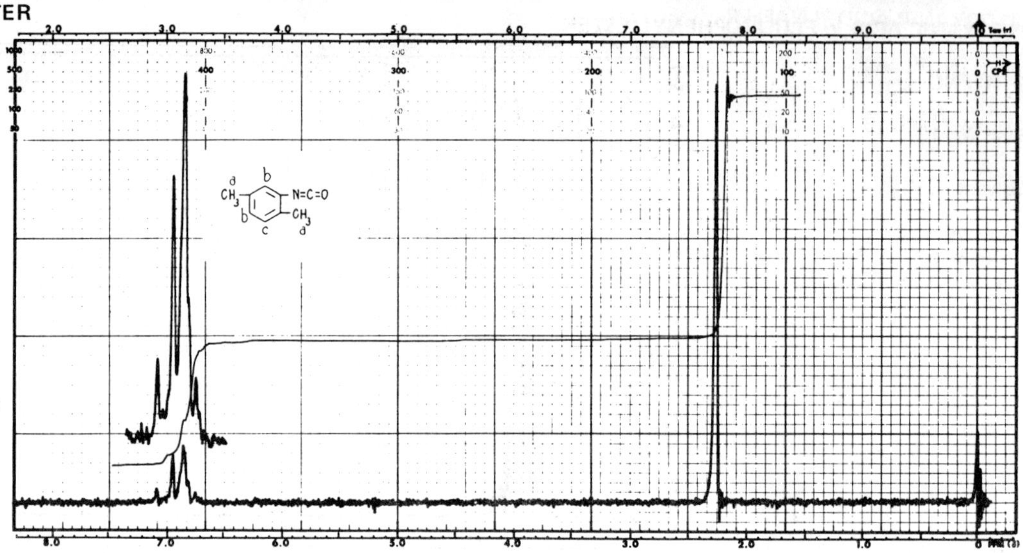

919

© 1978 Sadtler Research Laboratories, Inc.

ISOCYANIC ACID, 2,5-DICHLOROPHENYL ESTER

$C_7H_3Cl_2NO$

Mol. Wt. 188.01

M.P. 26-28°C

Solvent: $CDCl_3$

Assignments
a = ca 7.09
b = 7.26

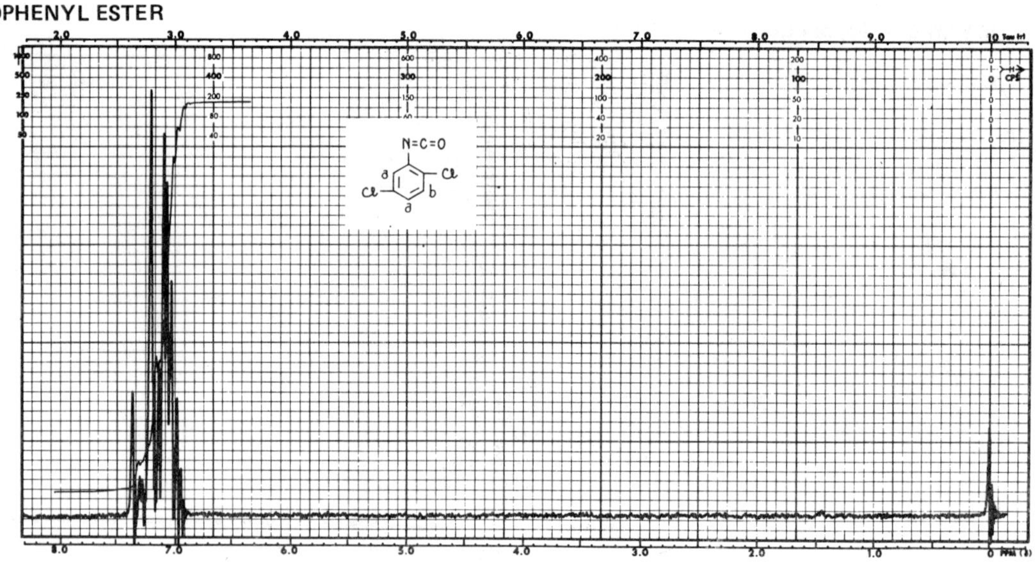

920

© 1978 Sadtler Research Laboratories, Inc.

ISOCYANIC ACID, 4-METHYL-m-PHENYLENE ESTER

$C_9H_6N_2O_2$

Mol. Wt. 174.16

M.P. 17-20°C

Solvent: CCl_4

Assignments
a = 2.29
b = 6.71
c = 6.79
d = 7.10

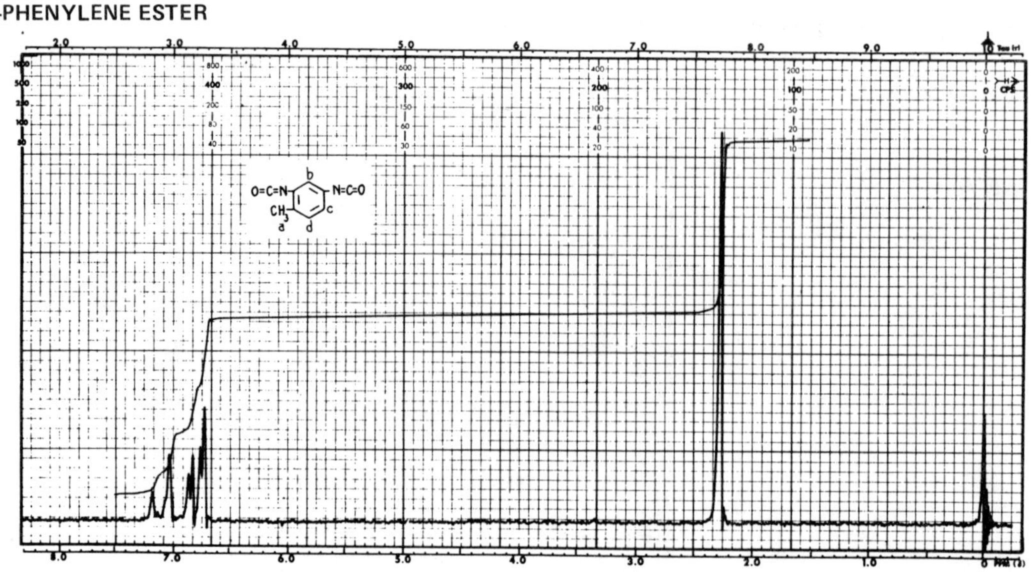

921

© 1978 Sadtler Research Laboratories, Inc.

ISOCYANIC ACID, 1-NAPHTHYL ESTER

$C_{11}H_7NO$

Mol. Wt. 169.18

M.P. 3-5°C

Solvent: $CDCl_3$

Assignments

a = 7.00-7.90

b = 8.00

922

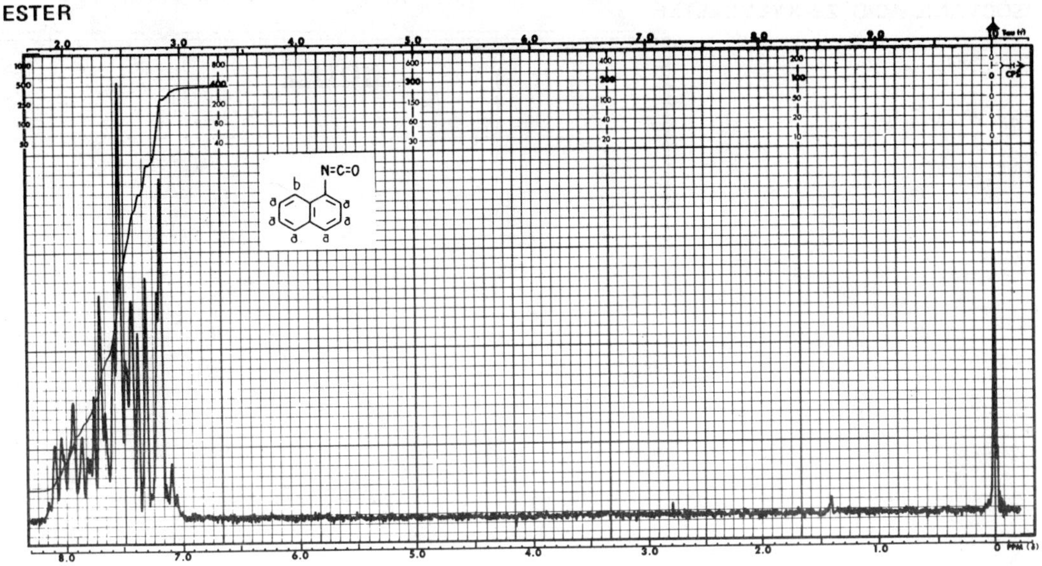

© 1978 Sadtler Research Laboratories, Inc.

NOTES

THE CARBODIIMIDES

Although relatively few compounds containing the Carbodiimide linkage are available, the indication is that the chemical shifts of both the aliphatic and aromatic groups bonded to it are similar to those of the Isocyanates (—N=C=O), the Benzylidenes (N=CH—Ø) and the Isothiocyanates (—N=C=S).

Chemical Shifts

Aliphatic Protons

HC–2	HC–1		Solvent	Spectrum
(1.22)	3.53	$(CH_3)_2$–CH–N=C=N–R	$CDCl_3$	923
(0.98)	- - - -	$(CH_3)_3$–C–N=C=N–R	$CDCl_3$	- - - -

Alicyclic Protons

Dicyclohexylcarbodiimide

HC–4,3,2	HC–1			
0.9–2.2	3.17	C_6–N=C=N–C_6	$CDCl_3$	924

Aromatic Protons

Spectra 927, 928 and 929 indicate that the Carbodiimide group is a weakly shielding group on the ortho aromatic hydrogens similar to a methyl substituent. Its presence in a molecule would be difficult to detect without the corresponding infrared data.

Syn-/Anti- Isomerism

Although both syn- and anti-forms can exist, most of the spectra included in this volume do not display the duplication of resonance bands expected in the spectra of such a mixture. It is assumed that the groups most often exist in the anti-form. One possible exception is noted when one compares the aromatic trityl resonance bands of spectrum 925 (a single broad band near 7.28 ppm) and the corresponding band of spectrum 926 (a complex, higher order series of multiplets in the range from 6.8 to 7.3 ppm).

DIISOPROPYLCARBODIIMIDE

$C_7H_{14}N_2$

Mol. Wt. 126.20

B.P. 145-147°C

Solvent: $CDCl_3$

Assignments
a = 1.22
b = 3.53

923

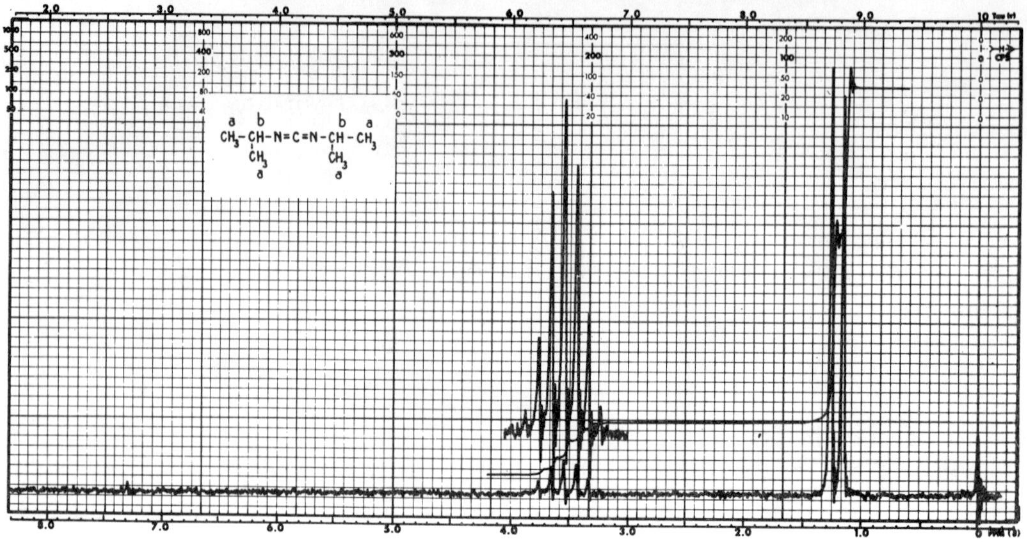

© 1978 Sadtler Research Laboratories, Inc.

DICYCLOHEXYLCARBODIIMIDE

$C_{13}H_{22}N_2$

Mol. Wt. 206.33

M.P. 33-34°C

B.P. 122-124°C/6mm (lit.)

Solvent: $CDCl_3$

Assignments
a = ca 0.89-2.13
b = 3.17

924

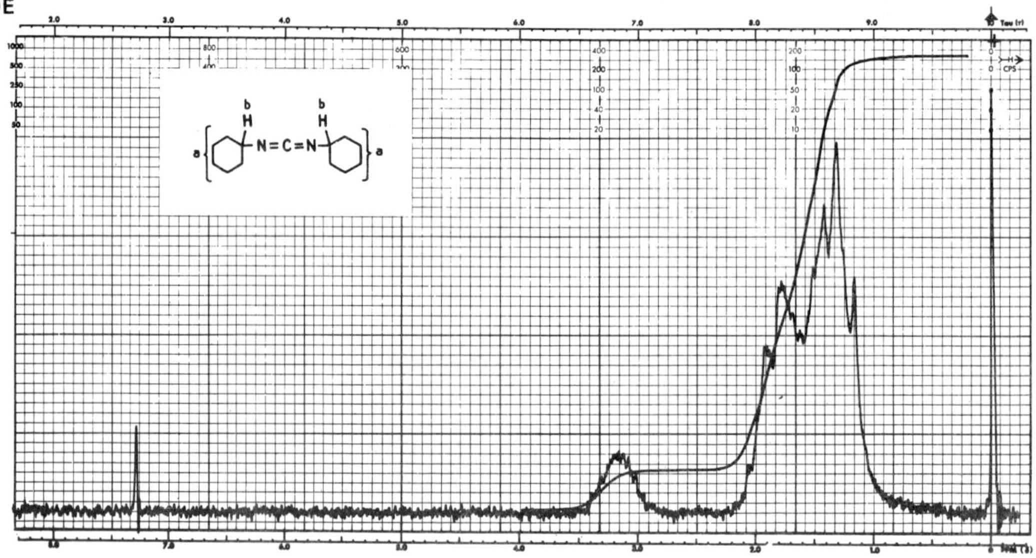

© 1978 Sadtler Research Laboratories, Inc.

tert-BUTYL(TRIPHENYLMETHYL)CARBODIIMIDE

$C_{24}H_{24}N_2$

Mol. Wt. 340.47

M.P. 71-73°C

Solvent: $CDCl_3$

Assignments
a = 0.98
b = ca 7.28

925

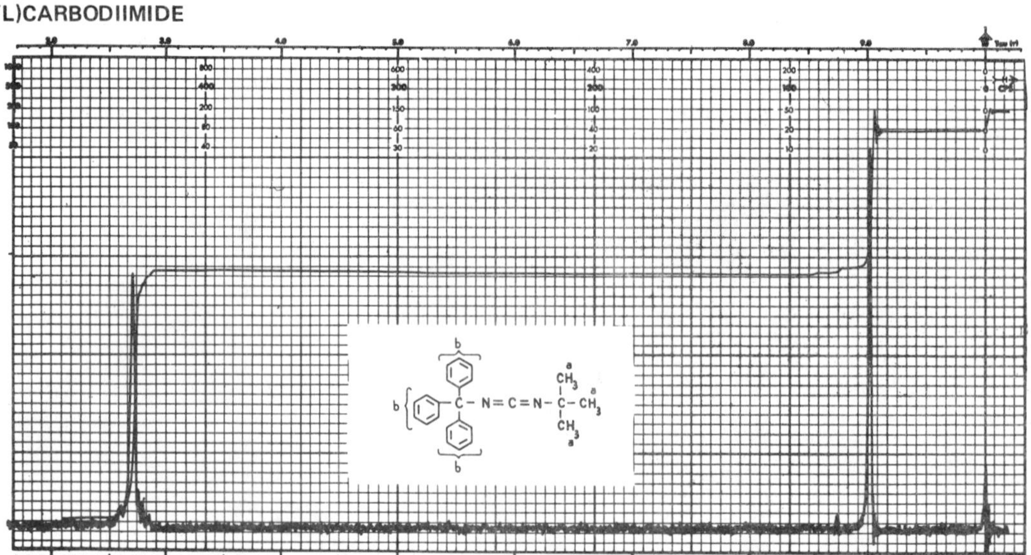

© 1978 Sadtler Research Laboratories, Inc.

DITRITYLCARBODIIMIDE

$C_{39}H_{30}N_2$

Mol. Wt. 526.69

M.P. 210°C (lit.)

Solvent: $CDCl_3$

Assignments
a = 6.83-7.33
no assignment made

926

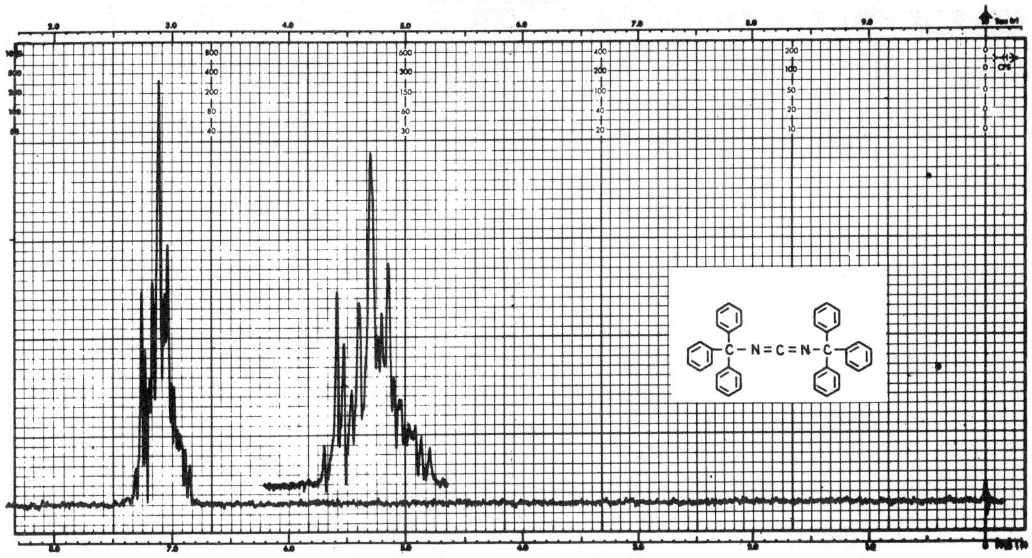

© 1978 Sadtler Research Laboratories, Inc.

DI-o-TOLYLCARBODIIMIDE

$C_{15}H_{14}N_2$

Mol. Wt. 222.29

B.P. 140°C/0.9mm

Solvent: CCl_4

Assignments
a = 2.34
b = 7.07

927

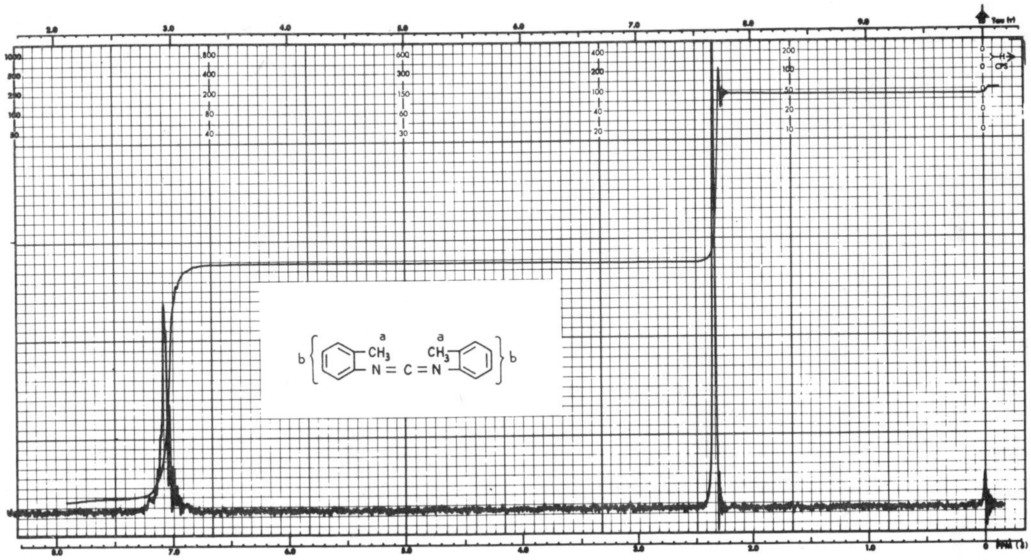

© 1978 Sadtler Research Laboratories, Inc.

DI-p-TOLYLCARBODIIMIDE

$C_{15}H_{14}N_2$

Mol. Wt. 222.29

Solvent: $CDCl_3$

Assignments
a = 2.30
b = 7.06

928

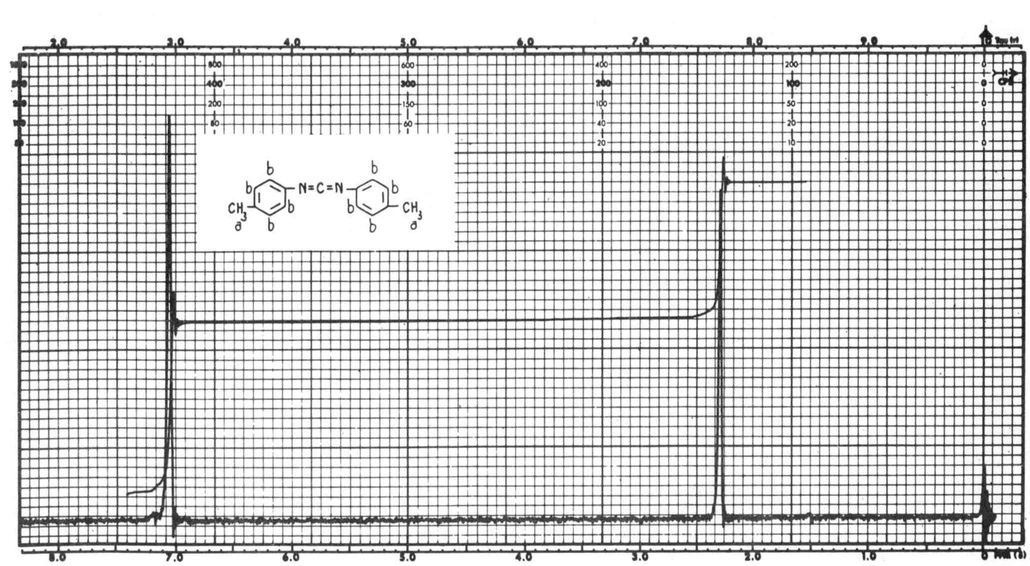

© 1978 Sadtler Research Laboratories, Inc.

BIS(2,6-DIETHYLPHENYL)CARBODIIMIDE

$C_{21}H_{26}N_2$

Mol. Wt. 306.45

B.P. 192-194°C/0.4mm

Solvent: $CDCl_3$

Assignments

a = 1.26

b = 2.80

c = 7.04

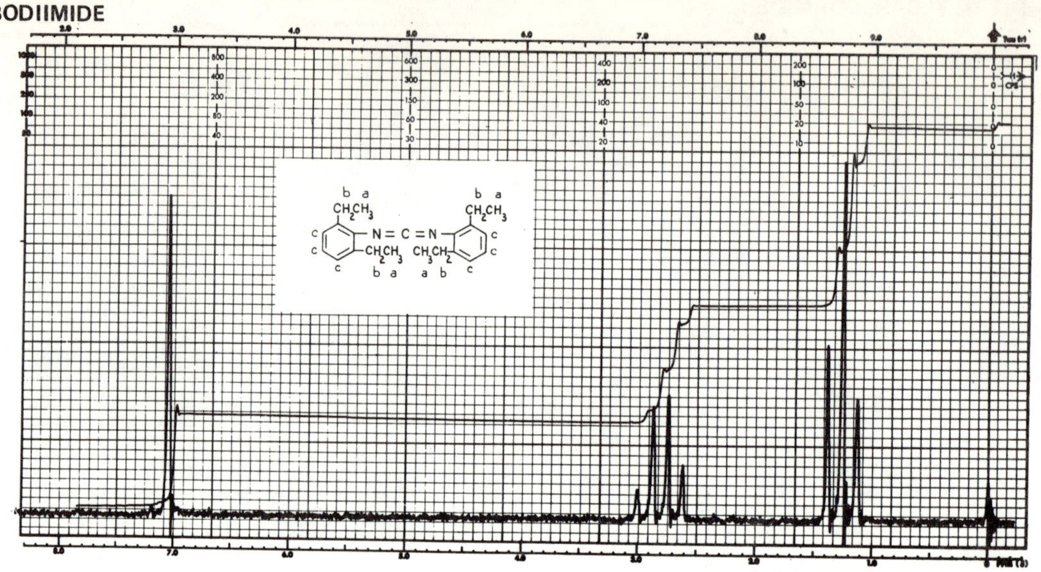

929

© 1978 Sadtler Research Laboratories, Inc.

THE ISOTHIOCYANATES

The esters of Isothiocyanic Acid (R–N=C=S) possess chemical shifts similar to those of the other –N=C= groups. The protons of adjacent aliphatic groups resonate in the range from 3.3 to 4.6 ppm.

The –N=C=S group exerts a weakly deshielding effect on the ortho aromatic protons.

Chemical Shifts

Aliphatic Protons

HC–4	HC–3	HC–2	HC–1		Solvent	Spectrum
			3.30	CH_3–N=C=S	$CDCl_3$	930
		1.37	3.55	CH_3–CH_2–N=C=S	$CDCl_3$	----
0.99	(1.2–2.1)		3.55	CH_3–$(CH_2)_2$–CH_2–N=C=S	CCl_4	931
		(1.45)	----	$(CH_3)_3$–C–N=C=S	CCl_4	932

Phenethylisothiocyanate

S=C=N	–CH_2	–CH_2	–X	Solvent	Spectrum
	3.48	2.79	–Ø	CCl_4	

Substituted Methylisothiocyanates

S=C=N–	CH_2	–X	Solvent	Spectrum
	3.48	–Cyclohexyl	$CDCl_3$	----
	4.15	–CH=CH_2	CCl_4	936
	4.59	–Ø	$CDCl_3$	937

Aromatic Protons

Phenyl Isothiocyanate

S=C=N–	ortho, meta, para	Solvent	Spectrum
	6.95–7.50	CCl_4	938

Para substituted aromatics

S=C=N–	ortho	meta	para	Solvent	Spectrum
	7.15	6.98	–F	CCl_4	941
	7.10	7.25	–Cl	$CDCl_3$	----
	7.08	7.45	–Br	$CDCl_3$	----
	7.25	8.02	–C(=O)–O–R_2	$CDCl_3$	----
	7.38	8.28	–NO_2	$CDCl_3$	----

ISOTHIOCYANIC ACID, METHYL ESTER

C_2H_3NS

Mol. Wt. 73.12

Solvent: $CDCl_3$

Assignments
a = 3.30

930

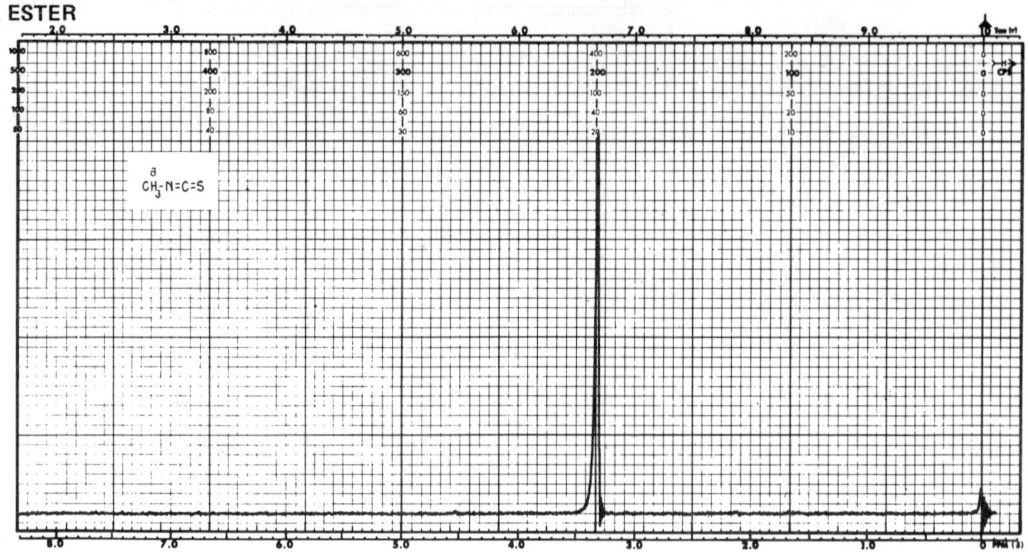

© 1978 Sadtler Research Laboratories, Inc.

ISOTHIOCYANIC ACID, BUTYL ESTER

C_5H_9NS

Mol. Wt. 115.20

B.P. 64-66°C/12mm

Solvent: CCl_4

Assignments
a = 0.99
b = 1.20-2.10
c = 3.55

931

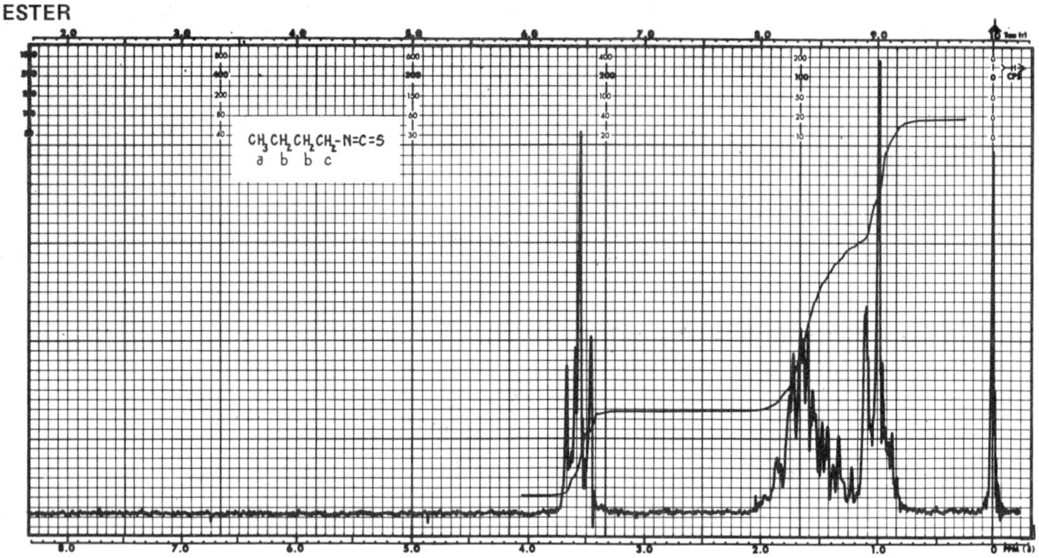

© 1978 Sadtler Research Laboratories, Inc.

---

tert-BUTYL ISOTHIOCYANATE

C_5H_9NS

Mol. Wt. 115.20

M.P. 8-11°C

Solvent: CCl_4

Assignments
a = 1.45

932

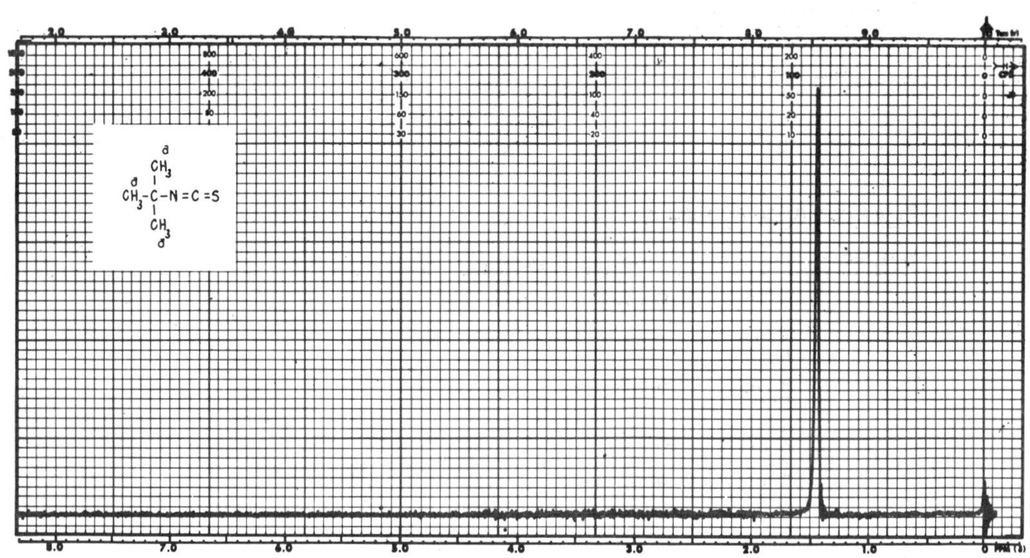

© 1978 Sadtler Research Laboratories, Inc.

ISOTHIOCYANIC ACID, HEPTYL ESTER

$C_8H_{15}NS$

Mol. Wt. 157.28

B.P. 107-109°C/10mm

Solvent: CCl_4

Assignments
a = 0.90
b = 1.10-1.60
c = 1.74
d = 3.51

933

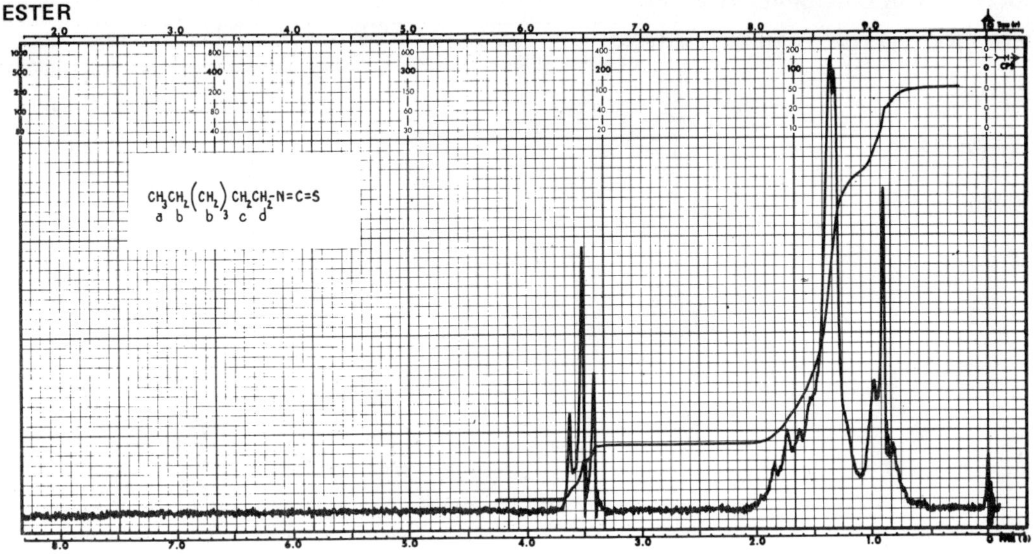

© 1978 Sadtler Research Laboratories, Inc.

ISOTHIOCYANIC ACID, OCTADECYL ESTER

$C_{19}H_{37}NS$

Mol. Wt. 311.58

M.P. 24-28°C

Solvent: $CDCl_3$

Assignments
a = 0.90
b = 1.10-1.90
c = 3.51

934

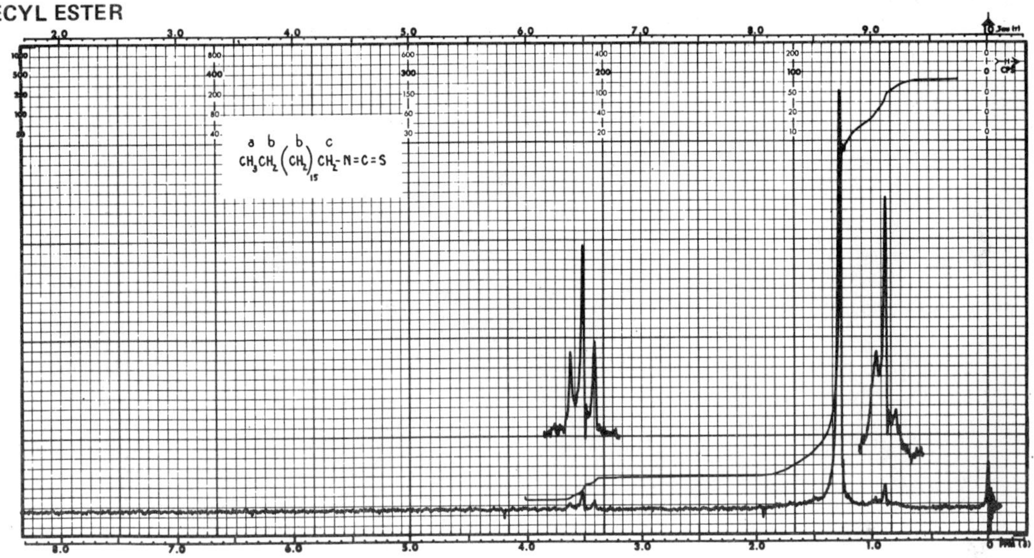

© 1978 Sadtler Research Laboratories, Inc.

ISOTHIOCYANIC ACID, 1-ADAMANTYL ESTER

$C_{11}H_{15}NS$

Mol. Wt. 193.31

M.P. 166.5-168°C

Solvent: $CDCl_3$

Assignments
a = 1.69
b = 2.02

935

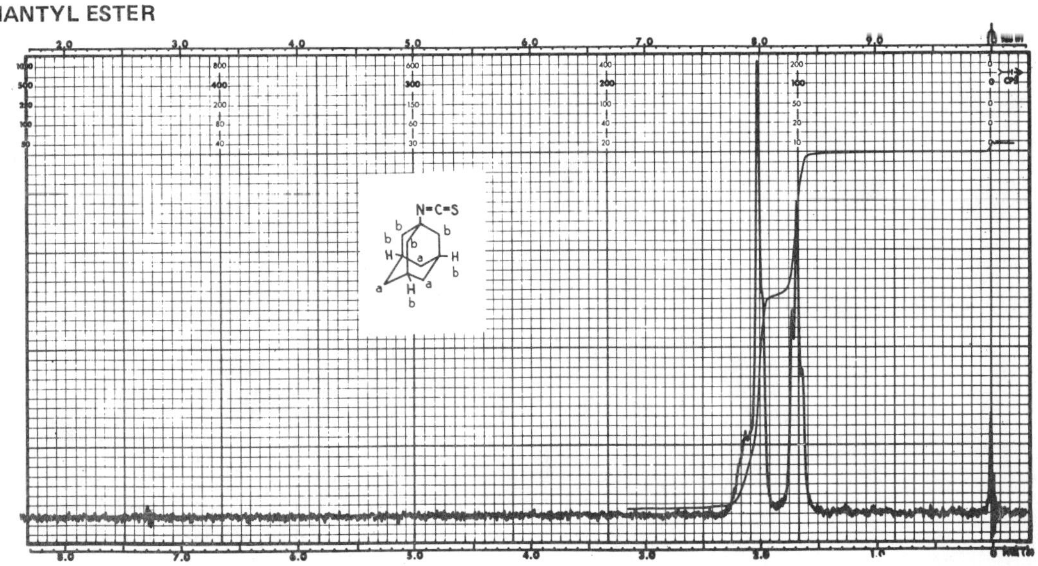

© 1978 Sadtler Research Laboratories, Inc.

ISOTHIOCYANIC ACID, ALLYL ESTER

C_4H_5NS

Mol. Wt. 99.16

B.P. 150-152°C

Solvent: CCl_4

Assignments
a = 4.15
b = 5.27
c = 5.35
d = ca 5.80
e = imp 3.51, 3.61

936

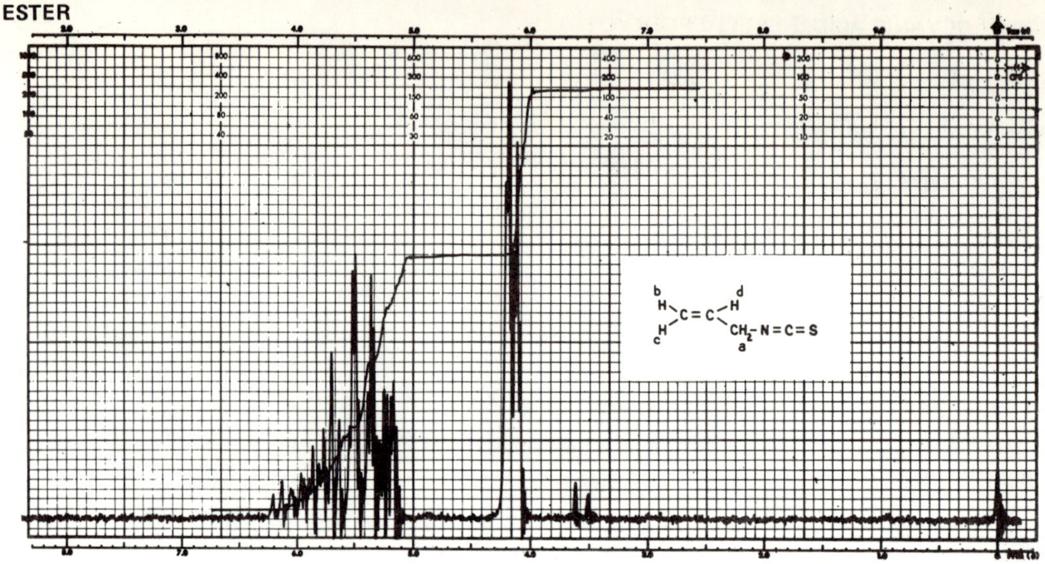

© 1978 Sadtler Research Laboratories, Inc.

ISOTHIOCYANIC ACID, BENZYL ESTER

C_8H_7NS

Mol. Wt. 149.22

B.P. 114-116°C/5mm

Solvent: $CDCl_3$

Assignments
a = 4.59
b = ca 7.29

937

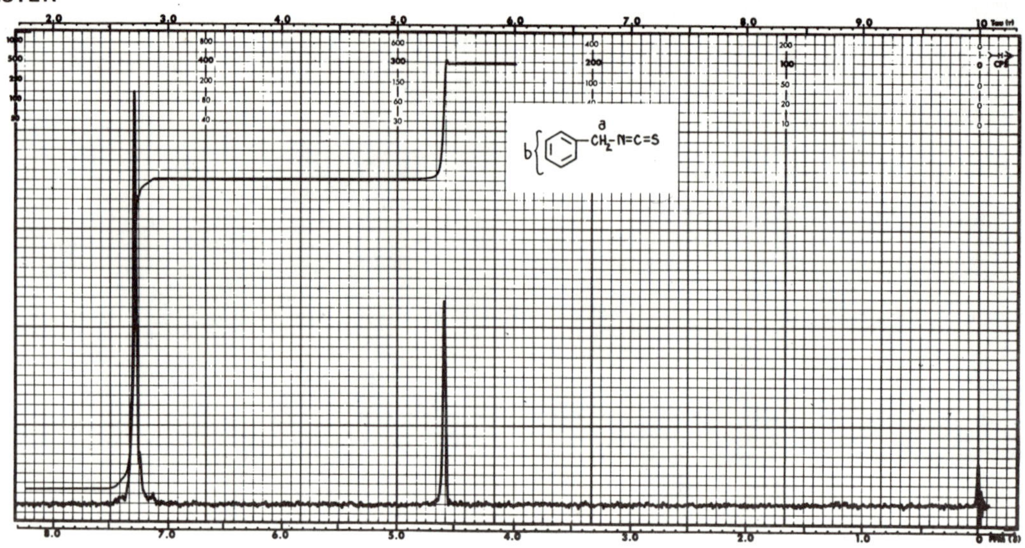

© 1978 Sadtler Research Laboratories, Inc.

ISOTHIOCYANIC ACID, PHENYL ESTER

C_7H_5NS

Mol. Wt. 135.19

B.P. 89-90°C/10mm

Solvent: CCl_4

Assignments
a = ca 7.20

938

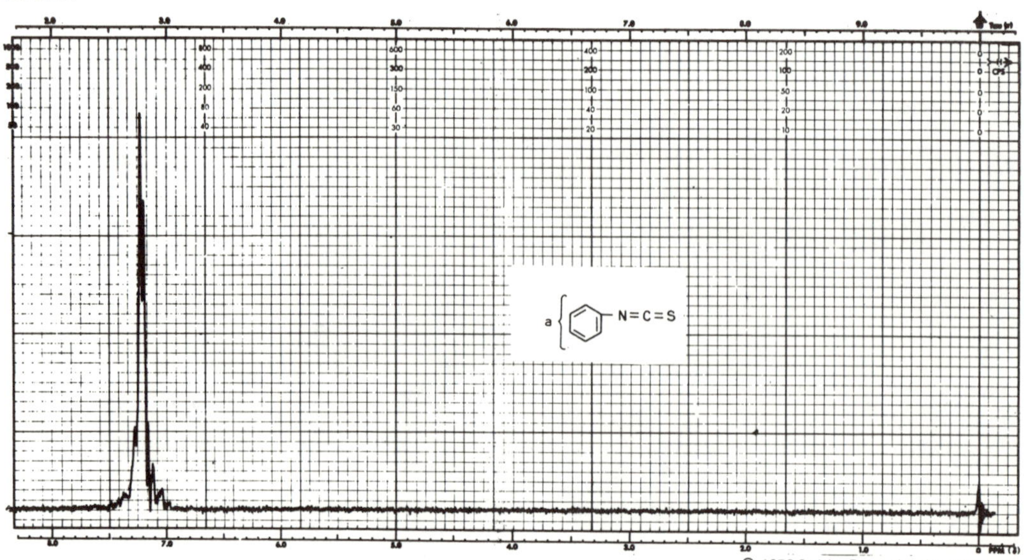

© 1978 Sadtler Research Laboratories, Inc.

ISOTHIOCYANIC ACID, *o*-FLUOROPHENYL ESTER

C_7H_4FNS

Mol. Wt. 153.18

Solvent: CCl_4

Assignments
a = 6.85-7.35

939

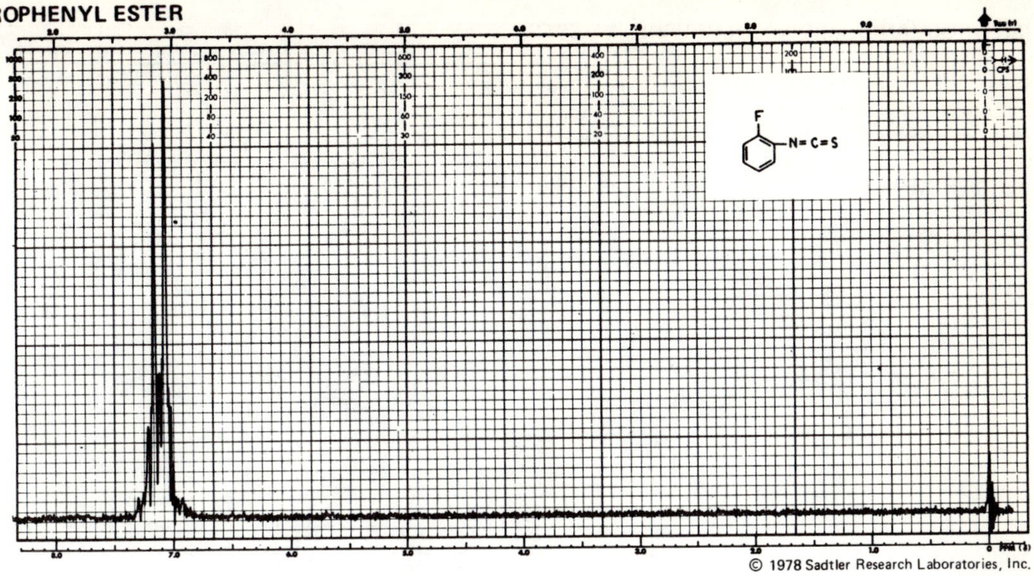

© 1978 Sadtler Research Laboratories, Inc.

ISOTHIOCYANIC ACID, m-FLUOROPHENYL ESTER

C_7H_4FNS

Mol. Wt. 153.18

Solvent: CCl_4

Assignments
a = 6.70-7.55

940

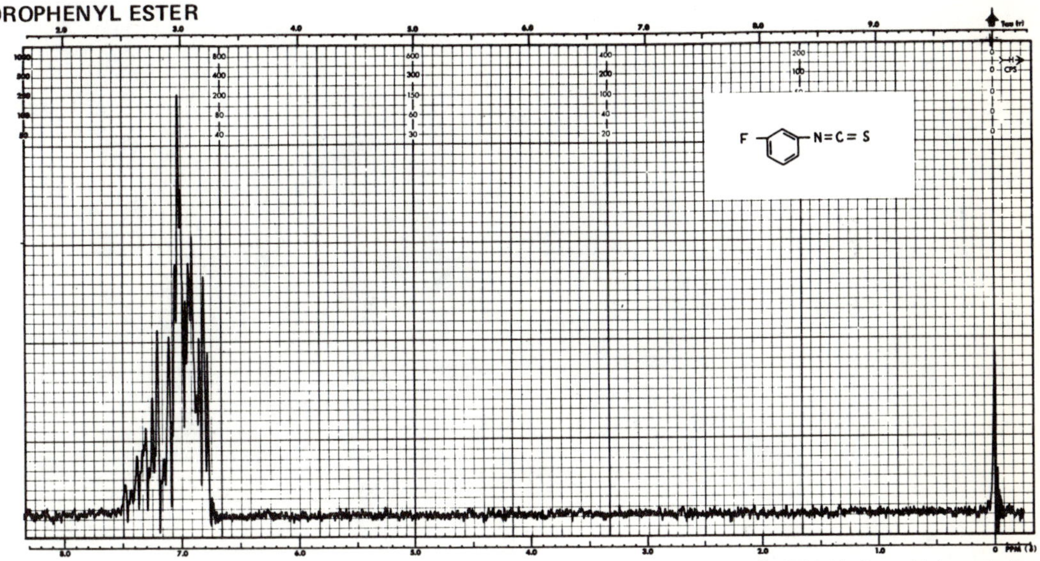

© 1978 Sadtler Research Laboratories, Inc.

ISOTHIOCYANIC ACID, p-FLUOROPHENYL ESTER

C_7H_4FNS

Mol. Wt. 153.18

M.P. 23.5-25.5°C

B.P. 60°C/1.3mm

Solvent: CCl_4

Assignments
a = 6.90
b = 7.15

941

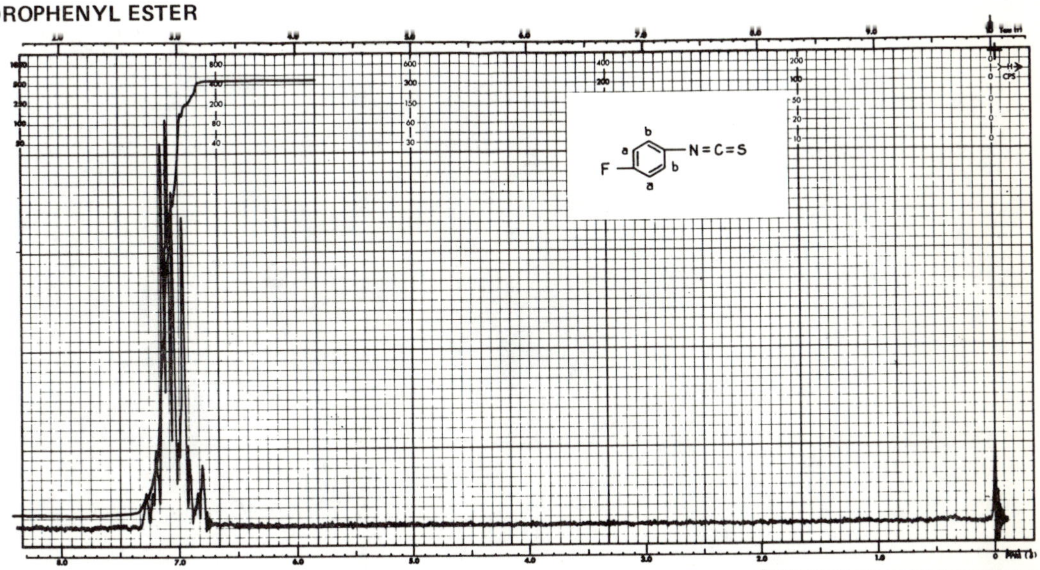

© 1978 Sadtler Research Laboratories, Inc.

ISOTHIOCYANIC ACID, o-CHLOROPHENYL ESTER

C_7H_4ClNS

Mol. Wt. 169.63

B.P. 90°C/1mm

Solvent: CCl_4

Assignments
a = ca 7.21

942

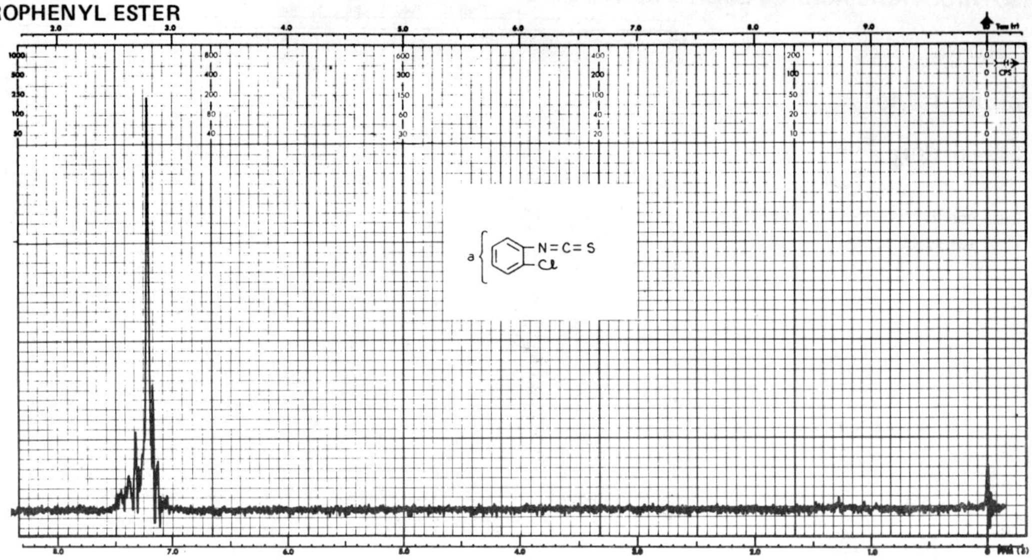

© 1978 Sadtler Research Laboratories, Inc.

A large amount of chemical shift data is available for this commercially important group of compounds. The Nitrile group is a weak deshielder of aliphatic and aromatic protons, similar to several other unsaturated carbon-carbon and carbon-nitrogen functional groups. Fortunately, the Nitrile group is easily characterized by its infrared absorption band allowing the NMR analyst to concentrate his energies on the proton groups in the molecules which are less easily defined by the infrared data.

Chemical Shifts

Aliphatic Protons

HC–4	HC–3	HC–2	HC–1		Solvent	Spectrum
			1.94	$CH_3–C≡N$	CCl_4	943
		1.25	2.34	$CH_3–CH_2–C≡N$	CCl_4	----
	1.07	1.67	2.27	$CH_3–CH_2–CH_2–C≡N$	CCl_4	944
		(1.30)	2.69	$(CH_3)_2–CH–C≡N$	CCl_4	945
0.97	(1.2–1.9)		2.30	$CH_3–(CH_2)_2–CH_2–C≡N$	CCl_4	946
	(1.08)	1.97	2.21	$(CH_3)_2–CH–CH_2–C≡N$	CCl_4	947
		(1.39)	----	$(CH_3)_3–C–C≡N$	CCl_4	948

2-Substituted Nitriloethanes

N≡C	–CH_2	–CH_2	–X	
	2.27	1.67	$–CH_3$	CCl_4
	2.62	2.62	$–C(=O)–O–CH_3$	CCl_4
	2.36	2.73	–Ø	$CDCl_3$
	2.78	2.78	–C≡N	$CDCl_3$
	2.55	2.96	$–NH–(CH_2)_2–C≡N$	$CDCl_3$
	2.49	2.99	$–NH_2$	$CDCl_3$
	2.83	3.20	$–NH_2$(TFA salt)	D_2O
	2.97	3.53	–Br	CCl_4
	2.80	3.70	–Cl	CCl_4
	2.67	3.73	$–O–(CH_2)_2–C≡N$	$CDCl_3$
	2.61	3.85	–OH	$CDCl_3$

Substituted Nitrilomethanes

N≡C	–CH_2	–X	
	2.34	$–CH_3$	CCl_4
	2.48	$–C(CH_3)_2–Ø$	CCl_4
	3.10	$–CH=CH_2$	CCl_4
	3.50	$–C(=O)–O–CH_3$	$CDCl_3$

Chemical Shifts continued

Aliphatic Protons continued

Substituted Nitrilomethanes continued

N≡C	$-CH_2$	$-X$	Solvent
	3.62	$-Ø$	CCl_4
	3.79	$-C(=O)-OH$	DMSO-d_6
	4.21	$-C≡N$	DMSO-d_6

Olefinic Protons

Although the Nitrile group deshields all of the vinyl protons, it has an abnormally strong deshielding effect upon the cis olefinic hydrogen. The chemical shifts of three olefinic nitriles are provided below.

Acrylonitrile

$H_2C=CH-C≡N$

cis	trans	geminal		Solvent	Spectrum
6.20	5.95	5.60	$-C≡N$	CCl_4	966

Methacrilonitrile

$H_2C=C(CH_3)-C≡N$

				Solvent	Spectrum
ca 5.79	1.98		$-C≡N$	CCl_4	967

Cinnamonitrile

$Ø-CH=CH-C≡N$

				Solvent	Spectrum
7.36	5.84		$-C≡N$	CCl_4	982

Aromatic Protons

Benzonitrile

N≡C–	ortho, meta, para		
	7.20–7.75	CCl_4	783

Para substituted Benzonitriles

N≡C–	ortho	meta	para	
	7.35	6.58	$-N(R_2)_2$	$CDCl_3$
	7.66	7.17	$-F$	$CDCl_3$
	7.47	7.23	$-CH_3$	CCl_4
	7.57	7.39	$-Cl$	$CDCl_3$
	7.90	7.90	$-Ø(4-C≡N)$	TFA

ACETONITRILE

C_2H_3N

Mol. Wt. 41.05

B.P. 81.6-82.0°C

Solvent: CCl_4

Assignments
a = 1.94

943

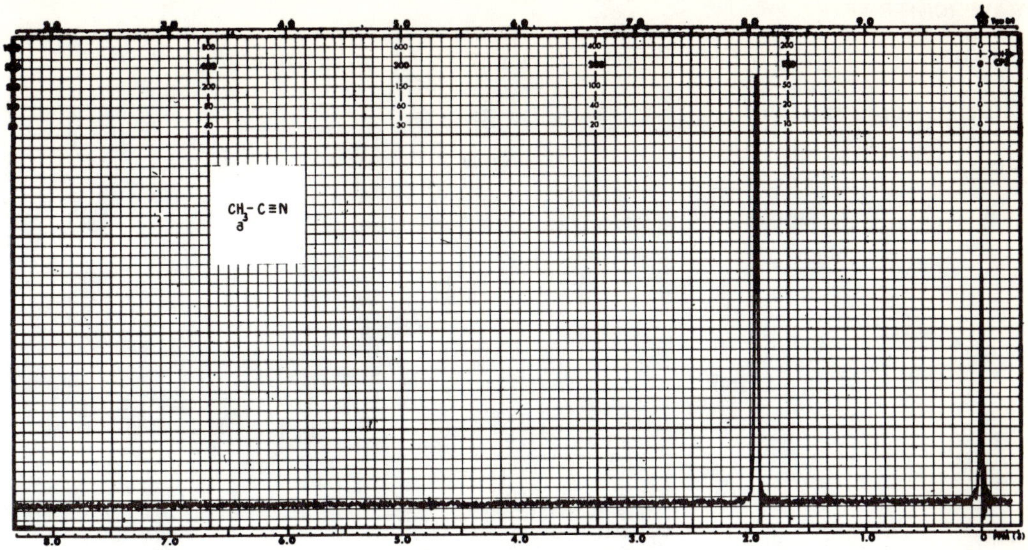

© 1978 Sadtler Research Laboratories, Inc.

BUTYRONITRILE

C_4H_7N

Mol. Wt. 69.10

B.P. 175.5°C

Solvent: CCl_4

Assignments
a = 1.07
b = 1.67
c = 2.27

944

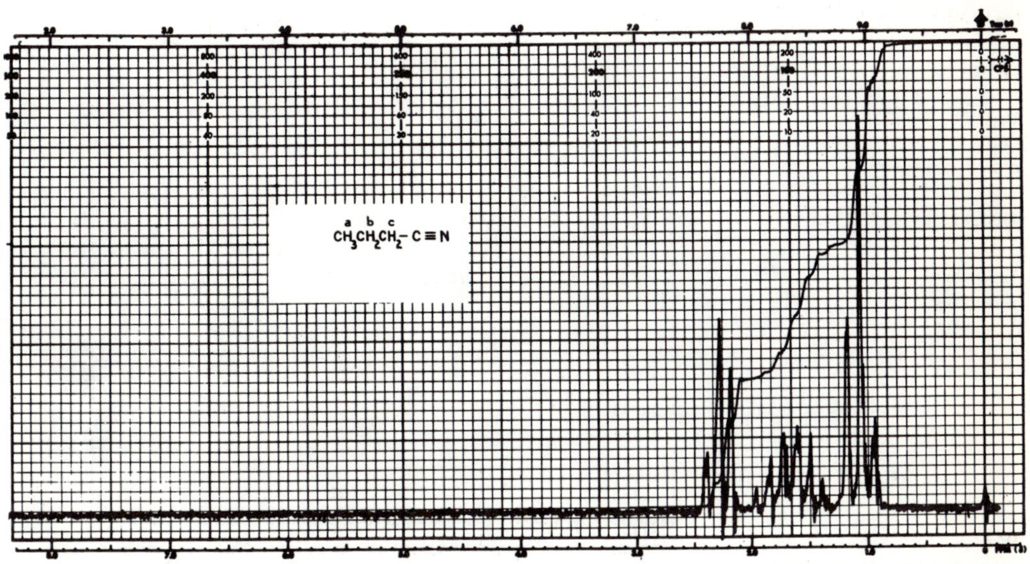

© 1978 Sadtler Research Laboratories, Inc.

ISOBUTYRONITRILE

C_4H_7N

Mol. Wt. 69.11

B.P. 102-104°C

Solvent: CCl_4

Assignments
a = 1.30
b = 2.69

945

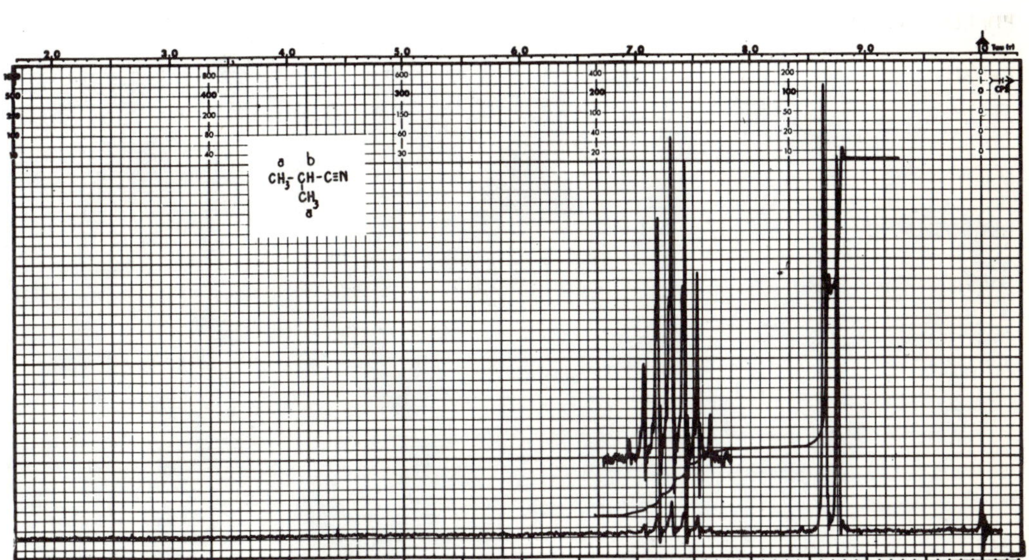

© 1978 Sadtler Research Laboratories, Inc.

VALERONITRILE

C_5H_9N

Mol. Wt. 83.13

Solvent: CCl_4

Assignments
a = 0.97
b = 1.59
c = 2.30

946

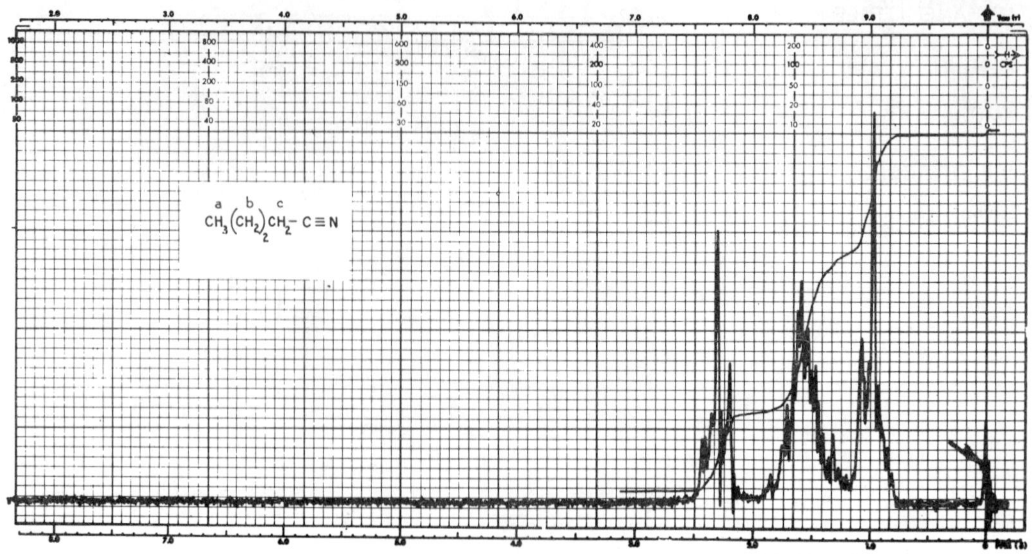

© 1978 Sadtler Research Laboratories, Inc.

ISOVALERONITRILE

C_5H_9N

Mol. Wt. 83.13

B.P. 127-128°C

Solvent: CCl_4

Assignments
a = 1.08
b = 1.97
c = 2.21

947

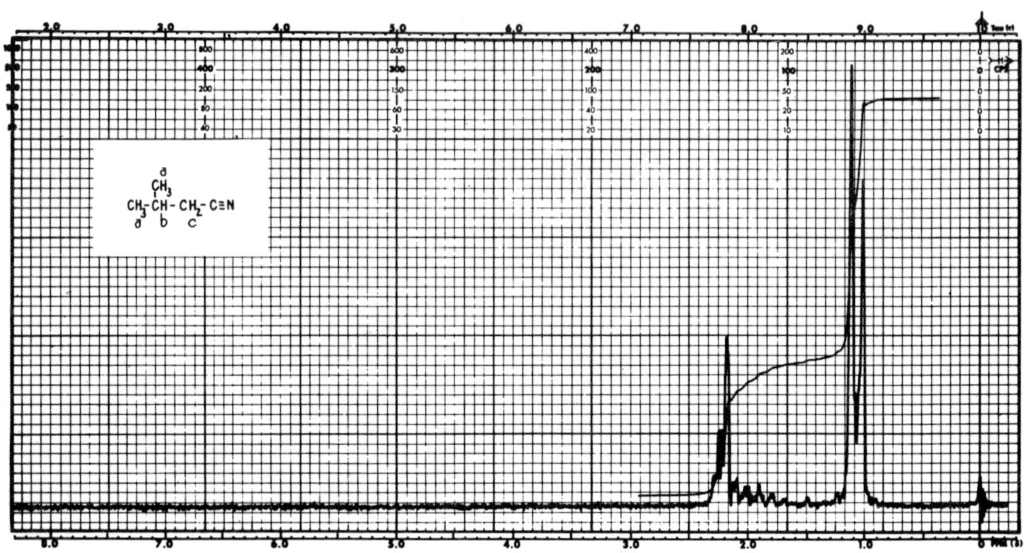

© 1978 Sadtler Research Laboratories, Inc.

PIVALONITRILE

C_5H_9N

Mol. Wt. 83.13

M.P. 15-16°C

B.P. 105-106°C

Solvent: CCl_4

Assignments
a = 1.39

948

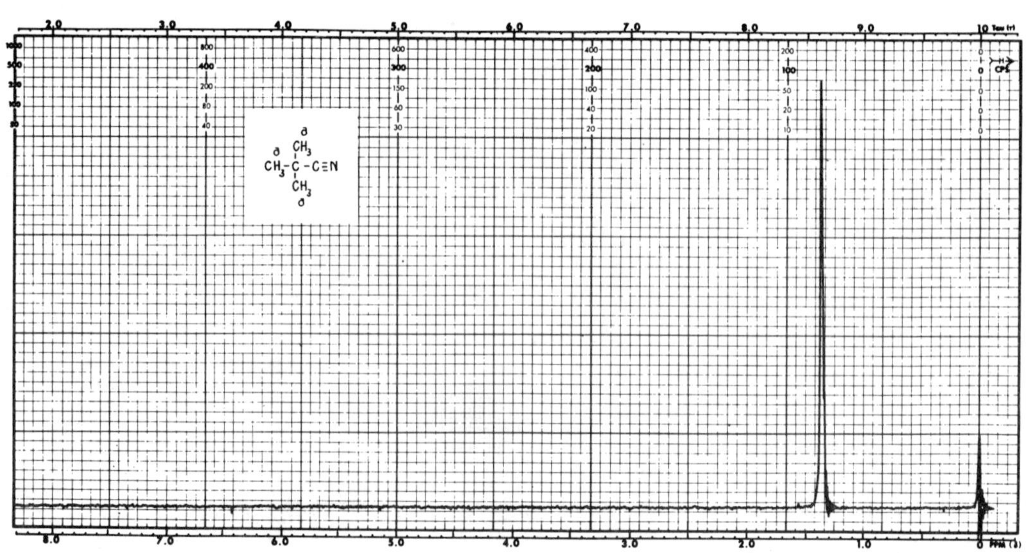

© 1978 Sadtler Research Laboratories, Inc.

HEXANENITRILE

$C_6H_{11}N$

Mol. Wt. 97.16

B.P. 161-164°C

Solvent: CCl_4

Assignments
a = 0.95
b = 1.10-1.60
c = 1.65
c = 2.30

949

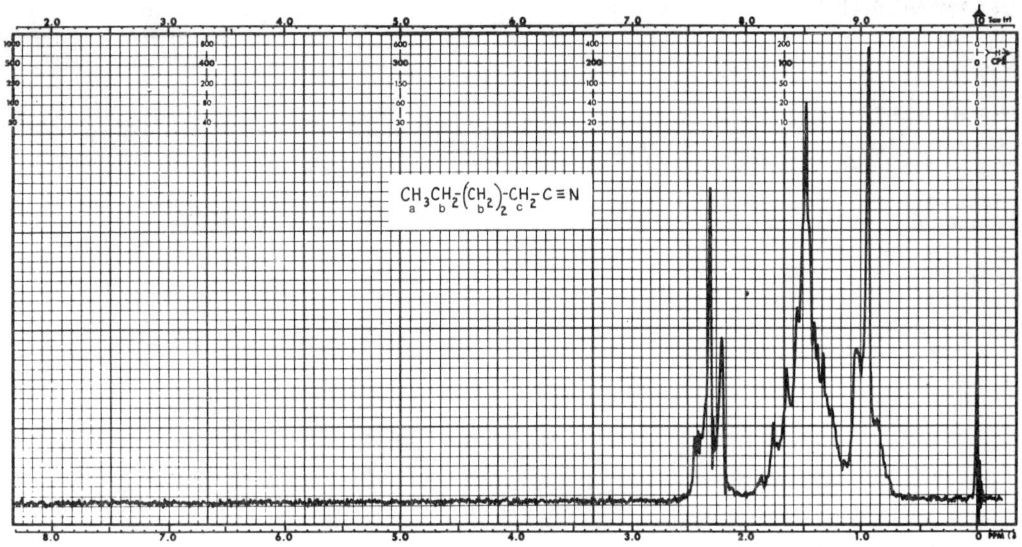

© 1978 Sadtler Research Laboratories, Inc.

HEPTANENITRILE

$C_7H_{13}N$

Mol. Wt. 111.19

B.P. 182-184°C

Solvent: CCl_4

Assignments
a = 0.92
b = 1.10-2.00
c = 2.29

950

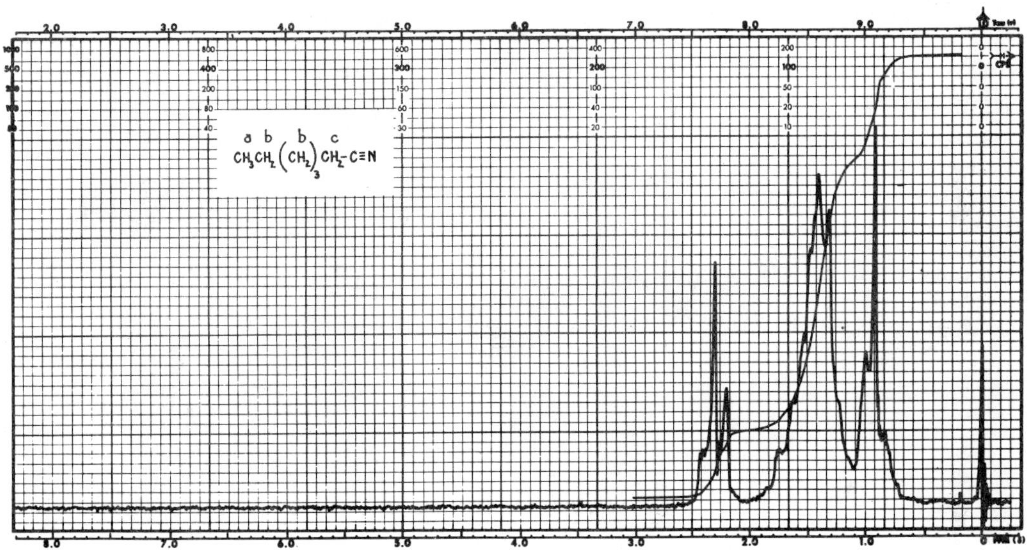

© 1978 Sadtler Research Laboratories, Inc.

OCTANENITRILE

$C_8H_{15}N$

Mol. Wt. 125.22

M.P. –45 to –46°C

Solvent: CCl_4

Assignments
a = 0.90
b = 1.10-1.90
c = 2.31

951

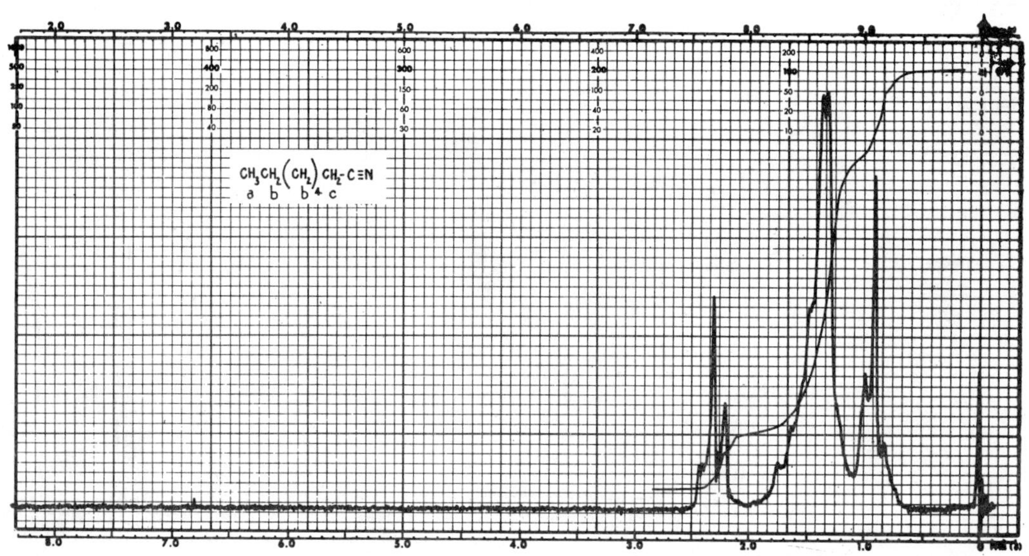

© 1978 Sadtler Research Laboratories, Inc.

DECANENITRILE

$C_{10}H_{19}N$

Mol. Wt. 153.27

M.P. −14 to −15°C

Solvent: CCl_4

Assignments
a = 0.90
b = 1.00-1.90
c = 2.29
d = 4.62 imp.

952

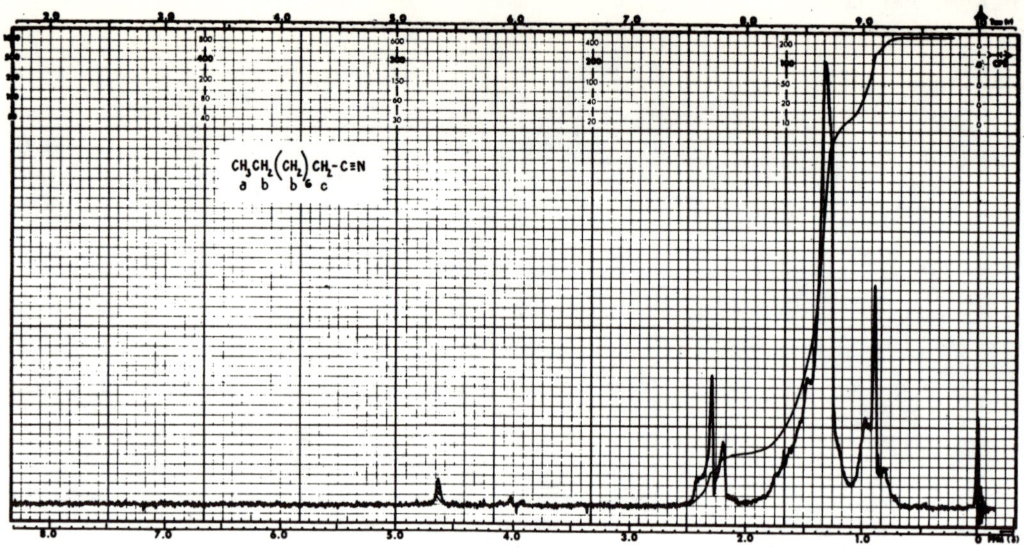

© 1978 Sadtler Research Laboratories, Inc.

PALMITONITRILE

$C_{16}H_{31}N$

Mol. Wt. 237.43

Solvent: $CDCl_3$

Assignments
a = 0.89
b = 1.10-1.50
c = 1.55
d = 2.31

953

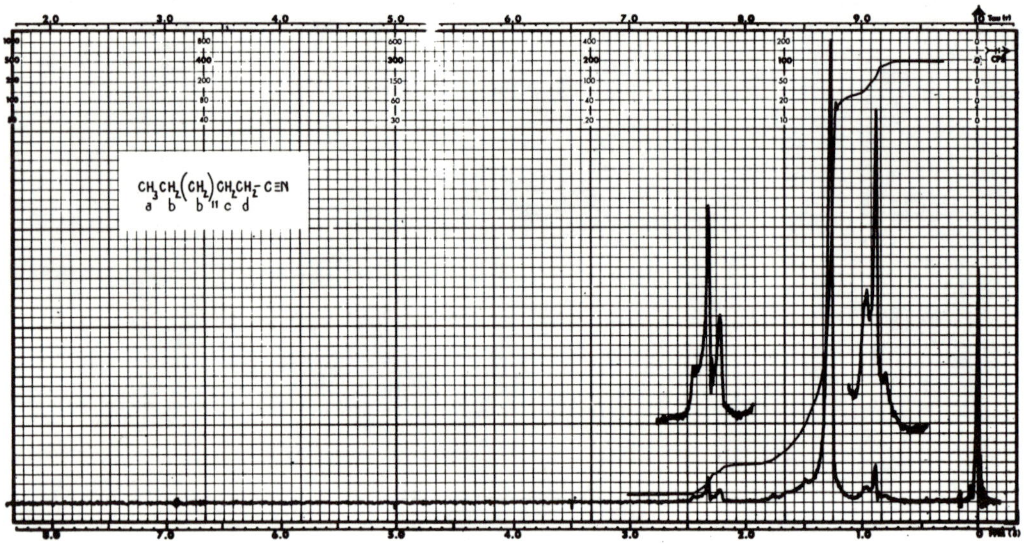

© 1978 Sadtler Research Laboratories, Inc.

3-CHLOROPROPIONITRILE

C_3H_4ClN

Mol. Wt. 89.52

B.P. 174-176°C

Solvent: CCl_4

Assignments
a = 2.80
b = 3.70

954

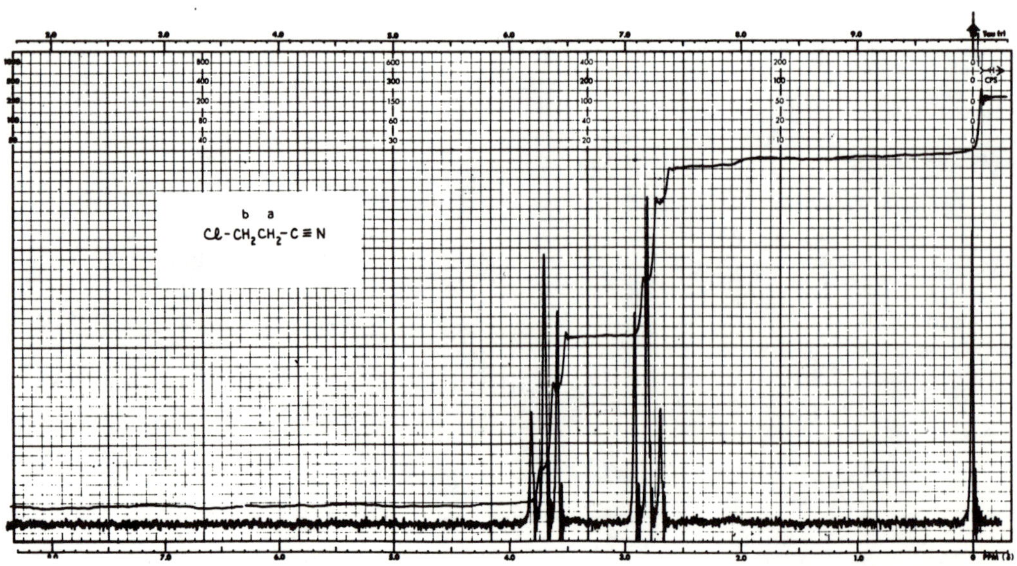

© 1978 Sadtler Research Laboratories, Inc.

4-CHLOROBUTYRONITRILE

C_4H_6ClN

Mol. Wt. 103.55

B.P. 74-77°C/10mm

Solvent: CCl_4

Assignments
a = 2.09
b = 2.51
c = 3.60

955

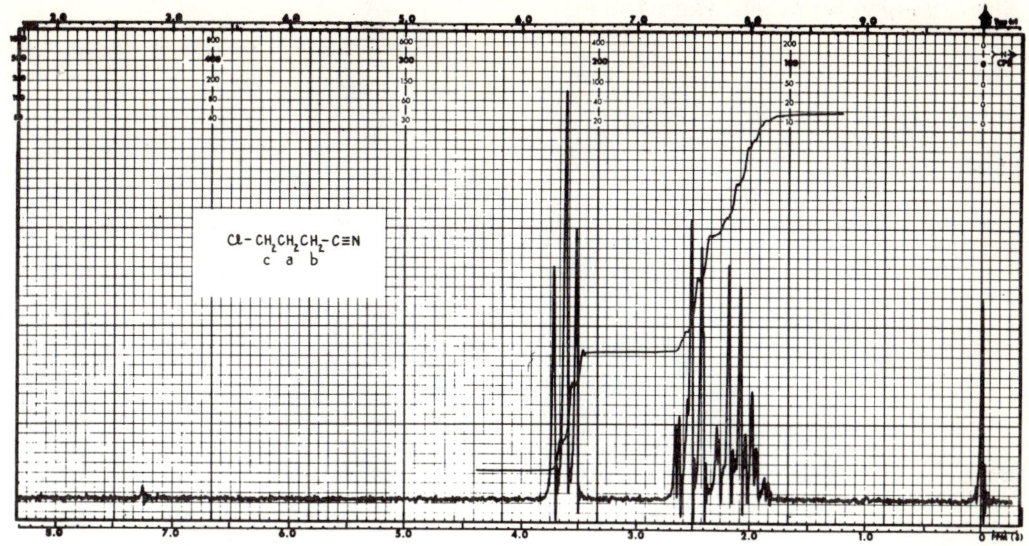

© 1978 Sadtler Research Laboratories, Inc.

3-BROMOPROPIONITRILE

C_3H_4BrN

Mol. Wt. 133.99

B.P. 91-93°C/25mm

Solvent: CCl_4

Assignments
a = 2.97
b = 3.53

956

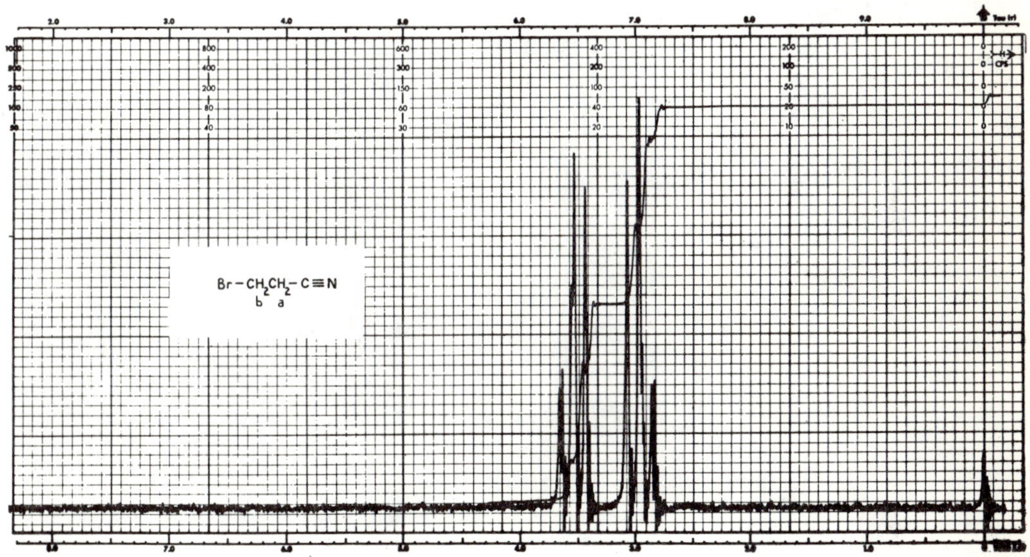

© 1978 Sadtler Research Laboratories, Inc.

7-BROMOHEPTANONITRILE

$C_7H_{12}BrN$

Mol. Wt. 190.10

Solvent: CCl_4

Assignments
a = 1.20-2.10
b = 2.33
c = 3.40

957

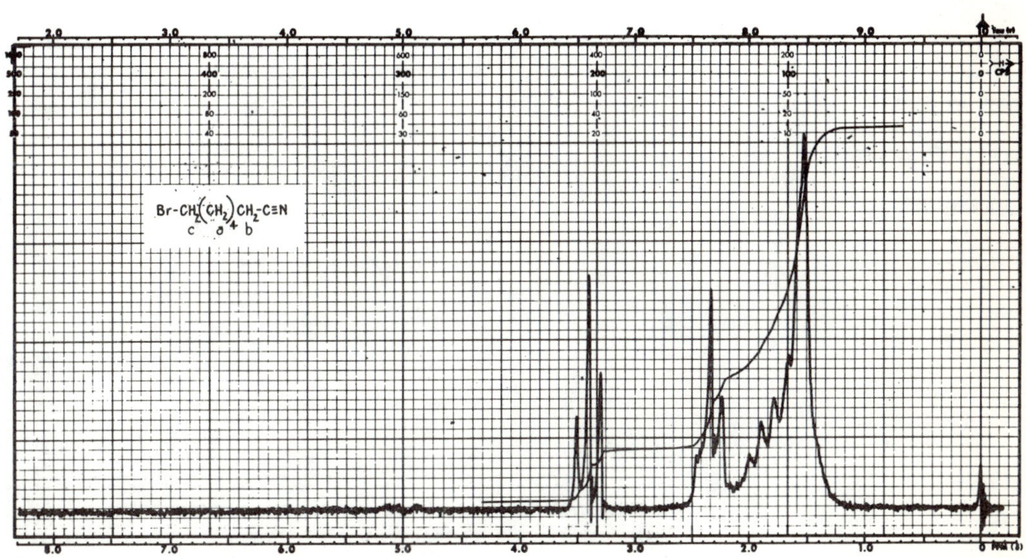

© 1978 Sadtler Research Laboratories, Inc.

4-(DIISOPROPYLAMINO)BUTYRONITRILE

$C_{10}H_{20}N_2$

Mol. Wt. 168.28

Solvent: CCl_4

Assignments
a = 1.00
b = 1.69
c = 2.30
d = 2.58
e = 3.00

958

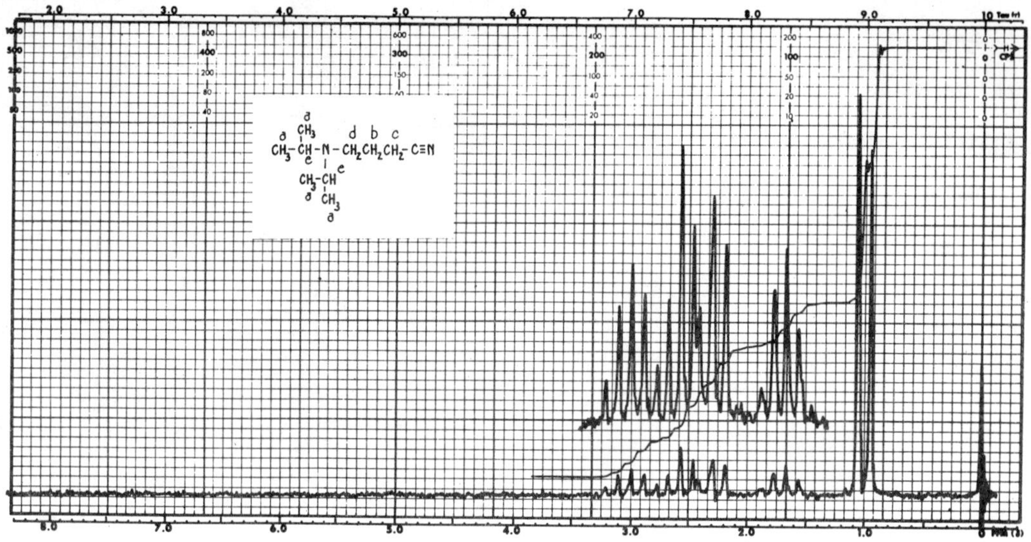

© 1978 Sadtler Research Laboratories, Inc.

4-(DIBUTYLAMINO)BUTYRONITRILE

$C_{12}H_{24}N_2$

Mol. Wt. 196.34

B.P. 91-94°C/0.8mm

Solvent: CCl_4

Assignments
a = 0.92
b = 1.12-2.08
c = 2.38
d = 2.48

959

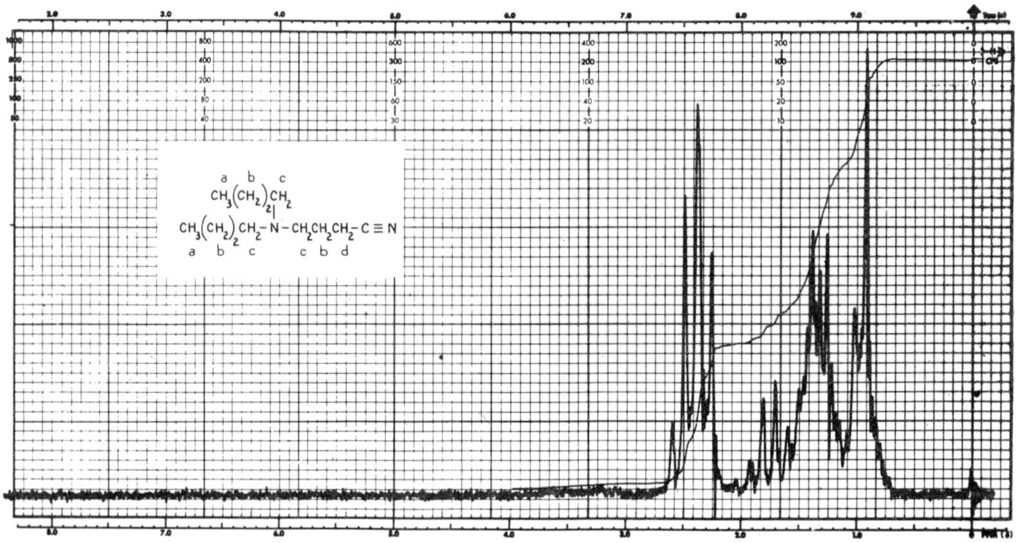

© 1978 Sadtler Research Laboratories, Inc.

3,3′-IMINODIPROPIONITRILE

$C_6H_9N_3$

Mol. Wt. 123.16

B.P. 173°C/10mm

Solvent: $CDCl_3$

Assignments
a = 1.80
b = 2.55
c = 2.96

960

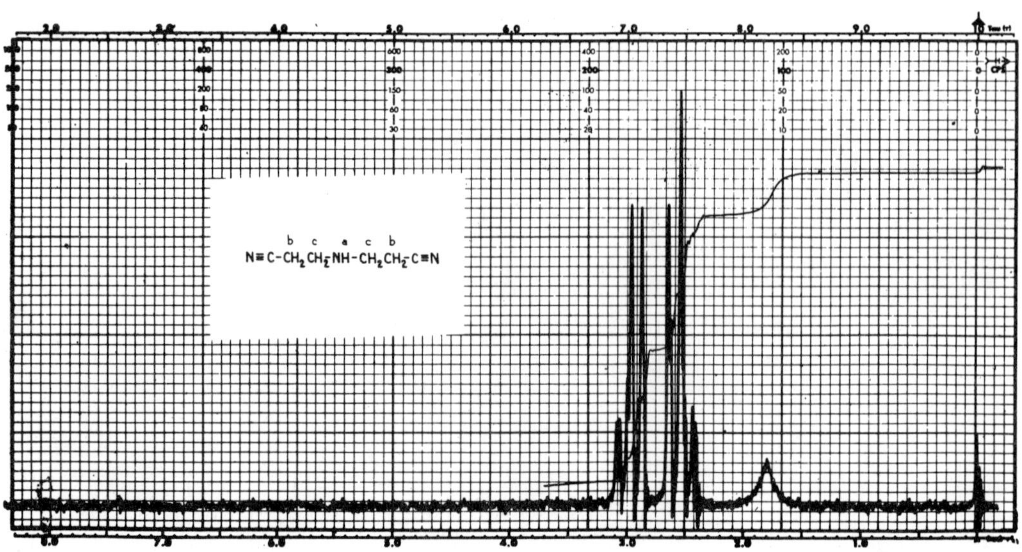

© 1978 Sadtler Research Laboratories, Inc.

MALONONITRILE

$C_3H_2N_2$

Mol. Wt. 66.06

Solvent: Dimethyl sulfoxide-d_6

Assignments
a = 4.21
b = 6.69, 6.81 imps.

961

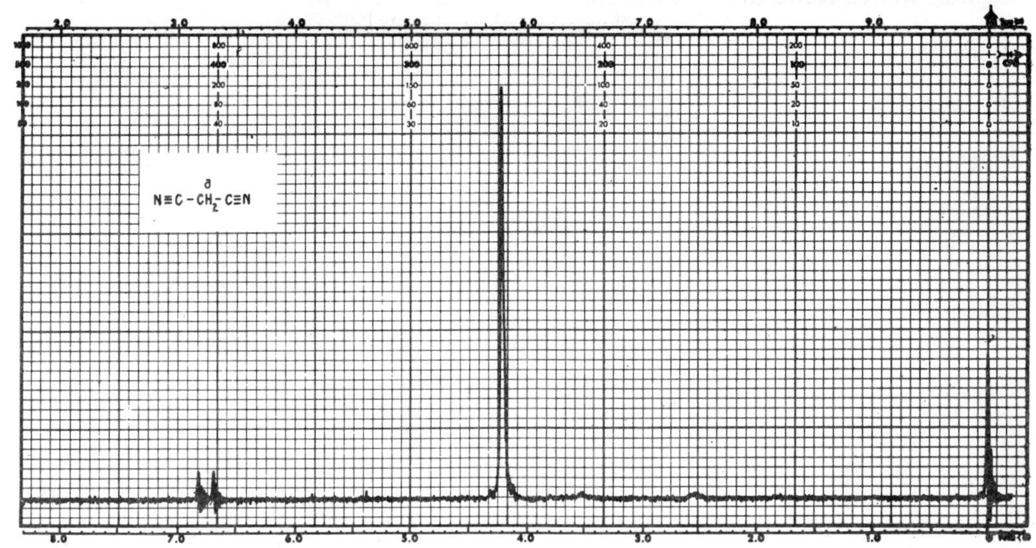

© 1978 Sadtler Research Laboratories, Inc.

SUCCINONITRILE

$C_4H_4N_2$

Mol. Wt. 80.09

M.P. 57-57.5°C

B.P. 265.7°C

Solvent: $CDCl_3$

Assignments
a = 2.78

962

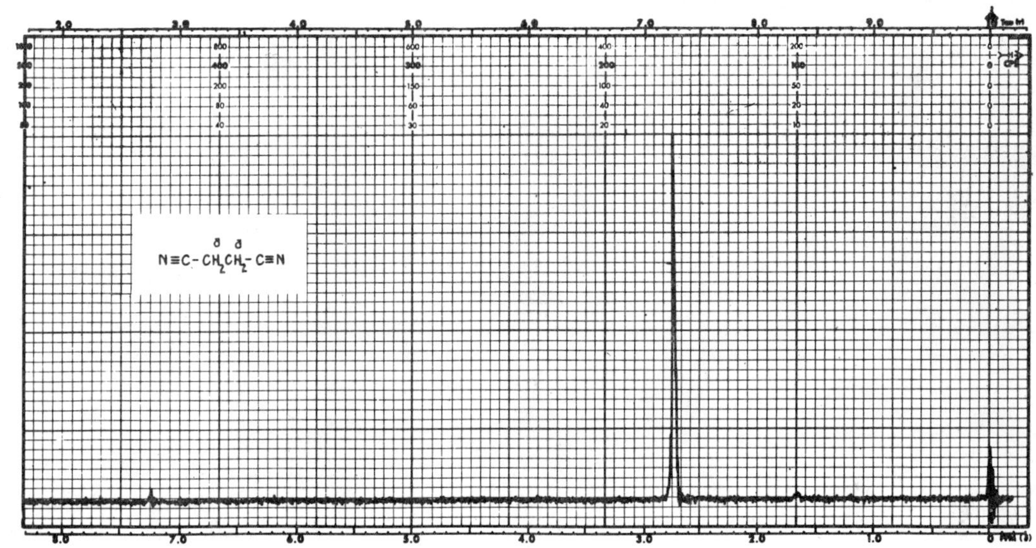

© 1978 Sadtler Research Laboratories, Inc.

PIMELONITRILE

$C_7H_{10}N_2$

Mol. Wt. 122.17

B.P. 169°C/13mm

Solvent: $CDCl_3$

Assignments
a = 1.69
b = 2.40

963

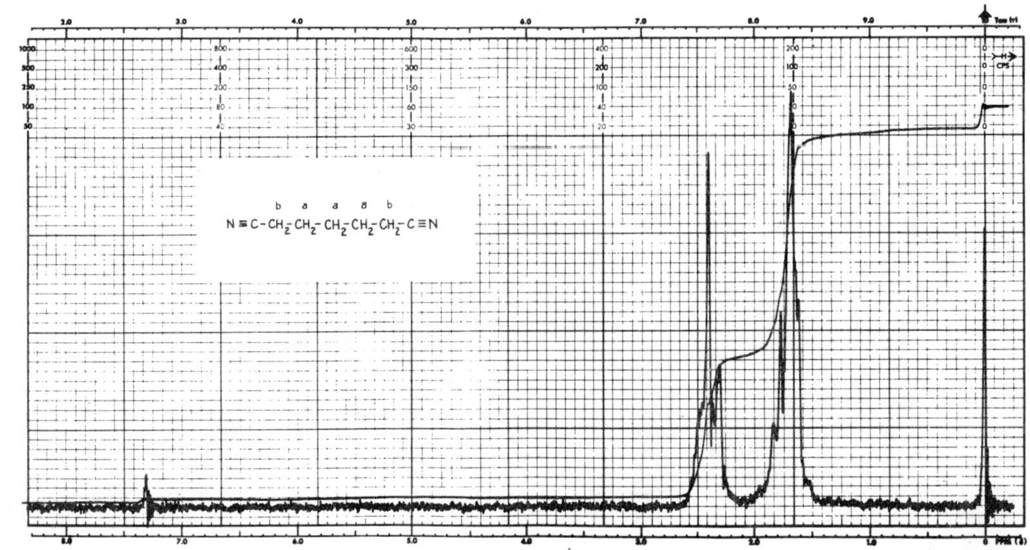

© 1978 Sadtler Research Laboratories, Inc.

CYCLOHEXANECARBONITRILE

$C_7H_{11}N$

Mol. Wt. 109.17

Solvent: CCl_4

Assignments
a = 1.10-2.10
b = 2.64

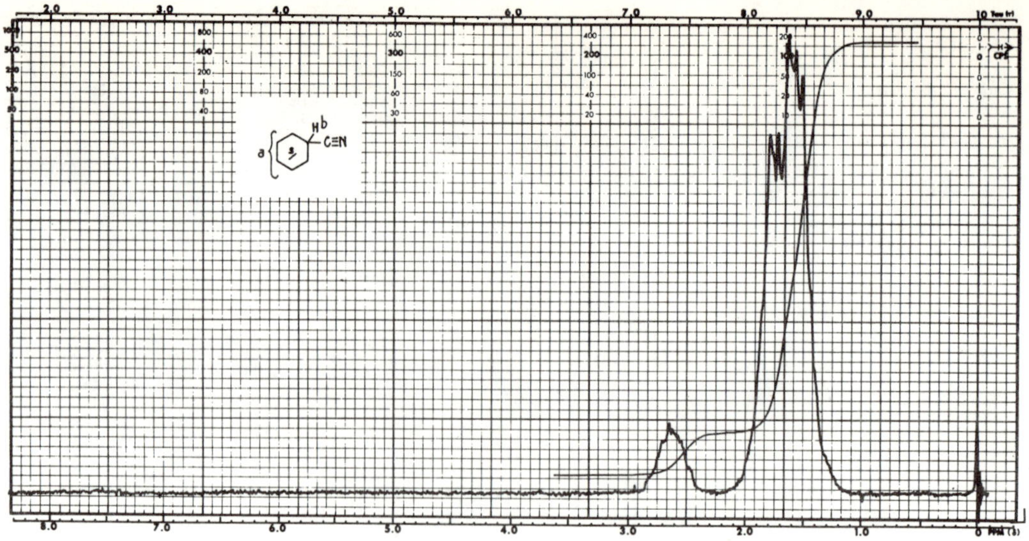

964

© 1978 Sadtler Research Laboratories, Inc.

trans-1,4-CYCLOHEXANEDICARBONITRILE

$C_8H_{10}N_2$

Mol. Wt. 134.18

M.P. 138-141°C

Solvent: $CDCl_3$

Assignments
a = 1.40-2.35
b = 2.71

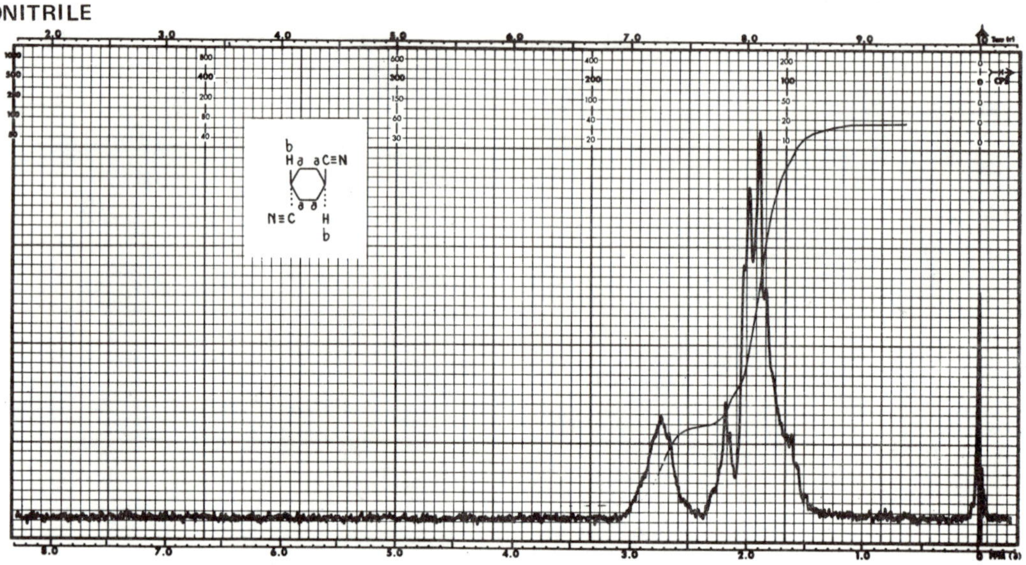

965

© 1978 Sadtler Research Laboratories, Inc.

ACRYLONITRILE

C_3H_3N

Mol. Wt. 53.06

Solvent: CCl_4

Assignments
a = 5.40-6.10
b = 6.22

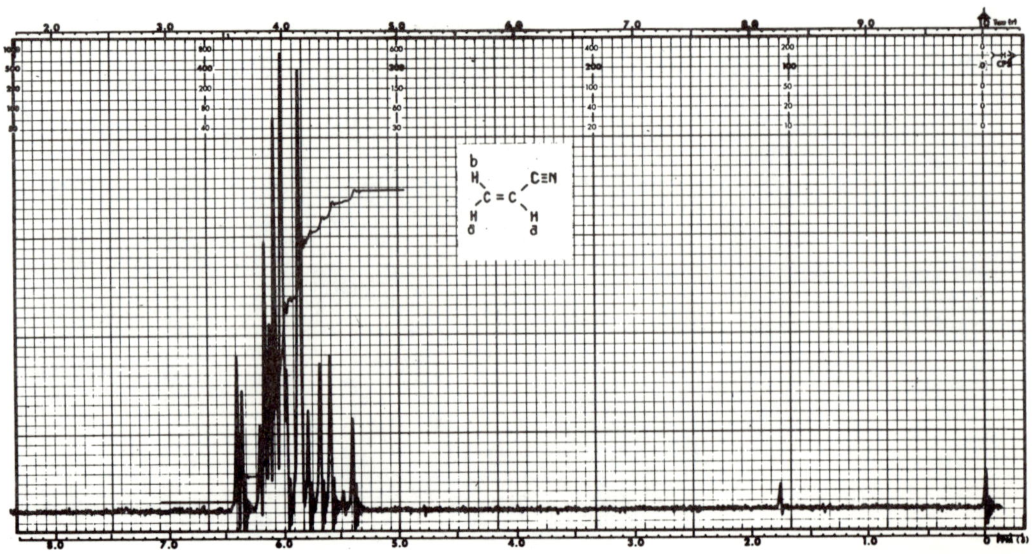

966

© 1978 Sadtler Research Laboratories, Inc.

METHACRYLONITRILE

C_4H_5N

Mol. Wt. 67.09

B.P. 90-91°C

Solvent: CCl_4

Assignments
a = 1.98
b = ca 5.79

967

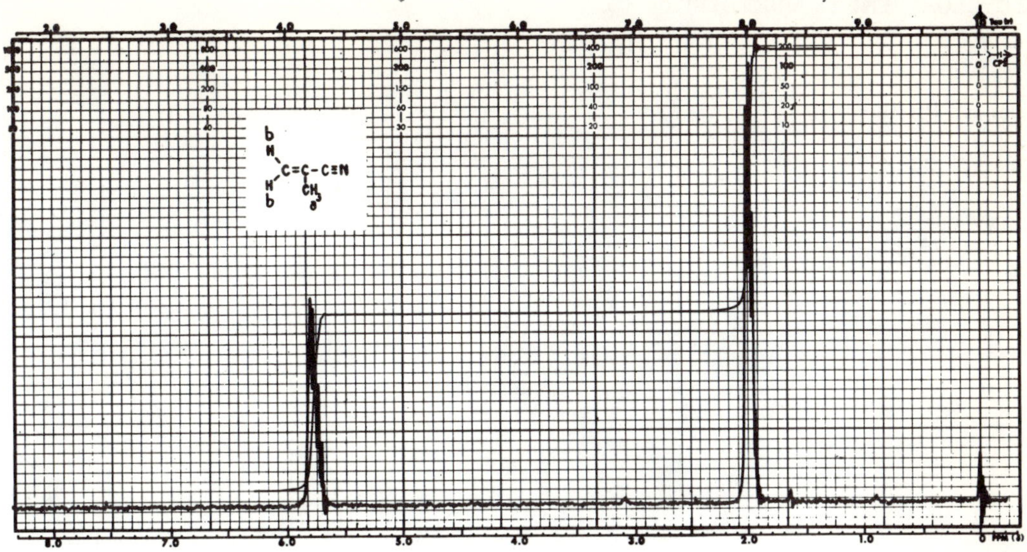

© 1978 Sadtler Research Laboratories, Inc.

3-BUTENENITRILE

C_4H_5N

Mol. Wt. 67.09

B.P. 117-119°C

Solvent: CCl_4

Assignments
a = 3.10
b = 5.10-5.65
c = 5.77

968

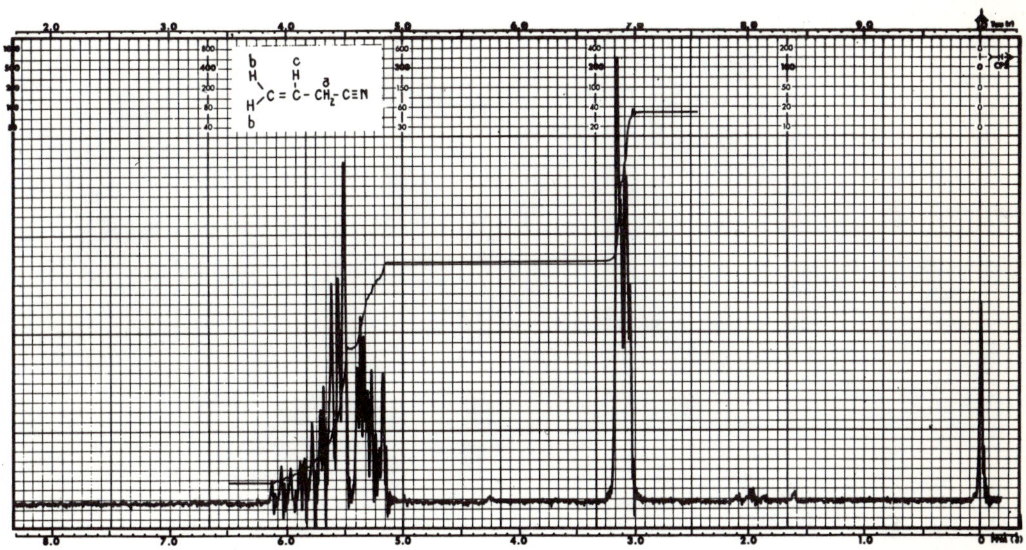

© 1978 Sadtler Research Laboratories, Inc.

(AMINOMETHYLENE)MALONONITRILE

$C_4H_3N_3$

Mol. Wt. 93.09

Solvent: DMSO-d_6

Assignments
a = 7.57
b = 7.63

969

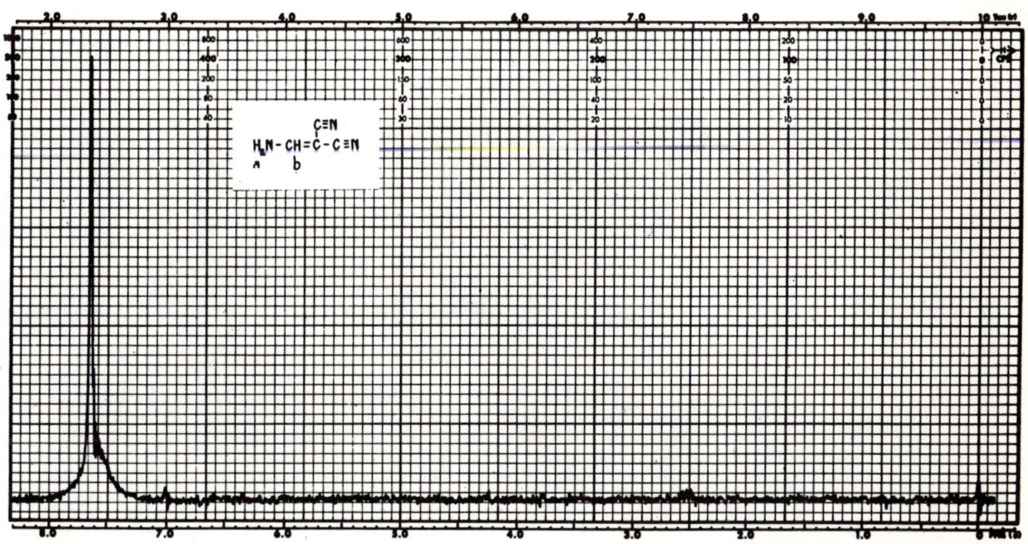

© 1978 Sadtler Research Laboratories, Inc.

3-CYCLOHEXENE-1-CARBONITRILE

C_7H_9N

Mol. Wt. 107.16

Solvent: CCl_4

Assignments
a = ca 2.10
b = ca 2.71
c = 5.65

970

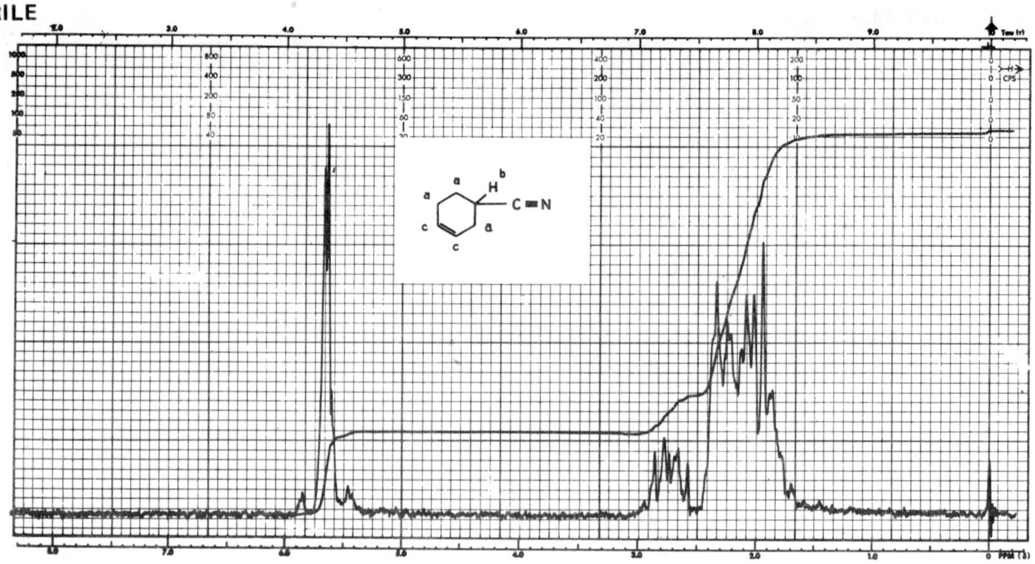

© 1978 Sadtler Research Laboratories, Inc.

1-CYCLOHEXENE-1-CARBONITRILE

C_7H_9N

Mol. Wt. 107.16

Solvent: CCl_4

Assignments
a = ca 1.68
b = ca 2.22
c = 6.58

971

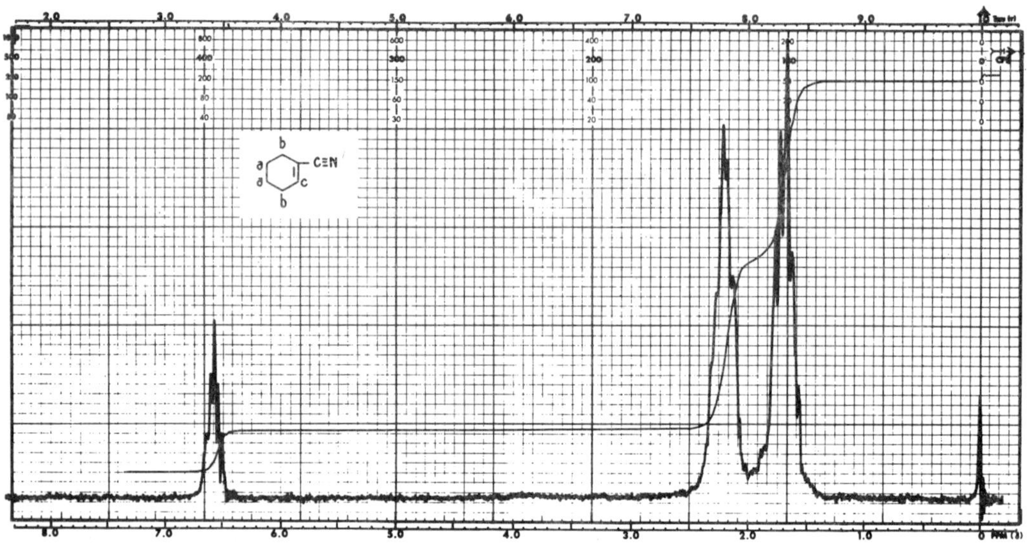

© 1978 Sadtler Research Laboratories, Inc.

PHENYLACETONITRILE

C_8H_7N

Mol. Wt. 117.15

B.P. 233.4°C

Solvent: CCl_4

Assignments
a = 3.62
b = 7.27

972

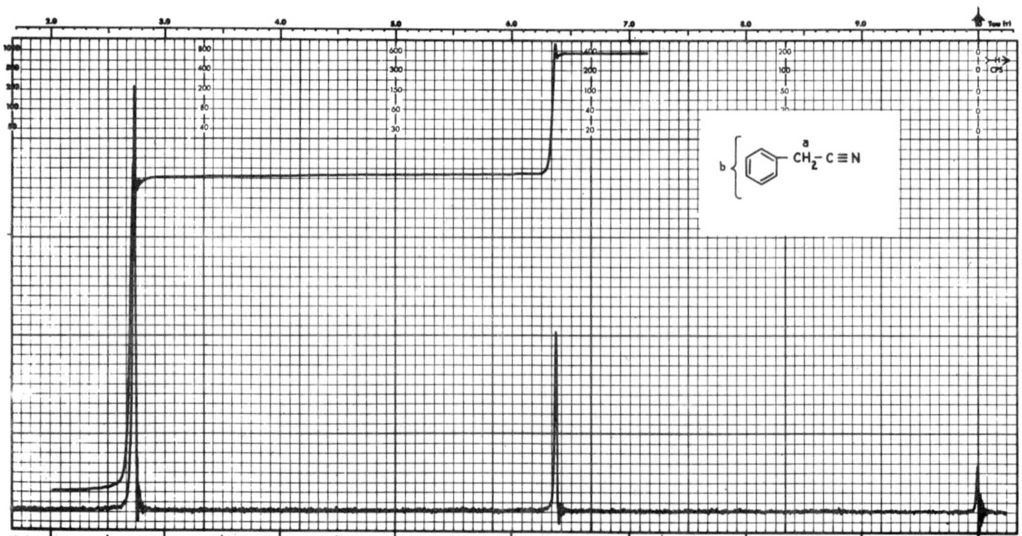

© 1978 Sadtler Research Laboratories, Inc.

HYDROCINNAMONITRILE

C_9H_9N

Mol. Wt. 131.18

Solvent: $CDCl_3$

Assignments
a = 2.36
b = 2.73
c = 7.16

973

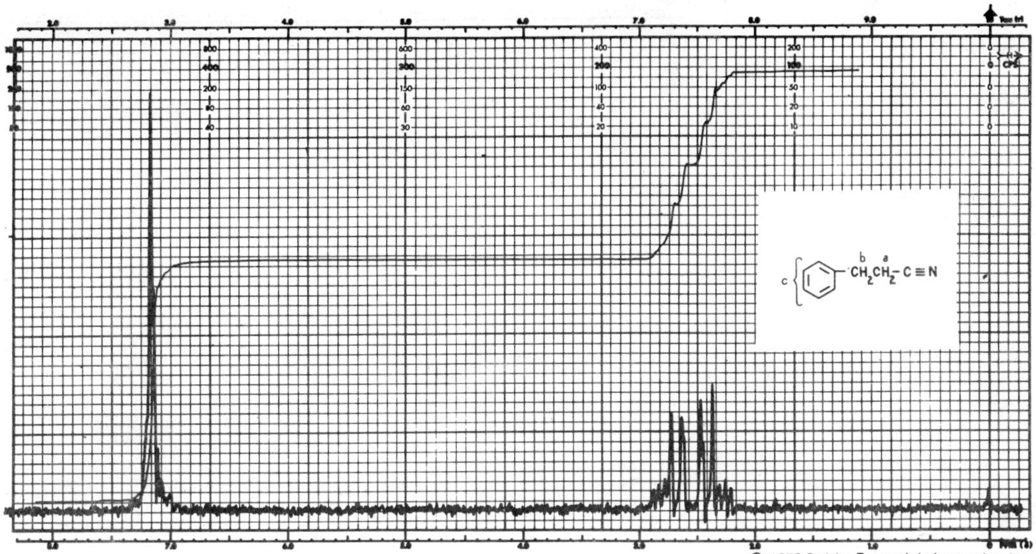

© 1978 Sadtler Research Laboratories, Inc.

3-METHYL-3-PHENYLBUTYRONITRILE

$C_{11}H_{13}N$

Mol. Wt. 159.23

B.P. 146°C/20mm

Solvent: CCl_4

Assignments
a = 1.47
b = 2.48
c = 7.22

974

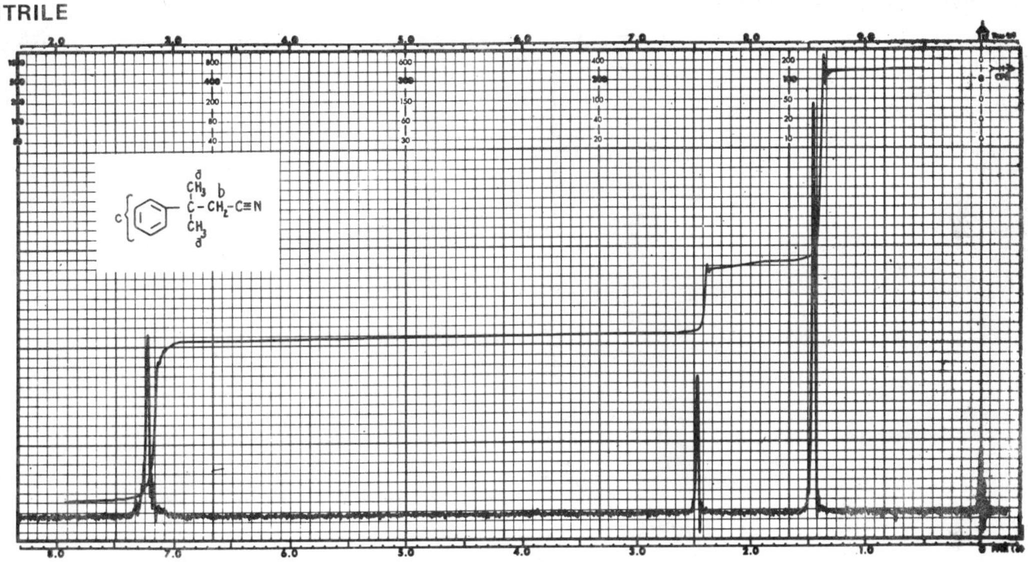

© 1978 Sadtler Research Laboratories, Inc.

N,N-DIMETHYL-2-PHENYLGLYCINONITRILE

$C_{10}H_{12}N_2$

Mol. Wt. 160.22

B.P. 76-78°C/0.8mm

Solvent: $CDCl_3$

Assignments
a = 2.33
b = 4.81
c = 7.20-7.60

975

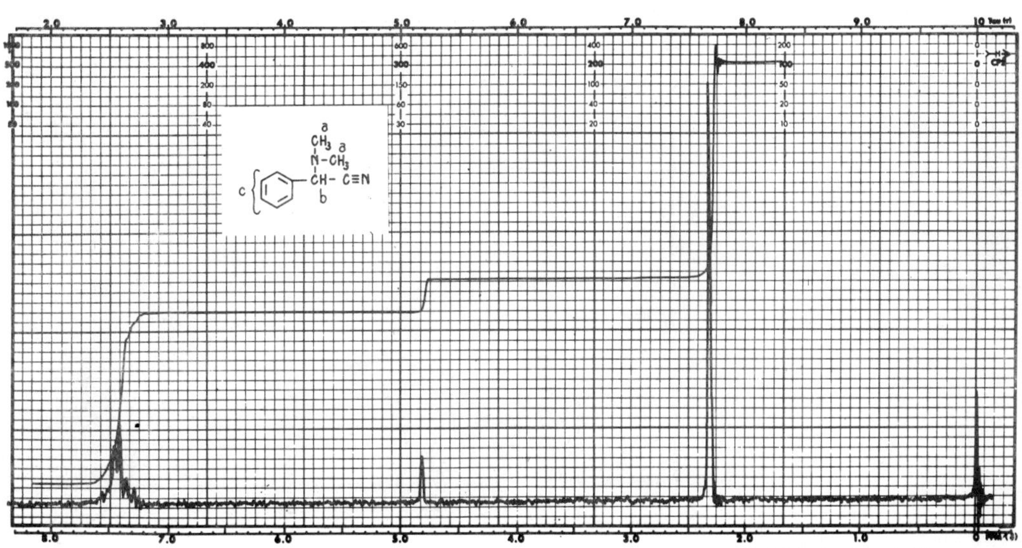

© 1978 Sadtler Research Laboratories, Inc.

3,3'-(PHENYLIMINO)DIPROPIONITRILE

$C_{12}H_{13}N_3$

Mol. Wt. 199.26

Solvent: DMSO-d_6

Assignments
a = 2.69
b = 3.69
c = 6.71
d = 6.79
e = 7.19
f = 2.49 DMSO-d_5
g = 3.10 H_2O

976

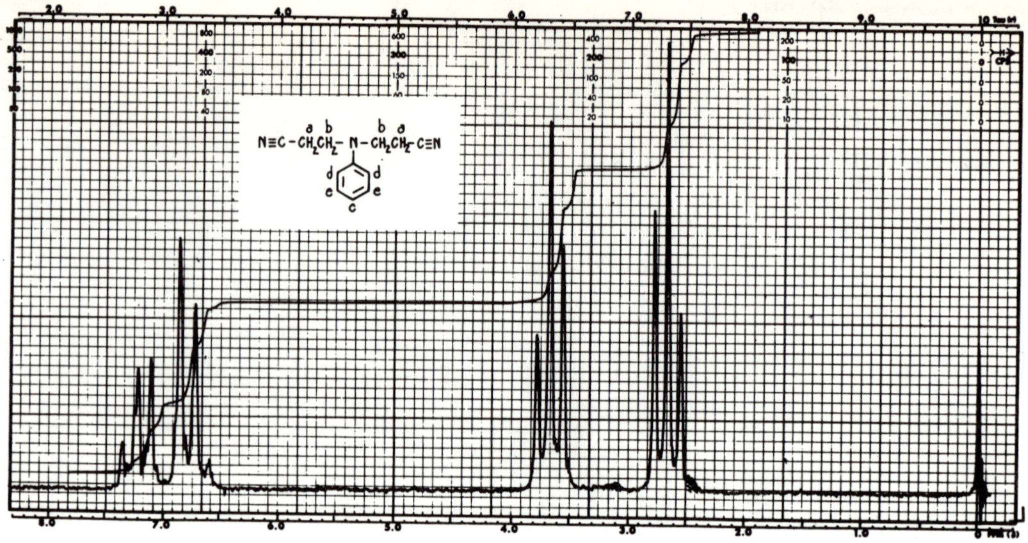

© 1978 Sadtler Research Laboratories, Inc.

2,2-DIPHENYLBUTYRONITRILE

$C_{16}H_{15}N$

Mol. Wt. 221.30

Solvent: $CDCl_3$

Assignments
a = 1.07
b = 2.41
c = ca 7.38
d = 7.78 imp.

977

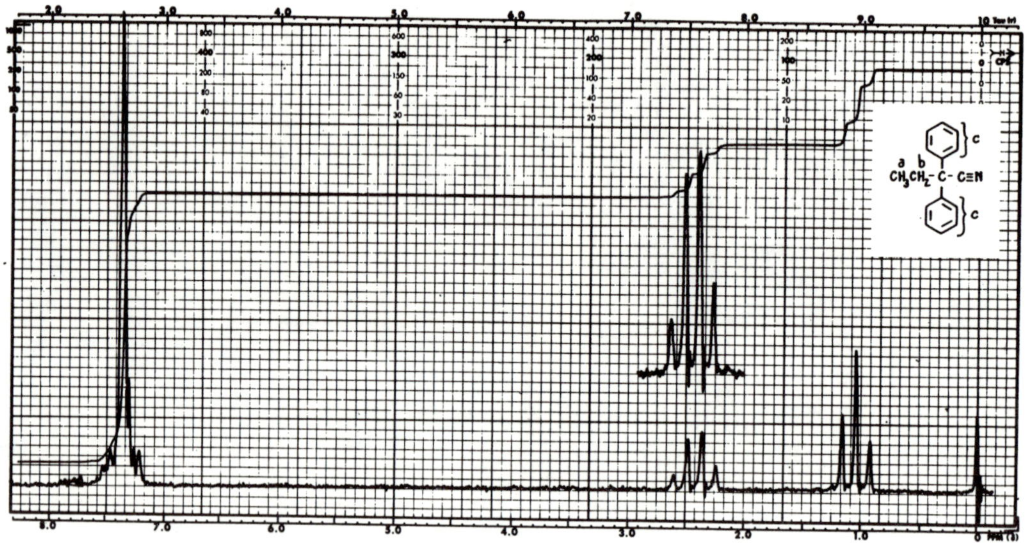

© 1978 Sadtler Research Laboratories, Inc.

2,3,4-TRIPHENYLBUTYRONITRILE

$C_{22}H_{19}N$

Mol. Wt. 297.60

M.P. 88-92°C

Solvent: $CDCl_3$

Assignments
a = 3.25
b = 4.08
c = 7.05
d = 7.20

978

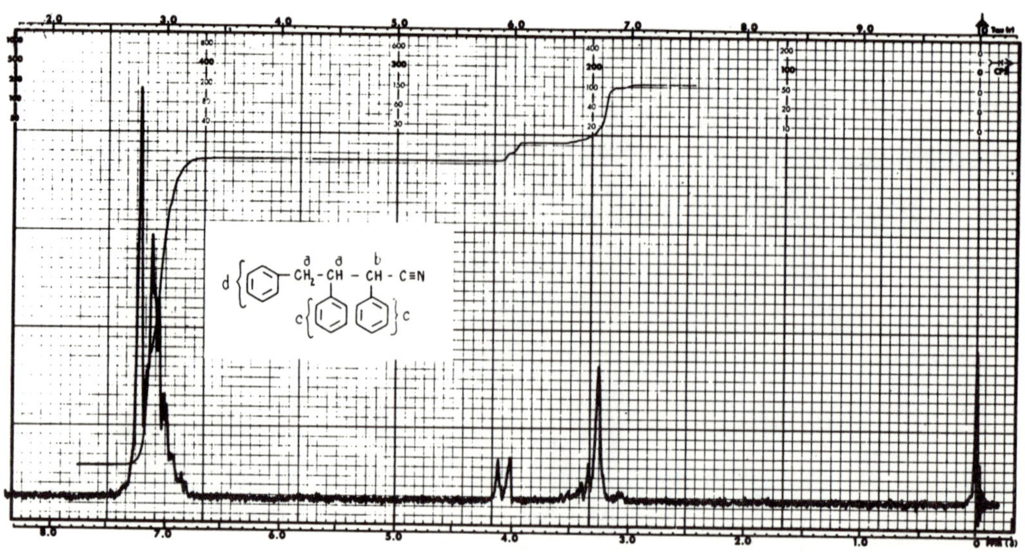

© 1978 Sadtler Research Laboratories, Inc.

4-CYANO-4-PHENYLHEPTANEDINITRILE

$C_{14}H_{13}N_3$

Mol. Wt. 223.28

Solvent: DMSO-d_6

Assignments
a = ca 2.34
b = ca 2.49
c = ca 7.49

979

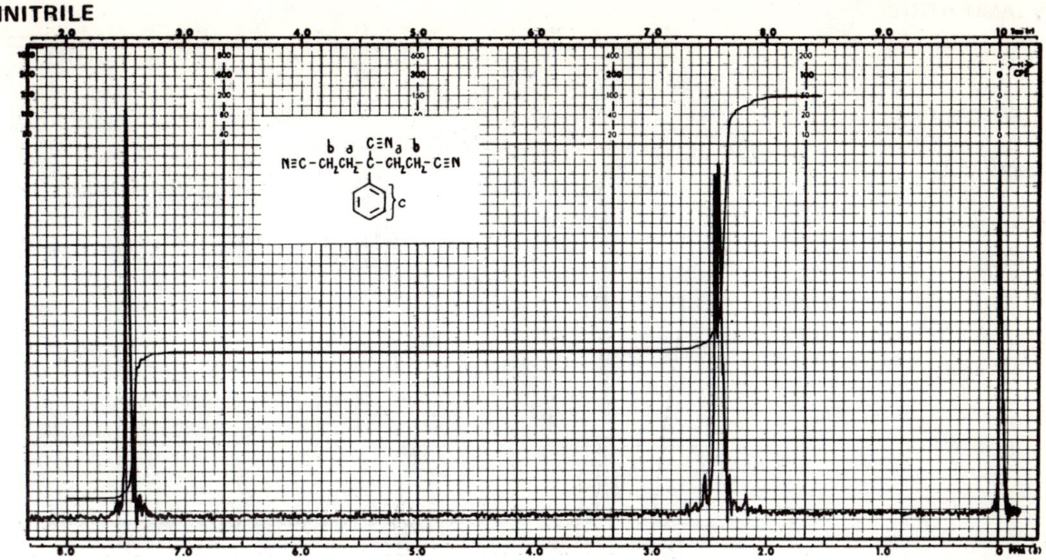

© 1978 Sadtler Research Laboratories, Inc.

(o-FLUOROPHENYL)ACETONITRILE

C_8H_6FN

Mol. Wt. 135.14

B.P. 122-126°C/10mm (lit.)

Solvent: CCl_4

Assignments
a = 3.62
b = 6.79-7.52

980

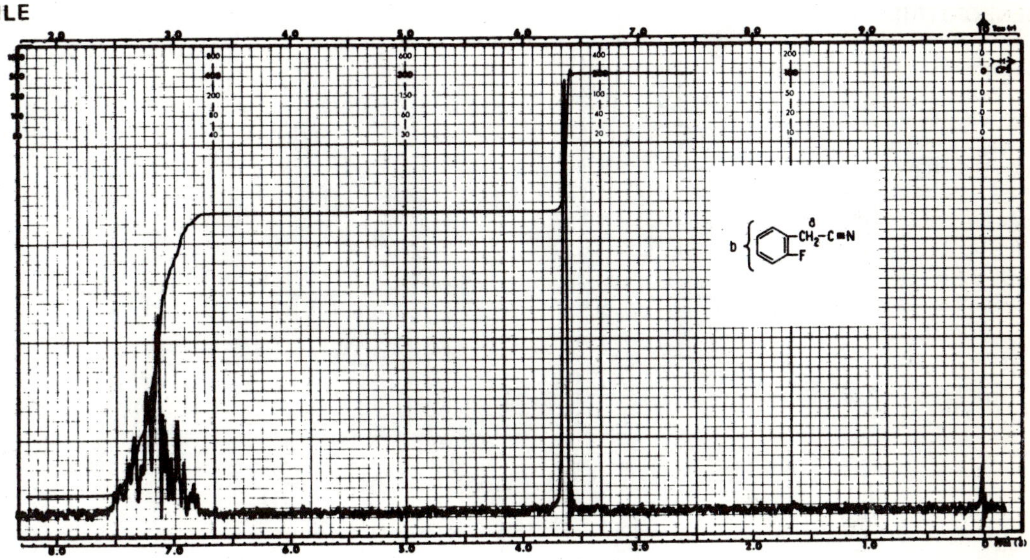

© 1978 Sadtler Research Laboratories, Inc.

(p-CHLOROPHENYL)ACETONITRILE

C_8H_6ClN

Mol. Wt. 151.60

M.P. 28-30°C (lit.)

Solvent: CCl_4

Assignments
a = 3.62
b = 7.24

981

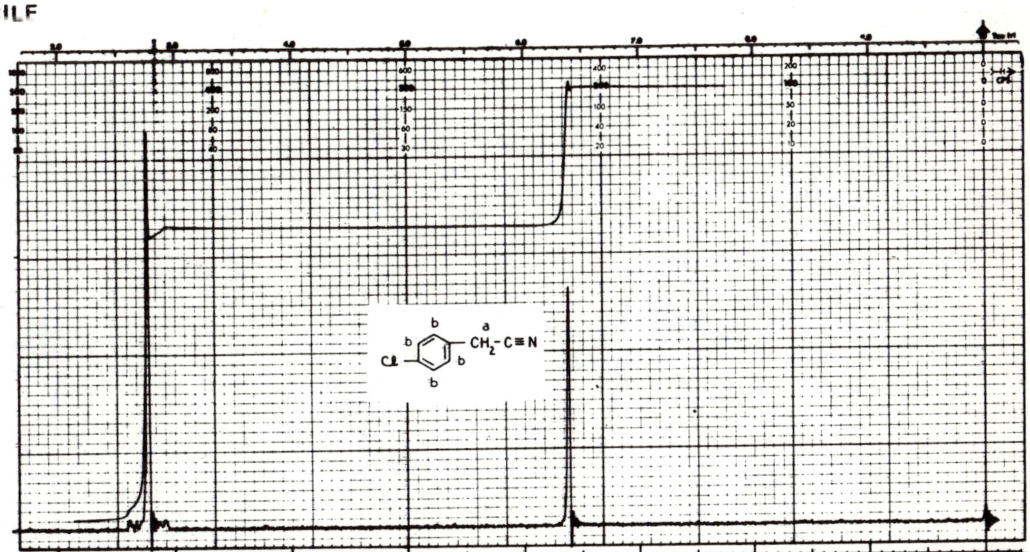

© 1978 Sadtler Research Laboratories, Inc.

CINNAMONITRILE

C_9H_7N

Mol. Wt. 129.16

Solvent: CCl_4

Assignments
a = 5.84
b = 7.36
c = 7.41

982

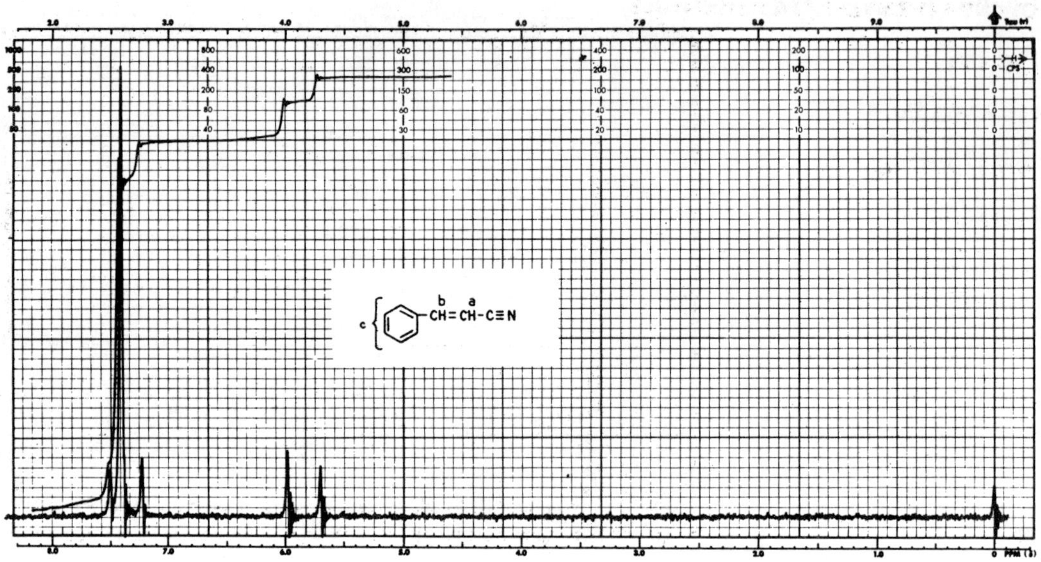

© 1978 Sadtler Research Laboratories, Inc.

BENZONITRILE

C_7H_5N

Mol. Wt. 103.12

B.P. 188-189°C

Solvent: CCl_4

Assignments
a = ca 7.50

983

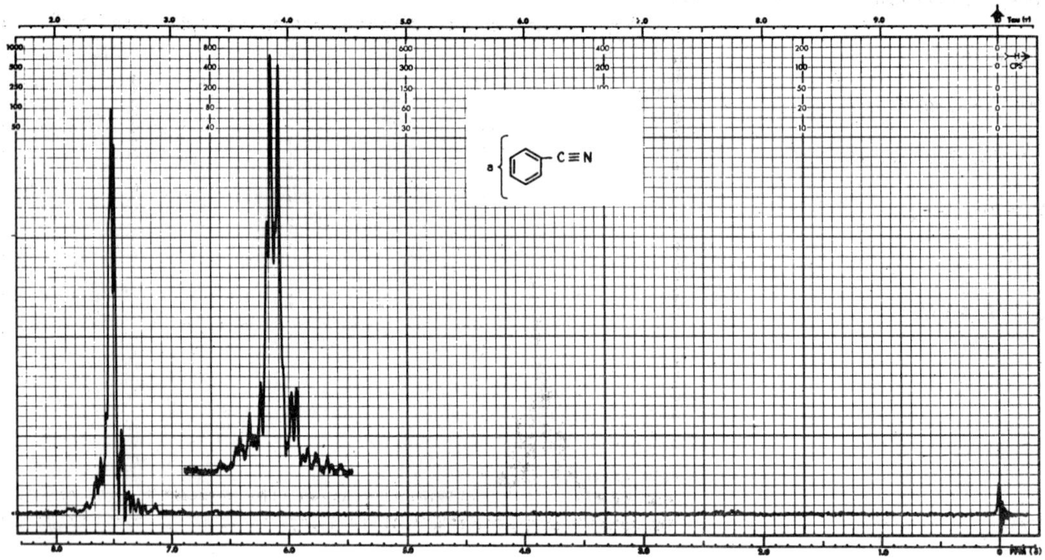

© 1978 Sadtler Research Laboratories, Inc.

o-TOLUNITRILE

C_8H_7N

Mol. Wt. 117.14

B.P. 93-94°C/18mm

Solvent: CCl_4

Assignments
a = 2.51
b = 7.02-7.66

984

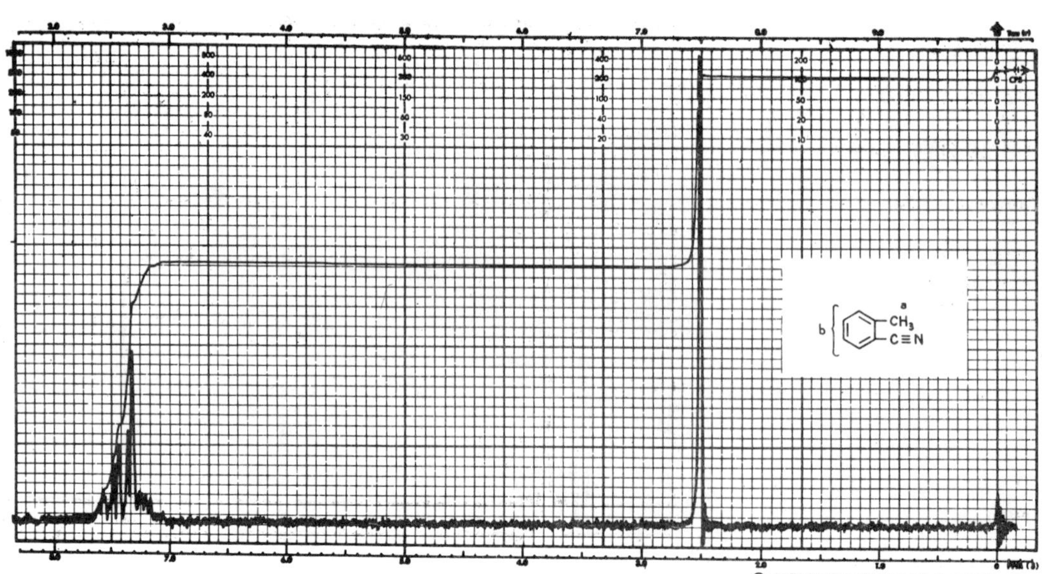

© 1978 Sadtler Research Laboratories, Inc.

m-TOLUNITRILE

C_8H_7N

Mol. Wt. 117.14

B.P. 89-90°C/11mm

Solvent: CCl_4

Assignments
a = 2.39
b = ca 7.37

985

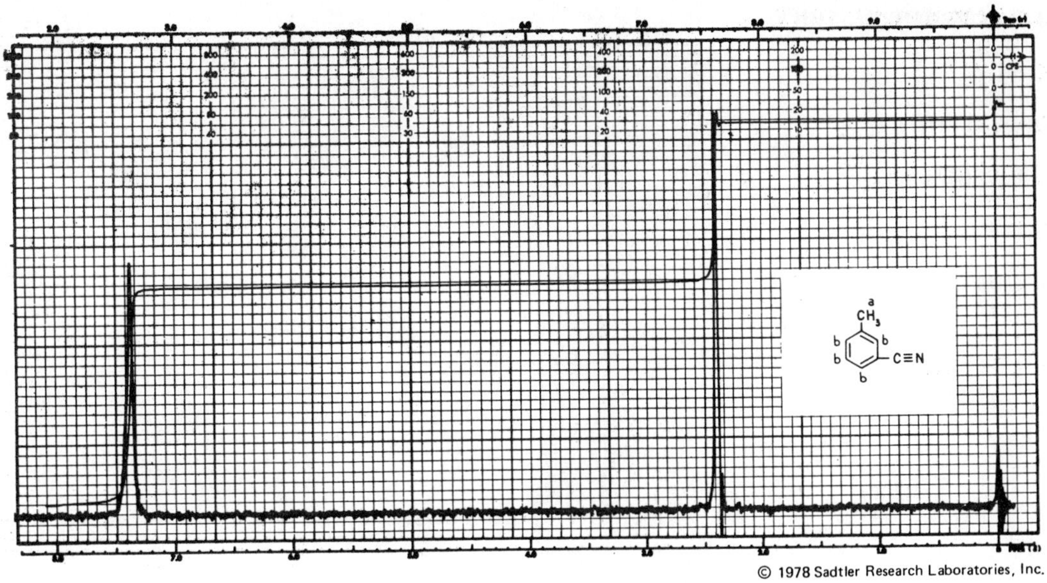

© 1978 Sadtler Research Laboratories, Inc.

p-TOLUNITRILE

C_8H_7N

Mol. Wt. 117.14

M.P. 25-27°C

Solvent: CCl_4

Assignments
a = 2.41
b = 7.23
c = 7.47

986

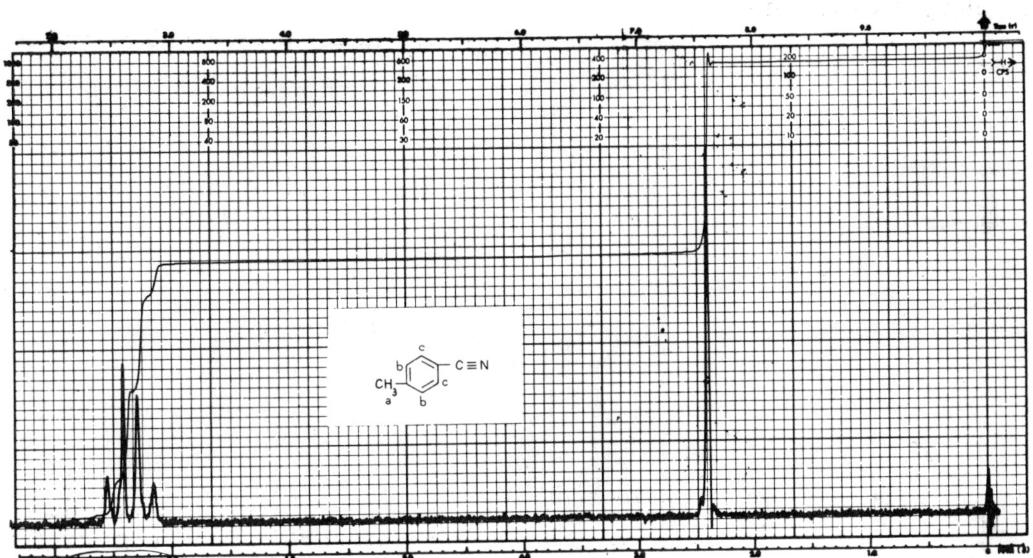

© 1978 Sadtler Research Laboratories, Inc.

o-FLUOROBENZONITRILE

C_7H_4FN

Mol. Wt. 121.11

B.P. 20°C/21mm (lit.)

Solvent: $CDCl_3$

987

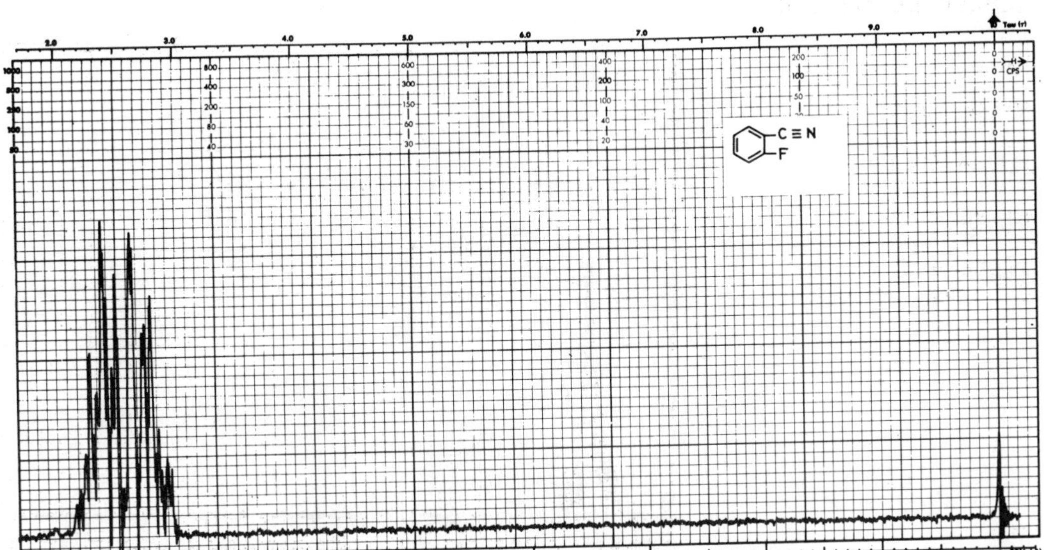

© 1978 Sadtler Research Laboratories, Inc.

p-FLUOROBENZONITRILE

C_7H_4FN

Mol. Wt. 121.11

M.P. 32-34°C (lit.)

Solvent: $CDCl_3$

Assignments
a = 7.17
b = 7.66
c = 1.68 imp.

988

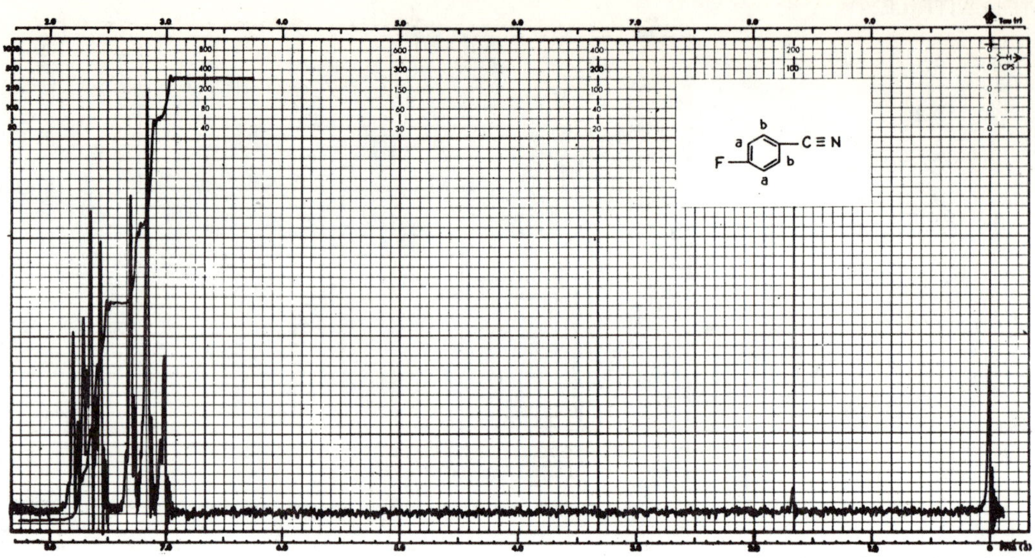

© 1978 Sadtler Research Laboratories, Inc.

2,6-DIFLUOROBENZONITRILE

$C_7H_3F_2N$

Mol. Wt. 139.11

Solvent: DMSO-d_6

Assignments
a = 7.31
b = 7.87
c = 2.55 DMSO-d_5
d = 3.22 H_2O

989

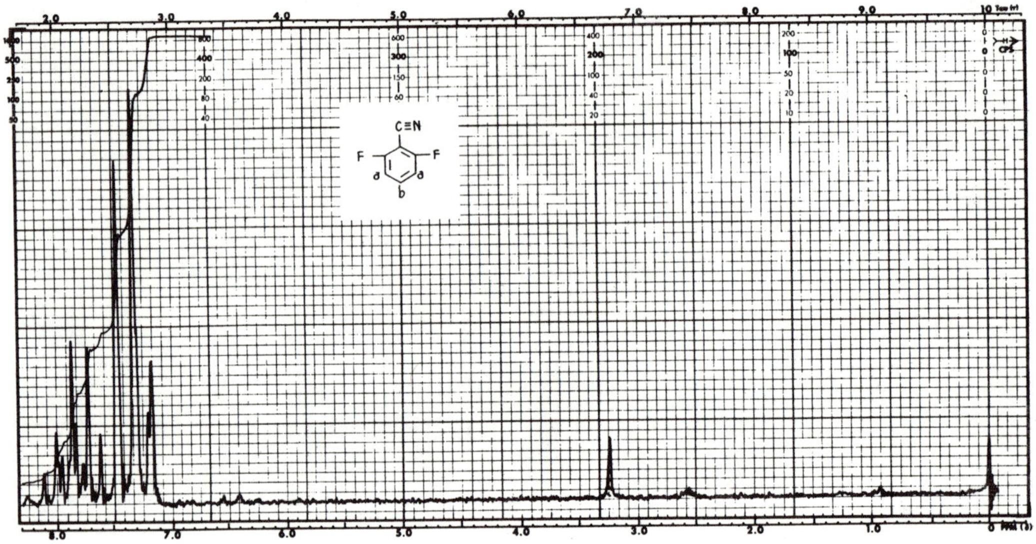

© 1978 Sadtler Research Laboratories, Inc.

---

o-CHLOROBENZONITRILE

C_7H_4ClN

Mol. Wt. 137.57

M.P. 44-46°C (lit.)

Solvent: $CDCl_3$

Assignments
a = 7.29-7.85

990

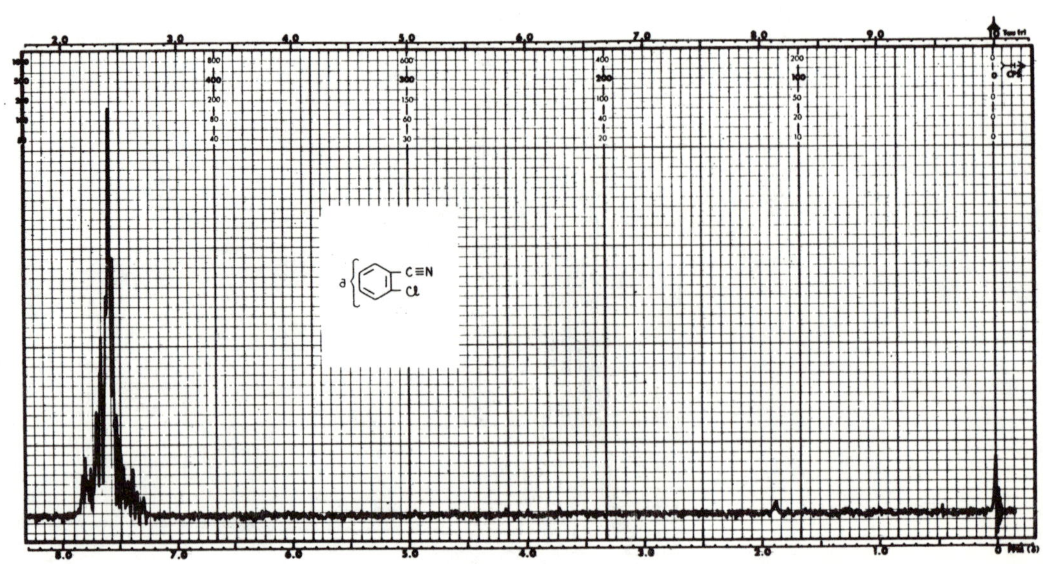

© 1978 Sadtler Research Laboratories, Inc.

p-CHLOROBENZONITRILE

C_7H_4ClN

Mol. Wt. 137.58

M.P. 91-92°C

Solvent: $CDCl_3$

Assignments
a = 7.39
b = 7.57

991

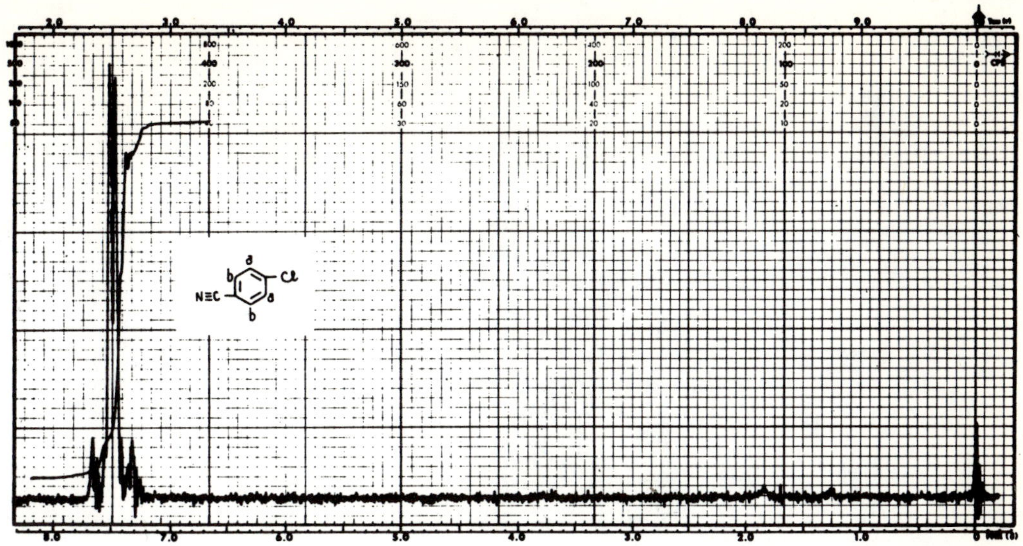

© 1978 Sadtler Research Laboratories, Inc.

o-BROMOBENZONITRILE

C_7H_4BrN

Mol. Wt. 182.03

Solvent: $CDCl_3$

Assignments
a = 7.33-7.90

992

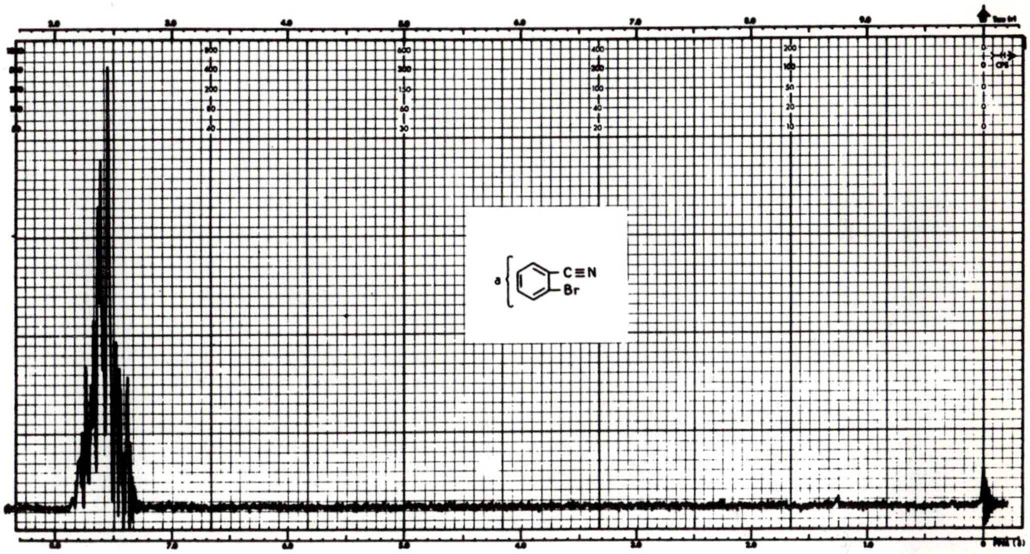

© 1978 Sadtler Research Laboratories, Inc.

m-BROMOBENZONITRILE

C_7H_4BrN

Mol. Wt. 182.03

Solvent: $CDCl_3$

Assignments
a = 7.40
b = 7.50-7.95

993

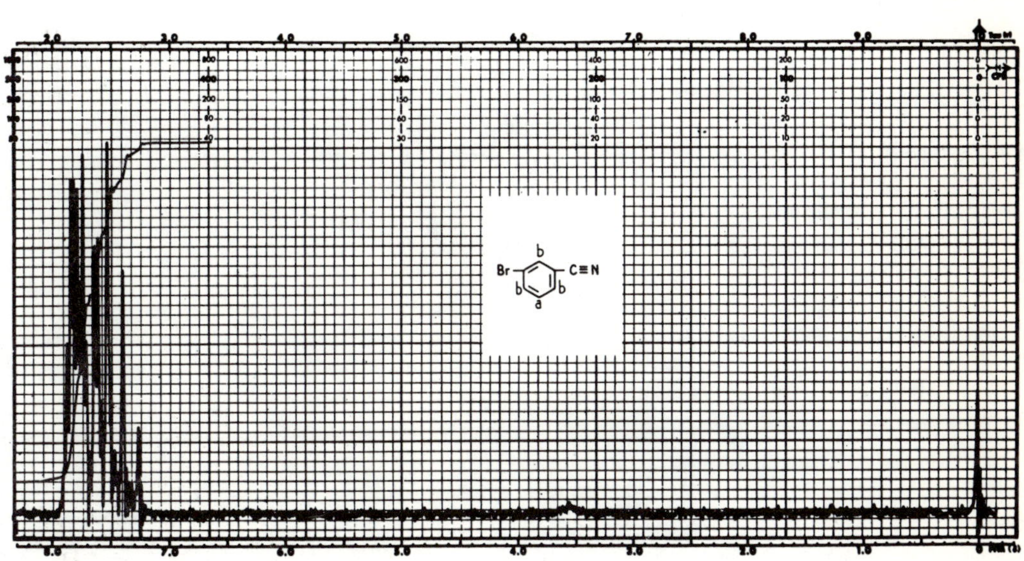

© 1978 Sadtler Research Laboratories, Inc.

m-AMINOBENZONITRILE

$C_7H_6N_2$

Mol. Wt. 118.14

Solvent: Polysol-d

Assignments
a = 5.29
b = 6.70-7.05
c = 7.19
d = 2.53 DMSO-d_5

994

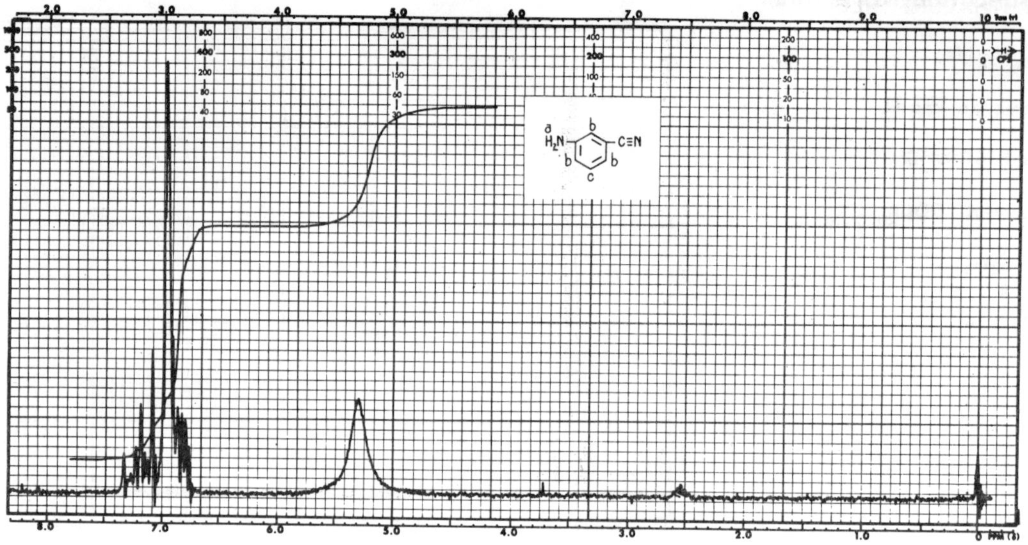

© 1978 Sadtler Research Laboratories, Inc.

p-(DIETHYLAMINO)BENZONITRILE

$C_{11}H_{14}N_2$

Mol. Wt. 174.25

Solvent: $CDCl_3$

Assignments
a = 1.28
b = 3.36
c = 6.58
d = 7.35
e = 7.67 imp.

995

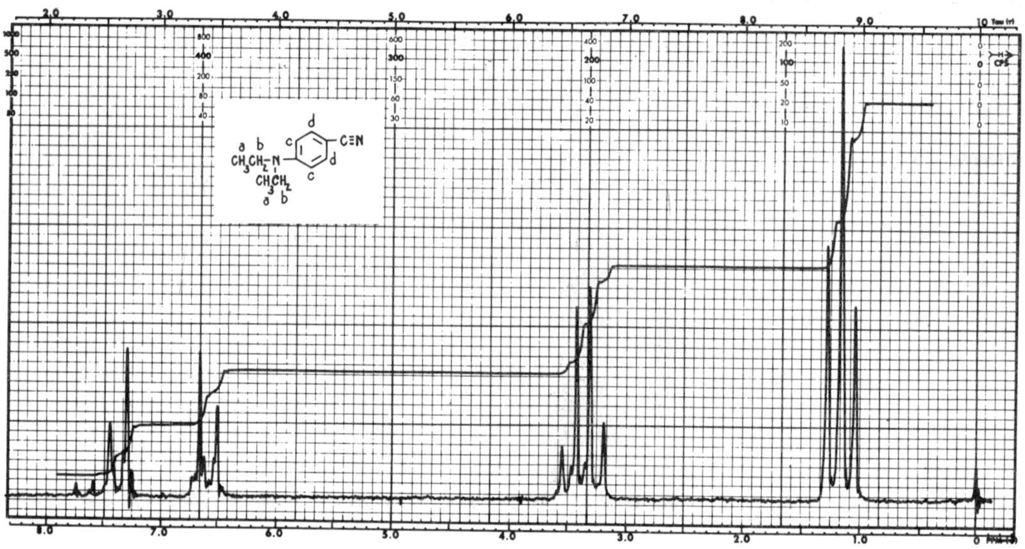

© 1978 Sadtler Research Laboratories, Inc.

4,4'-BIPHENYLDICARBONITRILE

$C_{14}H_8N_2$

Mol. Wt. 204.23

M.P. 235-238°C

Solvent: Trifluoroacetic acid

Assignments
a = 7.90

996

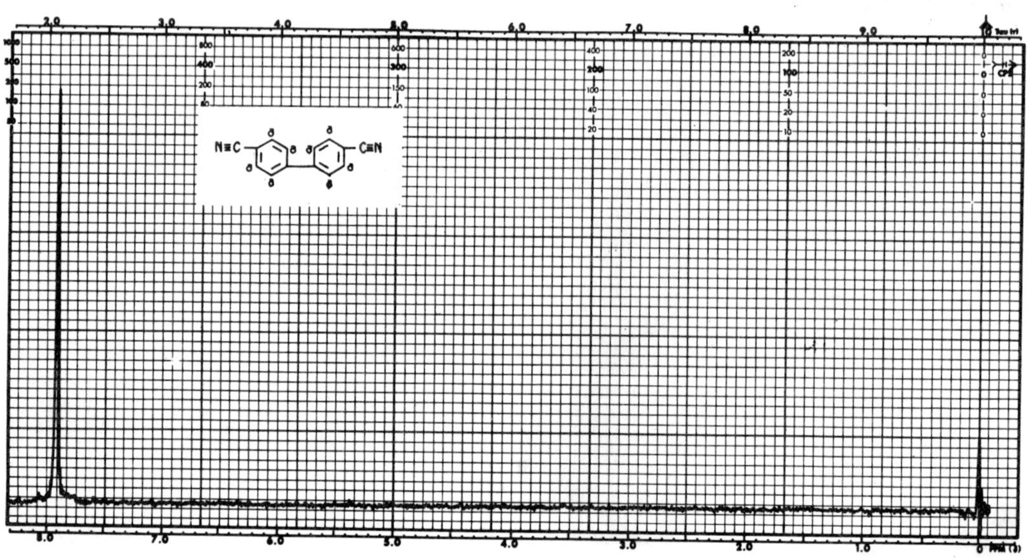

© 1978 Sadtler Research Laboratories, Inc.

PHTHALONITRILE

$C_8H_4N_2$

Mol. Wt. 128.13

M.P. 140°C

Solvent: Trifluoroacetic acid

Assignments
a = 8.00

997

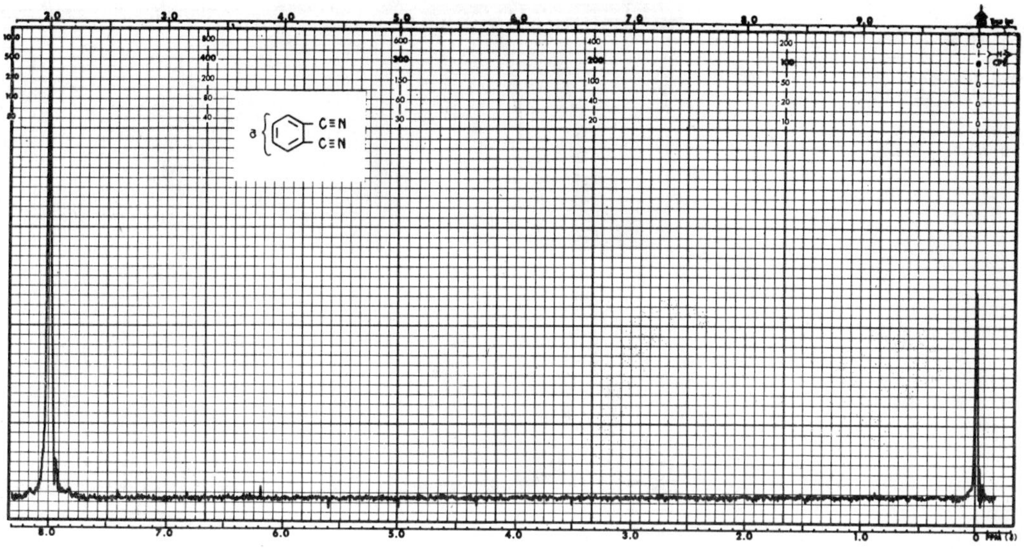

© 1978 Sadtler Research Laboratories, Inc.

2-NAPHTHALENEACETONITRILE

$C_{12}H_9N$

Mol. Wt. 167.21

M.P. 82-84°C

Solvent: Polysol-d

Assignments
a = 4.02
b = 7.30-7.65
c = 7.70-8.00
d = 2.52 DMSO-d_5
e = 3.26 H_2O

998

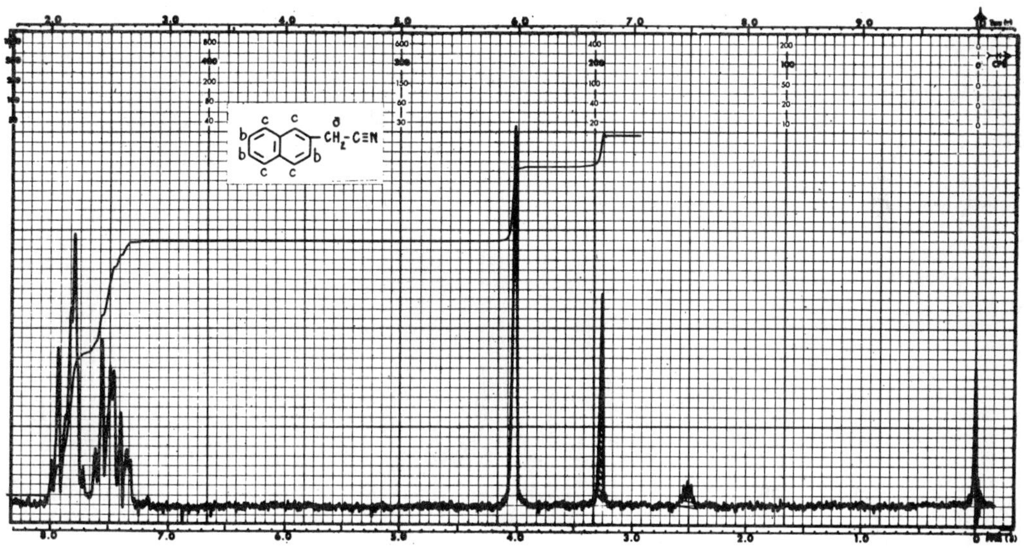

© 1978 Sadtler Research Laboratories, Inc.

1-NAPHTHONITRILE

$C_{11}H_7N$

Mol. Wt. 153.19

M.P. 34-36°C

Solvent: $CDCl_3$

Assignments
a = 7.33
b = 7.40-8.08
c = 8.11

999

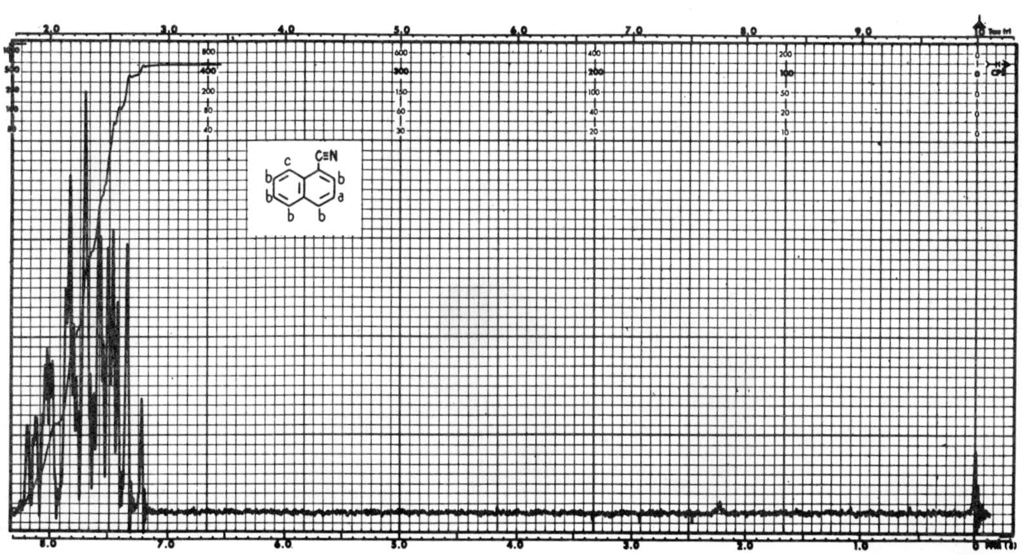

© 1978 Sadtler Research Laboratories, Inc.

2-NAPHTHONITRILE

$C_{11}H_7N$

Mol. Wt. 153.18

Solvent: $CDCl_3$

Assignments
a = 7.36-7.97
b = 8.10

1000

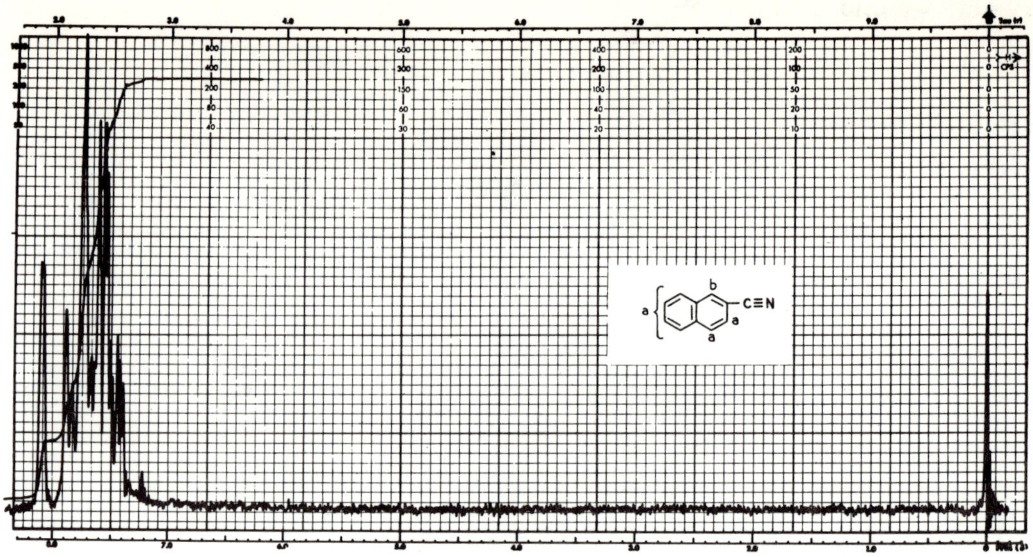

© 1978 Sadtler Research Laboratories, Inc.

9-PHENANTHRENECARBONITRILE

$C_{15}H_9N$

Mol. Wt. 203.25

Solvent: $CDCl_3$

Assignments
a = 7.50-8.00
b = 8.20
c = ca 8.25
d = ca 8.63

1001

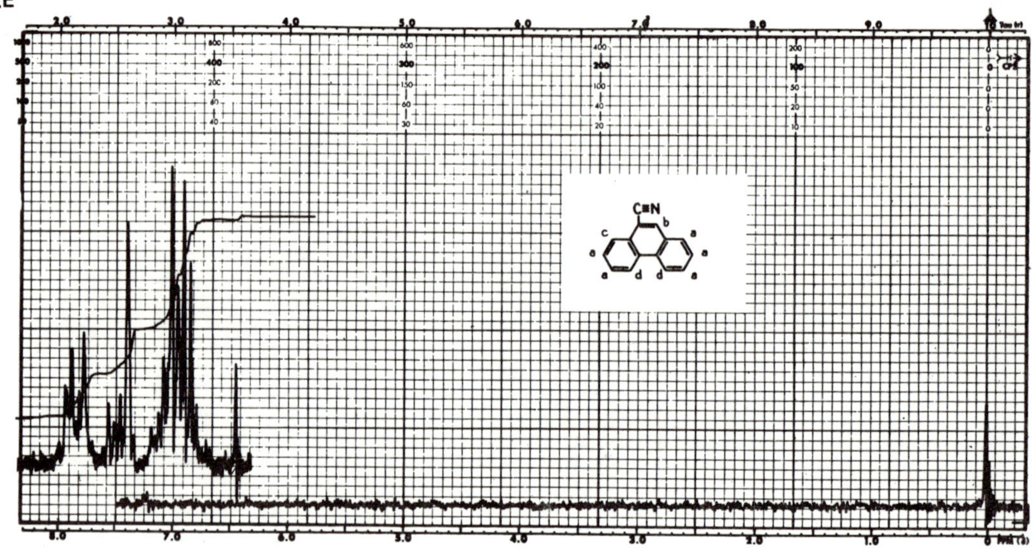

© 1978 Sadtler Research Laboratories, Inc.

NICOTINONITRILE

$C_6H_4N_2$

Mol. Wt. 104.11

M.P. 50-52°C

Solvent: $CDCl_3$

Assignments
a = 7.55
b = 8.11
c = 8.89
d = 8.93

1002

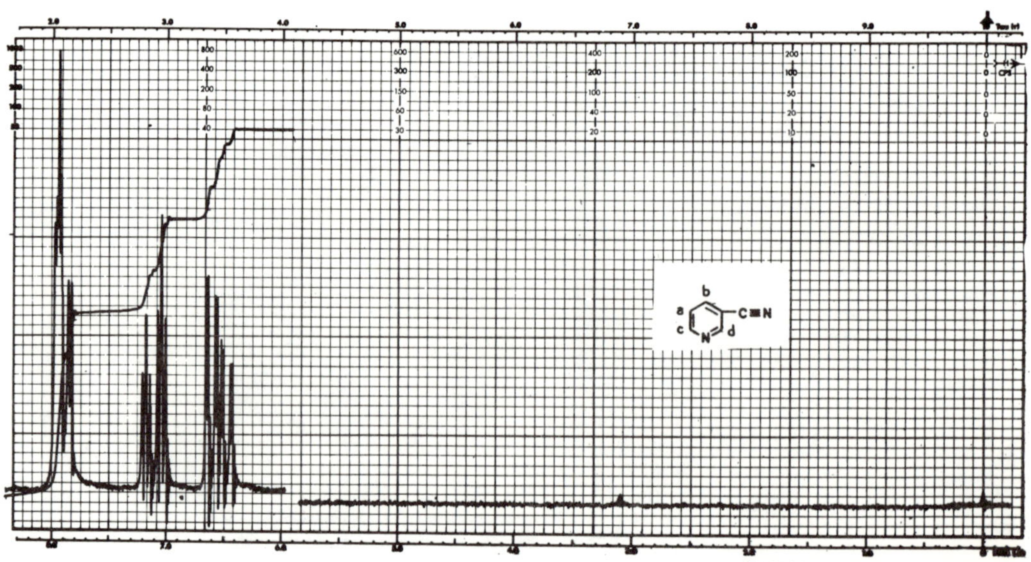

© 1978 Sadtler Research Laboratories, Inc.

THE CYANAMIDS

Although the Cyanamid group is of commercial importance, relatively few hydrocarbon derivatives are available for study. The chemical shifts of the aliphatic groups bonded to the Cyanamid group display intermediate deshielding similar to that of the corresponding amines.

Chemical Shifts

Aliphatic Protons

HC–2	HC–1		Solvent	Spectrum
	(2.88)	$(CH_3)_2$–N–C≡N	CCl_4	1033
(1.29)	----	$(CH_3)_3$–C–NH–C≡N	$CDCl_3$	----

Diallylcyanamide

HC–3,2	HC–1			
(5.0–6.2	3.60)	$(CH_2$=CH–CH_2–$)_2$–N–C≡N	CCl_4	1004

Dibenzylcyanamide

HC–1			
4.06	(Ø–$CH_2)_2$–N–N≡N	$CDCl_3$	1005

THE ISOCYANIDES

This rather rare functional group is one of the few that may display coupling (2n + 1) between the Nitrogen nucleus and adjacent hydrocarbon groups.

Chemical Shifts

Aliphatic Protons

Benzyl isocyanide

Ø	HC–1		Solvent	Spectrum
ca. 7.33	4.59	Ø–CH_2–N≡C	$CDCl_3$	----

Alicyclic Protons

Cyclohexyl isocyanide

HC–2,3,4,5,6	HC–1		
0.90–2.27	3.60	$CDCl_3$	1006

Coupling and Coupling Constants

$$J_{C\equiv N-CH_2} = 2.1 \text{ Hz}$$

DIMETHYLCYANAMIDE

$C_3H_6N_2$

Mol. Wt. 70.09

B.P. 160°C

Solvent: CCl_4

Assignments
a = 2.88

1003

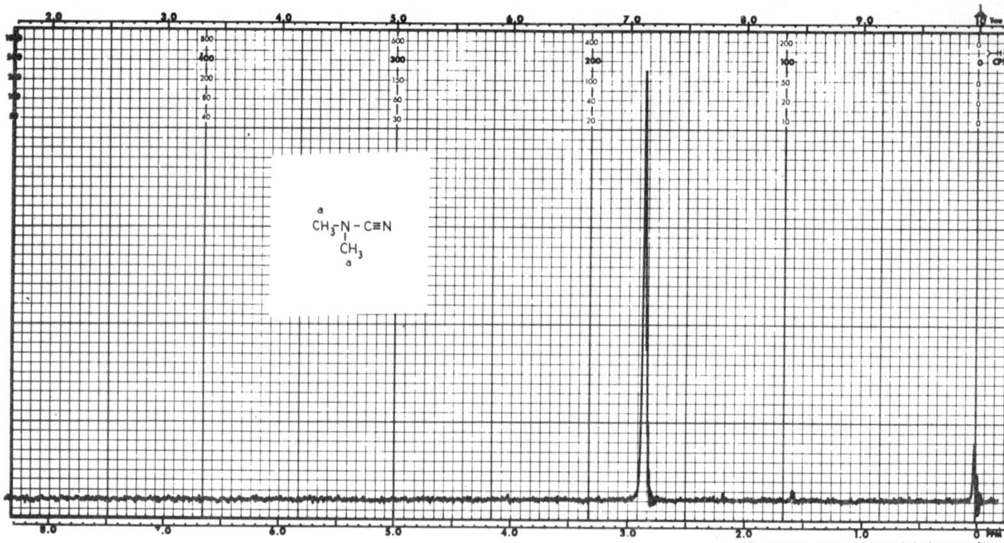

© 1978 Sadtler Research Laboratories, Inc.

DIALLYLCYANAMIDE

$C_7H_{10}N_2$

Mol. Wt. 122.17

B.P. 222°C

Solvent: CCl_4

Assignments
a = 3.60
b = 5.25
c = 5.33
d = ca 5.80

1004

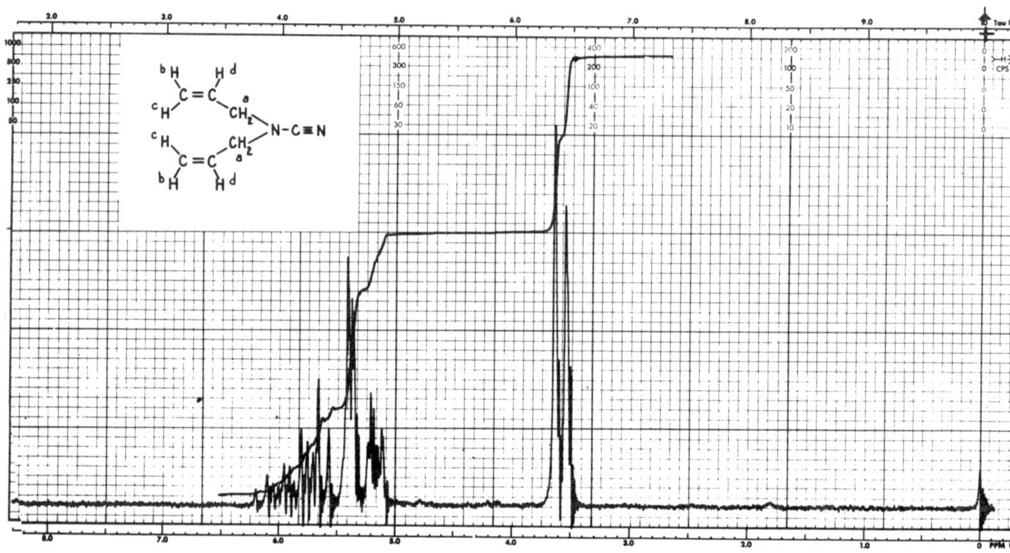

© 1978 Sadtler Research Laboratories, Inc.

DIBENZYLCYANAMIDE

$C_{15}H_{14}N_2$

Mol. Wt. 222.29

Solvent: $CDCl_3$

Assignments
a = 4.06
b = 7.30

1005

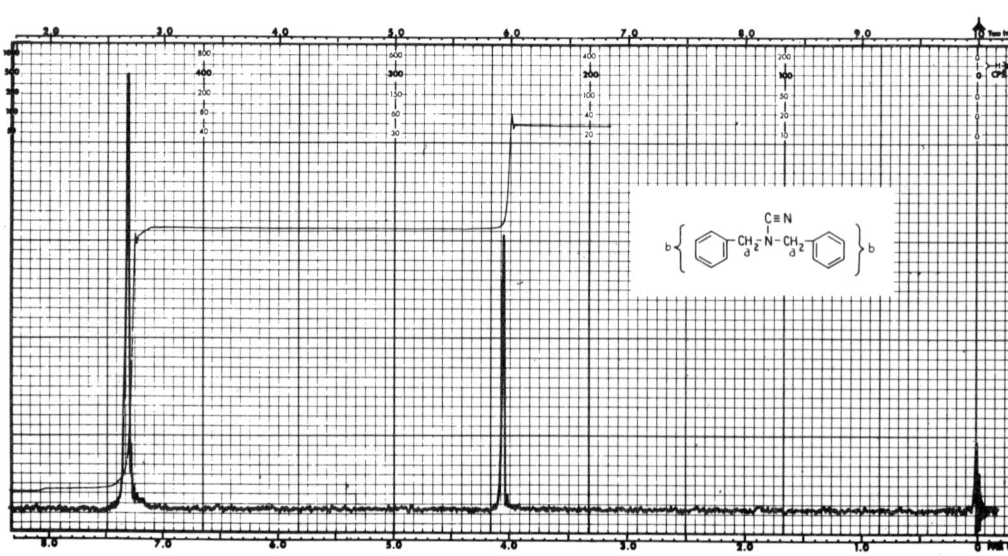

© 1978 Sadtler Research Laboratories, Inc.

CYCLOHEXYL ISOCYANIDE

$C_7H_{11}N$

Mol. Wt. 109.17

Solvent: $CDCl_3$

Assignments
a = 0.90-2.27
b = 3.60
c = 1.07, 2.53 imp.

1006

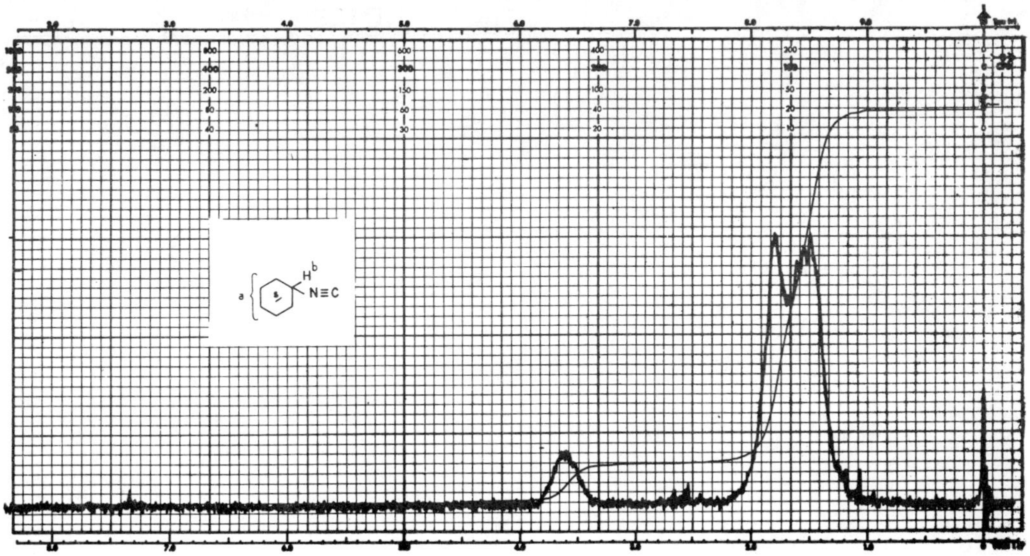

© 1978 Sadtler Research Laboratories, Inc.

NOTES

THE THIOCYANATES

The Thiocyanate group has an intermediate deshielding effect on adjacent aliphatic groups but little or no shielding/deshielding effect on the aromatic protons.

It is similar to many of the preceding groups in that it is difficult to determine the presence of this group in an NMR spectrum with any degree of certainty, without additional information such as elemental analysis data or infrared information.

Chemical Shifts

Aliphatic Protons

HC–4	HC–3	HC–2	HC–1		Solvent	Spectrum
			2.61	CH_3–S–C≡N	CCl_4	1007
		1.51	3.00	CH_3–CH_2–S–C≡N	$CDCl_3$	1008
0.99	(1.2–2.1)		2.97	CH_3–$(CH_2)_2$–CH_2–S–C≡N	CCl_4	1009

Substituted Methyl Thiocyanates

N≡C–S	–CH_2	–X	Solvent	Spectrum
	3.58	–CH=CH–Ø	$CDCl_3$	1019
	4.15	–Ø	$CDCl_3$	1017
	4.23	–Ø(4–NO_2)	$CDCl_3$	----
	4.40	–S–C≡N	$CDCl_3$	1014
	4.67	–C(=O)–Ø	$CDCl_3$	----
	4.92	–Cl	CCl_4	1013

Substituted Ethylthiocyanates

N≡C–S	–CH_2	–CH_2	–X	Solvent	Spectrum
	3.01	2.01	–$(CH_2)_2$–S–C≡N	$CDCl_3$	1016
	3.38	3.38	–S–C≡N	Polysol	1015
	3.11	3.78	–O$(CH_2)_2$–O–R_4	CCl_4	----

Aromatic Protons

Substituted Phenylisocyanates

C–1	C–2	C–3	C–4	C–5	C–6	Solvent
N≡C–S	6.92	6.71	NH_2	O–R_2	7.00	Polysol
	7.22	6.89	OH	CH_3	7.30	DMSO-d_6
	7.12	CH_3	OH	CH_3	7.12	$CDCl_3$
	7.30	CH_3	OH	6.89	7.22	DMSO-d_6
	7.35	6.68	OH	CH_3	CH_3	$CDCl_3$
	7.48	7.03	NH_2	7.41	C(=O)–R	DMSO-d_6
	7.92	7.61	Cl	8.01	C(=O)–R	$CDCl_3$
	7.93	8.12	C(=O)–R	8.36	S–C≡N	$CDCl_3$
	8.25	8.69	NO_2	8.99	NO_2	DMSO

THIOCYANIC ACID, METHYL ESTER

C_2H_3NS

Mol. Wt. 73.12

B.P. 131°C

Solvent: CCl_4

Assignments
a = 2.61

1007

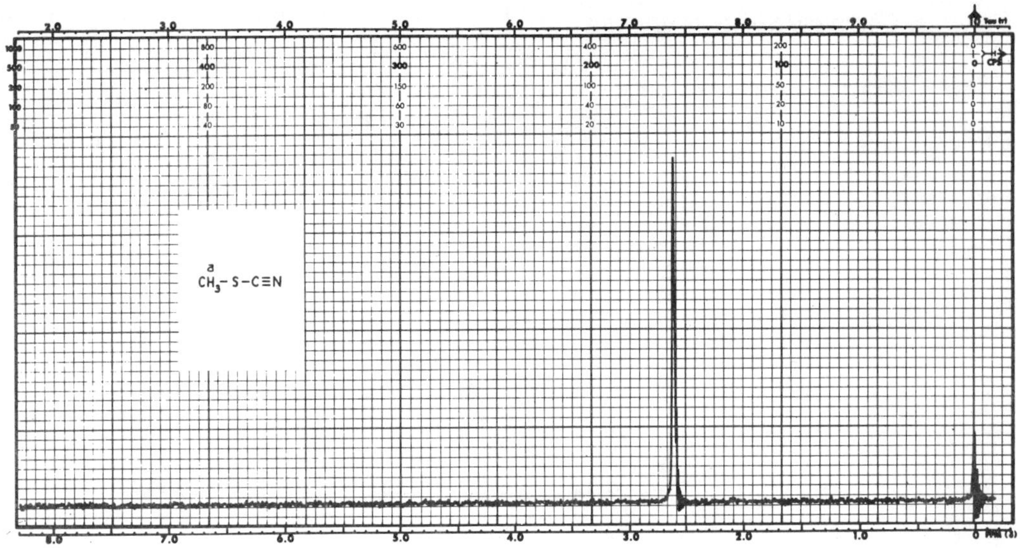

© 1978 Sadtler Research Laboratories, Inc.

THIOCYANIC ACID, ETHYL ESTER

C_3H_5NS

Mol. Wt. 87.14

B.P. 143-145°C

Solvent: $CDCl_3$

Assignments
a = 1.51
b = 3.00

1008

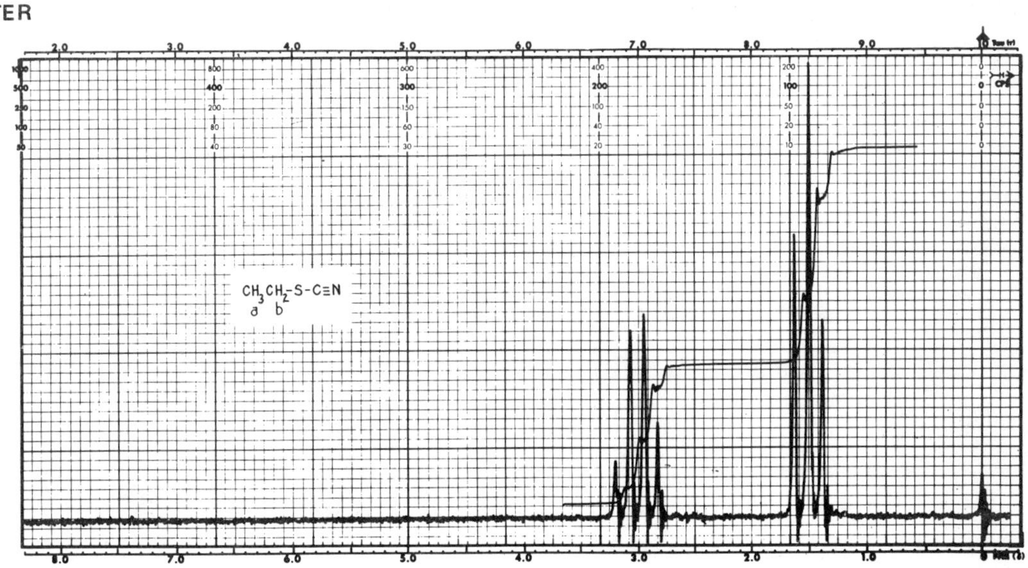

© 1978 Sadtler Research Laboratories, Inc.

THIOCYANIC ACID, BUTYL ESTER

C_5H_9NS

Mol. Wt. 115.20

Solvent: CCl_4

Assignments
a = 0.99
b = 1.16-2.13
c = 2.97

1009

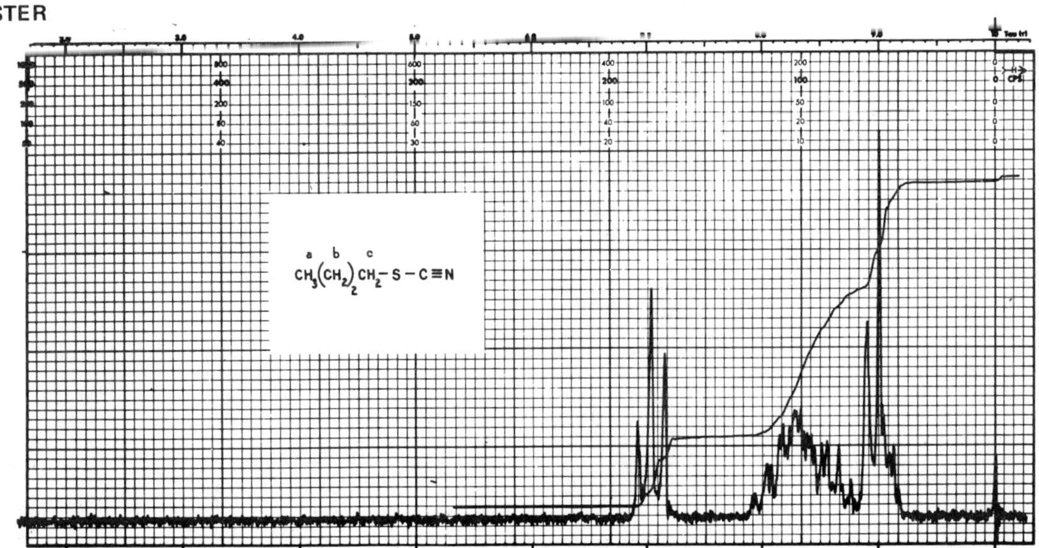

© 1978 Sadtler Research Laboratories, Inc.

THIOCYANIC ACID, OCTYL ESTER

$C_9H_{17}NS$

Mol. Wt. 171.31

Solvent: CCl_4

Assignments
a = 0.90
b = 1.33
c = 1.83
d = 2.93

1010

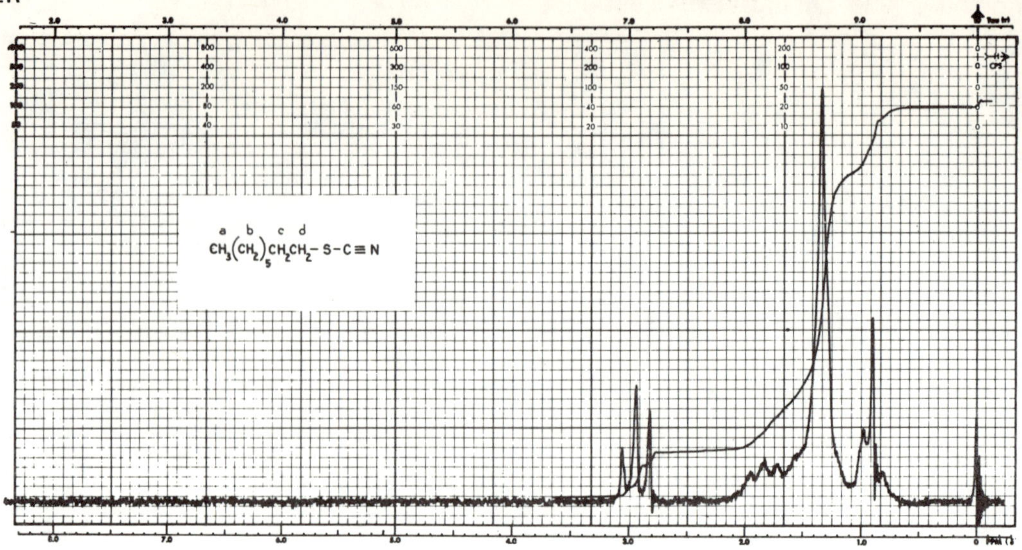

© 1978 Sadtler Research Laboratories, Inc.

THIOCYANIC ACID, DECYL ESTER

$C_{11}H_{21}NS$

Mol. Wt. 199.36

Solvent: CCl_4

Assignments
a = 0.89
b = 1.30
c = 1.82
d = 2.93

1011

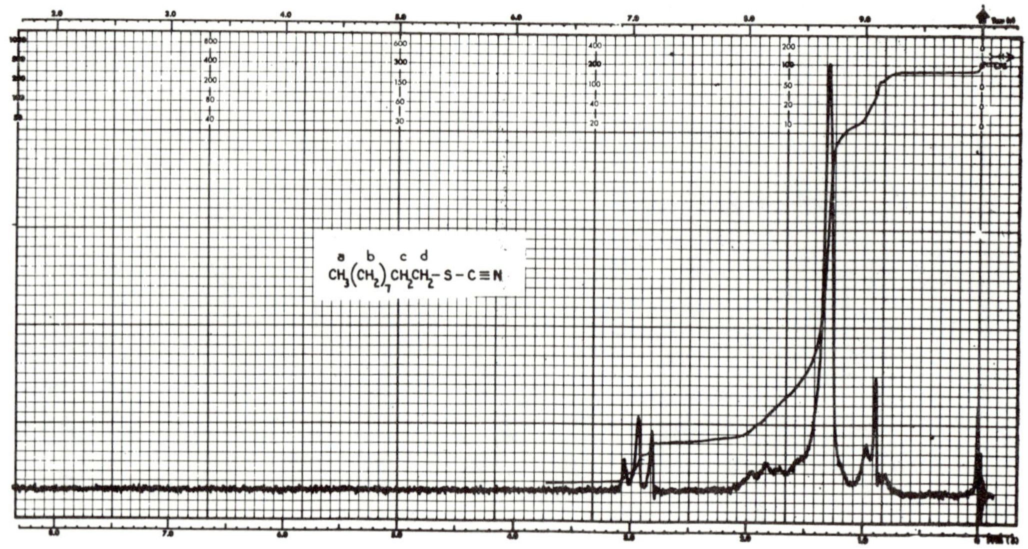

© 1978 Sadtler Research Laboratories, Inc.

THIOCYANIC ACID, DODECYL ESTER

$C_{13}H_{25}NS$

Mol. Wt. 227.41

B.P. 183-185°C/10mm

Solvent: CCl_4

Assignments
a = 0.90
b = ca 1.29
c = 1.81
d = 2.94

1012

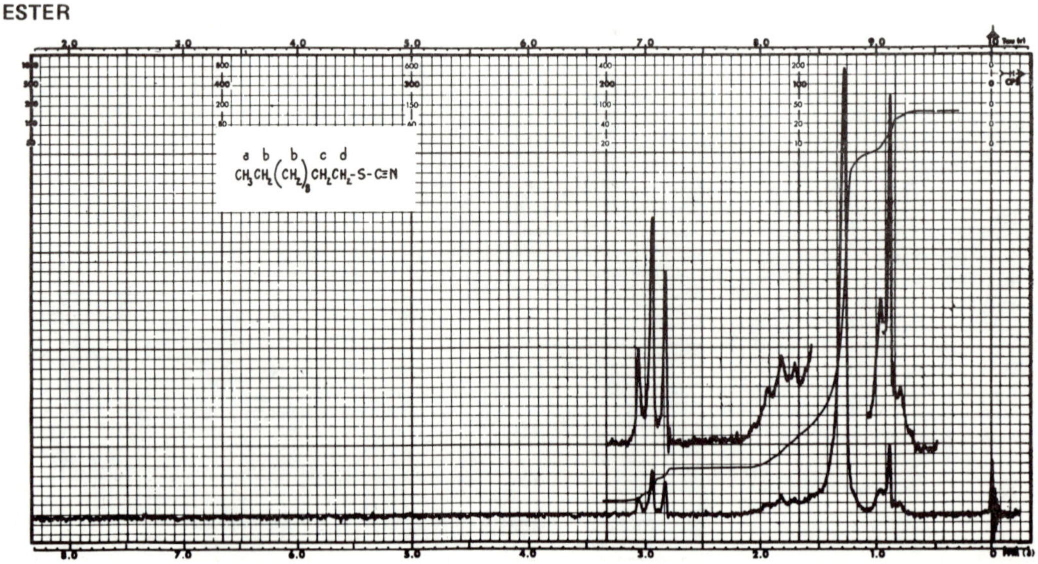

© 1978 Sadtler Research Laboratories, Inc.

THIOCYANIC ACID, CHLOROMETHYL ESTER

C_2H_2ClNS

Mol. Wt. 107.56

Solvent: CCl_4

Assignments
a = 4.92

1013

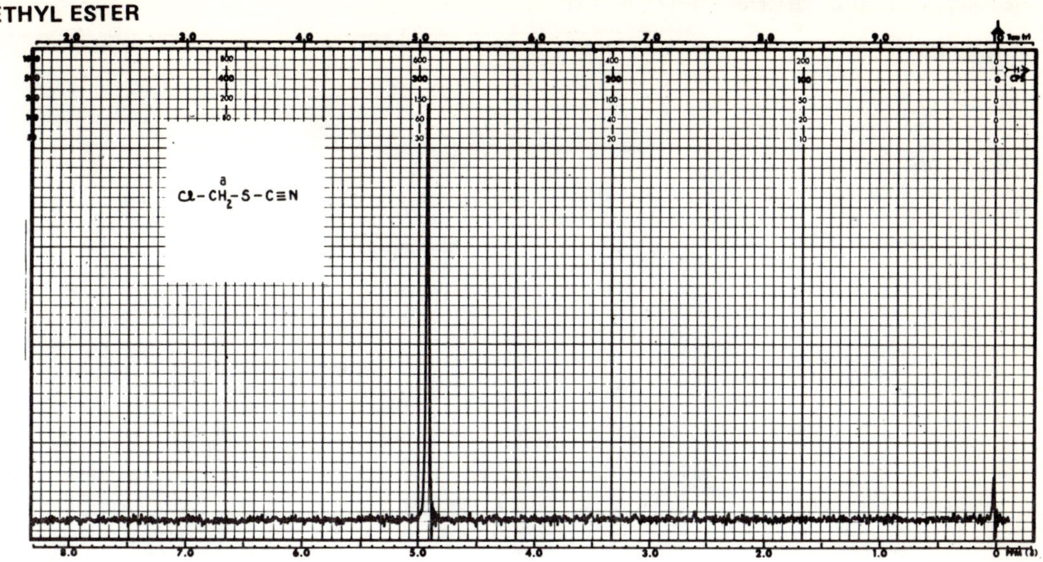

© 1978 Sadtler Research Laboratories, Inc.

THIOCYANIC ACID, METHYLENE ESTER

$C_3H_2N_2S_2$

Mol. Wt. 130.19

M.P. 104-106°C

Solvent: $CDCl_3$

Assignments
a = 4.40
b = 1.55 imp.

1014

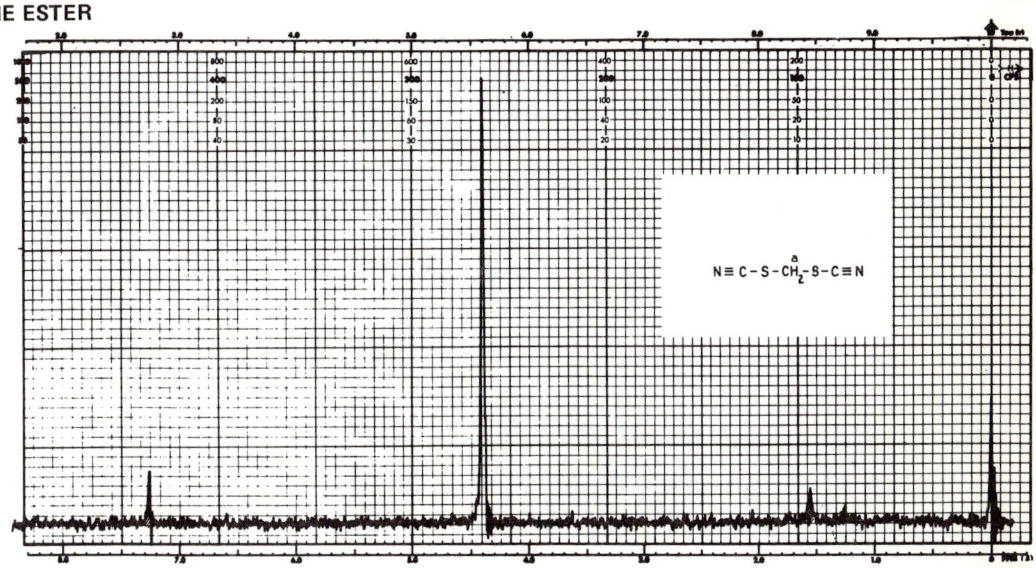

© 1978 Sadtler Research Laboratories, Inc.

THIOCYANIC ACID, ETHYLENE ESTER

$C_4H_4N_2S_2$

Mol. Wt. 144.22

M.P. 88-90°C

Solvent: Polysol-d

Assignments
a = 3.38

1015

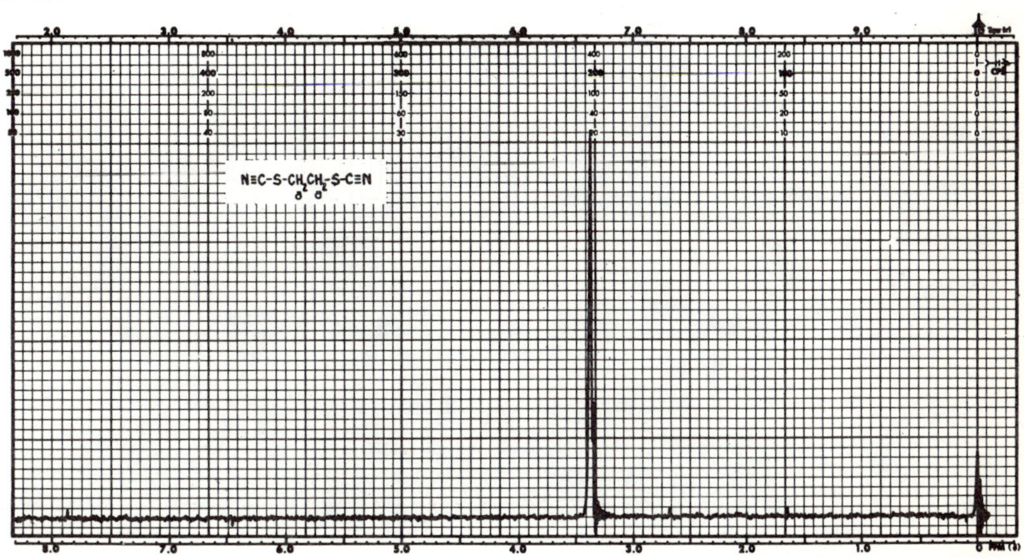

© 1978 Sadtler Research Laboratories, Inc.

THIOCYANIC ACID, TETRAMETHYLENE ESTER

$C_6H_8N_2S_2$

Mol. Wt. 172.27

B.P. 180-182°C/4mm

Solvent: $CDCl_3$

Assignments
a = 2.01
b = 3.01

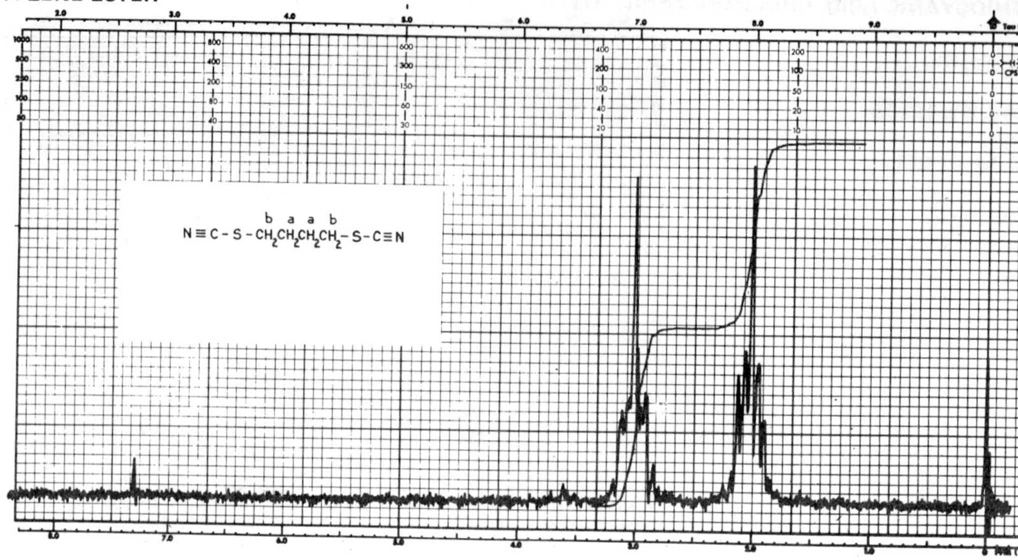

1016

© 1978 Sadtler Research Laboratories, Inc.

THIOCYANIC ACID, BENZYL ESTER

C_8H_7NS

Mol. Wt. 149.22

Solvent: $CDCl_3$

Assignments
a = 4.15
b = 7.38

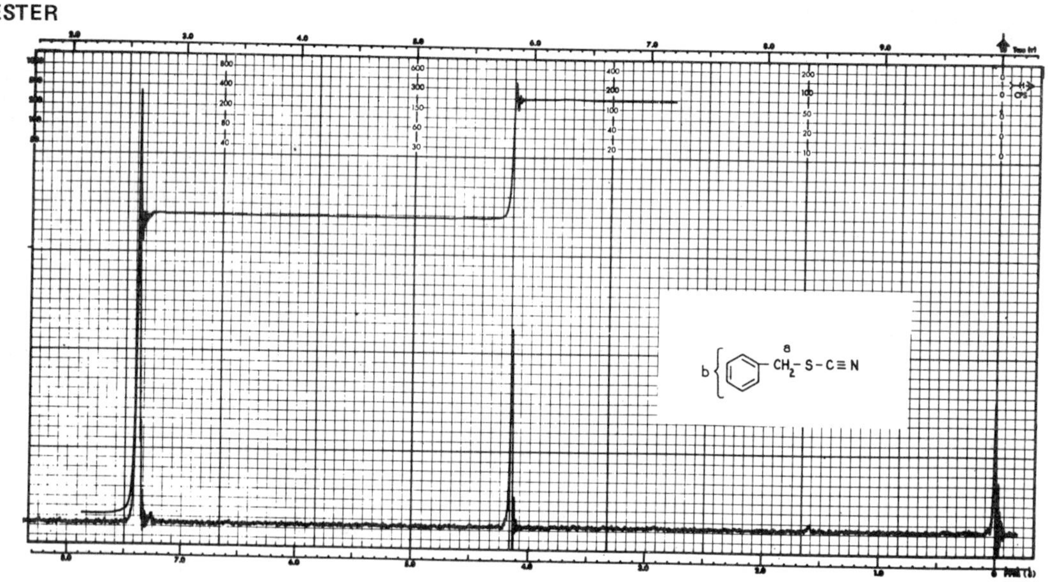

1017

© 1978 Sadtler Research Laboratories, Inc.

THIOCYANIC ACID, 2,6-DICHLOROBENZYL ESTER

$C_8H_5Cl_2NS$

Mol. Wt. 218.11

M.P. 40-42°C

Solvent: $CDCl_3$

Assignments
a = 4.51
b = ca 7.33

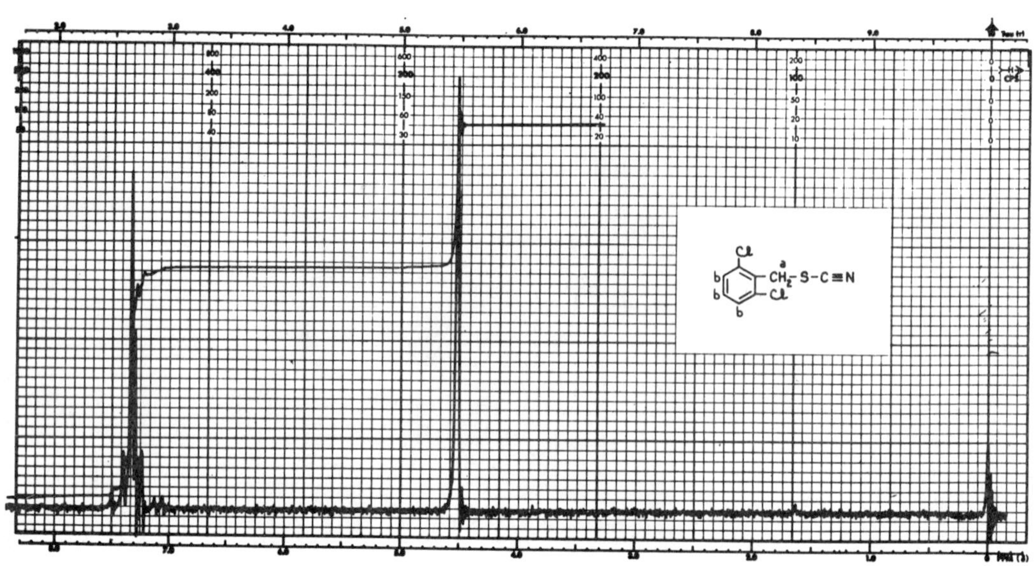

1018

© 1978 Sadtler Research Laboratories, Inc.

THIOCYANIC ACID, CINNAMYL ESTER

$C_{10}H_9NS$

Mol. Wt. 175.25

Solvent: $CDCl_3$

Assignments
a = 3.58
b = ca 6.10
c = 6.59
d = 7.25

1019

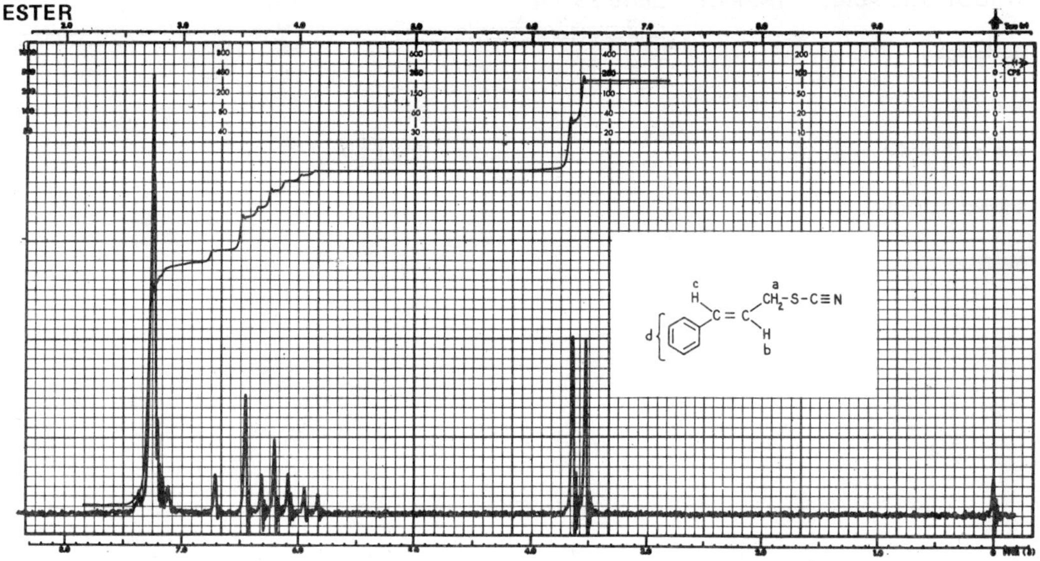

© 1978 Sadtler Research Laboratories, Inc.

THIOCYANIC ACID, 5-INDOLINYL ESTER

$C_9H_8N_2S$

Mol. Wt. 176.24

M.P. 60°C

Solvent: $CDCl_3$

Assignments
a = 2.99
b = 3.56
c = 4.18
d = 6.50
e = 7.18
f = 7.26

1020

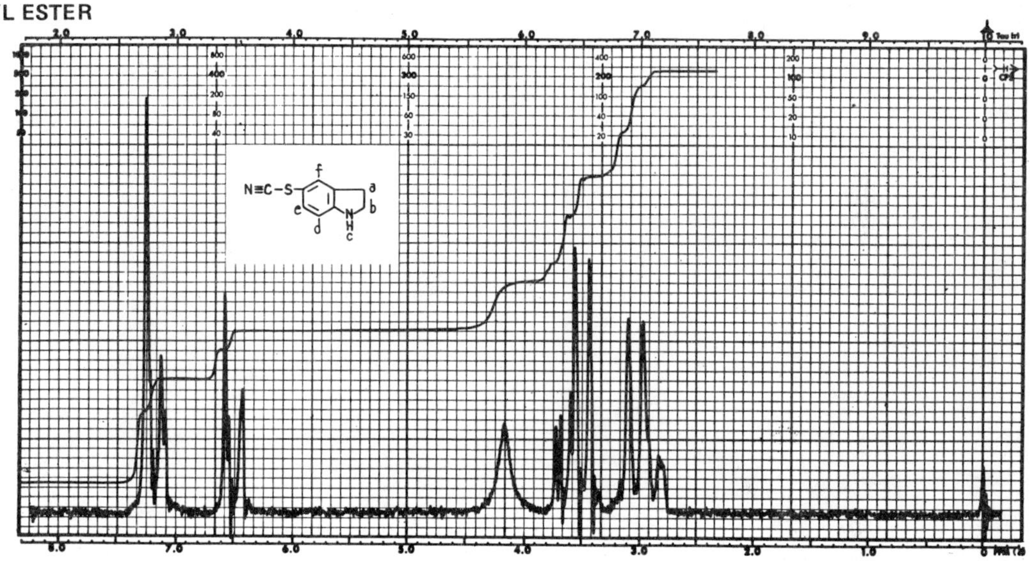

© 1978 Sadtler Research Laboratories, Inc.

NOTES

THE NITROSO COMPOUNDS

Little clear data is available for the aliphatic Nitroso compounds. The spectrum of 2-methyl-2-nitrosopropane (number 1021) presents a puzzle in that two bands are observed with an integration ratio of approximately 2:1. One explanation may be that there is restricted rotation about the C—N bond producing a different chemical shift for one of the t-Butyl methyl groups. Such restricted rotation may also be observed in the spectra of the aromatic compounds in that the protons ortho to the —N=O group are always slightly broadened in comparison to the other protons in the aromatic ring (spectra 1022, 1023, 1024).

The effect of the Nitroso group on the chemical shifts of the ortho aromatic compounds is that of a strongly deshielding group. A series of para substituted Nitroso phenyl compounds are presented below.

Chemical Shifts

Aromatic Protons

Para substituted nitroso benzenes

O=N—	ortho	meta	para	Solvent	Spectrum
	7.73	6.52	—O—Na	D_2O	- - - -
	7.70	6.60	$—N(CH_3)_2$	$CDCl_3$	1023
	7.67	6.63	—OH	Acetone	- - - -
	7.77	7.08	—NH—Ø	Polysol	1022
	7.79	6.65	$—N(R_2)_2$	$CDCl_3$	1024

2-METHYL-2-NITROSOPROPANE

C_4H_9NO

Mol. Wt. 87.12

Solvent: $CDCl_3$

Assignments
a = 1.21
b = 1.58

1021

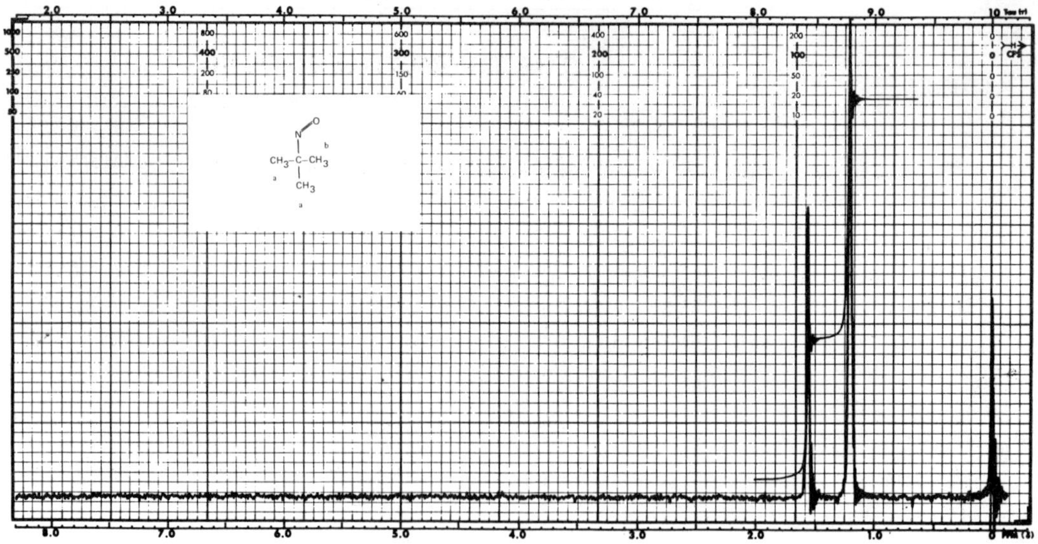

© 1978 Sadtler Research Laboratories, Inc.

4-NITROSODIPHENYLAMINE

$C_{12}H_{10}N_2O$

Mol. Wt. 198.23

M.P. 144-145°C

Solvent: Polysol-d

Assignments
a = 7.08
b = 7.10-7.50
c = 7.77
d = 3.60 H_2O

1022

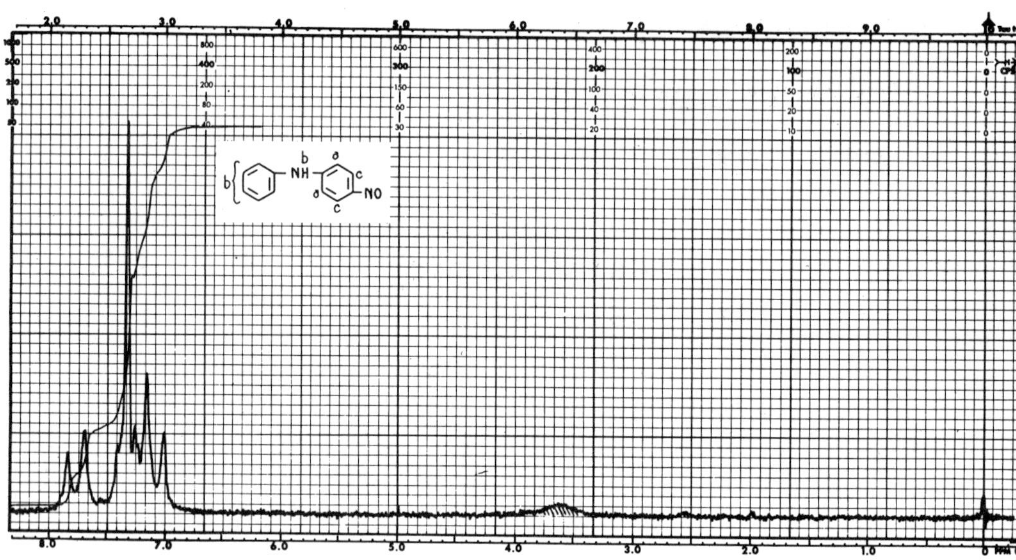

© 1978 Sadtler Research Laboratories, Inc.

N,N-DIMETHYL-p-NITROSOANILINE

$C_8H_{10}N_2O$

Mol. Wt. 150.15

M.P. 83-84°C

Solvent: $CDCl_3$

Assignments
a = 3.11
b = 6.60
c = 7.70

1023

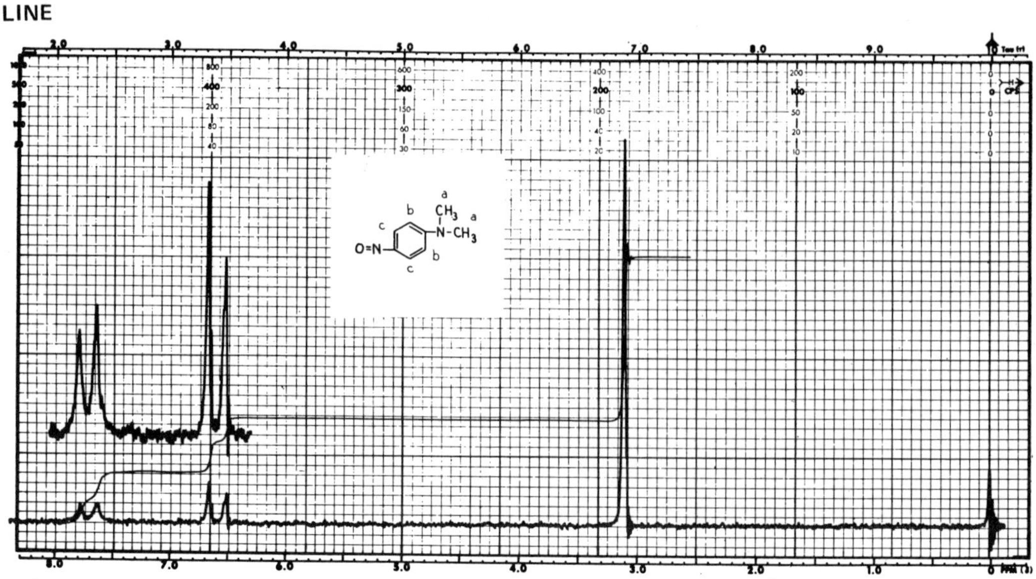

© 1978 Sadtler Research Laboratories, Inc.

N,N-DIETHYL-p-NITROSOANILINE

$C_{10}H_{14}N_2O$

Mol. Wt. 178.23

M.P. 82-83°C

Solvent: $CDCl_3$

Assignments

a = 1.23

b = 3.48

c = 6.65

d = ca 7.79

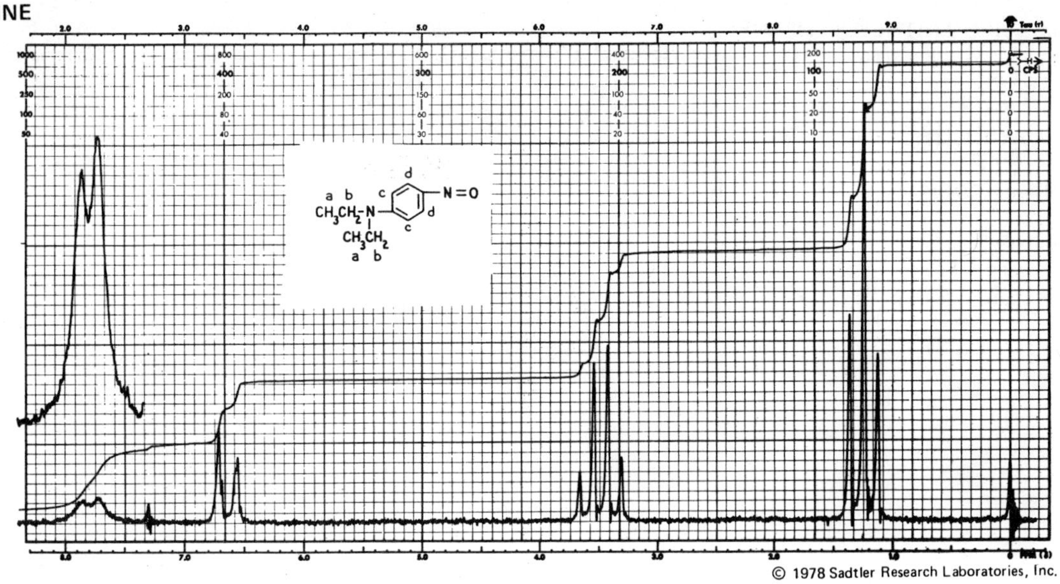

© 1978 Sadtler Research Laboratories, Inc.

1024

NOTES

Non-equivalence of similar groups bonded to the Nitrosoamine nitrogen nucleus is a characteristic of these compounds. Due to restricted rotation about the N—N bond the groups bonded to the amine nitrogen can reside either syn or anti to the nitroso oxygen atom producing a differentiation in their chemical shifts. The group syn to the oxygen atom usually resonates at higher field and may display a certain degree of broadening in comparison to the group in the anti position.

The differentiation in chemical shift decreases with distance from the amine nitrogen atom, i.e. the alpha groups differ in chemical shift by about 0.6 ppm, the beta groups by about 0.3 ppm and the gamma groups by about 0.1 ppm.

Spectrum 1031 is interesting in that only one form is observed. Apparently due to the bulk of the phenyl group it is always anti to the oxygen atom and the methyl group is syn.

Chemical Shifts

Aliphatic Protons

HC—3	HC—2	HC—1		Solvent	Spectrum
		(2.95)	$(CH_3)_2$—N—N=O (syn)	CCl_4	1025
		(3.71)	$(CH_3)_2$—N—N=O (anti)	CCl_4	1025
	(1.11	3.60)	$(CH_3$—$CH_2)_2$—N—N=O (syn)	$CDCl_3$	1026
	(1.41	4.12)	$(CH_3$—$CH_2)_2$—N—N=O (anti)	$CDCl_3$	1026
(0.88	1.50	3.48)	$(CH_3$—CH_2—$CH_2)_2$—N—N=O (syn)	CCl_4	1027
(0.97	1.80	4.06)	$(CH_3$—CH_2—$CH_2)_2$—N—N=O (anti)	CCl_4	1027

Aromatic Protons

As spectrum 1031 indicates, the n-nitroso group deshields all of the aromatic hydrogens of the anti ring forming a complex, higher order pattern centered at about 7.45 ppm. In spectrum 1032 one observes that two ortho hydrogens are slightly shielded and probably represent the ortho hydrogens of the ring syn to the N-nitroso oxygen atom.

N-NITROSODIMETHYLAMINE

$C_2H_6N_2O$

Mol. Wt. 74.08

B.P. 150-152°C

Solvent: CCl_4

Assignments
a = 2.95
b = 3.71

1025

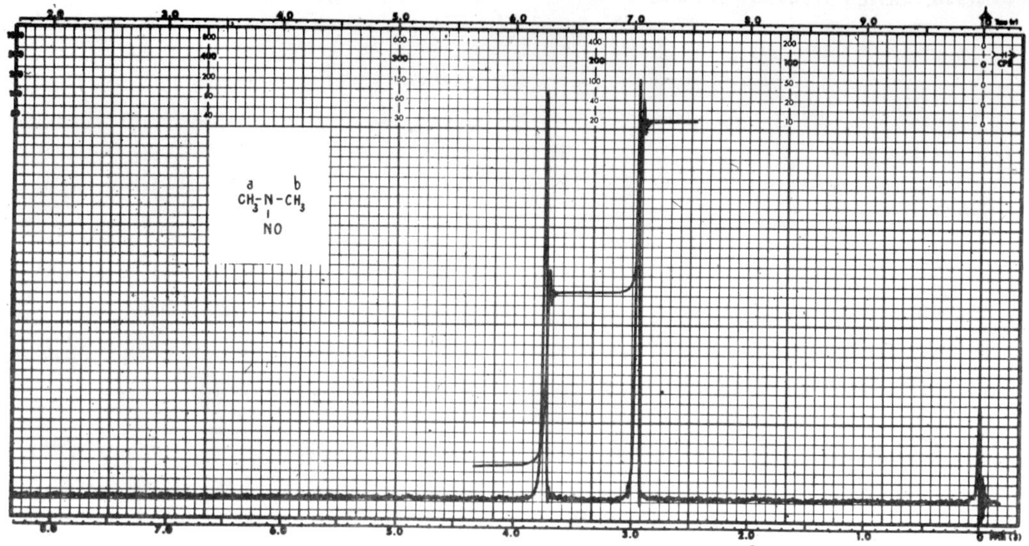

© 1978 Sadtler Research Laboratories, Inc.

N-NITROSODIETHYLAMINE

$C_4H_{10}N_2O$

Mol. Wt. 102.14

B.P. 59-60°C/8-9mm

Solvent: $CDCl_3$

Assignments
a = 1.11
b = 1.41
c = 3.60
d = 4.12

1026

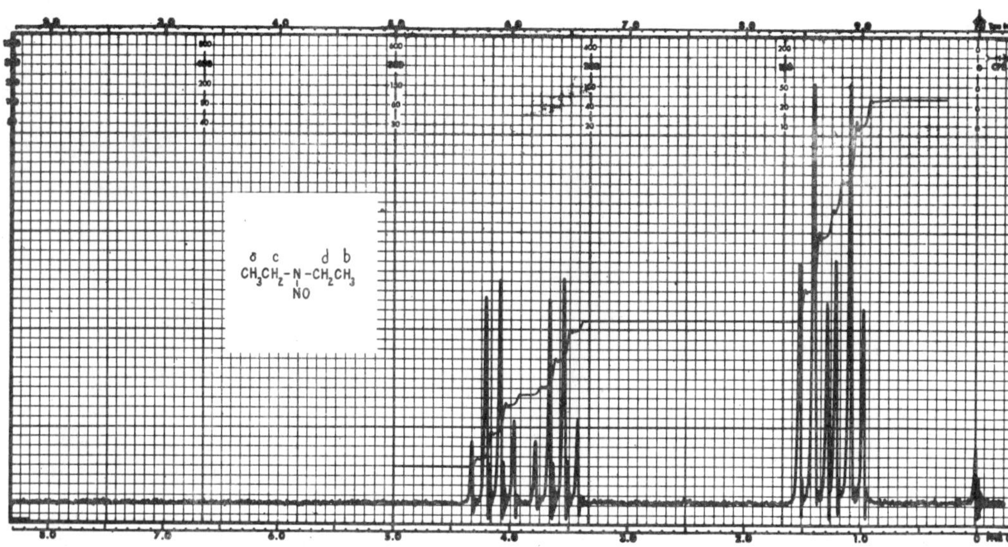

© 1978 Sadtler Research Laboratories, Inc.

N-NITROSODIPROPYLAMINE

$C_6H_{14}N_2O$

Mol. Wt. 130.19

B.P. 77-78°C/8mm

Solvent: CCl_4

Assignments
a = 0.88
b = 0.97
c = 1.00-2.05
d = 3.48
e = 4.06

1027

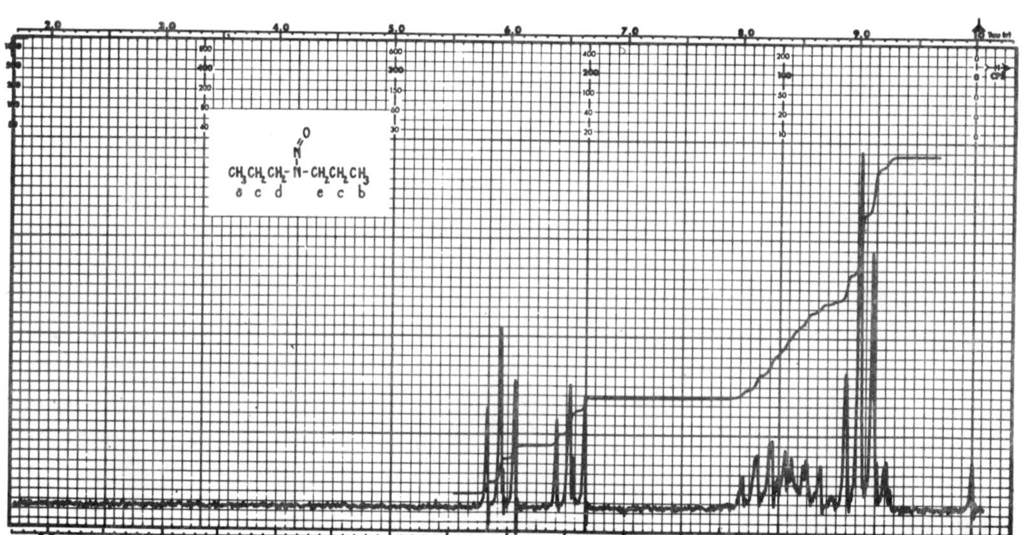

© 1978 Sadtler Research Laboratories, Inc.

N-NITROSODICYCLOHEXYLAMINE

$C_{12}H_{22}N_2O$

Mol. Wt. 210.32

M.P. 105-108°C

Solvent: $CDCl_3$

Assignments
a = 0.91-2.20
b = 3.75 or c
c = 4.83 or b

1028

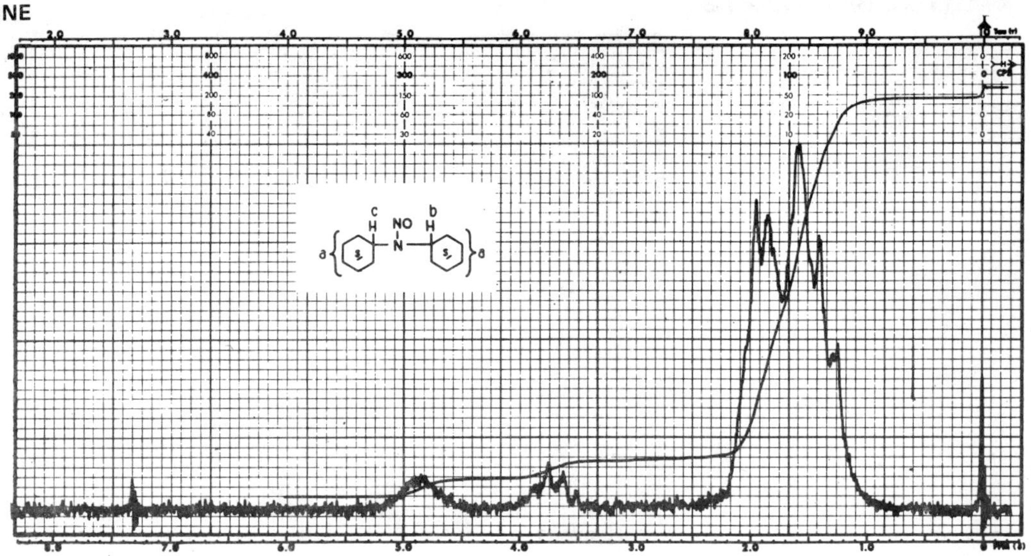

© 1978 Sadtler Research Laboratories, Inc.

1-NITROSOPYRROLIDINE

$C_4H_8N_2O$

Mol. Wt. 100.12

B.P. 74°C/4mm

Solvent: D_2O

Assignments
a = 2.08
b = 3.52
c = 4.21
d = 4.51 HDO

1029

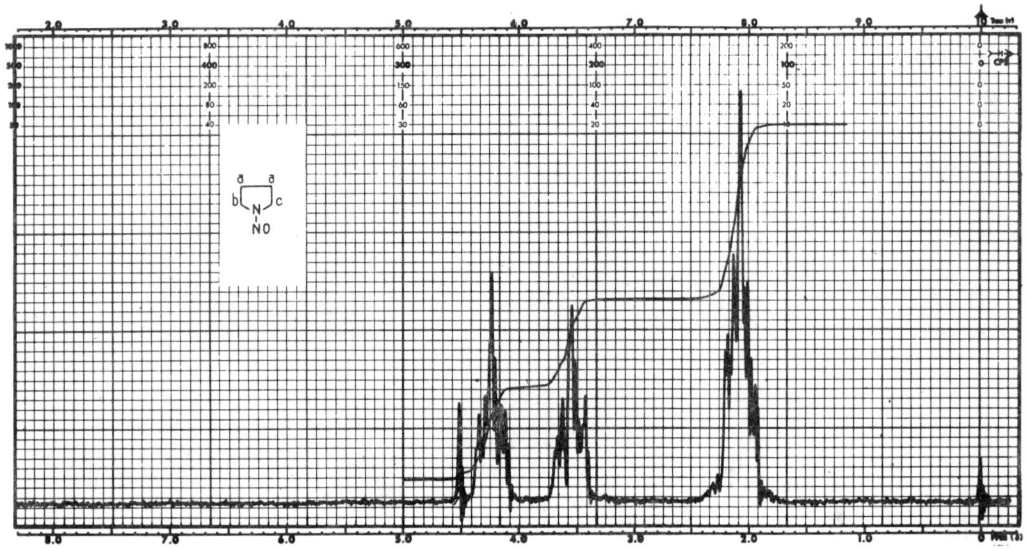

© 1978 Sadtler Research Laboratories, Inc.

1-NITROSOPIPERIDINE

$C_5H_{10}N_2O$

Mol. Wt. 114.15

B.P. 109-110°C/23mm

Solvent: $CDCl_3$

Assignments
a = 1.30-2.00
b = 3.78
c = 4.19

1030

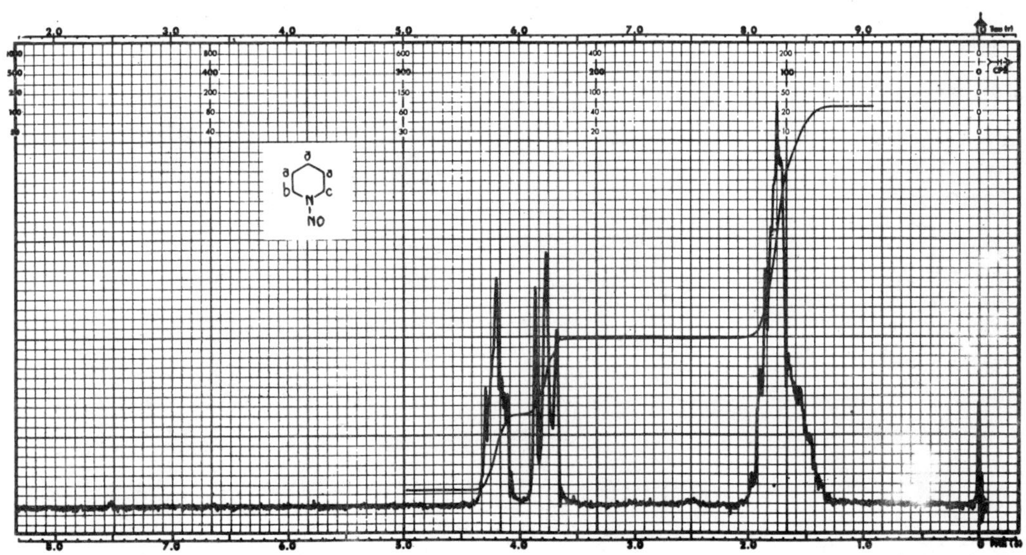

© 1978 Sadtler Research Laboratories, Inc.

N-METHYL-N-NITROSOANILINE

$C_7H_8N_2O$

Mol. Wt. 136.15

M.P. 14-15°C

Solvent: $CDCl_3$

Assignments
a = 3.39
b = 7.20-7.60

1031

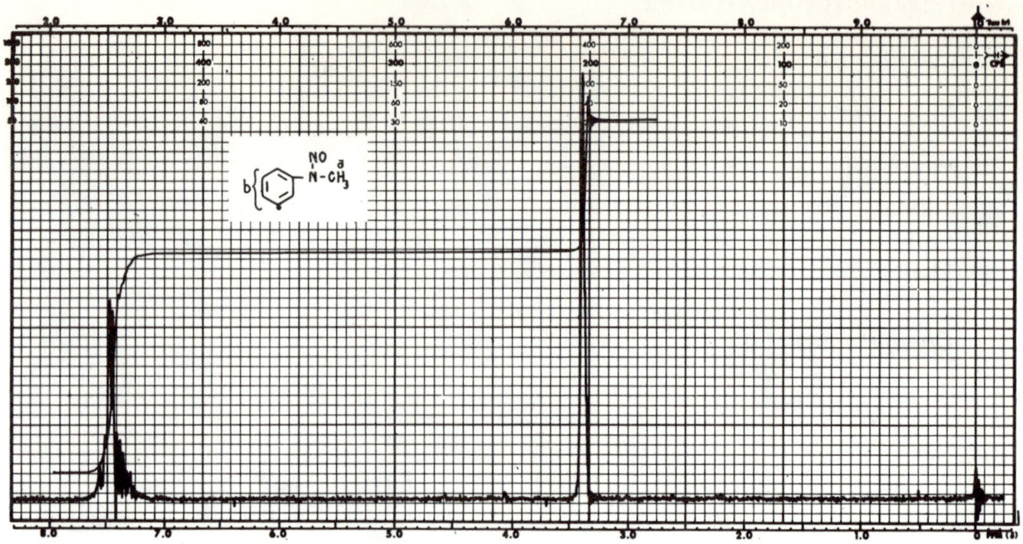

© 1978 Sadtler Research Laboratories, Inc.

N-NITROSODIPHENYLAMINE

$C_{12}H_{10}N_2O$

Mol. Wt. 198.23

M.P. 65-66°C

Solvent: $CDCl_3$

Assignments
a = 7.01
b = 7.15-7.58

1032

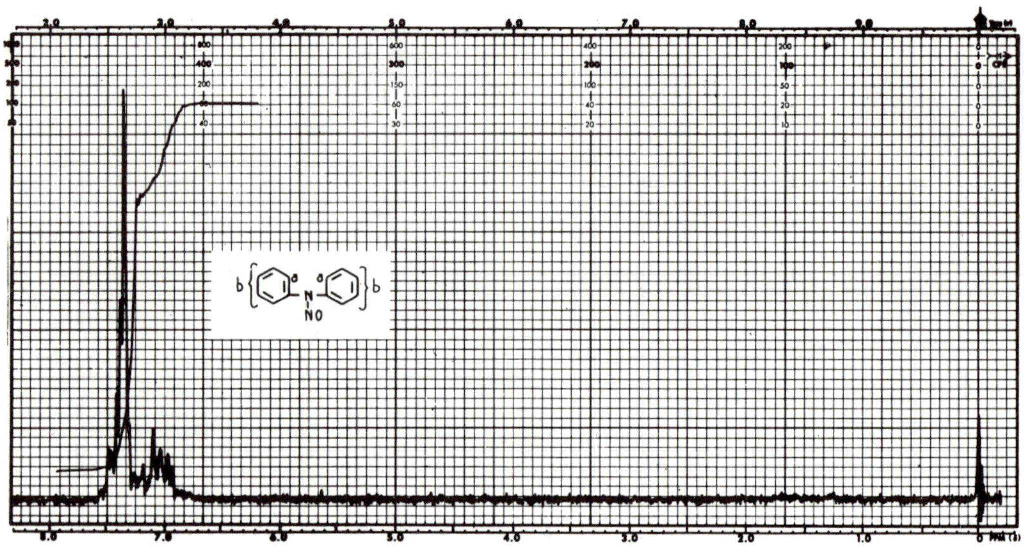

© 1978 Sadtler Research Laboratories, Inc.

THE NITRITES

The Nitrite group is one of the most strongly deshielding substituents in its effect on the alpha aliphatic groups. Methylene groups are deshielded to about 4.5 ppm and methines to about 5.5 ppm. Such extremes of chemical shift are characteristic of only a few substituents making the identification of a Nitrite compound a relatively easy matter.

Chemical Shifts

Aliphatic Protons

HC–3	HC–2	HC–1		Solvent	Spectrum
0.97	1.72	4.61	$CH_3–CH_2–CH_2–O–N=O$	CCl_4	1033
	(1.40)	5.59	$(CH_3)_2–CH–O–N=O$	CCl_4	- - - -
	(1.57)	- - - -	$(CH_3)_3–C–O–N=O$	$CDCl_3$	1036
(0.98)	1.98	4.45	$(CH_3)_2–CH–CH_2–O–N=O$	CCl_4	1035

Coupling and Coupling Constants

The nitrite compounds display no unusual coupling nor coupling constants. The normal vicinal aliphatic proton coupling of about 7 Hz is observed.

Solubility and Solvent Effects

The aliphatic nitrites are readily soluble in the halogenated NMR solvents; carbon tetrachloride and deuterochloroform. No special solvent effects have been noted.

PROPYL NITRITE

$C_3H_7NO_2$

Mol. Wt. 89.09

Solvent: CCl_4

Assignments
a = 0.97
b = 1.72
c = 4.61

1033

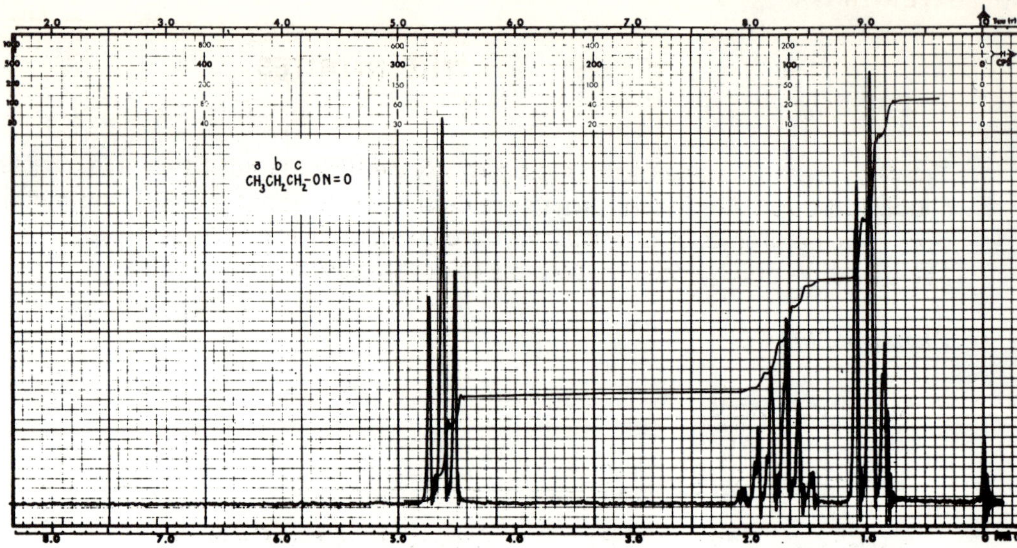

© 1978 Sadtler Research Laboratories, Inc.

---

sec-BUTYL NITRITE

$C_4H_9NO_2$

Mol. Wt. 103.12

Solvent: CCl_4

Assignments
a = 0.96
b = 1.39
c = 1.73
d = 5.40

1034

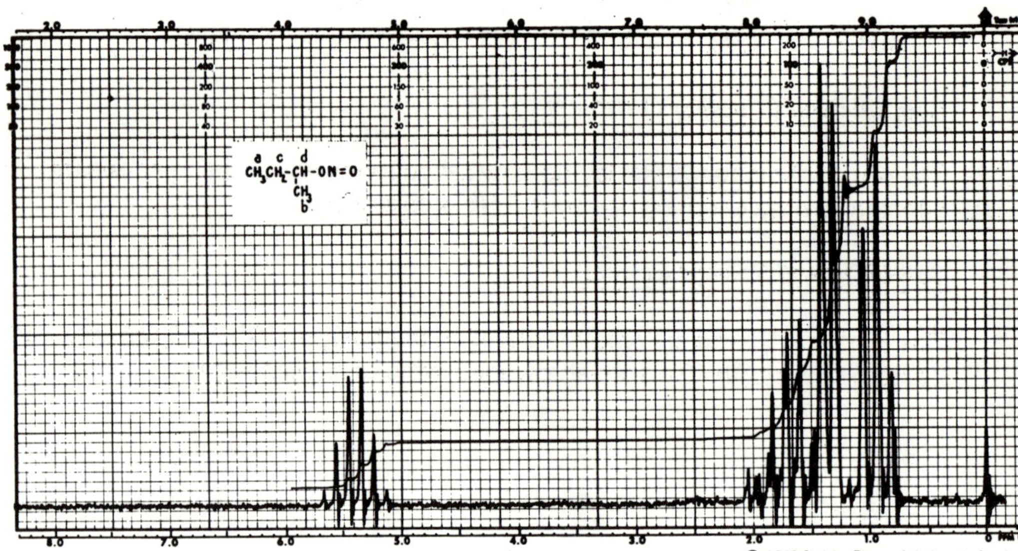

© 1978 Sadtler Research Laboratories, Inc.

ISOBUTYL NITRITE

$C_4H_9NO_2$

Mol. Wt. 103.12

Solvent: CCl_4

Assignments
a = 0.98
b = 1.98
c = 4.45
d = 2.09 imp.

1035

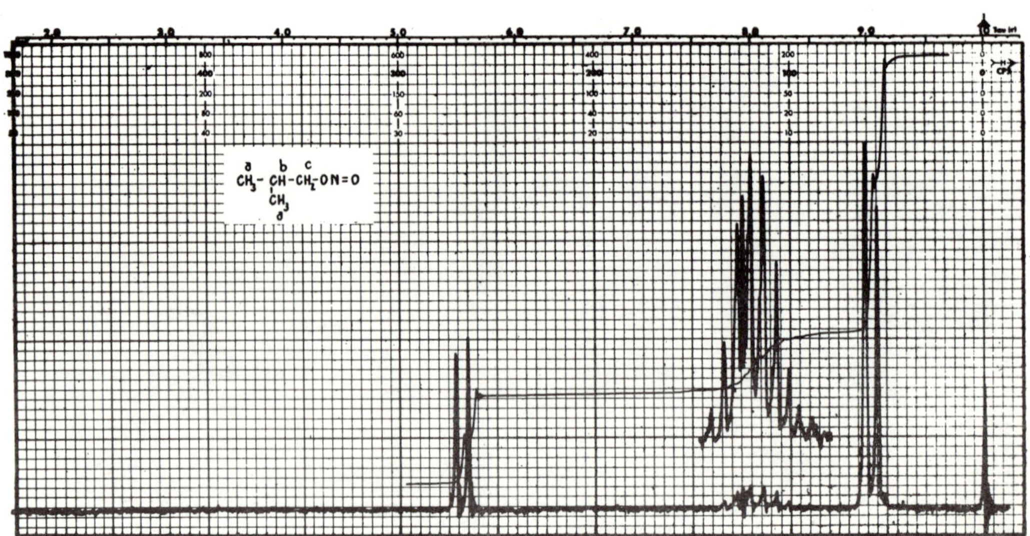

© 1978 Sadtler Research Laboratories, Inc.

tert-BUTYL NITRITE

$C_4H_9NO_2$

Mol. Wt. 103.12

B.P. 62-64°C

Solvent: $CDCl_3$

Assignments
a = 1.57

1036

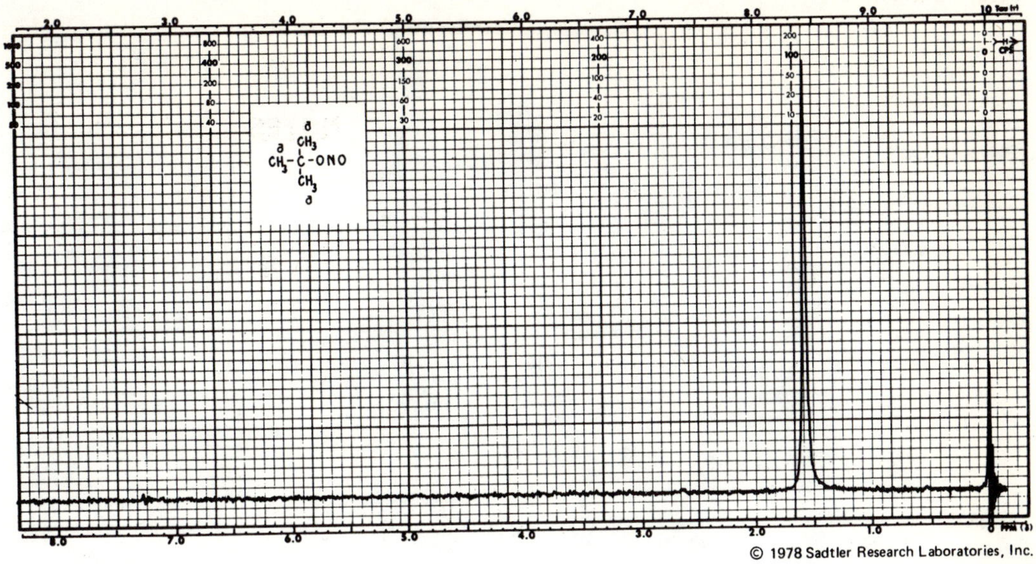

© 1978 Sadtler Research Laboratories, Inc.

ISOPENTYL NITRITE

$C_5H_{11}NO_2$

Mol. Wt. 117.15

B.P. 96.5-98.5°C

Solvent: $CDCl_3$

Assignments
a = 0.97
b = 1.00-2.20
c = 4.70
d = 4.51 imp.

1037

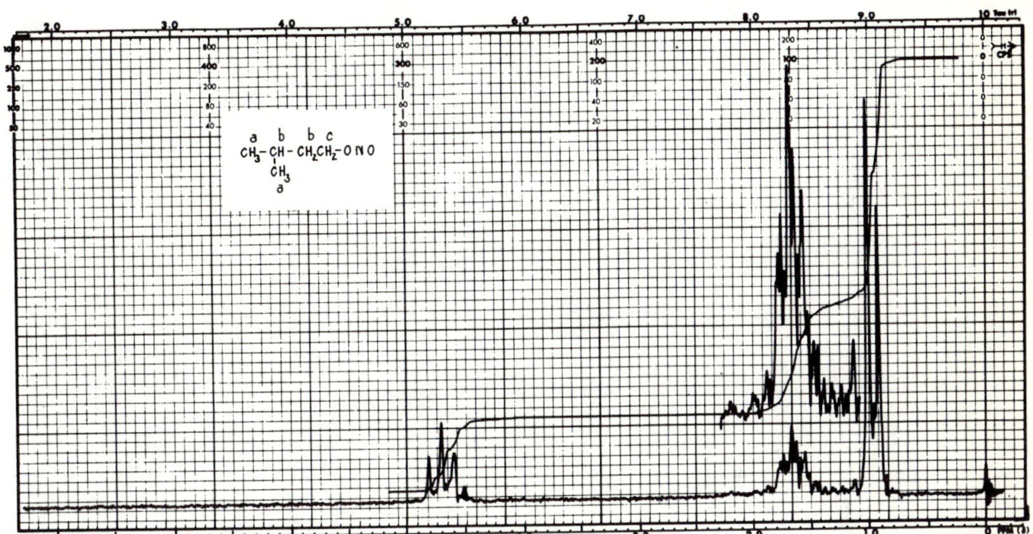

© 1978 Sadtler Research Laboratories, Inc.

OCTYL NITRITE

$C_8H_{17}NO_2$

Mol. Wt. 159.23

B.P. 52-53°C/6mm

Solvent: CCl_4

Assignments
a = 0.91
b = 1.10-1.60
c = 1.74
d = 4.62

1038

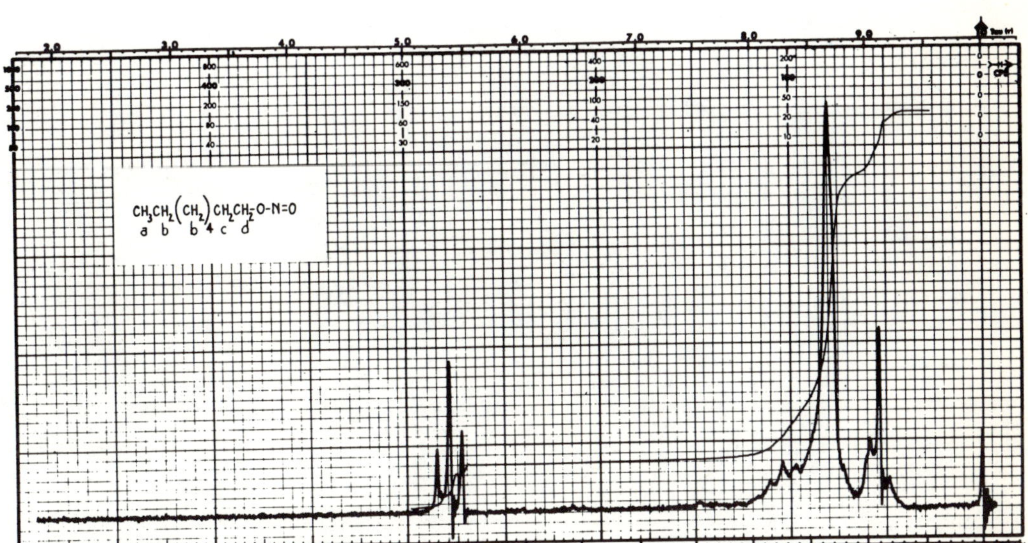

© 1978 Sadtler Research Laboratories, Inc.

NOTES

The Nitro functional group is one of the very few substituents which strongly deshield both adjacent aliphatic groups and the ortho aromatic protons. The group imparts no other distinguishing characteristics to the proton NMR spectrum.

Chemical Shifts

Aliphatic Protons

HC–3	HC–2	HC–1		Solvent	Spectrum
		4.22	$CH_3–NO_2$	CCl_4	1039
	1.55	4.40	$CH_3–CH_2–NO_2$	CCl_4	1040
1.01	2.00	4.31	$CH_3–CH_2–CH_2–NO_2$	CCl_4	1041
	(1.57)	4.65	$(CH_3)_2–CH–NO_2$	$CDCl_3$	1042
	(1.61)	----	$(CH_3)_3–C–NO_2$	CCl_4	1043

Aromatic Protons

Nitrobenzene

meta/para	ortho		Solvent	Spectrum
7.3–7.8	8.20	$–NO_2$ ·	CCl_4	----

Para Substituted Nitrobenzenes

$O_2N–$ ortho	meta	para	Solvent
7.84	6.19	–O–Na	DMSO
8.09	6.67	$–NH_2$	DMSO-d_6
8.12	6.91	$–O–CH_3$	CCl_4
8.15	7.00	–OH	Acetone
8.05	7.09	–NH–Ø	Acetone
8.20	7.20	–F	CCl_4
8.10	7.30	$–CH_3$	$CDCl_3$
8.19	7.32	–N=C=S	$CDCl_3$
8.09	7.46	–Cl	$CDCl_3$
8.03	7.62	–Br	$CDCl_3$
7.78	7.62	$–C(=O)–NH_2$	DMSO
8.26	7.82	–Ø	Acetone

NITROMETHANE

CH_3NO_2

Mol. Wt. 61.04

B.P. 101.5/765mm

Solvent: CCl_4

Assignments
a = 4.22

1039

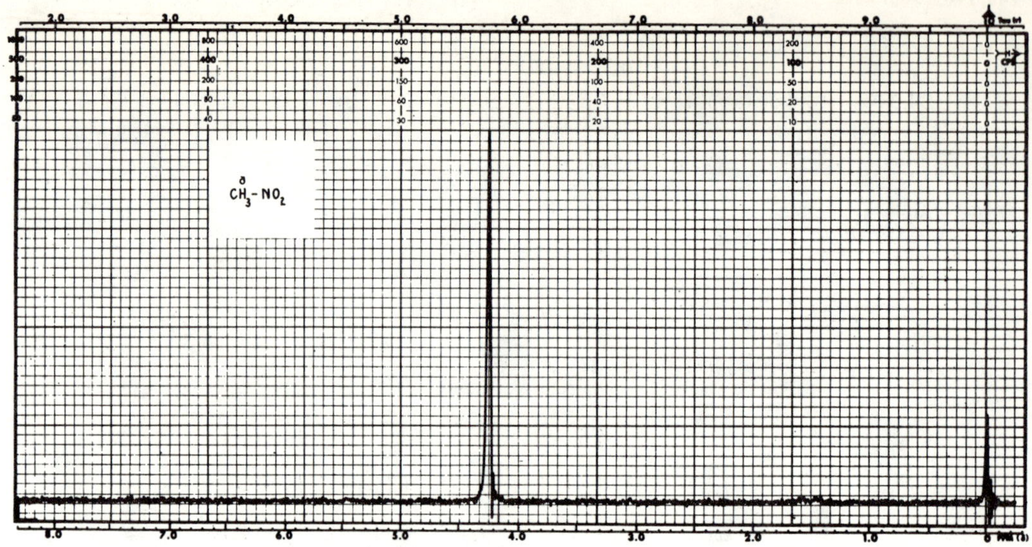

© 1978 Sadtler Research Laboratories, Inc.

NITROETHANE

$C_2H_5NO_2$

Mol. Wt. 75.05

Solvent: CCl_4

Assignments
a = 1.55
b = 4.40
c = 1.48 imp.

1040

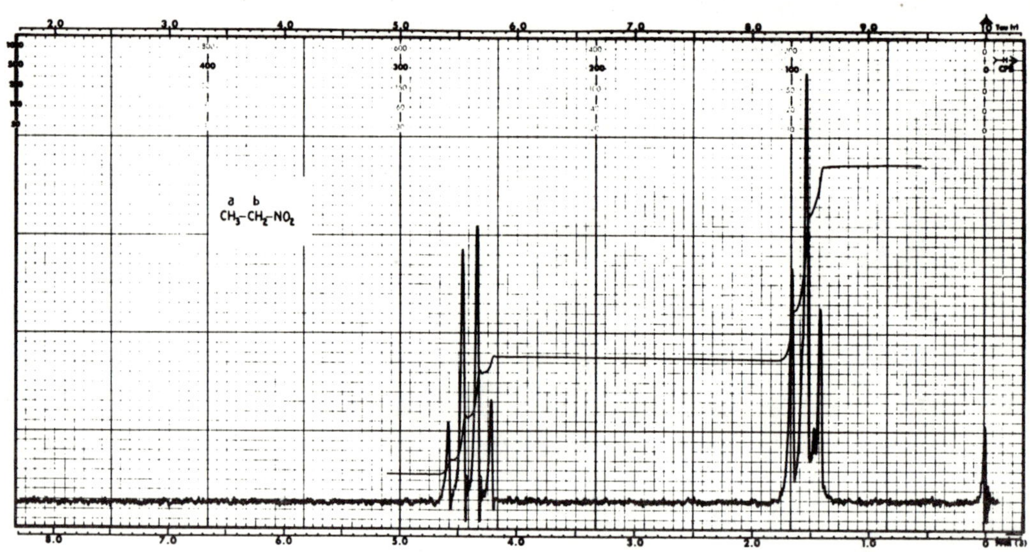

© 1978 Sadtler Research Laboratories, Inc.

1-NITROPROPANE

$C_3H_7NO_2$

Mol. Wt. 89.09

Solvent: CCl_4

Assignments
a = 1.01
b = 2.00
c = 4.31
d = 1.48, 1.59 imps.

1041

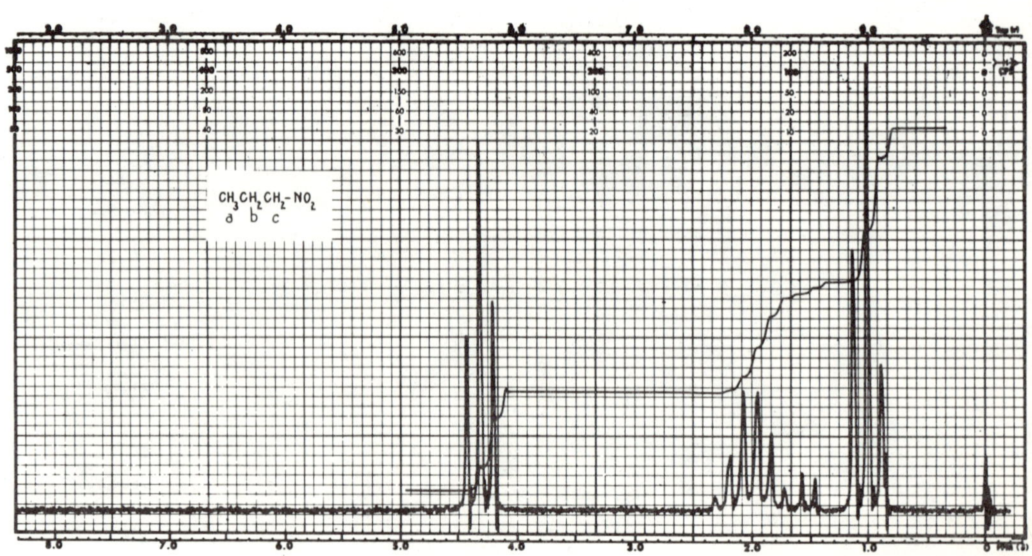

© 1978 Sadtler Research Laboratories, Inc.

2-NITROPROPANE

$C_3H_7NO_2$

Mol. Wt. 89.09

B.P. 120°C

Solvent: $CDCl_3$

Assignments
a = 1.57
b = 4.65

1042

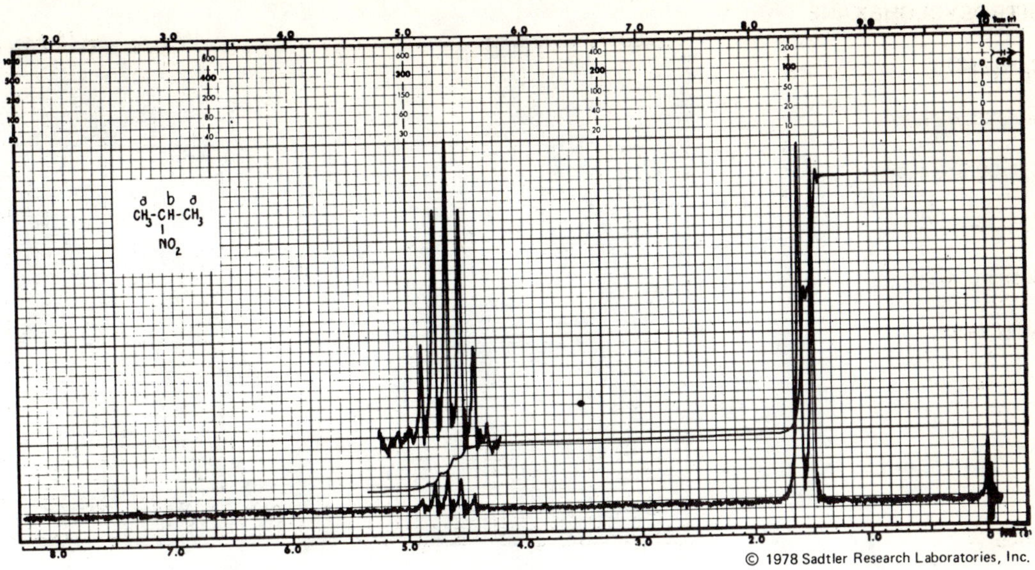

© 1978 Sadtler Research Laboratories, Inc.

2-METHYL-2-NITROPROPANE

$C_4H_9NO_2$

Mol. Wt. 103.12

M.P. 24-26°C

Solvent: CCl_4

Assignments
a = 1.61

1043

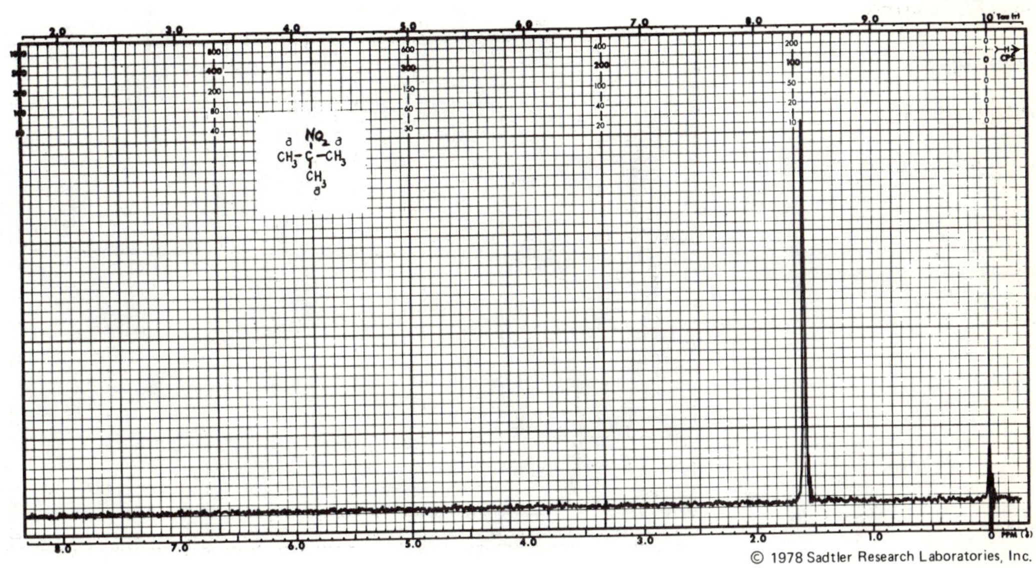

© 1978 Sadtler Research Laboratories, Inc.

1-CHLORO-1-NITROETHANE

$C_2H_4ClNO_2$

Mol. Wt. 109.51

B.P. 122-128.5°C

Solvent: CCl_4

Assignments
a = 2.02
b = 5.88

1044

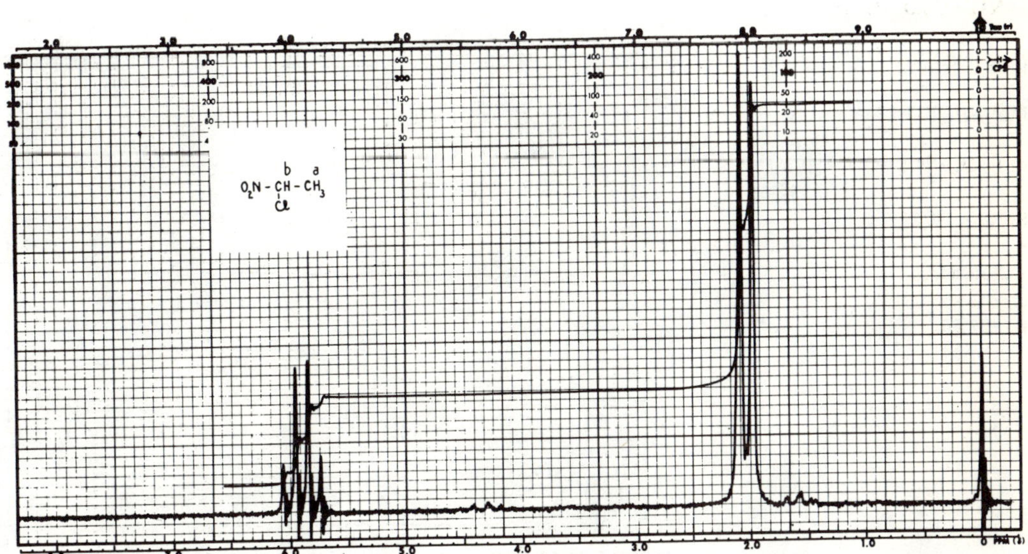

© 1978 Sadtler Research Laboratories, Inc.

NITROCYCLOHEXANE

$C_6H_{11}NO_2$

Mol. Wt. 129.16

B.P. 20.5°C (dec.)

F.P. 35-69°C

Solvent: CCl_4

Assignments
a = 1.00-2.50
b = 4.29

1045

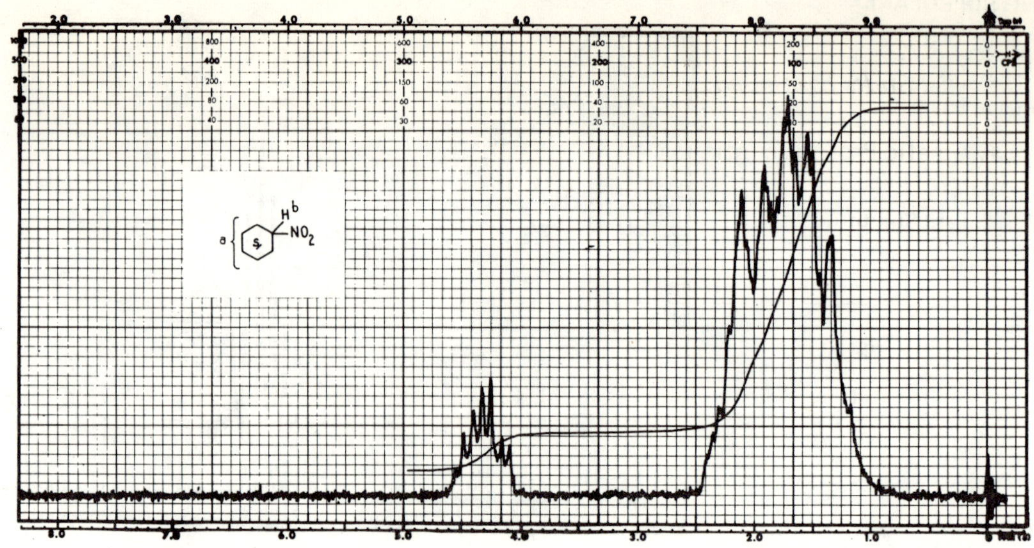

© 1978 Sadtler Research Laboratories, Inc.

o-NITROTOLUENE

$C_7H_7NO_2$

Mol. Wt. 137.14

B.P. 110-112°C/20mm (lit.)

Solvent: CCl_4

Assignments
a = 2.57
b = ca 7.33
c = 7.89

1046

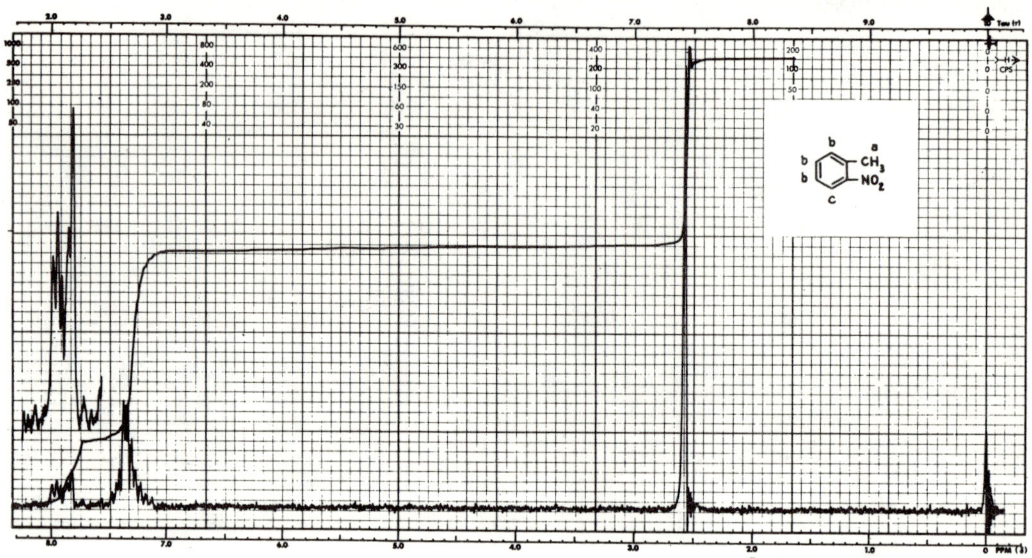

© 1978 Sadtler Research Laboratories, Inc.

m-NITROTOLUENE

$C_7H_7NO_2$

Mol. Wt. 137.13

M.P. 15.5°C (lit.)

Solvent: CCl_4

Assignments
a = 2.45
b = 7.38
c = 7.94

1047

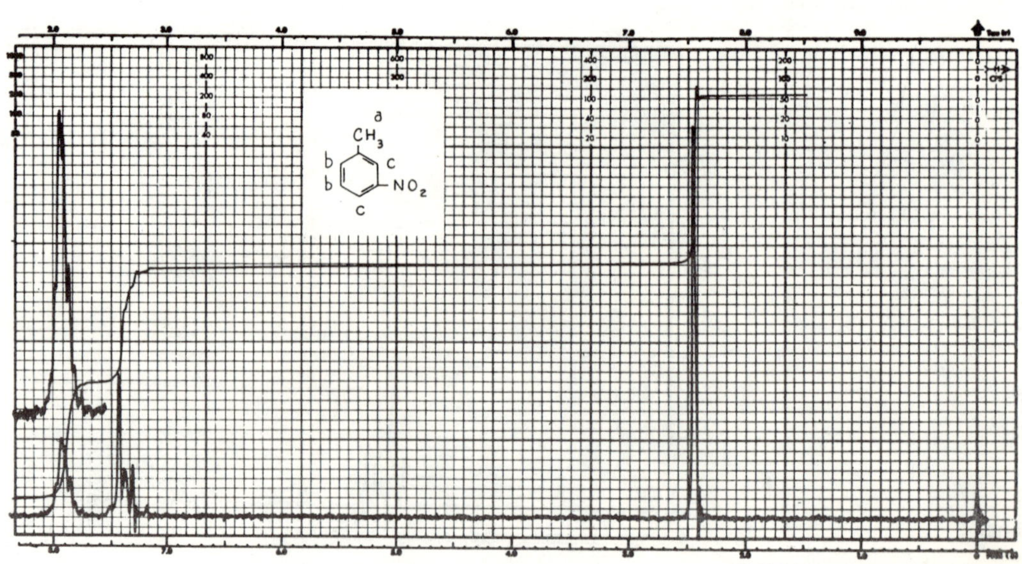

© 1978 Sadtler Research Laboratories, Inc.

p-NITROTOLUENE

$C_7H_7NO_2$

Mol. Wt. 137.14

M.P. 51-52°C

Solvent: $CDCl_3$

Assignments
a = 2.45
b = 7.30
c = 8.10

1048

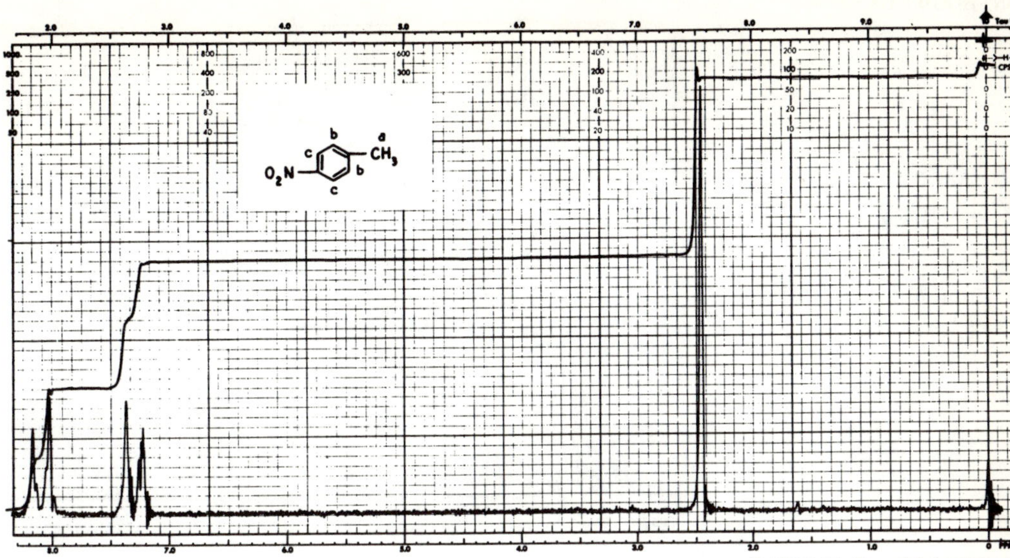

© 1978 Sadtler Research Laboratories, Inc.

1-ETHYL-2-NITROBENZENE

$C_8H_9NO_2$

Mol. Wt. 151.17

B.P. 228°C

Solvent: CCl_4

Assignments
a = 1.26
b = 2.88
c = 7.10-7.60
d = 7.75

1049

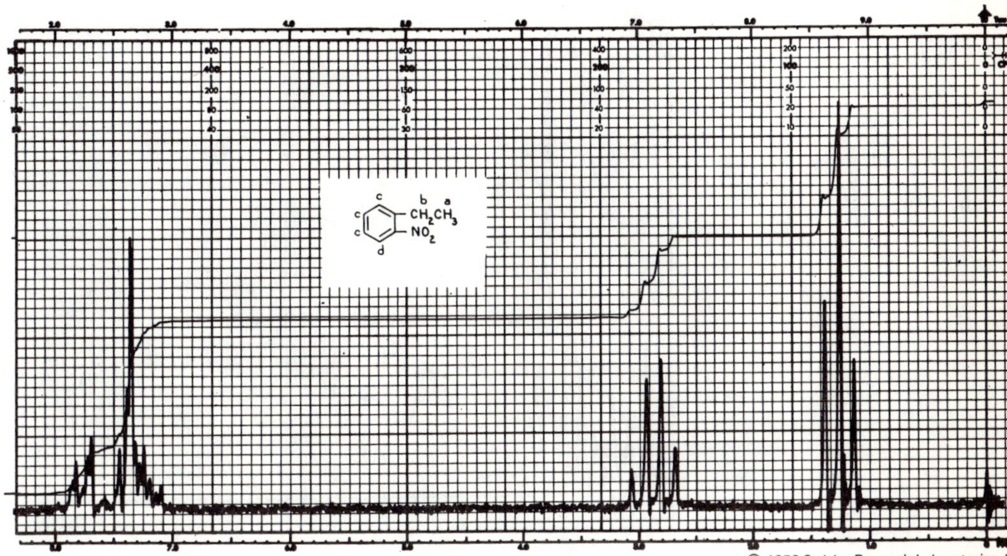

© 1978 Sadtler Research Laboratories, Inc.

1-ETHYL-4-NITROBENZENE

$C_8H_9NO_2$

Mol. Wt. 151.17

Solvent: CCl_4

Assignments
a = 1.27
b = 2.74
c = 7.30
d = 8.03

1050

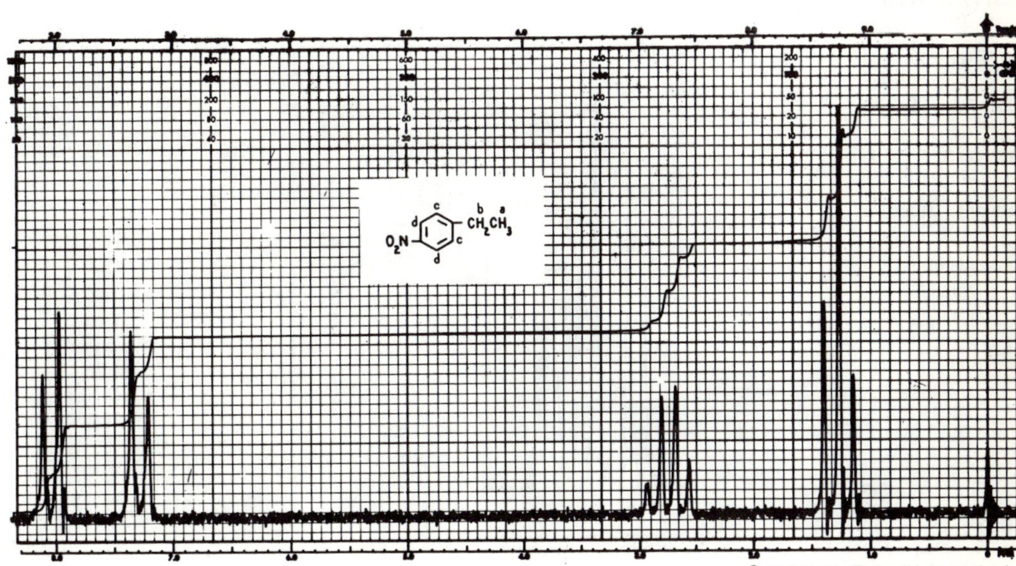

© 1978 Sadtler Research Laboratories, Inc.

2-NITROBIPHENYL

$C_{12}H_9NO_2$

Mol. Wt. 199.21

B.P. 325°C

Solvent: $CDCl_3$

Assignments
a = 7.20-7.70
b = 7.81

1051

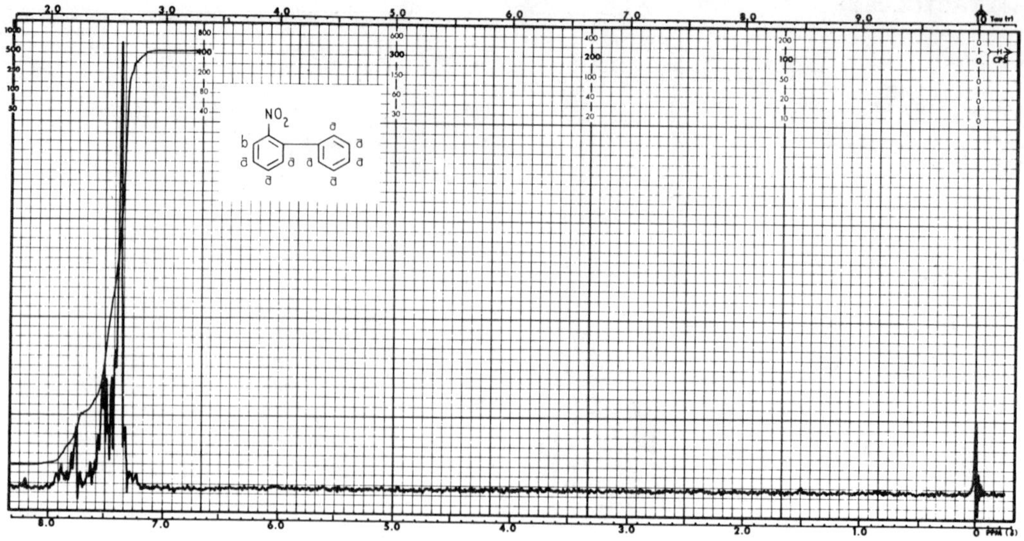

© 1978 Sadtler Research Laboratories, Inc.

3-NITROBIPHENYL

$C_{12}H_9NO_2$

Mol. Wt. 199.21

M.P. 58-60°C

Solvent: $CDCl_3$

Assignments
a = 7.20-7.65
b = 7.79
c = 8.08
d = 8.30

1052

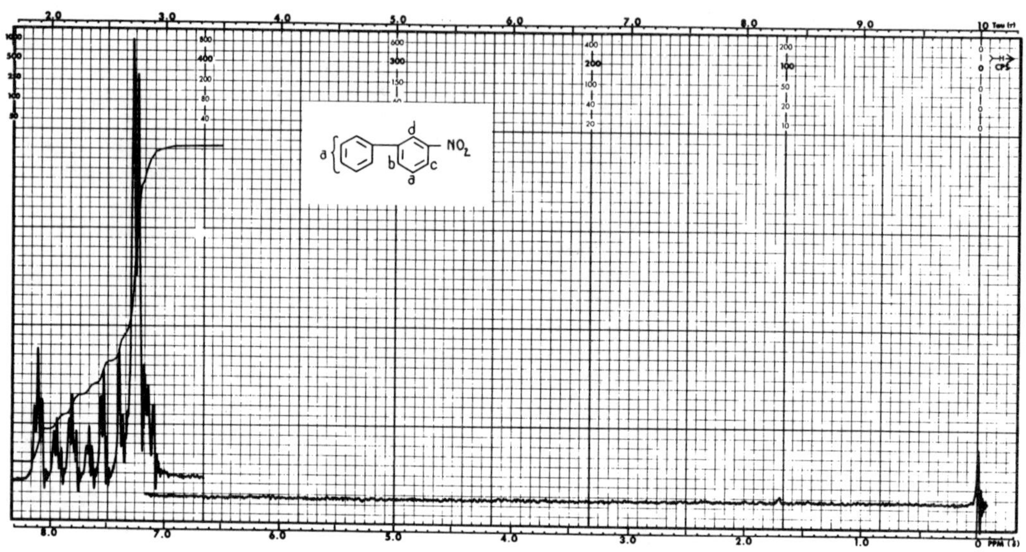

© 1978 Sadtler Research Laboratories, Inc.

4-NITROBIPHENYL

$C_{12}H_9NO_2$

Mol. Wt. 199.21

M.P. 113°C (lit.)

Solvent: Acetone

Assignments
a = 7.32-7.89
b = 7.82
c = 8.26

1053

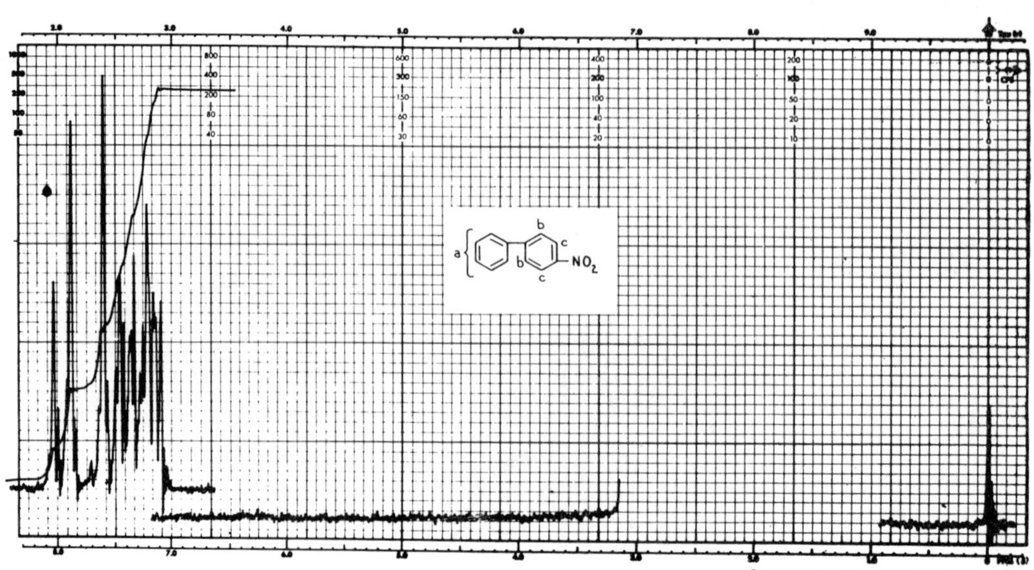

© 1978 Sadtler Research Laboratories, Inc.

1-FLUORO-2-NITROBENZENE

$C_6H_4FNO_2$

Mol. Wt. 141.10

B.P. 105-108°C/18mm

Solvent: CCl_4

Assignments

a = 7.10-7.50

b = 7.66

c = 8.03

1054

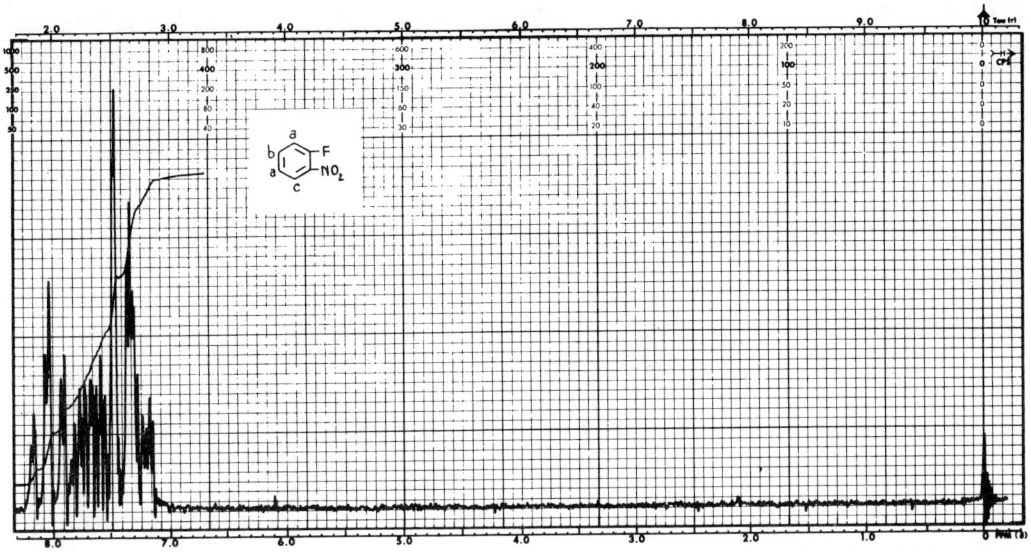

© 1978 Sadtler Research Laboratories, Inc.

1-FLUORO-4-NITROBENZENE

$C_6H_4FNO_2$

Mol. Wt. 141.10

M.P. 25-27°C (lit.)

Solvent: CCl_4

Assignments

a = 7.20

b = 8.20

1055

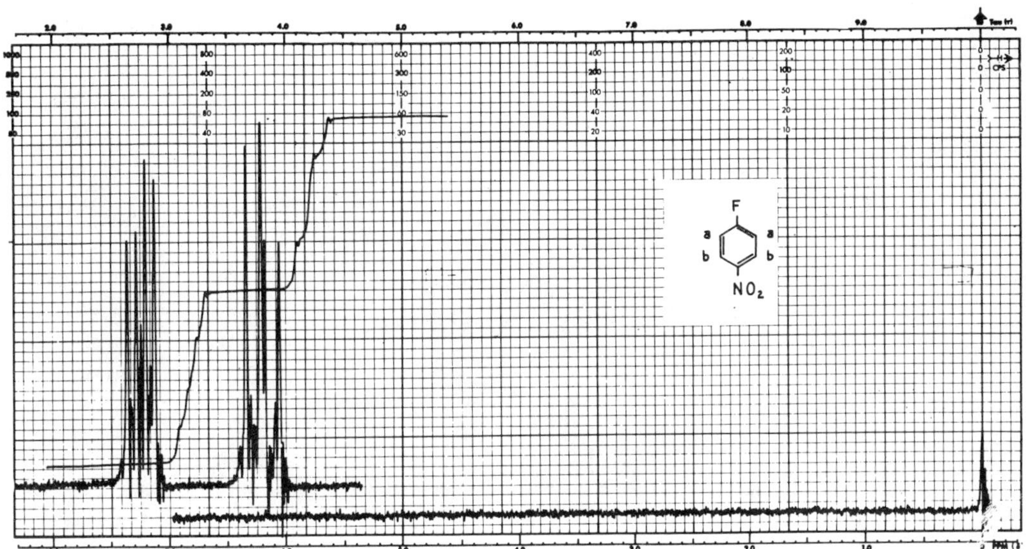

© 1978 Sadtler Research Laboratories, Inc.

1-CHLORO-2-NITROBENZENE

$C_6H_4ClNO_2$

Mol. Wt. 157.56

M.P. 31.7°C

Solvent: $CDCl_3$

Assignments

a = 7.19-7.59

b = 7.79

1056

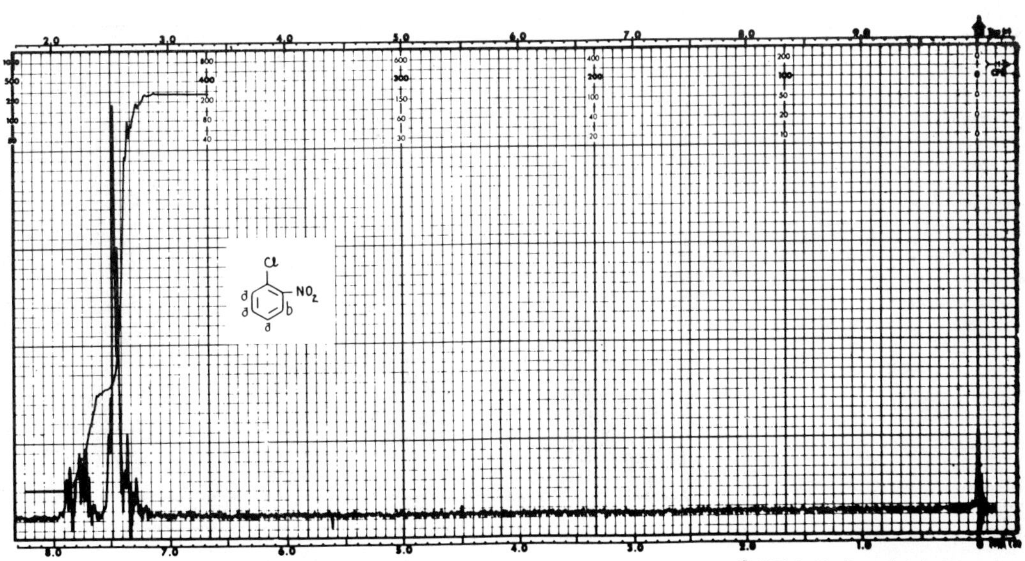

© 1978 Sadtler Research Laboratories, Inc.

1-CHLORO-3-NITROBENZENE

$C_6H_4ClNO_2$

Mol. Wt. 157.56

M.P. 43.9°C

Solvent: $CDCl_3$

Assignments
a = 7.53
b = 7.67
c = ca 8.10

1057

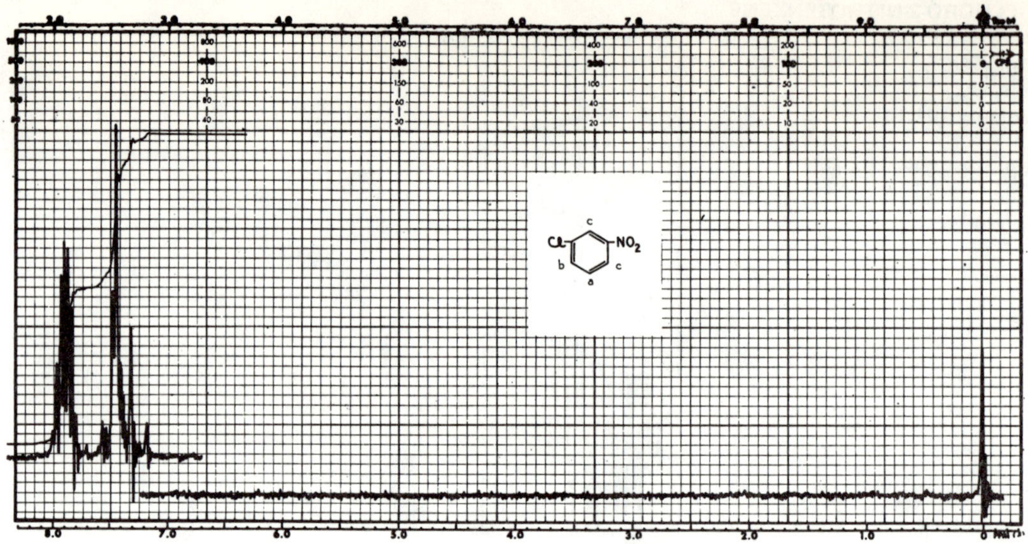

© 1978 Sadtler Research Laboratories, Inc.

---

p-NITROCHLOROBENZENE

$C_6H_4ClNO_2$

Mol. Wt. 157.56

F.P. –82.5°C

Solvent: $CDCl_3$

Assignments
a = 7.46
b = 8.09

1058

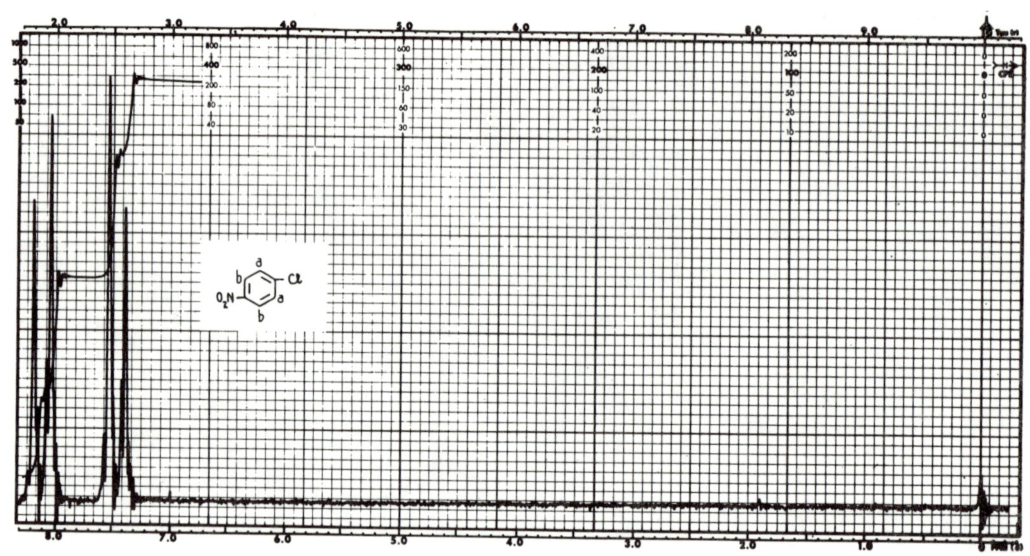

© 1978 Sadtler Research Laboratories, Inc.

1-BROMO-2-NITROBENZENE

$C_6H_4BrNO_2$

Mol. Wt. 202.03

M.P. 41-42°C

Solvent: $CDCl_3$

Assignments
a = 7.23-7.95

1059

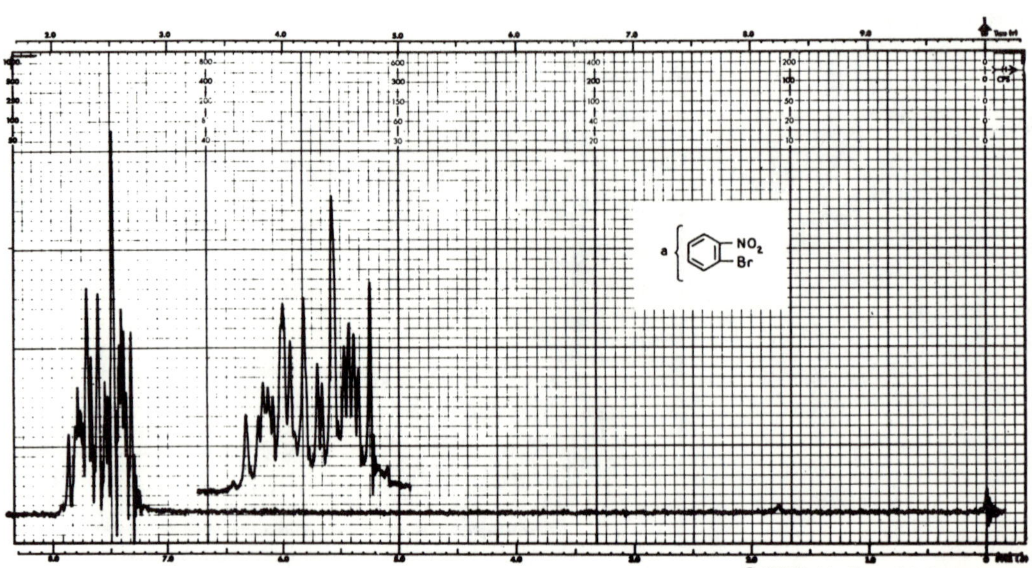

© 1978 Sadtler Research Laboratories, Inc.

1-IODO-3-NITROBENZENE

$C_6H_4INO_2$

Mol. Wt. 249.01

Solvent: Dimethyl sulfoxide

Assignments
a = 7.50
b = ca 8.21
c = 8.49

1060

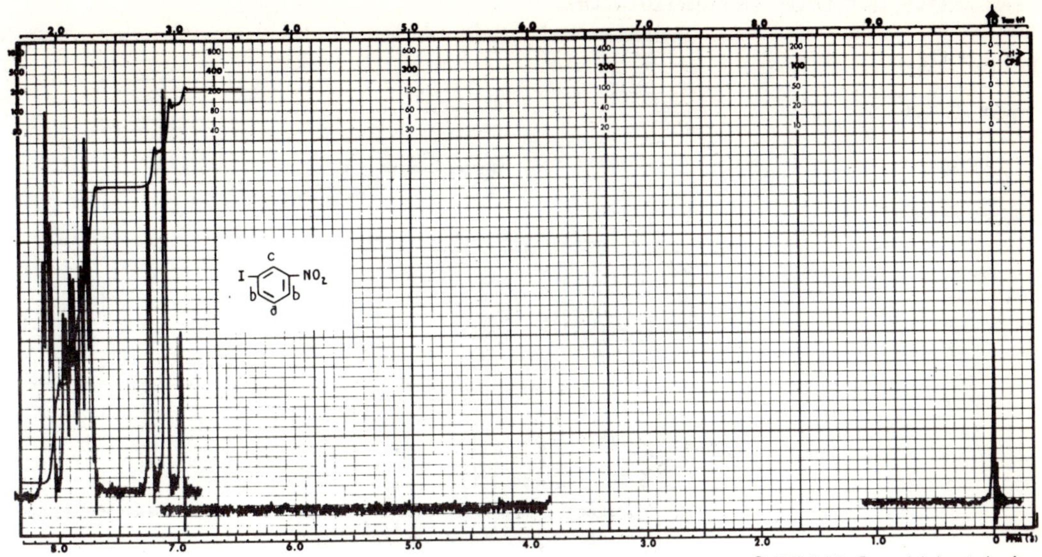

© 1978 Sadtler Research Laboratories, Inc.

1-IODO-4-NITROBENZENE

$C_6H_4INO_2$

Mol. Wt. 249.01

Solvent: Dimethyl sulfoxide

Assignments
a = ca 8.01

1061

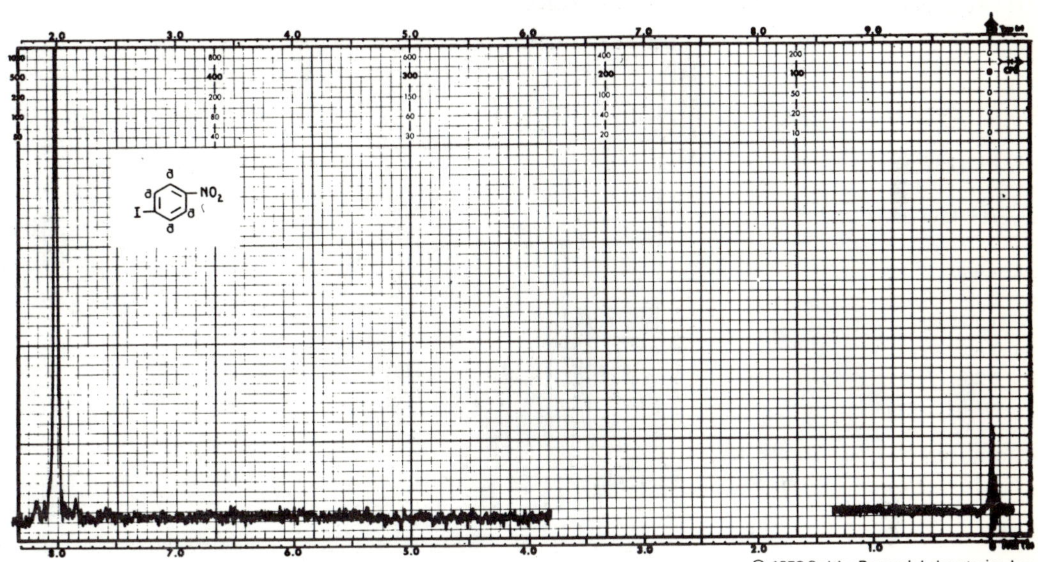

© 1978 Sadtler Research Laboratories, Inc.

N-ETHYL-o-NITROANILINE

$C_8H_{10}N_2O_2$

Mol. Wt. 166.18

B.P. 168°C/20mm

Solvent: CCl_4

Assignments
a = 1.38
b = 3.30
c = 6.53
d = 6.75
e = 7.34
f = 7.84
g = 8.06

1062

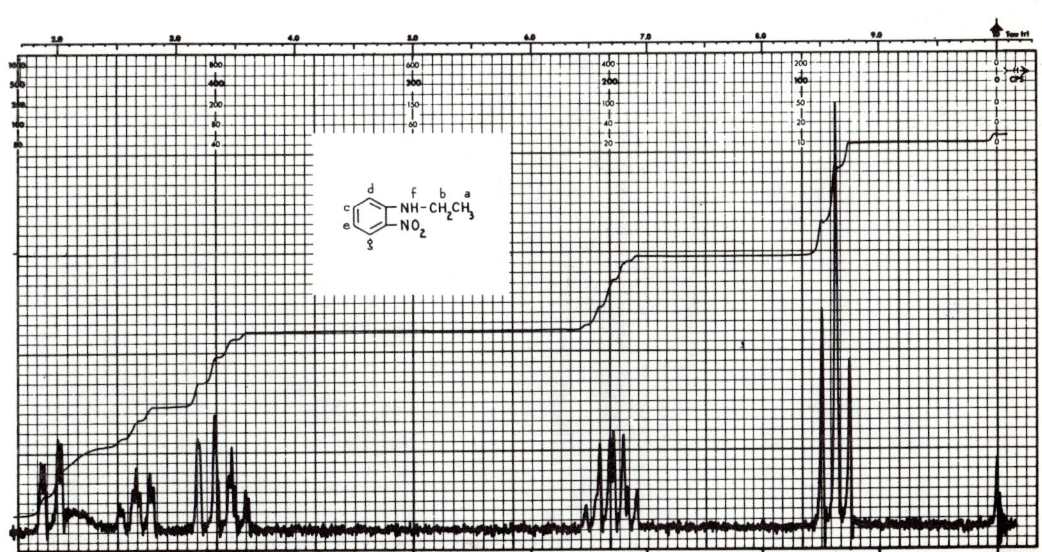

© 1978 Sadtler Research Laboratories, Inc.

2-PENTANONE, (p-NITROPHENYL)HYDRAZONE

$C_{11}H_{15}N_3O_2$

Mol. Wt. 221.26

M.P. 117°C

Solvent: Polysol-d

Assignments

a = 0.95
b = 1.59
c = ca 1.95
d = 2.27
e = 7.12
f = 8.01
g = 9.52

1063

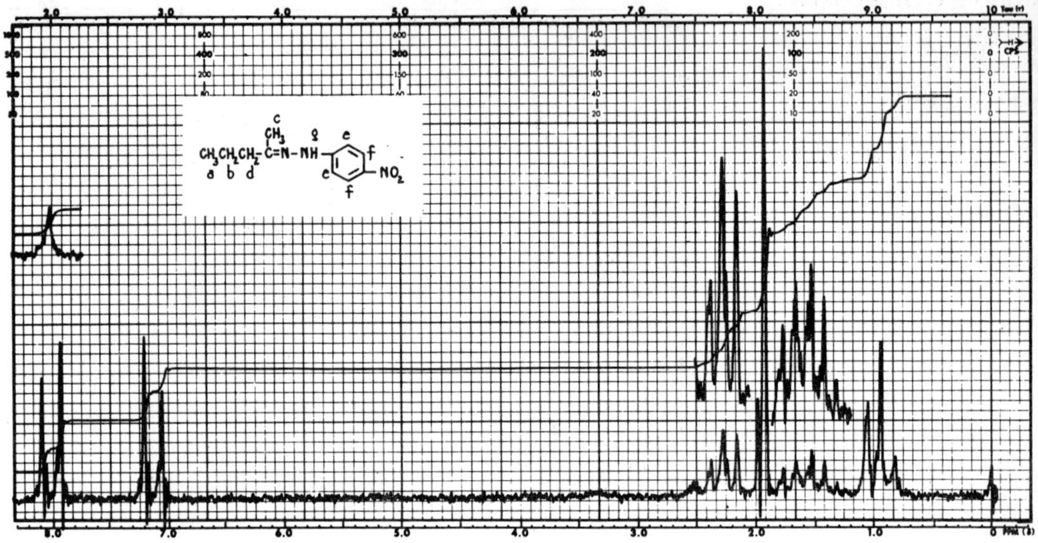

© 1978 Sadtler Research Laboratories, Inc.

BENZALDEHYDE, (p-NITROPHENYL)HYDRAZONE

$C_{13}H_{11}N_3O_2$

Mol. Wt. 241.25

M.P. 195-197°C

Solvent: Polysol-d

Assignments

a = 7.19
b = 7.20-7.50
c = 7.69
d = 8.06
e = 8.10
f = 11.14
g = 2.52 DMSO-d_5
h = 3.70 H_2O

1064

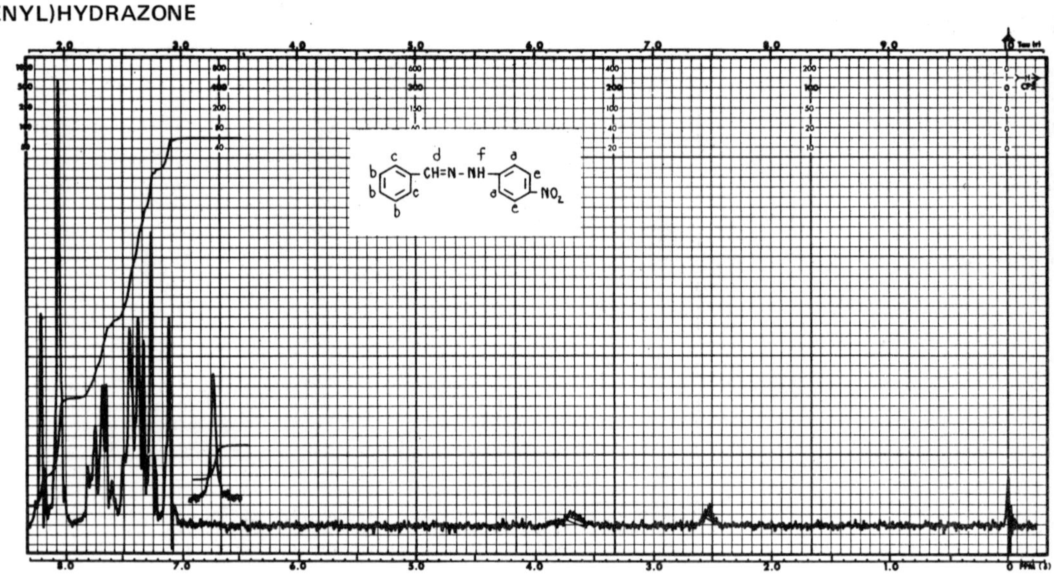

© 1978 Sadtler Research Laboratories, Inc.

4-NITROAZOBENZENE

$C_{12}H_9N_3O_2$

Mol. Wt. 227.22

Solvent: $CDCl_3$

Assignments

a = 7.40-7.65
b = ca 7.92
c = 8.32

1065

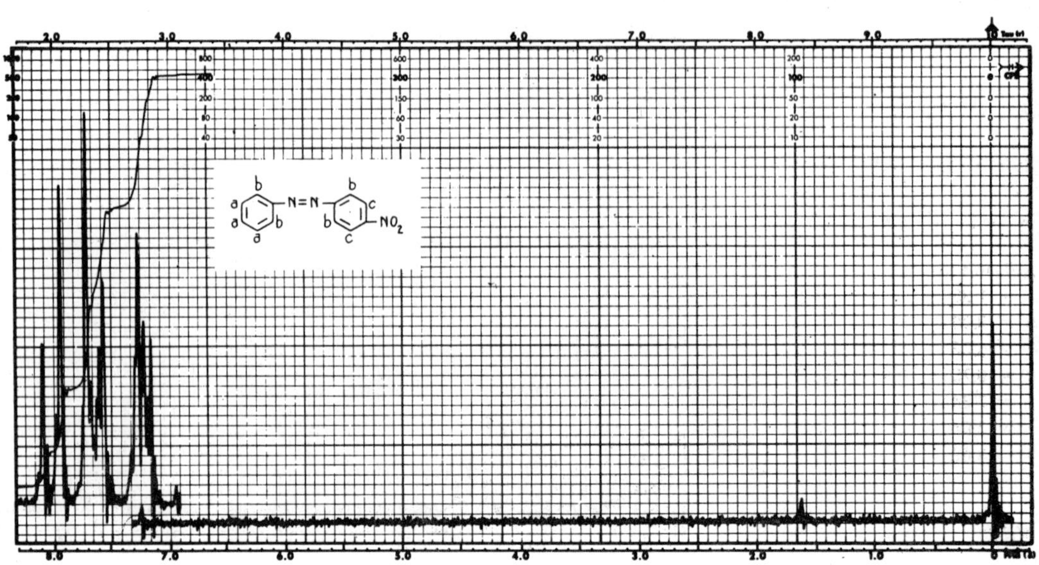

© 1978 Sadtler Research Laboratories, Inc.

o-DINITROBENZENE

$C_6H_4N_2O_4$

Mol. Wt. 168.11

Solvent: Dimethyl sulfoxide-D_6

Assignments
a = 8.08
b = 8.26
c = 2.59 DMSO-D_5

1066

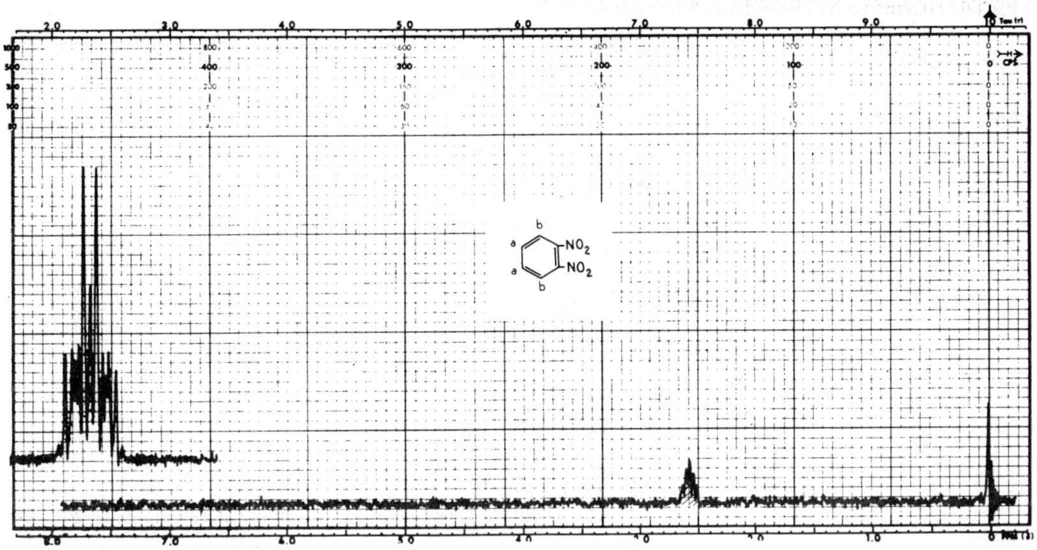

© 1978 Sadtler Research Laboratories, Inc.

m-DINITROBENZENE

$C_6H_4N_2O_4$

Mol. Wt. 168.11

M.P. 117-118°C

Solvent: $CDCl_3$

Assignments
a = 7.89
b = 8.59
c = 9.01

1067

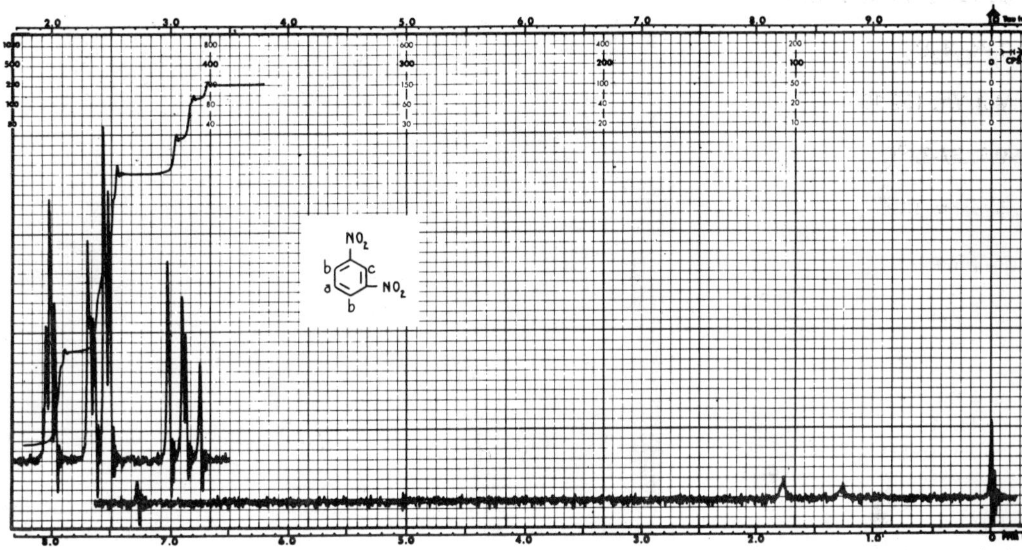

© 1978 Sadtler Research Laboratories, Inc.

4-NITRO-m-XYLENE

$C_8H_9NO_2$

Mol. Wt. 151.17

Solvent: $CDCl_3$

Assignments
a = 2.40
b = 2.58
c = 7.10
d = 7.13
e = 7.84

1068

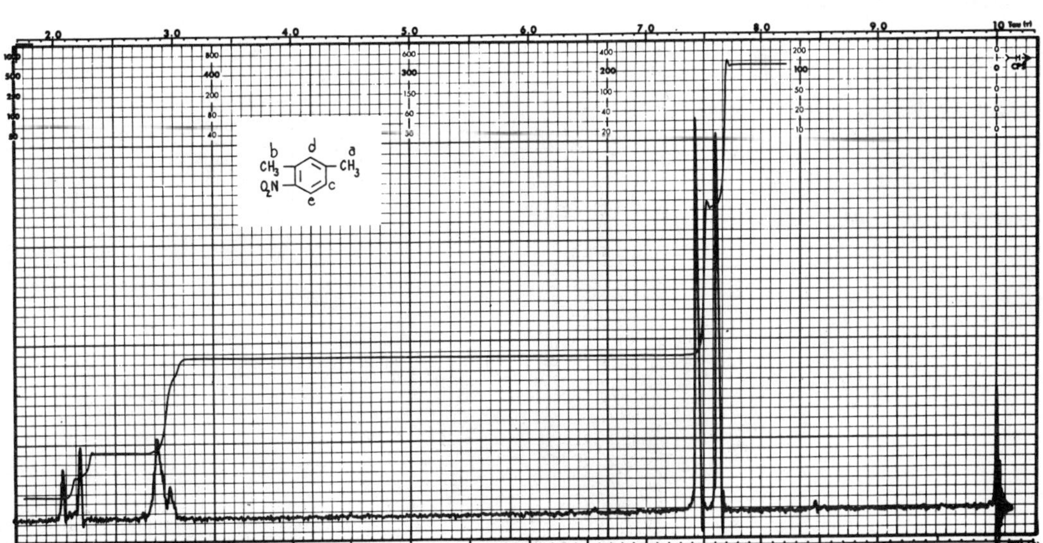

© 1978 Sadtler Research Laboratories, Inc.

1,2-DICHLORO-4-NITROBENZENE

$C_6H_3Cl_2NO_2$

Mol. Wt. 192.00

M.P. 40-42°C

Solvent: CCl_4

Assignments
a = 7.63
b = 8.08
c = 8.30

1069

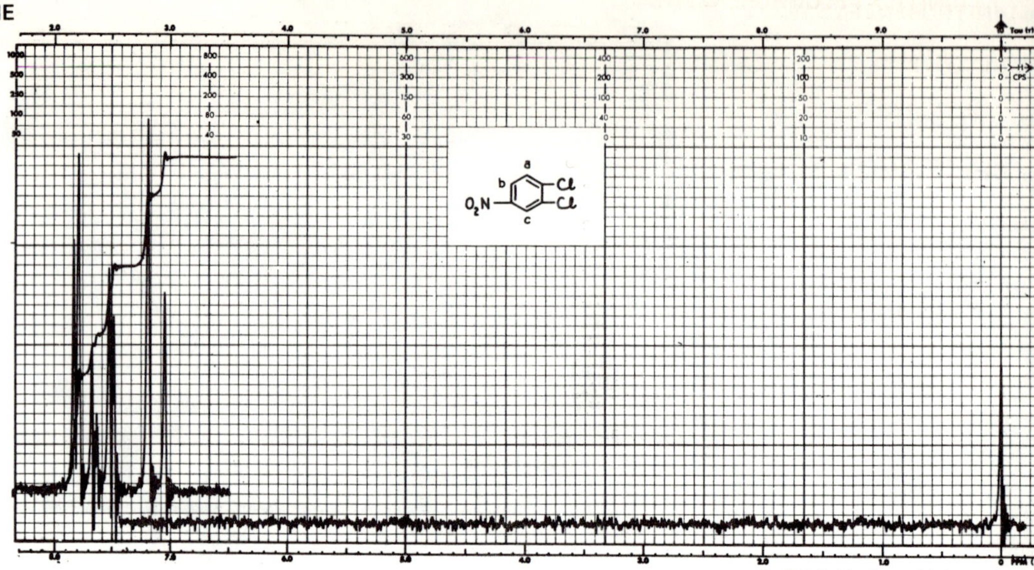

© 1978 Sadtler Research Laboratories, Inc.

3-NITRO-p-TOLUIDINE

$C_7H_8N_2O_2$

Mol. Wt. 152.15

M.P. 77-79°C

Solvent: $CDCl_3$

Assignments
a = 2.41
b = 3.80
c = 6.75
d = 7.03
e = 7.19

1070

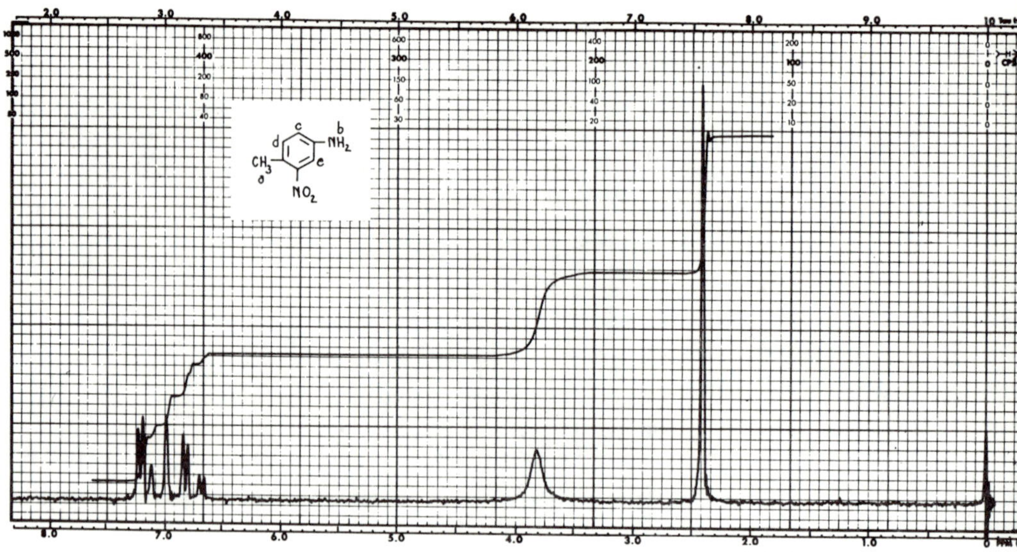

© 1978 Sadtler Research Laboratories, Inc.

4-NITRO-m-TOLUIDINE

$C_7H_8N_2O_2$

Mol. Wt. 152.15

M.P. 135°C

Solvent: Polysol-d

Assignments
a = 2.51
b = 6.12
c = 6.43
d = 6.50
e = 7.89

1071

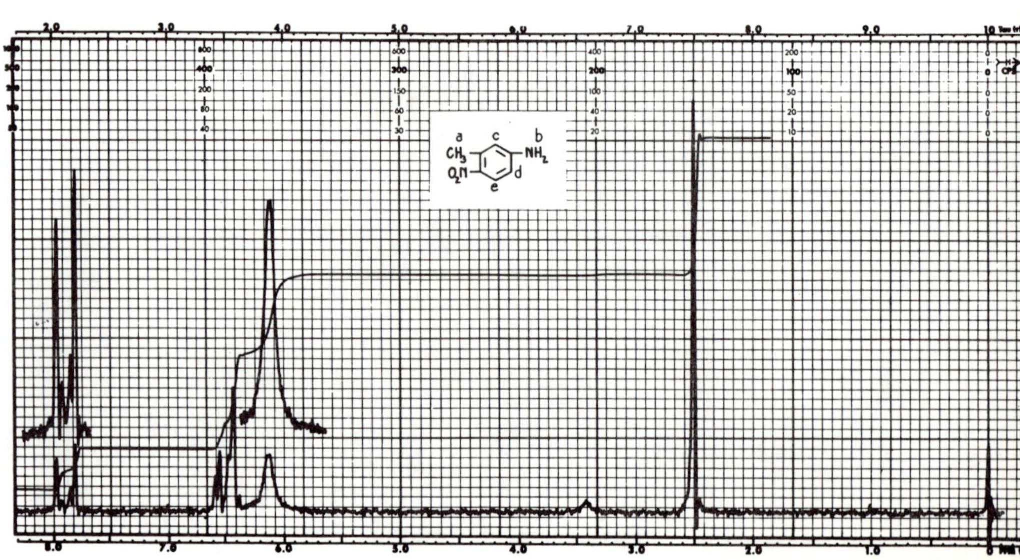

© 1978 Sadtler Research Laboratories, Inc.

2,4-DINITRO-1-FLUOROBENZENE

$C_6H_3FN_2O_4$

Mol. Wt. 186.10

Solvent: $CDCl_3$

Assignments
a = 7.49
b = 8.60
c = 8.93

1072

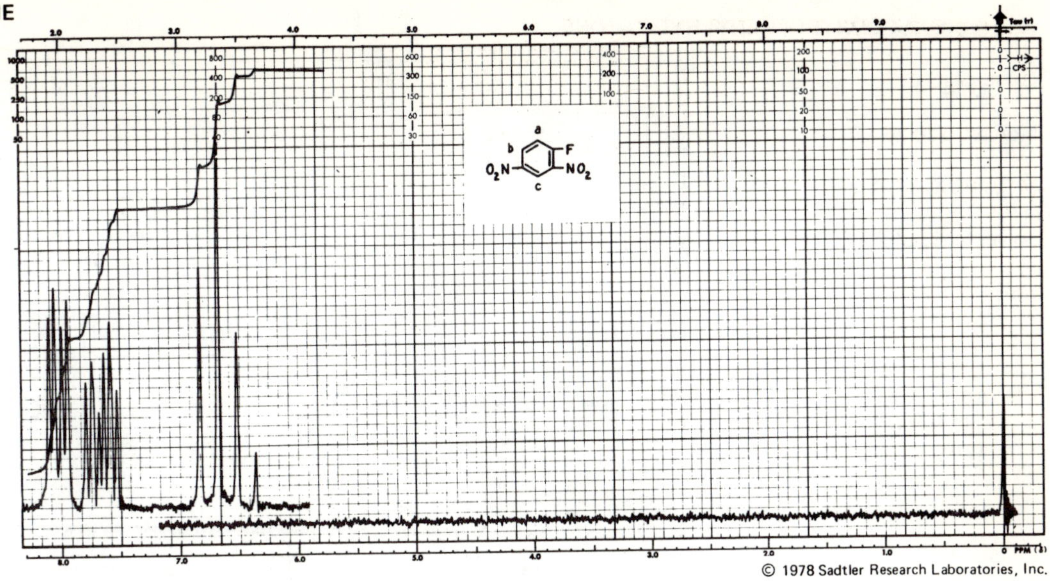

© 1978 Sadtler Research Laboratories, Inc.

1-BROMO-2,4-DINITROBENZENE

$C_6H_3BrN_2O_4$

Mol. Wt. 247.01

M.P. 70-72°C

Solvent: $CDCl_3$

Assignments
a = 8.09
b = 8.39
c = 8.70

1073

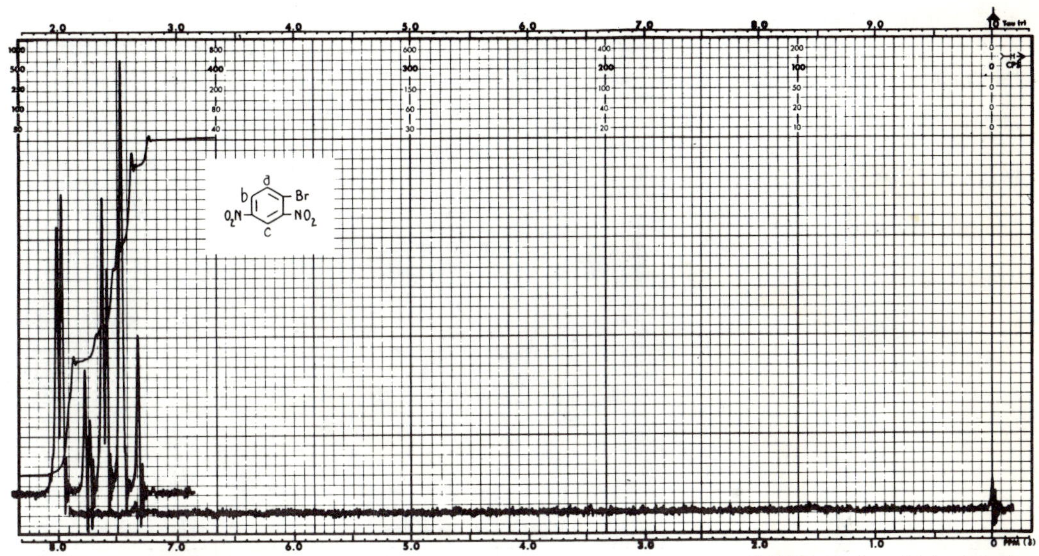

© 1978 Sadtler Research Laboratories, Inc.

N,N-DIMETHYL-3,4-DINITROANILINE

$C_8H_9N_3O_4$

Mol. Wt. 211.18

M.P. 175-176°C

Solvent: Dimethyl sulfoxide-d_6

Assignments
a = 3.11
b = 6.90
c = 7.11
d = 8.05
e = 2.50 DMSO-d_5

1074

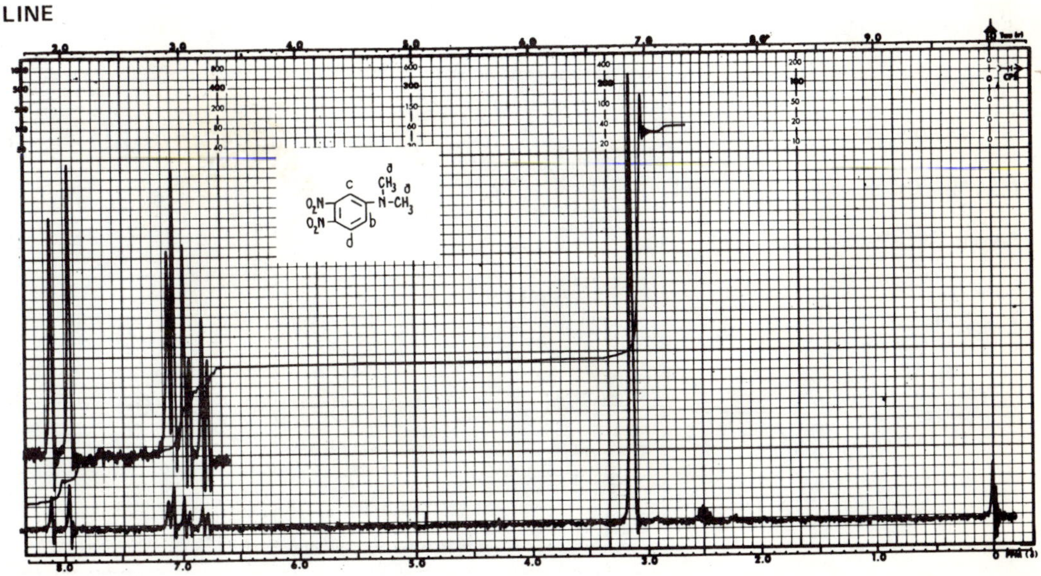

© 1978 Sadtler Research Laboratories, Inc.

ACETONE, 2,4-DINITROPHENYLHYDRAZONE

$C_9H_{10}N_4O_4$

Mol. Wt. 238.20

M.P. 126°C

Solvent: $CDCl_3$

Assignments

a = 2.09 or b
b = 2.18 or a
c = 7.94
d = 8.30
e = 9.11
f = 11.00

1075

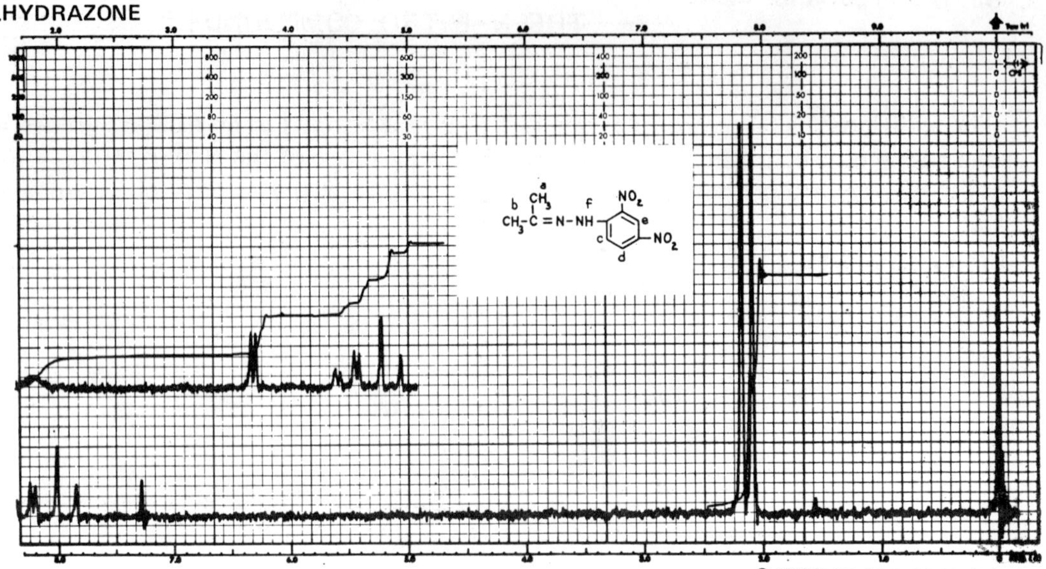

© 1978 Sadtler Research Laboratories, Inc.

1,3,5-TRINITROBENZENE

$C_6H_3N_3O_6$

Mol. Wt. 213.11

M.P. 121-122°C

Solvent: DMSO-d_6

Assignments

a = 9.12
b = 2.46 DMSO-d_5
c = 2.79 H_2O

1076

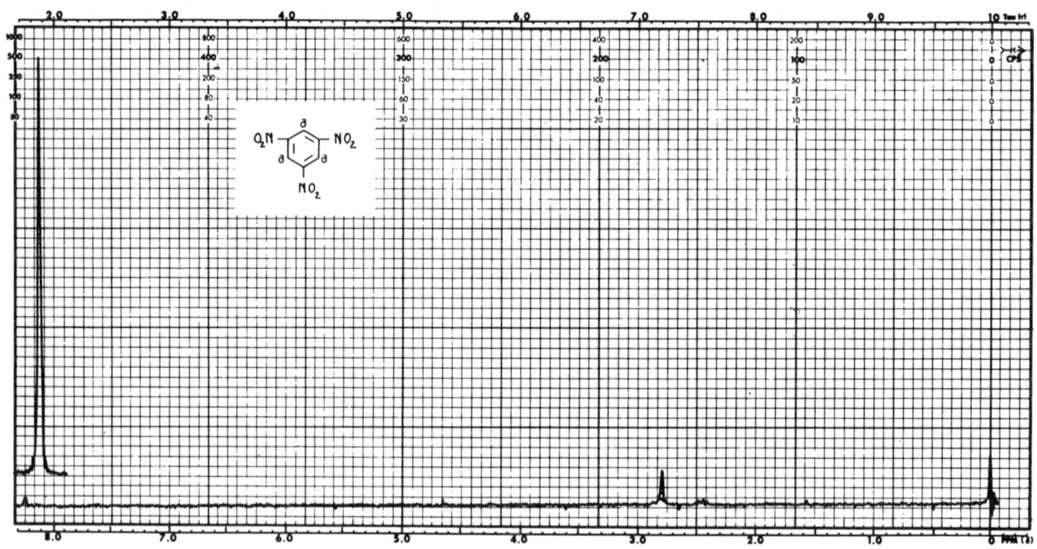

© 1978 Sadtler Research Laboratories, Inc.

4-NITRO-2,6-XYLIDINE

$C_8H_{10}N_2O_2$

Mol. Wt. 166.18

M.P. 164°C

Solvent: $CDCl_3$

Assignments

a = 2.21
b = 4.30
c = 7.88

1077

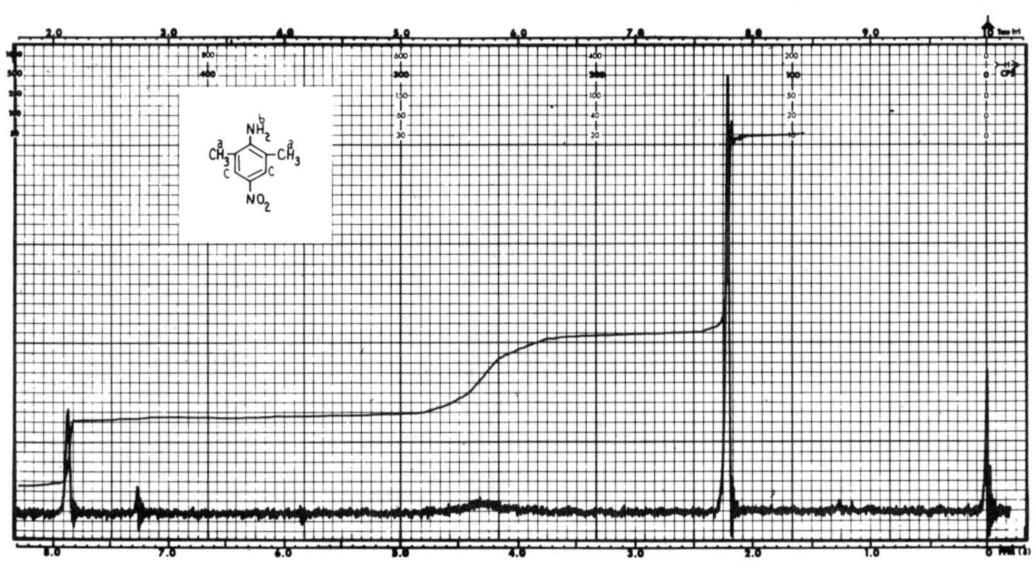

© 1978 Sadtler Research Laboratories, Inc.

THE N–NITRO COMPOUNDS

The addition of a Nitro group as one of the groups bonded to an Amine nitrogen atom increases the deshielding effect of the amine group. The presence of the N-Nitro group also decreases the relative solubility of these compounds in comparison to the corresponding amines.

Chemical Shifts

Aliphatic Protons

HC–3	HC–2	HC–1		Solvent	Spectrum
((0.90)	2.25	3.61)	$((CH_3)_2\text{–}CH\text{–}CH_2)_2\text{–}N\text{–}NO_2$	$CDCl_3$	1078
	(3.71	4.01)	$(O{=}C{=}N\text{–}CH_2\text{–}CH_2)_2\text{–}N\text{–}NO_2$	$CDCl_3$	- - - -
		3.65	$CH_3(\emptyset)\text{–}N\text{–}NO_2$	Polysol	- - - -

Alicyclic Protons

		4.01	1,4-Dinitropiperazine	DMSO-d_6	1079

THE NITRATES ($O\text{–}NO_2$)

The Nitrates which contain the $\text{–}O\text{–}NO_2$ substituent produce lower field shifts than the corresponding N-Nitro compounds (above). The effect of the $\text{–}O\text{–}NO_2$ group is not as strongly deshielding, however, as the –O–N=O group of the Nitrites previously discussed. Several comparisons of the Nitrate and Nitrite chemical shifts are presented below.

Chemical Shifts

Aliphatic Protonc

HC–3	HC–2	HC–1			
1.01	1.75	4.40	$CH_3\text{–}CH_2\text{–}CH_2\text{–}O\text{–}NO_2$	CCl_4	1080
0.97	1.72	4.61	$CH_3\text{–}CH_2\text{–}CH_2\text{–}O\text{–}N{=}O$	CCl_4	1033
	(1.37)	5.15	$(CH_3)_2\text{–}CH\text{–}O\text{–}NO_2$	CCl_4	1081
	(1.40)	5.59	$(CH_3)_2\text{–}CH\text{–}O\text{–}N{=}O$	CCl_4	- - - -
	1.74	4.47	$R_3\text{–}CH_2\text{–}CH_2\text{–}O\text{–}NO_2$	$CDCl_3$	1082

N-NITRODIISOBUTYLAMINE

$C_8H_{18}N_2O_2$

Mol. Wt. 174.24

M.P. 81.5-82.5°C

Solvent: $CDCl_3$

Assignments
a = 0.98
b = 1.70-2.64
c = 3.58

1078

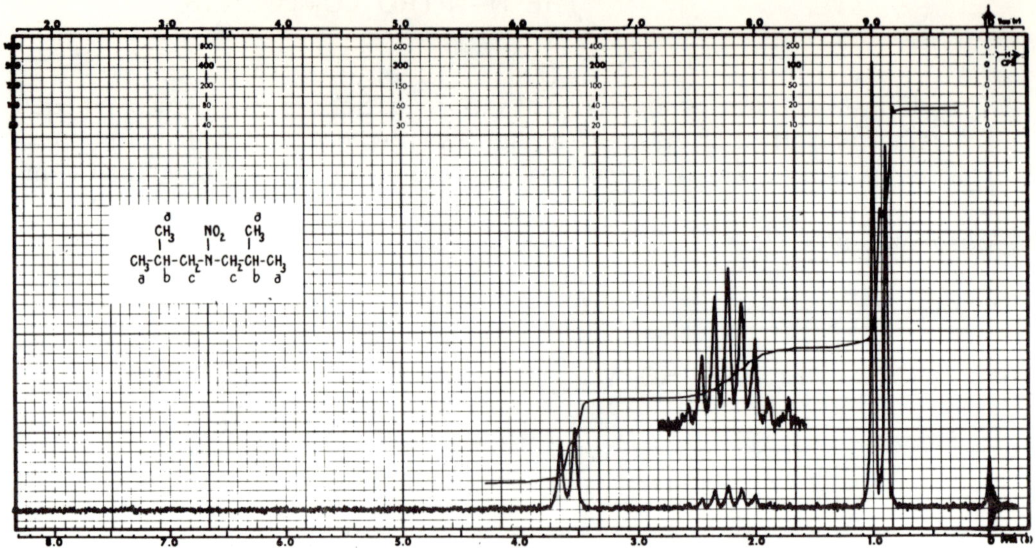

© 1978 Sadtler Research Laboratories, Inc.

1,4-DINITROPIPERAZINE

$C_4H_8N_4O_4$

Mol. Wt. 176.13

M.P. 207-212°C

Solvent: Dimethyl sulfoxide-D_6

Assignments
a = 4.10
b = 2.55 DMSO-D_5
c = 3.28 H_2O

1079

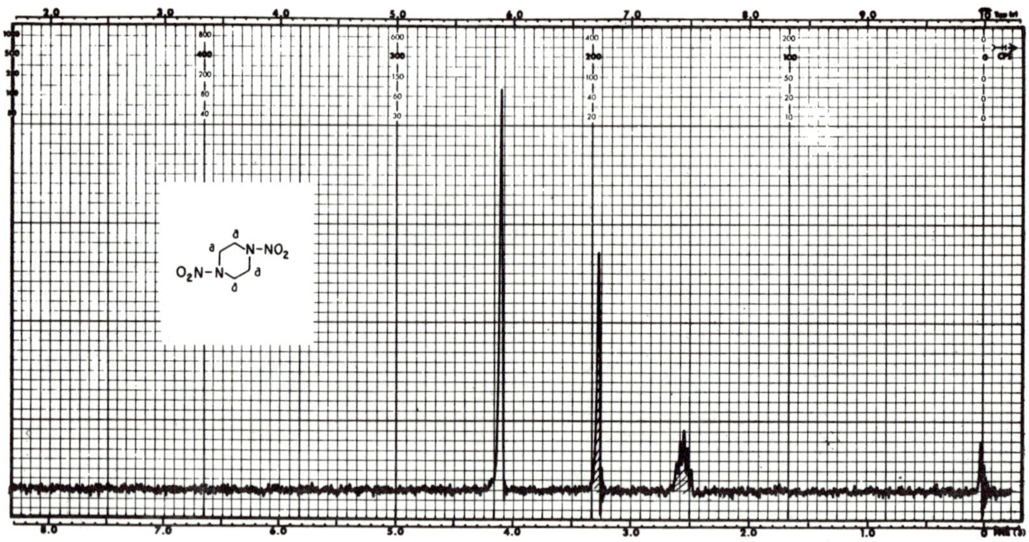

© 1978 Sadtler Research Laboratories, Inc.

PROPYL NITRATE

$C_3H_7NO_3$

Mol. Wt. 105.09

B.P. 108-110°C

Solvent: CCl_4

Assignments
a = 1.01
b = 1.75
c = 4.40

1080

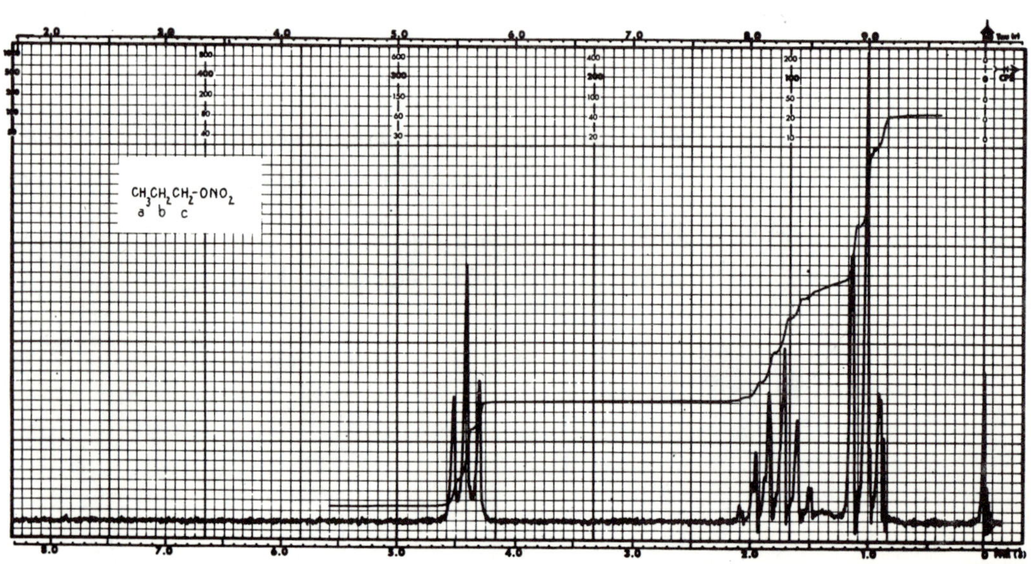

© 1978 Sadtler Research Laboratories, Inc.

ISOPROPYL NITRATE

$C_3H_7NO_3$

Mol. Wt. 105.09

B.P. 98-102°C

Solvent: CCl_4

Assignments
a = 1.37
b = 5.15

1081

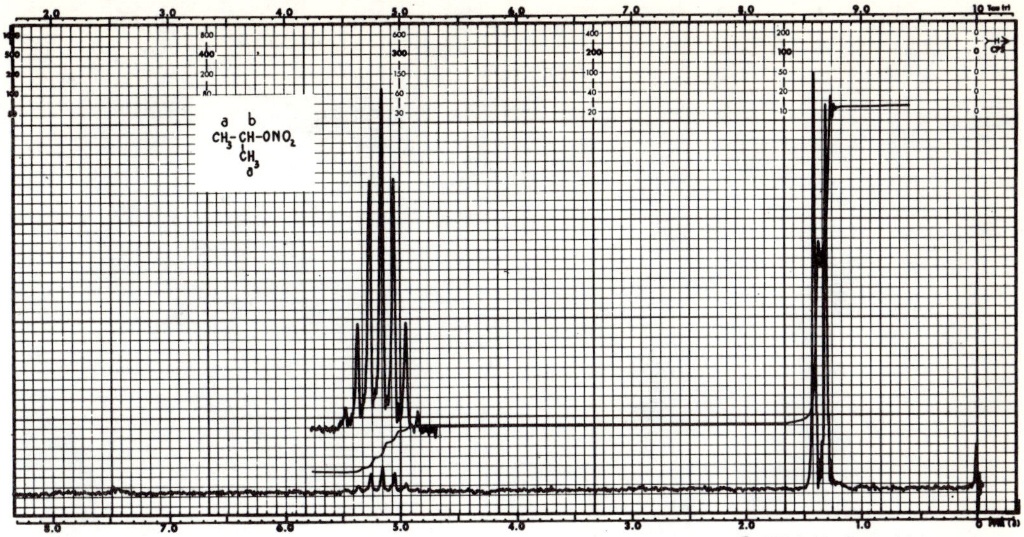

© 1978 Sadtler Research Laboratories, Inc.

NITRIC ACID, PENTYL ESTER

$C_5H_{11}NO_3$

Mol. Wt. 133.15

Solvent: $CDCl_3$

Assignments
a = 0.95
b = 1.10-1.60
c = 1.74
d = 4.47

1082

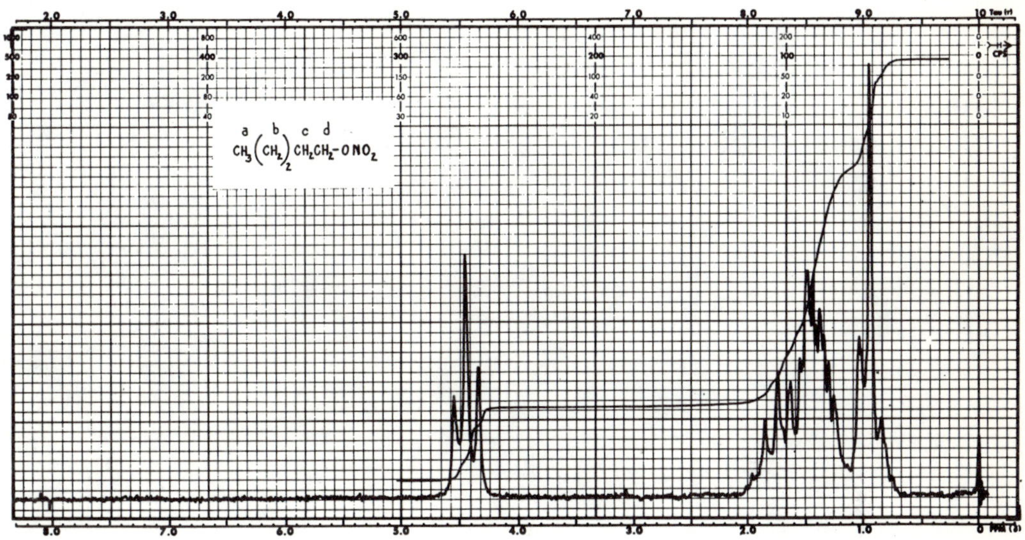

© 1978 Sadtler Research Laboratories, Inc.

NOTES

The outstanding characteristic of the Silicon compounds is the extremely high-field chemical shifts observed for aliphatic groups bonded to the Silicon nucleus. These aliphatic groups resonate at higher fields than any other group in corresponding molecular structures. Additionally, coupling between Si–H protons and adjacent aliphatic groups is observed as clear n+1 multiplets. The Silicon compounds are readily soluble in all the normal NMR solvents excluding D_2O.

The chemical shifts of groups bonded to the oxygen atom of the Silicon ethers (Siloxanes) are described later in the text with the other ether oxygen compounds.

Chemical Shifts

Aliphatic Protons

	CH_2	CH–1		Solvent	Spectrum
		(0.00)	$(CH_3)_4$–Si	----	----
		(0.03)	$(CH_3)_3$–Si–CH_2–CH=CH_2	CCl_4	1086
		(0.04)	$(CH_3)_3$–Si–Si–$(CH_3)_3$	CCl_4	1094
		(0.06)	$(CH_3)_3$–Si–CH=CH_2	C_2Cl_4	----
		(0.12)	$(CH_3)_3$–Si–CH_2–Cl	CCl_4	----
		(0.29)	$(CH_3)_3$–Si–Ø	$CDCl_3$	1089
		(0.40)	$(CH_3)_3$–Si–Cl	C_2Cl_4	----
		(0.48)	$(CH_3)_3$–Si–1–(imidazole)	$CDCl_3$	----
		(0.63)	$(CH_3)_2$–Si(Cl,Ø)	CCl_4	----
		0.77	CH_3–Si(Ø)$_3$	$CDCl_3$	----
		(0.80)	$(CH_3)_2$–Si(Cl)$_2$	$CDCl_3$	----
		1.17	CH_3–Si(Cl)$_3$	CCl_4	1092
	(0.92	0.56)	$(CH_3–CH_2)_4$–Si	CCl_4	1085
	1.10	1.30	CH_3–CH_2–Si(Ø)$_3$	$CDCl_3$	1091
0.90	(1.1–1.7)	0.90	$CH_3(CH_2)_4$–CH_2–Si(H)$_3$	$CDCl_3$	1083

Substituted Methyl Silanes

X–	CH_2	–Si(X,Y,Z)	
R_2–	0.67	–Si(H)$_3$	$CDCl_3$
R_5–	0.90	–Si(–O–CH_3)$_3$	CCl_4
CH_2=CH–	1.52	–Si(CH_3,CH_3–CH_2–CH=CH_2)	C_2Cl_4
CH_2=CH–	1.56	–Si$(CH_3)_3$	CCl_4
Ø–	2.59	–Si(Ø,Ø,–CH_2–Ø)	$CDCl_3$
Cl–	2.69	–Si$(CH_3)_3$	CCl_4
Ø–	2.91	–Si(Ø)$_3$	$CDCl_3$
Cl–	3.11	–Si(CH_3, Cl, Cl)	$CDCl_3$

Chemical Shifts continued

Silane Protons (Si–H)

		Solvent
3.52	H_3–Si–R_6	$CDCl_3$
4.03	H–Si(CH_2–Ø)$_3$	$CDCl_3$
4.14	H–Si(O–R_2)$_3$	CCl_4
4.46	H–Si(CH_3,CH_3,Si(H)(CH_3)$_2$)	CCl_4
4.92	H_2–Si(Ø)$_2$	$CDCl_3$
5.39	H–Si(Ø)$_3$ (4–O–CH_3)	$CDCl_3$
5.54	H–Si(CH_3,Cl,Cl)	CCl_4
5.79	H–Si(Ø)$_3$ (2–CH_3)	CCl_4

Olefinic Protons

The Silicon nucleus deshields all three of the vinyl protons producing a complex ABC pattern centered at about 5.9 ppm.

Aromatic Protons

Silicon substituents deshield the ortho aromatic hydrogens which resonate in the range 7.5–7.8 ppm depending on the other groups attached to the Silicon nucleus. The chemical shifts of a few representative aromatic silanes are provided below.

Phenyl Silanes

para, meta, ortho	–Si(X,Y,Z)	
7.10–7.60	–Si(Ø,R_{12},R_{12}	CCl_4
7.10–7.60	–Si(Ø,Ø,–CH_2–Ø)	$CDCl_3$
7.20–7.65	–Si(CH_3)$_3$	$CDCl_3$
7.10–7.70	–Si(Ø)$_3$	$CDCl_3$

para, meta	ortho	–Si(X,Y,Z)	
7.1–7.5	7.48	–Si(O–CH_3)$_3$	CCl_4
7.1–7.4	7.48	–Si(Ø,Ø,R_2)	$CDCl_3$
7.1–7.4	7.54	–Si(H,Ø,Ø)	$CDCl_3$
7.1–7.4	7.56	–Si(H,H,Ø)	$CDCl_3$
7.1–7.5	7.60	–Si(O–R_2)$_3$	CCl_4
7.0–7.4	7.62	–Si(Ø,Ø,Cl)	$CDCl_3$
7.2–7.7	7.81	–Si(Cl)$_3$	$CDCl_3$

Chemical Shifts continued

Aromatic Protons continued

Para Substituted Phenylsilanes

X–	ortho	meta	para	Solvent
CH_3–O–	6.85	7.40	–Si(CH_3,Ø,Ø)	$CDCl_3$
CH_3–O–	6.83	7.41	–Si(H,Ø,Ø)	$CDCl_3$
Cl–	7.35	7.51	–Si(CH_3,CH_3,Cl)	CCl_4
CH_3(OH)–CH–	7.33	7.62	–Si$(CH_3)_3$	$CDCl_3$

Coupling and Coupling Constants

Coupling between the Silane protons and adjacent aliphatic groups is observed as clear n+1 multiplets (spectrum 1083).

$$J_{H-Si-CH_2} = 3.1–3.9 \text{ Hz}$$

The isotope Silicon-29 has a natural abundance of 4.7% and possesses a spin of 1/2. These isotope sidebands can often be observed in the spectra of the Silanes if the noise level of the baseline is sufficiently low to allow their definition.

$J_{^{29}Si-H_3}$	=	184 Hz	(spectrum 1083)
$J_{^{29}Si-H_2}$	=	199 Hz	(spectrum 1084)

HEXYLSILANE

$C_6H_{16}Si$

Mol. Wt. 116.28

Solvent: $CDCl_3$

Assignments
a = ca 0.90
b = 1.00-1.70
c = 3.52

1083

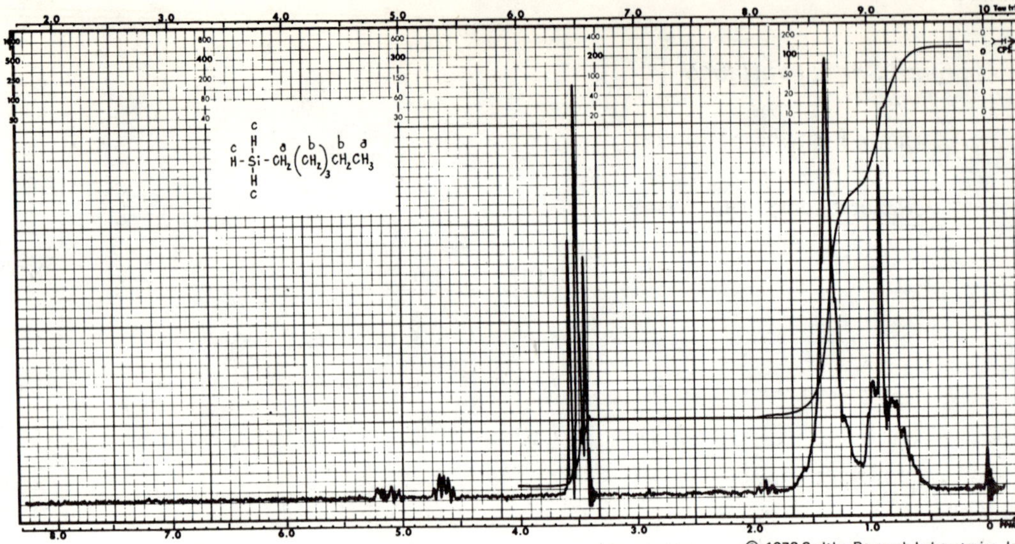

© 1978 Sadtler Research Laboratories, Inc.

DIPHENYLSILANE

$C_{12}H_{12}Si$

Mol. Wt. 184.32

B.P. 95-97°C/13mm

Solvent: $CDCl_3$

Assignments
a = 4.92
b = 7.15-7.43
c = 7.56

1084

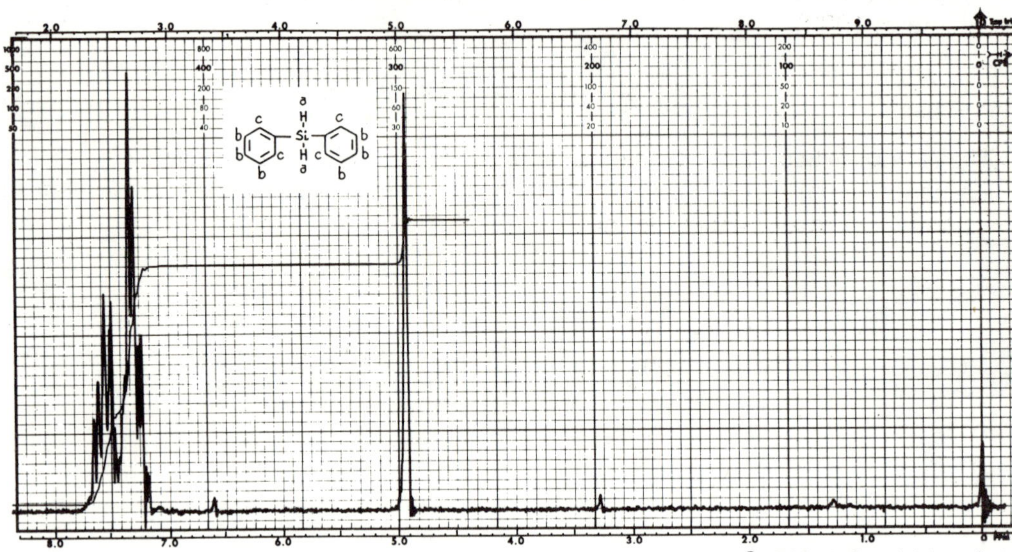

© 1978 Sadtler Research Laboratories, Inc.

TETRAETHYLSILANE

$C_8H_{20}Si$

Mol. Wt. 144.33

B.P. 155°C

Solvent: CCl_4

Assignments
a = 0.59
b = 0.91

1085

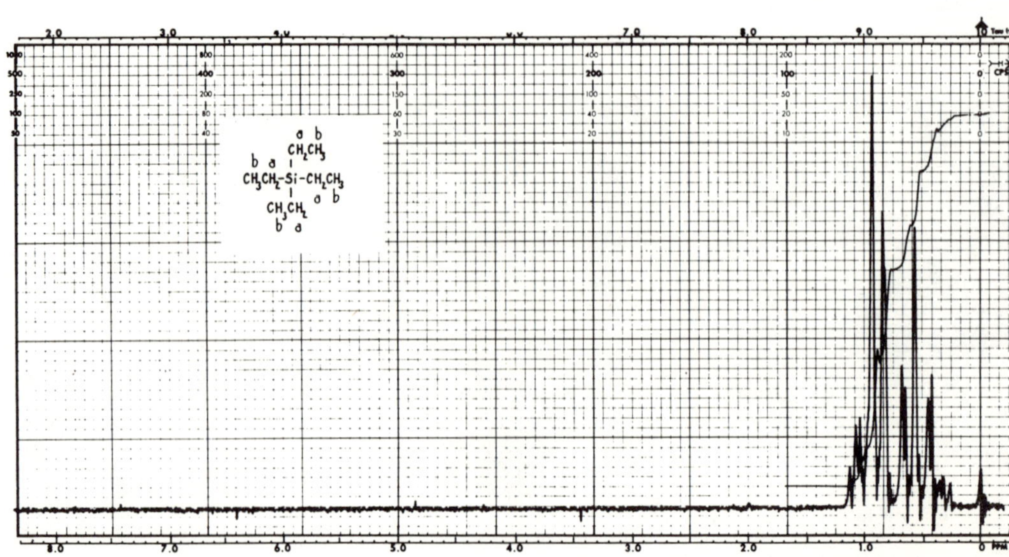

© 1978 Sadtler Research Laboratories, Inc.

ALLYLTRIMETHYLSILANE

$C_6H_{14}Si$

Mol. Wt. 114.26

B.P. 85-86°C

Solvent: CCl_4

Assignments
a = 0.03
b = 1.56
c = ca 4.86
d = 5.83

1086

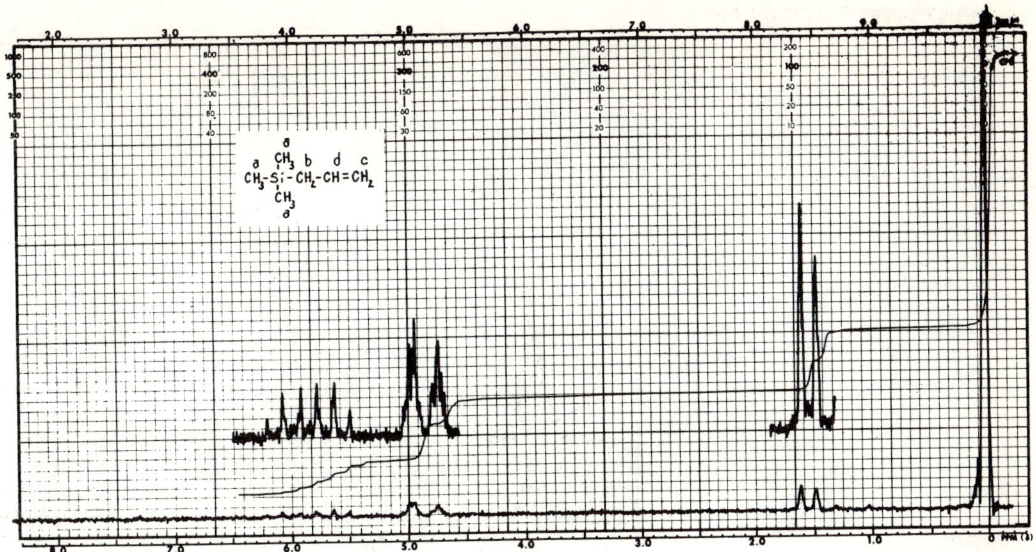

© 1978 Sadtler Research Laboratories, Inc.

DIALLYLDIMETHYLSILANE

$C_8H_{16}Si$

Mol. Wt. 140.30

Solvent: C_2Cl_4 (Acetone at 2.05 as internal reference)

Assignments
a = 0.01
b = 1.52
c = ca 4.79
d = ca 5.69
e = 2.05 Acetone

1087

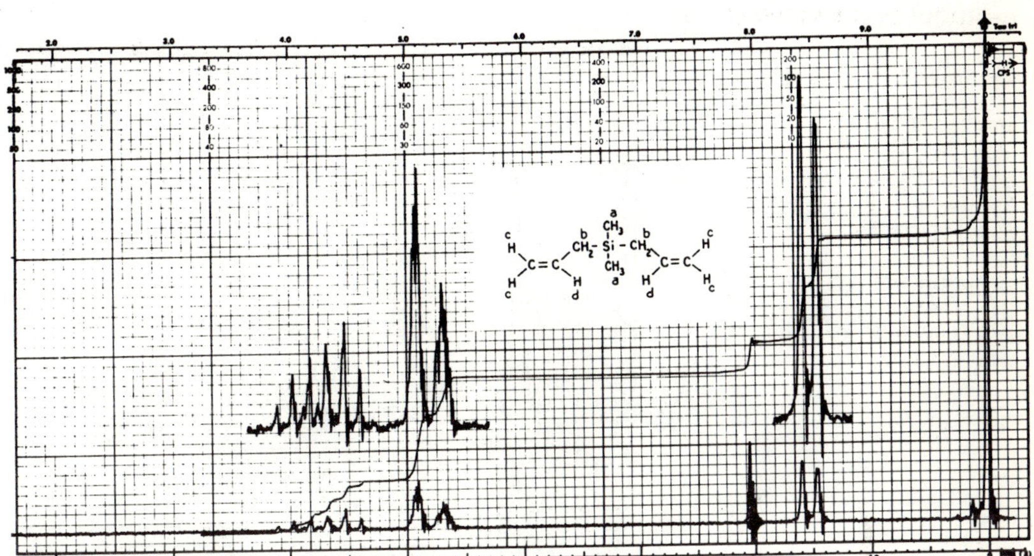

© 1978 Sadtler Research Laboratories, Inc.

DIBENZYLDIPHENYLSILANE

$C_{26}H_{24}Si$

Mol. Wt. 364.57

M.P. 63°C

Solvent: $CDCl_3$

Assignments
a = 2.59
b = 6.66-7.09
c = ca 7.28

1088

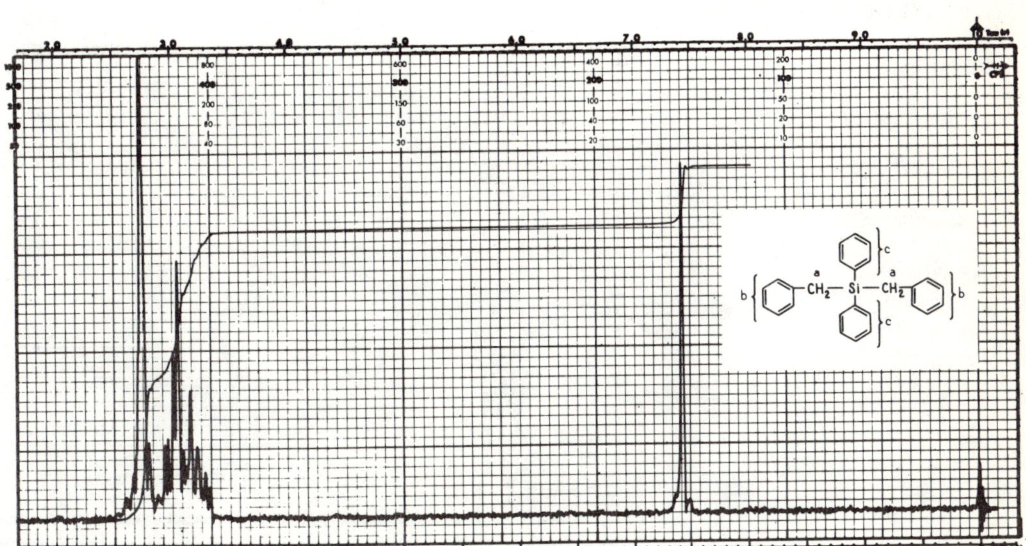

© 1978 Sadtler Research Laboratories, Inc.

PHENYLTRIMETHYLSILANE

$C_9H_{14}Si$

Mol. Wt. 150.29

B.P. 169-170°C

Solvent: $CDCl_3$

Assignments
a = 0.29
b = 7.20-7.65

1089

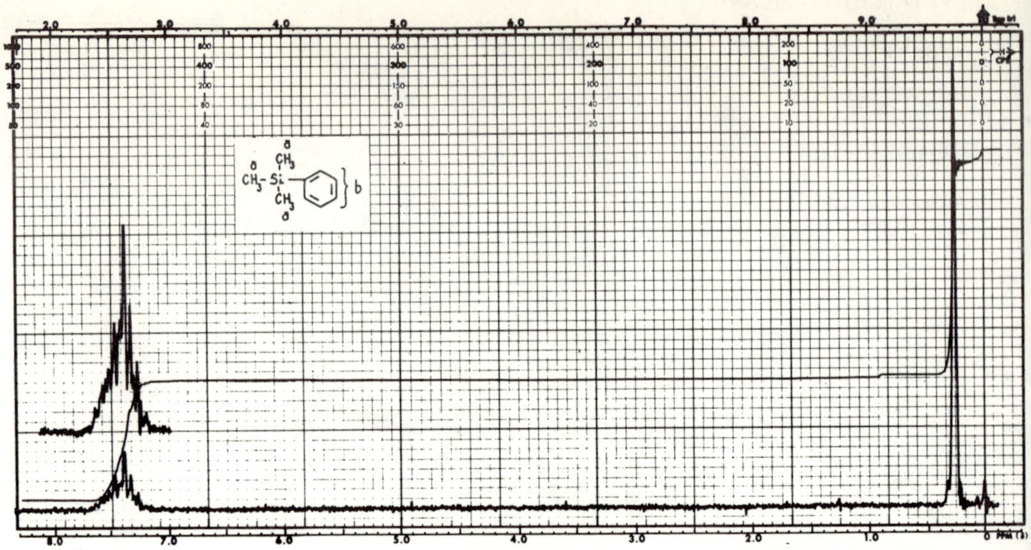

© 1978 Sadtler Research Laboratories, Inc.

DIDODECYLDIPHENYLSILANE

$C_{36}H_{60}Si$

Mol. Wt. 520.90

M.P. −40°C

Solvent: CCl_4

Assignments
a = 0.89
b = 0.90-1.60
c = 7.10-7.60

1090

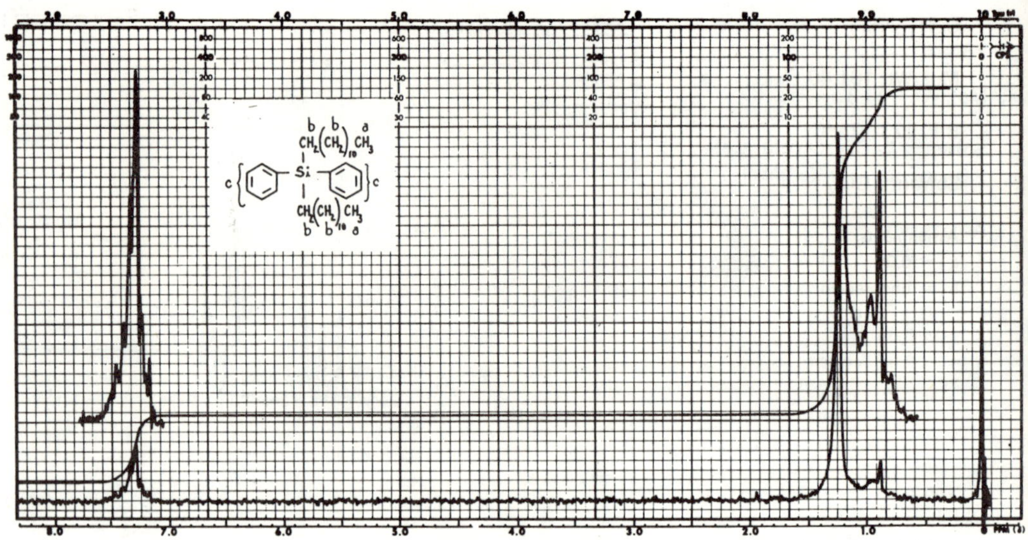

© 1978 Sadtler Research Laboratories, Inc.

ETHYLTRIPHENYLSILANE

$C_{20}H_{20}Si$

Mol. Wt. 288.47

M.P. 76°C

Solvent: $CDCl_3$

Assignments
a = 1.10
b = 1.22
c = 7.09-7.37
d = 7.48

1091

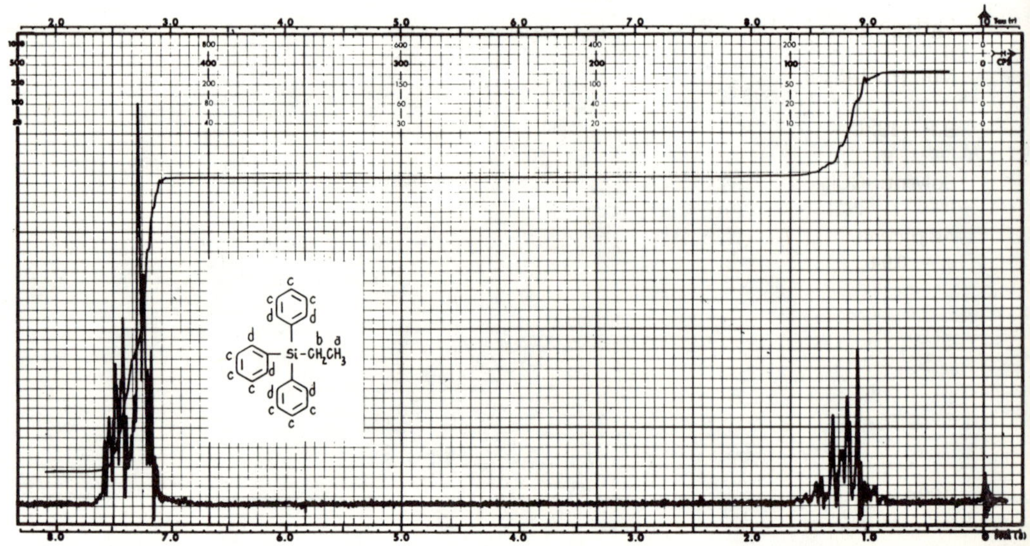

© 1978 Sadtler Research Laboratories, Inc.

METHYLTRICHLOROSILANE

CH_3CL_3Si

Mol. Wt. 149.48

Solvent: CCl_4

Assignments
a = 1.17

1092

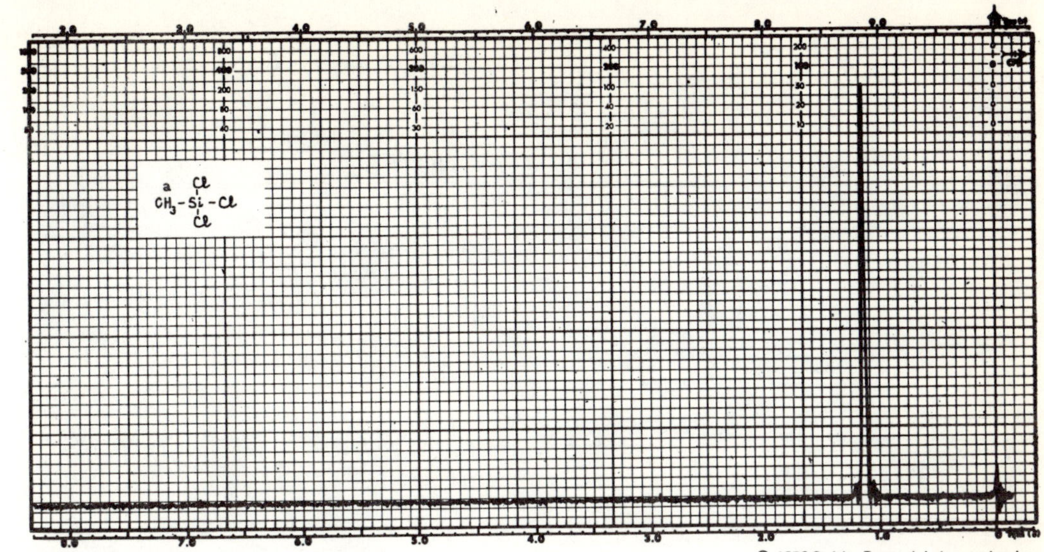

© 1978 Sadtler Research Laboratories, Inc.

1,1,1,3,3,3-HEXAMETHYLDISILAZANE

$C_6H_{19}NSi_2$

Mol. Wt. 161.40

B.P. 123-125°C

Solvent: CCl_4

Assignments
a = 0.04
b = unobserved

1093

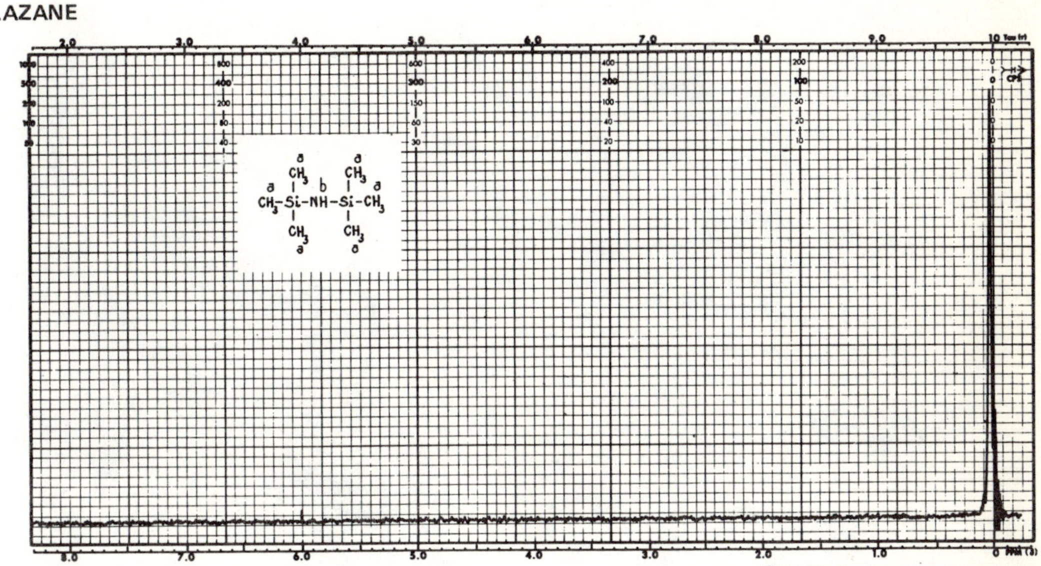

© 1978 Sadtler Research Laboratories, Inc.

HEXAMETHYLDISILANE

$C_6H_{18}Si_2$

Mol. Wt. 146.38

B.P. 113°C

Solvent: CCl_4

Assignments
a = 0.04

1094

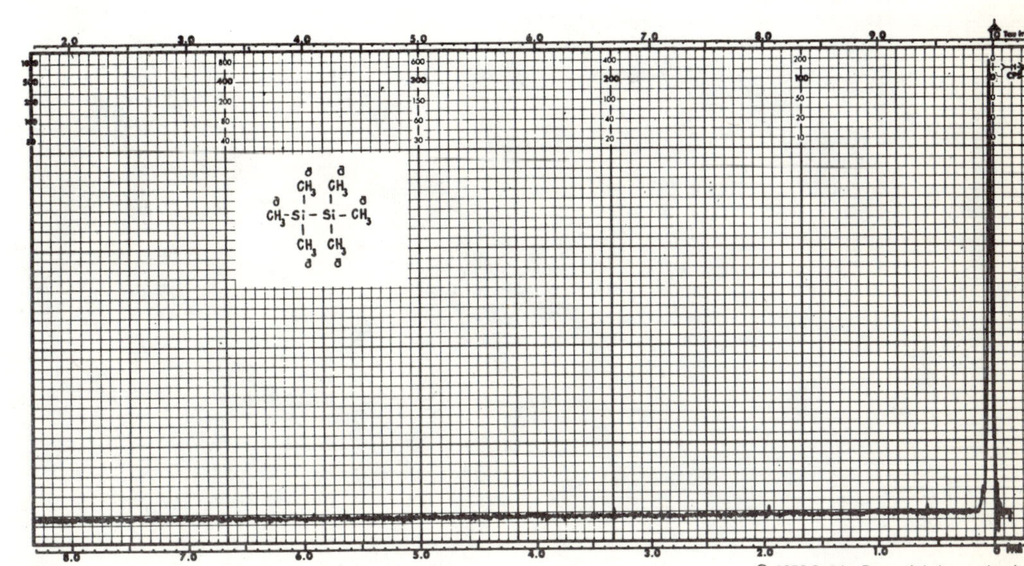

© 1978 Sadtler Research Laboratories, Inc.

NOTES

THE PHOSPHINES

For the sake of comparison, this group of compounds includes not only the Phosphines, but also several other tri-valent Phosphorus compounds. The Phosphine Phosphorus nucleus is a very weakly deshielding substituent in its effect on adjacent aliphatic groups, similar in effect to another aliphatic group (CH_3, CH_2, CH). Its effect on the aromatic protons varies, depending on the other groups bonded to the Phosphorus nucleus, from a weakly deshielding to a strongly deshielding group. Coupling between the Phosphorus atom and adjacent protons is usually present but often difficult to see clearly due to overlap with non-coupled protons.

Chemical Shifts

Aliphatic Protons

HC–4	HC–3	HC–2	HC–1		Solvent	Spectrum
0.94	(1.1–1.9)		1.1–1.9	$CH_3-(CH_2)_2-CH_2-P(R_4)_2$	$CDCl_3$	1095
		1.05	1.99	$CH_3-CH_2-P(\emptyset)_2$	$CDCl_3$	1096
			1.53	$CH_3-P(\emptyset)_2$	$CDCl_3$	----

Aromatic Protons

Phenyl Phosphines

para,meta,ortho	–X	Solvent
6.99–7.44	$-P(R_2,\emptyset)$	$CDCl_3$
7.12–7.59	$-P(O-CH_3)_2$	CCl_4
7.00–7.60	$-P(H,\emptyset)$	CCl_4
7.10–7.60	$-P(R_3,\emptyset)$	$CDCl_3$

para/meta	ortho	–X	Solvent
7.1–7.5	7.60	$-P(Cl,\emptyset)$	$CDCl_3$
7.1–7.6	7.83	$-P(Cl)_2$	$CDCl_3$
7.3–7.7	8.19	$-P(O-CH_3,\emptyset)$	CCl_4

Coupling and Coupling Constants

The Phosphorus nucleus of the Phosphine compounds couples weakly, if at all, with an alpha aliphatic group, strongly with the beta aliphatic group, and weakly with the gamma group. Certain anomalies appear to exist in the data presented below for those compounds in which two phosphine groups are present. Due to the small number of compounds available for analysis, the data is presented as it was deduced from the spectra via first order analysis.

Coupling and Coupling Constants continued

$CH_3{-}CH_2{-}P(\emptyset)_2$ $\quad J_{^{31}P-CH_2} = 0$ Hz, $\quad J_{^{31}P-C-CH_3} = 17$ Hz

$CH_3{-}P(\emptyset)_2$ $\quad J_{^{31}P-CH_3} = 4.0$ Hz

$N{\equiv}C{-}CH_2{-}CH_2{-}P(CH_2CH_2{-}CN)_2$ $\quad J_{^{31}P-CH_2} = 0$ Hz

$J_{^{31}P-C-CH_2} = 9.5$ Hz

$(\emptyset)_2P{-}CH_2{-}CH_2{-}P(\emptyset)_2$ $\quad J_{P-CH_2} = 4.5$ Hz, $\quad J_{P-C-CH_2} = 4.5$ Hz (?)

$(\emptyset)P{-}CH_2{-}P(\emptyset)_2$ $\quad J_{P-CH_2} =$ ca 1 Hz

Aromatic Protons

$\emptyset{-}P(Cl)_2$ $\quad J_{ortho} = 8.5$ Hz

$J_{P-C-C-CH_3} = 1.2$ Hz (spectrum 1098)

TRIBUTYLPHOSPHINE

$C_{12}H_{27}P$

Mol. Wt. 202.32

B.P. 150°C/50mm

Solvent: $CDCl_3$

Assignments
a = 0.94
b = 1.00-1.90
c = 2.31 imp.

1095

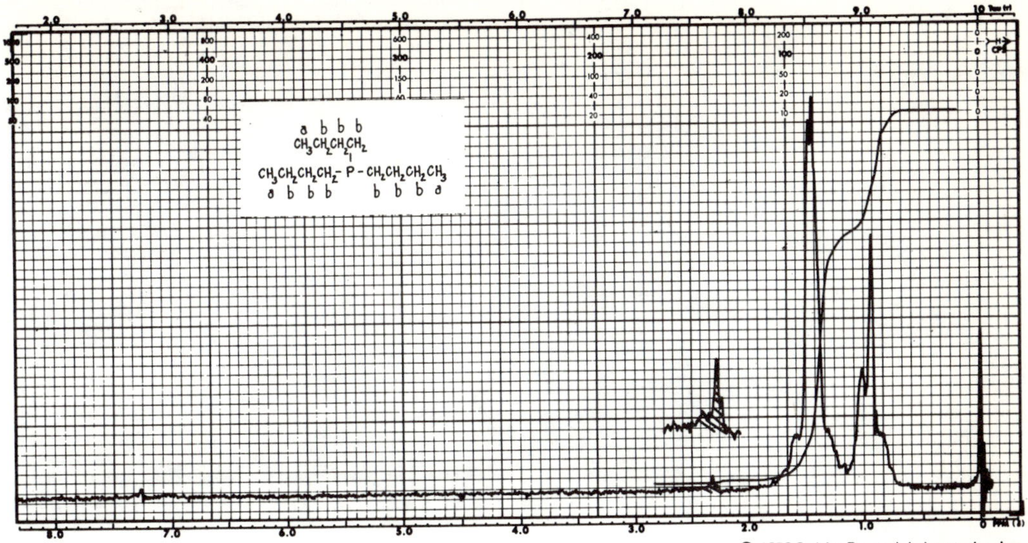

© 1978 Sadtler Research Laboratories, Inc.

DIPHENYLETHYLPHOSPHINE

$C_{14}H_{15}P$

Mol. Wt. 214.25

B.P. 130-132°C/1mm

Solvent: $CDCl_3$

Assignments
a = 1.05
b = 1.99
c = 6.99-7.44

1096

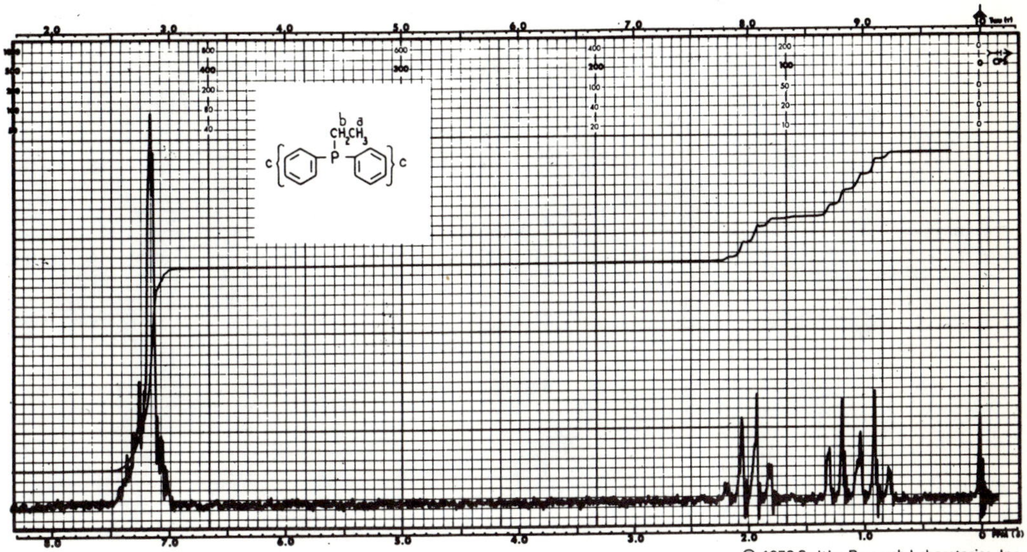

© 1978 Sadtler Research Laboratories, Inc.

TRIMETHYLENEBIS[DIPHENYLPHOSPHINE]

$C_{27}H_{26}P_2$

Mol. Wt. 412.46

M.P. 55-57°C

Solvent: $CDCl_3$

Assignments
a = 1.70
b = 2.21
c = 7.10-7.60

1097

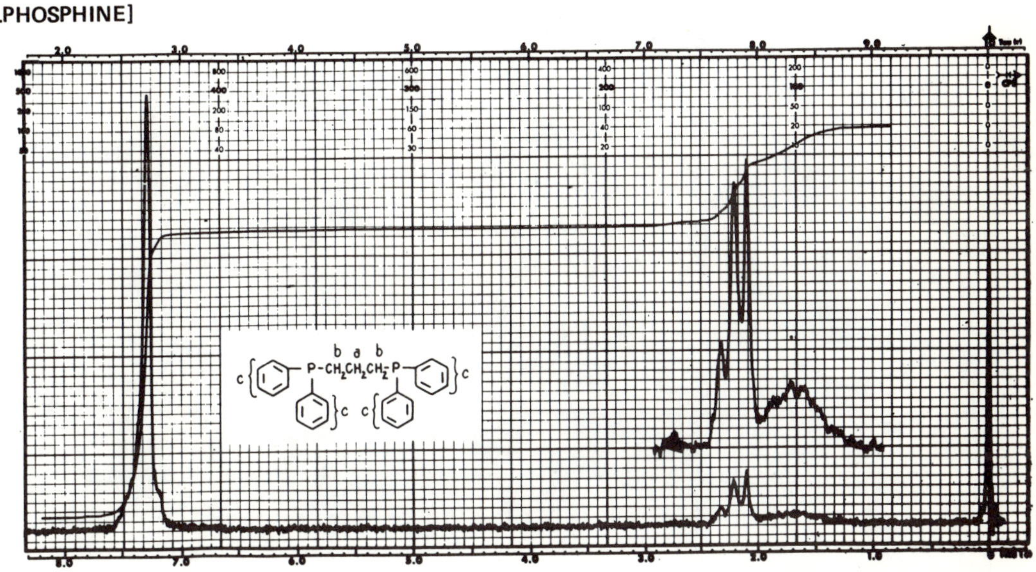

© 1978 Sadtler Research Laboratories, Inc.

TRI-o-TOLYLPHOSPHINE

$C_{21}H_{21}P$

Mol. Wt. 304.37

M.P. 120°C

Solvent: $CDCl_3$

Assignments
a = 2.39
b = 6.64-7.30

1098

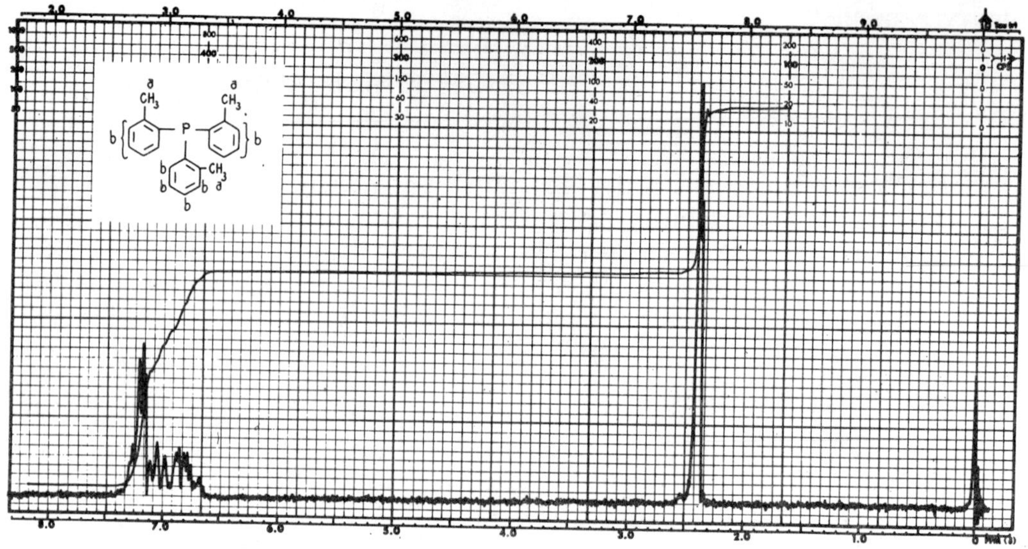

© 1978 Sadtler Research Laboratories, Inc.

TRI-p-TOLYLPHOSPHINE

$C_{21}H_{21}P$

Mol. Wt. 304.37

M.P. 142°C

Solvent: $CDCl_3$

Assignments
a = 2.21
b = 7.00-7.40

1099

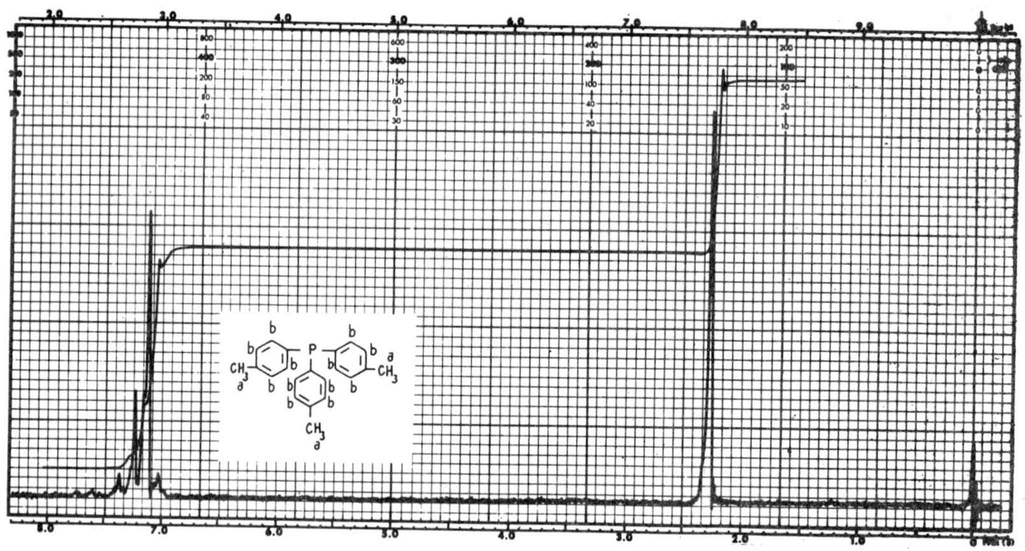

© 1978 Sadtler Research Laboratories, Inc.

TRIS(p-CHLOROPHENYL)PHOSPHINE

$C_{18}H_{12}Cl_3P$

Mol. Wt. 365.63

M.P. 90-93°C

Solvent: $CDCl_3$

Assignments
a = 7.04-7.50

1100

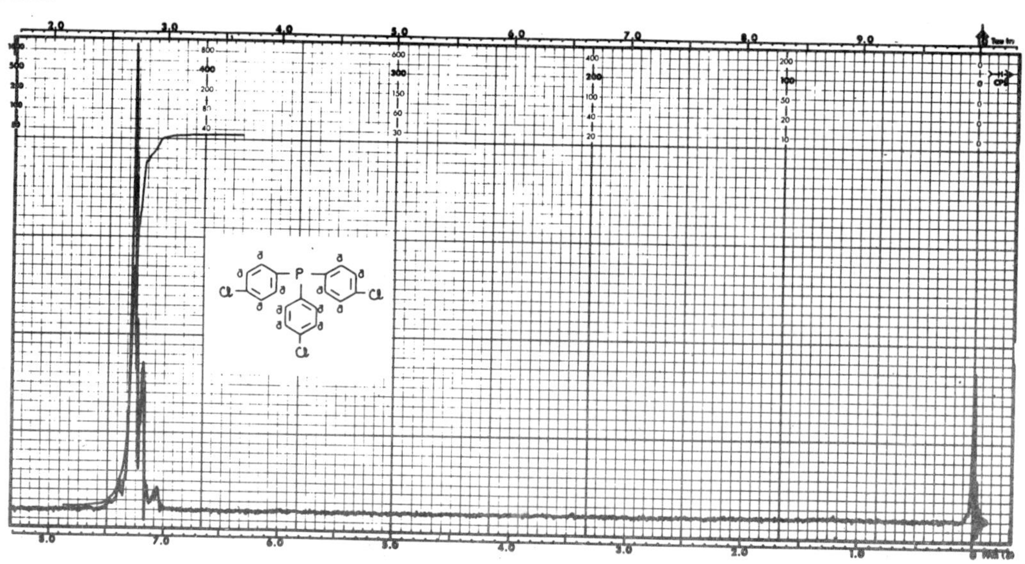

© 1978 Sadtler Research Laboratories, Inc.

THE PHOSPHONIUM COMPOUNDS

The Phosphorus nucleus of the Phosphonium compounds is a much more strongly deshielding group than that of the Phosphines. Clear coupling to the alpha aliphatic groups is observed which makes this group of compounds more easily identifiable than the Phosphines. Although soluble in deutero-chloroform, the spectra as a group suffer from a slightly higher noise level than the other Phosphorus containing compounds due to their lower solubility.

Chemical Shifts

Aliphatic Protons

HC–4	HC–3	HC–2	HC–1			Solvent	Spectrum
		(1.33	2.52)	$(CH_3{-}CH_2)_4{-}P^+$	($^-$I)	$CDCl_3$	1101
0.94	(1.1–2.0)		2.55	$CH_3{-}(CH_2)_2{-}CH_2{-}P^+(R)_4$	($^-$Cl)	$CDCl_3$	----
			(2.74)	$(CH_3)_2{-}P^+(\emptyset,R)$	($^-$Br)	$CDCl_3$	----
	(1.16	1.71	3.08)	$(CH_3{-}CH_2{-}CH_2)_2{-}P^+(\emptyset,R)$	($^-$Br)	$CDCl_3$	----
			3.27	$CH_3{-}P^+(\emptyset)_3$	($^-$Br)	$CDCl_3$	1102
0.89	(1.3–1.9)		3.69	$CH_3{-}(CH_2)_2{-}CH_2{-}P^+(\emptyset)_3$	($^-$Br)	$CDCl_3$	----
	1.24	1.79	3.70	$CH_3{-}CH_2{-}CH_2{-}P^+(\emptyset)_3$	($^-$Br)	$CDCl_3$	----

Substituted Methanes

(X,Y,Z)	$-P^+$	$-CH_2$	$-X$		
$(R_4)_3$	$-P^+-$	4.13	$-C(=O){-}O{-}R_2$	($^-$Br)	$CDCl_3$
$(R_4)_3$	$-P^+-$	4.43	$-\emptyset$	($^-$Cl)	$CDCl_3$
$(\emptyset,(R_3)_2)$	$-P^+-$	4.58	$-C(=O){-}O{-}R_2$	($^-$Br)	$CDCl_3$
$(\emptyset,(CH_3)_2)$	$-P^+-$	4.77	$-C(=O){-}O{-}R_2$	($^-$Br)	$CDCl_3$

Aromatic Protons

The Phosphonium group deshields all of the aromatic hydrogens producing a complex, higher order band at low field (spectrum 1102).

para,meta,ortho	$-X$			
7.62–8.08	$-P^+(\emptyset,\emptyset,CH_3)$	($^-$Br)	$CDCl_3$	1102
7.50–8.10	$-P^+(\emptyset)_3$	($^-$I)	$CDCl_3$	----
7.50–8.20	$-P^+(\emptyset,\emptyset,R)$	($^-$Br)	$CDCl_3$	----

Coupling and Coupling Constants

$J_{P^+-CH_3}$	=	13.6 Hz	(spectrum 1102)
$J_{P^+-CH_2}$	=	13.0–15.1 Hz	
$J_{P^+-C-CH_3}$	=	18.2 Hz	(spectrum 1101)

TETRAETHYLPHOSPHONIUM IODIDE

$C_8H_{20}IP$

Mol. Wt. 274.13

M.P. 270°C

Solvent: $CDCl_3$

Assignments
a = 1.33
b = 2.52

1101

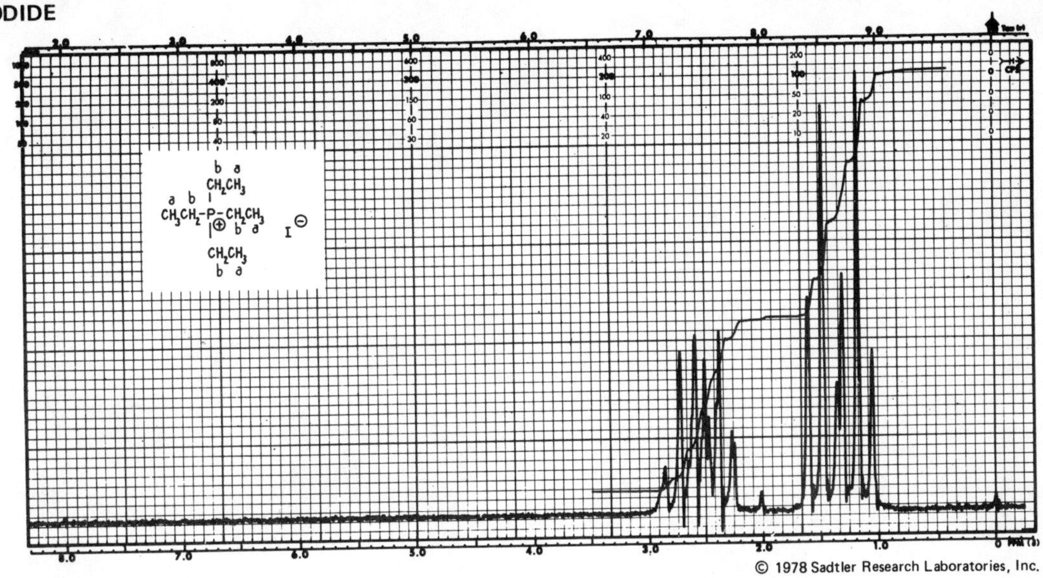

© 1978 Sadtler Research Laboratories, Inc.

METHYLTRIPHENYLPHOSPHONIUM BROMIDE

$C_{19}H_{18}BrP$

Mol. Wt. 357.23

M.P. 229-232°C (lit.)

Solvent: $CDCl_3$

Assignments
a = 3.27
b = 7.62-8.08

1102

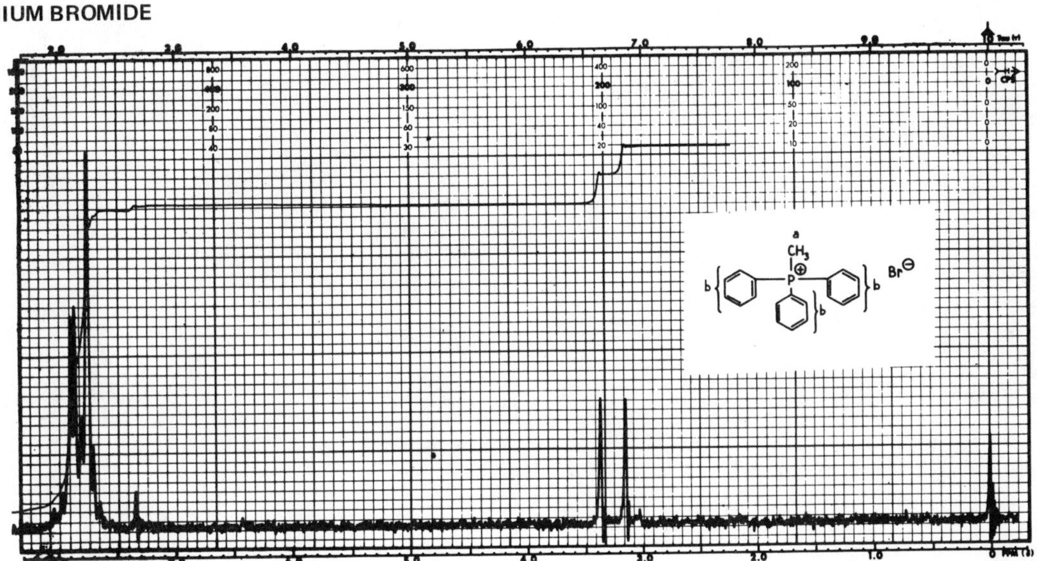

© 1978 Sadtler Research Laboratories, Inc.

NOTES

THE PHOSPHINE OXIDES

The Phosphine Oxide group produces chemical shifts similar to those of the Phosphines but with a significant increase in the magnitude of the coupling constants to the alpha hydrocarbon group. The long chain aliphatic Phosphine Oxides are often difficult to distinguish from the simple alkanes due to the weak carbon-like deshielding of the alpha methylene group.

Chemical Shifts

Aliphatic Protons

HC–13	HC–12 to HC–2	HC–1		Solvent	Spectrum
		(1.44)	$(CH_3)_2$–P(=O)–R_{13}	$CDCl_3$	1103
0.89	(1.1–1.5)	1.5 ca	CH_3–$(CH_2)_{11}$–CH_2–P(=O)(CH_3)	$CDCl_3$	1103

X–	CH_2–	P(=O)	(X, Y)	Solvent
R–	2.59	–P(=O)	$(CH_2–Ø)_2$	$CDCl_3$
Ø–	3.00	–P(=O)	$(CH_3)_2$	$CDCl_3$

Aromatic Protons

The Phosphine Oxide group is a moderately strong deshielding group in its effect on the ortho aromatic hydrogens. Coupling to these hydrogens, when it can be observed, is found to be about 12–13 Hz.

para/meta	ortho	–X	
7.3–7.7	7.72	–P(=O)(Ø, R)	$CDCl_3$

Coupling and Coupling Constants

$J_{P–CH_3}$ = 12–13 Hz

$J_{P–CH_2}$ = 10–12 Hz

DIMETHYLDODECYLPHOSPHINE OXIDE

$C_{14}H_{31}OP$

Mol. Wt. 246.39

M.P. 84-85°C

Solvent: $CDCl_3$

Assignments
a = 0.89
b = 1.27
c = 1.44
d = ca 1.60

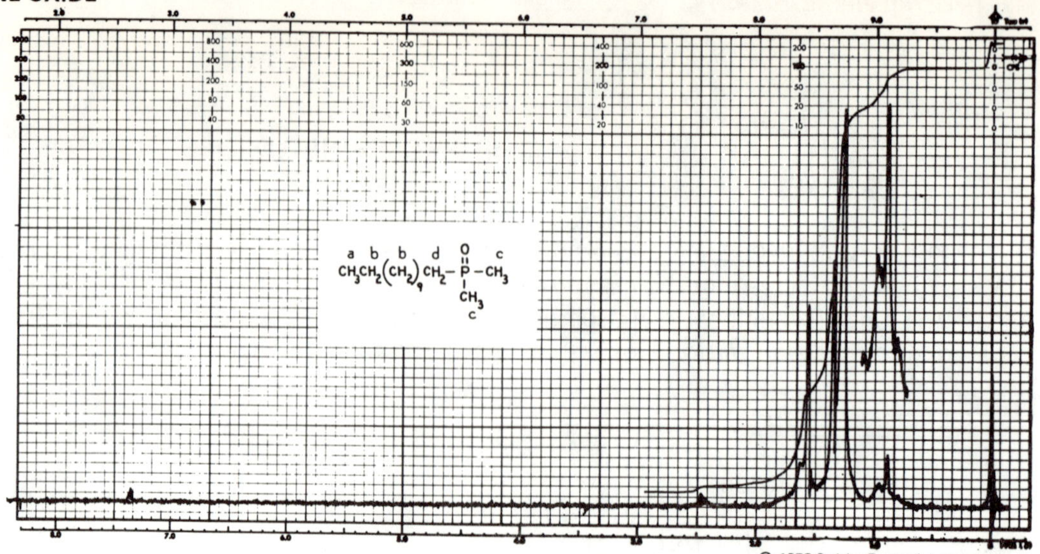

1103

© 1978 Sadtler Research Laboratories, Inc.

DIMETHYLHEXADECYLPHOSPHINE OXIDE

$C_{18}H_{39}OP$

Mol. Wt. 302.48

M.P. 93-94°C

Solvent: $CDCl_3$

Assignments
a = 0.89
b = 1.26
c = 1.41
d = ca 1.58

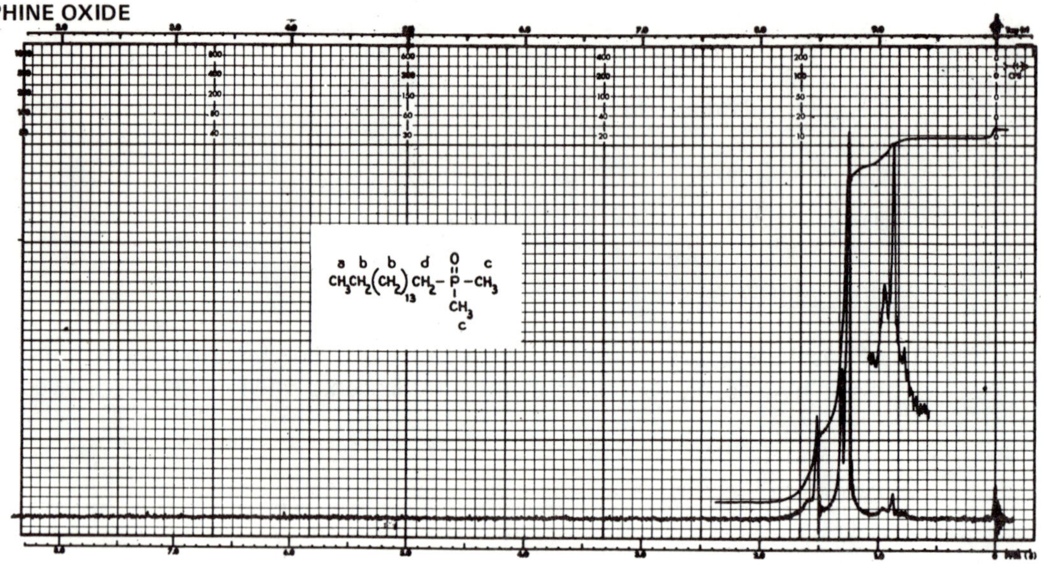

1104

© 1978 Sadtler Research Laboratories, Inc.

TRI-o-TOLYLPHOSPHINE OXIDE

$C_{21}H_{21}OP$

Mol. Wt. 320.37

M.P. 150-151°C

Solvent: $CDCl_3$

Assignments
a = 2.52
b = 6.87-7.62
c = 2.42 imp.

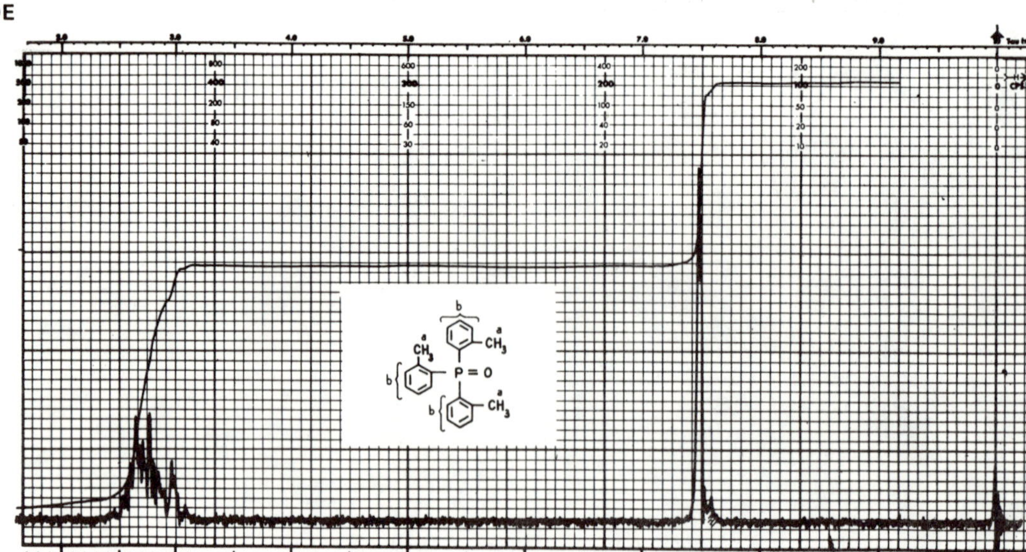

1105

© 1978 Sadtler Research Laboratories, Inc.

TRI-m-TOLYLPHOSPHINE OXIDE

$C_{21}H_{21}OP$

Mol. Wt. 320.37

M.P. 143-144°C

Solvent: $CDCl_3$

Assignments
a = 2.35
b = 7.25-7.75

1106

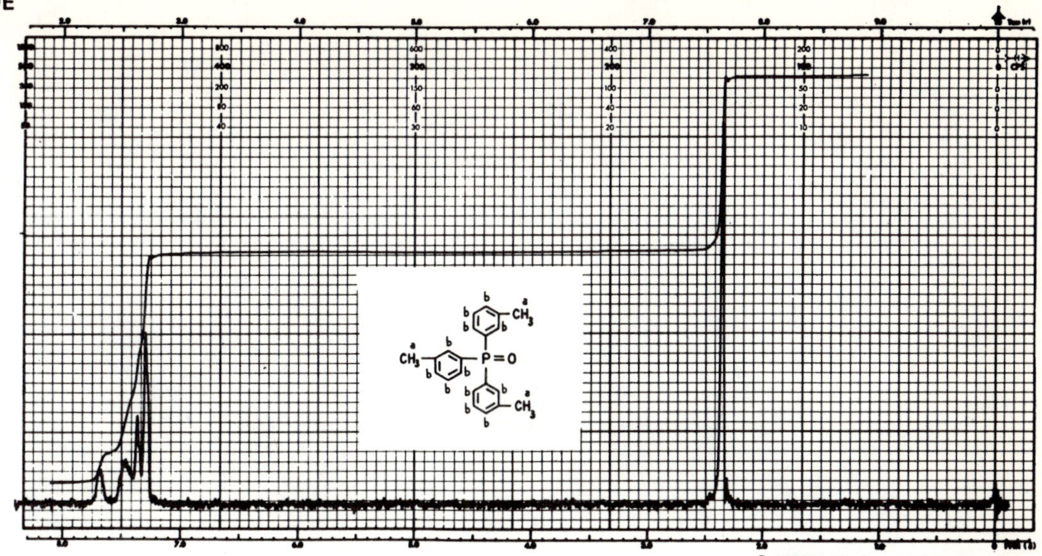

© 1978 Sadtler Research Laboratories, Inc.

NOTES

The Sulfide linkage has a weakly deshielding effect on adjacent aliphatic groups and a weak shielding effect on the ortho aromatic hydrogens. In both cases, the phenyl Sulfide group is a more strongly deshielding group than the corresponding aliphatic Sulfide linkage.

Except in the case of the heterocyclic molecules, the coupling constants are the same as those observed for the other substituents discussed thus far.

Chemical Shifts

Aliphatic Protons

HC–3	HC–2	HC–1		Solvent	Spectrum
		1.90	CH_3–S–CH_2–Ø	CCl_4	1158
		2.01	CH_3–S–R_{12}	CCl_4	1116
		2.04	CH_3–S–R_2	CCl_4	1107
		2.06	CH_3–S–CH_3	CCl_4	- - - -
		2.29	CH_3–S–Ø (2–NH_2)	CCl_4	- - - -
		2.39	CH_3–S–Ø	$CDCl_3$	- - - -
		2.45	CH_3–S–Ø (4–Br)	$CDCl_3$	1160
	1.19	2.32	CH_3–CH_2–S–CH_2–Ø	CCl_4	1159
	1.23	2.43	CH_3–CH_2–S–CH_3	CCl_4	1107
	1.23	2.47	CH_3–CH_2–S–R_2	CCl_4	1108
	1.23	2.68	CH_3–CH_2–S–CH_2–C(=O)–OH	CCl_4	- - - -
0.98	1.59	2.44	CH_3–CH_2–CH_2–S–R_3	CCl_4	1109
	(1.38)	3.12	$(CH_3)_2$–CH–S–R	TFA	- - - -
(0.99)	1.79	2.32	$(CH_3)_2$–CH–CH_2–S–R	CCl_4	1111
	(1.36)	- - - -	$(CH_3)_3$–C–S–C$(CH_3)_3$	CCl_4	- - - -

Substituted Methylsulfides

R–S–	CH_2	–X	
	2.32	–CH$(CH_3)_2$	CCl_4
	2.43	–CH_3	CCl_4
	3.09	–CH=CH_2	$CDCl_3$
	3.36	–C≡N	CCl_4
	3.77	–C(=O)–O–CH_3	$CDCl_3$
	3.80	–C(=O)–OH	$CDCl_3$
	3.80	–Ø	CCl_4
	4.83	–Cl	$CDCl_3$

Chemical Shifts continued

Aliphatic Protons continued

Substituted Methylsulfides continued

Ø–S–	CH_2	–X	Solvent	Spectrum
	2.10	$-Si(CH_3)_3$	CCl_4	
	3.49	–C(=O)–O–R	CCl_4	
	3.58	–C≡C–R	$CDCl_3$	
	3.59	–C(=O)–OH	$CDCl_3$	
	3.99	–Ø	$CDCl_3$	
	4.00	–C(=O)–R	Polysol	
	4.10	–C(=O)–Ø	$CDCl_3$	
	4.67	–NH–Ø	Polysol	

Substituted Ethylsulfides

R–S–	CH_2	$-CH_2$	–X	Solvent
	2.44	1.59	$-CH_3$	CCl_4
	2.52	1.69	$-(CH_2)_2-S-R_2$	$CDCl_3$
	2.55	2.66	–C(=O)–O–R	CCl_4
	2.71	2.71	$-S-R_2$	CCl_4
	2.75	2.88	–C≡N	$CDCl_3$
	2.61	3.63	–OH	CCl_4

Ø–S–	CH_2	$-CH_2$	–X	Solvent
	2.70	2.70	–S–Ø	Polysol
	3.10	2.58	–C(=O)–OH	Polysol
	3.11	2.78	$-C(=O)-CH_3$	Polysol
	3.07	2.90	–Ø	$CDCl_3$
	2.89	3.57	–OH	CCl_4

Aromatic Protons

Phenyl Sulfides

para/meta/ortho	–S–X	Solvent	Spectrum
6.90–7.30	$-S-CH_3$	$CDCl_3$	- - - -
7.00–7.43	–S–Ø	CCl_4	1169
7.10–7.70	–S–(4–pyridine)	$CDCl_3$	1172

Aromatic Protons continued

Para Substituted Phenylsulfides

CH_3–S– ortho	meta	para	Solvent	Spectrum
7.22	6.75	–OH	$CDCl_3$	
7.10	6.83	–NH–R	Polysol	
7.05	7.05	$-CH_3$	CCl_4	
7.11	7.39	–Br	$CDCl_3$	
7.21	7.55	–NH–C(=O)–R	DMSO-d_6	
7.24	7.81	$-C(=O)-CH_3$	Polysol	
7.27	7.89	–C(=O)–OH	DMSO-d_6	

Ø–S– ortho	meta	para	Solvent	Spectrum
7.33	6.63	$-N(CH_3)_2$	$CDCl_3$	
7.23	6.78	$-NH_2$	Polysol	
7.08	6.94	$-O-CH_3$	$CDCl_3$	
7.28	7.28	–Cl	Polysol	
7.21	7.07	$-CH_3$	$CDCl_3$	
7.47	7.47	$-NH_2$ (salt)	TFA	
7.44	7.97	$-NO_2$	$CDCl_3$	

The Thiophenes

Due to the small differentiation in chemical shift between the protons of the parent compound and the relatively large coupling constants involved, the multiplets that arise from the substituted Thiophenes are usually higher order in character. The coupling constants are unusual in that the J_{2-3} coupling constant is normally larger than the J_{3-4}, a situation that does not occur in the spin-spin interactions of the corresponding oxygen and nitrogen heterocyclics.

Thiophene

C–2,5	C–3,4	Solvent	Spectrum
6.90	7.10 ppm	CCl_4	1136

2-Substituted Thiophenes

C–5	C–4	C–3	C–2	Solvent	Spectrum
7.03	6.87	6.72	$-CH_3$	$CDCl_3$	- - - -
6.92	6.72	6.72	–Cl	$CDCl_3$	1138
7.12	6.79	6.98	–Br	CCl_4	1140

Chemical Shifts continued

The Thiophenes continued

2-Substituted Thiophenes continued

C–5	C–4	C–3	C–2	Solvent	Spectrum
7.19	6.72	7.27	–I	$CDCl_3$	1143
7.53	7.02	7.53	$-C(=O)-CH_3$	CCl_4	- - - -
7.60	7.10	7.60	$-C{\equiv}N$	$CDCl_3$	1144
7.54	7.07	7.62	$-C(=O)-R_2$	CCl_4	- - - -
7.62	7.07	7.71	$-SO_2-\emptyset$	$CDCl_3$	- - - -
7.67	7.12	7.67	$-C(=O)-H$	CCl_4	- - - -
7.56	7.07	7.69	$-C(=O)-O-Na$	D_2O	- - - -

3-Substituted Thiophenes

C–5	C–4	C–3	C–2		
7.11	6.75	$-C_3$	6.78	CCl_4	- - - -
7.05	6.77	$-CH_3$	6.77	CCl_4	1137
7.18	6.91	–Br	7.11	$CDCl_3$	1141

Coupling and Coupling Constants

J_{2-3} = 3–4 Hz

J_{3-4} = 4–5 Hz

J_{2-4} = 1–2 Hz

ETHYL METHYL SULFIDE

C_3H_8S

Mol. Wt. 76.16

B.P. 65-67°C

Solvent: CCl_4

Assignments
a = 1.23
b = 2.04
c = 2.43

1107

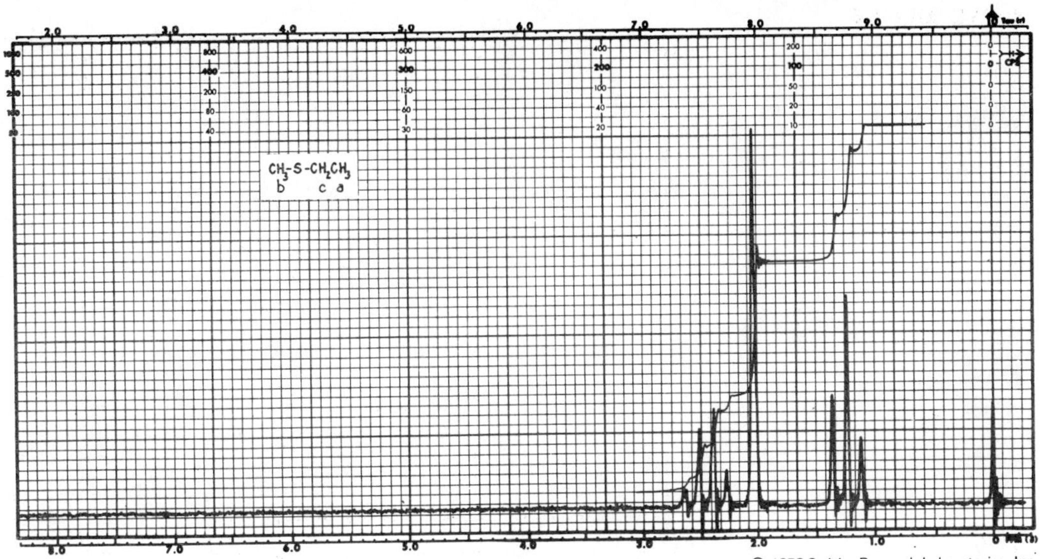

© 1978 Sadtler Research Laboratories, Inc.

ETHYL SULFIDE

$C_4H_{10}S$

Mol. Wt. 90.19

B.P. 90-92°C

Solvent: CCl_4

Assignments
a = 1.23
b = 2.47

1108

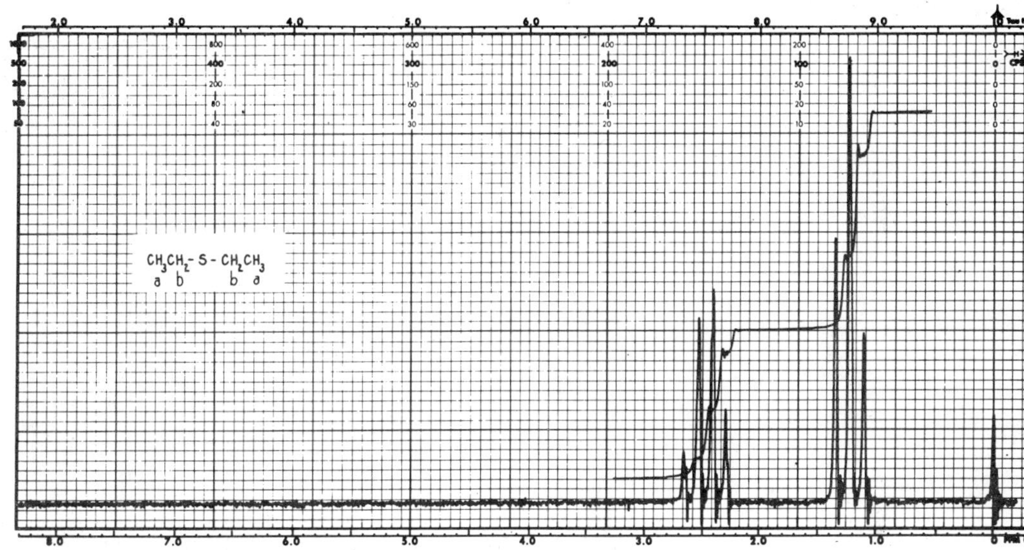

© 1978 Sadtler Research Laboratories, Inc.

PROPYL SULFIDE

$C_6H_{14}S$

Mol. Wt. 118.24

B.P. 142.5-144°C

Solvent: CCl_4

Assignments
a = 0.98
b = 1.59
c = 2.44

1109

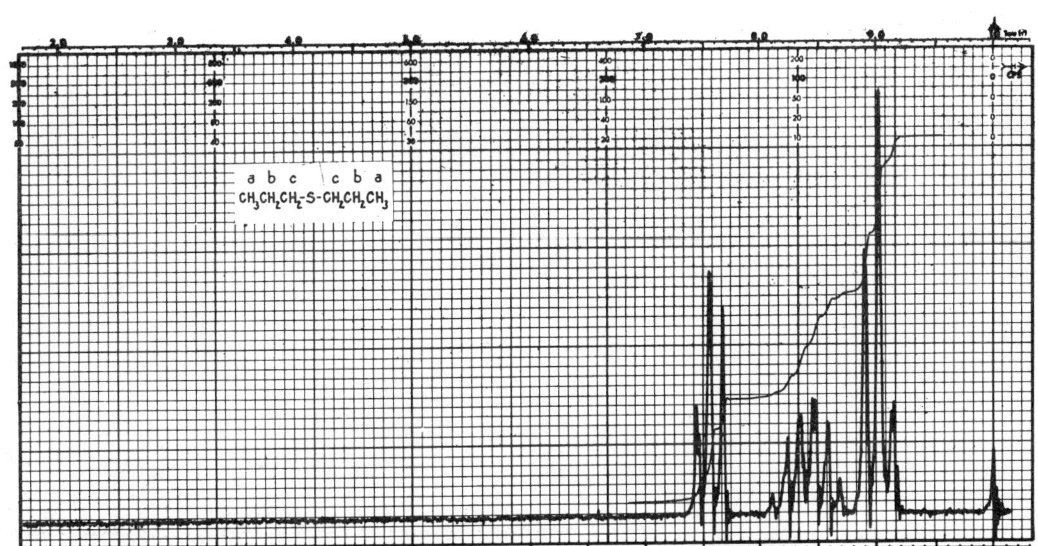

© 1978 Sadtler Research Laboratories, Inc.

BUTYL SULFIDE

$C_8H_{18}S$

Mol. Wt. 146.29

B.P. 182-186°C

Solvent: CCl_4

Assignments
a = 0.91
b = 1.47
c = 2.42

1110

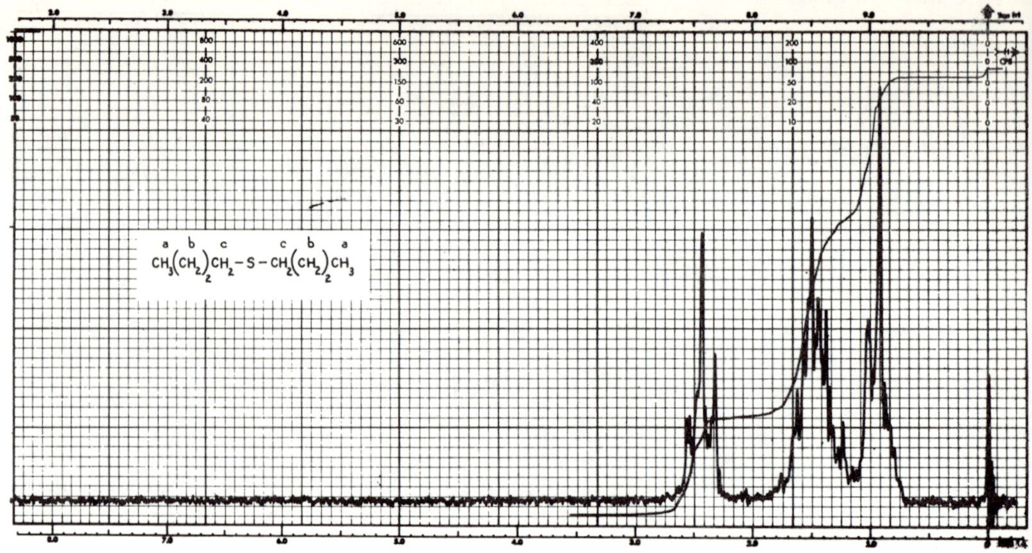

© 1978 Sadtler Research Laboratories, Inc.

ISOBUTYL SULFIDE

$C_8H_{18}S$

Mol. Wt. 146.30

B.P. 169.5-171.5°C

Solvent: CCl_4

Assignments
a = 0.99
b = 1.79
c = 2.32

1111

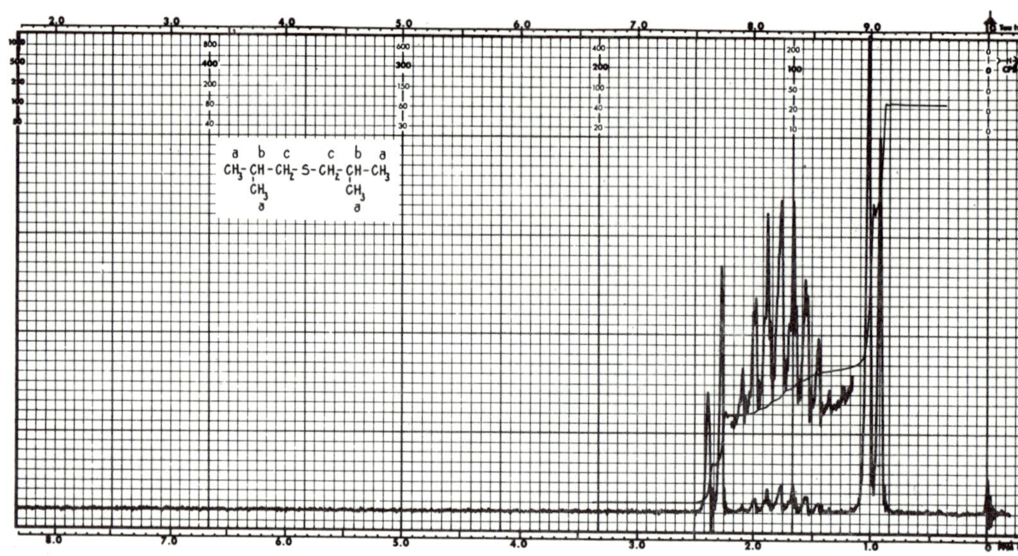

© 1978 Sadtler Research Laboratories, Inc.

ISOPENTYL SULFIDE

$C_{10}H_{22}S$

Mol. Wt. 174.25

B.P. 97-101°C/10mm

Solvent: $CDCl_3$

Assignments
a = 0.92
b = 1.20-2.00
c = 2.53

1112

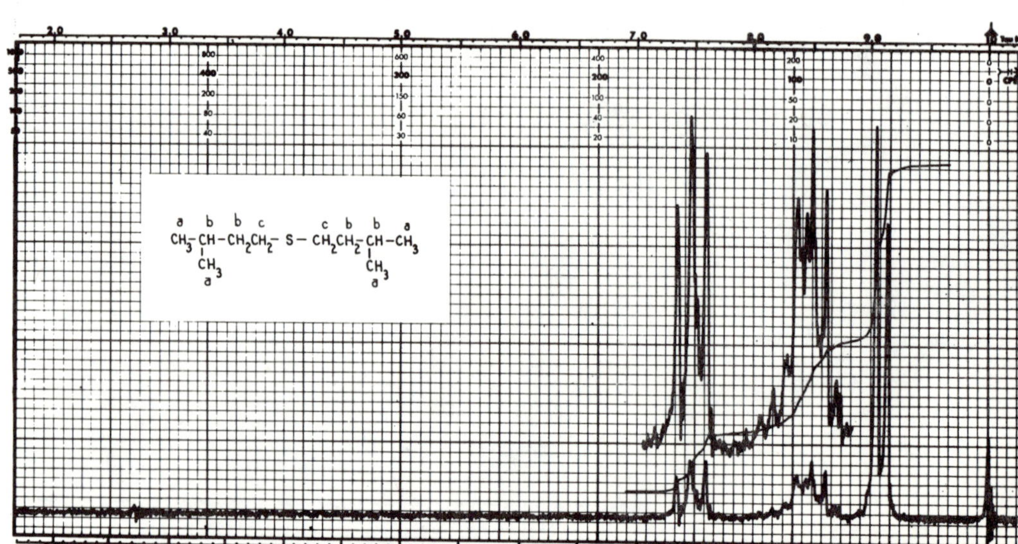

© 1978 Sadtler Research Laboratories, Inc.

BIS(2-ETHYLHEXYL) SULFIDE

$C_{16}H_{34}S$

Mol. Wt. 258.51

Solvent: CCl_4

Assignments
a = 0.87
b = 0.90
c = ca 1.31
d = 2.37

1113

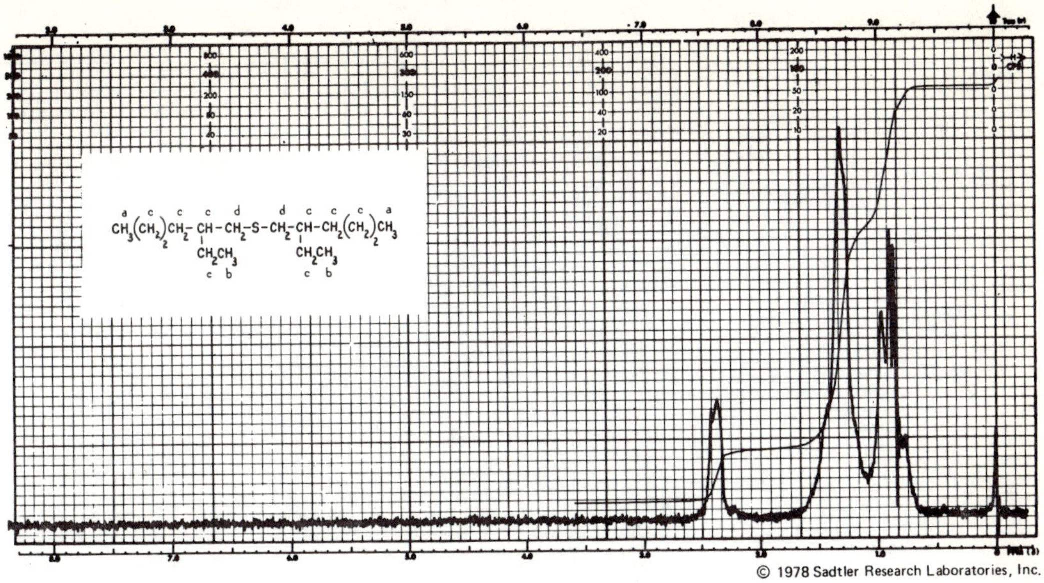

HEXYL SULFIDE

$C_{12}H_{26}S$

Mol. Wt. 202.41

B.P. 126-129°C/9mm

Solvent: CCl_4

Assignments
a = 0.91
b = 1.10-1.90
c = 2.43

1114

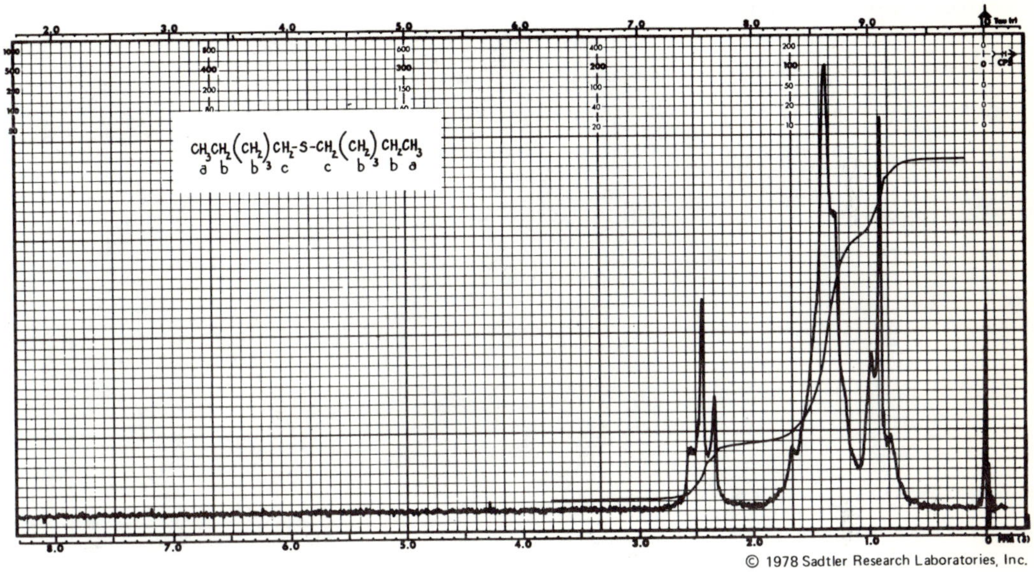

HEPTYL SULFIDE

$C_{14}H_{30}S$

Mol. Wt. 230.46

B.P. 143-145°C/8mm

Solvent: CCl_4

Assignments
a = 0.90
b = 1.10-1.90
c = 2.45

1115

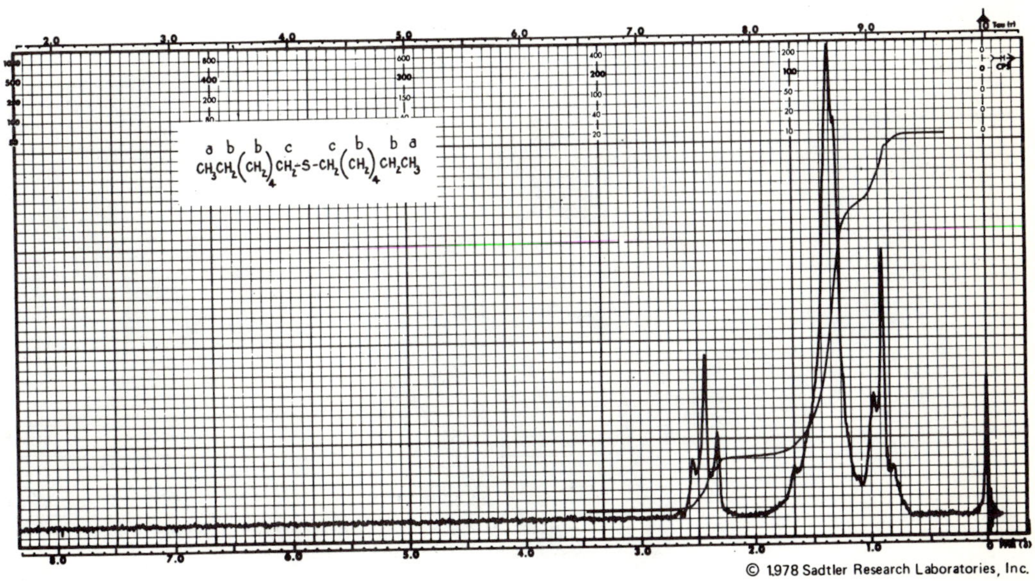

DODECYL METHYL SULFIDE

$C_{13}H_{28}S$

Mol. Wt. 216.43

B.P. 142-144°C/8mm

Solvent: CCl_4

Assignments
a = 0.90
b = 1.10-1.80
c = 2.01
d = 2.41

1116

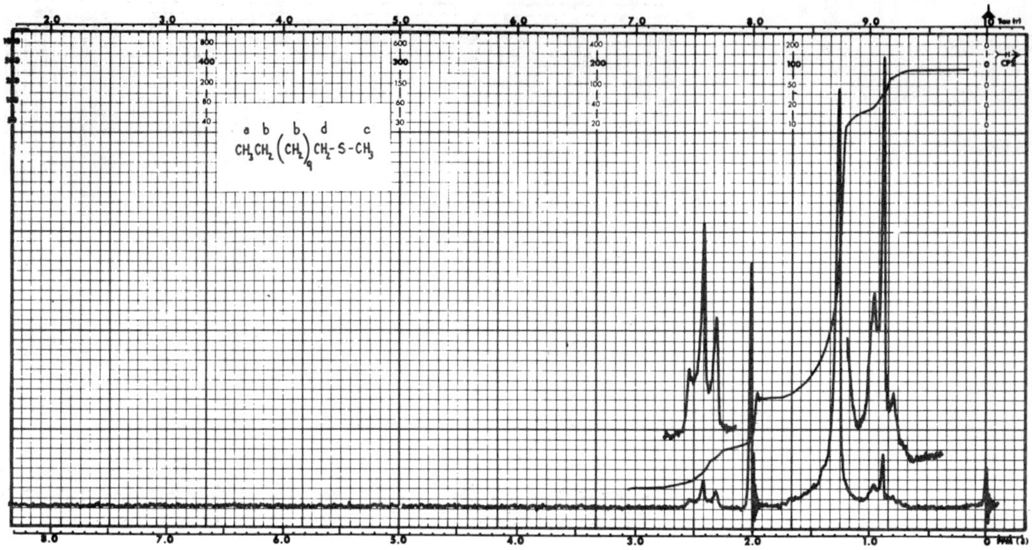

© 1978 Sadtler Research Laboratories, Inc.

DODECYL ETHYL SULFIDE

$C_{14}H_{30}S$

Mol. Wt. 230.46

B.P. 143-146°C/5mm

Solvent: CCl_4

Assignments
a = 0.90
b = 1.21
c = 1.10-1.80
d = 2.41
e = 2.43

1117

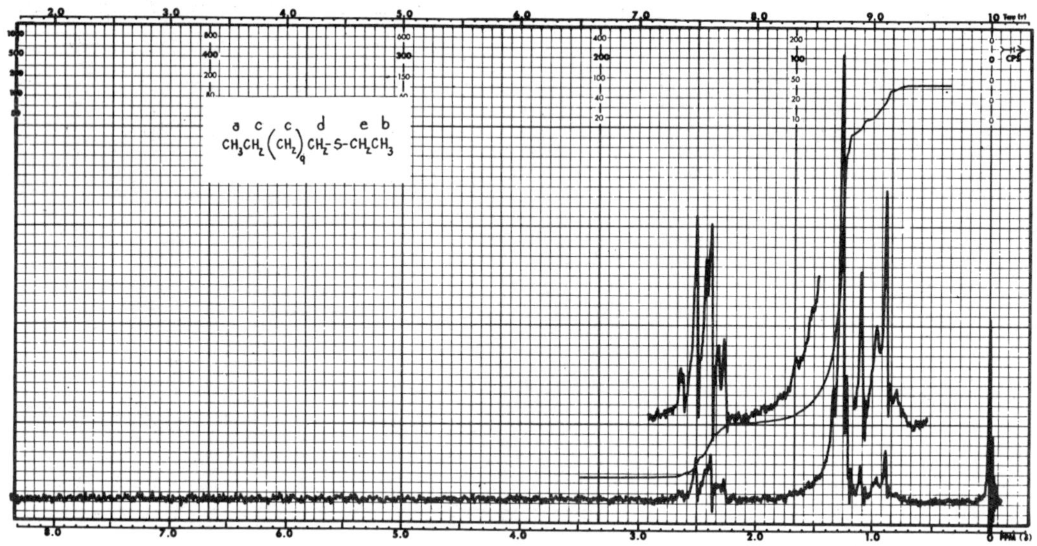

© 1978 Sadtler Research Laboratories, Inc.

DODECYL SULFIDE

$C_{24}H_{50}S$

Mol. Wt. 370.73

M.P. 40-41°C (lit.)

Solvent: CCl_4

Assignments
a = 0.89
b = 1.28
c = 2.41

1118

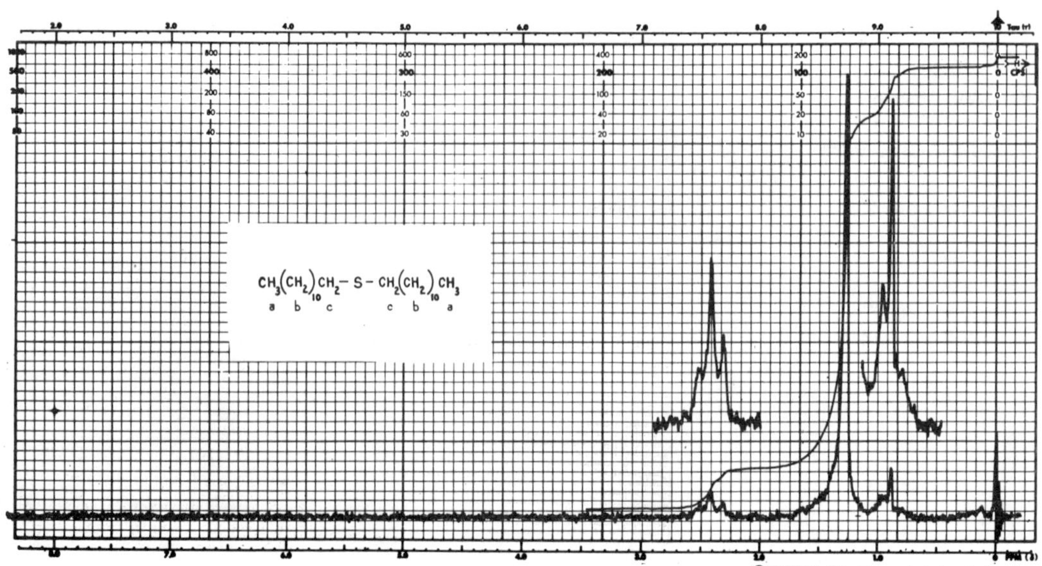

© 1978 Sadtler Research Laboratories, Inc.

ETHYL OCTADECYL SULFIDE

$C_{20}H_{42}S$

Mol. Wt. 314.62

M.P. 29-31°C

Solvent: $CDCl_3$

Assignments
a = 0.88
b = 1.24
c = ca 1.27
d = 2.49
e = 2.54

1119

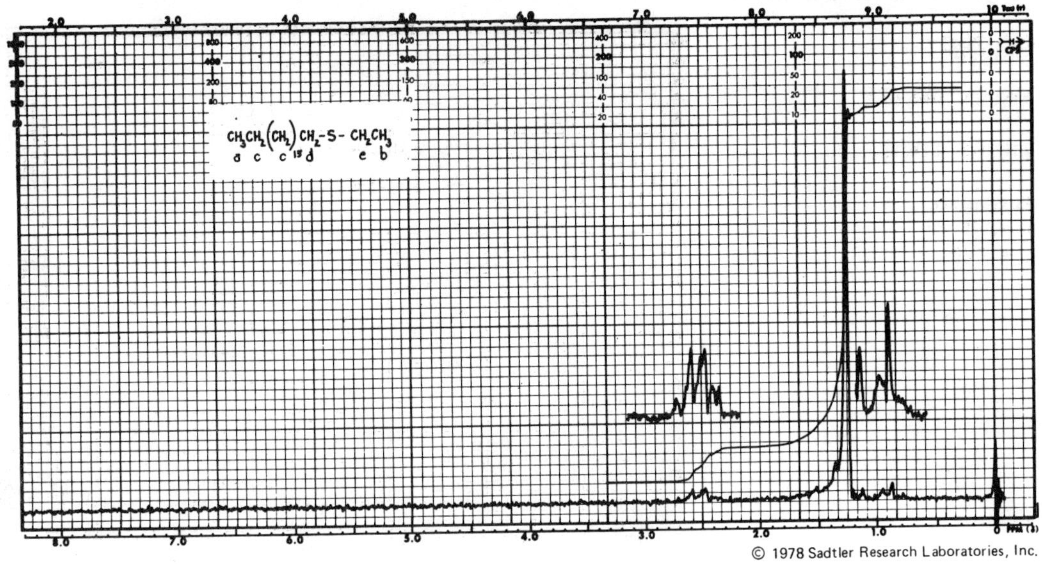

© 1978 Sadtler Research Laboratories, Inc.

BIS(CHLOROMETHYL) SULFIDE

$C_2H_4Cl_2S$

Mol. Wt. 131.02

Solvent: $CDCl_3$

Assignments
a = 4.83

1120

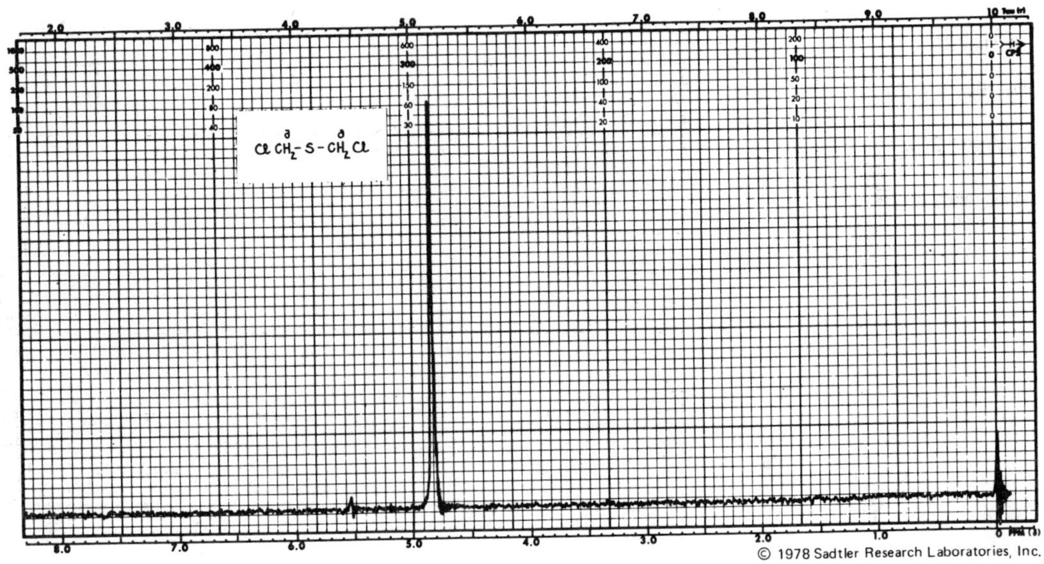

© 1978 Sadtler Research Laboratories, Inc.

3-(METHYLTHIO)PROPYLAMINE

$C_4H_{11}NS$

Mol. Wt. 105.20

B.P. 167-170°C

Solvent: D_2O

Assignments
a = 1.72
b = 2.10
c = 2.57
d = 2.68
e = 4.51 HDO

1121

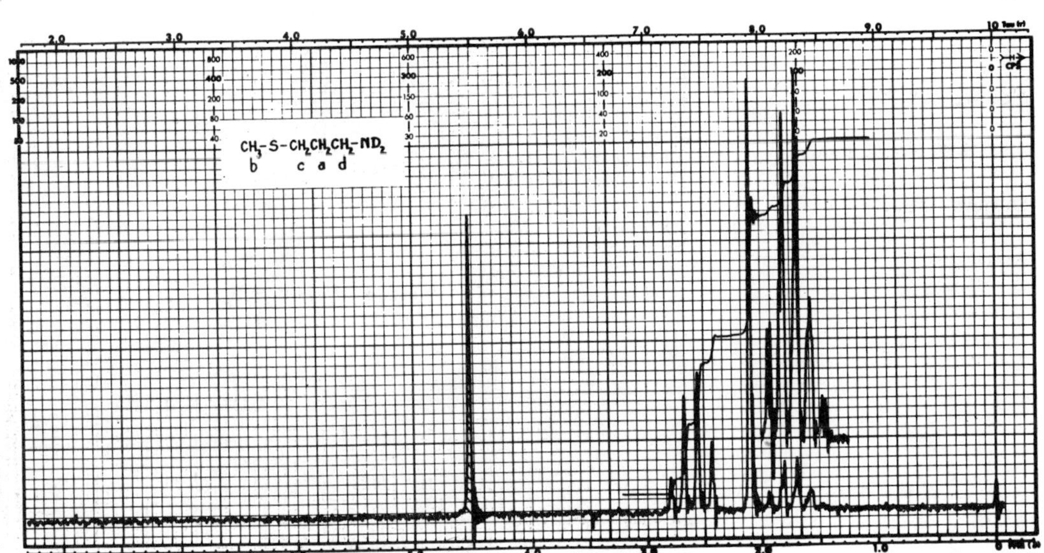

© 1978 Sadtler Research Laboratories, Inc.

3′,3′-THIODIPROPIONITRILE

$C_6H_8N_2S$

Mol. Wt. 140.20

Solvent: $CDCl_3$

Assignments
a = 2.75 or b
b = 2.88 or a

1122

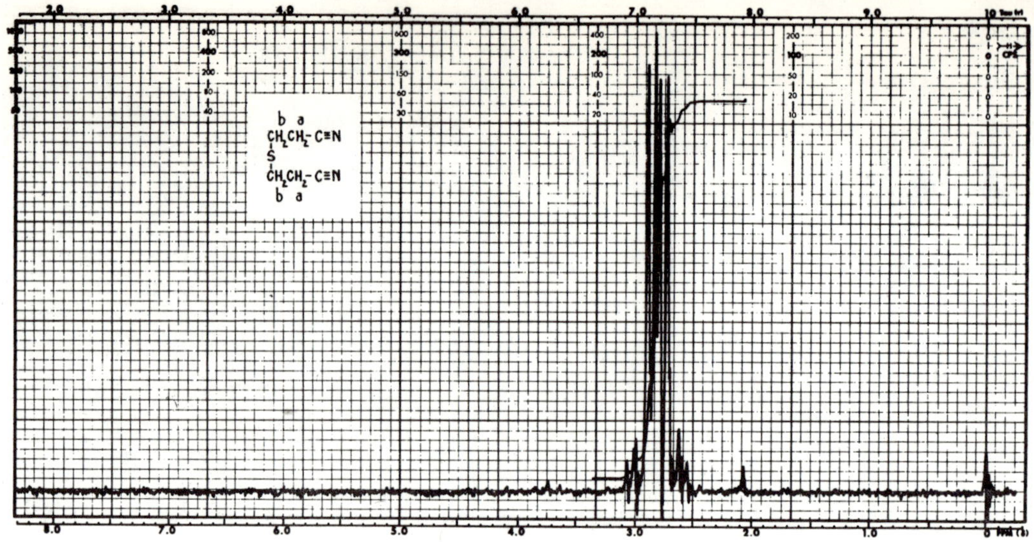

© 1978 Sadtler Research Laboratories, Inc.

1,2-BIS(ETHYLTHIO)ETHANE

$C_6H_{14}S_2$

Mol. Wt. 150.31

Solvent: CCl_4

Assignments
a = 1.28
b = 2.56
c = 2.71

1123

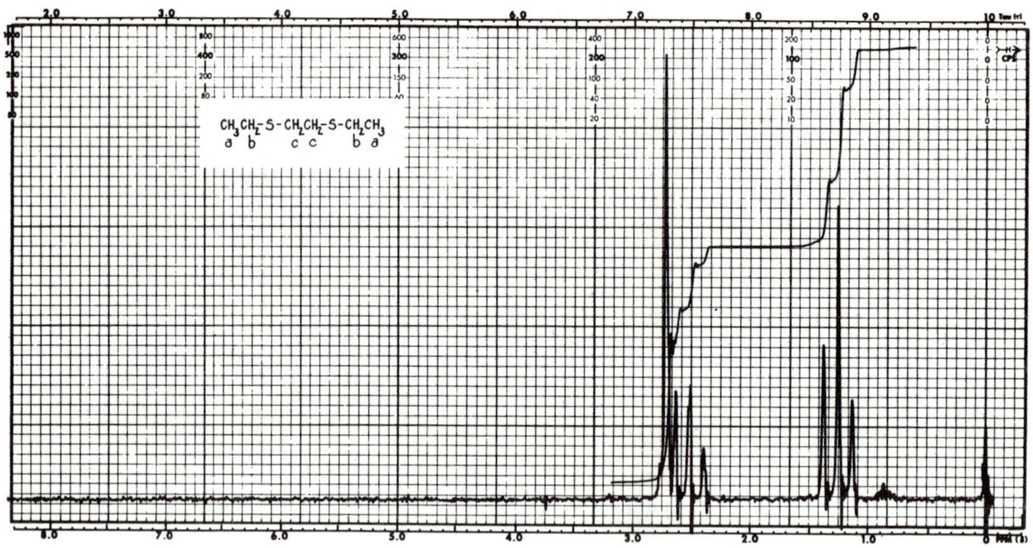

© 1978 Sadtler Research Laboratories, Inc.

1,2-BIS(PROPYLTHIO)ETHANE

$C_8H_{18}S_2$

Mol. Wt. 178.36

B.P. 91-92.5°C/0.3mm

Solvent: CCl_4

Assignments
a = 0.99
b = 1.60
c = 2.49
d = 2.61

1124

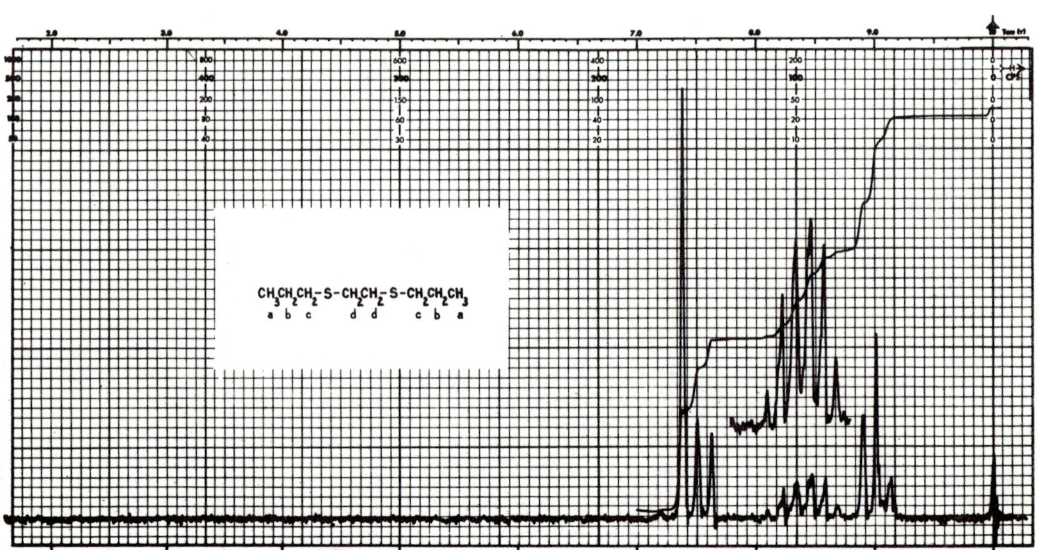

© 1978 Sadtler Research Laboratories, Inc.

1,4-BIS(ETHYLTHIO)BUTANE

$C_8H_{18}S_2$

Mol. Wt. 178.36

Solvent: $CDCl_3$

Assignments
a = 1.25
b = 1.69
c = 2.52

1125

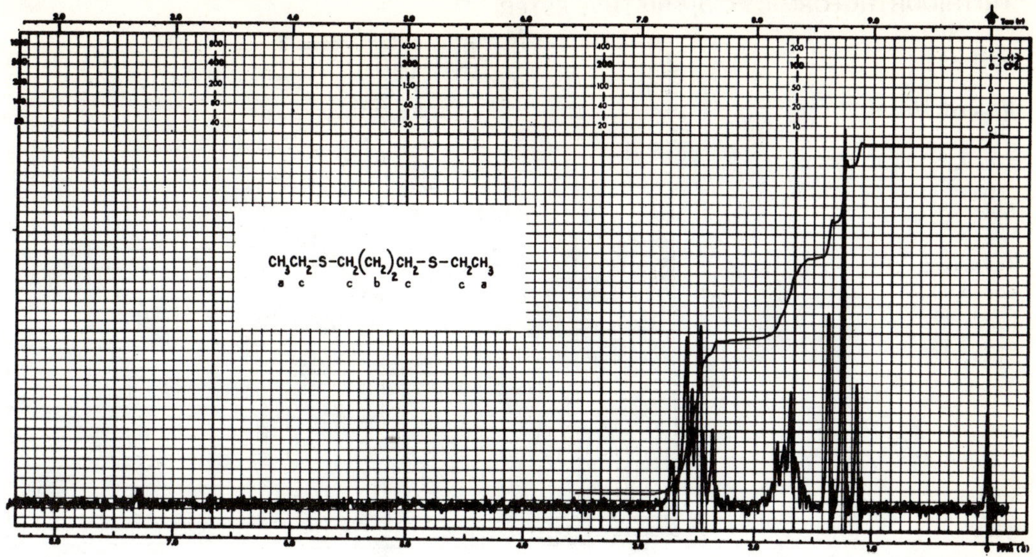

© 1978 Sadtler Research Laboratories, Inc.

1,4-BIS(PROPYLTHIO)BUTANE

$C_{10}H_{22}S_2$

Mol. Wt. 206.41

Solvent: CCl_4

Assignments
a = 0.96
b = 1.21-1.91
c = 2.43

1126

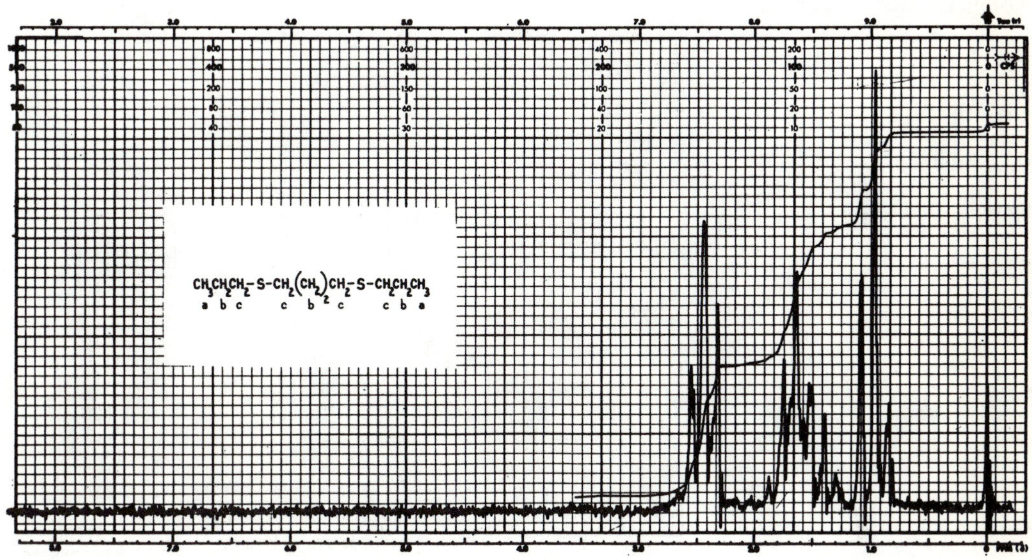

© 1978 Sadtler Research Laboratories, Inc.

1,6-BIS(METHYLTHIO)HEXANE

$C_8H_{18}S_2$

Mol. Wt. 178.36

B.P. 140°C/19mm (lit.)

Solvent: CCl_4

Assignments
a = 1.42
b = 2.00
c = 2.42

1127

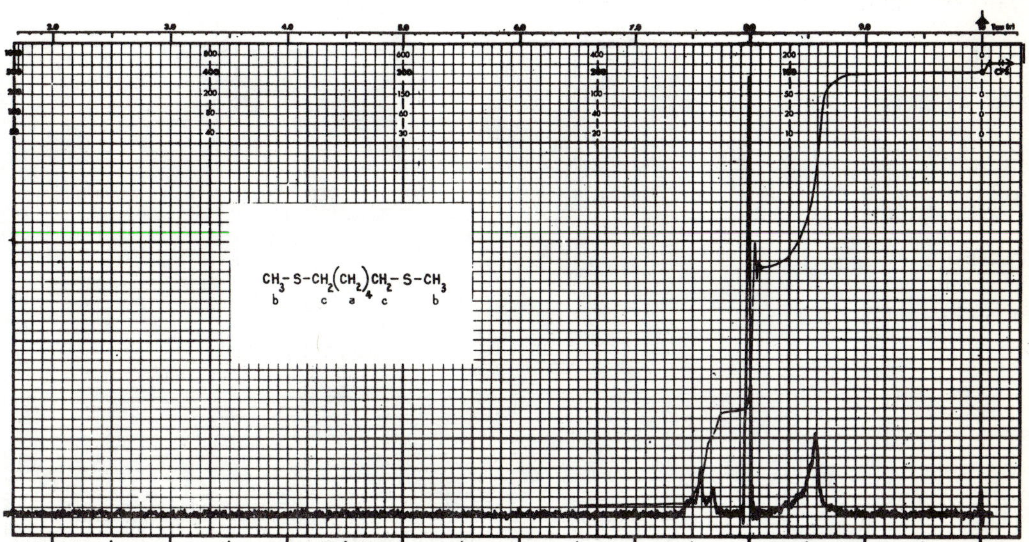

© 1978 Sadtler Research Laboratories, Inc.

TRITHIOORTHOFORMIC ACID, TRIETHYL ESTER

$C_7H_{16}S_3$

Mol. Wt. 196.40

B.P. 135-137°C/13mm

Solvent: CCl_4

Assignments
a = 1.29
b = 2.69
c = 4.80

1128

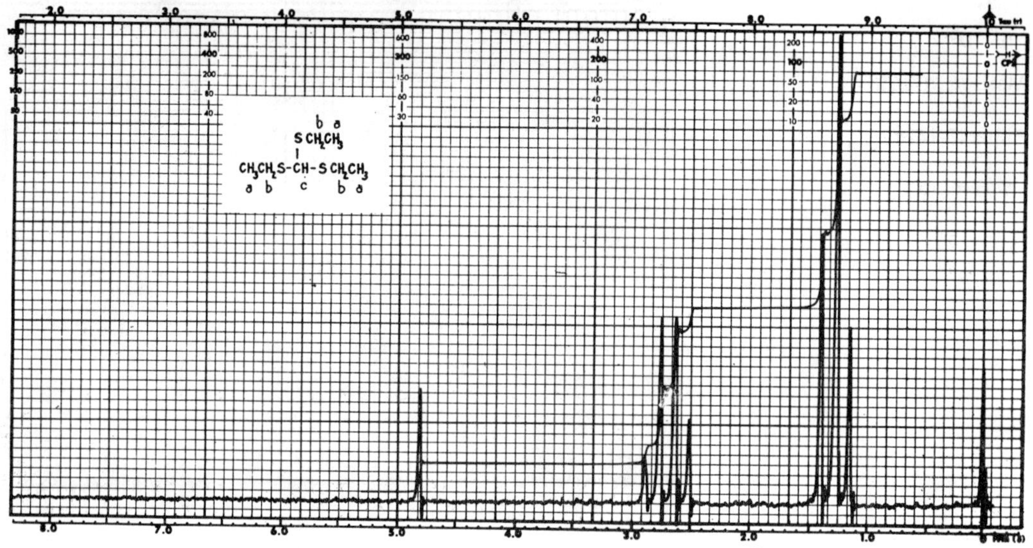

© 1978 Sadtler Research Laboratories, Inc.

TRIMETHYLENE SULFIDE

C_3H_6S

Mol. Wt. 74.15

B.P. 94-95°C

Solvent: $CDCl_3$

Assignments
a = 3.00
b = 3.23

1129

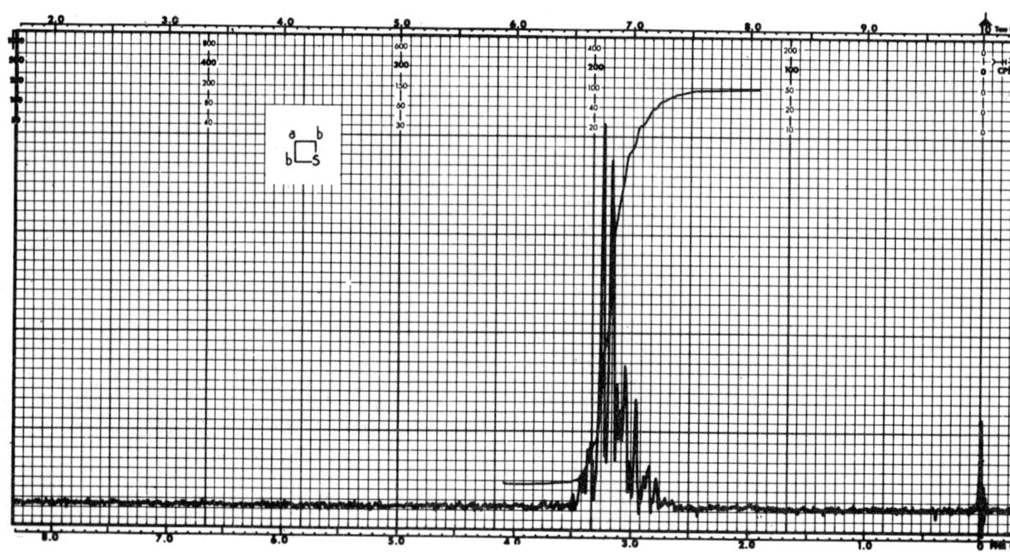

© 1978 Sadtler Research Laboratories, Inc.

TETRAHYDROTHIOPHENE

C_4H_8S

Mol. Wt. 88.17

B.P. 118-119°C

Solvent: CCl_4

Assignments
a = 1.90
b = 2.77

1130

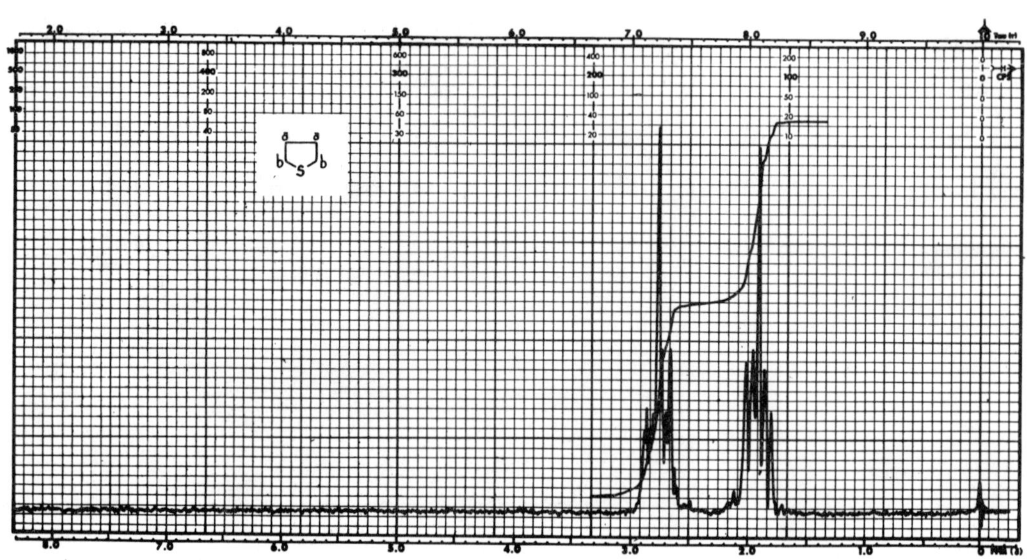

© 1978 Sadtler Research Laboratories, Inc.

TETRAHYDRO-2H-THIOPYRAN

$C_5H_{10}S$

Mol. Wt. 102.20

B.P. 140-142°C

Solvent: CCl_4

Assignments
a = 1.30-2.10
b = 2.58

1131

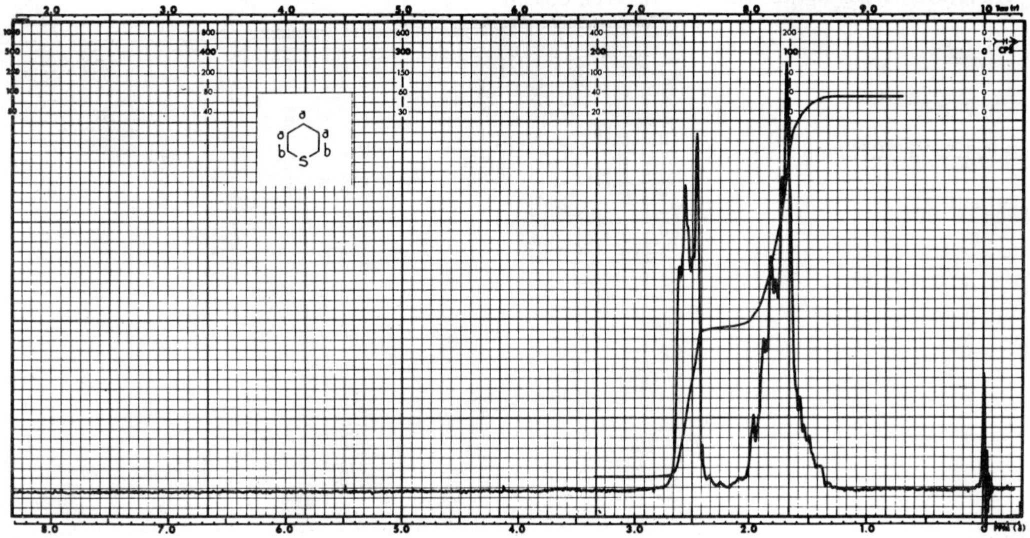

© 1978 Sadtler Research Laboratories, Inc.

---

m-DITHIANE

$C_4H_8S_2$

Mol. Wt. 120.24

M.P. 53-55°C

Solvent: Polysol-d

Assignments
a = 2.02
b = 2.83
c = 3.80

1132

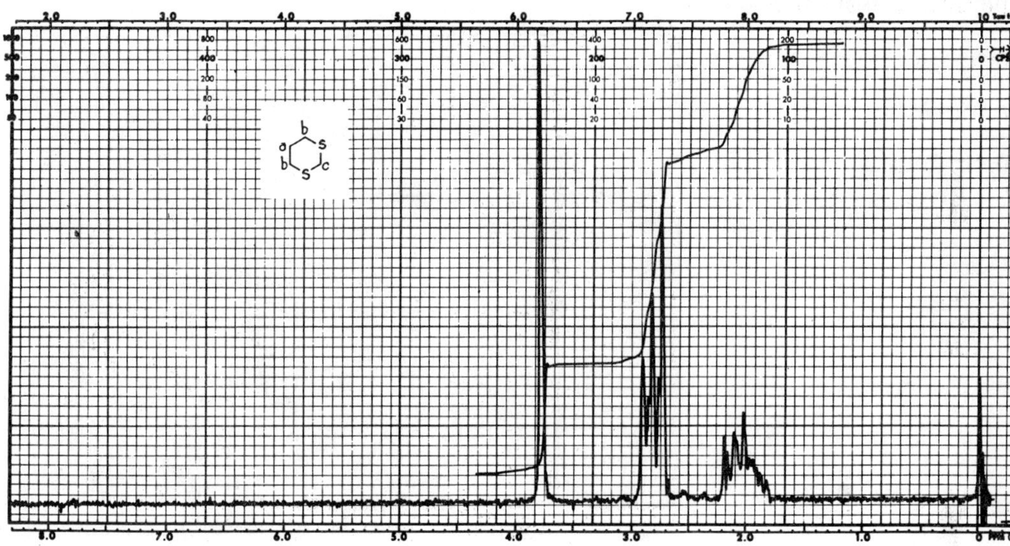

© 1978 Sadtler Research Laboratories, Inc.

---

p-DITHIANE

$C_4H_8S_2$

Mol. Wt. 120.24

M.P. 108°C

Solvent: CCl_4

Assignments
a = 2.85
b = 2.10 imp.

1133

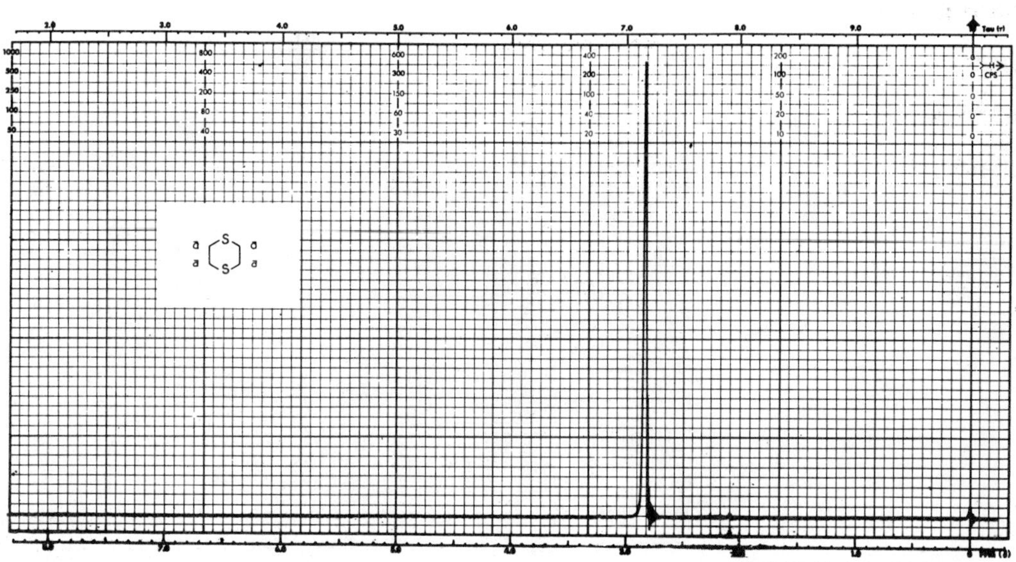

© 1978 Sadtler Research Laboratories, Inc.

2-METHYL-2-THIAZOLINE

C_4H_7NS

Mol. Wt. 101.17

B.P. 143-146°C

Solvent: CCl_4

Assignments
a = 2.12
b = 3.27
c = 4.11

1134

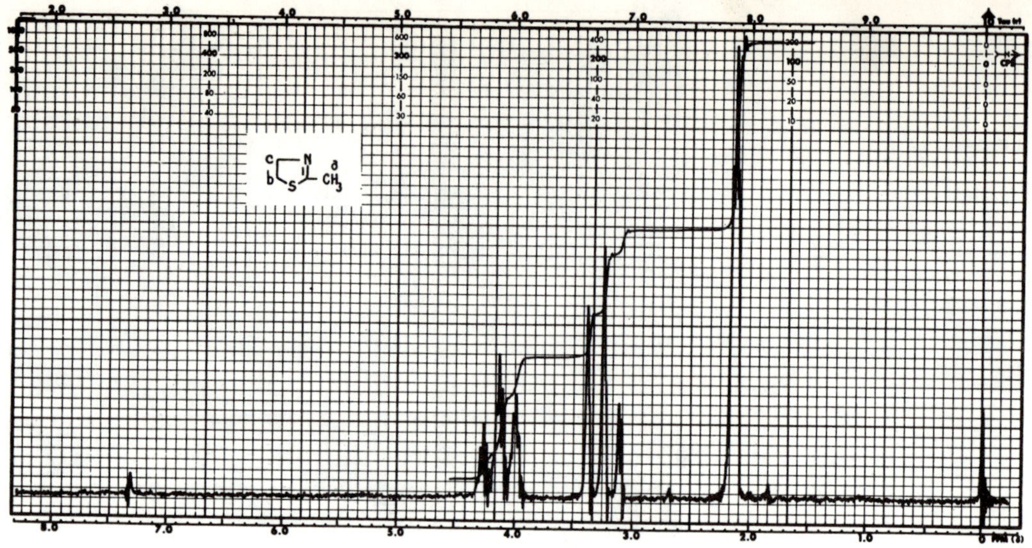

© 1978 Sadtler Research Laboratories, Inc.

2-(BENZYLAMINO)-2-THIAZOLINE

$C_{10}H_{12}N_2S$

Mol. Wt. 192.29

Solvent: $CDCl_3$

Assignments
a = 3.29
b = 4.00
c = 4.45
d = 4.62
e = 7.28

1135

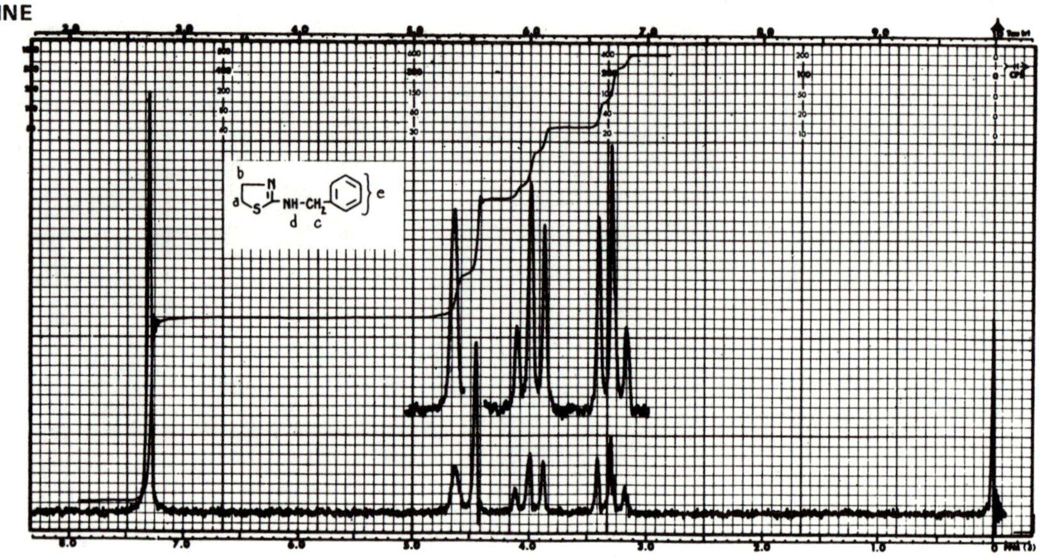

© 1978 Sadtler Research Laboratories, Inc.

THIOPHENE

C_4H_4S

Mol. Wt. 84.14

B.P. 84°C

Solvent: CCl_4

Assignments
a = 6.90
b = 7.10

1136

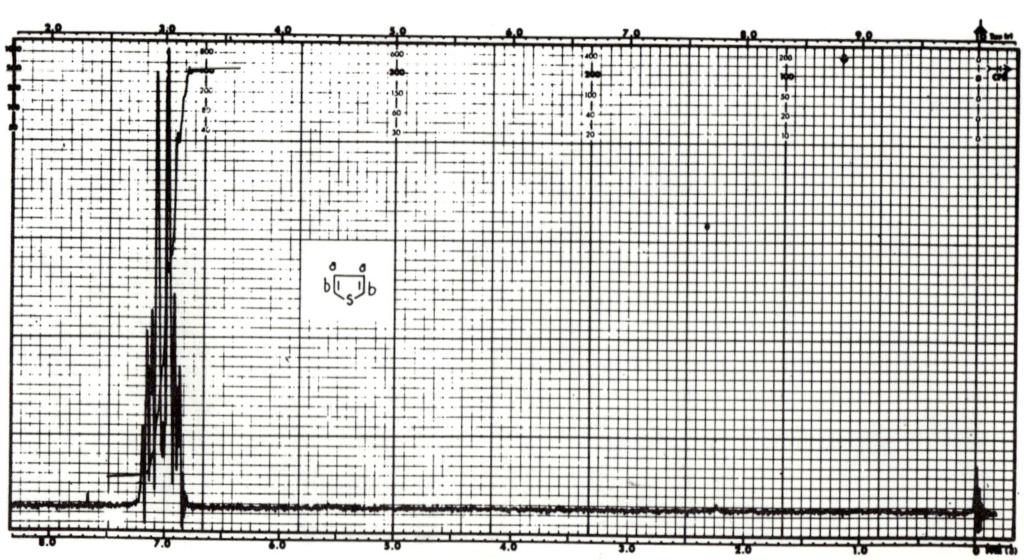

© 1978 Sadtler Research Laboratories, Inc.

3-METHYLTHIOPHENE

C_5H_6S

Mol. Wt. 98.17

B.P. 114°C/738mm

Solvent: CCl_4

Assignments
a = 2.22
b = ca 6.77
c = 7.05

1137

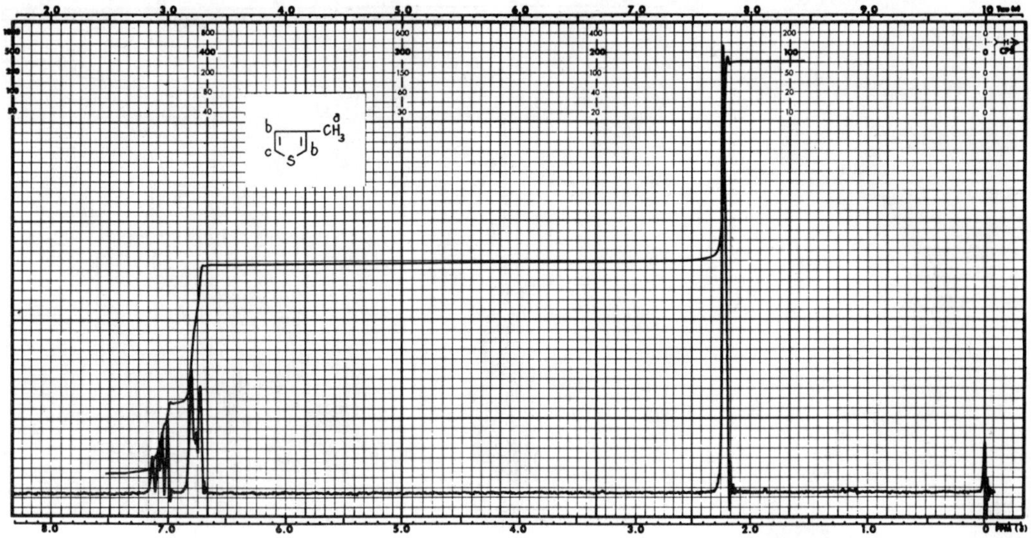

© 1978 Sadtler Research Laboratories, Inc.

2-CHLOROTHIOPHENE

C_4H_3ClS

Mol. Wt. 118.59

Solvent: $CDCl_3$

Assignments
a = 6.62-7.01

1138

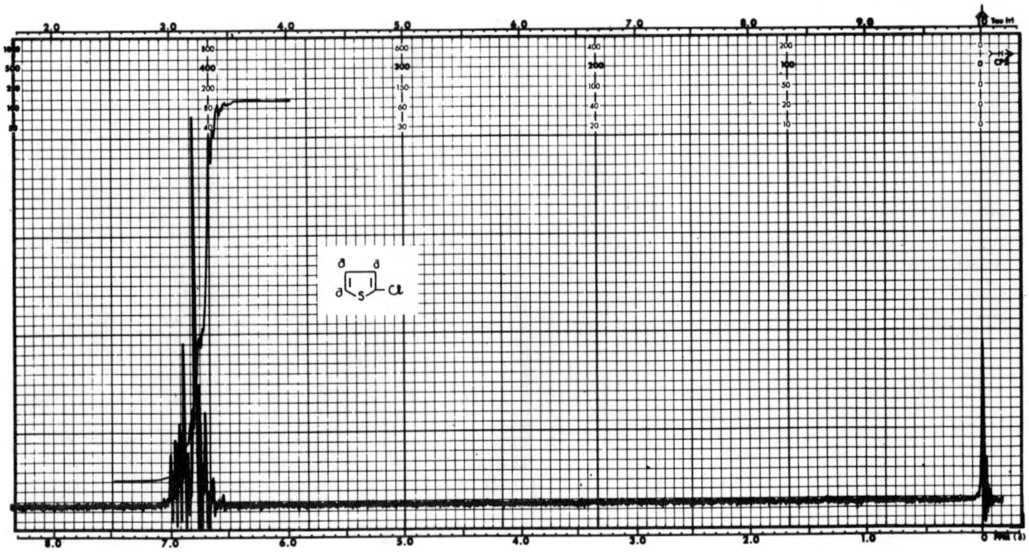

© 1978 Sadtler Research Laboratories, Inc.

2,5-DICHLOROTHIOPHENE

$C_4H_2Cl_2S$

Mol. Wt. 153.03

B.P. 160°C (lit.)

Solvent: CCl_4

Assignments
a = 6.64

1139

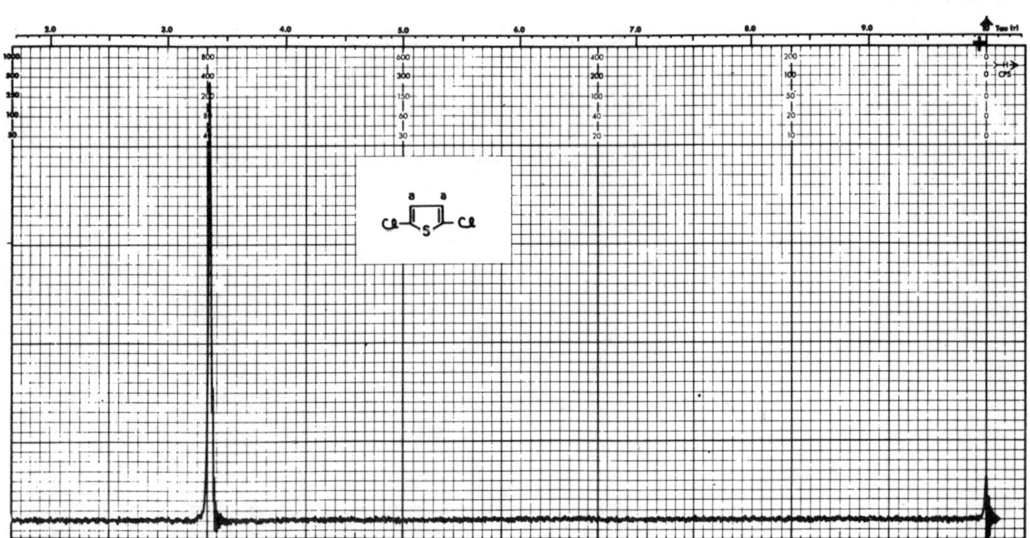

© 1978 Sadtler Research Laboratories, Inc.

2-BROMOTHIOPHENE

C_4H_3BrS

Mol. Wt. 163.04

B.P. 150-153°C

Solvent: CCl_4

Assignments
a = 6.79
b = 6.98 or c
c = 7.12 or b

1140

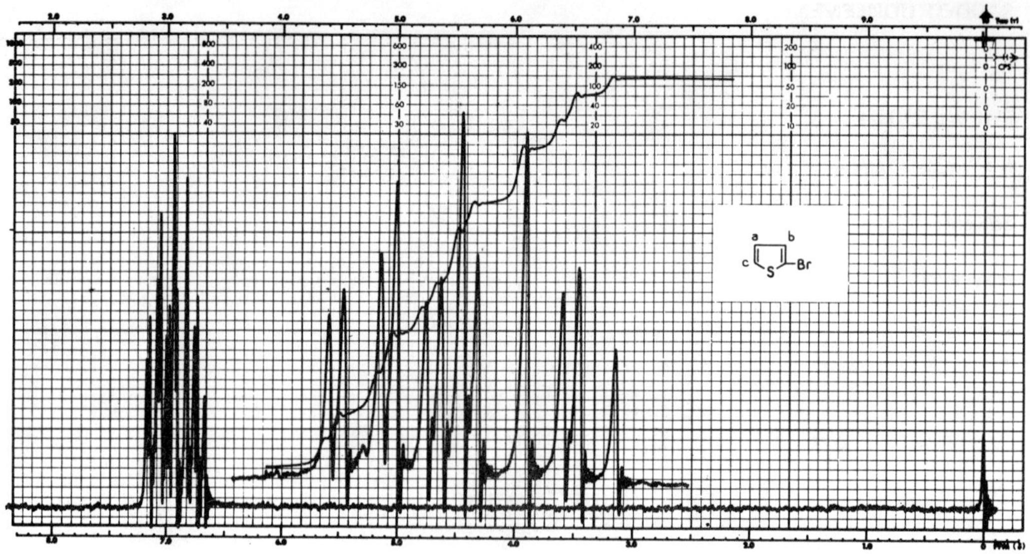

© 1978 Sadtler Research Laboratories, Inc.

3-BROMOTHIOPHENE

C_4H_3BrS

Mol. Wt. 163.04

B.P. 159-161°C

Solvent: $CDCl_3$

Assignments
a = 6.91
b = 7.11
c = 7.18

1141

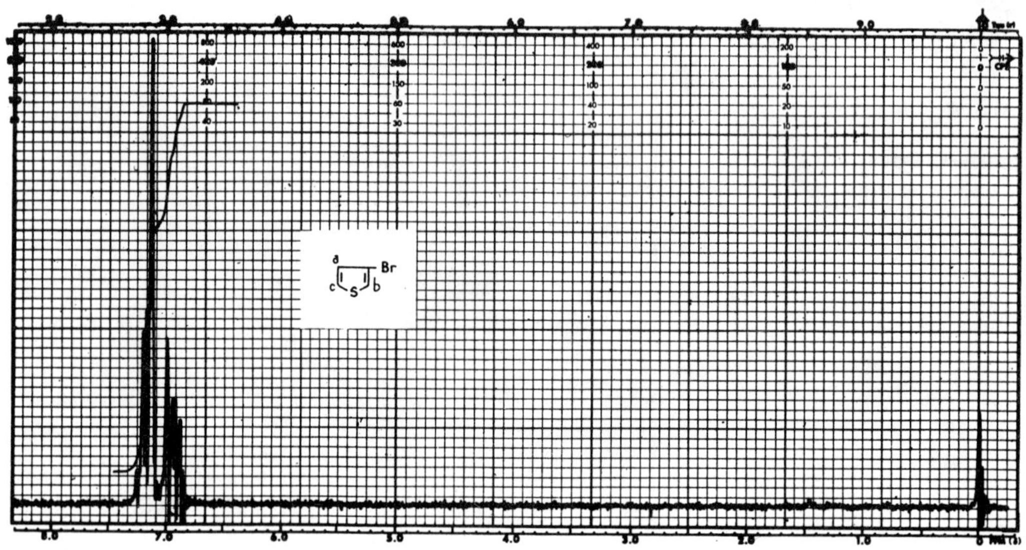

© 1978 Sadtler Research Laboratories, Inc.

2,5-DIBROMOTHIOPHENE

$C_4H_2Br_2S$

Mol. Wt. 241.94

B.P. 210.5-211.0°C

Solvent: CCl_4

Assignments
a = 6.71

1142

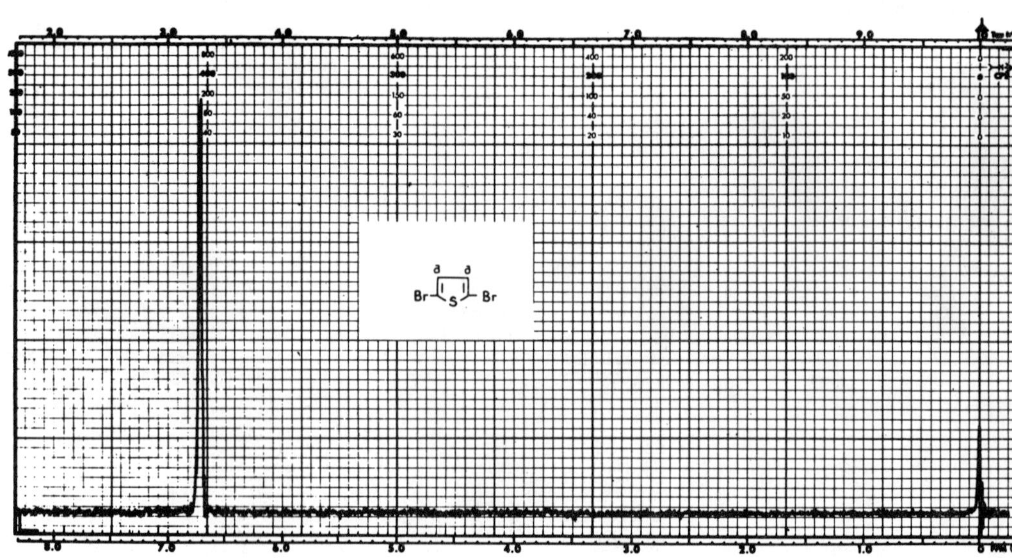

© 1978 Sadtler Research Laboratories, Inc.

2-IODOTHIOPHENE

C_4H_3IS

Mol. Wt. 210.04

B.P. 65-66°C/10mm

Solvent: $CDCl_3$

Assignments
a = 6.72
b = 7.19
c = 7.27

1143

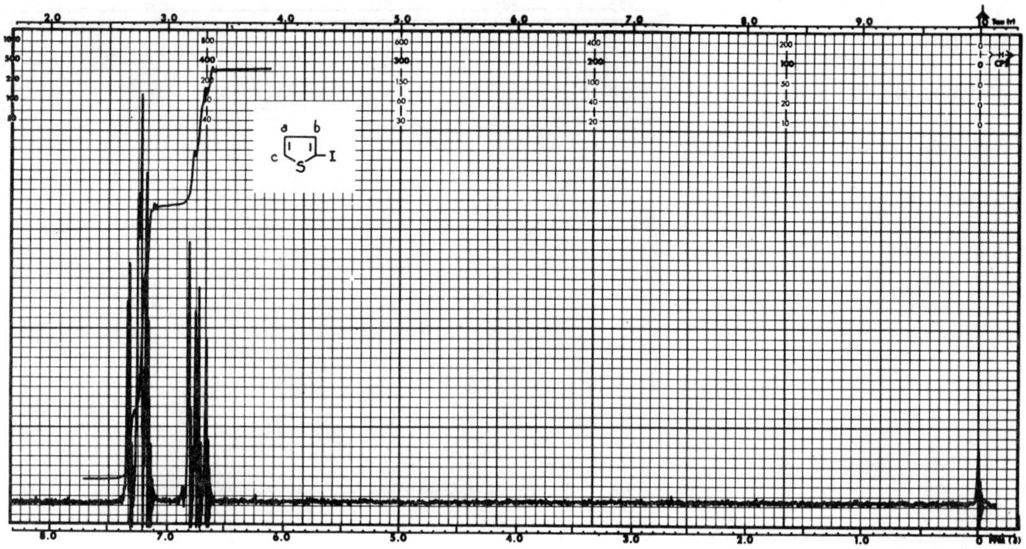

© 1978 Sadtler Research Laboratories, Inc.

2-THIOPHENECARBONITRILE

C_5H_3NS

Mol. Wt. 109.15

B.P. 75°/10mm

Solvent: CCl_4

Assignments
a = 7.10
b = ca 7.60

1144

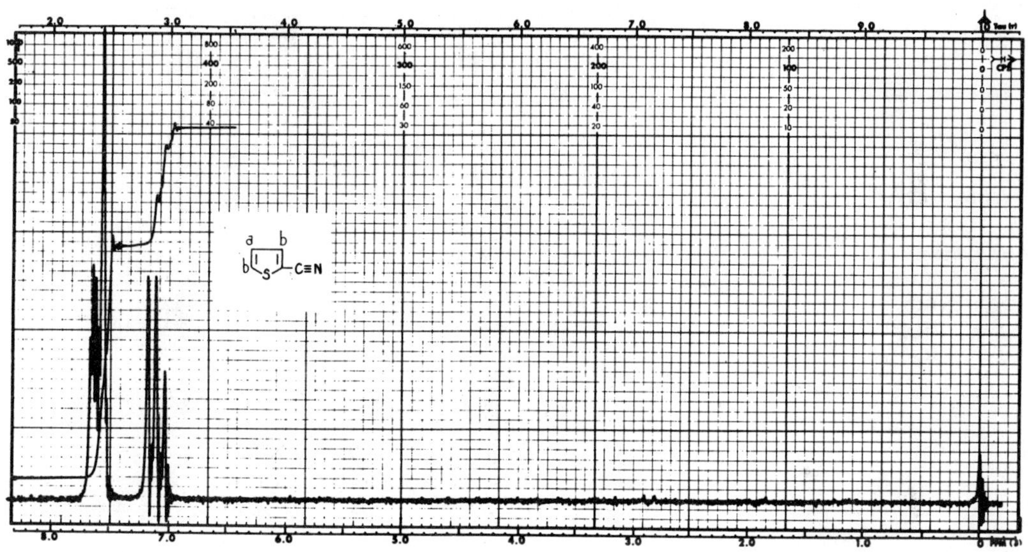

© 1978 Sadtler Research Laboratories, Inc.

2-THIOPHENECARBOXALDEHYDE, anti-OXIME

C_5H_5NOS

Mol. Wt. 127.17

M.P. 133-134°C

Solvent: Trifluoroacetic acid

Assignments
a = 7.46
b = 8.18
c = 8.35
d = 8.55
OH under TFA

1145

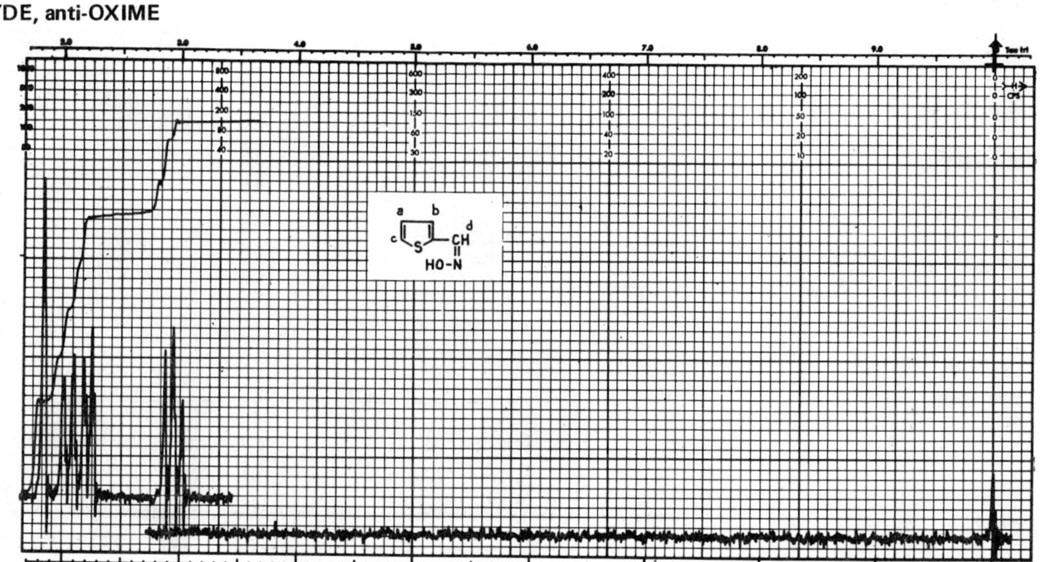

© 1978 Sadtler Research Laboratories, Inc.

3-NITROTHIOPHENE

$C_4H_3NO_2S$

Mol. Wt. 129.14

M.P. 75-76°C

Solvent: CCl_4

Assignments
a = 7.36 or b
b = 7.58 or a
c = 8.24

1146

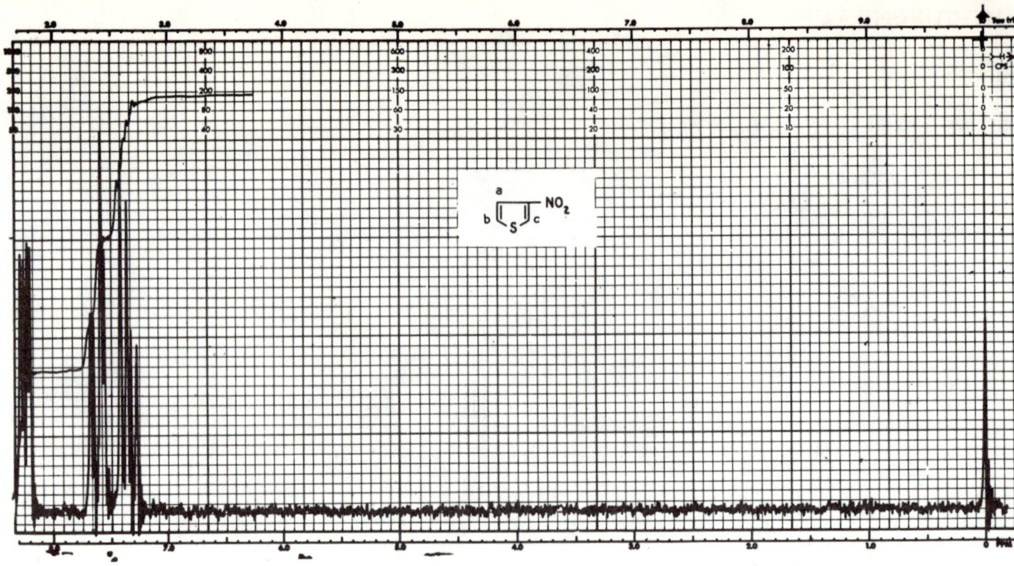

© 1978 Sadtler Research Laboratories, Inc.

2,2'-METHYLENEDITHIOPHENE

$C_9H_8S_2$

Mol. Wt. 180.29

M.P. 44-46°C

Solvent: $CDCl_3$

Assignments
a = 4.28
b = ca 6.82
c = 7.09
d = 4.10 imp.

1147

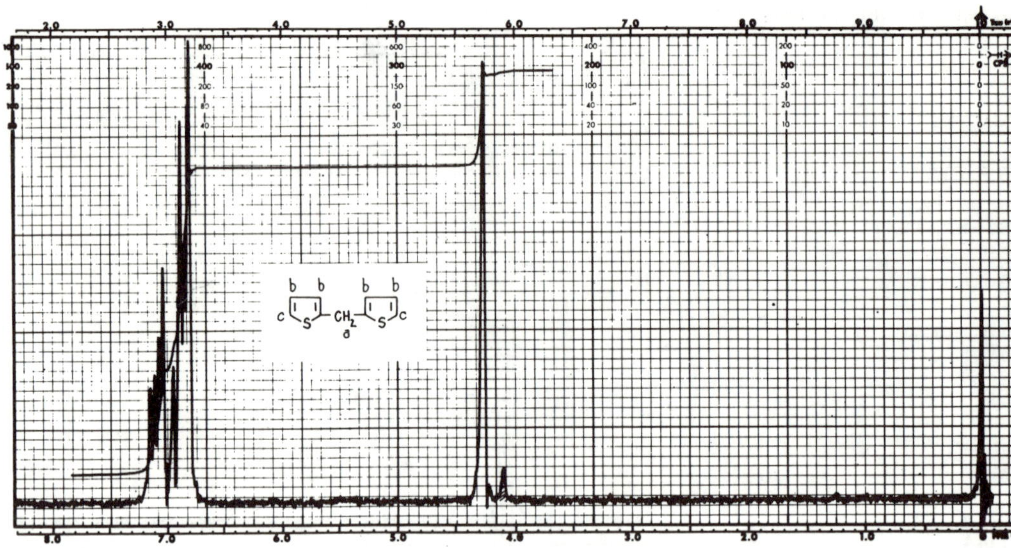

© 1978 Sadtler Research Laboratories, Inc.

3-METHYLISOTHIAZOLE

C_4H_5NS

Mol. Wt. 99.16

B.P. 134°C

M.P. –54°C

Solvent: CCl_4

Assignments
a = 2.50
b = 6.98
c = 8.52

1148

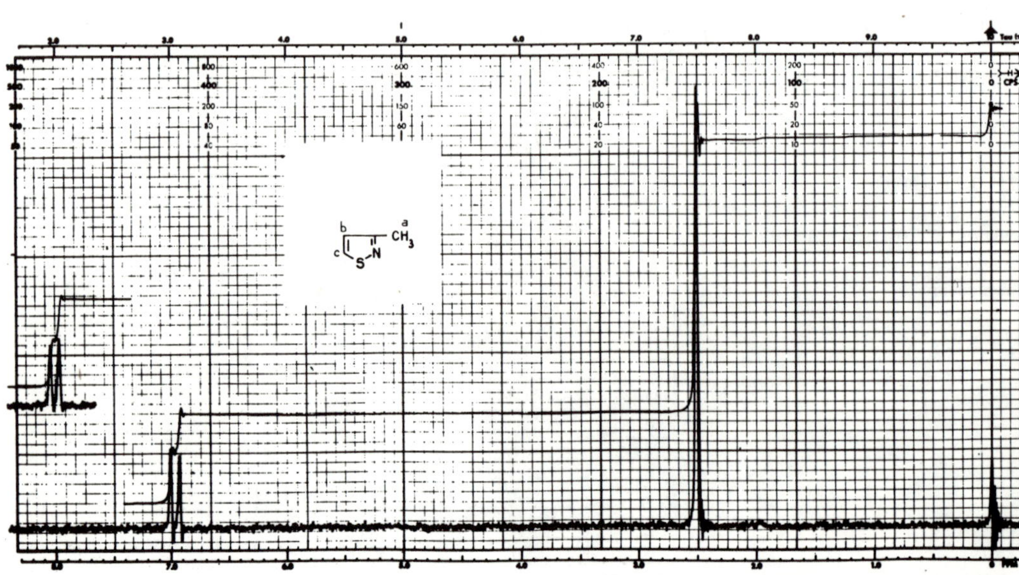

© 1978 Sadtler Research Laboratories, Inc.

4-METHYLISOTHIAZOLE

C_4H_5NS

Mol. Wt. 99.16

B.P. 146°C

M.P. −39°C

Solvent: CCl_4

Assignments
a = 2.32
b = 8.17

1149

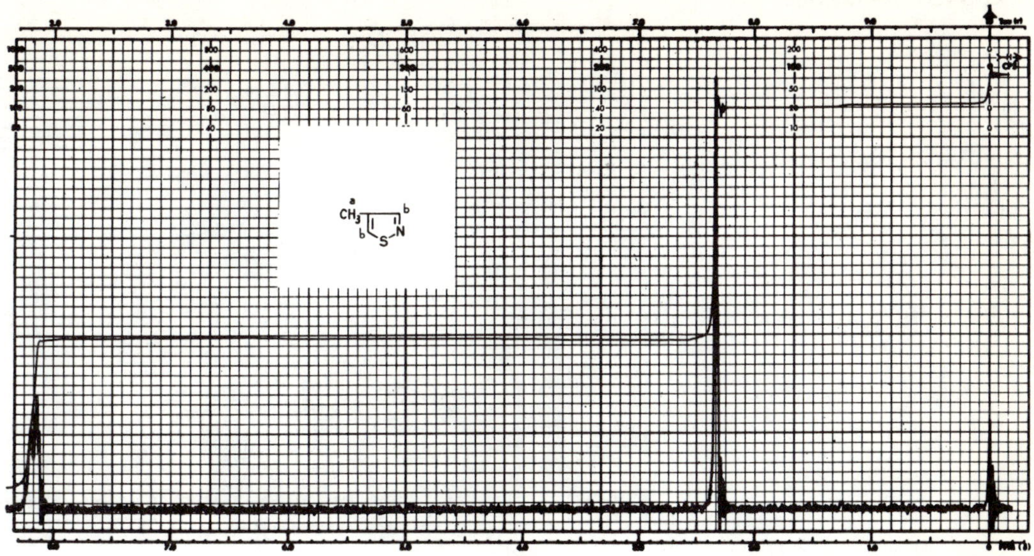

© 1978 Sadtler Research Laboratories, Inc.

4-PHENYLISOTHIAZOLE

C_9H_7NS

Mol. Wt. 161.23

M.P. 36-37°C

Solvent: CCl_4

Assignments
a = 7.10-7.64
b = 8.52 or c
c = 8.62 or b

1150

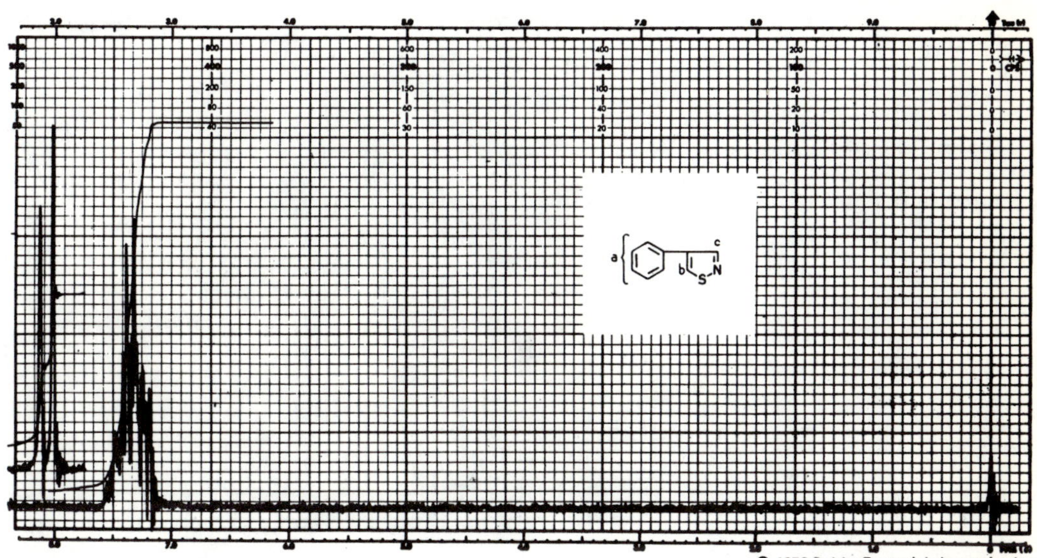

© 1978 Sadtler Research Laboratories, Inc.

4,5-DIPHENYL-2-METHYLTHIAZOLE

$C_{16}H_{13}NS$

Mol. Wt. 251.35

Solvent: $CDCl_3$

Assignments
a = 2.70
b = 7.29 or c
c = 7.10-7.65 or b

1151

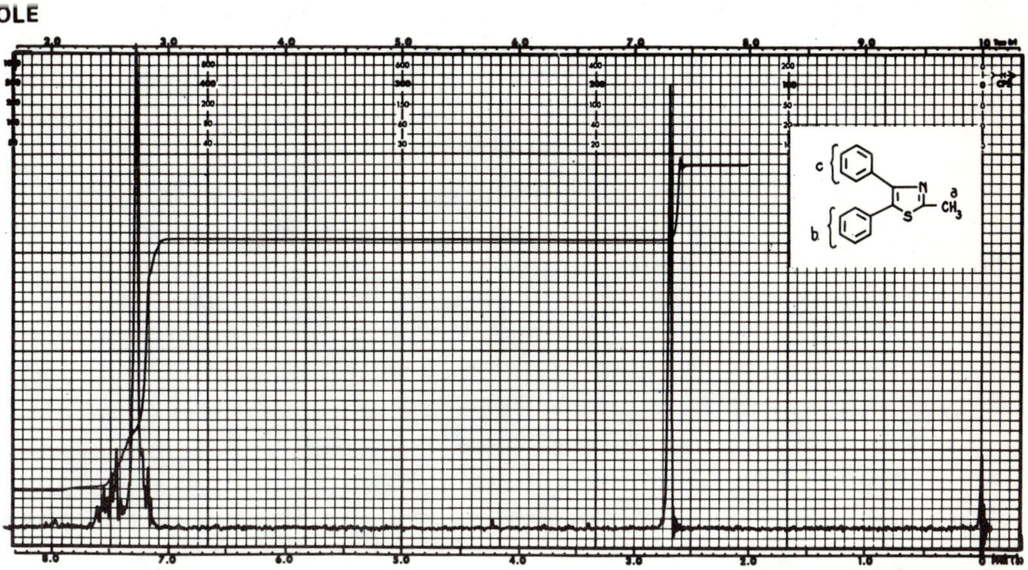

© 1978 Sadtler Research Laboratories, Inc.

2-BROMOTHIAZOLE

C_3H_2BrNS

Mol. Wt. 164.03

B.P. 55-56°C/10mm

Solvent: CCl_4

Assignments
a = 7.27
b = 7.56

1152

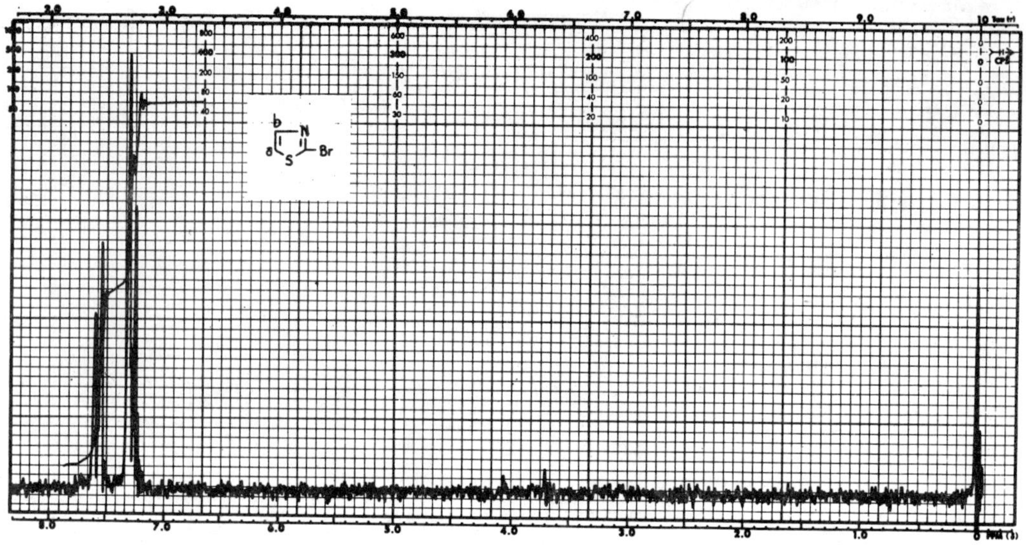

© 1978 Sadtler Research Laboratories, Inc.

2-AMINOTHIAZOLE

$C_3H_4N_2S$

Mol. Wt. 100.14

Solvent: Dimethyl sulfoxide-d_6

Assignments
a = 6.48
b = 6.87
c = 6.95
d = 2.50 DMSO-d_5
e = 3.82 H_2O

1153

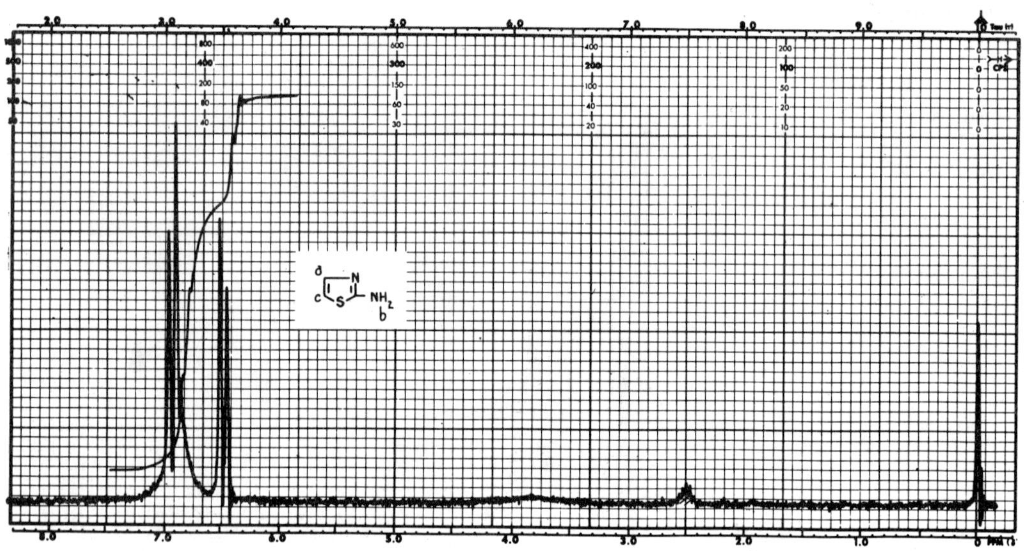

© 1978 Sadtler Research Laboratories, Inc.

2-AMINO-4-METHYLTHIAZOLE

$C_4H_6N_2S$

Mol. Wt. 114.17

Solvent: $CDCl_3$

Assignments
a = 2.15
b = 5.87
c = 5.97

1154

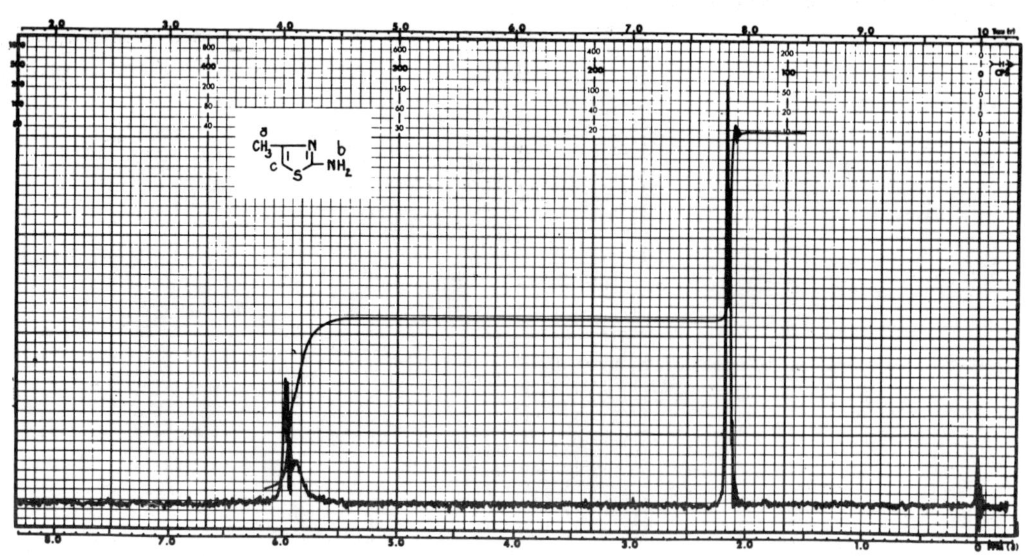

© 1978 Sadtler Research Laboratories, Inc.

2-AMINO-5-ETHYL-1,3,4-THIADIAZOLE

$C_4H_7N_3S$

Mol. Wt. 129.18

M.P. 200-202°C

Solvent: Trifluoroacetic acid

Assignments
a = 1.44
b = 2.99
c = 7.00-9.00

1155

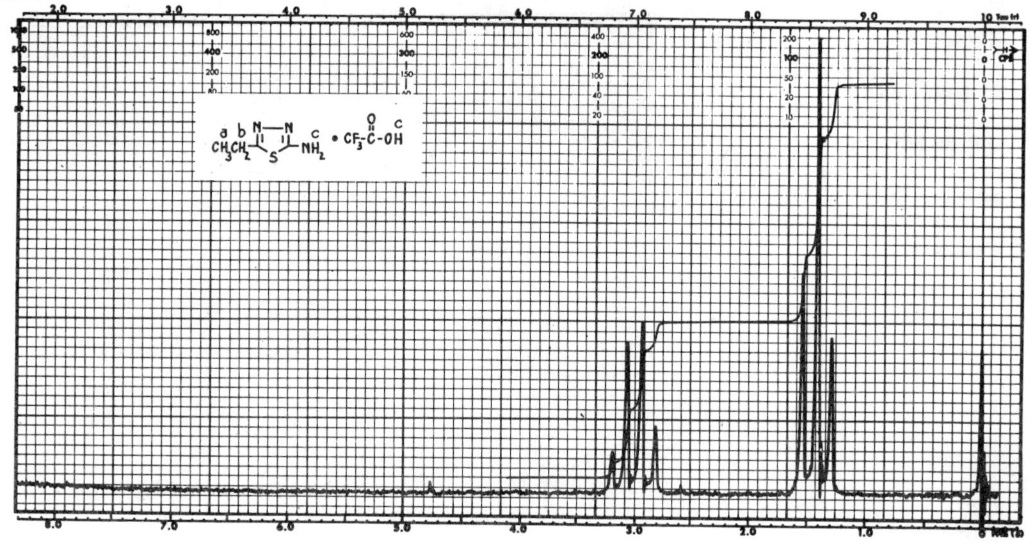

© 1978 Sadtler Research Laboratories, Inc.

2-(ETHYLAMINO)-1,3,4-THIADIAZOLE

$C_4H_7N_3S$

Mol. Wt. 129.18

M.P. 71-73°C

Solvent: Polysol-d

Assignments
a = 1.25
b = 3.39
c = 7.69
d = 8.50

1156

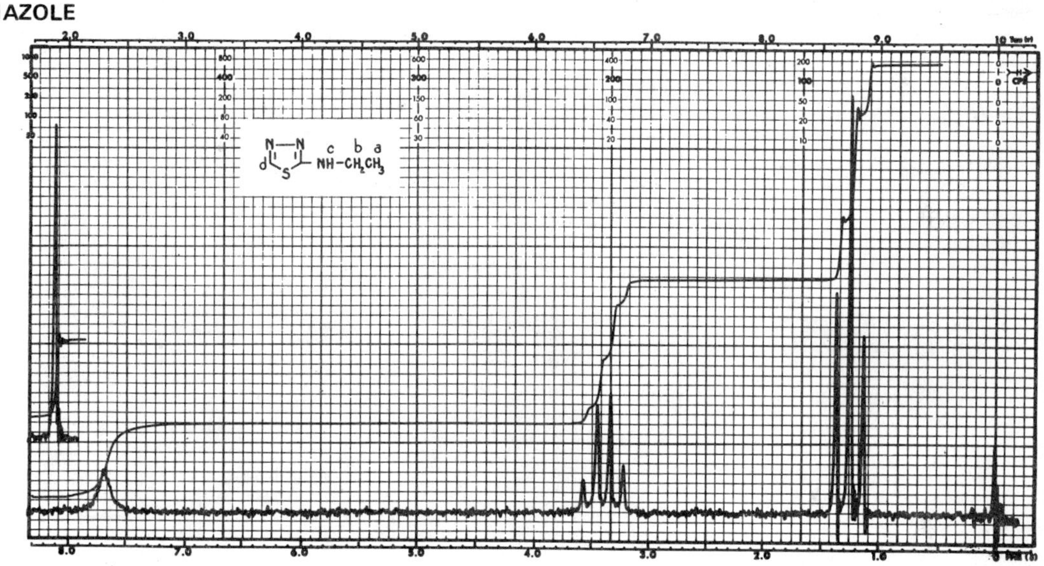

© 1978 Sadtler Research Laboratories, Inc.

2,5-DIPHENYL-p-DITHIIN

$C_{16}H_{12}S_2$

Mol. Wt. 268.40

M.P. 116-118°C

Solvent: $CDCl_3$

Assignments
a = 6.53
b = 7.20-7.48
c = 7.59

1157

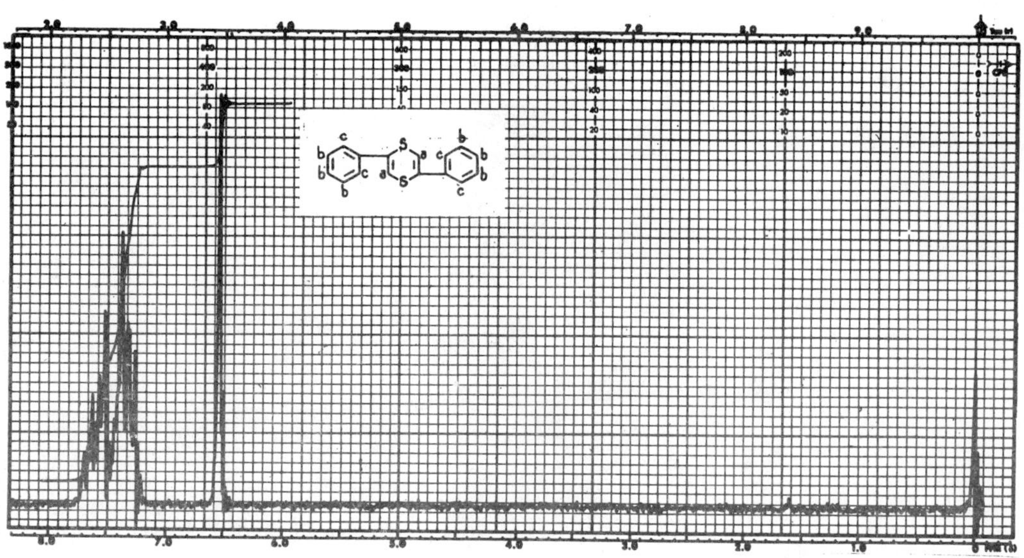

© 1978 Sadtler Research Laboratories, Inc.

BENZYL METHYL SULFIDE

$C_8H_{10}S$

Mol. Wt. 129.15

Solvent: CCl_4

Assignments
a = 1.90
b = 3.58
c = 7.20

1158

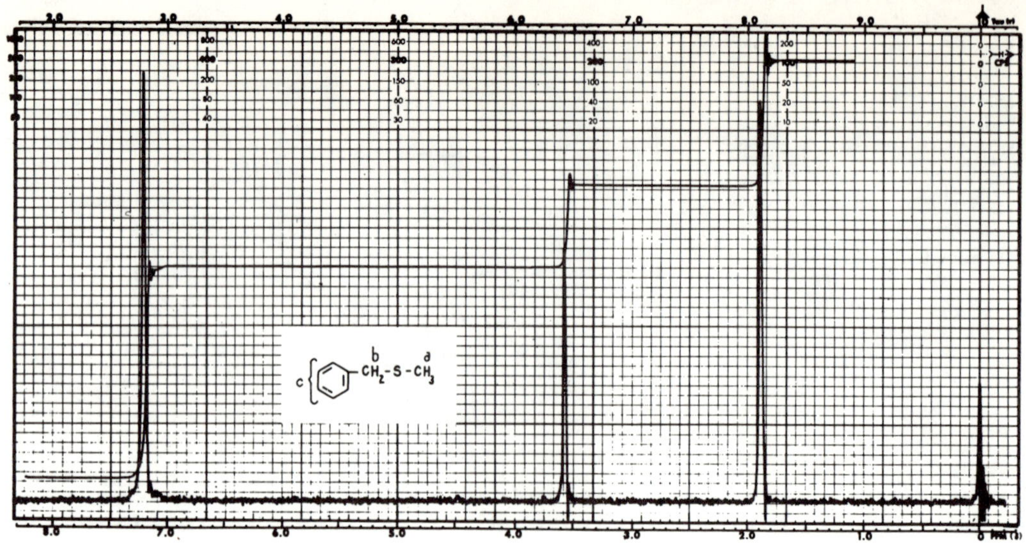

© 1978 Sadtler Research Laboratories, Inc.

BENZYL ETHYL SULFIDE

$C_9H_{12}S$

Mol. Wt. 152.26

Solvent: CCl_4

Assignments
a = 1.19
b = 2.32
c = 3.62
d = 7.21
e = 2.05 imp.

1159

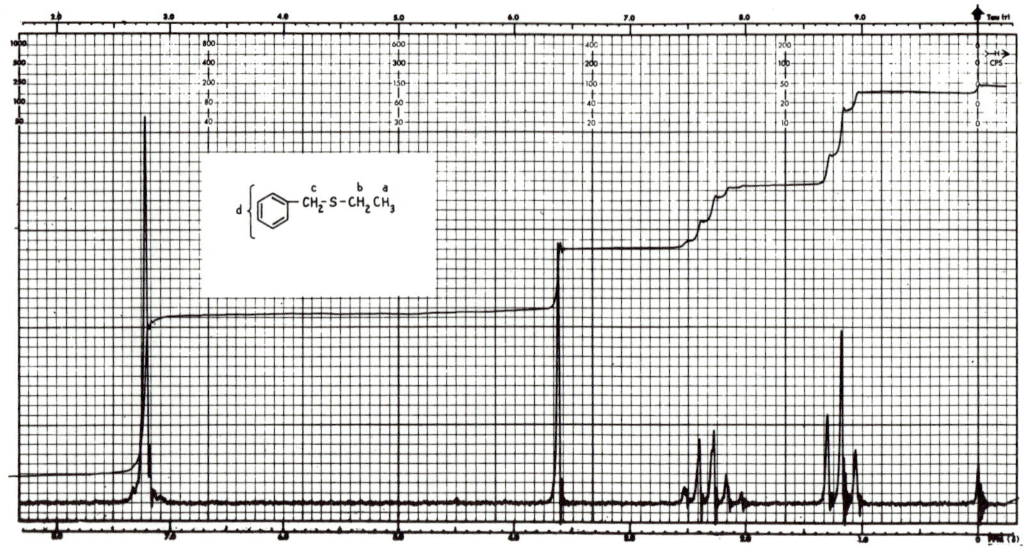

© 1978 Sadtler Research Laboratories, Inc.

p-BROMOPHENYL METHYL SULFIDE

C_7H_7BrS

Mol. Wt. 203.11

M.P. 39-41°C

Solvent: $CDCl_3$

Assignments
a = 2.45
b = 7.11
c = 7.39

1160

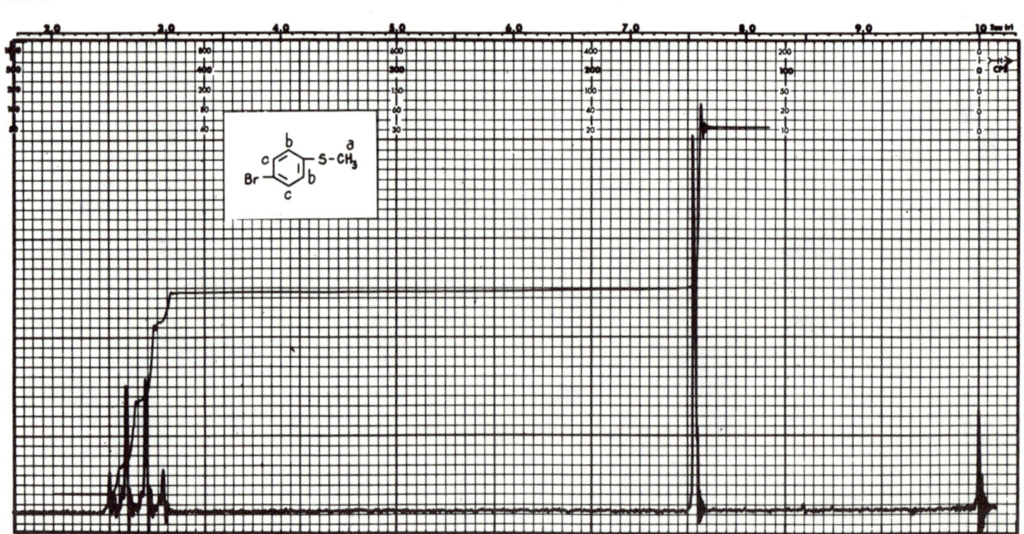

© 1978 Sadtler Research Laboratories, Inc.

o-(METHYLTHIO)ANILINE

C_7H_9NS

Mol. Wt. 139.22

B.P. 234°C

Solvent: CCl_4

Assignments
a = 2.29
b = 4.11
c = ca 6.55
d = 6.99
e = 7.28

1161

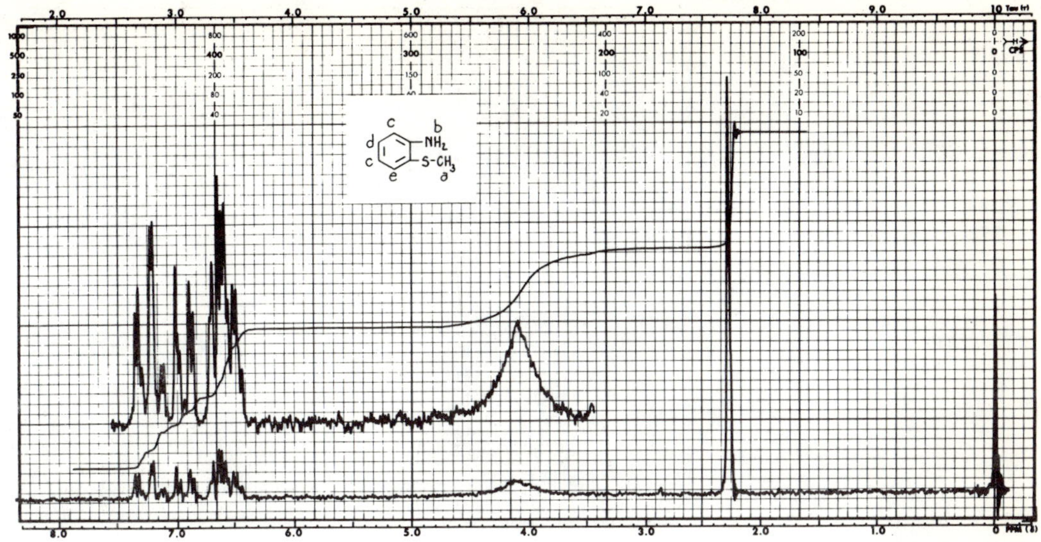

© 1978 Sadtler Research Laboratories, Inc.

BENZYL p-TOLYL SULFIDE

$C_{14}H_{14}S$

Mol. Wt. 214.33

Solvent: Polysol-d

Assignments
a = 2.21
b = 4.05
c = 7.01
d = 7.18
e = ca 7.19
f = 2.40 DMSO-d_5
g = 3.04 H_2O

1162

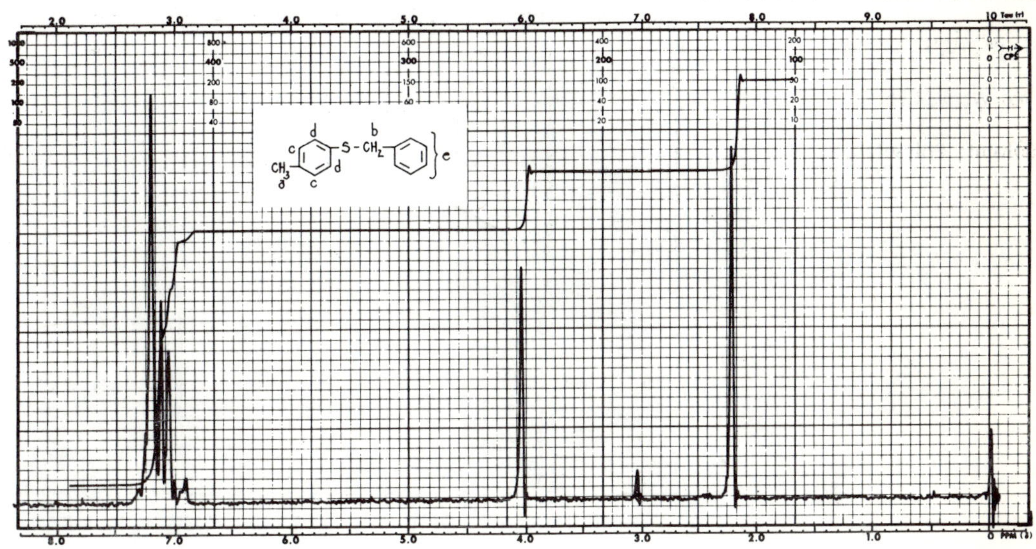

© 1978 Sadtler Research Laboratories, Inc.

PHENETHYL PHENYL SULFIDE

$C_{14}H_{14}S$

Mol. Wt. 214.33

Solvent: $CDCl_3$

Assignments
a = 2.90
b = 3.07
c = 7.17
d = 6.99-7.51

1163

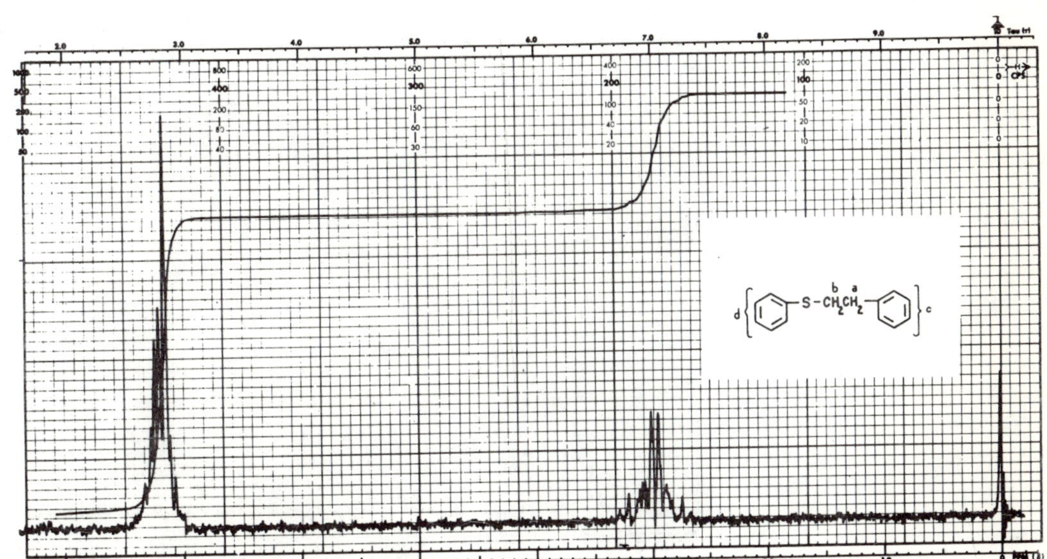

© 1978 Sadtler Research Laboratories, Inc.

PHENYL 3-PHENYLPROPYL SULFIDE

$C_{15}H_{16}S$

Mol. Wt. 228.36

Solvent: CCl_4

Assignments
a = 1.92
b = 2.69 or b
c = 2.80 or c
d = 7.12
e = 3.12 imp.
f = 3.22 imp.

1164

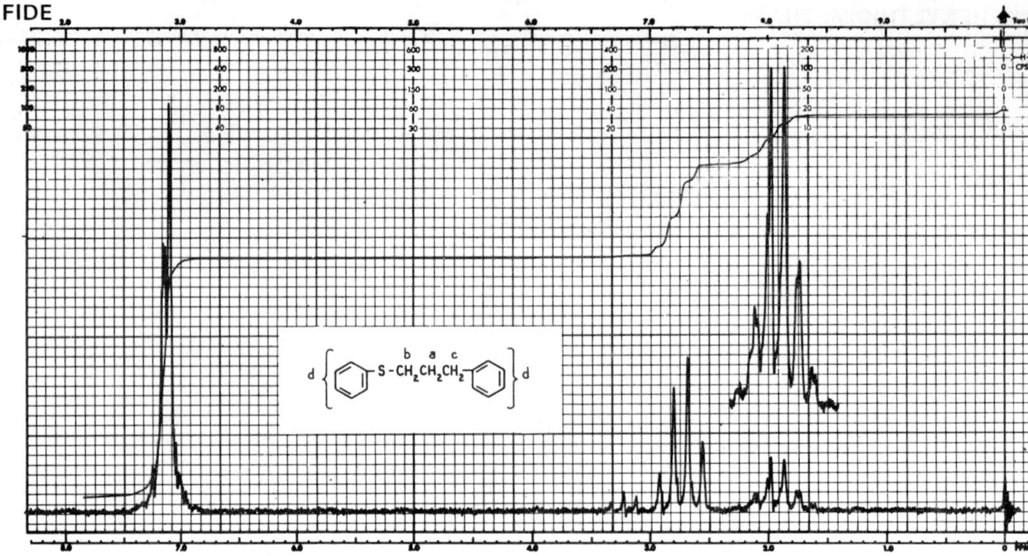

© 1978 Sadtler Research Laboratories, Inc.

BENZYL p-CHLOROPHENYL SULFIDE

$C_{13}H_{11}ClS$

Mol. Wt. 234.75

Solvent: Polysol-d

Assignments
a = 4.07
b = ca 7.20

1165

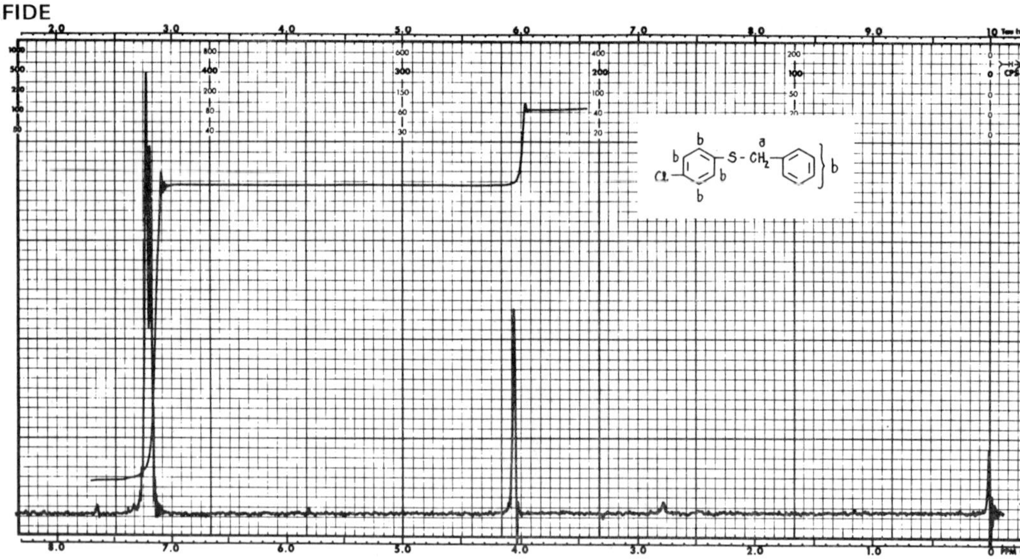

© 1978 Sadtler Research Laboratories, Inc.

2,2-DIPHENYLVINYL PHENYL SULFIDE

$C_{20}H_{16}S$

Mol. Wt. 288.41

M.P. 71-73°C

Solvent: $CDCl_3$

Assignments
a = 6.83
b = 7.10-7.55

1166

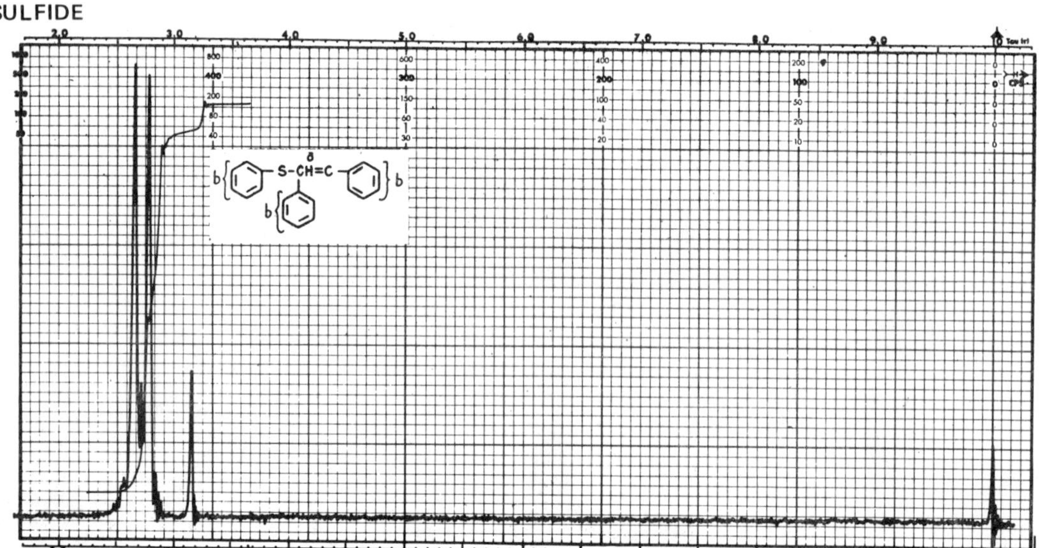

© 1978 Sadtler Research Laboratories, Inc.

BIS(PHENYLTHIO)METHANE

$C_{13}H_{12}S_2$

Mol. Wt. 232.37

Solvent: $CDCl_3$

Assignments
a = 4.25
b = 7.05-7.55

1167

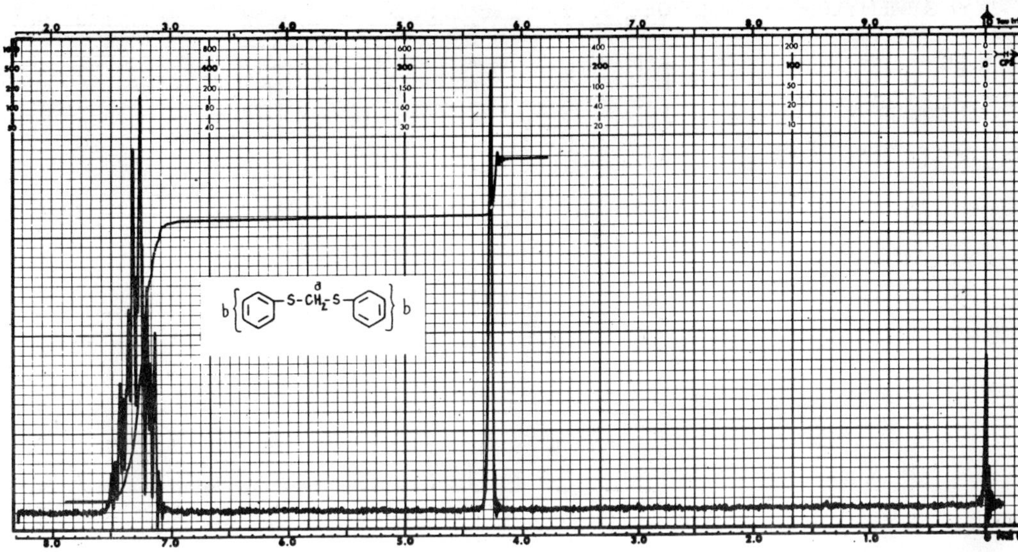

© 1978 Sadtler Research Laboratories, Inc.

TETRATHIOORTHOCARBONIC ACID, TETRAPHENYL ESTER

$C_{25}H_{20}S_4$

Mol. Wt. 448.69

M.P. 154-158°C

Solvent: $CDCl_3$

Assignments
a = ca 7.29
b = 7.65

1168

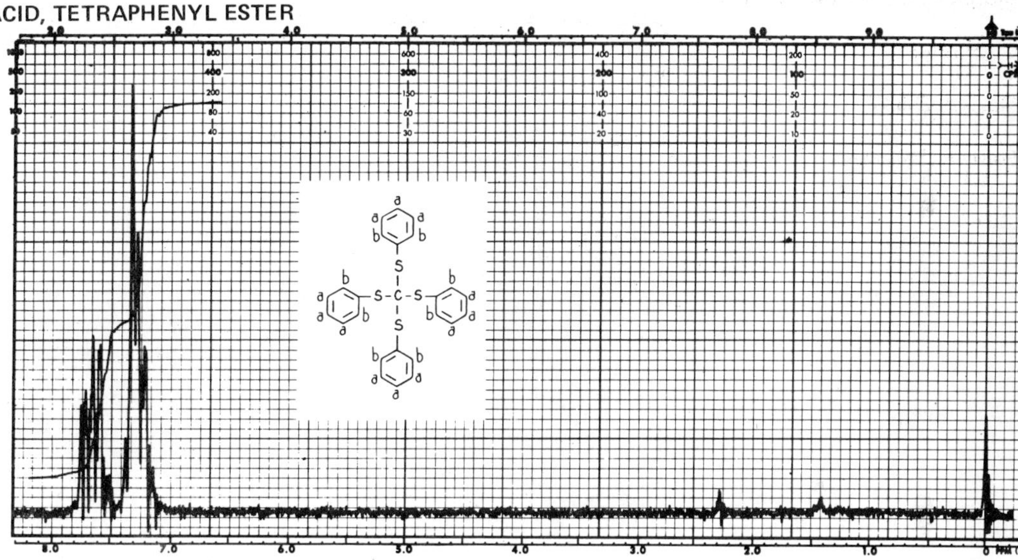

© 1978 Sadtler Research Laboratories, Inc.

PHENYL SULFIDE

$C_{12}H_{10}S$

Mol. Wt. 186.28

B.P. 96-100°C/0.1mm

Solvent: CCl_4

Assignments
a = 7.00-7.40

1169

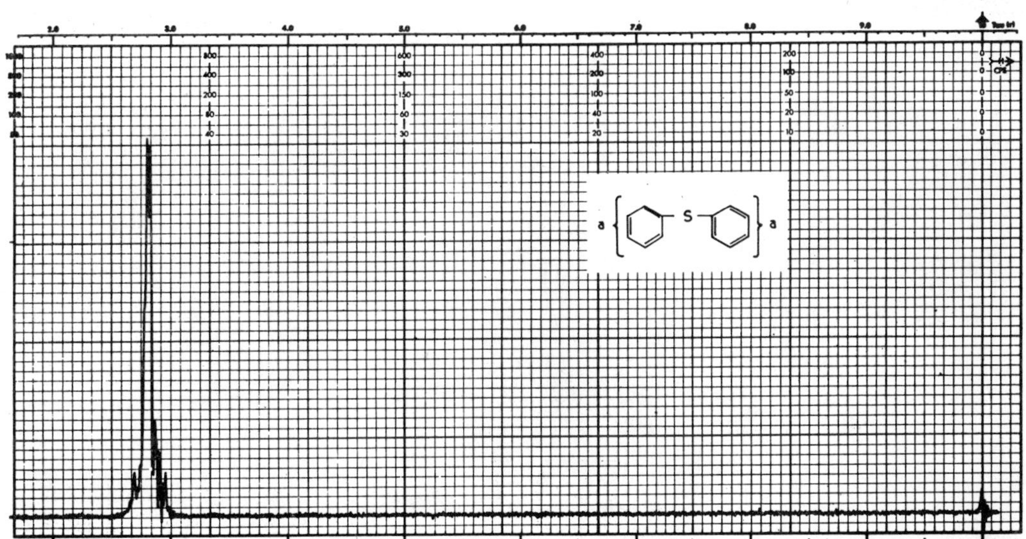

© 1978 Sadtler Research Laboratories, Inc.

BIS(p-TOLYL)SULFIDE

$C_{14}H_{14}S$

Mol. Wt. 214.33

M.P. 56.8-57.2^{o}C

Solvent: $CDCl_3$

Assignments
a = 2.30
b = 7.07
c = 7.21

1170

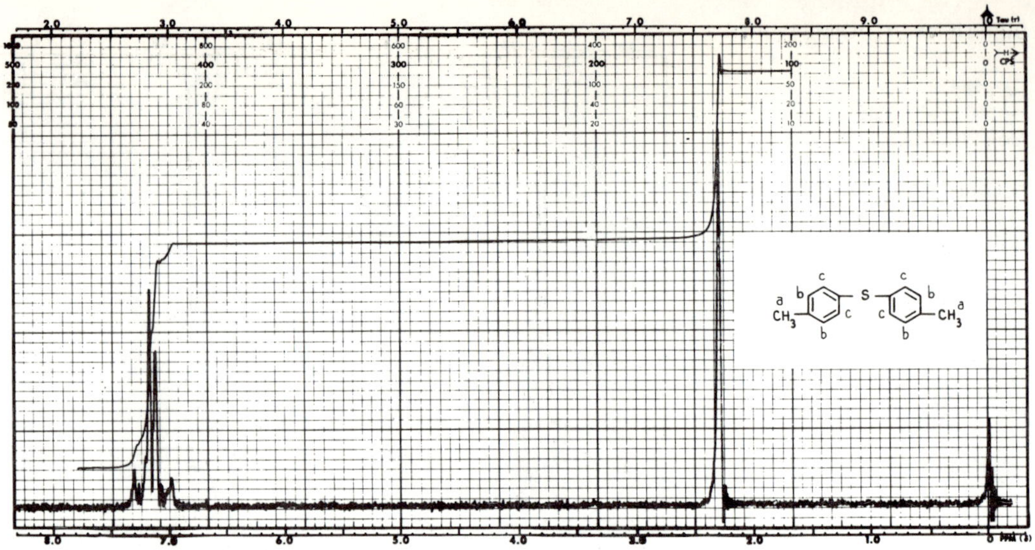

© 1978 Sadtler Research Laboratories, Inc.

BIS(p-CHLOROPHENYL) SULFIDE

$C_{12}H_8Cl_2S$

Mol. Wt. 255.17

Solvent: Polysol-d

Assignments
a = ca 7.28

1171

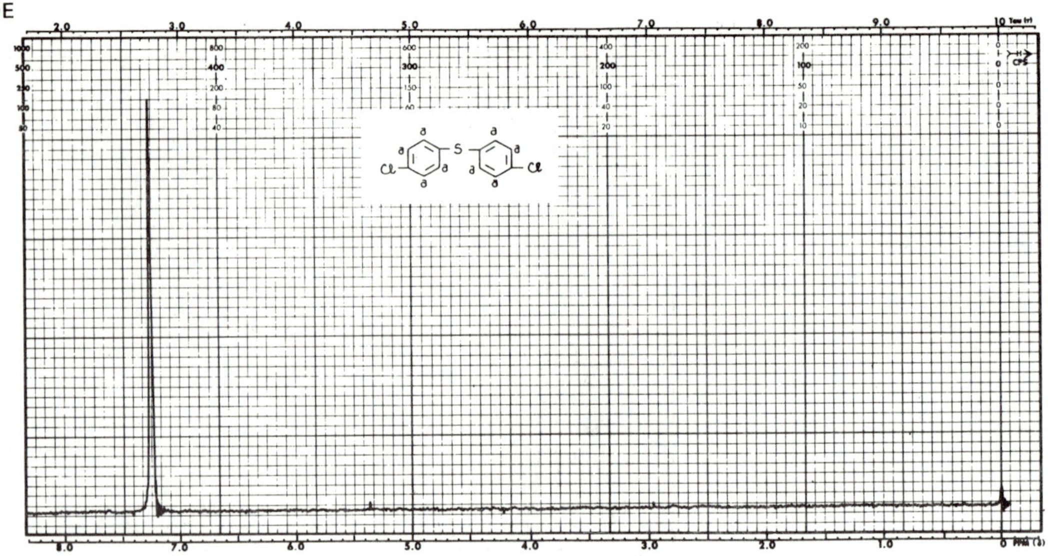

© 1978 Sadtler Research Laboratories, Inc.

4-(PHENYLTHIO)PYRIDINE

$C_{11}H_9NS$

Mol. Wt. 187.27

M.P. 0-5°C

Solvent: $CDCl_3$

Assignments
a = 6.92
b = 7.20-7.65
c = 8.33

1172

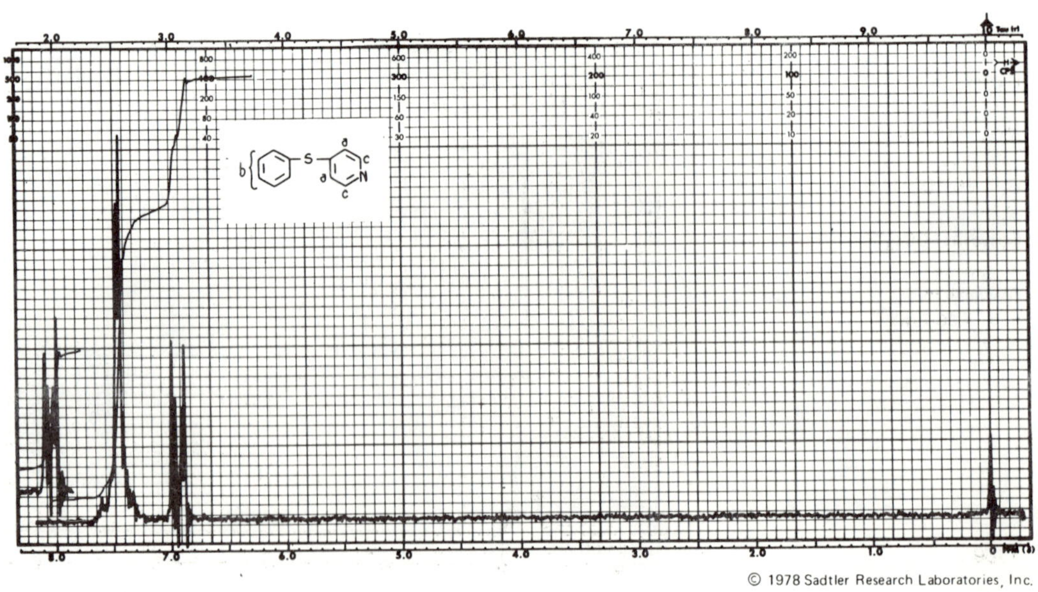

© 1978 Sadtler Research Laboratories, Inc.

BENZYL-2-NAPHTHYL SULFIDE

$C_{17}H_{14}S$

Mol. Wt. 250.37

M.P. 89.8-90.5°C

Solvent: $CDCl_3$

Assignments
a = 4.18
b = 7.28
c = 7.12-7.58
d = 7.60-7.85

1173

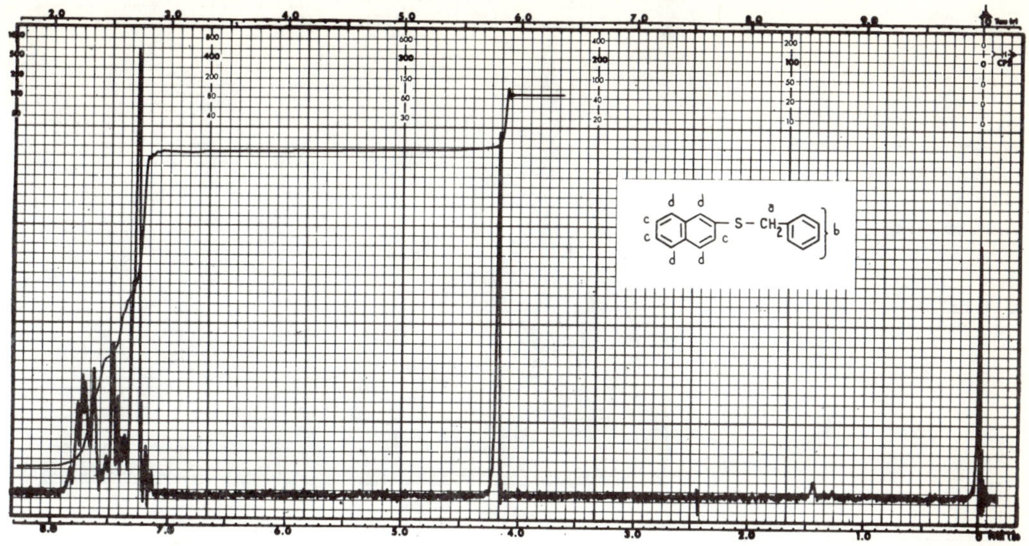

© 1978 Sadtler Research Laboratories, Inc.

2-(PHENYLTHIO)QUINOLINE

$C_{15}H_{11}NS$

Mol. Wt. 237.33

M.P. 43-48°C

Solvent: $CDCl_3$

Assignments
a = 6.91
b = 7.10-7.80
c = 7.98

1174

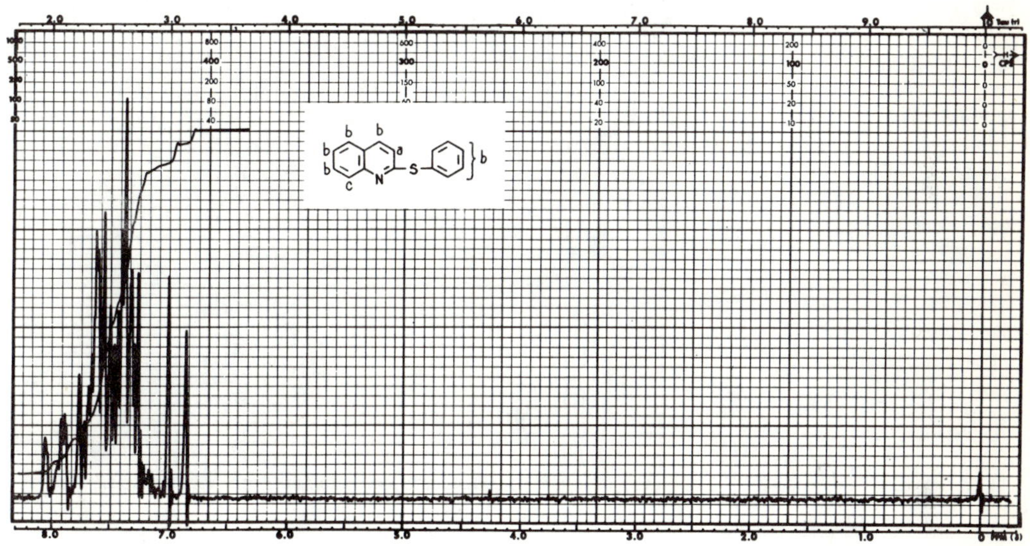

© 1978 Sadtler Research Laboratories, Inc.

2-METHYLBENZOTHIAZOLE

C_8H_7NS

Mol. Wt. 149.22

M.P. 13-15°C (lit.)

Solvent: $CDCl_3$

Assignments
a = 2.77
b = 7.08-7.53
c = 7.61-8.04

1175

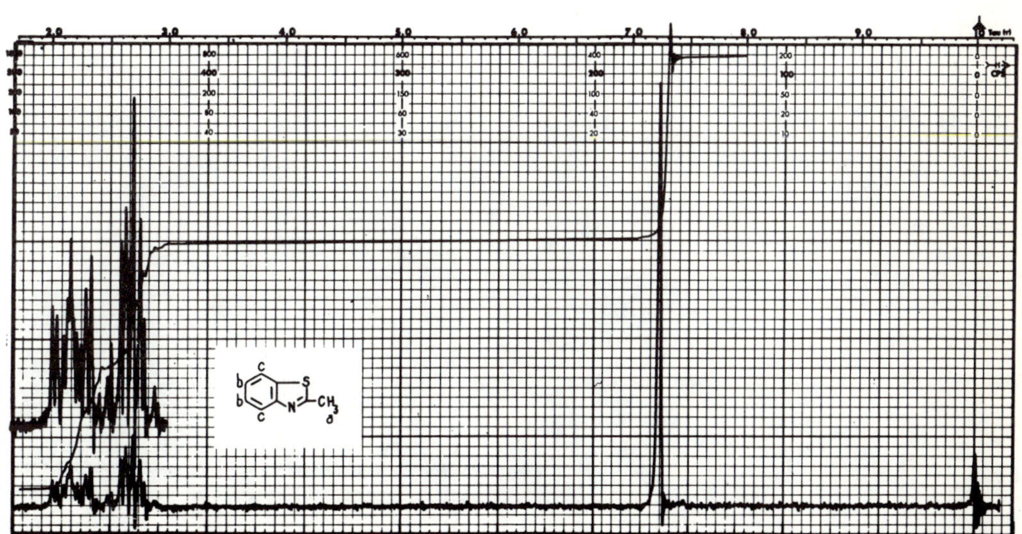

© 1978 Sadtler Research Laboratories, Inc.

2-PHENYLBENZOTHIAZOLE

$C_{13}H_9NS$

Mol. Wt. 211.29

M.P. 114°C

Solvent: $CDCl_3$

Assignments
a = 7.01-7.55
b = 7.67
c = ca 7.99

1176

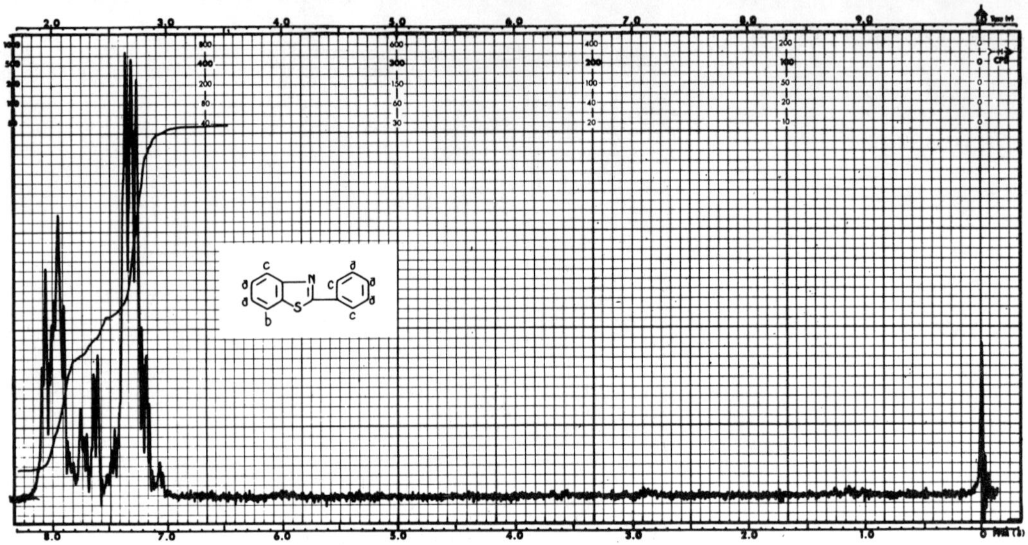

© 1978 Sadtler Research Laboratories, Inc.

5-CHLORO-2-METHYLBENZOTHIAZOLE

C_8H_6ClNS

Mol. Wt. 183.66

M.P. 68-70°C

Solvent: $CDCl_3$

Assignments
a = 2.71
b = 7.17
c = 7.55
d = 7.86

1177

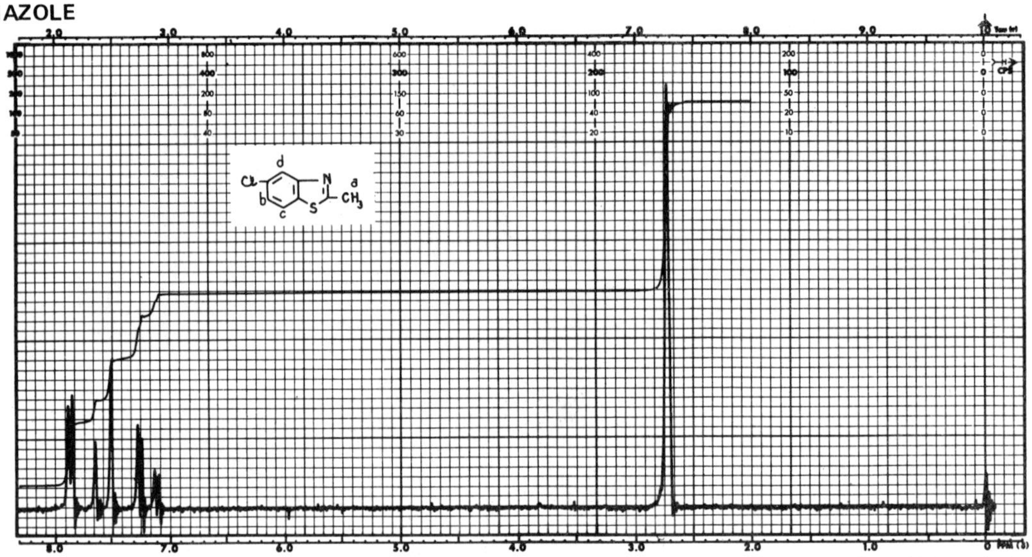

© 1978 Sadtler Research Laboratories, Inc.

2,1,3-BENZOTHIADIAZOLE

$C_6H_4N_2S$

Mol. Wt. 136.18

M.P. 43-45°C

Solvent: $CDCl_3$

Assignments
a = 7.41
b = 7.89

1178

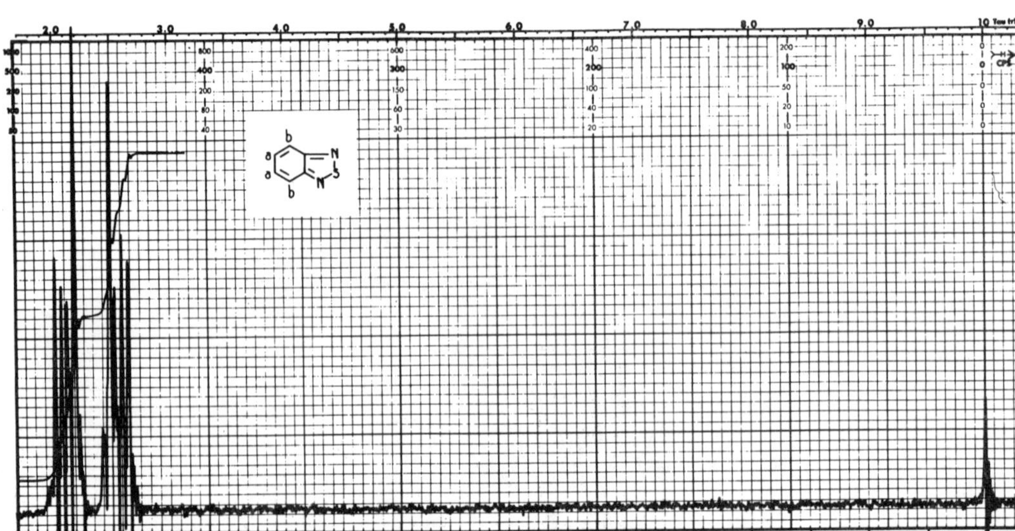

© 1978 Sadtler Research Laboratories, Inc.

3-DIBENZOTHIOPHENAMINE

$C_{12}H_9NS$

Mol. Wt. 199.28

M.P. 120-121°C

Solvent: Dimethyl sulfoxide

Assignments
a = 5.47
b = 6.88
c = 7.10-7.53
d = 7.61-8.16

1179

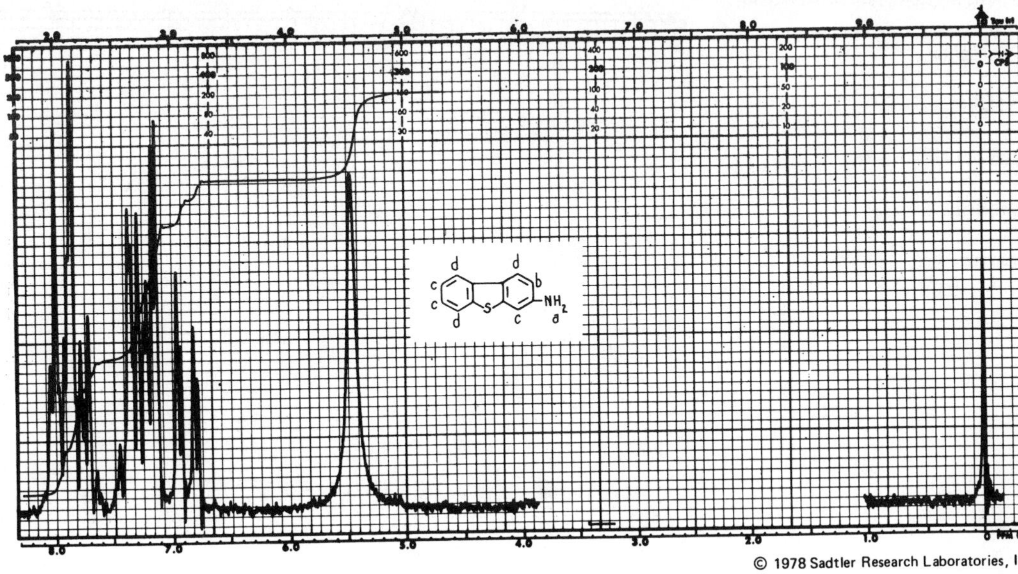

© 1978 Sadtler Research Laboratories, Inc.

3,7-DINITRODIBENZOTHIOPHENE

$C_{12}H_6N_2SO_4$

Mol. Wt. 274.26

Solvent: DMSO-d_6

Assignments
a = 8.11
b = 8.63
c = 8.89

1180

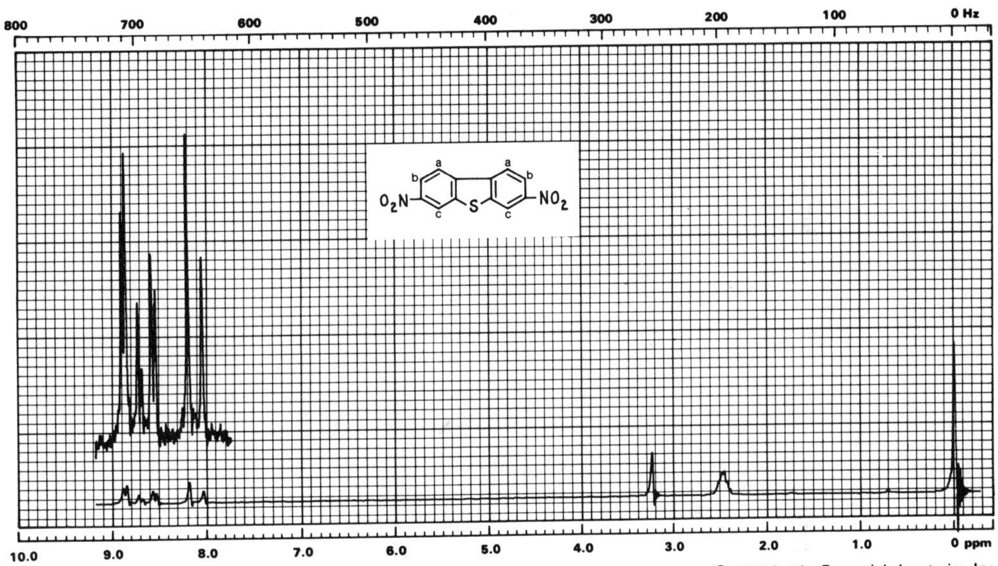

© 1978 Sadtler Research Laboratories, Inc.

PHENOTHIAZINE

$C_{12}H_9NS$

Mol. Wt. 199.26

Solvent: DMSO

Assignments
a = 6.52-7.15
b = 8.52
c = 3.62 imp.

1181

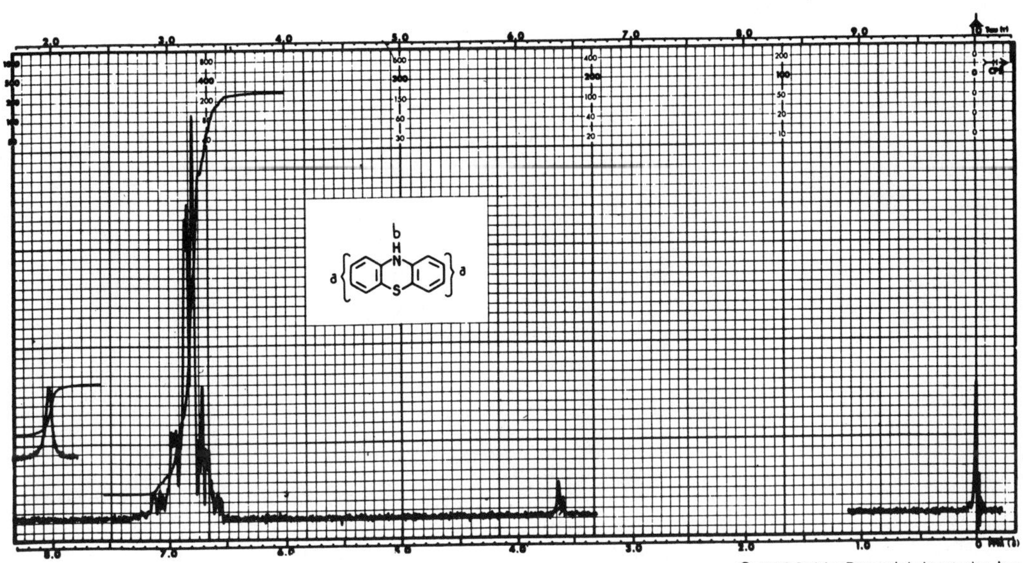

© 1978 Sadtler Research Laboratories, Inc.

3-CHLOROPHENOTHIAZINE

$C_{12}H_8ClNS$

Mol. Wt. 233.72

Solvent: DMSO-d_6

Assignments

a = 6.55-7.20

b = 8.61

c = 2.50 DMSO-d_5

d = 3.39 H_2O

1182

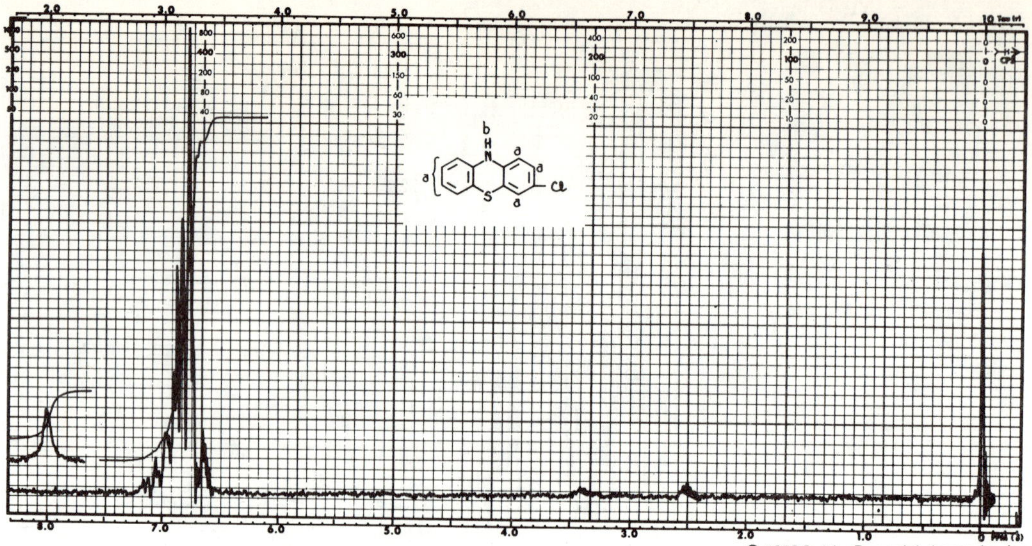

© 1978 Sadtler Research Laboratories, Inc.

THE DISULFIDES

The Disulfide group produces chemical shifts similar to those of the Sulfides. Its effect on the adjacent aliphatic groups is slightly more deshielding by about 0.1—0.4 ppm while its effect on the ortho aromatic protons is slightly less deshielding by about 0.2 ppm. Without prior knowledge that the element sulfur was present in the molecular formula of an unknown material, both linkages would be difficult to identify because of their weakly deshielding character, their lack of exchangeable protons and the fact that no unusual coupling constants are observed, with the exception of Thiophene as noted previously.

Chemical Shifts

Aliphatic Protons

HC–3	HC–2	HC–1		Solvent	Spectrum
		2.39	CH_3–S–S–CH_3	CCl_4	----
		2.33	CH_3–S–S–Ø	CCl_4	1192
	1.25	2.66	CH_3–CH_2–S–S–Ø	$CDCl_3$	1193
0.93	1.70	2.69	CH_3–CH_2–CH_2–S–S–R	$CDCl_3$	----
0.99	(1.2–2.2)	2.96	CH_3–$(CH_2)_2$–CH_2–S–S–R_4	$CDCl_3$	1183
(1.01)	1.92	2.55	$(CH_3)_2$–CH–CH_2–S–S–R	CCl_4	1185
	(1.30)	2.98	$(CH_3)_2$–CH–S–S–R	$CDCl_3$	----
	(1.30)		$(CH_3)_3$–C–S–S–R	CCl_4	1186

Substituted Methyldisulfides

R–S–S– CH_2	–X	
2.70	–CH_3	$CDCl_3$
3.29	–CH=CH_2	CCl_4
3.46	–Ø	CS_2
3.51	–C(=O)–O–R	CCl_4
3.70	–C(=O)–OH	TFA
3.70	–C(=O)–NH–R	Polysol

Substituted Ethyldisulfides

R–S–S– CH_2	–CH_2	–X	
2.69	1.70	–CH_3	$CDCl_3$
2.89	2.65	–C(=O)–OH	DMSO-d_6
3.07	3.39	–NH_2 (HCl)	D_2O
2.86	3.49	–NH–C(=O)–NH–Ø	Polysol

Aromatic Protons

Phenyl disulfides

	para/meta	ortho	Solvent
CH_3–S–S–	7.10–7.35	7.45	CCl_4
Ø–S–S–	7.00–7.35	7.48	$CDCl_3$

Para Substituted Diphenyl Disulfides

Ø–S–S– ortho	meta	para	
7.12	6.55	$-NH_2$	DMSO-d_6
7.48	6.86	$-O-CH_3$	$CDCl_3$
7.34	7.05	$-CH_3$	$CDCl_3$
7.38	7.20	–Cl	$CDCl_3$
7.39	7.39	–Br	$CDCl_3$
7.48	7.65	$-NH-C(=O)-CH_3$	DMSO-d_6
7.51	7.99	$-C(=O)-O-R_2$	$CDCl_3$
7.79	8.07	$-N^+(CH_3)_3(^-SO_3-CH_3)$	DMSO-d_6

BUTYL DISULFIDE

$C_8H_{18}S_2$

Mol. Wt. 178.36

Solvent: $CDCl_3$

Assignments
a = 0.99
b = 1.20-2.14
c = 2.96

1183

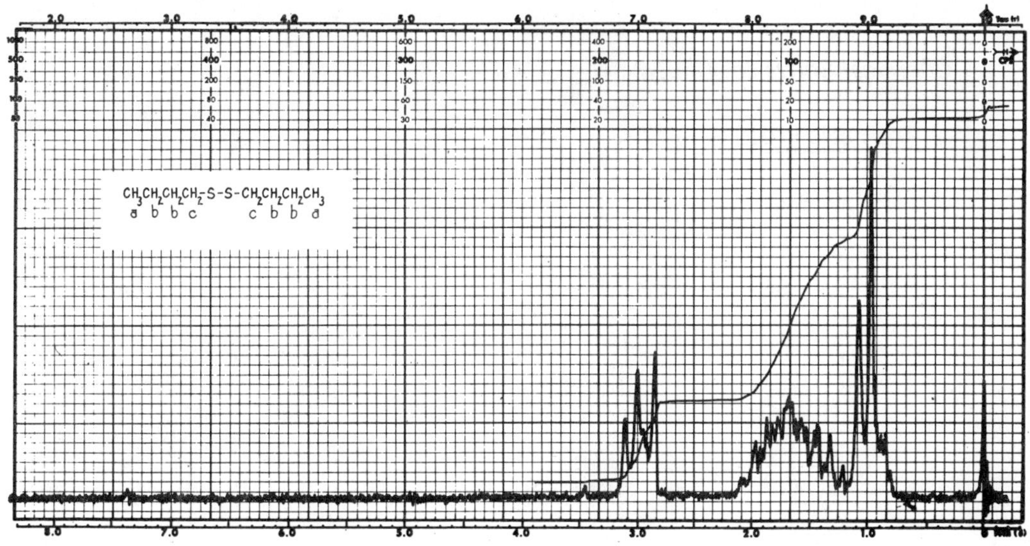

© 1978 Sadtler Research Laboratories, Inc.

sec-BUTYL DISULFIDE

$C_8H_{18}S_2$

Mol. Wt. 178.36

Solvent: CCl_4

Assignments
a = 0.99
b = 1.28
c = 1.60
d = 2.62

1184

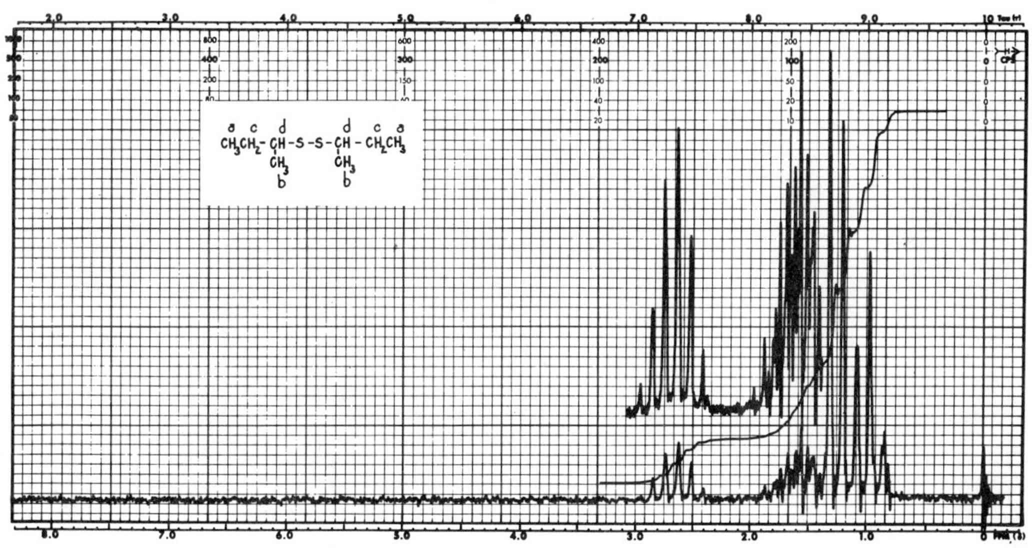

© 1978 Sadtler Research Laboratories, Inc.

ISOBUTYL DISULFIDE

$C_8H_{18}S_2$

Mol. Wt. 178.36

B.P. 84-86°C/10mm

Solvent: CCl_4

Assignments
a = 1.01
b = 1.92
c = 2.55

1185

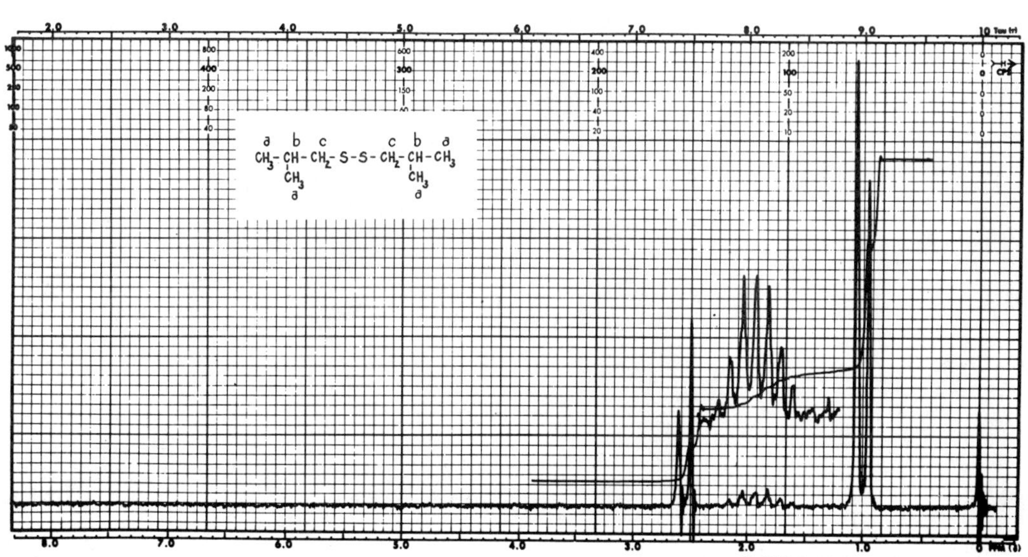

© 1978 Sadtler Research Laboratories, Inc.

tert-BUTYL DISULFIDE

$C_8H_{18}S_2$

Mol. Wt. 178.36

B.P. 198-204°C

Solvent: CCl_4

Assignments
a = 1.30

1186

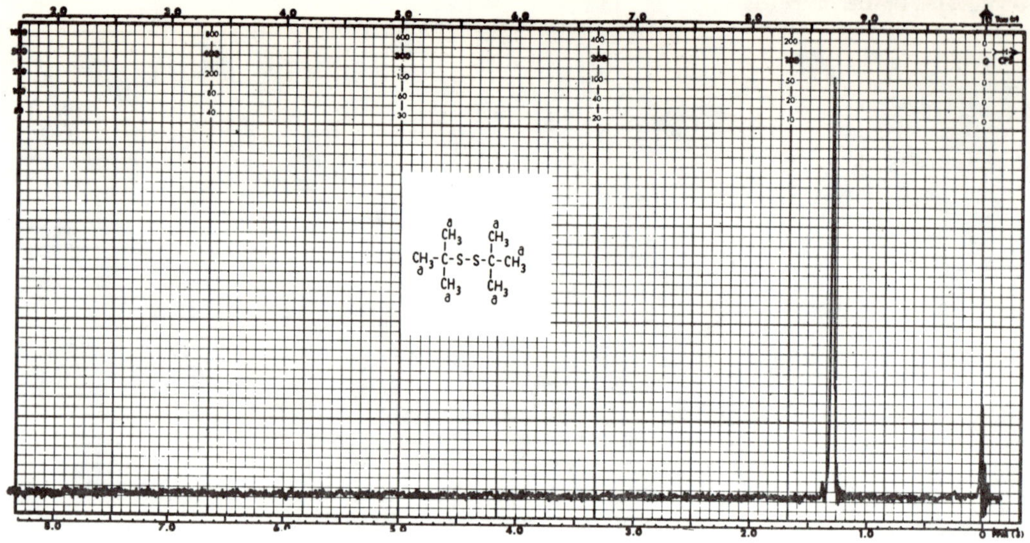

© 1978 Sadtler Research Laboratories, Inc.

PENTYL DISULFIDE

$C_{10}H_{22}S_2$

Mol. Wt. 206.41

Solvent: CCl_4

Assignments
a = 0.95
b = 1.10-1.60
c = ca 1.68
d = 2.64

1187

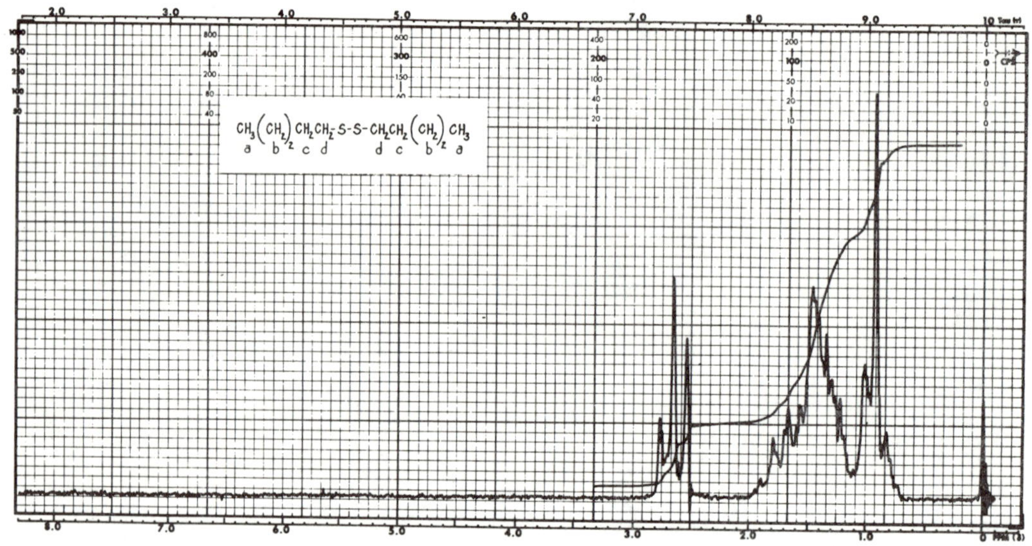

© 1978 Sadtler Research Laboratories, Inc.

ISOPENTYL DISULFIDE

$C_{10}H_{22}S_2$

Mol. Wt. 206.41

Solvent: CCl_4

Assignments
a = 0.95
b = 1.30-2.05
c = 2.69

1188

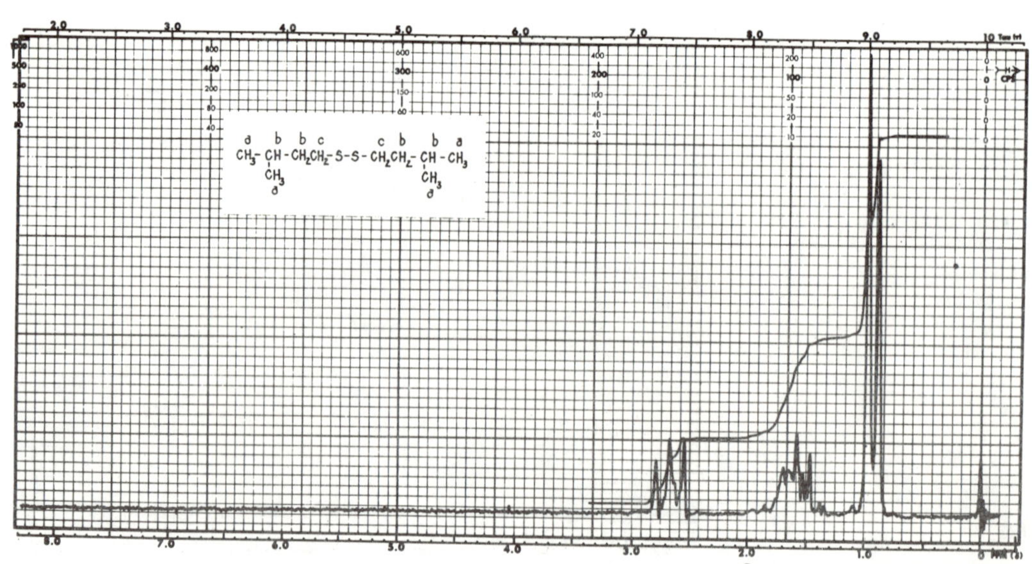

© 1978 Sadtler Research Laboratories, Inc.

DODECYL DISULFIDE

$C_{24}H_{50}S_2$

Mol. Wt. 402.79

Solvent: $CDCl_3$

Assignments
a = 0.89
b = ca 1.29
c = 1.68
d = 2.70

1189

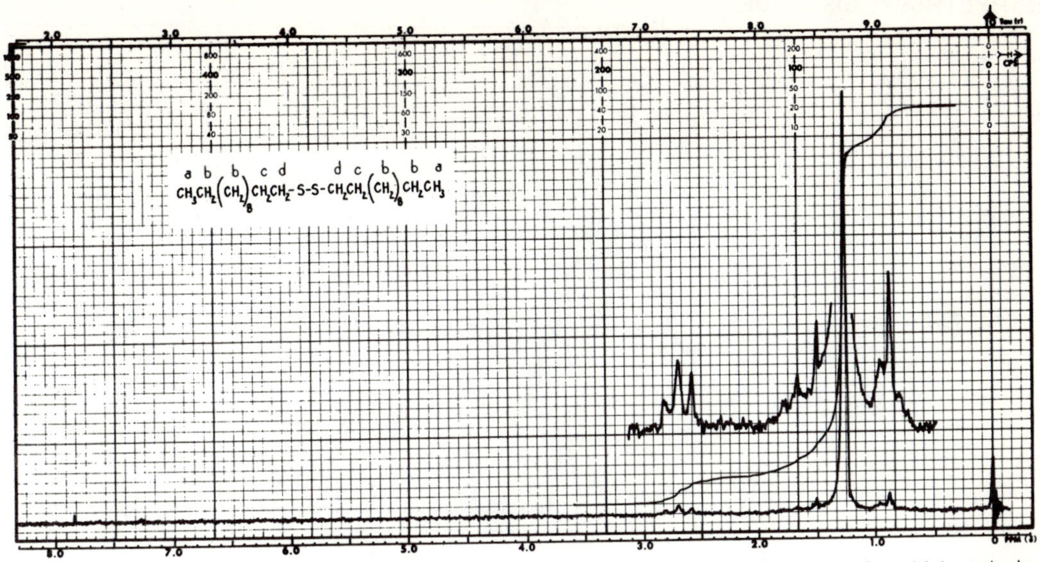

© 1978 Sadtler Research Laboratories, Inc.

ALLYL DISULFIDE

$C_6H_{10}S_2$

Mol. Wt. 146.27

Solvent: CCl_4

Assignments
a = 3.29
b = 5.05
c = 5.11
d = 5.72
e = 2.97, 3.08 imps.

1190

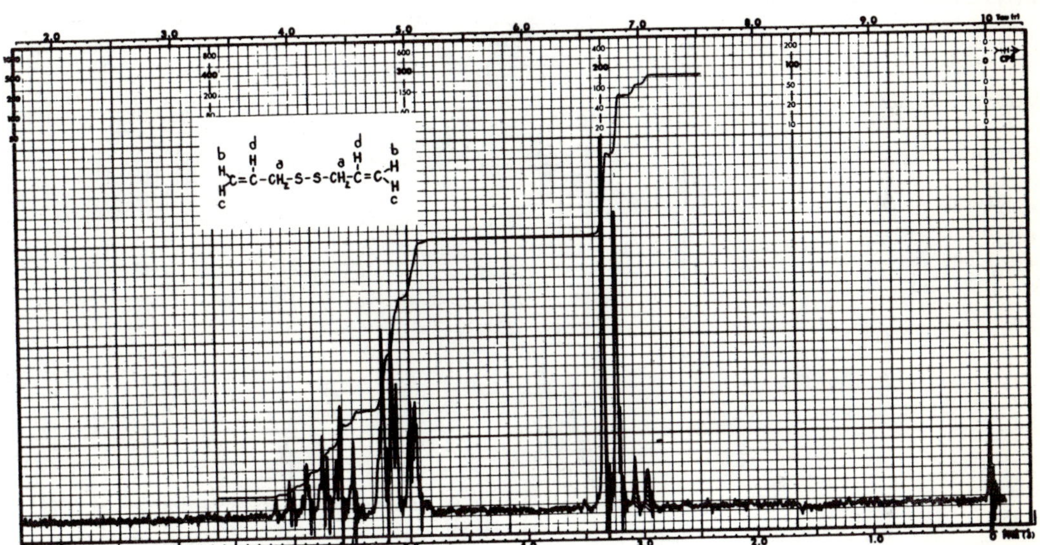

© 1978 Sadtler Research Laboratories, Inc.

BENZYL DISULFIDE

$C_{14}H_{14}S_2$

Mol. Wt. 246.40

M.P. 69-72°C

Solvent: $CDCl_3$

Assignments
a = 3.59
b = 7.25

1191

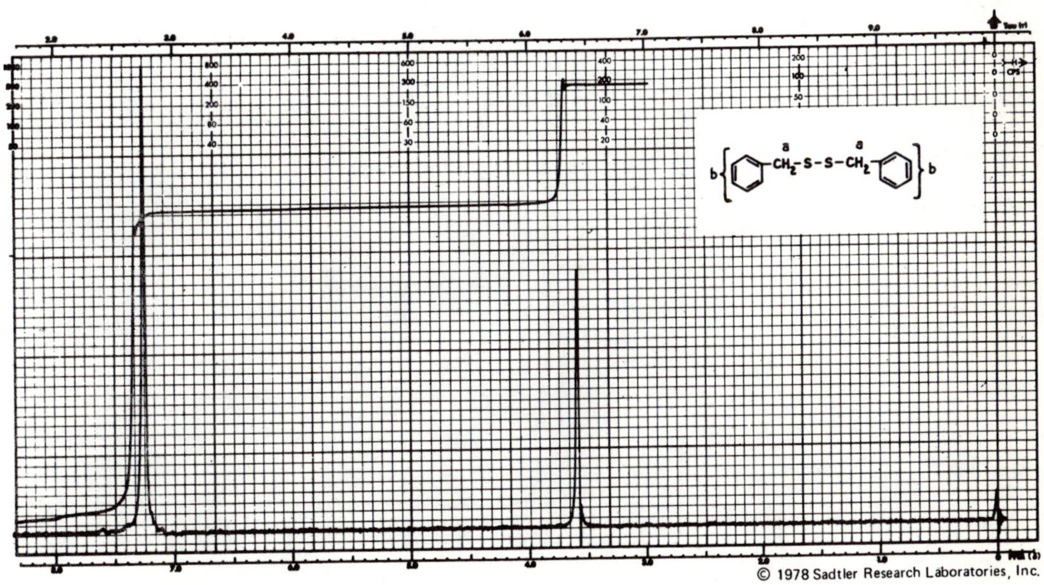

© 1978 Sadtler Research Laboratories, Inc.

METHYL PHENYL DISULFIDE

$C_7H_8S_2$

Mol. Wt. 156.27

B.P. 93-97°C/5mm

Solvent: CCl_4

Assignments
a = 2.33
b = 7.08-7.55

1192

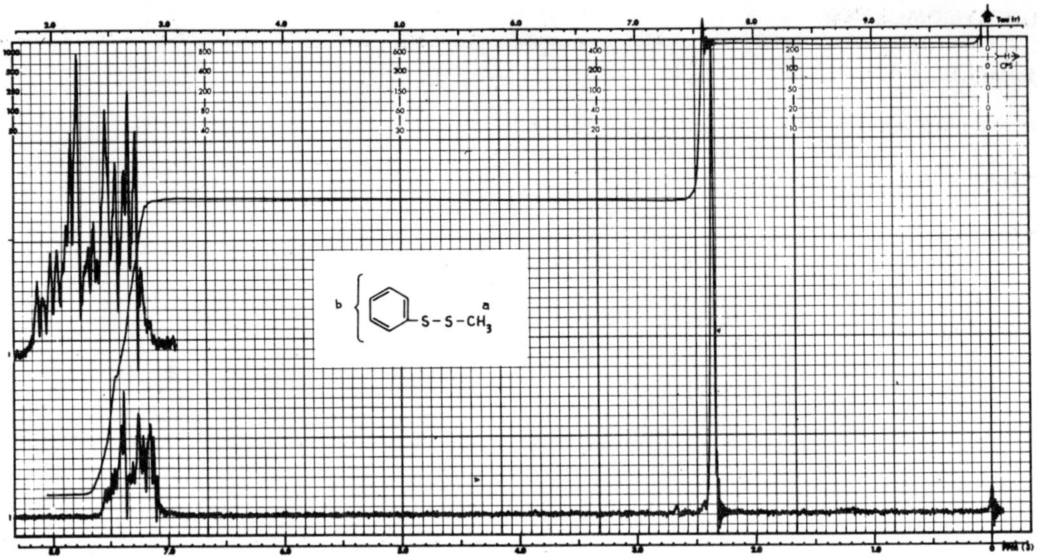

© 1978 Sadtler Research Laboratories, Inc.

ETHYL PHENYL DISULFIDE

$C_8H_{10}S_2$

Mol. Wt. 170.30

Solvent: CCl_4

Assignments
a = 1.25
b = 2.66
c = 7.10-7.50

1193

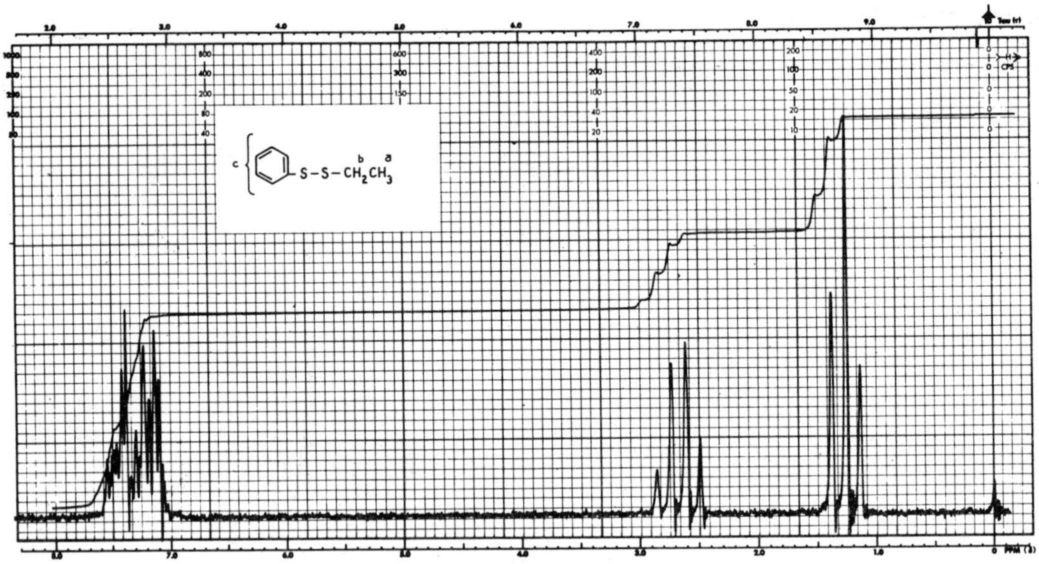

© 1978 Sadtler Research Laboratories, Inc.

PHENYL DISULFIDE

$C_{12}H_{10}S_2$

Mol. Wt. 218.34

M.P. 59-60°C

Solvent: $CDCl_3$

Assignments
a = 7.05-7.35
b = 7.46

1194

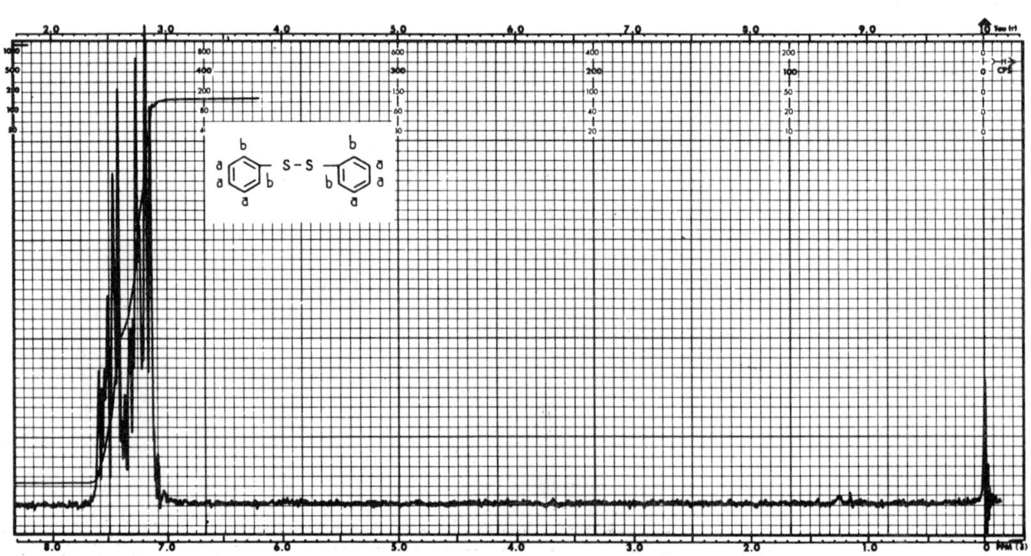

© 1978 Sadtler Research Laboratories, Inc.

o-TOLYL DISULFIDE

$C_{14}H_{14}S_2$

Mol. Wt. 246.40

Solvent: $CDCl_3$

Assignments
a = 2.24
b = ca 7.02

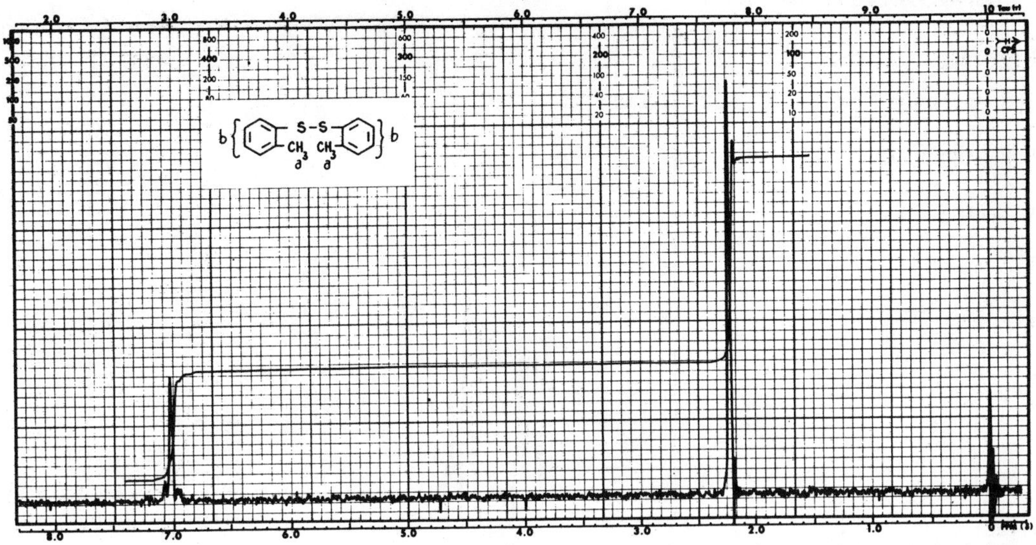

1195

© 1978 Sadtler Research Laboratories, Inc.

---

p-TOLYL DISULFIDE

$C_{14}H_{14}S_2$

Mol. Wt. 246.40

M.P. 43-46°C

Solvent: $CDCl_3$

Assignments
a = 2.30
b = 7.05
c = 7.34

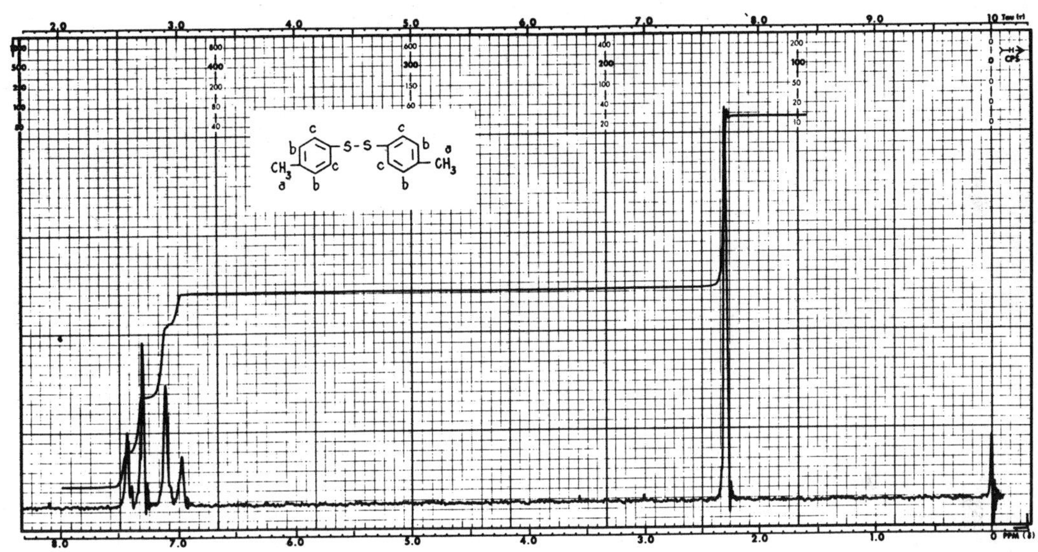

1196

© 1978 Sadtler Research Laboratories, Inc.

BIS(p-CHLOROPHENYL) DISULFIDE

$C_{12}H_8Cl_2S_2$

Mol. Wt. 287.23

M.P. 71-72°C

Solvent: $CDCl_3$

Assignments
a = 7.20
b = 7.38

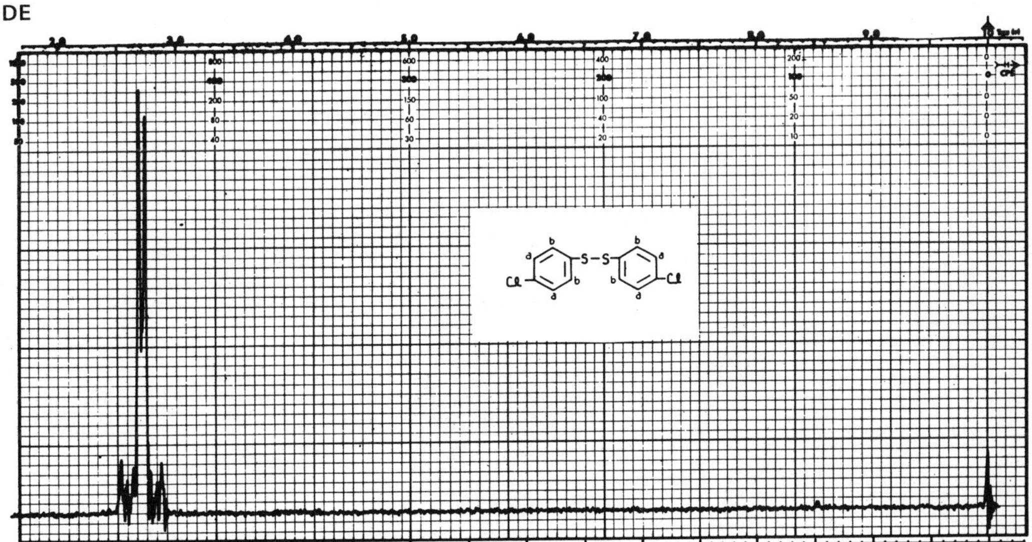

1197

© 1978 Sadtler Research Laboratories, Inc.

2,2-DITHIODIANILINE

$C_{12}H_{12}S_2$

Mol. Wt. 248.37

M.P. 91-92°C

Solvent: $CDCl_3$

Assignments
a = 4.24
b = 6.53
c = 6.60
d = 7.10
e = 7.16

1198

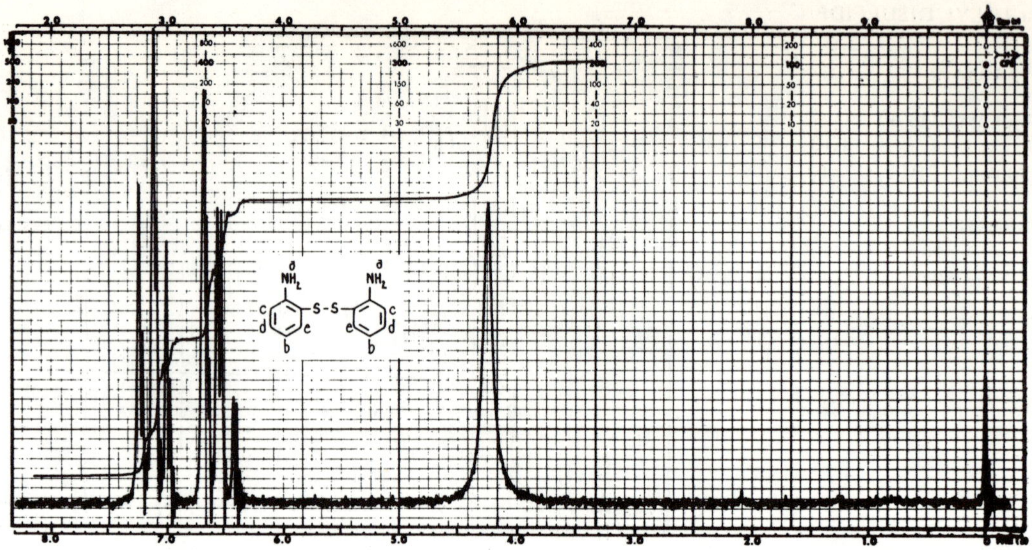

© 1978 Sadtler Research Laboratories, Inc.

The aliphatic Thiols are an especially easy group of compounds to characterize due to the clear coupling between the Thiol proton (—S—H) and adjacent aliphatic groups (except in D_2O solution). The aromatic Thiols are also relatively easy to characterize in that they contain an exchangeable proton which resonates at relatively high field (3.0—5.0 ppm) but the group does not strongly shield the ortho and para hydrogens as the aromatic amines and phenols do. The Thiol group is a weak to intermediate deshielding group in its effect on aliphatic protons but neither shields nor deshields the ortho aromatic hydrogens to any great extent.

Chemical Shifts

Aliphatic Protons

HC—3	HC—2	HC—1	—SH		Solvent	Spectrum
		2.07	1.24	CH_3—SH	$CDCl_3$	1199
	1.31	2.52	1.17	CH_3—CH_2—SH	CCl_4	1200
0.99	1.67	2.52	1.31	CH_3—CH_2—CH_2—SH	$CDCl_3$	1201
	(1.31)	3.07	1.40	$(CH_3)_2$—CH—SH	CCl_4	1202
	(1.41)	- - - -	1.62	$(CH_3)_3$—C—SH	CCl_4	- - - -

Substituted Methanethiols

HS	—CH_2	—X	
1.99	3.20	—C(=O)—O—R_8	CCl_4
2.22	3.31	—C(=O)—OH	$CDCl_3$
2.06	3.33	—C(=O)—O—$(CH_2)_2$—O—R	$CDCl_3$
2.06	3.41	—C(=O)—NH—Ø	$CDCl_3$
- - - -	3.51	—C(=O)—O—Na	D_2O
1.58	3.57	—Ø (4—Cl)	CCl_4
1.50	3.64	—Ø	CCl_4

Substituted Ethanethiols

HS	—CH_2	—CH_2	—X	
1.12	2.43	0.67	—Si(O—$CH_3)_3$	CCl_4
1.18	2.50	1.60	—$(CH_2)_6$—SH	CCl_4
1.19	2.49	1.60	—R_4	CCl_4
1.31	2.52	1.67	—CH_3	$CDCl_3$
1.18	2.51	1.69	—$(CH_2)_2$—SH	CCl_4
1.30	2.51	1.84	—CH_2—Ø	$CDCl_3$

Chemical Shifts continued

Aliphatic Protons continued

Substituted Ethanethiols continued

HS	$-CH_2$	$-CH_2$	$-X$	Solvent
1.27	2.67	1.89	$-CH_2-SH$	CCl_4
1.58	2.54	2.30	$-C(=O)-NH-R$	$CDCl_3$
----	2.71	2.71	$-SH$	D_2O
----	2.68	3.69	$-OH$	D_2O

Aromatic Protons

HS–	ortho/meta/para
3.19	ca 6.91 ppm (broad, single peak)

Para Substituted Benzenethiols

HS–	ortho	meta	para	
3.52	7.14	6.52	$-NH_2$	$CDCl_3$
3.37	7.30	6.81	$-O-CH_3$	$CDCl_3$
3.30	7.09	6.91	$-CH_3$	$CDCl_3$
3.40	7.11	7.11	$-Cl$	$CDCl_3$
3.16	7.13	7.13	$-C(CH_3)_3$	CCl_4
3.42	7.07	7.30	$-Br$	$CDCl_3$

Exchangeable Protons

The Thiol protons, being less active than the exchangeable hydrogens of the amines and alcohols, usually display coupling to the adjacent aliphatic groups in all solvents except D_2O. One exception to this fact occurs in the case in which another type of exchangeable is present in the molecule (spectra 1215 – 1218). The Thiol-aliphatic group coupling constant is similar to that observed for vicinal CH–CH coupling, J = 5–8 Hz.

These protons undergo deuteration only slowly upon the addition of a few drops of D_2O to a sample solution of $CDCl_3$ or CCl_4. In D_2O solution, however, they exchange completely and immediately.

METHANETHIOL

CH_4S

Mol. Wt. 48.11

Solvent: $CDCl_3$

Assignments
a = 1.24
b = 2.07

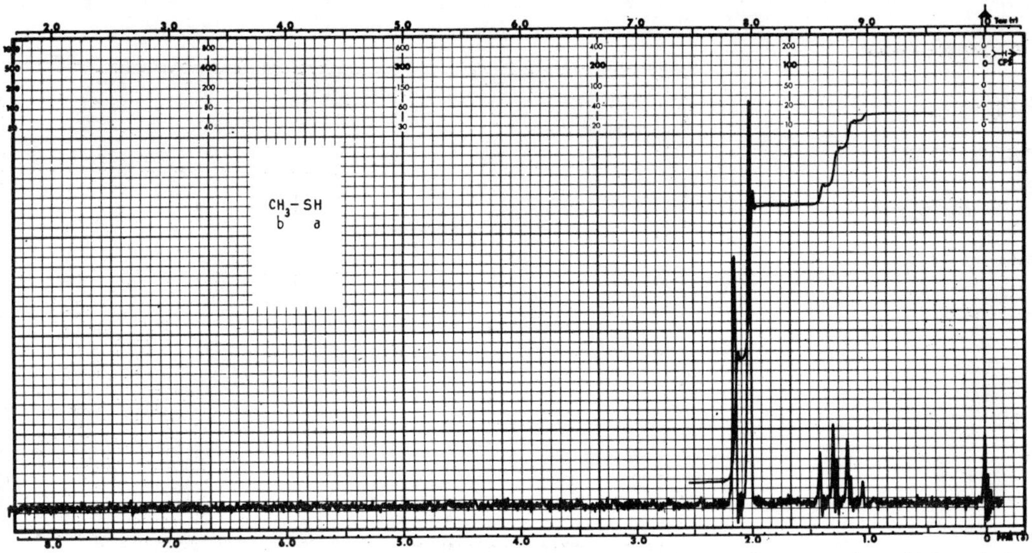

1199

© 1978 Sadtler Research Laboratories, Inc.

ETHANETHIOL

C_2H_6S

Mol. Wt. 62.13

B.P. 33.3°C

Solvent: CCl_4

Assignments
a = 1.17
b = 1.31
c = 2.52

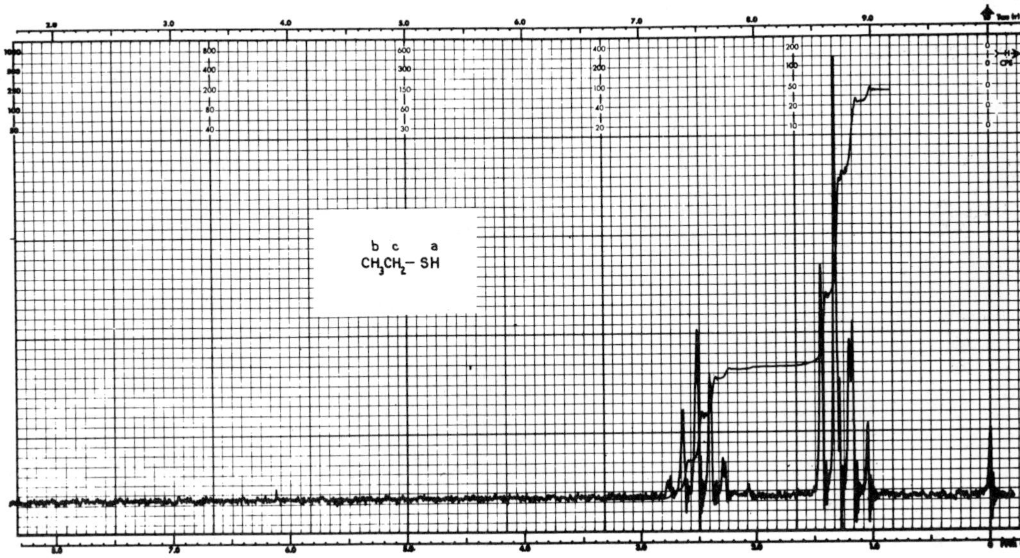

1200

© 1978 Sadtler Research Laboratories, Inc.

1-PROPANETHIOL

C_3H_8S

Mol. Wt. 76.17

B.P. 66-68°C

Solvent: $CDCl_3$

Assignments
a = 0.99
b = 1.31
c = 1.67
d = 2.52

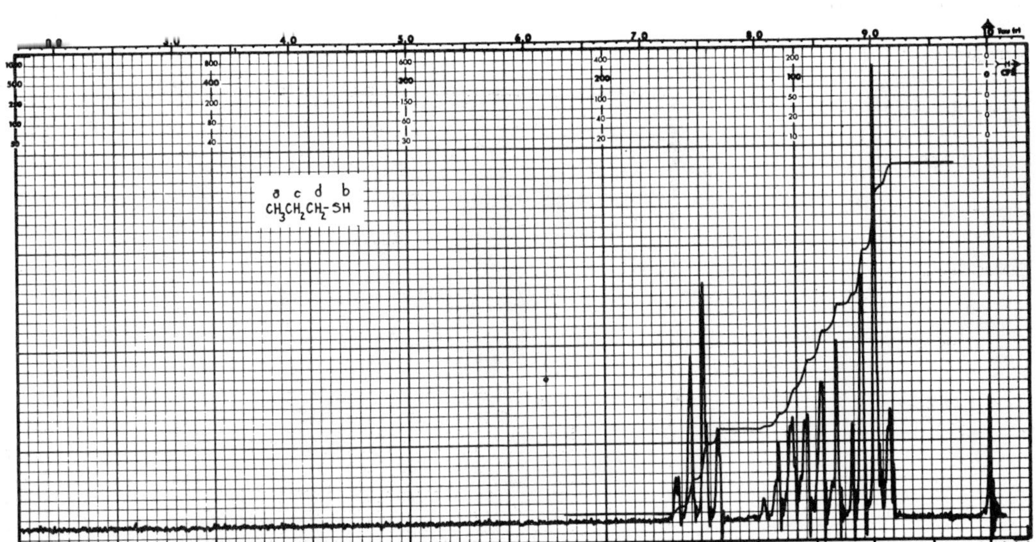

1201

© 1978 Sadtler Research Laboratories, Inc.

2-PROPANETHIOL

C_3H_8S

Mol. Wt. 79.16

B.P. 57-60°C

Solvent: CCl_4

Assignments
a = 1.31
b = 1.40
c = 3.07

1202

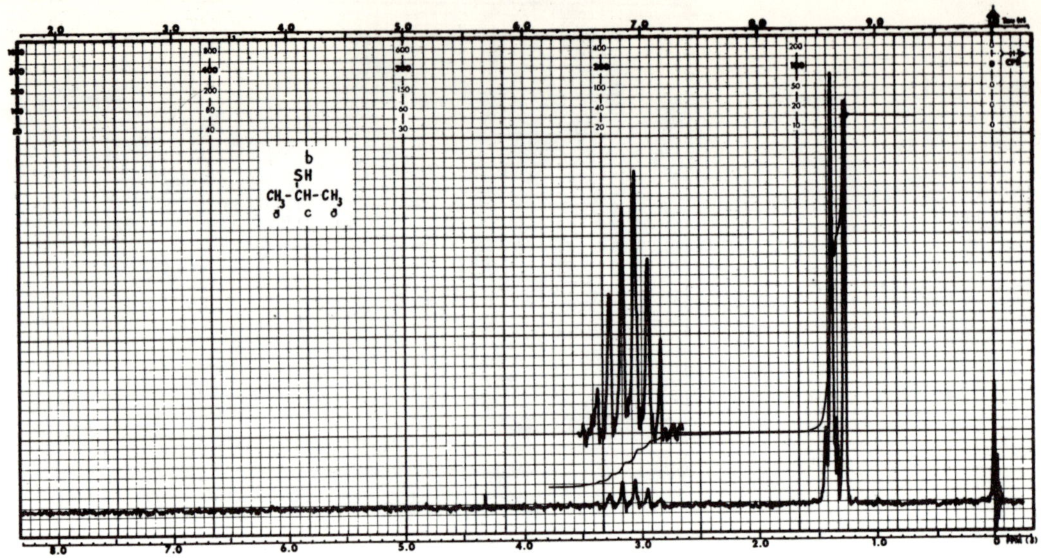

© 1978 Sadtler Research Laboratories, Inc.

2-BUTANETHIOL

$C_4H_{10}S$

Mol. Wt. 90.19

B.P. 73.3°C

Solvent: CCl_4

Assignments
a = 0.98
b = 1.24
c = 1.30
d = 1.50
e = 2.81

1203

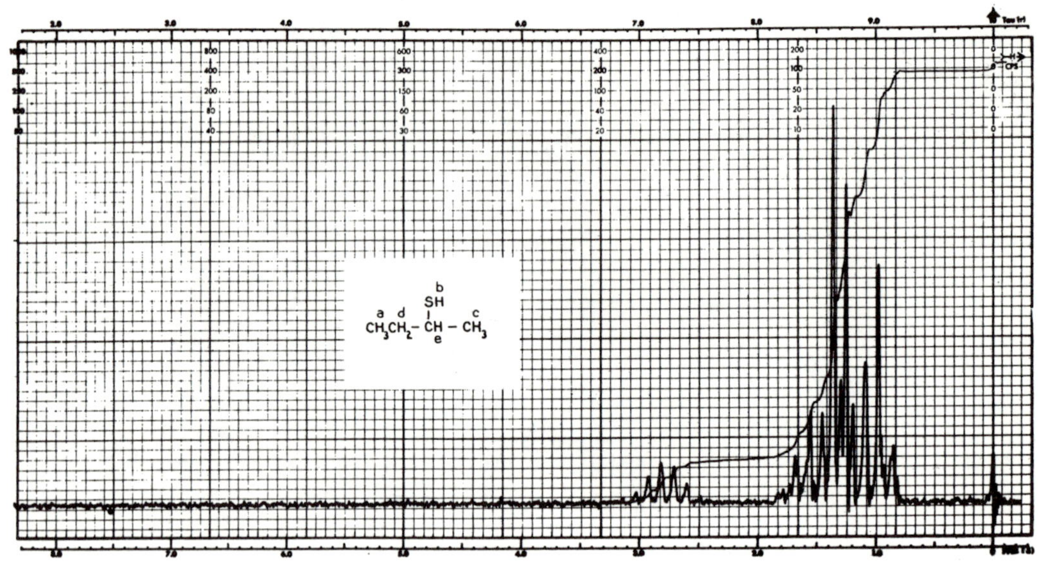

© 1978 Sadtler Research Laboratories, Inc.

1-PENTANETHIOL

$C_5H_{12}S$

Mol. Wt. 104.22

B.P. 123-126°C

Solvent: CCl_4

Assignments
a = 0.92
b = 1.14
c = 1.20-2.00
d = 2.49

1204

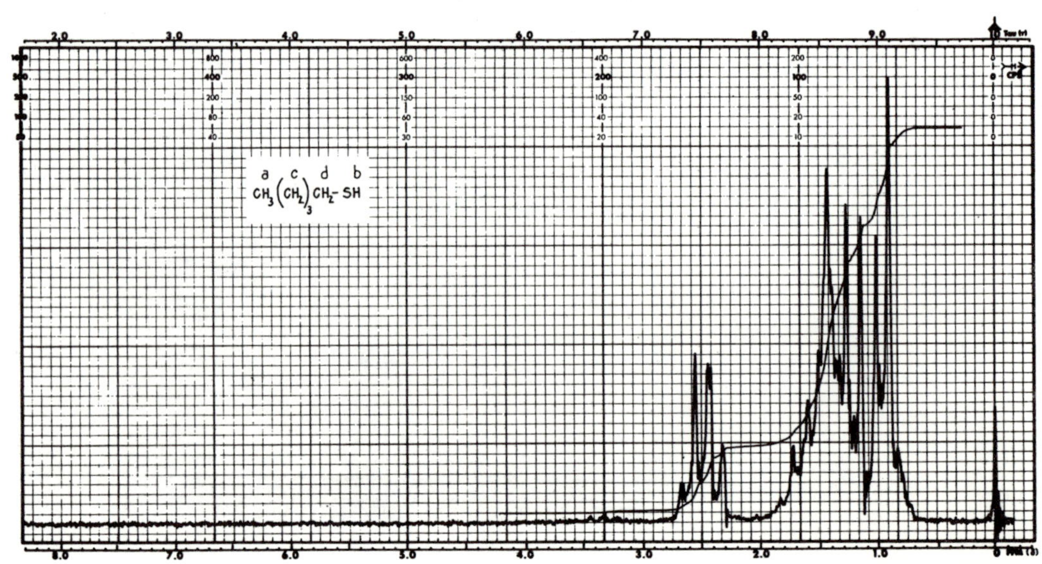

© 1978 Sadtler Research Laboratories, Inc.

3-METHYL-1-BUTANETHIOL

$C_5H_{12}S$

Mol. Wt. 104.22

B.P. 116-117°C (lit.)

Solvent: CCl_4

Assignments
a = 0.91
b = 1.11
c = 1.78-1.92
d = 2.49
e = 2.03 imp.

1205

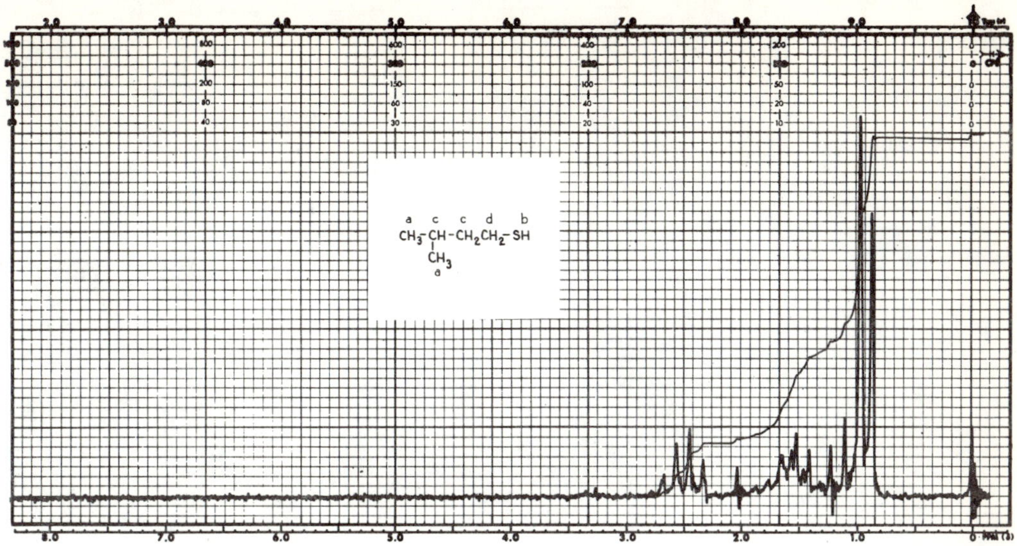

© 1978 Sadtler Research Laboratories, Inc.

1-HEXANETHIOL

$C_6H_{14}S$

Mol. Wt. 118.24

B.P. 149-152°C

Solvent: CCl_4

Assignments
a = 0.90
b = 1.19
c = 1.10-2.00
d = 2.49

1206

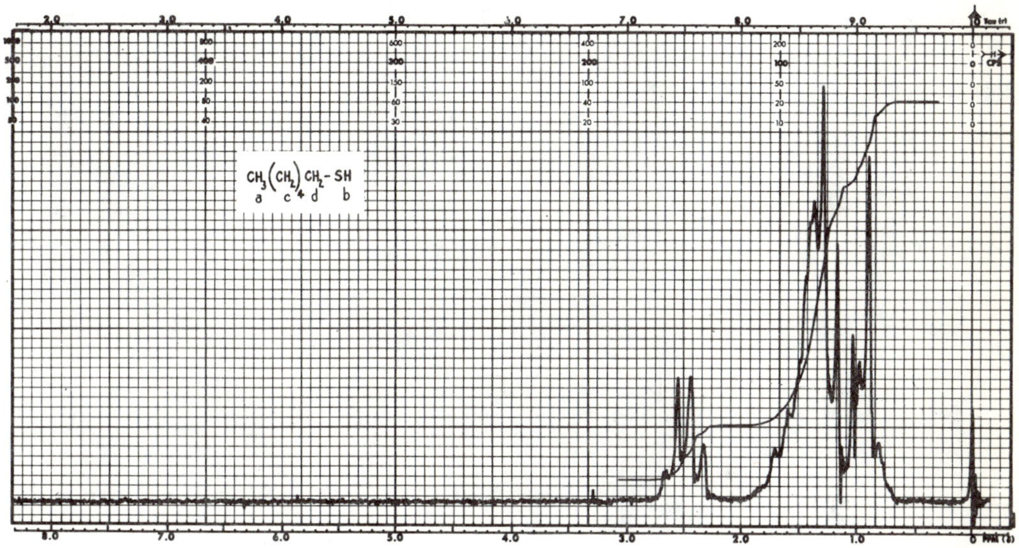

© 1978 Sadtler Research Laboratories, Inc.

2-ETHYL-1-HEXANETHIOL

$C_8H_{18}S$

Mol. Wt. 146.30

B.P. 90°C/3.5mm

Solvent: CCl_4

Assignments
a = 0.89
b = 0.92
c = 0.99
d = 1.32
e = 2.49

1207

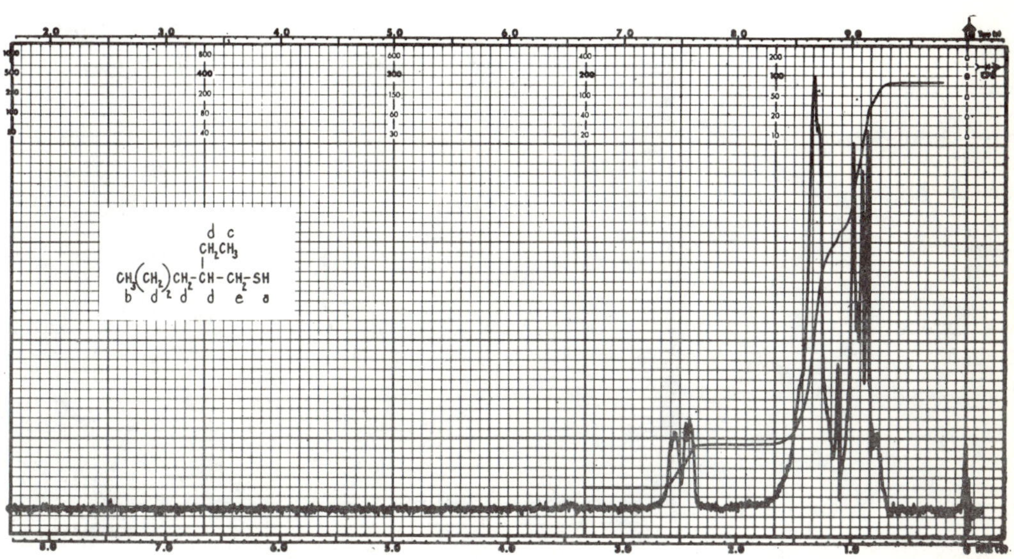

© 1978 Sadtler Research Laboratories, Inc.

1-HEPTANETHIOL

$C_7H_{16}S$

Mol. Wt. 132.26

B.P. 174-176°C

Solvent: $CDCl_3$

Assignments
a = 0.90
b = 1.29
c = 1.40
d = 2.48

1208

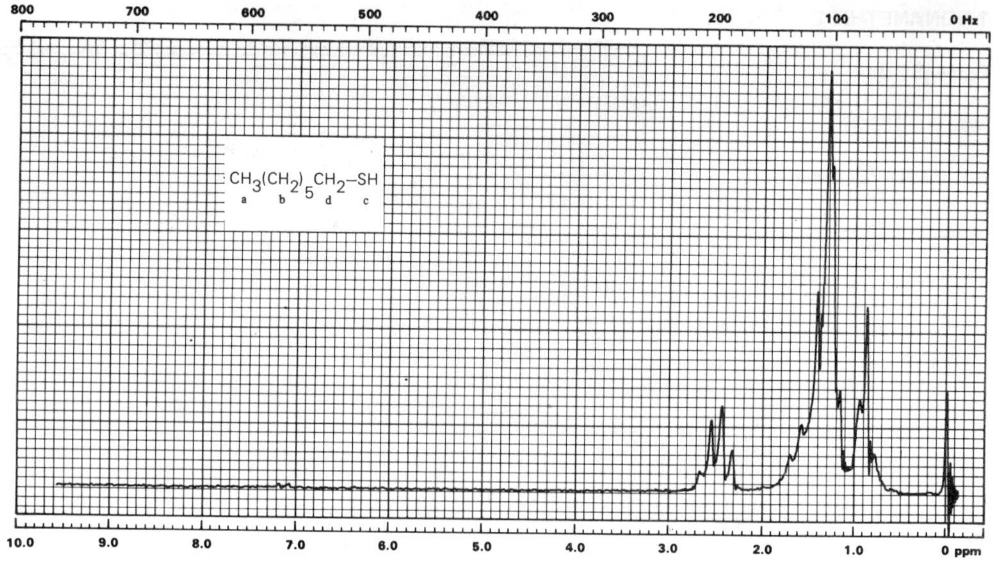

© 1978 Sadtler Research Laboratories, Inc.

1-OCTANETHIOL

$C_8H_{18}S$

Mol. Wt. 146.30

M.P. −45.2°C

B.P. 197-200°C

Solvent: CCl_4

Assignments
a = 0.90
b = 1.11
c = 1.15-1.85
d = 2.49

1209

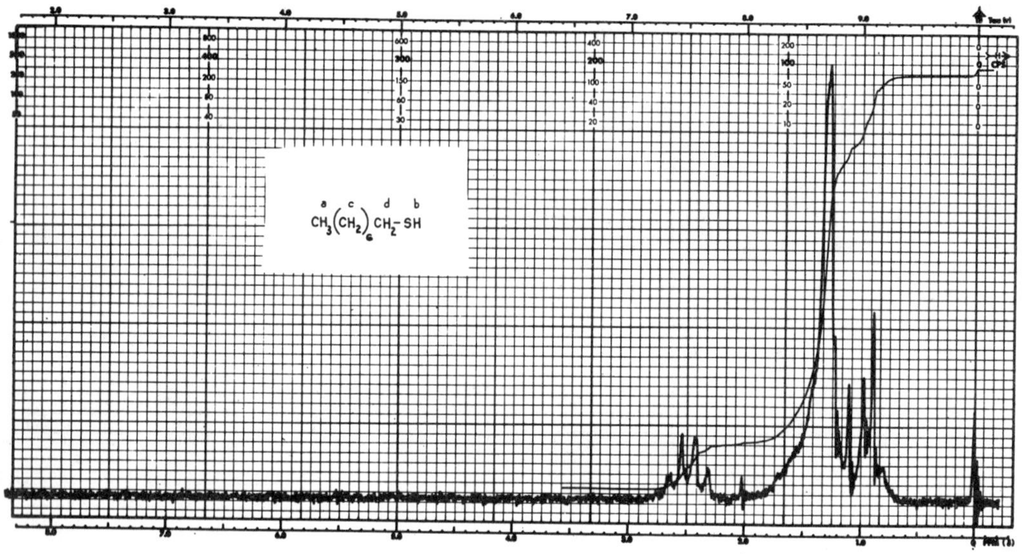

© 1978 Sadtler Research Laboratories, Inc.

2-OCTANETHIOL

$C_8H_{18}S$

Mol. Wt. 146.30

B.P. 75.5-76°C/18mm

Solvent: $CDCl_3$

Assignments
a = 0.90
b = 1.30
c = 1.00-2.00
d = 2.87

1210

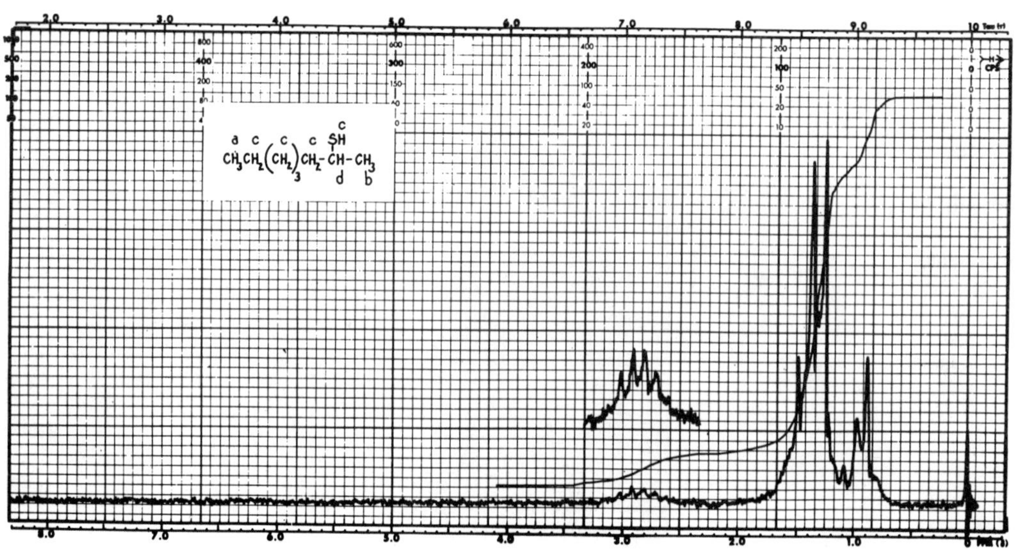

© 1978 Sadtler Research Laboratories, Inc.

1-NONANETHIOL

$C_9H_{20}S$

Mol. Wt. 160.32

B.P. 99-101°C/15mm

Solvent: $CDCl_3$

Assignments
a = 0.90
b = 1.10-1.50
c = 1.60
d = 2.53

1211

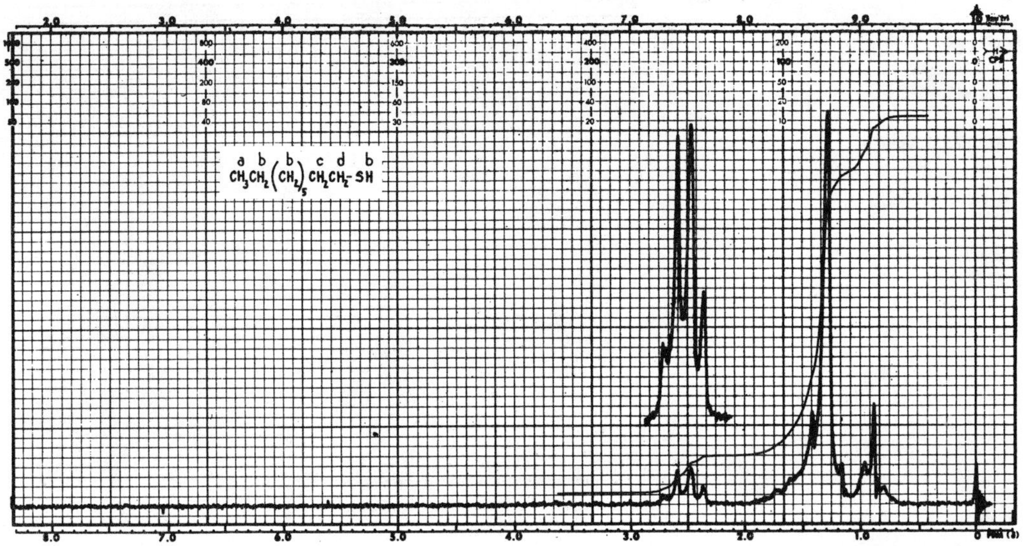

© 1978 Sadtler Research Laboratories, Inc.

1-DODECANETHIOL

$C_{12}H_{26}S$

Mol. Wt. 202.41

B.P. 146-148°C/15mm

Solvent: $CDCl_3$

Assignments
a = 0.90
b = 1.30
c = 1.46
d = 2.55

1212

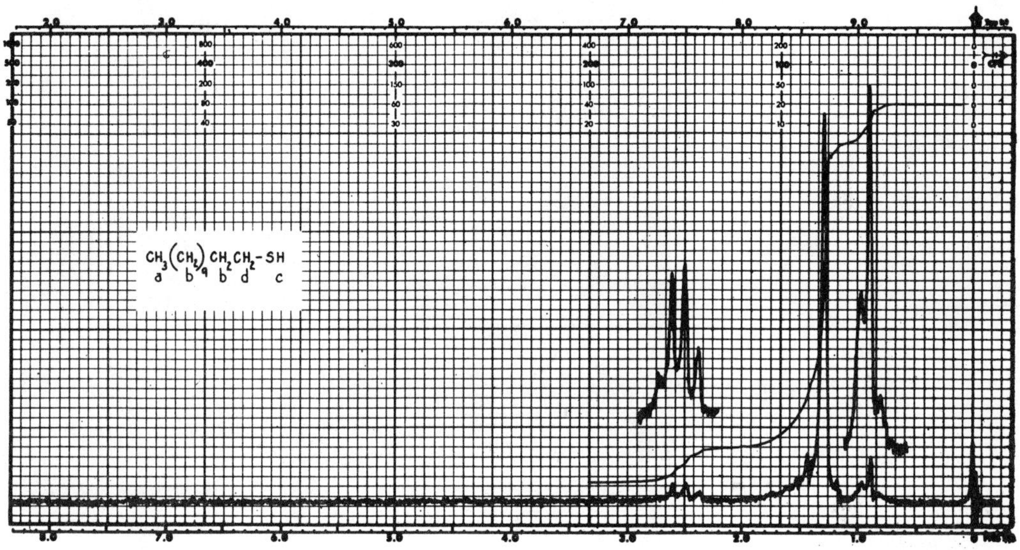

© 1978 Sadtler Research Laboratories, Inc.

1-HEXADECANETHIOL

$C_{16}H_{34}S$

Mol. Wt. 258.51

B.P. 169-170°C

Solvent: CCl_4

Assignments
a = 0.89
b = 1.20
c = 1.29
d = 2.43

1213

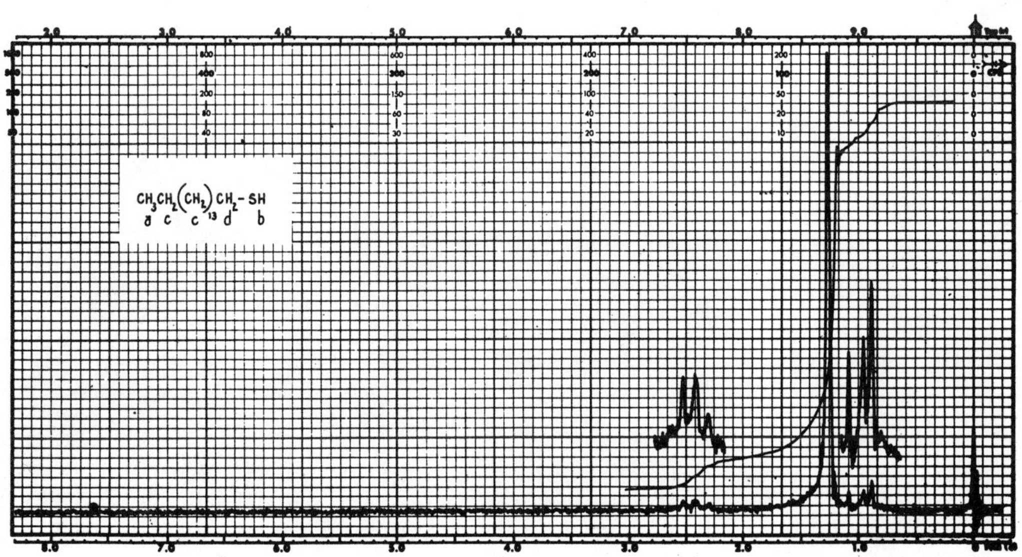

© 1978 Sadtler Research Laboratories, Inc.

1,3-DICHLORO-2-PROPANETHIOL

$C_3H_6Cl_2S$

Mol. Wt. 145.05

B.P. 104-106°C/88mm

Solvent: CCl_4

Assignments
a = 2.12
b = 3.22
c = 3.71
d = 3.92

1214

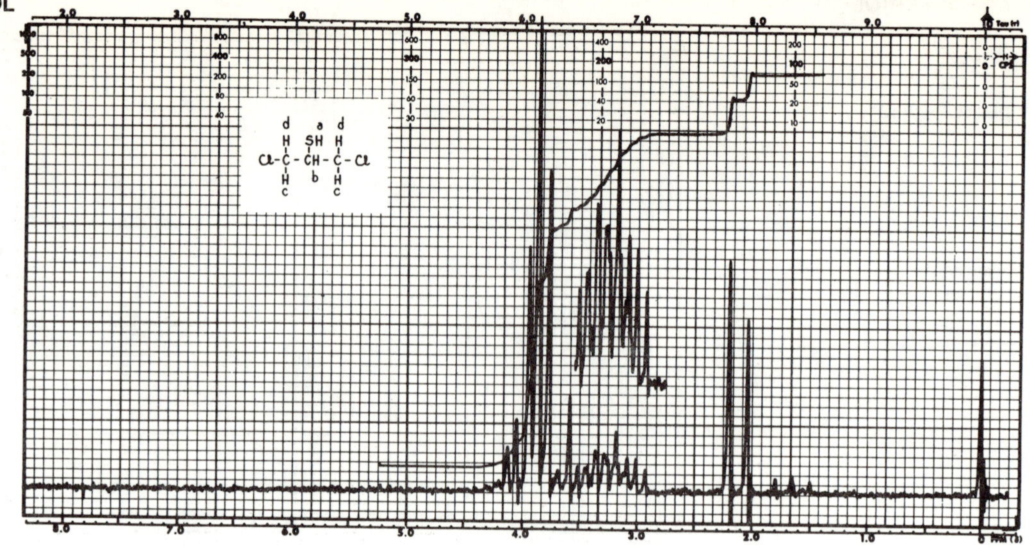

© 1978 Sadtler Research Laboratories, Inc.

2-(BUTYLAMINO)ETHANETHIOL

$C_6H_{15}NS$

Mol. Wt. 133.26

B.P. 112-115°C/10mm

Solvent: $CDCl_3$

Assignments
a = 0.93
b = 1.10-1.80
c = 1.61
d = 2.40-3.00

1215

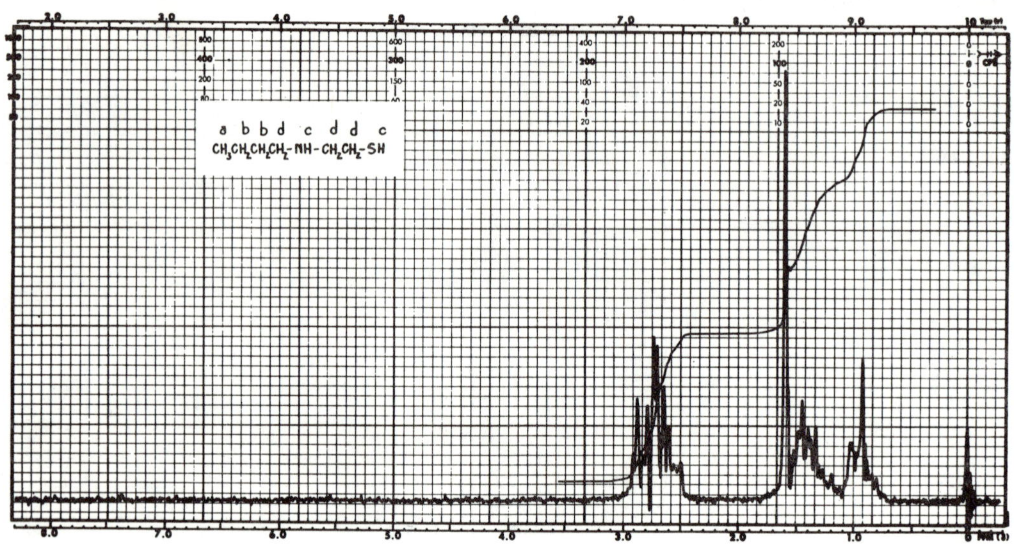

© 1978 Sadtler Research Laboratories, Inc.

2-(DECYLAMINO)ETHANETHIOL

$C_{12}H_{27}NS$

Mol. Wt. 217.42

B.P. 115-118°C/0.3mm

Solvent: $CDCl_3$

Assignments
a = 0.89
b = 1.10-1.90
c = 1.42
d = 2.40-3.10

1216

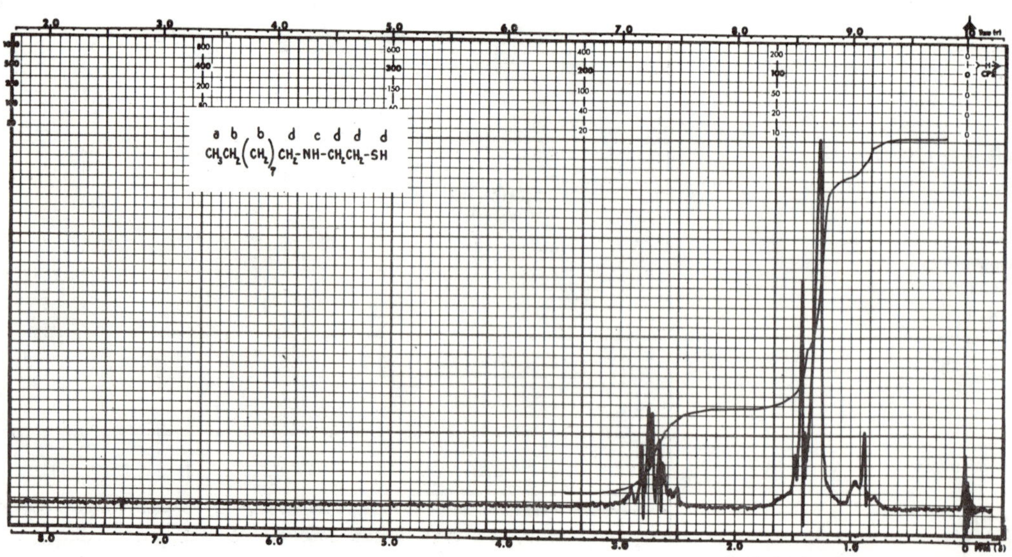

© 1978 Sadtler Research Laboratories, Inc.

2-(CYCLOHEXYLAMINO)ETHANETHIOL

$C_8H_{17}NS$

Mol. Wt. 159.30

B.P. 102-106°C/8mm

Solvent: $CDCl_3$

Assignments
a = 0.80-2.10
b = 1.83
c = 2.29
d = 2.50-3.10

1217

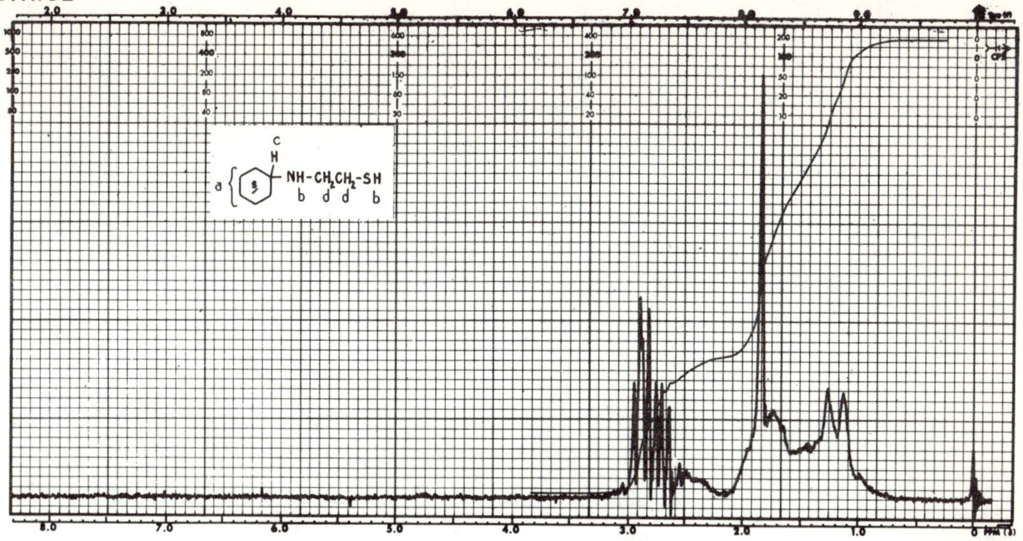

© 1978 Sadtler Research Laboratories, Inc.

2-(PHENETHYLAMINO)ETHANETHIOL

$C_{10}H_{15}NS$

Mol. Wt. 181.30

B.P. 130-135°C/1.25mm

Solvent: $CDCl_3$

Assignments
a = 1.58
b = 2.40-3.16
c = 7.21

1218

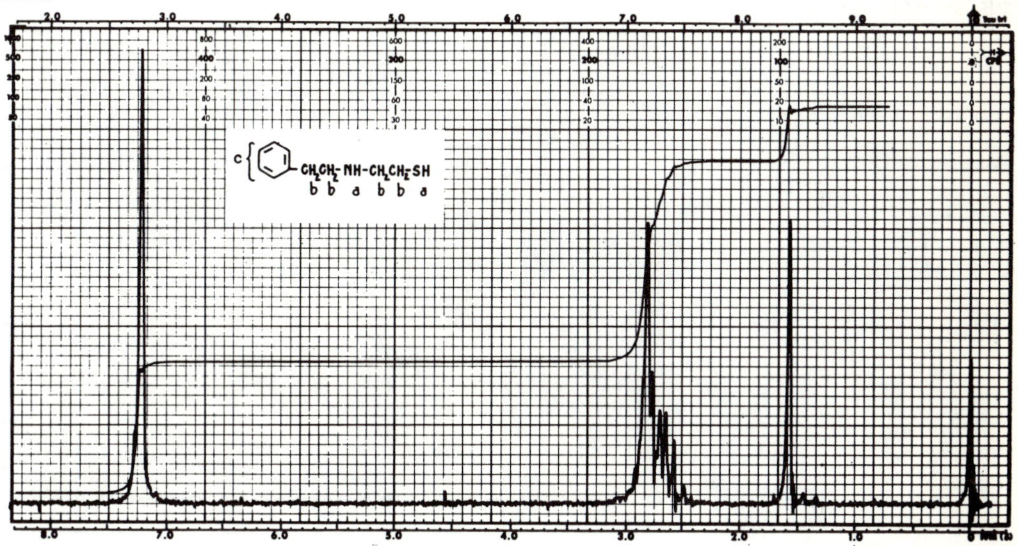

© 1978 Sadtler Research Laboratories, Inc.

1,2-PROPANEDITHIOL

$C_3H_8S_2$

Mol. Wt. 108.23

B.P. 41-43°C/11mm

Solvent: $CDCl_3$

Assignments
a = 1.41
b = 1.68
c = 1.86
d = 2.71
e = 3.00

1219

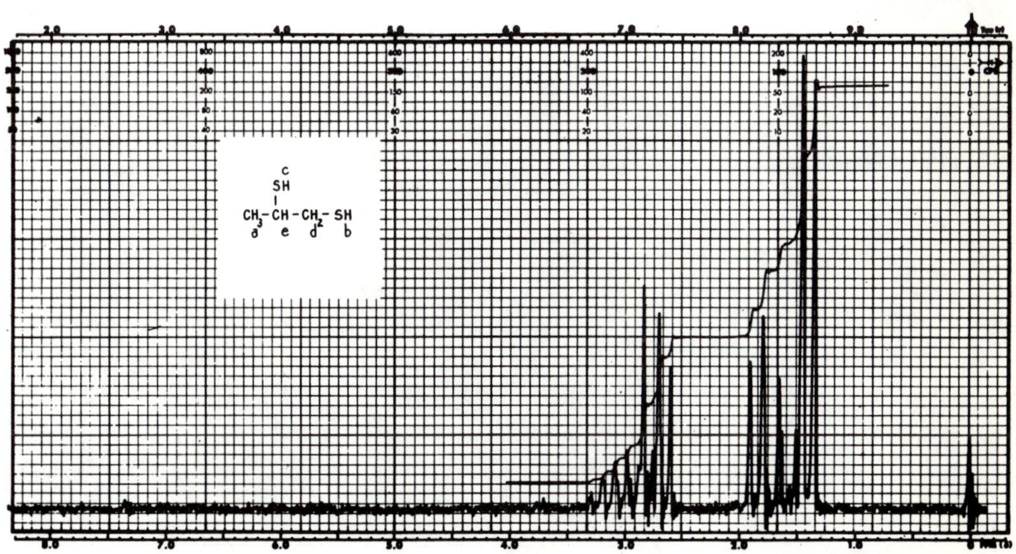

© 1978 Sadtler Research Laboratories, Inc.

1,4-BUTANEDITHIOL

$C_4H_{10}S_2$

Mol. Wt. 122.25

M.P. 65-67°C (lit.)

Solvent: CCl_4

Assignments
a = 1.18
b = 1.69
c = ca 2.51

1220

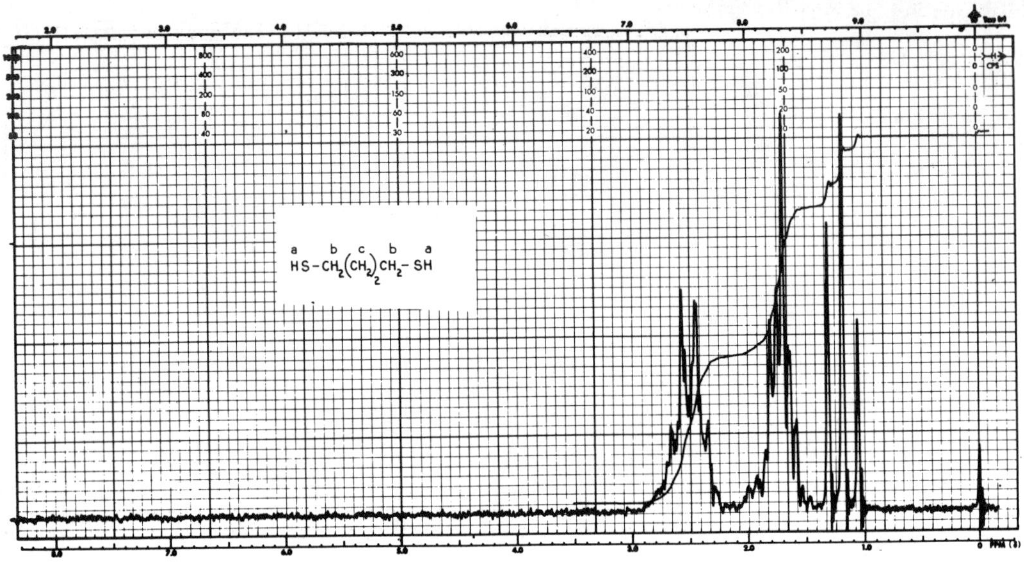

© 1978 Sadtler Research Laboratories, Inc.

1,8-OCTANEDITHIOL

$C_8H_{18}S_2$

Mol. Wt. 178.36

Solvent: CCl_4

Assignments
a = 1.18
b = 1.10-1.91
c = 2.50

1221

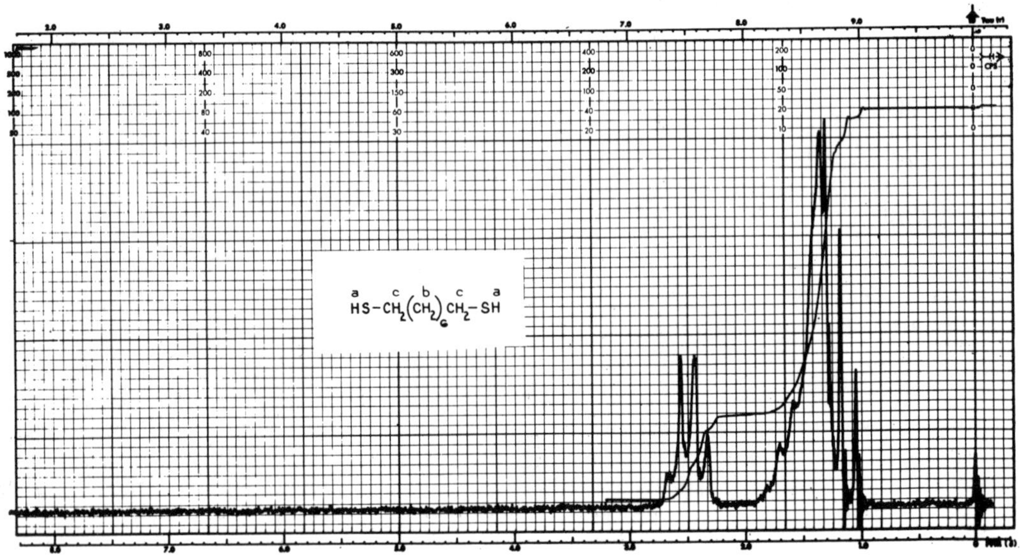

© 1978 Sadtler Research Laboratories, Inc.

1,9-NONANEDITHIOL

$C_9H_{20}S_2$

Mol. Wt. 192.39

B.P. 175-178°C/50mm (lit.)

Solvent: CCl_4

Assignments
a = 1.16
b = 1.21-1.98
c = 2.49

1222

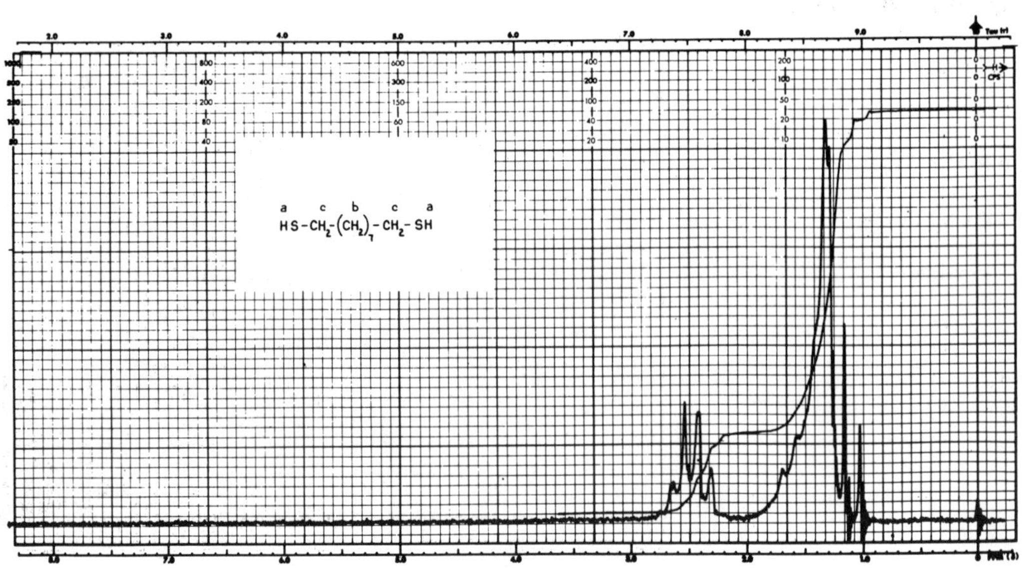

© 1978 Sadtler Research Laboratories, Inc.

2,6-DIMETHYL-2,6-OCTANEDITHIOL

$C_{10}H_{22}S_2$

Mol. Wt. 206.41

B.P. 64-65°C/0.2mm

Solvent: $CDCl_3$

Assignments
a = 0.98
b = 1.20-1.60
c = 1.31
d = 1.40
e = 1.57
f = 1.70

1223

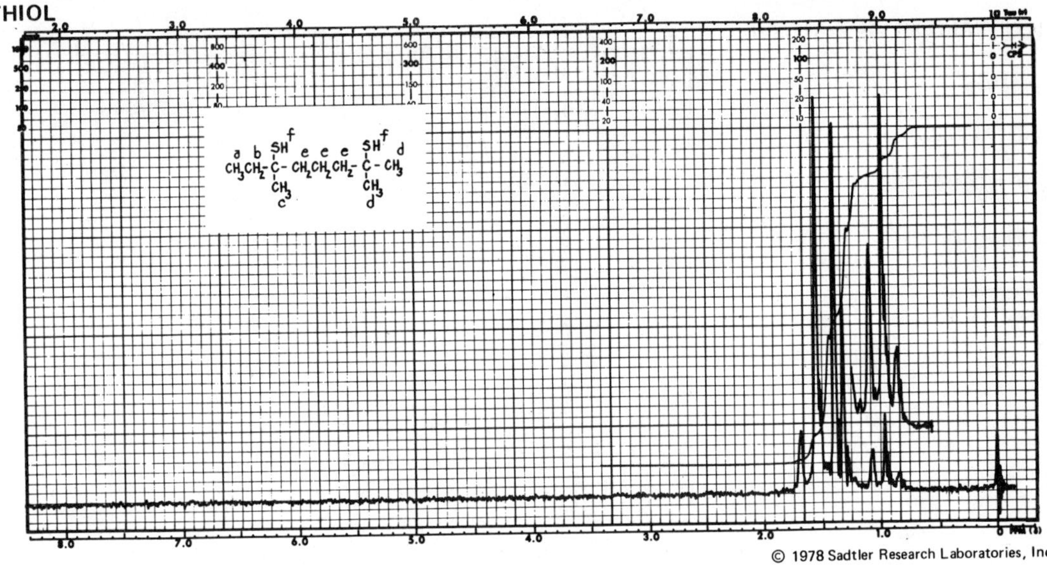

© 1978 Sadtler Research Laboratories, Inc.

α-TOLUENETHIOL

C_7H_8S

Mol. Wt. 124.20

B.P. 64-66°C/5mm

Solvent: CCl_4

Assignments
a = 1.50
b = 3.64
c = 7.22

1224

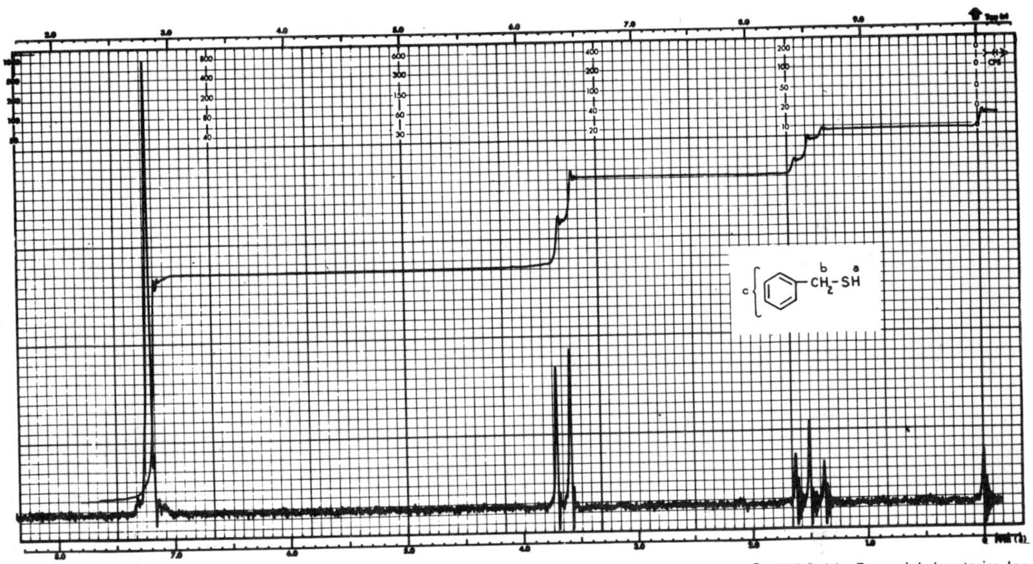

© 1978 Sadtler Research Laboratories, Inc.

p-CHLORO-α-TOLUENETHIOL

C_7H_7ClS

Mol. Wt. 158.65

B.P. 124°C/20mm

Solvent: CCl_4

Assignments
a = 1.58
b = 3.57
c = 7.14

1225

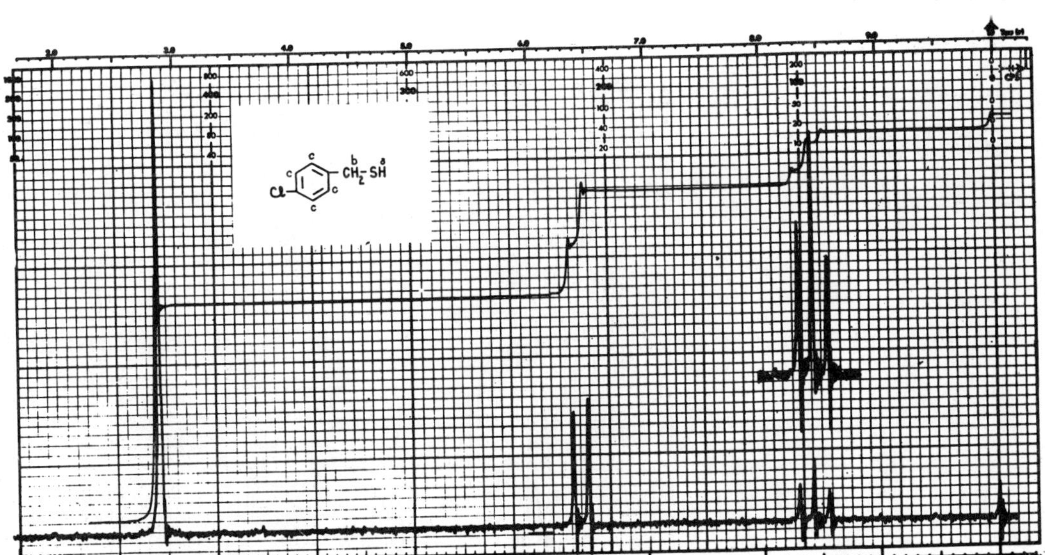

© 1978 Sadtler Research Laboratories, Inc.

TRIPHENYLMETHANETHIOL

$C_{19}H_{16}S$

Mol. Wt. 276.40

M.P. 106-107°C

Solvent: $CDCl_3$

Assignments
a = 3.05
b = ca 7.23

1226

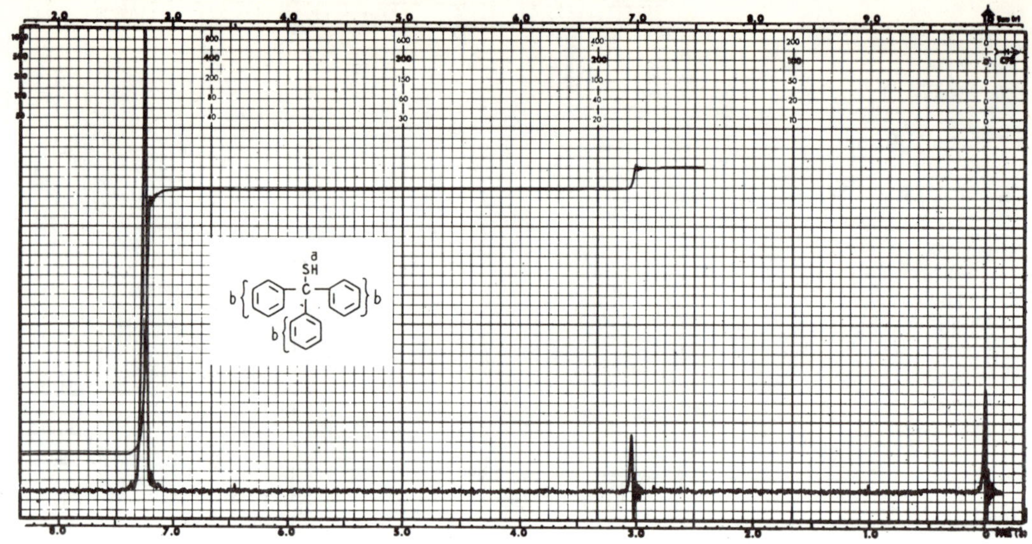

© 1978 Sadtler Research Laboratories, Inc.

4,5-DIMETHYL-o-XYLENE-α,α'-DITHIOL

$C_{10}H_{14}S_2$

Mol. Wt. 198.35

M.P. 67-69°C

Solvent: $CDCl_4$

Assignments
a = 1.81
b = 2.19
c = 3.75
d = 7.03

1227

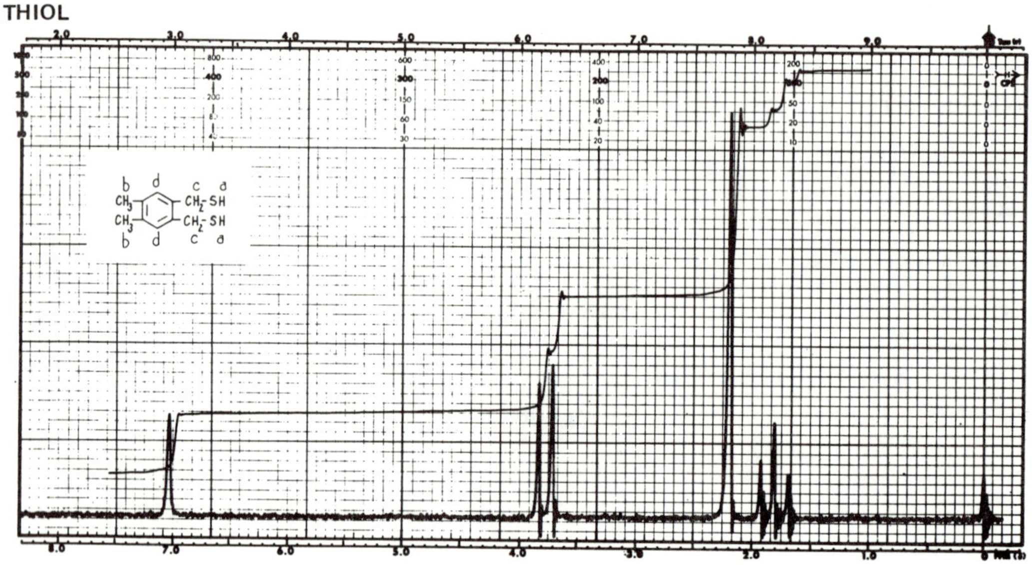

© 1978 Sadtler Research Laboratories, Inc.

o-TOLUENETHIOL

C_7H_8S

Mol. Wt. 124.21

Solvent: $CDCl_3$

Assignments
a = 2.32
b = 3.26
c = 6.92-7.40

1228

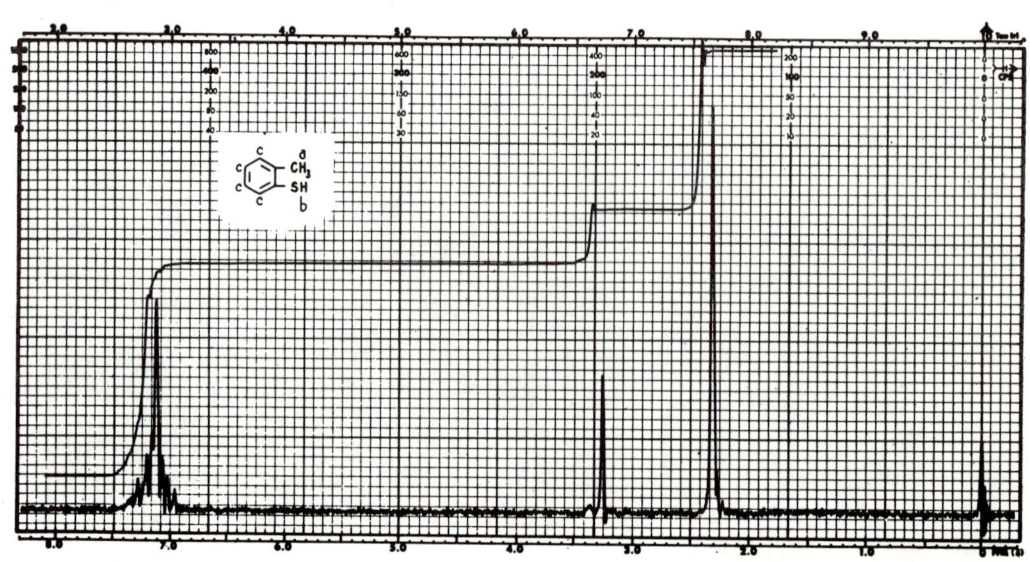

© 1978 Sadtler Research Laboratories, Inc.

m-TOLUENETHIOL

C_7H_8S

Mol. Wt. 124.21

Solvent: $CDCl_3$

Assignments
a = 2.30
b = 3.39
c = 6.81-7.33

1229

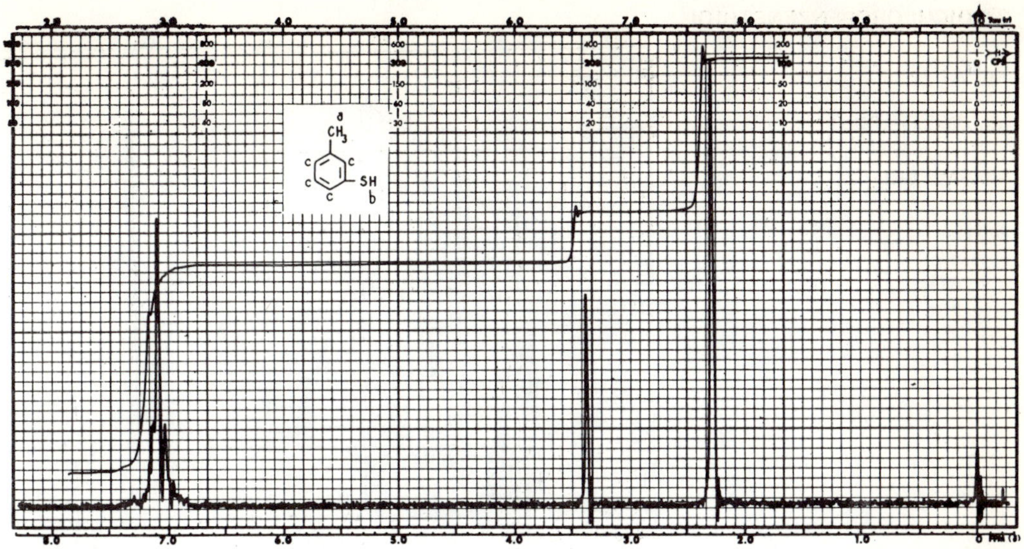

© 1978 Sadtler Research Laboratories, Inc.

---

p-TOLUENETHIOL

C_7H_8S

Mol. Wt. 124.20

M.P. 41-43°C

Solvent: $CDCl_3$

Assignments
a = 2.20
b = 3.30
c = 6.91
d = 7.09

1230

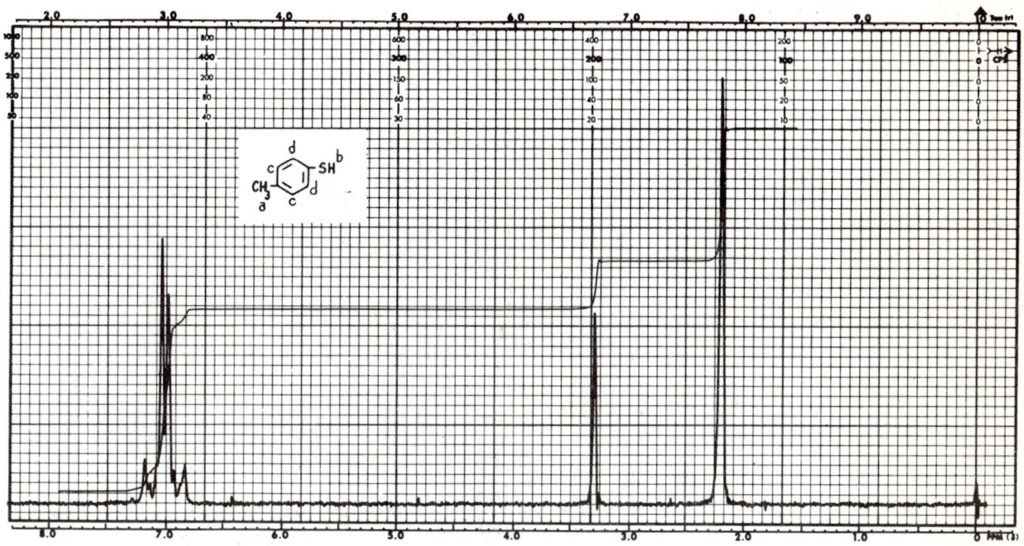

© 1978 Sadtler Research Laboratories, Inc.

---

p-tert-BUTYLBENZENETHIOL

$C_{10}H_{14}S$

Mol. Wt. 166.29

Solvent: CCl_4

Assignments
a = 1.28
b = 3.16
c = 7.13

1231

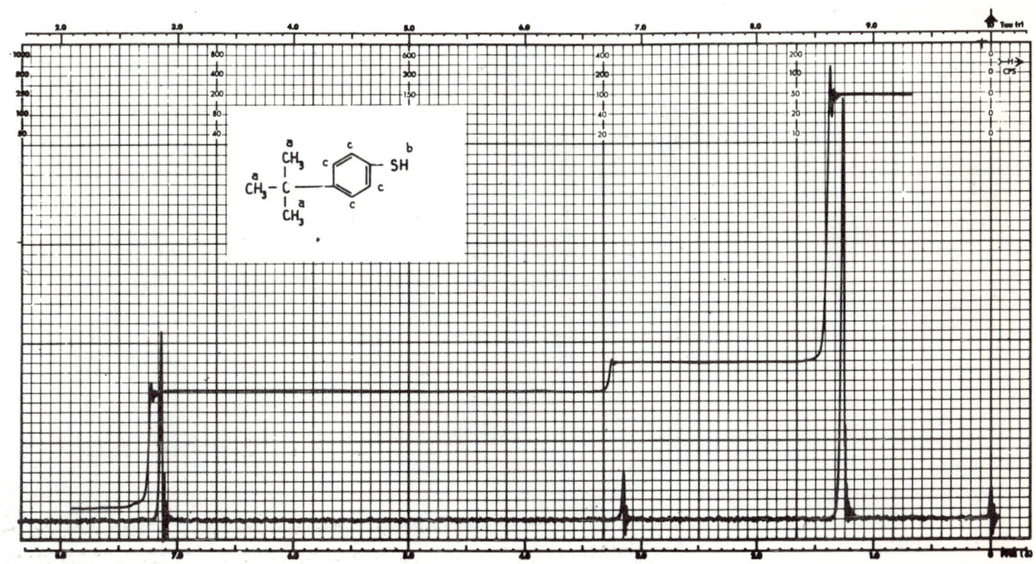

© 1978 Sadtler Research Laboratories, Inc.

2,5-DICHLOROBENZENETHIOL

$C_6H_4Cl_2S$

Mol. Wt. 179.07

Solvent: CCl_4

Assignments
a = 3.84
b = 6.99
c = 7.20
d = 7.28

1232

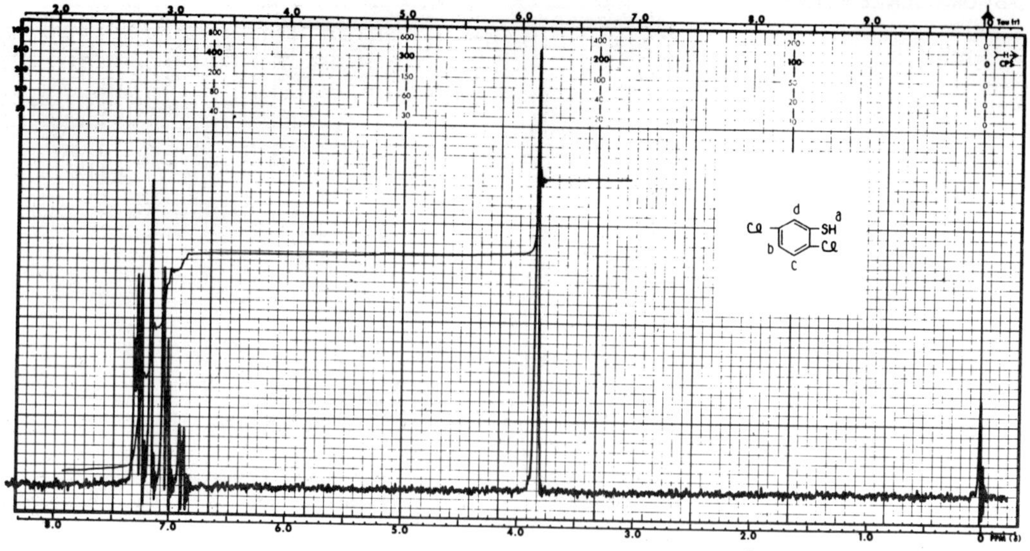

© 1978 Sadtler Research Laboratories, Inc.

3,4-DICHLOROBENZENETHIOL

$C_6H_4Cl_2S$

Mol. Wt. 179.07

Solvent: CCl_4

Assignments
a = 3.38
b = 7.00
c = 7.24
d = 7.29

1233

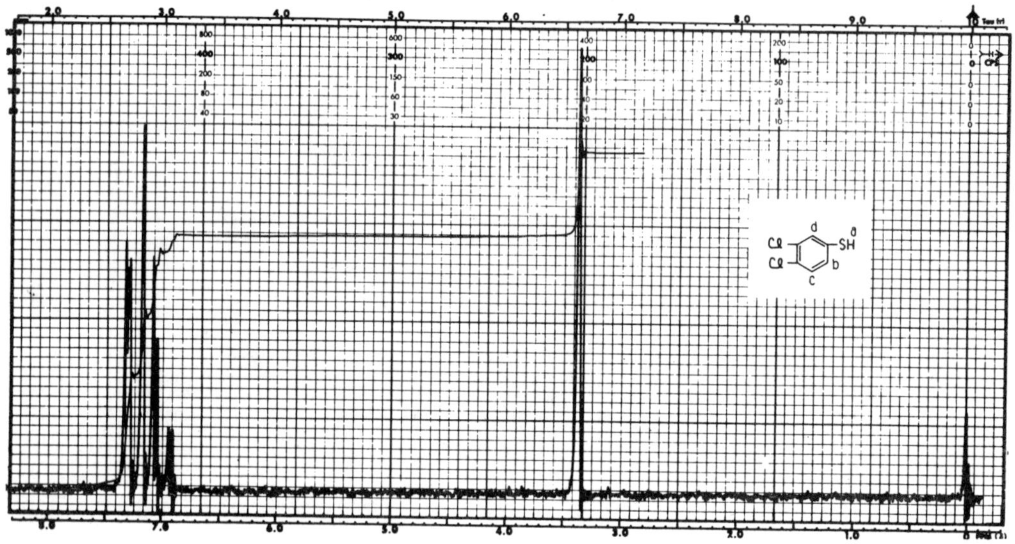

© 1978 Sadtler Research Laboratories, Inc.

2,4,5-TRICHLOROBENZENETHIOL

$C_6H_3Cl_3S$

Mol. Wt. 213.51

Solvent: $CDCl_3$

Assignments
a = 3.85
b = 7.35
c = 7.39
d = 7.60 imp.

1234

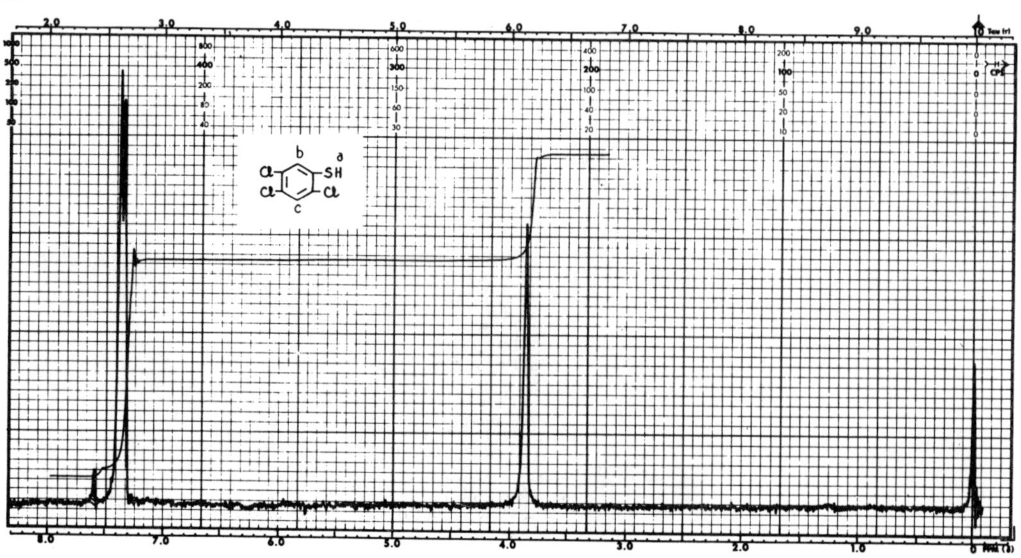

© 1978 Sadtler Research Laboratories, Inc.

p-BROMOBENZENETHIOL

C_6H_5BrS

Mol. Wt. 189.08

M.P. 74-76°C

B.P. 239°C

Solvent: $CDCl_3$

Assignments
a = 3.42
b = 7.07
c = 7.30

1235

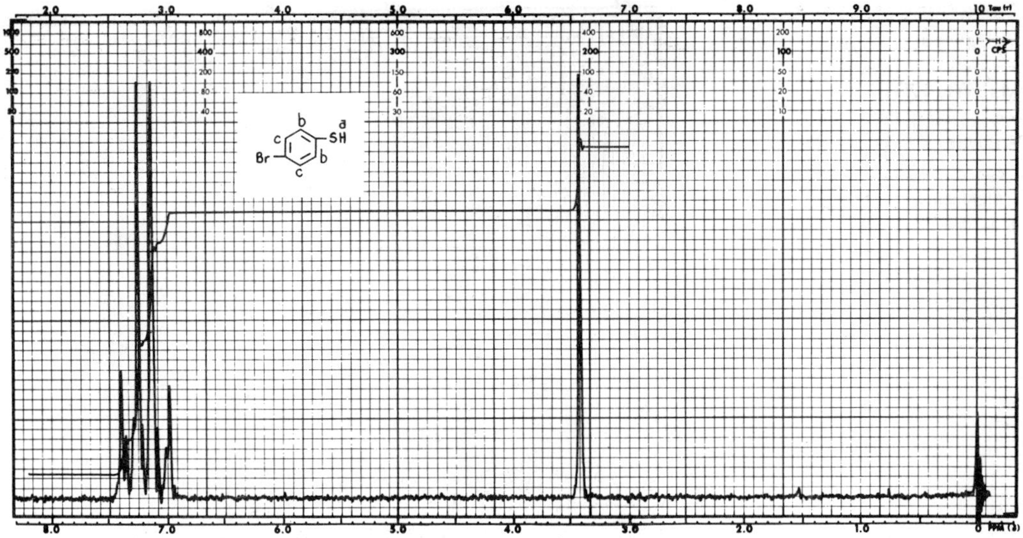

© 1978 Sadtler Research Laboratories, Inc.

---

o-AMINOBENZENETHIOL

C_6H_7NS

Mol. Wt. 125.19

B.P. 227.2°C (lit.)

Solvent: $CDCl_3$

Assignments
a = 3.68
b = 6.59
c = 6.64
d = 7.09
e = 7.32

1236

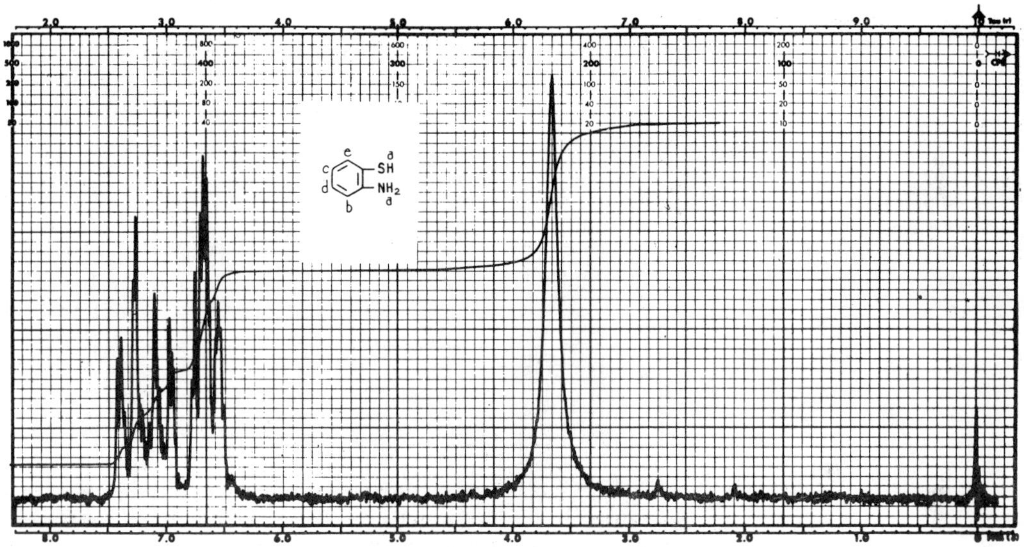

© 1978 Sadtler Research Laboratories, Inc.

---

m-AMINOBENZENETHIOL

C_6H_7NS

Mol. Wt. 125.19

Solvent: $CDCl_3$

Assignments
a = 3.50
b = 6.38
c = 6.45
d = 6.57
e = 6.97

1237

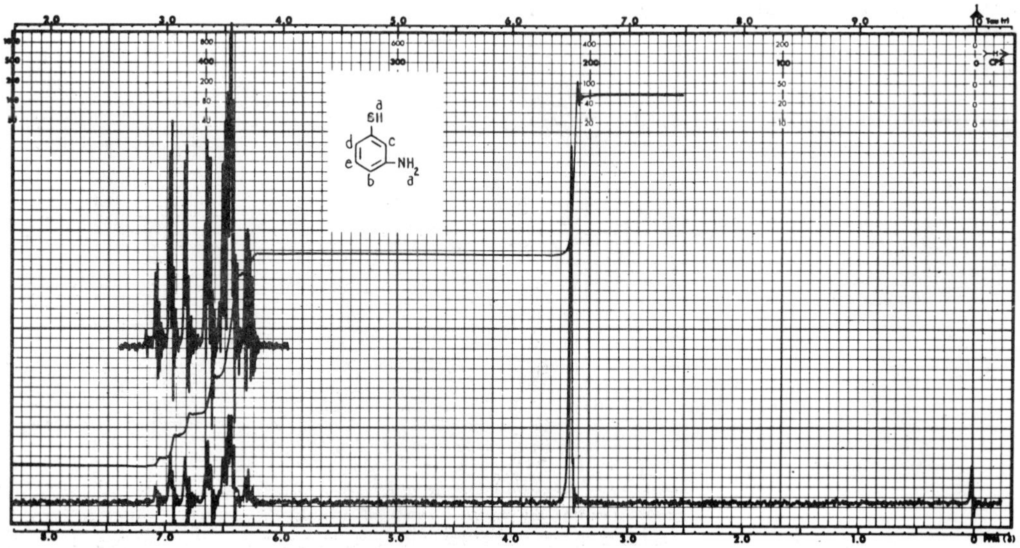

© 1978 Sadtler Research Laboratories, Inc.

4-AMINOBENZENETHIOL

C_6H_7NS

Mol. Wt. 125.19

Solvent: $CDCl_3$

Assignments
a = 3.52
b = 6.52
c = 7.14

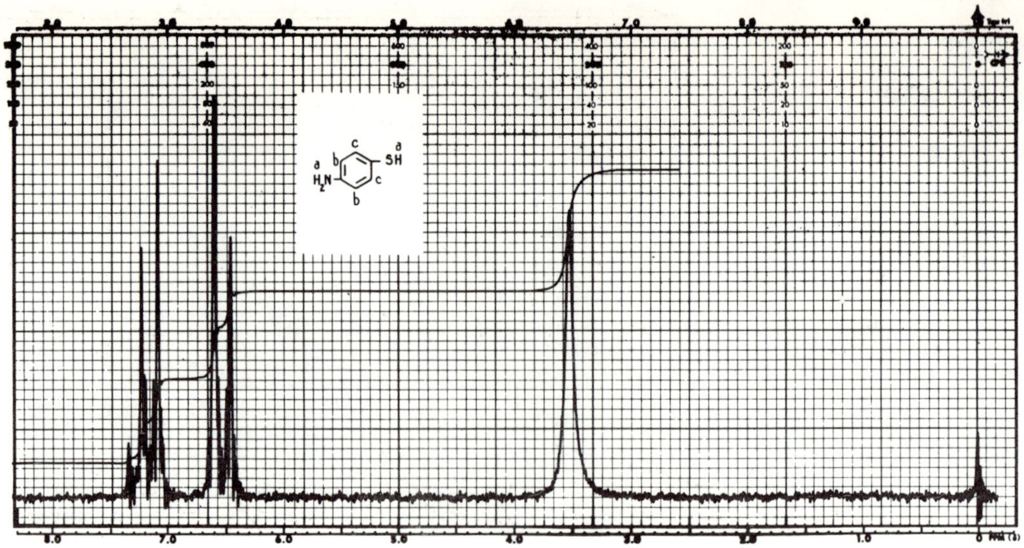

1238

© 1978 Sadtler Research Laboratories, Inc.

3,4-DITHIOLTOLUENE

$C_7H_8S_2$

Mol. Wt. 156.27

M.P. 28-30°C

Solvent: $CDCl_3$

Assignments
a = 2.17
b = 3.52
c = 3.69
d = 6.72
e = 7.05
f = 7.10

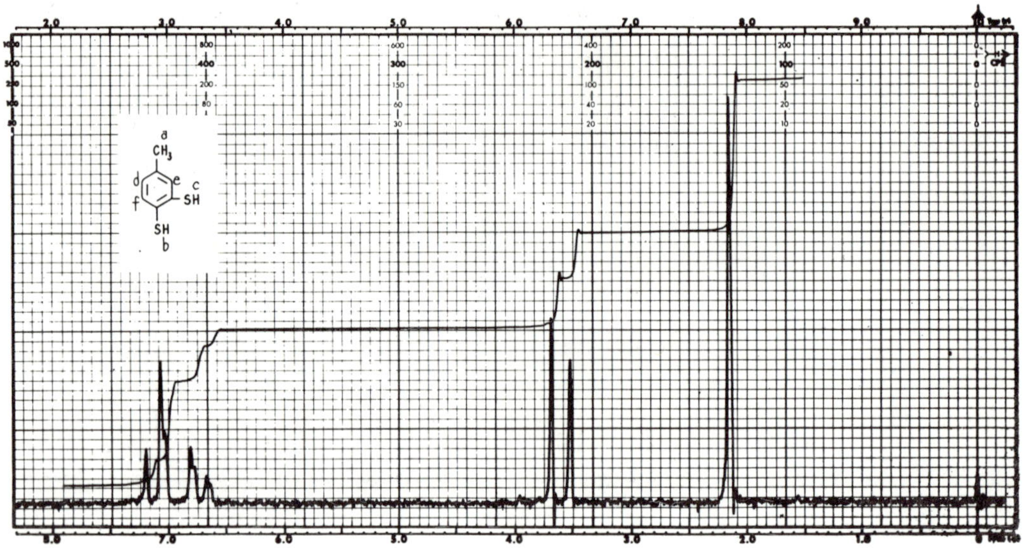

1239

© 1978 Sadtler Research Laboratories, Inc.

4-CHLORO-1,3-BENZENEDITHIOL

$C_6H_5ClS_2$

Mol. Wt. 176.69

Solvent: CCl_4

Assignments
a = 3.31 or b
b = 3.79 or a
c = 6.87
d = 7.11
e = 7.17

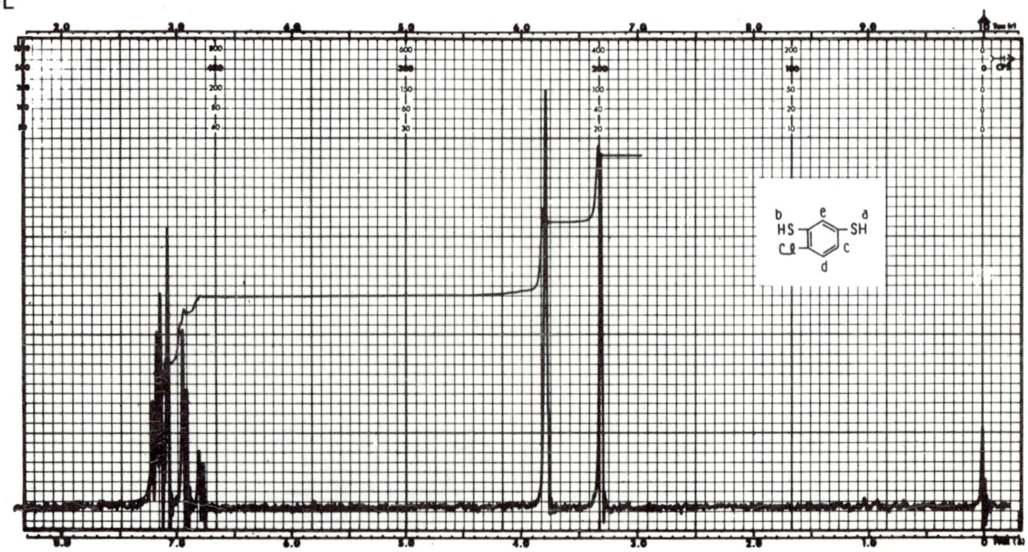

1240

© 1978 Sadtler Research Laboratories, Inc.

1-NAPHTHALENETHIOL

$C_{10}H_8S$

Mol. Wt. 160.24

B.P. 130-131°C/5mm

Solvent: $CDCl_3$

Assignments
a = 3.35
b = 7.02
c = 7.19-7.50
d = 7.57
e = 8.04

1241

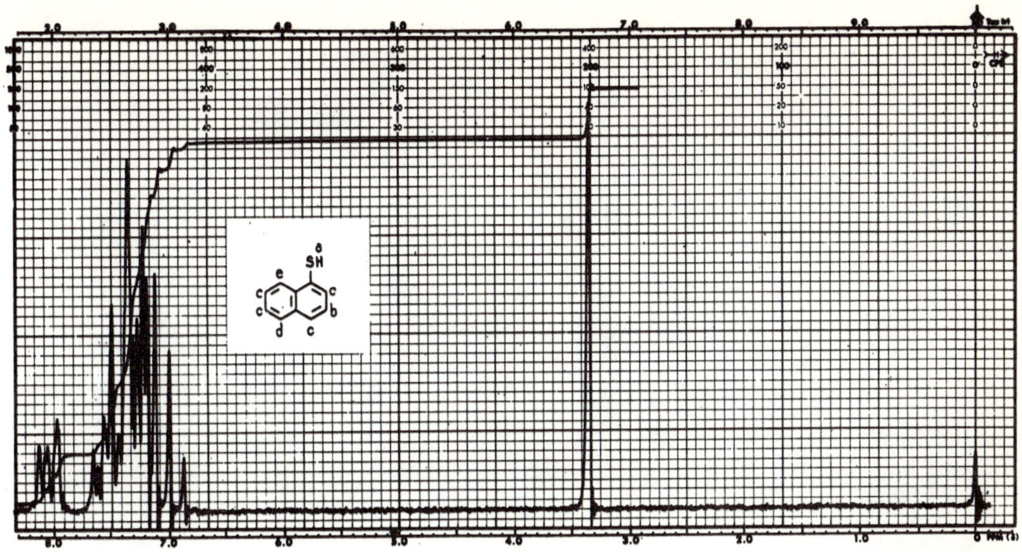

© 1978 Sadtler Research Laboratories, Inc.

2-NAPHTHALENETHIOL

$C_{10}H_8S$

Mol. Wt. 160.24

M.P. 81-83°C

Solvent: $CDCl_3$

Assignments
a = 3.45
b = 7.15
c = 7.15-7.75

1242

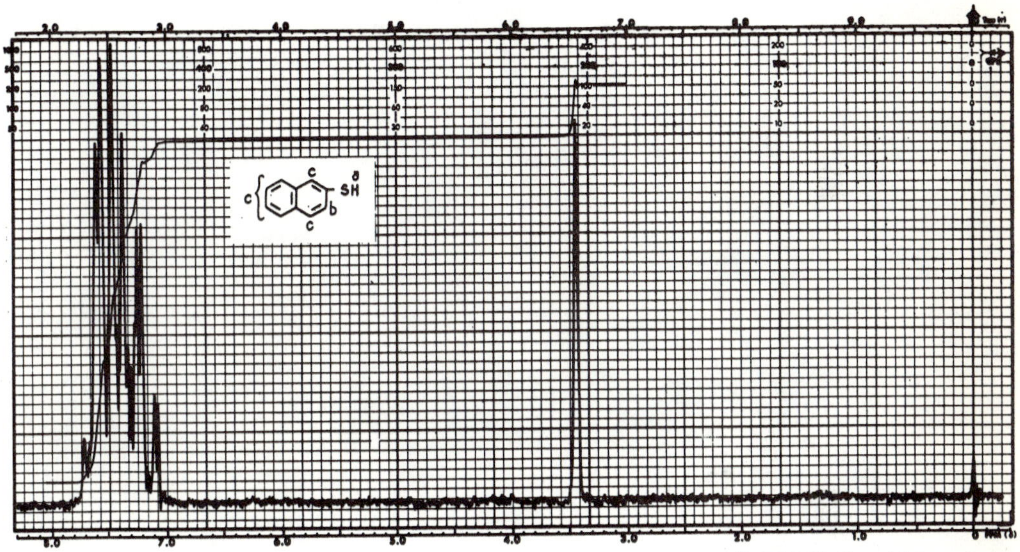

© 1978 Sadtler Research Laboratories, Inc.

NOTES

THE SULFOXIDES

The Sulfoxide group imparts no special features to its NMR spectrum. It is an intermediate to strong deshielder of adjacent aliphatic groups and a weak to intermediate deshielder of the ortho aromatic hydrogens.

Dimethyl Sulfoxide is well known as a solvent, with its deuterated form a commonly used material in NMR. All of the unsubstituted sulfoxides are readily soluble in the chlorinated solvents such as CCl_4 and $CDCl_3$.

Chemical Shifts

Aliphatic Protons

HC–4	HC–3	HC–2	HC–1		Solvent	Spectrum
			2.52	CH_3–S(=O)–CH_3	CCl_4	1243
		1.28	2.56	CH_3–CH_2–S(=O)–R_4	CCl_4	1244
	1.02	1.66	2.67	CH_3–CH_2–CH_2–S(=O)–Ø	CCl_4	1248
0.99	(1.1–2.0)		2.51	$CH_3(CH_2)_2$–CH_2–S(=O)–R_2	CCl_4	1244
0.96	(1.1–2.1)		2.68	$CH_3(CH_2)_2$–CH_2–S(=O)–R_4	$CDCl_3$	1245
		(1.22)	2.67	$(CH_3)_2$–CH–S(=O)–R_4	CCl_4	- - - -

Aromatic Protons

Phenyl Sulfoxides

	ortho/meta/para	Solvent	Spectrum
R_3–S(=O)–	7.30–7.70	CCl_4	1248
Ø–S(=O)–	7.25–7.80	$CDCl_3$	1251

Para Substituted Phenylsulfoxides

	ortho	meta	para	Solvent
Ø–S(=O)–	7.56	6.93	–O–CH_3	$CDCl_3$
CH_3–S(=O)–	7.86	7.36	–I	$CDCl_3$
R–S(=O)–	7.61	7.54	–Cl	$CDCl_3$
CH_3–S(=O)–	7.86	8.42	–NO_2	$CDCl_3$

METHYL SULFOXIDE

C_2H_6OS

Mol. Wt. 78.13

M.P. 17-19°C

B.P. 66-69°C/10mm

Solvent: CCl_4

Assignments
a = 2.52

1243

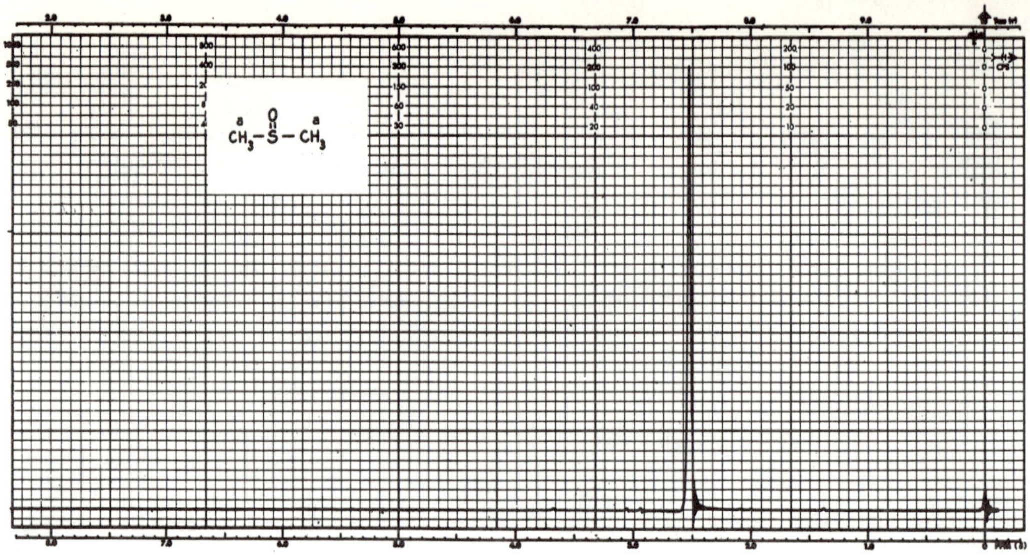

© 1978 Sadtler Research Laboratories, Inc.

BUTYL ETHYL SULFOXIDE

$C_6H_{14}OS$

Mol. Wt. 134.24

B.P. 63°C/0.6mm

Solvent: CCl_4

Assignments
a = 0.99
b = 1.28
c = ca 1.63
d = ca 2.56
e = 2.17 imp.

1244

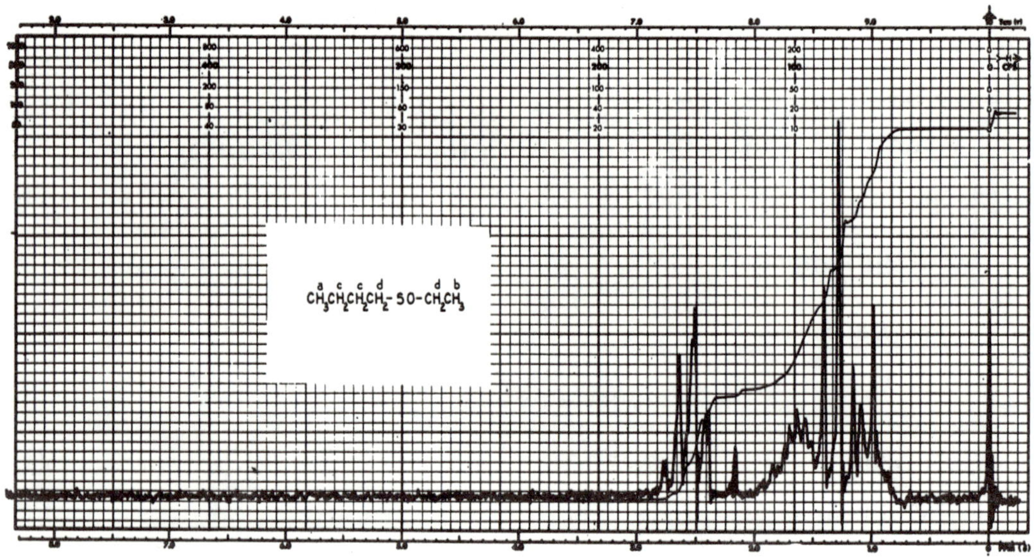

© 1978 Sadtler Research Laboratories, Inc.

BUTYL SULFOXIDE

$C_8H_{18}OS$

Mol. Wt. 162.30

Solvent: $CDCl_3$

Assignments
a = 0.96
b = 1.11-2.08
c = 2.68

1245

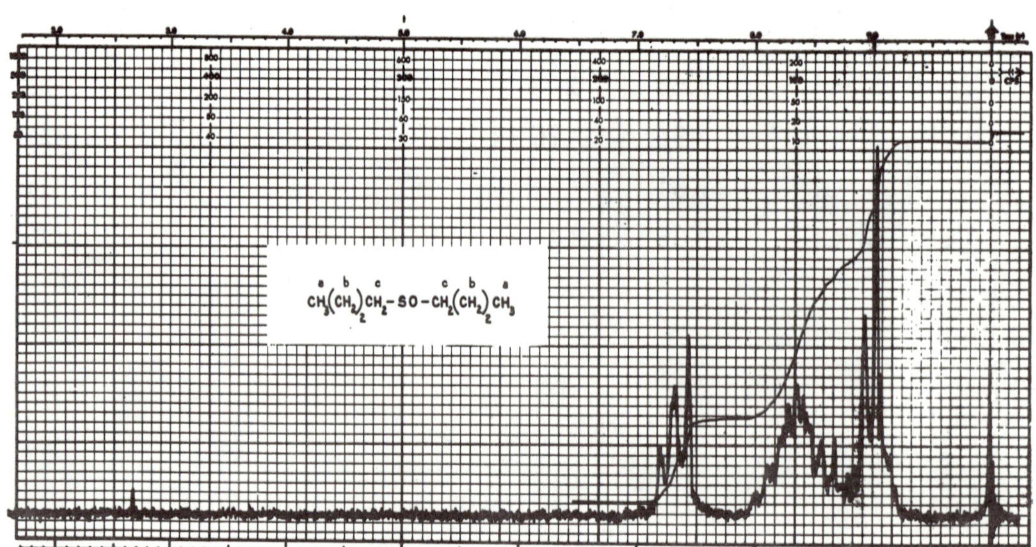

© 1978 Sadtler Research Laboratories, Inc.

ISOBUTYL SULFOXIDE

$C_8H_{18}OS$

Mol. Wt. 162.30

Solvent: CCl_4

Assignments
a = 1.08
b = 1.82-2.80

1246

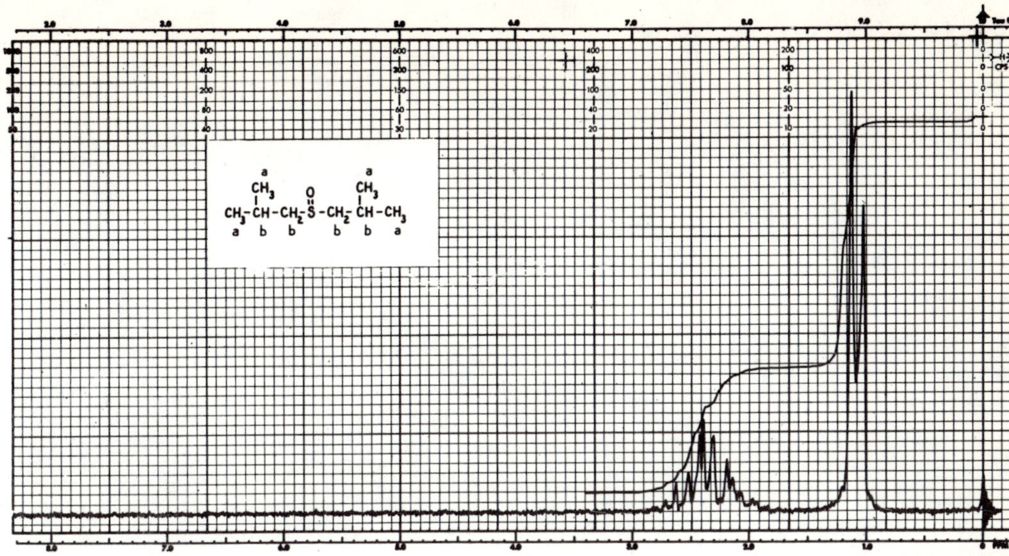

© 1978 Sadtler Research Laboratories, Inc.

DIBENZYL SULFOXIDE

$C_{14}H_{14}OS$

Mol. Wt. 230.33

M.P. 135-137°C

Solvent: $CDCl_3$

Assignments
a = 3.85
b = 7.29

1247

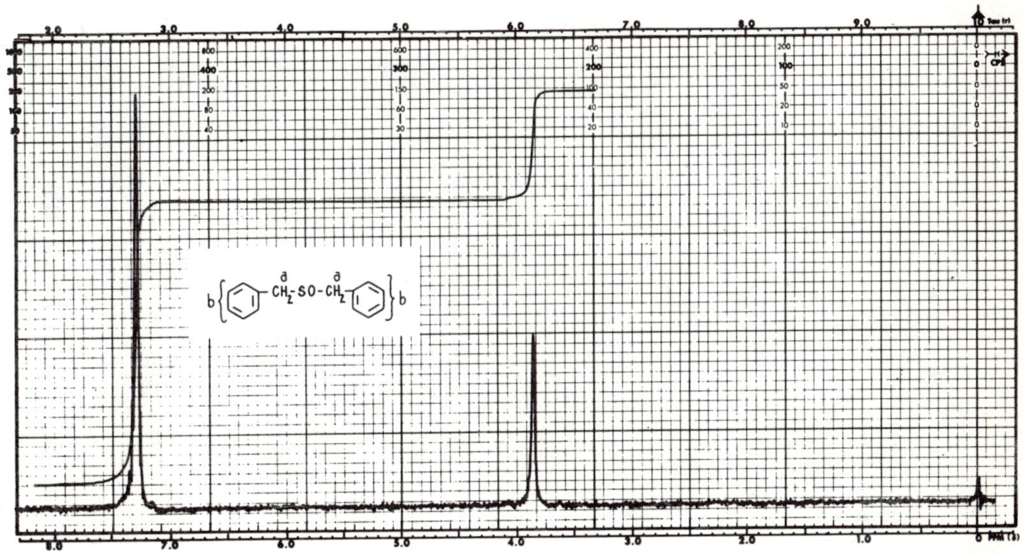

© 1978 Sadtler Research Laboratories, Inc.

PHENYL PROPYL SULFOXIDE

$C_9H_{12}OS$

Mol. Wt. 168.26

B.P. 80-83°C/0.02mm

Solvent: CCl_4

Assignments
a = 1.02
b = 1.66
c = 2.67
d = 7.30-7.70

1248

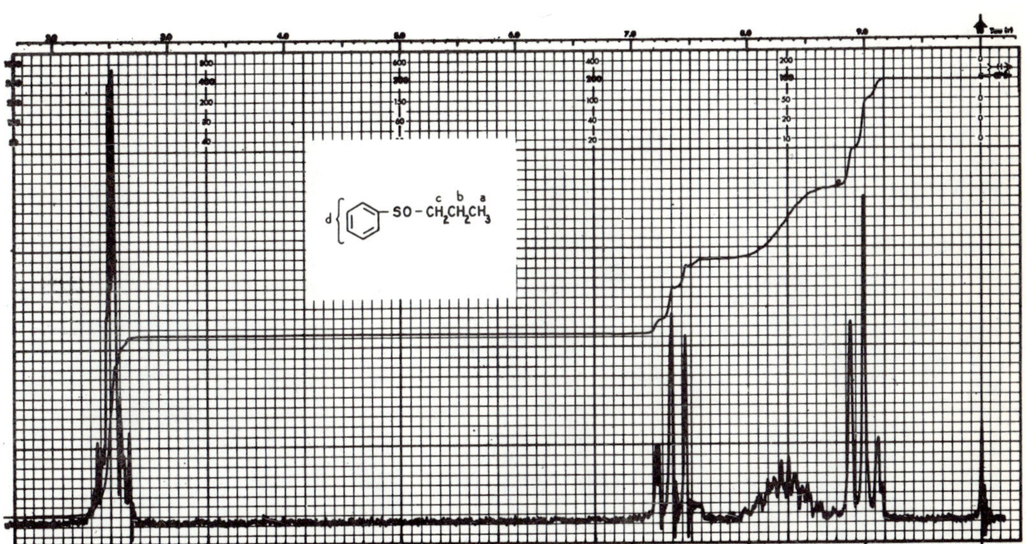

© 1978 Sadtler Research Laboratories, Inc.

p-IODOPHENYL METHYL SULFOXIDE

C_7H_7IOS

Mol. Wt. 266.10

M.P. 110-111°C

Solvent: $CDCl_3$

Assignments
a = 2.70
b = 7.36
c = 7.86

1249

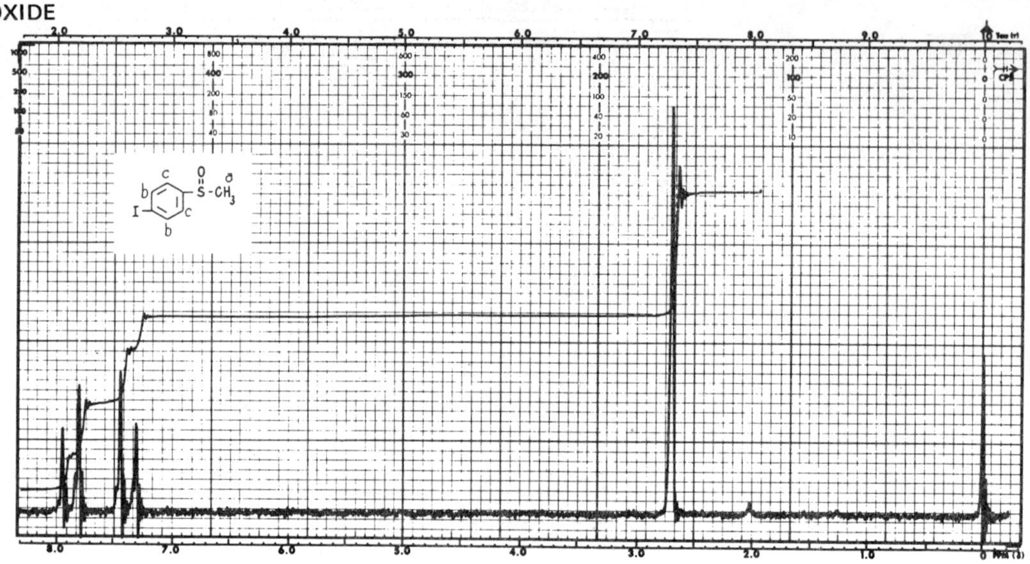

© 1978 Sadtler Research Laboratories, Inc.

p-CHLOROPHENYL 3-PENTYNYL SULFOXIDE

$C_{11}H_{11}ClOS$

Mol. Wt. 226.73

Solvent: $CDCl_3$

Assignments
a = 1.75
b = 2.55
c = 2.94
d = 7.53
e = 7.60

1250

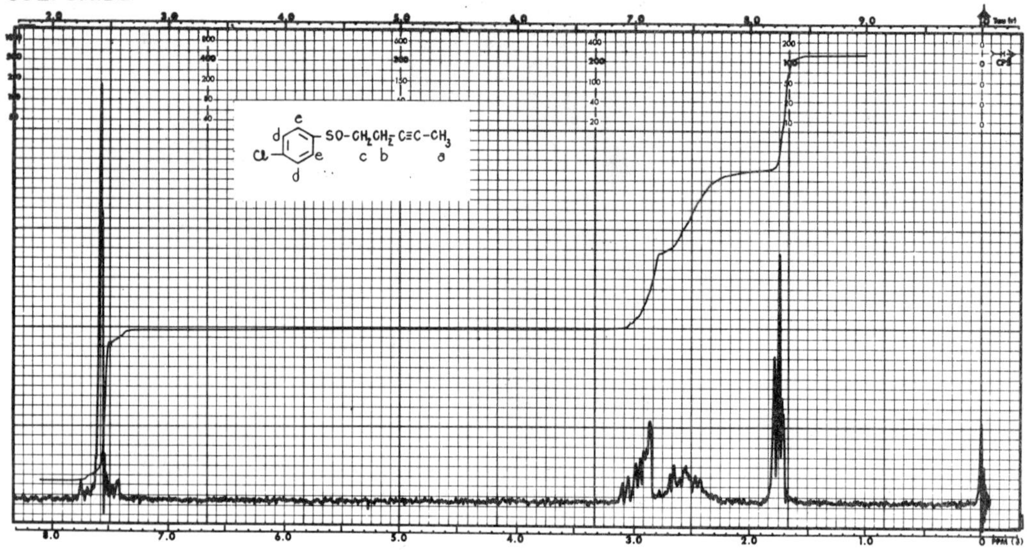

© 1978 Sadtler Research Laboratories, Inc.

PHENYL SULFOXIDE

$C_{12}H_{10}OS$

Mol. Wt. 202.28

M.P. 71-72°C

Solvent: $CDCl_3$

Assignments
a = 7.30-7.80
b = 1.99 imp.

1251

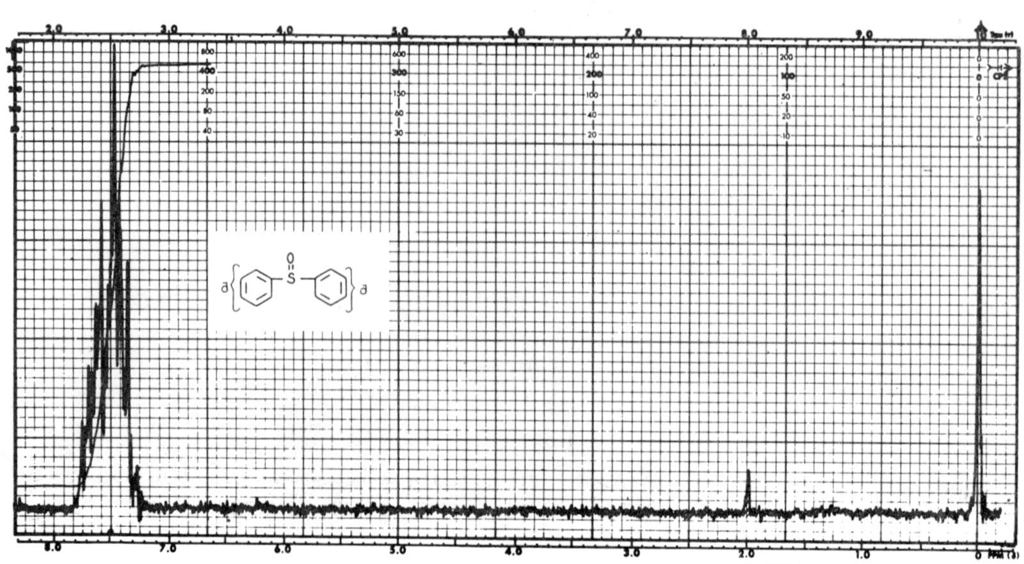

© 1978 Sadtler Research Laboratories, Inc.

THE SULFONES

In comparison to the Sulfoxides (—S(=O)—), the Sulfones (—S(=O)$_2$—) are a more strongly deshielding substituent in their effect on both the adjacent aliphatic groups and on the ortho aromatic protons. The relative deshielding effect of the Sulfur containing functional groups covered thus far are shown below.

1.90	CH_3—S—R
2.39	CH_3—S—S—R
2.52	CH_3—S(=O)—R
2.87	CH_3—S(=O)$_2$—R

The compounds containing the Sulfone group are somewhat less soluble in the chlorinated solvents than the sulfoxides but, are more soluble in solvents such as Acetone, DMSO-d_6, Polysol and D_2O.

Chemical Shifts

Aliphatic Protons

HC—4	HC—3	HC—2	HC—1		Solvent	Spectrum
			2.87	CH_3—S(=O)$_2$—CH_2—CH=CH—Ø	$CDCl_3$	----
			3.11	CH_3—S(=O)$_2$—CH_3	D_2O	1252
		1.26	2.97	CH_3—CH_2—S(=O)$_2$—CH=CH_2	CCl_4	1256
	1.09	1.81	2.91	CH_3—CH_2—CH_2—S(=O)$_2$—R_3	CCl_4	1253
0.97	(1.2—2.1)		2.95	CH_3—$(CH_2)_2$—CH_2—S(=O)$_2$—R_4	$CDCl_3$	1254

Substituted Methanesulfones

Ø—S(=O)$_2$ —CH_2	—X	Solvent	Spectrum
3.80	—CH=CH_2	$CDCl_3$	1261
4.30	—C≡C—R	Polysol	----
4.35	—C≡C—H	Polysol	1265
4.44	—Ø	Polysol	1267
4.60	—C(=O)—R	Polysol	----
4.88	—S(=O)$_2$—Ø	Polysol	1264

Olefinic Protons

The Sulfone group deshields all three vinyl protons, producing a complex, higher order pattern in the chemical shift range from 6.0—7.0 ppm (spectrum 1256). Analysis of these patterns suggests that the proton trans to the SO_2 group resonates at highest field (about 6.2 ppm), that the proton cis to the SO_2 group resonates at slightly lower field (about 6.4 ppm), and that the geminal proton resonates at lowest field (about 6.8 ppm).

Chemical Shifts continued

Aromatic Protons

The Sulfone substituted phenyl groups produce an aromatic pattern characteristic of a substituent which strongly deshields the ortho protons. The para and meta hydrogens overlap to produce a complex multiplet in the range from about 7.2–7.8 ppm while the ortho hydrogens appear as a higher order doublet of doublets at about 7.9 ppm (spectrum 1261).

Phenylsulfones

para/meta	ortho	–X	Solvent	Spectrum
7.2–7.7	7.89	$-S(=O)_2-CH_2-CH=CH_2$	$CDCl_3$	1261
7.2–7.8	8.02	$-S(=O)_2-CH(R_3)-C\equiv CH$	$CDCl_3$	- - - -

Para Substituted Phenylsulfones

$R-S(=O)_2-$	ortho	meta	para	
	7.85	7.11	$-O-CH_3$	Polysol
	7.78	7.34	$-CH_3$	$CDCl_3$
	7.81	7.50	–Cl	$CDCl_3$

Ø–$S(=O)_2-$	ortho	meta	para	
	7.79	6.91	$-O-CH_3$	$CDCl_3$
	7.95	7.18	–F	$CDCl_3$
	7.80	7.23	$-CH_3$	$CDCl_3$
	7.82	7.42	–Cl	$CDCl_3$
	7.82	7.65	–Br	$CDCl_3$
	8.08	8.30	$-NO_2$	$CDCl_3$

METHYL SULFONE

$C_2H_6O_2S$

Mol. Wt. 94.13

M.P. 109°C

B.P. 283°C/760mm

Solvent: D_2O

Assignments
a = 3.11
b = 4.53 HDO

1252

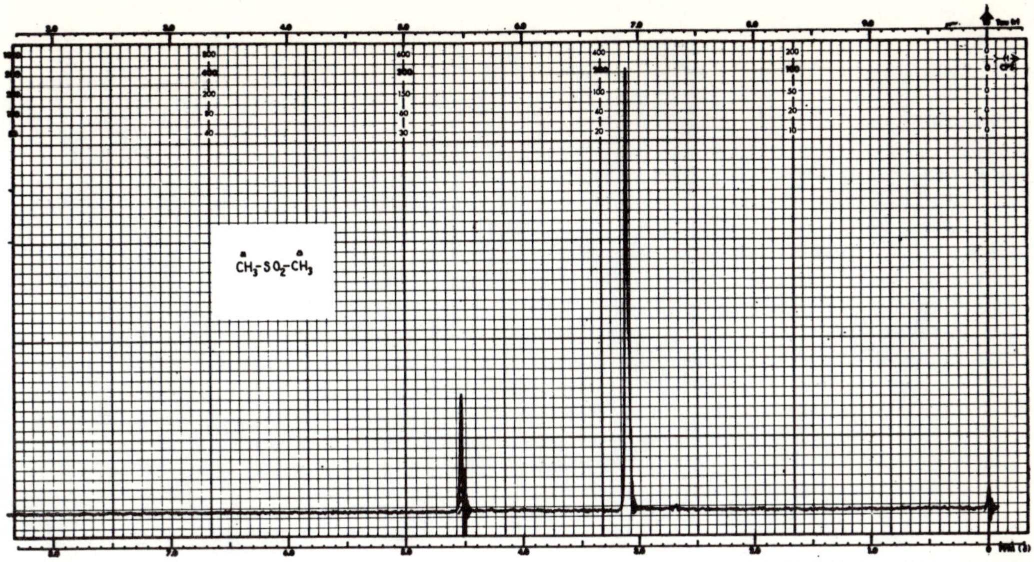

© 1978 Sadtler Research Laboratories, Inc.

PROPYL SULFONE

$C_6H_{14}O_2S$

Mol. Wt. 150.24

Solvent: CCl_4

Assignments
a = 1.09
b = 1.81
c = 2.91

1253

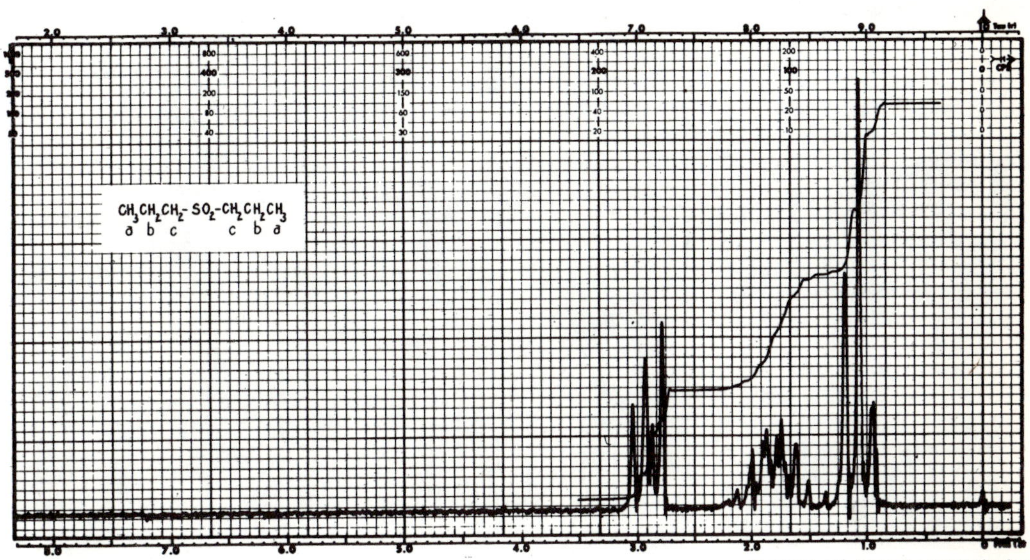

© 1978 Sadtler Research Laboratories, Inc.

BUTYL SULFONE

$C_8H_{18}O_2S$

Mol. Wt. 178.29

M.P. 44-46°C

Solvent: $CDCl_3$

Assignments
a = 0.97
b = 1.15-2.15
c = 2.95

1254

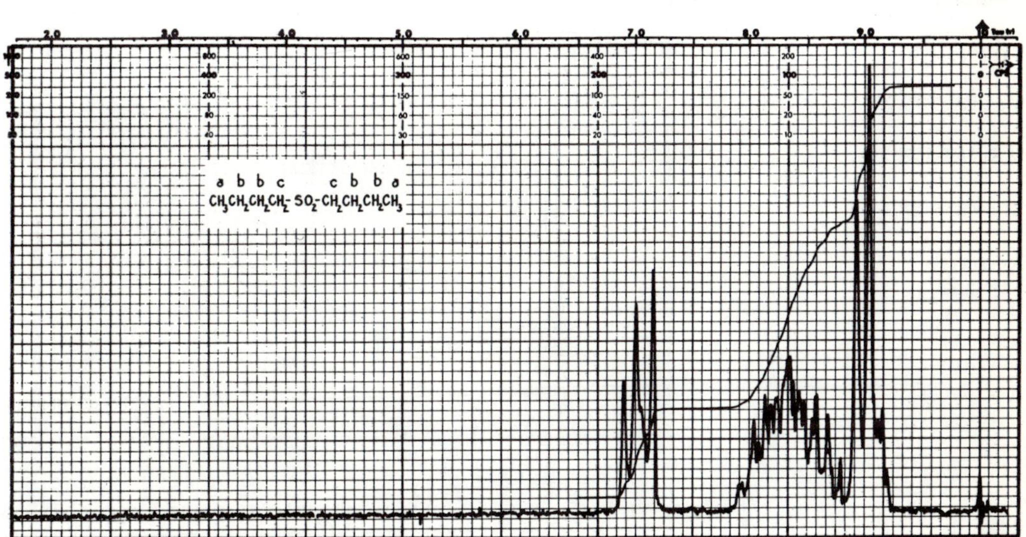

© 1978 Sadtler Research Laboratories, Inc.

PENTYL SULFONE

$C_{10}H_{22}O_2S$

Mol. Wt. 206.35

Solvent: CCl_4

Assignments
a = 0.98
b = 1.12-2.15
c = 2.84

1255

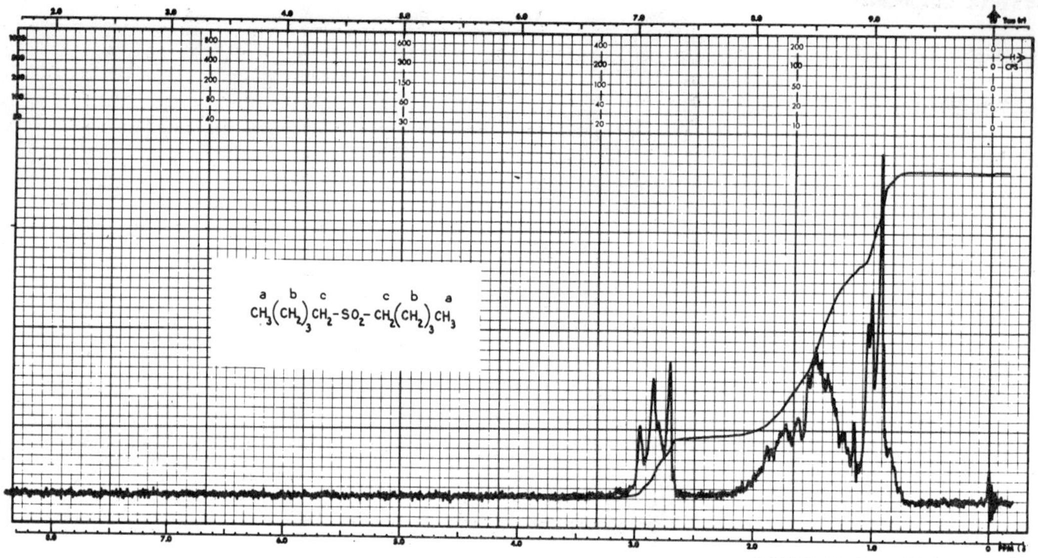

© 1978 Sadtler Research Laboratories, Inc.

ETHYL VINYL SULFONE

$C_4H_8O_2S$

Mol. Wt. 120.17

Solvent: CCl_4

Assignments
a = 1.26
b = 2.97
c = 6.00-7.00

1256

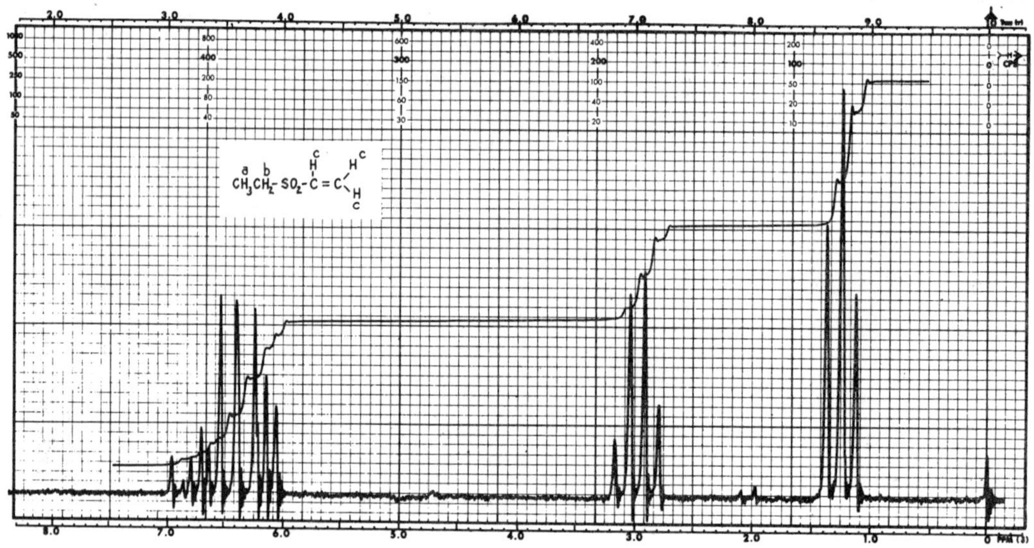

© 1978 Sadtler Research Laboratories, Inc.

trans-CINNAMYL METHYL SULFONE

$C_{10}H_{12}O_2S$

Mol. Wt. 196.27

M.P. 125-126°C

Solvent: $CDCl_3$

Assignments
a = 2.87
b = 3.86
c = 6.27
d = 6.72
e = ca 7.36
f = 1.72 imp.

1257

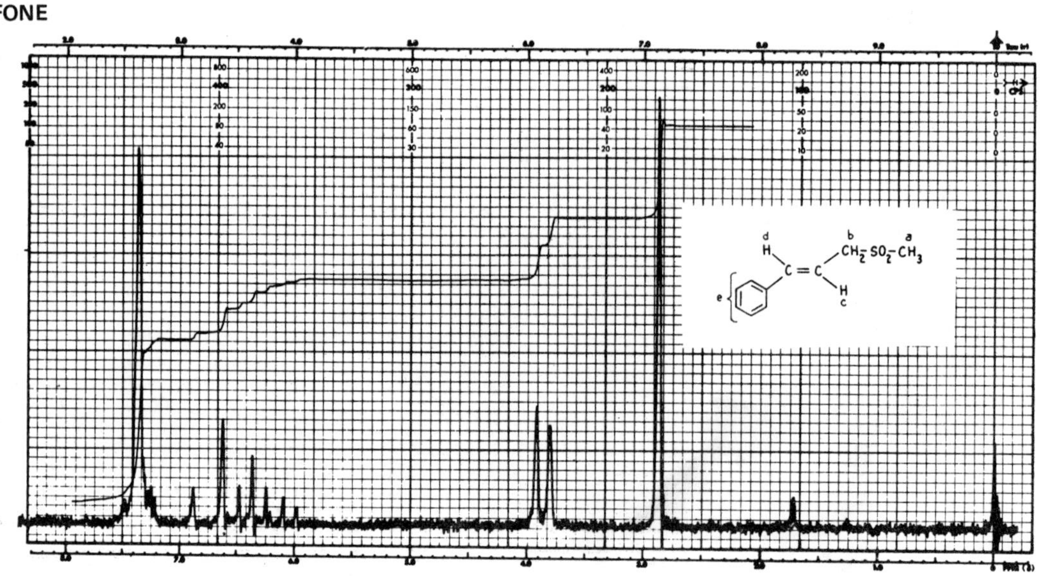

© 1978 Sadtler Research Laboratories, Inc.

STYRYL SULFONE

$C_{16}H_{14}O_2S$

Mol. Wt. 270.35

M.P. 99-100°C

Solvent: $CDCl_3$

Assignments
a = 6.82
b = ca 7.40
c = 7.60

1258

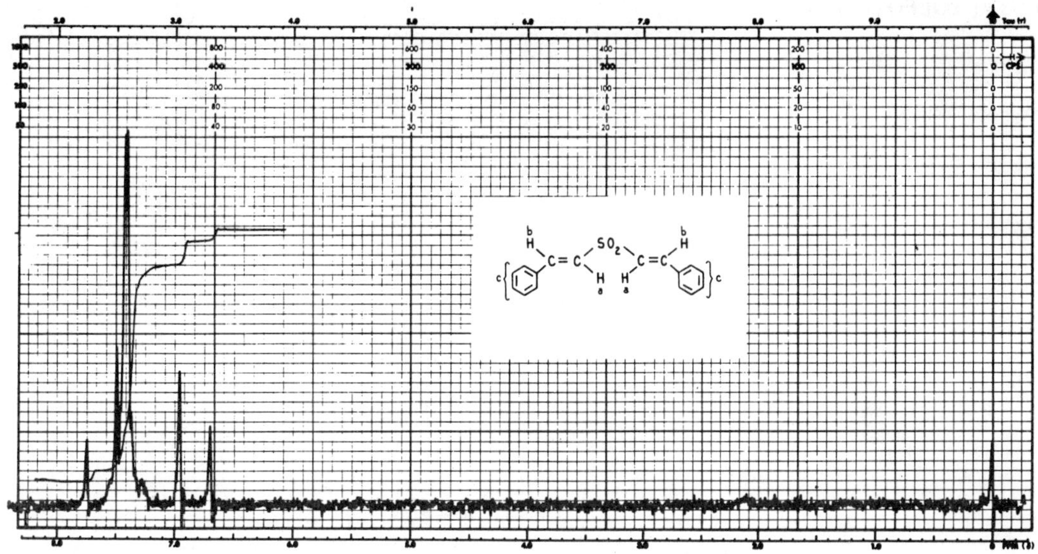

© 1978 Sadtler Research Laboratories, Inc.

2,5-DIHYDRO-2,4-DIMETHYLTHIOPHENE, 1,1-DIOXIDE

$C_6H_{10}O_2S$

Mol. Wt. 146.21

M.P. 40-43°C

Solvent: $CDCl_3$

Assignments
a = 1.39
b = 1.88
c = 3.64
d = 3.80
e = 5.63

1259

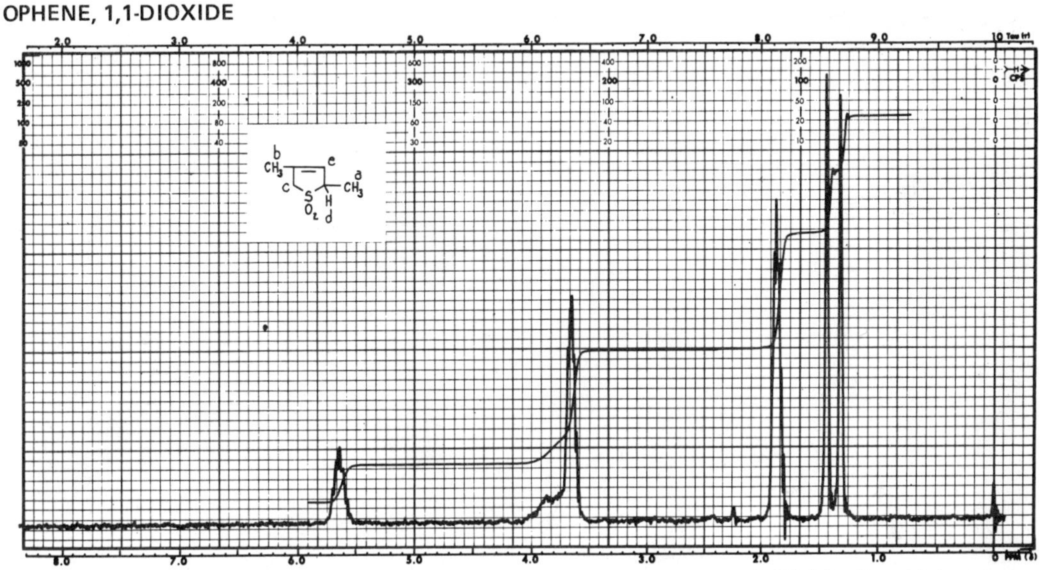

© 1978 Sadtler Research Laboratories, Inc.

3,4-DICHLOROTHIOPHENE, 1,1-DIOXIDE

$C_4H_2Cl_2O_2S$

Mol. Wt. 185.03

M.P. 112-113°C

Solvent: $CDCl_3$

Assignments
a = 6.75

1260

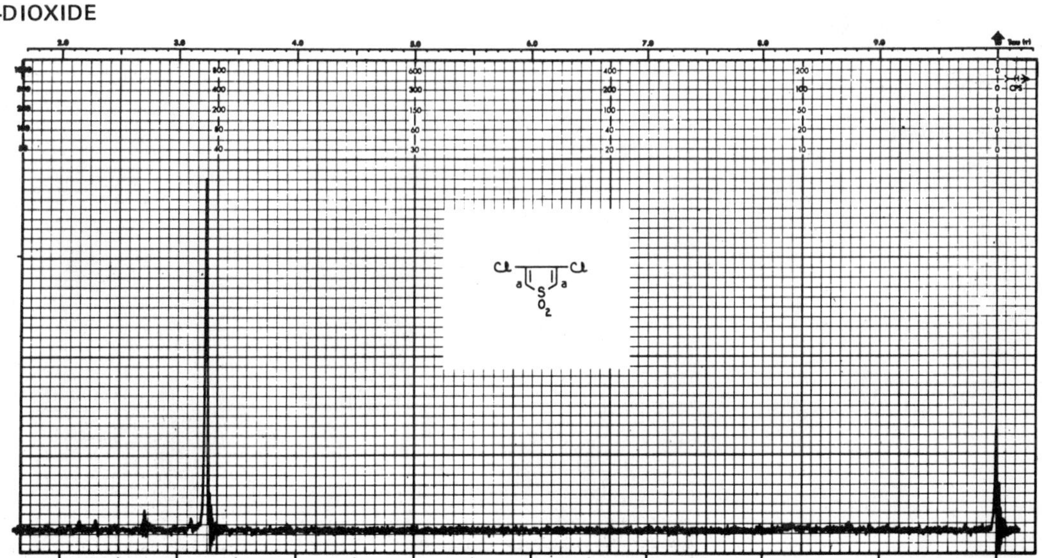

© 1978 Sadtler Research Laboratories, Inc.

ALLYL PHENYL SULFONE

$C_9H_{10}O_2S$

Mol. Wt. 182.24

Solvent: $CDCl_3$

Assignments

a = 3.80
b = 5.11
c = 5.30
d = 5.77
e = 7.20-7.75
f = 7.89
g = 0.99, 1.83 imps.

1261

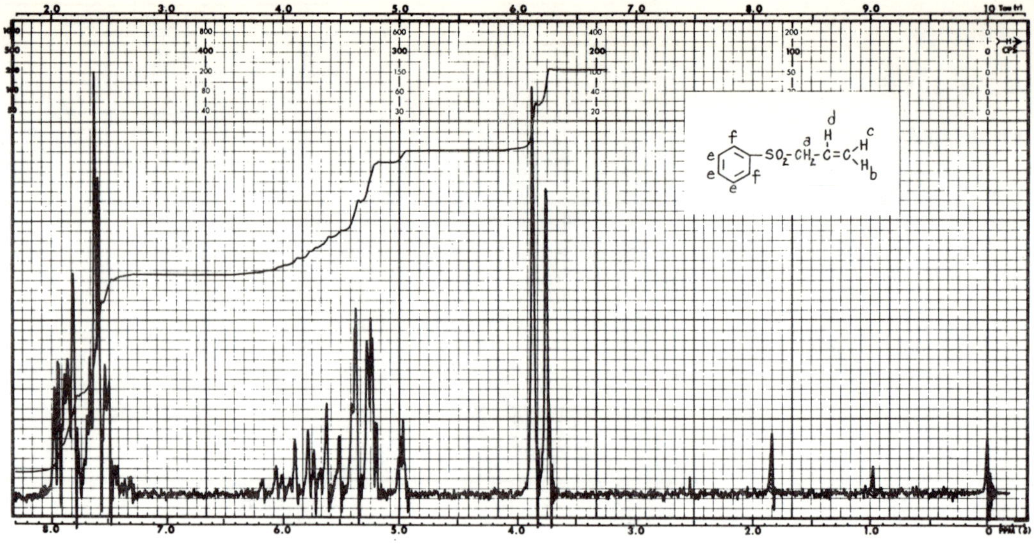

© 1978 Sadtler Research Laboratories, Inc.

ALLYL p-TOLYL SULFONE

$C_{10}H_{12}O_2S$

Mol. Wt. 196.27

Solvent: $CDCl_3$

Assignments

a = 2.40
b = 3.81
c = 5.12
d = 5.29
e = 5.79
f = 7.34
g = 7.78

1262

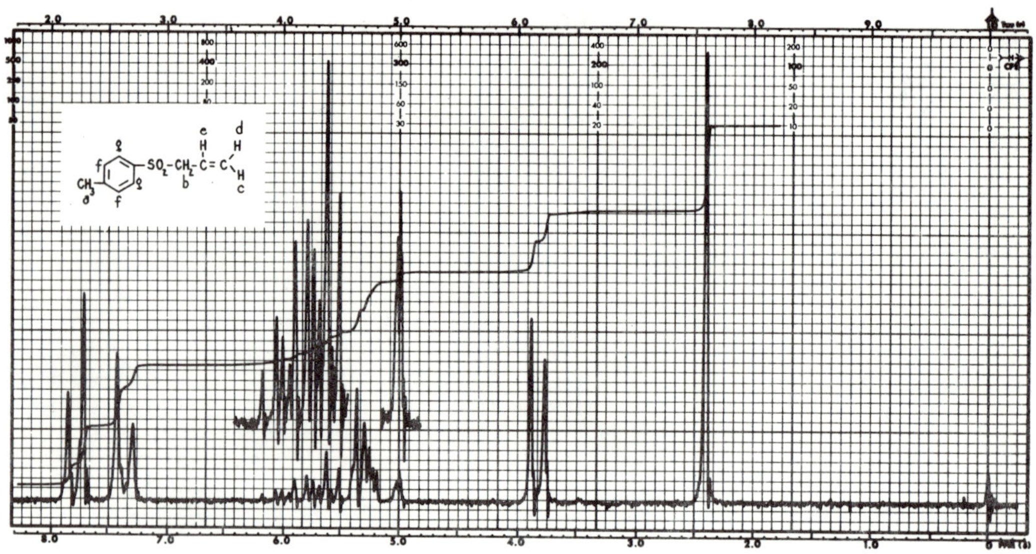

© 1978 Sadtler Research Laboratories, Inc.

ALLYL p-CHLOROPHENYL SULFONE

$C_9H_9ClO_2S$

Mol. Wt. 216.69

Solvent: $CDCl_3$

Assignments

a = 3.81
b = 5.11
c = 5.29
d = 5.76
e = 7.50
f = 7.81

1263

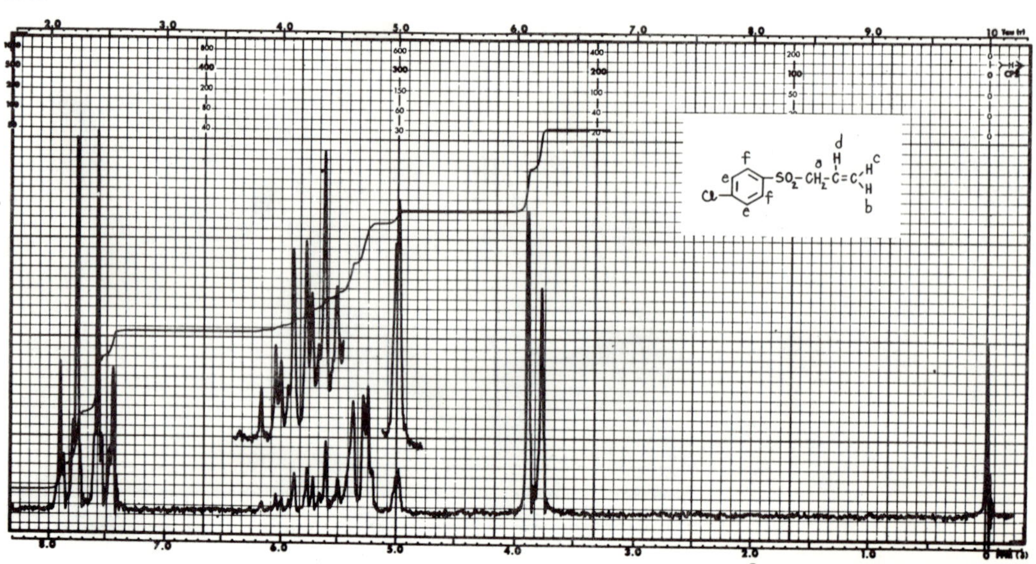

© 1978 Sadtler Research Laboratories, Inc.

BIS(TOLYLSULFONYL)METHANE

$C_{15}H_{16}O_4S_2$

Mol. Wt. 324.42

Solvent: Polysol-d

Assignments
a = 2.44
b = 4.88
c = 7.42
d = 7.81
e = 3.12 H_2O

1264

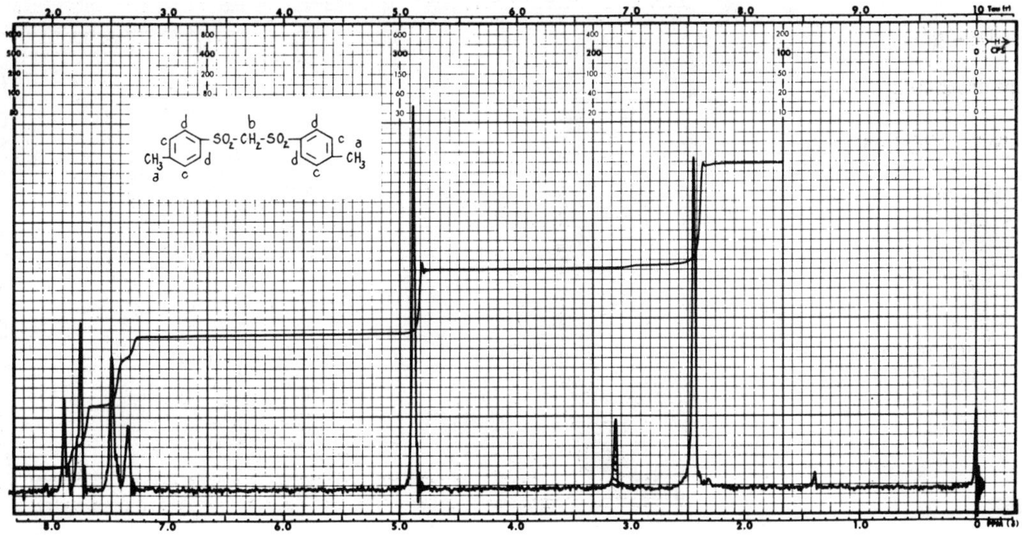

© 1978 Sadtler Research Laboratories, Inc.

PHENYL 2-PROPYNYL SULFONE

$C_9H_8O_2S$

Mol. Wt. 180.23

Solvent: Polysol-d

Assignments
a = 3.10
b = 4.35
c = 7.50-7.85
d = 8.01
e = 2.53 DMSO-d_5
f = 3.29 H_2O

1265

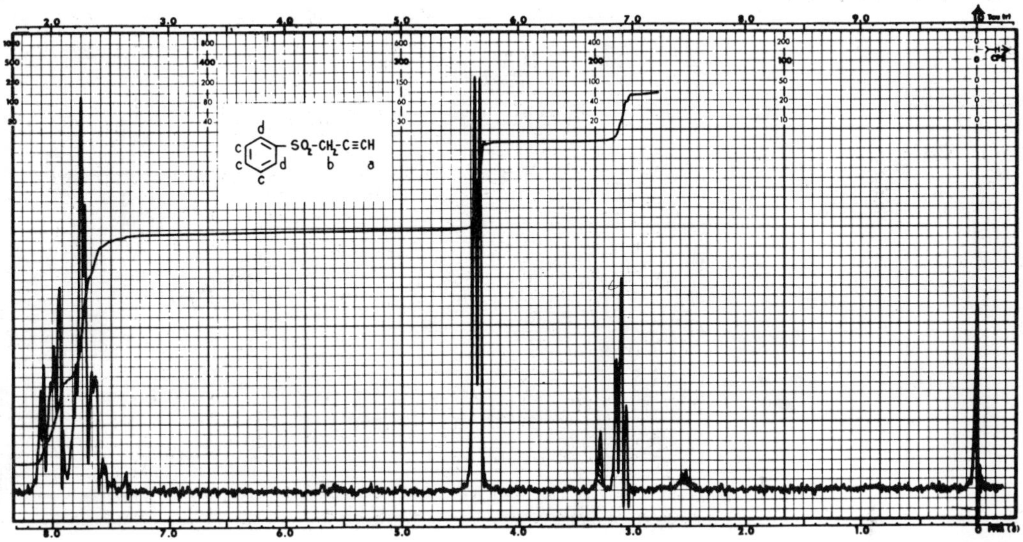

© 1978 Sadtler Research Laboratories, Inc.

3,4-DICHLOROPHENYL 2-PROPYNYL SULFONE

$C_9H_6Cl_2O_2S$

Mol. Wt. 249.12

Solvent: $CDCl_3$

Assignments
a = 2.49
b = 4.06
c = 7.70
d = 7.88
e = 8.10

1266

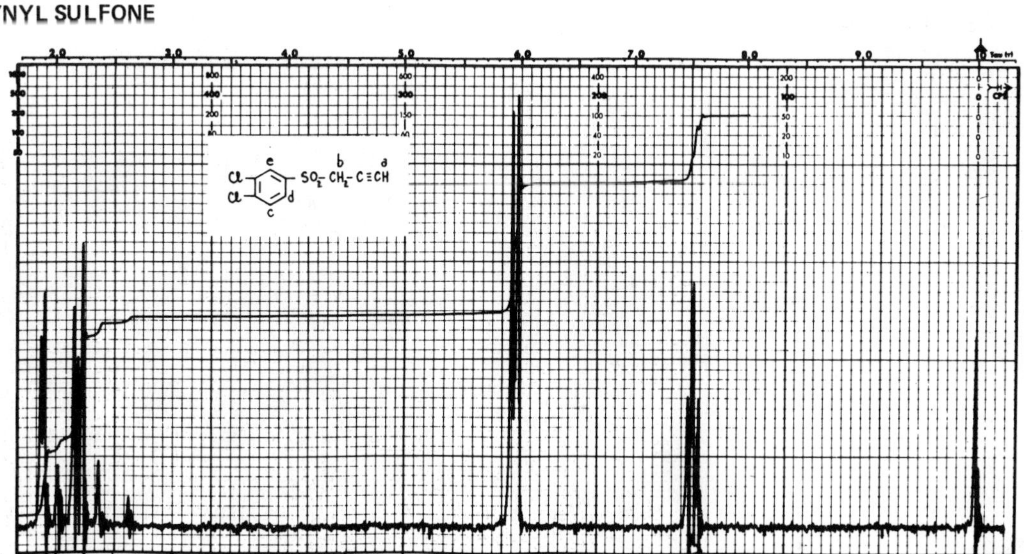

© 1978 Sadtler Research Laboratories, Inc.

BENZYL p-TOLYL SULFONE

$C_{14}H_{14}O_2S$

Mol. Wt. 246.33

Solvent: Polysol-d

Assignments
a = 2.39
b = 4.44
c = 7.00-7.30
d = 7.29
e = 7.54
f = 2.50 DMSO-d_5
g = 3.07 H_2O

1267

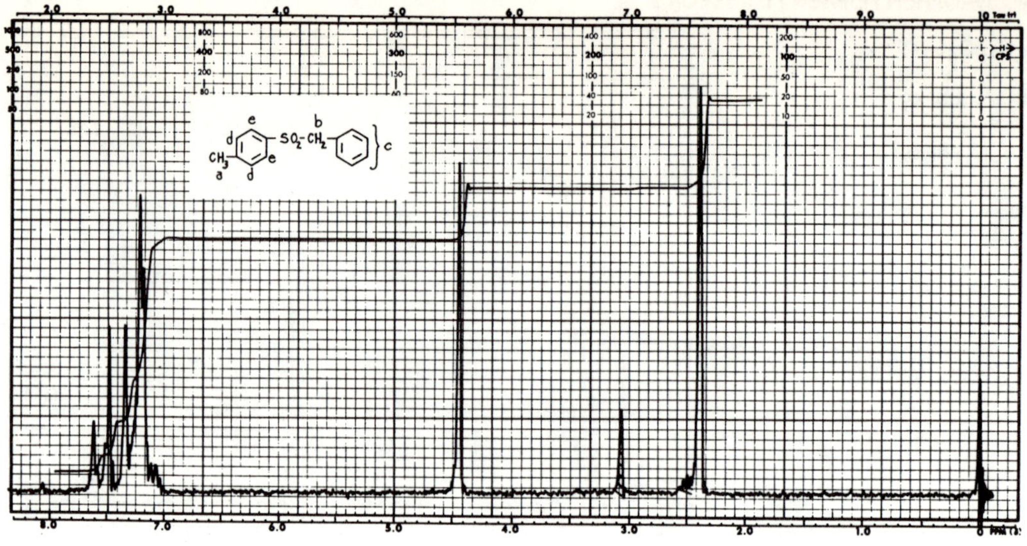

© 1978 Sadtler Research Laboratories, Inc.

PHENYL p-TOLYL SULFONE

$C_{13}H_{12}O_2S$

Mol. Wt. 232.30

Solvent: $CDCl_3$

Assignments
a = 2.30
b = 7.26
c = 7.30-7.60
d = 7.81
e = 7.90

1268

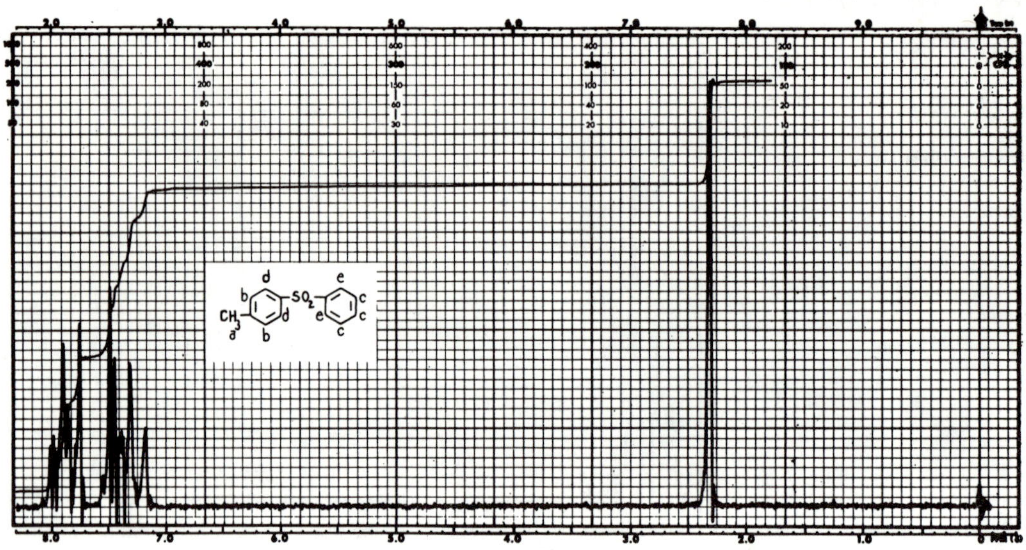

© 1978 Sadtler Research Laboratories, Inc.

p-TOLYL SULFONE

$C_{14}H_{14}O_2S$

Mol. Wt. 246.33

M.P. 158-159°C

Solvent: $CDCl_3$

Assignments
a = 2.31
b = 7.23
c = 7.80

1269

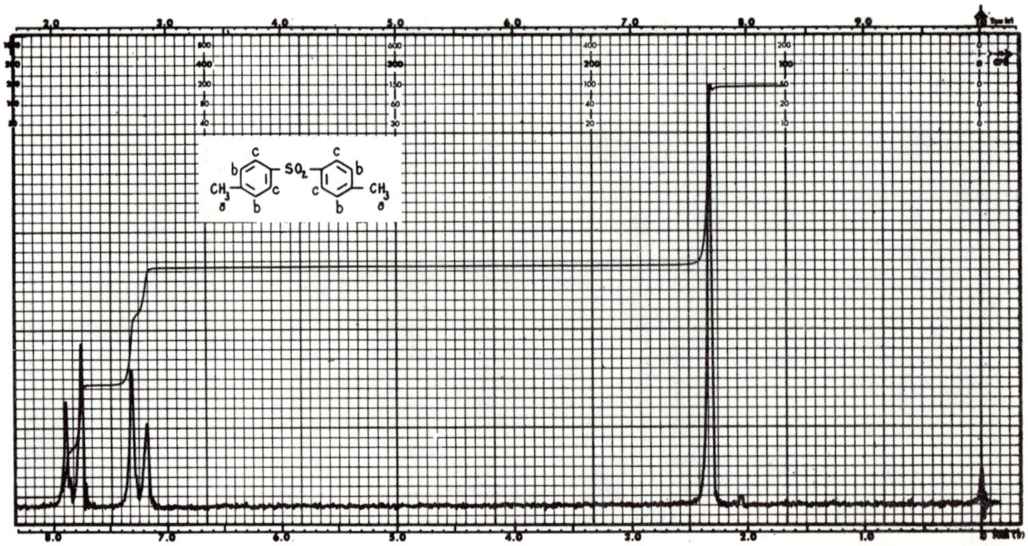

© 1978 Sadtler Research Laboratories, Inc.

p-FLUOROPHENYL PHENYL SULFONE

$C_{12}H_9FO_2S$

Mol. Wt. 236.27

M.P. 110-112°C

Solvent: $CDCl_3$

Assignments
a = 7.12
b = 7.20-7.65
c = ca 7.90

1270

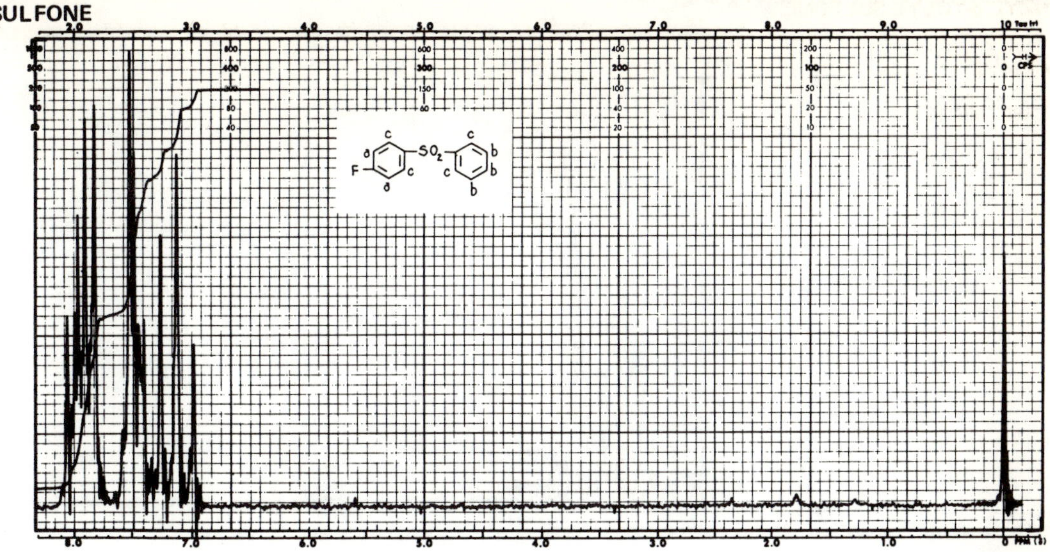

© 1978 Sadtler Research Laboratories, Inc.

BIS(p-FLUOROPHENYL) SULFONE

$C_{12}H_8F_2O_2S$

Mol. Wt. 254.26

M.P. 98-99°C

Solvent: $CDCl_3$

Assignments
a = 7.18
b = 7.95

1271

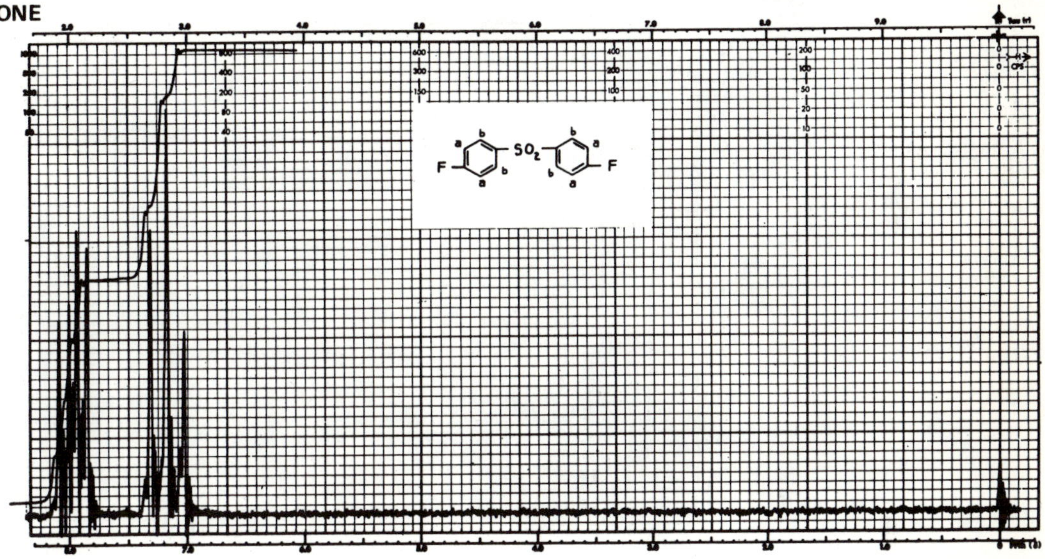

© 1978 Sadtler Research Laboratories, Inc.

PHENYL α,α,α-TRIFLUORO-m-TOLYL SULFONE

$C_{13}H_9F_3O_2S$

Mol. Wt. 286.28

M.P. 77-78°C

Solvent: $CDCl_3$

Assignments
No assignments made

1272

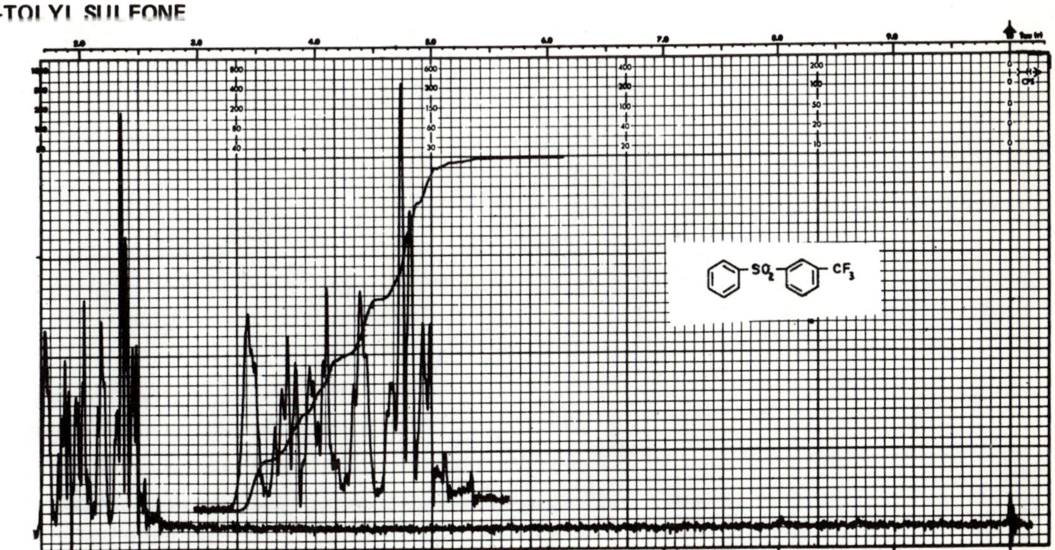

© 1978 Sadtler Research Laboratories, Inc.

2,5-DICHLOROPHENYL PHENYL SULFONE

$C_{12}H_8Cl_2O_2S$

Mol. Wt. 287.17

M.P. 145°C

Solvent: $CDCl_3$

Assignments
a = 7.25-7.70
b = 7.94
c = 8.32
d = 1.29 imp.

1273

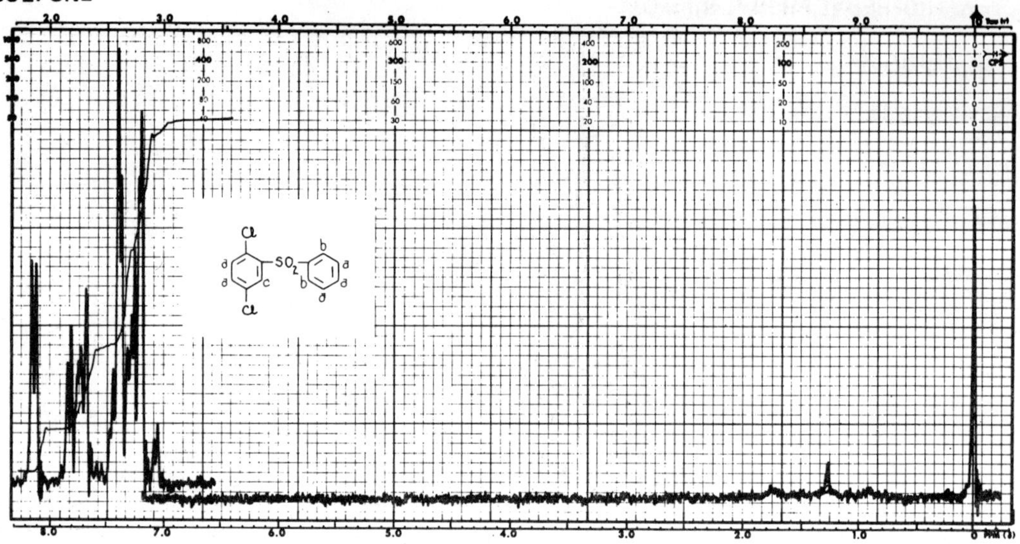

© 1978 Sadtler Research Laboratories, Inc.

BIS(p-CHLOROPHENYL) SULFONE

$C_{12}H_8Cl_2O_2S$

Mol. Wt. 287.17

M.P. 145.5-148.5°C

Solvent: $CDCl_3$

Assignments
a = 7.42
b = 7.82

1274

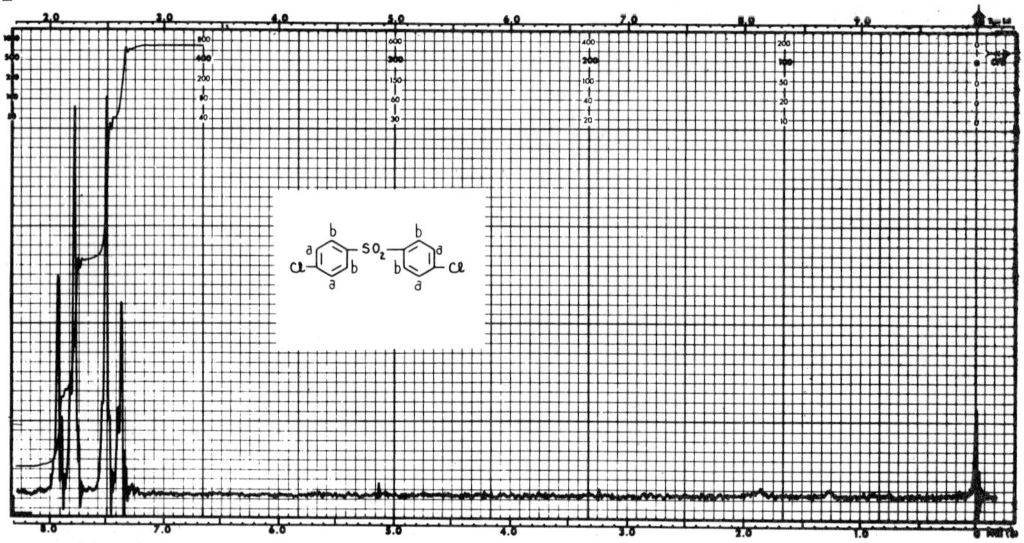

© 1978 Sadtler Research Laboratories, Inc.

BIS(p-BROMOPHENYL)SULFONE

$C_{12}H_8Br_2O_2S$

Mol. Wt. 376.08

M.P. 173°C

Solvent: $CDCl_3$

Assignments
a = 7.65
b = 7.82

1275

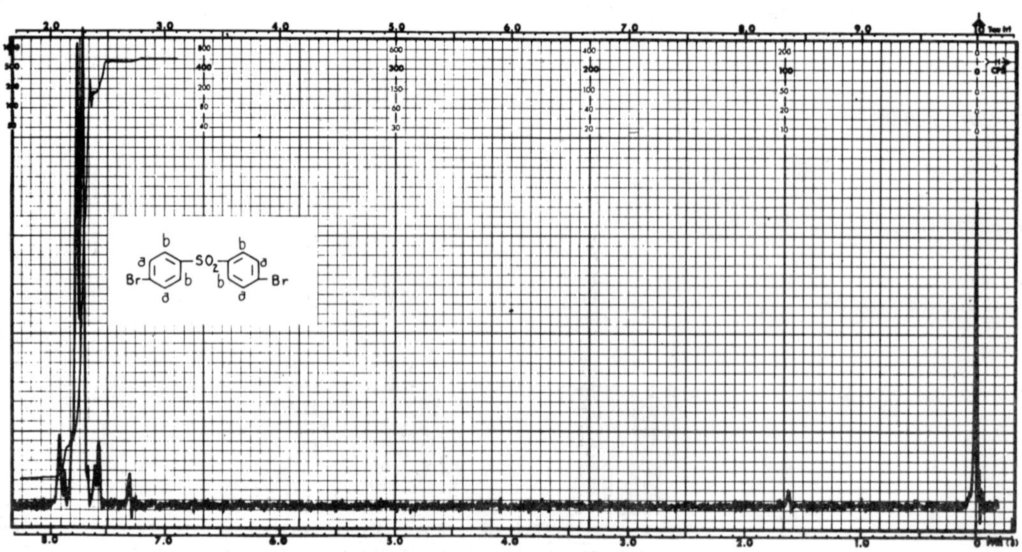

© 1978 Sadtler Research Laboratories, Inc.

p-NITROPHENYL PHENYL SULFONE

$C_{12}H_9NO_4S$

Mol. Wt. 263.27

M.P. 140-141°C

Solvent: $CDCl_3$

Assignments
a = 7.39-7.69
b = 7.92
c = 8.08
d = 8.30
e = 1.72 imp.

1276

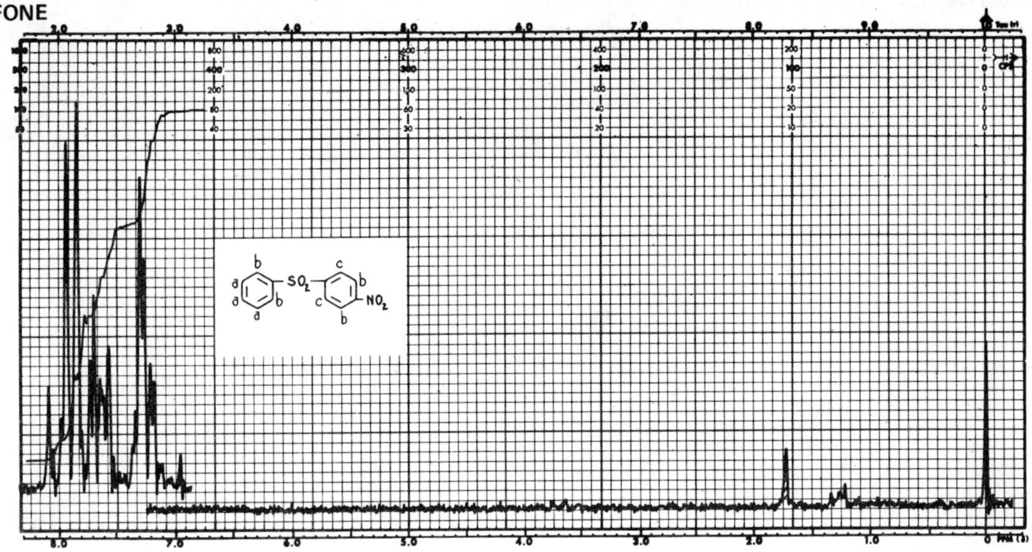

© 1978 Sadtler Research Laboratories, Inc.

BENZO[b] THIOPHENE, 1,1-DIOXIDE

$C_8H_6O_2S$

Mol. Wt. 166.20

M.P. 141-145°C

Solvent: $CDCl_3$

Assignments
a = 6.62
b = 7.16
c = 7.25-7.77
d = 1.79 imp.

1277

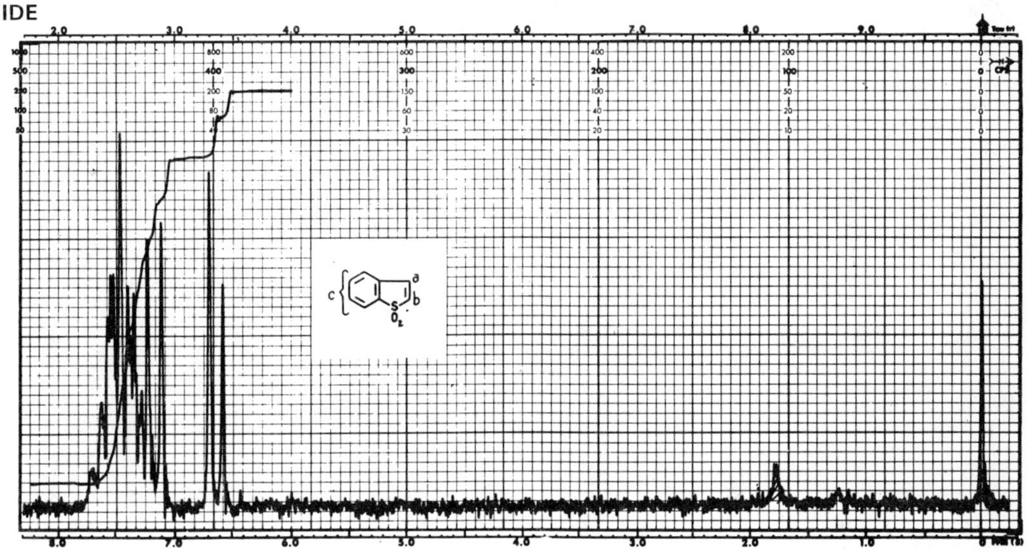

© 1978 Sadtler Research Laboratories, Inc.

2-DIBENZOTHIOPHENAMINE, 5,5-DIOXIDE

$C_{12}H_9NO_2S$

Mol. Wt. 231.28

M.P. 281°C

Solvent: Dimethyl sulfoxide

Assignments
a = 6.29
b = 6.80
c = 7.19
d = 7.62
e = 7.50-8.05

1278

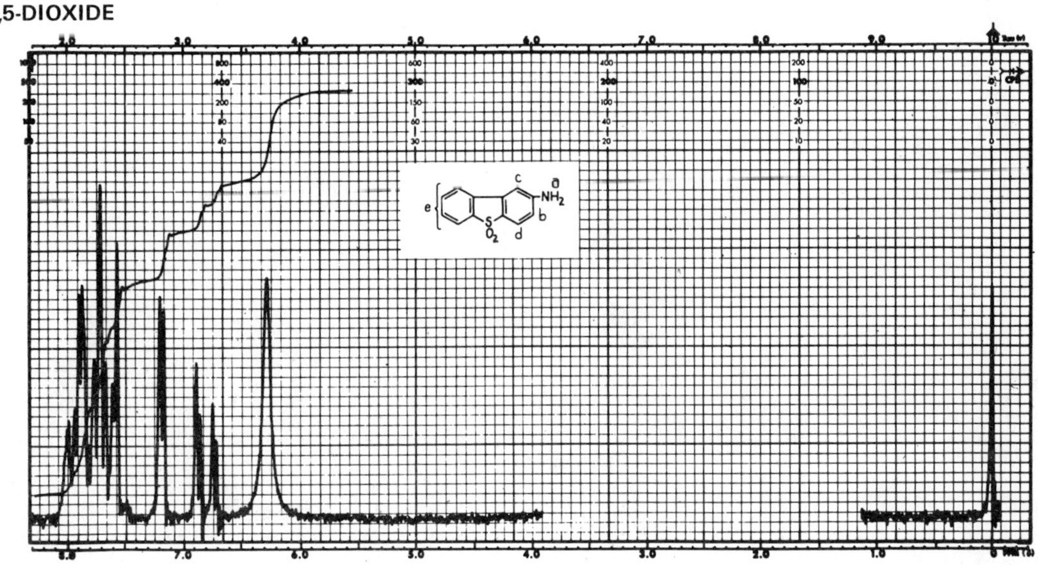

© 1978 Sadtler Research Laboratories, Inc.

NOTES

Due to the unavailability of the corresponding Bromine and Iodine analogs, this chapter presents chemical shift data only for the Sulfonyl Fluorides and Chlorides.

While the type of Halogen present appears to have a significant effect on the chemical shifts of aliphatic groups, the effect is much less pronounced in relation to the ortho aromatic hydrogens. For the sake of comparison, a series of methyl substituted SO_2 groups of various types is presented below.

CH_3–	SO_2–X	Solvent	Spectrum
2.82	CH_3–SO_2–OH	D_2O	----
3.03	CH_3–SO_2–NH_2	Polysol	1359
3.07	CH_3–SO_2–Ø	$CDCl_3$	----
3.11	CH_3–SO_2–CH_3	D_2O	1252
3.49	CH_3–SO_2–S–R	DMSO-d_6	----
3.65	CH_3–SO_2–Cl	CCl_4	----

It is interesting to note that when the Sulfonyl Chloride undergoes hydrolysis to form the corresponding Sulfonic acid, the methyl resonance is converted from the most strongly deshielded (3.65 ppm) to the least deshielded group of this series (2.82 ppm).

Chemical Shifts

Aliphatic Protons

HC–4	HC–3	HC–2	HC–1			
			3.65	CH_3–SO_2–Cl	CCl_4	----
		1.54	3.65	CH_3–CH_2–SO_2–Cl	CS_2	----
	1.17	2.09	3.68	CH_3–CH_2–CH_2–SO_2–Cl	CCl_4	1280
1.02	1.50	2.00	3.69	CH_3–CH_2–CH_2–CH_2–SO_2–Cl	CCl_4	1281

Substituted Methane Sulfonyl Halides

X–S(O_2)–	CH_2	–Y		
Cl–S(O_2)–	4.83	–Ø	$CDCl_3$	----
F–S(O_2)–	5.07	–Ø	Polysol	1279

Aromatic Protons

Both the Chlorine and Fluorine Sulfonyl compounds strongly deshield the ortho aromatic hydrogens. The chemical shifts observed for the two phenyl compounds are extremely similar. A comparison of the para substituted compounds indicates identical shifts for the two para substituted methyl compounds, higher field shifts for the chlorine substituted Sulfonyl Fluoride but lower field shifts for the corresponding carboxylic acid substituted Sulfonyl Fluoride compared to the corresponding Sulfonyl Chlorides.

Chemical Shifts continued

Aromatic Protons continued

Phenylsulfonyl Halides

	ortho	meta/para	Solvent	Spectrum
Cl–S(O$_2$)–Ø	8.03		CCl_4	3150
F–S(O$_2$)–Ø	8.00		CCl_4	7051

Para Substituted Phenylsulfonyl Halides

Cl–S(O$_2$)–	ortho	meta	para	
	7.98	7.07	–O–CH_3	$CDCl_3$
	7.90	7.41	–CH_3	$CDCl_3$
	8.05	7.67	–Cl	$CDCl_3$
	7.99	7.69	–I	$CDCl_3$
	7.90	7.77	–Br	$CDCl_3$
	8.17	8.08	–N=N–Ø	$CDCl_3$
	7.90	8.09	–C(=O)–OH	Polysol
	8.13	8.13	–SO_2–F	Polysol
	8.01	8.28	–NO_2	Polysol

F–S(O$_2$)–	ortho	meta	para	
	7.73	6.71	–NH_2	$CDCl_3$
	7.90	7.41	–CH_3	$CDCl_3$
	7.98	7.62	–Cl	$CDCl_3$
	8.01	8.01	–NH–C(=O)–CH_3	DMSO-d_6
	8.13	8.13	–S(O$_2$)–Cl	Polysol
	8.14	8.33	–C(=O)–OH	Polysol

Coupling and Coupling Constants

The only unusual coupling constant associated with the Sulfonyl Halides is that observed between the Sulfonyl Fluoride group and the adjacent aliphatic group:

$J_{F-S(O_2)-CH_2}$ = 4.5 Hz (spectrum 1279)

α-TOLUENESULFONYL FLUORIDE

$C_7H_7FO_2S$

Mol. Wt. 174.20

M.P. 93-94°C

Solvent: Polysol-d

Assignments
a = 5.07
b = ca 7.42

1279

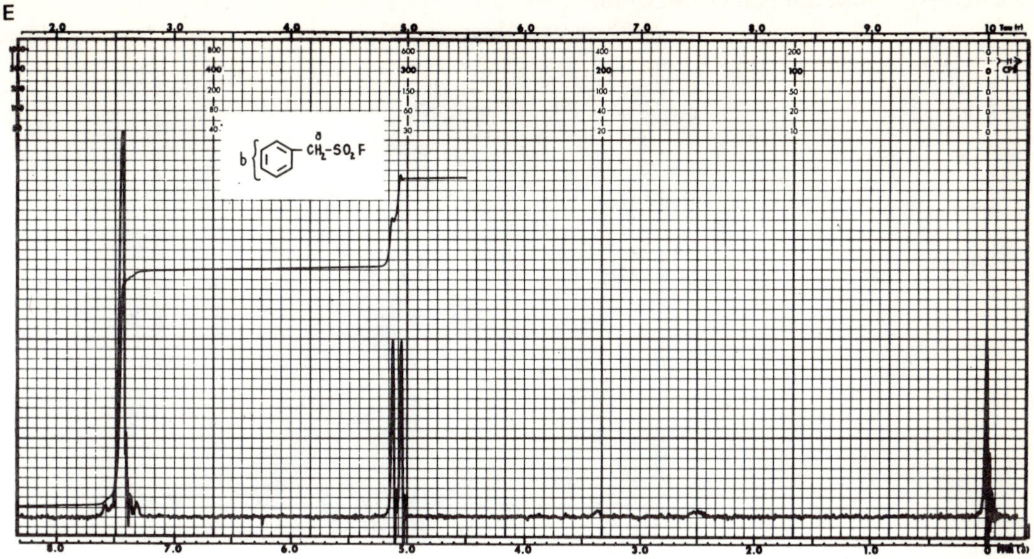

© 1978 Sadtler Research Laboratories, Inc.

1-PROPANESULFONYL CHLORIDE

$C_3H_7ClO_2S$

Mol. Wt. 142.61

B.P. 65-66°C/8mm

Solvent: CCl_4

Assignments
a = 1.17
b = 2.09
c = 3.68

1280

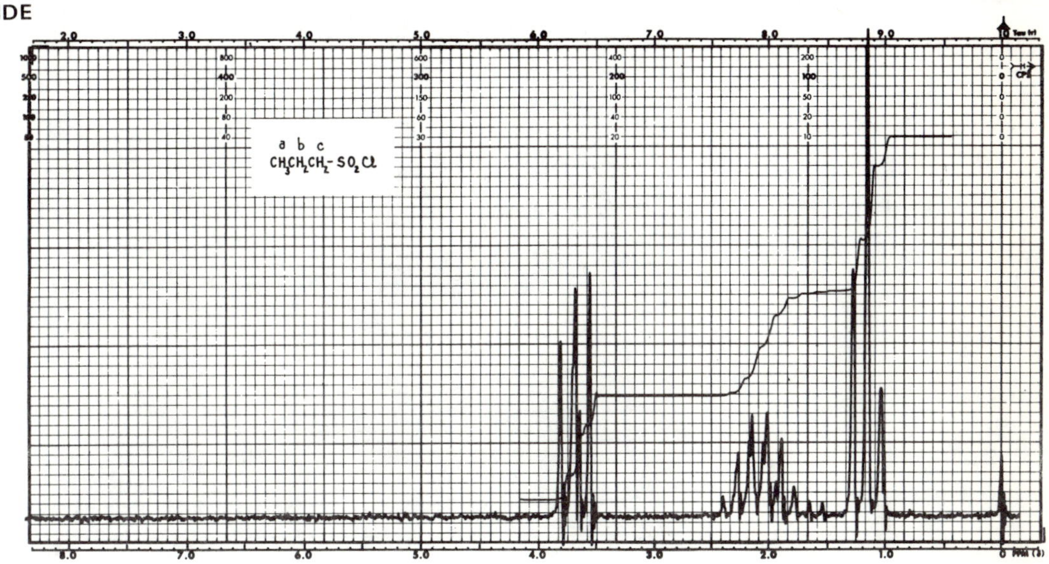

© 1978 Sadtler Research Laboratories, Inc.

1-BUTANESULFONYL CHLORIDE

$C_4H_9ClO_2S$

Mol. Wt. 156.63

B.P. 94-96°C/21mm

Solvent: CCl_4

Assignments
a = 1.02
b = 1.50
c = 2.01
d = 1.69

1281

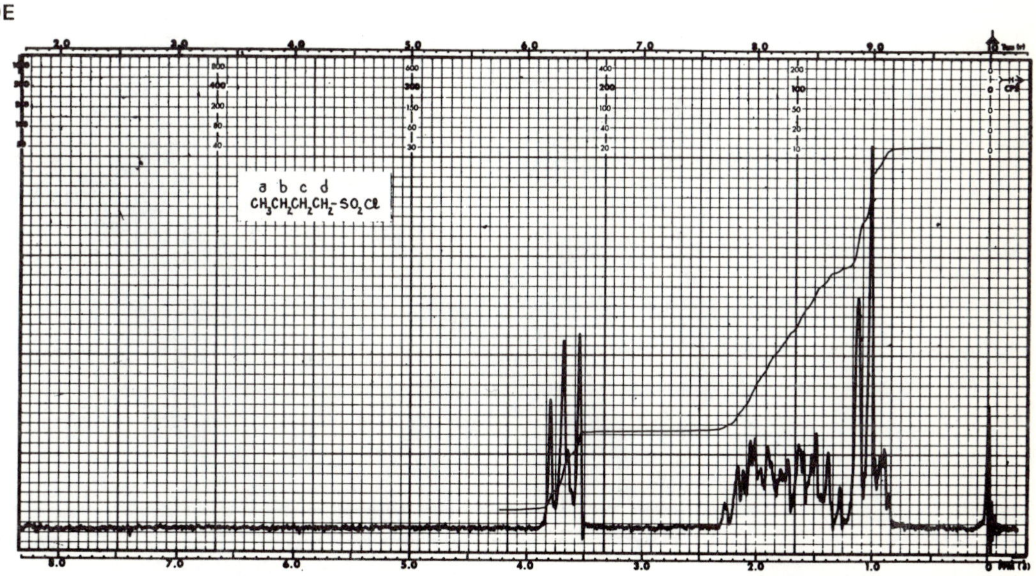

© 1978 Sadtler Research Laboratories, Inc.

2-CHLOROETHANESULFONYL CHLORIDE

$C_2H_4Cl_2O_2S$

Mol. Wt. 163.02

B.P. 76-78°C/9mm

Solvent: CCl_4

Assignments
a = 4.01

1282

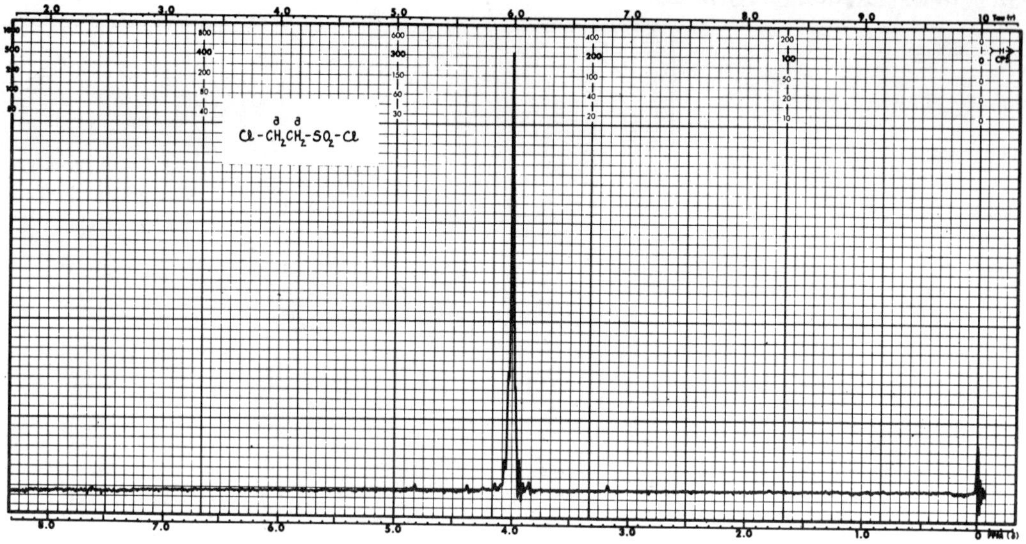

© 1978 Sadtler Research Laboratories, Inc.

p-TOLUENESULFONYL CHLORIDE

$C_7H_7ClO_2S$

Mol. Wt. 190.65

M.P. 66-68°C

Solvent: $CDCl_3$

Assignments
a = 2.49
b = 7.41
c = 7.90

1283

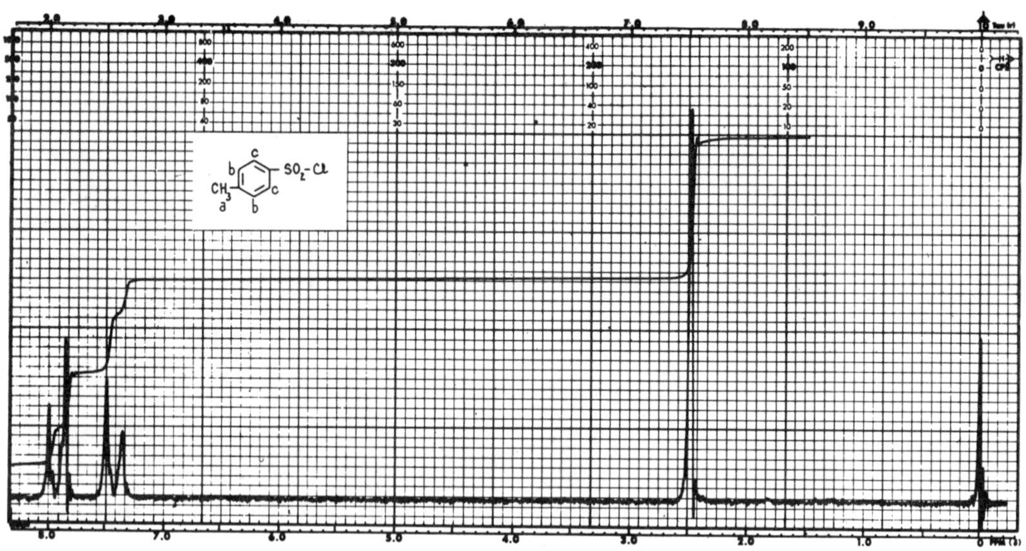

© 1978 Sadtler Research Laboratories, Inc.

2,5-XYLENESULFONYL CHLORIDE

$C_8H_9ClO_2S$

Mol. Wt. 204.68

M.P. 20.5-22°C

Solvent: CCl_4

Assignments
a = 2.41
b = 2.70
c = 7.29
d = 7.39
e = 7.81

1284

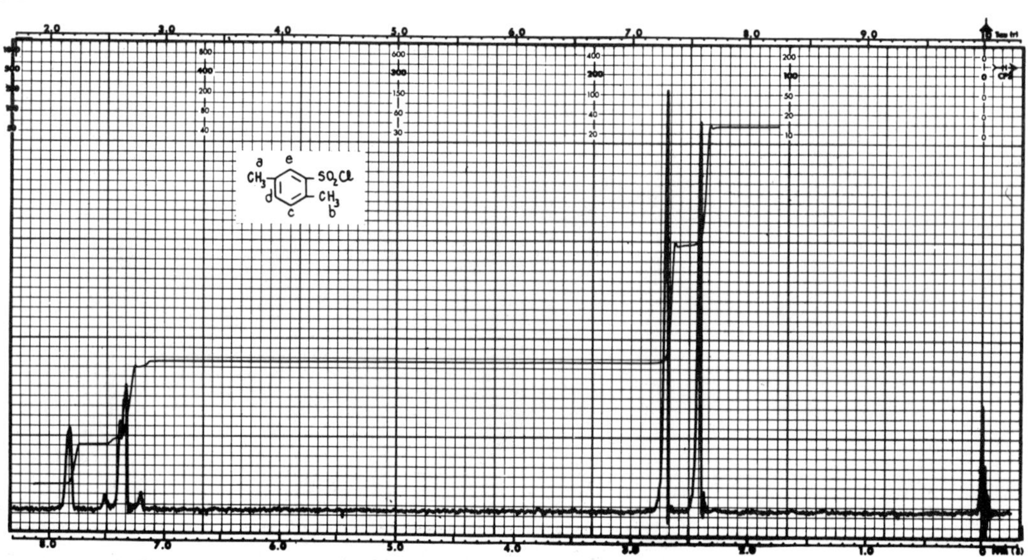

© 1978 Sadtler Research Laboratories, Inc.

o-NITROBENZENESULFONYL CHLORIDE

$C_6H_4ClNO_4S$

Mol. Wt. 221.62

M.P. 65-67°C

Solvent: $CDCl_3$

Assignments
a = 7.70-8.05
b = 8.29

1285

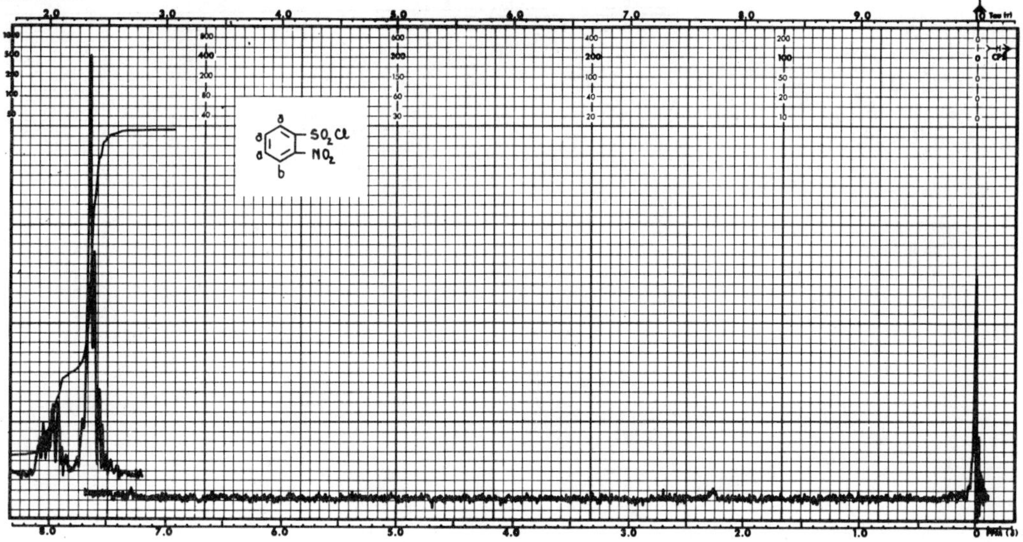

© 1978 Sadtler Research Laboratories, Inc.

1-NAPHTHALENESULFONYL CHLORIDE

$C_{10}H_7ClO_2S$

Mol. Wt. 226.68

M.P. 67-69°C

Solvent: Polysol-d

Assignments
a = 7.30-7.80
b = ca 7.91
c = 8.16

1286

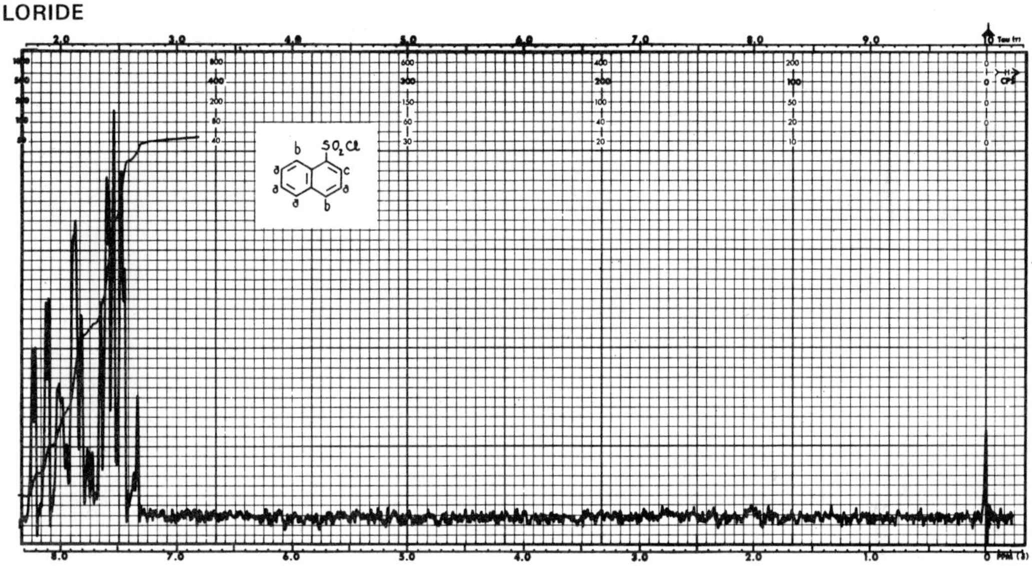

© 1978 Sadtler Research Laboratories, Inc.

NOTES

THE SULFONIC ACIDS

The Sulfonic acids are similar to the other sulfone ($-S(O_2)-$) compounds in their chemical shift effects, in that, they are a moderately strong deshielding group for adjacent aliphatic groups and a strong deshielding group for ortho aromatic hydrogens.

The lower molecular weight members of the series are soluble in both the chlorinated solvents as well as DMSO-d_6, polysol and D_2O. When an amine group is present in the molecule, they form an internal salt which makes these compounds soluble almost exclusively in D_2O or DMSO-d_6.

Chemical Shifts

Aliphatic Protons

HC–4	HC–3	HC–2	HC–1		Solvent	Spectrum
			2.82	CH_3-SO_2-OH	D_2O	----
		1.45	3.26	$CH_3-CH_2-SO_2-OH$	CCl_4	1287
	1.11	1.97	3.24	$CH_3-CH_2-CH_2-SO_2-OH$	$CDCl_3$	----
0.97	1.55	1.90	3.23	$CH_3-CH_2-CH_2-CH_2-SO_2-OH$	$CDCl_3$	----

Substituted Ethanesulfonic Acids

$HO-S(O_2)$	$-CH_2$	$-CH_2$	$-X$	
----	3.42	3.30	$-NH-CH_3$	D_2O
----	3.51	3.51	−2(pyridine)	D_2O
----	3.40	3.27	$-NH_2$	D_2O

Sulfonic Acid Protons

$HO-S(O_2)$	$-X$		
8.75	$-CH_2-CH(R)_2$	CCl_4	----
10.70	$-R_2$	CCl_4	1287
10.71	$-R_3$	$CDCl_3$	----
10.72	$-R_4$	$CDCl_3$	----

Aromatic Protons

The Sulfonic acid group strongly deshields the ortho aromatic hydrogens. The ortho protons resonate near 7.9 ppm as a distorted doublet of doublets while the meta and para hydrogens appear as a complex higher order band in the chemical shift range from 7.4–7.7 ppm.

Benzene sulfonic acid

	ortho	meta/para	
$HO-S(O_2)-$	7.93	7.4–7.7	D_2O

Chemical Shifts continued

Aromatic Protons continued

Para substituted benzene sulfonic acids

HO–S(O$_2$)–	ortho	meta	para	Solvent
	7.61	7.19	–CH_3	DMSO-d_6
	7.81	7.81	–Ø	DMSO-d_6

ETHANESULFONIC ACID

$C_2H_6O_3S$

Mol. Wt. 110.13

Solvent: CCl_4

Assignments
a = 1.45
b = 3.20
c = 10.70

1287

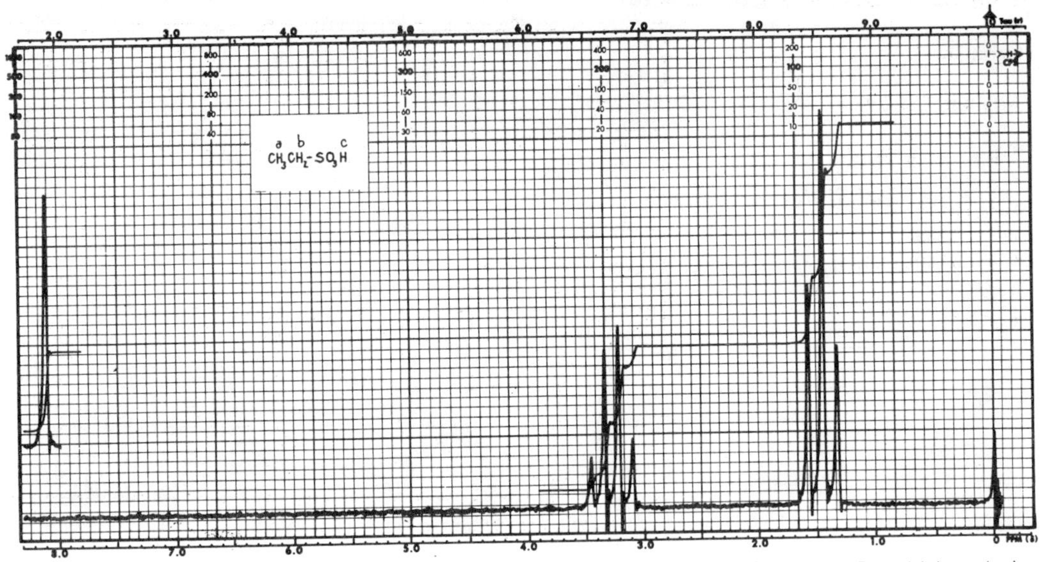

© 1978 Sadtler Research Laboratories, Inc.

N-METHYLTAURINE

$C_3H_9NO_3S$

Mol. Wt. 139.17

M.P. ∼ 242°C

Solvent: D_2O

Assignments
a = 2.79
b = 3.14-3.58
c = 4.58 HDO

1288

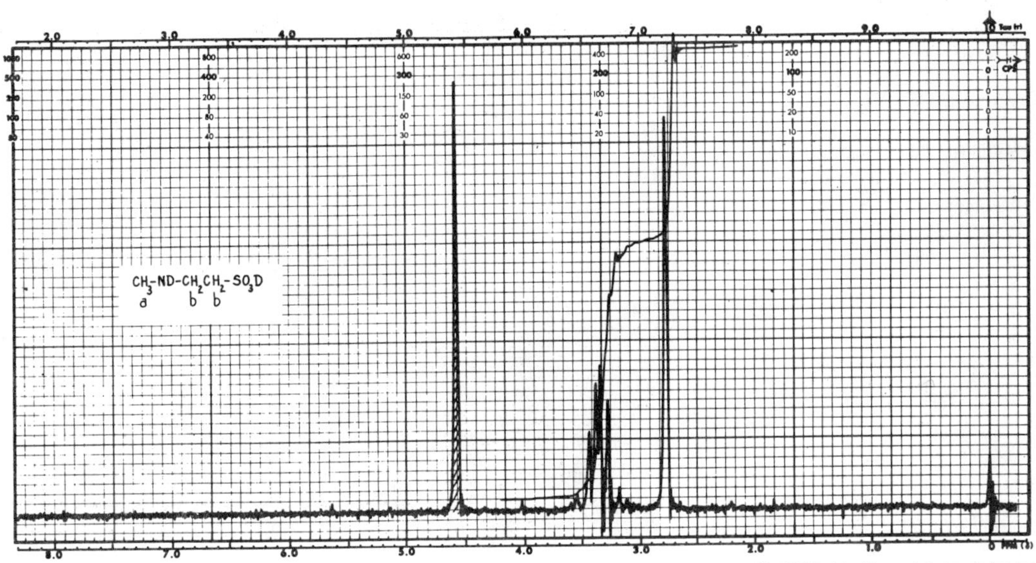

© 1978 Sadtler Research Laboratories, Inc.

4-BIPHENYLSULFONIC ACID

$C_{12}H_{10}O_3S$

Mol. Wt. 234.28

Solvent: Dimethyl sulfoxide

Assignments
a = 6.83 *
b = 7.39-8.00
c = ca 7.81
* H_2O present

1289

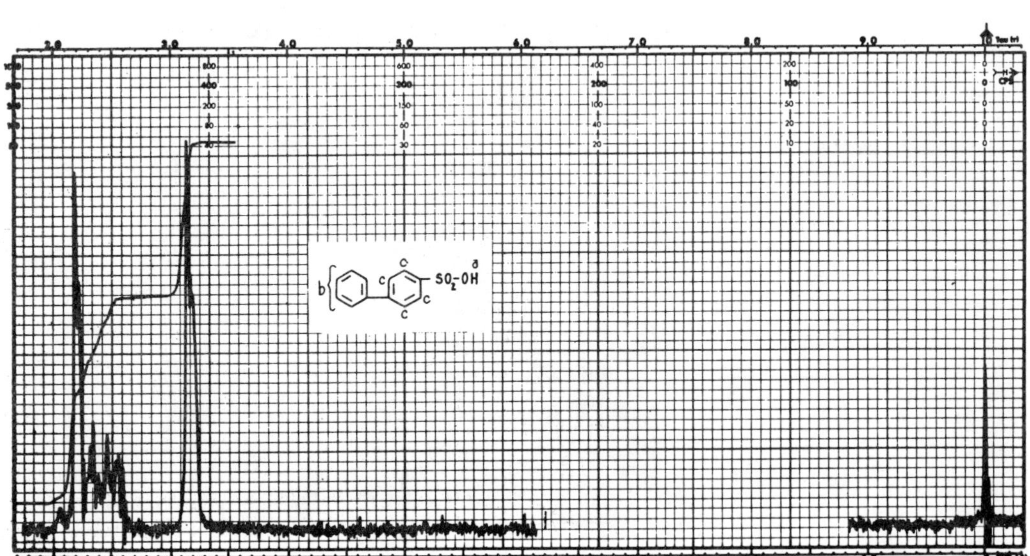

© 1978 Sadtler Research Laboratories, Inc.

6-AMINO-m-TOLUENESULFONIC ACID

$C_7H_9NO_3S$

Mol. Wt. 187.22

Solvent: DMSO-d_6

Assignments
a = 2.39
b = ca 7.40
c = 7.77
d = ca 9.20

1290

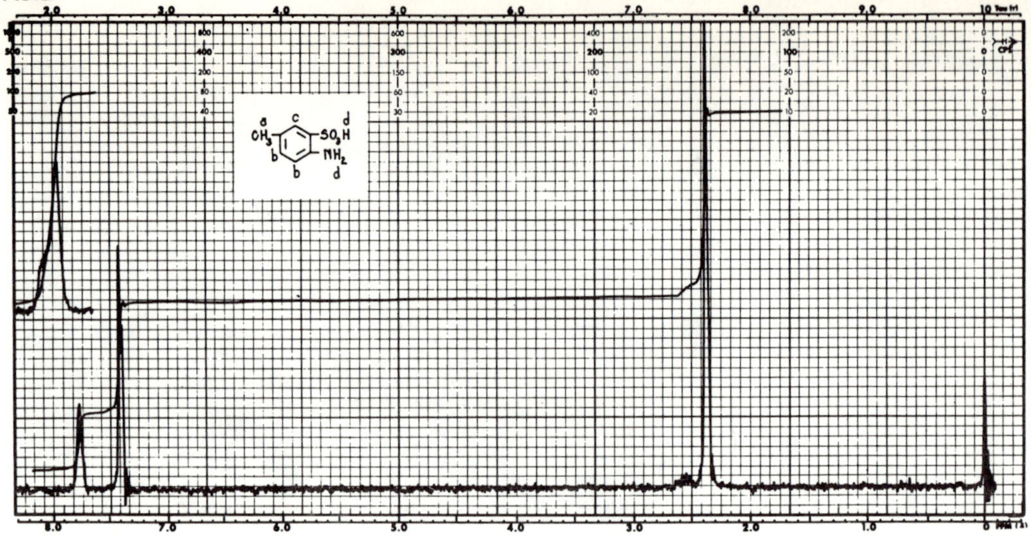

© 1978 Sadtler Research Laboratories, Inc.

3-AMINO-p-TOLUENESULFONIC ACID

$C_7H_9NO_3S$

Mol. Wt. 187.22

Solvent: Dimethyl sulfoxide-d_6

Assignments
a = 2.30
b = ca 7.15 *
c = 7.20
d = 7.52
e = 7.67
f = 2.49 DMSO-d_5
* Plus H_2O

1291

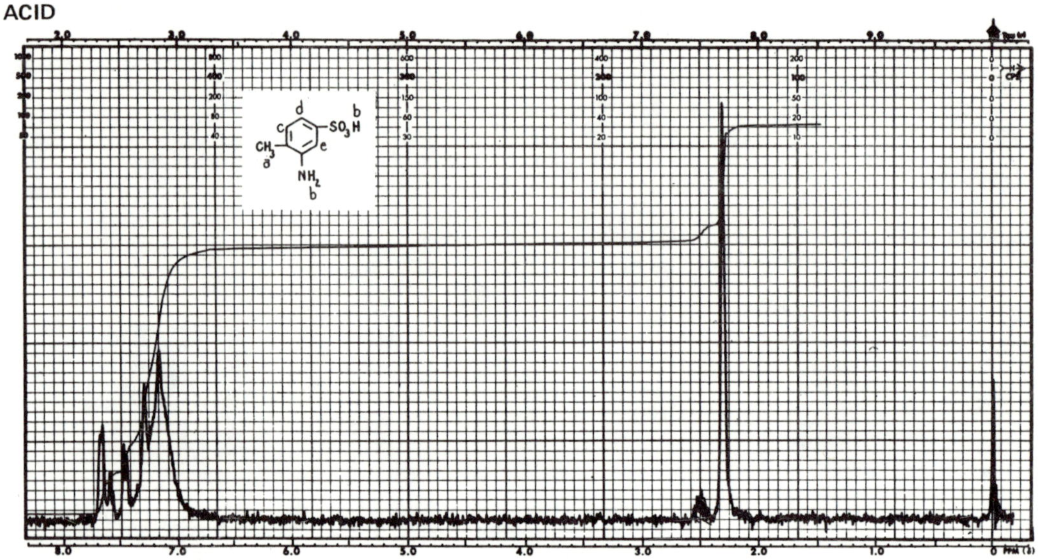

© 1978 Sadtler Research Laboratories, Inc.

1-NAPHTHALENESULFONIC ACID, DIHYDRATE

$C_{10}H_8O_3S{\cdot}2H_2O$

Mol. Wt. 244.26

Solvent: Trifluoroacetic acid

Assignments
a = 7.20-8.30
b = 8.51
c = unobserved *
* In exchange with solvent - OH group

1292

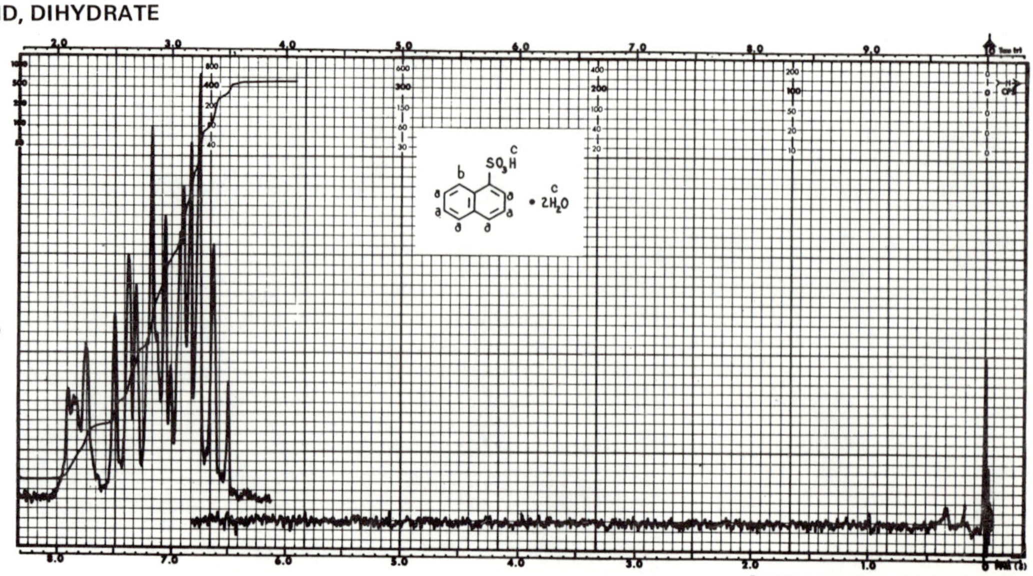

© 1978 Sadtler Research Laboratories, Inc.

2-AMINO-1-NAPHTHALENESULFONIC ACID

$C_{10}H_9NO_3S$

Mol. Wt. 223.25

Solvent: Dimethyl sulfoxide-D6

Assignments

a = 7.39
b = 7.63
c = 7.66
d = 7.87
e = 8.04
f = 9.06
g = 9.53
h = 2.50 DMSO-D5

1293

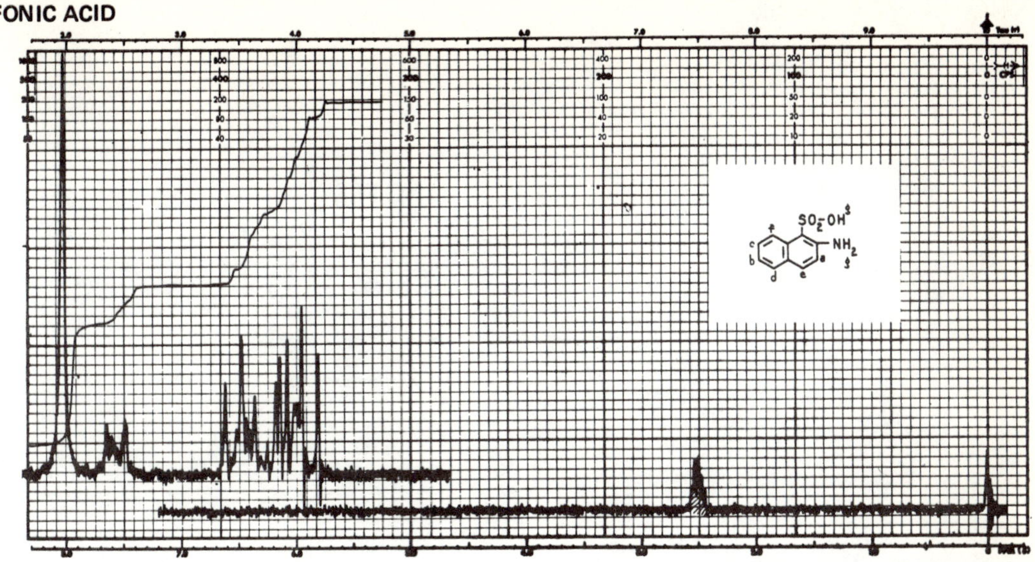

© 1978 Sadtler Research Laboratories, Inc.

NOTES

THE SALTS OF SULFONIC ACID

The water soluble metallic salts of sulfonic acid display aromatic chemical shifts very similar to those of the free acid with the aliphatic groups adjacent to the sulfonate group resonating at slightly higher field. Although of commercial importance, relatively small number of compounds are available for the preparation of their NMR spectra.

Chemical Shifts

Aliphatic Protons

HC–5	HC–3,4	HC–2	HC–1		Solvent	Spectrum
0.91	(1.2–1.6)	1.72	2.87	$CH_3(CH_2)_2–CH_2–CH_2–SO_3–Na$	D_2O	1327

2-Substituted Ethane Sulfonic Acid Salts

$Na–O_3S$ $–CH_2$	$–CH_2$	$–X$	Solvent	Spectrum
2.89	1.72	$–R_3$	D_2O	1294
3.15	2.96	–C≡N	D_2O	1297
3.27	3.27	$–SO_3–Na$	D_2O	1298
3.41	3.66	–Br	D_2O	1296
3.37	3.89	–Cl	D_2O	- - - -
3.15	3.95	–OH	D_2O	- - - -

Aromatic Protons

The aromatic chemical shifts of the sulfonic acid salts are similar to those of the parent acids. The shifts of benzene sulfonic acid, sodium salt are listed below followed by those of several para substituted derivatives.

Benzene sulfonic acid, sodium salt

para/meta	ortho			
7.4–7.7	7.02	$SO_3–Na$	D_2O	1300

Para substituted benzene sulfonic acid salts

X–	ortho	meta	para	
$H_2N–$	6.79	7.62	$–SO_3–Na$	D_2O
HO–	6.80	7.72	$–SO_3–K$	D_2O
$R_{14}–$	7.06	7.67	$–SO_3–Na$	D_2O
Ø–NH–	7.10	7.91	$–SO_3–Na$	D_2O
$CH_3–$	7.28	7.81	$–SO_3–NH_4$	D_2O
Cl–	7.45	7.83	$–SO_3–Na$	D_2O

Chemical Shifts continued

Aromatic Protons continued

Para substituted benzene sulfonic acid salts continued

X–	ortho	meta	para	Solvent
Br–	7.68	7.79	$-SO_3-Na$	D_2O
$CH_3-C(=O)-$	7.88	7.88	$-SO_3-Na$	D_2O
HO–C(=O)–	8.09	8.09	$-SO_3-K$	D_2O

1-PENTANESULFONIC ACID, SODIUM SALT, HYDRATE

$C_5H_{11}NaO_3S{\cdot}H_2O$

Mol. Wt. 192.21

Solvent: D_2O

Assignments
a = 0.91
b = 1.10-2.10
c = 2.89
d = 4.45 HDO

1294

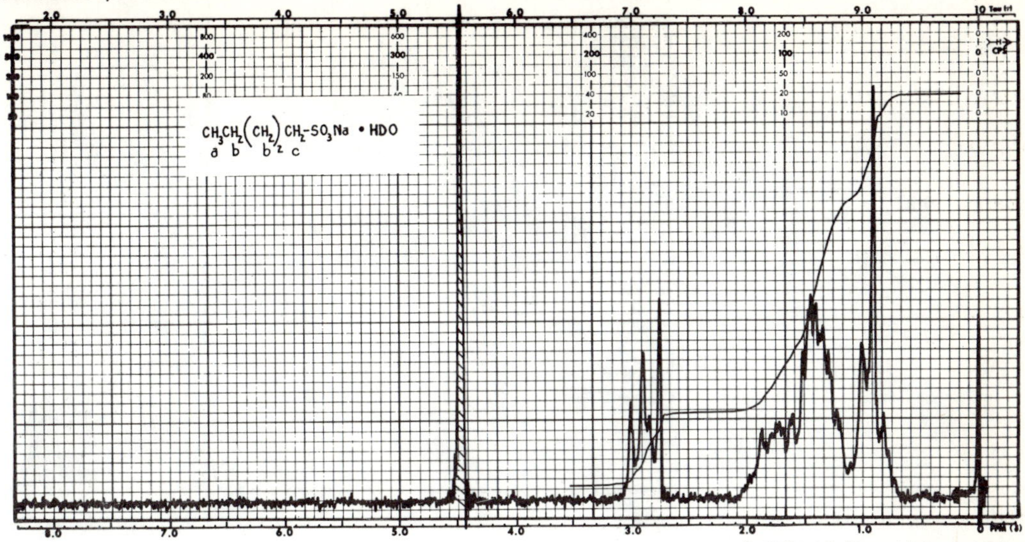

© 1978 Sadtler Research Laboratories, Inc.

1-HEPTANESULFONIC ACID, SODIUM SALT, HYDRATE

$C_7H_{15}NaO_3S{\cdot}H_2O$

Mol. Wt. 220.26

Solvent: D_2O

Assignments
a = 0.89
b = 1.10-1.60
c = 1.71
d = 2.81
e = 4.52 HDO

1295

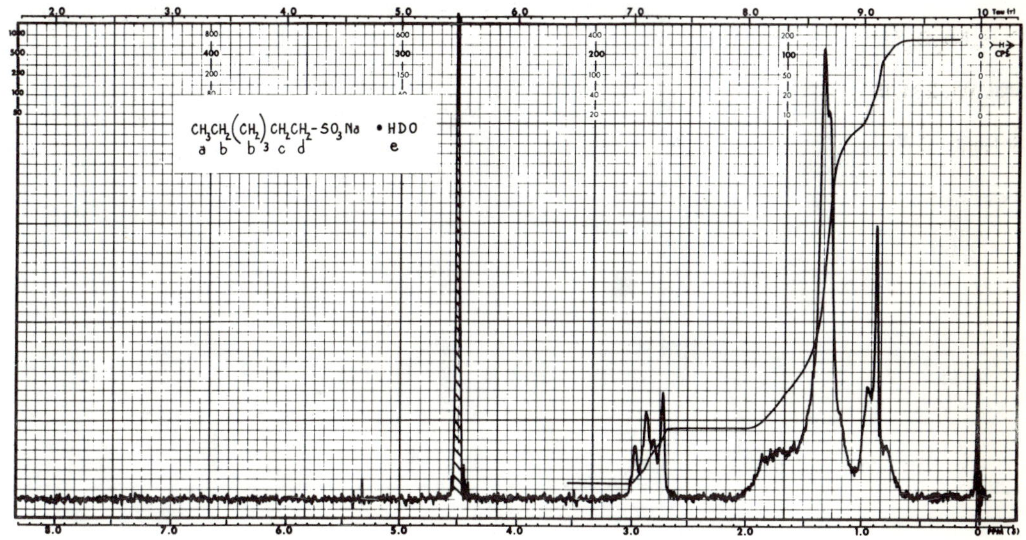

© 1978 Sadtler Research Laboratories, Inc.

2-BROMOETHANESULFONIC ACID, SODIUM SALT

$C_2H_4BrNaO_3S$

Mol. Wt. 211.02

Solvent: D_2O

Assignments
a = 3.41
b = 3.66
c = 4.42 HDO

1296

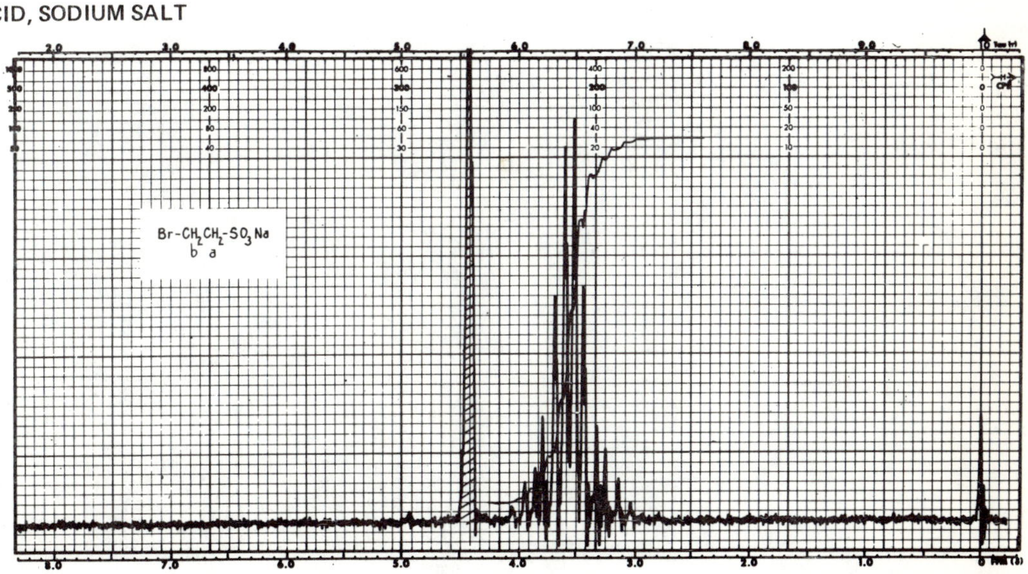

© 1978 Sadtler Research Laboratories, Inc.

2-CYANOETHANESULFONIC ACID, SODIUM SALT

$C_3H_4NNaO_3S$

Mol. Wt. 157.12

M.P. 243-244°C

Solvent: D_2O

Assignments
a = ca 2.96
b = ca 3.15
c = 4.32 HDO

1297

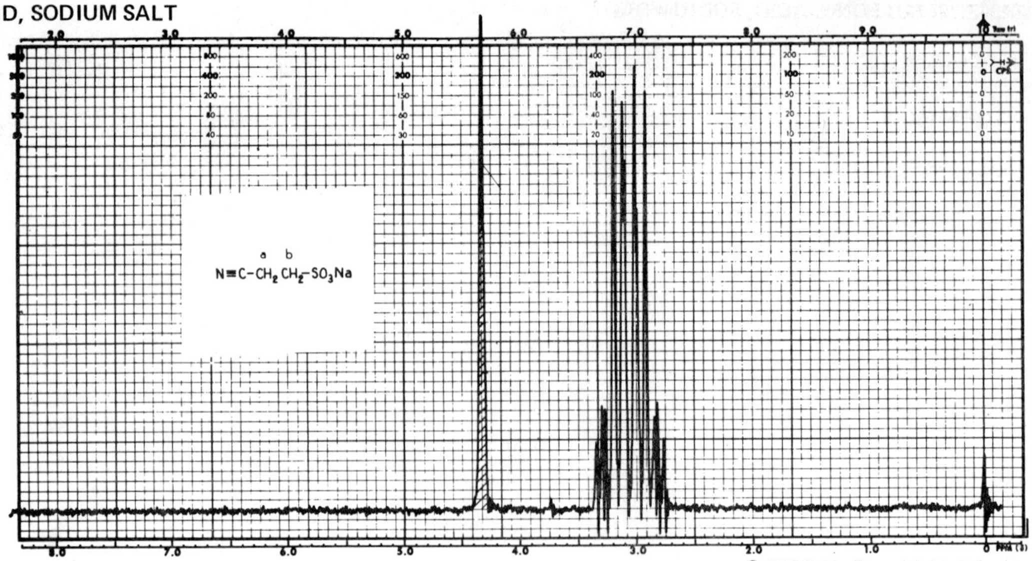

© 1978 Sadtler Research Laboratories, Inc.

1,2-ETHANEDISULFONIC ACID, DISODIUM SALT

$C_2H_4Na_2O_6S_2$

Mol. Wt. 234.18

Solvent: D_2O

Assignments
a = 3.27
b = 4.44 HDO

1298

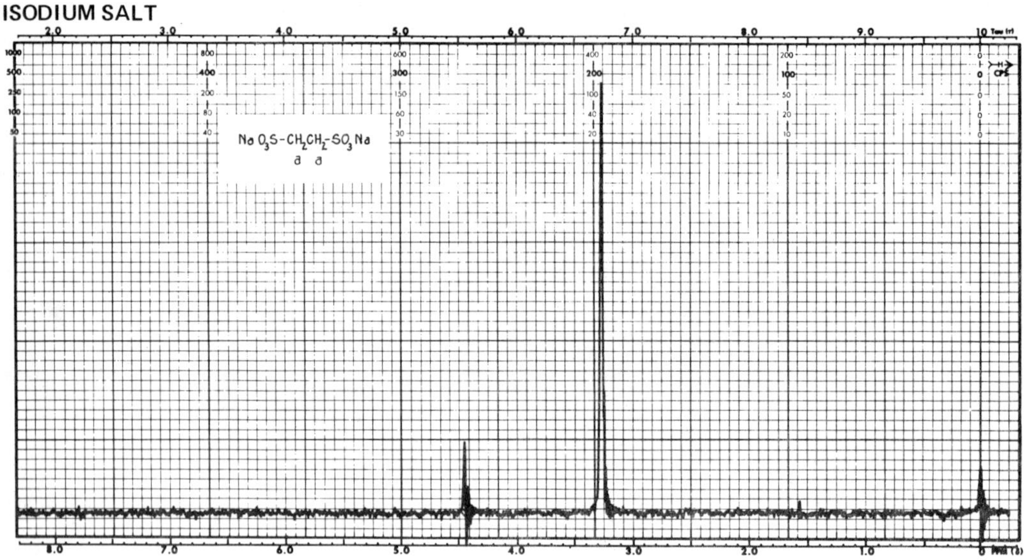

© 1978 Sadtler Research Laboratories, Inc.

2-PROPENE-1-SULFONIC ACID, SODIUM SALT

$C_3H_5NaO_3S$

Mol. Wt. 144.13

Solvent: D_2O

Assignments
a = 3.68
b = ca 5.34
c = 6.00
d = 4.51 HDO

1299

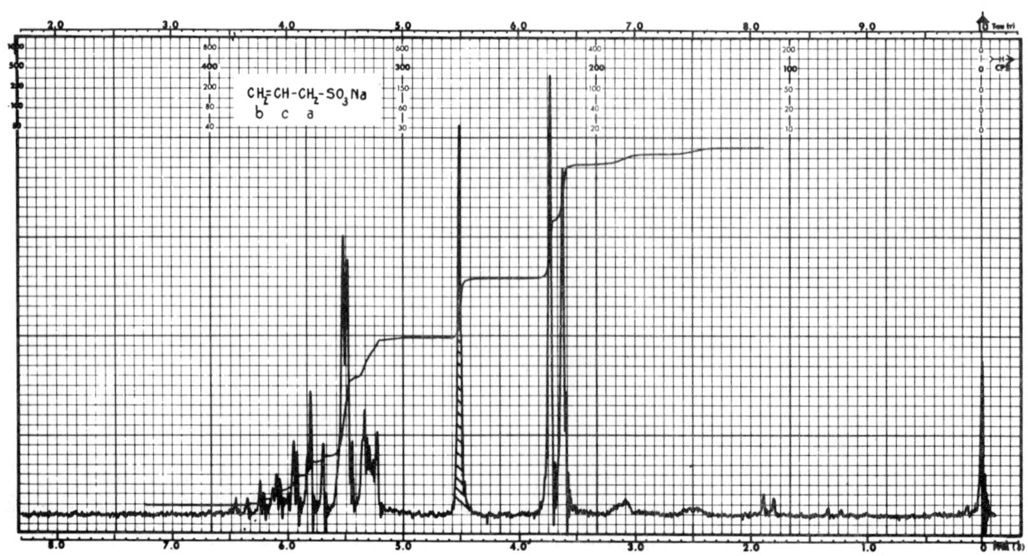

© 1978 Sadtler Research Laboratories, Inc.

BENZENESULFONIC ACID, SODIUM SALT

$C_6H_5NaO_3S$

Mol. Wt. 180.16

M.P. 450°C (dec.)

Solvent: D_2O

Assignments
a = 7.58
b = 7.92
c = 4.62 HDO

1300

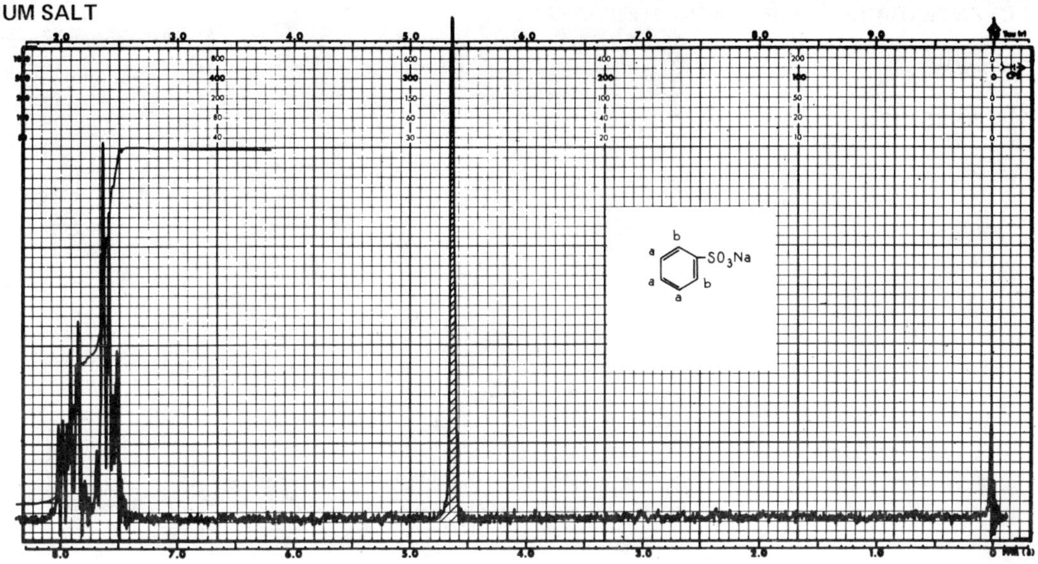

© 1978 Sadtler Research Laboratories, Inc.

2,4-XYLENESULFONIC ACID, SODIUM SALT

$C_8H_9NaO_3S$

Mol. Wt. 208.21

Solvent: D_2O

Assignments
a = 2.24
b = 2.65
c = 7.00-7.20
d = 7.87
e = 4.62 HDO

1301

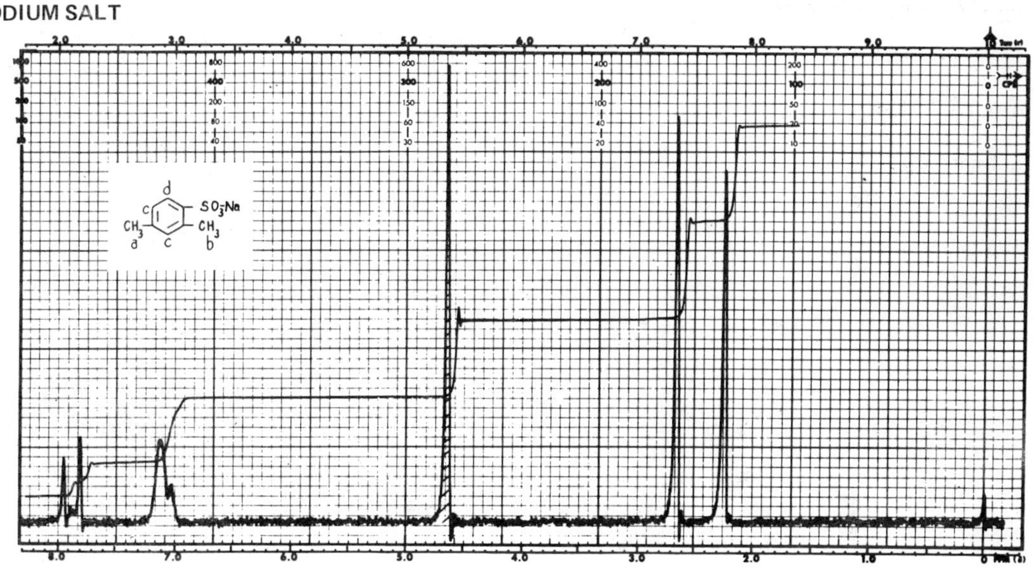

© 1978 Sadtler Research Laboratories, Inc.

p-TOLUENESULFONIC ACID, BARIUM SALT

$C_{14}H_{14}BaO_6S_2$

Mol. Wt. 479.73

Solvent: D_2O

Assignments
a = 2.37
b = 7.32
c = 7.76
d = 4.11 HDO
e = 2.65 imp.

1302

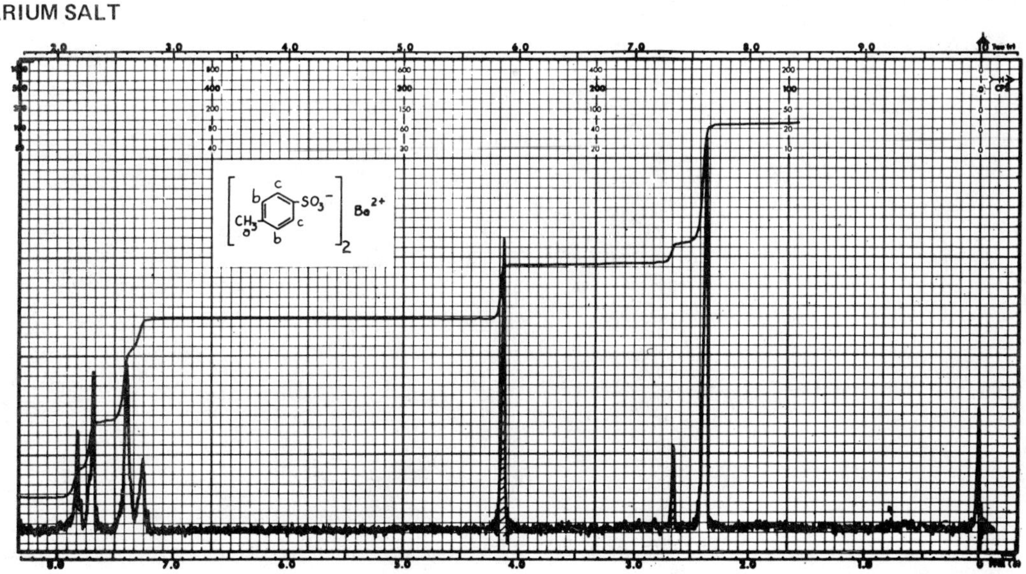

© 1978 Sadtler Research Laboratories, Inc.

p-BROMOBENZENESULFONIC ACID, SODIUM SALT

$C_6H_4BrNaO_3S$

Mol. Wt. 259.05

Solvent: D_2O

Assignments
a = 7.68
b = 7.79
c = 4.48 HDO

1303

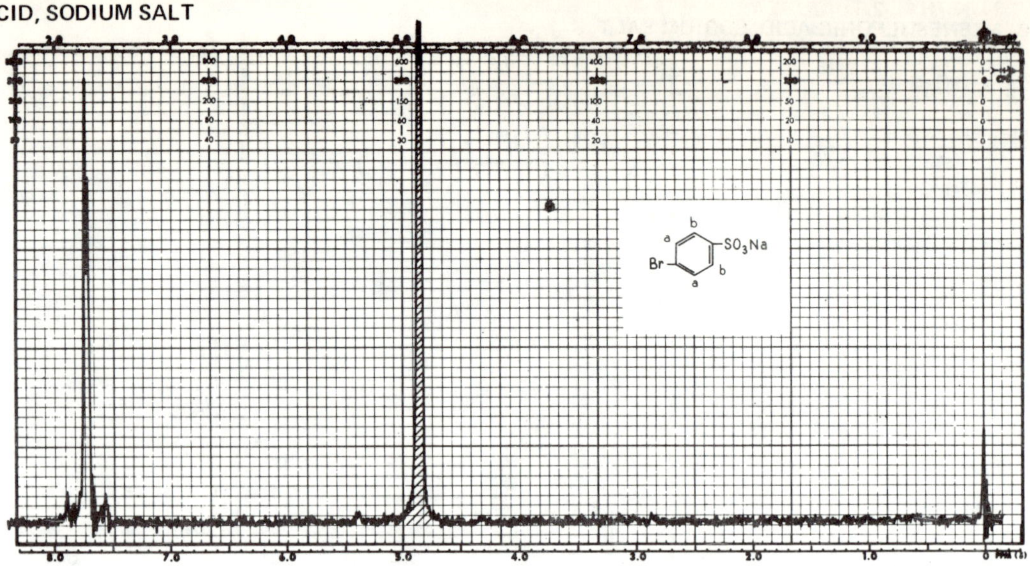

© 1978 Sadtler Research Laboratories, Inc.

SULFANILIC ACID, SODIUM SALT, DIHYDRATE

$C_6H_6NNaO_3S \cdot 2H_2O$

Mol. Wt. 231.20

Solvent: D_2O

Assignment
a = 6.79
b = 7.62
c = 4.10 HDO

1304

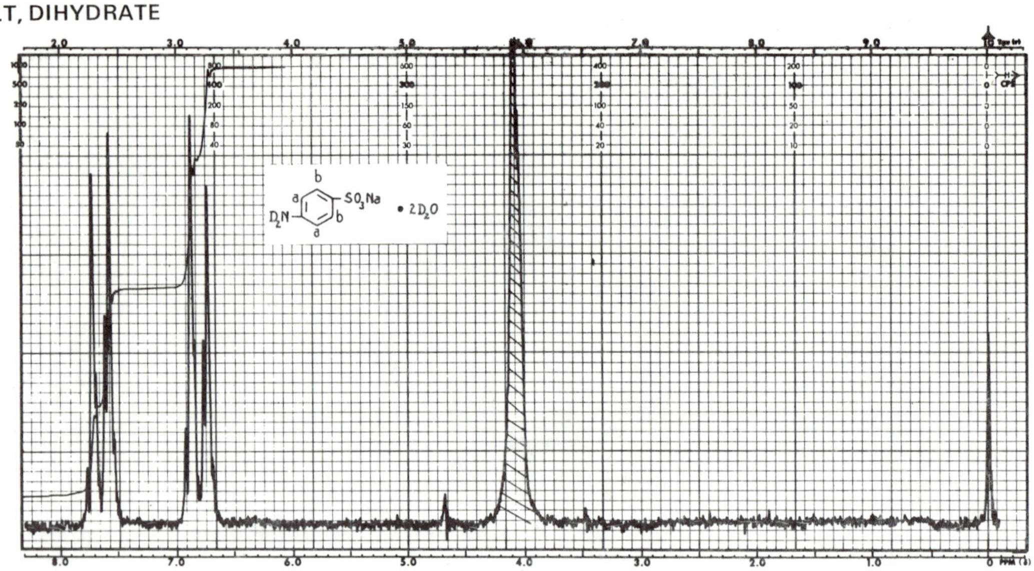

© 1978 Sadtler Research Laboratories, Inc.

p-(PHENYLAZO)BENZENESULFONIC ACID, SODIUM SALT

$C_{12}H_9N_2NaO_3S$

Mol. Wt. 284.27

Solvent: Dimethyl sulfoxide

Assignments
a = 7.50-7.75
b = 7.80-8.07

1305

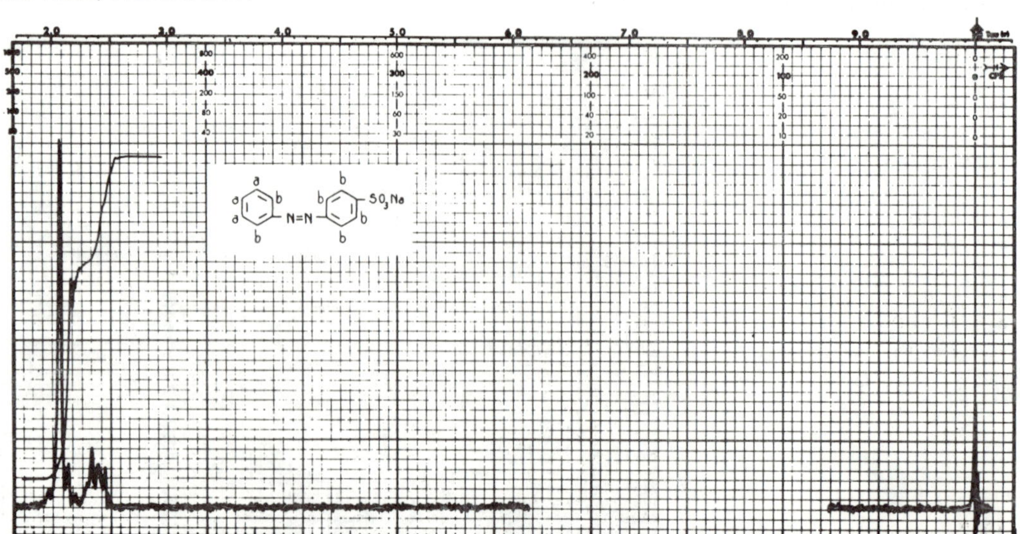

© 1978 Sadtler Research Laboratories, Inc.

2-ANILINO-5-NITROBENZENESULFONIC ACID, SODIUM SALT

$C_{12}H_9N_2NaO_5S$

Mol. Wt. 316.27

Solvent: Dimethyl sulfoxide

Assignments
a = 7.21
b = 7.10-7.95
c = 8.08
d = 8.51
e = 9.24

1306

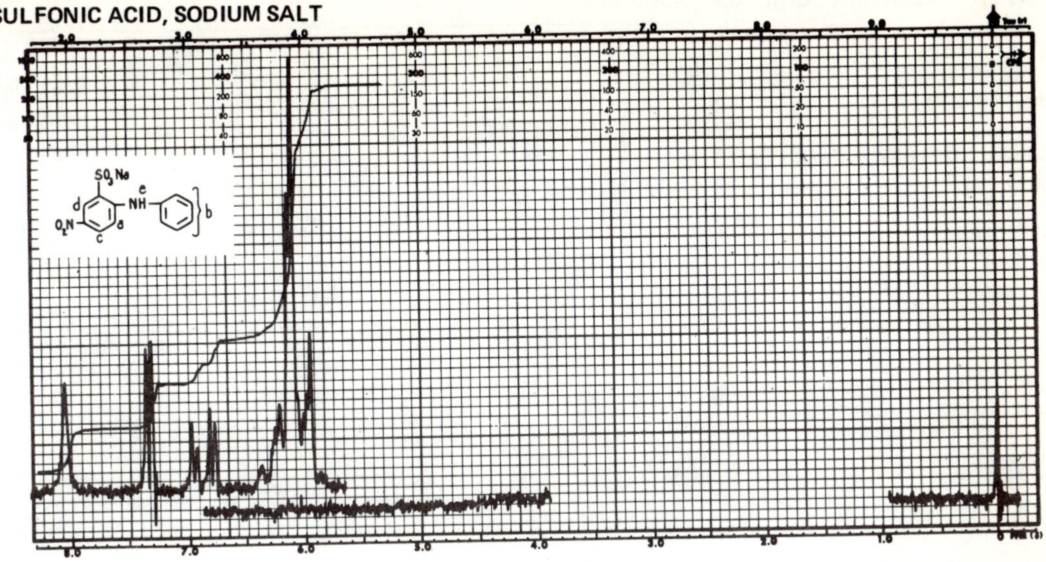

© 1978 Sadtler Research Laboratories, Inc.

2,4-DINITROBENZENESULFONIC ACID, SODIUM SALT

$C_6H_3NaN_2O_7S$

Mol. Wt. 270.15

Solvent: D_2O

Assignments
a = 8.29
b = 8.57
c = 8.62
d = 4.11 HDO

1307

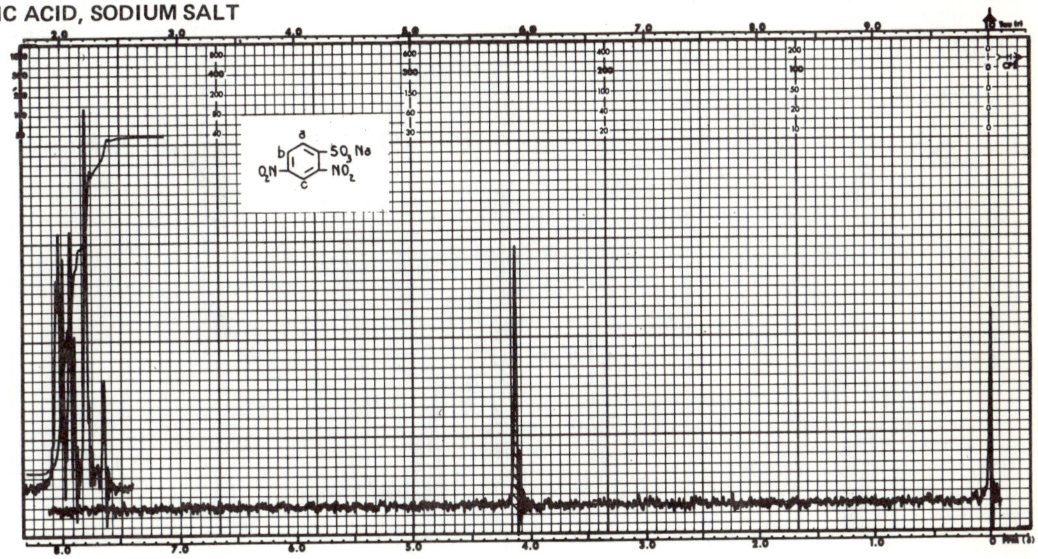

© 1978 Sadtler Research Laboratories, Inc.

1,3,5-BENZENETRISULFONIC ACID, TRISODIUM SALT, TRIHYDRATE

$C_6H_3Na_+O_9S_3 \cdot 3H_2O$

Mol. Wt. 438.29

Solvent: D_2O

Assignments
a = 8.45
b = 4.48 HDO
c = 2.71 imp.

1308

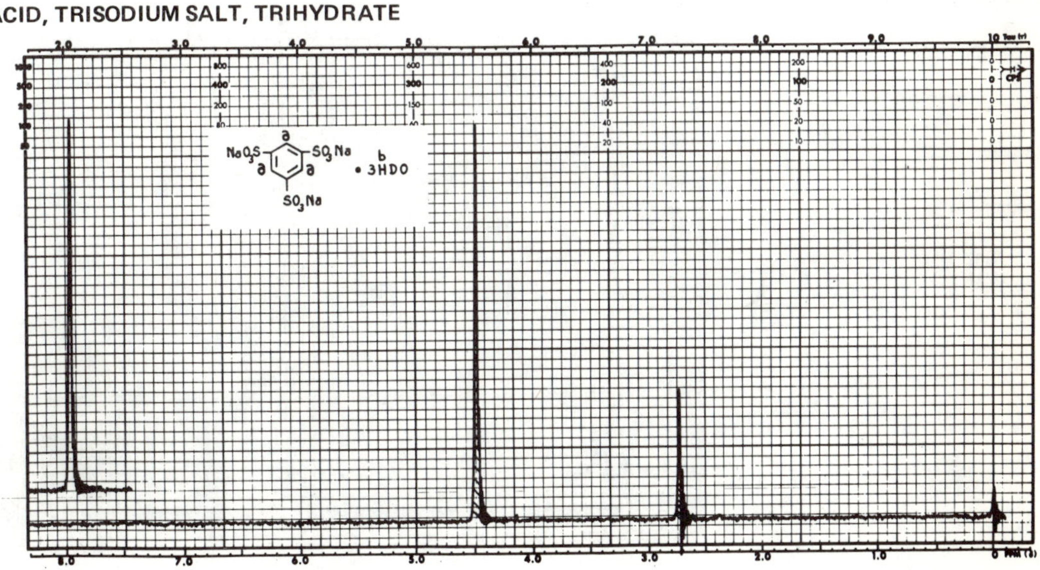

© 1978 Sadtler Research Laboratories, Inc.

2-NAPHTHALENESULFONIC ACID, SODIUM SALT

$C_{10}H_7NaO_3S$

Mol. Wt. 230.22

Solvent: Dimethyl sulfoxide-D_6

Assignments
a = 7.53
b = ca 7.89
c = 8.28
d = 2.51 DMSO-D_5

1309

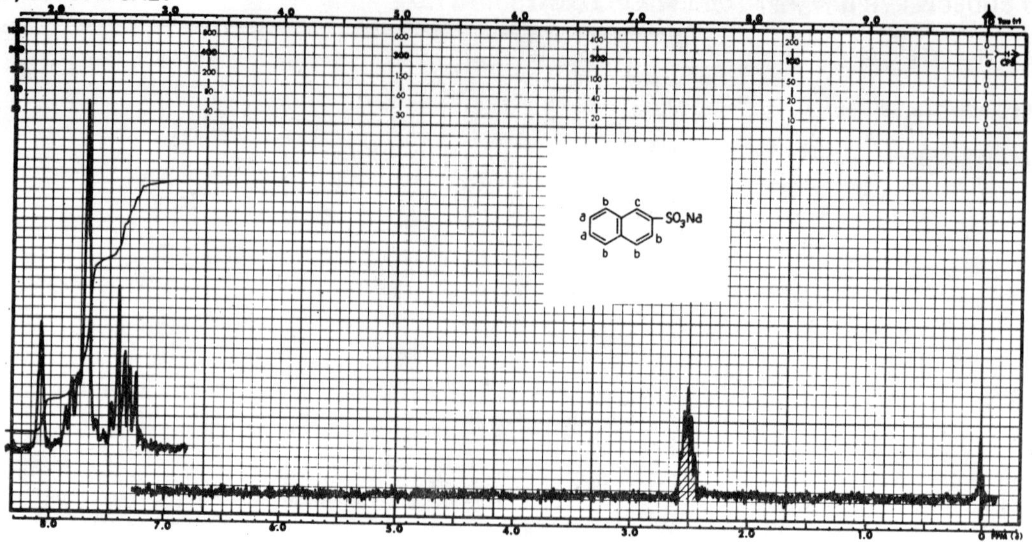

© 1978 Sadtler Research Laboratories, Inc.

4-AMINO-1-NAPHTHALENESULFONIC ACID, SODIUM SALT, HEMIHYDRATE

$C_{10}H_8NNaO_3S \cdot ½H_2O$

Mol. Wt. 254.24

Solvent: D_2O

Assignments
a = 6.76
b = 7.41
c = 7.55
d = 7.76
e = 8.09
f = 8.83
g = 4.30 HDO

1310

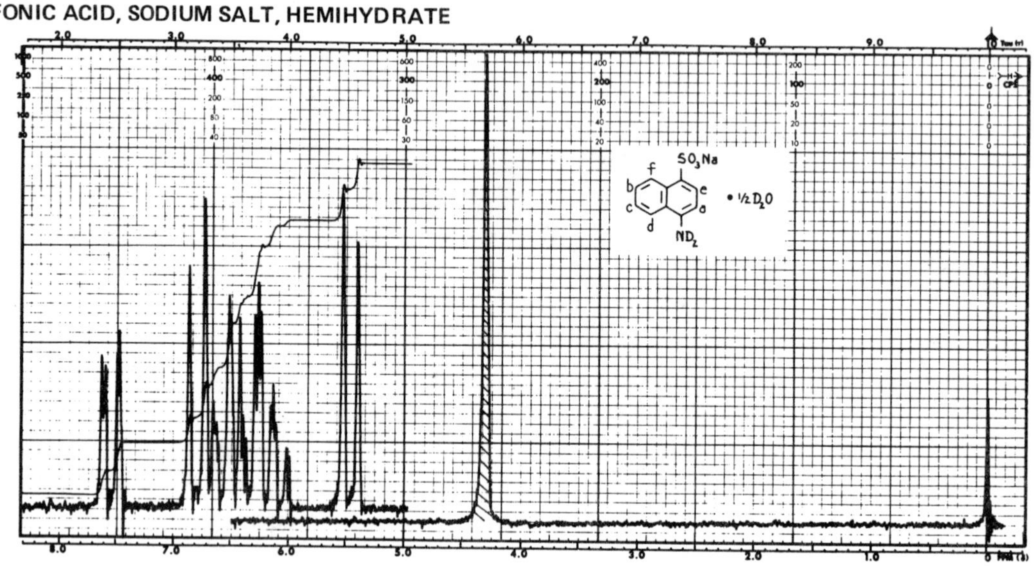

© 1978 Sadtler Research Laboratories, Inc.

2,7-NAPHTHALENEDISULFONIC ACID, DISODIUM SALT

$C_{10}H_6NaO_6S_2$

Mol. Wt. 332.26

Solvent: D_2O

Assignments
a = 7.79
b = 7.99
c = 8.40
d = 4.73 HDO

1311

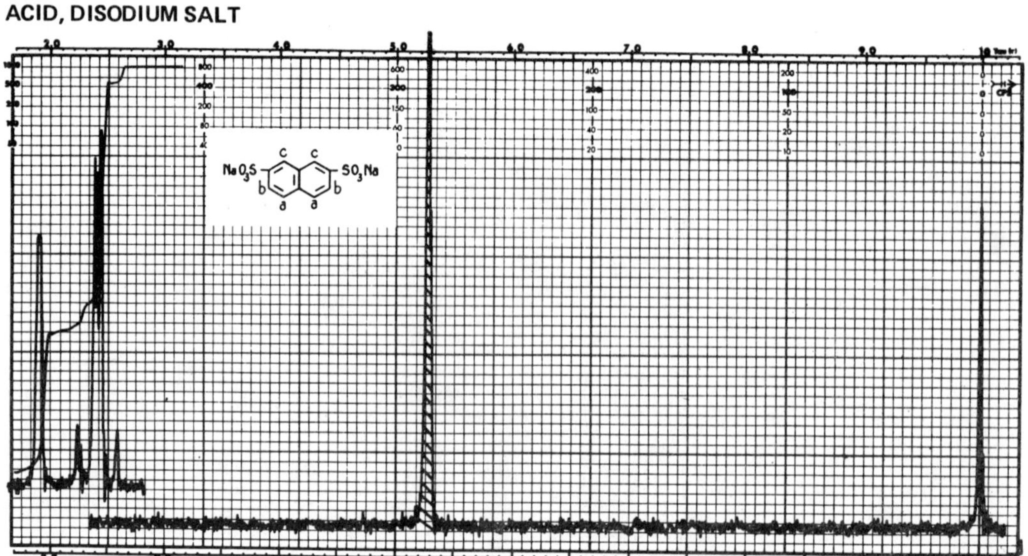

© 1978 Sadtler Research Laboratories, Inc.

1-DODECYLPYRIDINIUM p-TOLUENESULFONATE

$C_{24}H_{37}NO_3S$

Mol. Wt. 419.63

M.P. 138-140°C

Solvent: $CDCl_3$

Assignments

a =	0.89	h =	7.99
b =	ca 1.21	i =	8.38
c =	1.82	j =	9.16
d =	2.31	k =	3.13 imp.
e =	4.61		
f =	7.12		
g =	7.75		

1312

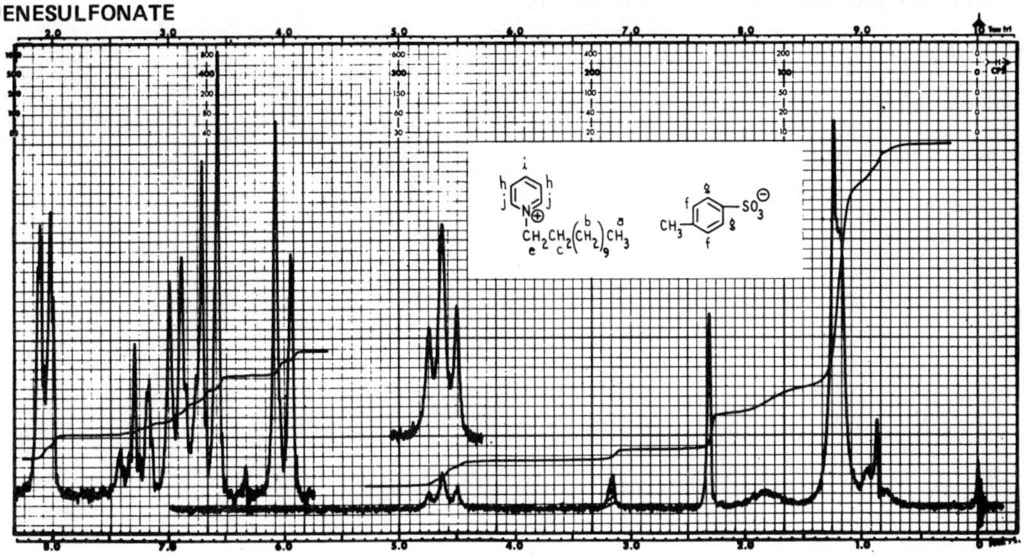

© 1978 Sadtler Research Laboratories, Inc.

PHENYLTRIMETHYLAMMONIUM, BENZENESULFONATE

$C_{15}H_{19}NO_3S$

Mol. Wt.293.39

M.P. 192-194°C

Solvent: Polysol-d

Assignments
- a = 3.70
- b = 7.20-7.45
- c = 7.50-7.70
- d = 7.77
- e = 8.01
- f = 2.52 DMSO-d_5

1313

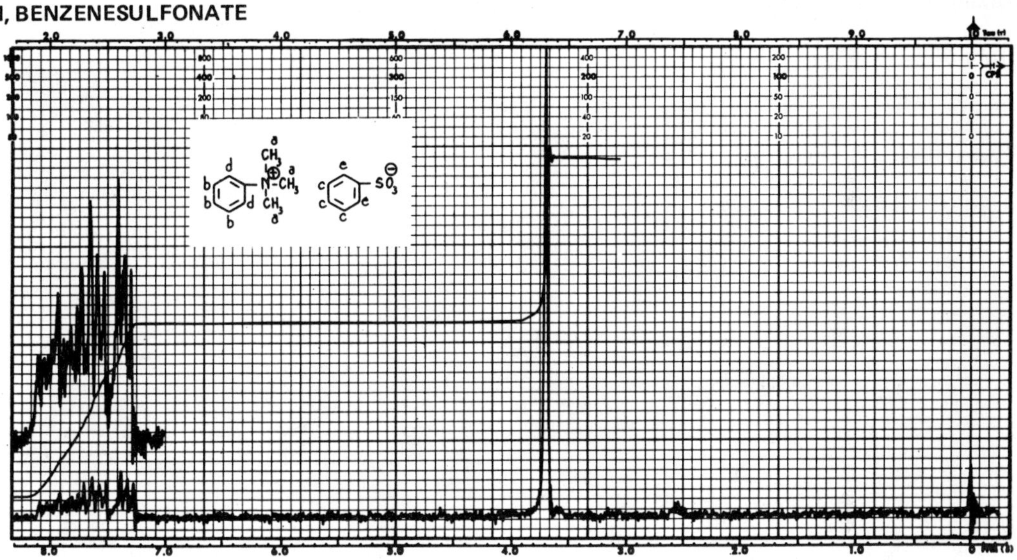

© 1978 Sadtler Research Laboratories, Inc.

METHYL POTASSIUM SULFATE

CH_3KO_4S

Mol. Wt. 150.20

Solvent: D_2O

Assignments
- a = 3.80
- b = 4.08 HDO

1314

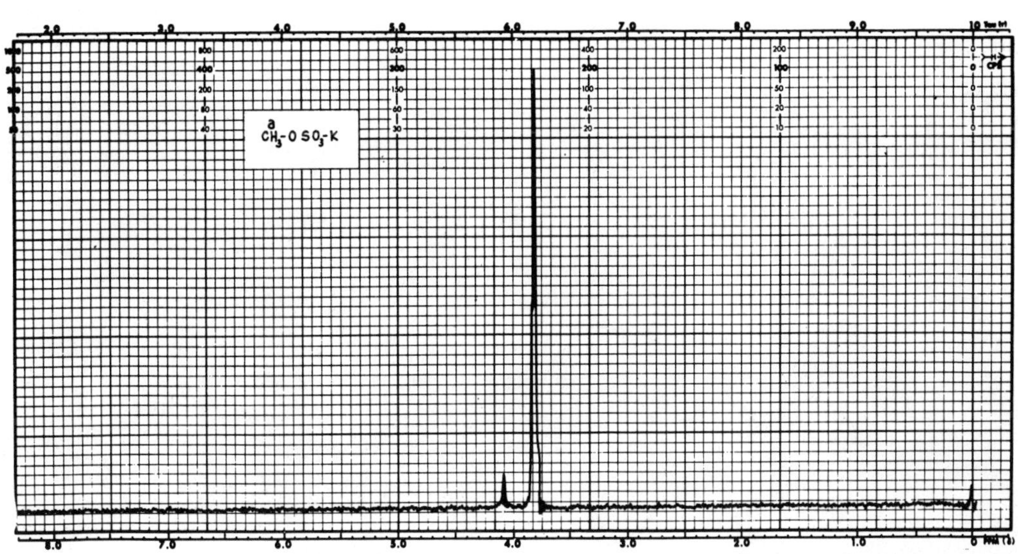

© 1978 Sadtler Research Laboratories, Inc.

* Belongs with group starting with spectrum 1326.

NOTES

ESTERS OF SULFONIC ACID

The sulfonic acid functional group produces two distinct sets of chemical shifts for the adjacent aliphatic groups. As noted previously, the groups bonded to the acid side of the linkage are weakly deshielded but those bonded to the alcohol side are very strongly deshielded. This situation is analogous to that which is encountered with the esters of carboxylic acids. The sulfonic acid esters, unlike the free acids and their salts, are readily soluble in $CDCl_3$ and CCl_4, and relatively insoluble in deuterium oxide.

Chemical Shifts

Aliphatic Protons

HC–3	HC–2	HC–1		Solvent	Spectrum
		2.94	CH_3–SO_2–O–R_2	CCl_4	----
		3.00	CH_3–SO_2–O–R_4	$CDCl_3$	1315
	1.39	3.09	CH_3–CH_2–SO_2–O–CH_3	CCl_4	1316
		3.85	CH_3–O–S(O_2)–R_2	CCl_4	1316
		3.89	CH_3–O–S(O_2)–Ø	$CDCl_3$	----
	1.25	4.01	CH_3–CH_2–O–S(O_2)–Ø	CCl_4	----
	1.39	4.21	CH_3–CH_2–O–S(O_2)–CH_3	CCl_4	----
0.89	1.59	3.99	CH_3–CH_2–CH_2–O–S(O_2)–Ø	$CDCl_3$	----
0.89	1.91	3.71	$(CH_3)_2$–CH–CH_2–O–S(O_2)–Ø	CCl_4	----

Aromatic Protons

In relation to the aromatic hydrogens, the effect of the sulfonic acid ester linkage is opposite to that observed for the aliphatic groups, i.e. the oxygen side of the linkage weakly shields the ortho protons while the sulfur side strongly deshields them.

para/meta/ortho		
6.9–7.5	–O–S(O_2)–Ø	$CDCl_3$
7.0–7.5	–O–S(O_2)–CH=CH_2	CCl_4

para/meta	ortho		
7.3–7.7	7.83	–S(O_2)–O–R_3	$CDCl_3$

METHANESULFONIC ACID, BUTYL ESTER

$C_5H_{12}O_3S$

Mol. Wt. 152.21

B.P. 102-103°C/5mm

Solvent: $CDCl_3$

Assignments
a = 0.97
b = 1.10-2.10
c = 3.00
d = 4.23

1315

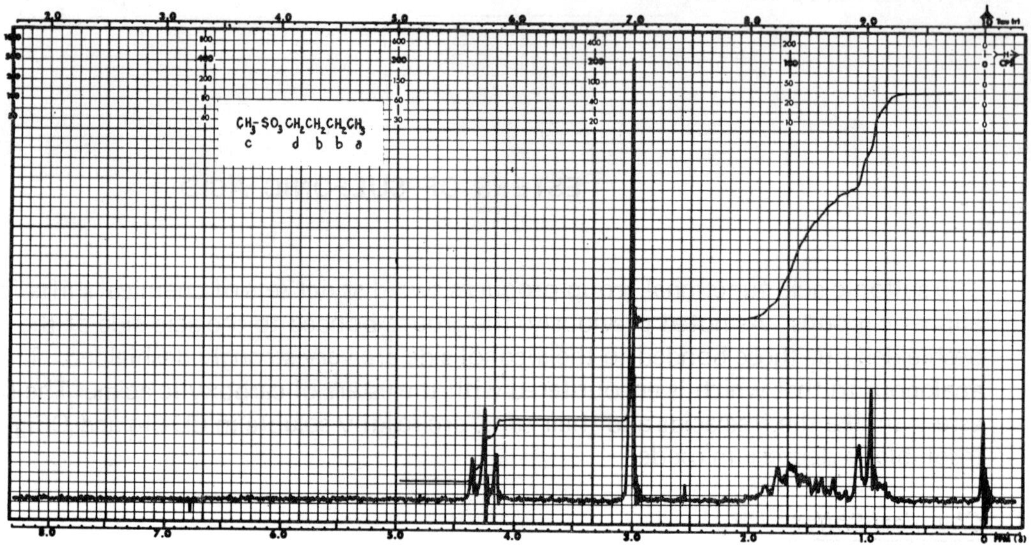

© 1978 Sadtler Research Laboratories, Inc.

ETHANESULFONIC ACID, METHYL ESTER

$C_3H_8O_3S$

Mol. Wt. 124.16

B.P. 97-98°C/16mm

Solvent: CCl_4

Assignments
a = 1.39
b = 3.09
c = 3.85

1316

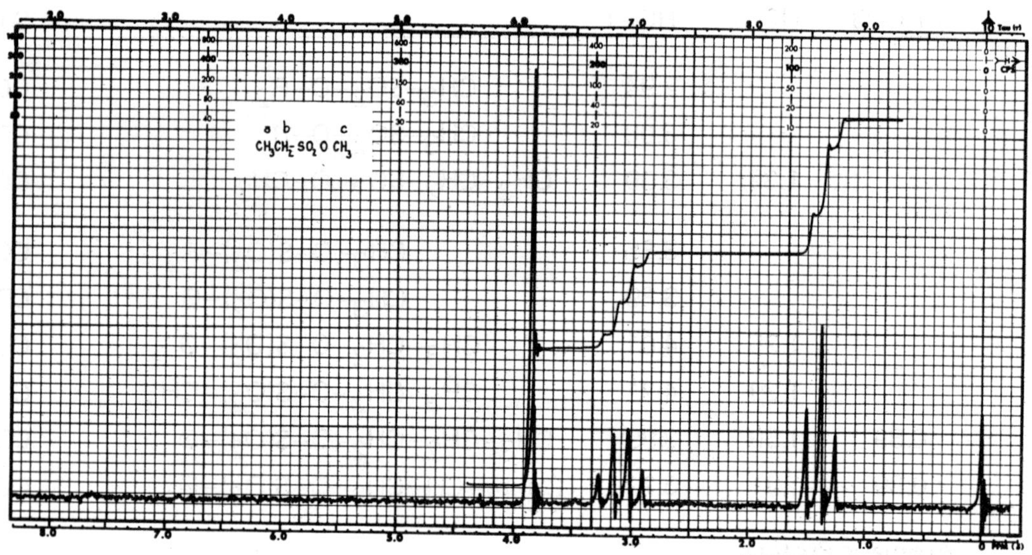

© 1978 Sadtler Research Laboratories, Inc.

METHANESULFONIC ACID, DECYL ESTER

$C_{11}H_{24}O_3S$

Mol. Wt. 236.37

M.P. 18.5°C

Solvent: $CDCl_3$

Assignments
a = 0.89
b = ca 1.28
c = 2.97
d = 4.20

1317

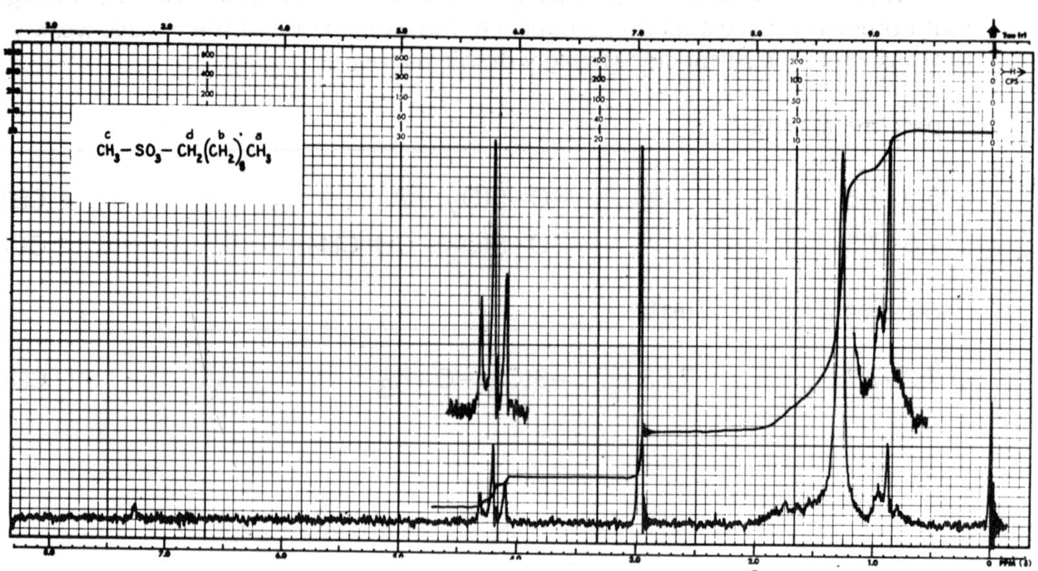

© 1978 Sadtler Research Laboratories, Inc.

1,2-OXATHIOLANE, 2,2-DIOXIDE

$C_3H_6O_3S$

Mol. Wt. 122.14

M.P. 30-33°C

B.P. 180°C/30mm

Solvent: $CDCl_3$

Assignments
a = 2.61
b = 3.23
c = 4.50

1318

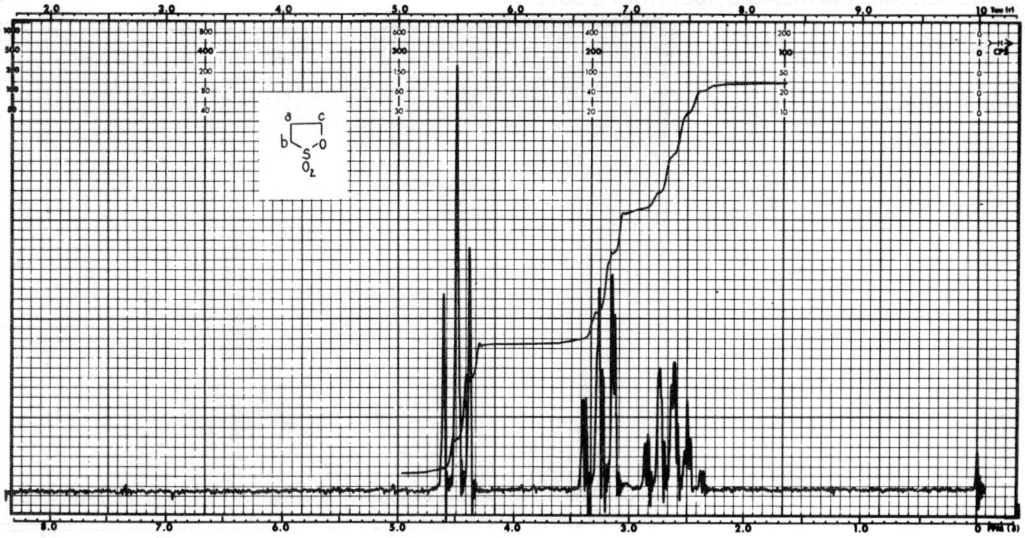

© 1978 Sadtler Research Laboratories, Inc.

1,2-OXATHIANE, 2,2-DIOXIDE

$C_4H_8O_3S$

Mol. Wt. 136.17

Solvent: $CDCl_3$

Assignments
a = 1.60-2.50
b = 3.17
c = 4.55
d = 1.44, 1.57 imps.

1319

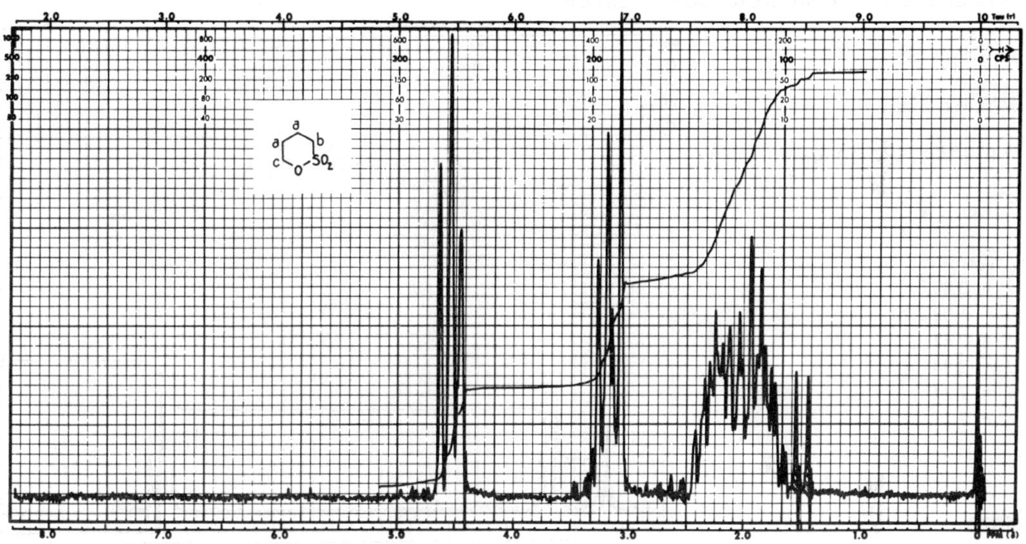

© 1978 Sadtler Research Laboratories, Inc.

METHANESULFONIC ACID, PHENYL ESTER

$C_7H_8O_3S$

Mol. Wt. 172.21

M.P. 60-62°C

Solvent: $CDCl_3$

Assignments
a = 3.07
b = ca 7.33

1320

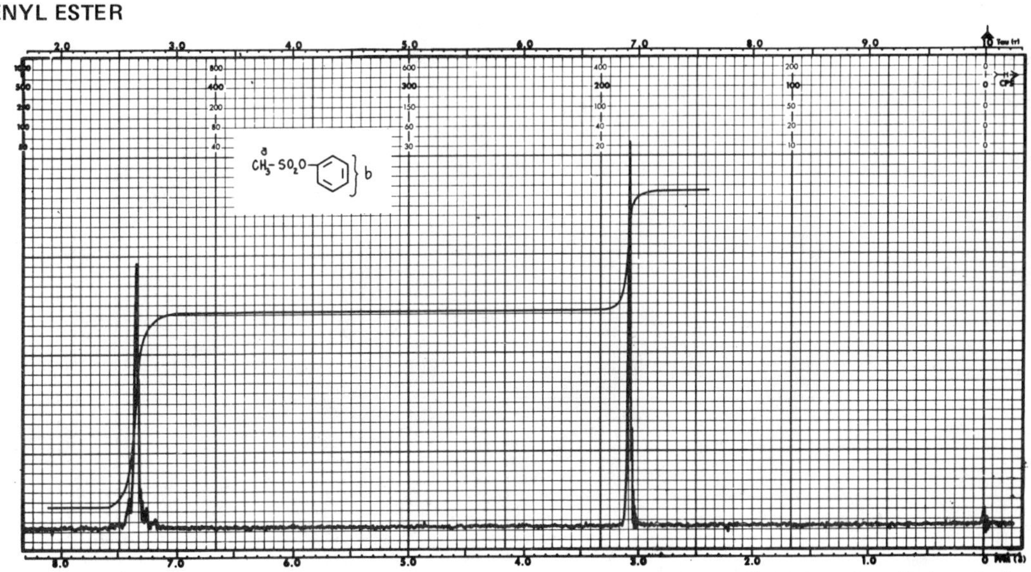

© 1978 Sadtler Research Laboratories, Inc.

p-TOLUENESULFONIC ACID, ISOPROPYL ESTER

$C_{10}H_{14}O_3S$

Mol. Wt. 214.29

M.P. 15-20°C

Solvent: CCl_4

Assignments

a = 1.23

b = 2.40

c = 4.66

d = 7.28

e = 7.20

1321

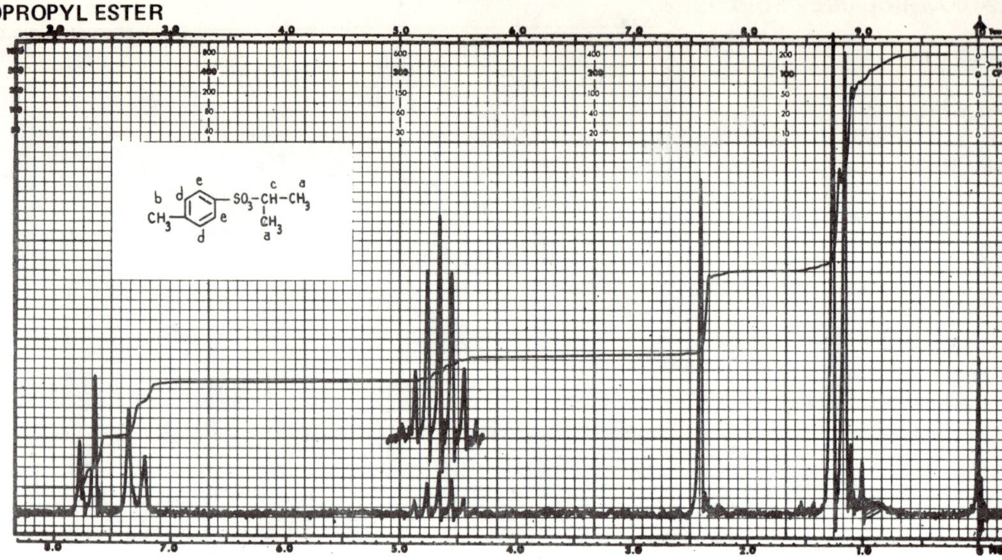

© 1978 Sadtler Research Laboratories, Inc.

p-TOLUENESULFONIC ACID, BUTYL ESTER

$C_{11}H_{16}O_3S$

Mol. Wt. 228.31

B.P. 174-175°C/10mm

Solvent: CCl_4

Assignments

a = 0.88

b = 1.08-2.70

c = 2.42

d = 3.94

e = 7.27

f = 7.71

1322

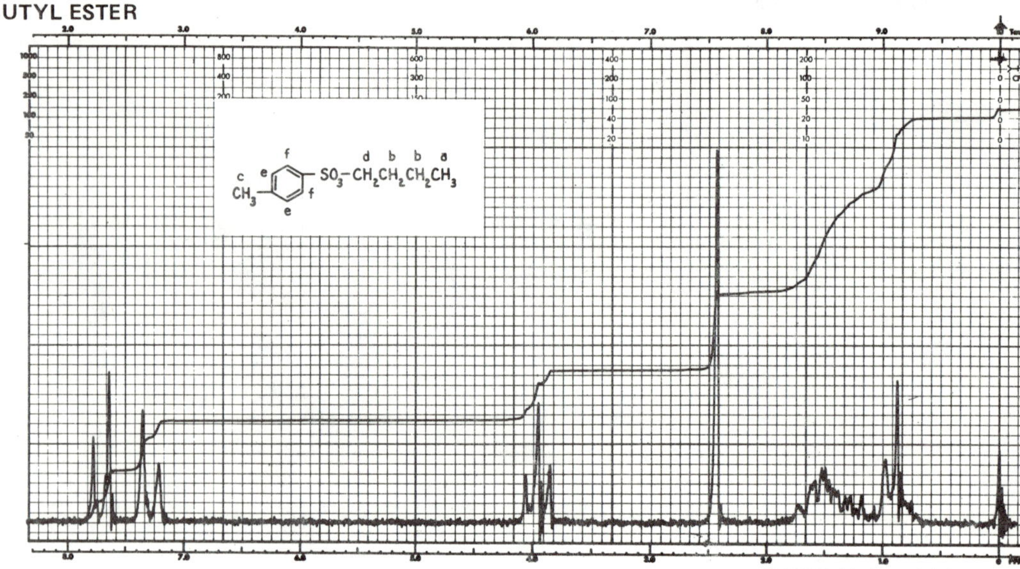

© 1978 Sadtler Research Laboratories, Inc.

THE ESTERS OF SULFUROUS ACID

The esters of Sulfurous acid are a small group of compounds for which only a few aliphatic derivatives are available commercially. As with all of the esters, the oxygen atom adjacent to the alpha carbon group has a strong deshielding effect upon the protons bonded to it.

Chemical Shifts

Aliphatic Protons

HC–5	HC–4,3	HC–2	HC–1		Solvent	Spectrum
			3.59	CH_3–O–S(=O)–O–CH_3	CCl_4	1323
		1.34	4.04	CH_3–CH_2–O–S(=O)–O–R_2	CCl_4	1324
	0.96	1.67	3.92	CH_3–CH_2–CH_2–O–S(=O)–O–R_3	$CDCl_3$	----
0.92	(1.2–2.0)		3.94	CH_3–$(CH_2)_3$–CH_2–O–S(=O)–O–R_5	CCl_4	1325

Coupling and Coupling Constants

No unusual couplings nor coupling constants have been noted for the esters of Sulfurous acid. The aliphatic patterns are similar in appearance to those of similarly strong deshielding groups such as the ether group.

Solubility and Solvent Effects

The Sulfites are readily soluble in the chlorinated solvents such as carbon tetrachloride and deuterochloroform.

Characterization

Because of the relatively narrow range of chemical shifts observed for the esters of the various sulfur containing acids (see table below) the functional group of such compounds is best characterized through the analysis of their Infrared spectra.

CH_3–		Acid		
3.59	CH_3–O–S(=O)–O–CH_3	Sulfurous acid	CCl_4	1323
3.73	CH_3–O–$S(O_2)$–O–K	Sulfuric acid	D_2O	1314
3.73	CH_3–O–$S(O_2)$–Ø	Benzenesulfonic acid	$CDCl_3$	----
3.88	CH_3–O–$S(O_2)$–CH_3	Methanesulfonic acid	CCl_4	----

SULFUROUS ACID, DIMETHYL ESTER

$C_2H_6O_3S$

Mol. Wt. 110.13

B.P. 126-127°C

Solvent: CCl_4

Assignments
a = 3.59

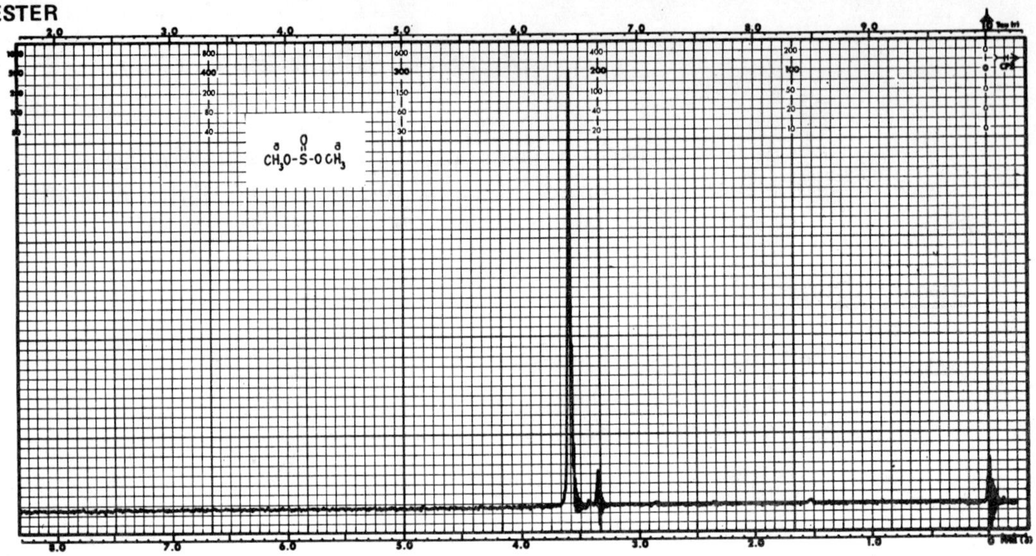

1323

© 1978 Sadtler Research Laboratories, Inc.

ETHYL SULFITE

$C_4H_{10}O_3S$

Mol. Wt. 138.19

Solvent: CCl_4

Assignments
a = 1.34
b = 4.04

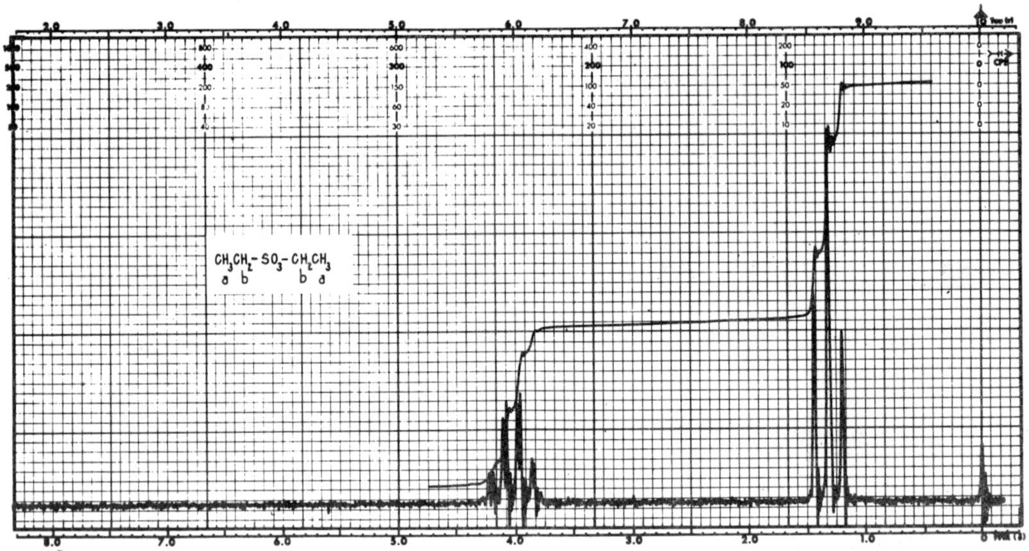

1324

© 1978 Sadtler Research Laboratories, Inc.

PENTYL SULFITE

$C_{10}H_{22}O_3S$

Mol. Wt. 222.35

B.P. 132-134°C/15mm

Solvent: CCl_4

Assignments
a = 0.92
b = 1.15-2.00
c = 3.94

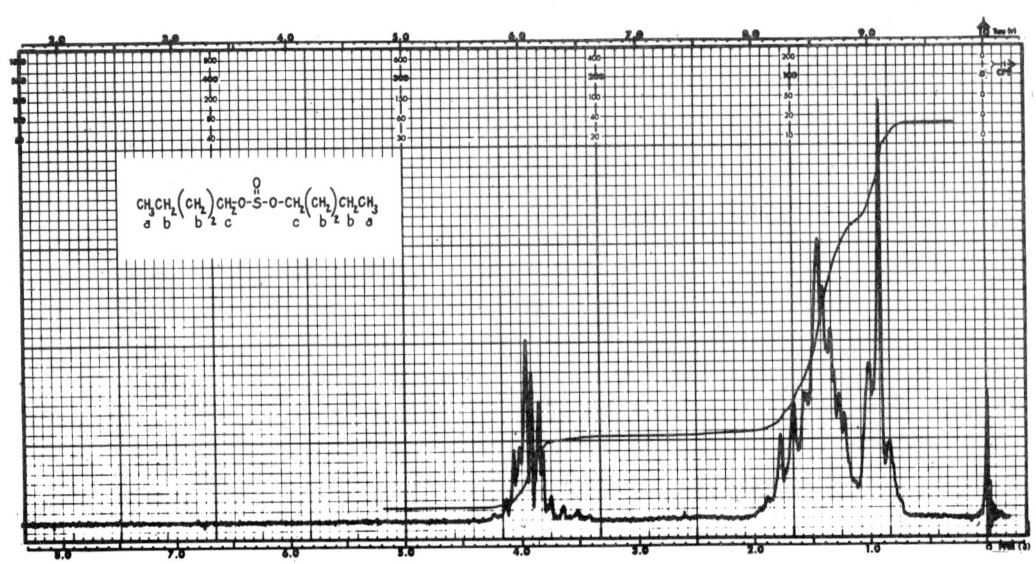

1325

© 1978 Sadtler Research Laboratories, Inc.

THE SALTS OF SULFURIC ACID

The mono-salts of Sulfuric acid display a strong deshielding effect on the aliphatic groups of the ester portion of the molecule. This deshielding effect is similar to that noted for the esters of Sulfurous acid (page 581).

Chemical Shifts

Aliphatic Protons

Methyl Esters

CH_3–	X	Solvent	Spectrum
3.71	$-O-S(O_2)-O^-$ $(^+N(CH_3)_4)$	D_2O	
3.72	$-O-S(O_2)-O^-$ $(^+S(R)_3)$	D_2O	
3.73	$-O-S(O_2)-O-Na$	D_2O	
3.95	$-O-S(O_2)-O^-$ $(^+N(R)_4)$	TFA	

$(CH_2)_n$–	HC–2	HC–1			
		3.73	$CH_3-O-S(O_2)-O-K$	D_2O	1314
	1.31	4.12	$CH_3-CH_2-O-S(O_2)-O-K$	Polysol	1326
$(1.1-1.6)_2$	1.69	4.06	$CH_3(CH_2)_2-CH_2-CH_2-O-S(O_2)-O-Na$	D_2O	1327
$(1.1-1.5)_5$	1.61	4.01	$CH_3(CH_2)_5-CH_2-CH_2-O-S(O_2)-O-Na$	D_2O	----
$(1.1-1.5)_7$	1.61	3.46	$CH_3(CH_2)_7-CH_2-CH_2-O-S(O_2)-O-Na$	Polysol	----
$(1.1-1.6)_{11}$	1.61	3.82	$CH_3(CH_2)_{10}-CH_2-CH_2-O-S(O_2)-O-Na$	Polysol	1328

Solubility and Solvent Effects

The presence of the mono-salt function makes the mono-esters much more soluble in solvents such as Polysol, DMSO-d_6 and D_2O than the corresponding di-esters (page 585). The rather wide divergence in chemical shift noted for the compounds examined in Polysol solution most probably arises from the varying amounts of H_2O which are often present in such solutions. Based upon the information supplied in the table above, it can be inferred that the chemical shifts of the Sulfuric acid ester/salts appear at highest field in relatively dry Polysol solution shifting to lower field as the amount of H_2O increases and finally reach maximum deshielding when the solvent is 100% H_2O (D_2O).

SULFURIC ACID, ETHYL ESTER, POTASSIUM SALT

$C_2H_5KO_4S$

Mol. Wt. 104.23

M.P. 203-204.5°C

Solvent: D_2O

Assignments
a = 1.31
b = 4.12
c = 4.51 HDO

1326

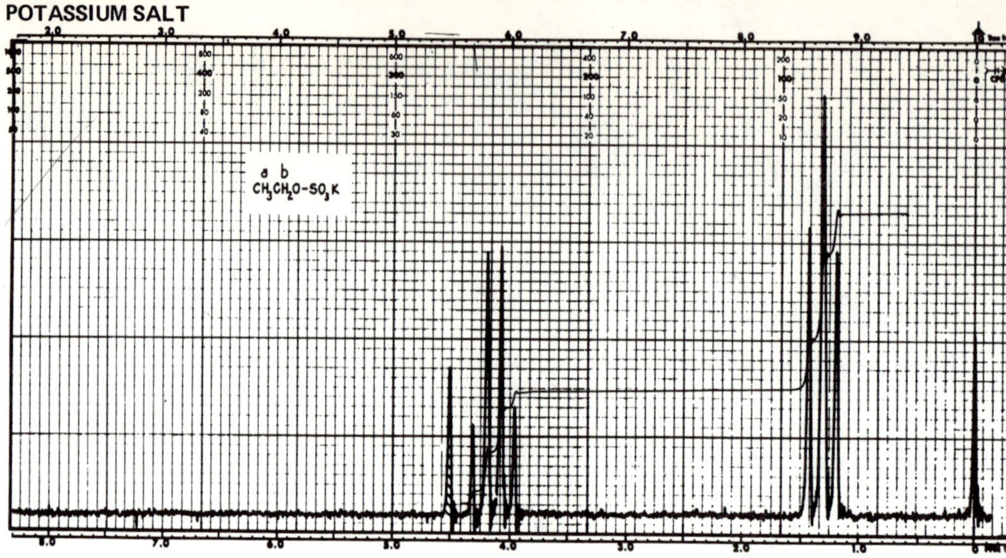

© 1978 Sadtler Research Laboratories, Inc.

SULFURIC ACID, MONOPENTYL ESTER, SODIUM SALT

$C_5H_{11}NaO_4S$

Mol. Wt. 190.19

Solvent: D_2O

Assignments
a = 0.92
b = 1.00-1.60
c = 1.69
d = 4.06
e = 4.32 HDO

1327

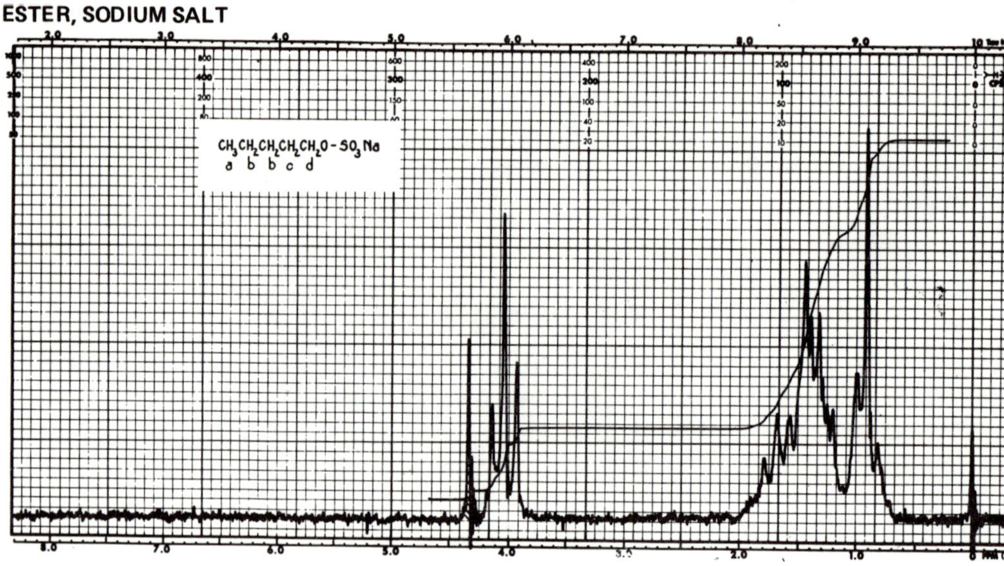

© 1978 Sadtler Research Laboratories, Inc.

SULFURIC ACID, MONOTETRADECYL ESTER, SODIUM SALT

$C_{14}H_{29}NaO_4S$

Mol. Wt. 316.44

Solvent: Polysol-d

Assignments
a = 0.89
b = 1.10-1.90
c = 3.82

1328

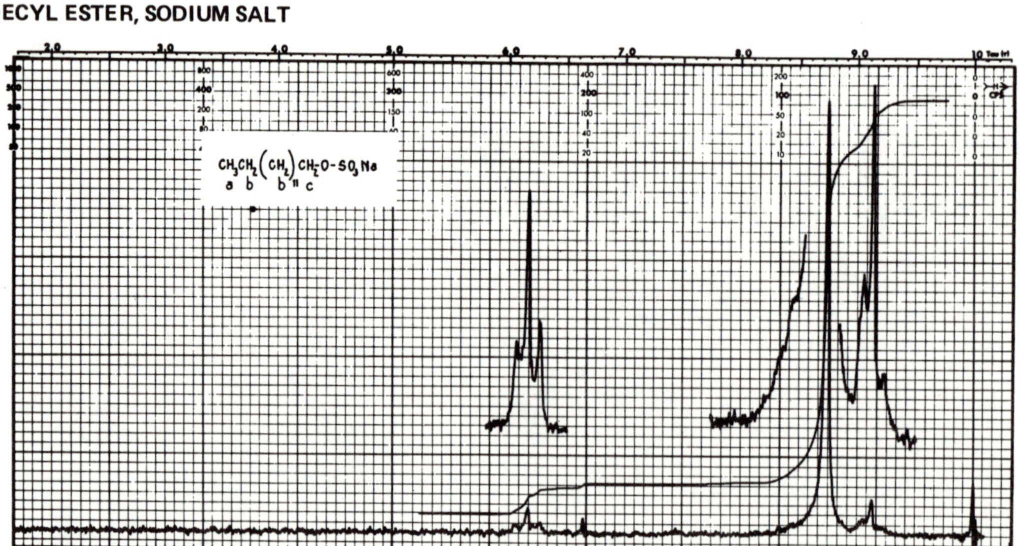

© 1978 Sadtler Research Laboratories, Inc.

THE DIESTERS OF SULFURIC ACID

The diesters of Sulfuric acid exhibit chemical shifts similar to those of the ester/salts with minor variations due to the different solvents employed. The diesters are found to be much more soluble in the chlorinated hydrocarbons than the more polar ester/salts.

Chemical Shifts

Aliphatic Protons

HC–3	HC–2	HC–1		Solvent	Spectrum
		3.87	CH_3–O–S(O_2)–O–CH_3	CCl_4	----
	1.45	4.28	CH_3–CH_2–O–S(O_2)–O–R_2	CCl_4	----
1.00	1.75	4.15	CH_3–CH_2–CH_2–O–S(O_2)–O–R_3	CCl_4	1329

Chlorinated Sulfate Esters

HC–2	HC–1			
3.80	4.52	Cl–CH_2–CH_2–O–S(O_2)–O–R	$CDCl_3$	1330
----	5.79	Cl–CH_2–O–S(O_2)–O–R	CCl_4	----

Coupling and Coupling Constants

The diesters of Sulfuric acid do not display any additional or different couplings from the protons of the other normal aliphatic groups. Vicinal coupling is observed (J_{HC-CH} = 6–8 Hz), but longer range coupling if present is too small in magnitude to be detected.

SULFURIC ACID, DIPROPYL ESTER

$C_6H_{14}O_4S$

Mol. Wt. 184.24

B.P. 116-118°C/16mm

Solvent: CCl_4

Assignments
a = 1.00
b = 1.75
c = 4.15

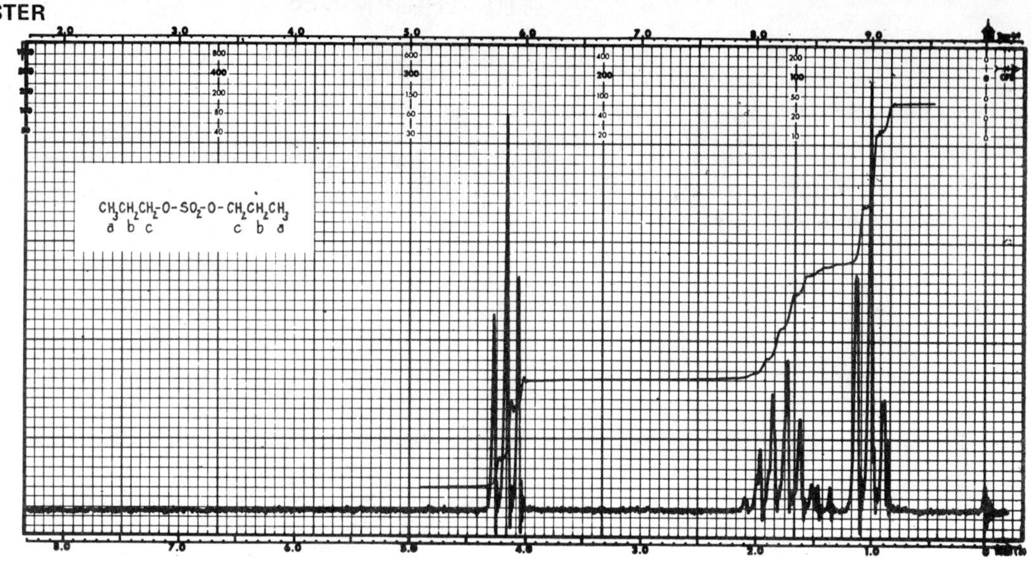

1329

© 1978 Sadtler Research Laboratories, Inc.

SULFURIC ACID, BIS(2-CHLOROETHYL) ESTER

$C_4H_8Cl_2O_4S$

Mol. Wt. 223.08

B.P. 117-119°C/1mm

Solvent: $CDCl_3$

Assignments
a = 3.80
b = 4.52

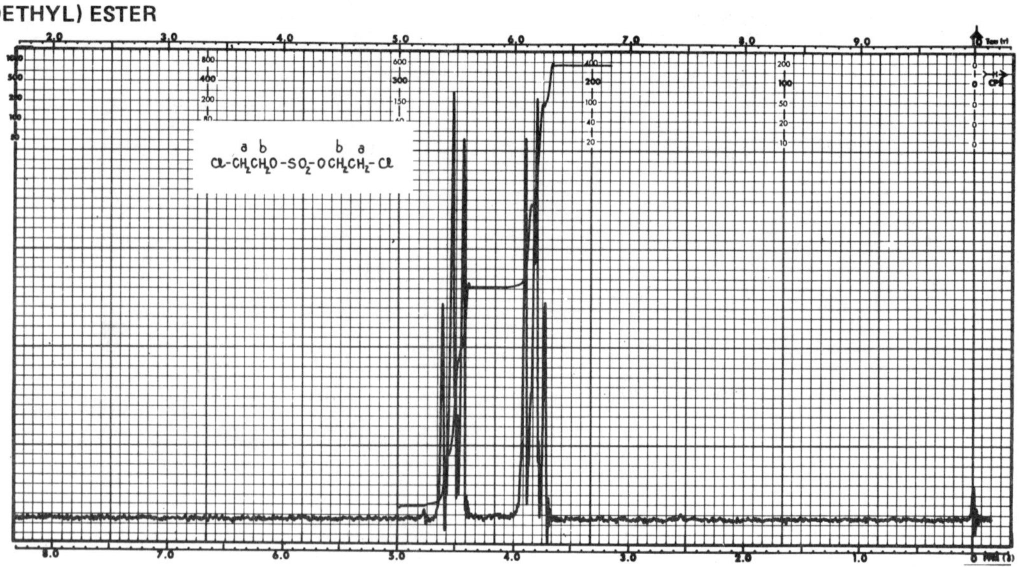

1330

© 1978 Sadtler Research Laboratories, Inc.

The Thioamide functional group is interesting in that a wide variety of chemical shifts may be observed depending upon the substituents and the side of the group to which they are bonded. Non-equivalence is commonly observed for the primary amide protons and also in the case in which two different groups are bonded to the tertiary amide nitrogen atom.

The Thioamides tend to be less soluble in the chlorinated solvents than most of the previously studied groups but are usually quite soluble in Polysol or DMSO-d_6.

Chemical Shifts

Aliphatic Protons

		Solvent	Spectrum
2.40 ppm	CH_3–C(=S)–NH_2	DMSO-d_6	1331
2.62 ppm	CH_3–C(=S)–$N(R)_2$	$CDCl_3$	1337
3.13 ppm	CH_3–N(R)–C(=S)–H	$CDCl_3$	----
(3.27, 3.30) ppm	$(CH_3)_2$–N–C(=S)–H	$CDCl_3$	1336
(3.31, 3.48) ppm	$(CH_3)_2$–N–C(=S)–R	$CDCl_3$	1337
3.34 ppm	CH_3–N(R)–C(=S)–H	$CDCl_3$	----

Aromatic Protons

The Thioamide carbon of this functional group has a moderately strong deshielding effect on the ortho aromatic hydrogens. They appear as a distorted doublet-of-doublets near 7.85 ppm while the para and meta protons produce a complex, higher order band at higher field.

	ortho	meta/para	
R–NH–C(=S)–Ø	7.81	7.3–7.6 ppm	----
NH_2–C(=S)–Ø	7.89	7.1–7.5 ppm	1333

Exchangeable Protons

The Thioamide protons usually appear as very broad bands at low field. They are often non-equivalent and thus may appear separated in chemical shift by 1–2 ppm.

H_2N–C(=S)	–X		
ca. 8.45	–CH_2–Ø	Polysol	1332
9.20	–CH_3	DMSO-d_6	1331
9.32, 9.65	–Ø	DMSO-d_6	1333

Thioformaldehyde Protons

The Thioformaldehyde protons resonate at very low field as a sharp to slightly broadened single peak. If the nitrogen atom is substituted by two different groups, it is possible for two stable forms to exist in which one group is syn to the aldehydic proton and the other group anti, and vice versa. Clear coupling between the aldehydic proton and aliphatic groups bonded to the nitrogen atom is usually not observed.

H–C(=S)–		Solvent	Spectrum
9.19	$N(CH_3)_2$	$CDCl_3$	1336
9.32, 9.46	$N(CH_3)–CH_2–Ø$	$CDCl_3$	- - - -
9.53	$N(CH_3)–C(CH_3)_3$	$CDCl_3$	- - - -

THIOACETAMIDE

C_2H_5NS

Mol. Wt. 75.13

M.P. 113-114^oC

Solvent: Dimethyl sulfoxide-d_6

Assignments

a = 2.40
b = 9.20
c = 2.59 DMSO-d_5
d = 3.54 H_2O

1331

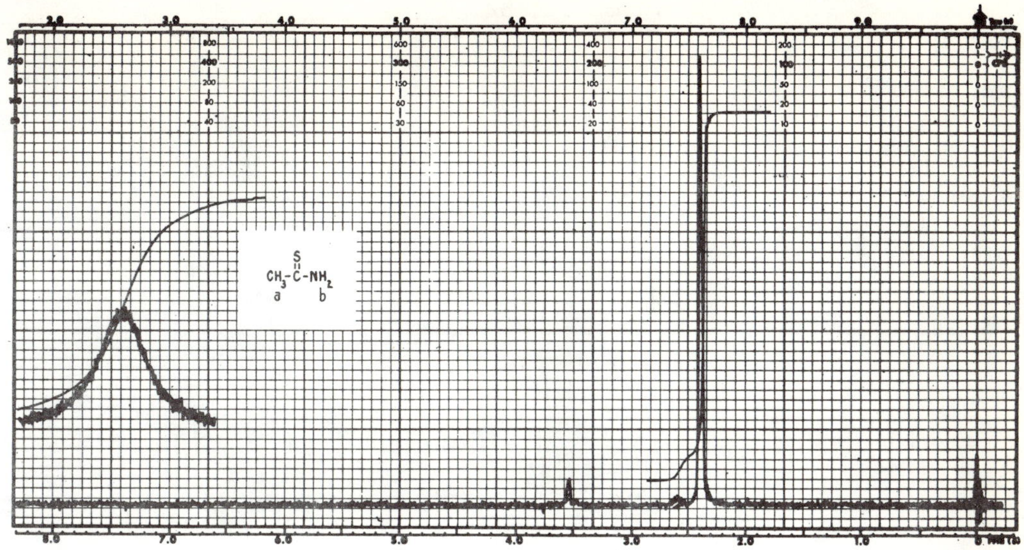

© 1978 Sadtler Research Laboratories, Inc.

2-PHENYLTHIOACETAMIDE

C_8H_9NS

Mol. Wt. 151.23

Solvent: Polysol-d

Assignments

a = 3.92
b = 7.05-7.45
c = ca 8.45

1332

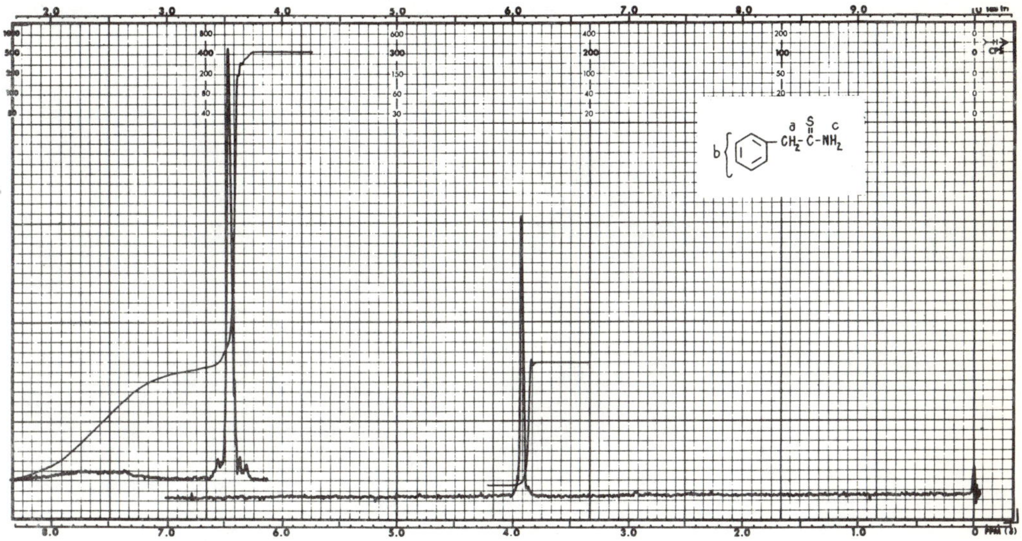

© 1978 Sadtler Research Laboratories, Inc.

THIOBENZAMIDE

C_7H_7NS

Mol. Wt. 137.20

M.P. 116-117^oC

Solvent: Polysol-d

Assignments

a = 7.10-7.50
b = 7.89
c = 9.32
d = 9.65
e = 2.50 DMSO-d_5

1333

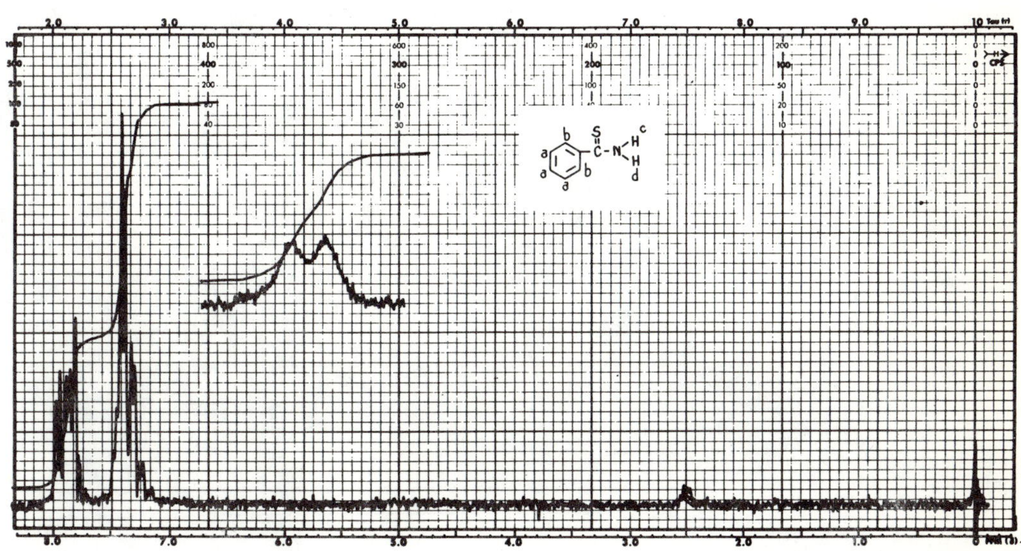

© 1978 Sadtler Research Laboratories, Inc.

6-METHYLTHIOPICOLINAMIDE

$C_7H_8N_2S$

Mol. Wt. 152.22

Solvent: Polysol-d

Assignments
a = 2.53
b = 7.29
c = 7.71
d = 8.40
e = ca 9.79

1334

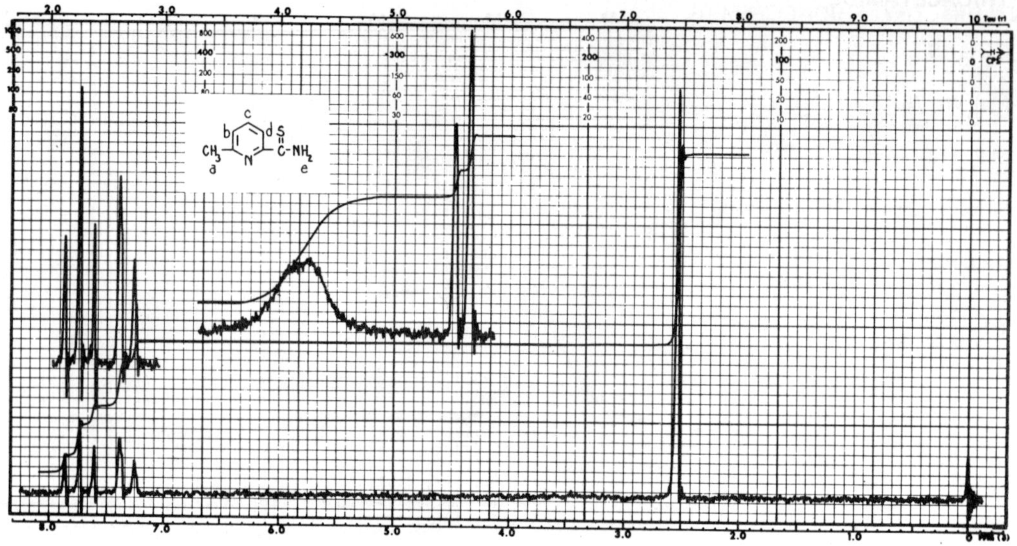

© 1978 Sadtler Research Laboratories, Inc.

THIOBENZANILIDE

$C_{13}H_{11}NS$

Mol. Wt. 213.30

M.P. 100-102°C

Solvent: $CDCl_3$

Assignments
a = 7.10-7.70
b = 7.70
c = 9.22
d = 1.66 imp.

1335

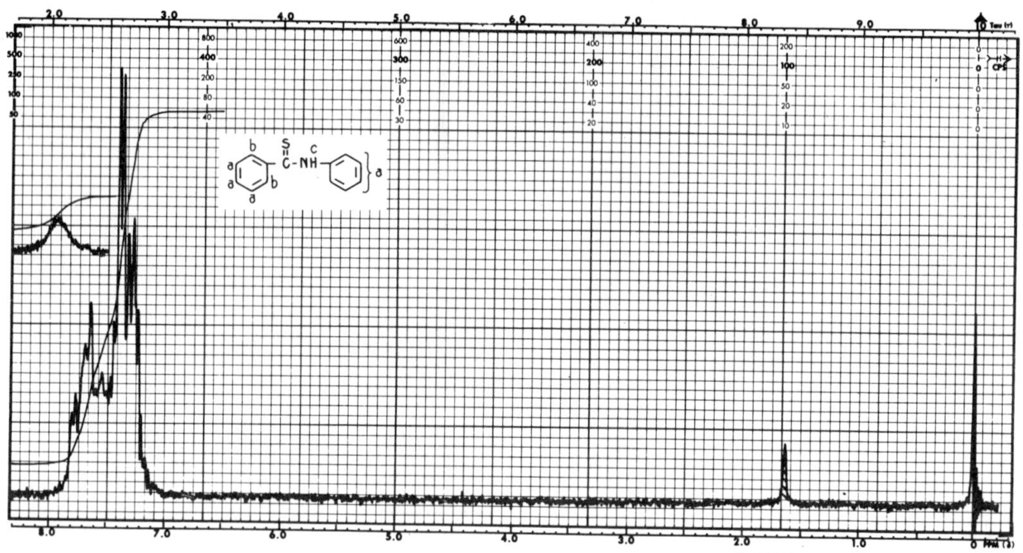

© 1978 Sadtler Research Laboratories, Inc.

N,N-DIMETHYLTHIOFORMAMIDE

C_3H_7NS

Mol. Wt. 89.16

B.P. 58-60°C/1mm

Solvent: $CDCl_3$

Assignments
a = 3.27
b = 3.30
c = 9.19

1336

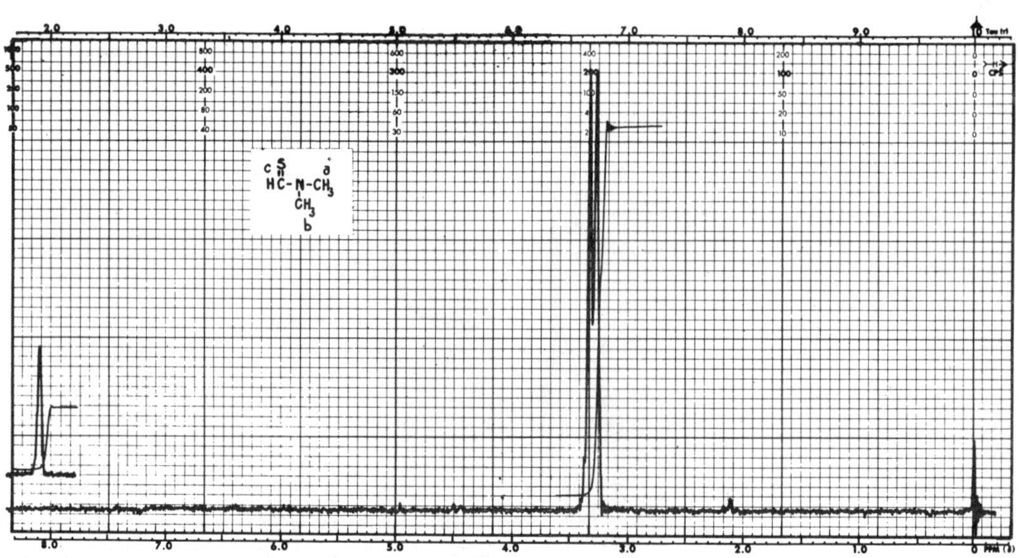

© 1978 Sadtler Research Laboratories, Inc.

N,N-DIMETHYLTHIOACETAMIDE

C_4H_9NS

Mol. Wt. 103.19

M.P. 73-75°C

Solvent: $CDCl_3$

Assignments
a = 2.62
b = 3.31
c = 3.48

1337

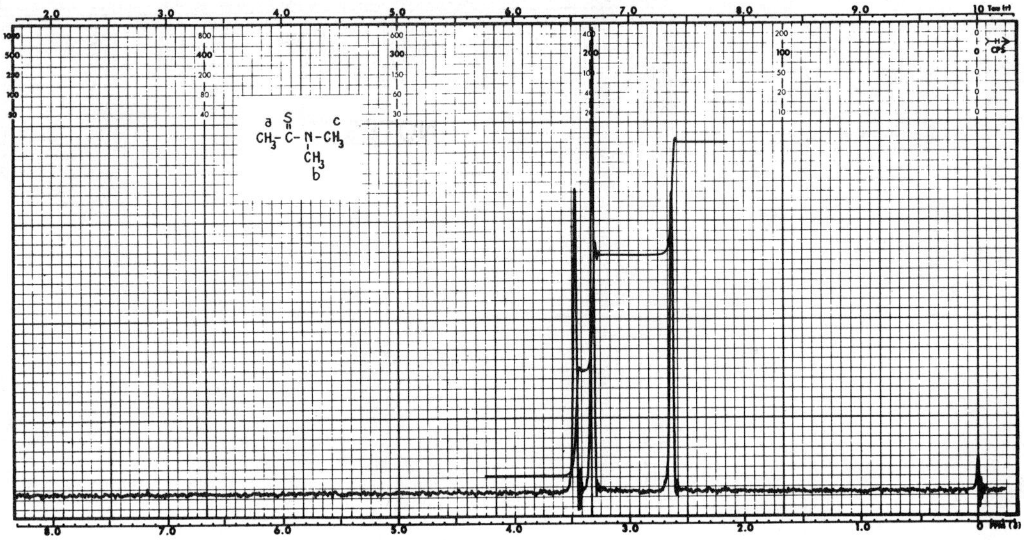

© 1978 Sadtler Research Laboratories, Inc.

NOTES

THE THIOUREAS

The Thioureas have a moderately strong deshielding effect on adjacent aliphatic groups and they usually display clear coupling to them in the case of the secondary Thiourea linkage (J = 4–5 Hz). The presence of the C(=S) thiocarbonyl group greatly reduces the shielding effect of the NH group on ortho aromatic hydrogens in comparison to the effect noted for the secondary amines. The chemical shift of the various NH hydrogens varies widely depending primarily on the type of substitution present in the molecule. In the case of the primary thiourea protons, the two hydrogens bonded to the nitrogen atom may be non-equivalent leading to different chemical shifts for each proton.

Chemical Shifts

Aliphatic Protons

HC–3	HC–2	HC–1		Solvent	Spectrum
		2.80	CH_3–NH–C(=S)–NH_2	DMSO-d_6	1338
		(3.29)	$(CH_3)_2$–N–C(=S)–NH–Ø	Polysol	1354
	1.04	3.28	CH_3–CH_2–NH–C(=S)–NH_2	DMSO-d_6	1339
	(1.20	3.52)	$(CH_3–CH_2)_2$–N–C(=S)–$N(R_2)_2$	$CDCl_3$	1356
	1.22	3.52	CH_3–CH_2–NH–C(=S)–NH–R_2	CCl_4	1344
	1.19	3.66	CH_3–CH_2–NH–C(=S)–NH–Ø	$CDCl_3$	1348
(0.92	1.64	3.61)	$(CH_3–CH_2–CH_2)_2$–N–C(=S)–NH–R	Polysol	- - - -
	(1.24)	4.32	$(CH_3)_2$–CH–NH–C(=S)–NH–R	$CDCl_3$	1345
(0.95)	1.92	3.30	$(CH_3)_2$–CH–CH_2–NH–C(=S)–NH–R	Polysol	1346

Aromatic Protons

The aromatic protons bonded to the nitrogen nuclei of the thiourea group are neither strongly shielded nor deshielded. They appear in the spectrum as a complex band in the chemical shift range from about 6.9 to 7.7 ppm. The shape and complexity of this higher order pattern is quite sensitive to the presence and type of other substituents bonded to the thiourea linkage. For example compare the appearance of the aromatic resonance bands in spectra 1348, 1349, 1350 and 1351.

The chemical shift ranges for three thiourea substituted phenyl groups is provided below.

para/meta/ortho			
6.90–7.60	Ø–NH–C(=S)–NH_2	DMSO-d_6	1340
7.00–7.70	Ø–NH–C(=S)–NH–R_2	$CDCl_3$	1348
7.10–7.60	$(Ø)_2$–N–C(=S)–NH_2	DMSO	1343

Exchangeable Protons

Although the chemical shifts of the exchangeable protons of the thioureas vary markedly with solvent, temperature and the presence of H_2O in solution, the tables below indicate a trend in shift with the type of substitution of the thiourea nitrogen atoms. It is noted that the primary amide protons resonate at highest field (H_2N–C(=S)–), that alkyl substituted secondary amide protons resonate at slightly lower field (R–NH–C(=S)–), and that the phenyl substituted secondary groups resonate at lowest field (Ø–NH–C(=S)–).

H_2H–C(=S)–		Solvent
5.0–8.0	–NH–CH_3	DMSO-d_6
6.31	–N(Ø)–R_2	Polysol
6.88	–NH–Ø (2–CH_3)	DMSO-d_6
6.93	–NH–R_2	DMSO-d_6
7.00	–NH–1(naphthalene)	Polysol
7.11	–NH_2	DMSO
7.32	–NH–Ø	DMSO-d_6

X	–NH–C(=S)	–Y	
$(CH_3)_2$–CH–	6.10	–NH–R	$CDCl_3$
R_2–	6.19	–NH–Ø	$CDCl_3$
R_2–	6.67	–NH–R_2	CCl_4
R_{10}–	7.10	–NH–R_{10}	Polysol
CH_3–	7.50	–NH_2	DMSO-d_6
R_2–	7.50	–NH_2	DMSO-d_6
Ø–	7.97	–NH–Ø	$CDCl_3$
Ø–	8.04	–NH–Ø (4–CH_3)	$CDCl_3$
Ø–	8.62	–NH–R_2	$CDCl_3$
Ø–	9.13	–NH–CH$(CH_3)_2$	Polysol
Ø	9.57	–NH_2	Polysol
Ø–	9.66	–NH_2	DMSO-d_6

1-METHYL-2-THIOUREA

$C_2H_6N_2S$

Mol. Wt. 90.15

Solvent: Dimethyl sulfoxide-d_6

Assignments
a = 2.80
b = 5.00-8.00
c = 2.55 DMSO-d_5

1338

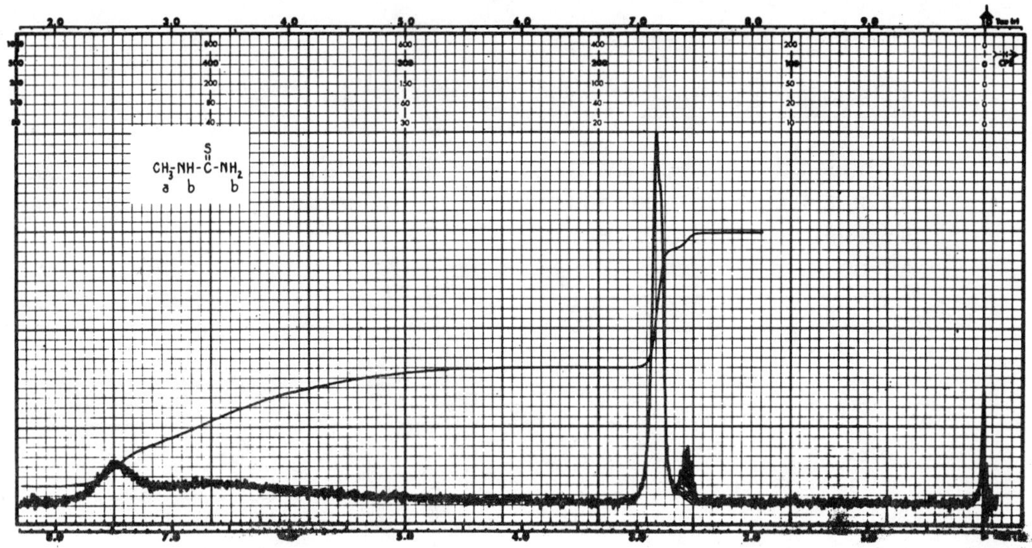

© 1978 Sadtler Research Laboratories, Inc.

1-ETHYL-2-THIOUREA

$C_3H_8N_2S$

Mol. Wt. 104.17

M.P. 110-111°C

Solvent: Dimethyl sulfoxide-d_6

Assignments
a = 1.04
b = 3.28
c = 6.93
d = 7.50
e = 2.52 DMSO-D_5

1339

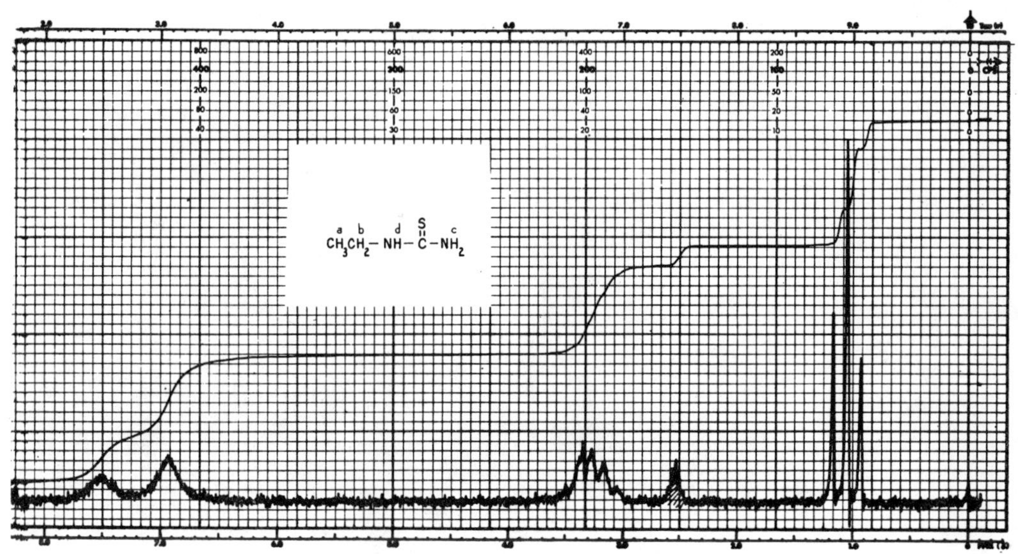

© 1978 Sadtler Research Laboratories, Inc.

1-PHENYL-2-THIOUREA

$C_7H_8N_2S$

Mol. Wt. 152.22

M.P. 153-154°C

Solvent: Dimethyl sulfoxide

Assignments
a = 6.90-7.60
b = 7.38
c = 9.66

1340

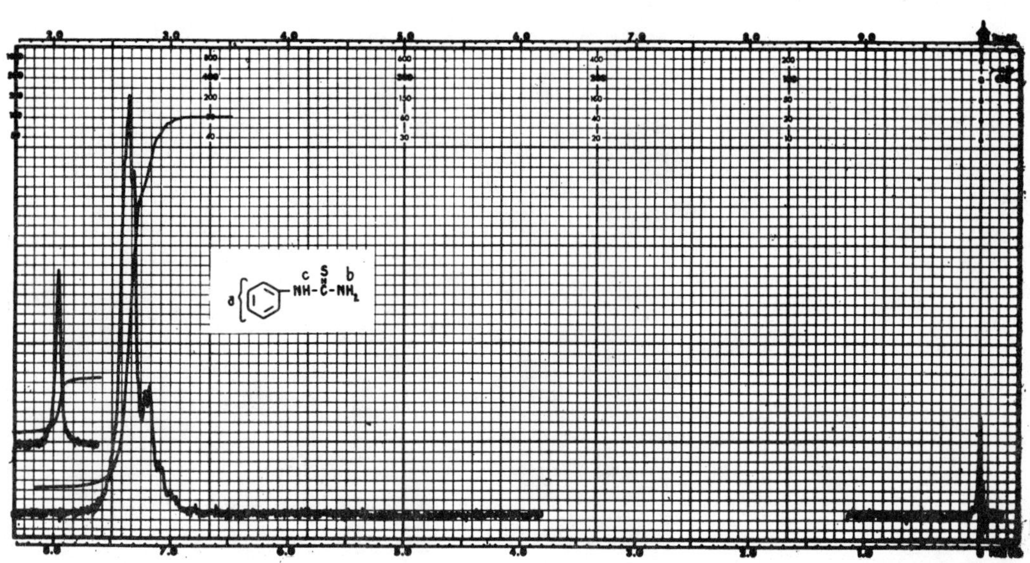

© 1978 Sadtler Research Laboratories, Inc.

2-THIO-1-o-TOLYLUREA

$C_8H_{10}N_2S$

Mol. Wt. 166.25

Solvent: DMSO-d_6

Assignments

a = 2.24

b = 6.88

c = ca 7.19

d = 8.96

1341

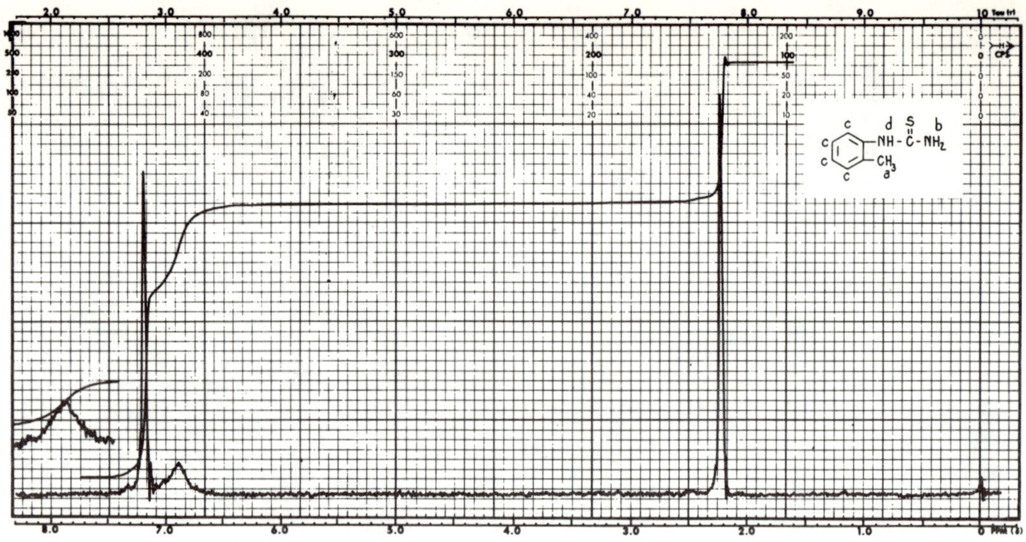

© 1978 Sadtler Research Laboratories, Inc.

1-(1-NAPHTHYL)-2-THIOUREA

$C_{11}H_{10}N_2S$

Mol. Wt. 202.28

M.P. 190-192°C

Solvent: Polysol-d

Assignments

a = 7.00

b = 7.20-7.65

c = 7.65-8.20

d = 9.61

e = 2.42 DMSO-d_5

1342

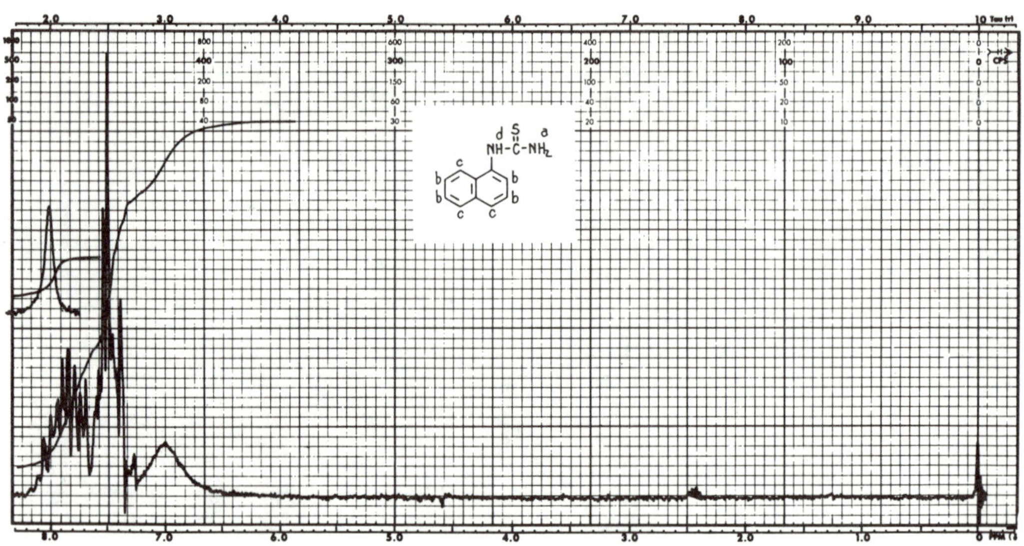

© 1978 Sadtler Research Laboratories, Inc.

1,1-DIPHENYL-2-THIOUREA

$C_{13}H_{12}N_2S$

Mol. Wt. 228.32

M.P. 219°C (dec.)

Solvent: Dimethyl sulfoxide

Assignments

a = 7.10-7.60

1343

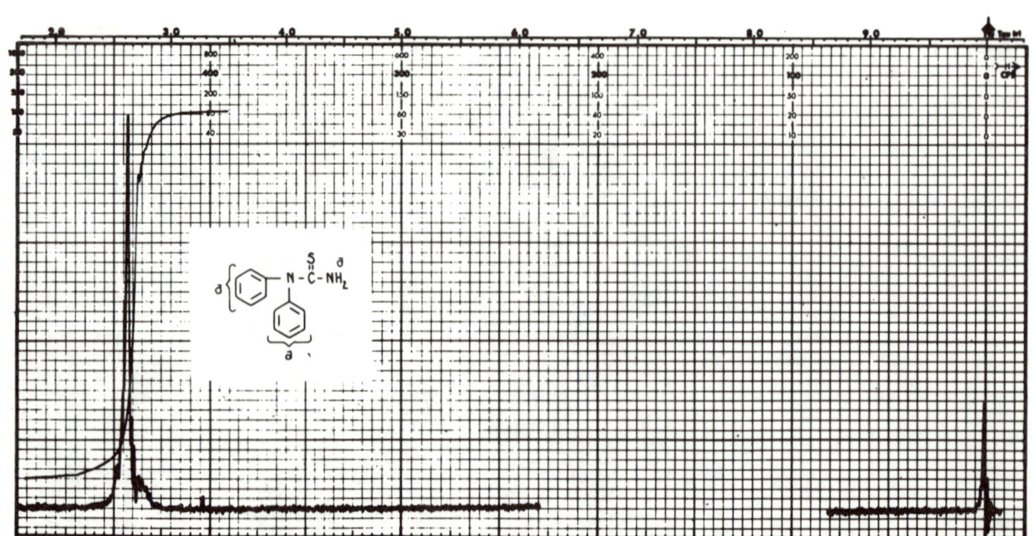

© 1978 Sadtler Research Laboratories, Inc.

1,3-DIETHYL-2-THIOUREA

$C_5H_{12}N_2S$

Mol. Wt. 132.22

M.P. 77-78°C

Solvent: CCl_4

Assignments
a = 1.22
b = 3.51
c = 6.67
Trace of acid added

1344

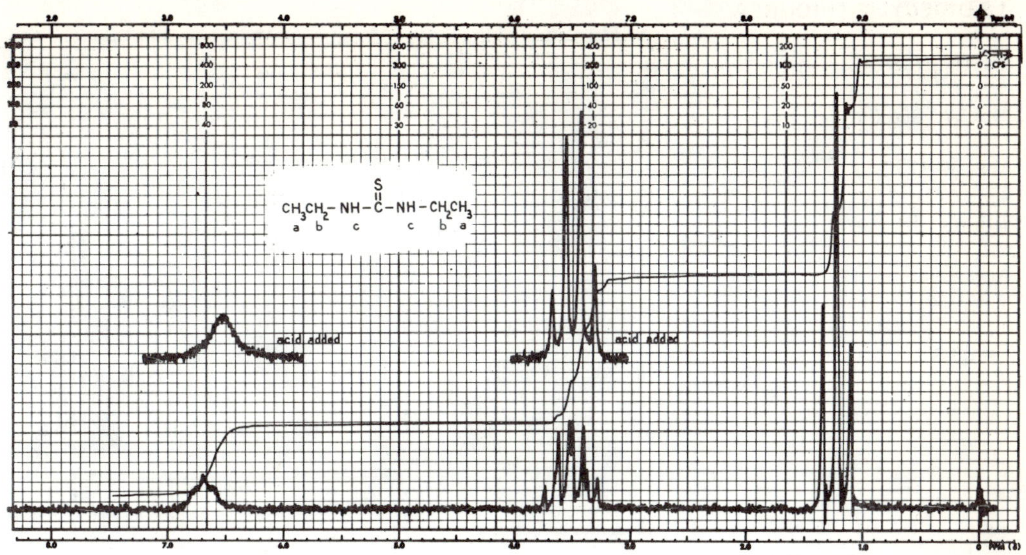

© 1978 Sadtler Research Laboratories, Inc.

1,3-DIISOPROPYL-2-THIOUREA

$C_7H_{16}N_2S$

Mol. Wt. 160.28

Solvent: $CDCl_3$

Assignments
a = 1.24
b = 4.32
c = 6.10

1345

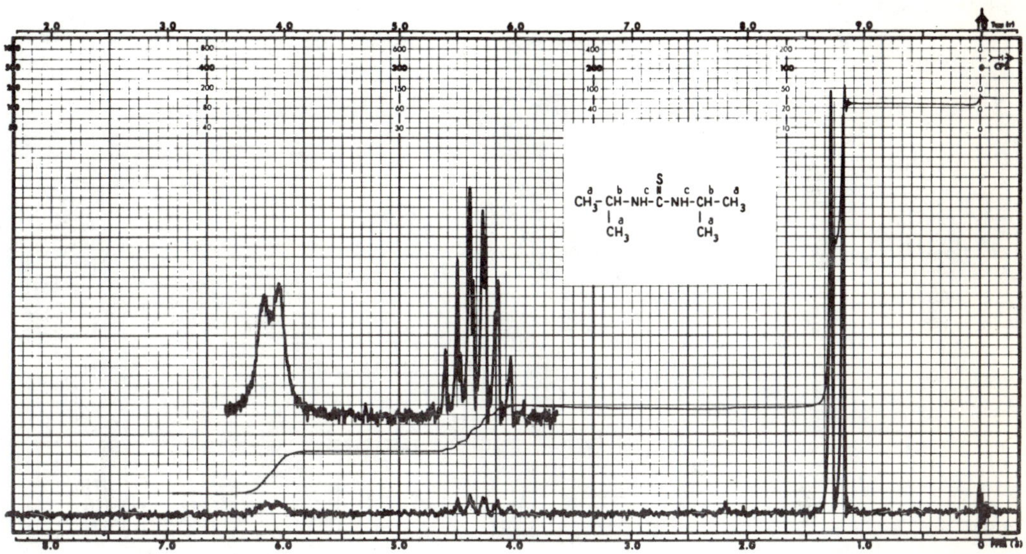

© 1978 Sadtler Research Laboratories, Inc.

1,3-DIISOBUTYL-2-THIOUREA

$C_9H_{20}N_2S$

Mol. Wt. 188.34

Solvent: $CDCl_3$

Assignments
a = 0.95
b = 1.92
c = 3.30
d = 6.55

1346

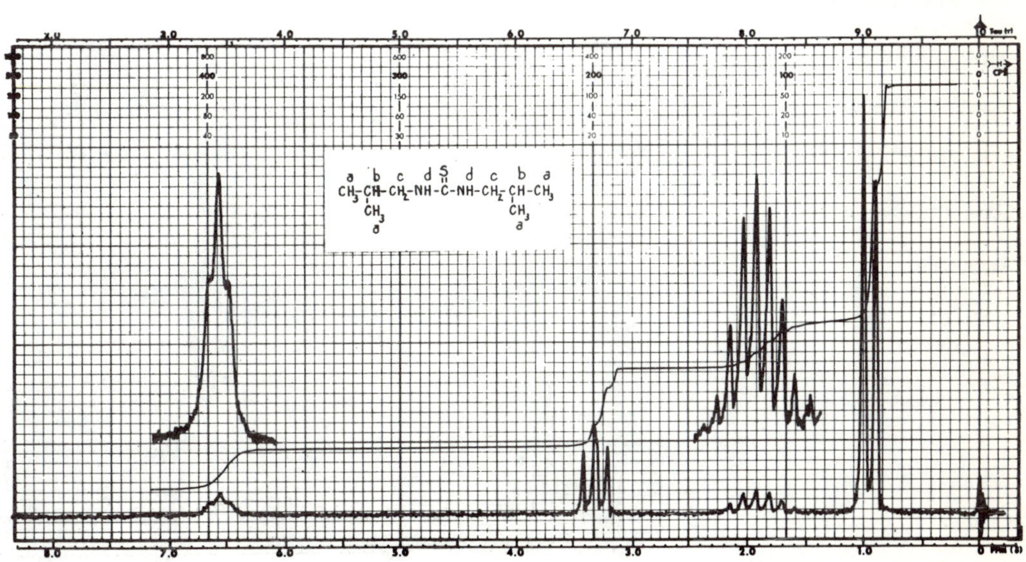

© 1978 Sadtler Research Laboratories, Inc.

1,3-DIDECYL-2-THIOUREA

$C_{21}H_{44}N_2S$

Mol. Wt. 356.66

M.P. 69-71°C

Solvent: Polysol-d

Assignments
a = 0.89
b = 1.10-1.80
c = 3.39
d = 7.10

1347

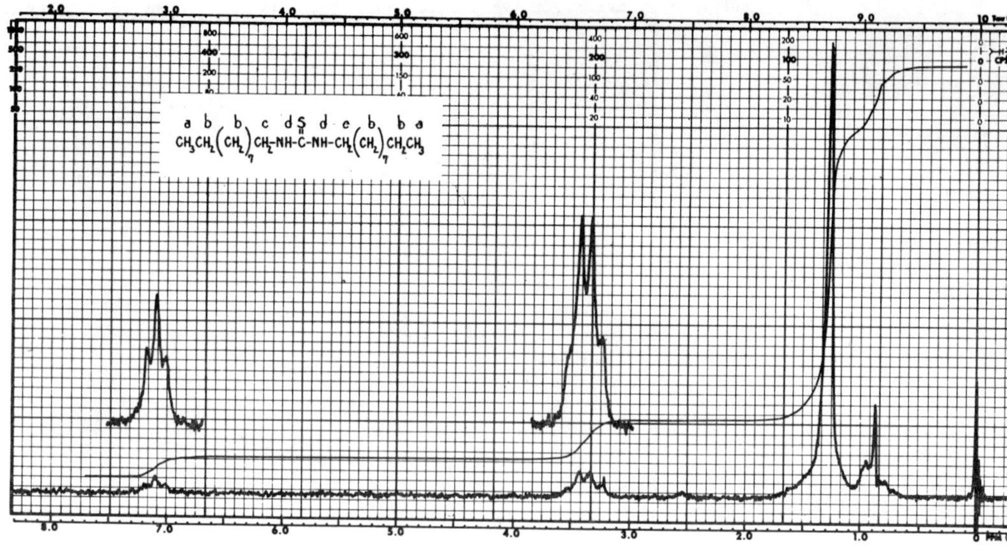

© 1978 Sadtler Research Laboratories, Inc.

1-ETHYL-3-PHENYL-2-THIOUREA

$C_9H_{12}N_2S$

Mol. Wt. 180.27

M.P. 102-104°C

Solvent: $CDCl_3$

Assignments
a = 1.19
b = 3.66
c = 6.19
d = 7.00-7.70
e = 8.62

1348

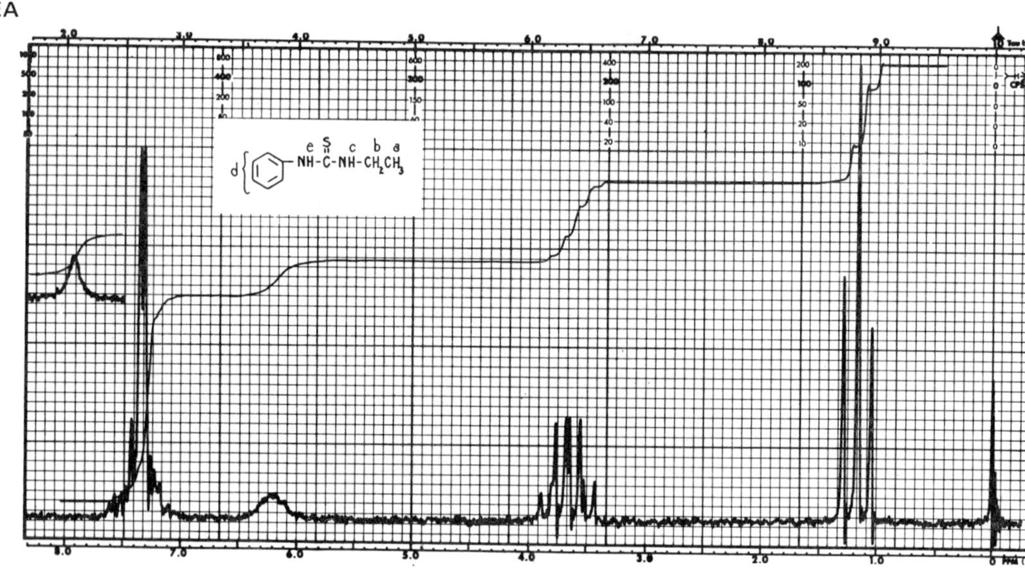

© 1978 Sadtler Research Laboratories, Inc.

1-ISOPROPYL-3-PHENYL-2-THIOUREA

$C_{10}H_{14}N_2S$

Mol. Wt. 194.30

M.P. 101°C

Solvent: Polysol-d

Assignments
a = 1.18
b = 4.51
c = 6.80-7.55
d = 9.13
e = 7.99 imp.

1349

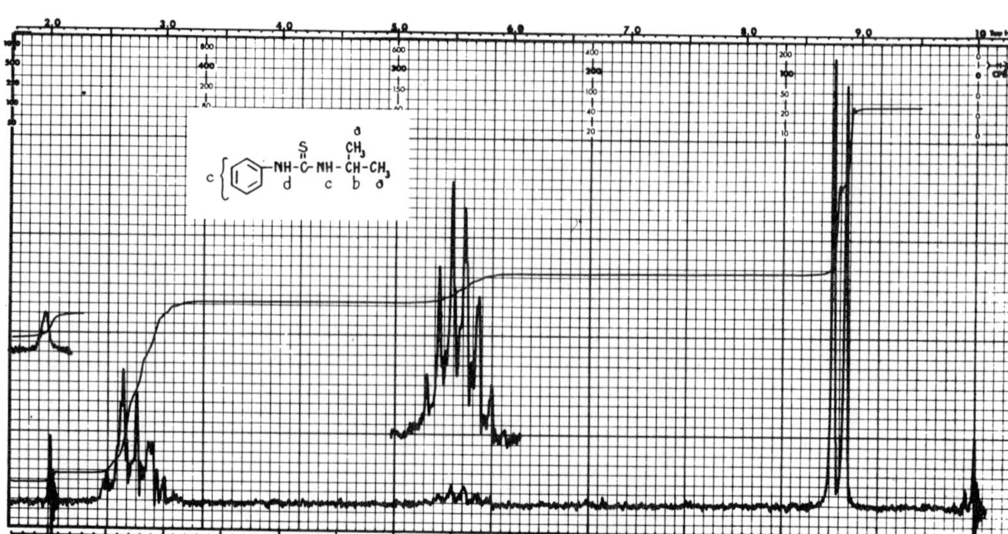

© 1978 Sadtler Research Laboratories, Inc.

1-ALLYL-3-PHENYL-2-THIOUREA

$C_{10}H_{12}N_2S$

Mol. Wt. 192.28

M.P. 98°C

Solvent: $CDCl_3$

Assignments
a = 4.22
b = 5.10
c = 5.52-6.20
d = ca 6.20
e = 7.01-7.59
f = 8.49
g = 1.89 imp.

1350

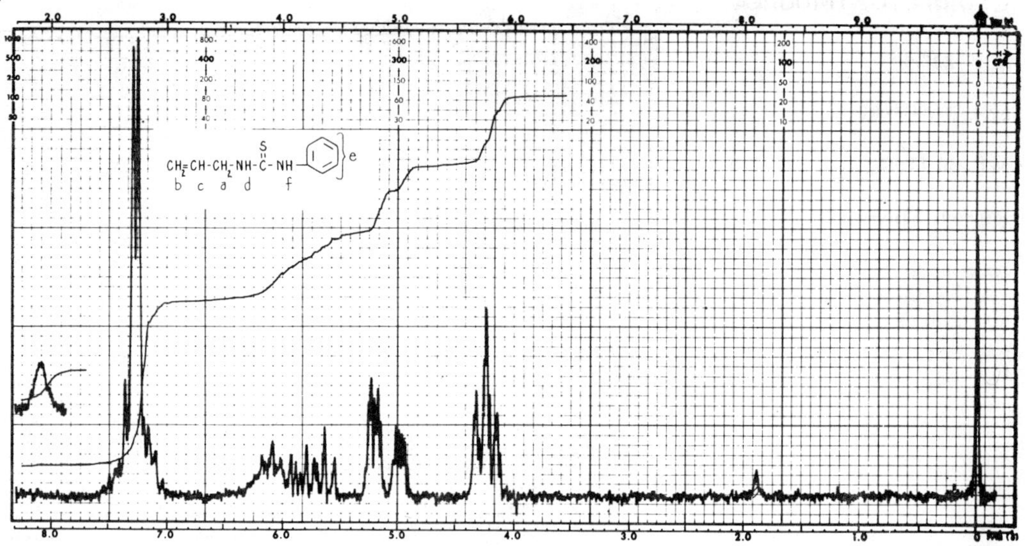

© 1978 Sadtler Research Laboratories, Inc.

THIOCARBANILIDE

$C_{13}H_{12}N_2S$

Mol. Wt. 228.32

M.P. 154.5-155.5°C

Solvent: $CDCl_3$

Assignments
a = 7.38
b = ca 7.97

1351

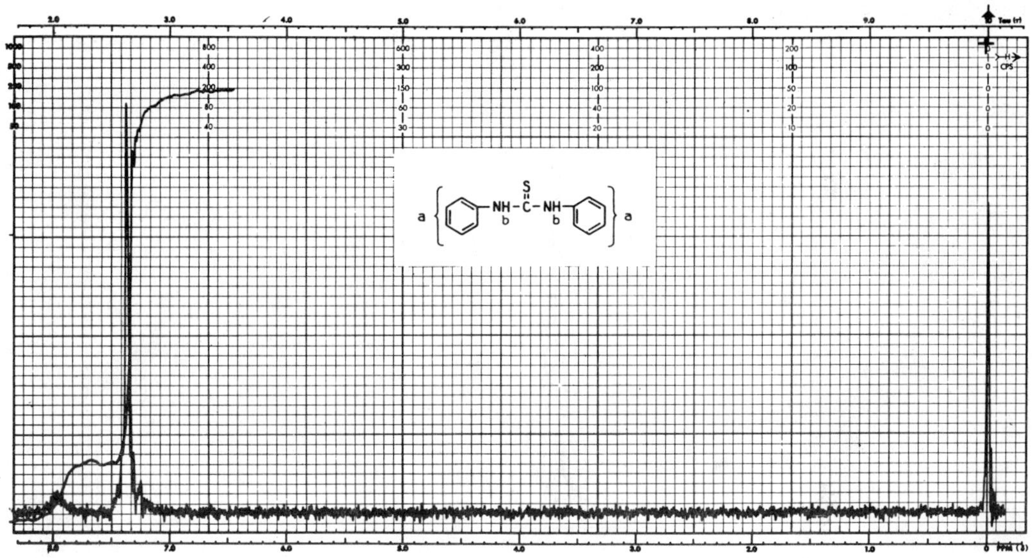

© 1978 Sadtler Research Laboratories, Inc.

1-(o-CHLOROPHENYL)-3-PHENYL-2-THIOUREA

$C_{13}H_{11}ClN_2S$

Mol. Wt. 262.76

M.P. 153-155°C

Solvent: Dimethyl sulfoxide

Assignments
a = 6.90-7.80
b = 9.60

1352

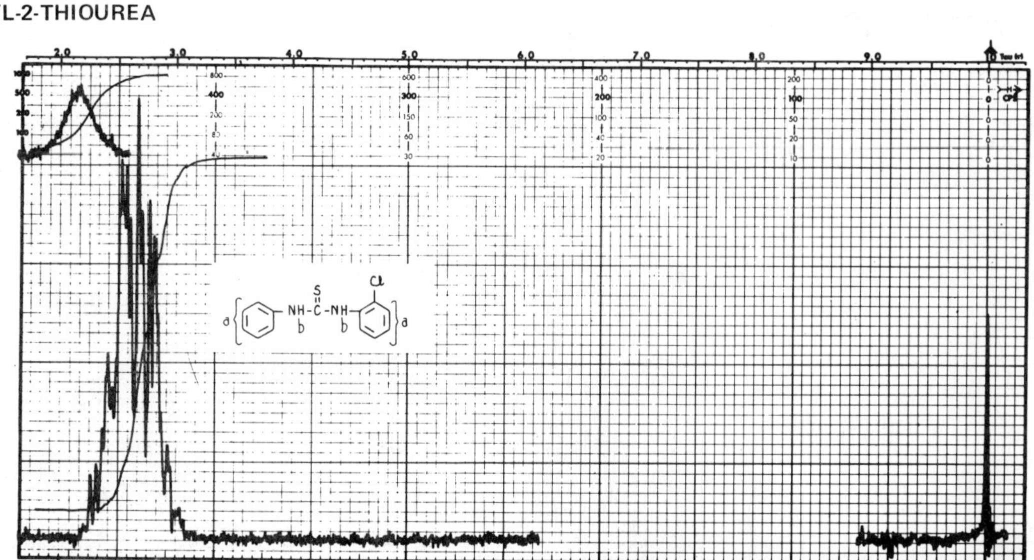

© 1978 Sadtler Research Laboratories, Inc.

1,1-DIMETHYL-4-(2-NORBORNYL)-3-THIOSEMICARBAZIDE

$C_{10}H_{19}N_3S$

Mol. Wt. 213.35

M.P. 152-153°C

Solvent: Polysol-d

Assignments
a = 0.80-2.10
b = ca 2.31
c = 2.51
d = 4.05
e = 7.19
f = 8.31

1353

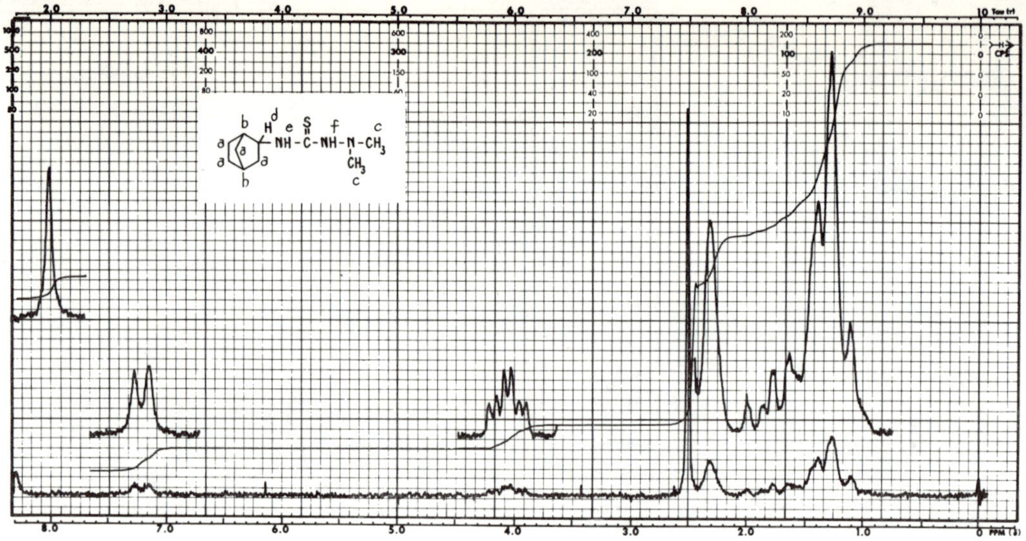

© 1978 Sadtler Research Laboratories, Inc.

3-(o-CHLOROPHENYL)-1,1-DIMETHYL-2-THIOUREA

$C_9H_{11}ClN_2S$

Mol. Wt. 214.72

M.P. 146°C

Solvent: Polysol-d

Assignments
a = 3.29
b = 7.00-7.60
c = 8.78

1354

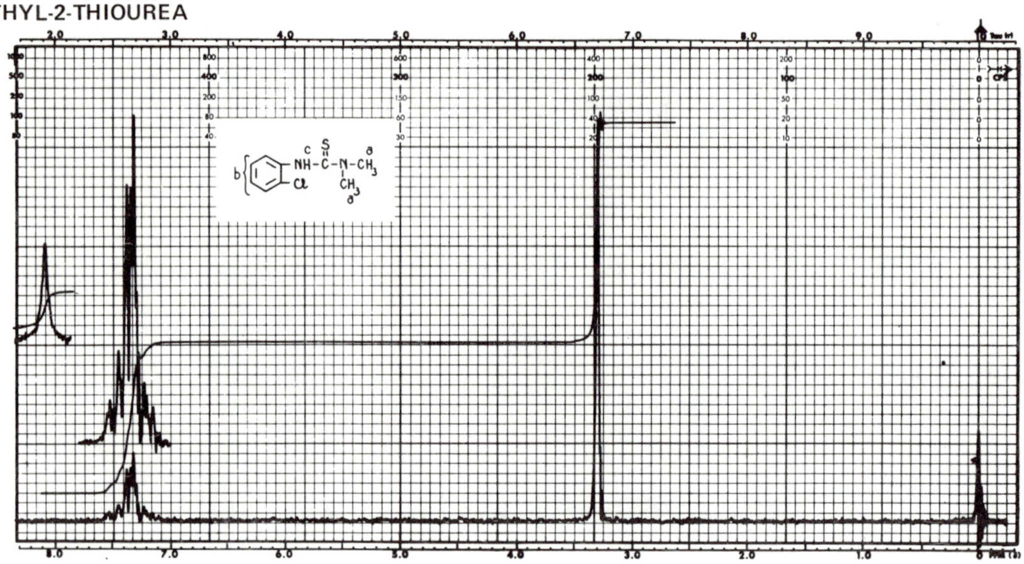

© 1978 Sadtler Research Laboratories, Inc.

THIO-1-PIPERIDINECARBOXY-m-TOLUIDIDE

$C_{13}H_{18}N_2S$

Mol. Wt. 234.37

M.P. 102°C

Solvent: $CDCl_3$

Assignments
a = 1.63
b = 2.32
c = 3.73
d = 6.70-7.40

1355

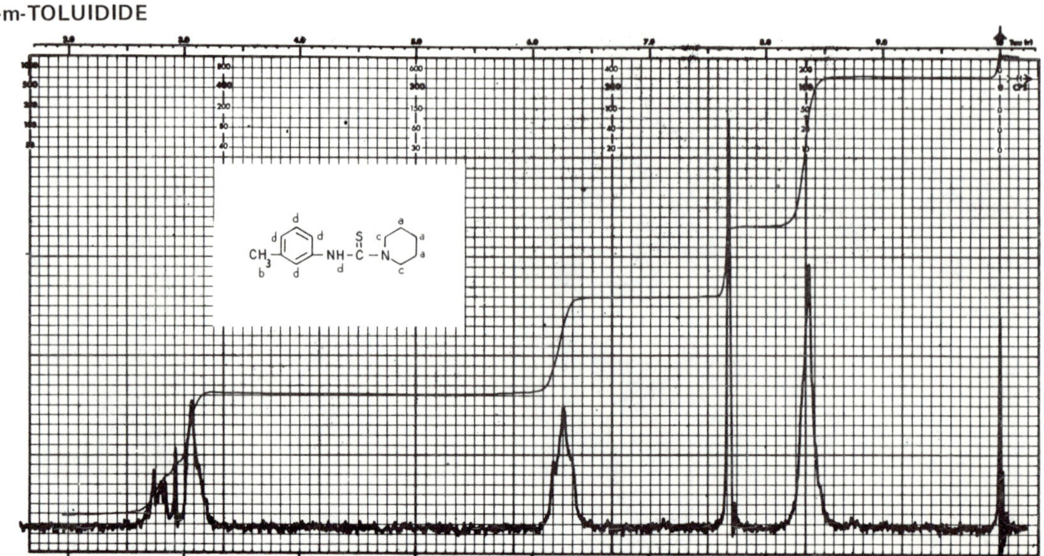

© 1978 Sadtler Research Laboratories, Inc.

1,1,3,3-TETRAETHYL-2-THIOUREA

$C_9H_{20}N_2S$

Mol. Wt. 188.34

B.P. 126-128°C/10mm

Solvent: $CDCl_3$

Assignments
a = 1.20
b = 3.52

1356

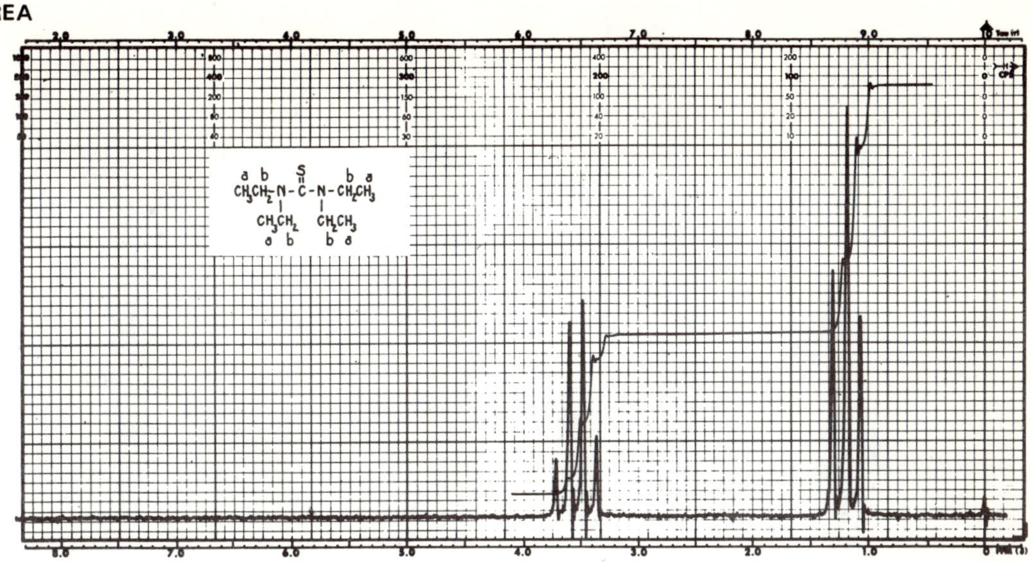

© 1978 Sadtler Research Laboratories, Inc.

1H-NAPHTH[2,3-d] IMIDAZOLE-2(3H)-THIONE

$C_{11}H_8N_2S$

Mol. Wt. 200.32

M.P. 305°C (dec.)

Solvent: Dimethyl sulfoxide-d_6

Assignments
a = 7.39
b = 7.60
c = 7.95
d = 12.62
e = 2.55 DMSO-d_5
f = 3.21 H_2O

1357

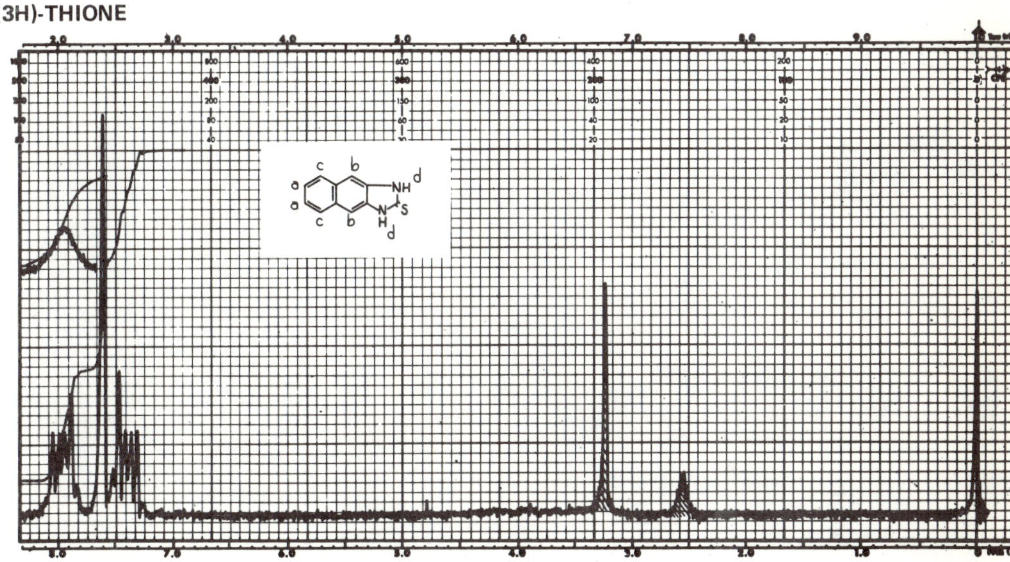

© 1978 Sadtler Research Laboratories, Inc.

1,5-DIHEPTYLIDENE-3-THIOCARBOHYDRAZIDE

$C_{15}H_{30}N_4S$

Mol. Wt. 298.50

M.P. 119°C

Solvent: $CDCl_3$

Assignments
a = 0.90
b = 1.34
c = 2.30
d = 7.42
e = 10.10

1358

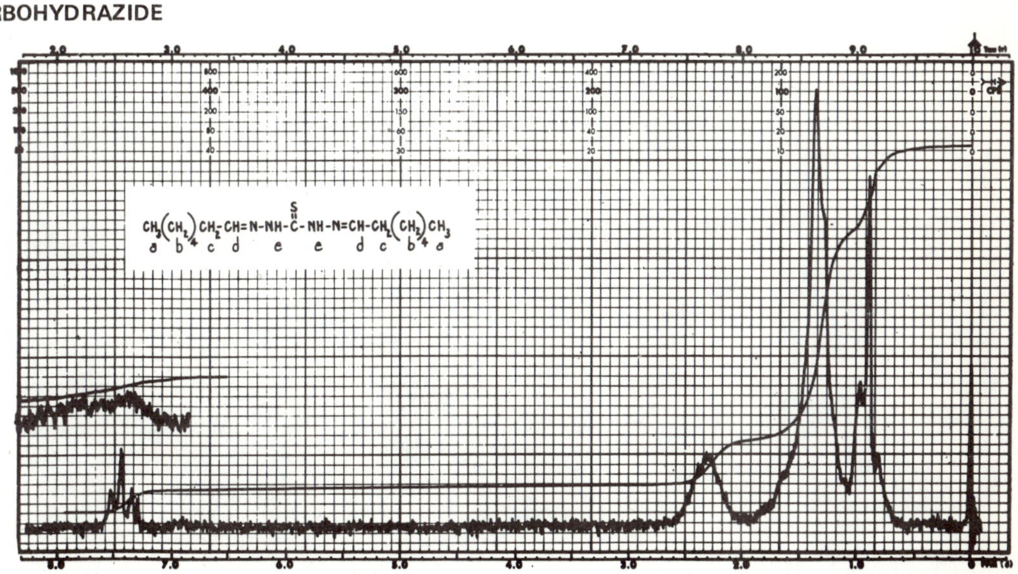

© 1978 Sadtler Research Laboratories, Inc.

NOTES

THE SULFONAMIDES

The Sulfonamides produce two distinct sets of chemical shifts. The protons of hydrocarbon groups bonded to the nitrogen side of the linkage resonate at higher field than the corresponding protons of groups bonded to the SO_2 side. In addition to the characteristic chemical shifts thus produced, the sulfonamides usually display clear coupling between an aliphatic group and the NH proton adjacent to it.

These compounds are generally more soluble in DMSO-d_6 and Polysol than in the chlorinated solvents. There appears to be a distinct deshielding of the NH protons in DMSO-d_6 and Polysol in comparison to similar protons in $CDCl_3$.

Chemical Shifts

Aliphatic Protons

HC–2	HC–1		Solvent	Spectrum
	2.60	CH_3–NH–SO_2–CH_2–Ø	Polysol	----
	(2.69)	$(CH_3)_2$N–SO_2–Ø	$CDCl_3$	1375
	(2.83)	$(CH_3)_2$N–SO_2–$N(R)_2$	$CDCl_3$	----
	(2.85)	$(CH_3)_2$N–SO_2–CH_3	$CDCl_3$	1373
1.08	2.97	CH_3–CH_2–NH–SO_2–CH_2–Ø	DMSO-d_6	----
(1.13)	3.40	$(CH_3)_2$–CH–NH–SO_2–CH_2–Ø	Polysol	----
(1.20)	----	$(CH_3)_3$–C–NH–SO_2–Ø	$CDCl_3$	----

HC–2	HC–1			
	2.78	CH_3–SO_2–$N(CH_3)_2$	$CDCl_3$	1373
	3.00	CH_3–SO_2–NH–Ø	$CDCl_3$	----
	3.03	CH_3–SO_2–NH_2	Polysol	1359
1.33	3.10	CH_3–CH_2–SO_2–NH–Ø	$CDCl_3$	1368
1.40	3.18	CH_3–CH_2–SO_2–NH_2	$CDCl_3$	----
1.40	3.47	CH_3–CH_2–SO_2–NH–C(=O)–CH_3	$CDCl_3$	----

Aromatic Protons

The SO_2 side of the sulfonamide linkage is a strong deshielding group in its effect on the ortho aromatic hydrogens. These protons usually resonate in the range from 7.5–7.9 ppm. The nitrogen side of the sulfonamide group is a weakly shielding substituent on all the aromatic protons and they appear as a broad single peak near 7.1 ppm.

A selection of Phenyl Sulfonamide compounds is presented below.

Aromatic Protons continued

	ortho/meta/para	Solvent	Spectrum
X-SO_2–NH–Ø	ca. 7.12	$CDCl_3$	1370
X–NH–SO_2–Ø	7.87, 7.5–7.8	DMSO-d_6	1371

Para Substituted sulfonamides

	ortho	meta	para	
Ø–SO_2–NH–	7.19	6.88	–O–SO_2–Ø	Polysol
Ø–SO_2–NH–	6.95	6.95	–CH_3	$CDCl_3$
Ø–SO_2–NH–	6.98	7.09	–R_2	TFA
R–SO_2–NH–	7.21	7.21	–Cl	DMSO-d_6
Ø–SO_2–NH–	7.01	7.31	–Br	$CDCl_3$
Ø–SO_2–NH–	7.01	7.54	–I	Acetone
Ø–SO_2–NH–	7.26	7.73	–SO_2–NH_2	Polysol
R–SO_2–NH–	7.40	8.03	–C(=O)–OH	Acetone

H_2N–SO_2–	ortho	meta	para	
	7.50	6.64	–NH_2	DMSO-d_6
	7.75	6.98	–O–CH_3	Polysol
	8.07	7.37	–F	Acetone
	7.90	7.57	–Cl	Acetone
	7.89	7.89	–NH–C(=O)–Ø	Polysol
	7.99	8.11	–C(=O)–OH	Polysol
	8.10	8.43	–NO_2	DMSO-d_6

Exchangeable Protons

Primary Sulfonamides

H_2N–SO_2	–X		
5.24	–R_2	$CDCl_3$	----
6.50	–CH=CH–Ø	Polysol	1360
6.89	–Ø (4–NH_2)	DMSO-d_6	1365
7.11	–1(naphthalene)	Polysol	----
7.11	–Ø (4–N=CH–Ø)	DMSO-d_6	1366
7.21	–Ø (2–CH_3)	DMSO-d_6	1362

Coupling and Coupling Constants

Coupling between aliphatic groups and the adjacent NH proton is usually observed. The coupling constant is similar in magnitude to normal vicinal CH—CH coupling, J_{CH-NH} = 6–8 Hz.

METHANESULFONAMIDE

CH_3NO_2S

Mol. Wt. 95.12

M.P. 91-93°C

Solvent: Polysol-d

Assignments
a = 3.03
b = 6.38

1359

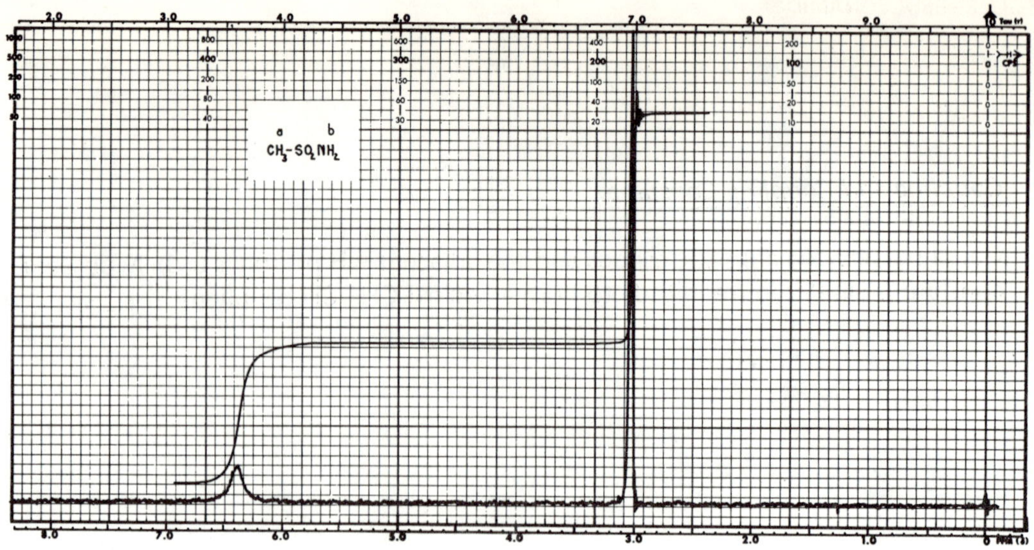

© 1978 Sadtler Research Laboratories, Inc.

trans, 2-PHENYLETHENESULFONAMIDE

$C_8H_9NO_2S$

Mol. Wt. 183.23

Solvent: Polysol-d

Assignments
a = ca 6.50
b = 7.10
c = 7.20-7.70
d = 7.41
e = 3.00-5.00 H_2O
f = 2.78 imp.

1360

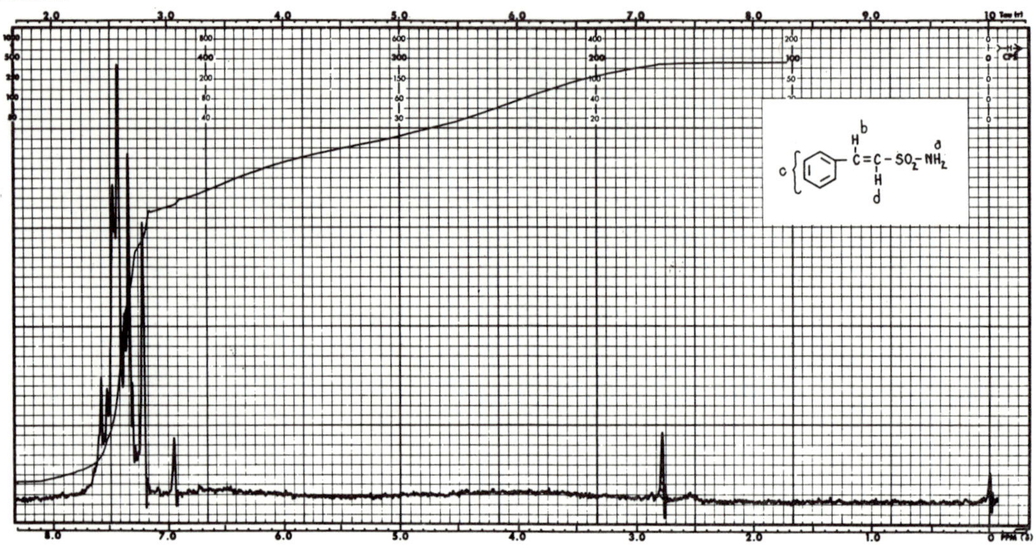

© 1978 Sadtler Research Laboratories, Inc.

BENZENESULFONAMIDE

$C_6H_7NO_2S$

Mol. Wt. 157.19

M.P. 152-154°C

Solvent: Trifluoroacetic acid

Assignments
a = ca 7.62
b = ca 7.95
NH_2 unobserved

1361

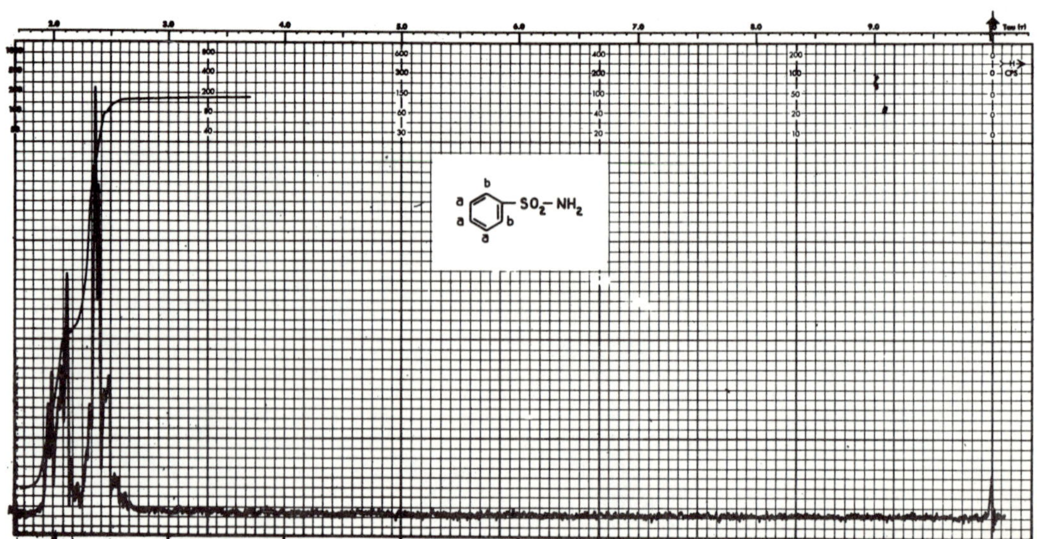

© 1978 Sadtler Research Laboratories, Inc.

o-TOLUENESULFONAMIDE

$C_7H_9NO_2S$

Mol. Wt. 171.22

M.P. 150-153°C

Solvent: Dimethyl sulfoxide-D_6

Assignments
a = 2.65
b = 7.21
c = 7.25-7.50
d = 7.91
e = 2.49 DMSO-D_5
f = 3.31 H_2O

1362

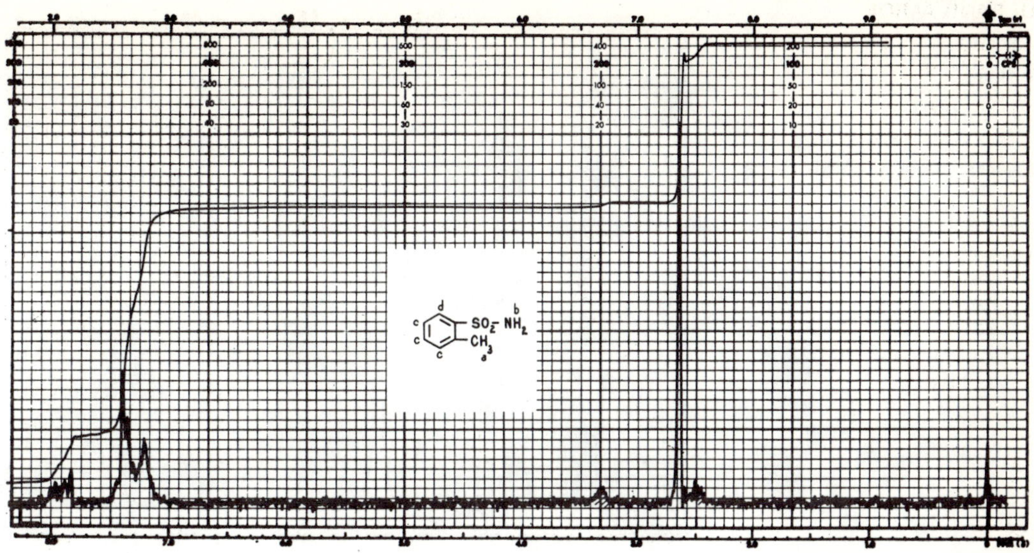

© 1978 Sadtler Research Laboratories, Inc.

p-TOLUENESULFONAMIDE

$C_7H_9NO_2S$

Mol. Wt. 171.22

M.P. 157°C (lit.)

Solvent: Trifluoroacetic acid

Assignments
a = 2.46
b = ca 4.03
c = 7.37
d = 7.84

1363

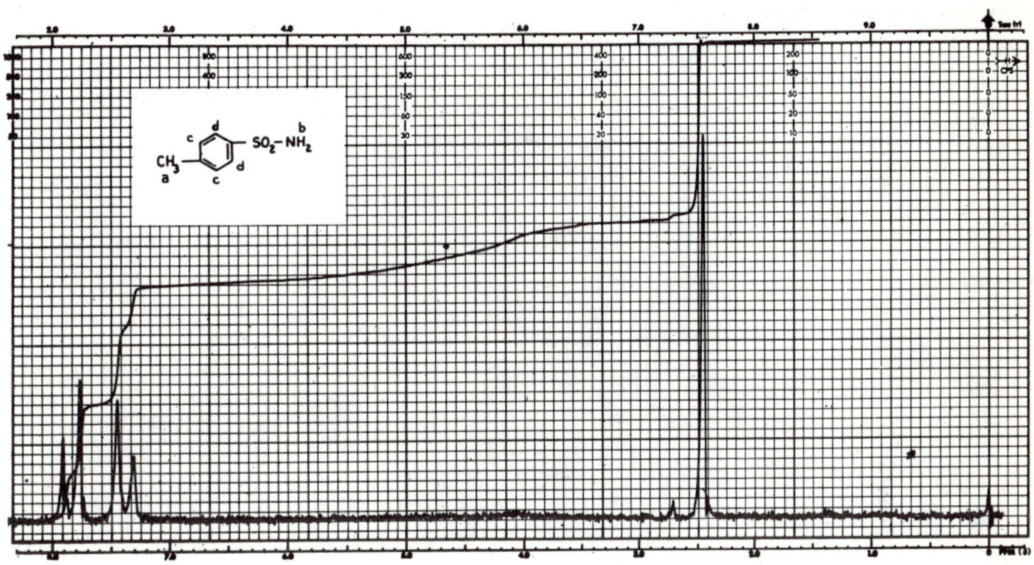

© 1978 Sadtler Research Laboratories, Inc.

2-MESITYLENESULFONAMIDE

$C_9H_{13}NO_2S$

Mol. Wt. 199.27

M.P. 140-144°C

Solvent: Trifluoroacetic acid

Assignments
a = 2.31
b = 2.64
c = 7.04
NH unobserved

1364

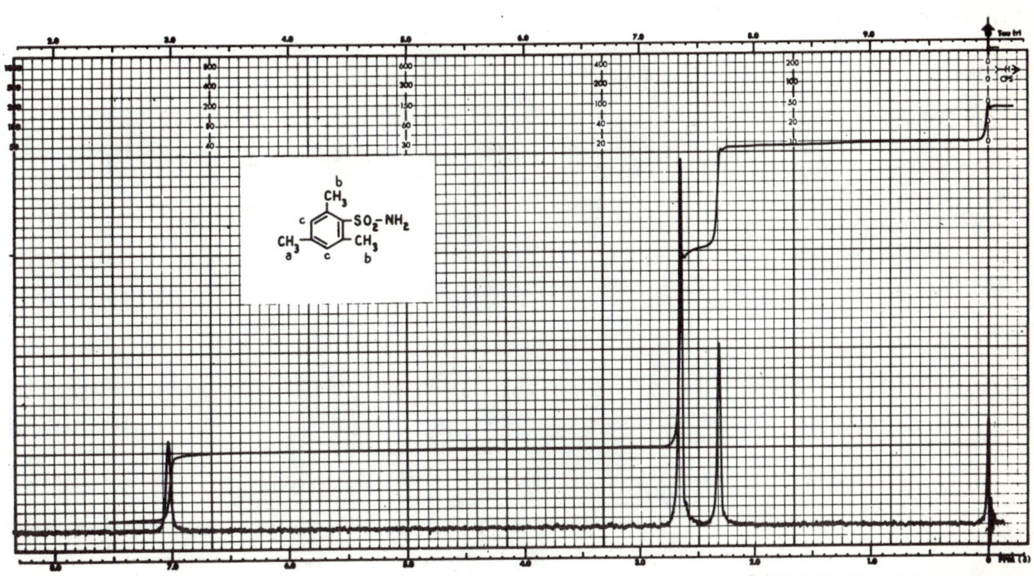

© 1978 Sadtler Research Laboratories, Inc.

SULFANILAMIDE

$C_6H_8N_2O_2S$

Mol. Wt. 172.21

Solvent: Dimethyl sulfoxide-d_6

Assignments
a = 5.69
b = 6.64
c = 6.89
d = 7.50
e = 2.49 DMSO-d_5
f = 3.49 H_2O

1365

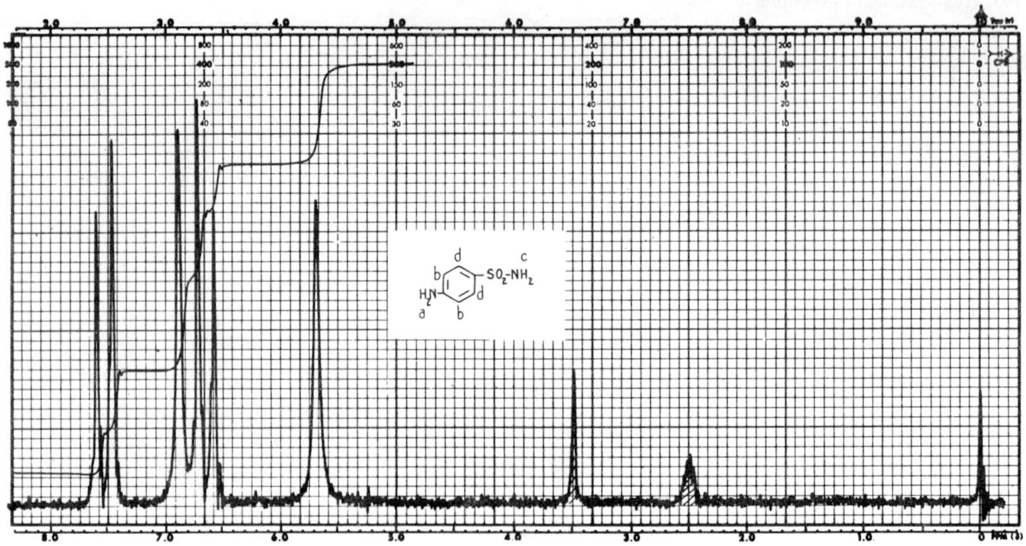

© 1978 Sadtler Research Laboratories, Inc.

N^4-BENZYLIDENESULFANILAMIDE

$C_{13}H_{12}N_2O_2S$

Mol. Wt. 260.32

M.P. 188-190°C

Solvent: DMSO-d_6

Assignments
a = 7.11
b = 7.37
c = 7.30-7.60
d = 7.95
e = 7.97
f = 8.59
g = 2.49 DMSO-d_5
h = 3.28 H_2O
i = 6.71 SSB

1366

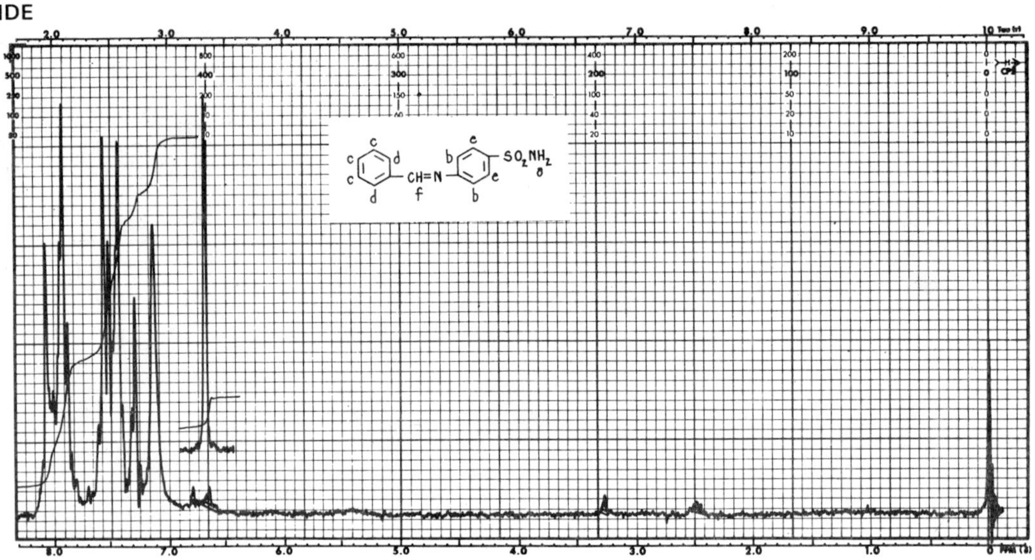

© 1978 Sadtler Research Laboratories, Inc.

2-NAPHTHALENESULFONAMIDE

$C_{10}H_9NO_2S$

Mol. Wt. 207.25

M.P. 213-216°C

Solvent: Polysol-d

Assignments
a = 7.11
b = ca 7.51
c = ca 7.85
d = ca 7.92
e = 8.45

1367

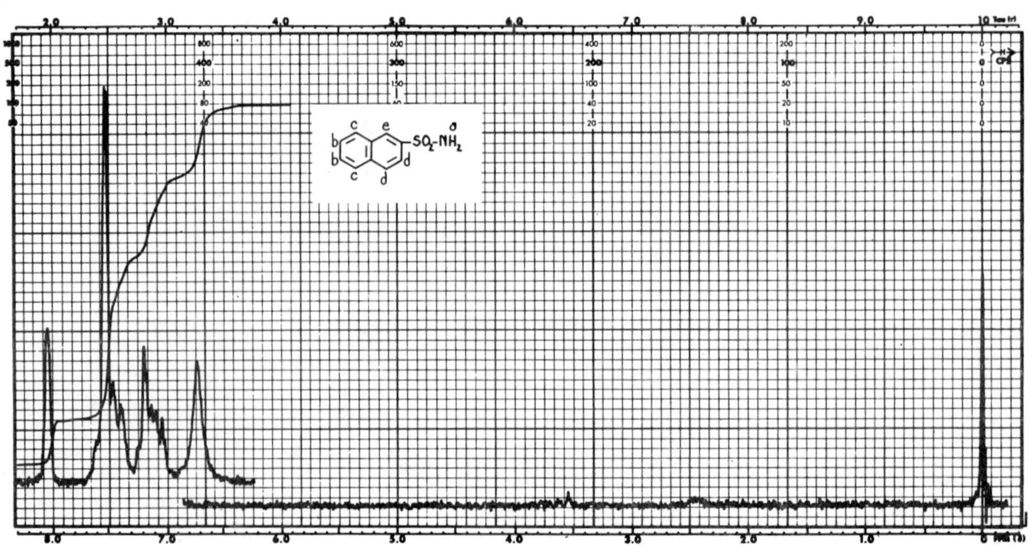

© 1978 Sadtler Research Laboratories, Inc.

ETHANESULFONO-p-TOLUIDIDE

$C_9H_{13}NO_2S$

Mol. Wt. 199.27

M.P. 81°C

Solvent: $CDCl_3$

Assignments

a = 1.33
b = 2.30
c = 3.10
d = ca 7.16
e = 7.36

1368

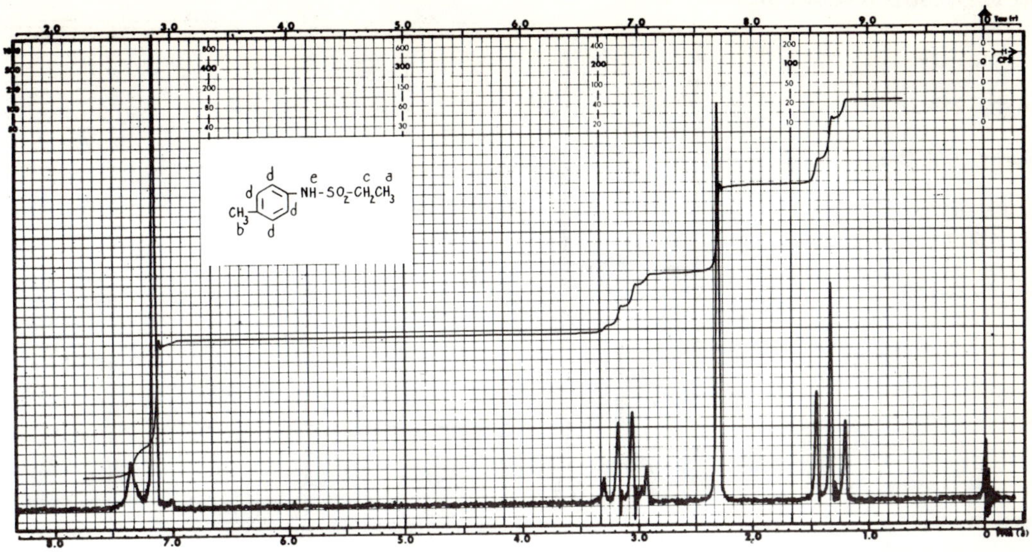

© 1978 Sadtler Research Laboratories, Inc.

N-(2-HYDROXYETHYL)-N-(3-HYDROXYPROPYL)-p-TOLUENESULFONAMIDE

$C_{12}H_{19}NO_4S$

Mol. Wt. 273.36

Solvent: $CDCl_3$

Assignments

a = 1.81
b = 2.40
c = 3.24
d = 3.30-4.20
e = 7.30
f = 7.69

1369

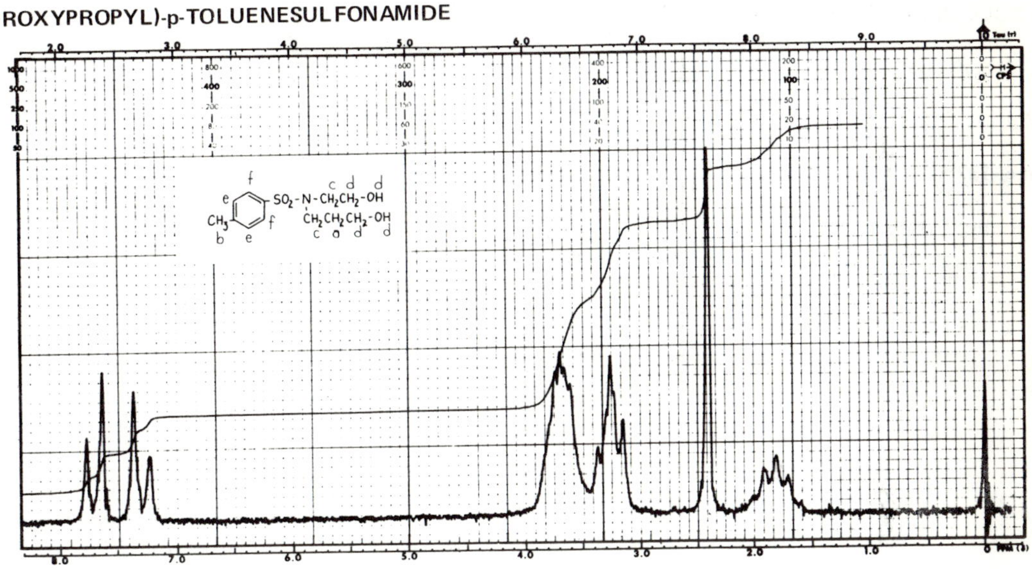

© 1978 Sadtler Research Laboratories, Inc.

p-TOLUENESULFONANILIDE

$C_{13}H_{13}NO_2S$

Mol. Wt. 247.32

M.P. 100-101°C

Solvent: $CDCl_3$

Assignments

a = 2.29
b = ca 7.12
c = 7.15
d = 7.65
e = 7.71

1370

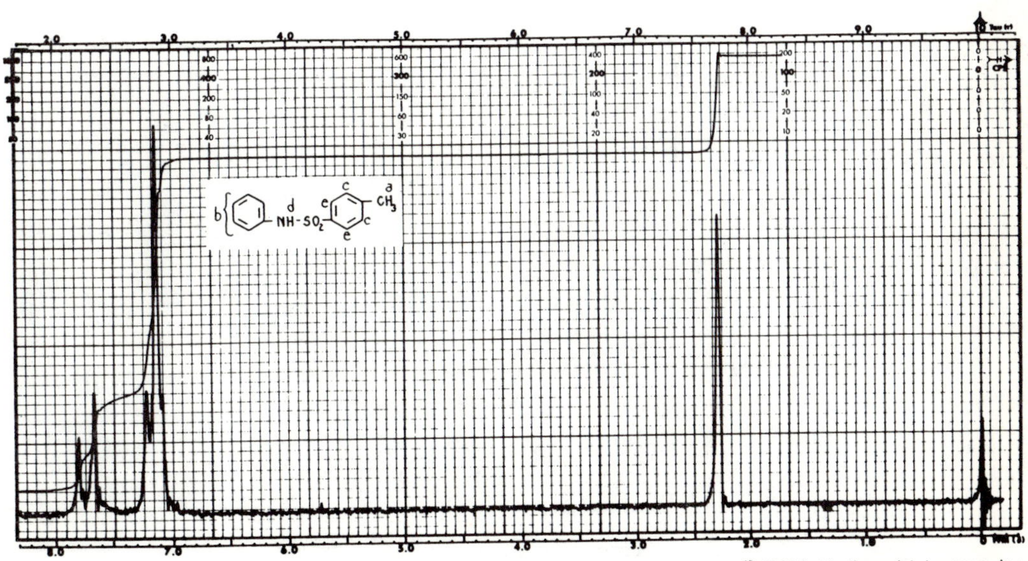

© 1978 Sadtler Research Laboratories, Inc.

o-BENZENESULFONOTOLUIDIDE

$C_{13}H_{13}NO_2S$

Mol. Wt. 247.32

Solvent: DMSO-d_6

Assignments

a = 2.11
b = ca 7.19
c = 7.50-7.80
d = 7.87
e = 9.36

1371

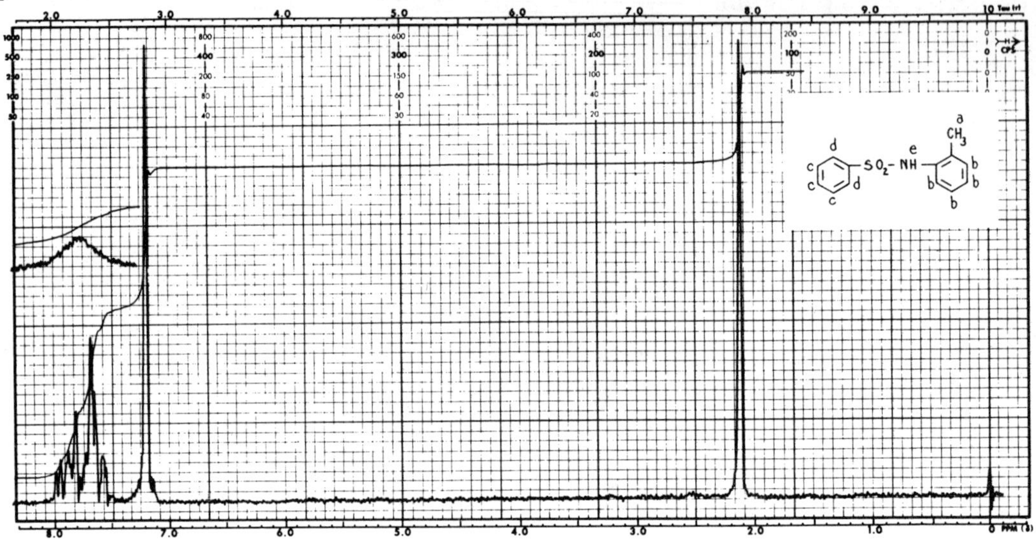

© 1978 Sadtler Research Laboratories, Inc.

p-TOLUENESULFONO-2',6'-XYLIDIDE

$C_{15}H_{17}NO_2S$

Mol. Wt. 275.37

M.P. 137°C

Solvent: $CDCl_3$

Assignments

a = 2.02
b = 2.38
c = 6.56
d = ca 7.01
e = 7.19
f = 7.60

1372

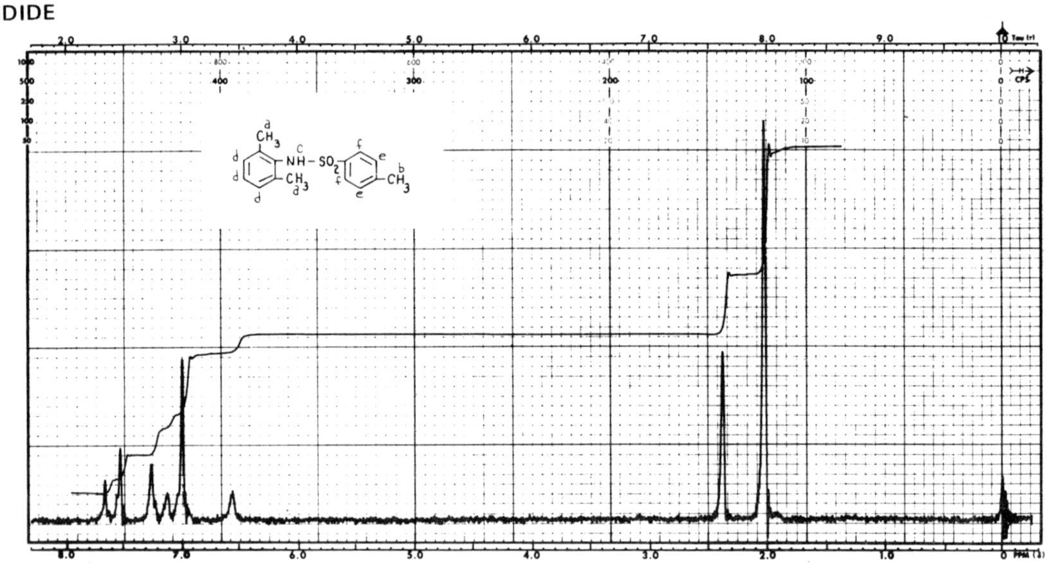

© 1978 Sadtler Research Laboratories, Inc.

N,N-DIMETHYLMETHANESULFONAMIDE

$C_3H_9NO_2S$

Mol. Wt. 123.17

Solvent: $CDCl_3$

Assignments

a = 2.78
b = 2.85

1373

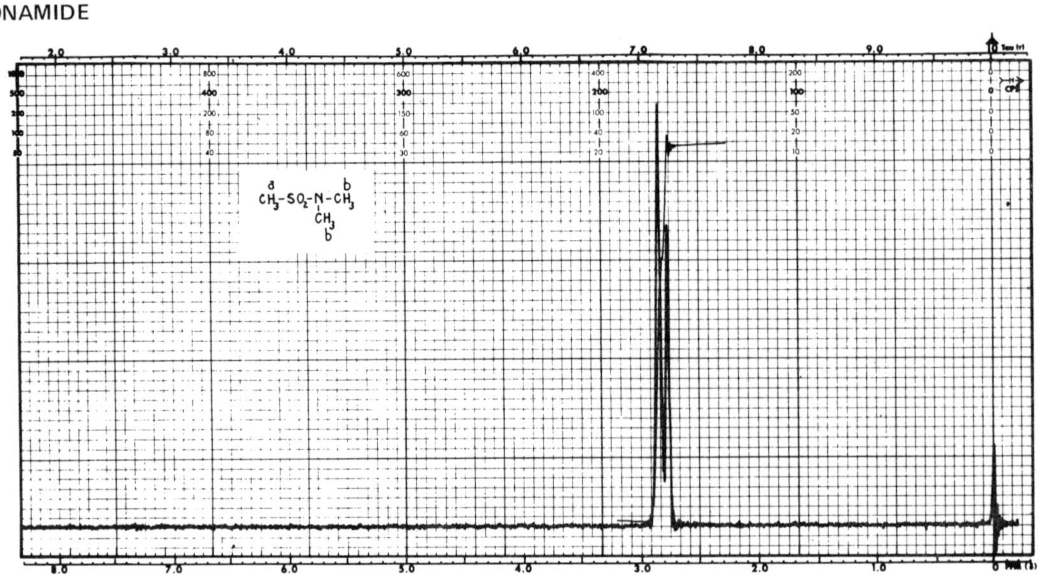

© 1978 Sadtler Research Laboratories, Inc.

N,N,N',N'-TETRAMETHYL-1,4-PIPERAZINEDISULFONAMIDE

$C_8H_{20}N_4O_4S_2$

Mol. Wt. 300.40

M.P. 217-218°C

Solvent: $CDCl_3$

Assignments
a = 2.83
b = 3.31

1374

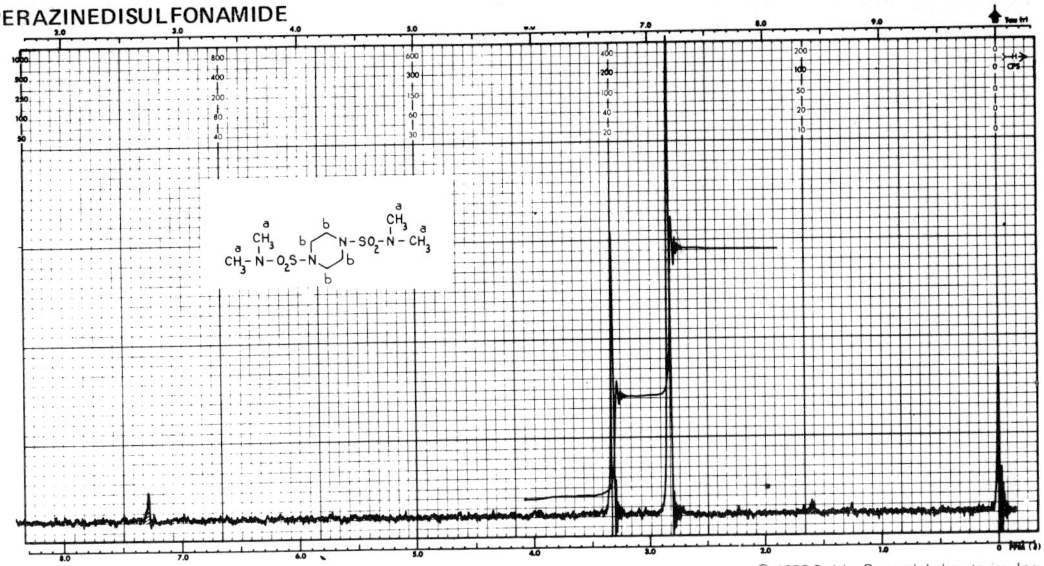

© 1978 Sadtler Research Laboratories, Inc.

N,N-DIMETHYL-p-TOLUENESULFONAMIDE

$C_9H_{13}NO_2S$

Mol. Wt. 199.27

M.P. 80-82°C

Solvent: $CDCl_3$

Assignments
a = 2.43
b = 2.69
c = 7.34
d = 7.65

1375

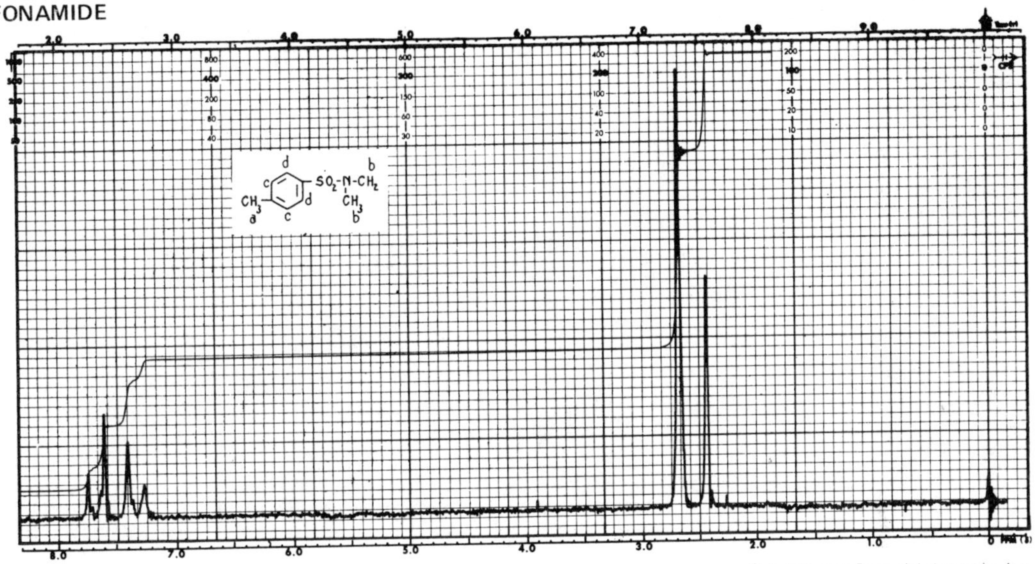

© 1978 Sadtler Research Laboratories, Inc.

1,3-BIS(PHENYLSULFONYL)IMIDAZOLIDINE

$C_{15}H_{16}N_2O_4S_2$

Mol. Wt. 352.43

Solvent: Polysol-d

Assignments
a = 3.27
b = 4.56
c = 7.30-8.00

1376

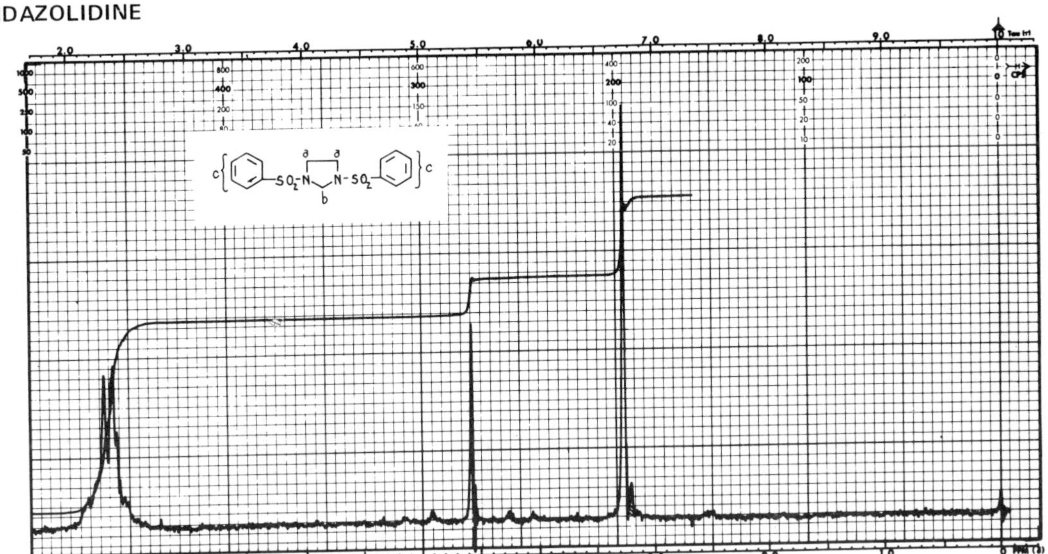

© 1978 Sadtler Research Laboratories, Inc.

N,N'-DICYCLOHEXYLSULFAMIDE

$C_{12}H_{24}N_2O_2S$

Mol. Wt. 260.40

M.P. 155-157°C

Solvent: Polysol-d

Assignments
a = 0.70-2.20
b = 3.00
c = 6.28
d = 2.53 DMSO-d_5

1377

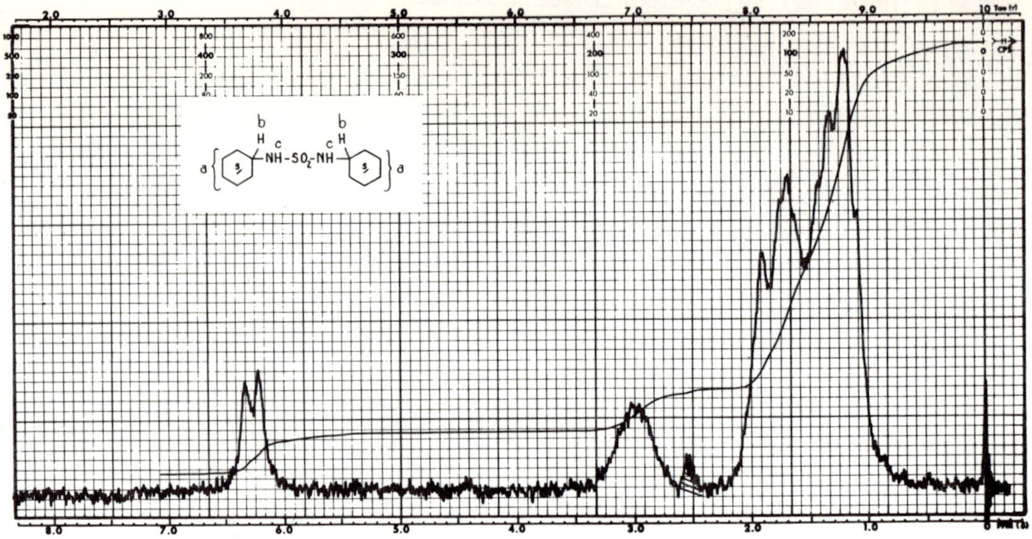

© 1978 Sadtler Research Laboratories, Inc.

N,N'-BIS(p-BROMOPHENYL)SULFAMIDE

$C_{12}H_{10}Br_2N_2O_2S$

Mol. Wt. 406.11

M.P. 122-123°C

Solvent: Polysol-d

Assignments
a = 7.09
b = 7.33
c = 10.11
d = 2.54 DMSO-d_5

1378

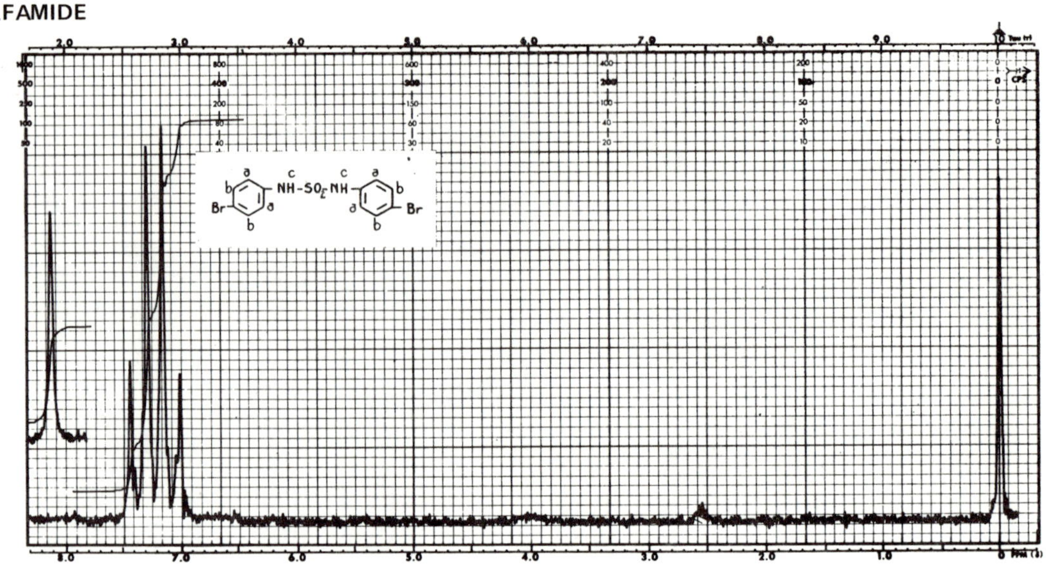

© 1978 Sadtler Research Laboratories, Inc.

THE ALIPHATIC AND OLEFINIC ETHERS

Because the compounds containing an ether linkage are of such commercial importance and because such a large number of compounds are available, the chemical shifts of this group have been divided into five separate sections; Aliphatic and Olefinic, Alicyclic, Aromatic, Heterocyclic and, the Silicon and Phosphorus Ethers.

Chemical Shifts

Aliphatic Protons

The aliphatic groups bonded to the ether linkage are moderately strongly deshielded. In addition, the aliphatic groups bonded to an olefinic ether linkage are more strongly deshielded than those of an aliphatic ether substituent.

HC–4	HC–3	HC–2	HC–1		Solvent	Spectrum
			3.11	CH_3–O–C$(CH_3)_3$	CCl_4	1396
			3.22	CH_3–O–R_5	CCl_4	- - - -
			3.30	CH_3–O–R_4	$CDCl_3$	1380
			3.72	CH_3–O–CH=CH–C(=O)–CH_3	CCl_4	- - - -
		1.13	3.38	CH_3–CH_2–O–R_2	CCl_4	1379
		1.23	3.71	CH_3–CH_2–O–CH=CH–CH_3	CCl_4	1408
	0.93	1.60	3.37	CH_3–CH_2–CH_2–O–R_3	$CDCl_3$	- - - -
		(1.12)	3.51	$(CH_3)_2$–CH–O–R_3	CCl_4	- - - -
0.91	(1.1–1.8)		3.37	$CH_3(CH_2)_2$–CH_2–O–CH_3	$CDCl_3$	1380
0.91	(1.1–1.8)		3.60	$CH_3(CH_2)_2$–CH_2–O–CH=CH_2	CCl_4	1404
	(0.89)	1.80	3.09	$(CH_3)_2$–CH–CH_2–O–R_2	CCl_4	1383
	(0.97)	1.90	3.38	$(CH_3)_2$–CH–CH_2–O–CH=CH_2	CCl_4	1405
		(1.12)	- - - -	$(CH_3)_3$–C–O–CH_3	CCl_4	1396

Substituted Methyl Ethers

CH_3 –O	–CH_2	–X	
3.30	3.37	–R_3	$CDCl_3$
3.43	3.89	–C(=O)–NH_2	$CDCl_3$
3.38	3.90	–C(=O)–O–CH_3	CCl_4
3.30	3.99	–C≡C–H	CCl_4
3.47	4.02	–C(=O)–CH_3	$CDCl_3$
3.46	4.14	–C≡N	CCl_4
3.49	4.30	–C(=O)–Cl	CCl_4
3.29	4.35	–Ø	$CDCl_3$
3.23	4.40	–O–CH_3	$CDCl_3$

Aliphatic Protons continued

Substituted Methyl Ethyl Ethers

CH_3	–O	–CH_2	–CH_2	–X	Solvent	Spectrum
3.38		3.54	2.52	–C≡N	CCl_4	
3.26		3.53	2.73	–Ø	$CDCl_3$	
3.37		3.42	2.86	–NH_2	$CDCl_3$	
3.39		3.65	3.40	–Br	CCl_4	
3.38		3.51	3.71	–OH	$CDCl_3$	

Olefinic Protons

In regard to the vinyl protons, the ether linkage is a strongly deshielding substituent in its effect on the chemical shift of the proton attached to the alpha carbon (the geminal hydrogen), but is a strong shielding group in its effect on the cis and trans protons.

Vinyl Ethers

	CH_2	=CH	–O–R		
cis	trans	gem			
4.01	3.84	6.32	–O–R_4	CCl_4	1404
4.08	3.89	6.35	–O–CH_2CH_2–$N(CH_3)_2$	CCl_4	1406

Substituted Vinyl Ethers

R–O–	CH	=CH	–X	
	5.82	4.27	–CH_3 (cis)	CCl_4
	6.12	4.68	–CH_3 (trans)	CCl_4
	7.54	5.42	–C(=O)–NH–Ø	$CDCl_3$
	7.50	5.52	–C(=O)–CH_3	CCl_4
	7.76	6.68	–C(=O)–1(thiophene)	Polysol
	7.79	6.72	–C(=O)–Ø	Polysol

Vinyl Coupling and Coupling Constants

Because of the clear separation in chemical shifts produced by ether substituents on the vinyl protons, the various coupling constants are often clearly displayed.

Spectrum 1406

Geminal	$J_{H_2C=C}$	=	1.7 Hz
Cis	$J_{H-C=C-H}$	=	7.0 Hz
Trans	$J_{H-C=C-H}$	=	14.5 Hz

Spectrum 1408

Geminal	J_{CH_3-C-H}	=	6.9 Hz
Cis	$J_{CH_3-C=C-H}$	=	1.6 Hz
Trans	$J_{CH_3-C=C-H}$	=	1.6 Hz

ETHYL ETHER

$C_4H_{10}O$

Mol. Wt. 74.12

Solvent: CCl_4

Assignments
a = 1.13
b = 3.38
c = 1.60 imp.

1379

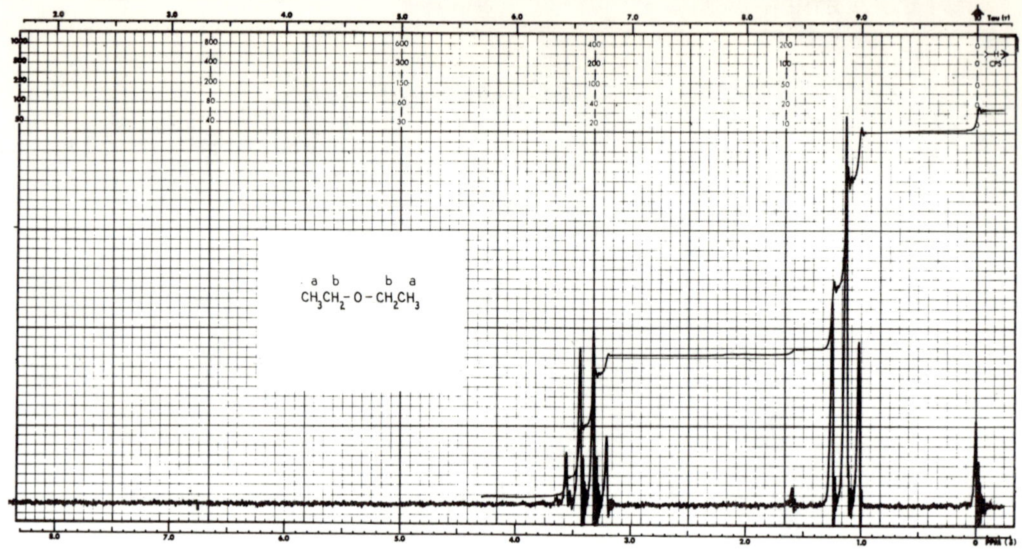

© 1978 Sadtler Research Laboratories, Inc.

BUTYL METHYL ETHER

$C_5H_{12}O$

Mol. Wt. 88.15

B.P. 69-71°C/760mm

Solvent: $CDCl_3$

Assignments
a = 0.91
b = 1.10-1.75
c = 3.30
d = 3.37

1380

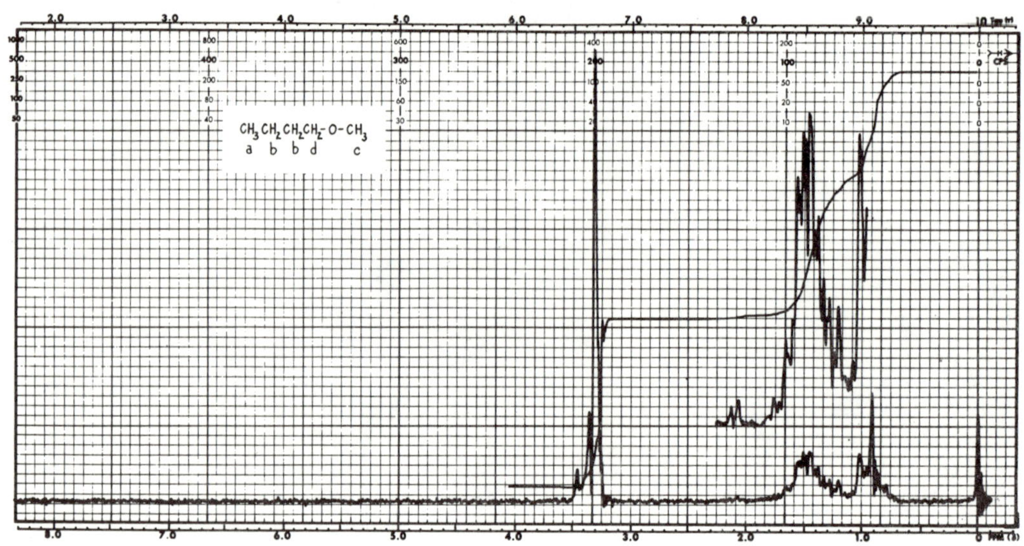

© 1978 Sadtler Research Laboratories, Inc.

PROPYL ETHER

$C_6H_{14}O$

Mol. Wt. 102.18

B.P. 88-90°C

Solvent: $CDCl_3$

Assignments
a = 0.93
b = 1.60
c = 3.37

1381

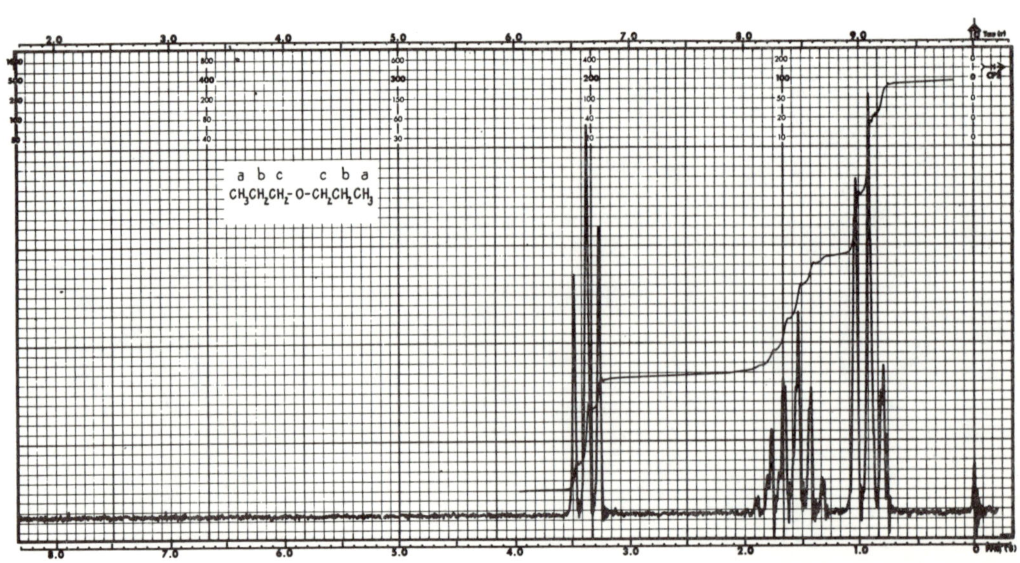

© 1978 Sadtler Research Laboratories, Inc.

BUTYL ETHYL ETHER

$C_6H_{14}O$

Mol. Wt. 102.18

B.P. 90-92°C

Solvent: CCl_4

Assignments
a = 0.93
b = 1.16
c = 1.20-1.80
d = 3.35
e = 3.39

1382

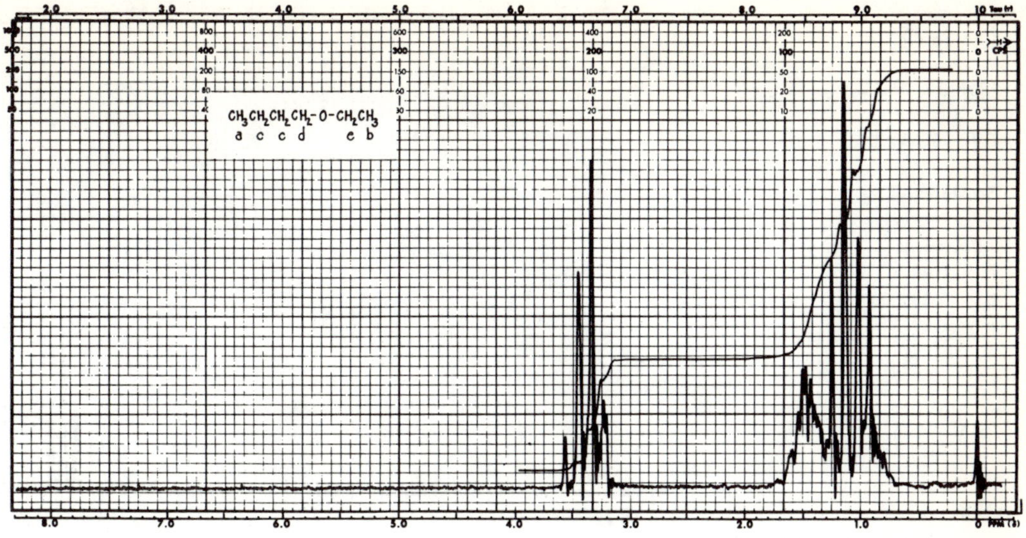

© 1978 Sadtler Research Laboratories, Inc.

ETHYL ISOBUTYL ETHER

$C_6H_{14}O$

Mol. Wt. 102.18

Solvent: CCl_4

Assignments
a = 0.89
b = 1.15
c = 1.80
d = 3.09
e = 3.38

1383

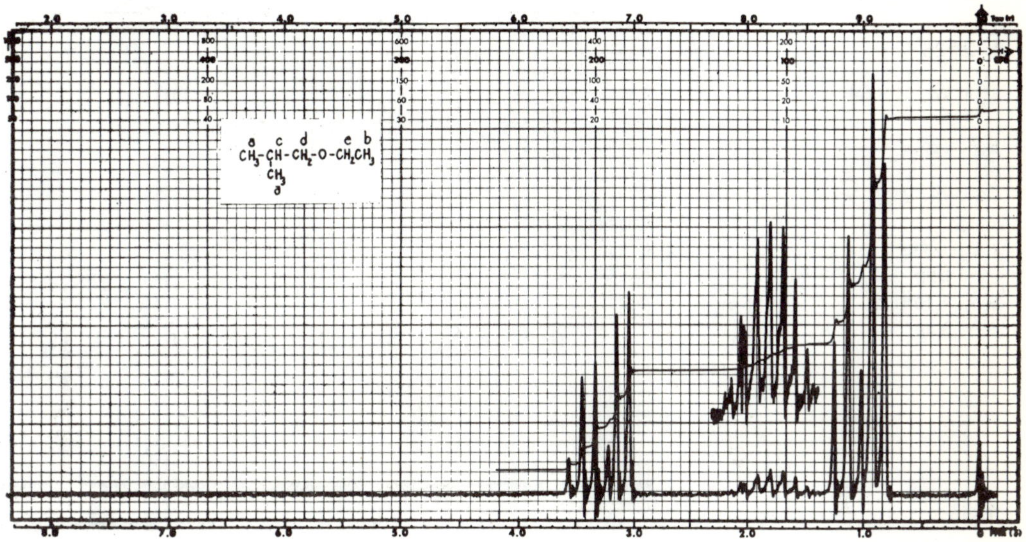

© 1978 Sadtler Research Laboratories, Inc.

BUTYL ETHER

$C_8H_{18}O$

Mol. Wt. 130.23

B.P. 140°C

Solvent: CCl_4

Assignments
a = 0.90
b = 1.10-1.70
c = 3.29

1384

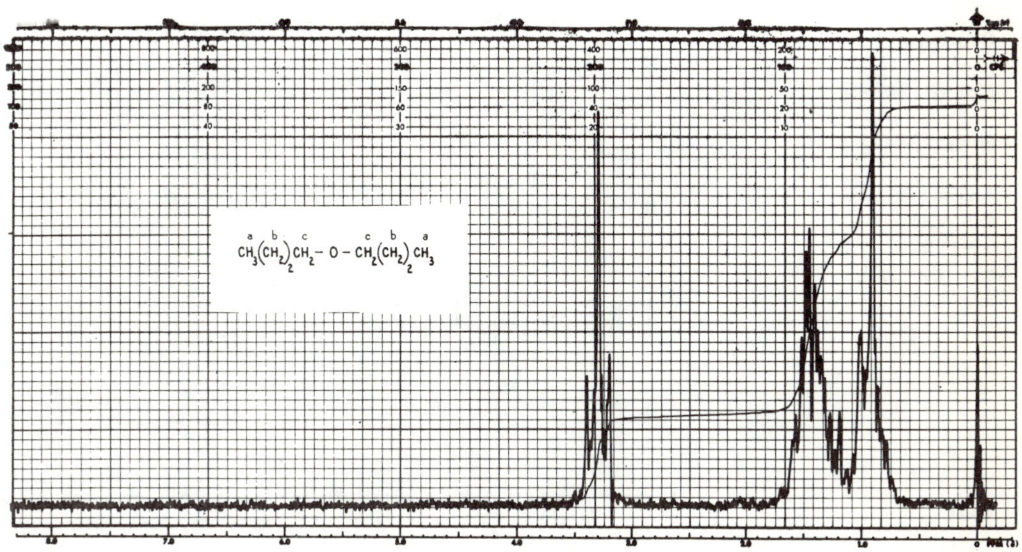

© 1978 Sadtler Research Laboratories, Inc.

ISOPENTYL ETHER

$C_{10}H_{22}O$

Mol. Wt. 158.29

B.P. 170-172°C

Solvent: CCl_4

Assignments
a = 0.90
b = 1.20-2.00
c = 3.32

1385

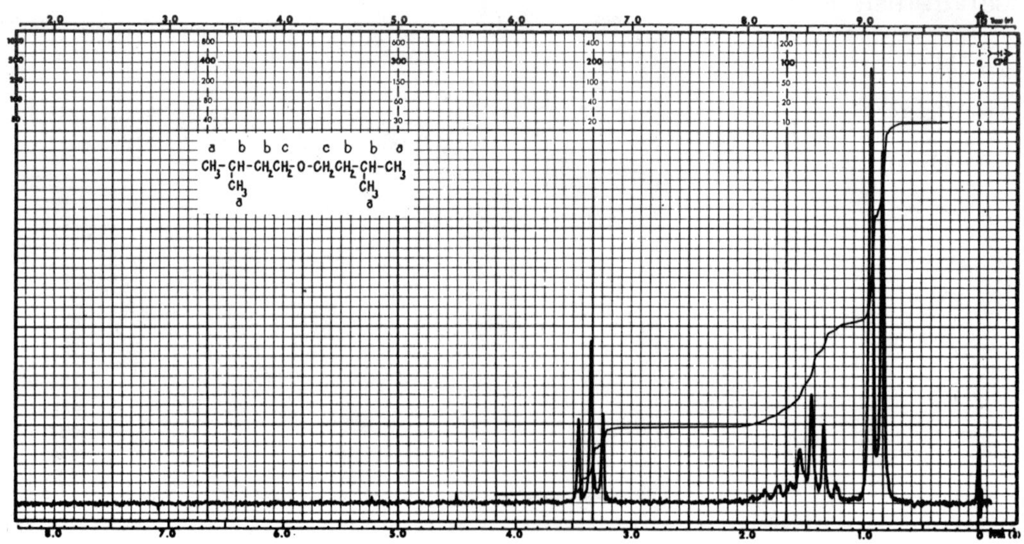

© 1978 Sadtler Research Laboratories, Inc.

HEXYL ETHER

$C_{12}H_{26}O$

Mol. Wt. 186.34

B.P. 226.2°C

Solvent: CCl_4

Assignments
a = 0.90
b = 1.10-1.70
c = 3.29

1386

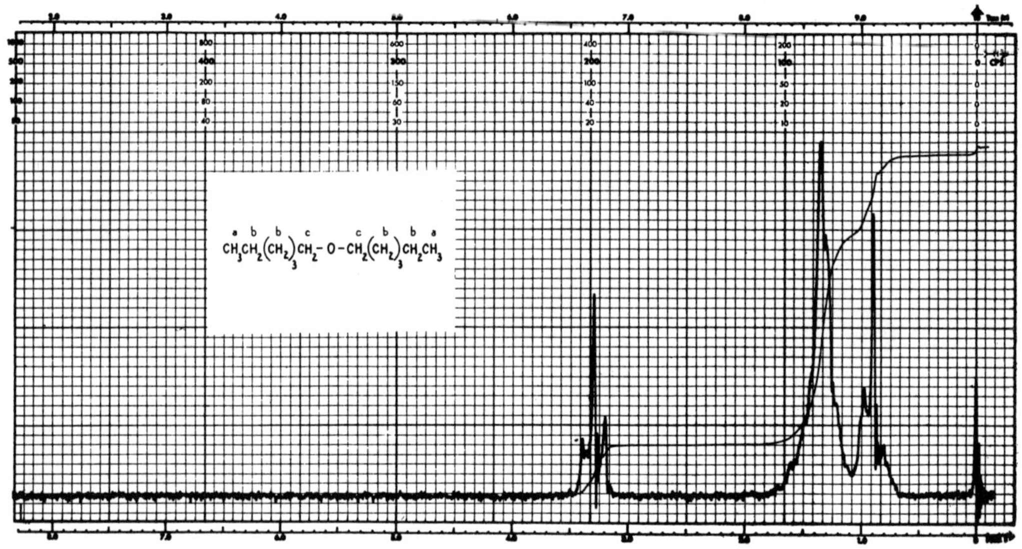

© 1978 Sadtler Research Laboratories, Inc.

BIS(2-ETHYLHEXYL) ETHER

$C_{16}H_{34}O$

Mol. Wt. 242.45

Solvent: CCl_4

Assignments
a = 0.88
b = 0.90
c = 1.09-1.62
d = 3.19

1387

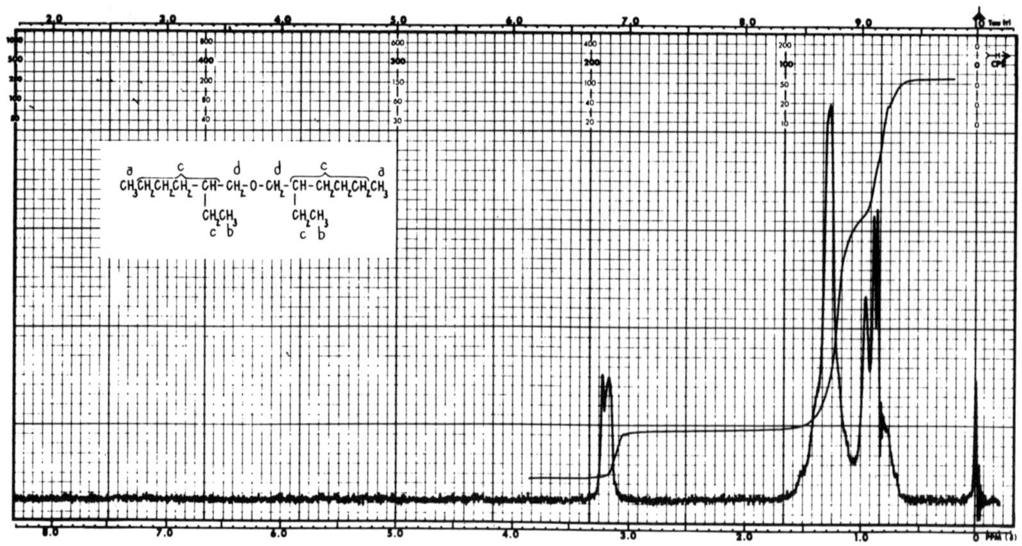

© 1978 Sadtler Research Laboratories, Inc.

OCTYL ETHER

$C_{16}H_{34}O$

Mol. Wt. 242.45

B.P. 143-144°C/2mm

Solvent: CCl_4

Assignments
a = 0.91
b = 1.31
c = 3.31

1388

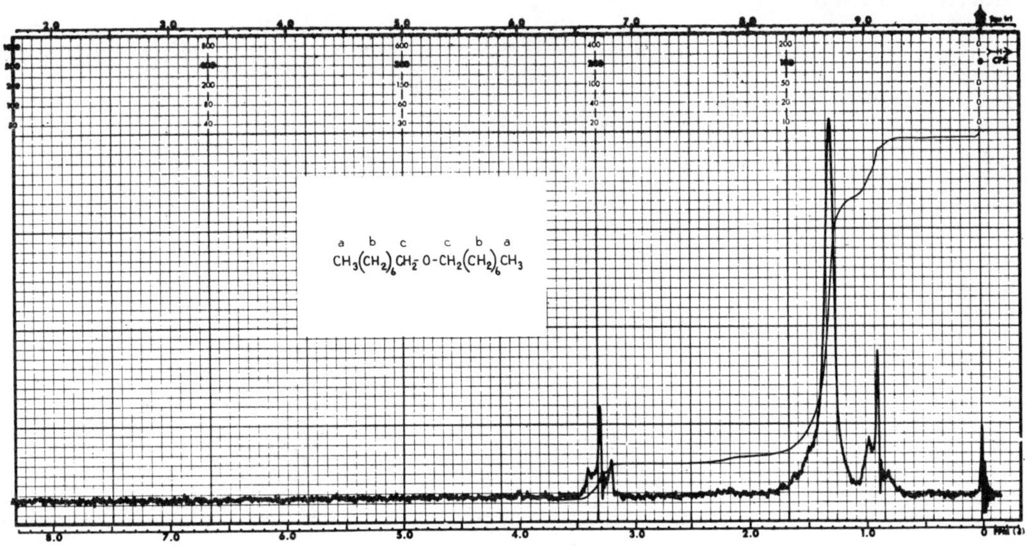

© 1978 Sadtler Research Laboratories, Inc.

DECYL ETHER

$C_{20}H_{42}O$

Mol. Wt. 298.56

M.P. 16°C

Solvent: CCl_4

Assignments
a = 0.89
b = 1.10-1.70
c = 3.60

1389

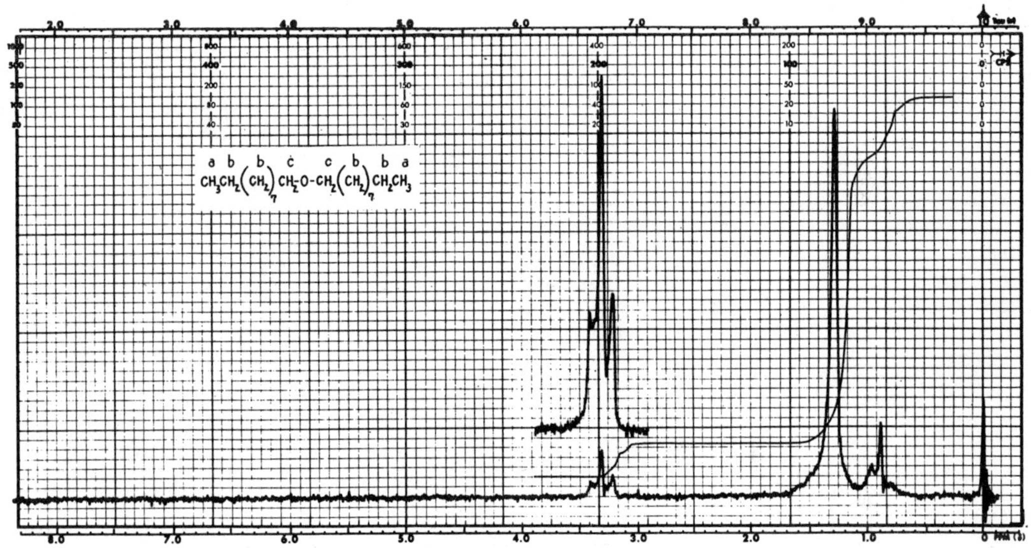

© 1978 Sadtler Research Laboratories, Inc.

ETHYL OCTADECYL ETHER

$C_{20}H_{42}O$

Mol. Wt. 298.56

Solvent: $CDCl_3$

Assignments
a = 0.89
b = 1.19
c = ca 1.25
d = 3.40
e = 3.46

1390

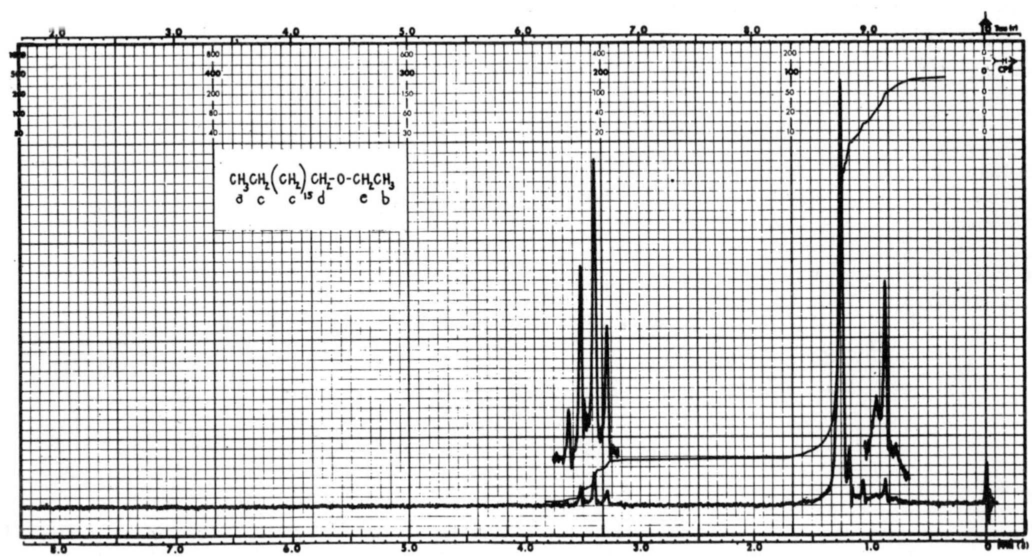

© 1978 Sadtler Research Laboratories, Inc.

HEXADECYL ETHER

$C_{32}H_{66}O$

Mol. Wt. 466.88

M.P. 49-51°C (lit.)

Solvent: $CDCl_3$

Assignments
a = 0.90
b = 1.29
c = 3.40

1391

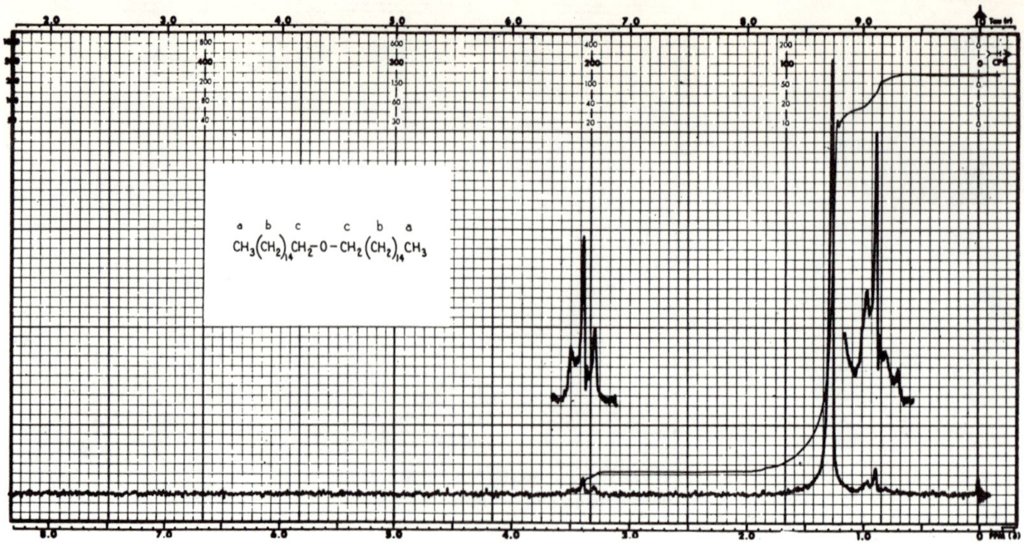

© 1978 Sadtler Research Laboratories, Inc.

OCTADECYL ETHER

$C_{36}H_{74}O$

Mol. Wt. 522.99

M.P. 60-62°C

Solvent: $CDCl_3$

Assignments
a = 0.89
b = 1.27
c = 3.38

1392

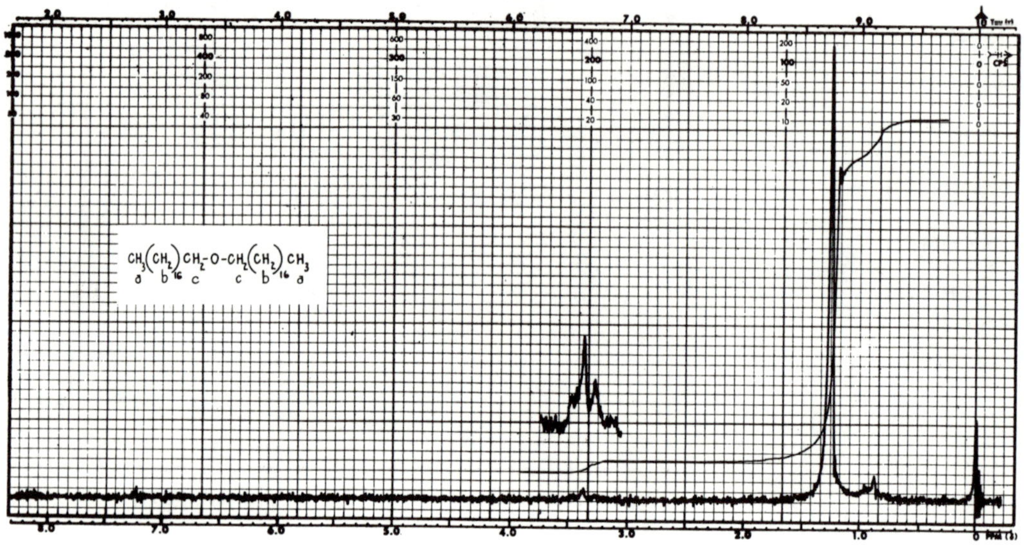

© 1978 Sadtler Research Laboratories, Inc.

2-CHLORO-1,1,2-TRIFLUOROETHYL) ETHYL ETHER

$C_4H_6ClF_3O$

Mol. Wt. 162.54

Solvent: CCl_4

Assignments
a = 1.35
b = 4.05
c = 5.98

1393

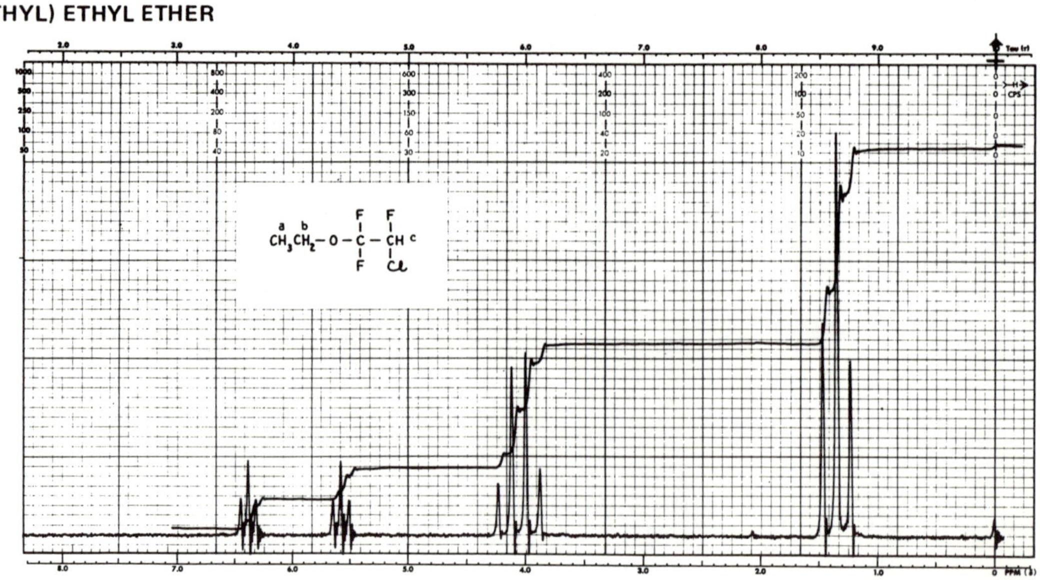

© 1978 Sadtler Research Laboratories, Inc.

ISOPROPYL PROPYL ETHER

$C_6H_{14}O$

Mol. Wt. 102.18

B.P. 79-81°C

Solvent: CCl_4

Assignments
a = 0.91
b = 1.12
c = 1.50
d = 3.31
e = 3.51

1394

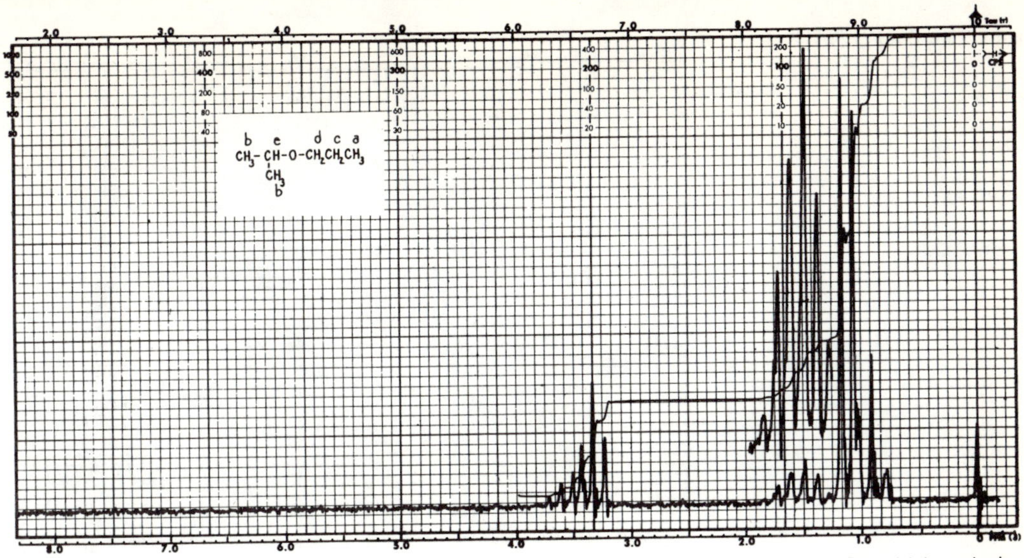

© 1978 Sadtler Research Laboratories, Inc.

ISOPROPYL ETHER

$C_6H_{14}O$

Mol. Wt. 102.18

B.P. 67-67.6°C

Solvent: CCl_4

Assignments
a = 1.05
b = 3.52

1395

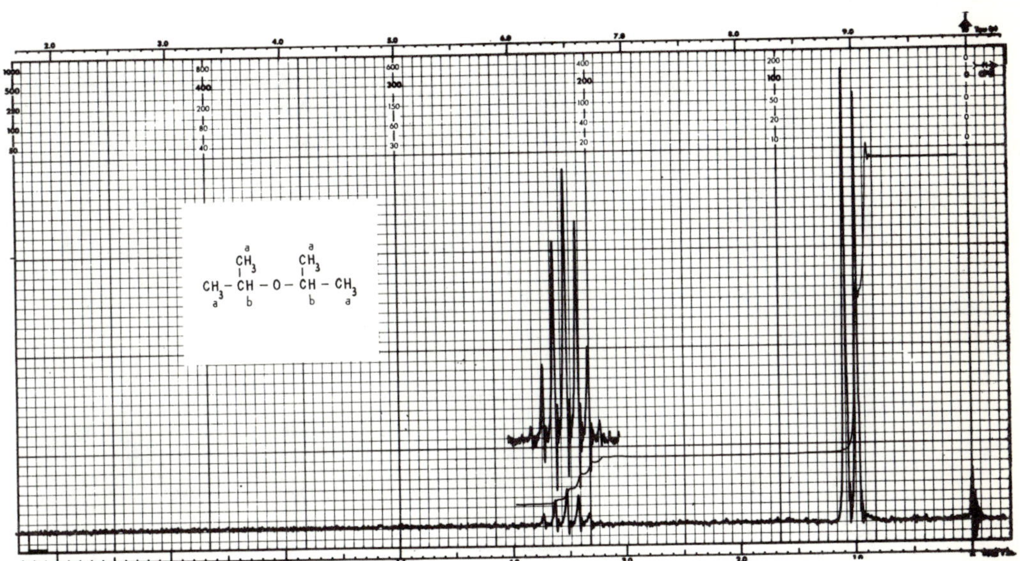

© 1978 Sadtler Research Laboratories, Inc.

tert-BUTYL METHYL ETHER

$C_5H_{12}O$

Mol. Wt. 88.15

B.P. 53-56°C

Solvent: CCl_4

Assignments
a = 1.12
b = 3.11

1396

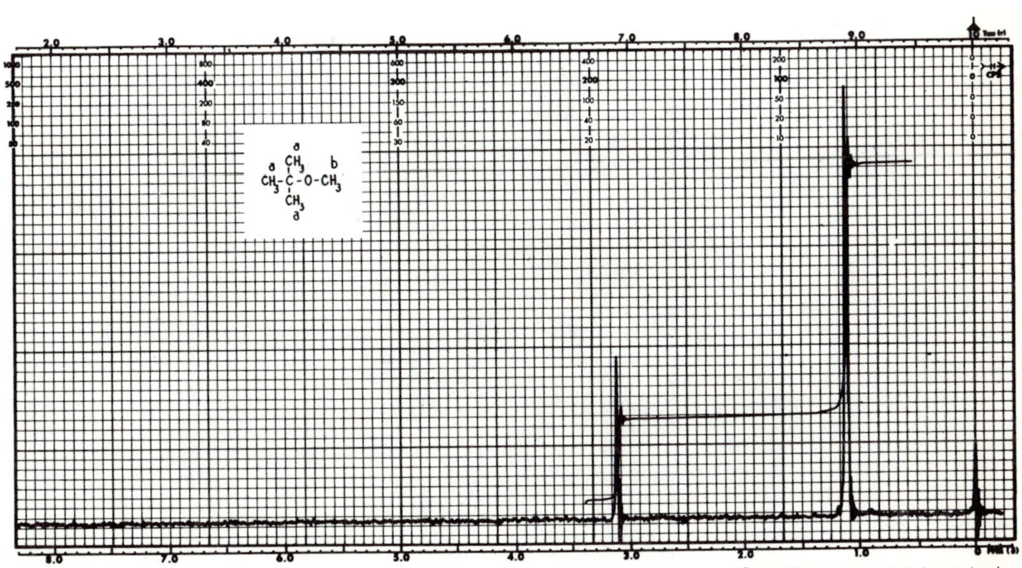

© 1978 Sadtler Research Laboratories, Inc.

tert-BUTYL ETHYL ETHER

$C_6H_{14}O$

Mol. Wt. 102.18

M.P. 70-72°C

Solvent: CCl_4

Assignments

a = 1.11

b = 1.13

c = 3.32

1397

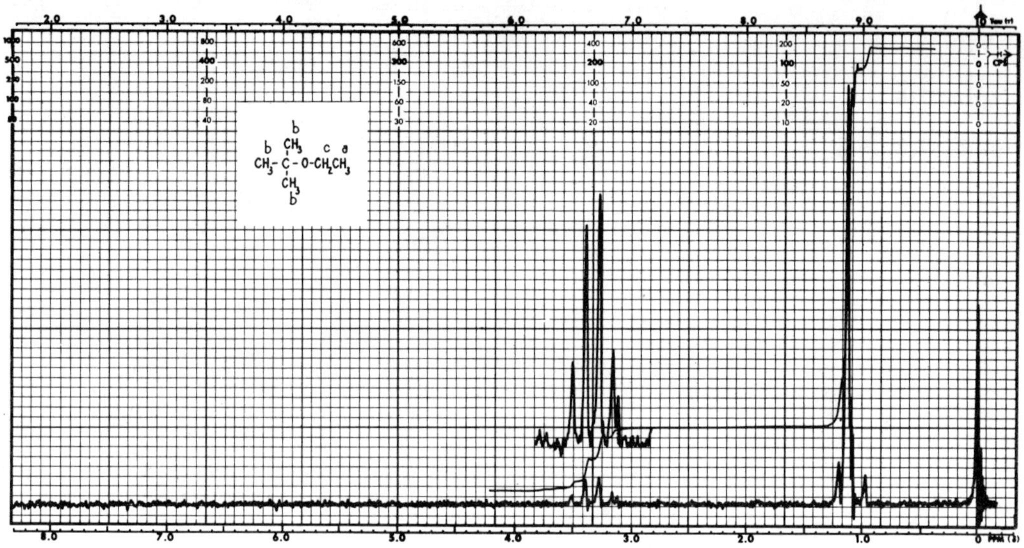

© 1978 Sadtler Research Laboratories, Inc.

tert-BUTYL ISOPROPYL ETHER

$C_7H_{16}O$

Mol. Wt. 116.20

B.P. 87-88°C

Solvent: CCl_4

Assignments

a = 1.04

b = 1.12

c = 3.70

1398

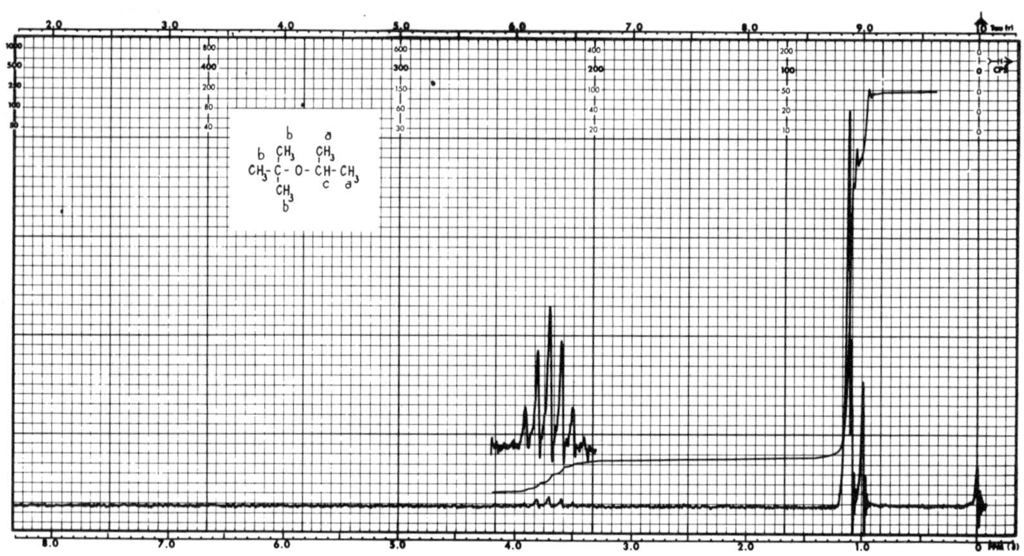

© 1978 Sadtler Research Laboratories, Inc.

BIS(CHLOROMETHYL) ETHER

$C_2H_4Cl_2O$

Mol. Wt. 114.96

B.P. 58-60°C

Solvent: $CDCl_3$

Assignments

a = 5.51

1399

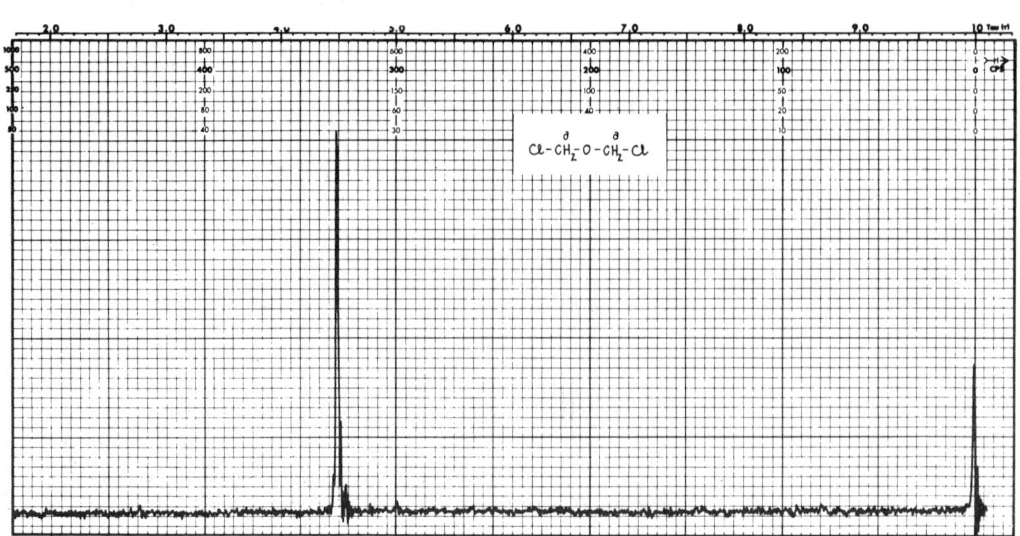

© 1978 Sadtler Research Laboratories, Inc.

3-ISOPROPOXYPROPIONITRILE

$C_6H_{11}NO$

Mol. Wt. 113.16

B.P. 65-65.5°C/10mm

Solvent: $CDCl_3$

Assignments
a = 1.18
b = 2.56
c = 3.63

1400

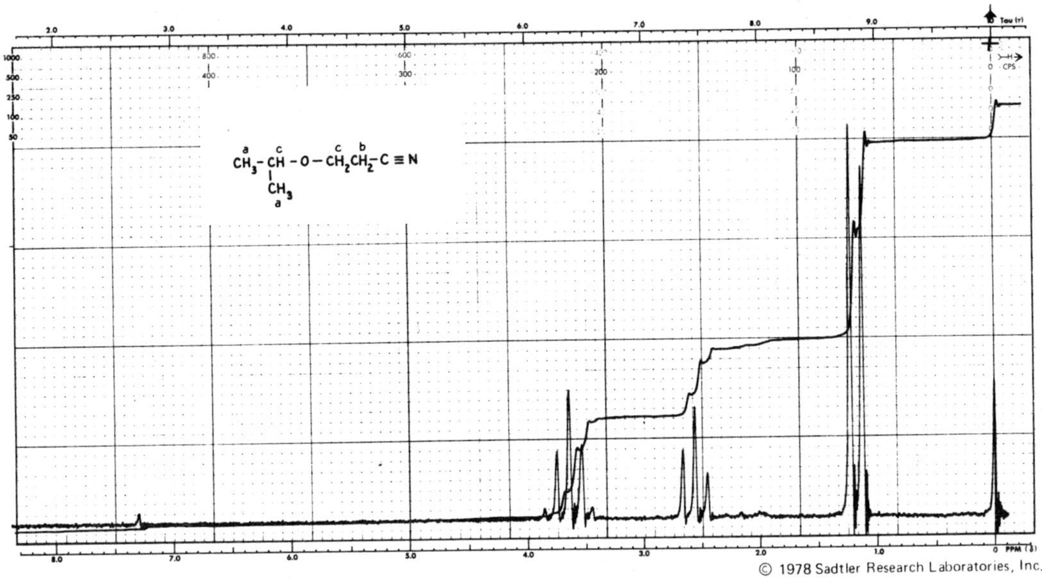

© 1978 Sadtler Research Laboratories, Inc.

2-ETHOXYETHANETHIOL

$C_4H_{10}OS$

Mol. Wt. 106.19

B.P. 126-128°C

Solvent: $CDCl_3$

Assignments
a = 1.20
b = 1.60
c = 2.69
d = 3.52
e = 3.59

1401

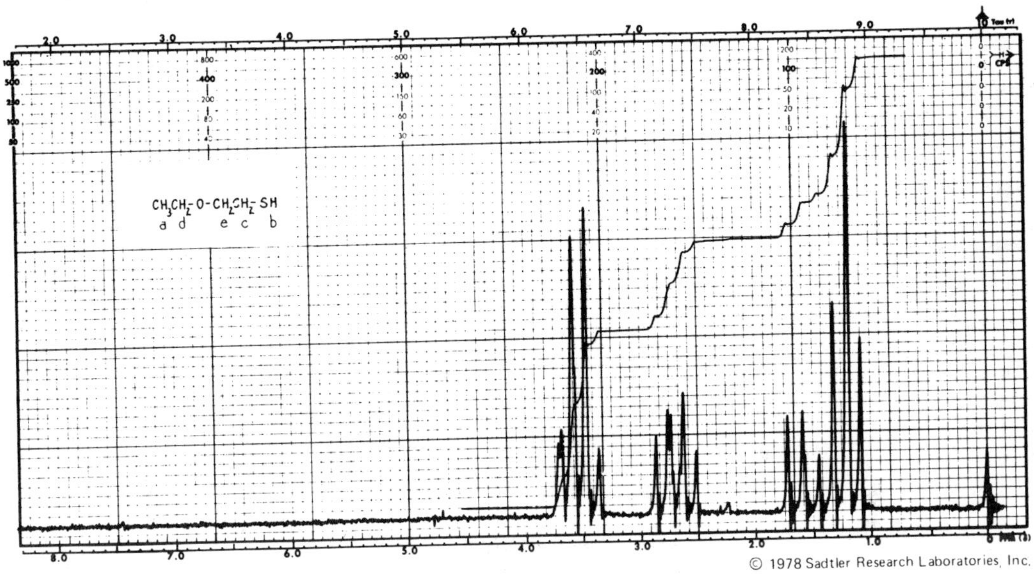

© 1978 Sadtler Research Laboratories, Inc.

2,2'-OXYDIETHANETHIOL

$C_4H_{10}OS_2$

Mol. Wt. 138.25

M.P. 64-65°C/2mm

B.P. 71-72°C/4mm

Solvent: CCl_4

Assignments
a = 1.60
b = 2.70
c = 3.55

1402

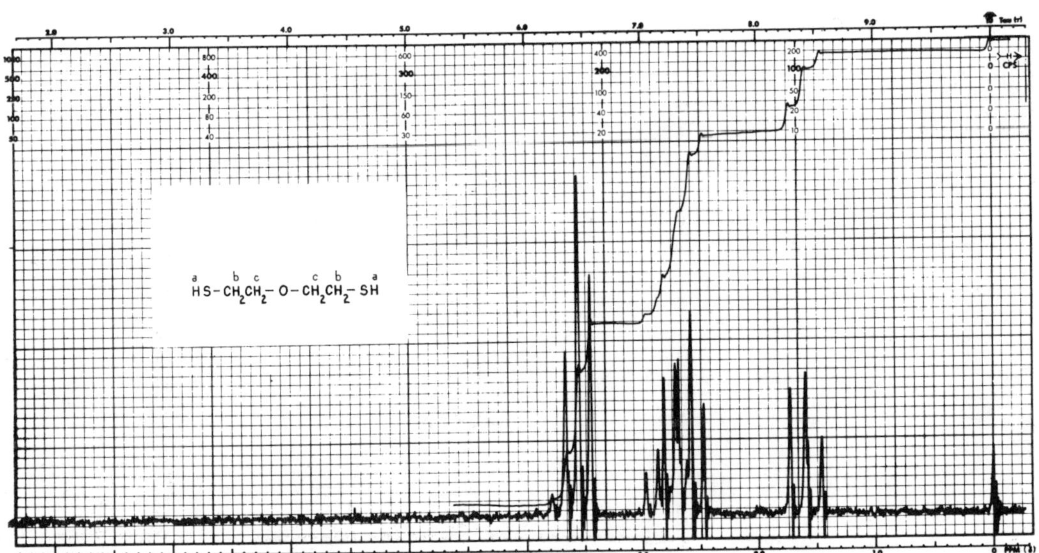

© 1978 Sadtler Research Laboratories, Inc.

ALLYL ETHYL ETHER

$C_5H_{10}O$

Mol. Wt. 86.13

B.P. 63-65°C

Solvent: $CDCl_3$

Assignments
a = 1.20
b = 3.49
c = 3.97
d = 5.16
e = 5.21
f = 5.94

1403

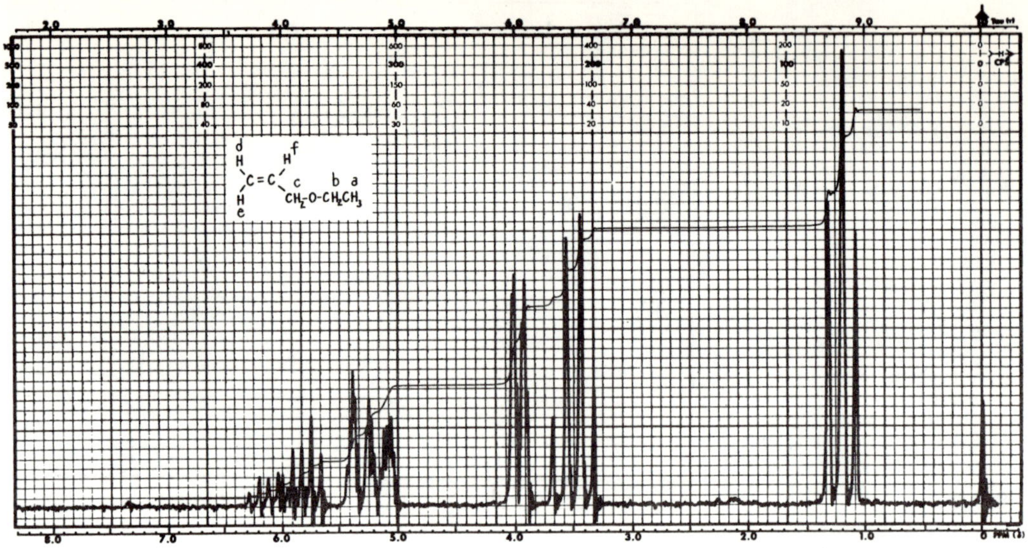

© 1978 Sadtler Research Laboratories, Inc.

BUTYL VINYL ETHER

$C_6H_{12}O$

Mol. Wt. 100.16

M.P. 94.1°C

Solvent: CCl_4

Assignments
a = 0.91
b = 1.10-1.80
c = 3.60
d = 3.84
e = 4.01
f = 6.32

1404

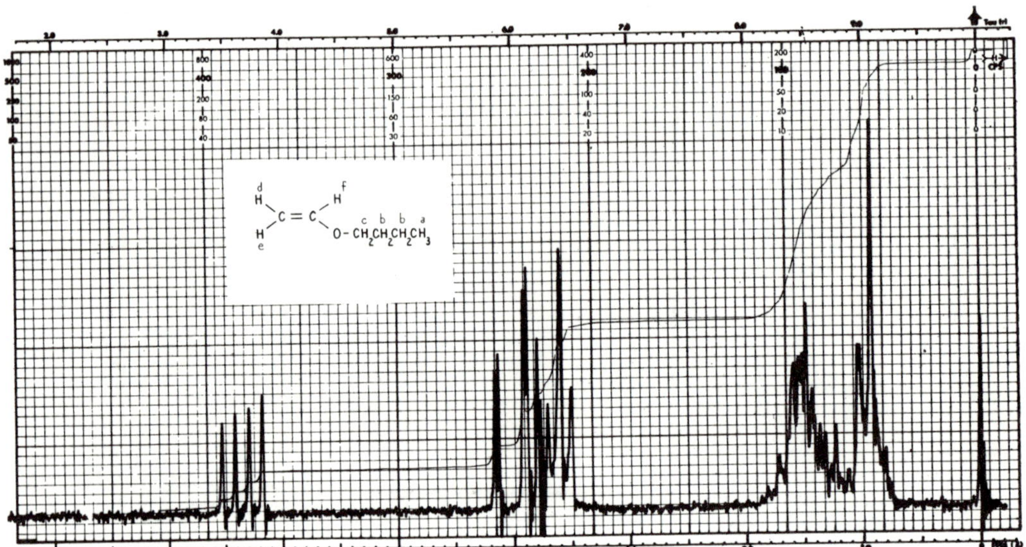

© 1978 Sadtler Research Laboratories, Inc.

ISOBUTYL VINYL ETHER

$C_6H_{12}O$

Mol. Wt. 100.16

B.P. 83.3°C

Solvent: CCl_4

Assignments
a = 0.97
b = 1.90
c = 3.38
d = 3.86
e = 4.03
f = 6.37

1405

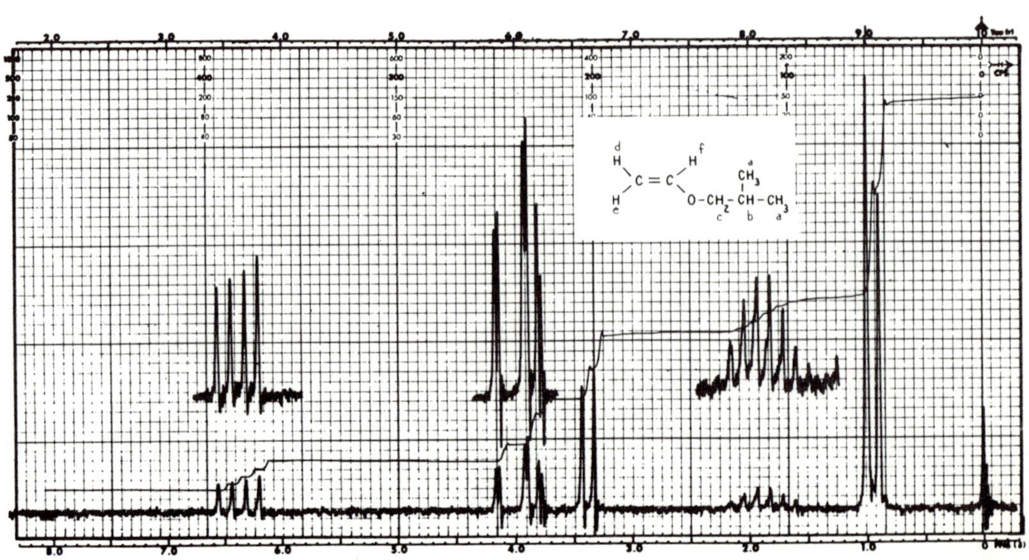

© 1978 Sadtler Research Laboratories, Inc.

N,N-DIMETHYL-2-(VINYLOXY)ETHYLAMINE

$C_6H_{13}NO$

Mol. Wt. 115.18

B.P. 123°C

Solvent: CCl_4

Assignments
a = 2.21
b = 2.51
c = 3.70
d = 3.89
e = 4.08
f = 6.35

1406

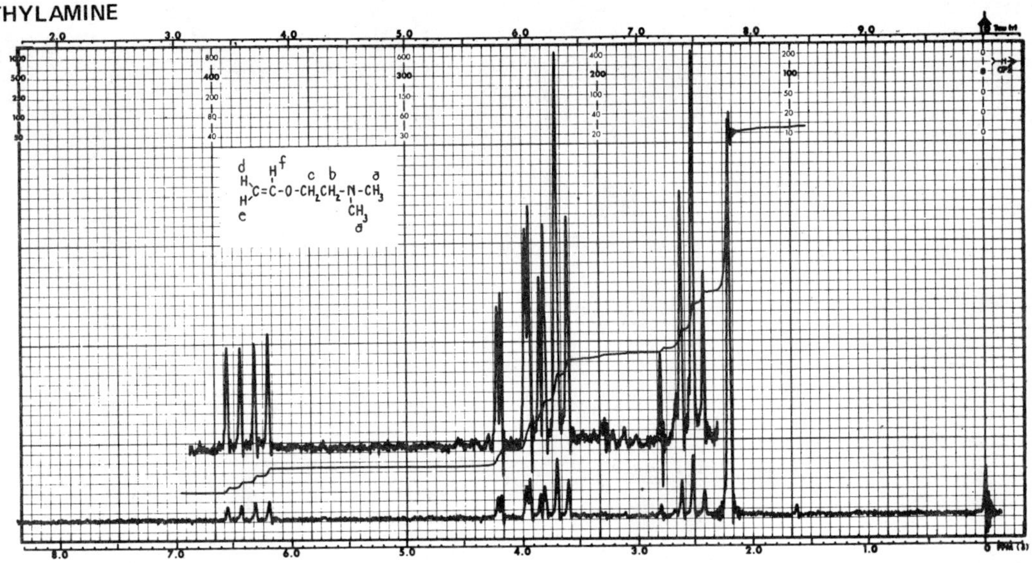

© 1978 Sadtler Research Laboratories, Inc.

ALLYL ETHER

$C_6H_{10}O$

Mol. Wt. 98.15

B.P. 93-94°C

Solvent: $CDCl_3$

Assignments
a = 3.90
b = 5.11
c = 5.19
d = 5.82

1407

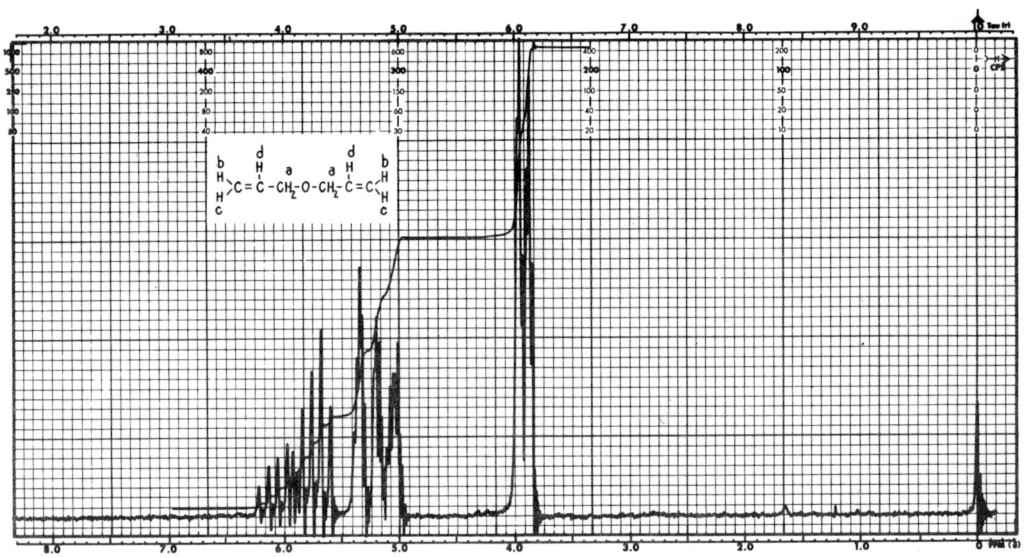

© 1978 Sadtler Research Laboratories, Inc.

ETHYL PROPENYL ETHER

$C_5H_{10}O$

Mol. Wt. 86.13

Solvent: CCl_4

Assignments
a = 1.23
b = 1.53
c = 3.71
d = 4.27
e = 4.68
f = 5.82
g = 6.12

1408

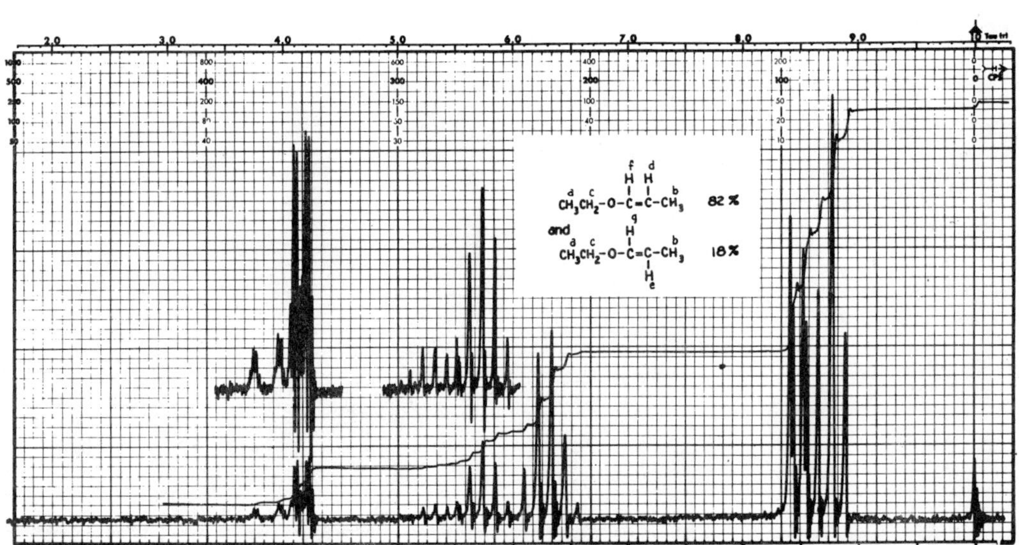

© 1978 Sadtler Research Laboratories, Inc.

METHYLAL

$C_3H_8O_2$

Mol. Wt. 76.10

B.P. 42-43.5°C

Solvent: CCl_4

Assignments
a = 3.23
b = 4.40

1409

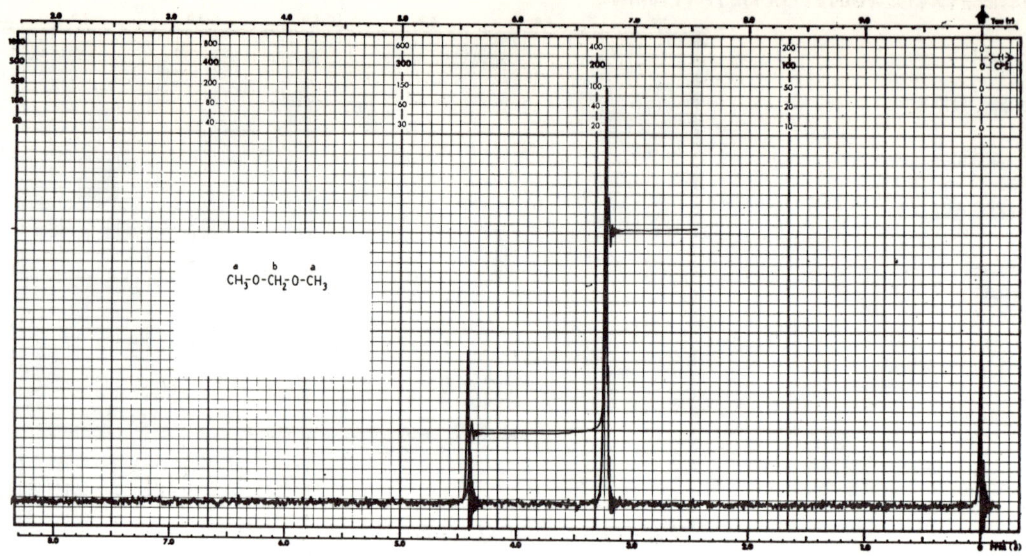

© 1978 Sadtler Research Laboratories, Inc.

DIETHOXYMETHANE

$C_5H_{12}O_2$

Mol. Wt. 104.15

B.P. 85-88°C

Solvent: $CDCl_3$

Assignments
a = 1.16
b = 3.51
c = 4.55
d = 4.62 imp.

1410

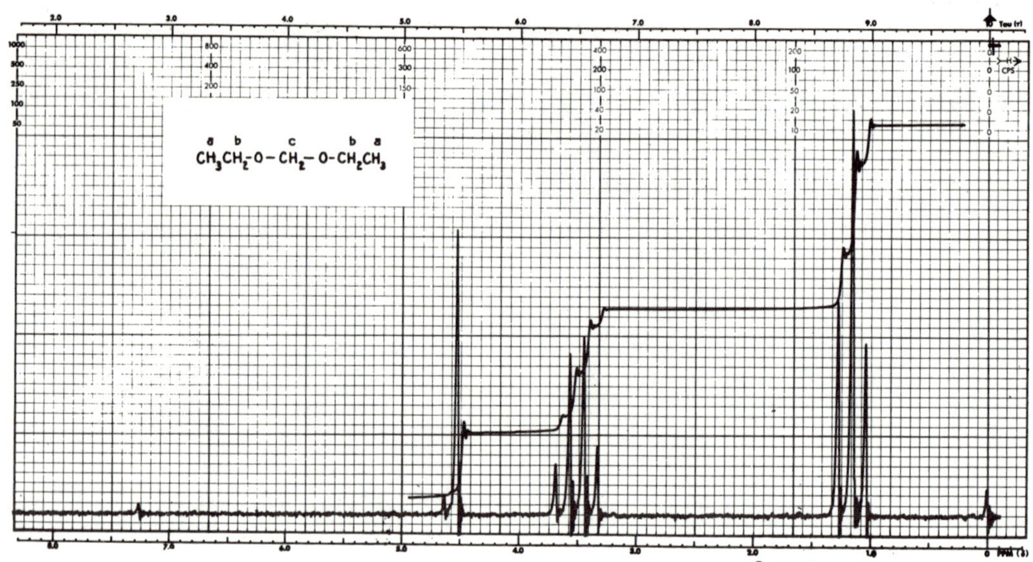

© 1978 Sadtler Research Laboratories, Inc.

BIS(2-CHLOROETHOXY)METHANE

$C_5H_{10}Cl_2O_2$

Mol. Wt. 173.04

B.P. 218.1°C

Solvent: CCl_4

Assignments
a = 3.68
b = 4.70

1411

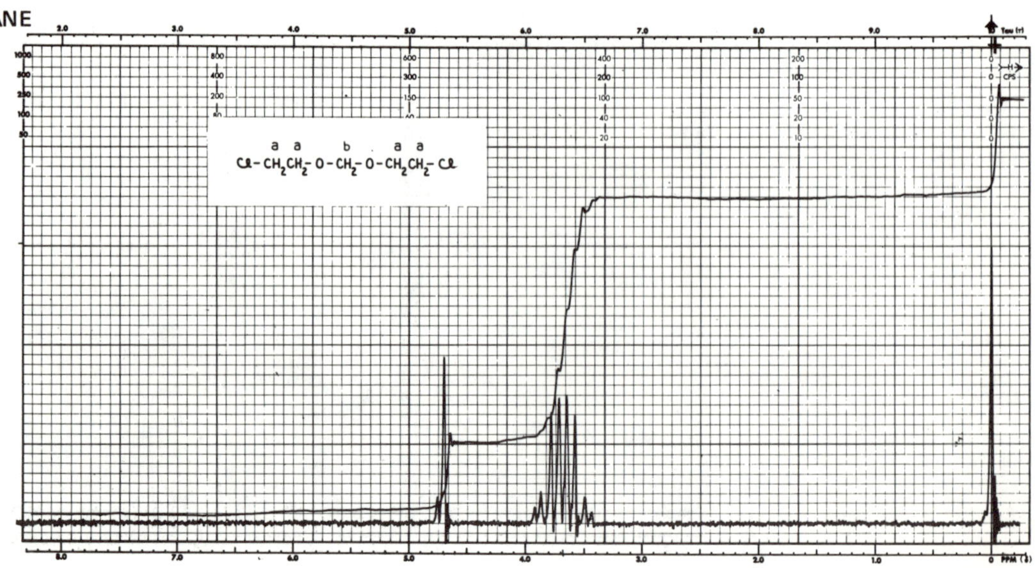

© 1978 Sadtler Research Laboratories, Inc.

1,2-DIETHOXYETHANE

$C_6H_{14}O$

Mol. Wt. 102.18

B.P. 121°C

Solvent: CCl_4

Assignments
a = 1.17
b = 3.42
c = 3.47

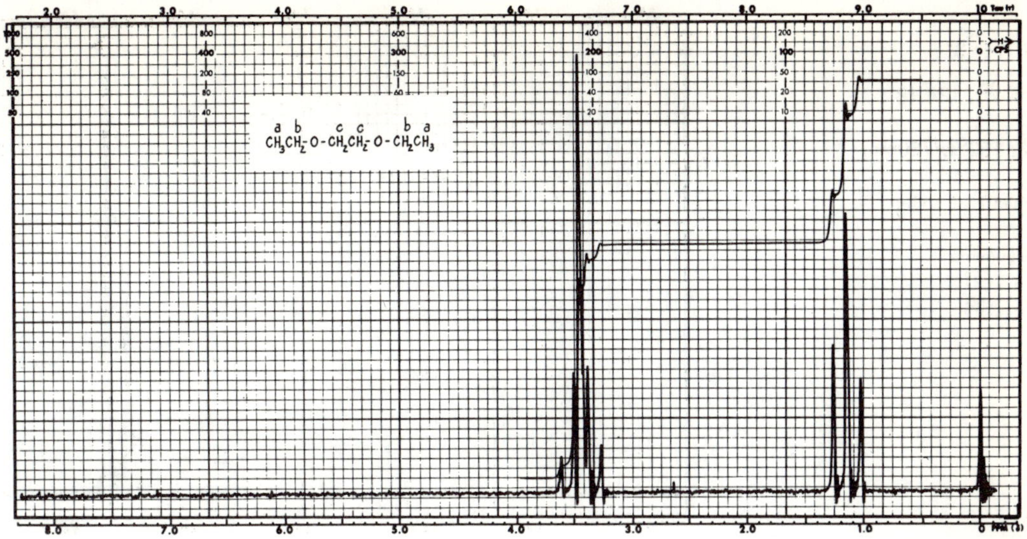

1412

© 1978 Sadtler Research Laboratories, Inc.

1,2-DIBUTOXYETHANE

$C_{10}H_{22}O_2$

Mol. Wt. 174.29

Solvent: CCl_4

Assignments
a = 0.91
b = 1.10-1.70
c = 3.39
d = 3.48

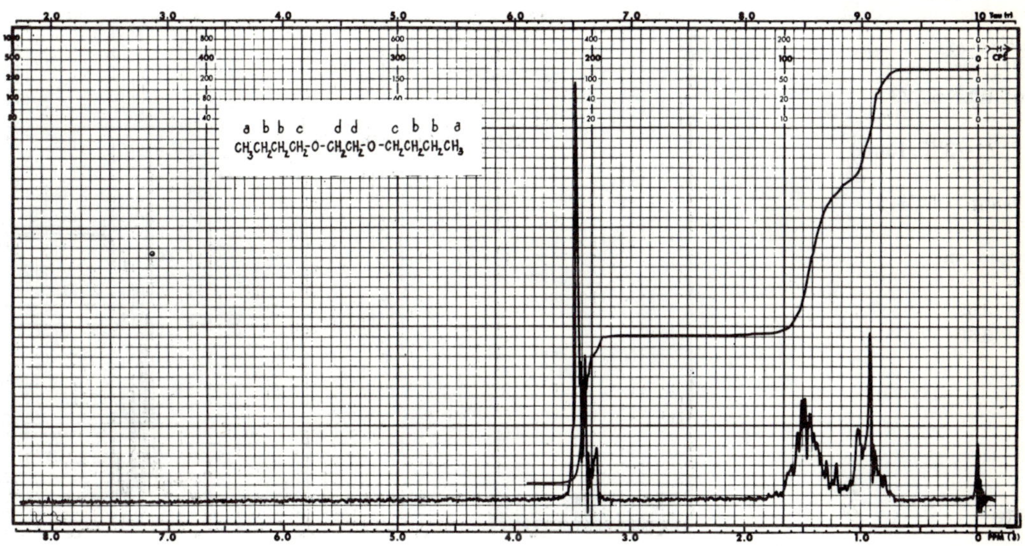

1413

© 1978 Sadtler Research Laboratories, Inc.

2-BUTOXYETHYL VINYL ETHER

$C_8H_{16}O_2$

Mol. Wt. 144.22

Solvent: CCl_4

Assignments
a = 0.95
b = 1.10-1.72
c = 3.30-3.80
d = 3.91
e = 4.09
f = 6.38

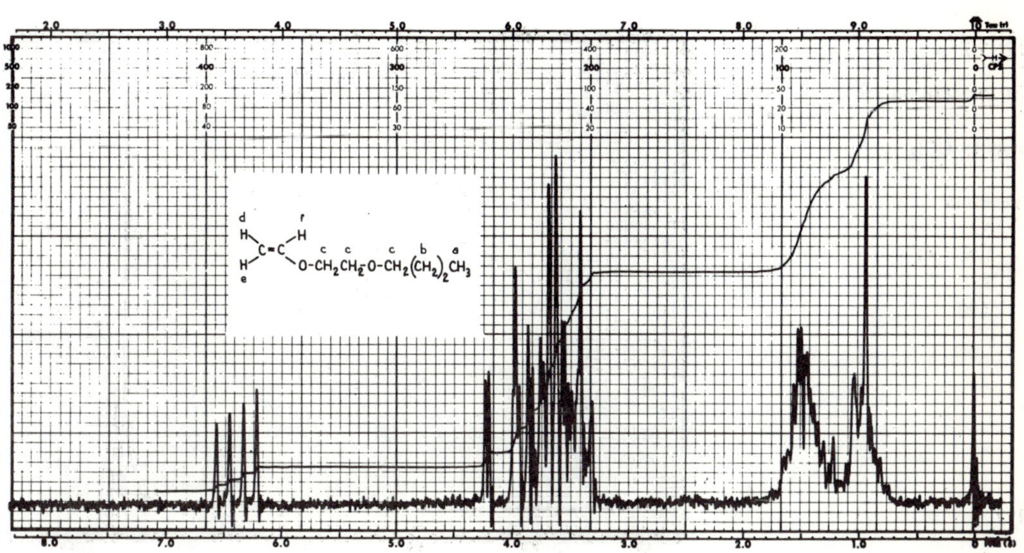

1414

© 1978 Sadtler Research Laboratories, Inc.

BIS(2-METHOXYETHYL) ETHER

$C_6H_{14}O_3$

Mol. Wt. 134.18

B.P. 162°C/760mm

Solvent: $CDCl_3$

Assignments
a = 3.31
b = 3.49

1415

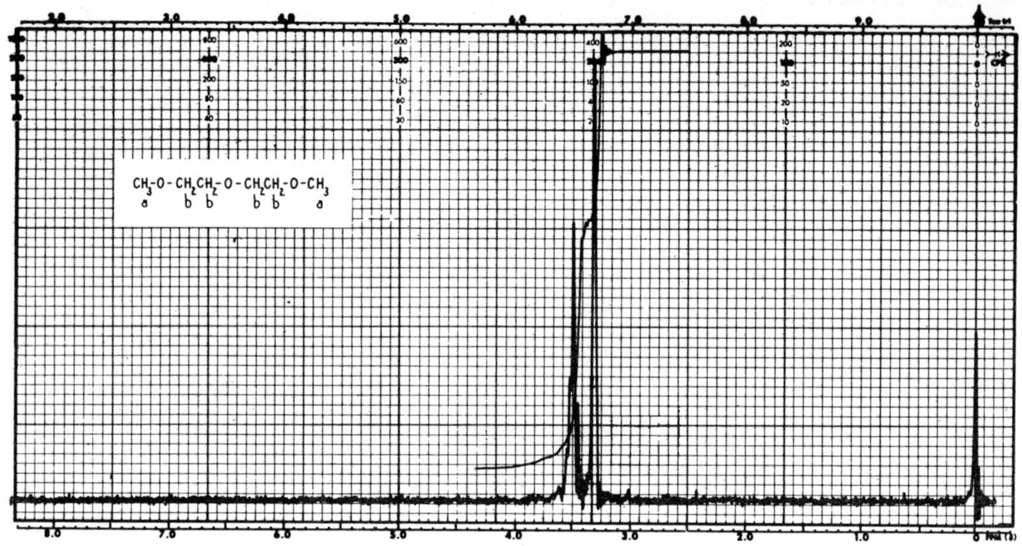

© 1978 Sadtler Research Laboratories, Inc.

BIS(2-ETHOXYETHYL) ETHER

$C_8H_{18}O_3$

Mol. Wt. 162.23

B.P. 71-73°C/10mm

Solvent: $CDCl_3$

Assignments
a = 1.21
b = 3.55
c = 3.63

1416

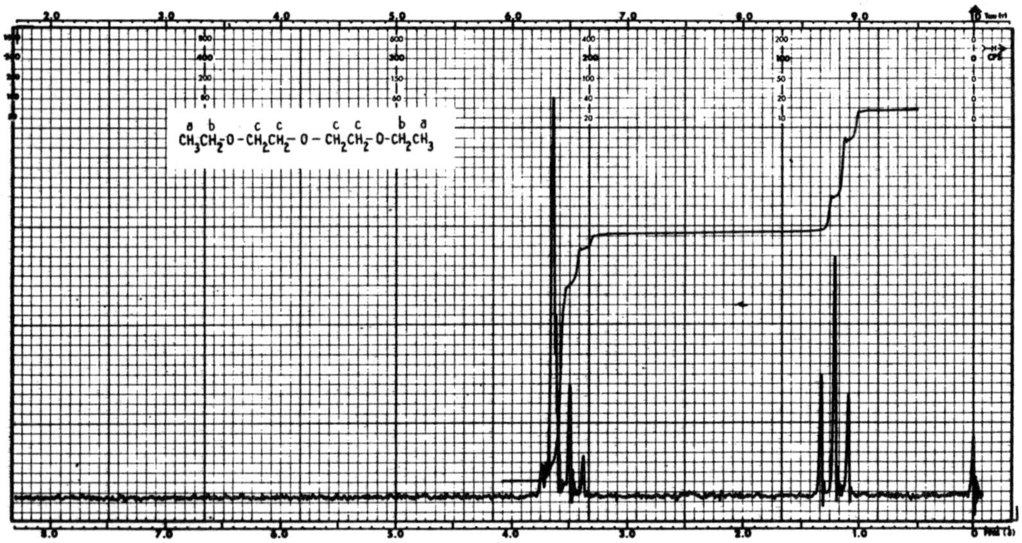

© 1978 Sadtler Research Laboratories, Inc.

BIS(2-BUTOXYETHYL) ETHER

$C_{12}H_{26}O_3$

Mol. Wt. 218.34

Solvent: CCl_4

Assignments
a = 0.91
b = 1.20-1.70
c = 3.40
d = 3.50

1417

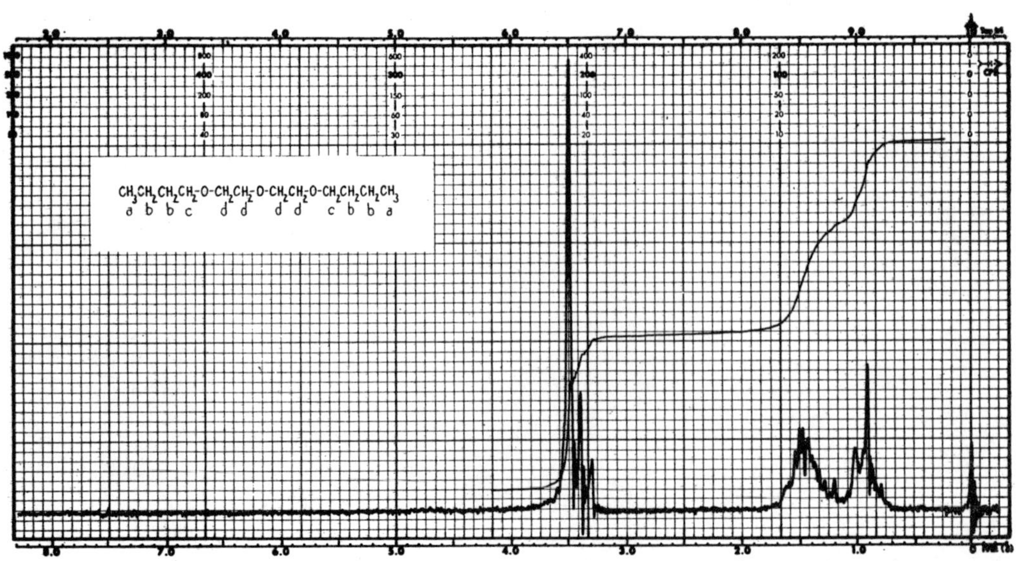

© 1978 Sadtler Research Laboratories, Inc.

BIS[2-(VINYLOXY)ETHYL] ETHER

$C_8H_{14}O_3$

Mol. Wt. 158.20

B.P. 81-82°C/10mm

Solvent: $CDCl_3$

Assignments
a = ca 3.79
b = 4.00
c = 4.16
d = 6.48

1418

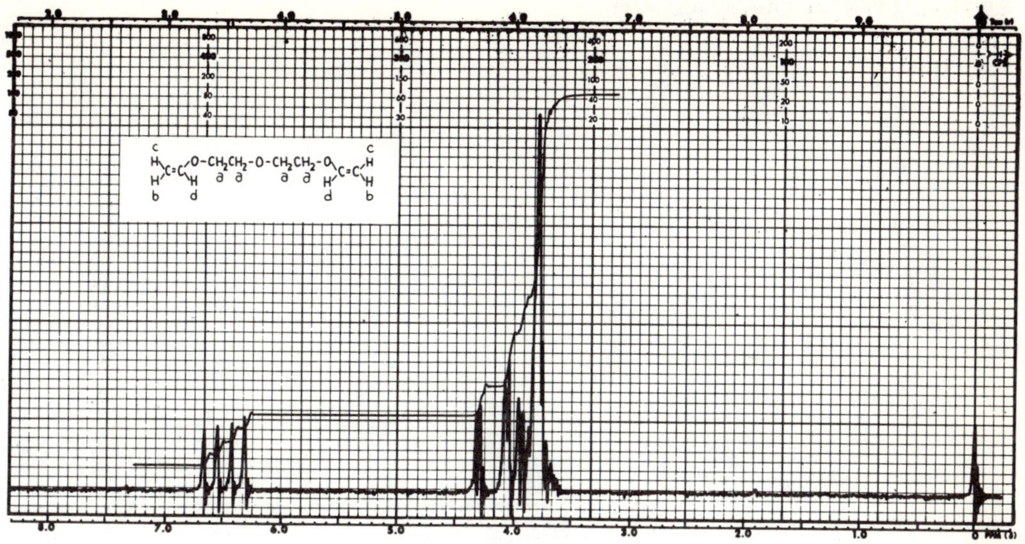

© 1978 Sadtler Research Laboratories, Inc.

2,5,8,11-TETRAOXADODECANE

$C_8H_{18}O_4$

Mol. Wt. 178.23

B.P. 216°C/760mm

F.P. −45°C

Solvent: CCl_4

Assignments
a = 3.29
b = ca 3.46
c = 3.50

1419

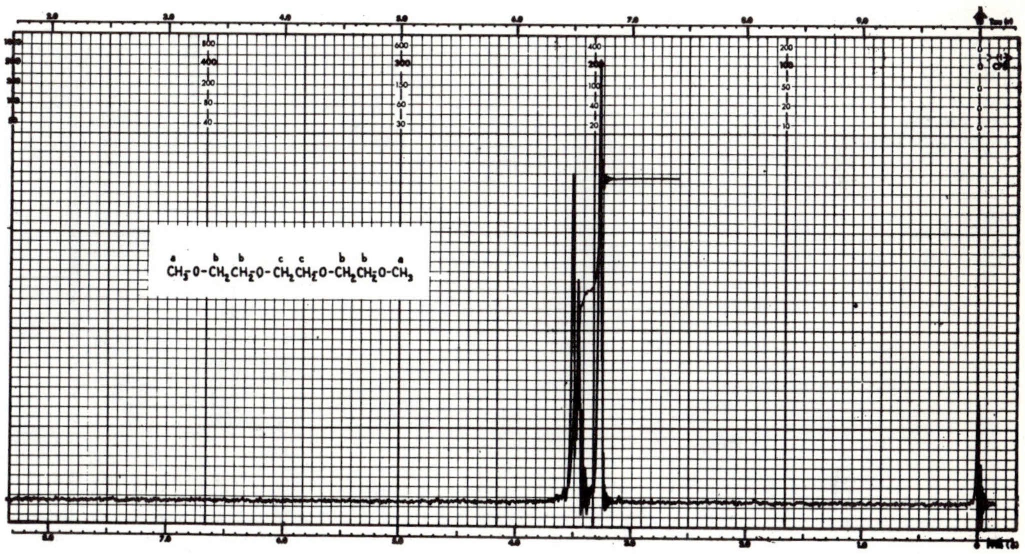

© 1978 Sadtler Research Laboratories, Inc.

2,5,8,11,14 PENTOXAPENTADECANE

$C_{10}H_{22}O_5$

Mol. Wt. 222.28

B.P. 275.3°C/760mm

F.P. −27°C

Solvent: CCl_4

Assignments
a = 3.30
b = ca 3.48
c = 3.52

1420

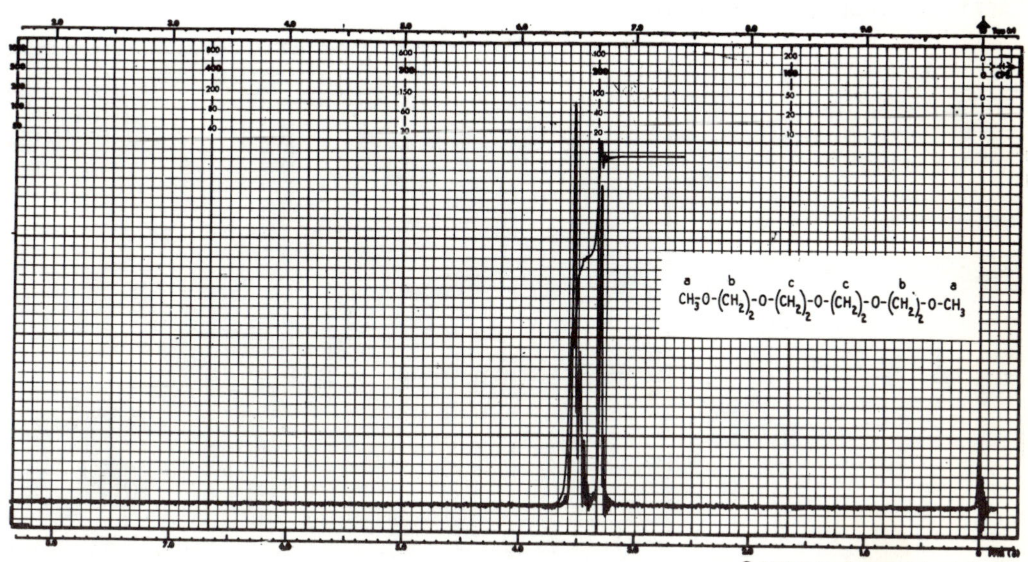

© 1978 Sadtler Research Laboratories, Inc.

5,8,11,14,17-PENTAOXAHENEICOSANE

$C_{16}H_{34}O_5$

Mol. Wt. 306.45

Solvent: CCl_4

Assignments
a = 0.92
b = 1.48
c = 3.41
d = ca 3.55

1421

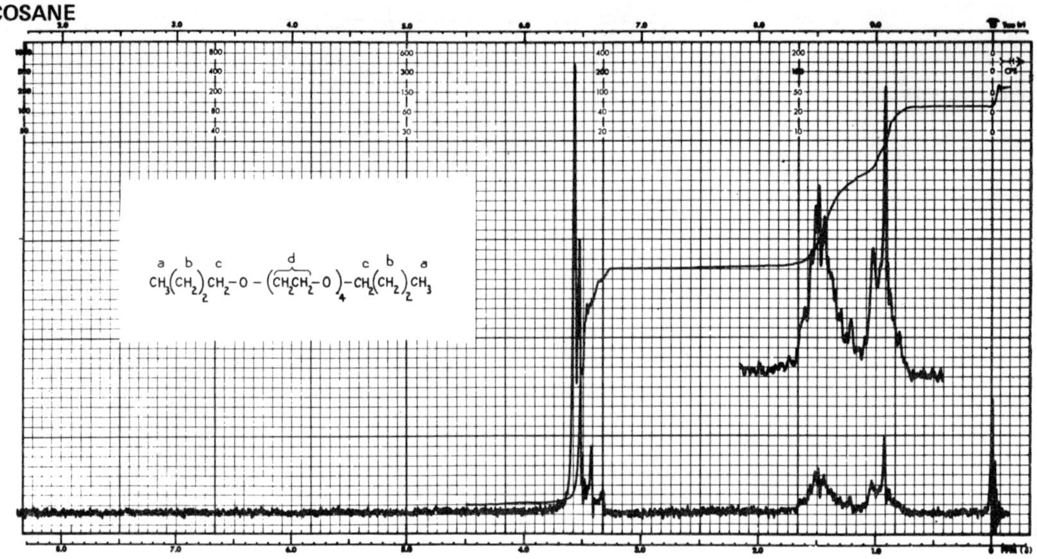

© 1978 Sadtler Research Laboratories, Inc.

ACETALDEHYDE, DIMETHYL ACETAL

$C_4H_{10}O_2$

Mol. Wt. 90.12

Solvent: CCl_4

Assignments
a = 1.20
b = 3.21
c = 4.50

1422

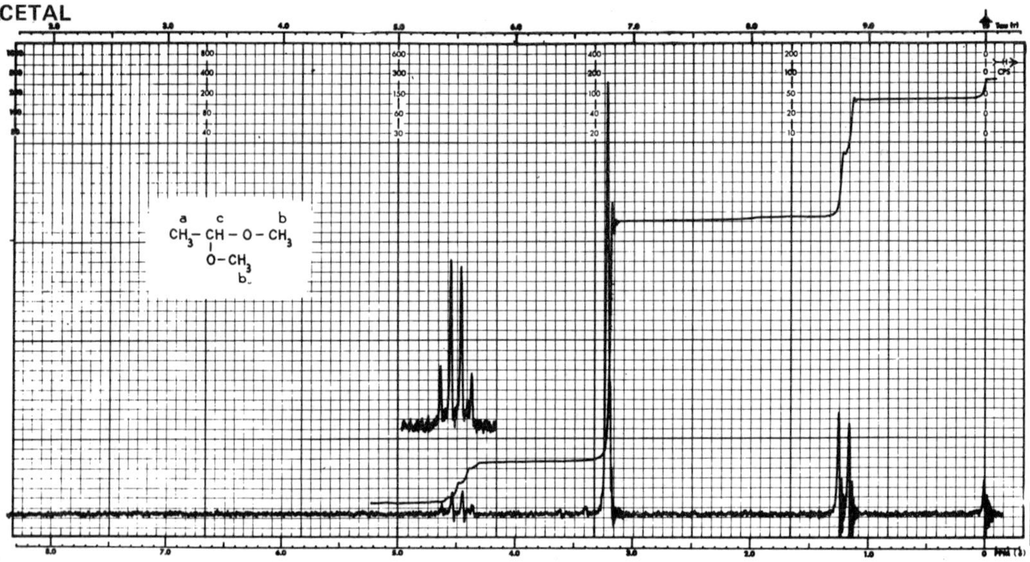

© 1978 Sadtler Research Laboratories, Inc.

ACETALDEHYDE, DIETHYL ACETAL

$C_6H_{14}O_2$

Mol. Wt. 118.18

B.P. 100-103°C

Solvent: CCl_4

Assignments
a = 1.14
b = 1.21
c = 3.36
d = 3.52
e = 4.59

1423

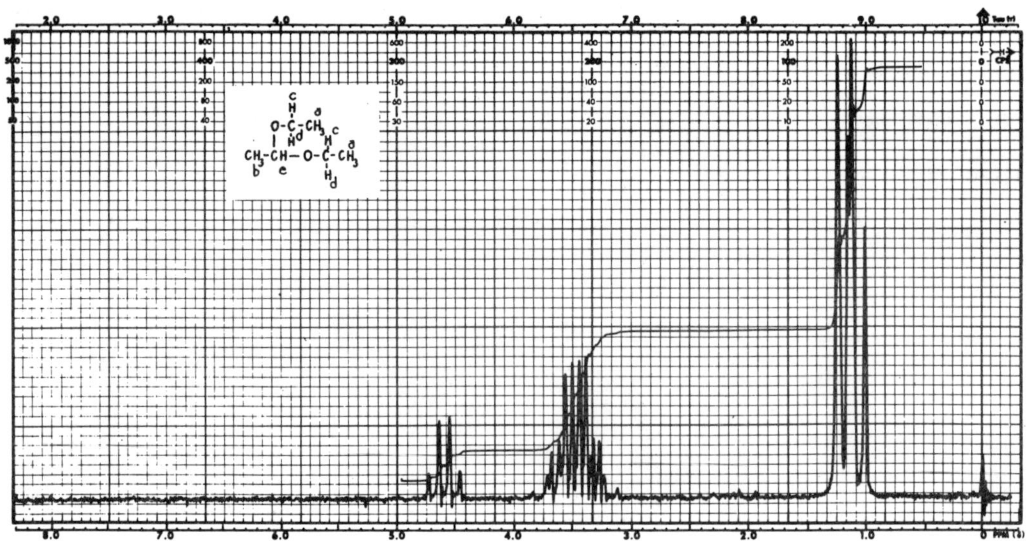

© 1978 Sadtler Research Laboratories, Inc.

ACETALDEHYDE, DIBUTYL ACETAL

$C_{10}H_{22}O_2$

Mol. Wt. 174.29

B.P. 40°C/2mm

Solvent: CCl_4

Assignments
a = 0.95
b = 1.22
c = 1.30-1.69
d = 3.40
e = 4.57

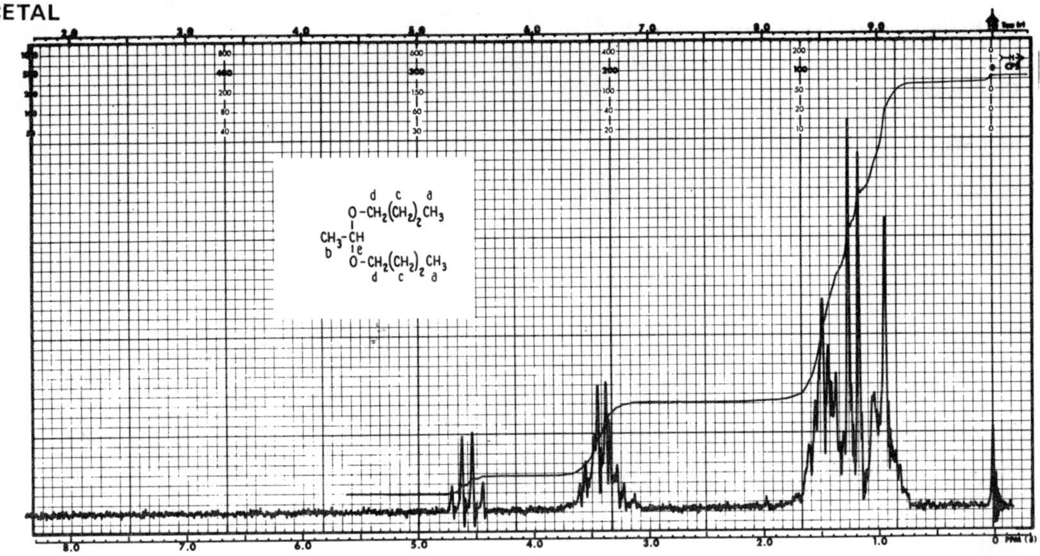

1424

© 1978 Sadtler Research Laboratories, Inc.

ACROLEIN, DIETHYL ACETAL

$C_7H_{14}O_2$

Mol. Wt. 130.19

Solvent: CCl_4

Assignments
a = 1.16
b = 3.42
c = 3.50
d = 4.79
e = 5.00-5.50
f = 5.67

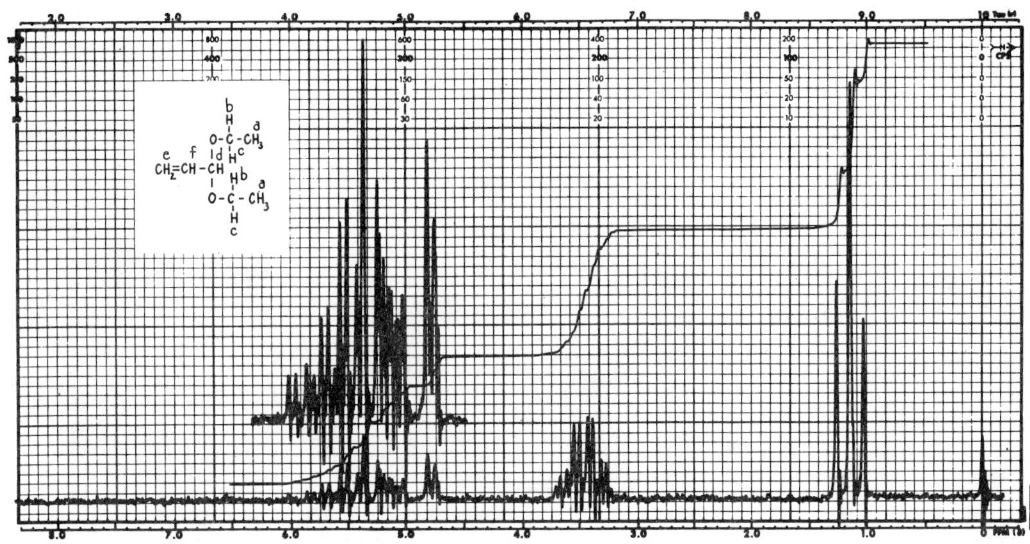

1425

© 1978 Sadtler Research Laboratories, Inc.

BROMOACETALDEHYDE, DIETHYL ACETAL

$C_6H_{13}BrO_2$

Mol. Wt. 179.08

B.P. 68-69°C/16mm

Solvent: CCl_4

Assignments
a = 1.19
b = 3.28
c = ca 3.58
d = 4.60

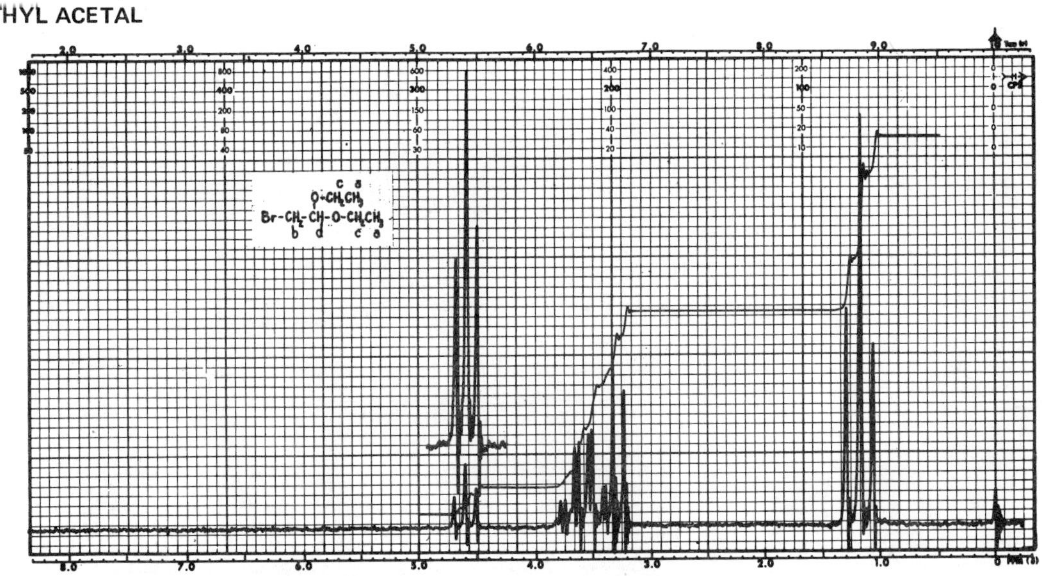

1426

© 1978 Sadtler Research Laboratories, Inc.

AMINOACETALDEHYDE, DIMETHYL ACETAL

$C_4H_{11}NO_2$

Mol. Wt. 105.14

B.P. 135-139°C/25mm

Solvent: CCl_4

Assignments
a = 1.69
b = 2.69
c = 3.30
d = 4.23

1427

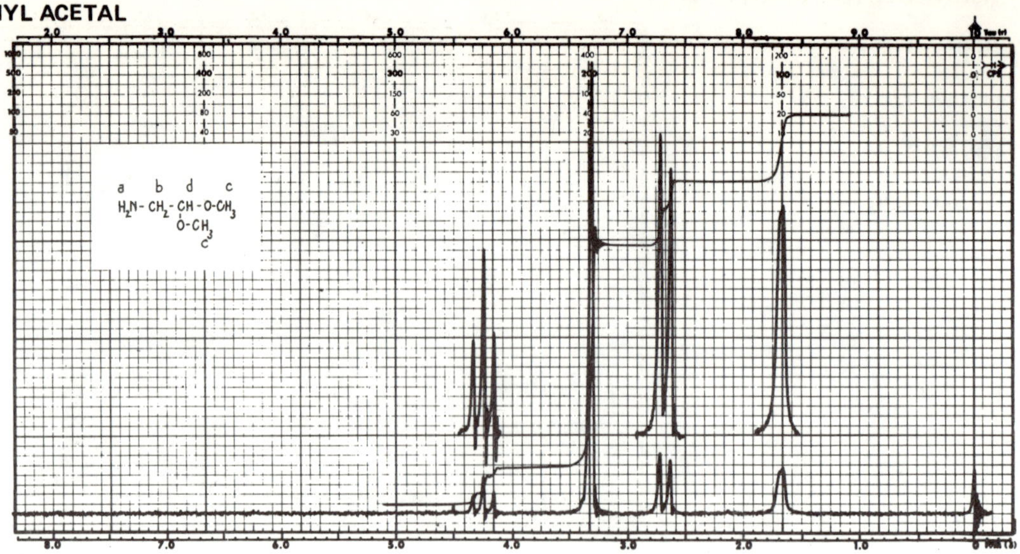

© 1978 Sadtler Research Laboratories, Inc.

AMINOACETALDEHYDE, DIETHYL ACETAL

$C_6H_{15}NO_2$

Mol. Wt. 133.19

B.P. 162-163°C

Solvent: $CDCl_3$

Assignments
a = 1.23
b = 2.76
c = 3.20-3.90
d = 4.40

1428

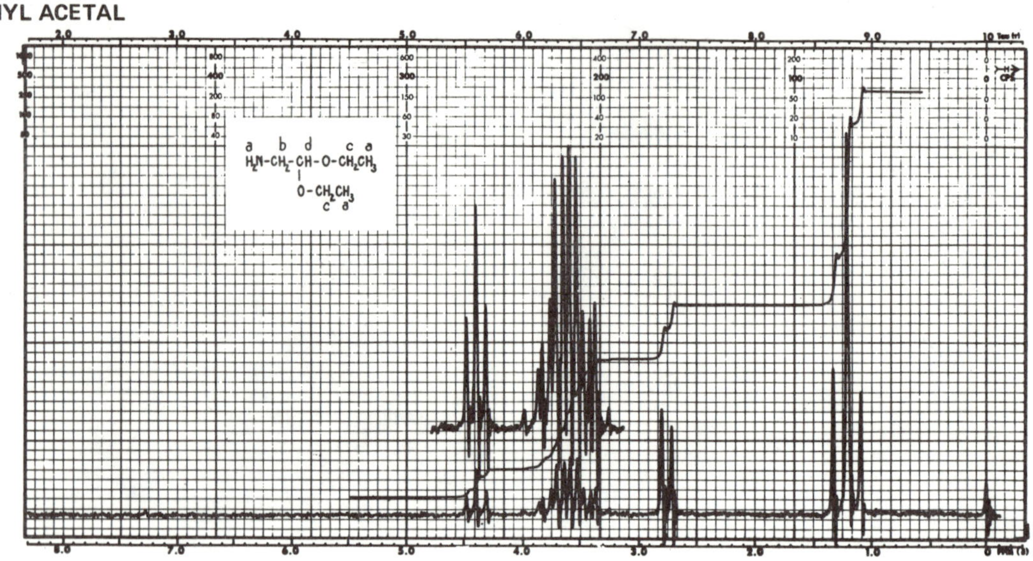

© 1978 Sadtler Research Laboratories, Inc.

MALONALDEHYDE, BIS(DIMETHYL ACETAL)

$C_7H_{16}O_4$

Mol. Wt. 164.20

B.P. 66-67°C/12mm

Solvent: CCl_4

Assignments
a = 1.80
b = 3.23
c = 4.37

1429

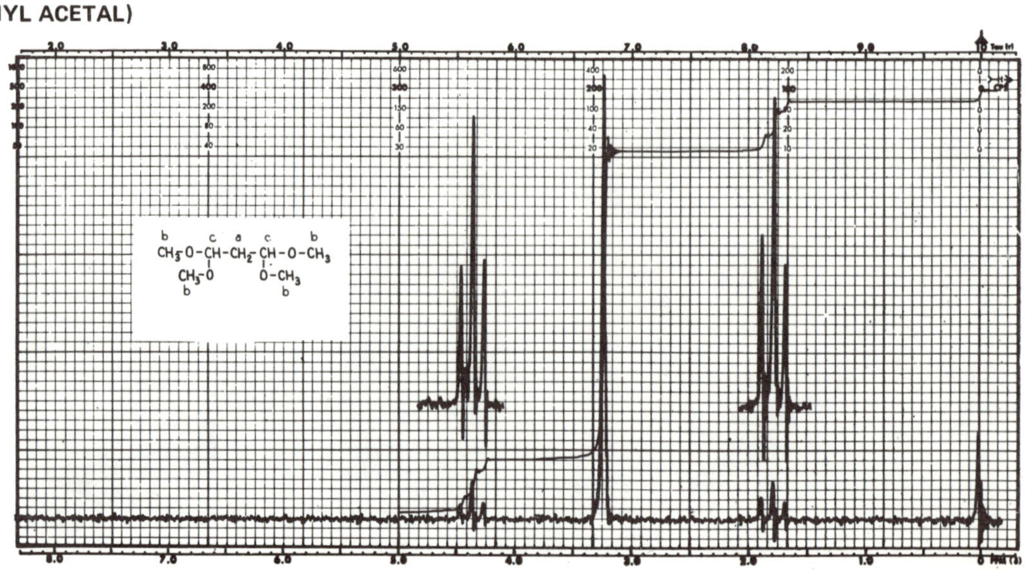

© 1978 Sadtler Research Laboratories, Inc.

(METHYLIMINO)DIACETALDEHYDE, BIS(DIMETHYLACETAL)

$C_9H_{21}NO_4$

Mol. Wt. 207.27

Solvent: CCl_4

Assignments
a = 2.34
b = 2.52
c = 3.30
d = 4.40

1430

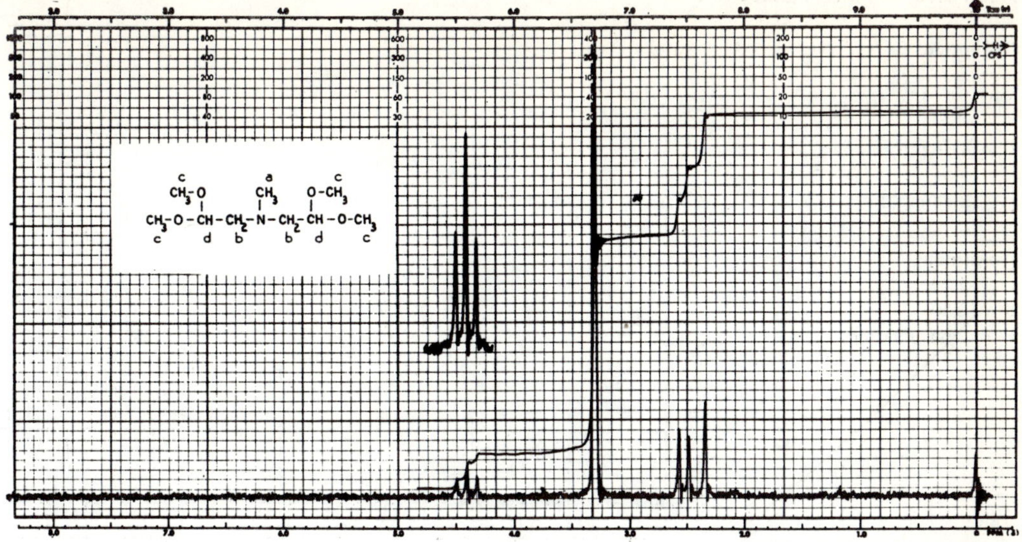

© 1978 Sadtler Research Laboratories, Inc.

N-METHYL-2,2,2',2'-TETRAETHOXYDIETHYLAMINE

$C_{13}H_{29}NO_4$

Mol. Wt. 263.38

Solvent: $CDCl_3$

Assignments
a = 1.20
b = 2.40
c = 2.64
d = 3.32-3.90
e = 4.60
f = 3.38 imp.

1431

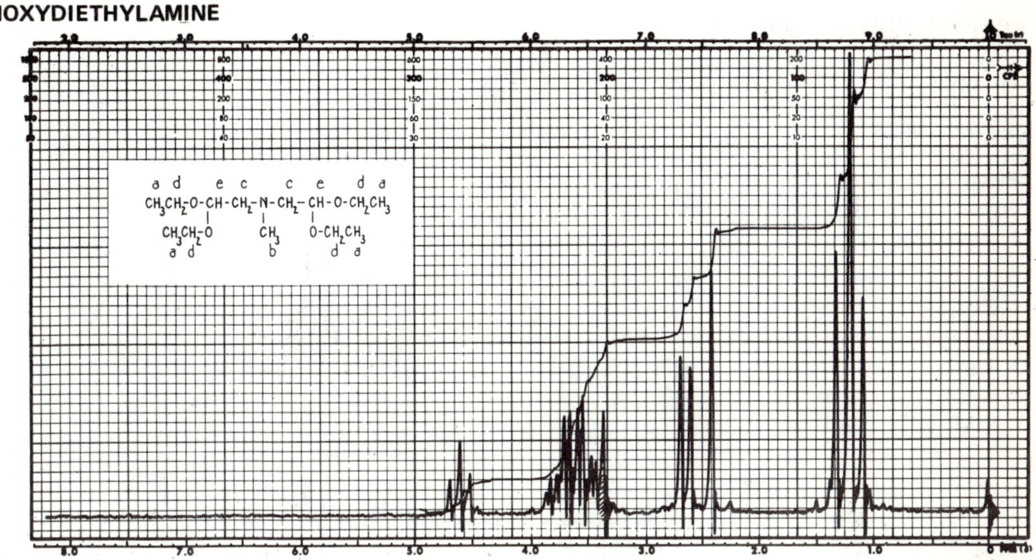

© 1978 Sadtler Research Laboratories, Inc.

METALDEHYDE

$(C_2H_4O)_n$

Mol. Wt. $(44.05)_n$

M.P. 246°C

Solvent: $CDCl_3$

Assignments
a = 1.37
b = 4.91

1432

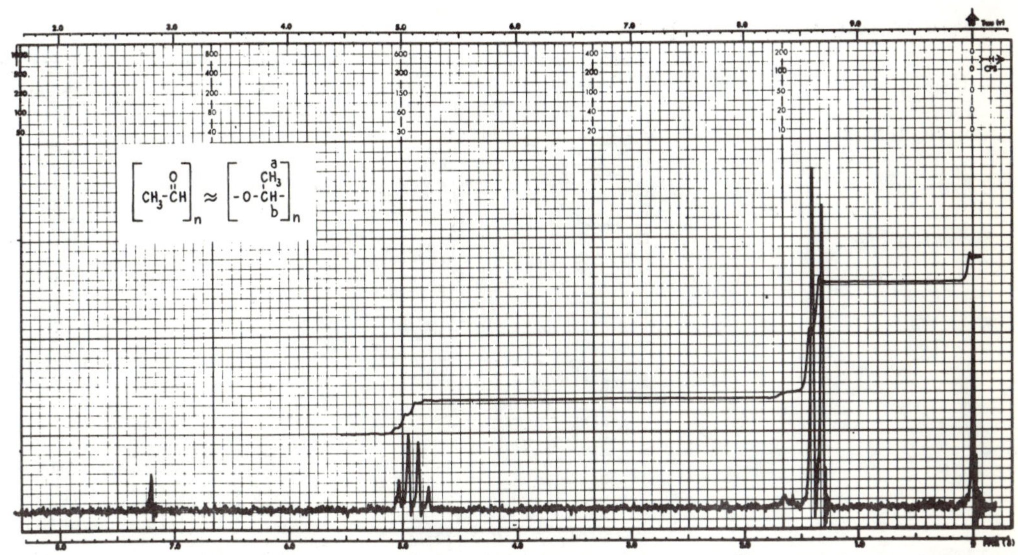

© 1978 Sadtler Research Laboratories, Inc.

ACETONE, DIMETHYL ACETAL

$C_5H_{12}O_2$

Mol. Wt. 104.15

B.P. 43°C/200mm

Solvent: CCl_4

Assignments
a = 1.23
b = 3.10
c = 2.08, 3.39 imp.

1433

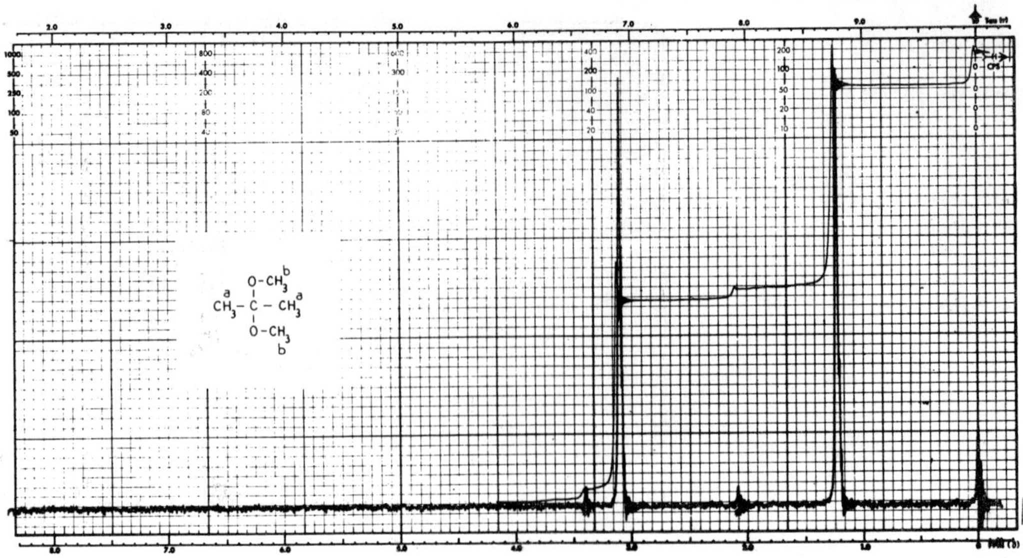

© 1978 Sadtler Research Laboratories, Inc.

ACETONE, DIALLYL ACETAL

$C_9H_{16}O_2$

Mol. Wt. 156.23

B.P. 59°C/20mm

Solvent: CCl_4

Assignments
a = 1.32
b = 3.89
c = 5.01
d = 5.16
e = 5.80

1434

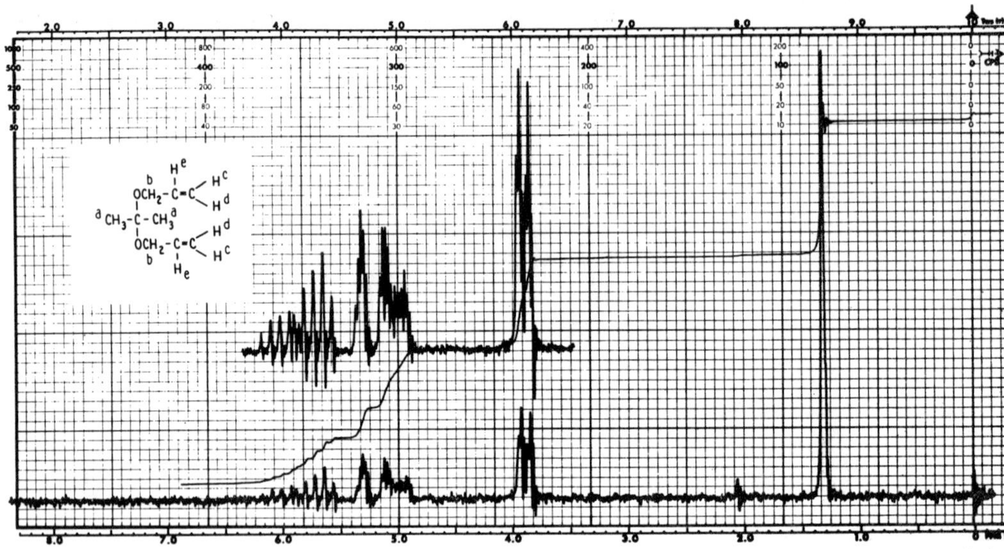

© 1978 Sadtler Research Laboratories, Inc.

ORTHOFORMIC ACID, TRIMETHYL ESTER

$C_4H_{10}O_3$

Mol. Wt. 106.12

B.P. 97-99°C

Solvent: CCl_4

Assignments
a = 3.22
b = 4.80

1435

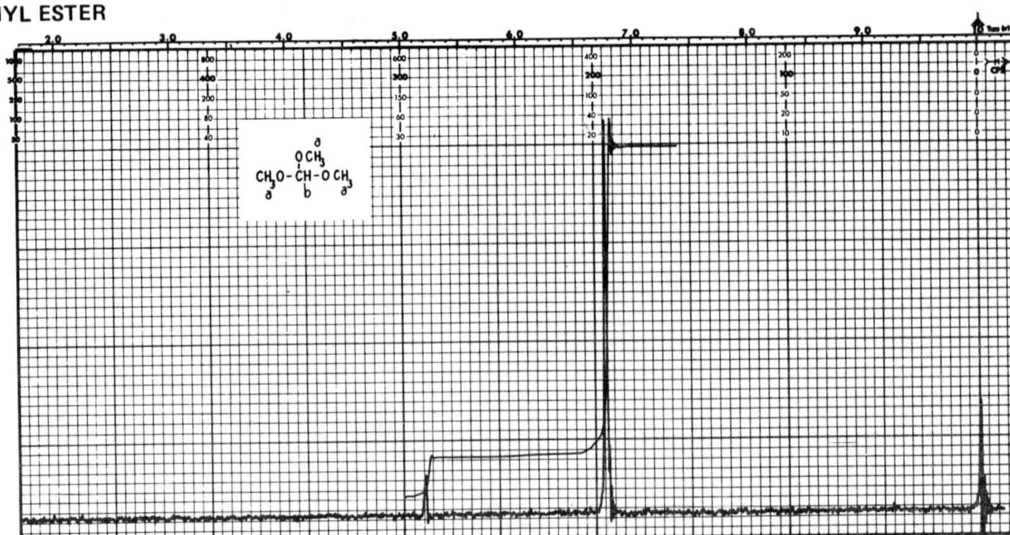

© 1978 Sadtler Research Laboratories, Inc.

ORTHOFORMIC ACID, TRIETHYL ESTER

$C_7H_{16}O_3$

Mol. Wt. 148.20

Solvent: CCl_4

Assignments
a = 1.19
b = 3.52
c = 5.00

1436

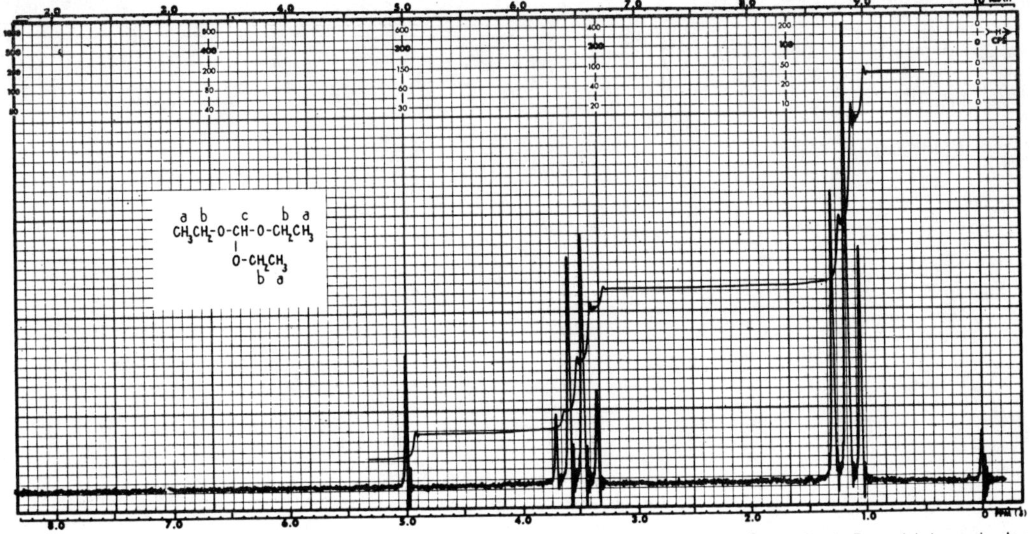

© 1978 Sadtler Research Laboratories, Inc.

ORTHROACETIC ACID, TRIMETHYL ESTER

$C_5H_{12}O_3$

Mol. Wt. 120.15

B.P. 105-108°C

Solvent: CCl_4

Assignments
a = 1.31
b = 3.19

1437

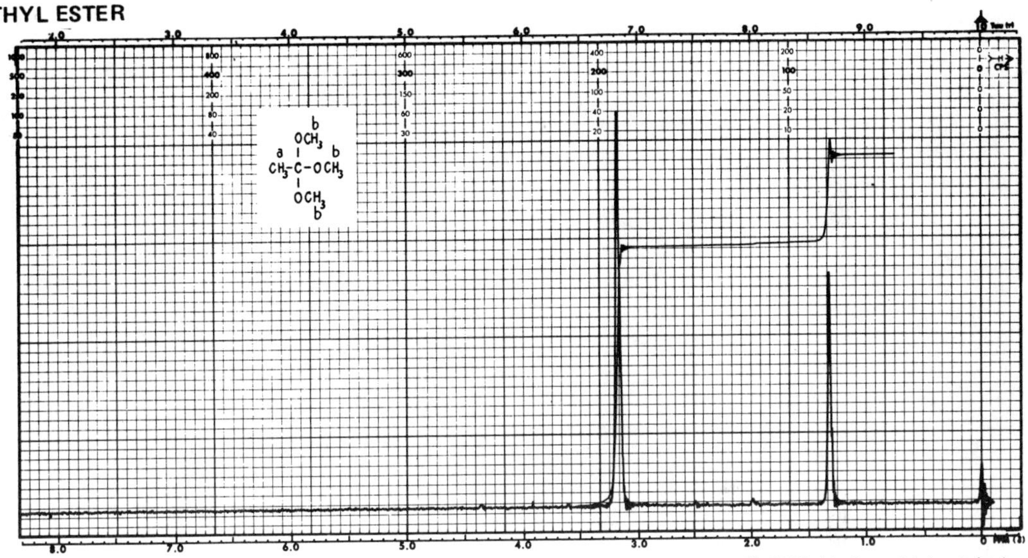

© 1978 Sadtler Research Laboratories, Inc.

NOTES

THE ALICYCLIC ETHERS

The broadening of multiplets due to the lack of rotation about the carbon-carbon bonds of the protons in the ring and the absence of terminal methyl groups are often sufficient evidence to characterize the nmr spectra of the alicyclic compounds. The cyclic ethers are all readily soluble in the chlorinated solvents CCl_4 and $CDCl_3$.

The three ring protons of the epoxide group are non-equivalent and appear as three distinct multiplets in the chemical shift range from 2.3 to 3.8 ppm delta. The two protons bonded to C–1 resonate at higher field than the proton attached to C–2. The appearance and chemical shifts of these bands are readily recognizable and quite characteristic of this group (spectra 1437 – 1447).

The chemical shifts of the parent compounds of the group are presented below.

CH_2–5	CH_2–4	CH_2–3	CH_2–2	CH_2–1		Solvent	Spectrum
		4.62	2.64	4.62	Trimethylene oxide	CCl_4	- - - -
	3.71	1.81	1.81	3.71	Tetrahydrofuran	$CDCl_3$	1455
3.56	1.59	1.59	1.59	3.56	Tetrahydropyran	CCl_4	- - - -

Chemical Shifts

The chemical shifts for a series of epoxy ring protons are listed below.

(cis) CH–1	(trans) CH–1	CH–2	–X	
2.31	2.55	2.69	$-R_2$	CCl_4
2.28	2.56	2.74	$-R_{10}$	CCl_4
2.23	2.59	2.80	$-CH_3$	CCl_4
2.46	2.63	2.98	$-CH_2-O-R$	CCl_4
2.58	2.79	3.17	$-CH_2-Cl$	CCl_4
2.60	2.90	3.30	$-CH{=}CH_2$	$CDCl_3$
2.58	2.84	3.40	$-CH_2-Br$	CCl_4
2.61	2.96	3.69	–Ø	CCl_4

Coupling and Coupling Constants

Due to the high degree of strain in the three membered ring, the coupling constants between the three hydrogens of the epoxide group are observed to be somewhat smaller than normally expected. In spectrum 1447, the coupling constants appear to be J = 5, 4 and 3 Hz for the geminal, cis and trans couplings.

PROPYLENE OXIDE

C_3H_6O

Mol. Wt. 58.08

Solvent: CCl_4

Assignments
a = 1.24
b = 2.23
c = 2.59
d = 2.80

1438

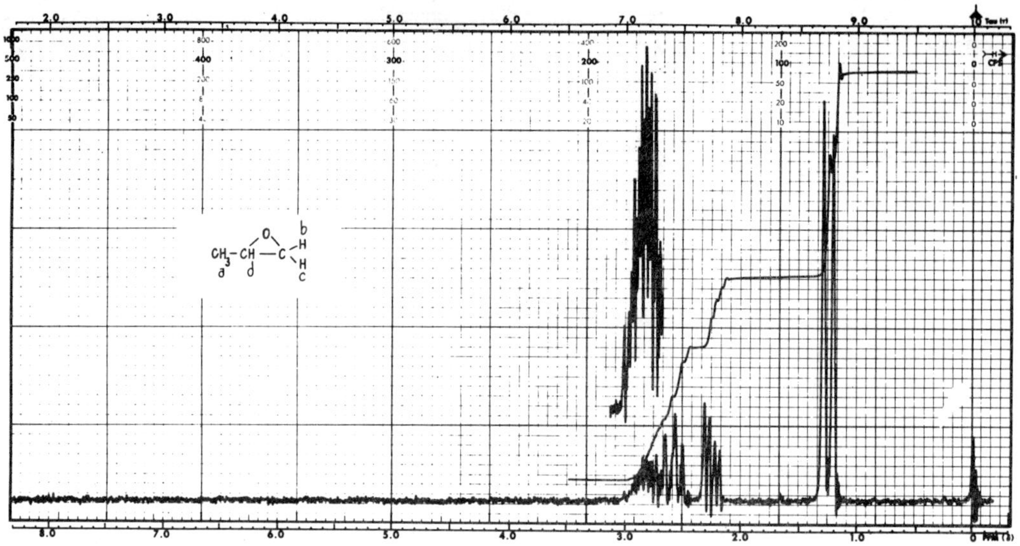

© 1978 Sadtler Research Laboratories, Inc.

1,2-EPOXYBUTANE

C_4H_8O

Mol. Wt. 72.11

Solvent: CCl_4

Assignments
a = 0.98
b = 1.50
c = 2.31
d = 2.55
e = 2.69

1439

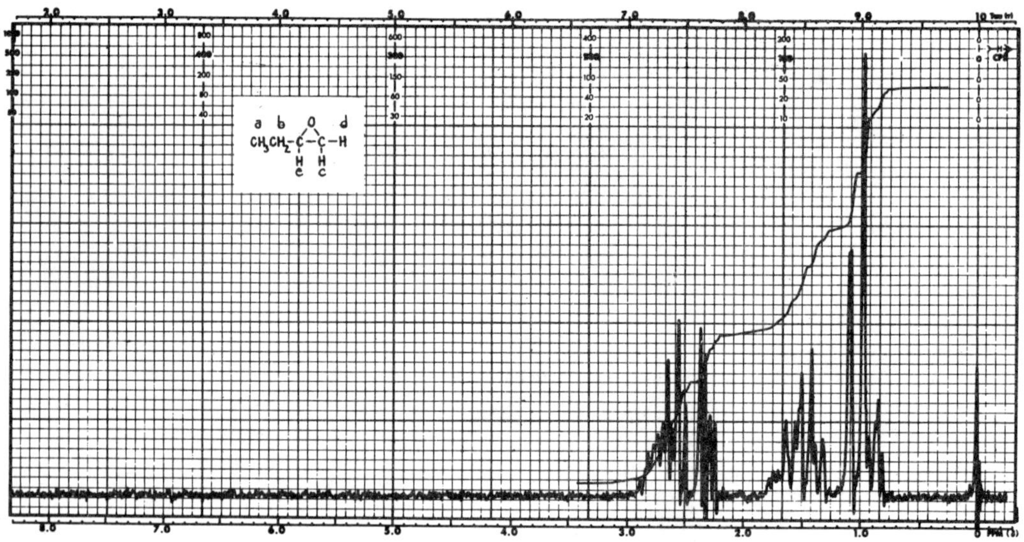

© 1978 Sadtler Research Laboratories, Inc.

1,2-EPOXYOCTANE

$C_8H_{16}O$

Mol. Wt. 128.22

Solvent: CCl_4

Assignments
a = 0.91
b = 1.10-1.80
c = 2.29
d = 2.56
e = 2.70

1440

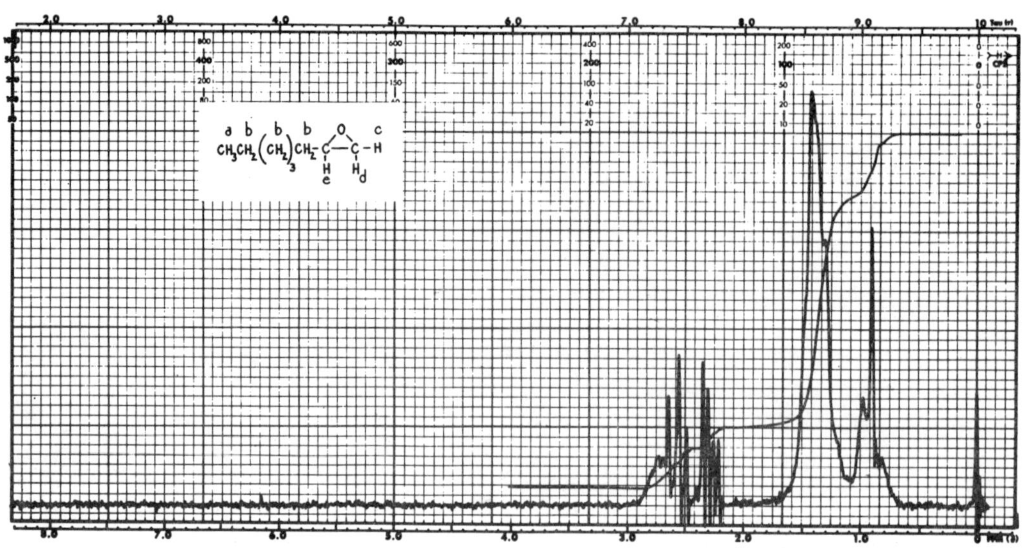

© 1978 Sadtler Research Laboratories, Inc.

1,2-EPOXYDECANE

$C_{10}H_{20}O$

Mol. Wt. 156.27

Solvent: CCl_4

Assignments
a = 0.90
b = 1.10-1.60
c = 2.29
d = 2.54
e = 2.71

1441

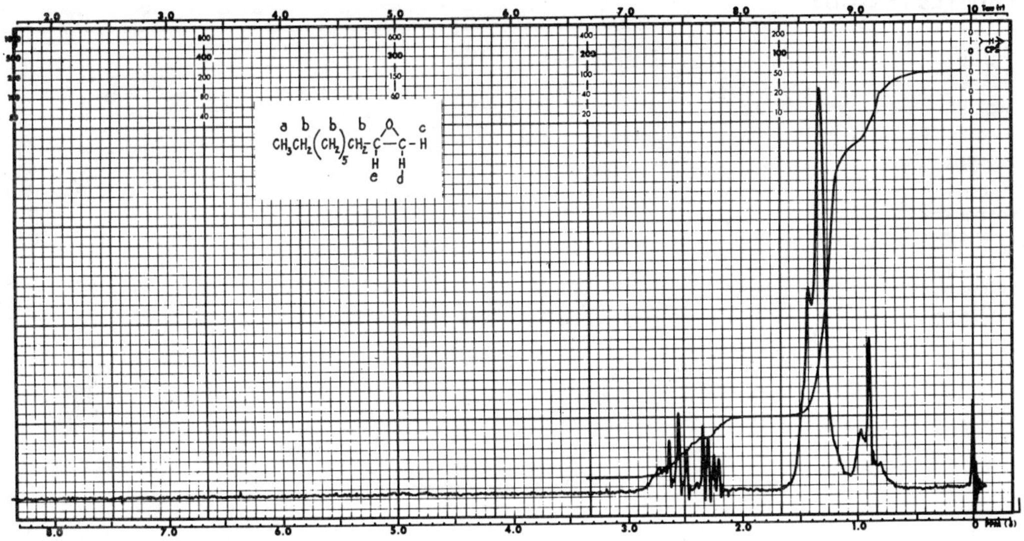

© 1978 Sadtler Research Laboratories, Inc.

1,2-EPOXYDODECANE

$C_{12}H_{24}O$

Mol. Wt. 184.32

Solvent: CCl_4

Assignments
a = 0.93
b = 1.18-1.55
c = 2.28
d = 2.56
e = 2.74
f = 2.00 imp.

1442

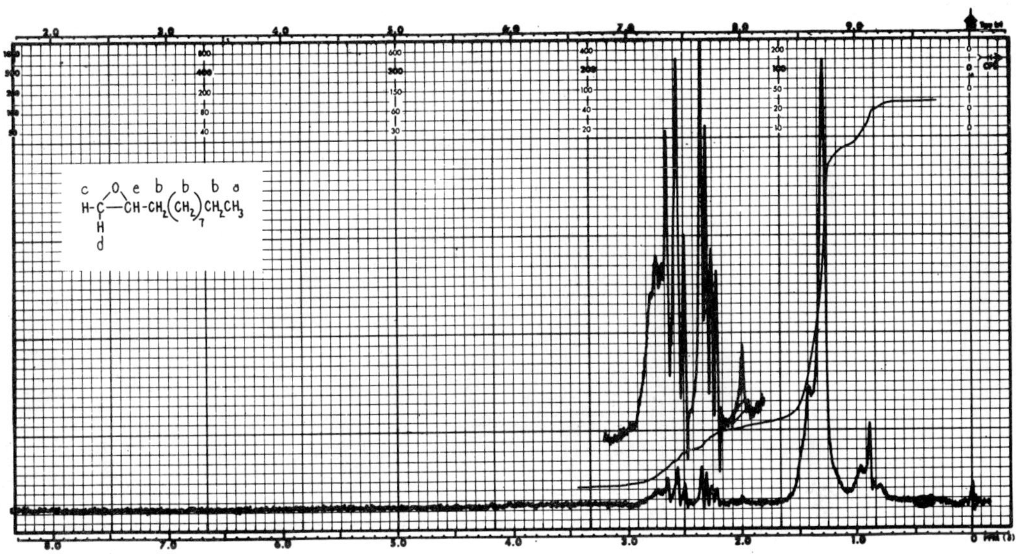

© 1978 Sadtler Research Laboratories, Inc.

EPICHLOROHYDRIN

C_3H_5ClO

Mol. Wt. 92.53

M.P. −25.6°C

B.P. 117°C/760mm (lit.)

Solvent: CCl_4

Assignments
a = 2.58 or b
b = 2.79 or a
c = 3.17
d = 3.48 or e
e = 3.56 or d

1443

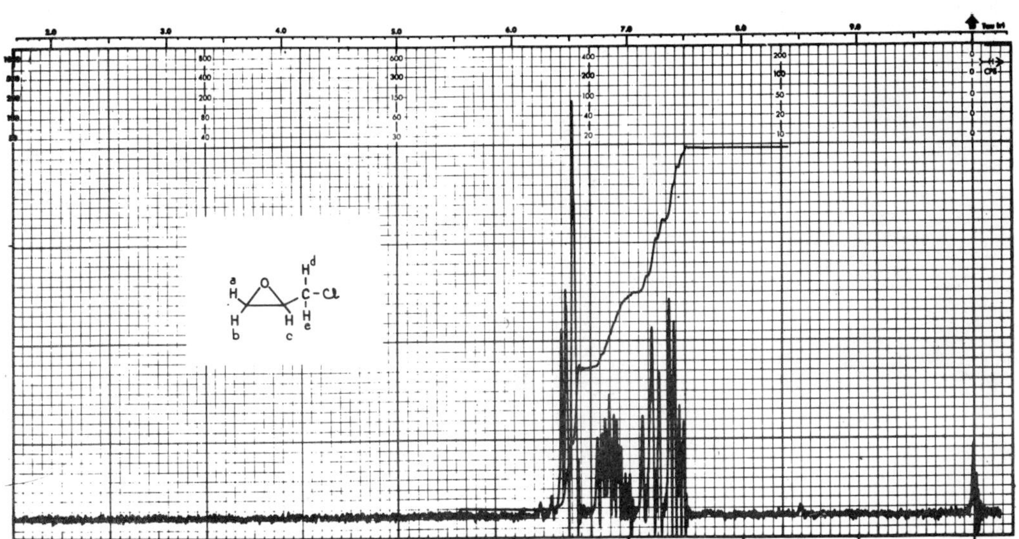

© 1978 Sadtler Research Laboratories, Inc.

1-BROMO-2,3-EPOXYPROPANE

C_3H_5BrO

Mol. Wt. 136.98

Solvent: CCl_4

Assignments
a = 2.58
b = 2.84
c = 3.22
d = ca 3.40

1444

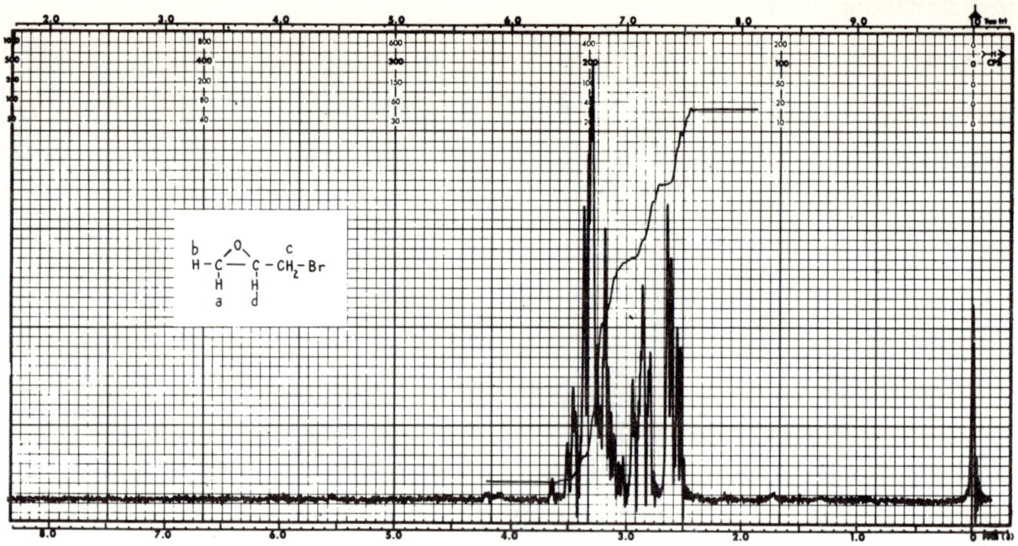

© 1978 Sadtler Research Laboratories, Inc.

1,2-EPOXY-3-ISOPROPOXYPROPANE

$C_6H_{12}O_2$

Mol. Wt. 116.16

B.P. 43-45°C/15mm

Solvent: CCl_4

Assignments
a = 1.12
b = 2.46
c = 2.63
d = 2.98
e = 3.15-3.90

1445

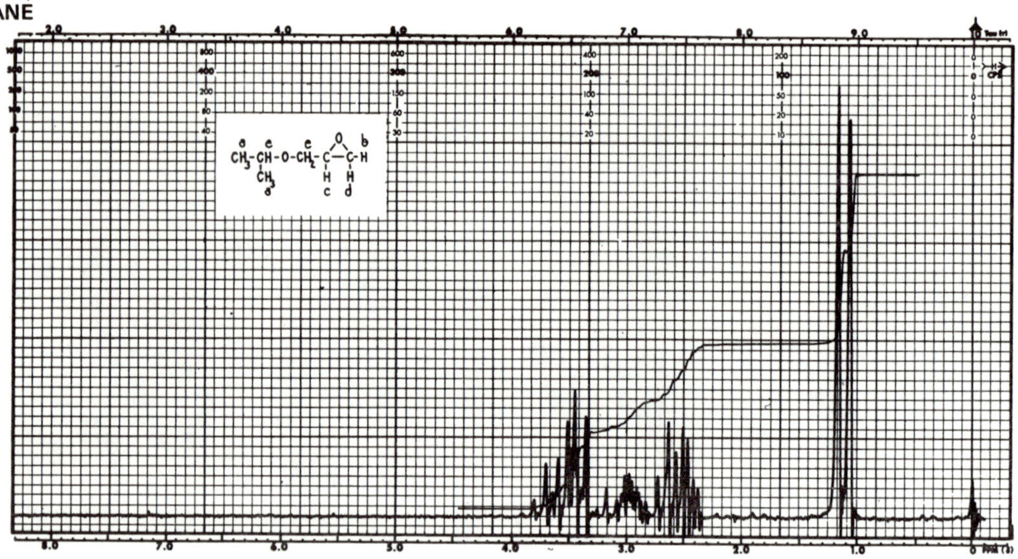

© 1978 Sadtler Research Laboratories, Inc.

3,4-EPOXY-1-BUTENE

C_4H_6O

Mol. Wt. 70.09

B.P. 65-66°C

Solvent: $CDCl_3$

Assignments
a = 2.60
b = 2.90
c = 3.30
d = 5.10-5.70

1446

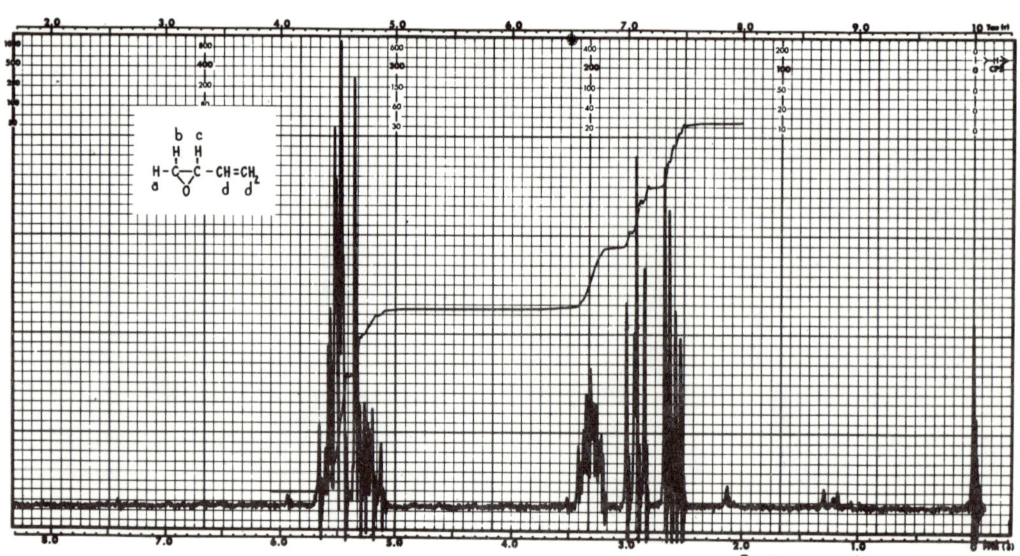

© 1978 Sadtler Research Laboratories, Inc.

(EPOXYETHYL) BENZENE

C_8H_8O

Mol. Wt. 120.15

B.P. 73-74°C/10mm

Solvent: CCl_4

Assignments
a = 2.61
b = 2.96
c = 3.69
d = 7.21

1447

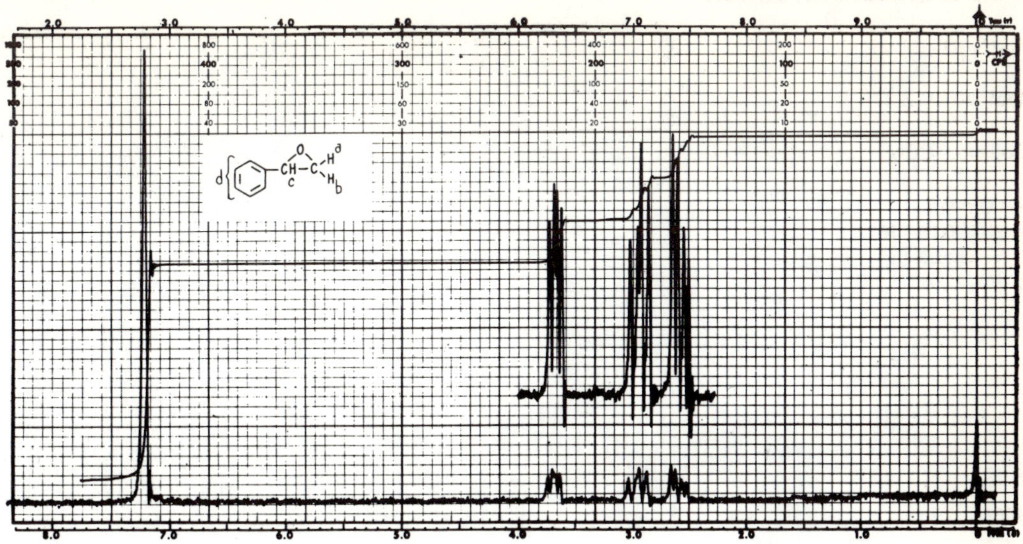

© 1978 Sadtler Research Laboratories, Inc.

7-OXABICYCLO[4.1.0] HEPTANE

$C_6H_{10}O$

Mol. Wt. 98.15

B.P. 130-132°C

Solvent: CCl_4

Assignments
a = 0.91-1.60
b = ca 1.82
c = 2.94

1448

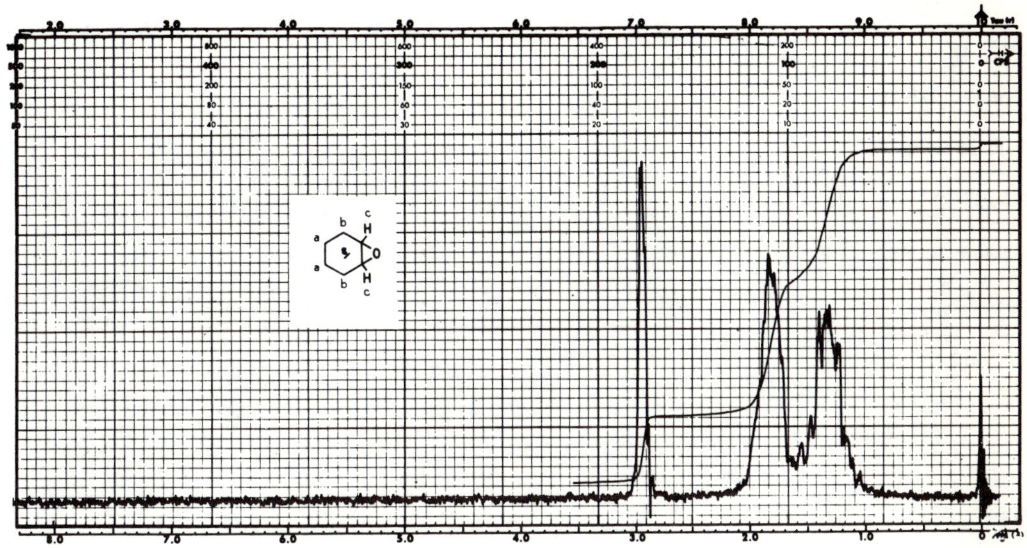

© 1978 Sadtler Research Laboratories, Inc.

2,3-EPOXYPINANE

$C_{10}H_{16}O$

Mol. Wt. 152.24

B.P. 84-86°C/24mm

Solvent: CCl_4

Assignments
a = 0.97
b = 1.30
c = 1.51-2.11
d = 2.89

1449

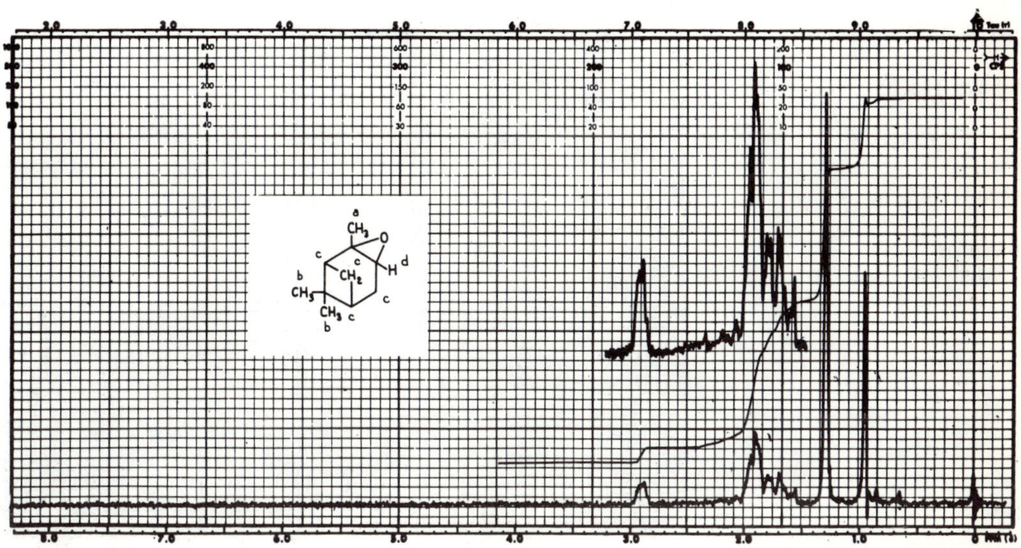

© 1978 Sadtler Research Laboratories, Inc.

8-OXABICYCLO[5.1.0] OCTANE

$C_7H_{12}O$

Mol. Wt. 112.17

Solvent: CCl_4

Assignments
a = 0.70-1.70
b = ca 1.93
c = 2.91

1450

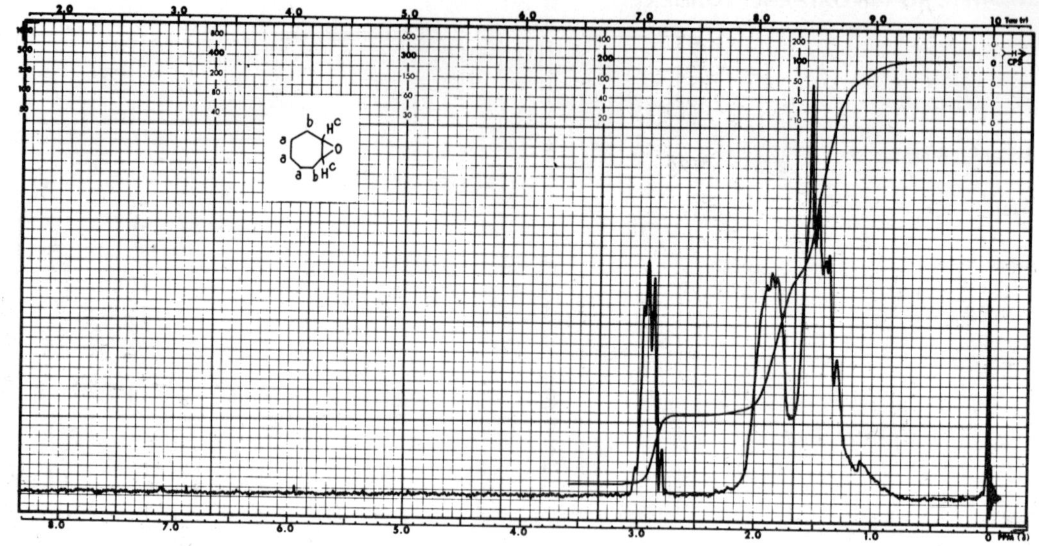

© 1978 Sadtler Research Laboratories, Inc.

9-OXABICYCLO[6.1.0] NONANE

$C_8H_{14}O$

Mol. Wt. 126.20

M.P. 53-56°C

B.P. 55°C/5mm

Solvent: $CDCl_3$

Assignments
a = 0.80-2.03
b = ca 2.78

1451

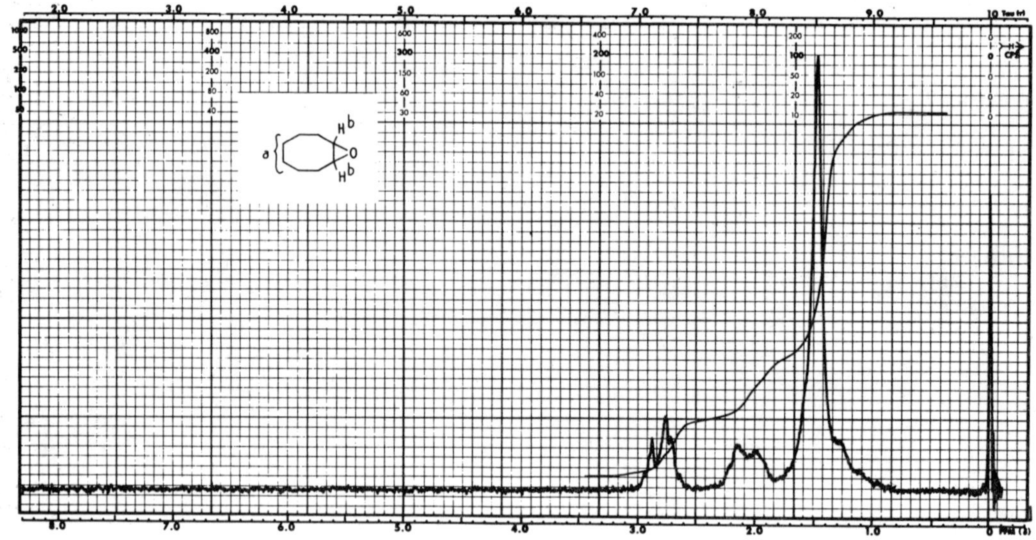

© 1978 Sadtler Research Laboratories, Inc.

5,10-DIOXATRICYCLO[7.1.0.0^{4,6}] DECANE

$C_8H_{12}O_2$

Mol. Wt. 140.18

B.P. 70°C/0.04mm

Solvent: CCl_4

Assignments
a = 1.45-2.02
b = ca 2.84

1452

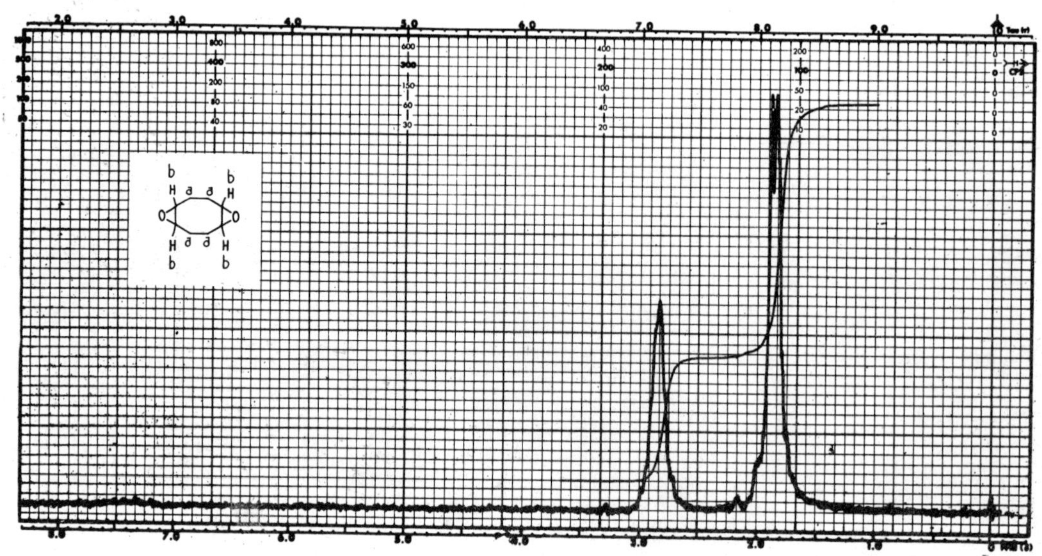

© 1978 Sadtler Research Laboratories, Inc.

6,6a-DIHYDRO-1aH-OXIRENO[a] INDENE

C_9H_8O

Mol. Wt. 132.16

B.P. 98°C/6mm

Solvent: CCl_4

Assignments
a = 2.71
b = 3.09
c = 3.87
d = 4.01
e = 6.85-7.50

1453

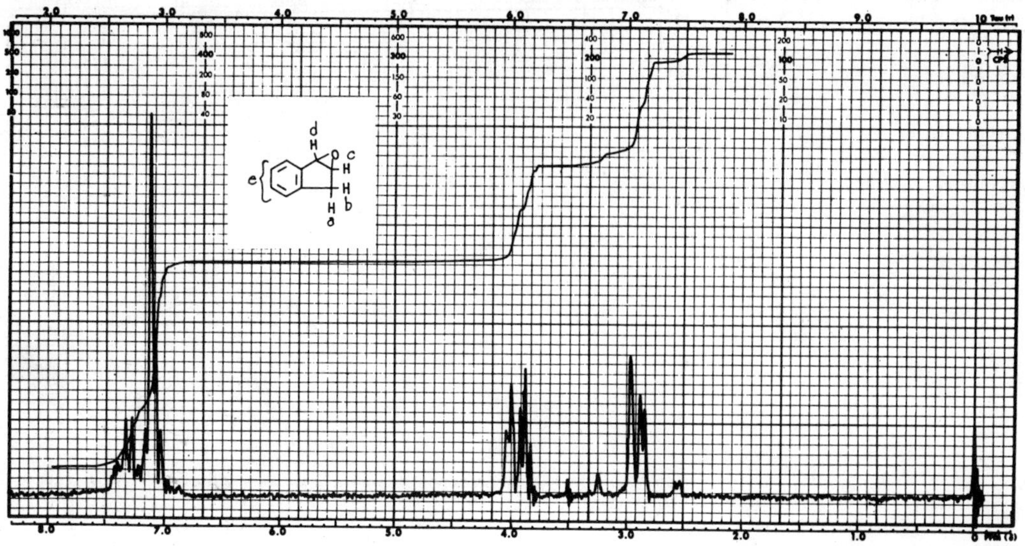

© 1978 Sadtler Research Laboratories, Inc.

2-CHLORO-1,3-EPOXYPROPANE

C_3H_5ClO

Mol. Wt. 92.53

B.P. 104°C

Solvent: CCl_4

Assignments
a = 4.45-5.00

1454

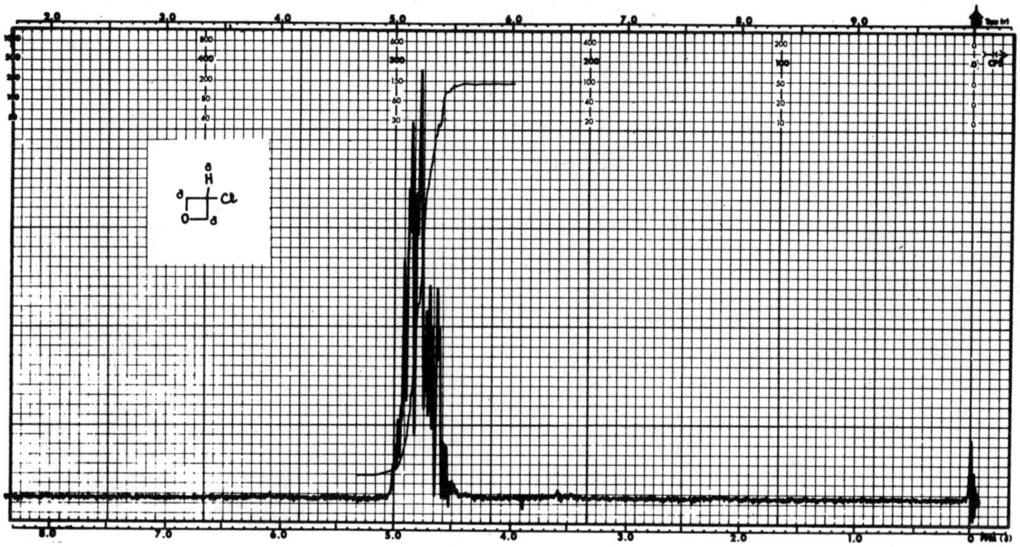

© 1978 Sadtler Research Laboratories, Inc.

TETRAHYDROFURAN

C_4H_8O

Mol. Wt. 72.11

B.P. 65-66.5°C

Solvent: $CDCl_3$

Assignments
a = 1.81
b = 3.71

1455

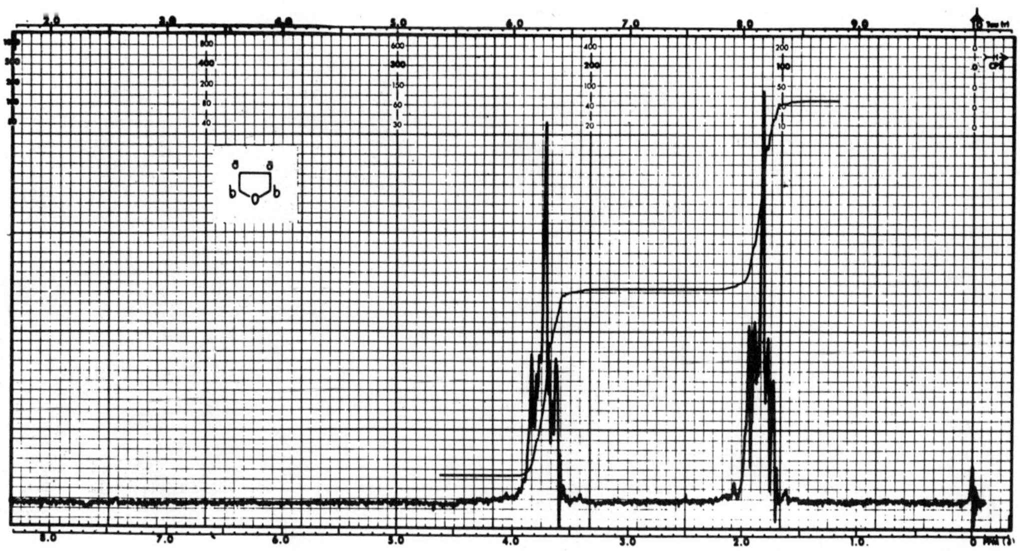

© 1978 Sadtler Research Laboratories, Inc.

2-METHYLTETRAHYDROFURAN

$C_5H_{10}O$

Mol. Wt. 86.13

B.P. 78-80°C

Solvent: CCl_4

Assignments
a = 1.13
b = 1.20-2.20
c = 3.30-4.00

1456

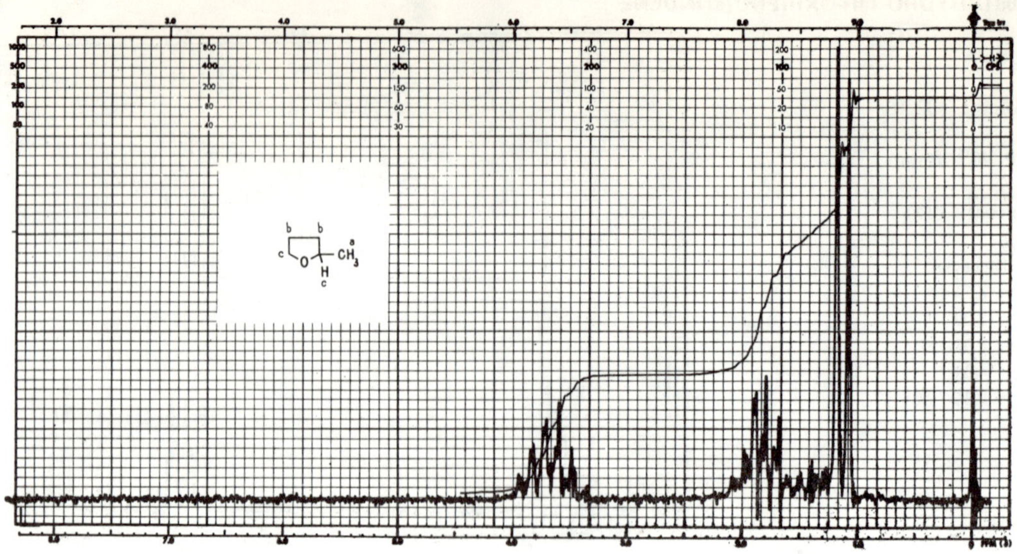

© 1978 Sadtler Research Laboratories, Inc.

2-(METHOXYMETHYL)TETRAHYDROFURAN

$C_6H_{12}O_2$

Mol. Wt. 116.16

M.P. < −78°C

B.P. 140°C

Solvent: CCl_4

Assignments
a = 1.60-2.12
b = 3.31
c = 3.50-4.08

1457

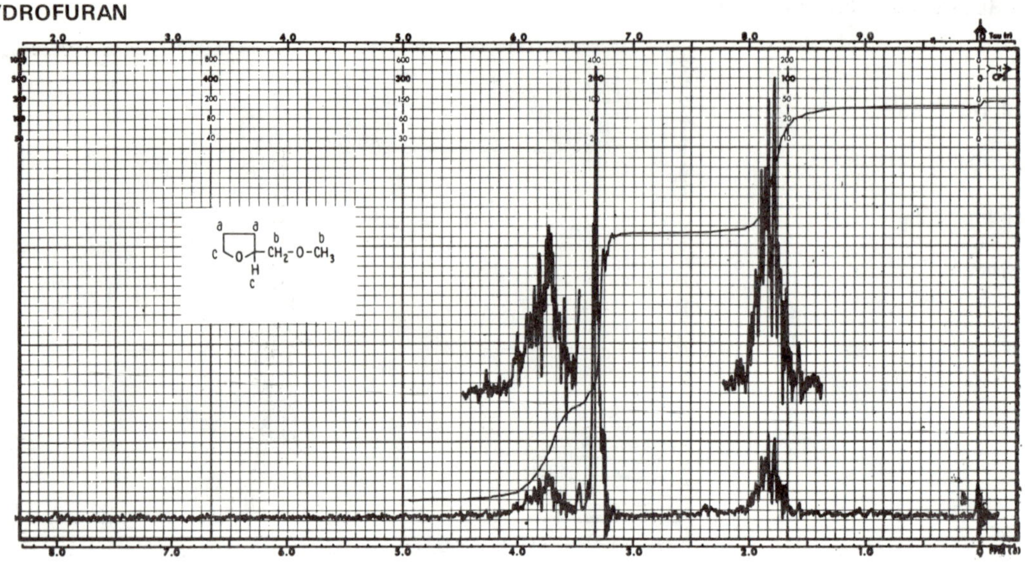

© 1978 Sadtler Research Laboratories, Inc.

1,3-DIOXOLANE

$C_3H_6O_2$

Mol. Wt. 74.08

B.P. 74-75°C

Solvent: CCl_4

Assignments
a = 3.79
b = 4.80

1458

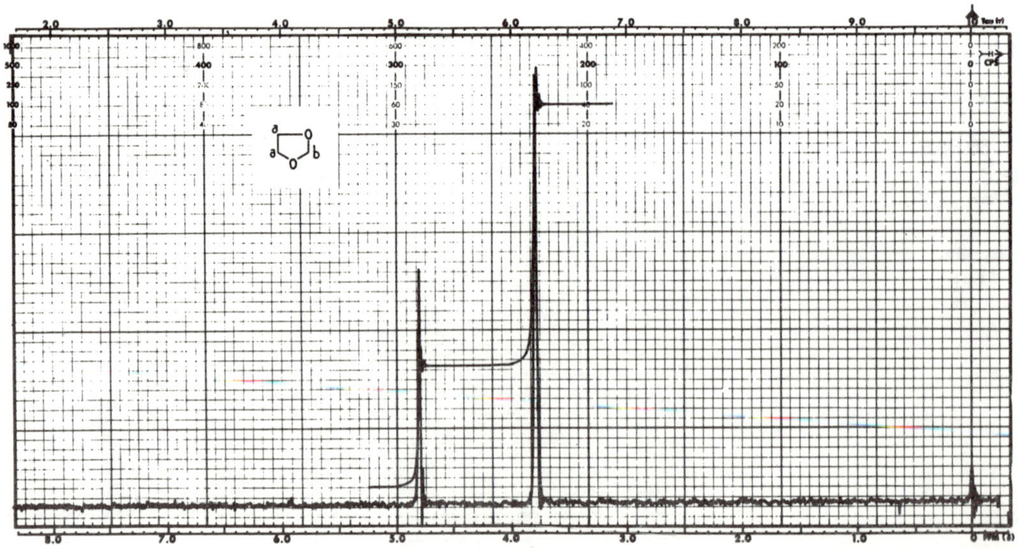

© 1978 Sadtler Research Laboratories, Inc.

2-ETHYL-2-METHYL-1,3-DIOXOLANE

$C_6H_{12}O_2$

Mol. Wt. 116.16

Solvent: CCl_4

Assignments
a = 0.88
b = 1.20
c = 1.55
d = 3.82

1459

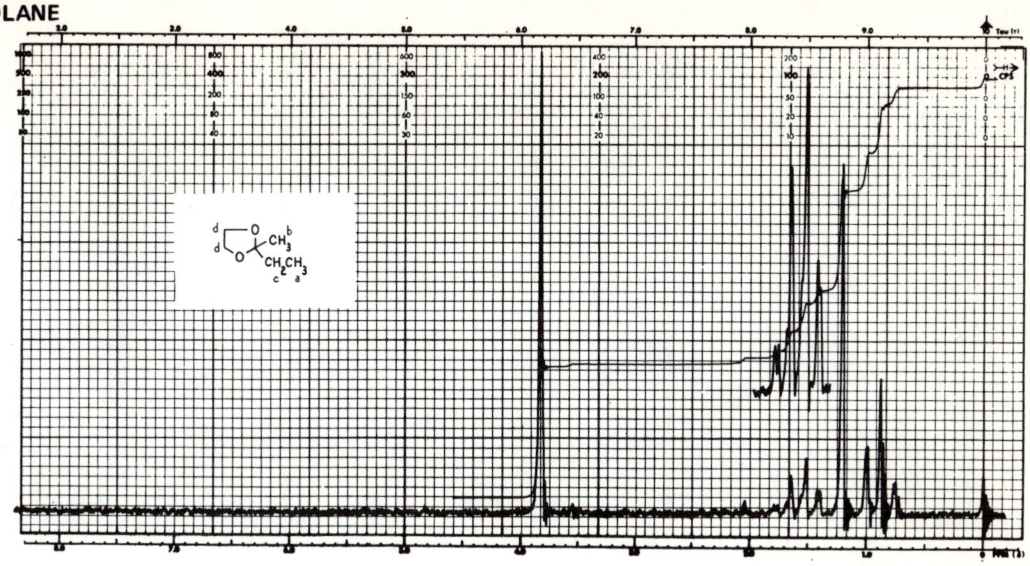

© 1978 Sadtler Research Laboratories, Inc.

FURFURYL 2-PYRANYL ETHER

$C_{10}H_{18}O_3$

Mol. Wt. 186.25

Solvent: CCl_4

Assignments
a = 1.17-1.70
b = 1.70-2.40
c = 3.15-4.10
d = 4.56

1460

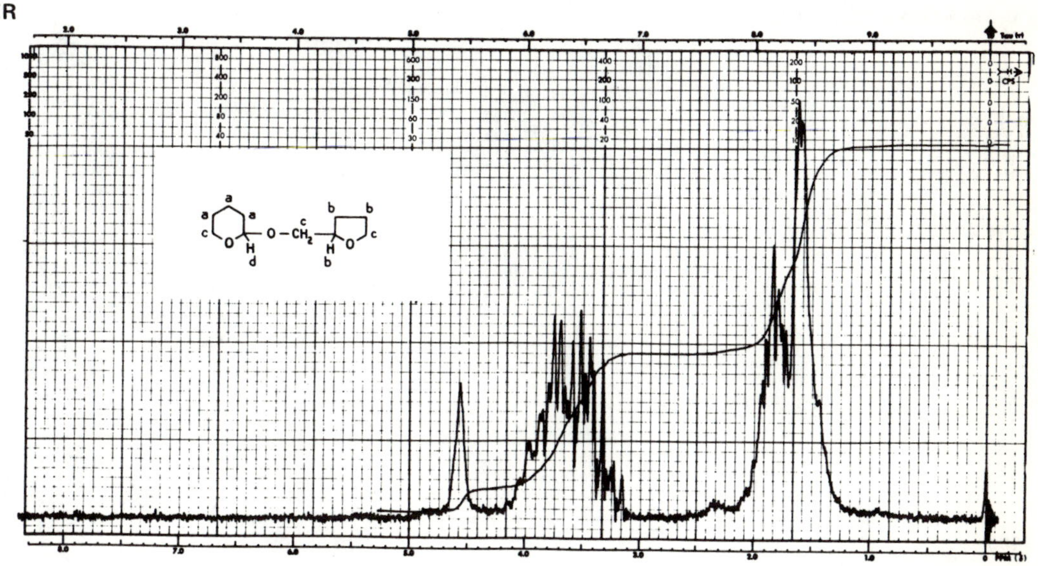

© 1978 Sadtler Research Laboratories, Inc.

3,6-DIHYDRO-2H-PYRAN

C_5H_8O

Mol. Wt. 84.12

Solvent: CCl_4

Assignments
a = 2.08
b = 3.69
c = 4.01
d = 5.65 or e
e = 5.80 or d

1461

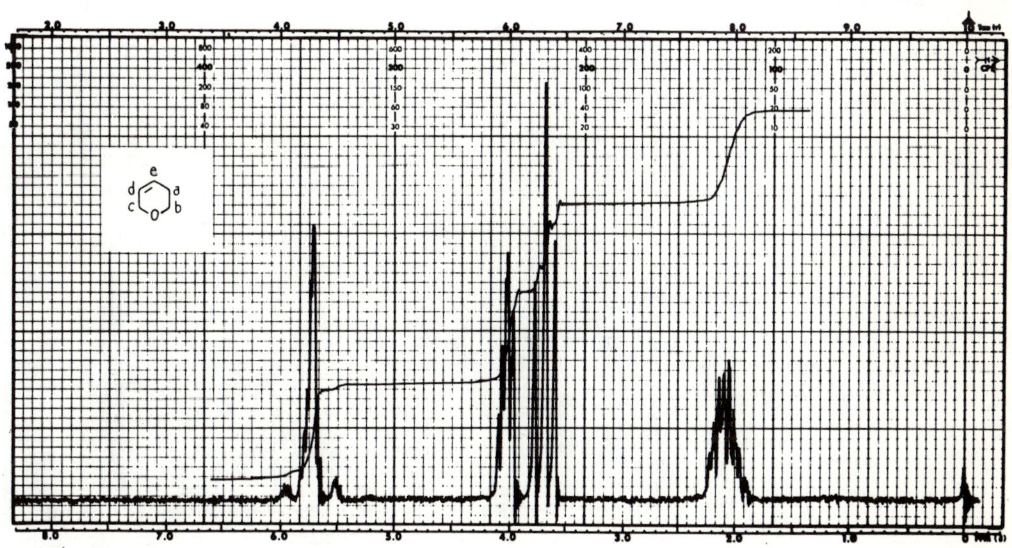

© 1978 Sadtler Research Laboratories, Inc.

MORPHOLINE

C_4H_9NO

Mol. Wt. 87.12

B.P. 128.6°C

Solvent: CCl_4

Assignments
a = 1.78
b = 2.72
c = 3.52

1462

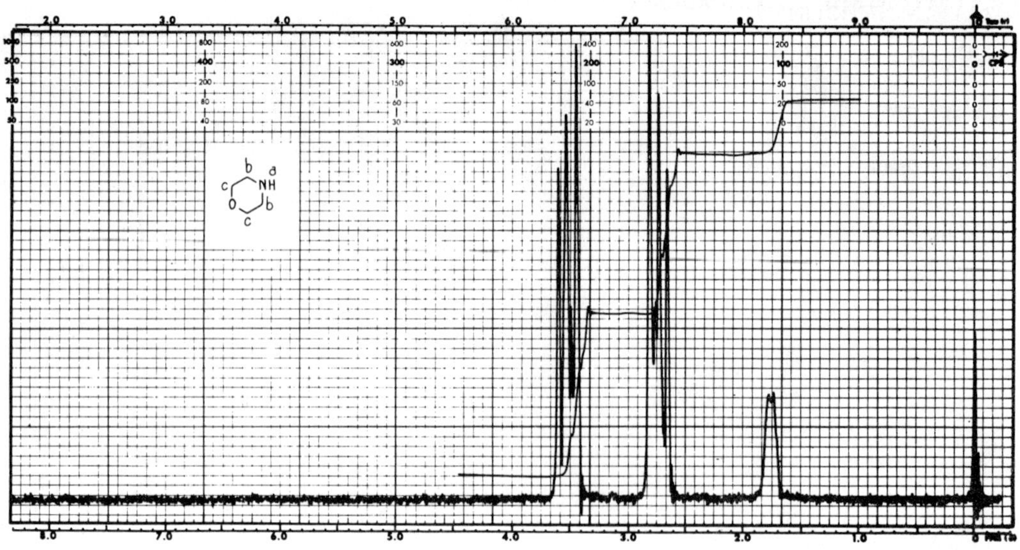

© 1978 Sadtler Research Laboratories, Inc.

2,6-DIMETHYLMORPHOLINE

$C_6H_{13}NO$

Mol. Wt. 115.18

B.P. 146.6°C

Solvent: $CDCl_3$

Assignments
a = 1.05-1.30
b = 2.20-3.20
c = 3.30-4.00
Mixture of isomers NH
is unobserved

1463

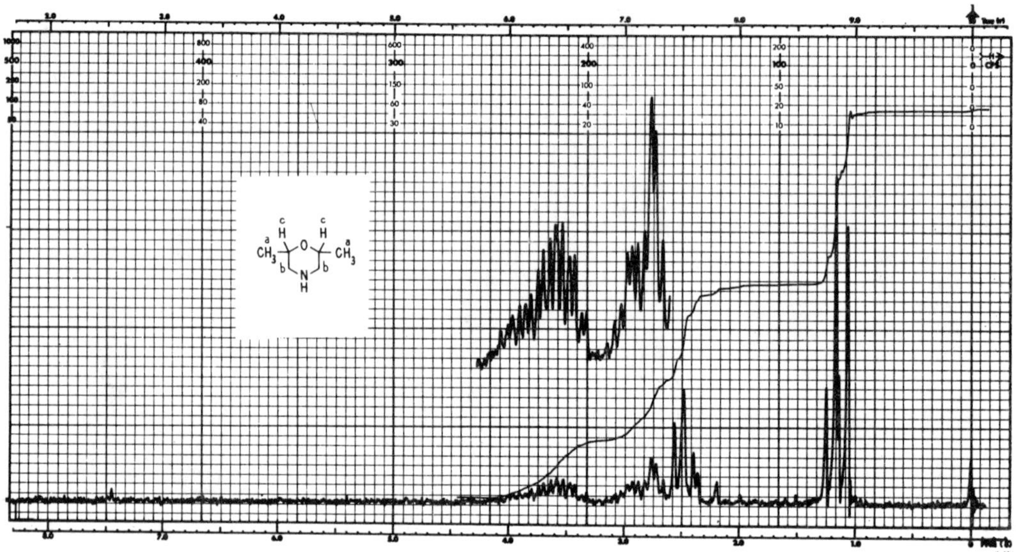

© 1978 Sadtler Research Laboratories, Inc.

4-ETHYLMORPHOLINE

$C_6H_{13}NO$

Mol. Wt. 115.18

B.P. 135-138°C

Solvent: CCl_4

Assignments
a = 1.02
b = ca 2.34
c = 3.57

1464

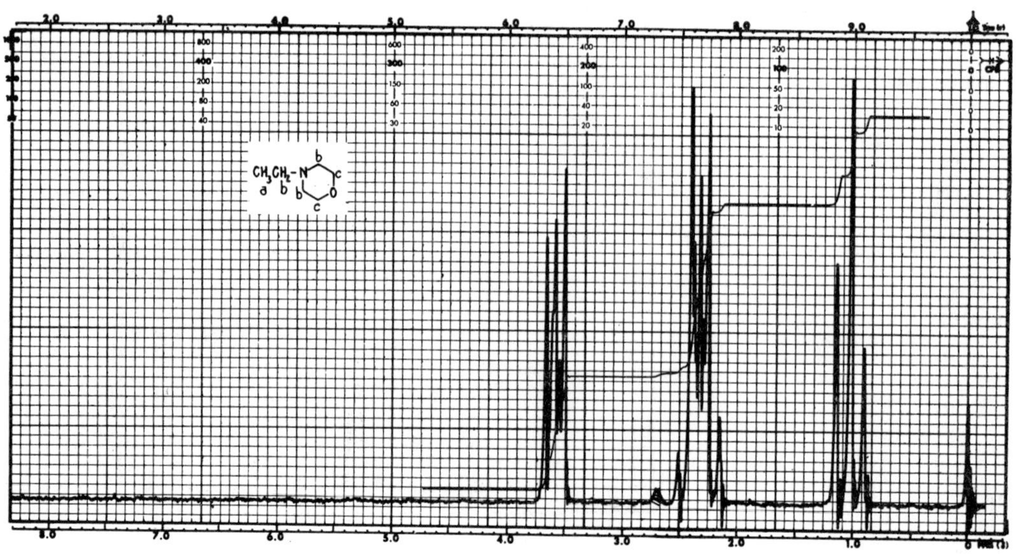

© 1978 Sadtler Research Laboratories, Inc.

4-ALLYLMORPHOLINE

$C_7H_{13}NO$

Mol. Wt. 127.19

B.P. 156-158°C/768mm

Solvent: $CDCl_3$

Assignments
a = 2.45
b = 3.00
c = 3.70
d = 5.15
e = 5.19
f = 5.86
g = 2.10, 3.23 imps.

1465

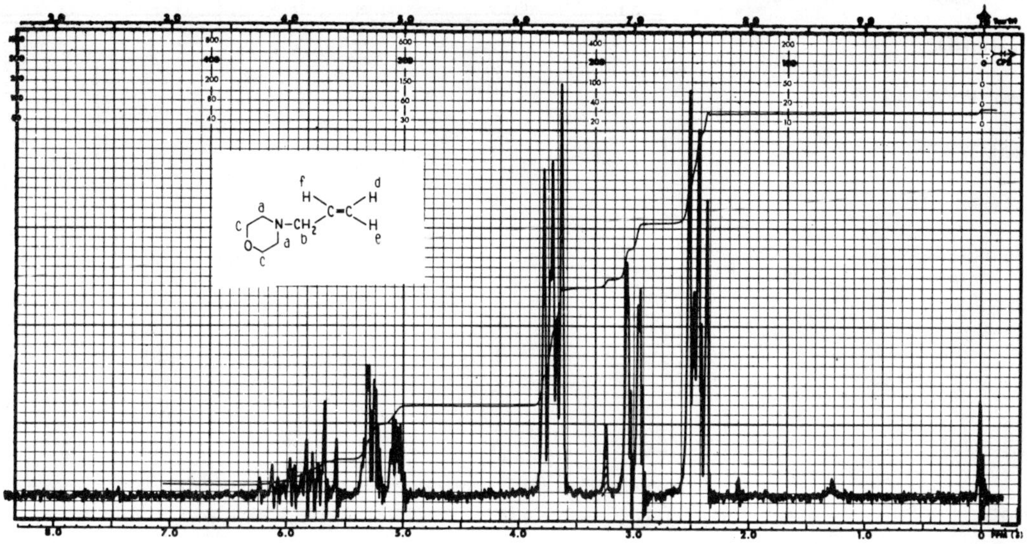

© 1978 Sadtler Research Laboratories, Inc.

4-(3-AMINOPROPYL)MORPHOLINE

$C_7H_{16}N_2O$

Mol. Wt. 144.23

B.P. 223°C

Solvent: CCl_4

Assignments
a = ca 1.25
b = 1.54
c = 2.35
d = 2.70
e = 3.58

1466

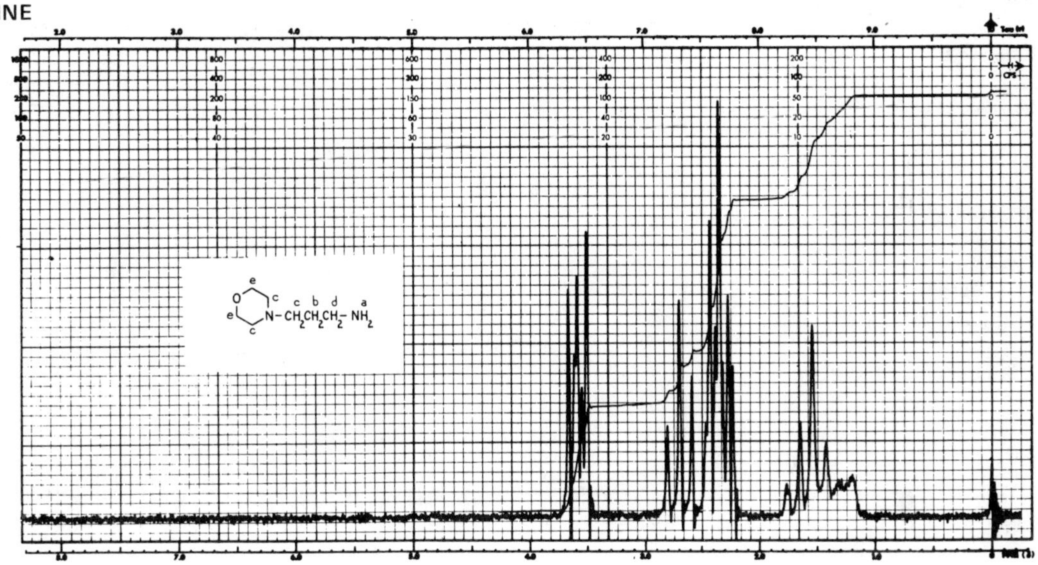

© 1978 Sadtler Research Laboratories, Inc.

α,α-DIMETHYL-4-MORPHOLINEACETONITRILE

$C_8H_{14}N_2O$

Mol. Wt. 154.21

B.P. 69-70°C/1.5mm

Solvent: $CDCl_3$

Assignments
a = 1.47
b = 2.61
c = 3.70

1467

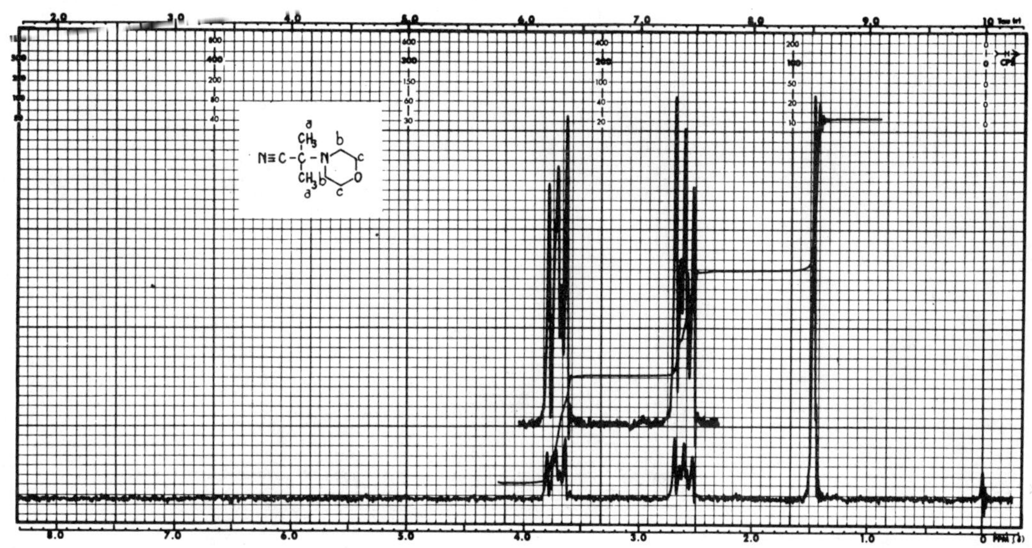

© 1978 Sadtler Research Laboratories, Inc.

4-m-TOLYLMORPHOLINE

$C_{11}H_{15}NO$

Mol. Wt. 177.25

M.P. 31-36°C

Solvent: $CDCl_3$

Assignments
a = 2.30
b = 3.05
c = 3.79
d = 6.50-6.85
e = 7.12

1468

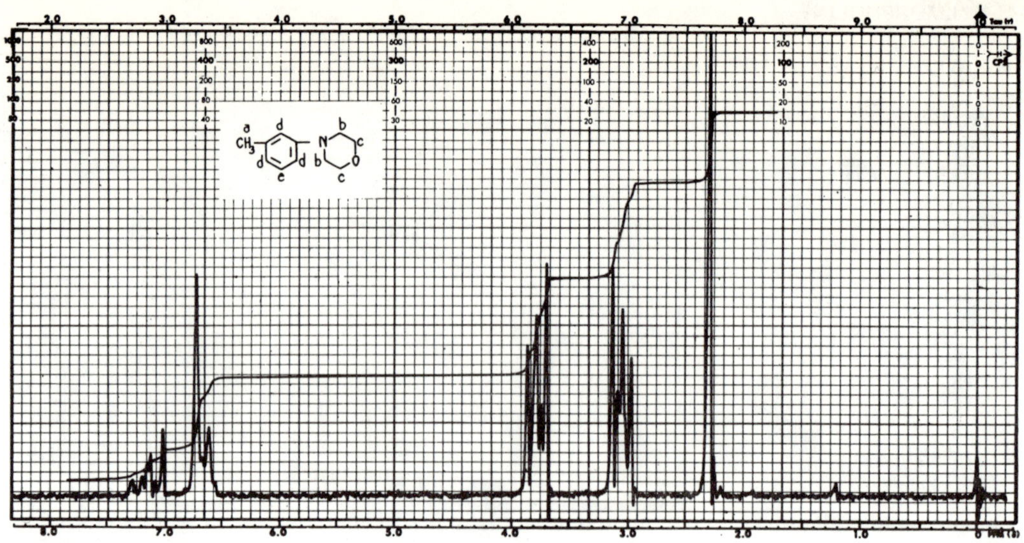

© 1978 Sadtler Research Laboratories, Inc.

MORPHOLINE, HYDROBROMIDE

$C_4H_9NO \cdot HBr$

Mol. Wt. 168.04

Solvent: D_2O

Assignments
a = 3.33
b = 3.98
c = 4.59 HDO

1469

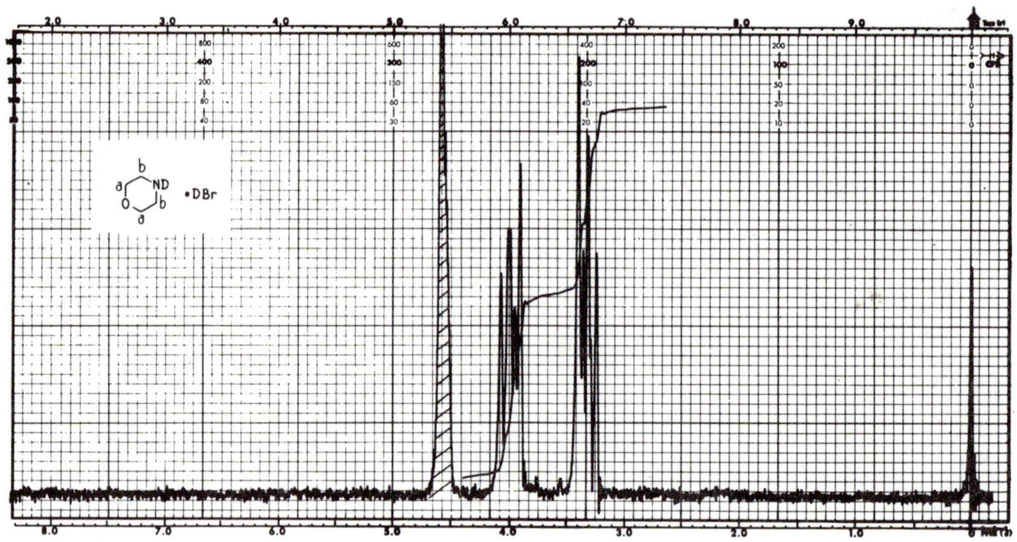

© 1978 Sadtler Research Laboratories, Inc.

1,4-OXATHIANE

C_4H_8OS

Mol. Wt. 104.17

B.P. 145-148°C

Solvent: CCl_4

Assignments
a = 2.55
b = 3.81

1470

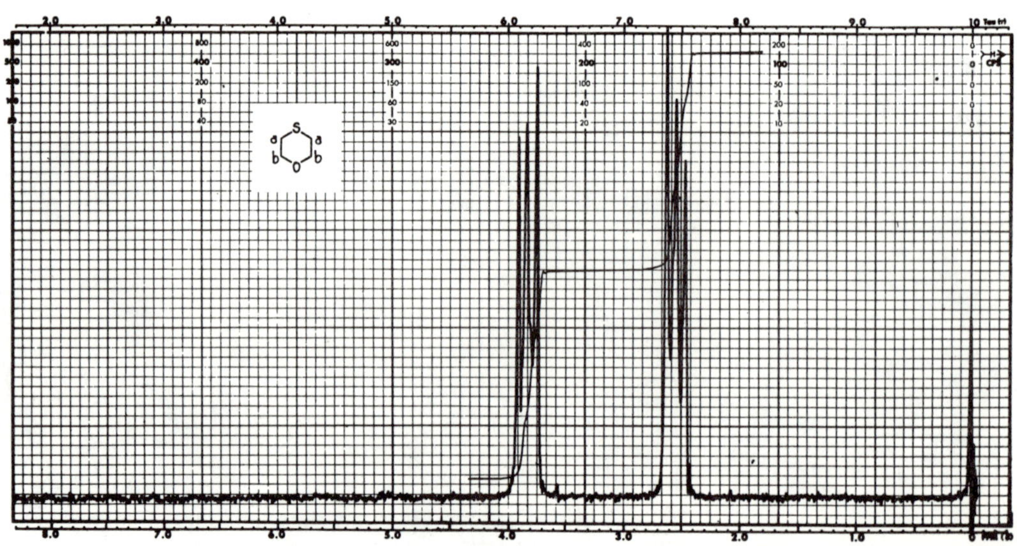

© 1978 Sadtler Research Laboratories, Inc.

1,4-DIOXANE

$C_4H_8O_2$

Mol. Wt. 88.11

M.P. 9.5-10.5°C

B.P. 101.1°C (lit.)

Solvent: CCl_4

Assignments

a = 3.63

1471

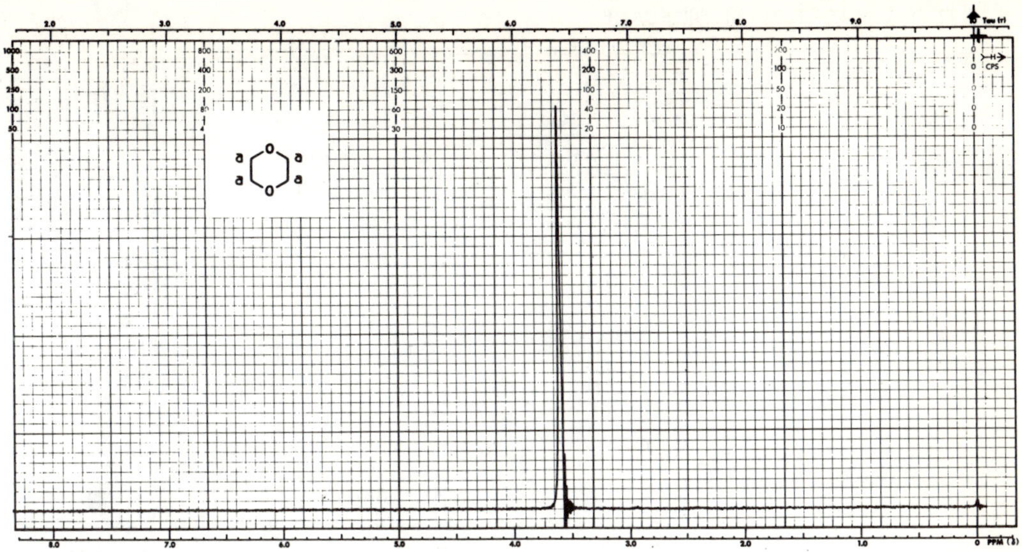

© 1978 Sadtler Research Laboratories, Inc.

5,5-DIMETHYL-2-HEXYL-m-DIOXANE

$C_{12}H_{24}O_2$

Mol. Wt. 200.32

Solvent: CCl_4

Assignments

a = 0.70 (Axial)
b = 0.90
c = 1.17 (Equatorial)
d = 1.10-1.70
e = 3.28
f = 3.51
g = 4.28

1472

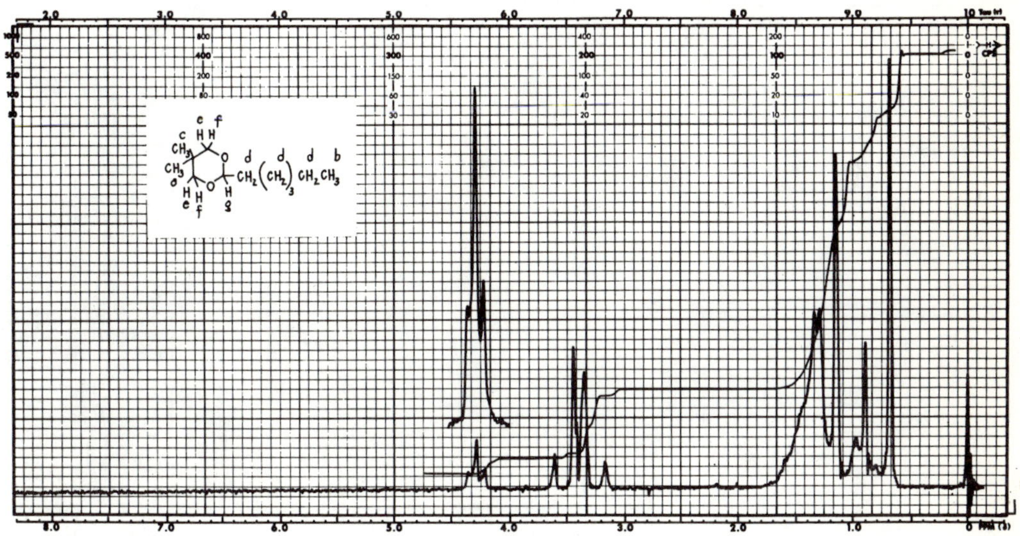

© 1978 Sadtler Research Laboratories, Inc.

2,5-BIS(MORPHOLINOMETHYL)-p-DIOXANE

$C_{14}H_{26}N_2O_4$

Mol. Wt. 286.37

Solvent: $CDCl_3$

Assignments

a = 2.00-2.70
b = 3.30-4.00
c = 3.29 imp.

1473

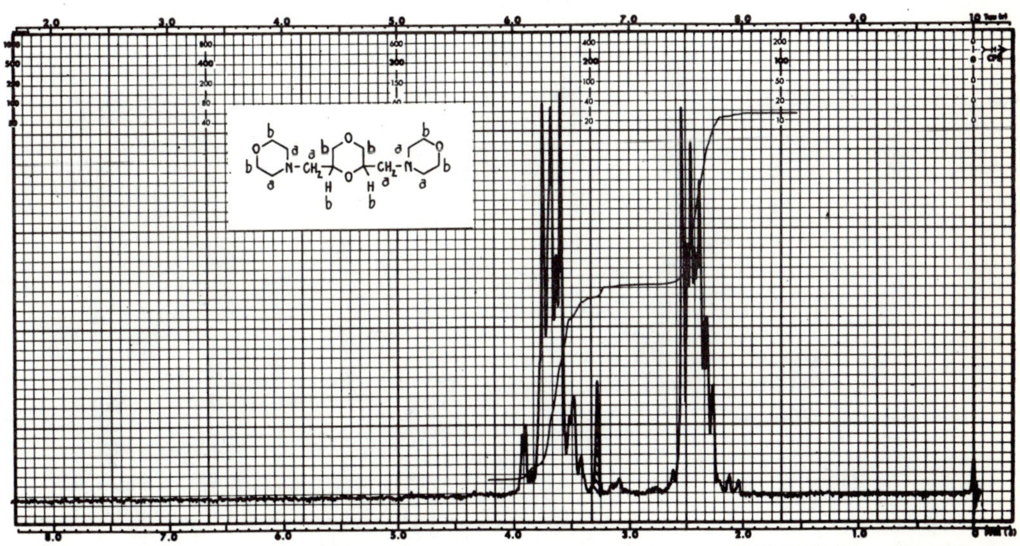

© 1978 Sadtler Research Laboratories, Inc.

2,4,8,10-TETRAOXASPIRO[5,5] UNDECANE

$C_7H_{12}O_4$

Mol. Wt. 160.17

M.P. 47-49°C

Solvent: $CDCl_3$

Assignments
a = 3.76
b = 4.80

1474

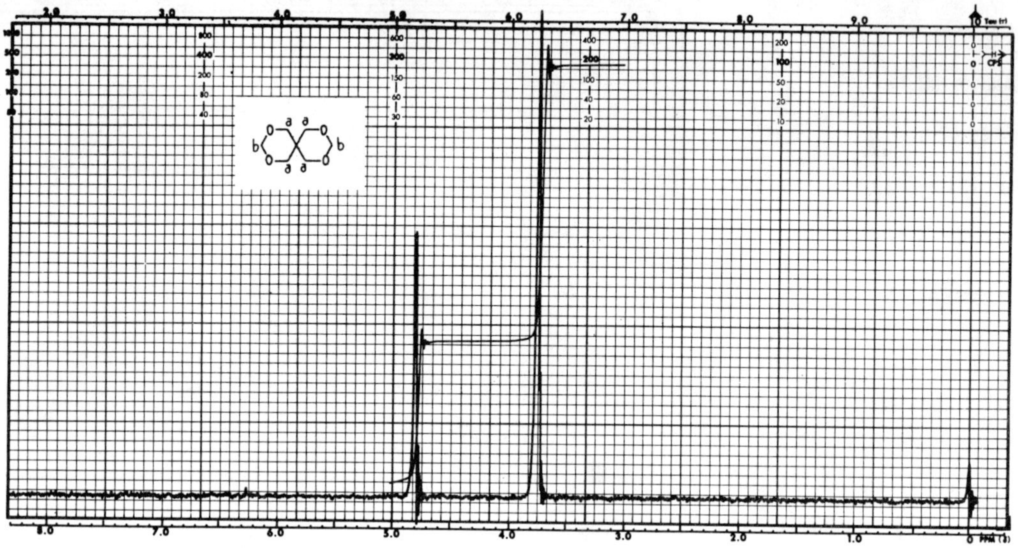

© 1978 Sadtler Research Laboratories, Inc.

3,9-DIMETHYL-2,4,8,10-TETRAOXASPIRO[5,5] UNDECANE

$C_9H_{16}O_4$

Mol. Wt. 188.23

M.P. 39-42°C

Solvent: $CDCl_3$

Assignments
a = 1.30
b = 3.19-3.69
c = 4.52
d = 4.58

1475

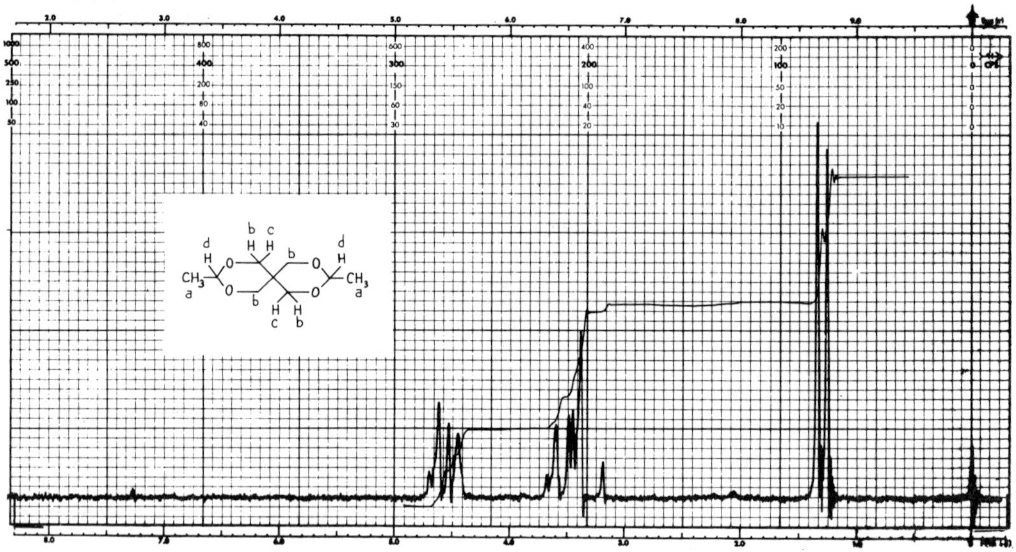

© 1978 Sadtler Research Laboratories, Inc.

3,9-DIETHYL-2,4,8,10-TETRAOXASPIRO[5.5] UNDECANE

$C_{11}H_{20}O_4$

Mol. Wt. 216.28

B.P. 258-259°C

Solvent: CCl_4

Assignments
a = 0.90
b = 1.55
c = 3.09-3.82
d = 4.29
e = 4.55

1476

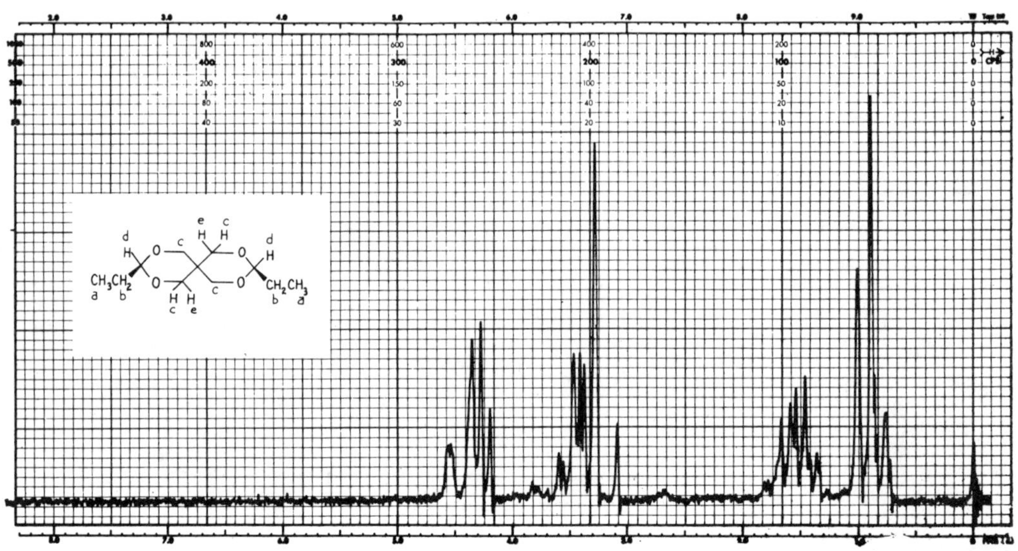

© 1978 Sadtler Research Laboratories, Inc.

s-TRIOXANE

$C_3H_6O_3$

Mol. Wt. 90.08

M.P. 61°C

B.P. 115°C

Solvent: Dimethyl sulfoxide-D_6

Assignments
a = 5.12
b = 2.51 DMSO-D_5
c = 3.32 H_2O

1477

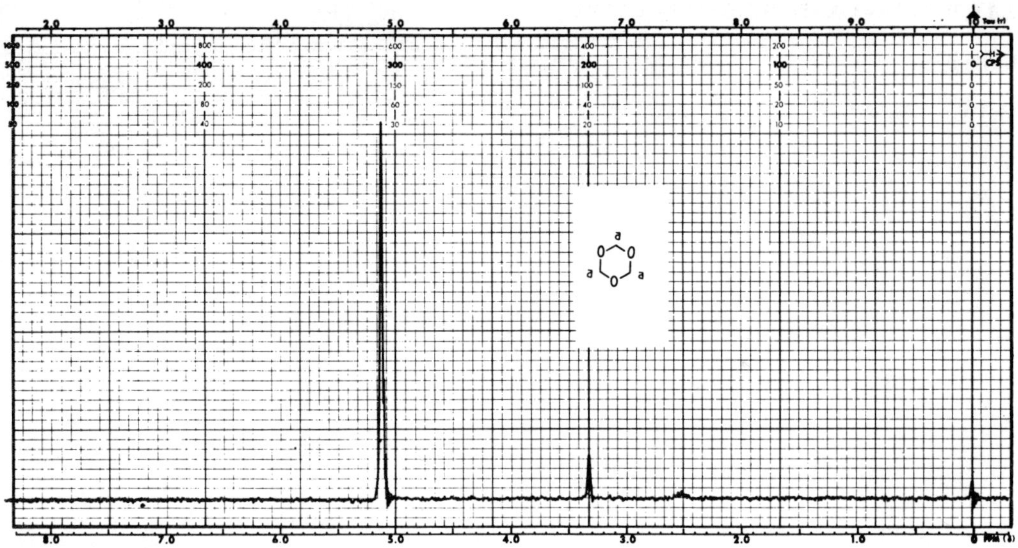

© 1978 Sadtler Research Laboratories, Inc.

1,3-DIOXEPANE

$C_5H_{10}O_2$

Mol. Wt. 102.13

B.P. 117-119°C

Solvent: CCl_4

Assignments
a = 1.65
b = 3.67
c = 4.62

1478

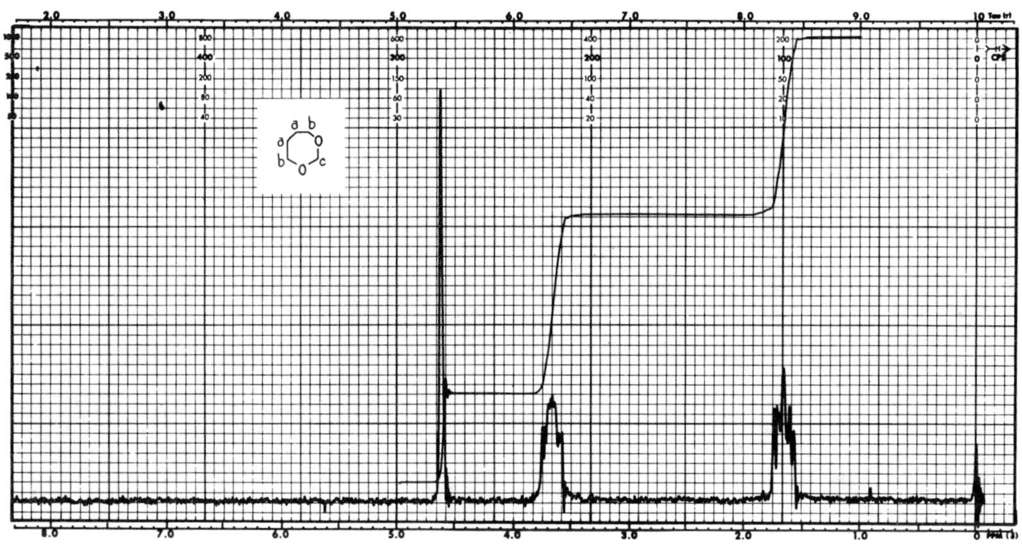

© 1978 Sadtler Research Laboratories, Inc.

NOTES

The NMR spectra of the phenyl ethers are often easily characterized by means of the low field shifts observed for aliphatic groups bonded to the phenoxy moiety and from the high field shifts observed for the ortho and para protons.

The compounds are normally soluble in CCl_4, $CDCl_3$ and DMSO-d_6. The spectra of the aromatic ethers possess no unusual features with the exception of the chemical shifts noted above.

Chemical Shifts

Aliphatic Protons

HC–4	HC–3	HC–2	HC–1		Solvent	Spectrum
			3.70	CH_3–O–Ø	CCl_4	1488
		1.31	3.89	CH_3–CH_2–O–Ø	CCl_4	1489
	0.90	1.71	3.85	CH_3–CH_2–CH_2–O–Ø	$CDCl_3$	1490
0.99	1.45	1.70	3.87	CH_3–CH_2–CH_2–CH_2–O–Ø	CCl_4	- - - -

1-Substituted Methyl Phenyl Ethers

Ø–O	–CH_2	–X	
	3.89	–CH_3	CCl_4
	4.36	–C(=O)–CH_3	CCl_4
	4.47	–C(=O)–NH_2	$CDCl_3$
	4.80	–C(=O)–Cl	CCl_4
	4.81	–C≡C–H	DMSO-d_6
	4.92	–C≡N	$CDCl_3$
	5.18	–C(=O)–Ø	$CDCl_3$

2-Substituted Ethyl Phenyl Ethers

Ø–O	–CH_2	–CH_2	–X	
	3.85	1.71	–CH_3	$CDCl_3$
	4.21	2.79	–C(=O)–OH	$CDCl_3$
	3.89	2.99	–NH_2	$CDCl_3$
	4.19	3.52	–Br	$CDCl_3$
	4.00	3.59	–O–CH_3	CCl_4
	4.05	3.63	–Cl	CCl_4
	4.06	4.29	–O–C(=O)–CH_3	CCl_4
	4.08	4.39	–O–C(=O)–H	$CDCl_3$

Chemical Shifts continued

Aromatic Protons

Methyl Phenyl Ether

CH_3–O–Ø	ortho	meta	para	Solvent	Spectrum
	6.78	7.19	6.80	CCl_4	1488

Diphenyl Ether

Ø–O–Ø	ortho/meta/para	Solvent	Spectrum
	6.78–7.40	$CDCl_3$	1541

Para Substituted Methyl Phenyl Ethers

CH_3–O–	ortho	meta	para	Solvent
3.75	6.79	6.79	–$N(CH_3)_2$	$CDCl_3$
3.71	6.79	6.79	–O–CH_3	$CDCl_3$
3.71	6.80	6.90	–O–Ø	$CDCl_3$
3.61	6.65	7.08	–CH=CH–CH_3	CCl_4
3.77	6.79	7.09	–CH_3	$CDCl_3$
3.70	6.72	7.18	–Cl	CCl_4
3.71	6.69	7.29	–Br	CCl_4
3.76	6.81	7.30	–SH	$CDCl_3$
3.76	6.93	7.49	–Ø	$CDCl_3$
3.74	6.68	7.53	–I	$CDCl_3$
3.65	6.81	7.73	–CH=N–Ø	$CDCl_3$
3.79	6.91	7.79	–SO_2–Ø	$CDCl_3$
3.89	6.91	8.12	–NO_2	CCl_4

Para Substituted Diphenyl Ethers

Ø–O–	ortho	meta	para	Solvent
	6.85	6.55	–NH_2	$CDCl_3$
	6.91	6.77	–OH	$CDCl_3$
	6.90	6.80	–O–CH_3	$CDCl_3$
	6.75	7.29	–Br	CCl_4
	7.02	7.39	–Cl	$CDCl_3$
	7.03	7.80	–C(=O)–Ø	$CDCl_3$
	6.98	8.07	–C(=O)–OH	$CDCl_3$
	7.10	8.27	–NO_2	CCl_4

BENZYL METHYL ETHER

$C_8H_{10}O$

Mol. Wt. 122.17

B.P. 58-60°C/10mm

Solvent: $CDCl_3$

Assignments

a = 3.29
b = 4.35
c = ca 7.25

1479

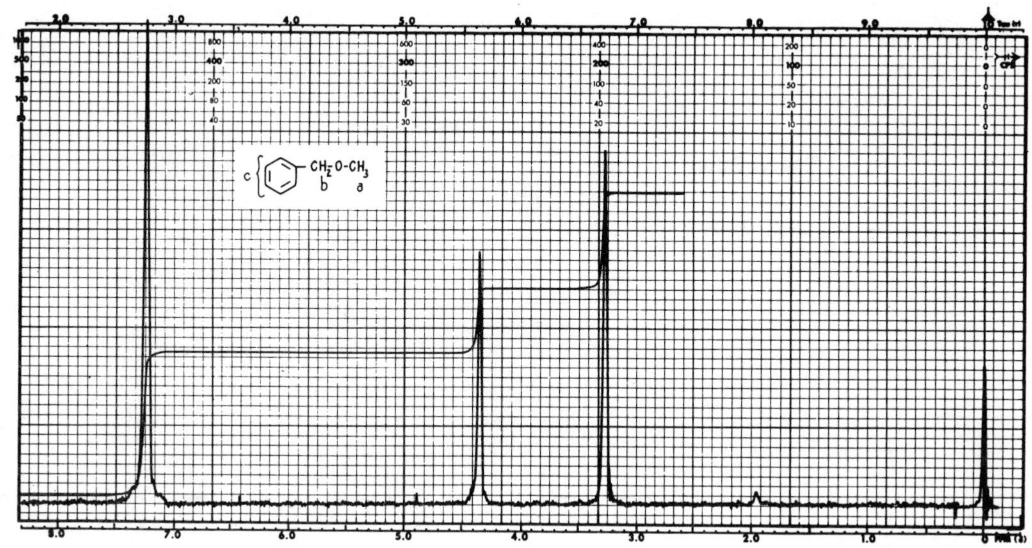

© 1978 Sadtler Research Laboratories, Inc.

4-ETHOXY-4-PHENYLBUTYRONITRILE

$C_{12}H_{15}NO$

Mol. Wt. 189.26

Solvent: $CDCl_3$

Assignments

a = 1.13
b = 1.97
c = ca 2.39
d = 3.31
e = 4.30
f = 7.25

1480

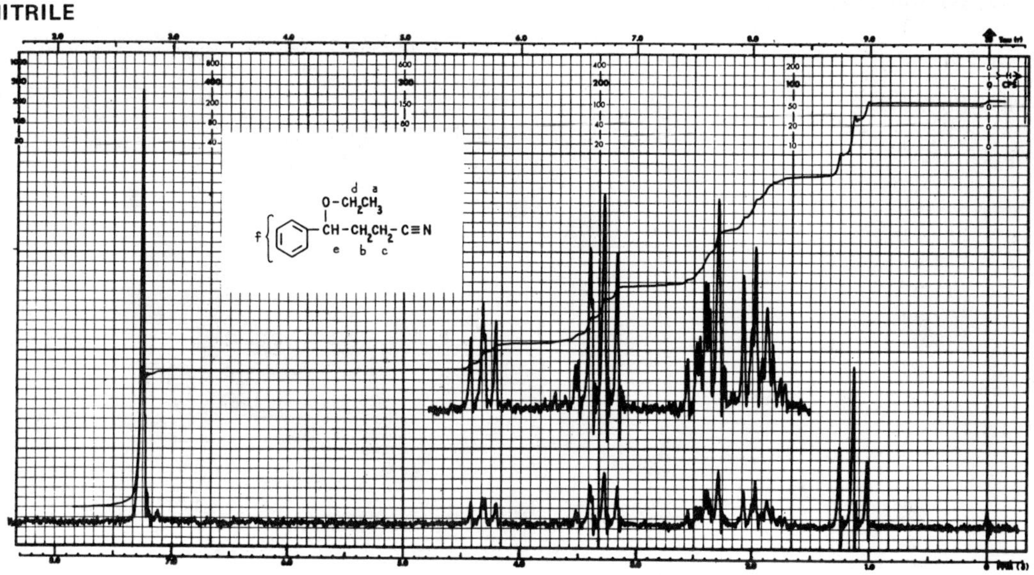

© 1978 Sadtler Research Laboratories, Inc.

ETHYL TRITYL ETHER

$C_{21}H_{20}O$

Mol. Wt. 288.39

M.P. 83°C (lit.)

Solvent: $CDCl_3$

Assignments

a = 1.16
b = 3.12
c = ca 7.12
d = 6.90-7.30
e = ca 7.43
f = 5.49 imp.

1481

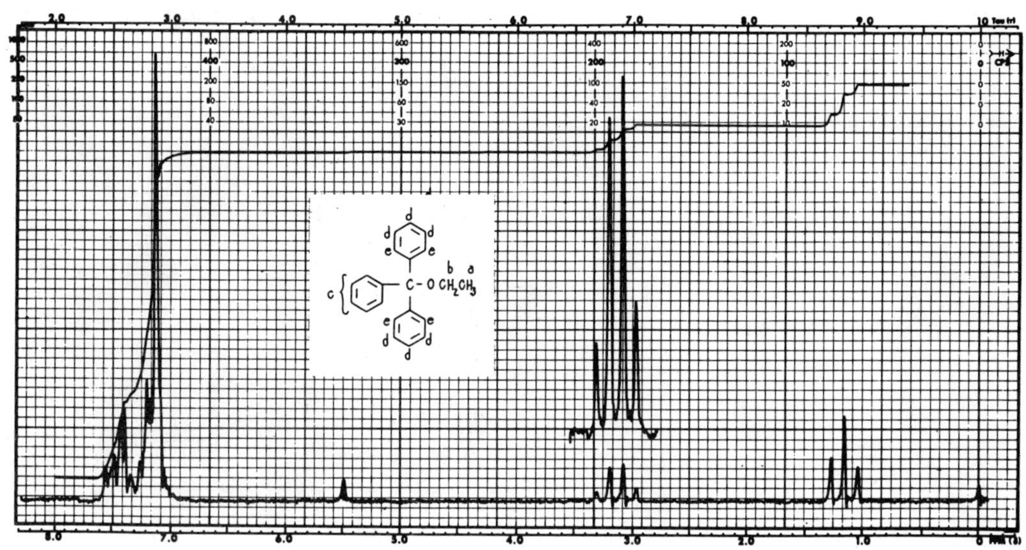

© 1978 Sadtler Research Laboratories, Inc.

BENZYL ETHER

$C_{14}H_{14}O$

Mol.Wt. 198.25

B.P. 295-298°C

M.P. 4-5°C

Solvent: CCl_4

Assignments
a = 4.47
b = 7.25

1482

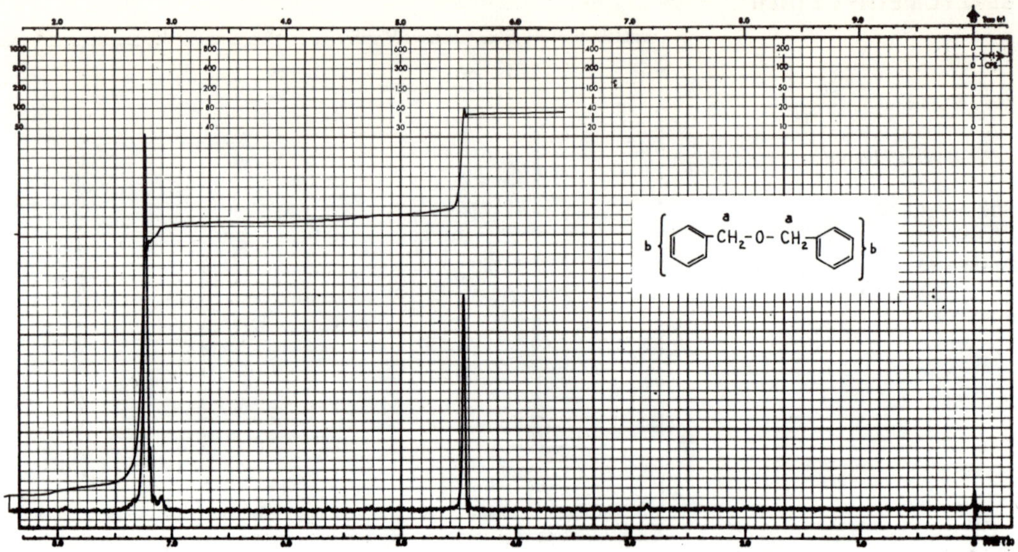

© 1978 Sadtler Research Laboratories, Inc.

α-METHYLBENZYL ETHER

$C_{16}H_{18}O$

Mol. Wt. 226.32

Solvent: CCl_4

Assignments
a = 1.34 or b
b = 1.36 or a
c = 4.15 or d
d = 4.45 or c
e = 7.16 or f
f = 7.21 or e

1483

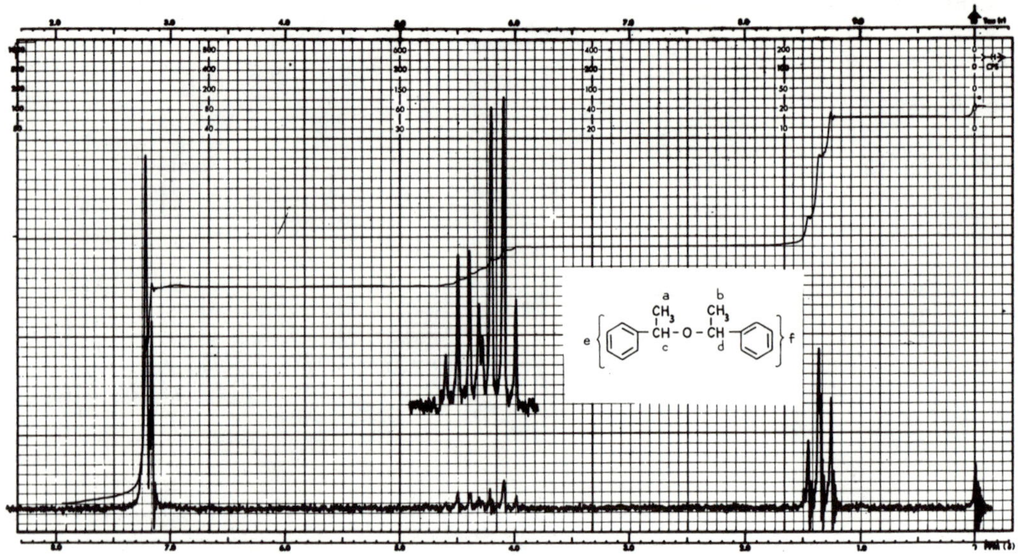

© 1978 Sadtler Research Laboratories, Inc.

METHYL PHENETHYL ETHER

$C_9H_{12}O$

Mol. Wt. 136.20

B.P. 195°C/760 mm

Solvent: $CDCl_3$

Assignments
a = 2.73
b = 3.26
c = 3.53
d = 7.19

1484

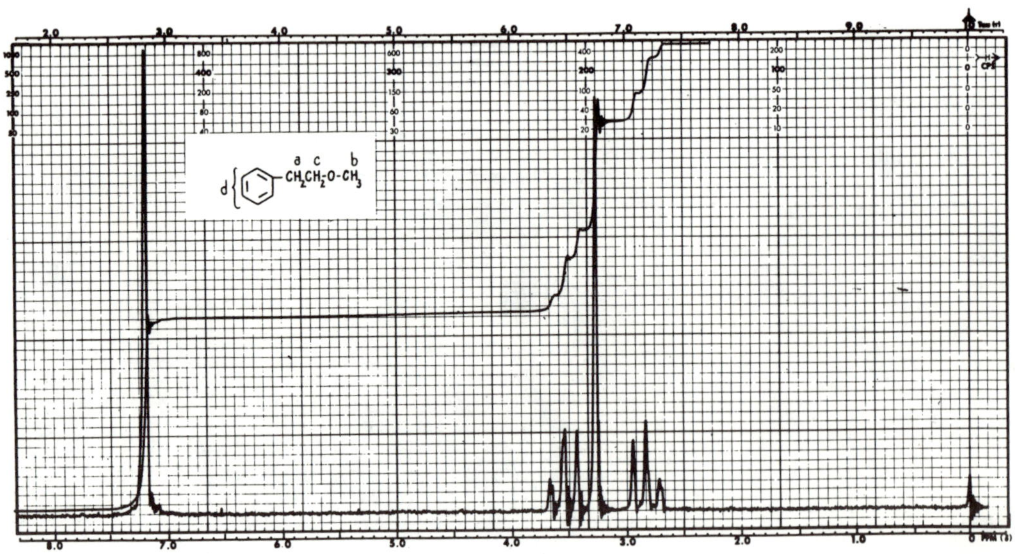

© 1978 Sadtler Research Laboratories, Inc.

ACETALDEHYDE, DIPHENETHYL ACETAL

$C_{18}H_{22}O_2$

Mol. Wt. 270.37

Solvent: CCl_4

Assignments
a = 1.19
b = 2.71
c = ca 3.49
d = 4.55
e = ca 7.10

1485

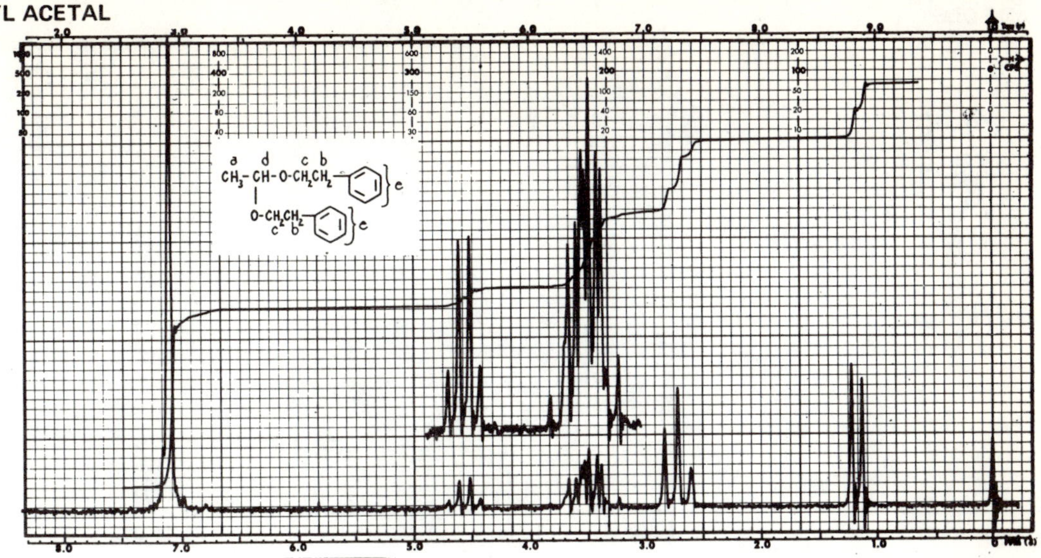

© 1978 Sadtler Research Laboratories, Inc.

DIMETHOXYDIPHENYLMETHANE

$C_{15}H_{16}O_2$

Mol. Wt. 228.29

M.P. 105-107°C

B.P. 288-290°C

Solvent: $CDCl_3$

Assignments
a = 3.11
b = 7.10-7.40
c = 7.49

1486

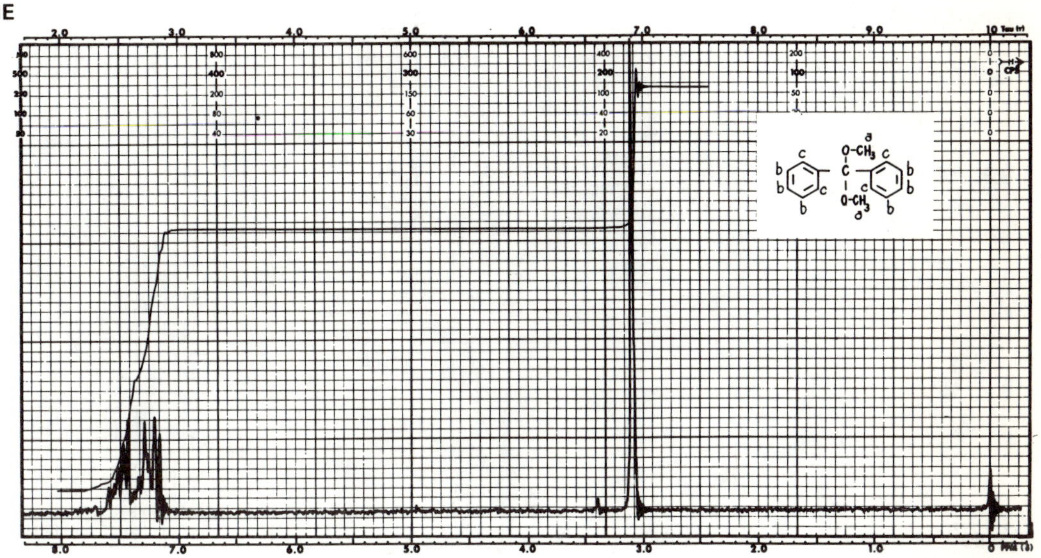

© 1978 Sadtler Research Laboratories, Inc.

ORTHOBENZOIC ACID, TRIMETHYL ESTER

$C_{10}H_{14}O_3$

Mol. Wt. 182.22

B.P. 87-88°C/7mm

Solvent: CCl_4

Assignments
a = 3.09
b = 7.20-7.60

1487

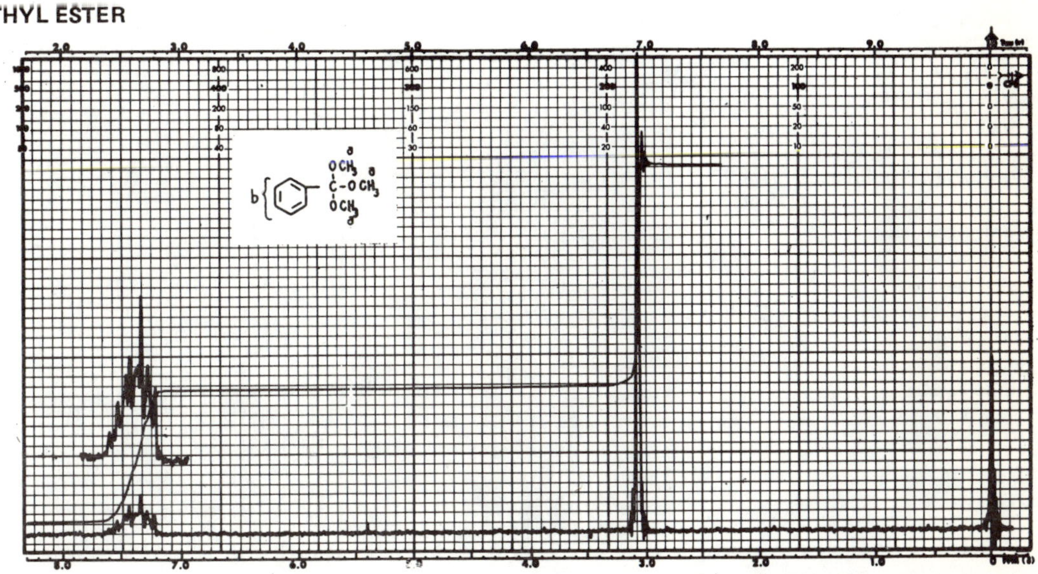

© 1978 Sadtler Research Laboratories, Inc.

ANISOLE

C_7H_8O

Mol. Wt. 108.14

Solvent: CCl_4

Assignments
a = 3.70
b = 6.80
c = 7.19

1488

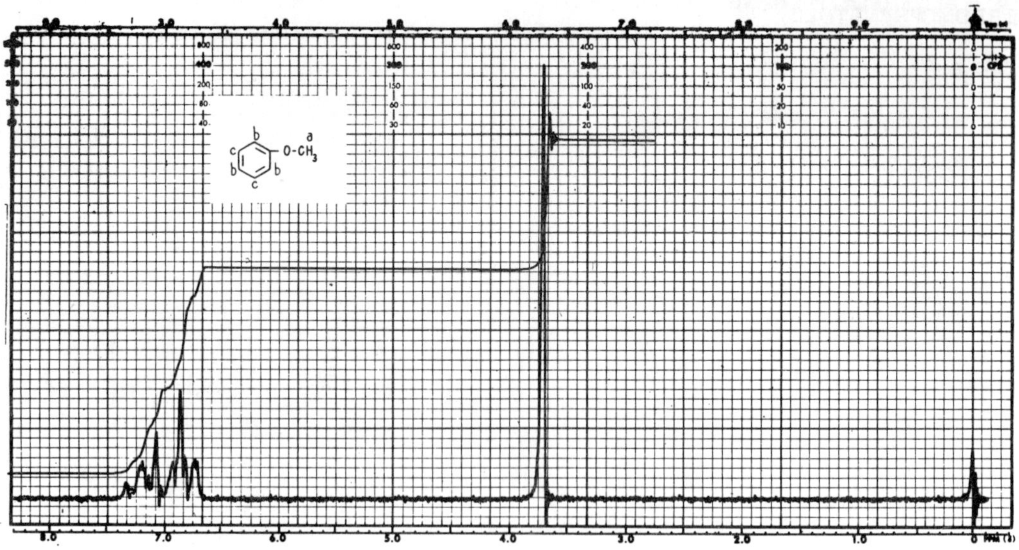

© 1978 Sadtler Research Laboratories, Inc.

PHENETOLE

$C_8H_{10}O$

Mol. Wt. 122.17

B.P. 169-171°C

Solvent: CCl_4

Assignments
a = 1.31
b = 3.89
c = 6.79
d = 6.81
e = 7.15

1489

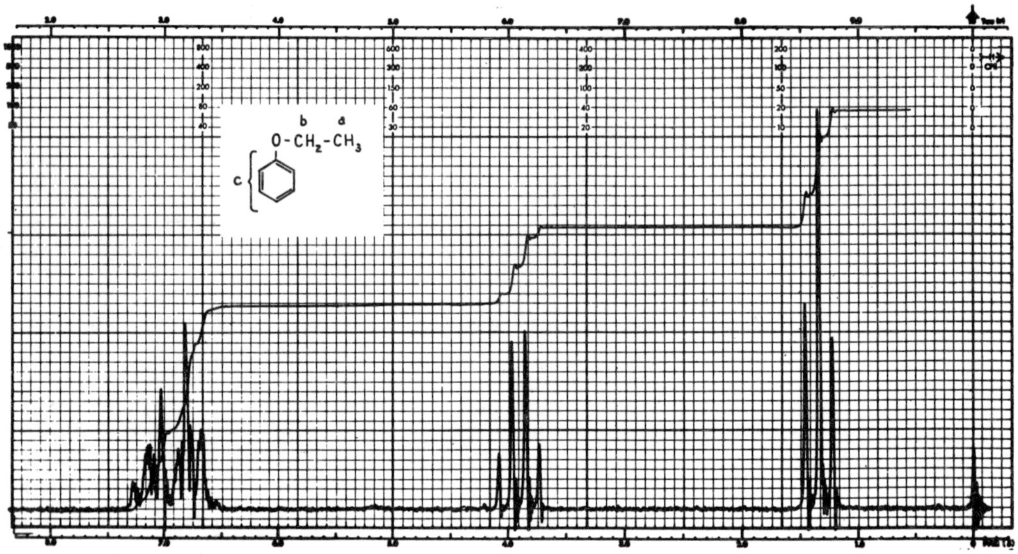

© 1978 Sadtler Research Laboratories, Inc.

PHENYL PROPYL ETHER

$C_9H_{12}O$

Mol. Wt. 136.20

B.P. 96-97°C/35mm

Solvent: $CDCl_3$

Assignments
a = 0.90
b = 1.71
c = 3.85
d = ca 6.93
e = 7.22

1490

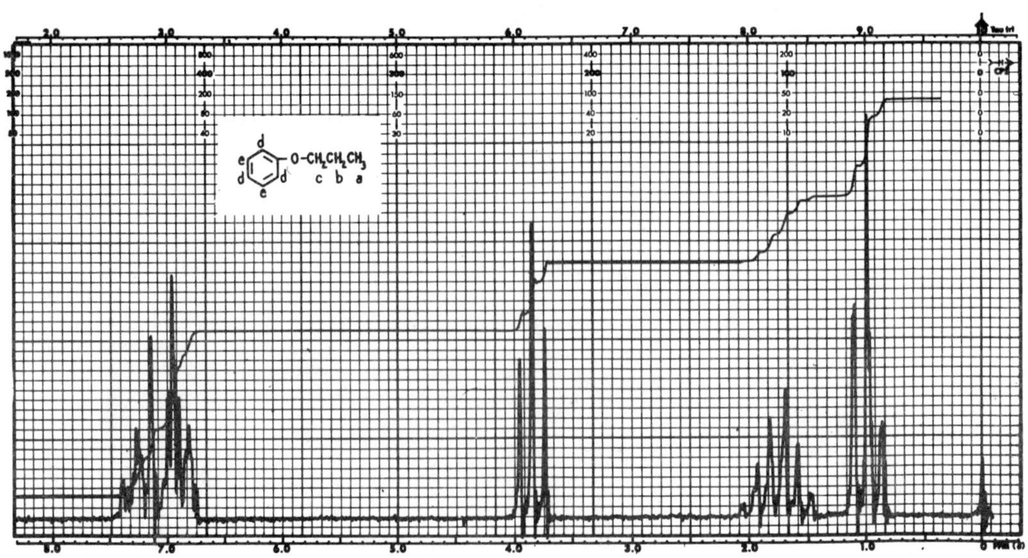

© 1978 Sadtler Research Laboratories, Inc.

β-BROMOPHENETOLE

C_8H_9BrO

Mol. Wt. 201.07

M.P. 30-31°C

Solvent: $CDCl_3$

Assignments
a = 3.52
b = 4.19
c = 6.88
d = 6.96
e = 7.28

1491

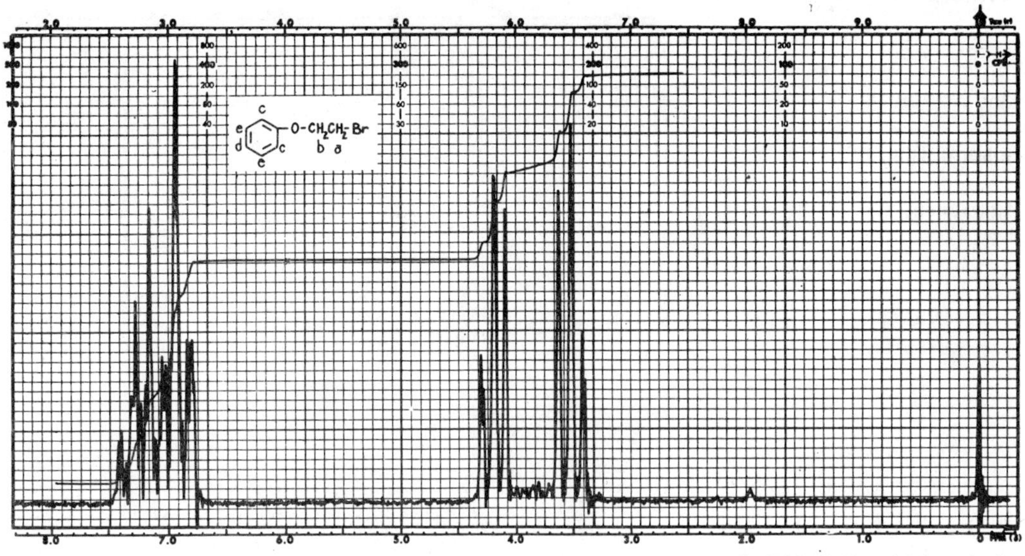

© 1978 Sadtler Research Laboratories, Inc.

(2-METHOXYETHOXY)BENZENE

$C_9H_{12}O_2$

Mol. Wt. 152.20

B.P. 219°C

Solvent: CCl_4

Assignments
a = 3.30
b = 3.59
c = 4.00
d = ca 6.79
e = 7.15

1492

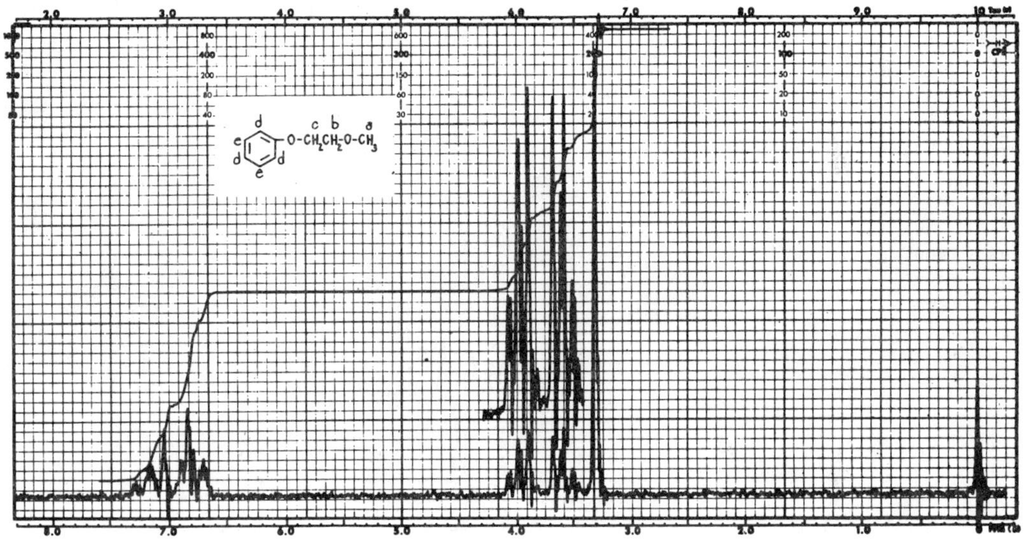

© 1978 Sadtler Research Laboratories, Inc.

ORTHOFORMIC ACID, DIETHYL PHENYL ESTER

$C_{11}H_{16}O_3$

Mol. Wt. 196.25

B.P. 103-104.5°C/10mm

Solvent: CCl_4

Assignments
a = 1.19
b = 3.68
c = 5.60
d = 6.65-7.40

1493

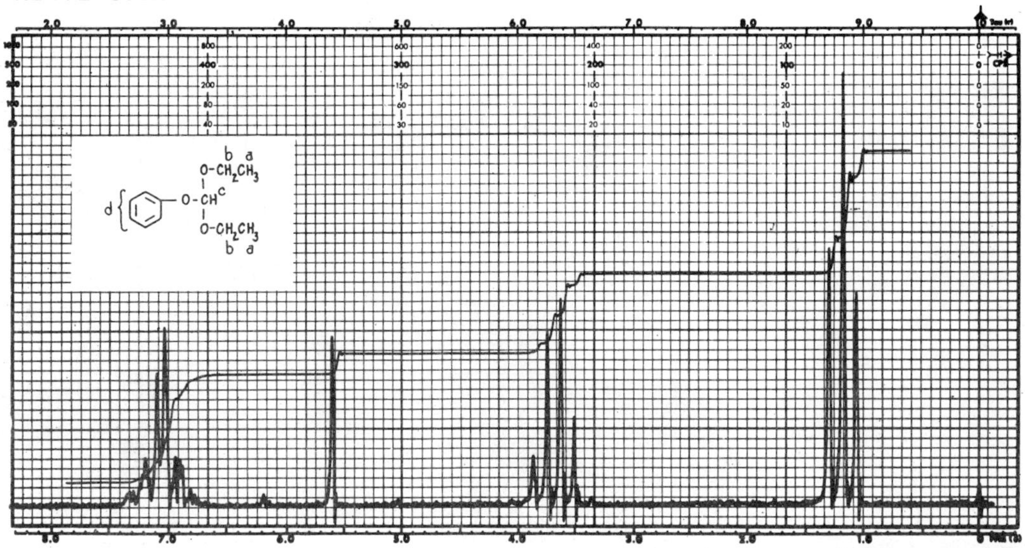

© 1978 Sadtler Research Laboratories, Inc.

o-METHYLANISOLE

$C_8H_{10}O$

Mol. Wt. 122.17

B.P. 170-172°C

Solvent: CCl_4

Assignments
a = 2.18
b = 3.73
c = 6.50-7.20

1494

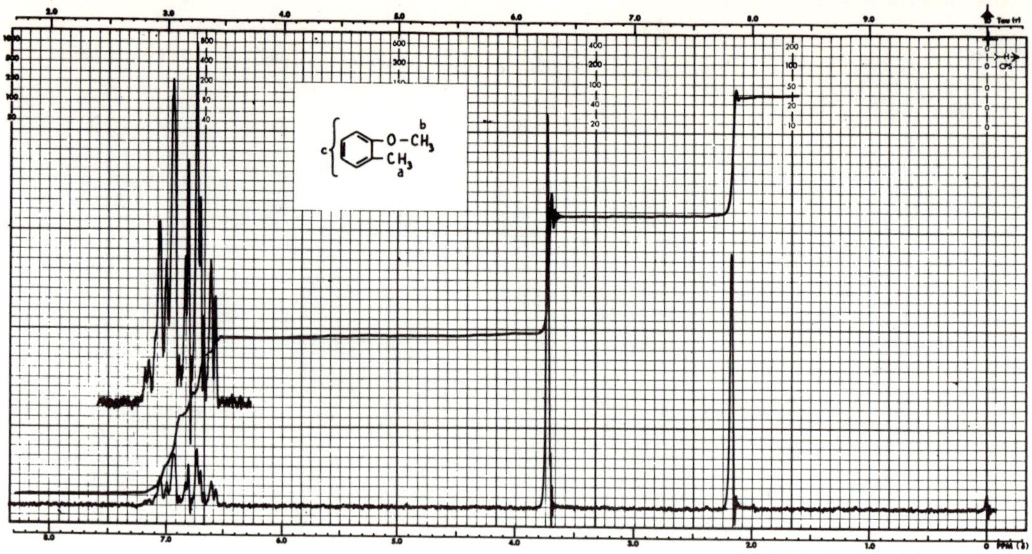

© 1978 Sadtler Research Laboratories, Inc.

m-METHYLANISOLE

$C_8H_{10}O$

Mol. Wt. 122.17

B.P. 102-103°C/100mm

Solvent: CCl_4

Assignments
a = 2.28
b = 3.68
c = 6.60
d = 7.00

1495

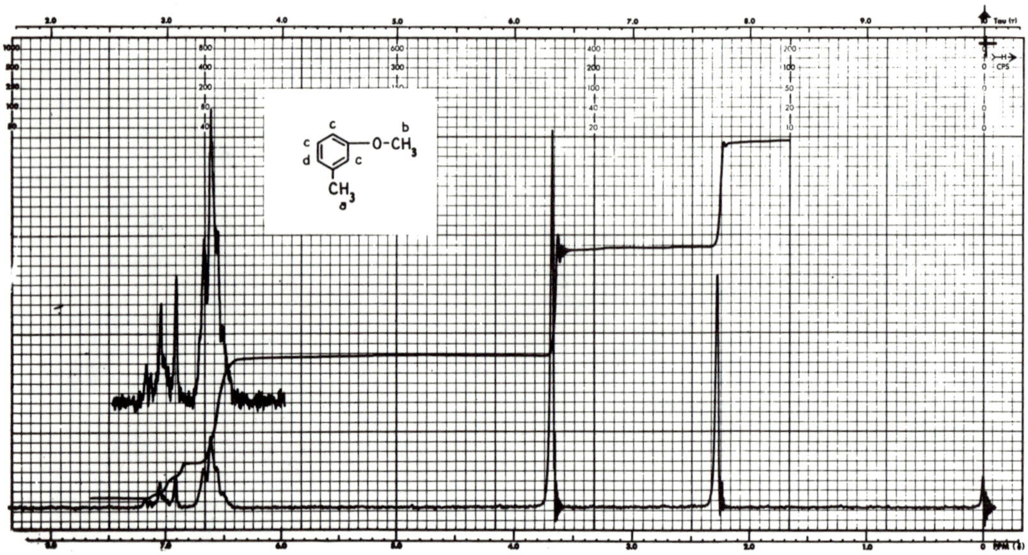

© 1978 Sadtler Research Laboratories, Inc.

p-METHYLANISOLE

$C_8H_{10}O$

Mol. Wt. 122.16

B.P. 173-174°C

Solvent: $CDCl_3$

Assignments
a = 2.28
b = 3.77
c = 6.79
d = 7.09

1496

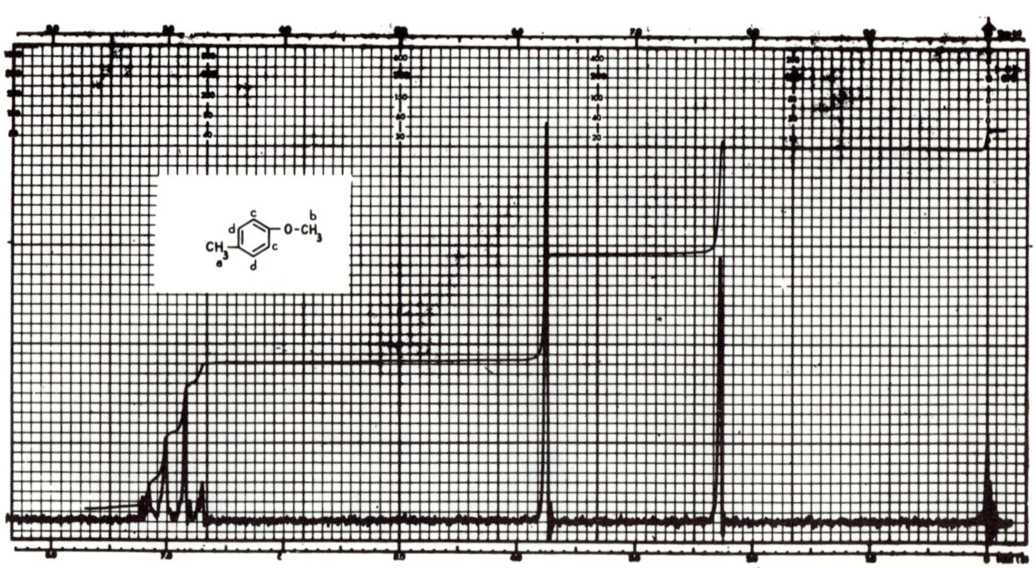

© 1978 Sadtler Research Laboratories, Inc.

p-ETHYLANISOLE

$C_9H_{12}O$

Mol. Wt. 136.20

B.P. 66-67°C/6mm

Solvent: CCl_4

Assignments

a = 1.20
b = 2.53
c = 3.69
d = 6.71
e = 7.03

1497

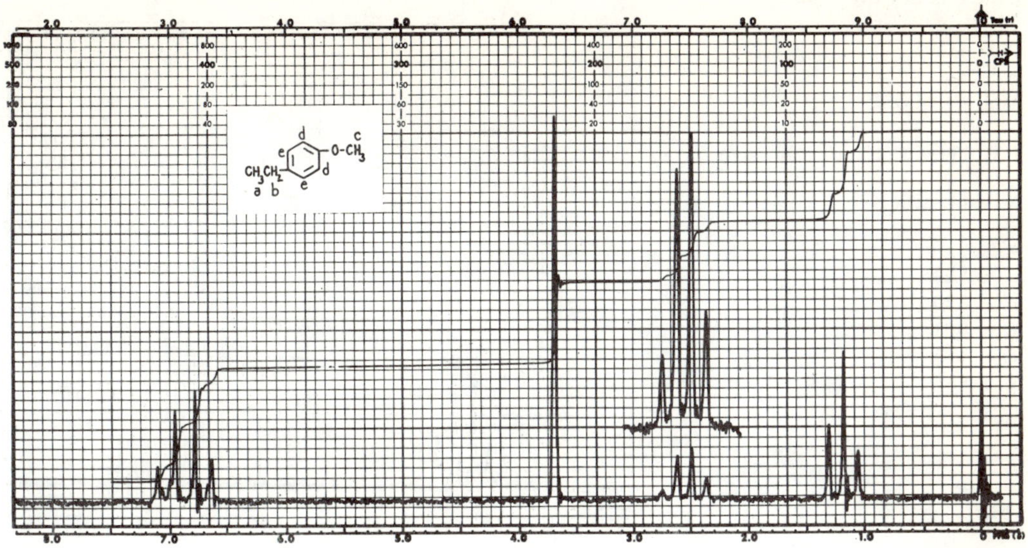

© 1978 Sadtler Research Laboratories, Inc.

o-METHYLPHENETOLE

$C_9H_{12}O$

Mol. Wt. 136.19

B.P. 55-56°C/5mm

Solvent: CCl_4

Assignments

a = 1.36
b = 2.18
c = 3.92
d = 6.52-7.18

1498

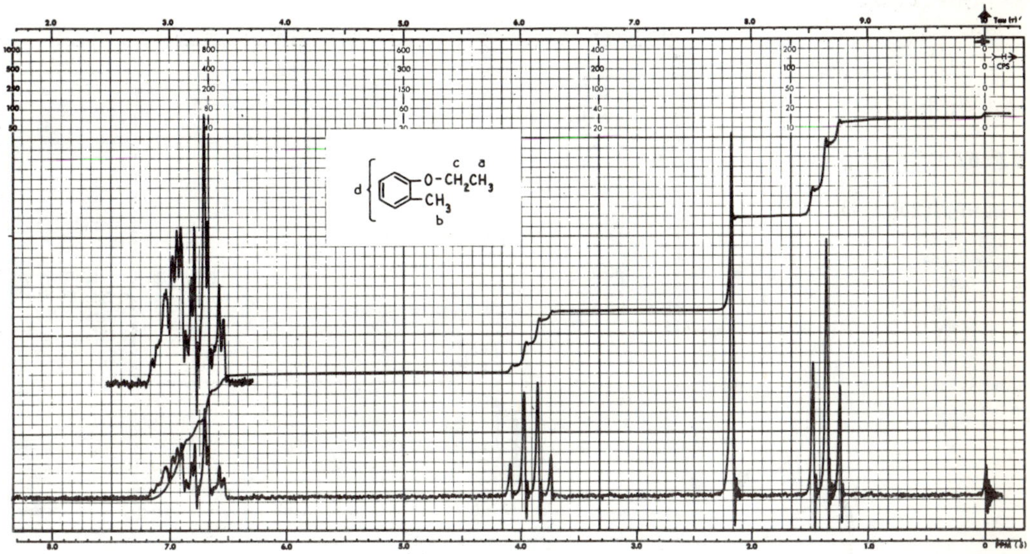

© 1978 Sadtler Research Laboratories, Inc.

m-METHYLPHENETOLE

$C_9H_{12}O$

Mol. Wt. 136.20

B.P. 63-64°C/5mm

Solvent: CCl_4

Assignments

a = 1.30
b = 2.24
c = 3.81
d = 6.50
e = 6.59
f = 7.00

1499

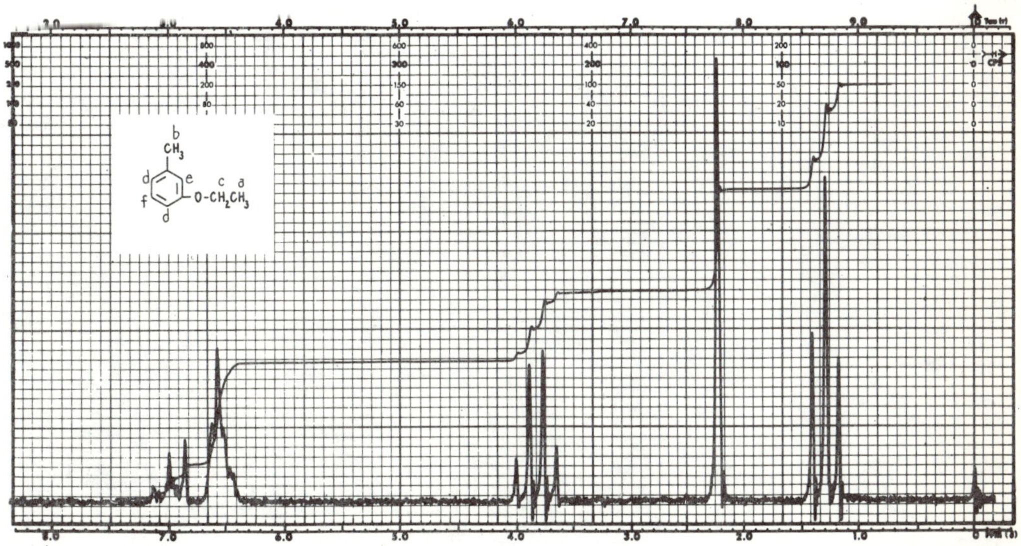

© 1978 Sadtler Research Laboratories, Inc.

p-METHYLPHENETOLE

$C_9H_{12}O$

Mol. Wt. 136.19

B.P. 63-64°C/10mm

Solvent: CCl_4

Assignments
a = 1.36
b = 2.26
c = 3.91
d = 6.65
e = 6.95

1500

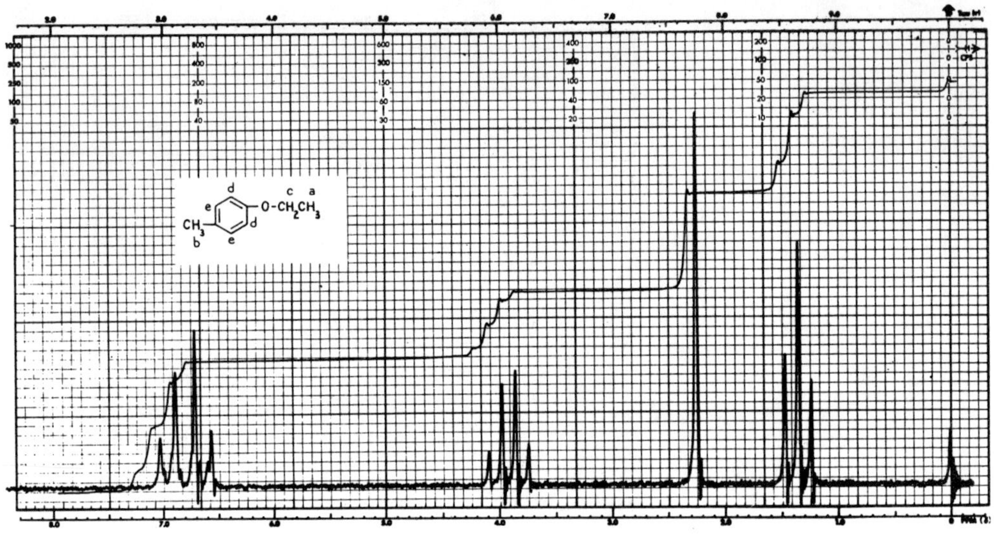

© 1978 Sadtler Research Laboratories, Inc.

p-PROPYLANISOLE

$C_{10}H_{14}O$

Mol. Wt. 150.22

B.P. 67°C/2mm

Solvent: CCl_4

Assignments
a = 0.90
b = 1.10-1.92
c = 2.50
d = 3.64
e = 6.71
f = 6.98
g = 1.28 imp.

1501

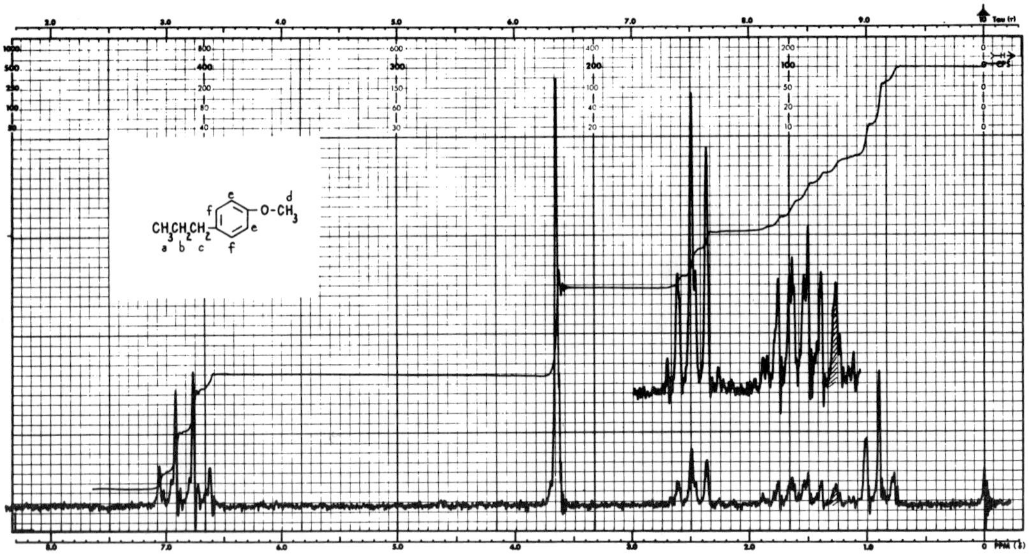

© 1978 Sadtler Research Laboratories, Inc.

4-CHLOROBUTYL m-TOLYL ETHER

$C_{11}H_{15}ClO$

Mol. Wt. 198.69

Solvent: CCl_4

Assignments
a = 1.89
b = 2.31
c = 3.52
d = 3.88
e = 6.40-6.80
f = 7.06

1502

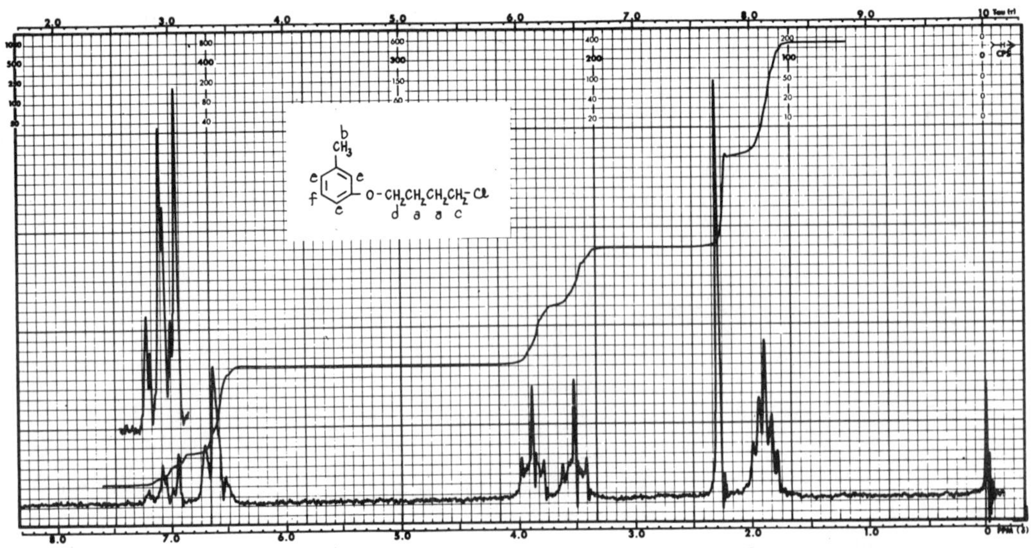

© 1978 Sadtler Research Laboratories, Inc.

p-(2-BROMOETHYL)ANISOLE

$C_9H_{11}BrO$

Mol. Wt. 215.10

Solvent: CCl_4

Assignments
a = 3.03
b = 3.41
c = 3.71
d = 6.71
e = 7.03

1503

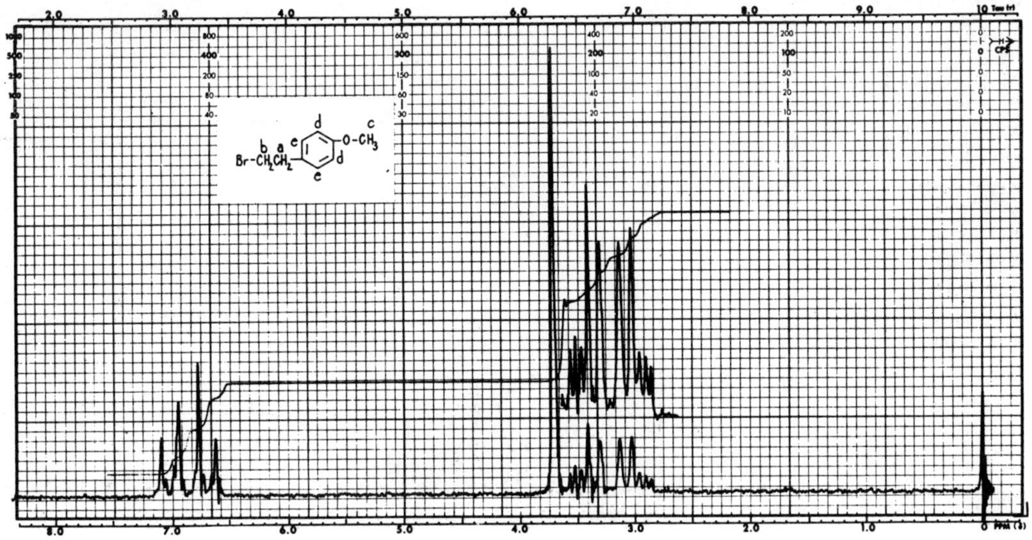

© 1978 Sadtler Research Laboratories, Inc.

---

o-METHOXYBENZYLAMINE

$C_8H_{11}NO$

Mol. Wt. 137.18

Solvent: $CDCl_3$

Assignments
a = 1.72
b = ca 3.88
c = 6.84
d = 6.91
e = 7.19
f = ca 7.23
g = 7.29 imp. ($CHCl_3$)

1504

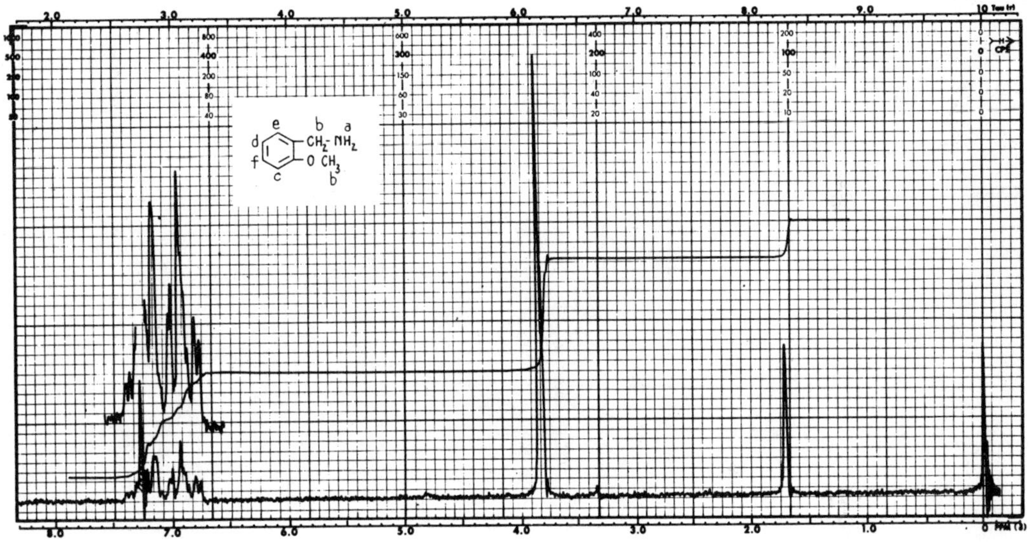

© 1978 Sadtler Research Laboratories, Inc.

---

m-METHOXYBENZYLAMINE

$C_8H_{11}NO$

Mol. Wt. 137.18

Solvent: $CDCl_3$

Assignments
a = 1.50
b = 1.74
c = 3.79
d = ca 6.74
e = 6.81
f = 7.21

1505

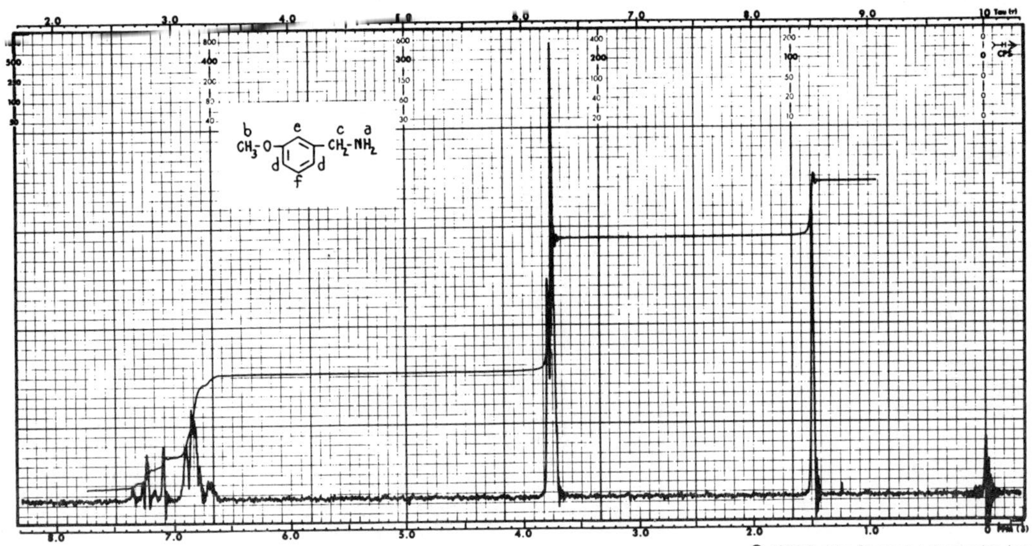

© 1978 Sadtler Research Laboratories, Inc.

p-PROPENYLANISOLE

$C_{10}H_{12}O$

Mol. Wt. 148.21

Solvent: CCl_4

Assignments
a = 1.78
b = 3.61
c = 5.91
d = 6.21
e = 6.65
f = 7.08

1506

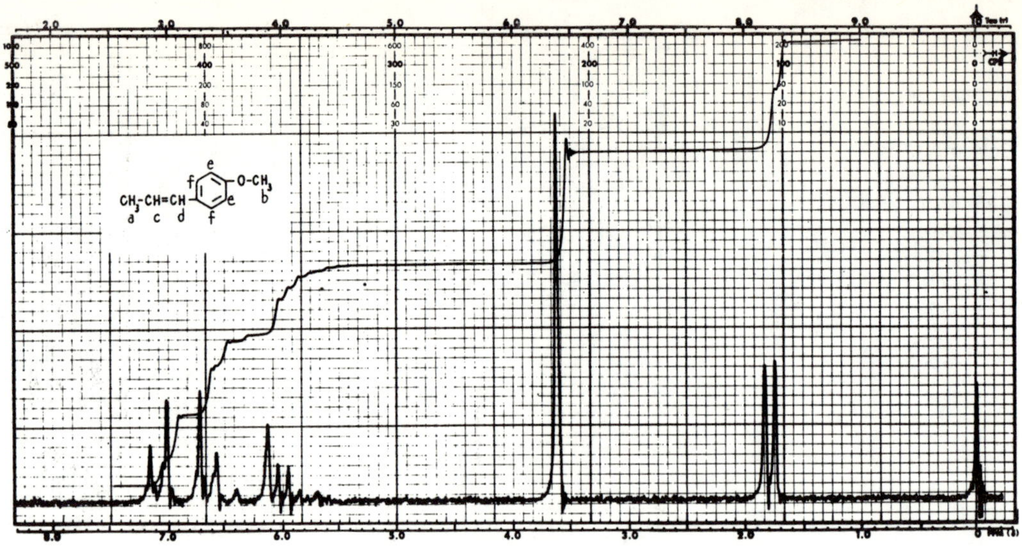

© 1978 Sadtler Research Laboratories, Inc.

---

o-PHENYLANISOLE

$C_{13}H_{12}O$

Mol. Wt. 184.24

M.P. 28-30°C

Solvent: CCl_4

Assignments
a = 3.53
b = 6.75
c = 6.90-7.55

1507

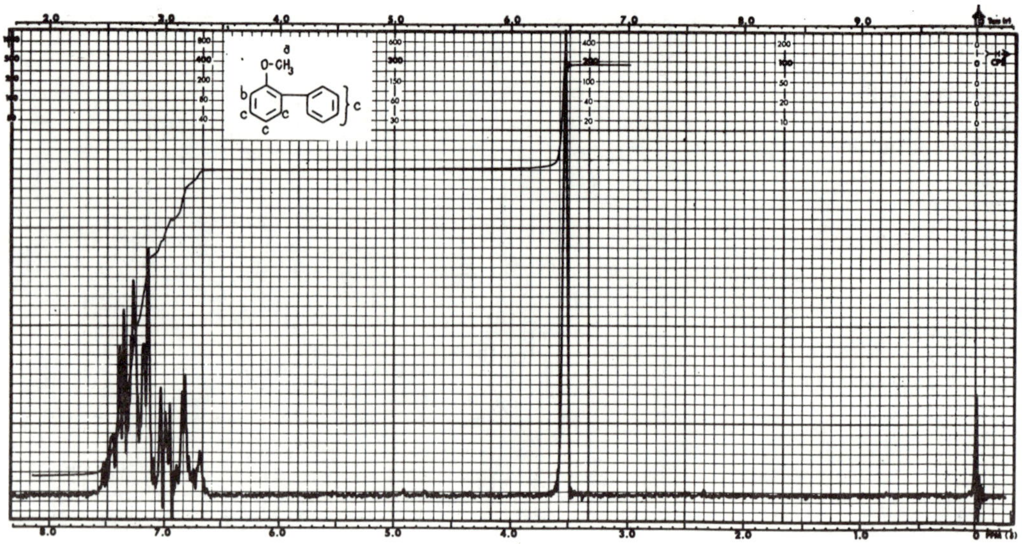

© 1978 Sadtler Research Laboratories, Inc.

---

p-PHENYLANISOLE

$C_{13}H_{12}O$

Mol. Wt. 184.24

M.P. 87-89°C

Solvent: $CDCl_3$

Assignments
a = 3.76
b = 6.93
c = 7.20-7.70
d = 7.49

1508

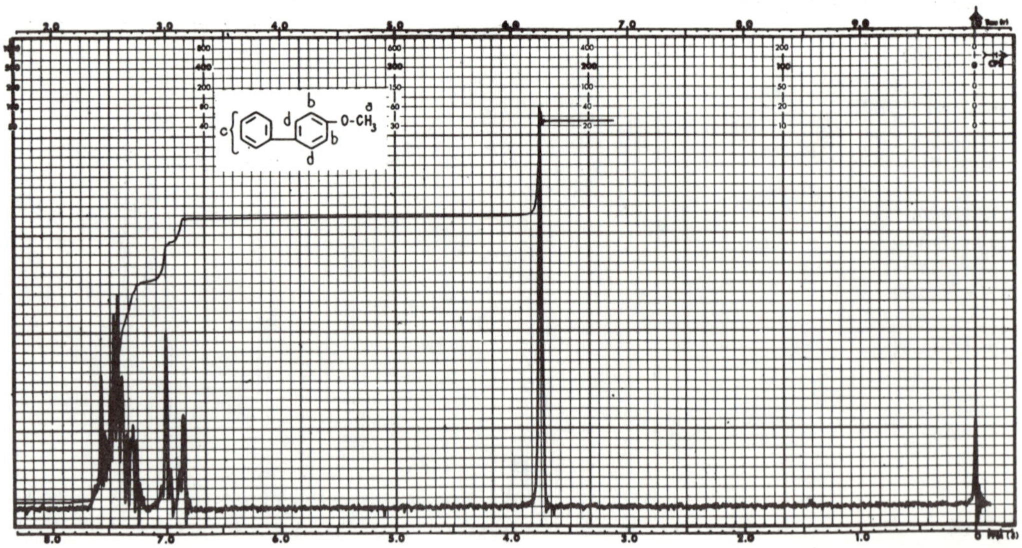

© 1978 Sadtler Research Laboratories, Inc.

o-CHLOROANISOLE

C_7H_7ClO

Mol. Wt. 142.59

B.P. 199.5-200°C

Solvent: $CDCl_3$

Assignments
a = 3.87
b = 6.85
c = 6.91
d = 7.10-7.50

1509

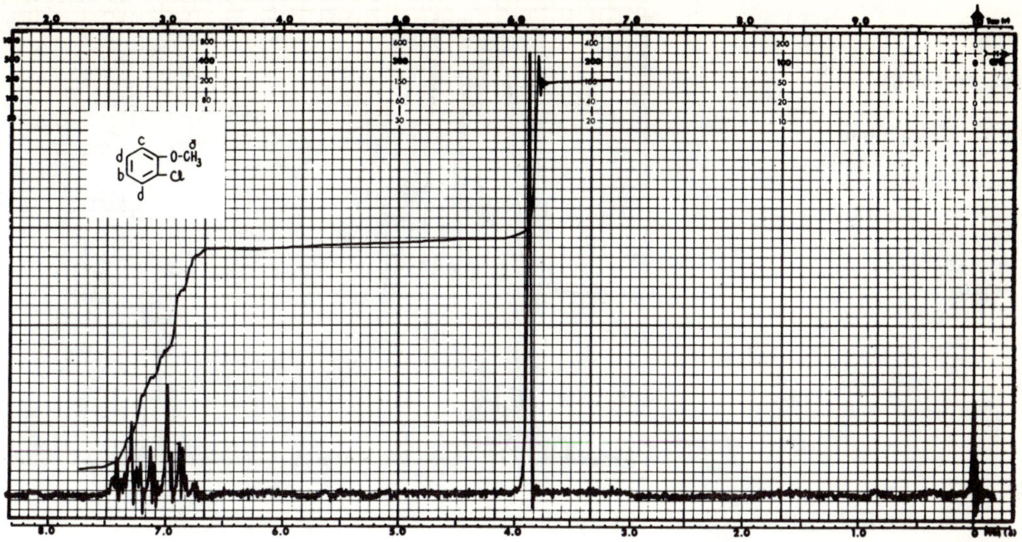

© 1978 Sadtler Research Laboratories, Inc.

---

m-CHLOROANISOLE

C_7H_7ClO

Mol. Wt. 142.59

B.P. 193°C

Solvent: CCl_4

Assignments
a = 3.68
b = 6.55-6.95
c = 7.11

1510

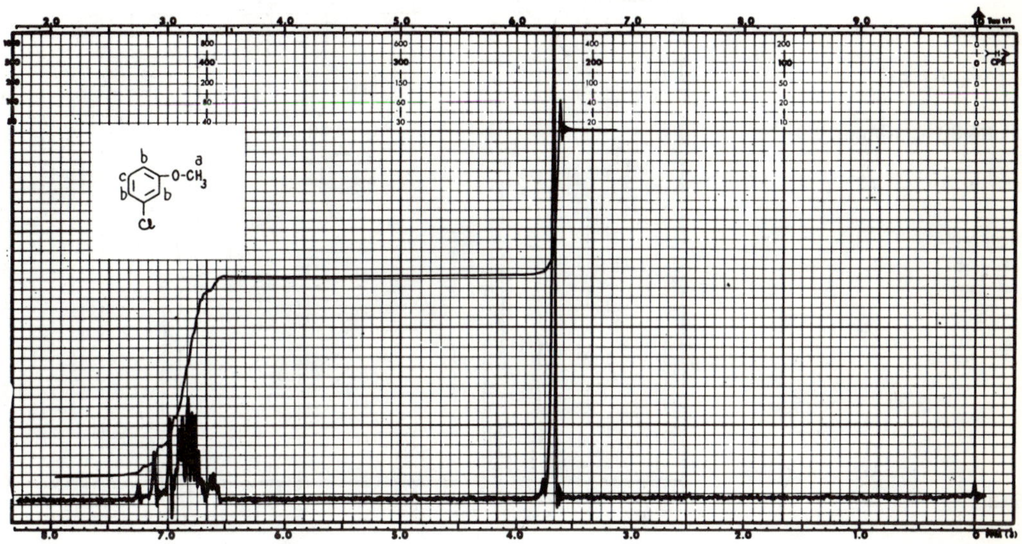

© 1978 Sadtler Research Laboratories, Inc.

---

p-CHLOROANISOLE

C_7H_7ClO

Mol. Wt. 142.59

B.P. 198°C

Solvent: CCl_4

Assignments
a = 3.70
b = 6.72
c = 7.18

1511

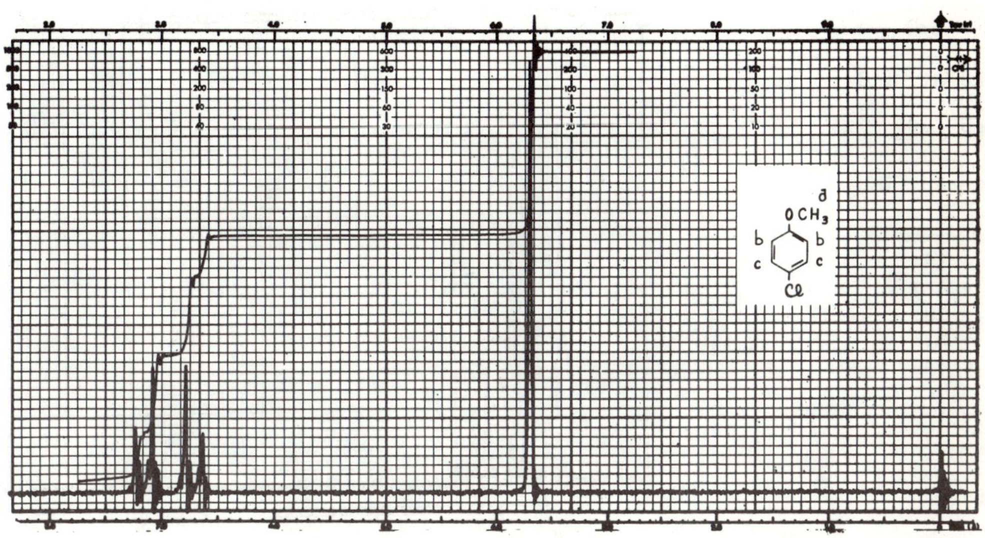

© 1978 Sadtler Research Laboratories, Inc.

o-CHLOROPHENETOLE

C_8H_9ClO

Mol. Wt. 156.61

B.P. 82-83°C/10mm

Solvent: CCl_4

Assignments
a = 1.42
b = 4.02
c = 6.60-6.90
d = 7.00
e = 7.24

1512

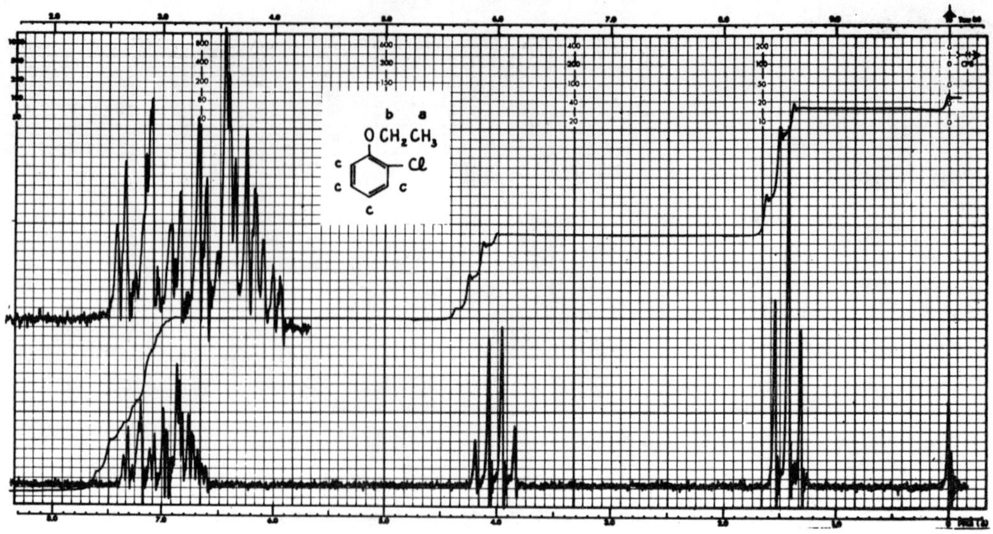

© 1978 Sadtler Research Laboratories, Inc.

---

p-CHLOROPHENETOLE

C_8H_9ClO

Mol. Wt. 156.61

B.P. 84-86°C/10mm

Solvent: CCl_4

Assignments
a = 1.35
b = 3.90
c = 6.68
d = 7.10

1513

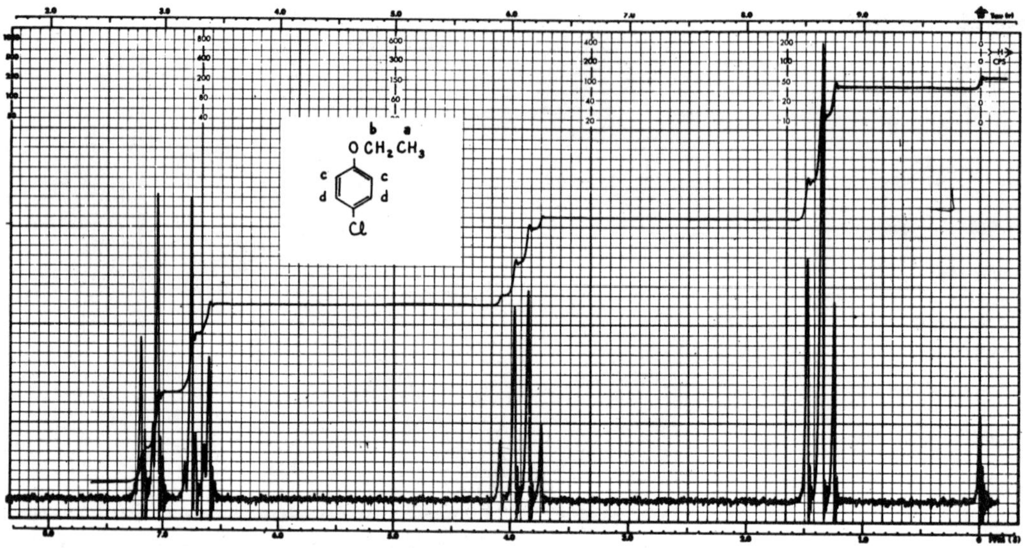

© 1978 Sadtler Research Laboratories, Inc.

---

o-BROMOANISOLE

C_7H_7BrO

Mol. Wt. 187.04

M.P. 2-4°C

Solvent: $CDCl_3$

Assignments
a = 3.84
b = 6.80
c = 6.88
d = 7.24
e = 7.49

1514

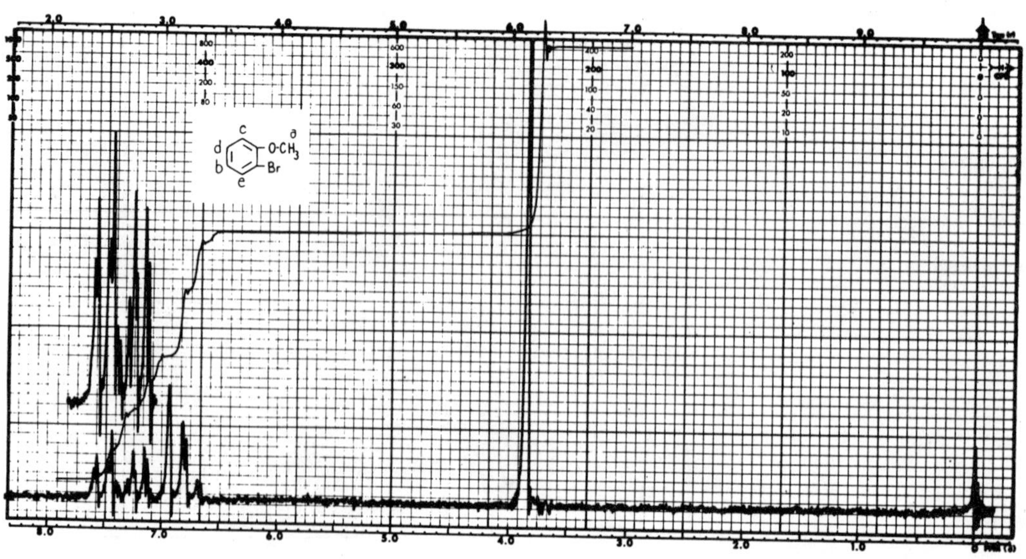

© 1978 Sadtler Research Laboratories, Inc.

m-BROMOANISOLE

C_7H_7BrO

Mol. Wt. 187.04

Solvent: $CDCl_3$

Assignments
a = 3.69
b = 6.77
c = 6.90-7.20

1515

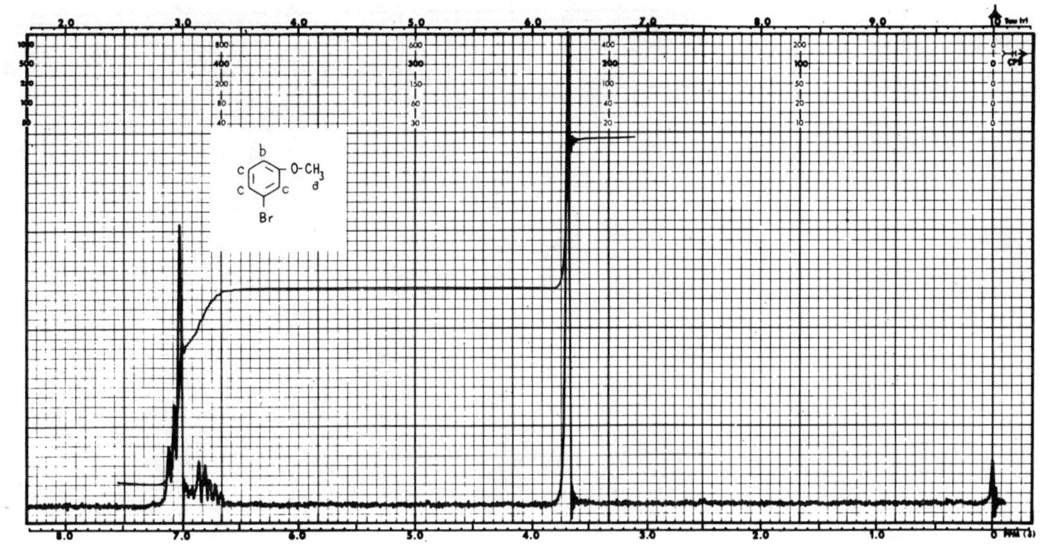

© 1978 Sadtler Research Laboratories, Inc.

---

p-BROMOANISOLE

C_7H_7BrO

Mol. Wt. 187.04

M.P. 12-13°C

Solvent: CCl_4

Assignments
a = 3.71
b = 6.69
c = 7.29

1516

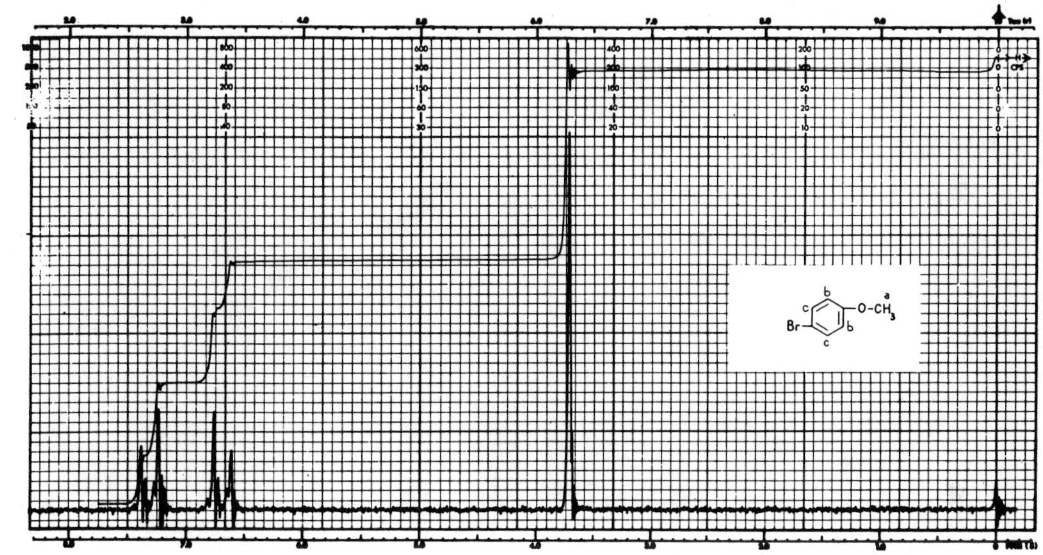

© 1978 Sadtler Research Laboratories, Inc.

---

o-BROMOPHENETOLE

C_8H_9BrO

Mol. Wt. 201.07

B.P. 122-125°C/22mm

Solvent: CCl_4

Assignments
a = 1.40
b = 3.99
c = 6.71
d = 6.76
e = 7.13
f = 7.46

1517

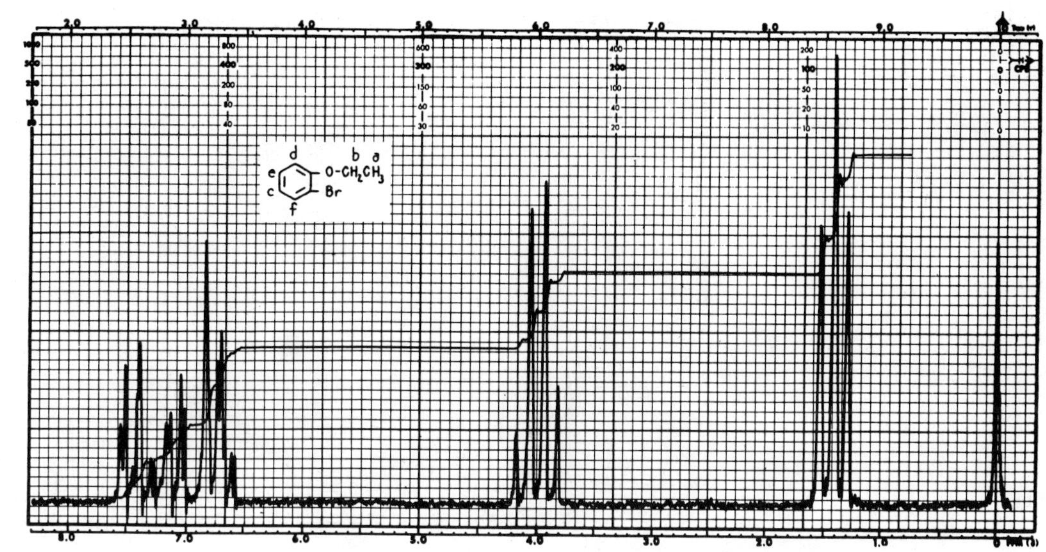

© 1978 Sadtler Research Laboratories, Inc.

p-BROMOPHENETOLE

C_8H_9BrO

Mol. Wt. 201.07

M.P. 10-12°C

Solvent: $CDCl_3$

Assignments
a = 1.35
b = 3.90
c = 6.71
d = 7.31

1518

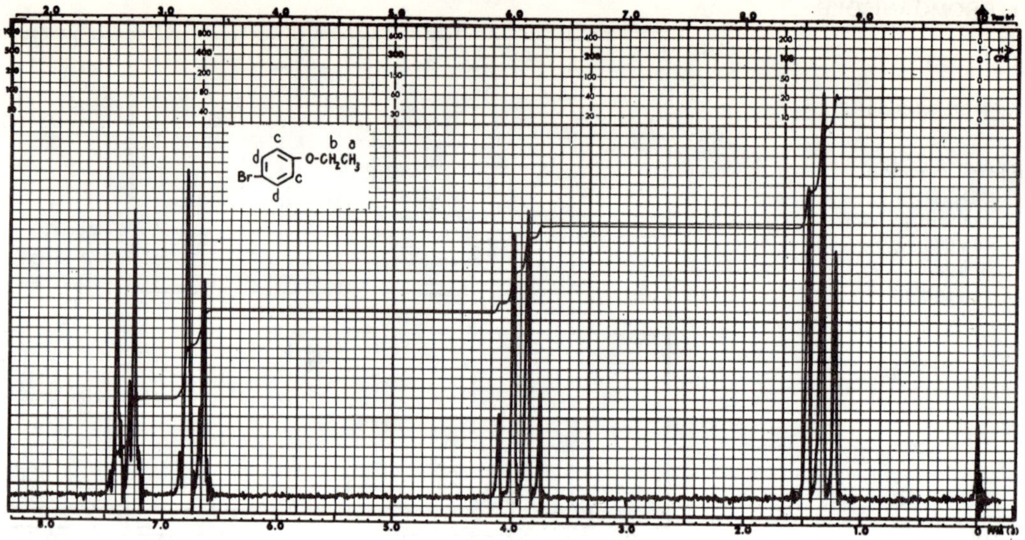

© 1978 Sadtler Research Laboratories, Inc.

---

o-IODOANISOLE

C_7H_7IO

Mol. Wt. 234.04

B.P. 239-241°C

Solvent: $CDCl_3$

Assignments
a = 3.71
b = 6.65
c = 6.70
d = 7.20
e = 7.71

1519

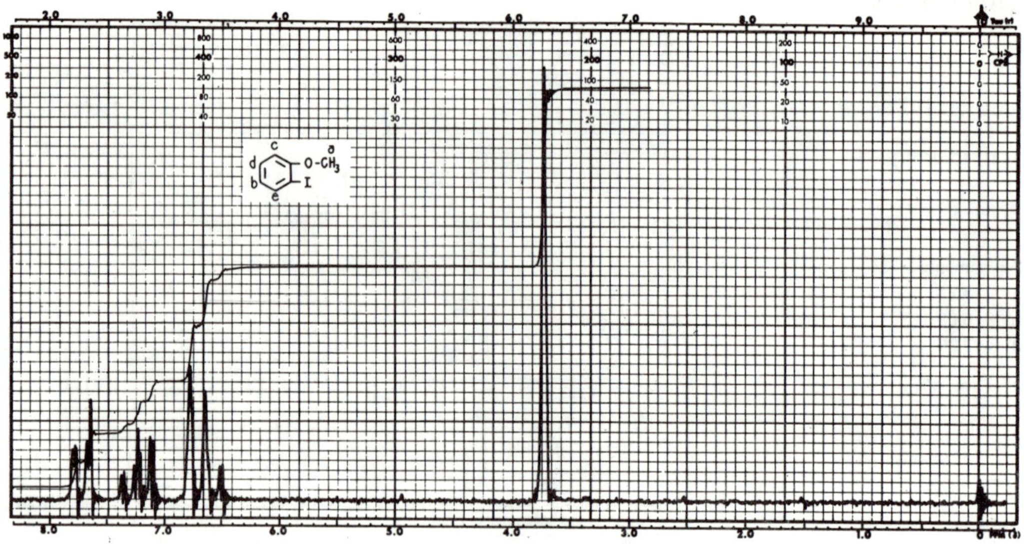

© 1978 Sadtler Research Laboratories, Inc.

---

m-IODOANISOLE

C_7H_7IO

Mol. Wt. 234.04

B.P. 109-110°C/10mm

Solvent: CCl_4

Assignments
a = 3.70
b = 6.79
c = 6.91
d = 7.19
e = 7.21

1520

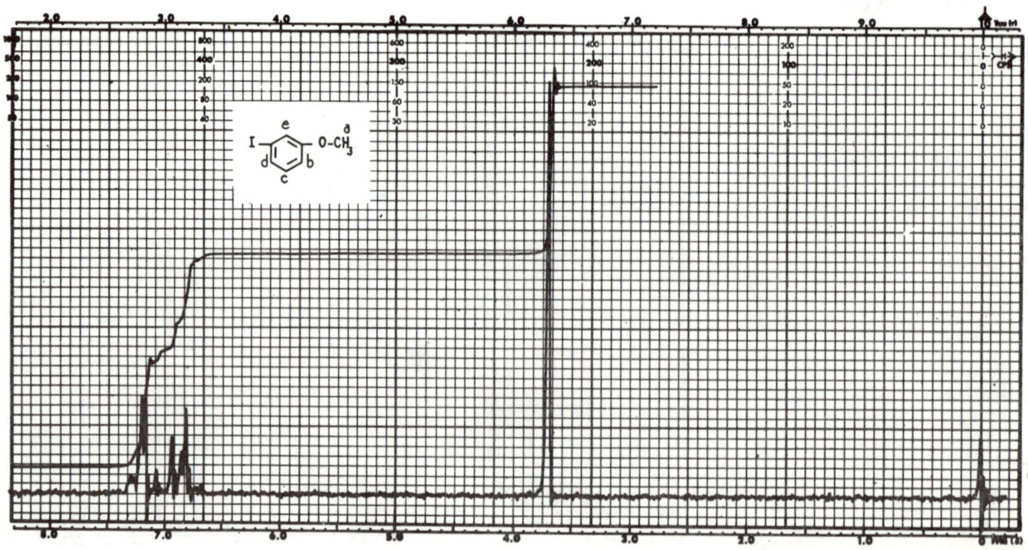

© 1978 Sadtler Research Laboratories, Inc.

p-IODOANISOLE

C_7H_7IO

Mol. Wt. 234.04

M.P. 51-52°C

Solvent: $CDCl_3$

Assignments
a = 3.74
b = 6.68
c = 7.53

1521

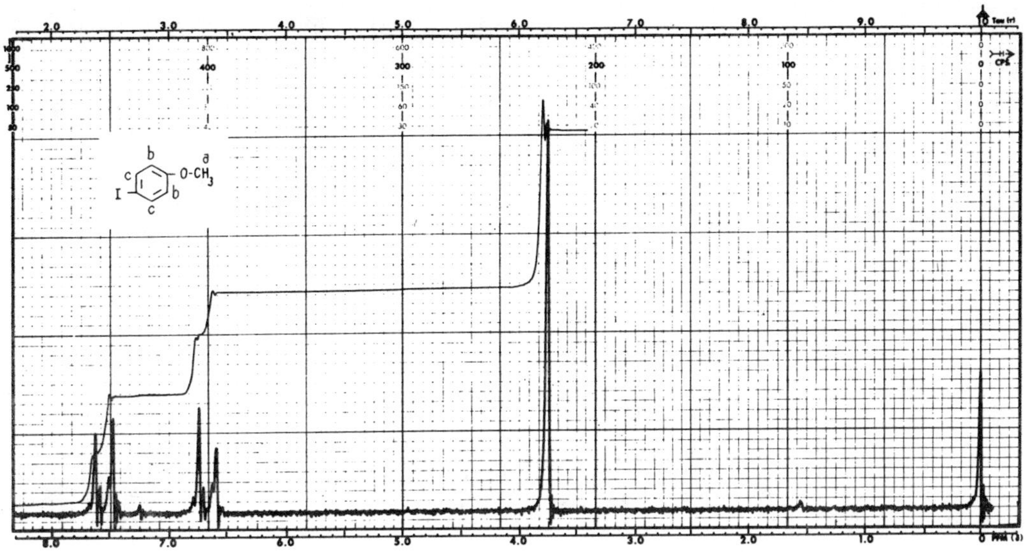

© 1978 Sadtler Research Laboratories, Inc.

m-PHENETIDINE

$C_8H_{11}NO$

Mol. Wt. 137.18

B.P. 124-126°C/7mm

Solvent: $CDCl_3$

Assignments
a = 1.29
b = 3.62
c = 3.82
d = ca 6.20
e = 6.98

1522

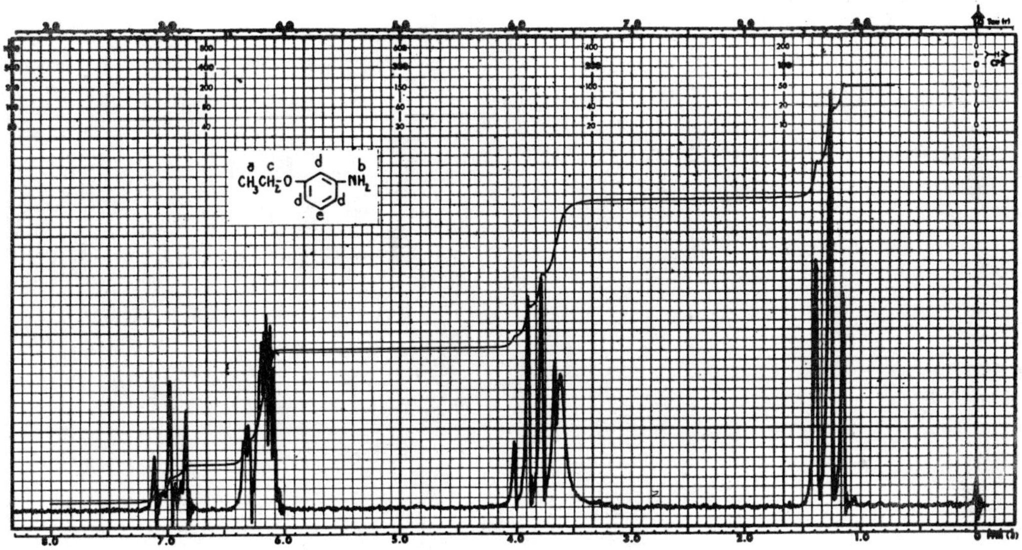

© 1978 Sadtler Research Laboratories, Inc.

N-METHYL-p-ANISIDINE

$C_8H_{11}NO$

Mol. Wt. 137.18

Solvent: Polysol-d

Assignments
a = 2.63
b = 3.61
c = 4.47
d = 6.45
e = 6.69

1523

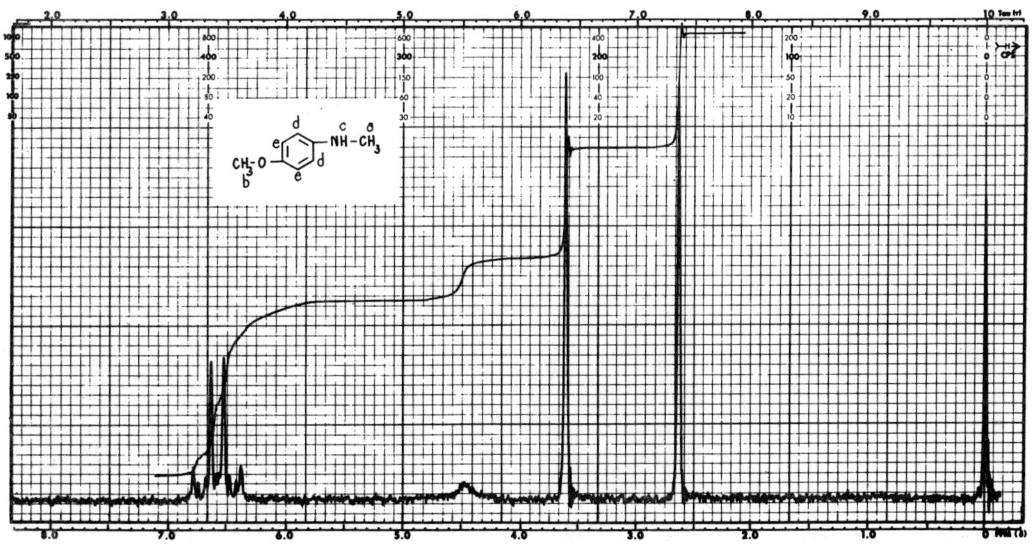

© 1978 Sadtler Research Laboratories, Inc.

N,N-DIMETHYL-p-ANISIDINE

$C_9H_{13}NO$

Mol. Wt. 151.21

M.P. 47°C

Solvent: $CDCl_3$

Assignments
a = 2.85
b = 3.75
c = 6.79

1524

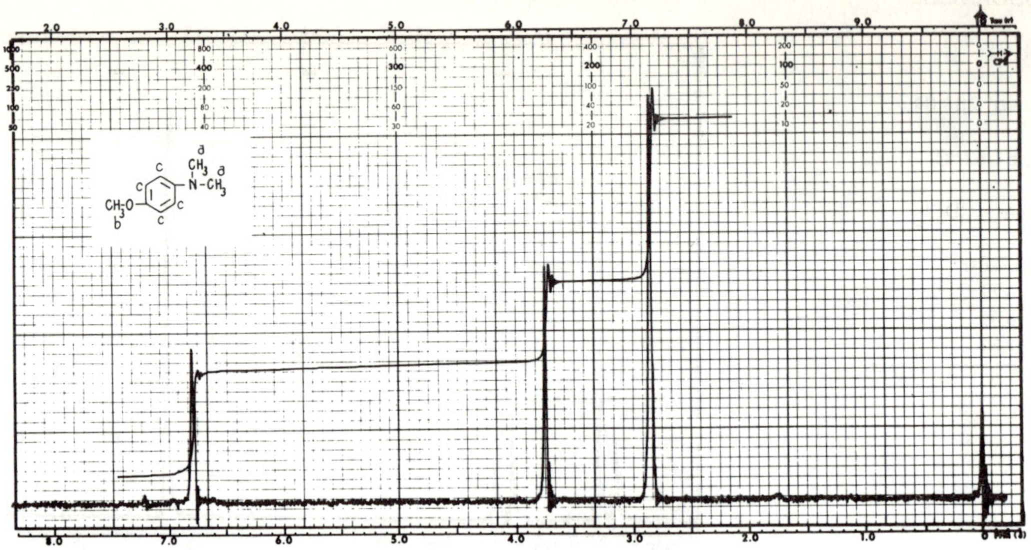

© 1978 Sadtler Research Laboratories, Inc.

N,N-DIETHYL-m-PHENETIDINE

$C_{12}H_{19}NO$

Mol. Wt. 193.29

B.P. 145-147°C/13mm

Solvent: CCl_4

Assignments
a = 1.13
b = 1.38
c = 3.29
d = 3.94
e = 6.00-6.30
f = 6.98

1525

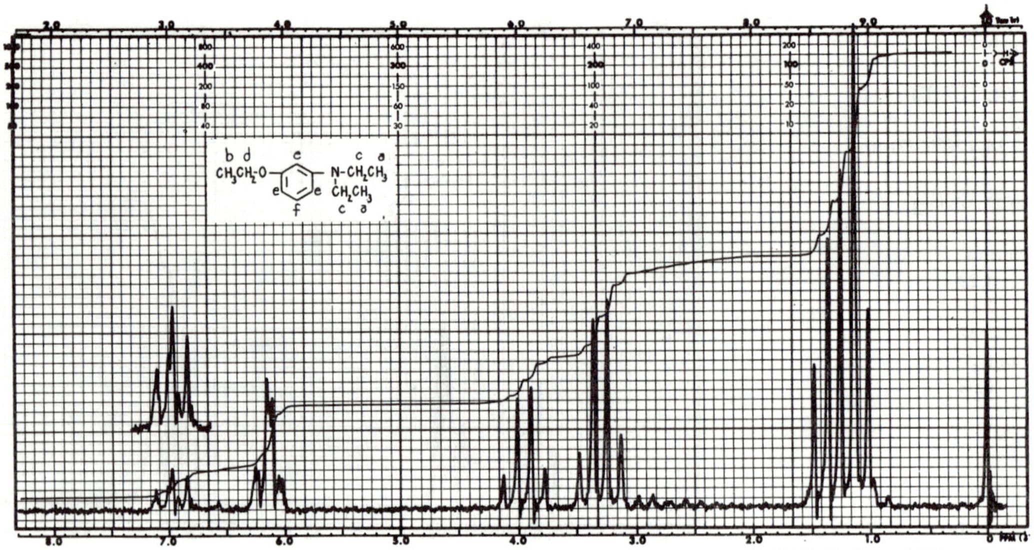

© 1978 Sadtler Research Laboratories, Inc.

o-ANISIDINE, HYDROCHLORIDE

$C_7H_9NO \cdot HCl$

Mol. Wt. 159.62

Solvent: Polysol-d

Assignments
a = 3.97
b = 7.00
c = 7.12
d = 7.39
e = 7.65
f = 9.80

1526

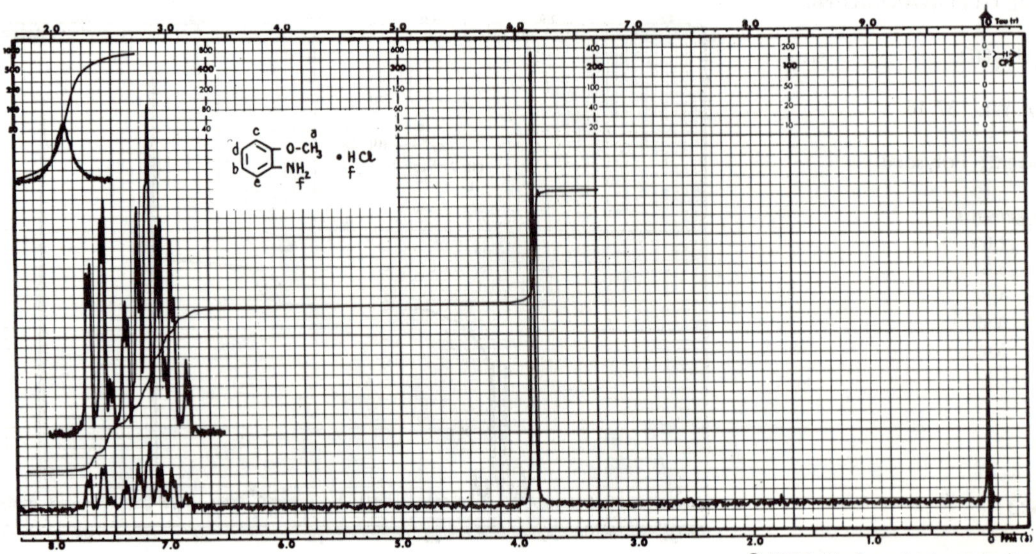

© 1978 Sadtler Research Laboratories, Inc.

(o-METHOXYPHENYL)HYDRAZINE, MONOHYDROCHLORIDE

$C_7H_{10}N_2O \cdot HCl$

Mol. Wt. 174.63

Solvent: Polysol-d

Assignments
a = 3.84
b = 6.65-7.05
c = 7.23
d = 7.40

1527

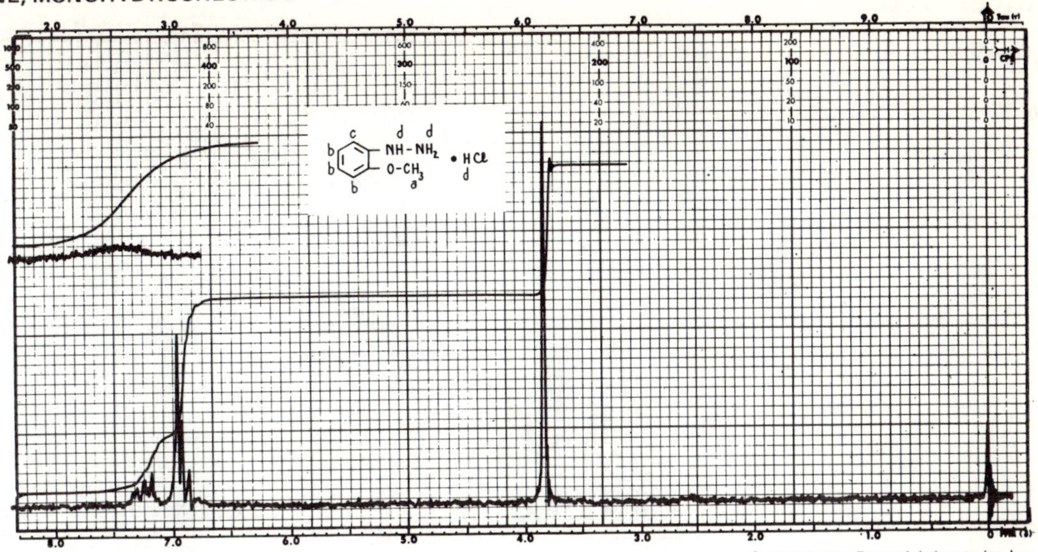

© 1978 Sadtler Research Laboratories, Inc.

N-(-METHOXYBENZYLIDENE)ANILINE

$C_{14}H_{13}NO$

Mol. Wt. 211.27

M.P. 63-64°C

Solvent: $CDCl_3$

Assignments
a = 3.65
b = 6.81
c = 7.00-7.40
d = 7.73
e = 8.23

1528

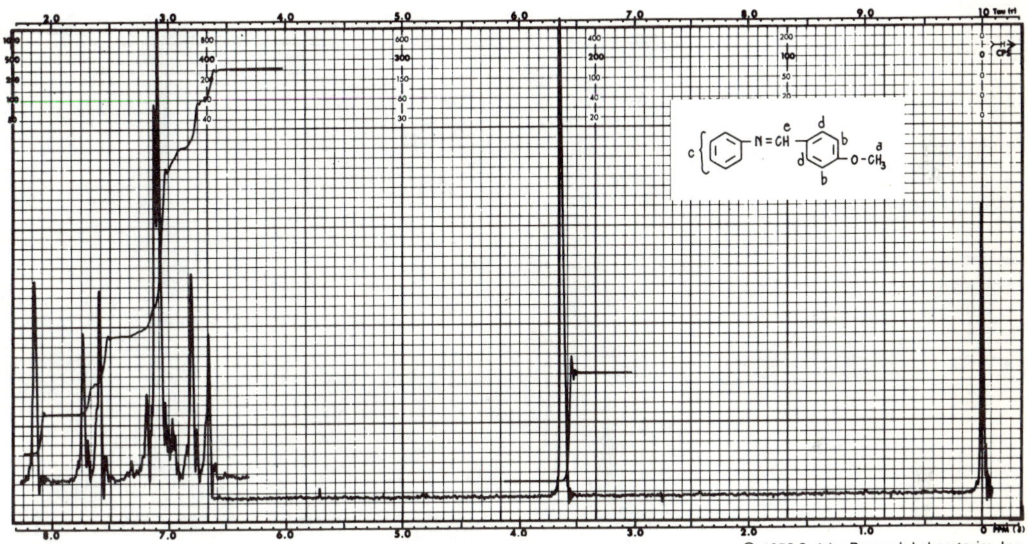

© 1978 Sadtler Research Laboratories, Inc.

o-ANISALDEHYDE, OXIME

$C_8H_9NO_2$

Mol. Wt. 151.17

Solvent: $CDCl_3$

Assignments
a = 3.82
b = 6.87
c = 6.93
d = 7.31
e = 7.64
f = 8.49
g = 9.34

1529

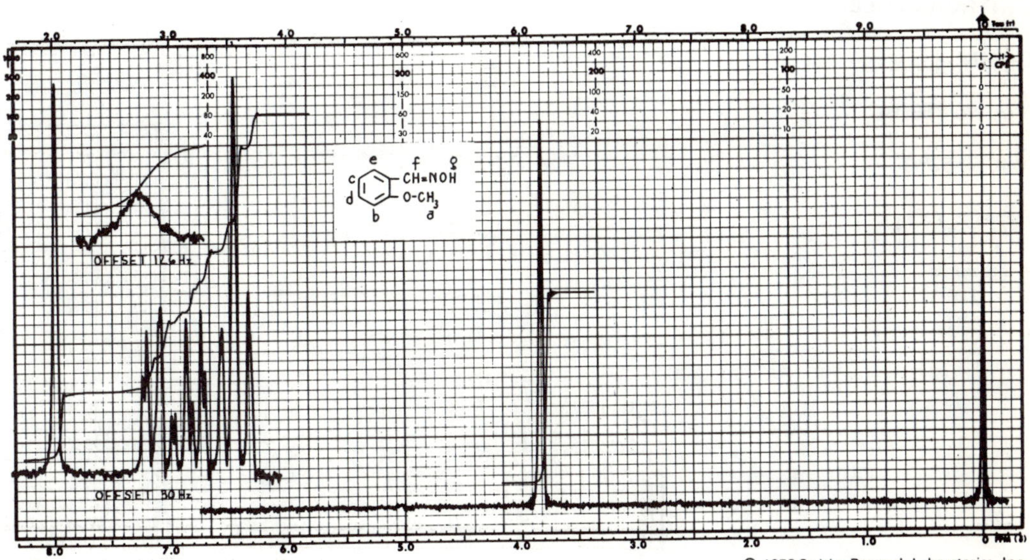

© 1978 Sadtler Research Laboratories, Inc.

2-NITROANISOLE

$C_7H_7NO_3$

Mol. Wt. 153.14

M.P. 9-10°C

Solvent: CCl_4

Assignments
a = 3.89
b = 6.83-7.85

1530

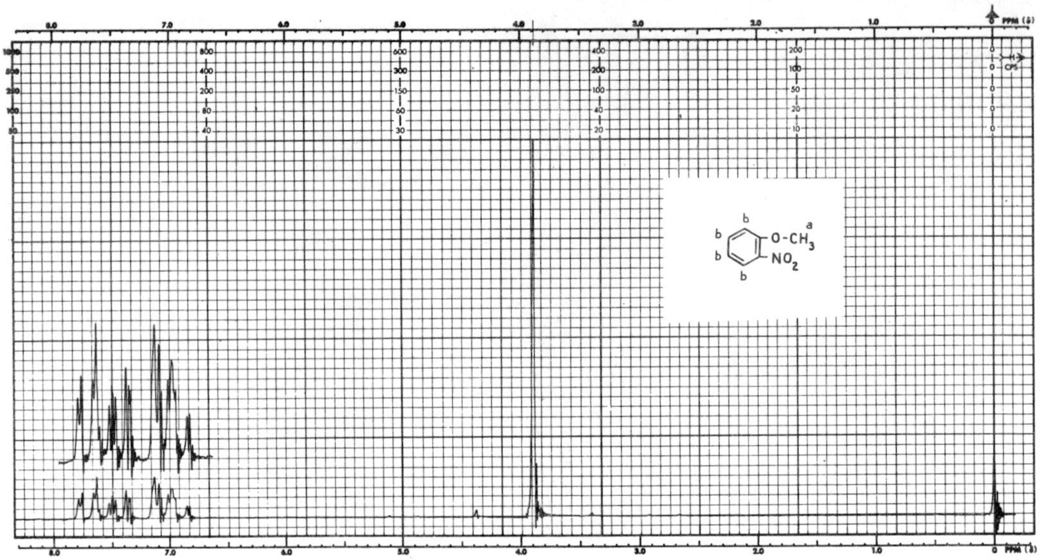

© 1978 Sadtler Research Laboratories, Inc.

---

m-NITROANISOLE

$C_7H_7NO_3$

Mol. Wt. 153.14

M.P. 36-38°C

Solvent: $CDCl_3$

Assignments
a = 3.86
b = 7.21
c = 7.42
d = 7.69
e = 7.76

1531

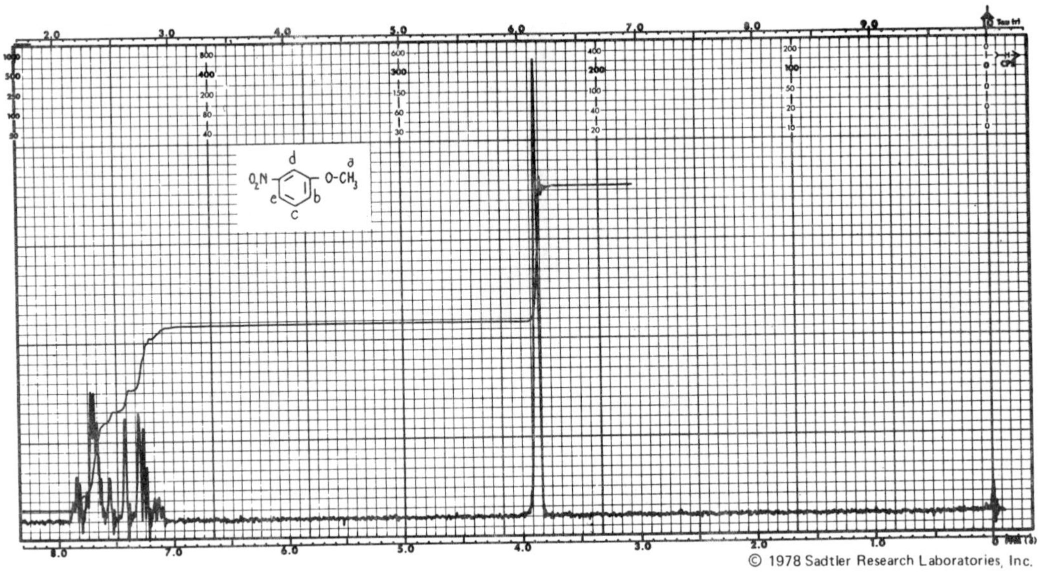

© 1978 Sadtler Research Laboratories, Inc.

---

p-NITROANISOLE

$C_7H_7NO_3$

Mol. Wt. 153.14

M.P. 52-53°C

Solvent: CCl_4

Assignments
a = 3.89
b = 6.91
c = 8.12

1532

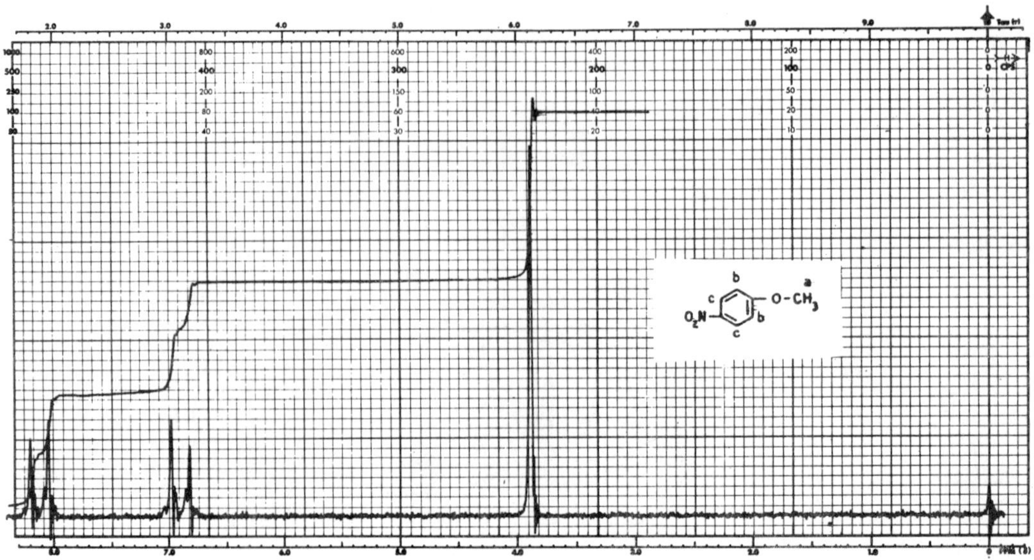

© 1978 Sadtler Research Laboratories, Inc.

p-NITROPHENETOLE

$C_8H_9NO_3$

Mol. Wt. 167.17

M.P. 57-58°C

Solvent: $CDCl_3$

Assignments
a = 1.42
b = 4.11
c = 6.91
d = 8.10

1533

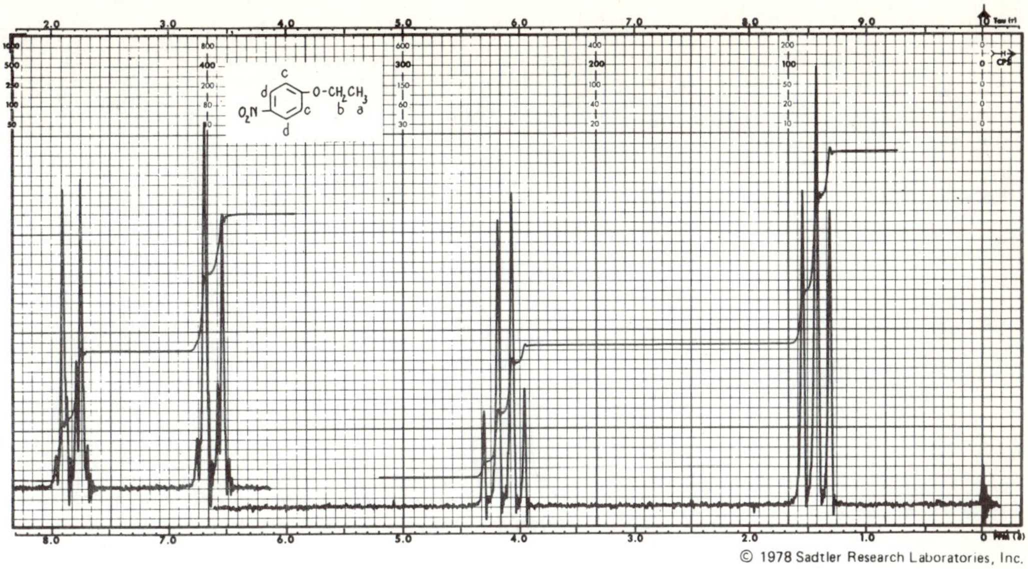

© 1978 Sadtler Research Laboratories, Inc.

1-BUTOXY-2-NITROBENZENE

$C_{10}H_{13}NO_3$

Mol. Wt. 195.22

Solvent: CCl_4

Assignments
a = 0.96
b = 1.19-2.08
c = 4.06
d = 6.96
e = 7.03
f = 7.46
g = 7.70

1534

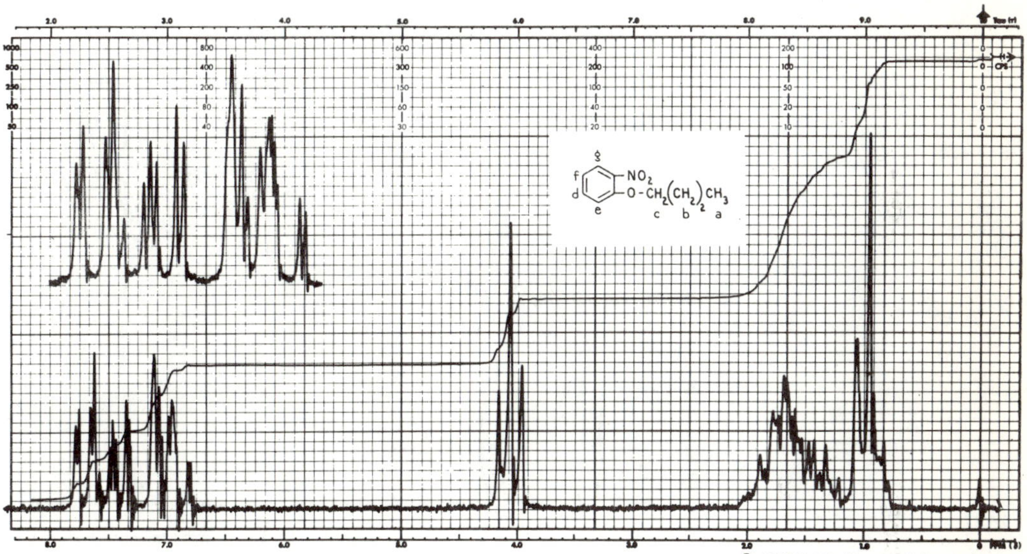

© 1978 Sadtler Research Laboratories, Inc.

m-METHOXYBENZENETHIOL

C_7H_8OS

Mol. Wt. 140.15

Solvent: $CDCl_3$

Assignments
a = 3.43
b = 3.71
c = 6.55-6.95
d = 7.10

1535

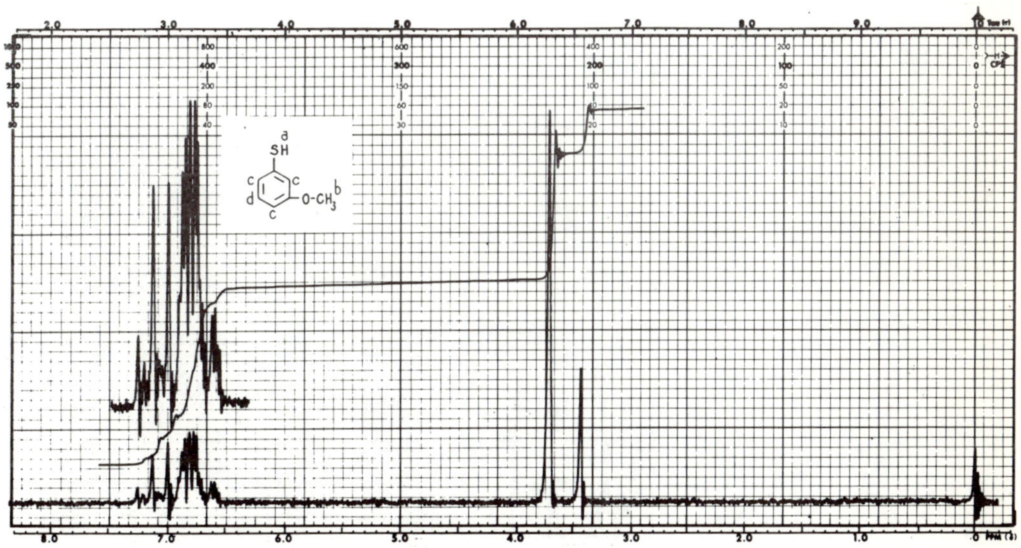

© 1978 Sadtler Research Laboratories, Inc.

p-METHOXYBENZENETHIOL

C_7H_8OS

Mol. Wt. 140.21

B.P. 102-104°C/13mm (lit.)

Solvent: $CDCl_3$

Assignments
a = 3.37
b = 3.76
c = 6.81
d = 7.30

1536

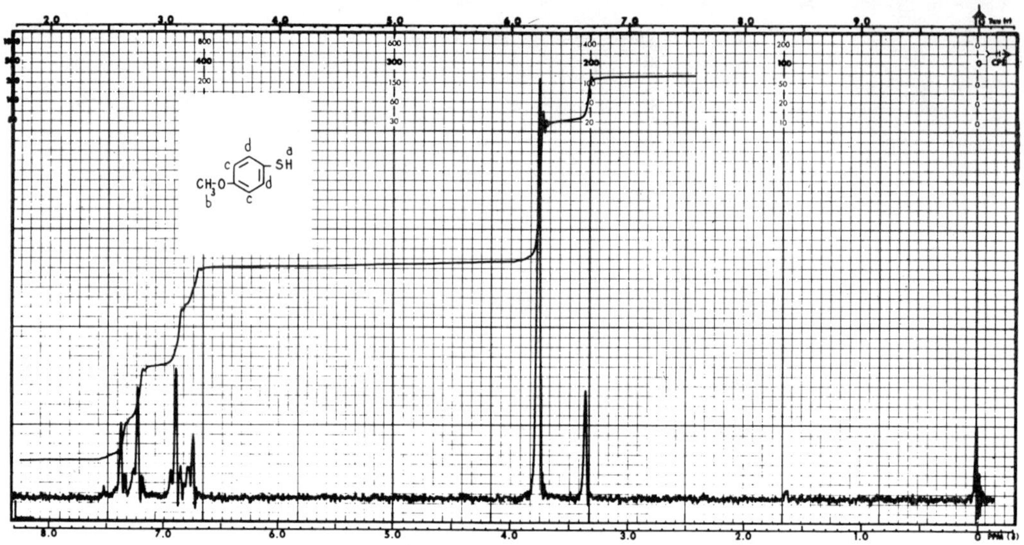

© 1978 Sadtler Research Laboratories, Inc.

4,4′-SULFONYL DIANISOLE

$C_{14}H_{14}O_4S$

Mol. Wt. 278.33

Solvent: $CDCl_3$

Assignments
a = 3.79
b = 6.91
c = 7.79
d = 1.80 imp.

1537

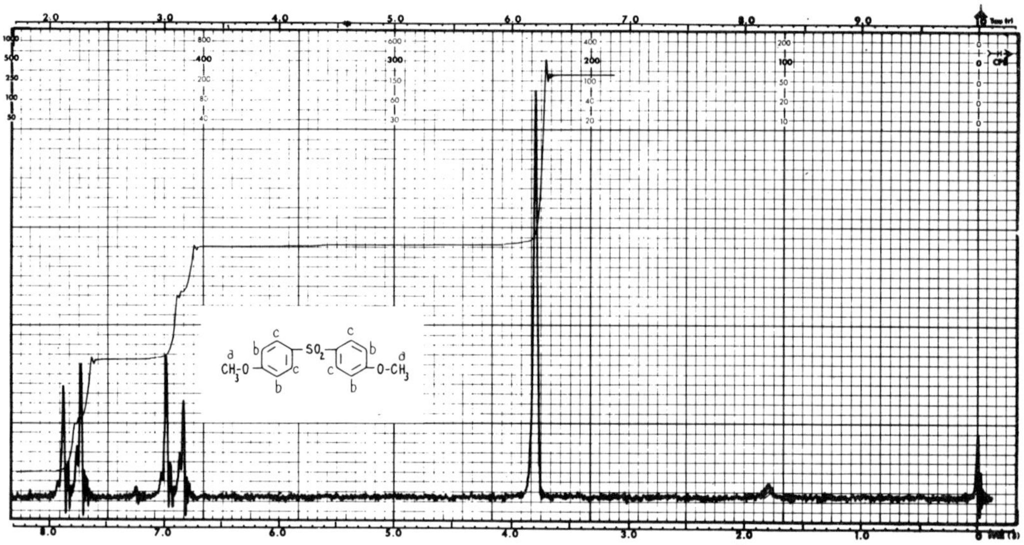

© 1978 Sadtler Research Laboratories, Inc.

p-METHOXYBENZENESULFONYL CHLORIDE

$C_7H_7ClO_3S$

Mol. Wt. 206.65

M.P. 40-43°C

Solvent: $CDCl_3$

Assignments
a = 3.92
b = 7.07
c = 7.98

1538

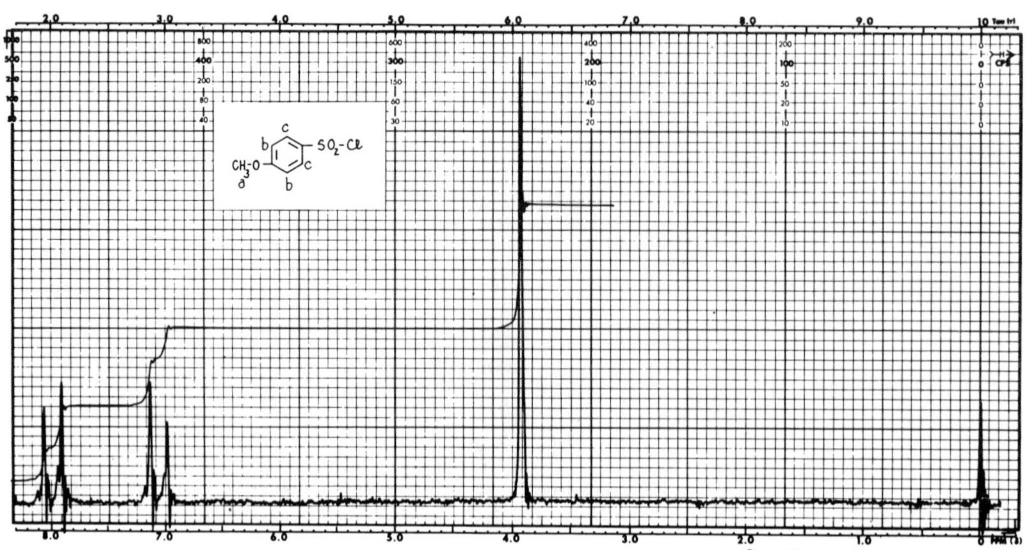

© 1978 Sadtler Research Laboratories, Inc.

2,2-BIS[p-(ALLYLOXY)PHENYL] PROPANE

$C_{21}H_{24}O_2$

Mol. Wt. 308.42

Solvent: $CDCl_3$

Assignments
a = 1.63
b = 4.48
c = 5.22
d = 5.35
e = 6.06
f = 6.81
g = 7.13

1539

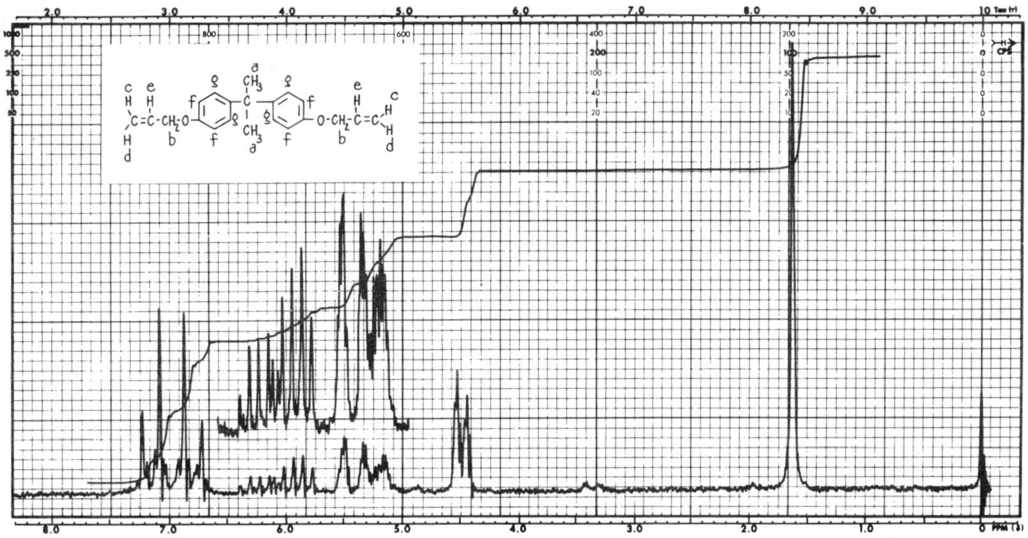

© 1978 Sadtler Research Laboratories, Inc.

1,2-EPOXY-3-(m-TOLYLOXY)PROPANE

$C_{10}H_{12}O_2$

Mol. Wt. 164.20

B.P. 122-123°C/8mm (lit.)

Solvent: CCl_4

Assignments
a = 2.31
b = 2.65
c = 3.15
d = 3.95
e = 6.45-7.25

1540

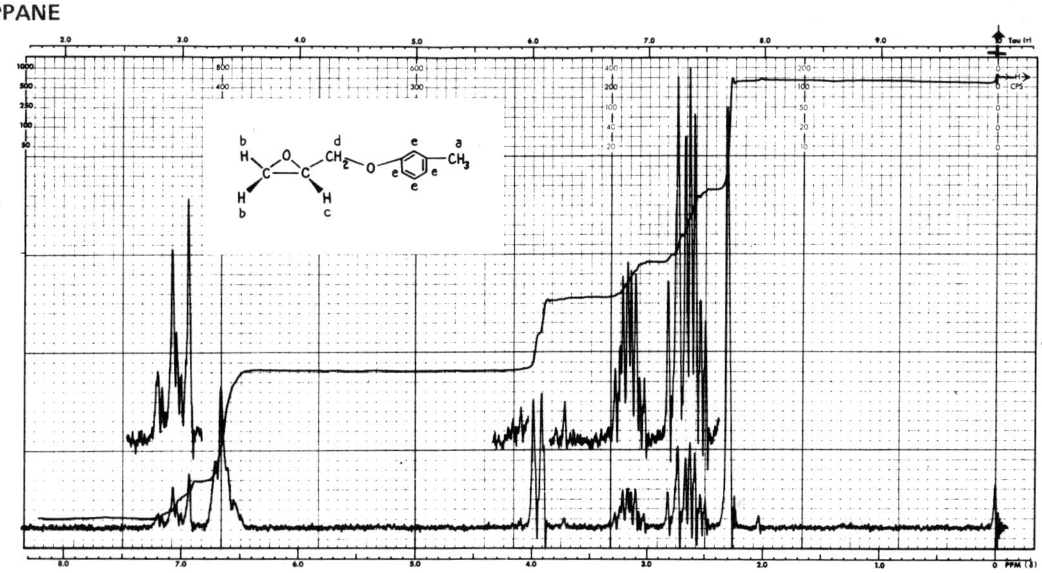

© 1978 Sadtler Research Laboratories, Inc.

PHENYL ETHER

$C_{12}H_{10}O$

Mol. Wt. 170.21

M.P. 25-27°C

Solvent: $CDCl_3$

Assignments
a = 6.78-7.40

1541

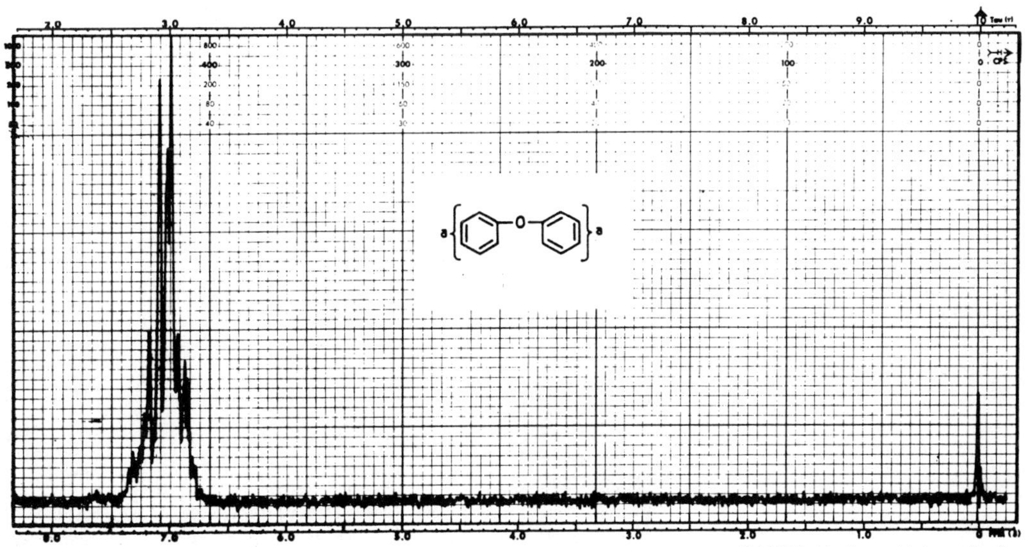

© 1978 Sadtler Research Laboratories, Inc.

p-BROMOPHENYL PHENYL ETHER

$C_{12}H_9BrO$

Mol. Wt. 249.11

B.P. 142-145°C/5mm

Solvent: CCl_4

Assignments
a = 6.75
b = 6.80-7.30
c = 7.29

1542

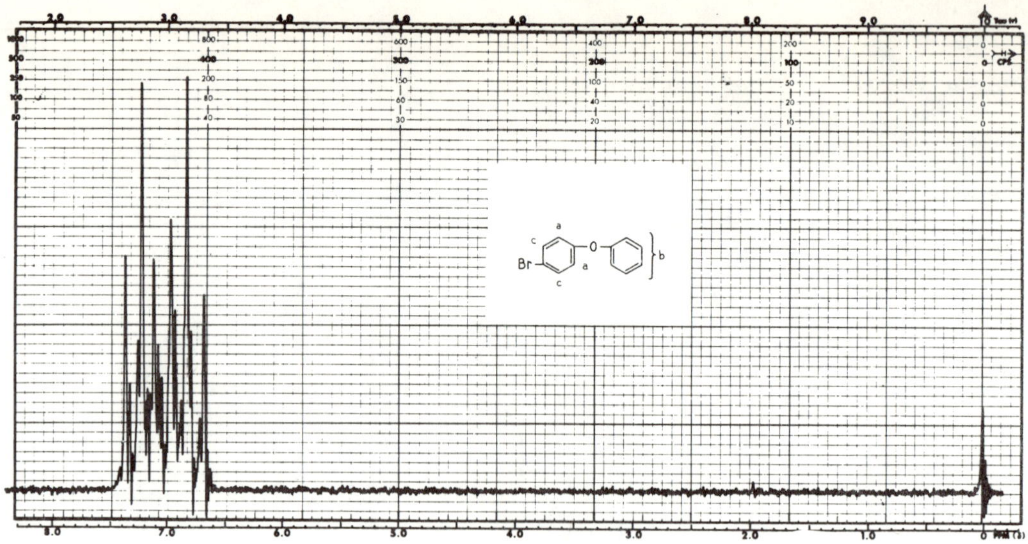

© 1978 Sadtler Research Laboratories, Inc.

o-NITROPHENYL PHENYL ETHER

$C_{12}H_9NO_3$

Mol. Wt. 215.21

M.P. 55-58°C

Solvent: $CDCl_3$

Assignments
a = 6.80-7.60
b = 7.86

1543

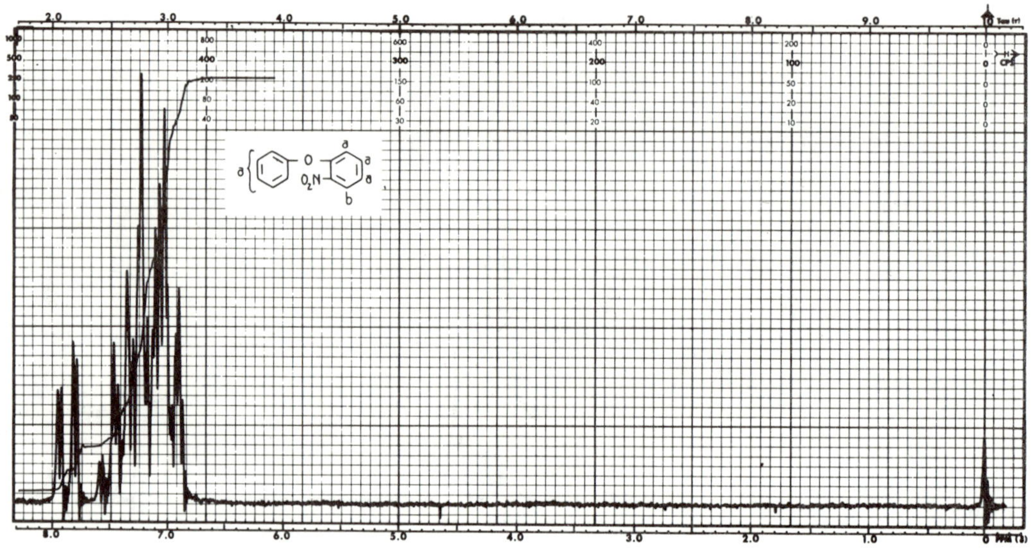

© 1978 Sadtler Research Laboratories, Inc.

m-NITROPHENYL PHENYL ETHER

$C_{12}H_9NO_3$

Mol. Wt. 215.21

B.P. 194-196°C/11mm

Solvent: $CDCl_3$

Assignments
a = 6.90-7.65
b = 7.81
c = 7.90

1544

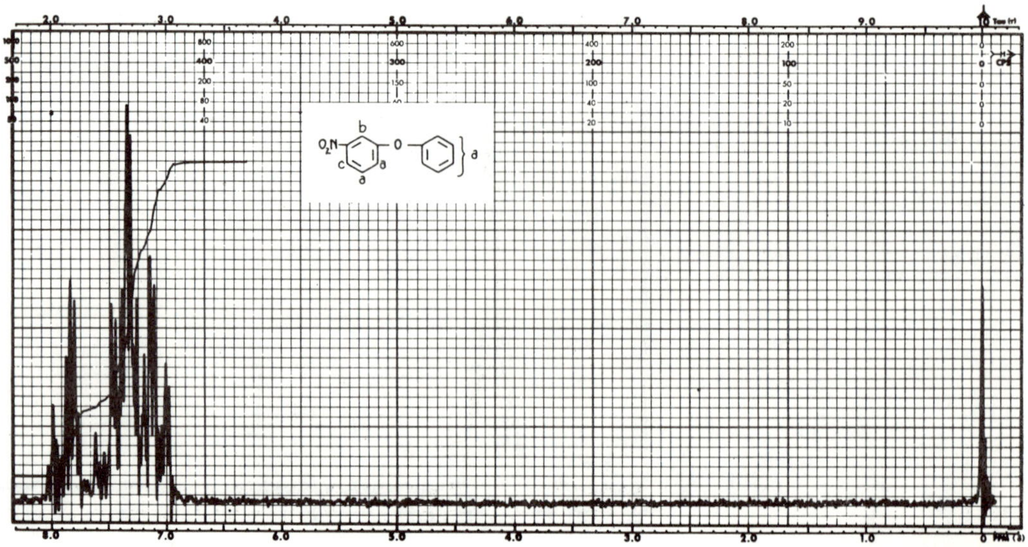

© 1978 Sadtler Research Laboratories, Inc.

VERTATROLE

$C_8H_{10}O_2$

Mol. Wt. 138.17

M.P. 22.5^{o}C

B.P. 207°C (lit.)

Solvent: CCl_4

Assignments
a = 3.72
b = 6.76

1545

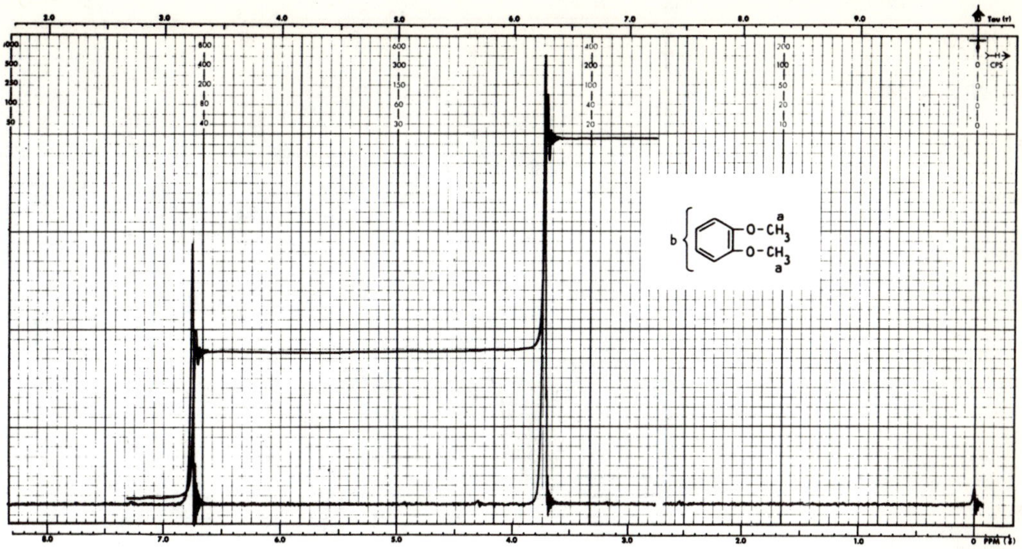

© 1978 Sadtler Research Laboratories, Inc.

---

m-DIMETHOXYBENZENE

$C_8H_{10}O_2$

Mol. Wt. 138.17

Solvent: CCl_4

Assignments
a = 3.29
b = 5.90-7.40

1546

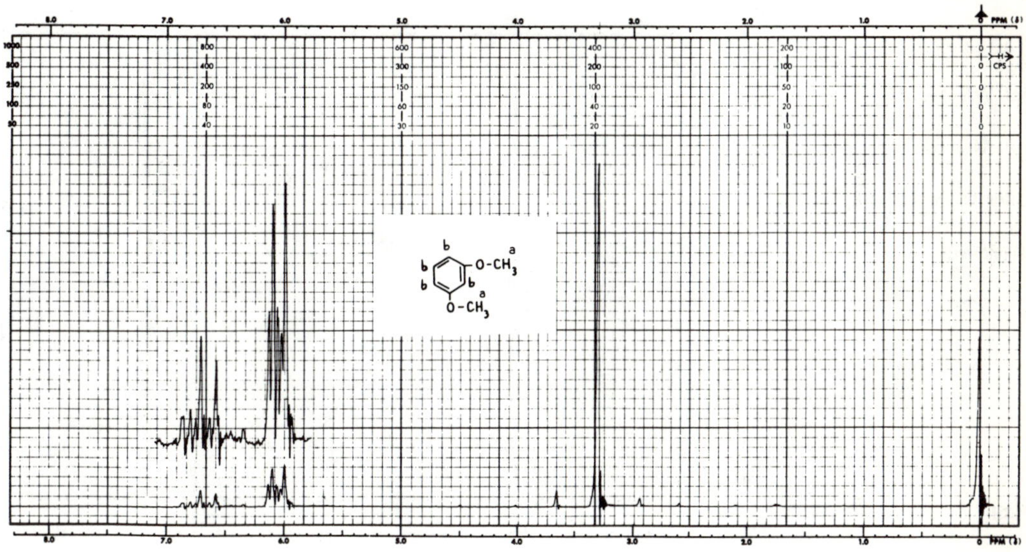

© 1978 Sadtler Research Laboratories, Inc.

1,4-DIMETHOXYBENZENE

$C_8H_{10}O_2$

Mol. Wt. 138.17

Solvent: $CDCl_3$

Assignments
a = 3.71
b = 6.79

1547

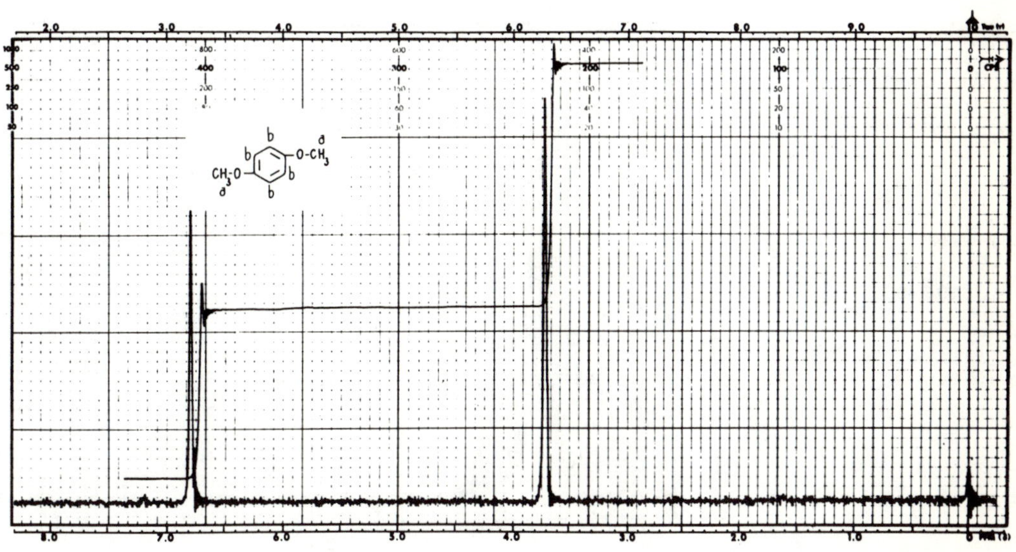

© 1978 Sadtler Research Laboratories, Inc.

1-METHOXY-4-PHENOXYBENZENE

$C_{13}H_{12}O_2$

Mol. Wt. 200.24

M.P. 5-9°C

Solvent: $CDCl_3$

Assignments
a = 3.71
b = 6.80
c = 6.90
d = 6.80-7.40

1548

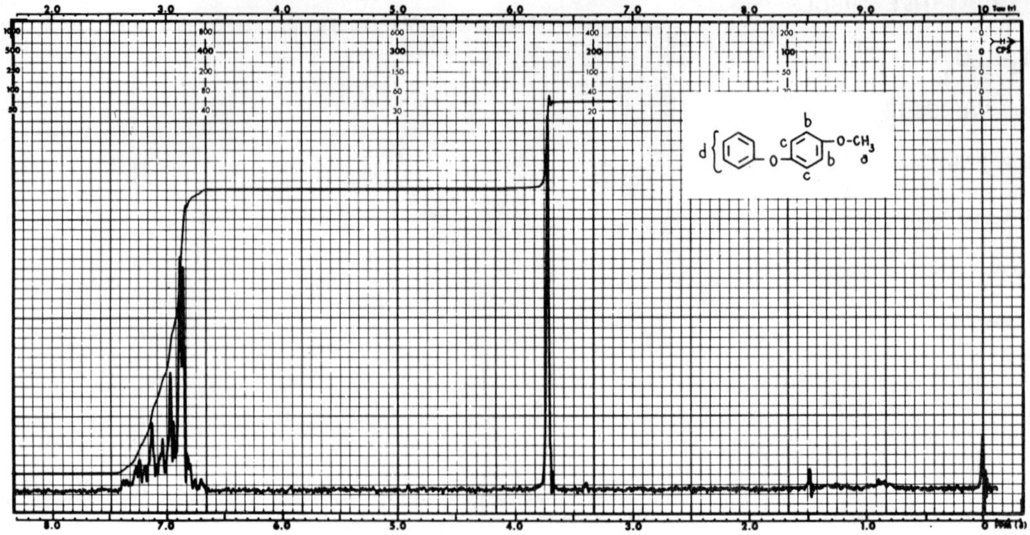

© 1978 Sadtler Research Laboratories, Inc.

4,4'-OXYDIANISOLE

$C_{14}H_{14}O_3$

Mol. Wt. 230.27

M.P. 102-104°C

Solvent: Polysol-d

Assignments
a = 3.75
b = 6.90

1549

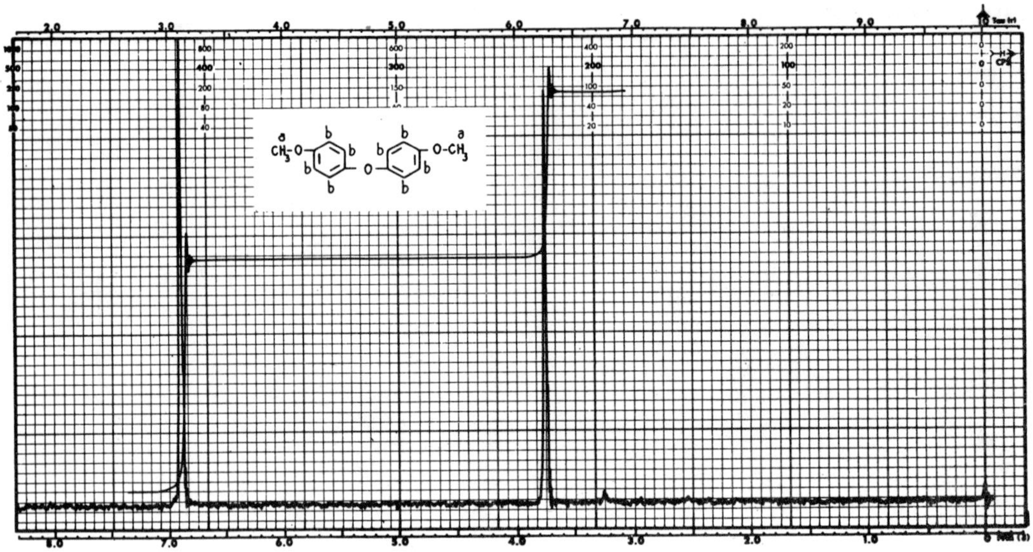

© 1978 Sadtler Research Laboratories, Inc.

m-DIPHENOXYBENZENE

$C_{18}H_{14}O_2$

Mol. Wt. 262.31

M.P. 60-62°C

Solvent: $CDCl_3$

Assignments
a = ca 6.70
b = 6.90-7.40

1550

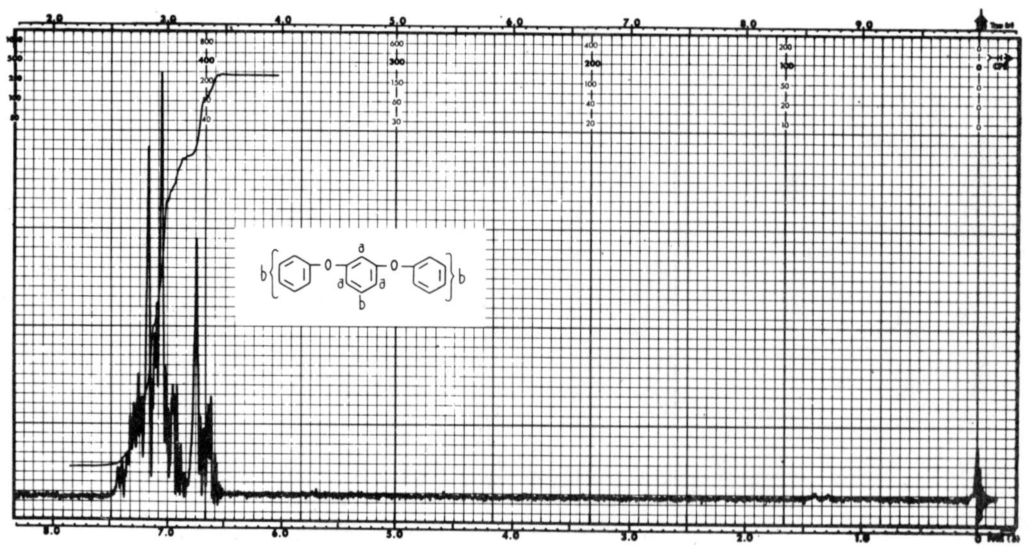

© 1978 Sadtler Research Laboratories, Inc.

2,3-DIMETHYLANISOLE

$C_9H_{12}O$

Mol. Wt. 136.20

B.P. 195°C

Solvent: $CDCl_3$

Assignments
a = 2.13
b = 2.27
c = 3.79
d = 6.68
e = 6.76
f = 7.07

1551

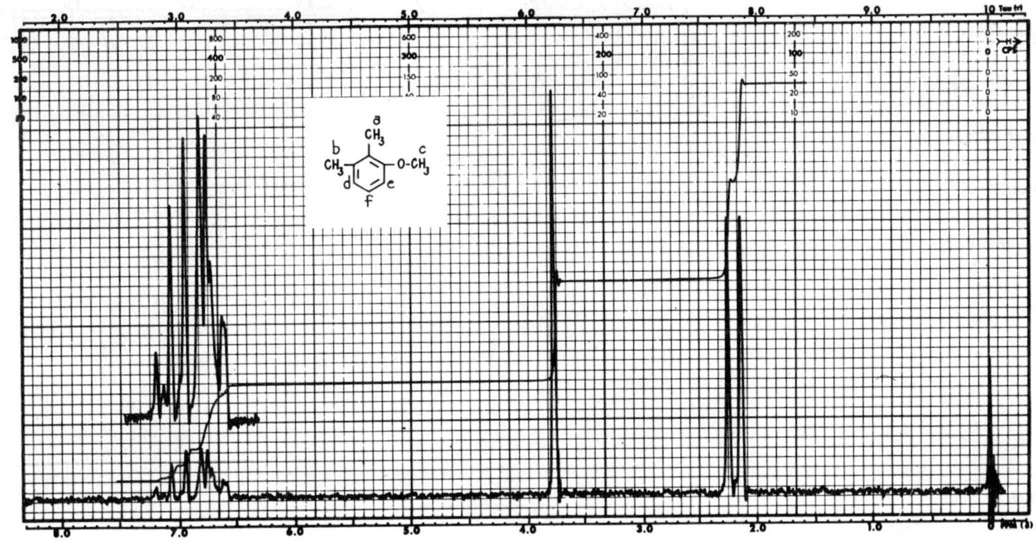

© 1978 Sadtler Research Laboratories, Inc.

2,4-DIMETHYLANISOLE

$C_9H_{12}O$

Mol. Wt. 136.20

B.P. 191°C

Solvent: $CDCl_3$

Assignments
a = 2.19
b = 2.24
c = 3.79
d = 6.69
e = ca 6.91

1552

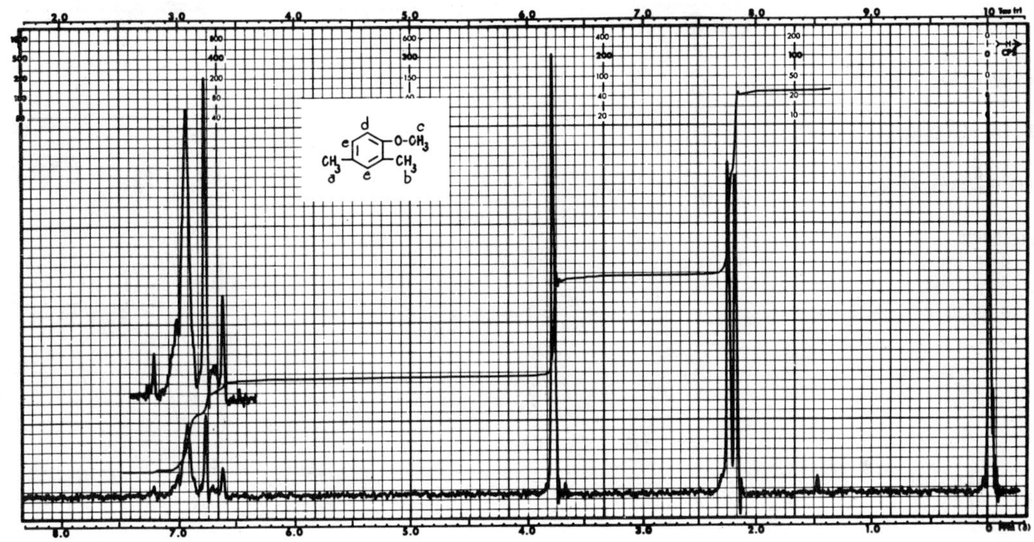

© 1978 Sadtler Research Laboratories, Inc.

2,6-DICHLOROANISOLE

$C_7H_6Cl_2O$

Mol. Wt. 177.03

Solvent: CCl_4

Assignments
a = 3.84
b = 6.87
c = 7.20

1553

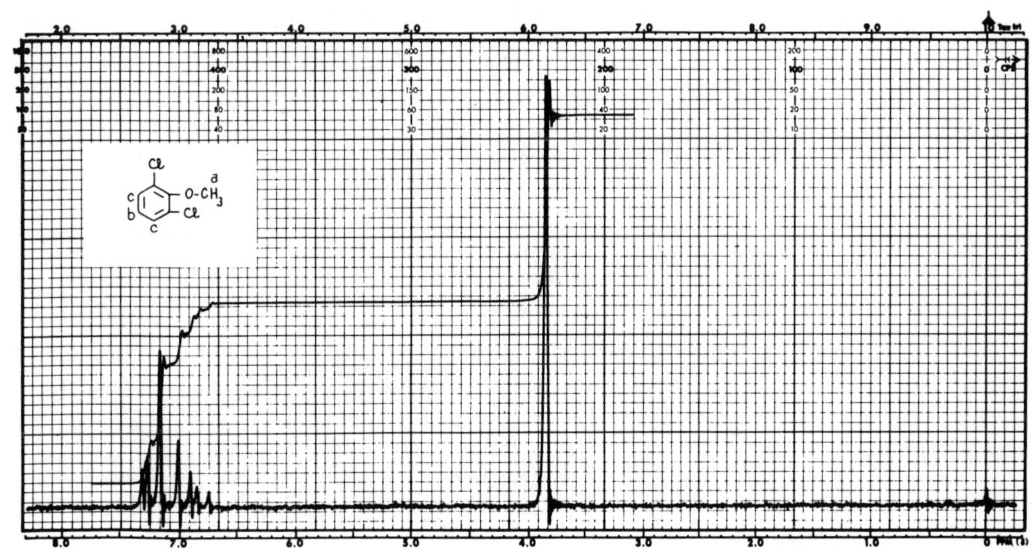

© 1978 Sadtler Research Laboratories, Inc.

p-CHLORO-o-ANISIDINE

C_7H_8ClNO

Mol. Wt. 157.60

M.P. 54°C (lit.)

Solvent: $CDCl_3$

Assignments
a = 3.72
b = 3.78
c = 6.60

1554

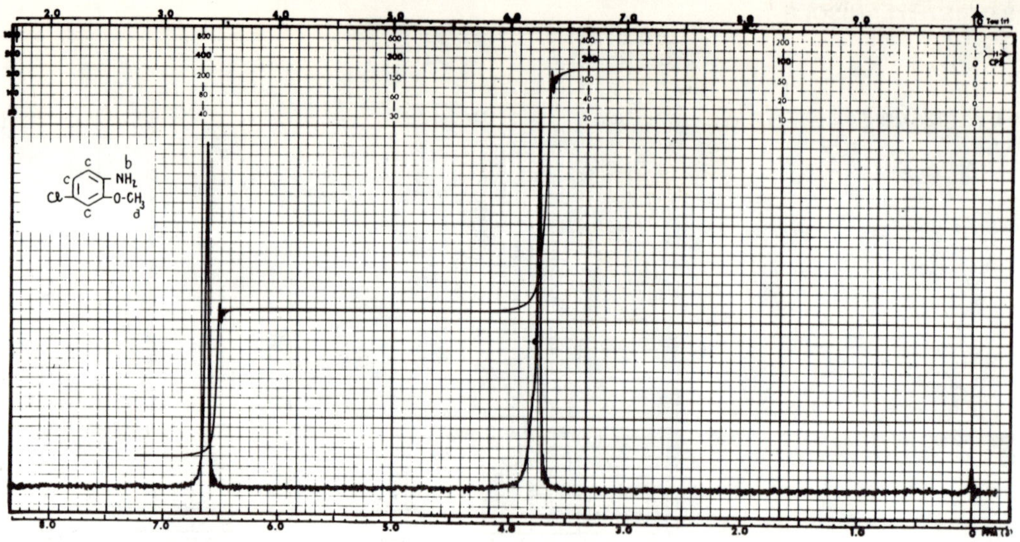

© 1978 Sadtler Research Laboratories, Inc.

2,4-DINITROANISOLE

$C_7H_6N_2O_5$

Mol. Wt. 198.14

M.P. 94-96°C

Solvent: Polysol-d

Assignments
a = 4.19
b = 7.60
c = 8.50
d = 8.65

1555

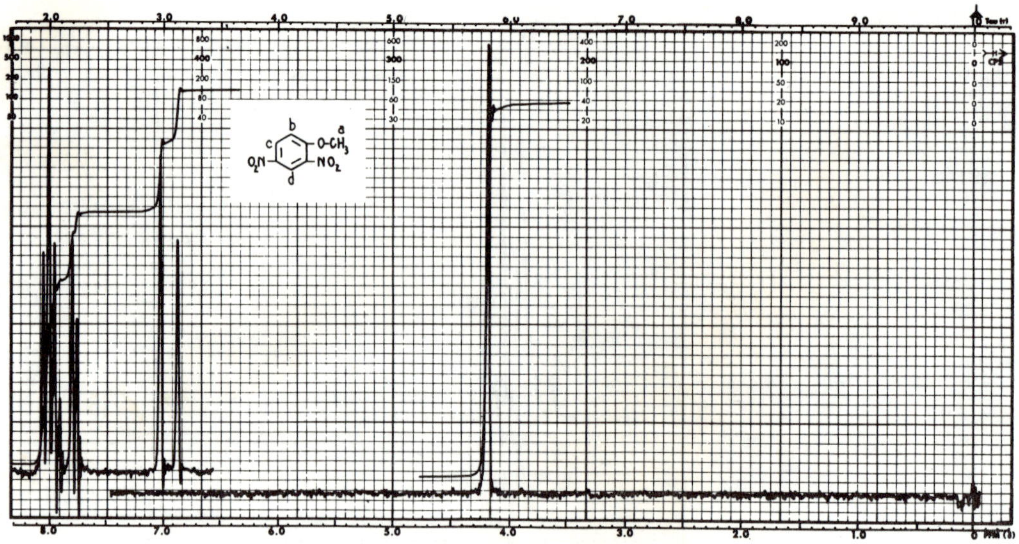

© 1978 Sadtler Research Laboratories, Inc.

2,6-DINITROANISOLE

$C_7H_6N_2O_5$

Mol. Wt. 198.14

M.P. 118°C

Solvent: $CDCl_3$

Assignments
a = 4.04
b = 7.31
c = 8.00

1556

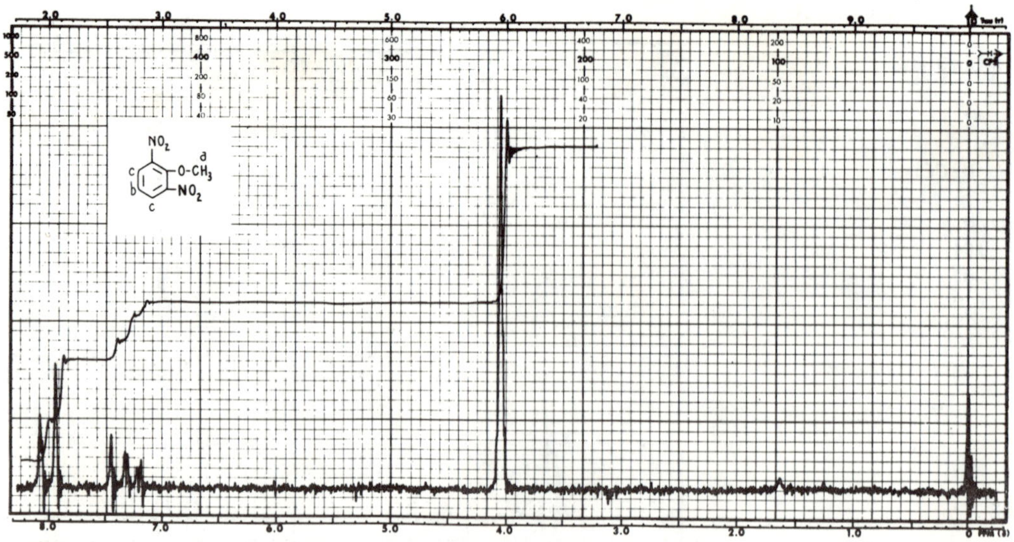

© 1978 Sadtler Research Laboratories, Inc.

2,4-DINITROPHENETOLE

$C_8H_8N_2O_5$

Mol. Wt. 212.16

M.P. 83-85.5°C

Solvent: $CDCl_3$

Assignments
a = 1.54
b = 4.37
c = 7.25
d = 8.43
e = 8.69
f = 1.70 imp.

1557

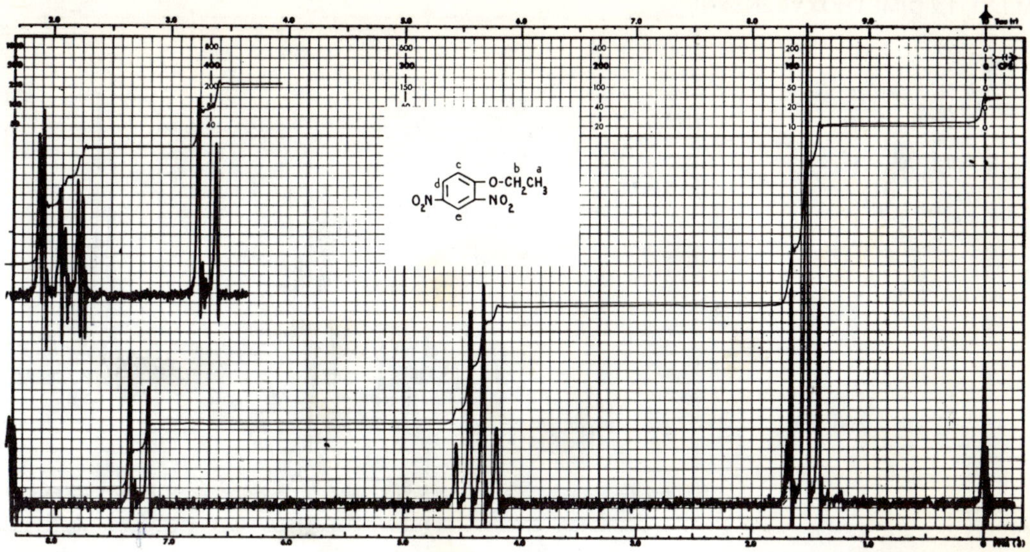

© 1978 Sadtler Research Laboratories, Inc.

2,5-DIMETHOXYTOLUENE

$C_9H_{12}O_2$

Mol. Wt. 152.19

M.P. 19-20°C

Solvent: CCl_4

Assignments
a = 2.18
b = 3.62
c = 3.69
d = ca 6.59

1558

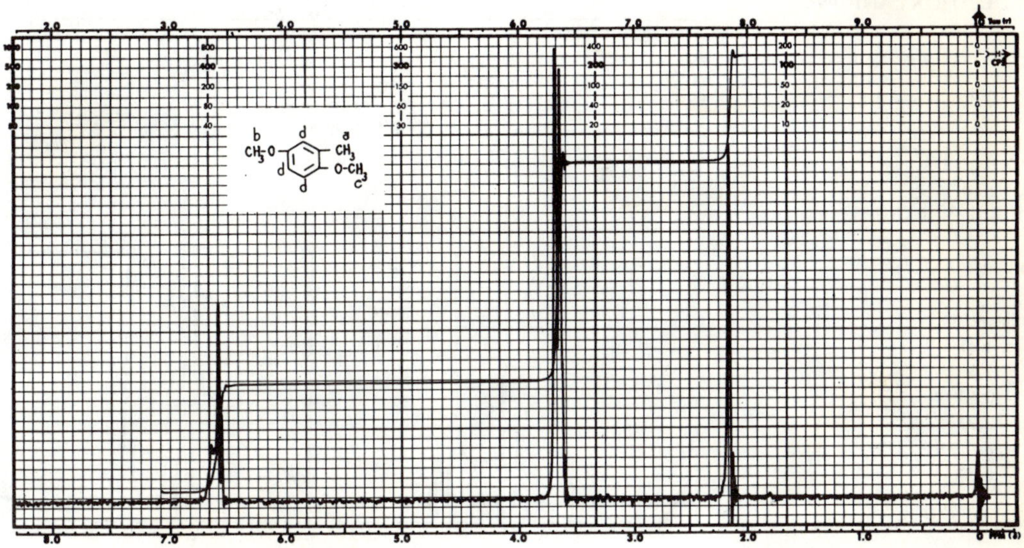

© 1978 Sadtler Research Laboratories, Inc.

1,2-DIMETHOXY-4-PROPENYLBENZENE

$C_{11}H_{14}O_2$

Mol. Wt. 178.23

M.P. 145-147°C/15mm

Solvent: $CDCl_3$

Assignments
a = 1.86
b = 3.84
c = 3.87
d = 5.91
e = 6.39
f = ca 6.85

1559

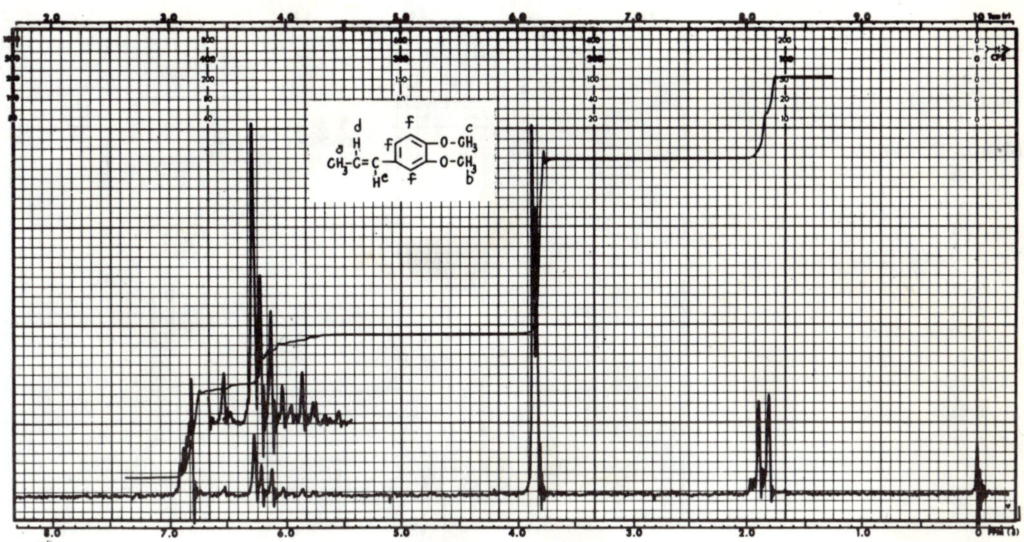

© 1978 Sadtler Research Laboratories, Inc.

4-ALLYL-1,2-DIMETHOXYBENZENE

$C_{11}H_{14}O_2$

Mol. Wt. 178.23

Solvent: CCl_4

Assignments
a = 3.25
b = 3.71
c = ca 5.02
d = 5.87
e = 6.60

1560

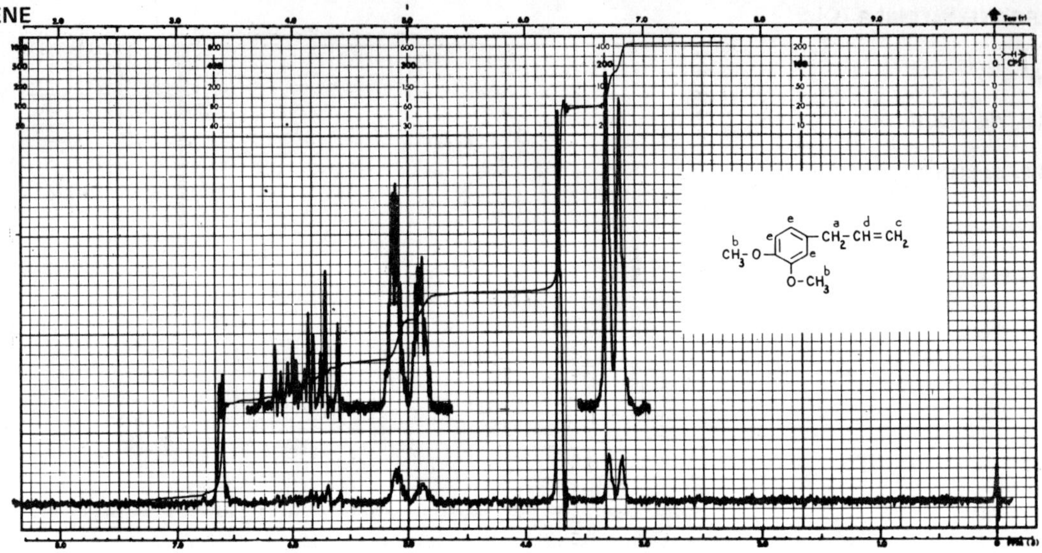

© 1978 Sadtler Research Laboratories, Inc.

3,4-DIETHOXYANILINE

$C_{10}H_{15}NO_2$

Mol. Wt. 181.24

M.P. 47-49°C

B.P. 132-135°C/0.45mm

Solvent: $CDCl_3$

Assignments
a = 1.31
b = 1.36
c = 3.43
d = 3.97
e = ca 6.18
f = 6.70

1561

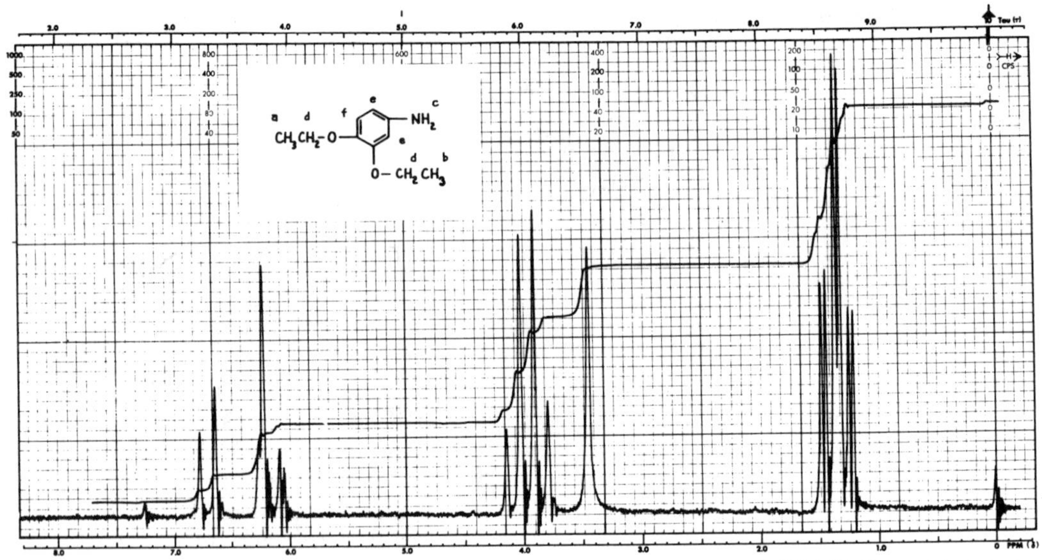

© 1978 Sadtler Research Laboratories, Inc.

2,5-DIBUTOXYANILINE

$C_{14}H_{23}NO_2$

Mol. Wt. 237.34

Solvent: $CDCl_3$

Assignments
a = 0.95
b = 1.15-2.08
c = 3.72
d = 3.82
e = 3.89
f = 6.19
g = 6.26
h = 6.65
i = 6.79 imp.

1562

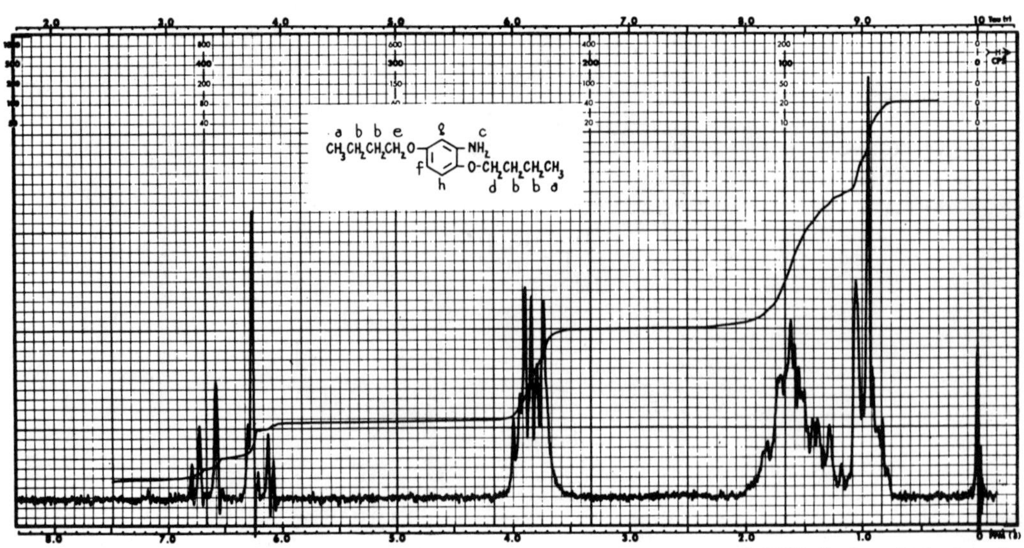

© 1978 Sadtler Research Laboratories, Inc.

VERATRONITRILE

$C_9H_9NO_2$

Mol. Wt. 163.18

Solvent: $CDCl_3$

Assignments
a = 3.83
b = 3.87
c = 6.85
d = 7.02
e = 7.20

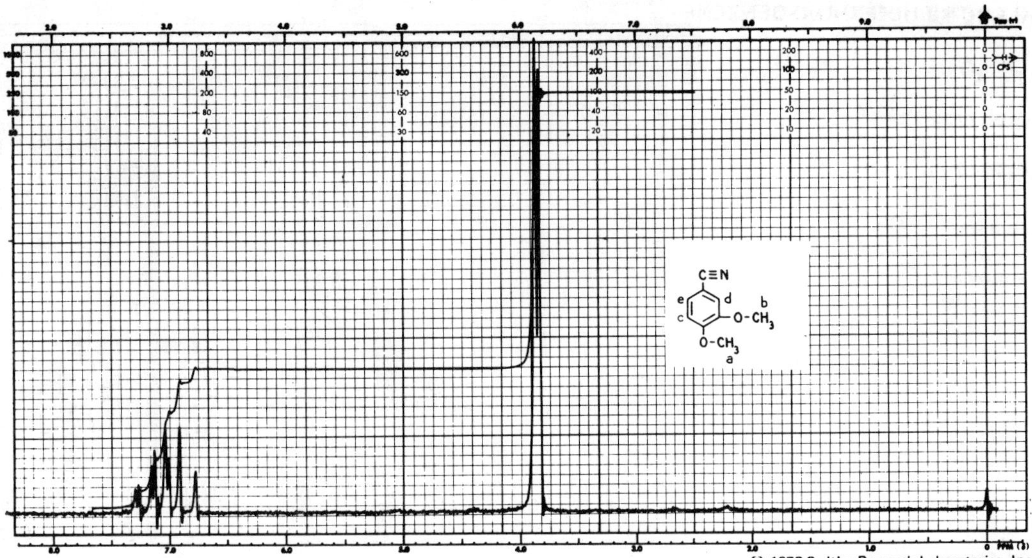

1563

© 1978 Sadtler Research Laboratories, Inc.

1,3-DIMETHOXY-4-NITROBENZENE

$C_8H_9NO_4$

Mol. Wt. 183.17

Solvent: $CDCl_3$

Assignments
a = 3.89
b = 3.92
c = 6.49
d = 6.58
e = 7.98

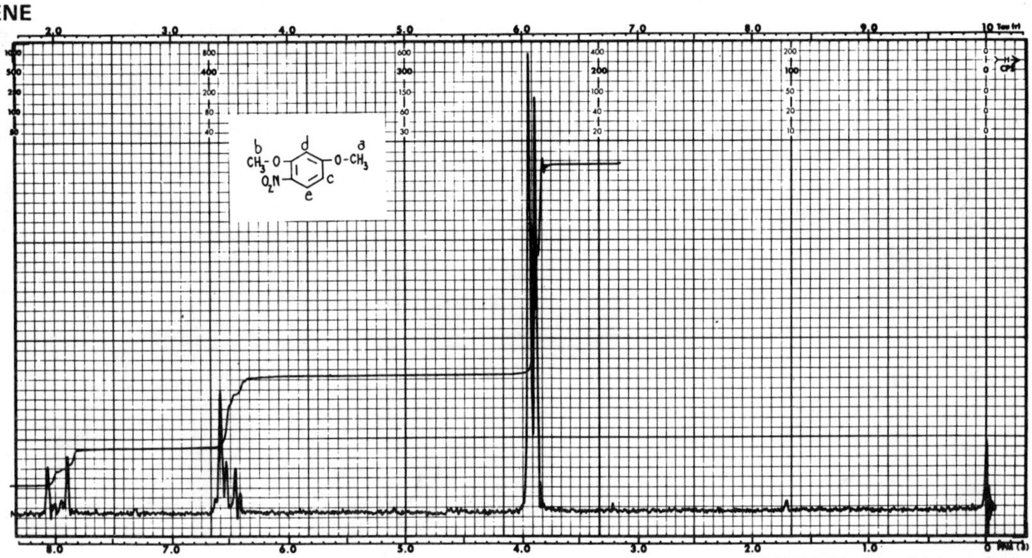

1564

© 1978 Sadtler Research Laboratories, Inc.

1,2,3-TRIMETHOXYBENZENE

$C_9H_{12}O_3$

Mol. Wt. 168.19

M.P. 43-45°C

Solvent: $CDCl_3$

Assignments
a = 3.86
b = 6.58
c = 7.00

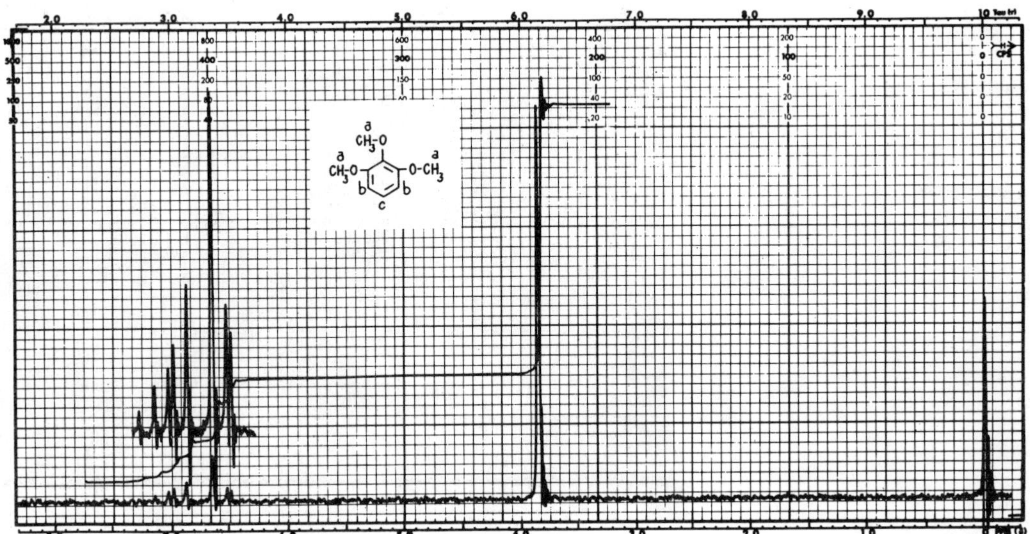

1565

© 1978 Sadtler Research Laboratories, Inc.

2,4,6-TRICHLOROANISOLE

$C_7H_5Cl_3O$

Mol. Wt. 211.48

M.P. 57-58°C

Solvent: $CDCl_3$

Assignments
a = 3.88
b = 7.29

1566

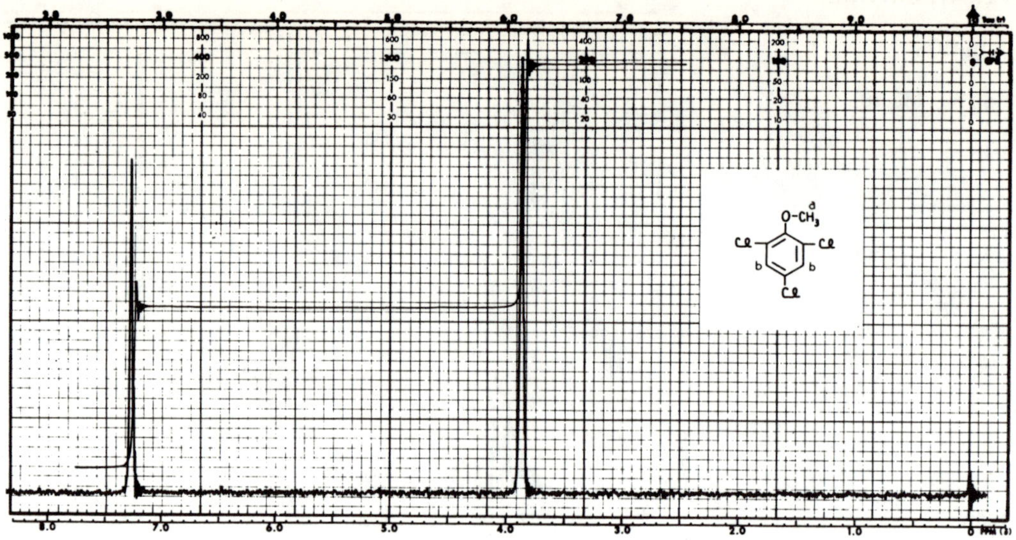

© 1978 Sadtler Research Laboratories, Inc.

2,4,6-TRIBROMOANISOLE

$C_7H_5Br_3O$

Mol. Wt. 344.84

M.P. 85-87°C

Solvent: CCl_4

Assignments
a = 3.81
b = 7.60

1567

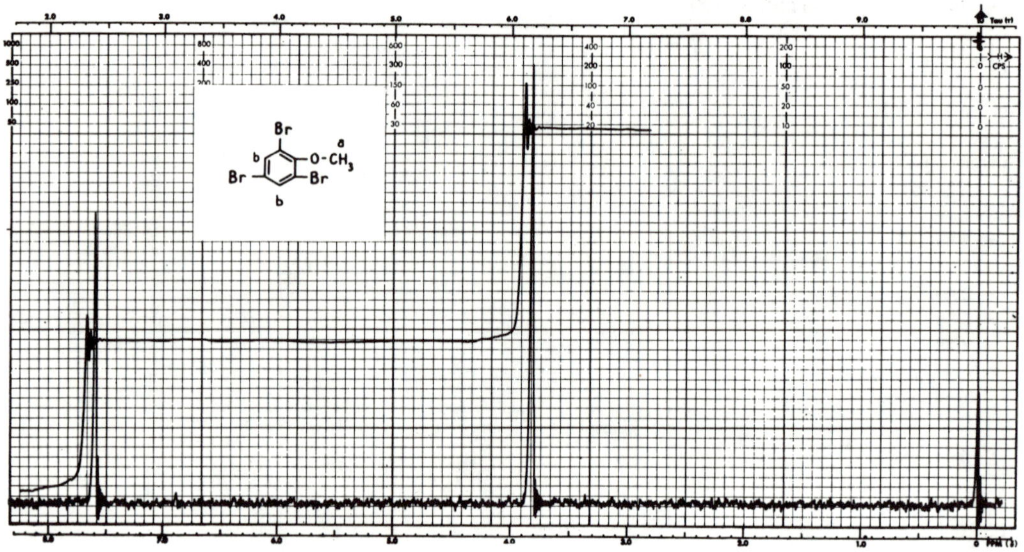

© 1978 Sadtler Research Laboratories, Inc.

2,4,6-TRINITROANISOLE

$C_7H_5N_3O_7$

Mol. Wt. 243.13

Solvent: $CDCl_3$

Assignments
a = 4.19
b = 8.93

1568

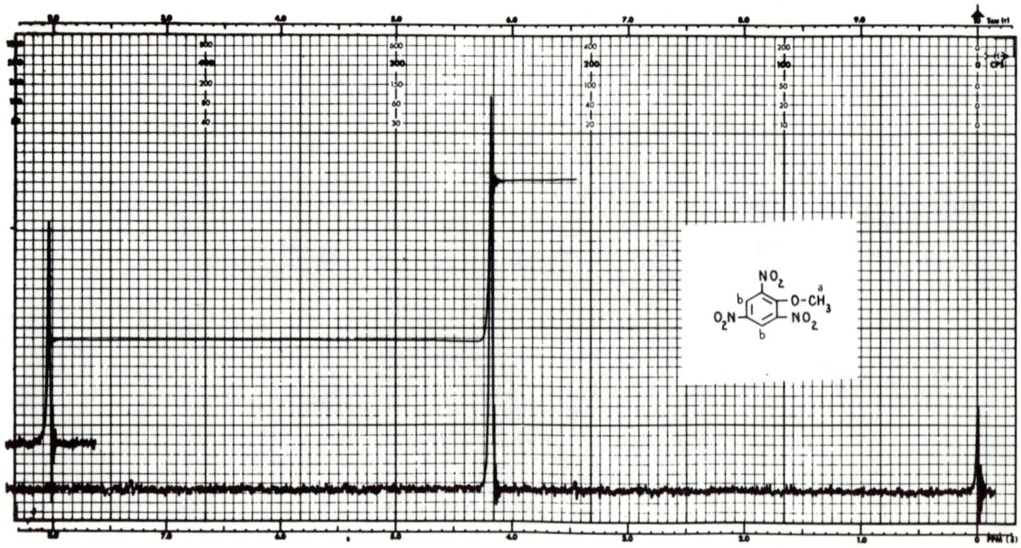

© 1978 Sadtler Research Laboratories, Inc.

1-METHOXYNAPHTHALENE

$C_{11}H_{10}O$

Mol. Wt. 158.20

M.P. 124°C/4mm

Solvent: $CDCl_3$

Assignments
a = 3.72
b = 6.51
c = 6.99-7.50
d = 7.60
e = 8.12

1569

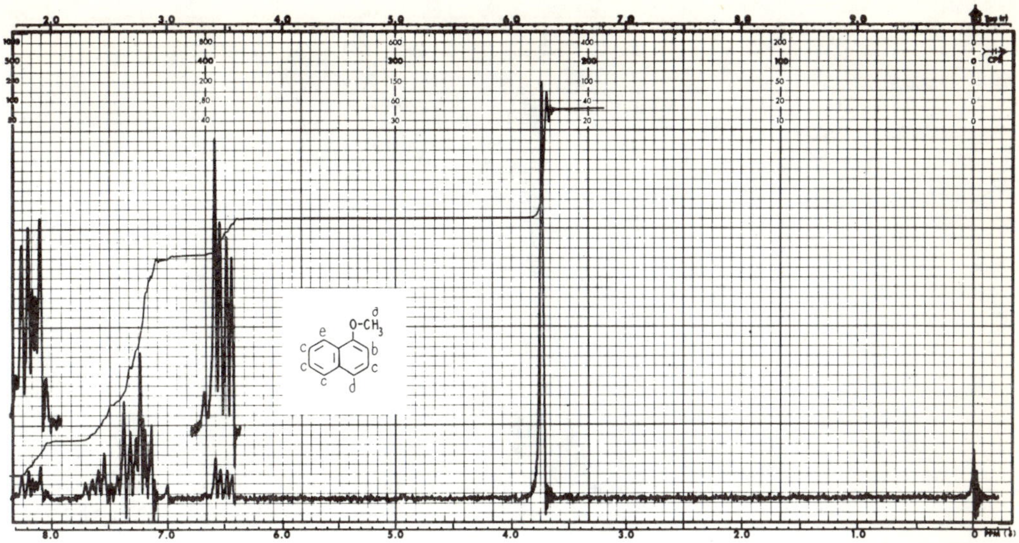

© 1978 Sadtler Research Laboratories, Inc.

1-ETHOXYNAPHTHALENE

$C_{12}H_{12}O$

Mol. Wt. 172.23

B.P. 143-145°C/12mm

Solvent: $CDCl_3$

Assignments
a = 1.37
b = 3.95
c = 6.60
d = 7.10-7.55
e = 7.70
f = 8.33

1570

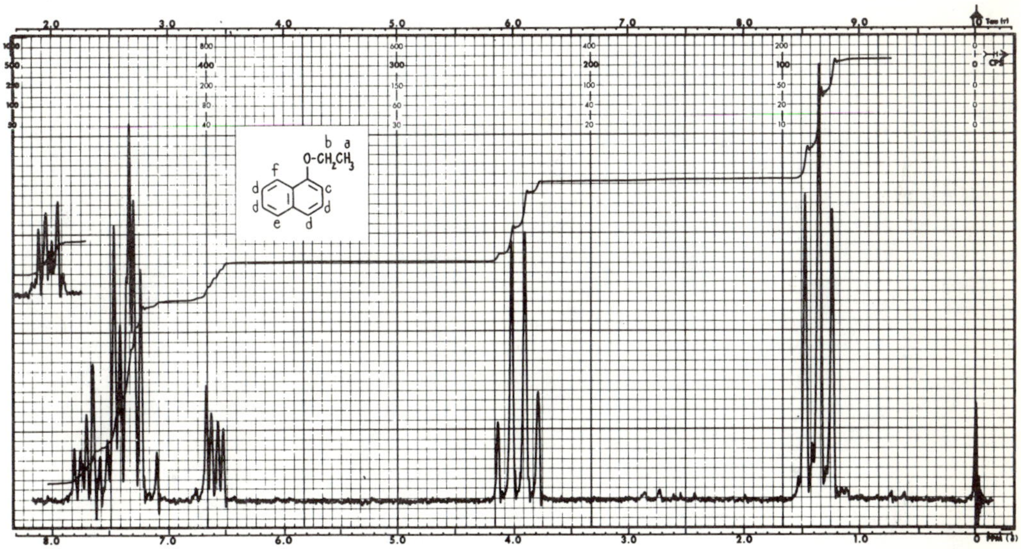

© 1978 Sadtler Research Laboratories, Inc.

2-ETHOXYNAPHTHALENE

$C_{12}H_{12}O$

Mol. Wt. 172.22

M.P. 37.5°C

Solvent: CCl_4

Assignments
a = 1.41
b = 4.02
c = ca 6.94
d = 7.00-7.50
e = ca 7.61

1571

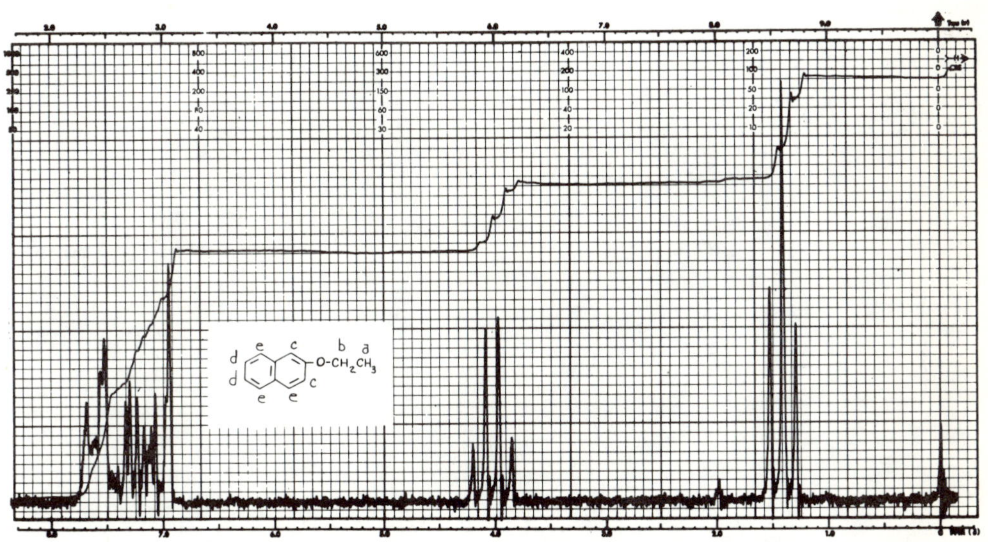

© 1978 Sadtler Research Laboratories, Inc.

2-(BENZYLOXY)NAPHTHALENE

$C_{17}H_{14}O$

Mol. Wt. 234.30

M.P. 99.5-100.5°C

Solvent: $CDCl_3$

Assignments

a = 5.10

b = 7.14-7.53

c = 7.71

1572

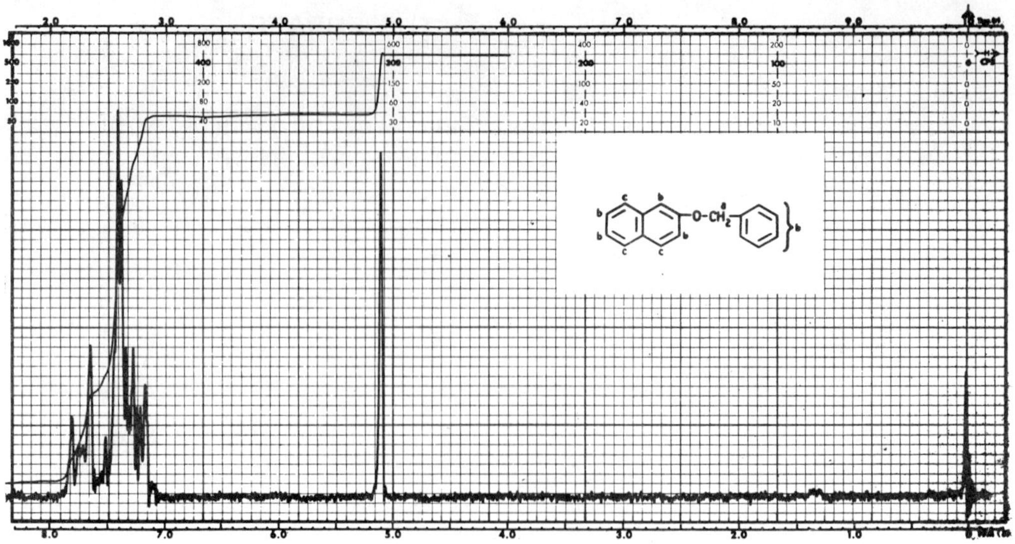

© 1978 Sadtler Research Laboratories, Inc.

Furan and its derivatives are the major heteroaromatic group of the ether compounds. Their NMR spectra display characteristic chemical shifts and coupling constants producing spin-spin coupling patterns similar to those of the pyrroles and thiophenes. The oxygen atom in the ring strongly deshields the hydrogens on the adjacent carbons (C–2 and C–5) but shields the protons bonded to positions C–3 and C–4.

Chemical Shifts

Furan

HC–5	HC–4	HC–3	HC–2	Solvent	Spectrum
7.37	6.30	6.30	7.37	CCl_4	1573

2-Substituted Furans

HC–5	HC–4	HC–3	C–2	Solvent
6.85	6.23	5.11	$-O-CH_3$	$CDCl_3$
7.15	6.13	5.83	$-CH_3$	CCl_4
7.30	6.25	6.14	$-CH_2-SH$	$CDCl_3$
7.82	7.29	6.67	$-C(=O)-NH_2$	DMSO-d_6
7.59	6.58	6.92	$-CH=CH-NO_2$	$CDCl_3$
7.64	6.59	7.05	$-C(=O)-O-K$	D_2O
7.60	6.55	7.11	$-C\equiv N$	CCl_4
7.82	6.70	7.52	$-C(=O)-Cl$	CCl_4

Coupling and Coupling Constants

The furan coupling constants are much smaller in magnitude than the corresponding ortho and meta coupling constants of the benzene derivatives. It is characteristic of the furans and the other heteroaromatic compounds that the "ortho" couplings, J_{2-3} and J_{3-4} are not the same.

$$J_{2-3} = 1.7\text{–}2.0 \text{ Hz}$$
$$J_{3-4} = 3.0\text{–}4.0 \text{ Hz}$$
$$J_{2-4} = 0.7\text{–}1.0 \text{ Hz}$$

Solubility and Solvent Effects

Excluding the solubility limitations imposed by the substituents that may be bonded to the furan ring system, the compounds are readily soluble in the chlorinated solvents normally utilized as NMR solvents (CCl_4 and $CDCl_3$).

UNIV. COLL. N.W. BANGOR LIBRARY

FURAN

C_4H_4O

Mol. Wt. 68.08

B.P. 31-33°C

Solvent: CCl_4

Assignments
a = 6.30
b = 7.37

1573

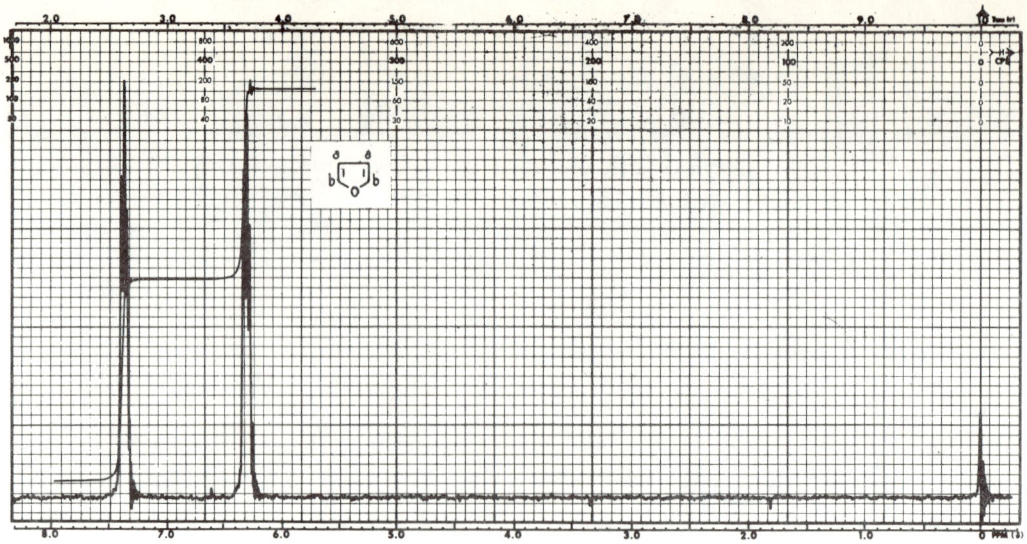

© 1978 Sadtler Research Laboratories, Inc.

2-METHYLFURAN

C_5H_6O

Mol. Wt. 82.10

B.P. 63-65°C

Solvent: CCl_4

Assignments
a = 2.22
b = 5.83
c = 6.13
d = 7.15

1574

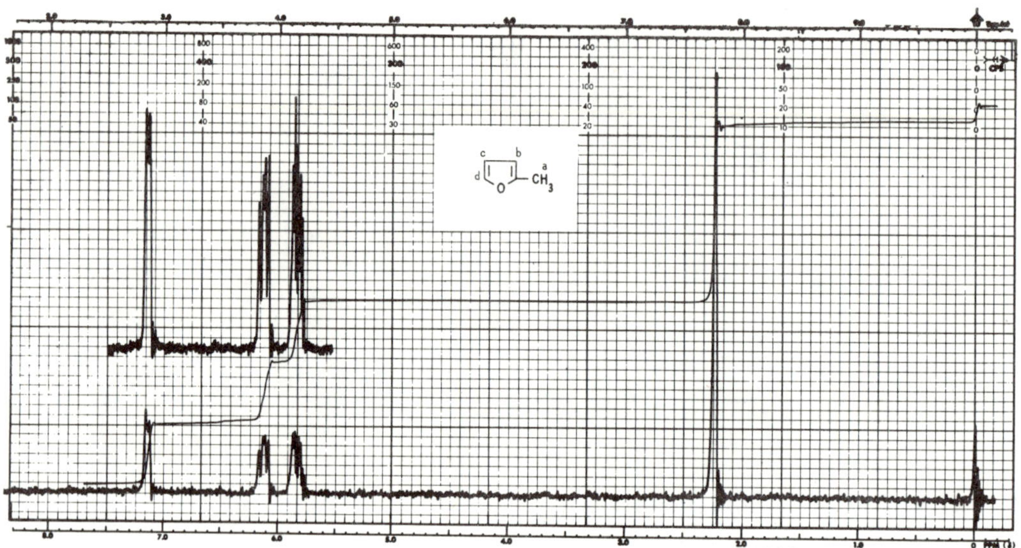

© 1978 Sadtler Research Laboratories, Inc.

2,5-DIMETHYLFURAN

C_6H_8O

Mol. Wt. 96.13

B.P. 92-94°C

Solvent: $CDCl_3$

Assignments
a = 2.23
b = 5.81

1575

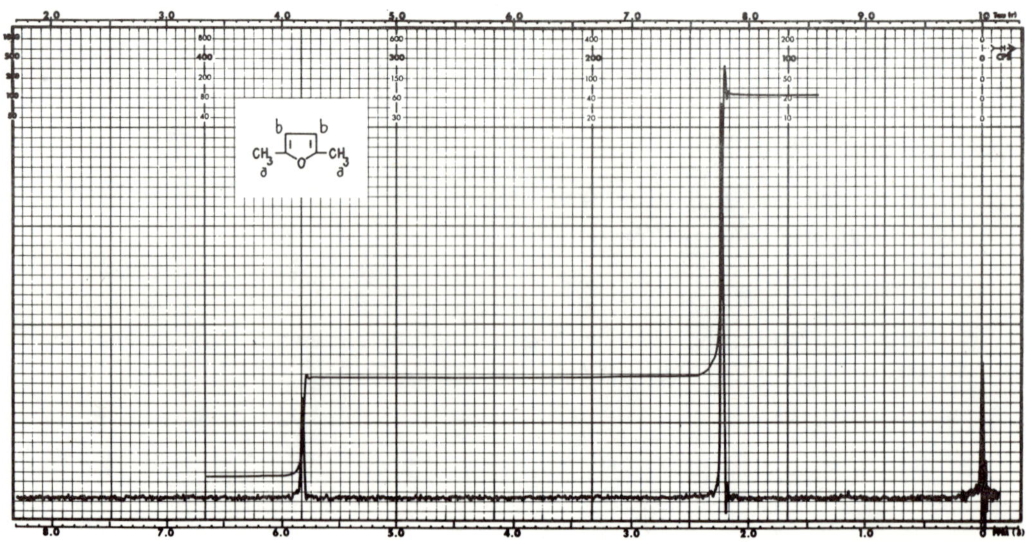

© 1978 Sadtler Research Laboratories, Inc.

2-FURONITRILE

C_5H_3NO

Mol. Wt. 93.09

Solvent: CCl_4

Assignments
a = 6.55
b = 7.11
c = 7.60

1576

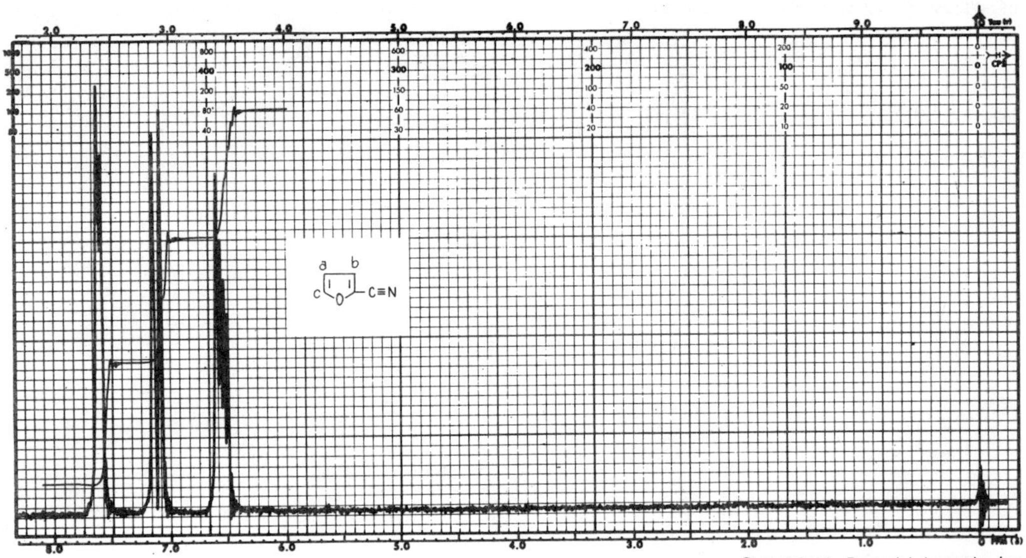

© 1978 Sadtler Research Laboratories, Inc.

2-FURANMETHANETHIOL

C_5H_6OS

Mol. Wt. 114.17

Solvent: $CDCl_3$

Assignments
a = 1.90
b = 3.70
c = 6.14
d = 6.25
e = 7.30
f = 1.29 imp.

1577

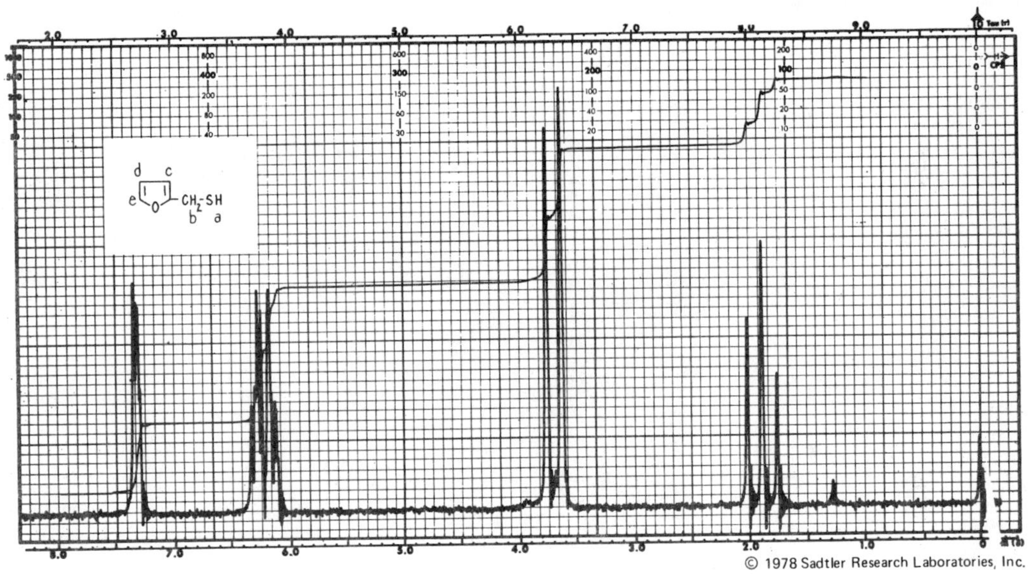

© 1978 Sadtler Research Laboratories, Inc.

2-(2-NITROVINYL)FURAN

$C_6H_5NO_3$

Mol. Wt. 139.11

Solvent: $CDCl_3$

Assignments
a = 6.58
b = 6.92
c = 7.48
d = 7.59
e = 7.78

1578

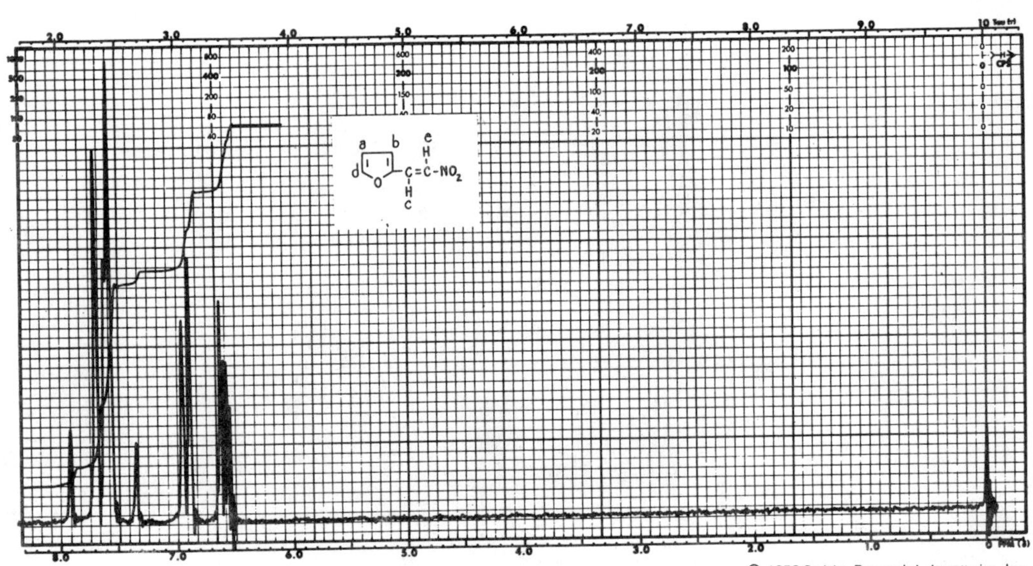

© 1978 Sadtler Research Laboratories, Inc.

3-METHYLISOXAZOLE

C_4H_5NO

Mol. Wt. 83.09

B.P. 118°C

Solvent: CCl_4

Assignments
a = 2.27
b = 6.13
c = 8.39

1579

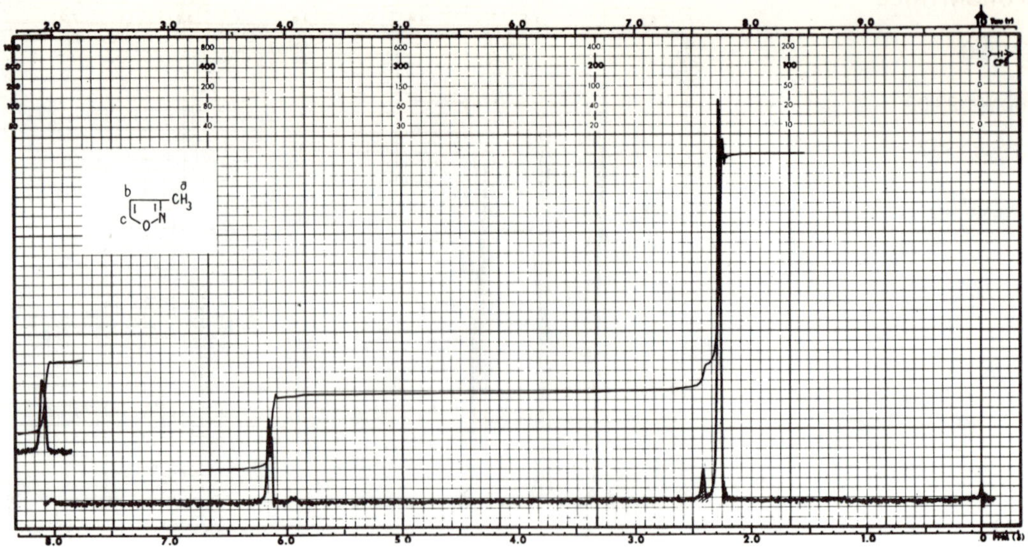

© 1978 Sadtler Research Laboratories, Inc.

2-(m-CHLOROPHENYL)-5-PHENYLOXAZOLE

$C_{15}H_{10}ClNO$

Mol. Wt. 255.71

Solvent: $CDCl_3$

Assignments
a = 7.20-7.80
b = 7.39
c = 7.95
d = 8.07

1580

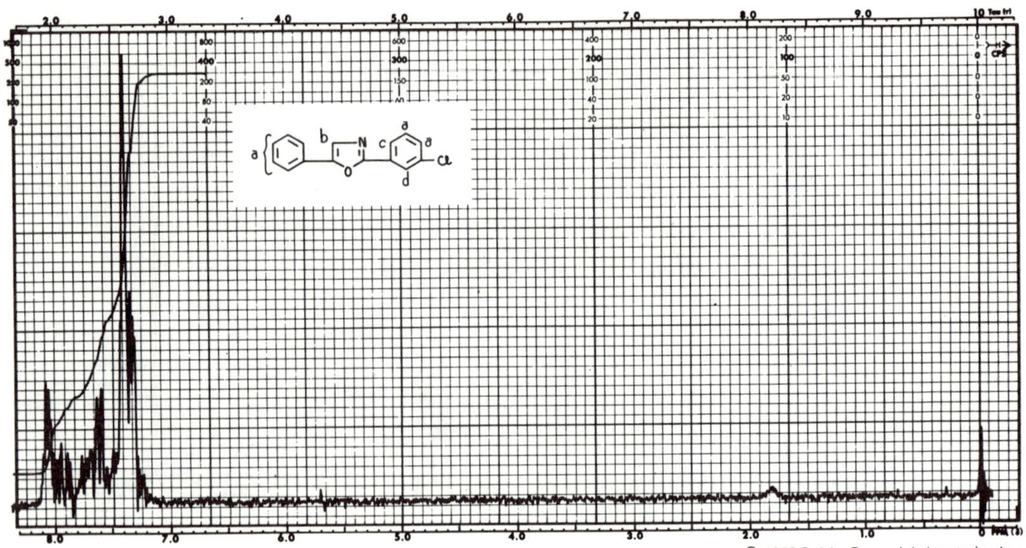

© 1978 Sadtler Research Laboratories, Inc.

2-(HEXADECYLOXY)PYRIDINE

$C_{21}H_{37}NO$

Mol. Wt. 319.54

Solvent: $CDCl_3$

Assignments
a = 0.90
b = 1.30
c = 1.80
d = 4.30
e = 6.80
f = 7.58
g = 8.18

1581

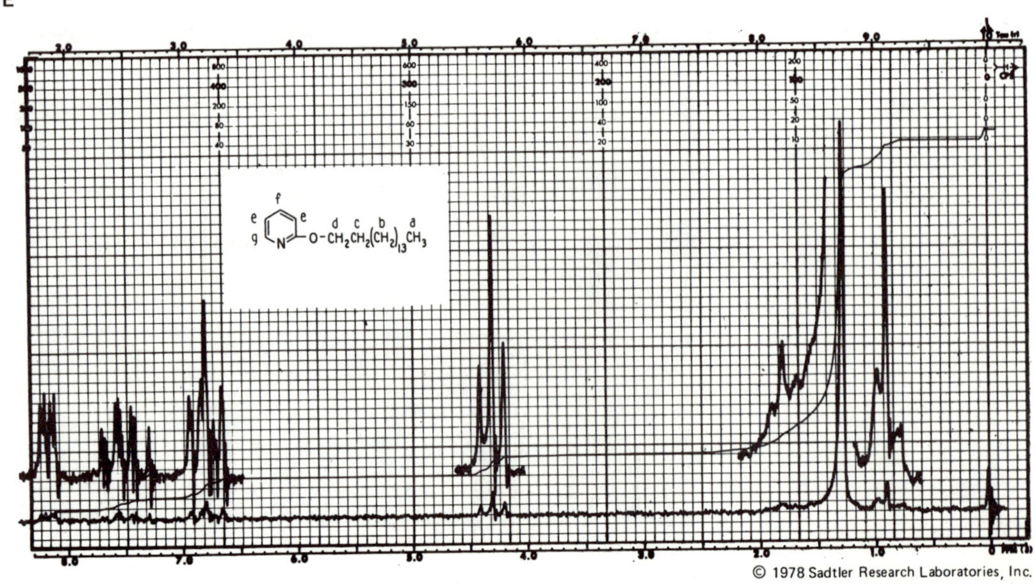

© 1978 Sadtler Research Laboratories, Inc.

METHOXYPYRAZINE

$C_5H_6N_2O$

Mol. Wt. 110.12

Solvent: $CDCl_3$

Assignments
a = 3.97
b = 8.10
c = 8.21

1582

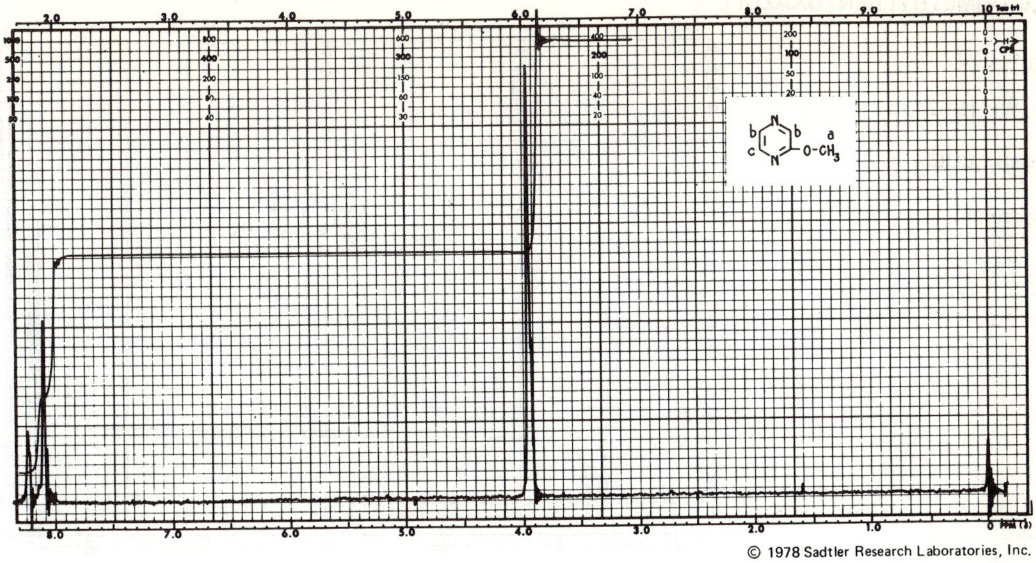

© 1978 Sadtler Research Laboratories, Inc.

3-METHOXYPHENOTHIAZINE

$C_{13}H_{11}NOS$

Mol. Wt. 229.30

Solvent: DMSO-d_6

Assignments
a = 3.70
b = 6.35
c = 6.39
d = 6.60-7.15
e = 8.34

1583

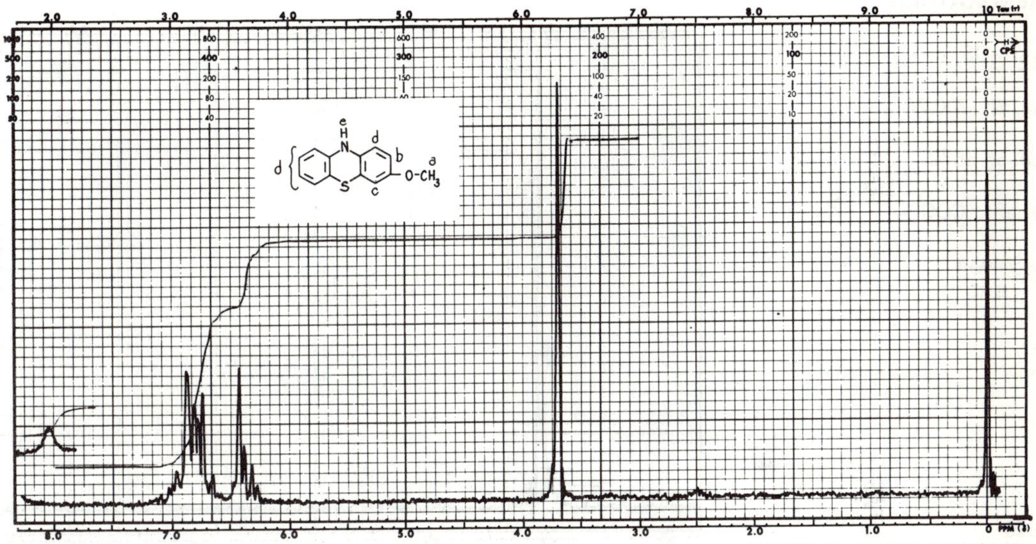

© 1978 Sadtler Research Laboratories, Inc.

BENZOFURAN

C_8H_6O

Mol. Wt. 118.14

B.P. 173-175°C

Solvent: CCl_4

Assignments
a = 6.59
b = 7.00-7.30
c = 7.30-7.60
d = 7.47

1584

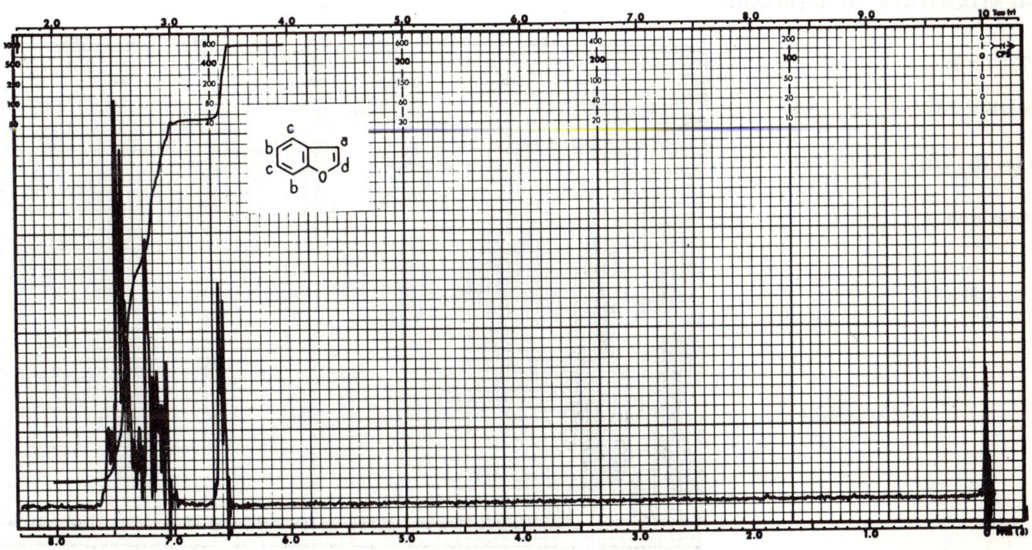

© 1978 Sadtler Research Laboratories, Inc.

2,5,6-TRIMETHYLBENZOXAZOLE

$C_{10}H_{11}NO$

Mol. Wt. 161.21

Solvent: $CDCl_3$

Assignments
a = 2.32
b = 2.56
c = 7.20 or d
d = 7.38 or c

1585

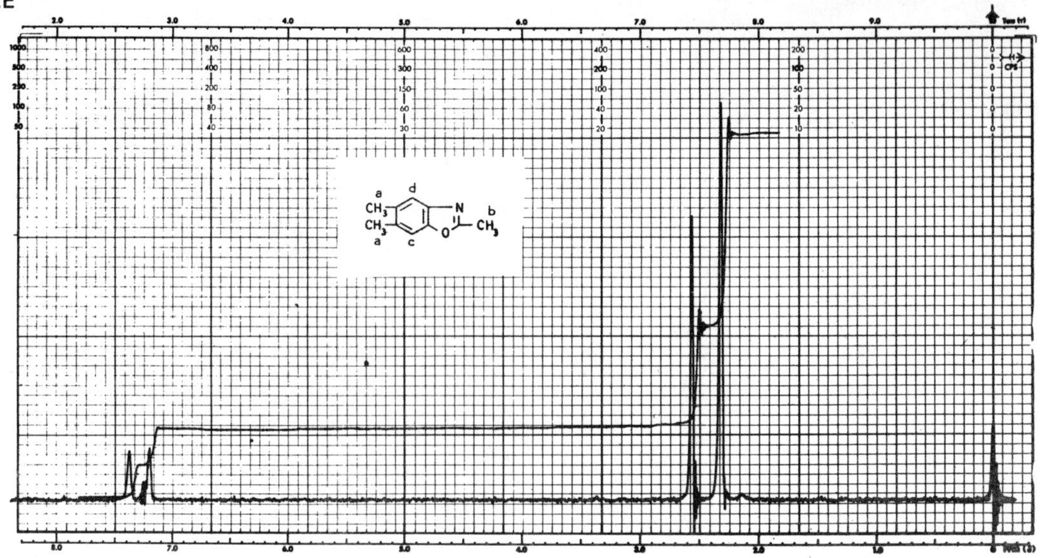

© 1978 Sadtler Research Laboratories, Inc.

PHENOXATHIIN

$C_{12}H_8OS$

Mol. Wt. 200.26

M.P. 56-58°C

Solvent: $CDCl_3$

Assignments
a = 6.92

1586

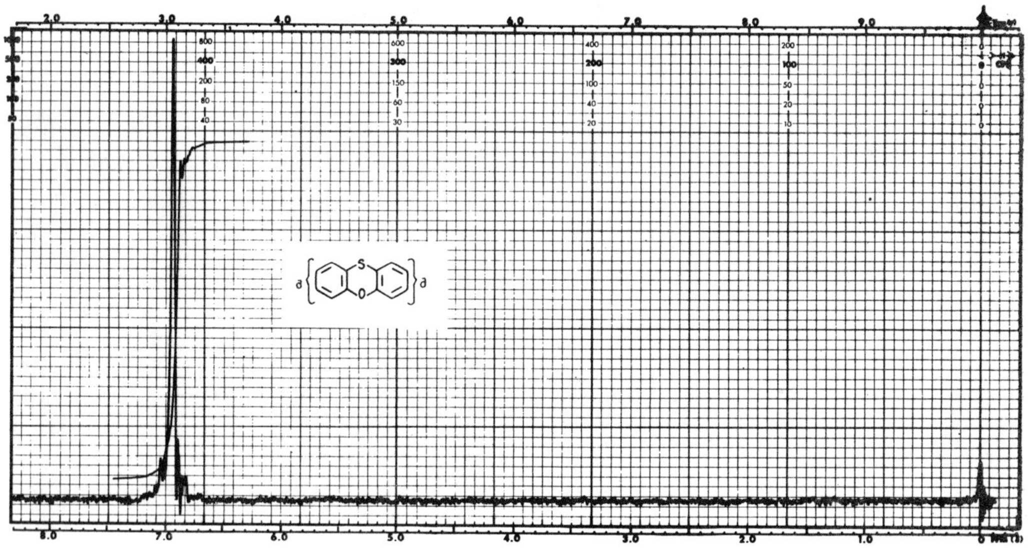

© 1978 Sadtler Research Laboratories, Inc.

PHENOXATHIIN, 10,10-DIOXIDE

$C_{12}H_8O_3S$

Mol. Wt. 232.26

Solvent: Trifluoroacetic acid

Assignments
a = 7.32
b = 7.40
c = 7.61
d = 8.03

1587

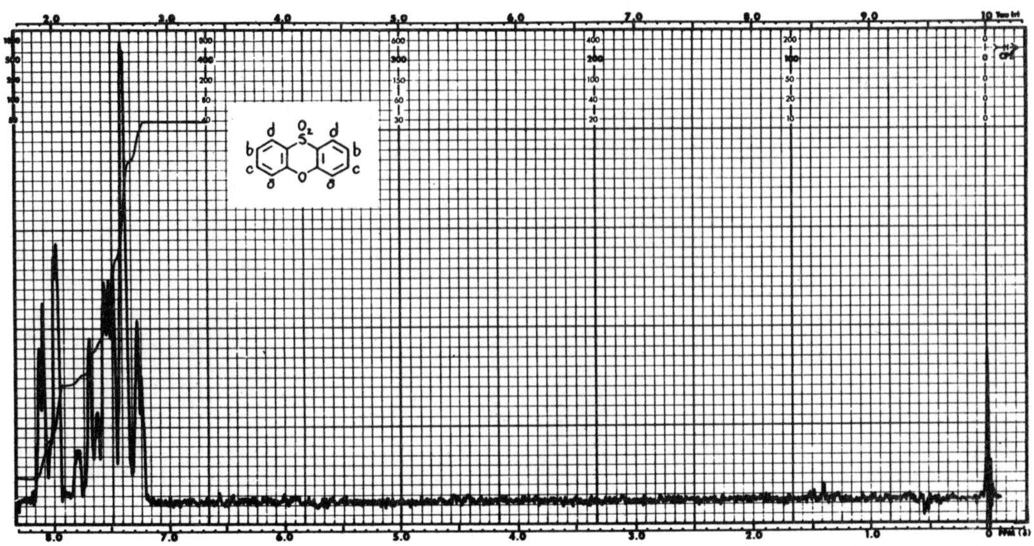

© 1978 Sadtler Research Laboratories, Inc.

1,3-BENZODIOXOLE

$C_7H_6O_2$

Mol. Wt. 122.12

B.P. 62°C/13mm

Solvent: CCl_4

Assignments
a = 5.89
b = 6.78

1588

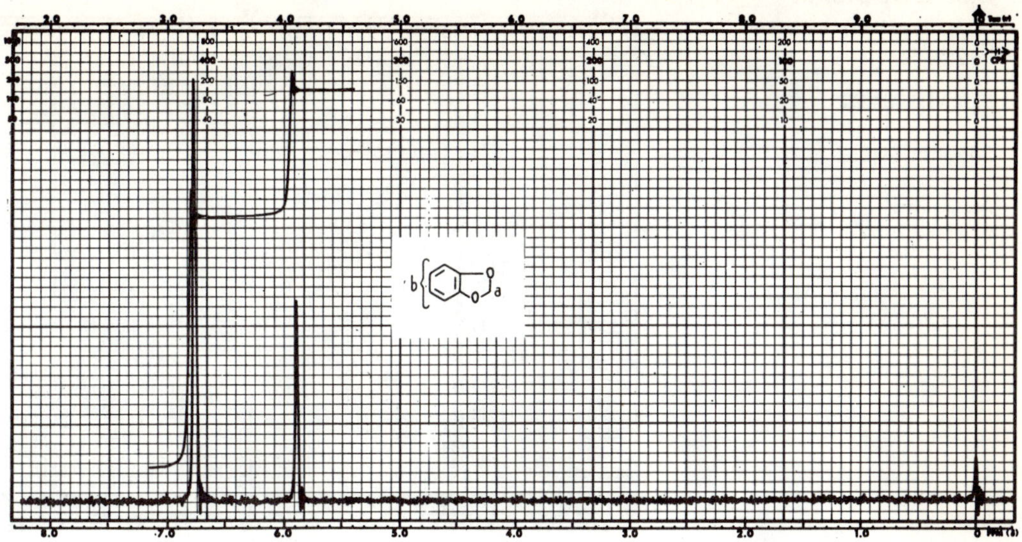

© 1978 Sadtler Research Laboratories, Inc.

ISOCHROMAN

$C_9H_{10}O$

Mol. Wt. 134.18

Solvent: CCl_4

Assignments
a = 2.69
b = 3.79
c = 4.60
d = 6.60-7.15

1589

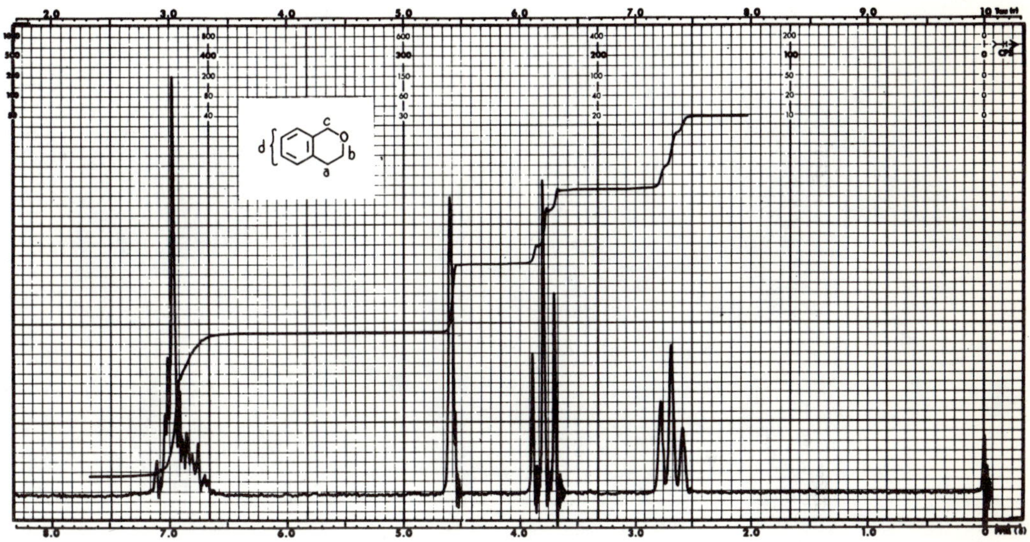

© 1978 Sadtler Research Laboratories, Inc.

1,4-BENZODIOXAN

$C_8H_8O_2$

Mol. Wt. 136.15

M.P. 87-90°C

Solvent: $CDCl_3$

Assignments
a = 4.23
b = 6.84

1590

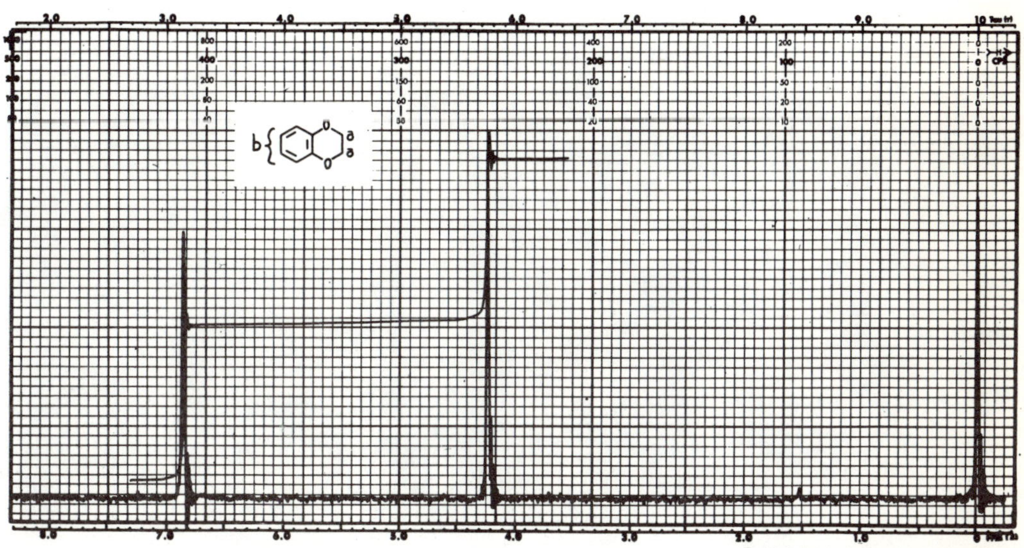

© 1978 Sadtler Research Laboratories, Inc.

DIBENZO-p-DIOXIN

$C_{12}H_8O_2$

Mol. Wt. 184.20

Solvent: $CDCl_3$

Assignments
a = ca 6.79

1591

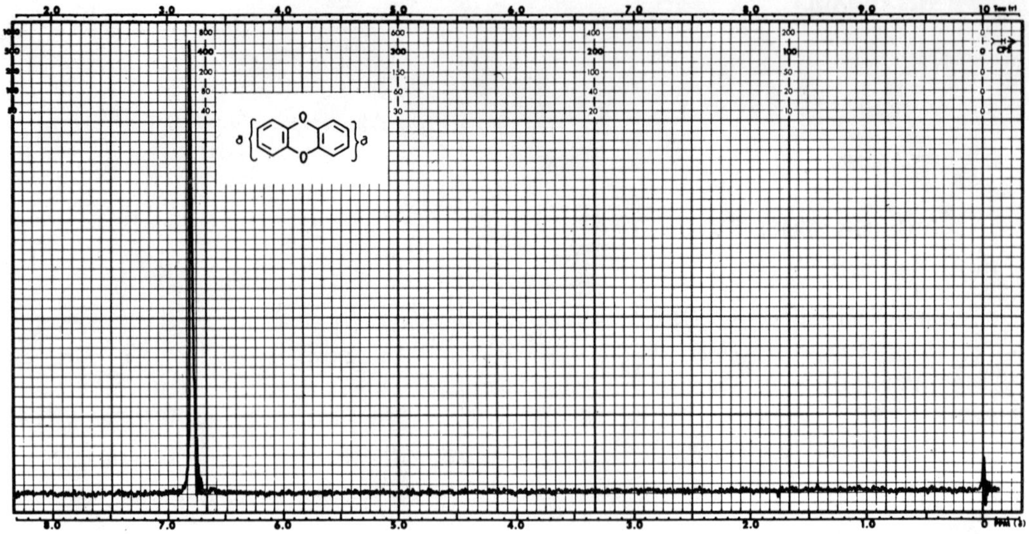

© 1978 Sadtler Research Laboratories, Inc.

MISCELLANEOUS ETHERS

THE SILICON ETHERS

The NMR spectra of the silicon ethers appear quite similar to those of the aliphatic ethers. The presence of the silicon nucleus can normally be detected only when a hydrocarbon group is bonded directly to it.

The silicon ethers are readily soluble in carbon tetrachloride and deuterochloroform.

Chemical Shifts

Aliphatic Protons

HC–4	HC–3	HC–2	HC–1		Solvent	Spectrum
			3.52	$(CH_3{-}O)_4{-}Si$	CCl_4	1598
			3.56	$(CH_3{-}O)_3{-}Si{-}\emptyset$	CCl_4	----
			3.57	$(CH_3{-}O)_3{-}Si{-}R_3$	$CDCl_3$	----
		1.13	3.59	$CH_3{-}CH_2{-}O{-}Si(CH_3)_3$	CCl_4	1593
		1.20	3.77	$(CH_3{-}CH_2{-}O)_3{-}Si{-}CH{=}CH_2$	CCl_4	----
		1.22	3.81	$(CH_3{-}CH_2{-}O)_3{-}Si{-}CH_3$	$CDCl_3$	1595
		1.15	4.17	$((CH_3)_2{-}CH{-}O)_3{-}Si{-}CH{=}CH_2$	CCl_4	----
		1.17	4.21	$((CH_3)_2{-}CH{-}Q)_3{-}Si{-}NH{-}Si(O{-}R)_3$	CCl_4	----
0.94	1.35	1.55	3.71	$(CH_3{-}CH_2{-}CH_2{-}CH_2{-}O)_4{-}Si$	CCl_4	1599

THE PHOSPHORUS ETHERS

The esters of phosphorus acid ($P{-}(O{-}R)_3$) possess chemical shifts characteristic of the oxygen substituent and in addition display additional coupling across the oxygen linkage to the phosphorus atom. This coupling to the first aliphatic group is usually similar to that of three bond proton-proton coupling (J = 6–8 Hz). The magnitude of the coupling constant between phosphorus atom and the second aliphatic group is usually too small to be clearly observed.

Chemical Shifts

Aliphatic Protons

HC–2	HC–1		Solvent	Spectrum
	3.41	$(CH_3–O)_3–P$	CCl_4	----
1.21	3.80	$(CH_3–CH_2–O)_3–P$	CCl_4	----

Aromatic Protons

Triphenyl Phosphite

$P–(\emptyset)_3$	ortho/meta/para	Solvent	Spectrum
	6.90–7.50	$CDCl_3$	1608

Coupling and Coupling Constants

$J_{P–O–CH_2}$ = 6–8 Hz $J_{P–O–C–CH_3}$ = 0–1 Hz

HYPOCHLOROUS ACID, tert-BUTYL ESTER

C_4H_9ClO

Mol. Wt. 108.57

Solvent: $CDCl_3$

Assignments
a = 1.33

1592

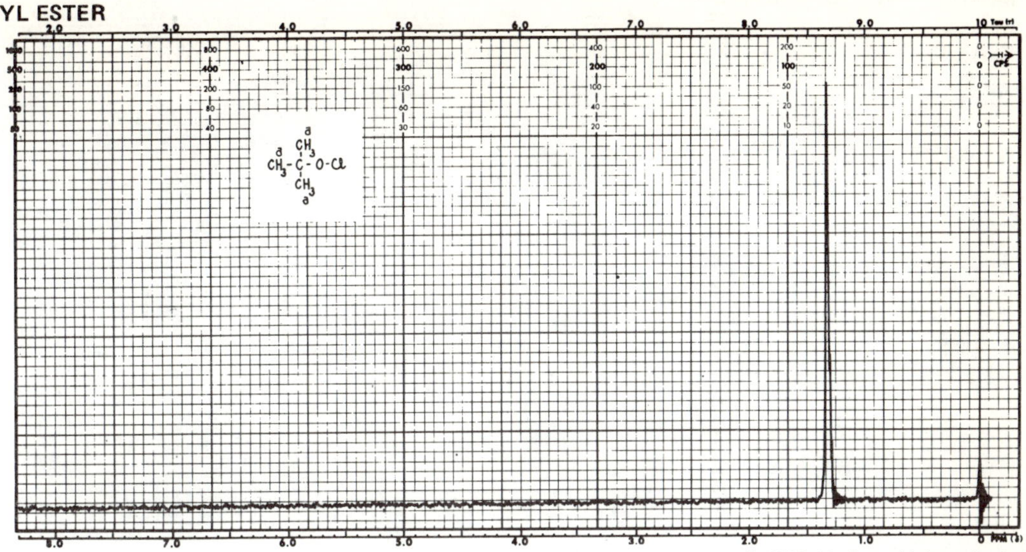

© 1978 Sadtler Research Laboratories, Inc.

ETHOXYTRIMETHYLSILANE

$C_5H_{14}OSi$

Mol. Wt. 118.25

B.P. 75-76°C

Solvent: CCl_4

Assignments
a = 0.08
b = 1.13
c = 3.59

1593

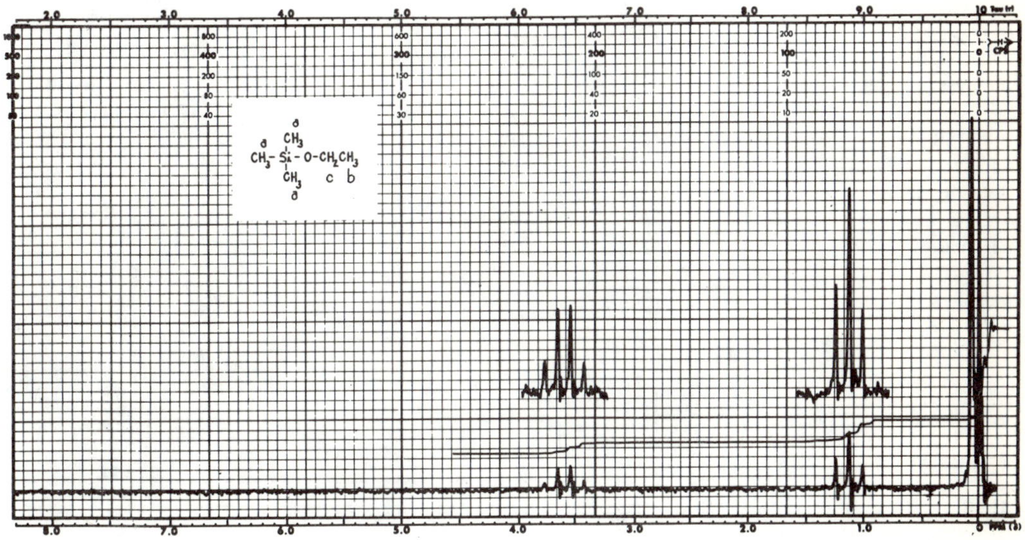

© 1978 Sadtler Research Laboratories, Inc.

DIMETHOXYDIPHENYLSILANE

$C_{14}H_{16}O_2Si$

Mol. Wt. 244.37

B.P. 191°C/53mm

Solvent: $CDCl_3$

Assignments
a = 3.61
b = 7.15-7.45
c = 7.61

1594

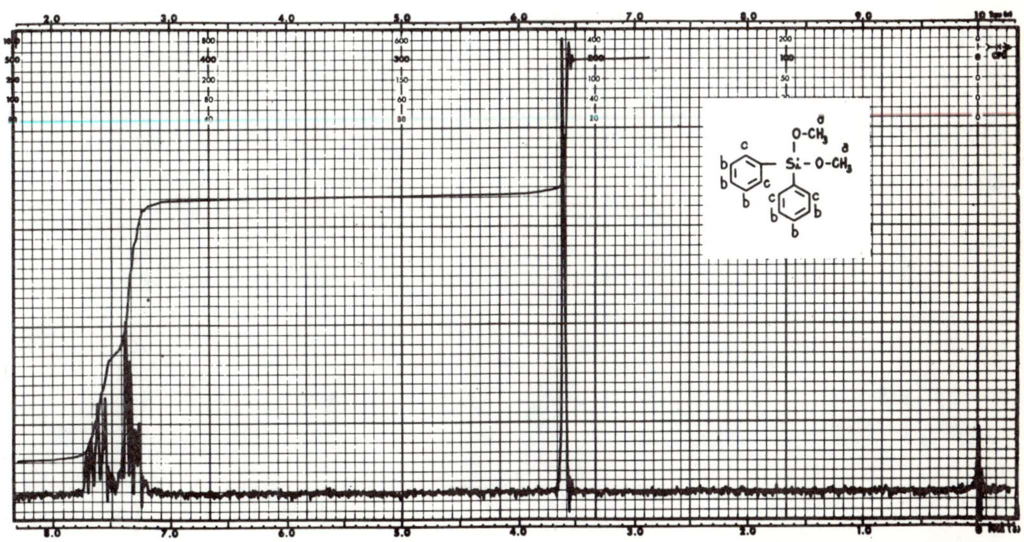

© 1978 Sadtler Research Laboratories, Inc.

METHYLTRIETHOXYSILANE

$C_7H_{18}O_3Si$

Mol. Wt. 178.31

B.P. 143°C

Solvent: $CDCl_3$

Assignments
a = 0.13
b = 1.22
c = 3.81

1595

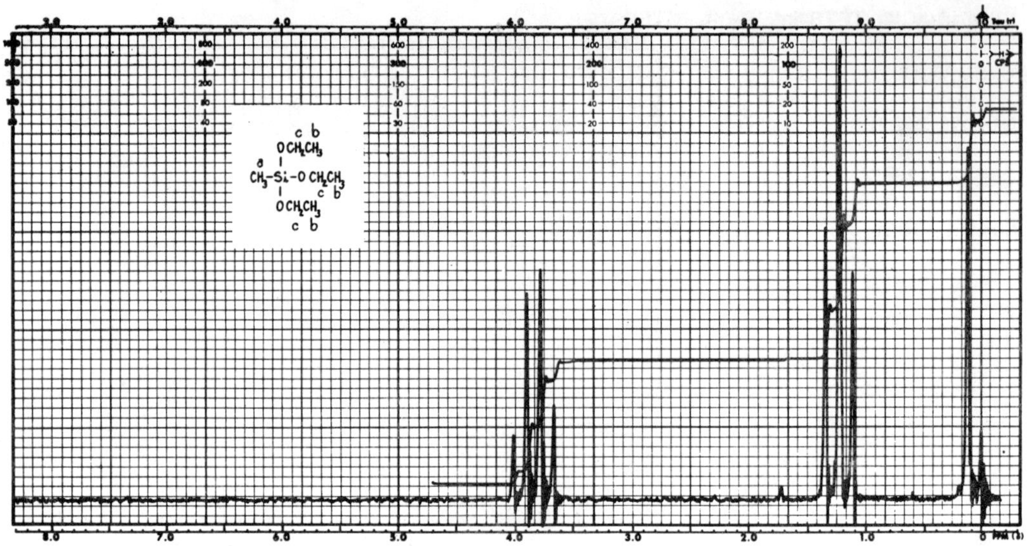

© 1978 Sadtler Research Laboratories, Inc.

ETHYLTRIETHOXYSILANE

$C_8H_{20}O_3Si$

Mol. Wt. 192.33

B.P. 158.5°C

Solvent: CCl_4

Assignments
a = 0.53
b = 0.98
c = 1.19
d = 3.76

1596

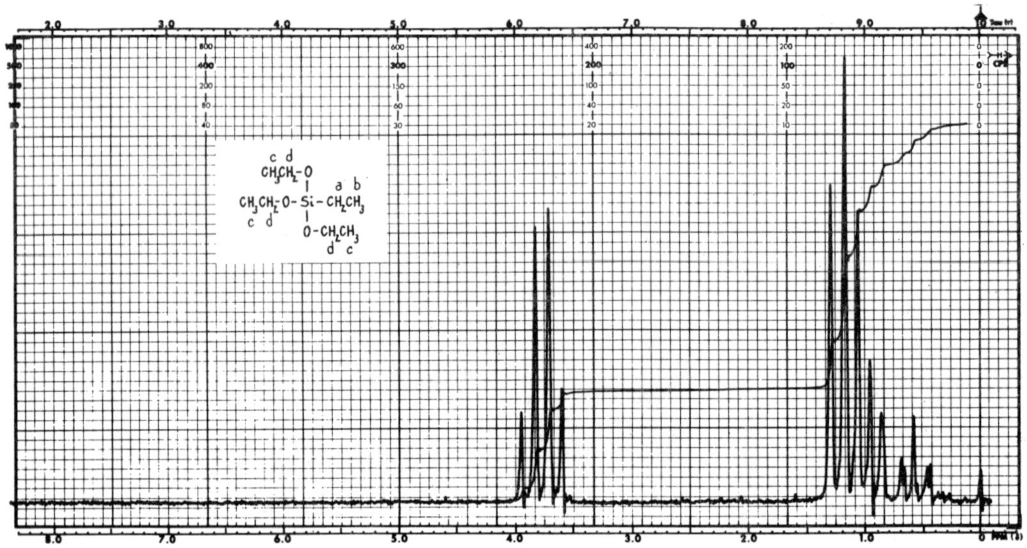

© 1978 Sadtler Research Laboratories, Inc.

3-(TRIETHOXYSILYL)PROPYLAMINE

$C_9H_{23}NO_3Si$

Mol. Wt. 221.37

B.P. 122-123°C/31mm

Solvent: $CDCl_3$

Assignments
a = 0.64
b = 1.22
c = 1.40-1.90
d = 2.68
e = 3.80
f = 3.47, 3.58 imps.
g = 3.69, 7.49 imps.

1597

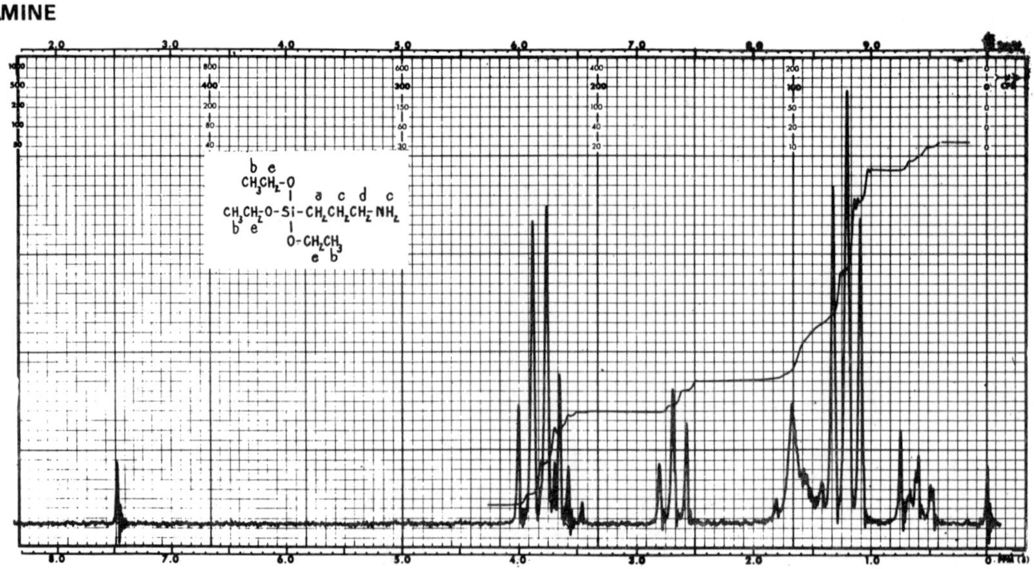

© 1978 Sadtler Research Laboratories, Inc.

SILICIC ACID, TETRAMETHYL ESTER

$C_4H_{12}O_4Si$

Mol. Wt. 152.22

B.P. 121-122°C

Solvent: $CDCl_3$

Assignments
a = 3.60

1598

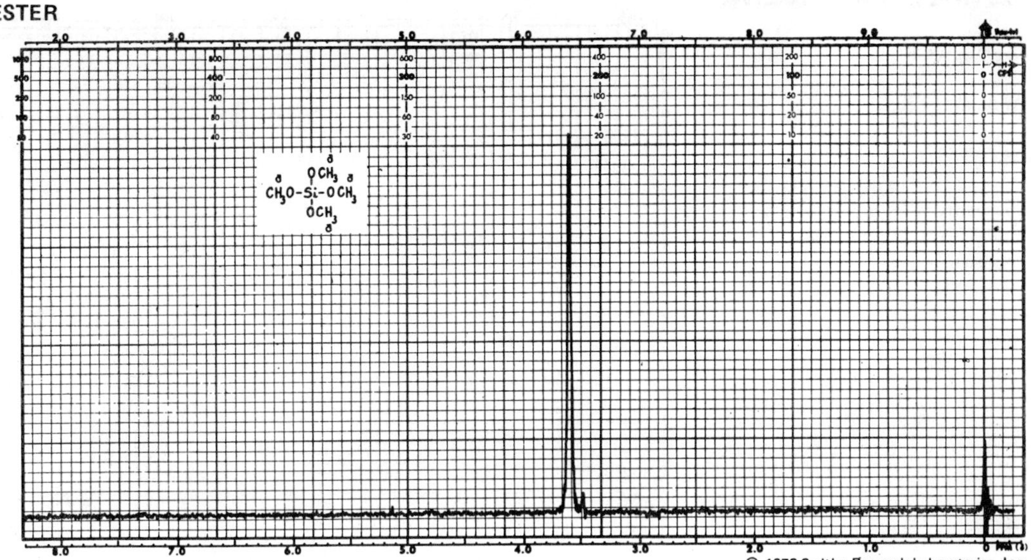

© 1978 Sadtler Research Laboratories, Inc.

BUTYL SILICATE

$C_{16}H_{36}O_4Si$

Mol. Wt. 320.55

Solvent: CCl_4

Assignments
a = 0.94
b = 1.00-1.70
c = 3.71

1599

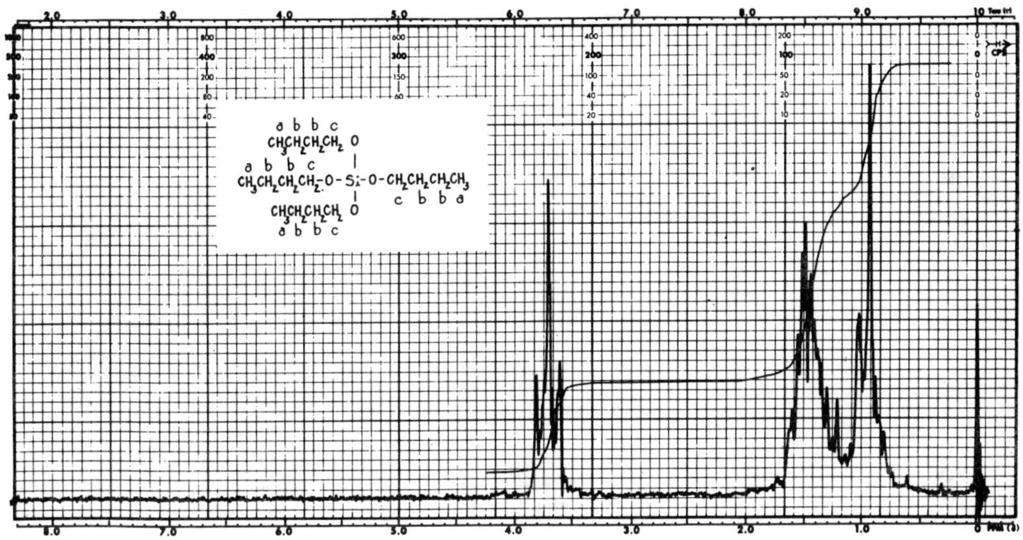

© 1978 Sadtler Research Laboratories, Inc.

PENTYL SILICATE

$C_{20}H_{44}O_4Si$

Mol. Wt. 376.65

B.P. 148°C/3mm

Solvent: CCl_4

Assignments
a = 0.90
b = 1.38
c = 3.67

1600

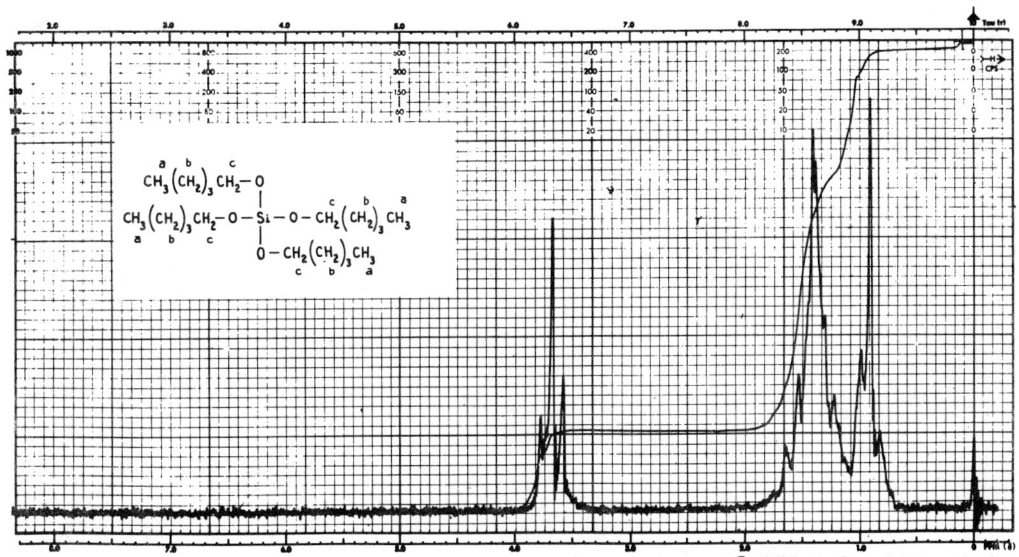

© 1978 Sadtler Research Laboratories, Inc.

(2-ETHYLBUTYL) SILICATE

$C_{24}H_{52}O_4Si$

Mol. Wt. 436.77

Solvent: CCl_4

Assignments
a = 0.92
b = 1.34
c = 3.66

1601

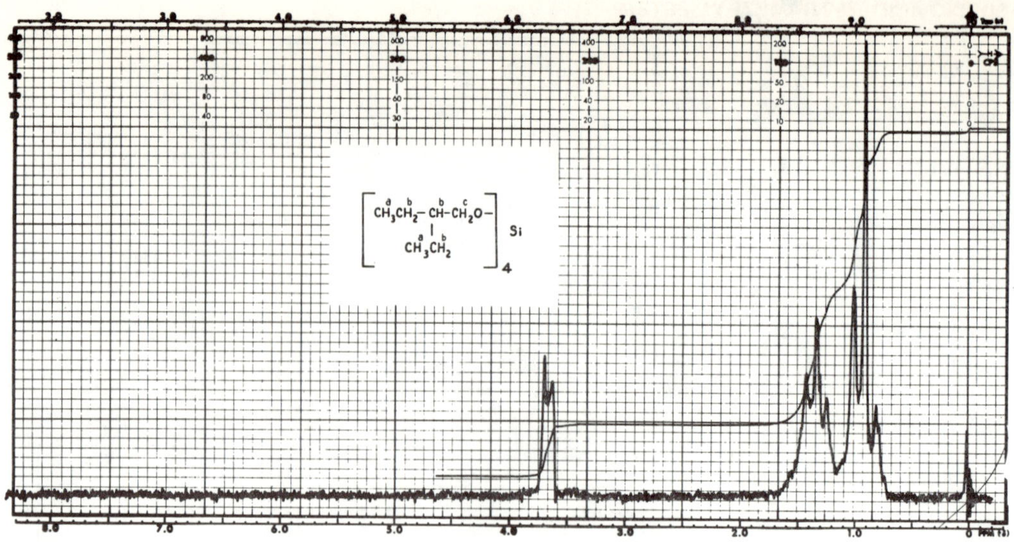

© 1978 Sadtler Research Laboratories, Inc.

SILICIC ACID, TETRAKIS(2-ETHYLHEXYL) ESTER

$C_{32}H_{68}O_4Si$

Mol. Wt. 544.98

Solvent: CCl_4

Assignments
a = ca 0.92
b = 1.10-1.60
c = 3.67

1602

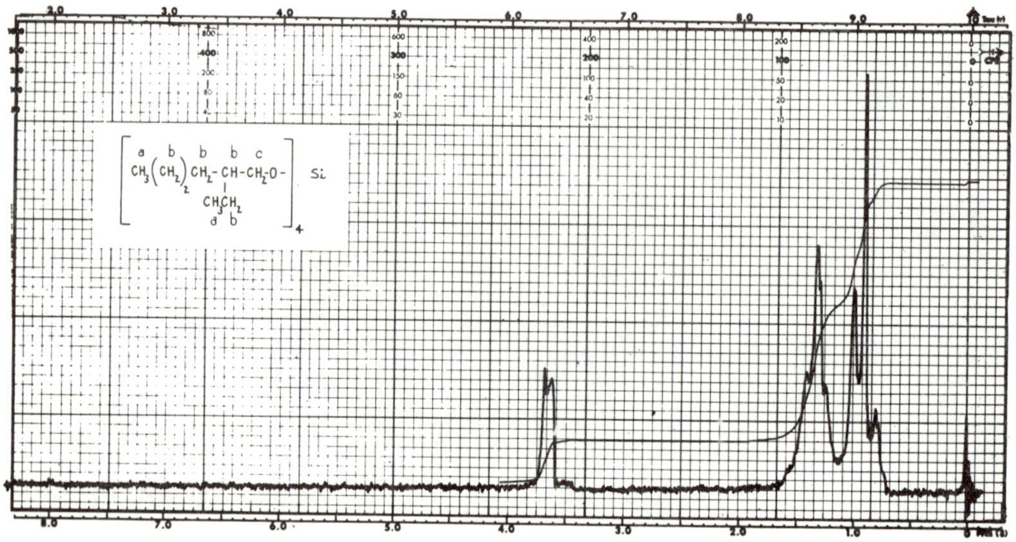

© 1978 Sadtler Research Laboratories, Inc.

HEXAMETHYLDISILOXANE

$C_6H_{18}OSi_2$

Mol. Wt. 162.38

Solvent: CCl_4

Assignments
a = 0.04

1603

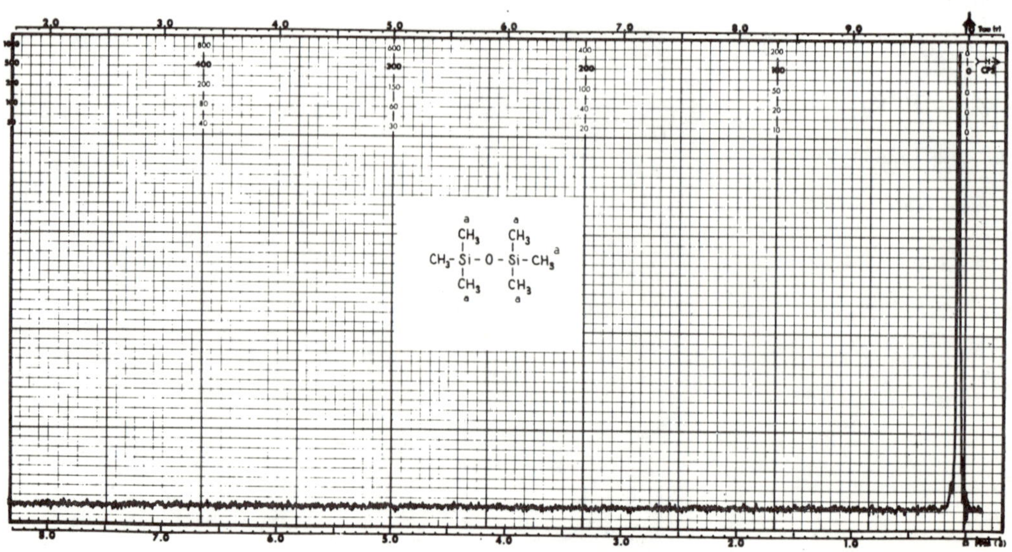

© 1978 Sadtler Research Laboratories, Inc.

1,3-BIS(CHLOROMETHYL)-1,1,3,3-TETRAMETHYLDISILOXANE

$C_6H_{16}Cl_2OSi_2$

Mol. Wt. 231.27

Solvent: $CDCl_3$

Assignments
a = 0.23
b = 2.75

1604

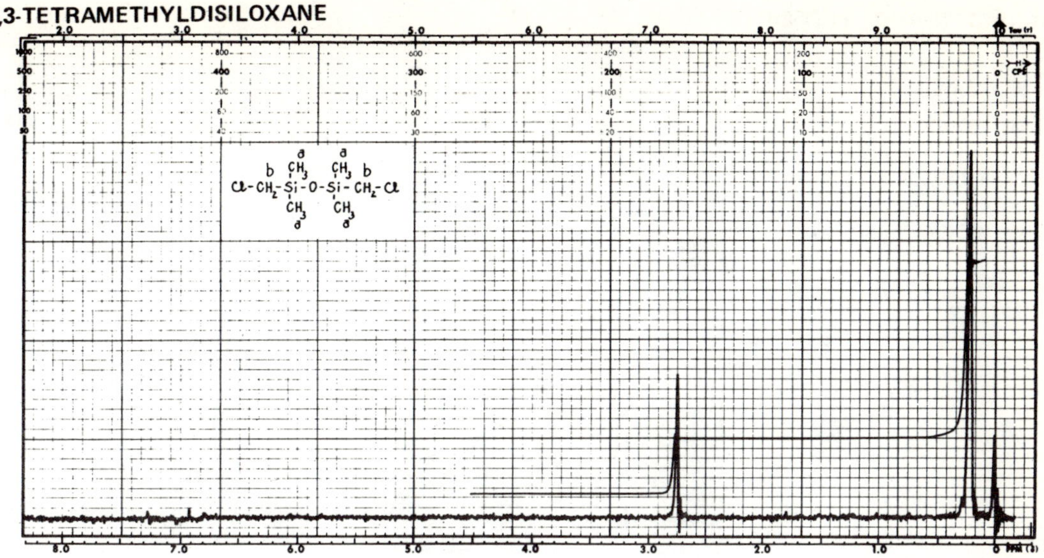

© 1978 Sadtler Research Laboratories, Inc.

OCTAMETHYLCYCLOTETRASILOXANE

$C_8H_{24}O_4Si_4$

Mol. Wt. 296.62

M.P. 17.4-17.6°C

B.P. 175°C

Solvent: CCl_4

Assignments
a = 0.08

1605

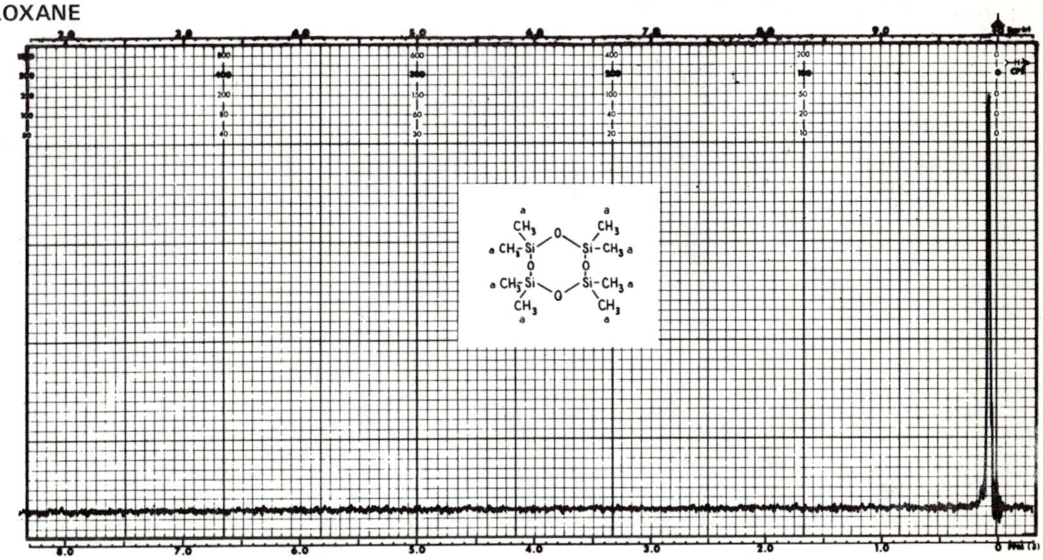

© 1978 Sadtler Research Laboratories, Inc.

TRI(2-ETHYLHEXYL)PHOSPHITE

$C_{24}H_{51}O_3P$

Mol. Wt. 418.65

B.P. 148-150°C/0.15mm

Solvent: CCl_4

Assignments
a = 0.92
b = 1.32
c = 3.62

1606

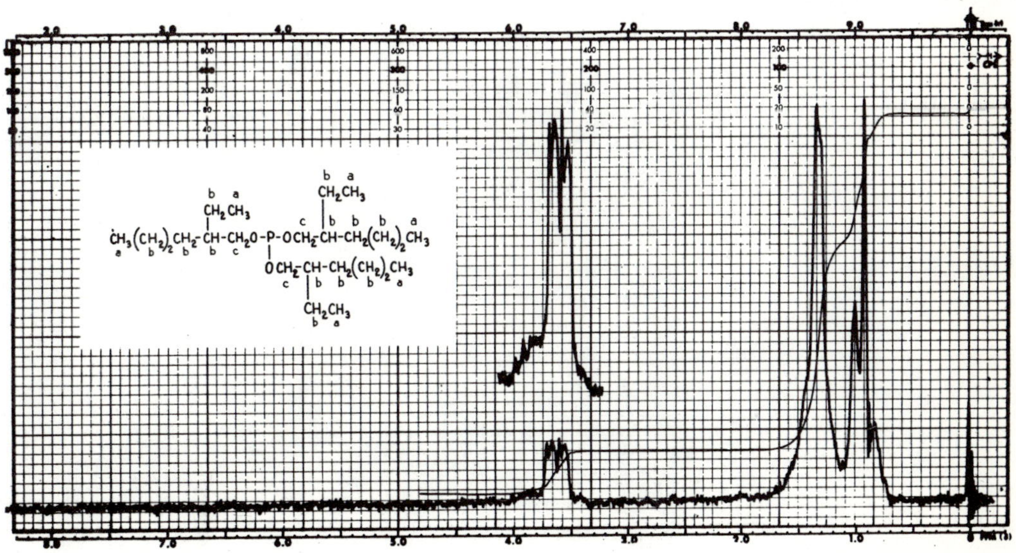

© 1978 Sadtler Research Laboratories, Inc.

2-CHLOROETHANOL, PHOSPHITE (3:1)

$C_6H_{12}Cl_3O_3P$

Mol. Wt. 269.49

B.P. 112-115°C/2mm

Solvent: $CDCl_3$

Assignments
a = 3.67
b = 4.10

1607

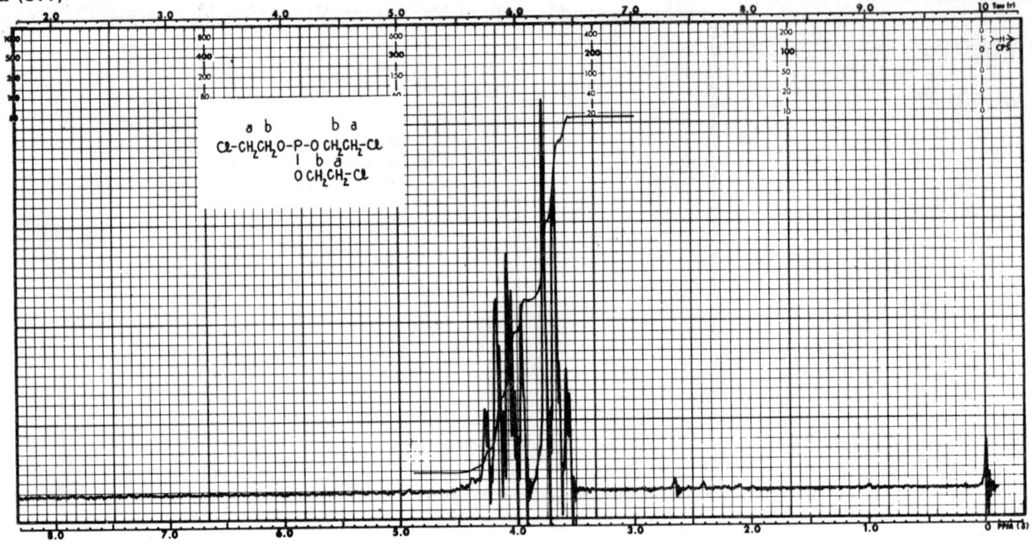

© 1978 Sadtler Research Laboratories, Inc.

PHOSPHOROUS ACID, TRIPHENYL ESTER

$C_{18}H_{15}O_3P$

Mol. Wt. 310.29

M.P. 22-24°C

Solvent: $CDCl_3$

Assignments
a = 6.90-7.50

1608

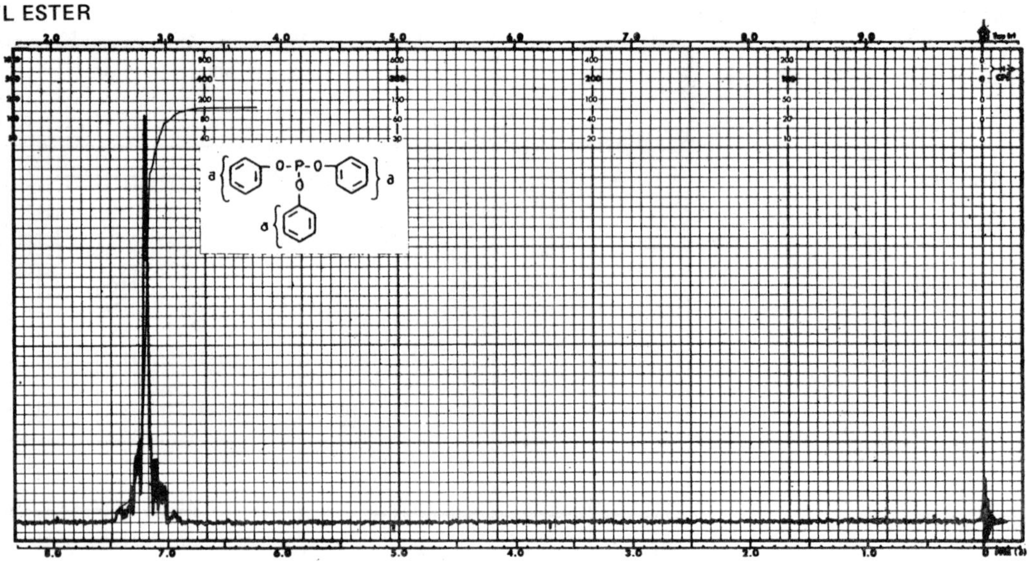

© 1978 Sadtler Research Laboratories, Inc.

DI-tert-BUTYL PEROXIDE

$C_8H_{18}O_2$

Solvent: CCl_4

Assignments
a = 1.19

1609

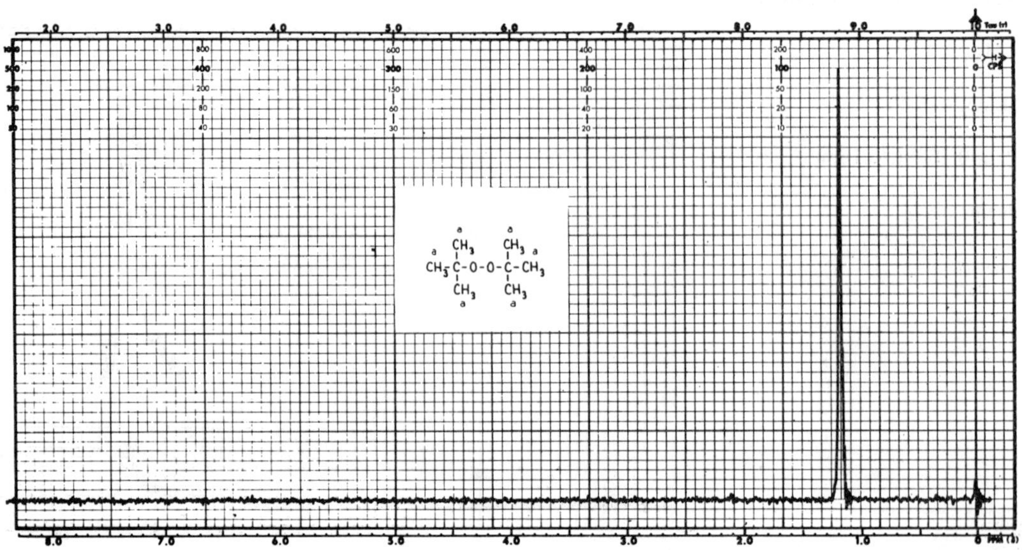

© 1978 Sadtler Research Laboratories, Inc.

1,4-EPIDIOXY-p-MENTH-2-ENE

$C_{10}H_{16}O_2$

Mol. Wt. 168.24

Solvent: CCl_4

Assignments
a = 0.96
b = 1.29
c = 1.30-2.15
d = 6.28
e = 6.40

1610

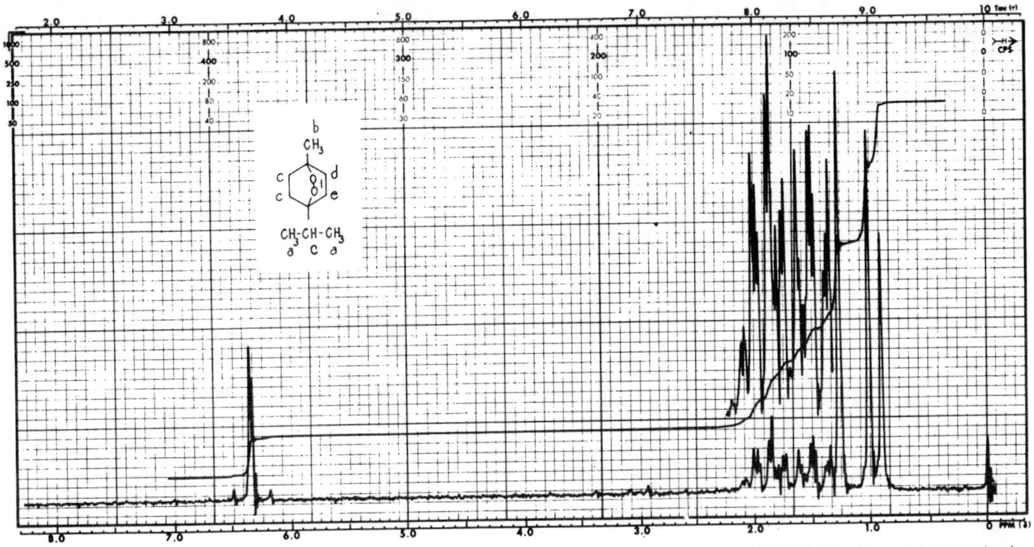

© 1978 Sadtler Research Laboratories, Inc.

NOTES

Because of the importance of the alcohols as solvents, their utilization as the starting material for the synthesis of many other compounds, and the large number of compounds available, the alcohols have been subdivided into four separate sections in this text. This chapter dealing with the primary alcohols is followed by that of the secondary alcohols, the tertiary alcohols and finally, the phenols.

The primary alcohols characteristically produce NMR spectra containing a methylene group in the chemical shift range from 3.3 to 5.4 ppm and one exchangeable proton which normally resonates over the range from 1.0 to 6.0 ppm. Both groups may be significantly broadened by partial coupling with each other. This coupling and the attendant broadening is easily eliminated by the addition of either acid or D_2O to the sample solution.

Chemical Shifts

Aliphatic Protons

HC–4	HC–3	HC–2	HC–1		Solvent	Spectrum
			3.34	CH_3–OH	CCl_4	1611
		1.17	3.58	CH_3–CH_2–OH	CCl_4	1612
	0.94	1.49	3.50	CH_3–CH_2–CH_2–OH	CCl_4	1613
	(0.89)	1.67	3.27	$(CH_3)_2$–CH–CH_2–OH	CCl_4	1615
	(0.87)	----	3.20	$(CH_3)_2$–C–CH_2–OH	CCl_4	----
0.91	1.35	1.55	3.52	CH_3–CH_2–CH_2–CH_2–OH	CCl_4	1614
(0.95)	----	1.50	3.71	$(CH_3)_3$–C–CH_2–CH_2–OH	$CDCl_3$	----
(0.91)	1.38	1.72	3.52	$(CH_3)_2$–CH–CH_2–CH_2–OH	CCl_4	1618

The Hydroxy Group

As with the other exchangeable protons, the chemical shift of the hydroxyl groups varies with concentration, temperature, solvent and the presence of impurities such as acid, base of H_2O. The trend for hydroxyl groups to resonate at lower field as their concentration in solution increases, can be illustrated by the selection of straight chain alcohols listed below. Note that as the molecular weight of the compound decreases with decreasing chain length, the chemical shift of the hydroxyl resonance increases proportionately.

HO–	X	Solvent	Spectrum
1.62	$-R_{20}$	$CDCl_3$	1634
2.05	$-R_{18}$	$CDCl_3$	1633
3.24	$-R_{11}$	CCl_4	1631
3.67	$-R_6$	CCl_4	1619
4.11	$-R_4$	CCl_4	1614
4.40	$-R_2$	CCl_4	1612

The Hydroxyl Group continued

Substituted Methanols

HO	$-CH_2$	$-X$	Solvent
3.49	3.07	−1(Adamantane)	Polysol
3.48	3.32	−Cyclohexane	CCl_4
4.40	3.58	$-CH_3$	CCl_4
- - - -	3.62	$-(CH_2-OH)_3$	D_2O
4.46	3.90	$-CF_3$	$CDCl_3$
4.11	3.92	$-CH=CH-CH_3$	CCl_4
4.60	4.05	$-CH=CH_2$	CCl_4
- - - -	4.06	$-C(=O)-NH_2$	D_2O
3.76	4.07	$-C(=O)-O-CH_3$	CCl_4
3.99	4.11	$-C\equiv C-CH_3$	CCl_4
4.01	4.23	$-C\equiv C-H$	$CDCl_3$
3.80	4.27	$-C(=O)-CH_3$	$CDCl_3$
- - - -	4.29	$-C(=O)-OH$	D_2O
3.10	4.41	−Ø	CCl_4
- - - -	4.46	$-SO_3-Na$	D_2O
3.32	4.50	−2(Furan)	$CDCl_3$
4.45	4.61	$-SO_2-$Ø	$CDCl_3$
6.27	4.71	−4(Pyridine)	$CDCl_3$
5.02	4.78	−2(Pyridine)	$CDCl_3$
3.63	4.86	$-C(=O)-$Ø	$CDCl_3$

2-Substituted Ethanols

HO	$-CH_2$	$-CH_2$	$-X$	
3.71	3.50	1.49	$-CH_3$	CCl_4
3.52	3.45	2.50	$-N(R_2)_2$	CCl_4
2.57	3.65	2.37	$-CH=CH_2$	$CDCl_3$
3.67	3.67	2.38	$-C\equiv C-H$	CCl_4
3.99	3.80	2.58	$-C\equiv N$	$CDCl_3$
- - - -	3.69	2.68	$-SH$	D_2O
2.82	3.54	2.73	$-NH_2$	$CDCl_3$
2.44	3.71	2.77	−Ø	$CDCl_3$
3.41	3.69	3.01	−1(Naphthalene)	CCl_4

The Hydroxyl Group continued

2-Substituted Ethanols continued

HO	$-CH_2$	$-CH_2$	$-X$	Solvent
----	3.93	3.12	$-SO_3-K$	D_2O
----	3.85	3.18	$-NH_2(HCl)$	D_2O
4.29	3.85	3.45	$-Br$	CCl_4
2.49	3.71	3.51	$-O-CH_3$	$CDCl_3$
4.61	4.05	3.56	$-N^+(CH_3)_3\,(Cl^-)$	$CDCl_3$
4.50	3.79	3.63	$-Cl$	$CDCl_3$

Coupling and Coupling Constants

Clearly defined coupling between the hydroxyl group and the adjacent hydrocarbon group is usually not observed in solutions of the alcohols in CCl_4 and $CDCl_3$. It appears that in these solvents, the hydroxyl protons are exchanging at an intermediate rate resulting in a broadening of both resonance bands such as is observed in spectra 1614, 1616 and 1626. In spectrum 1618, relatively clear coupling is observed (J_{CH_2-OH} = 5 Hz) indicating a much slower rate of exchange.

In solutions of alcohols in Acetone and DMSO, clear coupling between the hydroxyl protons and adjacent hydrocarbon groups is the rule rather than the exception and it appears to result from the presence of the small amount of water that is usually present in these solvents.

The addition of a small amount of D_2O or acid will remove any coupling or broadening that appears in the NMR spectra of the alcohols. This procedure was used to produce the very narrow lines that appear in spectra 1611, 1641 and 1689. The exchange rate can be increased by heating the sample solutions, resulting in sharp single peaks for the hydroxyl resonance, such as that of spectrum 1625.

The hydroxyl protons often interchange with other types of exchangeable protons present in the same molecular structure (spectra 1642 and 1730).

METHANOL

CH_4O

Mol. Wt. 32.04

B.P. 64.7°C

Solvent: CCl_4

Assignments
a = 3.34
b = 4.11

1611

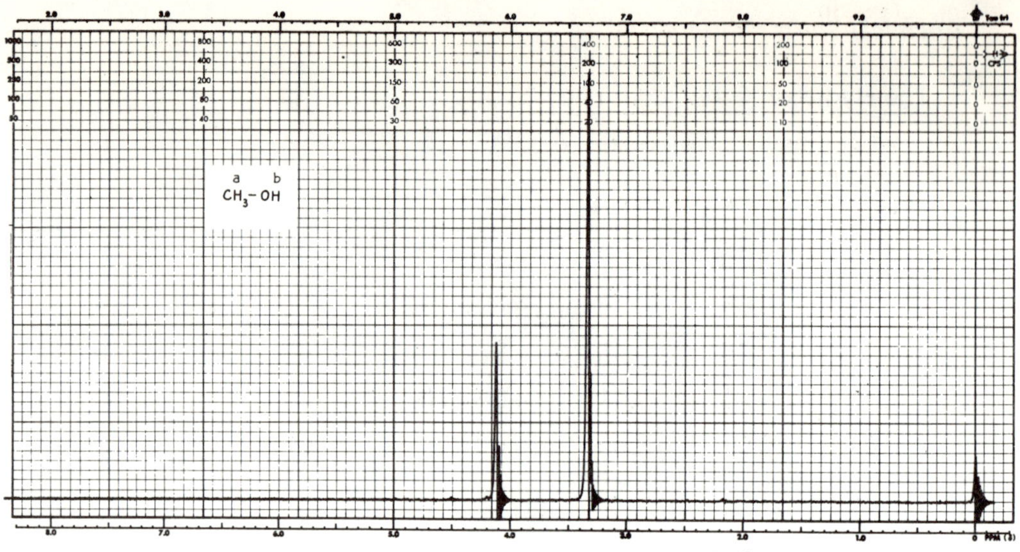

© 1978 Sadtler Research Laboratories, Inc.

ETHANOL

C_2H_6O

Mol. Wt. 46.07

B.P. 78.5°C

Solvent: CCl_4

Assignments
a = 1.17
b = 3.58
c = 4.40
d = 3.32 imp.

1612

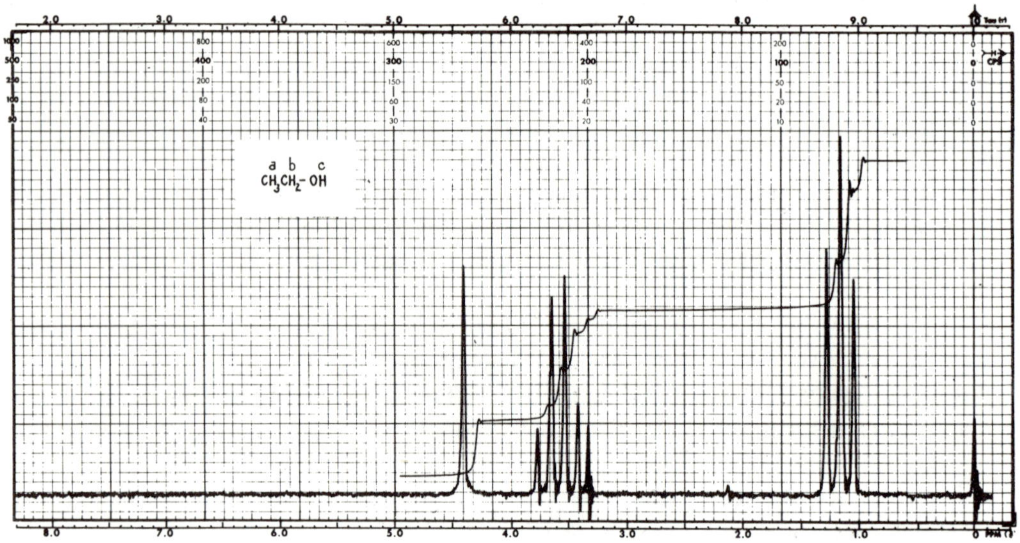

© 1978 Sadtler Research Laboratories, Inc.

PROPYL ALCOHOL

C_3H_8O

Mol. Wt. 60.09

B.P. 97.3-97.4°C

Solvent: CCl_4

Assignments
a = 0.94
b = 1.49
c = 3.50
d = 3.71

1613

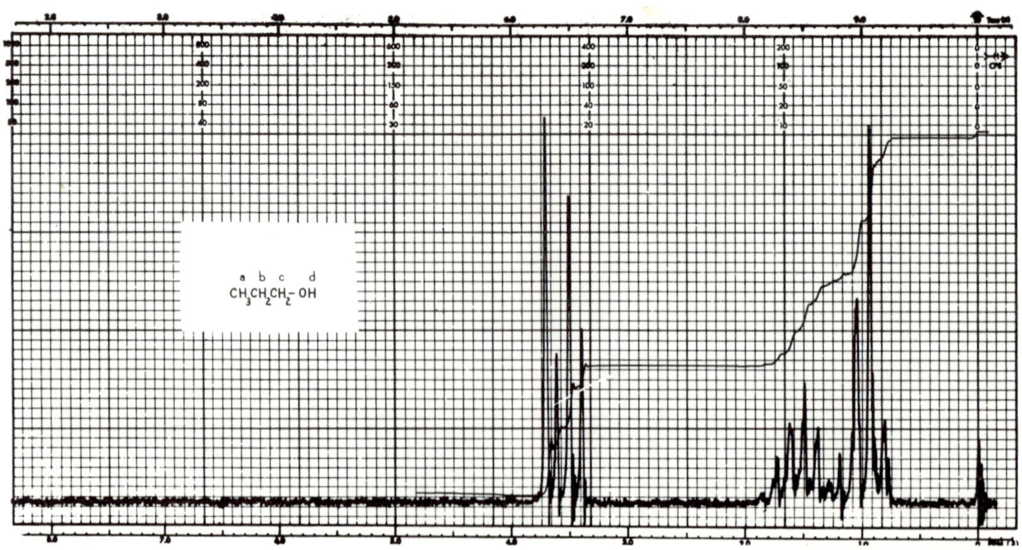

© 1978 Sadtler Research Laboratories, Inc.

BUTYL ALCOHOL

$C_4H_{10}O$

Mol. Wt. 74.12

M.P. --89.2°C

B.P. 117.7°C/760mm

Solvent: CCl_4

Assignments
a = 0.91
b = 1.10-1.70
c = 3.52
d = 4.11
Trace of D_2O added

1614

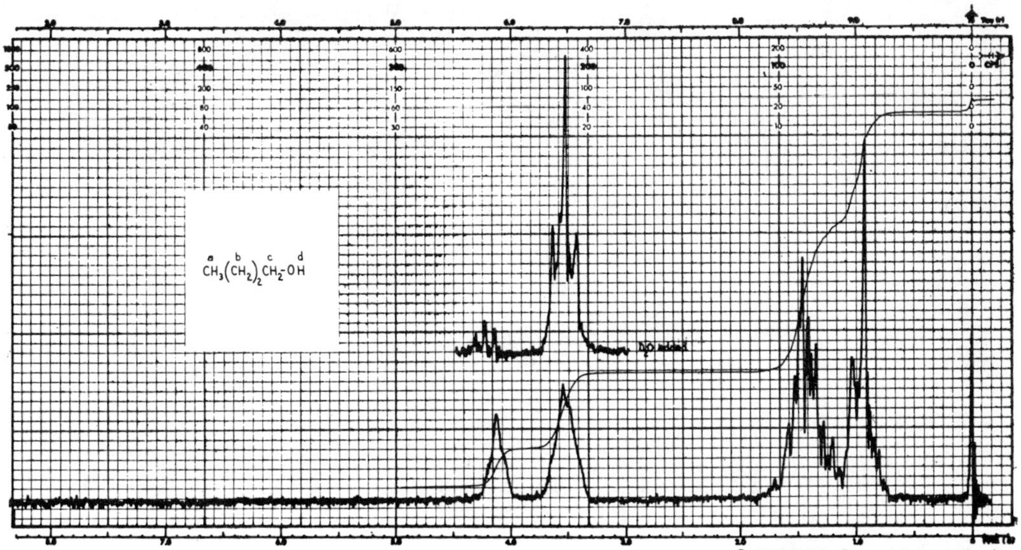

© 1978 Sadtler Research Laboratories, Inc.

ISOBUTYL ALCOHOL

$C_4H_{10}O$

Mol. Wt. 74.12

B.P. 107.4°C

Solvent: CCl_4

Assignments
a = 0.89
b = 1.67
c = 3.27
d = 3.98
Trace of D_2O added

1615

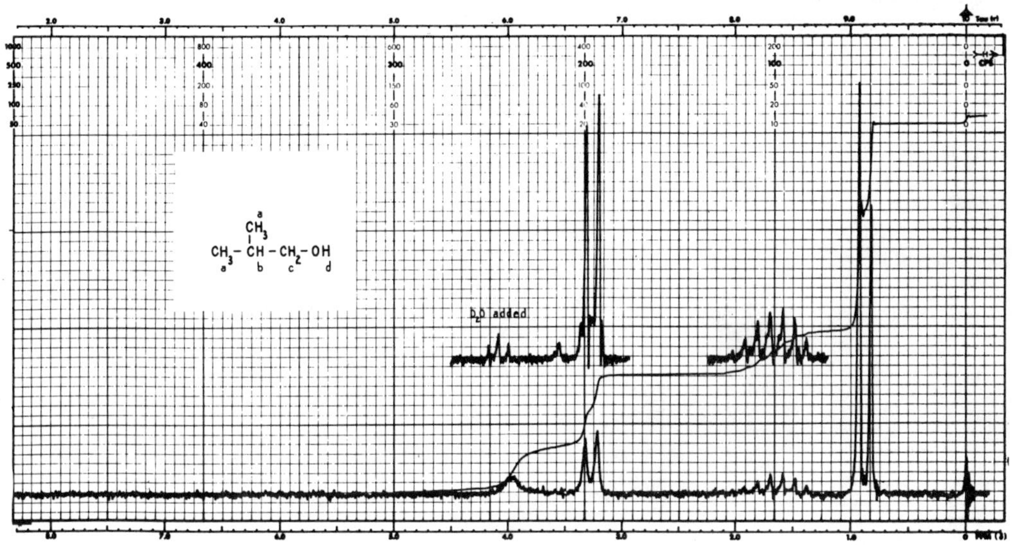

© 1978 Sadtler Research Laboratories, Inc.

1-PENTANOL

$C_5H_{12}O$

Mol. Wt. 88.15

B.P. 136-138°C

Solvent: $CDCl_3$

Assignments
a = 0.91
b = 1.09-1.80
c = 3.50
d = 4.41

1616

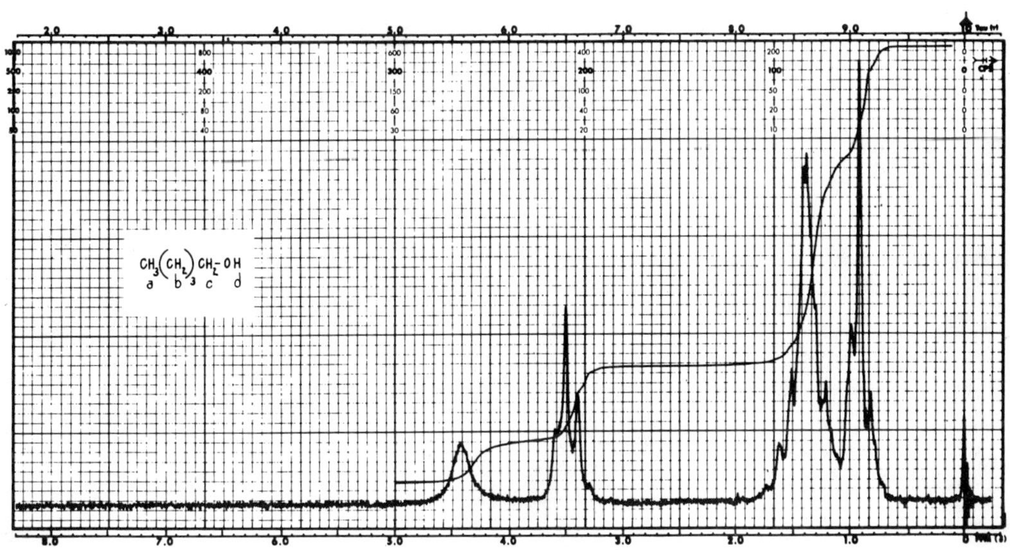

© 1978 Sadtler Research Laboratories, Inc.

2-METHYL-1-BUTANOL

$C_5H_{12}O$

Mol. Wt. 88.15

B.P. 126-128°C

Solvent: $CDCl_3$

Assignments
a = 0.88
b = 0.92
c = 1.00-1.90
d = 2.47
e = 3.48

1617

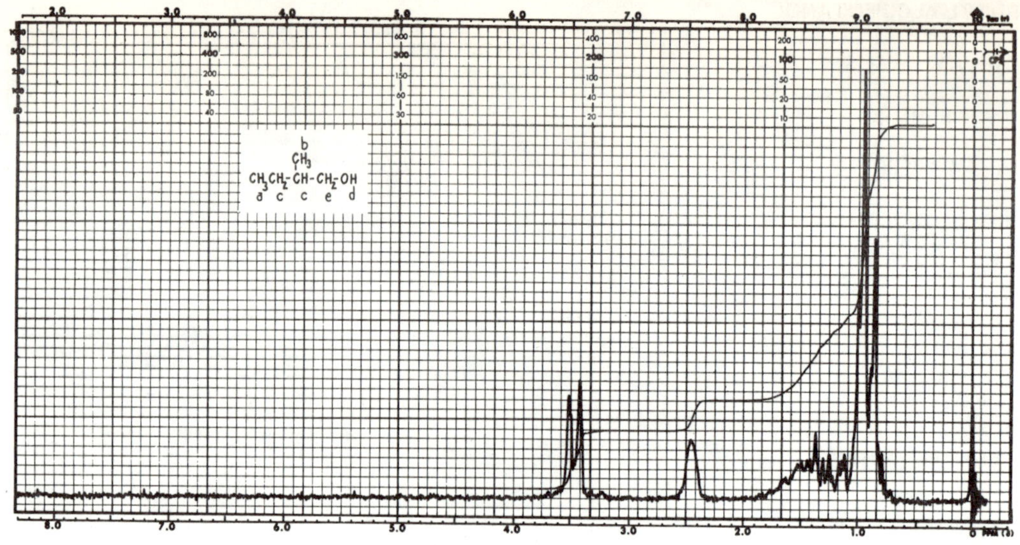

© 1978 Sadtler Research Laboratories, Inc.

3-METHYL-1-BUTANOL

$C_5H_{12}O$

Mol. Wt. 88.15

B.P. 131.4°C/760mm

Solvent: CCl_4

Assignments
a = 0.91
b = 1.10-1.90
c = 3.52
d = 4.11

1618

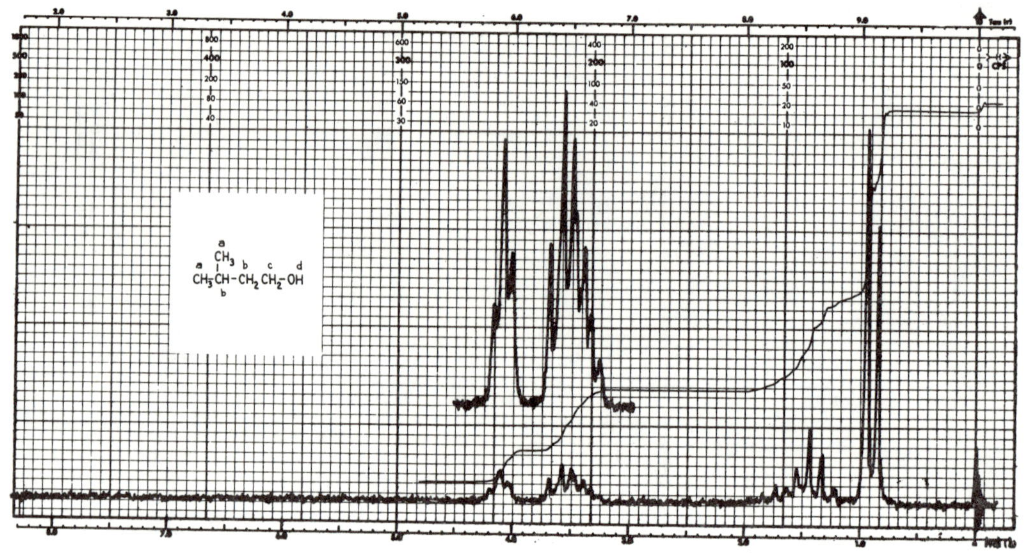

© 1978 Sadtler Research Laboratories, Inc.

HEXYL ALCOHOL

$C_6H_{14}O$

Mol. Wt. 102.17

B.P. 157.2°C

Solvent: CCl_4

Assignments
a = 0.90
b = ca 1.35
c = 3.50
d = 3.67
Trace of acid added for OH

1619

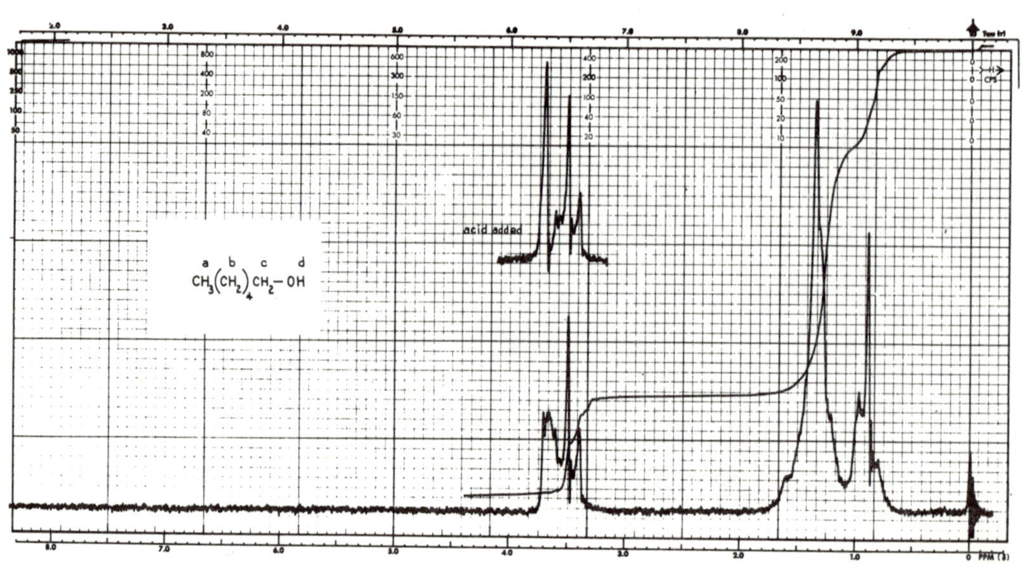

© 1978 Sadtler Research Laboratories, Inc.

3-METHYL-1-PENTANOL

$C_6H_{14}O$

Mol. Wt. 102.18

B.P. 151-152°C

Solvent: CCl_4

Assignments
a = 0.87
b = 0.95
c = 1.00-1.90
d = 3.56
e = 3.79

1620

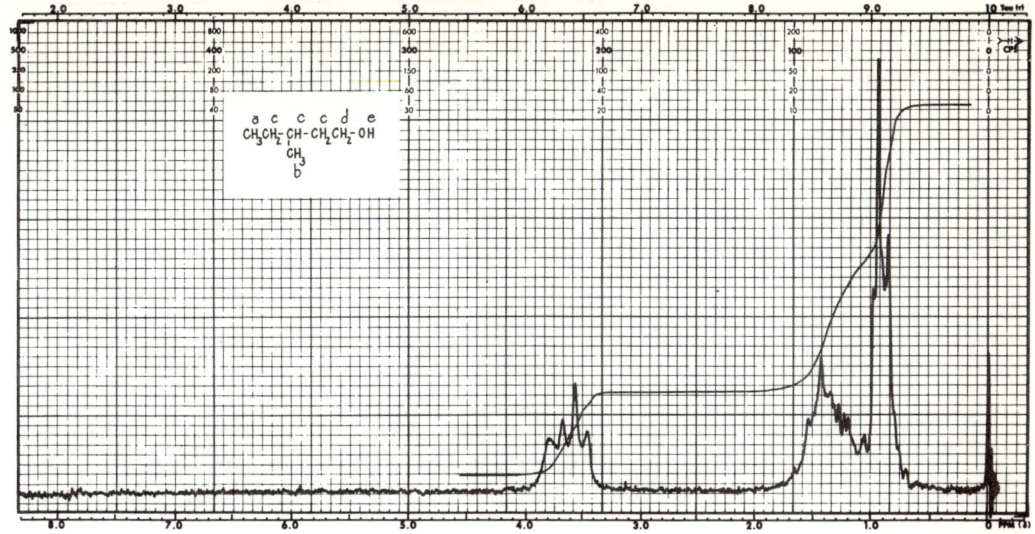

© 1978 Sadtler Research Laboratories, Inc.

4-METHYL-1-PENTANOL

$C_6H_{14}O$

Mol. Wt. 102.18

B.P. 160-165°C

Solvent: CCl_4

Assignments
a = 0.91
b = 1.00-1.90
c = 3.50
d = 4.13

1621

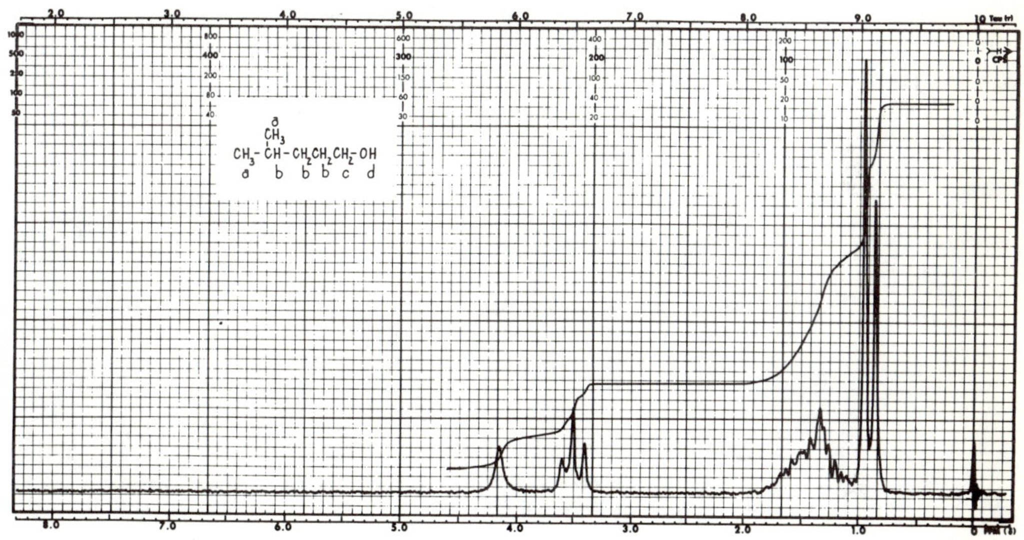

© 1978 Sadtler Research Laboratories, Inc.

2-ETHYLBUTANOL

$C_6H_{14}O$

Mol. Wt. 102.17

B.P. 148.2-149.1°C

Solvent: CCl_4

Assignments
a = 0.88
b = 1.10-1.60
c = 2.91
d = 3.43

1622

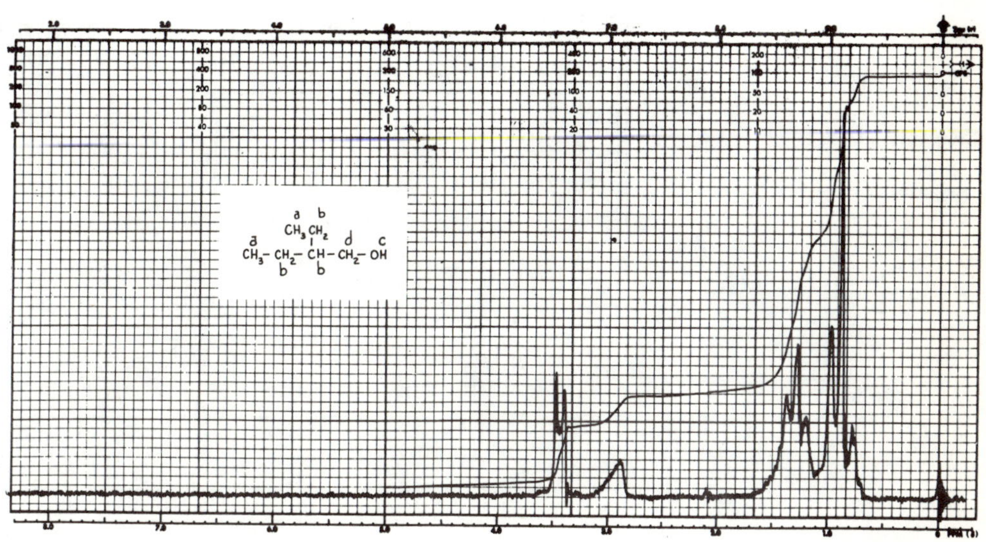

© 1978 Sadtler Research Laboratories, Inc.

1-HEPTANOL

$C_7H_{16}O$

Mol. Wt. 116.20

B.P. 172-176°C (lit.)

Solvent: CCl_4

Assignments
a = 0.89
b = ca 1.33
c = 3.40
d = 3.50
Trace of acid added for OH

1623

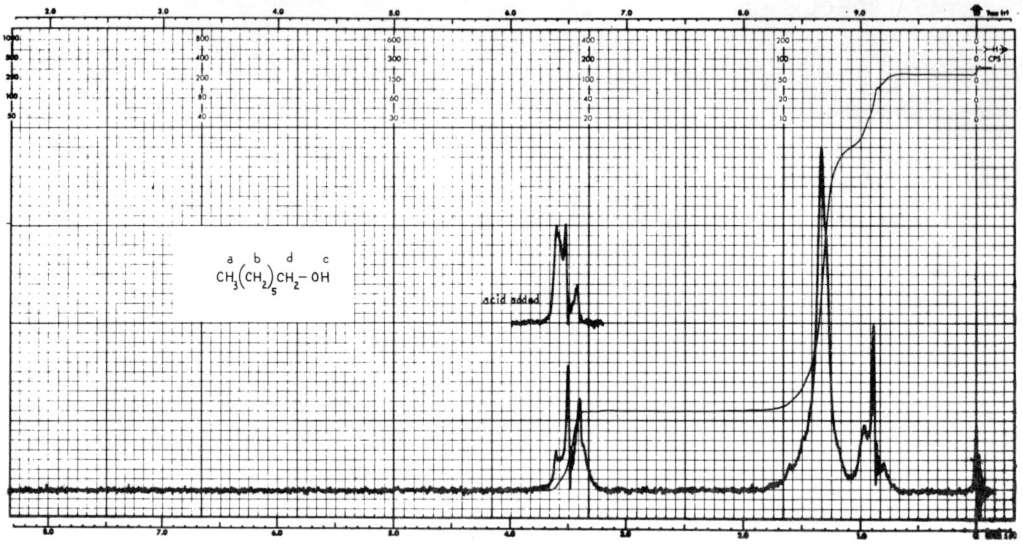

© 1978 Sadtler Research Laboratories, Inc.

2,2-DIMETHYL-1-PENTANOL

$C_7H_{16}O$

Mol. Wt. 116.20

Solvent: $CDCl_3$

Assignments
a = 0.85
b = 0.89
c = 1.00-1.40
d = 2.59
e = 3.27

1624

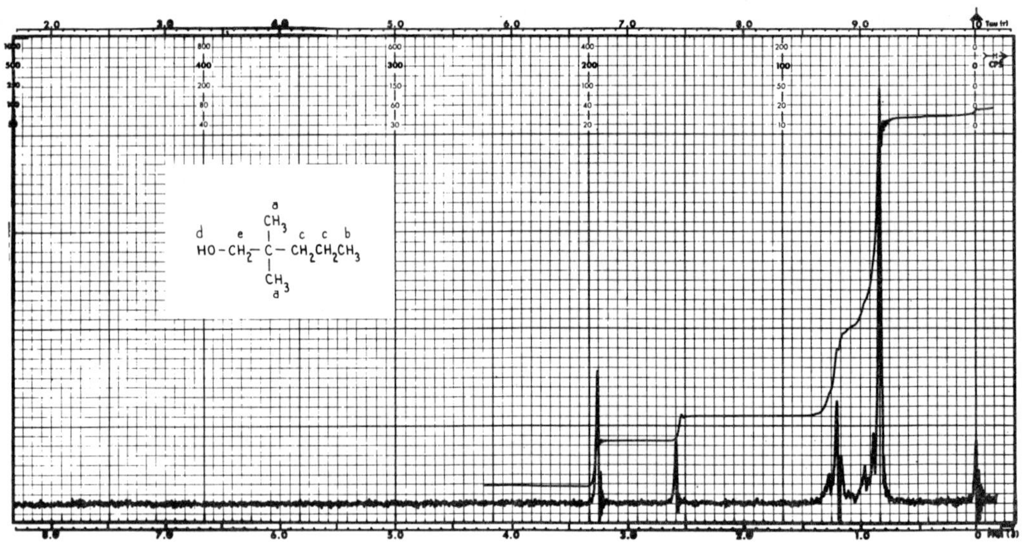

© 1978 Sadtler Research Laboratories, Inc.

5-METHYL-1-HEXANOL

$C_7H_{16}O$

Mol. Wt. 116.20

Solvent: $CDCl_3$

Assignments
a = 0.90
b = 1.10-1.80
c = 2.85
d = 3.60

1625

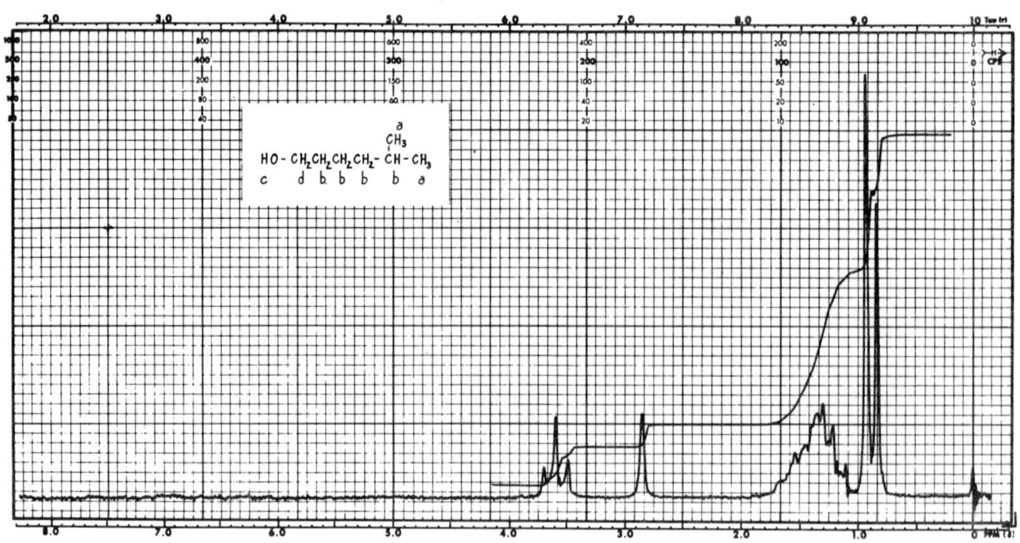

© 1978 Sadtler Research Laboratories, Inc.

OCTYL ALCOHOL

$C_8H_{18}O$

Mol. Wt. 130.23

B.P. 195°C

Solvent: CCl_4

Assignments
a = 0.88
b = ca 1.29
c = 3.49
d = 3.90

1626

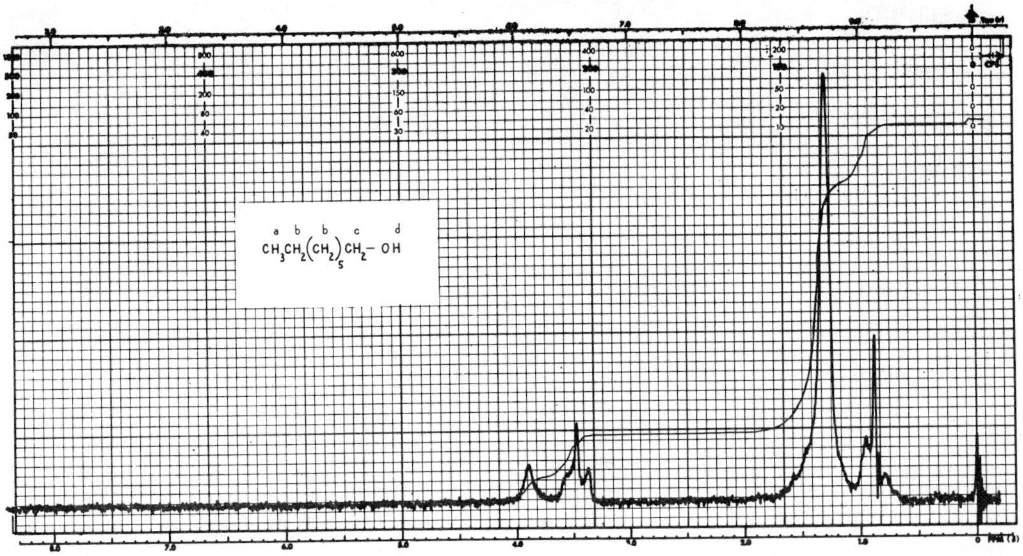

© 1978 Sadtler Research Laboratories, Inc.

2-ETHYL-4-METHYL-1-PENTANOL

$C_8H_{18}O$

Mol. Wt. 130.23

Solvent: CCl_4

Assignments
a = 0.90
b = 0.91
c = 1.04-1.83
d = 1.97
e = 3.43

1627

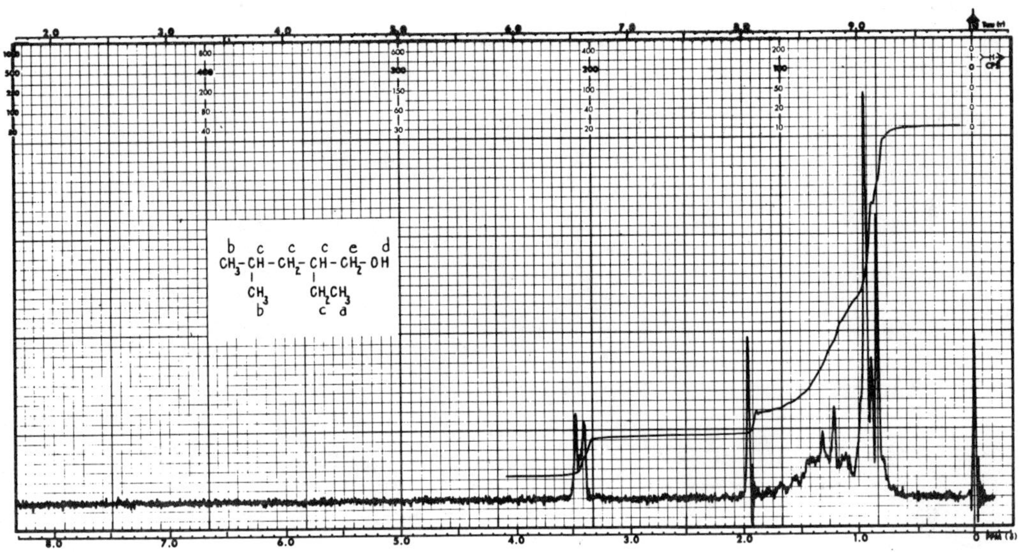

© 1978 Sadtler Research Laboratories, Inc.

1-NONANOL

$C_9H_{20}O$

Mol. Wt. 144.25

B.P. 213°C

Solvent: CCl_4

Assignments
a = 0.89
b = 1.30
c = 2.33
d = 3.55

1628

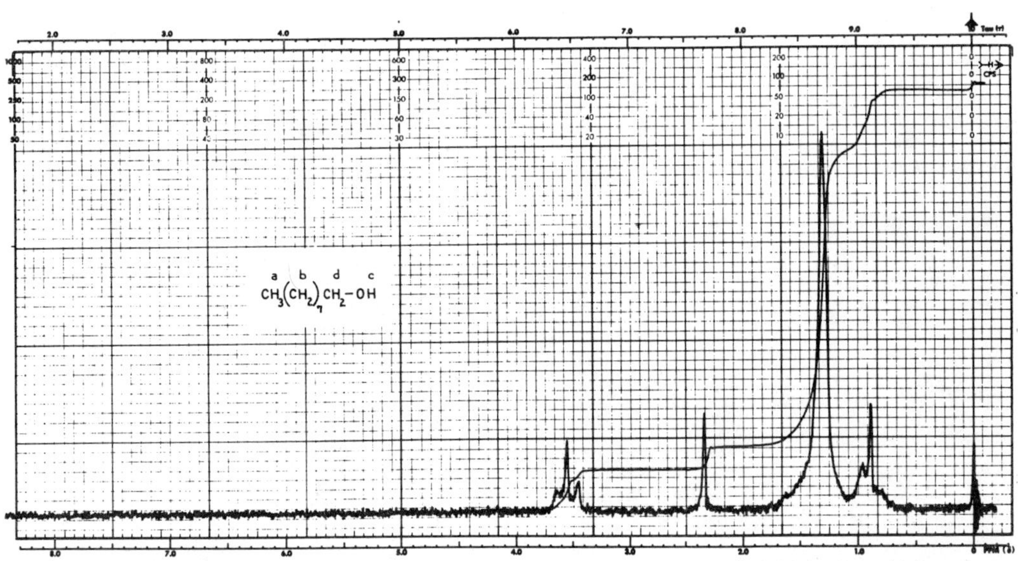

© 1978 Sadtler Research Laboratories, Inc.

3,5,5-TRIMETHYL-1-HEXANOL

$C_9H_{20}O$

Mol. Wt. 144.26

B.P. 193-195°C

Solvent: $CDCl_3$

Assignments
a = 0.89
b = 0.95-1.70
c = 1.74
d = 3.65

1629

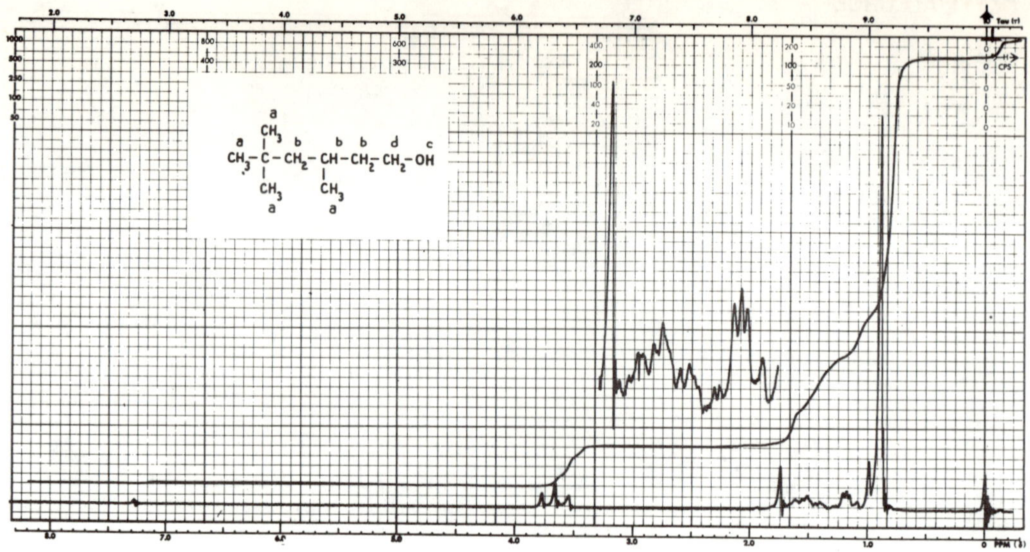

© 1978 Sadtler Research Laboratories, Inc.

1-DECANOL

$C_{10}H_{22}O$

Mol. Wt. 158.29

M.P. 5.5-6.5°C

B.P. 231°C

Solvent: CCl_4

Assignments
a = 0.90
b = 1.30
c = 3.50

1630

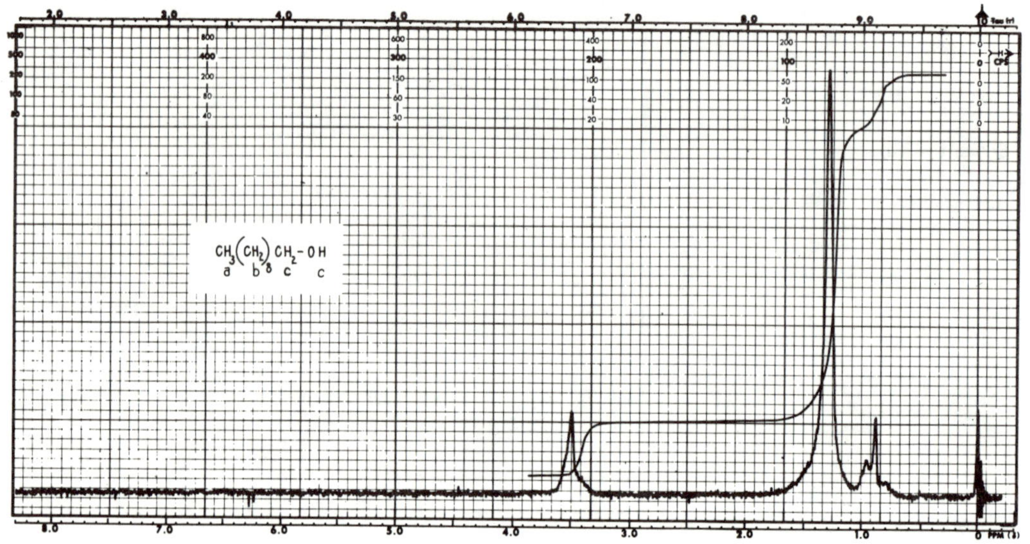

© 1978 Sadtler Research Laboratories, Inc.

1-UNDECANOL

$C_{11}H_{24}O$

Mol. Wt. 172.31

B.P. 109-110°C/5mm

Solvent: CCl_4

Assignments
a = 0.90
b = 1.30
c = 3.24
d = 3.51

1631

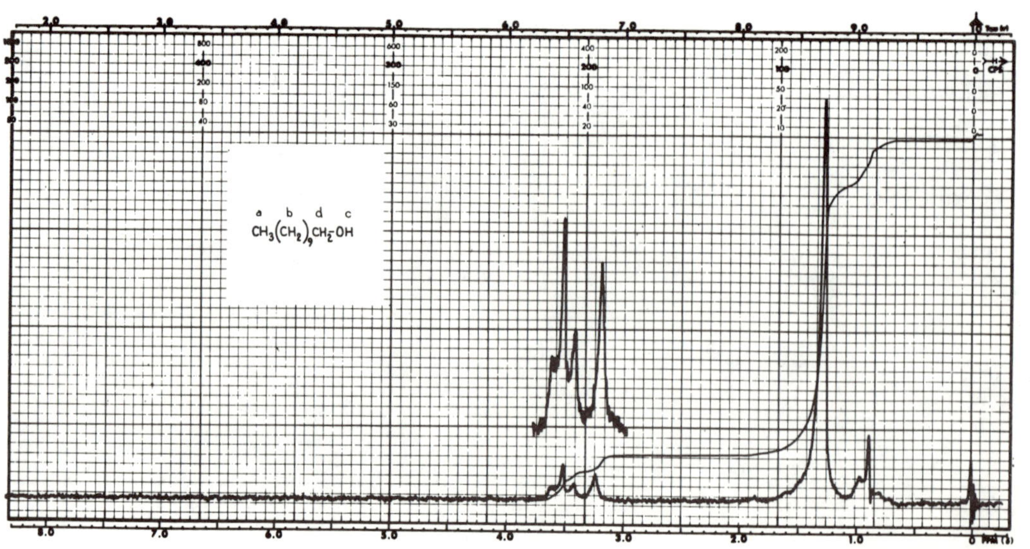

© 1978 Sadtler Research Laboratories, Inc.

1-TETRADECANOL

$C_{14}H_{30}O$

Mol. Wt. 214.38

B.P. 167°C/15mm

Solvent: $CDCl_3$

Assignments
a = 0.88
b = 1.28
c = 1.64
d = 3.61

1632

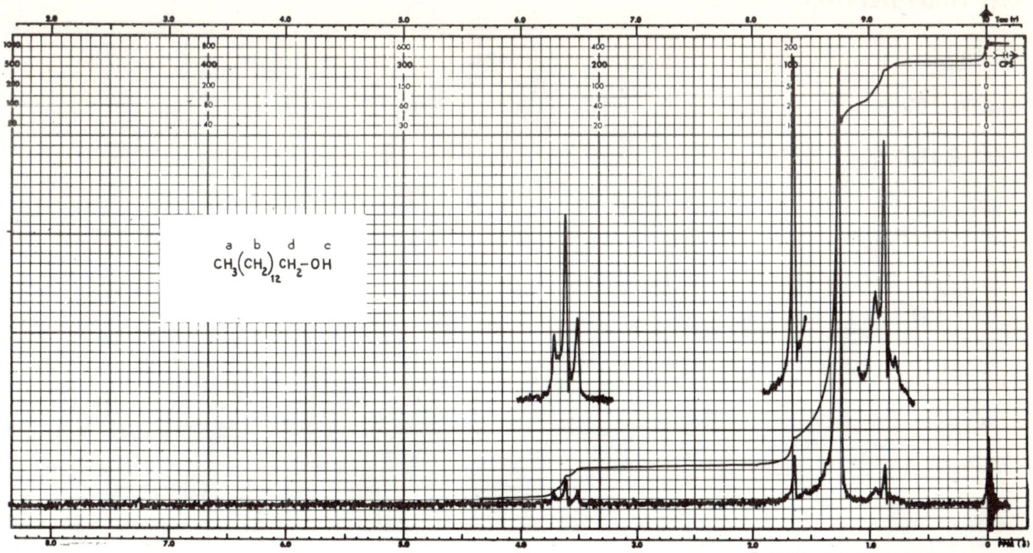

© 1978 Sadtler Research Laboratories, Inc.

1-OCTADECANOL

$C_{18}H_{38}O$

Mol. Wt. 270.50

M.P. 59°C

B.P. 210.5°C/15mm

Solvent: $CDCl_3$

Assignments
a = 0.88
b = ca 1.25
c = 2.05
d = 3.59

1633

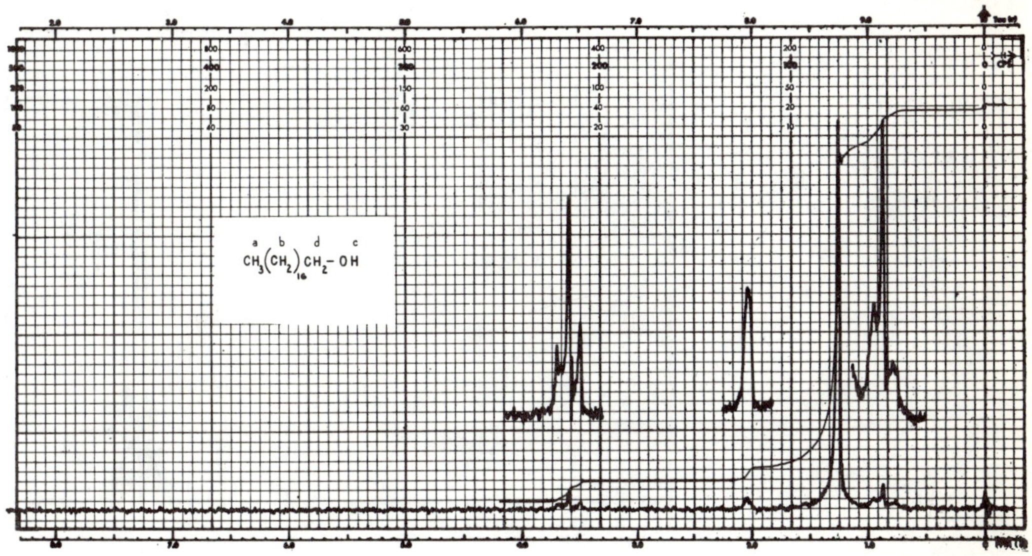

© 1978 Sadtler Research Laboratories, Inc.

1-EICOSANOL

$C_{20}H_{42}O$

Mol. Wt. 298.56

M.P. 65.3°C (lit.)

Solvent: $CDCl_3$

Assignments
a = 0.89
b = 1.26
c = 1.62
d = 3.62

1634

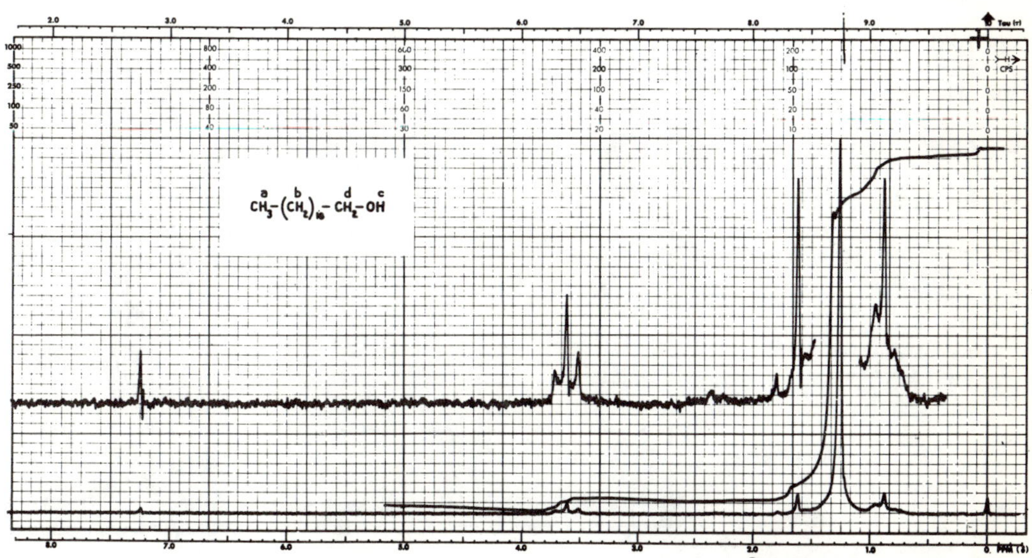

© 1978 Sadtler Research Laboratories, Inc.

1-TETRACOSANOL

$C_{24}H_{50}O$

Mol. Wt. 354.67

Solvent: Polysol-d

Assignments
a = 0.89
b = ca 1.29
c = 2.71
d = 3.48

1635

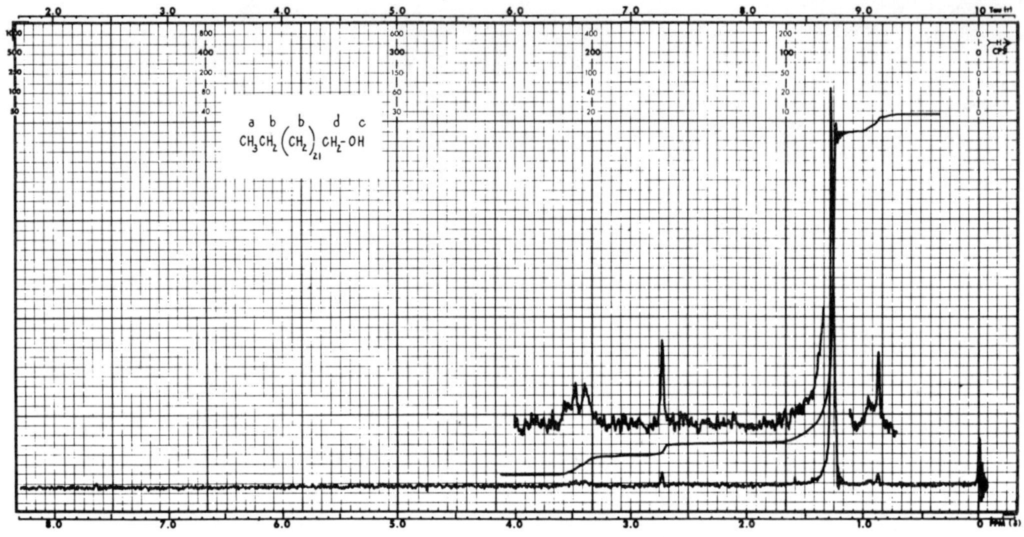

© 1978 Sadtler Research Laboratories, Inc.

2-CHLOROETHANOL

C_2H_5ClO

Mol. Wt. 80.51

B.P. 128°C

Solvent: $CDCl_3$

Assignments
a = 3.63 or b
b = 3.79 or a
c = 4.50

1636

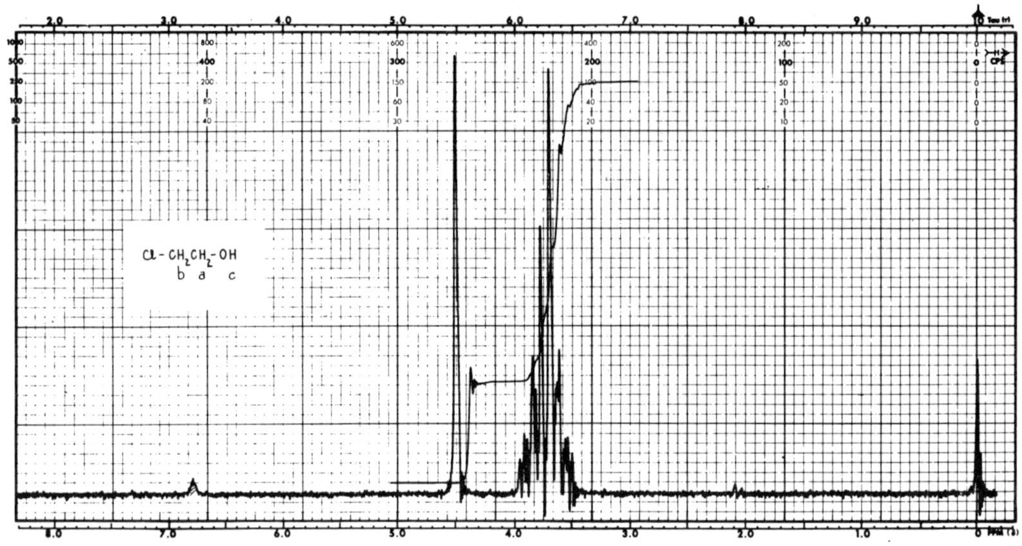

© 1978 Sadtler Research Laboratories, Inc.

3-CHLORO-1-PROPANOL

C_3H_7ClO

Mol. Wt. 94.54

B.P. 66-68°C/18mm

Solvent: CCl_4

Assignments
a = 1.97
b = 3.62
c = 3.72
Trace of D_2O added to base scan

1637

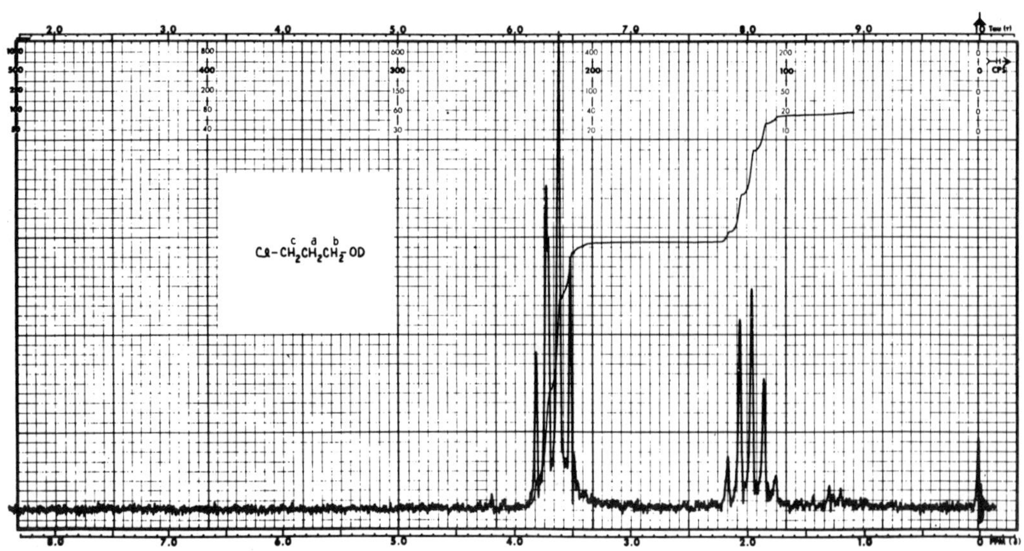

© 1978 Sadtler Research Laboratories, Inc.

2,2-DICHLOROETHANOL

$C_2H_4Cl_2O$

Mol. Wt. 114.96

B.P. 47-50°C/11mm

Solvent: CCl_4

Assignments
a = 3.97
b = 4.19
c = 5.76

1638

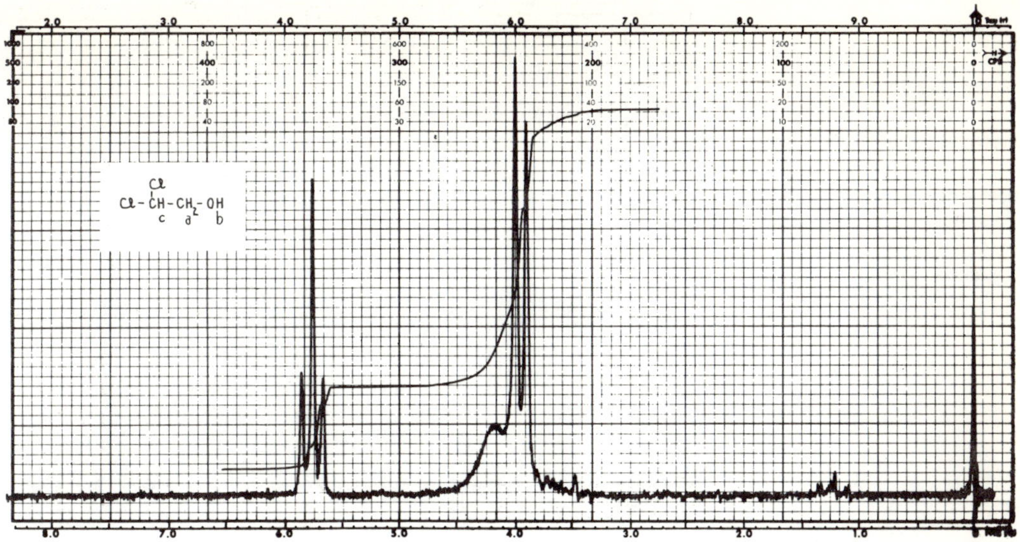

© 1978 Sadtler Research Laboratories, Inc.

2-BROMOETHANOL

C_2H_5BrO

Mol. Wt. 124.97

B.P. 48-50°C

Solvent: CCl_4

Assignments
a = 3.45
b = 3.85
c = 4.29

1639

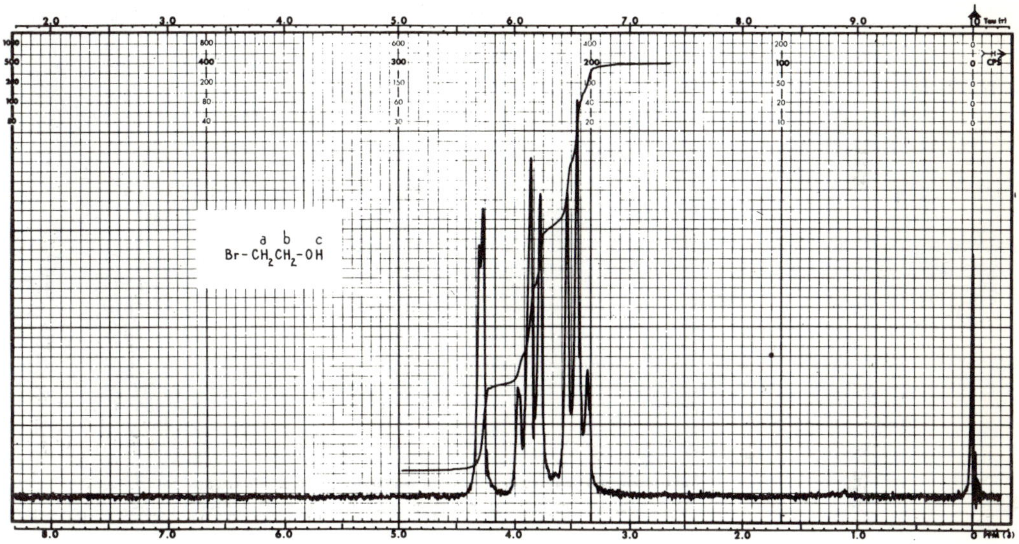

© 1978 Sadtler Research Laboratories, Inc.

3-BROMO-1-PROPANOL

C_3H_7BrO

Mol. Wt. 139.00

B.P. 81-85°C/23mm

Solvent: $CDCl_3$

Assignments
a = 2.10
b = 3.51
c = 3.72
d = 4.21

1640

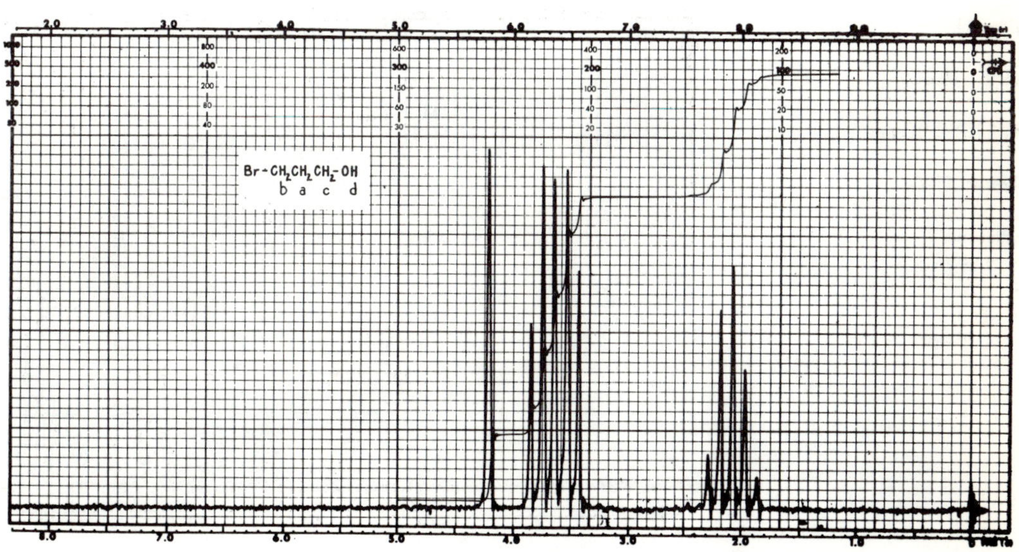

© 1978 Sadtler Research Laboratories, Inc.

2,2,2-TRIBROMOETHANOL

$C_2H_3Br_3O$

Mol. Wt. 282.77

Solvent: $CDCl_3$

Assignments
a = 3.28
b = 4.28
Trace of acid added

1641

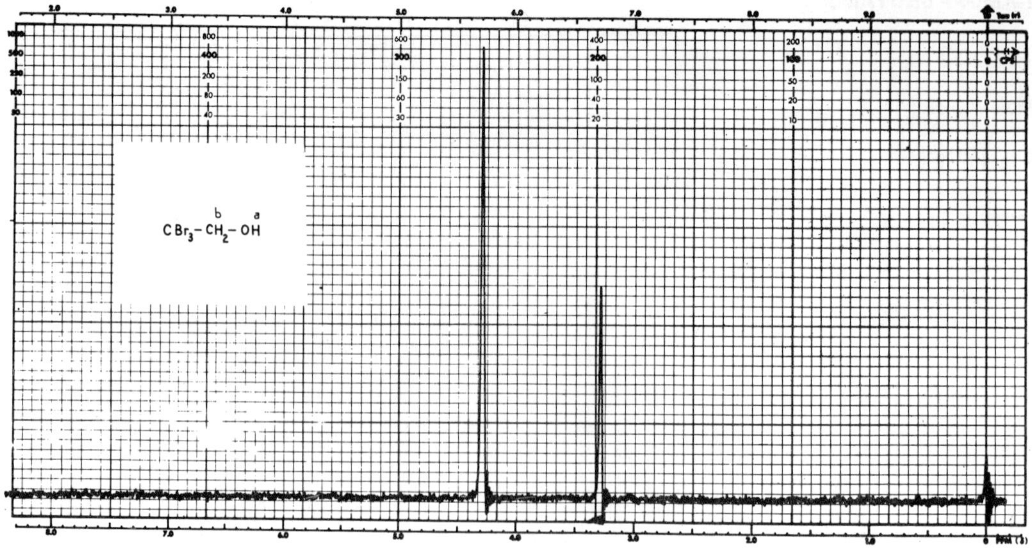

© 1978 Sadtler Research Laboratories, Inc.

2-AMINOETHANOL

C_2H_7NO

Mol. Wt. 61.08

B.P. 170.5°C

Solvent: $CDCl_3$

Assignments
a = 2.73
b = 2.82
c = 3.54

1642

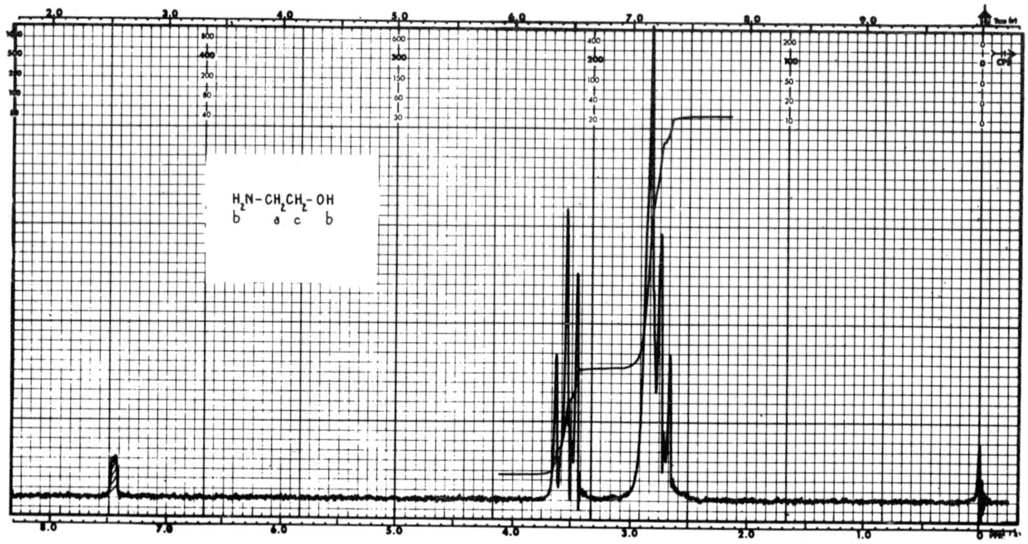

© 1978 Sadtler Research Laboratories, Inc.

3-AMINO-1-PROPANOL

C_3H_9NO

Mol. Wt. 75.11

B.P. 115°C/50mm

Solvent: D_2O

Assignments
a = 1.68
b = 2.70
c = 3.63
d = 4.62 HDO

1643

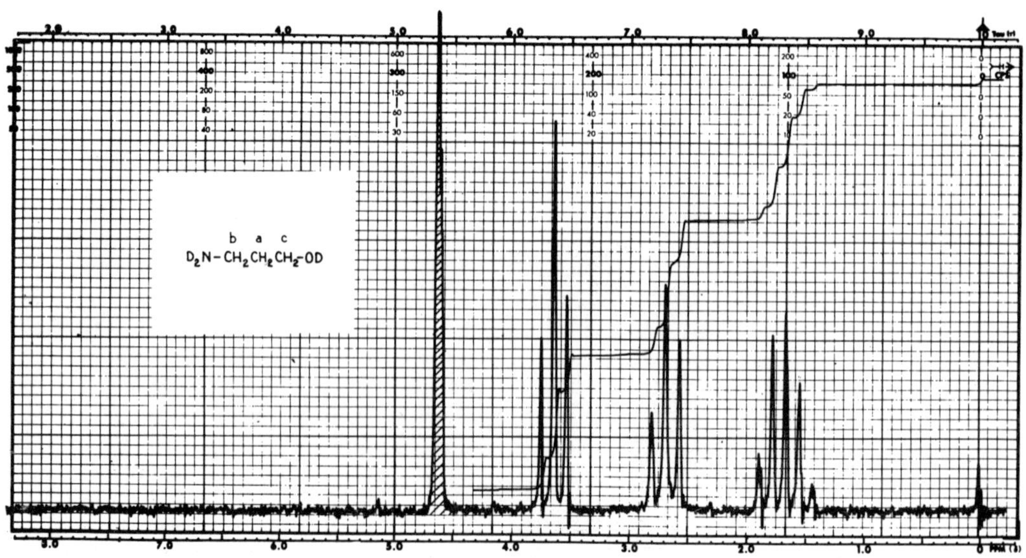

© 1978 Sadtler Research Laboratories, Inc.

2-AMINO-1-BUTANOL

$C_4H_{11}NO$

Mol. Wt. 89.14

B.P. 117-178°C

Solvent: $CDCl_3$

Assignments
a = 0.95
b = ca 1.36
c = 2.72
d = 3.02
e = 3.28
f = 3.62
Trace of D_2O added

1644

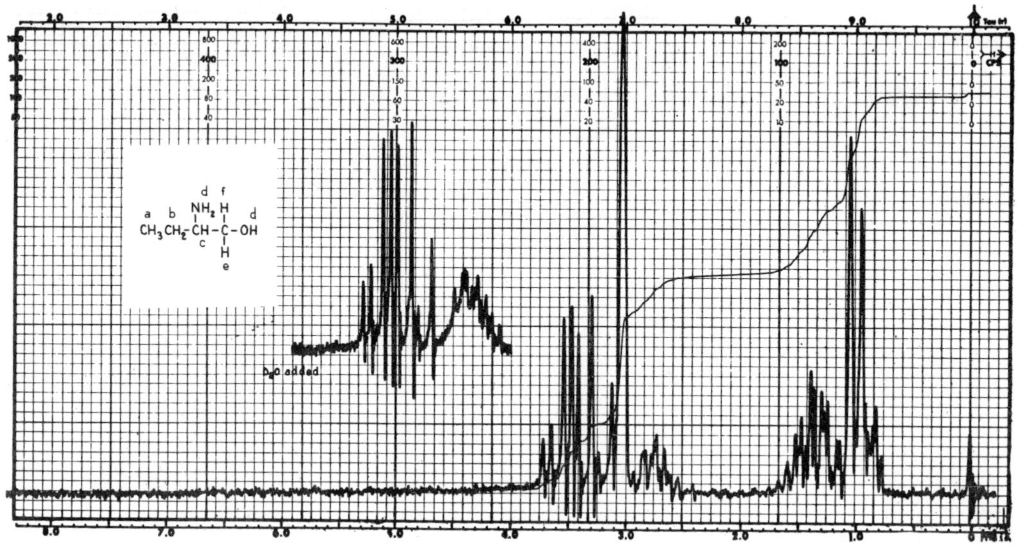

© 1978 Sadtler Research Laboratories, Inc.

3-AMINO-3-METHYL-1-BUTANOL

$C_5H_{13}NO$

Mol. Wt. 103.17

Solvent: $CDCl_3$

Assignments
a = 1.22
b = 1.56
c = 2.71
d = 3.81
e = 7.32 imp. ($CHCl_3$)

1645

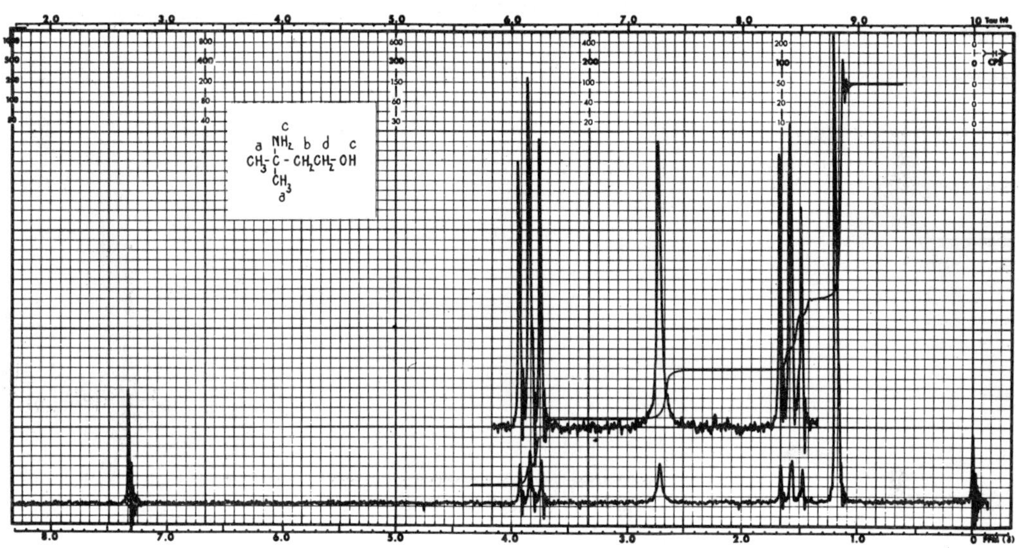

© 1978 Sadtler Research Laboratories, Inc.

2-[(2-AMINOETHYL)AMINO] ETHANOL

$C_4H_{12}N_2O$

Mol. Wt. 104.15

B.P. 243.7°C

Solvent: $CDCl_3$

Assignments
a = 2.50-2.90
b = 3.61

1646

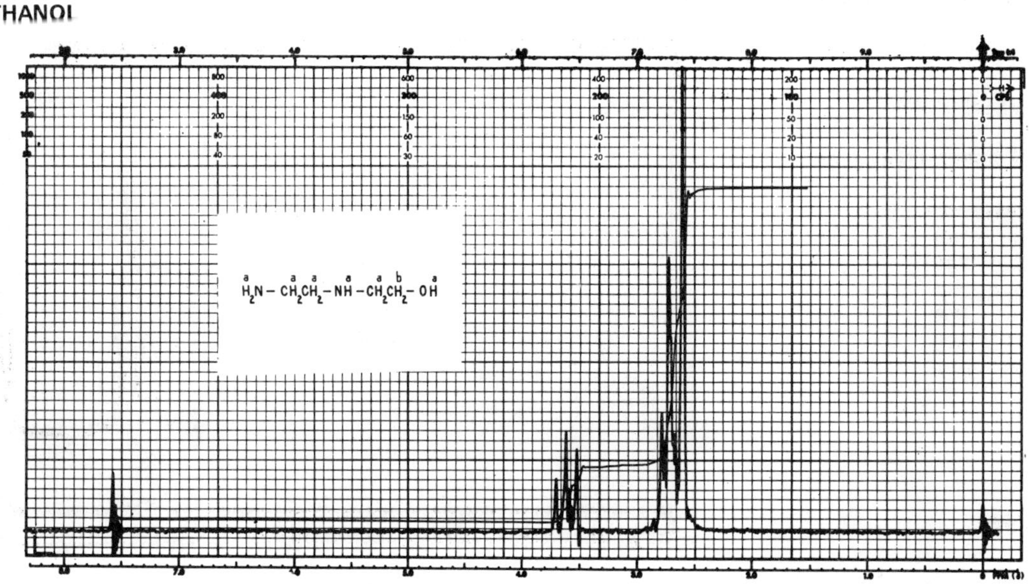

© 1978 Sadtler Research Laboratories, Inc.

2-(ISOPROPYLAMINO)ETHANOL

$C_5H_{13}NO$

Mol. Wt. 103.17

B.P. 69-74°C/10mm

Solvent: CCl_4

Assignments
a = 1.08
b = 2.66
c = 2.85
d = 3.19
e = 3.58

1647

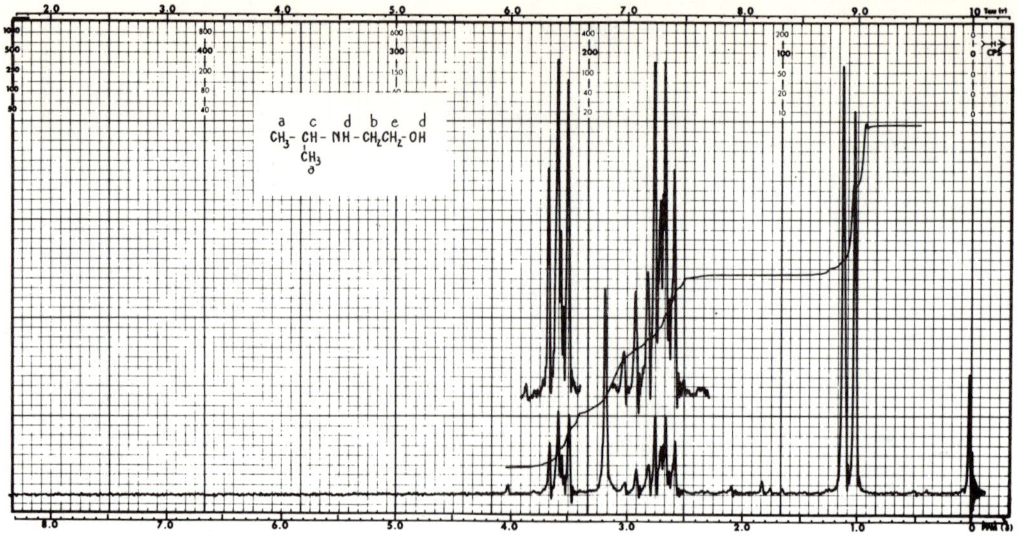

© 1978 Sadtler Research Laboratories, Inc.

2-(DIMETHYLAMINO)-1-PROPANOL

$C_5H_{13}NO$

Mol. Wt. 103.17

B.P. 145-155.5°C/760mm (lit.)

Solvent: D_2O

Assignments
a = 1.00
b = 2.20
c = 2.65
d = 3.46
e = 3.62
f = 4.62 HDO

1648

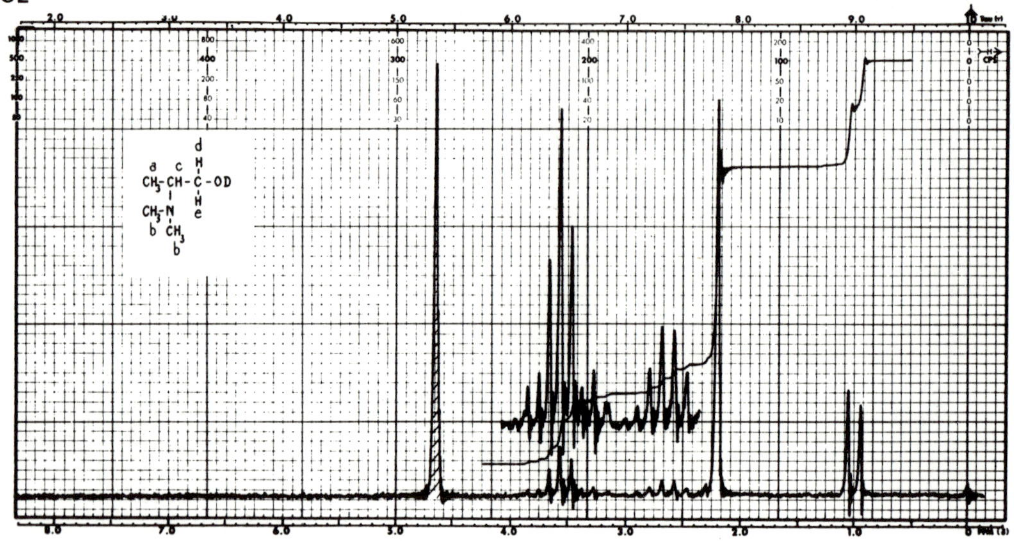

© 1978 Sadtler Research Laboratories, Inc.

3-(DIMETHYLAMINO)-1-PROPANOL

$C_5H_{13}NO$

Mol. Wt. 103.17

Solvent: CCl_4

Assignments
a = 1.60
b = 2.20
c = 2.41
d = 3.59
e = 3.74

1649

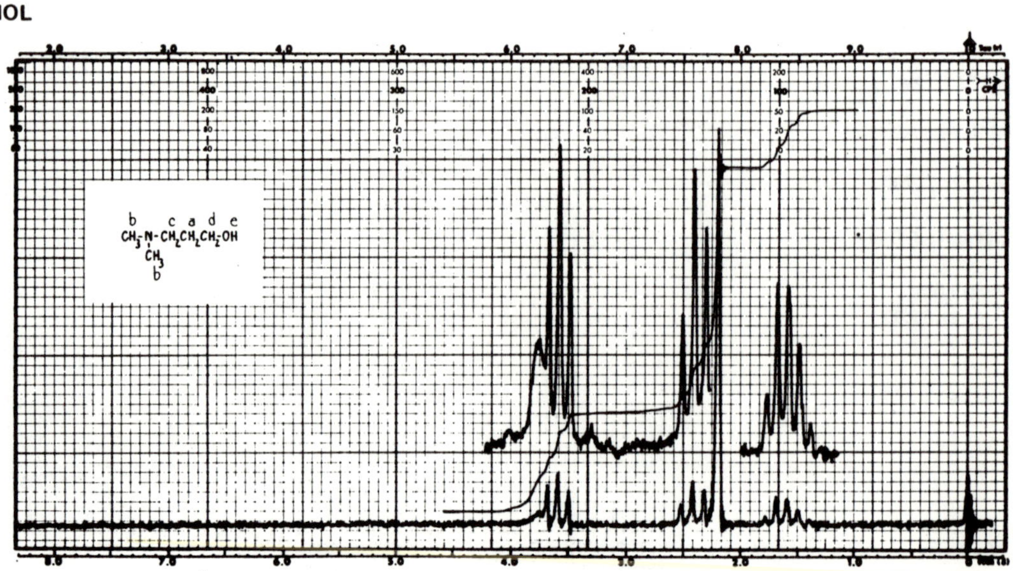

© 1978 Sadtler Research Laboratories, Inc.

2-(DIETHYLAMINO)ETHANOL

$C_6H_{15}NO$

Mol. Wt. 117.19

B.P. 158-165°C

Solvent: CCl_4

Assignments
a = 1.00
b = 2.50
c = 2.52
d = 3.45
e = 3.52

1650

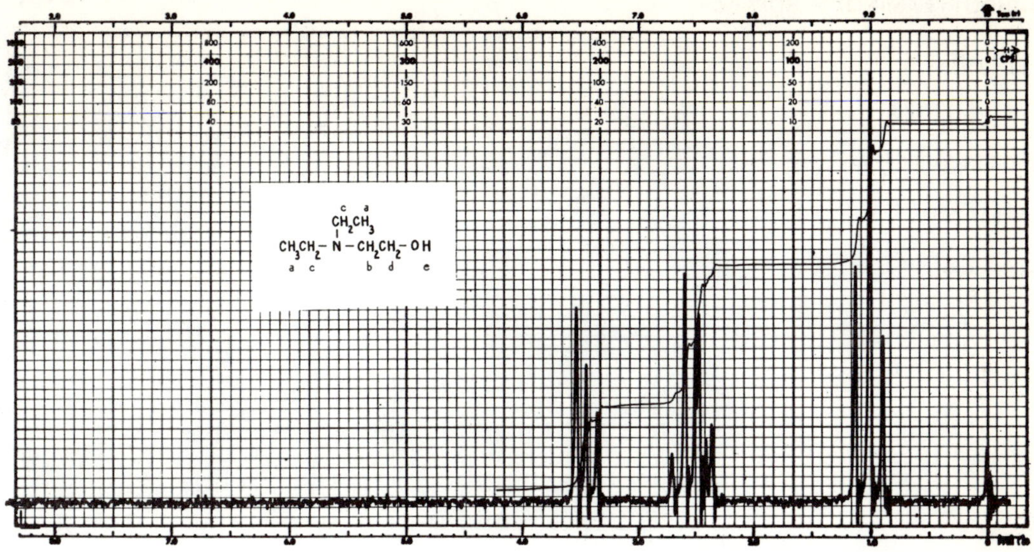

© 1978 Sadtler Research Laboratories, Inc.

2-DIETHYLAMINO-1-PROPANOL

$C_7H_{17}NO$

Mol. Wt. 131.22

B.P. 57-59°C/20mm

Solvent: CCl_4

Assignments
a = 0.88-1.28
b = 2.12-2.89
c = 3.60

1651

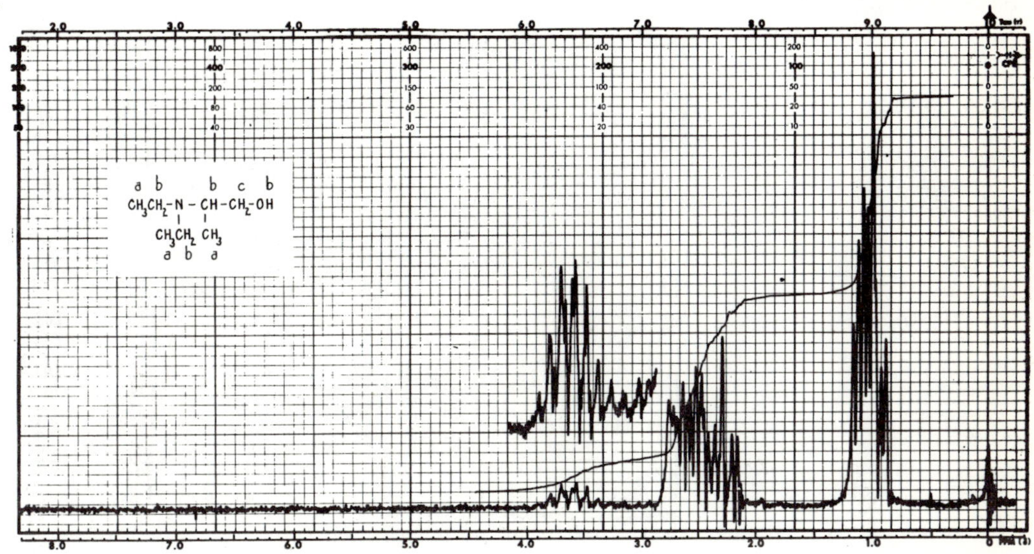

© 1978 Sadtler Research Laboratories, Inc.

2-(DIBUTYLAMINO)ETHANOL

$C_{10}H_{23}NO$

Mol. Wt. 173.30

B.P. 110-114°C/18mm

Solvent: $CDCl_3$

Assignments
a = 0.91
b = 1.10-1.80
c = 2.48
d = 2.57
e = 3.54
f = 3.64

1652

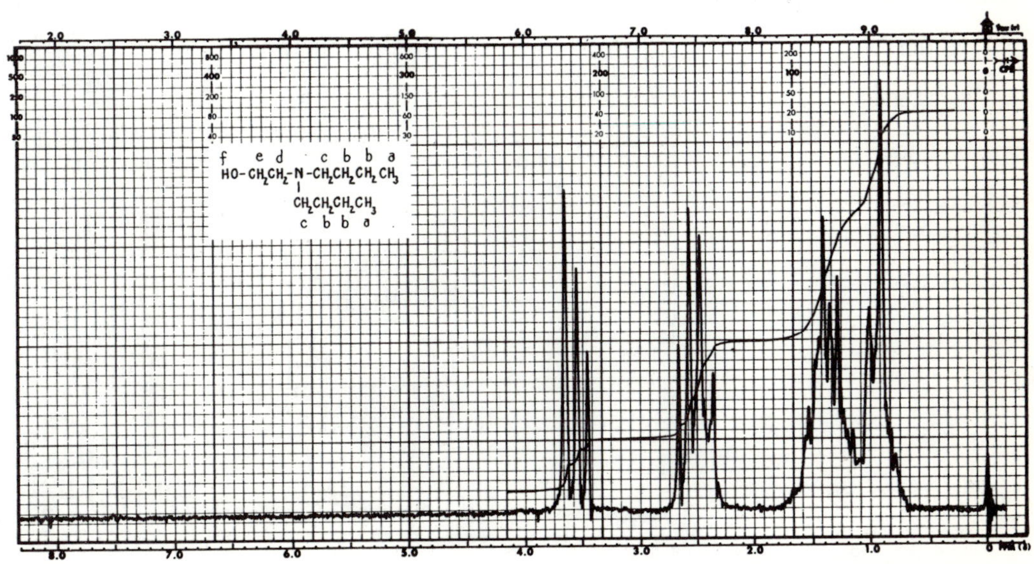

© 1978 Sadtler Research Laboratories, Inc.

3-(DIBUTYLAMINO)-1-PROPANOL

$C_{11}H_{25}NO$

Mol. Wt. 187.33

B.P. 110-116°C/8mm

Solvent: $CDCl_3$

Assignments
a = 0.92
b = 1.10-1.60
c = 1.69
d = 2.42
e = 2.63
f = 3.78
g = 5.19

1653

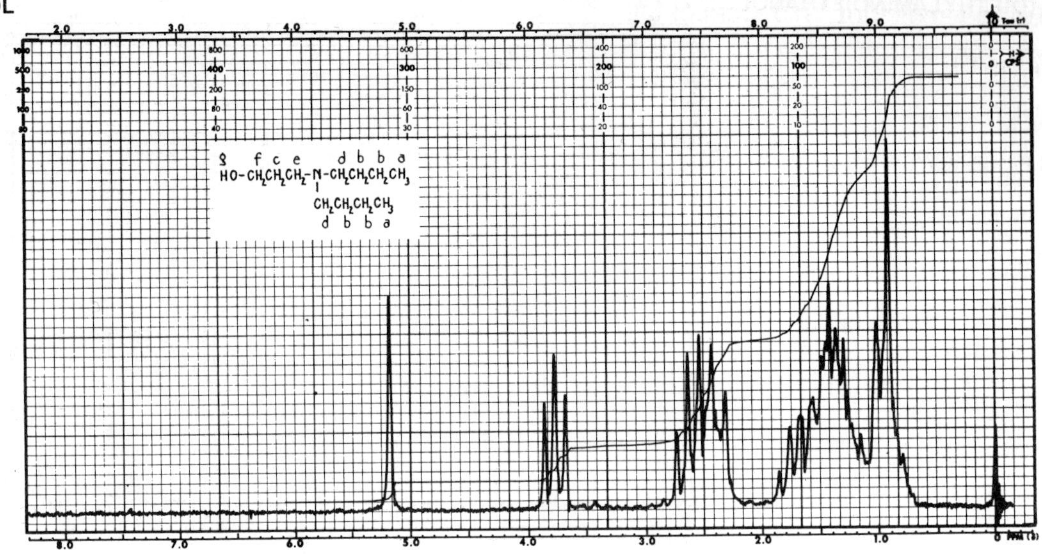

© 1978 Sadtler Research Laboratories, Inc.

2-HYDRAZINOETHANOL

$C_2H_8N_2O$

Mol. Wt. 76.10

Solvent: D_2O

Assignments
a = 2.93
b = 3.74
c = 4.78 HDO
d = 1.30 imp.

1654

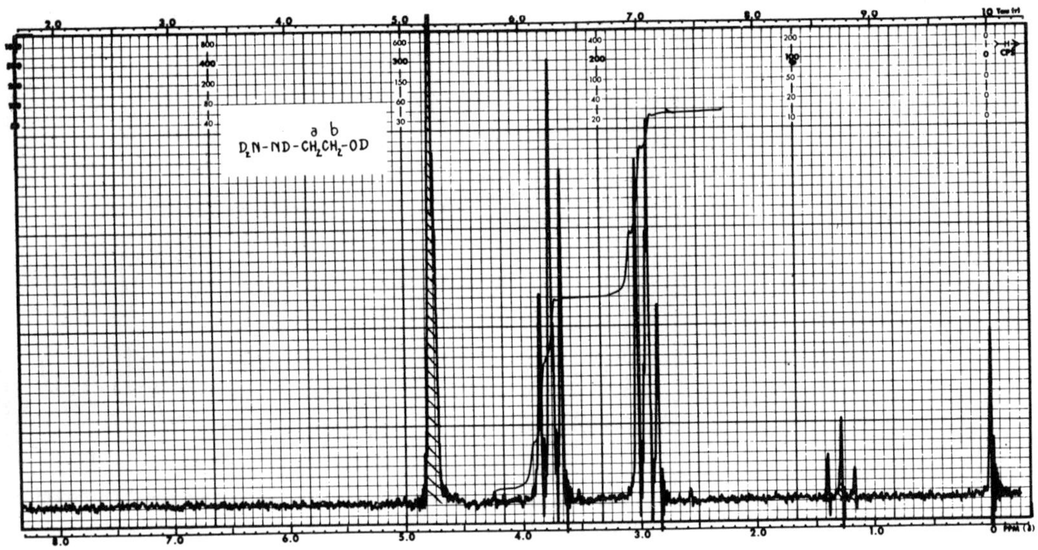

© 1978 Sadtler Research Laboratories, Inc.

2-AMINOETHANOL, HYDROCHLORIDE

$C_2H_7NO \cdot HCl$

Mol. Wt. 97.55

Solvent: D_2O

Assignments
a = 3.18
b = 3.85
c = 5.25 HDO

1655

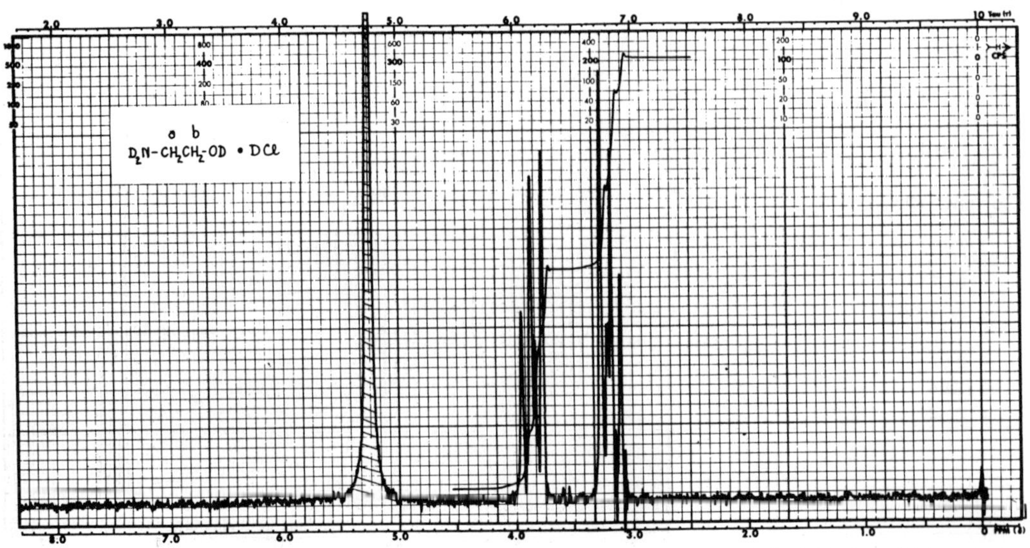

© 1978 Sadtler Research Laboratories, Inc.

3-AMINO-1-PROPANOL, HYDROCHLORIDE

$C_3H_9NO \cdot HCl$

Mol. Wt. 111.57

Solvent: D_2O

Assignments
a = 1.91
b = 3.17
c = 3.73
d = 4.54 HDO

1656

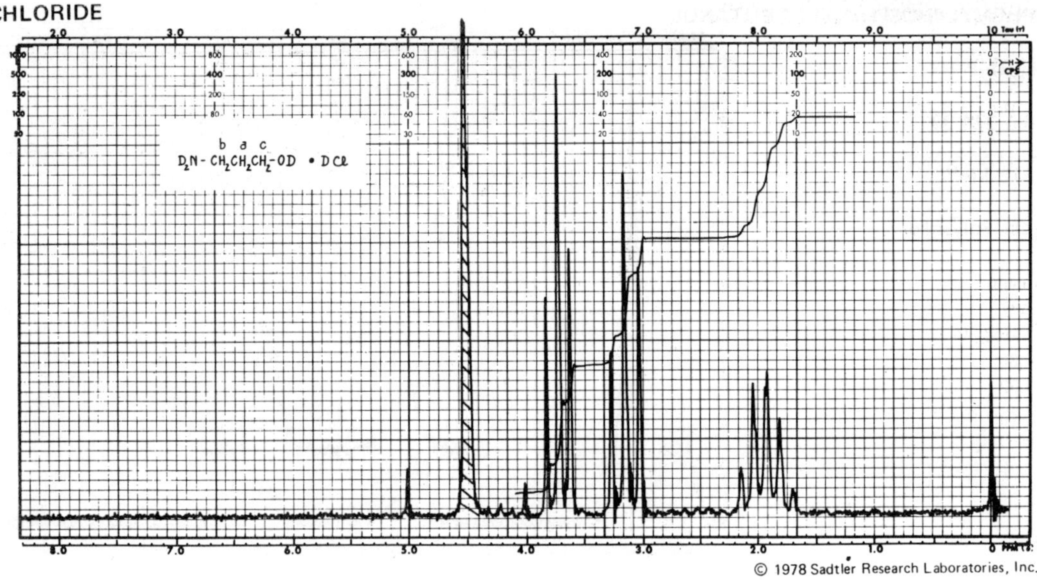

© 1978 Sadtler Research Laboratories, Inc.

CHOLINE CHLORIDE

$C_5H_{14}ClNO$

Mol. Wt. 139.63

M.P. 228-229°C

Solvent: $CDCl_3$

Assignments
a = 3.23
b = 3.56
c = 4.05
d = 4.61

1657

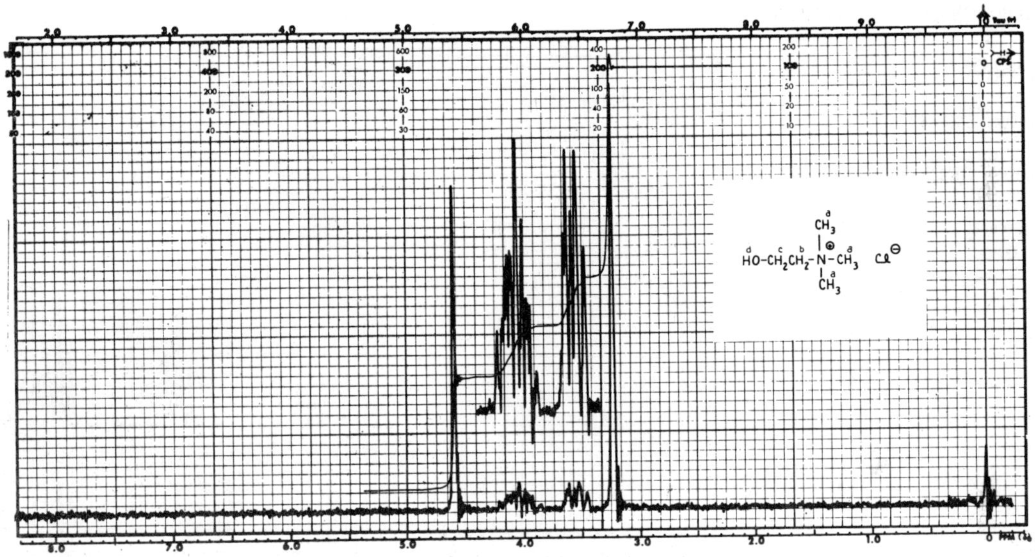

© 1978 Sadtler Research Laboratories, Inc.

HYDRACRYLONITRILE

C_3H_5NO

Mol. Wt. 71.05

B.P. 92-94°C/5mm

Solvent: $CDCl_3$

Assignments
a = 2.58
b = 3.80
c = 3.99

1658

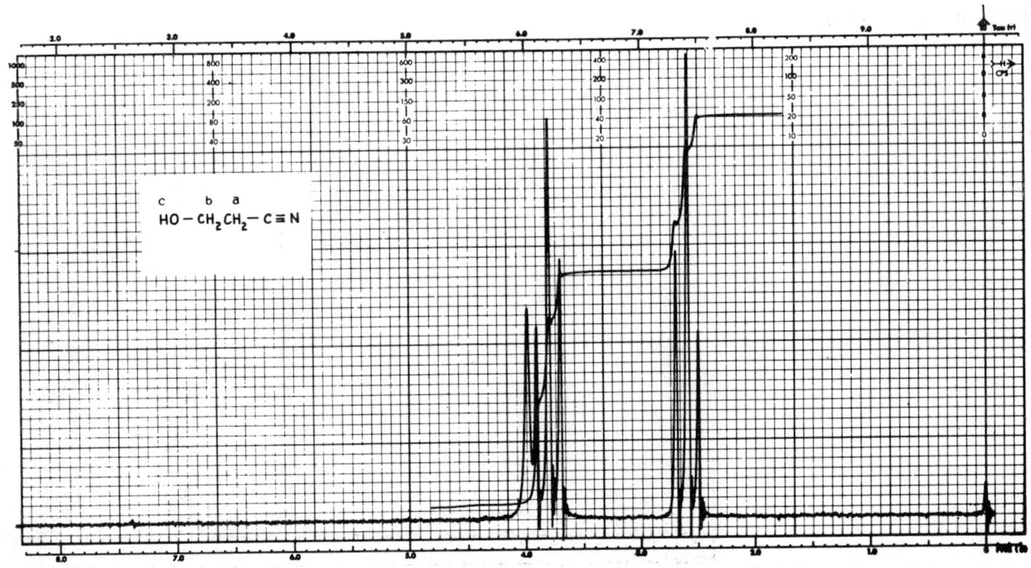

© 1978 Sadtler Research Laboratories, Inc.

1-(DIPHENYLPHOSPHINYL)-1-BUTANOL

$C_{16}H_{19}O_2P$

Mol. Wt. 274.30

M.P. 112-115°C

Solvent: Polysol-d

Assignments
a = 0.87
b = 1.10-2.00
c = 4.36
d = 5.52
e = 7.25-7.62
f = ca 7.89
g = 2.48 DMSO-d_5

1659

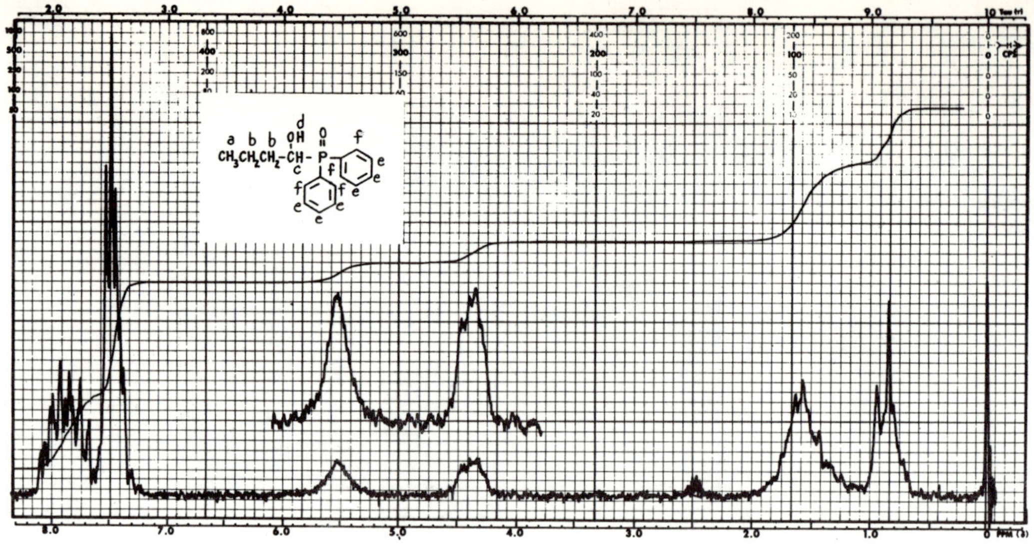

© 1978 Sadtler Research Laboratories, Inc.

2-(DODECYLTHIO)ETHANOL

$C_{14}H_{30}OS$

Mol. Wt. 246.46

Solvent: $CDCl_3$

Assignments
a = 0.89
b = 1.10-2.00
c = 2.07
d = 2.52
e = 2.71
f = 3.71

1660

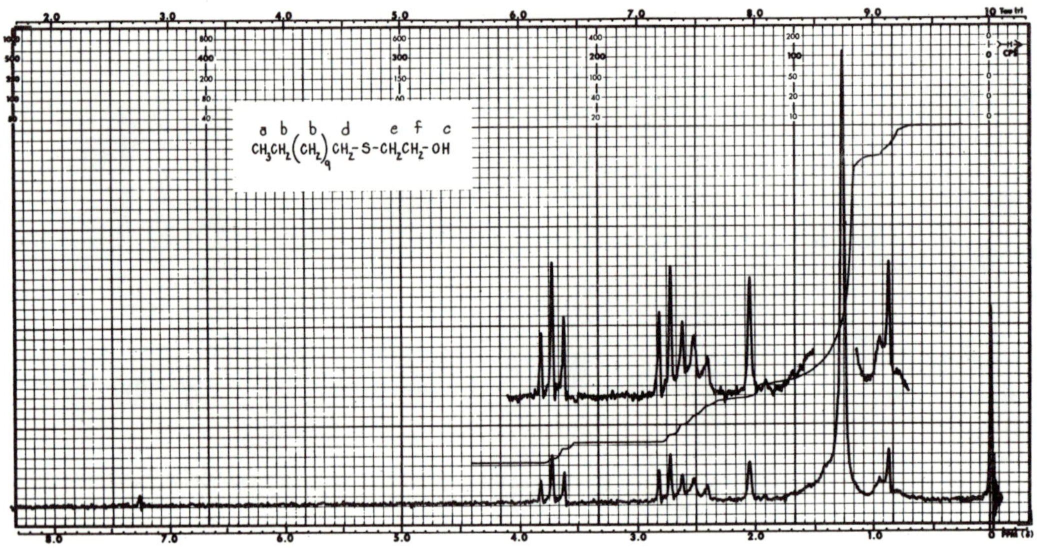

© 1978 Sadtler Research Laboratories, Inc.

2-MERCAPTOETHANOL

C_2H_6OS

Mol. Wt. 78.13

B.P. 50-52°C/10mm

Solvent: D_2O

Assignments
a = 2.68
b = 3.69
c = 4.56 HDO
d = 1.91 imp.

1661

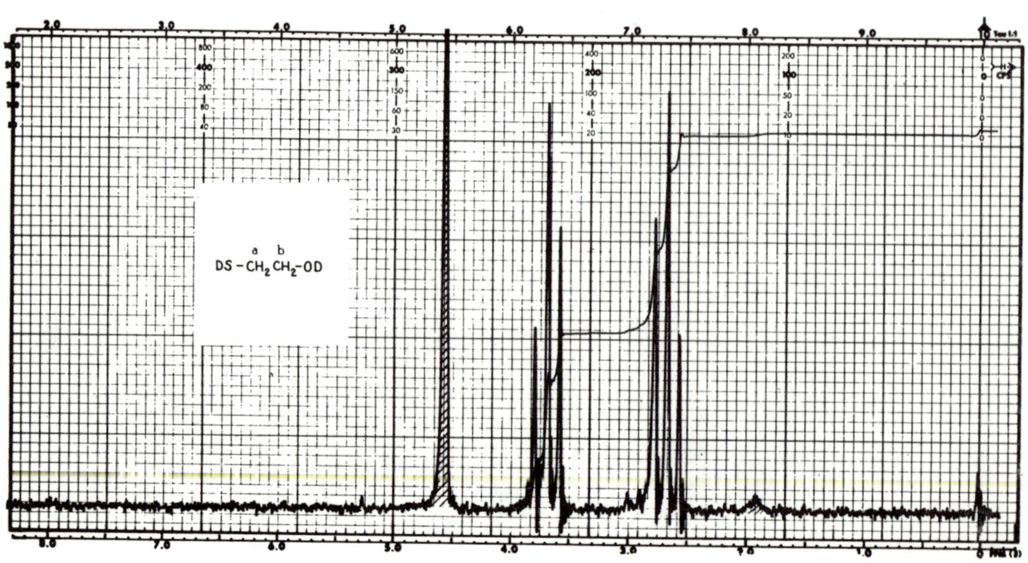

© 1978 Sadtler Research Laboratories, Inc.

2,3-DIMERCAPTO-1-PROPANOL

$C_3H_8OS_2$

Mol. Wt. 124.22

Solvent: $CDCl_3$

Assignments
a = 1.79
b = 1.85
c = 2.83
d = ca 3.00
e = 3.33
f = 3.77

1662

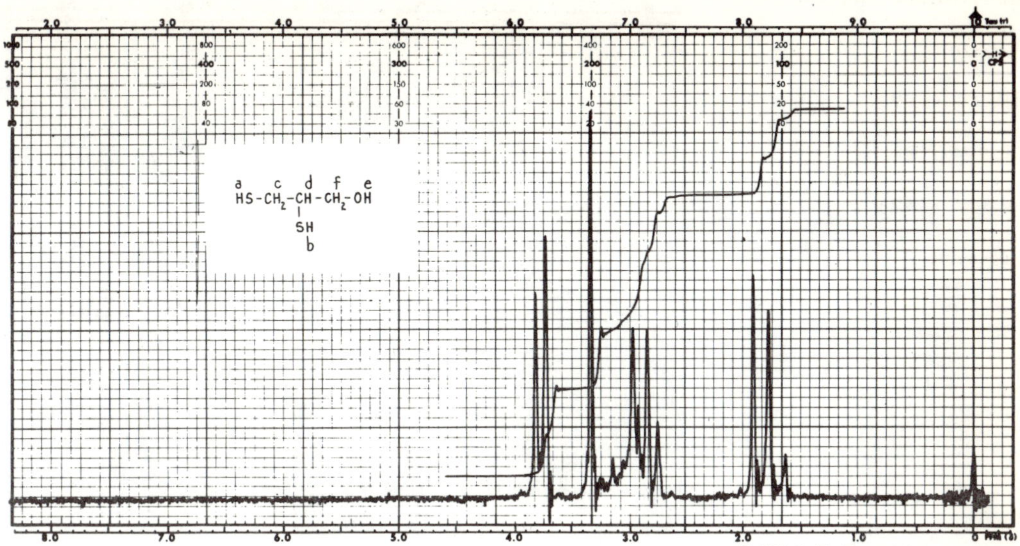

© 1978 Sadtler Research Laboratories, Inc.

2-HYDROXYETHANESULFONIC ACID, MONOPOTASSIUM SALT

$C_2H_5KO_4S$

Mol. Wt. 164.23

Solvent: D_2O

Assignments
a = 3.12
b = 3.93
c = 4.37 HDO
d = 3.66 imp.

1663

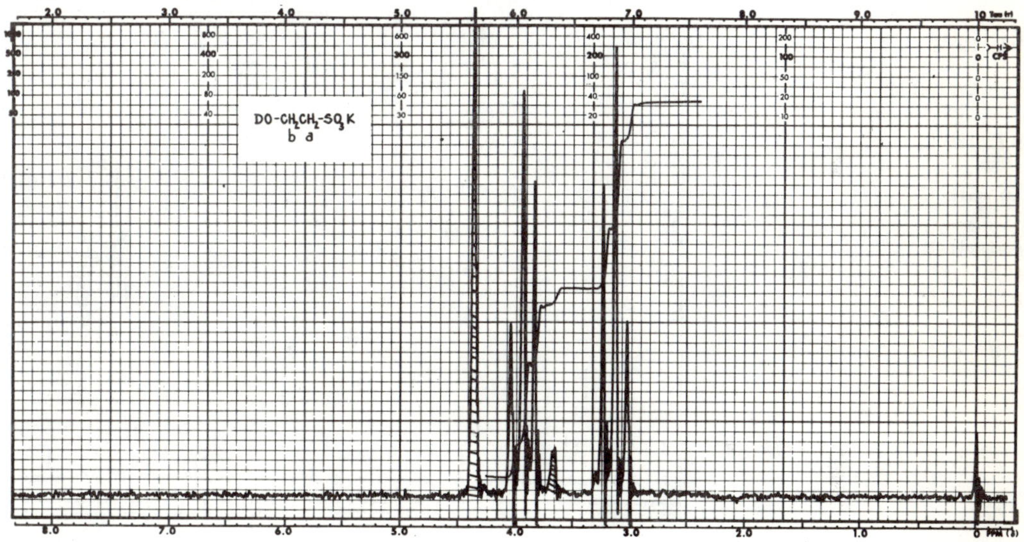

© 1978 Sadtler Research Laboratories, Inc.

1-(2-HYDROXYETHYL)-2-IMIDAZOLIDINETHIONE

$C_5H_{10}N_2OS$

Mol. Wt. 146.21

M.P. 136.5-137.5°C (lit.)

Solvent: Trifluoroacetic acid

Assignments
a = 3.94
b - ca 4.11
c = 4.70
OH unobserved

1664

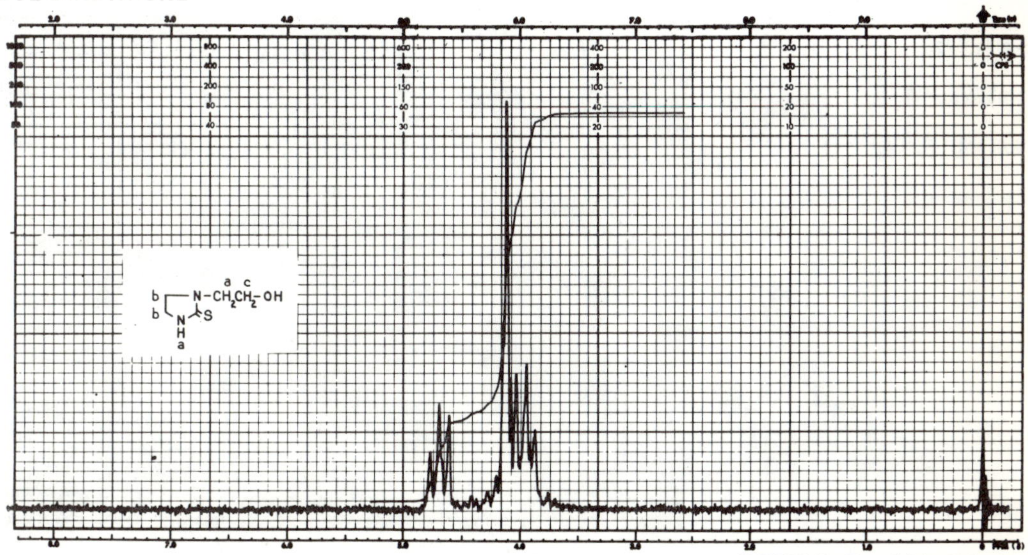

© 1978 Sadtler Research Laboratories, Inc.

2-METHOXYETHANOL

$C_3H_8O_2$

Mol. Wt. 76.09

B.P. 124.5°C

Solvent: $CDCl_3$

Assignments
a = 2.49
b = 3.38
c = 3.51 or d
d = 3.71 or c

1665

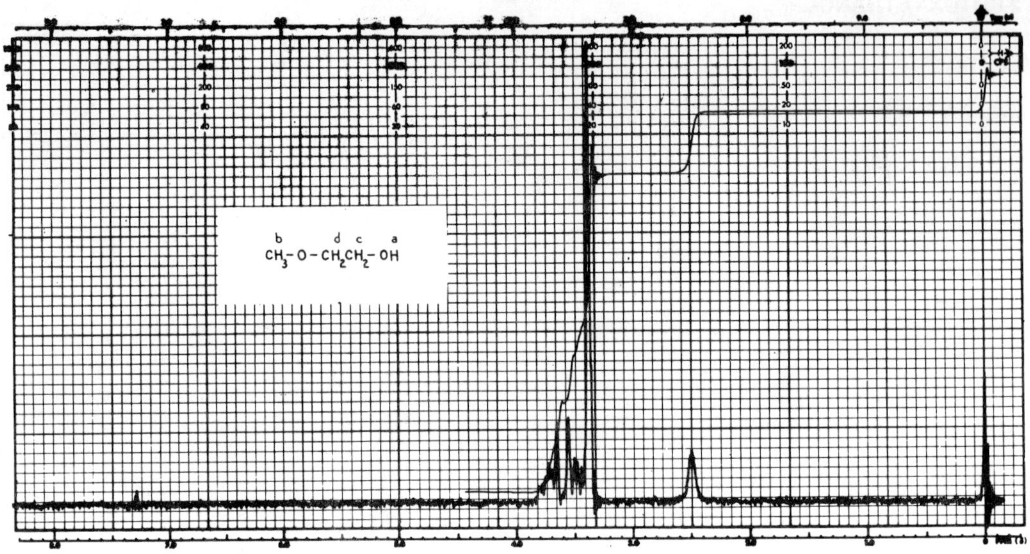

© 1978 Sadtler Research Laboratories, Inc.

2-ETHOXYETHANOL

$C_4H_{10}O_2$

Mol. Wt. 90.12

B.P. 136°C

Solvent: CCl_4

Assignments
a = 1.18
b = ca 3.51

1666

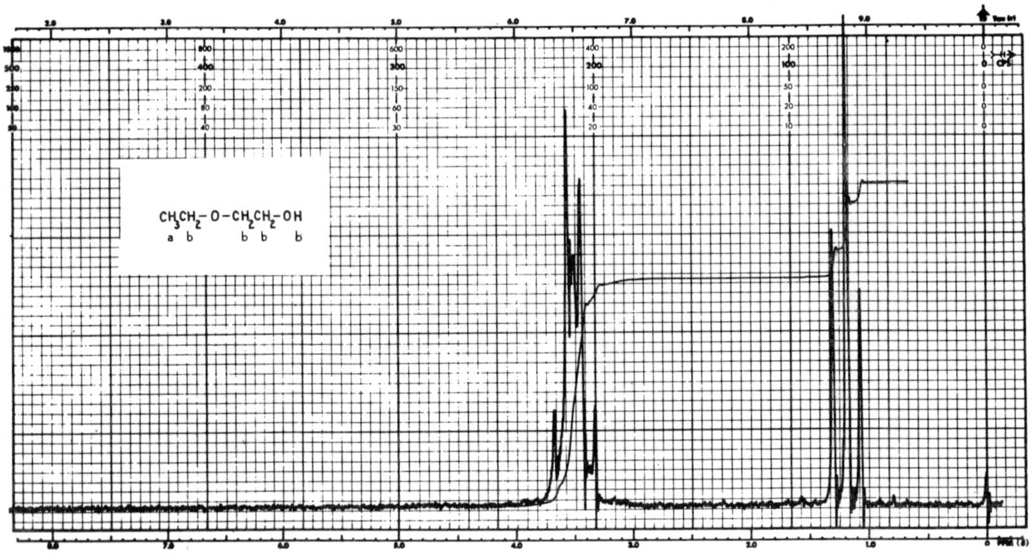

© 1978 Sadtler Research Laboratories, Inc.

2-PROPOXYETHANOL

$C_5H_{12}O_2$

Mol. Wt. 104.15

Solvent: $CDCl_3$

Assignments
a = 0.93
b = 1.59
c = 3.43
d = 3.30-3.85

1667

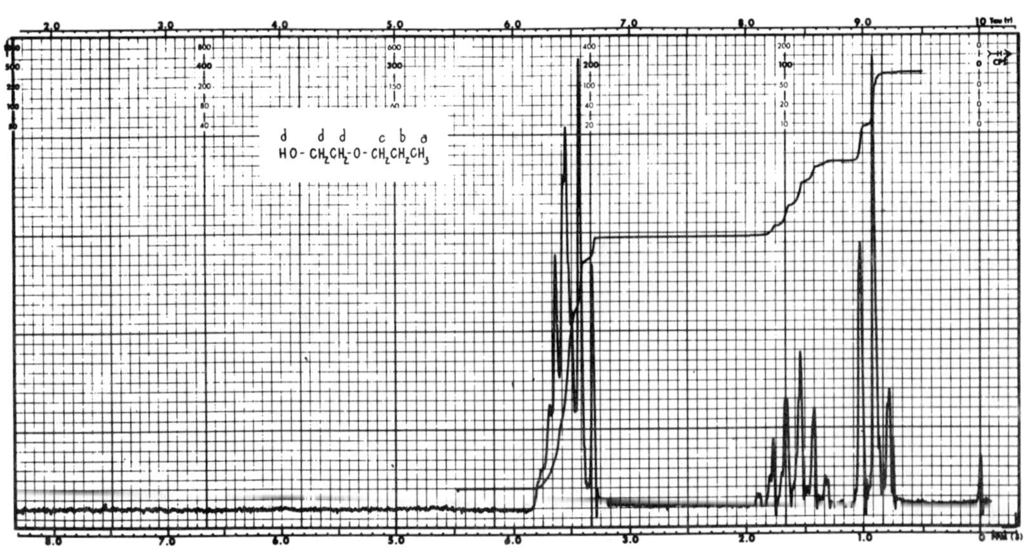

© 1978 Sadtler Research Laboratories, Inc.

2-BUTOXYETHANOL

$C_6H_{14}O_2$

Mol. Wt. 118.18

B.P. 171.2°C (lit.)

Solvent: CCl_4

Assignments
a = 0.93
b = 0.65-1.90
c = 3.25
d = 3.30-3.74

1668

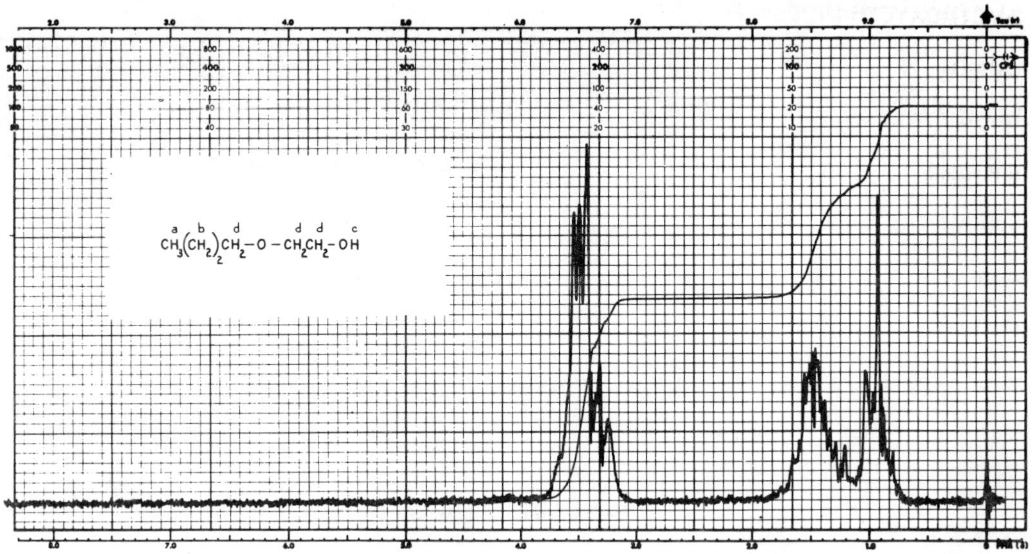

© 1978 Sadtler Research Laboratories, Inc.

2-(HEXYLOXY)ETHANOL

$C_8H_{18}O_2$

Mol. Wt. 146.23

B.P. 198-206°C

Solvent: CCl_4

Assignments
a = 0.92
b = 1.00-1.80
c = 2.75
d = 3.20-3.80

1669

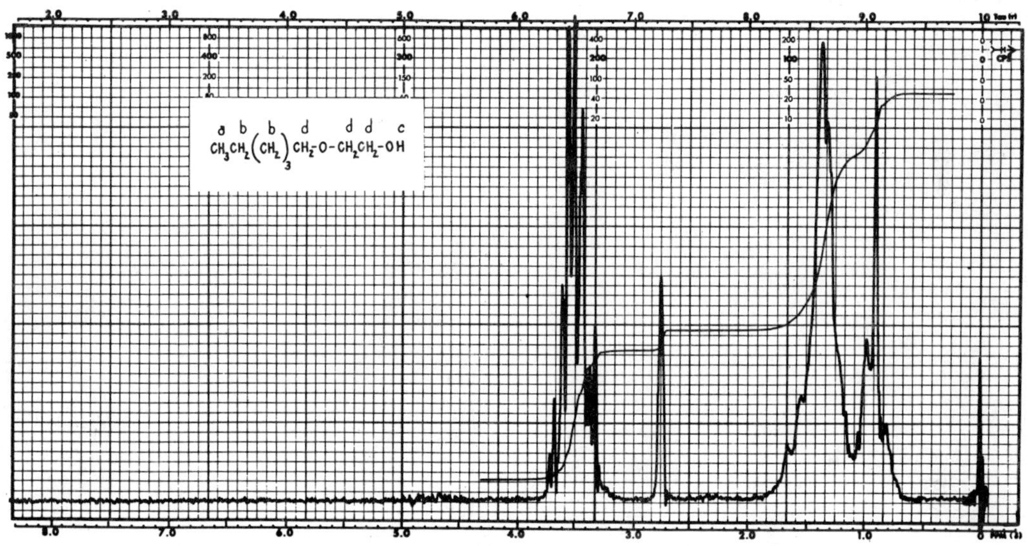

© 1978 Sadtler Research Laboratories, Inc.

2-(DODECYLOXY)ETHANOL

$C_{14}H_{30}O_2$

Mol. Wt. 230.39

B.P. 143°C

Solvent: $CDCl_3$

Assignments
a = 0.88
b = 1.10-1.85
c = 2.10
d = 3.30-3.80

1670

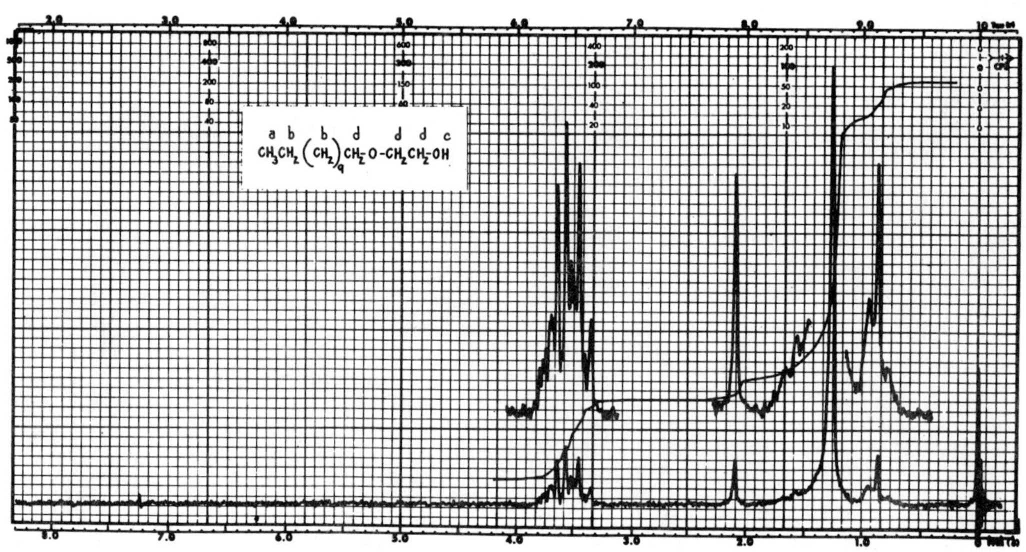

© 1978 Sadtler Research Laboratories, Inc.

2-(2-CHLOROETHOXY)ETHANOL

$C_4H_9ClO_2$

Mol. Wt. 124.57

B.P. 79-81°C/5mm

Solvent: $CDCl_3$

Assignments
a = 3.50-3.90

1671

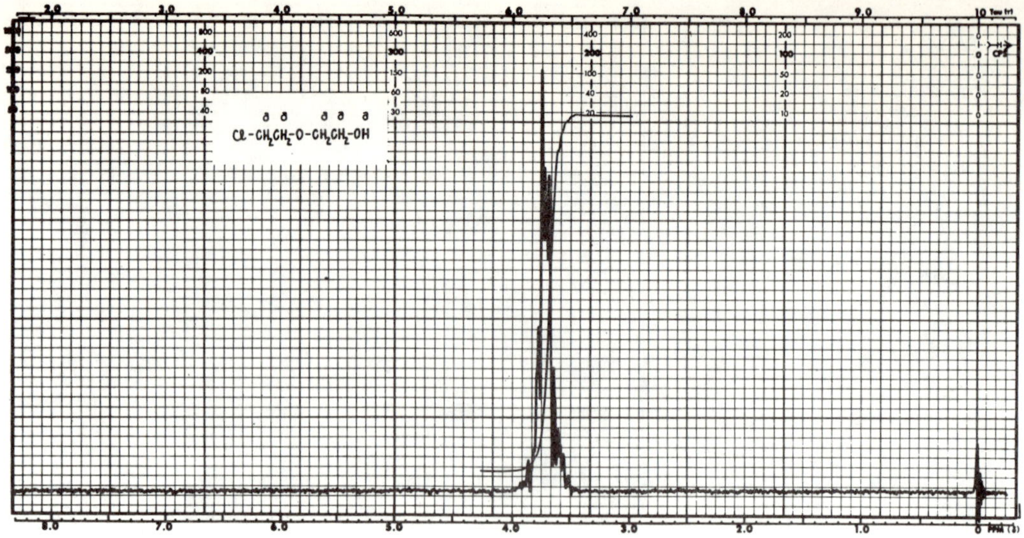

© 1978 Sadtler Research Laboratories, Inc.

2-(2-AMINOETHOXY)ETHANOL

$C_4H_{11}NO_2$

Mol. Wt. 105.14

Solvent: $CDCl_3$

Assignments
a = 2.63
b = 2.83
c = 3.52
d = 3.40-3.80
e = 7.53 imp.

1672

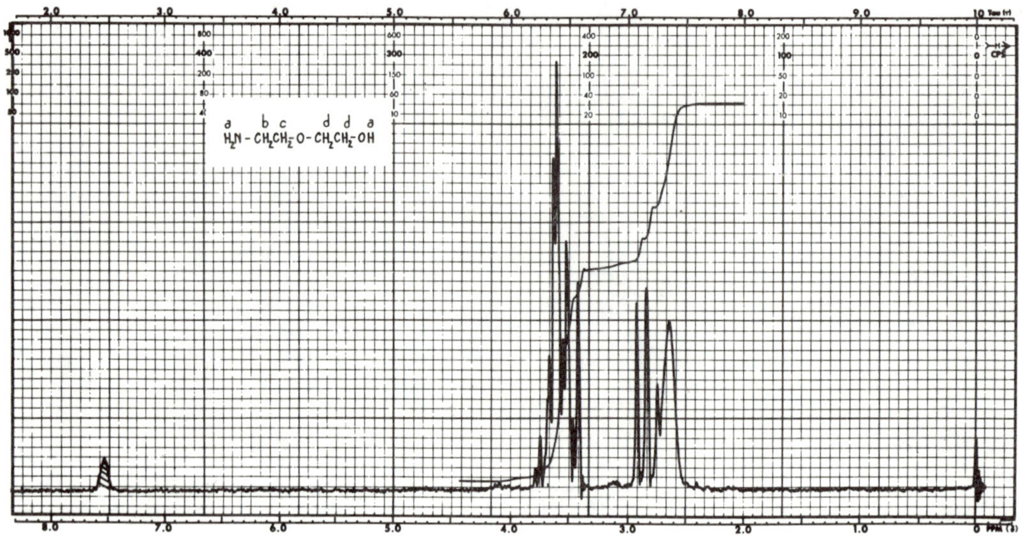

© 1978 Sadtler Research Laboratories, Inc.

2-(2-METHOXYETHOXY)ETHANOL

$C_5H_{12}O_3$

Mol. Wt. 120.15

B.P. 193.5°C (lit.)

Solvent: $CDCl_3$

Assignments
a = 3.27
b = 3.40
c = 3.63

1673

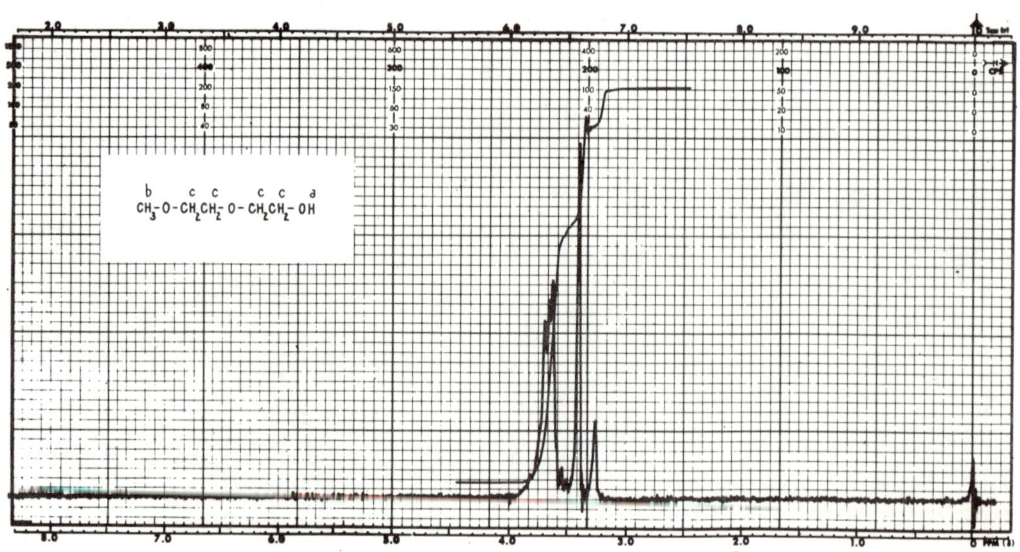

© 1978 Sadtler Research Laboratories, Inc.

2-(2-ETHOXYETHOXY)ETHANOL

$C_6H_{14}O_3$

Mol. Wt. 134.18

B.P. 90-92°C/11mm

Solvent: CCl_4

Assignments
a = 1.20
b = 3.12
c = 3.50
d = 3.56

1674

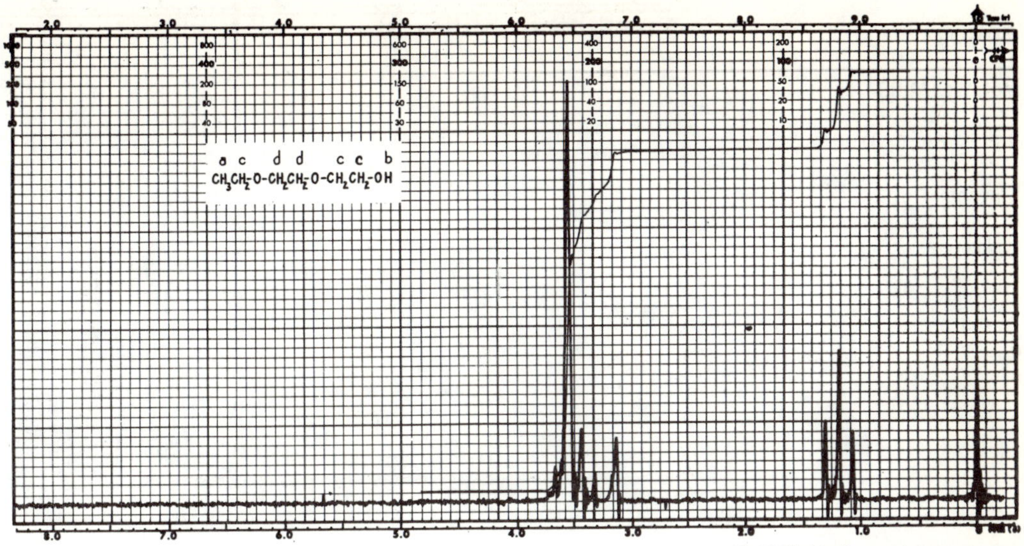

© 1978 Sadtler Research Laboratories, Inc.

2-(2-BUTOXYETHOXY)ETHANOL

$C_8H_{18}O_3$

Mol. Wt. 162.23

B.P. 231°C

Solvent: $CDCl_3$

Assignments
a = 0.90
b = 1.06-1.78
c = 3.07
d = 3.44
e = ca 3.60

1675

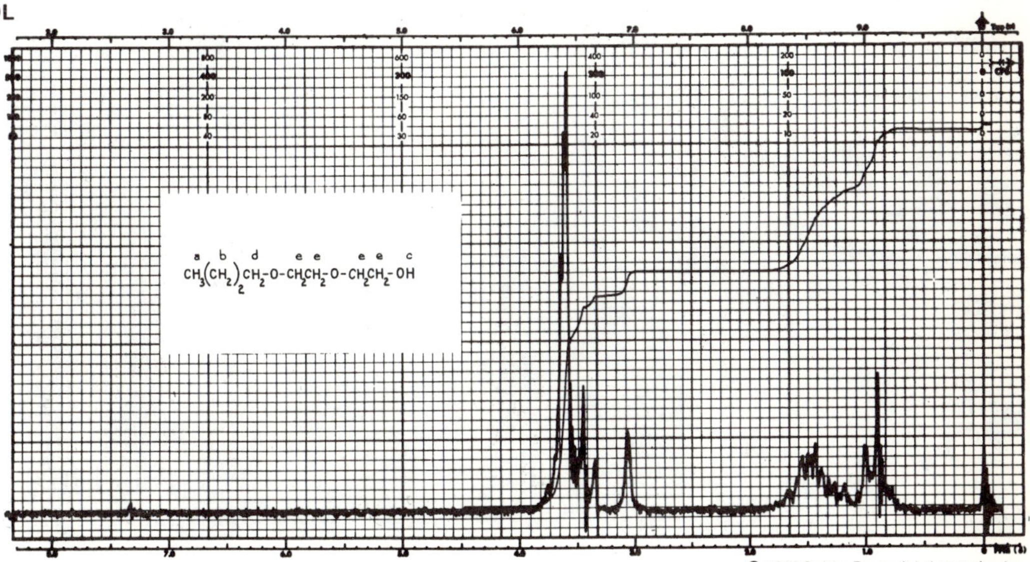

© 1978 Sadtler Research Laboratories, Inc.

2,2'-(2-BUTYNYLENEDIOXY)DIETHANOL

$C_8H_{14}O_4$

Mol. Wt. 174.20

Solvent: $CDCl_3$

Assignments
a = 3.40-4.10
b = 4.23

1676

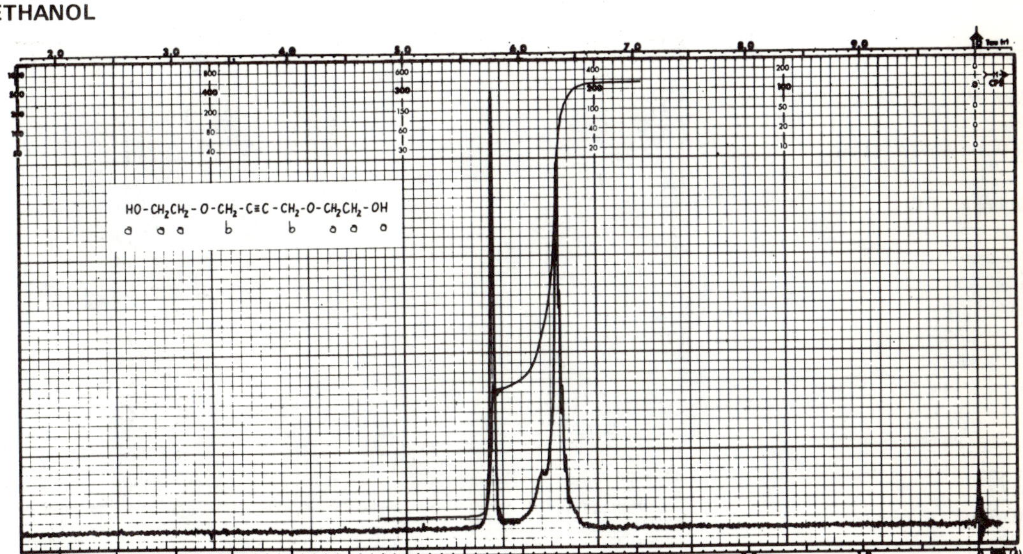

© 1978 Sadtler Research Laboratories, Inc.

2,3-EPOXY-1-PROPANOL

$C_3H_6O_2$

Mol. Wt. 74.08

B.P. 53-56°C/10mm

Solvent: D_2O

Assignments
a = 2.79
b = 2.91
c = 3.27
d = 3.49
e = 3.92
f = 4.60 HDO

1677

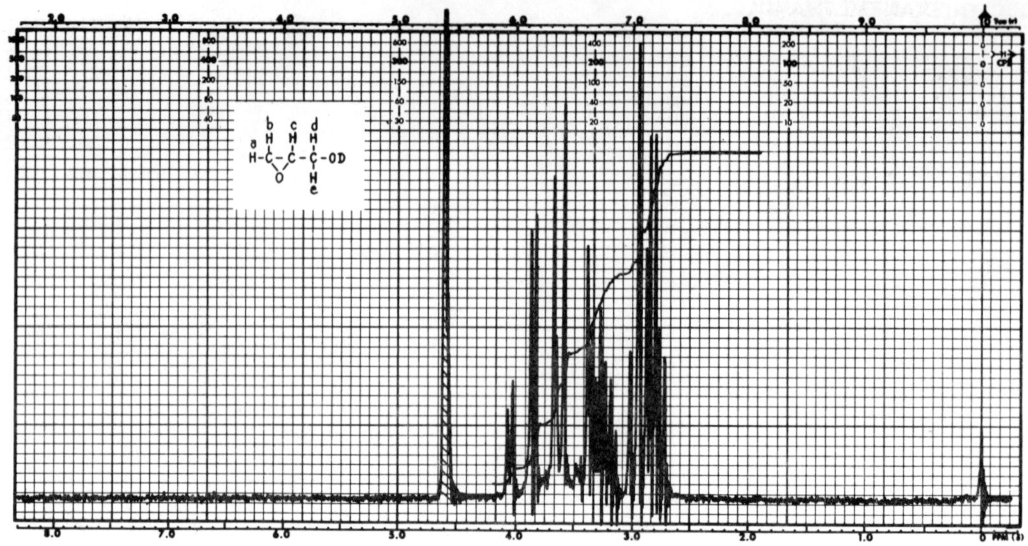

© 1978 Sadtler Research Laboratories, Inc.

2-THIOPHENETHANOL

C_6H_8OS

Mol. Wt. 128.19

B.P. 107-109°C/13mm

Solvent: CCl_4

Assignments
a = 2.91
b = 3.67
c = 3.78
d = 6.71
e = 6.81
f = 7.00

1678

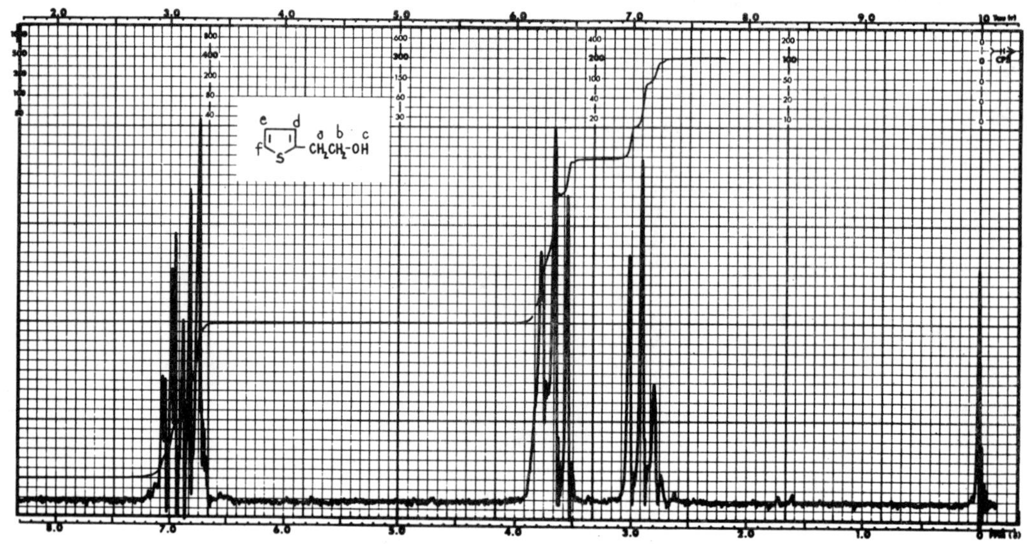

© 1978 Sadtler Research Laboratories, Inc.

2,2-DIMETHYL-1,3-DIOXOLANE-4-METHANOL

$C_6H_{12}O_3$

Mol. Wt. 132.16

B.P. 68-69°C/15mm

Solvent: CCl_4

Assignments
a = 1.31
b = 1.37
c = 3.46
d = 3.53
e = 3.60-4.30
f = 2.11 imp.
D_2O added in upper scan

1679

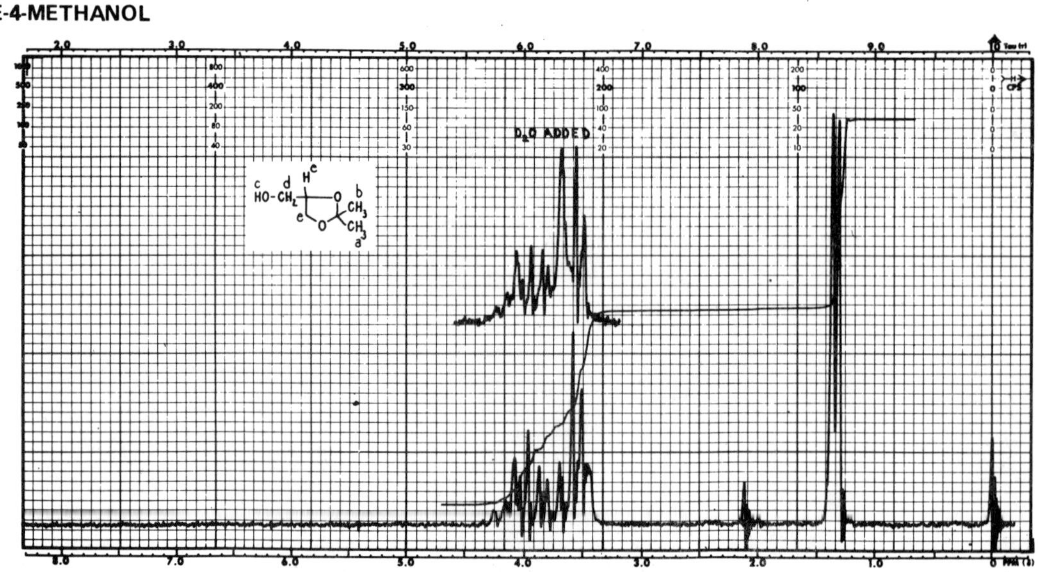

© 1978 Sadtler Research Laboratories, Inc.

CYCLOHEXANEMETHANOL

$C_7H_{14}O$

Mol. Wt. 114.19

B.P. 182-185°C

Solvent: CCl_4

Assignemtns
a = 0.25-2.82
b = 3.32
c = 3.48

1680

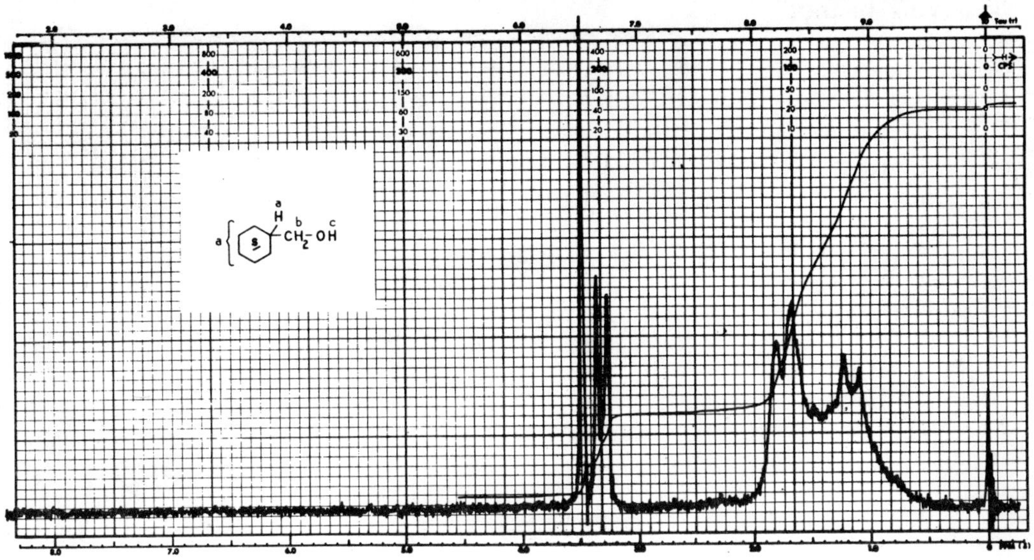

© 1978 Sadtler Research Laboratories, Inc.

4-METHYLCYCLOHEXANEMETHANOL

$C_8H_{16}O$

Mol. Wt. 128.22

B.P. 89-90°C/10mm

Solvent: CCl_4

Assignments
a = 0.92
b = 0.70-2.00
c = 3.32
d = 4.05

1681

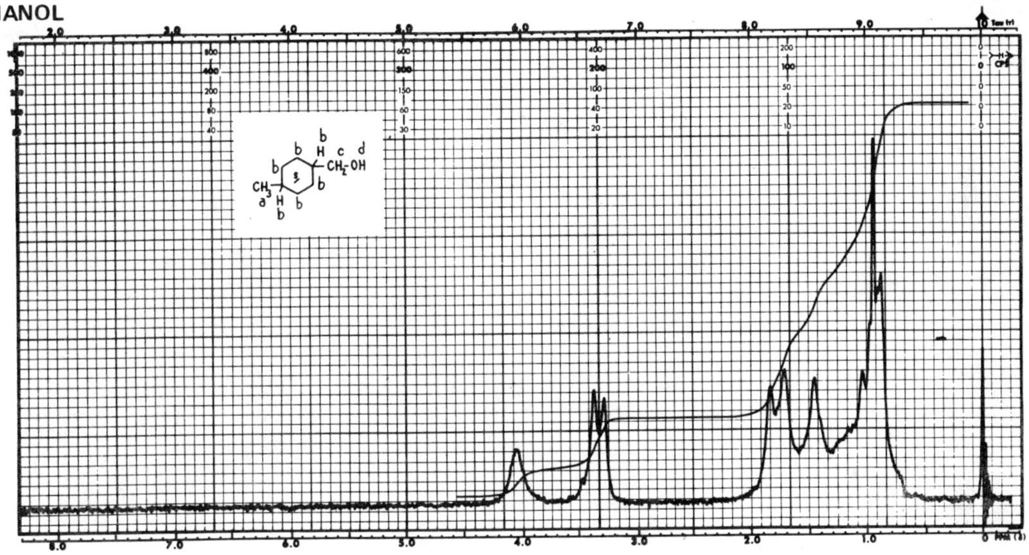

© 1978 Sadtler Research Laboratories, Inc.

CYCLOHEXANEETHANOL

$C_8H_{16}O$

Mol. Wt. 128.21

B.P. 121-122°C

Solvent: CCl_4

Assignments
a = 0.50-2.00
b = 3.49
c = 3.56

1682

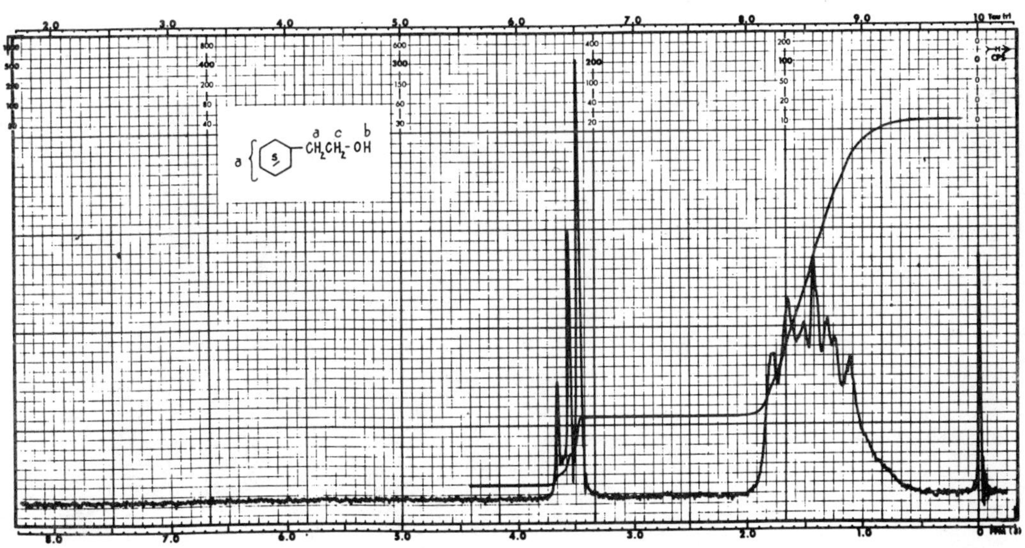

© 1978 Sadtler Research Laboratories, Inc.

3-PIPERIDINEMETHANOL

$C_6H_{13}NO$

Mol. Wt. 115.18

M.P. 60-62°C

Solvent: $CDCl_3$

Assignments
a = 0.85-2.00
b = 2.10-2.70 or c
c = 2.80-3.30 or b
d = 3.37
e = 3.67

1683

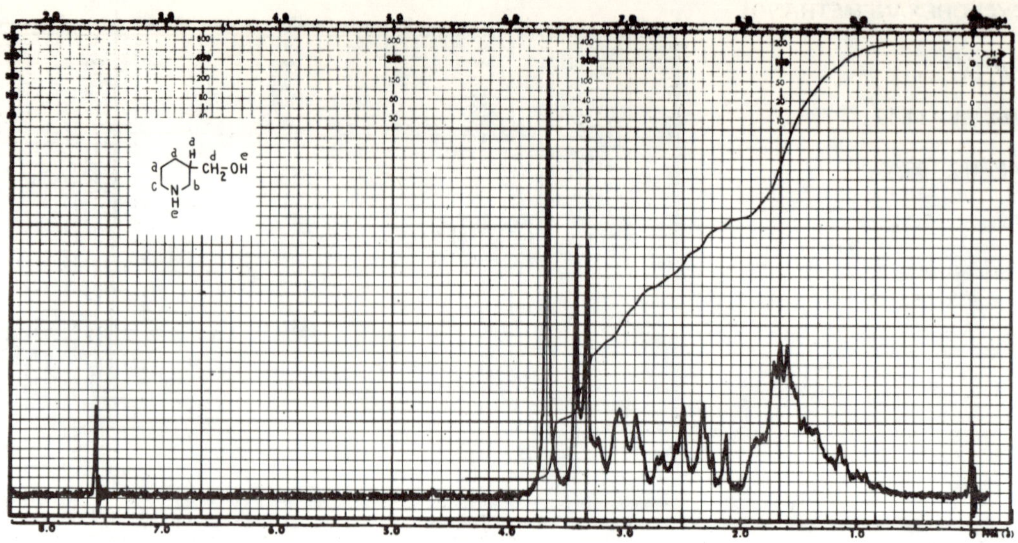

© 1978 Sadtler Research Laboratories, Inc.

1-ADAMANTANEMETHANOL

$C_{11}H_{18}O$

Mol. Wt. 166.27

Solvent: Polysol-d

Assignments
a = ca 1.51
b = ca 1.72
c = 1.93
d = 3.07
e = 3.49
f = 7.19 imp.

1684

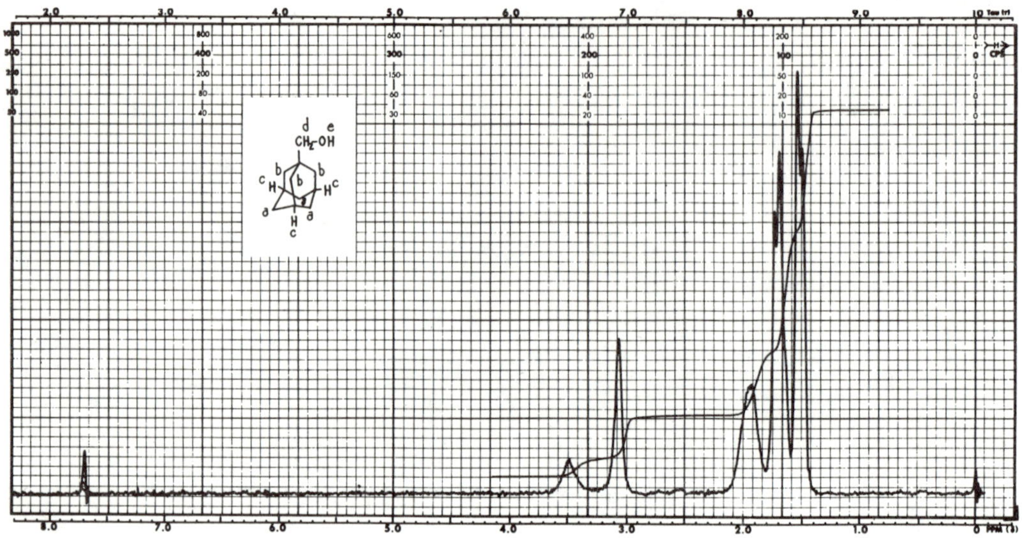

© 1978 Sadtler Research Laboratories, Inc.

ALLYL ALCOHOL

C_3H_6O

Mol. Wt. 58.08

M.P. 96.9°C

Solvent: CCl_4

Assignments
a = 4.05
b = 4.60
c = 5.07
d = 5.21
e = 5.88

1685

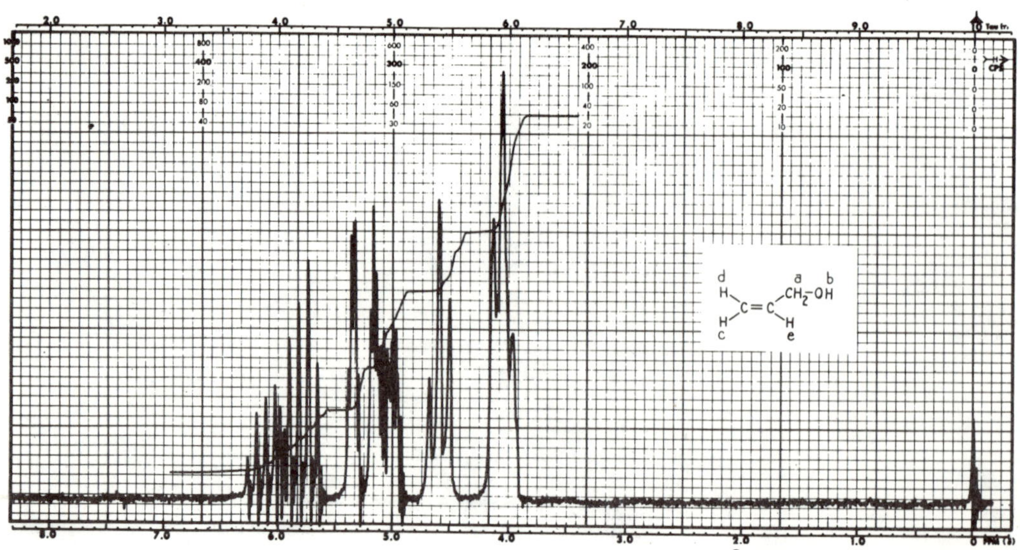

© 1978 Sadtler Research Laboratories, Inc.

2-BUTEN-1-OL

C_4H_8O

Mol. Wt. 72.11

Solvent: CCl_4

Assignments
a = 1.69
b = 3.92
c = 4.11
d = ca 5.57

1686

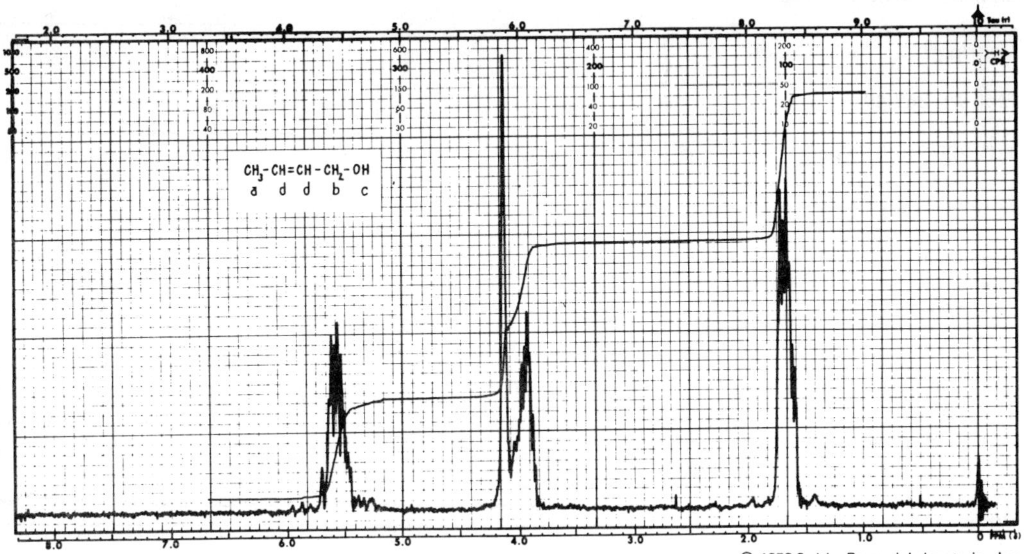

© 1978 Sadtler Research Laboratories, Inc.

3-BUTEN-1-OL

C_4H_8O

Mol. Wt. 72.11

Solvent: $CDCl_3$

Assignments
a = 2.37
b = 2.57
c = 3.65
d = 5.10
e = 5.14
f = 5.80

1687

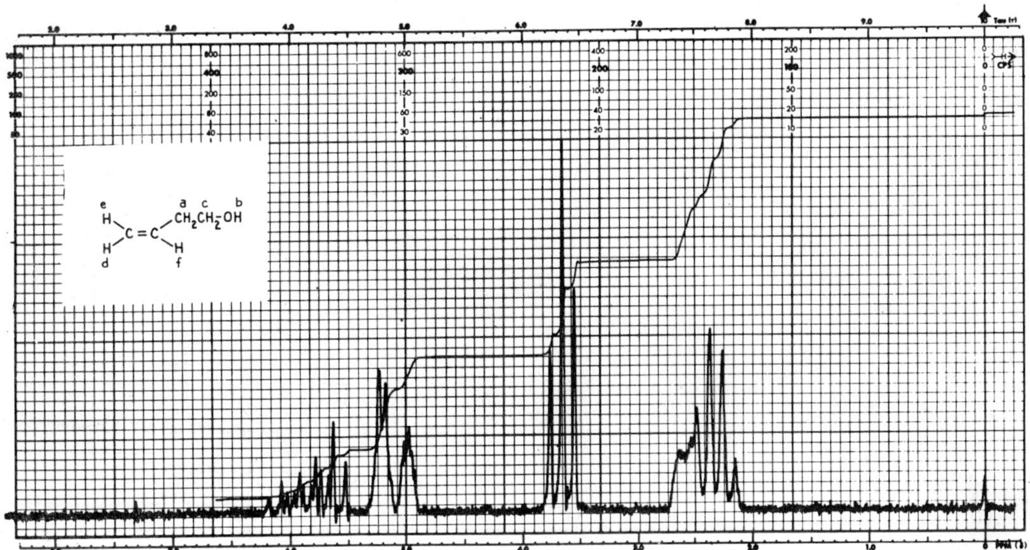

© 1978 Sadtler Research Laboratories, Inc.

cis-3-HEXEN-1-OL

$C_6H_{12}O$

Mol. Wt. 100.16

Solvent: CCl_4

Assignments
a = 0.95
b = 1.80-2.45
c = 3.52
d = 4.21
e = 5.05-5.70

1688

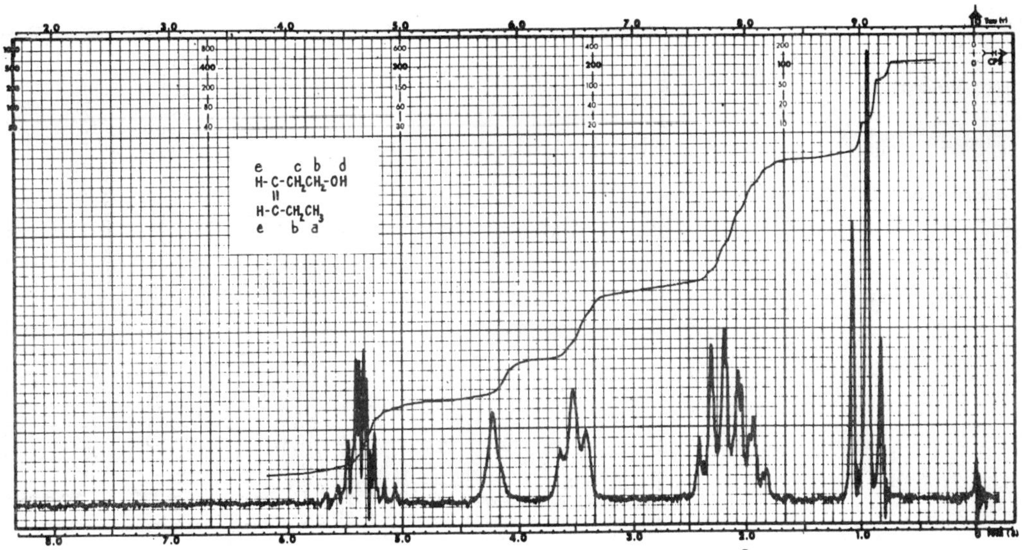

© 1978 Sadtler Research Laboratories, Inc.

cis-9-OCTADECEN-1-OL

$C_{18}H_{36}O$

Mol. Wt. 268.49

B.P. 205-210°C/15mm

Solvent: CCl_4

Assignments
a = 0.88
b = 1.29
c = 1.97
d = 2.87
e = 3.49
f = 5.25
Trace of acid added for OH

1689

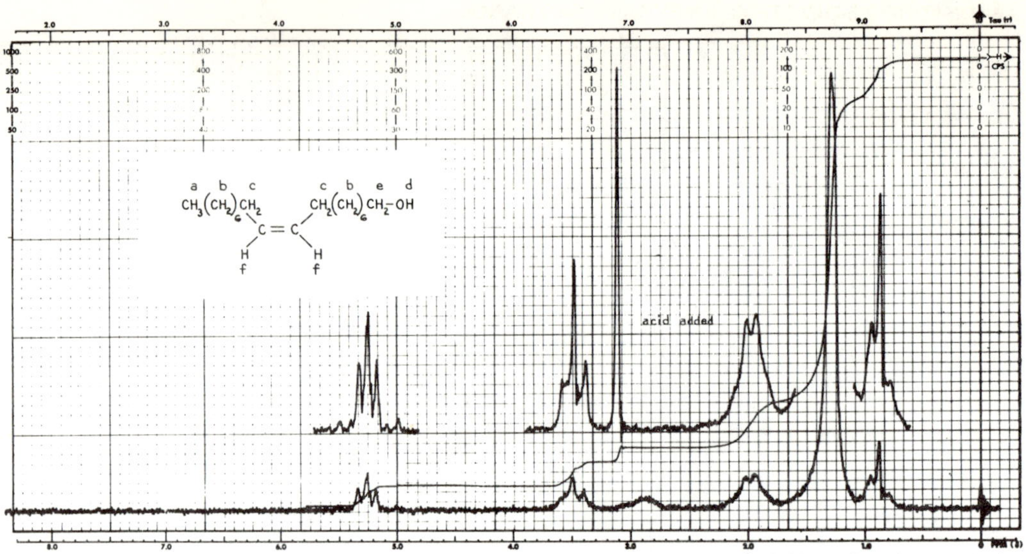

© 1978 Sadtler Research Laboratories, Inc.

PHYTOL

$C_{20}H_{40}O$

Mol. Wt. 296.54

B.P. 202-204°C/10mm

Solvent: CCl_4

Assignments
a = ca 0.89
b = 1.00-1.60
c = ca 1.67
d = 1.96
e = 2.30
f = 4.02
g = 5.35

1690

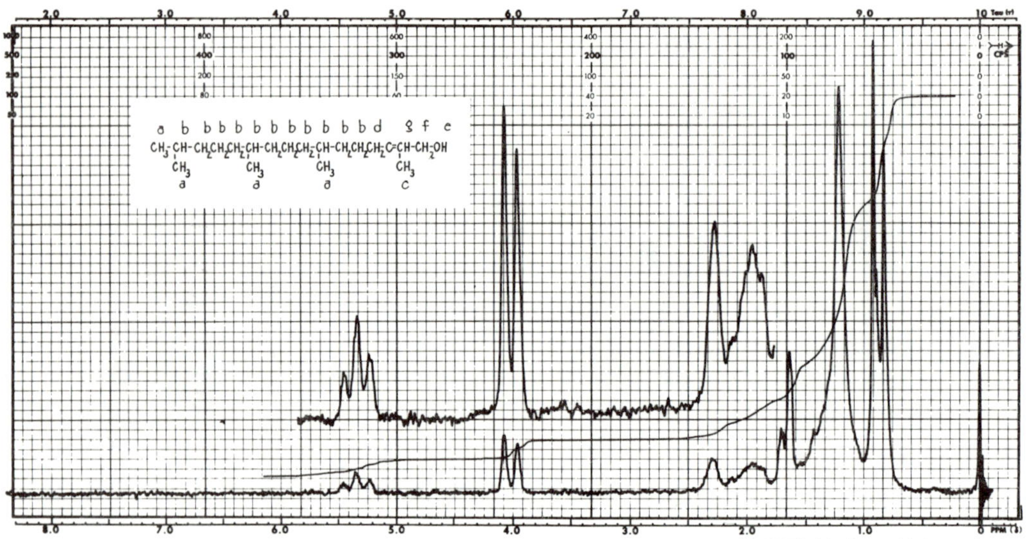

© 1978 Sadtler Research Laboratories, Inc.

2-(ALLYLOXY)ETHANOL

$C_5H_{10}O$

Mol. Wt. 102.13

Solvent: CCl_4

Assignments
a = 3.30-3.75
b = 3.97
c = 5.10
d = 5.16
e = 5.81

1691

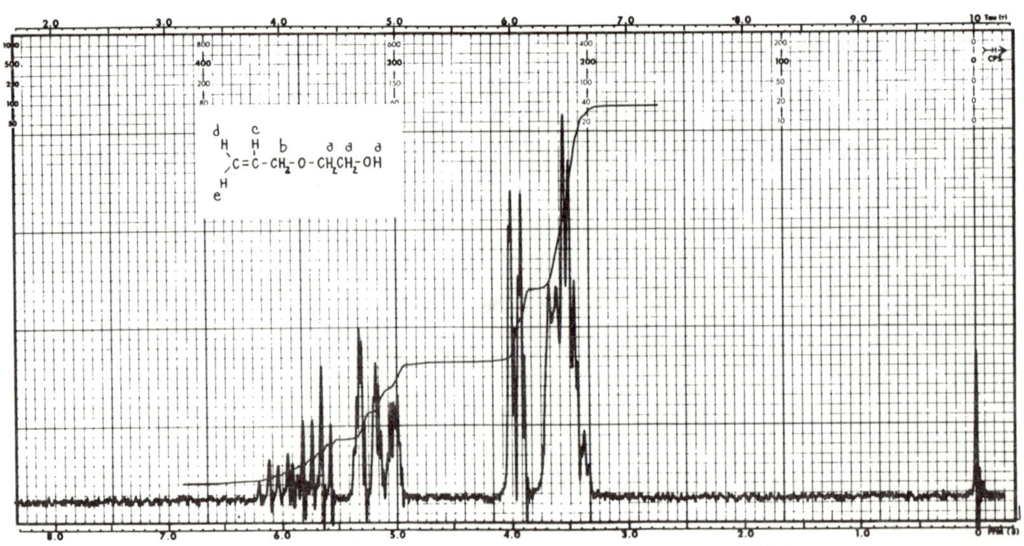

© 1978 Sadtler Research Laboratories, Inc.

3,4-DIHYDRO-2,5-DIMETHYL-2H-PYRAN-2-METHANOL

$C_8H_{14}O_2$

Mol. Wt. 142.20

B.P. 50-51°C/1mm

Solvent: $CDCl_3$

Assignments
a = 1.16
b = 1.56
c = 1.00-1.90
d = ca 3.49
e = 6.09

1692

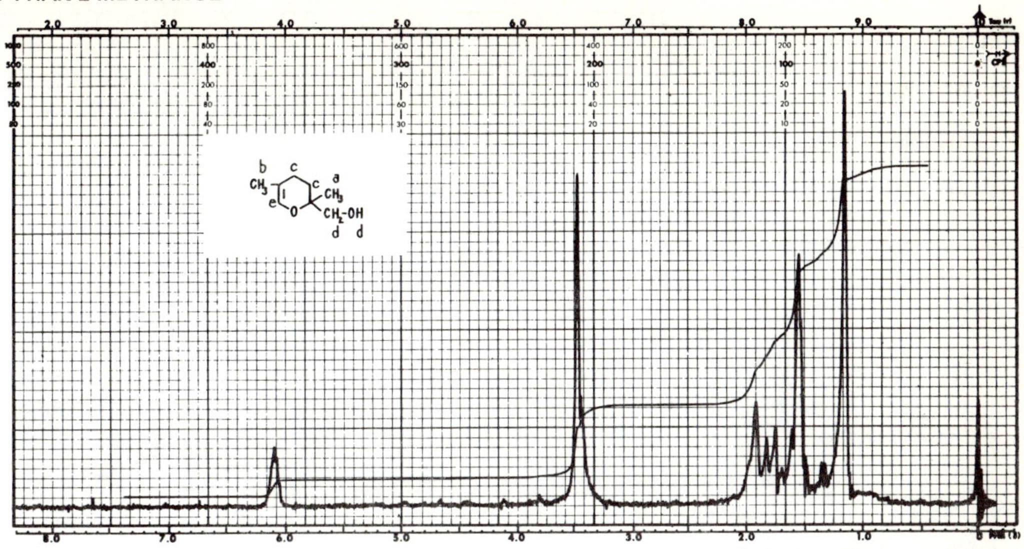

© 1978 Sadtler Research Laboratories, Inc.

2-PROPYN-1-OL

C_3H_4O

Mol. Wt. 56.06

B.P. 115°C

Solvent: $CDCl_3$

Assignments
a = 2.54
b = 4.01
c = 4.23

1693

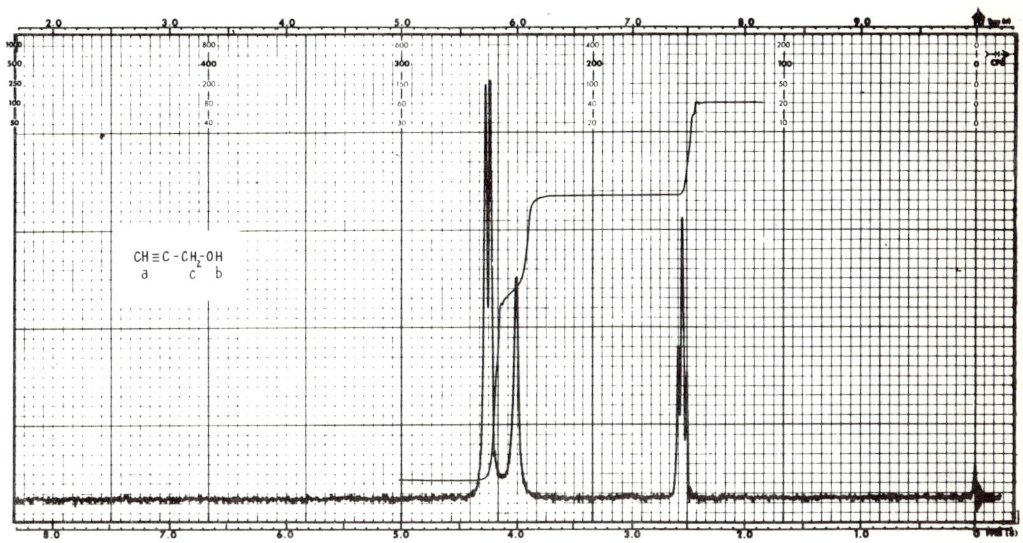

© 1978 Sadtler Research Laboratories, Inc.

2-BUTYN-1-OL

C_4H_6O

Mol. Wt. 70.09

Solvent: CCl_4

Assignments
a = 1.83
b = 3.99
c = 4.11

1694

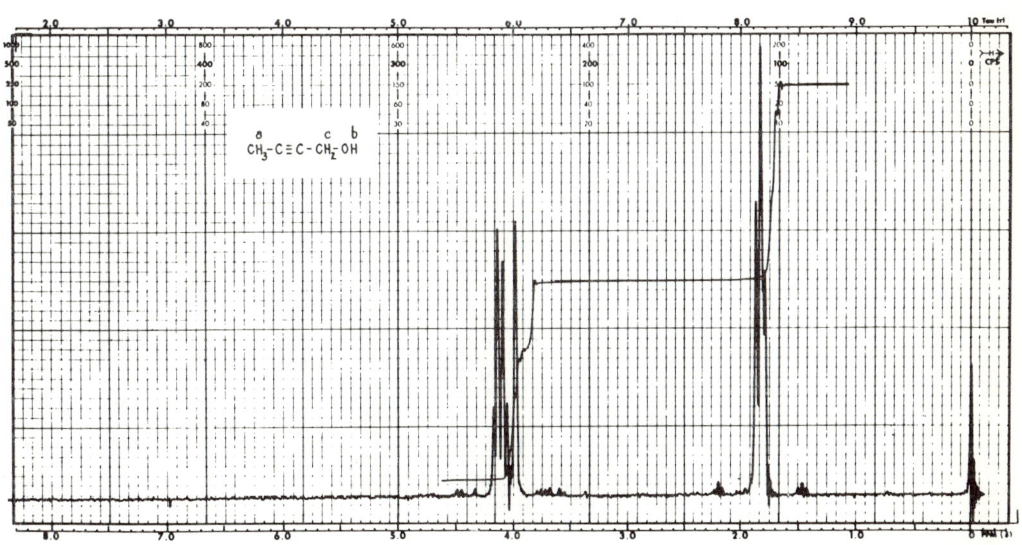

© 1978 Sadtler Research Laboratories, Inc.

3-BUTYN-1-OL

C_4H_6O

Mol. Wt. 70.09

Solvent: CCl_4

Assignments
a = 1.96
b = 2.38
c = 3.67

1695

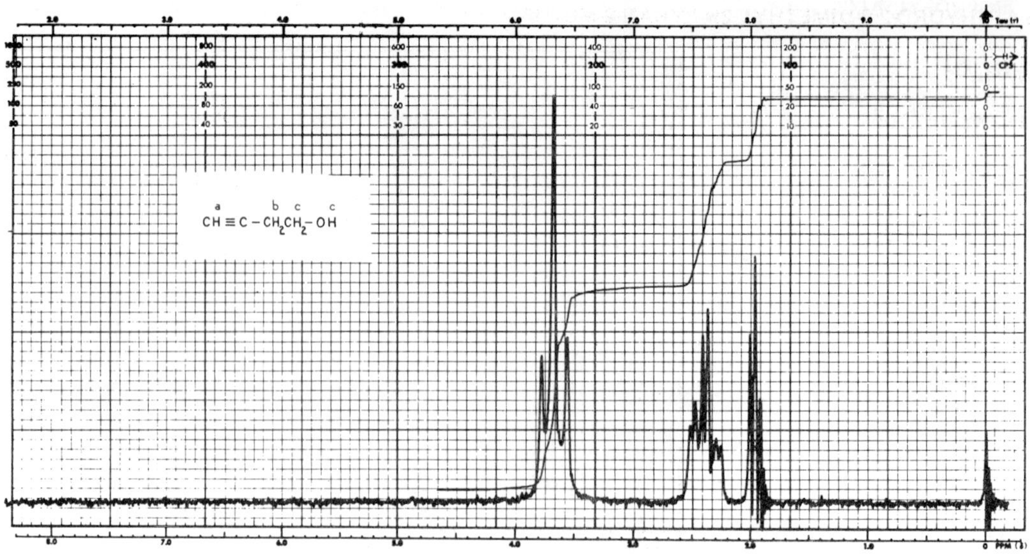

© 1978 Sadtler Research Laboratories, Inc.

2-PENTYN-1-OL

C_5H_8O

Mol. Wt. 84.12

Solvent: CCl_4

Assignments
a = 1.15
b = 2.23
c = 4.01
d = 4.17
e = 1.66, 4.79 imps.

1696

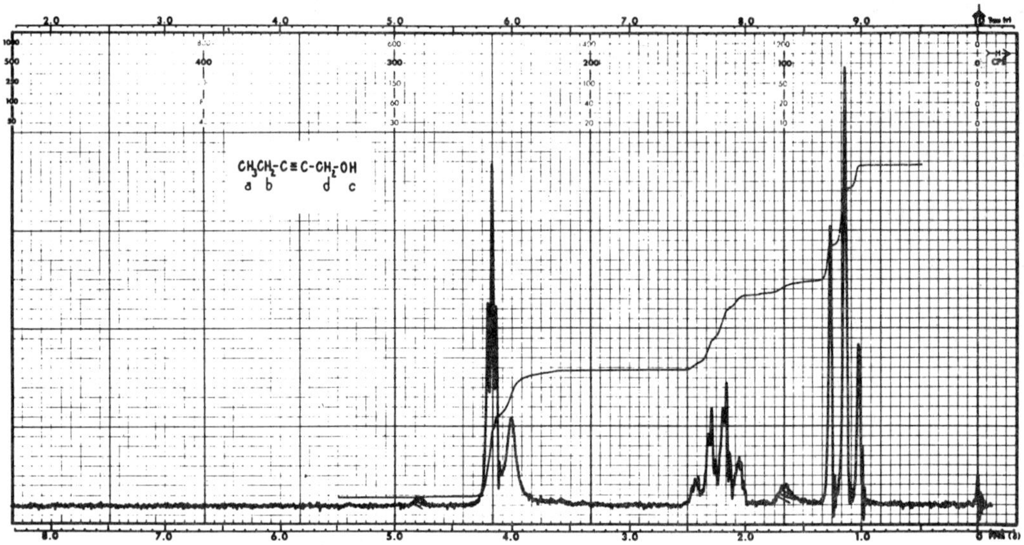

© 1978 Sadtler Research Laboratories, Inc.

2-NONYN-1-OL

$C_9H_{16}O$

Mol. Wt. 140.23

Solvent: CCl_4

Assignments
a = 0.92
b = 1.10-1.80
c = 2.20
d = ca 3.61
e = 4.19

1697

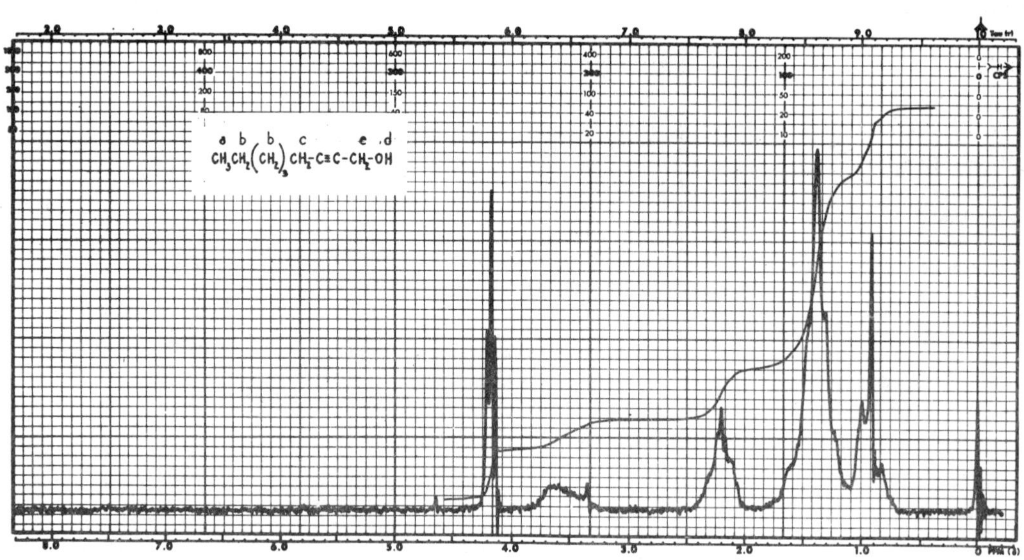

© 1978 Sadtler Research Laboratories, Inc.

3-NONYN-1-OL

$C_9H_{16}O$

Mol. Wt. 140.23

Solvent: CCl_4

Assignments
a = 0.92
b = 1.10-1.85
c = 2.15
d = 2.38
e = 3.31
f = 3.60

1698

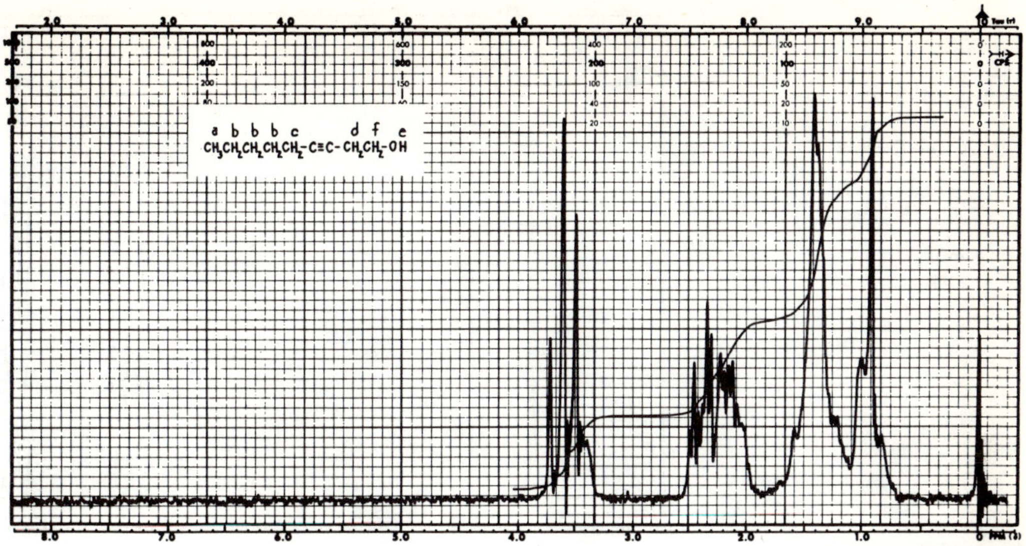

© 1978 Sadtler Research Laboratories, Inc.

3-DECYN-1-OL

$C_{10}H_{18}O$

Mol. Wt. 154.25

Solvent: CCl_4

Assignments
a = 0.90
b = 1.10-1.80
c = 2.12
d = 2.32
e = ca 3.30
f = 3.60

1699

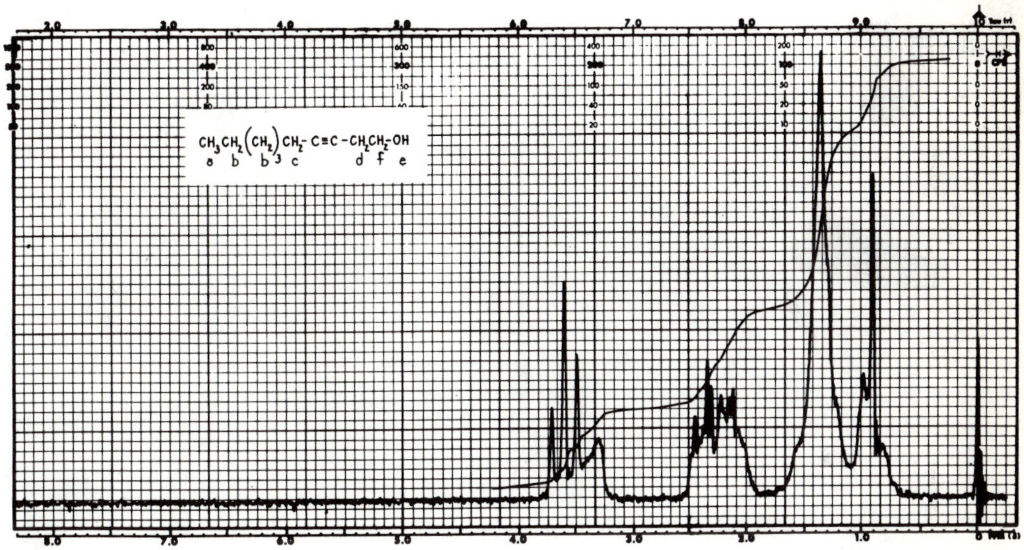

© 1978 Sadtler Research Laboratories, Inc.

BENZYL ALCOHOL

C_7H_8O

Mol. Wt. 108.14

B.P. 203-205°C

Solvent: CCl_4

Assignments
a = 3.10
b = 4.41
c = 7.19

1700

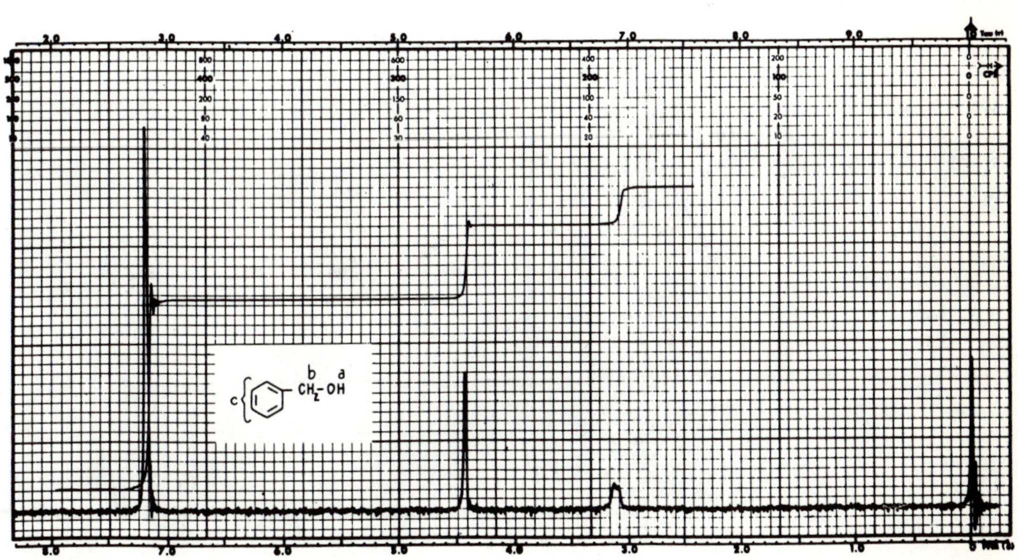

© 1978 Sadtler Research Laboratories, Inc.

PHENETHYL ALCOHOL

$C_8H_{10}O$

Mol. Wt. 122.16

B.P. 88-90°C/5mm

Solvent: $CDCl_3$

Assignments
a = 2.44
b = 2.77
c = 3.71
d = 7.19

1701

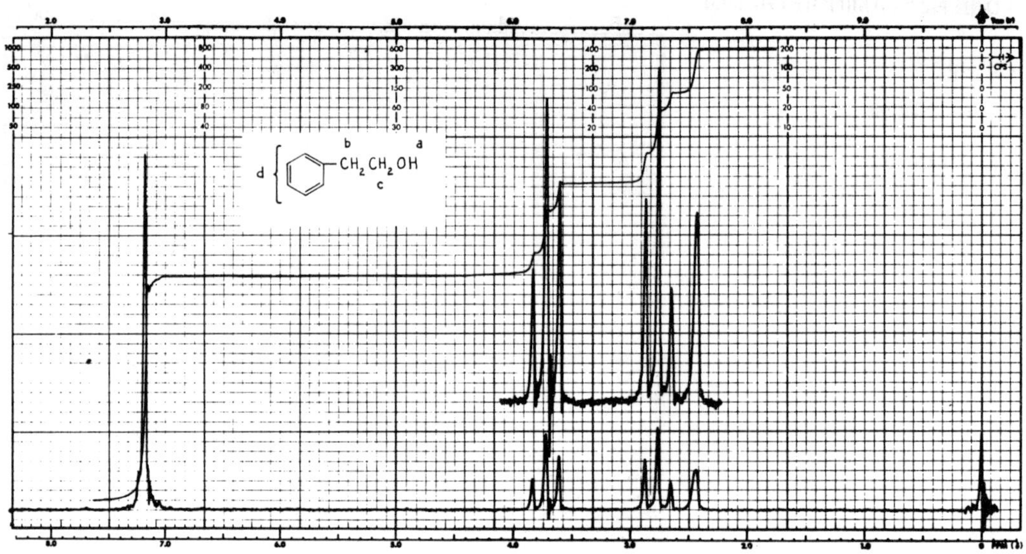

© 1978 Sadtler Research Laboratories, Inc.

3-PHENYL-1-PROPANOL

$C_9H_{12}O$

Mol. Wt. 136.18

B.P. 235.6°C (lit.)

Solvent: CCl_4

Assignments
a = 1.82
b = 2.62
c = 3.51
d = ca 3.63
e = 7.10

1702

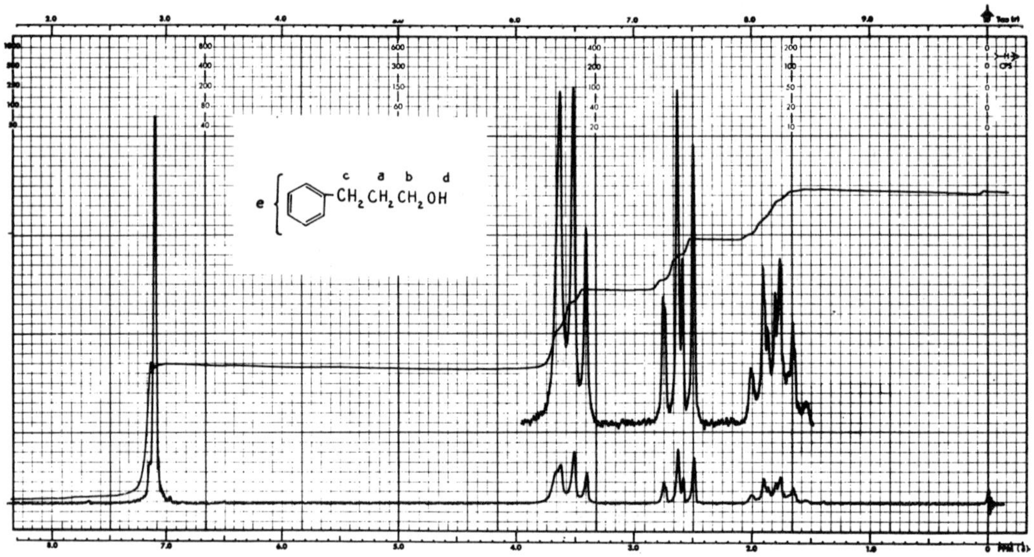

© 1978 Sadtler Research Laboratories, Inc.

2-[(N-BENZYL-N-METHYL)AMINO] ETHANOL

$C_{10}H_{15}NO$

Mol. Wt. 165.24

Solvent: $CDCl_3$

Assignments
a = 2.22
b = 2.59
c = 2.89
d = 3.55
e = 3.61
f = 7.30

1703

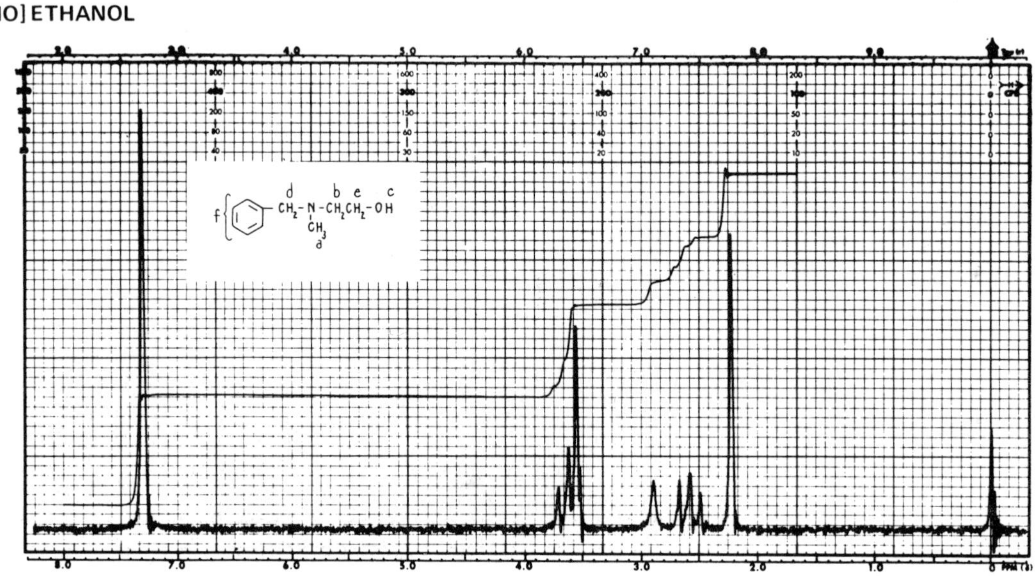

© 1978 Sadtler Research Laboratories, Inc.

2-(DIBENZYLAMINO)ETHANOL

$C_{16}H_{19}NO$

Mol. Wt. 241.34

Solvent: $CDCl_3$

Assignments
a = 2.31
b = 2.63
c = 3.54
d = 3.61
e = ca 7.29

1704

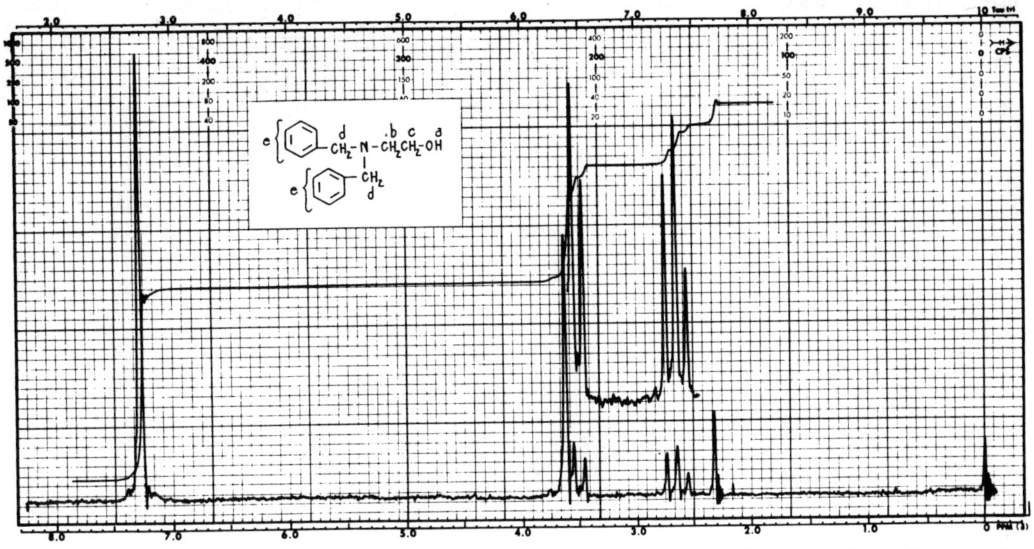

© 1978 Sadtler Research Laboratories, Inc.

2-(BENZYLOXY)ETHANOL

$C_9H_{12}O_2$

Mol. Wt. 152.19

B.P. 131-133°C/13mm

Solvent: CCl_4

Assignments
a = 2.59
b = 3.60 or c
c = 3.70 or b
d = 4.54
e = 7.33

1705

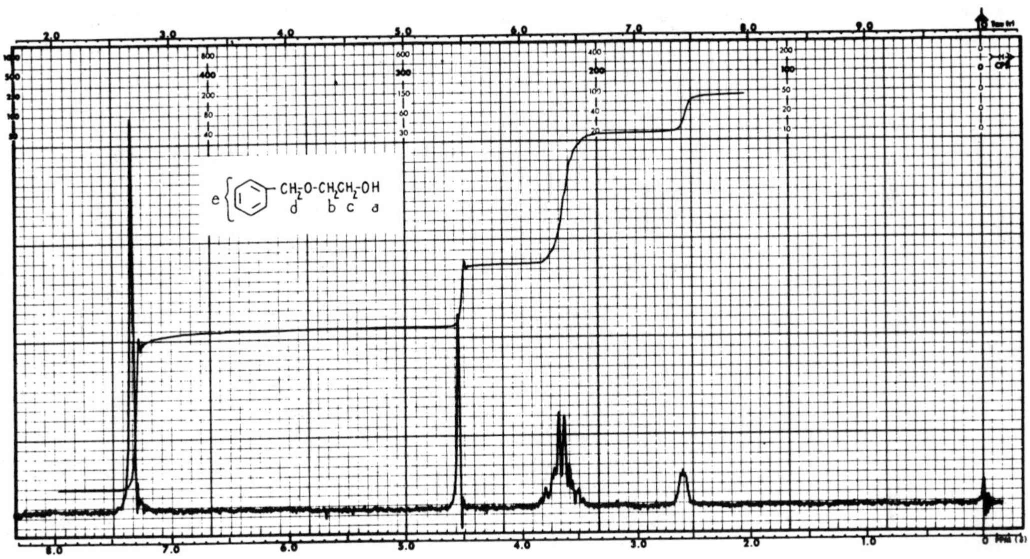

© 1978 Sadtler Research Laboratories, Inc.

2-(N-ETHYLANILINO)ETHANOL

$C_{10}H_{15}NO$

Mol. Wt. 165.24

Solvent: $CDCl_3$

Assignments
a = 1.09
b = 2.91
c = 3.30
d = 3.33
e = 3.62
f = 6.64
g = 6.72
h = 7.18

1706

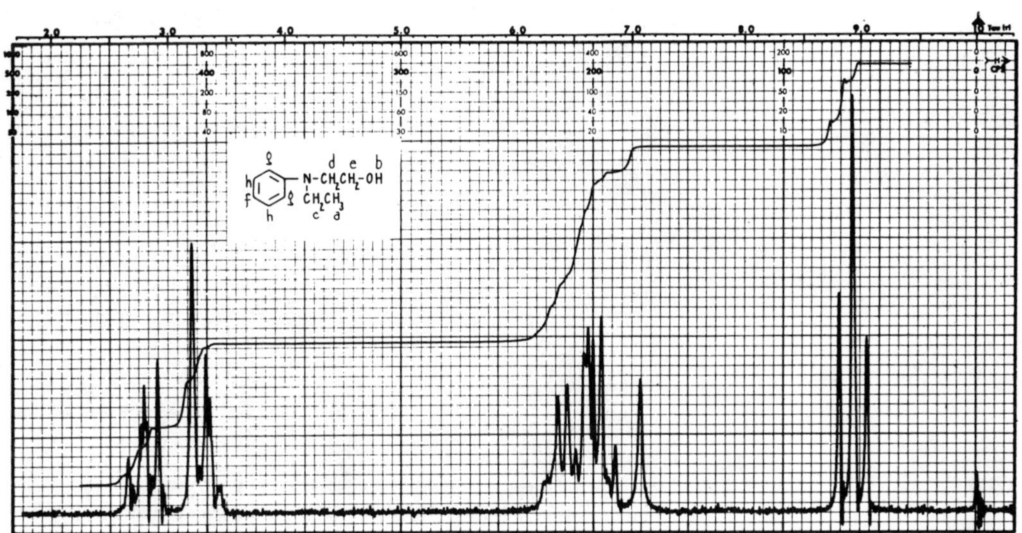

© 1978 Sadtler Research Laboratories, Inc.

2-[(p-CHLOROBENZYLIDENE)AMINO] ETHANOL

$C_9H_{10}ClNO$

Mol. Wt. 183.64

M.P. ∼70°C

Solvent: $CDCl_3$

Assignments

a = 3.40-4.00
b = 7.26
c = 7.55
d = 8.12

1707

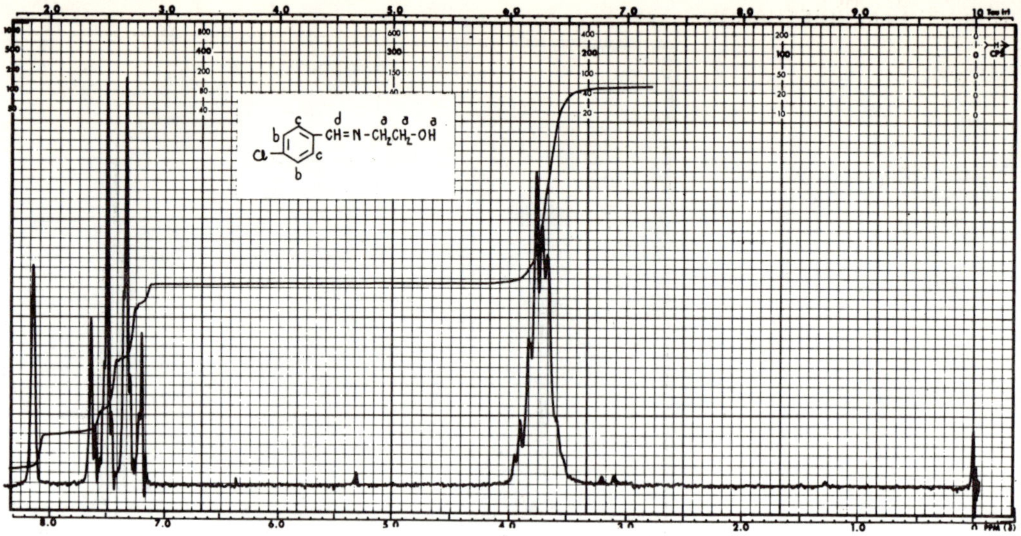

© 1978 Sadtler Research Laboratories, Inc.

2-(p-TOLYLTHIO)ETHANOL

$C_9H_{12}OS$

Mol. Wt. 168.26

B.P. 158-160°C/10mm

Solvent: CCl_4

Assignments

a = 2.22
b = 2.89
c = 3.28
d = 3.57
e = 6.92
f = 7.14

1708

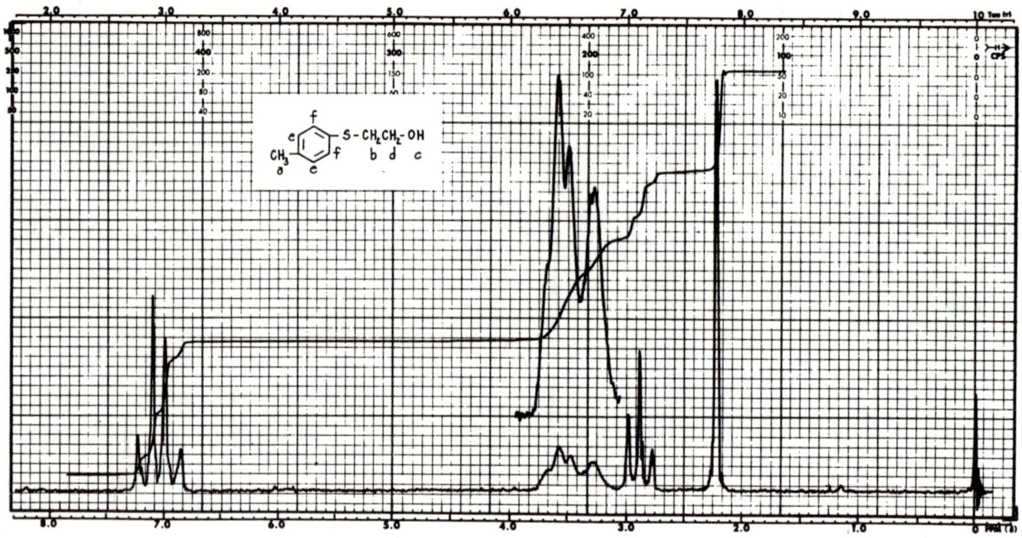

© 1978 Sadtler Research Laboratories, Inc.

(p-TOLYLSULFONYL)METHANOL

$C_8H_{10}O_3S$

Mol. Wt. 186.23

Solvent: $CDCl_3$

Assignments

a = 2.42
b = 4.45
c = 4.61
d = 7.32
e = 7.79

1709

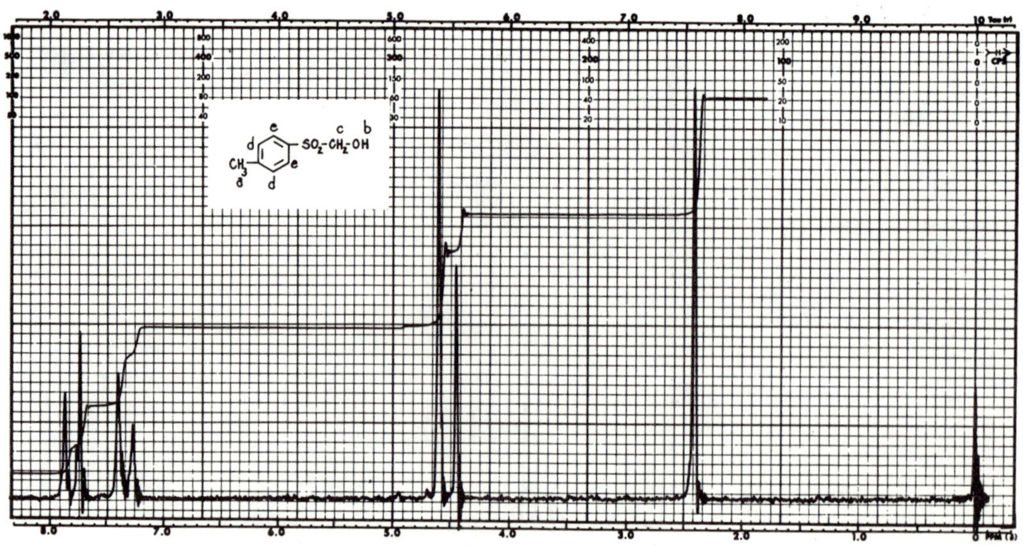

© 1978 Sadtler Research Laboratories, Inc.

2-PHENOXYETHANOL

$C_8H_{10}O_2$

Mol. Wt. 138.17

B.P. 103-105^oC/5mm

Solvent: CCl_4

Assignments
a = 3.50-4.00
b = 6.71
c = 6.74
d = 7.10

1710

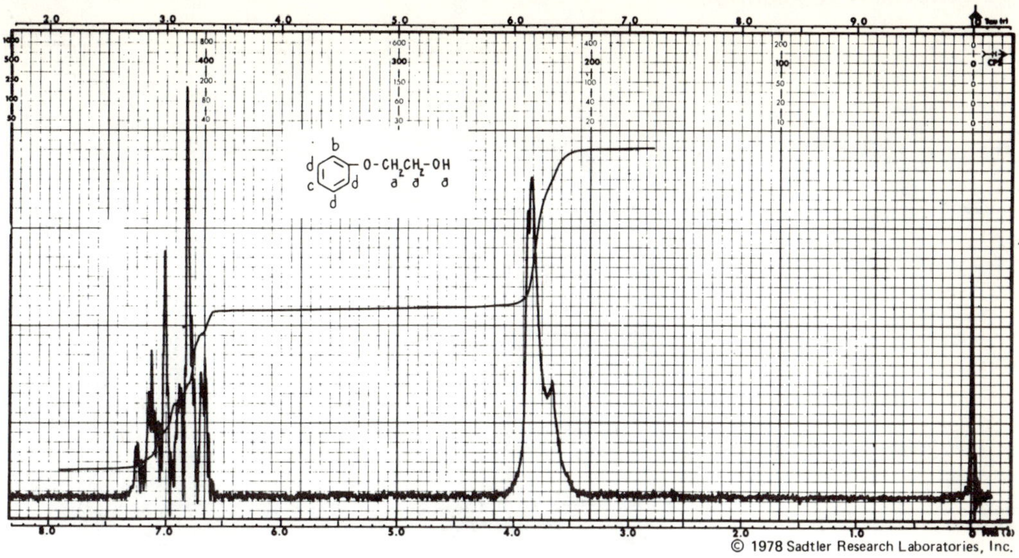

© 1978 Sadtler Research Laboratories, Inc.

CINNAMYL ALCOHOL

$C_9H_{10}O$

Mol. Wt. 134.17

M.P. 31-33^oC

Solvent: $CDCl_3$

Assignments
a = 2.33
b = 4.22
c = 6.31
d = 6.57
e = 7.28

1711

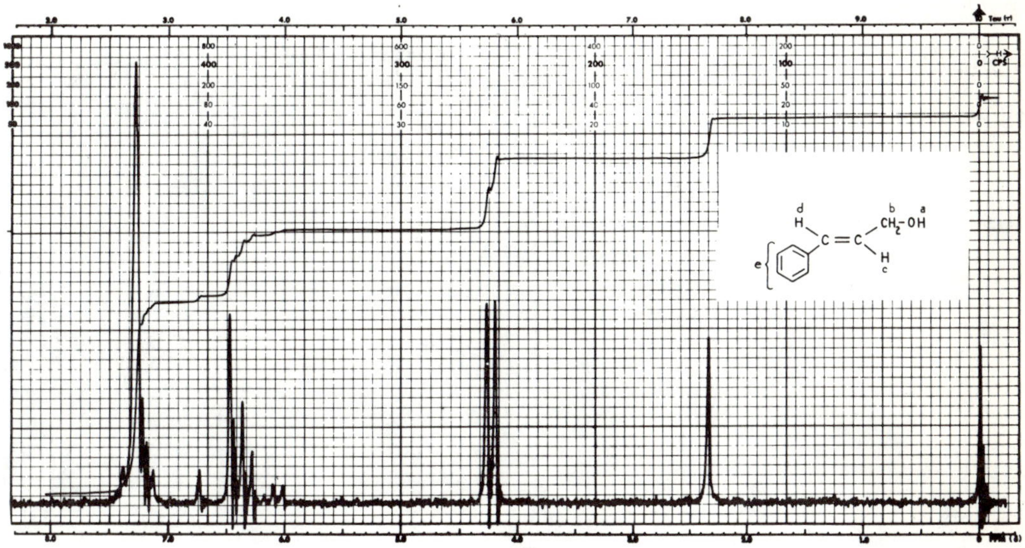

© 1978 Sadtler Research Laboratories, Inc.

3-PHENYL-2-BUTEN-1-OL

$C_{10}H_{12}O$

Mol. Wt. 148.21

Solvent: $CDCl_3$

Assignments
a = 1.81
b = 3.38
c = 4.09
d = 6.46
e = ca 7.18

1712

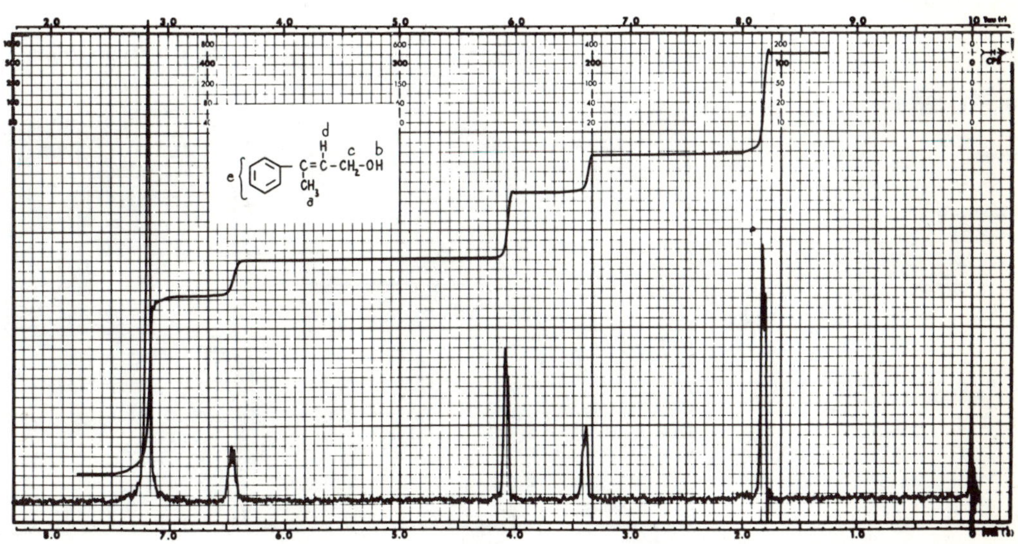

© 1978 Sadtler Research Laboratories, Inc.

p-METHYLBENZYL ALCOHOL

$C_8H_{10}O$

Mol. Wt. 122.17

M.P. 60-61°C

Solvent: CCl_4

Assignments
a = 2.28
b = 2.94
c = 4.37
d = 7.01

1713

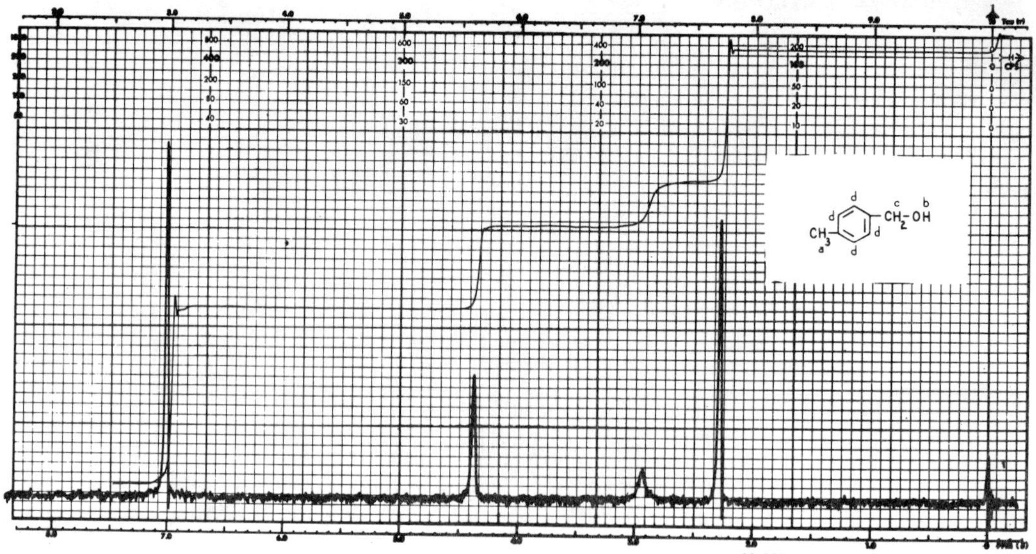

© 1978 Sadtler Research Laboratories, Inc.

---

p-CYMEN-7-OL

$C_{10}H_{14}O$

Mol. Wt. 150.22

B.P. 127-129°C/14mm

Solvent: CCl_4

Assignments
a = 1.21
b = 2.85
c = 4.05
d = 4.35
e = 7.07

1714

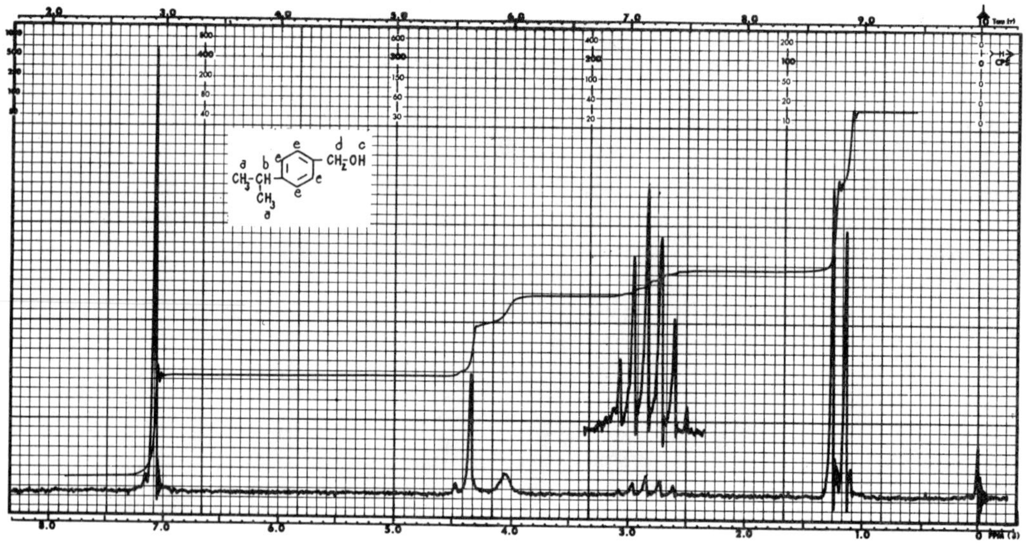

© 1978 Sadtler Research Laboratories, Inc.

---

o-FLUOROBENZYL ALCOHOL

C_7H_7FO

Mol. Wt. 126.13

Solvent: CCl_4

Assignments
a = 4.51
b = 4.46
c = 6.65-7.45

1715

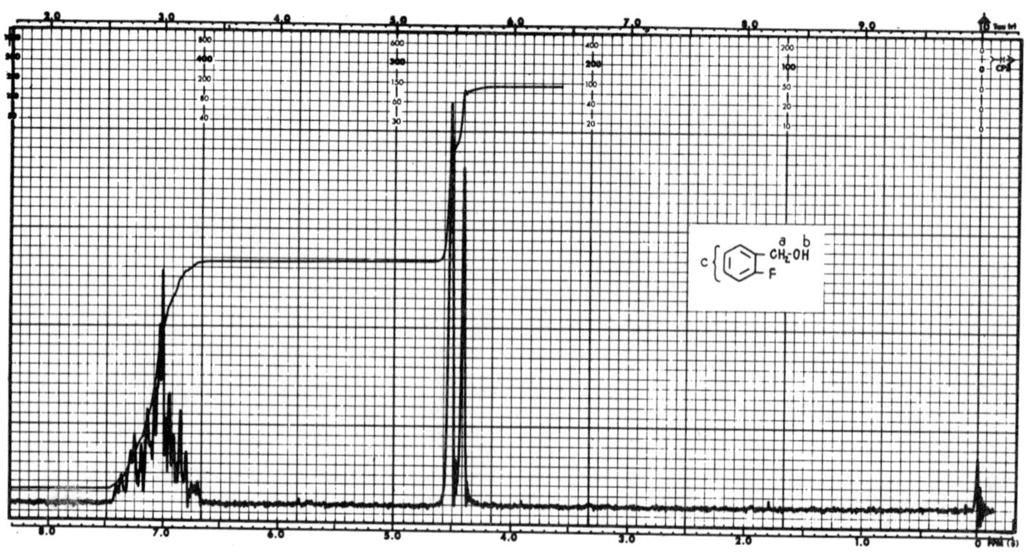

© 1978 Sadtler Research Laboratories, Inc.

p-FLUOROBENZYL ALCOHOL

C_7H_7FO

Mol. Wt. 126.13

M.P. 23°C

B.P. 76-78°C/13mm

Solvent: $CDCl_3$

Assignments
a = 4.40
b = 4.88
c = 6.90
d = 7.12

1716

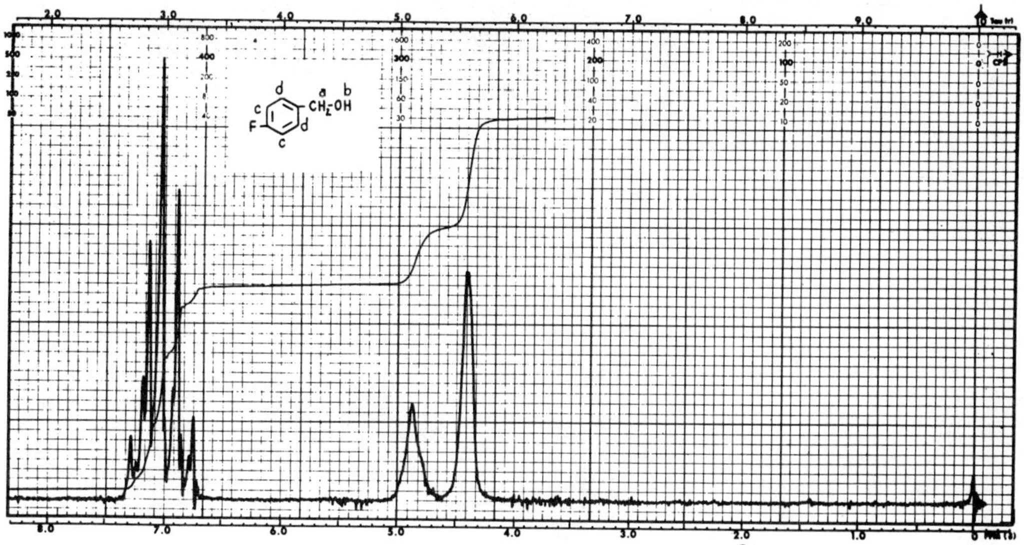

© 1978 Sadtler Research Laboratories, Inc.

o-CHLOROBENZYL ALCOHOL

C_7H_7ClO

Mol. Wt. 142.59

M.P. 69-71°C

Solvent: $CDCl_3$

Assignments
a = 3.30
b = 4.65
c = 7.02-7.51

1717

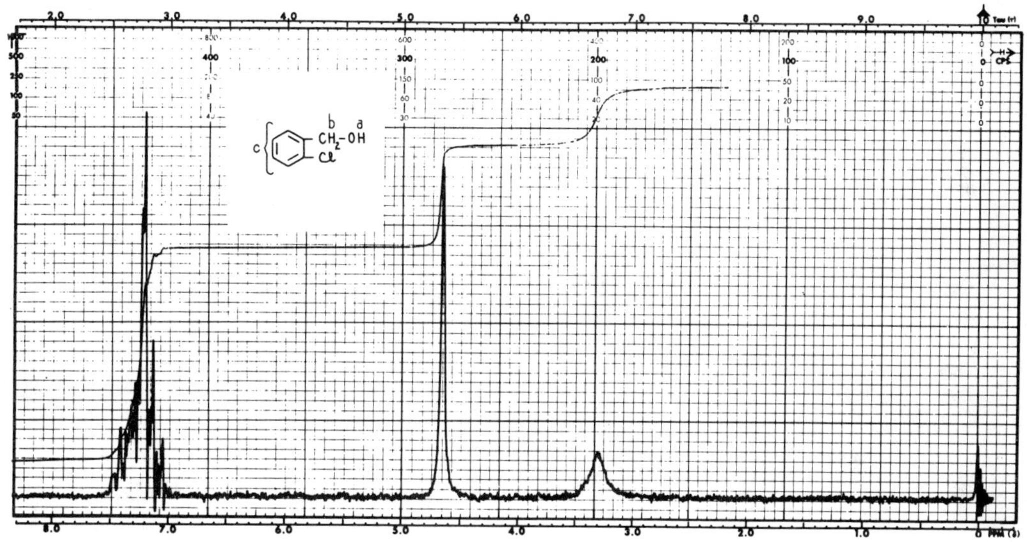

© 1978 Sadtler Research Laboratories, Inc.

p-AMINOPHENETHYL ALCOHOL

$C_8H_{11}NO$

Mol. Wt. 137.18

M.P. 105-108°C

Solvent: Polysol-d

Assignments
a = 2.63
b = 3.64
c = 4.05
d = 6.53
e = 6.91

1718

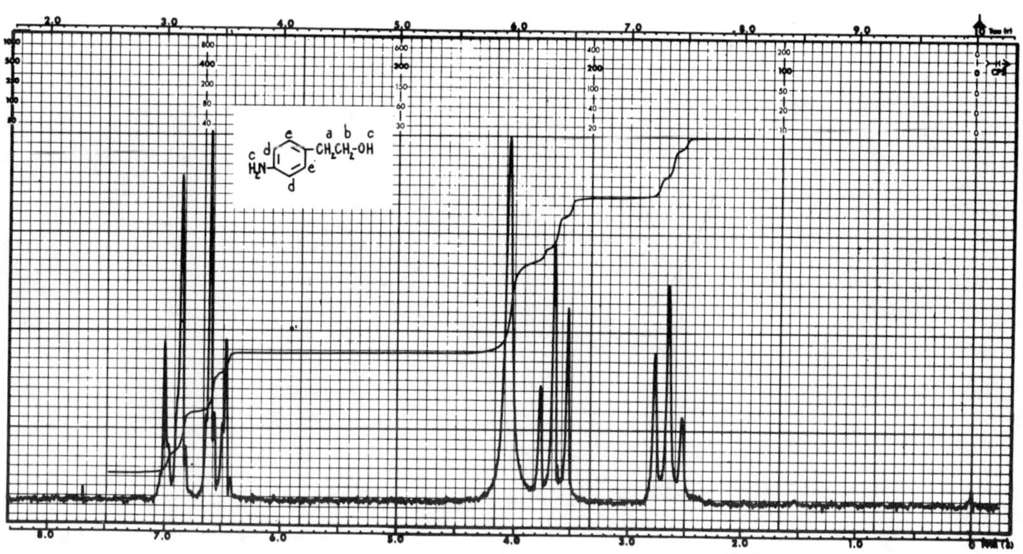

© 1978 Sadtler Research Laboratories, Inc.

o-NITROBENZYL ALCOHOL

$C_7H_7NO_3$

Mol. Wt. 153.14

M.P. 70-72°C

B.P. 270°C

Solvent: $CDCl_3$

Assignments
a = 3.12
b = 4.99
c = 7.25-7.90
d = 8.05

1719

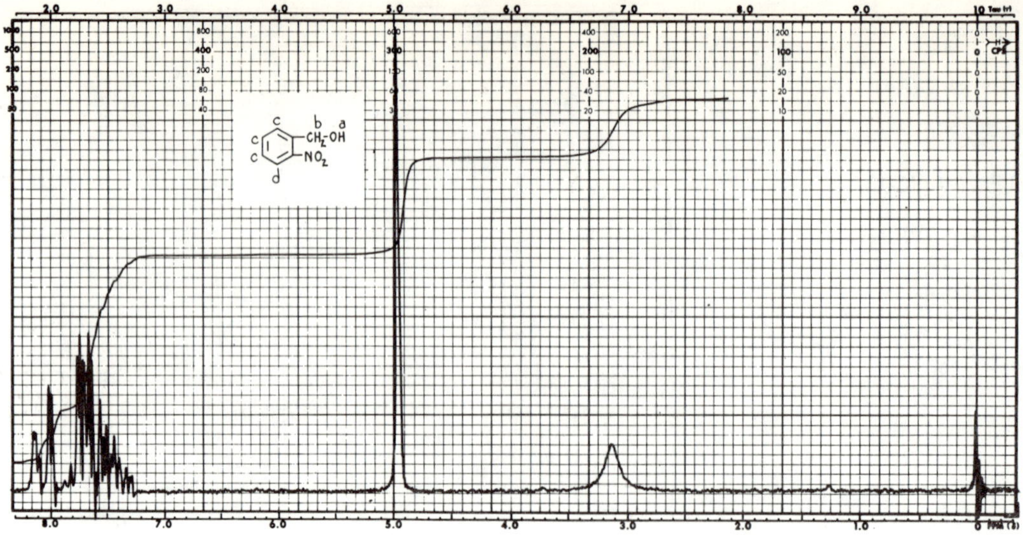

© 1978 Sadtler Research Laboratories, Inc.

m-NITROBENZYL ALCOHOL

$C_7H_7NO_3$

Mol. Wt. 153.14

M.P. 30-32°C

Solvent: $CDCl_3$

Assignments
a = 4.20
b = 4.78
c = 7.48
d = 7.64
e = 8.05
f = 8.11

1720

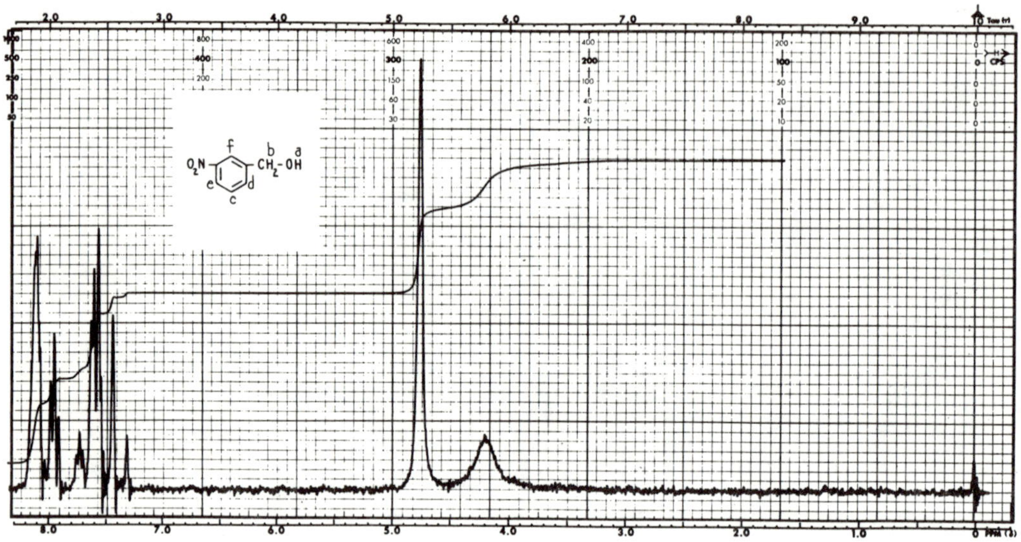

© 1978 Sadtler Research Laboratories, Inc.

p-NITROBENZYL ALCOHOL

$C_7H_7NO_3$

Mol. Wt. 153.14

Solvent: Acetone

Assignments
a = 4.79
b = 7.62
c = 8.18
OH is unobserved

1721

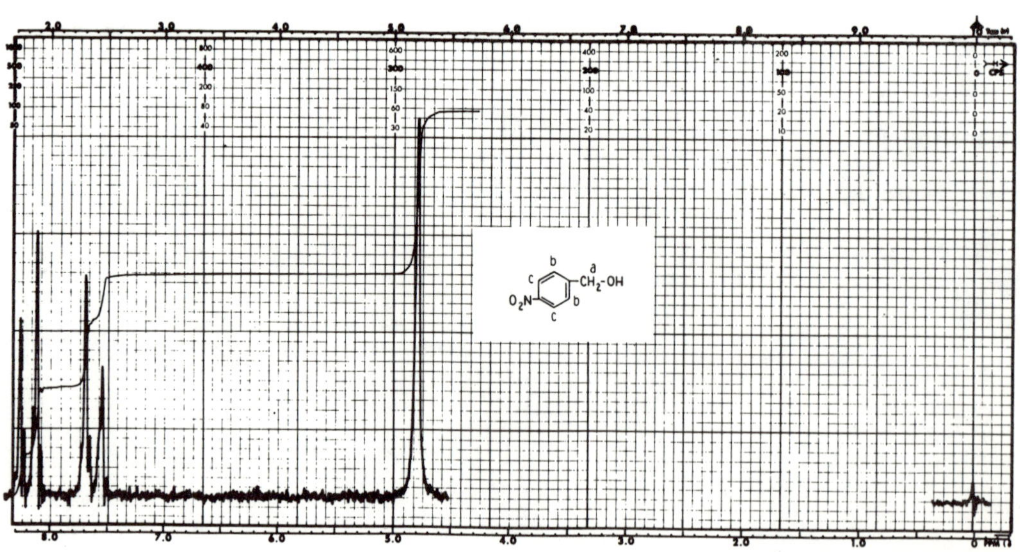

© 1978 Sadtler Research Laboratories, Inc.

o-METHOXYBENZYL ALCOHOL

$C_8H_{10}O_2$

Mol. Wt. 138.17

B.P. 290°C

Solvent: $CDCl_3$

Assignments

a = 3.11
b = 3.69
c = 4.60
d = 6.79
e = 6.85
f = 7.19
g = 7.24

1722

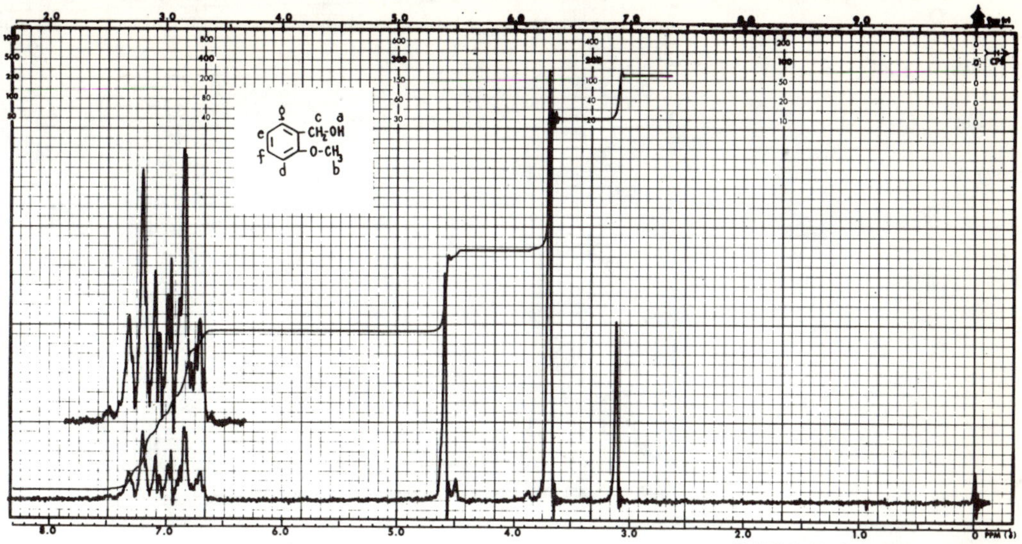

© 1978 Sadtler Research Laboratories, Inc.

p-METHOXYBENZYL ALCOHOL

$C_8H_{10}O_2$

Mol. Wt. 138.17

M.P. 23-25°C

Solvent: CCl_4

Assignments

a = 3.09
b = 3.69
c = 4.40
d = 6.73
e = 7.10

1723

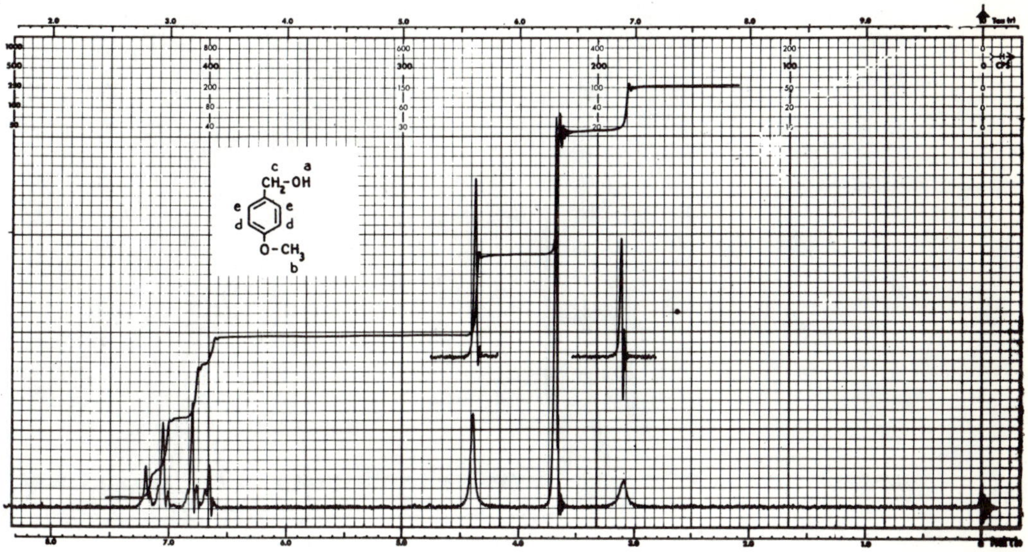

© 1978 Sadtler Research Laboratories, Inc.

3-(p-NITROPHENYL)-2-PROPEN-1-OL

$C_9H_9NO_3$

Mol. Wt. 179.18

M.P. 127-129°C

Solvent: Polysol-d

Assignments

a = 4.29
b = 4.42
c = ca 6.61
d = 7.49
e = 8.07

1724

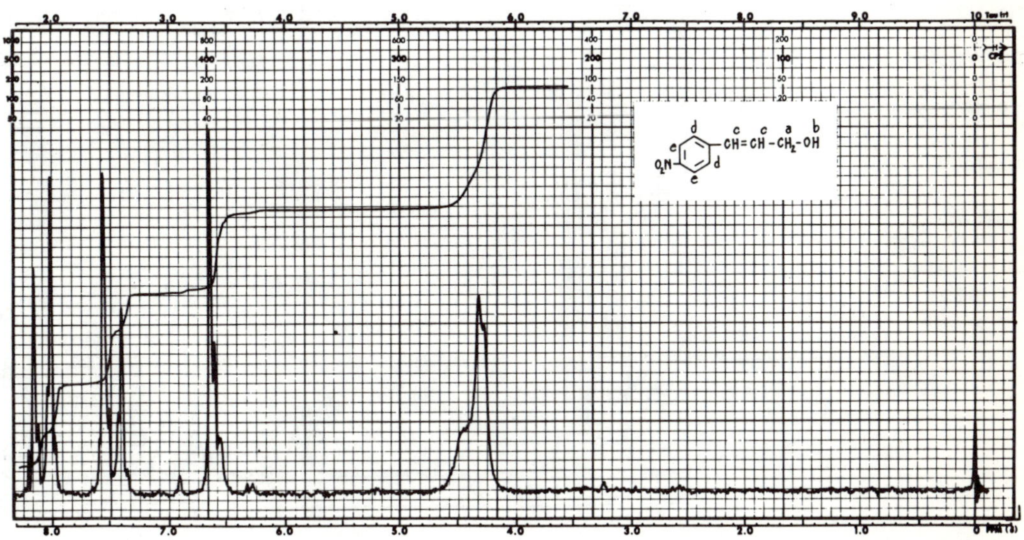

© 1978 Sadtler Research Laboratories, Inc.

1-NAPHTHALENEETHANOL

$C_{12}H_{12}O$

Mol. Wt. 172.23

Solvent: CCl_4

Assignments
a = 3.09
b = 3.41
c = 3.69
d = 7.00-8.00

1725

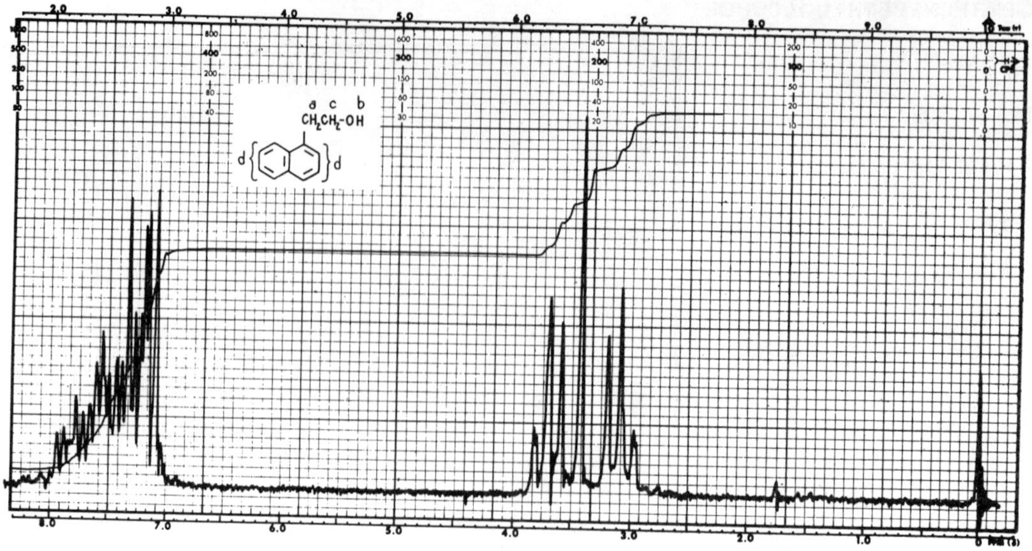

© 1978 Sadtler Research Laboratories, Inc.

2-(2-NAPHTHYLOXY)ETHANOL

$C_{12}H_{12}O_2$

Mol. Wt. 188.23

Solvent: $CDCl_3$

Assignments
a = 2.49
b = 3.98
c = 4.09
d = 7.04-7.60
e = ca 7.70

1726

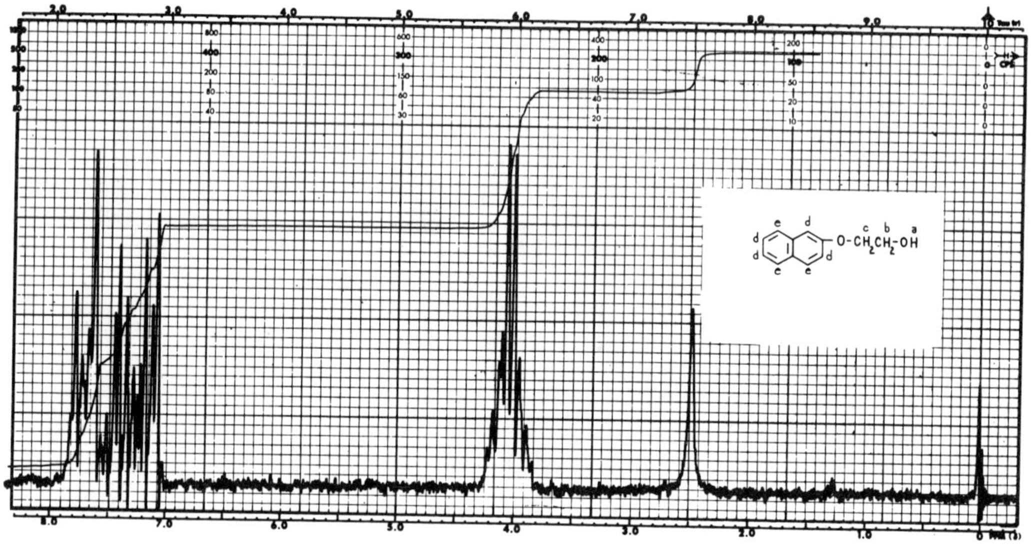

© 1978 Sadtler Research Laboratories, Inc.

FLUORENE-9-METHANOL

$C_{14}H_{12}O$

Mol. Wt. 196.25

M.P. 102-104°C

Solvent: Polysol-d

Assignments
a = 3.60-4.20
b = 4.93
c = ca 7.29
d = ca 7.69

1727

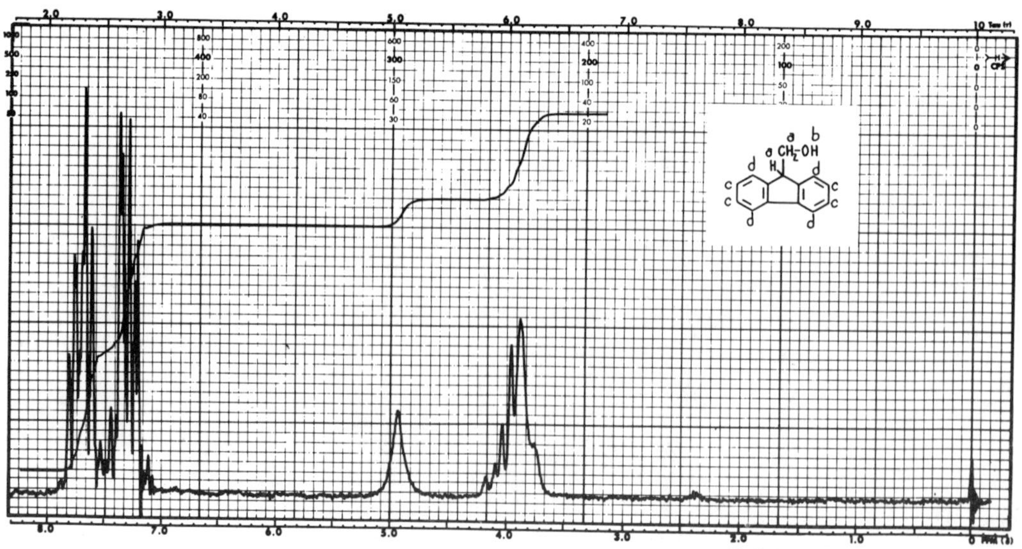

© 1978 Sadtler Research Laboratories, Inc.

2-PYRIDINEMETHANOL

C_6H_7NO

Mol. Wt. 109.13

B.P. 112-113°C/16mm

Solvent: $CDCl_3$

Assignments
a = 4.78
b = 5.02
c = 7.15
d = 7.35
e = 7.69
f = 8.50

1728

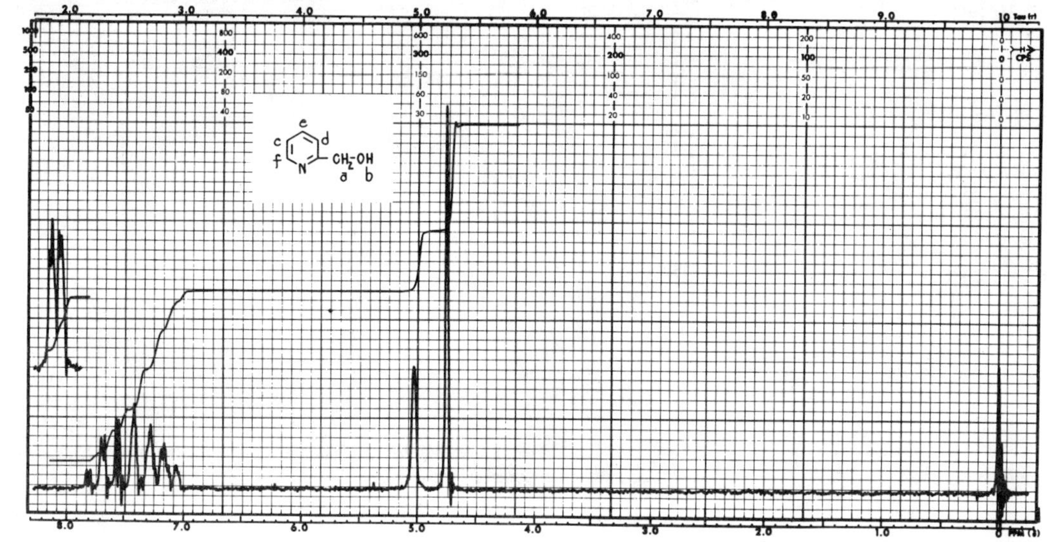

© 1978 Sadtler Research Laboratories, Inc.

4-PYRIDINEMETHANOL

C_6H_7NO

Mol. Wt. 109.13

M.P. 51-55°C

Solvent: $CDCl_3$

Assignments
a = 4.71
b = 6.27*
c = 7.32
d = 8.41
e = 2.48 imp.
*H_2O present

1729

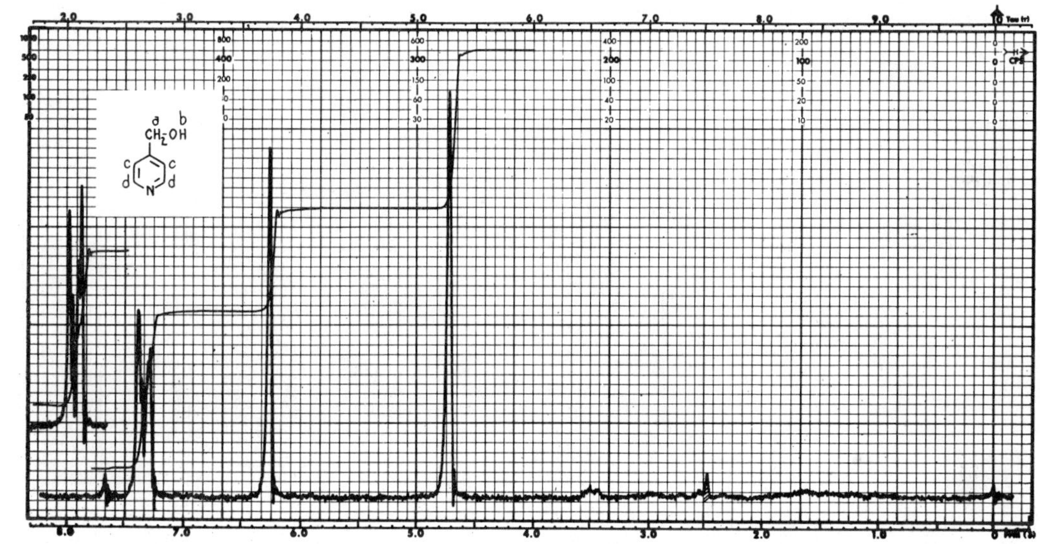

© 1978 Sadtler Research Laboratories, Inc.

2-BENZIMIDAZOLEMETHANOL

$C_8H_8N_2O$

Mol. Wt. 148.17

M.P. 170-172°C

Solvent: Dimethyl sulfoxide-d_6

Assignments
a = 4.95
b = 7.19
c = 7.66
d = 8.70

1730

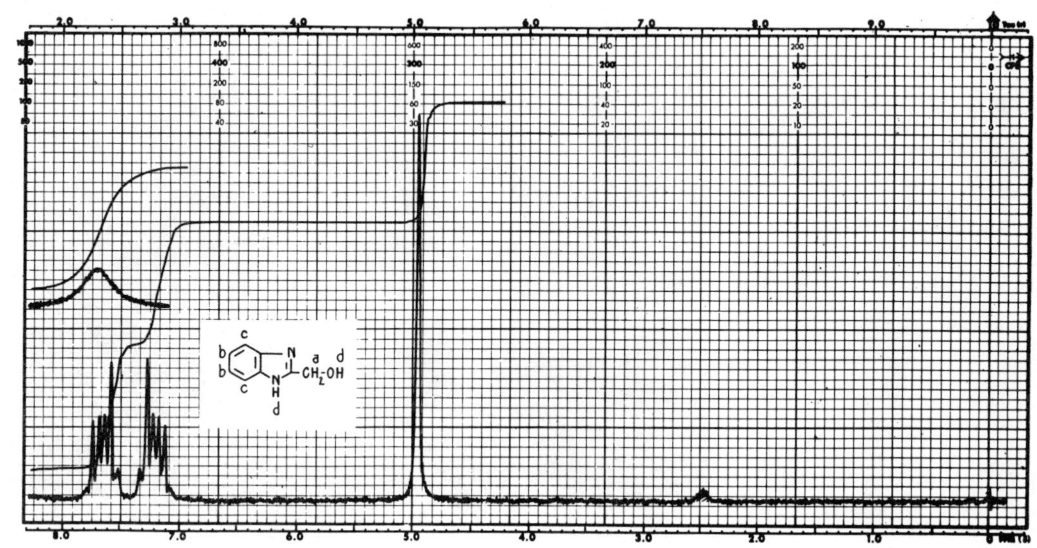

© 1978 Sadtler Research Laboratories, Inc.

3-QUINOLINEMETHANOL

$C_{10}H_9NO$

Mol. Wt. 159.19

M.P. 83.5-84°C

Solvent: $CDCl_3$

Assignments
a = 4.80
b = 5.49
c = 7.22-7.80
d = 7.98
e = 8.03
f = 8.75

1731

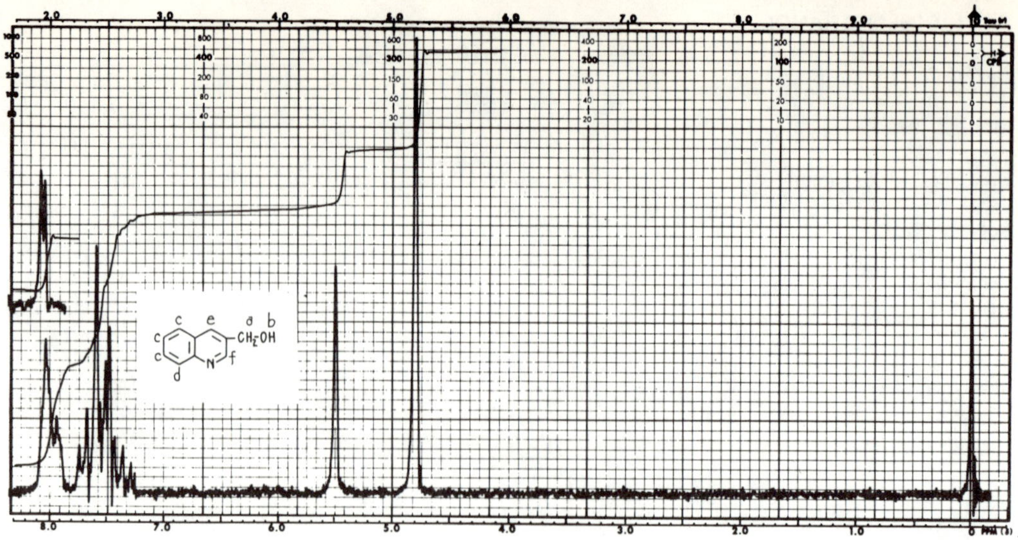

© 1978 Sadtler Research Laboratories, Inc.

5-PHENYL-2-PYRIMIDINEETHANOL

$C_{12}H_{12}N_2O$

Mol. Wt. 200.24

M.P. 92-97°C

Solvent: $CDCl_3$

Assignments
a = 3.22
b = 3.96
c = 4.11
d = ca 7.41
e = 8.79

1732

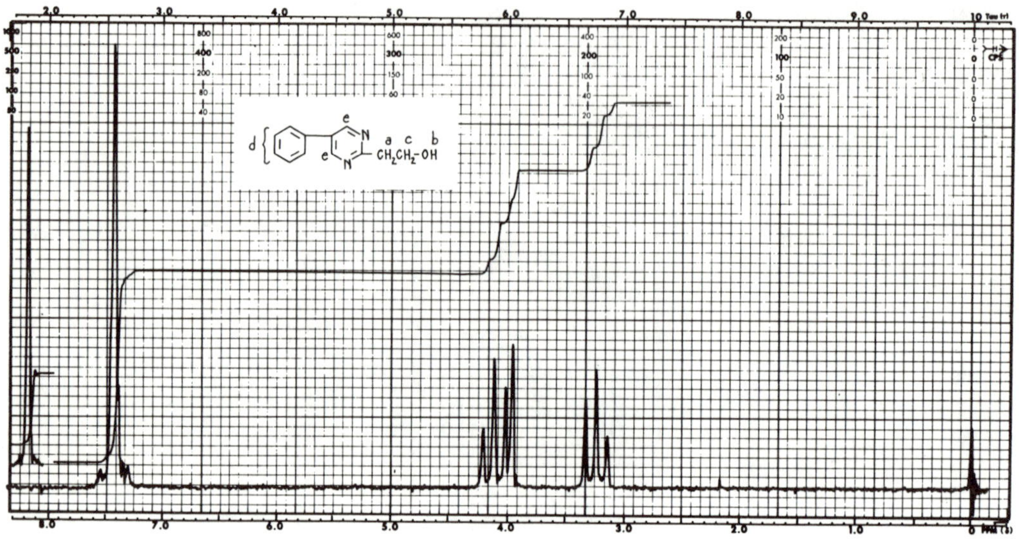

© 1978 Sadtler Research Laboratories, Inc.

FURFURYL ALCOHOL

$C_5H_6O_2$

Mol. Wt. 98.10

B.P. 75-77°C/17mm

Solvent: $CDCl_3$

Assignments
a = 3.30
b = 4.50
c = ca 6.29
d = 7.38

1733

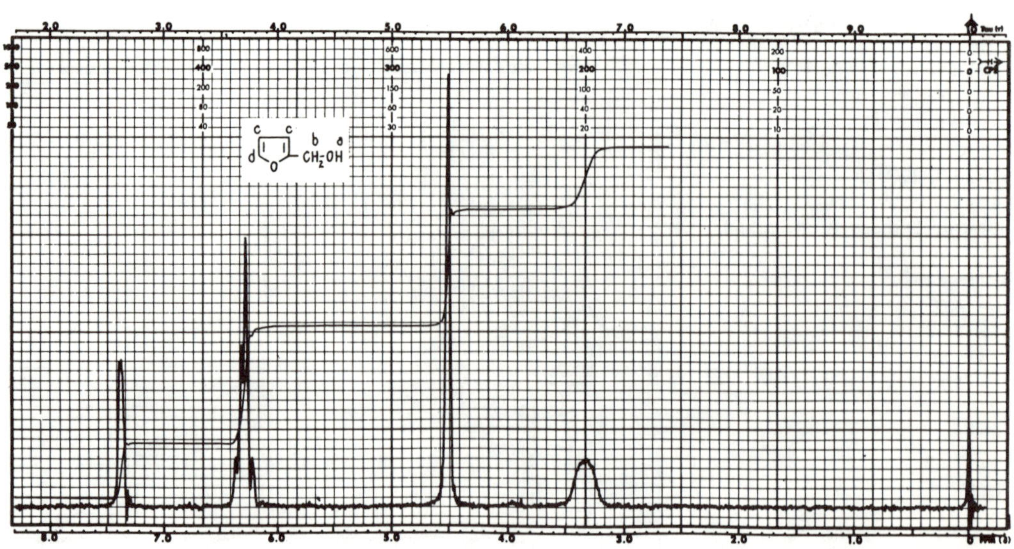

© 1978 Sadtler Research Laboratories, Inc.

PIPERONYL ALCOHOL

$C_8H_8O_3$

Mol. Wt. 152.15

M.P. 51°C (lit.)

Solvent: $CDCl_3$

Assignments

a = 2.83

b = 4.43

c = 5.85

d = ca 6.71

1734

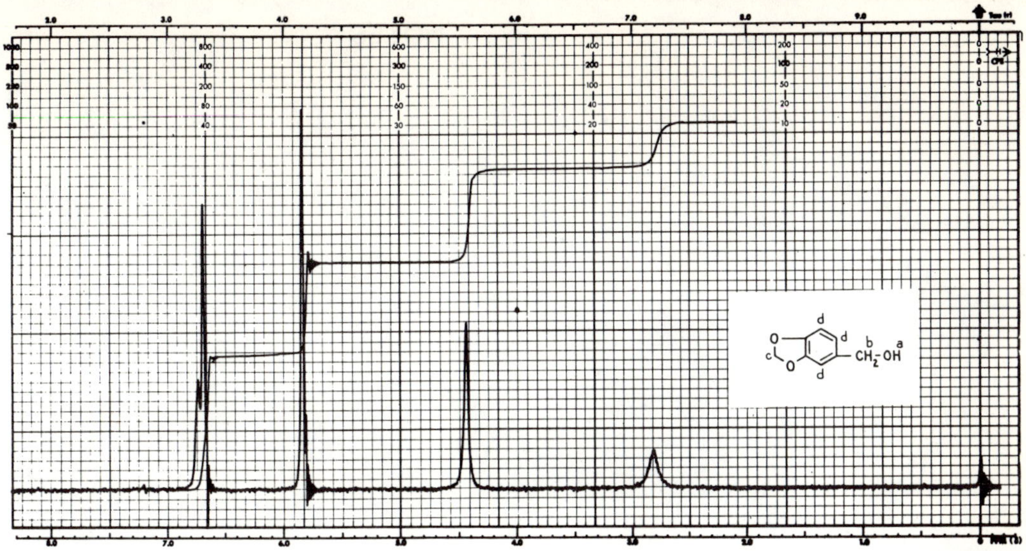

© 1978 Sadtler Research Laboratories, Inc.

NOTES

THE SECONDARY ALCOHOLS

The methine proton adjacent to the hydroxyl group of the secondary alcohols is very strongly deshielded and may appear as any of a wide variety of multiplets depending upon the aliphatic groups bonded to it. The highest degree of multiplicity that is observed is octet produced by the two methyl groups of isopropanol with additional coupling to the hydroxyl group.

Chemical Shifts

Aliphatic Protons

HC–3	HC–2	HC–1	–OH		Solvent	Spectrum
	(1.29)	3.99	2.90	$(CH_3)_2{-}CH{-}OH$	CCl_4	1735
(0.90	1.39)	3.36	3.30	$(CH_3{-}CH_2)_2{-}CH{-}OH$	CCl_4	1737
(0.92	1.62)	3.01	1.64	$((CH_3)_2{-}CH)_2{-}CH{-}OH$	$CDCl_3$	1741
		5.59	2.28	$(\emptyset)_2{-}CH{-}OH$	CCl_4	1790
	(3.59)	4.03	3.27	$(Br{-}CH_2)_2{-}CH{-}OH$	$CDCl_3$	1751

Alicyclic Protons

$(CH_2)_n$	HC–1	–OH			
1.1–2.5	4.16	5.48	Cyclobutanol	CCl_4	1763
1.3–2.1	4.21	3.58	Cyclopentanol	CCl_4	1764
0.8–2.5	3.49	4.20	Cyclohexanol	CCl_4	1766
1.2–2.3	3.80	2.69	Cycloheptanol	$CDCl_3$	1777

Disubstituted Methanols

HO	–CH	–X(Y)	
2.77	3.11	$-CH_3(-C_3)$	$CDCl_3$
1.95	3.18	$-R_2(-C_6)$	CCl_4
2.59	3.38	$-CH_3(-C(CH_3)_3)$	CCl_4
3.05	3.43	$-CH_3(-C_6)$	CCl_4
2.90	3.99	$-CH_3(-CH_3)$	CCl_4
3.63	4.03	$-CH_3(-CF_3)$	CCl_4
3.88	4.20	$-CH_3(-CH{=}CH_2)$	CCl_4
3.73	4.22	$-CH_3(-C(=O)-CH_3)$	$CDCl_3$
3.41	4.30	$-CH_3(-C(Cl)_3)$	$CDCl_3$
- - - -	4.40	$-CH_3(-C(=O)-OH)$	D_2O

Disubstituted Methanols continued

HO	–CH	–X(Y)	Solvent
4.08	4.50	$-CH_3(-C\equiv C-H)$	CCl_4
4.05	4.60	$-CH_3(-C\equiv N)$	$CDCl_3$
2.76	4.65	$-CH_3(-\emptyset)$	CCl_4
- - - -	4.94	$-\emptyset(-C(=O)-O-K)$	D_2O
4.60	5.91	$-\emptyset(-C(=O)-\emptyset)$	$CDCl_3$

ISOPROPYL ALCOHOL

C_3H_8O

Mol. Wt. 60.09

B.P. 81.5-82.5°C

Solvent: CCl_4

Assignments
a = 1.29
b = 2.90
c = 3.99

1735

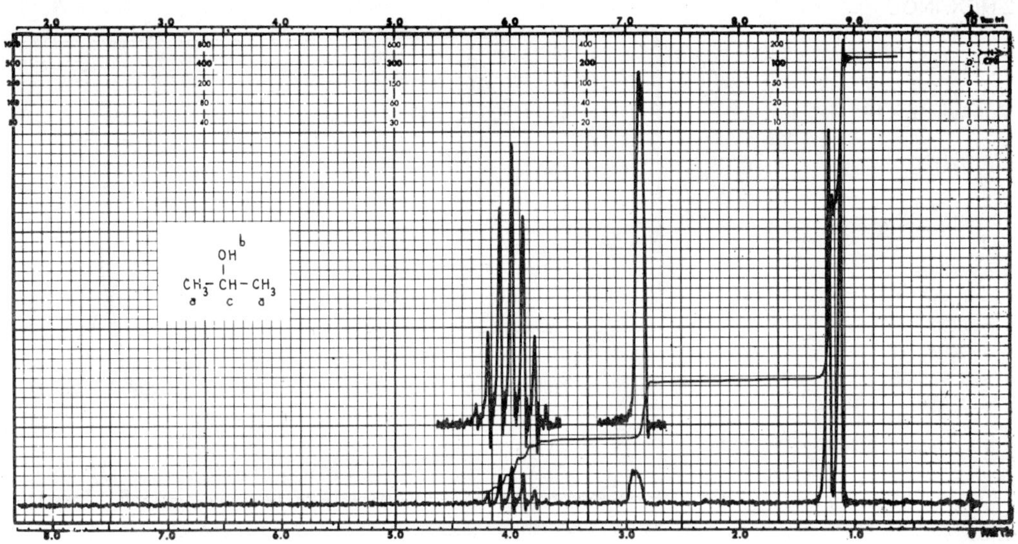

© 1978 Sadtler Research Laboratories, Inc.

2-BUTANOL

$C_4H_{10}O$

Mol. Wt. 74.12

B.P. 98-98.5°C

Solvent: CCl_4

Assignments
a = 0.90
b = 1.11
c = 1.38
d = 3.62
e = 3.99

1736

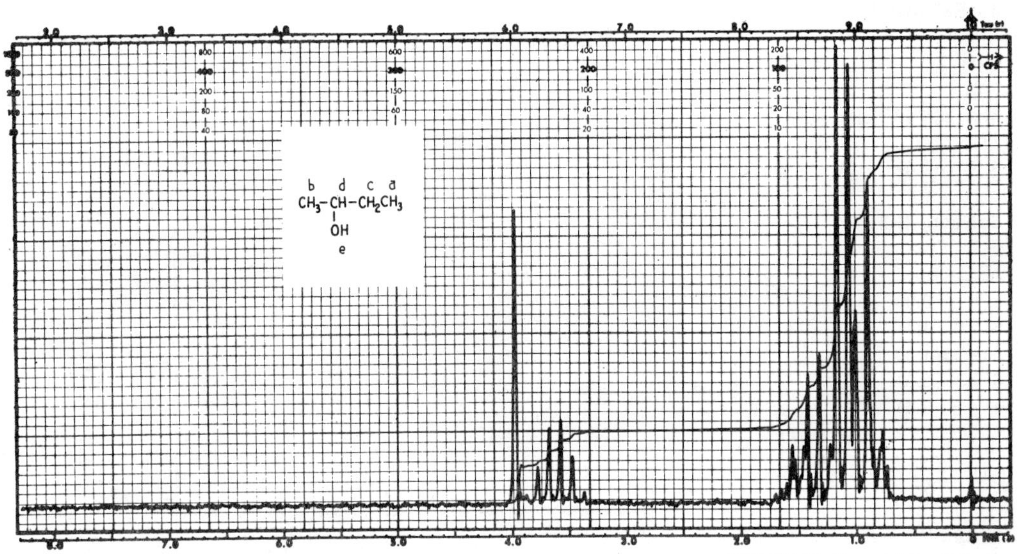

© 1978 Sadtler Research Laboratories, Inc.

3-PENTANOL

$C_5H_{12}O$

Mol. Wt. 88.15

B.P. 114°C/760mm

Solvent: CCl_4

Assignments
a = 0.90
b = 1.39
c = 3.30
d = 3.36

1737

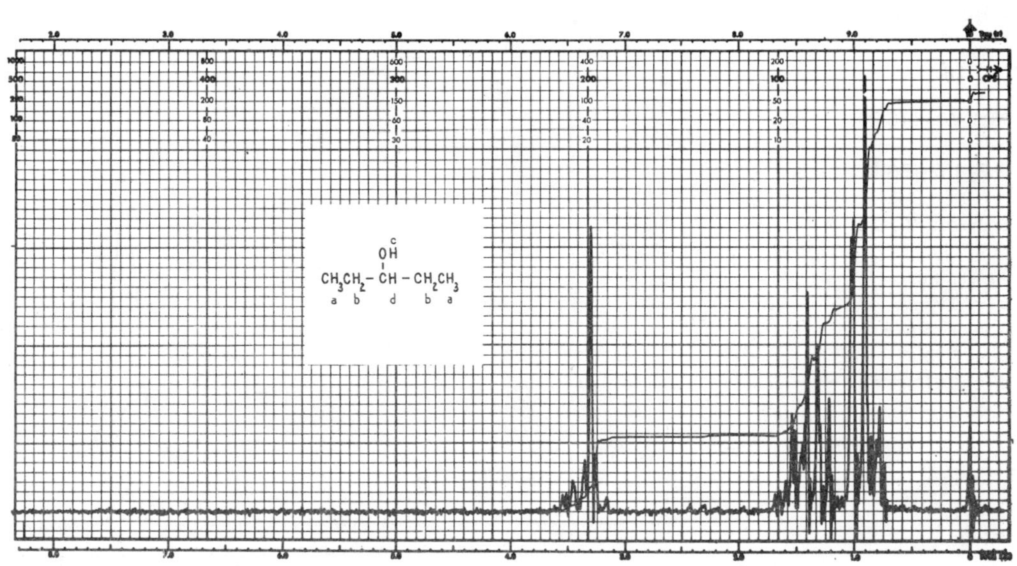

© 1978 Sadtler Research Laboratories, Inc.

2-HEXANOL

$C_6H_{14}O$

Mol. Wt. 102.18

B.P. 140°C

Solvent: CCl_4

Assignments
a = 0.91
b = 0.98
c = 1.18-1.71
d = 2.81
e = 3.42

1738

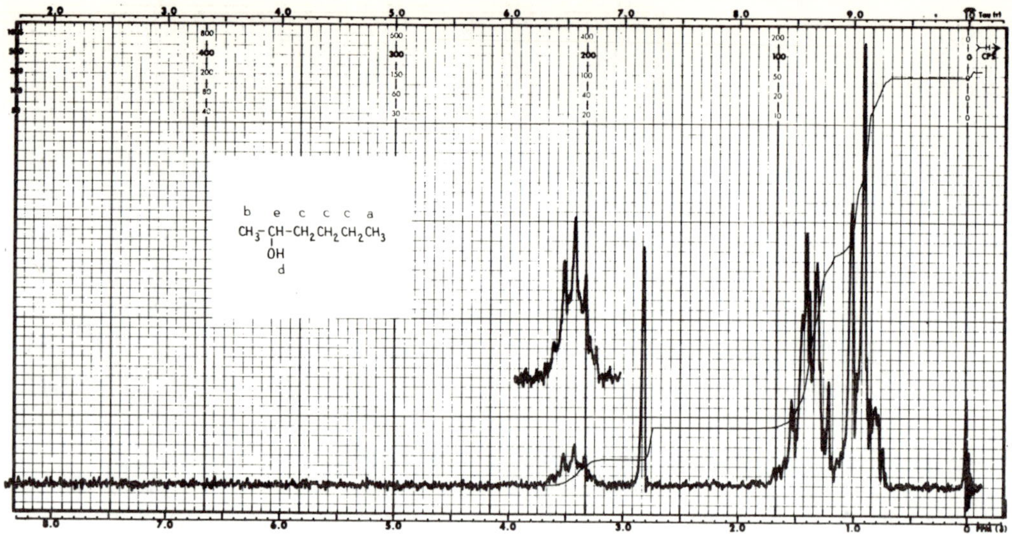

© 1978 Sadtler Research Laboratories, Inc.

2-HEPTANOL

$C_7H_{16}O$

Mol. Wt. 116.20

B.P. 160.4°C

Solvent: CCl_4

Assignments
a = 0.89
b = 1.11
c = ca 1.30
d = 3.39
e = 3.61

1739

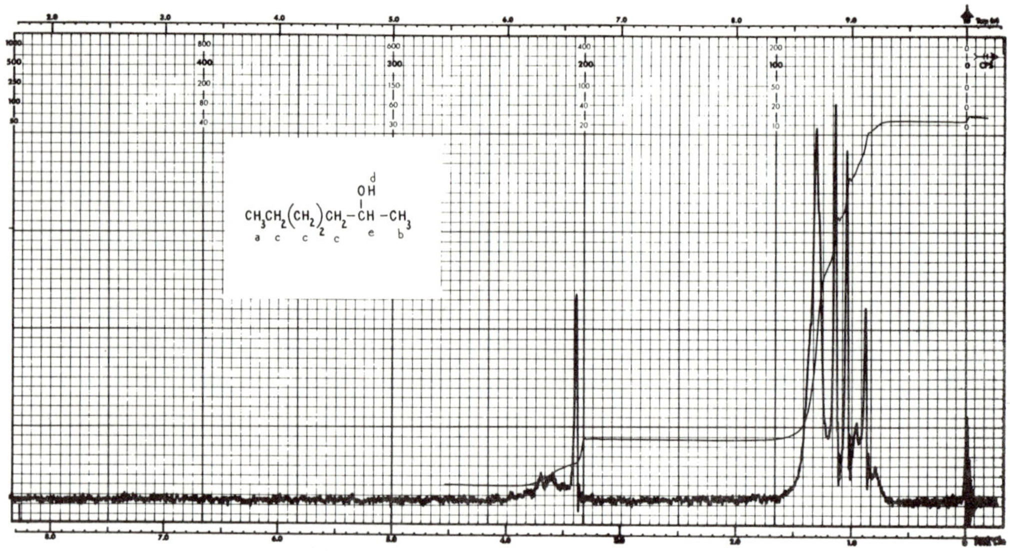

© 1978 Sadtler Research Laboratories, Inc.

4-HEPTANOL

$C_7H_{16}O$

Mol. Wt. 116.20

B.P. 152-156°C

Solvent: $CDCl_3$

Assignments
a = 0.93
b = 1.20-1.70
c = 2.18
d = 3.61

1740

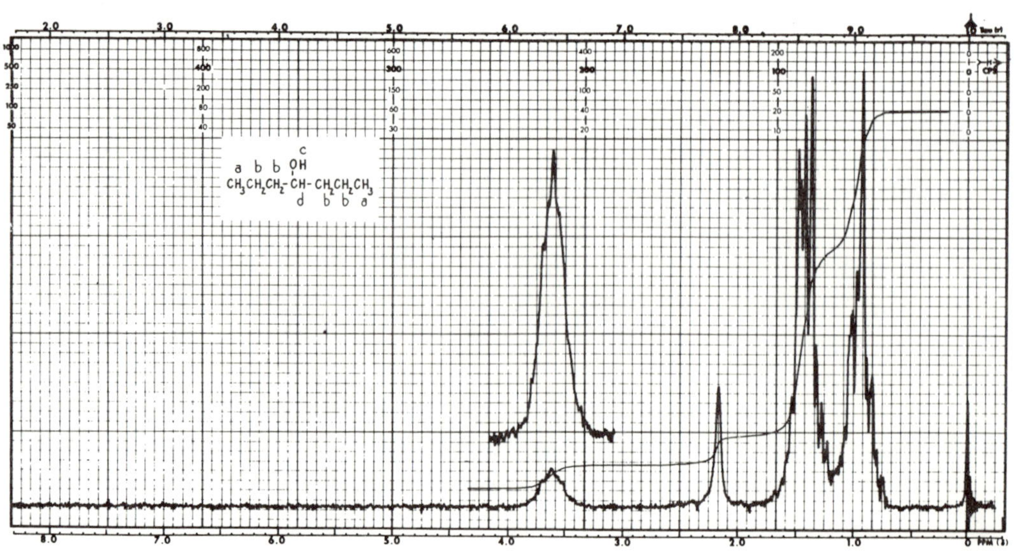

© 1978 Sadtler Research Laboratories, Inc.

2,4-DIMETHYL-3-PENTANOL

$C_7H_{16}O$

Mol. Wt. 116.20

B.P. 140°C

Solvent: $CDCl_3$

Assignments
a = 0.92
b = 1.64
c = 1.40-2.10
d = 3.01

1741

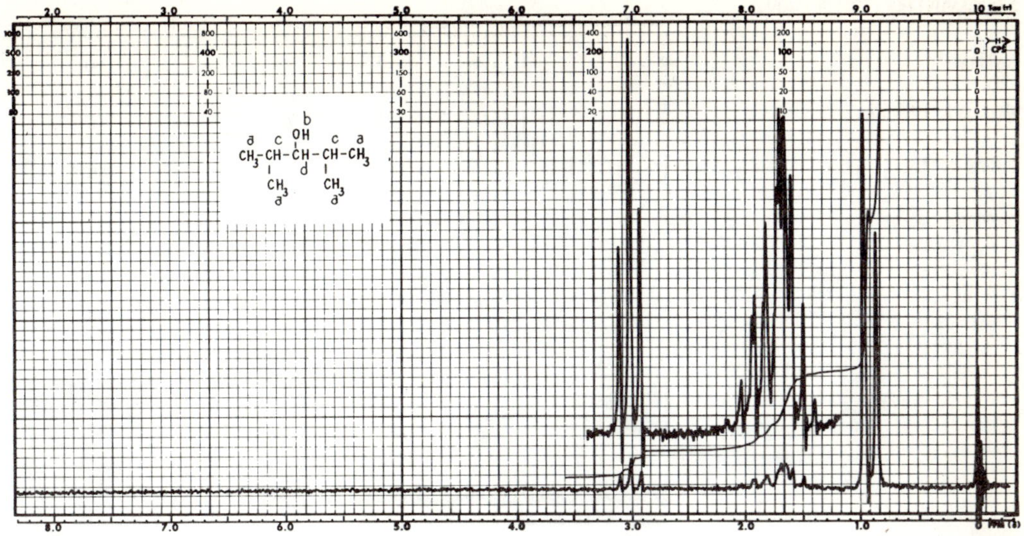

© 1978 Sadtler Research Laboratories, Inc.

(+)-2-OCTANOL

$C_8H_{18}O$

Mol. Wt. 130.23

B.P. 175°C

Solvent: CCl_4

Assignments
a = 0.90
b = 1.12
c = 1.31
d = 3.11
e = 3.68

1742

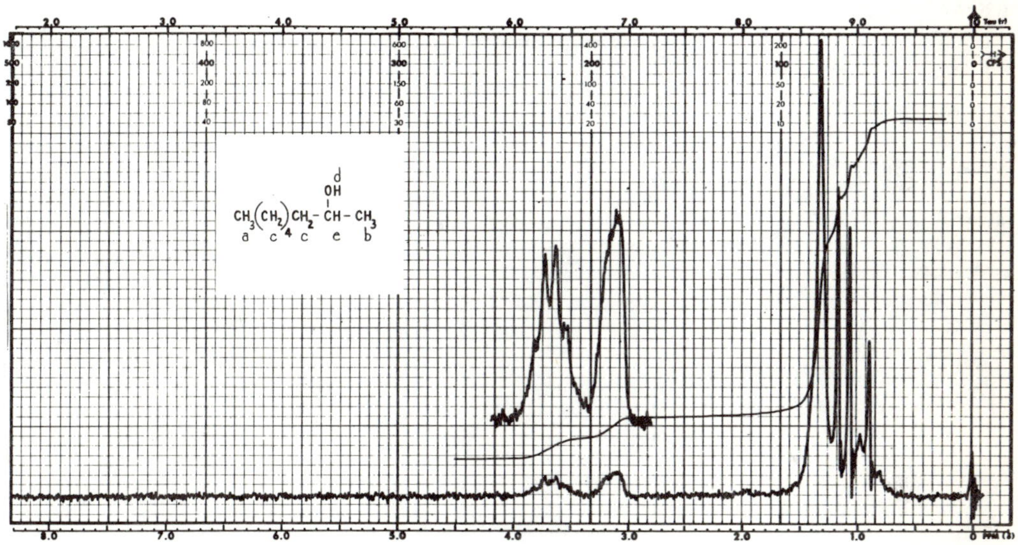

© 1978 Sadtler Research Laboratories, Inc.

4-ETHYL-3-HEXANOL

$C_8H_{18}O$

Mol. Wt. 130.23

Solvent: CCl_4

Assignments
a = 0.90
b = 0.94
c = 1.00-1.70
d = 2.05
e = 3.42

1743

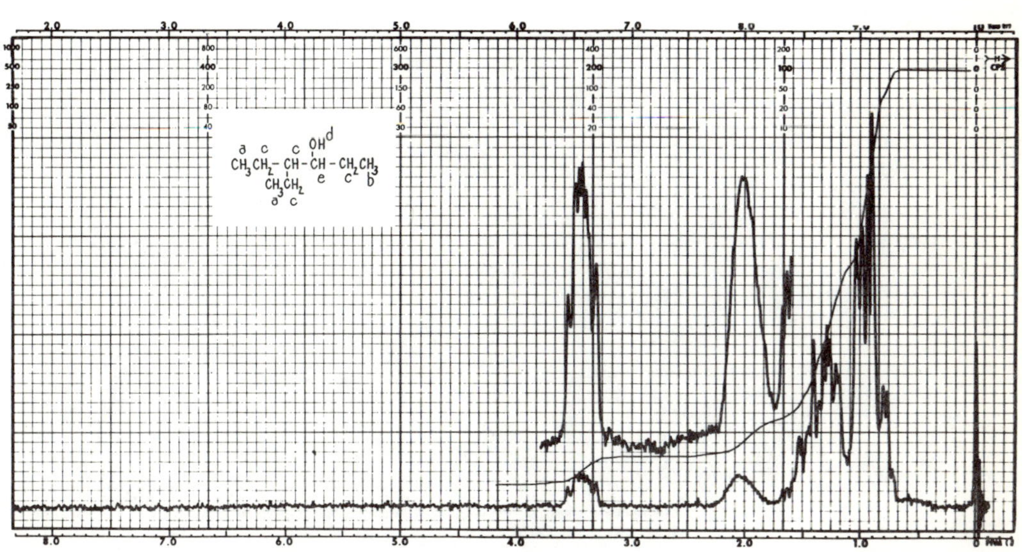

© 1978 Sadtler Research Laboratories, Inc.

2,2-DIMETHYL-3-HEXANOL

$C_8H_{18}O$

Mol. Wt. 130.23

Solvent: CCl_4

Assignments
a = 0.85
b = 1.32
c = ca 1.37
d = 3.13

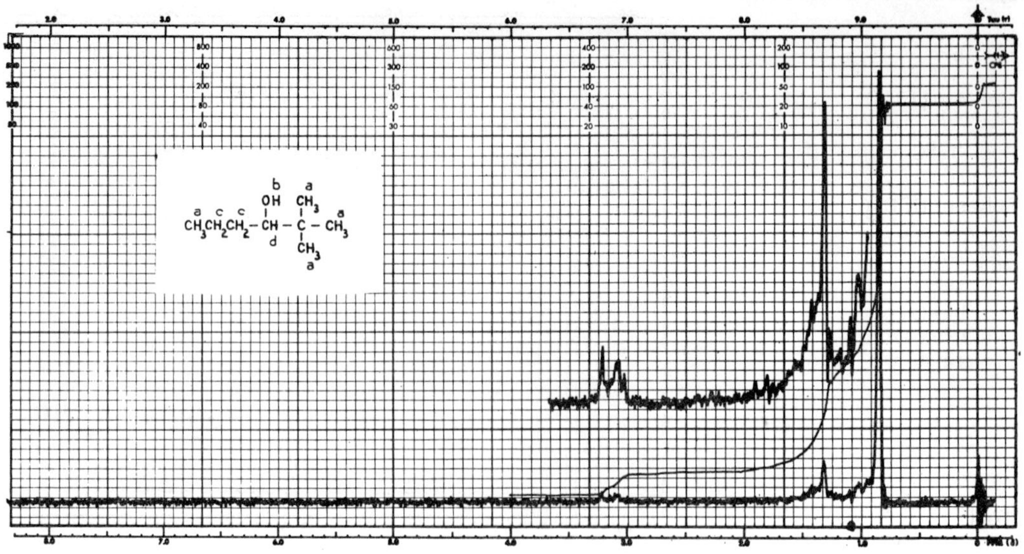

© 1978 Sadtler Research Laboratories, Inc.

1744

6-ETHYL-3-OCTANOL

$C_{10}H_{22}O$

Mol. Wt. 158.29

Solvent: $CDCl_3$

Assignments
a = 0.91
b = 0.94
c = 1.10-1.80
d = 2.71
e = 3.49

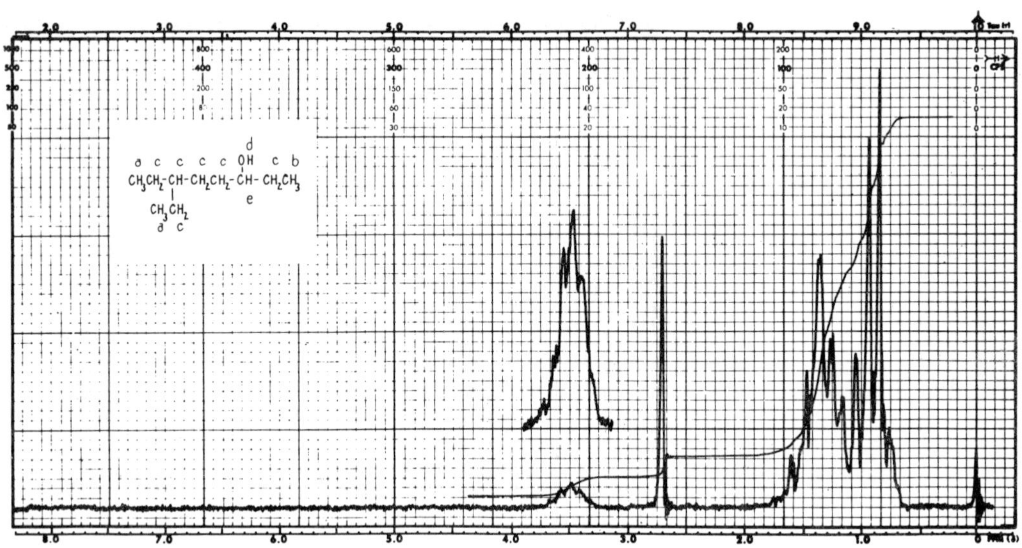

© 1978 Sadtler Research Laboratories, Inc.

1745

5-ETHYL-2-NONANOL

$C_{11}H_{24}O$

Mol. Wt. 172.31

Solvent: CCl_4

Assignments
a = ca 0.88
b = ca 1.12
c = ca 1.25
d = 2.42
e = 3.65

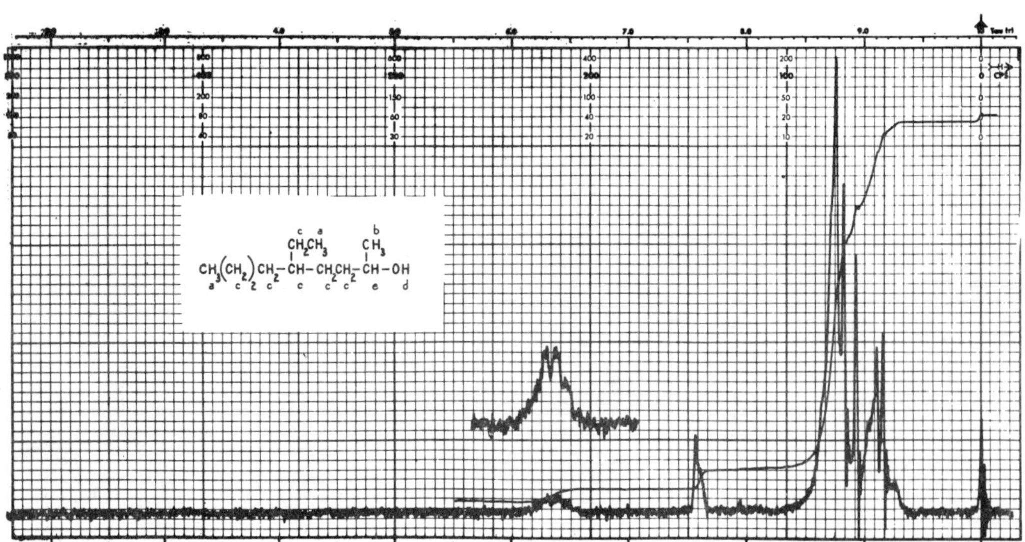

© 1978 Sadtler Research Laboratories, Inc.

1746

6-ETHYL-3-DECANOL

$C_{12}H_{26}O$

Mol. Wt. 186.34

Solvent: CCl_4

Assignments
a = 0.85
b = 0.91
c = 1.25
d = 2.97
e = 3.41
Trace of acid added for OH

1747

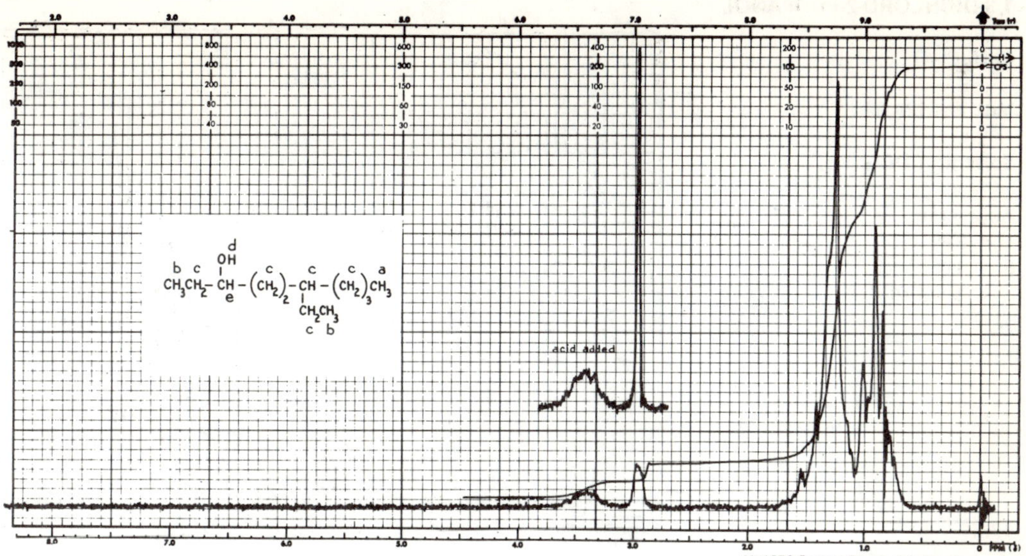

© 1978 Sadtler Research Laboratories, Inc.

2-TRIDECANOL

$C_{13}H_{28}O$

Mol. Wt. 200.37

Solvent: CCl_4

Assignments
a = 0.89
b = 1.10
c = 1.25
d = 1.67
e = 3.67
Trace of acid added for OH

1748

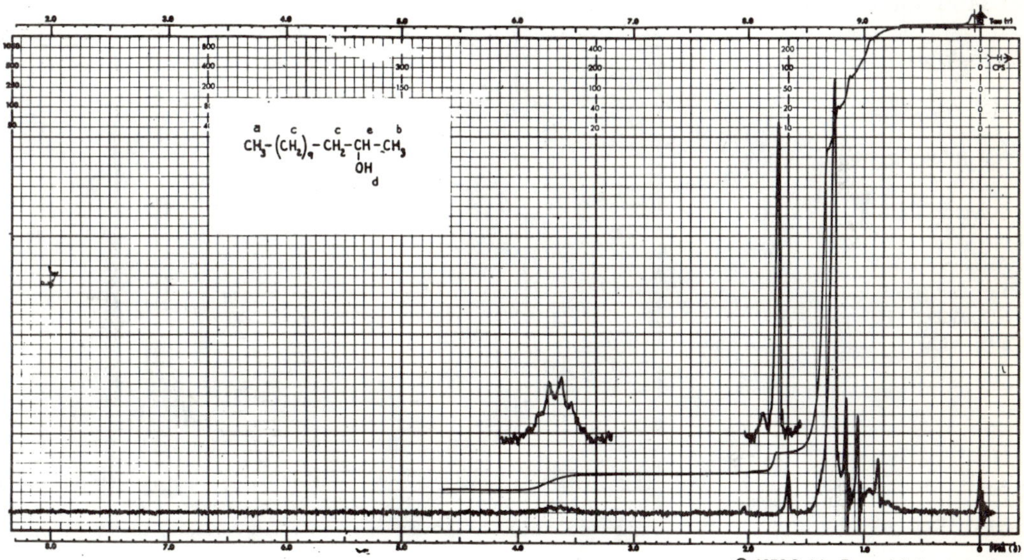

© 1978 Sadtler Research Laboratories, Inc.

3,9-DIETHYL-6-TRIDECANOL

$C_{17}H_{36}O$

Mol. Wt. 256.48

Solvent: CCl_4

Assignments
a = ca 0.88
b = 1.10-1.70
c = 2.02
d = 3.46

1749

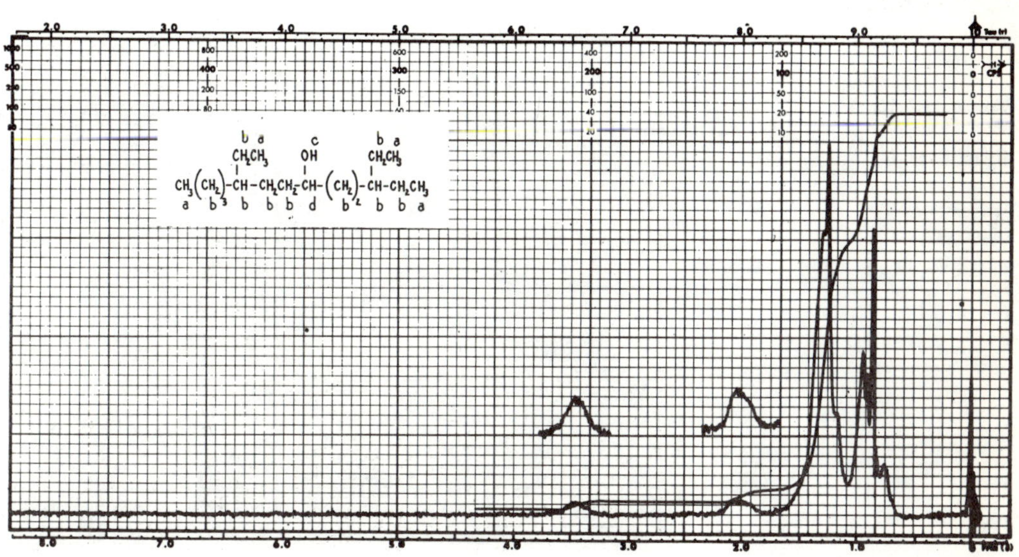

© 1978 Sadtler Research Laboratories, Inc.

1,3-DICHLORO-2-PROPANOL

$C_3H_6Cl_2O$

Mol. Wt. 128.99

Solvent: $CDCl_3$

Assignments
a = 3.69
b = 4.05
c = 3.40-4.40

1750

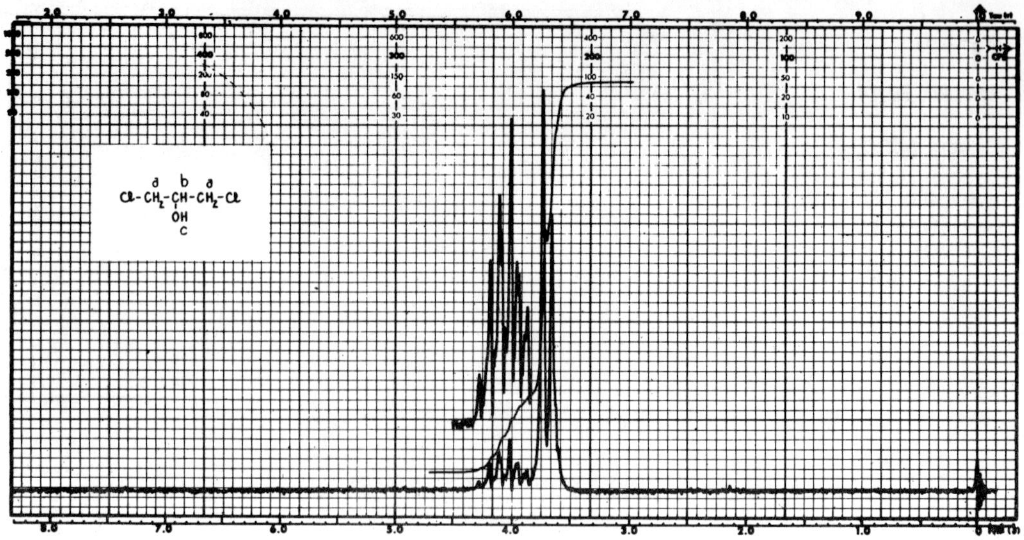

© 1978 Sadtler Research Laboratories, Inc.

1,3-DIBROMO-2-PROPANOL

$C_3H_6Br_2O$

Mol. Wt. 217.90

B.P. 88-90°C/10mm

Solvent: $CDCl_3$

Assignments
a = 3.27
b = 3.59
c = 4.03

1751

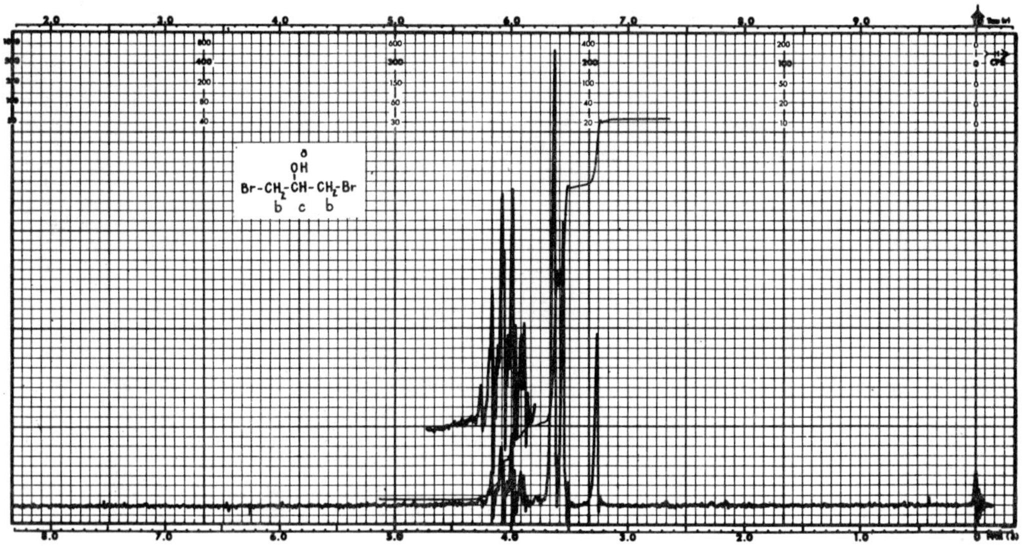

© 1978 Sadtler Research Laboratories, Inc.

1,3-DIAMINO-2-PROPANOL

$C_3H_{10}N_2O$

Mol. Wt. 90.13

Solvent: D_2O

Assignments
a = 2.50
b = 2.62
c = 3.53
d = 4.42 HDO

1752

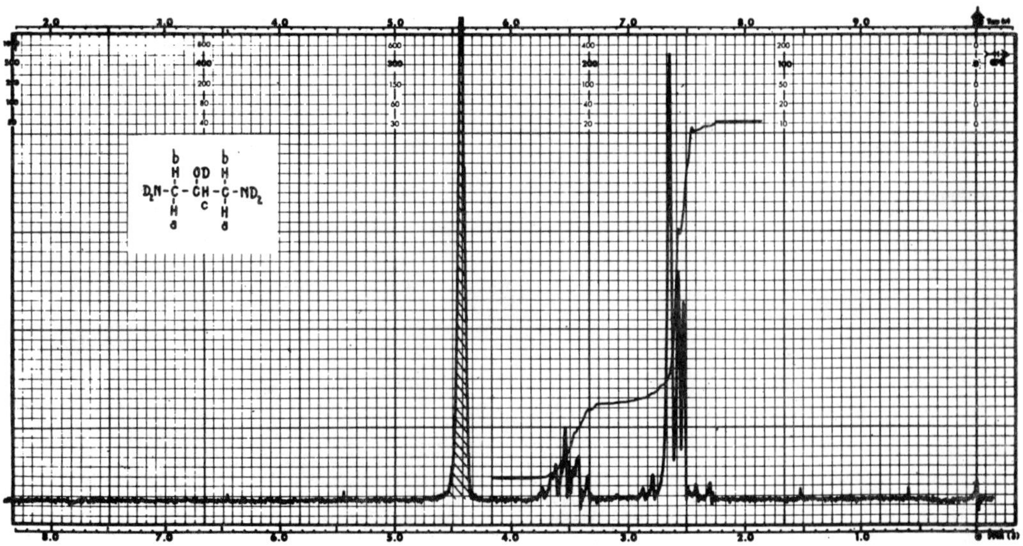

© 1978 Sadtler Research Laboratories, Inc.

3-AMINO-2-BUTANOL (ERYTHRO-THREO MIXTURE)

$C_4H_{11}NO$

Mol. Wt. 89.14

Solvent: $CDCl_3$

Assignments

a = 0.93-1.25

b = 2.22

c = 2.45-3.90

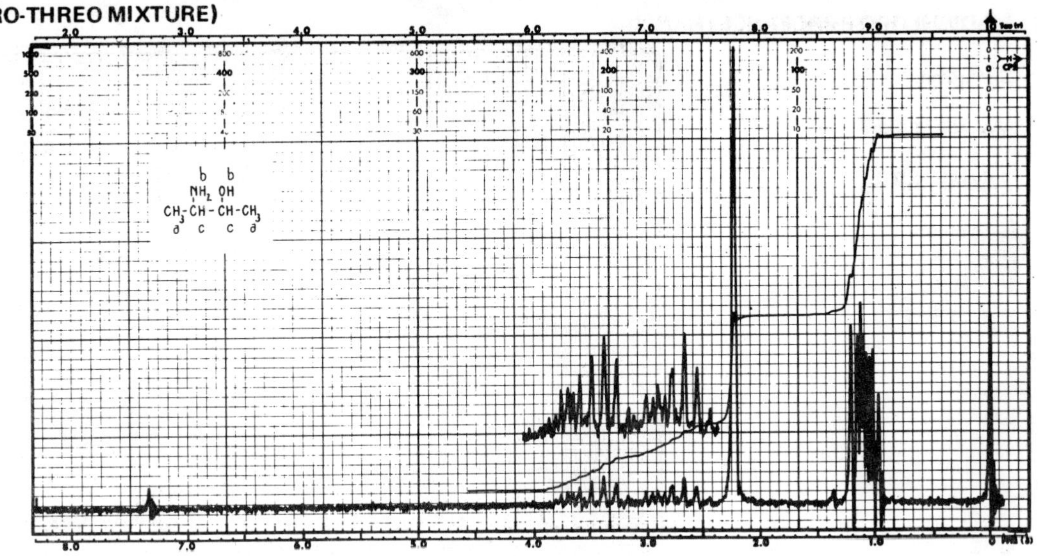

1753

© 1978 Sadtler Research Laboratories, Inc.

2-(DIMETHYLAMINO)-1-METHYLETHANOL

$C_5H_{13}NO$

Mol. Wt. 103.17

Solvent: $CDCl_3$

Assignments

a = 1.11

b = 2.17

c = 2.29

d = 3.19

e = 3.79

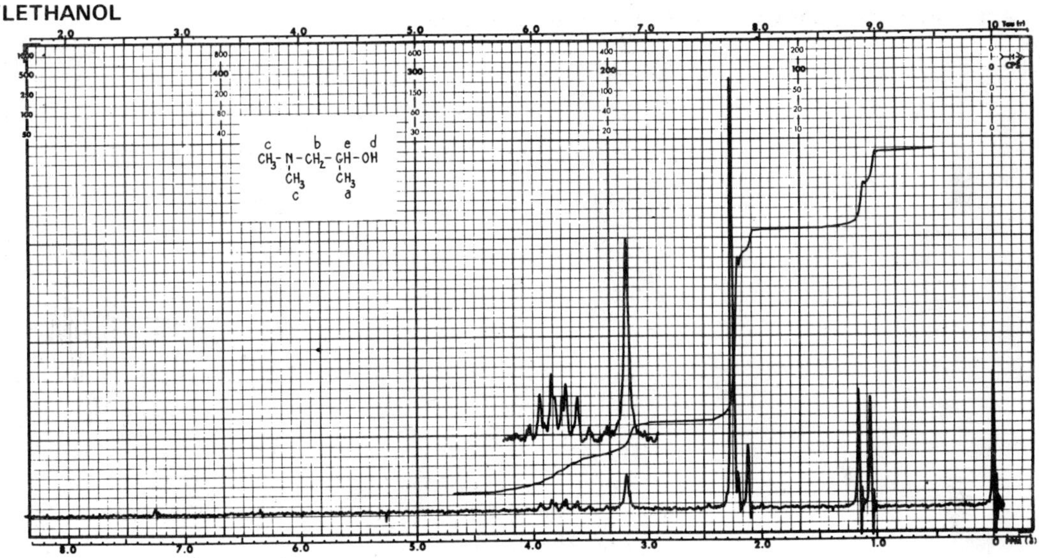

1754

© 1978 Sadtler Research Laboratories, Inc.

1-(DIBUTYLAMINO)-2-PROPANOL

$C_{11}H_{25}NO$

Mol. Wt. 187.33

Solvent: CCl_4

Assignments

a = 0.93

b = 1.02

c = ca 1.38

d = 2.18

e = 2.28

f = ca 2.42

g = 3.20

h = 3.59

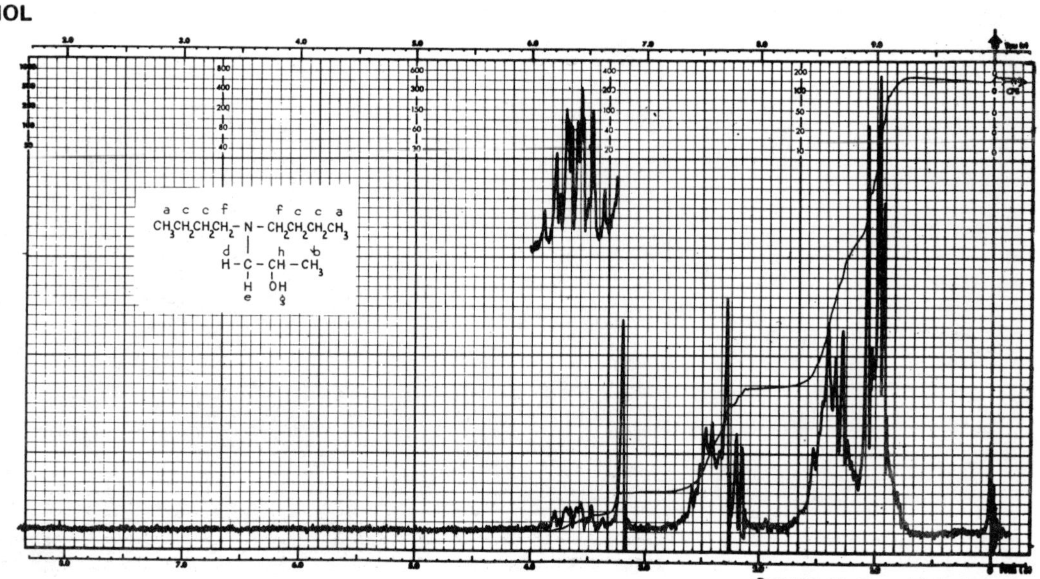

1755

© 1978 Sadtler Research Laboratories, Inc.

α-METHYL-1-PYRROLIDINEETHANOL

$C_7H_{15}NO$

Mol. Wt. 129.20

Solvent: $CDCl_3$

Assignments
a = 1.13
b = 1.76
c = 1.90-2.9
d = 3.40
e = 3.79

1756

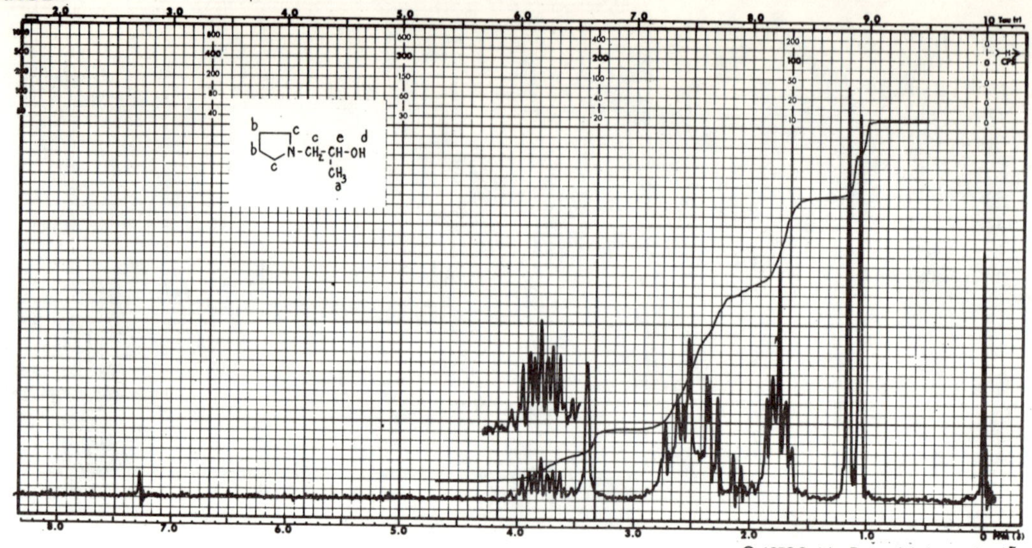

© 1978 Sadtler Research Laboratories, Inc.

4-(TETRADECYLAMINO)-2-BUTANOL, HYDROCHLORIDE

$C_{18}H_{39}NO \cdot HCl$

Mol. Wt. 321.98

Solvent: $CDCl_3$

Assignments
a = 0.89
b = ca 1.26
c = 1.91
d = 3.06
e = 4.08

1757

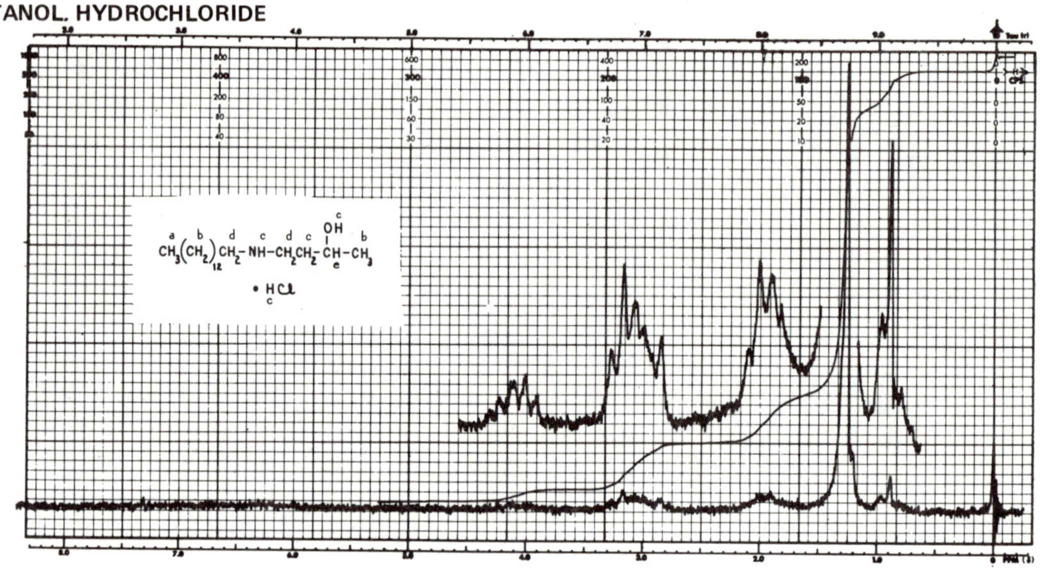

© 1978 Sadtler Research Laboratories, Inc.

LACTONITRILE

C_3H_5NO

Mol. Wt. 71.08

B.P. 80-81°C/10mm

Solvent: $CDCl_3$

Assignments
a = 1.59
b = 4.05
c = 4.60

1758

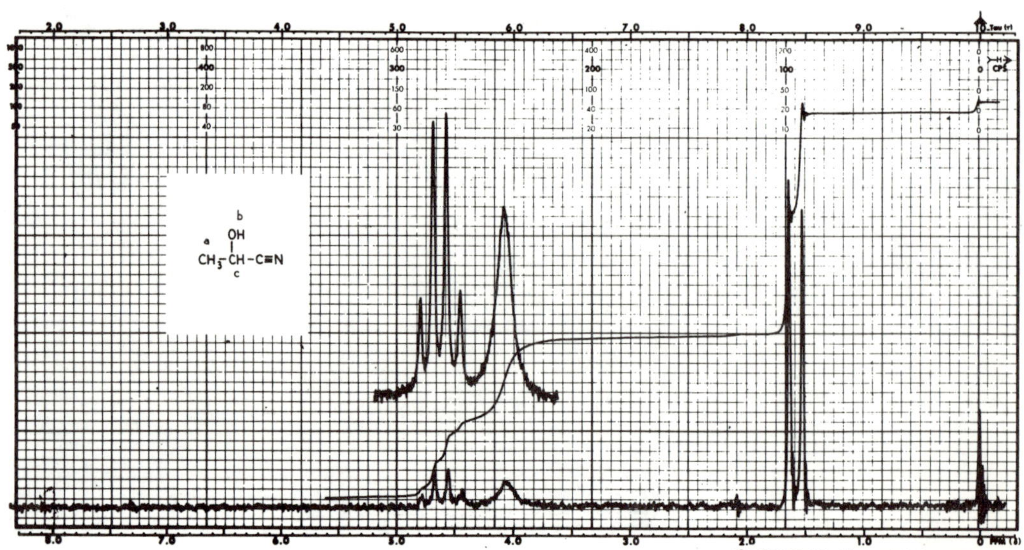

© 1978 Sadtler Research Laboratories, Inc.

4-METHOXY-4-METHYL-2-PENTANOL

$C_7H_{16}O_2$

Mol. Wt. 132.20

Solvent: CCl_4

Assignments
a = 1.03
b = 1.18
c = 1.21
d = 1.39-2.11
e = 3.18
f = 3.40
g = 3.98

1759

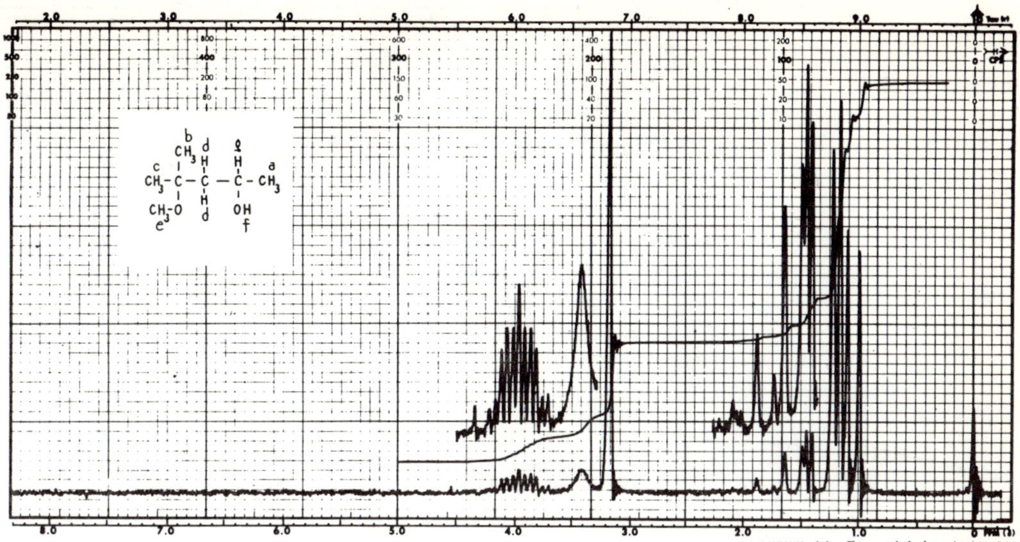

© 1978 Sadtler Research Laboratories, Inc.

1-[1-(2-METHOXY-1-METHYLETHOXY)-1-METHYLETHOXY]-2-PROPANOL

$C_{10}H_{22}O_4$

Mol. Wt. 206.28

Solvent: CCl_4

Assignments
a = 1.09
b = 3.30
c = 2.81-4.09

1760

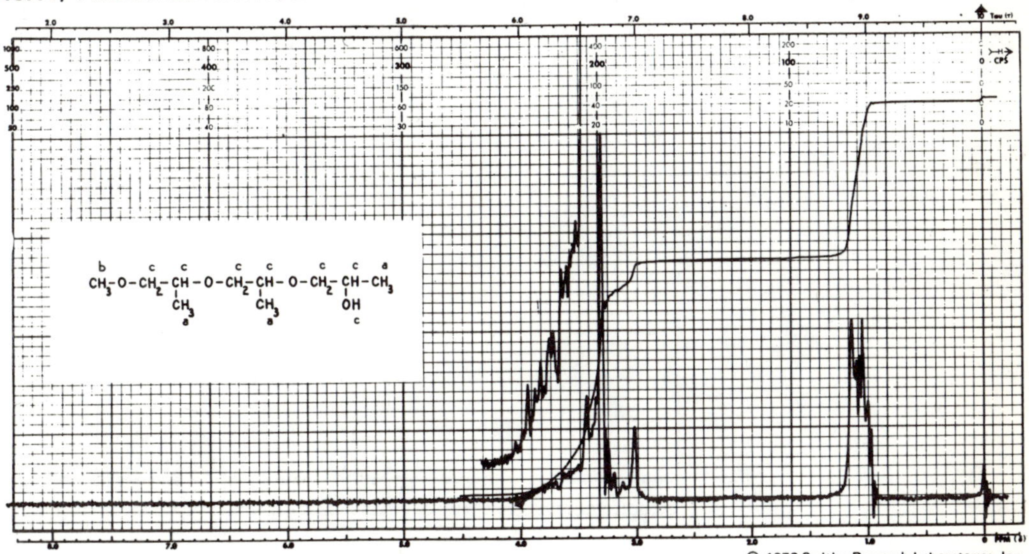

© 1978 Sadtler Research Laboratories, Inc.

a-METHYLCYCLOPROPANEMETHANOL

$C_5H_{10}O$

Mol. Wt. 86.13

B.P. 120-122°C

Solvent: $CDCl_3$

Assignments
a = 0.10-0.80
b = 0.86
c = 1.29
d = 2.77
e = 3.11

1761

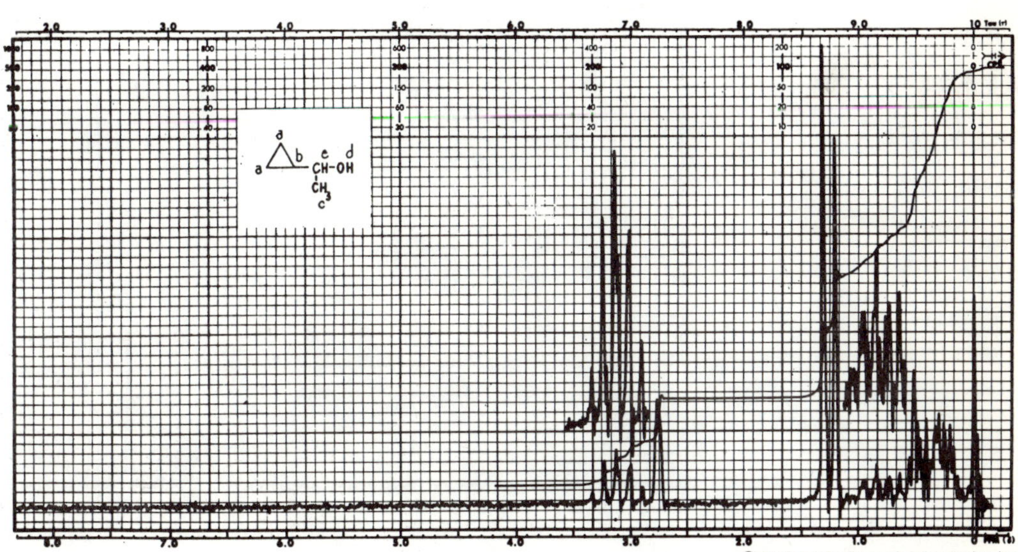

© 1978 Sadtler Research Laboratories, Inc.

α-METHYLCYCLOHEXANEMETHANOL

$C_8H_{16}O$

Mol. Wt. 128.22

Solvent: CCl_4

Assignments
a = 1.07
b = 1.76
c = 3.05
d = 3.43
e = 2.07 imp.

1762

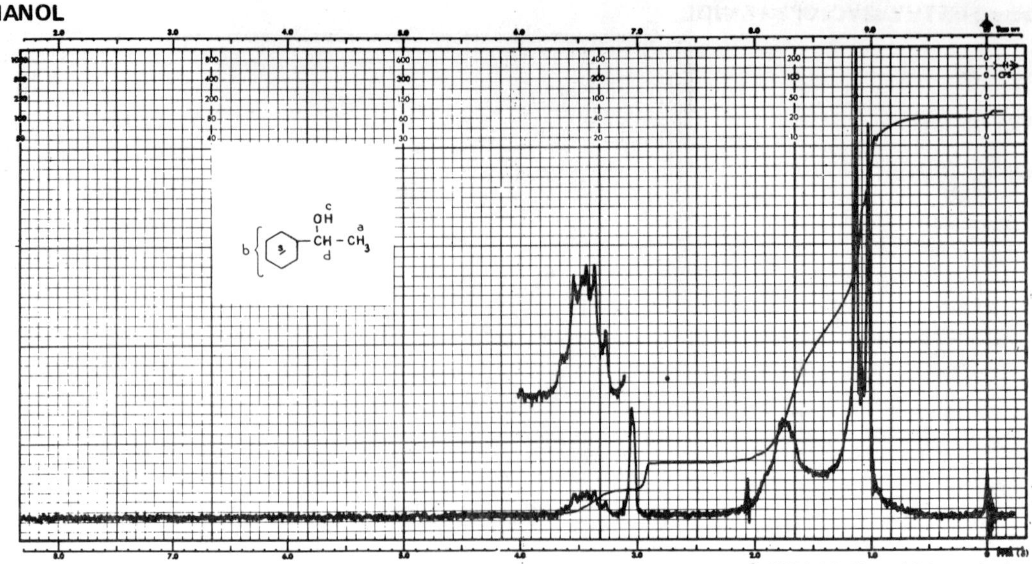

© 1978 Sadtler Research Laboratories, Inc.

CYCLOBUTANOL

C_4H_8O

Mol. Wt. 72.11

B.P. 123°C/733mm

Solvent: CCl_4

Assignments
a = 1.10-2.50
b = 4.16
c = 5.48

1763

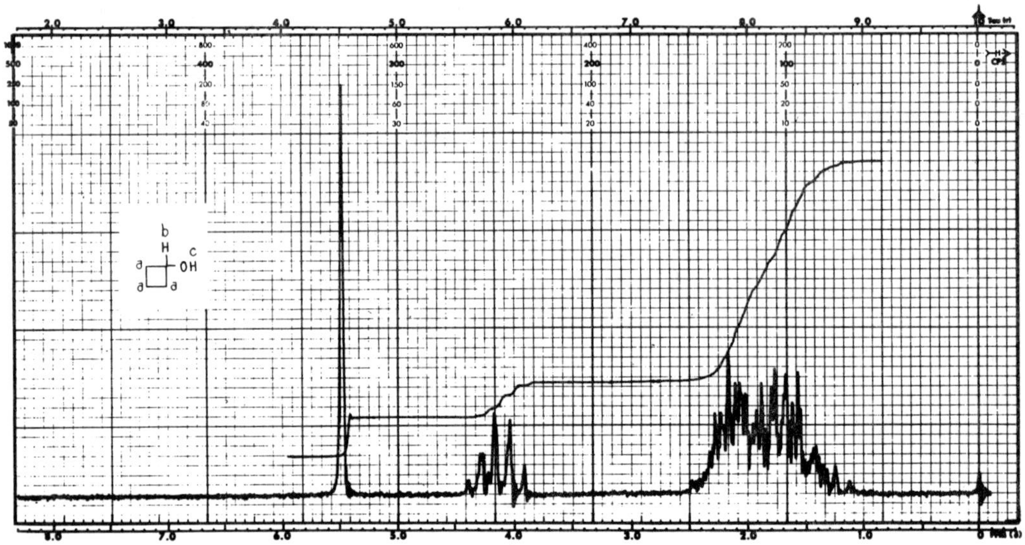

© 1978 Sadtler Research Laboratories, Inc.

CYCLOPENTANOL

$C_5H_{10}O$

Mol. Wt. 86.13

B.P. 139-140°C

Solvent: CCl_4

Assignments
a = ca 1.61
b = 3.58
c = 4.21

1764

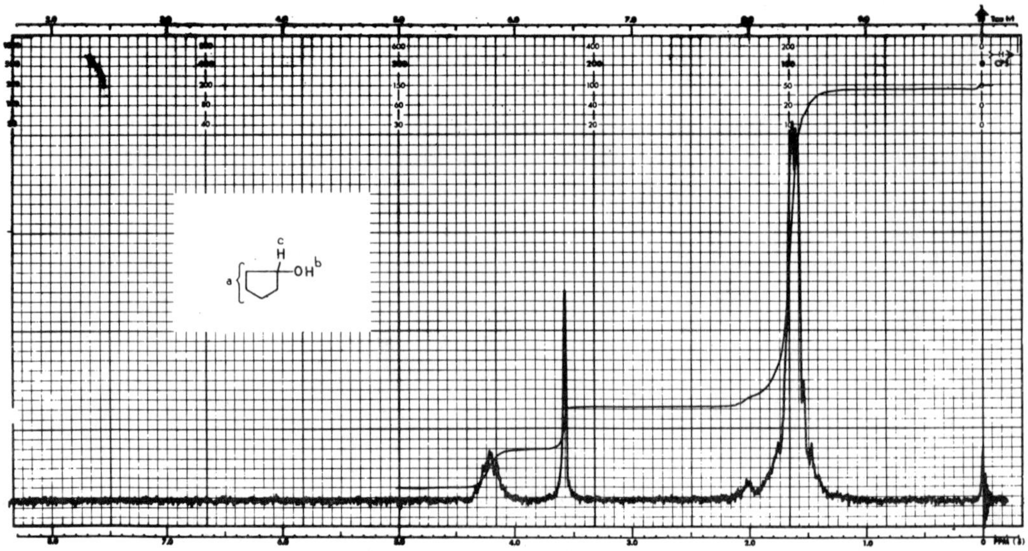

© 1978 Sadtler Research Laboratories, Inc.

trans-2-METHYLCYCLOPENTANOL

$C_6H_{12}O$

Mol. Wt. 100.16

Solvent: $CDCl_3$

Assignments
a = 0.96
b = 1.10-2.20
c = ca 2.47
d = 3.71

1765

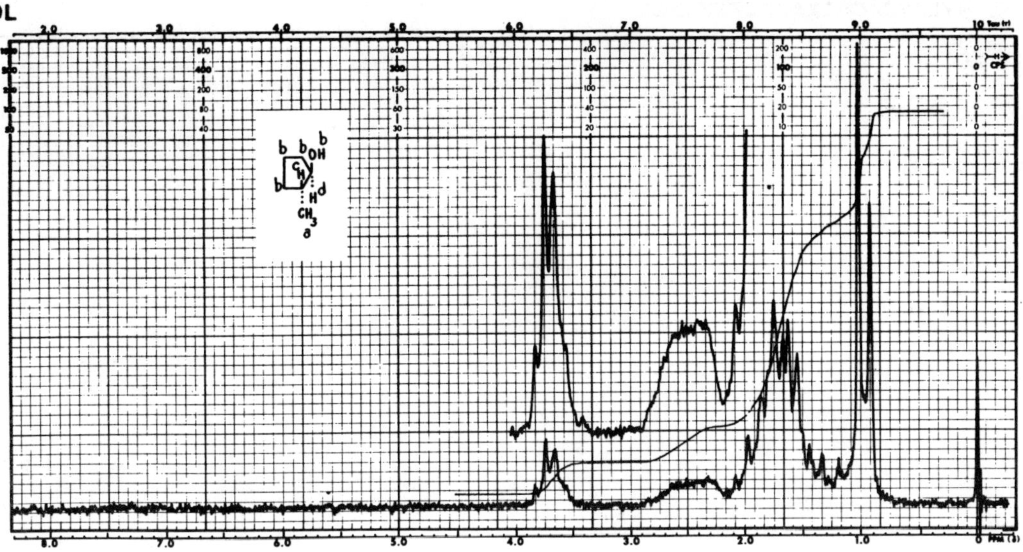

© 1978 Sadtler Research Laboratories, Inc.

CYCLOHEXANOL

$C_6H_{12}O$

Mol. Wt. 100.16

B.P. 160-161°C

Solvent: CCl_4

Assignments
a = 0.83-2.43
b = 3.49
c = 4.20

1766

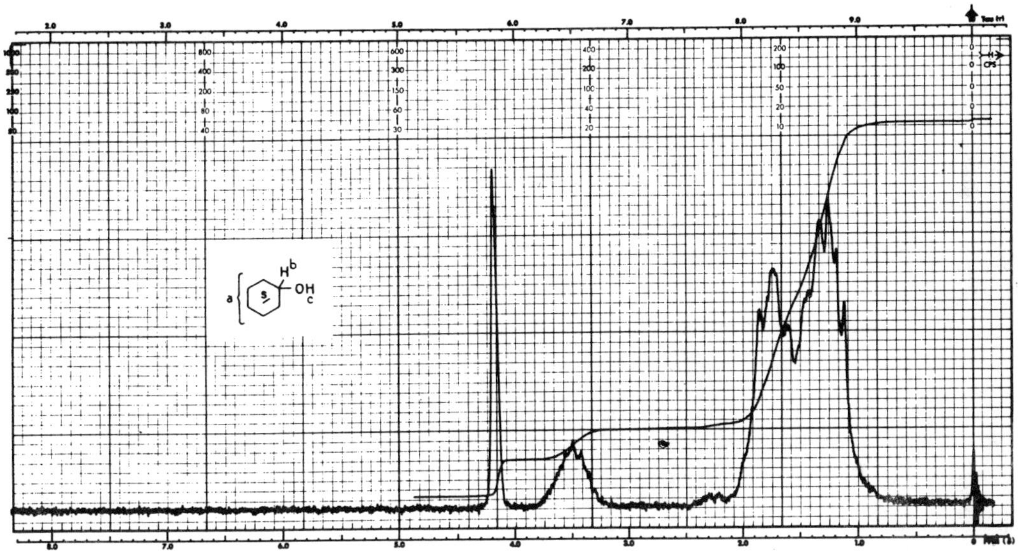

© 1978 Sadtler Research Laboratories, Inc.

---

trans-2-METHYLCYCLOHEXANOL

$C_7H_{14}O$

Mol. Wt. 114.19

Solvent: $CDCl_3$

Assignments
a = 1.01-1.03
b = 0.70-1.80
c = 1.86
d = 2.54
e = 3.12

1767

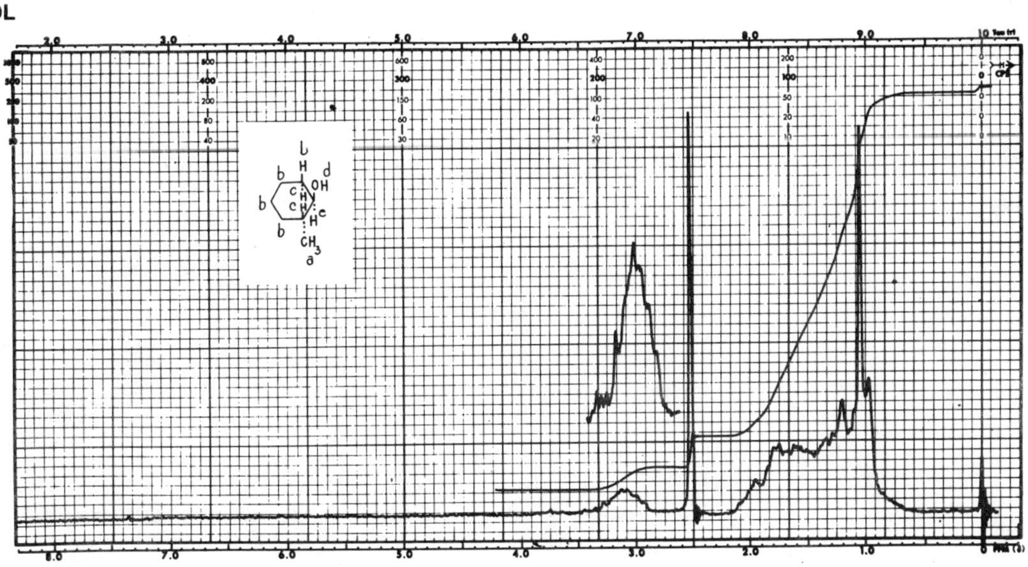

© 1978 Sadtler Research Laboratories, Inc.

trans-4-METHYLCYCLOHEXANOL

$C_7H_{14}O$

Mol. Wt. 114.19

B.P. 173-175°C

Solvent: $CDCl_3$

Assignments
a = 0.89
b = 1.00-2.20
c = 2.60
d = 3.51

1768

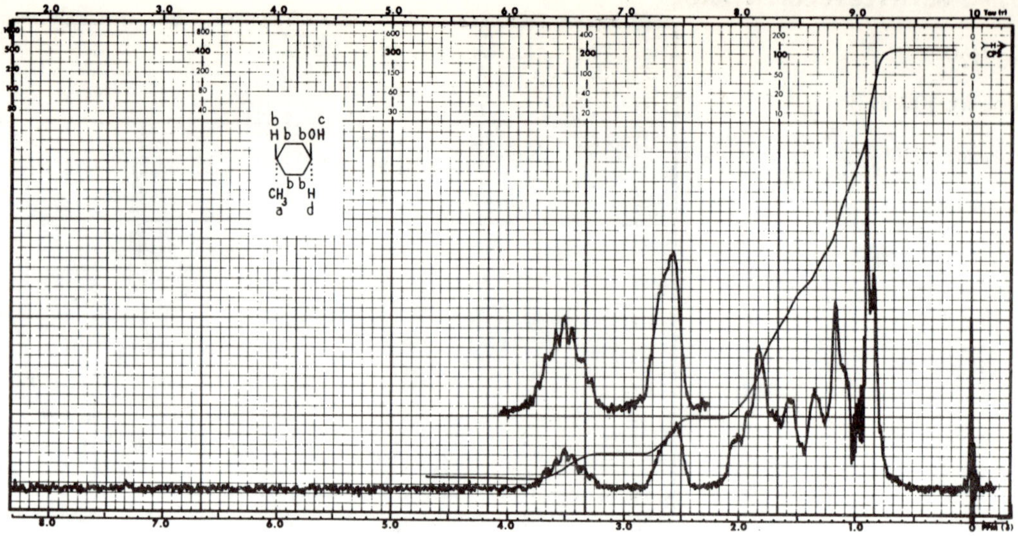

© 1978 Sadtler Research Laboratories, Inc.

cis-4-METHYLCYCLOHEXANOL

$C_7H_{14}O$

Mol. Wt. 114.19

B.P. 172-173°C

Solvent: $CDCl_3$

Assignments
a = 0.93
b = 1.10-2.00
c = 2.86
d = 3.89

1769

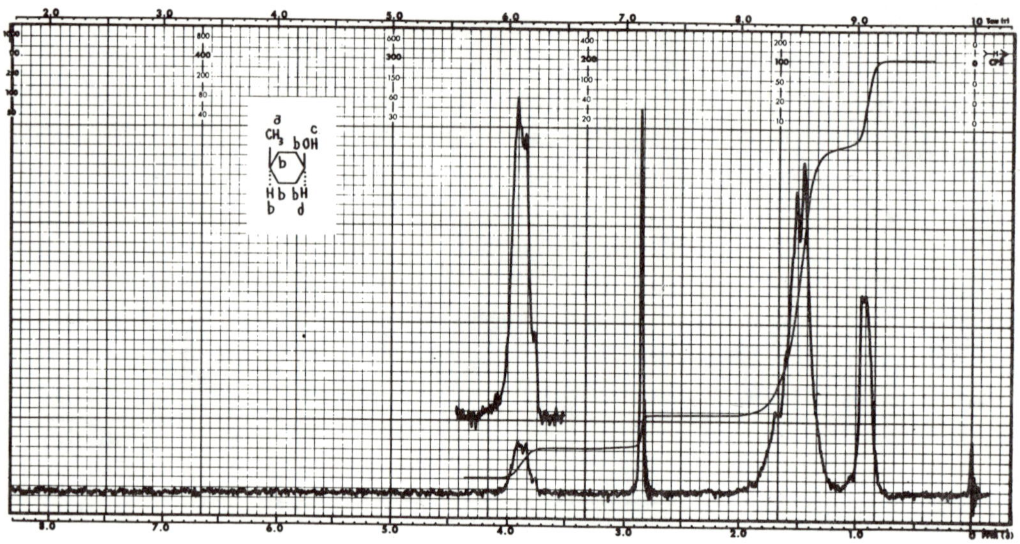

© 1978 Sadtler Research Laboratories, Inc.

2-ETHYLCYCLOHEXANOL (CIS AND TRANS)

$C_8H_{16}O$

Mol. Wt. 128.22

B.P. 74-79°C

Solvent: $CDCl_3$

Assignments
a = ca 1.00
b = 0.70-2.10
c = 2.39
d = 3.19
e = 3.87

1770

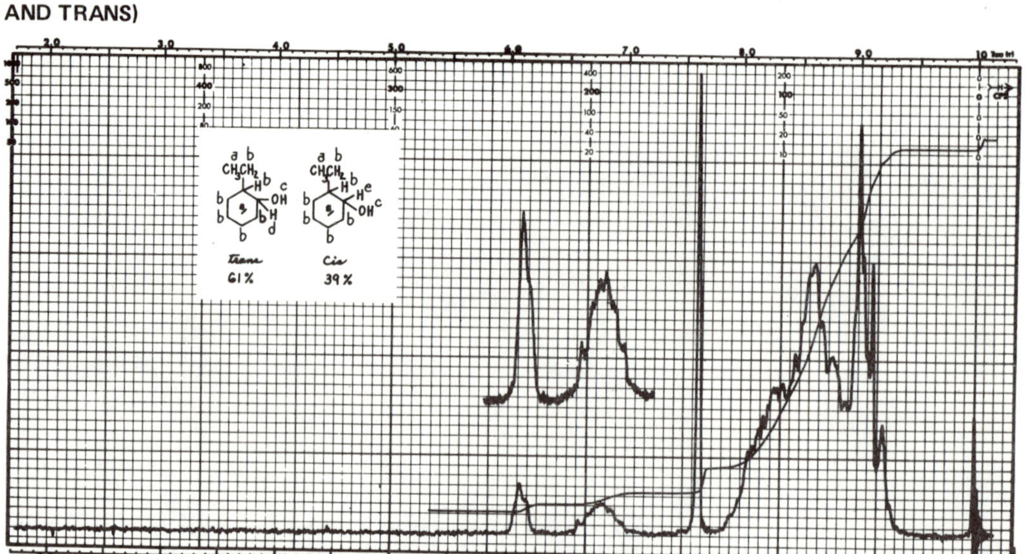

© 1978 Sadtler Research Laboratories, Inc.

2,4-DIMETHYLCYCLOHEXANOL

$C_8H_{16}O$

Mol. Wt. 128.22

B.P. 179°C

Solvent: $CDCl_3$

Assignments
a = 0.70-1.15
b = 0.50-2.10
c = 3.04 (Axial)
d = 3.11
e = 3.75 (Equatorial)

1771

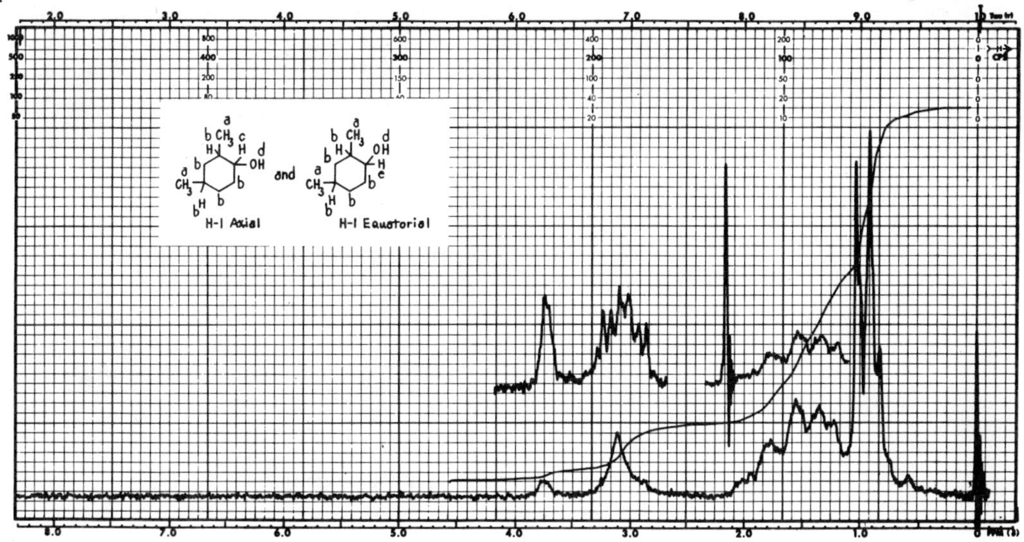

© 1978 Sadtler Research Laboratories, Inc.

MENTHOL

$C_{10}H_{20}O$

Mol. Wt. 156.27

M.P. 41-43°C

Solvent: $CDCl_3$

Assignments
a = 0.80
b = 0.91
c = 1.08-2.48
d = 2.23
e = 3.38

1772

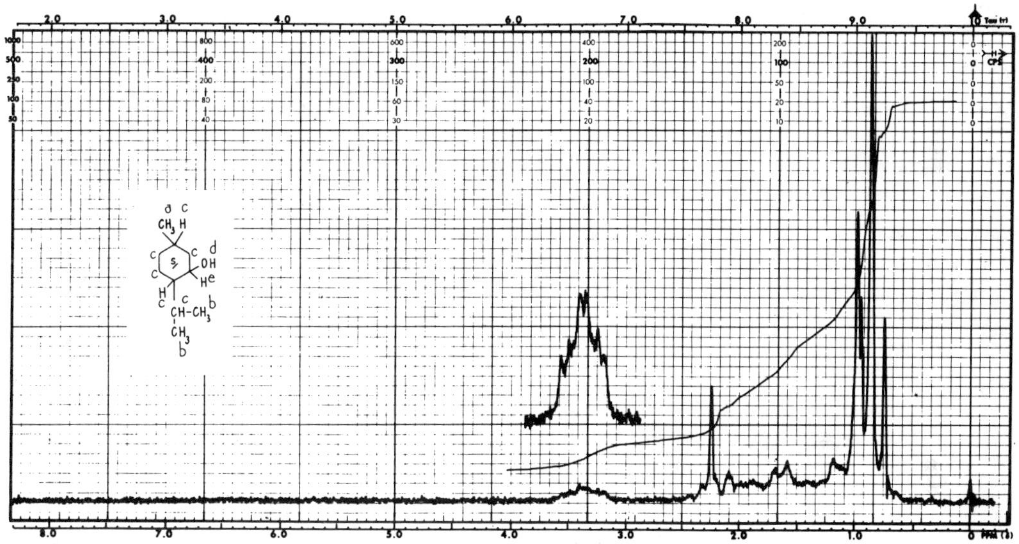

© 1978 Sadtler Research Laboratories, Inc.

2-CYCLOHEXYLCYCLOHEXANOL

$C_{12}H_{22}O$

Mol. Wt. 182.30

B.P. 126-128°C/7mm

Solvent: $CDCl_3$

Assignments
a = 0.50-2.20
b = 3.43
c = 4.12

1773

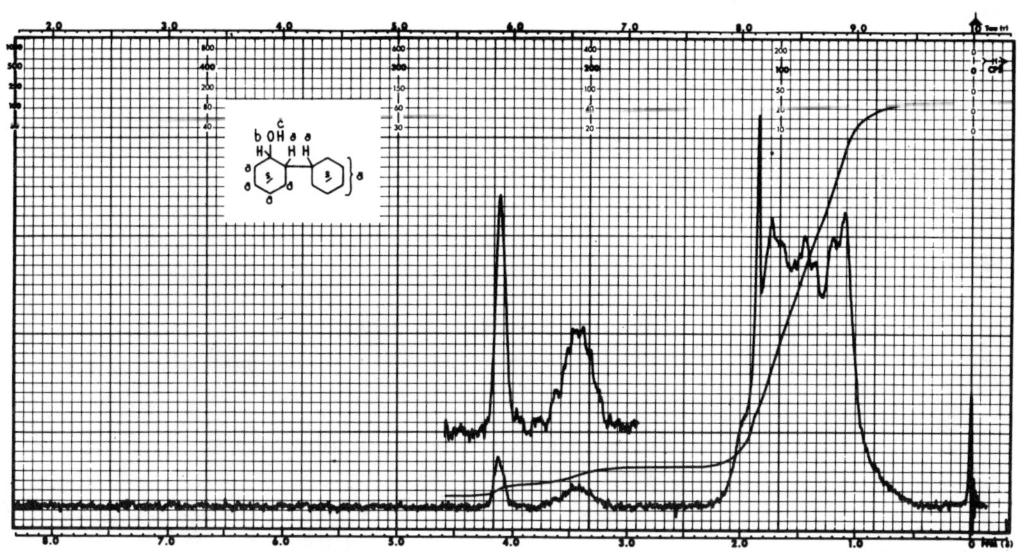

© 1978 Sadtler Research Laboratories, Inc.

trans-2-PHENYLCYCLOHEXANOL

$C_{12}H_{16}O$

Mol. Wt. 176.26

Solvent: $CDCl_3$

Assignments

a = 0.90-2.20
b = 1.89
c = 1.90-2.70
d = 3.52
e = ca 7.20

1774

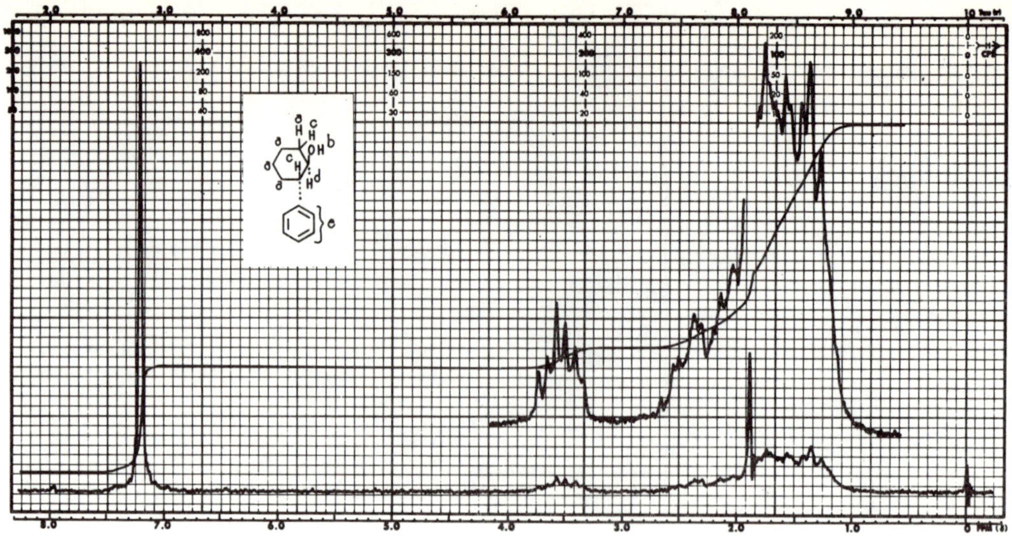

© 1978 Sadtler Research Laboratories, Inc.

cis-2-PHENYLCYCLOHEXANOL

$C_{12}H_{16}O$

Mol. Wt. 176.26

Solvent: CCl_4

Assignments

a = 0.80-2.90
b = 3.20-4.00
c = ca 7.18

1775

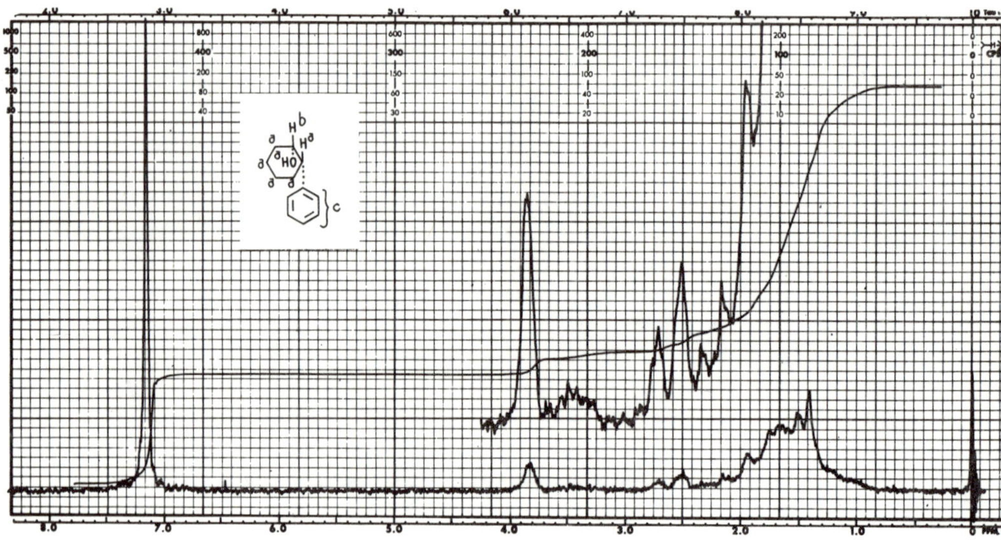

© 1978 Sadtler Research Laboratories, Inc.

2,2-DIPHENYLCYCLOHEXANOL

$C_{18}H_{20}O$

Mol. Wt. 252.36

M.P. 82-83°C

Solvent: $CDCl_3$

Assignments

a = 1.10-1.80
b = 1.49
c = ca 1.90
d = ca 2.37
e = 4.55
f = 6.90-7.60

1776

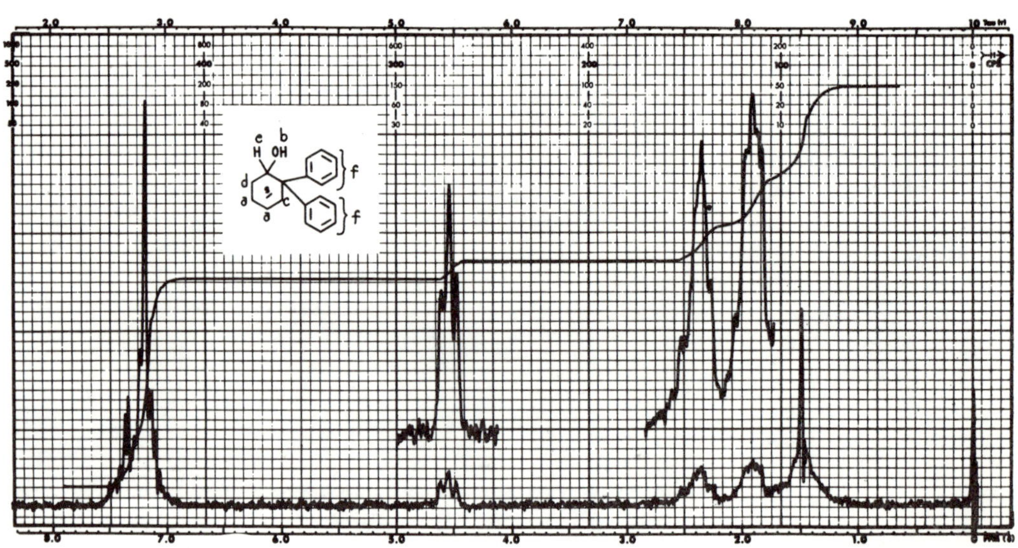

© 1978 Sadtler Research Laboratories, Inc.

CYCLOHEPTANOL

$C_7H_{14}O$

Mol. Wt. 114.19

B.P. 185°C

Solvent: $CDCl_3$

Assignments
a = 1.20-2.30
b = 2.69
c = 3.80

1777

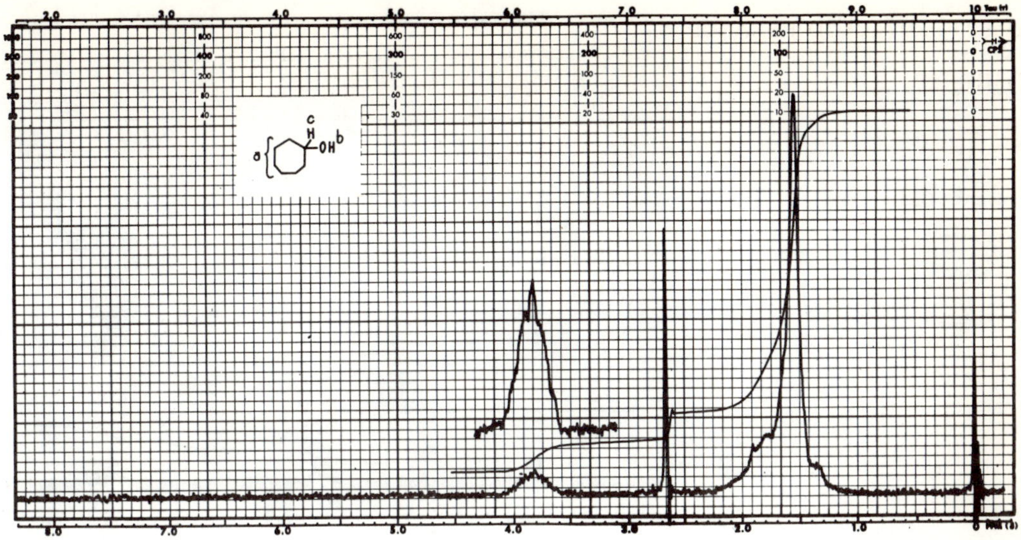

© 1978 Sadtler Research Laboratories, Inc.

DECAHYDRO-2-NAPHTHOL

$C_{10}H_{18}O_2$

Mol. Wt. 170.25

Solvent: Polysol-d

Assignments
a = 0.70-2.10
b = 2.48
c = 3.81

1778

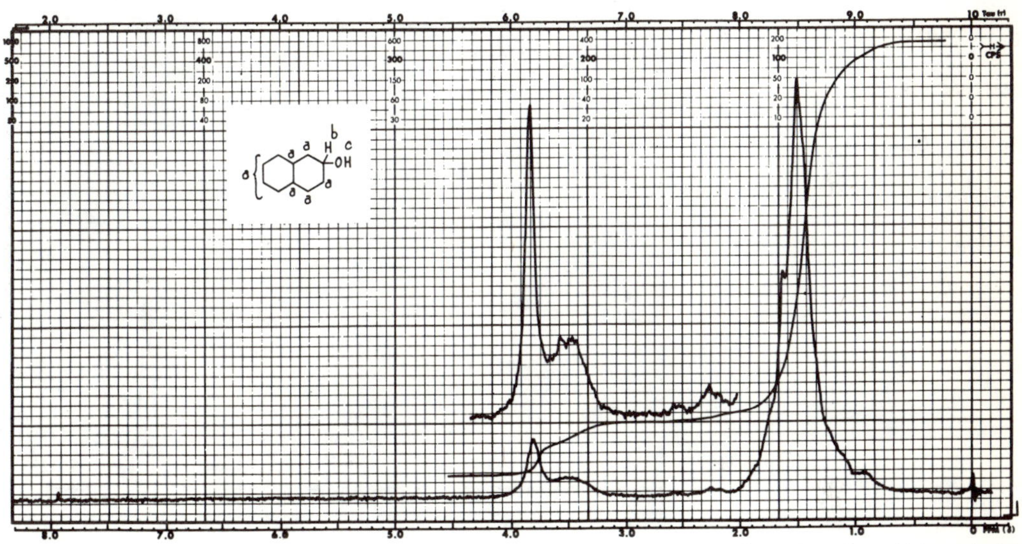

© 1978 Sadtler Research Laboratories, Inc.

BORNEOL

$C_{10}H_{18}O$

Mol. Wt. 154.25

M.P. 208-209°C

Solvent: $CDCl_3$

Assignments
a = 0.89
b = 1.00-2.60
c = 1.53
d = 4.00

1779

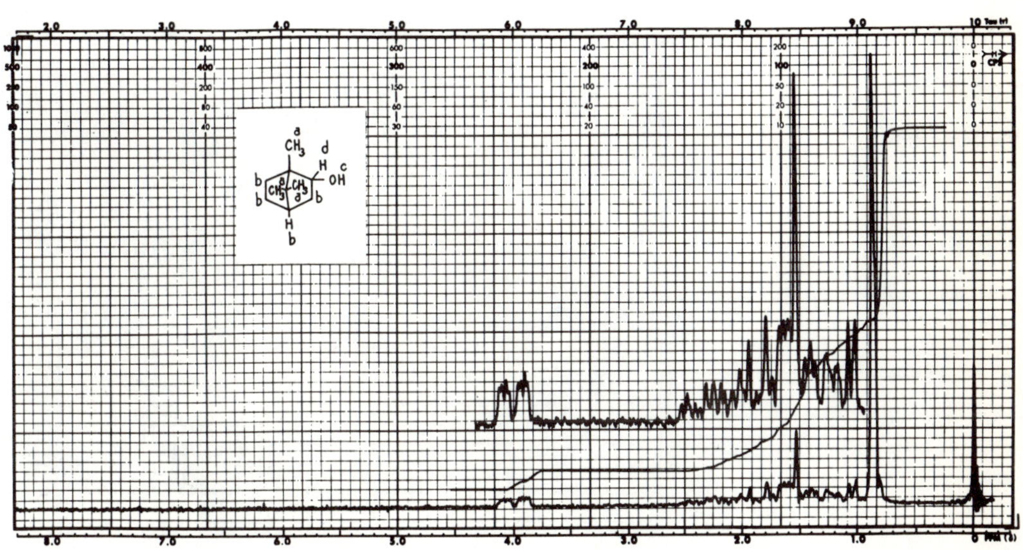

© 1978 Sadtler Research Laboratories, Inc.

CHOLESTROL

$C_{27}H_{46}O$

Mol. Wt. 386.35

M.P. 150°C

Solvent: $CDCl_3$

Assignments
a = 0.80
b = 0.88
c = 0.91
d = 0.80-2.40
e = 3.49
f = 5.31

1780

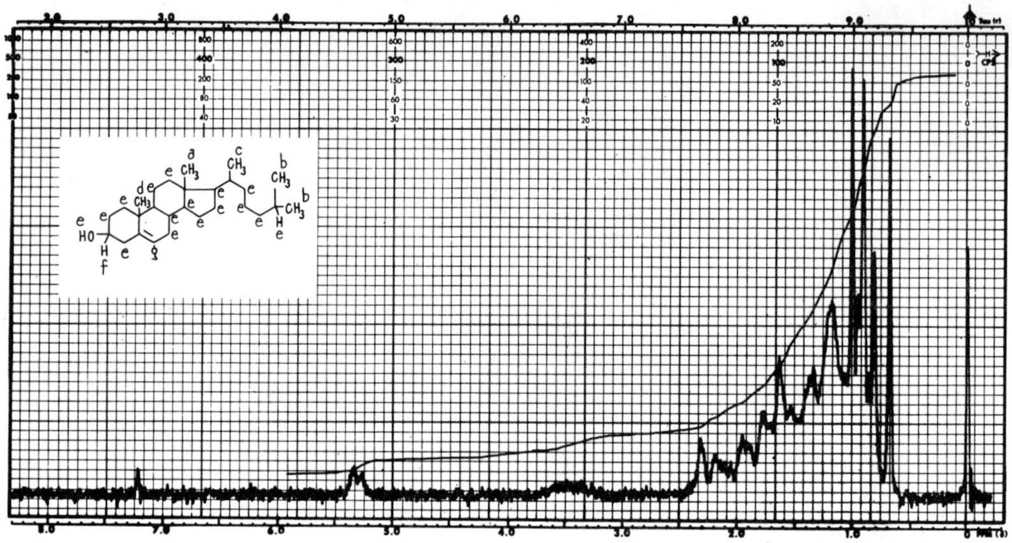

© 1978 Sadtler Research Laboratories, Inc.

TETRAHYDRO-2H-THIOPYRAN-3-OL

$C_5H_{10}OS$

Mol. Wt. 118.20

Solvent: $CDCl_3$

Assignments
a = 1.00-2.60
b = ca 2.63
c = 3.50-4.05

1781

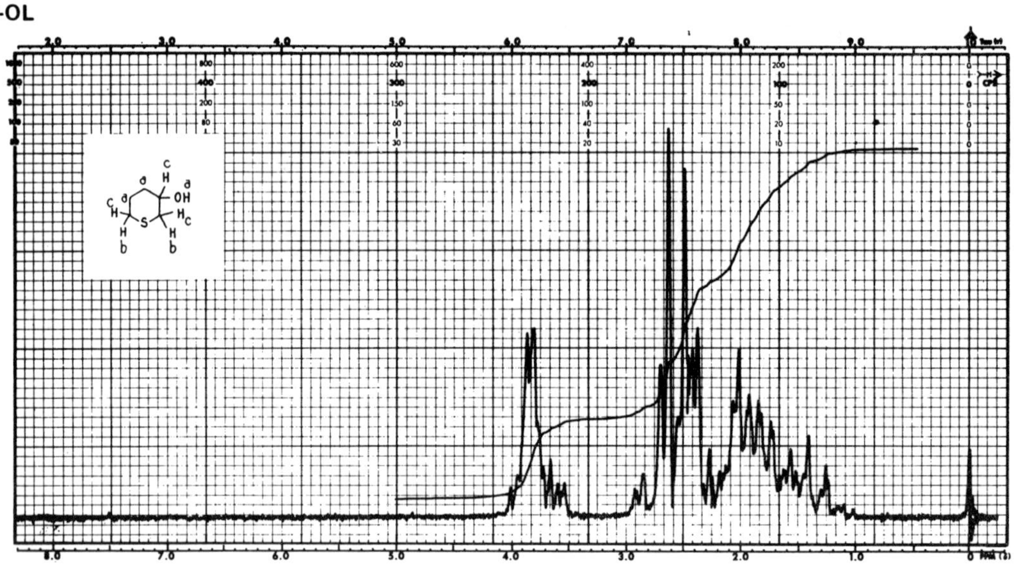

© 1978 Sadtler Research Laboratories, Inc.

3-METHYL-3-BUTEN-2-OL

$C_5H_{10}O$

Mol. Wt. 86.13

Solvent: CCl_4

Assignments
a = 1.20
b = 1.70
c = 3.09
d = 4.14
e = 4.70 or f
f = 4.89 or e
g = 3.91 imp.

1782

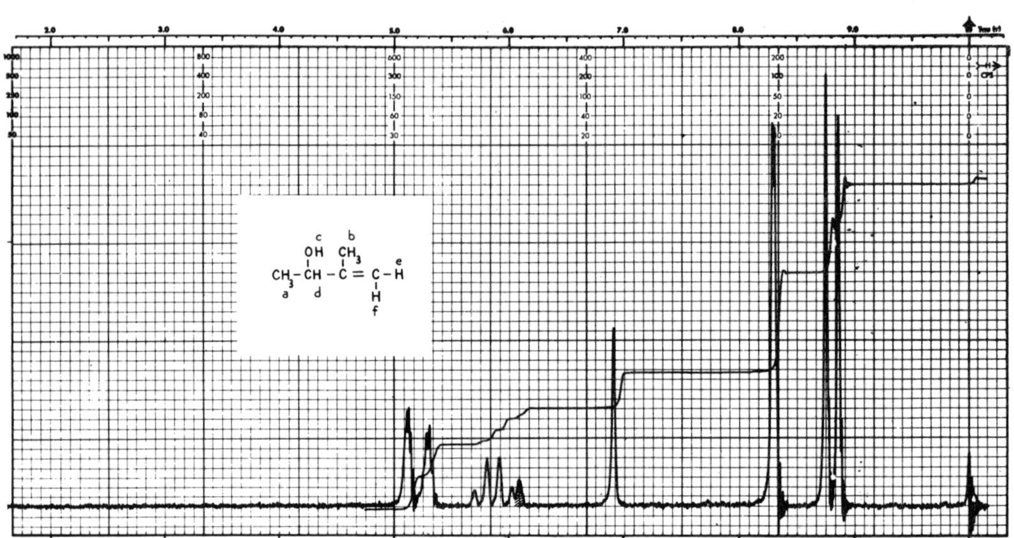

© 1978 Sadtler Research Laboratories, Inc.

1-HEPTEN-3-OL

$C_7H_{14}O$

Mol. Wt. 114.19

Solvent: CCl_4

Assignments

a = 0.90
b = ca 1.38
c = 2.78
d = 3.93
e = 4.93
f = 5.15
g = ca 5.78
h = 2.10 imp.

1783

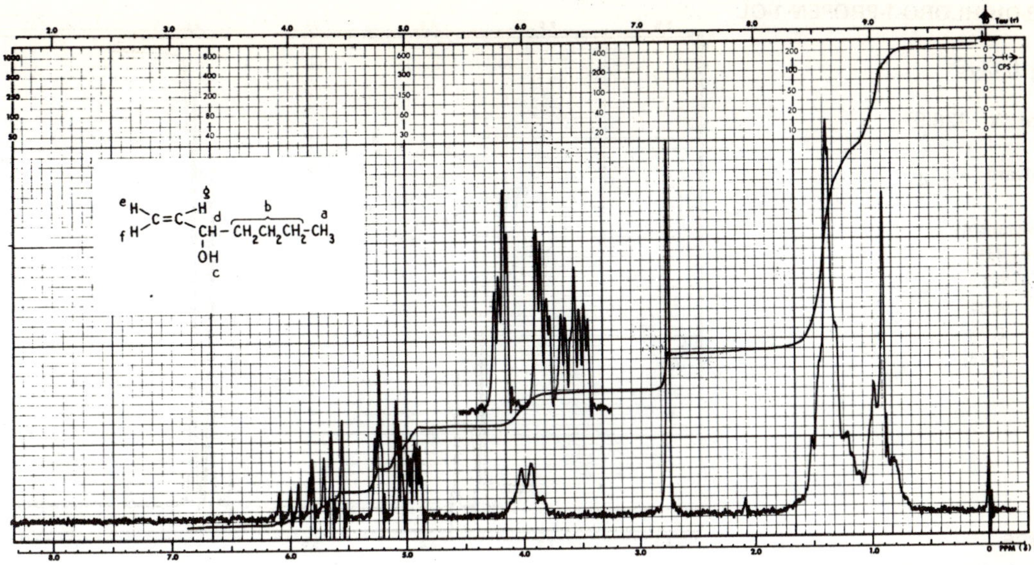

© 1978 Sadtler Research Laboratories, Inc.

1-HEPTEN-4-OL

$C_7H_{14}O$

Mol. Wt. 114.19

Solvent: $CDCl_3$

Assignments

a = 0.95
b = 1.10-1.65
c = 2.21
d = 2.43
e = 3.63
f = ca 5.06
g = 5.82

1784

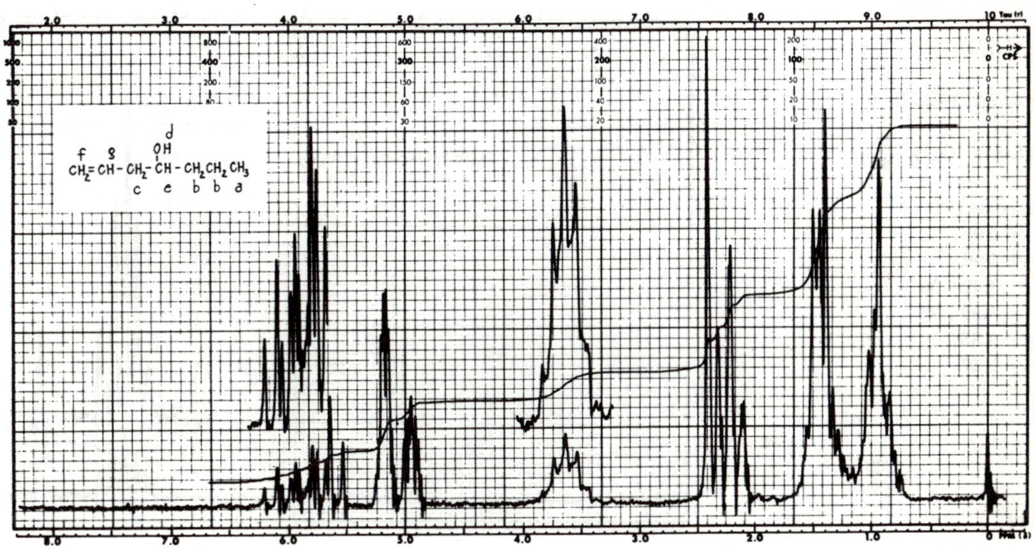

© 1978 Sadtler Research Laboratories, Inc.

1,5-HEXADIEN-3-OL

$C_6H_{10}O$

Mol. Wt. 98.15

Solent: $CDCl_3$

Assignments

a = 2.30
b = 3.15
c = 4.11
d = 4.85-5.38
e = ca 5.78

1785

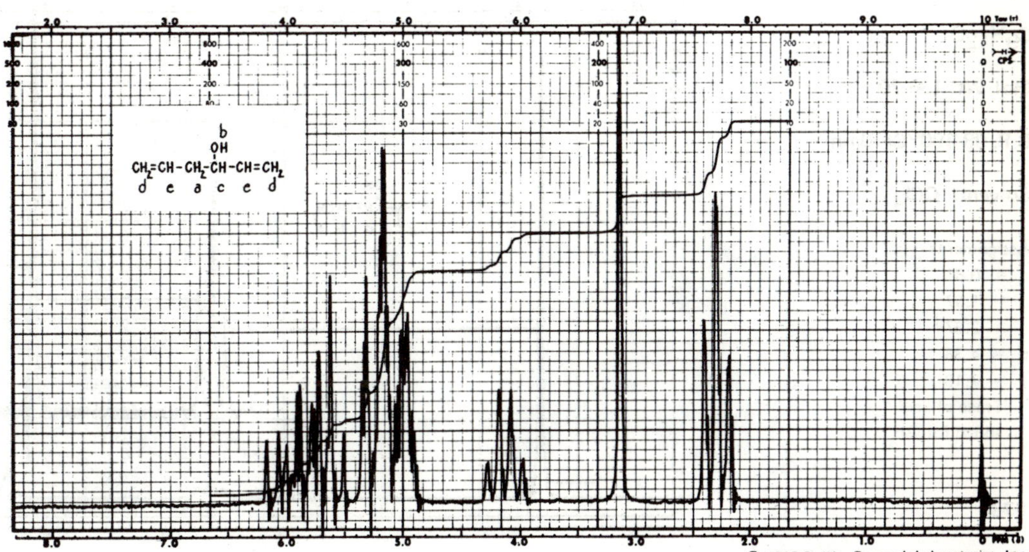

© 1978 Sadtler Research Laboratories, Inc.

2,3-DICHLORO-1-PROPEN-1-OL

$C_3H_4Cl_2O$

Mol. Wt. 126.97

Solvent: CCl_4

Assignments
a = 4.02
b = 4.13
c = 4.34
d = 6.23
e = 6.45

1786

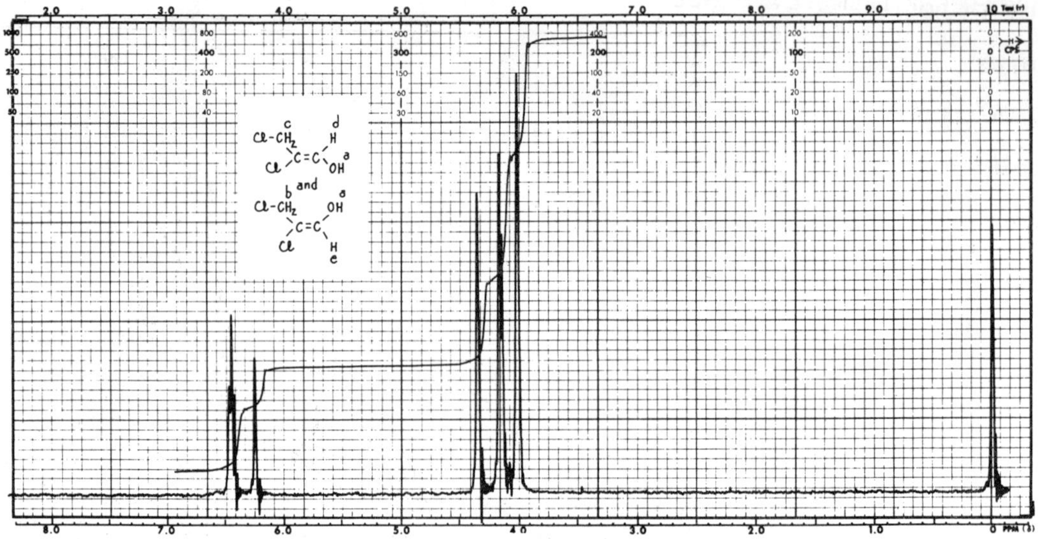

© 1978 Sadtler Research Laboratories, Inc.

5-NONYN-3-OL

$C_9H_{16}O$

Mol. Wt. 140.23

Solvent: CCl_4

Assignments
a = 0.95
b = 0.98
c = 1.10-1.90
d = 1.90-2.40
e = 2.86
f = 3.57

1787

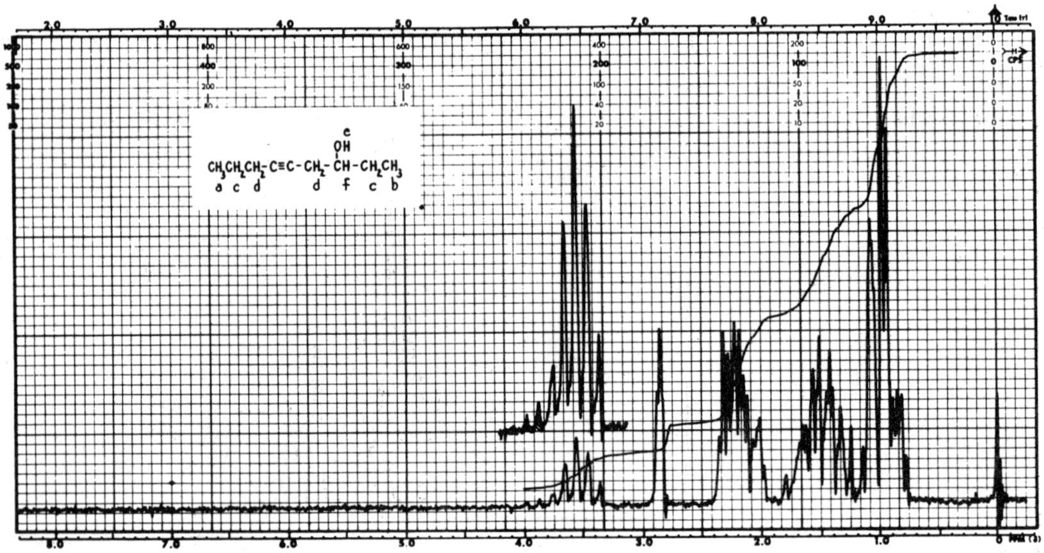

© 1978 Sadtler Research Laboratories, Inc.

α-ETHYLBENZYL ALCOHOL

$C_9H_{12}O$

Mol. Wt. 136.20

B.P. 98-100°C/10mm

Solvent: CCl_4

Assignments
a = 0.81
b = 1.57
c = 2.67
d = 4.38
e = 7.18
Trace of acid added for OH

1788

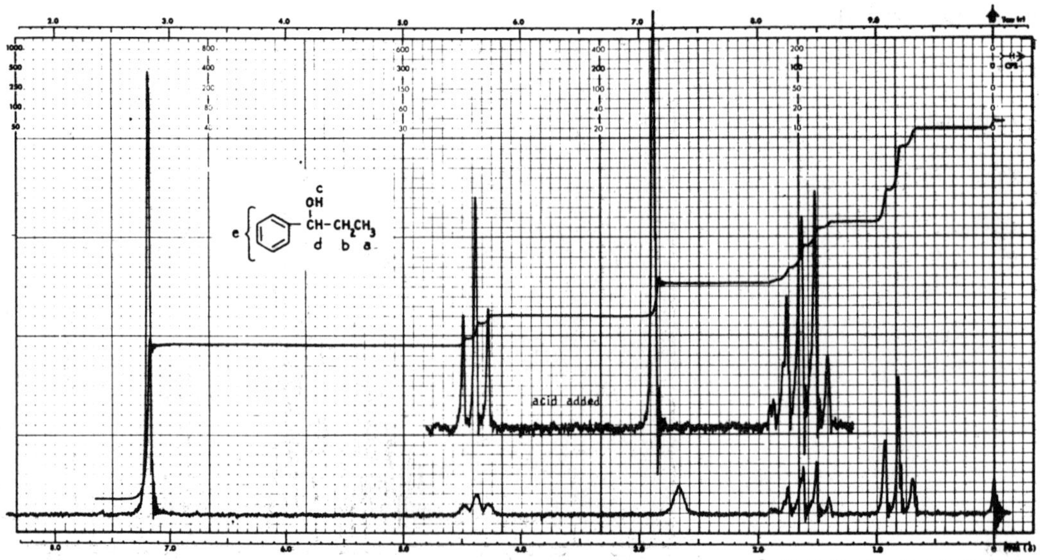

© 1978 Sadtler Research Laboratories, Inc.

α-CYCLOPROPYLBENZYL ALCOHOL

$C_{10}H_{12}O$

Mol. Wt. 148.21

B.P. 121°C/12mm

Solvent: CCl_4

Assignments
a = 0.90-0.64
b = 0.98
c = 3.22
d = 3.88
e = 7.27

1789

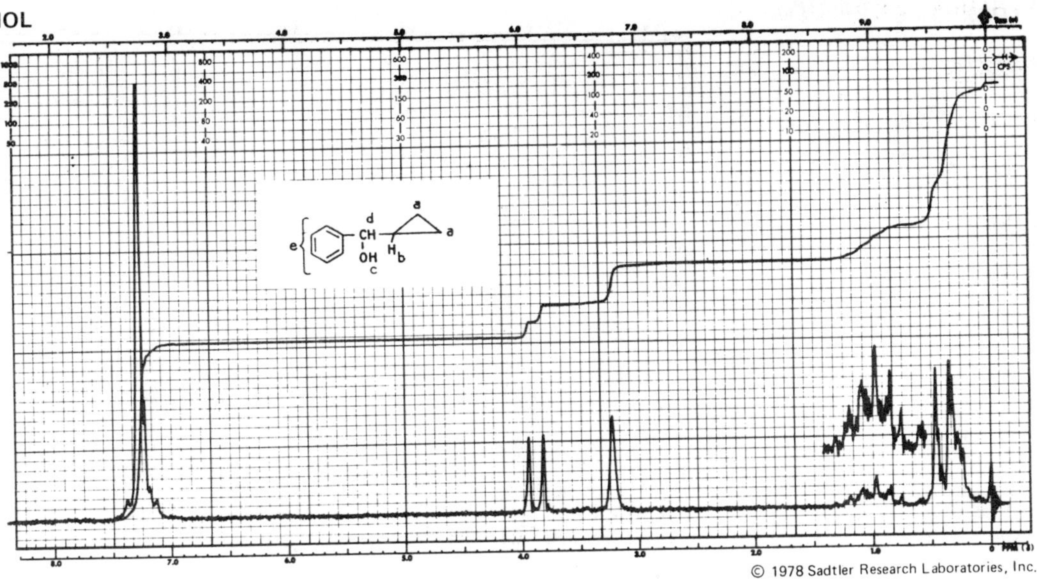

© 1978 Sadtler Research Laboratories, Inc.

BENZHYDROL

$C_{13}H_{12}O$

Mol. Wt. 184.24

M.P. 67-68°C

Solvent: CCl_4

Assignments
a = 2.28
b = 5.59
c = 7.18

1790

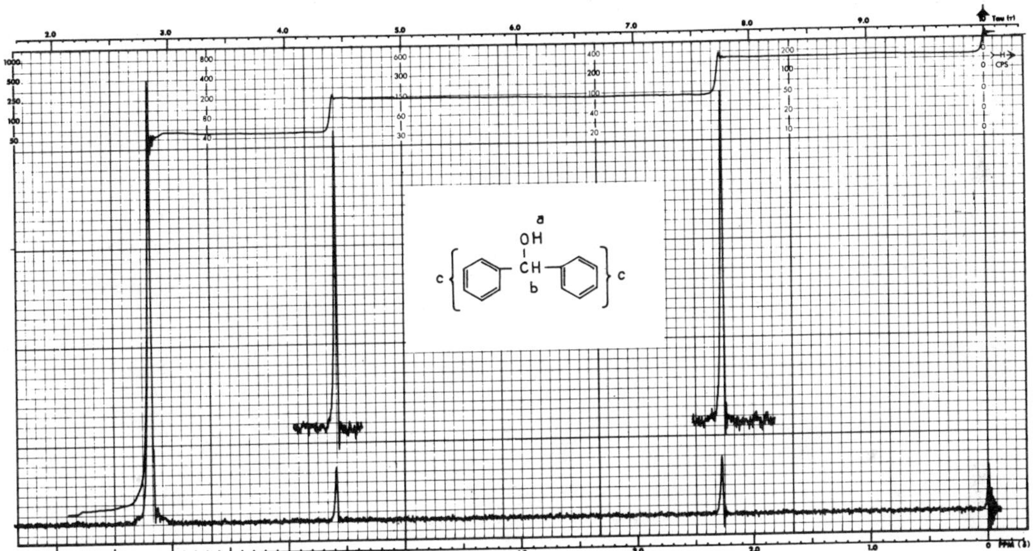

© 1978 Sadtler Research Laboratories, Inc.

1,2-DIPHENYLETHANOL

$C_{14}H_{14}O$

Mol. Wt. 198.27

M.P. 66-67°C

Solvent: $CDCl_3$

Assignments
a = 2.10
b = 2.97
c = 4.81
d = ca 7.21
e = ca 7.33

1791

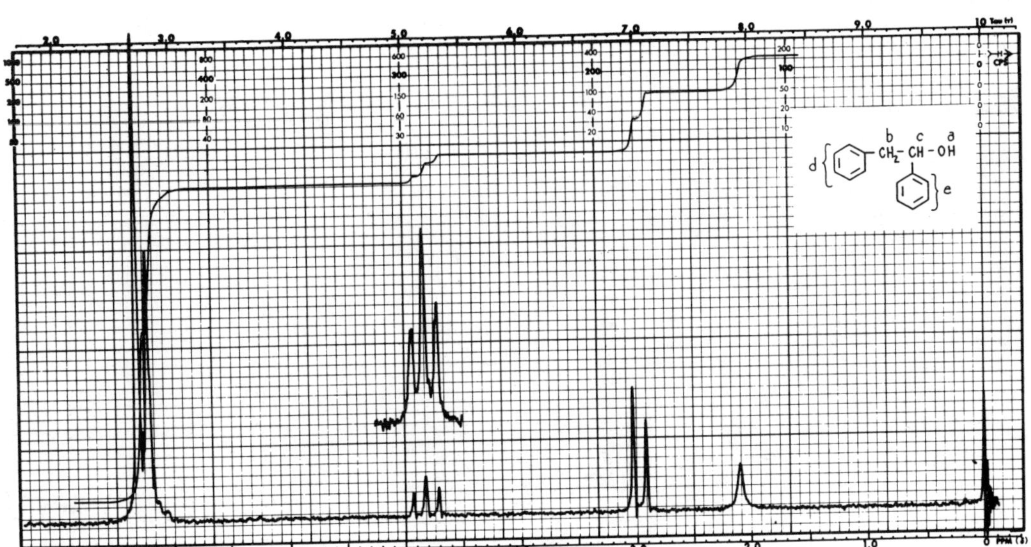

© 1978 Sadtler Research Laboratories, Inc.

1-PHENOXY-2-PROPANOL

$C_9H_{12}O_2$

Mol. Wt. 152.19

M.P. 13-17°C

Solvent: CCl_4

Assignments

a = 1.21
b = 3.41
c = 3.72
d = 4.08
e = 6.79
f = 6.83
g = 7.19

1792

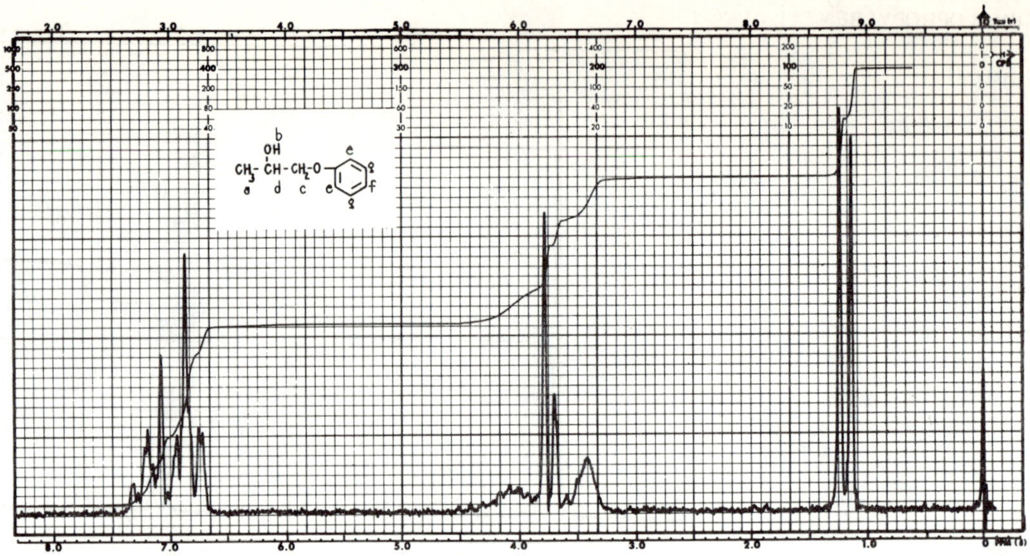

© 1978 Sadtler Research Laboratories, Inc.

2,5-DICHLORO-α-METHYLBENZYL ALCOHOL

$C_8H_8Cl_2O$

Mol. Wt. 191.06

Solvent: CCl_4

Assignments

a = 1.41
b = 2.82
c = 5.11
d = 7.15
e = 7.50

1793

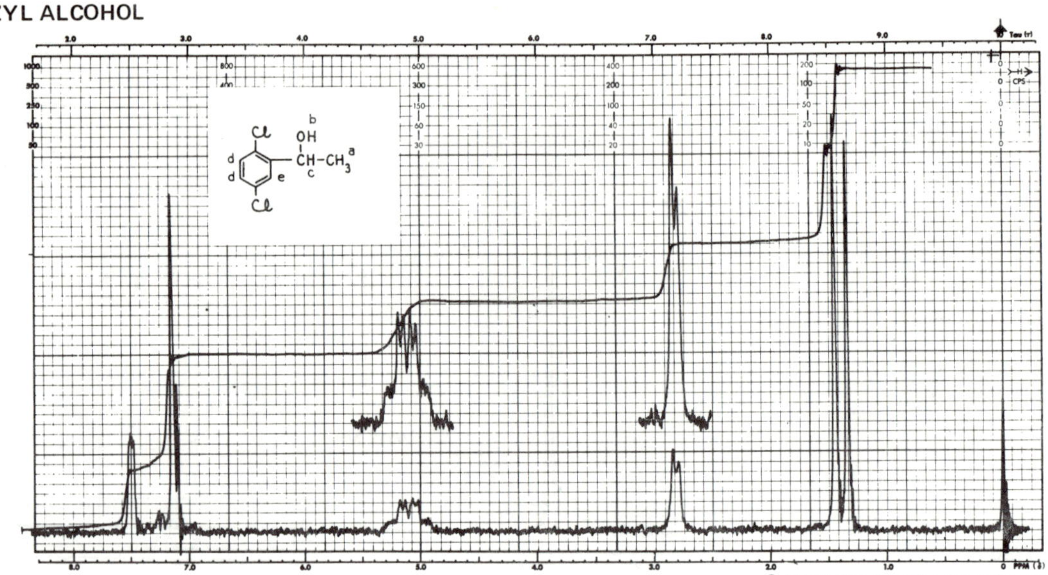

© 1978 Sadtler Research Laboratories, Inc.

QUININE

$C_{20}H_{24}N_2O_2$

Mol. Wt. 324.43

M.P. 177°C

Solvent: Polysol-d

a = 1.00-2.00		h = 7.31	
b = 2.00-3.70		i = ca 7.50	
c = 3.90		j = 7.93	
d = 4.89		k = 8.66	
e = 4.91			
f = 5.20-5.60			
g = 5.80			

1794

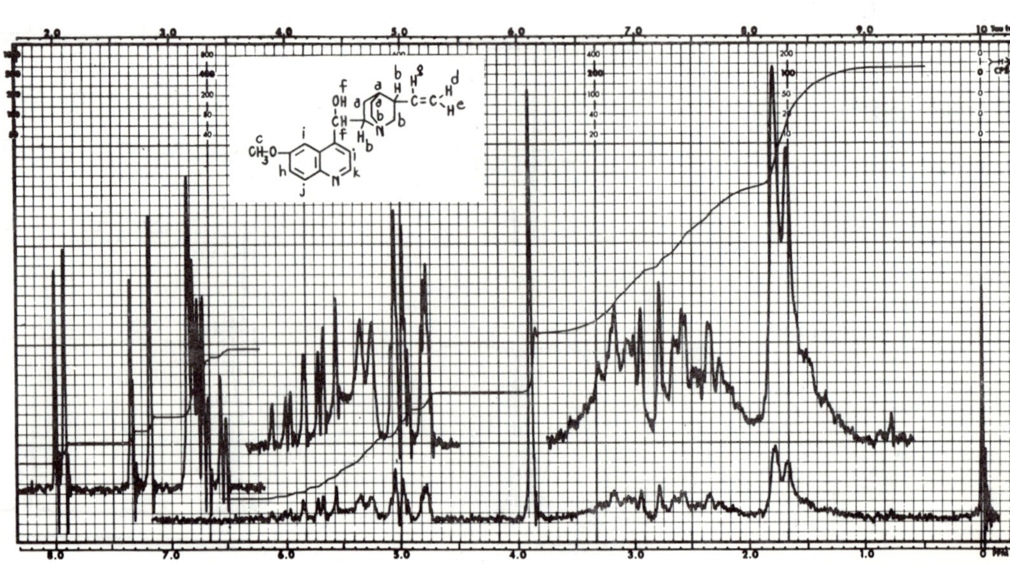

© 1978 Sadtler Research Laboratories, Inc.

THE TERTIARY ALCOHOLS

The tertiary alcohols are a difficult group of compounds to identify using only their NMR spectra. Their most characteristic feature is the presence of a single exchangeable proton which does not display any couplings since there are no protons on the adjacent carbon atom.

The phenols will be treated as a separate group in a later chapter. The series of 1-substituted-2-propanols listed below illustrates the relatively narrow range of chemical shifts (less than 1 ppm) that is observed for the hydrocarbon groups of the tertiary alcohols.

Chemical Shifts

Aliphatic Protons

1-Substituted-2-Propanols

HO	$-C-(CH_3)_2$	$-X$	Solvent	Spectrum
2.89	1.15	$-R_2$	CCl_4	1796
1.40	1.18	$-R_3$	CCl_4	1797
2.40	1.20	$-CH_3$	CCl_4	1795
2.14	1.23	$-CH{=}CH_2$	CCl_4	1812
2.34	1.51	−Ø	$CDCl_3$	1817
2.30	1.53	$-C{\equiv}C-H$	$CDCl_3$	1813
- - - -	1.54	$-SO_2-O-Na$	D_2O	1807
3.49	1.61	$-C{\equiv}N$	$CDCl_3$	1806
2.62	1.77	$-C(Br)_3$	$CDCl_3$	1805

Coupling and Coupling Constants

The coupling constants between protons on adjacent carbons (CH—CH) are similar to those of the corresponding structures of other types of substituents. Because none of the hydrocarbon groups are strongly deshielded by the hydroxyl group, complex higher order patterns at high field are to be expected.

Solubility and Solvent Effects

Except for compounds containing water soluble groups such as the sodium and potassium salts of organic acids, the simple tertiary alcohols are normally soluble in the chlorinated solvents, CCl_4 and $CDCl_3$.

The use of hydroscopic solvents such as acetone and DMSO should be avoided whenever possible because the relatively weak hydroxyl resonance may exchange with water in the solvent and its presence could go undetected.

tert-BUTYL ALCOHOL

$C_4H_{10}O$

Mol. Wt. 74.12

M.P. 24.5-25.5°C

B.P. 82.8°C

Solvent: CCl_4

Assignments
a = 1.20
b = 2.40

1795

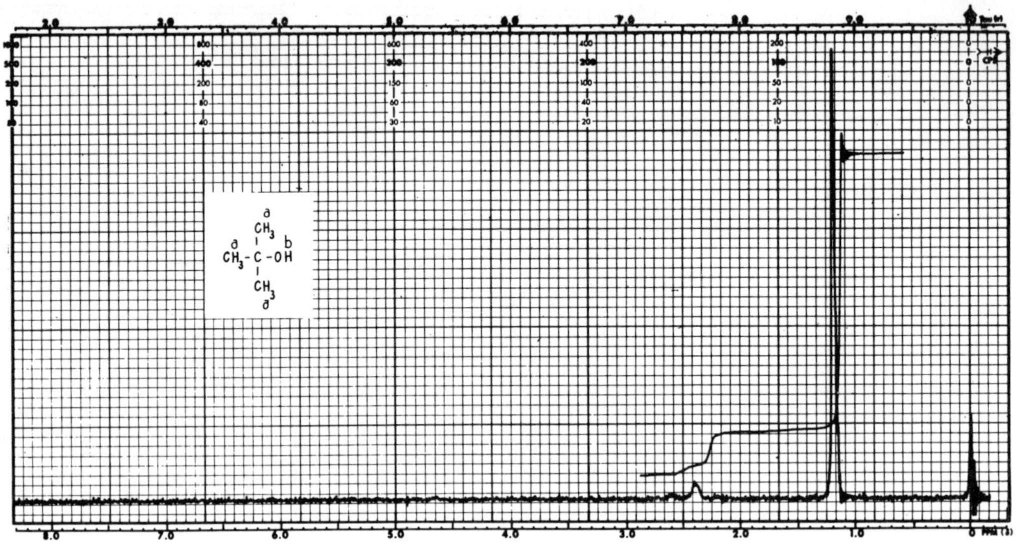

© 1978 Sadtler Research Laboratories, Inc.

2-METHYL-2-BUTANOL

$C_5H_{12}O$

Mol. Wt. 88.15

B.P. 102°C

Solvent: CCl_4

Assignments
a - 0.90
b = 1.15
c = 1.46
d = 2.89

1796

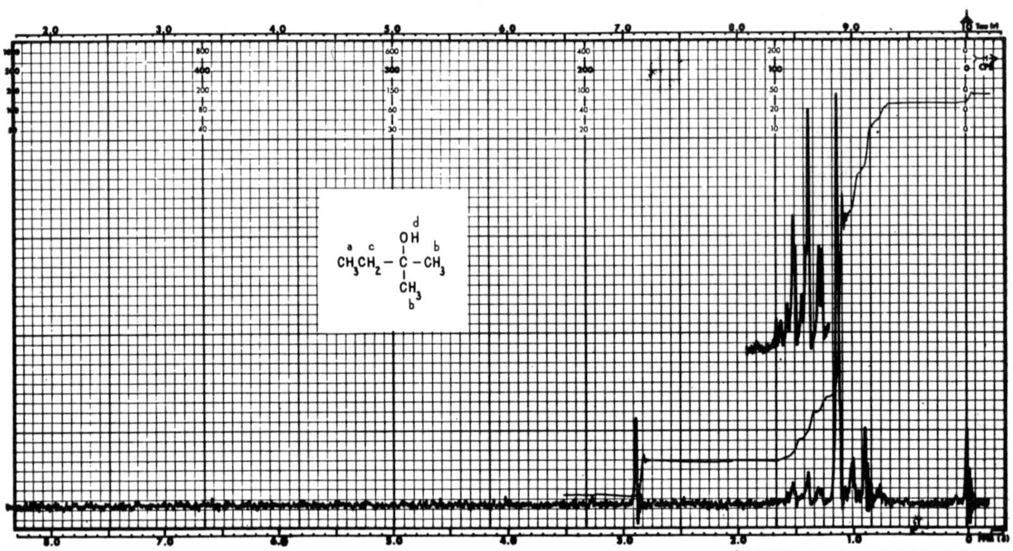

© 1978 Sadtler Research Laboratories, Inc.

2-METHYL-2-PENTANOL

$C_6H_{14}O$

Mol. Wt. 102.17

B.P. 122°C

Solvent: CCl_4

Assignments
a = 0.94
b = 1.18
c = 1.20-1.60

1797

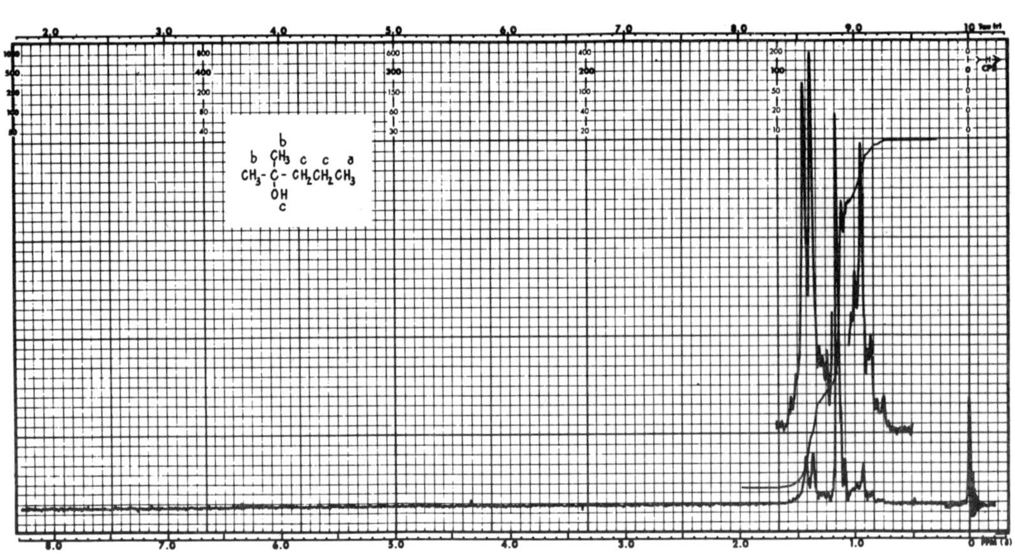

© 1978 Sadtler Research Laboratories, Inc.

2-METHYL-2-HEXANOL

$C_7H_{16}O$

Mol. Wt. 116.20

B.P. 58-60°C

Solvent: CCl_4

Assignments
a = 0.92
b = 1.17
c = ca 1.37
d = 1.78

1798

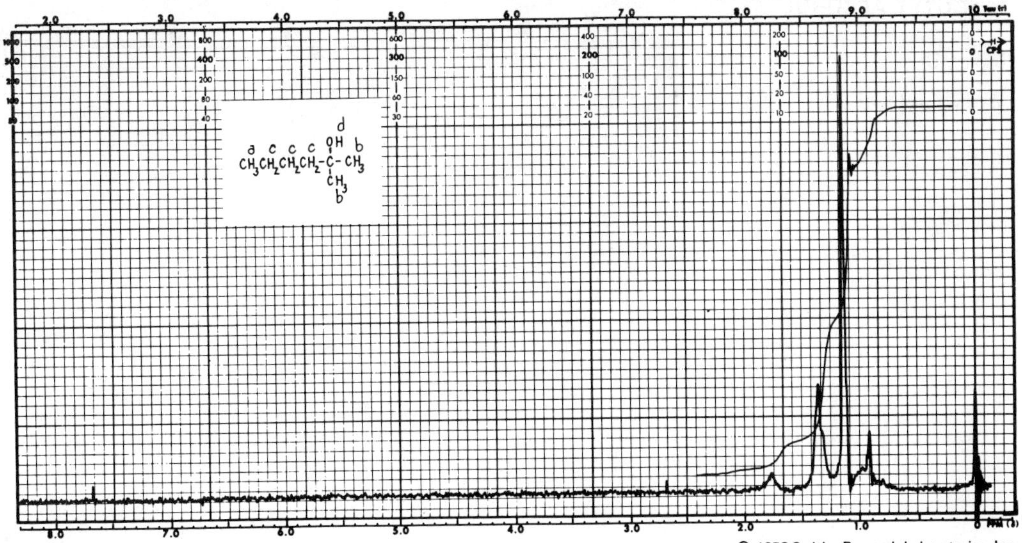

© 1978 Sadtler Research Laboratories, Inc.

2,5-DIMETHYL-2-HEXANOL

$C_8H_{18}O$

Mol. Wt. 130.23

Solvent: CCl_4

Assignments
a = 0.88
b = 1.12
c = 1.29
d = 1.46

1799

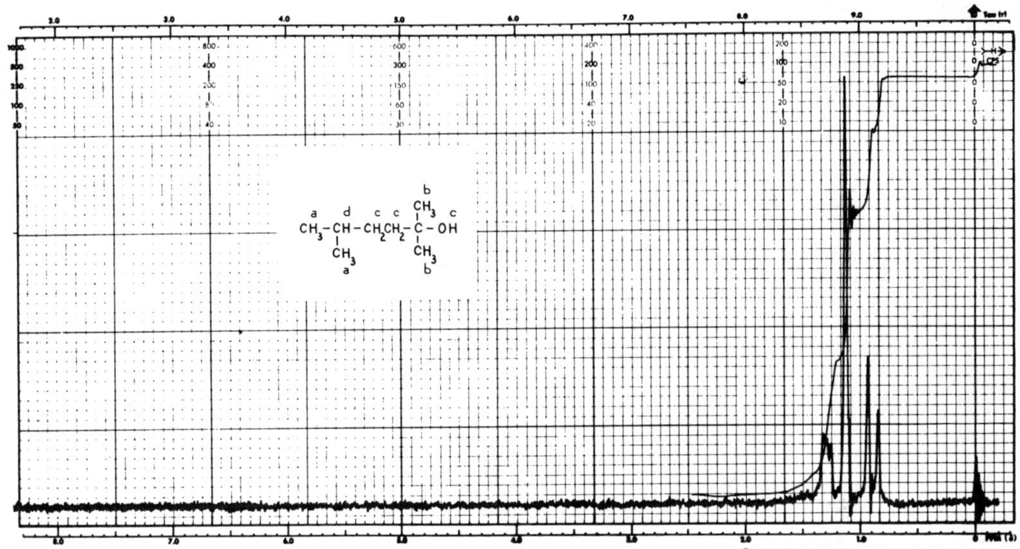

© 1978 Sadtler Research Laboratories, Inc.

2-METHYL-2-OCTANOL

$C_9H_{20}O$

Mol. Wt. 144.12

Solvent: $CDCl_3$

Assignments
a = 0.90
b = 1.21
c = 1.10-1.60
d = 1.89

1800

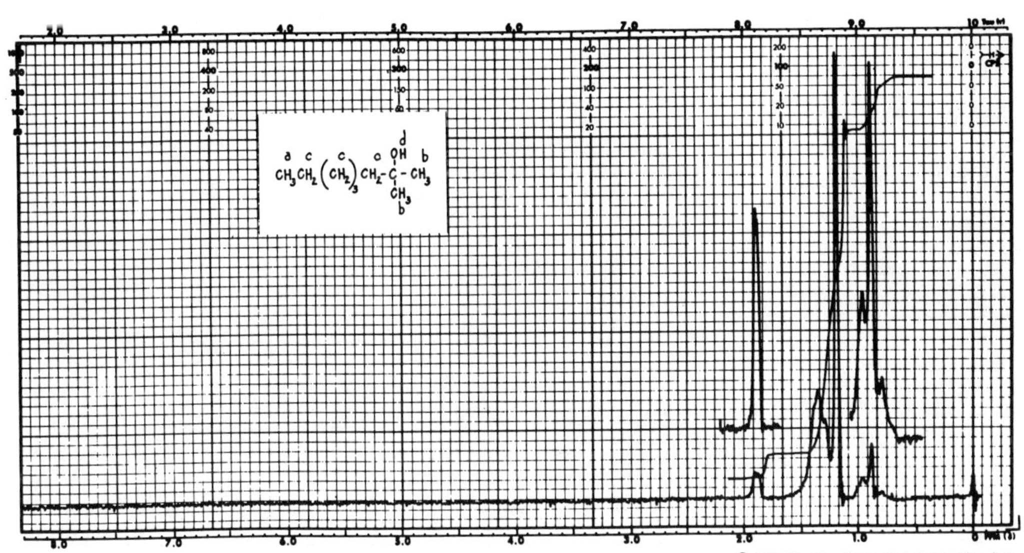

© 1978 Sadtler Research Laboratories, Inc.

3,6-DIMETHYL-3-HEPTANOL

$C_9H_{20}O$

Mol. Wt. 144.26

Solvent: CCl_4

Assignments
a = 0.90
b = 0.93
c = 1.12
d = 1.10-1.70
e = 2.00

1801

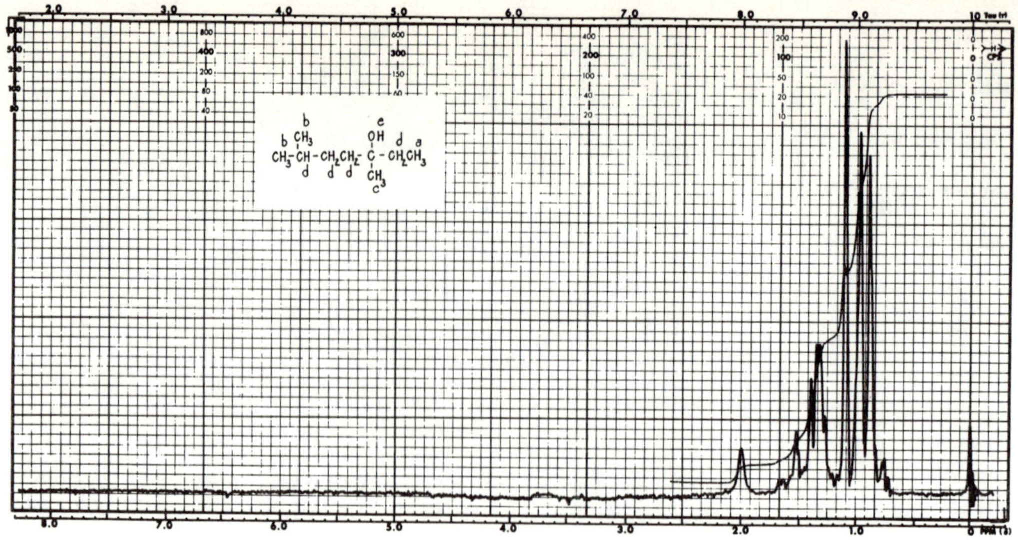

© 1978 Sadtler Research Laboratories, Inc.

3,7-DIMETHYL-3-OCTANOL

$C_{10}H_{22}O$

Mol. Wt. 158.29

Solvent: CCl_4

Assignments
a = ca 0.89
b = 0.90
c = 1.10
d = 1.13-1.78
e = 2.19

1802

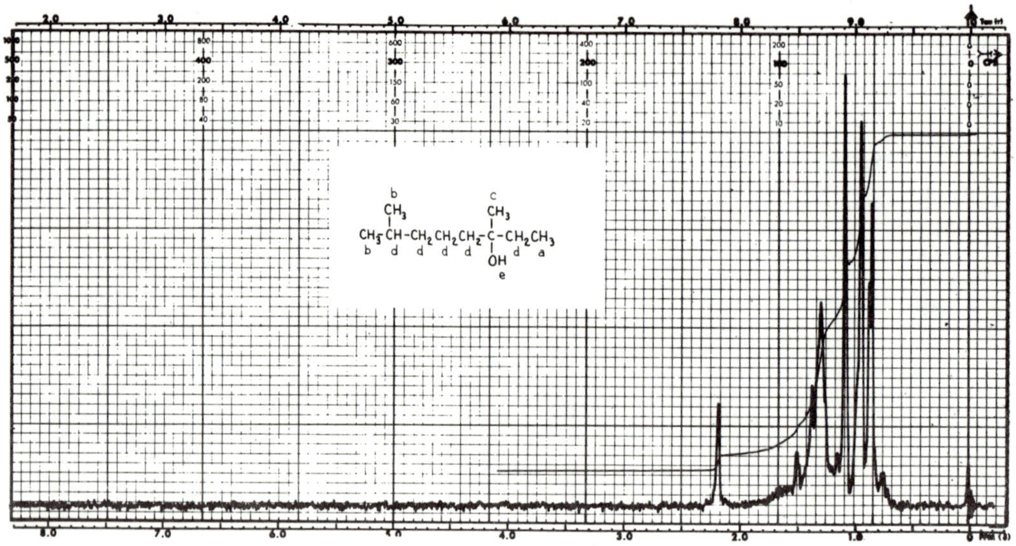

© 1978 Sadtler Research Laboratories, Inc.

3,4-DIETHYL-3-HEXANOL

$C_{10}H_{22}O$

Mol. Wt. 158.29

Solvent: CCl_4

Assignments
a = 0.89
b = 0.80-1.50
c = 1.49

1803

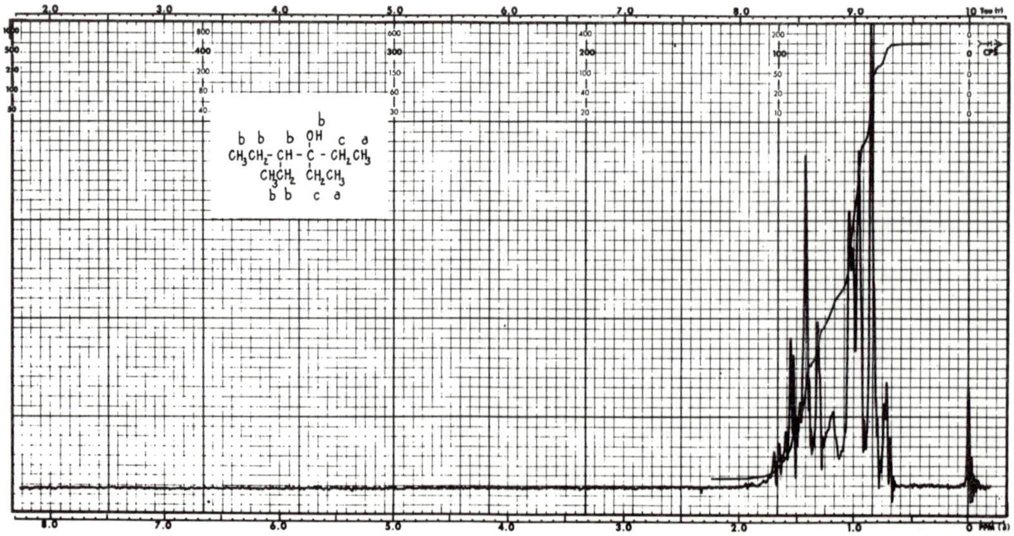

© 1978 Sadtler Research Laboratories, Inc.

14-TRIDECYL-14-HEPTACOSANOL

$C_{40}H_{82}O$

Mol. Wt. 579.10

M.P. 43.0-43.8°C

Solvent: $CDCl_3$

Assignments
a = 0.90
b = 1.16
c = 1.10-1.60

1804

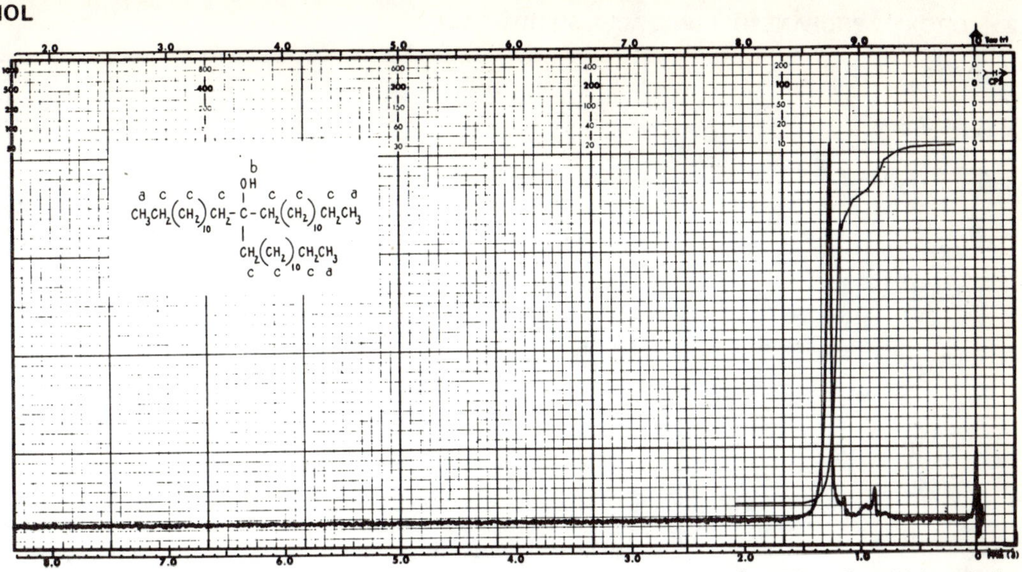

© 1978 Sadtler Research Laboratories, Inc.

2-METHYL-1,1,1-TRIBROMO-2-PROPANOL

$C_4H_7Br_3O$

Mol. Wt. 310.81

M.P. 175-177°C

Solvent: $CDCl_3$

Assignments
a = 1.77
b = 2.62

1805

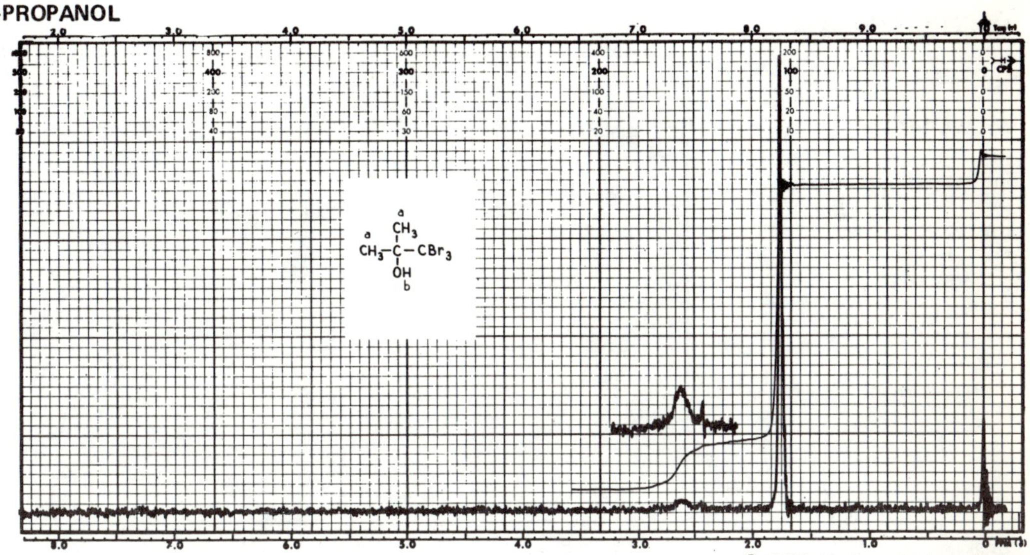

© 1978 Sadtler Research Laboratories, Inc.

2-METHYLLACTONITRILE

C_4H_7NO

Mol. Wt. 85.10

B.P. 75-76°C/20mm

Solvent: $CDCl_3$

Assignments
a = 1.61
b = 3.49

1806

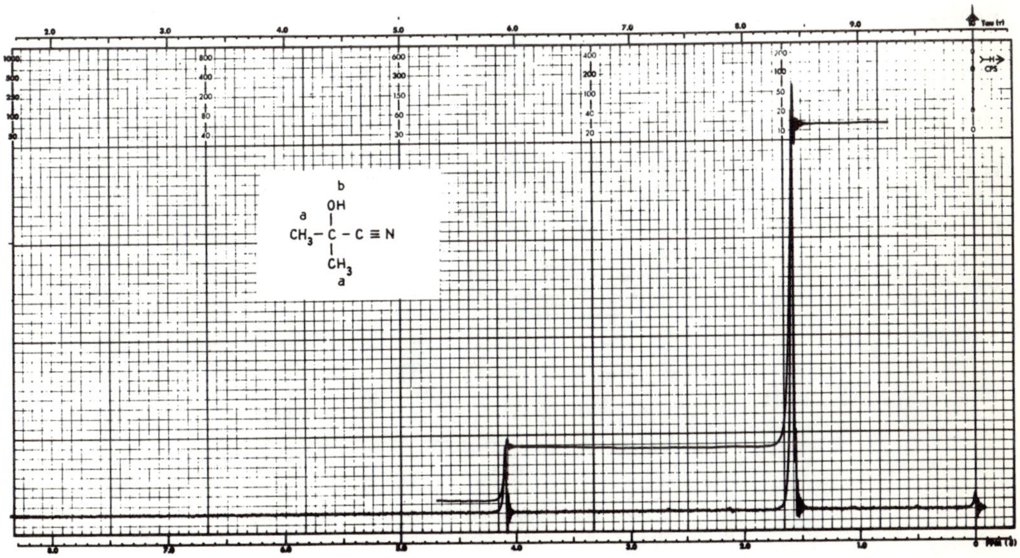

© 1978 Sadtler Research Laboratories, Inc.

2-HYDROXY-2-PROPANESULFONIC ACID, SODIUM SALT

$C_3H_7NaO_4S$

Mol. Wt. 162.14

Solvent: D_2O

Assignments
a = 1.54
b = 4.58 HDO
c = 2.24 imp.

1807

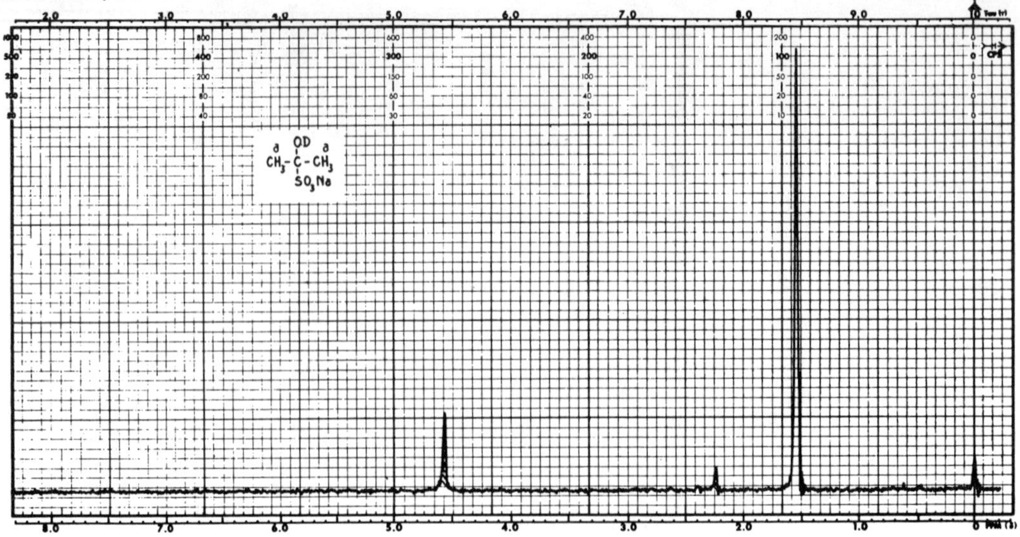

© 1978 Sadtler Research Laboratories, Inc.

p-MENTH-1-EN-8-OL

$C_{10}H_{18}O$

Mol. Wt. 154.25

Solvent: CCl_4

Assignments
a = 1.12
b = 1.24-2.29
c = 1.61
d = 5.30
e = 0.84 imp.
OH unobserved

1808

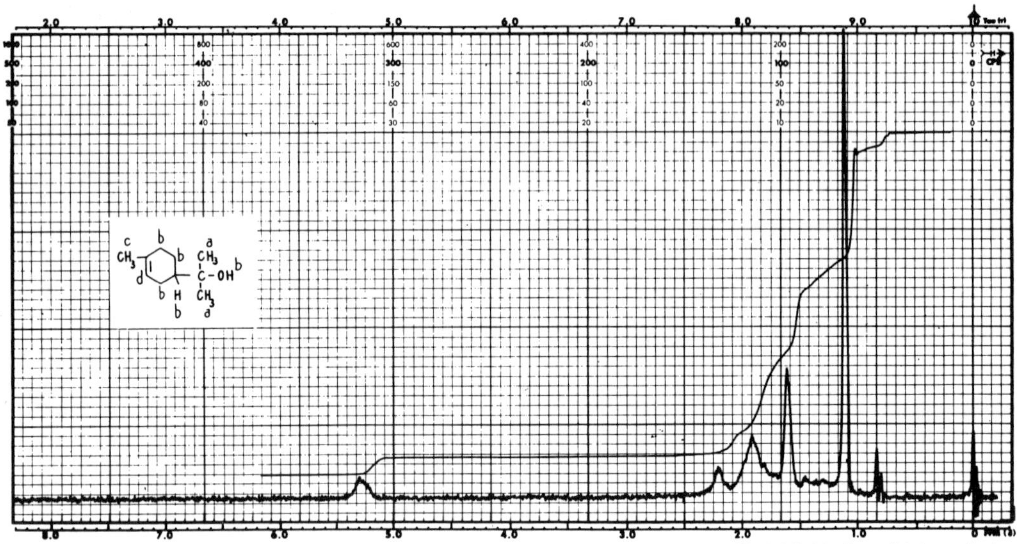

© 1978 Sadtler Research Laboratories, Inc.

1-METHYLCYCLOHEXANOL

$C_7H_{14}O$

Mol. Wt. 114.19

M.P. 26°C

B.P. 168°C/752mm

Solvent: $CDCl_3$

Assignments
a = 1.21
b = 1.33
c = ca 1.51

1809

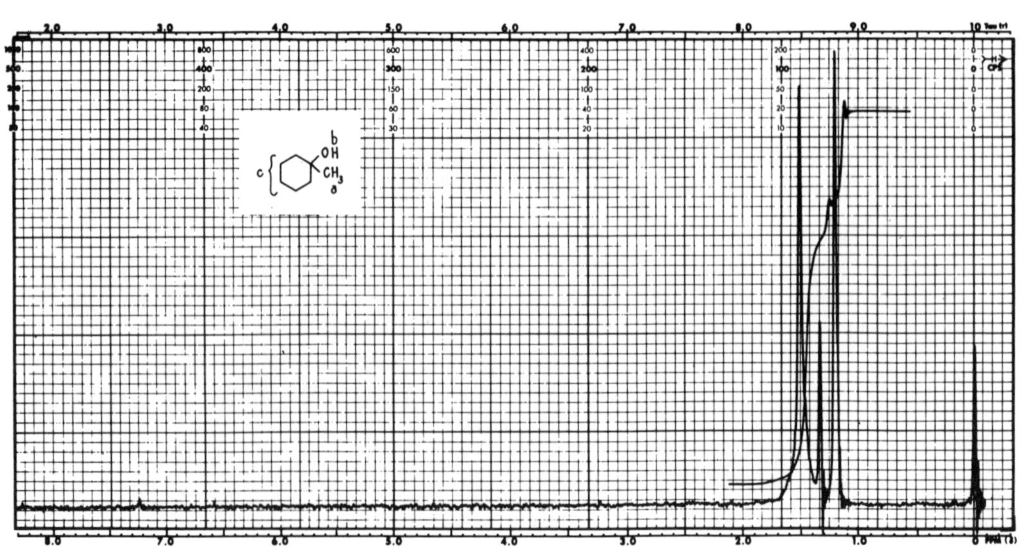

© 1978 Sadtler Research Laboratories, Inc.

1-ETHYLCYCLOHEXANOL

$C_8H_{16}O$

Mol. Wt. 128.22

M.P. 33-36°C

Solvent: CCl_4

Assignments
a = 0.89
b = ca 1.10
c = 1.34

1810

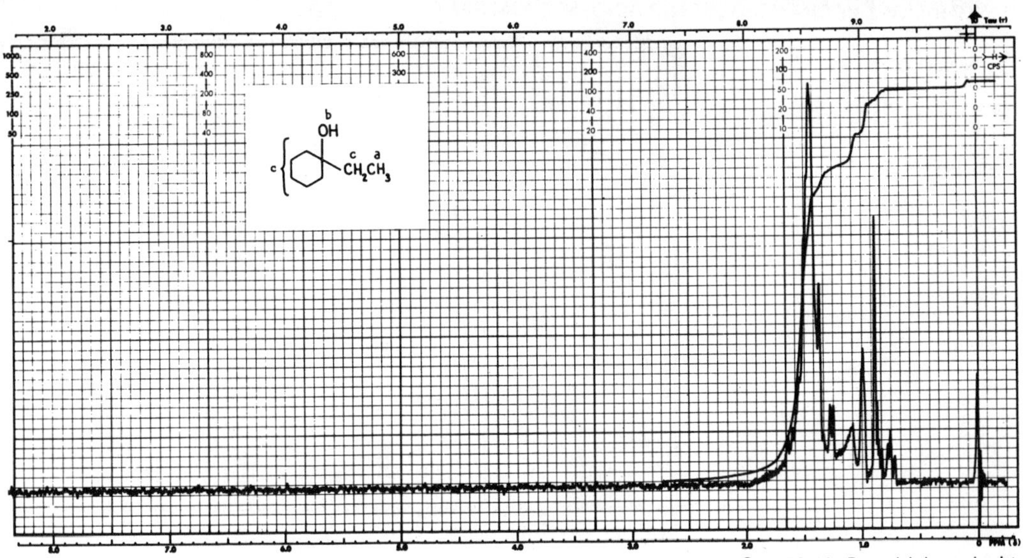

© 1978 Sadtler Research Laboratories, Inc.

1-METHYLCYCLOHEPTANOL

$C_8H_{16}O$

Mol. Wt. 128.22

Solvent: CCl_4

Assignments
a = 1.16
b = 1.54

1811

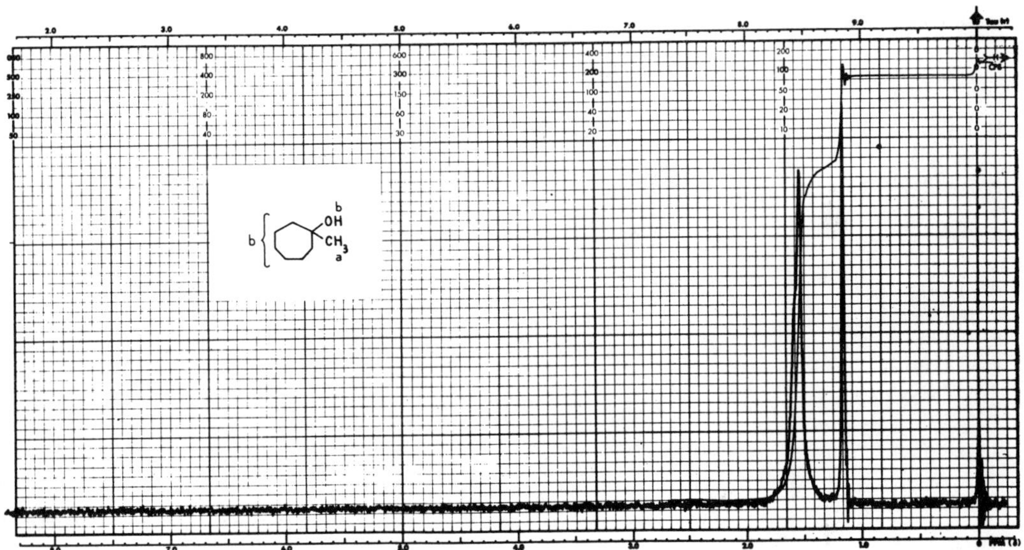

© 1978 Sadtler Research Laboratories, Inc.

2-METHYL-3-BUTEN-2-OL

$C_{22}H_{18}O_4$

Mol. Wt. 86.14

B.P. 96-97.5°C

Solvent: CCl_4

Assignments
a = 1.23
b = 2.14
c = 4.88
d = 5.11
e = 5.92

1812

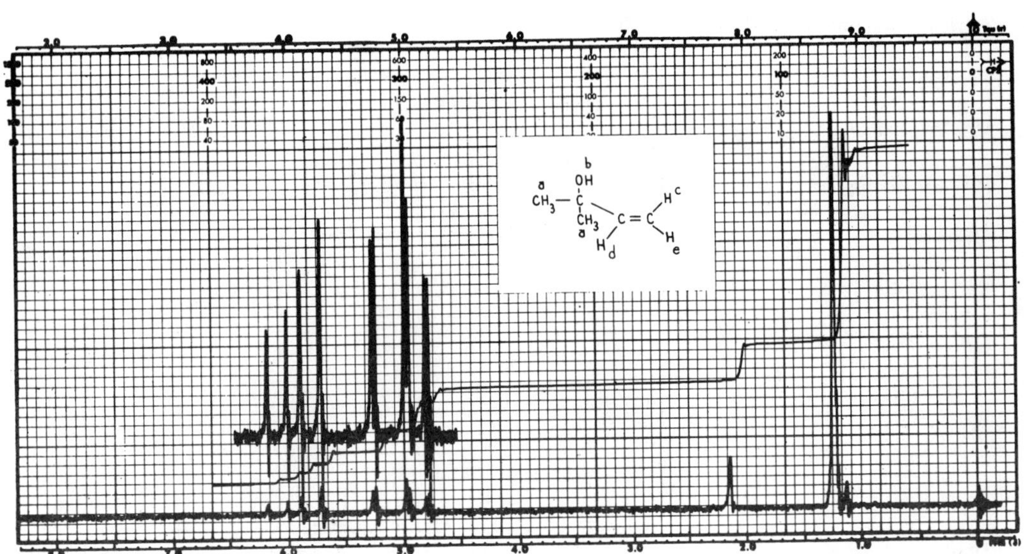

© 1978 Sadtler Research Laboratories, Inc.

2-METHYL-3-BUTYN-2-OL

C_5H_8O

Mol. Wt. 84.12

M.P. 2-3°C

Solvent: $CDCl_3$

Assignments
a = 1.53
b = 2.30
c = 2.41

1813

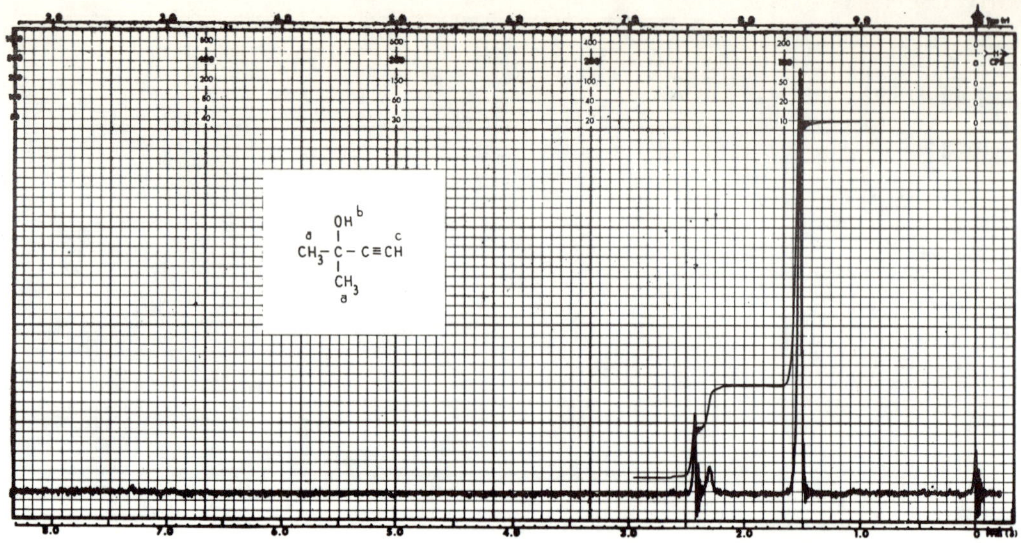

© 1978 Sadtler Research Laboratories, Inc.

3-METHYL-1-PENTYN-3-OL

$C_6H_{10}O$

Mol. Wt. 98.15

B.P. 121-122°C

Solvent: CCl_4

Assignments
a = 1.01
b = 1.41
c = 1.61
d = 2.30
e = 2.70

1814

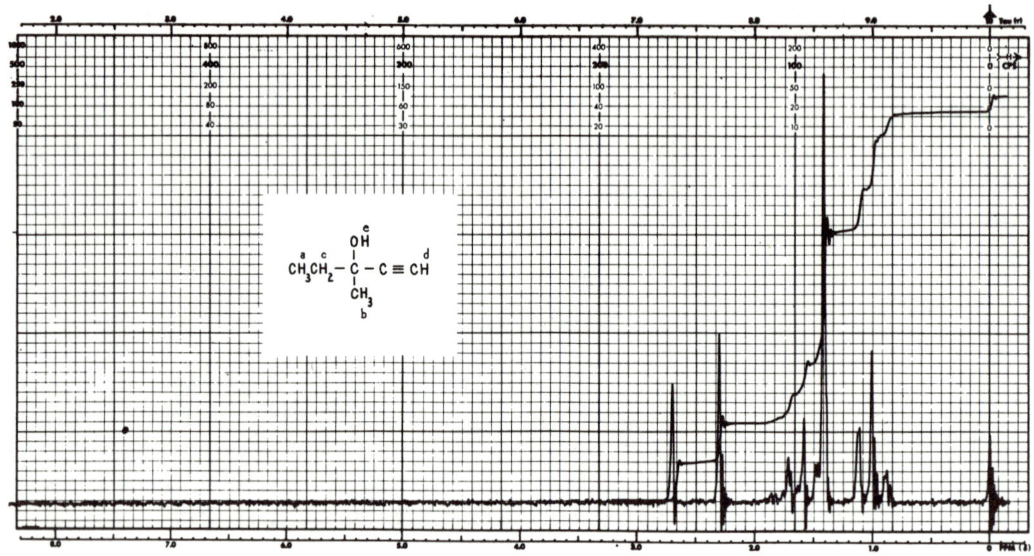

© 1978 Sadtler Research Laboratories, Inc.

2,4,7,9-TETRAMETHYL-5-DECYN-4,7-DIOL

$C_{14}H_{26}O_2$

Mol. Wt. 226.36

Solvent: $CDCl_3$

Assignments
a = 1.01
b = 1.49
c = 1.58
d = 1.93
e = ca 1.95

1815

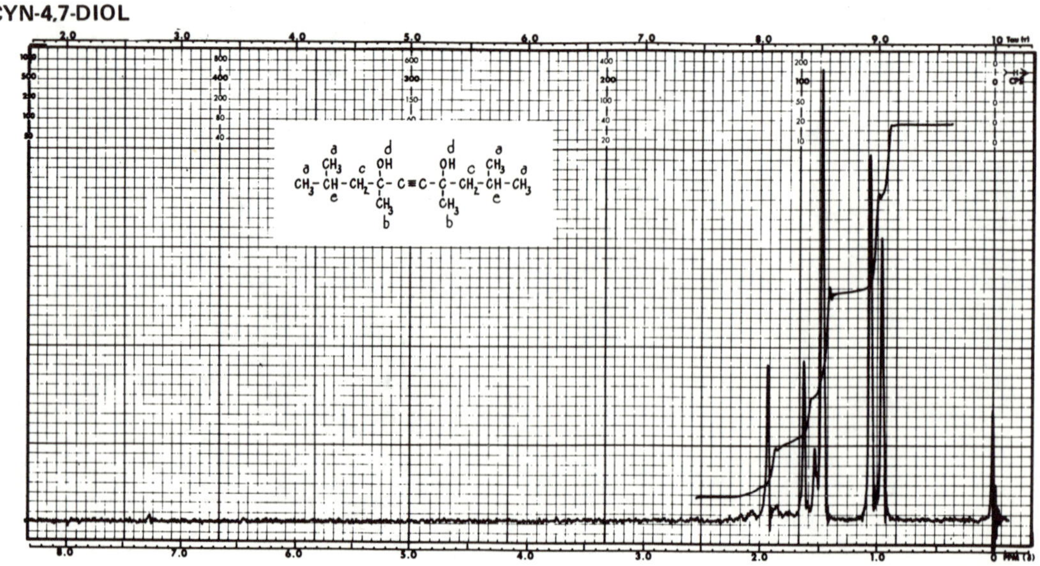

© 1978 Sadtler Research Laboratories, Inc.

1-ETHYNYLCYCLOHEXANOL

$C_8H_{12}O$

Mol. Wt. 124.18

M.P. 31-32°C

B.P. 79-81°C/21mm

Solvent: $CDCl_3$

Assignments
a = 1.00-2.20
b = 2.48
c = 3.08 *
*H_2O present

1816

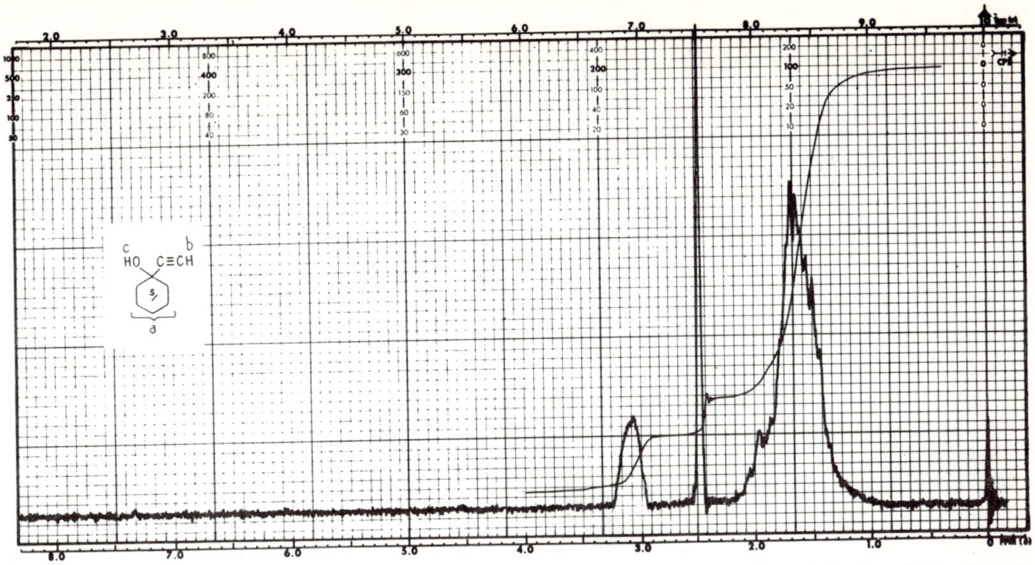

© 1978 Sadtler Research Laboratories, Inc.

***a,a*-DIMETHYLBENZYL ALCOHOL**

$C_9H_{12}O$

Mol. Wt. 136.20

B.P. 62-64°C/2mm

Solvent: $CDCl_3$

Assignments
a = 1.51
b = 2.34
c = 7.15-7.60

1817

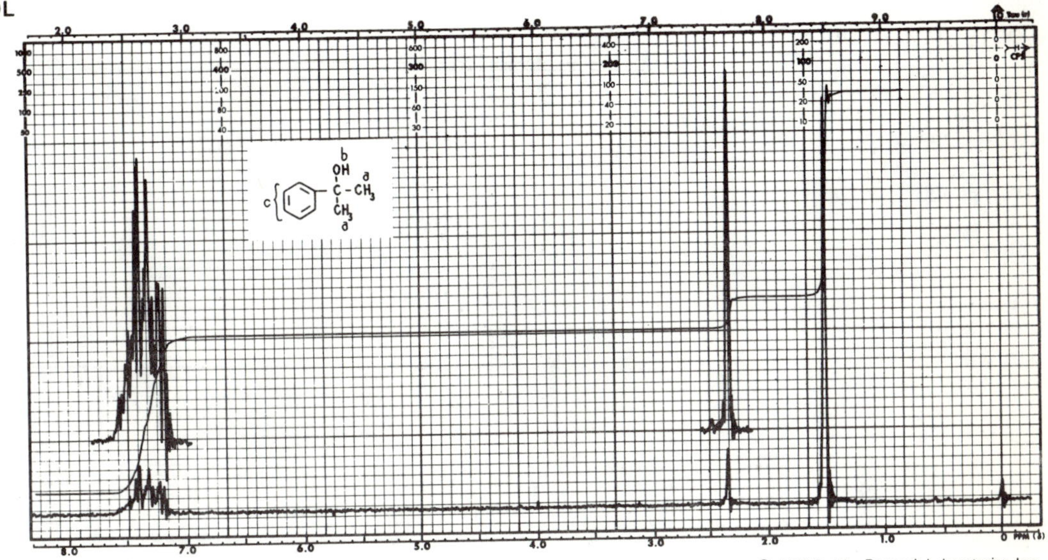

© 1978 Sadtler Research Laboratories, Inc.

***a,a*-DIETHYLBENZYL ALCOHOL**

$C_{11}H_{16}O$

Mol. Wt. 164.25

B.P. 80°C/2mm

Solvent: $CDCl_3$

Assignments
a = 0.77
b = 1.79
c = 1.82
d = 7.10-7.60

1818

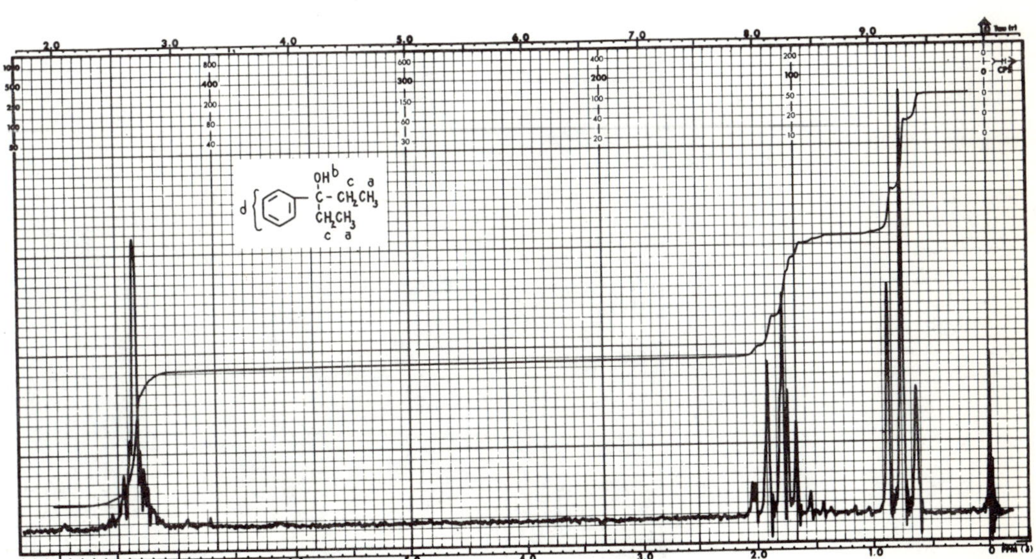

© 1978 Sadtler Research Laboratories, Inc.

2-PHENYL-3-BUTYN-2-OL

$C_{10}H_{10}O$

Mol. Wt. 146.18

M.P. 49°C

Solvent: $CDCl_3$

Assignments
a = 1.75
b = 2.63
c = 2.72
d = 7.36
e = 7.68
Trace of acid added for OH

1819

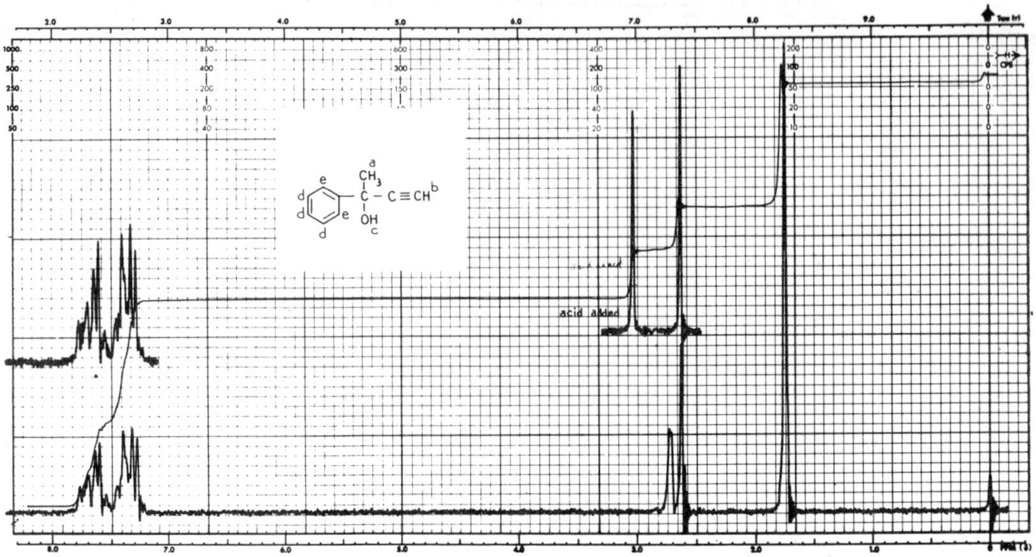

© 1978 Sadtler Research Laboratories, Inc.

α-**OCTYLBENZHYDROL**

$C_{21}H_{28}O$

Mol. Wt. 296.46

B.P. 194-200°C/0.02mm

Solvent: CCl_4

Assignments
a = 0.88
b = 1.00-1.60
c = 1.87
d = 2.15
e = 7.00-7.40

1820

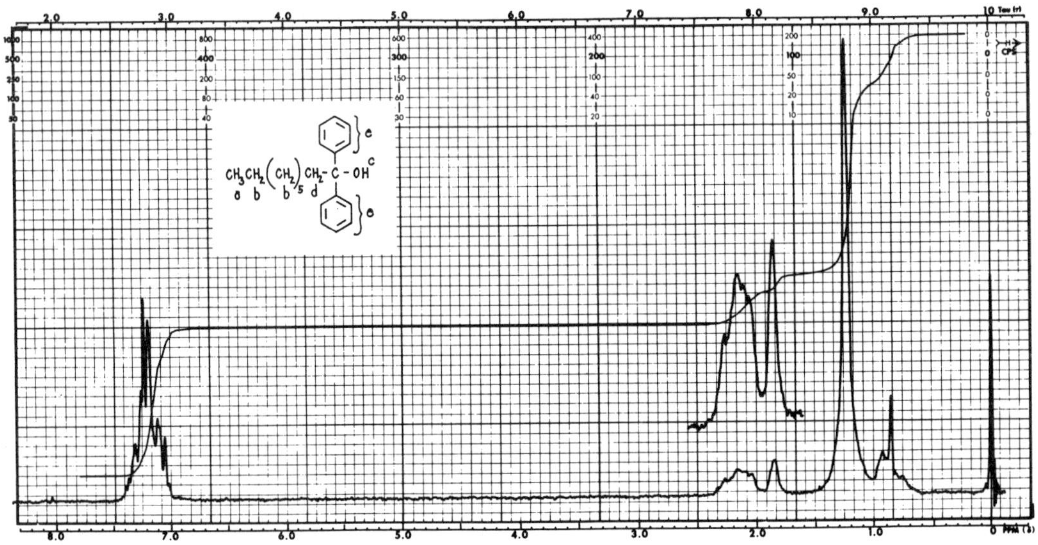

© 1978 Sadtler Research Laboratories, Inc.

1-(p-TOLYL)CYCLOHEXANOL

$C_{13}H_{18}O$

Mol. Wt. 190.29

Solvent: $CDCl_3$

Assignments
a = 1.30-2.10
b = 2.34
c = 2.36
d = 7.11
e = 7.36

1821

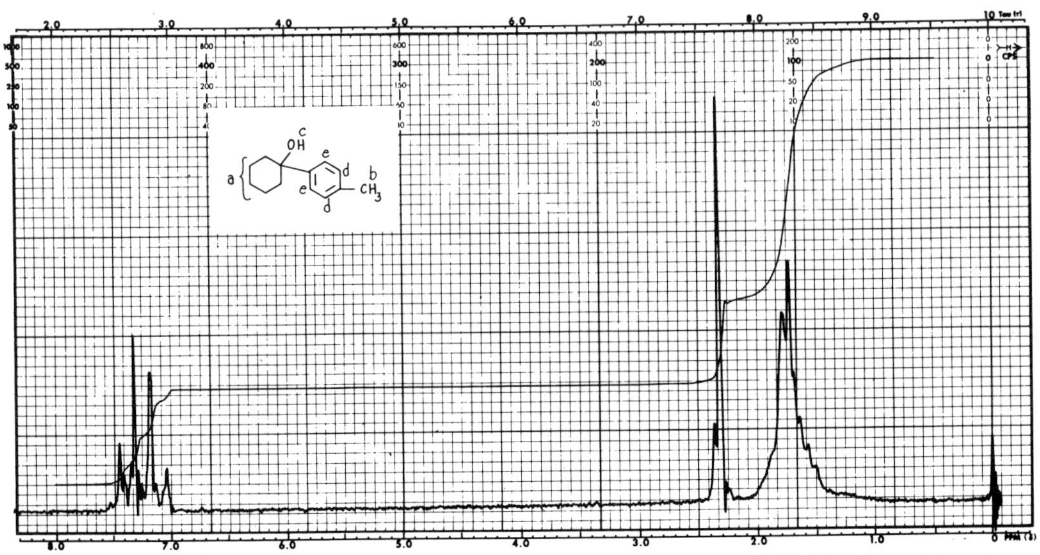

© 1978 Sadtler Research Laboratories, Inc.

TRIPHENYLMETHANOL

$C_{19}H_{16}O$

Mol. Wt. 260.34

M.P. 161-162°C

Solvent: $CDCl_3$

Assignments
a = ca 2.70
b = 7.19

1822

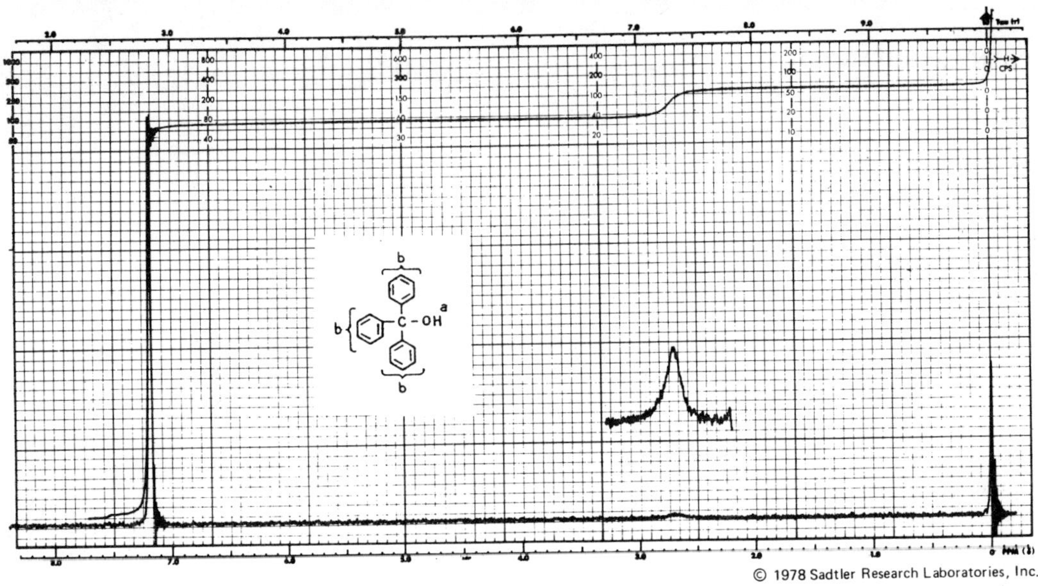

© 1978 Sadtler Research Laboratories, Inc.

NOTES

THE POLYOLS

The polyols are one of the groups of compounds for which the proton NMR integration ratios are most useful in determining their molecular structure. The polyols characteristically display the resonance of two or more exchangeable protons with three or more hydrocarbon groups resonating at low field in the range from 3 to 4 ppm. The straight chain polyols, and their cyclic counterparts, possess no methyl absorption bands at high field making their identification somewhat easier than the corresponding branched chain compounds.

A series of 1,n-diols are listed below.

Chemical Shifts

1,n-Alkanediols

	x	$(CH_2)_x$	$-(CH_2$	$-OH)_2$	Solvent	Spectrum
Ethylene Glycol	0	- - - -	3.68	- - - -	D_2O	1823
1,3-Propanediol	1	1.78	3.65	- - - -	D_2O	1824
1,4-Butanediol	2	1.43	3.39	4.34	DMSO-d_6	- - - -
1,5-Pentanediol	3	1.2–1.9	3.60	- - - -	D_2O	1825
1,6-Hexanediol	4	1.1–1.8	3.65	4.52	$CDCl_3$	1826
1,9-Nonanediol	7	1.1–1.8	3.60	3.05	$CDCl_3$	- - - -
1,16-Hexadecanediol	14	ca 1.32	3.40	3.92	DMSO-d_6	1827

Coupling and Coupling Constants

As with the primary and secondary alcohols, coupling between the hydroxyl group and adjacent methylene or methine groups may or may not be observed depending to a great extent on the solvent employed. In D_2O solution, of course, the hydroxyl proton will be replaced by deuterium and no coupling will be observed. In acetone and dimethylsulfoxide solutions coupling between the OH and adjacent aliphatic groups is usually clearly observed (J_{CH-OH} = 4–6 Hz) as in spectrum 1860. In carbon tetrachloride and deuterochloroform solutions, the coupling across the oxygen group of the polyols is usually not observed or at most, both groups may be badly broadened.

Solubility and Solvent Effects

With the exception of the very high molecular weight varieties, the polyols are readily soluble in one of the usually employed NMR solvents. The compounds containing approximately equal numbers of carbon and oxygen atoms are soluble in D_2O. At ratios of about three carbons per hydroxyl group, acetone or DMSO-d_6 will be found to be most effective. When the ratio of carbon to oxygen atoms exceeds 7 or 8, $CDCl_3$ and CCl_4 become more useful in dissolving such compounds containing large hydrocarbon fragments.

ETHYLENE GLYCOL

$C_2H_6O_2$

Mol. Wt. 62.07

M.P. −17°C

B.P. 197.4

Solvent: D_2O

Assignments
a = 3.68
b = 4.68 HDO

1823

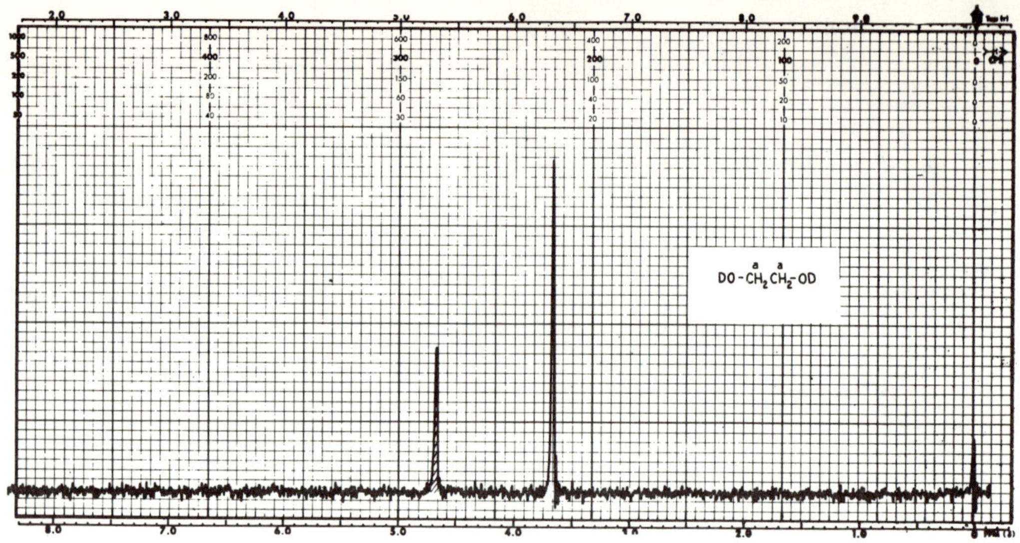

© 1978 Sadtler Research Laboratories, Inc.

1,3-PROPANEDIOL

$C_3H_8O_2$

Mol. Wt. 76.10

B.P. 214°C (lit.)

Solvent: D_2O

Assignments
a = 1.78
b = 3.65
c = 4.65 HDO

1824

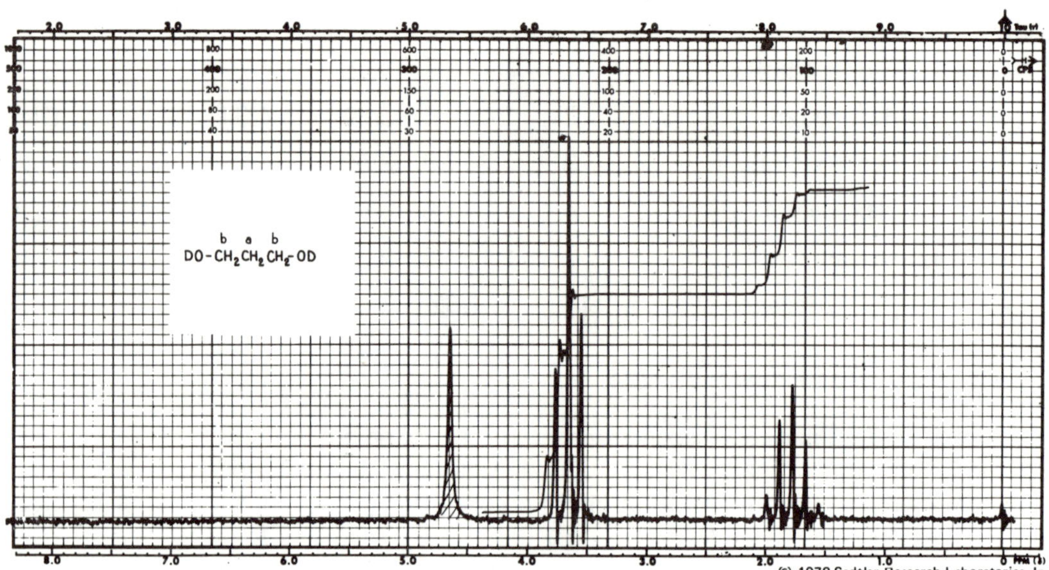

© 1978 Sadtler Research Laboratories, Inc.

1,5-PENTANEDIOL

$C_5H_{12}O_2$

Mol. Wt. 104.15

B.P. 239°C

Solvent: D_2O

Assignments
a = ca 1.49
b = 3.60
c = 4.63 HDO

1825

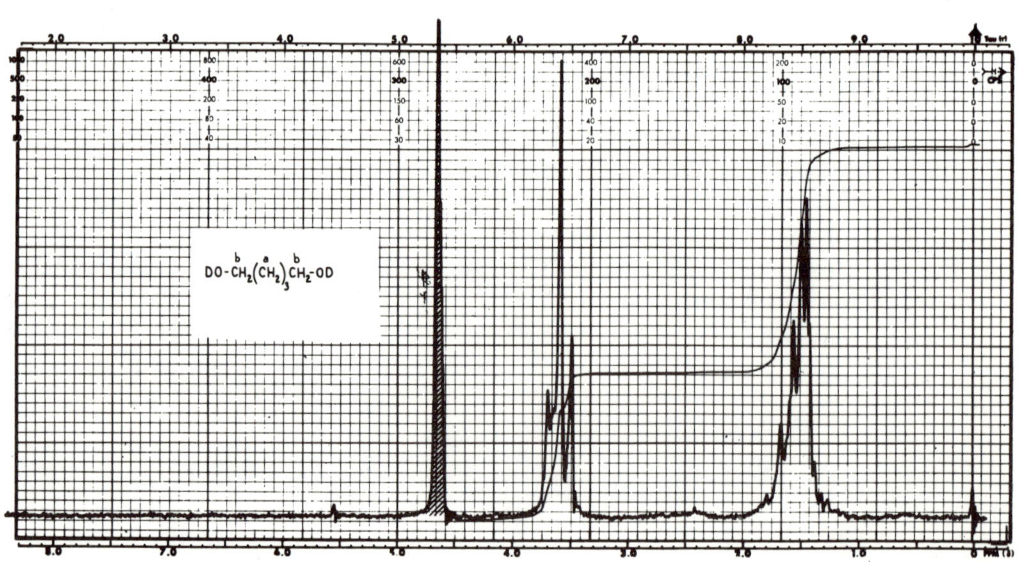

© 1978 Sadtler Research Laboratories, Inc.

1,6-HEXANEDIOL

$C_6H_{14}O_2$

Mol. Wt. 118.18

M.P. 40-42°C

Solvent: $CDCl_3$

Assignments
a = 1.10-1.80
b = 3.56
c = 4.52

1826

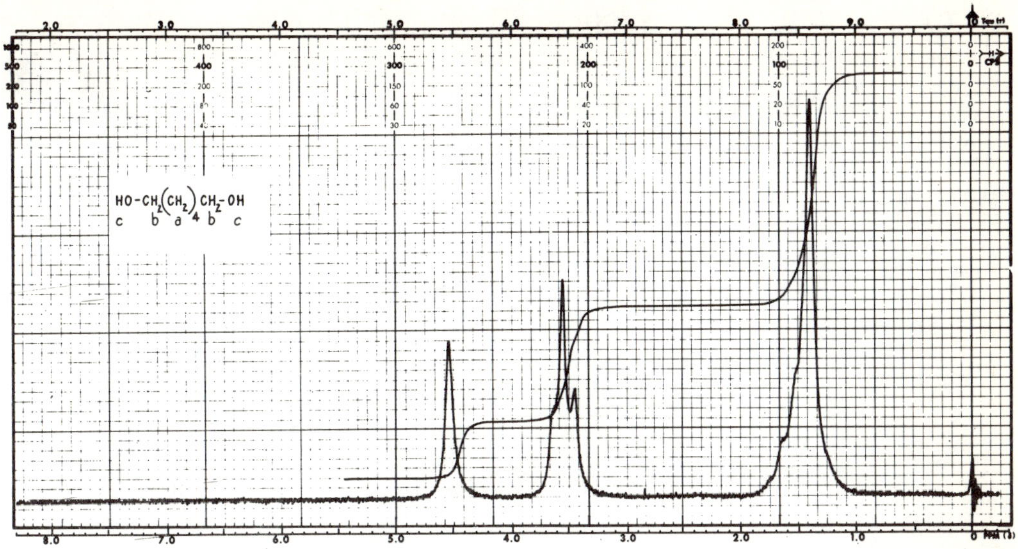

© 1978 Sadtler Research Laboratories, Inc.

1,16-HEXADECANEDIOL

$C_{16}H_{34}O_2$

Mol. Wt. 258.45

Solvent: DMSO-d_6

Assignments
a = ca 1.31
b = 3.40
c = 3.92
d = 2.50 DMSO-d_5
e = 3.00 H_2O

1827

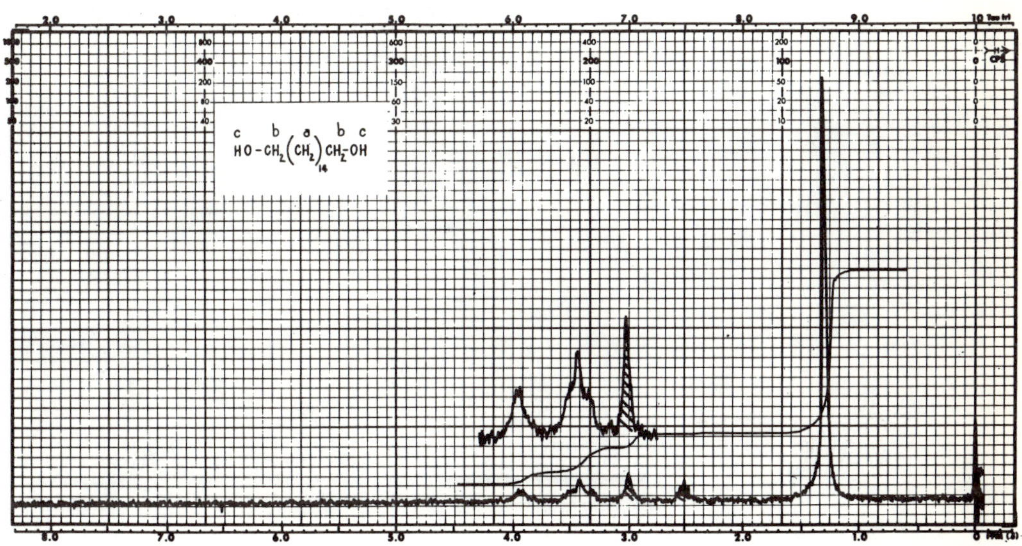

© 1978 Sadtler Research Laboratories, Inc.

1,2-PROPANEDIOL

$C_3H_8O_2$

Mol. Wt. 76.10

B.P. 187.2°C

Solvent: $CDCl_3$

Assignments
a = 1.13
b = 3.37
c = 3.50
d = 3.80
e = 4.38
Trace of acid added

1828

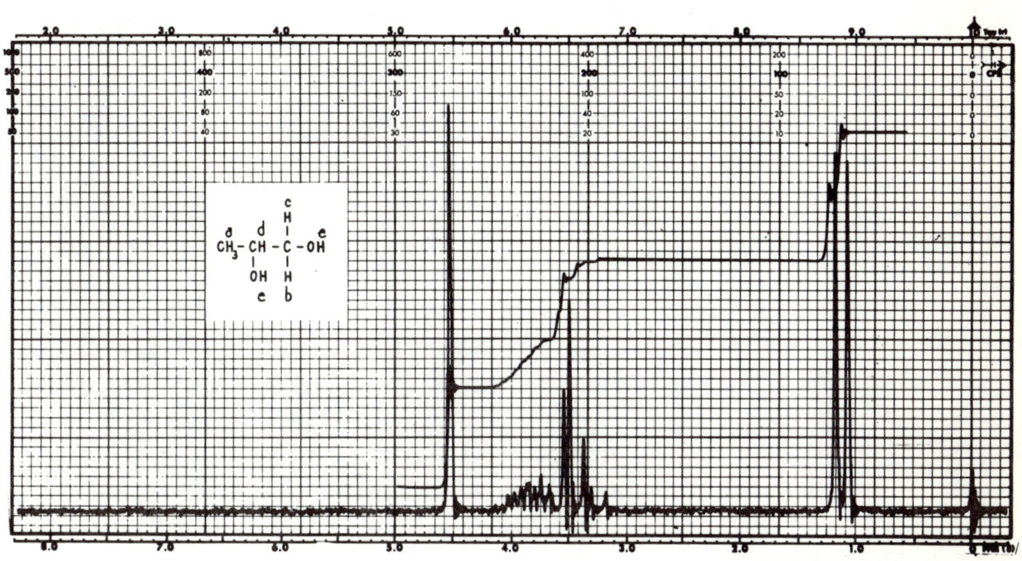

© 1978 Sadtler Research Laboratories, Inc.

1,2-BUTANEDIOL

$C_4H_{10}O_2$

Mol. Wt. 90.12

Solvent: $CDCl_3$

Assignments
a = 0.97
b = 1.10-1.80
c = 3.10-3.80
d = 4.30

1829

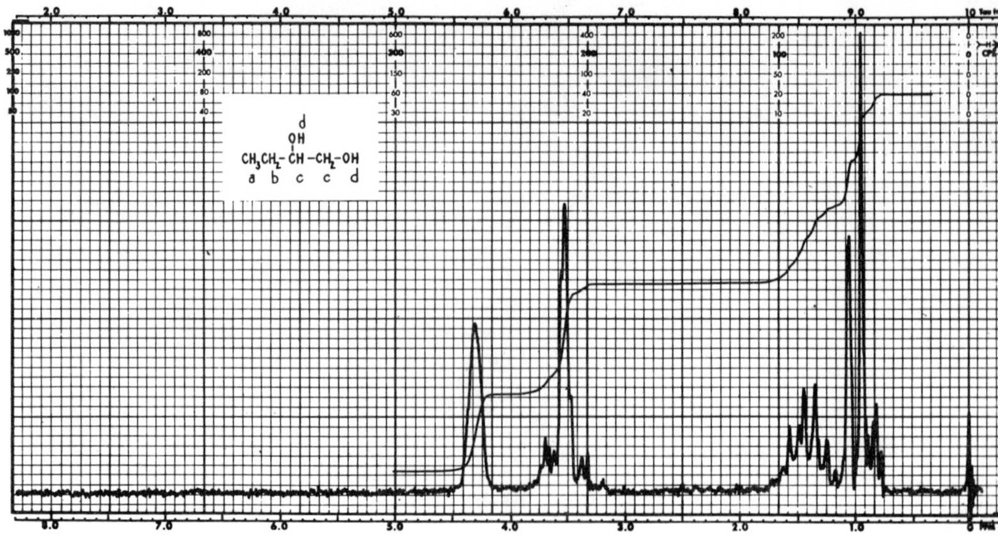

© 1978 Sadtler Research Laboratories, Inc.

2,4-PENTANEDIOL

$C_5H_{12}O_2$

Mol. Wt. 104.15

B.P. 198-201°C (lit.)

Solvent: CCl_4

Assignments
a = 1.19
b = 1.49
c = 3.95
d = 4.51-4.88

1830

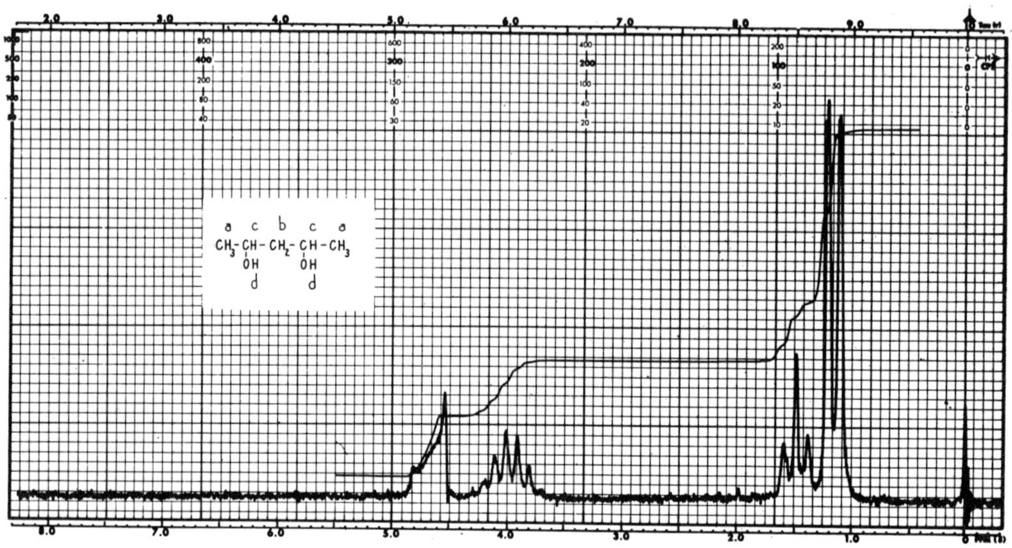

© 1978 Sadtler Research Laboratories, Inc.

2-METHYL-2,4-PENTANEDIOL

$C_6H_{14}O_2$

Mol. Wt. 118.18

B.P. 195-196°C

Solvent: CCl_4

Assignments
a = 1.12
b = 1.18
c = 1.23
d = 1.41
e = 1.55
f = 4.16
g = 4.72
h = 4.87
Trace of D_2O added

1831

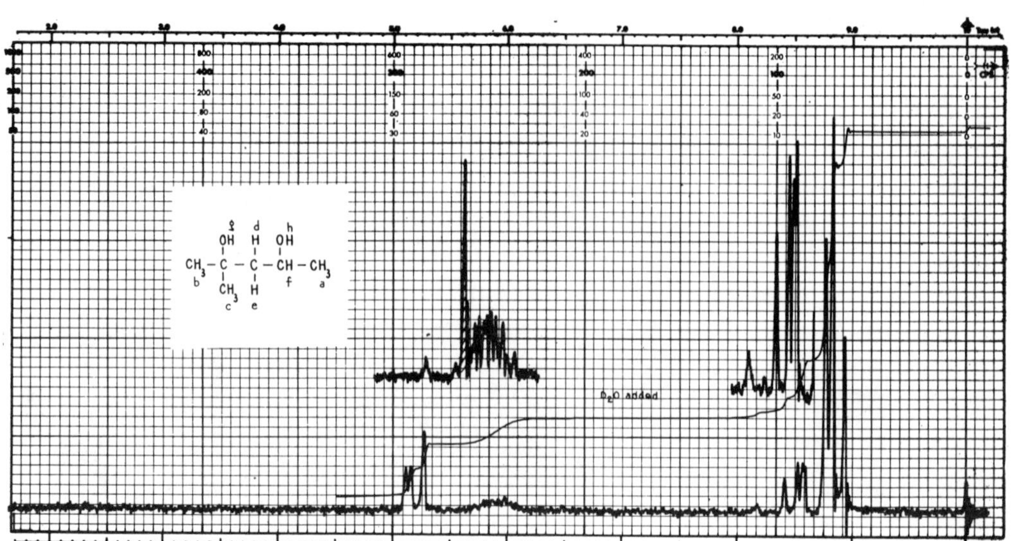

© 1978 Sadtler Research Laboratories, Inc.

3-METHYL-1,5-PENTANEDIOL

$C_6H_{14}O_2$

Mol. Wt. 118.18

Solvent: $CDCl_3$

Assignments
a = 0.92
b = 1.53
c = 3.47
d = 3.65
Trace of D_2O added

1832

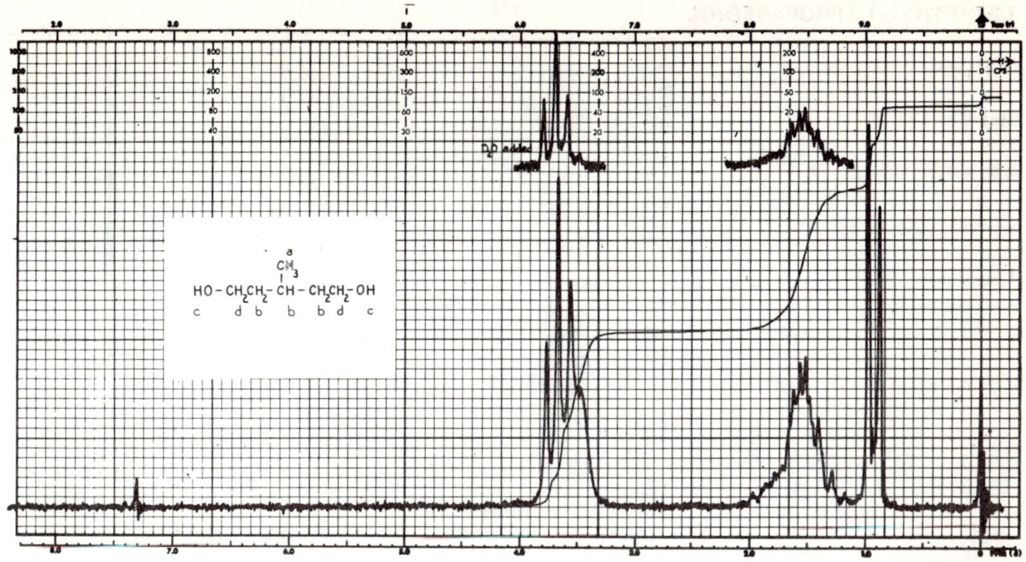

© 1978 Sadtler Research Laboratories, Inc.

2,5-HEXANEDIOL

$C_6H_{14}O_2$

Mol. Wt. 118.17

B.P. 220-222°C

Solvent: CCl_4

Assignments
a = 1.15
b = 1.48 or c
c = 1.55 or b
d = 3.75
e = 4.28

1833

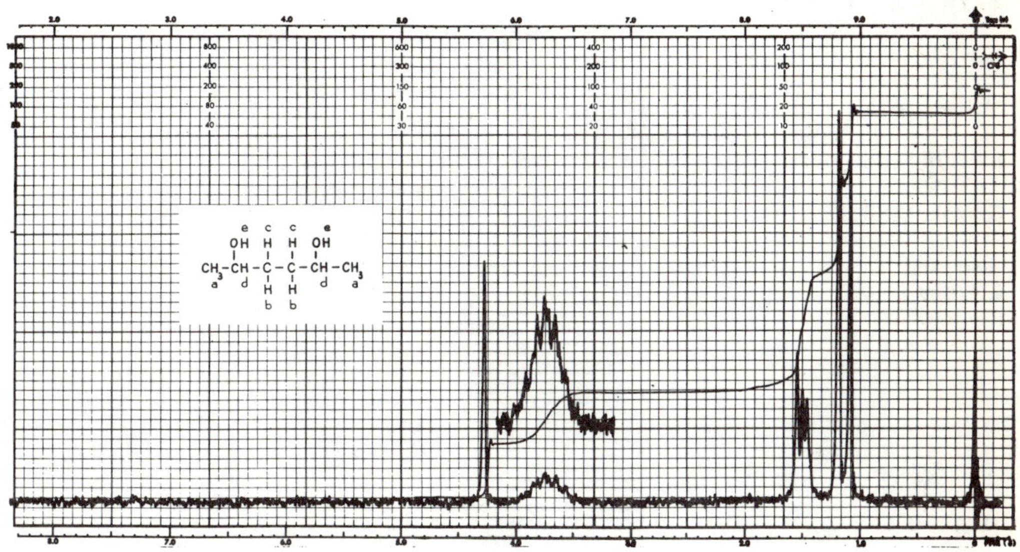

© 1978 Sadtler Research Laboratories, Inc.

2,3-DIMETHYL-2,3-BUTANEDIOL, HYDRATE

$C_6H_{14}O_2 \cdot H_2O$

Mol. Wt. 136.19

M.P. 46-48°C

Solvent: $CDCl_3$

Assignments
a = 1.26
b = 2.10

1834

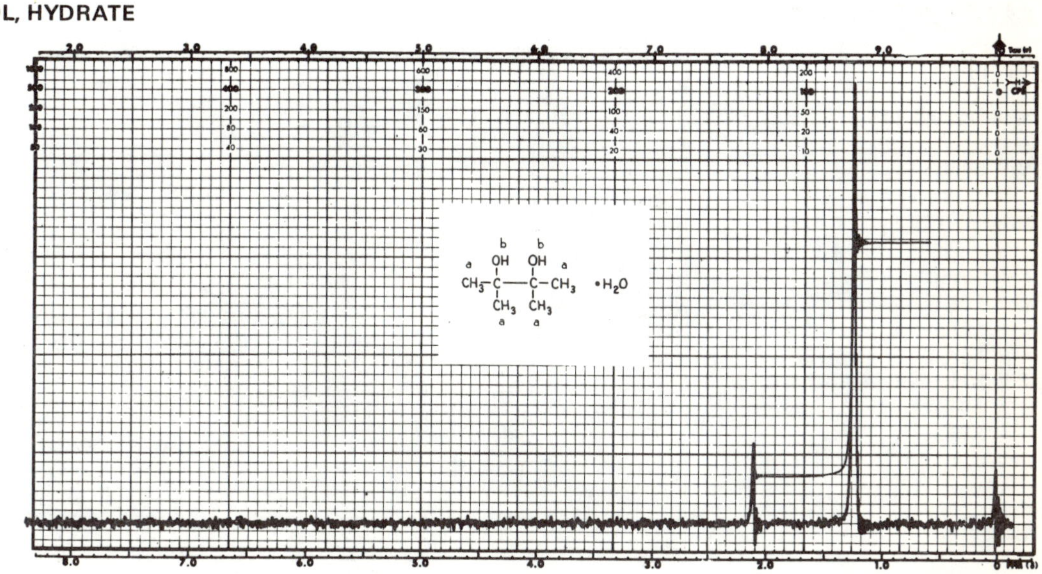

© 1978 Sadtler Research Laboratories, Inc.

2,2-DIETHYL-1,3-PROPANEDIOL

$C_7H_{16}O_2$

Mol. Wt. 132.20

B.P. 125°C/10mm

Solvent: $CDCl_3$

Assignments
a = 0.83
b = 1.29
c = 3.50
d = 4.09

1835

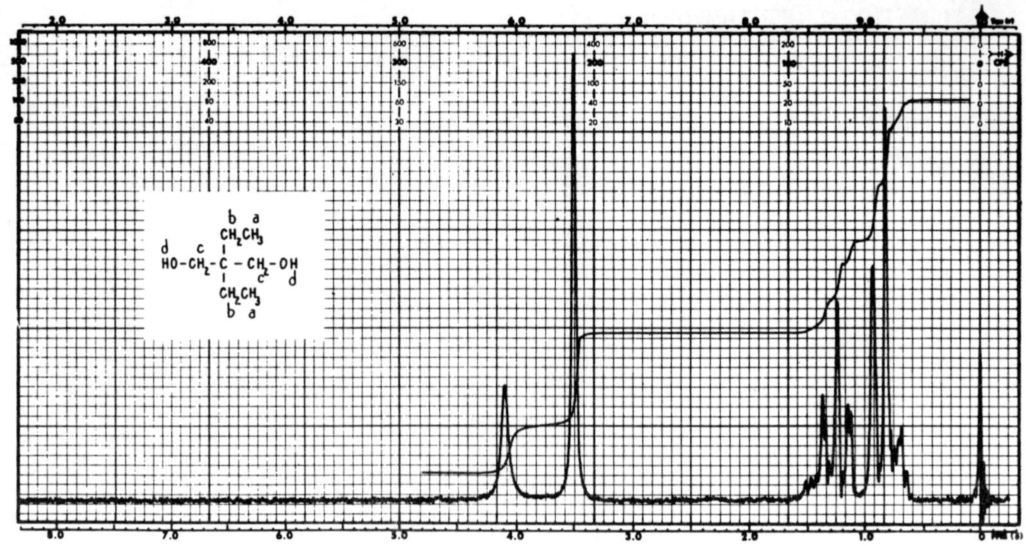

© 1978 Sadtler Research Laboratories, Inc.

2-ETHYL-1,3-HEXANEDIOL

$C_8H_{18}O_2$

Mol. Wt. 146.23

B.P. 129-131°C/10mm

Solvent: $CDCl_3$

Assignments
a = 0.95
b = 1.00-1.80
c = 3.40-4.20

1836

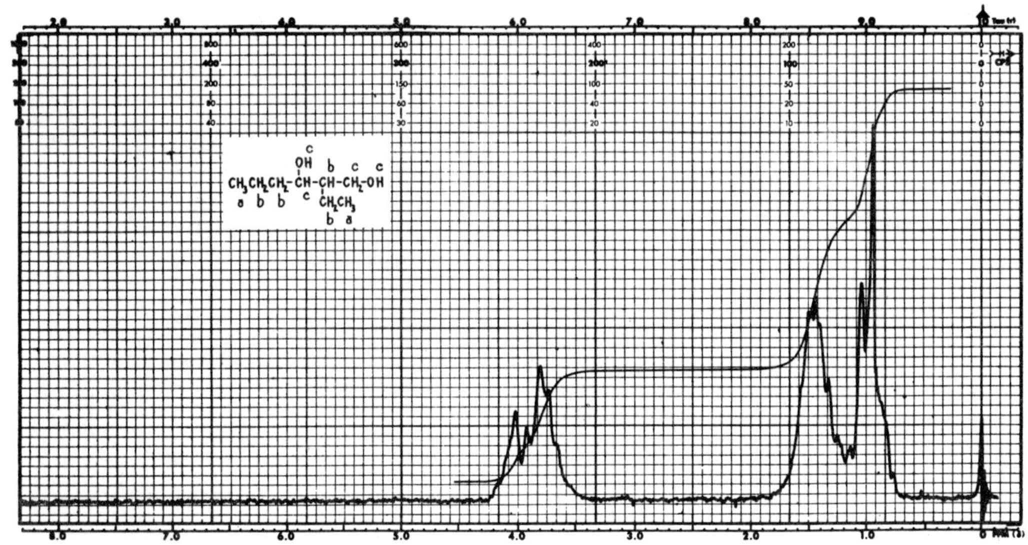

© 1978 Sadtler Research Laboratories, Inc.

2,4,4-TRIMETHYL-1,2-PENTANEDIOL

$C_8H_{18}O_2$

Mol. Wt. 146.23

M.P. 59°C

Solvent: $CDCl_3$

Assignments
a = 1.02
b = 1.27
c = 1.48
d = 2.15
e = 2.60
f = 3.39

1837

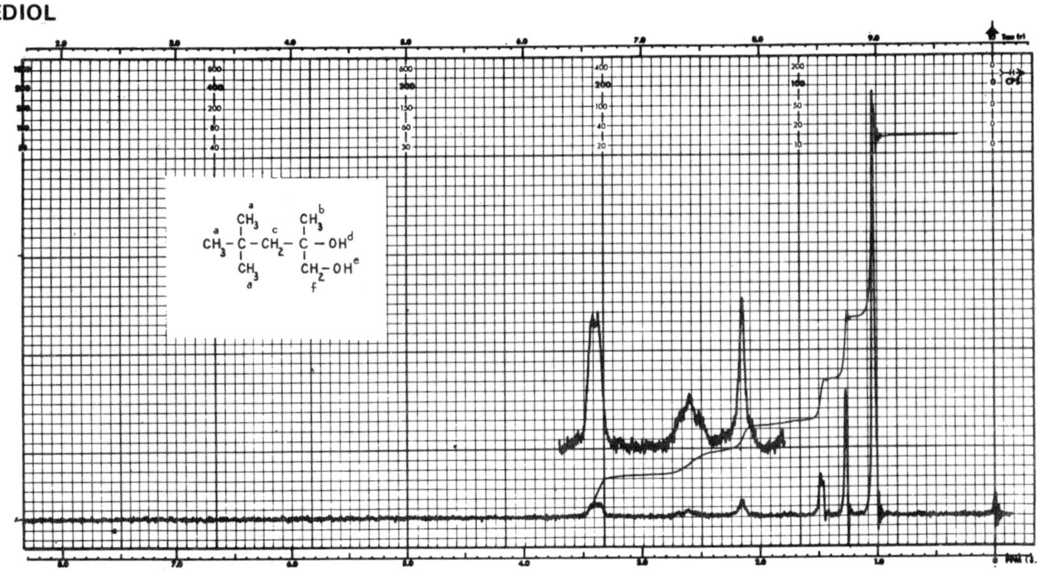

© 1978 Sadtler Research Laboratories, Inc.

2,4,4-TRIMETHYL-2,3-PENTANEDIOL

$C_8H_{18}O_2$

Mol. Wt. 146.23

M.P. 65°C

Solvent: $CDCl_3$

Assignments
a = 1.02
b = 1.30
c = 2.29
d = 2.58
e = 3.15

1838

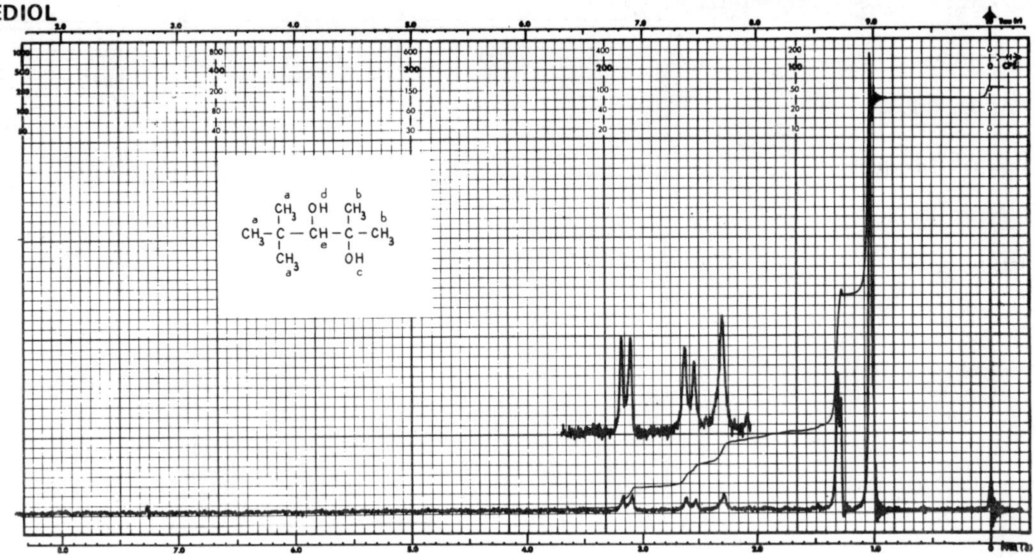

© 1978 Sadtler Research Laboratories, Inc.

2-BUTYL-2-ETHYL-1,3-PROPANEDIOL

$C_9H_{20}O_2$

Mol. Wt. 160.26

M.P. 39-44°C

Solvent: Acetone-D_6

Assignments
a = 0.83
b = 0.93
c = 1.00-1.60
d = 3.47
e = 3.71
f = 2.09 Acetone-D_5
g = 3.07 H_2O

1839

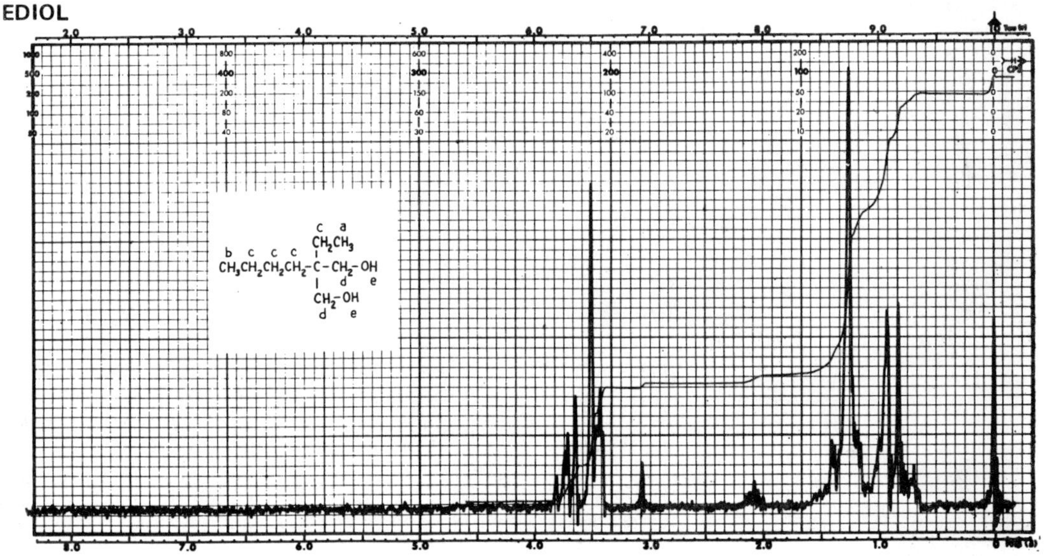

© 1978 Sadtler Research Laboratories, Inc.

3-CHLORO-1,2-PROPANEDIOL

$C_3H_7ClO_2$

Mol. Wt. 110.54

B.P. 115-117°C/11mm

Solvent: D_2O

Assignments
a = ca 3.66
b = 3.88
c = 4.64 HDO

1840

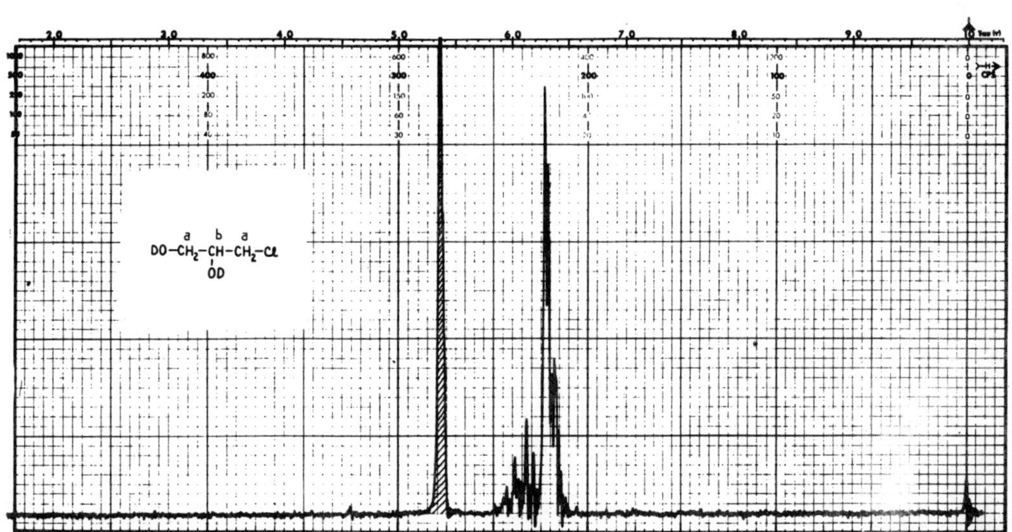

© 1978 Sadtler Research Laboratories, Inc.

2,2-BIS(BROMOMETHYL)-1,3-PROPANEDIOL

$C_5H_{10}Br_2O_2$

Mol. Wt. 261.95

Solvent: Polysol-d

Assignments
a = 3.56
b = 3.68
c = 3.93
d = 3.58 imp.

1841

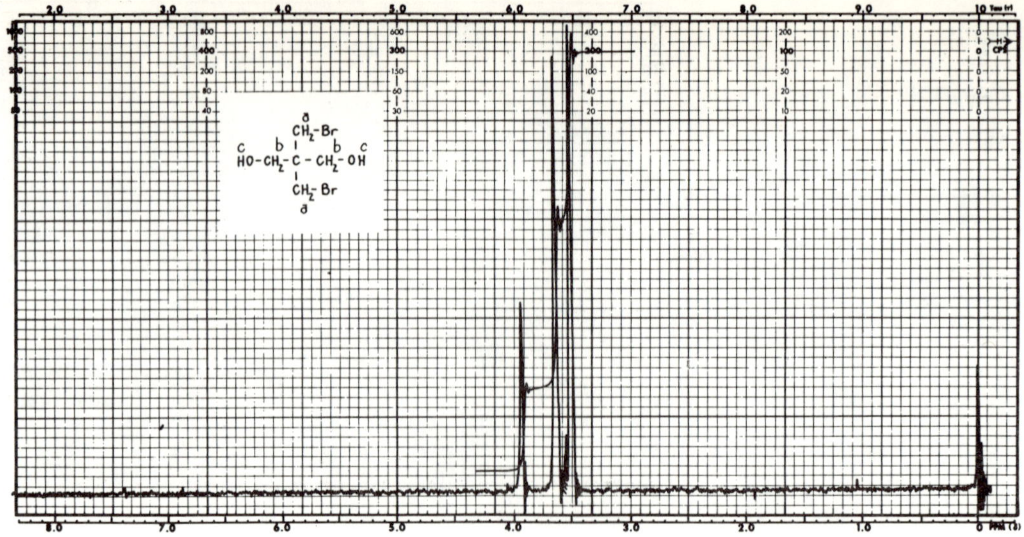

© 1978 Sadtler Research Laboratories, Inc.

2,2'-IMINODIETHANOL

$C_4H_{11}NO_2$

Mol. Wt. 105.14

M.P. 28°C

Solvent: D_2O

Assignments
a = 2.74
b = 3.69
c = 4.70 HDO

1842

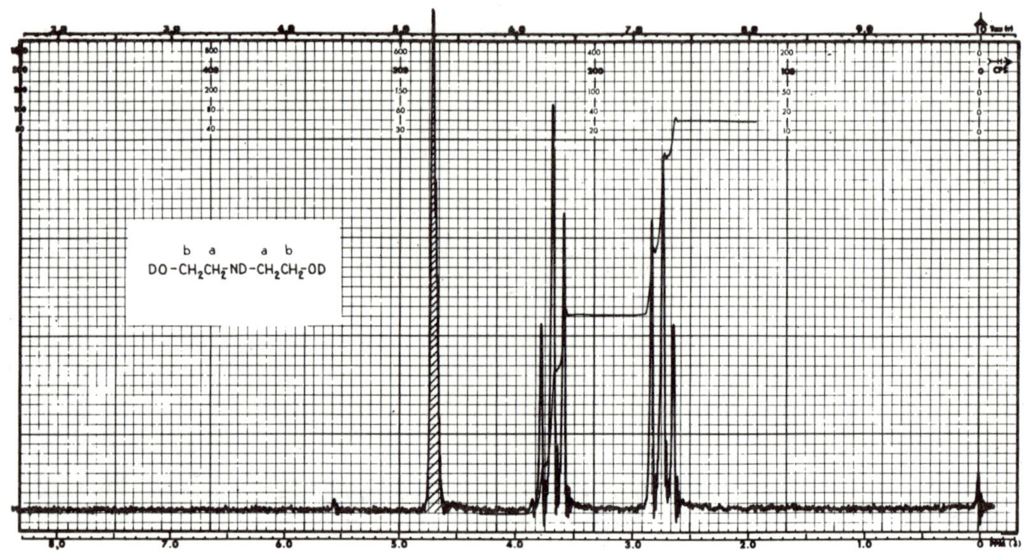

© 1978 Sadtler Research Laboratories, Inc.

2,2'-(ETHYLENEDIIMINO)DIETHANOL

$C_6H_{16}N_2O_2$

Mol. Wt. 148.21

B.P. 135-137°C/1mm

M.P. 96-99°C

Solvent: D_2O

Assignments
a = 2.72
b = 3.69
c = 4.69 HDO

1843

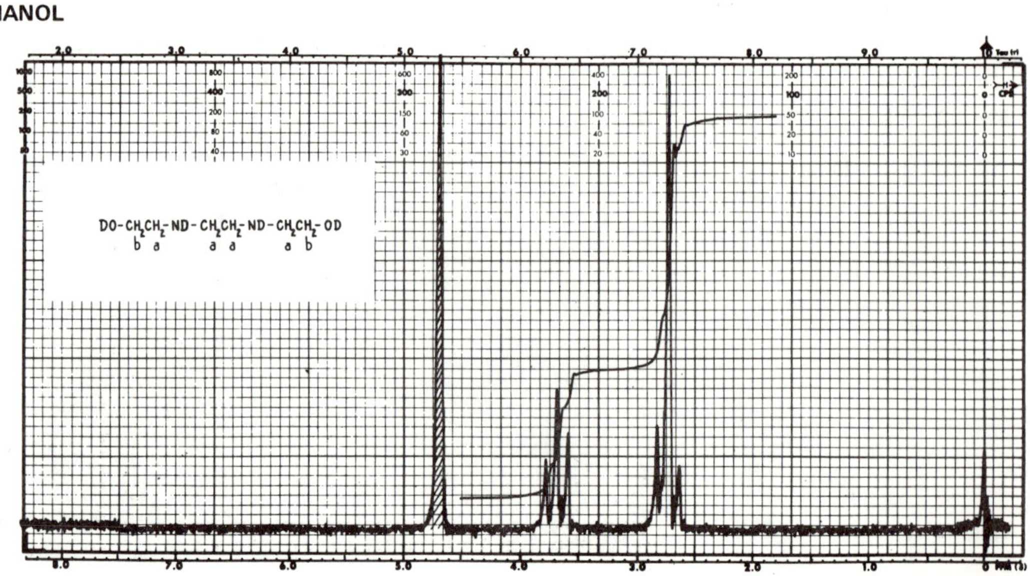

© 1978 Sadtler Research Laboratories, Inc.

3-AMINO-1,2-PROPANEDIOL

$C_3H_9NO_2$

Mol. Wt. 91.11

Solvent: D_2O

Assignments
a = 2.63
b = 3.30-3.90
c = 4.63 HDO

1844

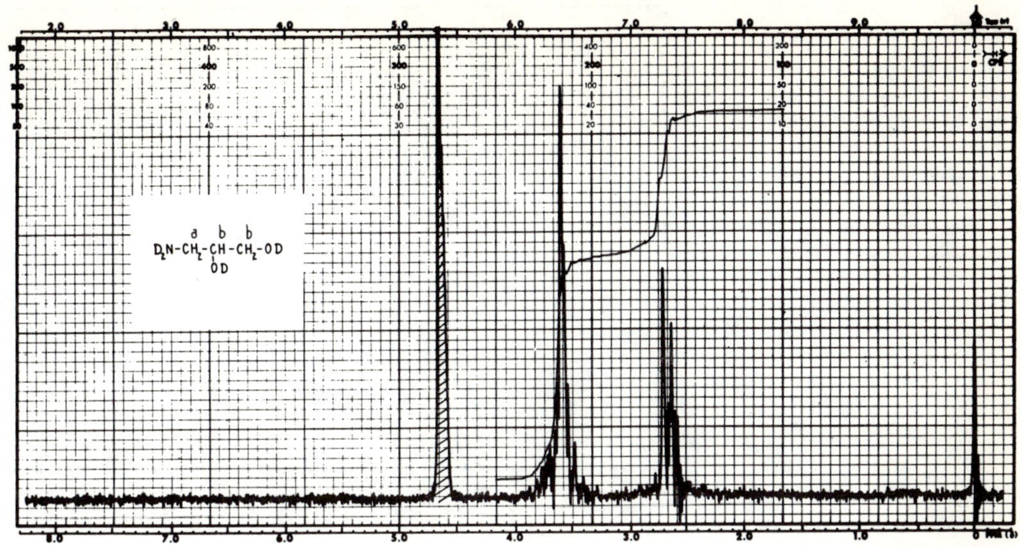

© 1978 Sadtler Research Laboratories, Inc.

DIETHYLENE GLYCOL

$C_4H_{10}O_3$

Mol. Wt. 106.12

B.P. 245°C

Solvent: $CDCl_3$

Assignments
a = 3.65
b = 4.22

1845

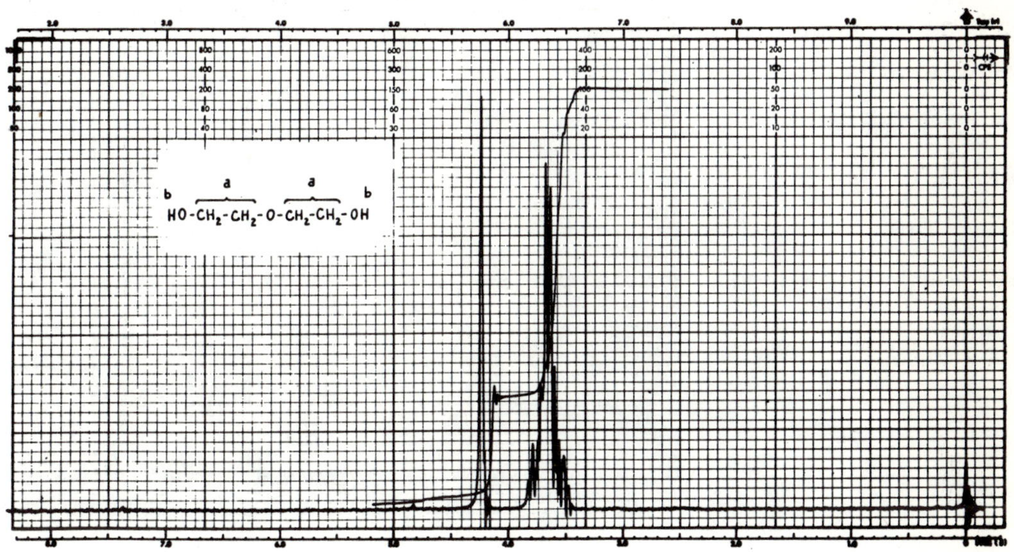

© 1978 Sadtler Research Laboratories, Inc.

TRIETHYLENE GLYCOL

$C_6H_{14}O_4$

Mol. Wt. 150.18

B.P. 284.4°C

Solvent: $CDCl_3$

Assignments
a = 3.48
b = 3.67

1846

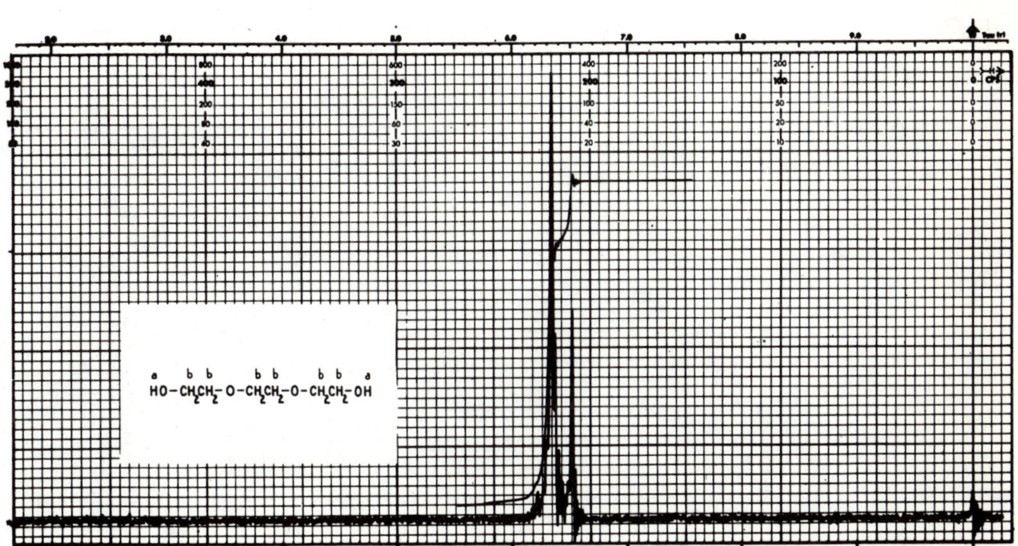

© 1978 Sadtler Research Laboratories, Inc.

TETRAETHYLENE GLYCOL

$C_8H_{18}O_5$

Mol. Wt. 194.23

Solvent: CCl_4

Assignments
a = 3.55
b = 3.60
OH is unobserved

1847

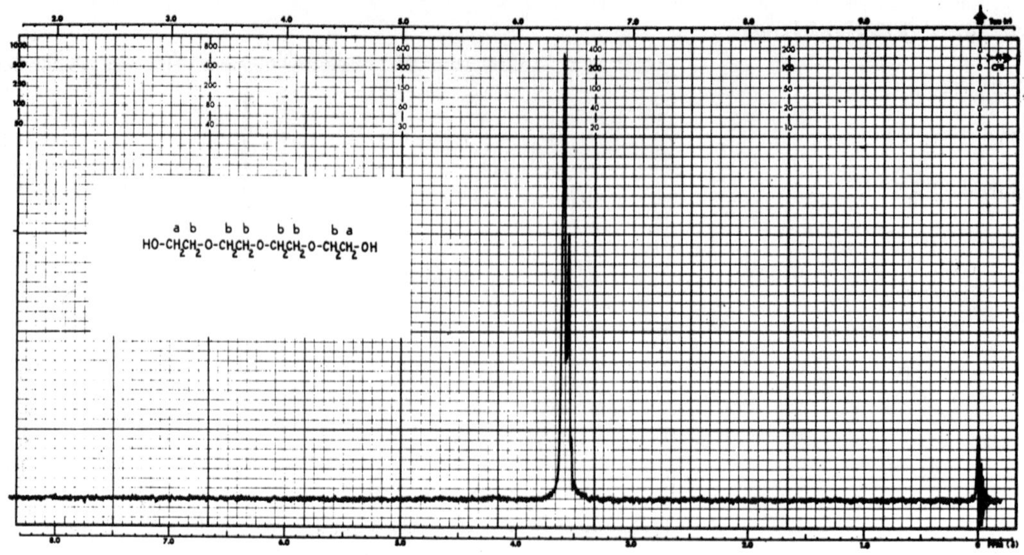

© 1978 Sadtler Research Laboratories, Inc.

GLYCEROL

$C_3H_8O_3$

Mol. Wt. 92.10

B.P. 182°C/2mm

M.P. 17.8°C

Solvent: D_2O

Assignments
a = 3.60
b = 3.62
c = 4.62 HDO

1848

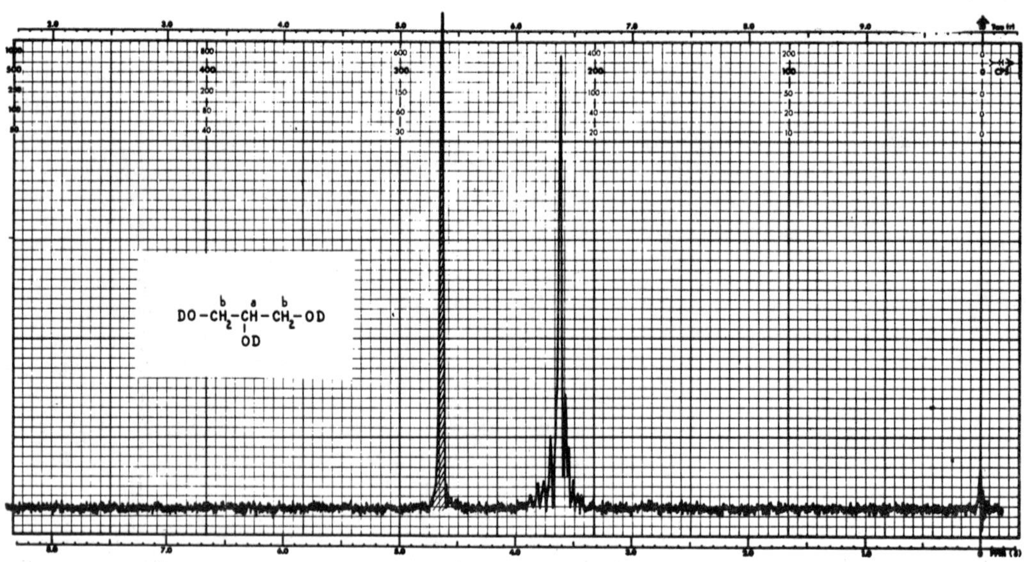

© 1978 Sadtler Research Laboratories, Inc.

1,4,7-HEPTANETRIOL

$C_7H_{16}O_3$

Mol. Wt. 148.20

B.P. 146°C/.05mm

Solvent: Polysol-d

Assignments
a = 1.10-2.00
b = 3.56
c = 4.53

1849

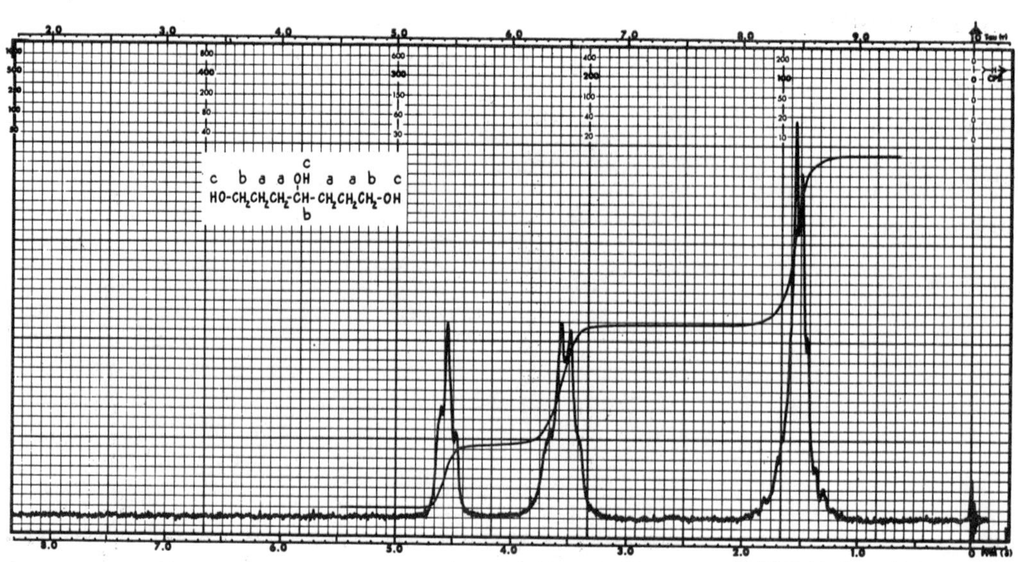

© 1978 Sadtler Research Laboratories, Inc.

2,2′,2″-NITRILOTRIETHANOL

$C_6H_{15}NO_3$

Mol. Wt. 149.19

B.P. 360°C

Solvent: D_2O

Assignments
a = 2.70
b = 3.64
c = 4.69 HDO

1850

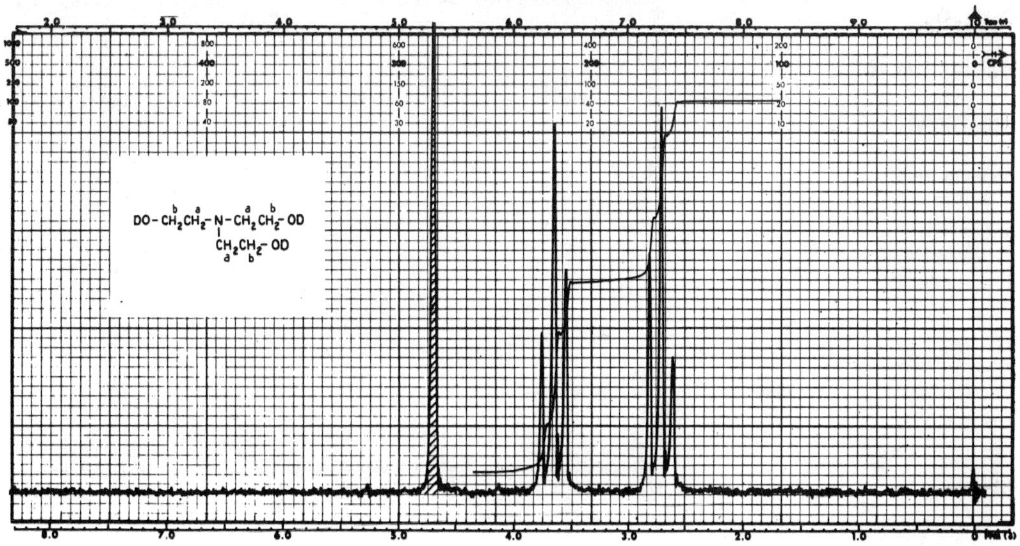

© 1978 Sadtler Research Laboratories, Inc.

1,1′,1″-NITRILOTRI-2-PROPANOL

$C_9H_{21}NO_3$

Mol. Wt. 191.27

B.P. 305.4°C

Solvent: $CDCl_3$

Assignments
a = 1.10
b = 2.48
c = ca 3.77
d = 4.70

1851

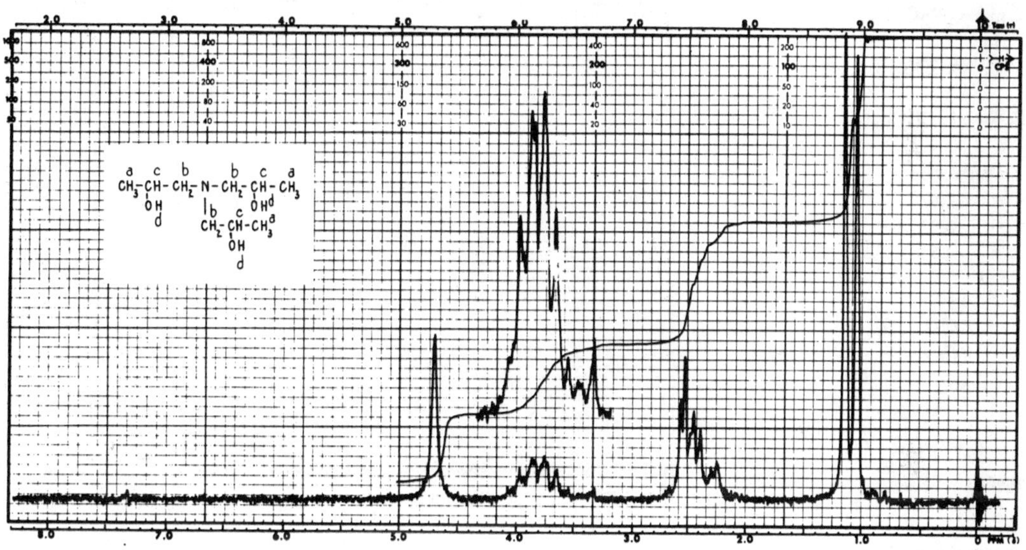

© 1978 Sadtler Research Laboratories, Inc.

2,2′,2″-NITRILOTRIETHANOL, HYDROCHLORIDE

$C_6H_{15}NO_3 \cdot HCl$

Mol. Wt. 185.65

M.P. 177-179°C

Solvent: D_2O

Assignments
a = 3.50
b = 3.97
c = 4.60 HDO

1852

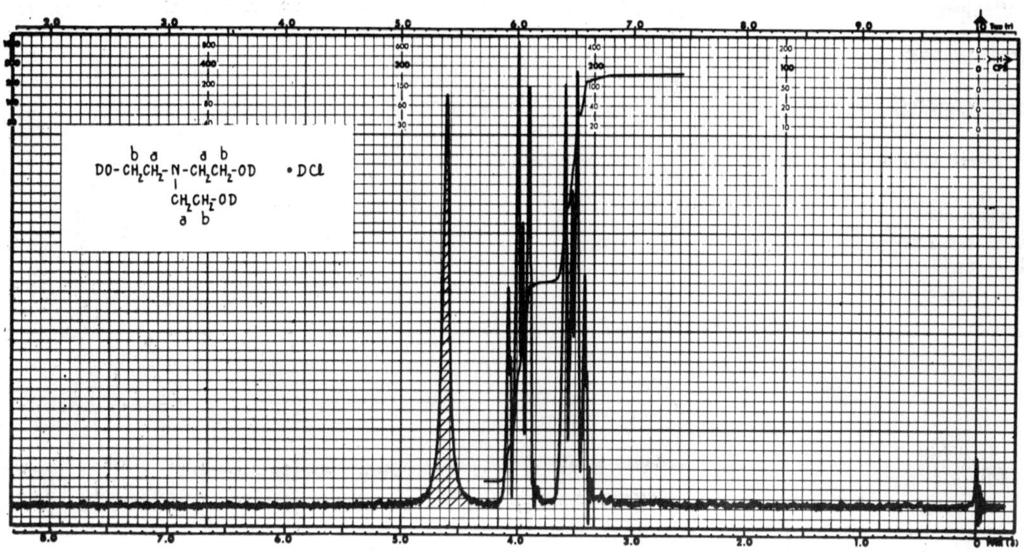

© 1978 Sadtler Research Laboratories, Inc.

PENTAERYTHRITOL

$C_5H_{12}O_4$

Mol. Wt. 136.15

M.P. 250°C

Solvent: D_2O

Assignments

a = 3.62

b = 4.71 HDO

1853

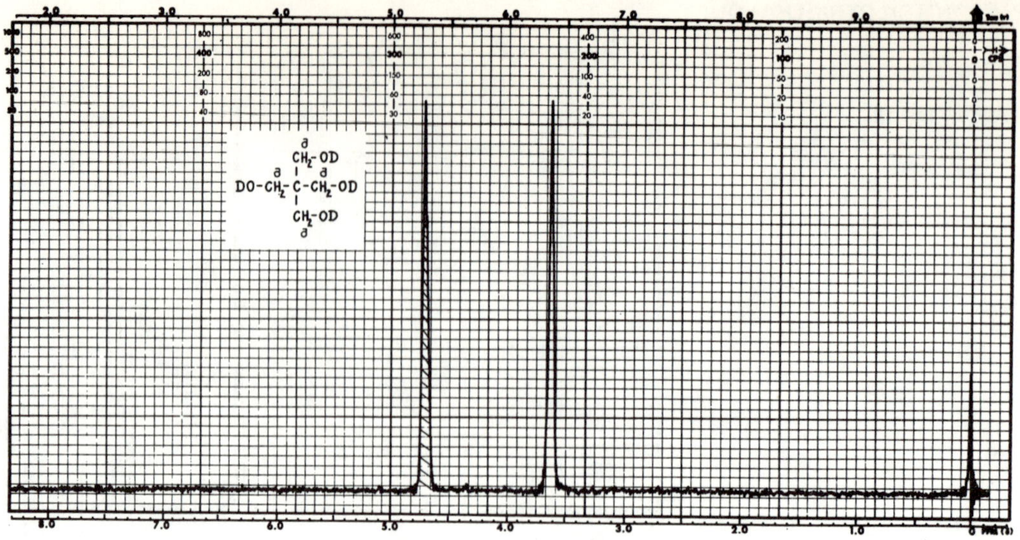

© 1978 Sadtler Research Laboratories, Inc.

2,2',2'',2''', (ETHYLENEDINITRILO)TETRAETHANOL

$C_{10}H_{24}N_2O_4$

Mol. Wt. 236.31

Solvent: $CDCl_3$

Assignments

a = ca 2.62

b = 3.63

c = 4.61

1854

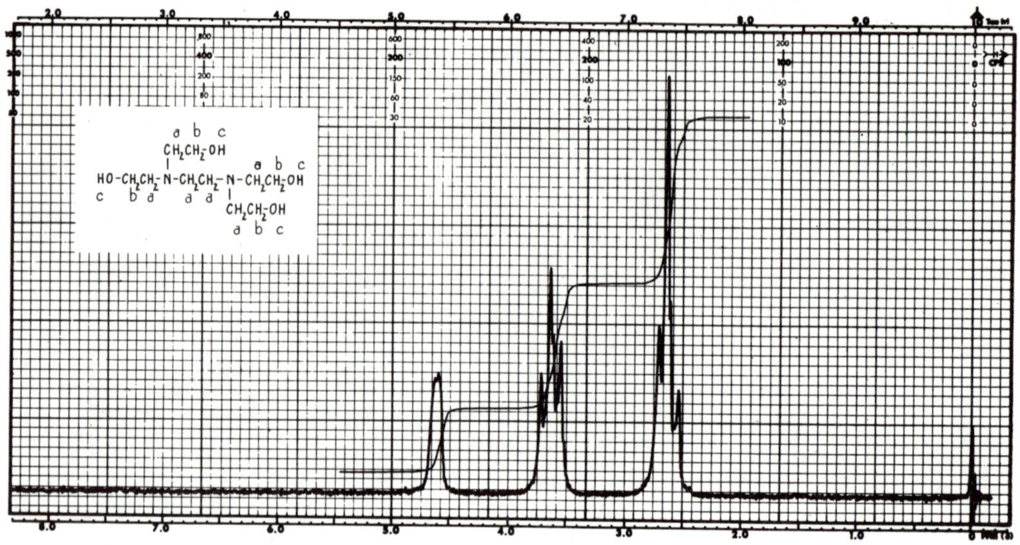

© 1978 Sadtler Research Laboratories, Inc.

cis-CYCLOHEXANE-1,2-DIOL

$C_6H_{12}O_2$

Mol. Wt. 116.16

M.P. 97-99°C

Solvent: $CDCl_3$

Assignments

a = 1.00-2.09

b = 3.20

c = 3.75

1855

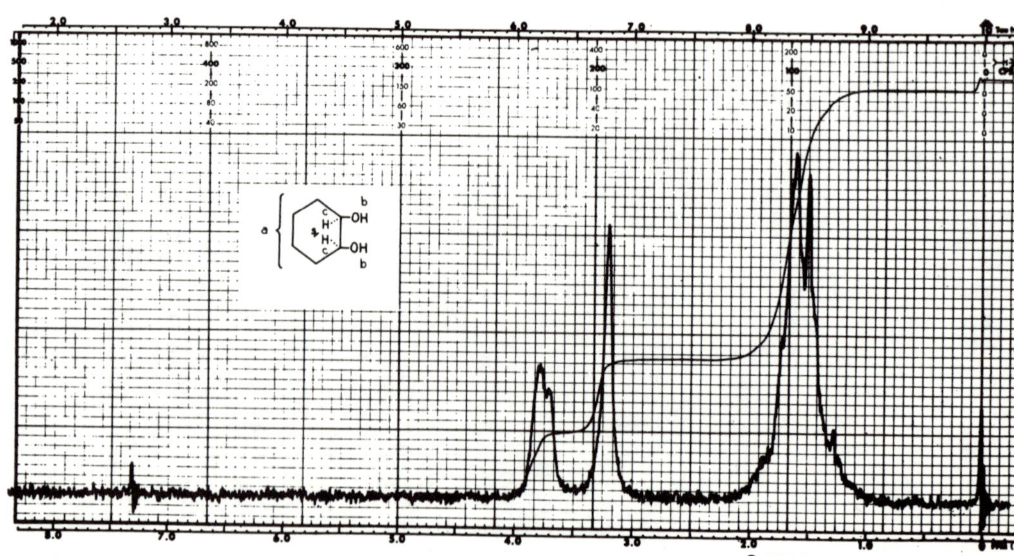

© 1978 Sadtler Research Laboratories, Inc.

1,4-CYCLOHEXANEDIOL

$C_6H_{12}O_2$

Mol. Wt. 116.16

Solvent: D_2O

Assignments
a = 1.00-2.20
b = 3.00-3.90
c = 4.67 HDO

1856

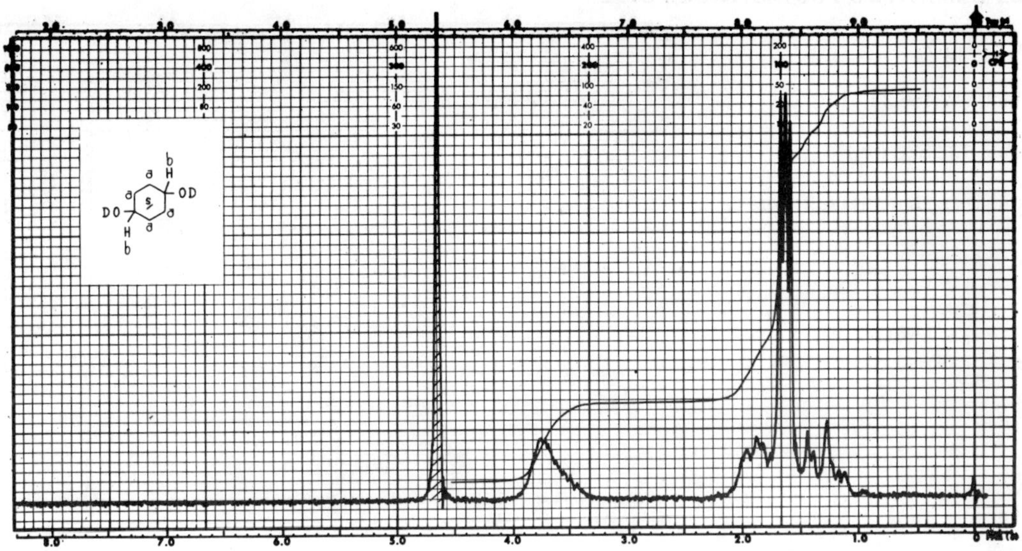

© 1978 Sadtler Research Laboratories, Inc.

---

cis-1,5-CYCLOOCTANEDIOL

$C_8H_{16}O_2$

Mol. Wt. 144.22

Solvent: D_2O

Assignments
a = 1.00-2.20
b = 3.77
c = 4.64 HDO

1857

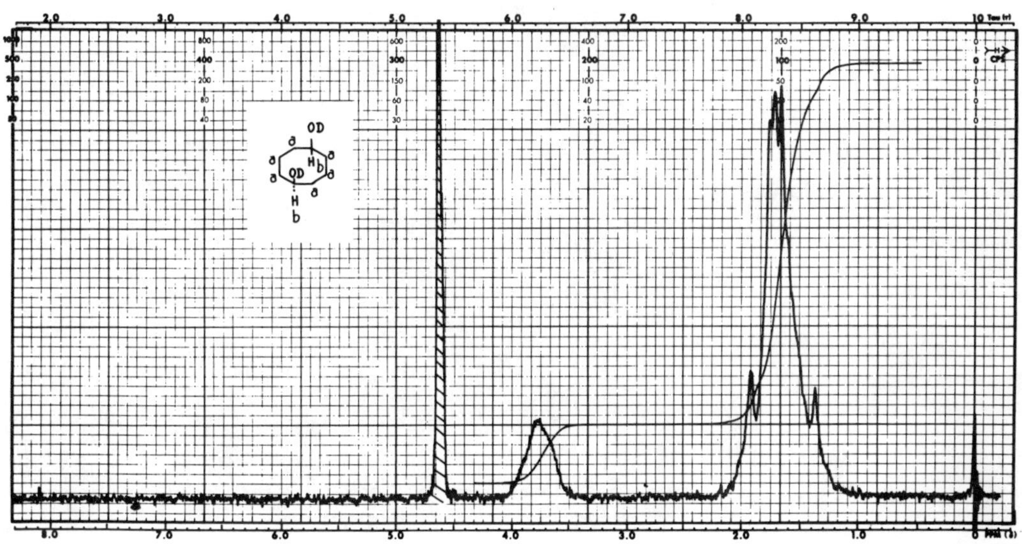

© 1978 Sadtler Research Laboratories, Inc.

2-BUTYNE-1,4-DIOL

$C_4H_6O_2$

Mol. Wt. 86.09

M.P. 50-56°C

Solvent: D_2O

Assignments
a = 4.29
b = 4.60 HDO

1858

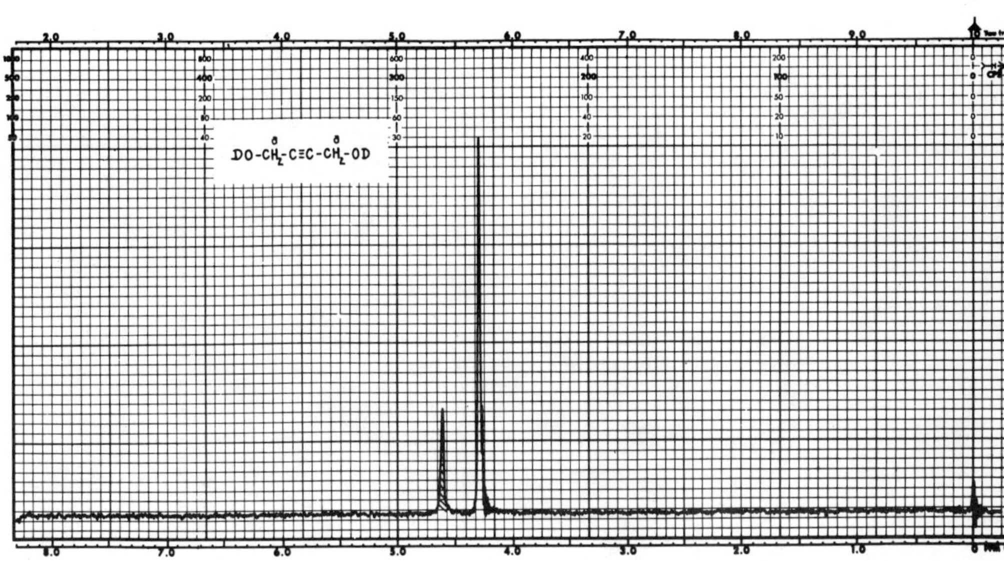

© 1978 Sadtler Research Laboratories, Inc.

2,7-DIMETHYL-3,5-OCTADIYNE-2,7-DIOL

$C_{10}H_{14}O_2$

Mol. Wt. 166.22

M.P. 132.5°C (lit.)

Solvent: Dimethyl sulfoxide-d_6

Assignments
a = 1.39
b = 5.47
c = 2.51 DMSO-d_5
d = 3.29 H_2O

1859

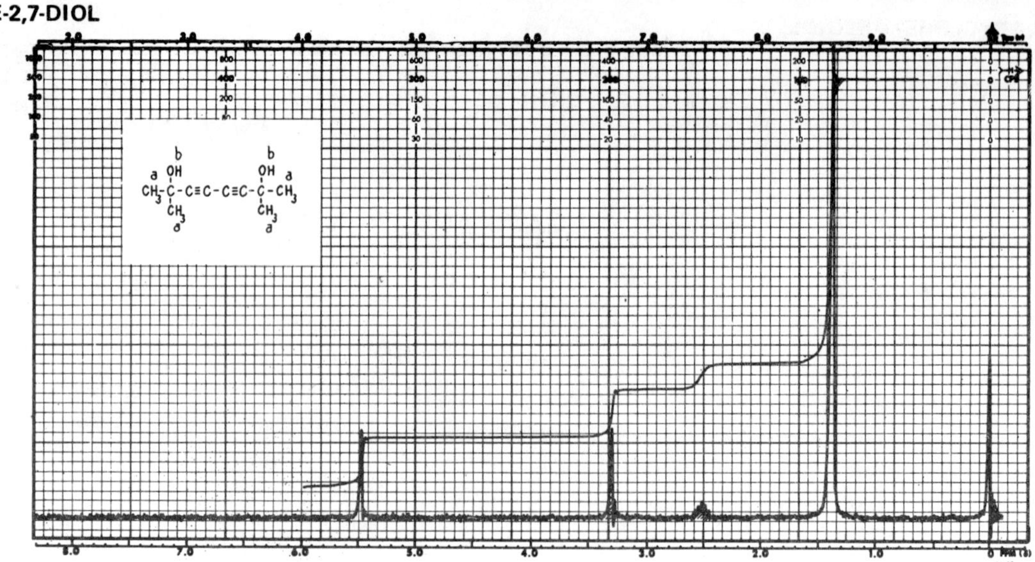

© 1978 Sadtler Research Laboratories, Inc.

p-XYLENE-*a,a'*-DIOL

$C_8H_{10}O_2$

Mol. Wt. 138.17

M.P. 118°C

Solvent: Dimethyl sulfoxide-D_6

Assignments
a = 4.51
b = 5.10
c = 7.27
d = 2.46 DMSO-D_5
e = 3.49 H_2O

1860

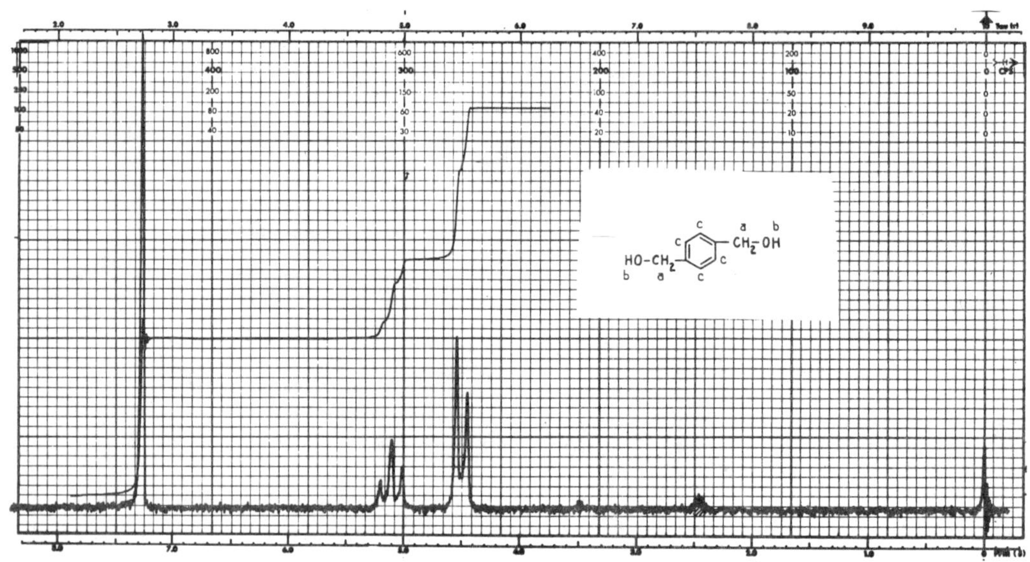

© 1978 Sadtler Research Laboratories, Inc.

ERYTHRITOL

$C_4H_{10}O_4$

Mol. Wt. 122.12

M.P. 121-122°C

Solvent: D_2O

Assignments
a = ca 3.74
b = 4.22 HDO

1861

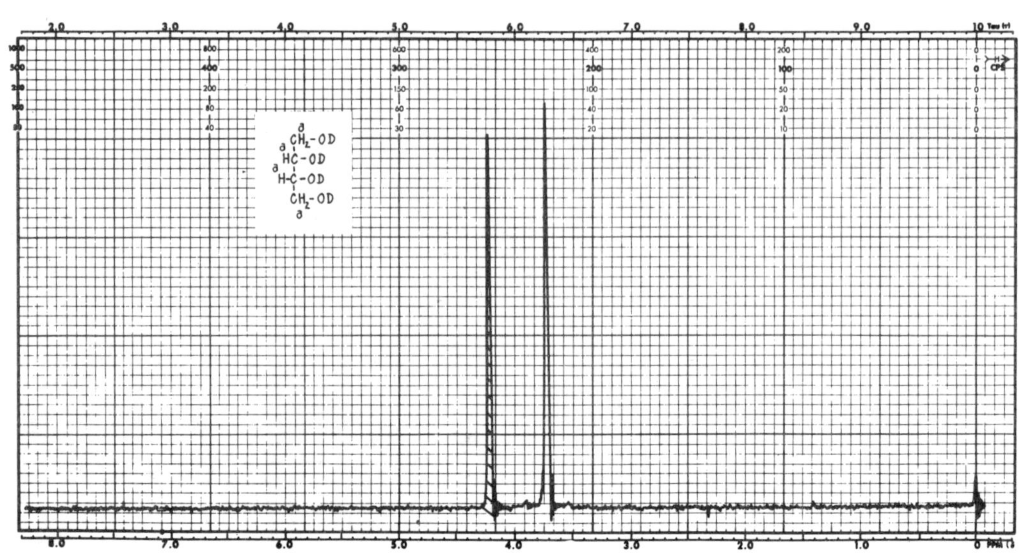

© 1978 Sadtler Research Laboratories, Inc.

SUGARS AND CARBOHYDRATES

This specialized group of polyols characteristically displays few resonance bands at high field. Their NMR spectra usually consist of a complex higher order pattern in the chemical shift range from 3–4 ppm, with the cyclic varieties displaying one or two additional doublets at lower field (4–6 ppm). These doublets represent the axial and equatorial protons bonded to the carbon nucleus adjacent to the ether linkage of the furanosides and pyranosides. All of the carbohydrates are readily soluble in deuterium oxide although they may dissolve only slowly. Proton NMR is useful in determining the relative percentages of alpha and beta forms, but the spectra are not otherwise readily interpretable except for direct comparison of the pattern with that of a reference compound.

Chemical Shifts

The chemical shift data for several selected pyranosides which appear in this volume are presented below.

	CH–2,3,4,5	Axial CH–1	Equat. CH–1	Solvent	Spectrum
Arabinopyranose	3.3–4.2	4.51	5.24	D_2O	1867
Glucose	3.0–4.1	4.60	5.20	D_2O	1871
Galactose	3.3–4.3	4.60	5.30	D_2O	1873
Xylose	3.1–4.2	4.68	5.28	D_2O	1868
Mannose	3.2–4.0	4.89	5.19	D_2O	1872

Coupling and Coupling Constants

Three types of coupling can occur in the spectra of the cyclic sugars; axial-axial, axial-equatorial and equatorial-equatorial. Because the size of the coupling constant varies with the dihedral angle between the coupled protons, the axial-axial interaction is significantly larger than the other two types. Ranges of observed coupling constants for these three couplings are given below.

$$J_{\text{axial-axial}} = 5\text{–}8 \text{ Hz}$$

$$J_{\text{axial-equatorial}} = 1\text{–}4 \text{ Hz}$$

$$J_{\text{equatorial-equatorial}} = 1\text{–}4 \text{ Hz}$$

XYLITOL

$C_5H_{12}O_5$

Mol. Wt. 152.15

Solvent: D_2O

Assignments

a = 3.40-3.80

b = 4.54 HDO

1862

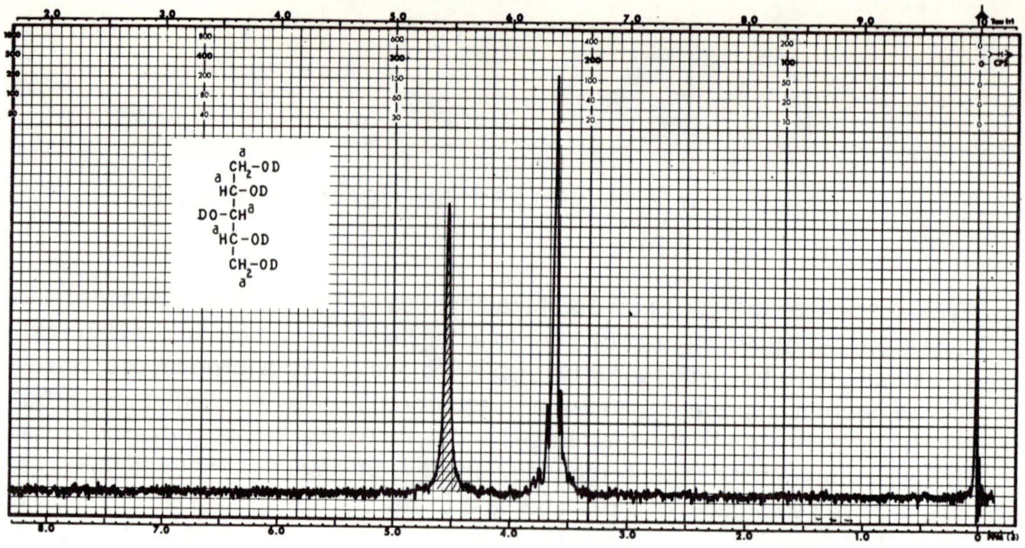

© 1978 Sadtler Research Laboratories, Inc.

GALACTITOL

$C_6H_{14}O_6$

Mol. Wt. 182.17

Solvent: D_2O

Assignments

a = 3.59-3.81

b = ca 3.96

c = 4.63 HDO

1863

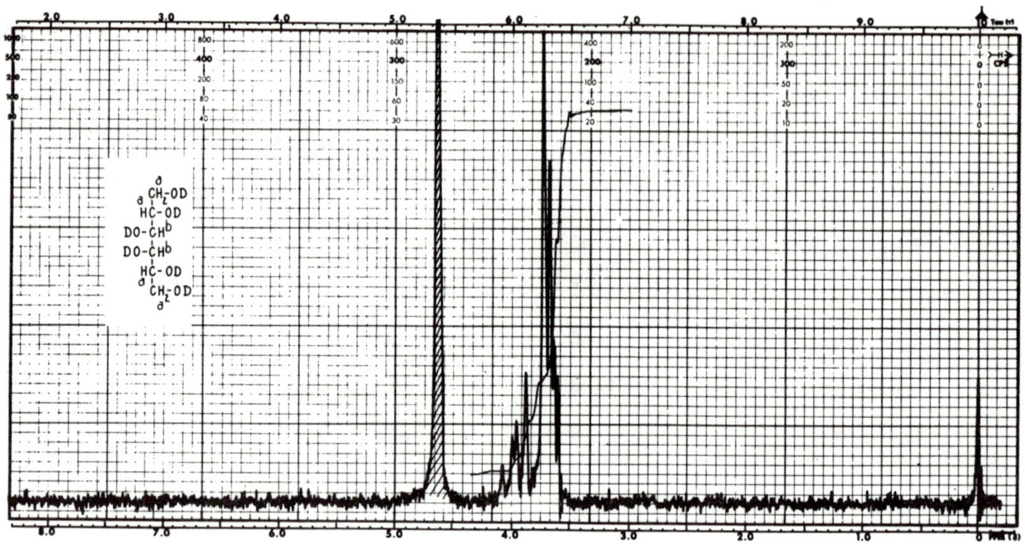

© 1978 Sadtler Research Laboratories, Inc.

D-GLUCITOL

$C_6H_{14}O_6$

Mol. Wt. 182.18

M.P. 99-101°C

Solvent: D_2O

Assignments

a = 3.49-4.05

b = 4.66 HDO

1864

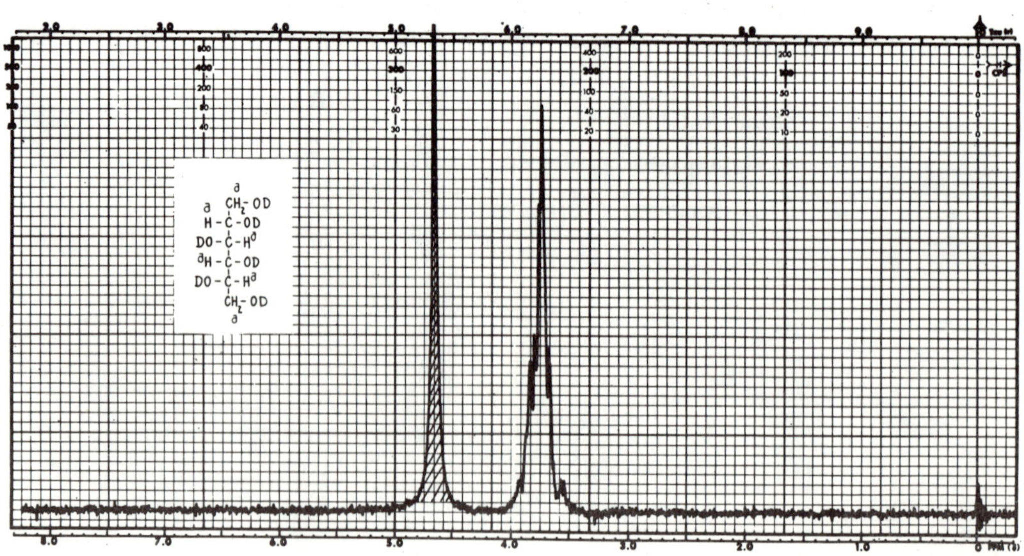

© 1978 Sadtler Research Laboratories, Inc.

D-(−)-MANNITOL

$C_6H_{14}O_6$

Mol. Wt. 182.17

Solvent: D_2O

Assignments
a = 3.79
b = 4.71 HDO

1865

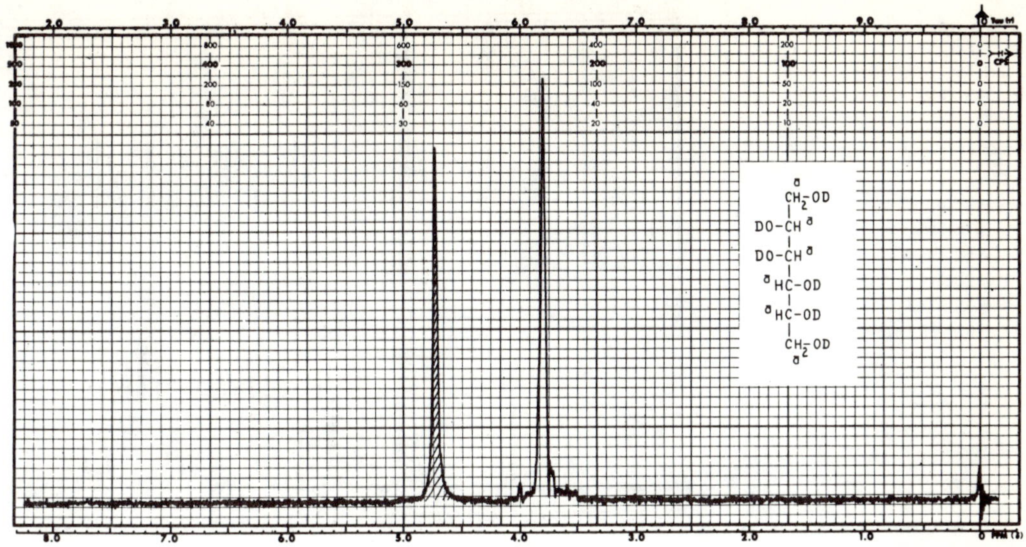

© 1978 Sadtler Research Laboratories, Inc.

D-RIBOSE

$C_5H_{10}O_5$

Mol. Wt. 150.13

M.P. 87°C

Solvent: D_2O

Assignments
a = 3.50-4.30
b = 4.86-5.16
c = 5.36 imp.
d = 5.58 HDO
e = 4.80, 6.37 sidebands

1866

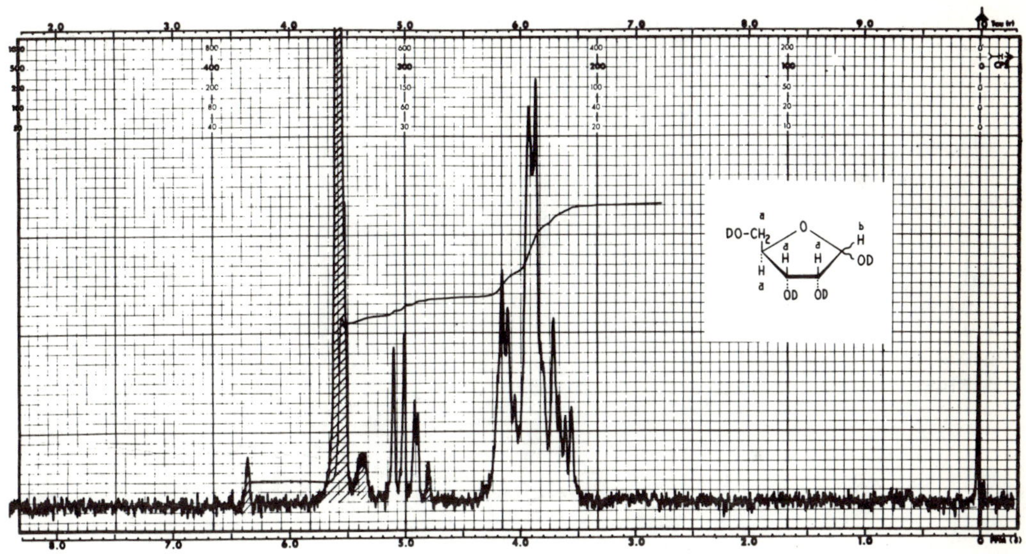

© 1978 Sadtler Research Laboratories, Inc.

L-ARABINOPYRANOSE

$C_5H_{10}O_5$

Mol. Wt. 150.13

M.P. 174°C (dec.)

Solvent: D_2O

Assignments
a = 3.30-4.20
b = 4.51
c = 5.24
d = 4.87 HDO
e = 5.78 imp.
Trace of acid added to shift HDO.

1867

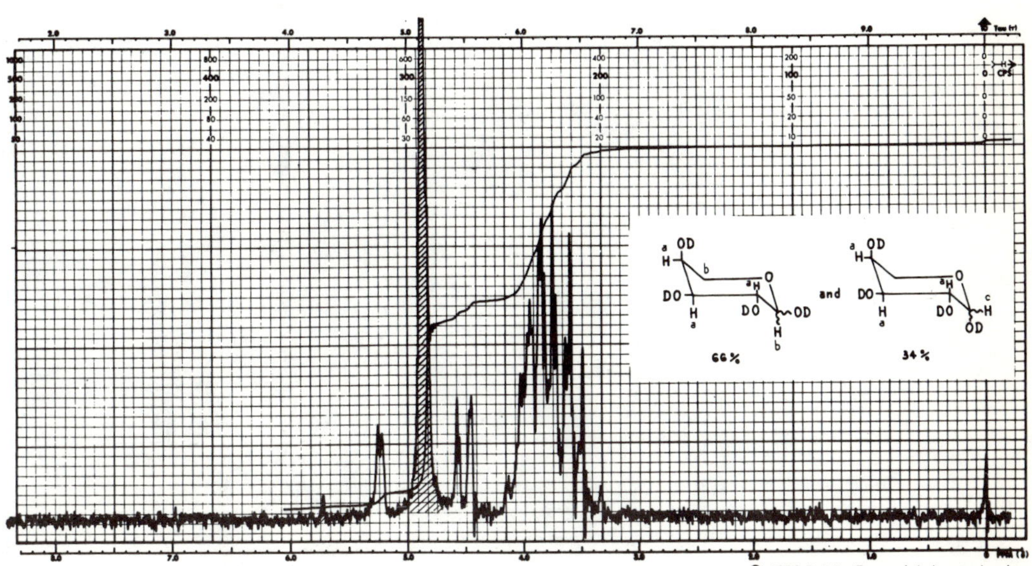

© 1978 Sadtler Research Laboratories, Inc.

(+)-XYLOSE

$C_5H_{10}O_5$

Mol. Wt. 150.13

Solvent: D_2O

Assignments
a = 3.16-4.19
b = 4.68
c = 5.28
d = 5.92 HDO
e = 5.03, 6.87 sidebands

1868

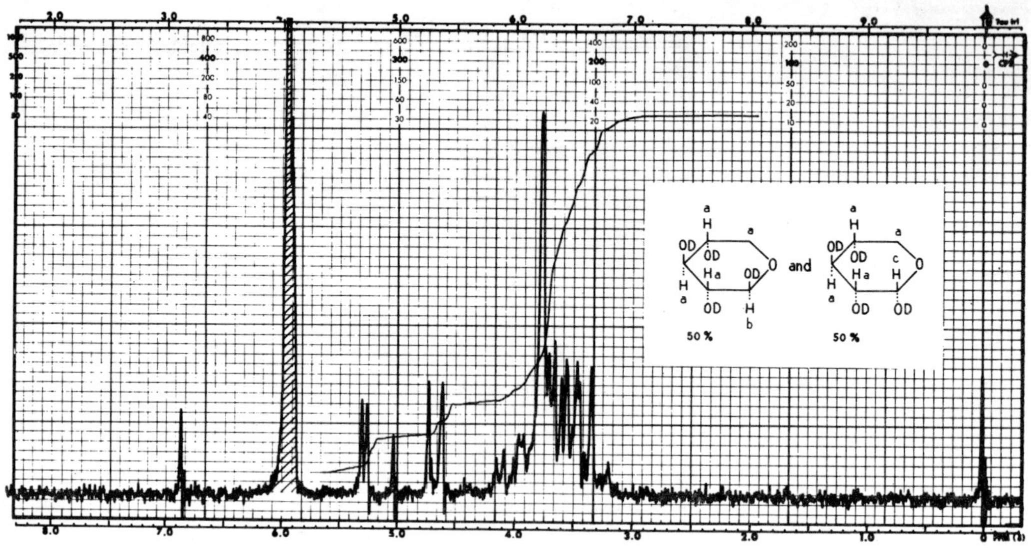

© 1978 Sadtler Research Laboratories, Inc.

L-(+)-RHAMNOSE, HYDRATE

$C_6H_{12}O_5 \cdot H_2O$

Mol. Wt. 182.17

M.P. 82-92°C

Solvent: D_2O

Assignments
a = 1.27
b = 3.20-4.10
c = 4.81
d = 5.10
e = 4.46 HDO

1869

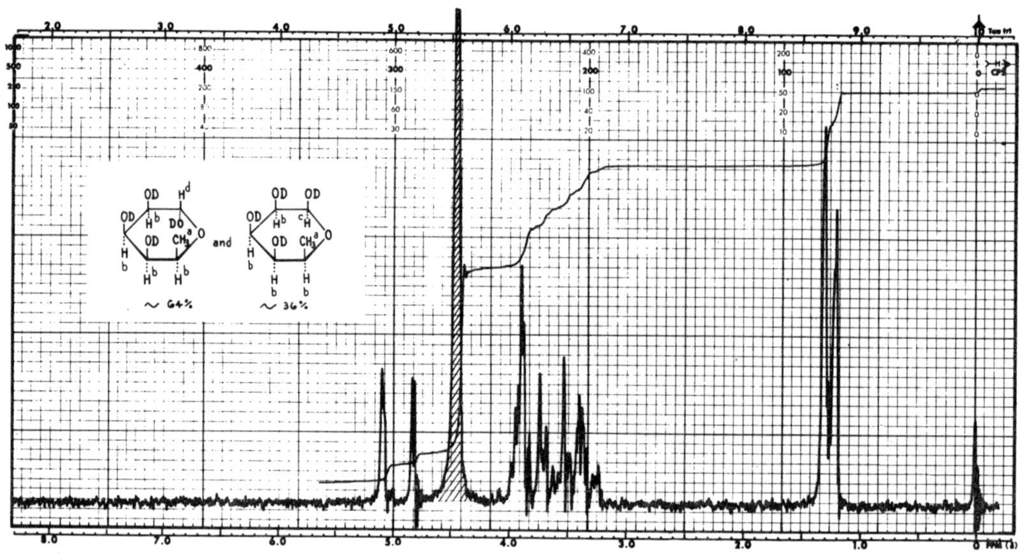

© 1978 Sadtler Research Laboratories, Inc.

D-(−)-FRUCTOSE

$C_6H_{12}O_6$

Mol. Wt. 180.16

Solvent: D_2O

Assignments
a = 3.50-4.20
b = 4.69 HDO

1870

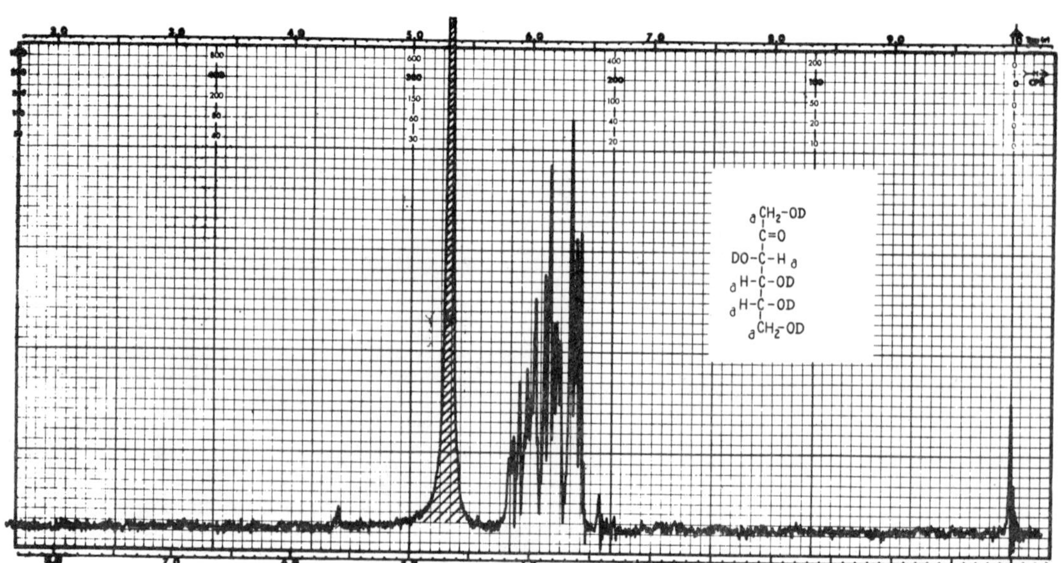

© 1978 Sadtler Research Laboratories, Inc.

D-GLUCOSE

$C_6H_{12}O_6$

Mol. Wt. 180.16

Solvent: D_2O

Assignments
a = 3.00-4.10
b = 4.60
c = 5.20
d = 4.21 HDO

1871

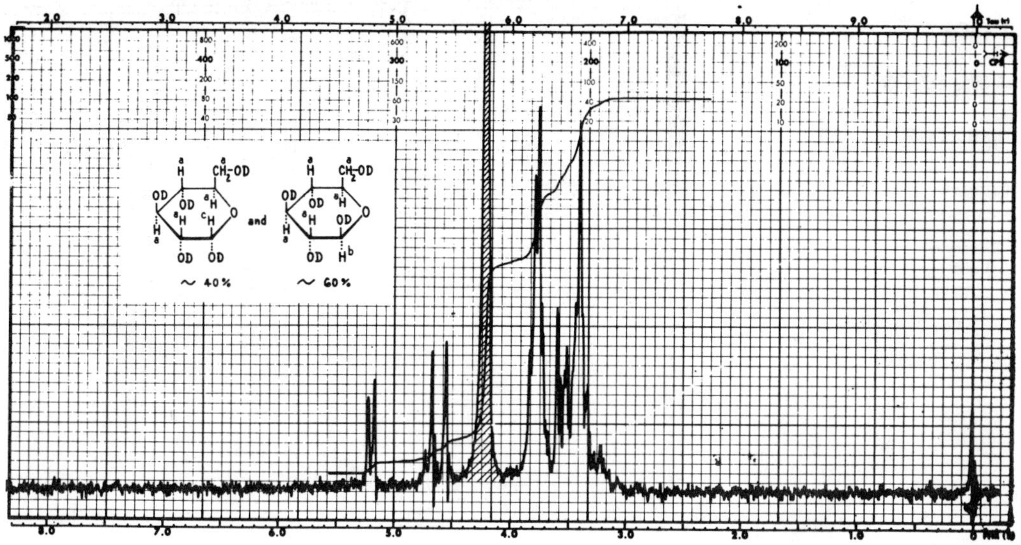

© 1978 Sadtler Research Laboratories, Inc.

D-(+)-MANNOSE

$C_6H_{12}O_6$

Mol. Wt. 180.16

M.P. 133-135°C

Solvent: D_2O

Assignments
a = 3.20-4.00
b = 4.89
c = 5.19
d = 4.41 HDO

1872

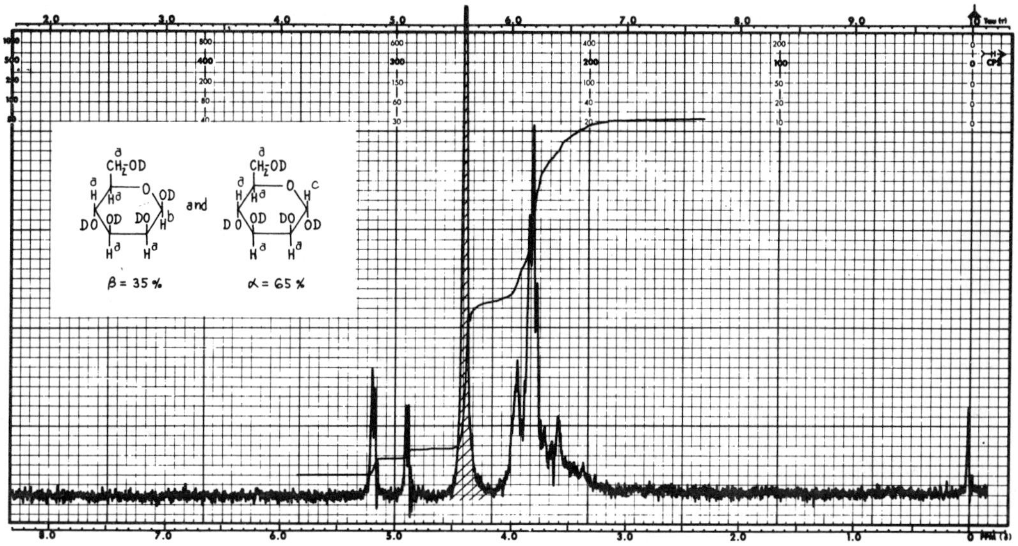

© 1978 Sadtler Research Laboratories, Inc.

d-GALACTOSE

$C_6H_{12}O_6$

Mol. Wt. 180.16

Solvent: D_2O

Assignments
a = 3.33-4.30
b = 4.60
c = 5.30
d = 5.52 HDO

1873

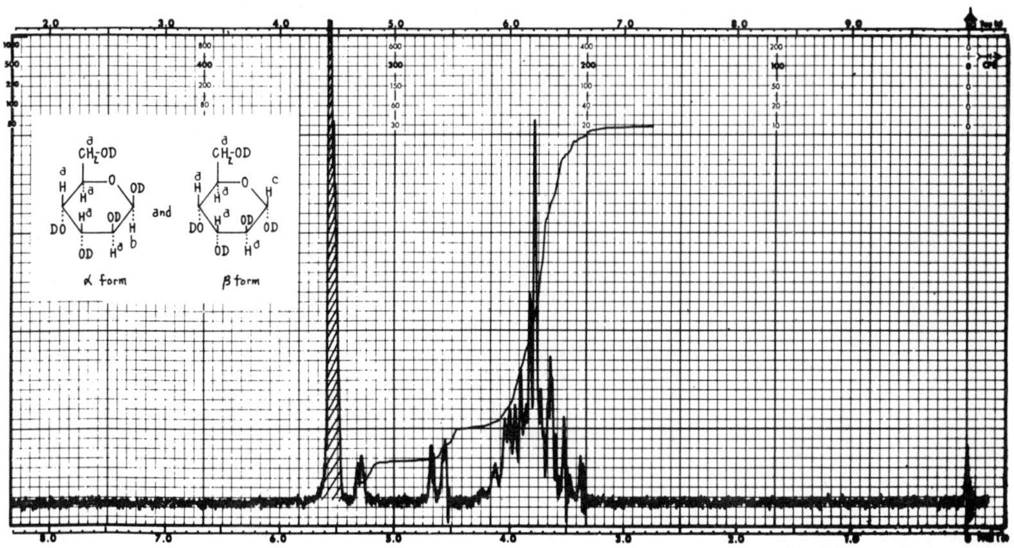

© 1978 Sadtler Research Laboratories, Inc.

SUCROSE

$C_{12}H_{22}O_{11}$

Mol. Wt. 342.30

M.P. 160-186°C (dec.)

Solvent: D_2O

Assignments
a = 3.35-4.33
b = 5.41
c = 4.68 HDO

1874

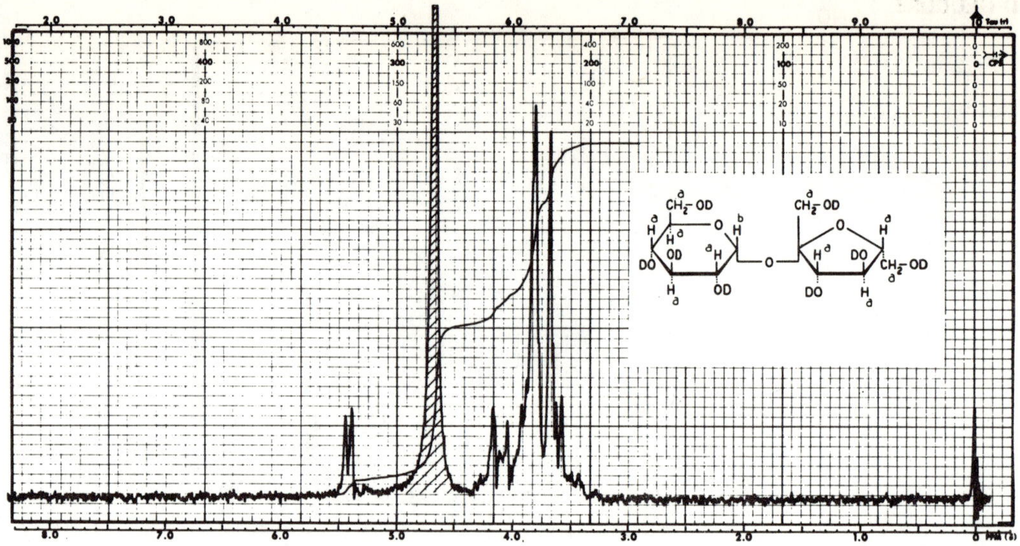

© 1978 Sadtler Research Laboratories, Inc.

The phenolic compounds characteristically display high field chemical shifts for the aromatic hydrogens ortho and para to the hydroxyl substituent. The single hydroxyl proton resonates at much lower field than the corresponding OH group of the alcohols but at higher field than that of the carboxylic acids.

Chemical Shifts

Aromatic Protons

Phenol

HO–	ortho	meta	para	Solvent	Spectrum
6.11	6.75	7.14	6.79	$CDCl_3$	1875

Para Substituted Phenols

HO–	ortho	meta	para		
5.77	6.62	6.62	$-O-CH_3$	CCl_4	- - - -
- - - -	6.79	6.79	–OH	D_2O	1912
6.45	6.69	6.92	$-CH_3$	$CDCl_3$	1878
6.53	6.79	6.91	–F	$CDCl_3$	1891
6.39	6.78	7.03	$-CH(CH_3)_2$	$CDCl_3$	1881
4.95	6.75	7.08	$-C_6$	$CDCl_3$	1887
6.33	6.71	7.10	–Cl	$CDCl_3$	1893
4.73	6.75	7.22	$-S-CH_3$	$CDCl_3$	- - - -
4.91	6.75	7.23	$-C(CH_3)_3$	$CDCl_3$	1883
5.20	6.69	7.30	–Br	$CDCl_3$	1895
9.49	6.87	7.43	–Ø	DMSO	1889
6.80	6.91	7.46	$-CF_3$	$CDCl_3$	- - - -
5.41	6.59	7.48	–I	$CDCl_3$	1898
6.84	6.88	7.58	–C(=O)–Ø	DMSO-d_6	- - - -
8.70	6.63	7.67	–N=O	Acetone	- - - -
10.50	6.93	7.72	$-SO_2-$Ø	DMSO-d_6	- - - -
- - - -	7.05	7.87	$-SO_2-O-Na$	D_2O	- - - -
9.76	7.07	8.11	–C(=O)–OH	Acetone	- - - -
9.32	7.06	8.15	$-NO_2$	Acetone	1905

Solubility and Solvent Effects

Phenol and the simple aliphatic substituted phenols are soluble in CCl_4 and $CDCl_3$. As indicated above in the table of chemical shifts, the presence of other functional groups may require the use of D_2O, Polysol, Acetone or DMSO-d_6.

Characterization

The presence of an exchangeable proton band at relatively low field and the relatively high field chemical shifts produced by the hydroxyl group on the ortho and para aromatic hydrogens makes the phenols a relatively simple group of compounds to characterize.

Only the anilines, produce similar chemical shifts, however, the resonance bands of the anilines are sensitive to the addition of acid to the sample solution while the phenols are not affected in the same manner.

PHENOL

C_6H_6O

Mol. Wt. 94.11

M.P. 42.5°C (lit.)

Solvent: $CDCl_3$

Assignments
a = 6.11
b = 6.65-7.32

1875

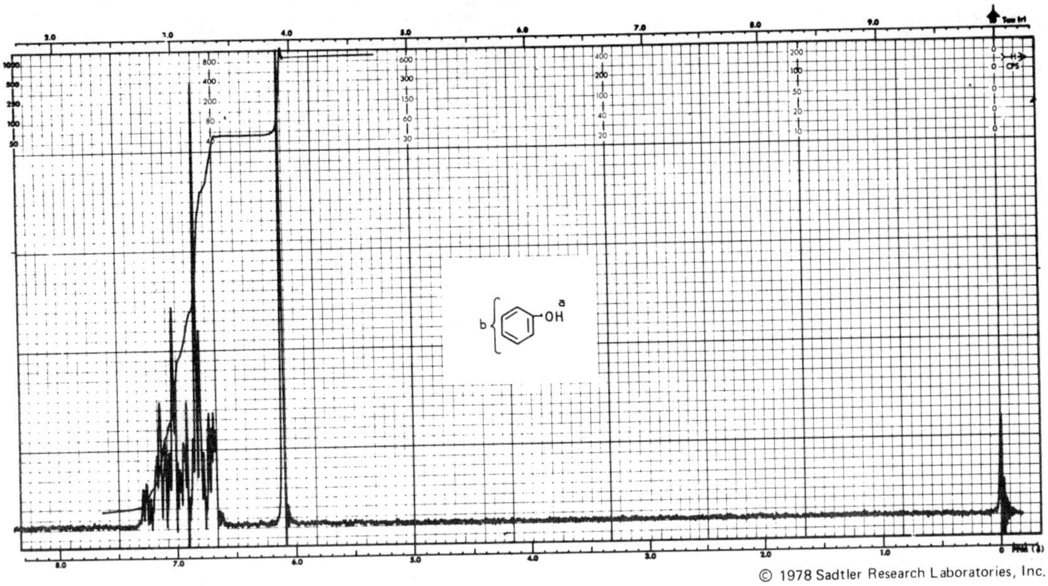

© 1978 Sadtler Research Laboratories, Inc.

---

o-CRESOL

C_7H_8O

Mol. Wt. 108.14

M.P. 30-31°C

Solvent: CCl_4

Assignments
a = 2.17
b = 5.18
c = 6.45-7.12

1876

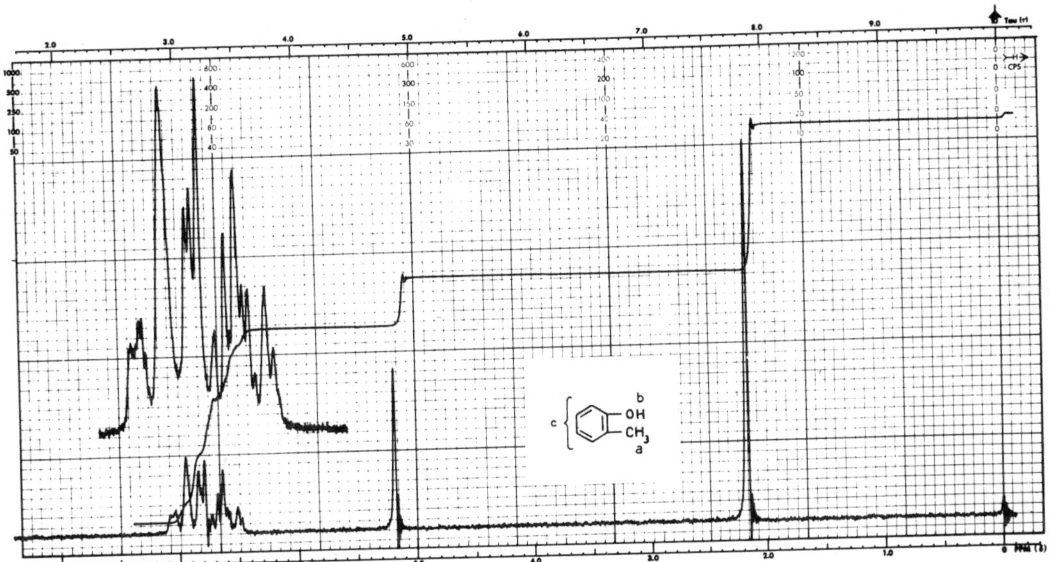

© 1978 Sadtler Research Laboratories, Inc.

---

m-CRESOL

C_7H_8O

Mol. Wt. 108.14

M.P. 10-12°C

Solvent: CCl_4

Assignments
a = 2.18
b = 6.51
c = 6.36-6.70
d = 6.94

1877

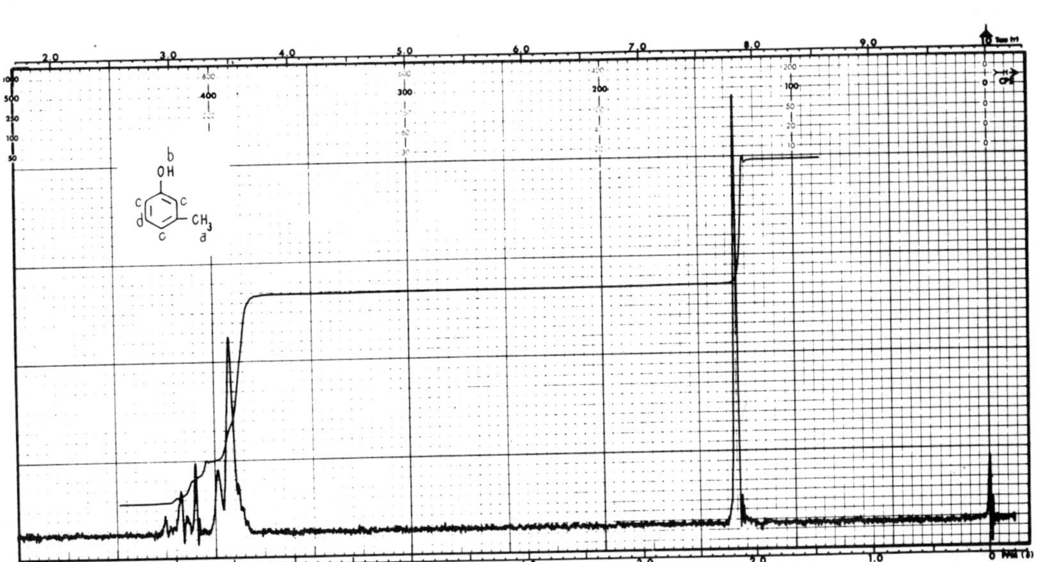

© 1978 Sadtler Research Laboratories, Inc.

p-CRESOL

C_7H_8O

Mol. Wt. 108.14

Solvent: $CDCl_3$

Assignments
a = 2.20
b = 6.45
c = 6.69
d = 6.92

1878

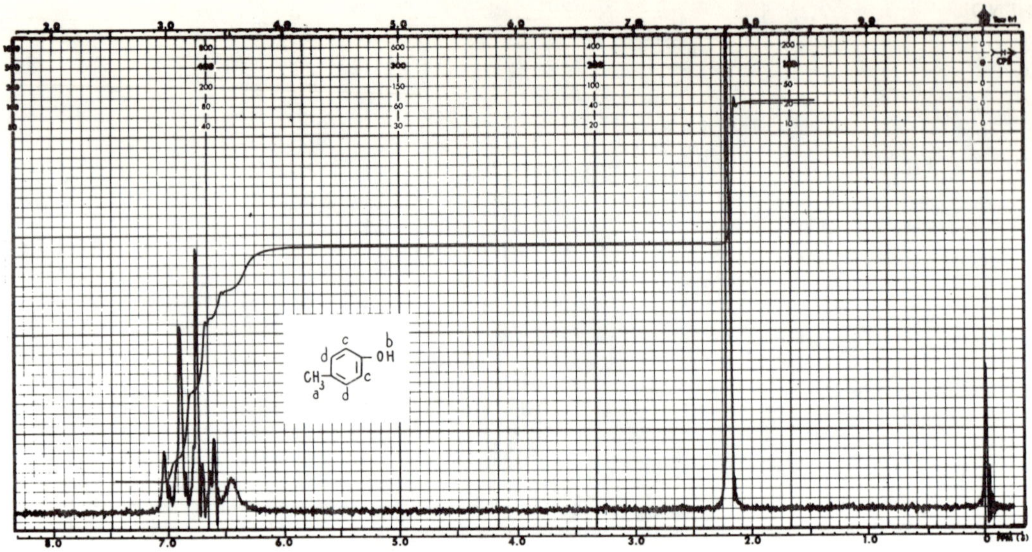

© 1978 Sadtler Research Laboratories, Inc.

---

o-ETHYLPHENOL

$C_8H_{10}O$

Mol. Wt. 122.17

B.P. 87-89°C/10mm

Solvent: CCl_4

Assignments
a = 1.17
b = 2.37
c = 5.52
d = 6.50-7.20

1879

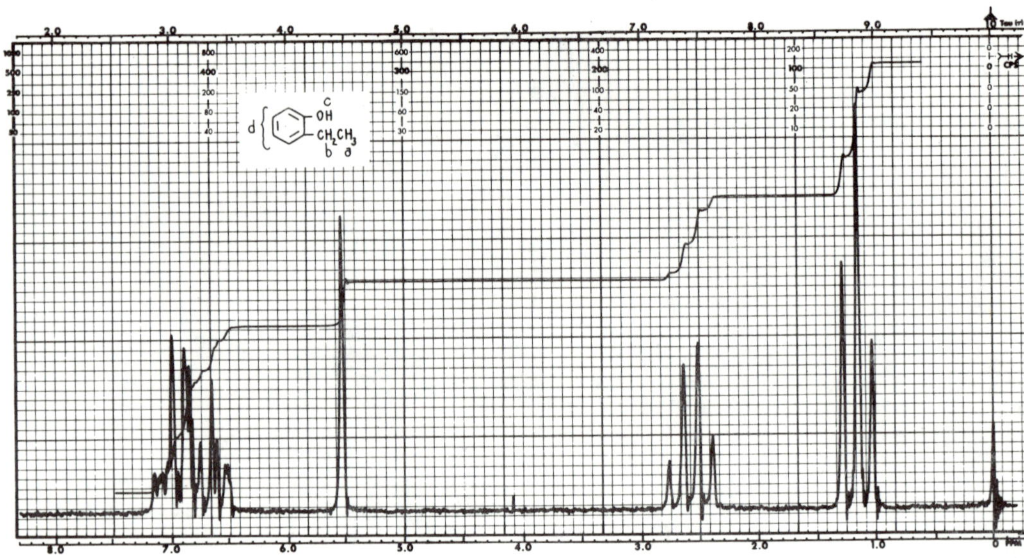

© 1978 Sadtler Research Laboratories, Inc.

3-ETHYLPHENOL

$C_8H_{10}O$

Mol. Wt. 122.17

M.P. −4°C

B.P. 214°C/752mm (lit.)

Solvent: $CDCl_3$

Assignments
a = 1.16
b = 2.55
c = 6.09
d = 6.56-6.90
e = 7.14

1880

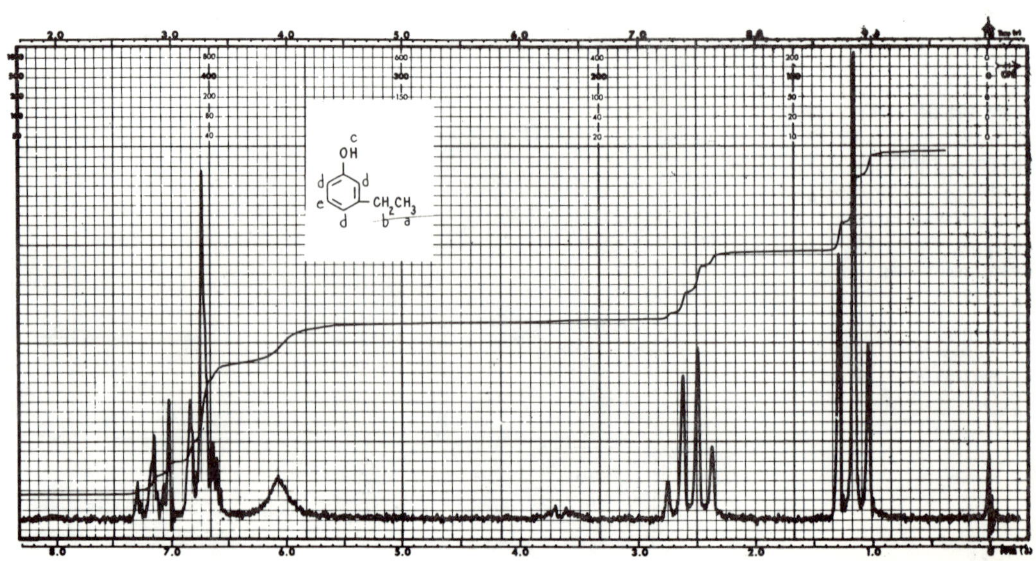

© 1978 Sadtler Research Laboratories, Inc.

p-ISOPROPYLPHENOL

$C_9H_{12}O$

Mol. Wt. 136.20

M.P. 61.5-62°C

Solvent: $CDCl_3$

Assignments
a = 1.20
b = 2.81
c = 6.39
d = 6.78
e = 7.03

1881

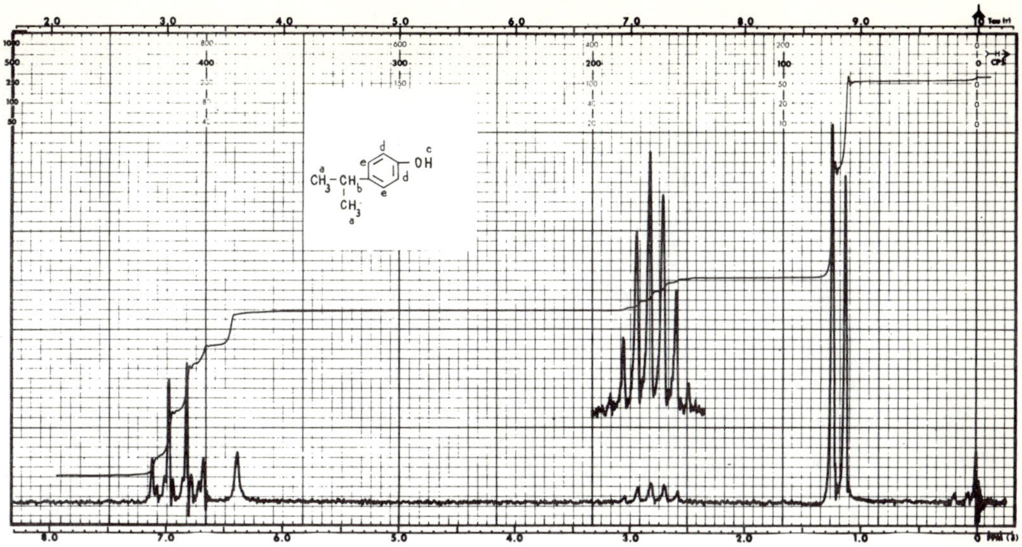

© 1978 Sadtler Research Laboratories, Inc.

o-tert-BUTYLPHENOL

$C_{10}H_{14}O$

Mol. Wt. 150.22

B.P. 224°C

M.P. −6.4°C

Solvent: CCl_4

Assignments
a = 1.40
b = 4.48
c = 6.49
d = 6.74
e = 6.87
f = 7.19

1882

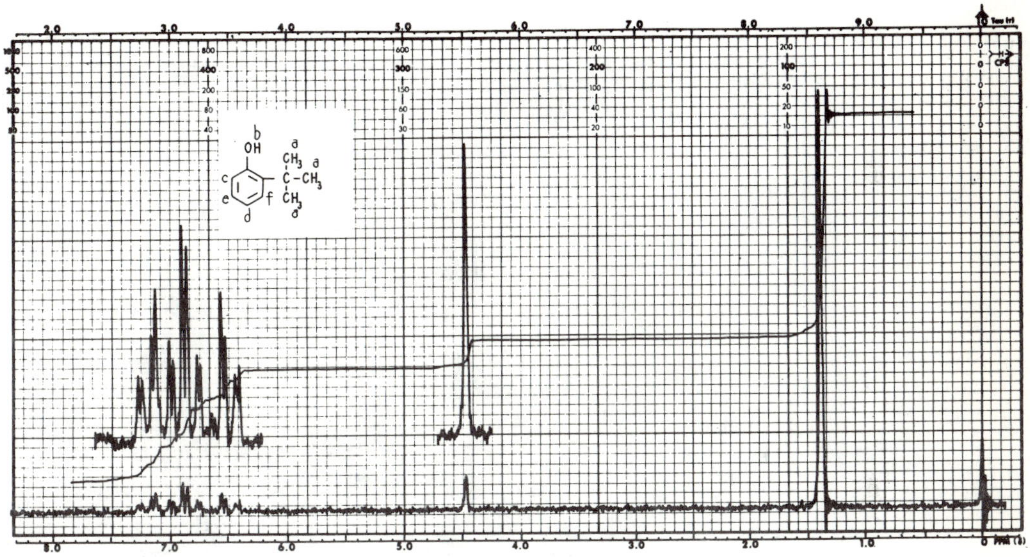

© 1978 Sadtler Research Laboratories, Inc.

p-tert-BUTYLPHENOL

$C_{10}H_{14}O$

Mol. Wt. 150.21

M.P. 99°C

B.P. 236-238°C/760mm

Solvent: $CDCl_3$

Assignments
a = 1.28
b = 4.91
c = 6.75
d = 7.23

1883

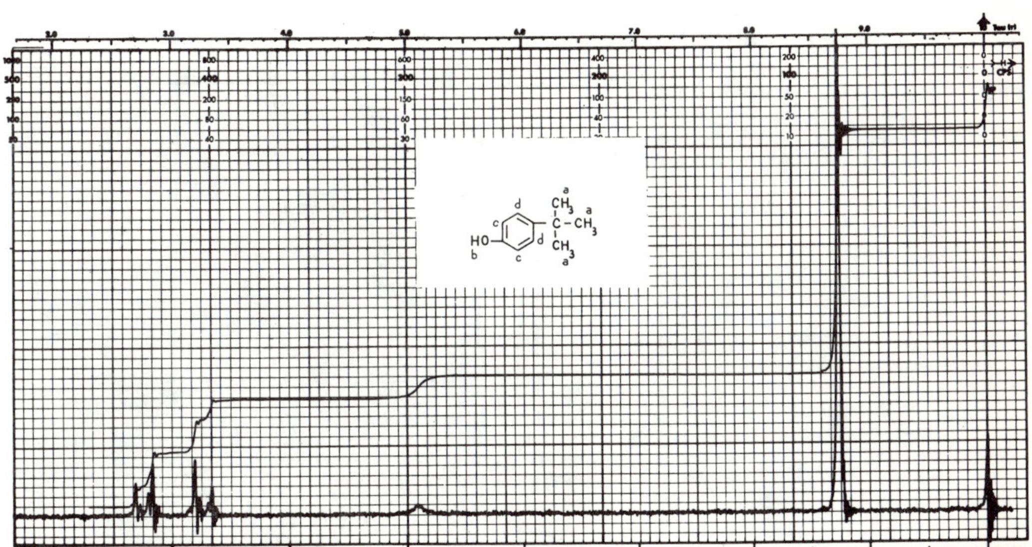

© 1978 Sadtler Research Laboratories, Inc.

p-PENTYLPHENOL

$C_{11}H_{16}O$

Mol. Wt. 164.25

M.P. 20-22°C

Solvent: CCl_4

Assignments
a = 0.89
b = 1.10-1.50
c = 1.52
d = 2.48
e = 6.69
f = 6.79
g = 6.93

1884

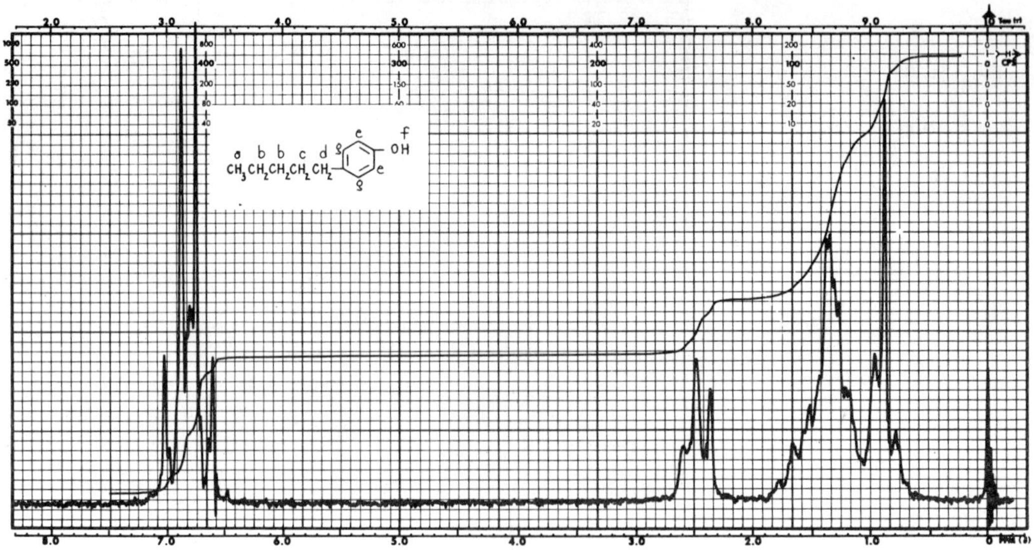

© 1978 Sadtler Research Laboratories, Inc.

p-tert-PENTYLPHENOL

$C_{11}H_{16}O$

Mol. Wt. 164.25

M.P. 89-93°C

Solvent: CCl_4

Assignments
a = 0.68
b = 1.23
c = 1.60
d = 5.46
e = 6.74
f = 7.17
Trace of acid added for OH

1885

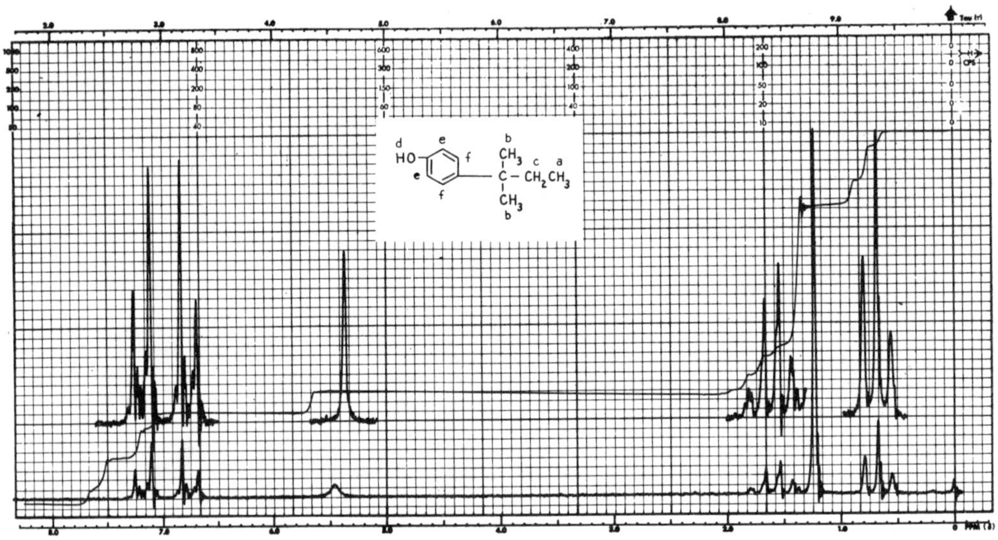

© 1978 Sadtler Research Laboratories, Inc.

p-(1,1,3,3-TETRAMETHYLBUTYL)PHENOL

$C_{14}H_{22}O$

Mol. Wt. 206.34

Solvent: $CDCl_3$

Assignments
a = 0.73
b = 1.32
c = 1.70
d = 4.67
e = 6.70
f = 7.17

1886

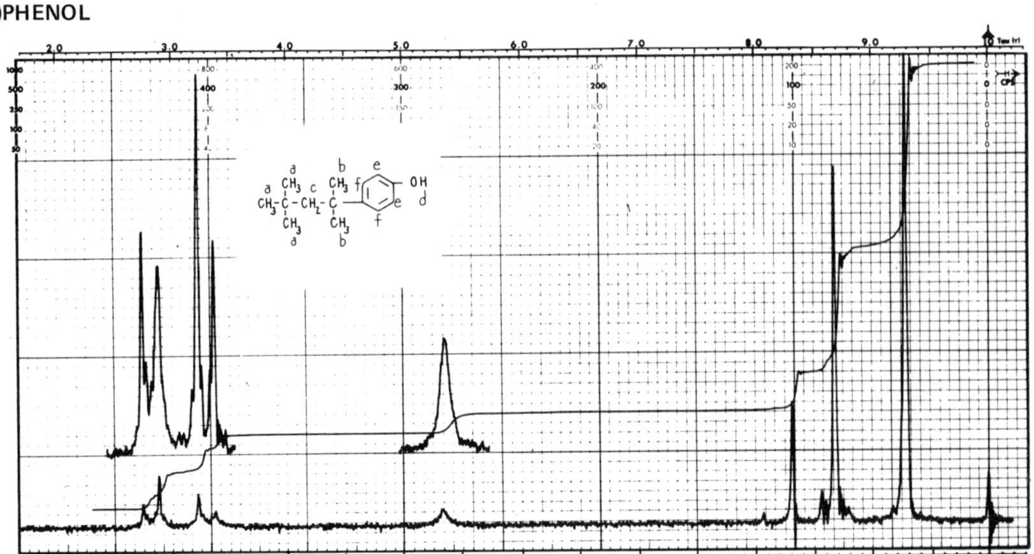

© 1978 Sadtler Research Laboratories, Inc.

p-CYCLOHEXYLPHENOL

$C_{12}H_{16}O$

Mol. Wt. 176.25

Solvent: $CDCl_3$

Assignments
a = 0.91-2.18
b = 2.47
c = 4.95
d = 6.75
e = 7.08

1887

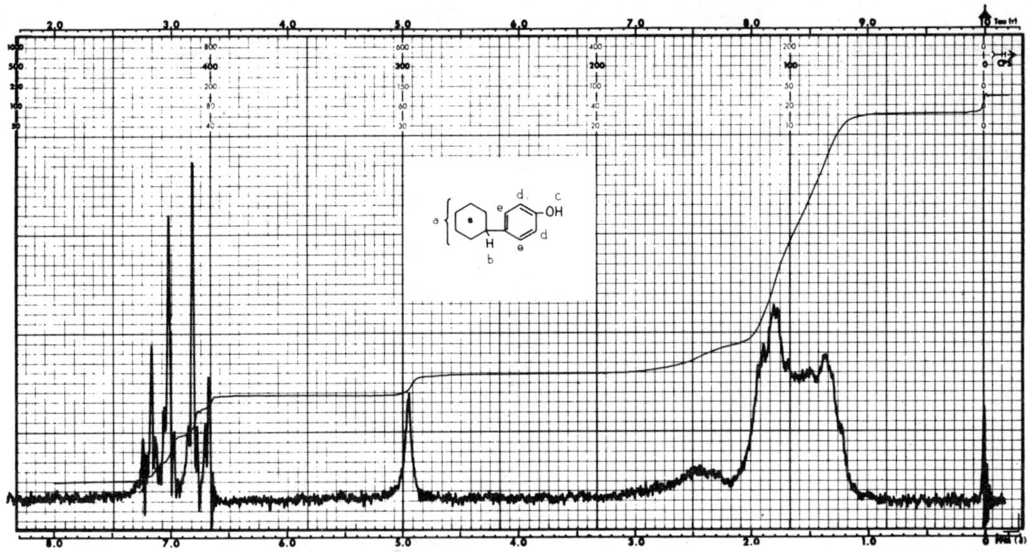

© 1978 Sadtler Research Laboratories, Inc.

---

p,p′-METHYLENEDIPHENOL

$C_{13}H_{12}O_2$

Mol. Wt. 200.24

Solvent: Dimethyl sulfoxide-D_6

Assignments
a = 3.70
b = 6.69
c = 7.00
d = 8.97
e = 2.49 DMSO-D_5
f = 3.32 H_2O

1888

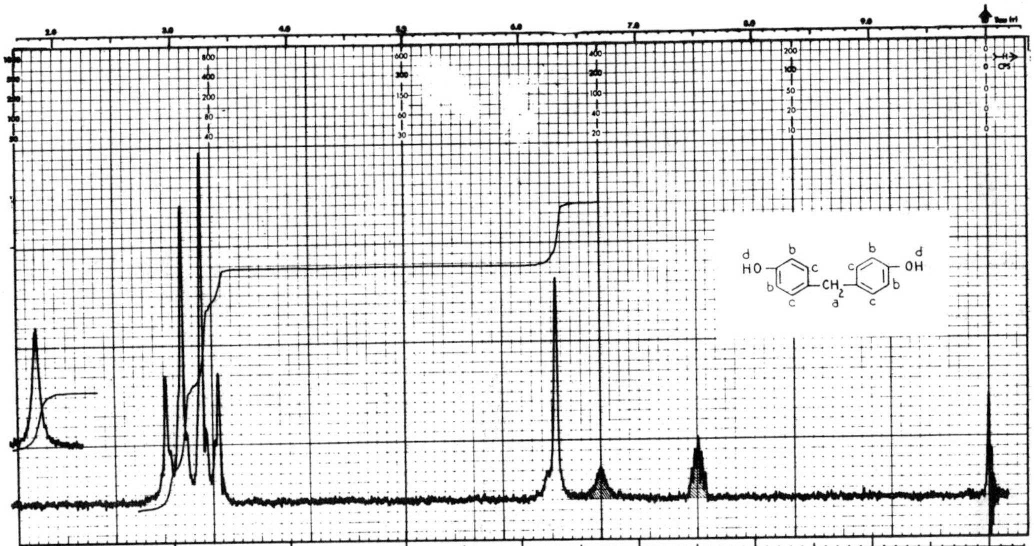

© 1978 Sadtler Research Laboratories, Inc.

---

p-PHENYLPHENOL

$C_{12}H_{10}O$

Mol. Wt. 170.21

Solvent: Dimethyl sulfoxide

Assignments
a = 6.87
b = 7.13-7.70
c = 7.43
d = 9.49

1889

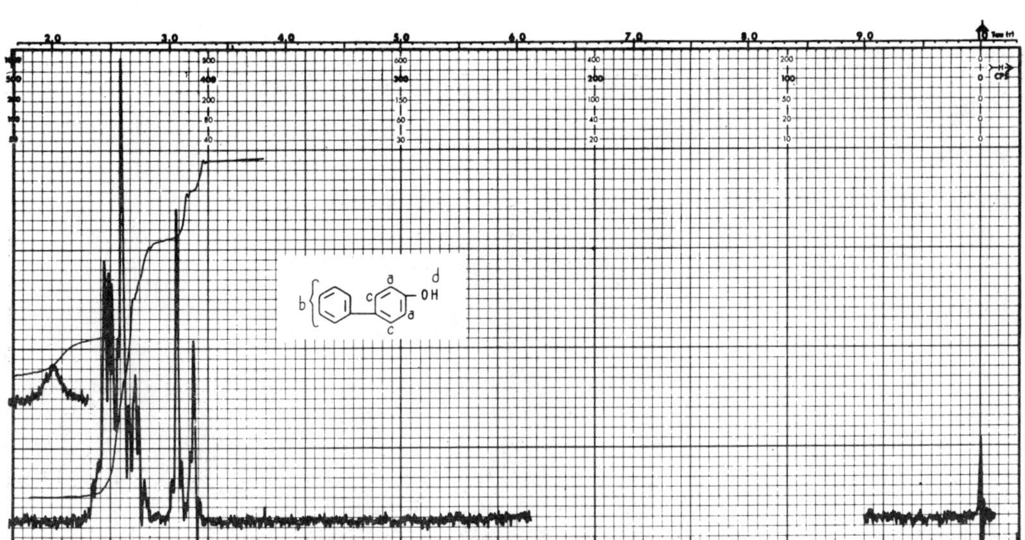

© 1978 Sadtler Research Laboratories, Inc.

o-FLUOROPHENOL

C_6H_5FO

Mol. Wt. 112.10

B.P. 149-151°C

Solvent: CCl_4

Assignments
a = 5.32
b = 6.50-7.20
c = 2.10-2.23 imp.

1890

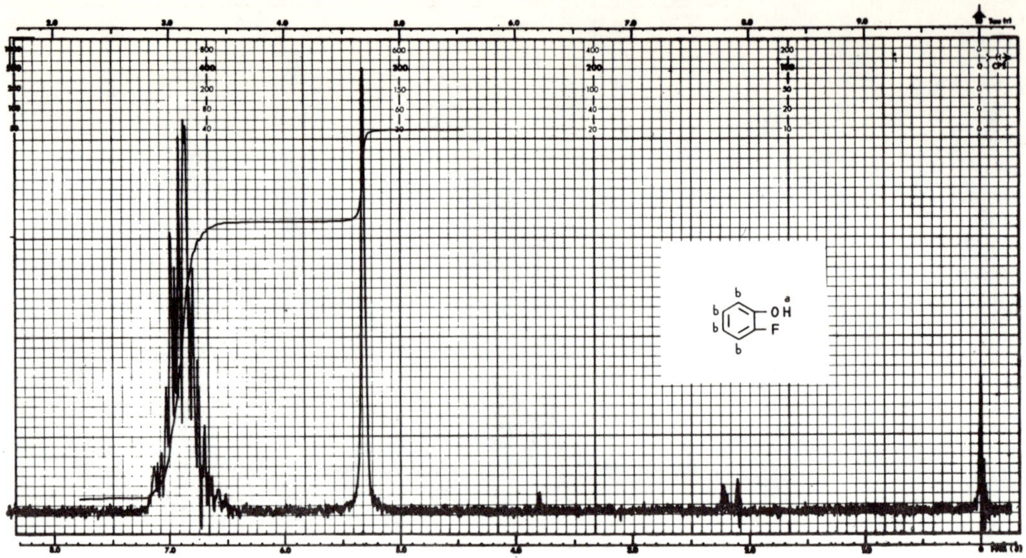

© 1978 Sadtler Research Laboratories, Inc.

p-FLUOROPHENOL

C_6H_5FO

Mol. Wt. 112.10

M.P. 47-49°C

Solvent: $CDCl_3$

Assignments
a = 6.53
b = 6.79
c = 6.91

1891

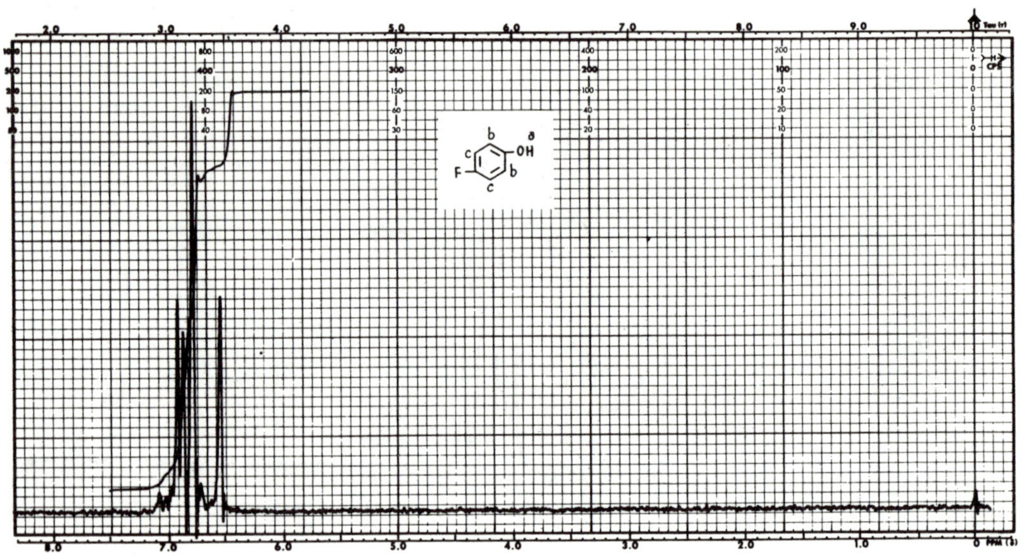

© 1978 Sadtler Research Laboratories, Inc.

o-CHLOROPHENOL

C_6H_5ClO

Mol. Wt. 128.56

M.P. 9-10°C

B.P. 62-63°C/11mm

Solvent: $CDCl_3$

Assignments
a = 5.37
b = 6.60-7.38
OH under TFA

1892

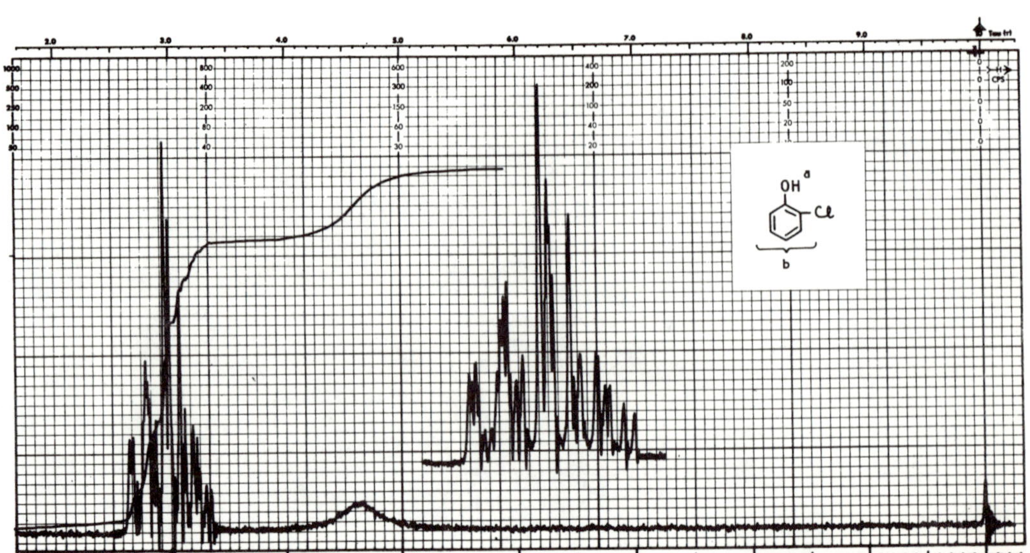

© 1978 Sadtler Research Laboratories, Inc.

p-CHLOROPHENOL

C_6H_5ClO

Mol. Wt. 128.56

M.P. 43-45°C

Solvent: $CDCl_3$

Assignments
a = 6.33
b = 6.71
c = 7.10

1893

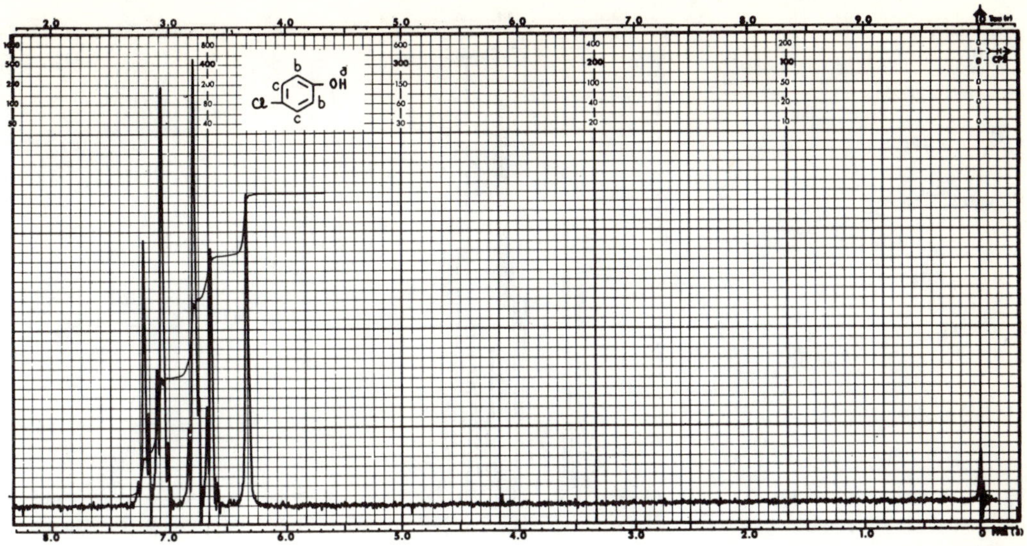

© 1978 Sadtler Research Laboratories, Inc.

o-BROMOPHENOL

C_6H_5BrO

Mol. Wt. 173.01

B.P. 195°C

Solvent: $CDCl_3$

Assignments
a = 5.08
b = 6.74
c = 6.98
d = 7.07
e = 7.43

1894

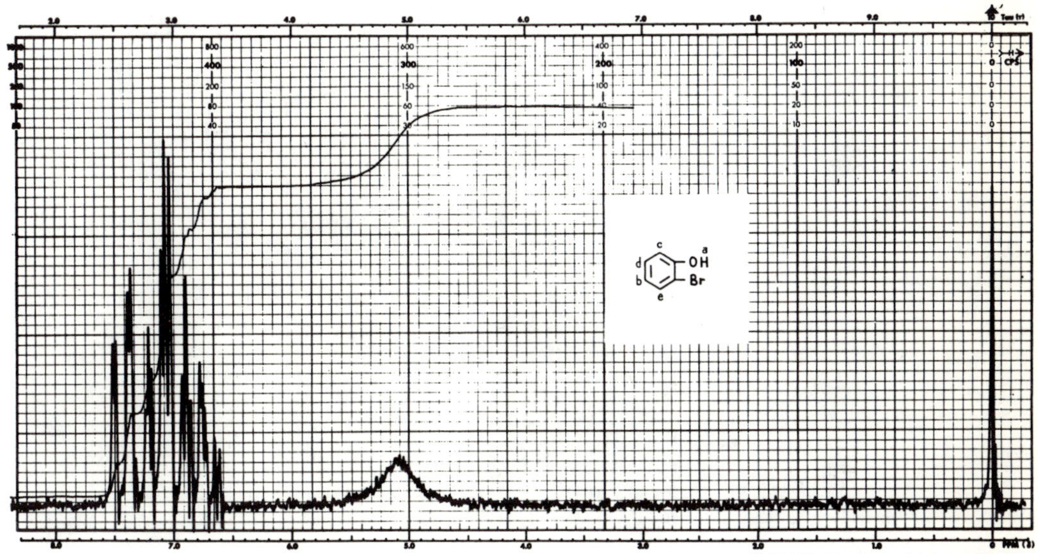

© 1978 Sadtler Research Laboratories, Inc.

p-BROMOPHENOL

C_6H_5BrO

Mol. Wt. 173.02

M.P. 64-66°C

Solvent: $CDCl_3$

Assignments
a = 5.20
b = 6.69
c = 7.30

1895

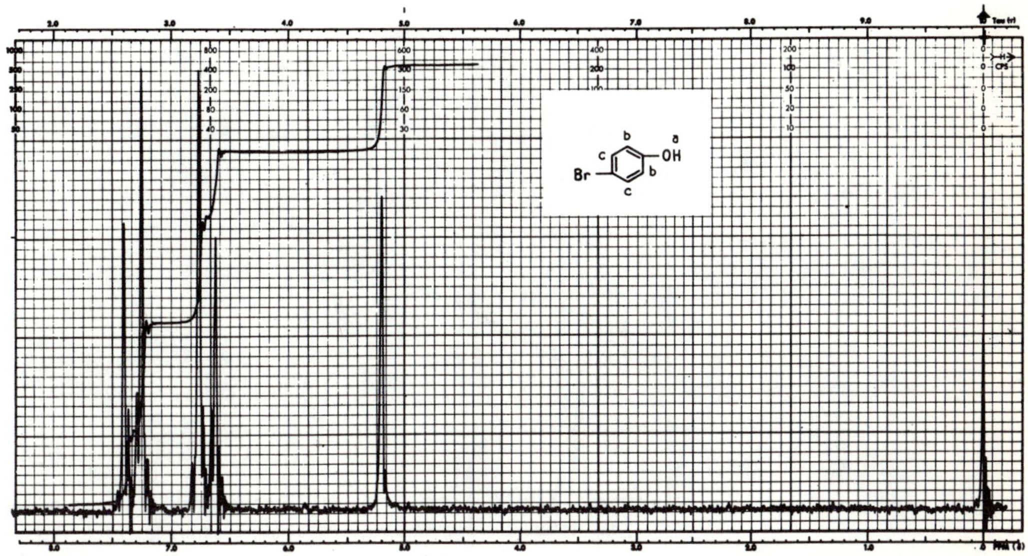

© 1978 Sadtler Research Laboratories, Inc.

o-IODOPHENOL

C_6H_5IO

Mol. Wt. 220.01

M.P. 41-43°C

Solvent: $CDCl_3$

Assignments

a = 5.55
b = 6.59
c = 6.98
d = 7.15
e = 7.58

1896

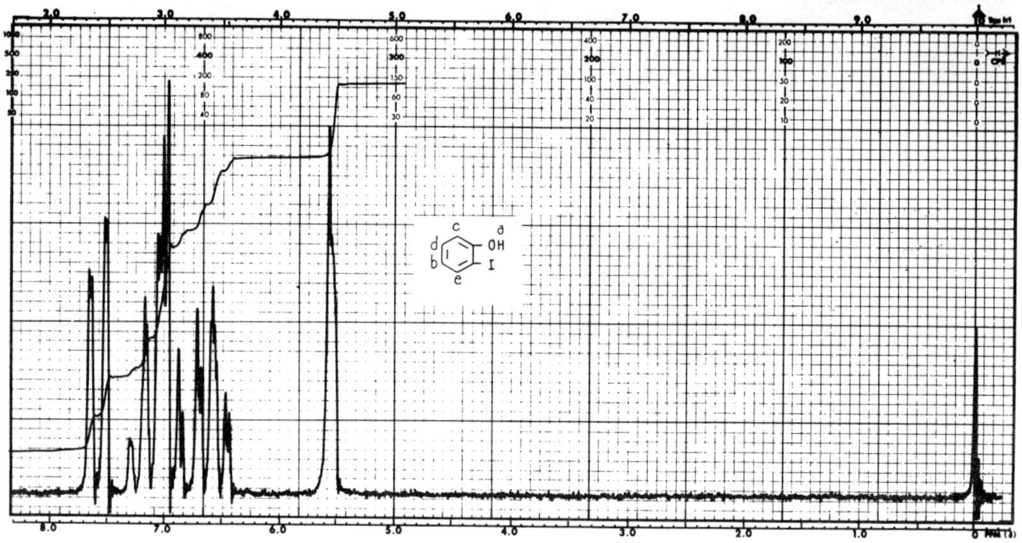

© 1978 Sadtler Research Laboratories, Inc.

---

m-IODOPHENOL

C_6H_5IO

Mol. Wt. 220.01

M.P. 40-42°C

Solvent: Polysol-d

Assignments

a = 6.70-7.00
b = 7.12
c = 7.28
d = 9.24

1897

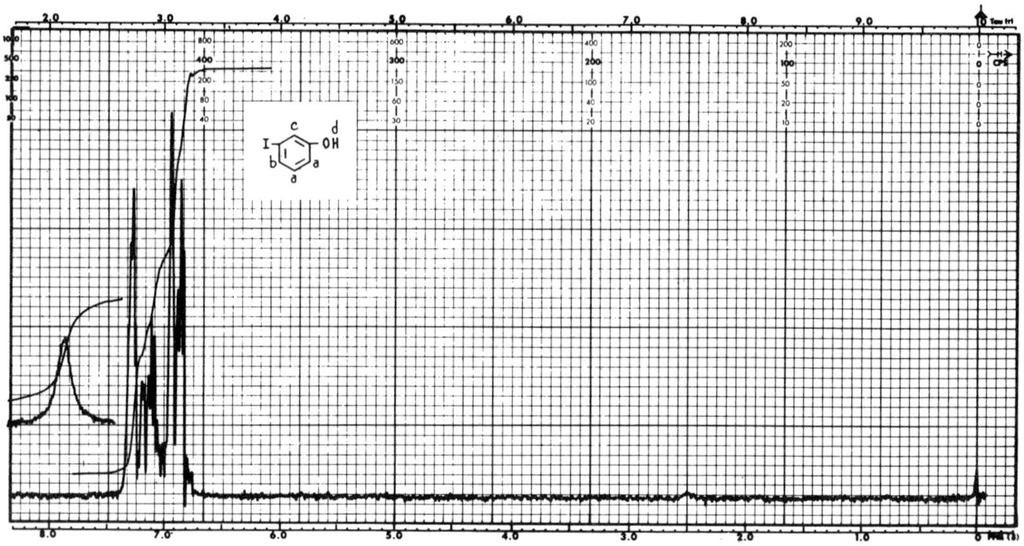

© 1978 Sadtler Research Laboratories, Inc.

---

p-IODOPHENOL

C_6H_5IO

Mol. Wt. 220.01

M.P. 94°C

Solvent: $CDCl_3$

Assignments

a = 5.41
b = 6.59
c = 7.48

1898

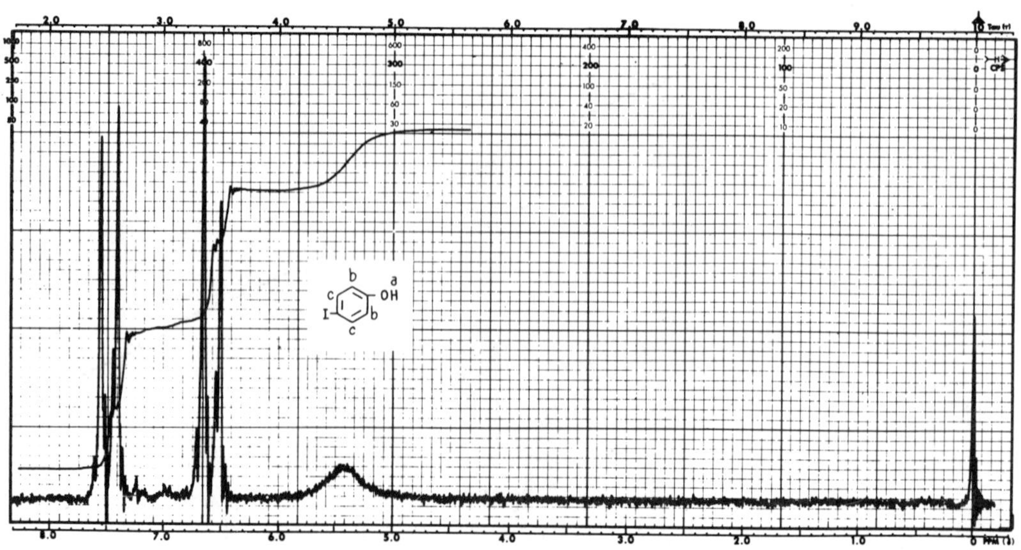

© 1978 Sadtler Research Laboratories, Inc.

m-AMINOPHENOL

C_6H_7NO

Mol. Wt. 109.13

M.P. 114°C

Solvent: Dimethyl sulfoxide

Assignments
a = 4.71
b = ca 5.99
c = 6.06
d = 6.78
e = 8.78

1899

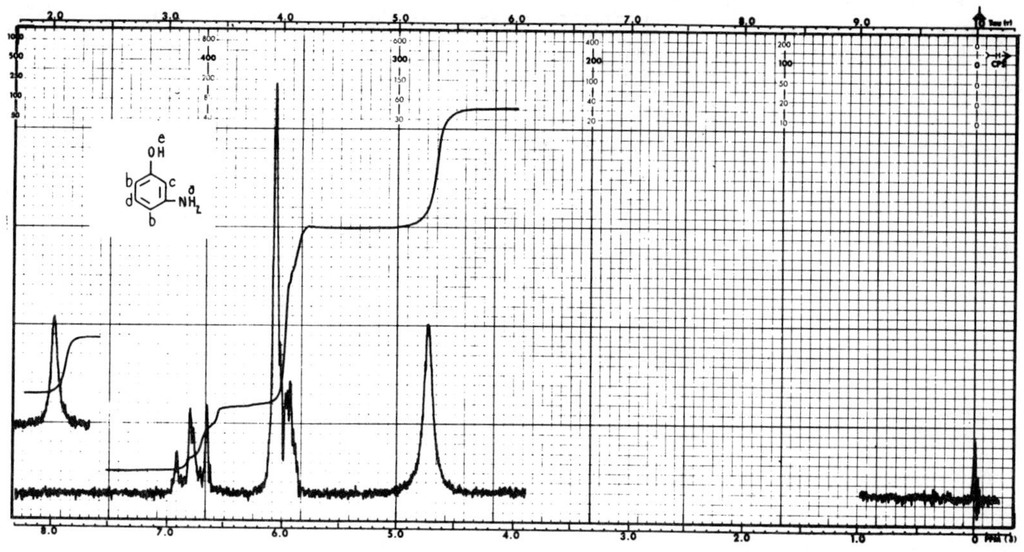

© 1978 Sadtler Research Laboratories, Inc.

---

m-DIMETHYLAMINOPHENOL

$C_8H_{11}NO$

Mol. Wt. 137.18

M.P. 85°C

Solvent: $CDCl_3$

Assignments
a = 2.85
b = 6.05
c = 6.10-6.50
d = 7.09

1900

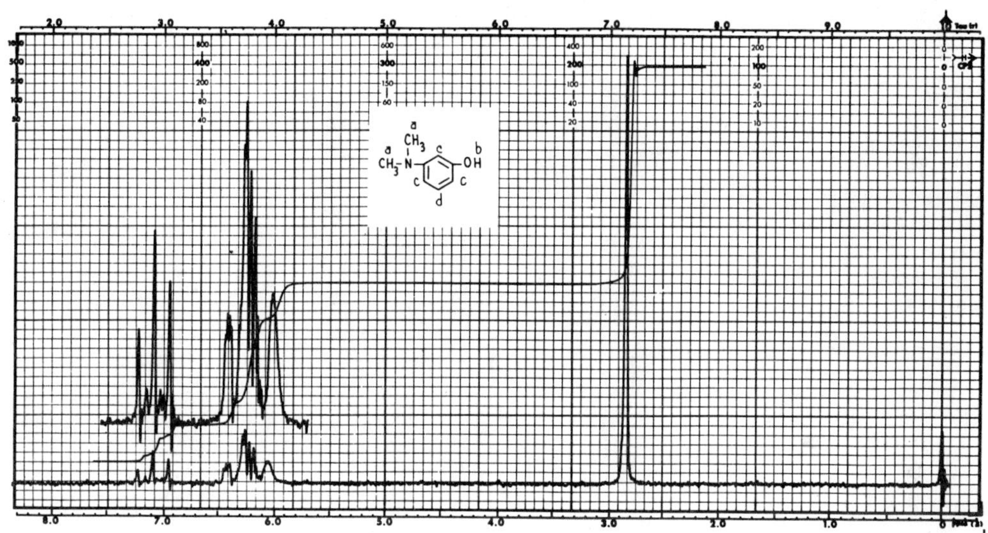

© 1978 Sadtler Research Laboratories, Inc

---

a,a-BIS[p-(DIMETHYLAMINO)PHENYL]-p-CRESOL

$C_{23}H_{26}N_2O$

Mol. Wt. 346.48

M.P. 166-168°C

Solvent: Polysol-d

Assignments
a = 2.89
b = 5.26
c = 6.63
d = 6.70
e = 7.00
f = 8.60

1901

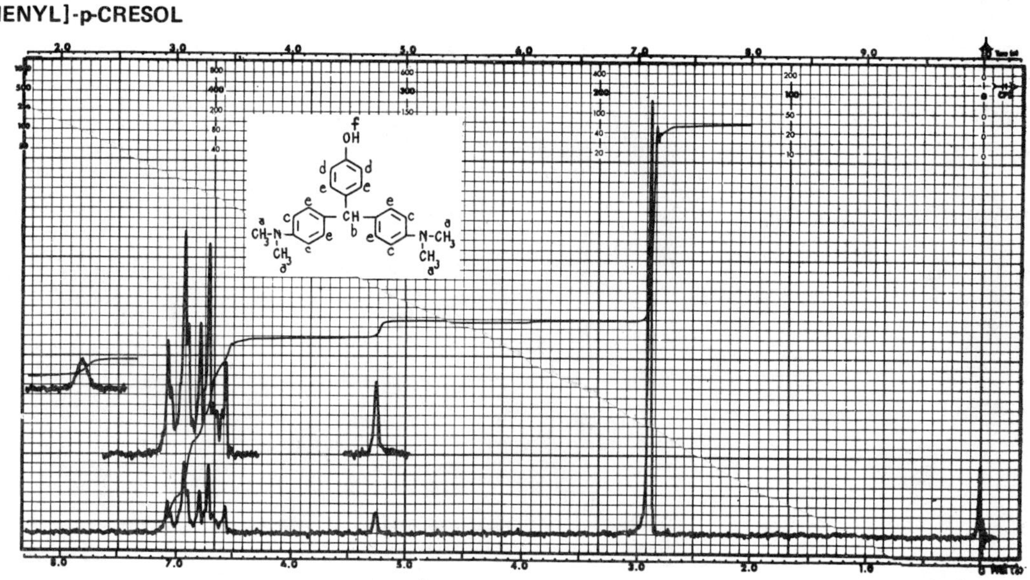

© 1978 Sadtler Research Laboratories, Inc.

p-HYDROXYHYDROCINNAMONITRILE

C_9H_9NO

Mol. Wt. 147.18

Solvent: $CDCl_3$

Assignments
a = 2.50
b = 2.75
c = 6.34
d = 6.77
e = 7.03

1902

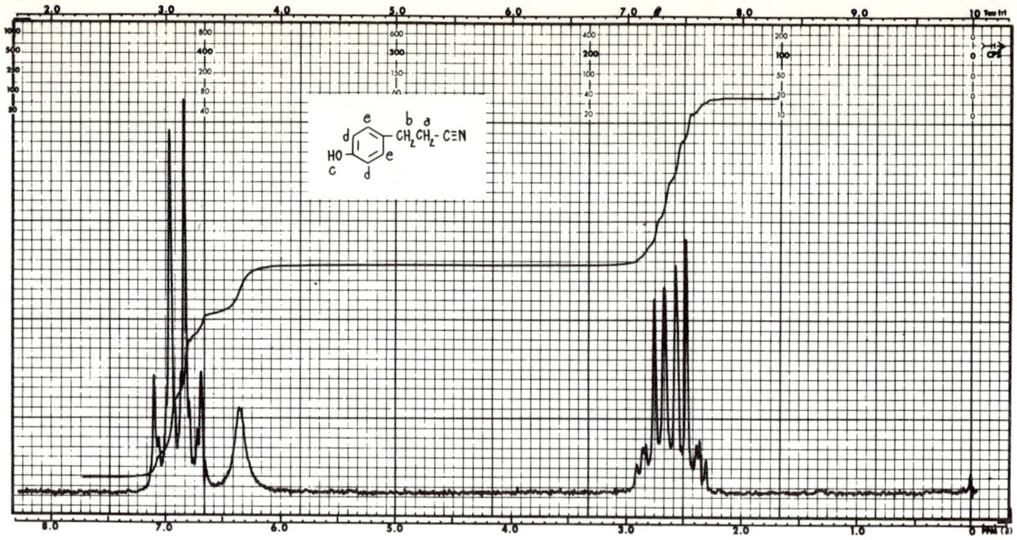

© 1978 Sadtler Research Laboratories, Inc.

---

o-NITROPHENOL

$C_6H_5NO_3$

Mol. Wt. 139.11

M.P. 44-45°C

Solvent: CCl_4

Assignments
a = 6.90
b = 7.09
c = 7.51
d = 8.00
e = 10.40

1903

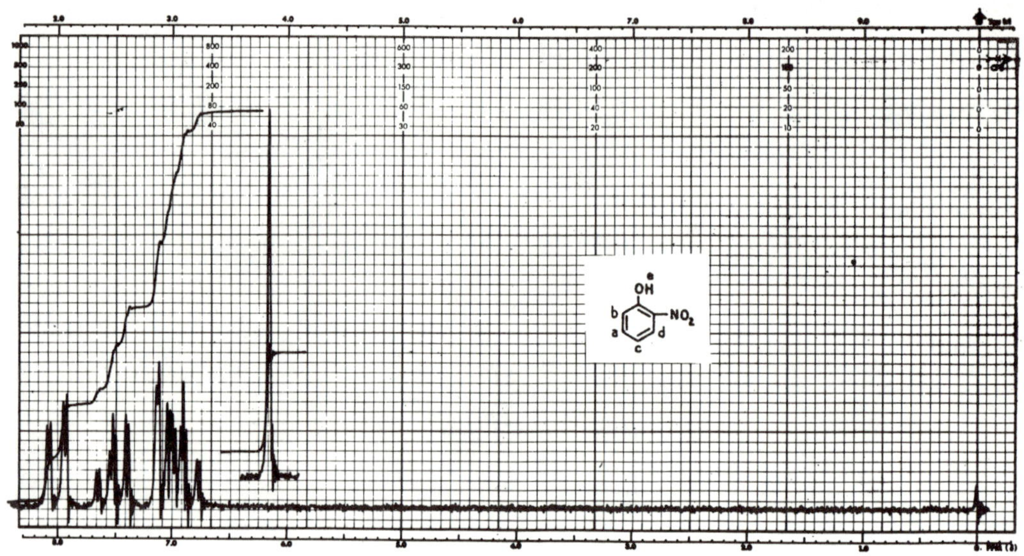

© 1978 Sadtler Research Laboratories, Inc.

---

m-NITROPHENOL

$C_6H_5NO_3$

Mol. Wt. 154.13

M.P. 142-143°C

Solvent: Dimethyl sulfoxide

Assignments
a = 7.19
b = 7.41
c = 7.57
d = 7.61
e = 10.31

1904

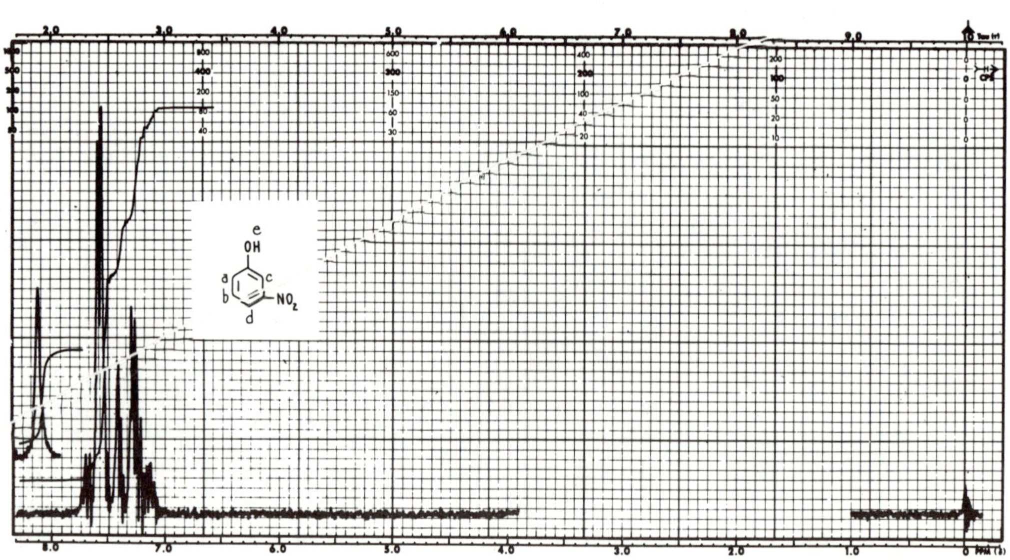

© 1978 Sadtler Research Laboratories, Inc.

p-NITROPHENOL

$C_6H_5NO_3$

Mol. Wt. 139.11

M.P. 112-113°C

Solvent: Acetone

Assignments
a = 7.06
b = 8.15
c = 9.32

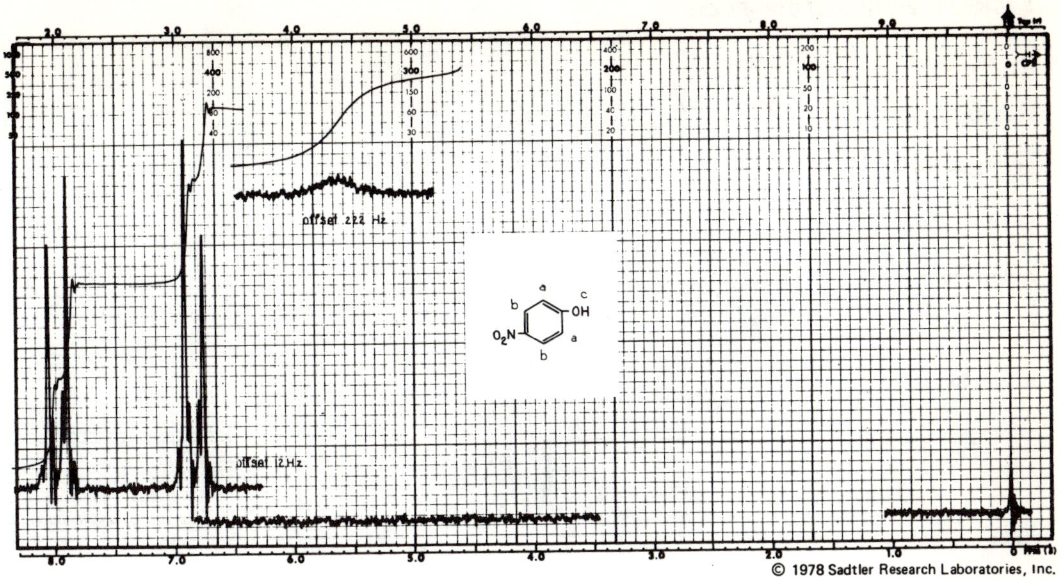

1905

© 1978 Sadtler Research Laboratories, Inc.

m-METHOXYPHENOL

$C_7H_8O_2$

Mol. Wt. 124.14

B.P. 244°C

Solvent: CCl_4

Assignments
a = 3.61
b = ca 6.34
c = 6.76
d = 7.00

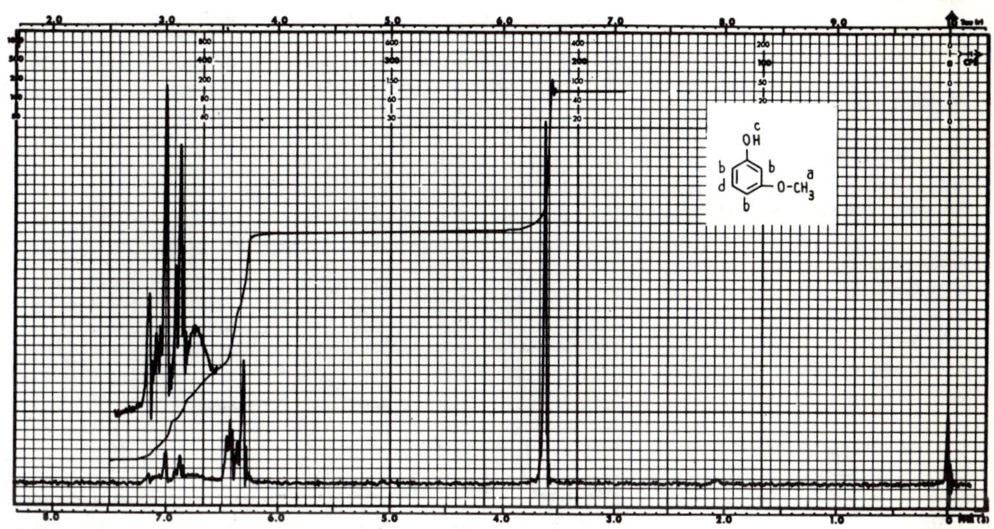

1906

© 1978 Sadtler Research Laboratories, Inc.

o-HYDROXYBENZYL ALCOHOL

$C_7H_8O_2$

Mol. Wt. 124.14

Solvent: Dimethyl sulfoxide-d_6

Assignments
a = 4.55
b = 4.69
c = 6.65=7.20
d = 7.31
e = 9.22
f = 2.49 DMSO-d_5
g = 3.80 H_2O

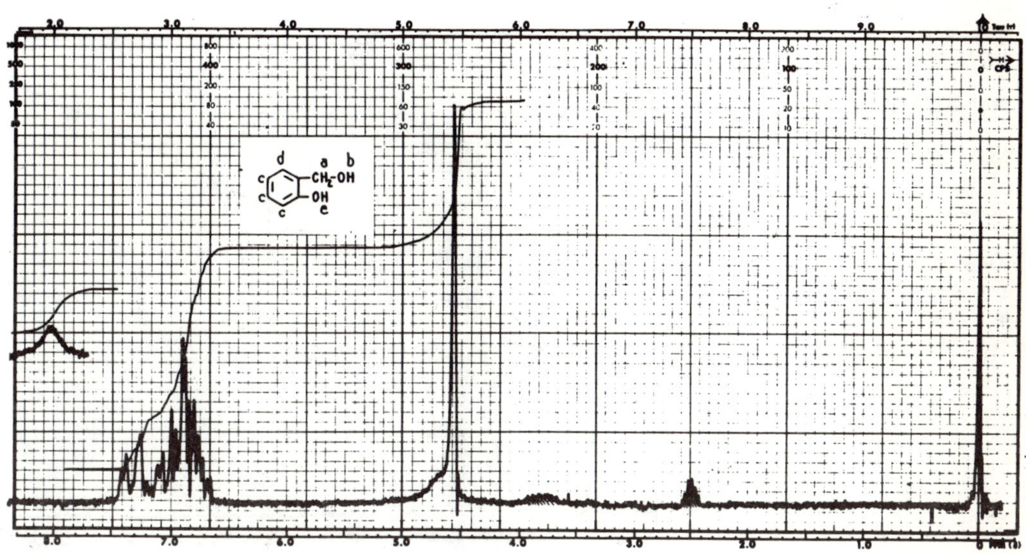

1907

© 1978 Sadtler Research Laboratories, Inc.

4,4'-ISOPROPYLIDENEDIPHENOL

$C_{15}H_{16}O_2$

Mol. Wt. 288.29

M.P. 156-158°C

Solvent: Polysol-d

Assignments
a = 1.57
b = 6.70
c = 7.04
d = 7.60

1908

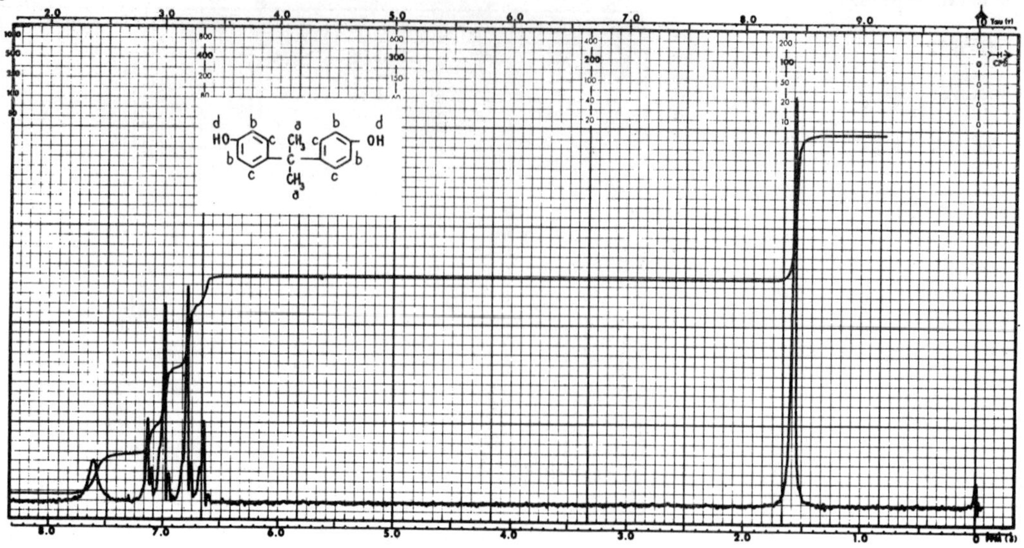

© 1978 Sadtler Research Laboratories, Inc.

---

p,p'-BIPHENOL

$C_{12}H_{10}O_2$

Mol. Wt. 186.21

M.P. 280°C

Solvent: Dimethyl sulfoxide-D_6

Assignments
a = 6.83
b = 7.43
c = 9.20
d = 2.50 DMSO-D_5

1909

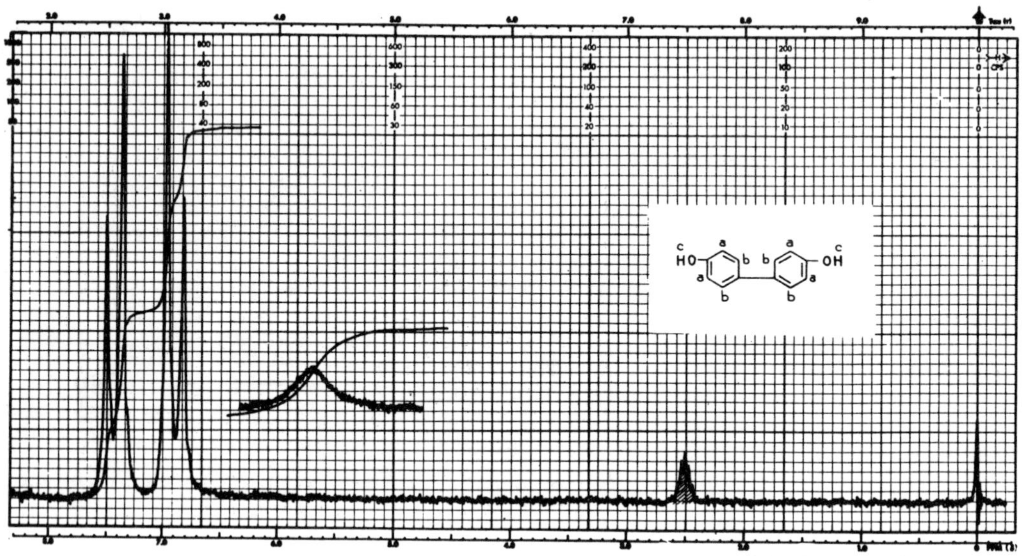

© 1978 Sadtler Research Laboratories, Inc.

PYROCATECHOL

$C_6H_6O_2$

Mol. Wt. 110.11

M.P. 104-106°C

Solvent: polysol-d

Assignments
a = 6.69
b = 6.83
c = 7.51

1910

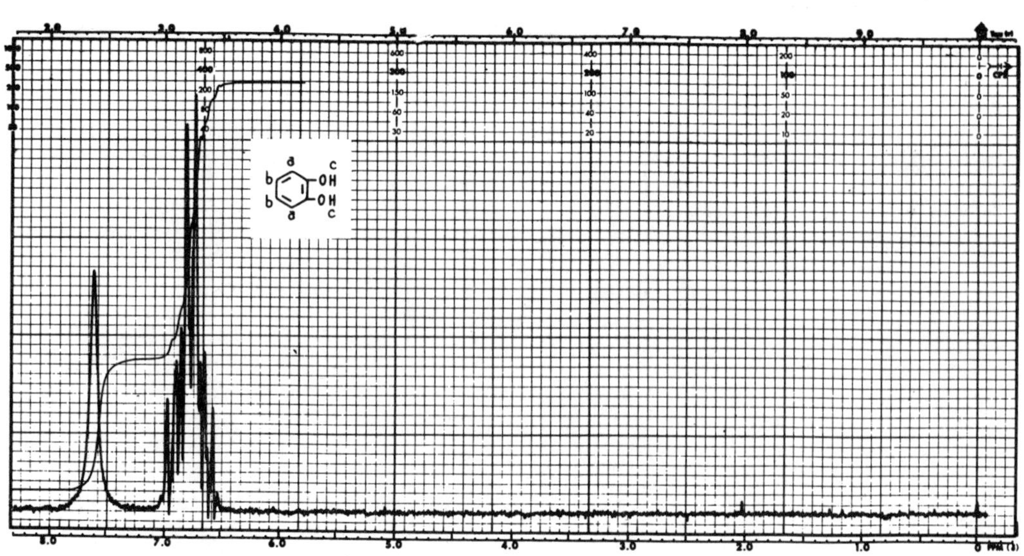

© 1978 Sadtler Research Laboratories, Inc.

RESORCINOL

$C_6H_6O_2$

Mol. Wt. 110.11

M.P. 109-111°C

Solvent: Acetone

Assignments
a = ca 6.49
b = 7.07
c = 8.16

1911

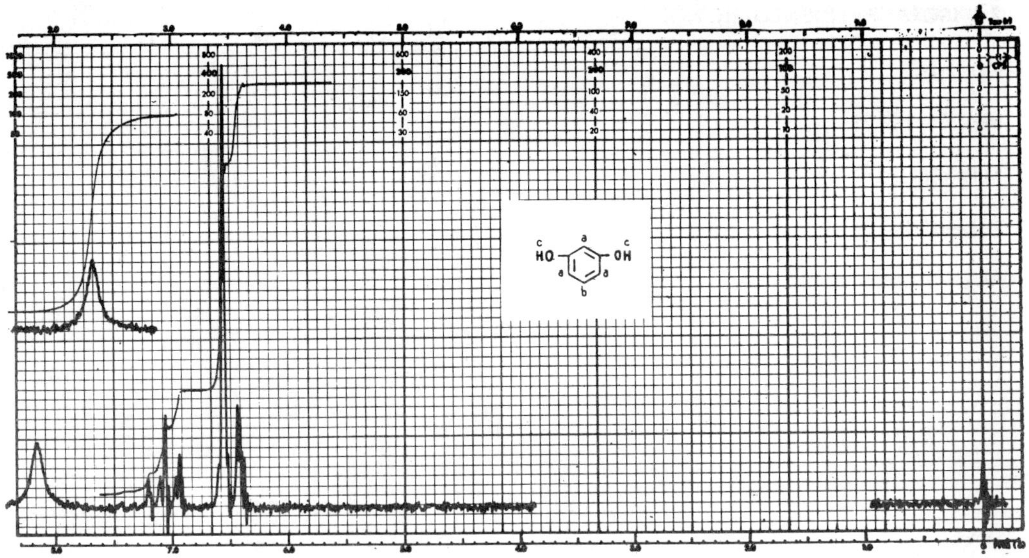

© 1978 Sadtler Research Laboratories, Inc.

HYDROQUINONE

$C_6H_6O_2$

Mol. Wt. 110.11

Solvent: D_2O

Assignments
a = 6.79
b = 4.68 HDO

1912

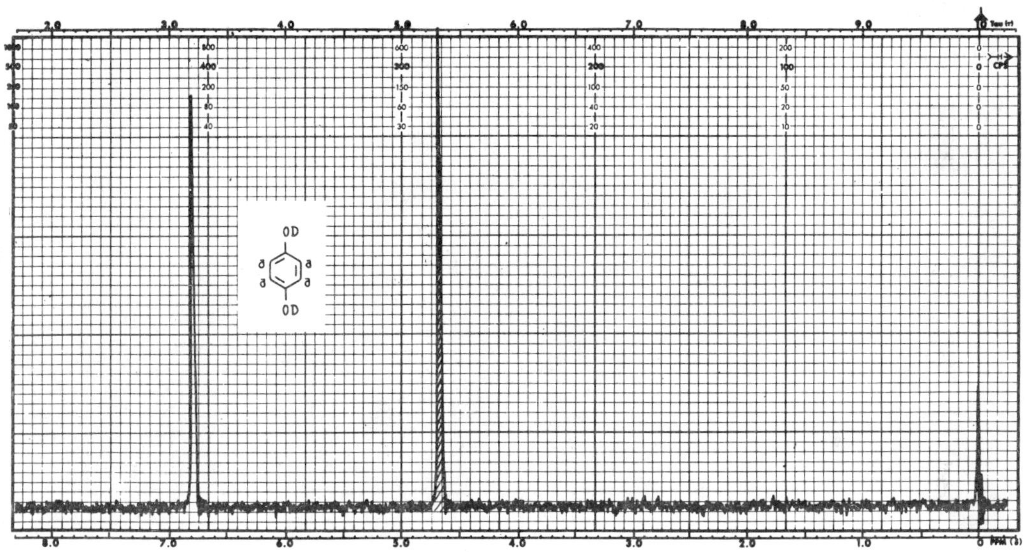

© 1978 Sadtler Research Laboratories, Inc.

2,3-XYLENOL

$C_8H_{10}O$

Mol. Wt. 122.17

M.P. 75°C

B.P. 218°C

Solvent: $CDCl_3$

Assignments
a = 2.14
b = 2.24
c = 4.84
d = 6.58
e = 6.69
f = 6.94

1913

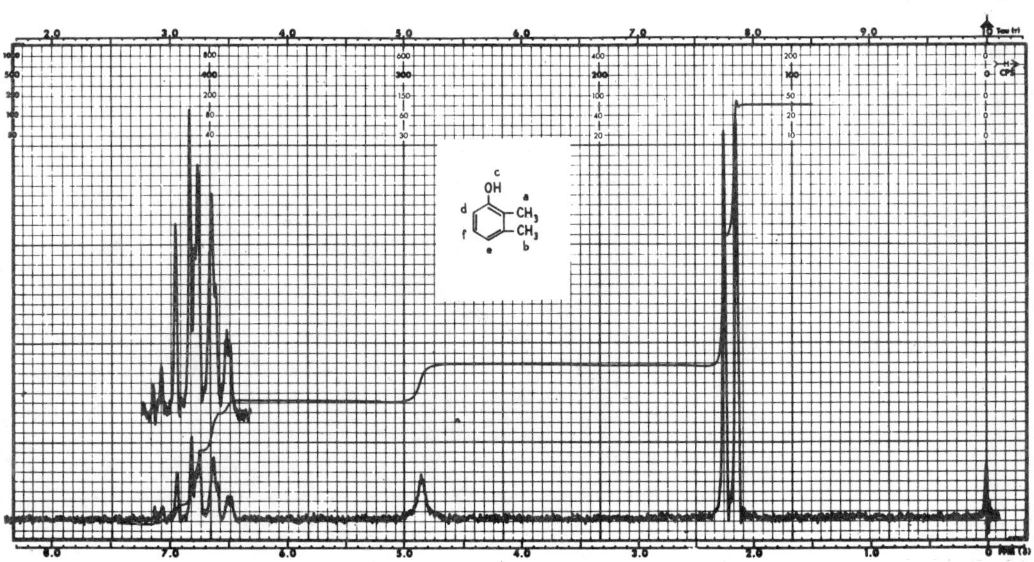

© 1978 Sadtler Research Laboratories, Inc.

3,4-XYLENOL

$C_8H_{10}O$

Mol. Wt. 122.17

M.P. 65°C

B.P. 228-229°C (lit.)

Solvent: CCl_4

Assignments
a = 2.12
b = 5.40
c = ca 6.46
d = 6.82
Trace of acid added for OH

1914

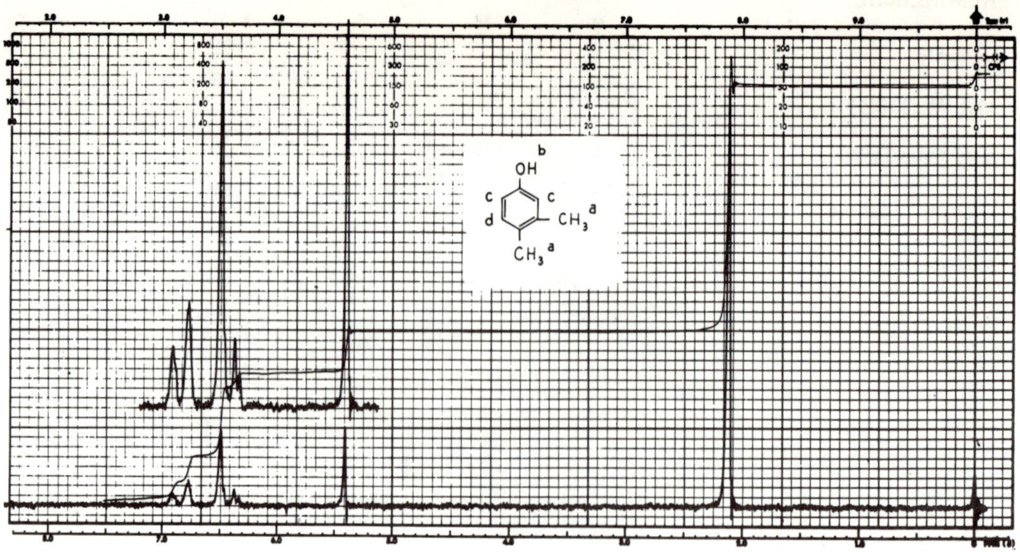

© 1978 Sadtler Research Laboratories, Inc.

3,5-XYLENOL

$C_8H_{10}O$

Mol. Wt. 122.16

M.P. 68°C

Solvent: CCl_4

Assignments
a = 2.18
b = 5.73
c = 6.33
d = 6.45

1915

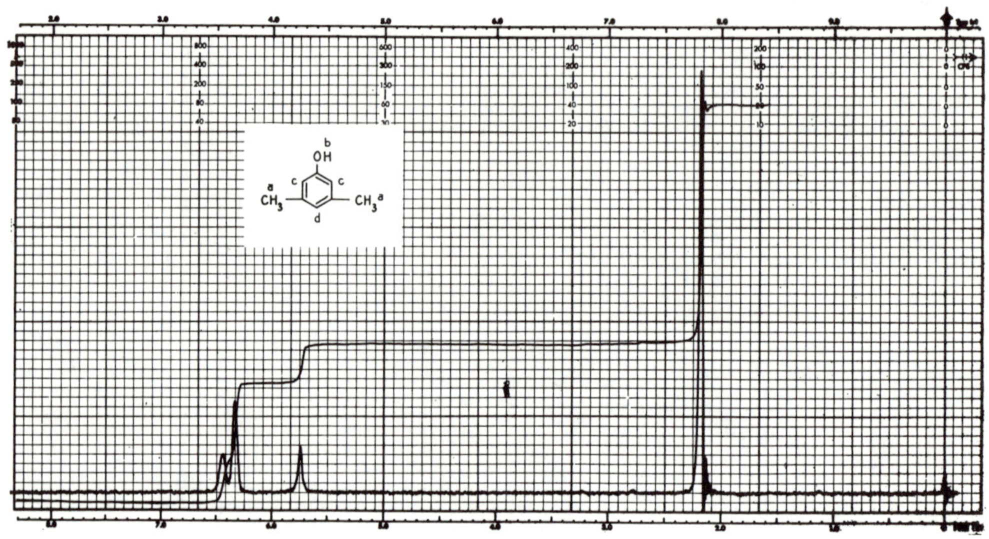

© 1978 Sadtler Research Laboratories, Inc.

2,6-XYLENOL

$C_8H_{10}O$

Mol. Wt. 122.17

Solvent: CCl_4

Assignments
a = 2.18
b = 4.39
c = 6.61
d = 6.82

1916

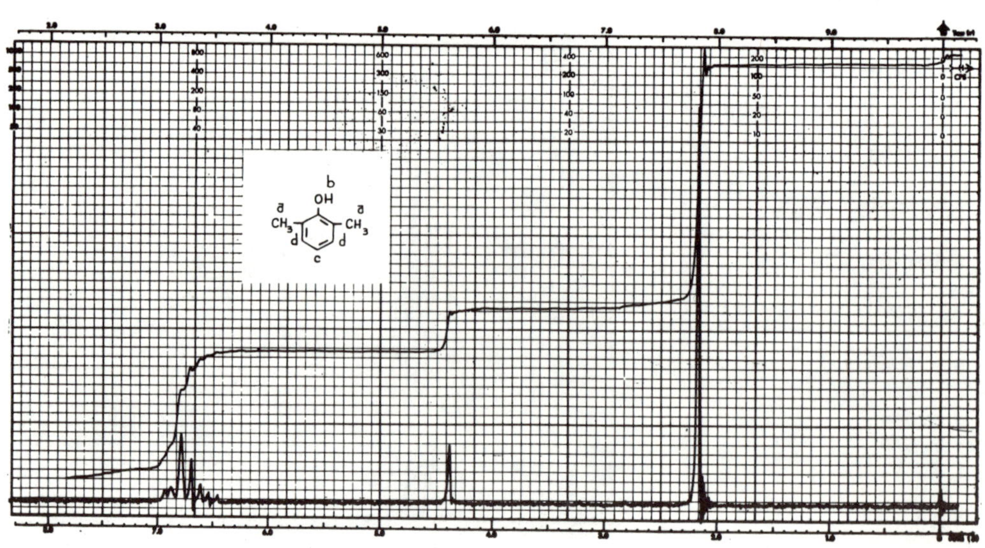

© 1978 Sadtler Research Laboratories, Inc.

THYMOL

$C_{10}H_{14}O$

Mol. Wt. 150.22

M.P. 50°C

Solvent: $CDCl_3$

Assignments
a = 1.22
b = 2.21
c = 3.15
d = 4.77
e = 6.53
f = 6.71
g = 7.07

1917

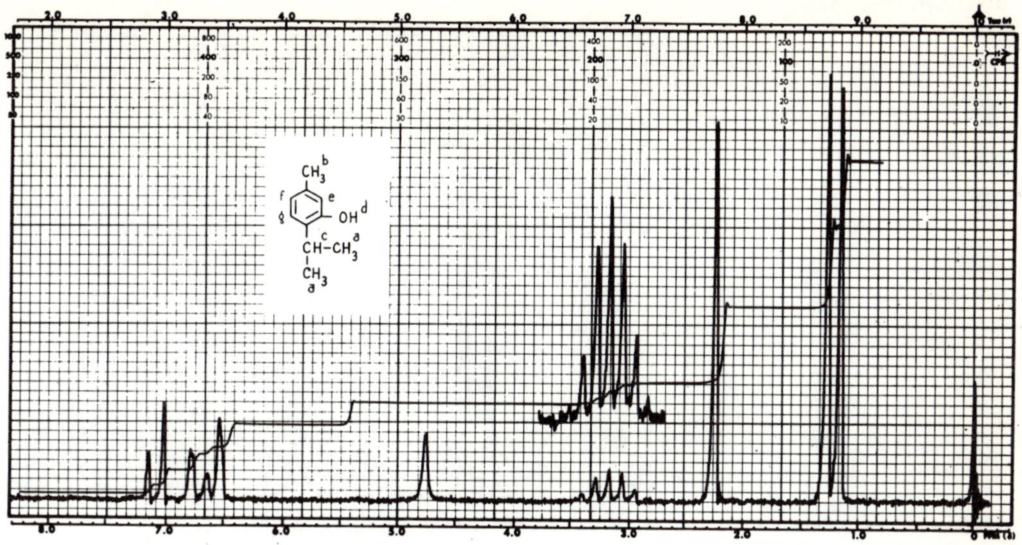

© 1978 Sadtler Research Laboratories, Inc.

6-ALLYL-o-CRESOL

$C_{10}H_{12}O$

Mol. Wt. 148.12

Solvent: CCl_4

Assignments
a = 2.12
b = 3.26
c = 4.79
d = ca 5.02
e = 5.84
f = 6.62
g = ca 6.80

1918

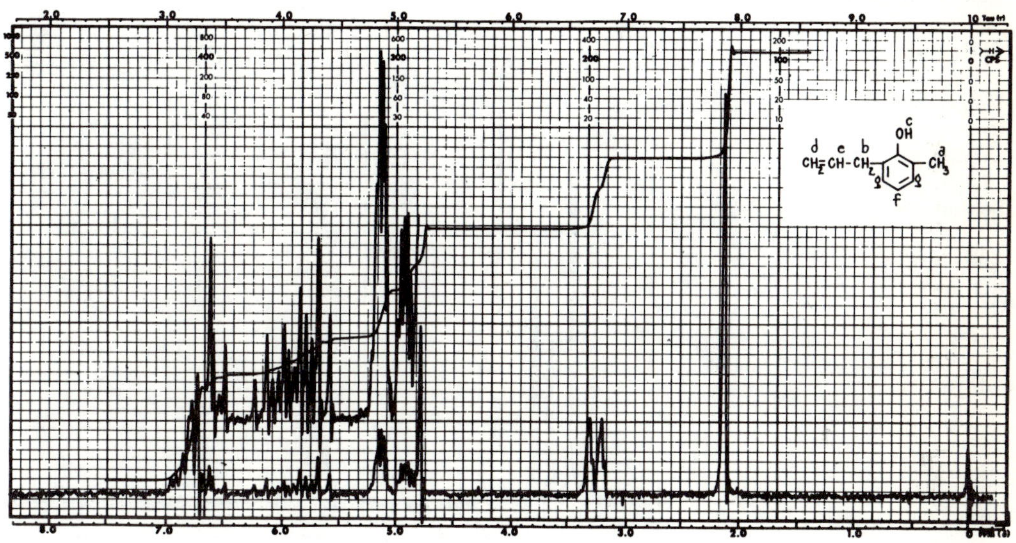

© 1978 Sadtler Research Laboratories, Inc.

2,4-DICHLOROPHENOL

$C_6H_4Cl_2O$

Mol. Wt. 163.00

M.P. 40-42°C

Solvent: $CDCl_3$

Assignments
a = 5.68
b = 6.90
c = 7.13
d = 7.28

1919

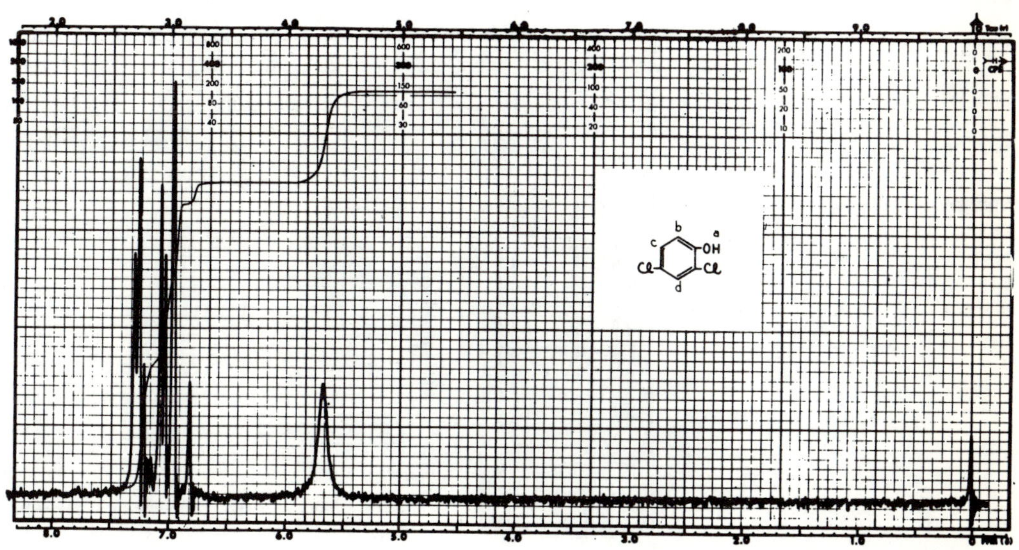

© 1978 Sadtler Research Laboratories, Inc.

3,5-DICHLOROPHENOL

$C_6H_4Cl_2O$

Mol. Wt. 163.00

M.P. 66-68°C

Solvent: $CDCl_3$

Assignments
a = 5.28
b = 6.71
c = 6.92

1920

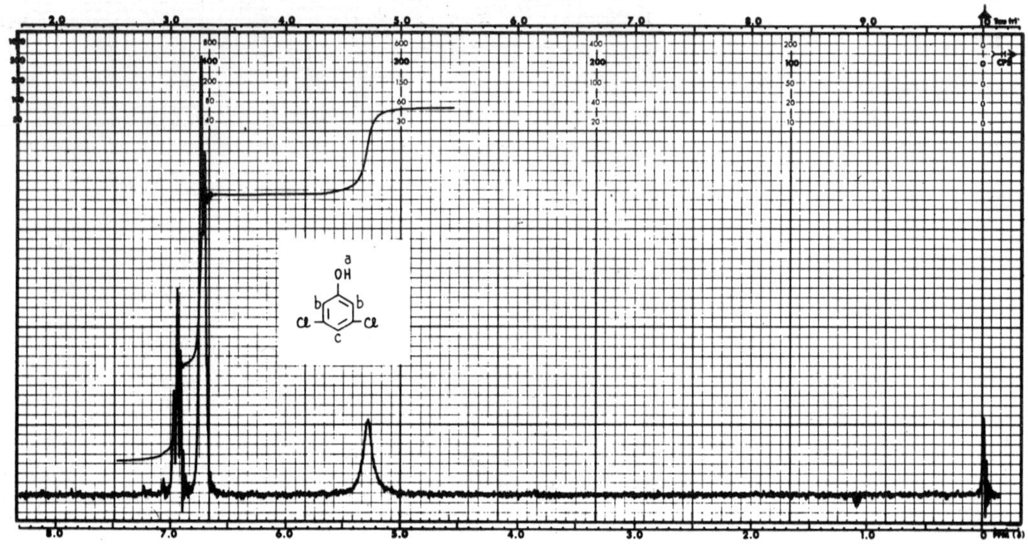

© 1978 Sadtler Research Laboratories, Inc.

2,4-DIBROMOPHENOL

$C_6H_4Br_2O$

Mol. Wt. 251.92

M.P. 38°C

Solvent: CCl_4

Assignments
a = 5.41
b = 6.91
c = 7.36
d = 7.62

1921

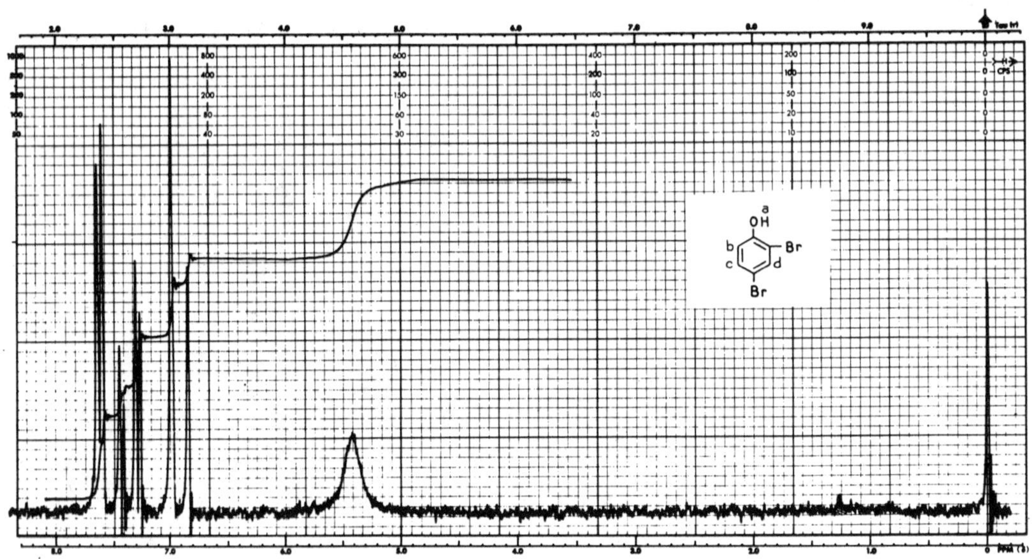

© 1978 Sadtler Research Laboratories, Inc.

2-CHLORO-6-NITROPHENOL

$C_6H_4ClNO_3$

Mol. Wt. 173.56

M.P. 70-71°C

Solvent: $CDCl_3$

Assignments
a = 7.01
b = 7.71
c = 8.05
d = 10.97

1922

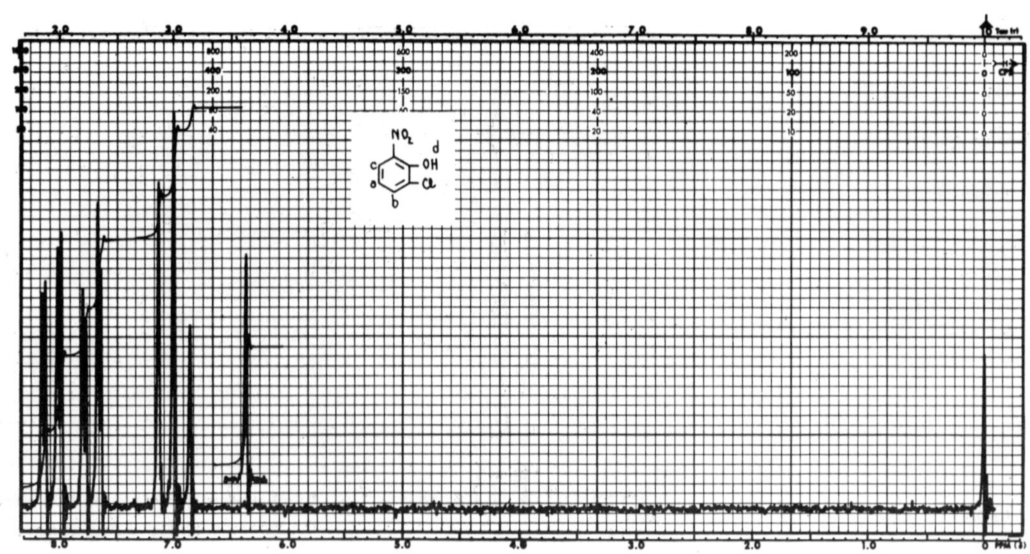

© 1978 Sadtler Research Laboratories, Inc.

4-ALLYL-2-METHOXYPHENOL

$C_{10}H_{12}O_2$

Mol. Wt. 164.21

Solvent: CCl_4

Assignments
a = 3.21
b = 3.70
c = 4.98
d = 5.40
e = 5.80
f = ca 6.53
g = 6.71
D_2O added on expansion

1923

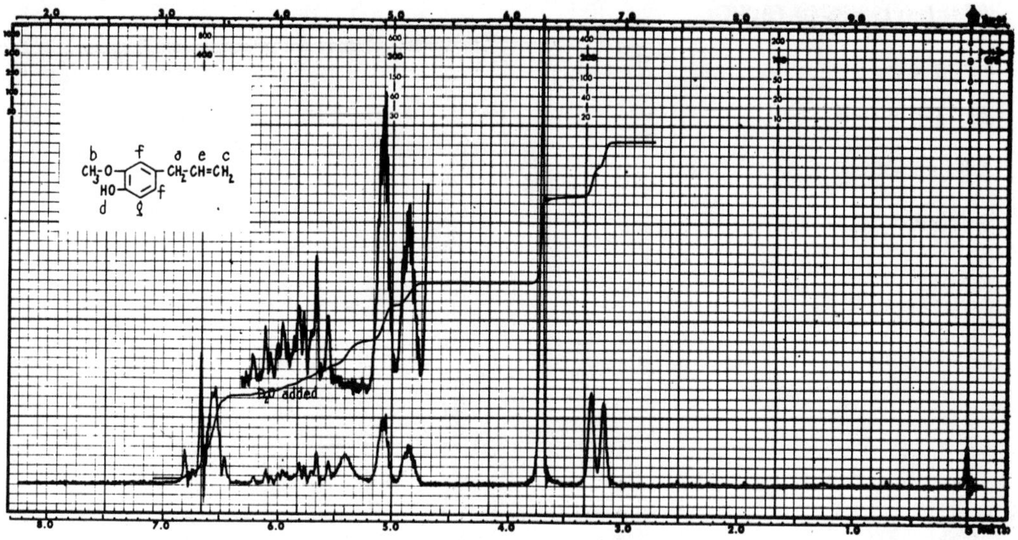

© 1978 Sadtler Research Laboratories, Inc.

4-ALLYL-2-METHOXYPHENOL

$C_{10}H_{12}O_2$

Mol. Wt. 164.21

Solvent: CCl_4

Assignments
a = 3.21
b = 3.70
c = 4.98
d = 5.40
e = 5.80
f = ca 6.53
g = 6.71
D_2O added on expansion

1924

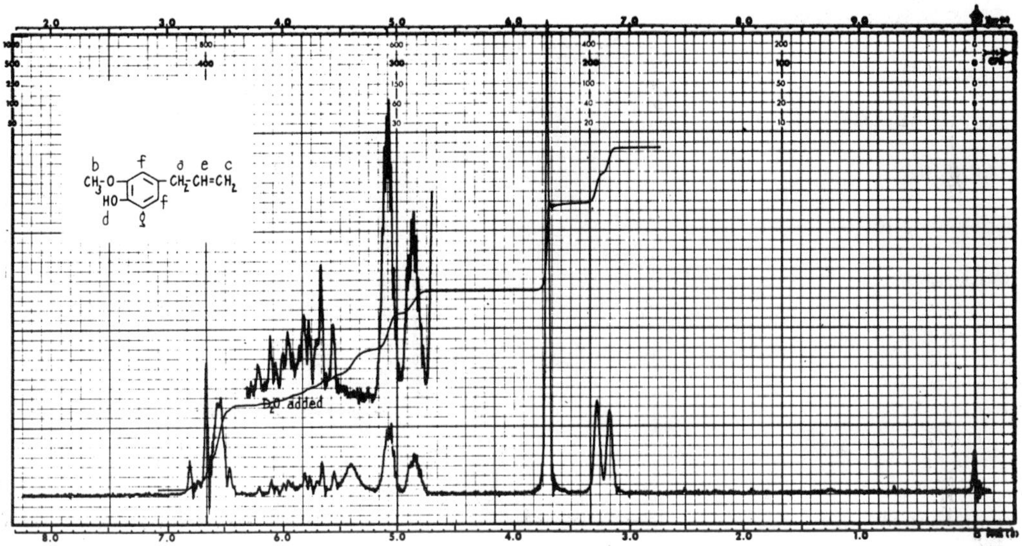

© 1978 Sadtler Research Laboratories, Inc.

3-METHYLPYROCATECHOL

$C_7H_8O_2$

Mol. Wt. 124.14

Solvent: $CDCl_3$

Assignments
a = 2.21
b = 5.78
c = 6.68

1925

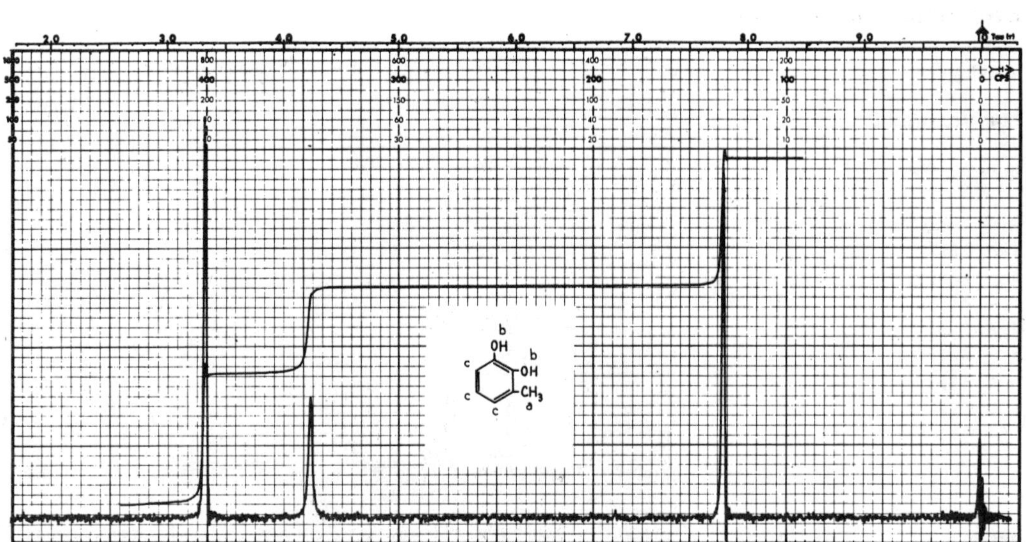

© 1978 Sadtler Research Laboratories, Inc.

4-METHYLPYROCATECHOL

$C_7H_8O_2$

Mol. Wt. 124.14

Solvent: $CDCl_3$

Assignments
a = 2.17
b = 5.87
c = 6.55
d = 6.61
e = 6.72

1926

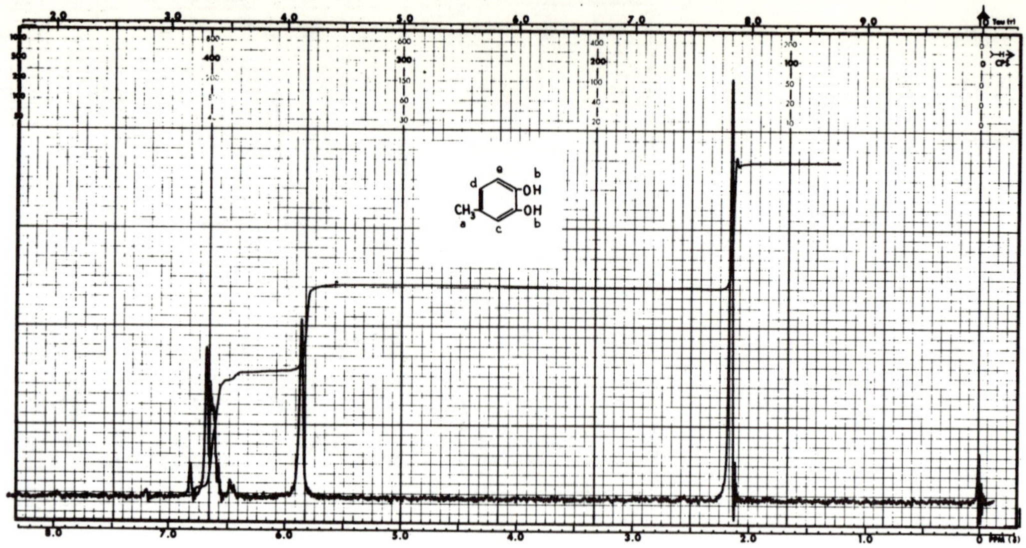

© 1978 Sadtler Research Laboratories, Inc.

4-HEXYLRESORCINOL

$C_{12}H_{18}O_2$

Mol. Wt. 194.28

M.P. 68-70°C

Solvent: $CDCl_3$

Assignments
a = 0.84
b = ca 1.29
c = 2.45
d = 5.85
e = 6.28
f = 6.32
g = 6.86
Trace of acid added for OH

1927

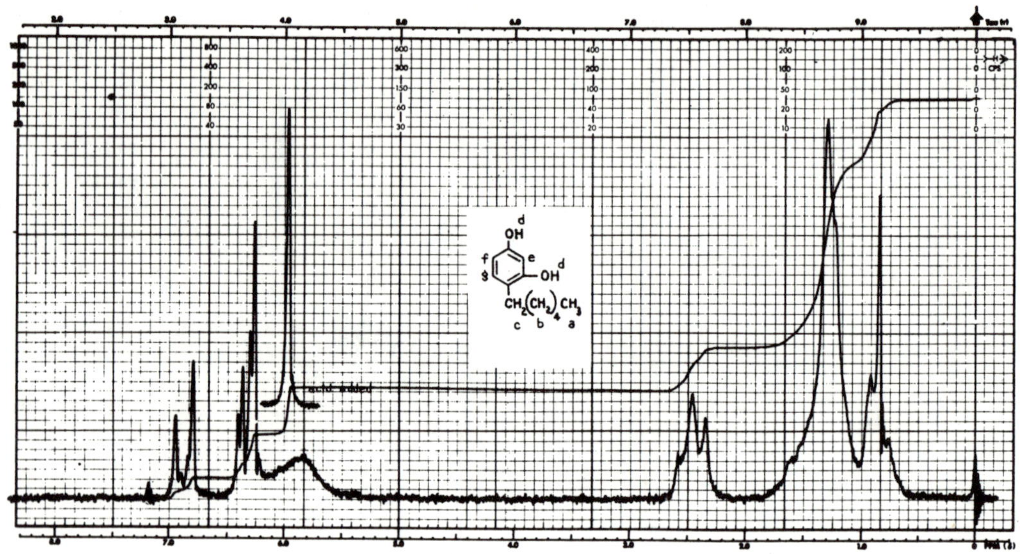

© 1978 Sadtler Research Laboratories, Inc.

4-HEXADECYLRESORCINOL

$C_{22}H_{38}O_2$

Mol. Wt. 334.55

Solvent: Polysol-d

Assignments
a = 0.89
b = 1.10-1.80
c = 2.50
d = 6.29
e = 6.37
f = 6.85
g = 7.70

1928

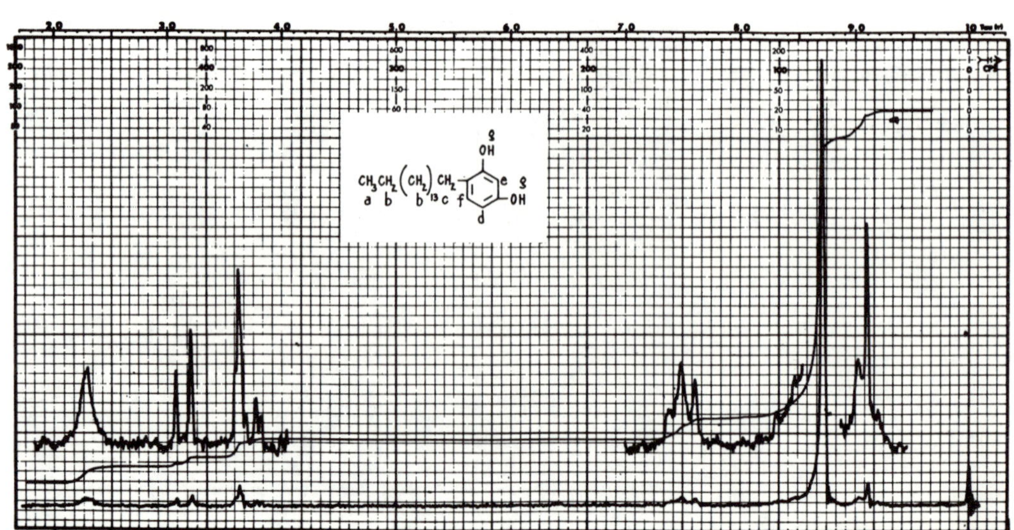

© 1978 Sadtler Research Laboratories, Inc.

4-BENZYLRESORCINOL

$C_{13}H_{12}O_2$

Mol. Wt. 200.24

Solvent: Dimethyl sulfoxide-d_6

Assignments
a = 3.80
b = 6.21
c = 6.39
d = 6.83
e = ca 7.19
f = 9.08
g = 2.49 DMSO-d_5
h = ca 3.52 H_2O

1929

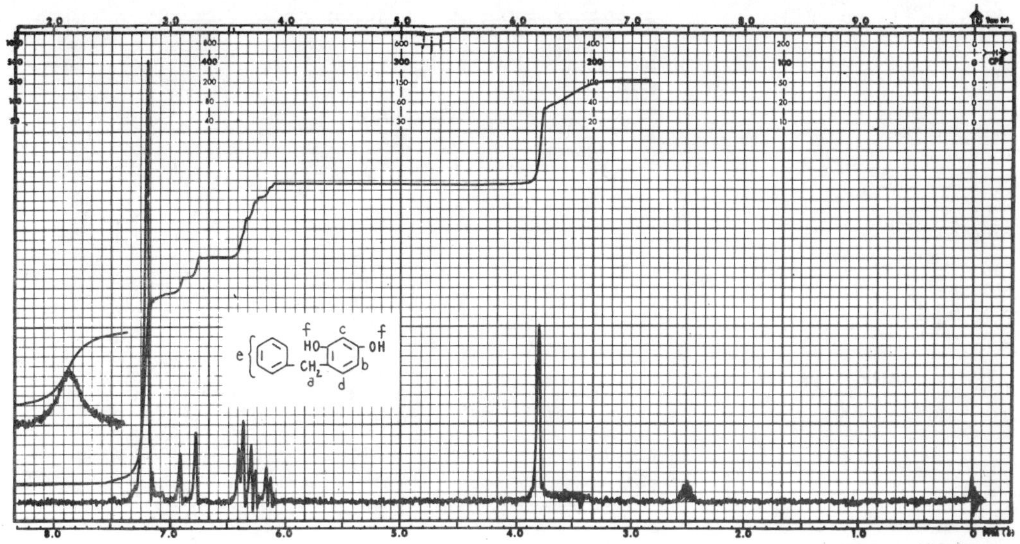

© 1978 Sadtler Research Laboratories, Inc.

3,4-BIPHENYLDIOL

$C_{12}H_{10}O_2$

Mol. Wt. 186.21

M.P. 137-138°C

Solvent: Polysol-d

Assignments
a = 6.97
b = 7.10-7.70
c = 8.86

1930

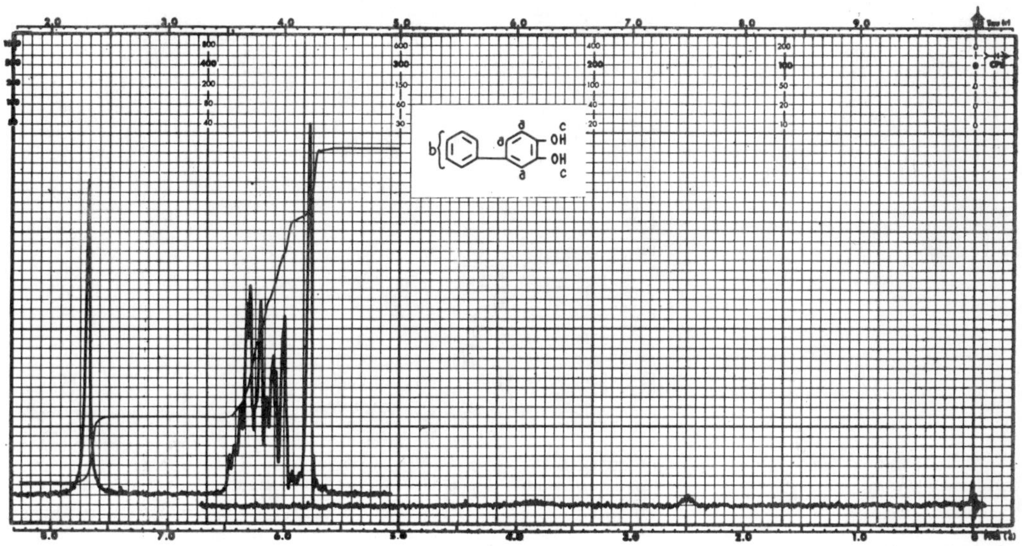

© 1978 Sadtler Research Laboratories, Inc.

PHENYLHYDROQUINONE

$C_{12}H_{10}O_2$

Mol. Wt. 186.21

M.P. 101.5-103°C

Solvent: $CDCl_3$

Assignments
a = 5.28
b = 6.40-6.85
c = 7.30

1931

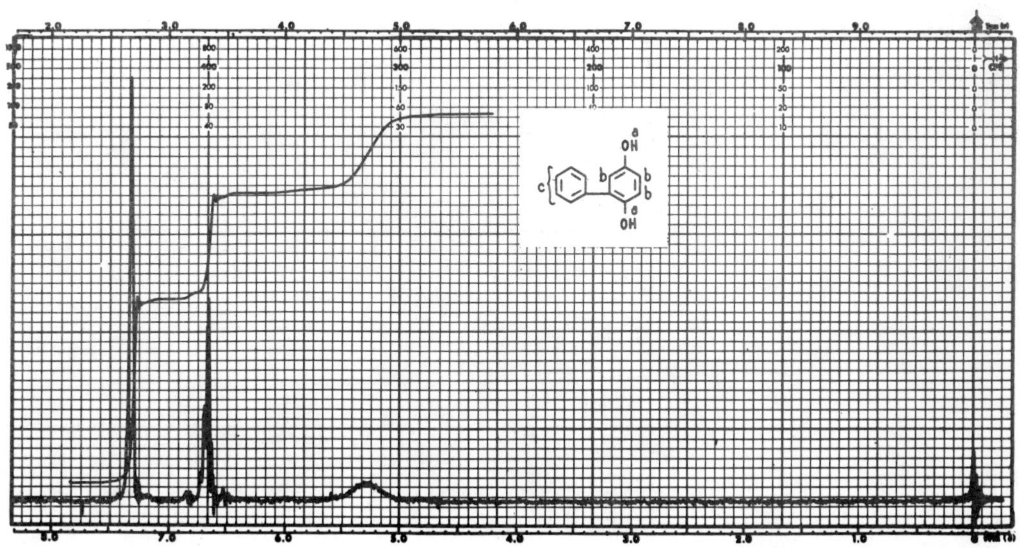

© 1978 Sadtler Research Laboratories, Inc.

2-ALLYL-3,5-XYLENOL

$C_{11}H_{14}O$

Mol. Wt. 162.23

M.P. 50.5°C

Solvent: $CDCl_3$

Assignments
a = 2.22
b = 3.36
c = ca 4.91
d = 5.00
e = 5.86
f = 6.43
g = 6.56

1932

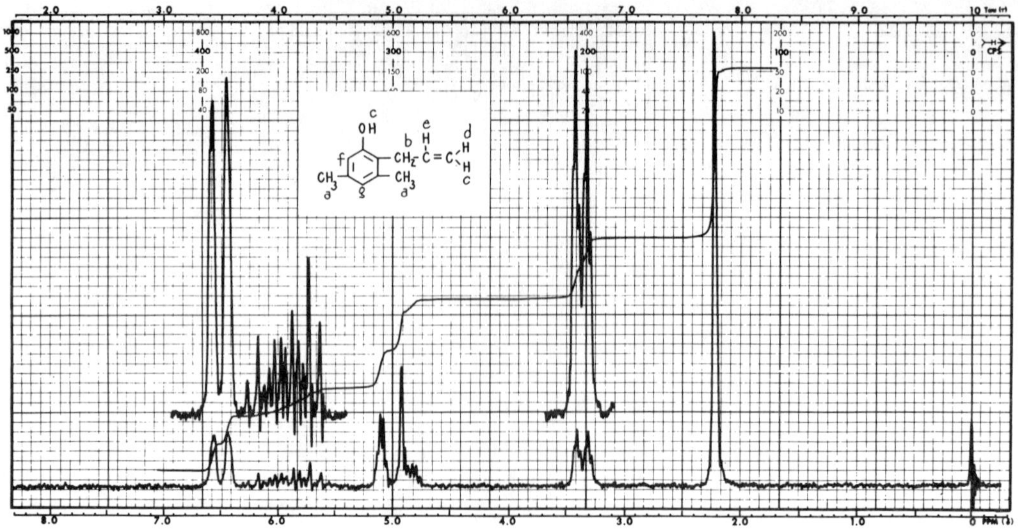

© 1978 Sadtler Research Laboratories, Inc.

2,2'-METHYLENEBIS(6-tert-BUTYL-4-ETHYLPHENOL)

$C_{25}H_{36}O_2$

Mol. Wt. 368.56

Solvent: $CDCl_3$

Assignments
a = 1.21
b = 1.40
c = 2.57
d = 3.90
e = ca 5.71
f = 7.00

1933

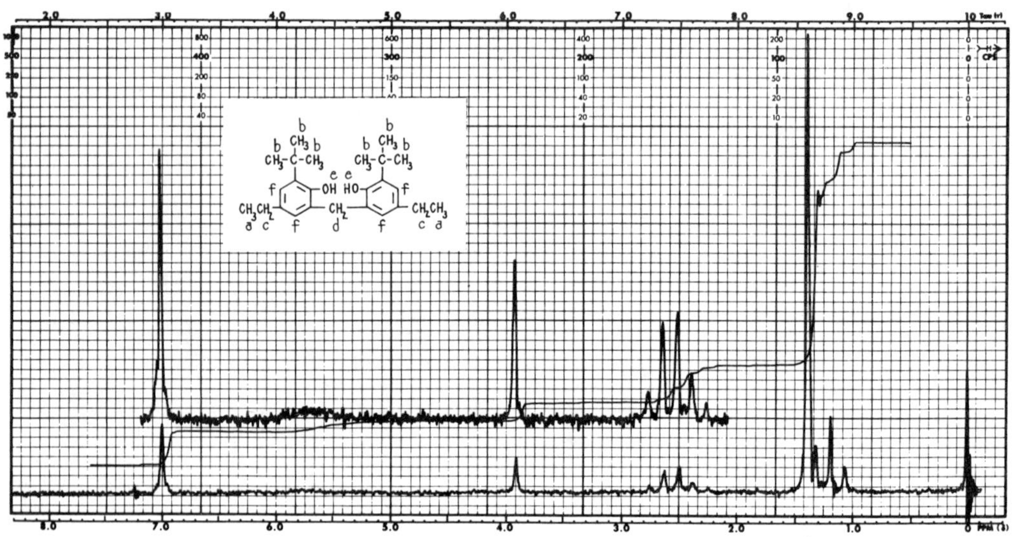

© 1978 Sadtler Research Laboratories, Inc.

2,2',6,6'-TETRA-tert-BUTYL-p,p'-BIPHENOL

$C_{28}H_{42}O_2$

Mol. Wt. 410.65

Solvent: $CDCl_3$

Assignments
a = 1.48
b = 5.13
c = 7.30

1934

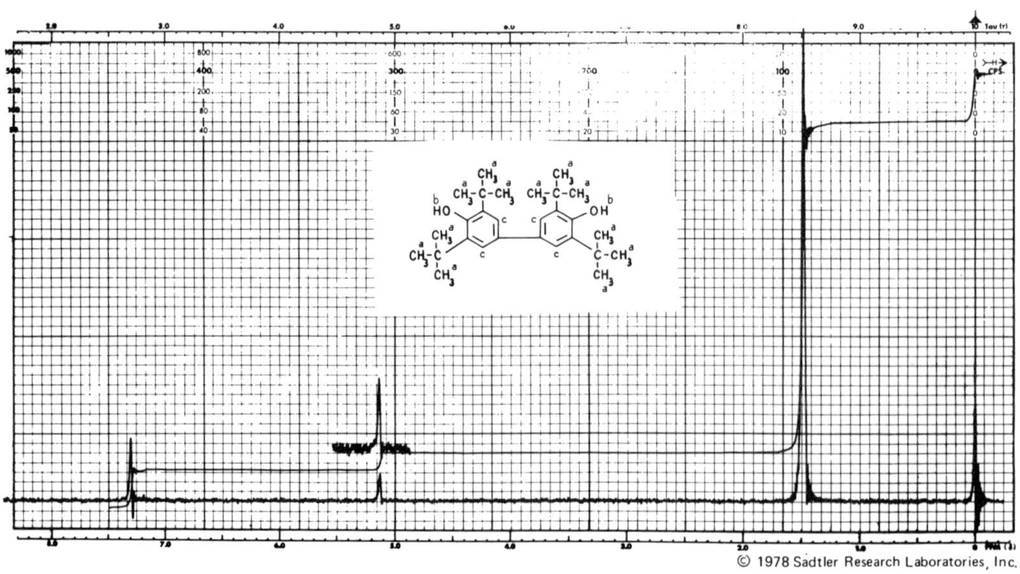

© 1978 Sadtler Research Laboratories, Inc.

4,4'-ISOPROPYLIDENEBIS(2,6-DIBROMOPHENOL)

$C_{15}H_{12}Br_4O_2$

Mol. Wt. 543.90

Solvent: $CDCl_3$

Assignments
a = 1.60
b = 5.81
c = 7.29

1935

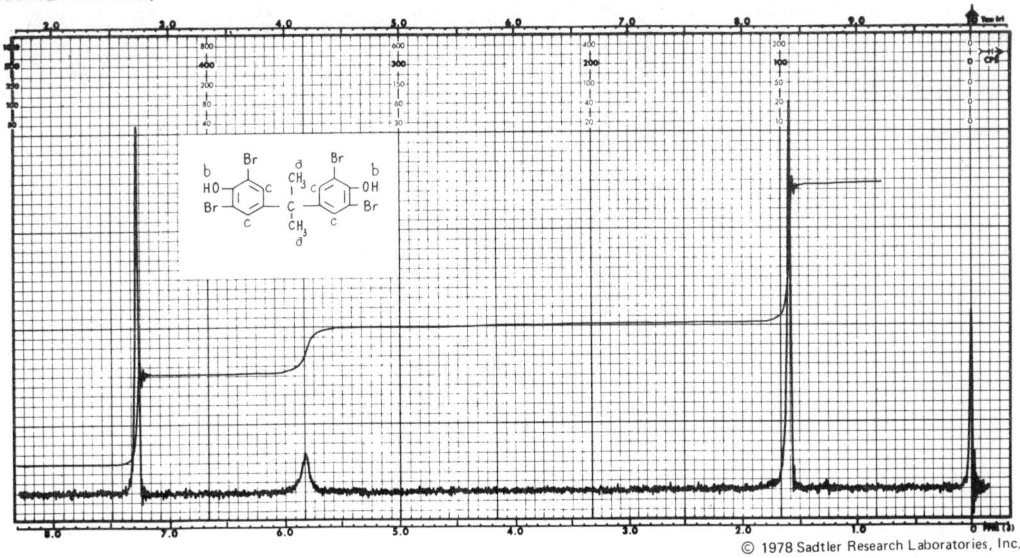

© 1978 Sadtler Research Laboratories, Inc.

---

a^2,a^6-BIS(5-tert-BUTYL-6-HYDROXY-m-TOLYL)MESITOL

$C_{31}H_{40}O_3$

Mol. Wt. 460.66

M.P. 171-172°C

Solvent: $CDCl_3$

Assignments
a = 1.38
b = 2.21
c = 3.81
d = 6.44
e = 6.92

1936

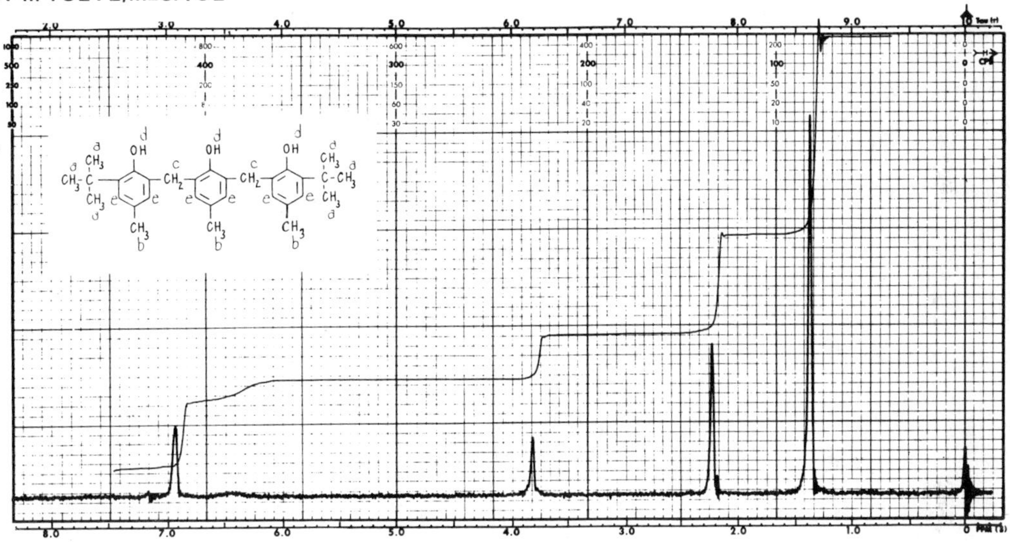

© 1978 Sadtler Research Laboratories, Inc.

2,4,6-TRIBROMOPHENOL

$C_6H_3Br_3O$

Mol. Wt. 330.82

M.P. 96°C

Solvent: CCl_4

Assignments
a = 5.71
b = 7.56
c = 6.90 imp.
d = 7.75 imp.

1937

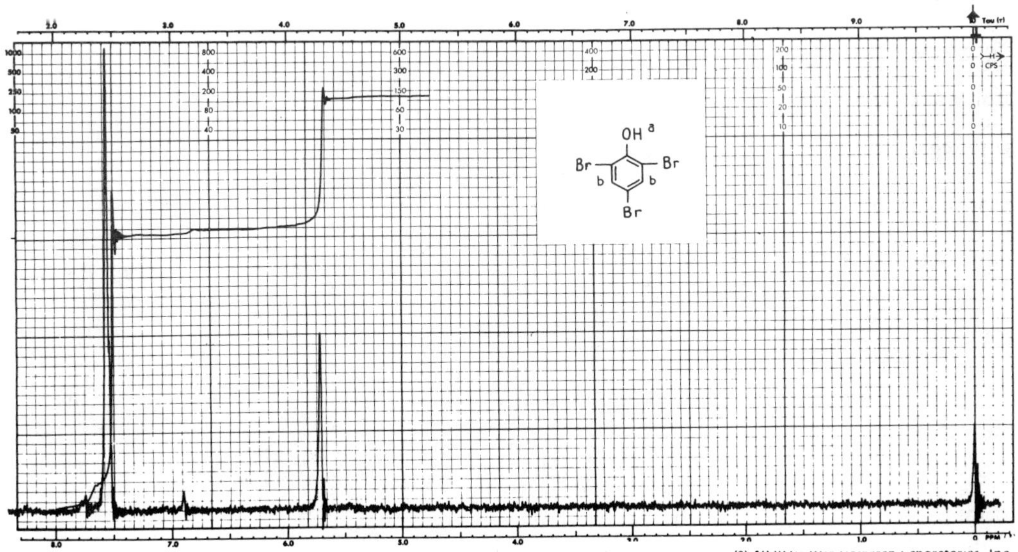

© 1978 Sadtler Research Laboratories, Inc.

2,4,6-TRIIODOPHENOL

$C_6H_3I_3O$

Mol. Wt. 471.80

M.P. 156-158°C

Solvent: Dimethyl sulfoxide

Assignments
a = ca 7.80 *
b = 8.02
* Plus H_2O

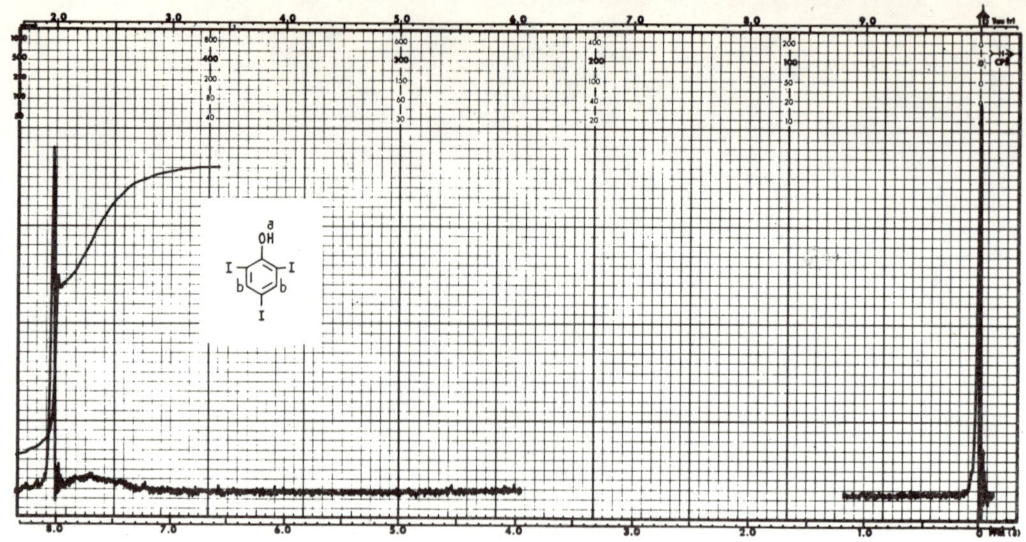

1938

© 1978 Sadtler Research Laboratories, Inc.

PICRIC ACID

$C_6H_3N_3O_7$

Mol. Wt. 229.11

M.P. 121.8°C (lit.)

Solvent: Acetone

Assignments
a = 6.73
b = 9.13
c = 2.05 acetone

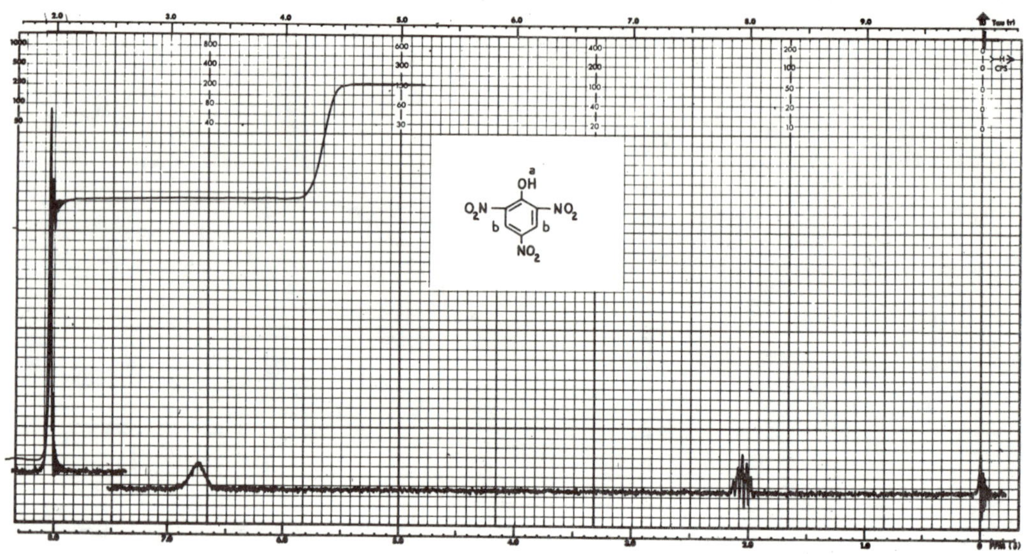

1939

© 1978 Sadtler Research Laboratories, Inc.

ANILINE, PICRATE

$C_6H_6N{\cdot}C_6H_3N_3O_7$

Mol. Wt. 321.23

M.P. 198°C

Solvent: Dimethyl sulfoxide

Assignments
a = ca 7.49
b = 7.91
c = 8.66

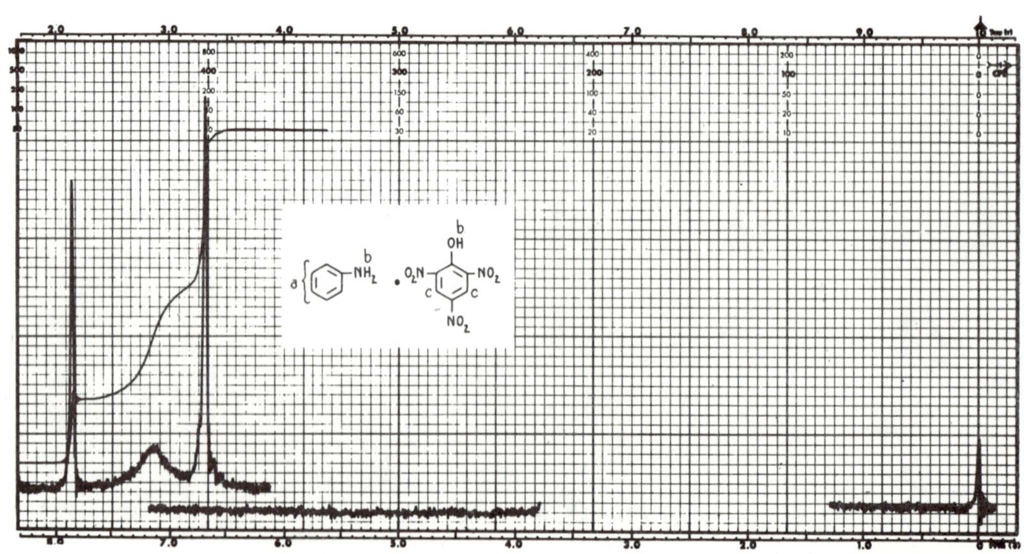

1940

© 1978 Sadtler Research Laboratories, Inc.

o-CHLOROANILINE, MONOPICRATE

$C_6H_6ClN{\cdot}C_6H_3N_3O_7$

Mol. Wt. 356.68

M.P. 177°C

Solvent: Dimethyl sulfoxide

Assignments
a = ca 7.43
b = 7.90 *
c = 8.69
* H_2O present

1941

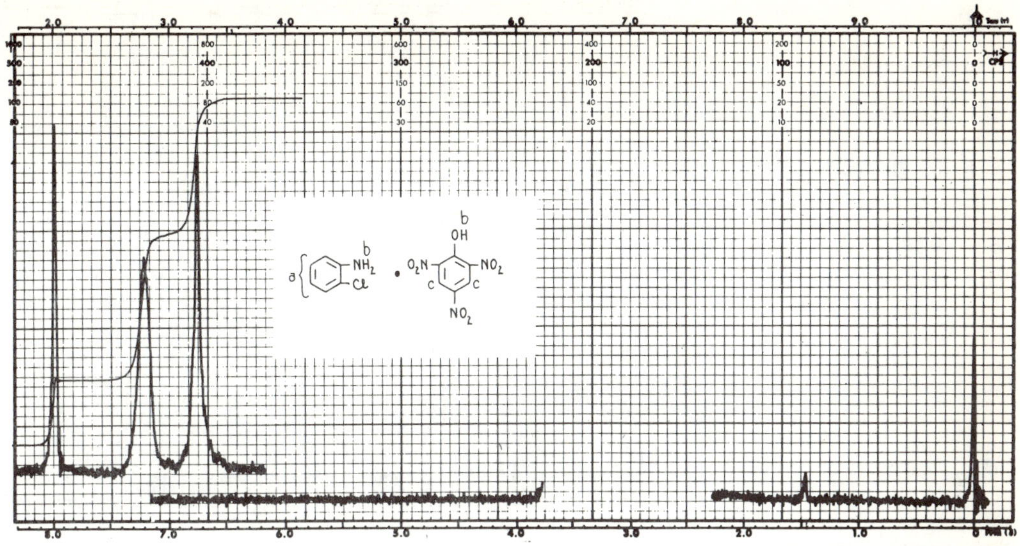

© 1978 Sadtler Research Laboratories, Inc.

PYRIDINE, PICRATE

$C_5H_5N{\cdot}C_6H_3N_3O_7$

Mol. Wt. 308.21

M.P. 161-167°C

Solvent: Polysol-d

Assignments
a = 8.06
b = 8.52
c = 8.54
d = 8.89
e = unobserved
f = 2.52 DMSO-d_5

1942

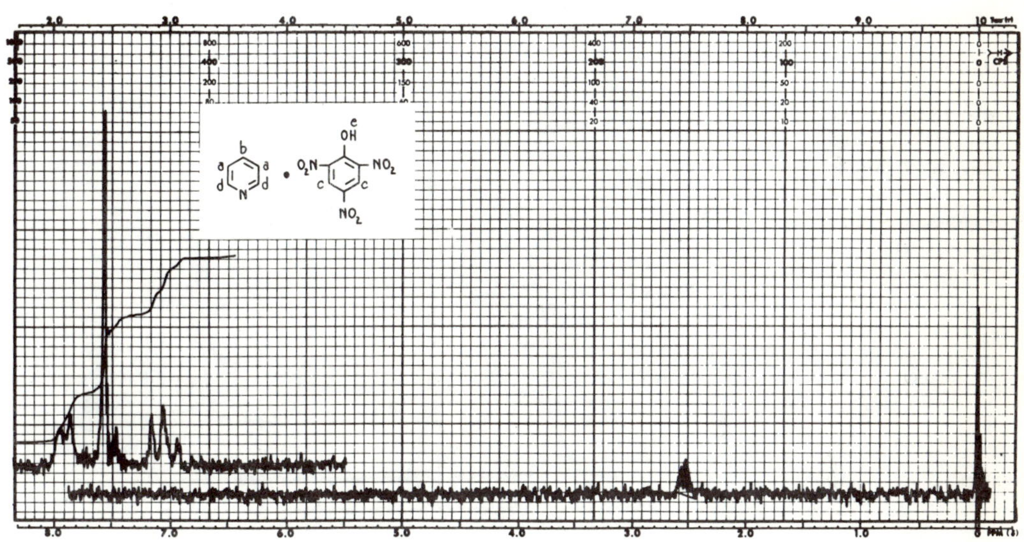

© 1978 Sadtler Research Laboratories, Inc.

2,6-DICHLOROHYDROQUINONE

$C_6H_4Cl_2O_2$

Mol. Wt. 179.00

Solvent: Polysol-d

Assignments
a = 6.76
b = 8.95
c = 9.37
d = 2.59 DMSO-d_5

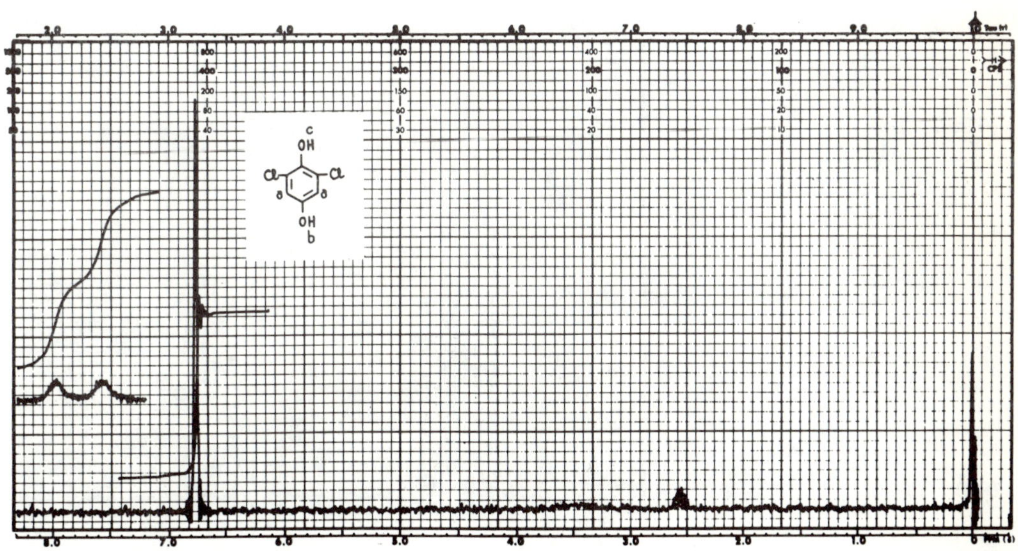

© 1978 Sadtler Research Laboratories, Inc.

2-NITRORESORCINOL

$C_6H_5NO_4$

Mol. Wt. 155.11

M.P. 82-83°C

Solvent: Polysol-d

Assignments
a = 6.54
b = 7.23
c = 10.35

1944

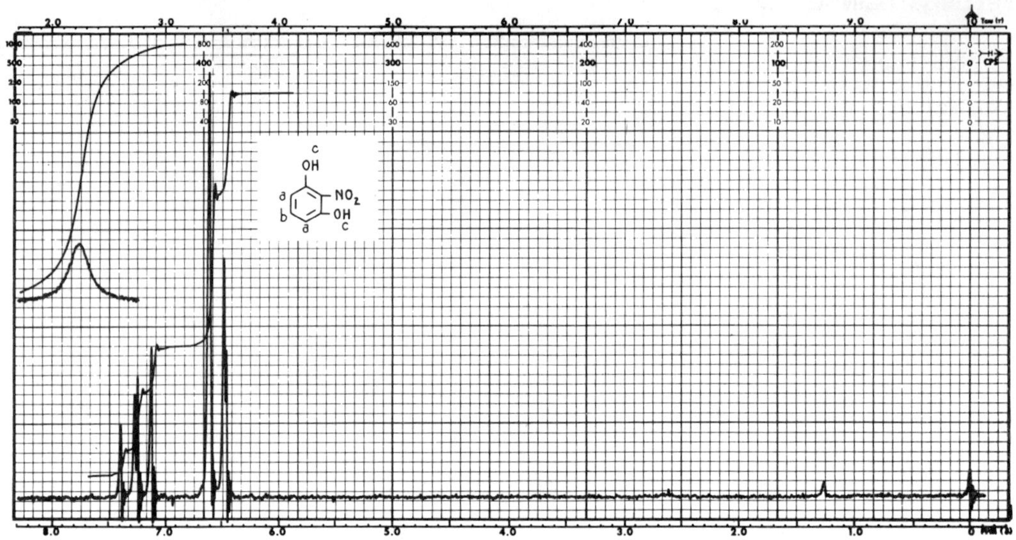

© 1978 Sadtler Research Laboratories, Inc.

PYROGALLOL

$C_6H_6O_3$

Mol. Wt. 126.11

M.P. 131-135°C

Solvent: Acetone

Assignments
a = ca 6.47
b = 7.54

1945

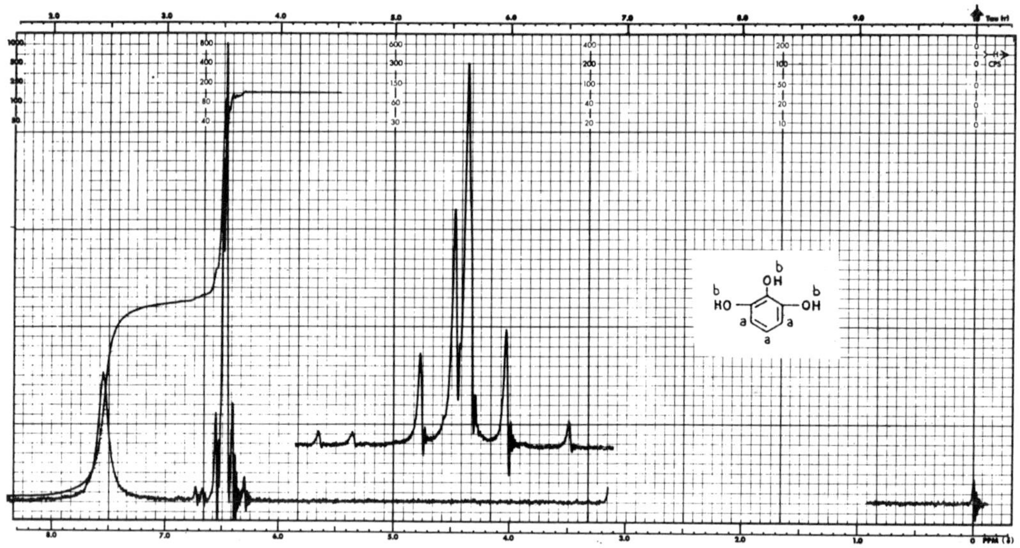

© 1978 Sadtler Research Laboratories, Inc.

1,2,4-BENZENETRIOL

$C_6H_6O_3$

Mol. Wt. 126.11

Solvent: D_2O

Assignments
a = 6.49
b = 6.62
c = 6.91
d = 5.00 HDO

1946

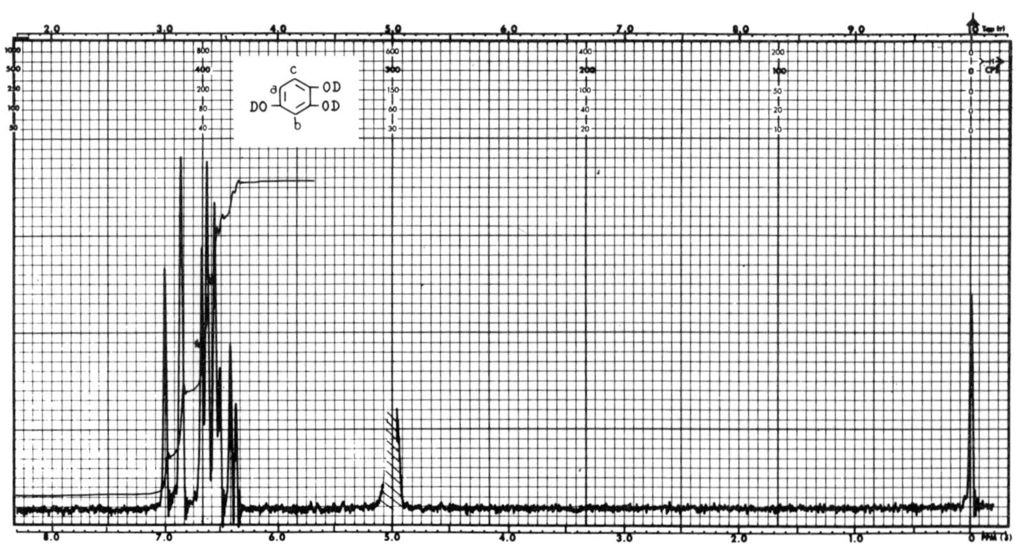

© 1978 Sadtler Research Laboratories, Inc.

PHLOROGLUCINOL

$C_6H_6O_3$

Mol. Wt. 126.11

M.P. 219°C (lit.)

Solvent: Acetone

Assignments
a = 5.99
b = 6.01

1947

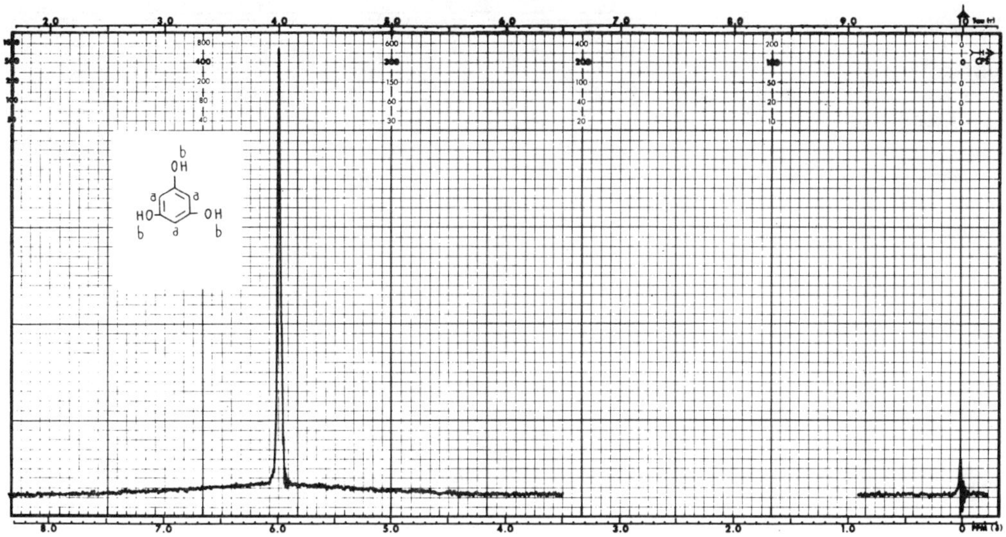

© 1978 Sadtler Research Laboratories, Inc.

4-INDANOL

$C_9H_{10}O$

Mol. Wt. 134.18

M.P. 47-49°C

Solvent: $CDCl_3$

Assignments
a = 1.92
b = ca 2.78
c = 6.15
d = 6.59
e = 6.75
f = 6.98

1948

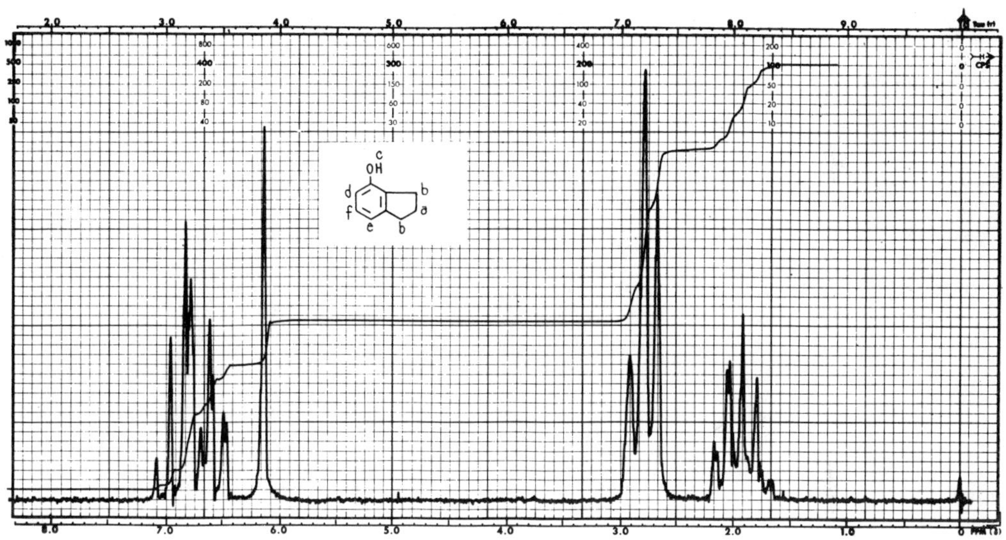

© 1978 Sadtler Research Laboratories, Inc.

2-FLUORENOL

$C_{13}H_{10}O$

Mol. Wt. 182.22

Solvent: Polysol-d

Assignments
a = 3.78
b = 6.82
c = 6.95-7.50
d = 7.56
e = 7.59
f = 8.99
g = 2.47 DMSO-d_5
h = 3.12 H_2O

1949

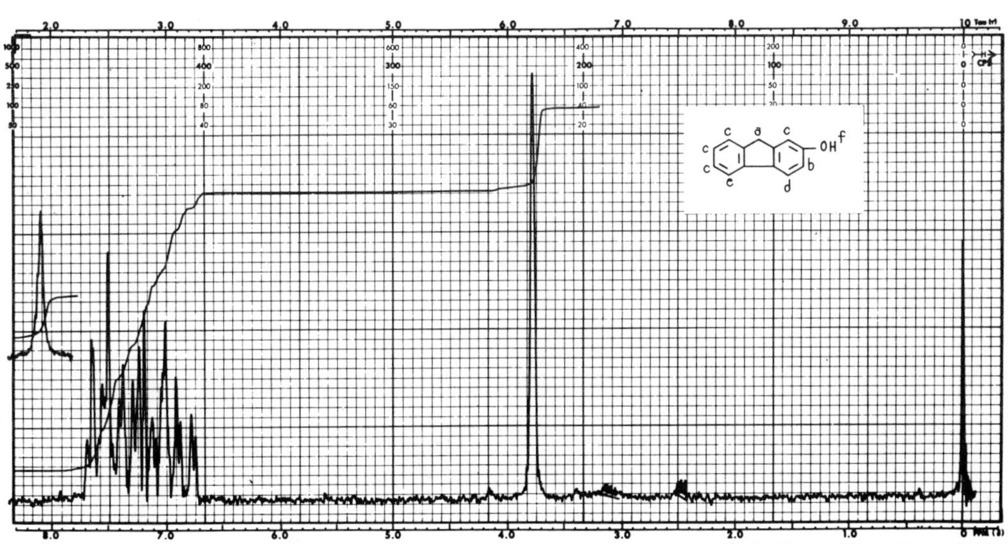

© 1978 Sadtler Research Laboratories, Inc.

1-NAPHTHOL

$C_{10}H_8O$

Mol. Wt. 144.17

M.P. 95.5-96.5°C

B.P. 288°C

Solvent: Acetone

Assignments
a = 7.01
b = 7.22-7.60
c = 7.72
d = 8.41
e = 8.92

1950

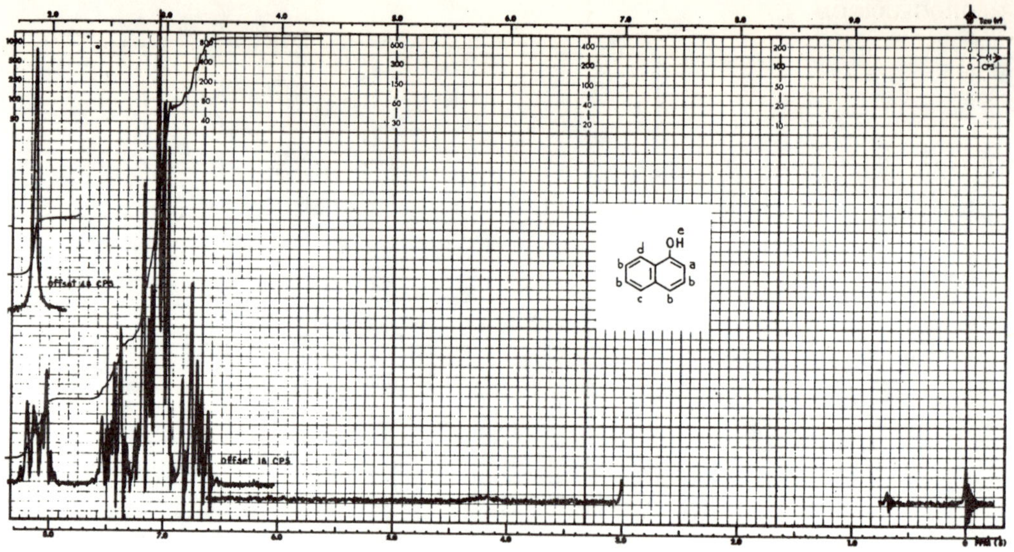

© 1978 Sadtler Research Laboratories, Inc.

2-NAPHTHOL

$C_{10}H_8O$

Mol. Wt. 144.17

M.P. 122-123°C

Solvent: Acetone

Assignments
a = 7.01-7.86
b = 8.35
Trace of acid
added for OH

1951

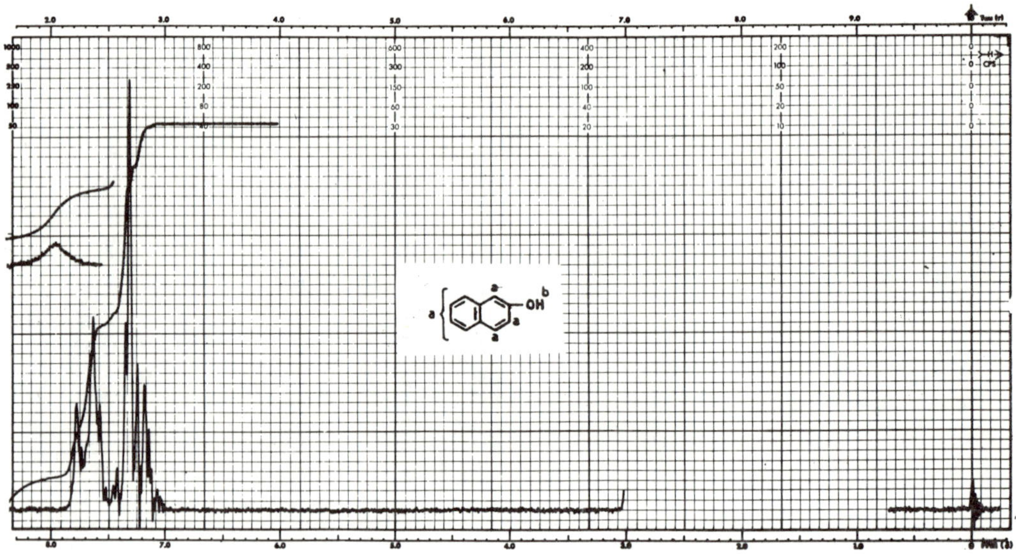

© 1978 Sadtler Research Laboratories, Inc.

2-METHYL-1-NAPHTHOL

$C_{11}H_{10}O$

Mol. Wt. 158.20

M.P. 246-248°C

Solvent: $CDCl_3$

Assignments
a = 2.21
b = 5.20
c = 7.06
d = 7.30
e = ca 7.32
f = 7.69
g = 8.06

1952

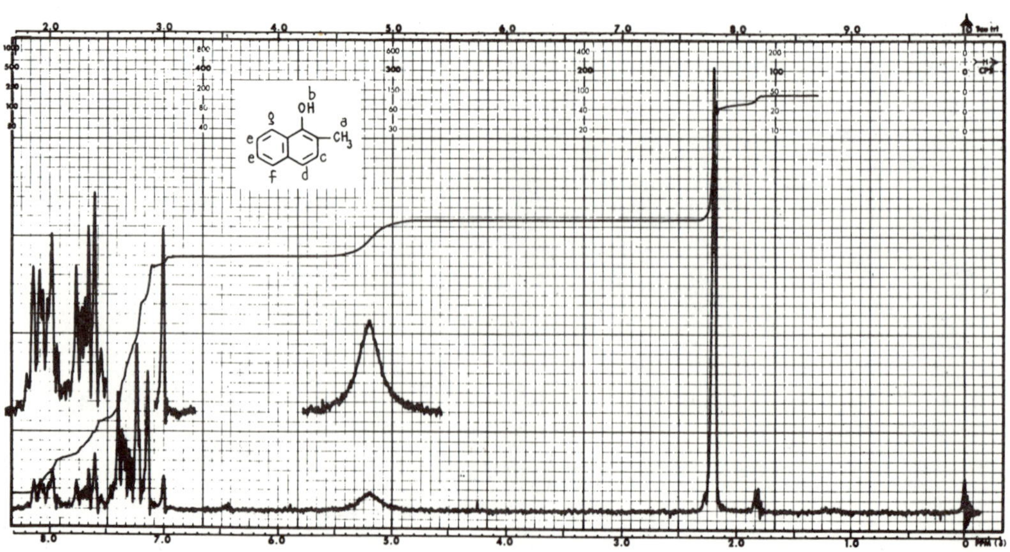

© 1978 Sadtler Research Laboratories, Inc.

6-AMINO-1-NAPHTHOL

$C_{10}H_9NO$

Mol. Wt. 159.19

Solvent: DMSO-d_6

Assignments
a = 4.71
b = 6.61
c = 6.87
d = 6.75-7.10
e = 7.17
f = 8.03

1953

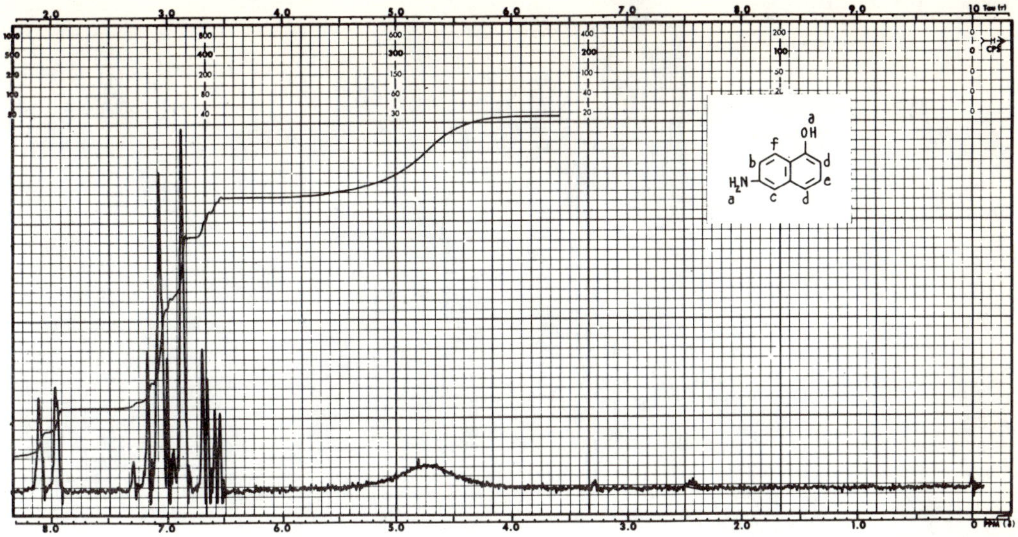

© 1978 Sadtler Research Laboratories, Inc.

2-HYDROXY-1-NAPHTHALENEPROPIONITRILE

$C_{13}H_{11}NO$

Mol. Wt. 197.24

Solvent: Polysol-d

Assignments
a = 2.64
b = 3.37
c = 7.05-7.45
d = 7.22
e = 7.61
f = 7.70
g = 7.81
h = ca 8.99

1954

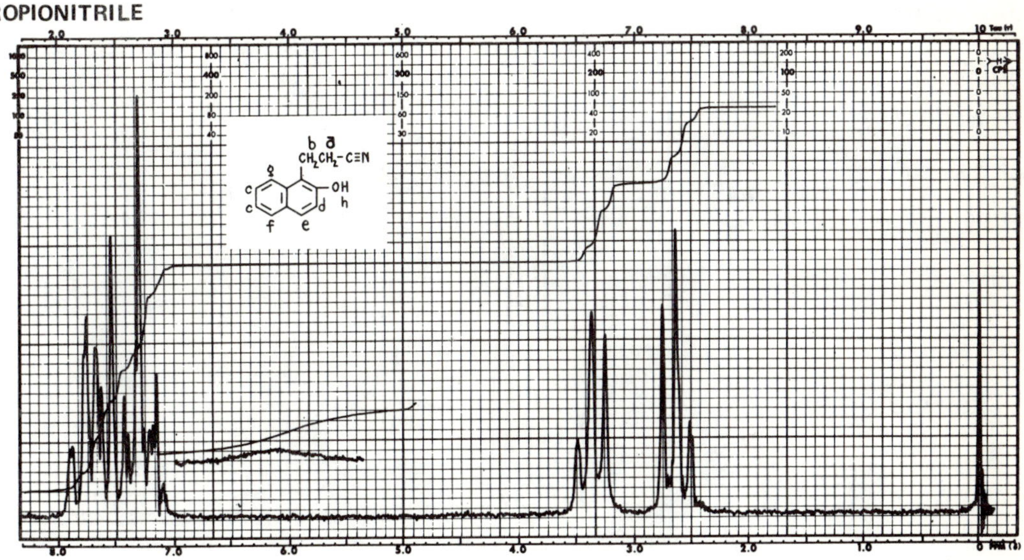

© 1978 Sadtler Research Laboratories, Inc.

1-NITROSO-2-NAPHTHOL

$C_{10}H_7NO_2$

Mol. Wt. 173.17

M.P. 106-109°C

Solvent: $CDCl_3$

Assignments
a = 6.56
b = 7.39-7.65
c = 7.67
d = 8.31
OH is unobserved

1955

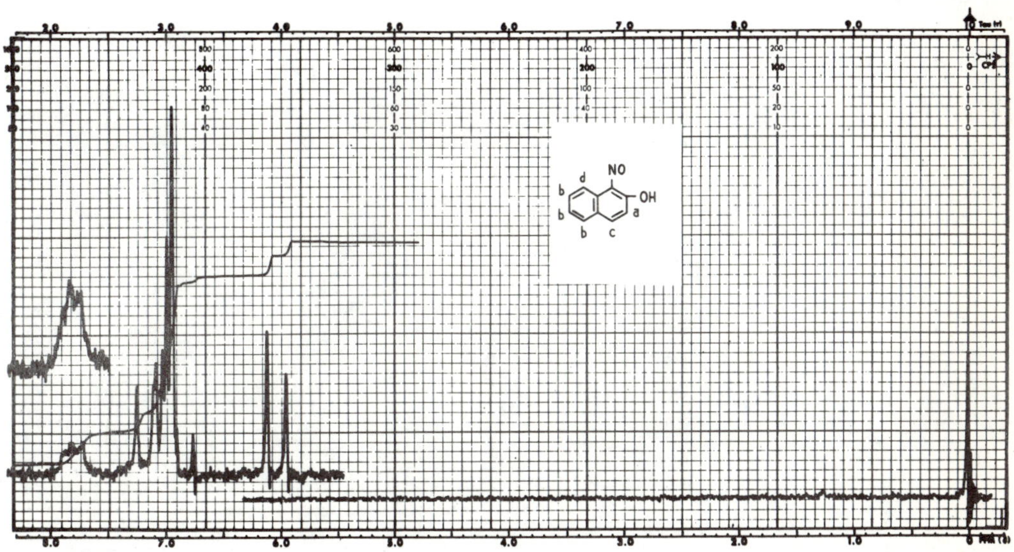

© 1978 Sadtler Research Laboratories, Inc.

1,3-NAPHTHALENEDIOL

$C_{10}H_8O_2$

Mol. Wt. 160.17

M.P. 120-125°C

Solvent: Acetone

Assignments
a = 6.70
b = 6.81
c = 7.08-7.85
d = 8.18
e = 8.48
f = 8.98
g = 6.30 imp.

1956

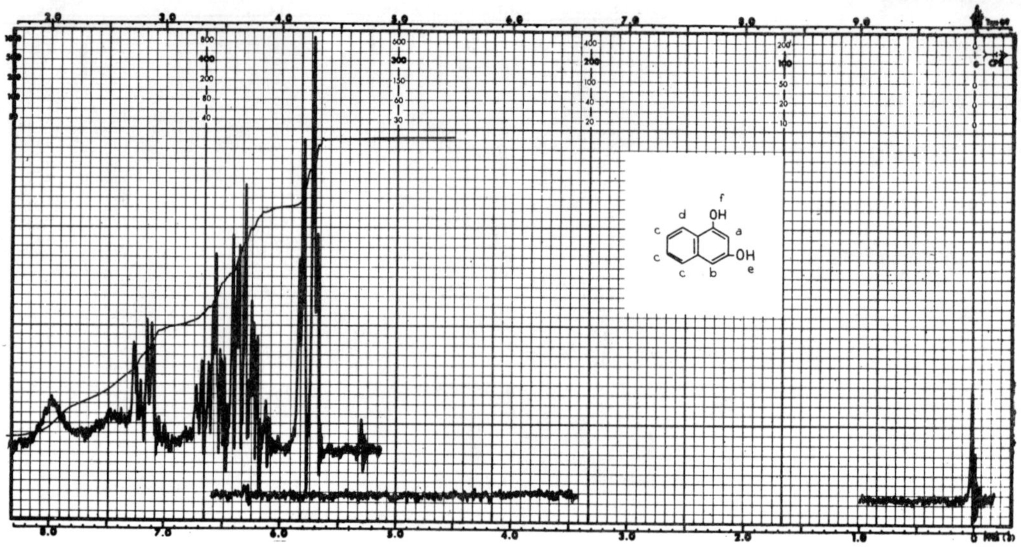

© 1978 Sadtler Research Laboratories, Inc.

1,5-NAPHTHALENEDIOL

$C_{10}H_8O_2$

Mol. Wt. 160.17

M.P. 256-260°C

Solvent: Dimethyl sulfoxide-D_6

Assignments
a = 6.89
b = 7.22
c = 7.66
d = 9.58
e = 2.50 DMSO-D_5

1957

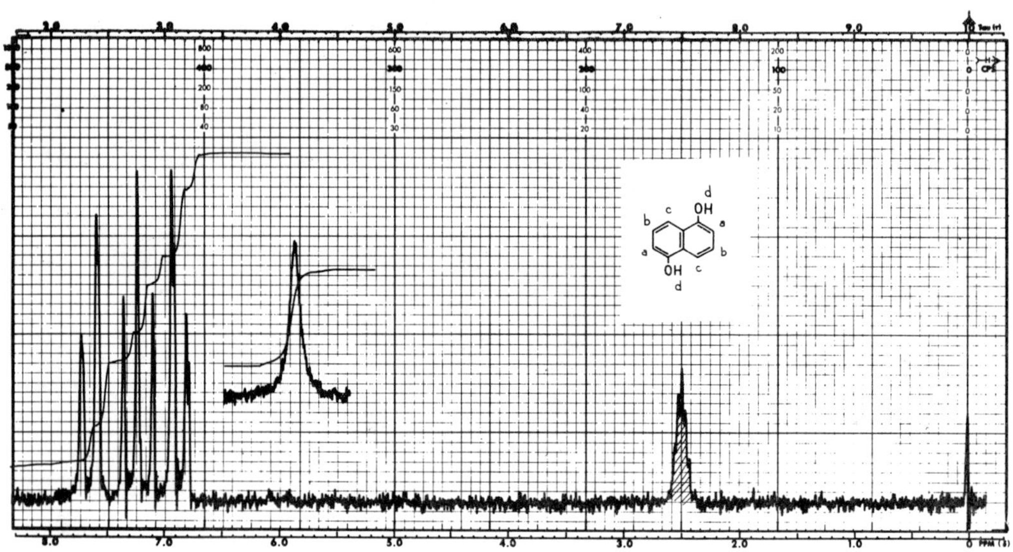

© 1978 Sadtler Research Laboratories, Inc.

2,7-NAPHTHALENEDIOL

$C_{10}H_8O_2$

Mol. Wt. 160.17

Solvent: Acetone

Assignments
a = 7.00
b = 7.10
c = 7.65
d = 8.32

1958

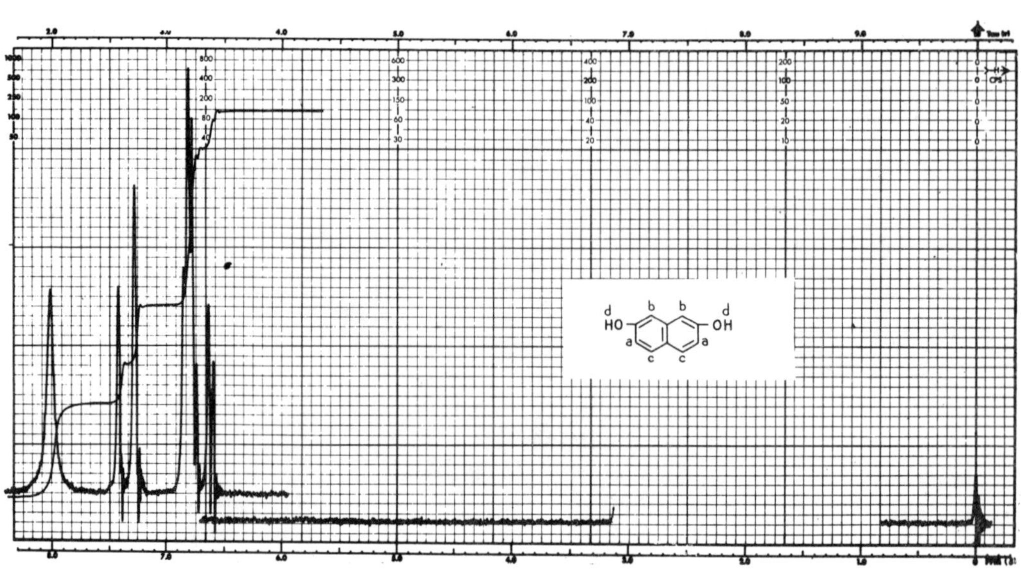

© 1978 Sadtler Research Laboratories, Inc.

3-PYRIDINOL

C_5H_5NO

Mol. Wt. 95.10

M.P. 126-129°C

B.P. 151-153°C/3mm

Solvent: Dimethyl sulfoxide-D_6

Assignments
a = 7.23
b = 8.09
c = 8.21
d = 9.93
e = 2.54 DMSO-D5

1959

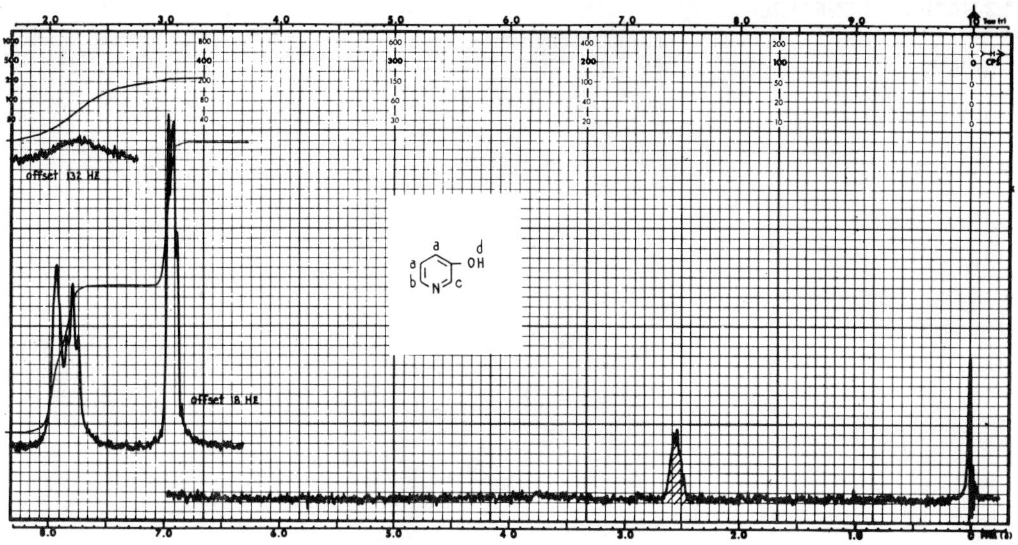

© 1978 Sadtler Research Laboratories, Inc.

2-PYRIMIDINOL

$C_4H_4N_2O$

Mol. Wt. 96.09

M.P. 170-174°C

Solvent: DMSO-d_6

Assignments
a = 6.40
b = 8.29
c = 10.30 *
d = 2.32 DMSO-d_5
* Plus H_2O

1960

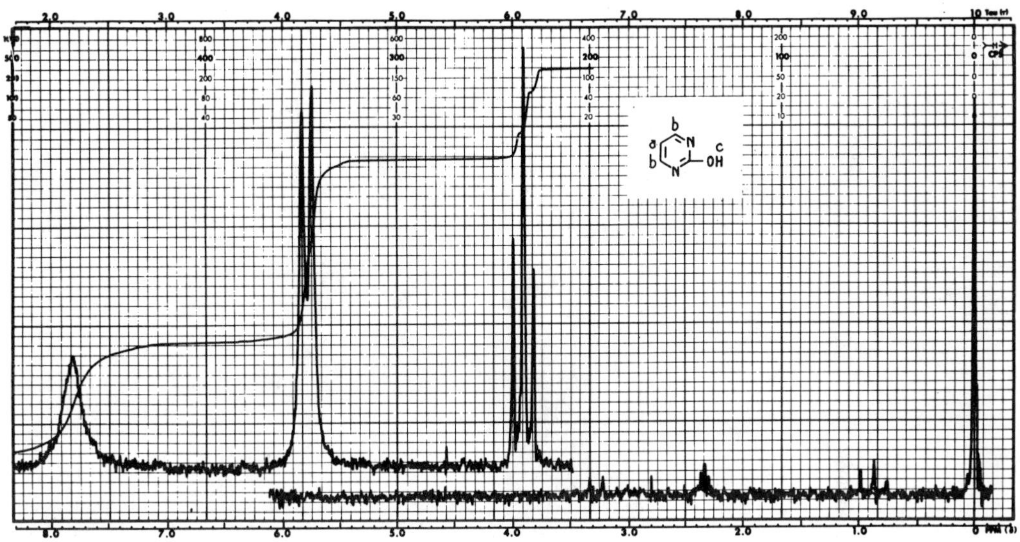

© 1978 Sadtler Research Laboratories, Inc.

CARBOSTYRYL

C_9H_7NO

Mol. Wt. 145.16

M.P. 196-197°C

Solvent: Dimethyl sulfoxide

Assignments
a = 6.57
b = 7.03-7.81
c = 7.91
d = 11.78

1961

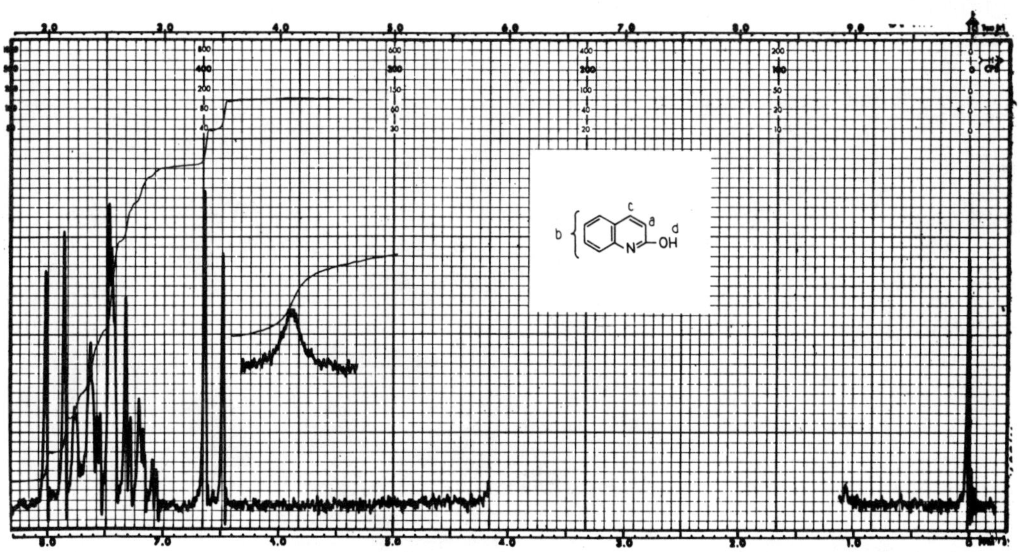

© 1978 Sadtler Research Laboratories, In

8-QUINOLINOL

C_9H_7NO

Mol. Wt. 145.15

M.P. 72.9-74.2°C

Solvent: $CDCl_3$

Assignments
a = 7.04-7.41
b = 7.44
c = 7.99
d = 8.61
e = 8.71

1962

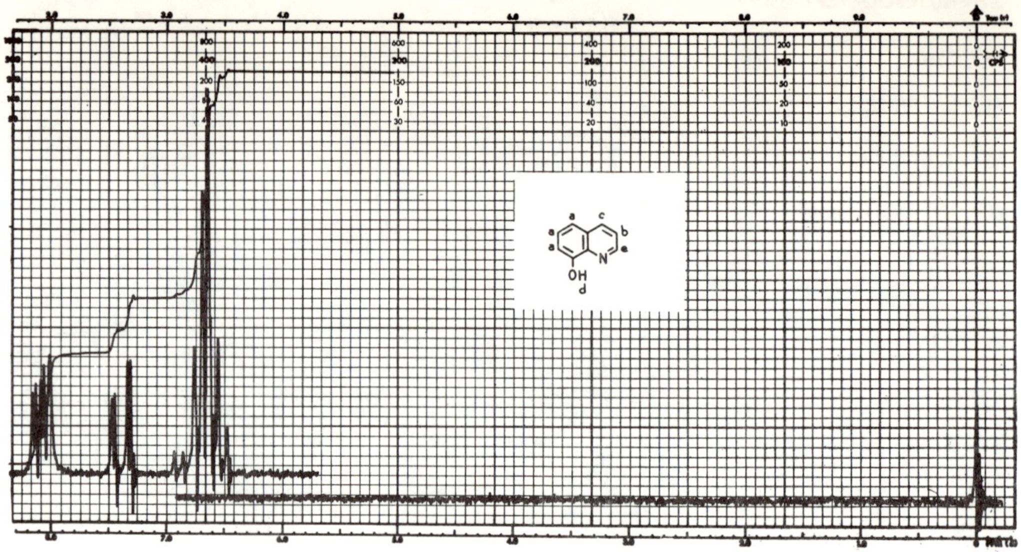

© 1978 Sadtler Research Laboratories, Inc.

4-METHYLCARBOSTYRIL

$C_{10}H_9NO$

Mol. Wt. 159.19

M.P. 221-223°C

Solvent: Trifluoroacetic acid

Assignments
a = 2.80
b = 7.11
c = 7.40-7.90
d = 8.04
e = unobserved*
* Under solvent band

1963

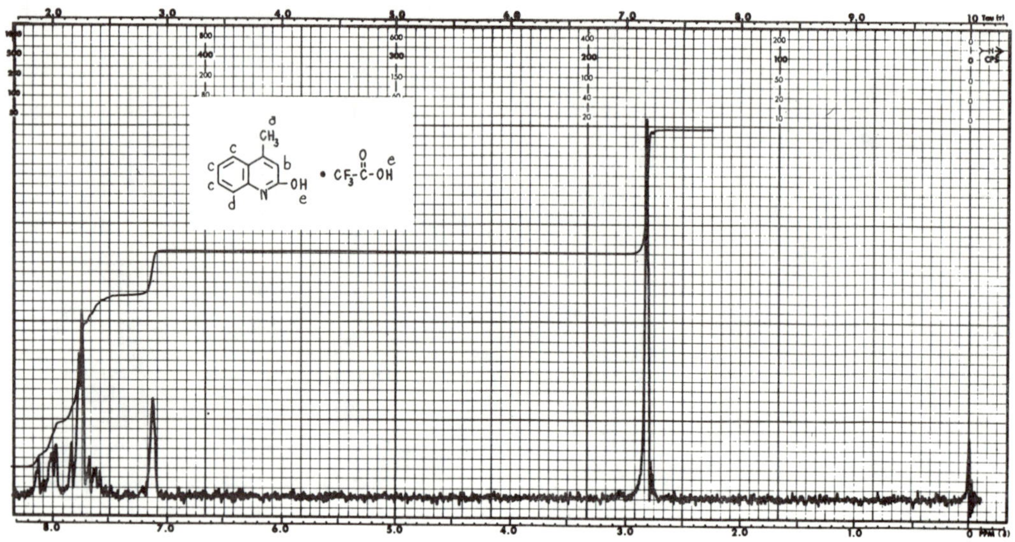

© 1978 Sadtler Research Laboratories, Inc.

4-NITROISOCARBOSTYRIL

$C_9H_6N_2O_3$

Mol. Wt. 190.16

M.P. 236-238°C (dec.)

Solvent: Dimethyl sulfoxide

Assignments
a = 7.62 or b
b = 7.85 or a
c = 8.29 or d
d = 8.50 or c
e = 8.52
OH unobserved

1964

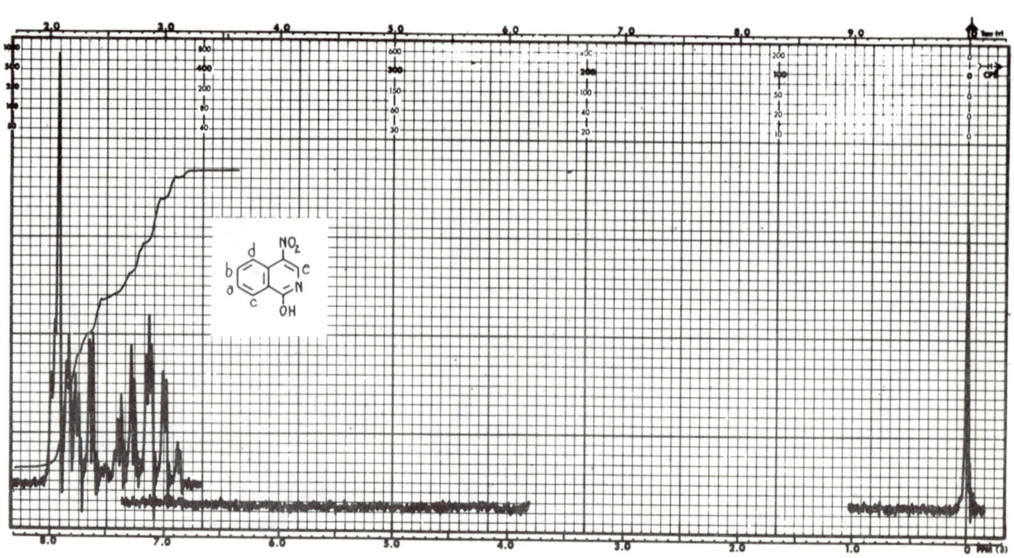

© 1978 Sadtler Research Laboratories, Inc.

PHOSPHONIC ACID, DIBUTYL ESTER

$C_8H_{19}O_3P$

Mol. Wt. 194.21

B.P. 118-119°C/7mm

Solvent: $CDCl_3$

Assignments

a = 0.95
b = 1.10-2.00
c = 4.03
d = ca 5.99

1965

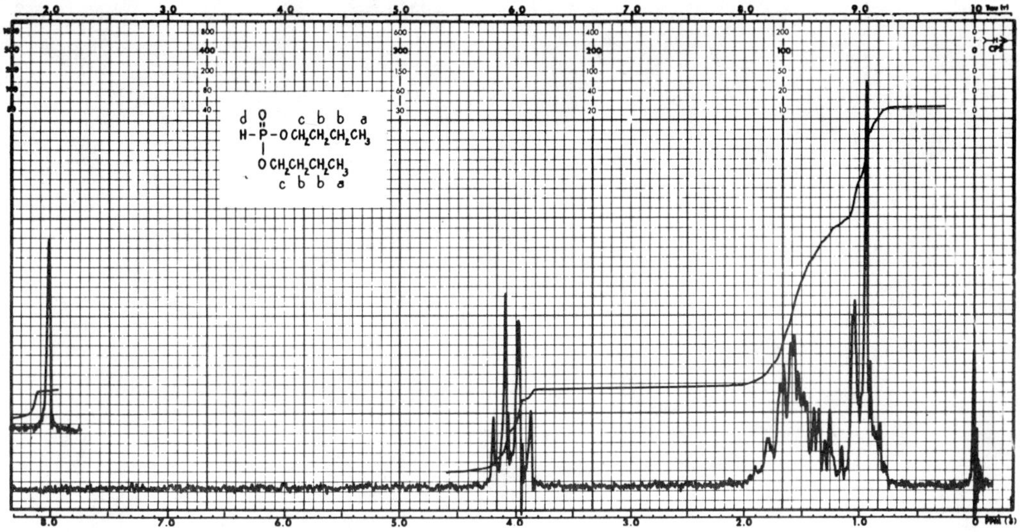

© 1978 Sadtler Research Laboratories, Inc.

TRI(2-ETHYLHEXYL)PHOSPHITE

$C_{24}H_{51}O_3P$

Mol. Wt. 418.65

B.P. 148-150°C/0.15mm

Solvent: CCl_4

Assignments

a = 0.92
b = 1.32
c = 3.62

1966

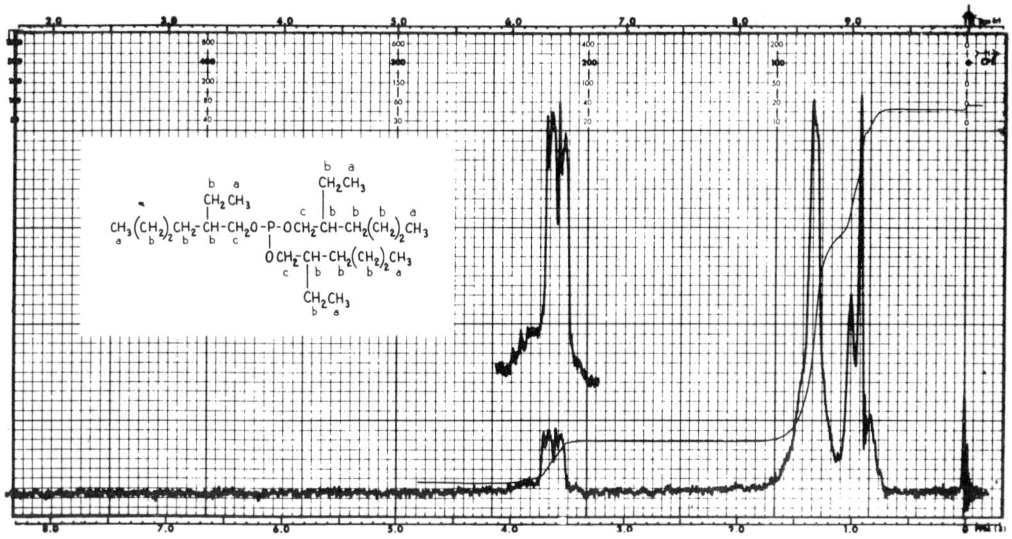

© 1978 Sadtler Research Laboratories, Inc.

NOTES

THE KETONES

This section deals with the NMR spectra of all the Ketones except the beta diketones which are covered in the next group of spectra beginning with spectrum 2137.

The Ketone linkage weakly deshields the protons of adjacent aliphatic groups but strongly deshields the ortho aromatic protons. The ketones are readily soluble in carbon tetrachloride and deuterochloroform.

Chemical Shifts

Aliphatic Protons

HC–3	HC–2	HC–1		Solvent	Spectrum
		2.05	$CH_3-C(=O)-R_2$	CCl_4	1968
		2.07	$CH_3-C(=O)-R_4$	CCl_4	1969
		2.11	$CH_3-C(=O)-CH_3$	- - - -	1967
		2.20	$CH_3-C(=O)-CH=CH_2$	CCl_4	2017
		2.40	$CH_3-C(=O)-C\equiv C-$Ø	$CDCl_3$	2044
		2.43	$CH_3-C(=O)-$Ø	CCl_4	2045
	1.03	2.29	$CH_3-CH_2-C(=O)-R_3$	CCl_4	1970
	0.99	2.40	$CH_3-CH_2-C(=O)-CH_3$	CCl_4	1968
	0.92	2.43	$CH_3-CH_2-C(=O)-R_5$	$CDCl_3$	1975
	1.18	2.94	$CH_3-CH_2-C(=O)-$Ø	$CDCl_3$	2046
0.90	1.58	2.31	$CH_3-CH_2-CH_2-C(=O)-R_3$	CCl_4	1972
0.95	1.72	2.82	$CH_3-CH_2-CH_2-C(=O)-$Ø	CCl_4	2047
	(1.05)	2.50	$(CH_3)_2-CH-C(=O)-R_6$	CCl_4	1982
	(1.00)	3.07	$(CH_3)_2-CH-C(=O)-C(CH_3)_3$	CCl_4	1977
	(1.18)	3.47	$(CH_3)_2-CH-C(=O)-$Ø	CCl_4	2048
	(1.11)	- - - -	$(CH_3)_3-C-C(=O)-CH(CH_3)_2$	CCl_4	1977
(0.91)	2.00	2.22	$(CH_3)_2-CH-CH_2-C(=O)-CH_3$	CCl_4	1971
(1.02)	- - - -	2.28	$(CH_3)_3-C-CH_2-C(=O)-CH_3$	CCl_4	1973

Alicyclic Protons

n	$(CH_2)_n$	HC–beta	HC–alpha	Ring		
0	- - - -	2.02	2.02	Cyclopentanone	CCl_4	1999
1	ca. 1.79	ca. 1.79	2.25	Cyclohexanone	CCl_4	2000
2	ca. 1.71	ca. 1.71	2.49	Cycloheptanone	$CDCl_3$	2009
3	ca. 1.47	1.82	2.31	Cyclooctanone	CCl_4	- - - -
8	ca. 1.29	1.62	2.36	Cyclotridecanone	CCl_4	2010
10	ca. 1.33	1.64	2.43	Cyclopentadecanone	$CDCl_3$	2011

Olefinic Protons

The carbonyl group of the ketones weakly deshields the geminal olefinic proton but strongly deshields the cis and trans hydrogens bonded to the beta carbon atom. Its effect is similar to that of the unsaturated carbon atom of the nitrile (—C≡N) functional group.

3-Buten-2-one

trans	cis	HC—1		Solvent	Spectrum
ca. 6.1	ca. 6.1	5.80	H_2C=CH—C(=O)—CH_3	CCl_4	2017

Substituted Olefinic Ketones

X—	CH	=CH	—C(=O)—R		
CH_3—O—	7.50	5.52	—C(=O)—CH_3	CCl_4	----
$(CH_3)_2$—N—	7.55	5.69	—C(=O)—Ø	$CDCl_3$	----
$(R)_2$—CH—	6.47	5.92	—C(=O)—CH_3	CCl_4	2021
CH_3—	6.70	6.00	—C(=O)—CH_3	CCl_4	2018
CH_3—	6.70	6.00	—C(=O)—Ø	CCl_4	----
Ø—	7.51	6.68	—C(=O)—R_4	$CDCl_3$	2041
Ø—	7.72	7.05	—C(=O)—Ø	$CDCl_3$	2043
Cl—	7.42	7.27	—C(=O)—Ø	$CDCl_3$	----
N≡C—S—	7.36	7.32	—C(=O)—Ø	$CDCl_3$	----
HO—C(=O)—	6.86	7.95	—C(=O)—Ø	$CDCl_3$	----
Ø—C(=O)—	7.95	7.95	—C(=O)—Ø	$CDCl_3$	----

Aromatic Protons

The Ketone functional group is one of the strongly deshielding groups in its effect on the ortho aromatic protons, deshielding them about 0.3 ppm in relation to the meta and para hydrogens. The ortho hydrogens of the phenyl ketones resonate at about 7.8 ppm while the meta and ortho hydrogens overlap to form a complex band in the range from 7.1—7.5 ppm (CCl_4 solution).

Phenyl Ketones

para/meta	ortho			
7.1—7.5	7.80	—C(=O)—CH_3	CCl_4	2045
7.2—7.6	7.92	—C(=O)—R_2	$CDCl_3$	2046
7.2—7.6	7.78	—C(=O)—Ø	$CDCl_3$	2098

Aromatic Protons continued

Para Substituted Acetophenones

para	meta	ortho	$-C(=O)-CH_3$	Solvent	Spectrum
CH_3-	7.09	7.69		CCl_4	2056
Br–	7.50	7.73		CCl_4	----
CH_3-CH_2-	7.17	7.75		CCl_4	2059
H_2N-	6.62	7.77		$CDCl_3$	----
R–NH–	6.54	7.80		$CDCl_3$	----
HO–	6.83	7.80		DMSO-d_6	----
Cl–	8.05	8.05		$CDCl_3$	2082
CH_3-S-	7.24	7.81		Polysol	2074
$(CH_3)_2-N-$	6.71	7.95		$CDCl_3$	----
CH_3-O-	6.98	7.97		$CDCl_3$	2075
F–	7.11	7.99		$CDCl_3$	----
$CH_3-C(=O)-$	8.05	8.05		$CDCl_3$	2082
Ø–	7.64	8.00		$CDCl_3$	----
O_2N-	8.32	8.19		$CDCl_3$	----

Para Substituted Benzophenones

para	meta	ortho	–C(=O)–Ø		
HO–	6.88	7.56		DMSO-d_6	----
Br–	7.60	7.60		$CDCl_3$	2102
CH_3-	7.22	7.60		$CDCl_3$	2100
$(CH_3)_2-N-$	6.65	7.73		$CDCl_3$	----
Ø–C(=O)–	7.92	7.92		$CDCl_3$	2107
HO–C(=O)–	8.39	7.99		TFA	2038

Coupling and Coupling Constants

No unusual couplings nor coupling constants are observed in the NMR spectra of the Ketones. The aliphatic three bond vicinal coupling J_{HC-CH} is 6–8 Hz, the aromatic ortho coupling constant varies from 8–9 Hz, and the olefinic coupling constants display the values listed below.

$J_{CH=CH}$ = 14–18 Hz (trans)

$J_{CH=CH}$ = 7–12 Hz (cis)

J_{H-C-H} = 1–4 Hz (geminal)

J_{CH_3-CH} = 5–7 Hz

$J_{CH_3-C=CH}$ = 0–2 Hz

ACETONE

C_3H_6O

Mol. Wt. 58.08

B.P. 56.5°C

Solvent: Neat

Assignments
a = 2.11

1967

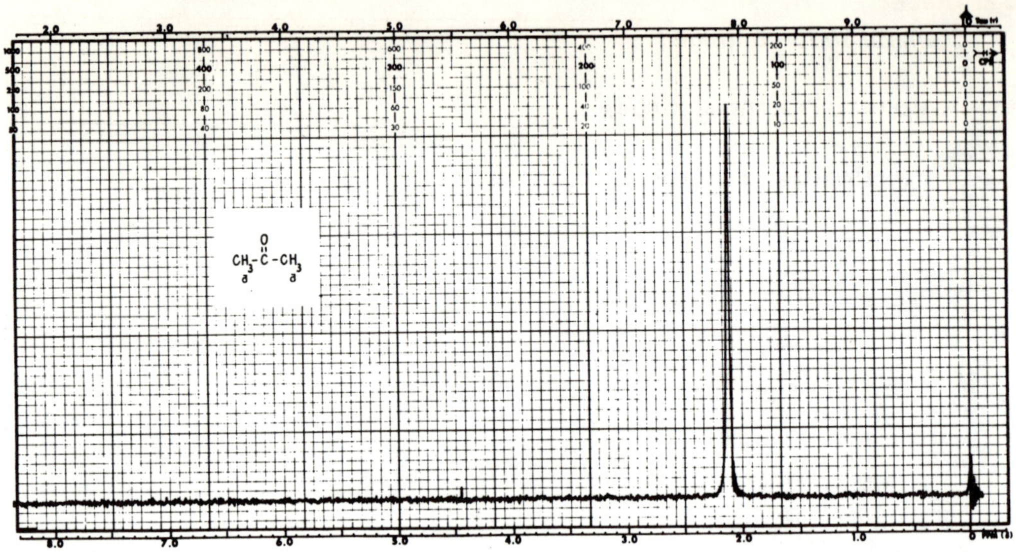

© 1978 Sadtler Research Laboratories, Inc.

2-BUTANONE

C_4H_8O

Mol. Wt. 72.11

B.P. 79-80°C

Solvent: CCl_4

Assignments
a = 0.99
b = 2.05
c = 2.40

1968

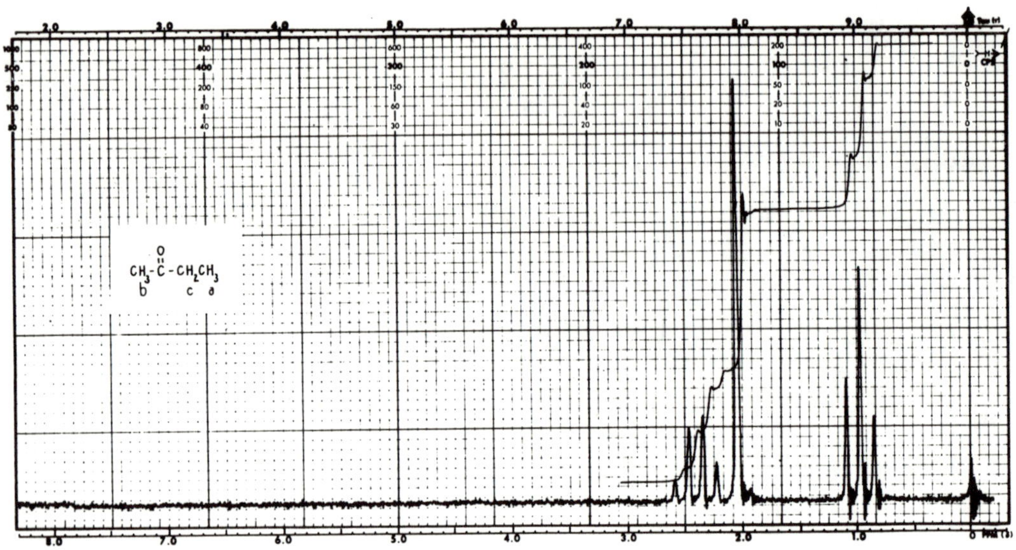

© 1978 Sadtler Research Laboratories, Inc.

2-HEXANONE

$C_6H_{12}O$

Mol. Wt. 100.16

B.P. 123°C

Solvent: CCl_4

Assignments
a = 1.40
b = 2.28
c = 3.37
d = 4.60
e = 4.62

1969

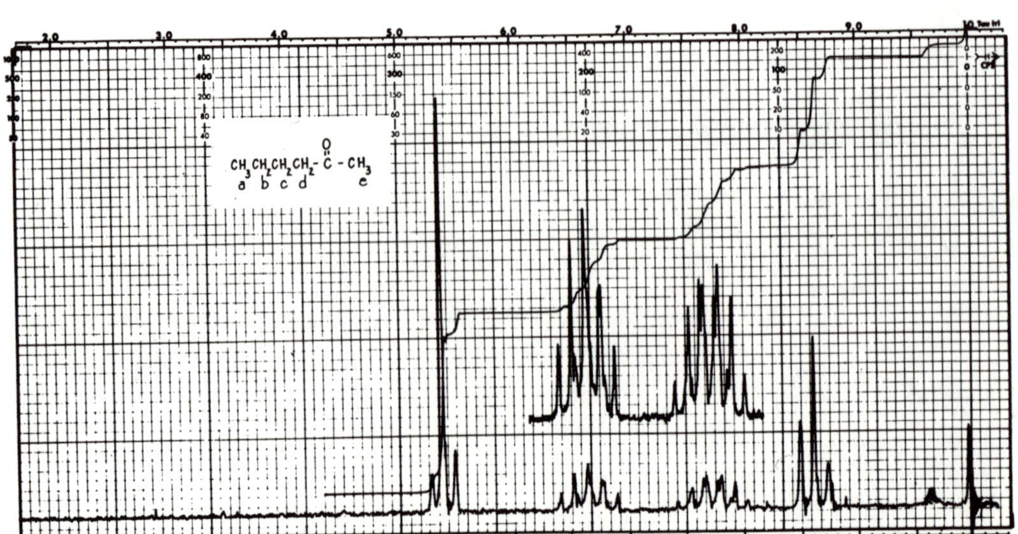

© 1978 Sadtler Research Laboratories, Inc.

3-HEXANONE

$C_6H_{12}O$

Mol. Wt. 100.16

B.P. 120-125°C

Solvent: CCl_4

Assignments
a = 0.92
b = 1.03
c = 1.59
d = 2.29
e = 2.37

1970

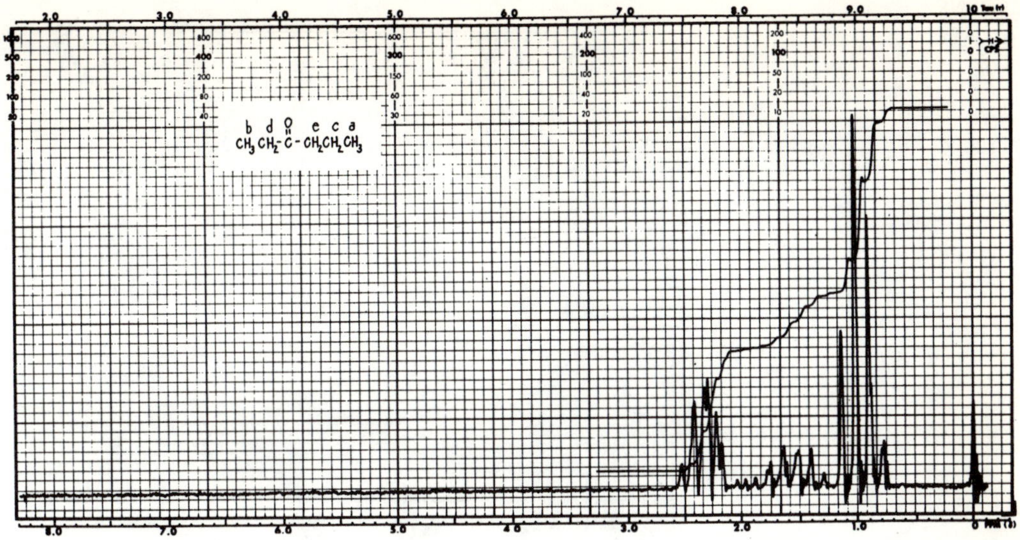

© 1978 Sadtler Research Laboratories, Inc.

4-METHYL-2-PENTANONE

$C_6H_{12}O$

Mol. Wt. 100.16

B.P. 114-116°C

Solvent: CCl_4

Assignments
a = 0.91
b = 2.01
c = 1.90-2.37

1971

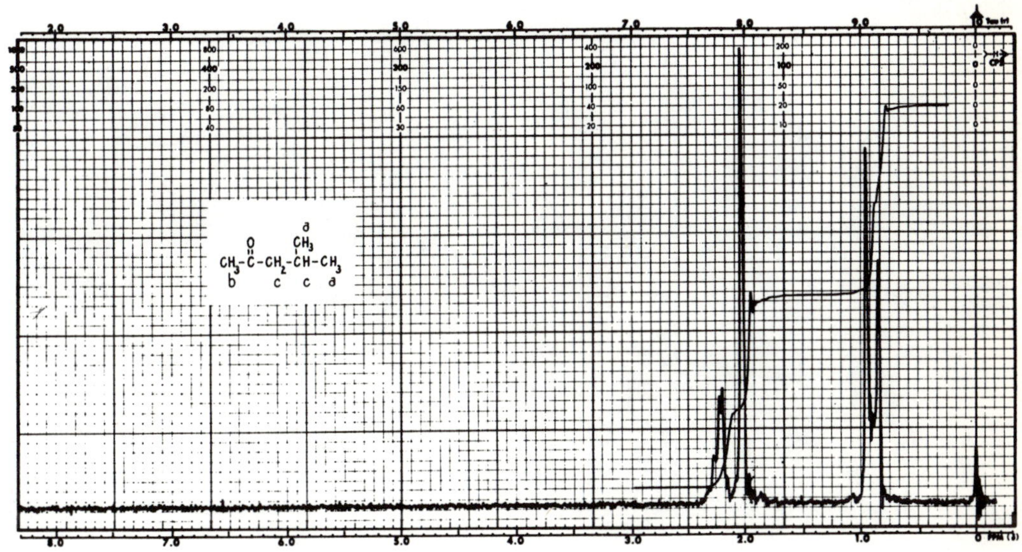

© 1978 Sadtler Research Laboratories, Inc.

4-HEPTANONE

$C_7H_{14}O$

Mol. Wt. 114.19

B.P. 144°C

Solvent: CCl_4

Assignments
a = 0.90
b = 1.58
c = 2.31

1972

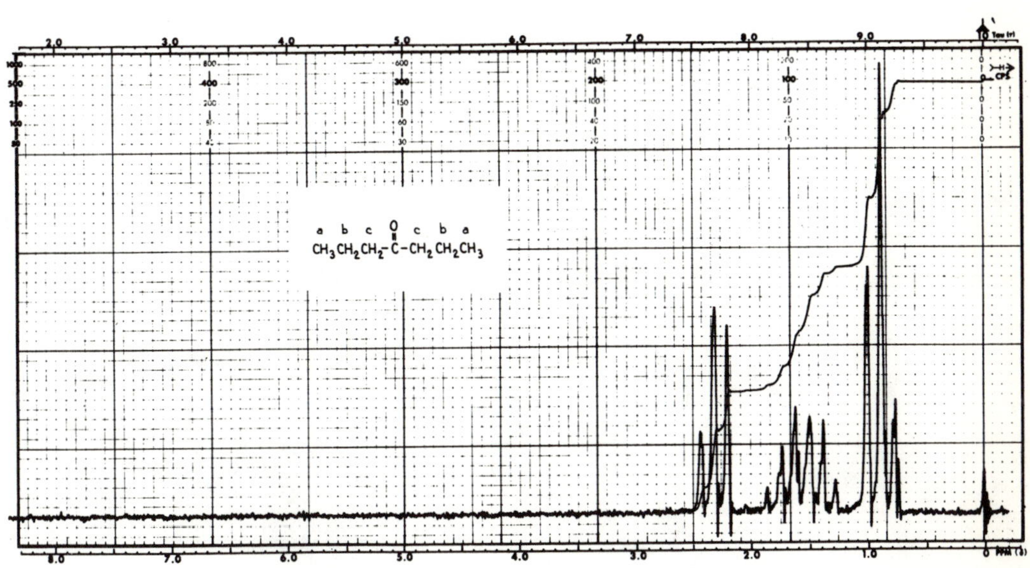

© 1978 Sadtler Research Laboratories, Inc.

4,4-DIMETHYL-2-PENTANONE

$C_7H_{14}O$

Mol. Wt. 114.19

B.P. 126°C

Solvent: CCl_4

Assignments
a = 1.02
b = 2.06
c = 2.28

1973

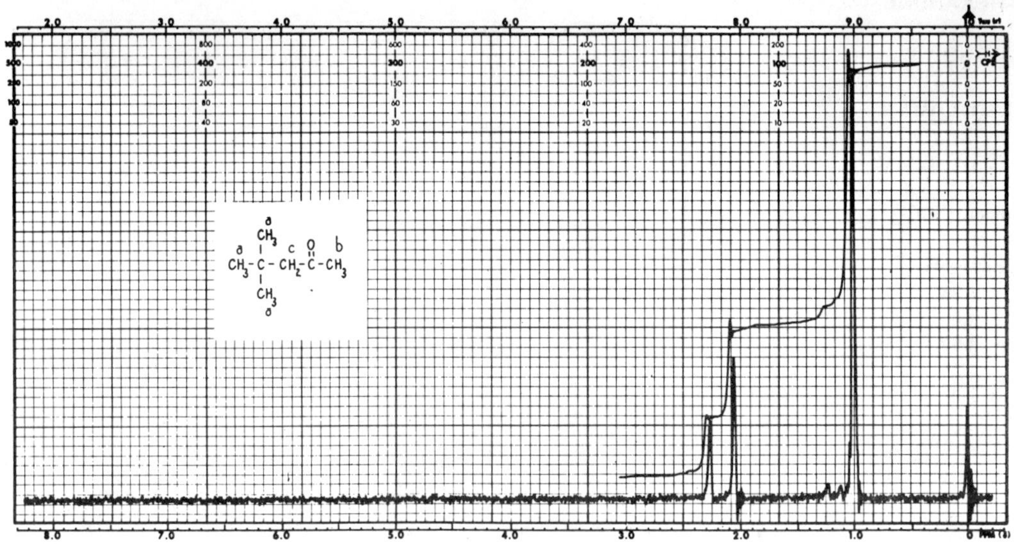

© 1978 Sadtler Research Laboratories, Inc.

2-OCTANONE

$C_8H_{16}O$

Mol. Wt. 128.22

Solvent: CCl_4

Assignments
a = 0.91
b = 1.12-1.72
c = 2.06
d = 2.36

1974

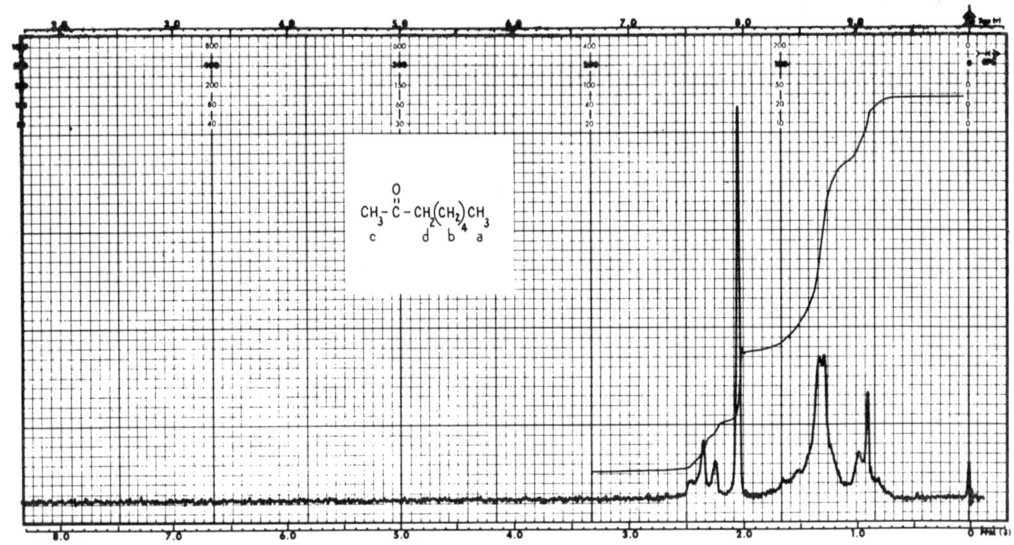

© 1978 Sadtler Research Laboratories, Inc.

3-OCTANONE

$C_8H_{16}O$

Mol. Wt. 128.22

B.P. 167-168°C

Solvent: $CDCl_3$

Assignments
a = 0.92
b = 1.07
c = 1.00-1.90
d = 2.38
e = 2.43

1975

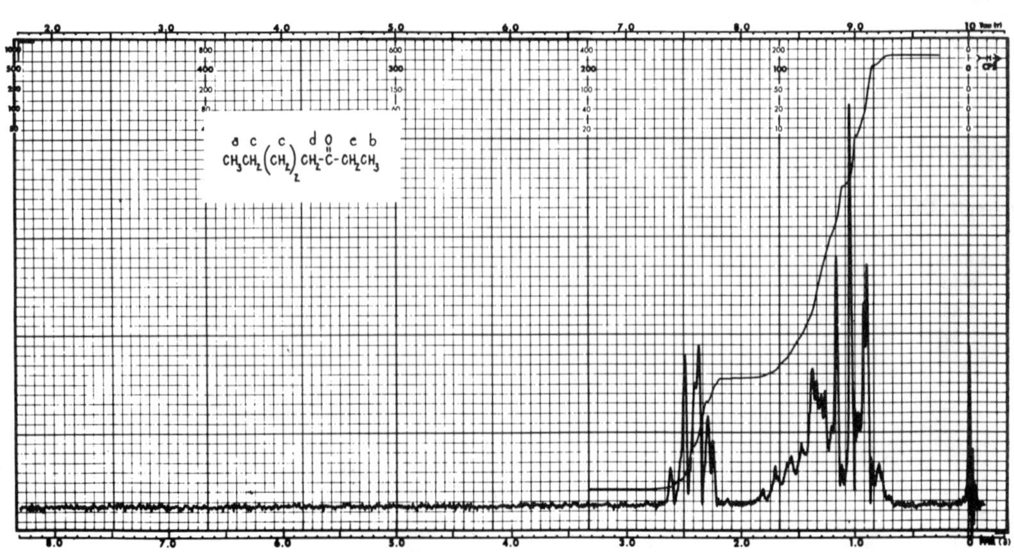

© 1978 Sadtler Research Laboratories, Inc.

6-METHYL-2-HEPTANONE

$C_8H_{16}O$

Mol. Wt. 128.22

Solvent: CCl_4

Assignments
a = 0.90
b = 1.02-1.90
c = 2.06
d = 2.34

1976

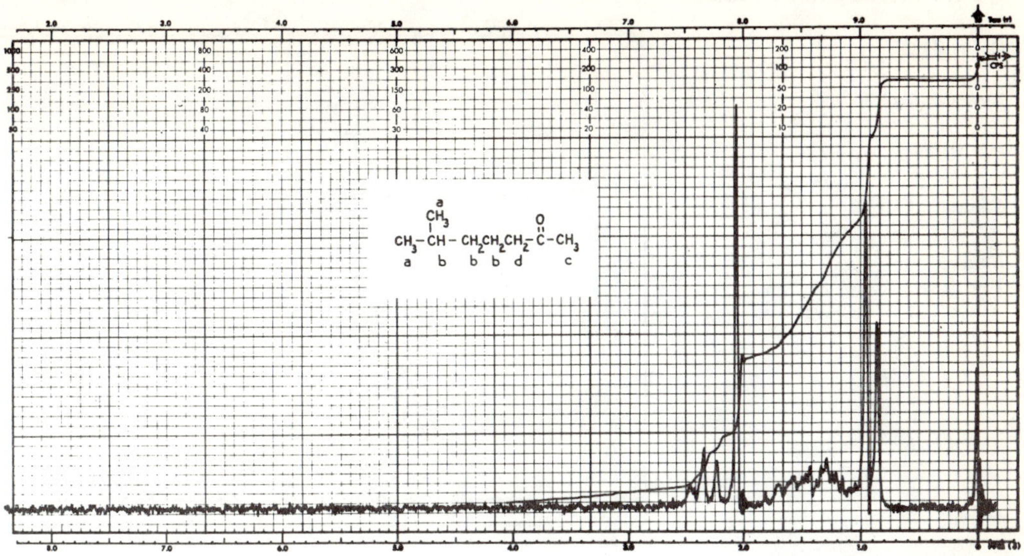

© 1978 Sadtler Research Laboratories, Inc.

2,2,4-TRIMETHYL-3-PENTANONE

$C_8H_{16}O$

Mol. Wt. 128.21

B.P. 136.2°C

Solvent: CCl_4

Assignments
a = 1.00
b = 1.11
c = 3.07
d = 0.90 imp.

1977

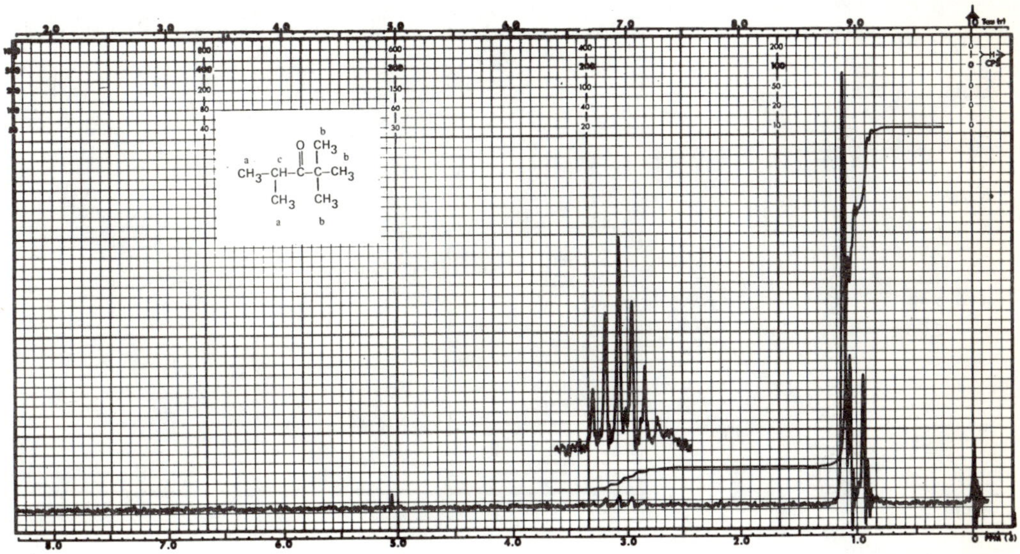

© 1970 Sadtler Research Laboratories, Inc.

5-NONANONE

$C_9H_{18}O$

Mol. Wt. 142.24

M.P. −7 to −5°C

Solvent: $CDCl_3$

Assignments
a = 0.91
b = 1.10-1.90
c = 2.40

1978

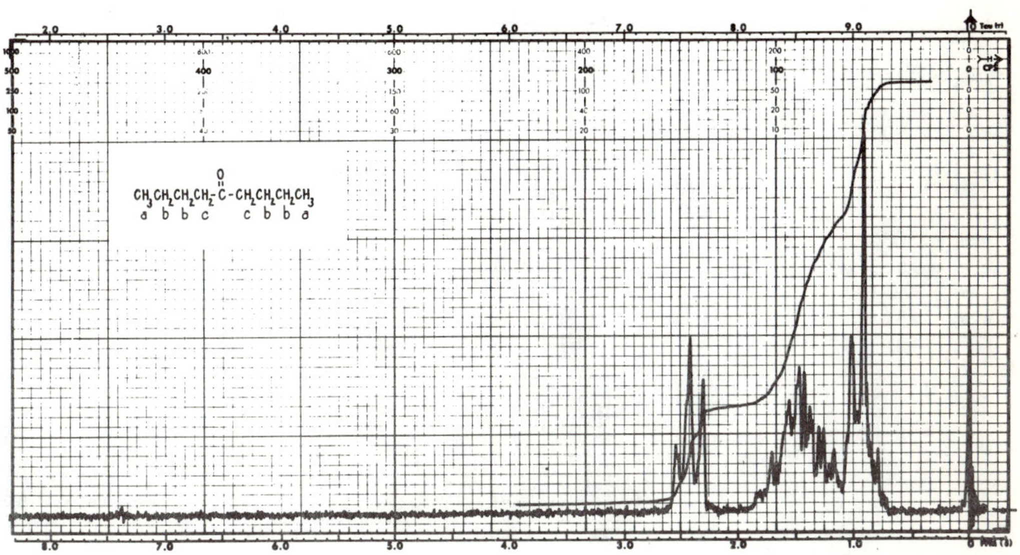

© 1978 Sadtler Research Laboratories, Inc.

2,6-DIMETHYL-4-HEPTANONE

$C_9H_{18}O$

Mol. Wt. 142.24

B.P. 168°C/760mm

Solvent: CCl_4

Assignments
a = 0.90
b = 1.96-2.34

1979

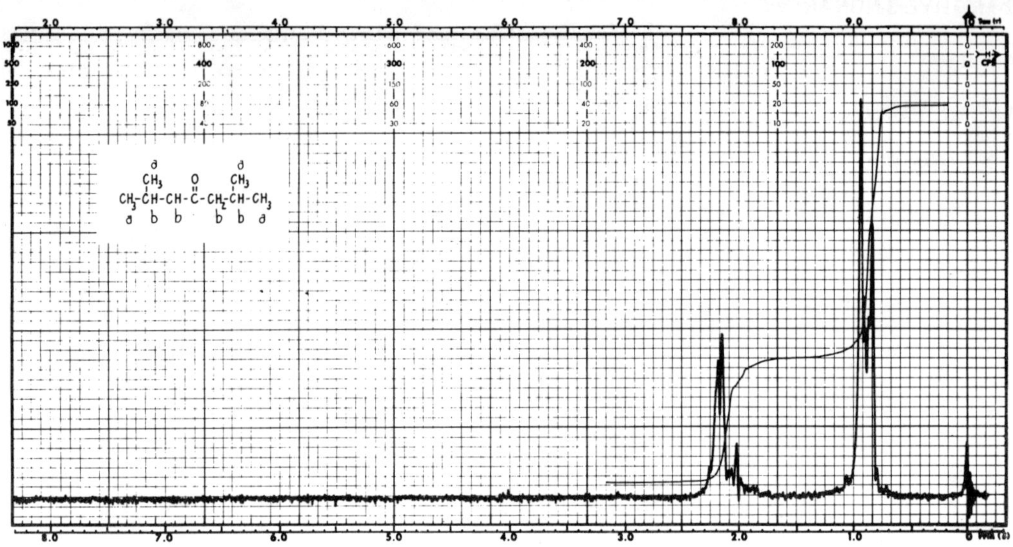

© 1978 Sadtler Research Laboratories, Inc.

3-DECANONE

$C_{10}H_{20}O$

Mol. Wt. 156.27

Solvent: CCl_4

Assignments
a = 0.88
b = 1.00
c = 1.28
d = 2.34
e = 2.36

1980

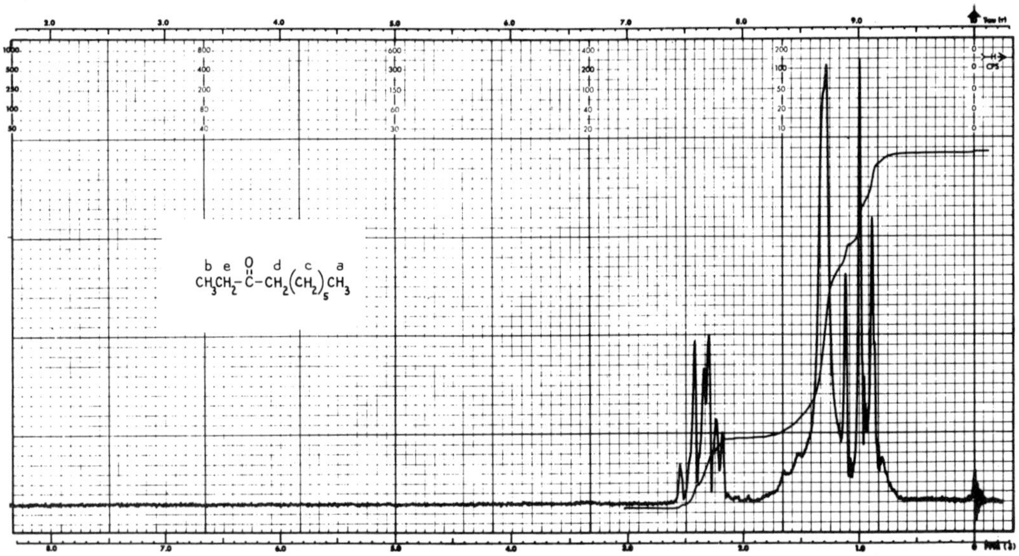

© 1978 Sadtler Research Laboratories, Inc.

4-DECANONE

$C_{10}H_{20}O$

Mol. Wt. 156.27

B.P. 207-209°C

Solvent: CCl_4

Assignments
a = ca 0.93
b = 1.00-1.50
c = 1.54
d = ca 2.30

1981

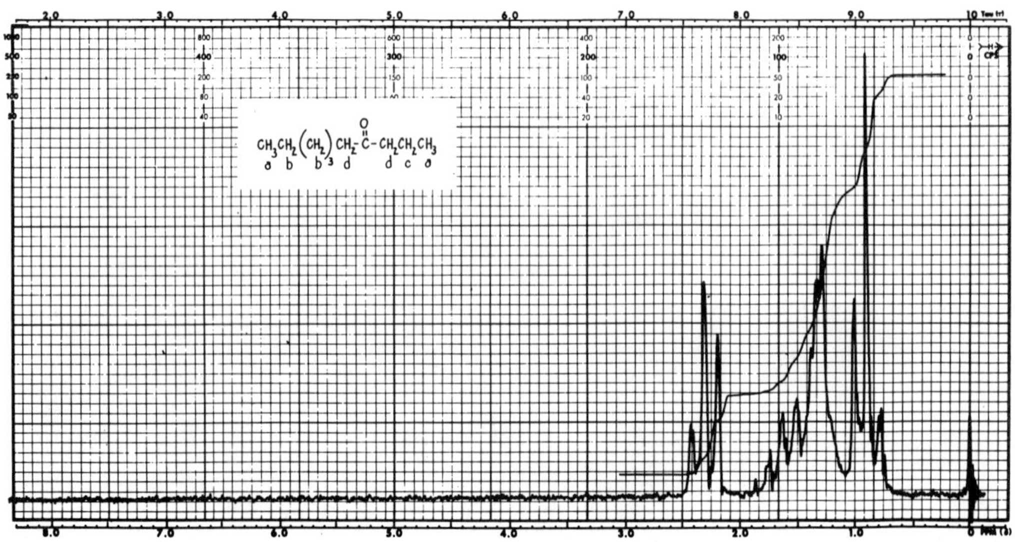

© 1978 Sadtler Research Laboratories, Inc.

2-METHYL-3-NONANONE

$C_{10}H_{20}O$

Mol. Wt. 156.27

B.P. 200-201°C

Solvent: CCl_4

Assignments
a = 0.90
b = 1.05
c = 1.00-1.90
d = 2.35
e = 2.50

1982

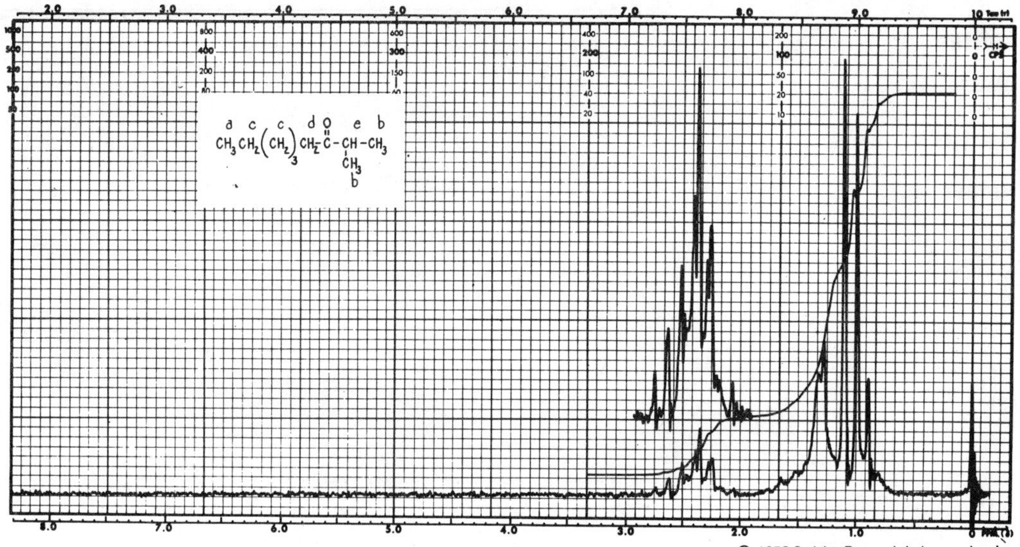

© 1978 Sadtler Research Laboratories, Inc.

6-UNDECANONE

$C_{11}H_{22}O$

Mol. Wt. 170.30

M.P. 15°C

B.P. 227°C

Solvent: CCl_4

Assignments
a = 0.91
b = 1.10-1.80
c = 2.31

1983

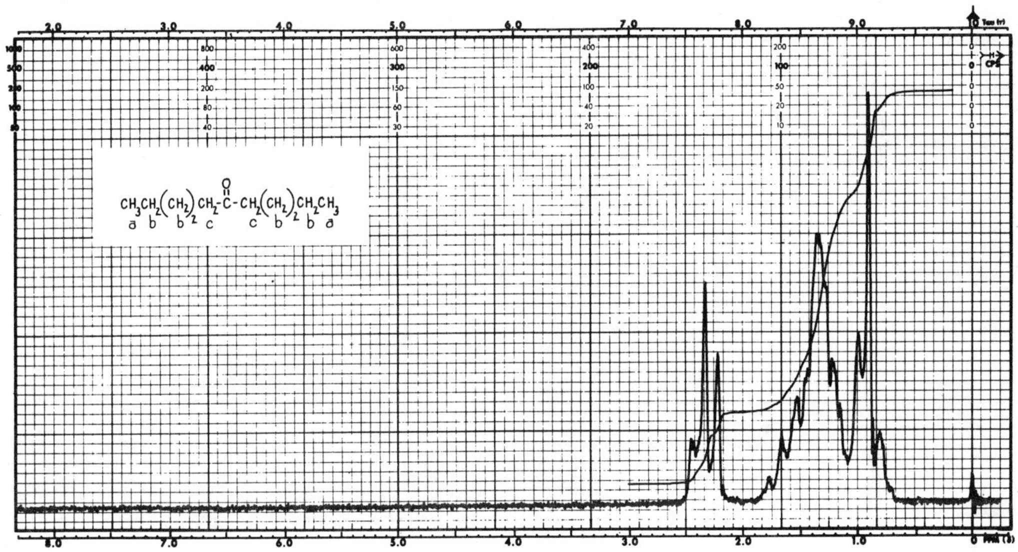

© 1978 Sadtler Research Laboratories, Inc.

4-DODECANONE

$C_{12}H_{24}O$

Mol. Wt. 184.32

Solvent: CCl_4

Assignments
a = 0.91
b = 1.10-1.50
c = 1.56
d = 2.30

1984

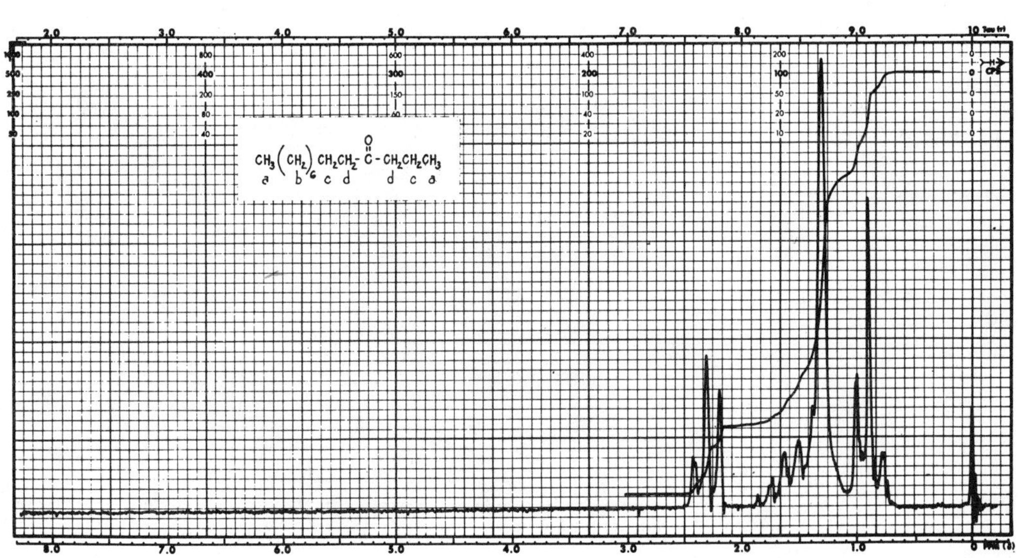

© 1978 Sadtler Research Laboratories, Inc.

6-DODECANONE

$C_{12}H_{24}O$

Mol. Wt. 184.32

Solvent: CCl_4

Assignments
a = 0.90
b = 1.10-1.85
c = 2.31

1985

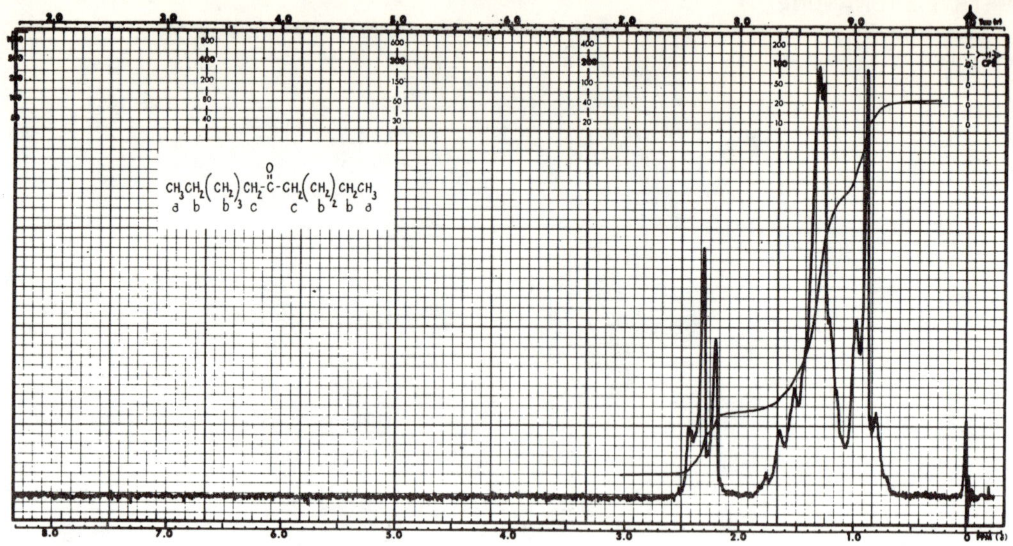

© 1978 Sadtler Research Laboratories, Inc.

6-TRIDECANONE

$C_{13}H_{26}O$

Mol. Wt. 198.35

Solvent: $CDCl_3$

Assignments
a = 0.90
b = 1.10-1.90
c = 2.40

1986

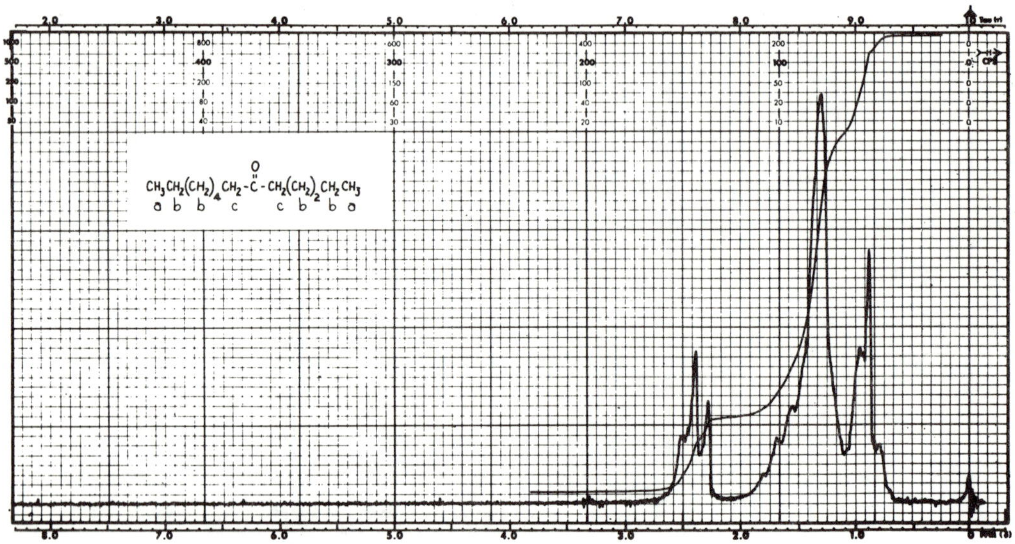

© 1978 Sadtler Research Laboratories, Inc.

7-TRIDECANONE

$C_{13}H_{26}O$

Mol. Wt. 198.35

M.P. 28-29°C

Solvent: $CDCl_3$

Assignments
a = 0.89
b = 1.10-1.80
c = 2.39

1987

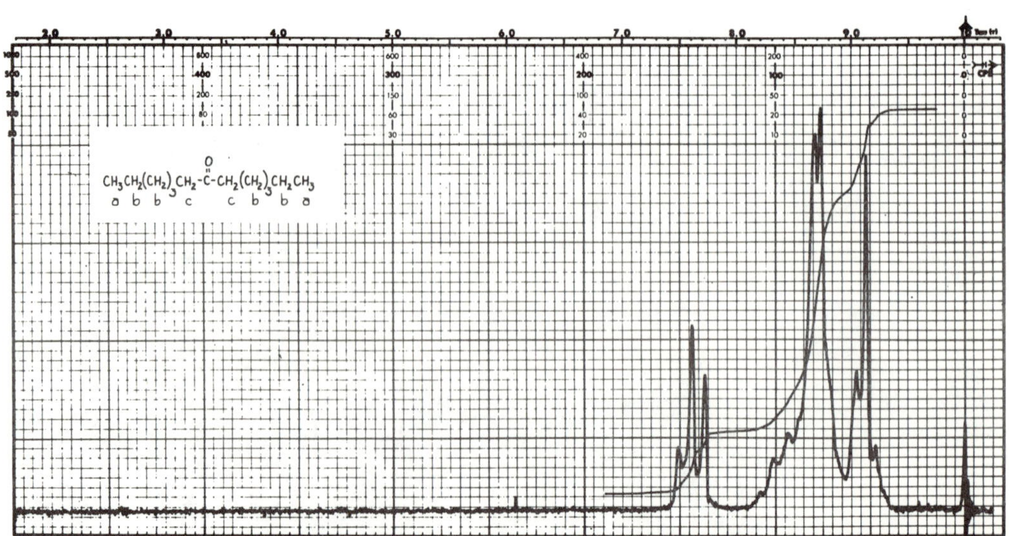

© 1978 Sadtler Research Laboratories, Inc.

3-TRIDECANONE

$C_{13}H_{26}O$

Mol. Wt. 198.35

Solvent: CCl_4

Assignments
a = 0.90
b = 1.01
c = 1.10-1.90
d = ca 2.34

1988

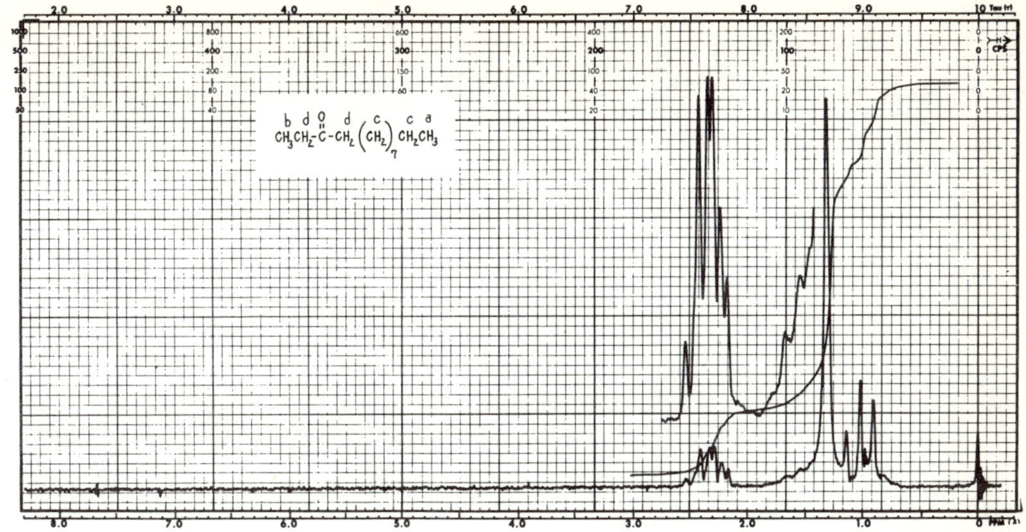

© 1978 Sadtler Research Laboratories, Inc.

2-TETRADECANONE

$C_{14}H_{28}O$

Mol. Wt. 212.38

M.P. 33-34°C

Solvent: $CDCl_3$

Assignments
a = 0.89
b = ca 1.29
c = 1.54
d = 2.13
e = 2.43

1989

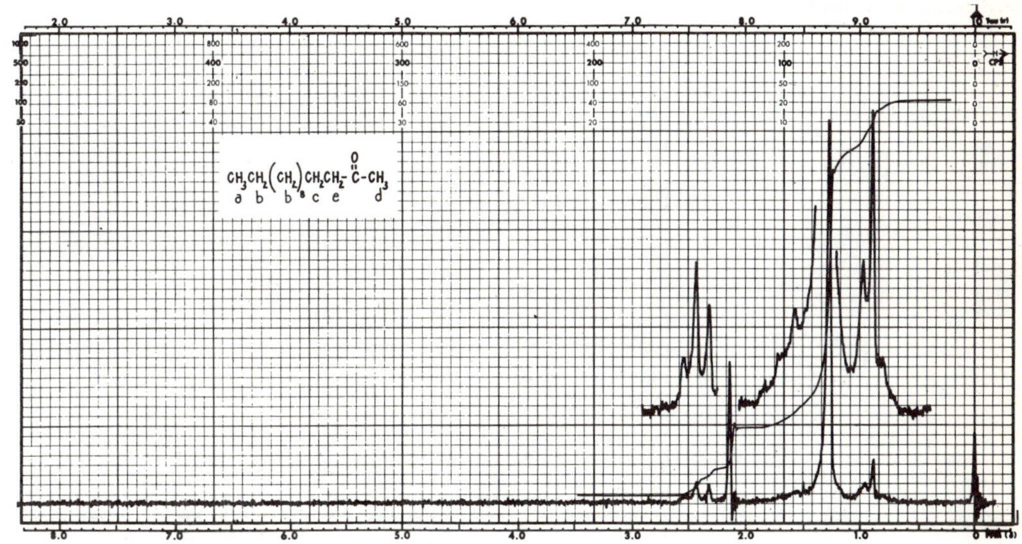

© 1978 Sadtler Research Laboratories, Inc.

1,3-DIFLUORO-2-PROPANONE

$C_3H_4F_2O$

Mol. Wt. 94.06

B.P. 101-102°C

Solvent: CCl_4

Assignments
a = 5.11

1990

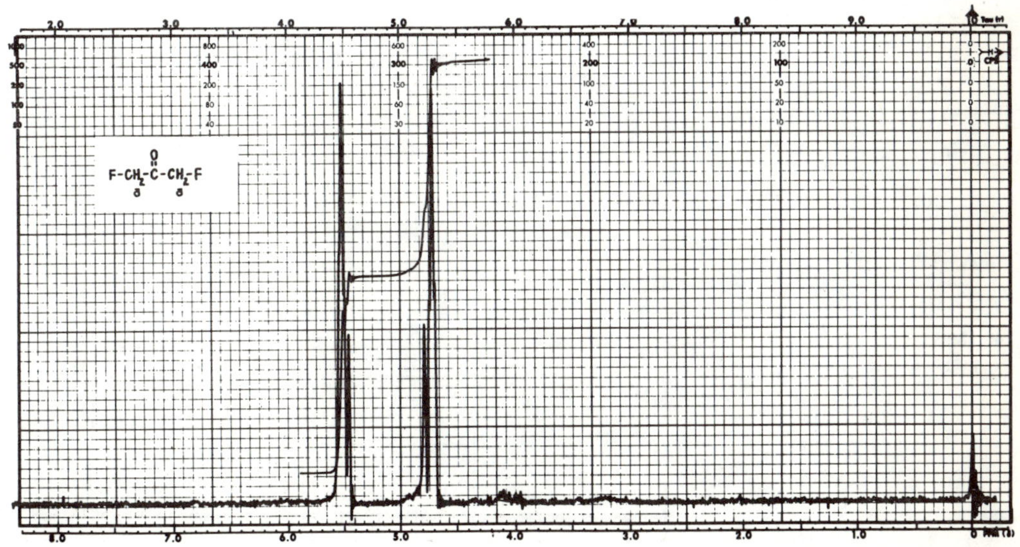

© 1978 Sadtler Research Laboratories, Inc.

3-CHLORO-2-BUTANONE

C_4H_7ClO

Mol. Wt. 106.55

Solvent: CCl_4

Assignments
a = 1.57
b = 2.29
c = 4.23

1991

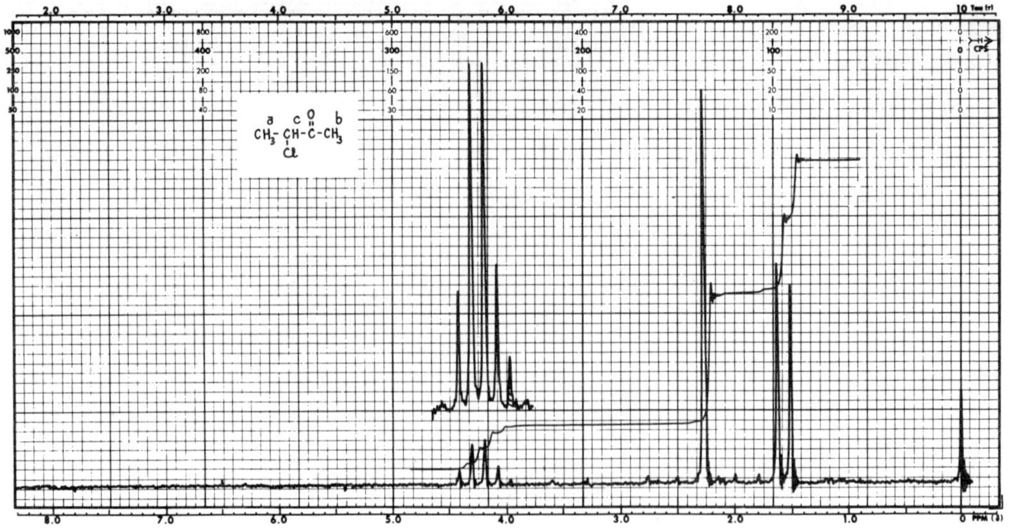

© 1978 Sadtler Research Laboratories, Inc.

3-BROMO-2-BUTANONE

C_4H_7BrO

Mol. Wt. 151.01

Solvent: CCl_4

Assignments
a = 1.71
b = 2.31
c = 4.39
d = 2.08 imp.

1992

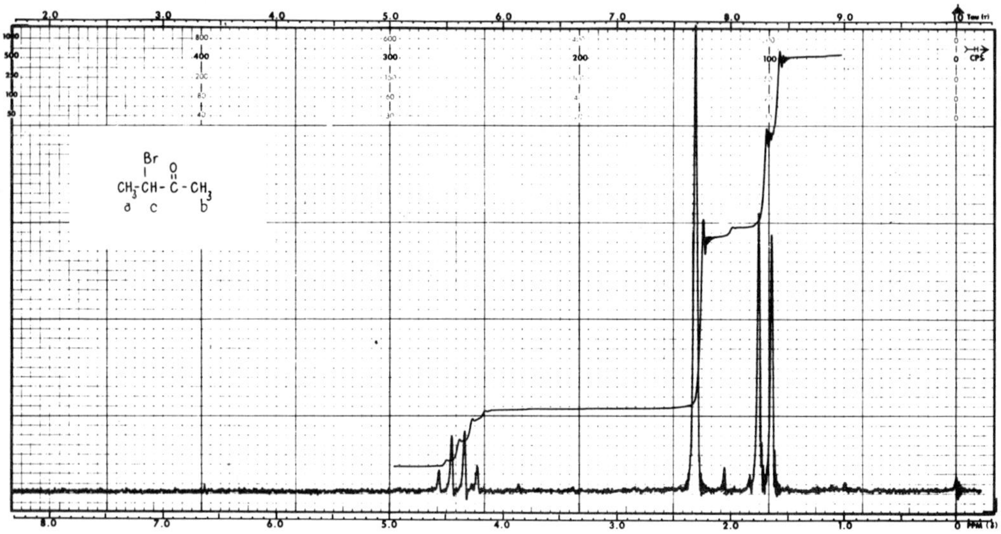

© 1978 Sadtler Research Laboratories, Inc.

3-HYDROXY-3-METHYL-2-BUTANONE

$C_5H_{10}O_2$

Mol. Wt. 102.13

B.P. 140°C

Solvent: $CDCl_3$

Assignments
a = 1.37
b = 2.23
c = 3.30
d = 1.13 imp.

1993

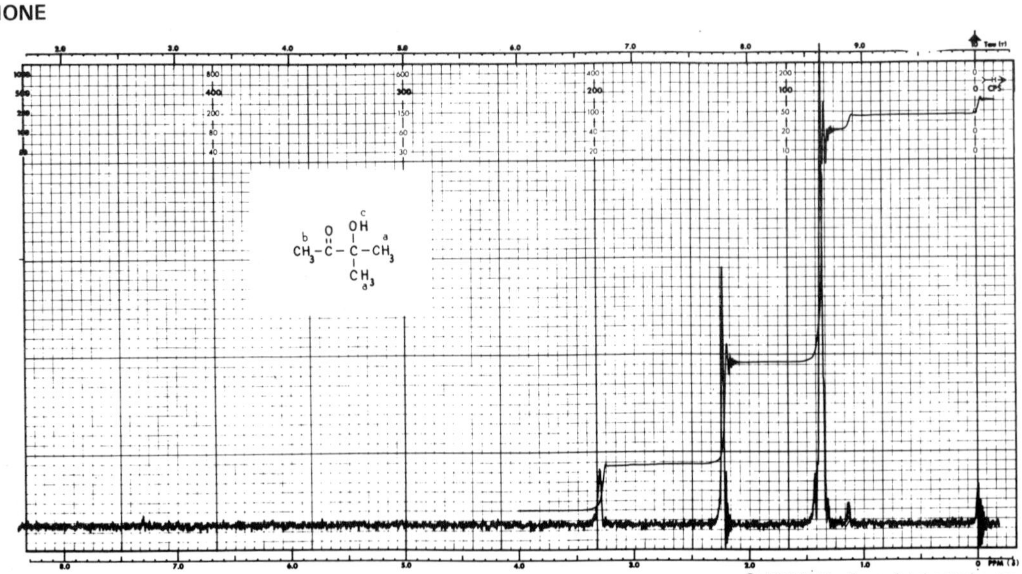

© 1978 Sadtler Research Laboratories, Inc.

4-HYDROXY-3-METHYL-2-BUTANONE

$C_5H_{10}O_2$

Mol. Wt. 102.13

B.P. 90-95°C/15mm

Solvent: $CDCl_3$

Assignments
a = 1.12
b = 2.21
c = 2.71
d = 3.11
e = 3.71

1994

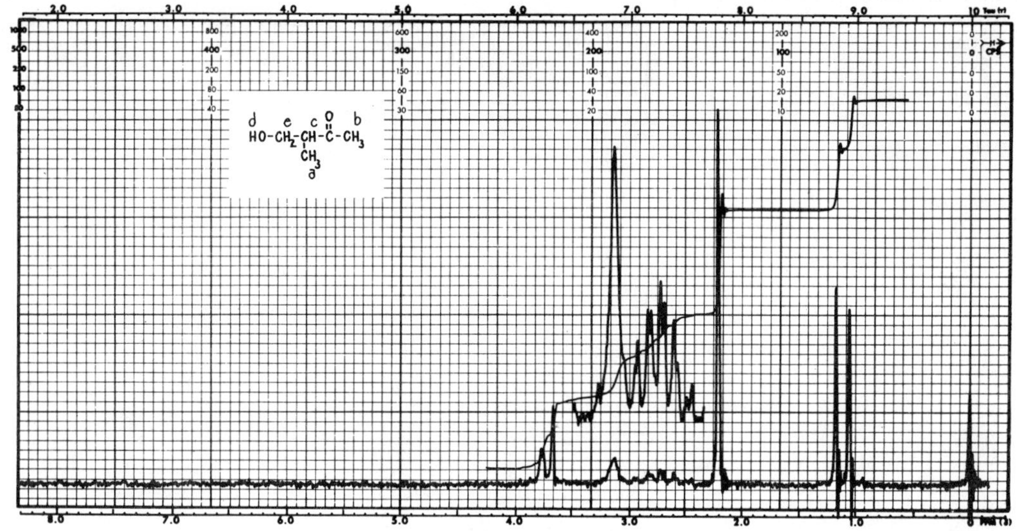

© 1978 Sadtler Research Laboratories, Inc.

2,5-HEXANEDIONE

$C_6H_{10}O_2$

Mol. Wt. 114.15

Solvent: CCl_4

Assignments
a = 2.12
b = 2.60

1995

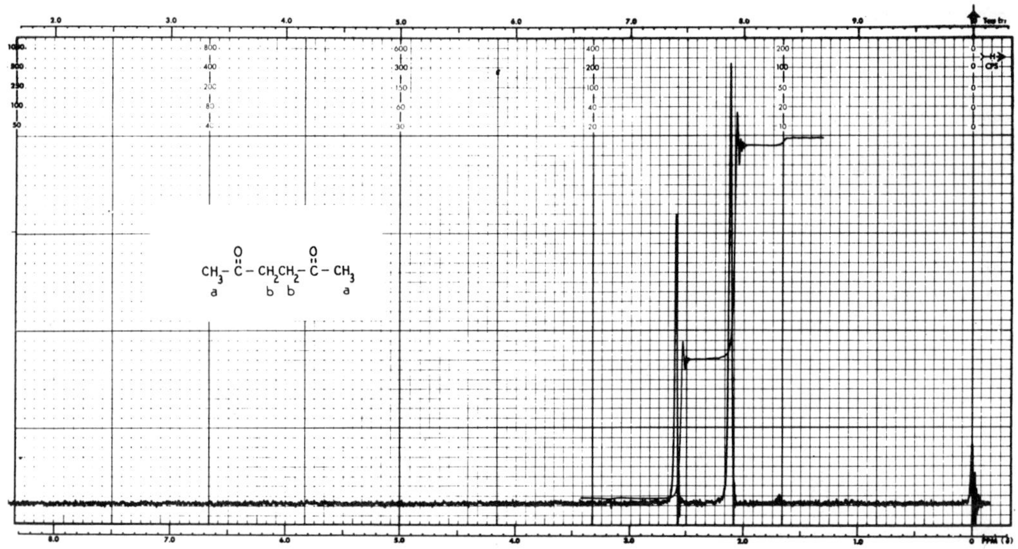

© 1978 Sadtler Research Laboratories, Inc.

CYCLOPROPYL KETONE

$C_7H_{10}O$

Mol. Wt. 110.16

Solvent: CCl_4

Assignments
a = 0.60-1.10
b = 2.00

1996

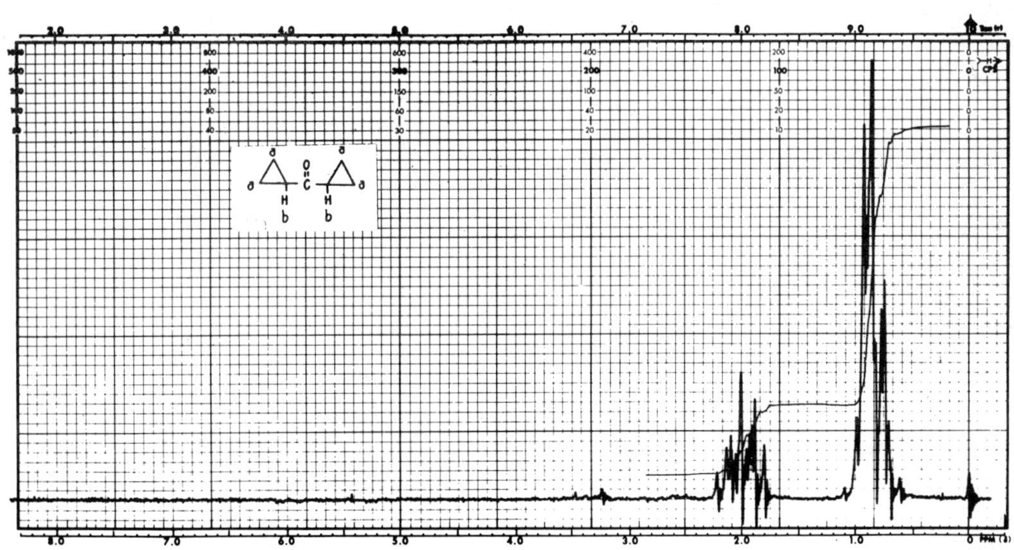

© 1978 Sadtler Research Laboratories, Inc.

1-CYCLOPENTYL-2-PROPANONE

$C_8H_{14}O$

Mol. Wt. 126.20

Solvent: CCl_4

Assignments
a = 0.90-2.60
b = 2.05
c = ca 2.38

1997

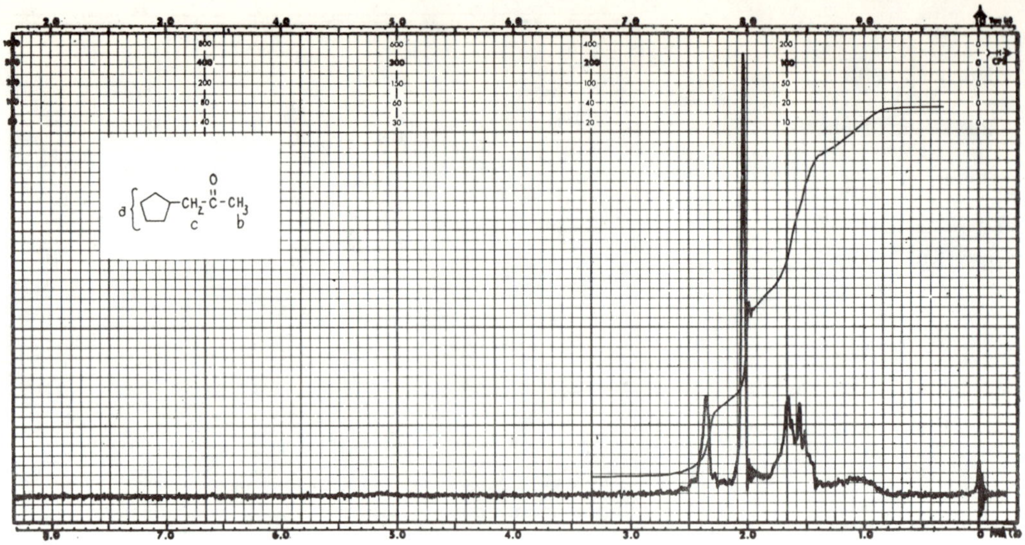

© 1978 Sadtler Research Laboratories, Inc.

METHYL 2-NORBORNYL KETONE

$C_9H_{14}O$

Mol. Wt. 138.21

Solvent: $CDCl_3$

Assignments
a = 1.00-2.00
b = ca 2.11
c = 2.00-3.10

1998

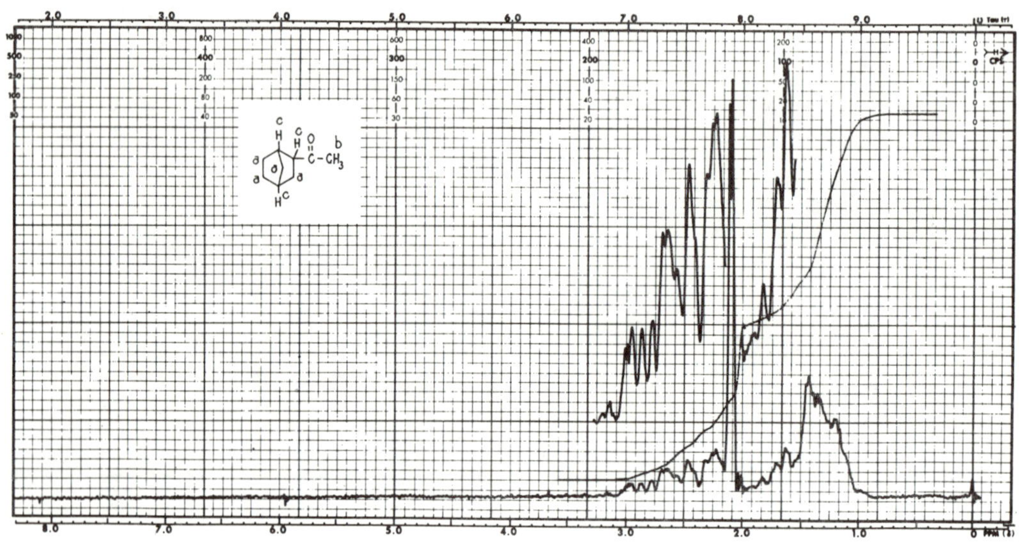

© 1978 Sadtler Research Laboratories, Inc.

CYCLOPENTANONE

C_5H_8O

Mol. Wt. 84.12

B.P. 129-131°C

Solvent: CCl_4

Assignments
a = ca 2.02

1999

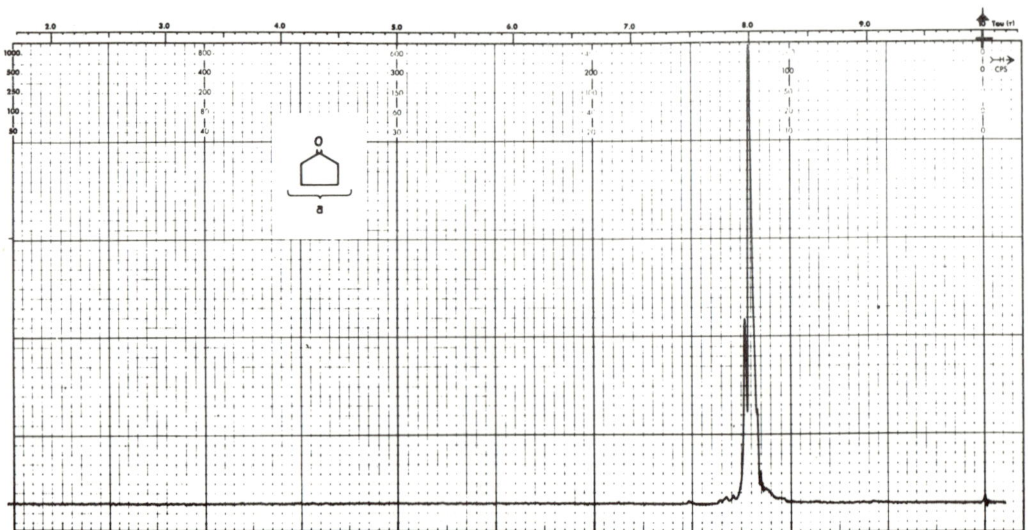

© 1978 Sadtler Research Laboratories, Inc.

CYCLOHEXANONE

$C_6H_{10}O$

Mol. Wt. 98.15

B.P. 154-156°C

Solvent: CCl_4

Assignments
a = 1.79
b = 2.25
c = 4.99 imp.

2000

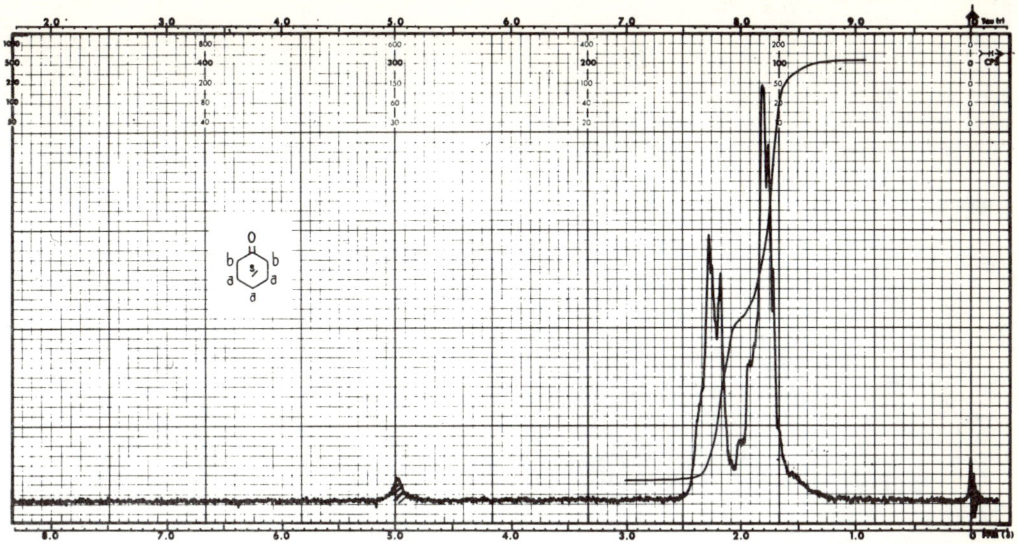

© 1978 Sadtler Research Laboratories, Inc.

2-METHYLCYCLOHEXANONE

$C_7H_{12}O$

Mol. Wt. 112.17

B.P. 161-163°C

Solvent: CCl_4

Assignments
a = 0.94
b = 1.00-2.10
c = ca 2.22

2001

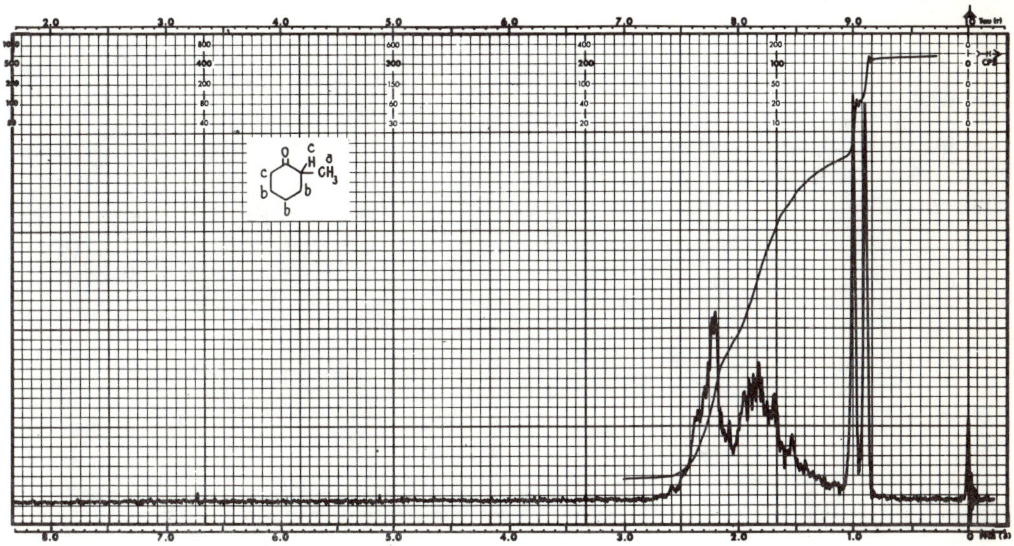

© 1978 Sadtler Research Laboratories, Inc.

3-METHYLCYCLOHEXANONE

$C_7H_{12}O$

Mol. Wt. 112.17

B.P. 167-169°C

Solvent: CCl_4

Assignments
a = 1.02
b = 1.10-2.40

2002

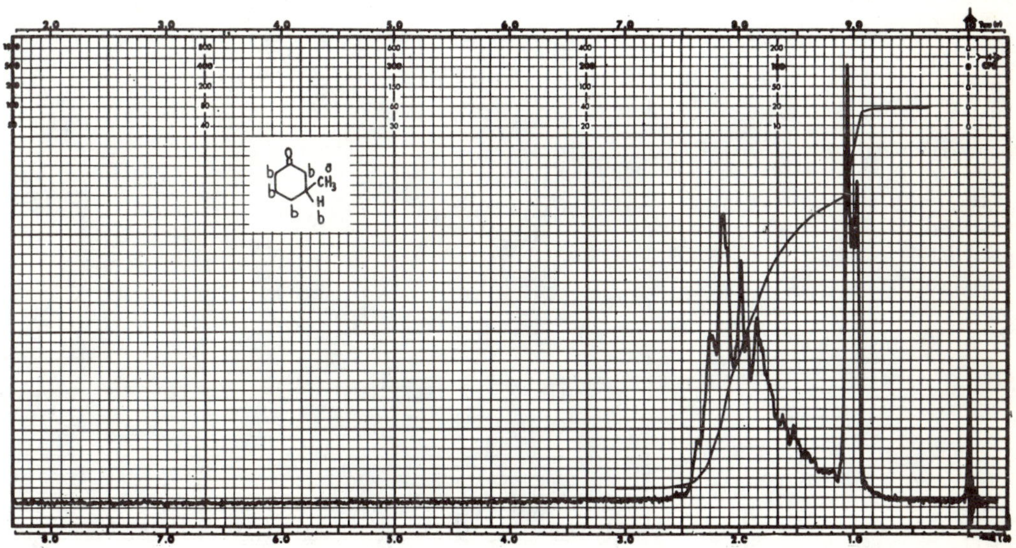

© 1978 Sadtler Research Laboratories, Inc.

4-METHYLCYCLOHEXANONE

$C_7H_{12}O$

Mol. Wt. 112.17

B.P. 169°C/760mm

Solvent: CCl_4

Assignments
a = 1.02
b = 1.18-2.28
c = ca 2.21

2003

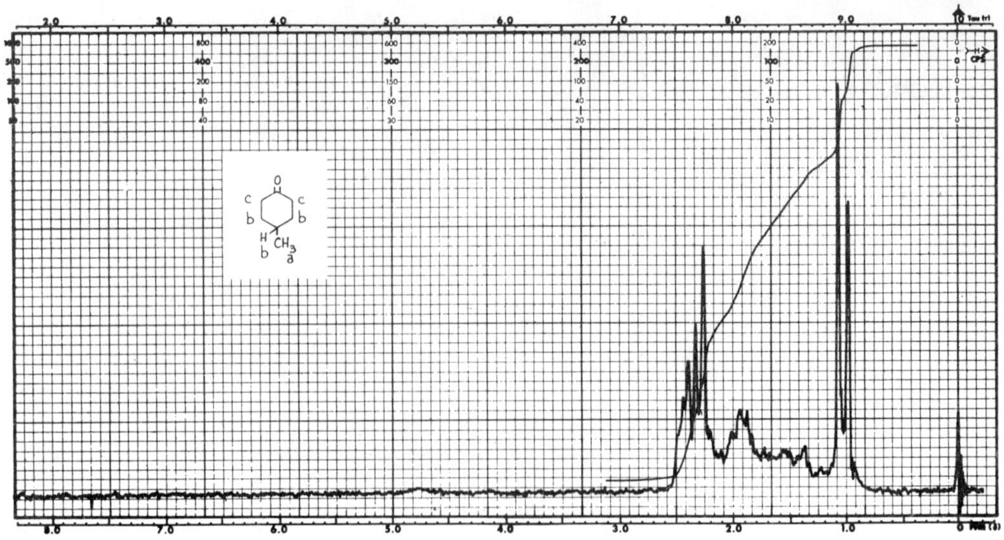

© 1978 Sadtler Research Laboratories, Inc.

4-tert-BUTYLCYCLOHEXANONE

$C_{10}H_{18}O$

Mol. Wt. 154.25

B.P. 90-92°C/9mm

Solvent: $CDCl_3$

Assignments
a = 0.95
b = 1.20-1.85
c = 1.85-2.60

2004

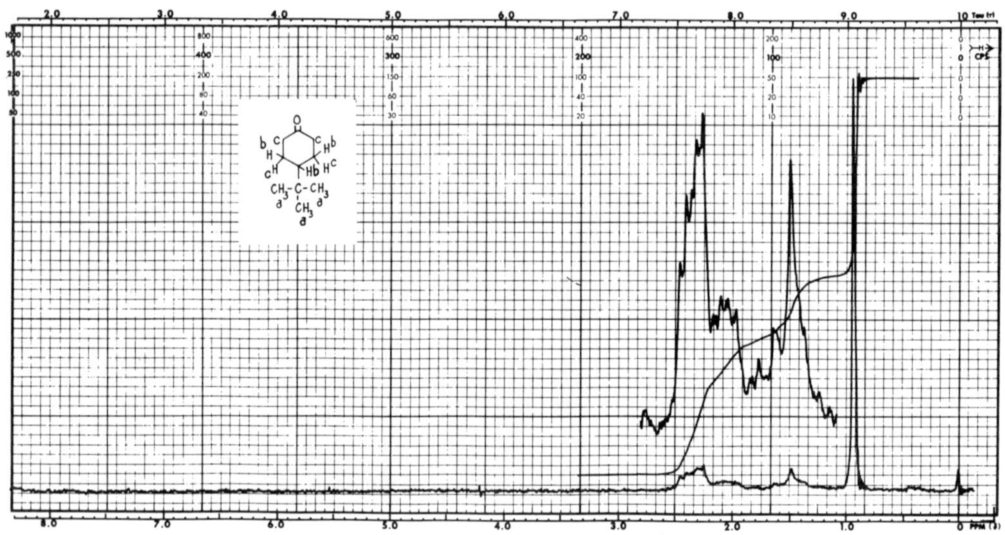

© 1978 Sadtler Research Laboratories, Inc.

5-HEXYL-3-METHYLCYCLOHEXANONE

$C_{13}H_{24}O$

Mol. Wt. 196.34

B.P. 276°C/760mm

Solvent: $CDCl_3$

Assignments
a = ca 0.99
b = 1.30
c = 1.10-2.60

2005

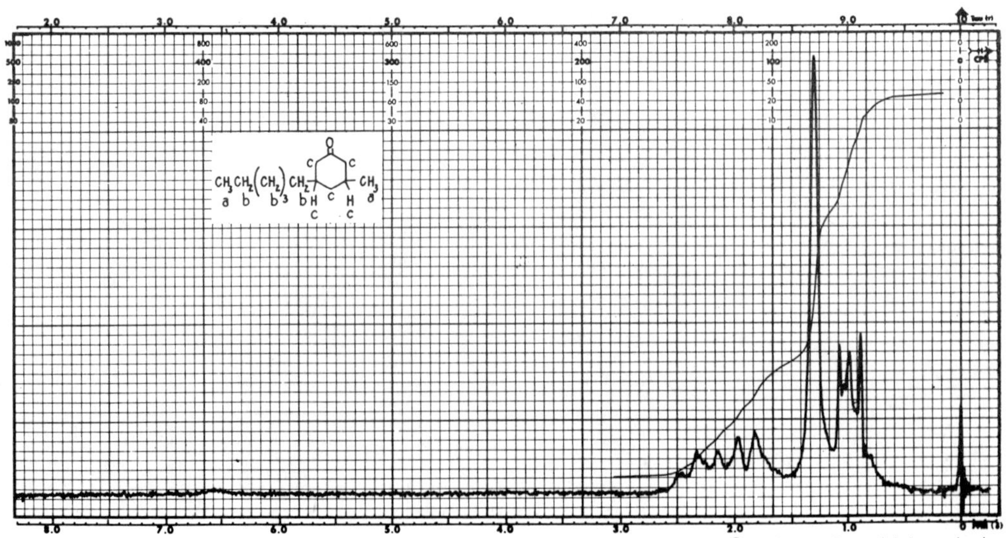

© 1978 Sadtler Research Laboratories, Inc.

(+)-1,3,3-TRIMETHYL-2-NORBORNANONE

$C_{10}H_{16}O$

Mol. Wt. 152.24

M.P. 3.5-3.6°C

B.P. 62-65°C/13mm

Solvent: CCl_4

Assignments
a = 1.01
b = 1.09
c = 1.20-2.20

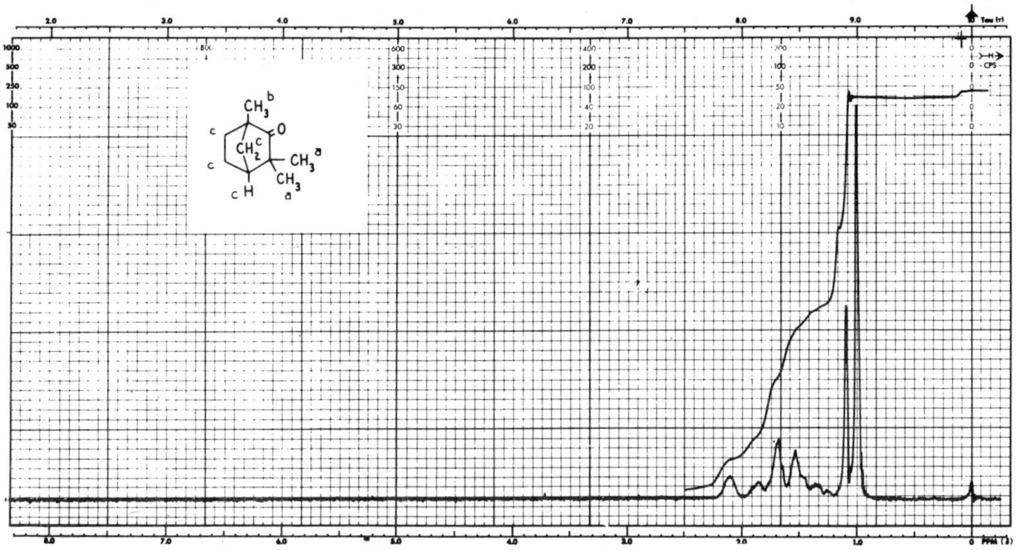

2006

DL-10-CAMPHORSULFONIC ACID, SODIUM SALT

$C_{10}H_{15}NaO_4S$

Mol. Wt. 254.28

Solvent: D_2O

Assignments
a = 0.89
b = 1.10
c = 1.30-2.70
d = 2.80
e = 3.27
f = 3.98 HDO

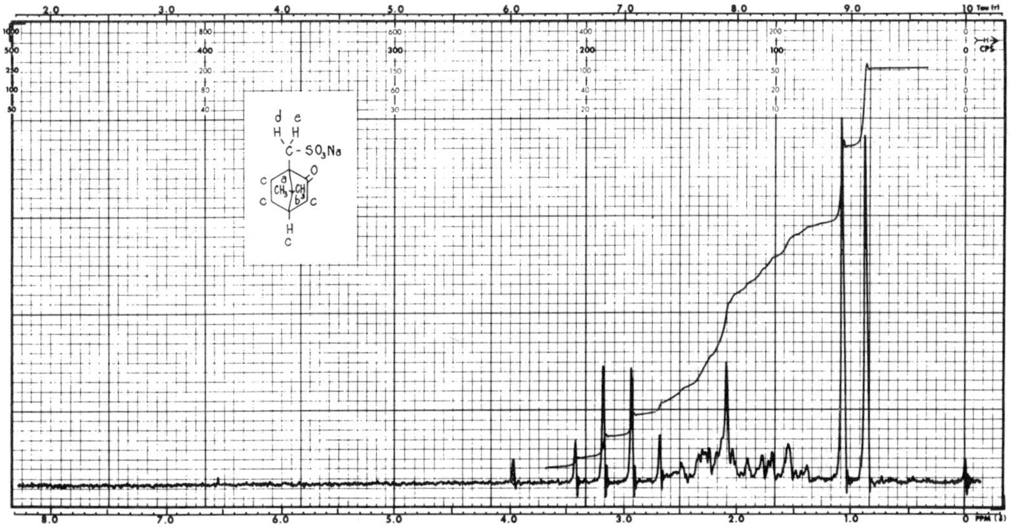

2007

5-OXOCAMPHOR

$C_{10}H_{14}O_2$

Mol. Wt. 166.22

Solvent: $CDCl_3$

Assignments
a = 0.93
b = 1.05
c = 1.11
d = 1.40-2.30
e = 2.61

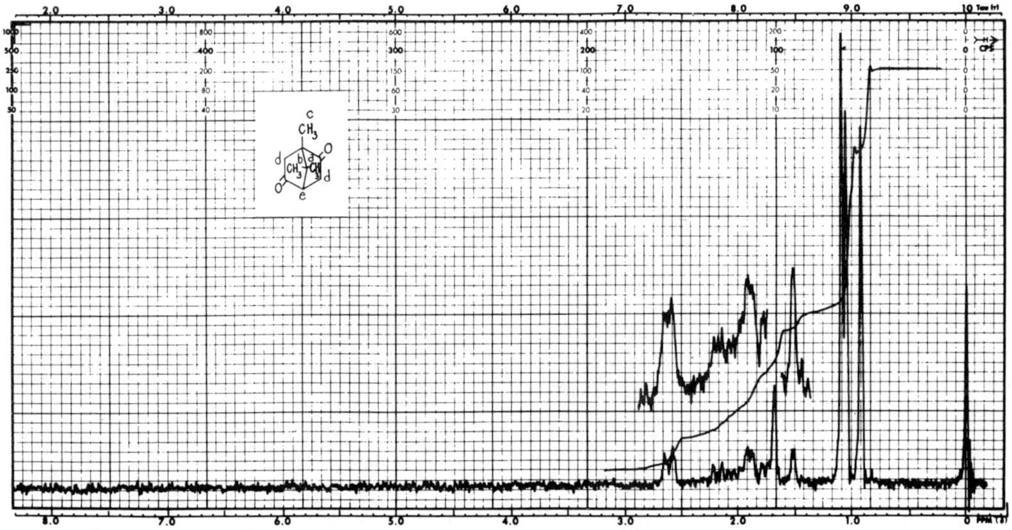

2008

CYCLOHEPTANONE

$C_7H_{12}O$

Mol. Wt. 112.17

B.P. 179°C

Solvent: $CDCl_3$

Assignments
a = ca 1.71
b = 2.49

2009

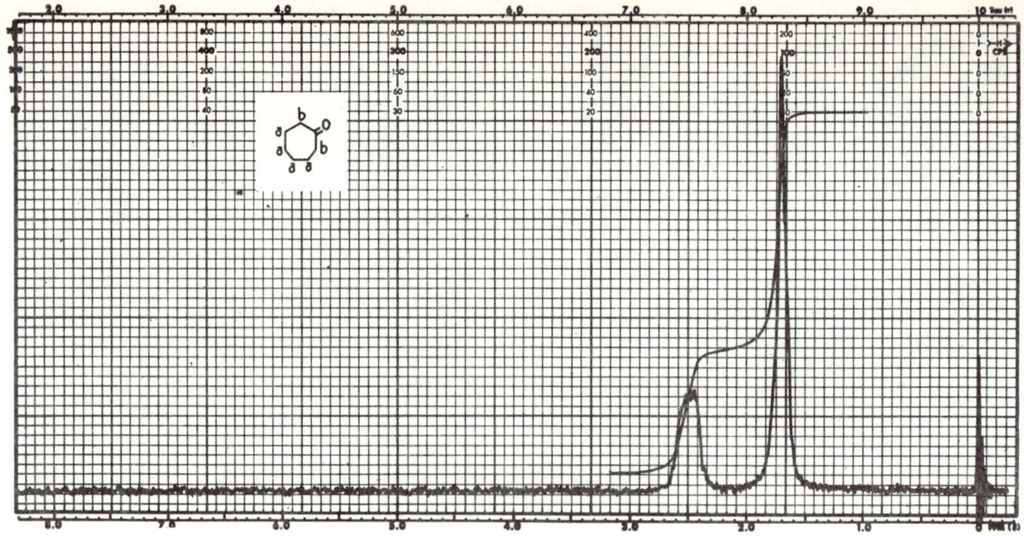

© 1978 Sadtler Research Laboratories, Inc.

CYCLOTRIDECANONE

$C_{13}H_{24}O$

Mol. Wt. 196.34

M.P. 25°C

Solvent: CCl_4

Assignments
a = 1.29
b = 1.40-1.95
c = 2.36

2010

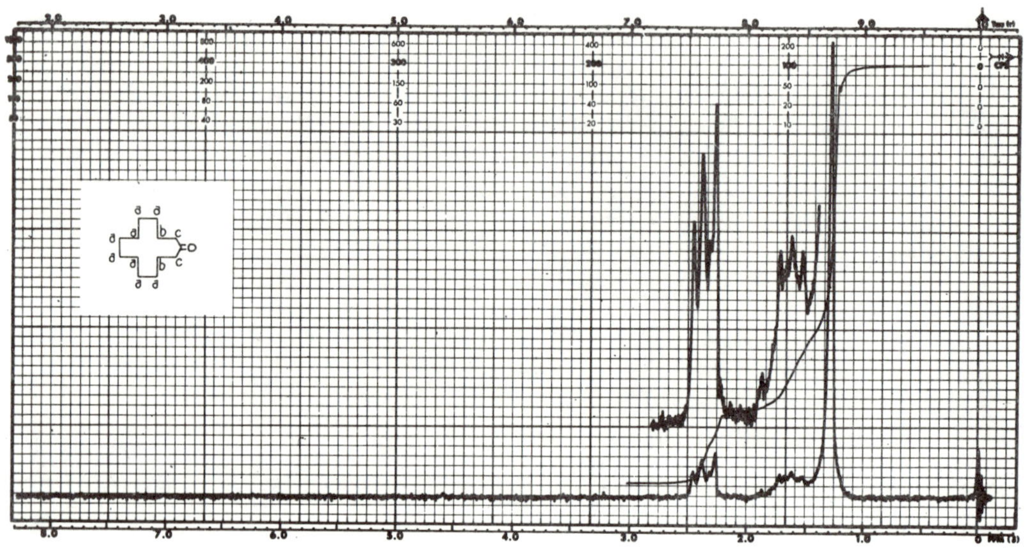

© 1978 Sadtler Research Laboratories, Inc.

CYCLOPENTACECANONE

$C_{15}H_{28}O$

Mol. Wt. 224.39

M.P. 61-63°C

B.P. 120°C/0.3mm

Solvent: $CDCl_3$

Assignments
a = ca 1.33
b = 1.64
c = 2.43

2011

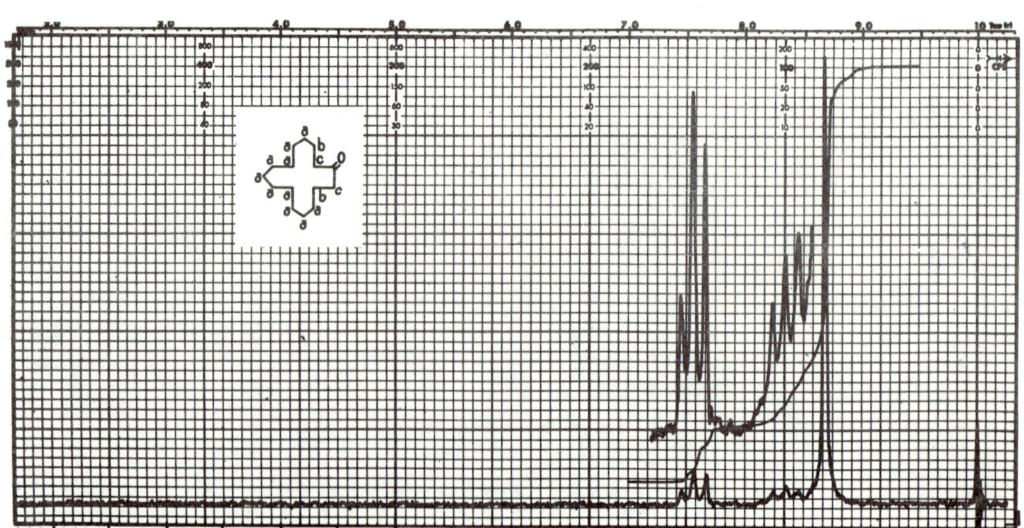

© 1978 Sadtler Research Laboratories, Inc.

2-ADAMANTANONE

$C_{10}H_{14}O$

Mol. Wt. 150.22

M.P. 250°C (dec.)

Solvent: $CDCl_3$

Assignments
a = 2.04
b = 2.54

2012

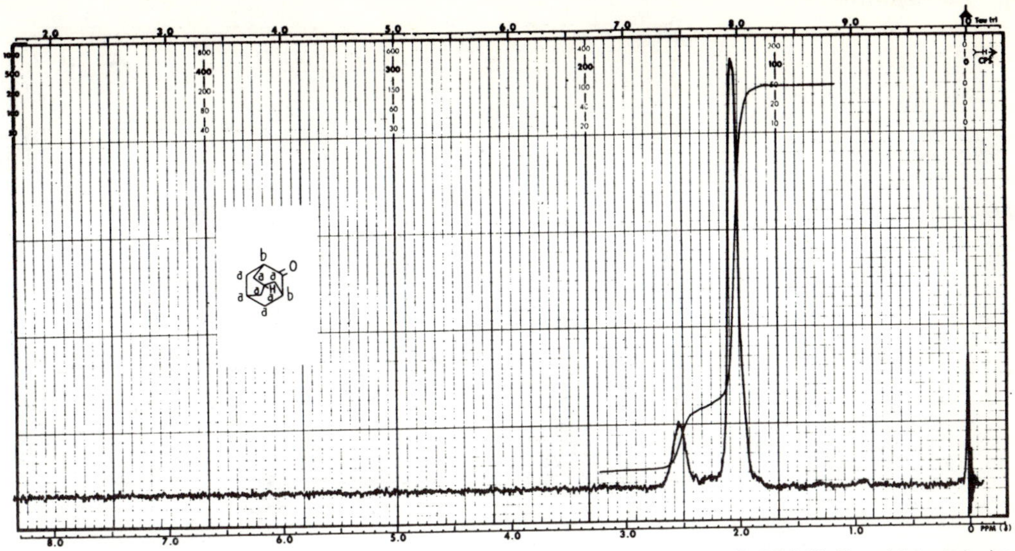

© 1978 Sadtler Research Laboratories, Inc.

2,2,6,6-TETRAMETHYL-4-PIPERIDONE, HYDROCHLORIDE

$C_9H_{17}NO \cdot HCl$

Mol. Wt. 191.70

Solvent: $CDCl_3$

Assignments
a = 1.74
b = 2.80
c = 7.90-11.20

2013

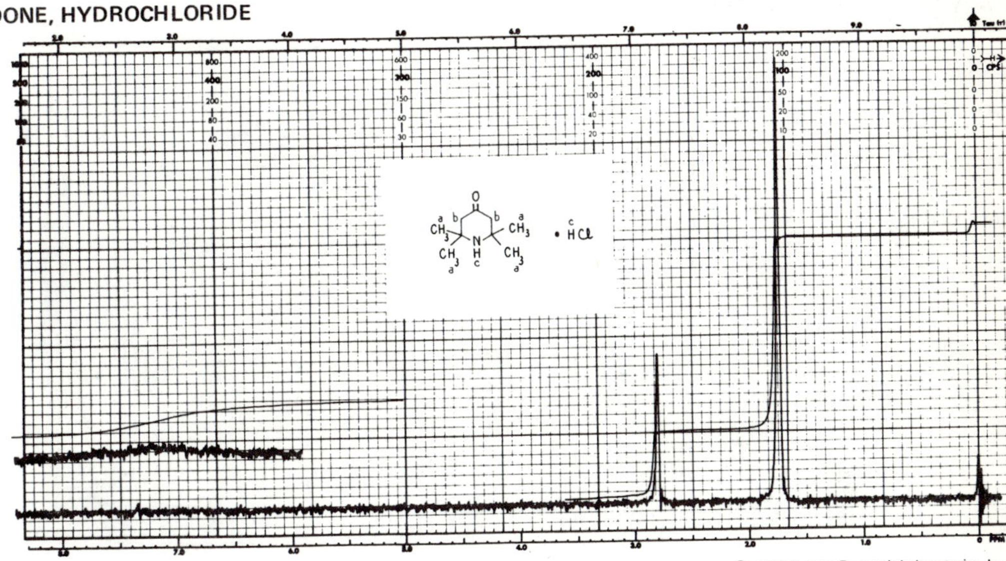

© 1978 Sadtler Research Laboratories, Inc.

1-PHENETHYL-4-PIPERIDONE

$C_3H_{17}NO$

Mol. Wt. 203.29

M.P. 57-60°C

Solvent: $CDCl_3$

Assignments
a = 2.43
b = ca 2.79
c = ca 7.22

2014

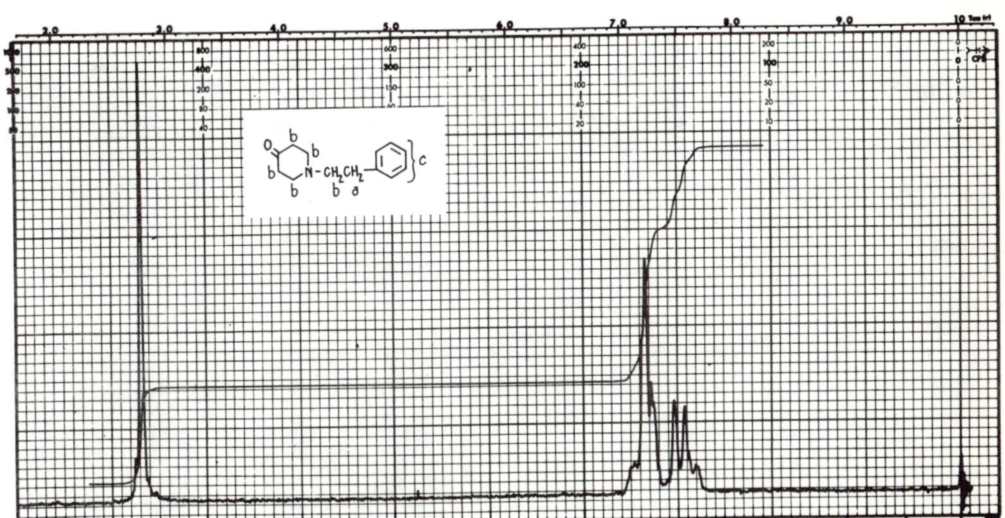

© 1978 Sadtler Research Laboratories, Inc.

1-BENZYL-4-PIPERIDONE

$C_{12}H_{15}NO$

Mol. Wt. 189.26

B.P. 134°C/7mm

Solvent: $CDCl_3$

Assignments
a = 2.41
b = 2.71
c = 3.60
d = ca 7.53

2015

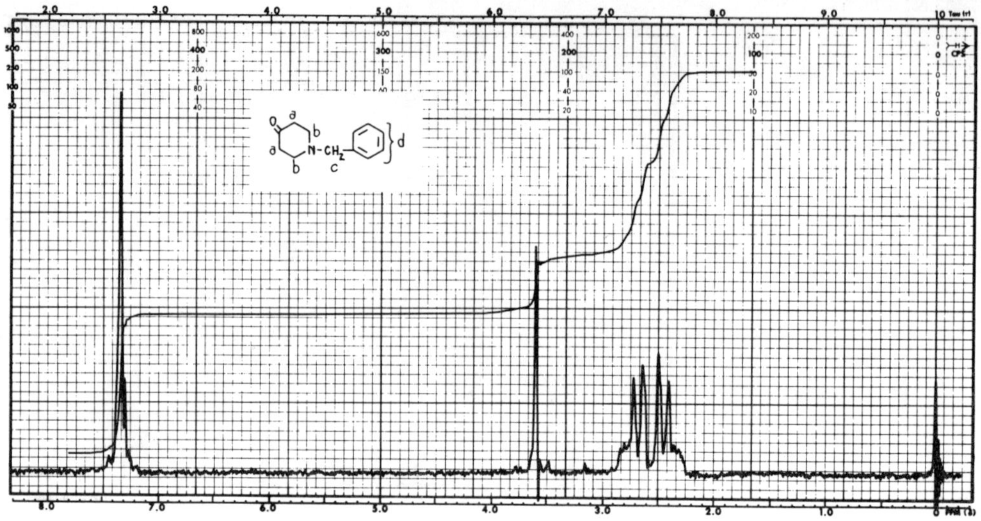

© 1978 Sadtler Research Laboratories, Inc.

TETRAHYDRO-4H-THIOPYRAN-4-ONE

C_5H_8OS

Mol. Wt. 116.18

M.P. 59-61°C

Solvent: $CDCl_3$

Assignments
a = 2.67
b = 2.98

2016

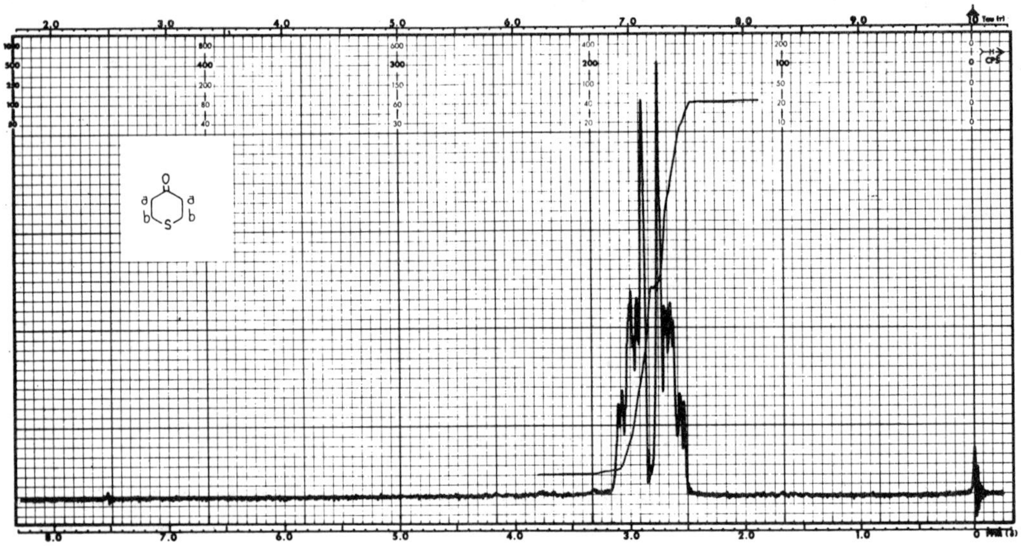

© 1978 Sadtler Research Laboratories, Inc.

3-BUTEN-2-ONE

C_4H_6O

Mol. Wt. 70.09

B.P. 83-87°C

Solvent: CCl_4

Assignments
a = 2.20
b = ca 5.82
c = ca 6.18
d = ca 6.28

2017

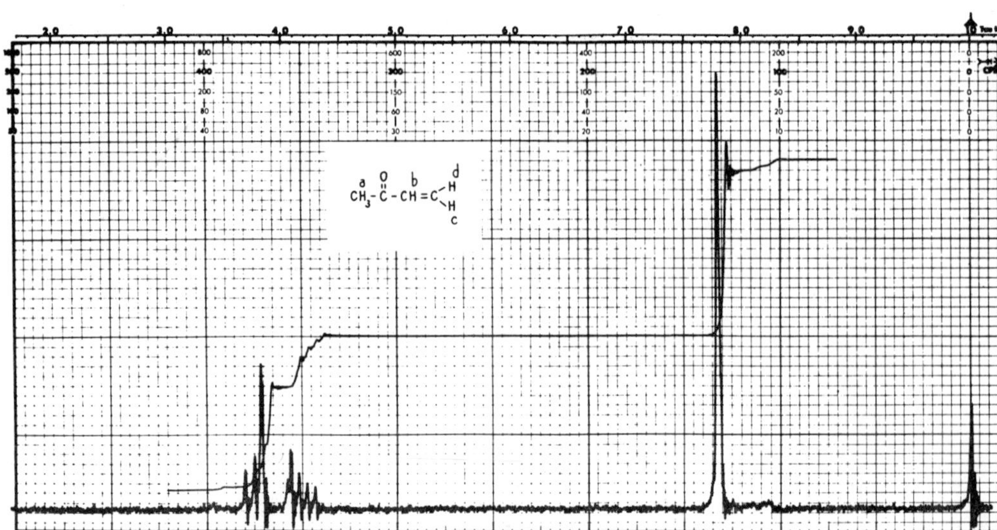

© 1978 Sadtler Research Laboratories, Inc.

3-PENTEN-2-ONE

C_5H_8O

Mol. Wt. 84.12

B.P. 121-124°C

Solvent: CCl_4

Assignments
a = 1.89
b = 2.12
c = 6.00
d = 6.70
e = 2.08, 6.03 imp.

2018

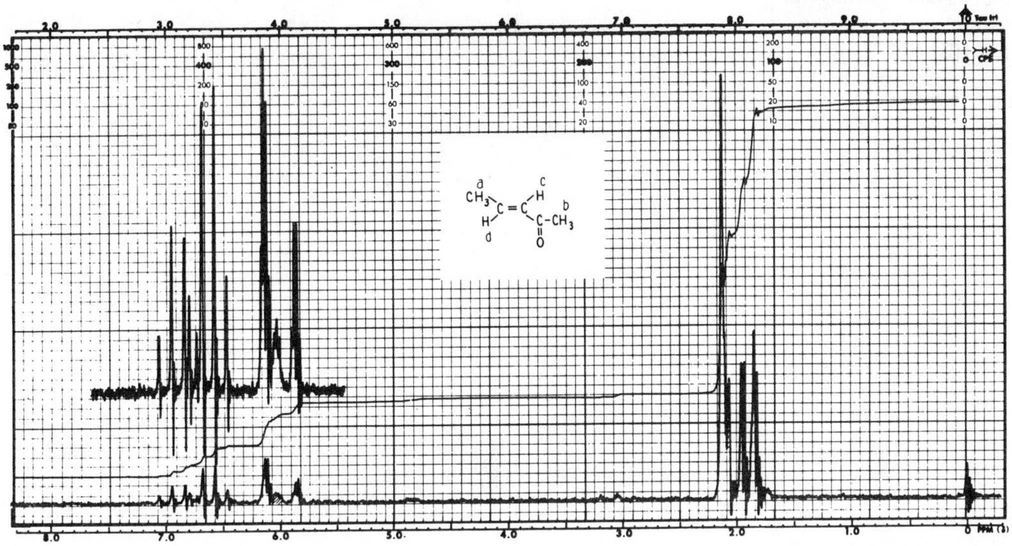

© 1978 Sadtler Research Laboratories, Inc.

5-HEXEN-2-ONE

$C_6H_{10}O$

Mol. Wt. 98.15

B.P. 129.5°C (lit.)

Solvent: CCl_4

Assignments
a = 2.09
b = ca 2.40
c = 4.91
d = 4.97
e = 5.75

2019

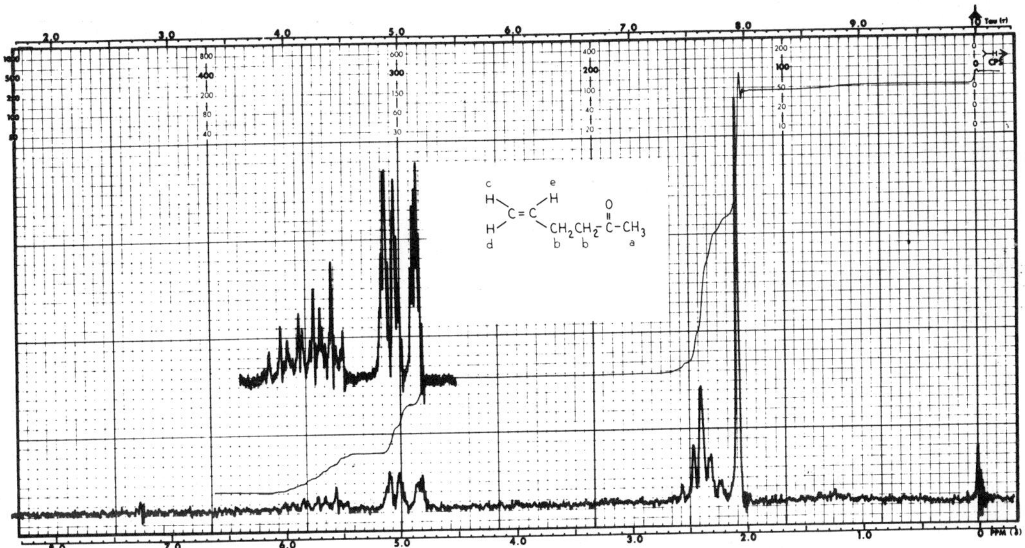

© 1978 Sadtler Research Laboratories, Inc.

4-METHYL-3-PENTEN-2-ONE

$C_6H_{10}O$

Mol. Wt. 98.15

B.P. 129°C

Solvent: $CDCl_3$

Assignments
a = 1.90
b = ca 2.15
c = 6.10

2020

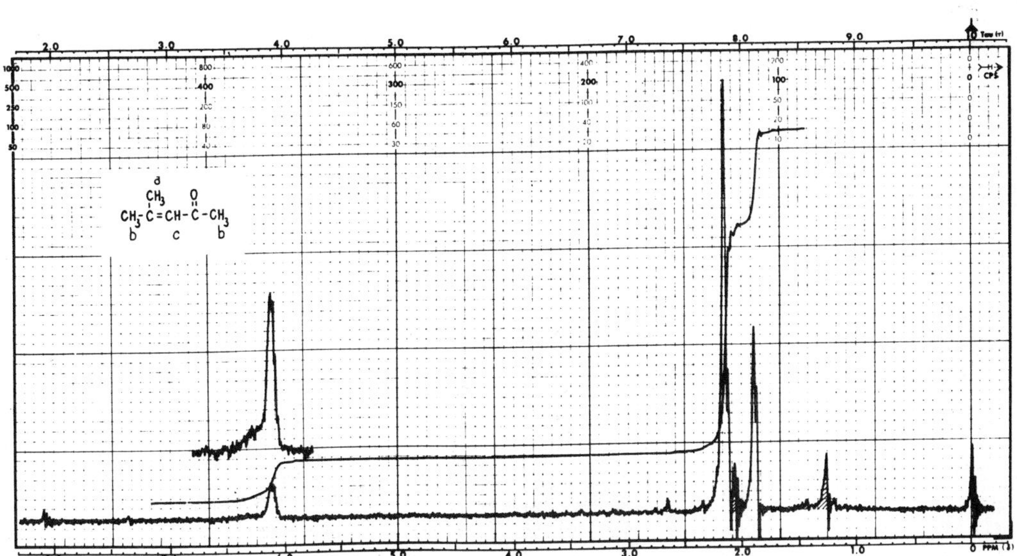

© 1978 Sadtler Research Laboratories, Inc.

5-ETHYL-3-HEPTEN-2-ONE

$C_9H_{16}O$

Mol. Wt. 140.23

Solvent: CCl_4

Assignments
a = 0.89
b = 1.50
c = ca 2.00
d = 2.14
e = 5.92
f = 6.47

2021

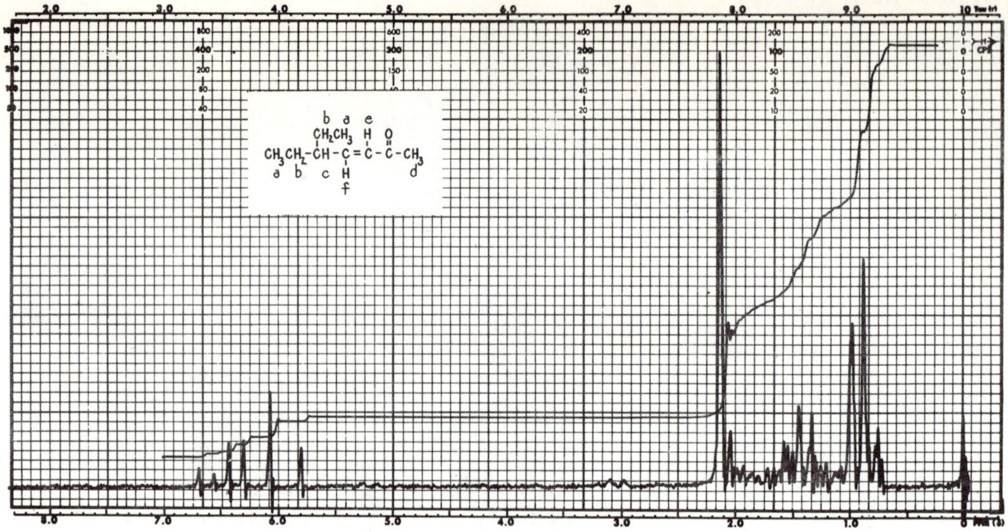

© 1978 Sadtler Research Laboratories, Inc.

1-CYCLOHEXEN-1-YL METHYL KETONE

$C_8H_{12}O$

Mol. Wt. 124.18

Solvent: CCl_4

Assignments
a = 1.60
b = 2.15
c = 2.19
d = 6.78

2022

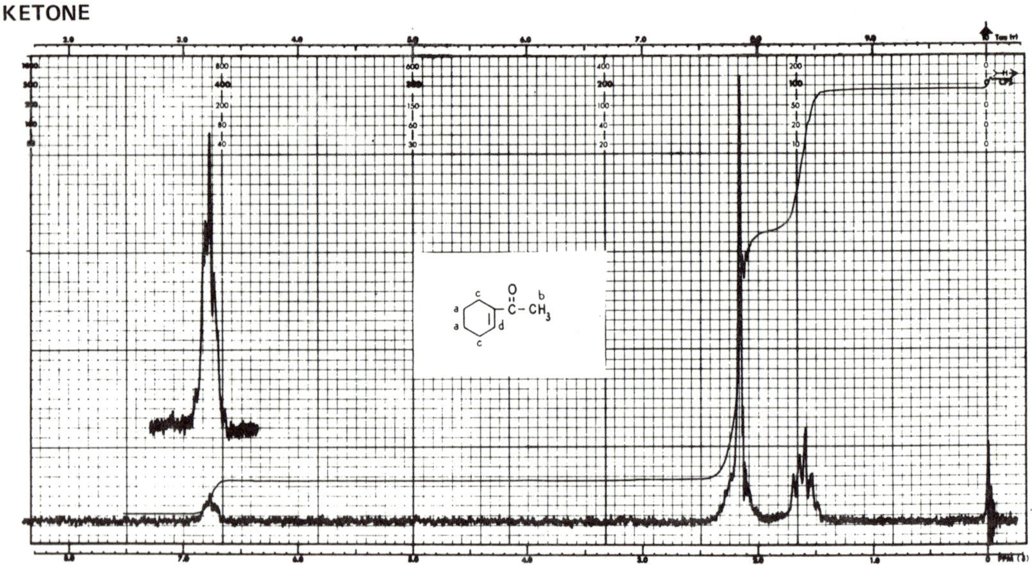

© 1978 Sadtler Research Laboratories, Inc.

FUROIN

$C_{10}H_8O_4$

Mol. Wt. 192.17

Solvent: Dimethyl sulfoxide

Assignments
a = 5.31 *
b = 5.79
c = ca 6.40
d = 6.64
e = ca 7.49
f = 7.89
*H_2O present

2023

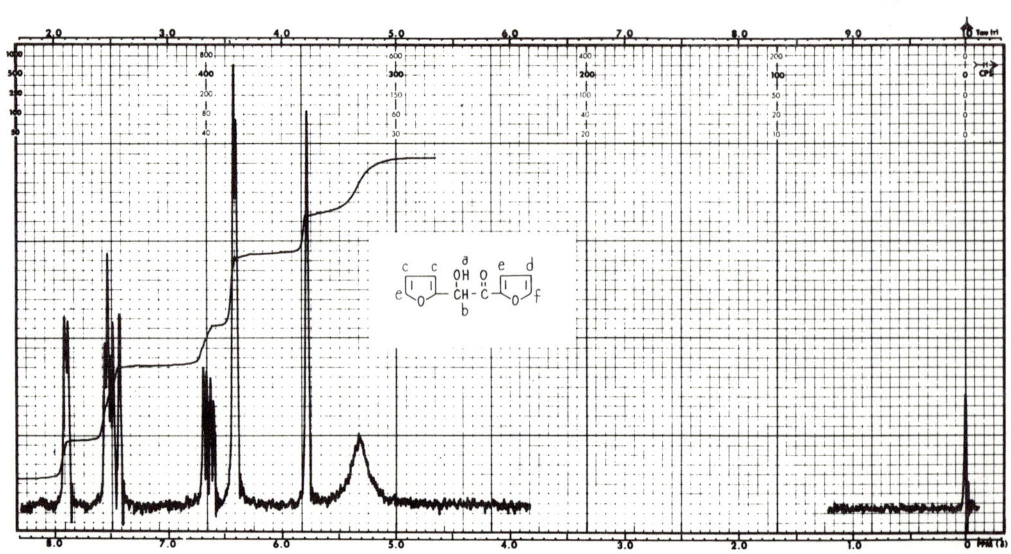

© 1978 Sadtler Research Laboratories, Inc.

3-CHLORO-2-CYCLOPENTEN-1-ONE

C_5H_5ClO

Mol. Wt. 116.55

B.P. 35°C/0.7mm

Solvent: CCl_4

Assignments
a = 2.49
b = 2.85
c = 6.11

2024

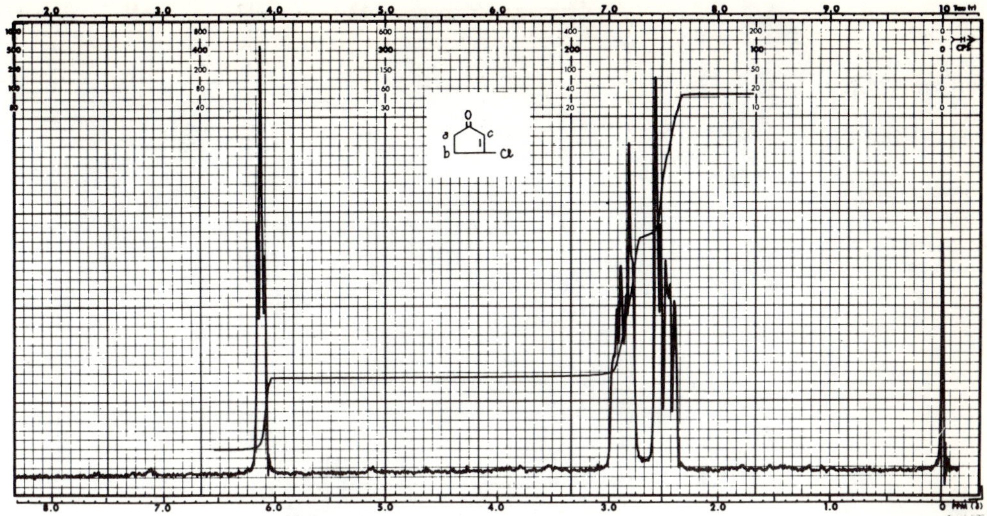

© 1978 Sadtler Research Laboratories, Inc.

2-CYCLOHEXEN-1-ONE

C_6H_8O

Mol. Wt. 96.13

B.P. 168°C/760mm

F.P. 53.5°C

Solvent: CCl_4

Assignments
a = 1.72-2.58
b = 5.88
c = 6.91

2025

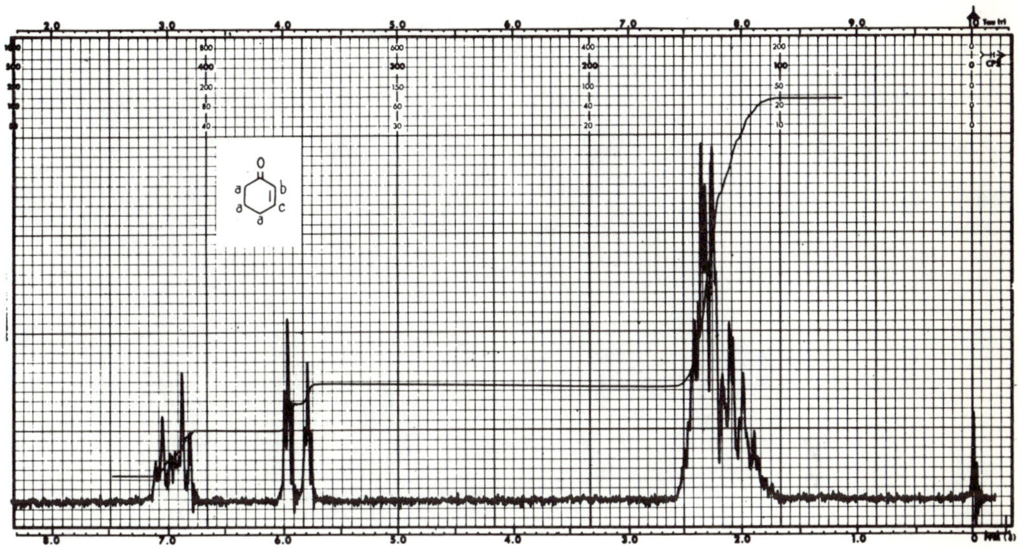

© 1978 Sadtler Research Laboratories, Inc.

3-CHLORO-2-CYCLOHEXEN-1-ONE

C_6H_7ClO

Mol. Wt. 130.58

B.P. 63°C/4mm

Solvent: CCl_4

Assignments
a = 1.70-2.50
b = 2.70
c = 6.07

2026

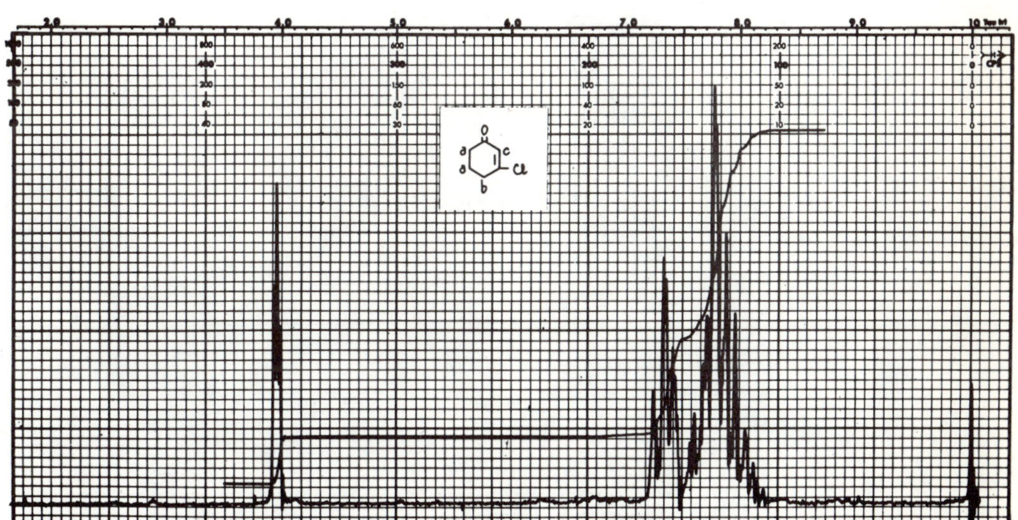

© 1978 Sadtler Research Laboratories, Inc.

3,5-DIMETHYL-2-CYCLOHEXEN-1-ONE

$C_8H_{12}O$

Mol. Wt. 124.18

B.P. 211-212°C

Solvent: CCl_4

Assignments
a = 1.07
b = 1.97
c = 1.70-2.60
d = 5.71

2027

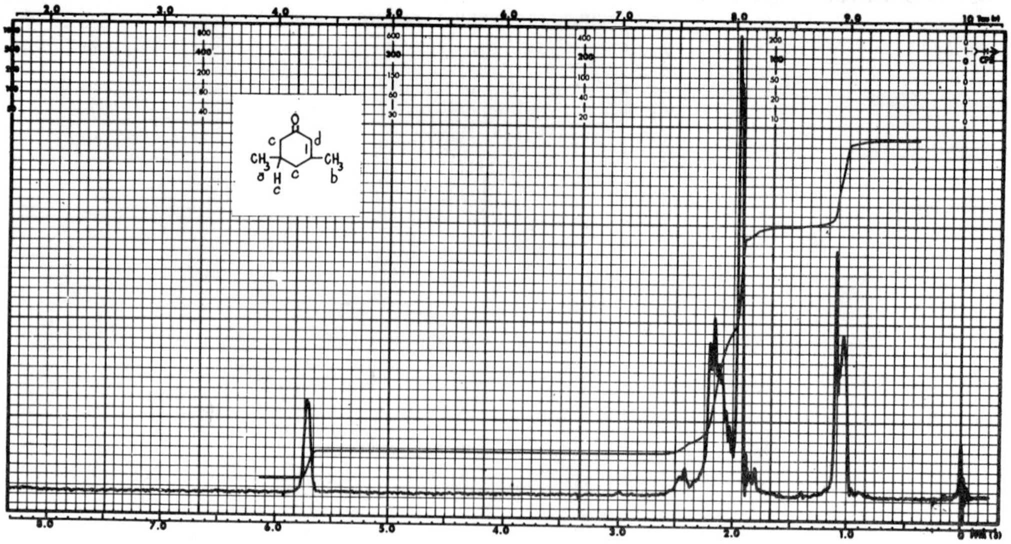

© 1978 Sadtler Research Laboratories, Inc.

3-CHLORO-2-METHYL-2-CYCLOHEXEN-1-ONE

C_7H_9ClO

Mol. Wt. 144.60

B.P. 62°C/2mm

Solvent: CCl_4

Assignments
a = 1.85
b = ca 2.01
c = 2.32
d = 2.71

2028

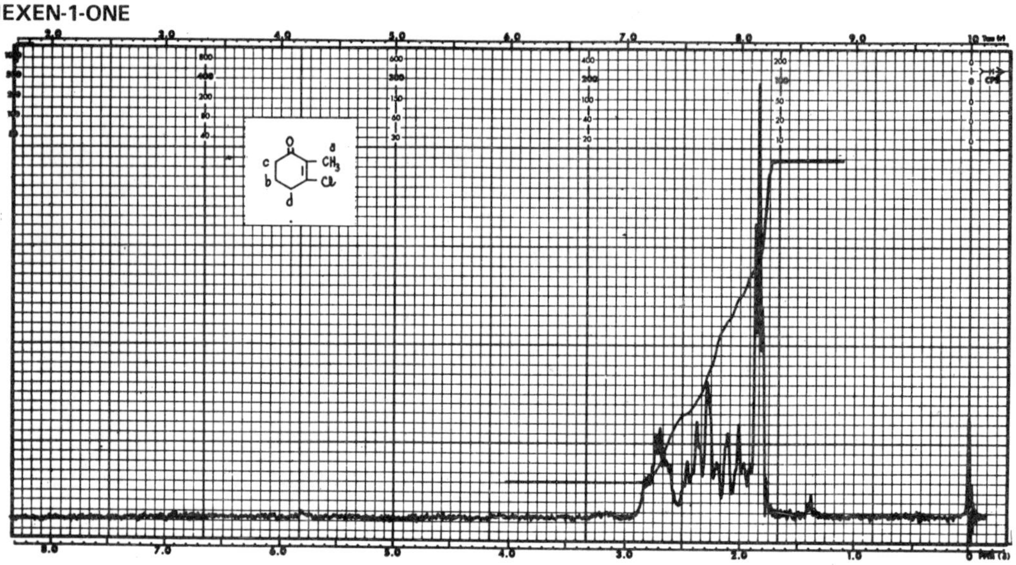

© 1978 Sadtler Research Laboratories, Inc.

3-CHLORO-5,5-DIMETHYL-2-CYCLOHEXEN-1-ONE

$C_8H_{11}ClO$

Mol. Wt. 158.63

B.P. 72°C/5mm

Solvent: CCl_4

Assignments
a = 1.11
b = 2.13
c = 2.49
d = 6.09

2029

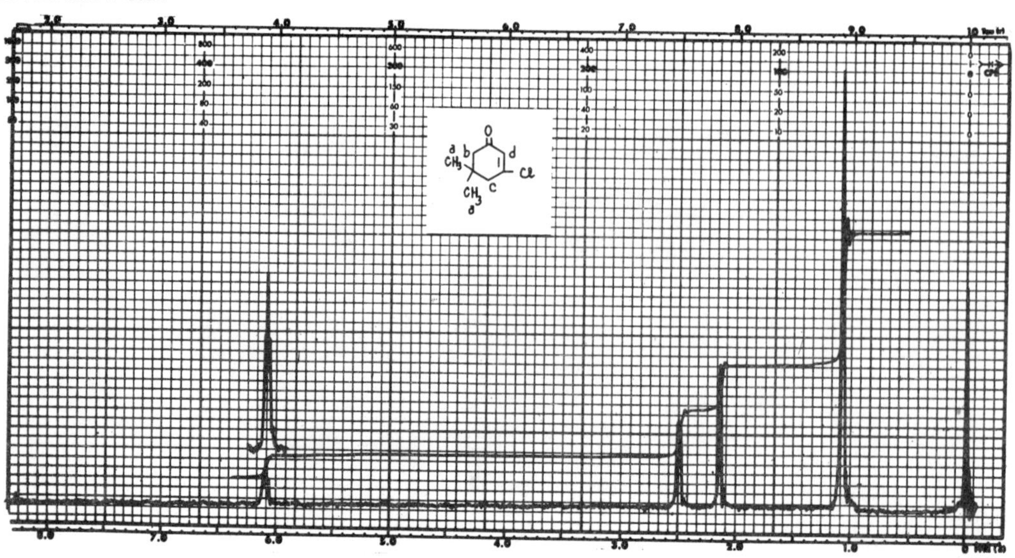

© 1978 Sadtler Research Laboratories, Inc.

p-TOLUQUINONE

$C_7H_6O_2$

Mol. Wt. 122.12

Solvent: $CDCl_3$

Assignments
a = 2.08
b = 6.63
c = 6.78

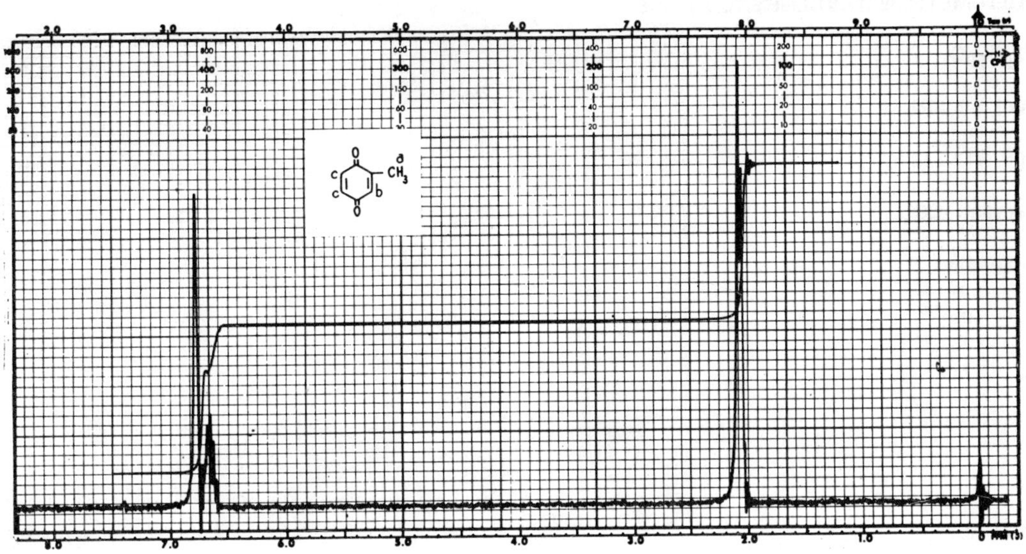

2030

© 1978 Sadtler Research Laboratories, Inc.

2,5-DI-tert-PENTYL-p-BENZOQUINONE

$C_{16}H_{24}O_2$

Mol. Wt. 248.37

M.P. 134-135°C

Solvent: $CDCl_3$

Assignments
a = 0.71
b = 1.21
c = 1.77
d = 6.40

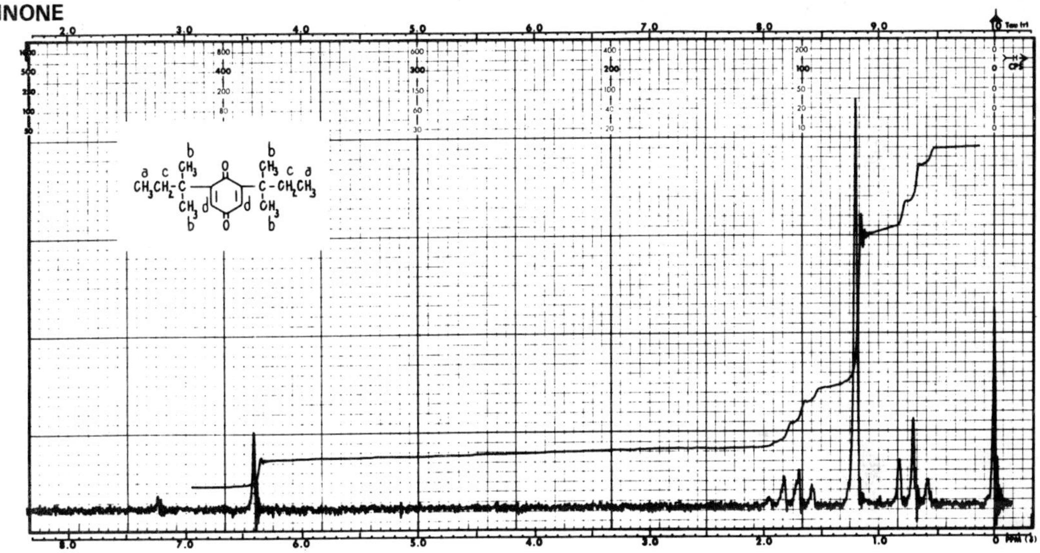

2031

© 1978 Sadtler Research Laboratories, Inc.

TETRAMETHYL-p-BENZOQUINONE

$C_{10}H_{12}O_2$

Mol. Wt. 164;21

M.P. 109.5-110°C

Solvent: $CDCl_3$

Assignments
a = 2.01

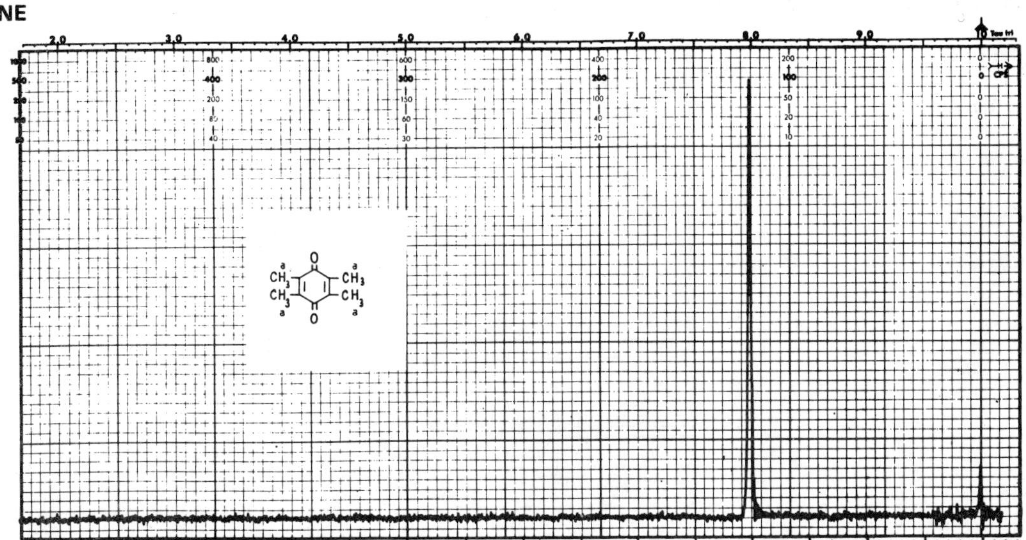

2032

© 1978 Sadtler Research Laboratories, Inc.

PHENYL-p-BENZOQUINONE

$C_{12}H_8O_2$

Mol. Wt. 184.20

M.P. 112-114°C

Solvent: $CDCl_3$

Assignments
a = 6.84
b = 7.47

2033

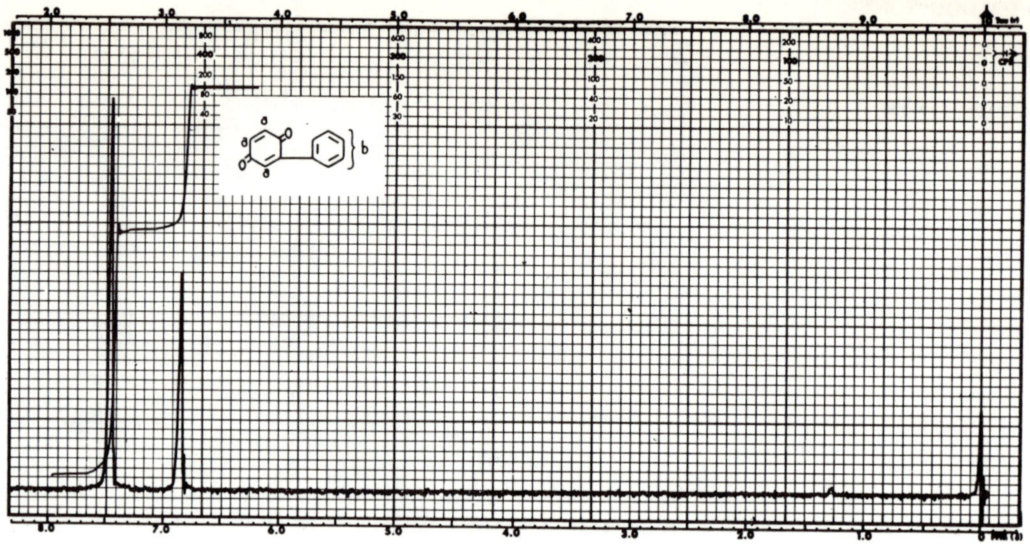

© 1978 Sadtler Research Laboratories, Inc.

1-PHENYL-2-BUTANONE

$C_{10}H_{12}O$

Mol. Wt. 148.21

B.P. 109-113°C/15mm

Solvent: CCl_4

Assignments
a = 0.93
b = 2.34
c = 3.53
d = 7.17

2034

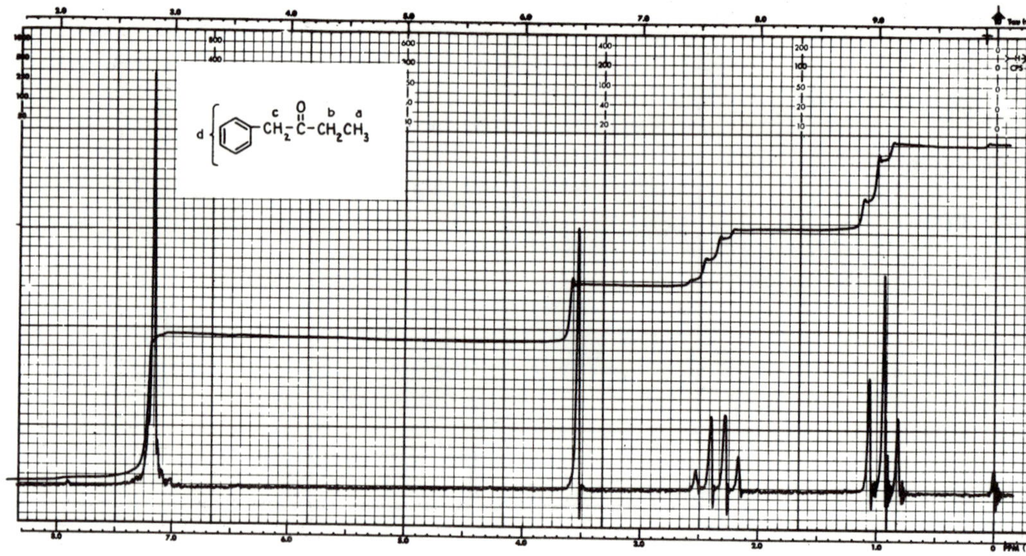

© 1978 Sadtler Research Laboratories, Inc.

3-METHYL-1-PHENYL-2-BUTANONE

$C_{11}H_{14}O$

Mol. Wt. 162.23

Solvent: CCl_4

Assignments
a = 1.01
b = 2.61
c = 3.61
d = ca 7.15

2035

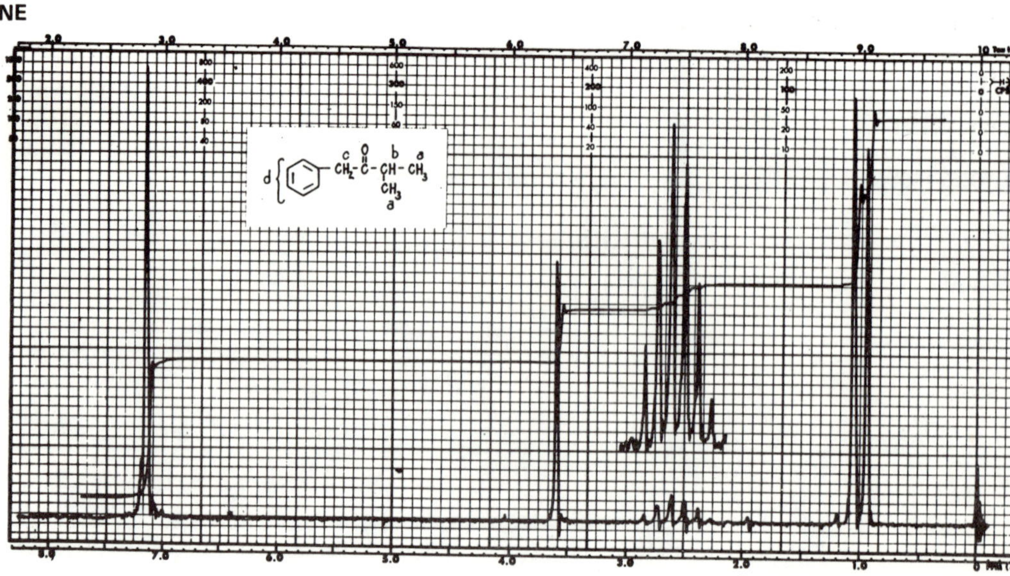

© 1978 Sadtler Research Laboratories, Inc.

4-PHENYL-2-BUTANONE

$C_{10}H_{12}O$

Mol. Wt. 148.21

Solvent: CCl_4

Assignments
a = 2.00
b = 2.64
c = 2.81
d = ca 7.19

2036

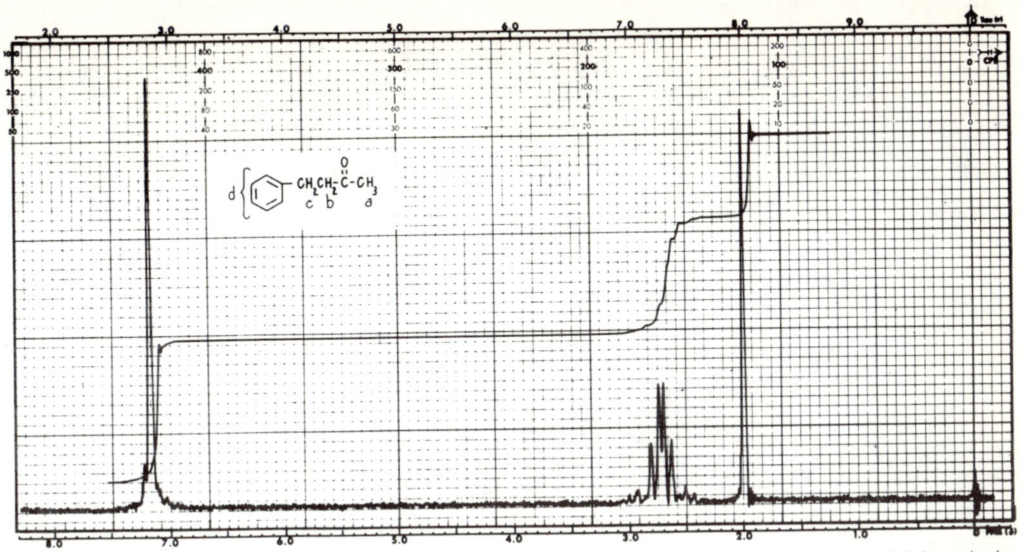

© 1978 Sadtler Research Laboratories, Inc.

3,4-DIBROMO-4-PHENYL-2-BUTANONE

$C_{10}H_{10}Br_2O$

Mol. Wt. 306.01

Solvent: $CDCl_3$

Assignments
a = 2.42
b = 4.91
c = 5.33
d = ca 7.35

2037

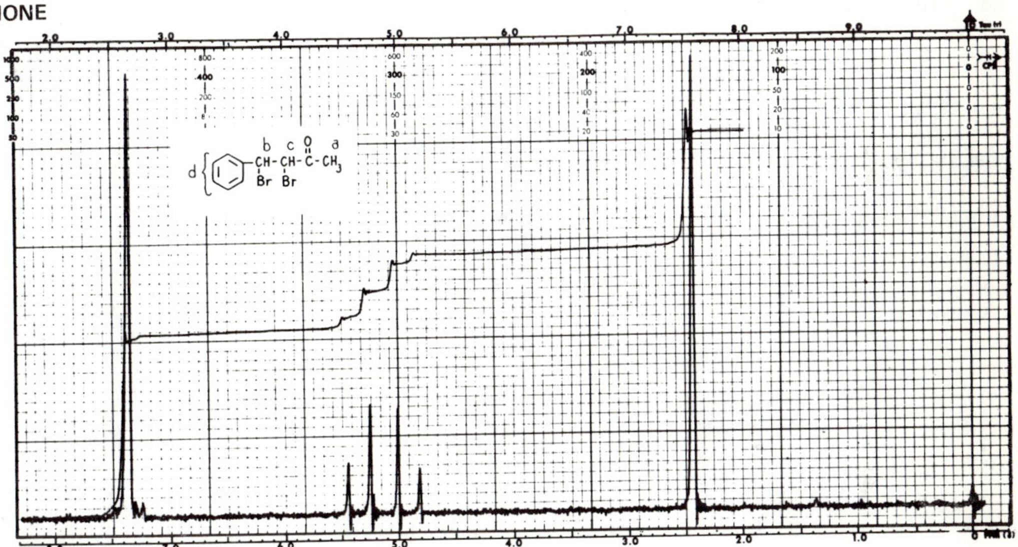

© 1978 Sadtler Research Laboratories, Inc.

4-(p-HYDROXYPHENYL)-2-BUTANONE

$C_{10}H_{12}O_2$

Mol. Wt. 164.21

M.P. 82-83°C (lit.)

Solvent: $CDCl_3$

Assignments
a = 2.11
b = 2.71
c = 2.77
d = 6.75
e = 6.99

2038

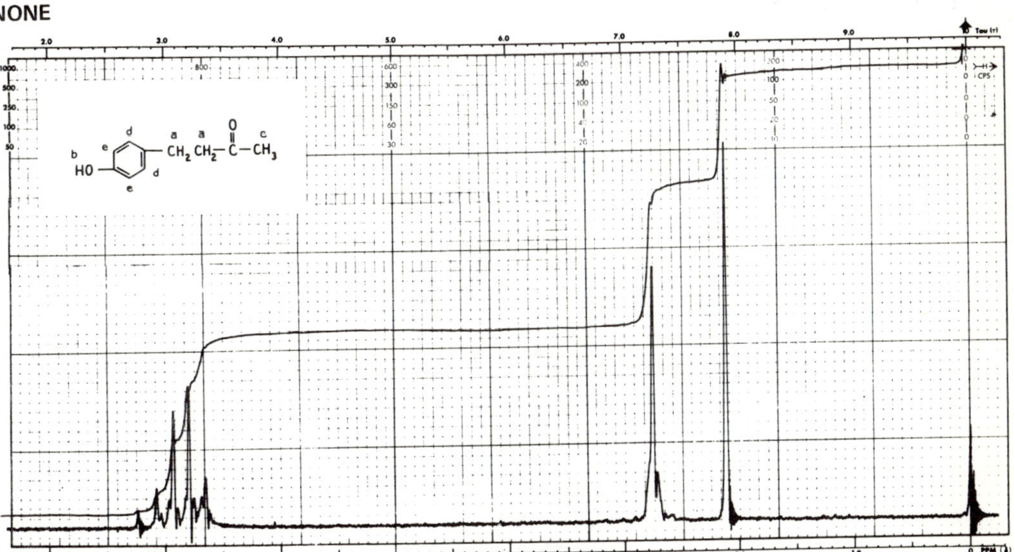

© 1978 Sadtler Research Laboratories, Inc.

1-[3,4-(METHYLENEDIOXY)PHENYL]-2-PROPANONE

$C_{10}H_{10}O_3$

Mol. Wt. 178.19

Solvent: CCl_4

Assignments
a = 2.09
b = 3.48
c = 5.85
d = 6.45-6.76

2039

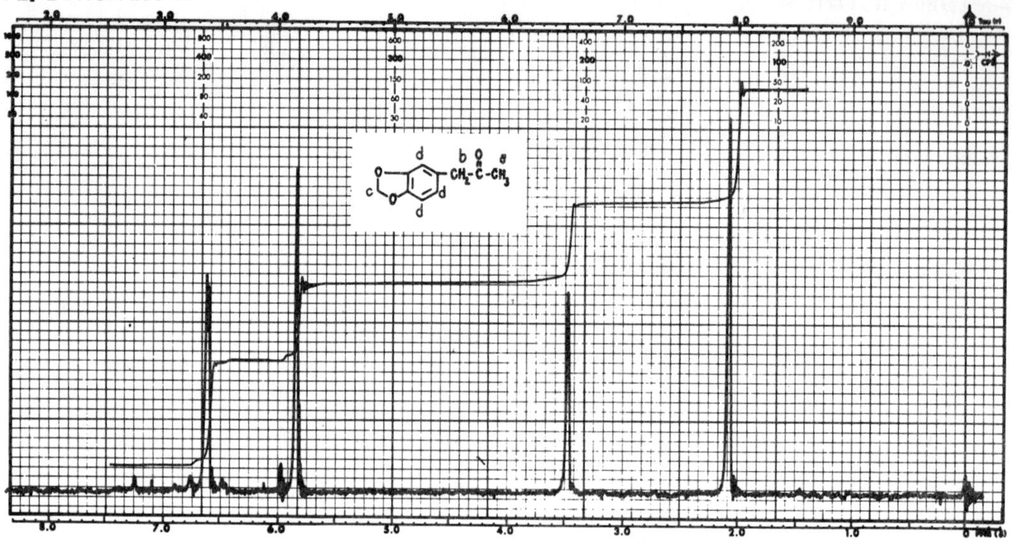

© 1978 Sadtler Research Laboratories, Inc.

1,3-DIPHENYL-2-PROPANONE

$C_{15}H_{14}O$

Mol. Wt. 210.28

M.P. 31-34°C

Solvent: $CDCl_3$

Assignments
a = 3.67
b = 7.00-7.40

2040

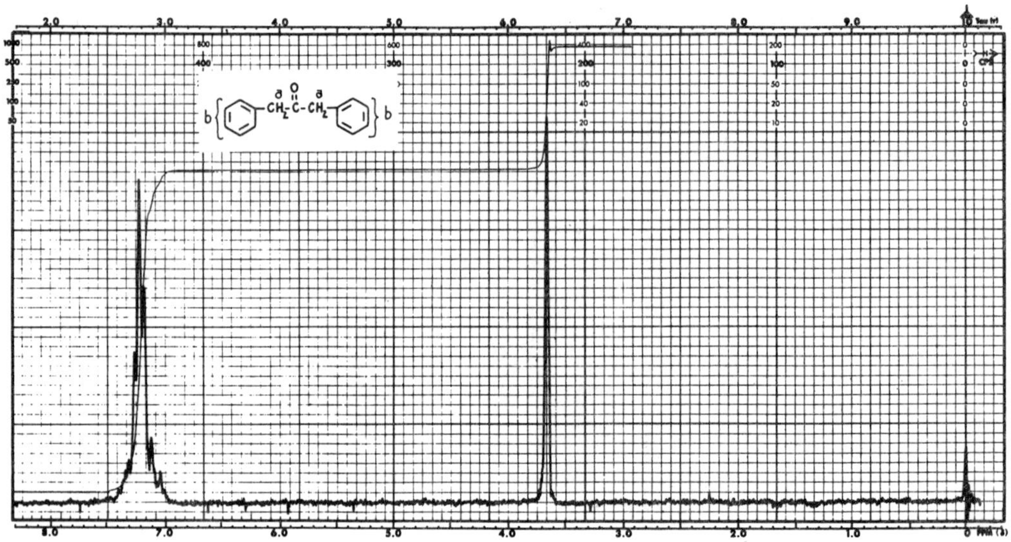

© 1978 Sadtler Research Laboratories, Inc.

1-PHENYL-1-HEPTEN-3-ONE

$C_{13}H_{16}O$

Mol. Wt. 188.27

M.P. 34-36°C

Solvent: $CDCl_3$

Assignments
a = 0.92
b = 1.00-2.00
c = 2.61
d = 6.68
e = 7.20-7.65
f = 7,51

2041

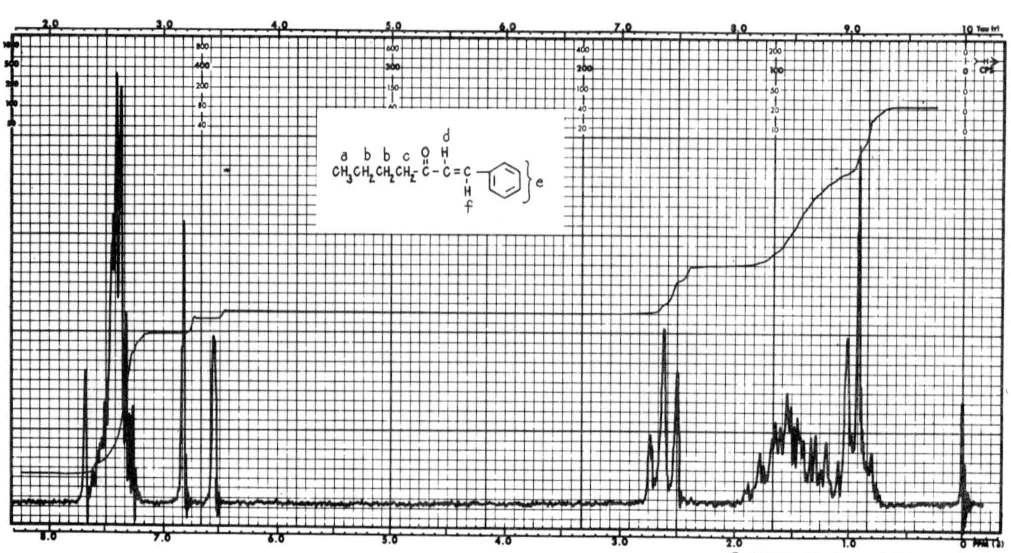

© 1978 Sadtler Research Laboratories, Inc.

4,4-DIPHENYL-3-BUTEN-2-ONE

$C_{16}H_{14}O$

Mol. Wt. 222.29

M.P. 33°C

Solvent: $CDCl_3$

Assignments
a = 1.89
b = 6.54
c = 7.00-7.50

2042

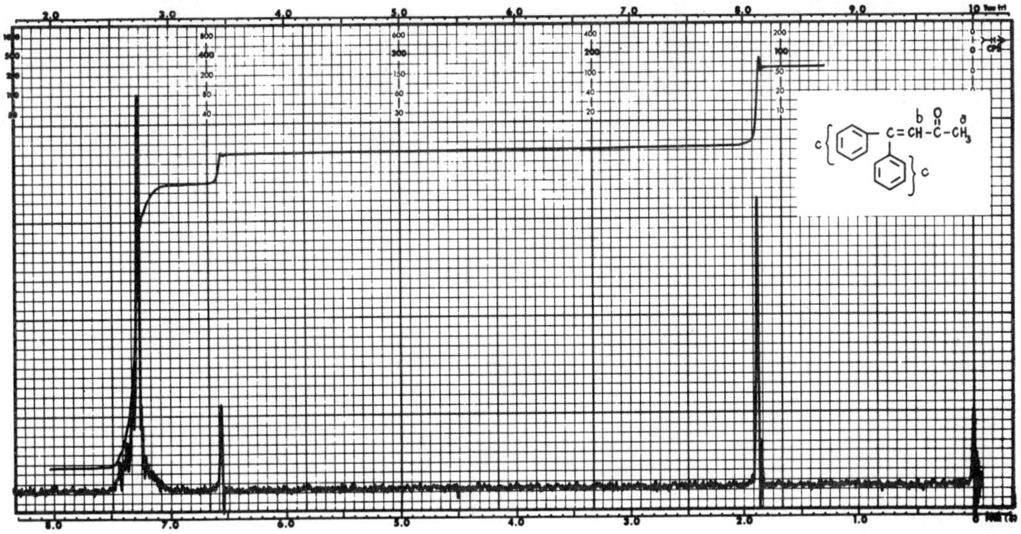

© 1978 Sadtler Research Laboratories, Inc.

1,5-DIPHENYL-1,4-PENTADIEN-3-ONE

$C_{17}H_{14}O$

Mol. Wt. 234.30

M.P. 111-112°C

Solvent: $CDCl_3$

Assignments
a = 7.05
b = 7.14-7.80
c = 7.72
d = 0.92, 1.04 imp.

2043

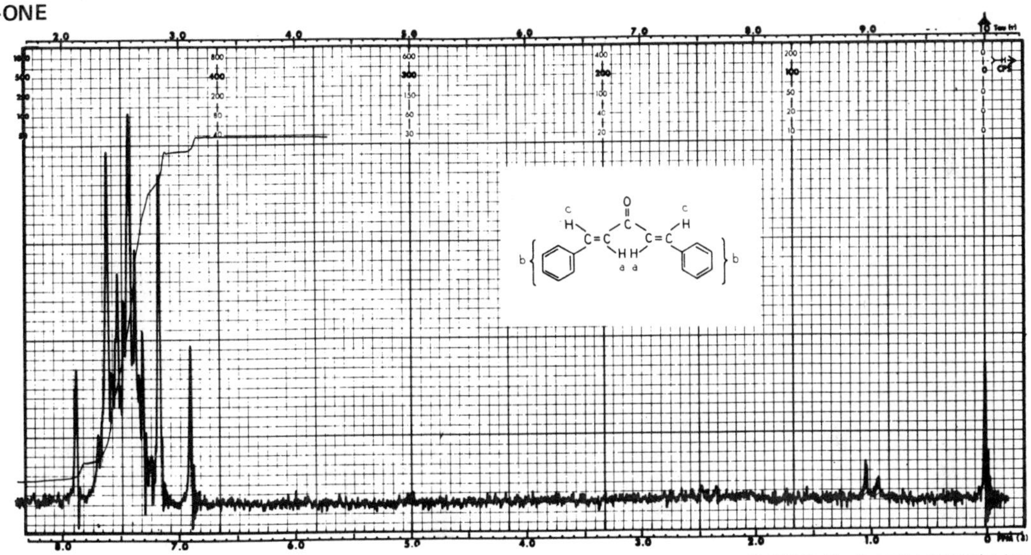

© 1978 Sadtler Research Laboratories, Inc.

4-PHENYL-3-BUTYN-2-ONE

$C_{10}H_8O$

Mol. Wt. 144.17

B.P. 75-76°C/0.8mm

Solvent: $CDCl_3$

Assignments
a = 2.40
b = 7.20-7.70

2044

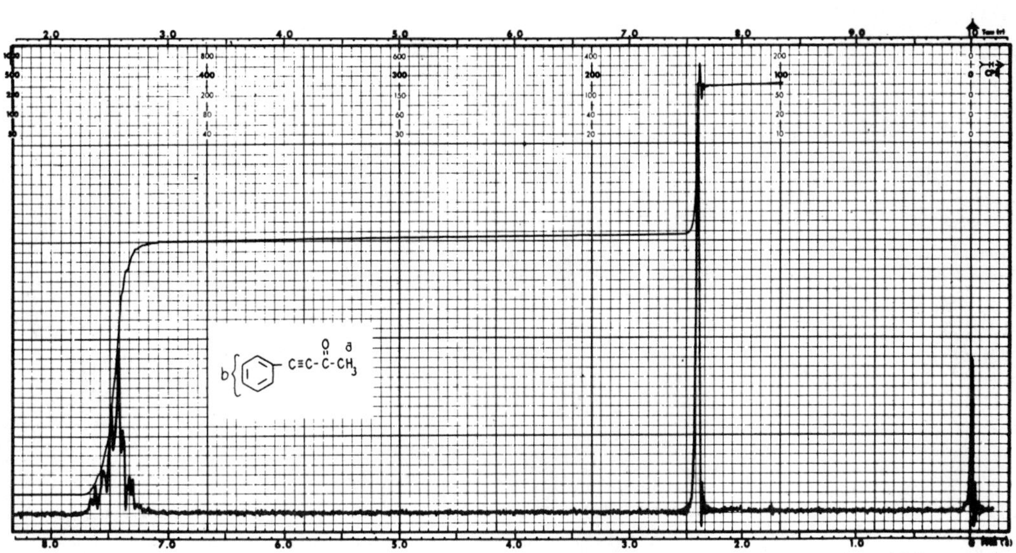

© 1978 Sadtler Research Laboratories, Inc.

ACETOPHENONE

C_8H_8O

Mol. Wt. 120.15

M.P. 19-20°C

Solvent: CCl_4

Assignments
a = 2.43
b = 7.15-7.50
c = 7.80

2045

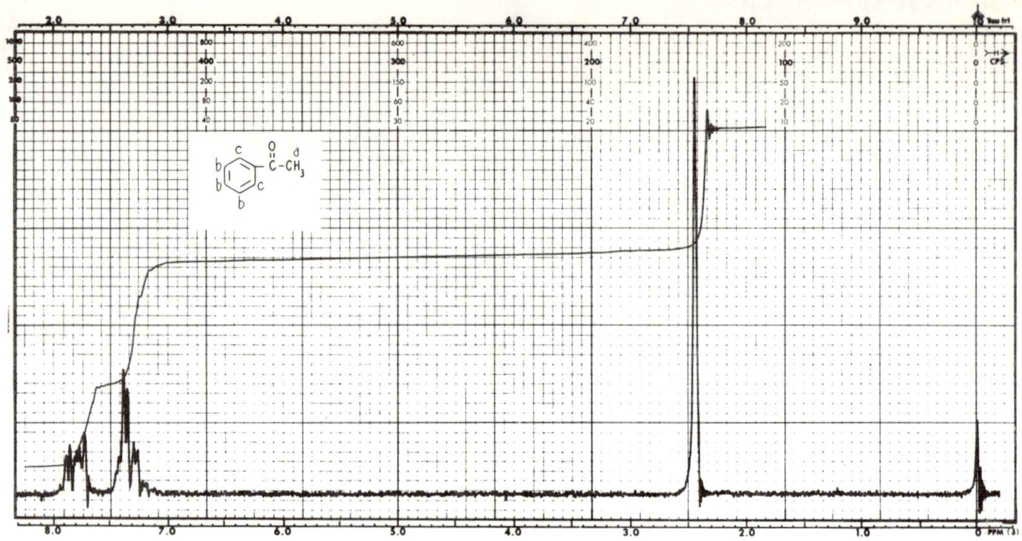

© 1978 Sadtler Research Laboratories, Inc.

PROPIOPHENONE

$C_9H_{10}O$

Mol. Wt. 134.18

B.P. 218°C

Solvent: $CDCl_3$

Assignments
a = 1.18
b = 2.94
c = ca 7.47
d = ca 7.92

2046

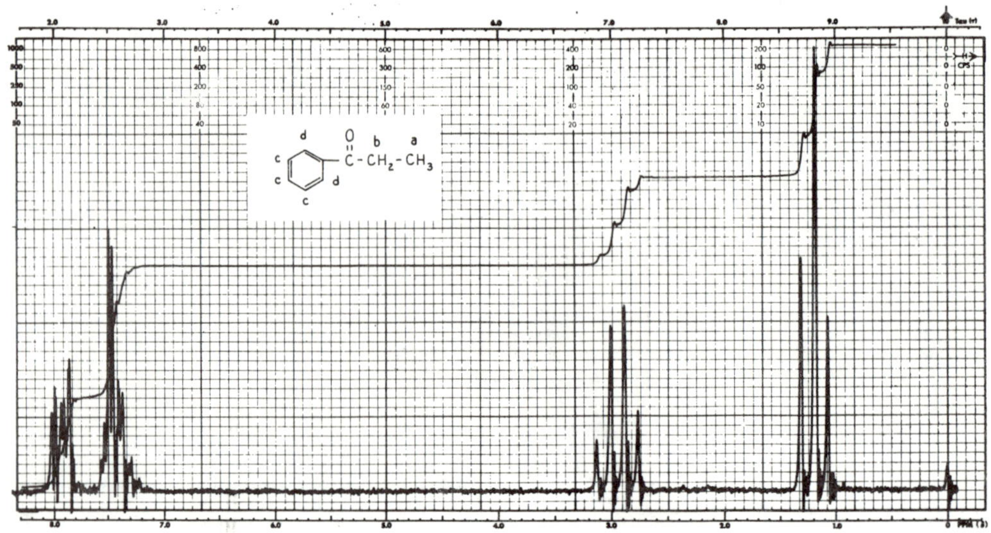

© 1978 Sadtler Research Laboratories, Inc.

BUTYROPHENONE

$C_{10}H_{12}O$

Mol. Wt. 148.21

B.P. 101-102°C/9mm

Solvent: CCl_4

Assignments
a = 0.95
b = 1.72
c = 2.82
d = 7.12-7.55
e = 7.87

2047

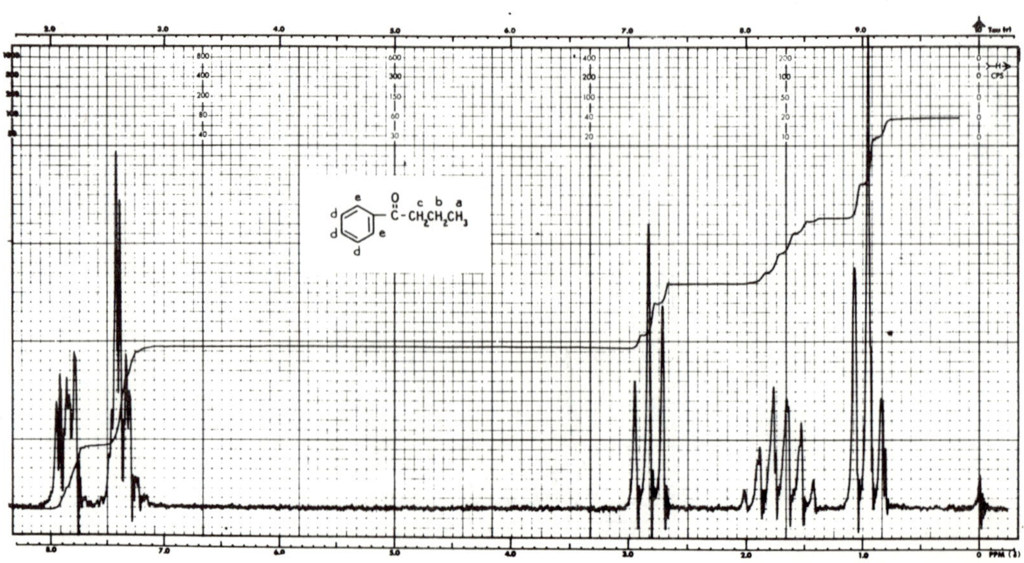

© 1978 Sadtler Research Laboratories, Inc.

ISOBUTYROPHENONE

$C_{10}H_{12}O$

Mol. Wt. 148.21

B.P. 91.5-93.5°C/7mm

Solvent: CCl_4

Assignments

a = 1.18
b = 3.47
c = 7.33
d = 7.85

2048

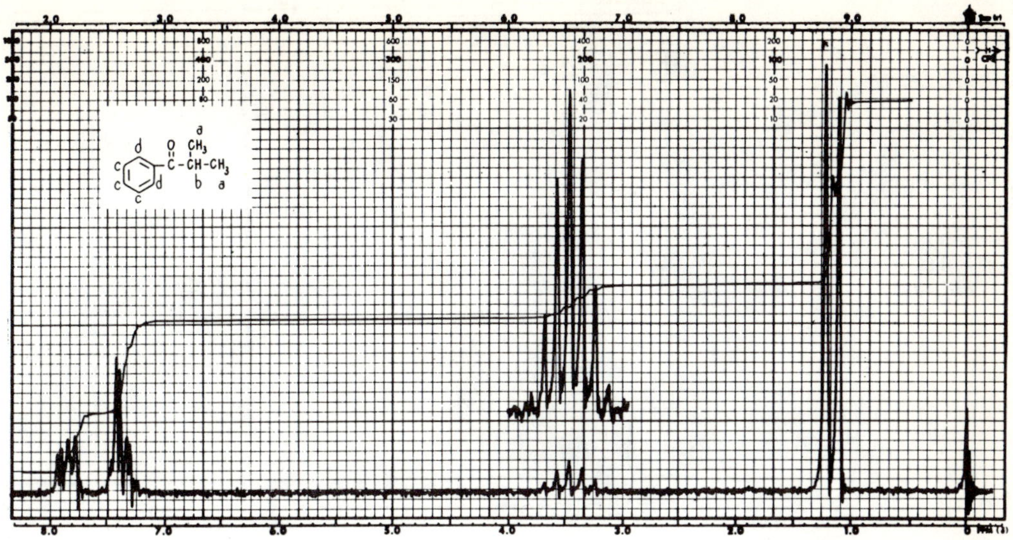

© 1978 Sadtler Research Laboratories, Inc.

HEPTANOPHENONE

$C_{13}H_{18}O$

Mol. Wt. 190.29

Solvent: CCl_4

Assignments

a = 0.89
b = 1.00-1.60
c = 1.69
d = 2.86
e = 7.15-7.55
f = 7.89

2049

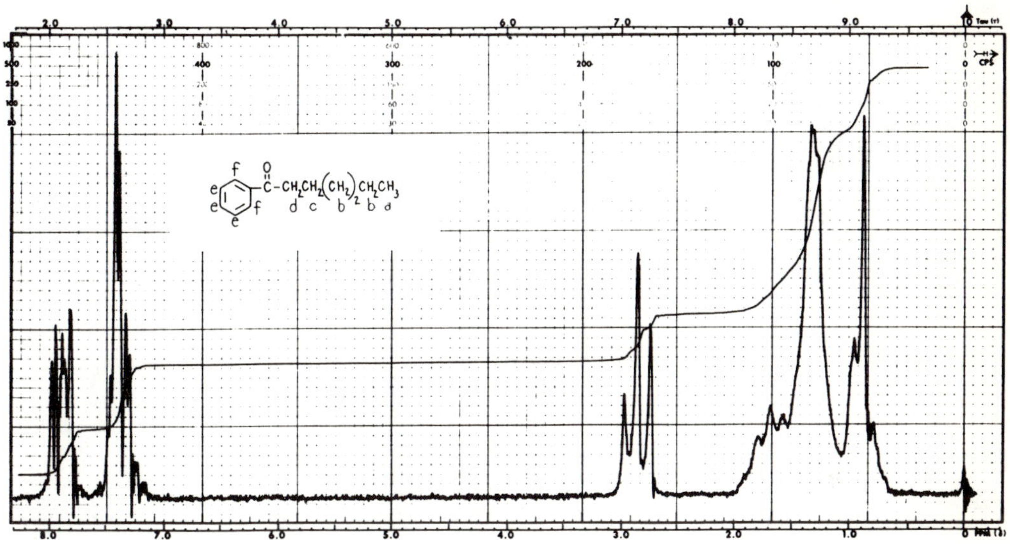

© 1978 Sadtler Research Laboratories, Inc.

MYRISTOPHENONE

$C_{20}H_{32}O$

Mol. Wt. 288.48

M.P. 53-55°C

Solvent: $CDCl_3$

Assignments

a = 0.89
b = ca 1.29
c = 1.72
d = 2.93
e = 7.20-7.60
f = 7.91

2050

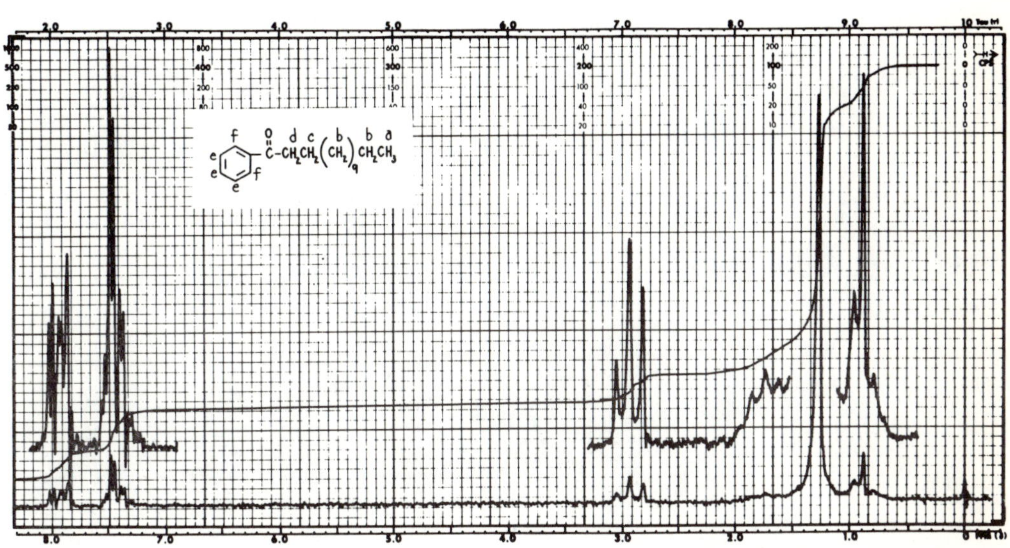

© 1978 Sadtler Research Laboratories, Inc.

2-CHLOROACETOPHENONE

C_8H_7ClO

Mol. Wt. 154.60

M.P. 54-56°C

Solvent: $CDCl_3$

Assignments
a = 4.66
b = ca 7.49
c = 7.90

2051

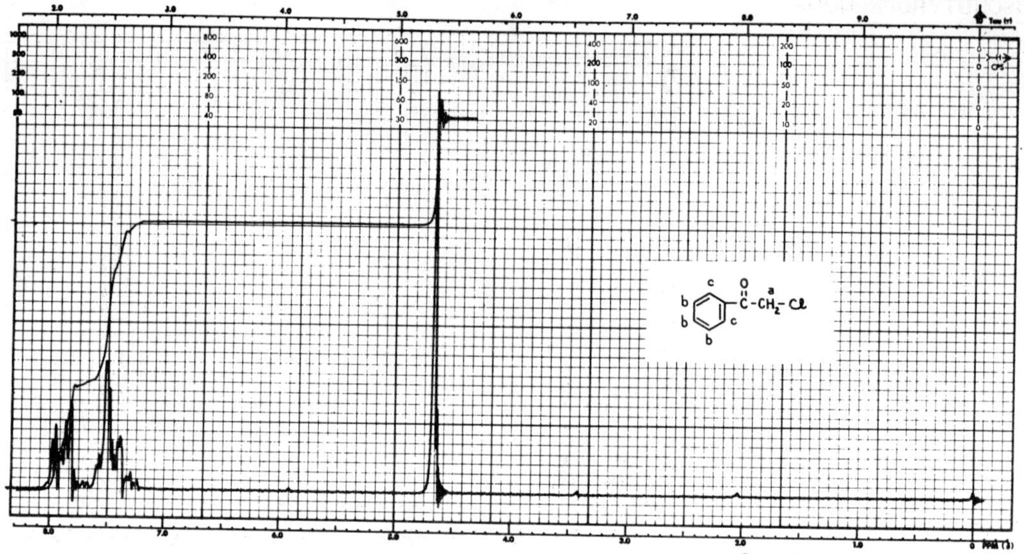

© 1978 Sadtler Research Laboratories, Inc.

3-CHLOROPROPIOPHENONE

C_9H_9ClO

Mol. Wt. 168.62

M.P. 40-44°C

Solvent: $CDCl_3$

Assignments
a = 3.39
b = 3.87
c = 7.19-7.69
d = 7.88

2052

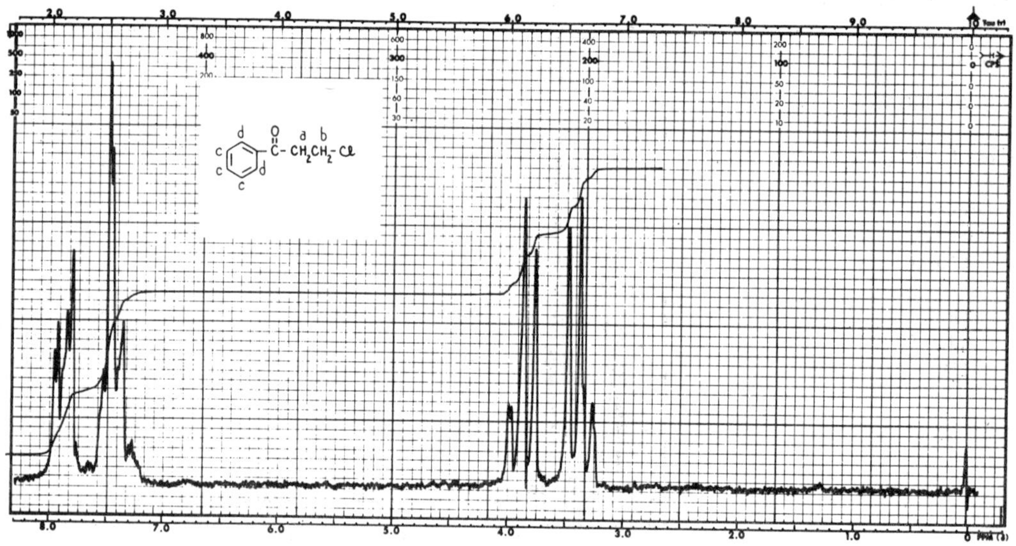

© 1978 Sadtler Research Laboratories, Inc.

2-BROMOACETOPHENONE

C_8H_7BrO

Mol. Wt. 199.05

M.P. 49-50°C

Solvent: $CDCl_3$

Assignments
a = 4.42
b = 7.20-7.60
c = 7.91

2053

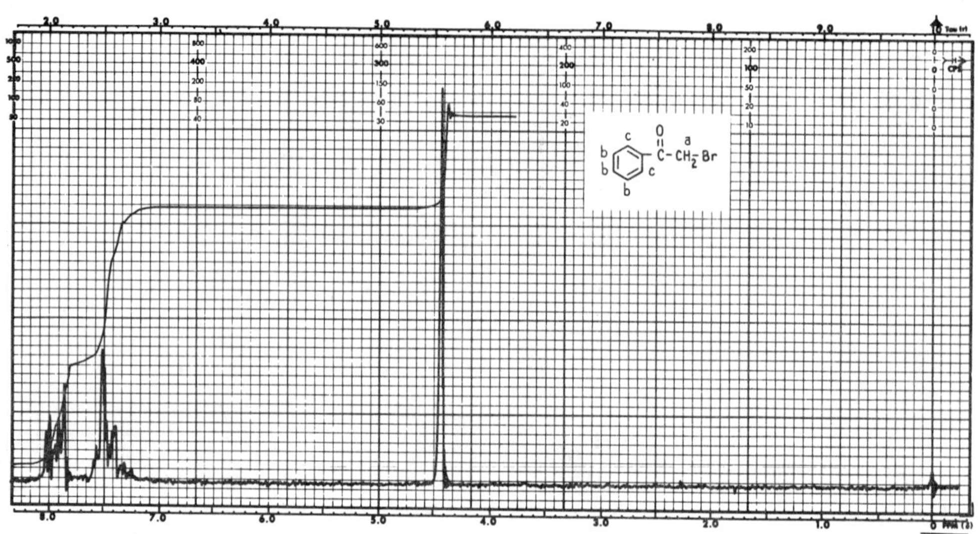

© 1978 Sadtler Research Laboratories, Inc.

BENZOYLACETONITRILE

C_9H_7NO

Mol. Wt. 145.16

M.P. 80-81.5°C

Solvent: $CDCl_3$

Assignments
a = 4.14
b = 7.30-7.70
c = 7.90

2054

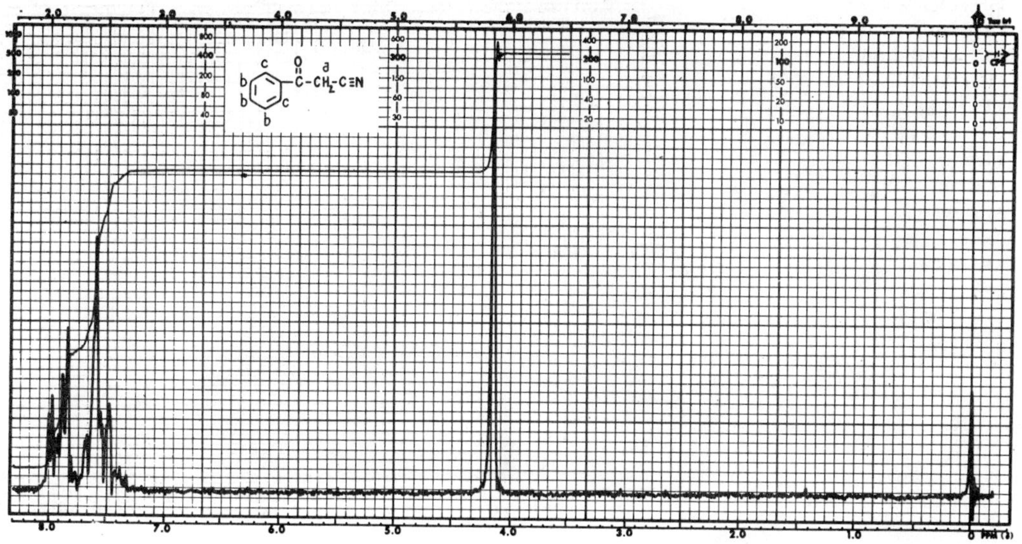

© 1978 Sadtler Research Laboratories, Inc.

2'-METHYLACETOPHENONE

$C_9H_{10}O$

Mol. Wt. 134.18

B.P. 214°C

Solvent: CCl_4

Assignments
a = 4.61
b = 5.43
c = 7.60-8.10
d = 9.13

2055

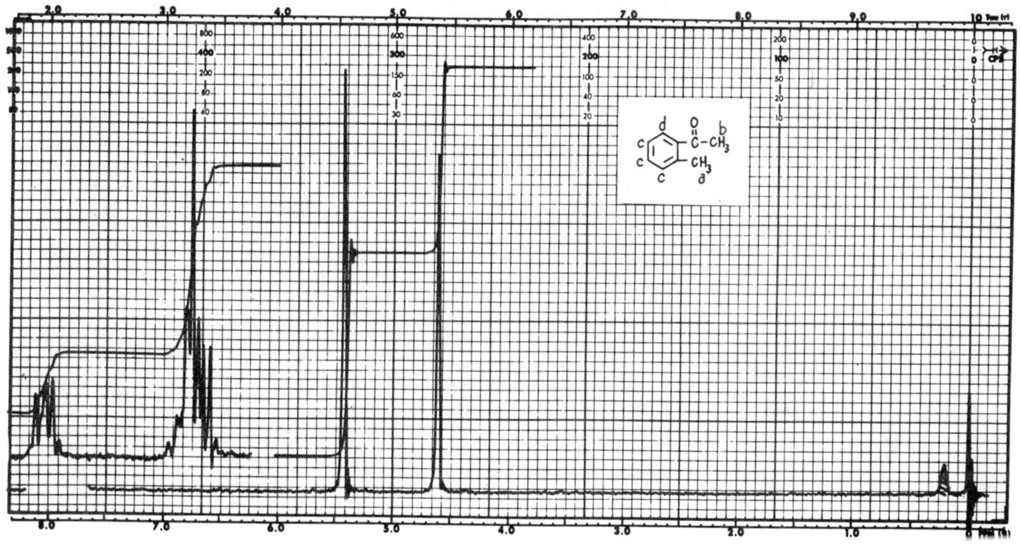

© 1978 Sadtler Research Laboratories, Inc.

4'-METHYLACETOPHENONE

$C_9H_{10}O$

Mol. Wt. 134.18

Solvent: CCl_4

Assignments
a = 2.32
b = 2.42
c = 7.09
d = 7.69
e = 1.49 imp.

2056

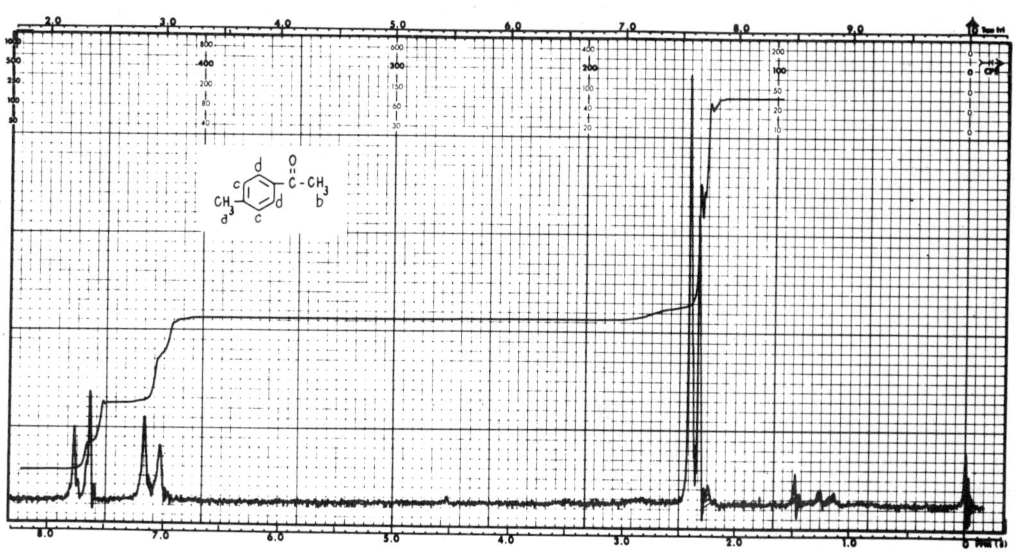

© 1978 Sadtler Research Laboratories, Inc.

2',5'-DIMETHYLACETOPHENONE

$C_{10}H_{12}O$

Mol. Wt. 148.21

B.P. 114-116°C/17mm

Solvent: CCl_4

Assignments
a = 2.33
b = ca 2.44
c = 7.07
d = 7.42

2057

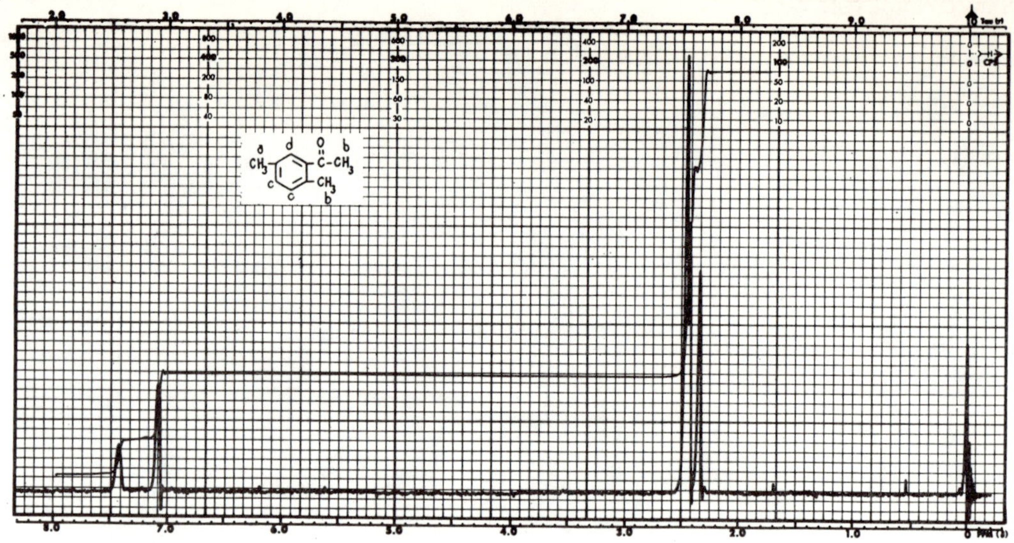

© 1978 Sadtler Research Laboratories, Inc.

2',3',5',6'-TETRAMETHYLACETOPHENONE

$C_{12}H_{16}O$

Mol. Wt. 176.26

M.P. 70-73°C

Solvent: $CDCl_3$

Assignments
a = 2.08
b = 2.91
c = 2.40
d = 6.90

2058

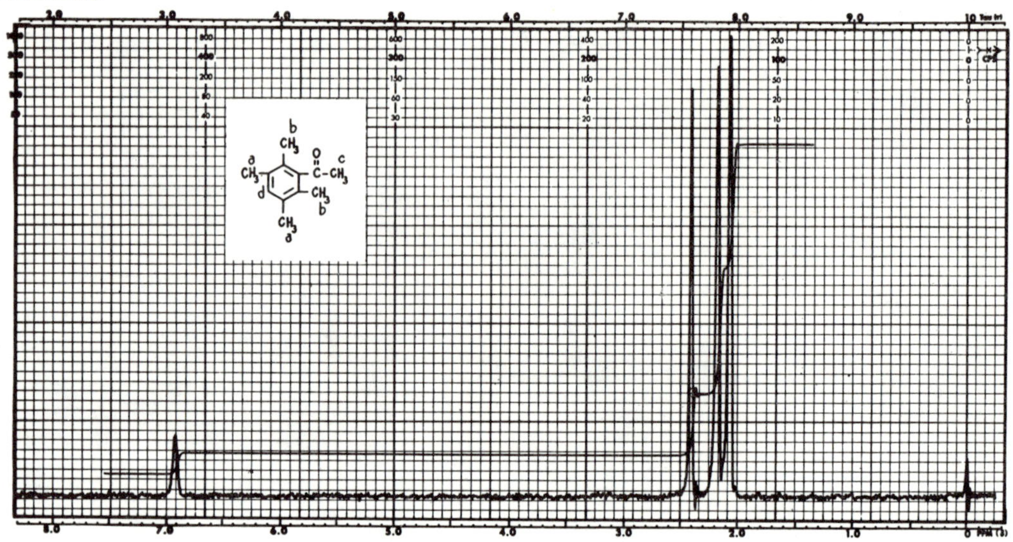

© 1978 Sadtler Research Laboratories, Inc.

4'-ETHYLACETOPHENONE

$C_{10}H_{12}O$

Mol. Wt. 148.21

B.P. 113-115°C/11mm

Solvent: CCl_4

Assignments
a = 1.22
b = 2.42
c = 2.65
d = 7.17
e = 7.75

2059

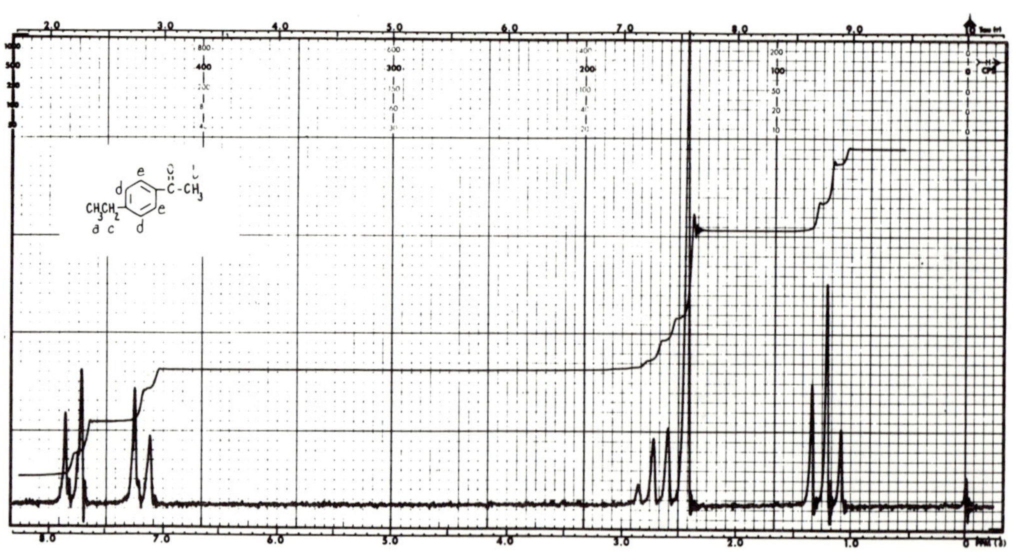

© 1978 Sadtler Research Laboratories, Inc.

4'-METHYLVALEROPHENONE

$C_{12}H_{16}O$

Mol. Wt. 176.26

Solvent: $CDCl_3$

Assignments

a = 0.96
b = 1.00-2.05
c = 2.40
d = 2.90
e = 7.21
f = 7.83

2060

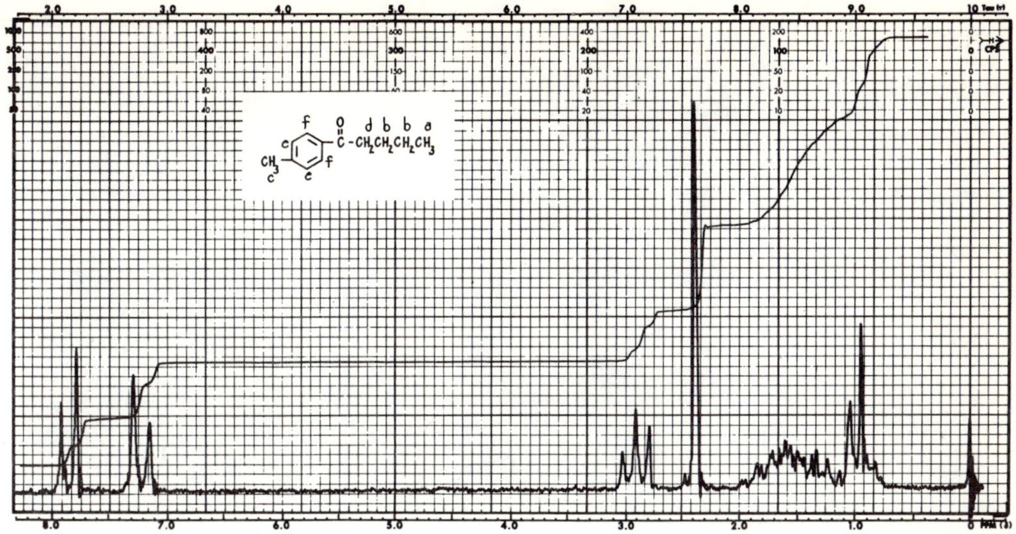

© 1978 Sadtler Research Laboratories, Inc.

2',4'-DIMETHYLPIVALOPHENONE

$C_{13}H_{18}O$

Mol. Wt. 190.29

Solvent: CCl_4

Assignments

a = 1.21
b = 2.13
c = 2.30
d = ca 6.91

2061

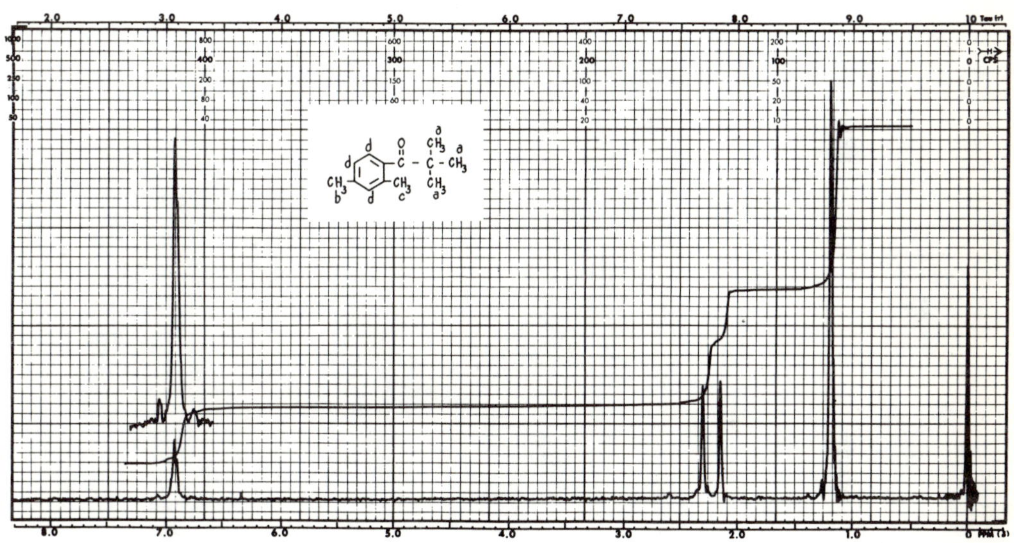

© 1978 Sadtler Research Laboratories, Inc.

2-BROMO-4'-METHYLACETOPHENONE

C_9H_9BrO

Mol. Wt. 213.08

M.P. 45-49°C

Solvent: $CDCl_3$

Assignments

a = 2.39
b = 4.39
c = 7.23
d = 7.83

2062

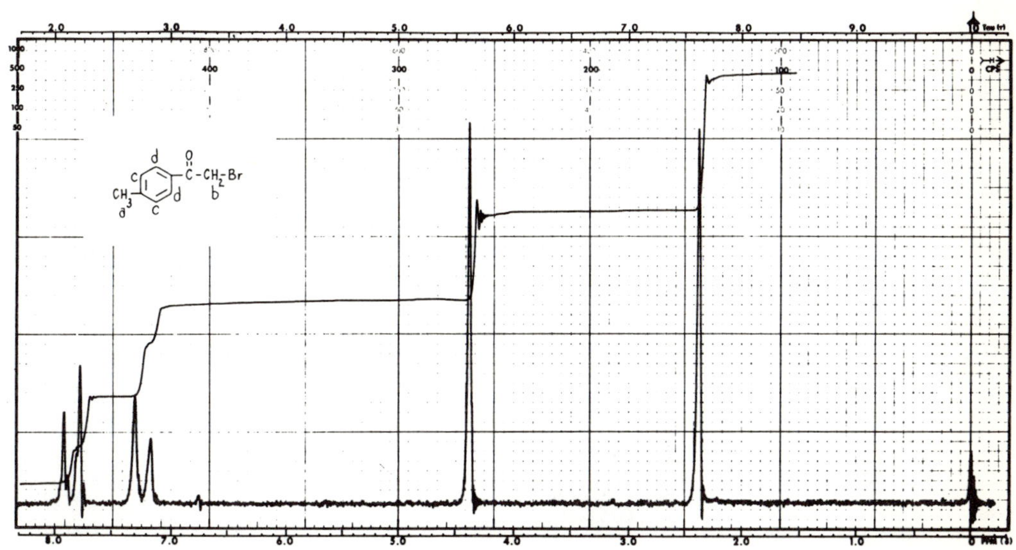

© 1978 Sadtler Research Laboratories, Inc.

2',4'-DIFLUOROACETOPHENONE

$C_8H_6F_2O$

Mol. Wt. 156.13

B.P. 60-61°C/7mm

Solvent: CCl_4

Assignments
a = 2.53
b = 6.60-7.10
c = 7.86

2063

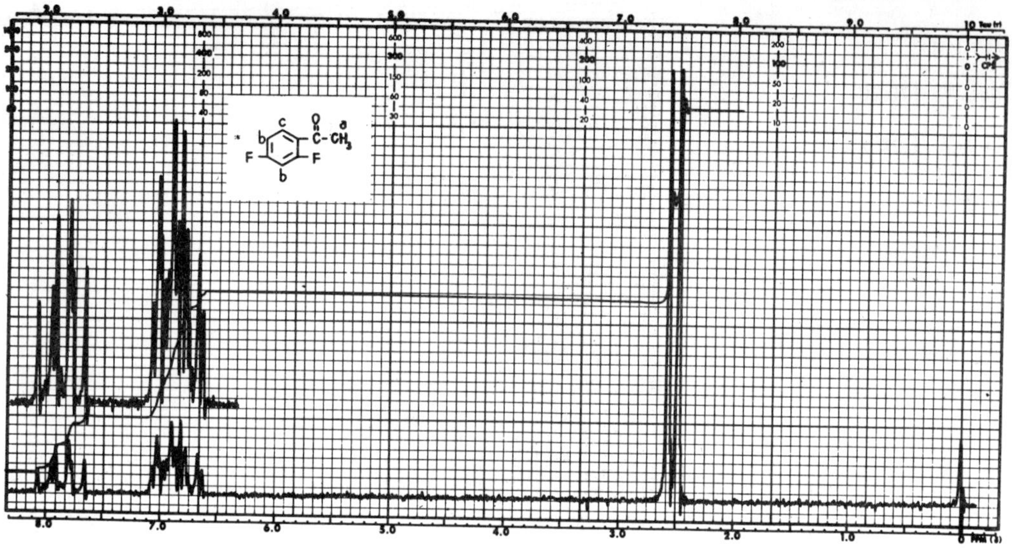

© 1978 Sadtler Research Laboratories, Inc.

2'-CHLOROACETOPHENONE

C_8H_7ClO

Mol. Wt. 154.60

B.P. 104-106°C/12mm

Solvent: CCl_4

Assignments
a = 2.55
b = 7.15-7.60

2064

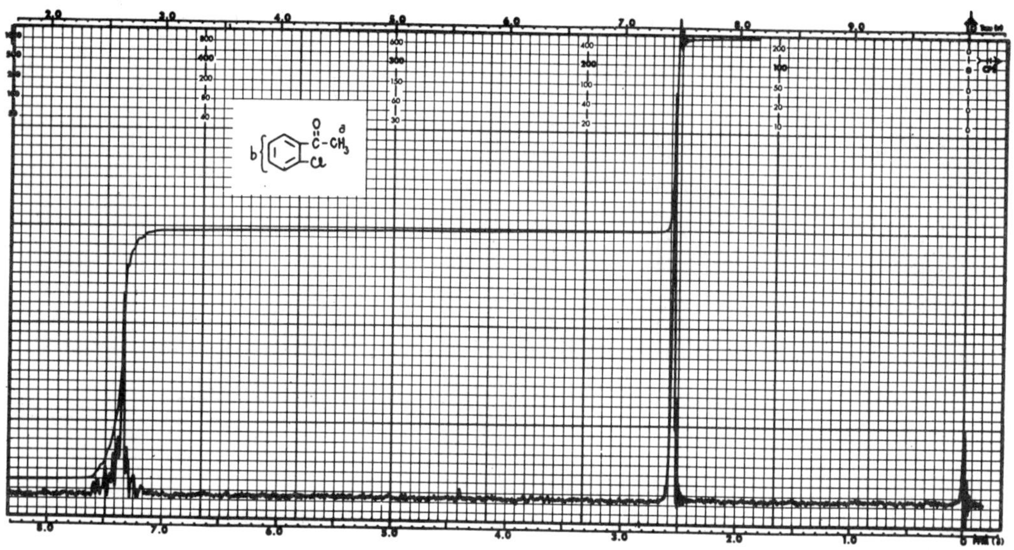

© 1978 Sadtler Research Laboratories, Inc.

3'-CHLOROACETOPHENONE

C_8H_7ClO

Mol. Wt. 154.60

B.P. 233-235°C

Solvent: CCl_4

Assignments
a = 2.50
b = 7.15-7.63
c = 7.78
d = 7.82

2065

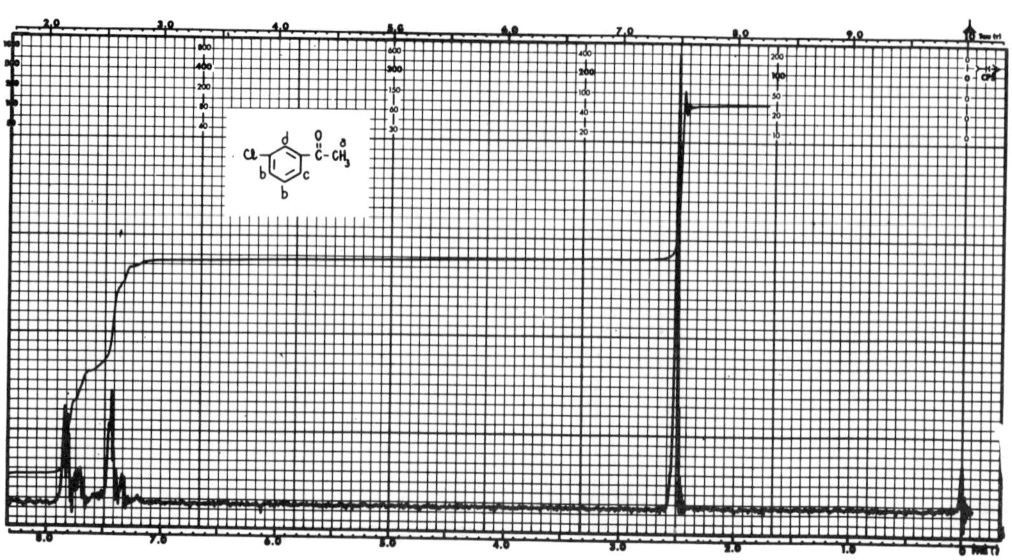

© 1978 Sadtler Research Laboratories, Inc.

4'-CHLOROVALEROPHENONE

$C_{11}H_{13}ClO$

Mol. Wt. 196.68

Solvent: $CDCl_3$

Assignments
a = 0.94
b = 1.00-2.00
c = 2.90
d = 7.39
e = 7.89

2066

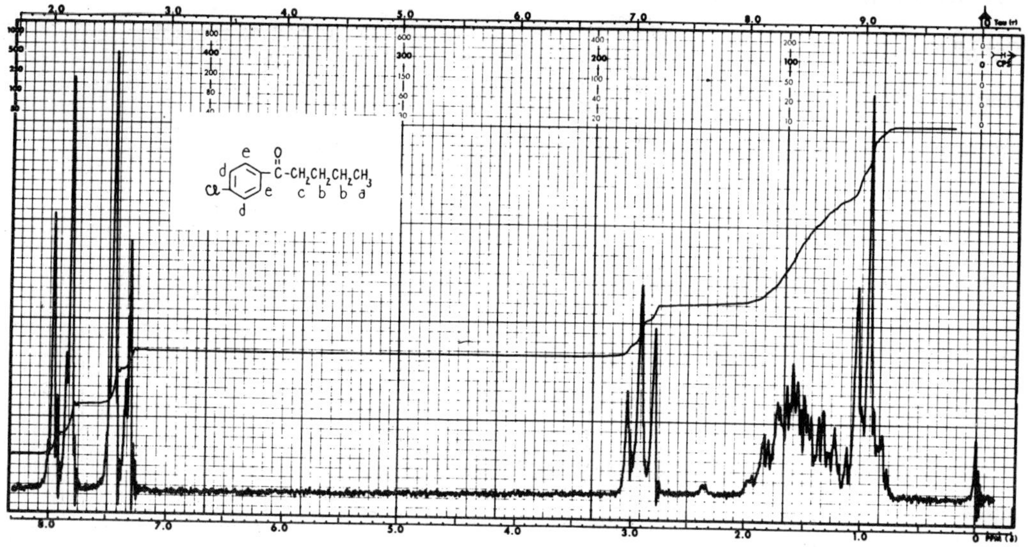

© 1978 Sadtler Research Laboratories, Inc.

2',5'-DICHLOROACETOPHENONE

$C_8H_6Cl_2O$

Mol. Wt. 189.04

M.P. 11-13°C

Solvent: CCl_4

Assignments
a = 2.59
b = 7.32
c = 7.49

2067

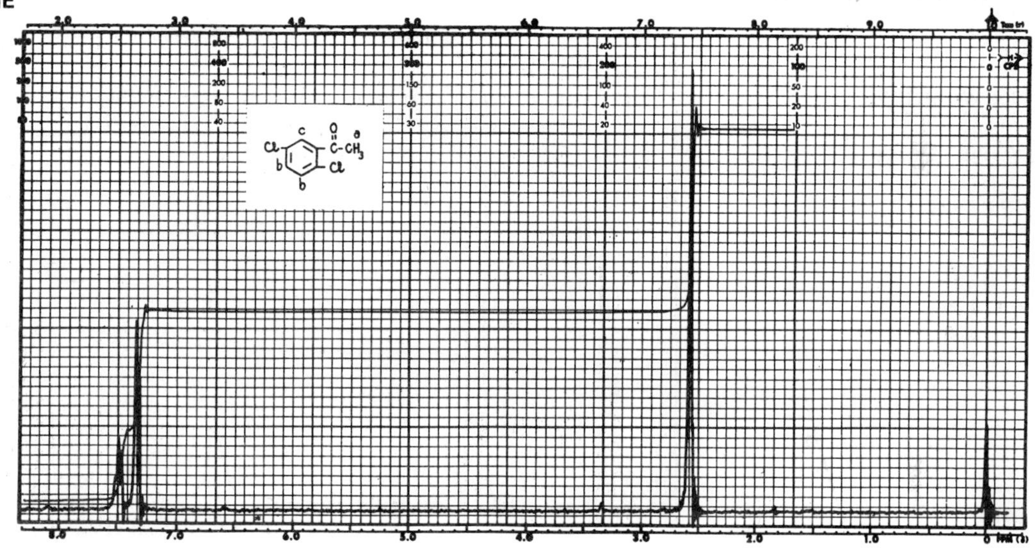

© 1978 Sadtler Research Laboratories, Inc.

3'-BROMOACETOPHENONE

C_8H_7BrO

Mol. Wt. 199.05

M.P. 13-15°C

Solvent: CCl_4

Assignments
a = 2.50
b = 7.28
c = 7.62
d = 7.80
e = 7.99

2068

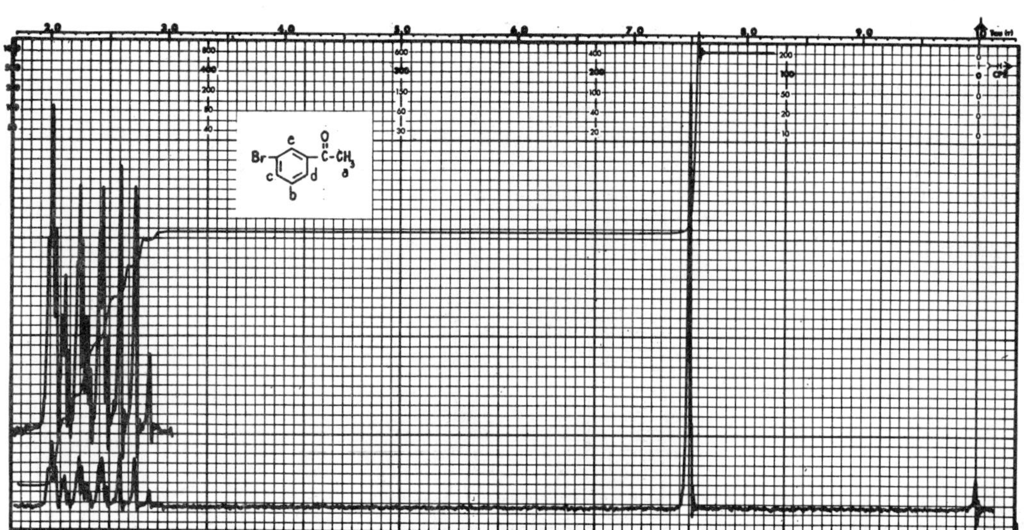

© 1978 Sadtler Research Laboratories, Inc.

3'-IODOACETOPHENONE

C_8H_7IO

Mol. Wt. 246.05

M.P. 26-29°C

Solvent: $CDCl_3$

Assignments
a = 2.50
b = 7.10
c = 7.79
d = 8.17

2069

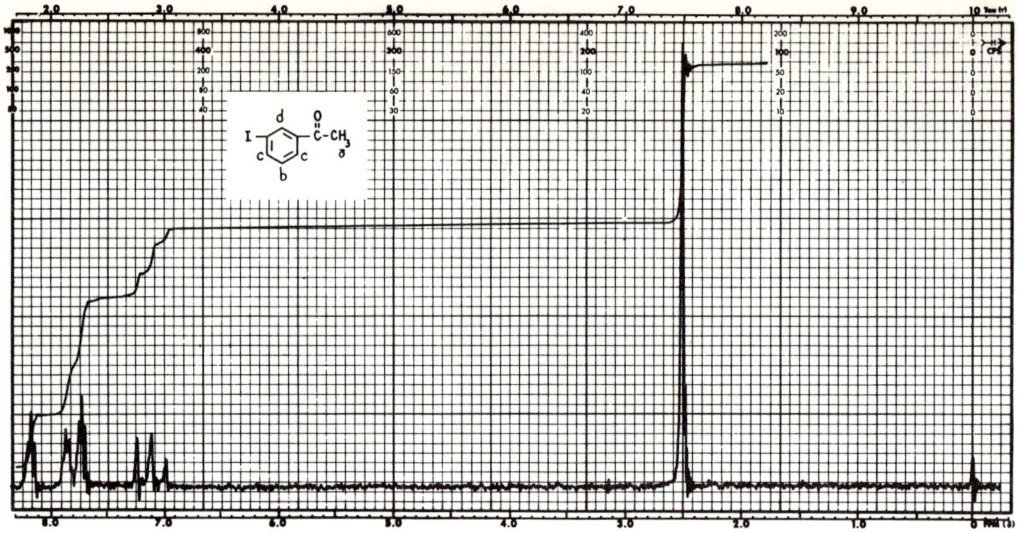

© 1978 Sadtler Research Laboratories, Inc.

4'-AMINOBUTYROPHENONE

$C_{10}H_{13}NO$

Mol. Wt. 163.22

Solvent: $CDCl_3$

Assignments
a = 0.98
b = 1.72
c = 2.81
d = 4.30
e = 6.58
f = 7.73

2070

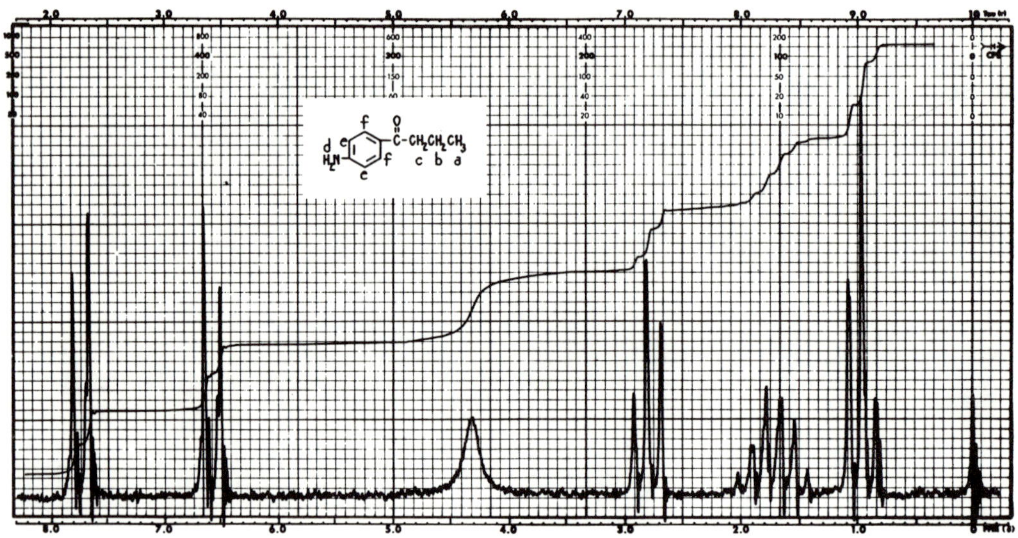

© 1978 Sadtler Research Laboratories, Inc.

4'-AMINOVALEROPHENONE

$C_{11}H_{15}NO$

Mol. Wt. 177.25

M.P. 71-73°C

Solvent: $CDCl_3$

Assignments
a = 0.93
b = 1.10-2.00
c = 2.82
d = 4.38
e = 6.59
f = 7.75

2071

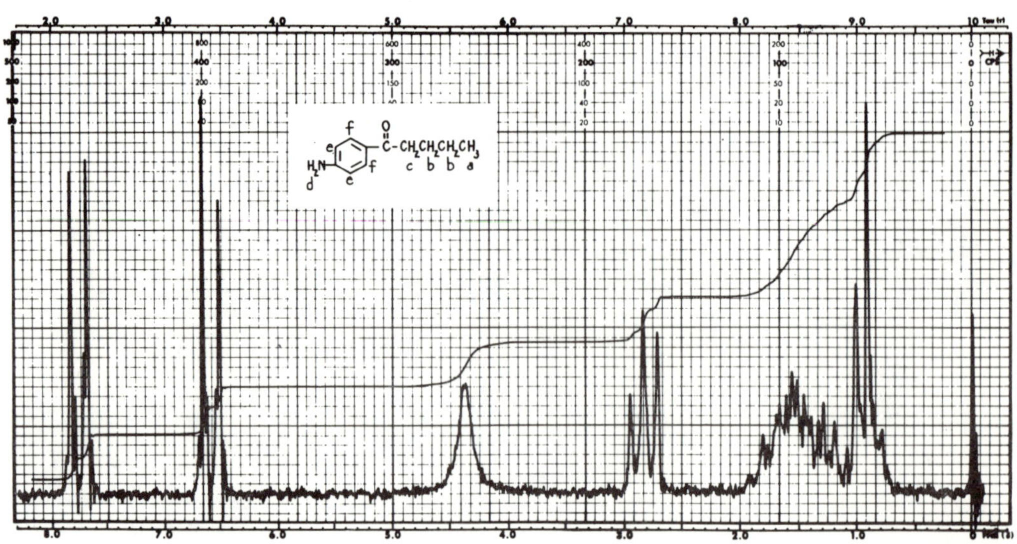

© 1978 Sadtler Research Laboratories, Inc.

2'-NITROACETOPHENONE

$C_8H_7NO_3$

Mol. Wt. 165.15

B.P. 158.5-159°C/16mm (lit.)

Solvent: CCl_4

Assignments
a = 2.46
b = 7.25-7.80
c = 7.91

2072

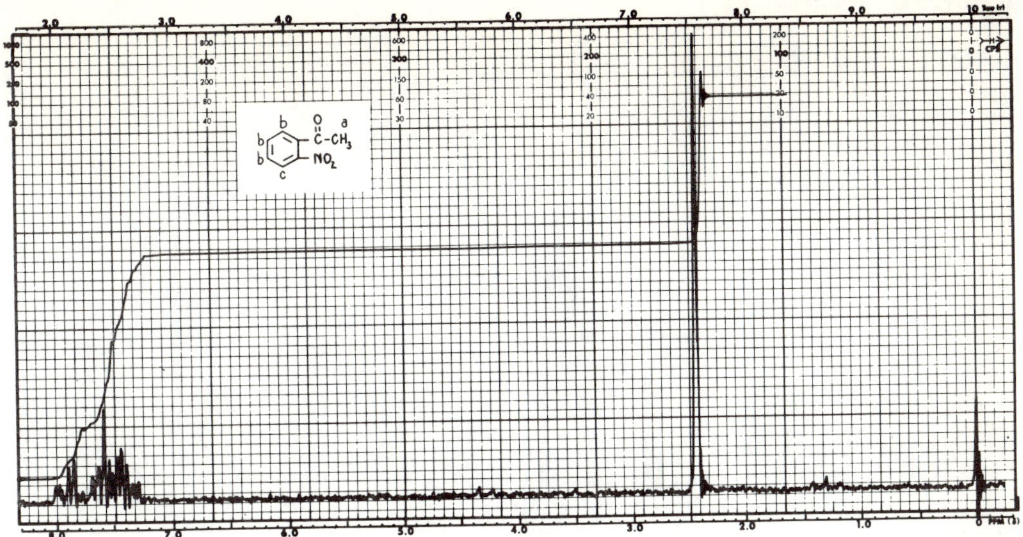

© 1978 Sadtler Research Laboratories, Inc.

3'-NITROACETOPHENONE

$C_8H_7NO_3$

Mol. Wt. 165.15

M.P. 77-79°C

Solvent: $CDCl_3$

Assignments
a = 2.68
b = 7.67
c = 8.28
d = 8.41
e = 8.72

2073

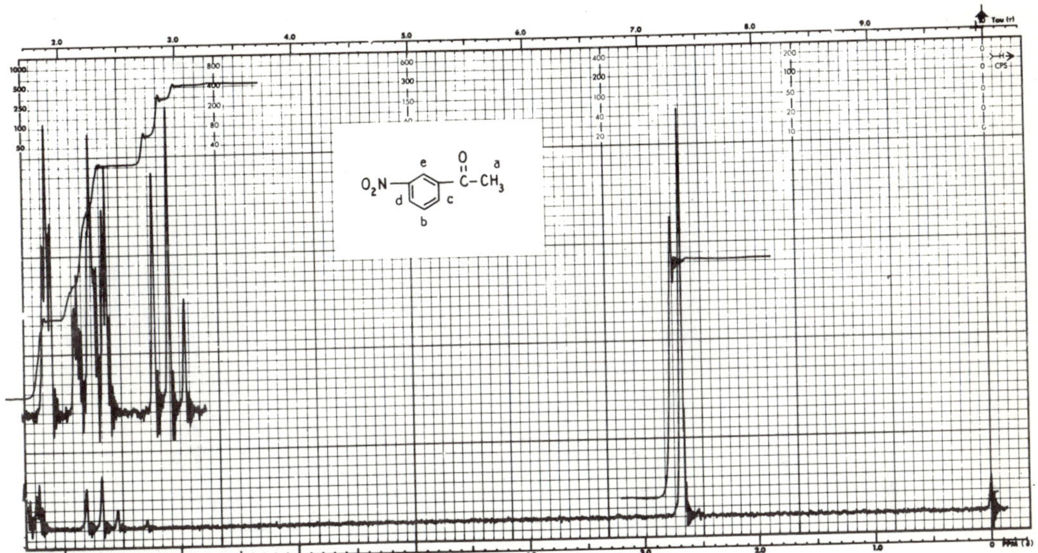

© 1978 Sadtler Research Laboratories, Inc.

4'-(METHYLTHIO)ACETOPHENONE

$C_9H_{10}OS$

Mol. Wt. 166.24

Solvent: Polysol-d

Assignments
a = 2.52
b = 7.24
c = 7.81

2074

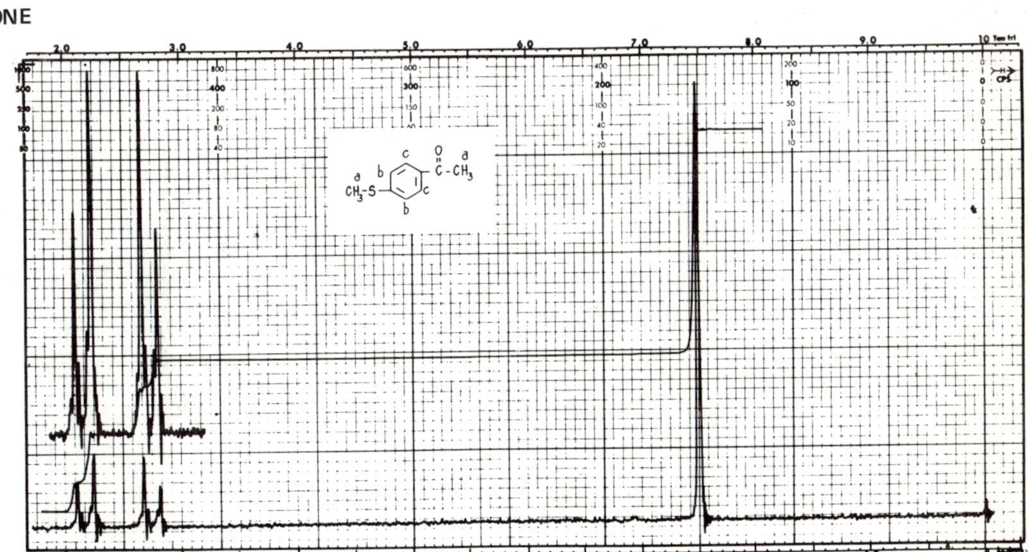

© 1978 Sadtler Research Laboratories, Inc.

4'-METHOXYACETOPHENONE

$C_9H_{10}O_2$

Mol. Wt. 150.18

Solvent: $CDCl_3$

Assignments
a = 2.56
b = 3.88
c = 6.98
d = 7.97

2075

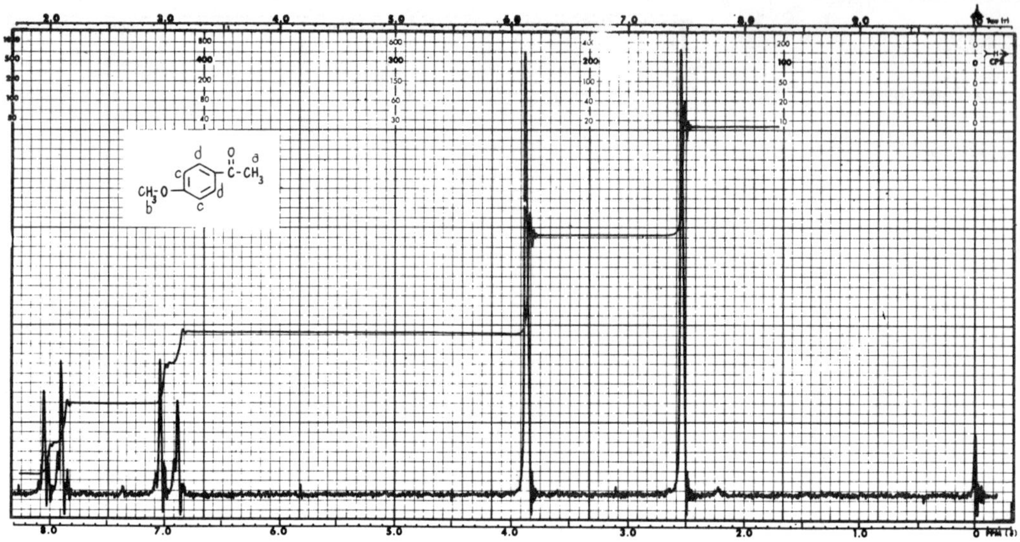

© 1978 Sadtler Research Laboratories, Inc.

4'-ETHOXYACETOPHENONE

$C_{10}H_{12}O_2$

Mol. Wt. 164.21

M.P. 33-36°C

Solvent: $CDCl_3$

Assignments
a = 1.42
b = 2.51
c = 4.09
d = 6.91
e = 7.91

2076

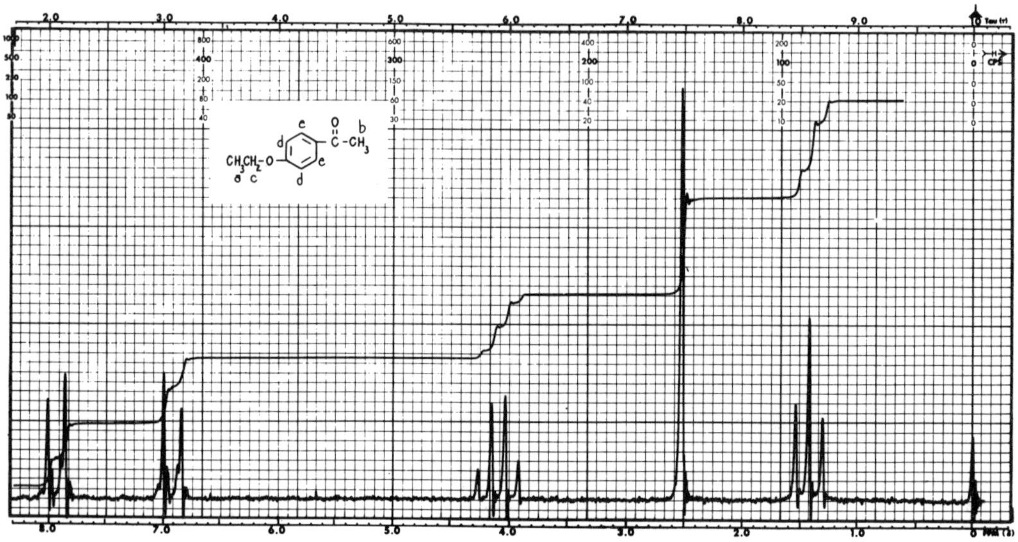

© 1978 Sadtler Research Laboratories, Inc.

4'-METHOXYBUTYROPHENONE

$C_{11}H_{14}O_2$

Mol. Wt. 178.23

Solvent: CCl_4

Assignments
a = 0.98
b = 1.72
c = 2.79
d = 3.80
e = 6.81
f = 7.80

2077

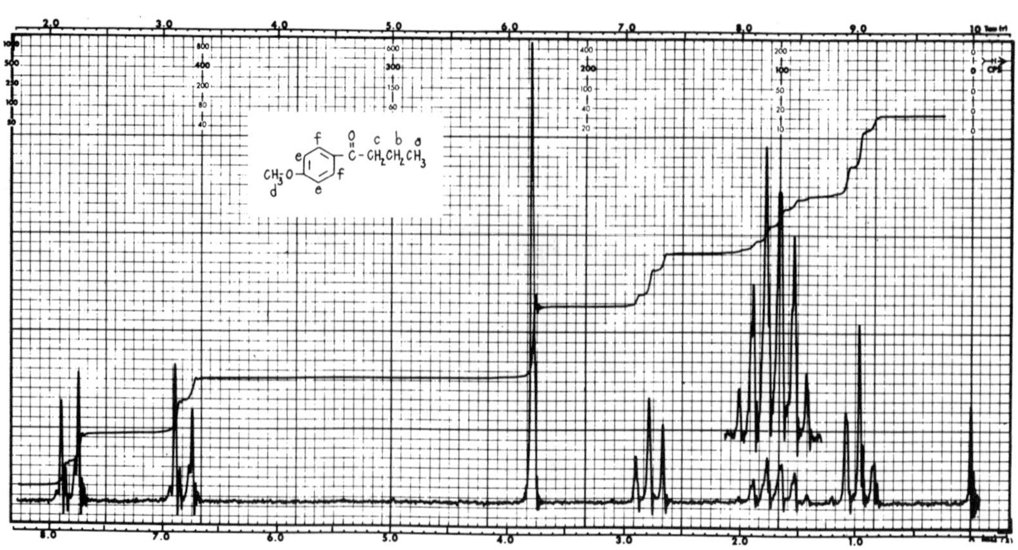

© 1978 Sadtler Research Laboratories, Inc.

4'-HYDROXY-2'-METHYLACETOPHENONE

$C_9H_{10}O_2$

Mol. Wt. 150.18

M.P. 127-129°C

B.P. 313°C

Solvent: DMSO-d_6

Assignments
a = 2.48
b = 6.67
c = 6.70
d = 7.70
e = 8.20-10.20

2078

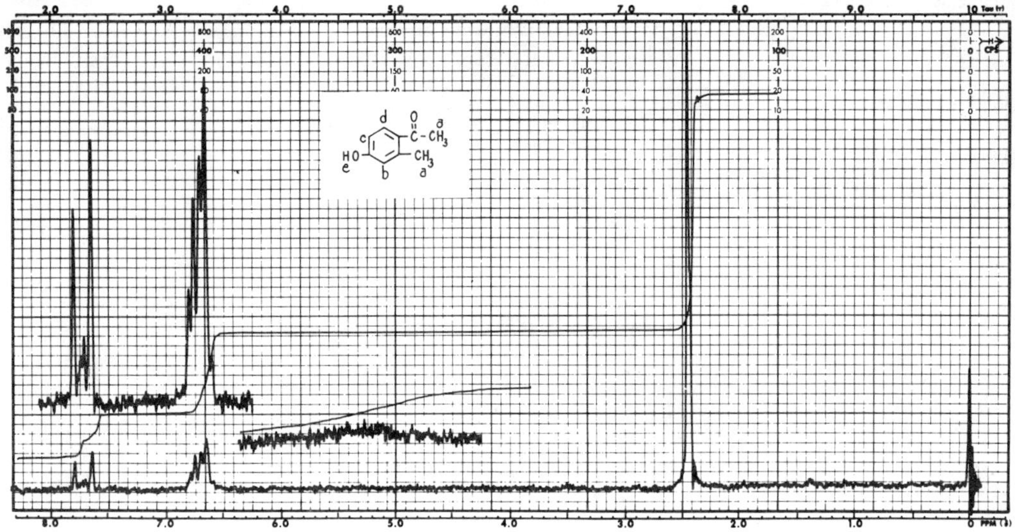

© 1978 Sadtler Research Laboratories, Inc.

5'-CHLORO-2'-HYDROXYACETOPHENONE

$C_8H_7ClO_2$

Mol. Wt. 170.60

Solvent: $CDCl_3$

Assignments
a = 2.62
b = 6.89
c = 7.38
d = 7.67
e = 12.02

2079

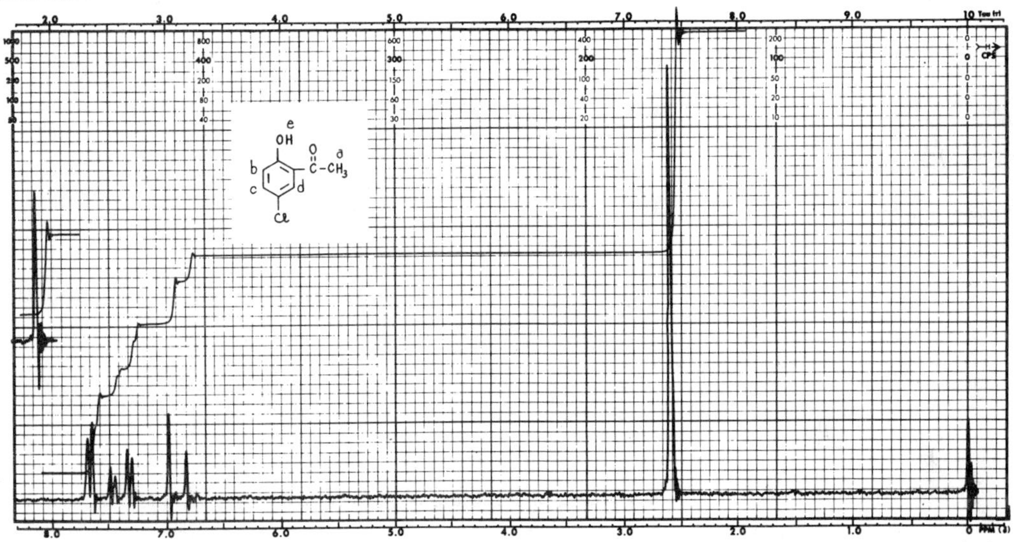

© 1978 Sadtler Research Laboratories, Inc.

2',5'-DIHYDROXYACETOPHENONE

$C_8H_8O_3$

Mol. Wt. 152.15

Solvent: Dimethyl sulfoxide-D_6

Assignments
a = 2.59
b = 6.80
c = 7.06
d = 7.24
e = 9.14
f = 11.40
g = 2.58 DMSO-D_5
h = 3.47 H_2O

2080

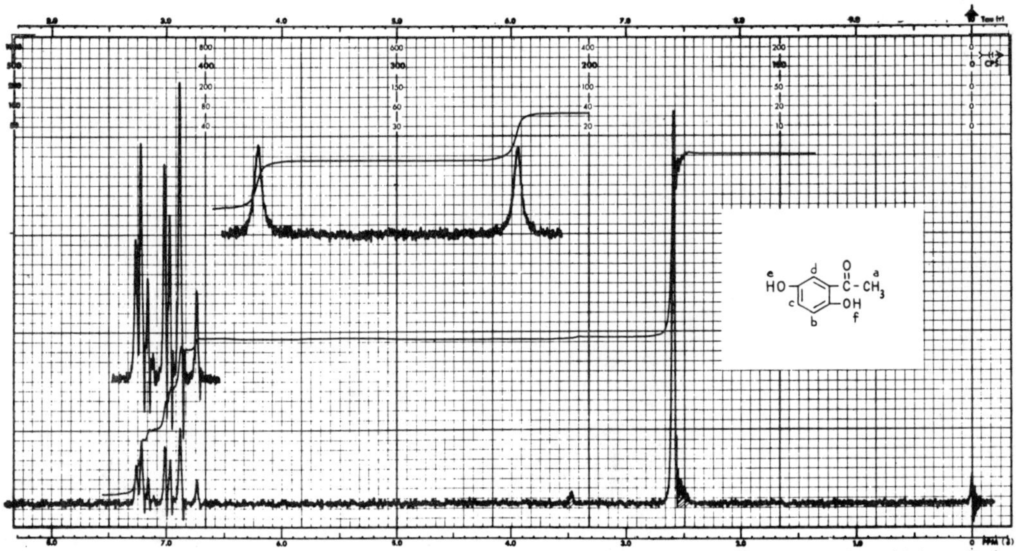

© 1978 Sadtler Research Laboratories, Inc.

m-DIACETYLBENZENE

$C_{10}H_{10}O_2$

Mol. Wt. 162.19

M.P. 31-32°C

Solvent: $CDCl_3$

Assignments
a = 2.64
b = 7.57
c = 8.13
d = 8.49

2081

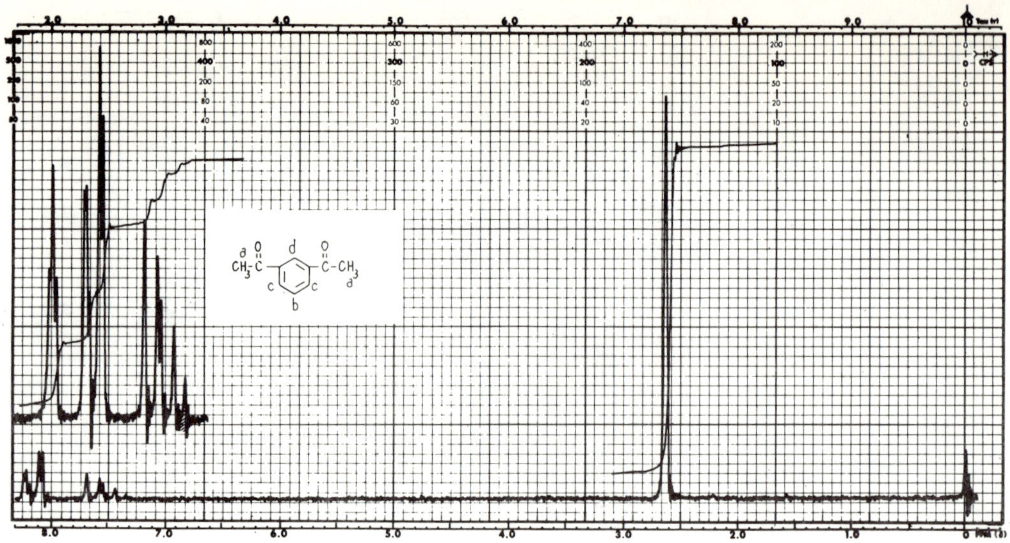

© 1978 Sadtler Research Laboratories, Inc.

---

p-DIACETYLBENZENE

$C_{10}H_{10}O_2$

Mol. Wt. 162.19

Solvent: $CDCl_3$

Assignments
a = 2.63
b = 8.05

2082

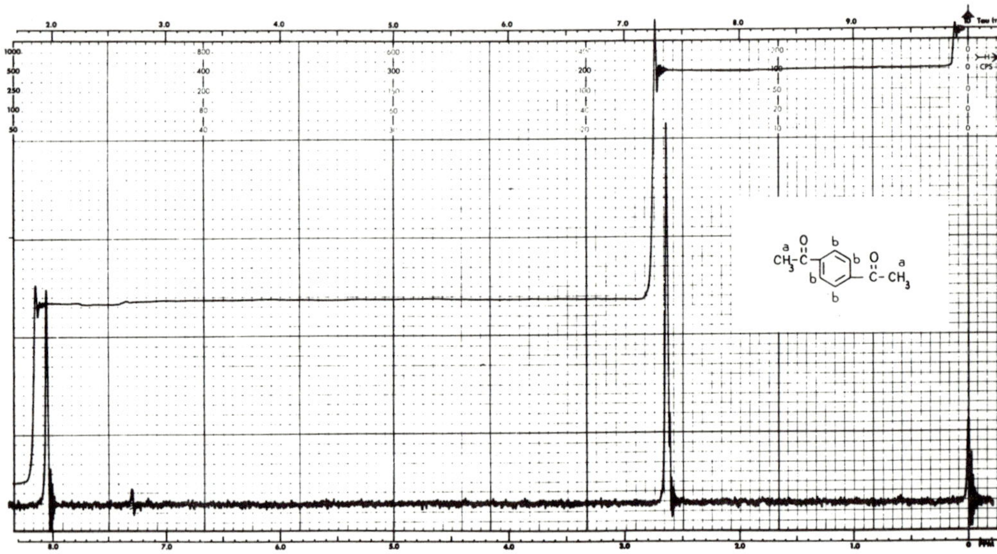

© 1978 Sadtler Research Laboratories, Inc.

CYCLOPENTYL PHENYL KETONE

$C_{12}H_{14}O$

Mol. Wt. 174.24

B.P. 106.5-107.5°C/2mm (lit.)

Solvent: CCl_4

Assignments
a = 1.38-2.15
b = 3.65
c = 7.16-7.60
d = 7.95

2083

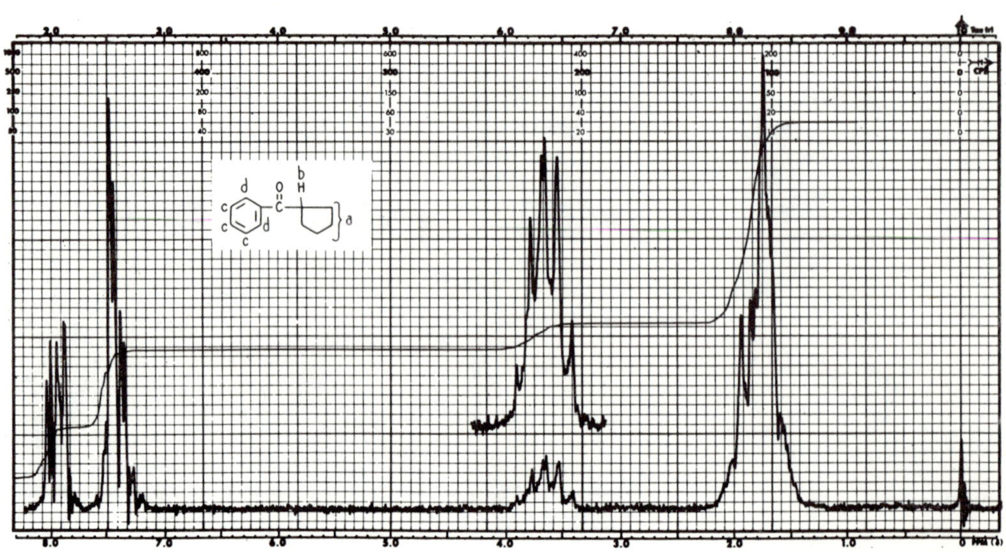

© 1978 Sadtler Research Laboratories, Inc.

CYCLOHEXYL PHENYL KETONE

$C_{13}H_{16}O$

Mol. Wt. 188.27

Solvent: $CDCl_3$

Assignments
a = 1.00-2.20
b = 3.25
c = 7.15-7.60
d = 7.91

2084

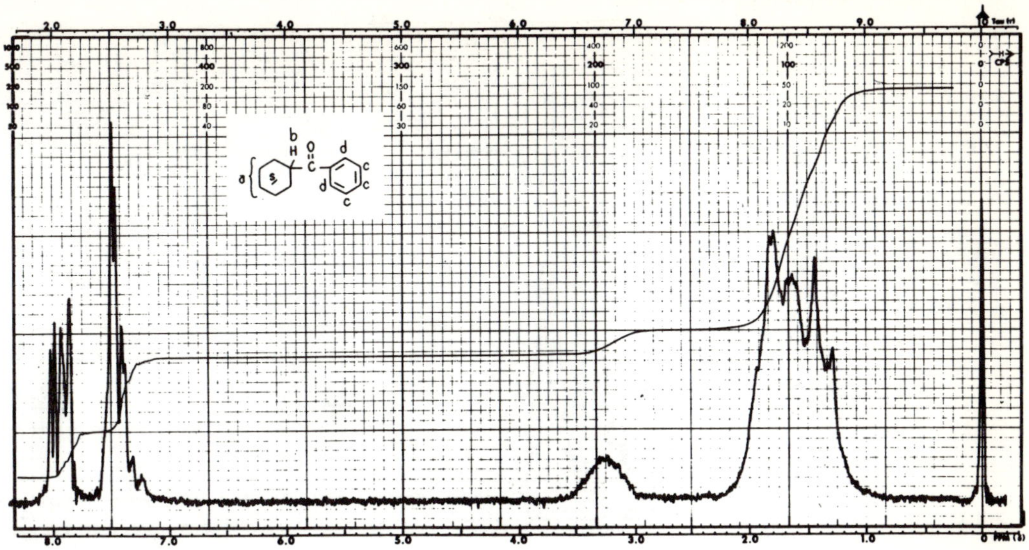

© 1978 Sadtler Research Laboratories, Inc.

2-PHENYLACETOPHENONE

$C_{14}H_{12}O$

Mol. Wt. 196.25

M.P. 55-56°C

Solvent: $CDCl_3$

Assignments
a = 4.21
b = 7.27
c = 7.30-7.60
d = 7.98

2085

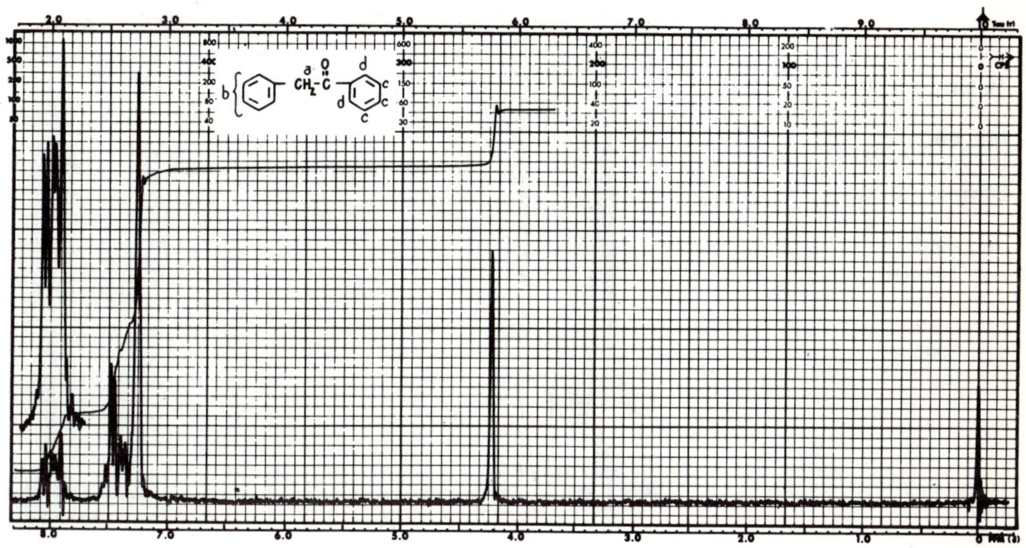

© 1978 Sadtler Research Laboratories, Inc.

2-(p-CHLOROPHENYL)ACETOPHENONE

$C_{14}H_{11}ClO$

Mol. Wt. 230.70

M.P. 134-136°C

Solvent: Polysol-d

Assignments
a = 4.27
b = ca 7.21
c = 7.25-7.60
d = 7.95

2086

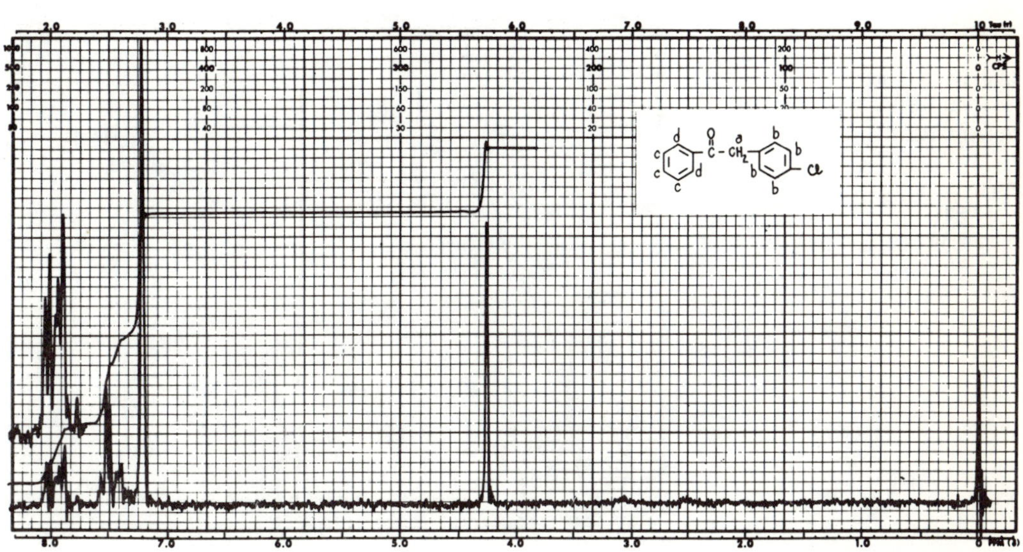

© 1978 Sadtler Research Laboratories, Inc.

2-BROMO-2-PHENYLACETOPHENONE

$C_{14}H_{11}BrO$

Mol. Wt. 275.15

M.P. 55-57°C

Solvent: $CDCl_3$

Assignments
a = 6.41
b = 7.20-7.70
c = 8.01

2087

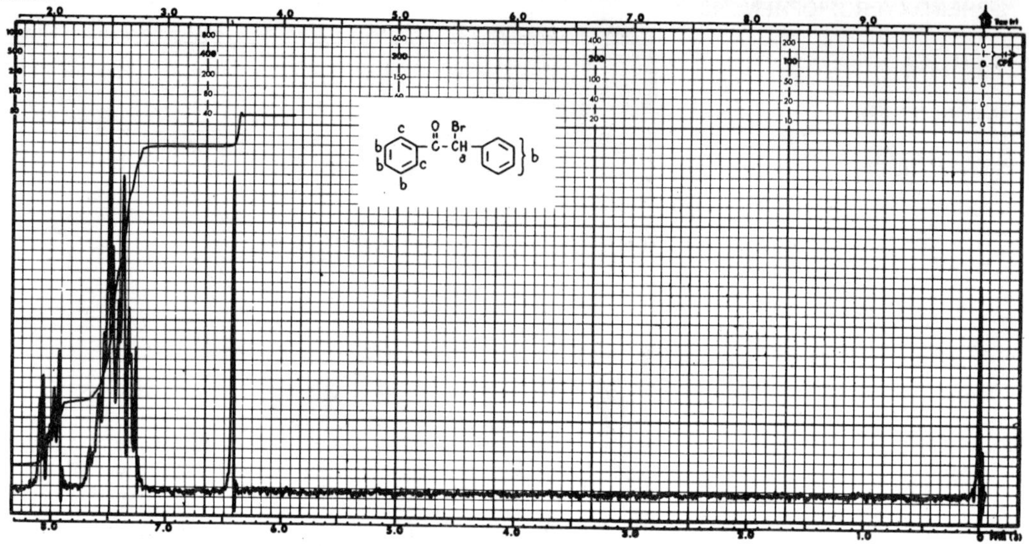

© 1978 Sadtler Research Laboratories, Inc.

3-PHENYLPROPIOPHENONE

$C_{15}H_{14}O$

Mol. Wt. 210.28

M.P. 69-71°C

Solvent: $CDCl_3$

Assignments
a = 3.08 or b
b = 3.20 or a
c = ca 7.20
d = 7.30-7.55
e = 7.90

2088

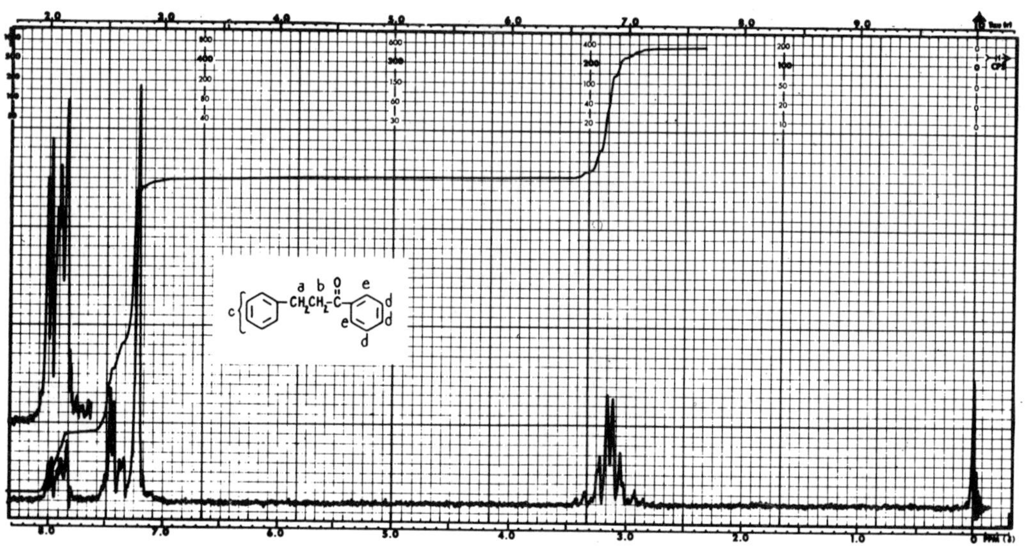

© 1978 Sadtler Research Laboratories, Inc.

2-PHENOXYACETOPHENONE

$C_{14}H_{12}O_2$

Mol. Wt. 212.25

Solvent: $CDCl_3$

Assignments
a = 5.21
b = 6.80=7.40
c = 7.30-7.65
d = 7.98

2089

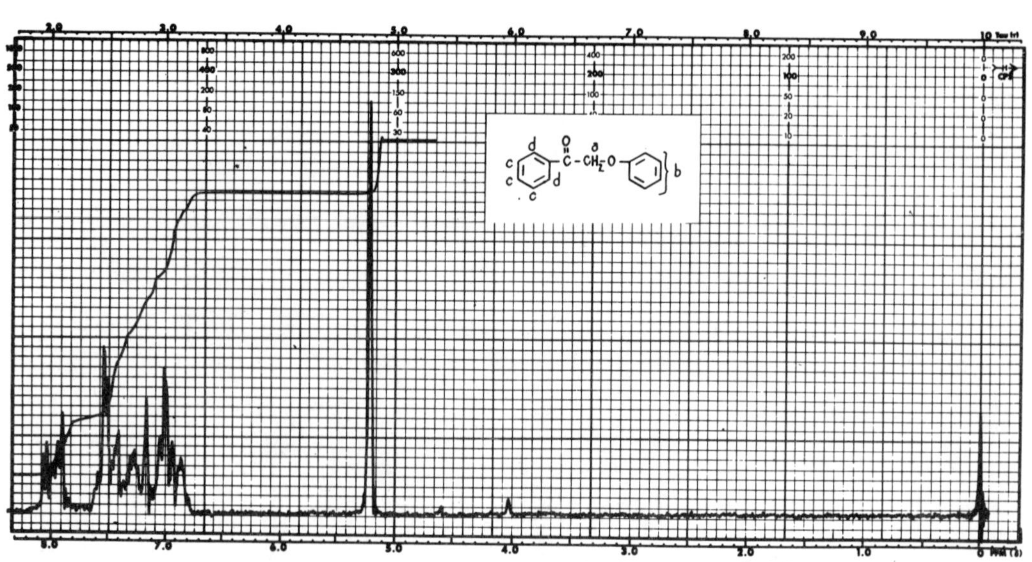

© 1978 Sadtler Research Laboratories, Inc.

2-PHENOXYPROPIOPHENONE

$C_{15}H_{14}O_2$

Mol. Wt. 226.28

Solvent: $CDCl_3$

Assignments
a = 1.64
b = 5.42
c = 6.70-7.30
d = 7.25-7.55
e = 8.03

2090

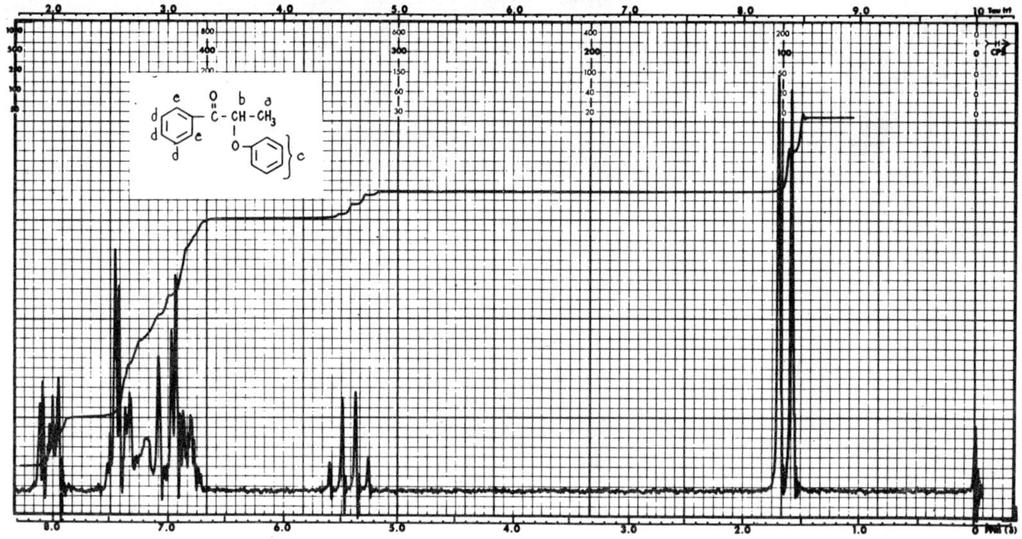

© 1978 Sadtler Research Laboratories, Inc.

3-(2-FURYL)ACRYLOPHENONE

$C_{13}H_{10}O_2$

Mol. Wt. 198.22

B.P. 178-181°C/8mm

Solvent: $CDCl_3$

Assignments
a = 6.52
b = 6.72
c = 7.55-7.85
d = 8.28
e = 8.93
f = 9.08
Shift Reagent Added

2091

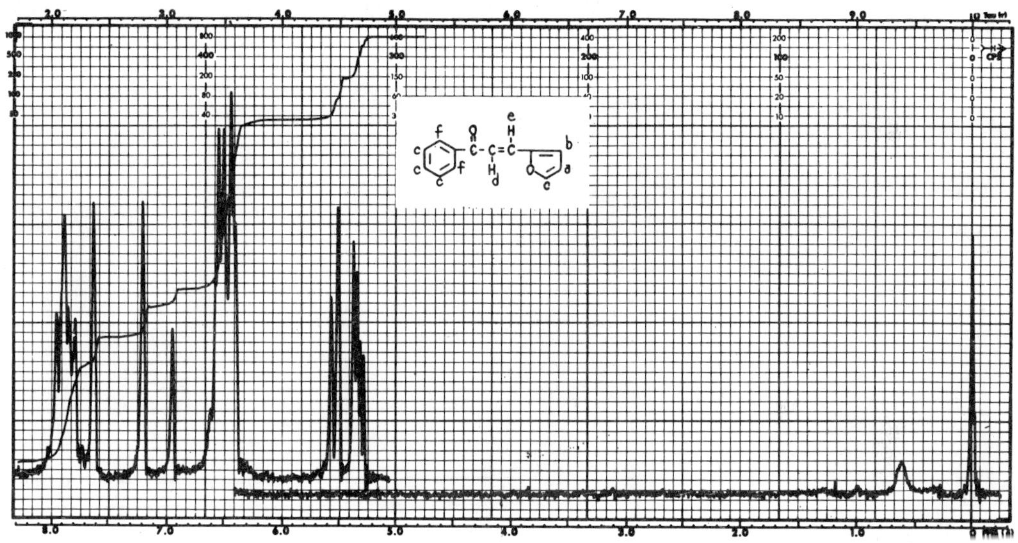

© 1978 Sadtler Research Laboratories, Inc.

CHALCONE

$C_{15}H_{12}O$

Mol. Wt. 208.26

M.P. 62°C (lit.)

Solvent: $CDCl_3$

Assignments
a = 7.20-7.70
b = 7.81
c = 8.02

2092

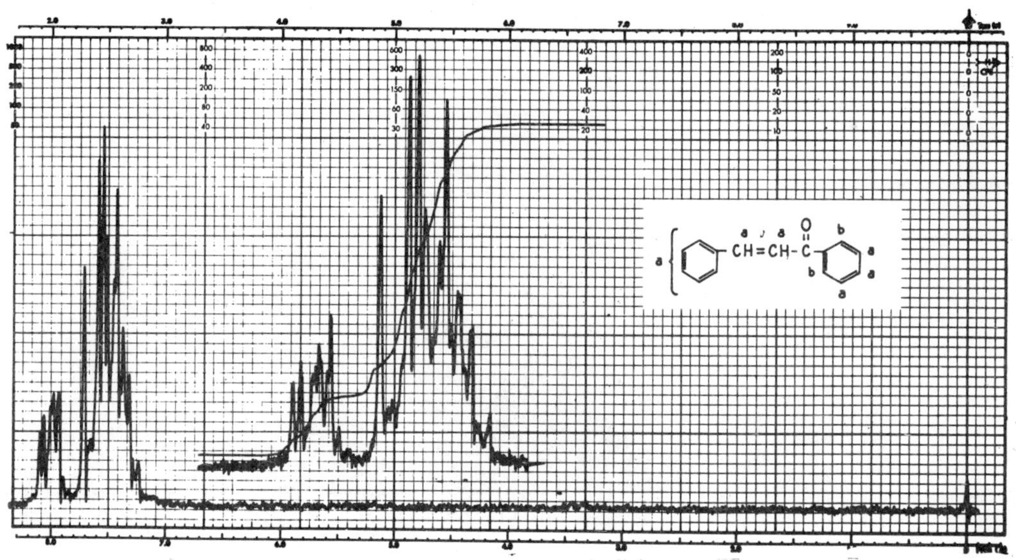

© 1978 Sadtler Research Laboratories, Inc.

4'-METHYLCHALCONE

$C_{16}H_{14}O$

Mol. Wt. 222.29

Solvent: $CDCl_3$

Assignments
a = 2.36
b = 7.15-7.75
c = 7.98

2093

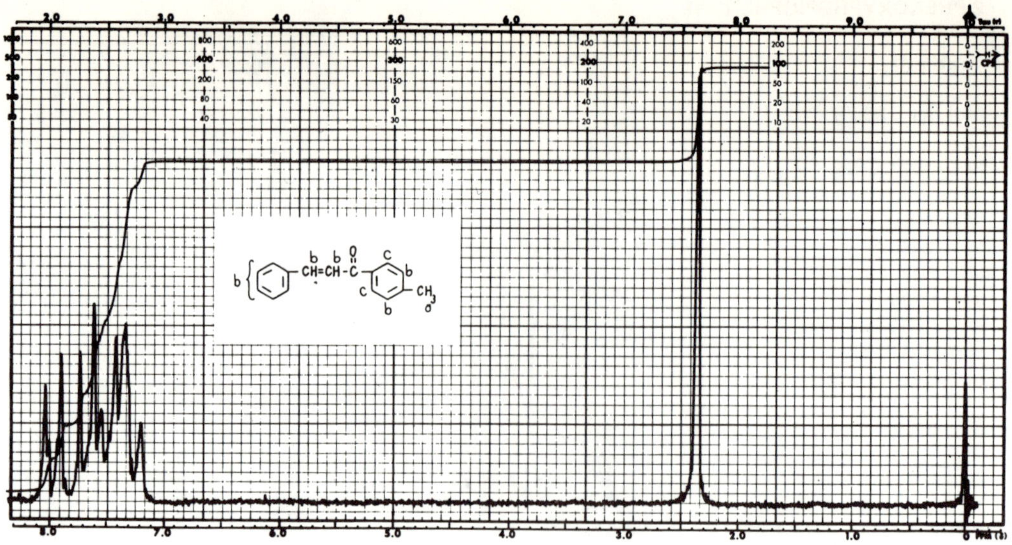

© 1978 Sadtler Research Laboratories, Inc.

4'-BROMOCHALCONE

$C_{15}H_{11}BrO$

Mol. Wt. 287.16

Solvent: $CDCl_3$

Assignments
a = 7.20-8.00

2094

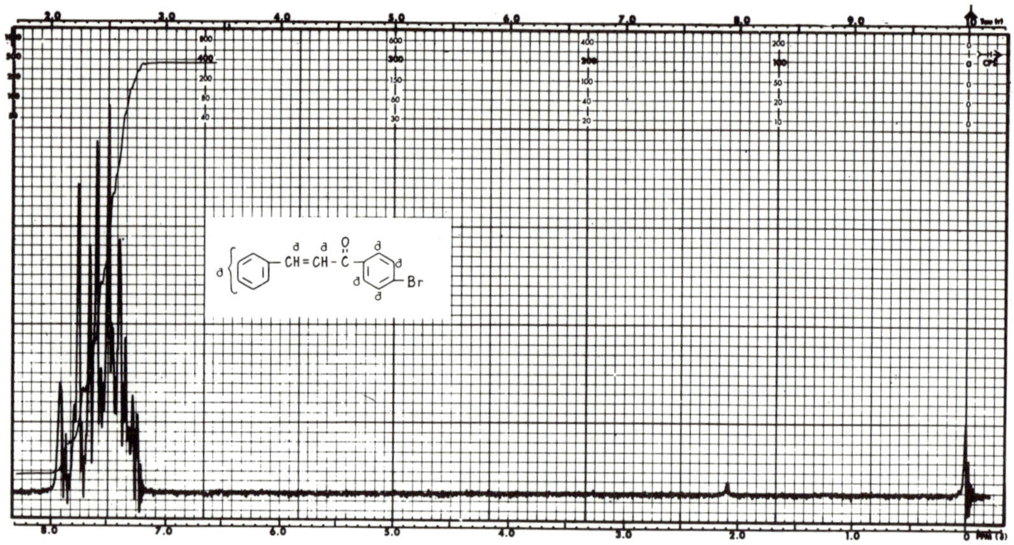

© 1978 Sadtler Research Laboratories, Inc.

3-NITROCHALCONE

$C_{15}H_{11}NO_3$

Mol. Wt. 253.26

M.P. 143-144°C

Solvent: Polysol-d

Assignments
a = 7.30-7.70
b = ca 7.79
c = 8.00-8.30
d = 8.52
e = 2.97 H_2O

2095

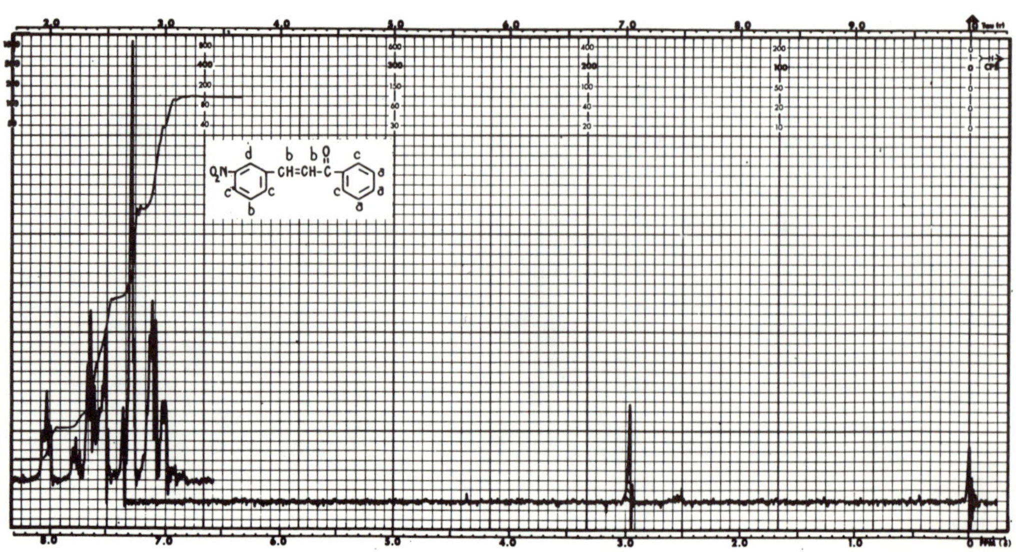

© 1978 Sadtler Research Laboratories, Inc.

cis-4′-METHOXYCHALCONE

$C_{16}H_{14}O_2$

Mol. Wt. 238.29

M.P. 104-106°C

Solvent: Polysol-d

Assignments
a = 3.89
b = 7.03
c = 7.30-7.55
d = ca 7.77
e = 8.11
f = 2.52 DMSO-d_5
g = 8.12 imp.

2096

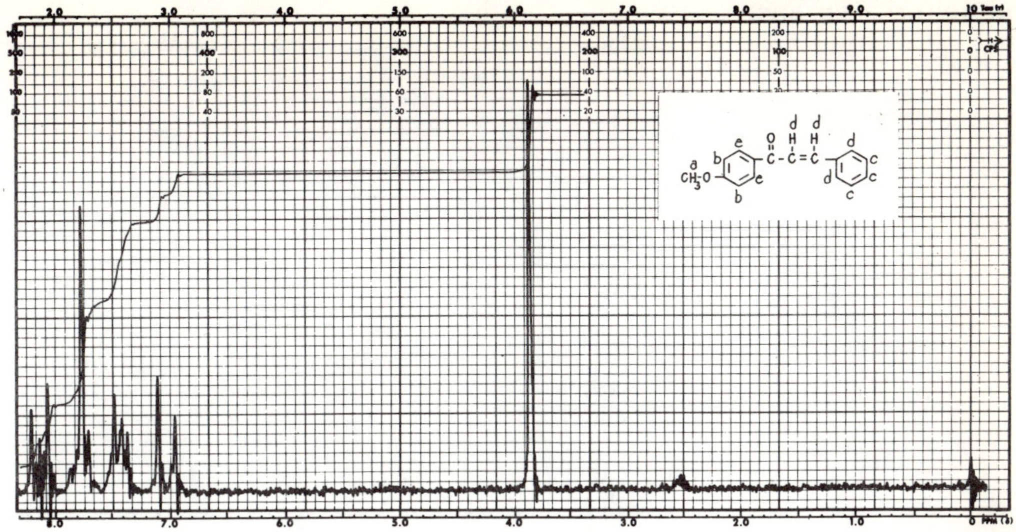

© 1978 Sadtler Research Laboratories, Inc.

---

trans-1,4-DIPHENYL-2-BUTENE-1,4-DIONE

$C_{16}H_{12}O_2$

Mol. Wt. 236.27

M.P. 109-110°C

Solvent: $CDCl_3$

2097

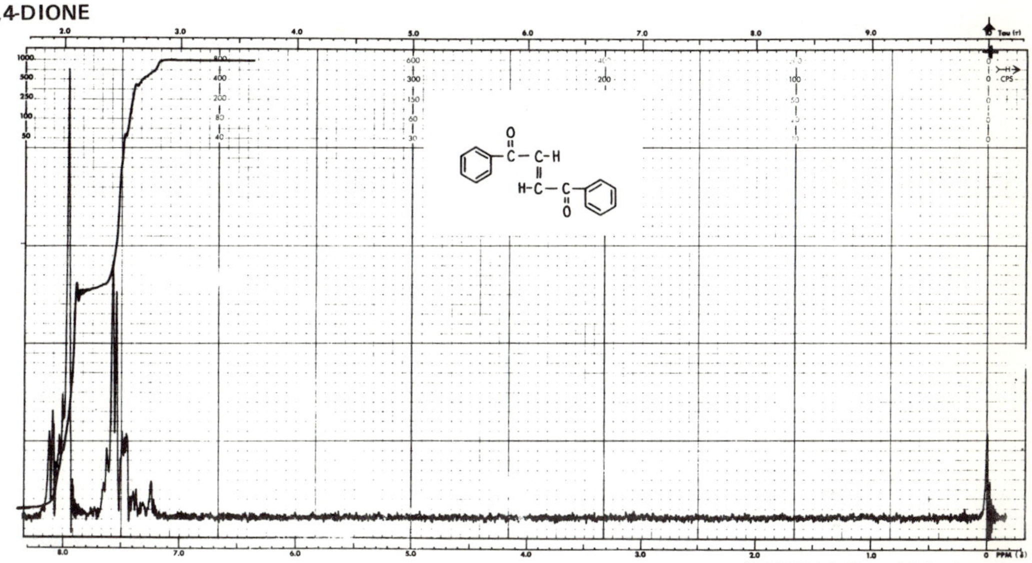

© 1978 Sadtler Research Laboratories, Inc.

α-BROMOTOLUENE

C_7H_7Br

Mol. Wt. 171.04

B.P. 197-199°C

Solvent: CCl_4

Assignments
a = 4.34
b = 7.22

2098

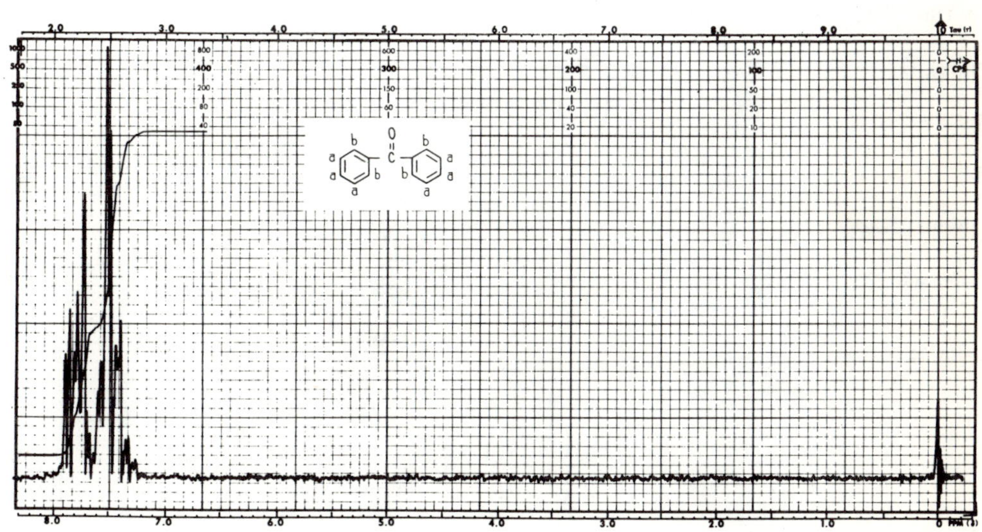

© 1978 Sadtler Research Laboratories, Inc.

2-METHYLBENZOPHENONE

$C_{14}H_{12}O$

Mol. Wt. 196.25

Solvent: CCl_4

Assignments
a = 2.29
b = ca 7.19
c = 7.20-7.50
d = 7.70

2099

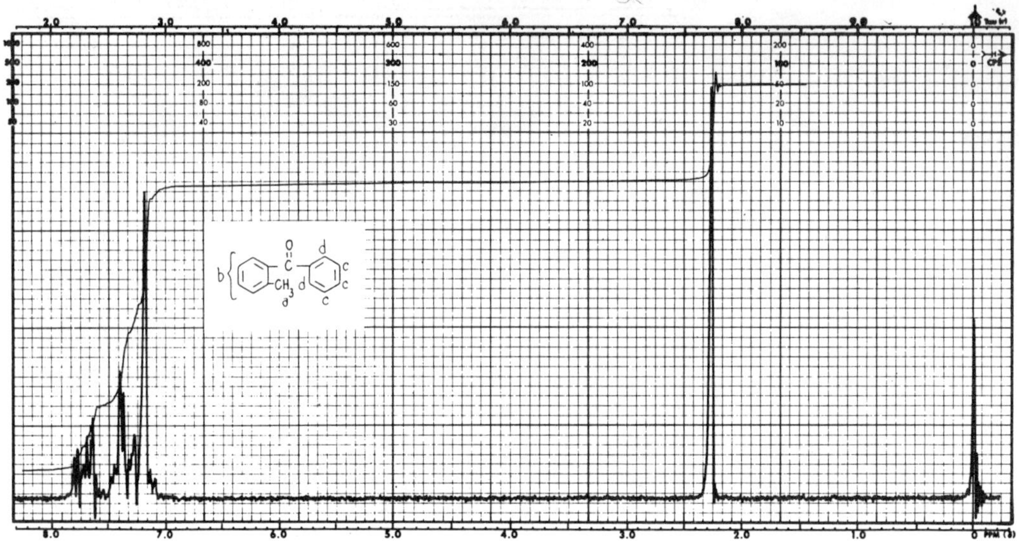

© 1978 Sadtler Research Laboratories, Inc.

4-METHYLBENZOPHENONE

$C_{14}H_{12}O$

Mol. Wt. 196.25

Solvent: $CDCl_3$

Assignments
a = 2.39
b = 7.22
c = 7.32-7.84

2100

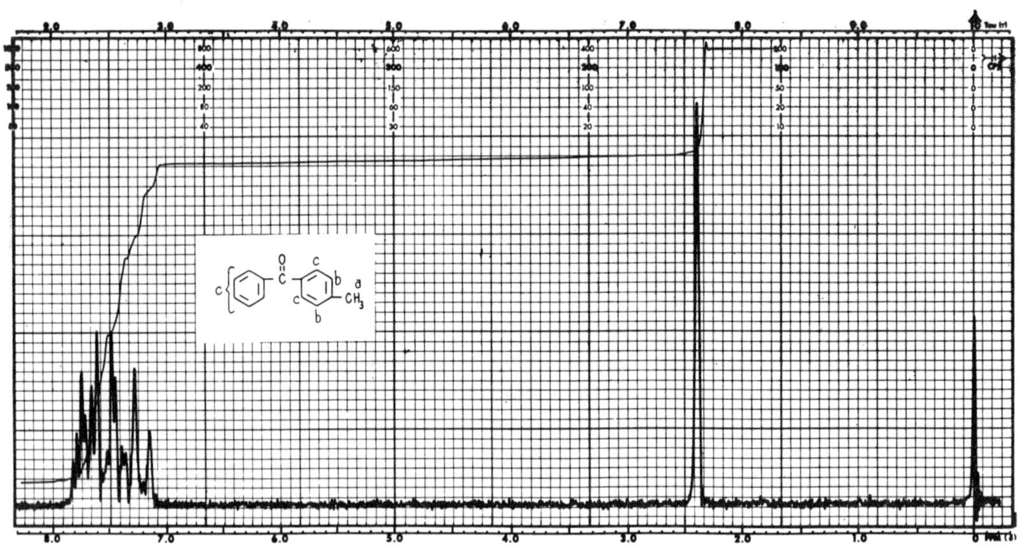

© 1978 Sadtler Research Laboratories, Inc.

4,4'-DIMETHYLBENZOPHENONE

$C_{15}H_{14}O$

Mol. Wt. 210.28

M.P. 95°C

Solvent: $CDCl_3$

Assignments
a = 2.36
b = 7.20
c = 7.64

2101

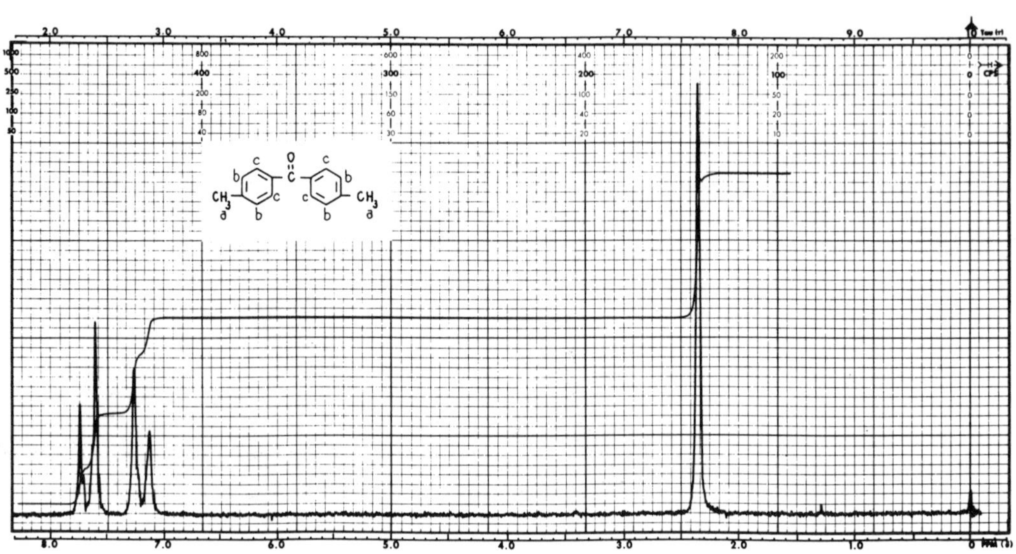

© 1978 Sadtler Research Laboratories, Inc.

4-BROMOBENZOPHENONE

$C_{13}H_9BrO$

Mol. Wt. 261.13

Solvent: $CDCl_3$

Assignments
a = 7.30-7.80
b = 7.60

2102

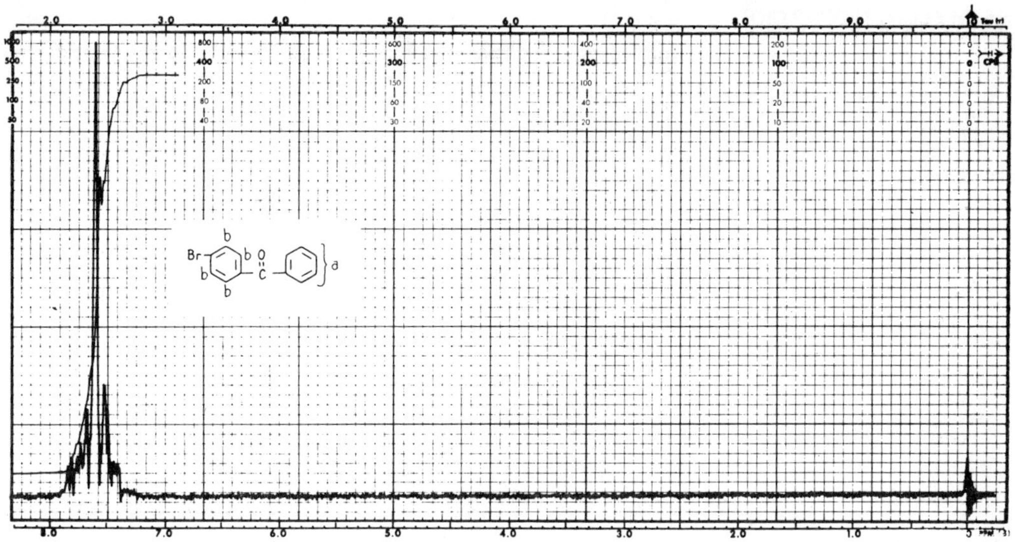

© 1978 Sadtler Research Laboratories, Inc.

3-AMINOBENZOPHENONE

$C_{13}H_{11}NO$

Mol. Wt. 197.24

Solvent: DMSO-d_6

Assignments
a = 5.39
b = 6.88
c = 6.70-7.30
d = 7.30-7.90
e = 2.50 DMSO-d_5
f = 3.47 H_2O

2103

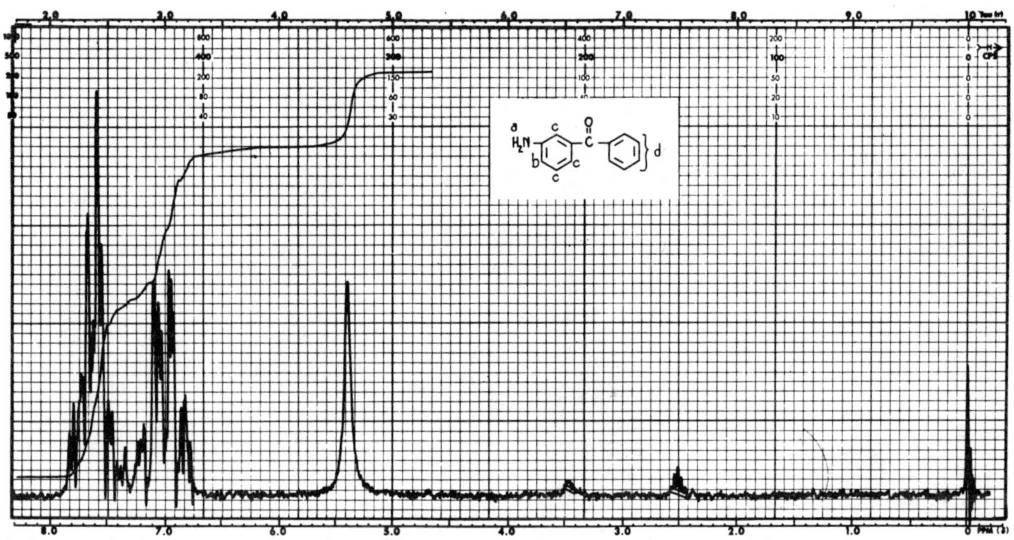

© 1978 Sadtler Research Laboratories, Inc.

4,4'-BIS(DIETHYLAMINO)BENZOPHENONE

$C_{21}H_{28}N_2O$

Mol. Wt. 324.47

M.P. 88-91°C

Solvent: $CDCl_3$

Assignments
a = 1.20
b = 3.41
c = 6.66
d = 7.75

2104

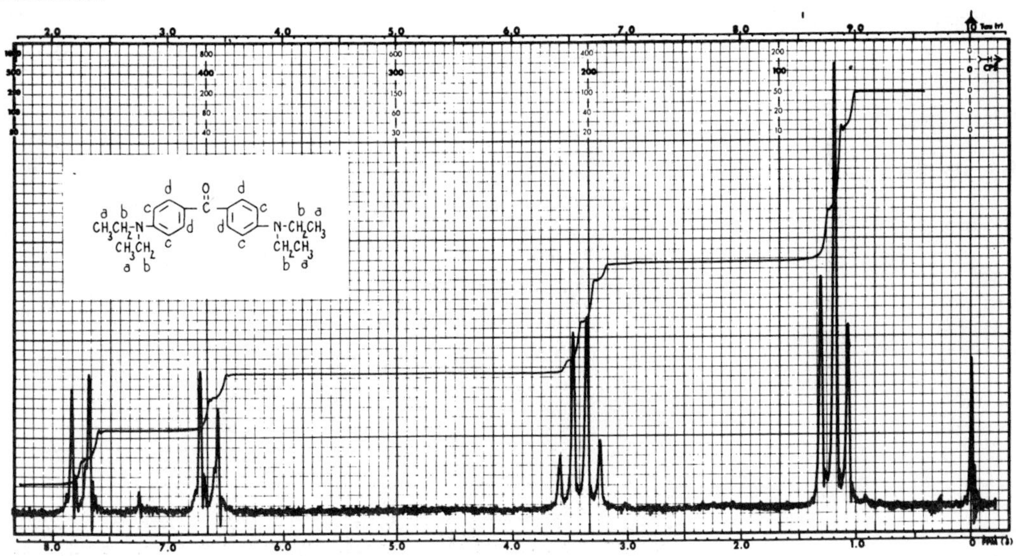

© 1978 Sadtler Research Laboratories, Inc.

2-HYDROXY-4'-METHYLBENZOPHENONE

$C_{14}H_{12}O_2$

Mol. Wt. 212.25

M.P. 40-41°C

Solvent: $CDCl_3$

Assignments

a = 2.38
b = 6.78
c = 6.99
d = 7.20
e = 7.34-7.69
f = 7.51
g = 12.00

2105

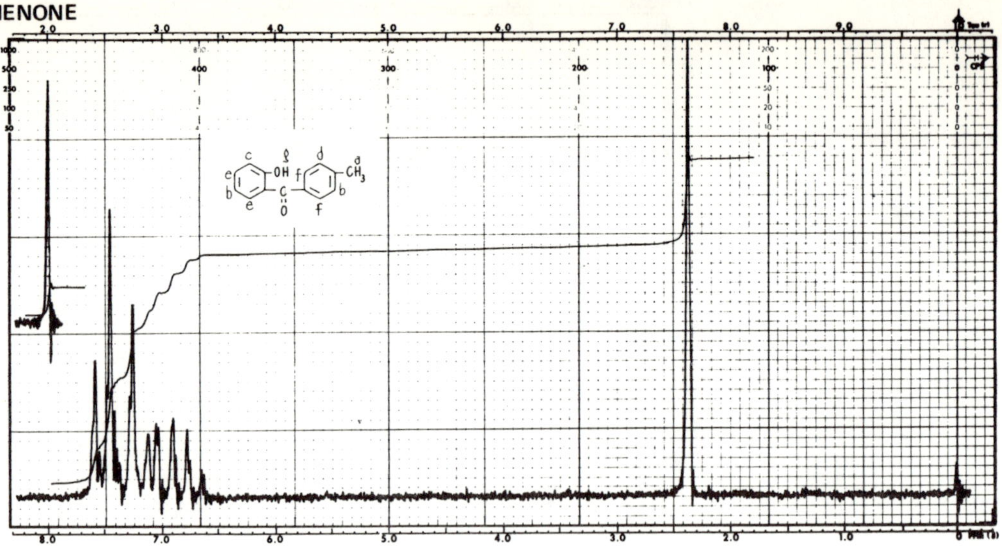

© 1978 Sadtler Research Laboratories, Inc.

m-DIBENZOYLBENZENE

$C_{20}H_{14}O_2$

Mol. Wt. 286.33

Solvent: Polysol-d

Assignments

a = 7.25-8.08
b = 8.11

2106

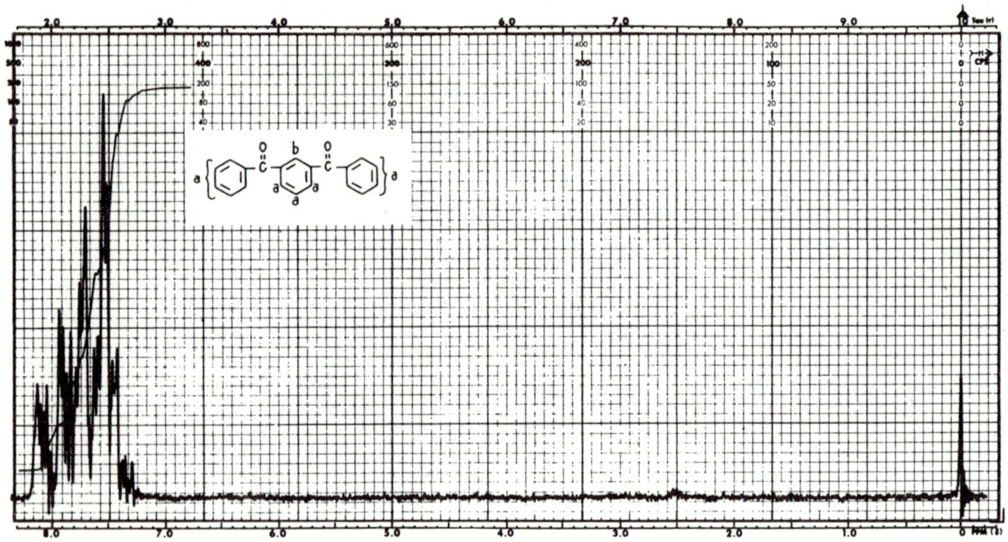

© 1978 Sadtler Research Laboratories, Inc.

p-DIBENZOYLBENZENE

$C_{20}H_{14}O_2$

Mol. Wt. 286.33

M.P. 158-159°C (lit.)

Solvent: $CDCl_3$

Assignments

a = 7.32-7.72
b = 7.88
c = 7.92

2107

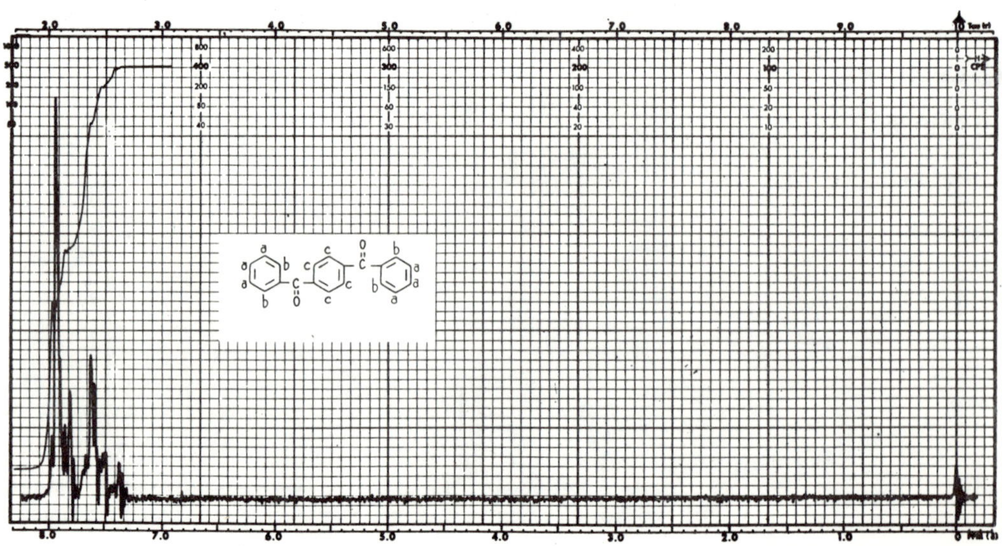

© 1978 Sadtler Research Laboratories, Inc.

5-INDANYL METHYL KETONE

$C_{11}H_{12}O$

Mol. Wt. 160.22

B.P. 132-136°C/7mm

Solvent: $CDCl_3$

Assignments
a = 2.03
b = 2.48
c = 2.89
d = 7.21
e = 7.70
f = 7.76

2108

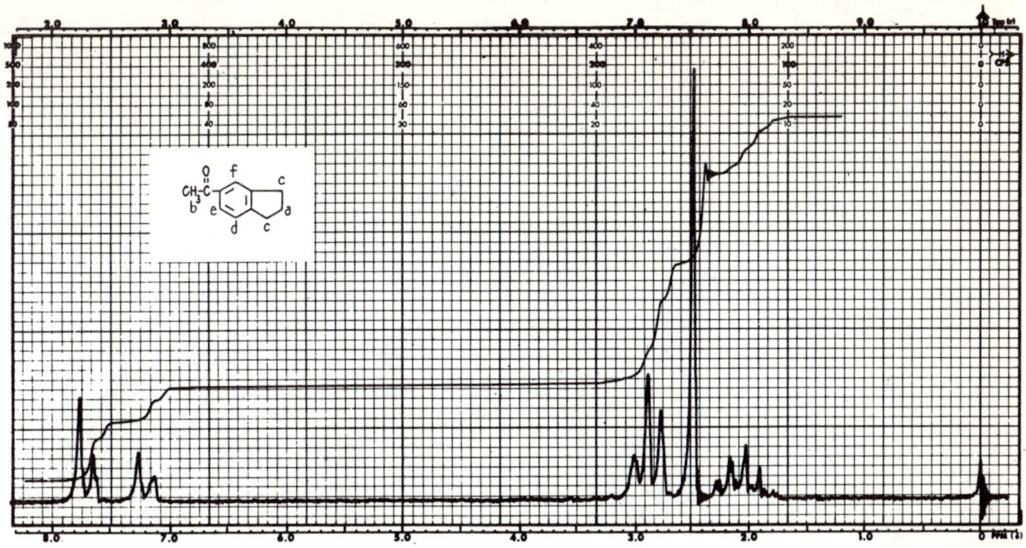

© 1978 Sadtler Research Laboratories, Inc.

1,4-BENZODIOXAN-6-YL METHYL KETONE

$C_{10}H_{10}O_3$

Mol. Wt. 178.19

Solvent: $CDCl_3$

Assignments
a = 2.49
b = 4.27
c = 6.82
d = ca 7.41

2109

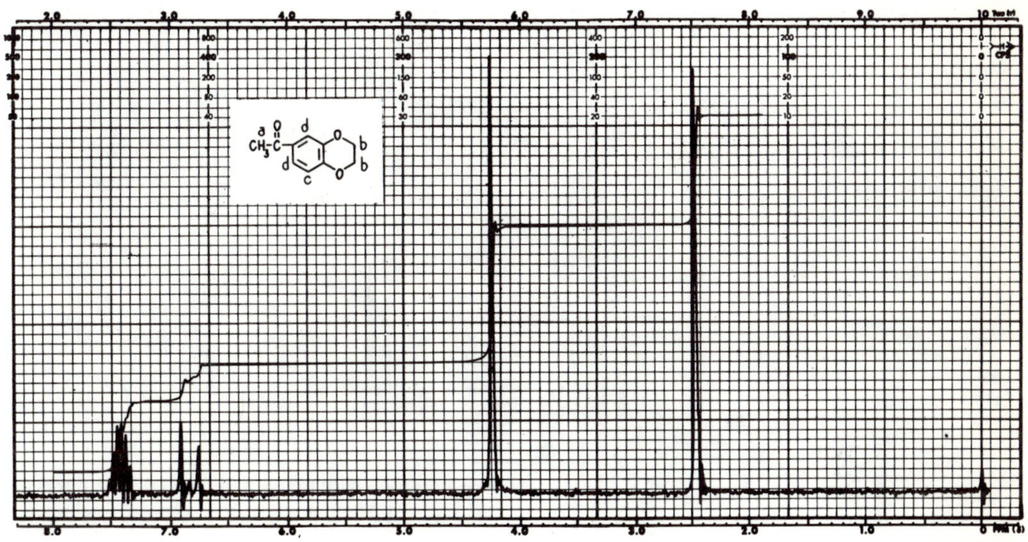

© 1978 Sadtler Research Laboratories, Inc.

METHYL 2-PYRIDYL KETONE

C_7H_7NO

Mol. Wt. 121.14

Solvent: CCl_4

Assignments
a = 2.62
b = 7.39
c = 7.81
d = 7.90
f = 8.71

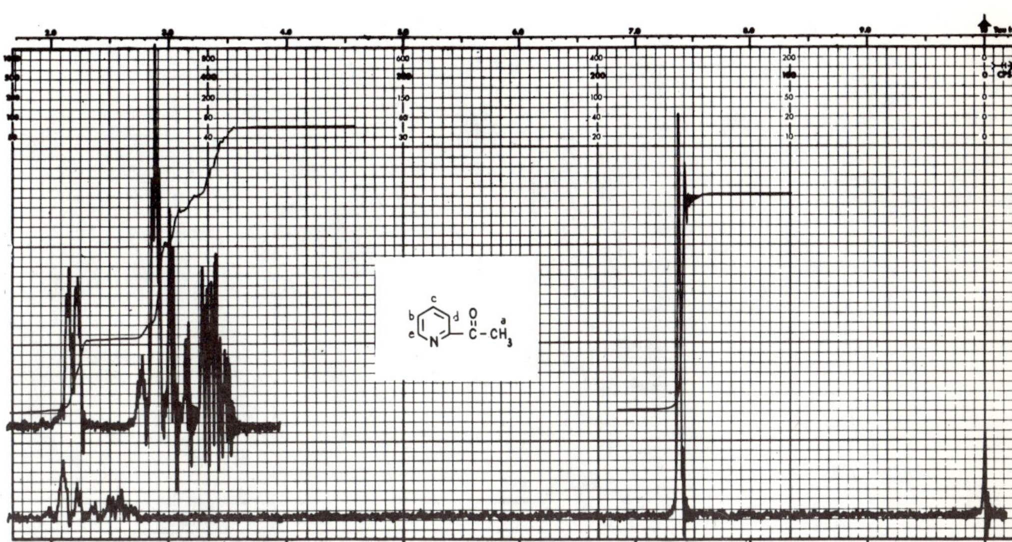

© 1978 Sadtler Research Laboratories, Inc.

METHYL 6-METHYL-2-PYRIDYL KETONE

C_8H_9NO

Mol. Wt. 135.17

Solvent: $CDCl_3$

Assignments
a = 2.60
b = 2.69
c = 7.30
d = 7.69
e = 7.81

2111

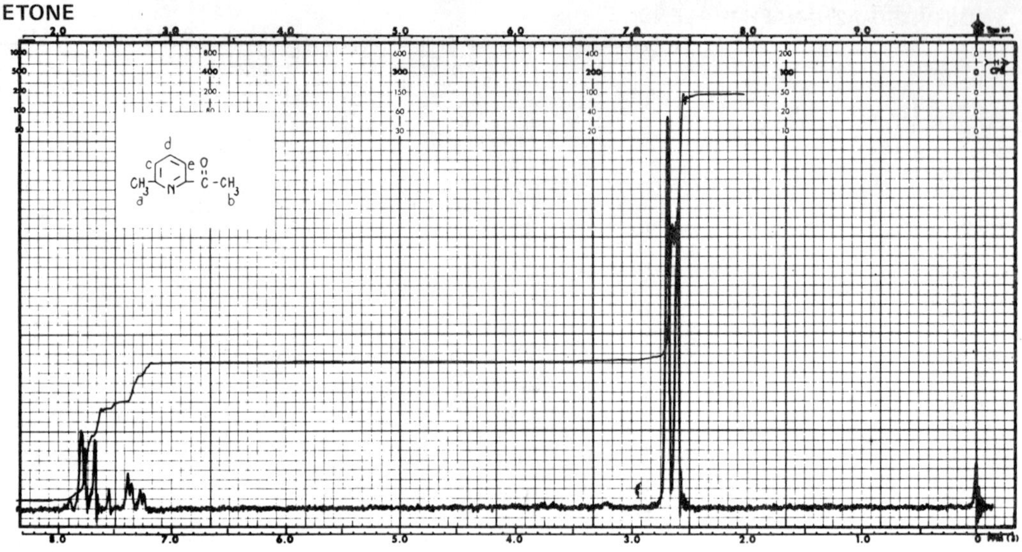

© 1978 Sadtler Research Laboratories, Inc.

PHENYL 3-PYRIDYL KETONE

$C_{12}H_9NO$

Mol. Wt. 183.21

M.P. 36-40°C

Solvent: $CDCl_3$

Assignments
a = 7.18-7.69
b = 7.80
c = 8.08
d = 8.78
e = 8.97
f = 3.32 imp.

2112

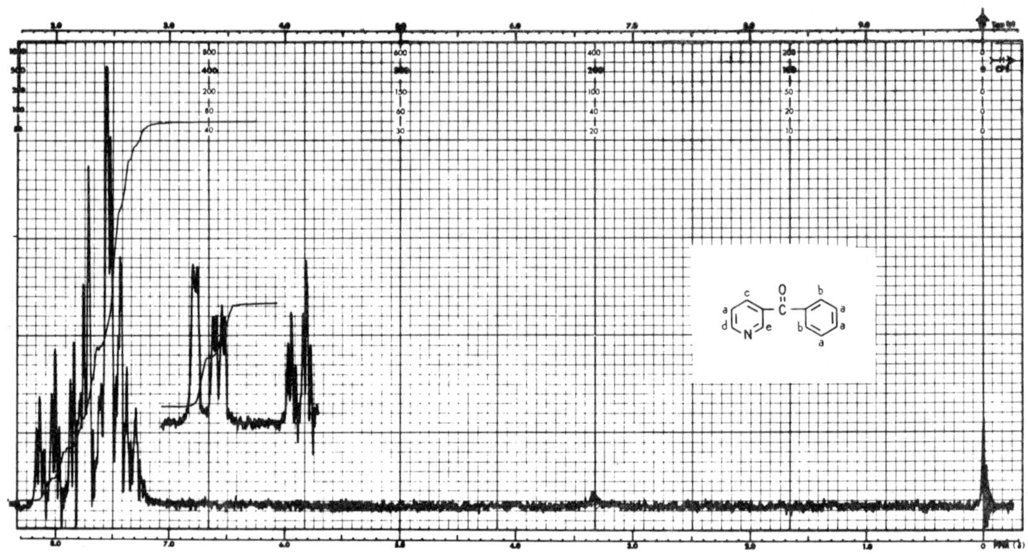

© 1978 Sadtler Research Laboratories, Inc.

2'-ACETONAPHTHONE

$C_{12}H_{10}O$

Mol. Wt. 170.21

M.P. 53-55°C

Solvent: CCl_4

Assignments
a = 2.60
b = 7.30-8.40

2113

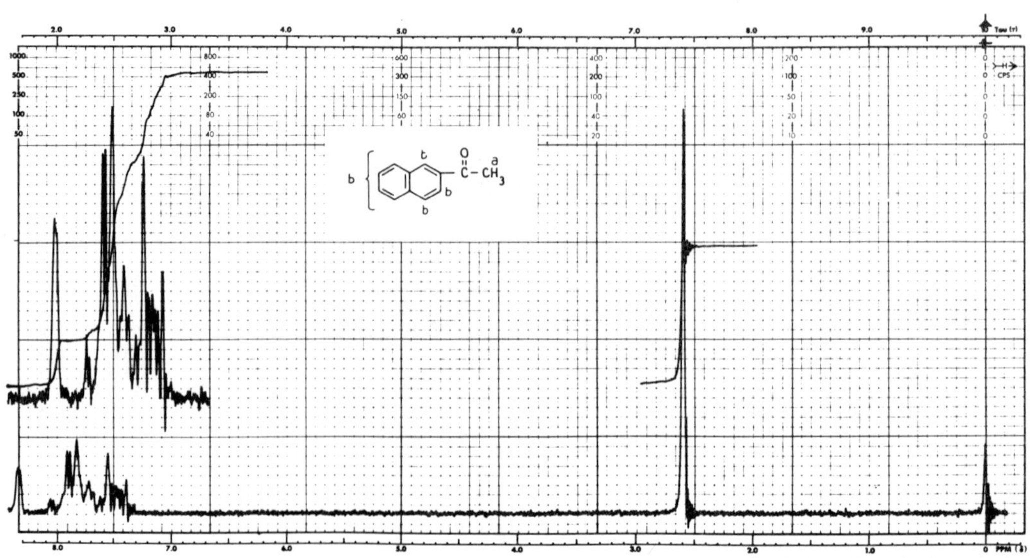

© 1978 Sadtler Research Laboratories, Inc.

3,4-DIHYDRO-1(2H)-NAPHTHALENONE

$C_{10}H_{10}O$

Mol. Wt. 146.19

Solvent: CCl_4

Assignments
a = 2.07
b = 2.52
c = 2.90
d = 7.00-7.50
e = 7.89

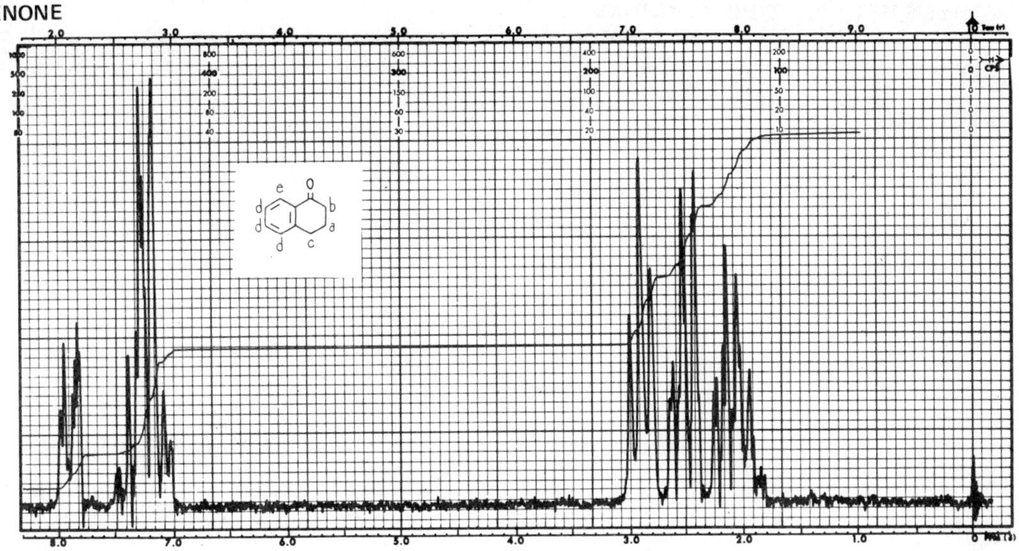

2114

© 1978 Sadtler Research Laboratories, Inc.

3,4-DIHYDRO-7-ETHYL-1(2H)NAPHTHALENONE

$C_{12}H_{14}O$

Mol. Wt. 174.24

B.P. 142-144°C/6mm

Solvent: $CDCl_3$

Assignments
a = 1.22
b = 2.10
c = 2.40-3.10
d = 7.14
e = 7.31
f = 7.89

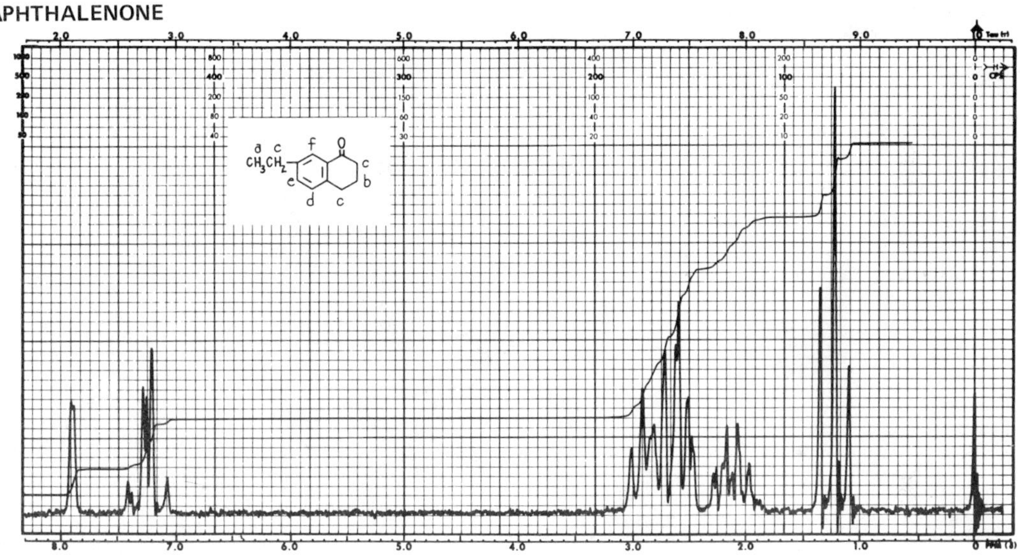

2115

© 1978 Sadtler Research Laboratories, Inc.

3,4-DIHYDRO-7-METHOXY-2(1H)-NAPHTHALENONE

$C_{11}H_{12}O_2$

Mol. Wt. 176.22

Solvent: $CDCl_3$

Assignments
a = 2.49
b = 2.97
c = 3.51
d = 3.78
e = 6.64
f = 6.71
g = 7.11
h = 3.87, 7.56 imps.

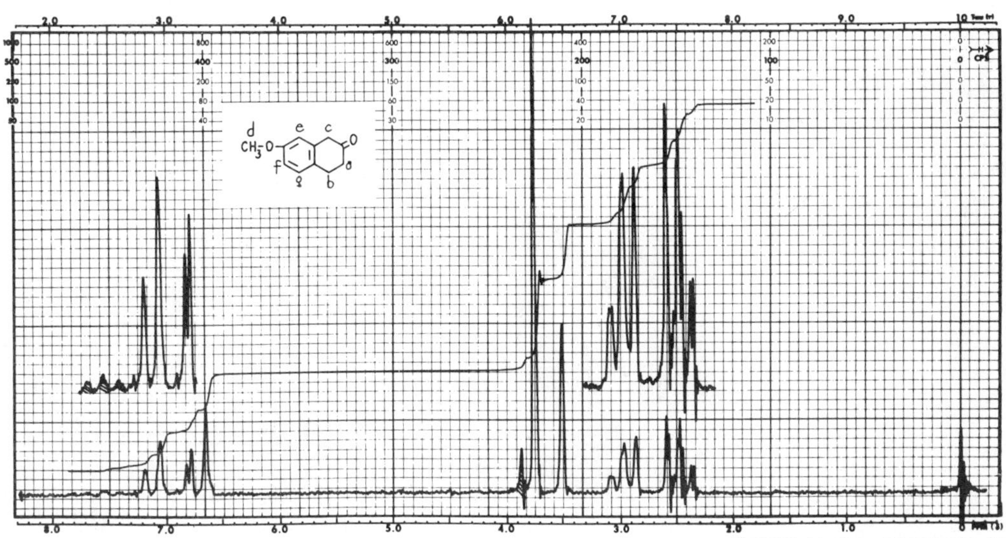

2116

© 1978 Sadtler Research Laboratories, Inc.

6-METHYL-4-CHROMANONE

$C_{10}H_{10}O_2$

Mol. Wt. 162.19

M.P. 33-35°C

Solvent: $CDCl_3$

Assignments

a = 2.30
b = 2.75
c = 4.48
d = 6.82
e = 7.28
f = 7.69

2117

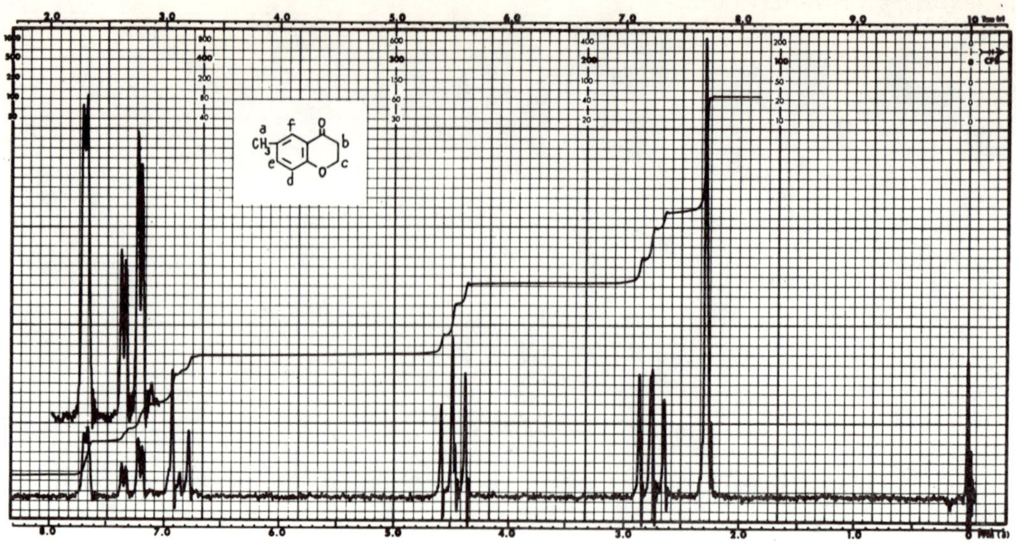

© 1978 Sadtler Research Laboratories, Inc.

6-CHLORO-4-CHROMANONE

$C_9H_7ClO_2$

Mol. Wt. 182.61

M.P. 102-104°C

Solvent: $CDCl_3$

Assignments

a = 2.77
b = 4.53
c = 6.89
d = 7.35
e = 7.74

2118

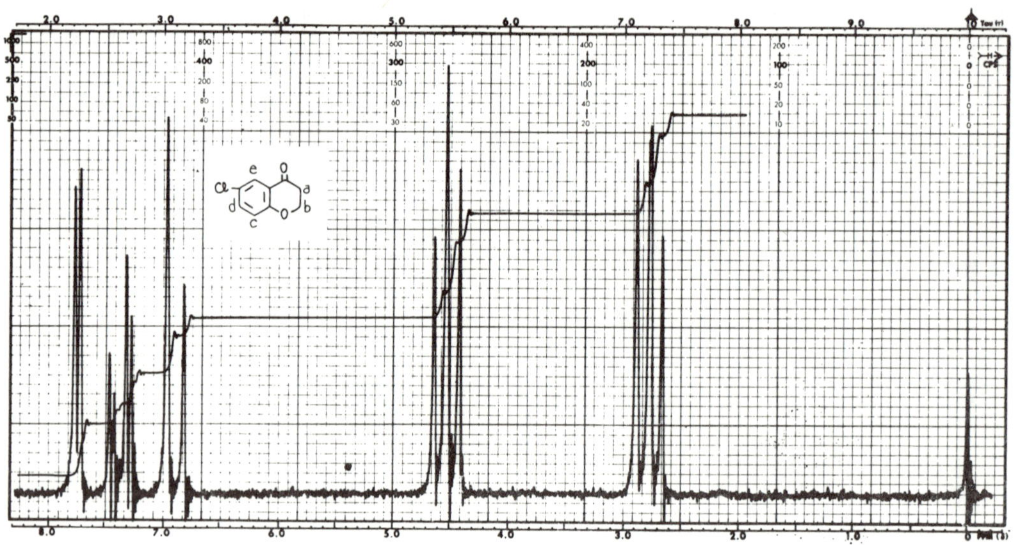

© 1978 Sadtler Research Laboratories, Inc.

2-METHYL-1,4-NAPHTHOQUINONE

$C_{11}H_8O_2$

Mol. Wt. 172.19

M.P. 104-106°C

Solvent: $CDCl_3$

Assignments

a = 2.15
b = 6.76
c = ca 7.63
d = ca 7.94

2119

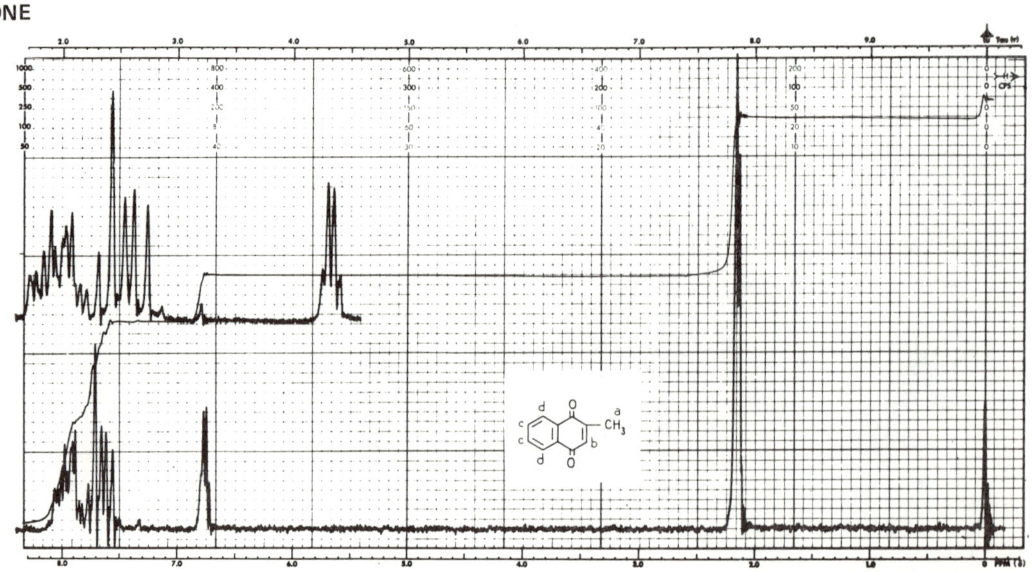

© 1978 Sadtler Research Laboratories, Inc.

2-tert-BUTYL-1,4-NAPHTHOQUINONE

$C_{14}H_{14}O_2$

Mol. Wt. 214.27

M.P. 73-76°C

Solvent: $CDCl_3$

Assignments
a = 1.39
b = 6.85
c = 7.59-7.87
d = 7.91-8.20

2120

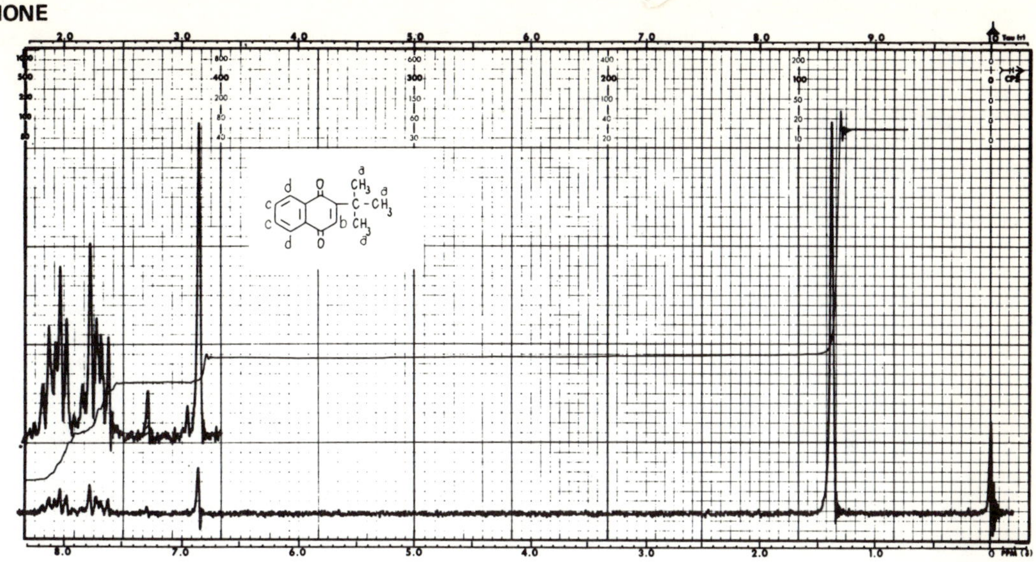

© 1978 Sadtler Research Laboratories, Inc.

2-CYCLOHEXYL-1,4-NAPHTHOQUINONE

$C_{16}H_{16}O_2$

Mol. Wt. 240.30

Solvent: $CDCl_3$

Assignments
a = 0.83-2.17
b = 2.90
c = 6.67
d = ca 7.68
e = ca 7.96

2121

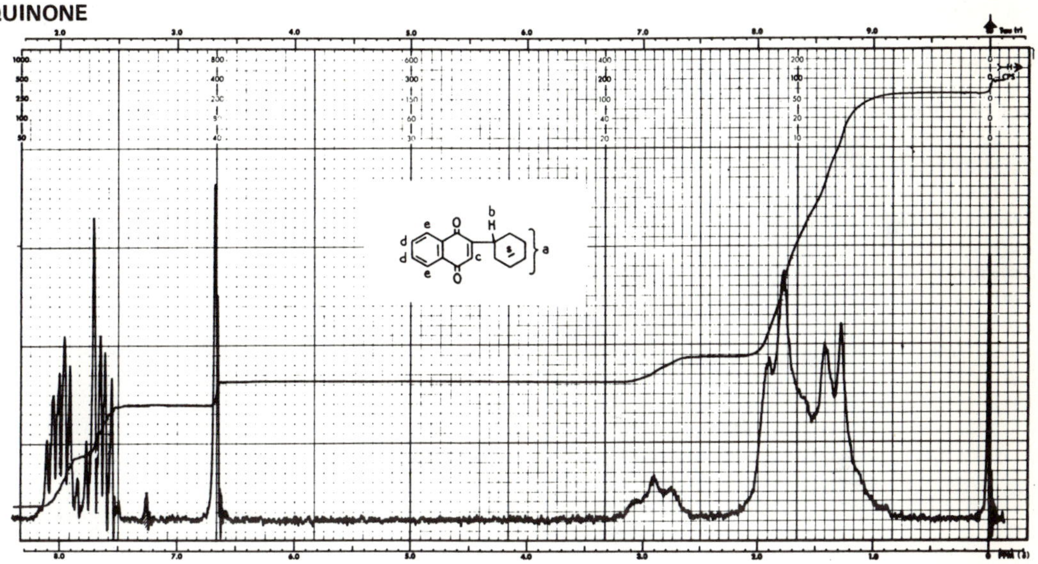

© 1978 Sadtler Research Laboratories, Inc.

9-FLUORENONE

$C_{13}H_8O$

Mol. Wt. 180.19

M.P. 83-83.5°C

Solvent: CCl_4

Assignments
a = 7.05-7.70

2122

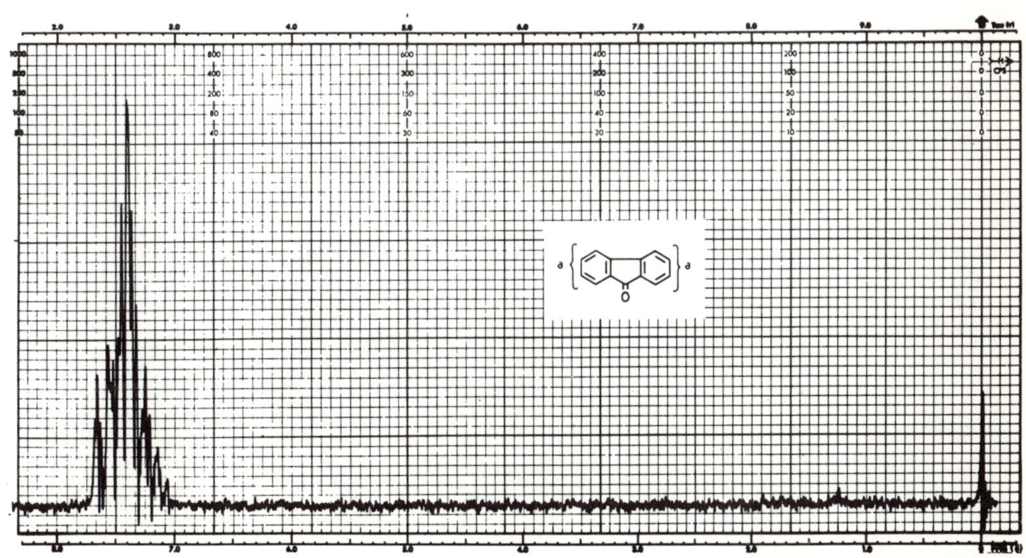

© 1978 Sadtler Research Laboratories, Inc.

2-FLUORO-9-FLUORENONE

$C_{13}H_7NO$

Mol. Wt. 198.20

Solvent: $CDCl_3$

Assignments
a = 7.01
b = 7.10-7.65

2123

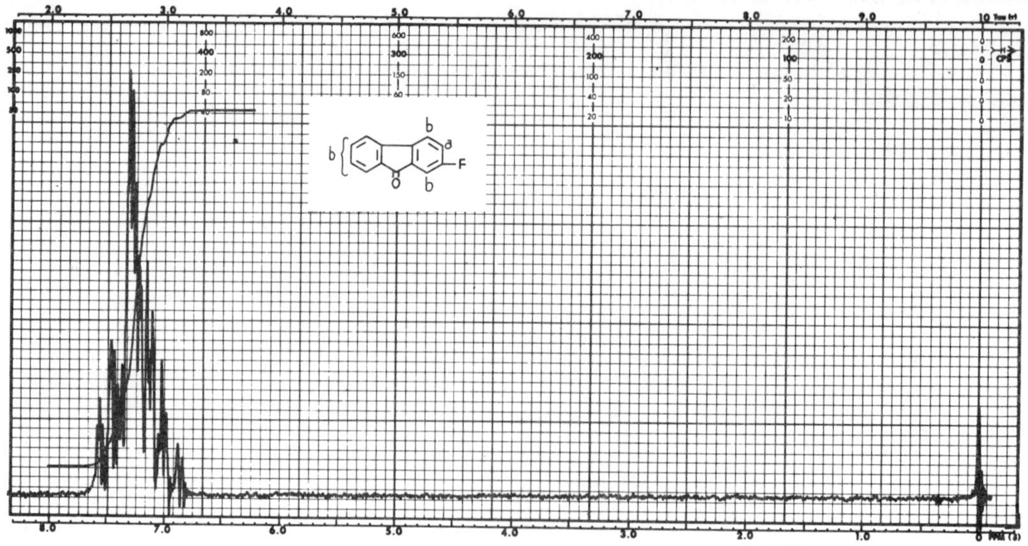

© 1978 Sadtler Research Laboratories, Inc.

ANTHRONE

$C_{14}H_{10}O$

Mol. Wt. 194.22

M.P. 154-155°C

Solvent: $CDCl_3$

Assignments
a = 4.31
b = ca 7.53
c = 7.88

2124

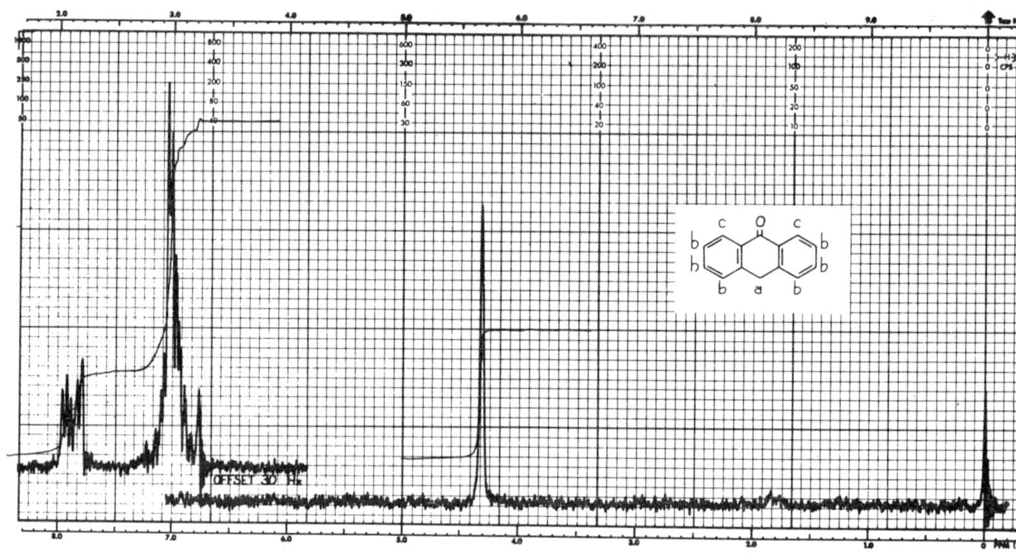

© 1978 Sadtler Research Laboratories, Inc.

5H-DIBENZO[a,d] CYCLOHEPTEN-5-ONE

$C_{15}H_{10}O$

Mol. Wt. 206.24

M.P. 88-89°C

Solvent: CCl_4

Assignments
a = 6.85
b = 7.40
c = 8.12

2125

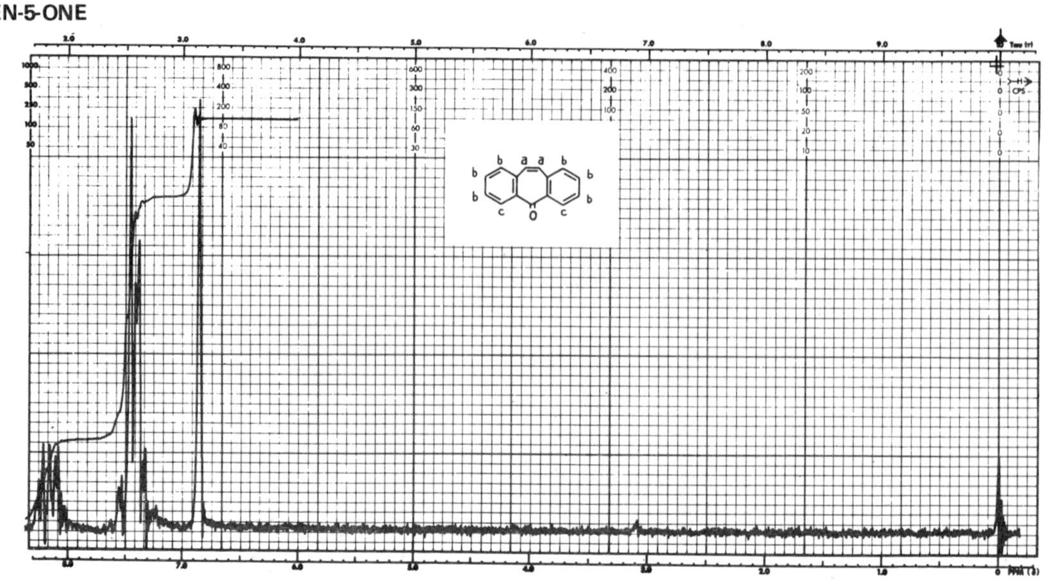

© 1978 Sadtler Research Laboratories, Inc.

2-METHYLANTHRAQUINONE

$C_{15}H_{10}O$

Mol. Wt. 222.25

M.P. 174-176°C

Solvent: CCl_4

Assignments
a = 2.45
b = 7.29-8.38

2126

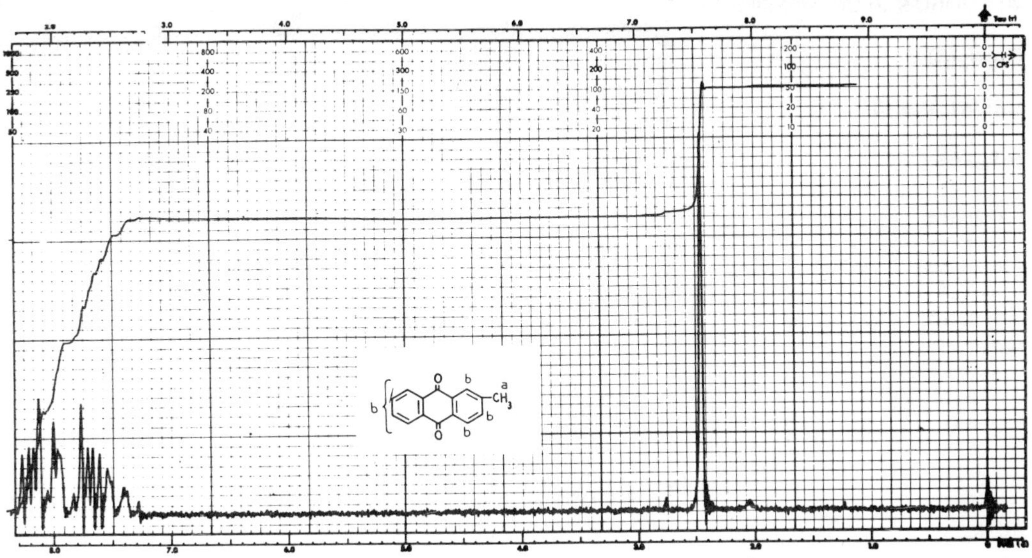

© 1978 Sadtler Research Laboratories, Inc.

1,4-DIMETHYLANTHRAQUINONE

$C_{16}H_{12}O_2$

Mol. Wt. 236.27

M.P. 133-135°C

Solvent: $CDCl_3$

Assignments
a = 2.66
b = 7.21
c = 7.61
d = 8.03
e = 2.24, 2.50 imps.

2127

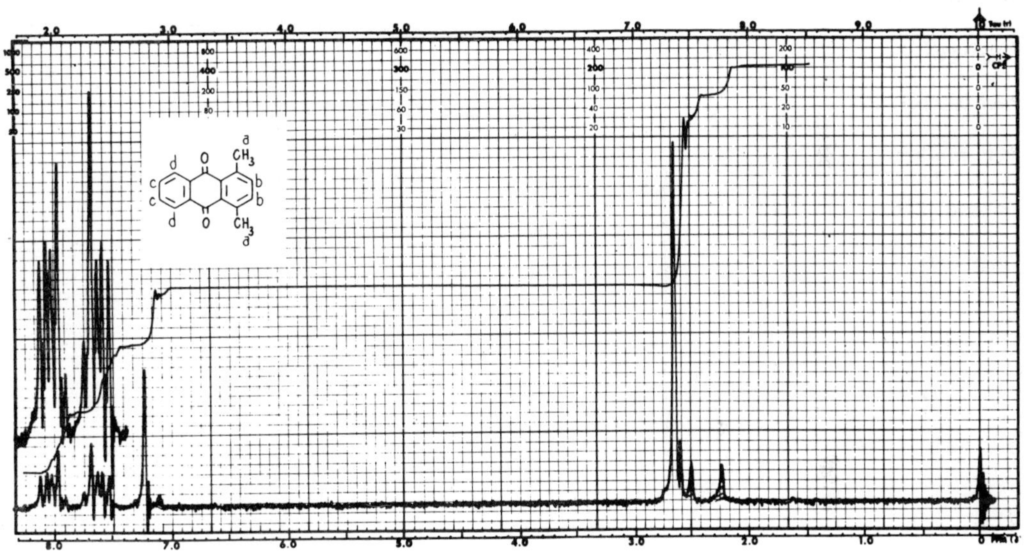

© 1978 Sadtler Research Laboratories, Inc.

2,3-DIMETHYLANTHRAQUINONE

$C_{16}H_{12}O_2$

Mol. Wt. 236.27

M.P. 211°C

Solvent: $CDCl_3$

Assignments
a = 2.38
b = 7.78
c = 8.00
d = 8.30

2128

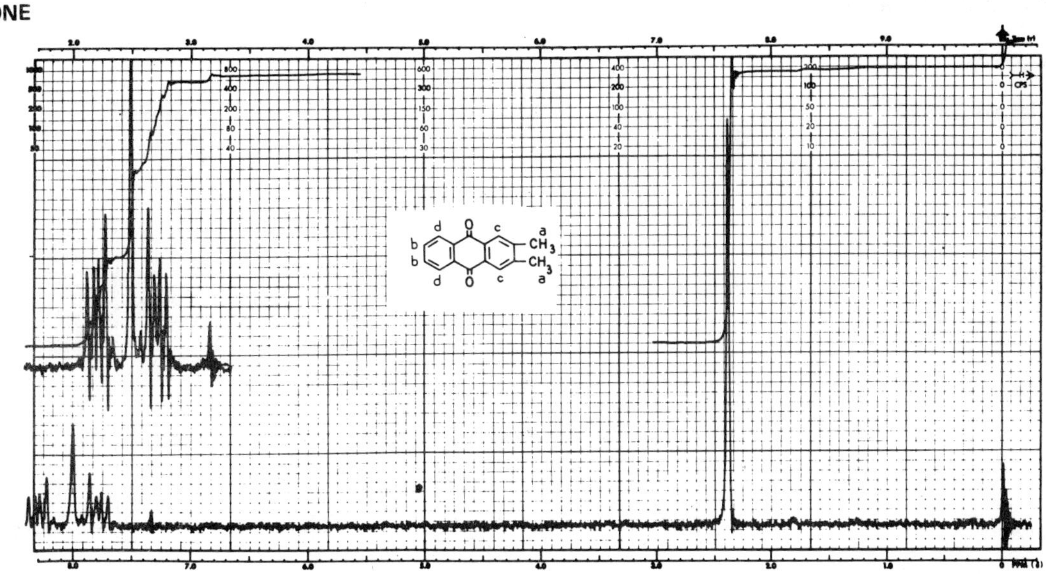

© 1978 Sadtler Research Laboratories, Inc.

3β-HYDROXYANDROSTEN-17-ONE

$C_{19}H_{28}O_2$

Mol. Wt. 288.43

M.P. 140-141°C (lit.)

Solvent: $CDCl_3$

Assignments
a = 0.90
b = 1.05
c = 1.01-2.10
d = 2.00-2.50
e = 2.92
f = 3.49
g = 5.34

2129

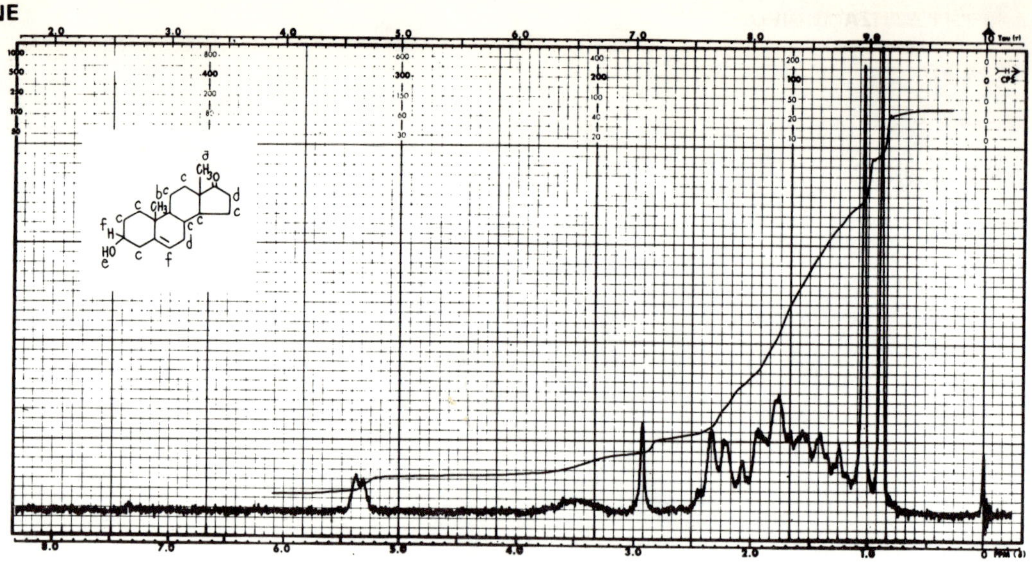

© 1978 Sadtler Research Laboratories, Inc.

CHOLEST-4-EN-3-ONE

$C_{27}H_{44}O$

Mol. Wt. 384.65

M.P. 77-79°C

Solvent: $CDCl_3$

Assignments
a = 0.72
b = 0.89
c = 0.91
d = 1.20
e = 0.80-2.10
f = 2.10-2.55
g = 5.70

2130

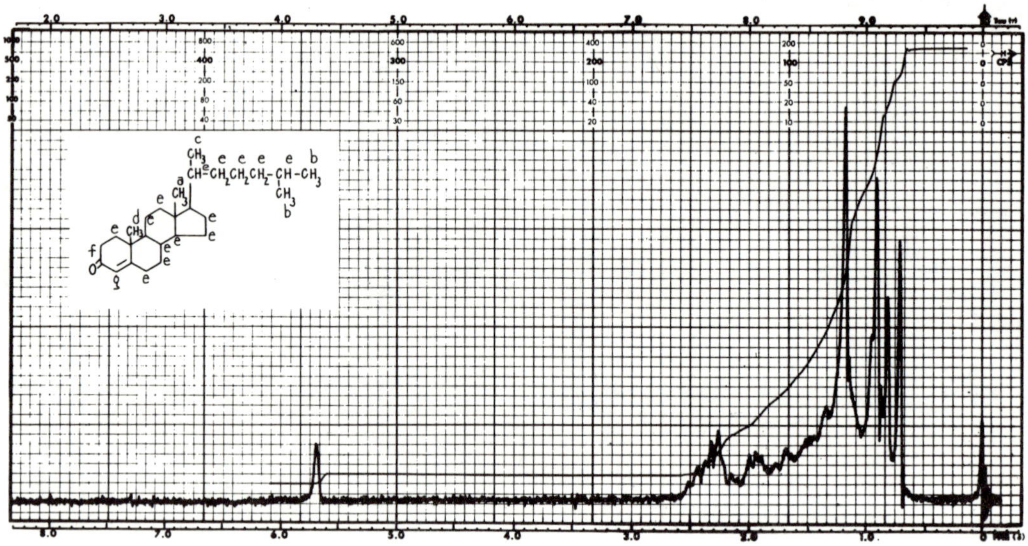

© 1978 Sadtler Research Laboratories, Inc.

2,3-PENTANEDIONE

$C_5H_8O_2$

Mol. Wt. 100.12

B.P. 107-109°C

Solvent: CCl_4

Assignments
a = 1.03
b = 2.26
c = 2.71

2131

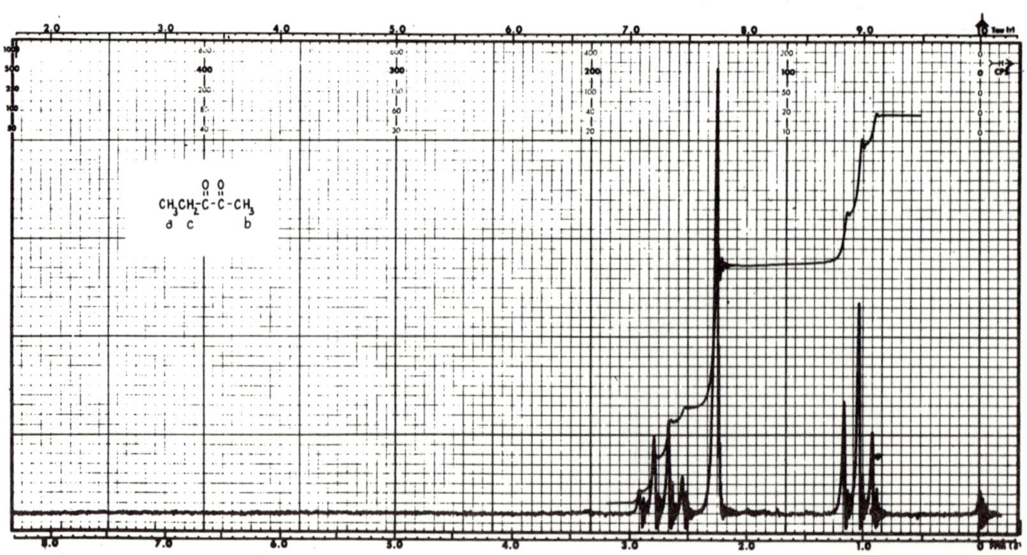

© 1978 Sadtler Research Laboratories, Inc.

d-2,3-BORNANEDIONE

$C_{10}H_{14}O_2$

Mol. Wt. 166.22

M.P. 201-203°C

Solvent: Polysol-d

Assignments
a = 0.91
b = 1.09
c = 1.40-2.40
d = 2.60

2132

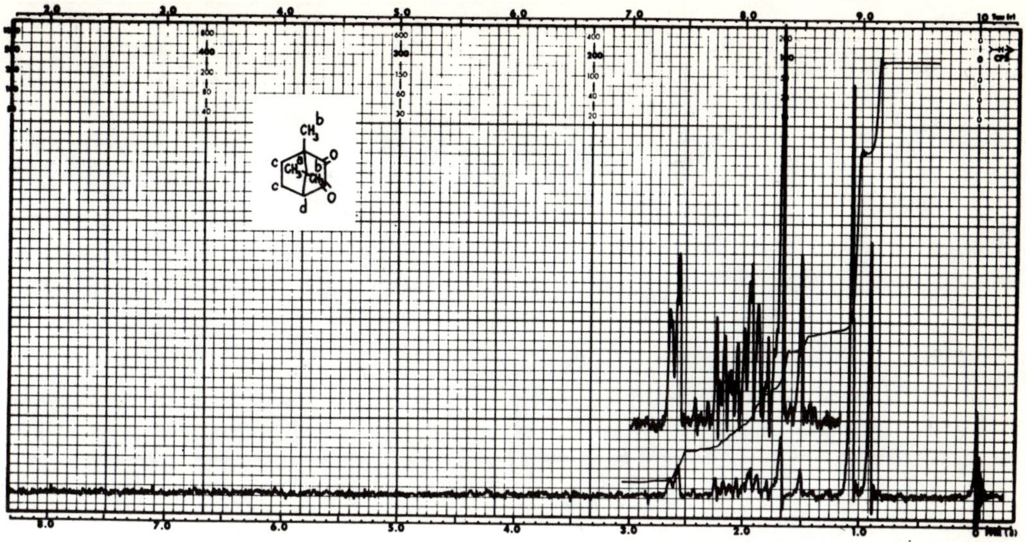

© 1978 Sadtler Research Laboratories, Inc.

FURIL

$C_{10}H_6O_4$

Mol. Wt. 190.16

M.P. 161-162°C

Solvent: $CDCl_3$

Assignments
a = 6.64
b = 7.63
c = 7.78

2133

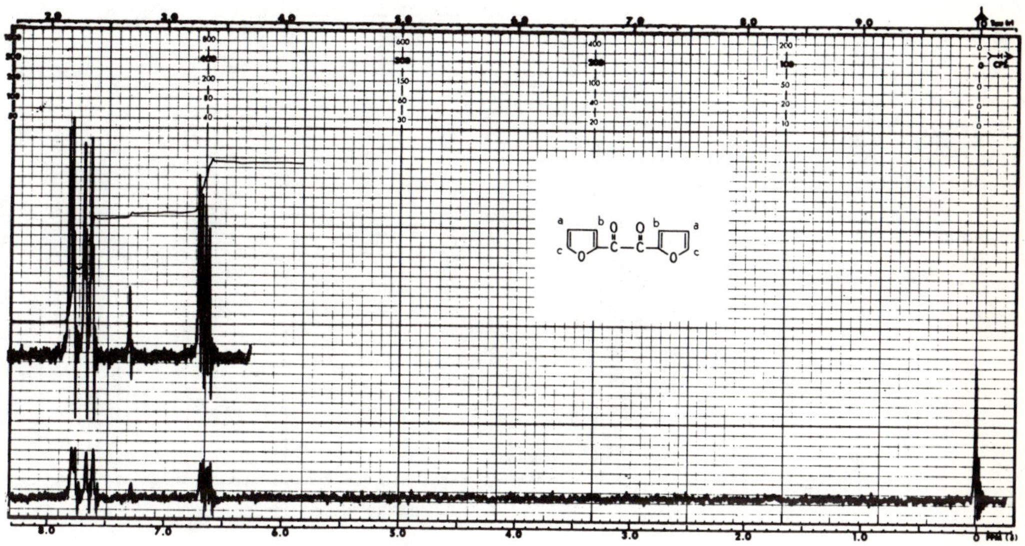

© 1978 Sadtler Research Laboratories, Inc.

BENZIL

$C_{14}H_{10}O_2$

Mol. Wt. 210.22

M.P. 95°C

B.P. 346-348°C

Solvent: CCl_4

Assignments
a = ca 7.21-7.65
b = ca 7.67-8.02

2134

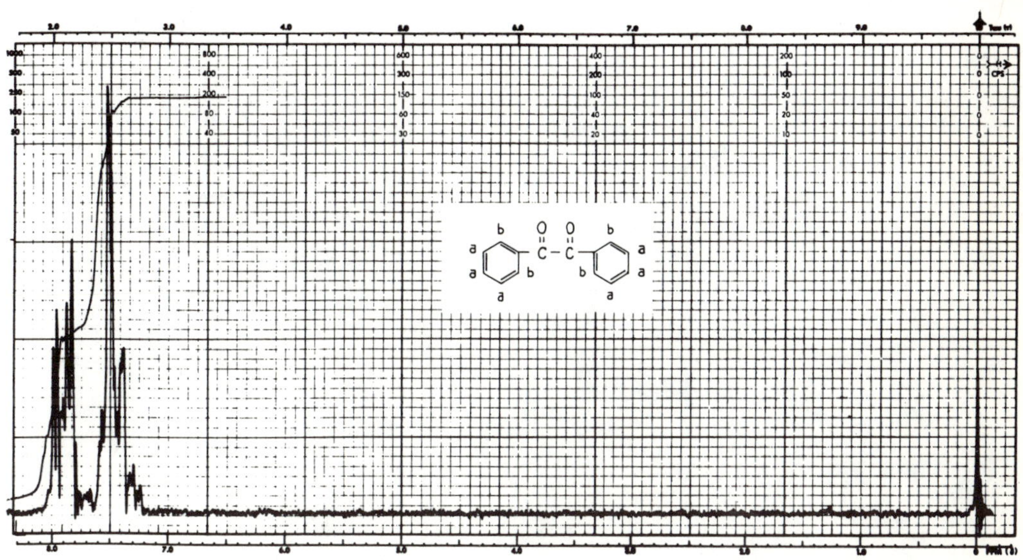

© 1978 Sadtler Research Laboratories, Inc.

p-ANISIL

$C_{16}H_{14}O_4$

Mol. Wt. 270.29

M.P. 131-132°C

Solvent: $CDCl_3$

Assignments
a = 3.83
b = 6.92
c = 7.94

2135

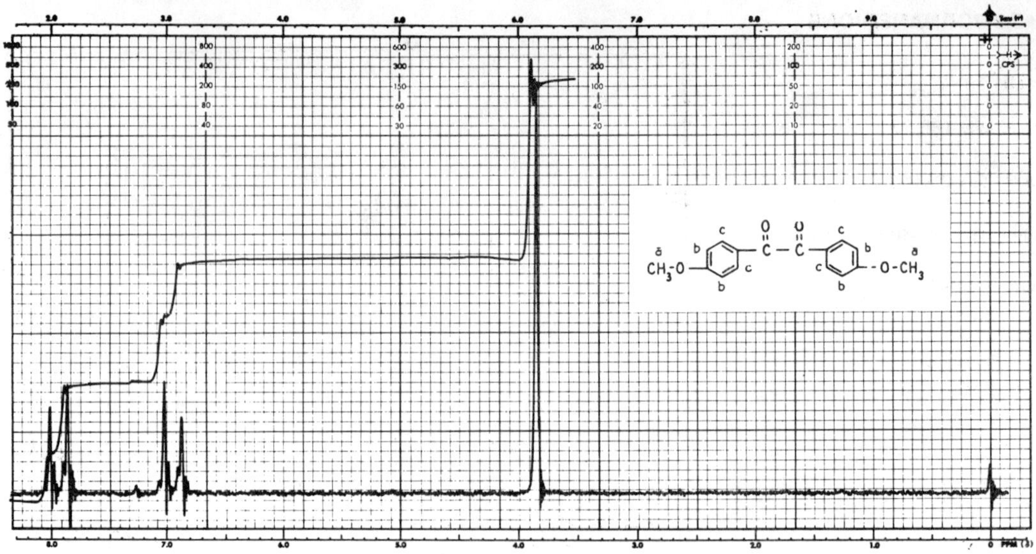

© 1978 Sadtler Research Laboratories, Inc.

ACENAPHTHENEQUINONE

$C_{12}H_6O_2$

Mol. Wt. 182.18

M.P. 259-262°C

Solvent: Trifluoroacetic acid

Assignments
a = 7.90
b = 8.10
c = 8.32

2136

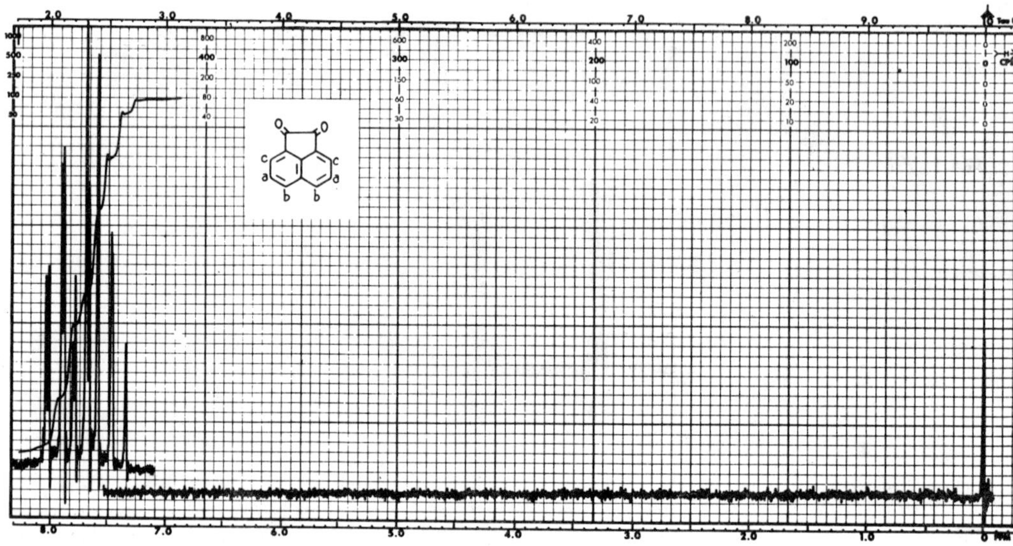

© 1978 Sadtler Research Laboratories, Inc.

The beta Diketones are unusual in that two distinct forms with different chemical shifts exist in solution. Their concentration changes with time, normally the keto form predominates when the sample is first dissolved but upon standing, the keto form increases until it becomes the form at higher concentration. The chemical shifts of aliphatic groups bonded to the diketone fragment are weakly deshielded with the aliphatic groups of the enol form resonating at slightly higher field than those of the keto form.

Because the enol hydroxyl group is often quite weak and broadened, it is often difficult to locate its resonance in the offset range below 10 ppm.

Chemical Shifts

Keto Form			Enol Form					
$CH_3-C(=O)-CH_2-C(=O)-R$			$CH_3-C(OH)=CH-C(=O)-R$				Solvent	Spectrum
1.98	3.32	$-CH_2-$Ø	1.88	15.0	5.29	$-CH_2-$Ø	CCl_4	2146
2.17	3.54	$-CH_3$	1.99	(?)	5.38	$-CH_3$	CCl_4	----
2.20	3.86	–2(Furan)	2.08	15.0	6.00	–2(Furan)	CCl_4	2141
2.24	4.03	–Ø	2.13	(?)	6.12	–Ø	$CDCl_3$	2147

Exchangeable Protons

The hydroxyl group formed during enolization resonates at lower field than nearly any other type of proton. The range of chemical shifts extends from about 11–20 ppm depending upon the structure of the beta diketone and the amount of H_2O present in the solution. The high field values are usually observed for solutions containing a relatively large percentage of H_2O.

Coupling and Coupling Constants

As with the mono-ketones discussed previously, coupling between groups on opposite sides of the carbonyl carbon is usually not observed. Similarly, coupling across the enolized carbonyl group (–C(OH)=CH) is not observed.

Solubility and Solvent Effects

The beta Diketones are readily soluble in carbon tetrachloride and deuterochloroform. Because these solvents are least likely to contain large amounts of water which could exchange with and mask the enol –OH resonance, their use is preferable under normal circumstances.

3-ETHYL-2,4-PENTANEDIONE

$C_7H_{12}O_2$

Mol. Wt. 128.17

Solvent: $CDCl_3$

Assignments

a = 0.91
b = 1.03
c = 1.85
d = ca 2.20
e = 2.29
f = 3.56
g = 16.60

2137

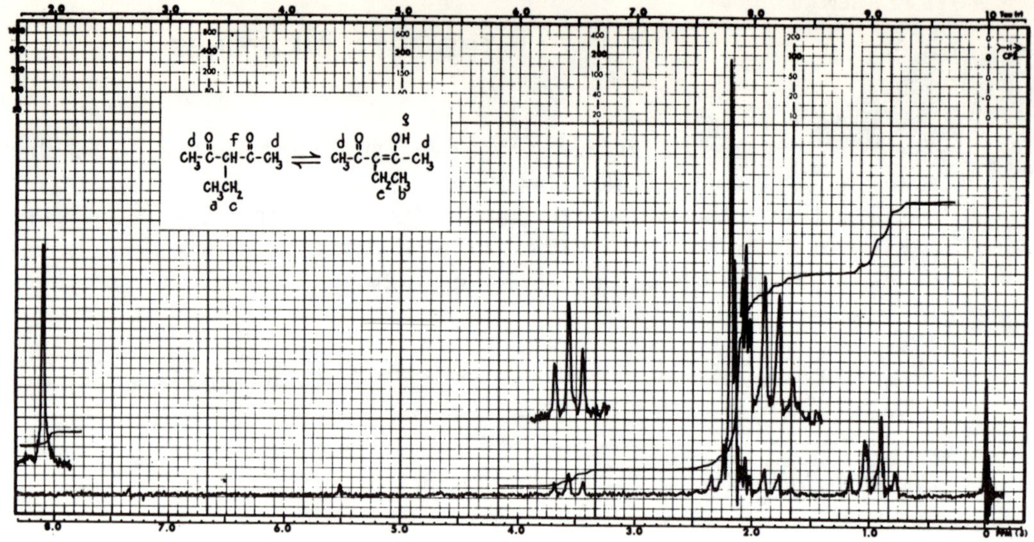

© 1978 Sadtler Research Laboratories, Inc.

3,5-HEPTANEDIONE

$C_7H_{12}O_2$

Mol. Wt. 128.17

B.P. 46-48°C/6mm

Solvent: CCl_4

Assignments

a = 1.00
b = 1.11
c = 2.26
d = 2.47
e = 3.47
f = 5.42
g = 15.16

2138

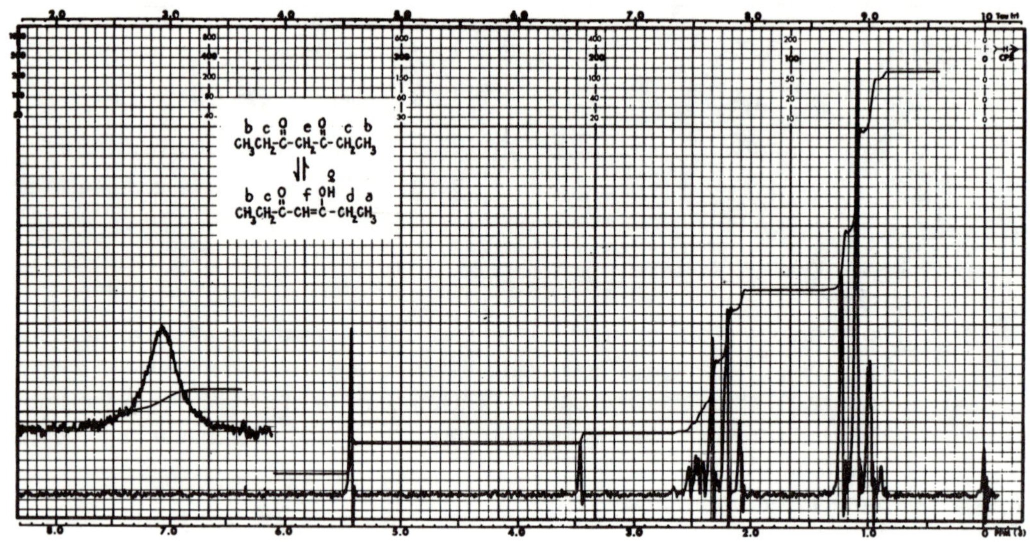

© 1978 Sadtler Research Laboratories, Inc.

4,6-NONANEDIONE

$C_9H_{16}O_2$

Mol. Wt. 156.23

B.P. 79-80°C/8mm

Solvent: CCl_4

Assignments

a = 0.95
b = ca 1.62
c = 2.21
d = 2.40
e = 3.41
f = 5.39
g = 15.24

2139

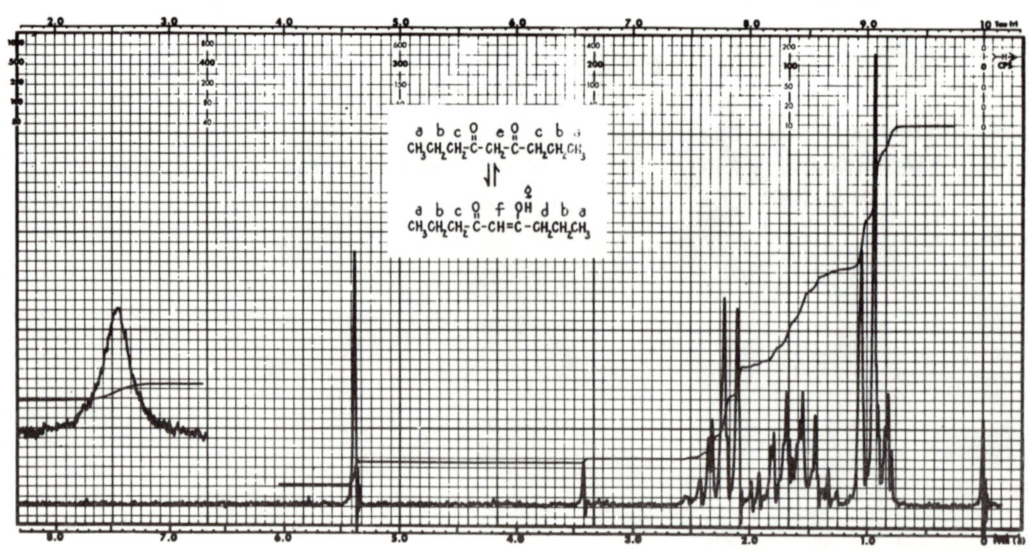

© 1978 Sadtler Research Laboratories, Inc.

3-ISOBUTYL-2,4-PENTANEDIONE (KETO & ENOL FORMS)

$C_9H_{16}O_2$

Mol. Wt. 156.23

Solvent: CCl_4

Assignments
a = ca 0.92
b = 1.10-2.10
c = ca 2.16
d = 3.71
e = unobserved

2140

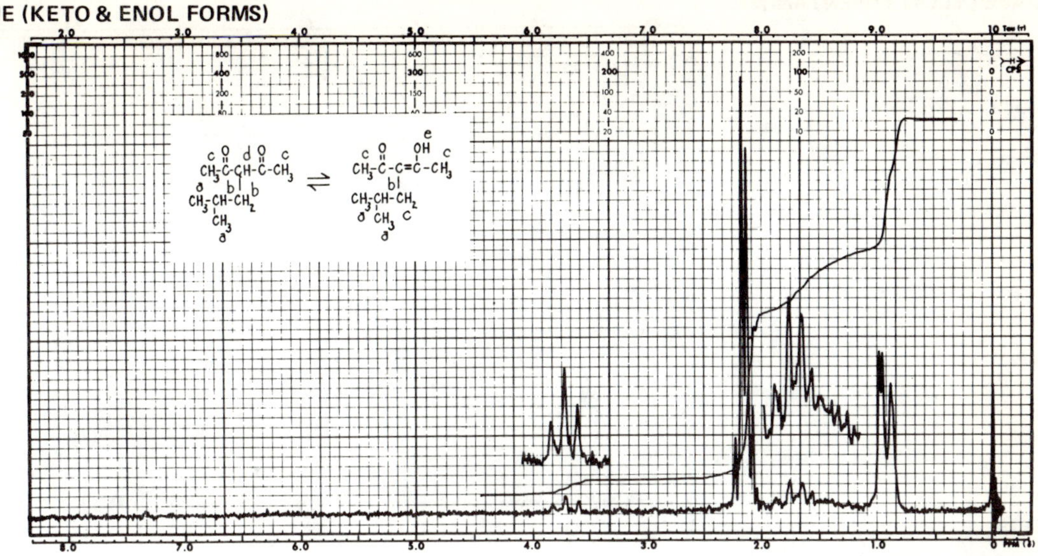

© 1978 Sadtler Research Laboratories, Inc.

1-(2-FURYL)-1,3-BUTANEDIONE

$C_8H_8O_3$

Mol. Wt. 152.15

B.P. 97-98°C/5mm

Solvent: CCl_4

Assignments
a = 2.08
b = 2.20
c = 3.86
d = 6.00
e = 6.46
f = 7.06
g = 7.49
h = 15.03

2141

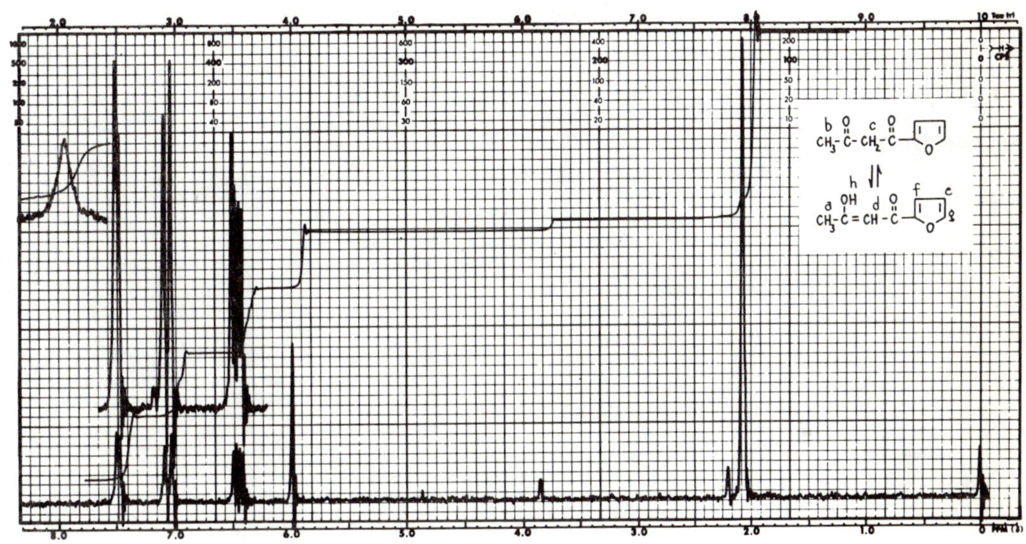

© 1978 Sadtler Research Laboratories, Inc.

2,2,4,4-TETRAMETHYL-1,3-CYCLOBUTANEDIONE

$C_8H_{12}O_2$

Mol. Wt. 140.18

Solvent: $CDCl_3$

Assignments
a = 1.32

2142

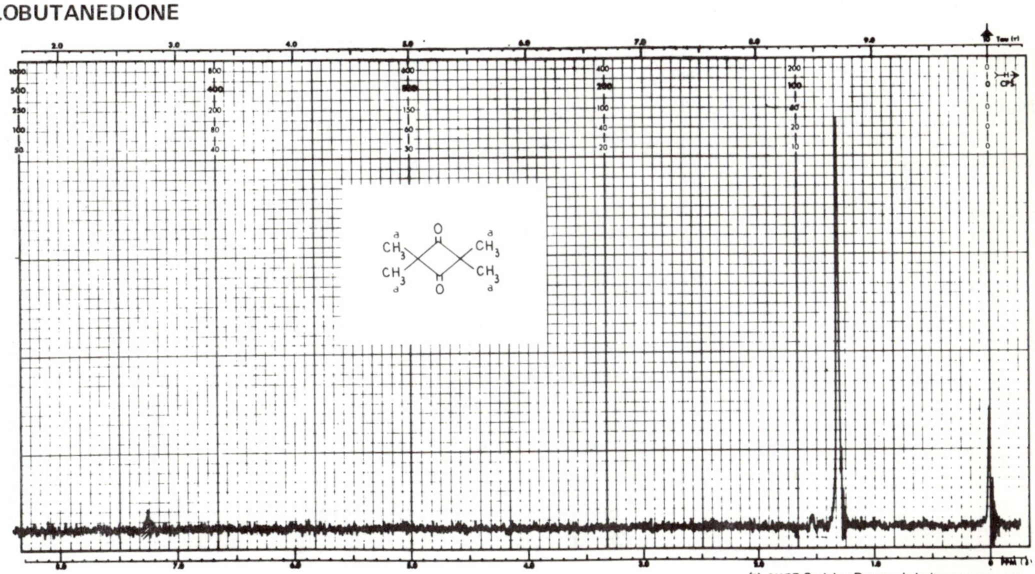

© 1978 Sadtler Research Laboratories, Inc.

2-ACETYLCYCLOPENTANONE

$C_7H_{10}O_2$

Mol. Wt. 126.16

Solvent: CCl_4

Assignments
a = 1.94
b = 2.21
c = 1.65-2.70
d = 3.35
e = 13.27

2143

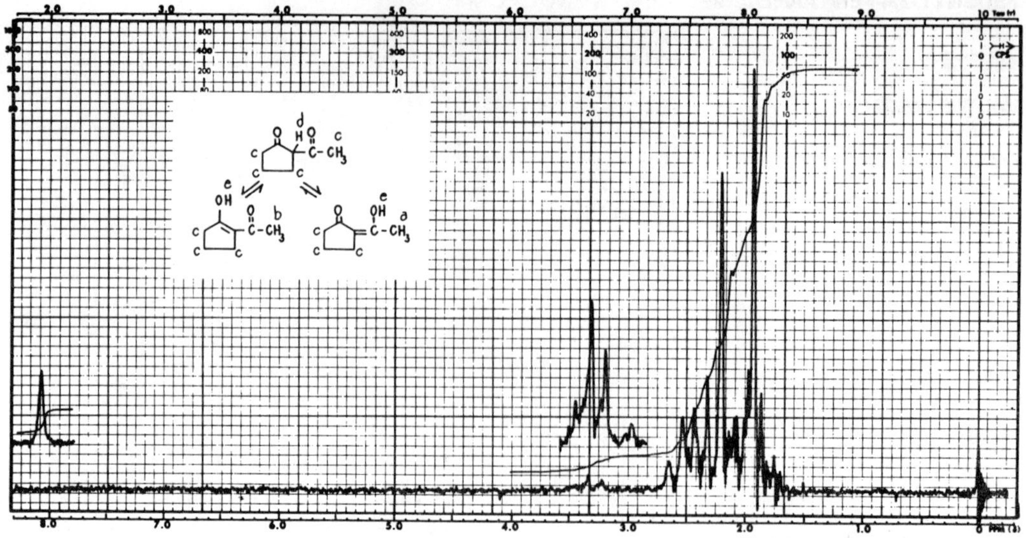

© 1978 Sadtler Research Laboratories, Inc.

5,5-DIMETHYL-1,3-CYCLOHEXANDIONE

$C_8H_{12}O_2$

Mol. Wt. 140.18

M.P. 148-150°C

Solvent: Polysol-d

Assignments
a = 1.09
b = 2.02
c = 5.41
d = 10.00

2144

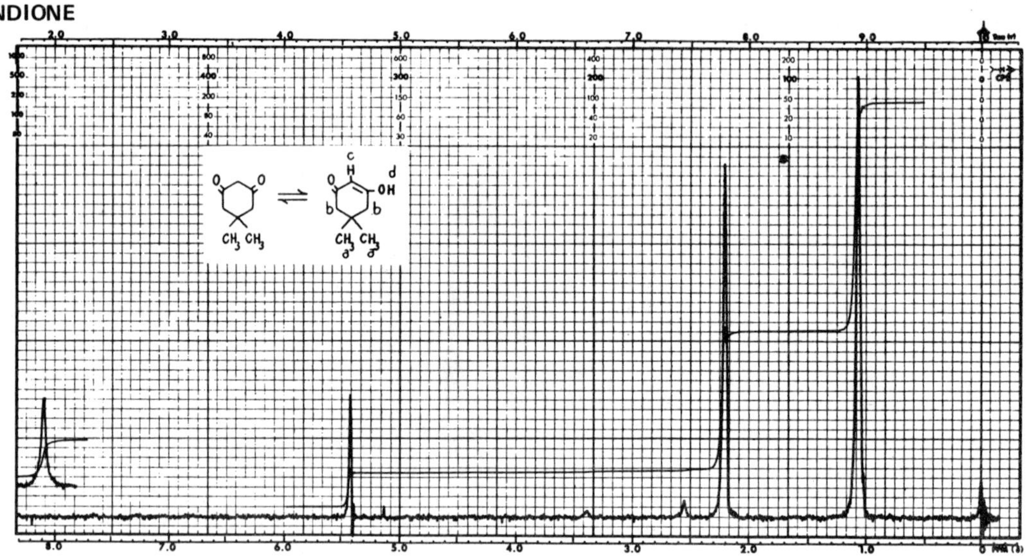

© 1978 Sadtler Research Laboratories, Inc.

2-ACETYL-1,3-CYCLOHEXANEDIONE

$C_8H_{10}O_3$

Mol. Wt. 154.17

Solvent: CCl_4

Assignments
a = 1.98
b = 2.39 or d
c = 2.51
d = 2.61 or b
e = unobserved

2145

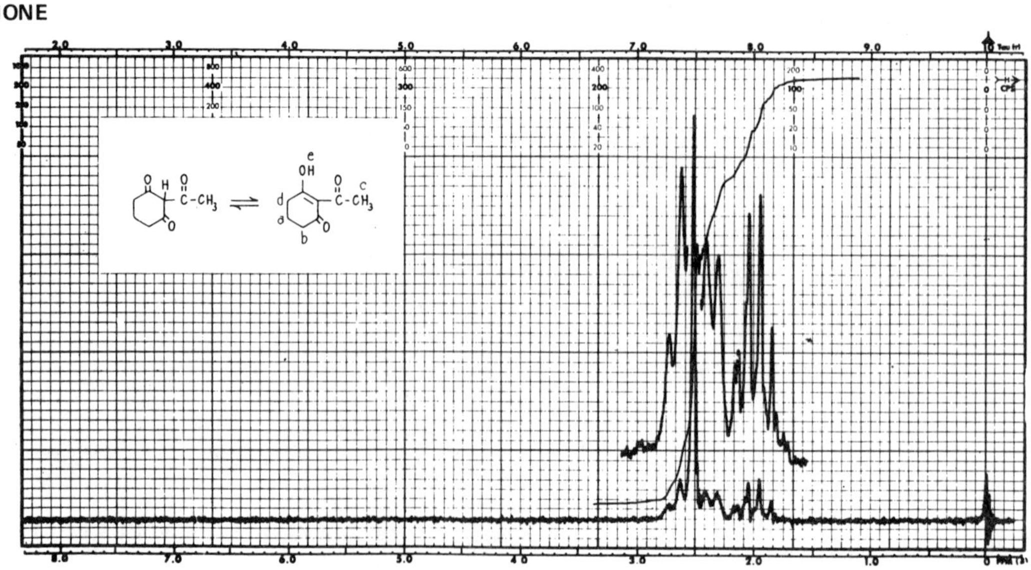

© 1978 Sadtler Research Laboratories, Inc.

1-PHENYL-2,4-PENTANEDIONE

$C_{11}H_{12}O_2$

Mol. Wt. 176.22

B.P. 142-145°C/15mm

Solvent: CCl_4

Assignments
a = 1.88
b = 1.98
c = 3.32
d = 3.43
e = 3.60
f = 5.29
g = ca 7.12
h = 15.02

2146

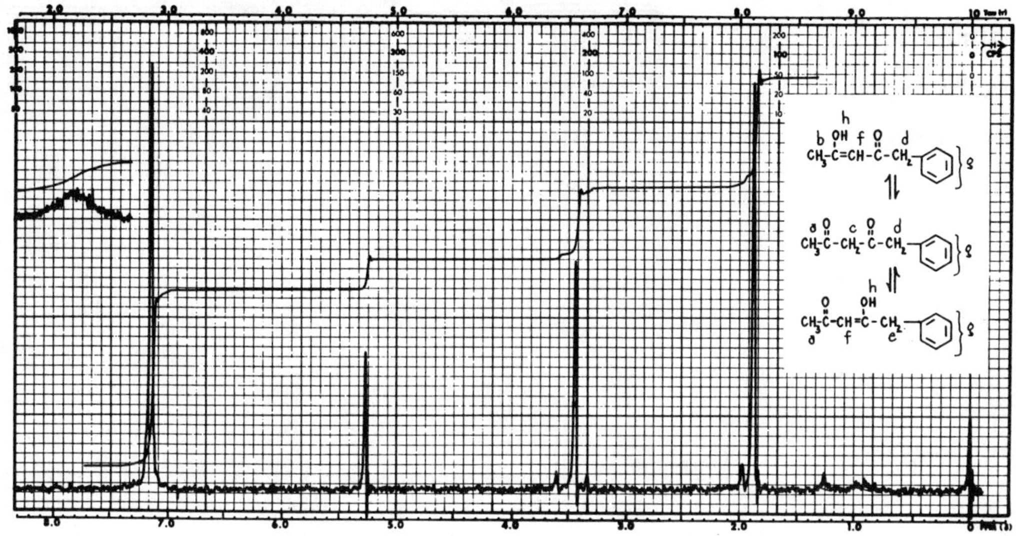

© 1978 Sadtler Research Laboratories, Inc.

1-PHENYL-1,3-BUTANEDIONE

$C_{10}H_{10}O_2$

Mol. Wt. 162.19

M.P. 57-58°C

Solvent: $CDCl_3$

Assignments
a = 2.13
b = 2.24
c = 4.03
d = 6.12
e = 7.20-7.50
f = 7.82
OH unobserved

2147

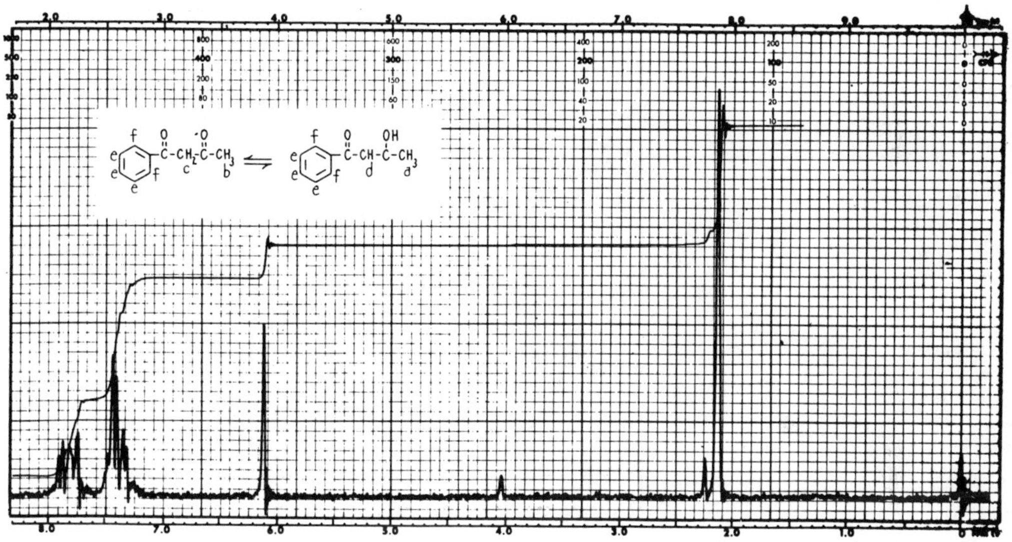

© 1978 Sadtler Research Laboratories, Inc.

1,5-DIPHENYL 1,3,5-PENTANETRIONE

$C_{17}H_{14}O_3$

Mol. Wt. 266.30

M.P. 111-113°C

Solvent: Polysol-d

Assignments
a = 4.20
b = 4.36
c = 6.28
d = 6.51
e = 7.20-7.65
f = 7.70-8.15
g = unobserved
h = 2.48 MDSO-d_5

2148

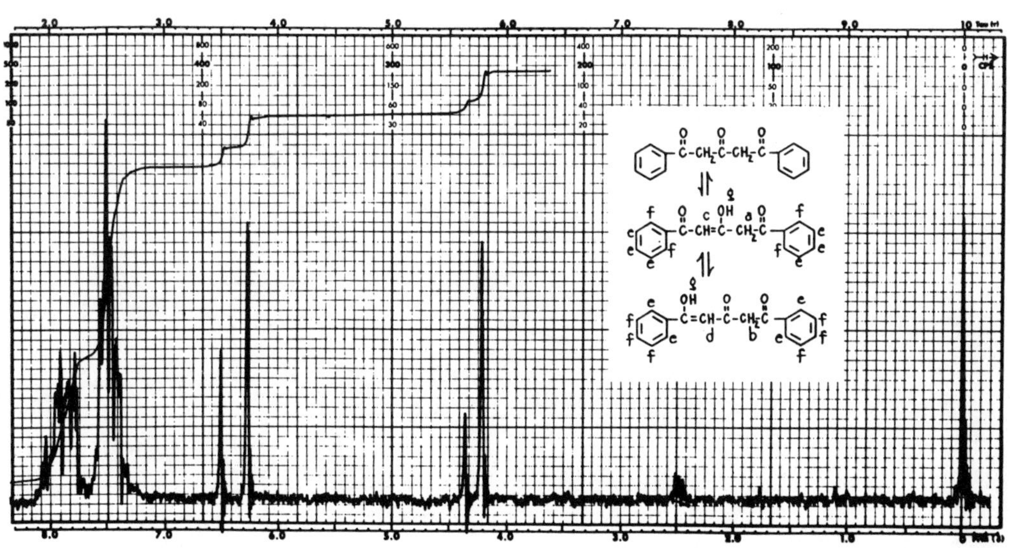

© 1978 Sadtler Research Laboratories, Inc.

5-CHLORO-1,3-INDANDIONE

$C_9H_5ClO_2$

Mol. Wt. 180.59

M.P. 137-138°C

Solvent: Trifluoroacetic acid

Assignments
a = 3.63
b = ca 8.07

2149

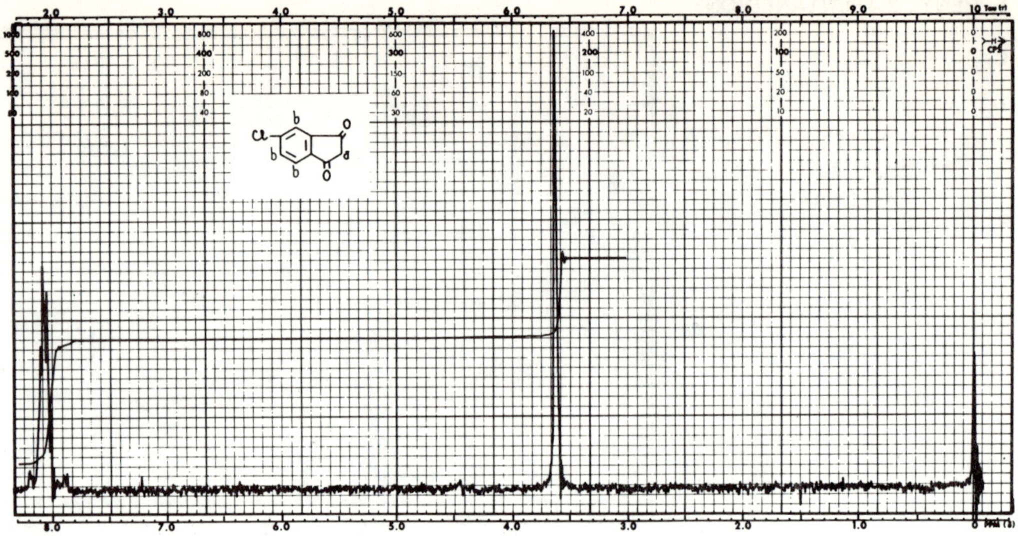

© 1978 Sadtler Research Laboratories, Inc.

2,2-DIHYDROXY-1,3-INDANDIONE

$C_9H_6O_4$

Mol. Wt. 178.15

Solvent: Dimethyl sulfoxide

Assignments
a = 7.72
b = 8.20

2150

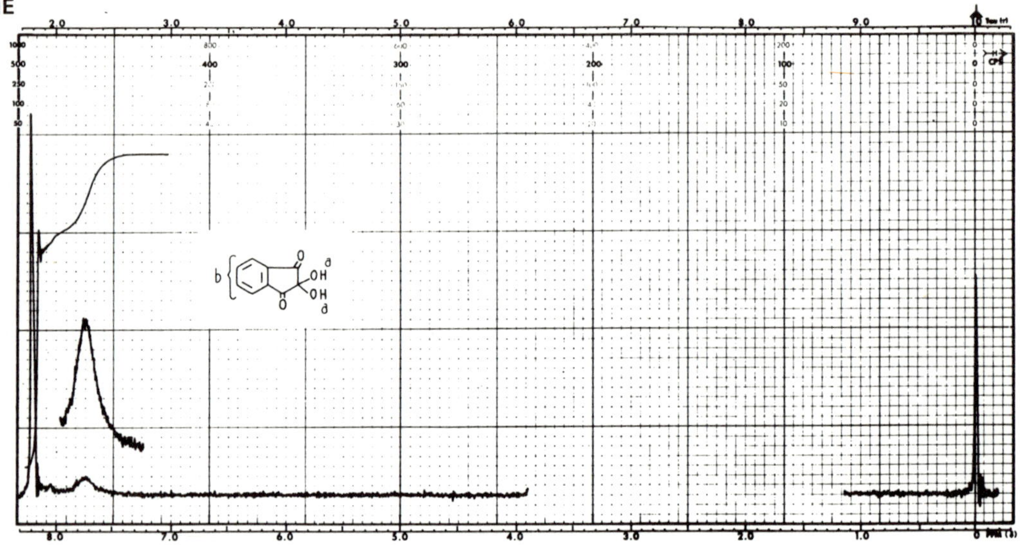

© 1978 Sadtler Research Laboratories, Inc.

The aldehydic functional group produces a characteristic band at very low field arising from the resonance of the proton bonded to the carbonyl group. Coupling between this proton and the adjacent aliphatic groups is usually observed in the NMR spectra of the aliphatic aldehydes.

The aldehydic group weakly deshields aliphatic protons but has a relatively strong deshielding effect on the ortho aromatic protons. The aldehydes oxidize easily and their NMR spectra often display impurity bands arising from the presence of the corresponding carboxylic acid.

Chemical Shifts

Aliphatic Protons

HC–4	HC–3	HC–2	HC–1		Solvent	Spectrum
			2.12	$CH_3-C(=O)-H$	CCl_4	----
		1.09	2.45	$CH_3-CH_2-C(=O)-H$	$CDCl_3$	----
	0.95	1.61	2.36	$CH_3-CH_2-CH_2-C(=O)-H$	CCl_4	2151
0.90	1.35	1.55	2.43	$CH_3-CH_2-CH_2-CH_2-C(=O)-H$	DMSO-d_6	----
		(1.12)	2.38	$(CH_3)_2-CH-C(=O)-H$	CCl_4	2152
	(0.98)	2.12	2.29	$(CH_3)_2-CH-CH_2-C(=O)-H$	$CDCl_3$	----

Aromatic Protons

Benzaldehyde

H–C(=O)–	ortho	meta/para	Solvent	Spectrum
9.94	7.79	7.2–7.6	CCl_4	2163

Para Substituted Benzaldehydes

H–C(=O)–	ortho	meta	para	Solvent	Spectrum
9.66	7.63	6.63	$-N(CH_3)_2$	$CDCl_3$	2170
9.78	7.69	6.89	$-O-R_3$	$CDCl_3$	----
9.80	7.71	6.92	$-O-R_5$	CCl_4	2174
9.82	7.72	6.92	$-O-CH_2-CH=CH_2$	CCl_4	2175
9.89	7.80	7.04	$-OH$	Acetone	----
9.91	7.80	7.19	$-O-C(=O)-CH_3$	CCl_4	----
10.00	7.91	7.21	$-F$	$CDCl_3$	----
9.97	7.76	7.31	$-CH_3$	$CDCl_3$	----
9.91	7.73	7.40	$-Cl$	$CDCl_3$	----

Aromatic Protons continued

Para Substituted Benzaldehydes continued

H–C(=O)–	ortho	meta	para	Solvent	Spectrum
10.00	7.88	7.67	–Ø	$CDCl_3$	- - - -
9.97	7.68	7.68	–Br	$CDCl_3$	2169
10.13	7.89	8.01	–C≡N	$CDCl_3$	- - - -
10.18	8.09	8.39	$-NO_2$	$CDCl_3$	2172

Aldehydic Protons

H–C(=O)–	X		
9.33	$-C(R_2)=CH-R_3$	CCl_4	2157
9.48	–2–Pyrrole	$CDCl_3$	2182
9.53	$-CH=CH-R_3$	$CDCl_3$	2156
9.57	$-C(CH_3)=CH-$Ø	CCl_4	2160
9.58	$-CH(CH_3)_2$	CCl_4	2152
9.63	–2–Furan	CCl_4	2185
9.66	–Cyclohexane	$CDCl_3$	2155
9.68	$-R_3$	CCl_4	2151
9.69	$-CH_3$	CCl_4	- - - -
9.94	–Ø	CCl_4	2163
9.97	–Ø (4–Br)	$CDCl_3$	2169
10.18	–Ø $(4-NO_2)$	$CDCl_3$	2172

As the table below indicates there is a general trend in the chemical shift of the aldehydic proton in relation to the type of group to which it is bonded. In general, the aldehydic protons bonded to vinyl or heteroaromatic groups resonate at slightly higher field than those bonded to aliphatic groups. At lowest field appear the benzaldehyde protons which are further differentiated in chemical shift by the deshielding effect of other substituents on the aromatic ring as indicated in the table of chemical shifts for para substituted benzaldehydes presented on the previous page.

Coupling and Coupling Constants

The aldehydic proton normally displays coupling to protons bonded to the carbon atom alpha to the carbonyl group. The coupling constants for such vinyl protons tend to be significantly larger than the corresponding coupling constant observed for aliphatic protons.

J =	7.6 Hz	H–C(=O)–CH=C	2162, 2156
J =	1.4–1.8 Hz	$H-C(=O)-CH_2-R$	2159, 2151
J =	2.5 Hz	$H-C(=O)-CH(R)_2$	- - - -

BUTYRALDEHYDE

C_4H_8O

Mol. Wt. 72.11

B.P. 73-75°C

Solvent: CCl_4

Assignments
a = 0.95
b = 1.61
c = 2.36
d = 9.68

2151

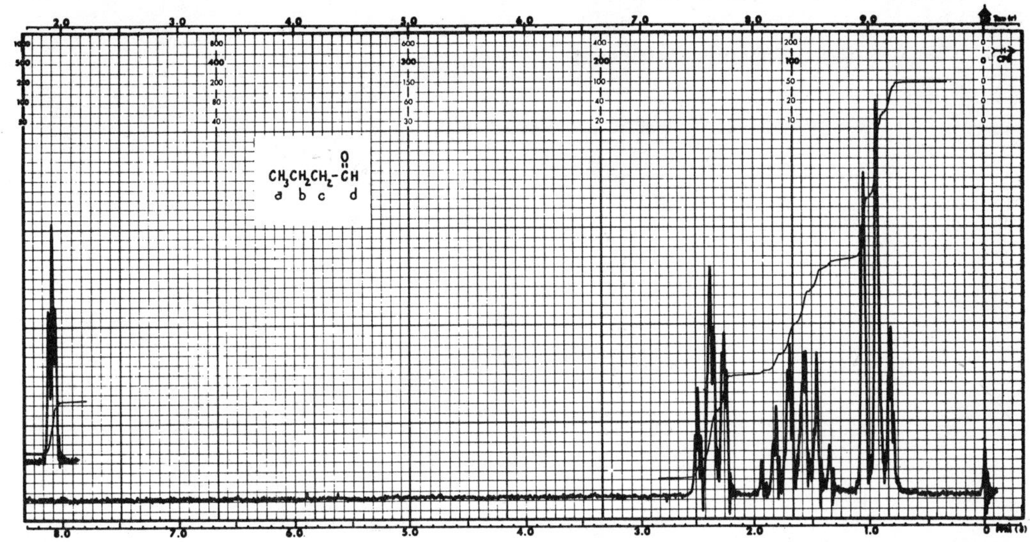

© 1978 Sadtler Research Laboratories, Inc.

ISOBUTYRALDEHYDE

C_4H_8O

Mol. Wt. 72.10

Solvent: CCl_4

Assignments
a = 1.12
b = 2.38
c = 9.58
d = 1.25 imp.

2152

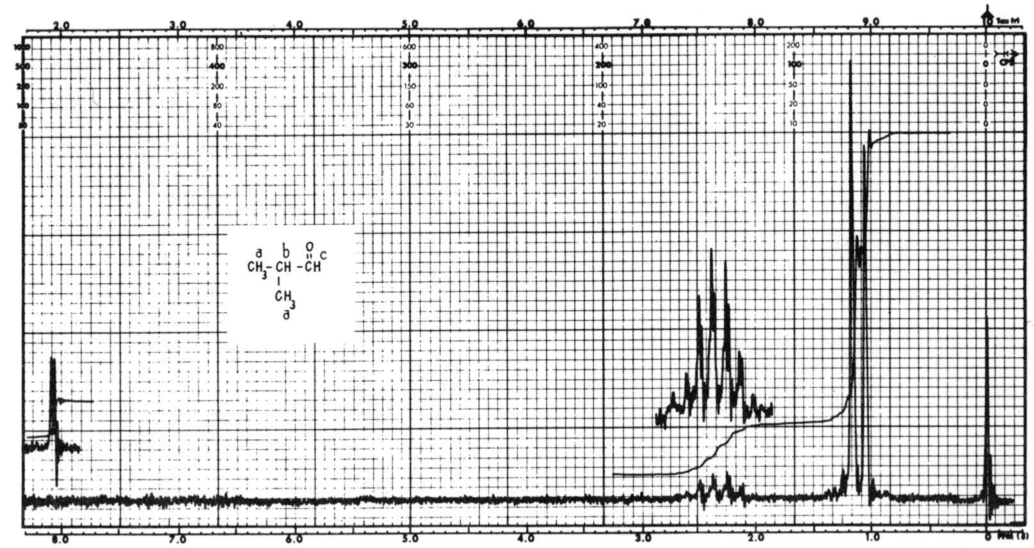

© 1978 Sadtler Research Laboratories, Inc.

HEPTANAL

$C_7H_{14}O$

Mol. Wt. 114.19

B.P. 54-56°C/20mm

Solvent: CCl_4

Assignments
a = 0.90
b = 1.34
c = 2.35
d = 4.78
e = 9.72

2153

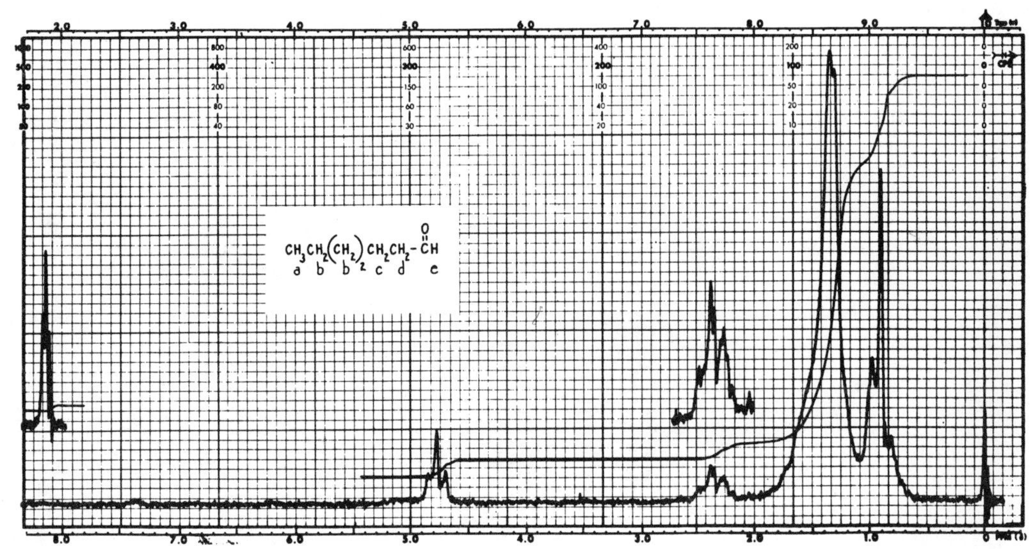

© 1978 Sadtler Research Laboratories, Inc.

DECANAL

$C_{10}H_{20}O$

Mol. Wt. 156.27

Solvent: CCl_4

Assignments
a = 0.91
b = 1.30
c = 2.34
d = 9.73

2154

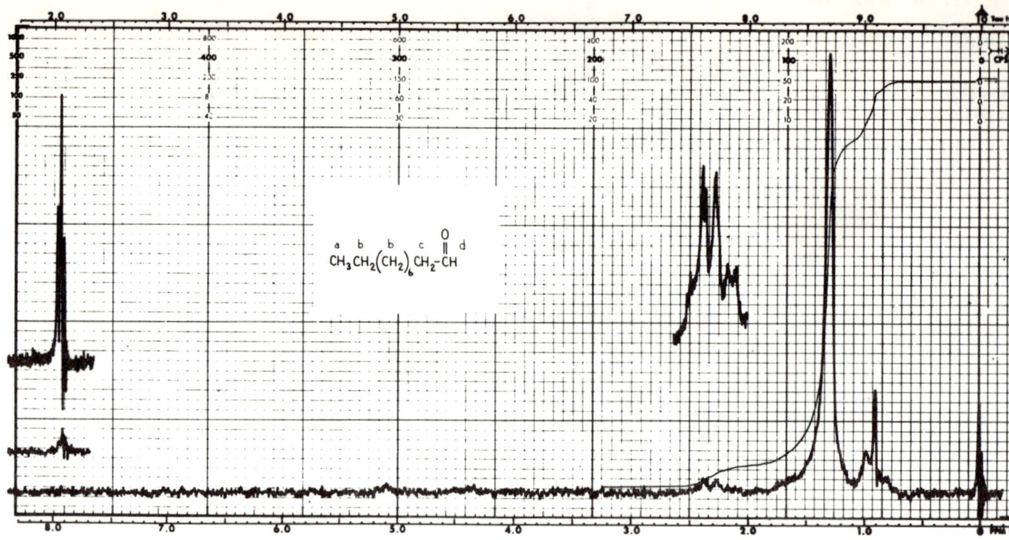

© 1978 Sadtler Research Laboratories, Inc.

CYCLOHEXANECARBOXALDEHYDE

$C_7H_{12}O$

Mol. Wt. 112.17

B.P. 161-163°C

Solvent: $CDCl_3$

Assignments
a = 0.90-2.60
b = 9.66

2155

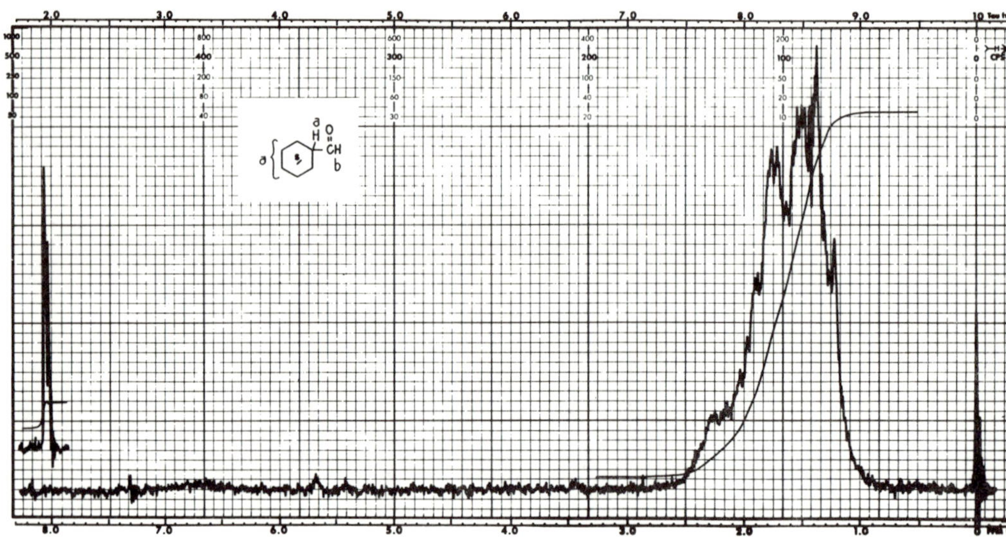

© 1978 Sadtler Research Laboratories, Inc.

---

trans-2-HEXENAL

$C_6H_{10}O$

Mol. Wt. 98.15

B.P. 47°C/17mm

Solvent: $CDCl_3$

Assignments
a = 0.98
b = 1.54
c = 2.33
d = 6.13
e = 6.87
f = 9.53

2156

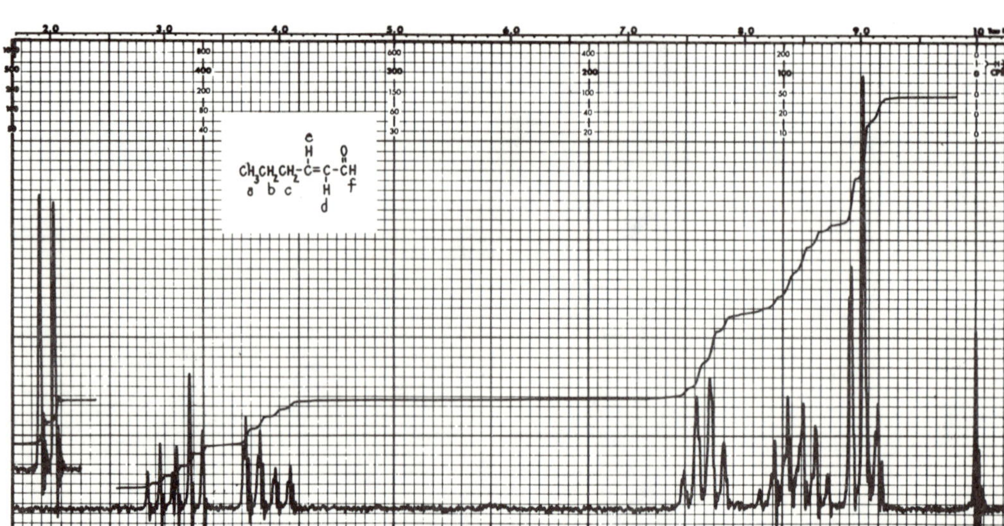

© 1978 Sadtler Research Laboratories, Inc.

2-ETHYL-2-HEXENAL

$C_8H_{14}O$

Mol. Wt. 126.20

B.P. 175°C

Solvent: CCl_4

Assignments
a = 0.98
b = 1.01
c = 1.58
d = 2.32
e = 6.37
f = 9.33

2157

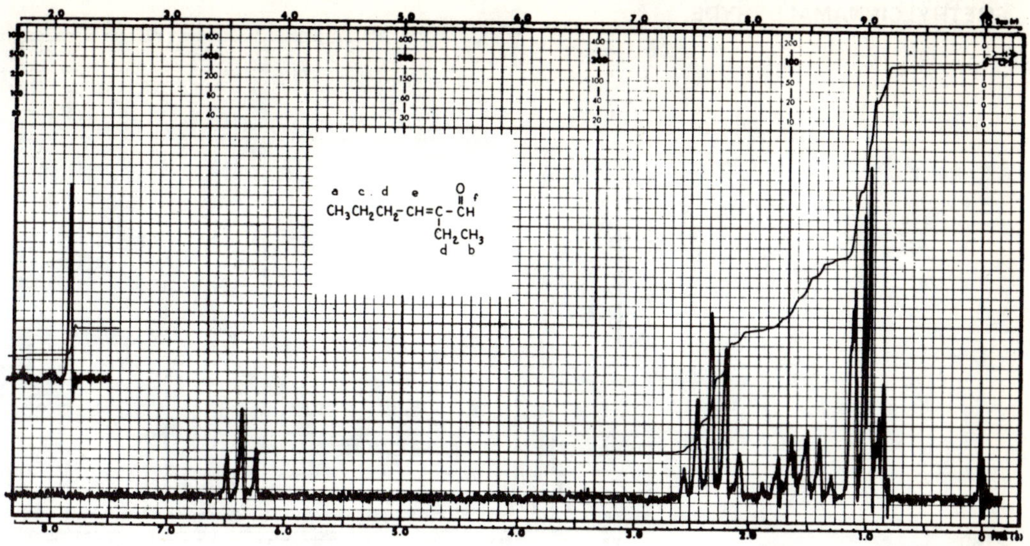

© 1978 Sadtler Research Laboratories, Inc.

3-CYCLOHEXENE-1-CARBOXALDEHYDE

$C_7H_{10}O$

Mol. Wt. 110.16

B.P. 164.5°C

Solvent: CCl_4

Assignments
a = 1.40-2.72
b = 5.68
c = 9.64

2158

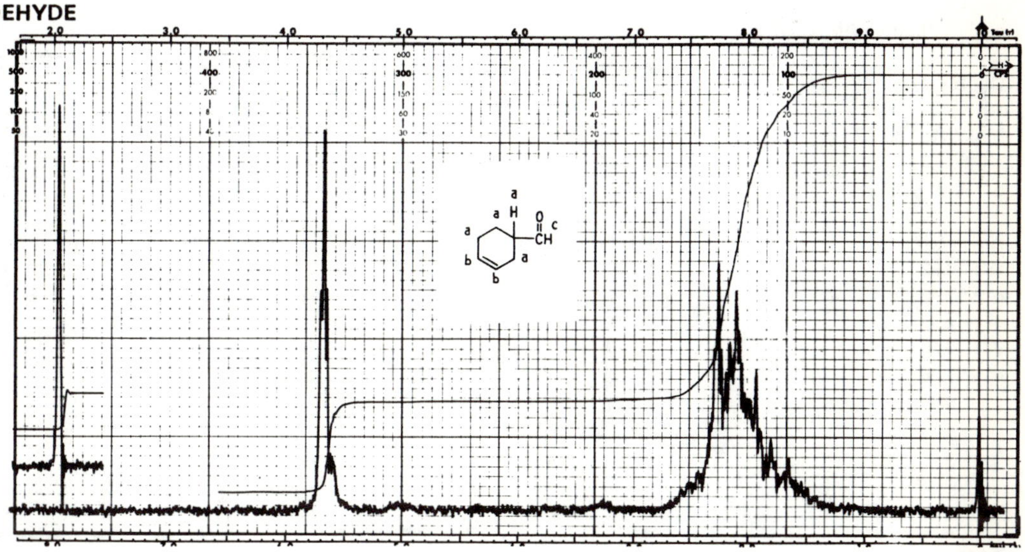

© 1978 Sadtler Research Laboratories, Inc.

HYDROCINNAMALDEHYDE

$C_9H_{10}O$

Mol. Wt. 134.18

Solvent: $CDCl_3$

Assignments
a = 2.69
b = 2.89
c = ca 7.20
d = 9.71

2159

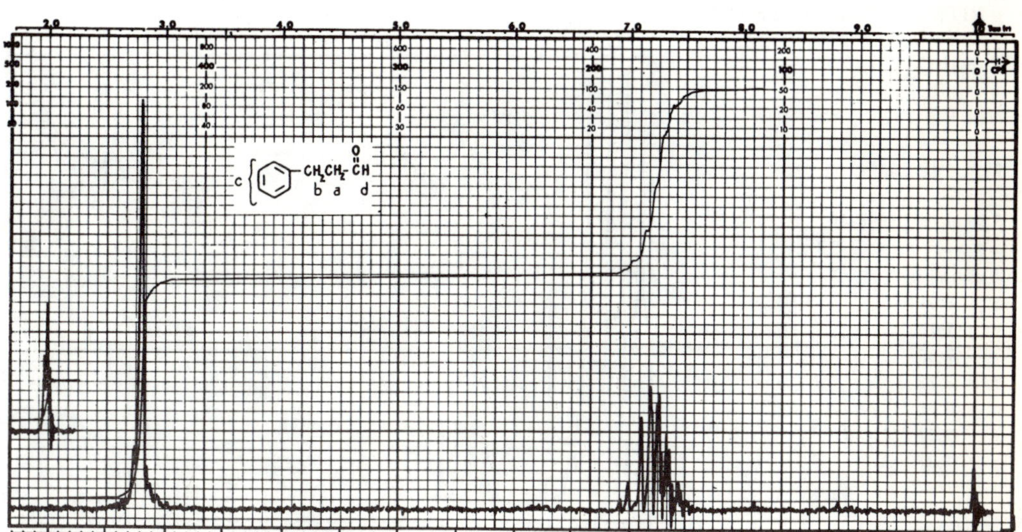

© 1978 Sadtler Research Laboratories, Inc.

α-METHYLCINNAMALDEHYDE.

$C_{10}H_{10}O$

Mol. Wt. 149.19

B.P. 118-120°C/10mm

Solvent: CCl_4

Assignments
a = 1.99
b = 7.19
c = 7.25-7.65
d = 9.57

2160

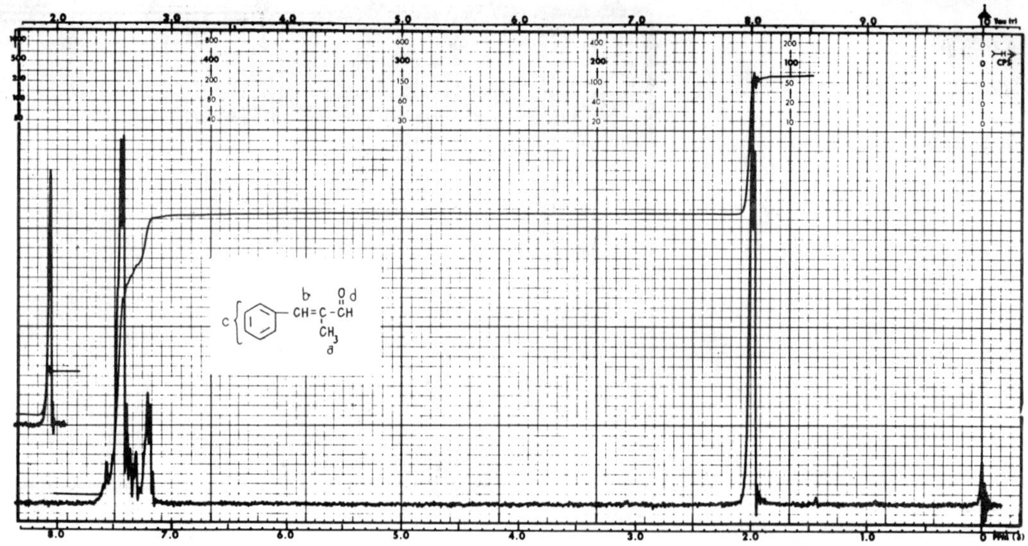

© 1978 Sadtler Research Laboratories, Inc.

α-BUTYLCINNAMALDEHYDE

$C_{13}H_{16}O$

Mol. Wt. 188.27

Solvent: CCl_4

Assignments
a = 0.92
b = 1.40
c = 2.48
d = 7.10
e = 7.38
f = 9.47

2161

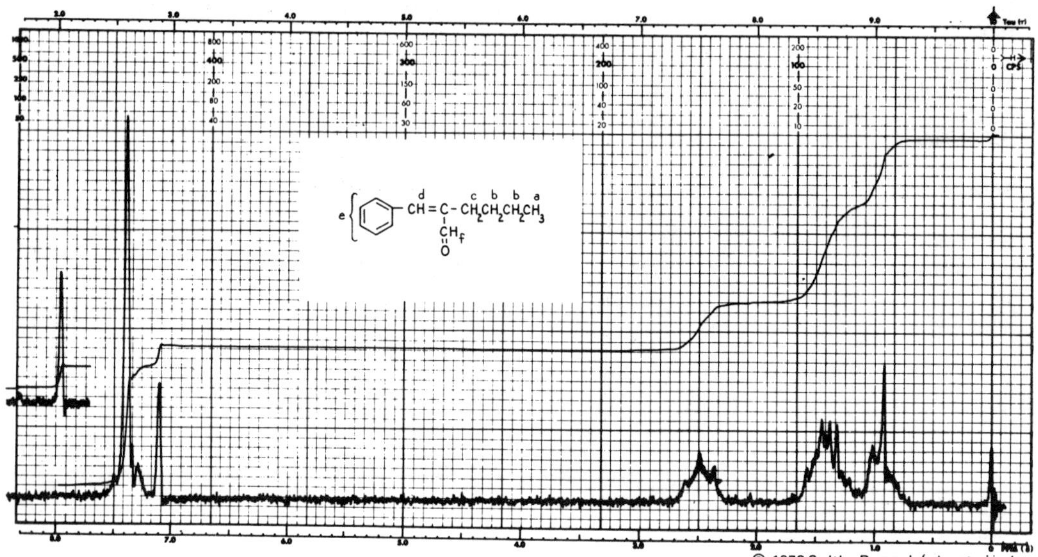

© 1978 Sadtler Research Laboratories, Inc.

p-NITROCINNAMALDEHYDE

$C_9H_7NO_3$

Mol. Wt. 177.16

M.P. 142-144°C

Solvent: DMSO-d_6

Assignments
a = 6.91
b = 7.79
c = 7.90
d = 8.23
e = 9.76
f = 2.57 DMSO-d_5
g = 3.37 H_2O

2162

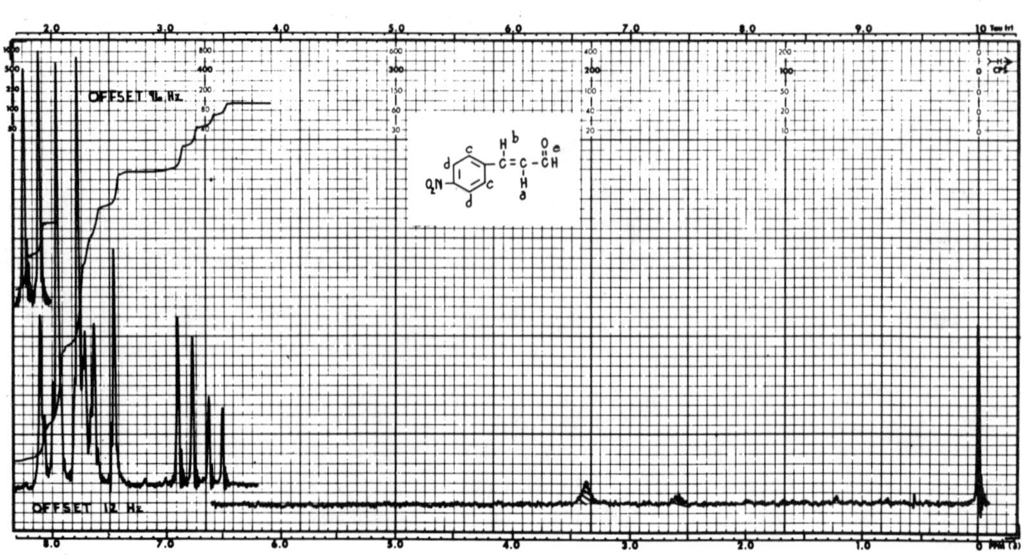

© 1978 Sadtler Research Laboratories, Inc.

BENZALDEHYDE

C_7H_6O

Mol. Wt. 106.13

B.P. 75-76°C/16mm

Solvent: CCl_4

Assignments
a = 7.20-7.60
b = 7.79
c = 9.94

2163

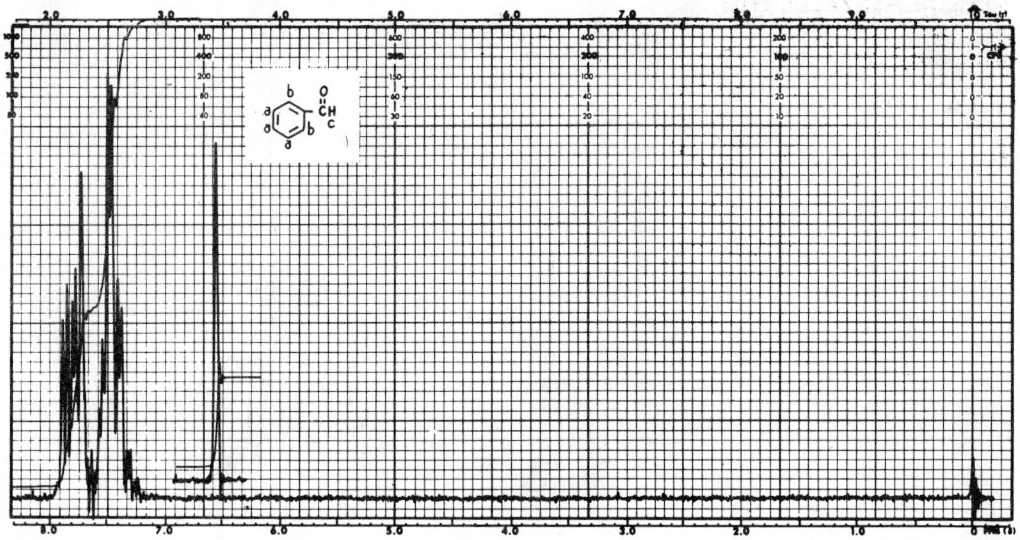

© 1978 Sadtler Research Laboratories, Inc.

---

o-TOLUALDEHYDE

C_8H_8O

Mol. Wt. 120.15

B.P. 85-87°C/14mm

Solvent: CCl_4

Assignments
a = 2.65
b = 7.31
c = 7.72
d = 10.22
e = 2.21 imp.

2164

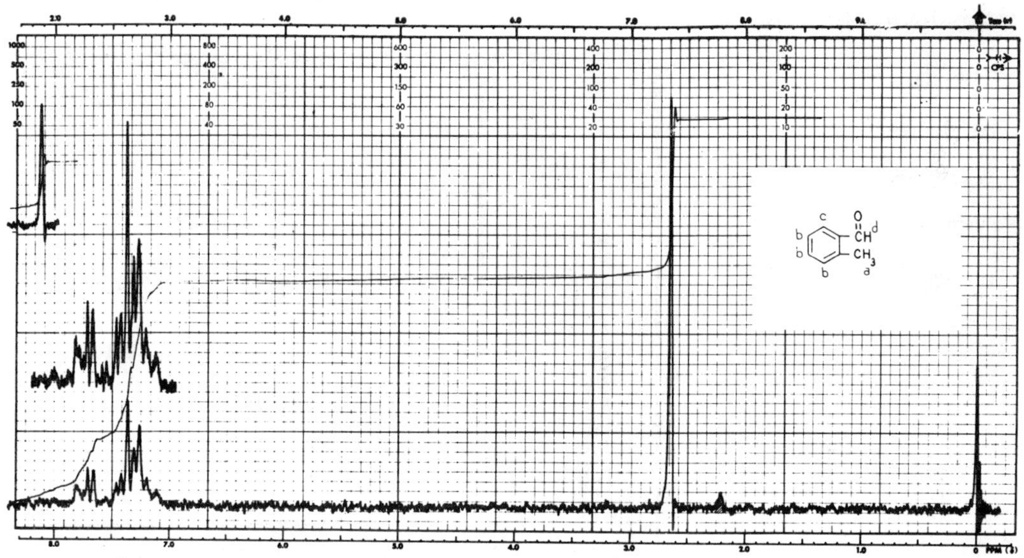

© 1978 Sadtler Research Laboratories, Inc.

---

m-TOLUALDEHYDE

C_8H_8O

Mol. Wt. 120.15

B.P. 80-82°C/11mm

Solvent: CCl_4

Assignments
a = 2.42
b = 7.37
c = 7.60
d = 9.91

2165

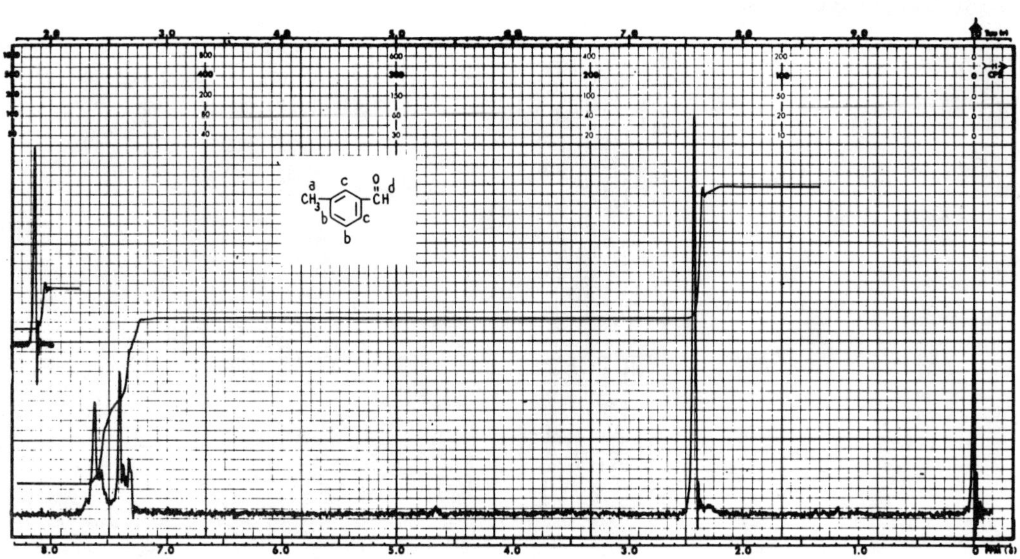

© 1978 Sadtler Research Laboratories, Inc.

o-CHLOROBENZALDEHYDE

C_7H_5ClO

Mol. Wt. 140.57

M.P. 10-12°C

B.P. 208°C/748mm

Solvent: CCl_4

Assignments
a = ca 7.40
b = ca 7.81
c = 10.48

2166

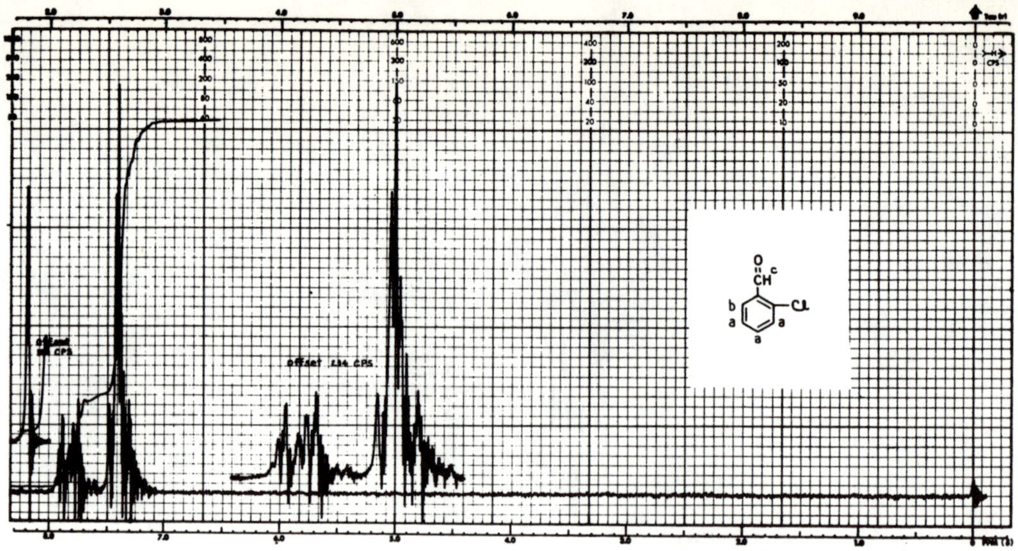

© 1978 Sadtler Research Laboratories, Inc.

---

m-CHLOROBENZALDEHYDE

C_7H_5ClO

Mol. Wt. 140.57

B.P. 213-214°C

Solvent: $CDCl_3$

Assignments
a = 7.60
b = 7.80
c = 7.85
d = 10.00
e = 2.11 imp.

2167

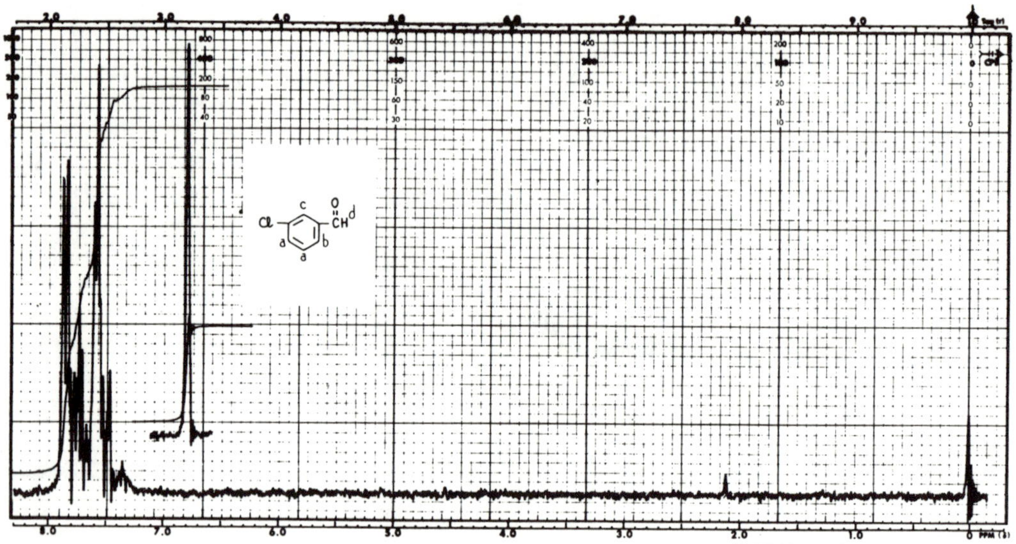

© 1978 Sadtler Research Laboratories, Inc.

---

o-BROMOBENZALDEHYDE

C_7H_5BrO

Mol. Wt. 185.03

M.P. 21-22°

B.P. 230°C

Solvent: CCl_4

Assignments
a = 7.20-7.70
b = 7.82
c = 10.30

2168

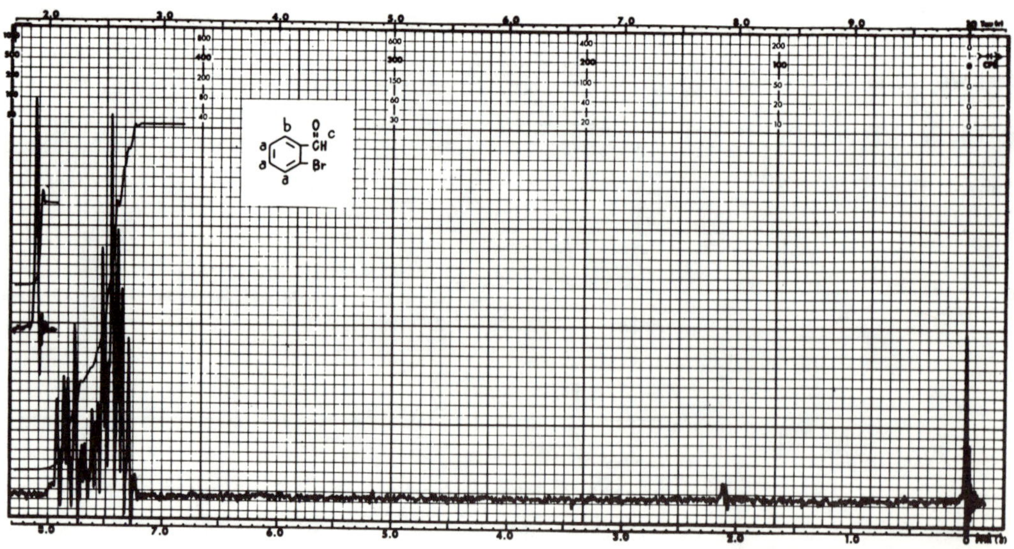

© 1978 Sadtler Research Laboratories, Inc.

p-BROMOBENZALDEHYDE

C_7H_5BrO

Mol. Wt. 185.03

Solvent: $CDCl_3$

Assignments
a = ca 7.68
b = 9.97

2169

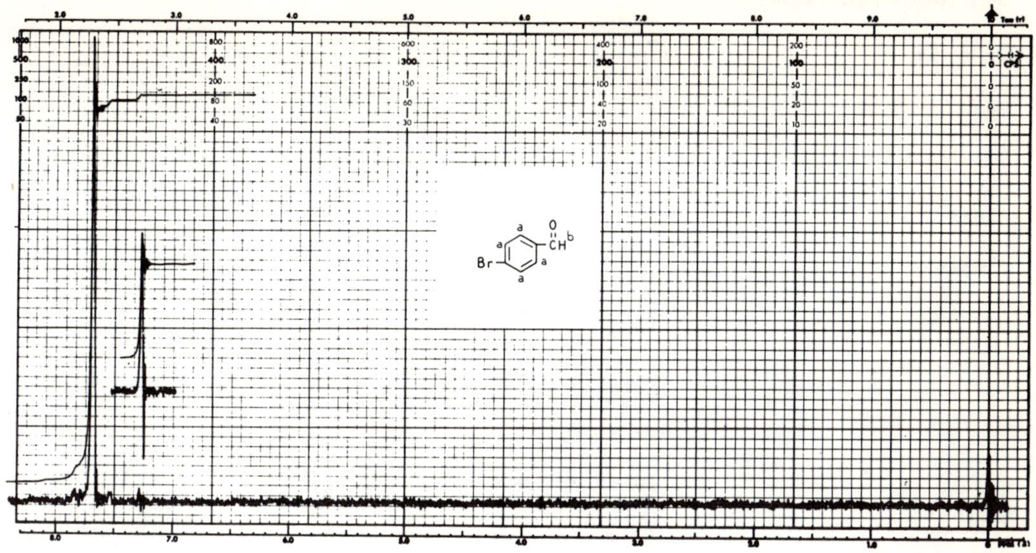

© 1978 Sadtler Research Laboratories, Inc.

---

p-(DIMETHYLAMINO)BENZALDEHYDE

$C_9H_{11}NO$

Mol. Wt. 149.19

M.P. 73°C

Solvent: $CDCl_3$

Assignments
a = 3.01
b = 6.63
c = 7.63
d = 9.66

2170

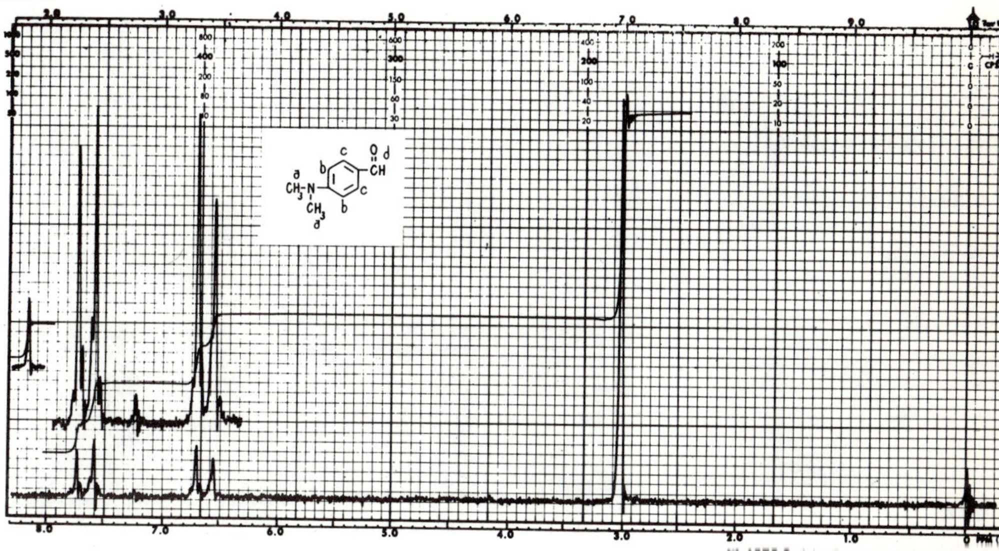

© 1978 Sadtler Research Laboratories, Inc.

---

m-NITROBENZALDEHYDE

$C_7H_5NO_3$

Mol. Wt. 151.12

M.P. 58°C

B.P. 164°C/23mm

Solvent: $CDCl_3$

Assignments
a = 7.85
b = 8.28 or c
c = 8.51 or b
d = 8.67
e = 10.19

2171

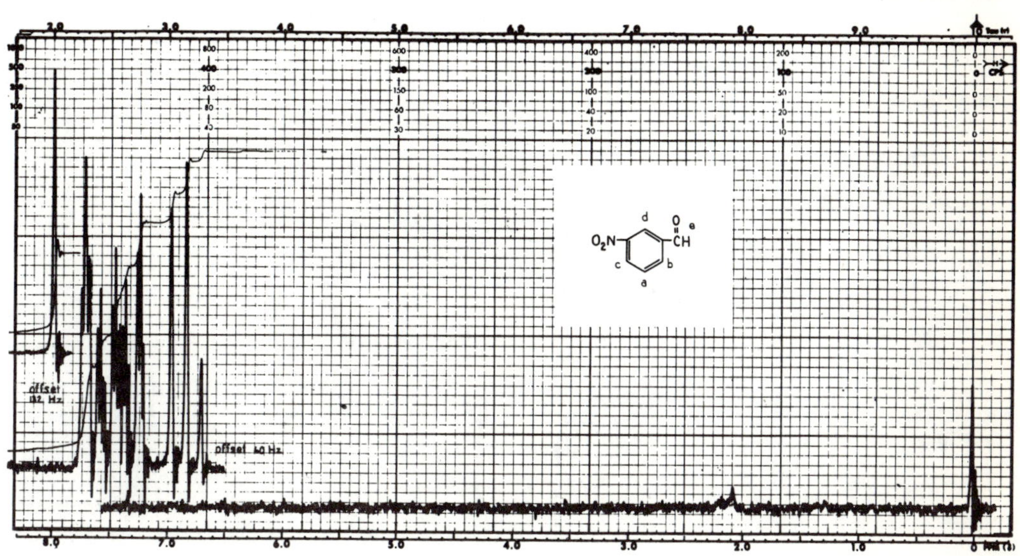

© 1978 Sadtler Research Laboratories, Inc.

p-NITROBENZALDEHYDE

$C_7H_5NO_3$

Mol. Wt. 151.12

M.P. 106-107°C (lit.)

Solvent: $CDCl_3$

Assignments

a = 8.09

b = 8.39

c = 10.18

2172

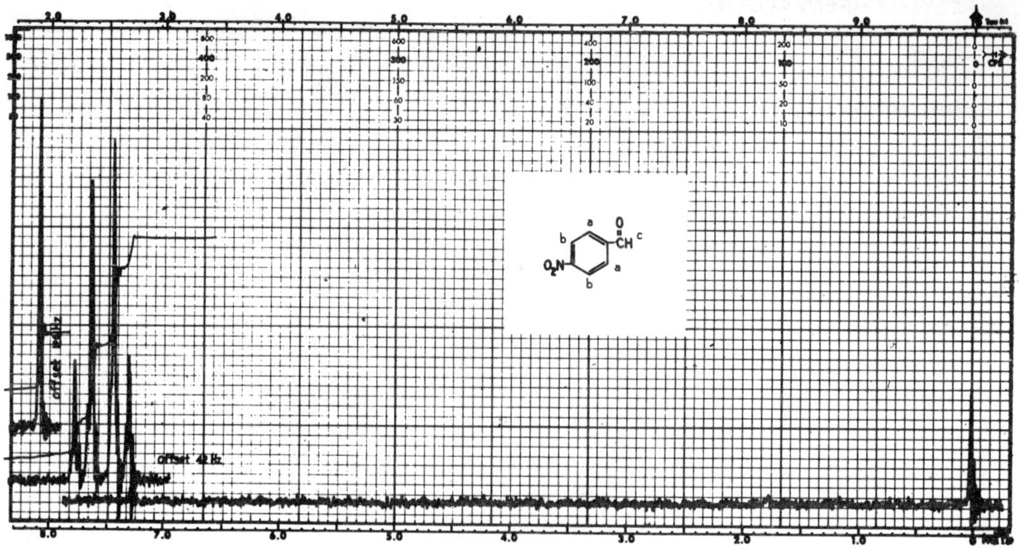

© 1978 Sadtler Research Laboratories, Inc.

---

p-PROPOXYBENZALDEHYDE

$C_{10}H_{12}O_2$

Mol. Wt. 164.21

B.P. 129-130°C/9mm

Solvent: CCl_4

Assignments

a = 1.04

b = 1.76

c = 3.93

d = 6.89

e = 7.69

f = 9.78

2173

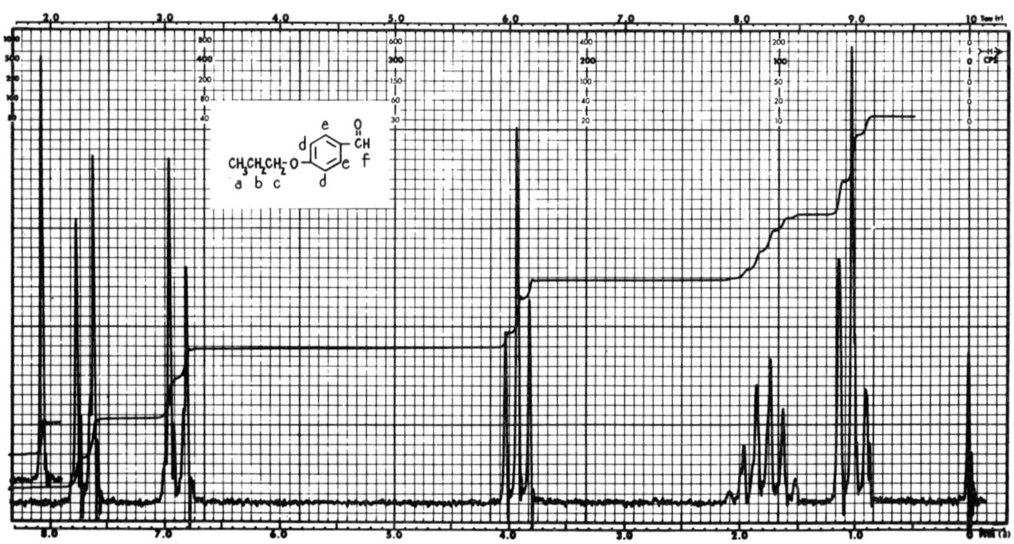

© 1978 Sadtler Research Laboratories, Inc.

---

p-(PENTYLOXY)BENZALDEHYDE

$C_{12}H_{16}O_2$

Mol. Wt. 192.26

M.P. 154-155°C

Solvent: CCl_4

Assignments

a = 0.93

b = 1.10-1.60

c = 1.80

d = 3.99

e = 6.92

f = 7.71

g = 9.80

2174

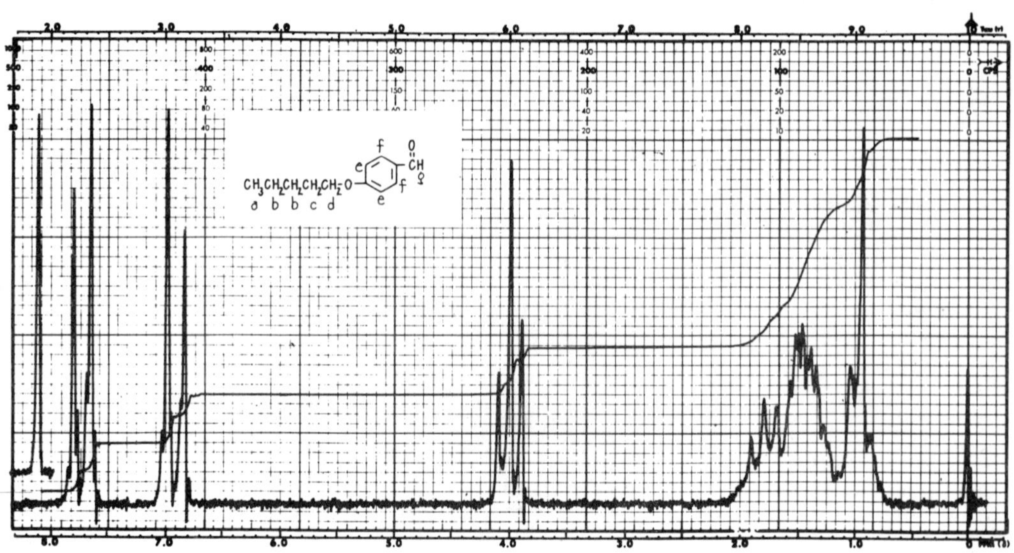

© 1978 Sadtler Research Laboratories, Inc.

p-(ALLYLOXY)BENZALDEHYDE

$C_{10}H_{10}O_2$

Mol. Wt. 162.19

B.P. 142°C/10mm (lit.)

Solvent: CCl_4

Assignments
a = 4.57
b = 5.28
c = 5.35
d = 5.94
e = 6.92
f = 7.72
g = 9.82

2175

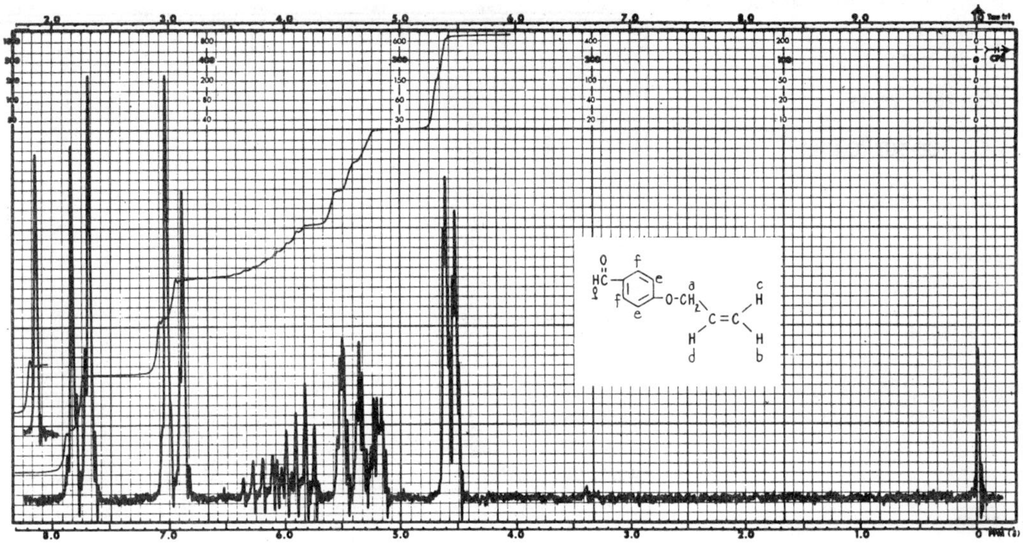

© 1978 Sadtler Research Laboratories, Inc.

2,4-DICHLOROBENZALDEHYDE

$C_7H_4Cl_2O$

Mol. Wt. 175.02

M.P. 68-72°C

Solvent: $CDCl_3$

Assignments
a = 7.31
b = 7.40
c = 7.80
d = 10.35

2176

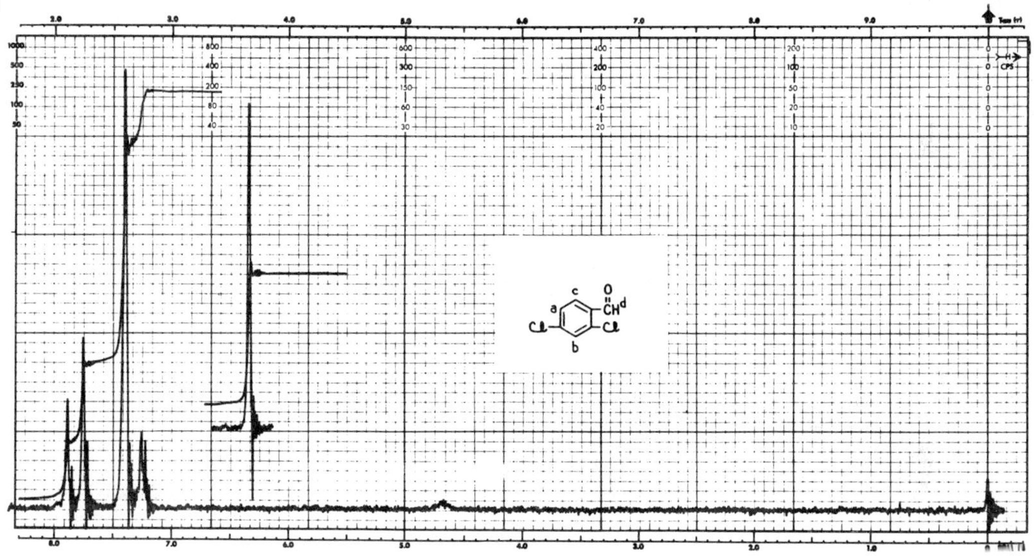

© 1978 Sadtler Research Laboratories, Inc.

3,4-DICHLOROBENZALDEHYDE

$C_7H_4Cl_2O$

Mol. Wt. 175.02

M.P. 42-44°C (lit.)

Solvent: CCl_4

Assignments
a = 7.58
b = 7.68
c = 7.89
d = 9.95

2177

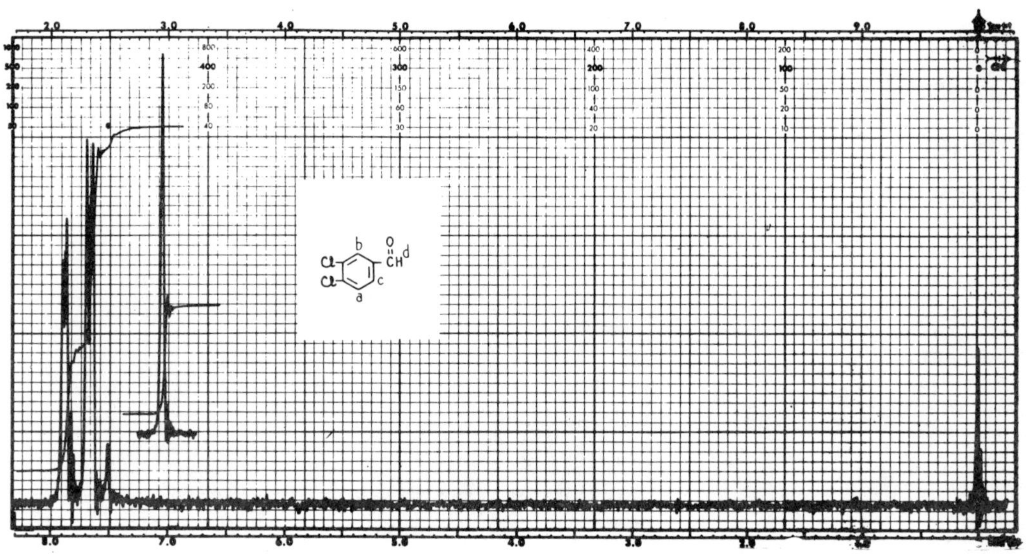

© 1978 Sadtler Research Laboratories, Inc.

6-HYDROXY-m-ANISALDEHYDE

$C_8H_8O_3$

Mol. Wt. 152.15

B.P. 103°C/2.5mm

Solvent: CCl_4

Assignments
a = 3.70
b = 6.79
c = 6.91
d = 7.02
e = 9.70
f = 10.50

2178

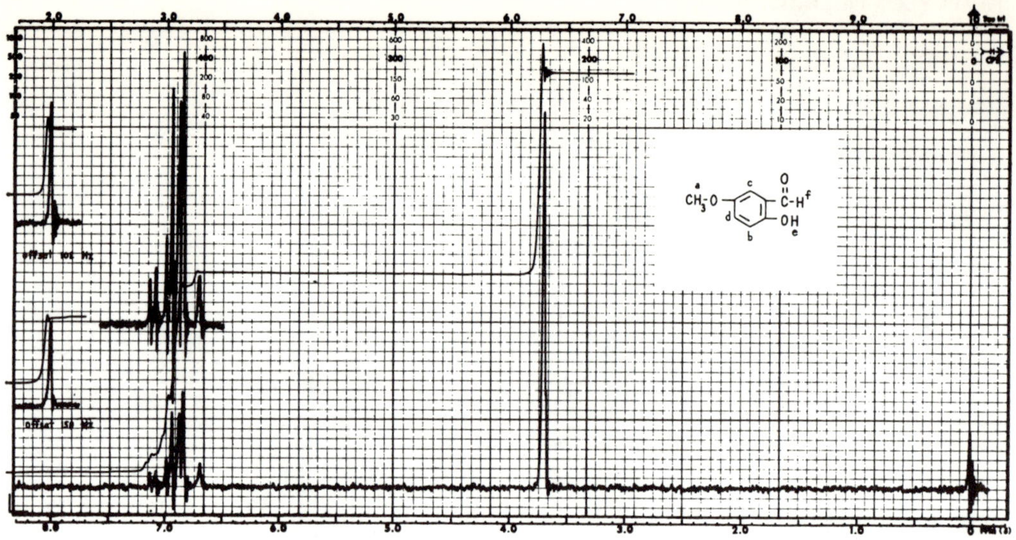

© 1978 Sadtler Research Laboratories, Inc.

5-INDANCARBOXALDEHYDE

$C_{10}H_{10}O$

Mol. Wt. 146.19

B.P. 255-257°C

Solvent: CCl_4

Assignments
a = 2.12
b = 2.91
c = 7.20
d = 7.52
e = 7.59
f = 9.82

2179

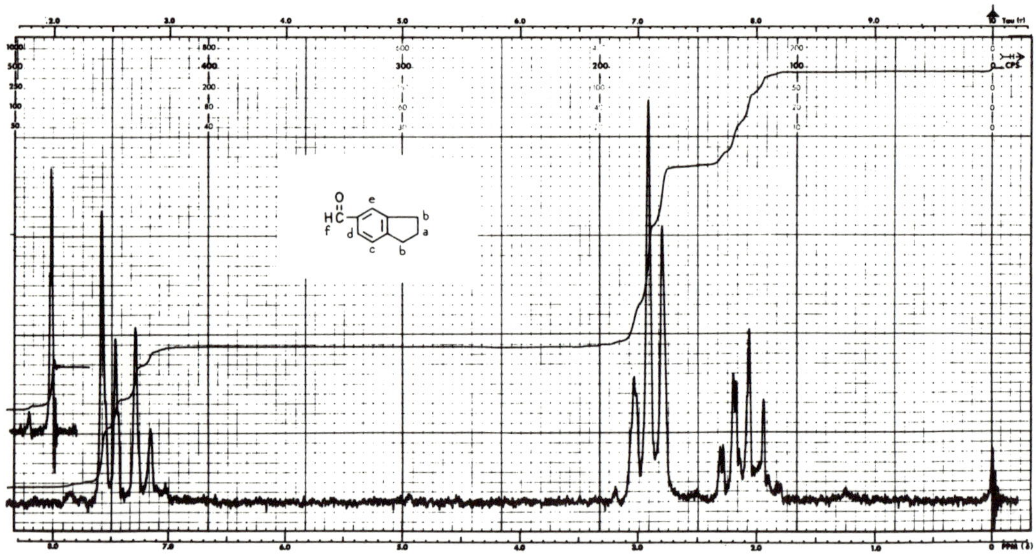

© 1978 Sadtler Research Laboratories, Inc.

FLUORENE-2-CARBOXALDEHYDE

$C_{14}H_{10}O$

Mol. Wt. 194.22

M.P. 85-86°C

Solvent: $CDCl_3$

Assignments
a = 3.72
b = 7.19-7.51
c = 7.53-7.91
d = 9.93

2180

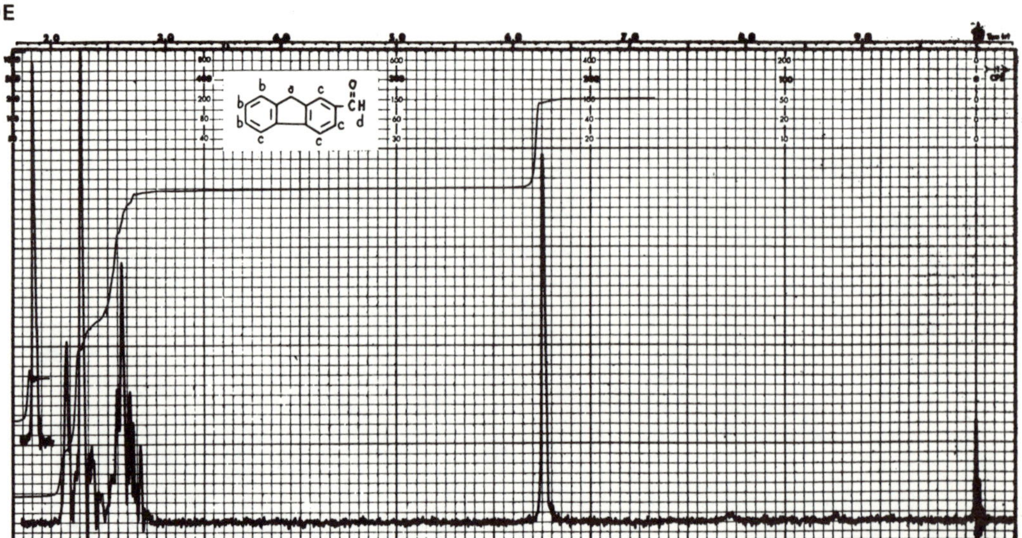

© 1978 Sadtler Research Laboratories, Inc.

3-PYRENECARBOXALDEHYDE

$C_{17}H_{10}O$

Mol. Wt. 230.27

M.P. 124°C

Solvent: $CDCl_3$

Assignments
a = 7.48-8.07
b = 9.96
c = 10.39

2181

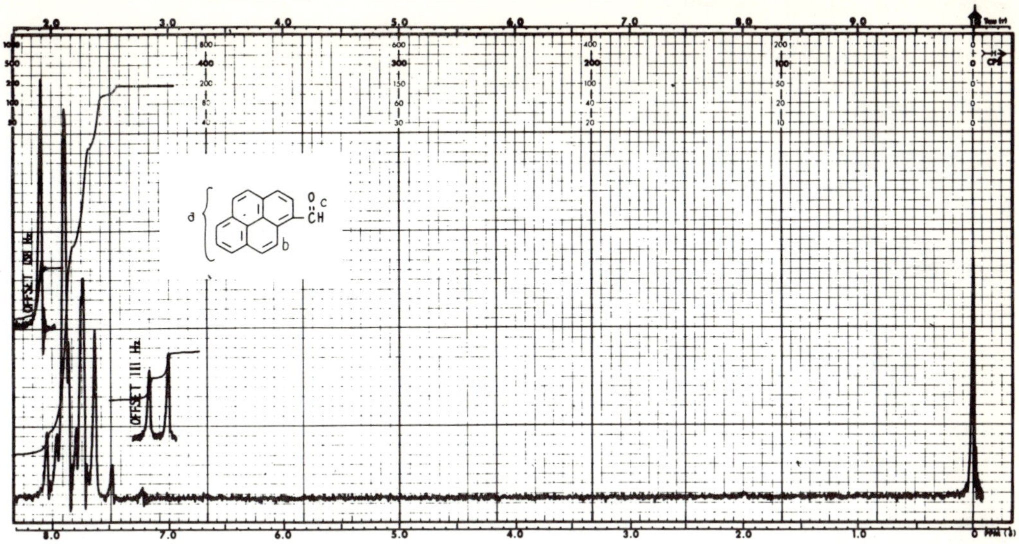

© 1978 Sadtler Research Laboratories, Inc.

PYRROLE-2-CARBOXALDEHYDE

C_5H_5NO

Mol. Wt. 95.10

M.P. 45-47°C

Solvent: $CDCl_3$

Assignments
a = 6.29
b = 6.97
c = 7.12
d = 9.48
e = 11.20

2182

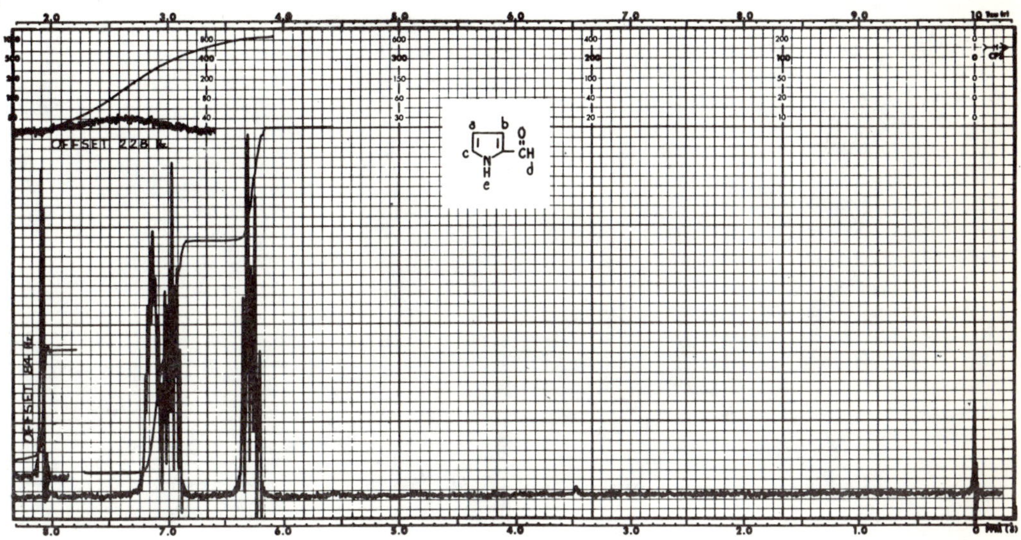

© 1978 Sadtler Research Laboratories, Inc.

3-QUINOLINECARBOXALDEHYDE

$C_{10}H_7NO$

Mol. Wt. 157.17

Solvent: $CDCl_3$

Assignments
a = 7.40-8.10
b = 8.15
c = 8.58
d = 9.31
e = 10.23

2183

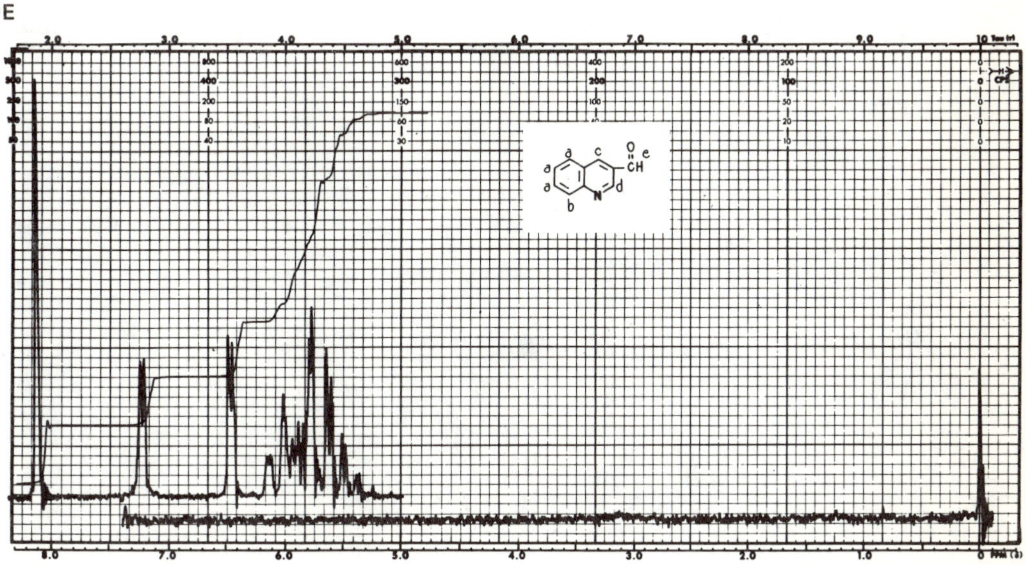

© 1978 Sadtler Research Laboratories, Inc.

5-METHYL-2-THIOPHENECARBOXALDEHYDE

C_6H_6OS

Mol. Wt. 126.18

Solvent: $CDCl_3$

Assignments
a = 2.54
b = 6.87
c = 7.59
d = 9.76

2184

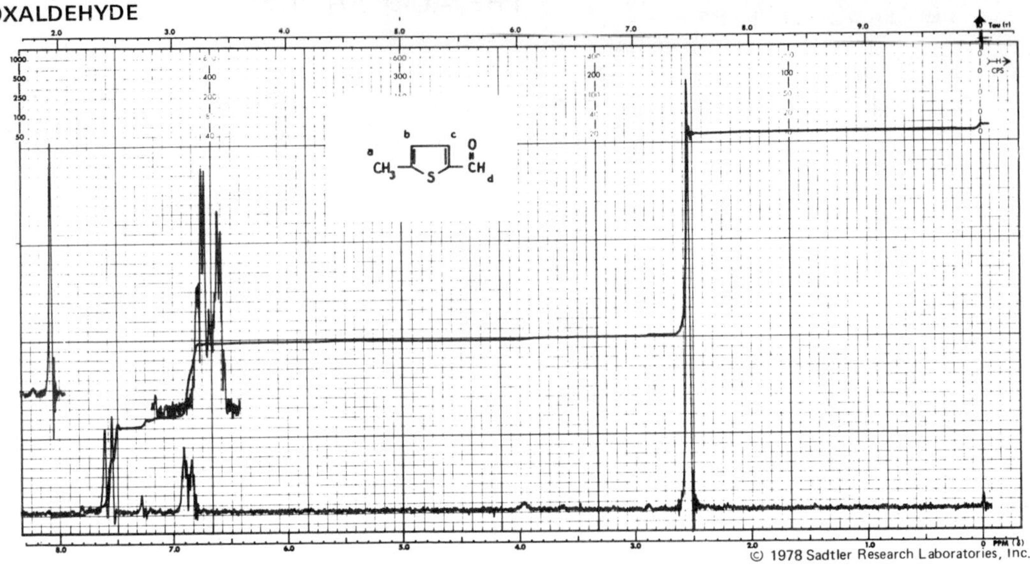

© 1978 Sadtler Research Laboratories, Inc.

2-FURALDEHYDE

$C_5H_4O_2$

Mol. Wt. 96.09

B.P. 41-42°C/5mm

Solvent: CCl_4

Assignments
a = 6.52
b = 7.11
c = 7.61
d = 9.63

2185

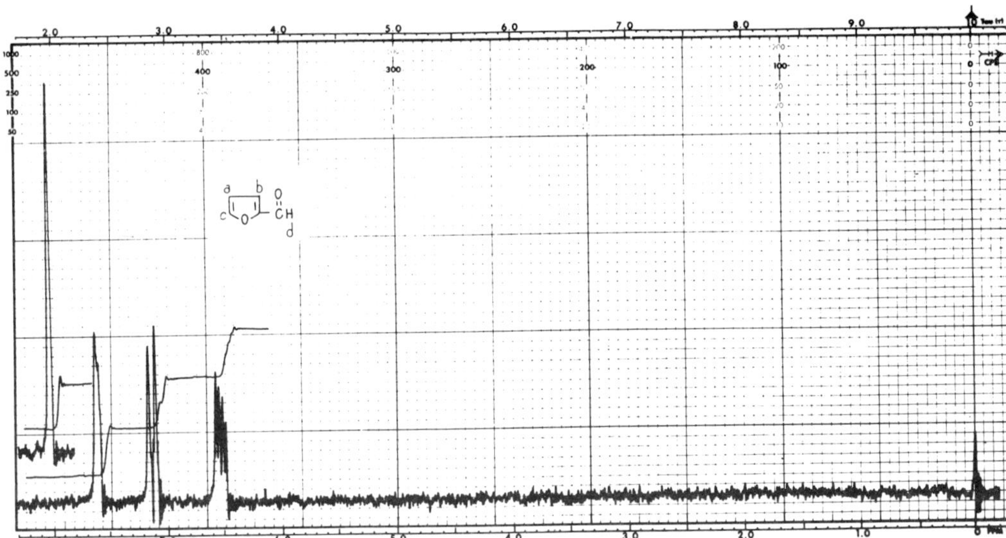

© 1978 Sadtler Research Laboratories, Inc.

PIPERONAL

$C_8H_6O_3$

Mol. Wt. 150.14

M.P. 35.3°C

Solvent: $CDCl_3$

Assignments
a = 6.01
b = 6.83
c = 7.24
d = 7.31
e = 9.72

2186

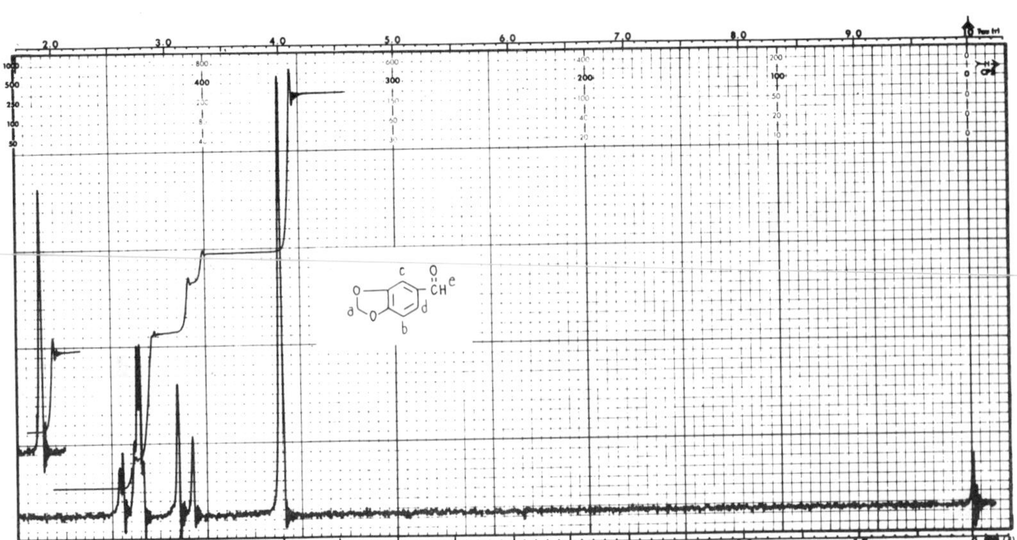

© 1978 Sadtler Research Laboratories, Inc.

Due to the limited availability of compounds containing Acid Fluoride, Acid Bromide or Acid Iodide groups, the spectra in this chapter and the following discussion will deal primarily with the NMR parameters encountered in the spectra of the Acid Chlorides. The table shown below indicates the relative deshielding effect of three of the Acetyl Halides and their hydrolysis product, Acetic acid.

CH_3–C(=O)–X		Solvent	Spectrum
2.06	–OH	CCl_4	
2.18	–I	CCl_4	
2.66	–Cl	CCl_4	
2.79	–Br	$CDCl_3$	

Chemical Shifts

Aliphatic Protons

HC–3	HC–2	HC–1			
		2.66	CH_3–C(=O)–Cl	CCl_4	----
	1.22	2.93	CH_3–CH_2–C(=O)–Cl	$CDCl_3$	2187
0.99	1.75	2.88	CH_3–CH_2–CH_2–C(=O)–Cl	$CDCl_3$	2188
	(1.21)	2.59	$(CH_3)_2$–CH–C(=O)–Cl	$CDCl_3$	----
(1.10)	----	2.79	$(CH_3)_3$–C–CH_2–C(=O)–Cl	CCl_4	2189

2-Substituted Propionyl Chlorides

Cl–C(=O)–	CH_2	–CH_2	–X		
	2.88	1.75	–CH_3	$CDCl_3$	2188
	2.88	1.77	–C_5	CCl_4	2203
	2.62	2.62	–C(=O)–O–R	CCl_4	----
	3.02	2.89	–Ø	CCl_4	2212
	3.28	3.28	–C(=O)–Cl	$CDCl_3$	2198
	3.51	3.51	–Br	CCl_4	2196

Substituted Acetyl Chlorides

Cl–C(=O)–	CH_2	–X		
	2.79	–$C(CH_3)_3$	CCl_4	2189
	2.93	–CH_3	$CDCl_3$	2187
	3.80	–C(=O)–O–R	CCl_4	----

Aliphatic Protons continued

Substituted Acetyl Chlorides continued

Cl–C(=O)–	CH_2	–X	Solvent	Spectrum
	4.02	–Ø	CCl_4	- - - -
	4.10	–Ø (4–Cl)	$CDCl_3$	2217
	4.80	–O–Ø	CCl_4	2215

Aromatic Protons

Benzoyl Halides

para/meta	ortho	–C(=O)–X	Solvent	Spectrum
7.2–7.7	7.94	–C(=O)–F	CCl_4	2219
7.2–7.7	8.02	–C(=O)–Br	CCl_4	2221
7.2–7.7	8.04	–C(=O)–Cl	CCl_4	2220

Para Substituted Benzoyl Chlorides

Cl–C(=O)–	ortho	meta	para	Solvent	Spectrum
	7.98	6.89	$-O-R_5$	CCl_4	- - - -
	8.02	6.90	$-O-CH_3$	CCl_4	- - - -
	7.96	7.24	$-CH_3$	CCl_4	2223
	8.02	7.46	–Cl	$CDCl_3$	2231
	8.06	7.52	$-C(CH_3)_3$	$CDCl_3$	2224
	7.94	7.61	–Br	$CDCl_3$	- - - -
	8.20	7.73	$-CF_3$	$CDCl_3$	2226
	8.22	7.81	–C≡N	$CDCl_3$	- - - -
	8.21	7.90	–N=N–Ø	$CDCl_3$	- - - -
	8.10	8.10	–C(=O)–O–R	CCl_4	- - - -
	8.26	8.26	–C(=O)–Cl	$CDCl_3$	- - - -
	8.38	8.38	$-NO_2$	Acetone	- - - -

Solubility and Solvent Effects

The Acid Halides are readily soluble in the chlorinated solvents carbon tetrachloride and deutero-chloroform. Due to the ease with which the Acid Halides hydrolyze to form the corresponding carboxylic acid, these solvents are preferable to those such as Polysol, Acetone or DMSO-d_6 which may contain traces of water.

PROPIONYL CHLORIDE

C_3H_5ClO

Mol. Wt. 92.53

B.P. 77-79°C

Solvent: $CDCl_3$

Assignments
a = 1.22
b = 2.93

2187

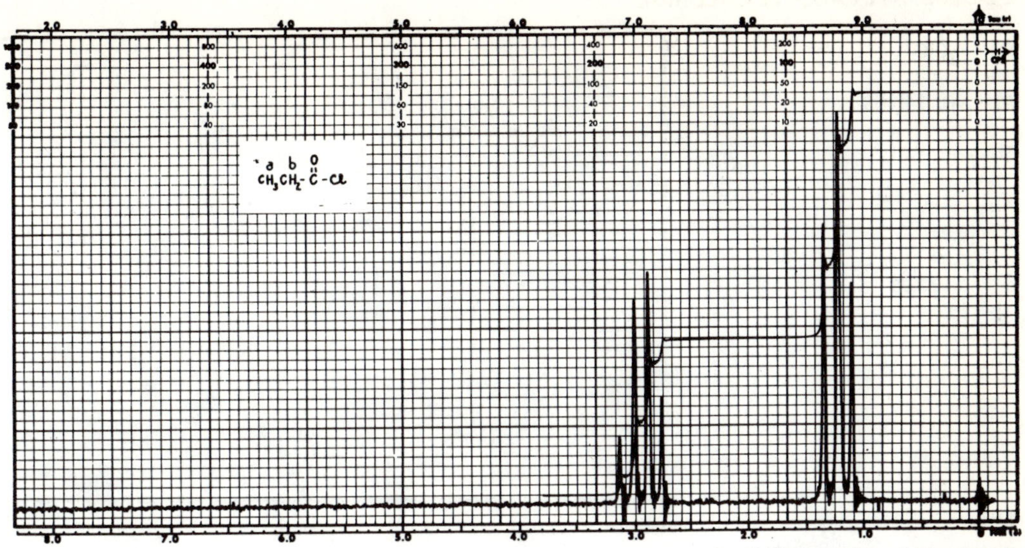

© 1978 Sadtler Research Laboratories, Inc.

BUTYRYL CHLORIDE

C_4H_7ClO

Mol. Wt. 106.55

B.P. 100-102°C

Solvent: $CDCl_3$

Assignments
a = 0.99
b = 1.75
c = 2.88

2188

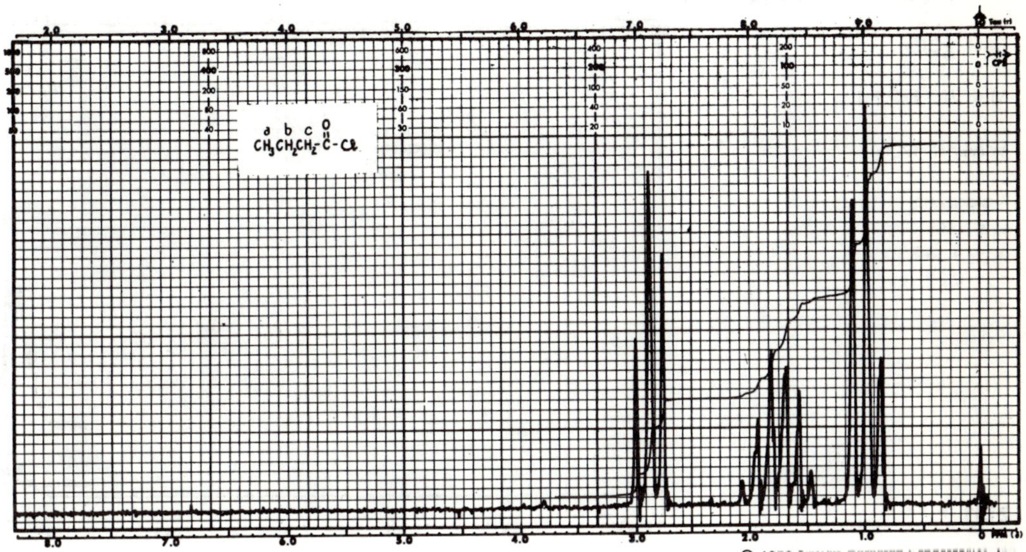

© 1978 Sadtler Research Laboratories, Inc.

3,3-DIMETHYLBUTYRYL CHLORIDE

$C_6H_{11}ClO$

Mol. Wt. 134.61

B.P. 127-129°C

Solvent: CCl_4

Assignments
a = 1.10
b = 2.79

2189

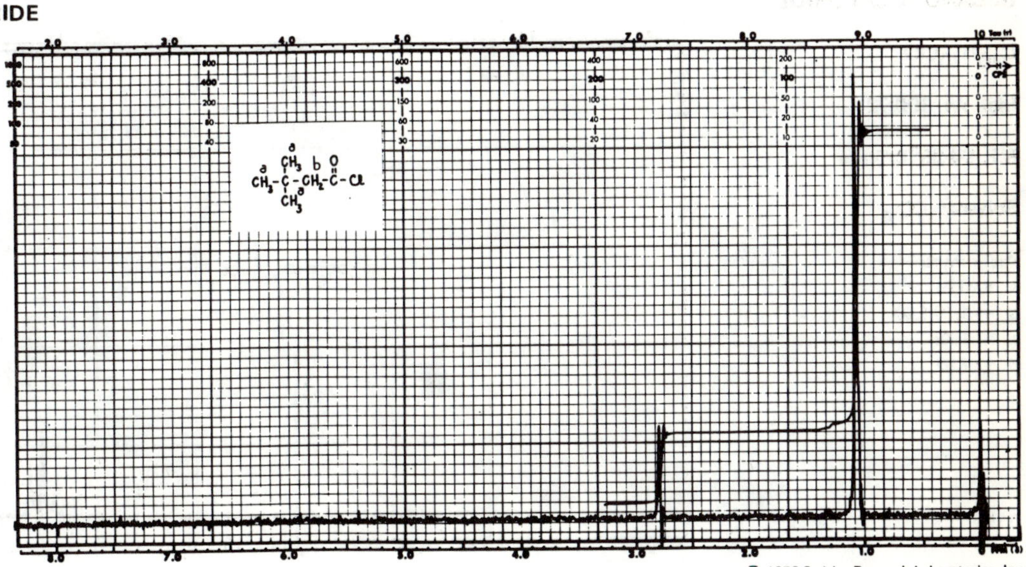

© 1978 Sadtler Research Laboratories, Inc.

2,2-DIMETHYLVALERYL CHLORIDE

$C_7H_{13}ClO$

Mol. Wt. 148.63

Solvent: CCl_4

Assignments
a = 0.95
b = 1.29
c = 1.10-2.00

2190

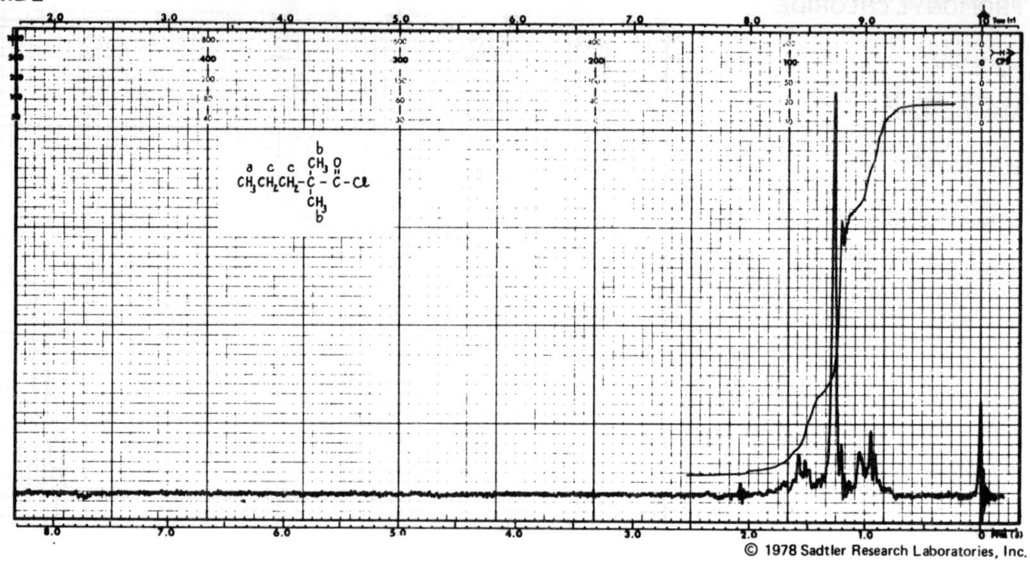

© 1978 Sadtler Research Laboratories, Inc.

2-PROPYLVALERYL CHLORIDE

$C_8H_{15}ClO$

Mol. Wt. 162.66

Solvent: CCl_4

Assignments
a = 0.95
b = 1.10-2.10
c = 2.72

2191

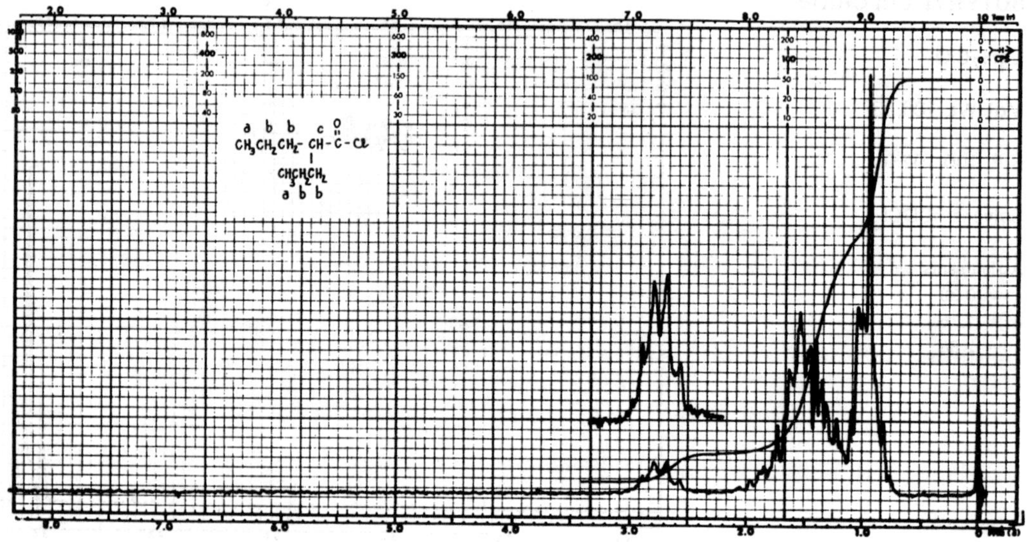

© 1978 Sadtler Research Laboratories, Inc.

DECANOYL CHLORIDE

$C_{10}H_{10}ClO$

Mol. Wt. 190.72

B.P. 94-96°C/5mm

Solvent: CCl_4

Assignments
a = 0.91
b = 1.10-1.60
c = 1.72
d = 2.87

2192

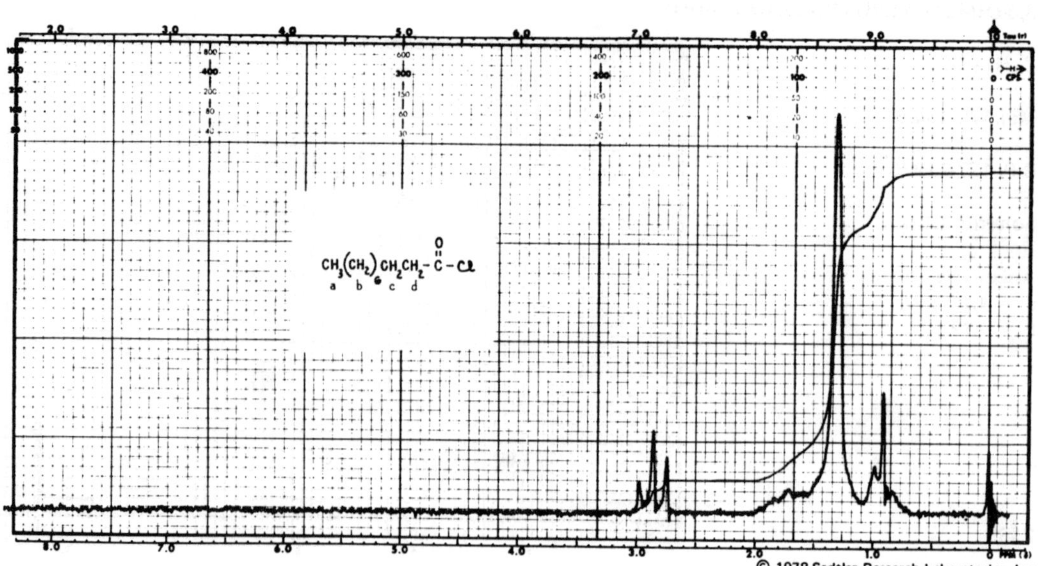

© 1978 Sadtler Research Laboratories, Inc.

LAUROYL CHLORIDE

$C_{12}H_{23}ClO$

Mol. Wt. 218.77

B.P. 87-89°C/0.5mm

Solvent: $CDCl_3$

Assignments

a = 0.89
b = 1.30
c = 1.70
d = 2.86

2193

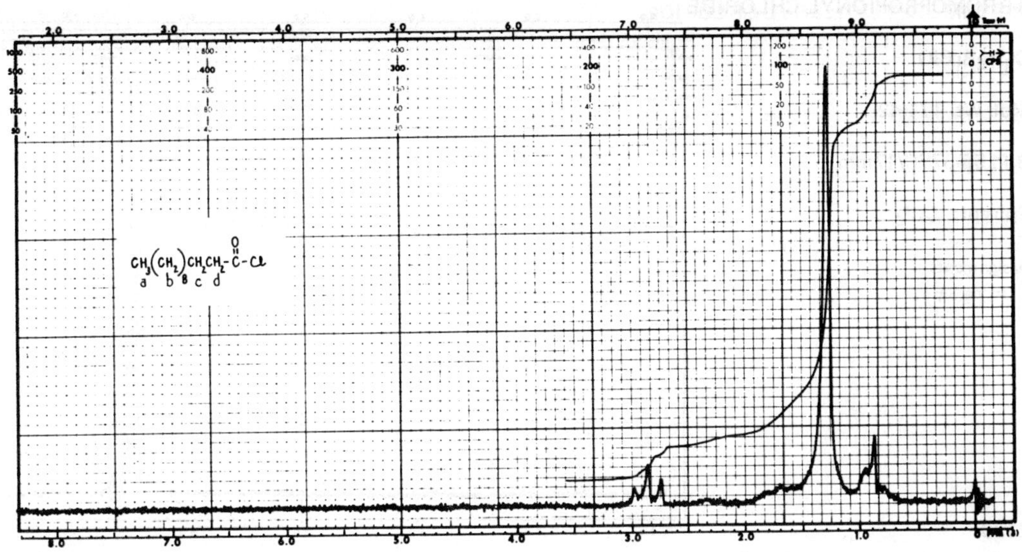

© 1978 Sadtler Research Laboratories, Inc.

STEAROYL CHLORIDE

$C_{18}H_{35}ClO$

Mol. Wt. 302.93

M.P. 19-21°C

Solvent: CCl_4

Assignments

a = 0.90
b = ca 1.28
c = 1.72
d = 2.85

2194

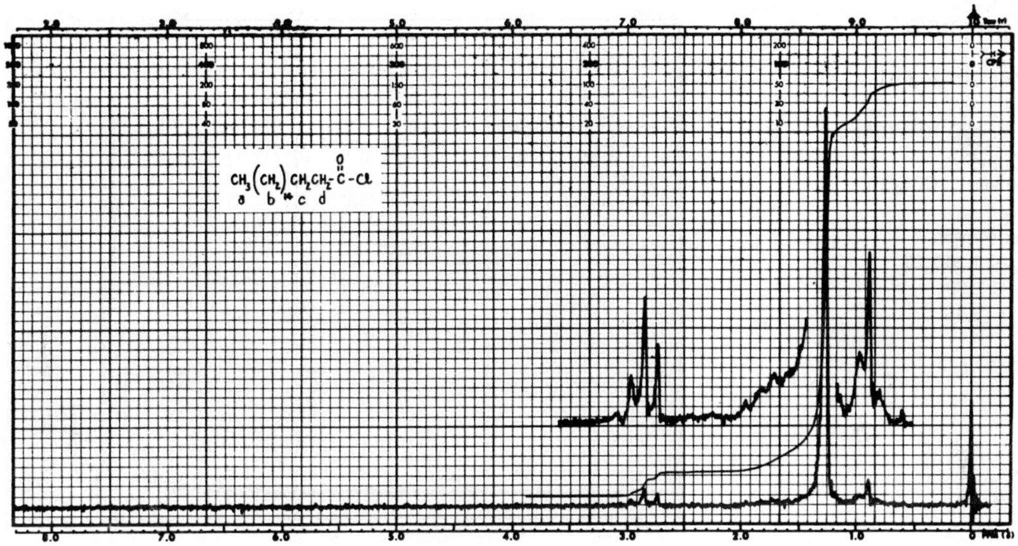

© 1978 Sadtler Research Laboratories, Inc.

4-CHLOROBUTYRYL CHLORIDE

$C_4H_6Cl_2O$

Mol. Wt. 141.00

Solvent: CCl_4

Assignments

a = 2.14
b = 3.09
c = 3.55

2195

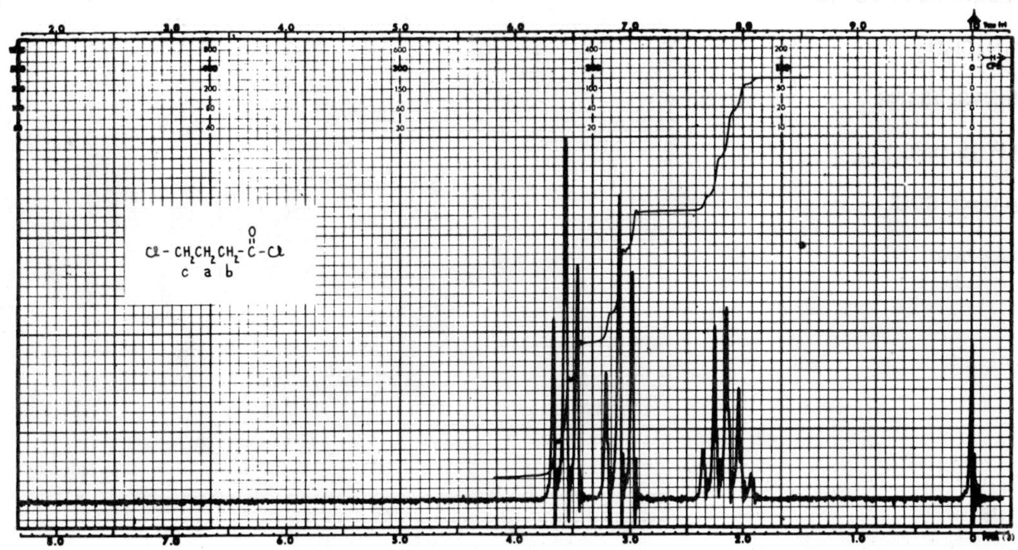

© 1978 Sadtler Research Laboratories, Inc.

3-BROMOPROPIONYL CHLORIDE

C_3H_4BrClO

Mol. Wt. 171.42

B.P. 55-57°C/17mm

Solvent: CCl_4

Assignments
a = ca 3.51

2196

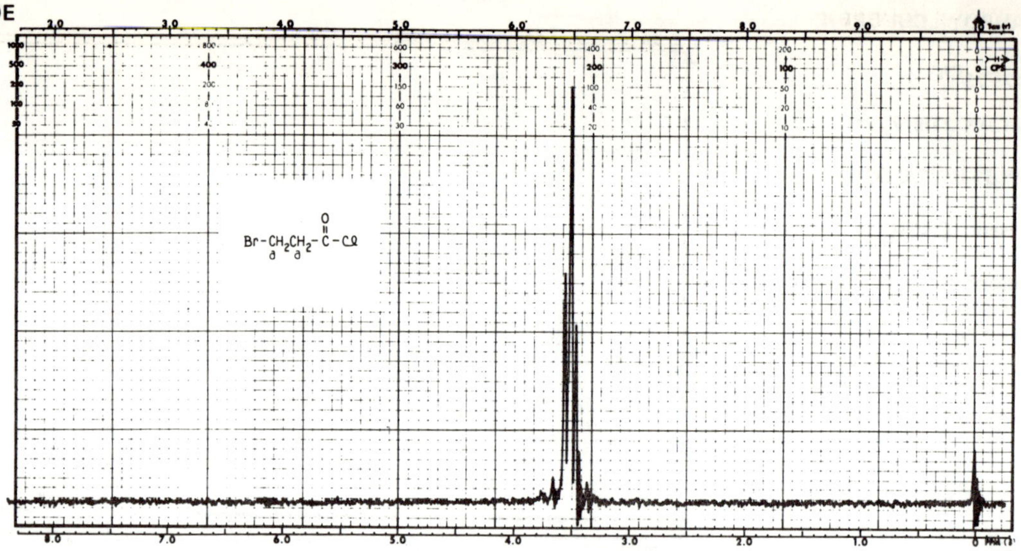

© 1978 Sadtler Research Laboratories, Inc.

6-BROMOHEXANOYL CHLORIDE

$C_6H_{10}BrClO$

Mol. Wt. 213.51

B.P. 100-101°C/6mm

Solvent: CCl_4

Assignments
a = 1.20-2.20
b = 2.92
c = 3.40

2197

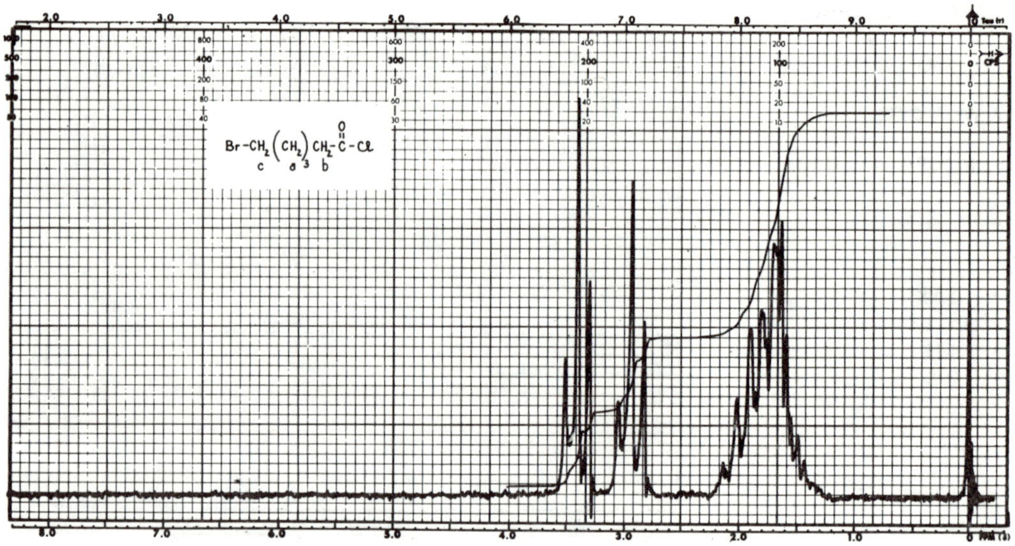

© 1978 Sadtler Research Laboratories, Inc.

SUCCINYL CHLORIDE

$C_4H_4Cl_2O_2$

Mol. Wt. 154.98

M.P. 15-18°C

Solvent: $CDCl_3$

Assignments
a = 3.28

2198

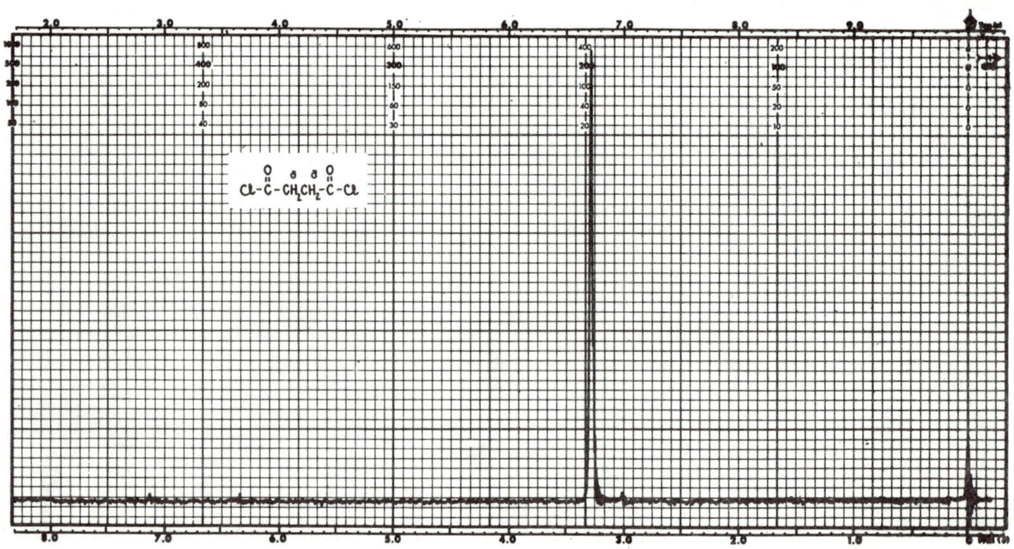

© 1978 Sadtler Research Laboratories, Inc.

ADIPOYL CHLORIDE

$C_6H_8Cl_2O_2$

Mol. Wt. 183.04

B.P. 99-101°C/5mm

Solvent: $CDCl_3$

Assignments
a = 1.77
b = 2.94

2199

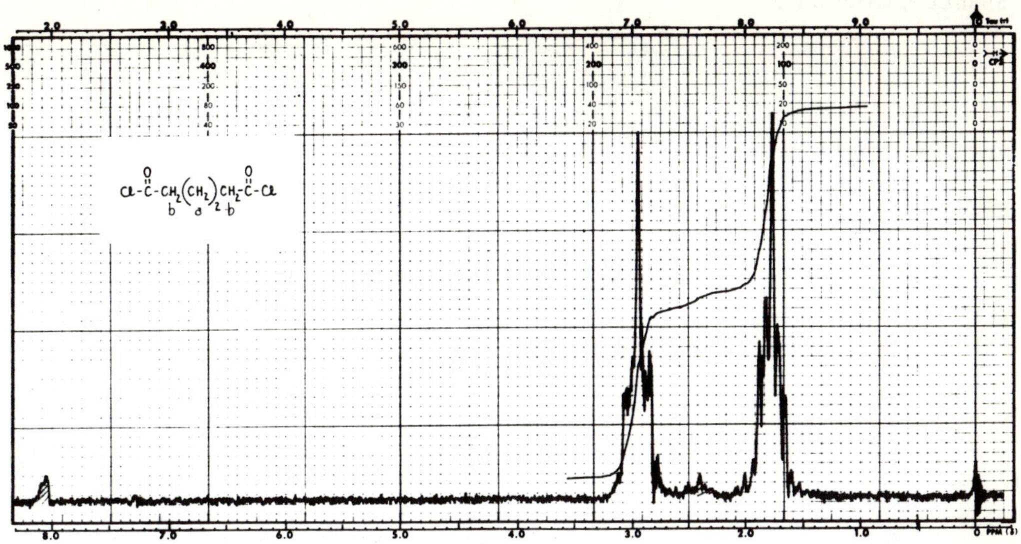

© 1978 Sadtler Research Laboratories, Inc.

PIMELOYL CHLORIDE

$C_7H_{10}Cl_2O_2$

Mol. Wt. 197.06

Solvent: CCl_4

Assignments
a = 1.20-2.05
b = 2.89

2200

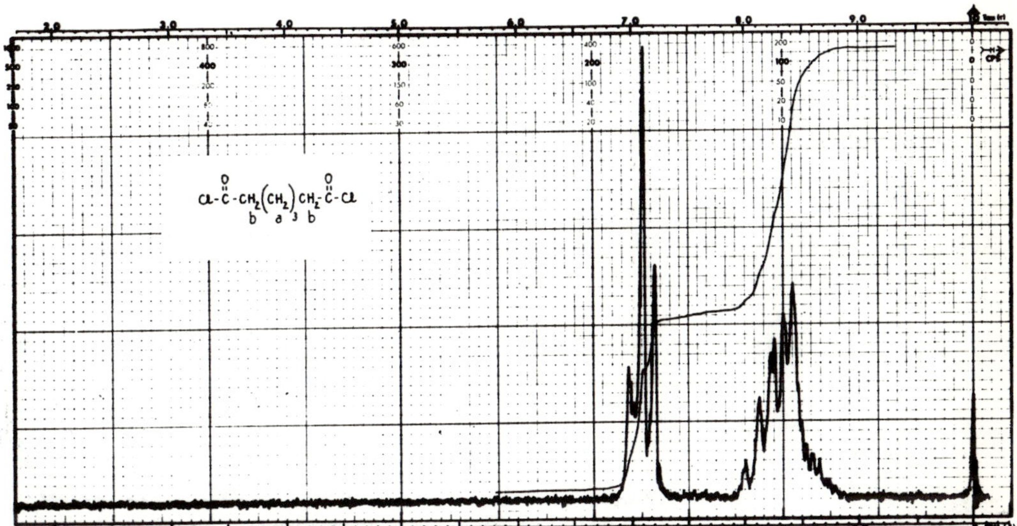

© 1978 Sadtler Research Laboratories, Inc.

AZELAOYL CHLORIDE

$C_9H_{14}Cl_2O_2$

Mol. Wt. 255.12

B.P. 180-183.5°C/35mm (lit.)

Solvent: $CDCl_3$

Assignments
a = 1.20-2.00
b = 2.90

2201

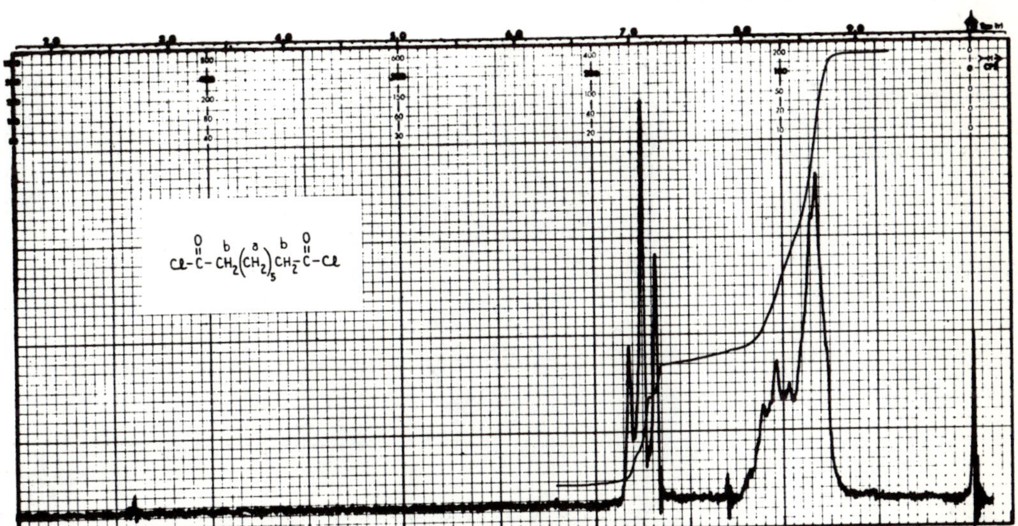

© 1978 Sadtler Research Laboratories, Inc.

SEBACOYL CHLORIDE

$C_{10}H_{16}Cl_2O_2$

Mol. Wt. 239.14

M.P. -7° to $-5^{o}C$

Solvent: CCl_4

Assignments
a = 1.37
b = 1.71
c = 2.88

2202

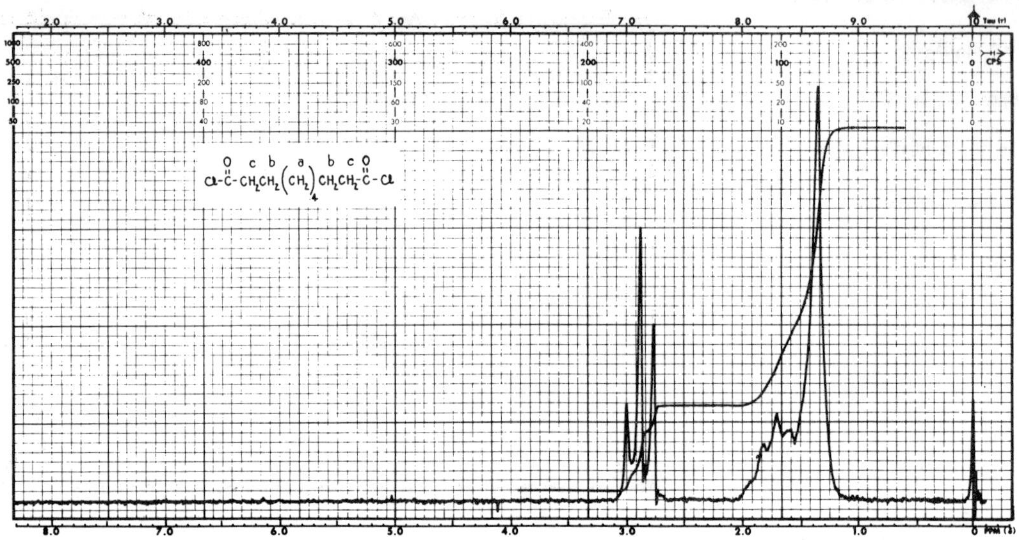

© 1978 Sadtler Research Laboratories, Inc.

3-CYCLOPENTANEPROPIONYL CHLORIDE

$C_8H_{13}ClO$

Mol. Wt. 160.65

B.P. $81\text{-}82^{o}C/10mm$

Solvent: CCl_4

Assignments
a = 0.85-2.11
b = 2.88

2203

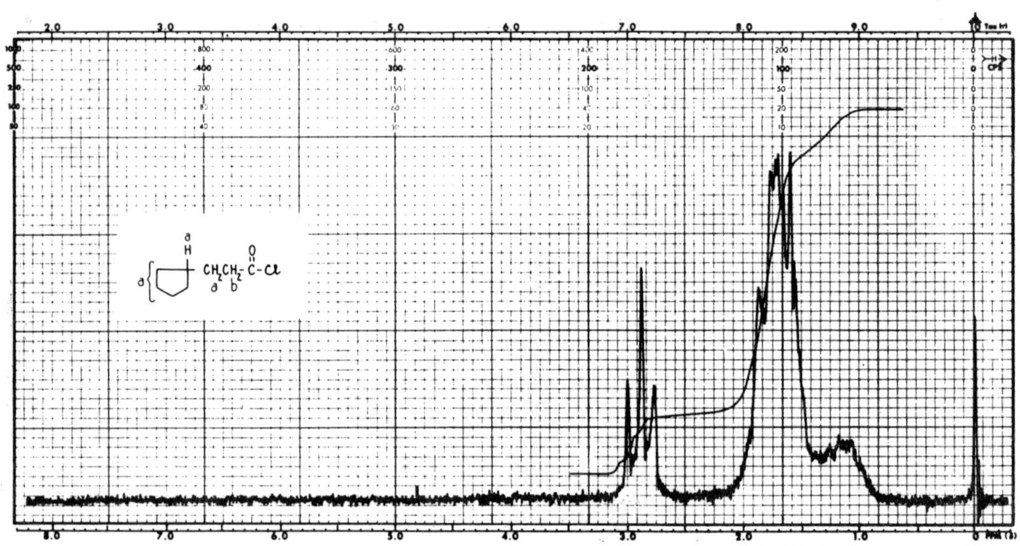

© 1978 Sadtler Research Laboratories, Inc.

CYCLOPENTANECARBONYL CHLORIDE

C_6H_9ClO

Mol. Wt. 132.59

Solvent: CCl_4

Assignments
a = 1.40-2.25
b = 3.19

2204

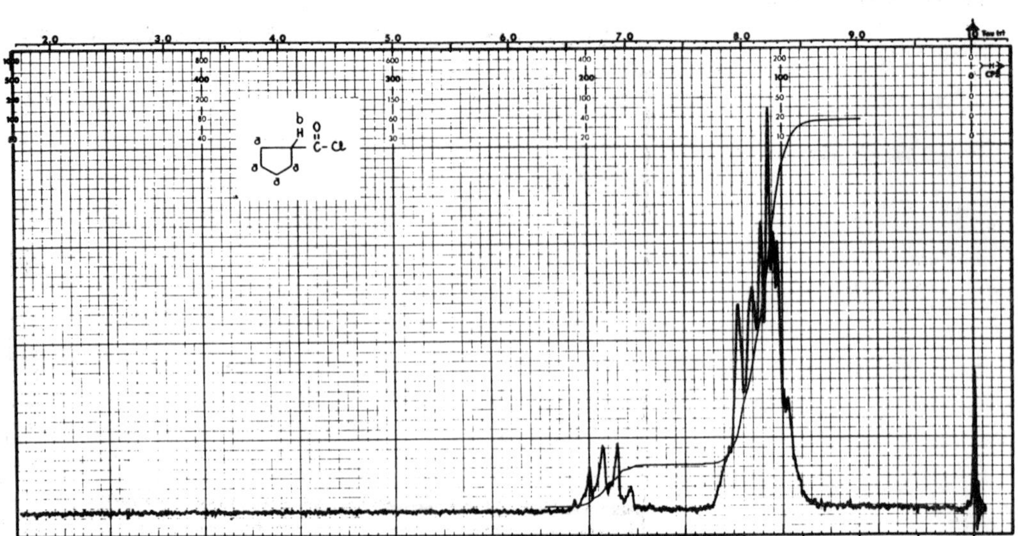

© 1978 Sadtler Research Laboratories, Inc.

CYCLOHEXANECARBONYL CHLORIDE

$C_7H_{11}ClO$

Mol. Wt. 146.62

B.P. 71-73°C/1mm

Solvent: CCl_4

Assignments
a = 1.00-2.45
b = 2.71

2205

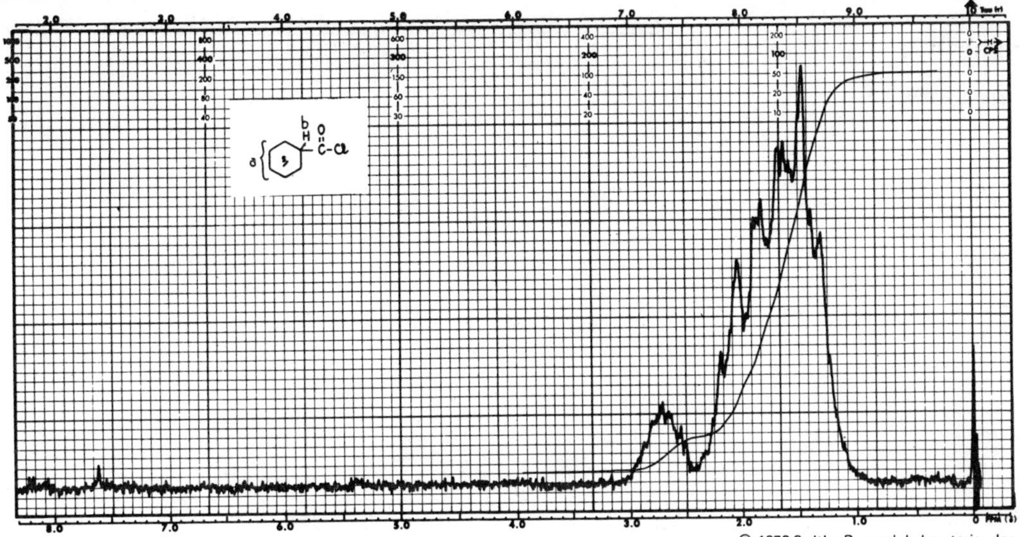

© 1978 Sadtler Research Laboratories, Inc.

10-UNDECENOYL CHLORIDE

$C_{11}H_{19}ClO$

Mol. Wt. 202.73

B.P. 120-122°C/10mm

Solvent: CCl_4

Assignments
a = 1.39
b = 1.78
c = 2.10
d = 2.92
e = 5.01
f = 5.07
g = 5.84

2206

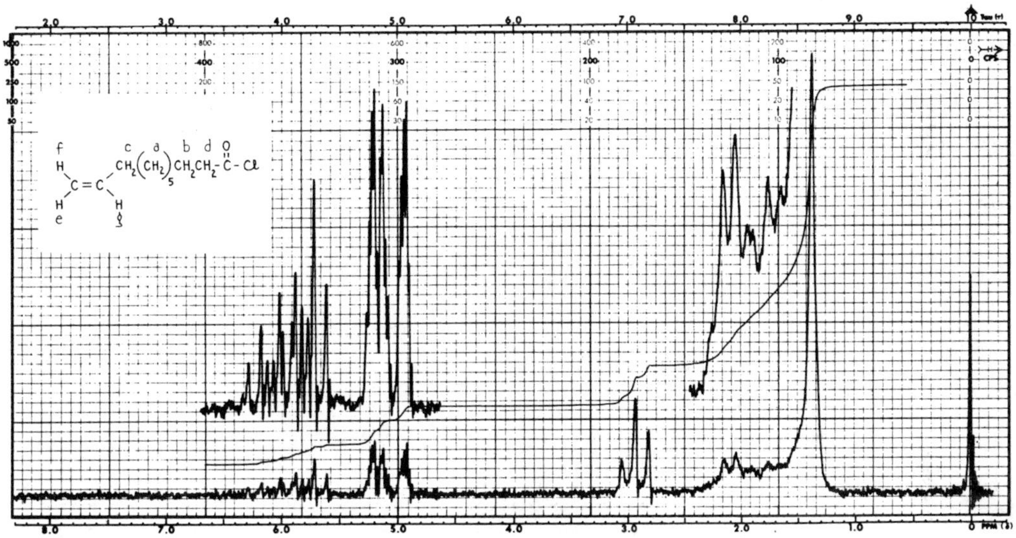

© 1978 Sadtler Research Laboratories, Inc.

OLEOYL CHLORIDE

$C_{10}H_{13}ClO$

Mol. Wt. 280.76

Solvent: CCl_4

Assignments
a = 0.90
b = 1.20-1.70
c = 1.70-2.40
d = 2.85
e = 5.30

2207

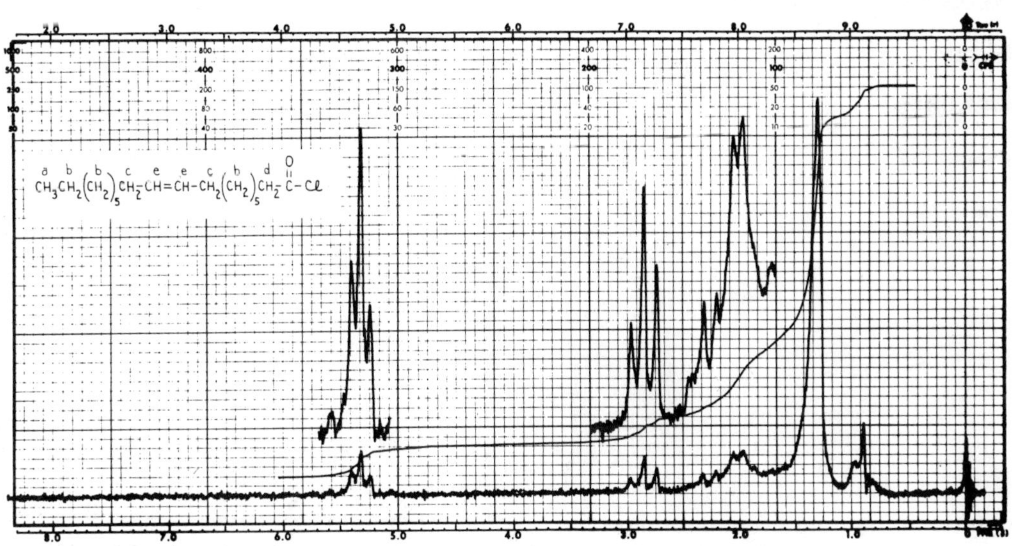

© 1978 Sadtler Research Laboratories, Inc.

SORBOYL CHLORIDE

C_6H_7ClO

Mol. Wt. 130.58

Solvent: CCl_4

Assignments
a = 1.92
b = 5.70-6.70
c = 7.38

2208

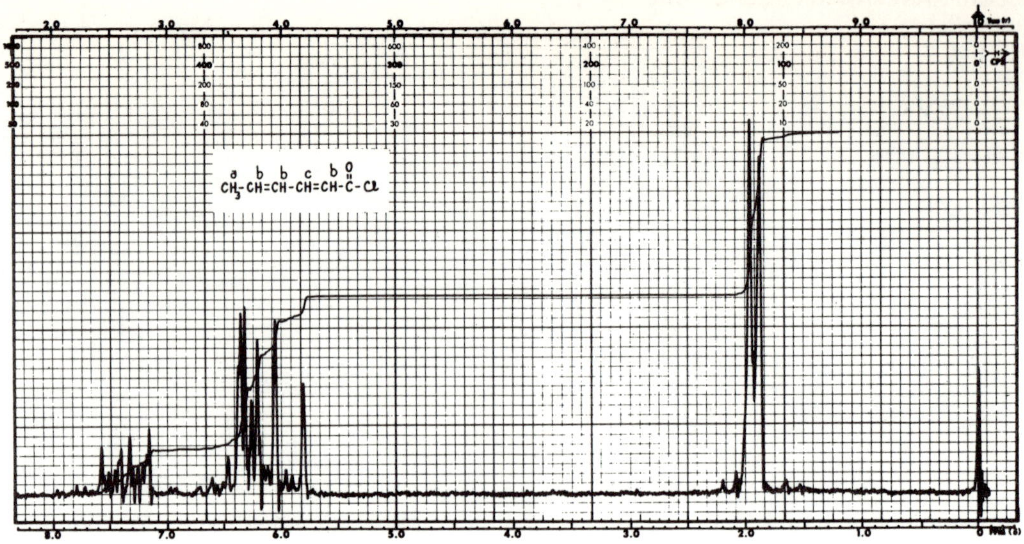

© 1978 Sadtler Research Laboratories, Inc.

FUMAROYL CHLORIDE

$C_4H_2Cl_2O_2$

Mol. Wt. 152.97

B.P. 159-161°C

Solvent: CCl_4

Assignments
a = 7.08

2209

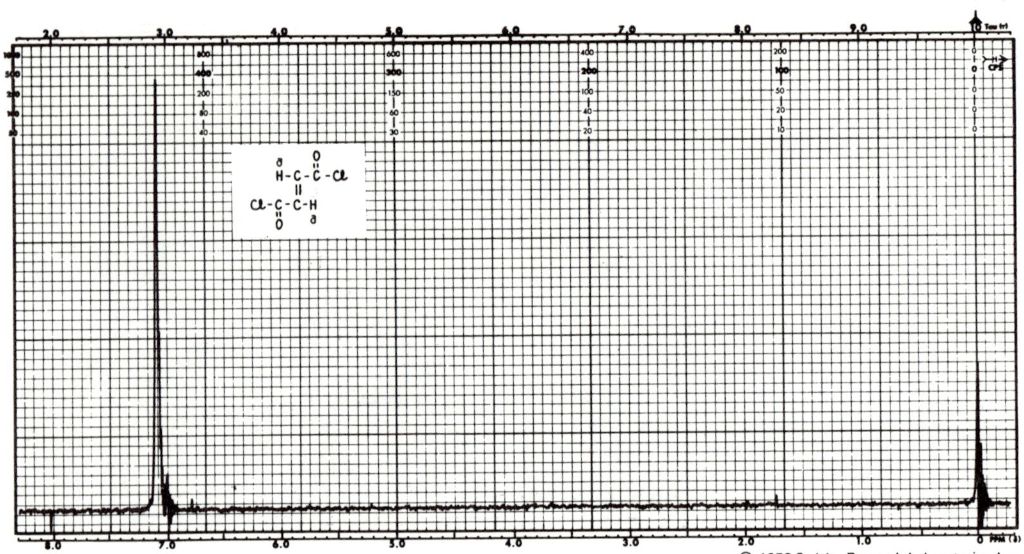

© 1978 Sadtler Research Laboratories, Inc.

1-CYCLOHEXENE-1-CARBONYL CHLORIDE

C_7H_9ClO

Mol. Wt. 144.60

Solvent: CCl_4

Assignments
a = ca 1.69
b = ca 2.32
c = 7.39

2210

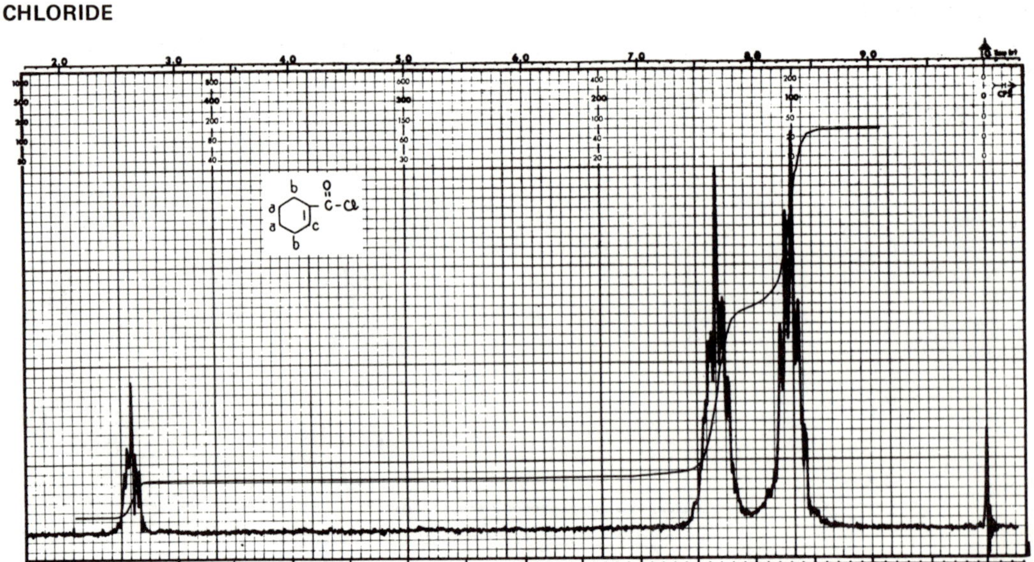

© 1978 Sadtler Research Laboratories, Inc.

3-CYCLOHEXENE-1-CARBONYL CHLORIDE

C_7H_9ClO

Mol. Wt. 144.60

B.P. 76-78°C/760mm (lit.)

Solvent: CCl_4

Assignments
a = 1.50-2.50
b = 2.98
c = ca 5.70

2211

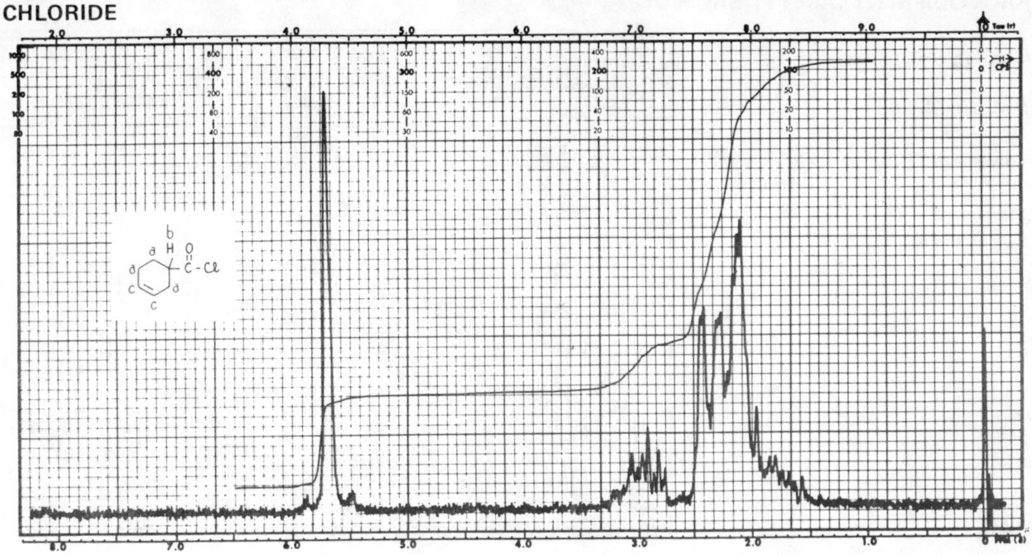

© 1978 Sadtler Research Laboratories, Inc.

HYDROCINNAMOYL CHLORIDE

C_9H_9ClO

Mol. Wt. 168.62

M.P. −7° to −5°C

Solvent: CCl_4

Assignments
a = 2.89
b = 3.02
c = 6.90-7.40

2212

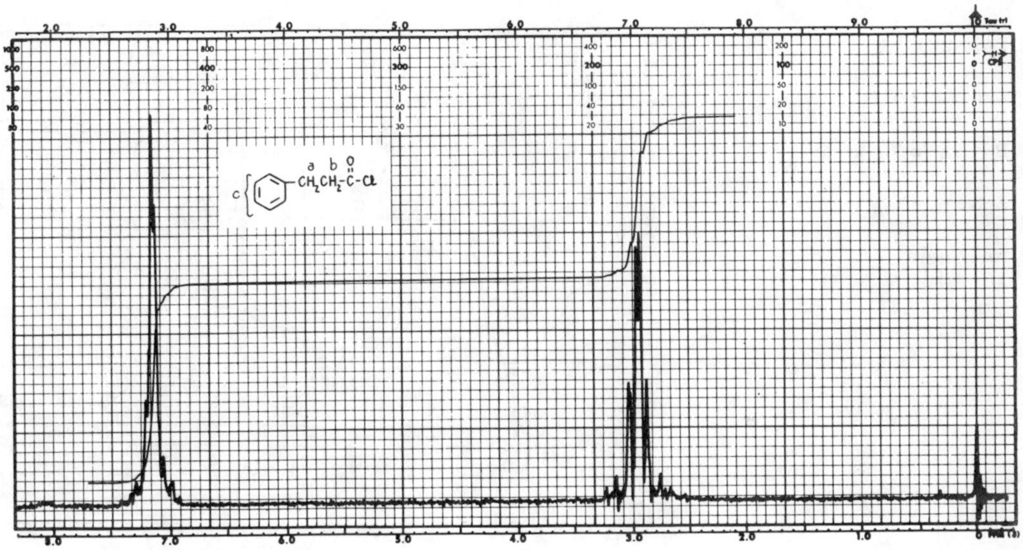

© 1978 Sadtler Research Laboratories, Inc.

CHLOROPHENYLACETYL CHLORIDE

$C_8H_6Cl_2O$

Mol. Wt. 189.04

B.P. 120°C/23mm

Solvent: CCl_4

Assignments
a = 5.52
b = 7.40

2213

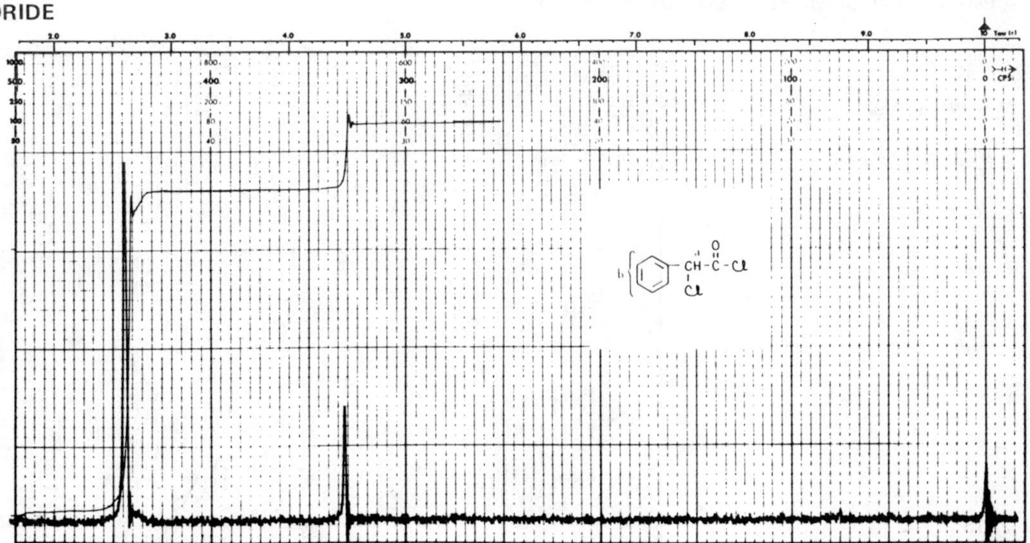

© 1978 Sadtler Research Laboratories, Inc.

BROMODIPHENYLACETYL BROMIDE

$C_{14}H_{10}Br_2O$

Mol. Wt. 354.05

Solvent: $CDCl_3$

Assignments
a = 7.10-7.60

2214

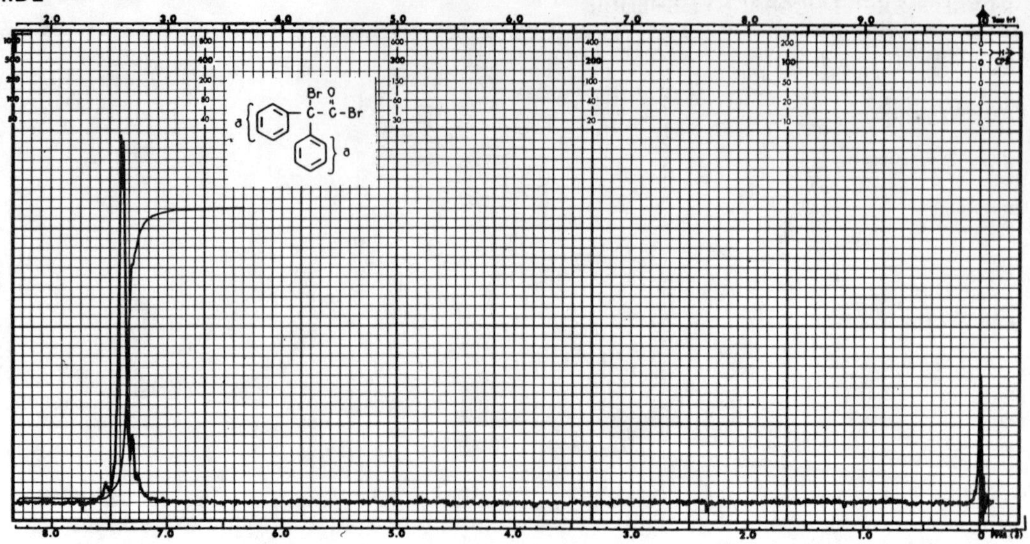

© 1978 Sadtler Research Laboratories, Inc.

PHENOXYACETYL CHLORIDE

$C_8H_7ClO_2$

Mol. Wt. 170.60

B.P. 96-98°C/16mm

Solvent: CCl_4

Assignments
a = 4.80
b = 6.70-7.40

2215

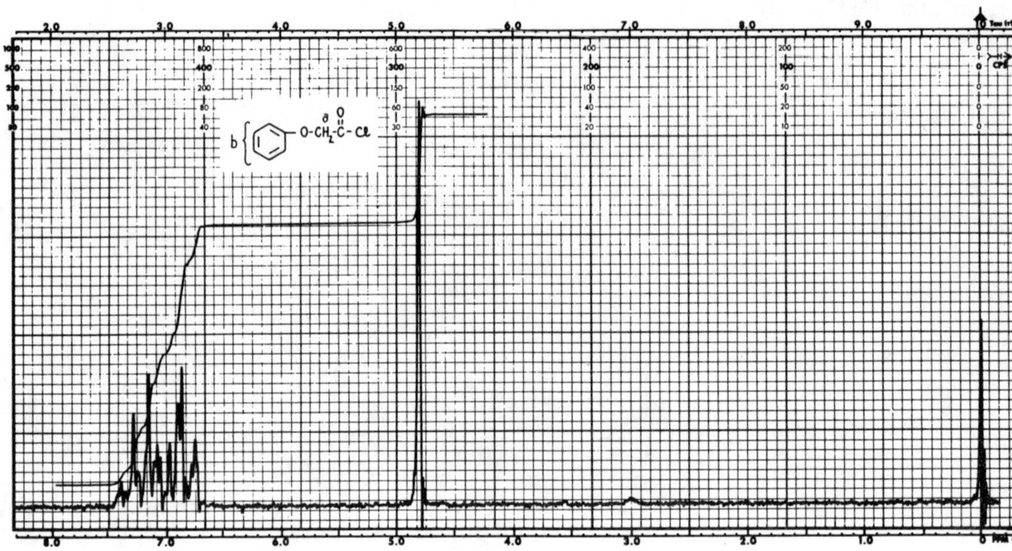

© 1978 Sadtler Research Laboratories, Inc.

2-PHENOXYPROPIONYL CHLORIDE

$C_9H_9ClO_2$

Mol. Wt. 184.62

B.P. 92-94°C/5mm

Solvent: CCl_4

Assignments
a = 1.71
b = 4.86
c = 6.82
d = 6.93
e = 7.29

2216

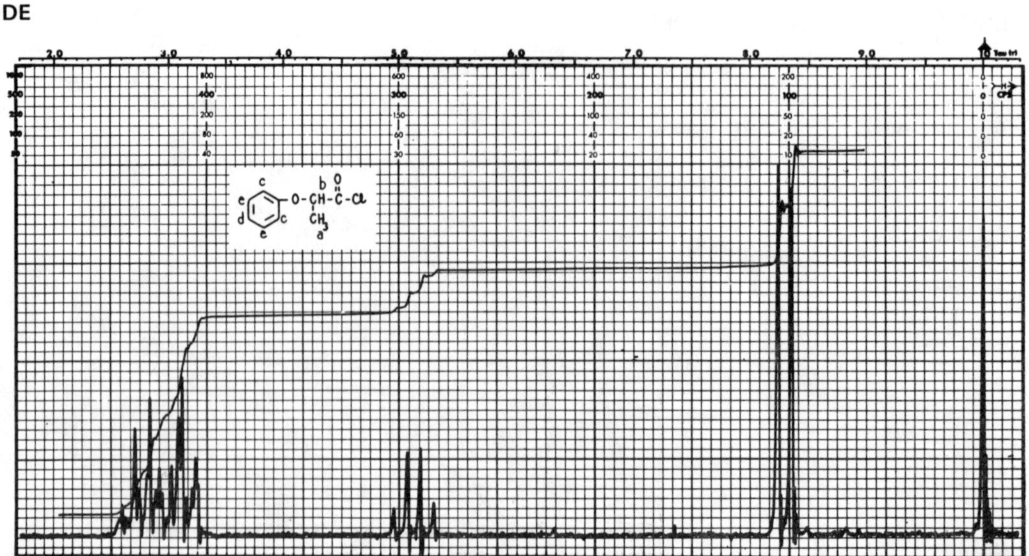

© 1978 Sadtler Research Laboratories, Inc.

(p-CHLOROPHENYL)ACETYL CHLORIDE

$C_8H_6Cl_2O$

Mol. Wt. 189.04

Solvent: $CDCl_3$

Assignments
a = 4.10
b = 7.18
c = 7.32

2217

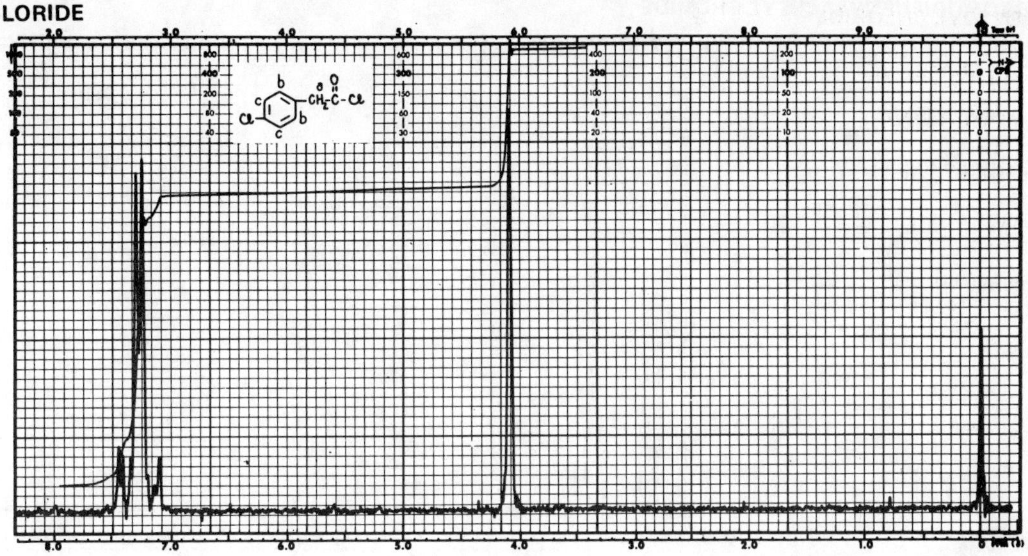

© 1978 Sadtler Research Laboratories, Inc.

4-(p-BROMOPHENYL)BUTYRYL CHLORIDE

$C_{10}H_{10}BrClO$

Mol. Wt. 261.55

B.P. 147-148°C/4mm (lit.)

Solvent: $CDCl_3$

Assignments
a = 1.87
b = 2.27 or c
c = 2.57 or b
d = 6.98
e = 7.33

2218

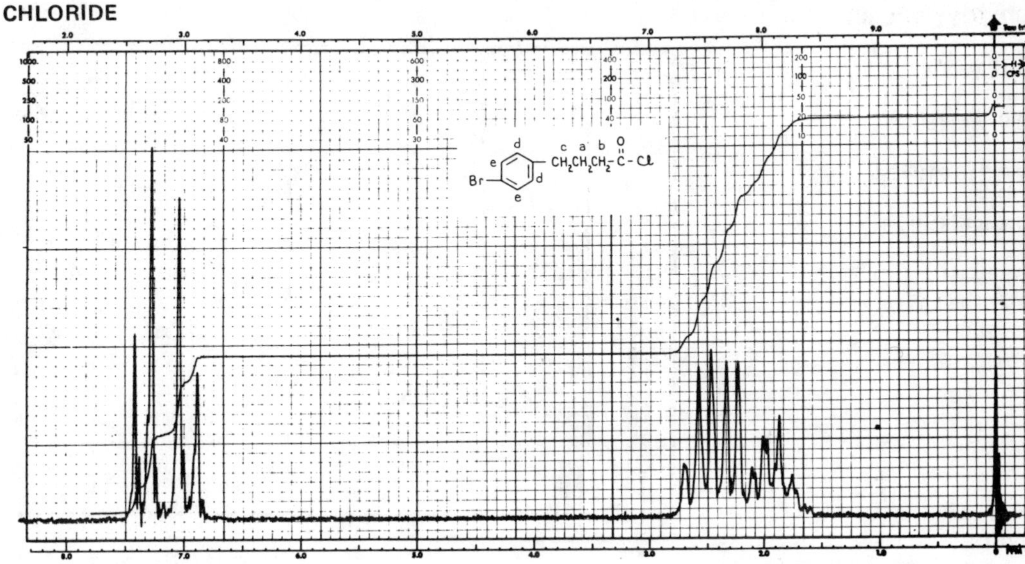

© 1978 Sadtler Research Laboratories, Inc.

BENZOYL FLUORIDE

C_7H_5FO

Mol. Wt. 124.12

B.P. 159-161°C

Solvent: CCl_4

Assignments
a = 7.25-7.80
b = 7.94

2219

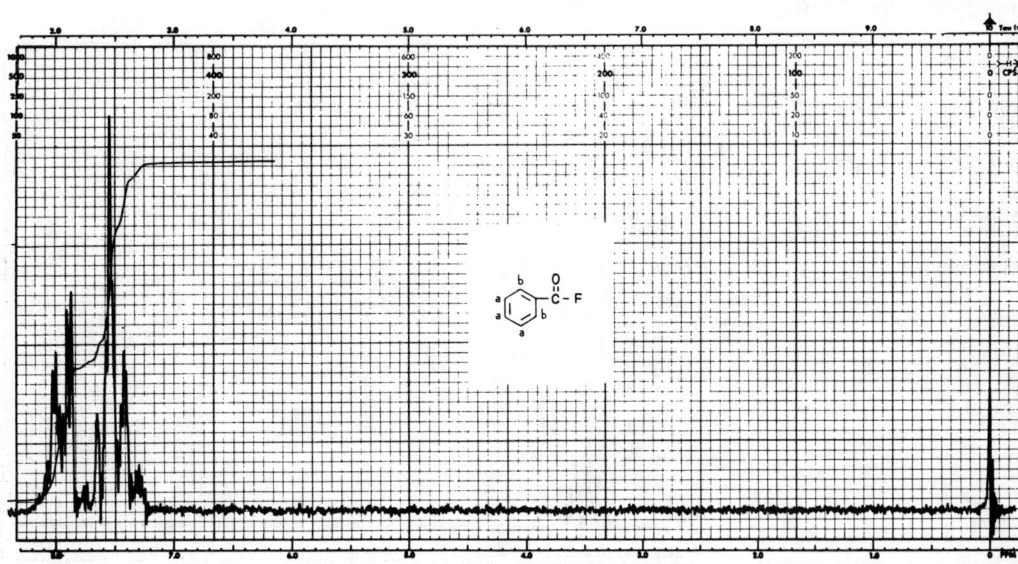

© 1978 Sadtler Research Laboratories, Inc.

BENZOYL CHLORIDE

C_7H_5ClO

Mol. Wt. 140.57

B.P. 198°C (lit.)

Solvent: CCl_4

Assignments
a = 7.24-7.78
b = 8.04

2220

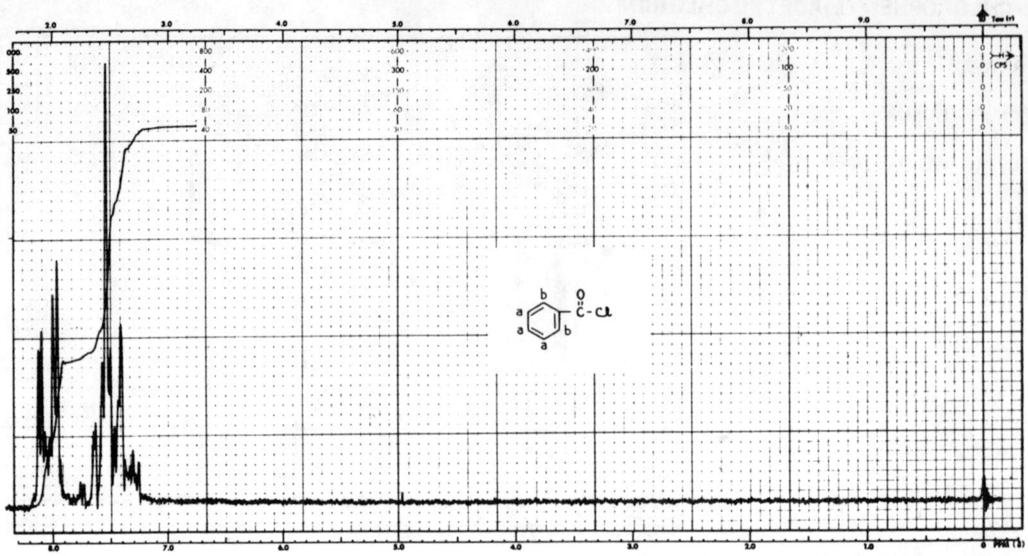

© 1978 Sadtler Research Laboratories, Inc.

BENZOYL BROMIDE

C_7H_5BrO

Mol. Wt. 185.20

B.P. 218-219°C

Solvent: CCl_4

Assignments
a = 7.28-7.90
b = 8.02

2221

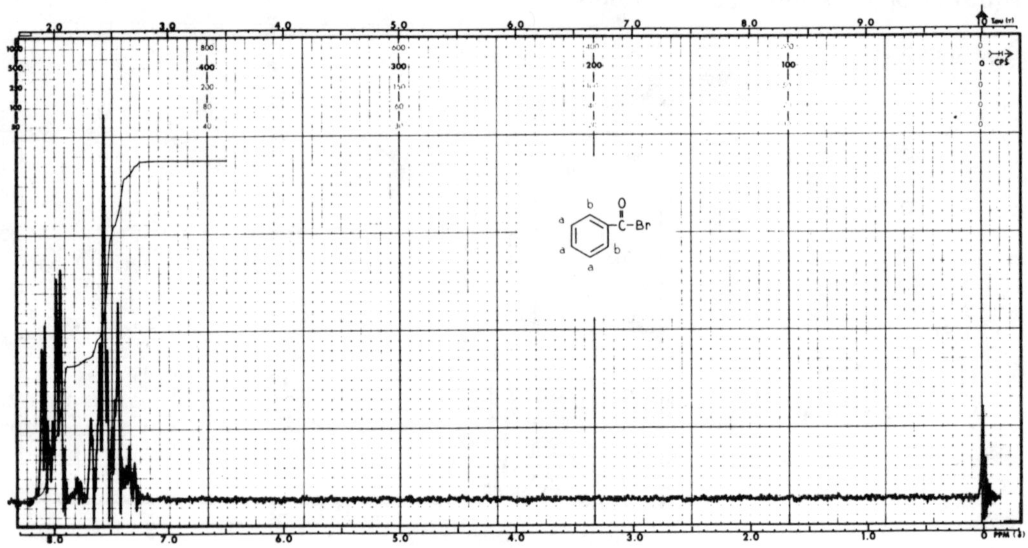

© 1978 Sadtler Research Laboratories, Inc.

---

o-TOLUOYL CHLORIDE

C_8H_7ClO

Mol. Wt. 154.60

B.P. 88-89°C/11mm

Solvent: CCl_4

Assignments
a = 2.51
b = 7.10-7.65
c = 8.19

2222

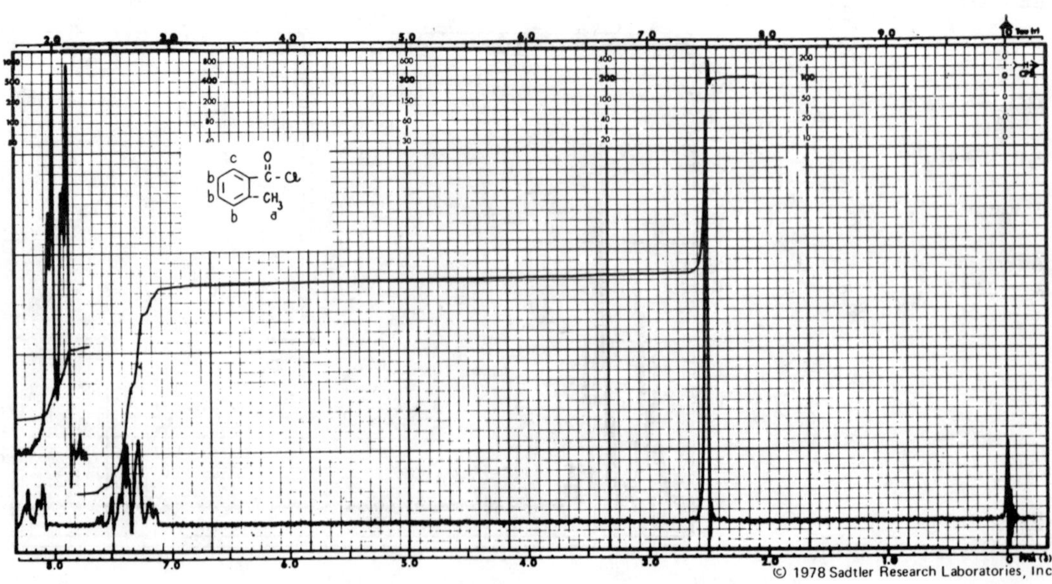

© 1978 Sadtler Research Laboratories, Inc.

p-TOLUOYL CHLORIDE

C_8H_7ClO

Mol. Wt. 154.60

B.P. 225-227°C

Solvent: CCl_4

Assignments
a = 2.44
b = 7.24
c = 7.96

2223

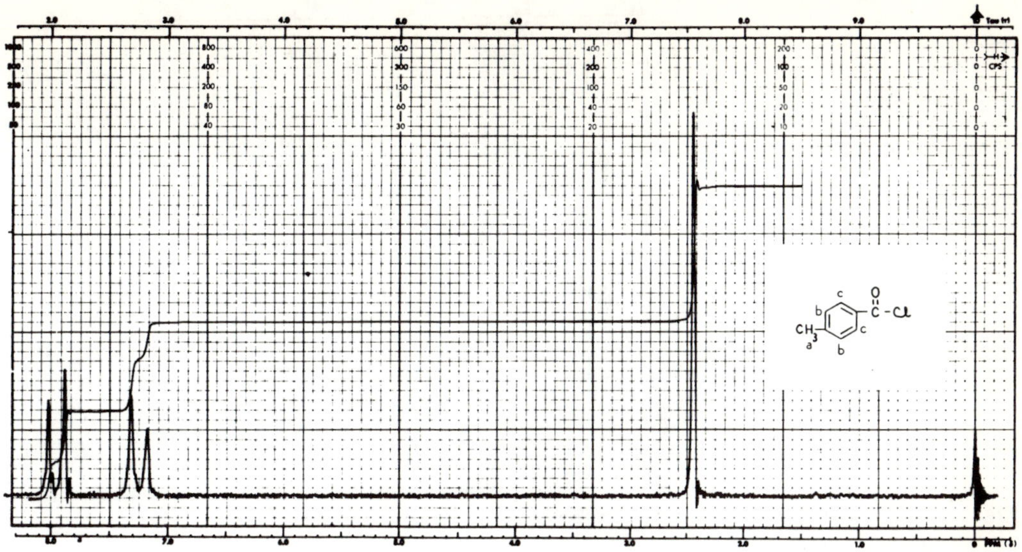

© 1978 Sadtler Research Laboratories, Inc.

p-tert-BUTYLBENZOYL CHLORIDE

$C_{11}H_{13}ClO$

Mol. Wt. 196.68

B.P. 135°C/20mm

Solvent: $CDCl_3$

Assignments
a = 1.39
b = 7.52
c = 8.06

2224

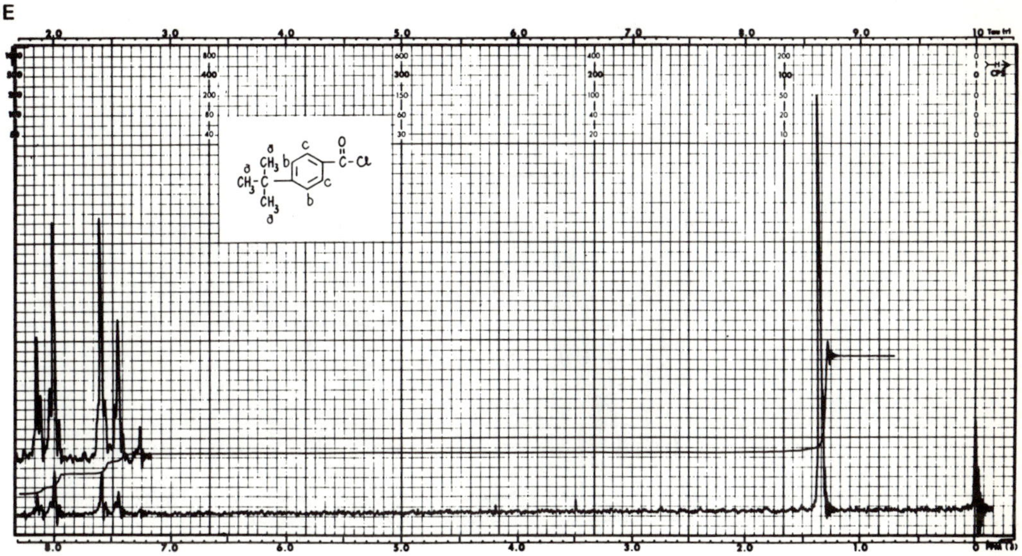

© 1978 Sadtler Research Laboratories, Inc.

***a,a,a*-TRIFLUORO-o-TOLUOYL CHLORIDE**

$C_8H_4ClF_3O$

Mol. Wt. 208.57

Solvent: CCl_4

Assignments
a = 7.59-8.12

2225

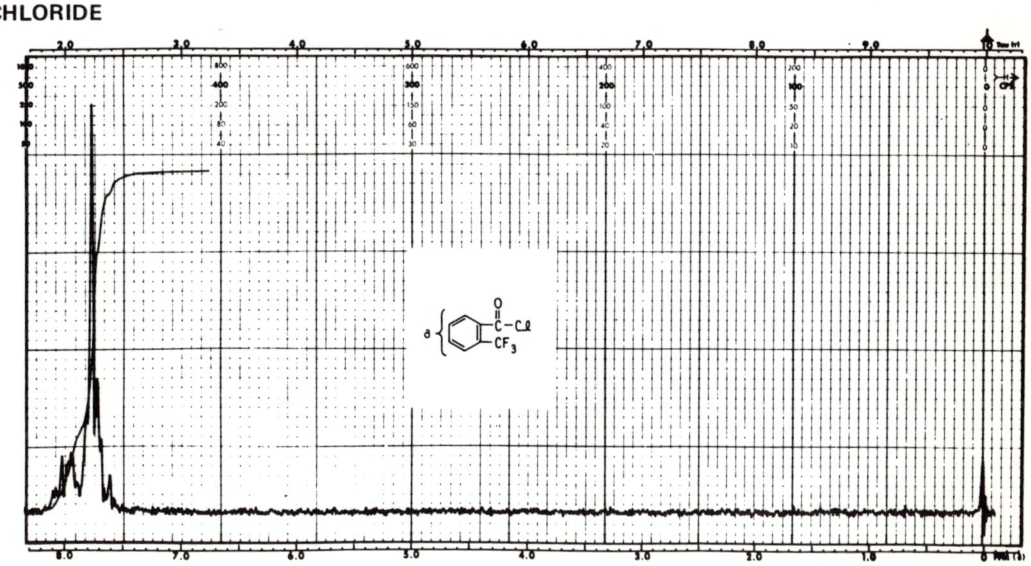

© 1978 Sadtler Research Laboratories, Inc.

α,α,α-TRIFLUORO-p-TOLUOYL CHLORIDE

$C_8H_4ClF_3O$

Mol. Wt. 208.57

B.P. 199-200°C/750mm (lit.)

Solvent: $CDCl_3$

Assignments
a = 7.73
b = 8.20

2226

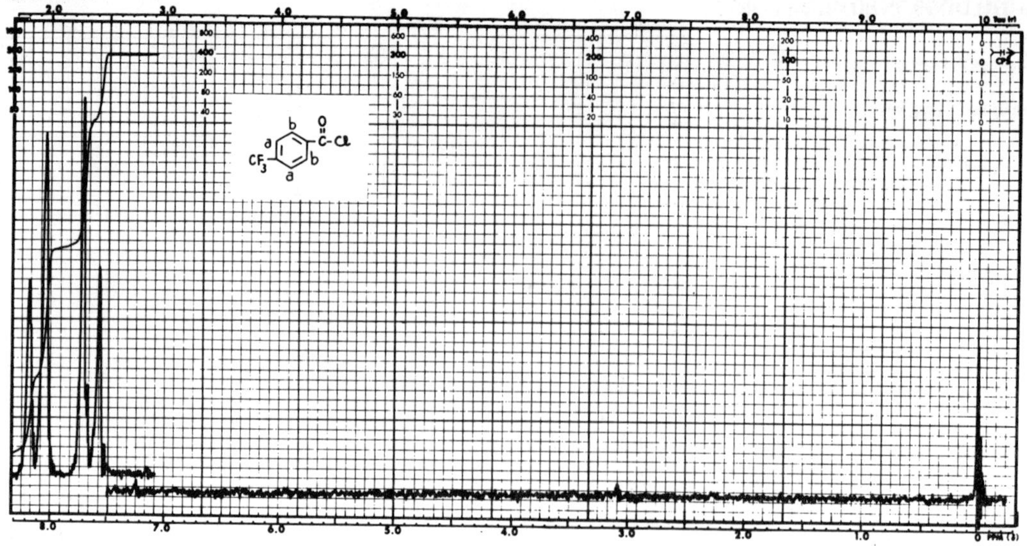

© 1978 Sadtler Research Laboratories, Inc.

o-FLUOROBENZOYL CHLORIDE

C_7H_4ClFO

Mol. Wt. 158.56

Solvent: CCl_4

Assignments
a = 7.17
b = 7.30-8.30

2227

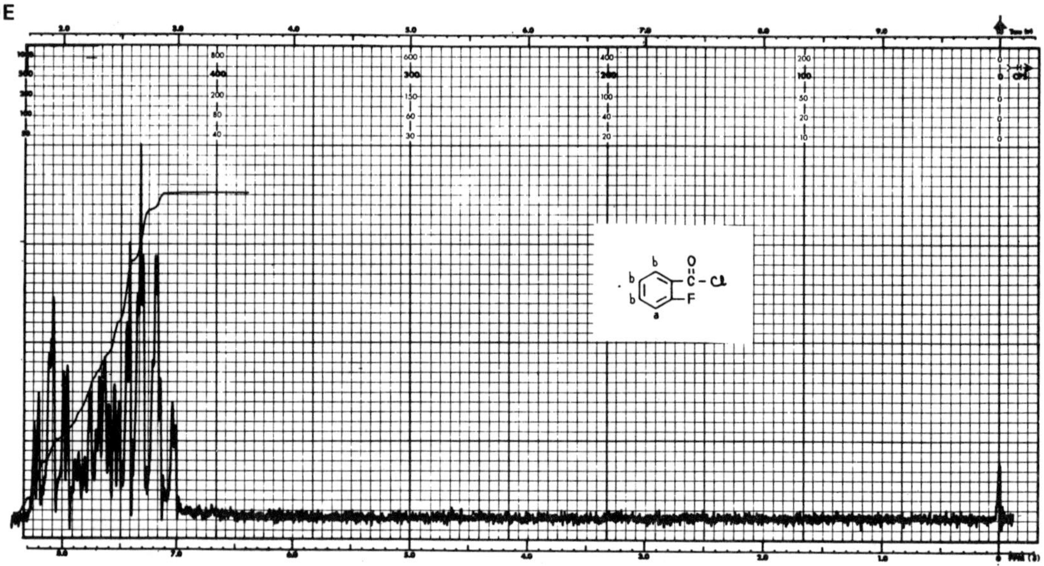

© 1978 Sadtler Research Laboratories, Inc.

m-FLUOROBENZOYL CHLORIDE

C_7H_4ClFO

Mol. Wt. 158.56

B.P. 189°C

Solvent: CCl_4

Assignments
a = 7.15-7.70
b = 7.82

2228

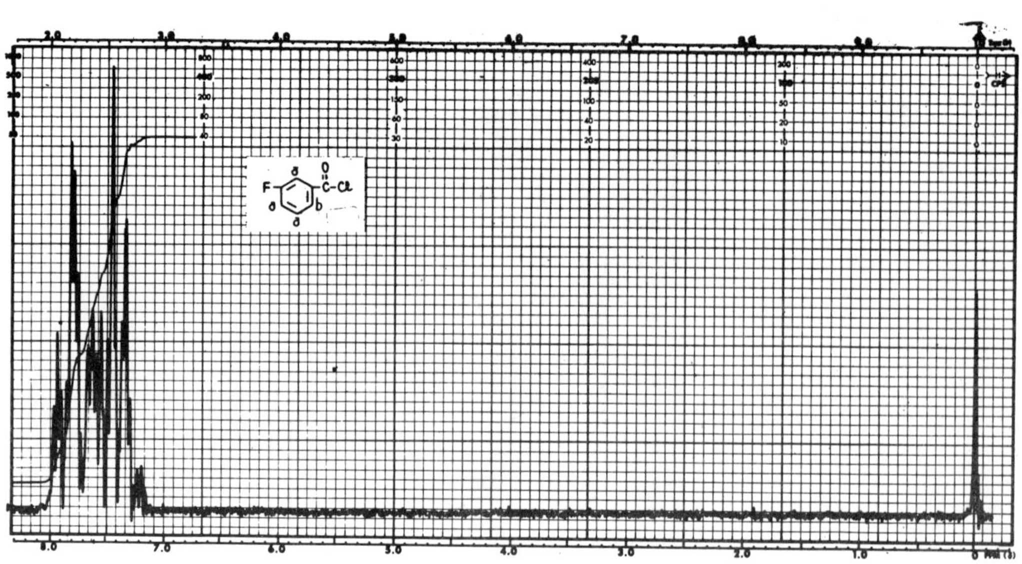

© 1978 Sadtler Research Laboratories, Inc.

o-CHLOROBENZOYL CHLORIDE

$C_7H_4Cl_2O$

Mol. Wt. 175.02

M.P. −4°C

B.P. 235-238°C

Solvent: CCl_4

Assignments
a = 7.21-7.58
b = 8.01

2229

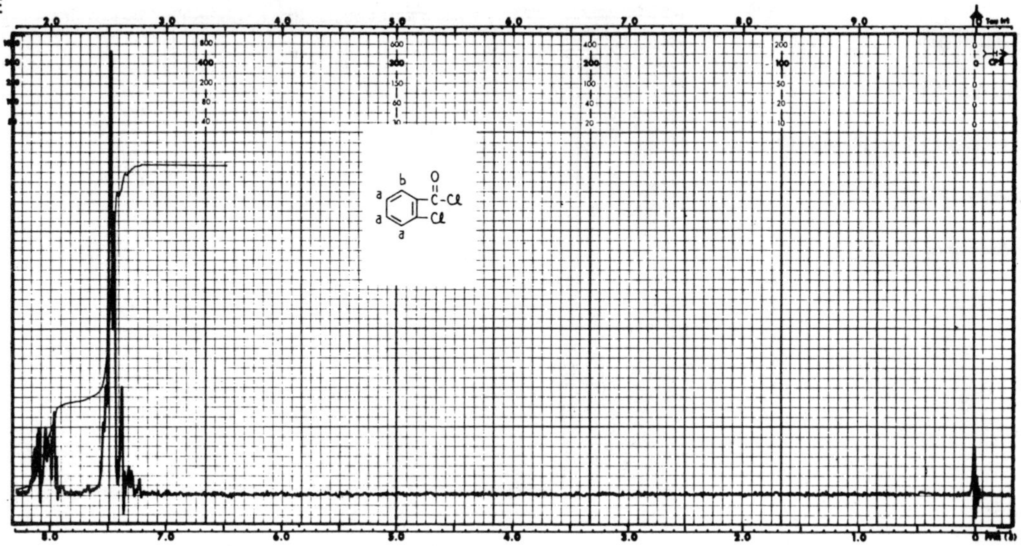

© 1978 Sadtler Research Laboratories, Inc.

3-CHLOROBENZOYL CHLORIDE

$C_7H_4Cl_2O$

Mol. Wt. 175.02

Solvent: CCl_4

Assignments
a = 7.41
b = 7.60
c = 7.98
d = 8.02

2230

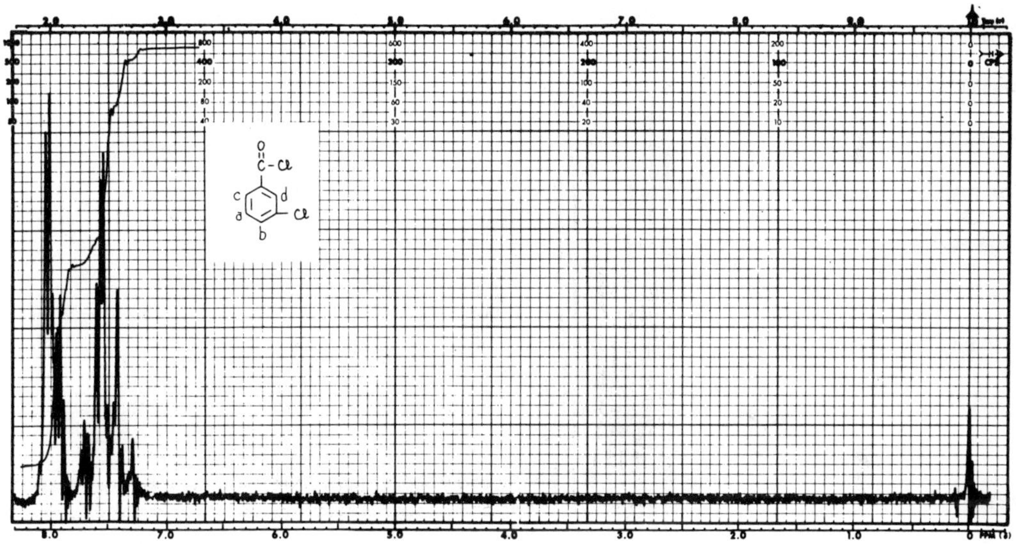

© 1978 Sadtler Research Laboratories, Inc.

p-CHLOROBENZOYL CHLORIDE

$C_7H_4Cl_2O$

Mol. Wt. 175.02

M.P. 12-14°C

Solvent: $CDCl_3$

Assignments
a = 7.46
b = 8.02

2231

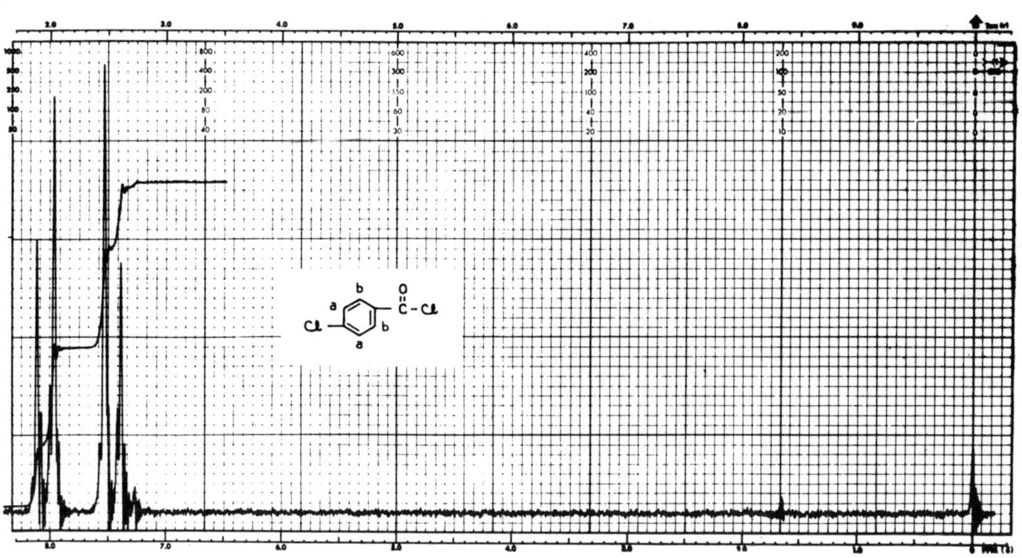

© 1978 Sadtler Research Laboratories, Inc.

o-BROMOBENZOYL CHLORIDE

C_7H_4BrClO

Mol. Wt. 219.47

B.P. 120-125°C/15mm

Solvent: CCl_4

Assignments
a = 7.23-7.83
b = 8.00

2232

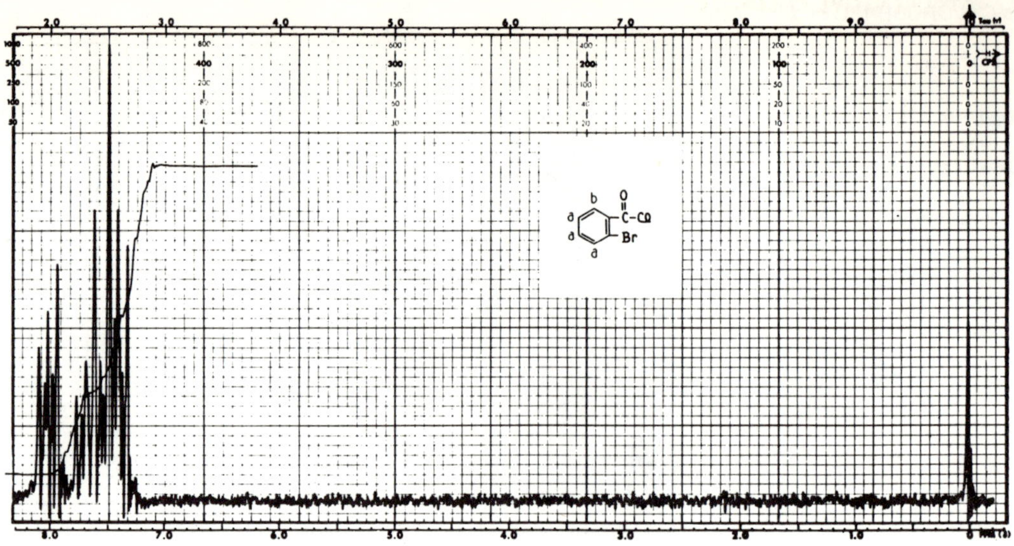

© 1978 Sadtler Research Laboratories, Inc.

---

m-BROMOBENZOYL CHLORIDE

C_7H_4BrClO

Mol. Wt. 219.47

B.P. 118-119°C/13mm

Solvent: CCl_4

Assignments
a = 7.39
b = 7.79
c = 8.03
d = 8.19

2233

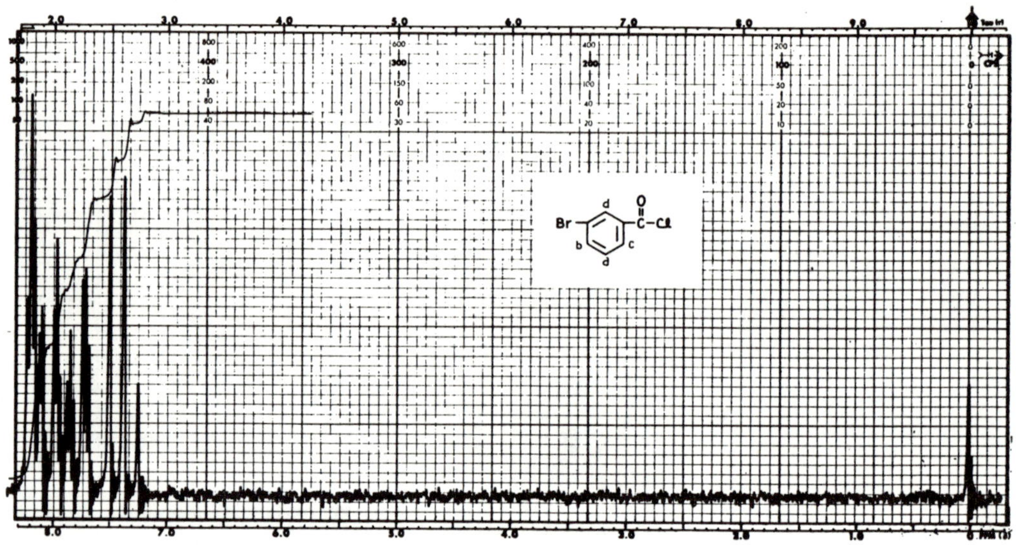

© 1978 Sadtler Research Laboratories, Inc.

---

m-BROMOBENZOYL BROMIDE

$C_7H_4Br_2O$

Mol. Wt. 263.93

B.P. 124-126°C/10mm

Solvent: CCl_4

Assignments
a = 7.38
b = 7.79
c = 7.98
d = 8.07

2234

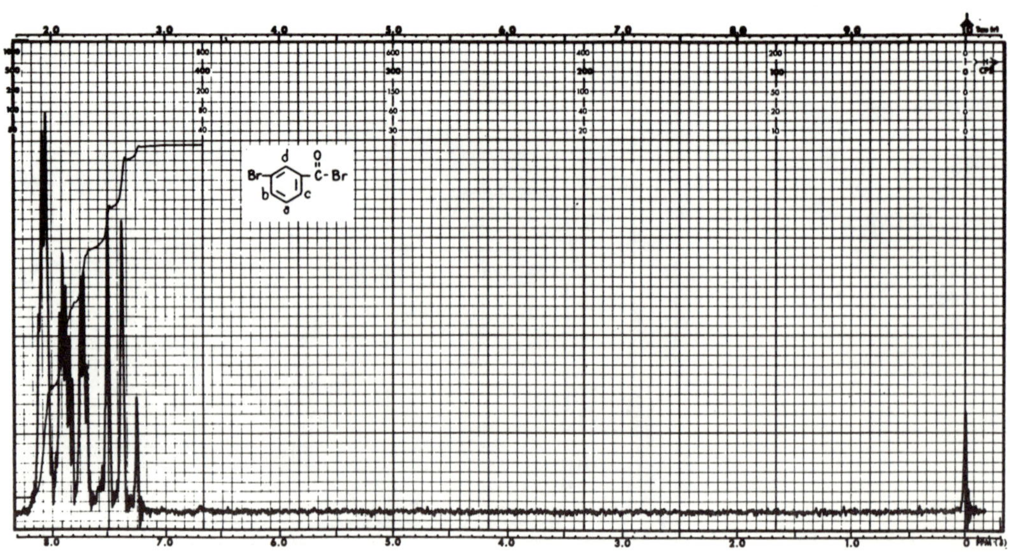

© 1978 Sadtler Research Laboratories, Inc.

o-NITROBENZOYL CHLORIDE

$C_7H_4ClNO_3$

Mol. Wt. 185.47

Solvent: CCl_4

Assignments
a = 7.70-7.99
b = 8.09

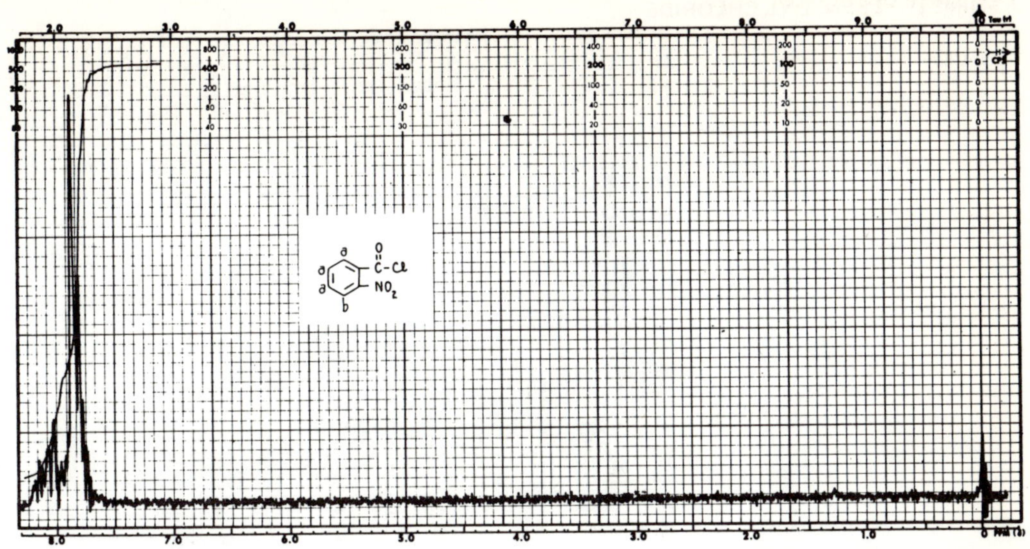

2235

© 1978 Sadtler Research Laboratories, Inc.

m-NITROBENZOYL CHLORIDE

$C_7H_4ClNO_3$

Mol. Wt. 185.57

B.P. 275-278°C

Solvent: $CDCl_3$

Assignments
a = 7.85
b = 8.49 or c
c = 8.61 or b
d = 8.94

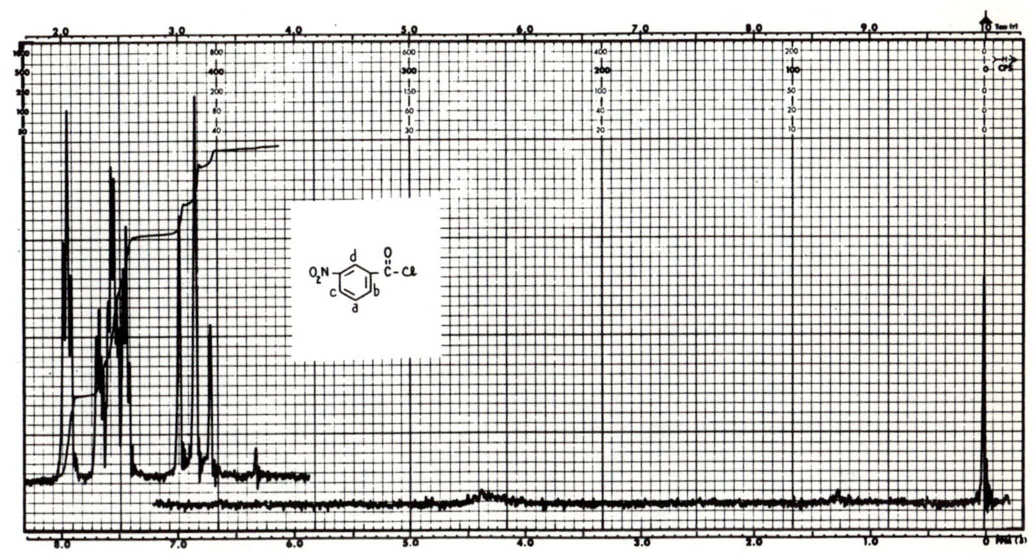

2236

© 1978 Sadtler Research Laboratories, Inc.

PHTHALOYL CHLORIDE

$C_8H_4Cl_2O_2$

Mol. Wt. 203.03

M.P. 10-13°C

B.P. 131-135°C/11mm

Solvent: CCl_4

Assignments
a = 7.80
b = 7.91

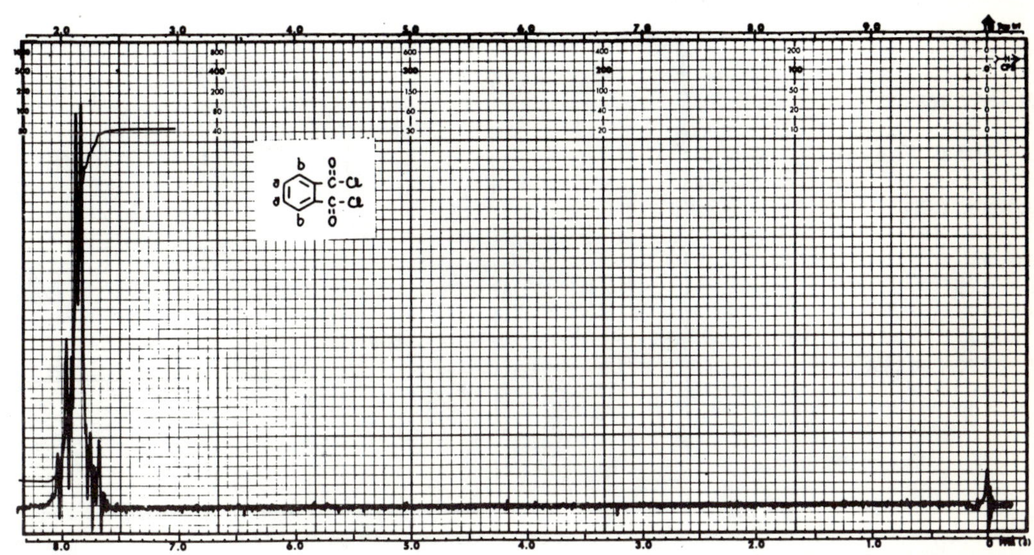

2237

© 1978 Sadtler Research Laboratories, Inc.

2,3-DIMETHYLBENZOYL CHLORIDE

C_9H_9ClO

Mol. Wt. 168.62

Solvent: CCl_4

Assignments
a = 2.33
b = 2.39
c = 7.10
d = 7.28
e = 7.84

2238

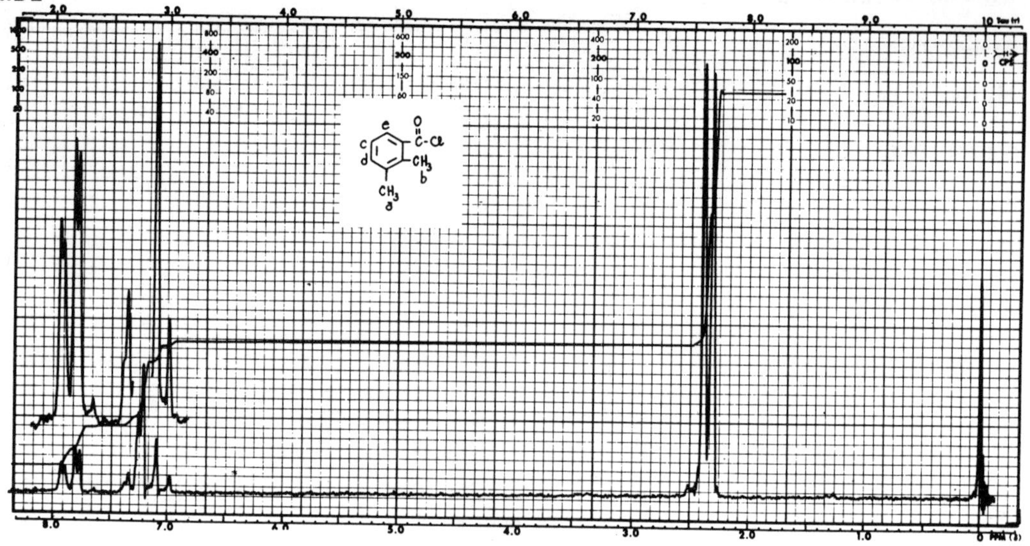

© 1978 Sadtler Research Laboratories, Inc.

3,4-DIMETHYLBENZOYL CHLORIDE

C_9H_9ClO

Mol. Wt. 168.62

Solvent: CCl_4

Assignments
a = ca 2.31
b = 7.13
c = ca 7.71

2239

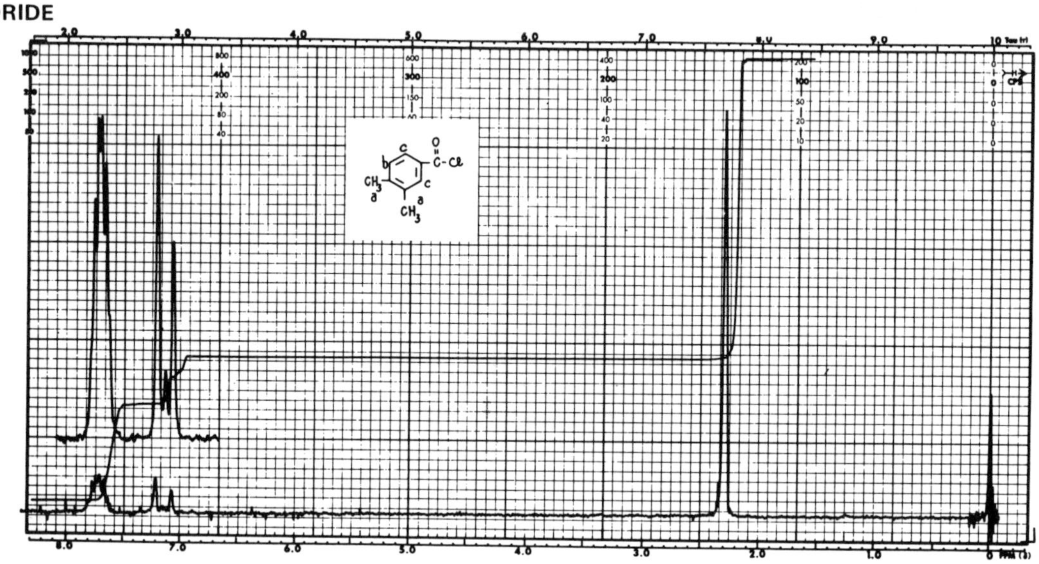

© 1978 Sadtler Research Laboratories, Inc.

3,5-DIMETHYLBENZOYL CHLORIDE

C_9H_9ClO

Mol. Wt. 168.62

B.P. 101-103°C/8mm

Solvent: CCl_4

Assignments
a = 2.35
b = 7.25
c = 7.67
d = 2.20, 2.40 imps.

2240

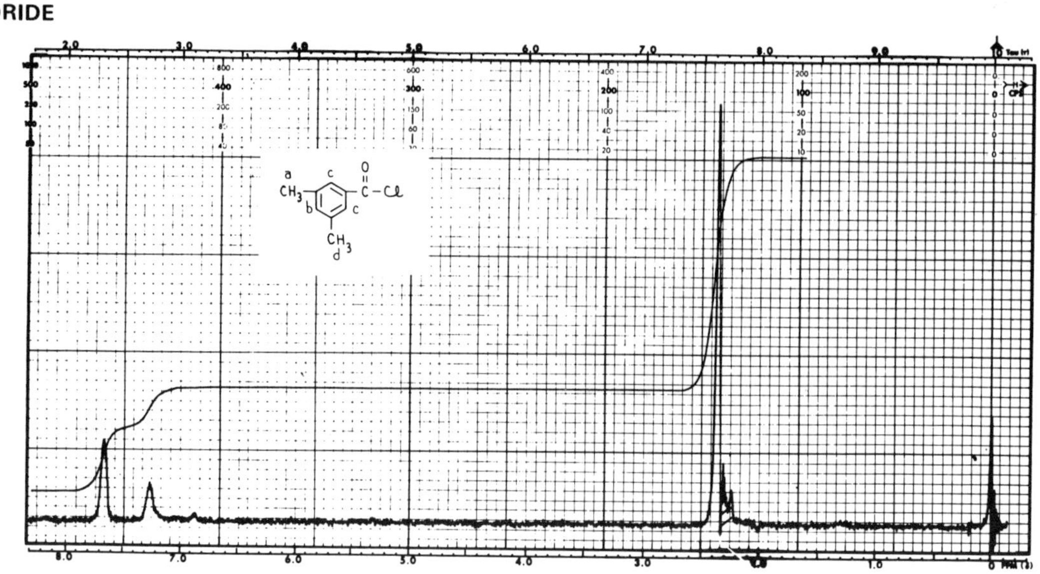

© 1978 Sadtler Research Laboratories, Inc.

3-NITRO-p-TOLUOYL CHLORIDE

$C_8H_6ClNO_3$

Mol. Wt. 199.59

Solvent: CCl_4

Assignments
a = 2.70
b = 7.60
c = 8.18
d = 8.49

2241

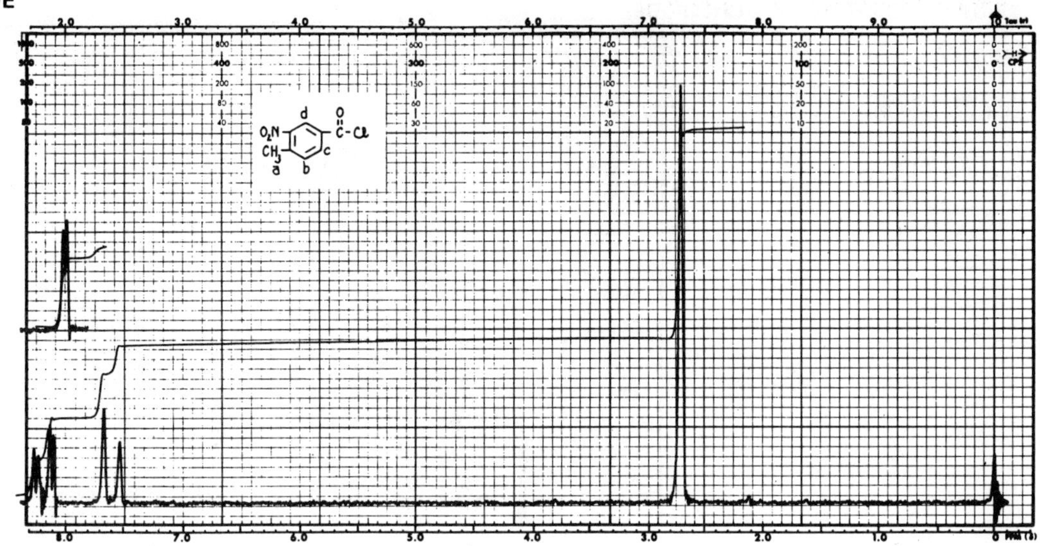

© 1978 Sadtler Research Laboratories, Inc.

2-CHLORO-4-NITROBENZOYL CHLORIDE

$C_7H_3Cl_2NO_3$

Mol. Wt. 220.01

Solvent: $CDCl_3$

Assignments
a = 8.15-8.45

2242

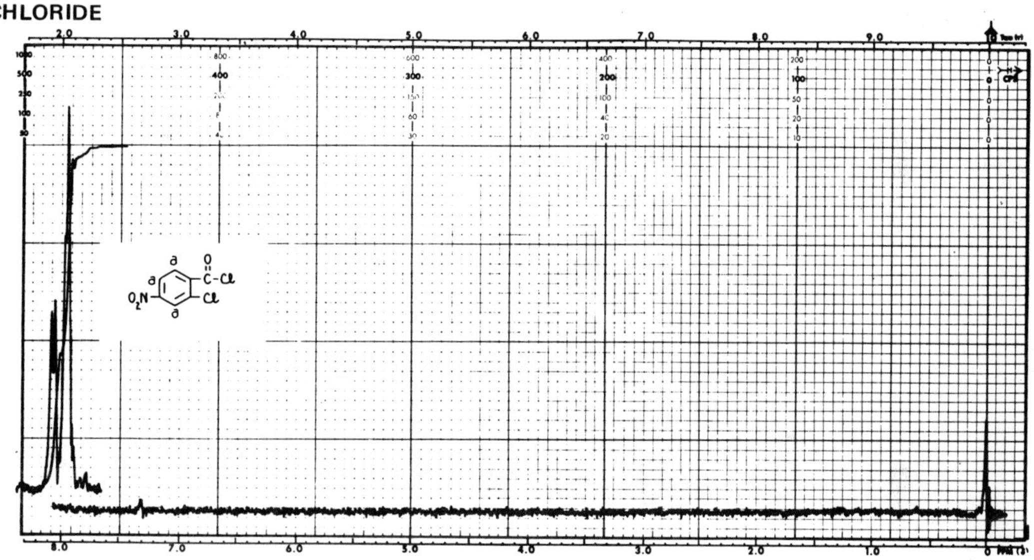

© 1978 Sadtler Research Laboratories, Inc.

2-CHLORO-5-NITROBENZOYL CHLORIDE

$C_7H_3Cl_2NO_3$

Mol. Wt. 220.01

Solvent: $CDCl_3$

Assignments
a = 7.80
b = 8.42
c = 8.90

2243

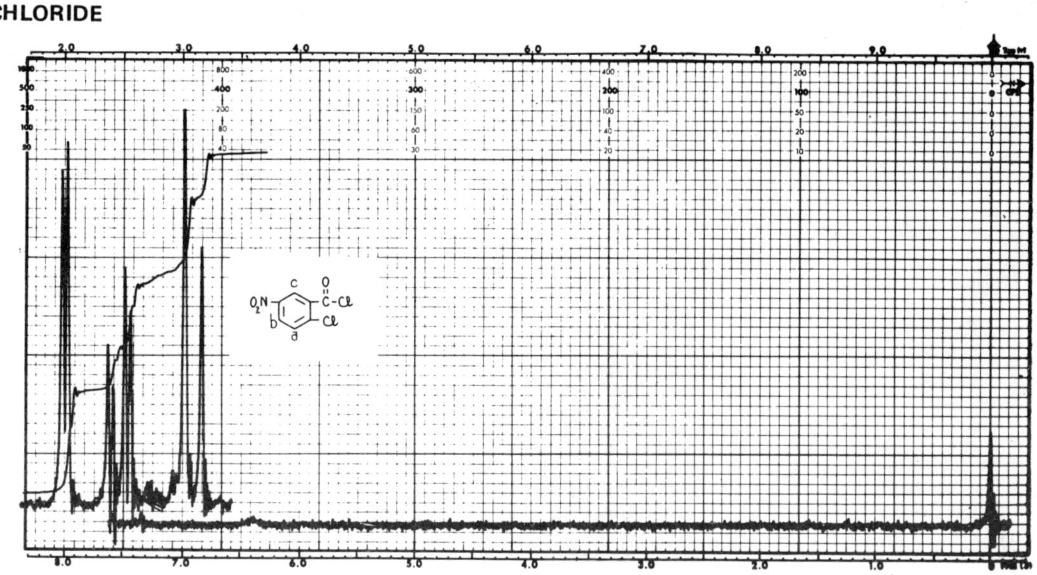

© 1978 Sadtler Research Laboratories, Inc.

3,5-DIMETHOXYBENZOYL CHLORIDE

$C_9H_9ClO_3$

Mol. Wt. 200.62

Solvent: CCl_4

Assignments
a = 3.80
b = 6.61
c = 7.12

2244

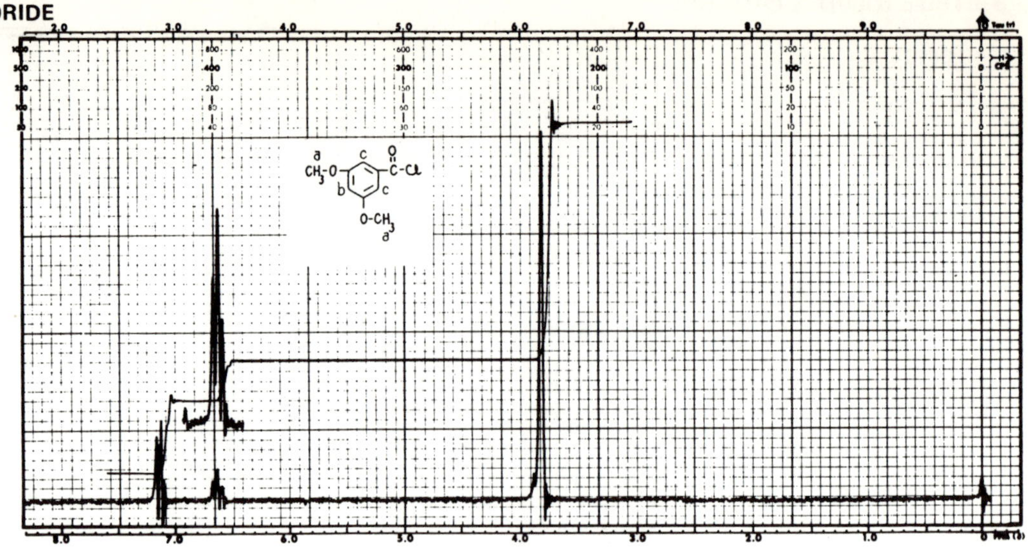

© 1978 Sadtler Research Laboratories, Inc.

1-NAPHTHOYL CHLORIDE

$C_{11}H_7ClO$

Mol. Wt. 190.63

M.P. 17-19°C

Solvent: CCl_4

Assignments
a = 7.20-7.80
b = 7.89
c = 8.39
d = 8.63

2245

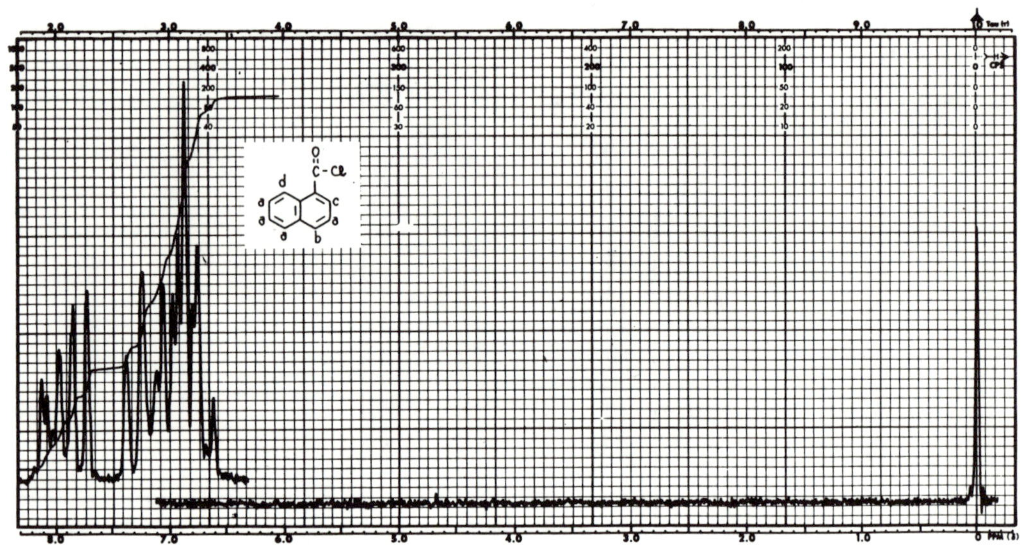

© 1978 Sadtler Research Laboratories, Inc.

2-FUROYL CHLORIDE

$C_5H_3ClO_2$

Mol. Wt. 130.53

B.P. 173-174°C

Solvent: CCl_4

Assignments
a = 6.70
b = 7.52
c = 7.82

2246

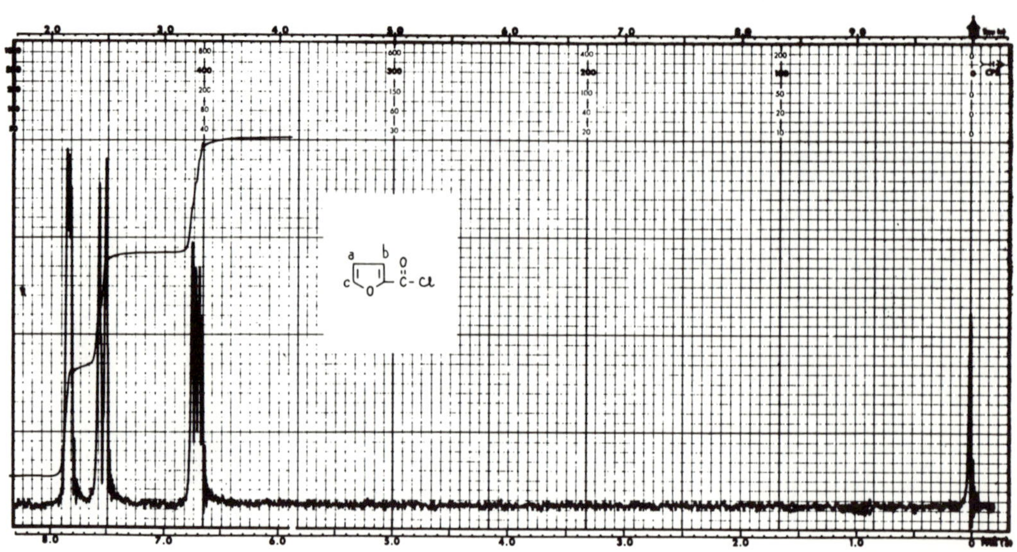

© 1978 Sadtler Research Laboratories, Inc.

The Anhydrides are similar to other carbonyl containing compounds in that, adjacent aliphatic groups are weakly deshielded while the ortho aromatic protons are strongly deshielded. Their NMR spectra are similar to those of the carboxylic acids which begin with spectrum 2471. Because the Anhydrides hydrolyze easily to form the corresponding carboxylic acid, the chemical shift range from 10–12 ppm should be checked carefully to determine the presence of the carboxylic acid –OH group which would indicate that hydrolysis has occurred and to what extent such decomposition products are present in solution.

Chemical Shifts

Aliphatic Protons

HC–4	HC–3	HC–2	HC–1		Solvent	Spectrum
			2.20	CH_3–C(=O)–O–C(=O)–R_1	CCl_4	2247
		1.17	2.43	CH_3–CH_2–C(=O)–O–C(=O)–R_2	CCl_4	2248
	1.01	1.69	2.40	CH_3–CH_2–CH_2–C(=O)–O–C(=O)–R_3	CCl_4	2249
		(1.24)	2.68	$(CH_3)_2$–CH–C(=O)–O–C(=O)–R	$CDCl_3$	- - - -
0.95	1.45	1.65	2.48	CH_3–CH_2–CH_2–CH_2–C(=O)–O–C(=O)–R_4	$CDCl_3$	2250
	(0.97	1.60)	2.29	$(CH_3–CH_2)_2$–CH–C(=O)–O–C(=O)–R	CCl_4	2253
		(1.25)	- - - -	$(CH_3)_3$–C–C(=O)–O–C(=O)–R	CCl_4	2251

Aromatic Protons

Benzoic Anhydride

meta/para	ortho		Solvent	Spectrum
7.2–7.7	8.11	–C(=O)–O–C(=O)–Ø	$CDCl_3$	2270

Coupling and Coupling Constants

The Anhydrides display no special couplings nor coupling constants, other than the usual aliphatic H–C–C–H coupling and the normal aromatic ortho and meta couplings that are commonly observed for most functional groups.

Solubility and Solvent Effects

The Anhydrides are readily soluble in carbon tetrachloride and deuterochloroform which are the solvents of preference. Solvents such as DMSO-d_6, Polysol and Acetone-d_6 could contain sufficient amounts of waιer to lead to the decomposition of the sample.

ACETIC ANHYDRIDE

$C_4H_6O_3$

Mol. Wt. 102.09

Solvent: CCl_4

Assignments
a = 2.20

2247

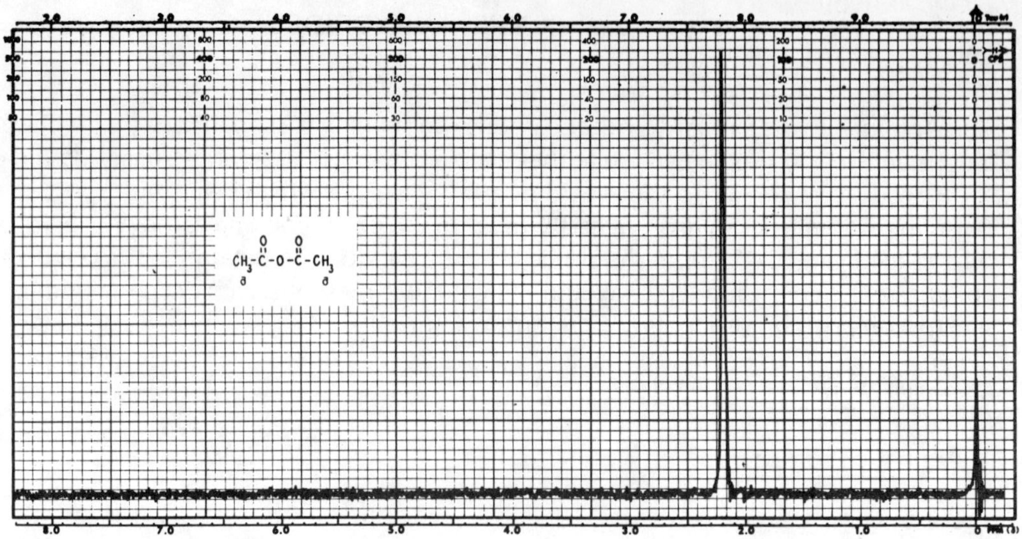

© 1978 Sadtler Research Laboratories, Inc.

PROPIONIC ANHYDRIDE

$C_6H_{10}O_3$

Mol. Wt. 130.14

B.P. 169°C

Solvent: CCl_4

Assignments
a = 1.17
b = 2.43

2248

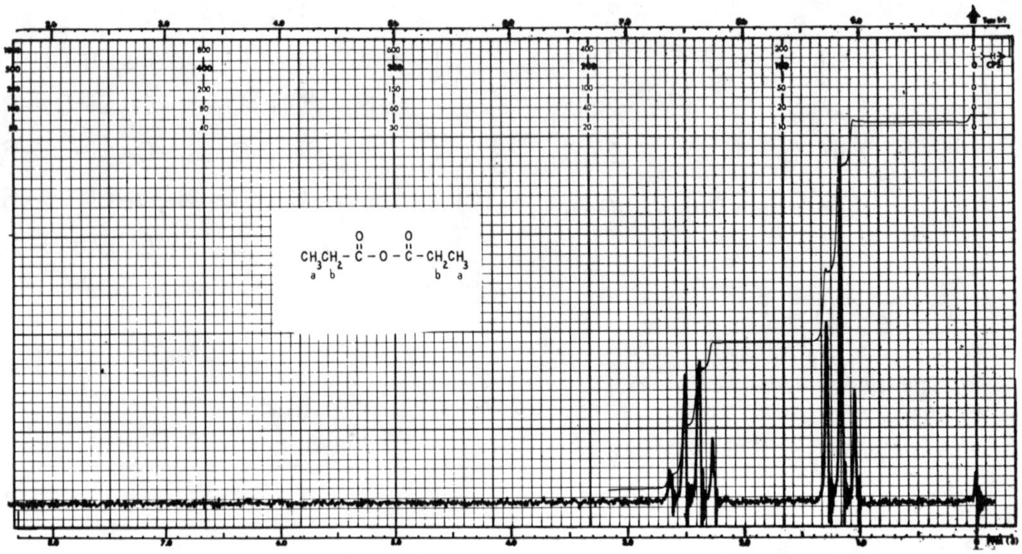

© 1978 Sadtler Research Laboratories, Inc.

BUTYRIC ANHYDRIDE

$C_8H_{14}O_3$

Mol. Wt. 158.20

B.P. 76-78°C/11mm

Solvent: CCl_4

Assignments
a = 1.01
b = 1.69
c = 2.40

2249

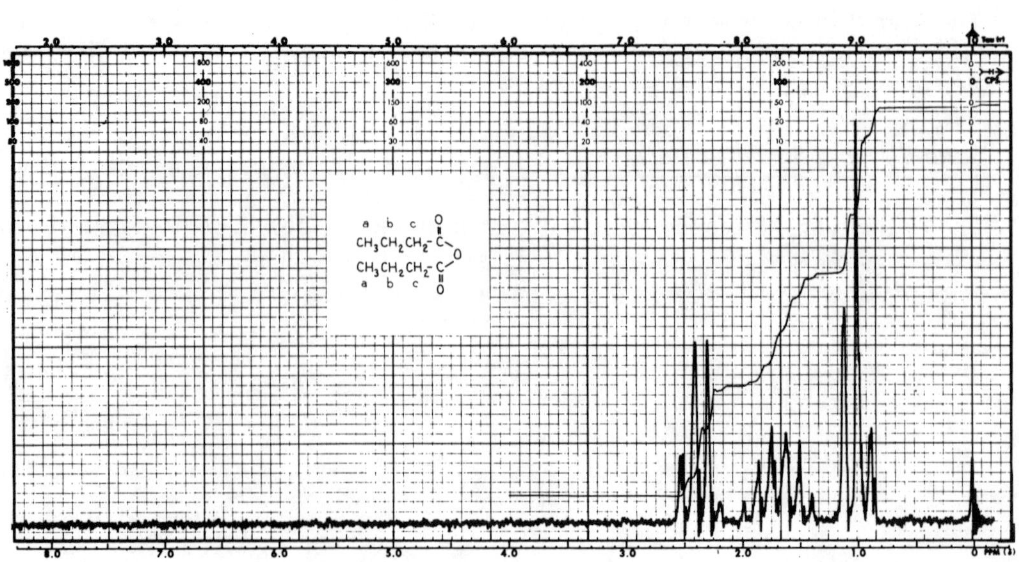

© 1978 Sadtler Research Laboratories, Inc.

VALERIC ANHYDRIDE

$C_{10}H_{18}O_3$

Mol. Wt. 186.25

Solvent: $CDCl_3$

Assignments
a = 0.95
b = 1.11-1.97
c = 2.48

2250

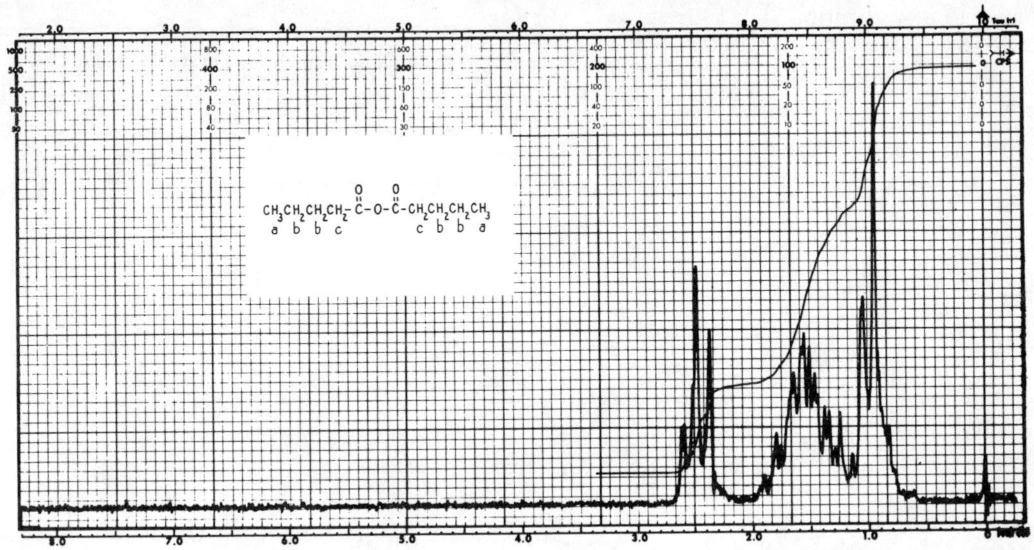

© 1978 Sadtler Research Laboratories, Inc.

PIVALIC ANHYDRIDE

$C_{10}H_{18}O_3$

Mol. Wt. 186.25

Solvent: CCl_4

Assignments
a = 1.25

2251

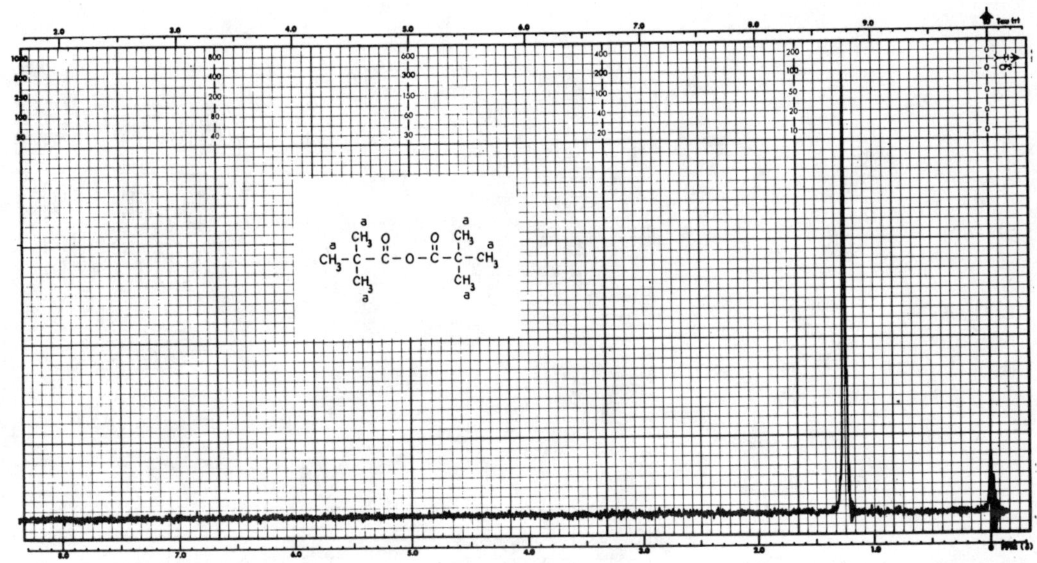

© 1978 Sadtler Research Laboratories, Inc.

HEXANOIC ANHYDRIDE

$C_{12}H_{22}O_3$

Mol. Wt. 214.31

M.P. −41°C

Solvent: $CDCl_3$

Assignments
a = 0.91
b = 1.10-1.55
c = 1.63
d = 2.37

2252

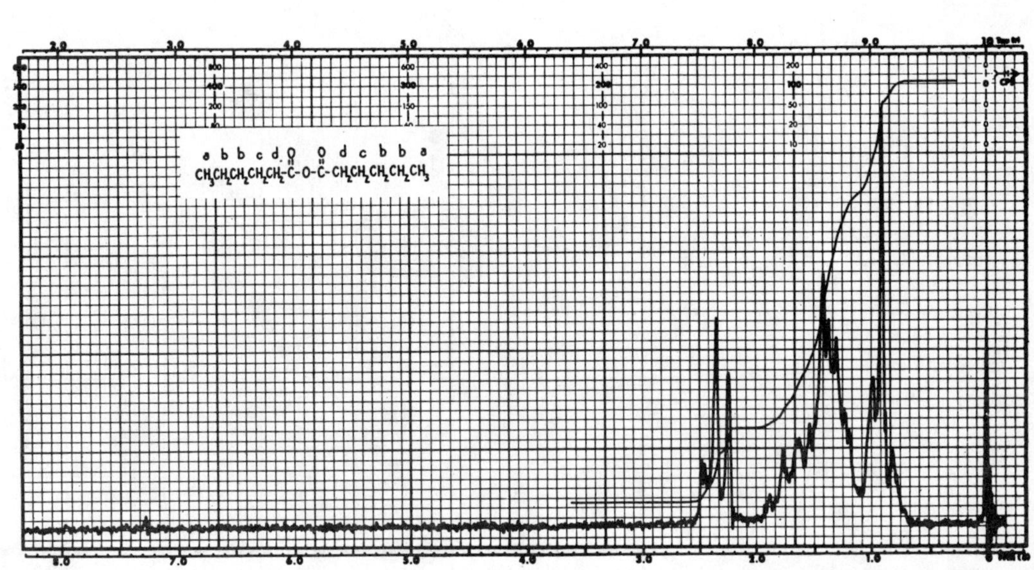

© 1978 Sadtler Research Laboratories, Inc.

2-ETHYLBUTYRIC ANHYDRIDE

$C_{12}H_{22}O_3$

Mol. Wt. 214.31

Solvent: CCl_4

Assignments
a - 0.97
b = 1.60
c = 2.29

2253

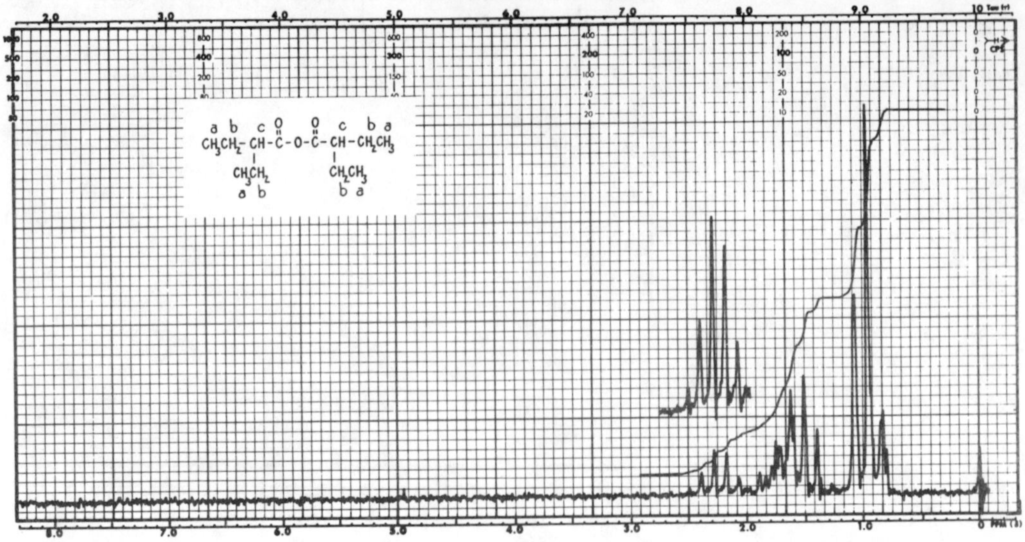

© 1978 Sadtler Research Laboratories, Inc.

HEPTANOIC ANHYDRIDE

$C_{14}H_{26}O_3$

Mol. Wt. 242.36

M.P. −13 to −11°C

Solvent: CCl_4

Assignments
a = 0.90
b = 1.10-1.60
c = 2.40

2254

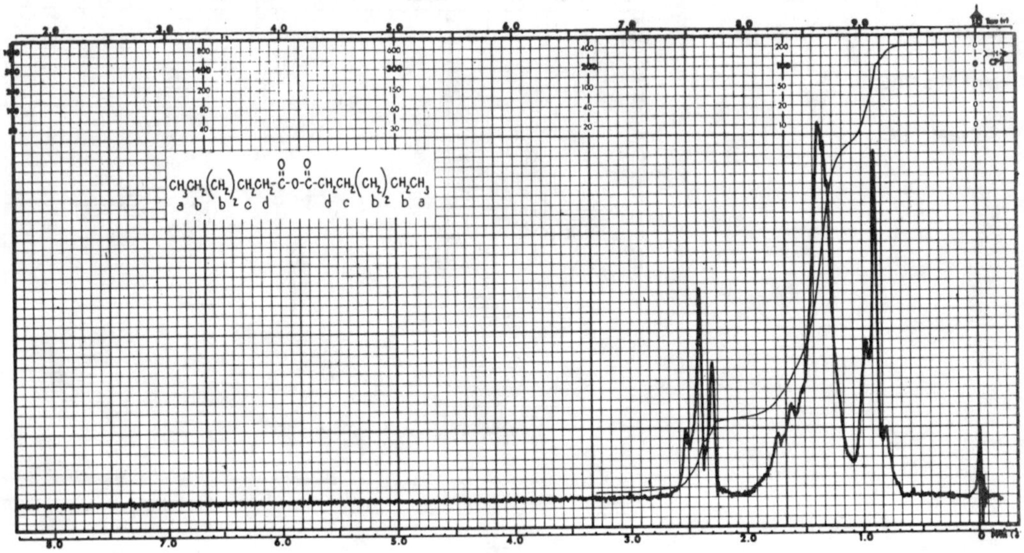

© 1978 Sadtler Research Laboratories, Inc.

SUCCINIC ANHYDRIDE

$C_4H_4O_3$

Mol. Wt. 100.07

M.P. 119-120°C

Solvent: $CDCl_3$

Assignments
a = 3.00
b = 1.62 imp.

2255

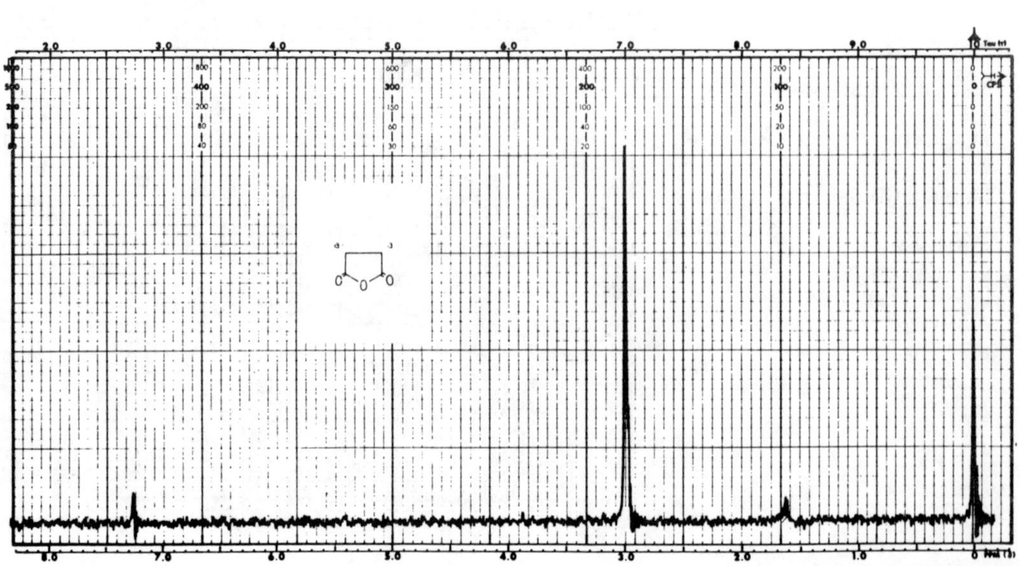

© 1978 Sadtler Research Laboratories, Inc.

METHYLSUCCINIC ANHYDRIDE

$C_5H_6O_3$

Mol. Wt. 114.10

M.P. 30-32°C

Solvent: $CDCl_3$

Assignments
a = 1.41
b = 2.30-3.55

2256

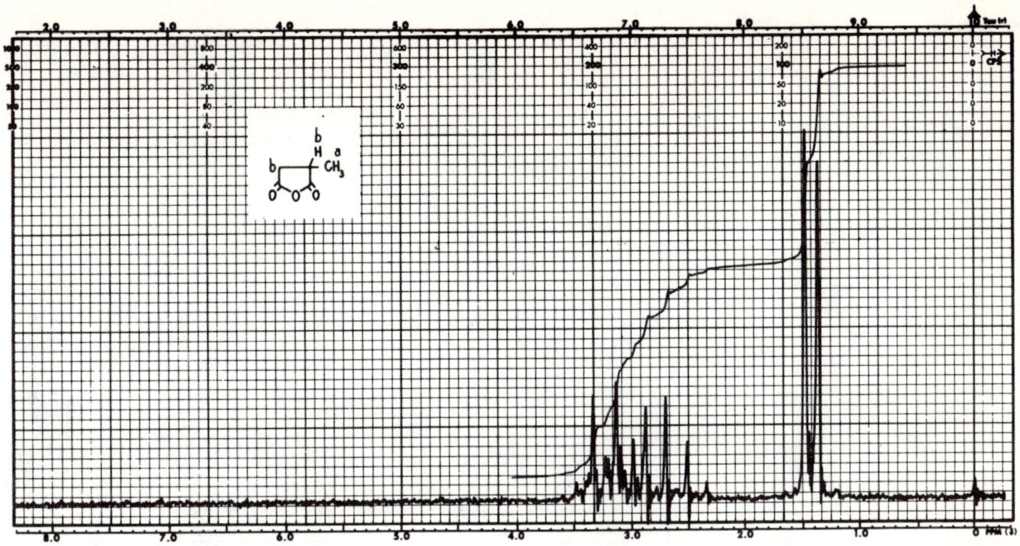

© 1978 Sadtler Research Laboratories, Inc.

(2-DODECENYL)SUCCINIC ANHYDRIDE

$C_{16}H_{26}O_3$

Mol. Wt. 266.38

M.P. 38-40°C

Solvent: $CDCl_3$

Assignments
a = 0.89
b = 1.30
c = 1.99
d = 2.27-3.40
e = ca 5.41

2257

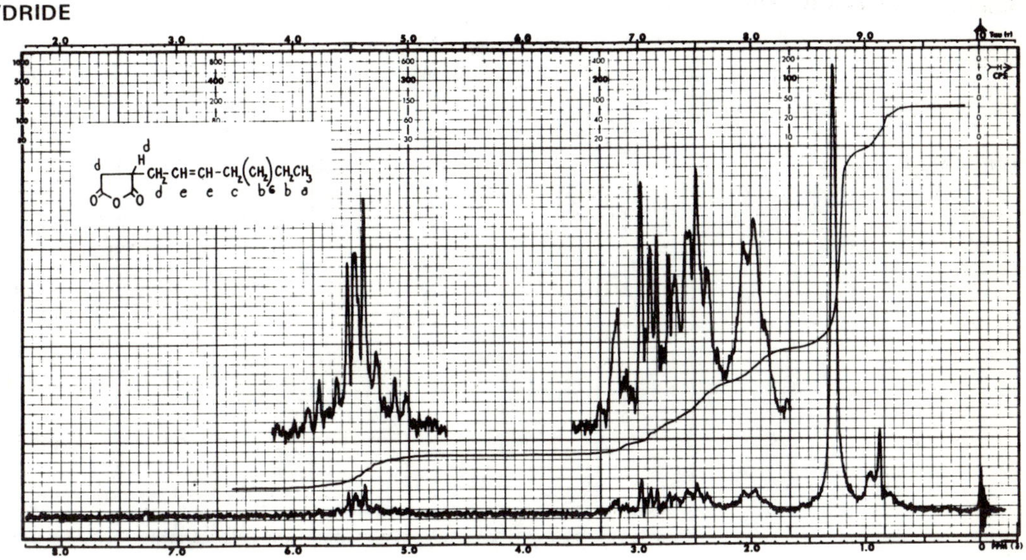

© 1978 Sadtler Research Laboratories, Inc.

METHYLENESUCCINIC ANHYDRIDE

$C_5H_4O_3$

Mol. Wt. 112.09

M.P. 66-68°C

Solvent: $CDCl_3$

Assignments
a = 3.62
b = 5.94 or c
c = 6.53 or b

2258

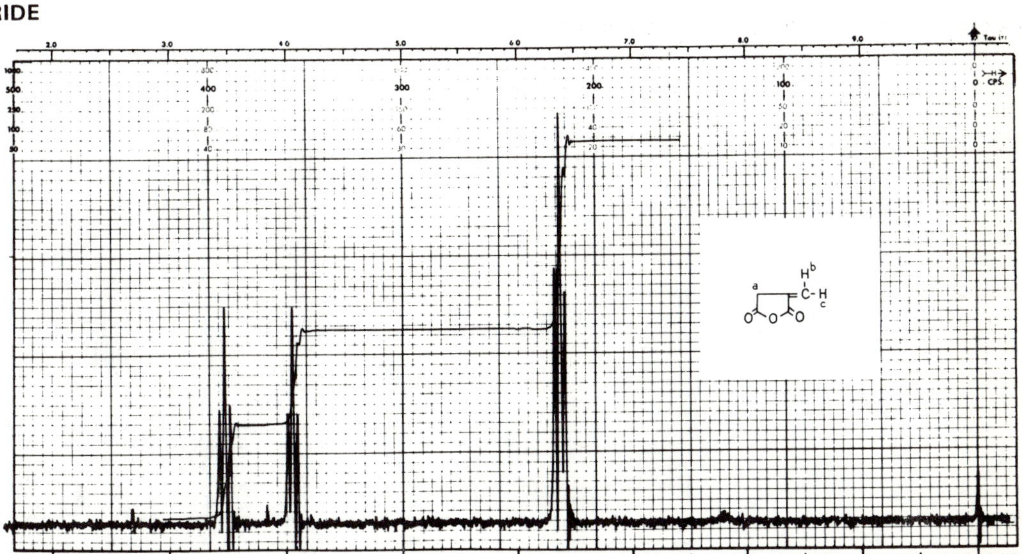

© 1978 Sadtler Research Laboratories, Inc.

3-METHYLGLUTARIC ANHYDRIDE

$C_6H_8O_3$

Mol. Wt. 128.13

Solvent: $CDCl_3$

Assignments
a = 1.12
b = 2.04-3.15

2259

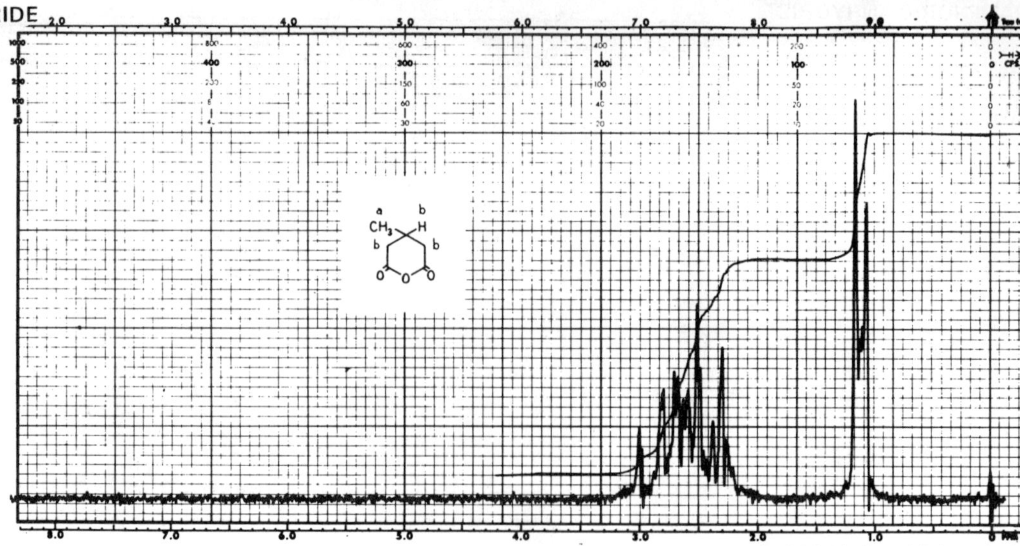

© 1978 Sadtler Research Laboratories, Inc.

2,2-DIMETHYLGLUTARIC ANHYDRIDE

$C_7H_{10}O_3$

Mol. Wt. 142.16

Solvent: $CDCl_3$

Assignments
a = 1.21
b = 1.92
c = 2.42

2260

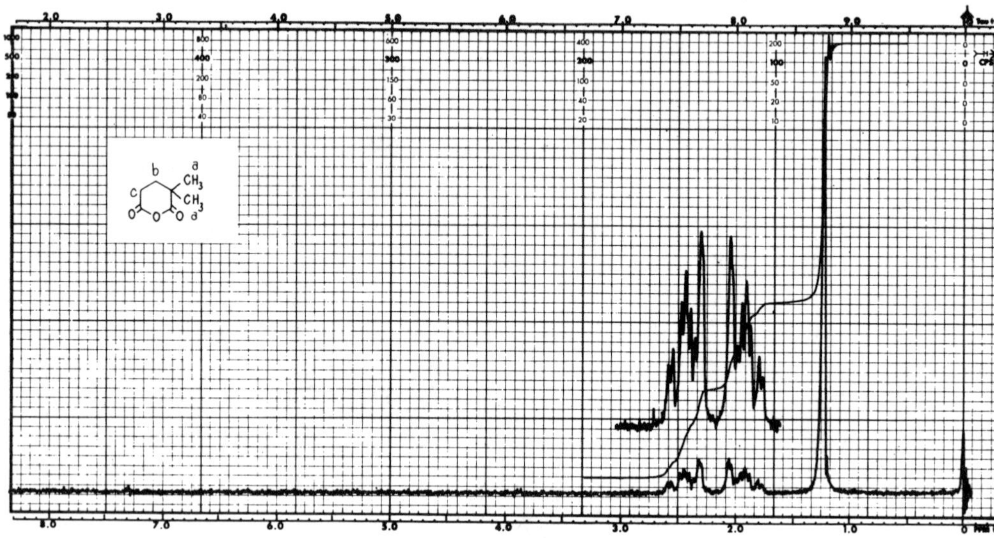

© 1978 Sadtler Research Laboratories, Inc.

1,2-CYCLOHEXANEDICARBOXYLIC ANHYDRIDE

$C_8H_{10}O_3$

Mol. Wt. 154.17

M.P. 33-35°C

Solvent: $CDCl_3$

Assignments
a = 1.20-2.30
b = 3.22

2261

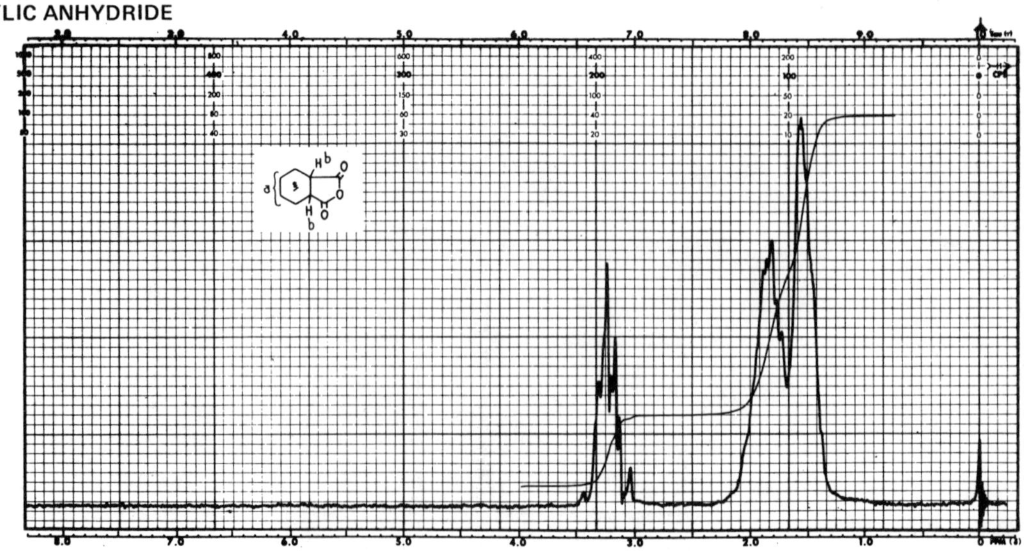

© 1978 Sadtler Research Laboratories, Inc.

DL-CAMPHORIC ANHYDRIDE

$C_{10}H_{14}O_3$

Mol. Wt. 182.22

M.P. 221°C

Solvent: $CDCl_3$

Assignments
a = 1.01
b = 1.10
c = 1.26
d = 1.80-2.60
e = 2.84

2262

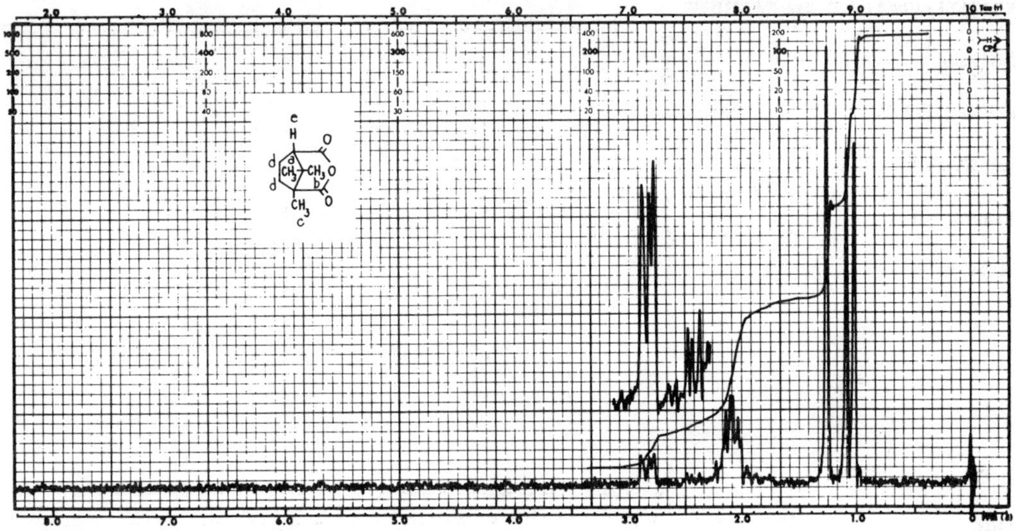

© 1978 Sadtler Research Laboratories, Inc.

MALEIC ANHYDRIDE

$C_4H_2O_3$

Mol. Wt. 98.06

M.P. 52.4-56°C

B.P. 199-200°C

Solvent: D_2O

Assignments
a = 6.41
b = 5.03 HDO

2263

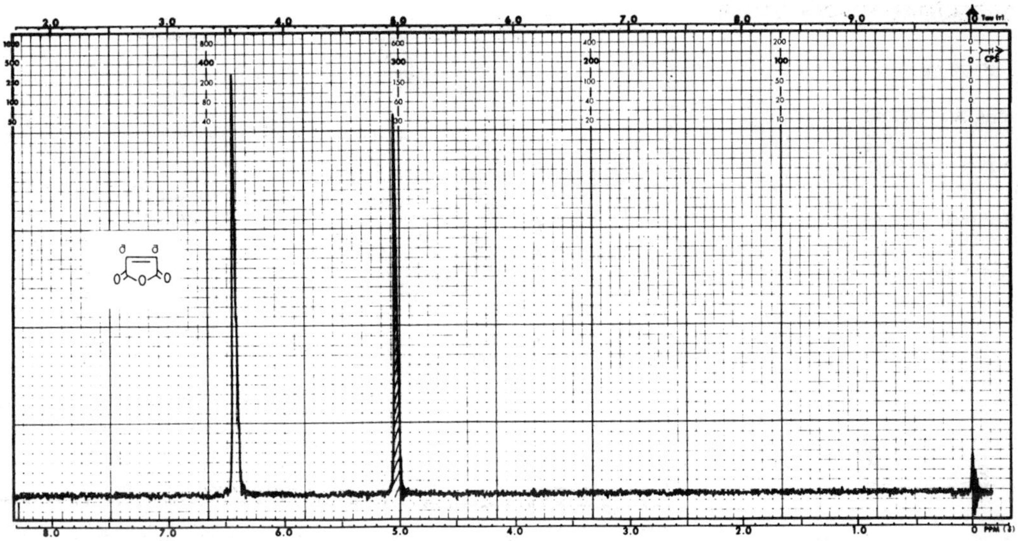

© 1978 Sadtler Research Laboratories, Inc.

CITRACONIC ANHYDRIDE

$C_5H_4O_3$

Mol. Wt. 112.09

Solvent: CCl_4

Assignments
a = 2.19
b = 6.60

2264

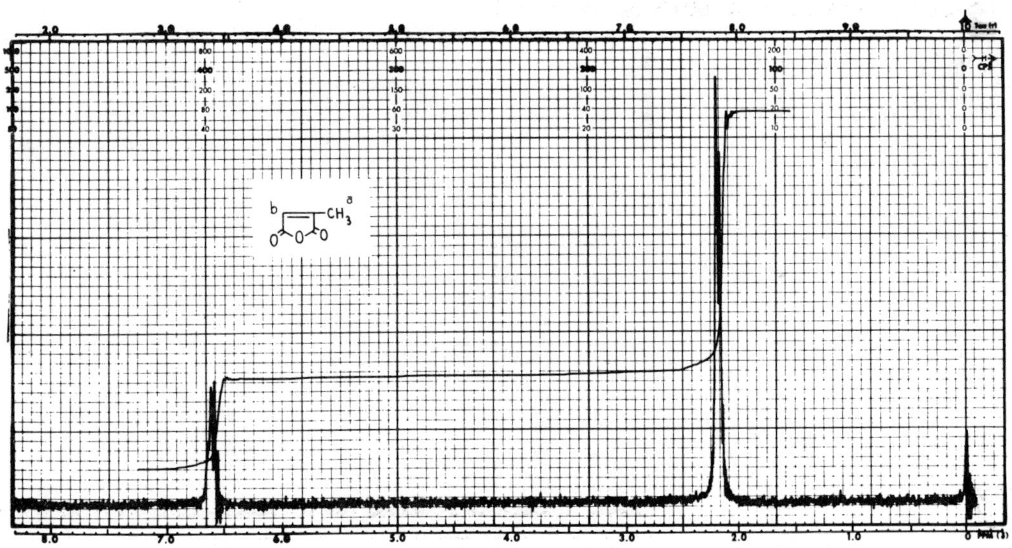

© 1978 Sadtler Research Laboratories, Inc.

DIMETHYLMALEIC ANHYDRIDE

$C_6H_6O_3$

Mol. Wt. 126.11

M.P. 88-90°C

Solvent: $CDCl_3$

Assignments
a = 2.09

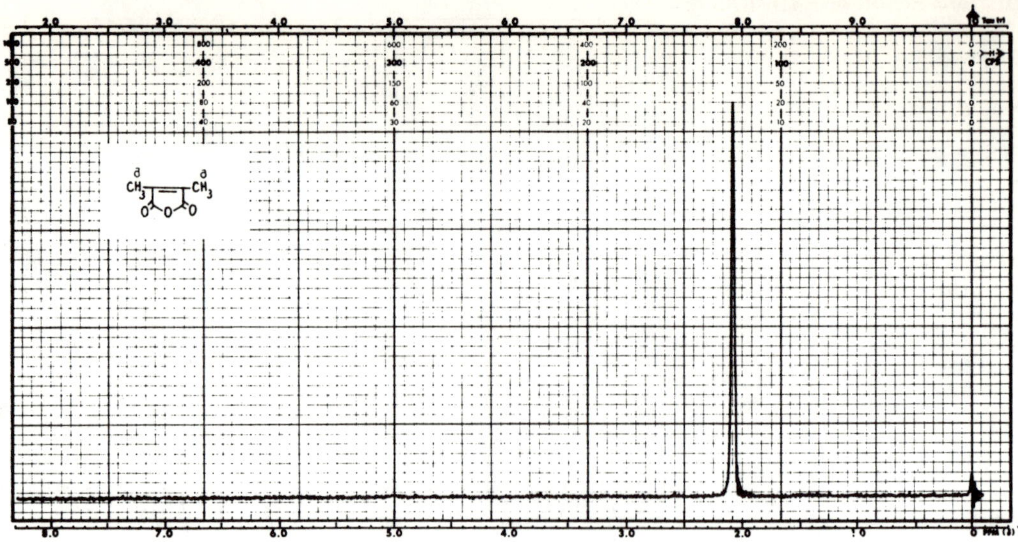

2265

© 1978 Sadtler Research Laboratories, Inc.

BROMOMALEIC ANHYDRIDE

C_4HBrO_3

Mol. Wt. 176.96

B.P. 211-212°C

Solvent: $CDCl_3$

Assignments
a = 7.21

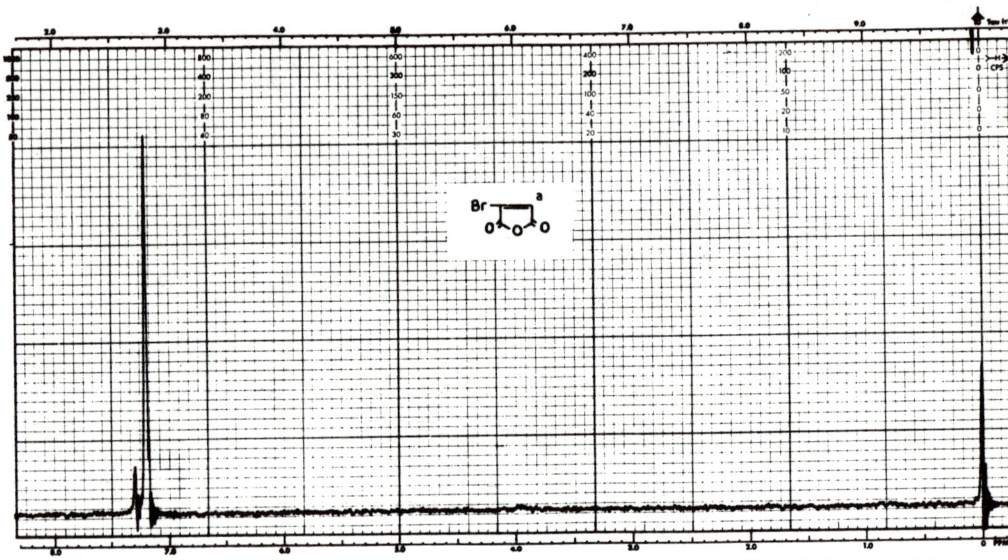

2266

© 1978 Sadtler Research Laboratories, Inc.

1-CYCLOHEXENE-1,2-DICARBOXYLIC ANHYDRIDE

$C_8H_8O_3$

Mol. Wt. 152.15

Solvent: $CDCl_3$

Assignments
a = 1.82
b = 2.45

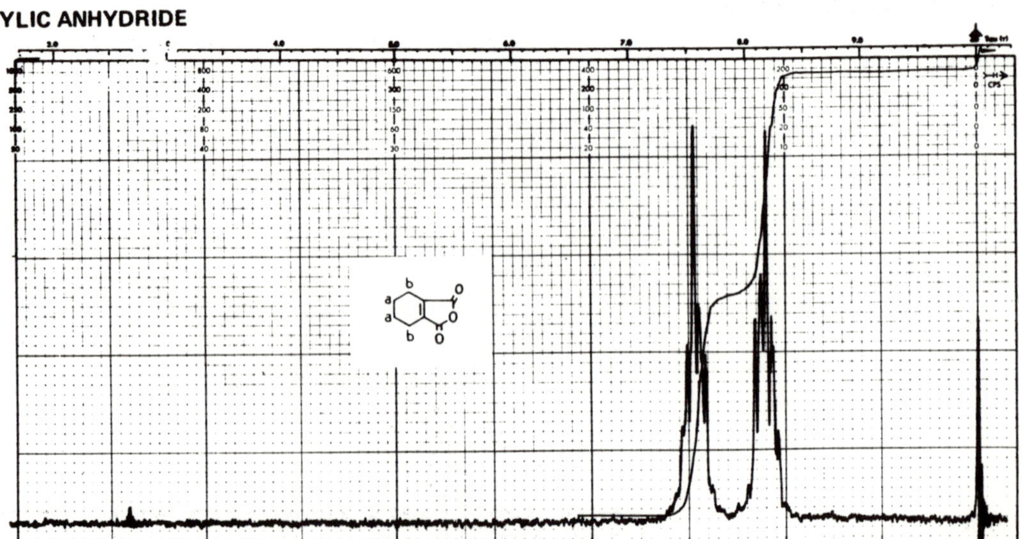

2267

© 1978 Sadtler Research Laboratories, Inc.

cis-4-CYCLOHEXENE-1,2-DICARBOXYLIC ANHYDRIDE

$C_8H_8O_3$

Mol. Wt. 152.15

M.P. 101-102°C

Solvent: $CDCl_3$

Assignments
a = 2.00-2.90
b = ca 3.48
c = ca 6.01
d = 5.69 imp.

2268

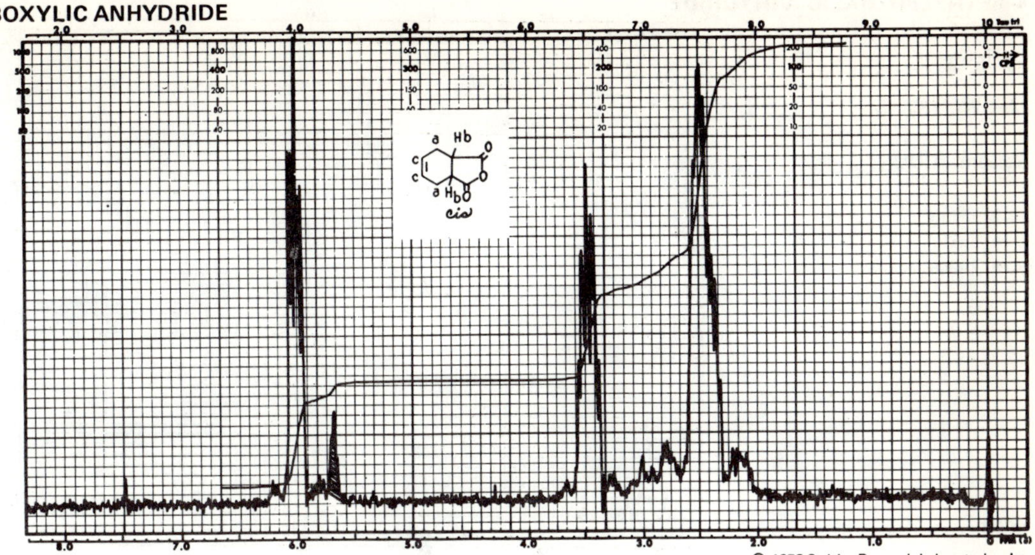

© 1978 Sadtler Research Laboratories, Inc.

CINNAMIC ANHYDRIDE

$C_{18}H_{14}O_3$

Mol. Wt. 278.31

M.P. 134-136°C

Solvent: $CDCl_3$

Assignments
a = 6.49
b = 7.41
c = 7.80

2269

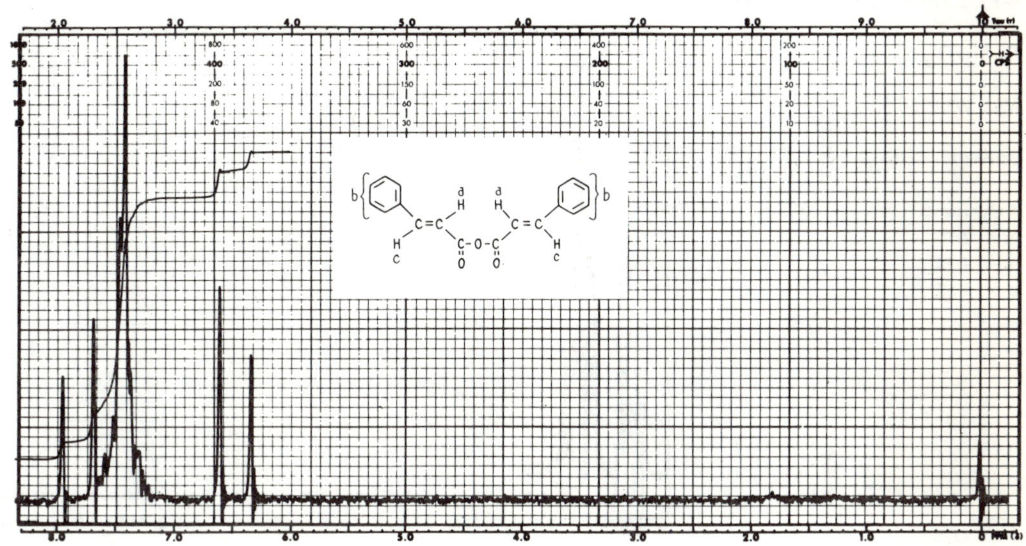

© 1978 Sadtler Research Laboratories, Inc.

BENZOIC ANHYDRIDE

$C_{14}H_{10}O_3$

Mol. Wt. 226.22

M.P. 42°C

Solvent: $CDCl_3$

Assignments
a =ca 7.57
b = 8.11

2270

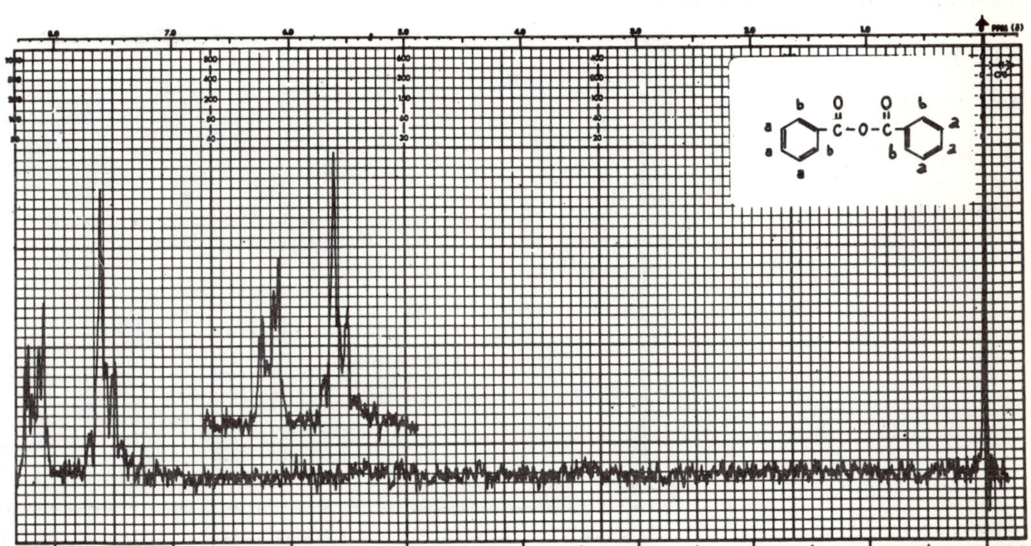

© 1978 Sadtler Research Laboratories, Inc.

4-METHYLPHTHALIC ANHYDRIDE

$C_9H_6O_3$

Mol. Wt. 162.15

Solvent: $CDCl_3$

Assignments
a = 2.63
b = 7.65-8.10

2271

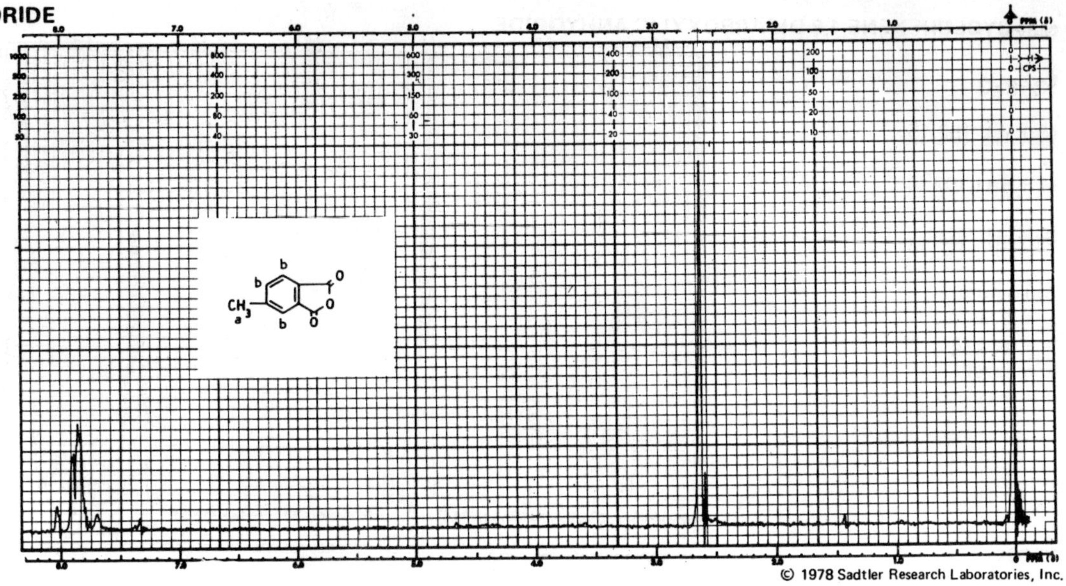

© 1978 Sadtler Research Laboratories, Inc.

4-FLUOROPHTHALIC ANHYDRIDE

$C_8H_3FO_3$

Mol. Wt. 166.11

Solvent: $CDCl_3$

Assignments
a = 7.72
b = 7.79
c = 8.22

2272

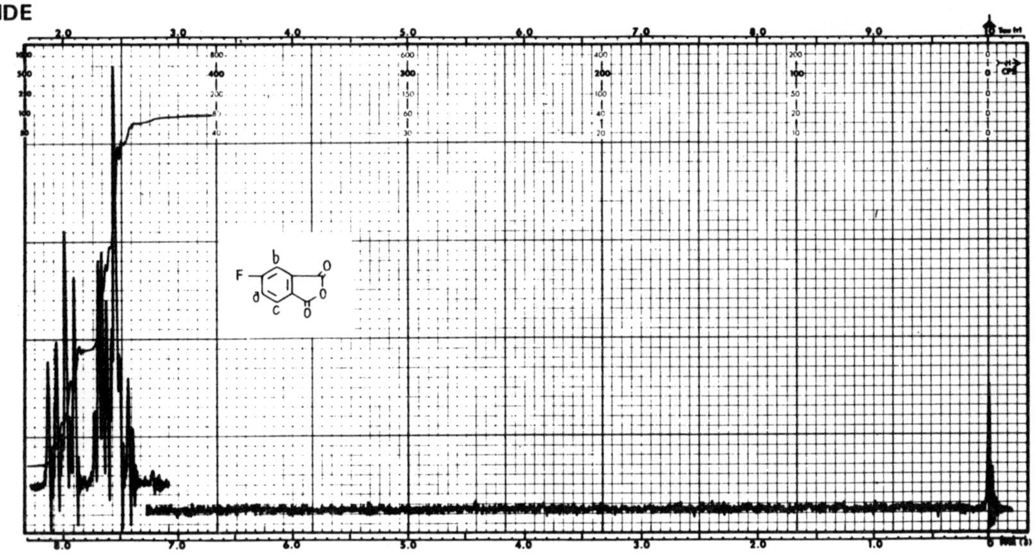

© 1978 Sadtler Research Laboratories, Inc.

1,2,4,5-BENZENETETRACARBOXYLIC 1,2:4,5-DIANHYDRIDE

$C_{10}H_2O_6$

Mol. Wt. 218.12

Solvent: DMSO-d_6

Assignments
a = 8.61

2273

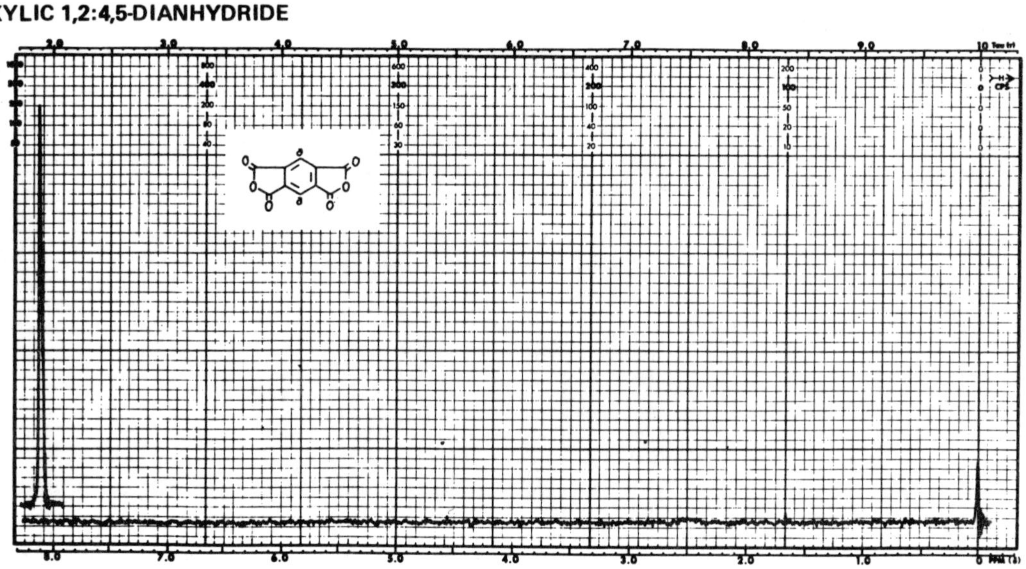

© 1978 Sadtler Research Laboratories, Inc.

NAPHTHALIC ANHYDRIDE

$C_{12}H_6O_3$

Mol. Wt. 198.18

M.P. 274-276°C

Solvent: Trifluoroacetic acid

Assignments
a = 7.95
b = 8.46
c = 8.71

2274

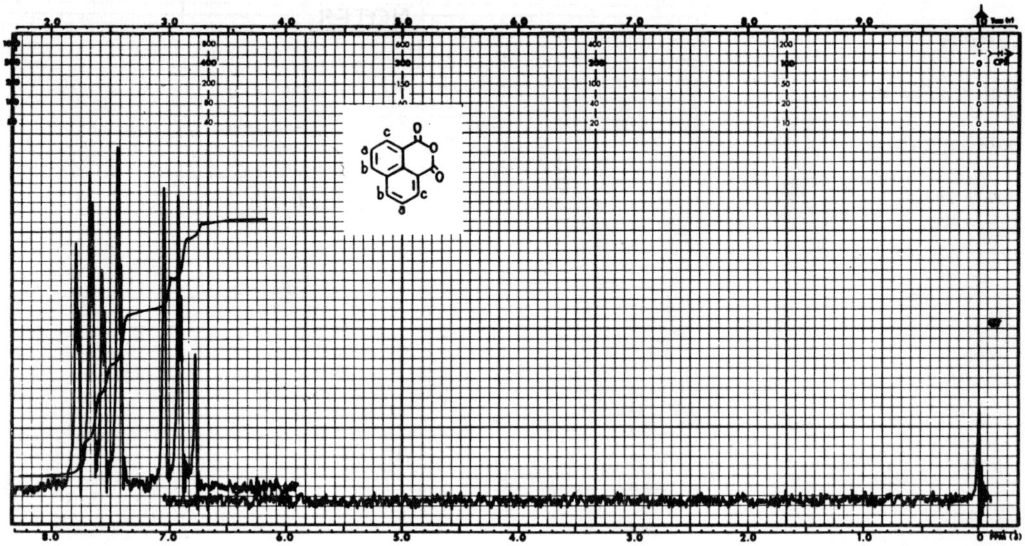

© 1978 Sadtler Research Laboratories, Inc.

NOTES

Although similar in chemical shifts to the other carbonyl containing compounds, the Primary Amides can be distinguished by the presence of one or two very broad bands at low field (5.5–8.9 ppm) arising from the resonance of the two $-NH_2$ protons. These bands are exchangeable and will not be observed if D_2O or TFA are used as the solvent. Due to the partial double bond character of the amide $-C(=O)-NH_2$ bond, the two NH protons may be non-equivalent resulting in two distinct but overlapping resonance bands.

Chemical Shifts

Aliphatic Protons

HC–3	HC–2	HC–1	$-C(=O)-NH_2$		Solvent	Spectrum
		1.98	- - - -	$CH_3-C(=O)-NH_2$	D_2O	2275
	1.12	2.23	6.0–7.5	$CH_3-CH_2-C(=O)-NH_2$	$CDCl_3$	2276
0.94	1.62	2.28	- - - -	$CH_3-CH_2-CH_2-C(=O)-NH_2$	D_2O	- - - -
	(1.19)	2.39	5.6–6.2	$(CH_3)_2-CH-C(=O)-NH_2$	$CDCl_3$	- - - -
(0.98)	- - - -	2.11	- - - -	$(CH_3)_3-C-CH_2-C(=O)-NH_2$	D_2O	- - - -

Substituted Acetamides

$H_2N-C(=O)-$	CH_2	$-X$	
6.0–7.5	2.23	$-CH_3$	$CDCl_3$
- - - -	2.28	$-CH_2-CH_3$	D_2O
7.0, 7.4	3.00	$-C(=O)-NH_2$	DMSO-d_6
5.1–5.7	3.56	–Ø	$CDCl_3$
7,3, 7.6	3.59	$-C\equiv N$	DMSO-d_6
5.5–8.0	3.89	$-O-CH_3$	$CDCl_3$
- - - -	4.18	–Cl	D_2O
- - - -	4.19	–1–Naphthalene	TFA

2-Substituted Propionamides

$H_2N-C(=O)-$	CH_2	$-CH_2$	$-X$	
6.5–7.1	2.43	2.43	–C(=O)–N(Piperazine)	DMSO-d_6
6.2, 6.9	2.59	2.59	–C(=O)–N(Pyrrolidine)	CCl_4
6.1–6.7	2.61	2.61	–C(=O)–N(Piperidine)	$CDCl_3$
6.6–7.3	2.48	2.62	$-S-CH_2-$Ø	DMSO-d_6
- - - -	3.00	3.00	$-C(=O)-NH_2$	TFA
5.7–6.7	2.54	3.12	$-S-C(=O)-CH_3$	$CDCl_3$
6.6–7.2	2.72	3.85	–Cl	Acetone-d_6
5.9, 6.4	2.74	4.38	–O–Ø	$CDCl_3$

Aromatic Protons

Benzamide

para/meta	ortho		Solvent	Spectrum
7.3–7.9	8.00	–C(=O)–NH_2	TFA	2288

Para Substituted Benzamides

H_2N–C(=O)–	ortho	meta	para		
6.5–7.0	7.62	6.60	–NH_2	Polysol	- - - -
7.1–7.7	7.90	6.94	–O–CH_3	Polysol	- - - -
7.1–7.8	7.87	7.61	–Br	DMSO-d_6	- - - -
- - - -	7.92	7.43	–CH_3	TFA	2290
6.8–7.6	7.62	7.78	–NO_2	DMSO-d_6	2292

Exchangeable Protons

As discussed previously, the two exchangeable Primary Amide protons resonate at low field as either one or two very broad bands. The table of chemical shifts provided below indicates that the aliphatic Primary Amides resonate at slightly higher field than the aromatic compounds. The chemical shift(s) of these protons vary over a relatively wide range of values due to their sensitivity to the concentration of the sample solution, the solvent employed and the temperature at which the solution was examined, in addition to any possible hydrogen bonding effects and other structural considerations.

H_2N–C(=O)	–X		
5.5–6.9	–R_7	$CDCl_3$	2279
5.8–6.9	–R_6	$CDCl_3$	2278
5.9, 6.5	–CH(Ø)–R_2	CCl_4	2286
6.0–7.5	–R_2	$CDCl_3$	2276
6.0–7.0	–Ø (3–CH_3)	$CDCl_3$	2289
6.8–7.5	–Ø (4–NO_2)	DMSO	2292
7.3–8.3	–2–Furan	DMSO	2294
7.6–8.9	–Ø (2–OH)	DMSO-d_6	2293

ACETAMIDE

C_2H_5NO

Mol. Wt. 59.07

M.P. 81°C

Solvent: D_2O

Assignments
a = 1.98
b = 4.61 HDO

2275

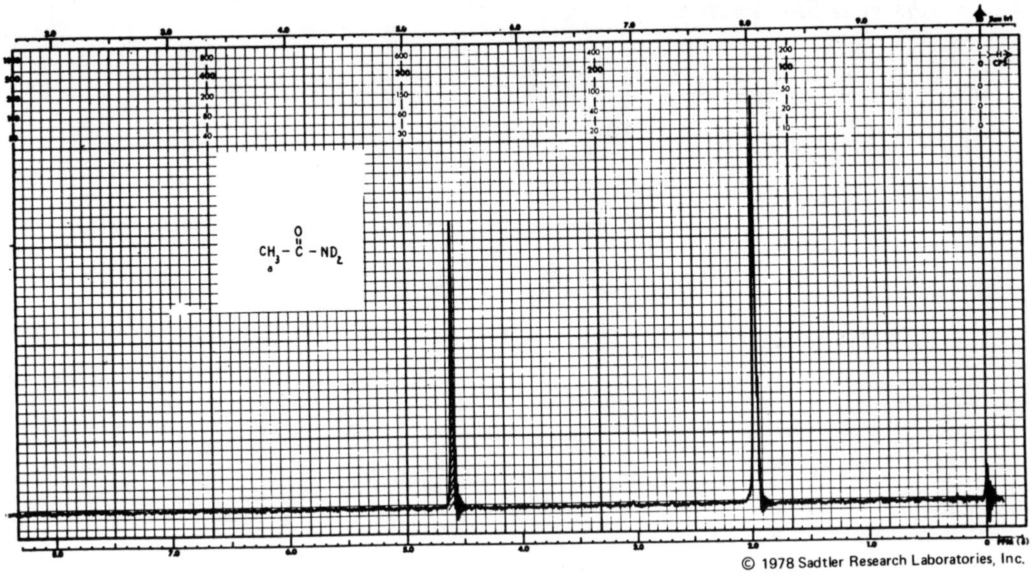

© 1978 Sadtler Research Laboratories, Inc.

PROPIONAMIDE

C_3H_7NO

Mol. Wt. 73.10

M.P. 76-78°C

Solvent: $CDCl_3$

Assignments
a = 1.12
b = 2.23
c = 6.00-7.50
d = 5.04 imp.

2276

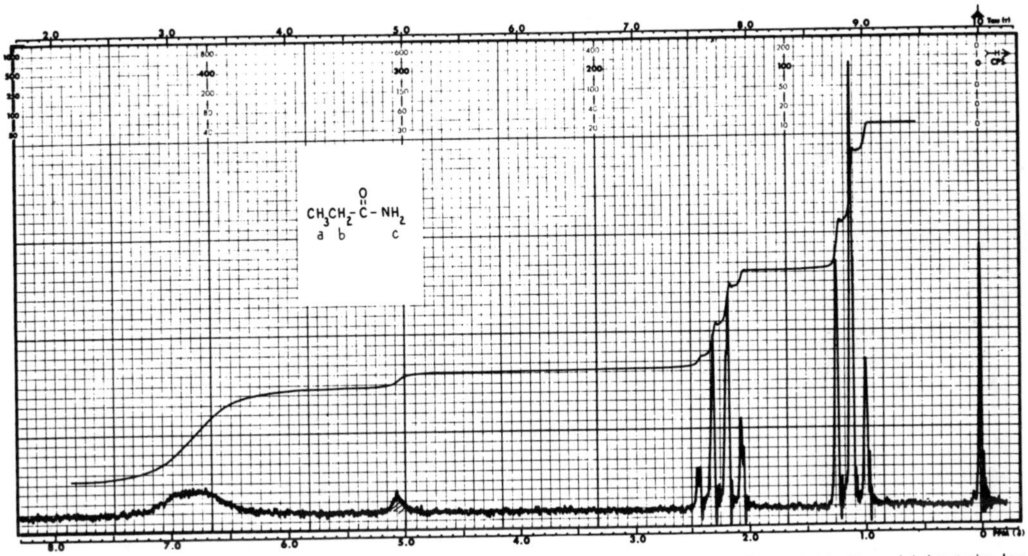

© 1978 Sadtler Research Laboratories, Inc.

HEXANAMIDE

$C_6H_{13}NO$

Mol. Wt. 115.18

M.P. 100-101°C

Solvent: Dimethyl sulfoxide-d_6

Assignments
a = 0.85
b = 1.00-1.80
c = 1.85-2.25
d = ca 6.65
e = ca 7.15
f = 2.15 DMSO-d_5
g = 3.33 H_2O

2277

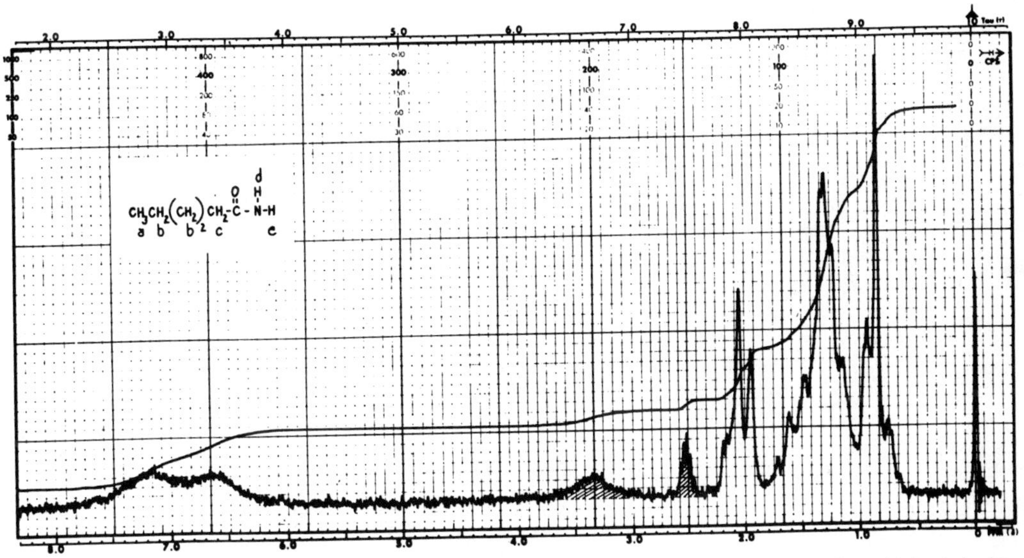

© 1978 Sadtler Research Laboratories, Inc.

HEPTANAMIDE

$C_7H_{15}NO$

Mol. Wt. 129.20

M.P. 96°C

Solvent: $CDCl_3$

Assignments
a = 0.89
b = 1.31
c = ca 1.60
d = 2.21
e = 6.25
f = 2.89 imp.

2278

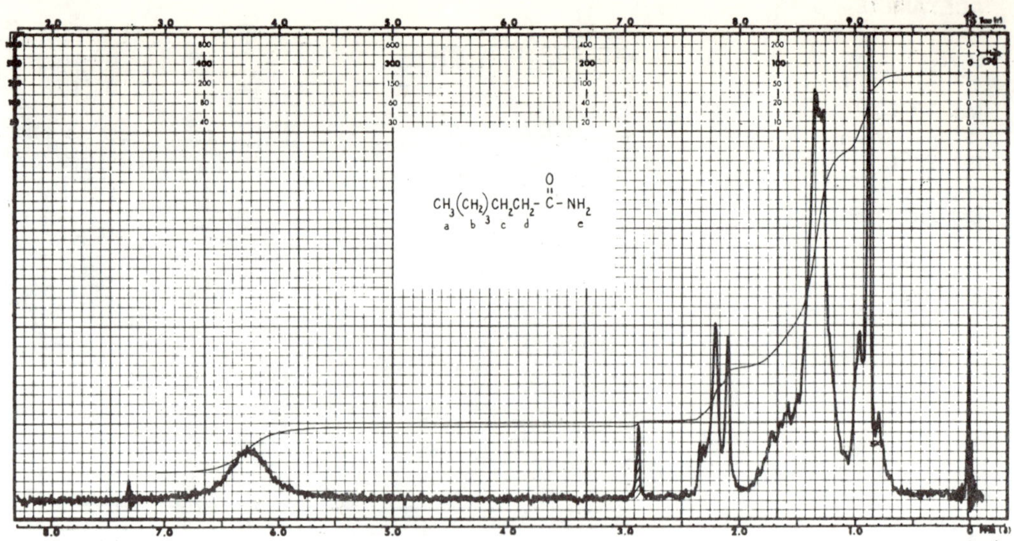

© 1978 Sadtler Research Laboratories, Inc.

OCTANAMIDE

$C_8H_{17}NO$

Mol. Wt. 143.23

Solvent: $CDCl_3$

Assignments
a = 0.90
b = ca 1.31
c = 1.60
d = 2.21
e = 6.11

2279

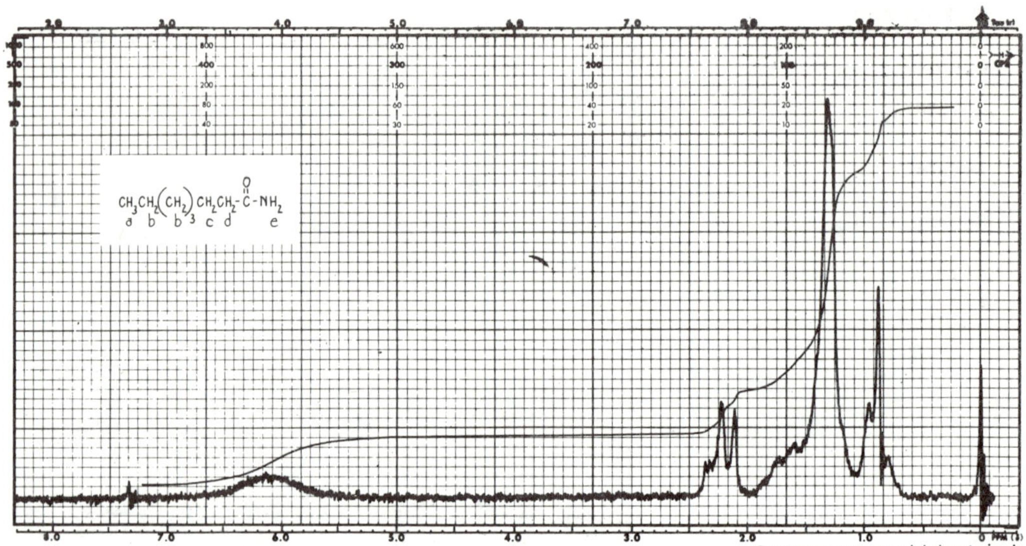

© 1978 Sadtler Research Laboratories, Inc.

MYRISTAMIDE

$C_{14}H_{29}NO$

Mol. Wt. 227.39

Solvent: Trifluoroacetic acid

Assignments
a = 0.92
b = 1.10-2.10
c = 2.63
d = unobserved *
*Under solvent band

2280

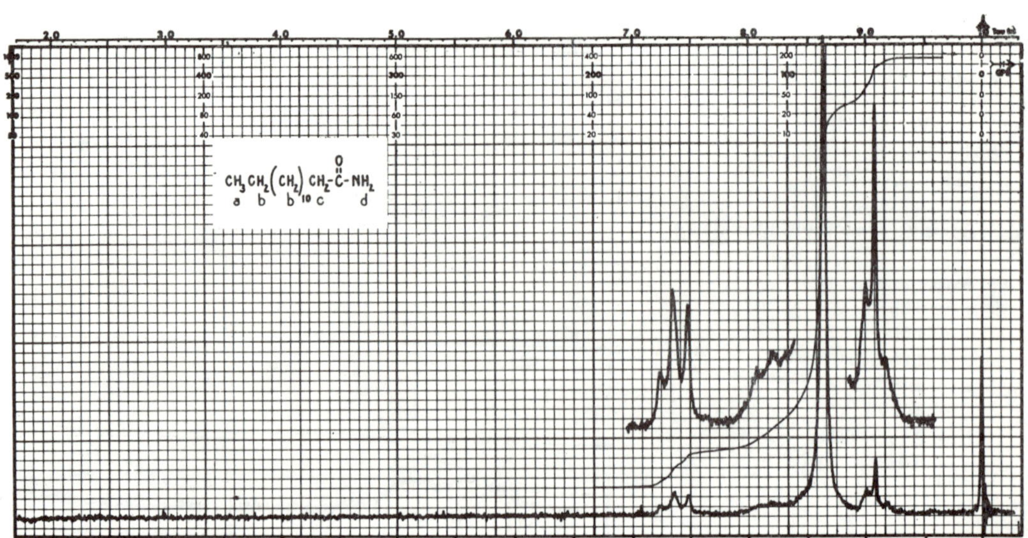

© 1978 Sadtler Research Laboratories, Inc.

2-CHLOROACETAMIDE

C_2H_4ClNO

Mol. Wt. 93.51

Solvent: D_2O

Assignments
a = 4.18
b = 4.62 HDO

2281

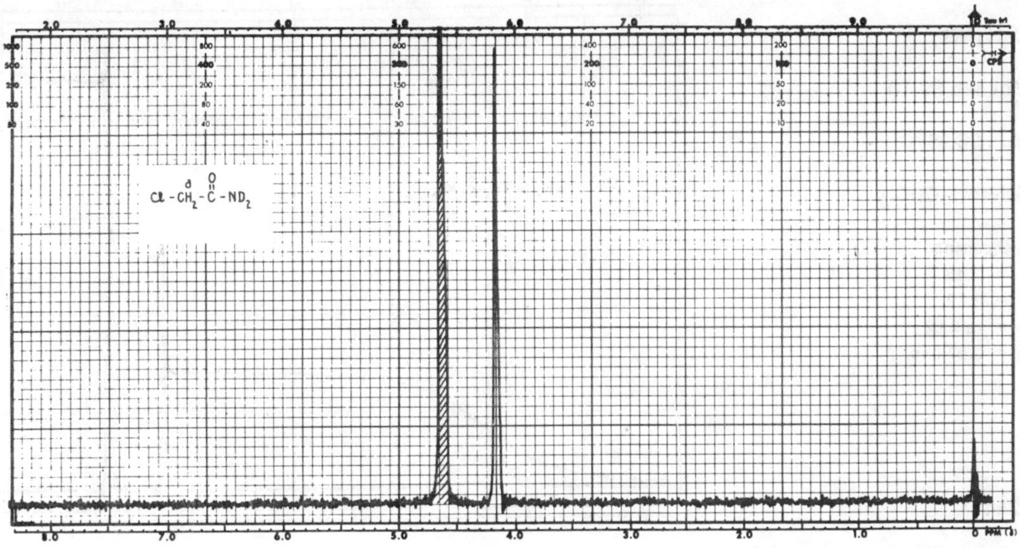

© 1978 Sadtler Research Laboratories, Inc.

2,2-DICHLOROACETAMIDE

$C_2H_3Cl_2NO$

Mol. Wt. 127.96

M.P. 97-99°C

Solvent: Acetone

Assignments
a = 6.31
b = ca 7.35

2282

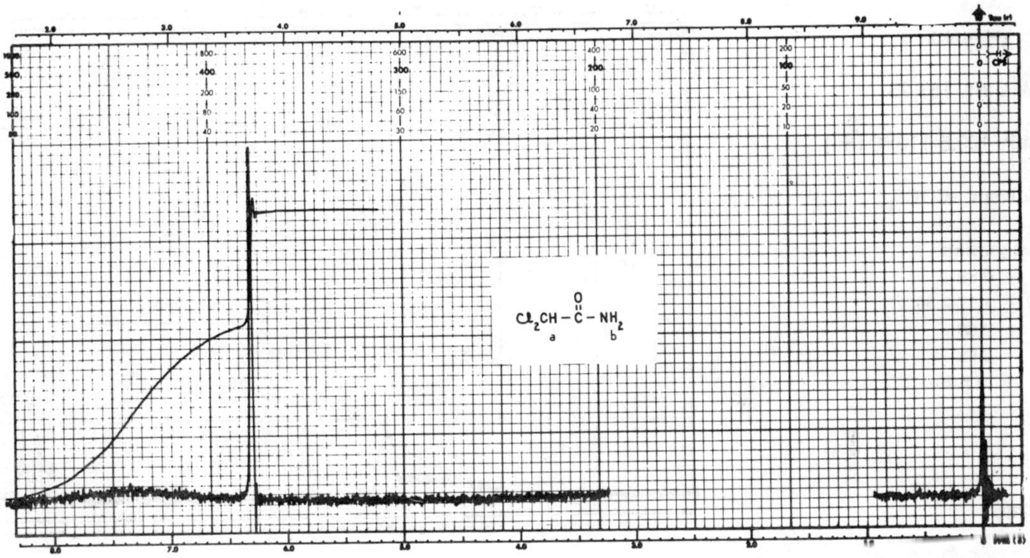

© 1978 Sadtler Research Laboratories, Inc.

LACTAMIDE

$C_3H_7NO_2$

Mol. Wt. 89.09

M.P. 74-76°C

Solvent: D_2O

Assignments
a = 1.38
b = 4.27
c = 4.68 HDO

2283

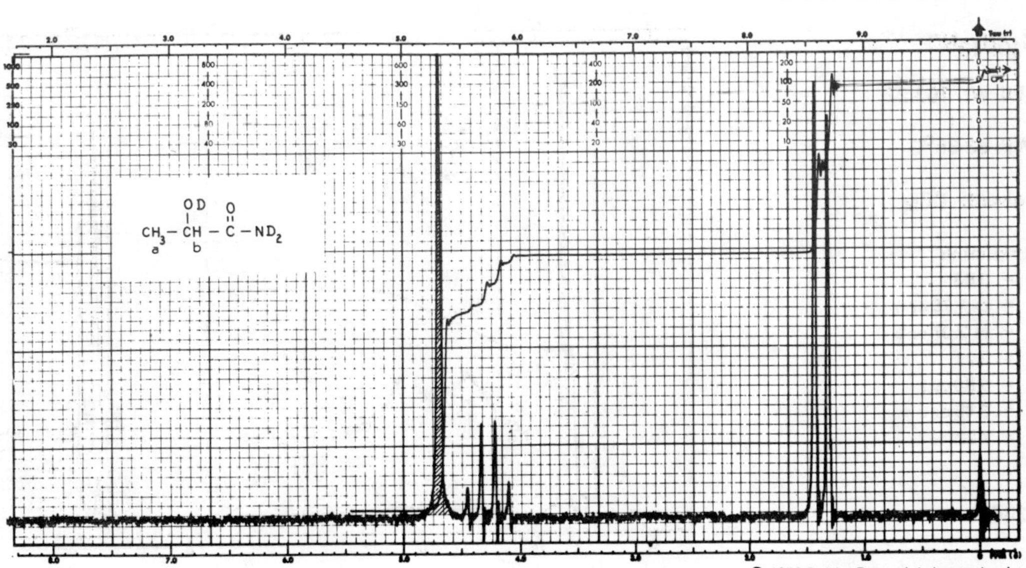

© 1978 Sadtler Research Laboratories, Inc.

2-METHOXYACETAMIDE

$C_3H_7NO_2$

Mol. Wt. 89.10

M.P. 96-99°C

Solvent: $CDCl_3$

Assignments
a = 3.43
b = 3.89
c = ca 6.62
d = 2.57 imp.

2284

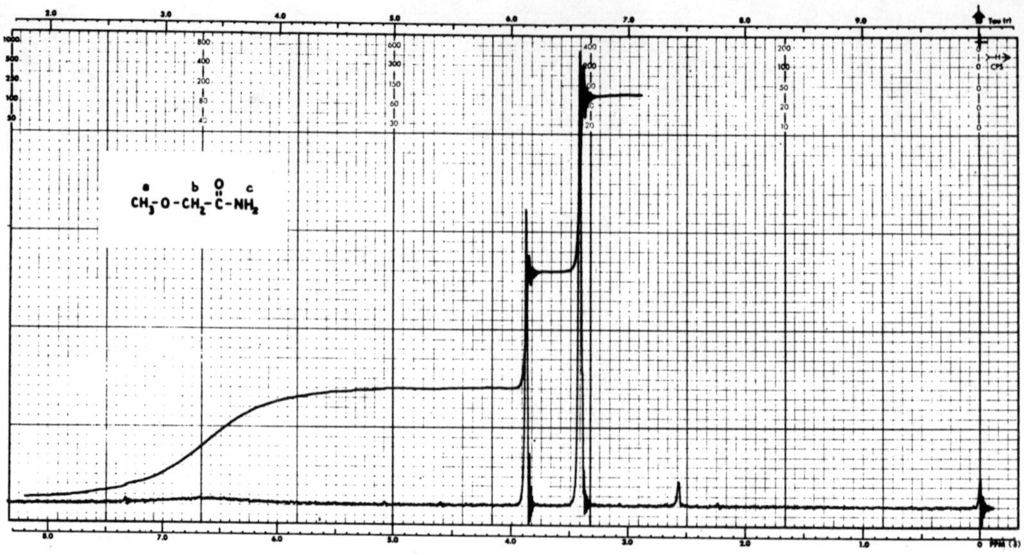

© 1978 Sadtler Research Laboratories, Inc.

α-ETHYL-1-CYCLOHEXENE-1-ACETAMIDE

$C_{10}H_{17}NO$

Mol. Wt. 167.25

M.P. 109.5-110°C (lit.)

Solvent: $CDCl_3$

Assignments
a = 0.88
b = 1.17-2.30
c = 2.68
d = 5.66
e = ca 6.20

2285

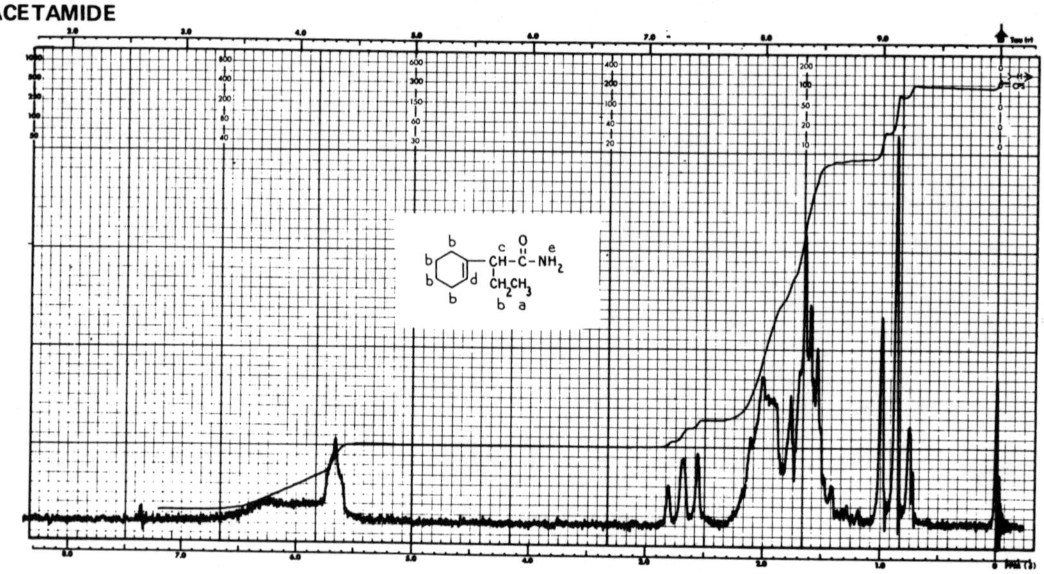

© 1978 Sadtler Research Laboratories, Inc.

2-PHENYLBUTYRAMIDE

$C_{10}H_{13}NO$

Mol. Wt. 163.22

M.P. 83-85°C

Solvent: CCl_4

Assignments
a = 0.83
b = ca 1.75 or c
c = ca 2.07 or b
d = 3.21
e = 5.86 or f
f = 6.45 or e
g = 7.22

2286

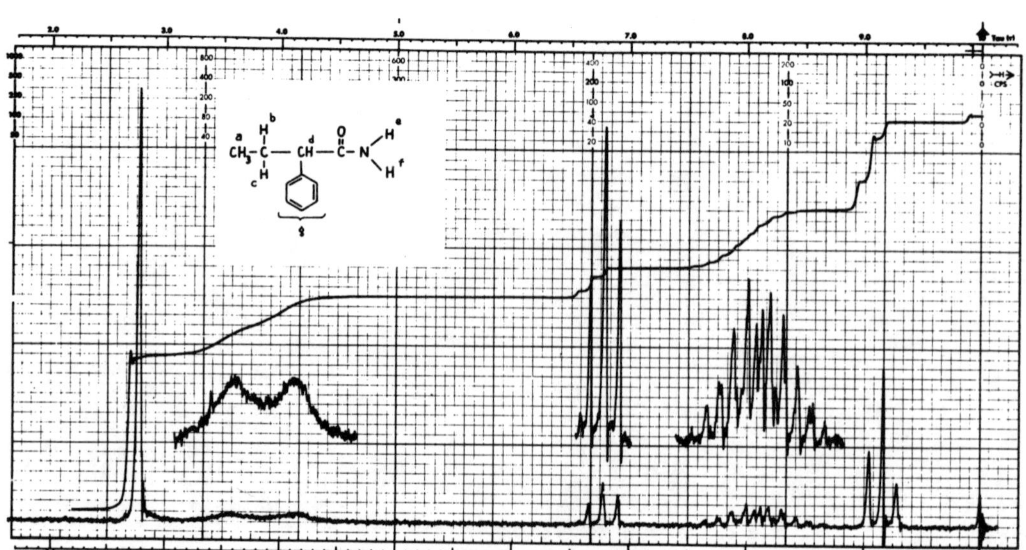

© 1978 Sadtler Research Laboratories, Inc.

1-NAPHTHALENEACETAMIDE

$C_{12}H_{11}NO$

Mol. Wt. 185.23

M.P. 183-184°C

Solvent: Trifluoroacetic acid

Assignments
a = 4.19
b = 7.20-7.65
c = ca 7.80
d = unobserved *
* Under solvent band

2287

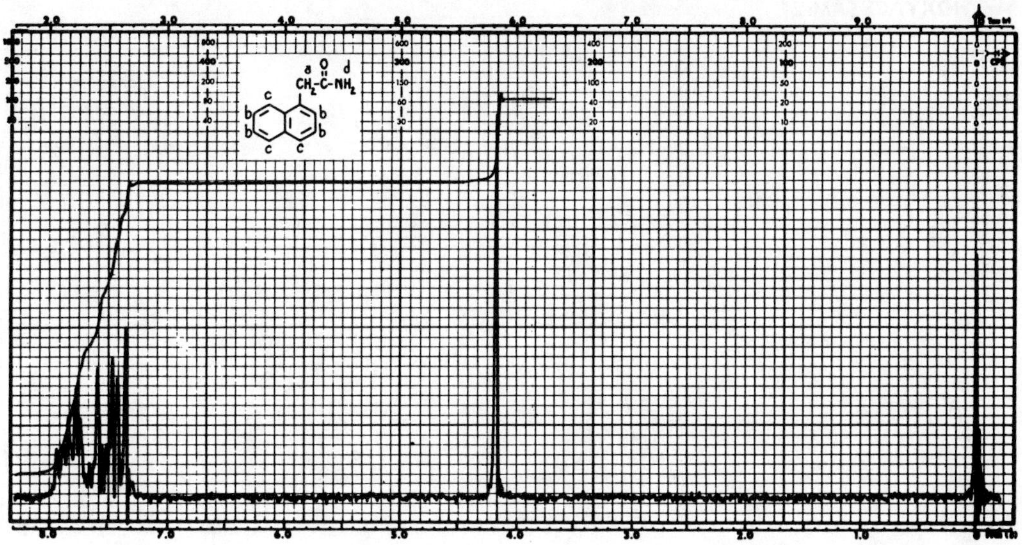

© 1978 Sadtler Research Laboratories, Inc.

BENZAMIDE

C_7H_7NO

Mol. Wt. 121.14

M.P. 130°C

Solvent: Trifluoroacetic acid

Assignments
a = 7.35-7.85
b = 8.00

2288

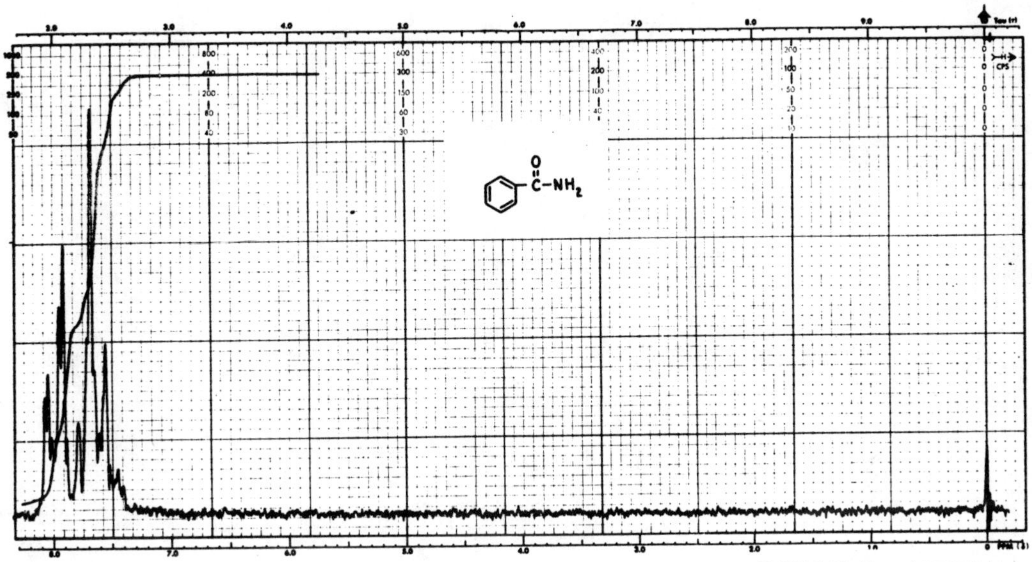

© 1978 Sadtler Research Laboratories, Inc.

m-TOLUAMIDE

C_8H_9NO

Mol. Wt. 135.17

M.P. 94-95°C

Solvent: $CDCl_3$

Assignments
a = 2.34
b = ca 6.50
c = 7.30
d = 7.63

2289

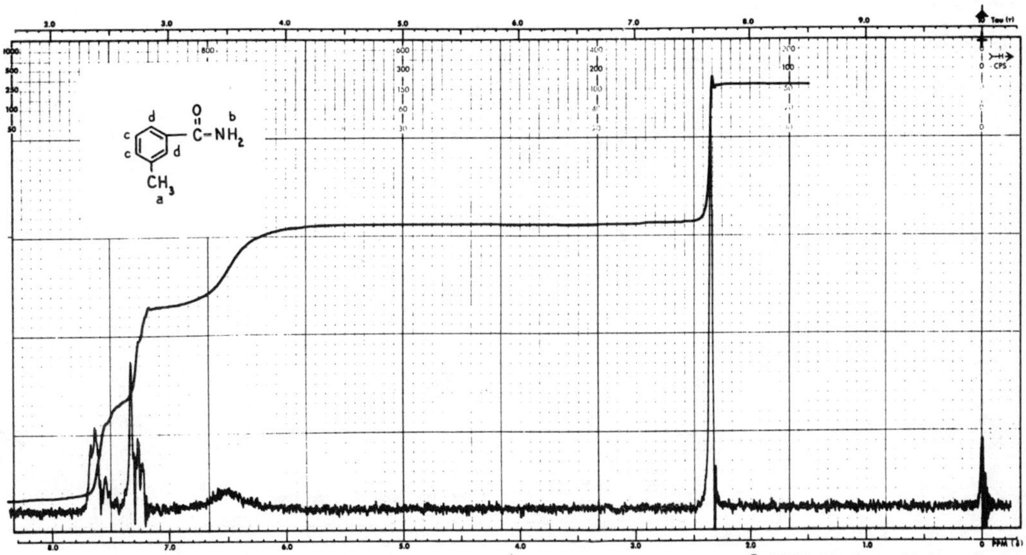

© 1978 Sadtler Research Laboratories, Inc.

p-TOLUAMIDE

C_8H_9NO

Mol. Wt. 135.17

M.P. 159-160°C

Solvent: Trifluoroacetic acid

Assignments
a = 2.49
b = 7.43
c = 7.92
d = ca 11.70

2290

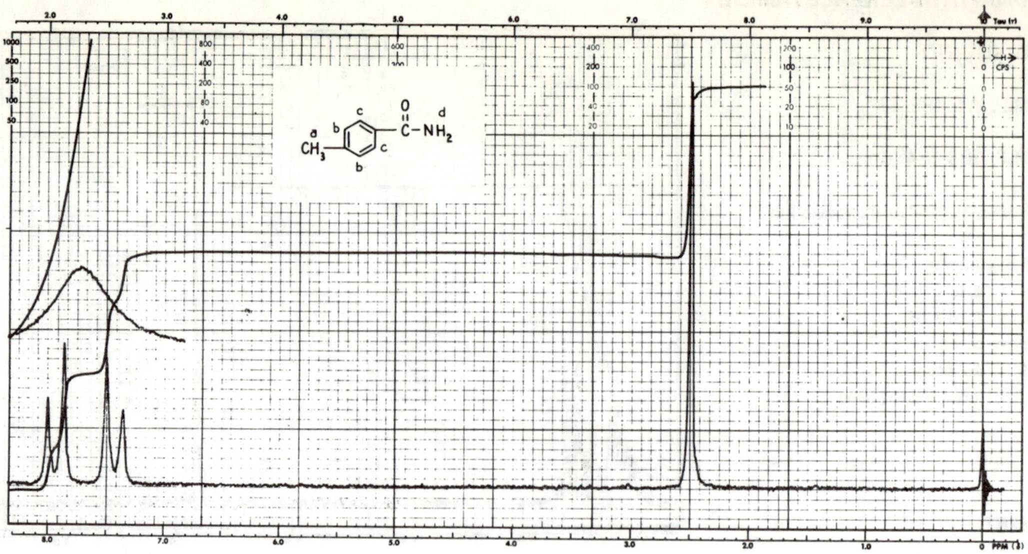

© 1978 Sadtler Research Laboratories, Inc.

o-BROMOBENZAMIDE

C_7H_6BrNO

Mol. Wt. 200.04

M.P. 160-161°C

Solvent: Trifluoroacetic acid

Assignments
a = ca 7.58

2291

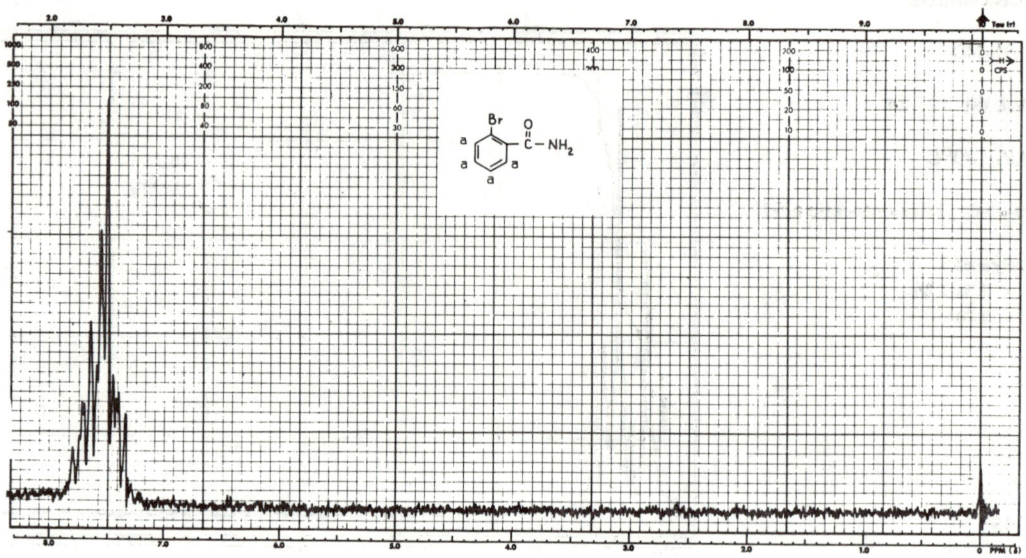

© 1978 Sadtler Research Laboratories, Inc.

p-NITROBENZAMIDE

$C_7H_6N_2O_3$

Mol. Wt. 166.13

M.P. 199-200°C

Solvent: Dimethyl sulfoxide

Assignments
a = 7.20
b = 7.62
c = 7.78

2292

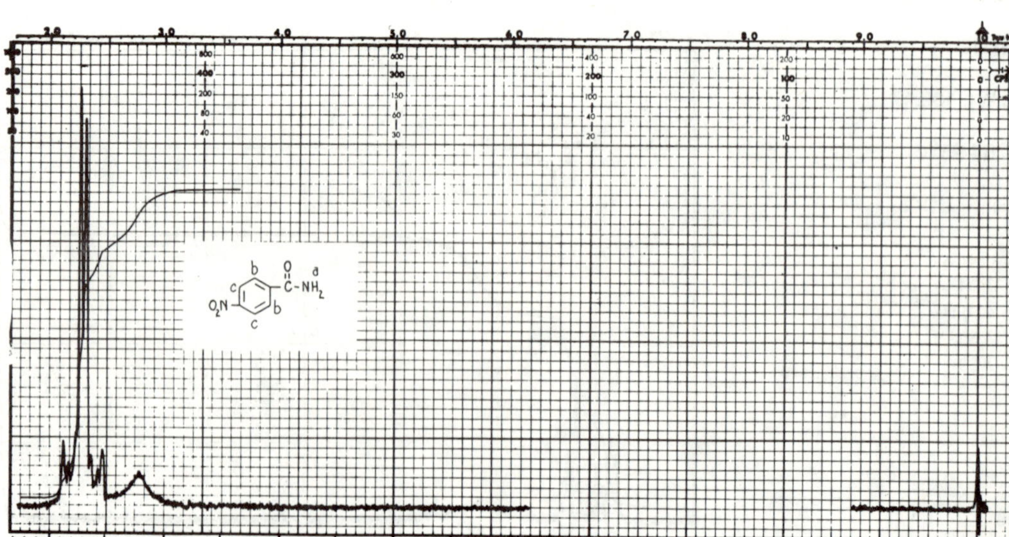

© 1978 Sadtler Research Laboratories, Inc.

SALICYLAMIDE

$C_7H_7NO_2$

Mol. Wt. 121.14

M.P. 140-142°C

Solvent: Dimethyl sulfoxide-D_6

Assignments
a = 6.96
b = 7.02
c = 7.50
d = 8.02
e = ca 8.20
f = 13.10
g = 2.60 DMSO-D_5

2293

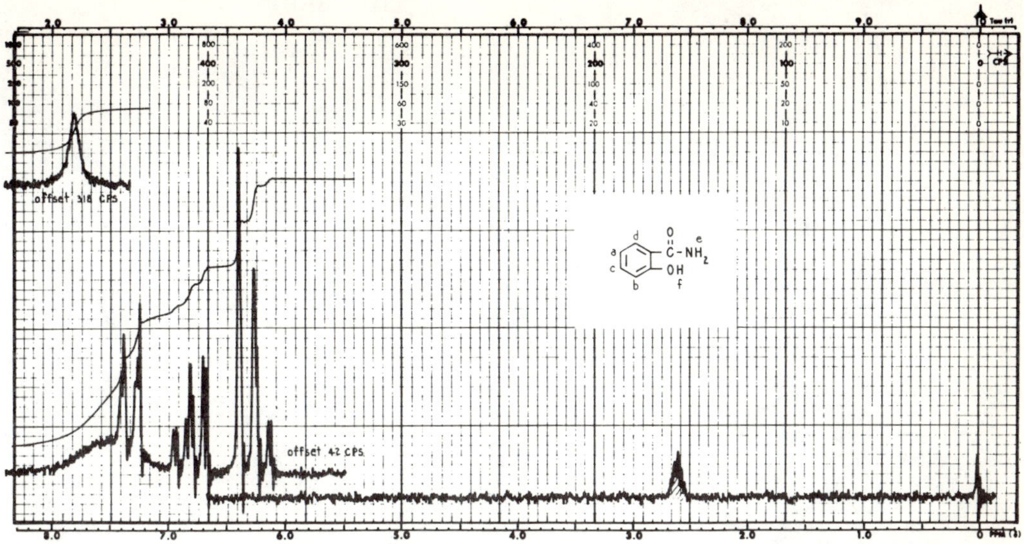

© 1978 Sadtler Research Laboratories, Inc.

2-FURAMIDE

$C_5H_5NO_2$

Mol. Wt. 111.10

M.P. 144.5-145.0°C

Solvent: Dimethyl sulfoxide

Assignments
a = 6.67
b = 7.29
c = 7.72
d = 7.82

2294

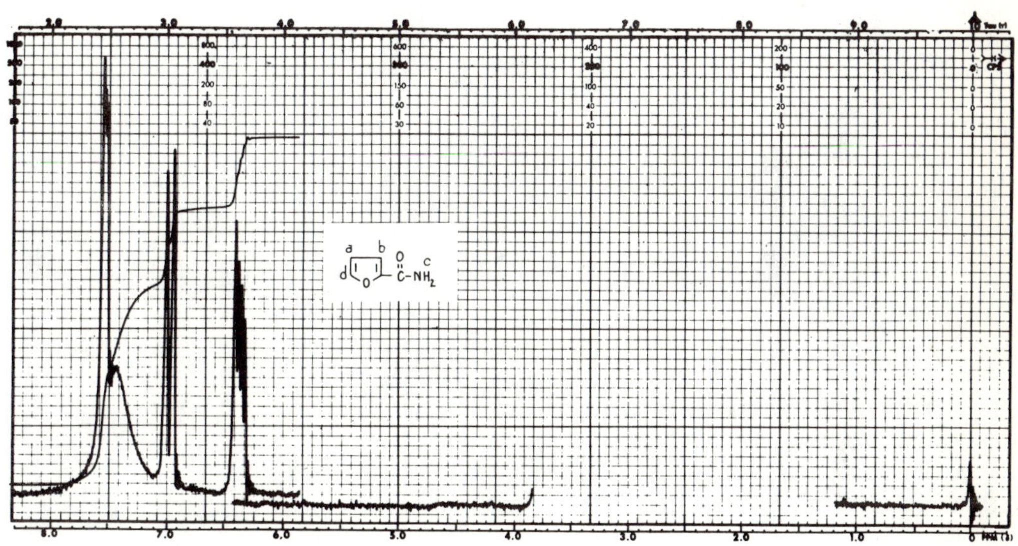

© 1978 Sadtler Research Laboratories, Inc.

NOTES

The NMR spectra of the Secondary Amides are usually more complex than the primary amides due to the presence of a substituent bonded to the amide nitrogen atom. These substituents produce a much wider range of chemical shifts for the amide proton which may, in addition, display coupling to aliphatic groups bonded to it. The chemical shifts of aliphatic groups bonded to the carbonyl side of this functional group are similar to those observed for the Primary Amides, while those groups bonded to the nitrogen side of the linkage resonate at slightly lower field than the corresponding amines (ca 0.4 ppm).

Chemical Shifts

Aliphatic Protons

HC–3	HC–2	HC–1		Solvent	Spectrum
		1.99	CH_3–C(=O)–NH–R_4	D_2O	2297
		2.27	CH_3–C(=O)–NH–Ø (2-CH_3)	$CDCl_3$	2321
		2.27	CH_3–C(=O)–NH–Ø (2–NO_2)	$CDCl_3$	2333
	1.17	2.24	CH_3–CH_2–C(=O)–NH–CH_3	$CDCl_3$	2296
	1.19	2.35	CH_3–CH_2–C(=O)–NH–Ø	$CDCl_3$	2315
0.92	1.64	2.29	CH_3–CH_2–CH_2–C(=O)–NH–Ø	$CDCl_3$	2316
	(1.12)	2.54	$(CH_3)_2$–CH–C(=O)–NH–Ø	Polysol	2317
	(1.20)	----	$(CH_3)_3$–C–C(=O)–NH–CH_3	$CDCl_3$	----
	(1.23)	----	$(CH_3)_3$–C–C(=O)–NH–Ø	Polysol	----
		2.66	CH_3–NH–C(=O)–CH_2–Ø	$CDCl_3$	2312
		2.79	CH_3–NH–C(=O)–C$(CH_3)_3$	$CDCl_3$	----
		2.80	CH_3–NH–C(=O)–R_2	$CDCl_3$	2296
		2.97	CH_3–NH–C(=O)–Ø	$CDCl_3$	2342
	1.10	3.21	CH_3–CH_2–NH–C(=O)–R	Polysol	----
	1.15	3.29	CH_3–CH_2–NH–C(=O)–H	D_2O	2295
	1.20	3.39	CH_3–CH_2–NH–C(=O)–Ø	$CDCl_3$	----
0.90	1.51	3.20	CH_3–CH_2–CH_2–NH–C(=O)–R_{11}	$CDCl_3$	2299
	(1.40)	----	$(CH_3)_3$–C–NH–C(=O)–CH=CH_2	$CDCl_3$	2304

Olefinic Protons

The olefinic protons of the Acrylamides display the same differentiation in chemical shifts noted for the other carbonyl containing groups, i.e. the protons bonded to the beta carbon are deshielded in comparison to the geminal proton which is bonded to the alpha carbon.

cis/trans	geminal			
5.9–6.3	5.52	–C(=O)–NH–C$(CH_3)_3$	$CDCl_3$	2304

Chemical Shifts continued

Aromatic Protons

Both sides of the Secondary Amide linkage deshield the ortho aromatic protons. The protons ortho to the NH group resonate near 7.5 ppm while the protons ortho to the carbonyl group resonate at slightly lower field near 7.8 ppm.

N-Substituted Benzamides

para/meta	ortho		Solvent	Spectrum
6.8–7.7	7.81	$-C(=O)-NH-CH_3$	$CDCl_3$	2342
7.2–7.6	7.74	$-C(=O)-NH-\emptyset$ (4-CH_3)	$CDCl_3$	2345

N-Phenyl Amides

para	meta	ortho		Solvent	Spectrum
7.20	7.05	7.52	$-NH-C(=O)-R_3$	$CDCl_3$	2316
7.25	7.12	7.54	$-NH-C(=O)-R_2$	$CDCl_3$	2315

Para Substituted Acetanilides

$CH_3-C(=O-NH-$	ortho	meta	para	Solvent	Spectrum
2.03	7.30	6.71	$-OH$	Polysol	2336
2.02	7.48	6.85	$-O-CH_3$	DMSO-d_6	2337
2.11	7.40	6.99	$-F$	$CDCl_3$	2327
2.05	7.45	7.03	$-CH_3$	DMSO-d_6	2323
2.07	7.32	7.04	$-CH_2-CH_3$	$CDCl_3$	2324
2.10	7.68	7.32	$-Cl$	DMSO-d_6	2330
2.10	7.40	7.28	$-C(CH_3)_3$	$CDCl_3$	2325
2.18	7.60	7.60	$-CF_3$	$CDCl_3$	----
2.10	7.45	7.61	$-Br$	DMSO-d_6	2331
2.11	7.47	7.61	$-CH=N-OH$	Polysol	2332
2.44	7.19	7.70	$-I$	TFA	----
2.48	7.82	8.35	$-NO_2$	TFA	----

Coupling and Coupling Constants

Clear coupling is normally observed between the NH group and the protons of adjacent hydrocarbon groups.

J_{HN-CH} = 4.8–5.2 Hz   Spectra 2296, 2298, 2299

N-ETHYLFORMAMIDE

C_3H_7NO

Mol. Wt. 73.10

B.P. 200-202°C

Solvent: D_2O

Assignments
a = 1.15
b = 3.29
c = 8.02
d = 4.65 HDO

2295

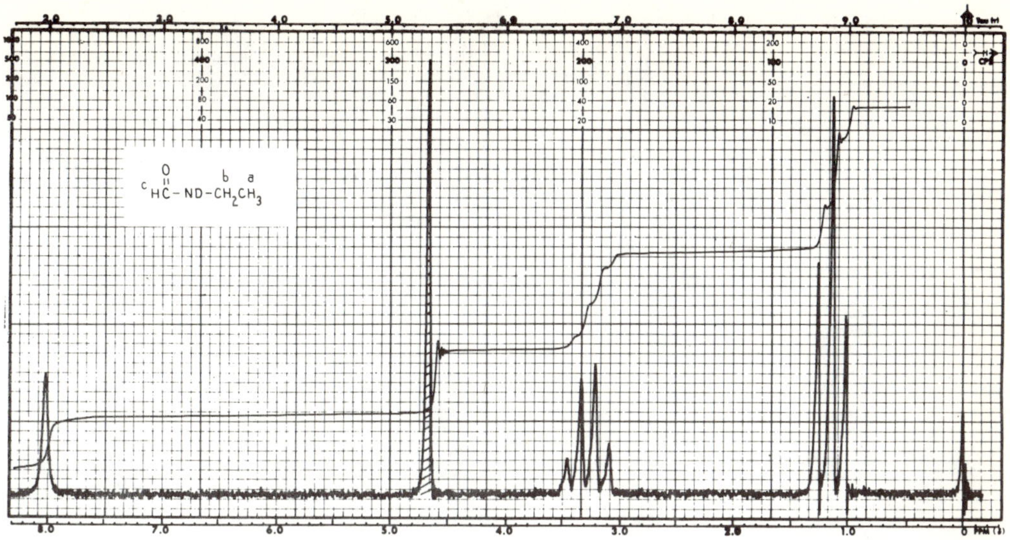

© 1978 Sadtler Research Laboratories, Inc.

N-METHYLPROPIONAMIDE

C_4H_9NO

Mol. Wt. 87.12

B.P. 86-88°C/6mm

Solvent: $CDCl_3$

Assignments
a = 1.17
b = 2.24
c = 2.80
d = 6.76

2296

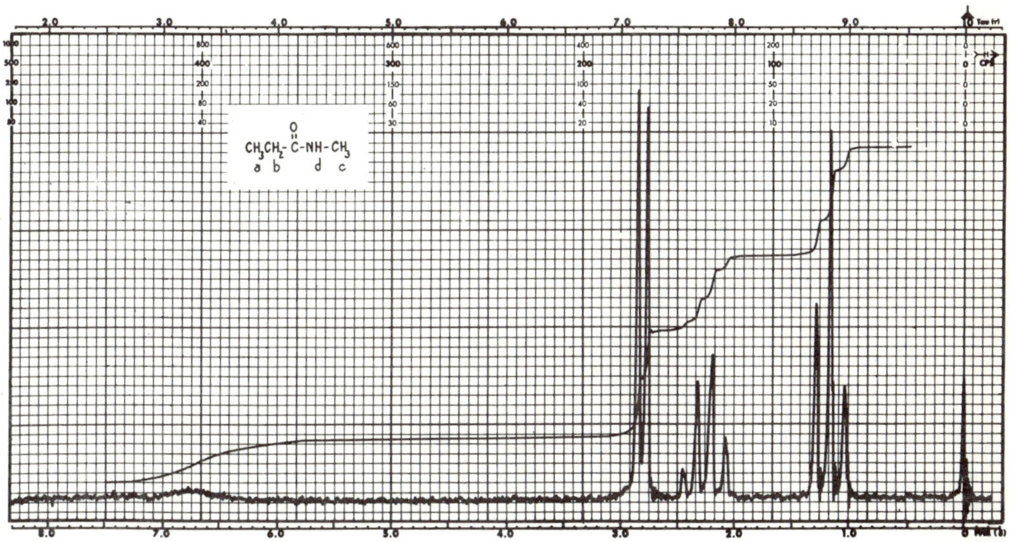

© 1978 Sadtler Research Laboratories, Inc.

N-BUTYLACETAMIDE

$C_6H_{13}NO$

Mol. Wt. 115.18

B.P. 116-118°C/10mm

Solvent: D_2O

Assignments
a = 0.90
b = 1.10-1.70
c = 1.99
d = 3.19
e = 2.09 imp.
f = 4.66 HDO

2297

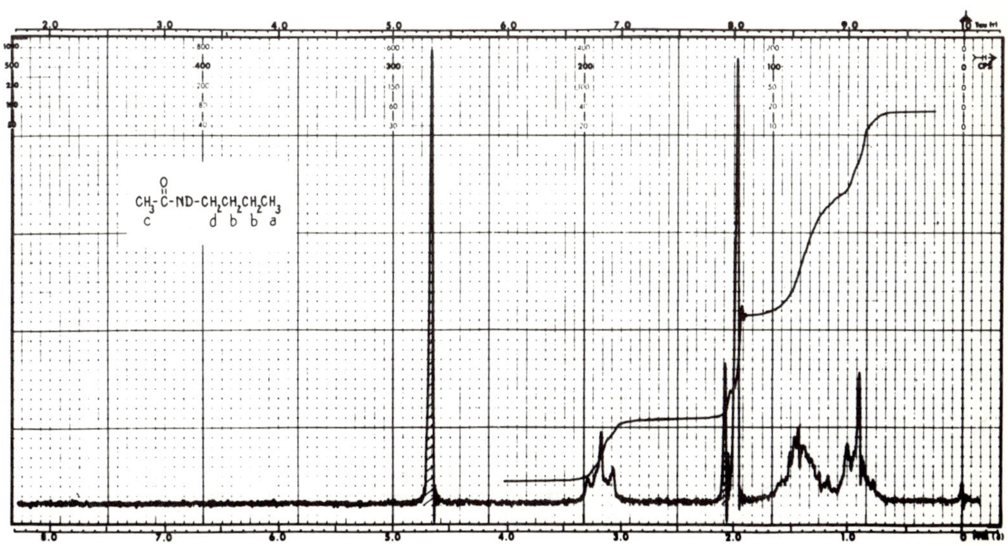

© 1978 Sadtler Research Laboratories, Inc.

N-METHYLDODECANAMIDE

$C_{13}H_{27}NO$

Mol. Wt. 213.37

Solvent: $CDCl_3$

Assignments
a = 0.90
b = 1.29
c = ca 1.62
d = 2.20
e = 2.80
f = 6.23

2298

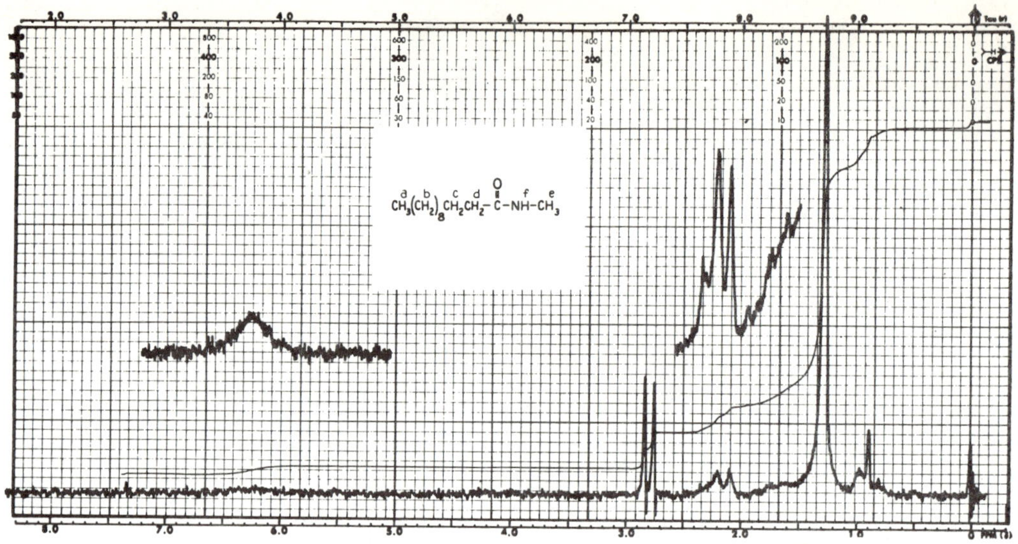

© 1978 Sadtler Research Laboratories, Inc.

N-PROPYLDODECANAMIDE

$C_{15}H_{31}NO$

Mol. Wt. 241.42

M.P. 58-60°C

Solvent: $CDCl_3$

Assignments
a = ca 0.90
b = 1.10-1.50
c = 1.51
d = 2.21
e = 3.20
f = 6.80

2299

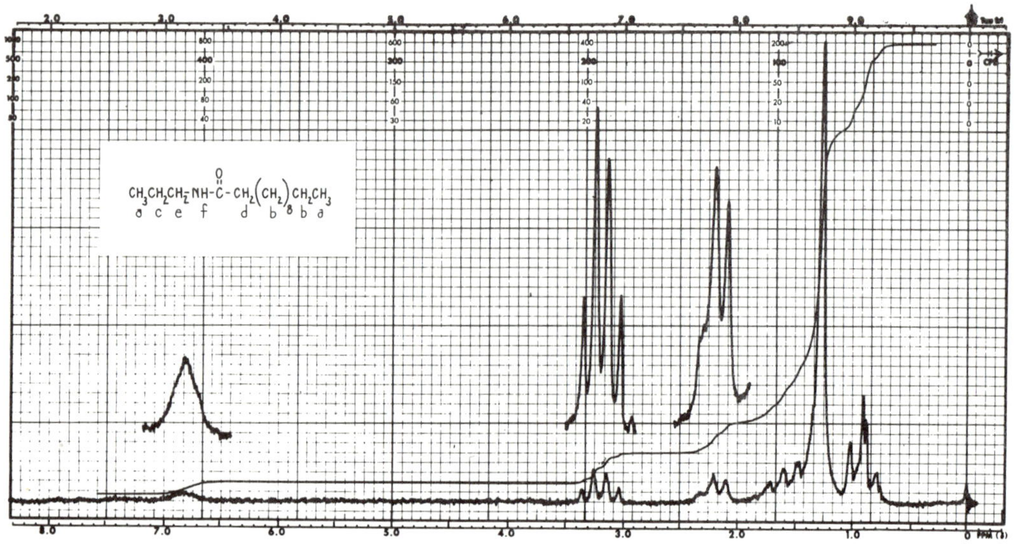

© 1978 Sadtler Research Laboratories, Inc.

N,N'-METHYLENEBISFORMAMIDE

$C_3H_6N_2O_2$

Mol. Wt. 102.09

M.P. 142-147°C

Solvent: Trifluoroacetic acid

Assignments
a = 5.01
b = 8.10
c = 8.38
d = 4.08 imp.

2300

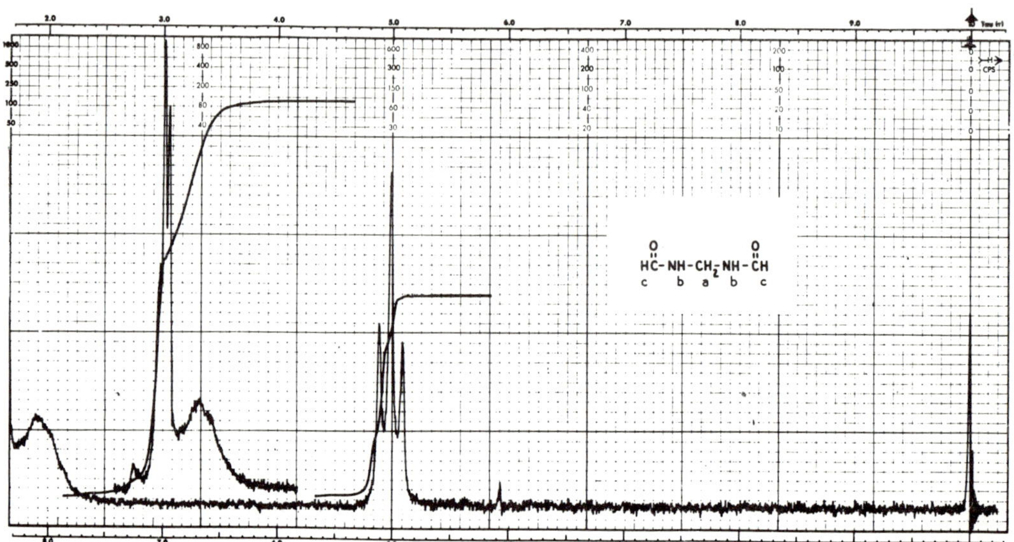

© 1978 Sadtler Research Laboratories, Inc.

N,N'-ETHYLENEBISACETAMIDE

$C_6H_{12}N_2O_2$

Mol. Wt. 144.17

Solvent: Polysol-d

Assignments
a = 1.90
b = 3.24
c = 7.36
d = 3.50 imp.

2301

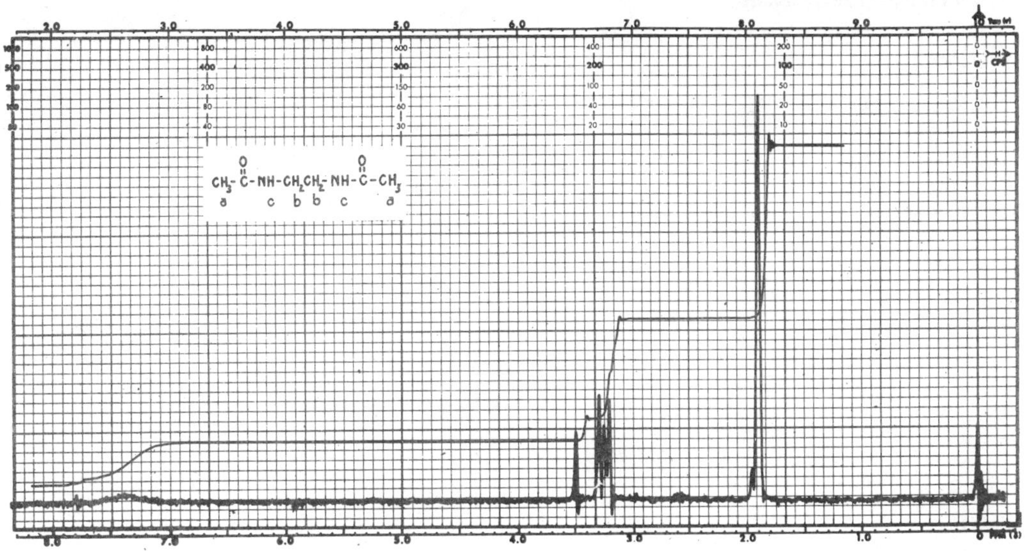

© 1978 Sadtler Research Laboratories, Inc.

N,N'-BIS(1,1-DIMETHYLPROPYL)GLUTARAMIDE

$C_{15}H_{30}N_2O_2$

Mol. Wt. 270.42

Solvent: $CDCl_3$

Assignments
a = 0.83
b = 1.29
c = 1.70
d = 1.92-2.61
e = 5.64
f = 2.52 imp.

2302

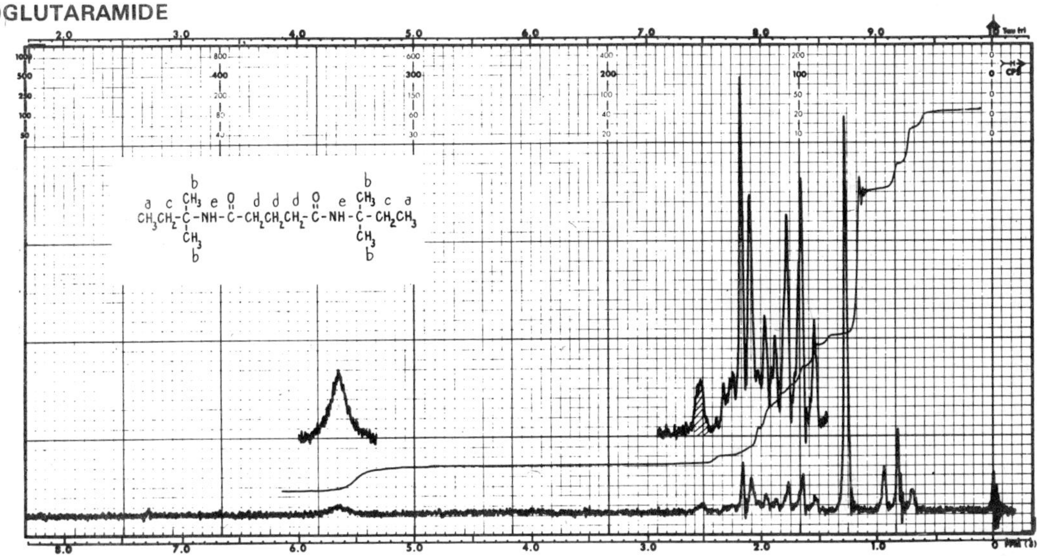

© 1978 Sadtler Research Laboratories, Inc.

2-PIPERIDONE

C_5H_9NO

Mol. Wt. 99.13

Solvent: CCl_4

Assignments
a = 1.77
b = 2.20
c = 3.23
d = 8.09
e = 4.30 imp.

2303

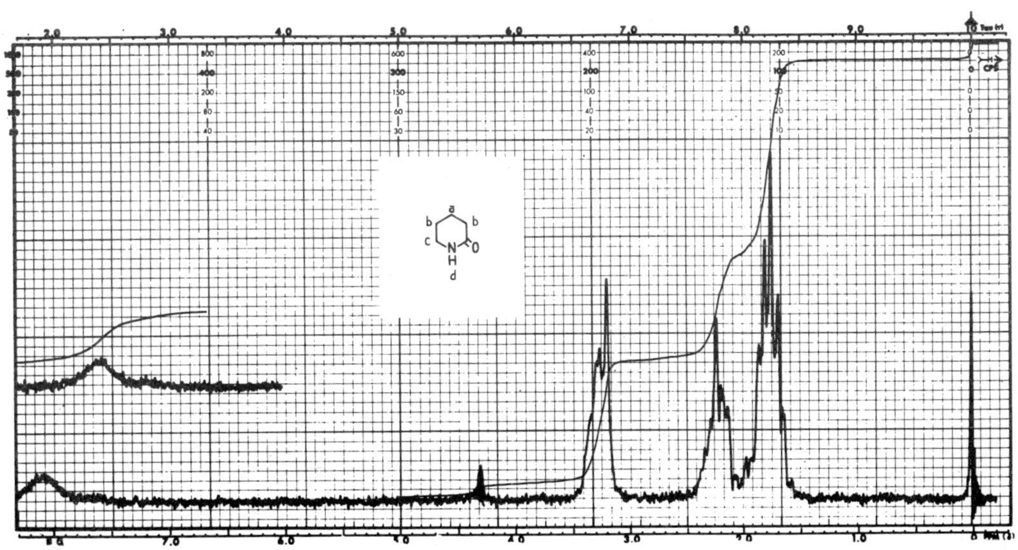

© 1978 Sadtler Research Laboratories, Inc.

N-tert-BUTYLACRYLAMIDE

$C_7H_{13}NO$

Mol. Wt. 127.19

M.P. 128-130°C

Solvent: $CDCl_3$

Assignments
a = 1.40
b = 5.40-6.50
c = 6.03

2304

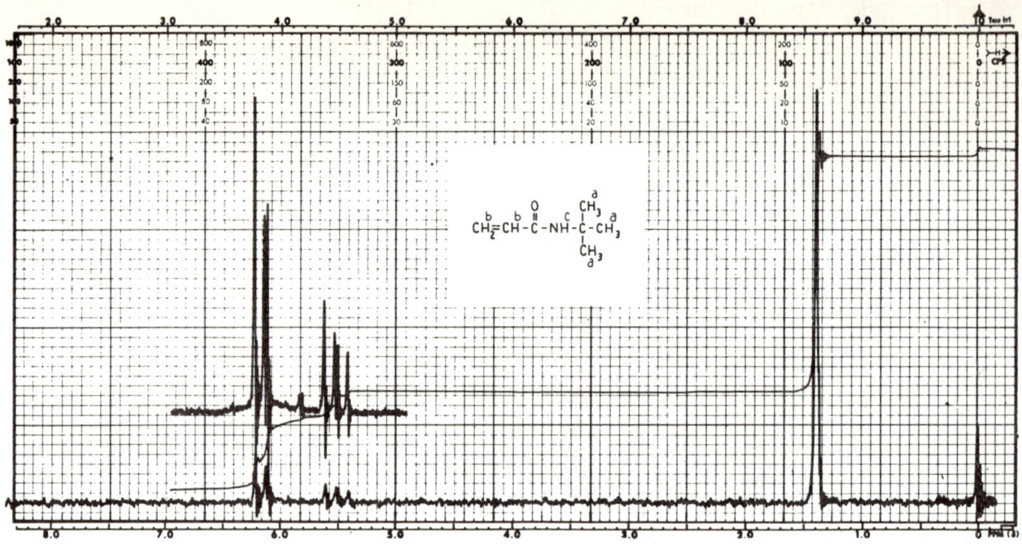

© 1978 Sadtler Research Laboratories, Inc.

N-(1,1,3,3-TETRAMETHYLBUTYL)-ACRYLAMIDE

$C_{11}H_{21}NO$

Mol. Wt. 183.30

M.P. 50-52°C

Solvent: $CDCl_3$

Assignments
a = 1.00
b = 1.47
c = 1.82
d = 5.51
e = 5.81
f = ca 6.09 or g
g = ca 6.19 or f

2305

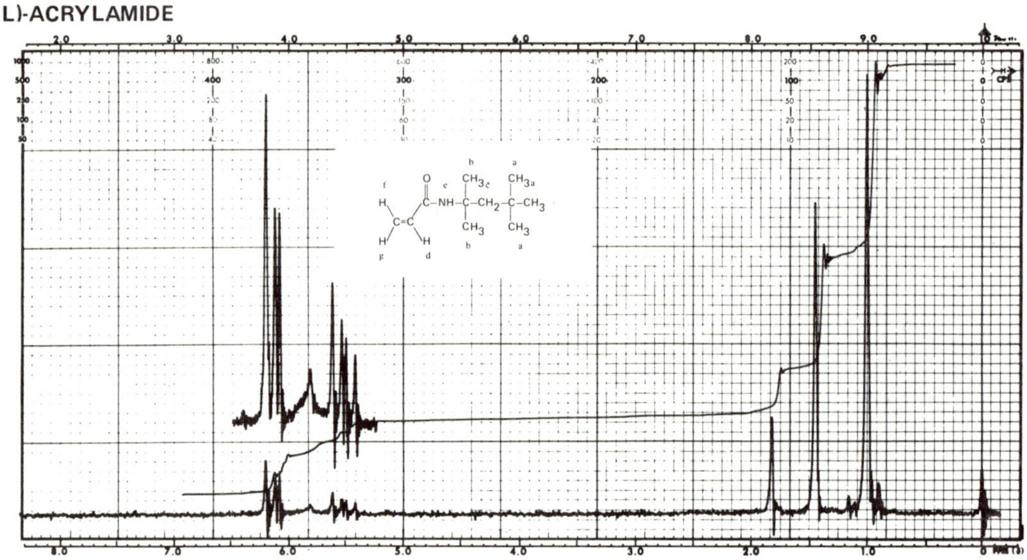

© 1978 Sadtler Research Laboratories, Inc.

N-ALLYLACRYLAMIDE

C_6H_9NO

Mol. Wt. 111.14

B.P. 75-78°C/0.5mm

Solvent: $CDCl_3$

Assignments
a = 3.91
b = 5.11
c = 5.18
d = ca 5.61
e = 5.88
f = ca 6.29
g = 7.29

2306

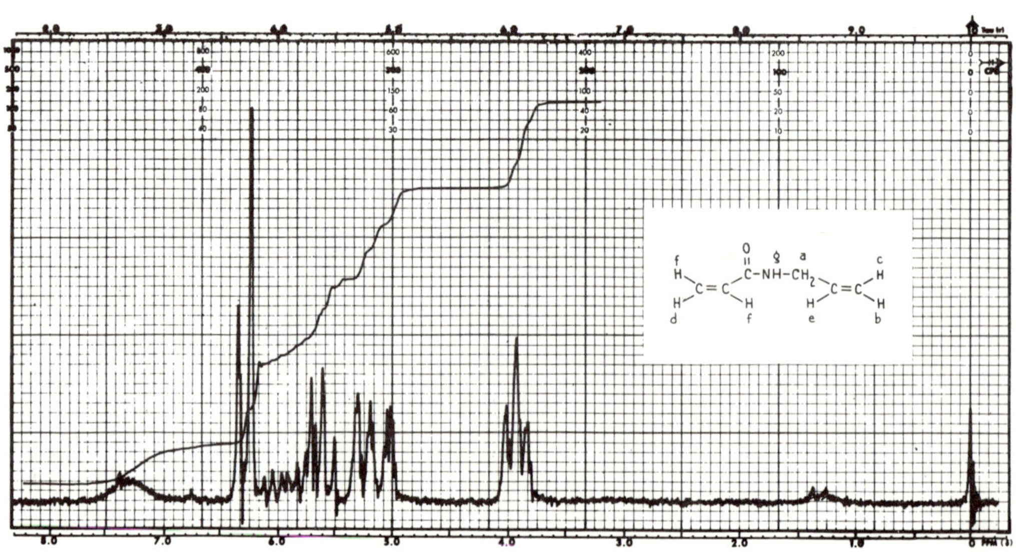

© 1978 Sadtler Research Laboratories, Inc.

N-(α-METHYLBENZYL)FORMAMIDE

$C_9H_{11}NO$

Mol. Wt. 149.19

M.P. 45°C

Solvent: CCl_4

Assignments
a = 1.36
b = 5.00
c = 7.19
d = 7.65
e = 7.84
f = 2.03 imp.

2307

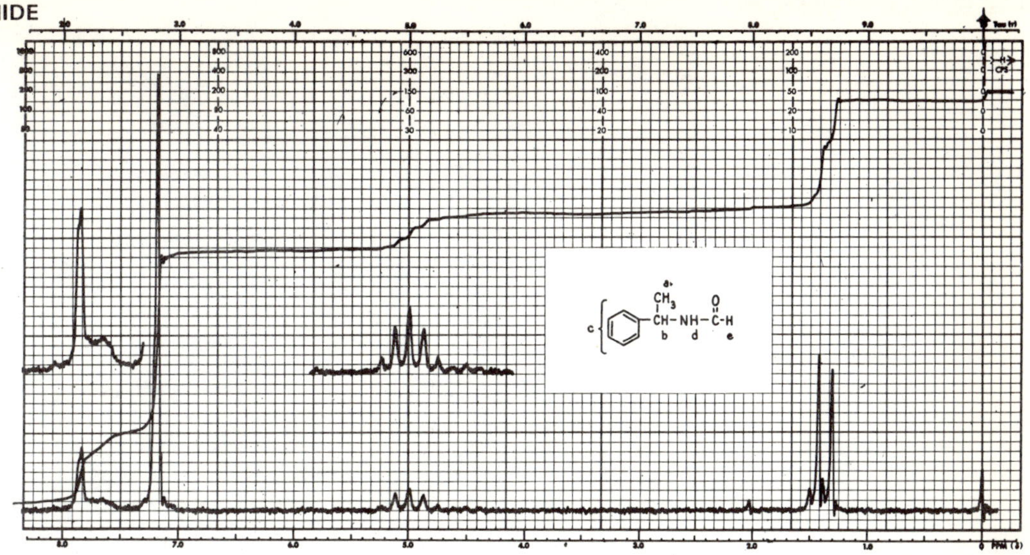

© 1978 Sadtler Research Laboratories, Inc.

N-(α-BENZYLPHENETHYL)FORMAMIDE

$C_{16}H_{17}NO$

Mol. Wt. 239.32

Solvent: DMSO-d_6

Assignments
a = 2.82
b = 4.32
c = 7.27
d = 6.70-7.80
e = 7.95

2308

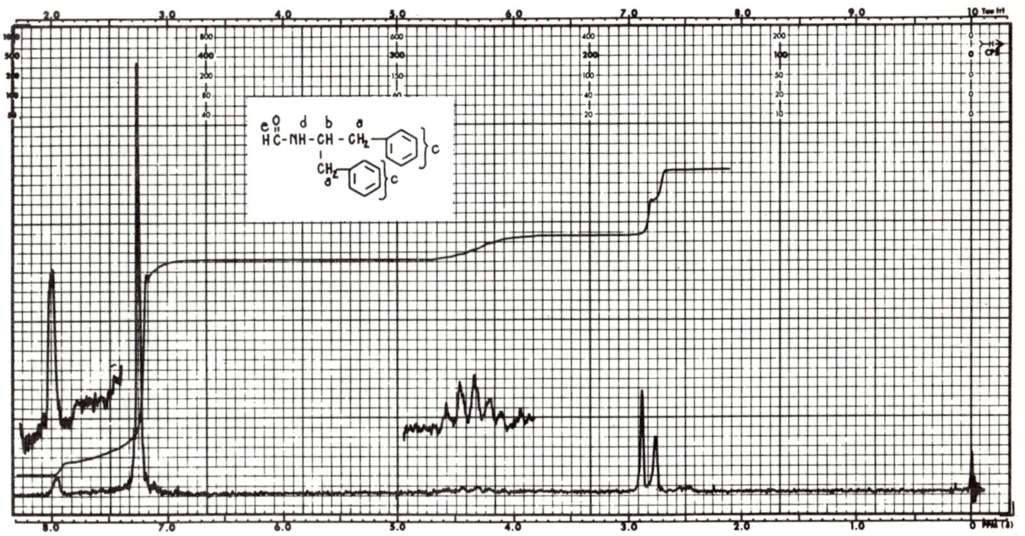

© 1978 Sadtler Research Laboratories, Inc

N-PHENETHYLACETAMIDE

$C_{10}H_{13}NO$

Mol. Wt. 163.22

M.P. 52-54°C

Solvent: $CDCl_3$

Assignments
a = 1.90
b = 2.79
c = 3.48
d = 6.58
e = ca 7.21

2309

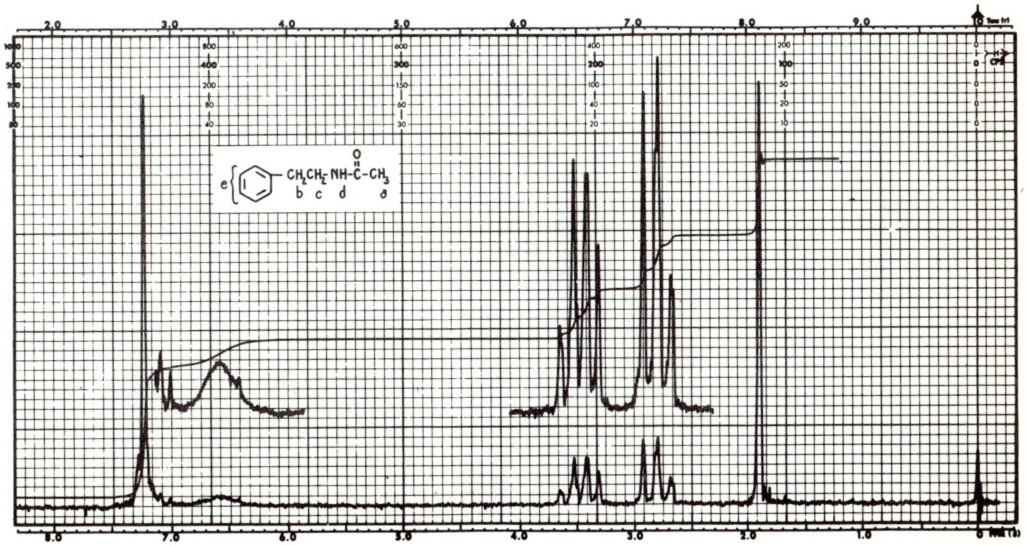

© 1978 Sadtler Research Laboratories, Inc.

N-(α-METHYLBENZYL)ACETAMIDE

$C_{10}H_{13}NO$

Mol. Wt. 163.22

M.P. 76°C

Solvent: $CDCl_3$

Assignments
a = 1.45
b = 1.90
c = 5.10
d = 6.68
e = 7.29

2310

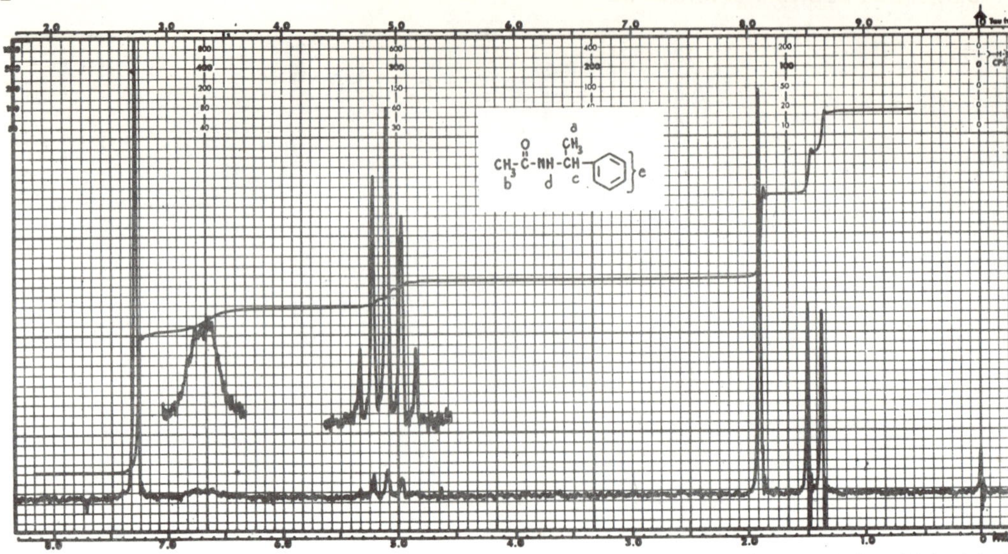

© 1978 Sadtler Research Laboratories, Inc.

N-(PHENETHYL)-2,2,2-TRIFLUOROACETAMIDE

$C_{10}H_{10}F_3NO$

Mol. Wt. 217.19

M.P. 55.5-56.5°C

Solvent: $CDCl_3$

Assignments
a = 2.83
b = 3.52
c = ca 7.22
d = 7.40

2311

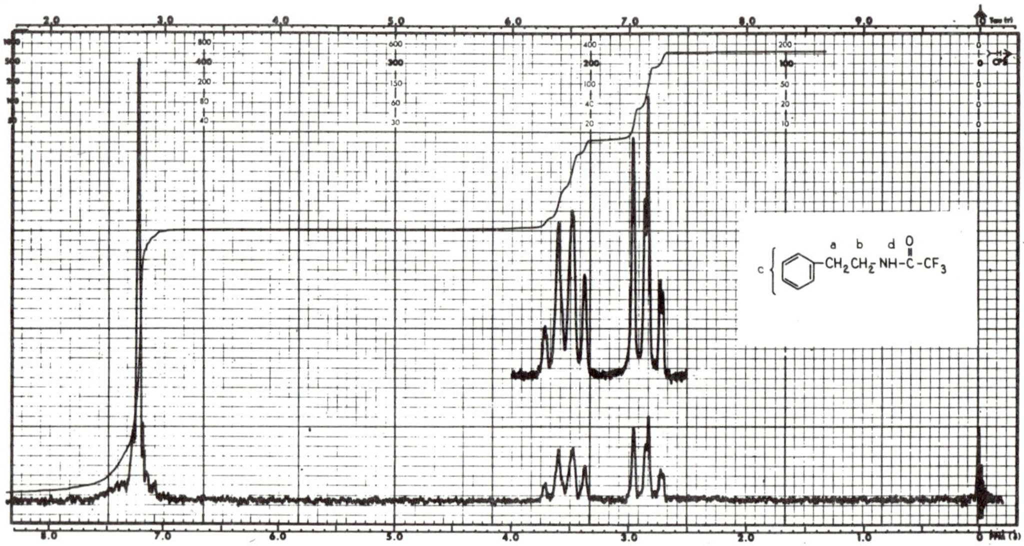

© 1978 Sadtler Research Laboratories, Inc.

N-METHYL-2-PHENYLACETAMIDE

$C_9H_{11}NO$

Mol. Wt. 149.20

M.P. 56-59°C

Solvent: $CDCl_3$

Assignments
a = 2.66
b = 3.48
c = 6.62
d = 7.21

2312

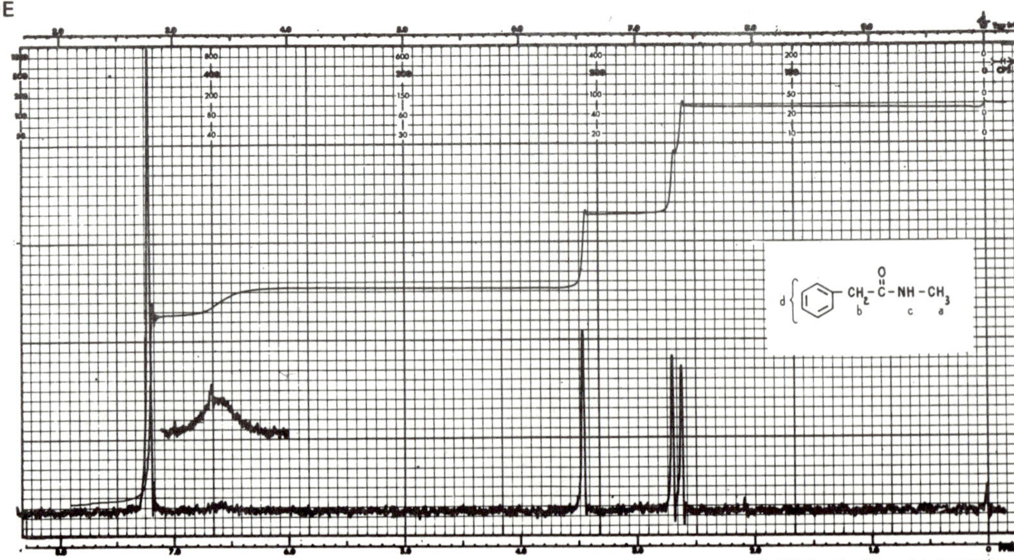

© 1978 Sadtler Research Laboratories, Inc.

N-BENZYL-3-CHLOROPROPIONAMIDE

$C_{10}H_{12}ClNO$

Mol. Wt. 197.67

Solvent: $CDCl_3$

Assignments
a = 2.56
b = 3.68
c = 4.34
d = ca 6.96
e = 7.25

2313

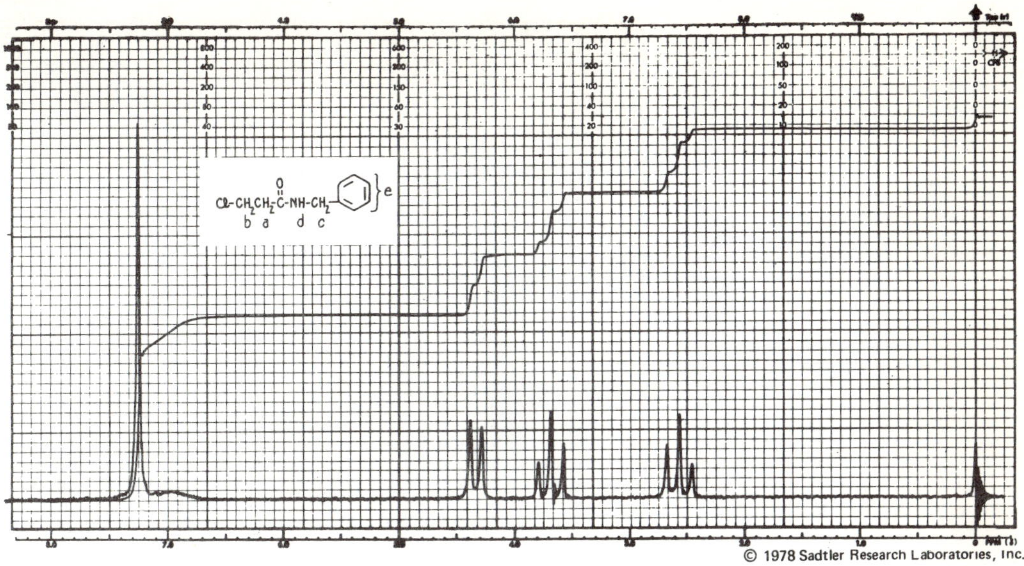

© 1978 Sadtler Research Laboratories, Inc.

N-BENZYL-2-PHENYLACETAMIDE

$C_{15}H_{15}NO$

Mol. Wt. 225.29

M.P. 119-120°C

Solvent: $CDCl_3$

Assignments
a = 3.55
b = 4.32
c = 5.95
d = ca 7.21

2314

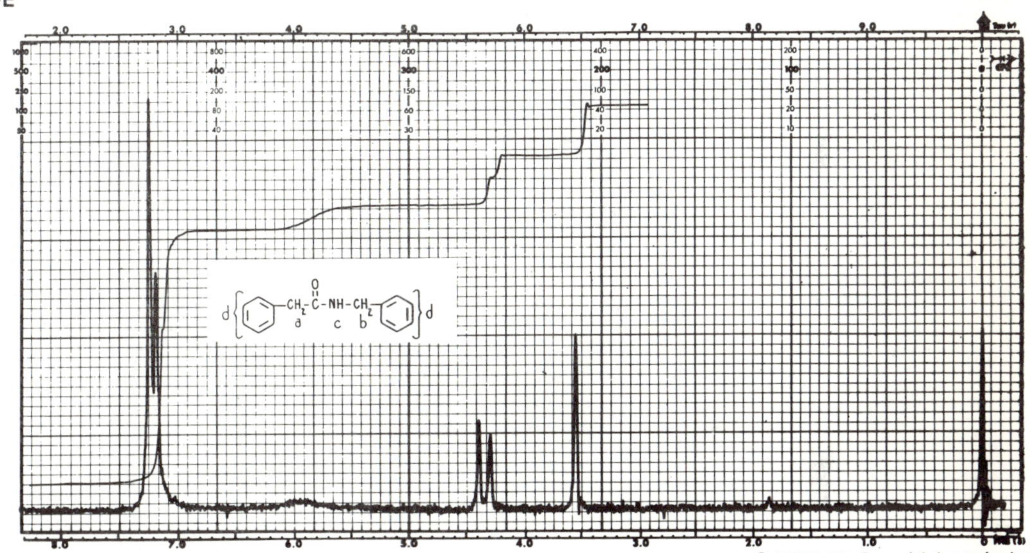

© 1978 Sadtler Research Laboratories, Inc.

PROPIONANILIDE

$C_9H_{11}NO$

Mol. Wt. 149.19

M.P. 105-107°C

Solvent: $CDCl_3$

Assignments
a = 1.19
b = 2.35
c = 6.90-7.42
d = 7.54
e = 8.38

2315

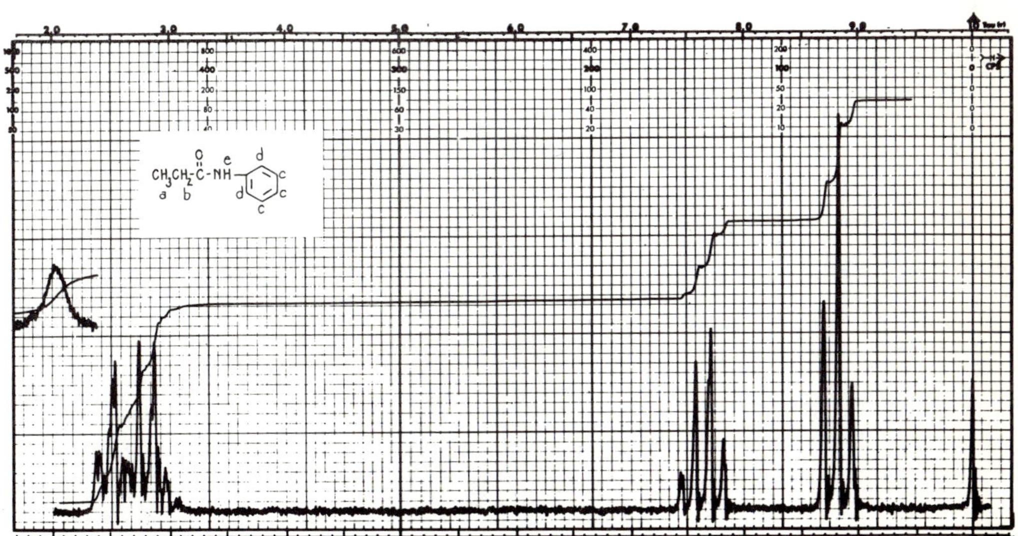

© 1978 Sadtler Research Laboratories, Inc.

BUTYRANILIDE

$C_{10}H_{13}NO$

Mol. Wt. 163.21

M.P. 93-95°C

Solvent: $CDCl_3$

Assignments
a = 0.92
b = 1.64
c = 2.29
d = 6.90-7.70
e = 8.45

2316

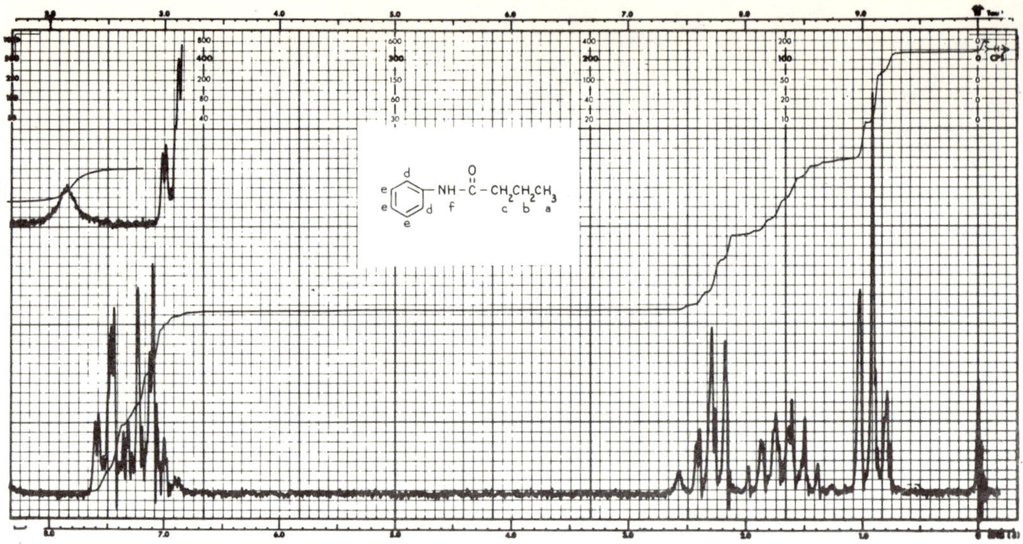

© 1978 Sadtler Research Laboratories, Inc.

ISOBUTYRANILIDE

$C_{10}H_{13}NO$

Mol. Wt. 163.22

M.P. 104°C

Solvent: Polysol-d

Assignments
a = 1.12
b = 2.54
c = 7.00
d = 7.21
e = 7.60
f = 9.58

2317

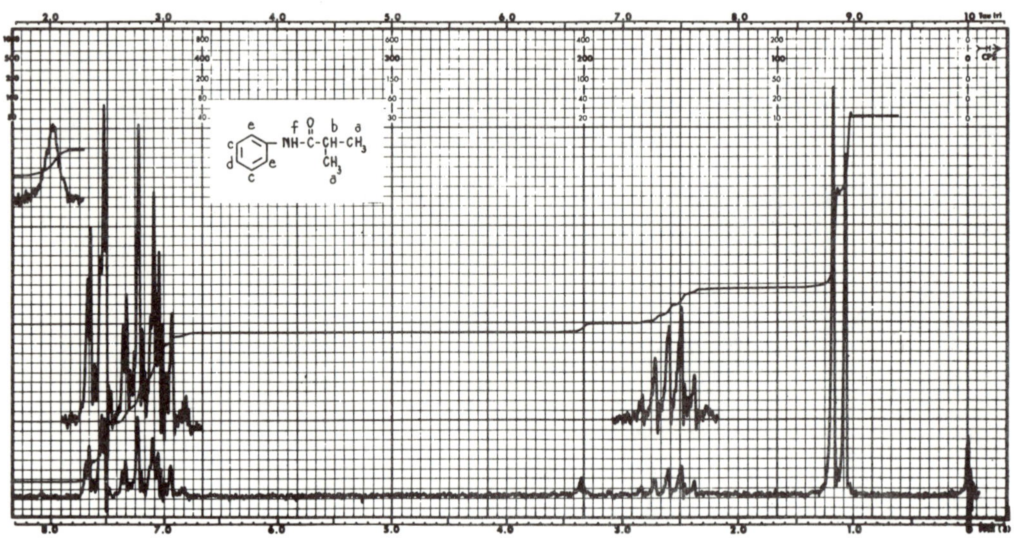

© 1978 Sadtler Research Laboratories, Inc.

2-CHLOROACETANILIDE

C_8H_8ClNO

Mol. Wt. 169.61

M.P. 134-135°C

Solvent: Dimethyl sulfoxide-d_6

Assignments
a = 4.27
b = 6.90-7.50
c = 7.61
d = 10.27
e = 2.50 DMSO-d_5
f = 4.04 imp.

2318

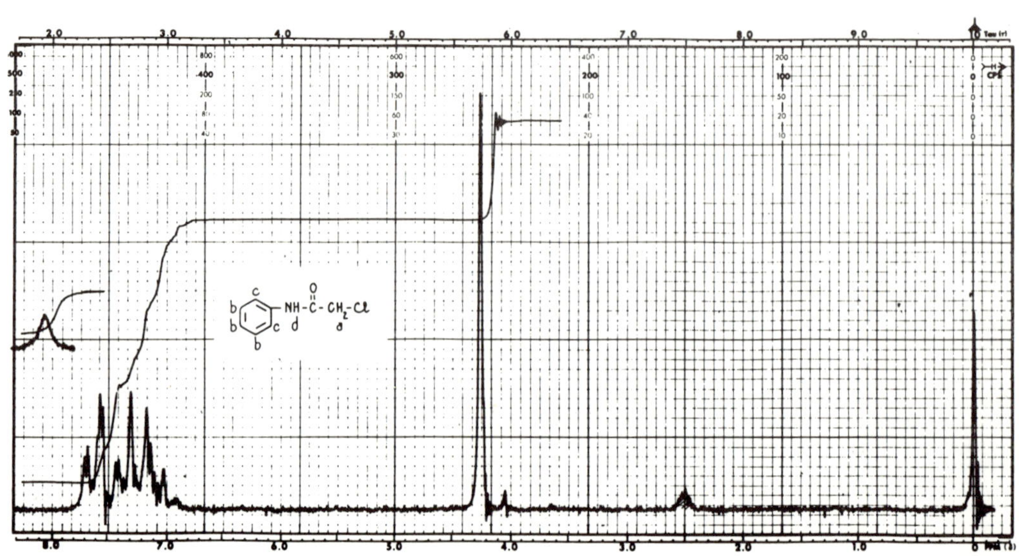

© 1978 Sadtler Research Laboratories, Inc.

ACETOACETANILIDE

$C_{10}H_{11}NO_2$

Mol. Wt. 171.21

M.P. 85°C

Solvent: Dimethyl sulfoxide-d_6

Assignments

a = 1.93
b = 2.21
c = 3.56
d = 5.21
e = 6.90-7.75
f = 10.05
g = 2.50 DMSO-d_5
h = 3.34 H_2O
OH unobserved

2319

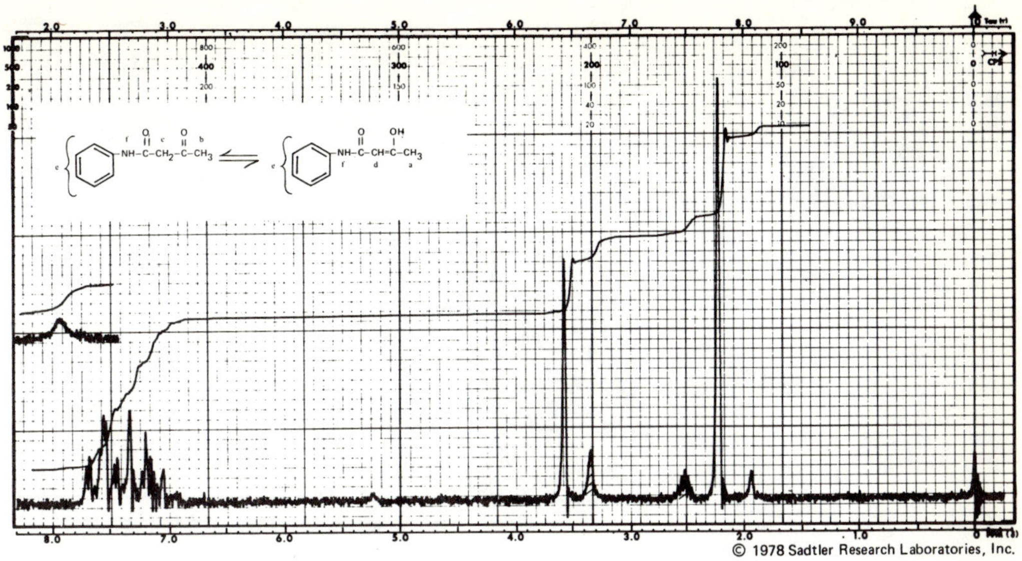

© 1978 Sadtler Research Laboratories, Inc.

MALONANILIDE

$C_{15}H_{14}N_2O_2$

Mol. Wt. 254.29

M.P. 229-230.5°C

Solvent: Dimethyl sulfoxide-D_6

Assignments

a = 3.51
b = 6.90-7.34
c = 7.55
d = 9.93
e = 2.49 DMSO-D_5
f = 3.21 H_2O

2320

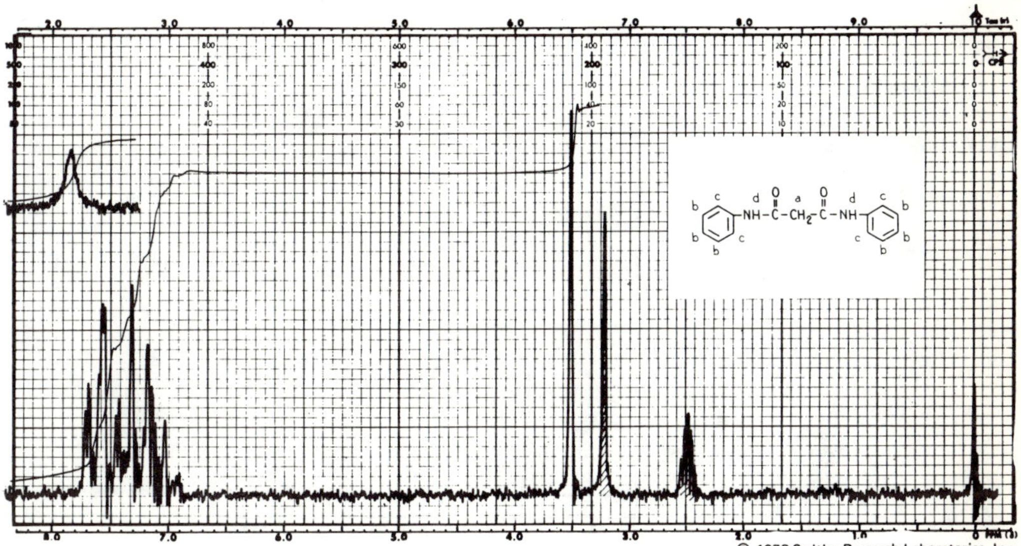

© 1978 Sadtler Research Laboratories, Inc.

o-ACETOTOLUIDIDE

$C_9H_{11}NO$

Mol. Wt. 149.19

M.P. 110-111°C

Solvent: $CDCl_3$

Assignments

a = 2.05
b = 2.15
c = 6.90-7.90

2321

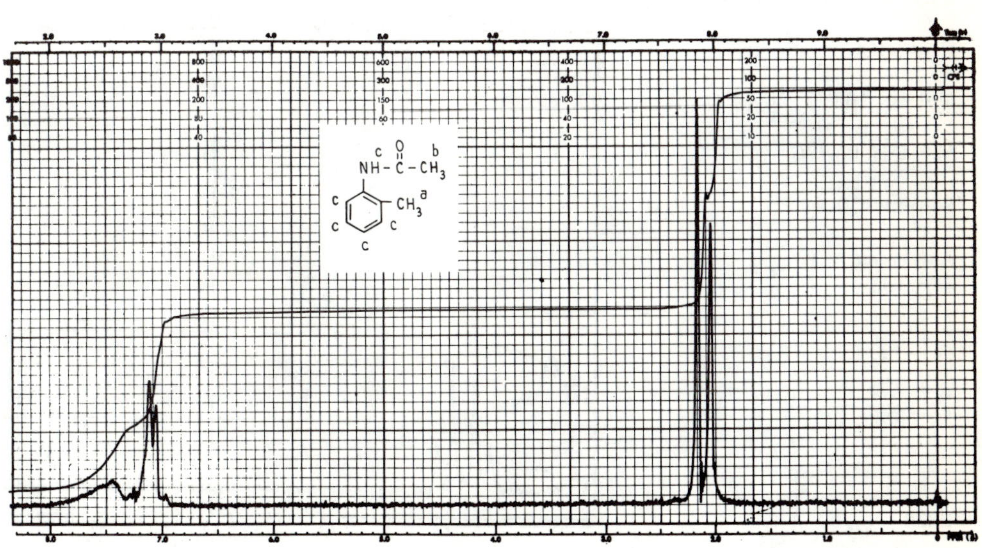

© 1978 Sadtler Research Laboratories, Inc.

m-ACETOTOLUIDIDE

$C_9H_{11}NO$

Mol. Wt. 149.19

M.P. 65-66^oC

Solvent: $CDCl_3$

Assignments
a = 2.06
b = 2.20
c = 6.82
d = 7.08
e = 7.25
f = 7.29
g = 8.72
h = 2.83 imp.

2322

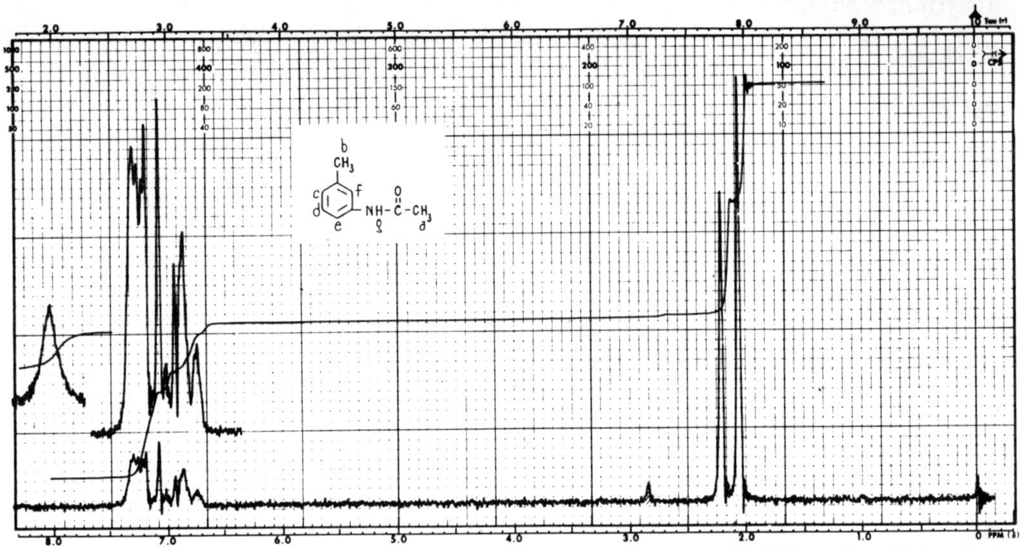

© 1978 Sadtler Research Laboratories, Inc.

ACETYL-p-TOLUIDINE

$C_9H_{11}NO$

Mol. Wt. 149.19

M.P. 148-149^oC

Solvent: Dimethyl sulfoxide-d_6

Assignments
a = 2.05
b = 2.22
c = 7.03
d = 7.45
e = 9.74
f = 2.49 DMSO-d_5

2323

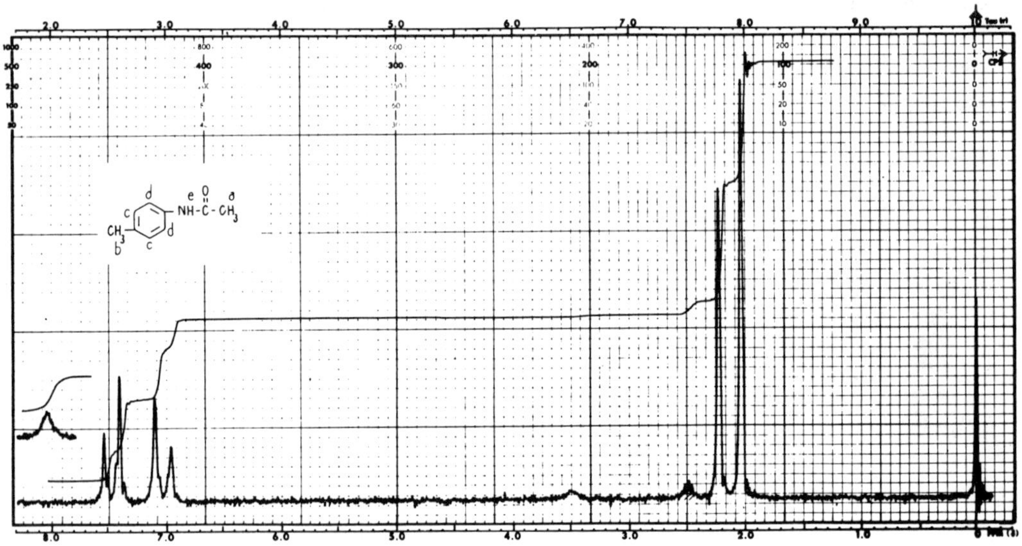

© 1978 Sadtler Research Laboratories, Inc.

---

p-ETHYLACETANILIDE

$C_{10}H_{13}NO$

Mol. Wt. 163.22

M.P. 90.1-90.7^oC

Solvent: $CDCl_3$

Assignments
a = 1.18
b = 2.07
c = 2.56
d = 7.04
e = 7.32
f = 8.15

2324

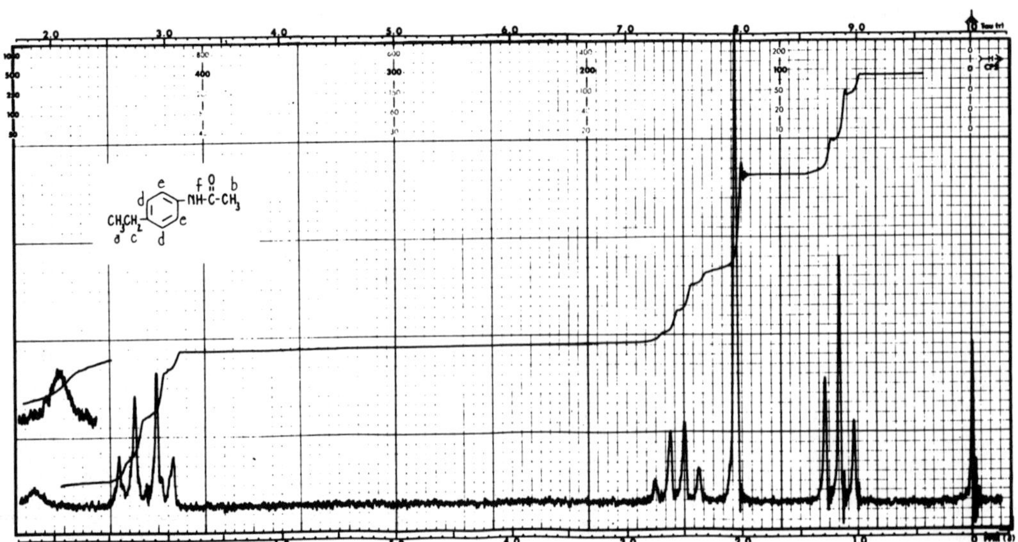

© 1978 Sadtler Research Laboratories, Inc.

p-tert-BUTYLACETANILIDE

$C_{12}H_{17}NO$

Mol. Wt. 191.28

Solvent: $CDCl_3$

Assignments
a = 1.29
b = 2.10
c = 7.28 or d
d = 7.40 or c
e = 7.80

2325

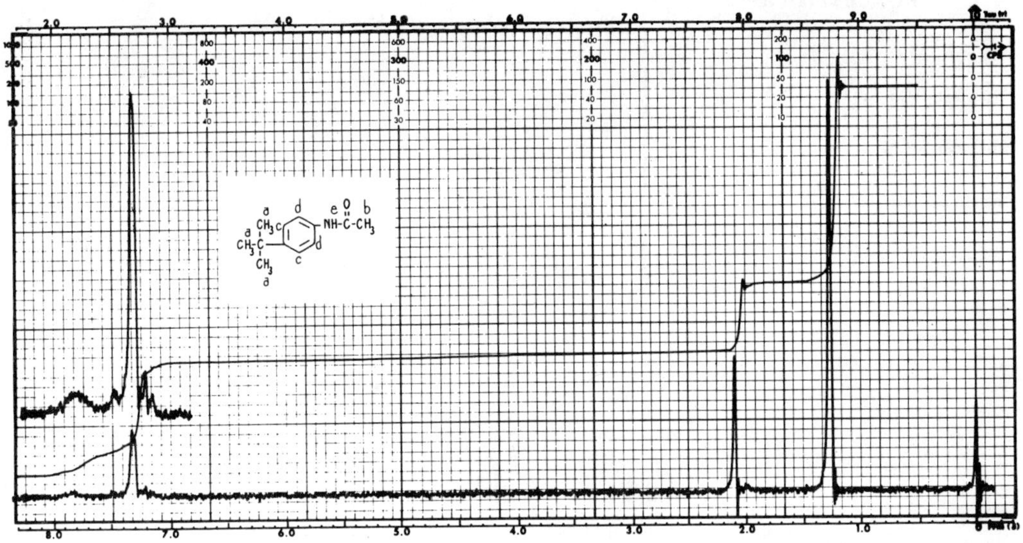

© 1978 Sadtler Research Laboratories, Inc.

3'-FLUOROACETANILIDE

C_8H_8FNO

Mol. Wt. 153.16

M.P. 88°C

Solvent: $CDCl_3$

Assignments
a = 2.17
b = 6.82
c = 7.22
d = 7.52
e = 8.50

2326

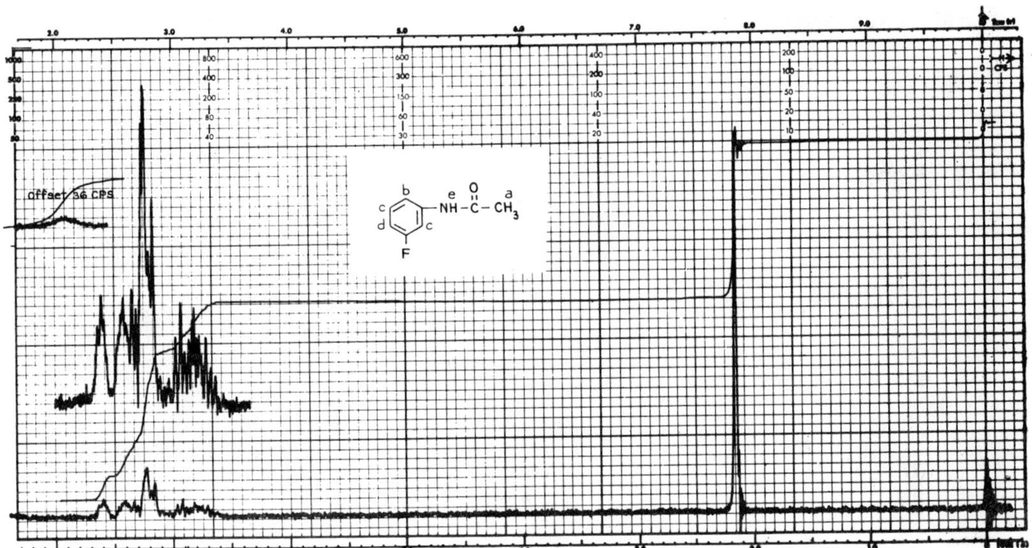

© 1978 Sadtler Research Laboratories, Inc.

4'-FLUOROACETANILIDE

C_8H_8FNO

Mol. Wt. 153.16

M.P. 152-153°C

Solvent: $CDCl_3$

Assignments
a = 2.11
b = 6.99
c = 7.40
d = ca 7.45

2327

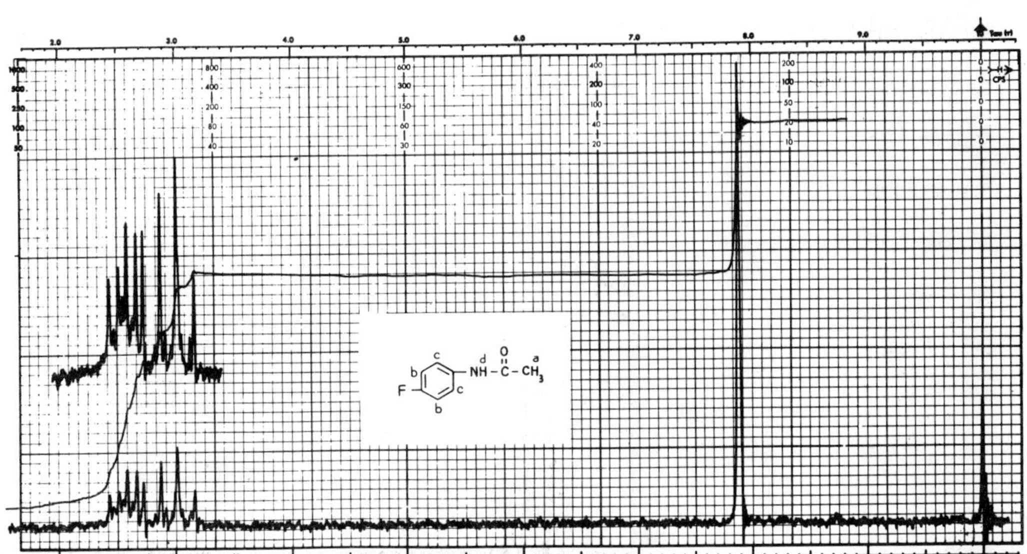

© 1978 Sadtler Research Laboratories, Inc.

2'-CHLOROACETANILIDE

C_8H_8ClNO

Mol. Wt. 169.61

M.P. 86-88°C

Solvent: $CDCl_3$

Assignments
a = 2.17
b = 6.78-7.40
c = 7.83
d = 8.15

2328

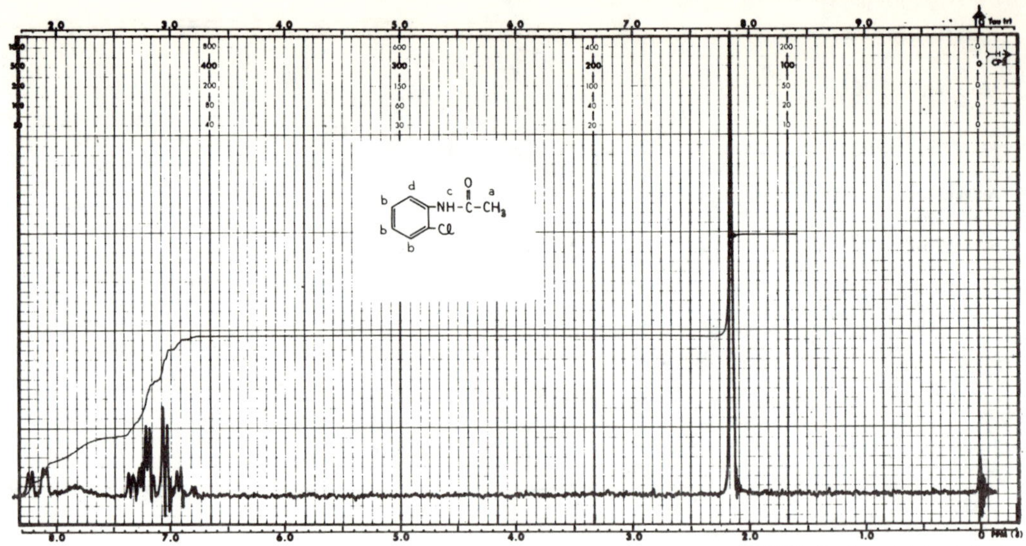

© 1978 Sadtler Research Laboratories, Inc.

3'-CHLOROACETANILIDE

C_8H_8ClNO

Mol. Wt. 169.61

Solvent: $CDCl_3$

Assignments
a = 2.16
b = 7.00-7.50
c = 7.70
d = 8.74

2329

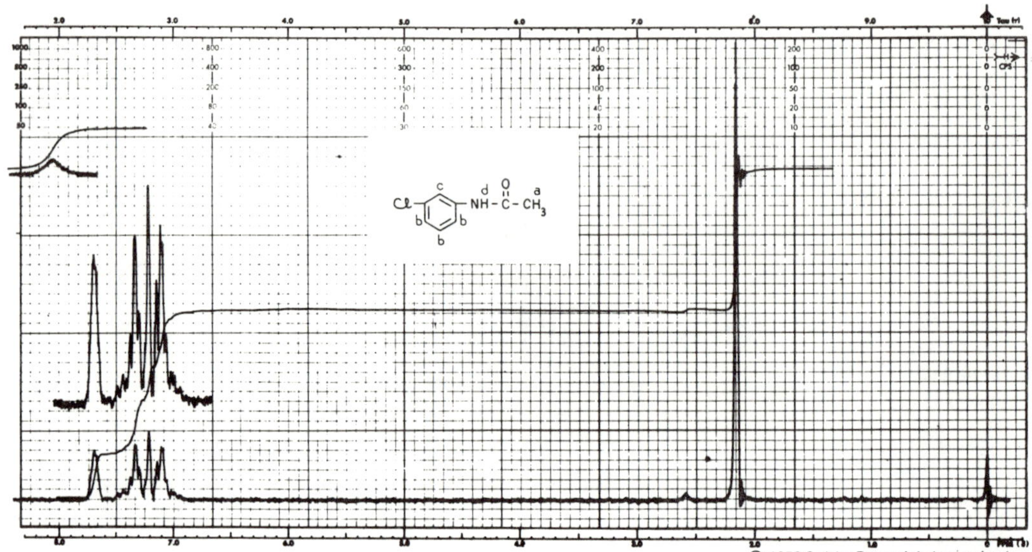

© 1978 Sadtler Research Laboratories, Inc.

4'-CHLOROACETANILIDE

C_8H_8ClNO

Mol. Wt. 169.61

M.P. 177-179°C

Solvent: Dimethyl sulfoxide-d_6

Assignments
a = 2.10
b = 7.32
c = 7.68
d = 10.03

2330

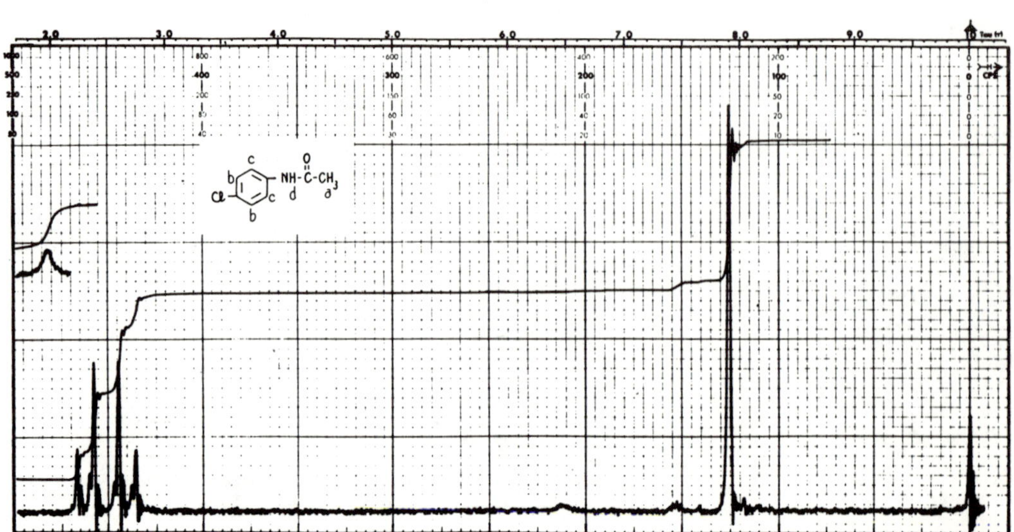

© 1978 Sadtler Research Laboratories, Inc.

4'-BROMOACETANILIDE

C_8H_8BrNO

Mol. Wt. 214.06

M.P. 165-167°C

Solvent: Dimethyl sulfoxide-D_6

Assignments
a = 2.10
b = 7.45 or c
c = 7.61 or b
d = 10.04
e = 2.57 DMSO-D_5
f = 3.48 H_2O

2331

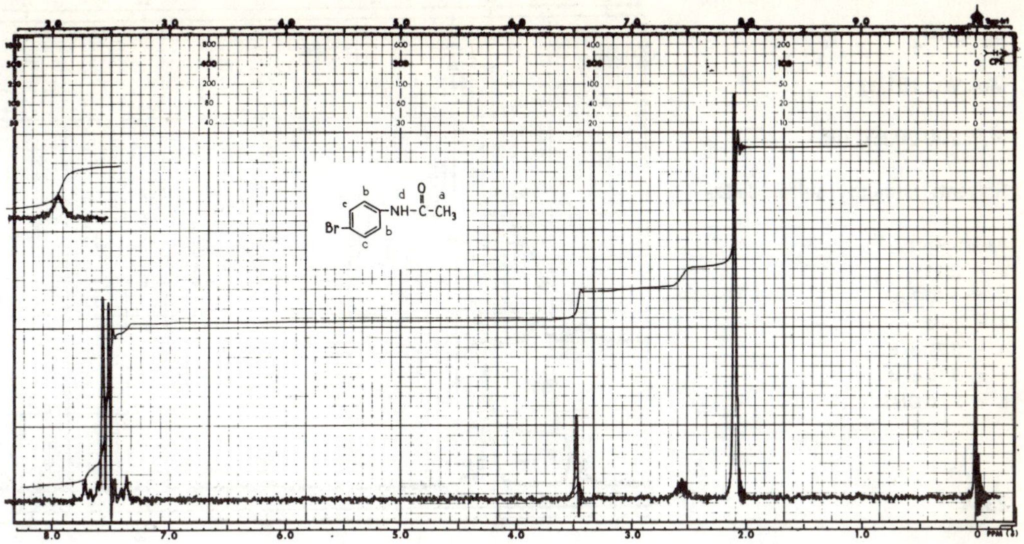

© 1978 Sadtler Research Laboratories, Inc.

4'-FORMYLACETANILIDE, 4'-OXIME

$C_9H_{10}N_2O_2$

Mol. Wt. 178.19

Solvent: Polysol-d

Assignments
a = 2.11
b = 7.47
c = 7.61
d = 8.04
e = 9.50

2332

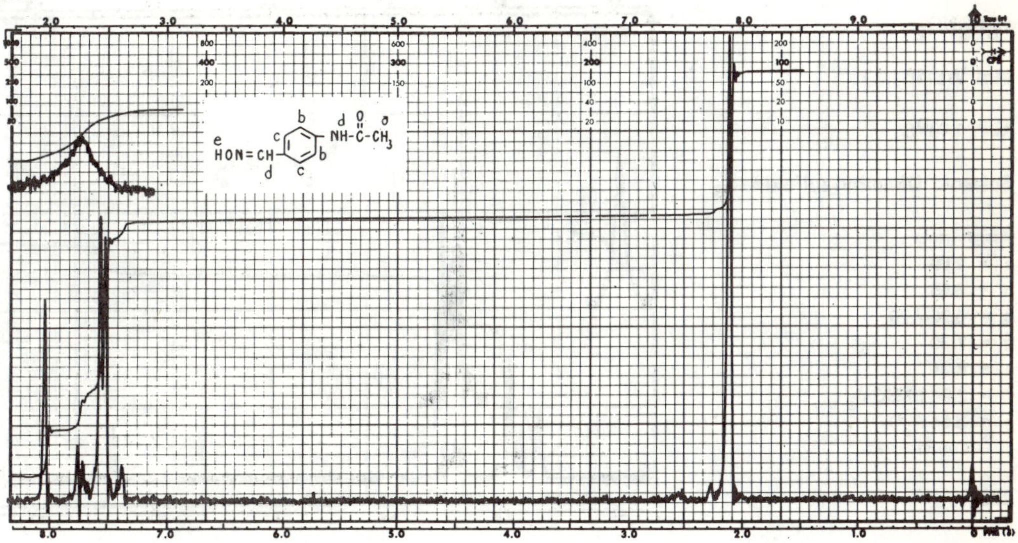

© 1978 Sadtler Research Laboratories, Inc.

2'-NITROACETANILIDE

$C_8H_8N_2O_3$

Mol. Wt. 180.16

M.P. 90-91°C

Solvent: $CDCl_3$

Assignments
a = 2.27
b = 7.11
c = 7.61
d = 8.15
e = 8.74
f = 10.27

2333

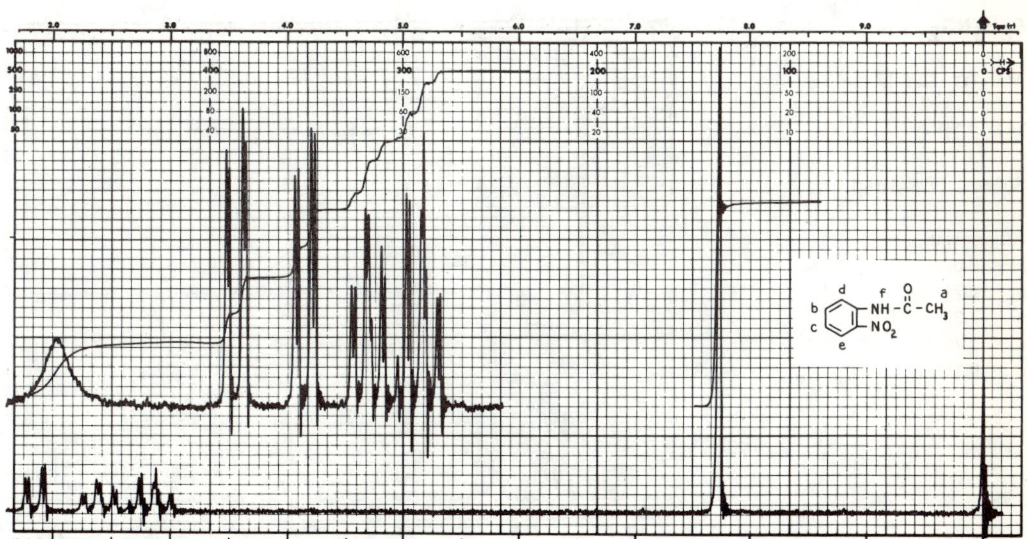

© 1978 Sadtler Research Laboratories, Inc.

2'-HYDROXYACETANILIDE

$C_8H_9NO_2$

Mol. Wt. 151.17

M.P. 200-203°C (lit.)

Solvent: Dimethyl sulfoxide-D_6

Assignments
a = 2.11
b = 6.60-7.08
c = 7.68
d = 9.37 or e
e = 9.75 or d
f = 2.50 DMSO-D_5

2334

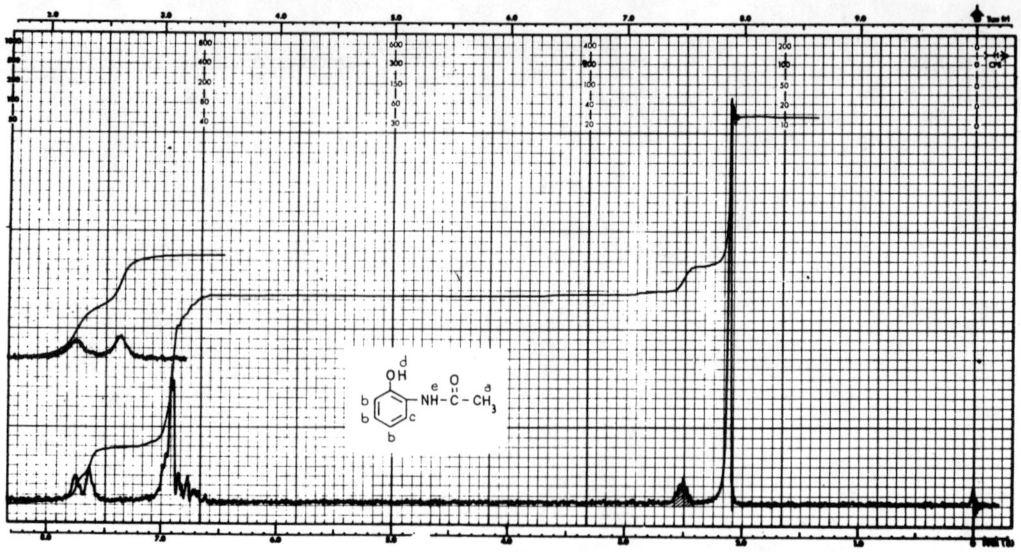

© 1978 Sadtler Research Laboratories, Inc.

3'-HYDROXYACETANILIDE

$C_8H_9NO_2$

Mol. Wt. 151.17

M.P. 148°C

Solvent: Dimethyl sulfoxide-d_6

Assignments
a = 2.09
b = 6.55
c = 6.95-7.20
d = 7.30
e = 9.34
f = 2.54 DMSO-d_5
g = 3.64 H_2O
OH unobserved

2335

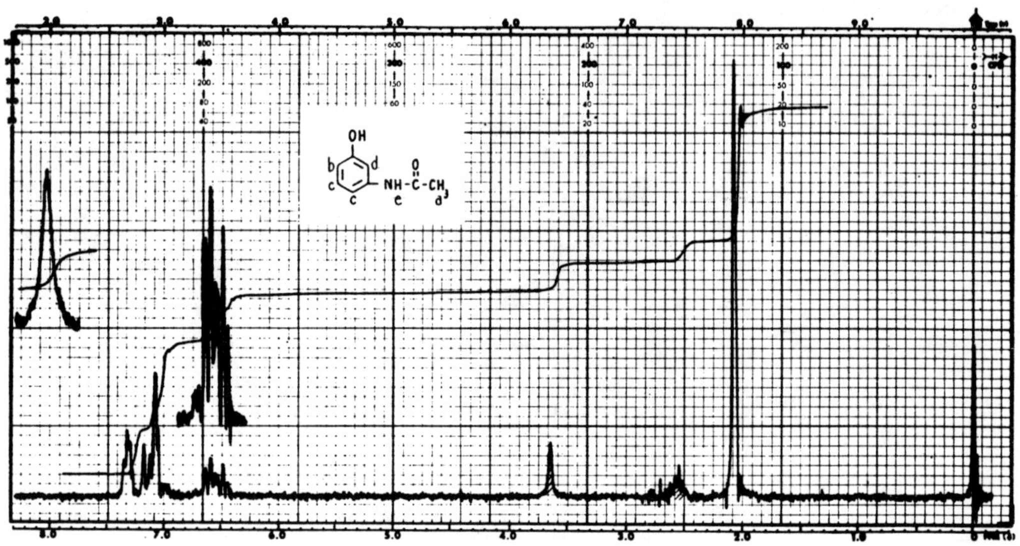

© 1978 Sadtler Research Laboratories, Inc.

4-HYDROXYACETANILIDE

$C_8H_9NO_2$

Mol. Wt. 151.17

M.P. 201°C

Solvent: Polysol-d

Assignments
a = 2.03
b = 6.71
c = 7.30
d = 8.76
e = 9.21

2336

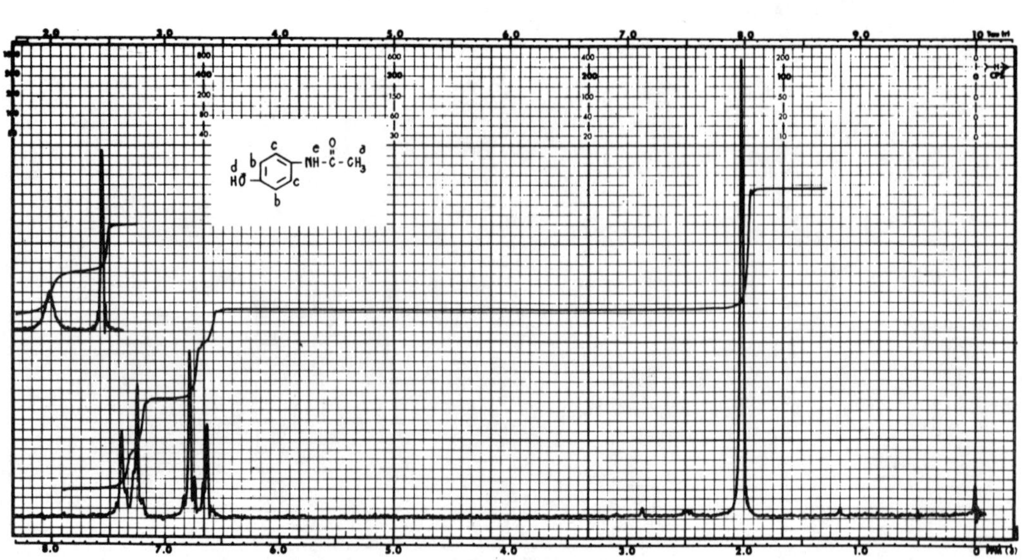

© 1978 Sadtler Research Laboratories, Inc.

p-ACETANISIDIDE

$C_9H_{11}NO_2$

Mol. Wt. 165.19

M.P. 129-131°C (lit.)

Solvent: Dimethyl sulfoxide-D_6

Assignments

a = 2.02
b = 3.70
c = 6.85
d = 7.48
e = 9.71
f = 2.52 DMSO-D_5
g = 3.40 H_2O

2337

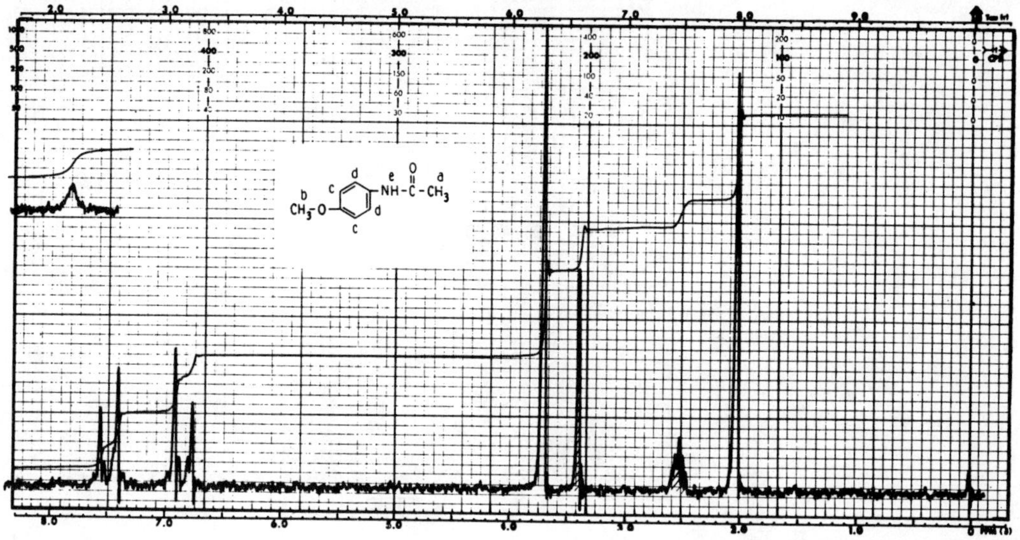

© 1978 Sadtler Research Laboratories, Inc.

2',4'-ACETOXYLIDIDE

$C_{10}H_{13}NO$

Mol. Wt. 163.22

Solvent: $CDCl_3$

Assignments

a = 1.99
b = 2.09 or c
c = 2.22 or b
d = 6.84
e = 6.90
f = 7.21
g = 7.96

2338

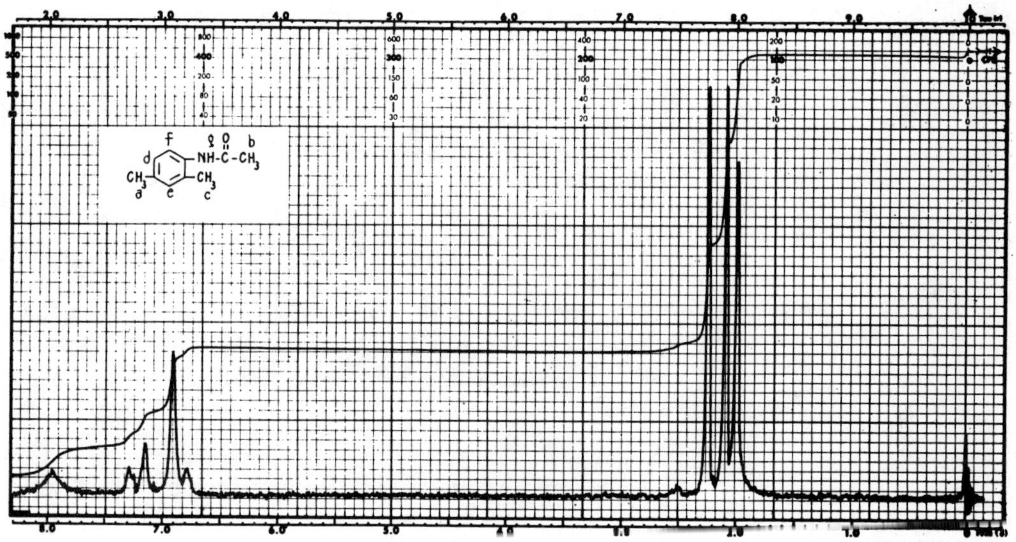

© 1978 Sadtler Research Laboratories, Inc.

2',5'-ACETOXYLIDIDE

$C_{10}H_{13}NO$

Mol. Wt. 163.22

Solvent: $CDCl_3$

Assignments

a = 2.07
b = 2.13 or c
c = 2.26 or b
d = 6.84
e = 7.01
f = 7.31
g = 7.62

2339

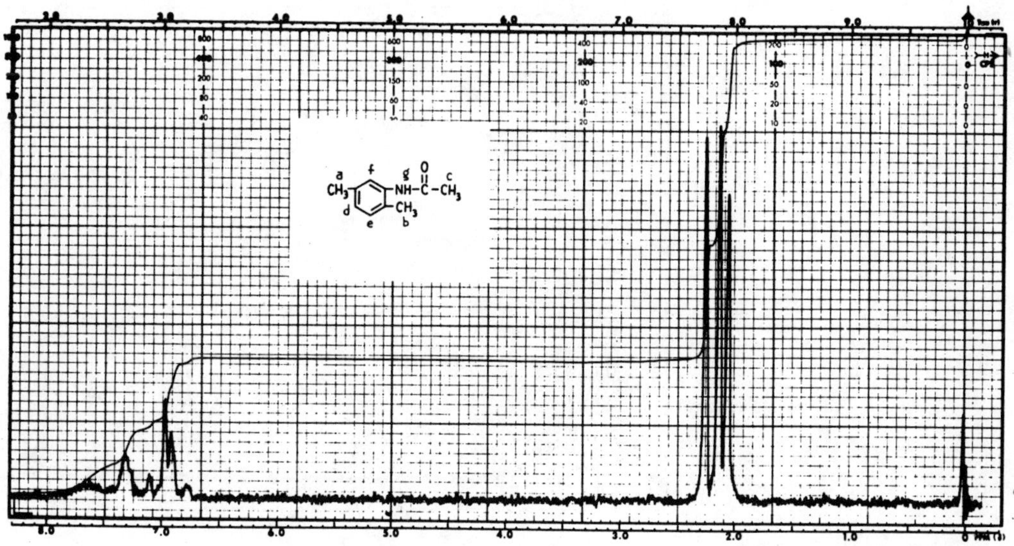

© 1978 Sadtler Research Laboratories, Inc.

N-(2-NAPHTHYL)ACETAMIDE

$C_{12}H_{11}NO$

Mol. Wt. 184.22

M.P. 131-132°C

Solvent: $CDCl_3$

Assignments
a = 2.11
b = 7.34
c = ca 7.61
d = 8.10
e = 8.24

2340

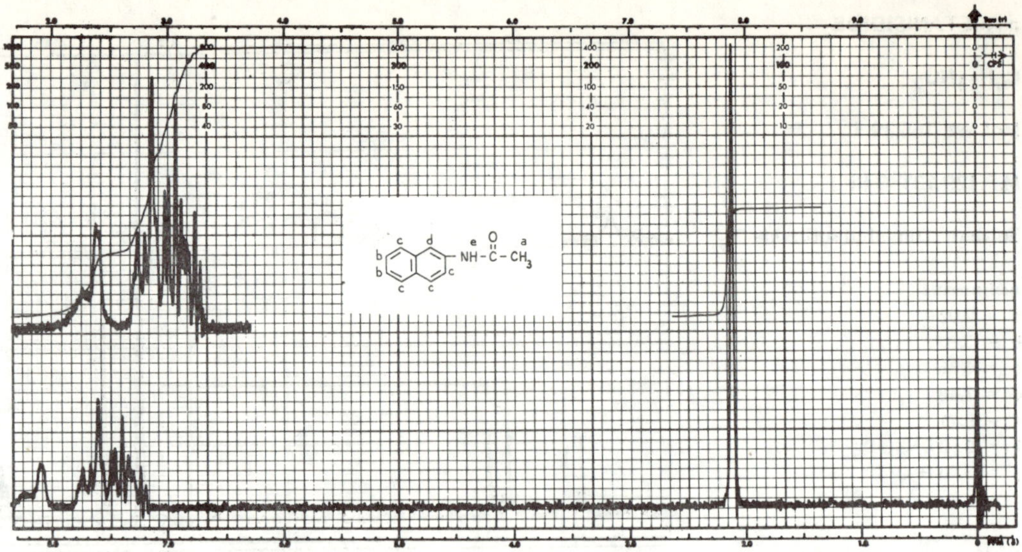

© 1978 Sadtler Research Laboratories, Inc.

2-ACETAMIDOPYRIDINE

$C_7H_8N_2O$

Mol. Wt. 136.15

Solvent: $CDCl_3$

Assignments
a = 2.20
b = 7.03
c = 7.70
d = ca 8.27
e = 9.94

2341

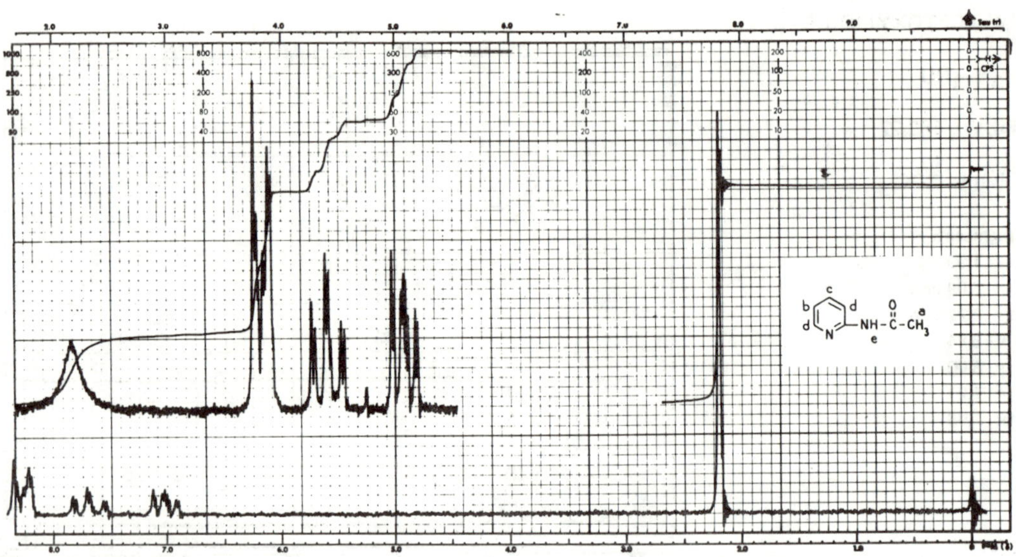

© 1978 Sadtler Research Laboratories, Inc.

N-METHYLBENZAMIDE

C_8H_9NO

Mol. Wt. 135.17

M.P. 79-82°C

Solvent: $CDCl_3$

Assignments
a = 2.97
b = 6.80-7.70
c = 7.81

2342

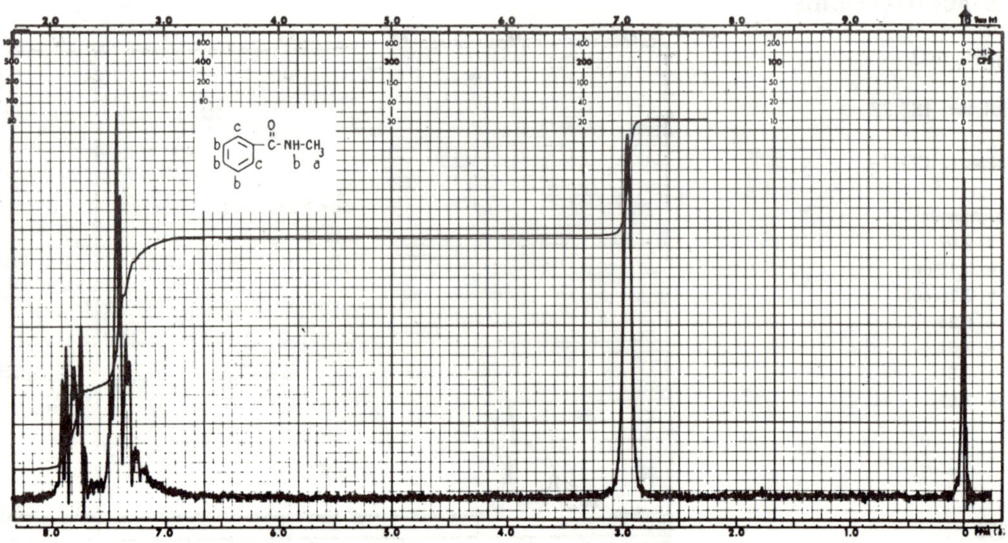

© 1978 Sadtler Research Laboratories, Inc.

BENZANILIDE

$C_{13}H_{11}NO$

Mol. Wt. 197.24

Solvent: Dimethyl sulfoxide

Assignments
a = 6.90-7.70
b = 7.81
c = 7.98
d = 10.21

2343

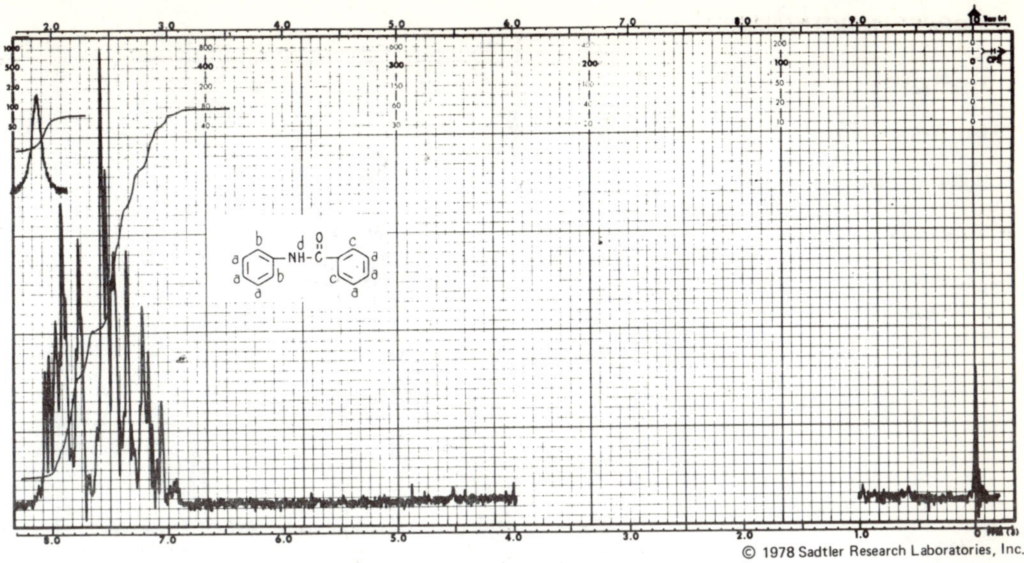

© 1978 Sadtler Research Laboratories, Inc.

---

o-BENZOTOLUIDIDE

$C_{14}H_{13}NO$

Mol. Wt. 211.27

M.P. 144°C

Solvent: $CDCl_3$

Assignments
a = 2.31
b = 7.00-7.50
c = 7.30-7.65
d = ca 7.89

2344

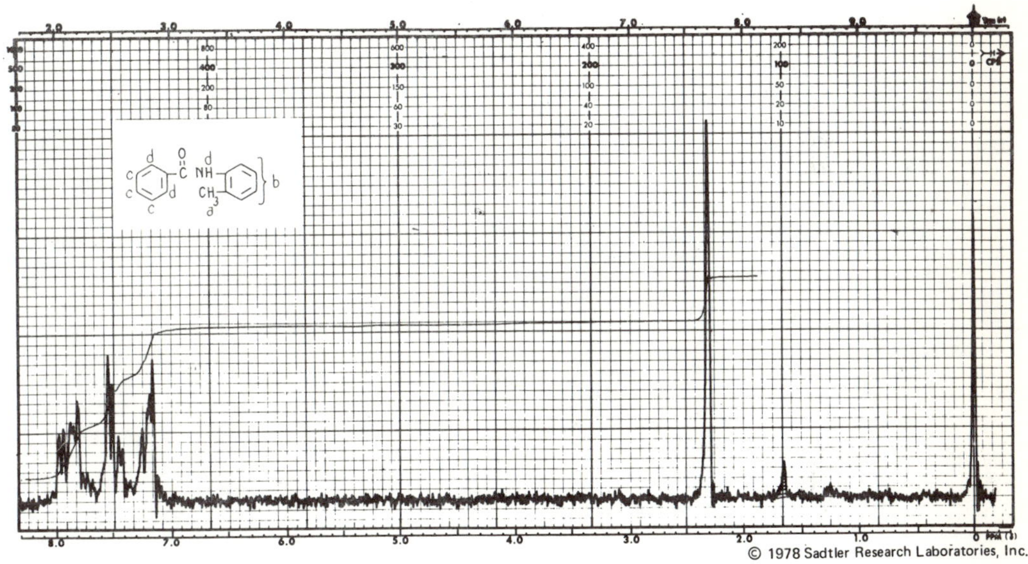

© 1978 Sadtler Research Laboratories, Inc.

---

p-BENZOTOLUIDIDE

$C_{14}H_{13}NO$

Mol. Wt. 211.27

M.P. 155-157°C

Solvent: $CDCl_3$

Assignments
a = 2.30
b = 7.08
c = 7.28-7.59
d = 7.76
e = 8.04

2345

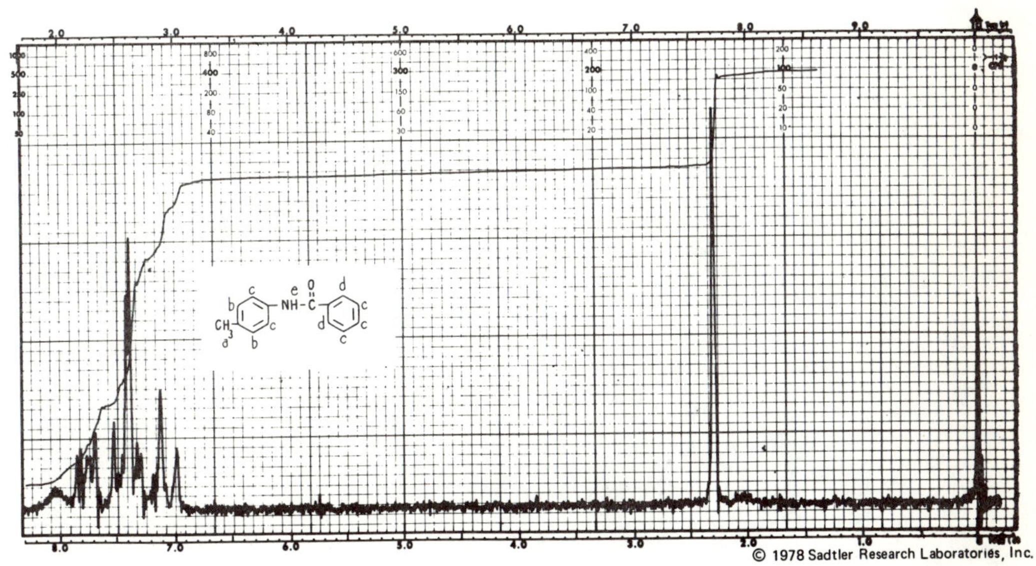

© 1978 Sadtler Research Laboratories, Inc.

N-BENZYLNICOTINAMIDE

$C_{13}H_{12}N_2O$

Mol. Wt. 212.25

Solvent: $CDCl_3$

Assignments

a = 4.54
b = 7.23
c = 7.26
d = 7.79
e = 8.09
f = 8.55
g = 8.92

2346

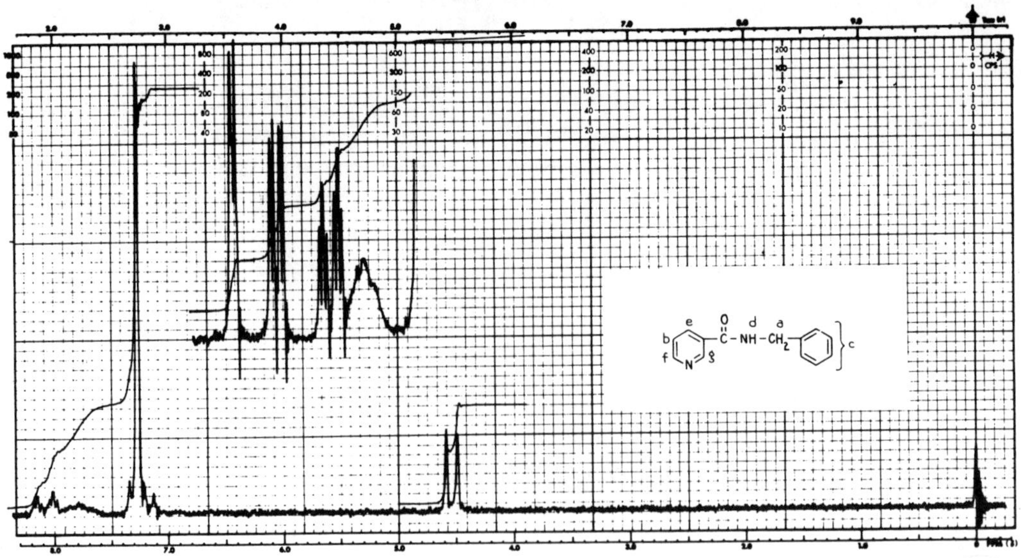

© 1978 Sadtler Research Laboratories, Inc.

A characteristic feature of the NMR spectra of the Tertiary Amides is the non-equivalence in chemical shift of similar groups bonded to the nitrogen nucleus. In the case of the N,N-dimethyl amides, the methyl groups range in appearance from a broad single peak (spectrum 2357) to two distinct bands separated by about 10 Hz (spectrum 2352). The chemical shifts of aliphatic groups bonded to the carbonyl side of the Amide linkage are similar to those observed for the Primary and Secondary Amides and thus are not covered in this discussion.

Chemical Shifts

Aliphatic Protons

HC–3	HC–2	HC–1		Solvent	Spectrum
		2.89	CH_3–N(CH_3)–C(=O)–R_{11}	CCl_4	2355
		(2.94)	CH_3–N(CH_3)–C(=O)–R_{11}	CCl_4	2355
		2.94	CH_3–N(CH_3)–C(=O)–R_2	D_2O	2352
		(3.11)	CH_3–N(CH_3)–C(=O)–R_2	D_2O	2352
		(2.99)	$(CH_3)_2$–N–C(=O)–CH(OH)–CH_3	CCl_4	2357
		(3.23)	CH_3–N(Ø)–C(=O)–CH_3	$CDCl_3$	2363
	1.12	3.31	CH_3–CH_2–N(R_2)–C(=O)–H	$CDCl_3$	2347
	(1.20	3.37)	CH_3–CH_2–N(R_2)–C(=O)–H	$CDCl_3$	2347
	1.11	3.36	CH_3–CH_2–N(R_2)–C(=O)–R_2	D_2O	2350
	(1.11	3.40)	CH_3–CH_2–N(R_2)–C(=O)–R_2	D_2O	2350
	1.11	3.81	CH_3–CH_2–N(Ø)–C(=O)–CH_3	$CDCl_3$	2363
0.89	1.58	3.17	CH_3–CH_2–CH_2–N(R_3)–C(=O)–R_9	$CDCl_3$	2354
(0.89	1.58	3.21)	CH_3–CH_2–CH_2–N(R_3)–C(=O)–R_9	$CDCl_3$	2354
0.88	1.55	3.27	CH_3–CH_2–CH_2–N(R_3)–C(=O)–CH_3	D_2O	----
(0.91	1.55	3.29)	CH_3–CH_2–CH_2–N(R_3)–C(=O)–CH_3	D_2O	----
0.89	1.56	3.69	CH_3–CH_2–CH_2–N(Ø)–C(=O)–CH_3	$CDCl_3$	2365
	(1.29	3.69)	($(CH_3)_2$–CH$)_2$–N–C(=O)–CH_3	CCl_4	2348

The aromatic pattern produced by the Tertiary Amide nitrogen atom depends primarily on whether there are one or two phenyl groups bonded to the nitrogen atom. The aromatic protons of the N-phenyl compounds appear as a complex band over the chemical shift range from 7.0–7.6 ppm while those of the N,N-diphenyl compounds appear as a single peak centered at about 7.3 ppm.

para/meta/ortho		Solvent	Spectrum
ca. 7.22	$(\emptyset)_2$–N–C(=O)–H	CCl_4	2367
ca. 7.33	$(\emptyset)_2$–N–C(=O)–CH_3	$CDCl_3$	----
7.0–7.6	Ø–N(CH_3)–C(=O)–CH_3	$CDCl_3$	2363
7.0–7.6	Ø–N(CH_3)–C(=O)–H	$CDCl_3$	2366
7.1–7.6	Ø–N(R_3)–C(=O)–CH_3	$CDCl_3$	2365

Para Substituted Tertiary Benzamides

para	meta	ortho		
CH_3–O–	7.11	7.53	–C(=O)–N(Piperazine)	D_2O
Cl–	7.71	7.98	–C(=O)–N(R)–CH_3	TFA
R–SO_2–	7.87	7.52	–C(=O)–N$(R_2)_2$	$CDCl_3$
R–SO_2–	7.87	7.52	–C(=O)–N(Piperazine)	$CDCl_3$
CH_3–SO_2–	7.92	7.52	–C(=O)–N$(R_2)_2$	$CDCl_3$
O_2N–	8.16	7.42	–C(=O)–N$(R)_2$	$CDCl_3$
O_2N–	8.22	7.50	–C(=O)–N$(R_2)_2$	Polysol
O_2N–	8.29	7.60	–C(=O)–N(Piperazine)	$CDCl_3$
O_2N–	8.62	8.29	–C(=O)–N(R)–CH_3	TFA

Solubility and Solvent Effects

The low molecular weight Tertiary Amides are often most soluble in D_2O or DMSO-d_6 while the higher molecular weight materials are usually more soluble in CCl_4 or $CDCl_3$. For the difficulty soluble material, Trifluoroacetic Acid is often effective although its use results in unusually low field chemical shifts for hydrocarbon groups adjacent to both the carbonyl side and the nitrogen side of the Tertiary Amide group.

N,N-DIETHYLFORMAMIDE

$C_5H_{11}NO$

Mol. Wt. 101.15

B.P. 177-179°C

Solvent: $CDCl_3$

Assignments
a = 1.12
b = 1.20
c = 3.31
d = 3.37
e = 8.08

2347

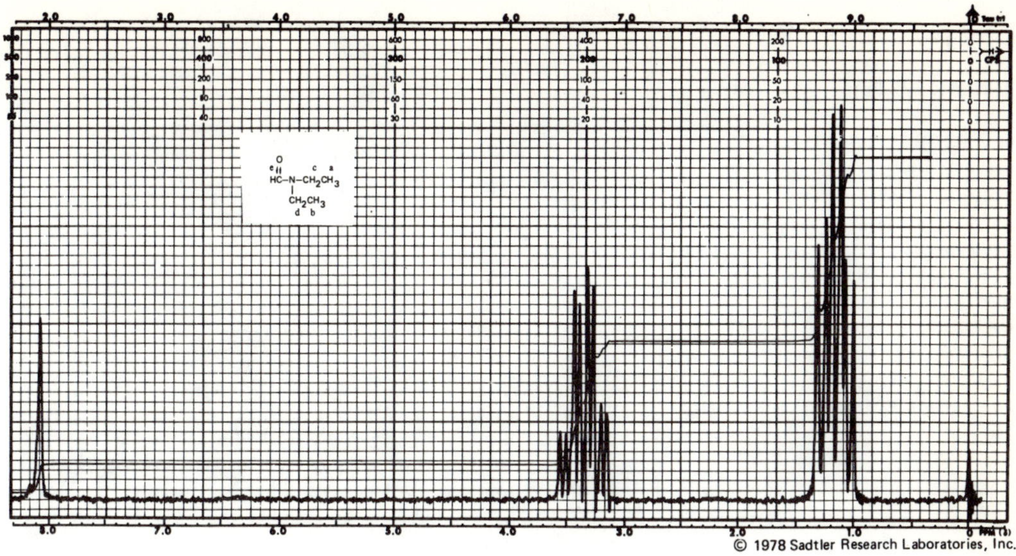

© 1978 Sadtler Research Laboratories, Inc.

N,N-DIISOPROPYLACETAMIDE

$C_8H_{17}NO$

Mol. Wt. 143.23

B.P. 71-73°C/6mm

Solvent: CCl_4

Assignments
a = 1.29
b = 1.97
c = 3.69

2348

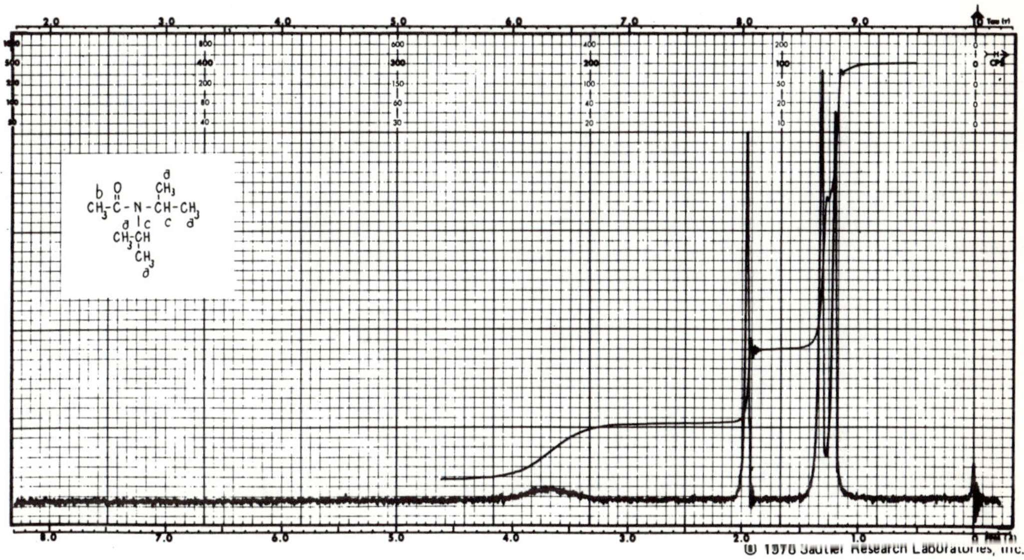

© 1978 Sadtler Research Laboratories, Inc.

N,N-DIBUTYLACETAMIDE

$C_{10}H_{21}NO$

Mol. Wt. 171.28

B.P. 114-116°C/10mm

Solvent: CCl_4

Assignments
a = 0.97
b = 1.41
c = 1.98
d = 3.22

2349

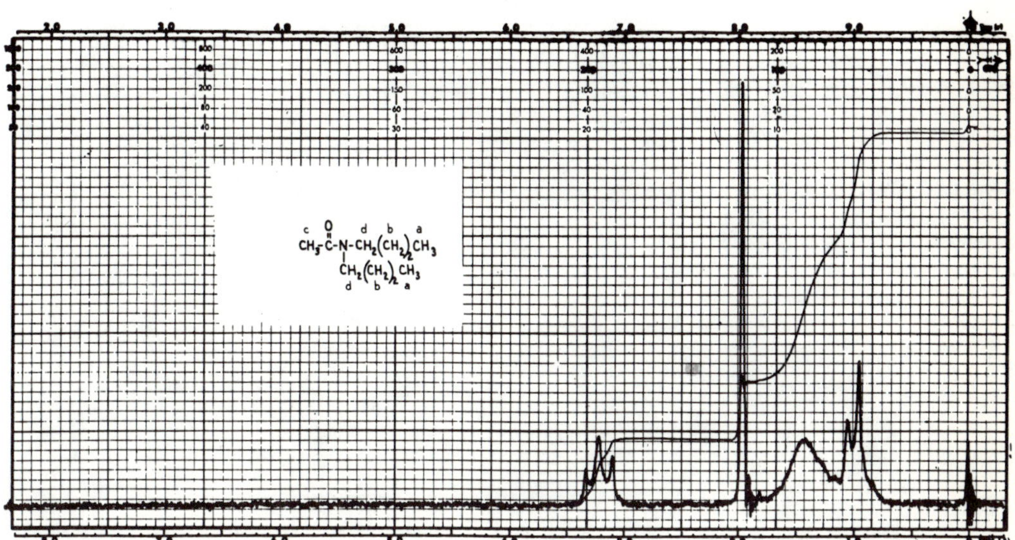

© 1978 Sadtler Research Laboratories, Inc.

N,N-DIETHYLPROPIONAMIDE

$C_7H_{15}NO$

Mol. Wt. 129.20

B.P. 68-69°C/5mm

Solvent: D_2O

Assignments
a = 1.11
b = 1.21
c = 2.41
d = 3.36
e = 3.40
f = 4.70 HDO

2350

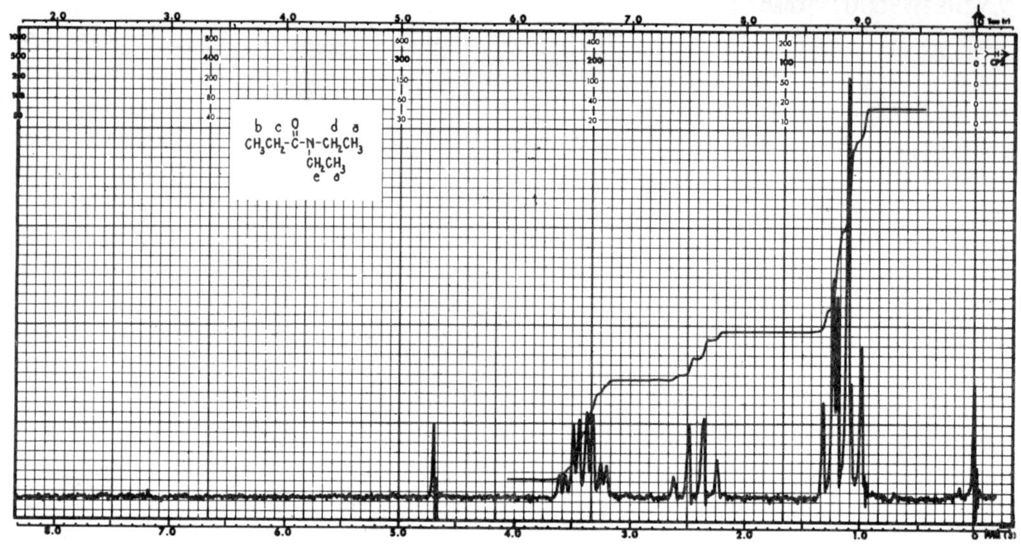

© 1978 Sadtler Research Laboratories, Inc.

N,N-DIBUTYLPROPIONAMIDE

$C_{11}H_{23}NO$

Mol. Wt. 185.31

B.P. 115-116°C/6mm

Solvent: $CDCl_3$

Assignments
a = 0.95
b = 1.16
c = 1.10-1.90
d = 2.31
e = 3.29

2351

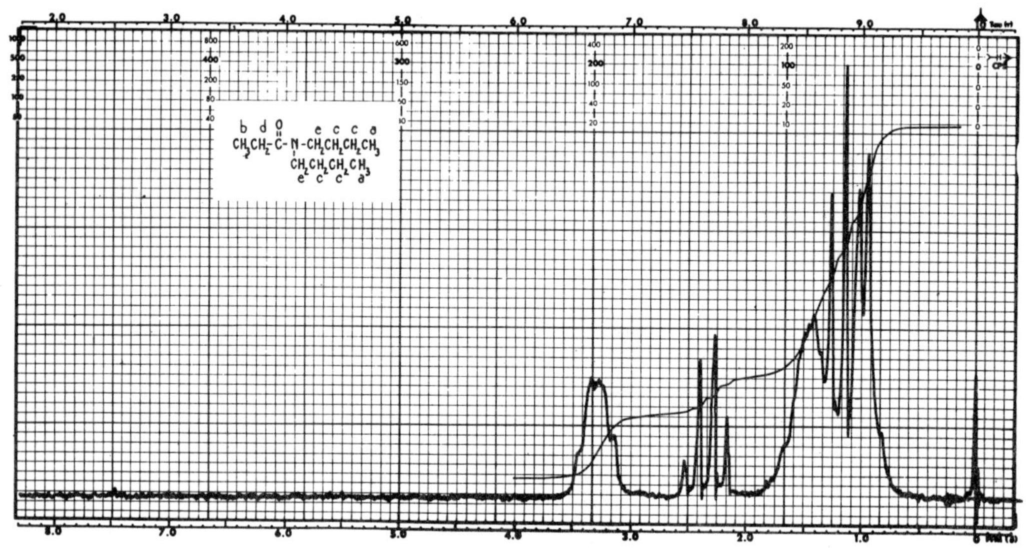

© 1978 Sadtler Research Laboratories, Inc.

N,N-DIMETHYLBUTYRAMIDE

$C_6H_{13}NO$

Mol. Wt. 115.18

B.P. 190-191°C

Solvent: D_2O

Assignments
a = 0.96
b = 1.59
c = 2.41
d = 2.94
e = 3.11
f = 4.69 HDO

2352

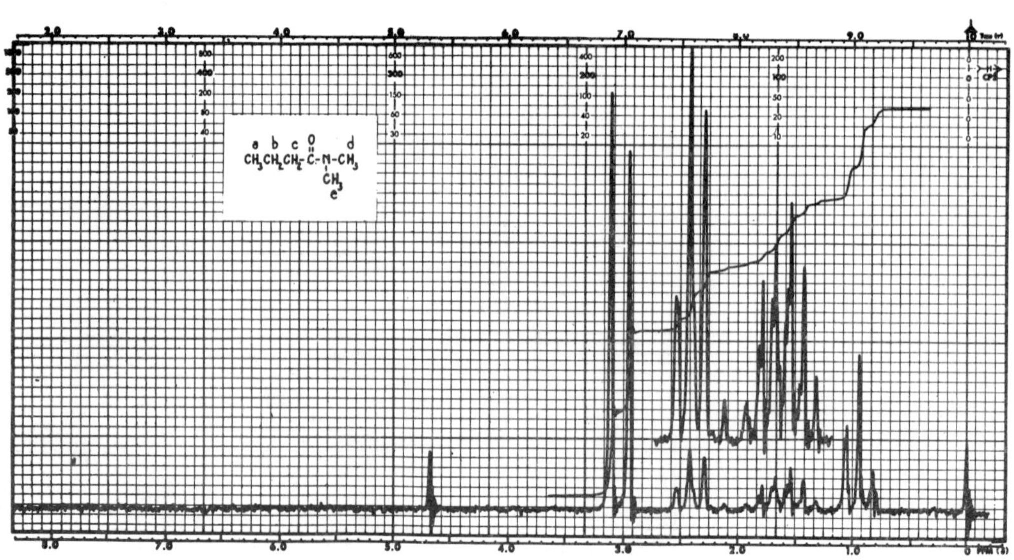

© 1978 Sadtler Research Laboratories, Inc.

N,N-DIETHYLBUTYRAMIDE

$C_8H_{17}NO$

Mol. Wt. 143.23

B.P. 73-74°C/6mm

Solvent: D_2O

Assignments
a = 0.85-1.35
b = 1.63
c = 2.38
d = 3.36
e = 3.41
f = 4.66 HDO

2353

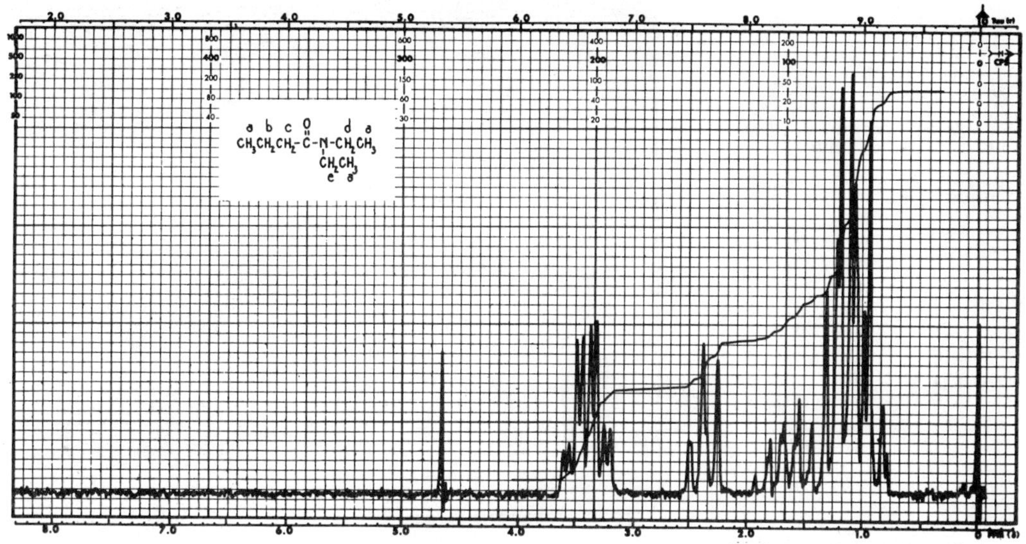

© 1978 Sadtler Research Laboratories, Inc.

N,N-DIPROPYLDECANAMIDE

$C_{16}H_{33}NO$

Mol. Wt. 255.45

B.P. 166-168°C/5mm

Solvent: $CDCl_3$

Assignments
a = ca 0.89
b = ca 1.30
c = 1.58
d = 2.29
e = 3.17
f = 3.21

2354

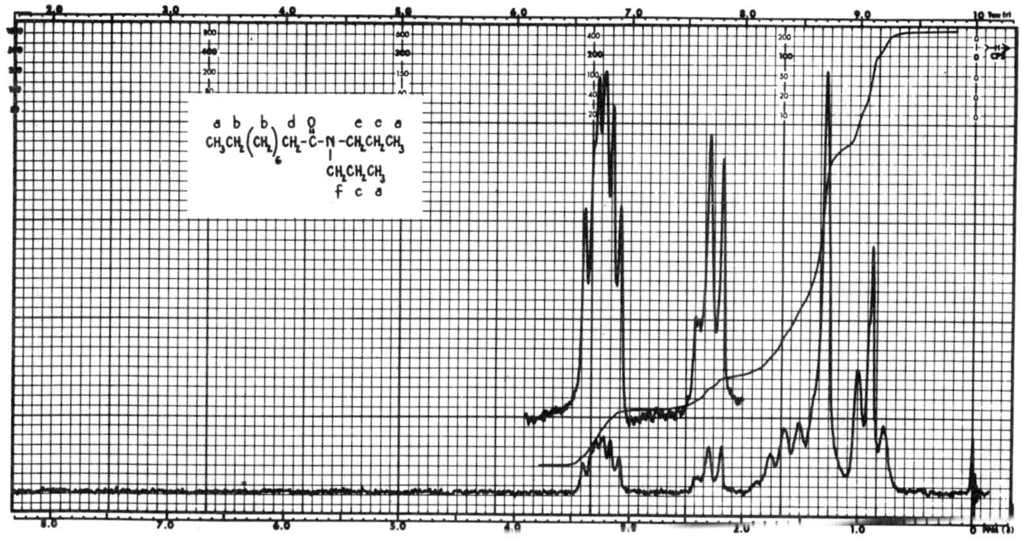

© 1978 Sadtler Research Laboratories, Inc.

N,N-DIMETHYLDODECANAMIDE

$C_{14}H_{29}NO$

Mol. Wt. 227.39

M.P. 15-16°C

Solvent: CCl_4

Assignments
a = 0.88
b = 1.28
c = 2.20
d = ca 2.91

2355

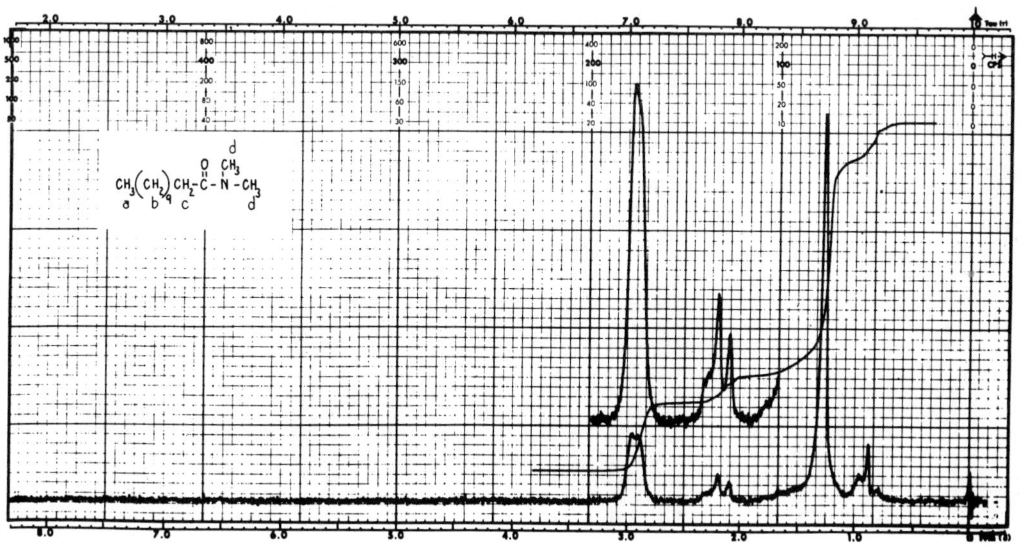

© 1978 Sadtler Research Laboratories, Inc.

N,N-DIETHYLLAURAMIDE

$C_{16}H_{33}NO$

Mol. Wt. 255.45

B.P. 166-167°C/2mm

Solvent: CCl_4

Assignments
a = 0.89
b = 1.11
c = 1.29
d = 2.19
e = 3.27

2356

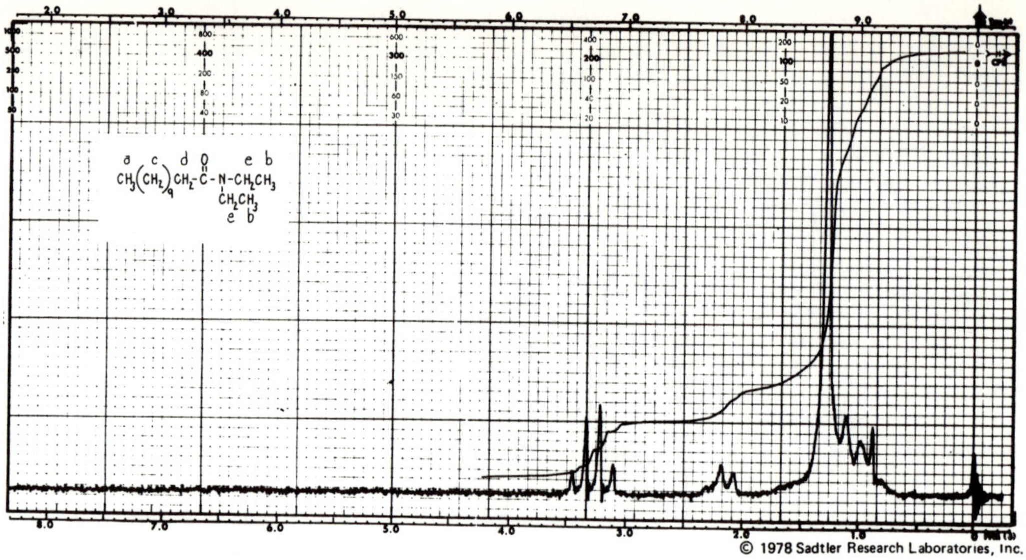

© 1978 Sadtler Research Laboratories, Inc.

N,N-DIMETHYLLACTAMIDE

$C_5H_{11}NO_2$

Mol. Wt. 117.15

B.P. 58-61°C/1mm

Solvent: CCl_4

Assignments
a = 1.22
b = 2.99
c = 3.83
d = 4.37
e = 3.69 imp.

2357

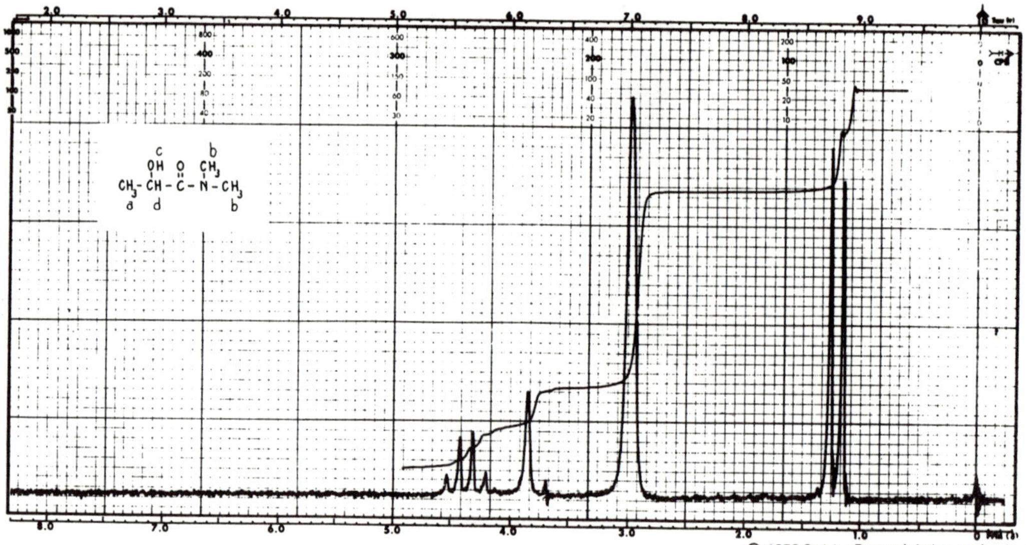

© 1978 Sadtler Research Laboratories, Inc.

N,N-DIETHYLACETOACETAMIDE

$C_8H_{15}NO_2$

Mol. Wt. 157.21

B.P. 124-126°C

Solvent: $CDCl_3$

Assignments
a = 1.12
b = ca 1.19
c = 1.92
d = 2.26
e = 3.10-3.60
f = 3.51
g = 5.10
h = unobserved

2358

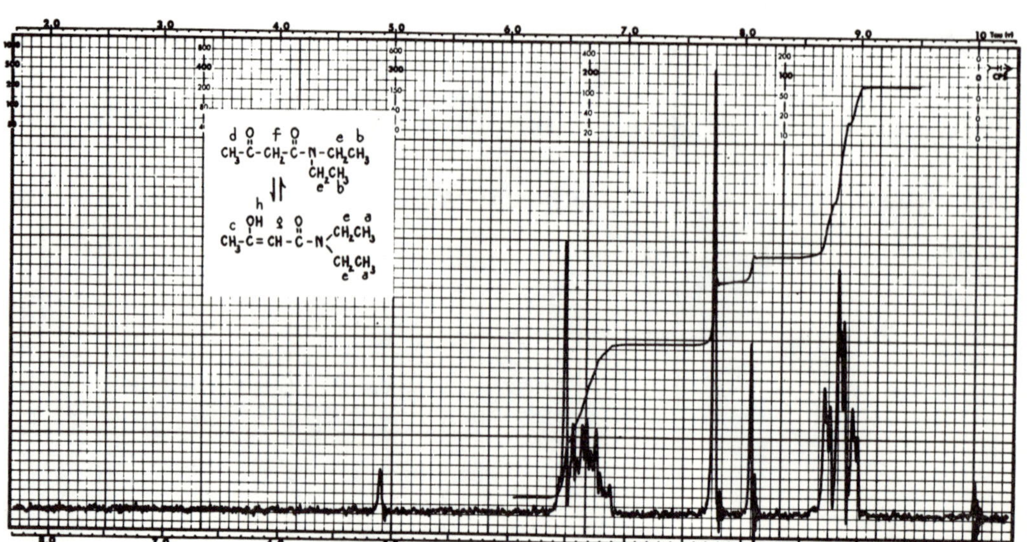

© 1978 Sadtler Research Laboratories, Inc.

N,N-DIMETHYL-1-ADAMANTANECARBOXAMIDE

$C_{13}H_{21}NO$

Mol. Wt. 207.32

M.P. 79°C

Solvent: $CDCl_3$

Assignments
a = 1.72
b = 2.02
c = 3.04

2359

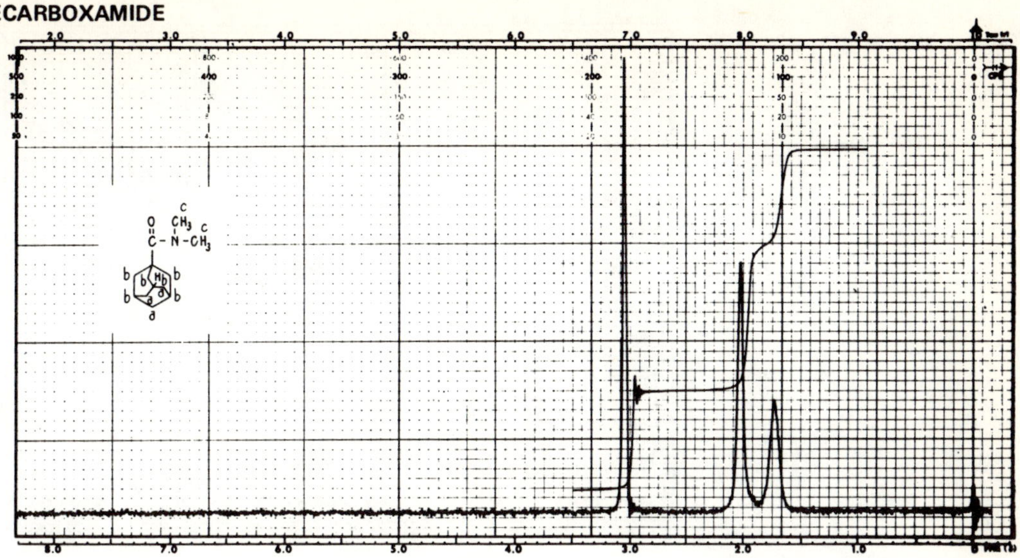

© 1978 Sadtler Research Laboratories, Inc.

4-ACETYLMORPHOLINE

$C_6H_{11}NO_2$

Mol. Wt. 129.16

M.P. −23.5°C

B.P. 89°C/1mm

Solvent: CCl_4

Assignments
a = 1.98
b = 3.45 or c
c = 3.53 or b

2360

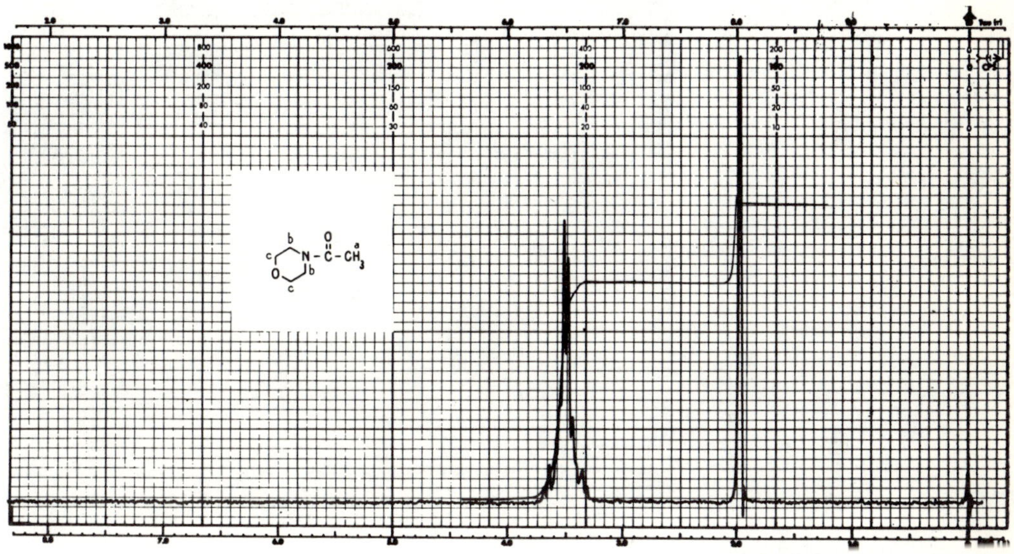

© 1978 Sadtler Research Laboratories, Inc.

N-ETHYLFORMANILIDE

$C_9H_{11}NO$

Mol. Wt. 149.19

B.P. 130-131°C/14mm

Solvent: $CDCl_3$

Assignments
a = 1.11
b = 3.81
c = 7.00-7.60
d = 8.30

2361

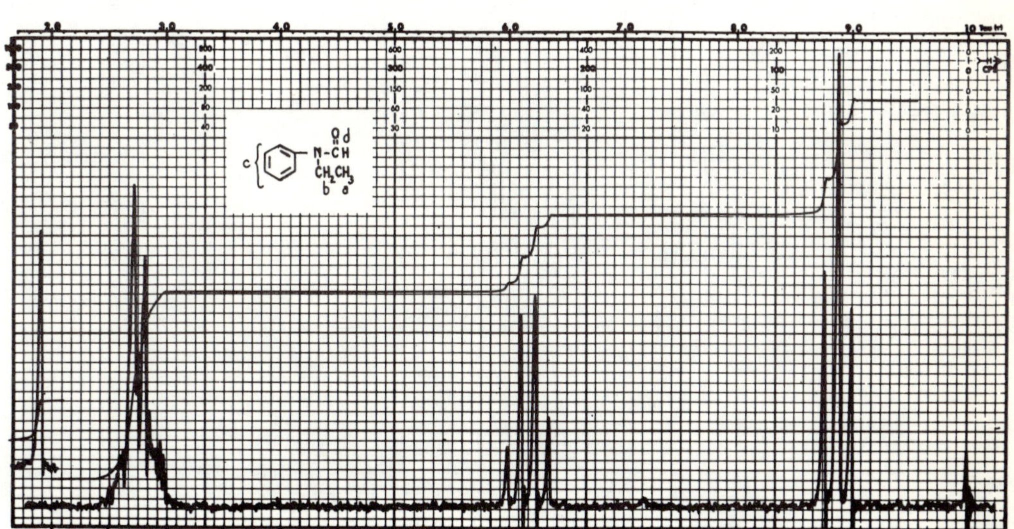

© 1978 Sadtler Research Laboratories, Inc.

N-ETHYL-p-FORMOTOLUIDIDE

$C_{10}H_{13}NO$

Mol. Wt. 163.22

B.P. 128-134°C/9mm

Solvent: $CDCl_3$

Assignments
a = 1.14
b = 2.36
c = 3.80
d = 7.02
e = 7.19
f = 8.27

2362

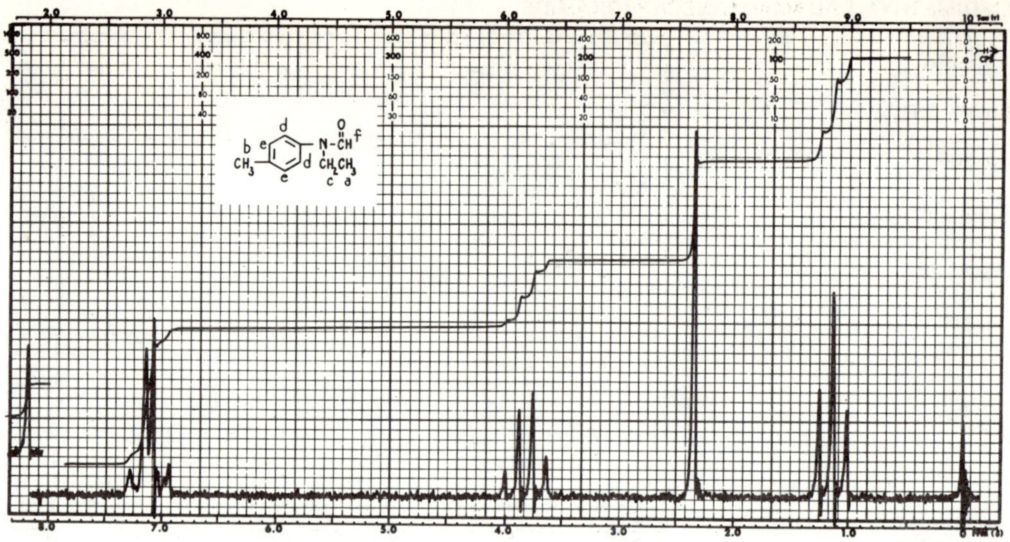

© 1978 Sadtler Research Laboratories, Inc.

N-METHYLACETANILIDE

$C_9H_{11}NO$

Mol. Wt. 149.19

M.P. 100-102°C

Solvent: $CDCl_3$

Assignments
a = 1.86
b = 3.23
c = 7.07-7.59

2363

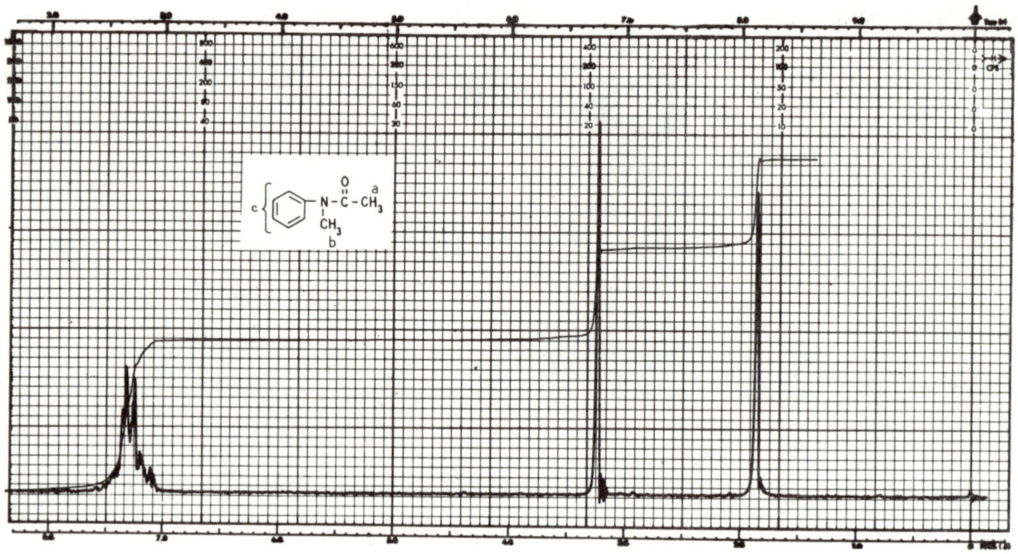

© 1978 Sadtler Research Laboratories, Inc.

N-ETHYLACETANILIDE

$C_{10}H_{13}NO$

Mol. Wt. 163.22

Solvent: CCl_4

Assignments
a = 1.08
b = 1.72
c = 3.69
d = 7.50

2364

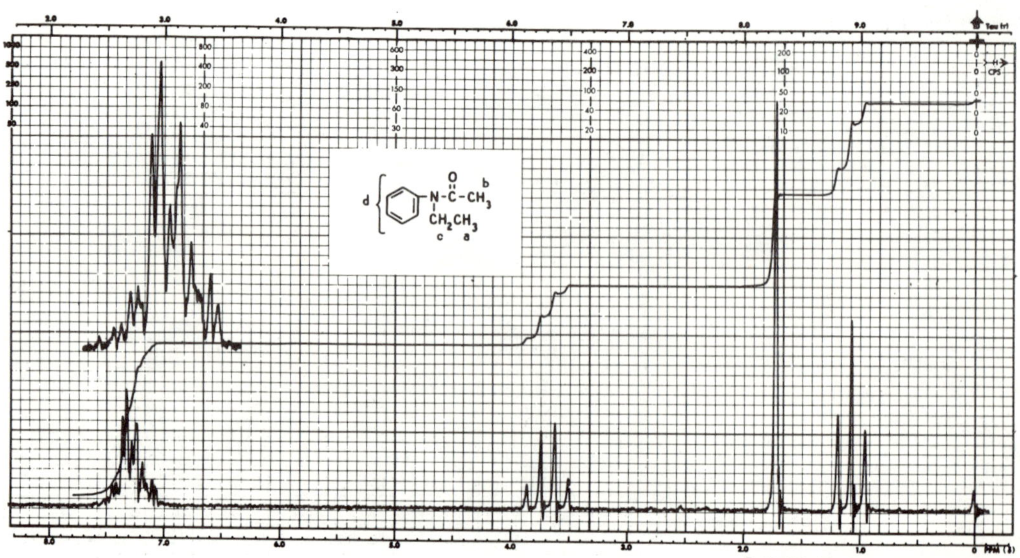

© 1978 Sadtler Research Laboratories, Inc.

N-PROPYLACETANILIDE

$C_{11}H_{15}NO$

Mol. Wt. 177.25

M.P. 48-50°C

Solvent: $CDCl_3$

Assignments
a = 0.89
b = 1.56
c = 1.82
d = 3.69
e = 7.07-7.60

2365

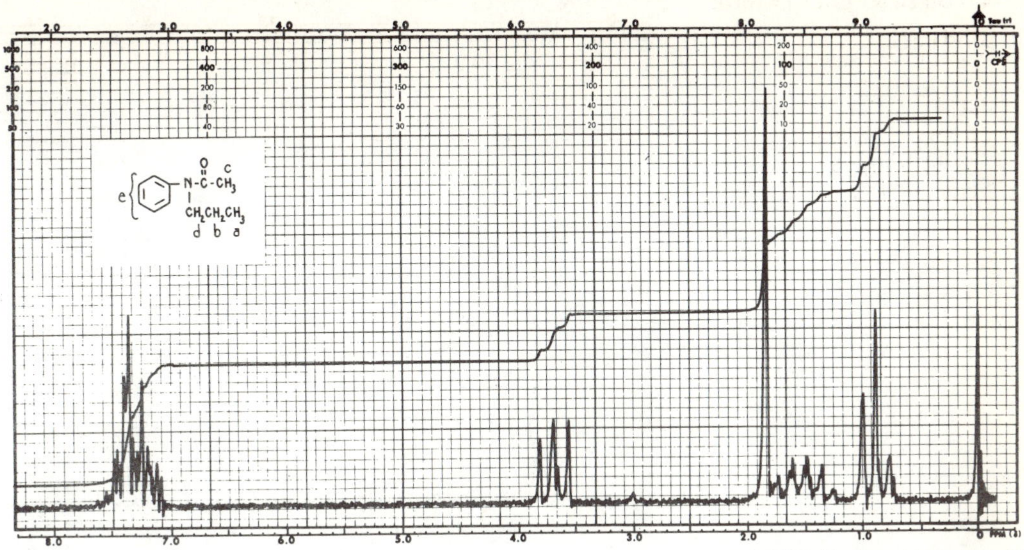

© 1978 Sadtler Research Laboratories, Inc.

N-BUTYLACETANILIDE

$C_{12}H_{17}NO$

Mol. Wt. 191.28

B.P. 140-142°C/9mm (lit.)

Solvent: CCl_4

Assignments
a = 0.89
b = 1.38
c = 1.72
d = 3.65
e = 7.02-7.51

2366

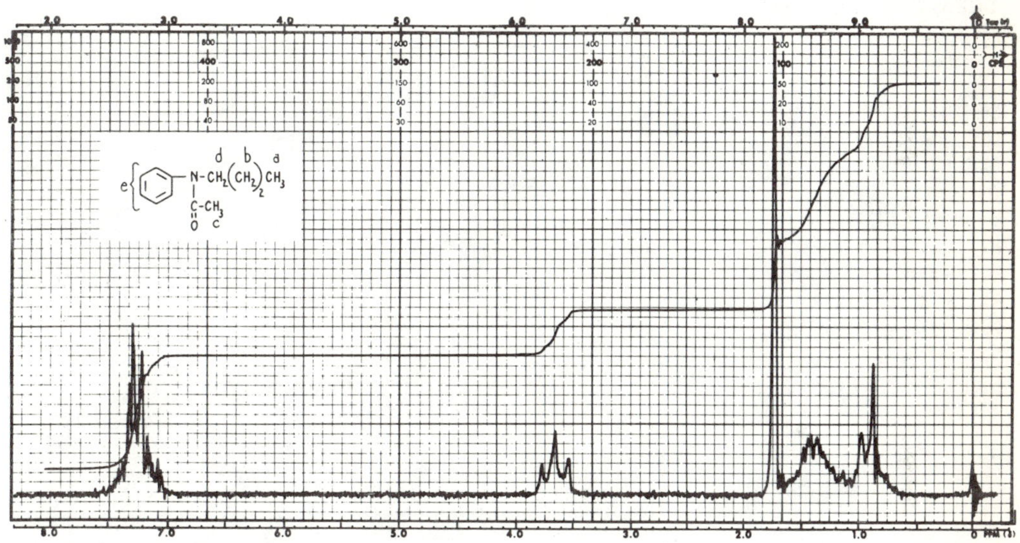

© 1978 Sadtler Research Laboratories, Inc.

DIPHENYLFORMAMIDE

$C_{13}H_{11}NO$

Mol. Wt. 197.24

M.P. 69-71°C

Solvent: CCl_4

Assignments
a = 7.22
b = 8.50
c = 2.02 imp.

2367

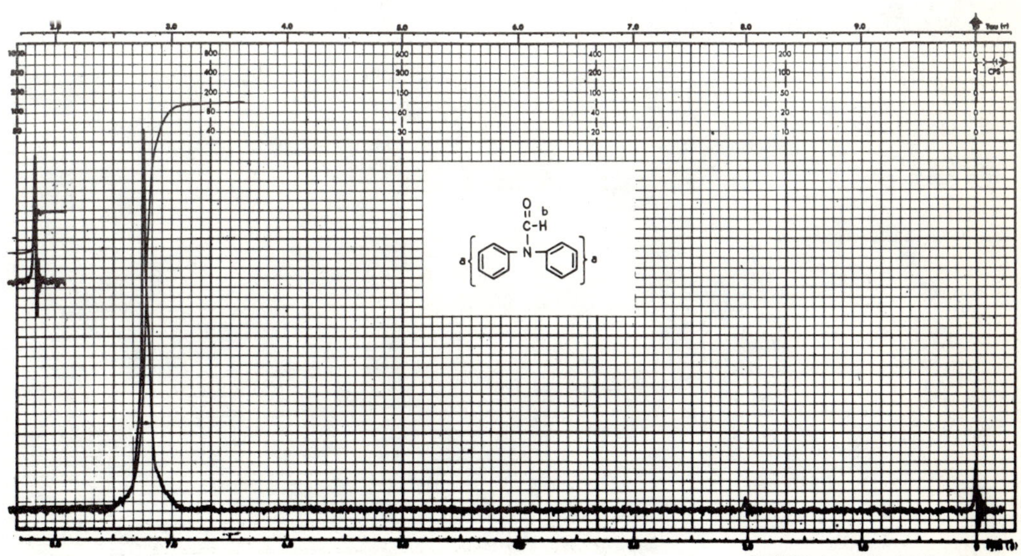

© 1978 Sadtler Research Laboratories, Inc.

N,N-DIPHENYLACETAMIDE

$C_{14}H_{13}NO$

Mol. Wt. 211.27

M.P. 99°C

Solvent: $CDCl_3$

Assignments
a = 2.06
b = 7.33
c = 1.57 imp.

2368

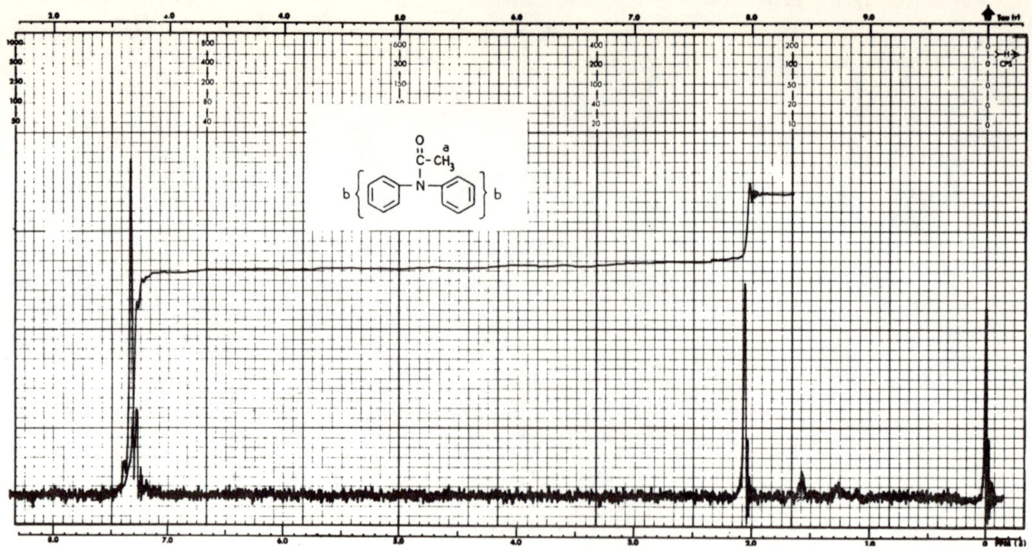

© 1978 Sadtler Research Laboratories, Inc.

1-BENZOYLPIPERIDINE

$C_{12}H_{15}NO$

Mol. Wt. 189.26

M.P. 47-49°C

Solvent: CCl_4

Assignments
a = 1.62
b = 3.47
c = 7.33

2369

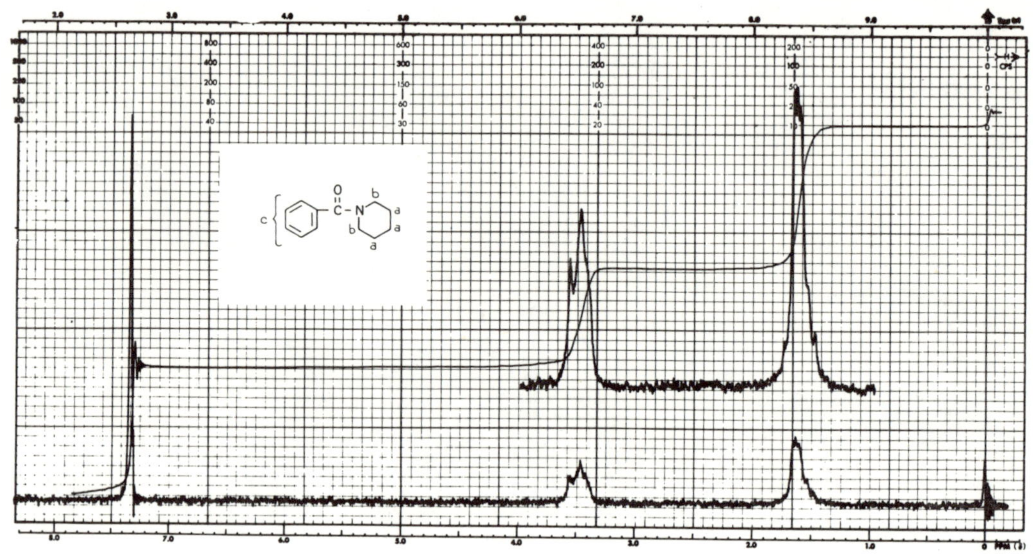

© 1978 Sadtler Research Laboratories, Inc.

N,N,N′,N′-TETRAETHYLPHTHALAMIDE

$C_{16}H_{24}N_2O_2$

Mol. Wt. 276.38

M.P. 37-39.5°C

Solvent: $CDCl_3$

Assignme nts
a = 1.09
b = 1.20
c = 3.28
d = 3.49
e = ca 7.32

2370

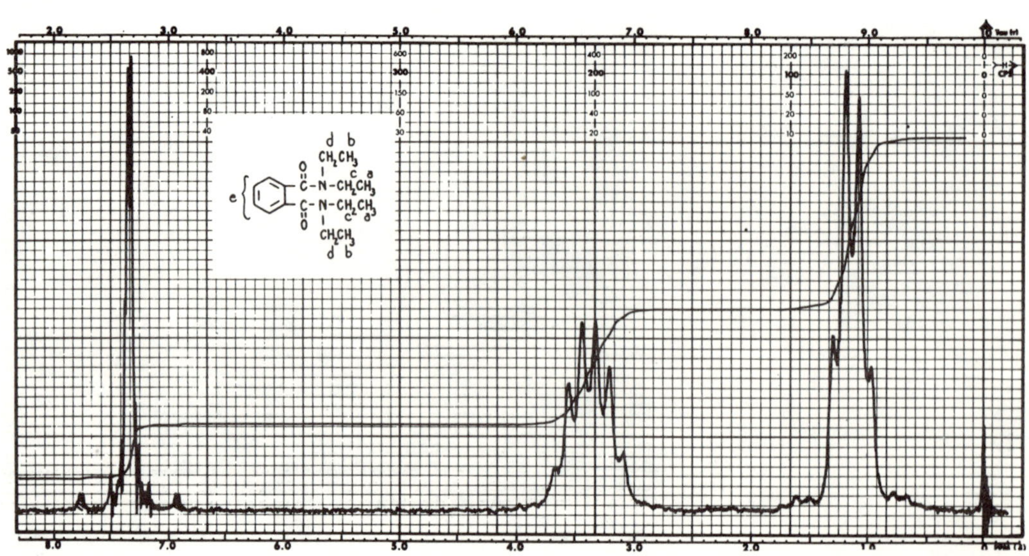

© 1978 Sadtler Research Laboratories, Inc.

3,4-DIHYDRO-1(2H)-QUINOLINECARBOXALDEHYDE

$C_{10}H_{11}NO$

Mol. Wt. 161.21

B.P. 165-168°C/14mm

Solvent: $CDCl_3$

Assignments
a = 1.92
b = 2.81
c = 3.80
d = 7.18
e = 8.81

2371

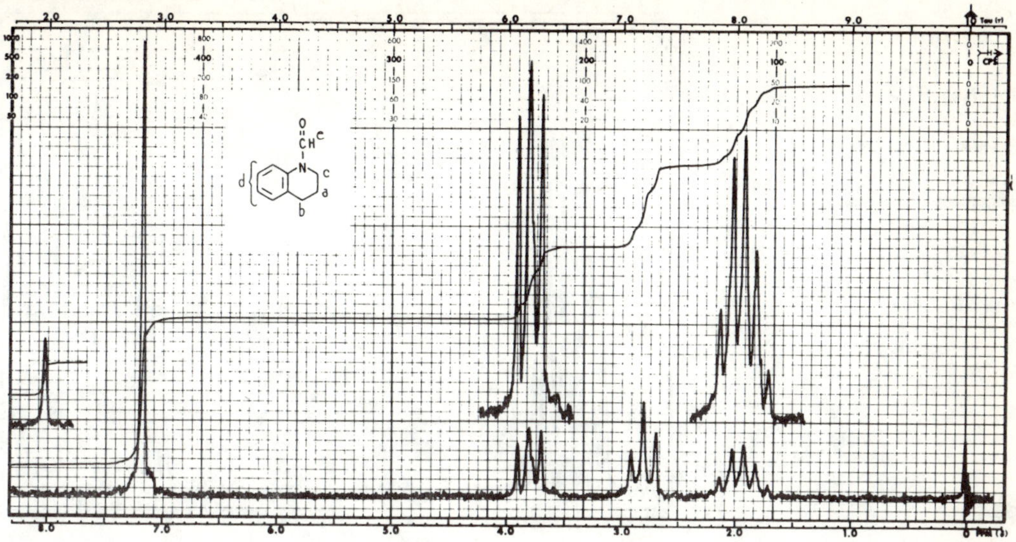

© 1978 Sadtler Research Laboratories, Inc.

1-ACETYL-1,2,3,4-TETRAHYDROQUINOLINE

$C_{11}H_{13}NO$

Mol. Wt. 175.23

B.P. 290-294°C

Solvent: $CDCl_3$

Assignments
a = 1.95
b = 2.21
c = 2.72
d = 3.79
e = 7.06-7.40

2372

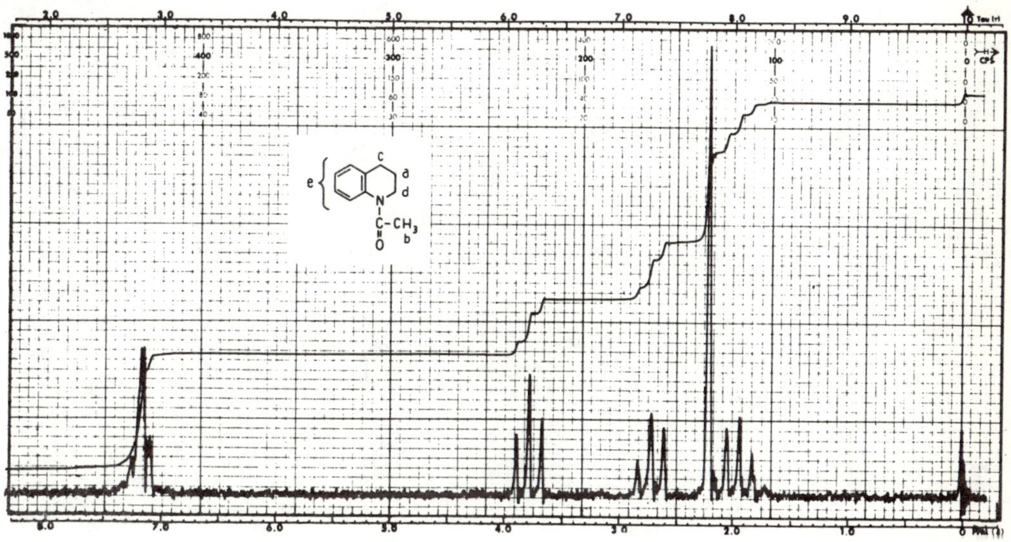

© 1978 Sadtler Research Laboratories, Inc.

9-ACETYLCARBAZOLE

$C_{14}H_{11}NO$

Mol. Wt. 209.25

M.P. 68-69°C

Solvent: $CDCl_3$

Assignments
a = 2.70
b = 7.15-7.55
c = 7.83 or d
d = 8.09 or c

2373

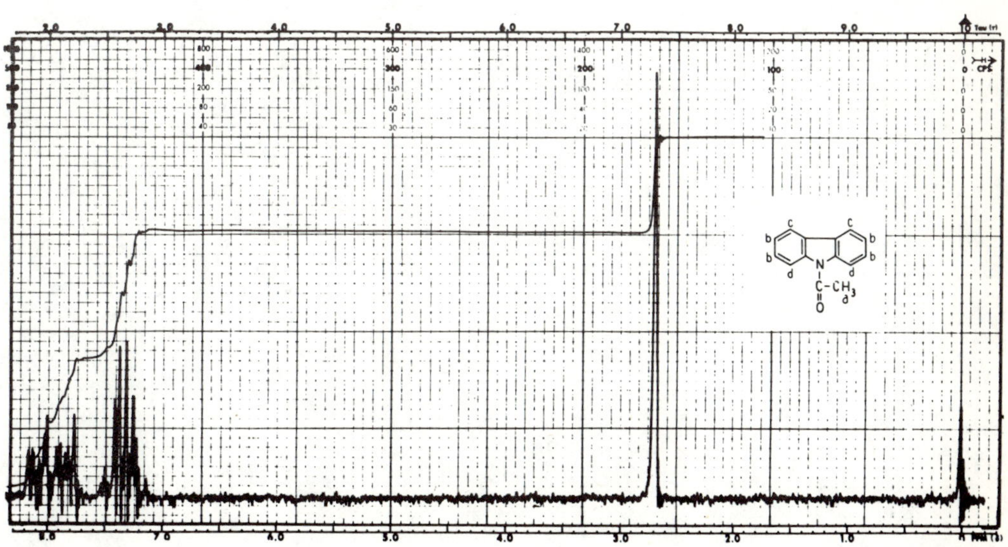

© 1978 Sadtler Research Laboratories, Inc.

1-PHENYL-2-PYRROLIDINONE

$C_{10}H_{11}NO$

Mol. Wt. 161.21

Solvent: $CDCl_3$

Assignments

a = 2.10

b = 2.55

c = 3.78

d = 6.90-7.80

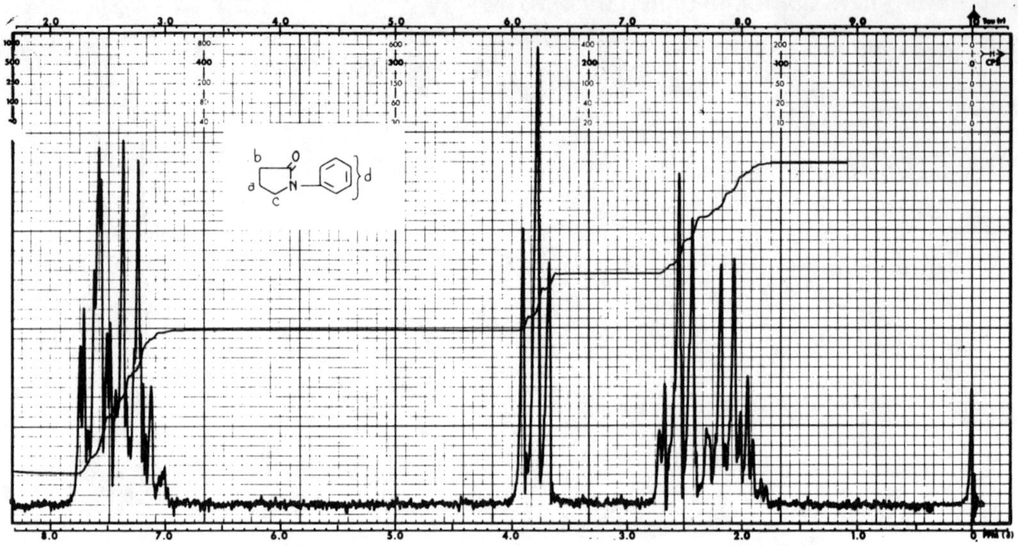

© 1978 Sadtler Research Laboratories, Inc.

2374

THE IMIDES

The Imides are nitrogen containing analogs of the Anhydrides. They contain the –C(=O)–NH–C(=O)– linkage and are usually cyclic in structure. As with the other carbonyl containing compounds, the protons of aliphatic groups alpha to the C(=O) group are weakly deshielded. The Imide NH proton resonates at low field (8–12 ppm) and is usually a very broad absorption band.

Chemical Shifts

Aliphatic Protons

HC–2	HC–1		Solvent	Spectrum
	2.30	CH_3–C(=O)–NH–C(=O)–CH_3	$CDCl_3$	----
	2.99	CH_3–N(Succinimide)	$CDCl_3$	----
	3.23	CH_3–N(Phthalimide)	TFA	2391
1.18	3.55	CH_3–CH_2–N(Maleiimide)	$CDCl_3$	2384
1.29	3.71	CH_3–CH_2–N(Phthalimide)	$CDCl_3$	----

Olefinic Protons

The ring olefinic protons of Maleiimide appear as a single sharp peak in the chemical shift range from 6.6 to 7.2 ppm. The olefinic protons of a vinyl group bonded to the Imide nitrogen atom are well differentiated in chemical shift as shown in spectrum 2395. The hydrogen in the geminal position resonates at lowest field (6.83 ppm) as a doublet of doublets due to coupling to the hydrogens cis and trans to it. These hydrogens resonate at higher field, the cis proton at 5.01 ppm as a 10 Hz doublet and the trans proton at 6.02 ppm as a 16 Hz doublet.

Aromatic Protons

Phenyl groups bonded to the Imide nitrogen atom usually appear as a single, relatively sharp band near 7.4 ppm.

The four aromatic hydrogens of the Phthalimides are observed as a symmetrical, higher-order series of bands centered at about 7.8 ppm.

Exchangeable Protons

The chemical shift of the Imide NH proton, as noted above resonates over a range of about 4 ppm at low field. The major determining factor producing this range is the type of ring system in which the Imide group is found. Generally the alicyclic systems, Succinimide and Glutarimide contain NH protons resonating at highest field, the Maleimides occupy the middle of the range, while the Phthalimides appear at the low field end of the range.

SUCCINIMIDE

$C_4H_5NO_2$

Mol. Wt. 99.09

M.P. 124-126°C

B.P. 228°C

Solvent: D_2O

Assignments
a = 2.79
b = 4.58 HDO

2375

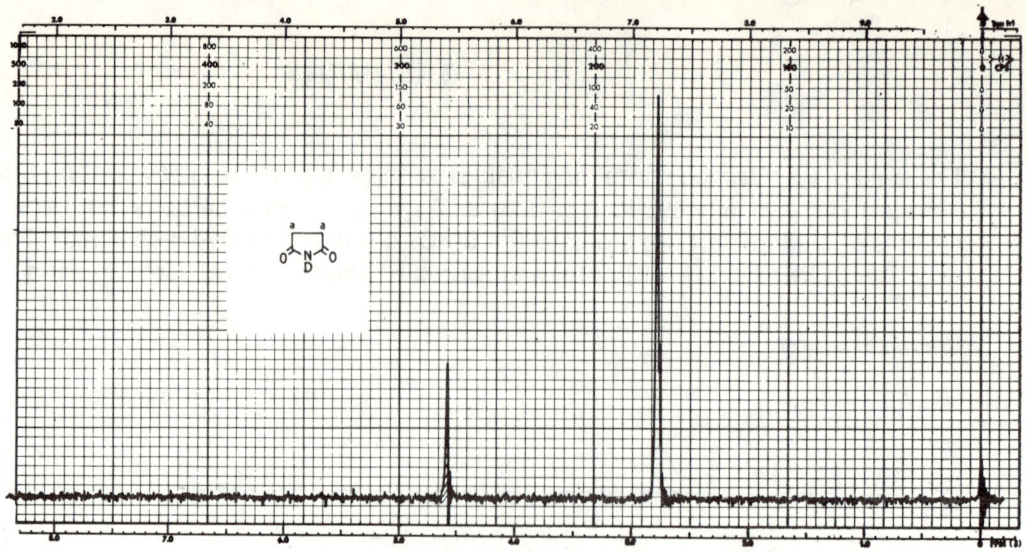

© 1978 Sadtler Research Laboratories, Inc.

2-METHYLSUCCINIMIDE

$C_5H_7NO_2$

Mol. Wt. 113.12

M.P. 68-70°C

Solvent: $CDCl_3$

Assignments
a = 1.31
b = 2.38
c = 2.60-3.30
d = 9.51

2376

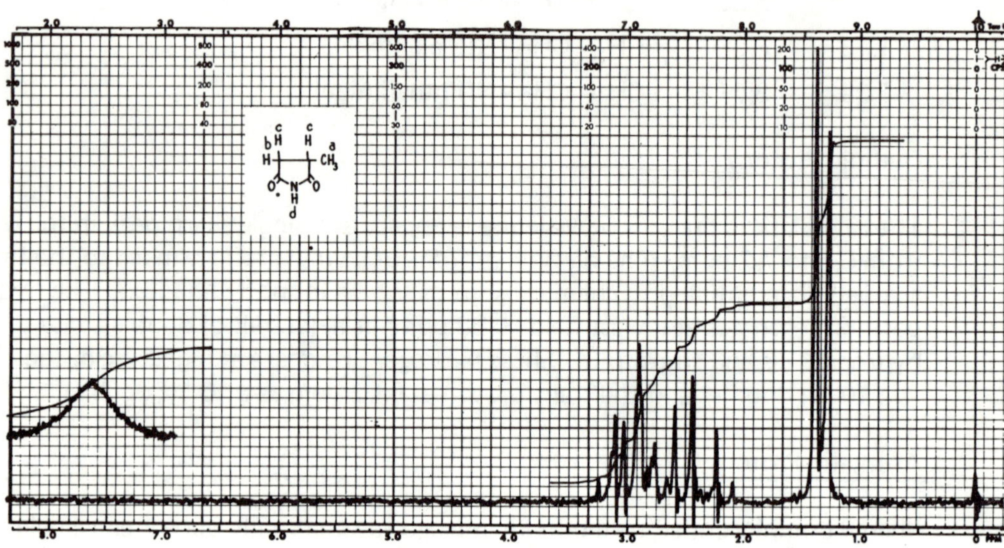

© 1978 Sadtler Research Laboratories, Inc.

GLUTARIMIDE

$C_5H_7NO_2$

Mol. Wt. 113.12

Solvent: Trifluoroacetic acid

Assignments
a = 2.12
b = 2.79
c = unobserved *
* Under solvent band

2377

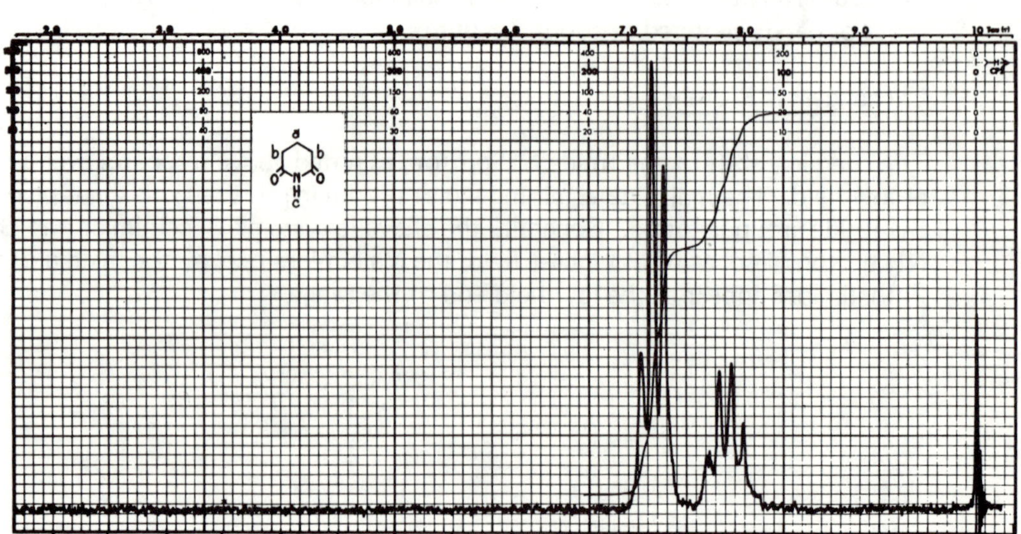

© 1978 Sadtler Research Laboratories, Inc.

3-AZASPIRO(5,5)UNDECANE-2,4-DIONE

$C_{10}H_{15}NO_2$

Mol. Wt. 181.24

Solvent: $CDCl_3$

Assignments
a = ca 1.51
b = 2.50
c = 8.42

2378

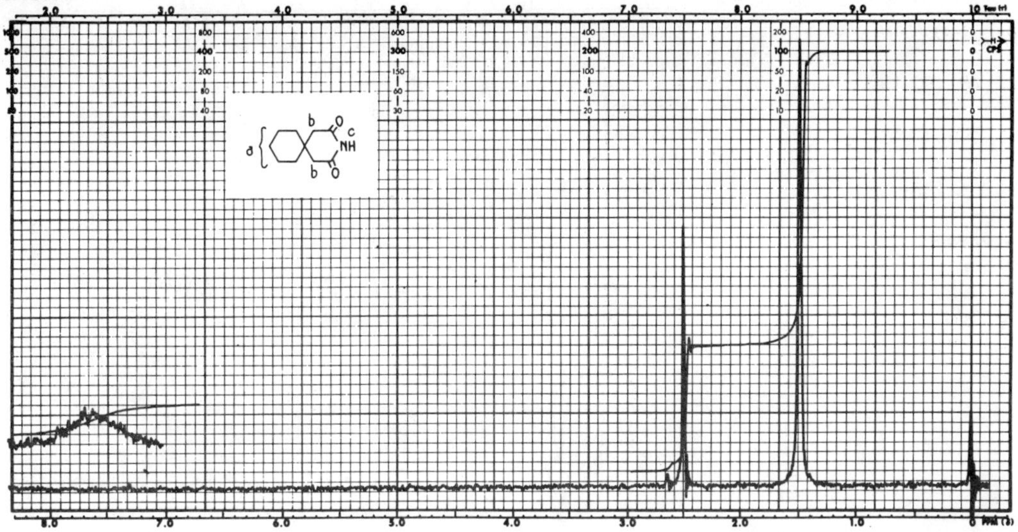

© 1978 Sadtler Research Laboratories, Inc.

MALEIMIDE

$C_4H_3NO_2$

Mol. Wt. 97.08

M.P. 91-93°C

Solvent: Polysol-d

Assignments
a = 6.79
b = 10.60

2379

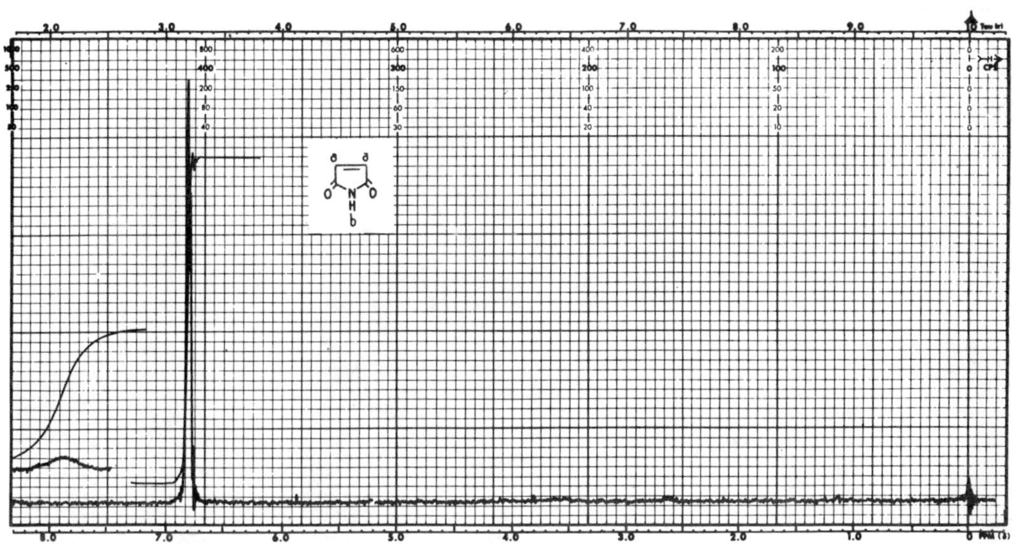

© 1978 Sadtler Research Laboratories, Inc.

PHTHALIMIDE

$C_8H_5NO_2$

Mol. Wt. 147.13

M.P. 235-237°C

Solvent: Dimethyl sulfoxide

Assignments
a = 7.86
b = 11.29

2380

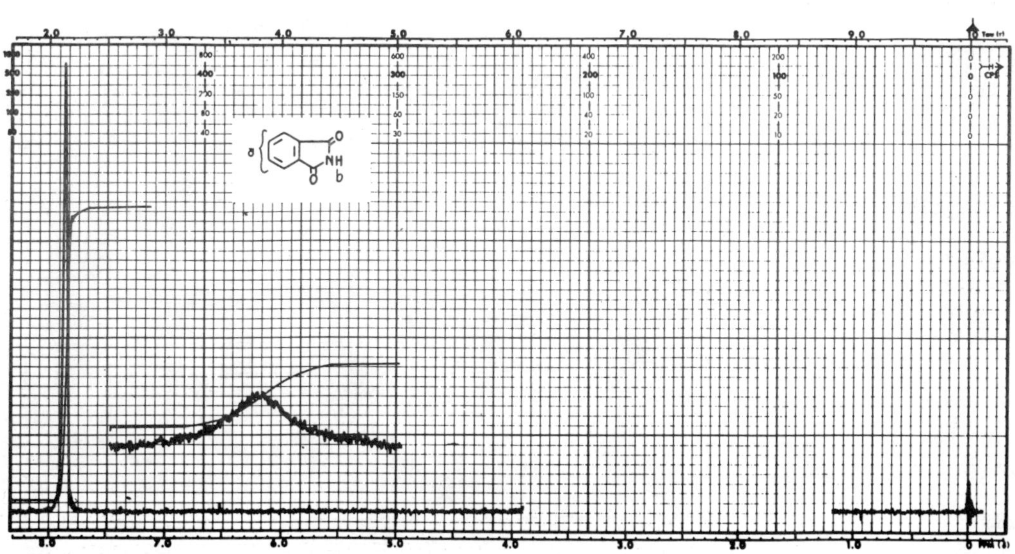

© 1978 Sadtler Research Laboratories, Inc.

4-NITROPHTHALIMIDE

$C_8H_4N_2O_4$

Mol. Wt. 192.13

M.P. 199-201°C

Solvent: Polysol-d

Assignments
a = 8.04
b = 8.49
c = 8.60
d = 11.33

2381

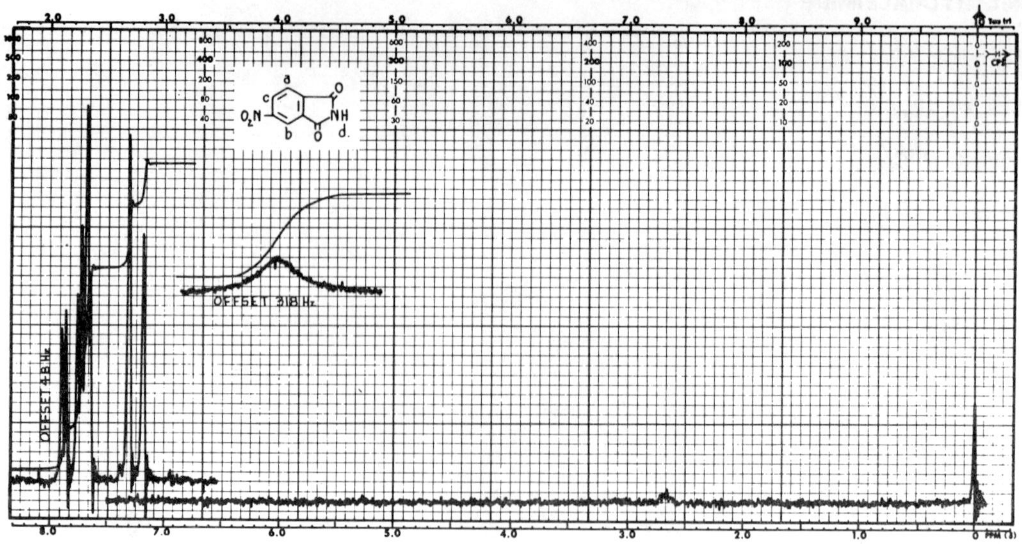

© 1978 Sadtler Research Laboratories, Inc.

N-PENTYLSUCCINIMIDE

$C_9H_{15}NO_2$

Mol. Wt. 169.23

B.P. 138-140°C/9mm

Solvent: $CDCl_3$

Assignments
a = 0.90
b = 1.10-1.90
c = 2.70
d = 3.49

2382

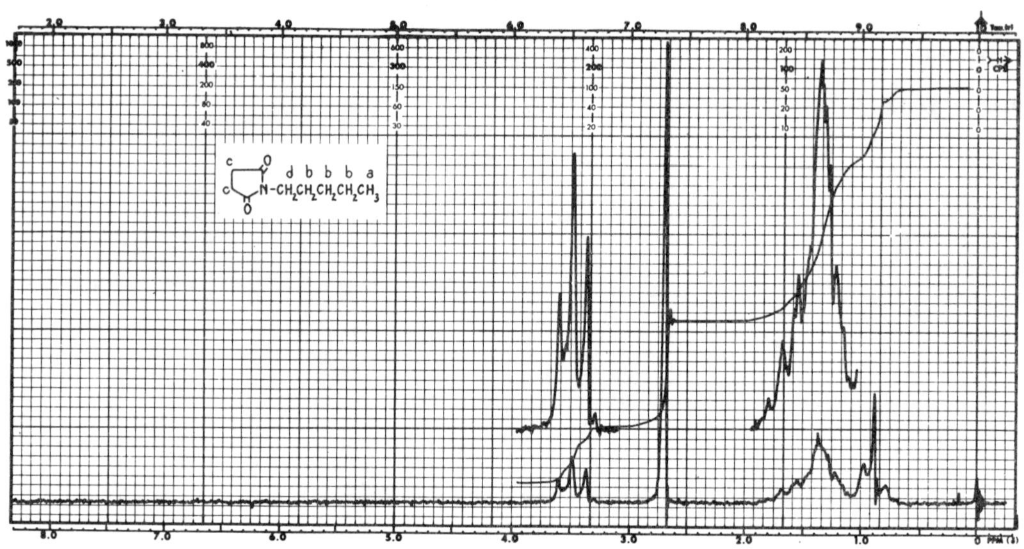

© 1978 Sadtler Research Laboratories, Inc.

1-ACETYL-2-OXOHEXAMETHYLENEIMINE

$C_8H_{13}NO_2$

Mol. Wt. 155.20

B.P. 132-133°C/19mm

Solvent: $CDCl_3$

Assignments
a = 1.77
b = 2.49
c = 2.70
d = 3.91

2383

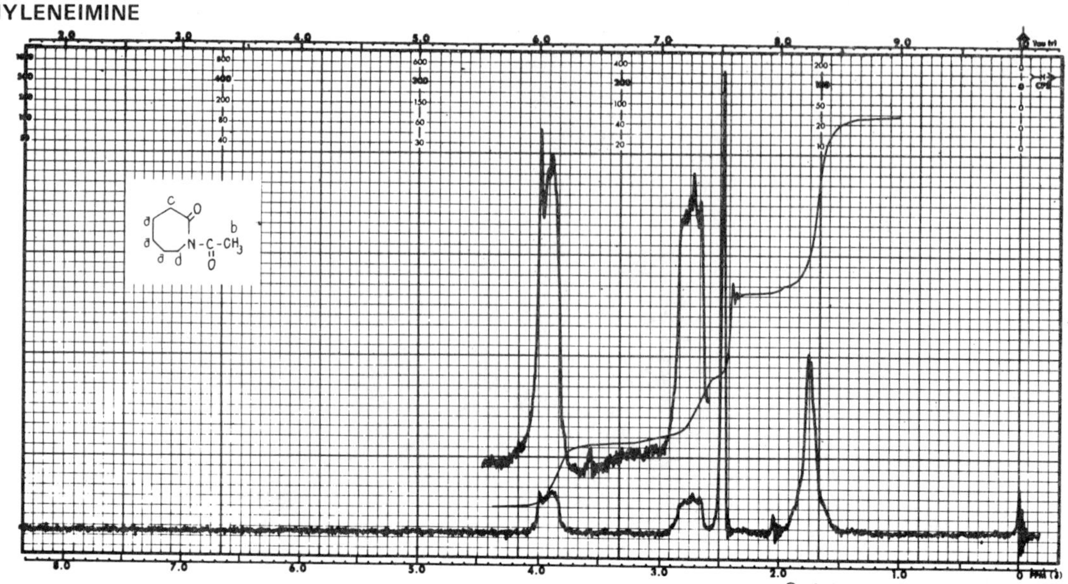

© 1978 Sadtler Research Laboratories, Inc.

N-ETHYLMALEIMIDE

$C_6H_7NO_2$

Mol. Wt. 125.13

M.P. 46-48°C

Solvent: $CDCl_3$

Assignments
a = 1.18
b = 3.55
c = 6.72

2384

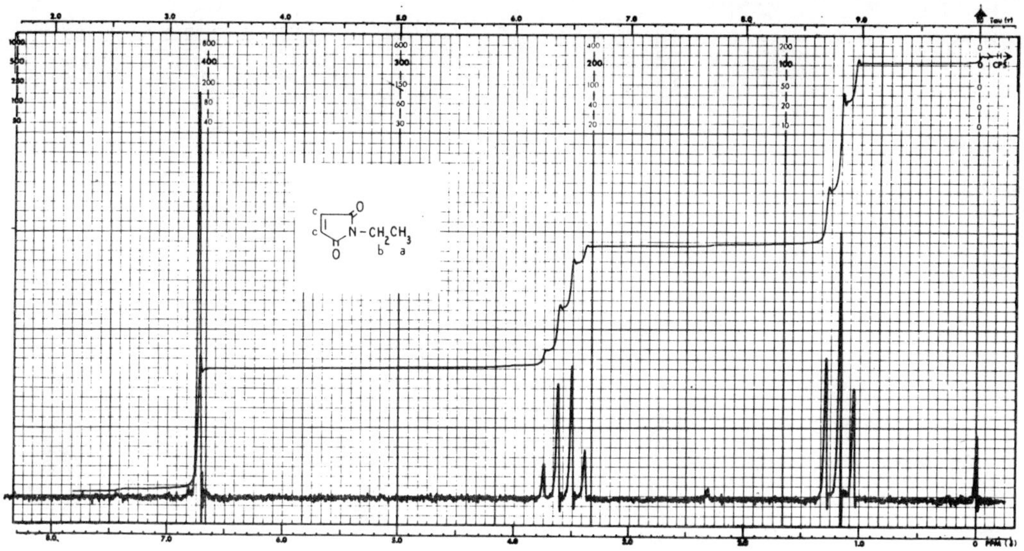

© 1978 Sadtler Research Laboratories, Inc.

N-PHENETHYLMALEIMIDE

$C_{12}H_{11}NO_2$

Mol. Wt. 201.23

M.P. 102°C

Solvent: Polysol-d

Assignments
a = 2.87
b = 3.69
c = 6.78
d = ca 7.20

2385

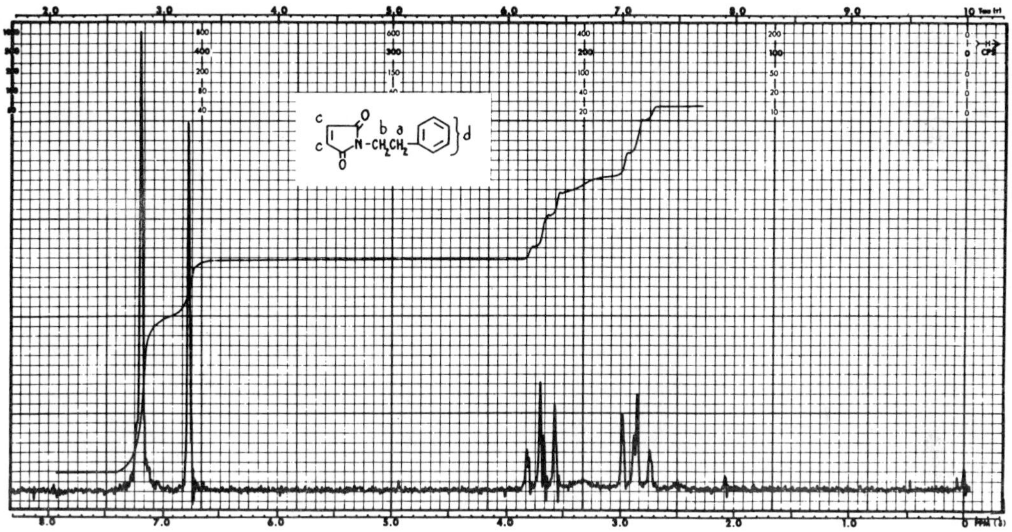

© 1978 Sadtler Research Laboratories, Inc.

N-BENZYLMALEIMIDE

$C_{11}H_9NO_2$

Mol. Wt. 187.20

M.P. 70°C

Solvent: $CDCl_3$

Assignments
a = 4.63
b = 6.63
c = ca 7.29

2386

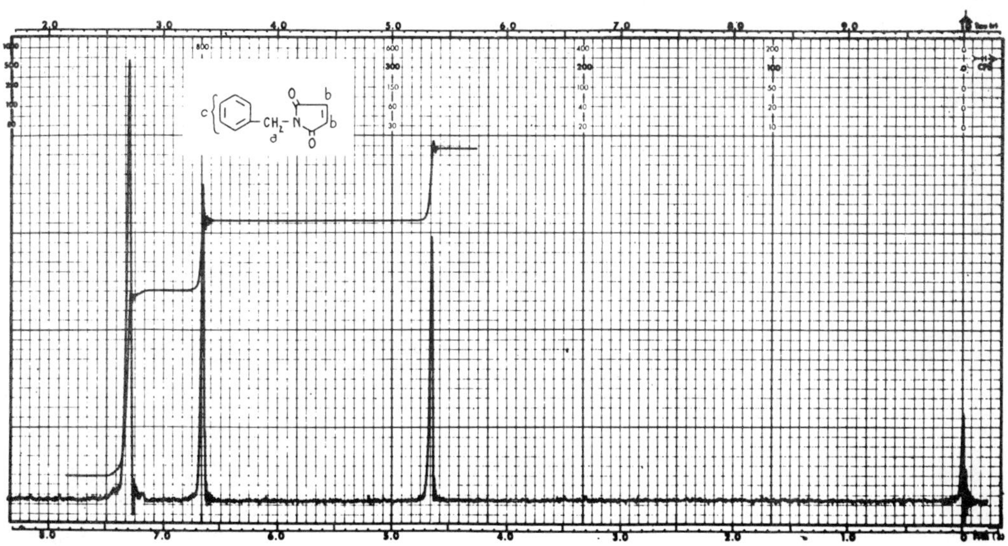

© 1978 Sadtler Research Laboratories, Inc.

N-(o-CHLOROPHENYL)MALEIMIDE

$C_{10}H_6ClNO_2$

Mol. Wt. 207.62

M.P. 74°C

Solvent: $CDCl_3$

Assignments
a = 6.71
b = 7.16-7.60

2387

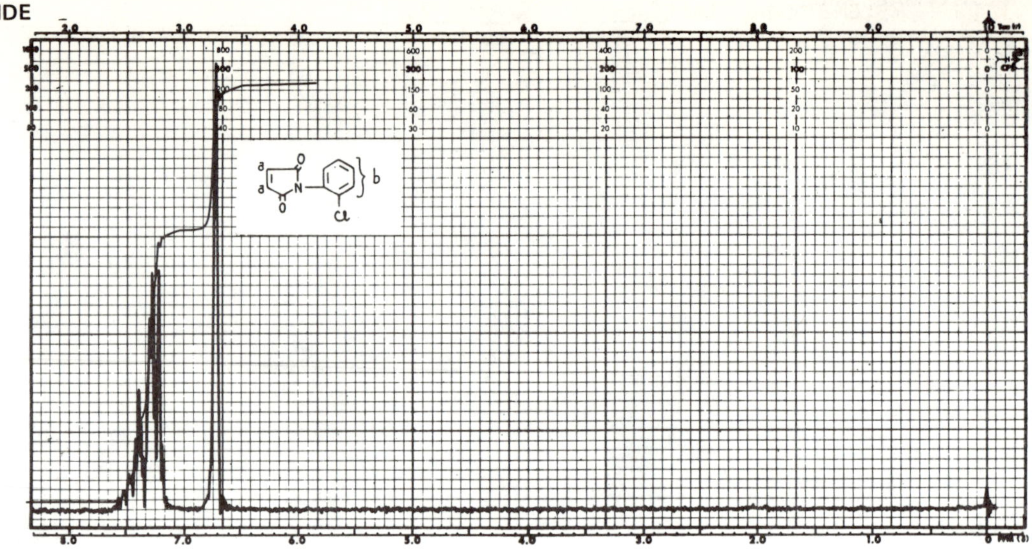

© 1978 Sadtler Research Laboratories, Inc.

N-(p-BROMOPHENYL)MALEIMIDE

$C_{10}H_6BrNO_2$

Mol. Wt. 252.07

M.P. 112°C

Solvent: Polysol-d

Assignments
a = 7.07
b = 7.31
c = 7.64

2388

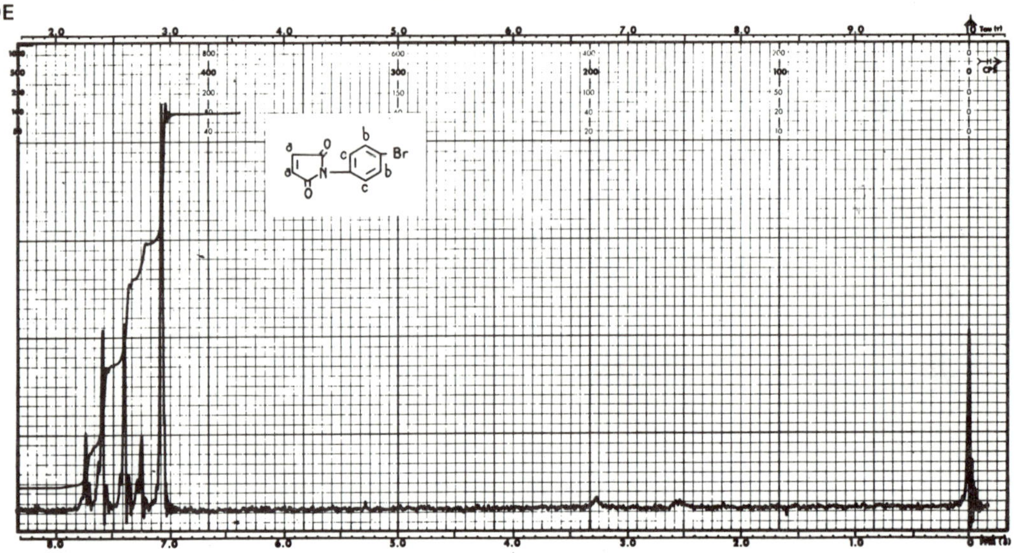

© 1978 Sadtler Research Laboratories, Inc.

N-(p-IODOPHENYL)MALEIMIDE

$C_{10}H_6INO_2$

Mol. Wt. 299.07

M.P. 150°C

Solvent: Trifluoroacetic acid

Assignments
a = 6.99
b = 7.06
c = 7.82

2389

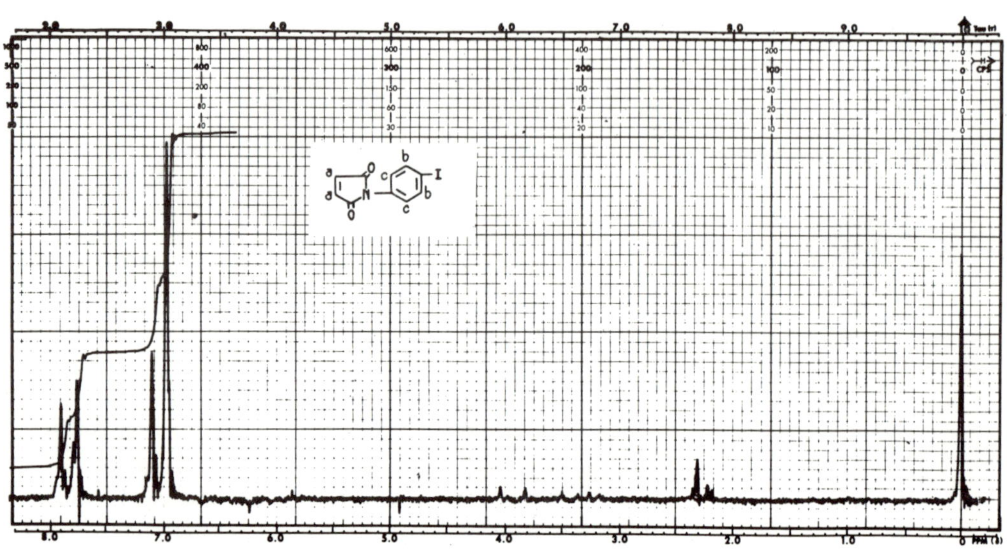

© 1978 Sadtler Research Laboratories, Inc.

N,N'-(m-PHENYLENE)DIMALEIMIDE

$C_{14}H_8N_2O_4$

Mol. Wt. 268.23

Solvent: DMSO-d_6

Assignments
a = 7.15
b = 7.25-7.70
c = 2.49 DMSO-d_5
d = 1.90, 6.08 imps.

2390

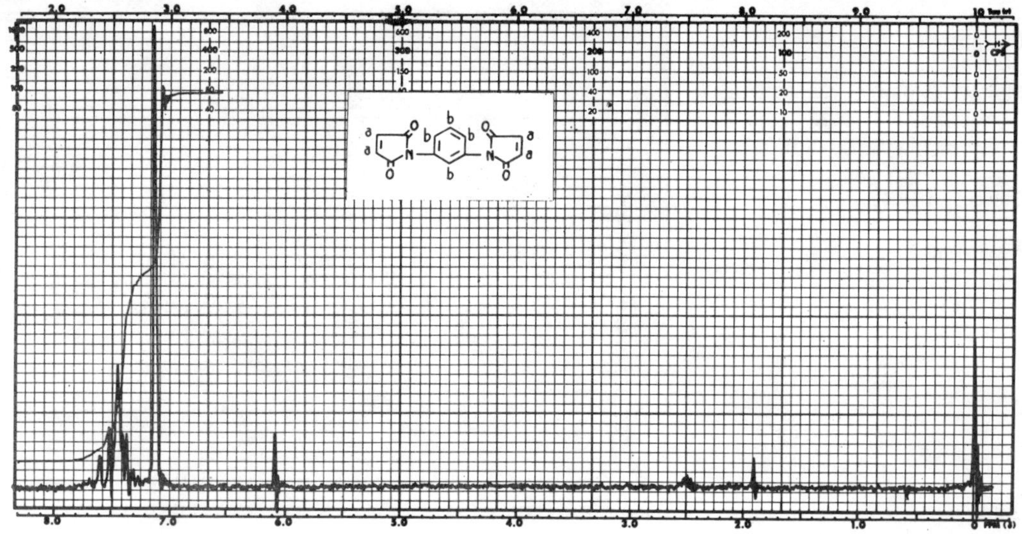

© 1978 Sadtler Research Laboratories, Inc.

N-METHYLPHTHALIMIDE

$C_9H_7NO_2$

Mol. Wt. 161.16

M.P. 132-135°C

Solvent: Trifluoroacetic acid

Assignments
a = 3.23
b = 7.84

2391

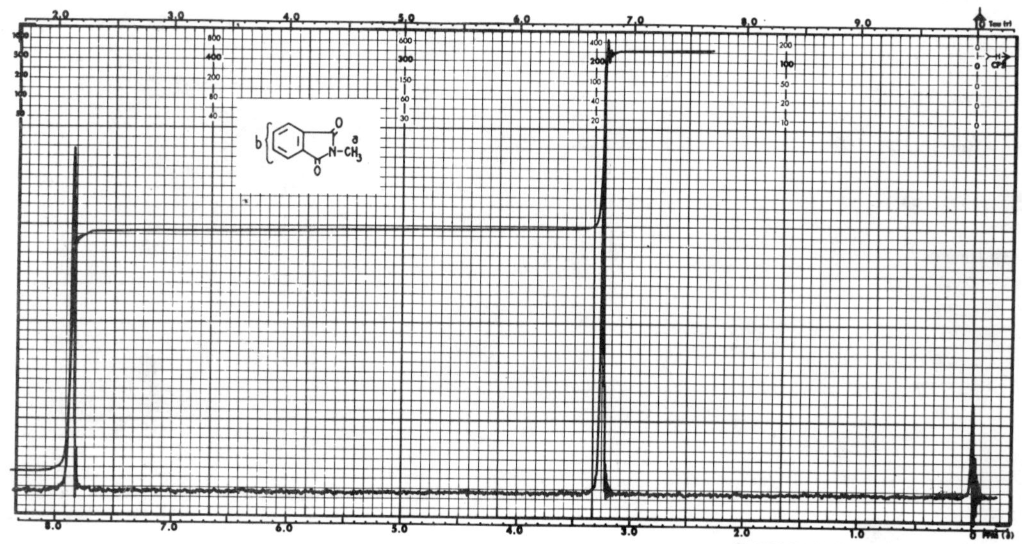

© 1978 Sadtler Research Laboratories, Inc.

N-(2-CHLOROETHYL)PHTHALIMIDE

$C_{10}H_8ClNO_2$

Mol. Wt. 209.63

Solvent: $CDCl_3$

Assignments
a = 3.80 or b
b = 4.06 or a
c = 7.76
d = 7.93

2392

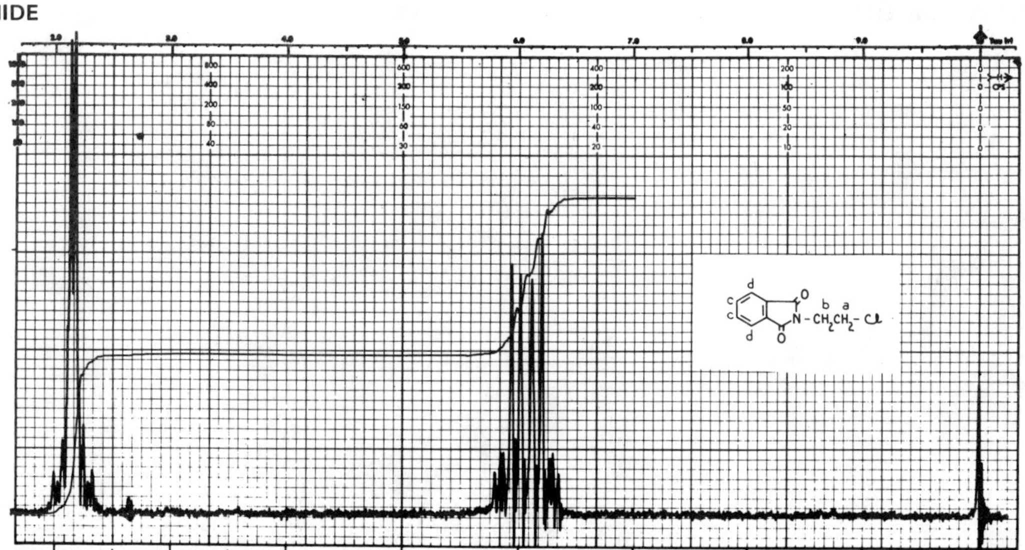

© 1978 Sadtler Research Laboratories, Inc.

N-(BROMOMETHYL)PHTHALIMIDE

$C_9H_6BrNO_2$

Mol. Wt. 240.06

M.P. 149-150°C

Solvent: $CDCl_3$

Assignments
a = 5.54
b = 7.84
c = 7.99
d = 5.30 imp.

2393

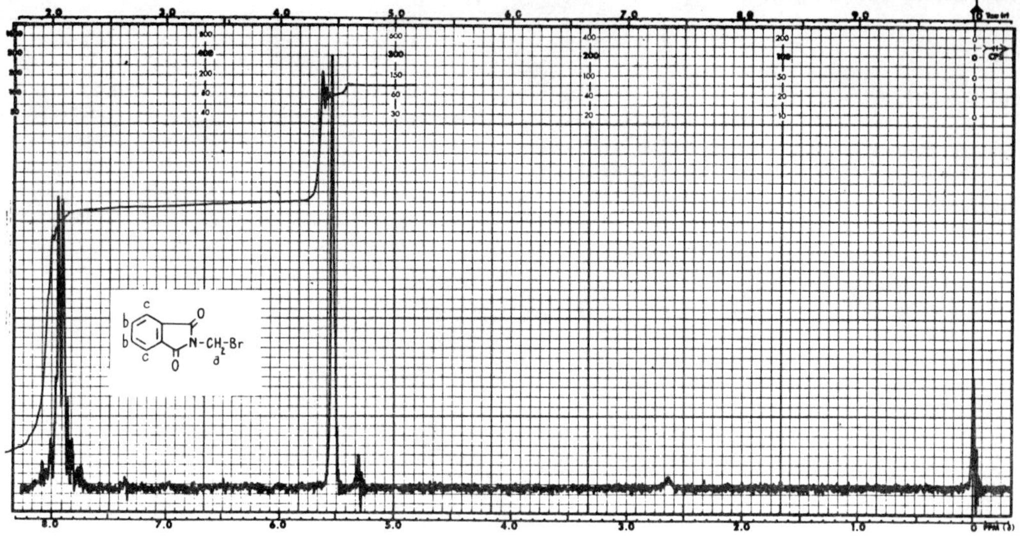

© 1978 Sadtler Research Laboratories, Inc.

N-(2,3-EPOXYPROPYL)PHTHALIMIDE

$C_{11}H_9NO_3$

Mol. Wt. 203.20

M.P. 98-100°C

Solvent: $CDCl_3$

Assignments
a = 2.58-2.95
b = 3.25
c = 3.81
d = 3.94
e = ca 7.82

2394

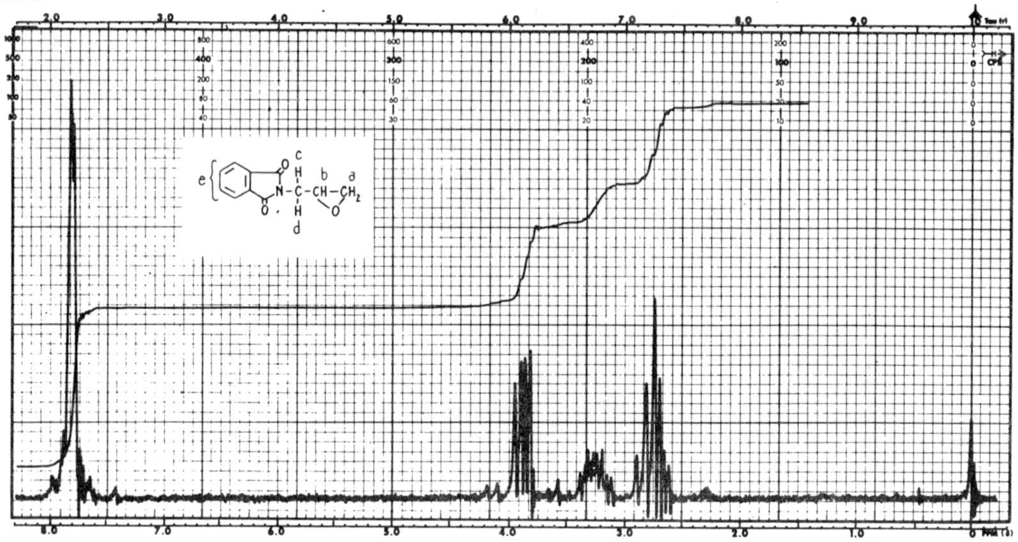

© 1978 Sadtler Research Laboratories, Inc.

N-VINYLPHTHALIMIDE

$C_{10}H_7NO_2$

Mol. Wt. 173.17

M.P. 82-86°C

Solvent: $CDCl_3$

Assignments
a = 5.01
b = 6.02
c = 6.83
d = 7.73
e = 7.82

2395

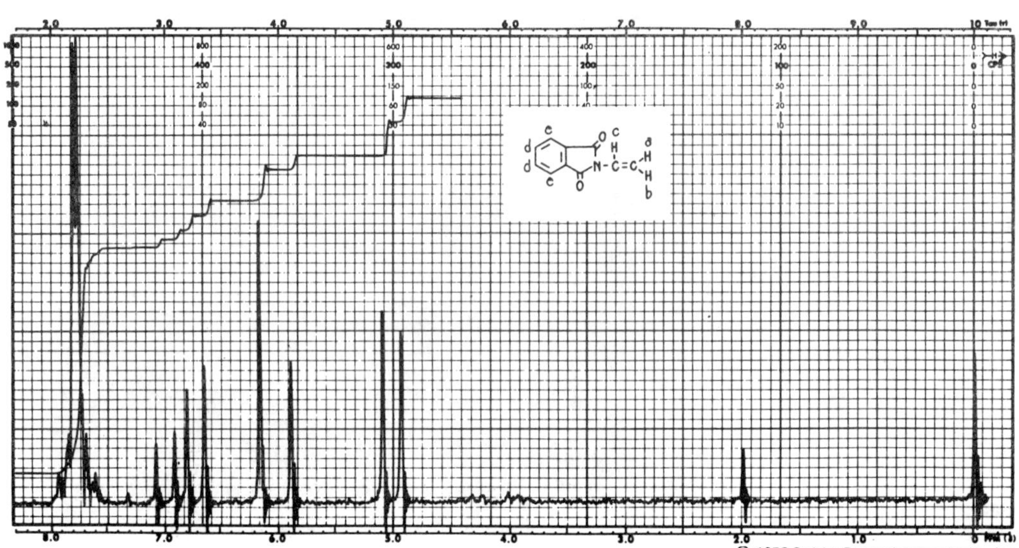

© 1978 Sadtler Research Laboratories, Inc.

N-BENZYLPHTHALIMIDE

$C_{15}H_{11}NO_2$

Mol. Wt. 237.26

Solvent: Dimethyl sulfoxide-d_6

Assignments
a = 4.78
b = 7.29
c = 7.81
d = 2.52 DMSO-d_5
e = 3.38 H_2O

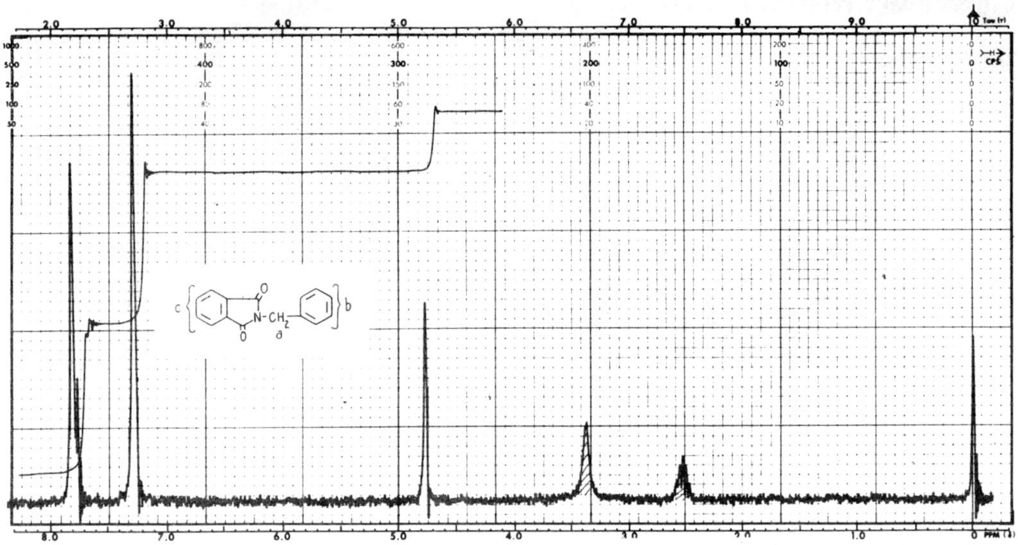

2396

© 1978 Sadtler Research Laboratories, Inc.

N-o-TOLYLPHTHALIMIDE

$C_{15}H_{11}NO_2$

Mol. Wt. 237.26

M.P. 182°C

Solvent: $CDCl_3$

Assignments
a = 2.21
b = 7.10-7.40
c = 7.74
d = 7.91
e = 1.74imp.

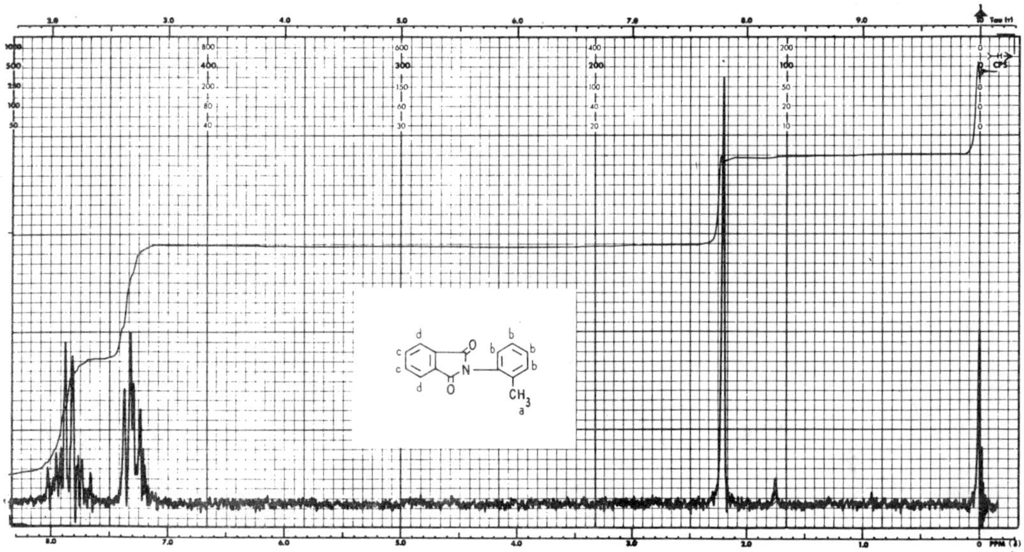

2397

© 1978 Sadtler Research Laboratories, Inc.

N-(p-METHOXYPHENYL)PHTHALIMIDE

$C_{15}H_{11}NO_3$

Mol. Wt. 253.26

M.P. 162°C

Solvent: $CDCl_3$

Assignments
a = 3.80
b = 6.97
c = 7.30
d = 7.71
e = 7.83
f = 1.83 imp.

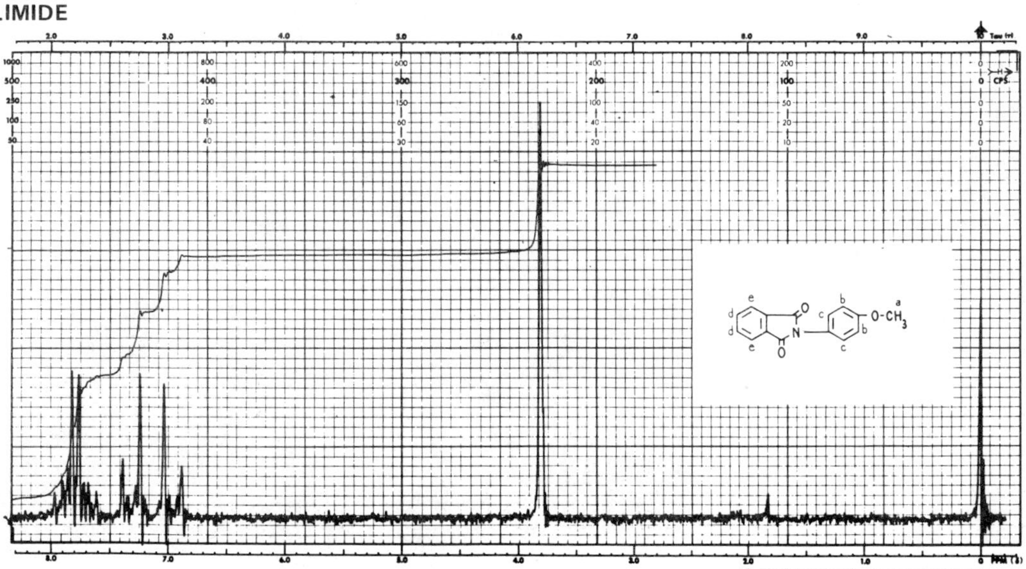

2398

© 1978 Sadtler Research Laboratories, Inc.

N,N'-HEXAMETHYLENEDIPHTHALIMIDE

$C_{22}H_{20}N_2O_4$

Mol. Wt. 376.42

M.P. 183-184°C

Solvent: $CDCl_3$

Assignments
a = 1.10-2.00
b = 3.69
c = 7.70
d = 7.79

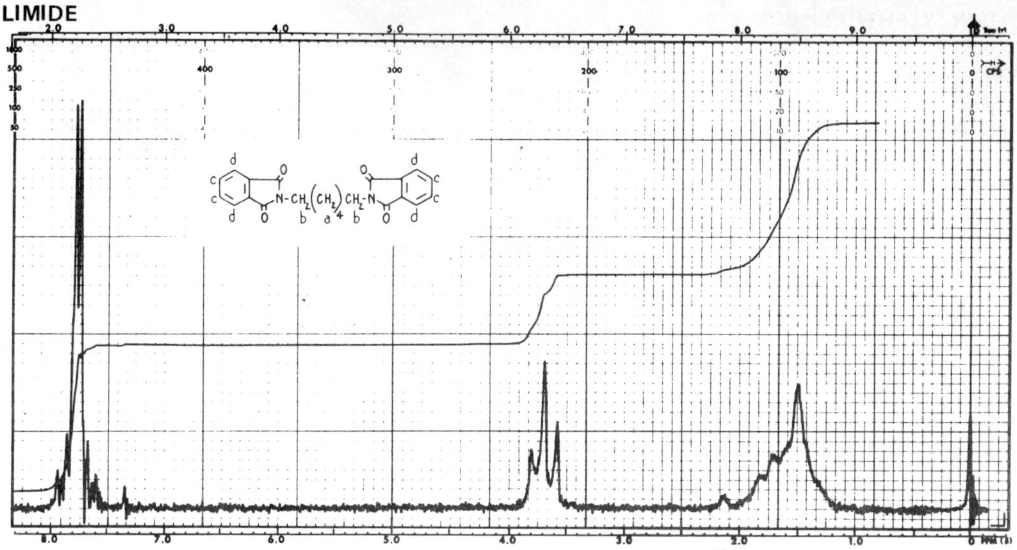

2399

© 1978 Sadtler Research Laboratories, Inc.

N,N'-[(BENZYLIMINO)DIMETHYLENE]DIPHTHALIMIDE

$C_{25}H_{19}N_3O_4$

Mol. Wt. 425.45

M.P. 187-188°C

Solvent: $CDCl_3$

Assignments
a = 4.02
b = 4.92
c = 6.90-7.30
d = 7.50-7.90
e = 1.88 imp.

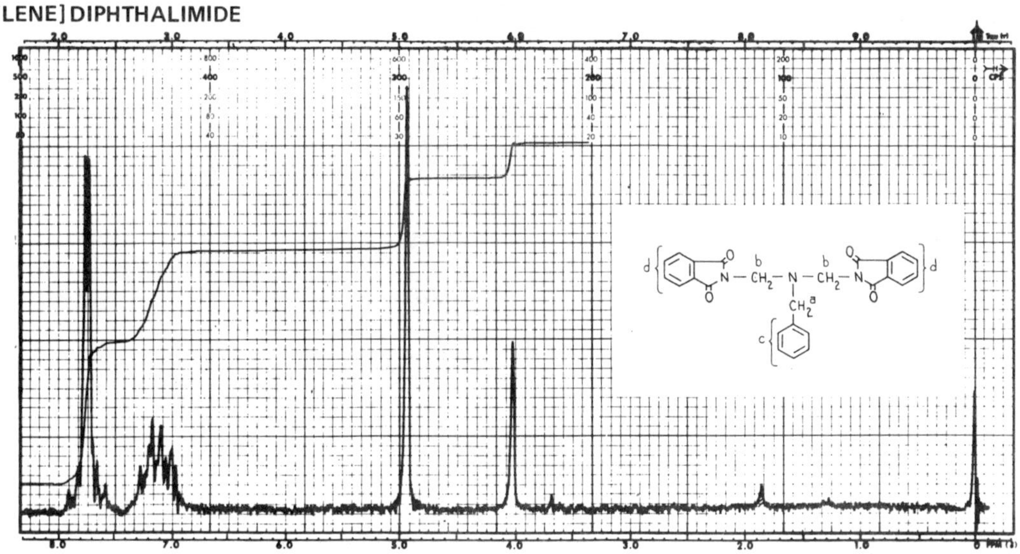

2400

© 1978 Sadtler Research Laboratories, Inc.

1,2,4,5-BENZENETETRACARBOXYLIC 1,2:4,5-DIIMIDE

$C_{10}H_4N_2O_4$

Mol. Wt. 216.15

M.P. > 320°C

Solvent: DMSO

Assignments
a = 8.07
b = 11.42

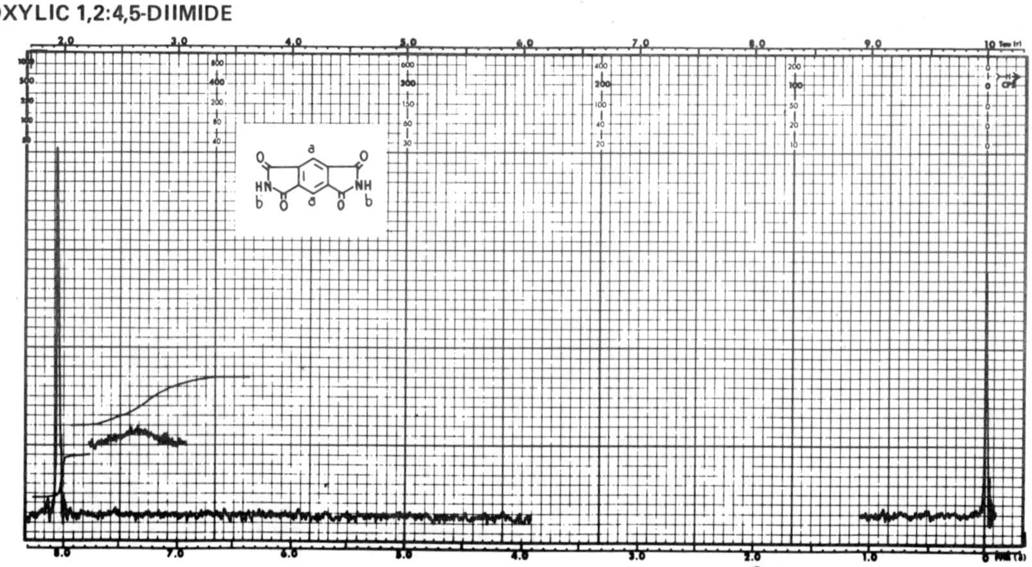

2401

© 1978 Sadtler Research Laboratories, Inc.

N-AMINOPHTHALIMIDE

$C_8H_6N_2O_2$

Mol. Wt. 162.15

Solvent: Dimethyl sulfoxide

Assignments

a = 4.93

b = ca 7.80

2402

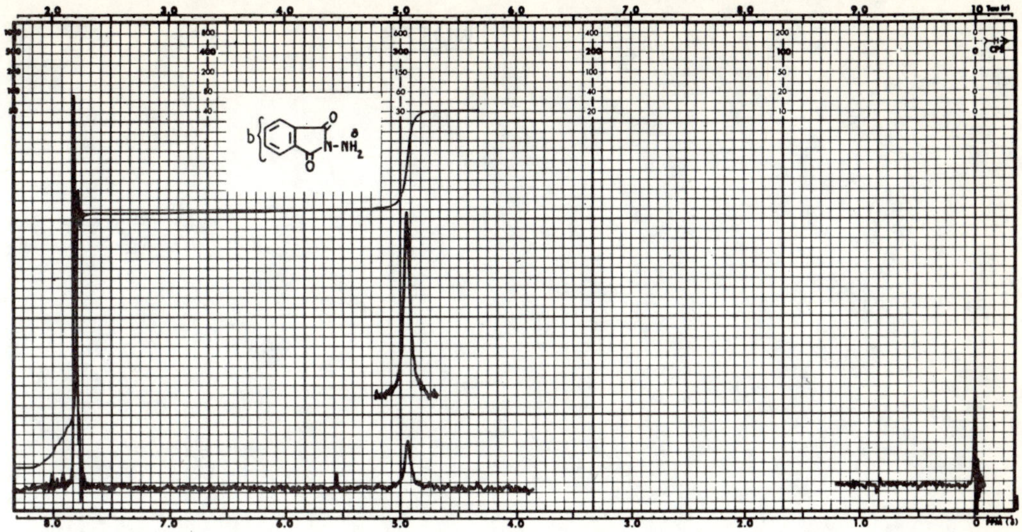

© 1978 Sadtler Research Laboratories, Inc.

NOTES

The Hydrazides are similar to the other Amide-like structures in the chemical shifts of protons bonded to carbons alpha to the carbonyl group. The most distinguishing feature of the Hydrazides is the very broad two or three proton band in the range from 3–6 ppm which represents the resonance of the $-NH_2$ exchangeable hydrogens. The NH proton may be in exchange with the NH_2 and may resonate in the same range or, if not in exchange, will appear at lower field (7–10 ppm).

Chemical Shifts

Aliphatic Protons

HC–3	HC–2	HC–1		Solvent	Spectrum
		1.93	CH_3–C(=O)–NH–NH_2	$CDCl_3$	2403
	1.11	2.21	CH_3–CH_2–C(=O)–NH–NH–Ø	Polysol	----
0.90	1.58	2.11	CH_3–CH_2–CH_2–C(=O)–NH–NH–C(=O)–R_3	DMSO-d_6	2413
		(2.66)	$(CH_3)_2$–N–NH–C(=O)–Ø	$CDCl_3$	2410

Aromatic Protons

Benzoic Acid, Hydrazides

para/meta	ortho			
7.2–7.6	7.88	Ø–C(=O)–NH–NH_2	DMSO	----
7.1–7.5	7.82	Ø–C(=O)–NH–N=C_6	$CDCl_3$	2411

Para Substituted Benzoic Acid, Hydrazides

H_2N–NH–C(=O)– ortho	meta	para		
7.72	7.23	$-CH_3$	$CDCl_3$	----
7.88	7.62	–Br	DMSO	2406
8.07	8.22	$-NO_2$	Polysol	----

Exchangeable Protons

Because the Hydrazides are soluble in solvents which may contain significant amounts of water, it is often difficult to characterize the exchangeable proton resonances with confidence. In general, it appears that the NH_2 group resonates at intermediate field and is often in exchange with any water which may be present in the solution producing an erroneous 3-hydrogen integration value. In addition, the C(=O)–NH proton may resonate at much lower field (7–9 ppm) as a very broad band which may be difficult to locate.

H_2N–	NH–C(=O)–		
4.5	7.7	–Ø (2–NH–CH_3)	DMSO-d_6
3.9	8.0	–Ø (2-naphthalene)	Polysol
4.5	9.1	–Ø (2–CH_2–OH)	DMSO-d_6

ACETIC ACID, HYDRAZIDE

$C_2H_6N_2O$

Mol. Wt. 74.08

Solvent: $CDCl_3$

Assignments
a = 1.93
b = 5.02

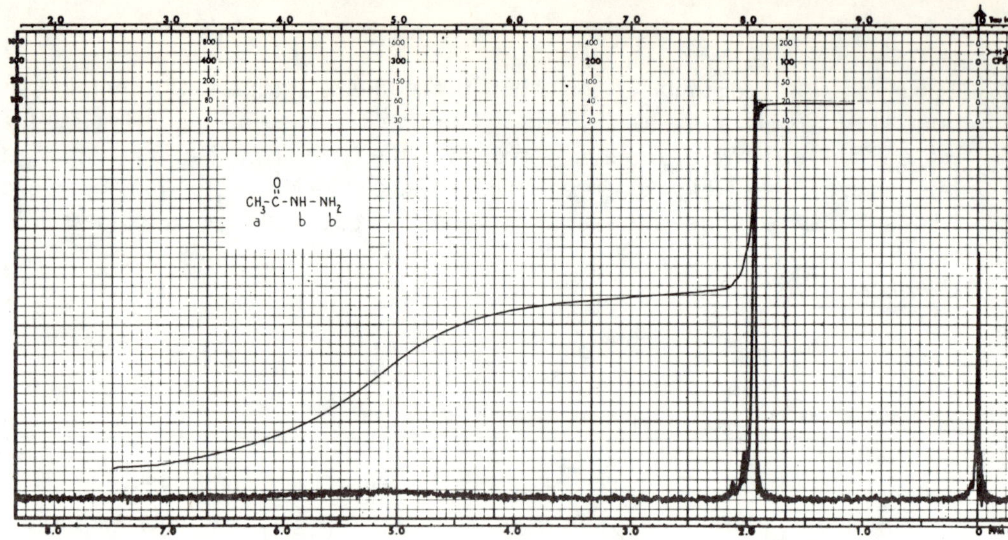

2403

© 1978 Sadtler Research Laboratories, Inc.

ADIPIC ACID, DIHYDRAZIDE

$C_6H_{14}N_4O_2$

Mol. Wt. 174.20

M.P. 181-183°C

Solvent: D_2O

Assignments
a = 1.61
b = 2.28
c = 4.69 HDO

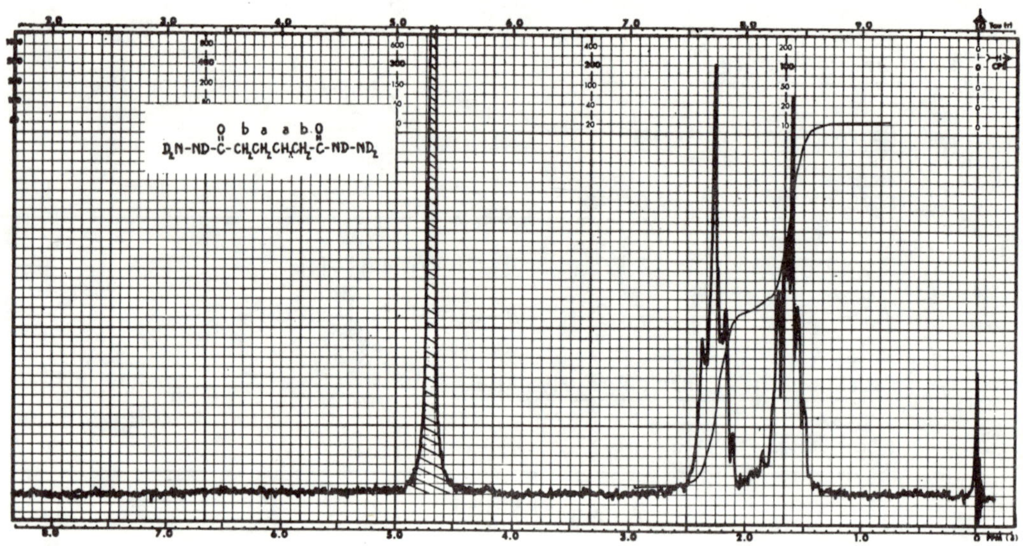

2404

© 1978 Sadtler Research Laboratories, Inc.

(p-AMINOPHENYL)ACETIC ACID, HYDRAZIDE

$C_8H_{11}N_3O$

Mol. Wt. 165.20

M.P. 160°C

Solvent: Polysol-d

Assignments
a = 3.28
b = 4.05
c = 6.51
d = 6.96
e = 8.38
f = 2.49 DMSO-d_5

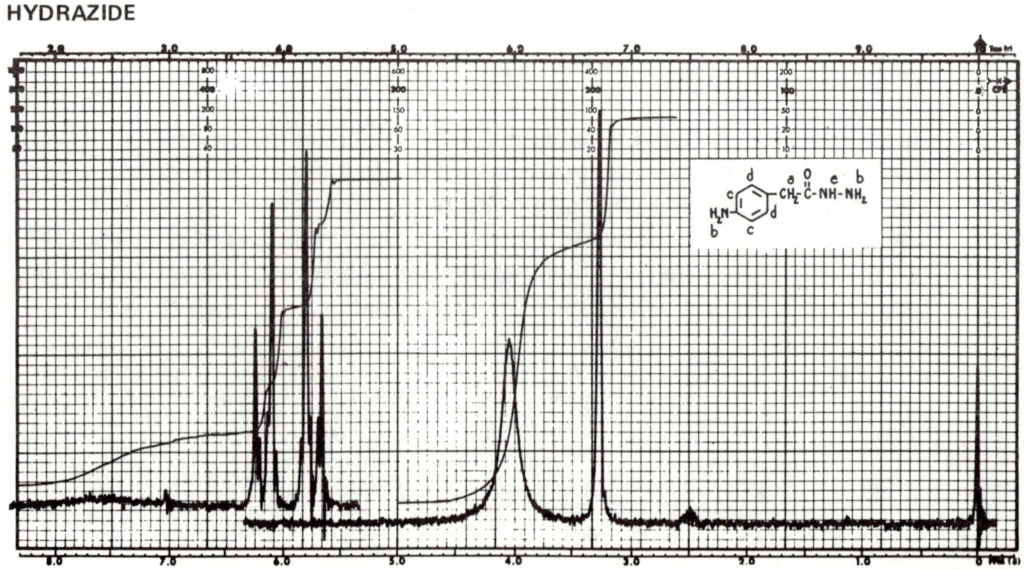

2405

© 1978 Sadtler Research Laboratories, Inc.

p-BROMOBENZOIC ACID, HYDRAZIDE

$C_7H_7BrN_2O$

Mol. Wt. 215.06

Solvent: Dimethyl sulfoxide

Assignments
a = 4.52 *
b = 7.62
c = 7.88
* H_2O present

2406

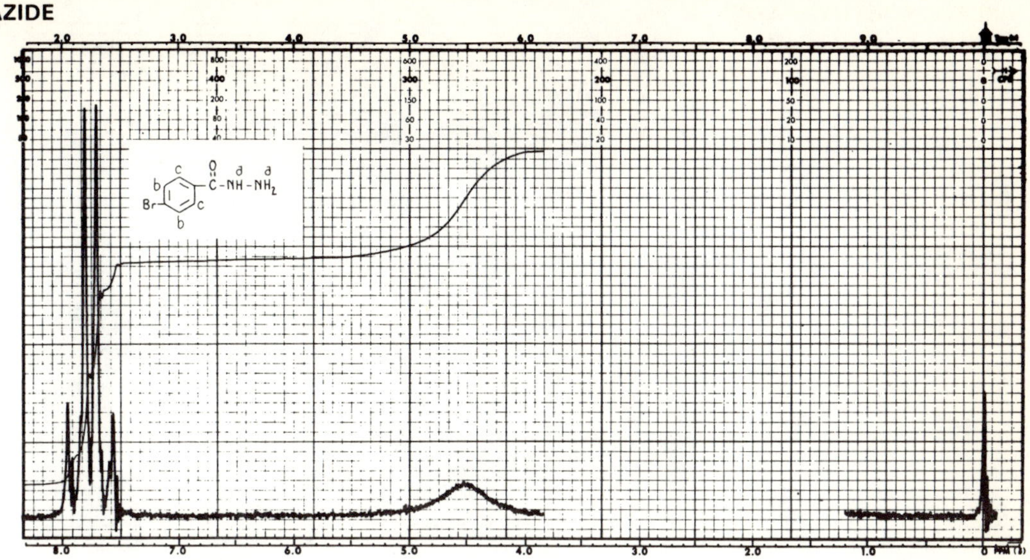

© 1978 Sadtler Research Laboratories, Inc.

o-NITROBENZOIC ACID, HYDRAZIDE

$C_7H_7N_3O_3$

Mol. Wt. 181.15

Solvent: DMSO-d_6

Assignments
a = ca 4.55
b = 7.40-7.90
c = 8.02
d = 2.49 DMSO-d_5

2407

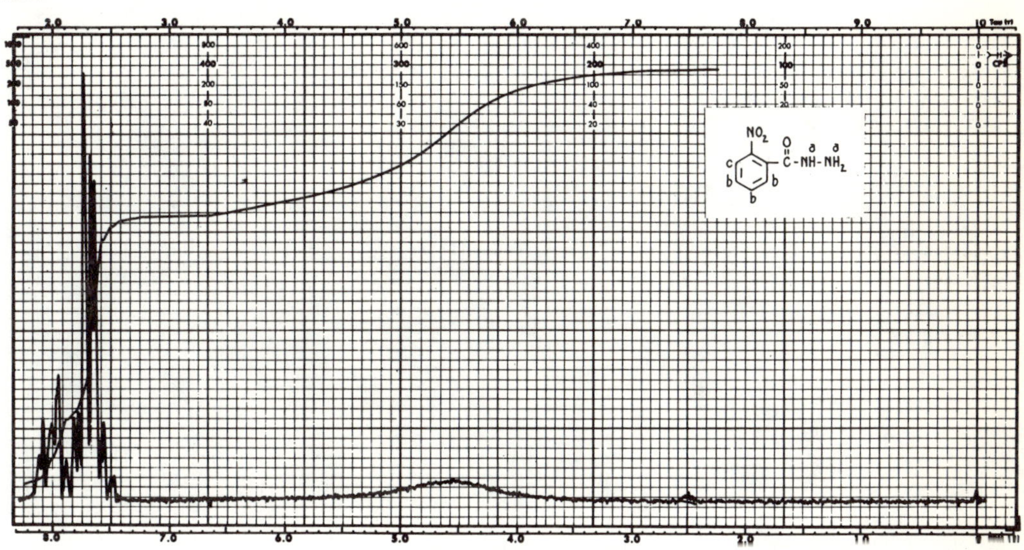

© 1978 Sadtler Research Laboratories, Inc.

PICOLINIC ACID, HYDRAZIDE

$C_6H_7N_3O$

Mol. Wt. 137.14

M.P. 98-100°C

Solvent: $CDCl_3$

Assignments
a = ca 4.60
b = 7.37
c = 7.76
d = 8.15
e = 8.53

2408

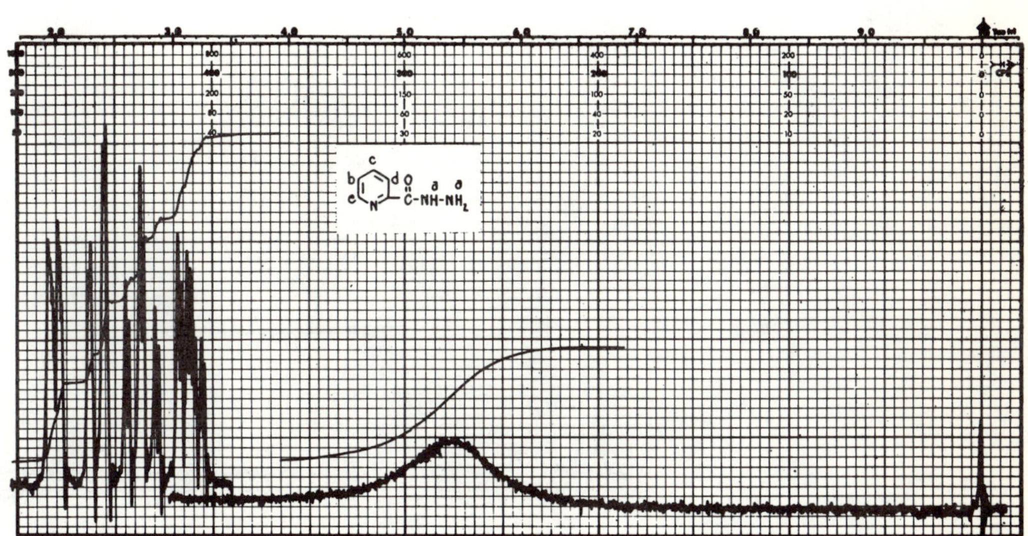

© 1978 Sadtler Research Laboratories, Inc.

NICOTINIC ACID, HYDRAZIDE

$C_6H_7N_3O$

Mol. Wt. 137.14

M.P. 160-163°C

Solvent: DMSO-D6

Assignments
a = 4.51
b = 7.46
c = 8.18
d = 8.70
e = 9.04
f = 2.53 DMSO-D_5
-NH is unobserved

2409

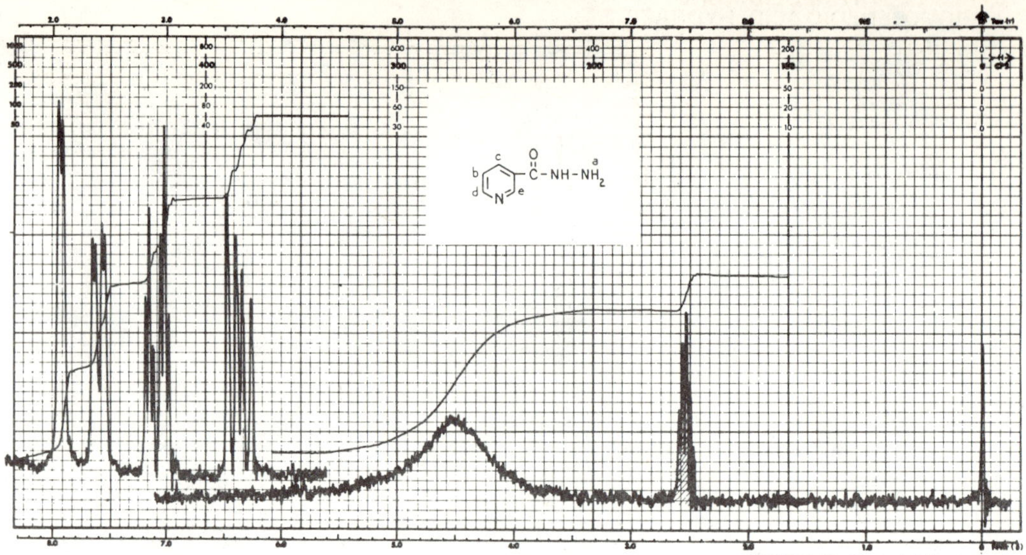

© 1978 Sadtler Research Laboratories, Inc.

p-CHLOROBENZOIC ACID, 2,2-DIMETHYLHYDRAZIDE

$C_9H_{11}ClN_2O$

Mol. Wt. 198.65

M.P. 132-134°C

Solvent: $CDCl_3$

Assignments
a = 2.66
b = 6.50-8.50
c = 7.30
d = 7.69

2410

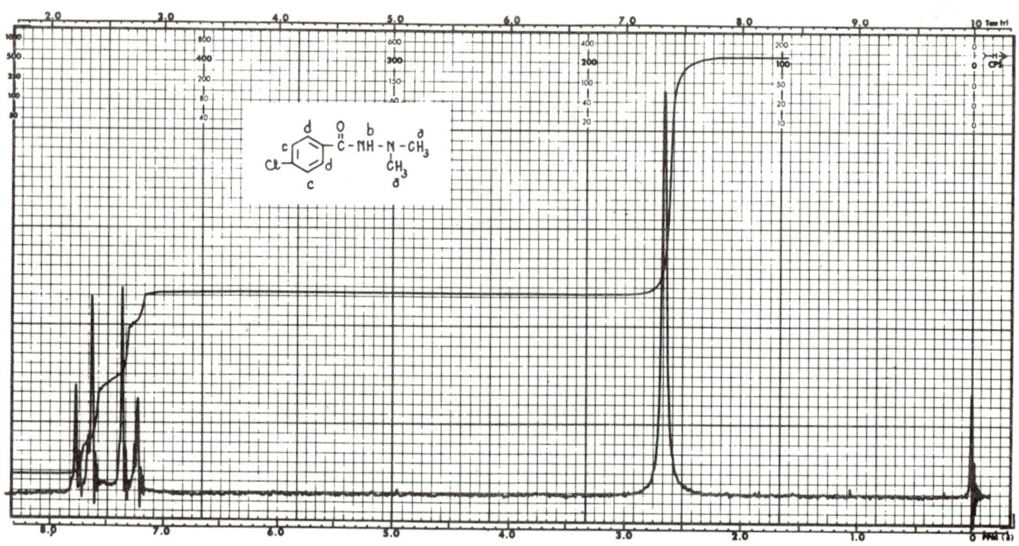

© 1978 Sadtler Research Laboratories, Inc.

BENZOIC ACID, CYCLOHEXYLIDENEHYDRAZIDE

$C_{13}H_{16}N_2O$

Mol. Wt. 216.29

M.P. 163-165°C

Solvent: $CDCl_3$

Assignments
a = ca 1.60
b = ca 2.38
c = 7.10-7.50
d = 7.82
e = 9.55

2411

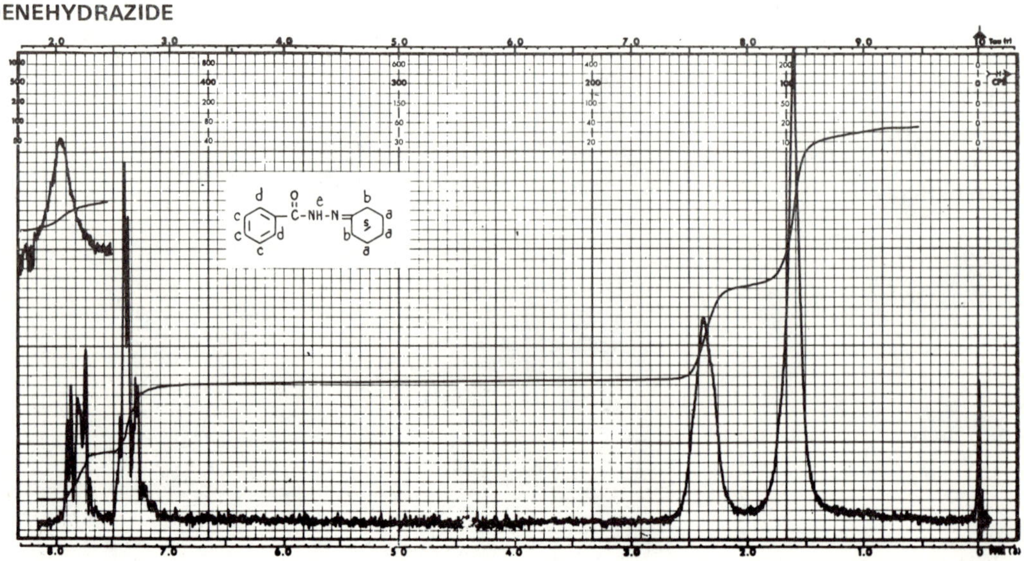

© 1978 Sadtler Research Laboratories, Inc.

5-PHENYL-3-PYRAZOLIDINONE

$C_9H_{10}N_2O$

Mol. Wt. 162.19

M.P. 104 -105°C

Solvent: Dimethyl sulfoxide-d_6

Assignments

a = 2.39 or b
b = 2.79 or a
c = 3.48
d = 4.62
e = 5.58
f = 7.40

2412

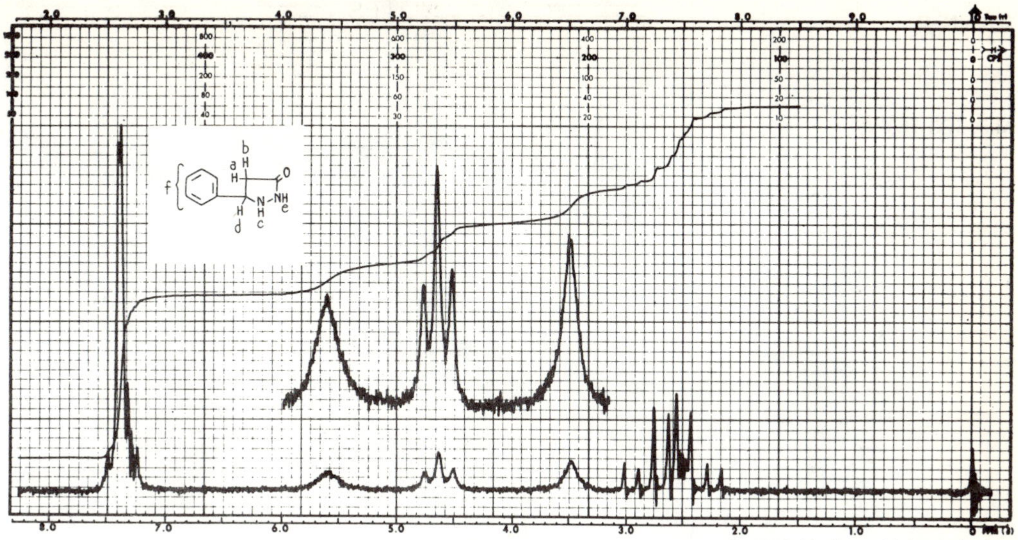

© 1978 Sadtler Research Laboratories, Inc.

1,2-DIBUTYRYLHYDRAZINE

$C_8H_{16}N_2O_2$

Mol. Wt. 172.23

M.P. 163-164°C

Solvent: Dimethyl sulfoxide-d_6

Assignments

a = 0.90
b = 1.58
c = 2.11
d = 9.60
e = 2.52 DMSO-d_5
f = 3.34 H_2O

2413

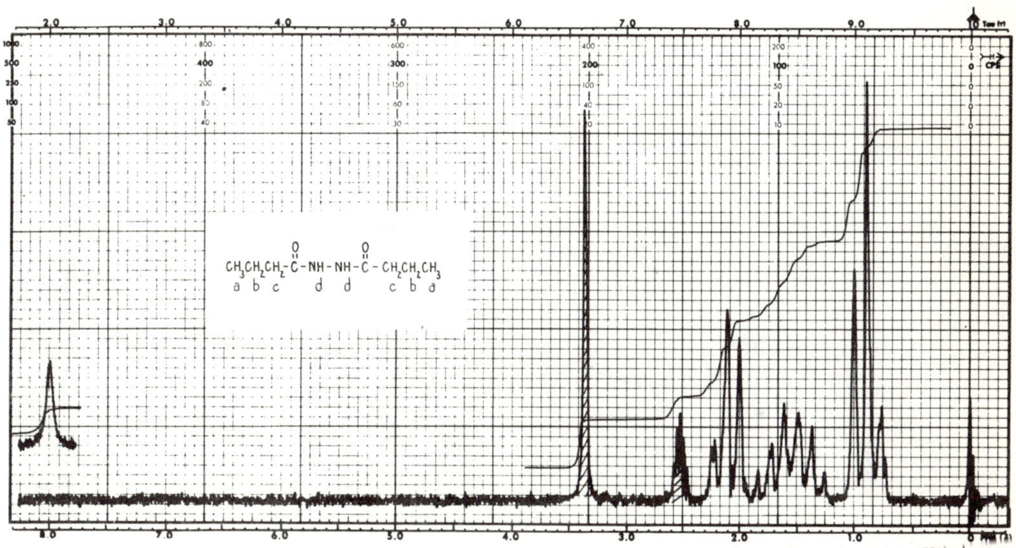

© 1978 Sadtler Research Laboratories, Inc.

1,2-DIBENZOYL-1,2-DIMETHYLHYDRAZINE

$C_{16}H_{16}N_2O$

Mol. Wt. 252.32

Solvent: $CDCl_3$

Assignments

a = 3.14
b = 7.00-7.60

2414

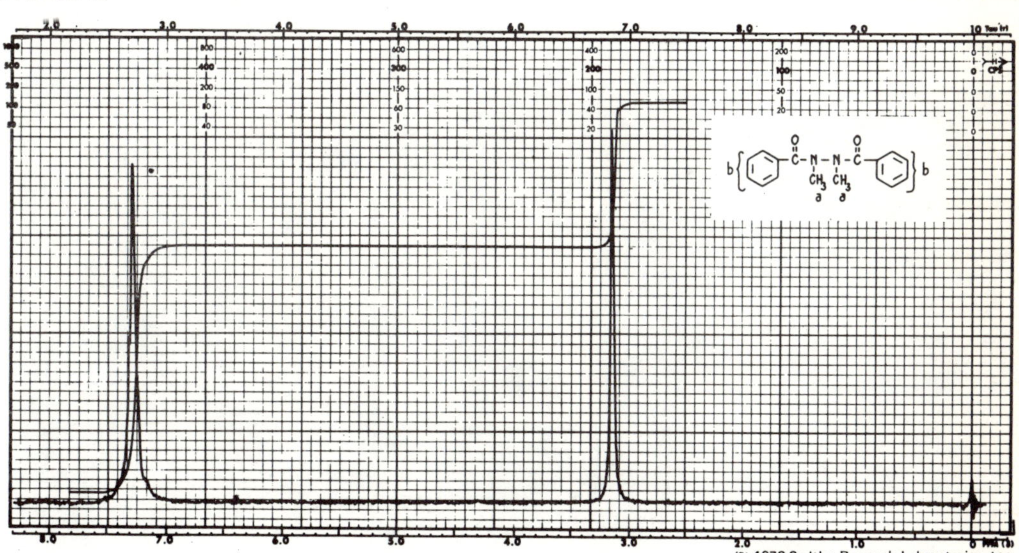

© 1978 Sadtler Research Laboratories, Inc.

NOTES

THE UREAS

The monosubstituted aliphatic Ureas characteristically display separate resonance bands for the two different types of NH protons. Coupling is usually observed between the NH and the protons of the adjacent hydrocarbon group (NH—CH_2). The chemical shift of aliphatic groups adjacent to the Urea nitrogen nucleus varies with the degree of substitution of the Urea moiety as well as the other types of substituents in the molecule.

Chemical Shifts

Aliphatic Protons

HC—4	HC—3	HC—2	HC—1		Solvent	Spectrum
			2.70	CH_3—NH—C(=O)—NH_2	D_2O	2416
			(2.75)	$(CH_3)_2$—N—C(=O)—N$(CH_3)_2$	CCl_4	2437
			2.78	CH_3—NH—C(=O)—N$(Ø)_2$	$CDCl_3$	2436
			(2.91)	$(CH_3)_2$—N—C(=O)—NH_2	$CDCl_3$	2426
			(2.97)	$(CH_3)_2$—N—C(=O)—NH—Ø	$CDCl_3$	2435
			3.16	CH_3—N(Ø)—C(=O)—N(Ø)—CH_3	$CDCl_3$	2440
		1.11	3.18	CH_3—CH_2—NH—C(=O)—NH_2	$CDCl_3$	- - - -
		1.22	3.50	CH_3—CH_2—NH—C(=O)—NH—R_2	$CDCl_3$	2428
		(1.09)	3.72	$(CH_3)_2$—CH—NH—C(=O)—NH_2	Polysol	2417
0.96	1.35	1.50	3.15	CH_3—CH_2—CH_2—CH_2—NH—C(=O)—NH—R_4	$CDCl_3$	2429
		(1.29)	- - - -	$(CH_3)_3$—C—NH—C(=O)—NH_2	Polysol	- - - -
	(0.90)	1.69	2.91	$(CH_3)_2$—CH—CH_2—NH—C(=O)—NH_2	Polysol	2418

Aromatic Protons

The 1-phenyl Ureas display an unusual type of aromatic shielding, i.e. the para proton is shielded while the ortho protons are deshielded (see table below). This shielding is not observed when two phenyl groups are bonded to the same nitrogen atom. In this case, a single peak is often observed near 7.3 ppm.

Phenyl Substituted Ureas

para	meta	ortho			
7.02	7.32	7.58	Ø—NH—C(=O)—NH—Ø	DMSO	- - - -
	6.8—7.5		Ø—NH—C(=O)—N$(CH_3)_2$	$CDCl_3$	2435
	ca. 7.25		$(Ø)_2$—N—C(=O—NH—CH_3	$CDCl_3$	2436
	ca. 7.27		$(Ø)_2$—N—C(=O)—NH_2	Polysol	2427

Exchangeable Protons

Substitution of a Urea molecule deshields the adjacent NH proton in comparison to the unsubstituted NH_2 group in the same molecule. A series of monosubstituted Ureas are presented below.

$H_2N-C(=O)$	$-NH$	$-X$	Solvent	Spectrum
5.21	5.63	$-C(CH_3)_2-R_2$	DMSO-d_6	2419
5.21	5.74	$-CH(CH_3)_2$	Polysol	2417
5.39	5.92	–Cyclohexane	Polysol	2421
5.22	6.00	$-CH_2-CH(CH_3)_2$	Polysol	2418
5.29	6.35	$-CH_2-Ø$	Polysol	2422
6.79	6.79	–H	Polysol	2415
5.53	8.19	$-Ø$ $(4-O-CH_3)$	Polysol	2425
5.79	8.39	$-Ø$ $(4-CH_3)$	DMSO-d_6	2423
5.88	8.62	$-Ø$ (4–Br)	DMSO-d_6	2424

Coupling and Coupling Constants

Coupling is usually observed between the NH proton and the hydrogens of adjacent hydrocarbon groups. The coupling constant varies over the range from 6–8 Hz (see spectra 2418, 2421, 2422, 2429 and 2430).

It is interesting to note that the type of non-equivalence so common in the spectra of N,N-dimethyl amides is not observed in the NMR spectra of the N,N-dimethyl ureas (spectra 2426, 2435 and 2437).

Solubility and Solvent Effects

The solubility of the Ureas varies primarily with the degree of substitution and the type of substituent(s). Urea and its monosubstituted derivatives are generally more soluble in solvents such as D_2O, DMSO-d_6 and Polysol. The trisubstituted and tetrasubstituted compounds, as well as those with large hydrocarbon groups tend to be more soluble in carbon tetrachloride or deuterochloroform.

There does not appear to be any unusual solvent effects when CCl_4, $CDCl_3$, DMSO-d_6 and Polysol are used. As with most compounds, D_2O exchanges with the labile hydrogens in the molecule resulting in the loss of their resonance bands, and the use of trifluoroacetic acid produces unusually low field chemical shifts.

UREA

CH_4N_2O

Mol. Wt. 60.06

M.P. 132-134°C

Solvent: Polysol-d

Assignments
a = 6.79

2415

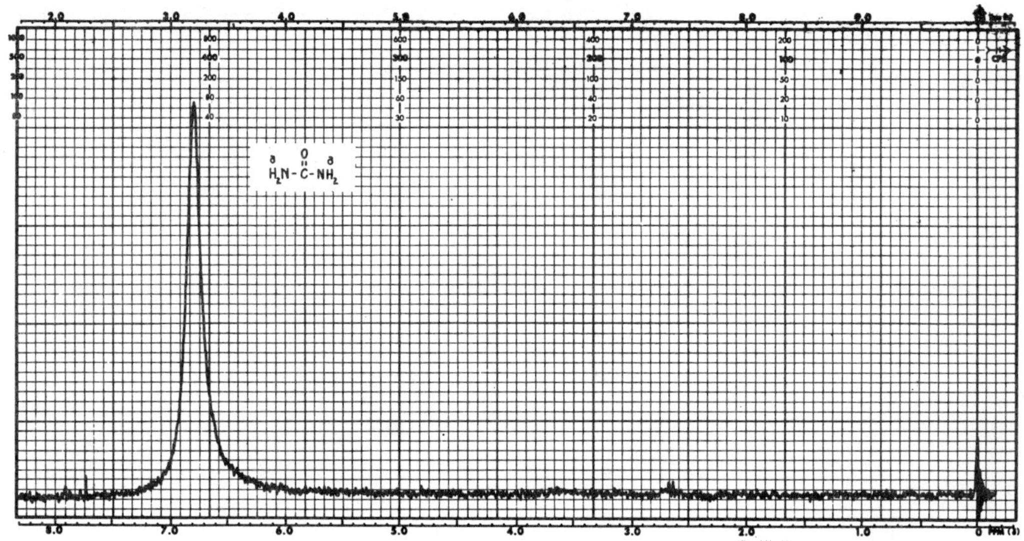

© 1978 Sadtler Research Laboratories, Inc.

METHYLUREA

$C_2H_6N_2O$

Mol. Wt. 74.08

M.P. 99-101°C

Solvent: D_2O

Assignments
a = 2.70
b = 4.67 HDO

2416

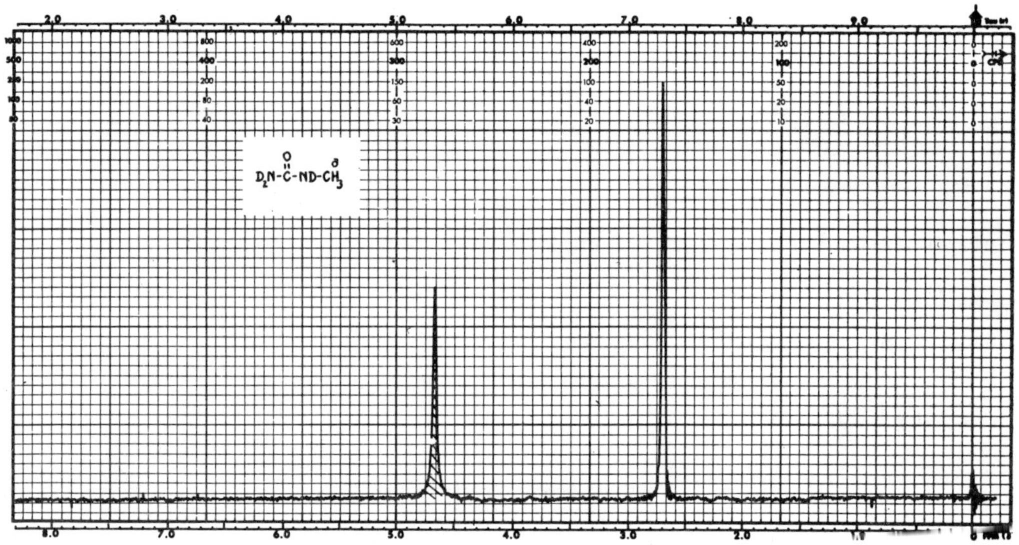

© 1978 Sadtler Research Laboratories, Inc.

ISOPROPYLUREA

$C_4H_{10}N_2O$

Mol. Wt. 102.14

M.P. 154-156°C

Solvent: Polysol-d

Assignments
a = 1.09
b = 3.72
c = ca 5.21
d = 5.74

2417

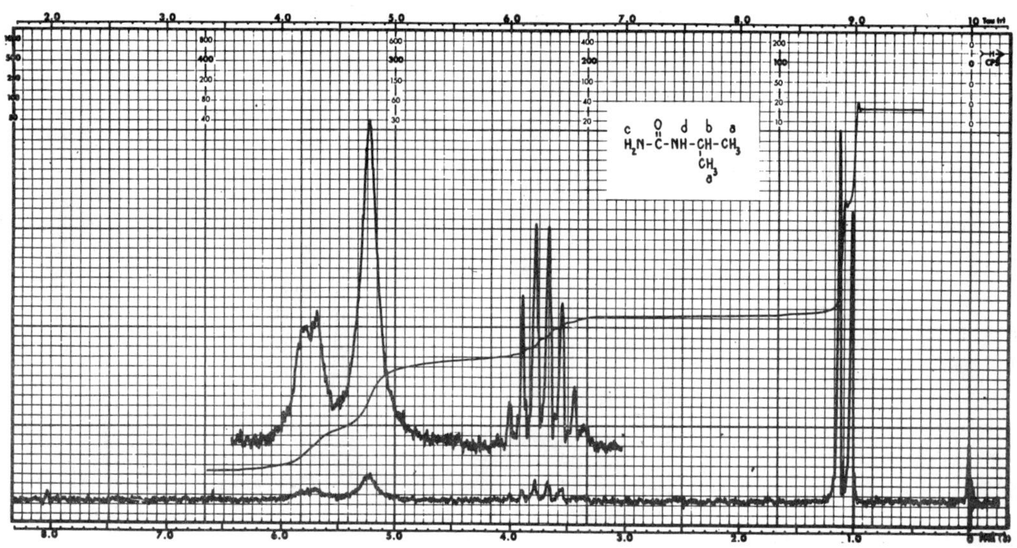

© 1978 Sadtler Research Laboratories, Inc.

ISOBUTYLUREA

$C_5H_{12}N_2O$

Mol. Wt. 116.16

Solvent: Polysol-d

Assignments
a = 0.90
b = 1.69
c = 2.91
d = 5.22
e = 6.00
f = 7.86 imp.

2418

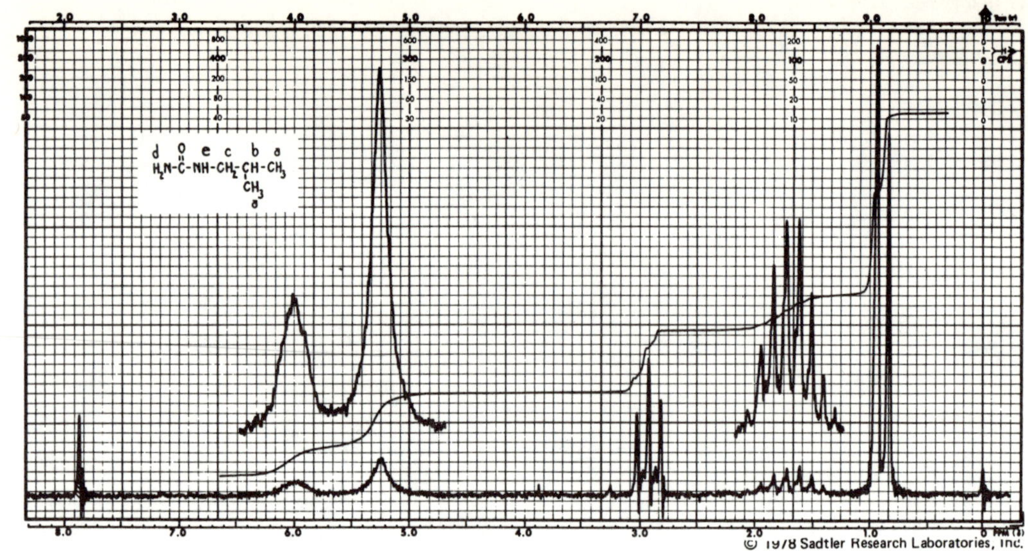

© 1978 Sadtler Research Laboratories, Inc.

tert-PENTYLUREA

$C_6H_{14}N_2O$

Mol. Wt. 130.19

M.P. 152-155°C

Solvent: Dimethyl sulfoxide-d_6

Assignments
a = 0.81
b = 1.17
c = 1.61
d = 5.21
e = 5.63
f = 2.52 DMSO-d_5
g = 3.38 H_2O

2419

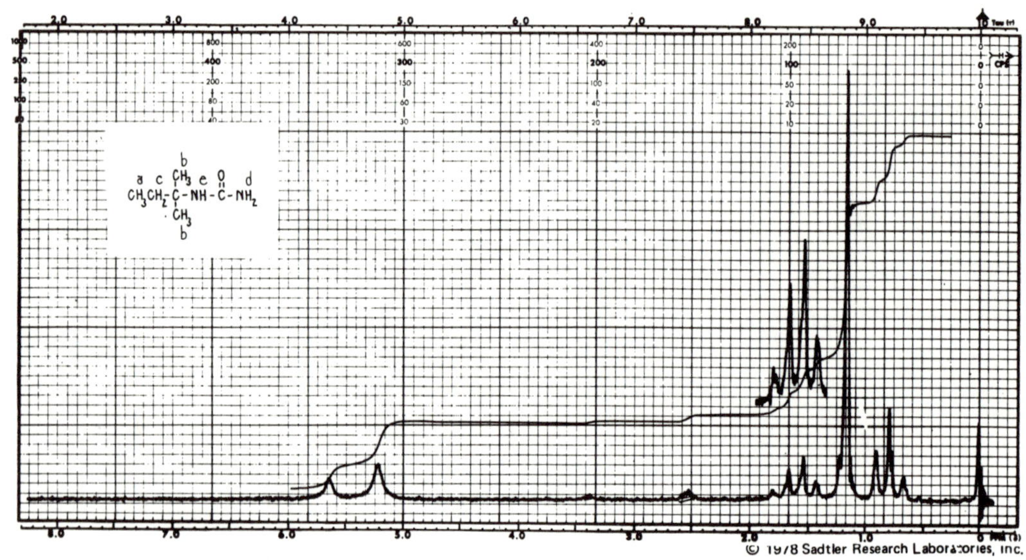

© 1978 Sadtler Research Laboratories, Inc.

1,1-ETHYLENEDIUREA

$C_4H_{10}N_4O_2$

Mol. Wt. 146.15

M.P. 190-191°C

Solvent: Dimethyl sulfoxide-d_6

Assignments
a = 3.04
b = 5.48
c = 6.01
d = 2.54 DMSO-d_5
e = 3.40 H_2O

2420

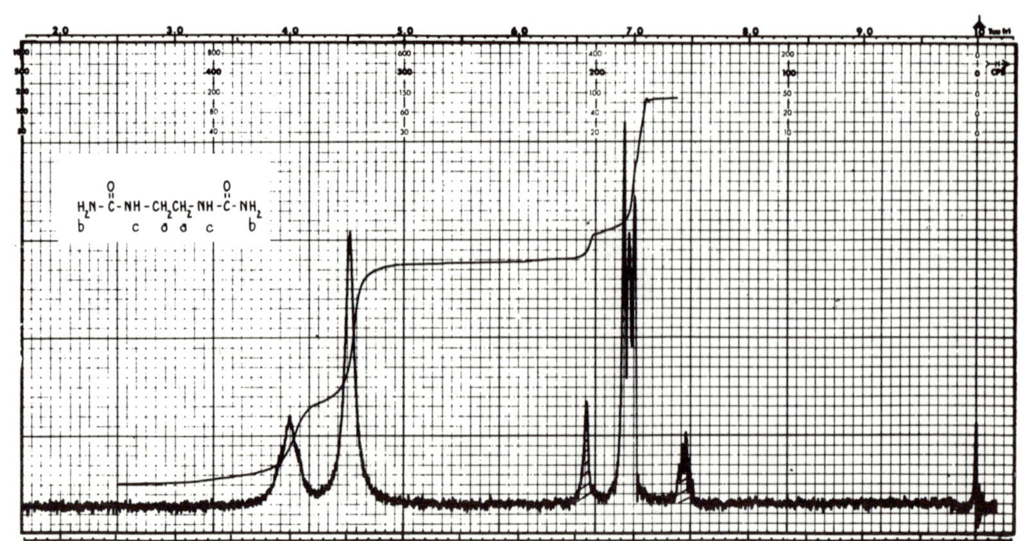

© 1978 Sadtler Research Laboratories, Inc.

CYCLOHEXYLUREA

$C_7H_{14}N_2O$

Mol. Wt. 142.20

Solvent: Polysol-d

Assignments
a = 0.70-2.10
b = 3.36
c = 5.39
d = 5.92
e = 2.53 DMSO-d_5

2421

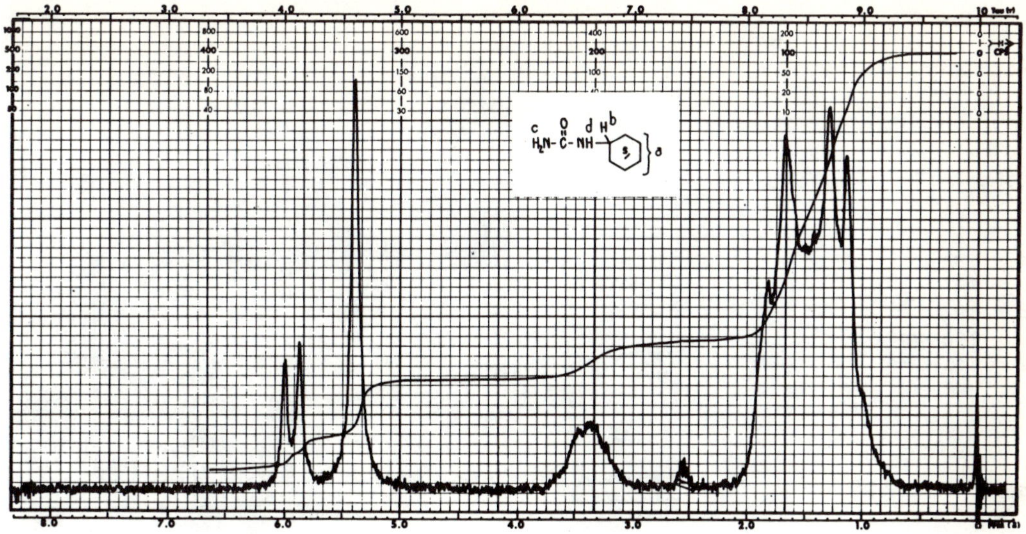

© 1978 Sadtler Research Laboratories, Inc.

BENZYLUREA

$C_8H_{10}N_2O$

Mol. Wt. 150.18

Solvent: Polysol-d

Assignments
a = 4.25
b = 5.29
c = 6.35
d = ca 7.25

2422

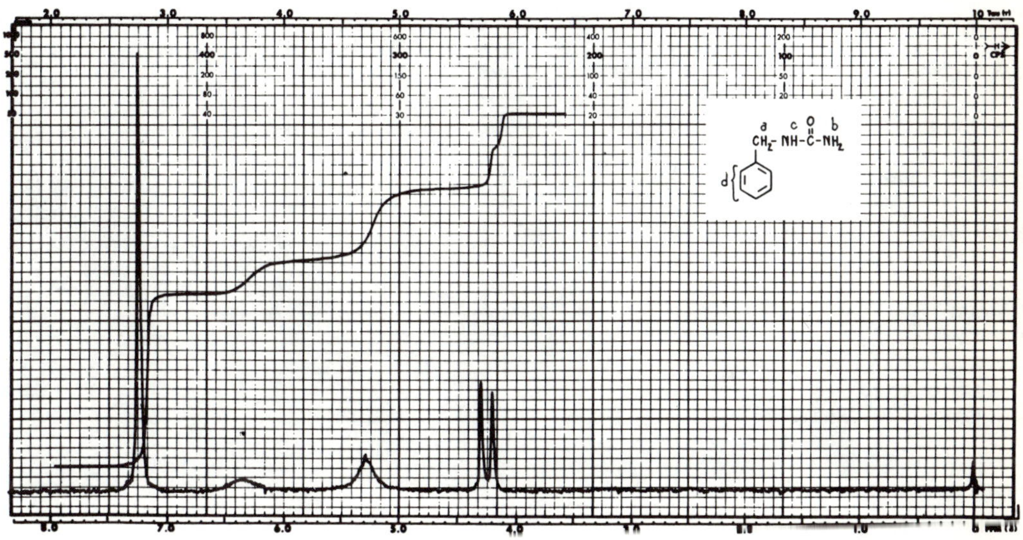

© 1978 Sadtler Research Laboratories, Inc.

---

p-TOLYLUREA

$C_8H_{10}N_2O$

Mol. Wt. 150.18

M.P. 185°C

Solvent: Dimethyl sulfoxide-d_6

Assignments
a = 2.21
b = 5.79
c = 7.01
d = 7.29
e = 8.39
f = 2.51 DMSO-d_5
g = 3.56 H_2O

2423

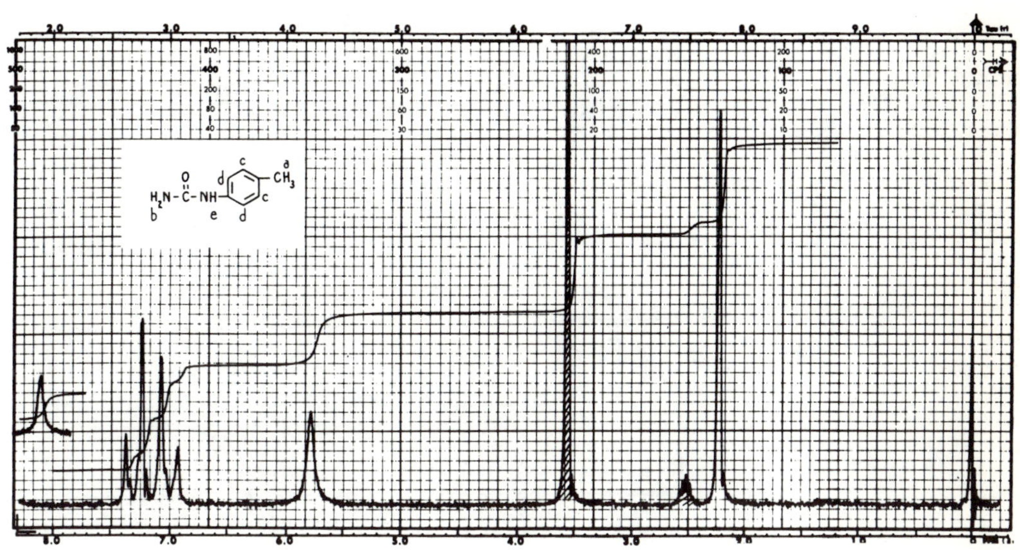

© 1978 Sadtler Research Laboratories, Inc.

(p-BROMOPHENYL)UREA

$C_7H_7BrN_2O$

Mol. Wt. 215.05

M.P. > 260°C (dec.)

Solvent: Dimethyl sulfoxide-d6

Assignments
a = 5.88
b = 7.37
c = 8.62
d = 2.50 DMSO-D_5
e = 3.38 H_2O

2424

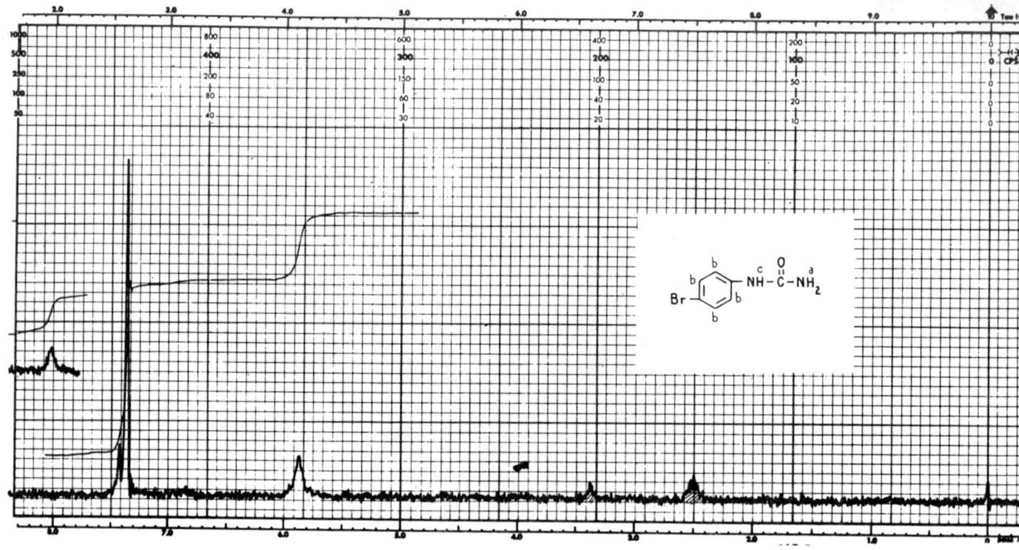

© 1978 Sadtler Research Laboratories, Inc.

(p-METHOXYPHENYL)UREA

$C_8H_{10}N_2O_2$

Mol. Wt. 166.18

M.P. 169-171°C

Solvent: Polysol-d

Assignments
a = 3.71
b = 5.53
c = 6.77
d = 7.29
e = 8.19
f = 2.52 DMSO-d_5
g = 3.30 H_2O

2425

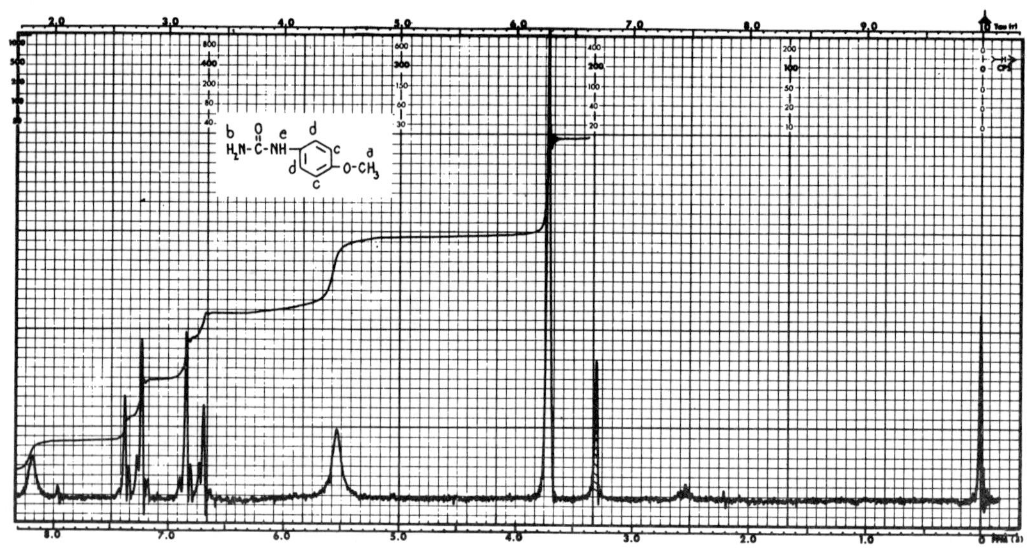

© 1978 Sadtler Research Laboratories, Inc.

1,1-DIMETHYLUREA

$C_3H_8N_2O$

Mol. Wt. 88.11

Solvent: $CDCl_3$

Assignments
a = 2.91
b = 4.38

2426

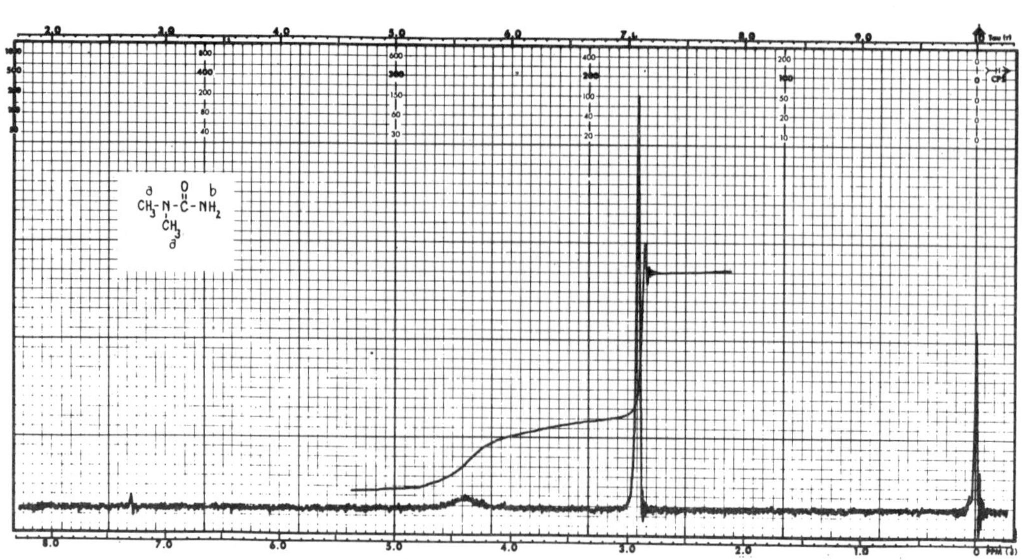

© 1978 Sadtler Research Laboratories, Inc.

1,1-DIPHENYLUREA

$C_{13}H_{12}N_2O$

Mol. Wt. 212.25

M.P. 188°C (dec.)

Solvent: Polysol-d

Assignments
a = ca 7.27
b = unobserved

2427

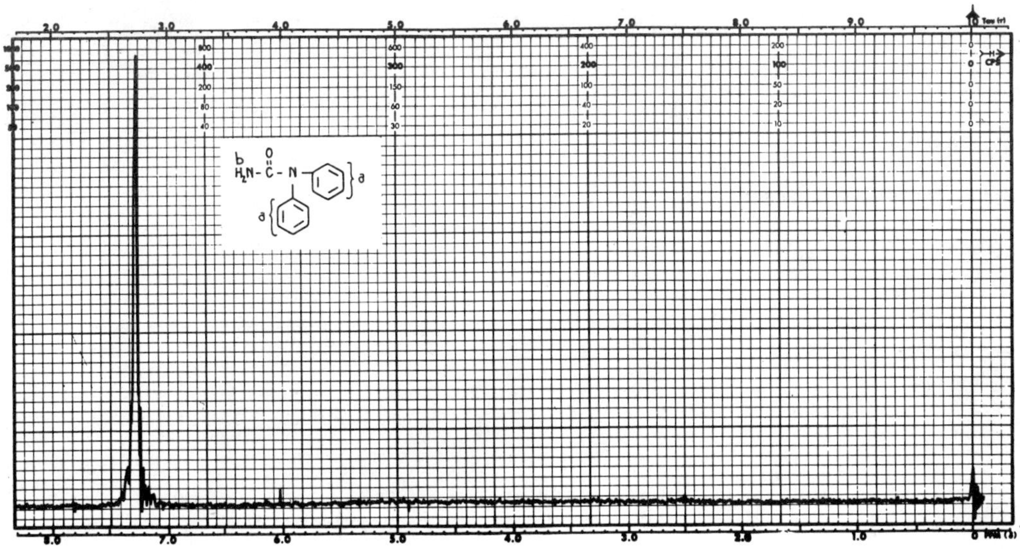

© 1978 Sadtler Research Laboratories, Inc.

1,3-DIETHYLUREA

$C_5H_{12}N_2O$

Mol. Wt. 116.16

M.P. 112°C

Solvent: $CDCl_3$

Assignments
a = 1.22
b = 3.50
c = 6.53

2428

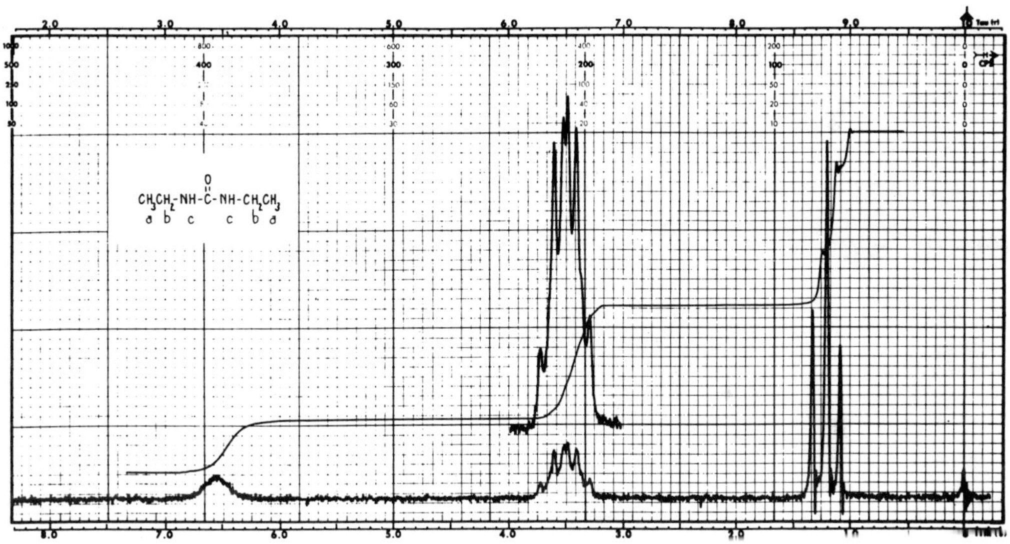

© 1978 Sadtler Research Laboratories, Inc.

1,3-DIBUTYLUREA

$C_9H_{20}N_2O$

Mol. Wt. 172.27

M.P. 72-74°C

Solvent: $CDCl_3$

Assignments
a = 0.96
b = 1.10-1.170
c = 3.15
d = 5.71

2429

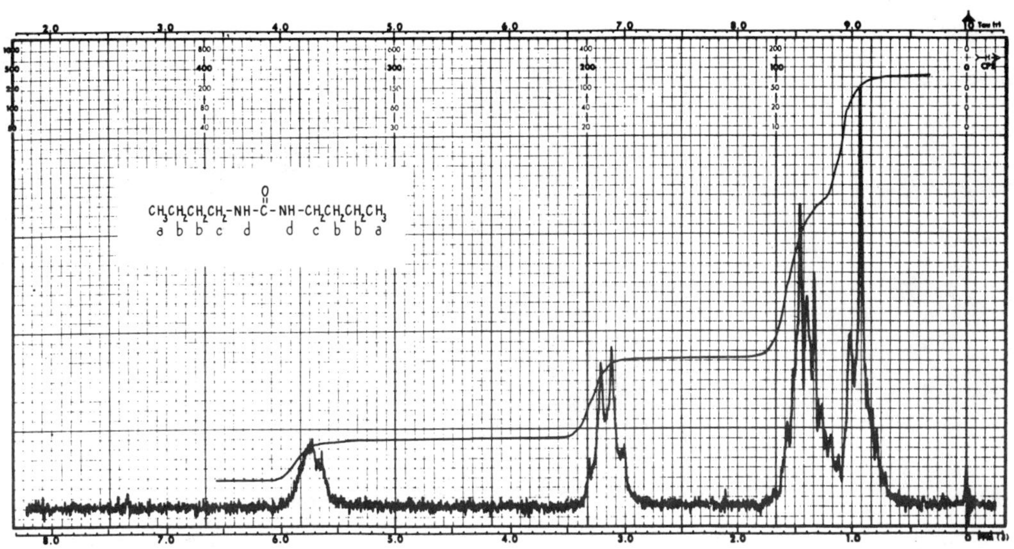

© 1978 Sadtler Research Laboratories, Inc.

1,3-BIS(HYDROXYMETHYL)UREA

$C_3H_8N_2O_3$

Mol. Wt. 102.11

M.P. 121-123^oC (dec.)

Solvent: Dimethyl sulfoxide-D_6

Assignments
a = 4.51
b = 5.27
c = 6.71
d = 2.55 DMSO-D_5
e = 3.50 H_2O

2430

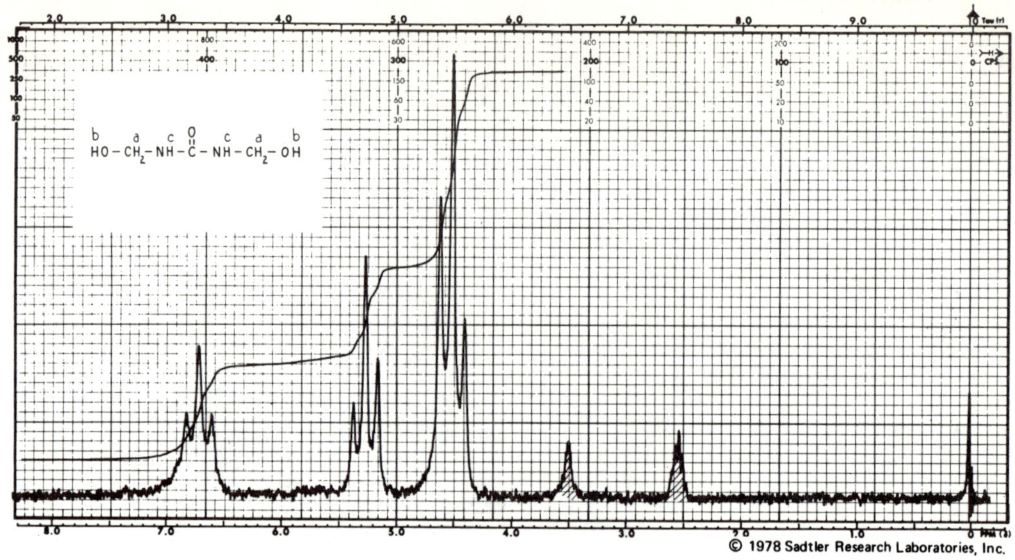

© 1978 Sadtler Research Laboratories, Inc.

2-IMIDAZOLIDINONE

$C_3H_6N_2O$

Mol. Wt. 86.09

M.P. 234^oC

Solvent: $CDCl_3$

Assignments
a = 3.50
b = 5.30

2431

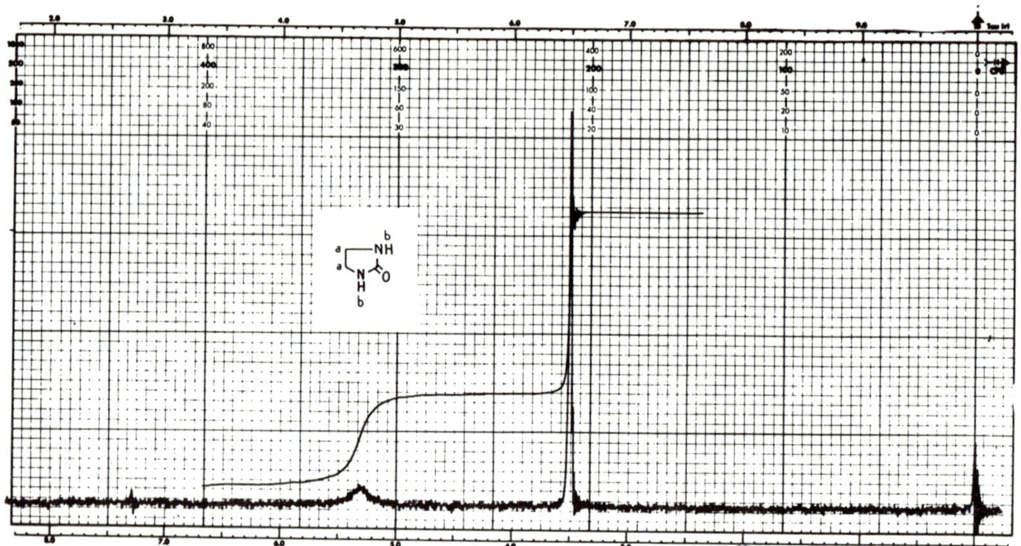

© 1978 Sadtler Research Laboratories, Inc.

1,3-DIBENZYLUREA

$C_{15}H_{16}N_2O$

Mol. Wt. 240.31

M.P. 167-170^oC

Solvent: Polysol-d

Assignments
a = 4.28
b = 6.16
c = ca 7.20

2432

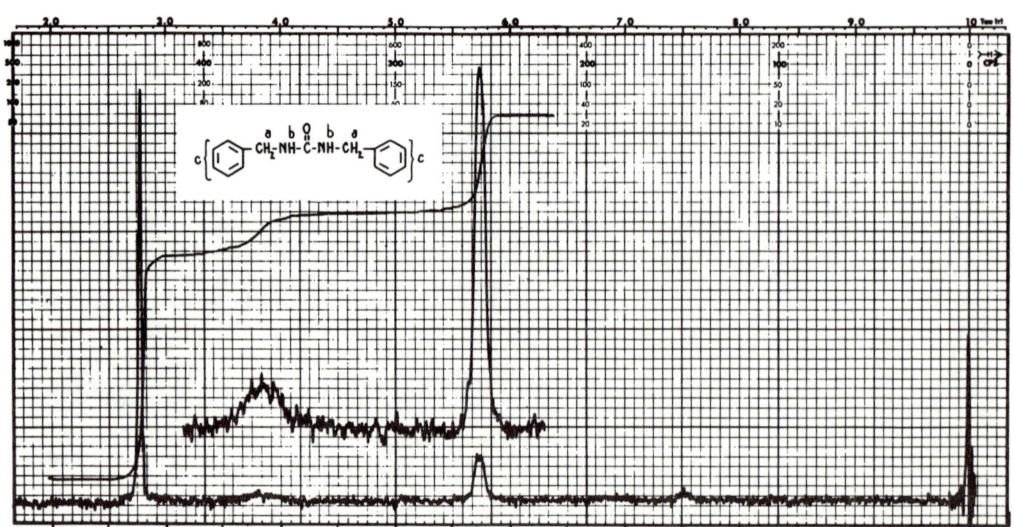

© 1978 Sadtler Research Laboratories, Inc.

1-BUTYL-2-IMIDAZOLIDINONE

$C_7H_{14}N_2O$

Mol. Wt. 142.20

B.P. 102°C/0.3mm

M.P. 36-39°C

Solvent: CCl_4

Assignments
a = 0.95
b = 1.10-1.65
c = 3.10
d = 3.38
e = 6.69

2433

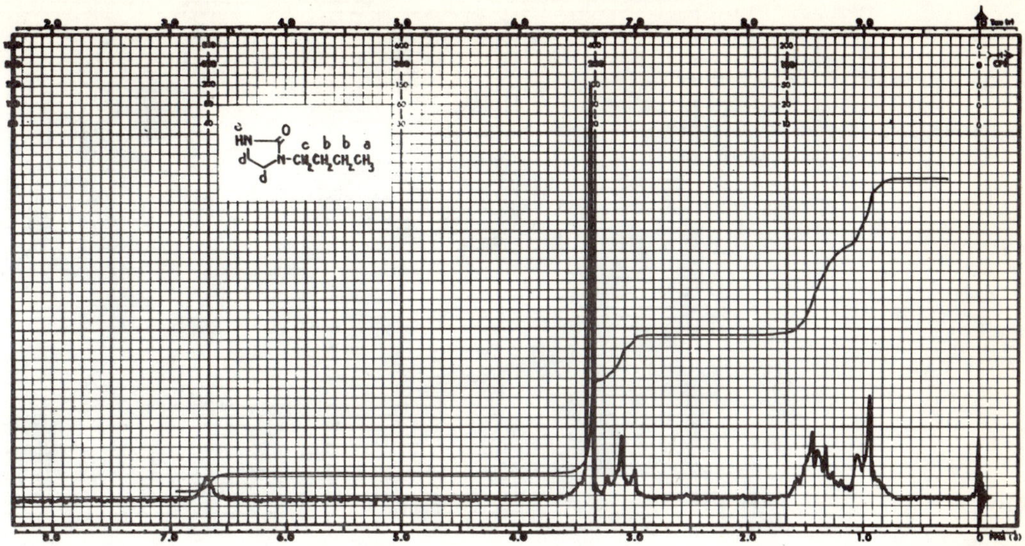

© 1978 Sadtler Research Laboratories, Inc.

1-ALLYL-2-IMIDAZOLIDINONE

$C_6H_{10}N_2O$

Mol. Wt. 126.16

B.P. 95°C/0.25mm

M.P. 56.5-58°C

Solvent: $CDCl_3$

Assignments
a = 3.39
b = 3.77
c = 5.19
d = 5.75
e = 6.42

2434

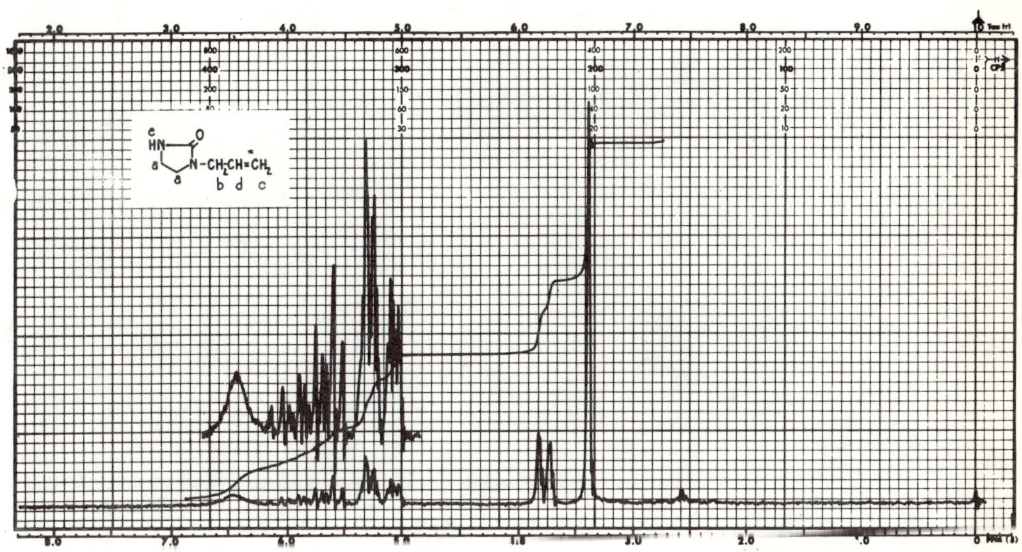

© 1978 Sadtler Research Laboratories, Inc.

1,1-DIMETHYL-3-PHENYLUREA

$C_9H_{12}N_2O$

Mol. Wt. 164.21

M.P. 131-132°C

Solvent: $CDCl_3$

Assignments
a = 2.97
b = 6.47
c = 7.06
d = 7.20-7.50

2435

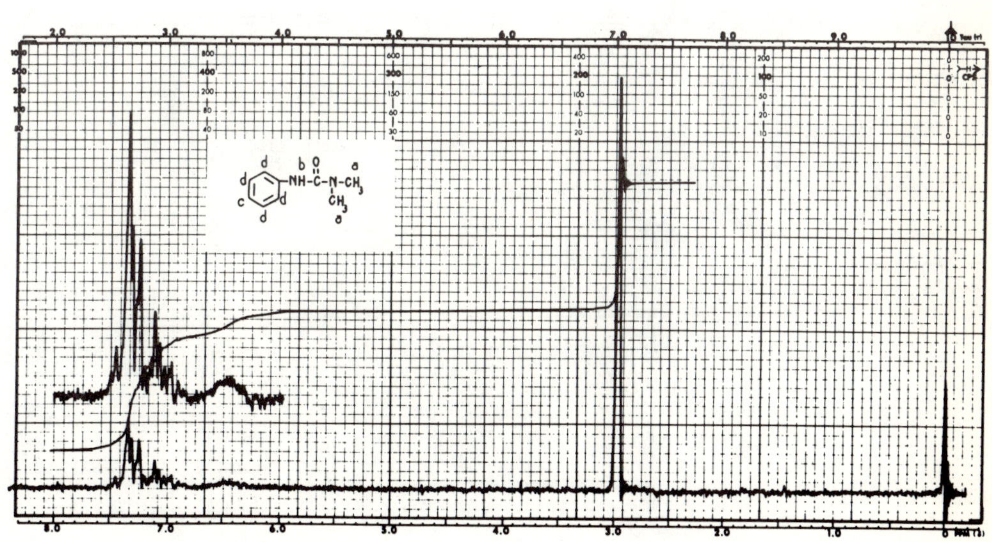

© 1978 Sadtler Research Laboratories, Inc.

1,1-DIPHENYL-3-METHYLUREA

$C_{14}H_{14}N_2O$

Mol. Wt. 226.28

M.P. 170-174°C

Solvent: $CDCl_3$

Assignments
a = 2.78
b = 4.48
c = 7.25
Trace of D_2O added

2436

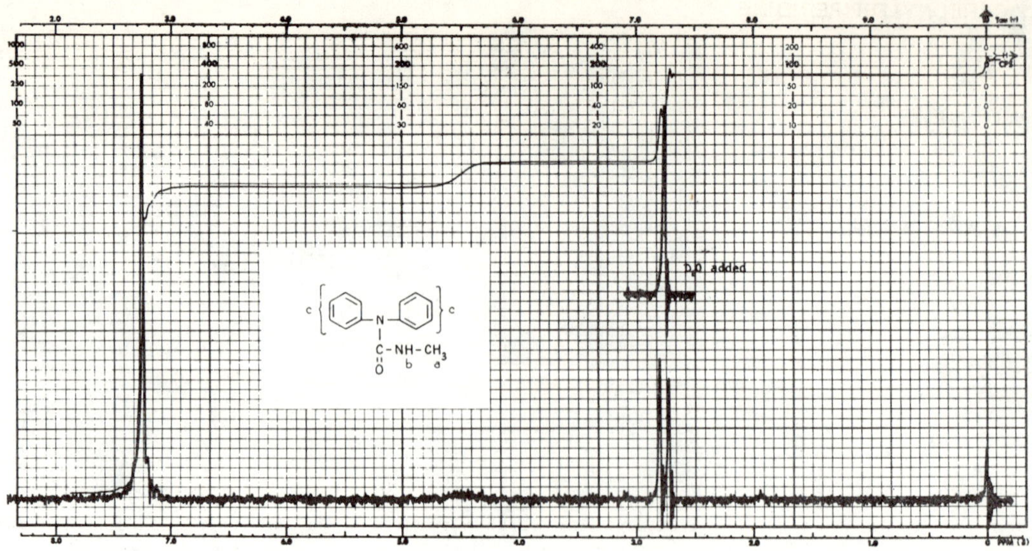

© 1978 Sadtler Research Laboratories, Inc.

TETRAMETHYLUREA

$C_5H_{12}N_2O$

Mol. Wt. 116.16

Solvent: CCl_4

Assignments
a = 2.75

2437

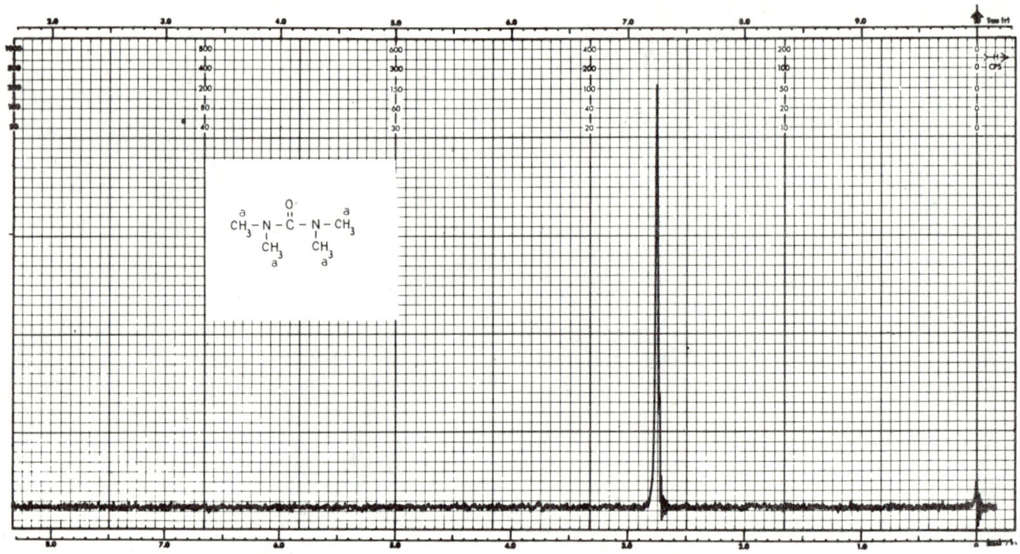

© 1978 Sadtler Research Laboratories, Inc.

TETRABUTYLUREA

$C_{17}H_{36}N_2O$

Mol. Wt. 284.49

B.P. 134-136°C/2mm

Solvent: CCl_4

Assignments
a = 0.94
b = 1.39
c = 3.02

2438

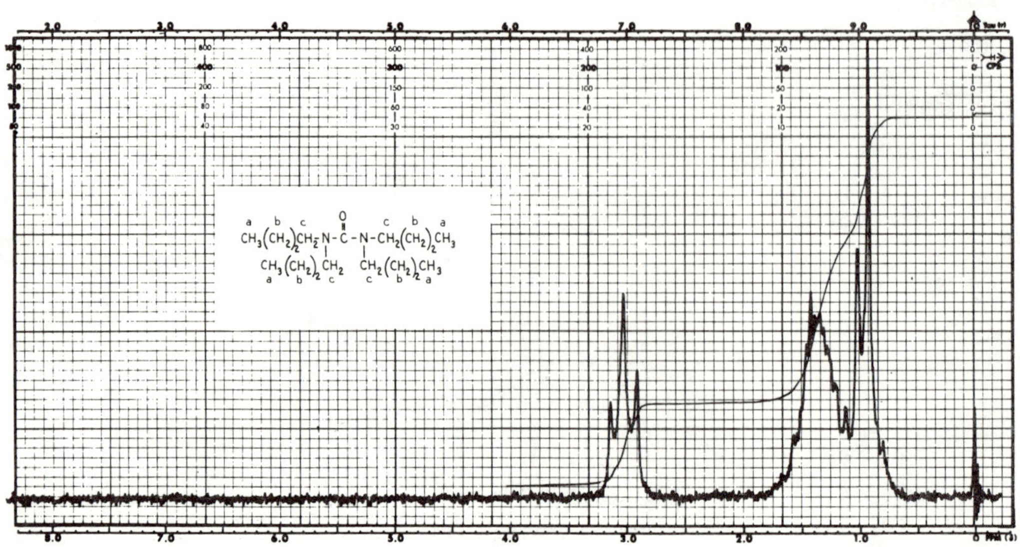

© 1978 Sadtler Research Laboratories, Inc.

1,1'-CARBONYLDIPIPERIDINE

$C_{11}H_{20}N_2O$

Mol. Wt. 196.29

M.P. 44-47°C

B.P. 296-298°C

Solvent: $CDCl_3$

Assignments
a = ca 1.60
b = 3.19

2439

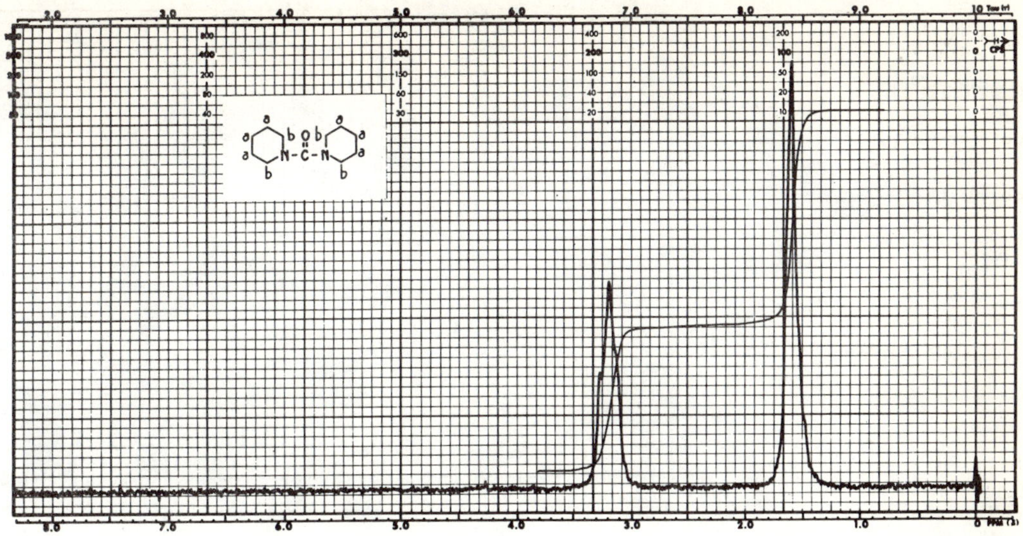

© 1978 Sadtler Research Laboratories, Inc.

N,N-DIMETHYLCARBANILIDE

$C_{15}H_{16}N_2O$

Mol. Wt. 240.31

M.P. 120-122°C

Solvent: $CDCl_3$

Assignments
a = 3.16
b = 6.68-7.18

2440

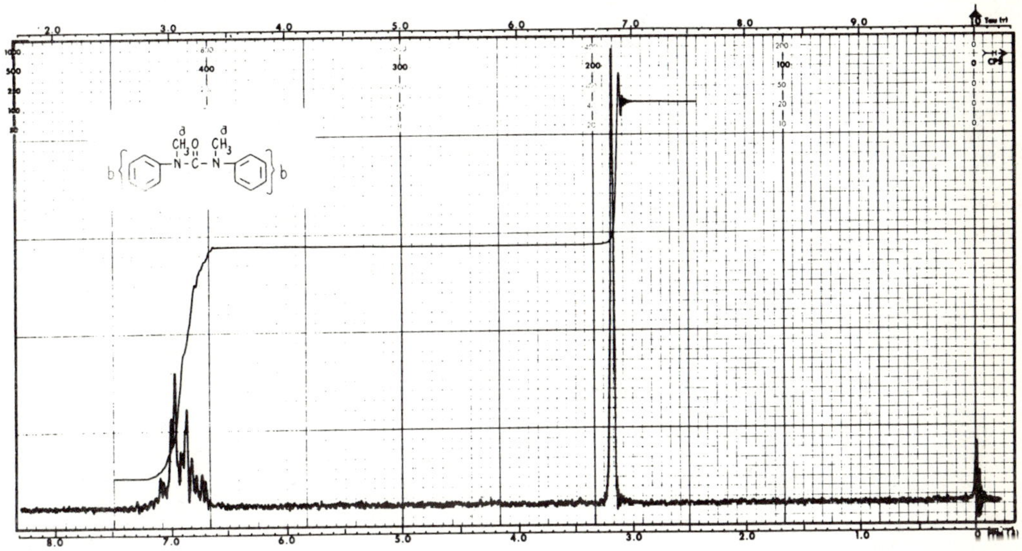

© 1978 Sadtler Research Laboratories, Inc.

1-PHENYL-2-BENZIMIDAZOLINONE

$C_{13}H_{10}N_2O$

Mol. Wt. 210.24

M.P. 203.5-204°C

Solvent: Dimethyl sulfoxide-D6

Assignments
a = ca 7.02
b = 7.51
c = 11.10
d = 2.50 DMSO-D5
e = 3.40 3.40 H_2O

2441

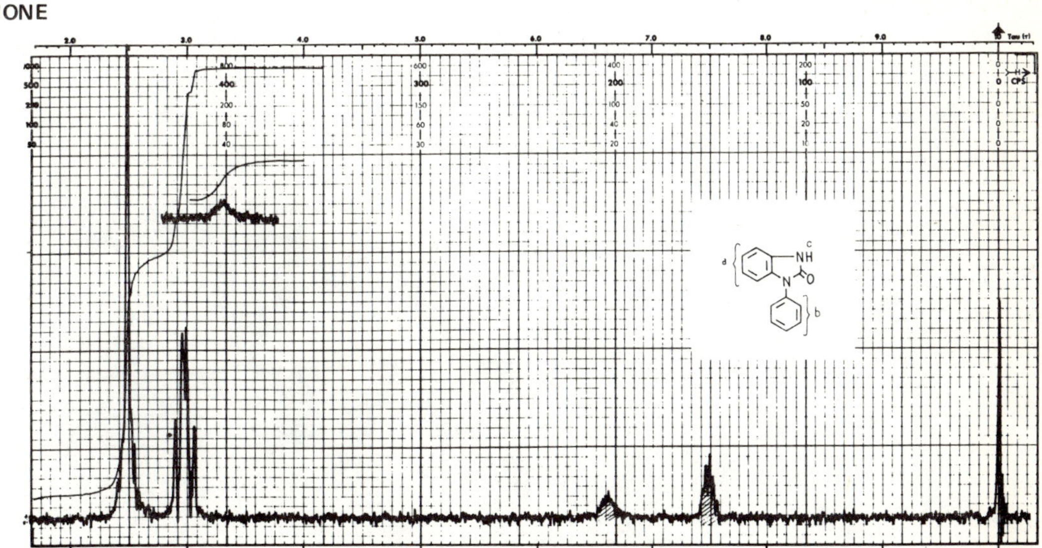

© 1978 Sadtler Research Laboratories, Inc.

5,5-DIMETHYLTETRAHYDRO-2(1H)-PYRIMIDINONE

$C_6H_{12}N_2O$

Mol. Wt. 128.18

M.P. 255-257°C

Solvent: Trifluoroacetic acid

Assignments
a = 1.13
b = 3.19
c = unobserved *
* Under solvent band

2442

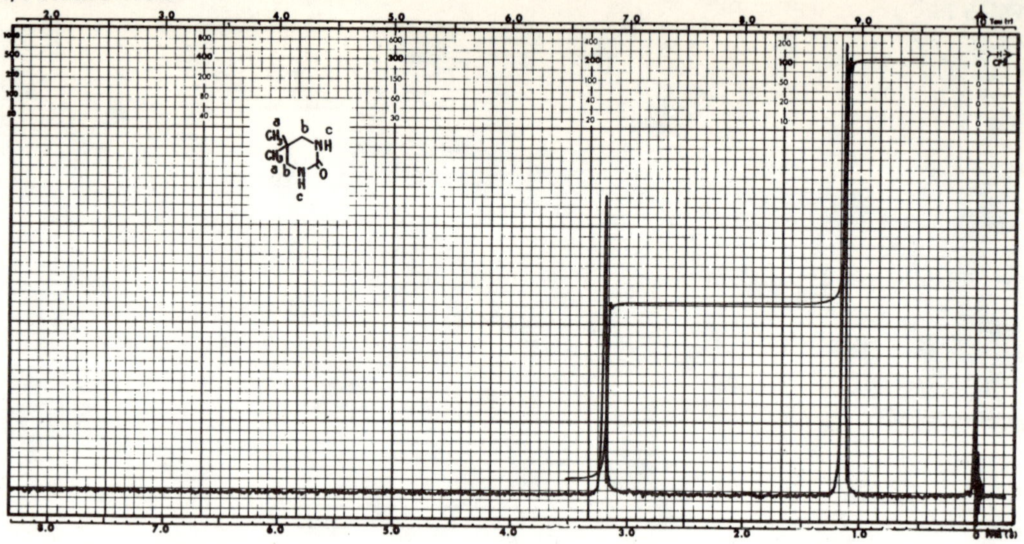

© 1978 Sadtler Research Laboratories, Inc.

THE HYDANTOINS AND URACILS

The Hydantoins and Uracils are cyclic structures containing the group —NH—C(=O—NH—C(=O)—. Hydantoin is a five membered ring while Uracil is a six membered ring. When present, the NH proton at position one (adjacent to one carbonyl) resonates at higher field than the NH proton at position three (adjacent to two carbonyl groups). Aliphatic groups bonded to the rings possess chemical shifts determined by their position on the ring.

THE HYDANTOINS

Chemical Shifts

Aliphatic Protons

		Group	Ring System	Solvent	Spectrum
	1.31	CH_3—	(5-Hydantoin)	Polysol	2444
	(1.38)	$(CH_3)_2$—	(5,5-Hydantoin, 3-Benzyl)	$CDCl_3$	2445
	1.68	CH_3—	(5-Hydantoin, 5-Phenyl)	DMSO-d_6	2447
	3.00	CH_3—	(3-Hydantoin, 5-Phenyl)	$CDCl_3$	2448
1.12	3.47	CH_3—CH_2—	(3-Hydantoin, 5-Phenyl)	$CDCl_3$	2449

Cyclic Protons

—NH—C(=O)	—NH—C(=O)	—CH—	X		
7.44	9.52	4.03	—CH_3	Polysol	2444
- - - -	- - - -	4.29	—H	TFA	2443
8.34	10.74	5.14	—Ø	DMSO-d_6	2446
6.94	7.92	5.39	—NH—C(=O)—NH_2	DMSO-d_6	2451

THE URACILS

Chemical Shifts

Aliphatic Protons

	Group	Ring System		
1.75	CH_3—	(1-Uracil, 3-Methyl)	DMSO-d_6	2453
2.05	CH_3—	(5-Uracil, 6-Methyl)	TFA	2455
2.38	CH_3—	(6-Uracil, 5-Methyl)	TFA	2455
3.30	CH_3—	(3-Uracil, 1-Methyl)	$CDCl_3$	2452
3.41	CH_3—	(1-Uracil, 3-Methyl)	$CDCl_3$	2452

Chemical Shifts continued

Cyclic Protons

—NH—C(=O)—	NH—C(=O)——	CH=	CH—	Solvent	Spectrum
----	----	$-CH_3$	7.21	DMSO	2453
$-CH_3$	$-CH_3$	5.71	7.29	$CDCl_3$	2452
10.55	10.55	—Br	7.71	DMSO	2456
9.10	9.10	—I	7.88	DMSO	2457
11.69	11.69	$-NO_2$	8.80	DMSO	2458
10.10	----	$-CH_3$	$-CH_3$	TFA	2455
10.19	----	6.02	$-CH_3$	TFA	2454

Solubility and Solvent Effects

The cyclic di-amides such as Hydantoin and Uracil are normally not soluble in Carbon Tetrachloride nor Deuterochloroform unless one or both of the nitrogen atoms are substituted by an aliphatic group. The compounds are usually readily soluble in Polysol, DMSO-d_6, Acetone and Trifluoroacetic Acid.

Trifluoroacetic Acid is usually the solvent of last choice since one or both of the NH resonance bands may not be observed due to overlap with the acid proton band at low field.

HYDANTOIN

$C_3H_4N_2O_2$

Mol. Wt. 100.08

M.P. 221-223°C

Solvent: Trifluoroacetic acid

Assignments
a = 4.29
b = unobserved *
* Under solvent band

2443

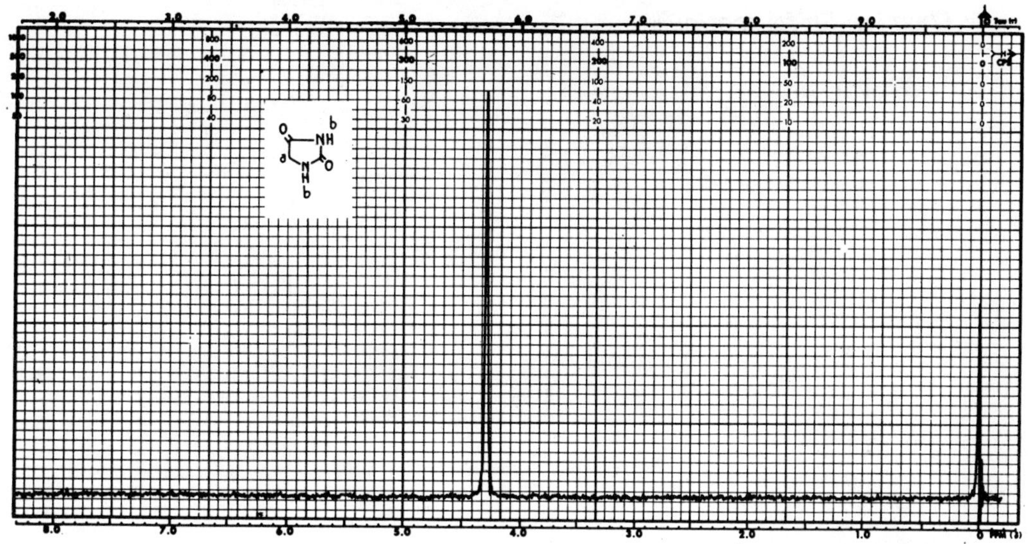

© 1978 Sadtler Research Laboratories, Inc.

5-METHYLHYDANTOIN

$C_4H_6N_2O_2$

Mol. Wt. 114.10

Solvent: Polysol-d

Assignments
a = 1.31
b = 4.03
c = 7.44
d = 9.52
e = 5.45, 6.23 imps.
f = 6.81 imp.

2444

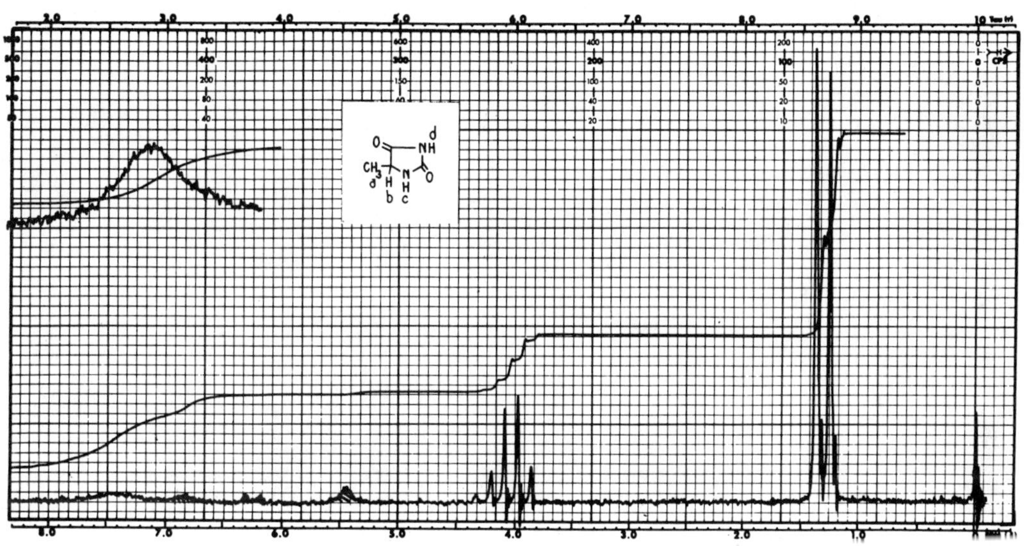

© 1978 Sadtler Research Laboratories, Inc.

3-BENZYL-5,5-DIMETHYLHYDANTOIN

$C_{12}H_{14}N_2O_2$

Mol. Wt. 218.26

M.P. 105-106°C

Solvent: $CDCl_3$

Assignments
a = 1.38
b = 4.64
c = 7.13
d = 7.30

2445

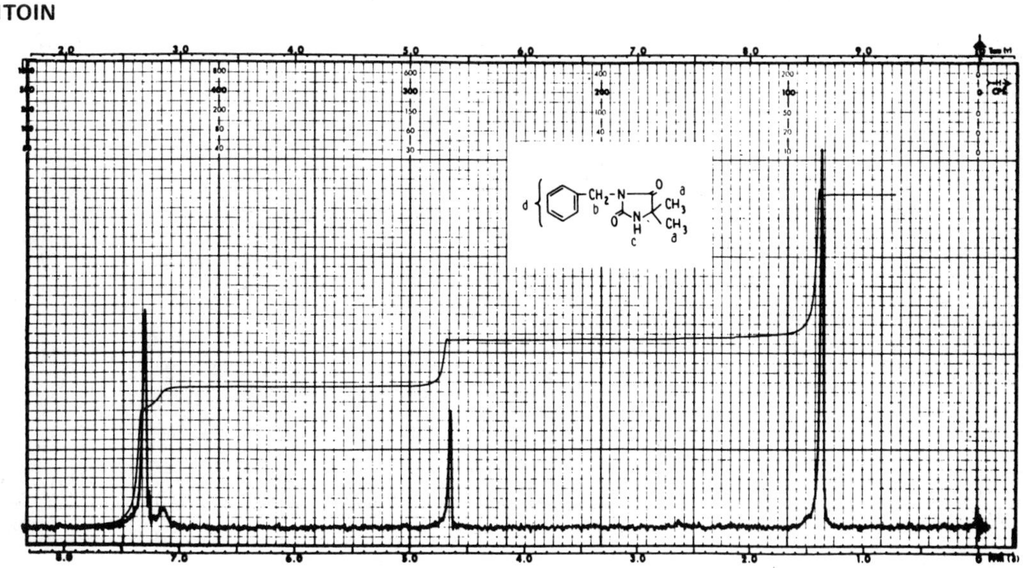

© 1978 Sadtler Research Laboratories, Inc.

5-PHENYLHYDANTOIN

$C_9H_8N_2O_2$

Mol. Wt. 176.18

M.P. 183-185°C (lit.)

Solvent: Dimethyl sulfoxide-D6

Assignments
a = 5.14
b = 7.33
c = 8.34
d = 10.74
e = 2.49 DMSO-D5

2446

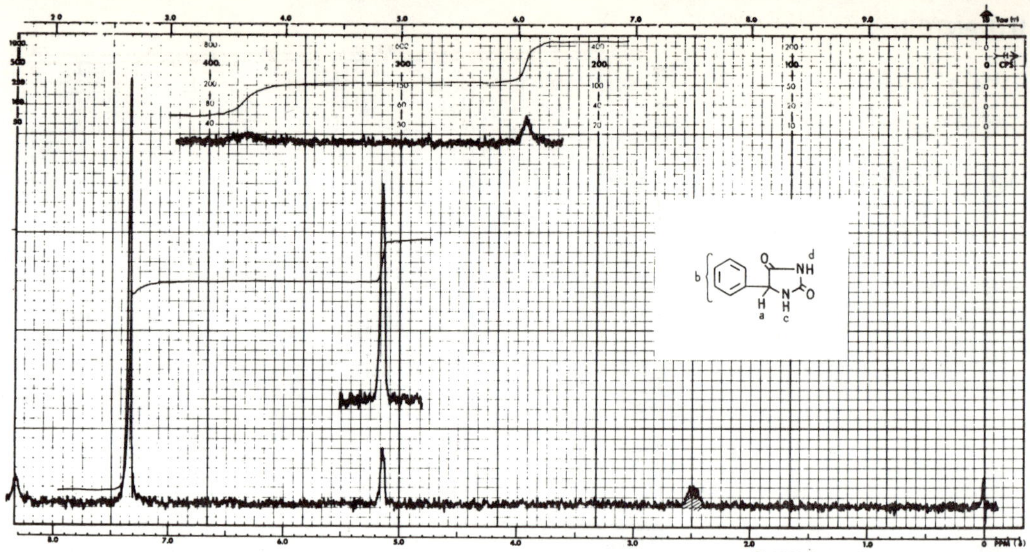

© 1978 Sadtler Research Laboratories, Inc.

5-METHYL-5-PHENYLHYDANTOIN

$C_{10}H_{10}N_2O_2$

Mol. Wt. 190.20

M.P. 244°C (lit.)

Solvent: Dimethyl sulfoxide-d_6

Assignments
a = 1.68
b = 7.25-7.60
c = 8.52
d = 2.50 DMSO-d_5
e = 3.30 H_2O
NH unobserved

2447

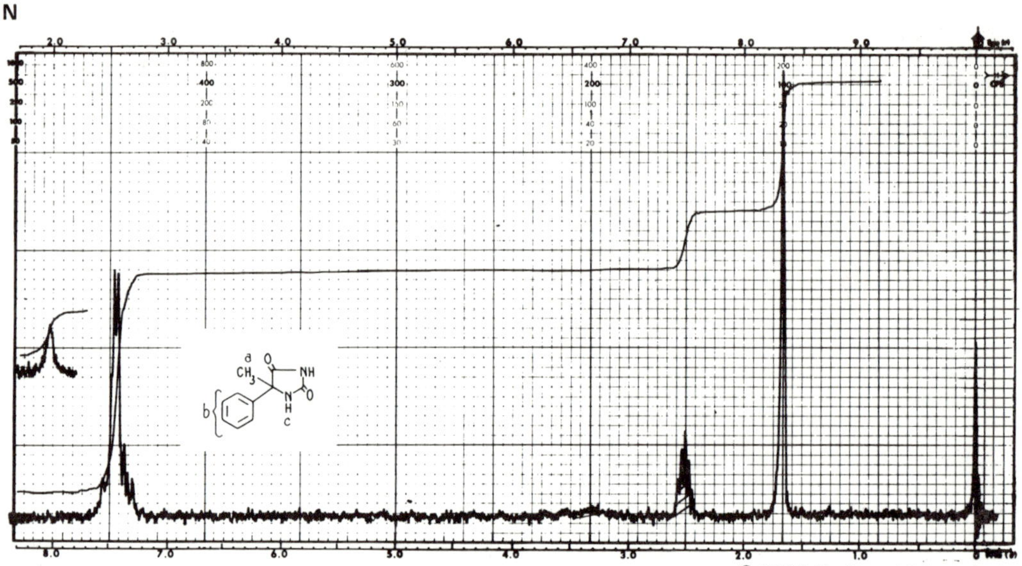

© 1978 Sadtler Research Laboratories, Inc.

3-METHYL-5-PHENYLHYDANTOIN

$C_{10}H_{10}N_2O_2$

Mol. Wt. 190.20

M.P. 163-166°C

Solvent: $CDCl_3$

Assignments
a = 3.00
b = 5.00
c = 6.61
d = 7.31
e = 1.80 imp.

2448

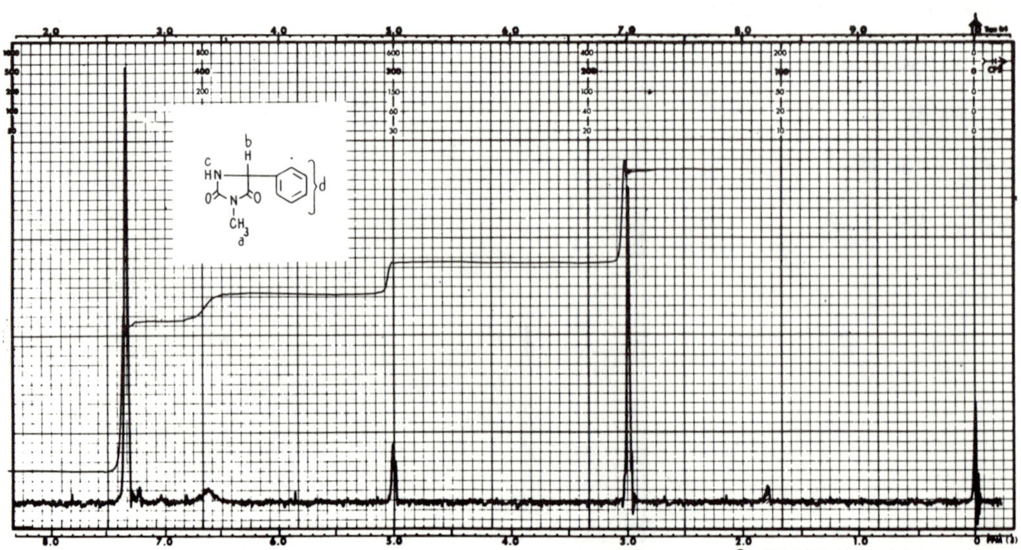

© 1978 Sadtler Research Laboratories, Inc.

3-ETHYL-5-PHENYLHYDANTOIN

$C_{11}H_{12}N_2O_2$

Mol. Wt. 204.23

M.P. 91-93°C

Solvent: $CDCl_3$

Assignments
a = 1.12
b = 3.47
c = 4.93
d = 7.33

2449

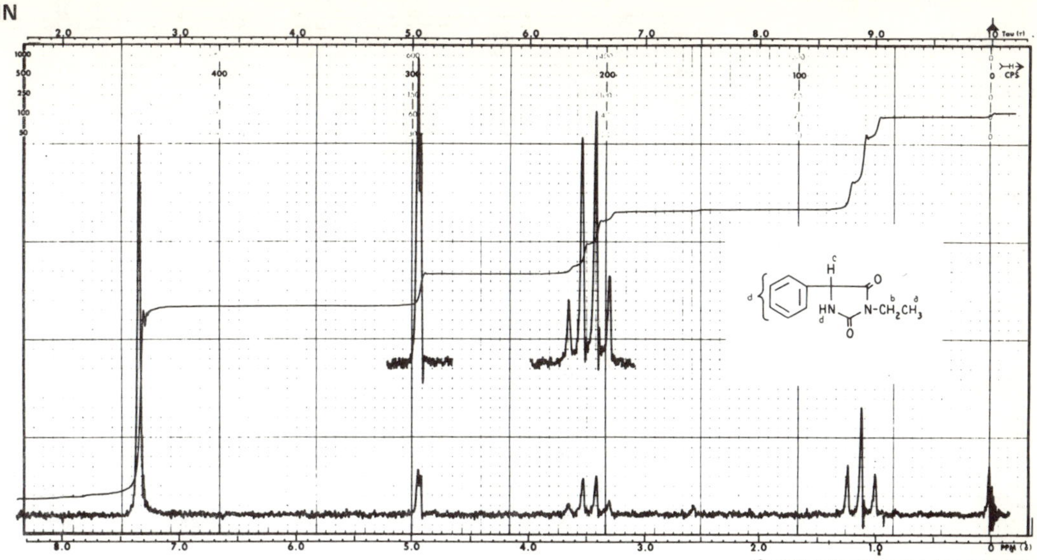

© 1978 Sadtler Research Laboratories, Inc.

1,3-DIPHENYLHYDANTOIN

$C_{15}H_{12}N_2O_2$

Mol. Wt. 252.28

Solvent: $CDCl_3$

Assignments
a = 4.22
b = 7.05-7.65

2450

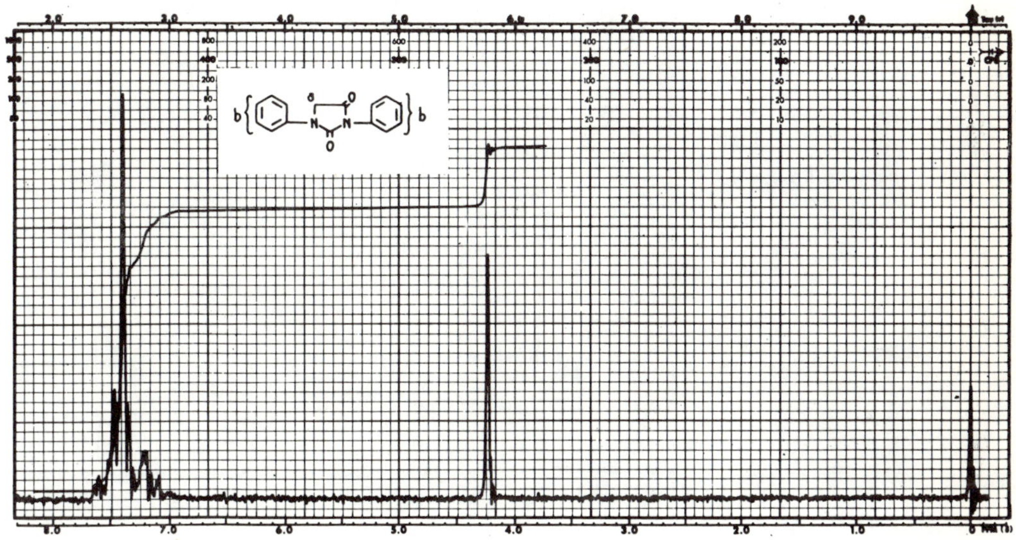

© 1978 Sadtler Research Laboratories, Inc.

ALLANTOIN

$C_4H_6N_4O_3$

Mol. Wt. 158.12

Solvent: DMSO-d_6

Assignments
a = 5.39
b = ca 5.70
c = 6.94
d = 7.92
e = 2.57 DMSO-d_5

2451

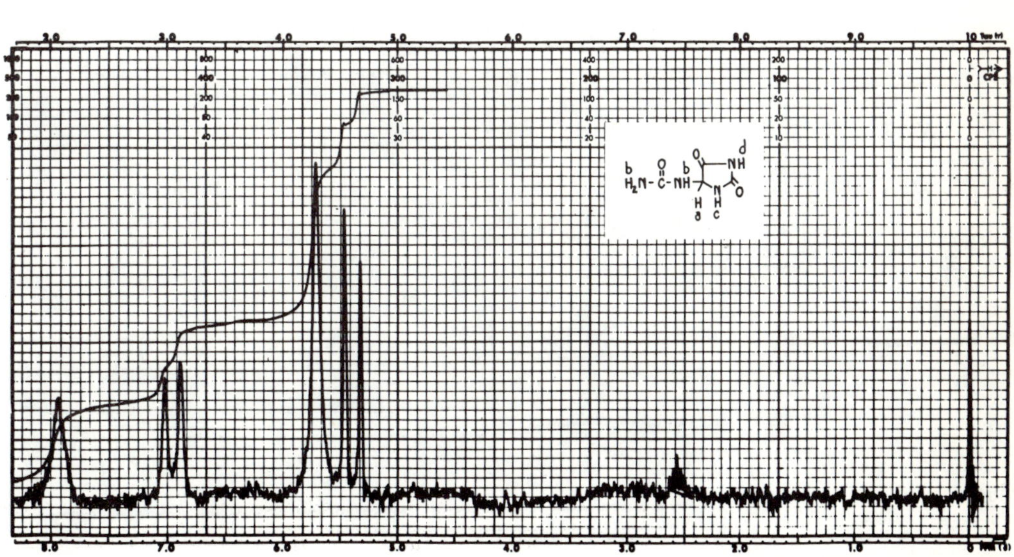

© 1978 Sadtler Research Laboratories, Inc.

1,3-DIMETHYLURACIL

$C_6H_8N_2O_2$

Mol. Wt. 140.14

Solvent: $CDCl_3$

Assignments
a = 3.30
b = 3.41
c = 5.71
d = 7.29

2452

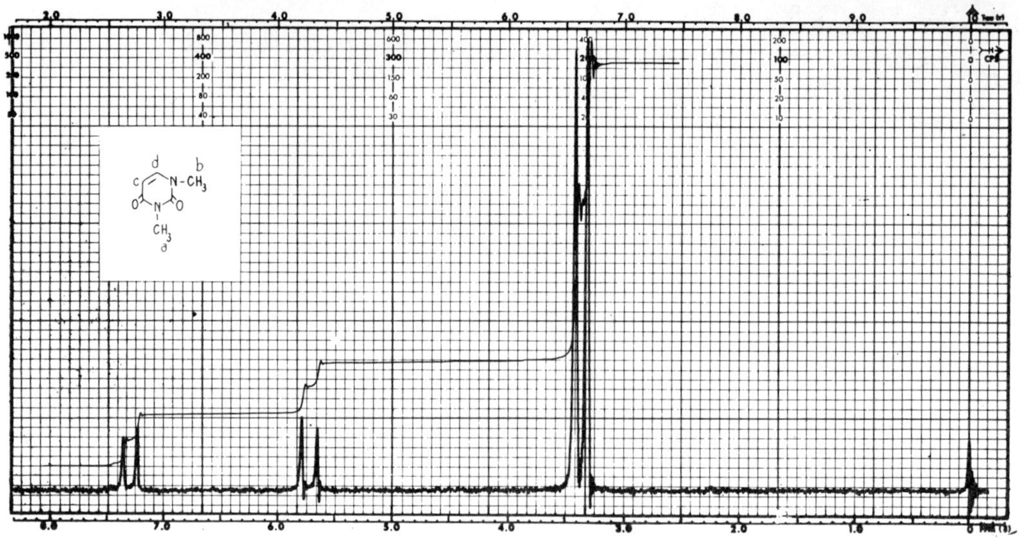

© 1978 Sadtler Research Laboratories, Inc.

THYMINE

$C_5H_6N_2O_2$

Mol. Wt. 126.12

M.P. 335-337°C

Solvent: Dimethyl sulfoxide-d_6

Assignments
a = 1.75
b = 7.21
c = 2.52 DMSO-d_5
d = 3.35 H_2O
NH's unobserved

2453

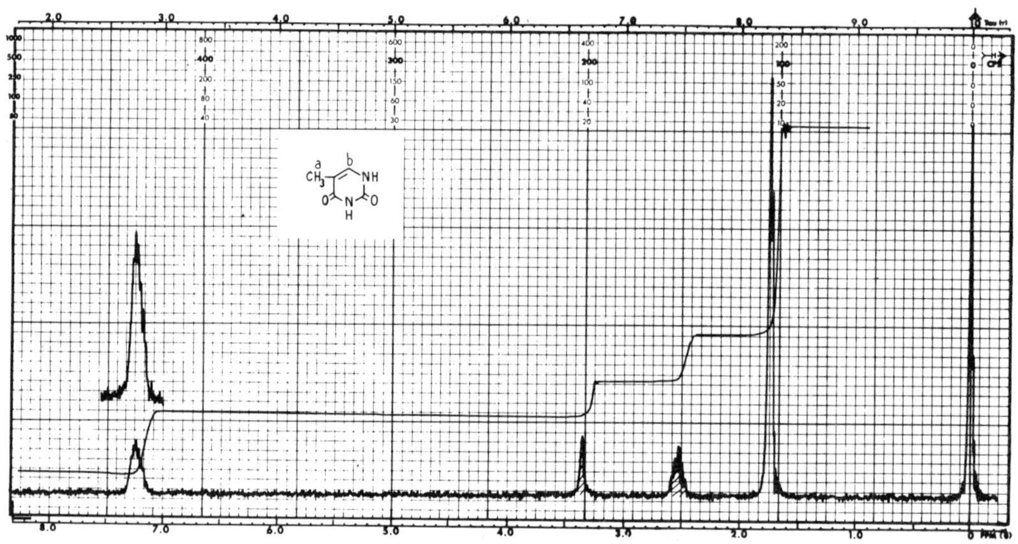

© 1978 Sadtler Research Laboratories, Inc.

6-METHYLURACIL

$C_5H_6N_2O_2$

Mol. Wt. 126.12

M.P. 320°C (dec.)

Solvent: Trifluoroacetic acid

Assignments
a = 2.39
b = 6.02
c = 10.19 or d
d = unobserved or c

2454

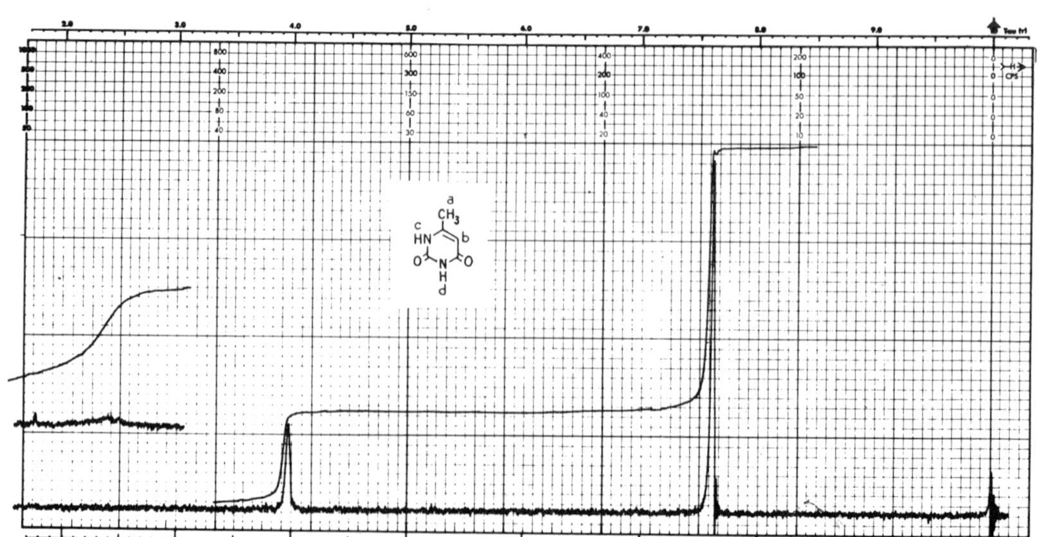

© 1978 Sadtler Research Laboratories, Inc.

5,6-DIMETHYLURACIL

$C_6H_8N_2O_2$

Mol. Wt. 140.14

Solvent: Trifluoroacetic acid

Assignments
a = 2.05
b = 2.38
c = 10.10
d = 6.01 imp.

2455

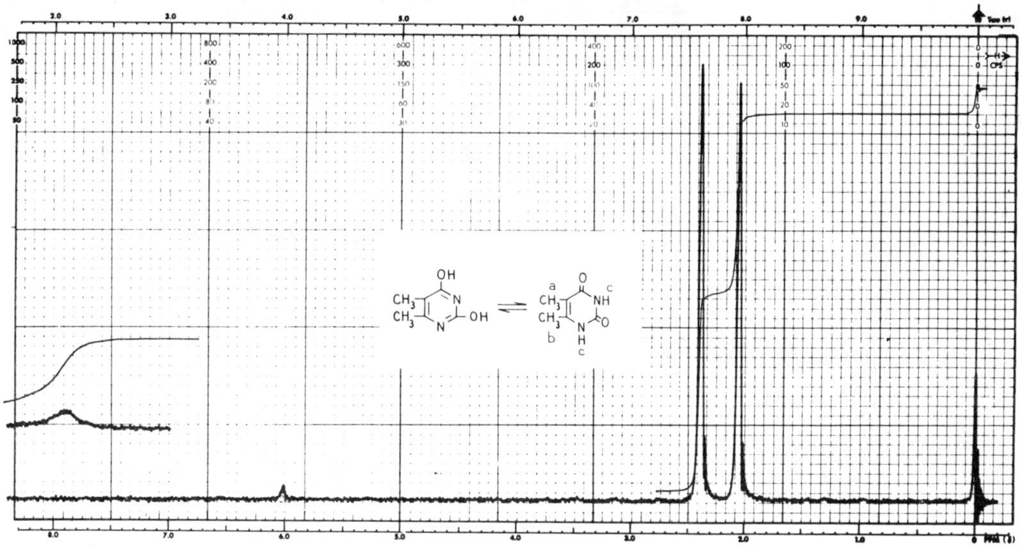

© 1978 Sadtler Research Laboratories, Inc.

5-BROMOURACIL

$C_4H_3BrN_2O_2$

Mol. Wt. 190.99

M.P. 305-310°C (dec.)

Solvent: Dimethyl sulfoxide

Assignments
a = 7.71
b = 10.65
c = 5.49, 7.31 imps.

2456

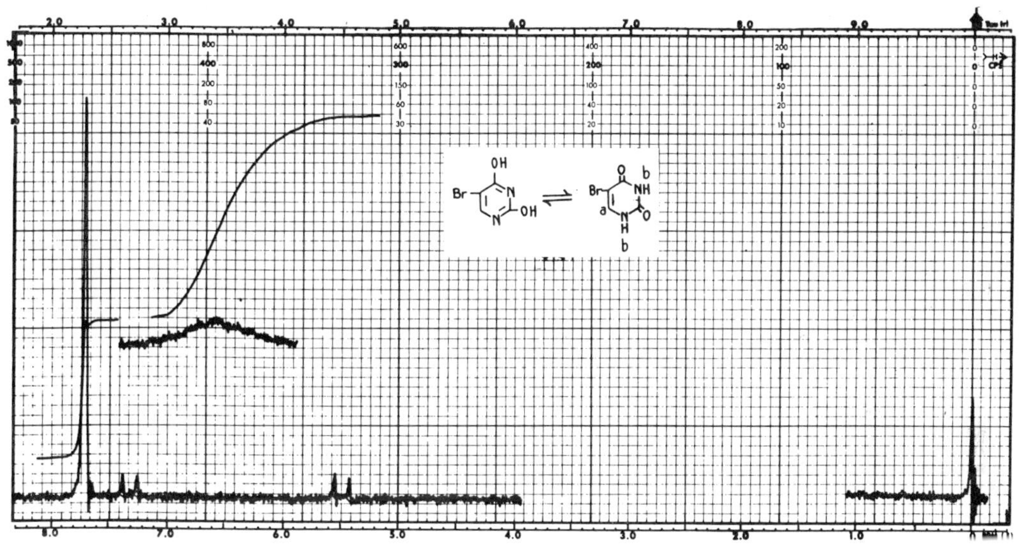

© 1978 Sadtler Research Laboratories, Inc.

5-IODOURACIL

$C_4H_3IN_2O_2$

Mol. Wt. 237.99

M.P. 263-266°C

Solvent: DMSO

Assignments
a = 7.88
b = 9.10 *
* Plus H_2O from solvent

2457

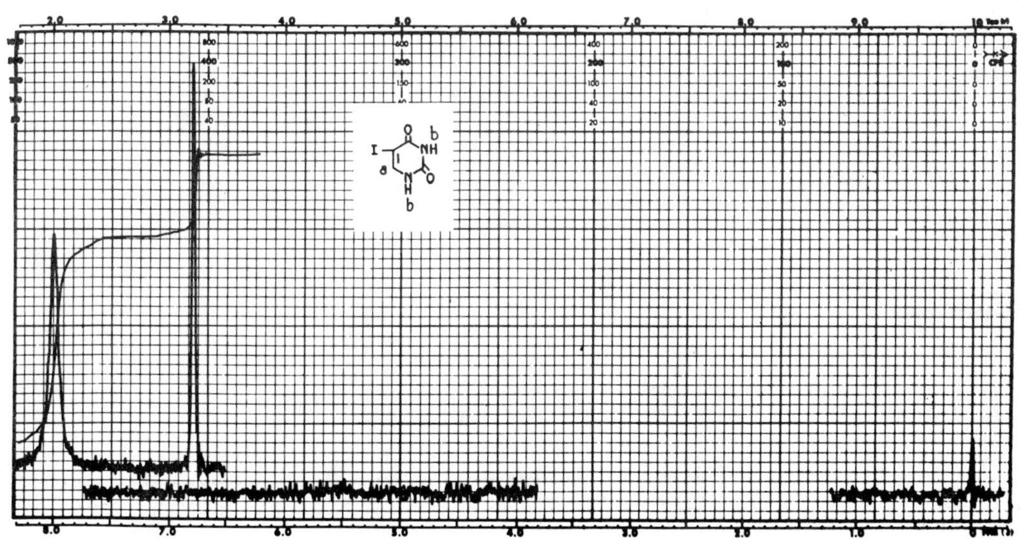

© 1978 Sadtler Research Laboratories, Inc.

5-NITROURACIL

$C_4H_3N_3O_4$

Mol. Wt. 157.09

Solvent: Dimethyl sulfoxide

Assignments
a = 8.80
b = 11.69

2458

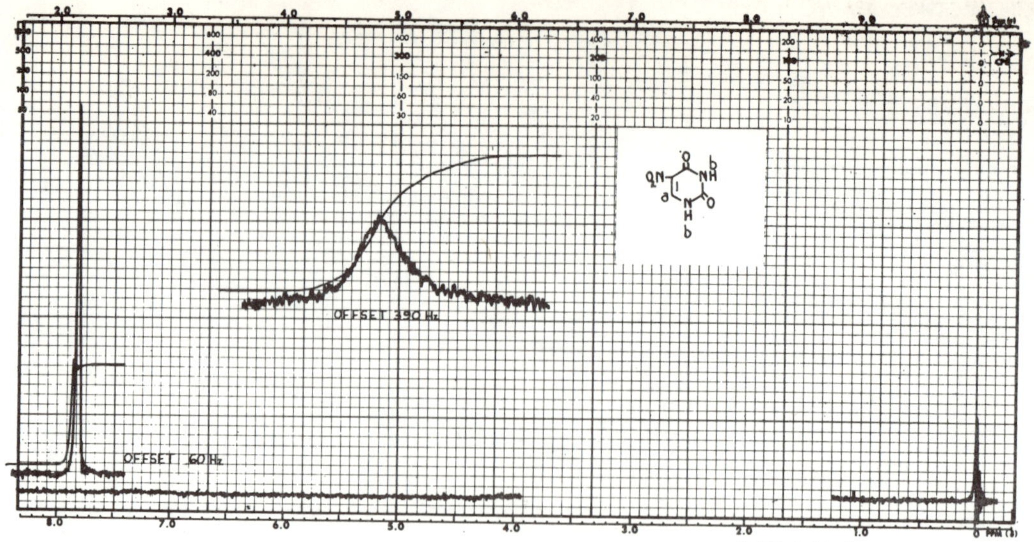

© 1978 Sadtler Research Laboratories, Inc.

5-(1-METHYLBUTYL)BARBITURIC ACID

$C_9H_{14}N_2O_3$

Mol. Wt. 198.22

M.P. 165-168°C

Solvent: Dimethyl sulfoxide-d_6

Assignments
a = 0.89
b = 1.00
c = 1.10-1.80
d = 2.33
e = unobserved
f = 2.52 DMSO-d_5
g = 3.32 H_2O

2459

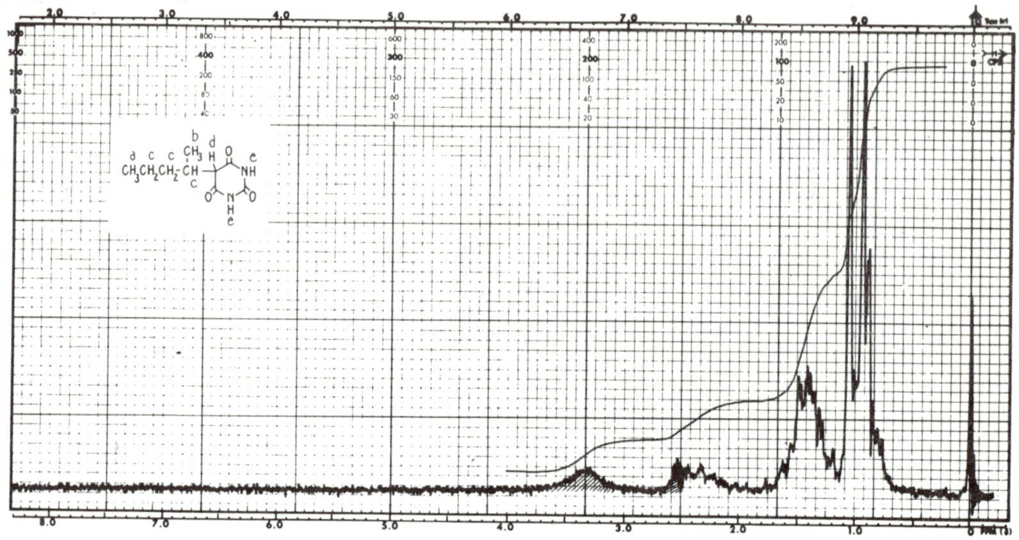

© 1978 Sadtler Research Laboratories, Inc.

5,5-DIETHYLBARBITURIC ACID

$C_8H_{12}N_2O_3$

Mol. Wt. 184.20

M.P. 186-188°C

Solvent: Trifluoroacetic acid

Assignments
a = 1.01
b = 2.22
NH's unobserved

2460

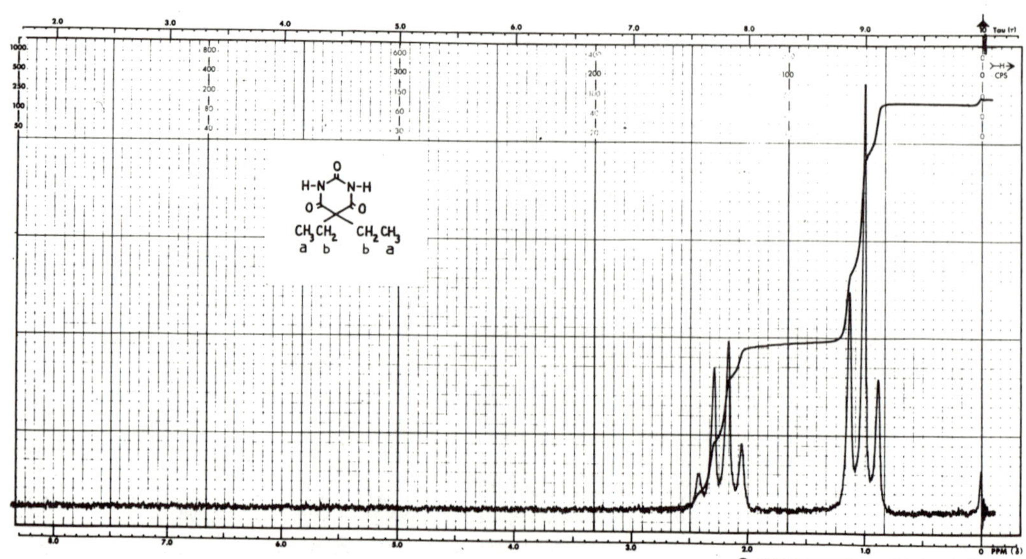

© 1978 Sadtler Research Laboratories, Inc.

5,5-DIALLYLBARBITURIC ACID

$C_{10}H_{12}N_2O_3$

Mol. Wt. 208.22

M.P. 174°C

Solvent: Dimethyl sulfoxide-d_6

Assignments
a = 2.60
b = 4.89-5.91
c = 11.38
d = 5.19 H_2O

2461

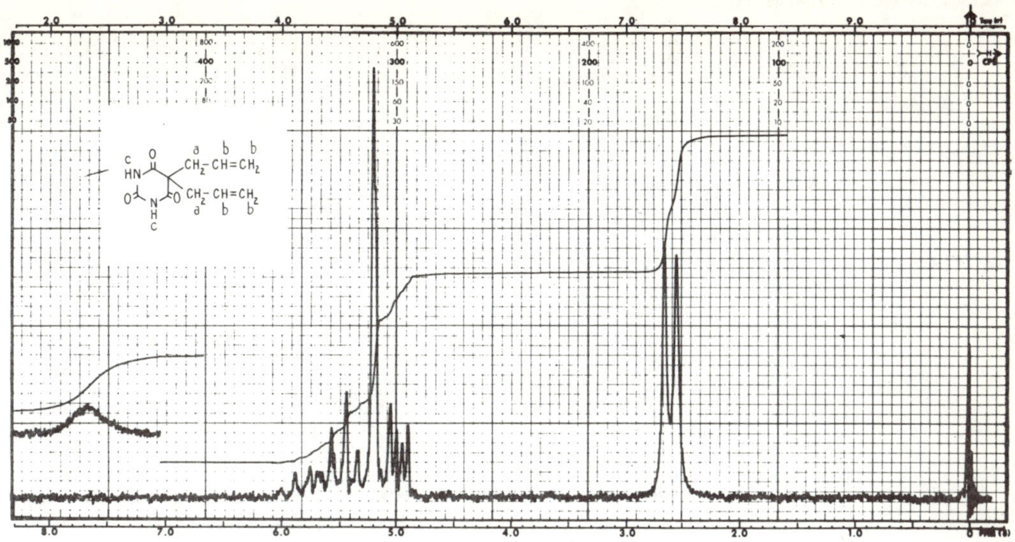

© 1978 Sadtler Research Laboratories, Inc.

5,5-DIBENZYLBARBITURIC ACID

$C_{18}H_{16}N_2O_3$

Mol. Wt. 308.34

M.P. 221-225°C

Solvent: Dimethyl sulfoxide-d_6

Assignments
a = 3.30 *
b = 6.95-7.40
c = 2.50 DMSO-d_5
* H_2O present
NH's unobserved

2462

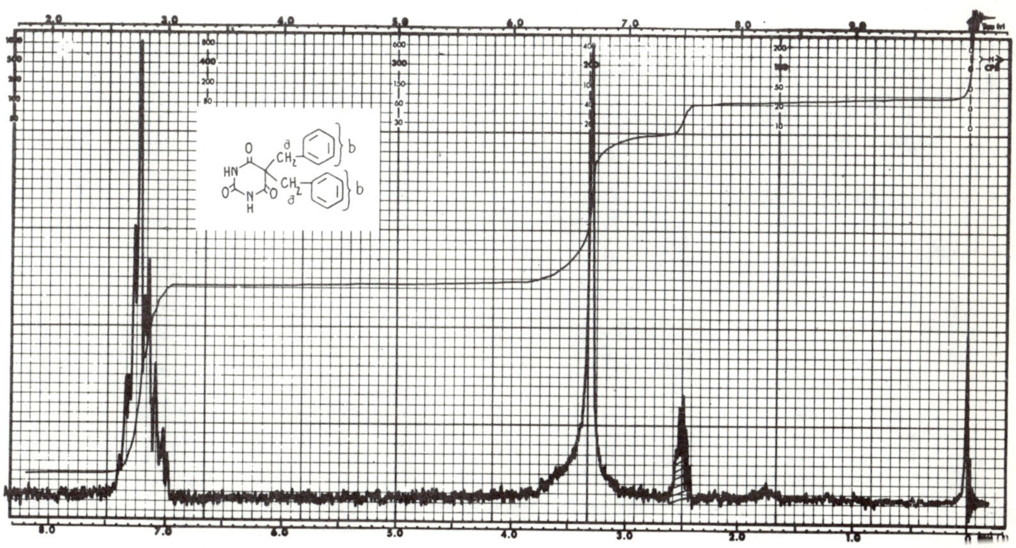

© 1978 Sadtler Research Laboratories, Inc.

5-BENZYLIDENEBARBITURIC ACID

$C_{11}H_8N_2O_3$

Mol. Wt. 216.20

M.P. 256°C

Solvent: Dimethyl sulfoxide

Assignments
a = 7.46
b = 8.08
c = 8.30

2463

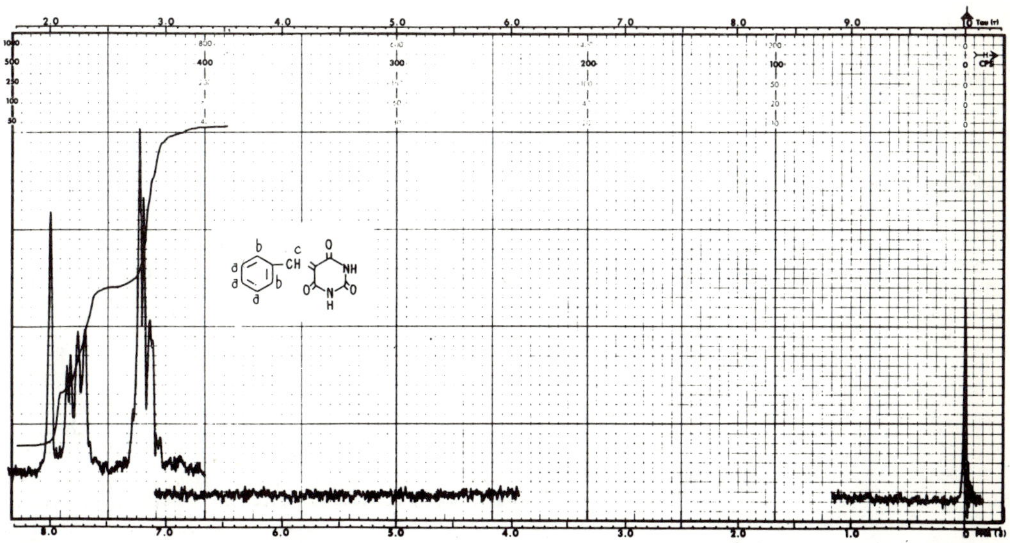

© 1978 Sadtler Research Laboratories, Inc.

5-ALLYL-5-PHENYLBARBITURIC ACID

$C_{13}H_{12}N_2O_3$

Mol. Wt. 244.25

M.P. 156-157.5°C

Solvent: Dimethyl sulfoxide-D_6

Assignments

a = 3.09
b = 4.96-6.13 *
c = 7.39
d = 11.64
e = 2.51 DMSO-D_5
* Contains H_2O from DMSO

2464

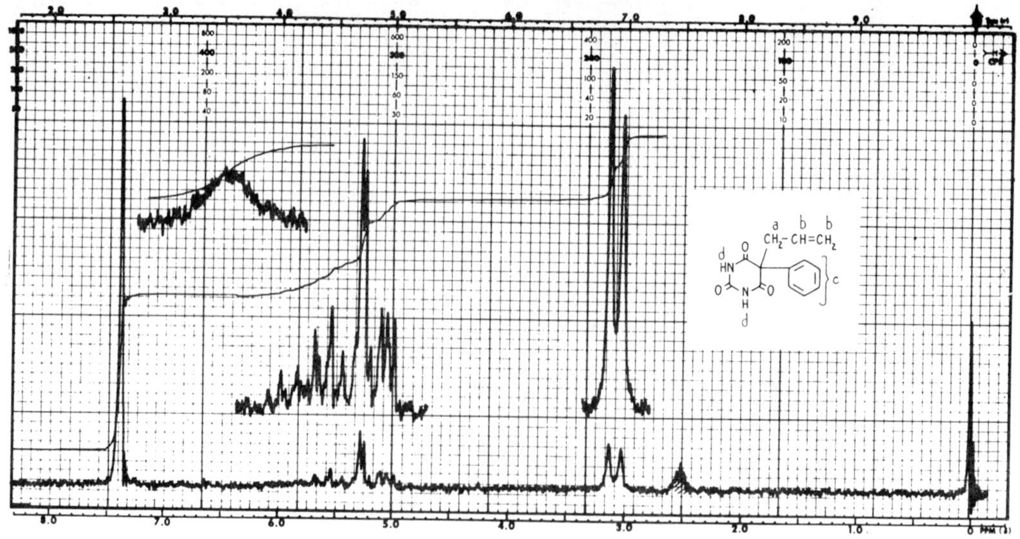

© 1978 Sadtler Research Laboratories, Inc.

TRIS(2-HYDROXYETHYL)-s-TRIAZINE-2,4,6(1H,3H,5H)-TRIONE

$C_9H_{15}N_3O_6$

Mol. Wt. 261.24

M.P. 134-136°C

Solvent: Dimethyl sulfoxide-D_6

Assignments

a = 3.60
b = 3.83
c = 4.54
d = 2.50 DMSO-D_6

2465

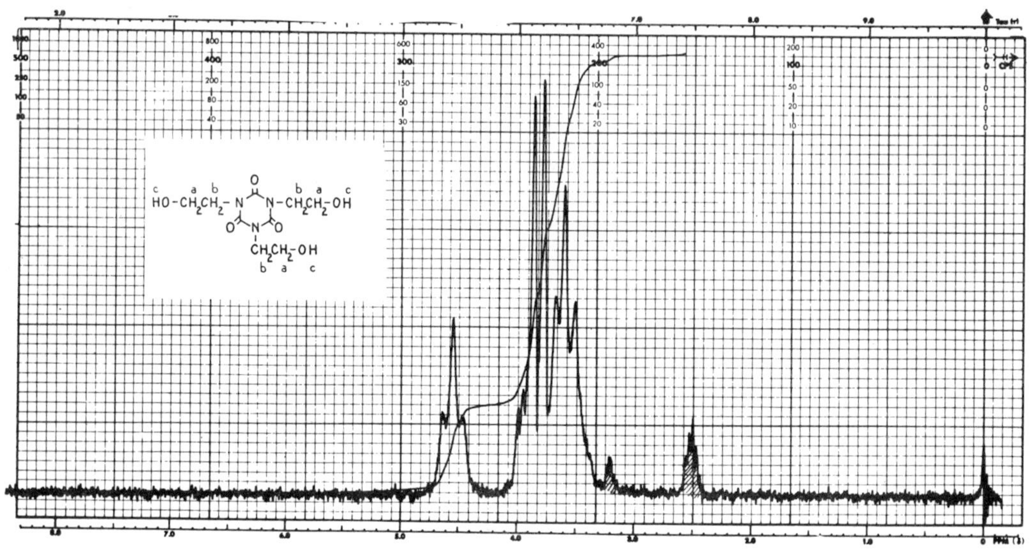

© 1978 Sadtler Research Laboratories, Inc.

1,3-DIPHENYL-5-ETHYL-s-TRIAZINE-2,4,6(1H,3H,5H)-TRIONE

$C_{17}H_{15}N_3O_3$

Mol. Wt. 309.33

M.P. 155-156°C

Solvent: $CDCl_3$

Assignments

a = 1.29
b = 4.02
c = 7.15-7.60

2466

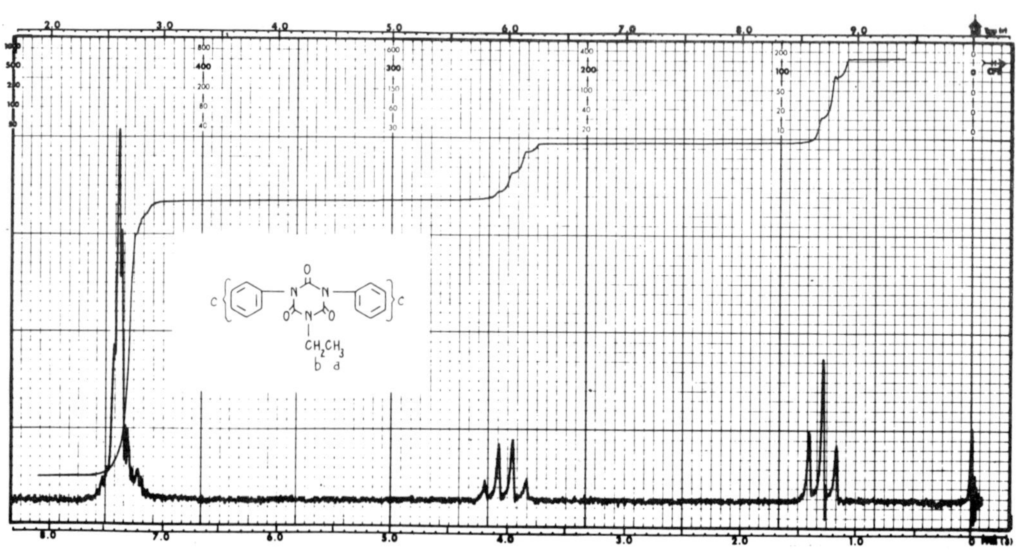

© 1978 Sadtler Research Laboratories, Inc.

1,4-DIPHENYLSEMICARBAZIDE

$C_{13}H_{13}N_3O$

Mol. Wt. 227.27

M.P. 175-178°C

Solvent: Dimethyl sulfoxide

Assignments
a = ca 6.82
b = 7.00-7.45
c = 7.60
d = 7,69
e = 8.18
f = 8.67

2467

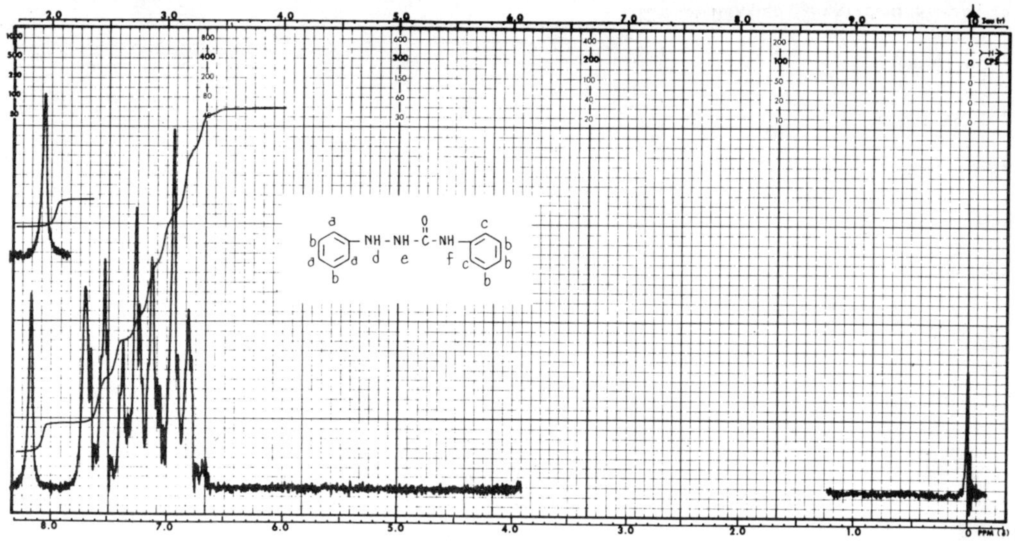

© 1978 Sadtler Research Laboratories, Inc.

4,4-DIPHENYLSEMICARBAZIDE

$C_{13}H_{13}N_3O$

Mol. Wt. 227.27

M.P. 155-156°C

Solvent: $CDCl_3$

Assignments
a = 3.50
b = 5.90

2468

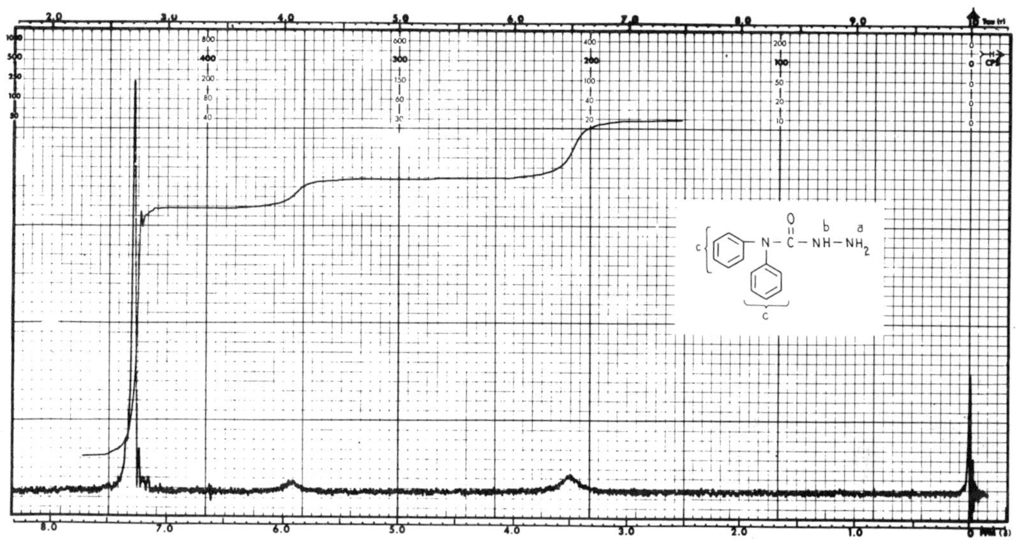

© 1978 Sadtler Research Laboratories, Inc.

3-METHYL-2-BUTANONE, SEMICARBAZONE

$C_6H_{13}N_3O$

Mol. Wt. 151.19

M.P. 110°C

Solvent: $CDCl_3$

Assignments
a = 1.09
b = 1.81
c = 2.48
d = 5.89
e = 8.58

2469

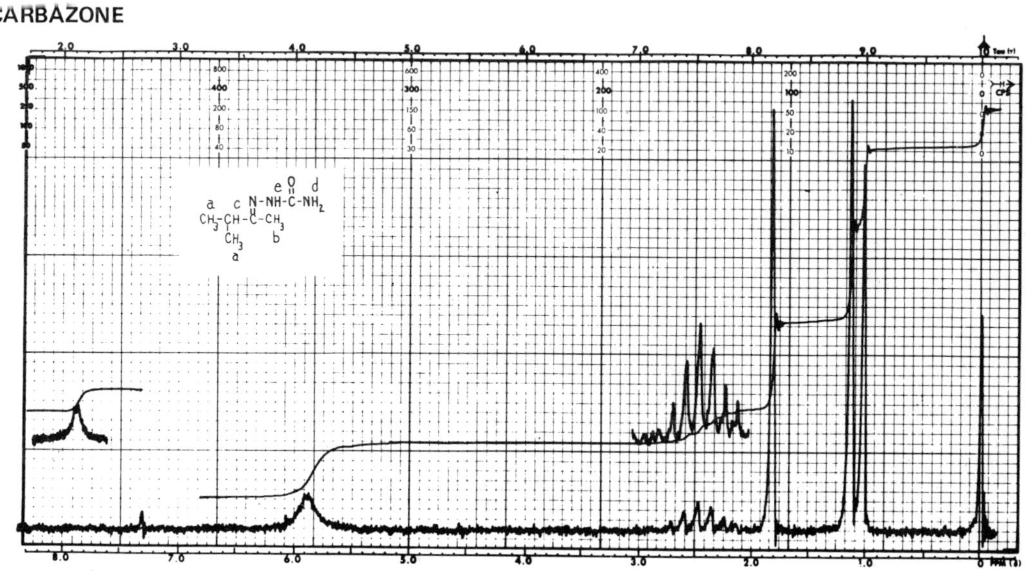

© 1978 Sadtler Research Laboratories, Inc.

2-HEPTANONE, SEMICARBAZONE

$C_8H_{17}N_3O$

Mol. Wt. 171.24

M.P.125-126°C

Solvent: $CDCl_3$

Assignments
a = 0.88
b = 1.06-1.68
c = 1.82
d = 2.20
e = 5.92
f = 8.76

2470

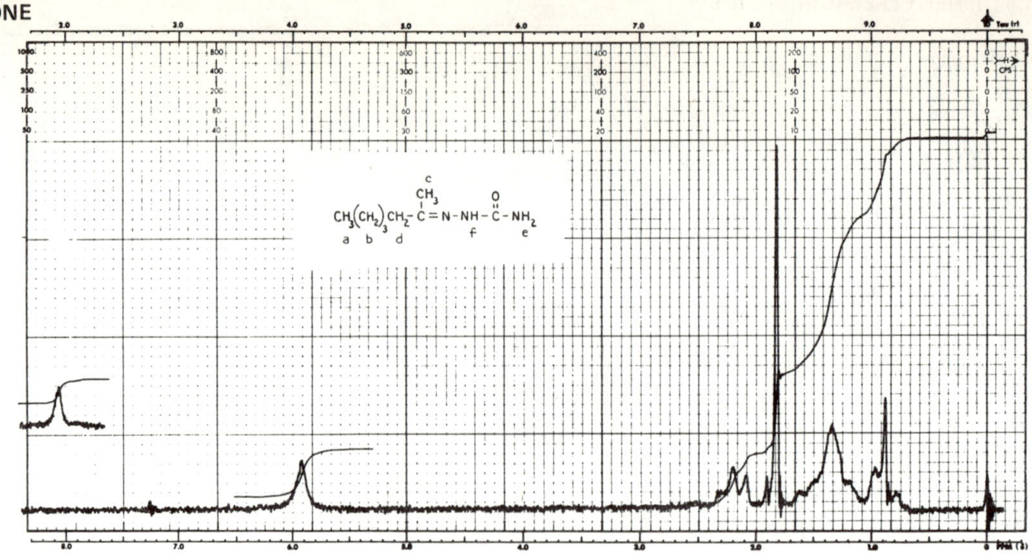

© 1978 Sadtler Research Laboratories, Inc.

The Carboxylic acid functional group (–C(=O)–OH) has a weakly deshielding effect on the protons of adjacent aliphatic groups but a strongly deshielding effect on the ortho aromatic protons.

A distinguishing feature of this group of compounds is the carboxylic acid –OH group which normally resonates at very low field (10–13 ppm). This band may appear at higher field when a significant amount of water is present in the solution.

Chemical Shifts

Aliphatic Protons

HC–3	HC–2	HC–1	–C(=O)–OH		Solvent	Spectrum
		2.06	11.90	CH_3–C(=O)–OH	CCl_4	2472
	1.14	2.37	10.49	CH_3–CH_2–C(=O)–OH	$CDCl_3$	2473
0.90	1.67	2.29	11.97	CH_3–CH_2–CH_2–C(=O)–OH	CCl_4	- - - -
	(1.20)	2.55	12.08	$(CH_3)_2$–CH–C(=O)–OH	CCl_4	2474
(0.98)	2.08	2.19	11.00	$(CH_3)_2$–CH–CH_2–C(=O)–OH	CCl_4	- - - -
	(1.21)	- - - -	12.01	$(CH_3)_3$–C–C(=O)–OH	$CDCl_3$	2476
(0.93	1.55)	2.21	12.08	$(CH_3–CH_2)_2$–CH–C(=O)–OH	CCl_4	- - - -

Substituted Acetic Acids

HO–C(=O)–	CH_2	–X		
10.49	2.37	–CH_3	$CDCl_3$	2473
11.80	2.43	–CH_2–C(=O)–OH	DMSO-d_6	2497
10.08	3.12	–CH=CH_2	$CDCl_3$	2516
11.28	3.25	–CH=CH–Ø	$CDCl_3$	2549
11.50	3.37	–C(=O)–OH	Polysol	2496
10.88	3.61	–Ø	$CDCl_3$	2539
10.80	3.71	–I	$CDCl_3$	2488
11.49	3.79	–2–Thiophene	$CDCl_3$	2512
11.75	3.79	–C≡N	DMSO-d_6	- - - -
10.59	3.92	–Br	$CDCl_3$	2485
11.22	4.05	–Cl	CCl_4	2483
- - - -	4.10	–O–CH_3	D_2O	2491
- - - -	4.29	–OH	D_2O	2492
8.78	4.67	–O–Ø	Acetone	2553

Alicyclic Protons

	n	$(CH_2)_n$	–CH	–C(=O)–OH	Solvent	Spectrum
Cyclopropane	2	(0.7–1.2)	1.58	11.72	CCl_4	- - - -
Cyclobutane	3	(1.6–2.7)	3.19	11.99	$CDCl_3$	2513
Cyclopentane	4	(1.2–2.2)	2.69	11.25	CCl_4	- - - -
Cyclohexane	5	(1.1–2.6)	2.25	12.00	CCl_4	2514
Cycloheptane	6	(1.1–2.2)	2.53	11.55	$CDCl_3$	- - - -

Olefinic Protons

The three olefinic protons of Acrylic Acid resonate as a complex higher-order pattern in the chemical shift range from 5.7–6.8 ppm. The most deshielded of these protons is the hydrogen that is cis to the carboxylic acid group, while the geminal proton resonates at higher field. This order of chemical shifts is the reverse of that observed for simple alkenes.

3-Substituted Acrylic acids

HO	–C(=O)– CH	=CH	–X			
12.28	6.66	6.66	–C(=O)–OH	(trans)	DMSO-d_6	2533
12.27	6.27	6.84	–Cl	(cis)	Polysol	2531
11.91	5.77	7.03	$-R_3$	(trans)	CCl_4	2520
11.79	5.82	7.04	$-CH_3$	(trans)	$CDCl_3$	2517
11.34	6.28	7.50	–Cl	(trans)	$CDCl_3$	- - - -
11.90	6.41	7.73	–Ø	(trans)	$CDCl_3$	- - - -
11.00	6.90	7.75	–C(=O)–Ø	(trans)	$CDCl_3$	- - - -

The Aromatic Protons

Benzoic Acid

para/meta	ortho	–C(=O)–OH		
7.2–7.8	8.14	12.82	CCl_4	2566

Para Substituted Benzoic Acids

HO–C(=O)–	ortho	meta	para		
6.55	7.83	6.76	$-NH_2$	Acetone	2578
10.95	7.95	6.90	$-O-R_3$	Polysol	2586
- - - -	7.91	7.00	$-O-CH_3$	DMSO-d_6	2584
9.76	8.11	7.07	–OH	Acetone	2588
12.02	8.09	7.20	–O–Ø (4–C(=O)–OH)	DMSO	2589

Para Substituted Benzoic Acids continued

HO–C(=O)–	ortho	meta	para	Solvent	Spectrum
9.72	8.02	7.30	–F	Acetone	2570
----	8.02	7.30	$-CH_3$	TFA	2567
8.27	8.00	7.56	–Cl	DMSO	2571
7.28	7.90	7.71	–Br	DMSO-d_6	2574
----	7.90	7.71	–I	DMSO-d_6	2577
9.41	8.11	8.11	–C(=O)–OH	DMSO-d_6	2592
----	8.21	8.30	$-NO_2$	DMSO-d_6	2582

Exchangeable Protons

The carboxylic acid protons which are extremely labile hydrogens exchange with many other types of labile hydrogens to produce an averaged chemical shift for the protons involved in the exchange.

Such chemical shift averaging is noted in spectra 2578 and 2588. In addition, they will be in exchange with any water present in the solution resulting in either higher-field chemical shifts than expected and/or very broad resonance bands covering several ppm. The latter case is much more noticeable in the NMR spectra of the Benzoic acids than in the spectra of the more soluble aliphatic compounds.

FORMIC ACID

CH_2O_2

Mol. Wt. 46.03

M.P. 8.6°C

B.P. 100.8°C

Solvent: D_2O

Assignments
a = 8.22
b = 5.08 HDO

2471

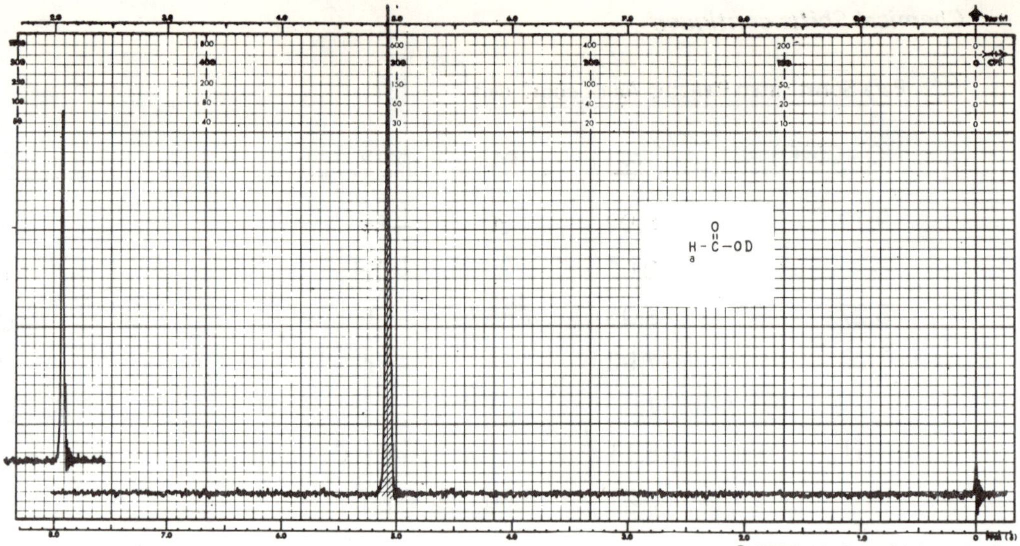

© 1978 Sadtler Research Laboratories, Inc.

ACETIC ACID

$C_2H_4O_2$

Mol. Wt. 60.05

Solvent: CCl_4

Assignments
a = 2.06
b = 11.90

2472

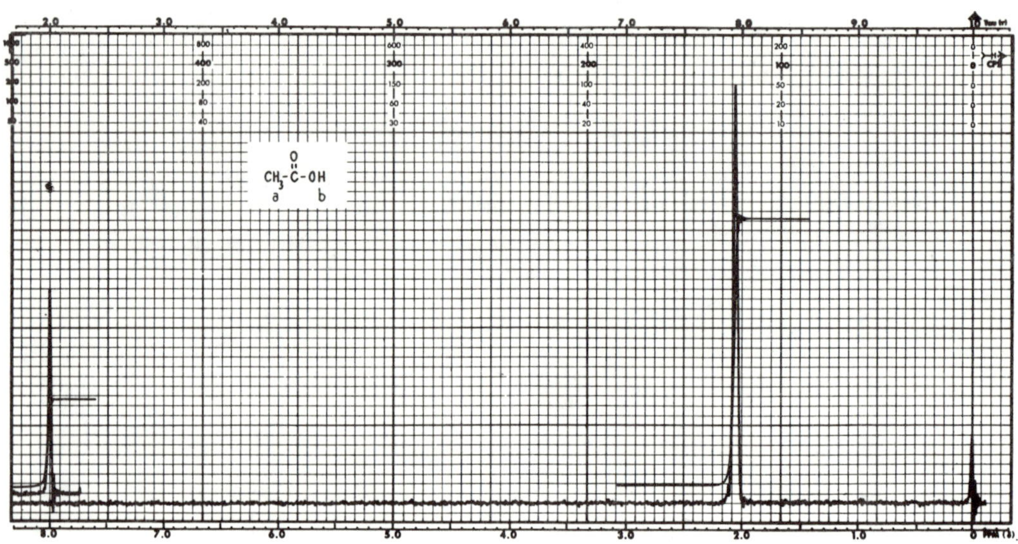

© 1978 Sadtler Research Laboratories, Inc.

PROPIONIC ACID

$C_3H_6O_2$

Mol. Wt. 74.08

B.P. 141°C

Solvent: $CDCl_3$

Assignments
a = 1.14
b = 2.37
c = 10.49

2473

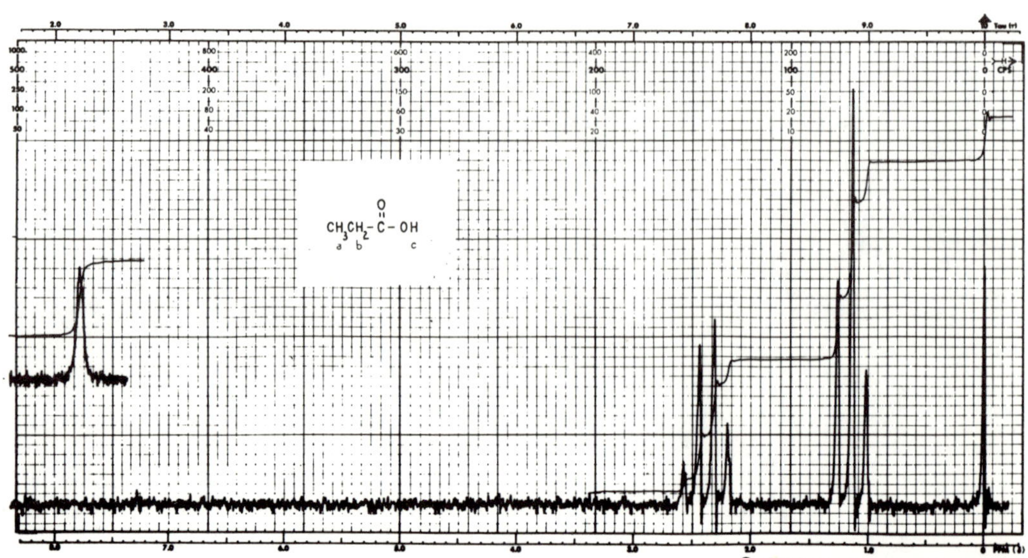

© 1978 Sadtler Research Laboratories, Inc.

ISOBUTYRIC ACID

$C_4H_8O_2$

Mol. Wt. 88.06

B.P. 154.4°C

F.P. −47°C

Solvent: CCl_4

Assignments
a = 1.20
b = 2.55
c = 12.08

2474

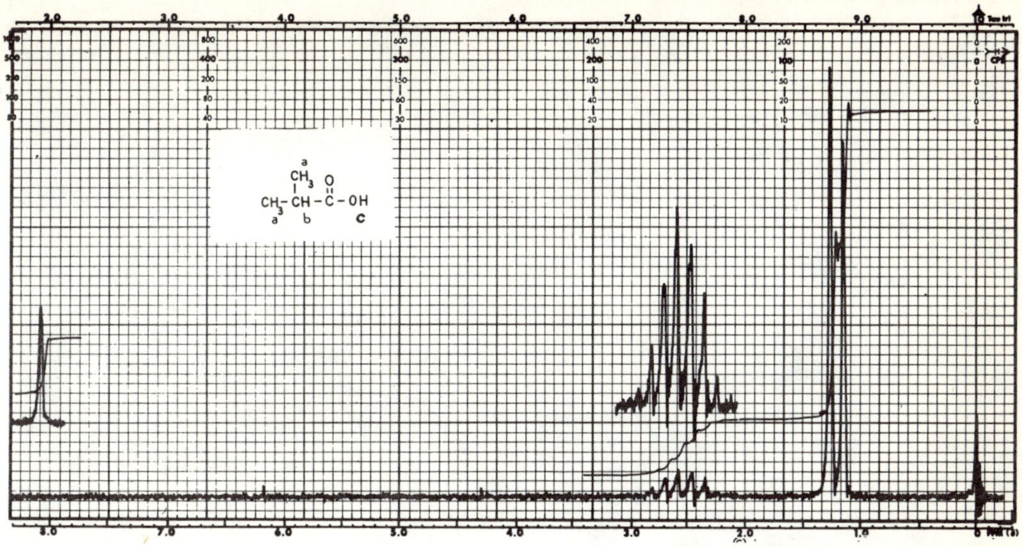

© 1978 Sadtler Research Laboratories, Inc.

2-METHYLBUTYRIC ACID

$C_5H_{10}O_2$

Mol. Wt. 102.13

B.P. 176.1°C

Solvent: $CDCl_3$

Assignments
a = 0.93
b = 1.17
c = ca 1.56
d = 2.36
e = 10.60

2475

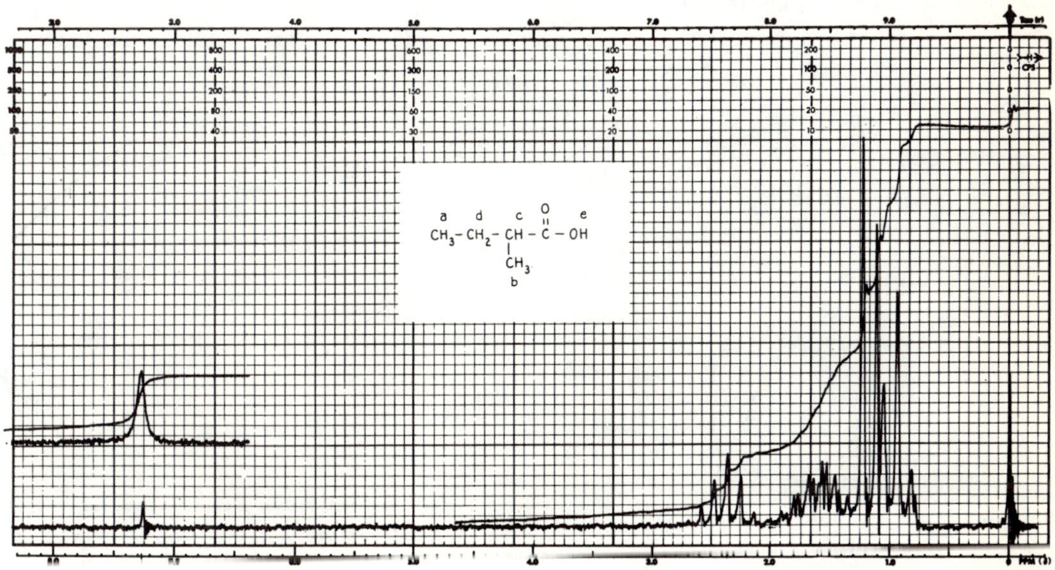

© 1978 Sadtler Research Laboratories, Inc.

PIVALIC ACID

$C_5H_{10}O_2$

Mol. Wt. 102.13

M.P. 35°C

Solvent: $CDCl_3$

Assignments
a = 1.21
b = 12.01

2476

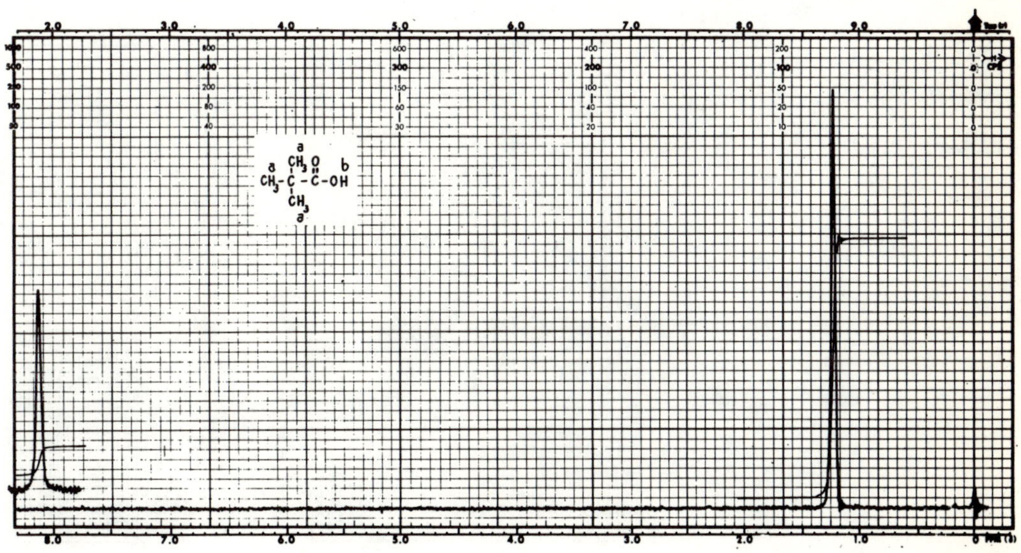

© 1978 Sadtler Research Laboratories, Inc.

2-ETHYLHEXANOIC ACID

$C_8H_{16}O_2$

Mol. Wt. 144.22

Solvent: CCl_4

Assignments
a = 0.92
b = 0.97
c = 1.16-1.90
d = 2.29
e = 12.09

2477

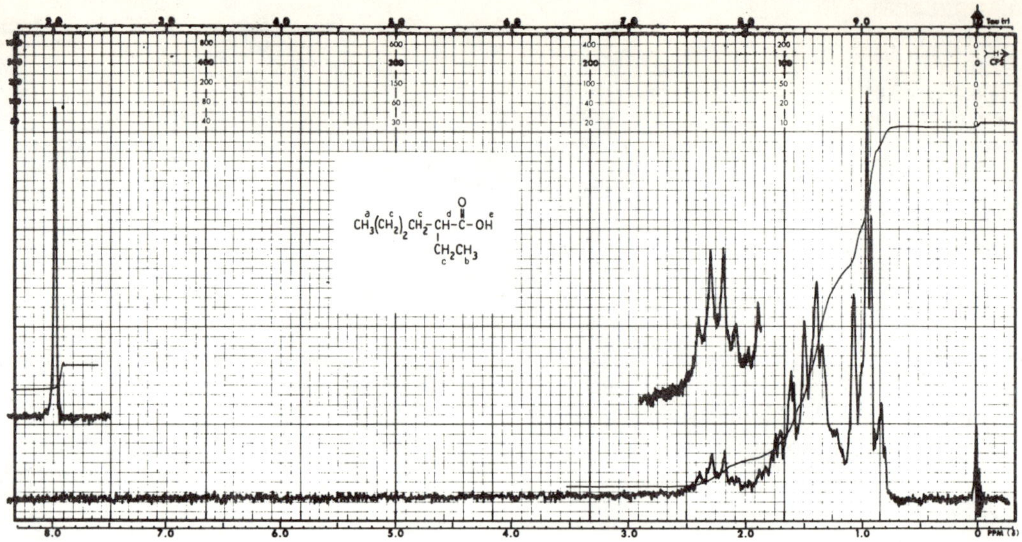

© 1978 Sadtler Research Laboratories, Inc.

2-ETHYL-4-METHYLVALERIC ACID

$C_8H_{16}O_2$

Mol. Wt. 144.22

Solvent: CCl_4

Assignments
a = 0.92
b = 0.99
c = 1.10-2.00
d = 2.38
e = 12.01

2478

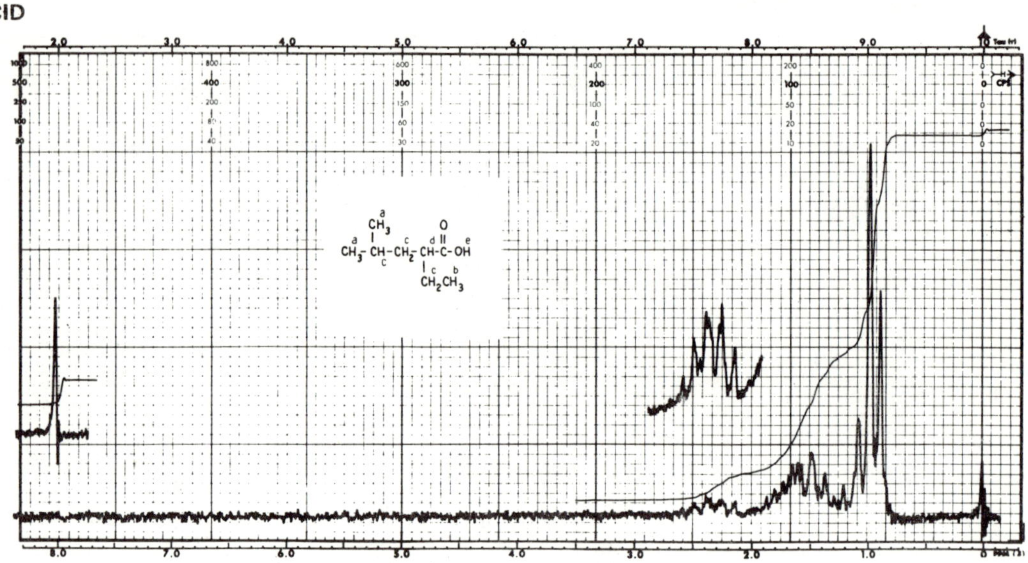

© 1978 Sadtler Research Laboratories, Inc.

2-PROPYLVALERIC ACID

$C_8H_{16}O_2$

Mol. Wt. 144.22

Solvent: CCl_4

Assignments
a = 0.91
b = 1.10-2.00
c = 2.33
d = 12.23

2479

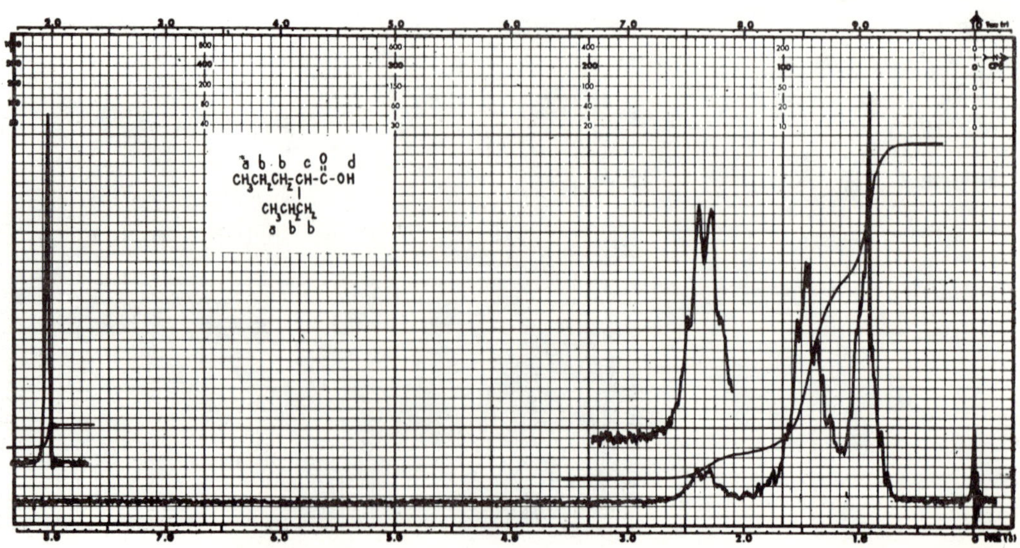

© 1978 Sadtler Research Laboratories, Inc.

LAURIC ACID

$C_{12}H_{24}O_2$

Mol. Wt. 200.32

M.P. 44-44.5°C

Solvent: $CDCl_3$

Assignments

a = 0.89
b = 1.29
c = 1.60
d = 2.33
e = 10.96

2480

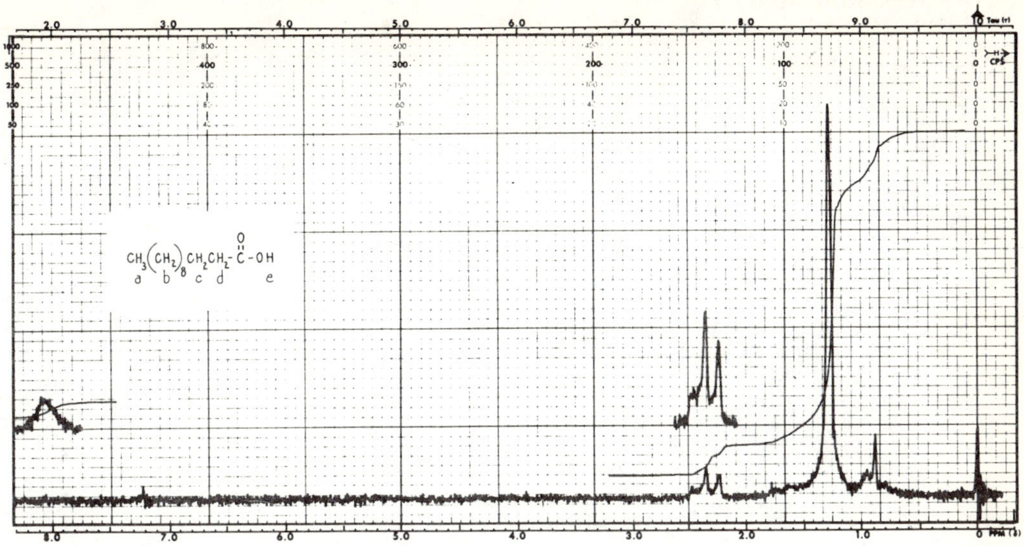

© 1978 Sadtler Research Laboratories, Inc.

2-PENTYLHEPTANOIC ACID

$C_{12}H_{24}O_2$

Mol. Wt. 200.32

Solvent: CCl_4

Assignments

a = 0.90
b = 1.10-2.00
c = 2.31
d = 11.50

2481

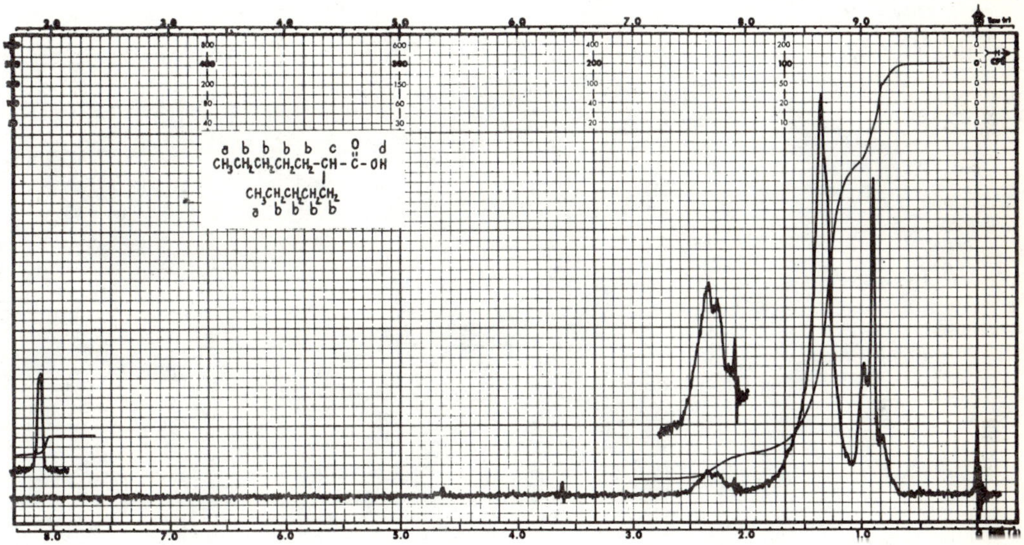

© 1978 Sadtler Research Laboratories, Inc.

HEXADECANOIC ACID

$C_{16}H_{32}O_2$

Mol. Wt. 256.43

Solvent: $CDCl_3$

Assignments

a = 0.90
b = ca 1.27
c = 1.70
d = 2.37
e = 7.90-8.90

2482

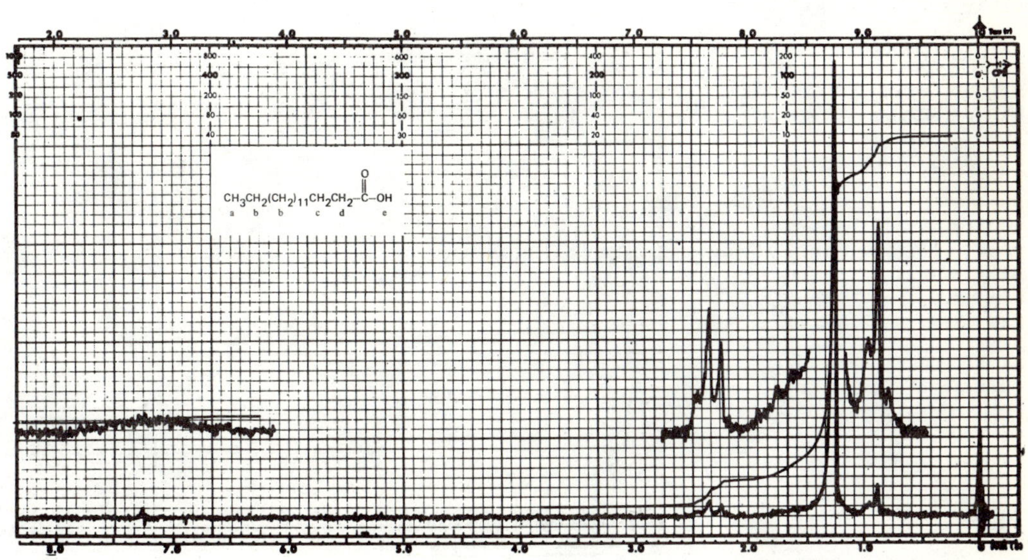

© 1978 Sadtler Research Laboratories, Inc.

CHLOROACETIC ACID

$C_2H_3ClO_2$

Mol. Wt. 94.50

M.P. 61.2°C

B.P. 189.5°C

Solvent: CCl_4

Assignments
a = 4.05
b = 11.22

2483

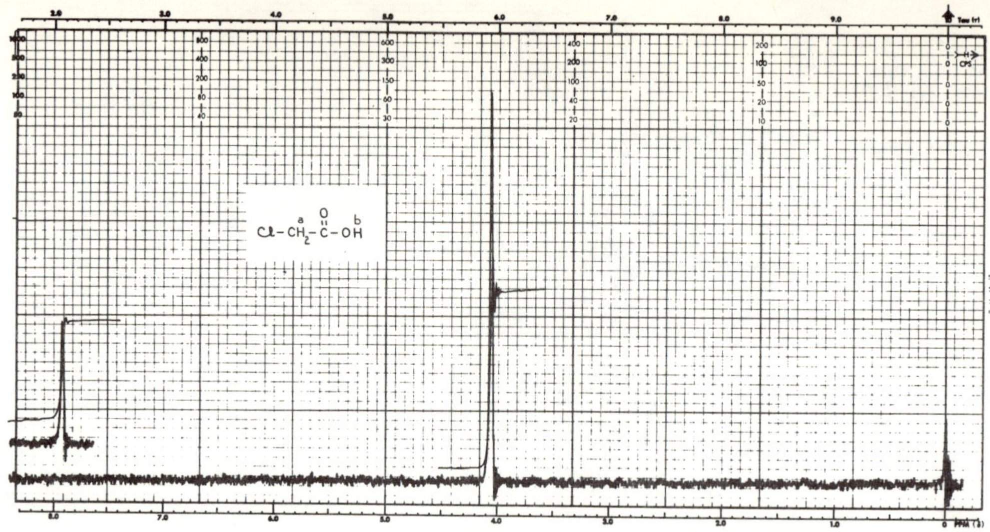

© 1978 Sadtler Research Laboratories, Inc.

2-CHLOROBUTYRIC ACID

$C_4H_7ClO_2$

Mol. Wt. 122.55

Solvent: $CDCl_3$

Assignments
a = 1.08
b = 2.04
c = 4.28
d = 11.57

2484

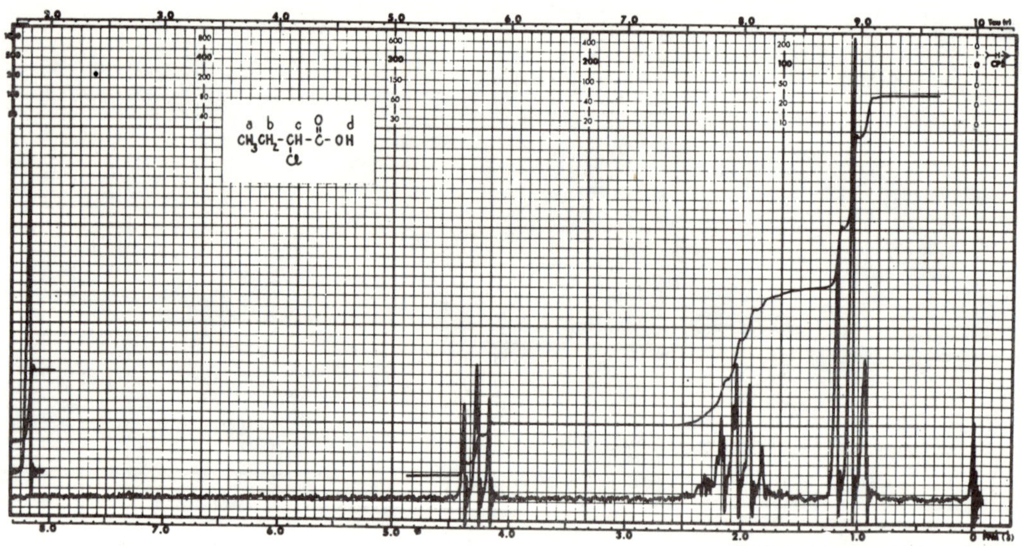

© 1978 Sadtler Research Laboratories, Inc.

BROMOACETIC ACID

$C_2H_3BrO_2$

Mol. Wt. 138.95

M.P. 48-50°C (lit.)

Solvent: $CDCl_3$

Assignments
a = 3.92
b = 10.59

2485

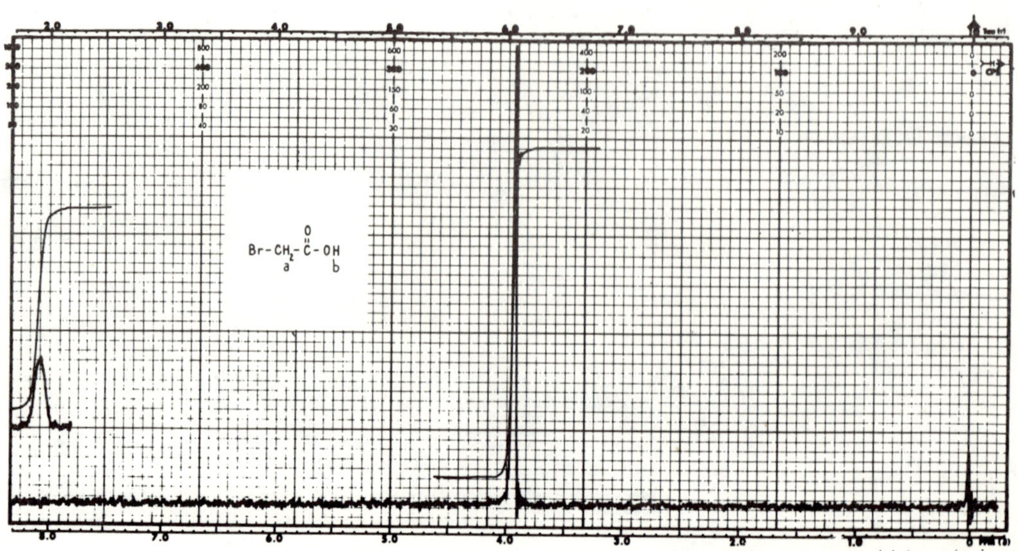

© 1978 Sadtler Research Laboratories, Inc.

2-BROMOVALERIC ACID

$C_5H_9BrO_2$

Mol. Wt. 181.04

B.P. 126-130°C/27mm

Solvent: $CDCl_3$

Assignments

a = 0.98
b = 1.49
c = 2.06
d = 4.29
e = 11.78

2486

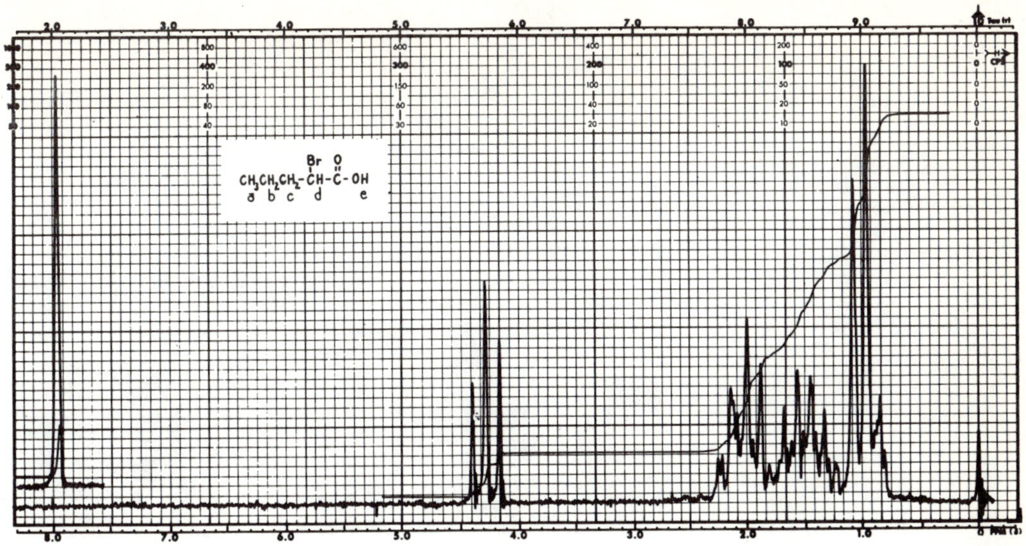

© 1978 Sadtler Research Laboratories, Inc.

2-BROMOHEPTANOIC ACID

$C_7H_{13}BrO_2$

Mol. Wt. 209.09

Solvent: CCl_4

Assignments

a = 0.90
b = 1.10-.175
c = 2.01
d = 4.19
e = 12.13

2487

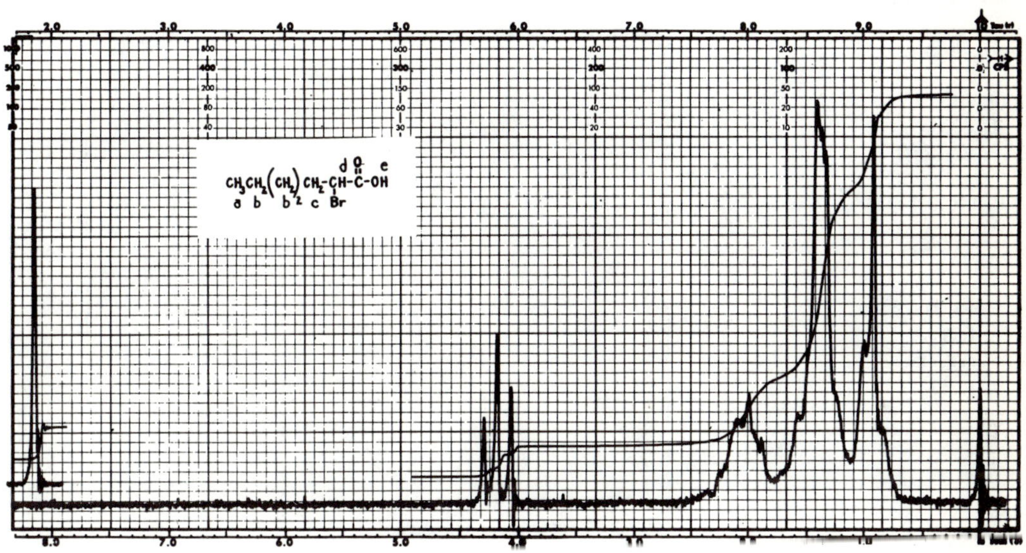

© 1978 Sadtler Research Laboratories, Inc.

IODOACETIC ACID

$C_2H_3IO_2$

Mol. Wt. 185.96

M.P. 80-82°C

Solvent: $CDCl_3$

Assignments

a = 3.71
b = 10.80

2488

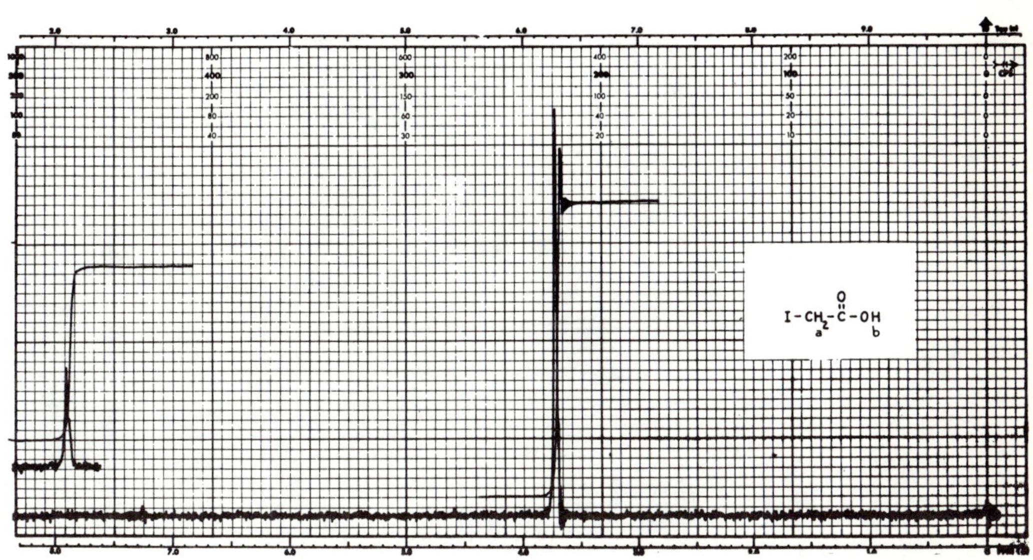

© 1978 Sadtler Research Laboratories, Inc.

2-MERCAPTOPROPIONIC ACID

$C_3H_6O_2S$

Mol. Wt. 106.14

M.P. 10-14°C

B.P. 94-96°C610mm

Solvent: CCl_4

Assignments
a = 1.52
b = 2.22
c = 3.50
d = 11.09

2489

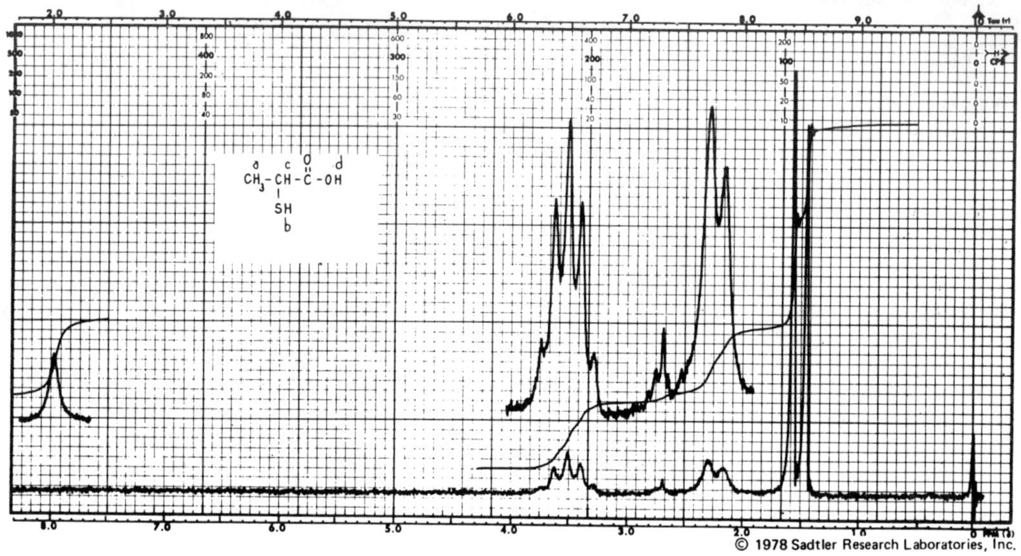

© 1978 Sadtler Research Laboratories, Inc.

3-MERCAPTOPROPIONIC ACID

$C_3H_6O_2S$

Mol. Wt. 106.14

M.P. 17-19°C

B.P. 110.5-111.5°C/15mm

Solvent: CCl_4

Assignments
a = 1.58
b = ca 2.74
c = 11.50

2490

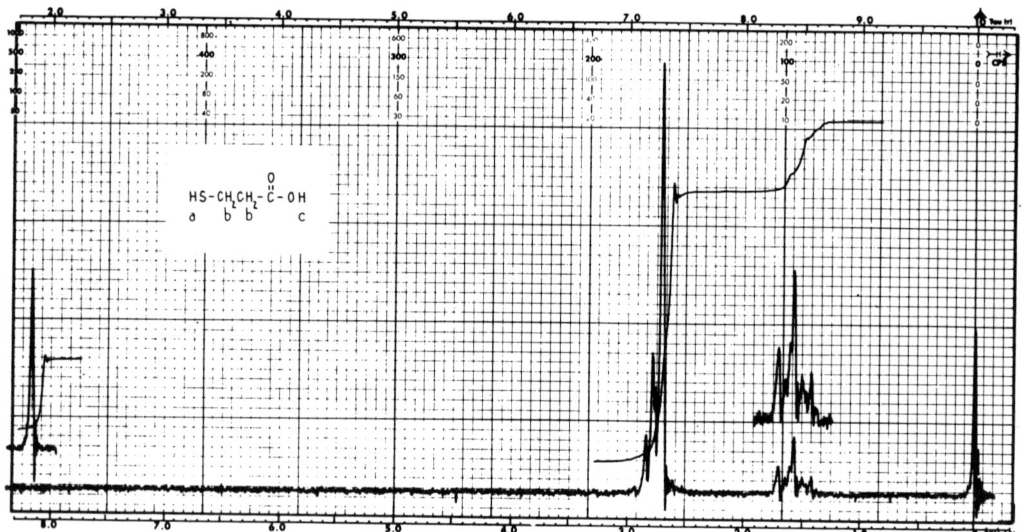

© 1978 Sadtler Research Laboratories, Inc.

METHOXYACETIC ACID

$C_3H_6O_3$

Mol. Wt. 90.08

B.P. 197-198°C/733mm (lit.)

F.P. 7.7°C (min.)

Solvent: D_2O

Assignments
a = 3.40
b = 4.10
c = 4.74 HDO

2491

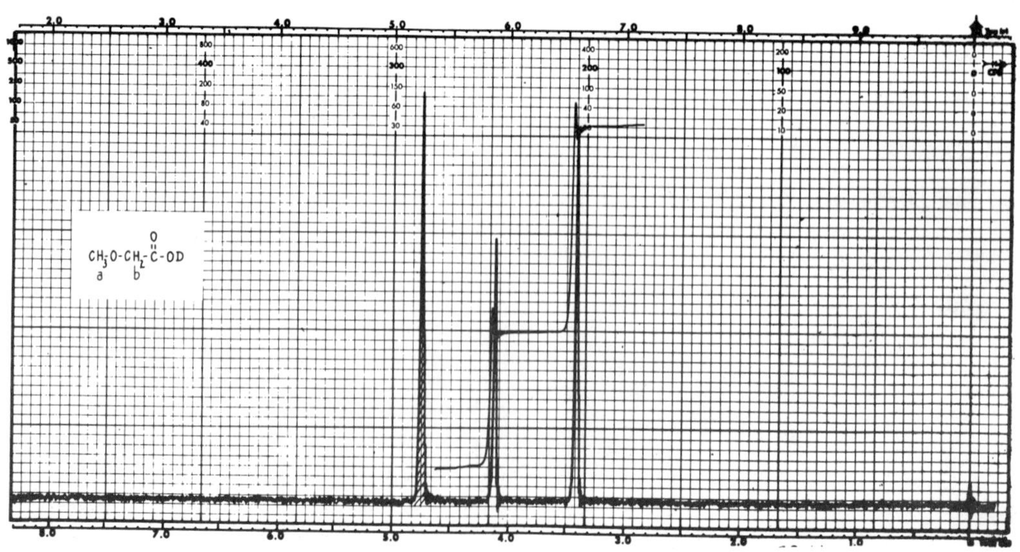

© 1978 Sadtler Research Laboratories, Inc.

GLYCOLIC ACID

$C_2H_4O_3$

Mol. Wt. 76.05

M.P. 76-77°C (lit.)

Solvent: D_2O

Assignments
a = 4.29
b = 5.35 HDO

2492

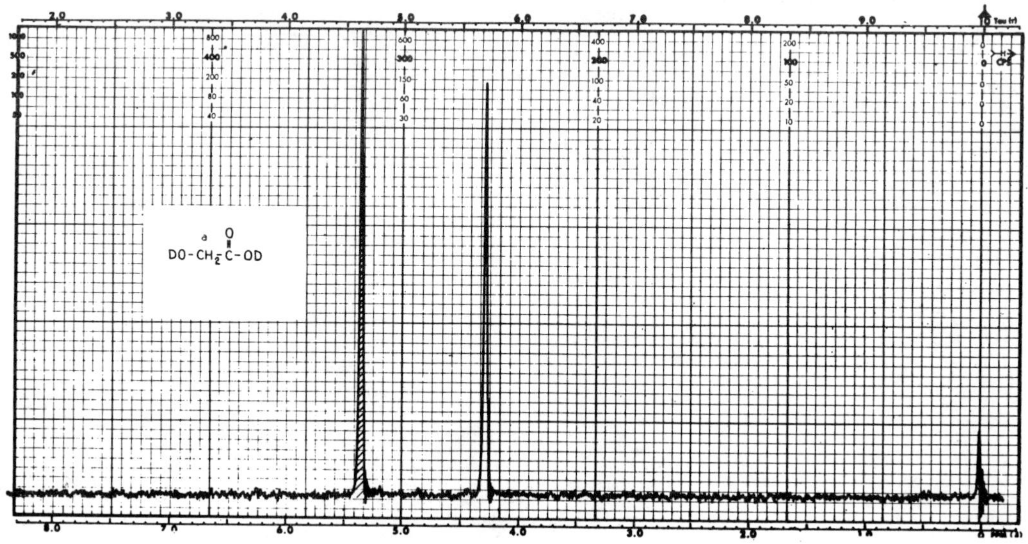

© 1978 Sadtler Research Laboratories, Inc.

LACTIC ACID

$C_3H_6O_3$

Mol. Wt. 90.08

Solvent: D_2O

Assignments
a = 1.43
b = 4.40
c = 5.09 HDO
d = 1.60 imp.

2493

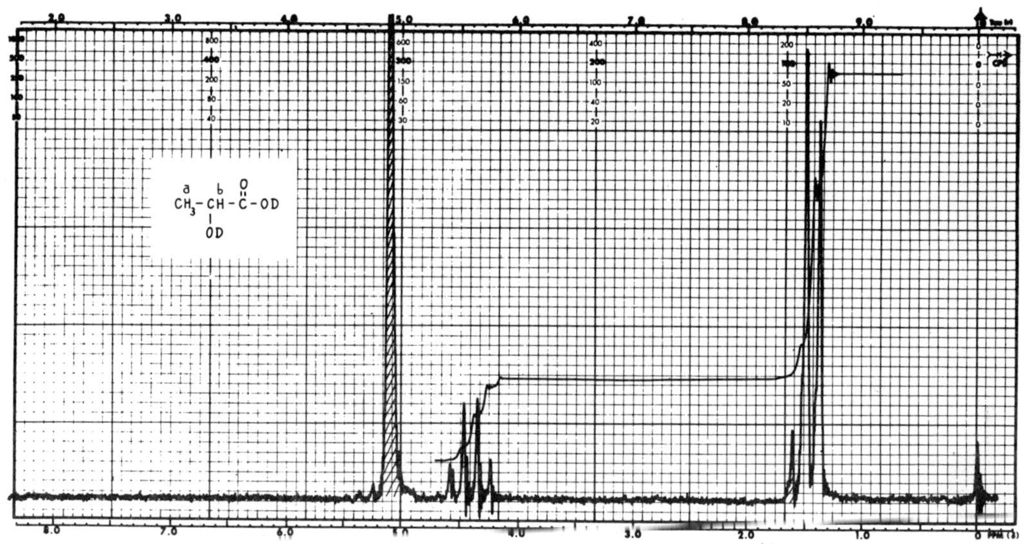

© 1978 Sadtler Research Laboratories, Inc.

LEVULINIC ACID

$C_5H_8O_3$

Mol. Wt. 116.12

M.P. 37°C

B.P. 246°C/760mm

Solvent: $CDCl_3$

Assignments
a = 2.17
b = ca 2.59
c = 2.70
d = 9.33

2494

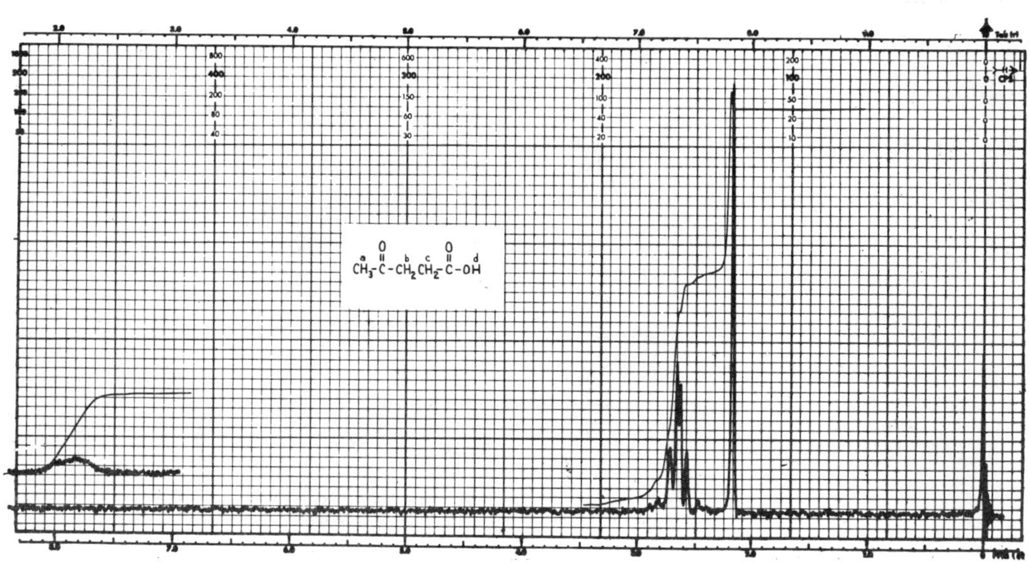

© 1978 Sadtler Research Laboratories, Inc.

9,12-DIOXOOCTADECANOIC ACID

$C_{18}H_{32}O_4$

Mol. Wt. 312.45

M.P. 99-100°C

Solvent: $CDCl_3$

Assignments
a = 0.89
b = 1.10-2.00
c = ca 2.40
d = 2.67
e = 10.03

2495

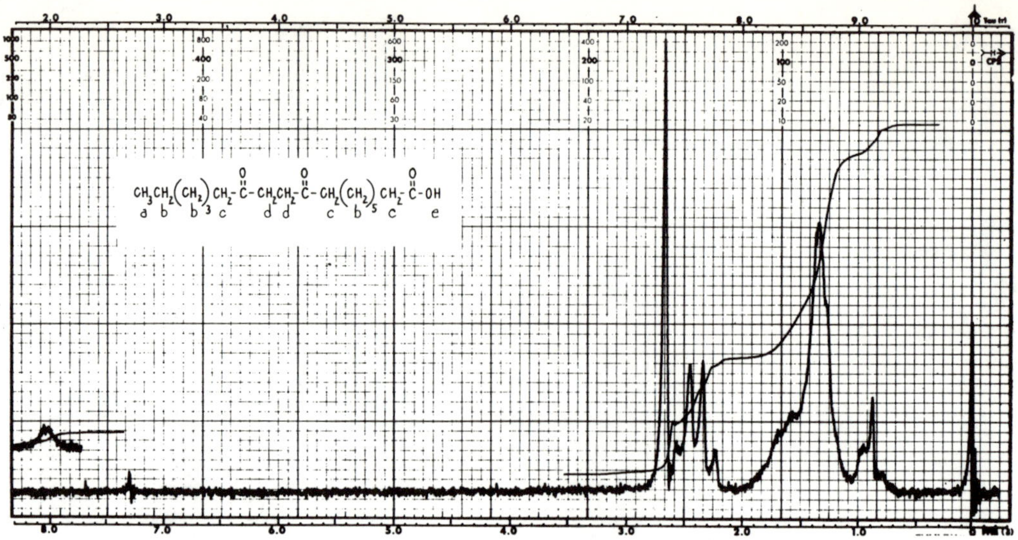

© 1978 Sadtler Research Laboratories, Inc.

MALONIC ACID

$C_3H_4O_4$

Mol. Wt. 104.06

M.P. 135-136°C

Solvent: Polysol-d

Assignments
a = 3.37
b = 11.50

2496

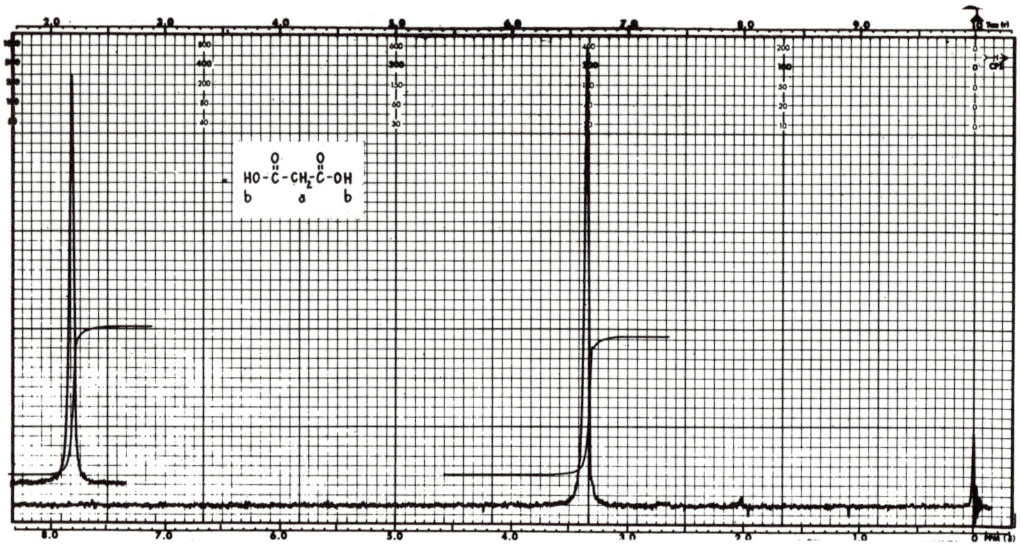

© 1978 Sadtler Research Laboratories, Inc.

SUCCINIC ACID

$C_4H_6O_4$

Mol. Wt. 118.09

M.P. 189-190°C (lit.)

Solvent: Dimethyl sulfoxide-D_6

Assignments
a = 2.43
b = ca 11.80
c = 2.52 DMSO-D_5

2497

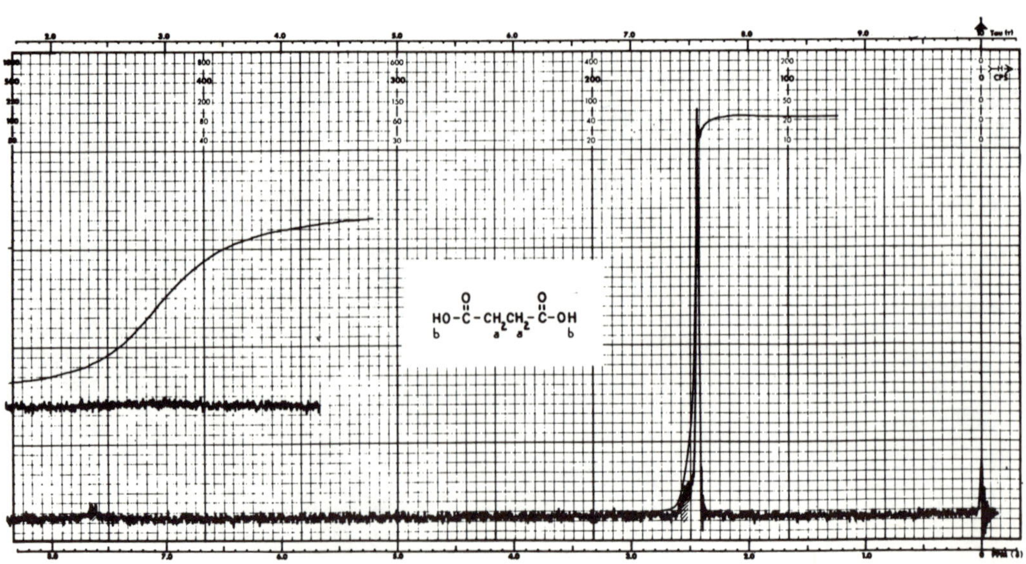

© 1978 Sadtler Research Laboratories, Inc.

GLUTARIC ACID

$C_5H_8O_4$

Mol. Wt. 132.11

M.P. 95-96°C

Solvent: DMSO-D6

Assignments
a = 1.80
b = 2.26
c = 11.74
d = 2.53 DMSO-D5

2498

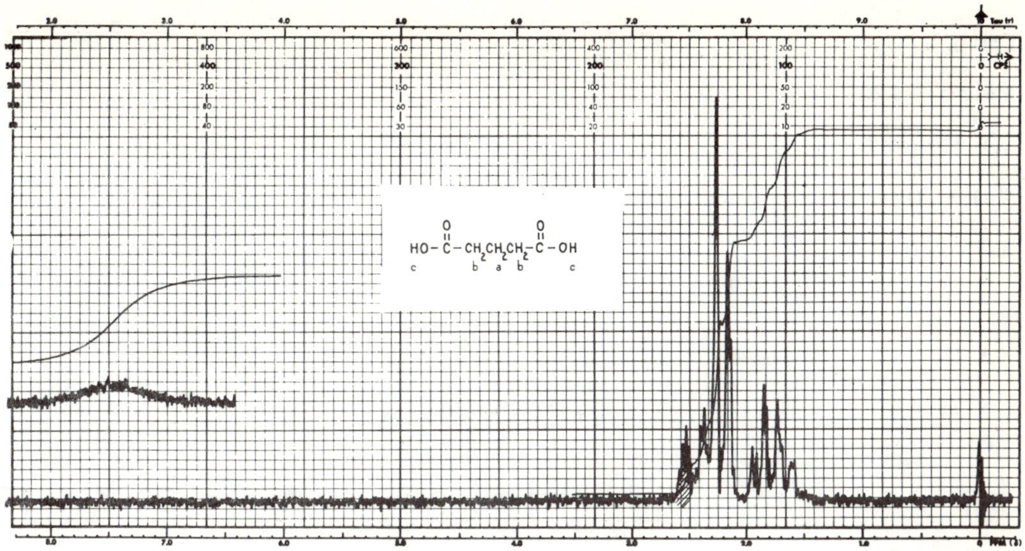

© 1978 Sadtler Research Laboratories, Inc.

ADIPIC ACID

$C_6H_{10}O_4$

Mol. Wt. 146.14

M.P. 152°C

B.P. 265°C/100mm

Solvent: DMSO-D6

Assignments
a = 1.52
b = 2.21
c = 11.32
d = 2.53 DMSO-D5

2499

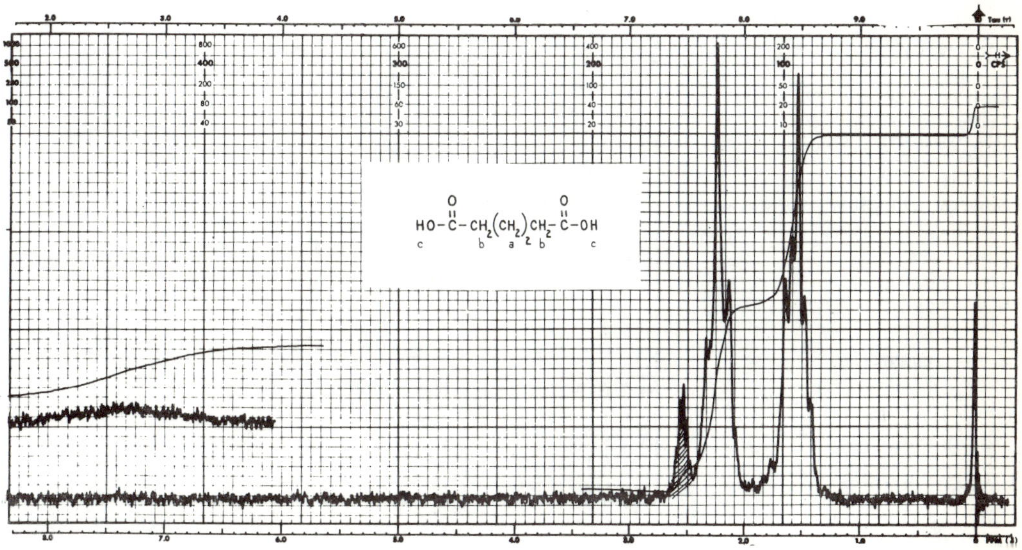

© 1978 Sadtler Research Laboratories, Inc.

3,3-DIMETHYLGLUTARIC ACID

$C_7H_{12}O_4$

Mol. Wt. 160.17

M.P. 101-103°C

Solvent: Polysol-d

Assignments
a = 1.10
b = 2.31
c = 9.74

2500

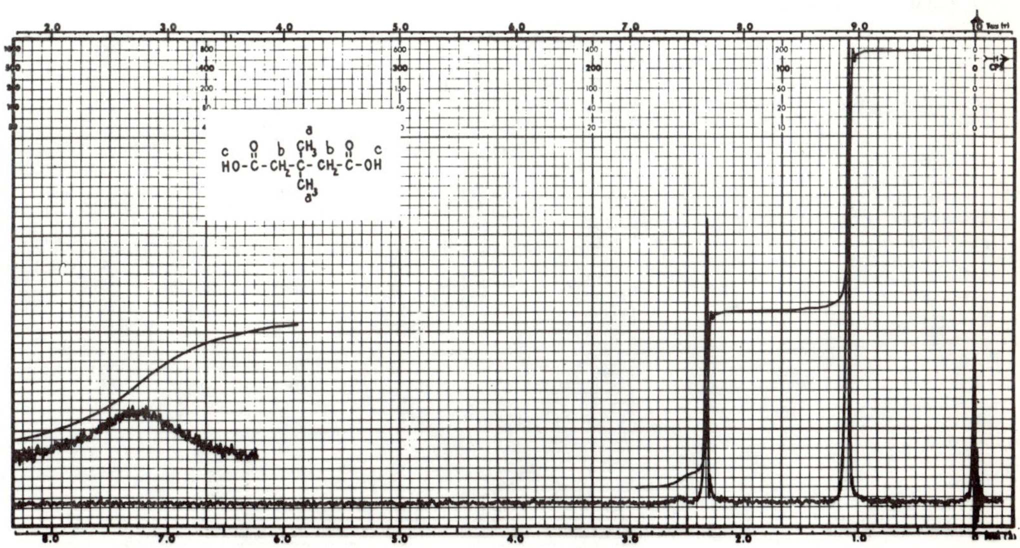

© 1978 Sadtler Research Laboratories, Inc.

SUBERIC ACID

$C_8H_{14}O_4$

Mol. Wt. 174.20

M.P. 139-141°C

Solvent: Dimethyl sulfoxide-D_6

Assignments

a = 1.35
b = 1.62
c = 2.20
d = 11.67
e = 2.52 DMSO-D_5
f = 3.61 imp.

2501

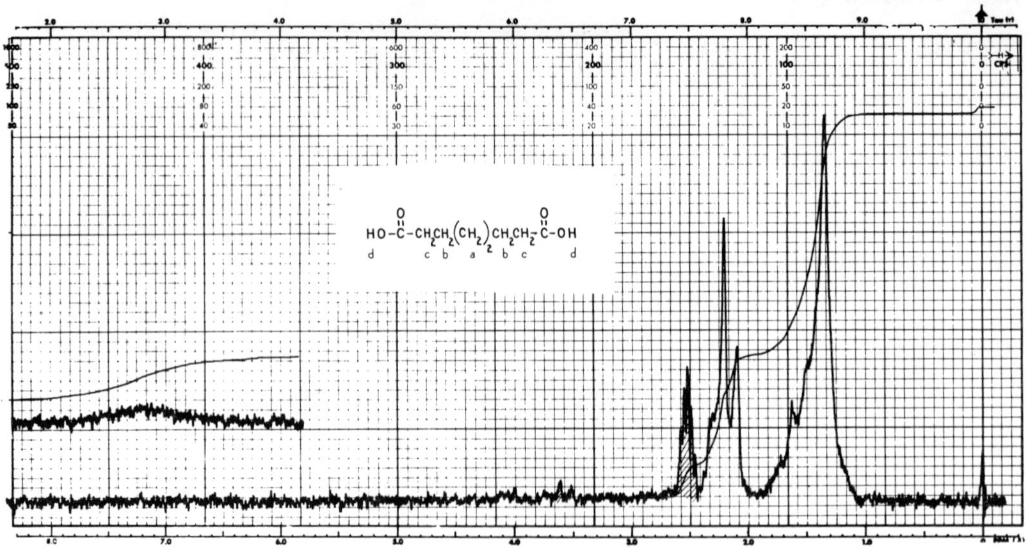

© 1978 Sadtler Research Laboratories, Inc.

SEBACIC ACID

$C_{10}H_{18}O_4$

Mol. Wt. 202.25

M.P. 131°C

Solvent: Dimethyl sulfoxide-d_6

Assignments

a = 1.02-1.88
b = 2.22
c = 11.05
d = 2.60 DMSO-d_5

2502

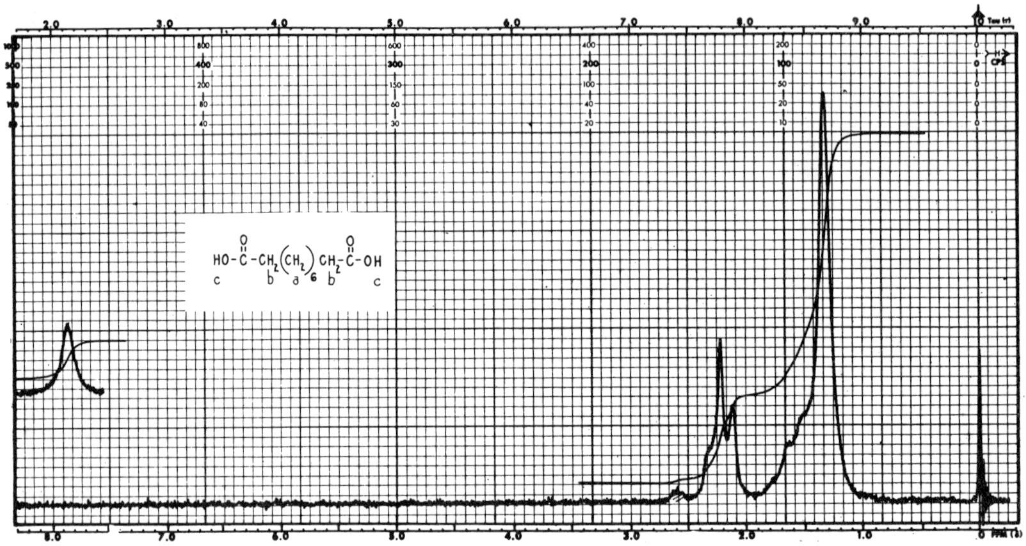

© 1978 Sadtler Research Laboratories, Inc.

HEXADECANEDIOIC ACID

$C_{16}H_{30}O_4$

Mol. Wt. 286.42

Solvent: Polysol-d

Assignments

a = 1.29
b = 1.58
c = 2.27
d = 8.97

2503

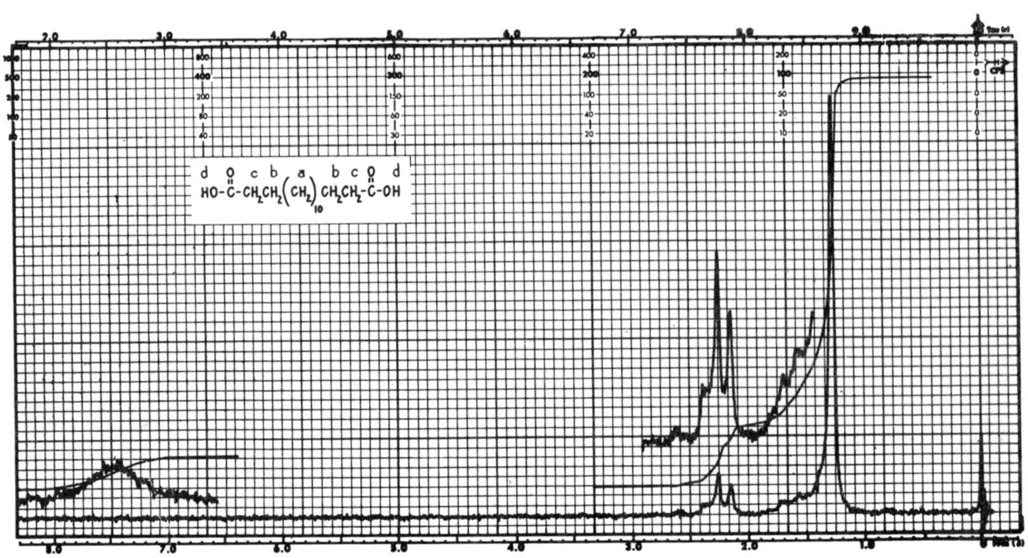

© 1978 Sadtler Research Laboratories, Inc.

HEXAFLUOROGLUTARIC ACID

$C_5H_2F_6O_4$

Mol. Wt. 240.06

M.P. 78-88°C

Solvent: Polysol-d

Assignments
a = 12.53

2504

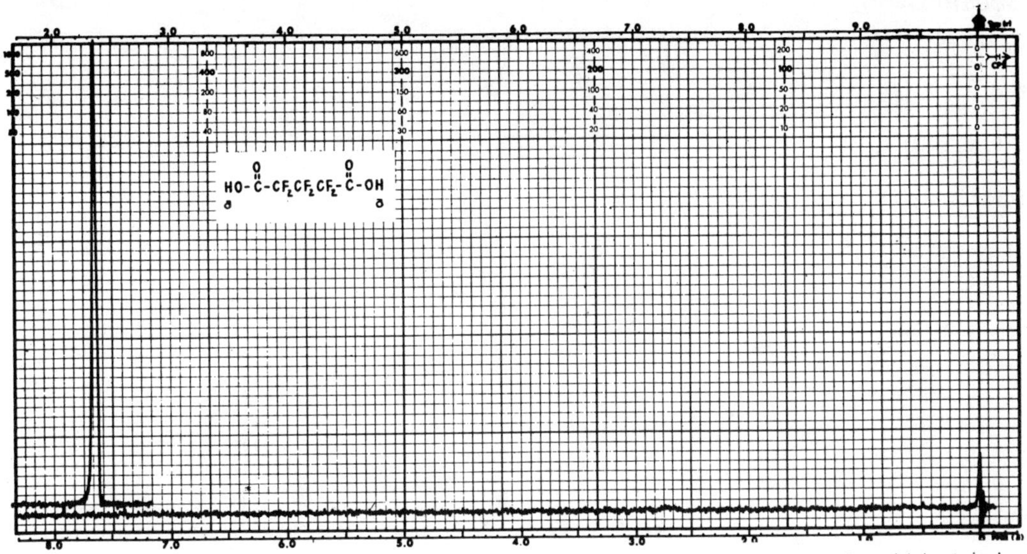

© 1978 Sadtler Research Laboratories, Inc.

3-CHLOROGLUTARIC ACID

$C_5H_7ClO_4$

Mol. Wt. 166.56

M.P. 125-126°C

Solvent: D_2O

Assignments
a = 2.83
b = 3.08
c = 4.69
d = 5.46 HDO
TFA added

2505

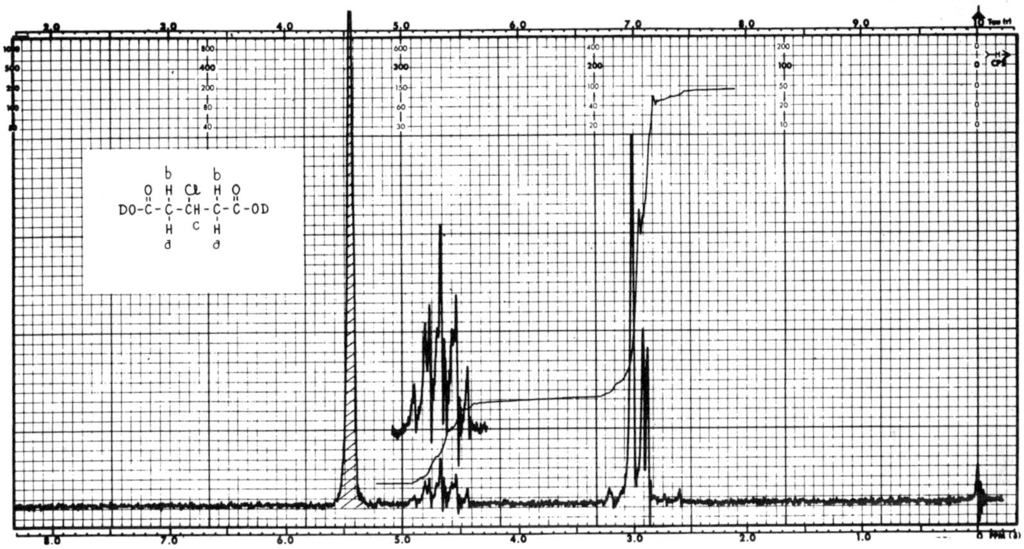

© 1978 Sadtler Research Laboratories, Inc.

TARTRONIC ACID

$C_3H_4O_5$

Mol. Wt. 120.06

M.P. 155-156°C

Solvent: D_2O

Assignments
a = 5.01
b = 5.45 HDO

2506

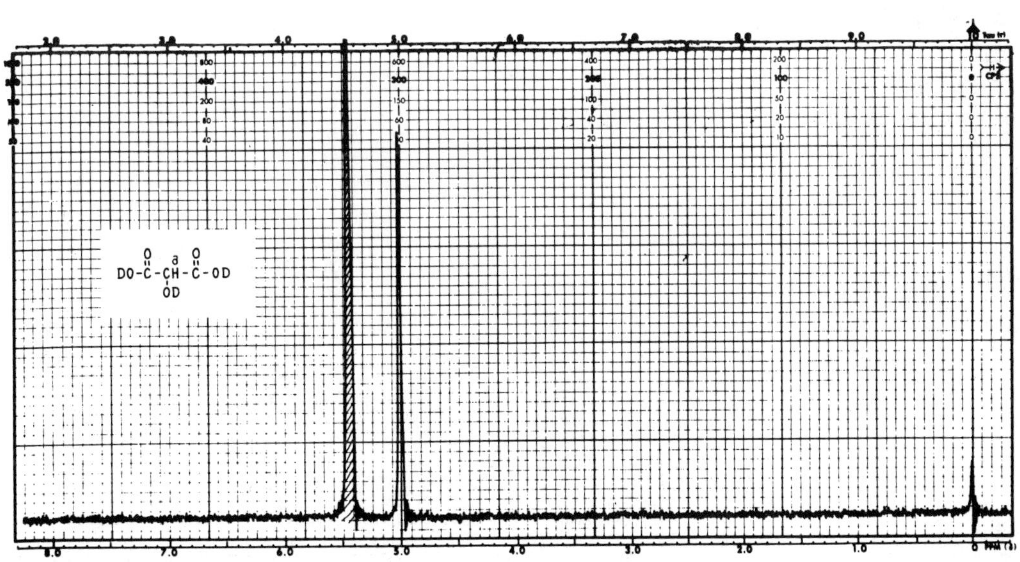

© 1978 Sadtler Research Laboratories, Inc.

DL-MALIC ACID

$C_4H_6O_5$

Mol. Wt. 134.09

M.P. 129-131°C

Solvent: D_2O

Assignments
a = 2.91
b = 4.63
c = 5.10 HDO

2507

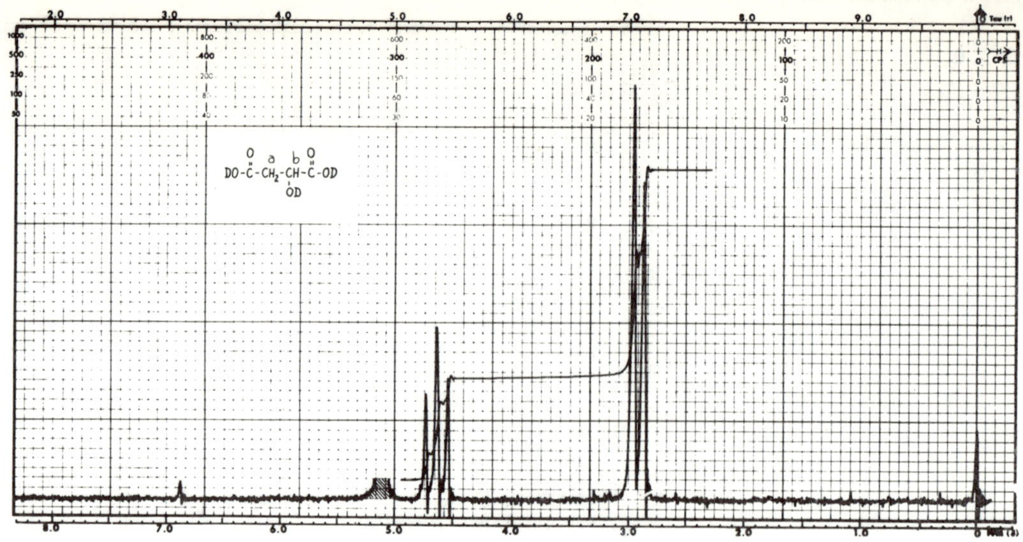

© 1978 Sadtler Research Laboratories, Inc.

L-(+)-TARTARIC ACID

$C_4H_6O_6$

Mol. Wt. 150.09

Solvent: D_2O

Assignments
a = 4.68
b = 4.47 HDO

2508

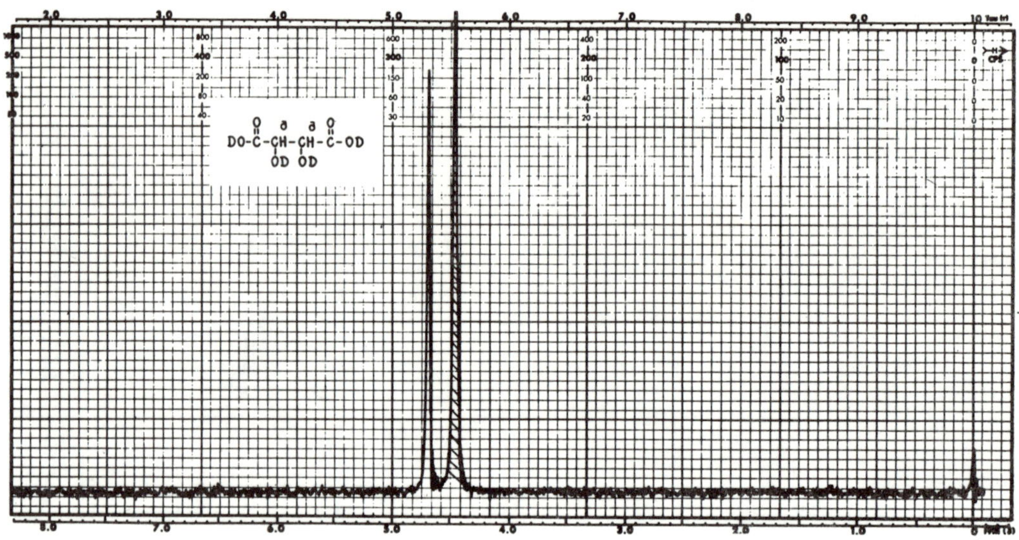

© 1978 Sadtler Research Laboratories, Inc.

DL-TARTARIC ACID

$C_4H_6O_6$

Mol. Wt. 150.09

M.P. 205°C (dec.)

Solvent: D_2O

Assignments
a = 4.79
b = 4.97 HDO

2509

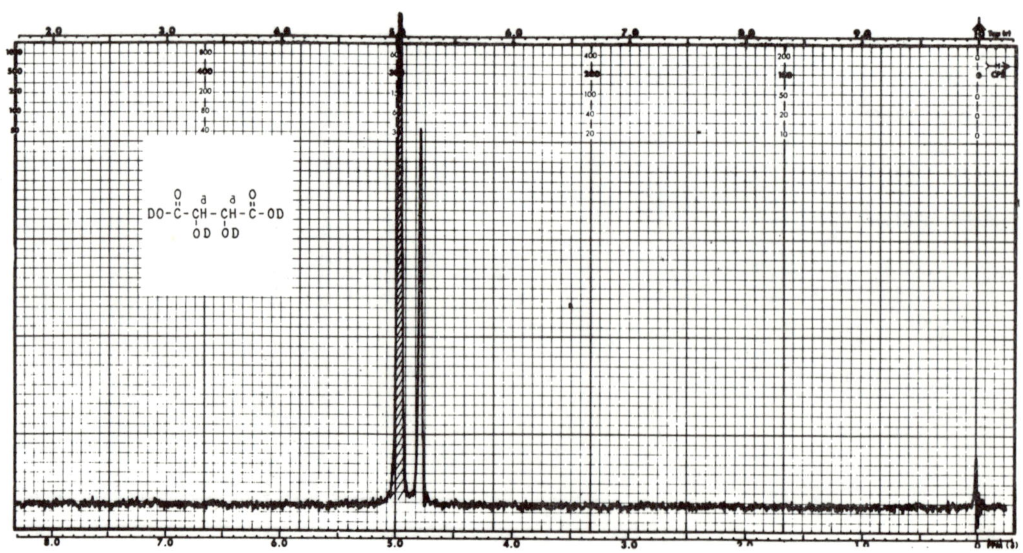

© 1978 Sadtler Research Laboratories, Inc.

1,2,3-PROPANETRICARBOXYLIC ACID

$C_6H_8O_6$

Mol. Wt. 176.13

Solvent: D_2O

Assignments
a = 2.78
b = 3.22
c = 4.89 HDO

2510

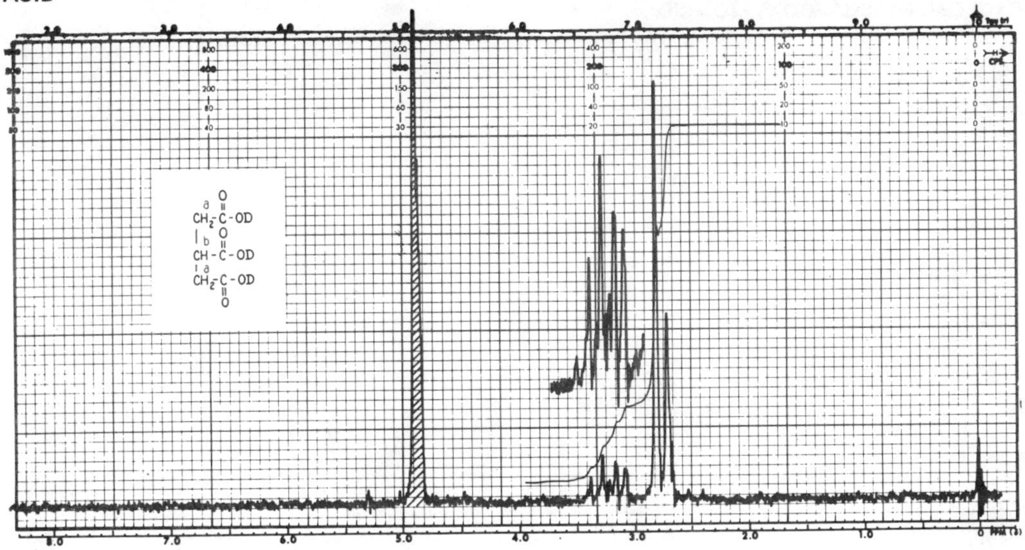

© 1978 Sadtler Research Laboratories, Inc.

CITRIC ACID (ANHYDROUS)

$C_6H_8O_7$

Mol. Wt. 192.13

M.P. 153°C

Solvent: D_2O

Assignments
a = 2.90
b = 3.08
c = 5.05 HDO

2511

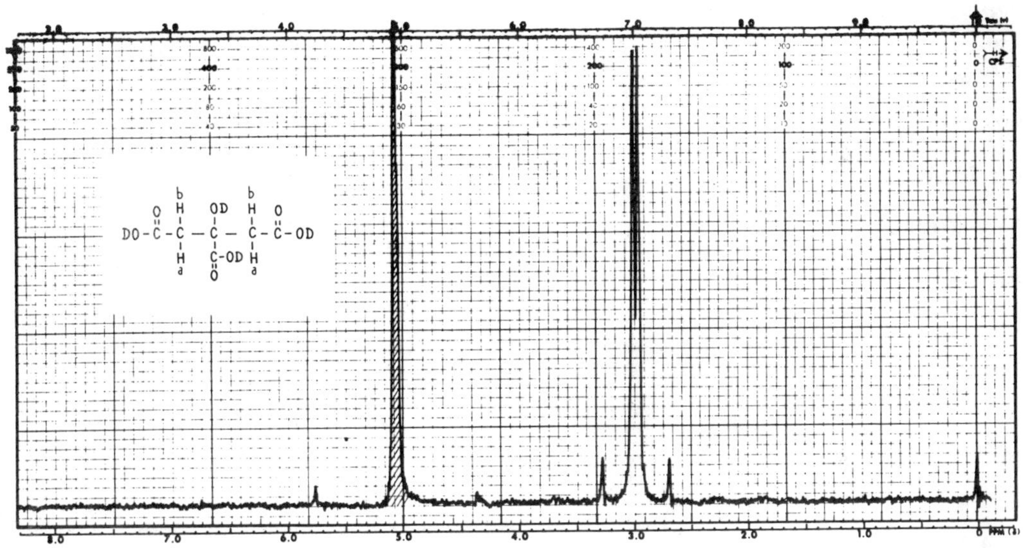

© 1978 Sadtler Research Laboratories, Inc.

2-THIOPHENEACETIC ACID

$C_6H_6O_2S$

Mol. Wt. 142.18

Solvent: $CDCl_3$

Assignments
a = 3.79
b = ca 6.89
c = 7.13
d = 11.49
e = 1.16 imp.

2512

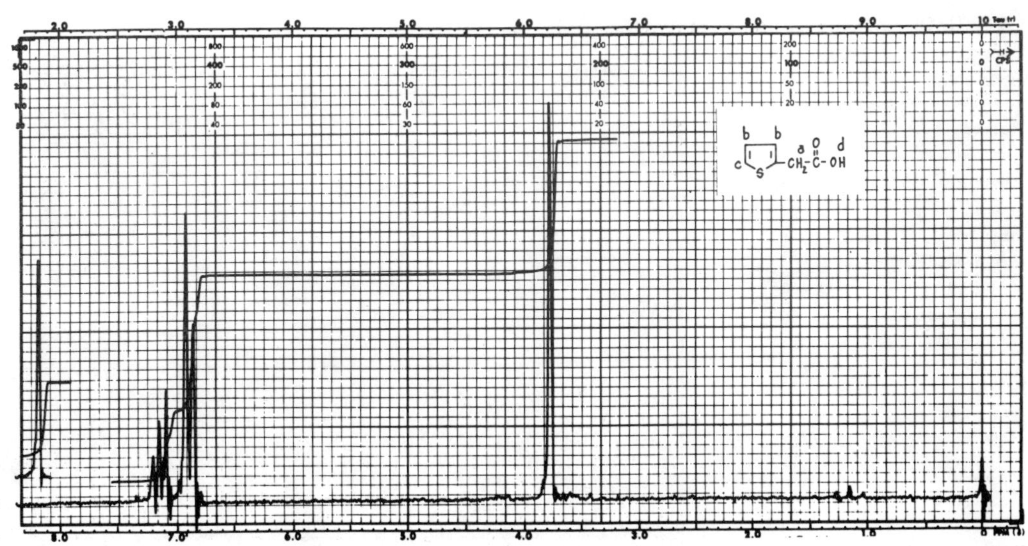

© 1978 Sadtler Research Laboratories, Inc.

CYCLOBUTANECARBOXYLIC ACID

$C_5H_8O_2$

Mol. Wt. 100.12

B.P. 195°C

Solvent: $CDCl_3$

Assignments
a = 1.60-2.70
b = 3.19
c = 11.99

2513

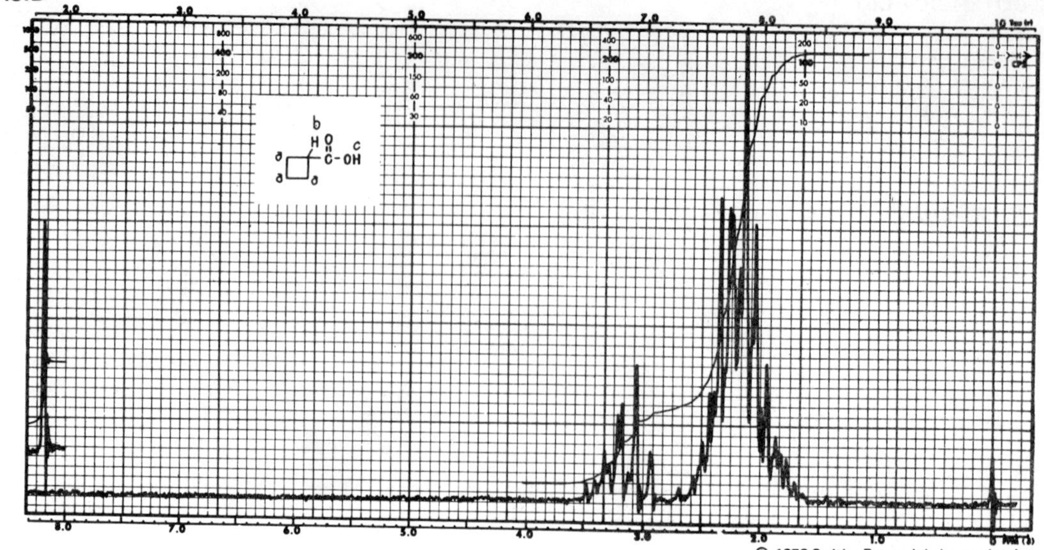

© 1978 Sadtler Research Laboratories, Inc.

CYCLOHEXANECARBOXYLIC ACID

$C_7H_{12}O_2$

Mol. Wt. 128.17

M.P. 29-31°C

Solvent: CCl_4

Assignments
a = 1.10-2.60
b = 12.00

2514

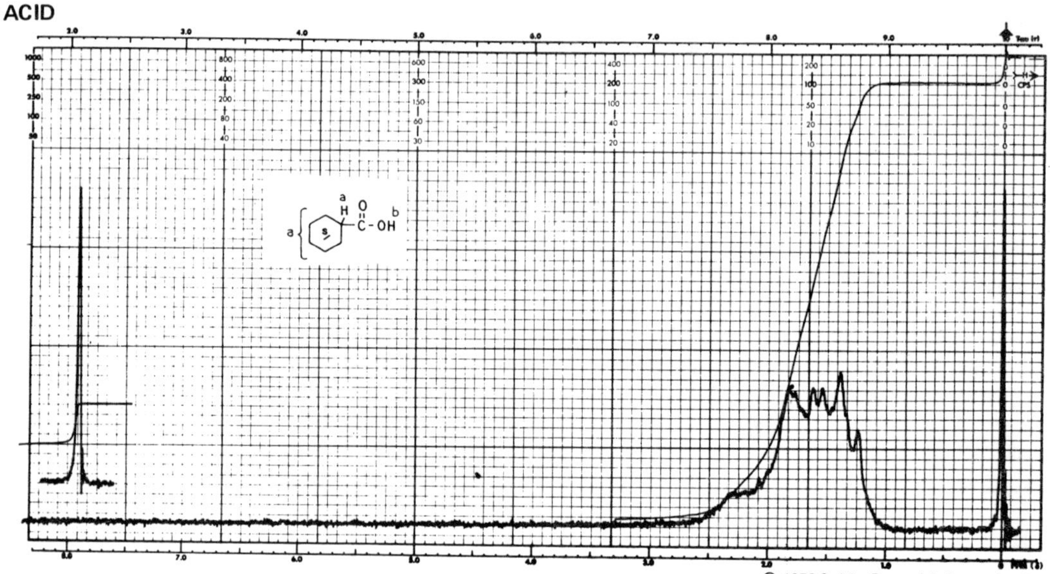

© 1978 Sadtler Research Laboratories, Inc.

1,4-CYCLOHEXANEDICARBOXYLIC ACID

$C_8H_{12}O_4$

Mol. Wt. 172.18

Solvent: Trifluoroacetic acid

Assignments
a = 1.30-2.30
b = 2.69
c = unobserved *
* In exchange with solvent

2515

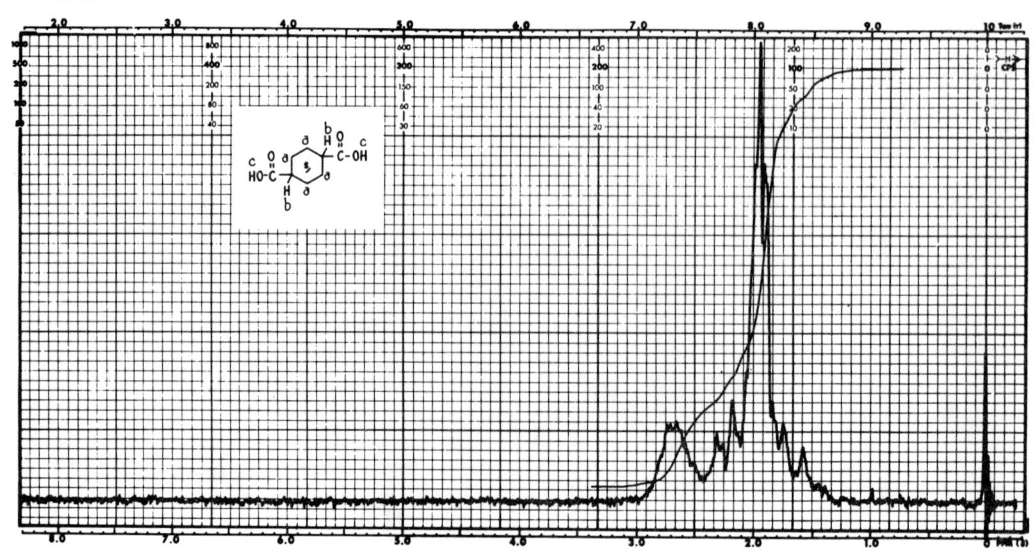

© 1978 Sadtler Research Laboratories, Inc.

3-BUTENOIC ACID

$C_4H_6O_2$

Mol. Wt. 86.09

B.P. 69-70°C/12mm

Solvent: $CDCl_3$

Assignments

a = 3.12
b = 5.12
c = 5.18
d = 5.87
e = 10.08
f = 1.83, 1.94 imps.

2516

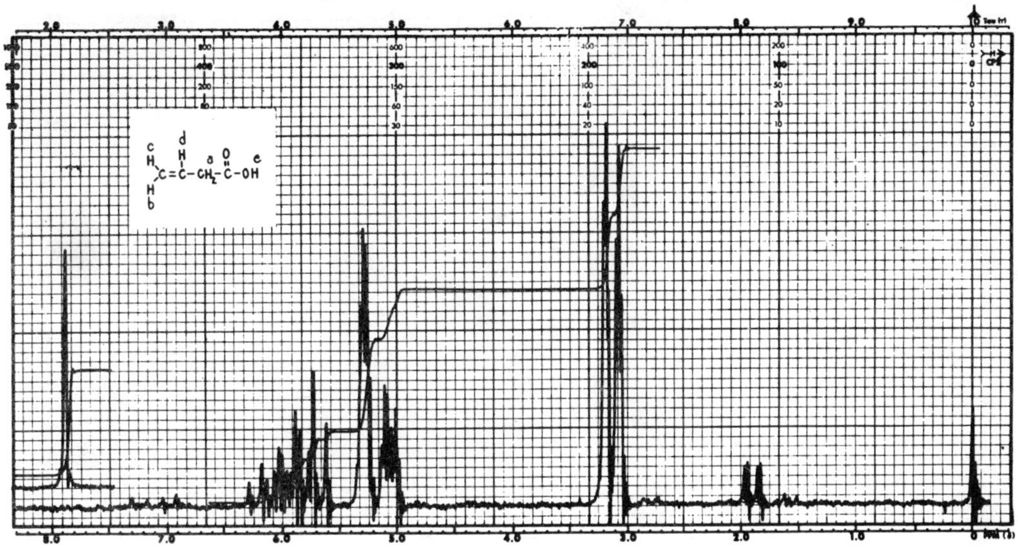

© 1978 Sadtler Research Laboratories, Inc.

CROTONIC ACID

$C_4H_6O_2$

Mol. Wt. 86.09

M.P. 70-72°C

B.P. 182-185°C

Solvent: $CDCl_3$

Assignments

a = 1.89
b = 5.82
c = 7.04
d = 11.79

2517

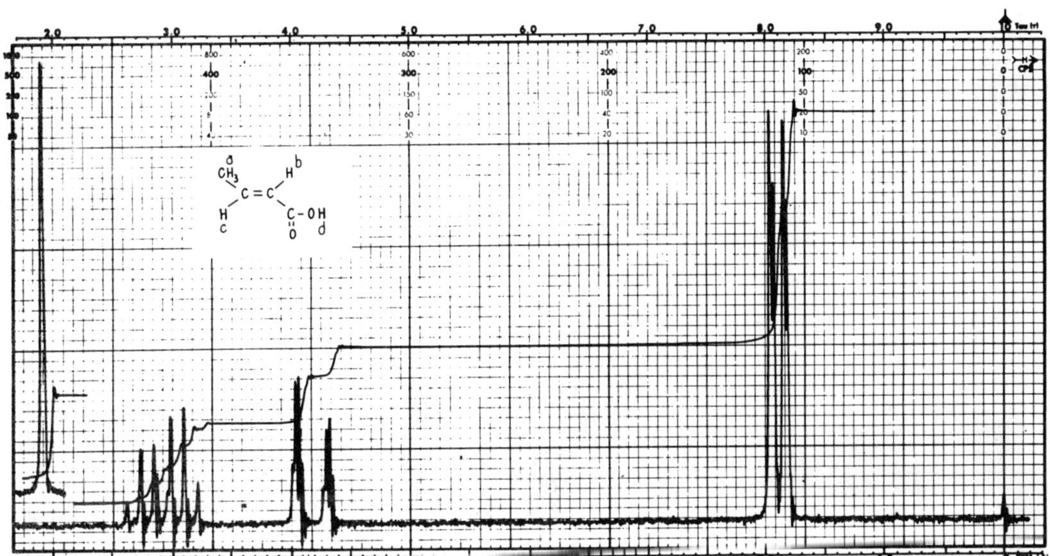

© 1978 Sadtler Research Laboratories, Inc.

4-PENTENOIC ACID

$C_5H_8O_2$

Mol. Wt. 110.12

Solvent: $CDCl_3$

Assignments

a = 2.20-2.70
b = 5.02
c = 5.08
d = 5.80
e = 10.81

2518

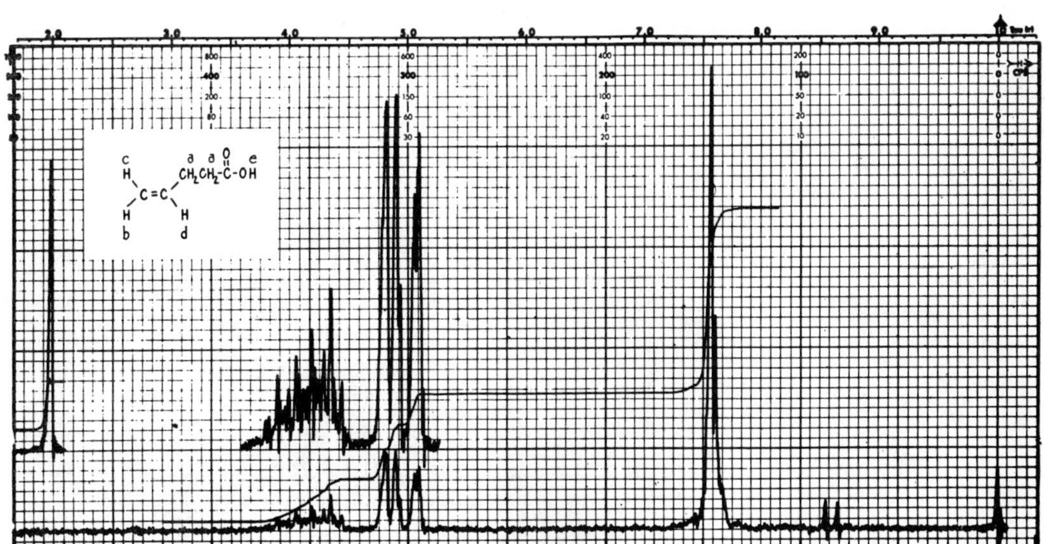

© 1978 Sadtler Research Laboratories, Inc.

3-METHYLCROTONIC ACID

$C_5H_8O_2$

Mol. Wt. 100.12

M.P. 64-67°C

Solvent: CCl_4

Assignments
a = 1.95
b = 2.19
c = 5.69
d = 12.22

2519

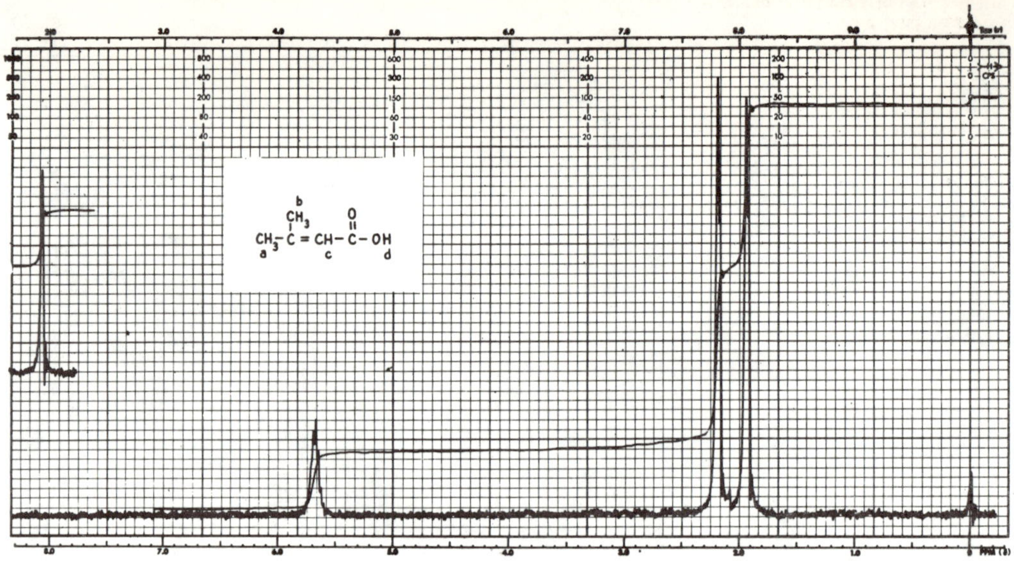

© 1978 Sadtler Research Laboratories, Inc.

trans-2-HEXENOIC ACID

$C_6H_{10}O_2$

Mol. Wt. 114.15

Solvent: CCl_4

Assignments
a = 0.98
b = 1.54
c = 2.21
d = 5.77
e = 7.03
f = 11.91

2520

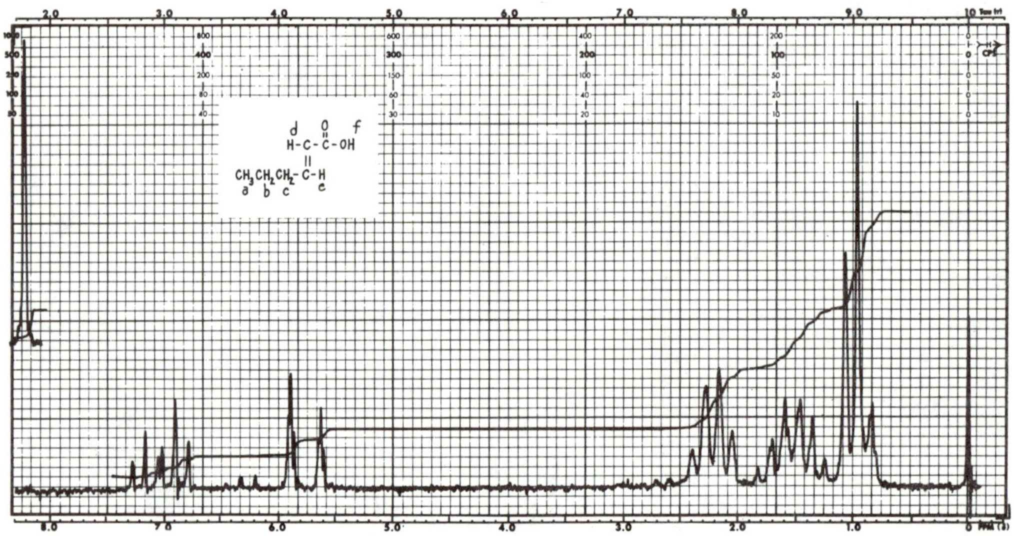

© 1978 Sadtler Research Laboratories, Inc.

3-HEPTENOIC ACID

$C_7H_{12}O_2$

Mol. Wt. 128.17

Solvent: CCl_4

Assignments
a = 0.90
b = 1.41
c = 2.02
d = 3.03
e = ca 5.54
f = 11.47

2521

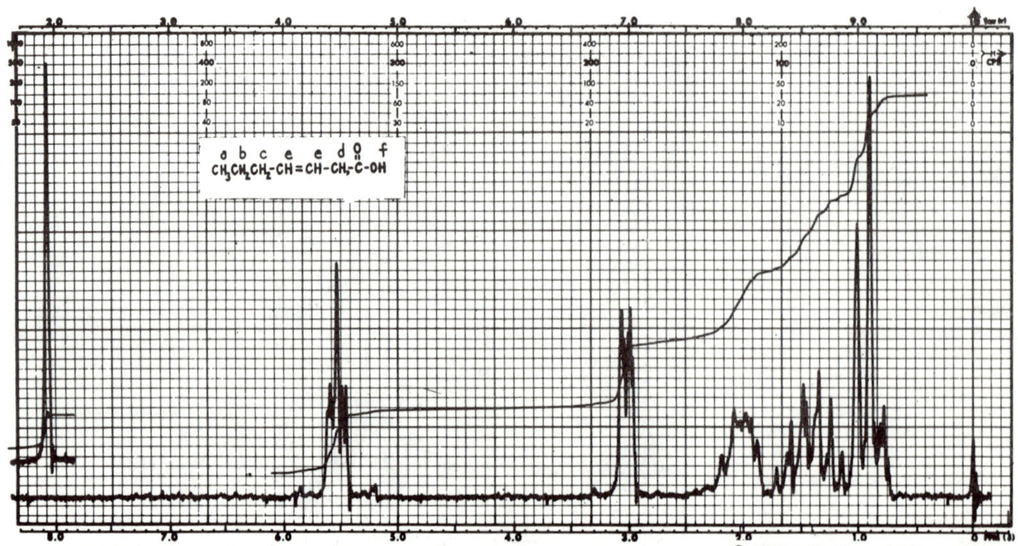

© 1978 Sadtler Research Laboratories, Inc.

4-METHYL-2-HEPTENOIC ACID

$C_8H_{14}O_2$

Mol. Wt. 142.20

Solvent: CCl_4

Assignments

a = 0.93
b = 1.08
c = 1.10-1.70
d = 2.36
e = 5.71
f = 6.93
g = 12.15

2522

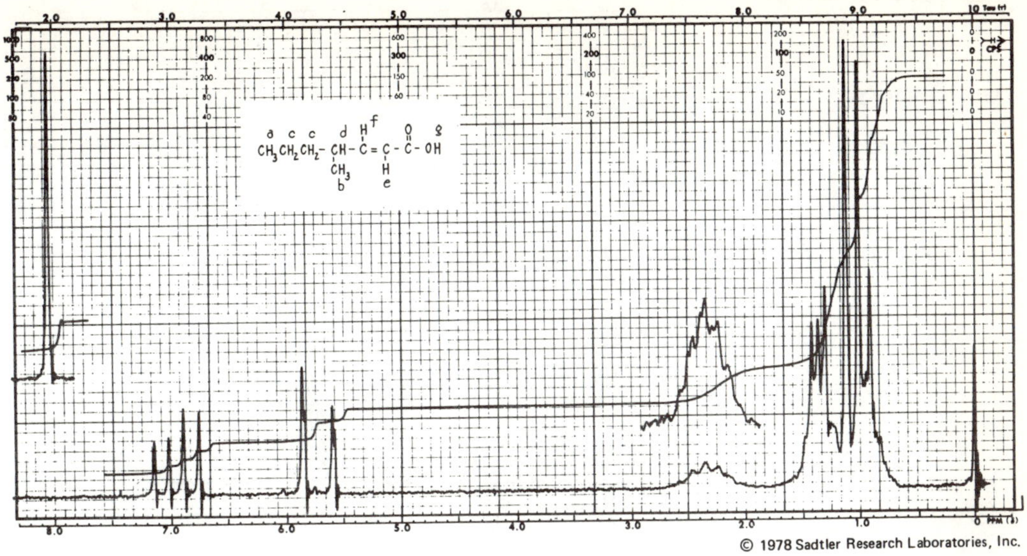

© 1978 Sadtler Research Laboratories, Inc.

2-ETHYL-2-HEXENOIC ACID

$C_8H_{14}O_2$

Mol. Wt. 142.20

B.P. 126-128°C/10mm

Solvent: CCl_4

Assignments

a = 0.96
b = 1.03
c = 1.50
d = 2.27
e = 6.88
f = 12.34

2523

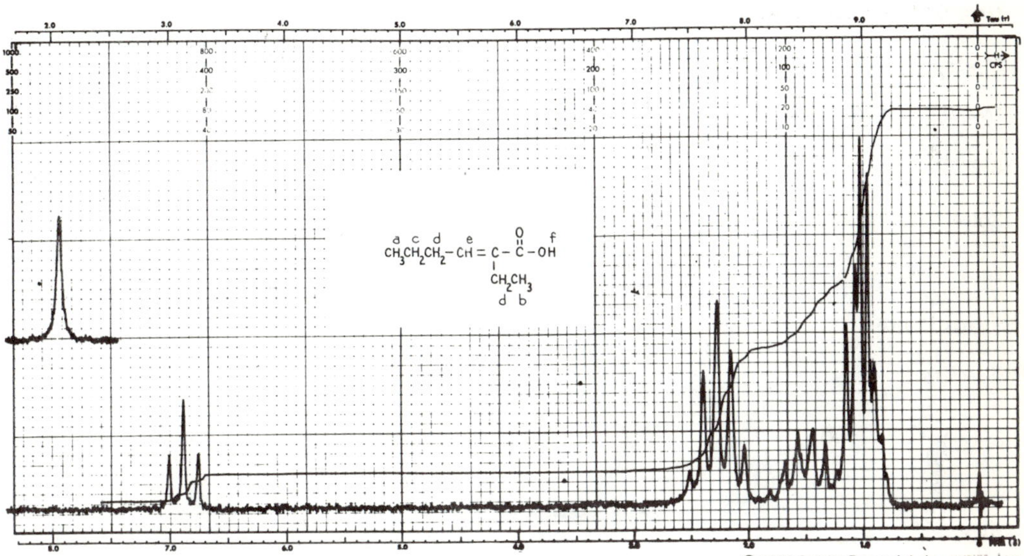

© 1978 Sadtler Research Laboratories, Inc.

2-DECENOIC ACID

$C_{10}H_{18}O_2$

Mol. Wt. 170.25

Solvent: CCl_4

Assignments

a = 0.89
b = 1.31
c = 2.20
d = 5.76
e = 7.04
f = 12.35

2524

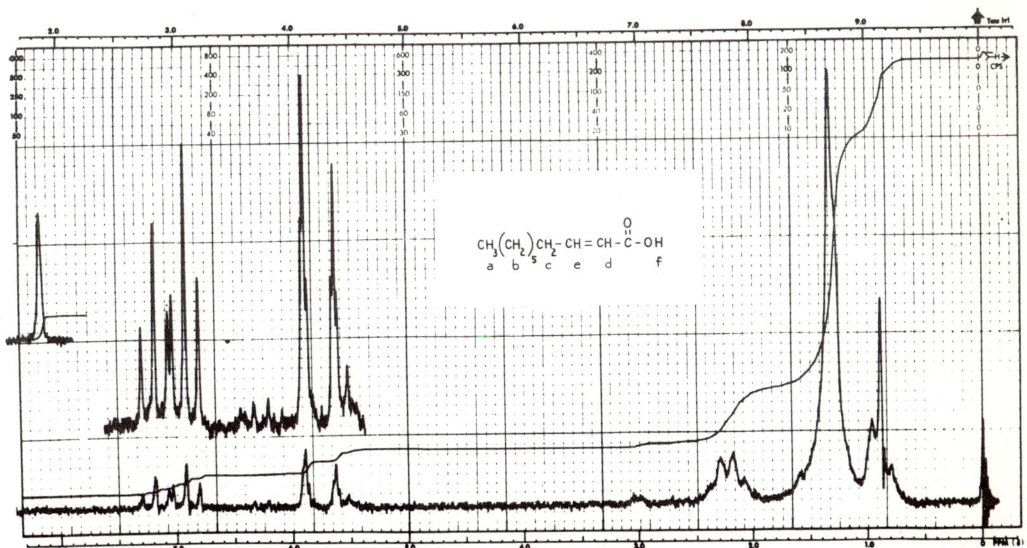

© 1978 Sadtler Research Laboratories, Inc.

trans-4-DECENOIC ACID

$C_{10}H_{18}O_2$

Mol. Wt. 170.25

Solvent: CCl_4

Assignments
a = 1.90
b = 1.00-1.70
c = 1.99
d = ca 2.37
e = 5.48
f = 11.80

2525

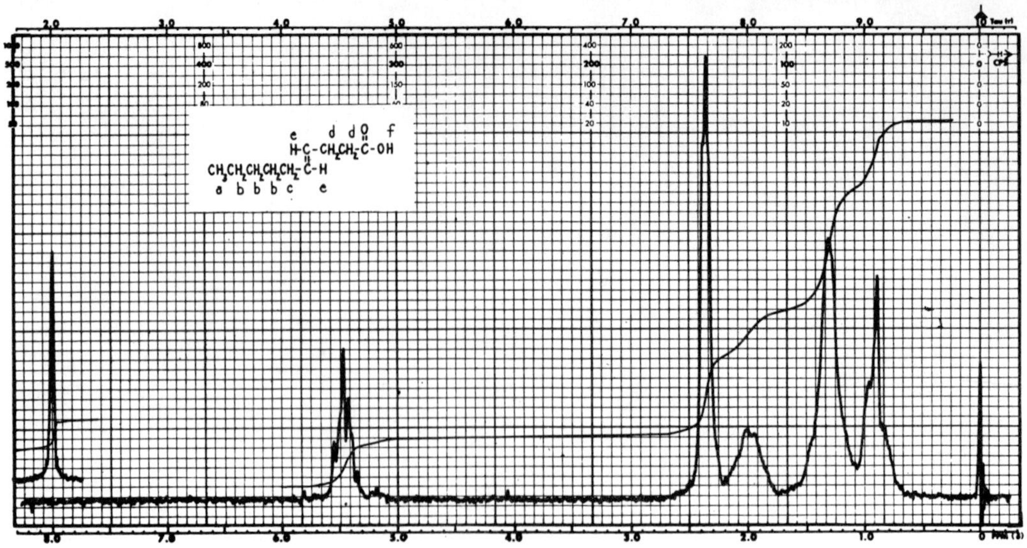

© 1978 Sadtler Research Laboratories, Inc.

trans-4-UNDECENOIC ACID

$C_{11}H_{20}O_2$

Mol. Wt. 184.28

B.P. 129-130°C/4mm

Solvent: CCl_4

Assignments
a = 0.90
b = 1.10-1.65
c = 2.00
d = ca 2.37
e = 5.47
f = 11.73

2526

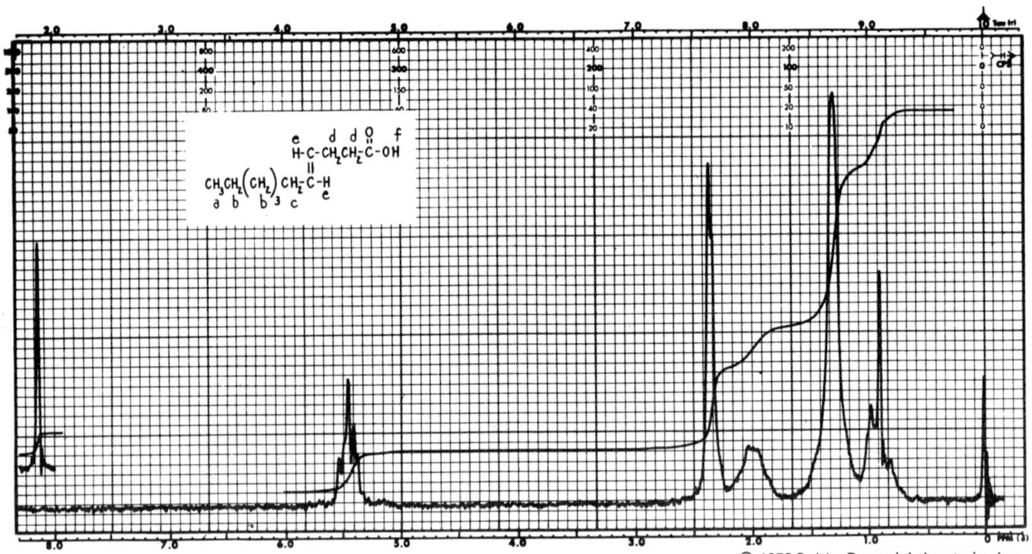

© 1978 Sadtler Research Laboratories, Inc.

trans-4-TRIDECENOIC ACID

$C_{13}H_{24}O_2$

Mol. Wt. 212.34

B.P. 146.5-147.5°C/3mm

M.P. 23.5-24.5°C

Solvent: CCl_4

Assignments
a = 0.89
b = 1.10-1.65
c = 1.97
d = 2.33
e = 5.43
f = 11.91

2527

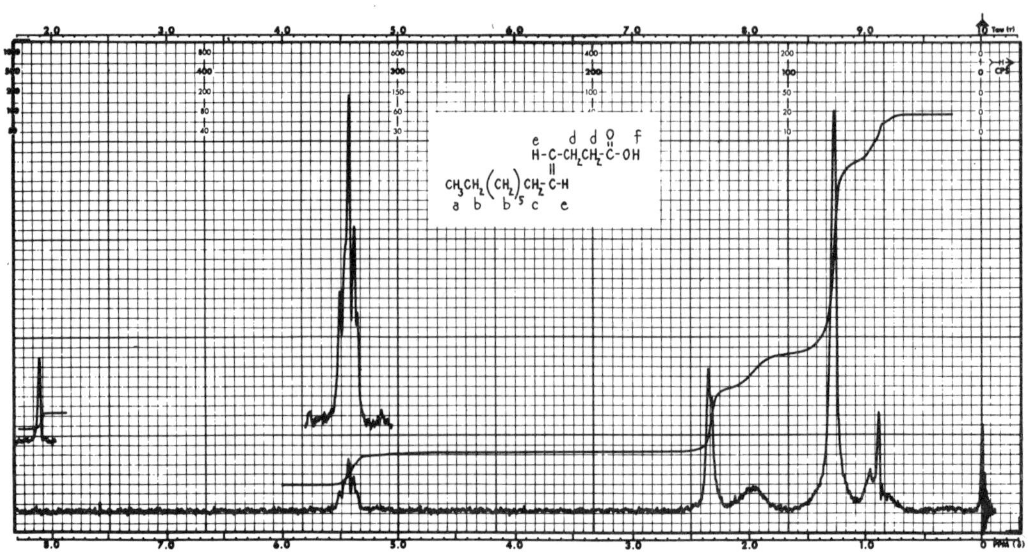

© 1978 Sadtler Research Laboratories, Inc.

OLEIC ACID

$C_{18}H_{34}O_2$

Mol. Wt. 282.45

M.P. 14°C

B.P. 286°C/100mm

Solvent: CCl_4

Assignments
a = 0.89
b = 1.30
c = 2.00
d = 2.30
e = 5.28
f = 11.86

2528

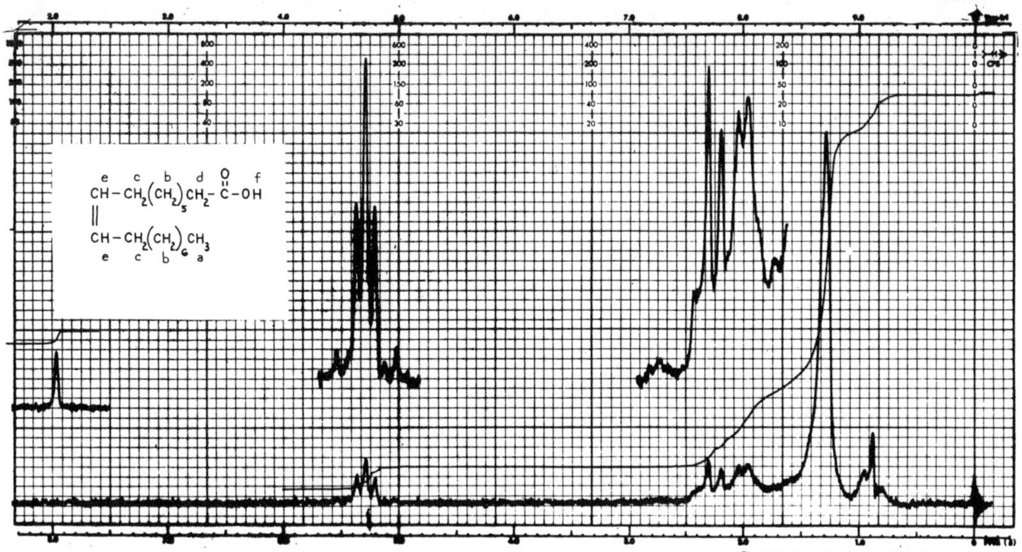

© 1978 Sadtler Research Laboratories, Inc.

ELAIDIC ACID

$C_{18}H_{34}O_2$

Mol. Wt. 282.47

Solvent: $CDCl_3$

Assignments
a = 0.90
b = 1.32
c = 1.98
d = 2.34
e = 5.39
OH is unobserved

2529

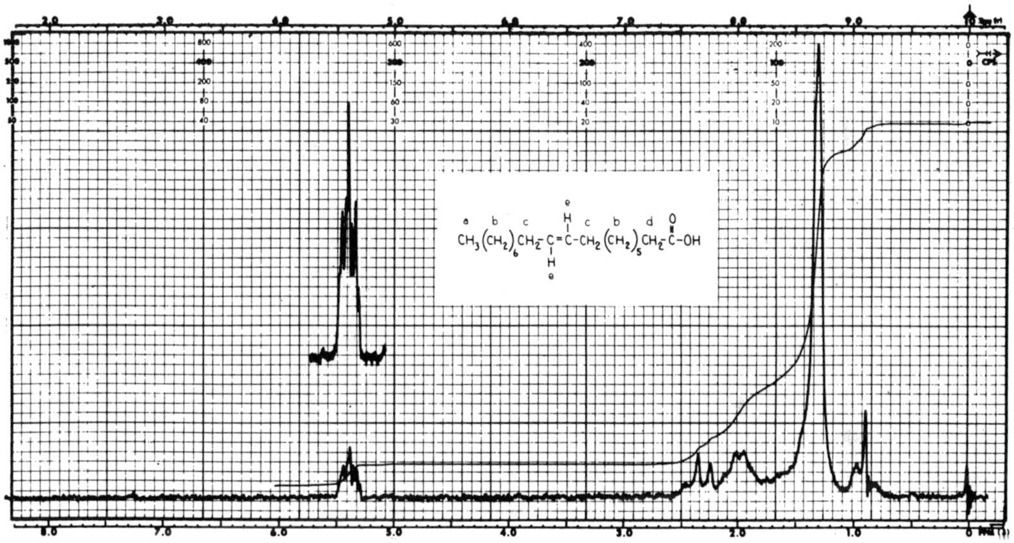

© 1978 Sadtler Research Laboratories, Inc.

SORBIC ACID

$C_6H_8O_2$

Mol. Wt. 112.12

B.P. 120°C/10mm

Solvent: $CDCl_3$

Assignments
a = 1.89
b = 5.73
c = 6.01-6.51
d = 7.30
e = 10.04

2530

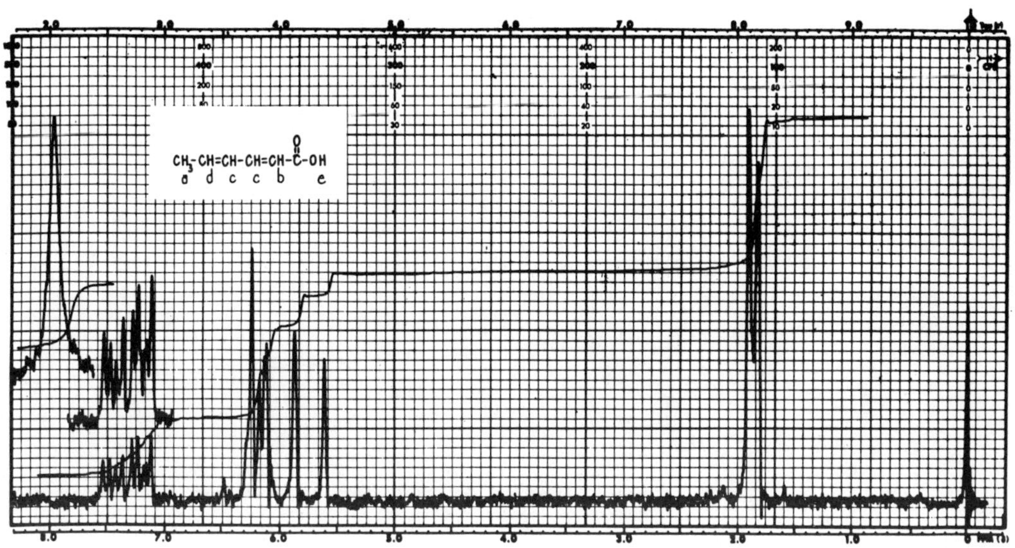

© 1978 Sadtler Research Laboratories, Inc.

cis-3-CHLOROACRYLIC ACID

$C_3H_3ClO_2$

Mol. Wt. 106.51

Solvent: Polysol-d

Assignments
a = 6.27
b = 6.84
c = 12.27

2531

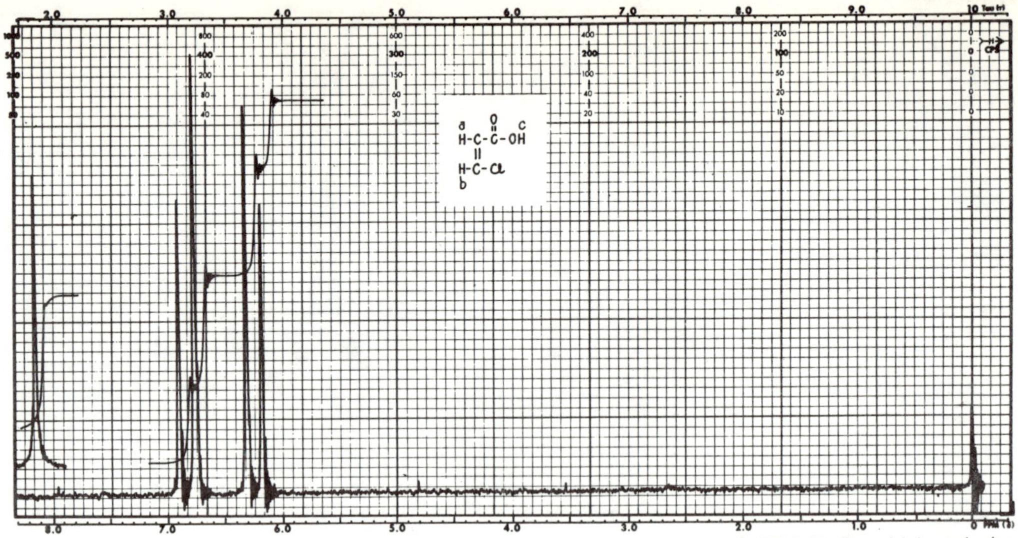

© 1978 Sadtler Research Laboratories, Inc.

RICINOLEIC ACID

$C_{18}H_{34}O_3$

Mol. Wt. 298.47

Solvent: CCl_4

Assignments
a = 0.90
b = 1.32
c = 1.79-2.47
d = 3.58
e = 5.40
f = 7.17

2532

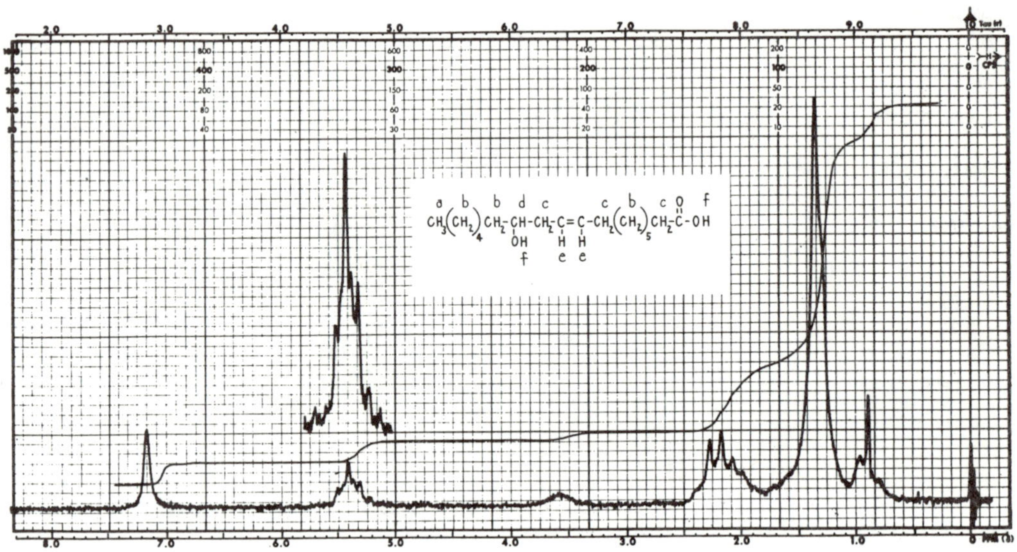

© 1978 Sadtler Research Laboratories, Inc.

FUMARIC ACID

$C_4H_4O_4$

Mol. Wt. 116.08

M.P. 288.7°C (in sealed tube)

Solvent: DMSO-D6

Assignments
a = 6.66
b = 12.28
c = 2.56 DMSO-D5

2533

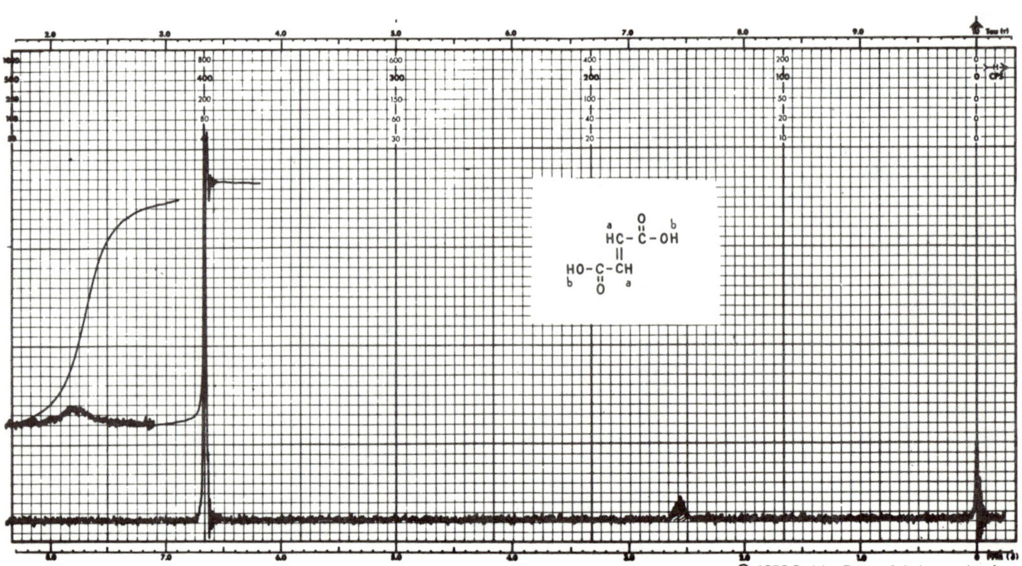

© 1978 Sadtler Research Laboratories, Inc.

MESACONIC ACID

$C_5H_6O_4$

Mol. Wt. 130.10

M.P. 200-202°C

Solvent: Trifluoroacetic acid

Assignments
a = 2.41
b = 7.04
c = 2.71, 2.95 imp.
OH's are unobserved

2534

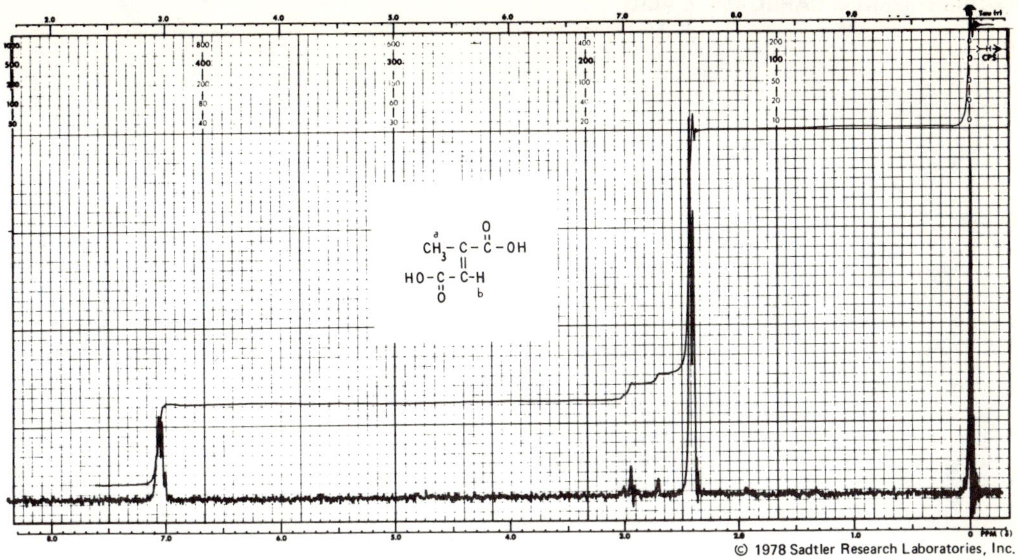

© 1978 Sadtler Research Laboratories, Inc.

1-CYCLOHEXENE-1-CARBOXYLIC ACID

$C_7H_{10}O_2$

Mol. Wt. 126.16

Solvent: CCl_4

Assignments
a = 1.40-1.90
b = 2.00-2.50
c = ca 7.09
d = 12.42

2535

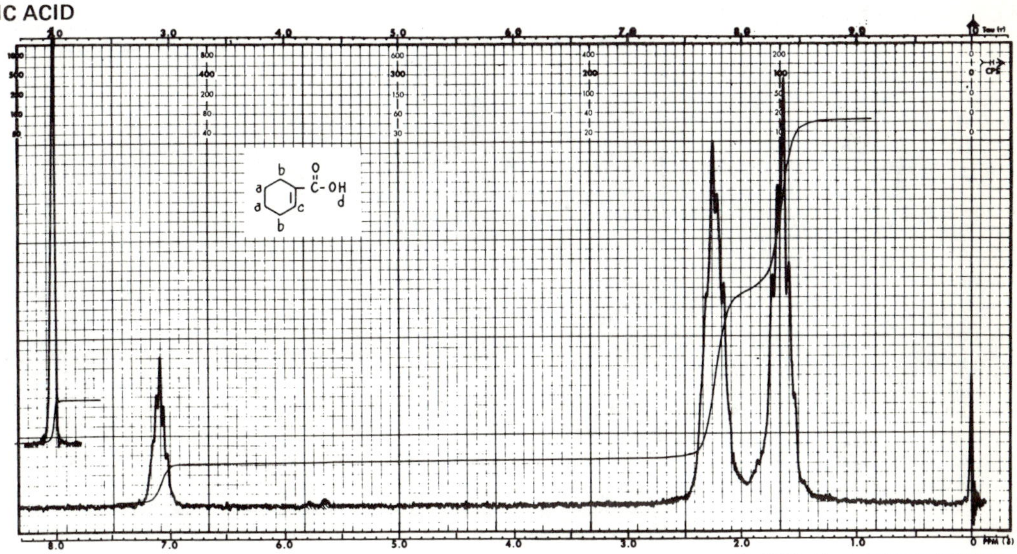

© 1978 Sadtler Research Laboratories, Inc.

4-CYCLOOCTENE-1-CARBOXYLIC ACID

$C_9H_{14}O_2$

Mol. Wt. 154.21

Solvent: $CDCl_3$

Assignments
a = 1.00-2.80
b = ca 5.67
c = 11.72

2536

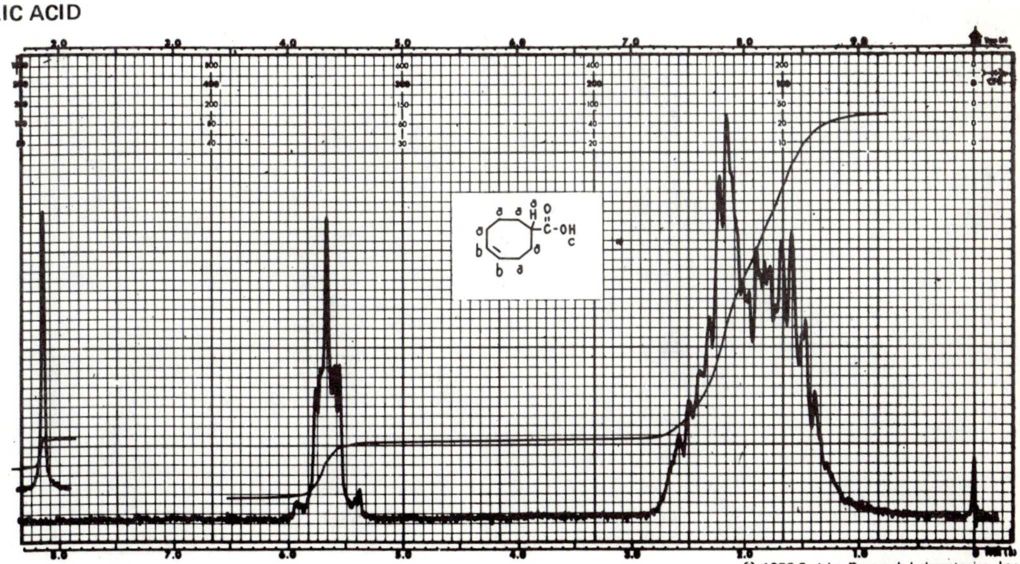

© 1978 Sadtler Research Laboratories, Inc.

5-NORBORNENE-2-CARBOXYLIC ACID

$C_8H_{10}O_2$

Mol. Wt. 138.17

Solvent: CCl_4

Assignments
a = 1.10-1.60
b = 1.50-2.40
c = 2.75-3.30
d = 5.70-6.40
e = 11.70

2537

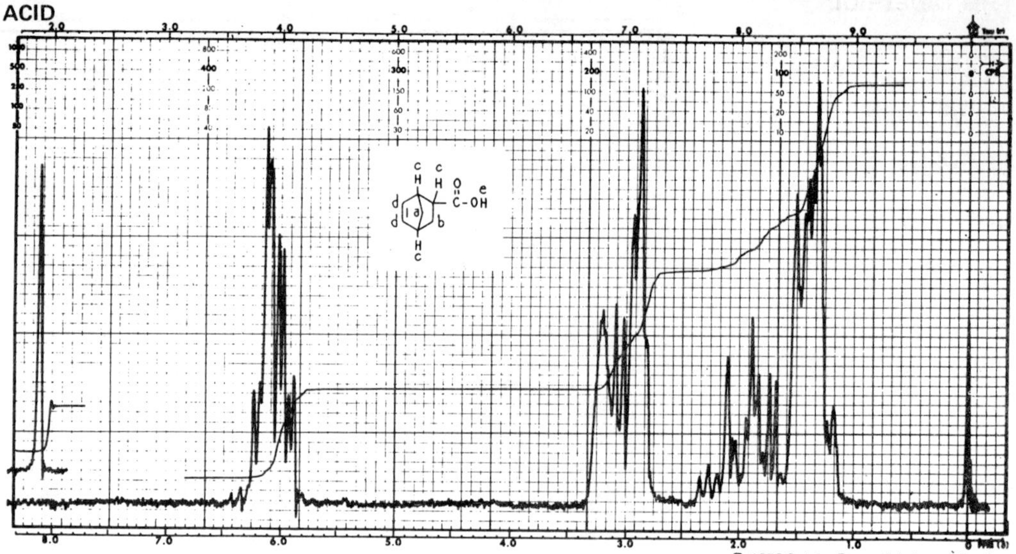

© 1978 Sadtler Research Laboratories, Inc.

4-CYCLOHEXENE-1,2-DICARBOXYLIC ACID

$C_8H_{10}O_4$

Mol. Wt. 170.17

M.P. 171-173°C

Solvent: Acetone-D_6

Assignments
a = ca 2.51
b = 3.09
c = 5.69
d = 9.17
e = 2.09 Acetone-D_5

2538

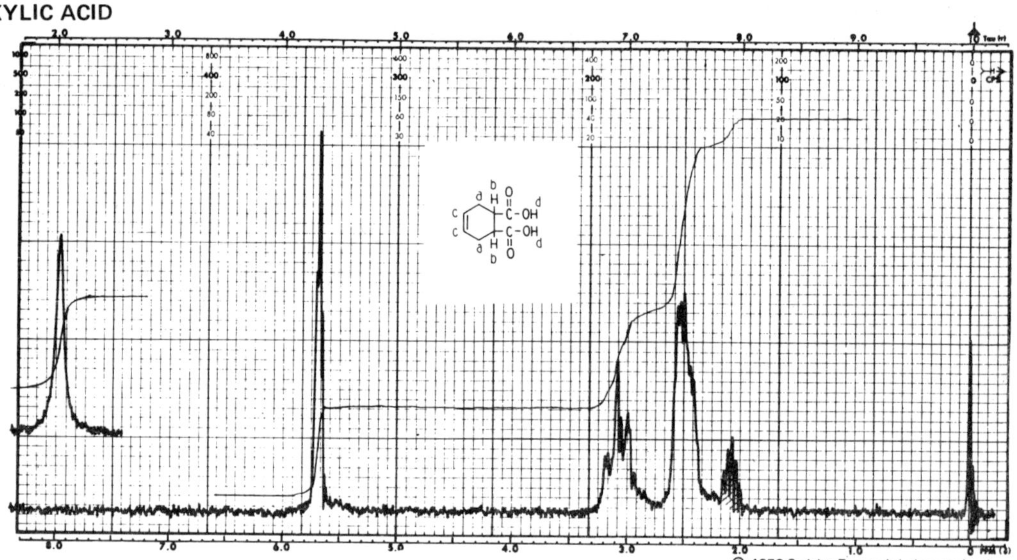

© 1978 Sadtler Research Laboratories, Inc.

PHENYLACETIC ACID

$C_8H_8O_2$

Mol. Wt. 136.14

M.P. 76-77°C

Solvent: $CDCl_3$

Assignments
a = 3.61
b = 7.27
c = 10.88

2539

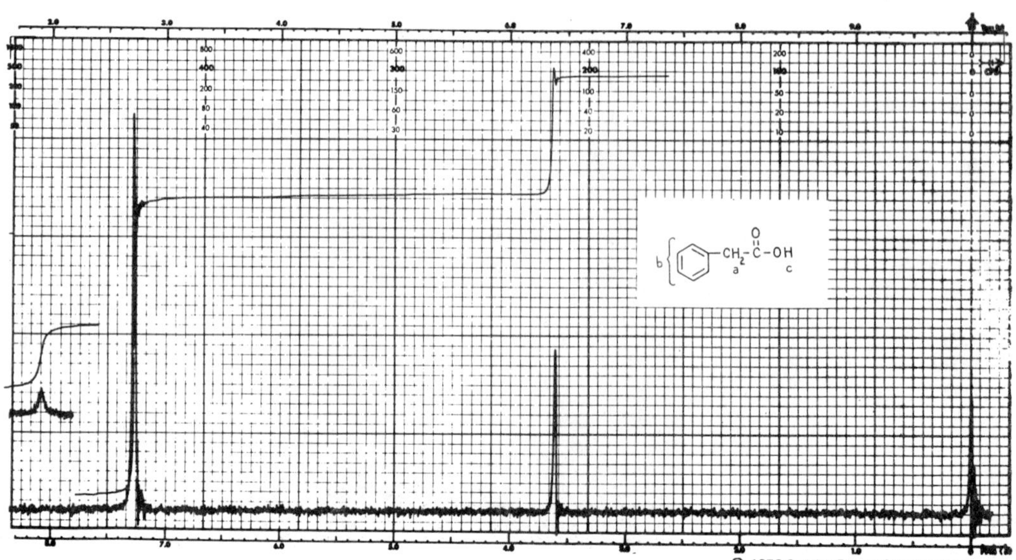

© 1978 Sadtler Research Laboratories, Inc.

1-HEPTANETHIOL

$C_7H_{16}S$

Mol. Wt. 132.26

B.P. 174-176°C

Solvent: $CDCl_3$

Assignments
a = 0.90
b = 1.29
c = 2.09
d = 2.53

2540

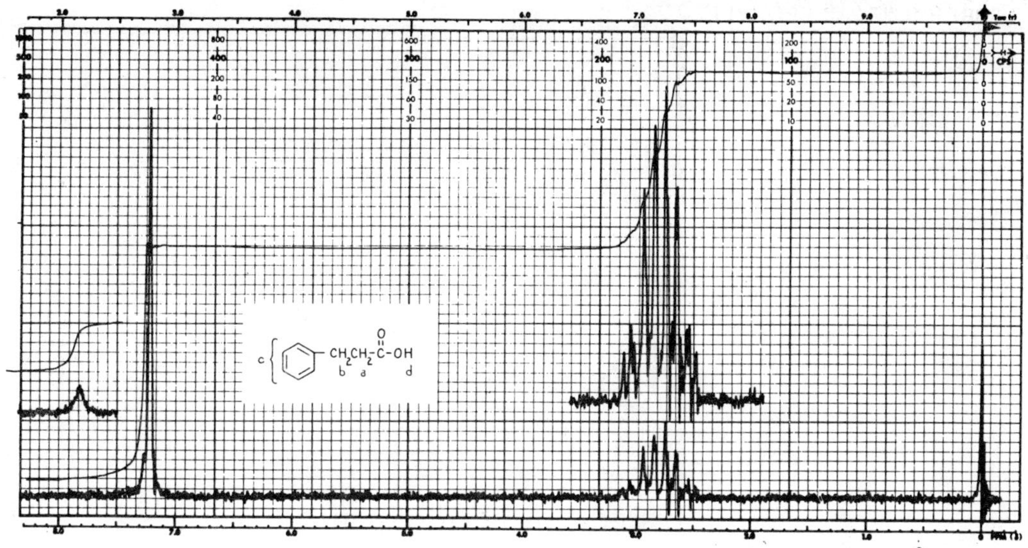

© 1978 Sadtler Research Laboratories, Inc.

2-PHENYLPROPIONIC ACID

$C_9H_{10}O_2$

Mol. Wt. 150.18

Solvent: CCl_4

Assignments
a = 1.47
b = 3.62
c = ca 7.20
d = 11.50

2541

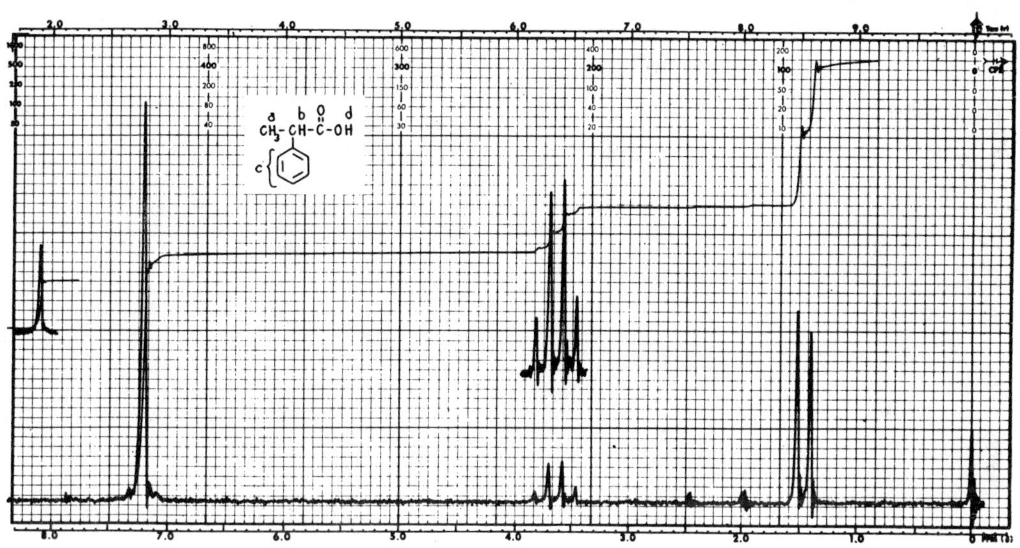

© 1978 Sadtler Research Laboratories, Inc.

2,5 DIPHENYLHEXANEDIOIC ACID

$C_{18}H_{18}O_4$

Mol. Wt. 298.34

M.P. 238°C

Solvent: Dimethyl sulfoxide

Assignments
a = 1.79
b = 3.52
c = 7.26
d = unobserved
e = 2.48 DMSO-d_5

2542

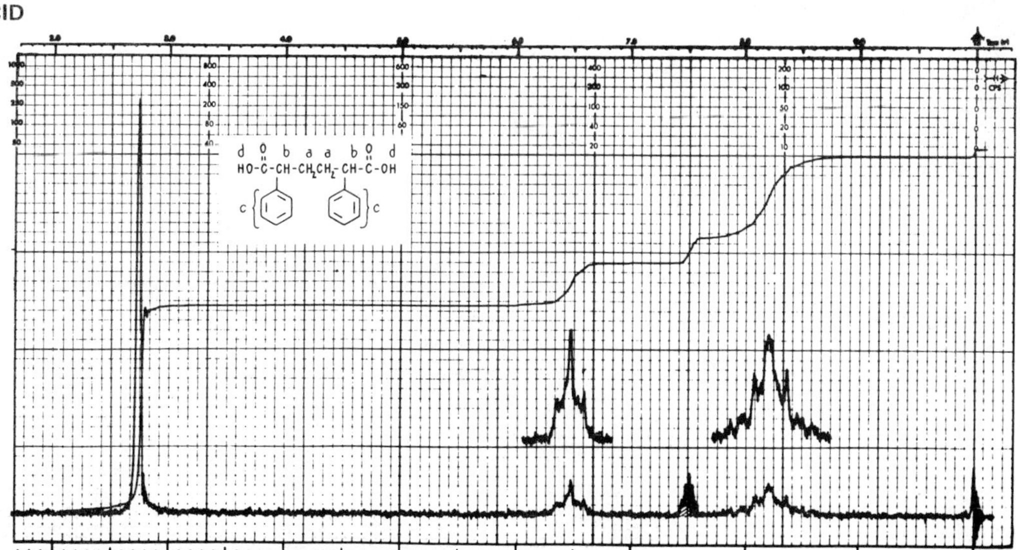

© 1978 Sadtler Research Laboratories, Inc.

3-PHENYL-3-PROPYLGLUTARIC ACID

$C_{14}H_{18}O_4$

Mol. Wt. 250.30

Solvent: DMSO-d_6

Assignments
a = 0.87
b = 1.20-1.40
c = 1.89
d = 2.99
e = ca 7.28
f = 10.59

2543

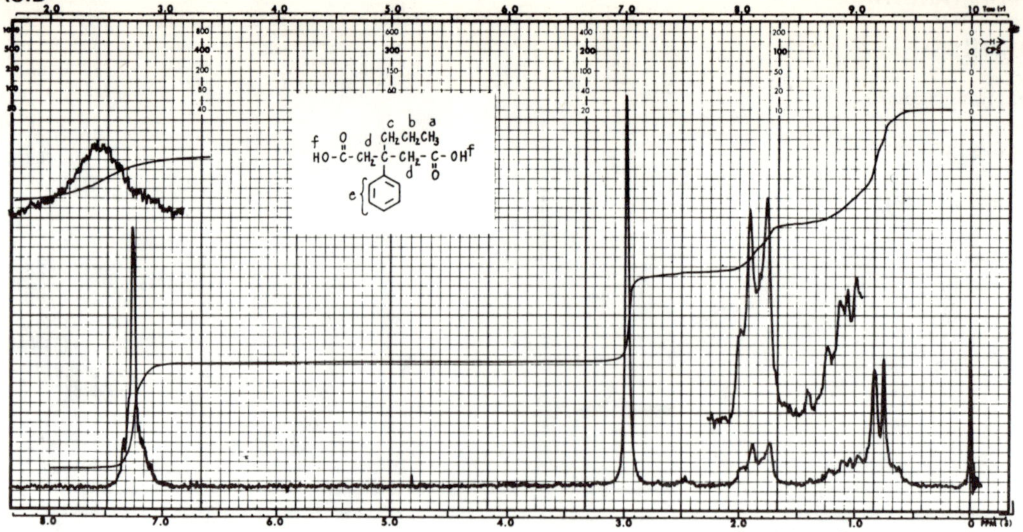

© 1978 Sadtler Research Laboratories, Inc.

α-BENZYLHYDROCINNAMIC ACID

$C_{16}H_{16}O_2$

Mol. Wt. 240.30

M.P. 88-89°C

Solvent: $CDCl_3$

Assignments
a = 2.60-3.30
b = ca 7.24
c = 10.58

2544

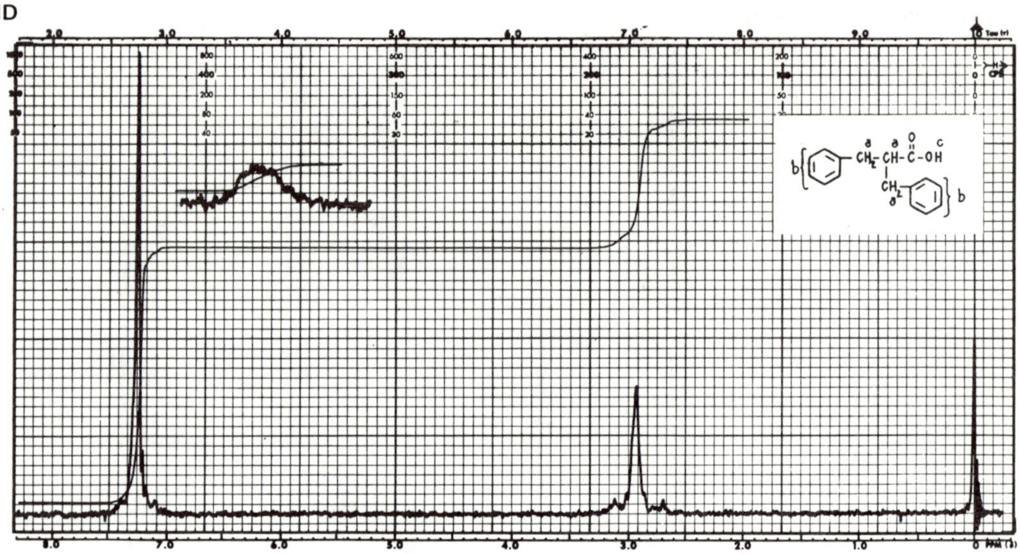

© 1978 Sadtler Research Laboratories, Inc.

2,2-DIPHENYLVALERIC ACID

$C_{17}H_{18}O_2$

Mol. Wt. 254.33

Solvent: $CDCl_3$

Assignments
a = 0.70-1.40
b = 2.31
c = 7.23
d = 10.48

2545

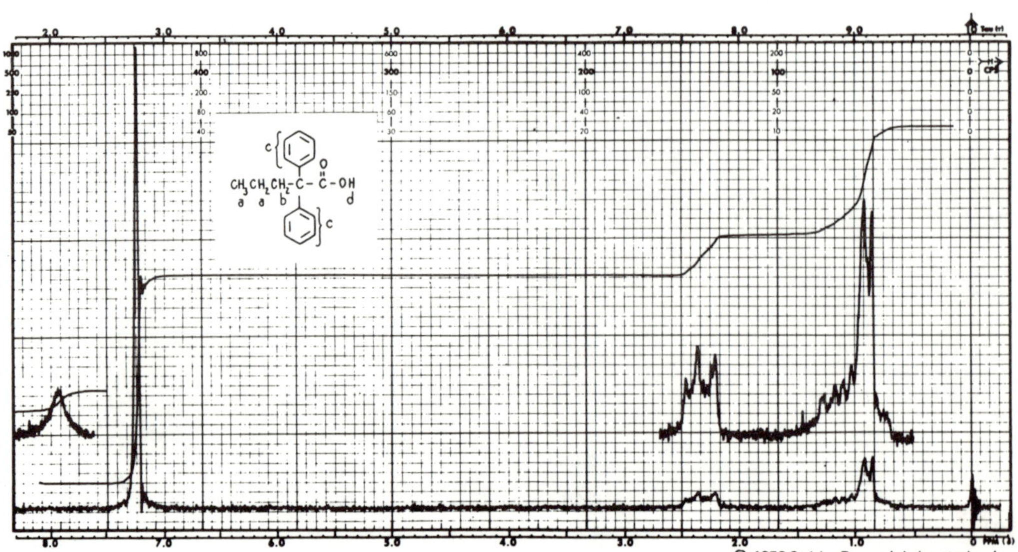

© 1978 Sadtler Research Laboratories, Inc.

DIPHENYLACETIC ACID

$C_{14}H_{12}O_2$

Mol. Wt. 212.24

M.P. 145-146°C

Solvent: $CDCl_3$

Assignments
a = 5.04
b = 7.32
c = 10.60

2546

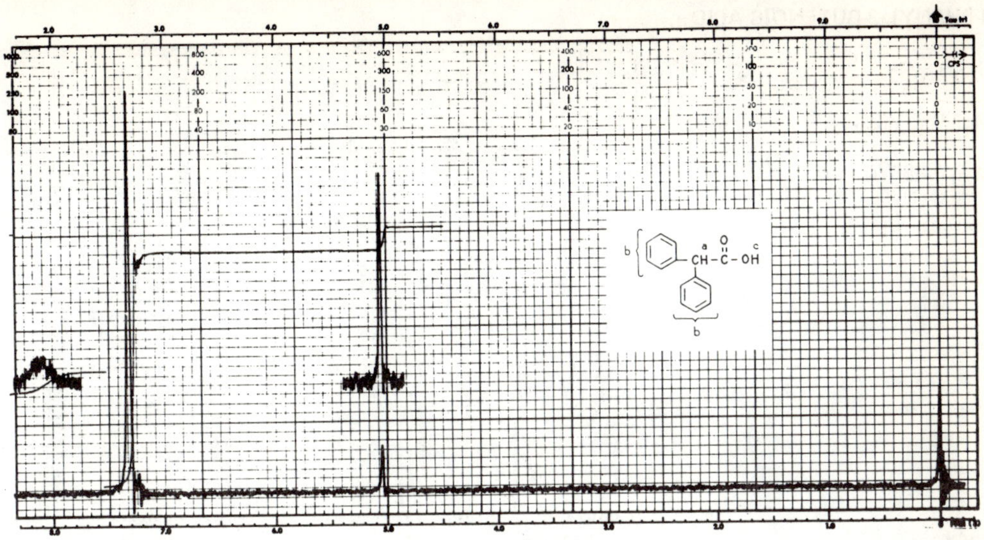

© 1978 Sadtler Research Laboratories, Inc.

TRIPHENYLACETIC ACID

$C_{20}H_{16}O_2$

Mol. Wt. 288.35

M.P. 270-273°C

Solvent: DMSO-d_6

Assignments
a = 6.90-7.50
b = unobserved
c = 2.50 DMSO-d_5

2547

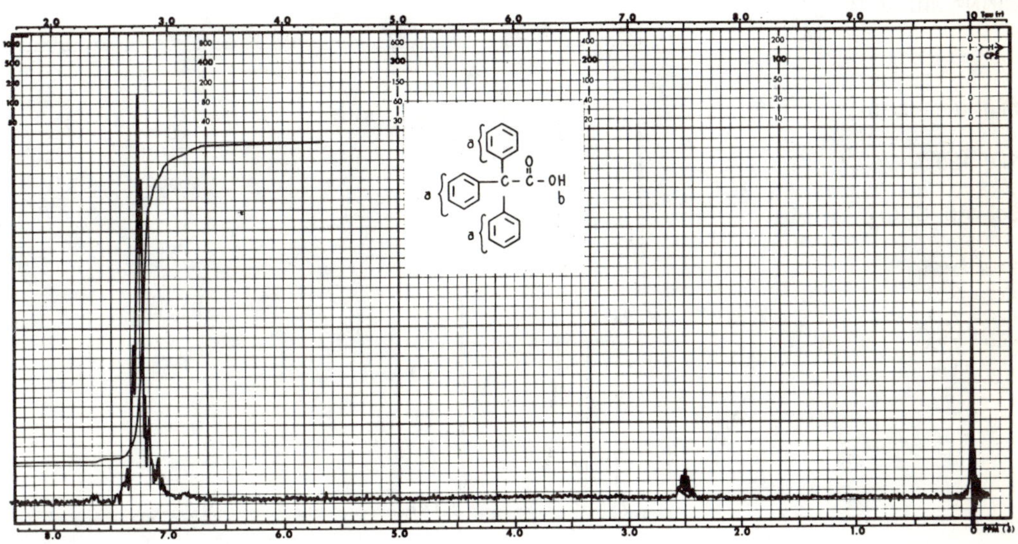

© 1978 Sadtler Research Laboratories, Inc.

(PHENYLIMINO)DIACETIC ACID

$C_{10}H_{11}NO_4$

Mol. Wt. 209.20

Solvent: Polysol-d

Assignments
a = 4.08
b = 6.45-6.85
c = 7.14
d = 11.68
e = 2.50 DMSO-d_5
f = 2.93 imp.

2548

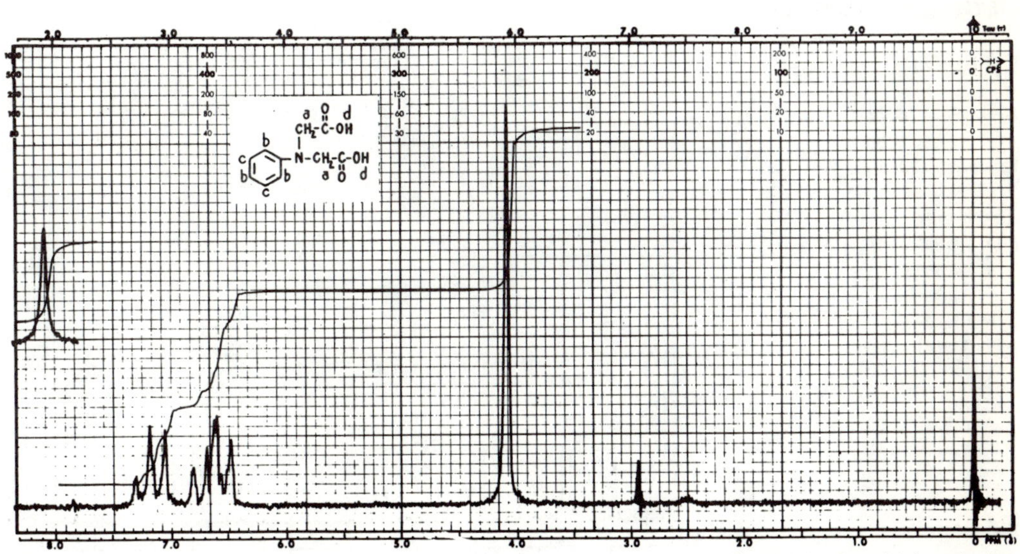

© 1978 Sadtler Research Laboratories, Inc.

4-PHENYL-3-BUTENOIC ACID

$C_{10}H_{10}O_2$

Mol. Wt. 162.19

M.P. 84-86°C

Solvent: $CDCl_3$

Assignments
a = 3.25
b = 6.14
c = 6.53
d = ca 7.30
e = 11.28

2549

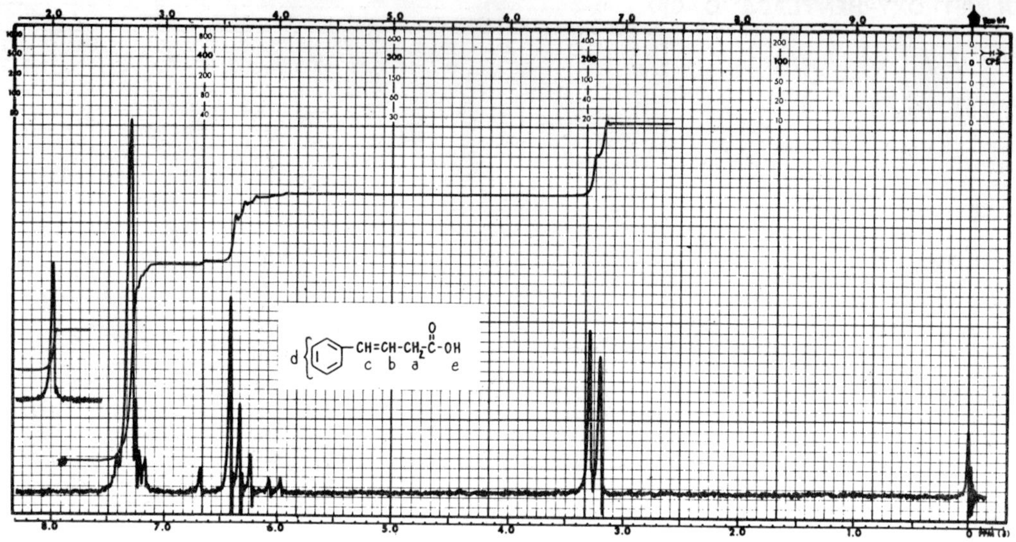

© 1978 Sadtler Research Laboratories, Inc.

CINNAMIC ACID

$C_9H_8O_2$

Mol. Wt. 148.16

M.P. 133°C

Solvent: $CDCl_3$

Assignments
a = 6.41
b = 7.17-7.69
c = 7.73
d = 11.90

2550

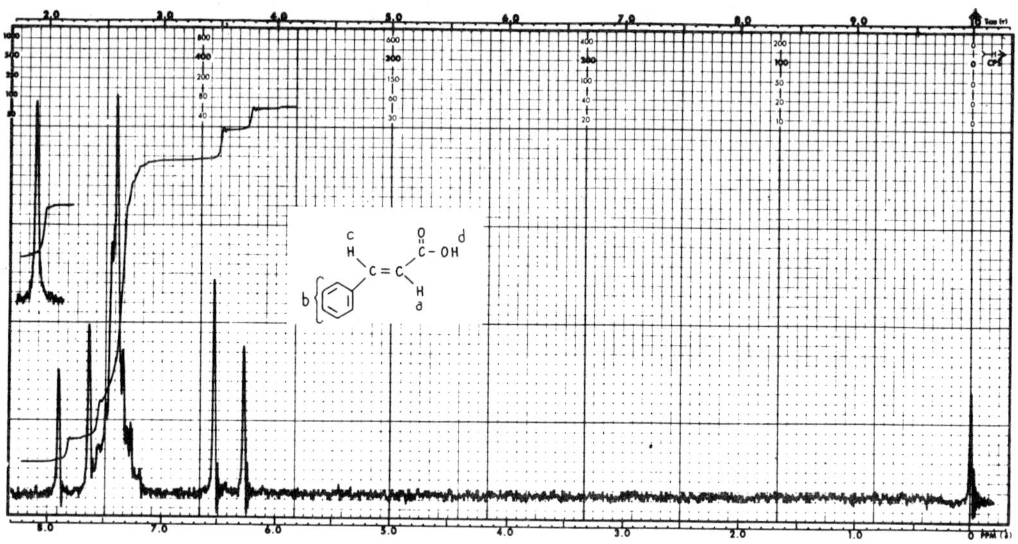

© 1978 Sadtler Research Laboratories, Inc.

α-ETHYLCINNAMIC ACID

$C_{11}H_{12}O_2$

Mol. Wt. 176.22

Solvent: $CDCl_3$

Assignments
a = 1.20
b = 2.58
c = 7.38
d = 7.80
e = 11.76

2551

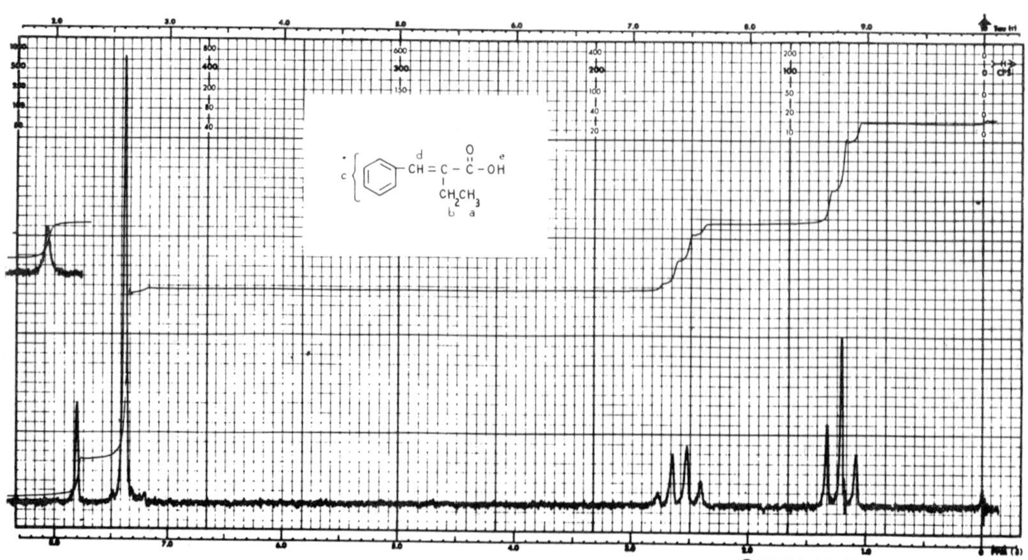

© 1978 Sadtler Research Laboratories, Inc.

DL-METHOXYPHENYLACETIC ACID

$C_9H_{10}O_3$

Mol. Wt. 166.18

M.P. 70-72°C

Solvent: $CDCl_3$

Assignments
a = 3.32
b = 4.78
c = 7.15-7.60
d = 11.58

2552

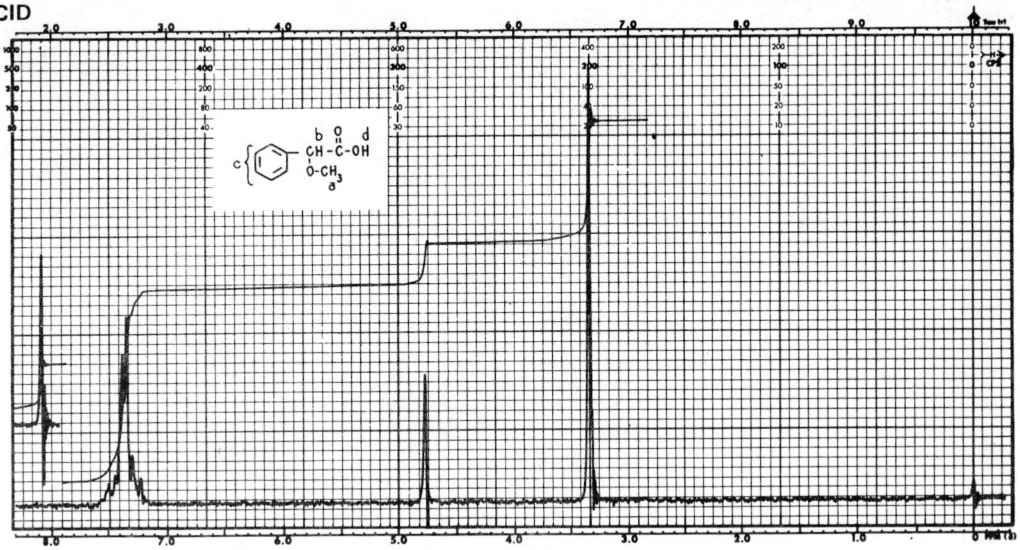

© 1978 Sadtler Research Laboratories, Inc.

PHENOXYACETIC ACID

$C_8H_8O_3$

Mol. Wt. 152.14

M.P. 98-99°C

Solvent: Acetone

Assignments
a = 4.67
b = ca 6.92
c = ca 7.25
d = 8.78

2553

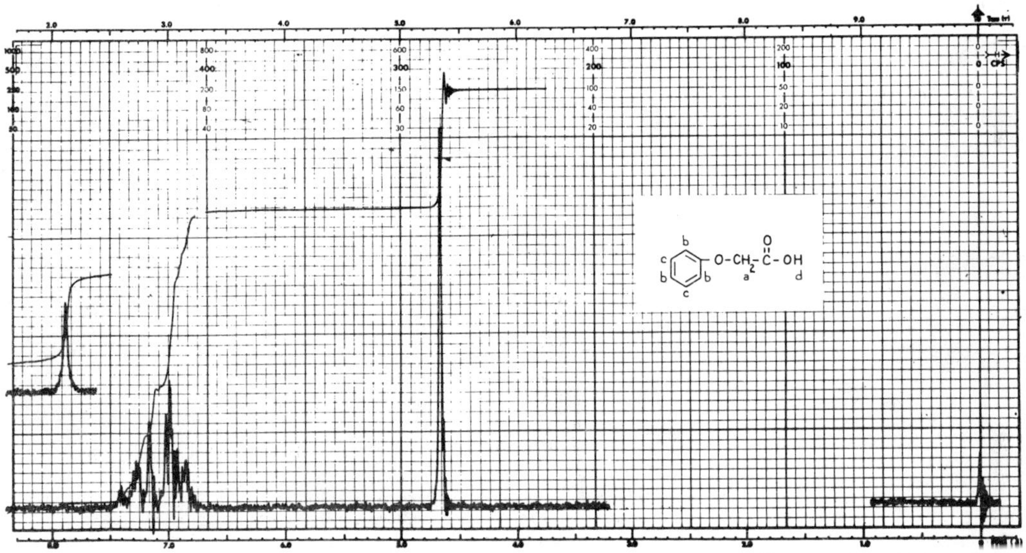

© 1978 Sadtler Research Laboratories, Inc.

11-PHENOXYUNDECANOIC ACID

$C_{17}H_{26}O_3$

Mol. Wt. 278.40

Solvent: $CDCl_3$

Assignments
a = ca 1.31
b = ca 1.70
c = 2.31
d = 3.92
e = ca 6.88
f = 7.22
g = 10.68

2554

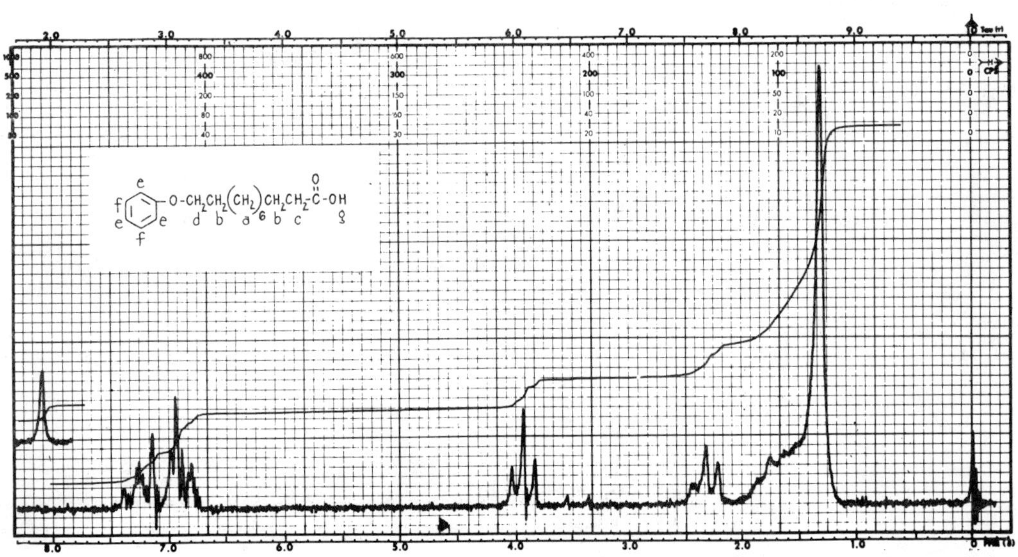

© 1978 Sadtler Research Laboratories, Inc.

D(–)MANDELIC ACID

$C_8H_8O_3$

Mol. Wt. 152.15

M.P. 132-135°C

Solvent: Dimethyl sulfoxide

Assignments
a = 5.05
b = 7.15-7.60
c = ca 7.72

2555

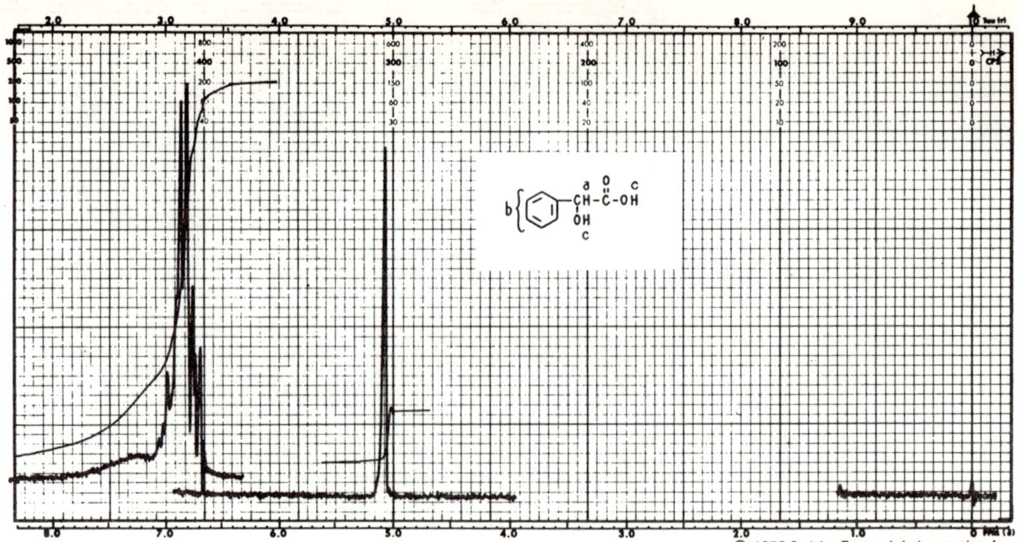

© 1978 Sadtler Research Laboratories, Inc.

BENZILIC ACID

$C_{14}H_{12}O_3$

Mol. Wt. 228.25

M.P. 146-149°C (lit.)

Solvent: Acetone

Assignments
a = 6.27 *
b = 7.23-7.72
* H_2O present

2556

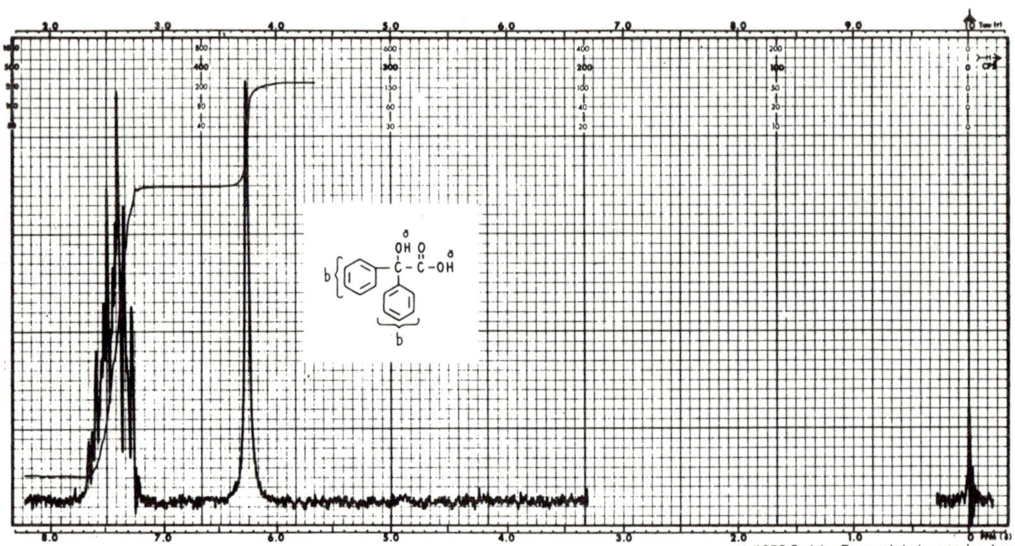

© 1978 Sadtler Research Laboratories, Inc.

---

p-TOLYLACETIC ACID

$C_9H_{10}O_2$

Mol. Wt. 150.17

M.P. 88-91°C

Solvent: $CDCl_3$

Assignments
a = 2.31
b = 3.57
c = 7.12
d = 9.07

2557

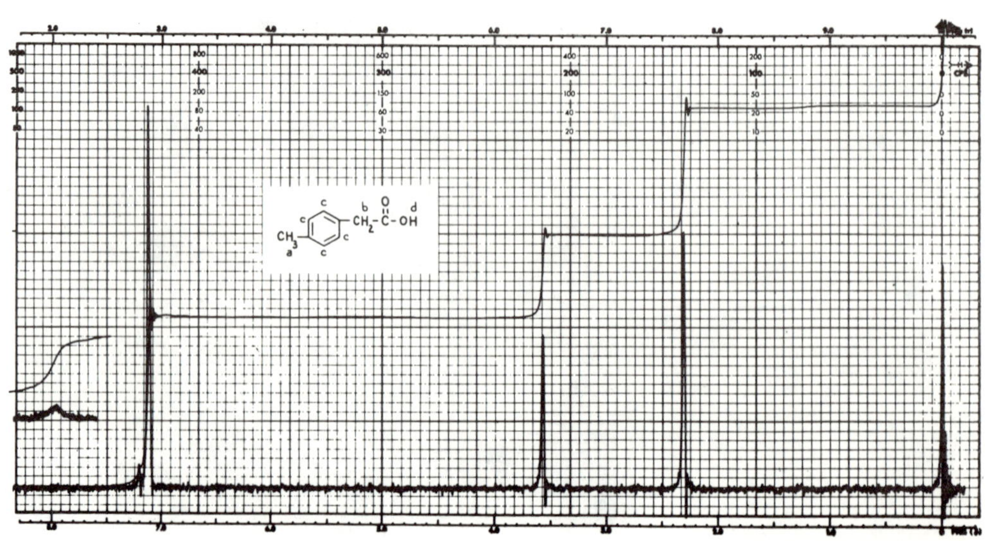

© 1978 Sadtler Research Laboratories, Inc.

(o-FLUOROPHENYL)ACETIC ACID

$C_8H_7FO_2$

Mol. Wt. 154.14

Solvent: $CDCl_3$

Assignments
a = 3.69
b = 6.81-7.45
c = 11.00

2558

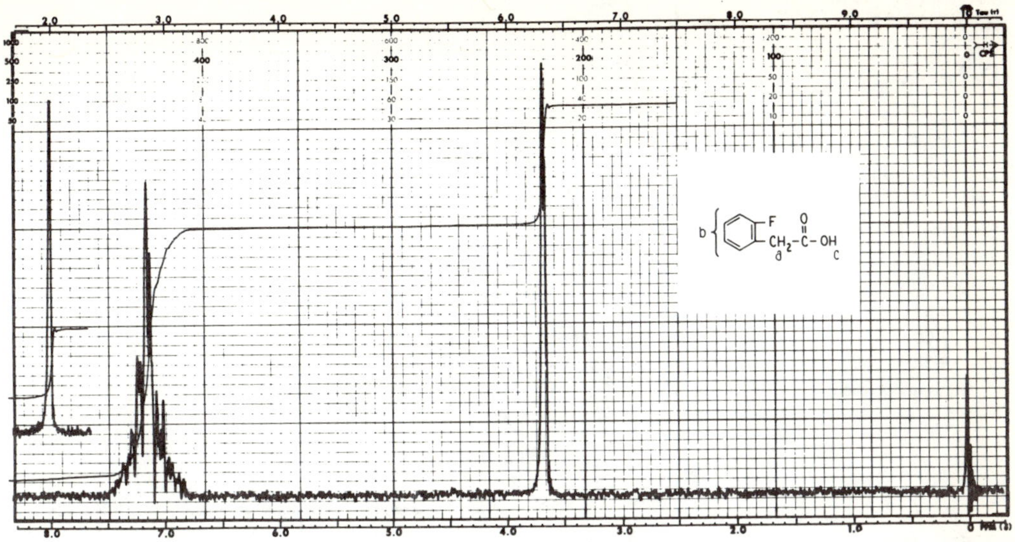

© 1978 Sadtler Research Laboratories, Inc.

(m-FLUOROPHENYL)ACETIC ACID

$C_8H_7FO_2$

Mol. Wt. 154.14

Solvent: $CDCl_3$

Assignments
a = 3.58
b = 6.70-7.50
c = 11.89

2559

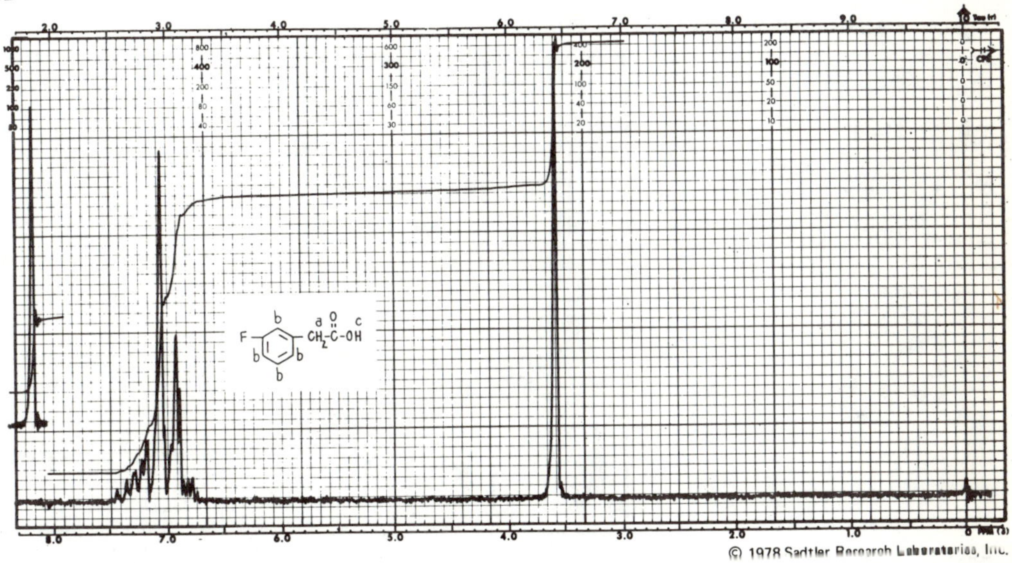

© 1978 Sadtler Research Laboratories, Inc.

4-(o-AMINOPHENYL)BUTYRIC ACID, HYDROCHLORIDE

$C_{10}H_{13}NO_2 \cdot HCl$

Mol. Wt. 215.68

Solvent: D_2O

Assignments
a = 2.00
b = 2.55
c = 2.87
d = 7.10-7.60
e = 4.36 HDO

2560

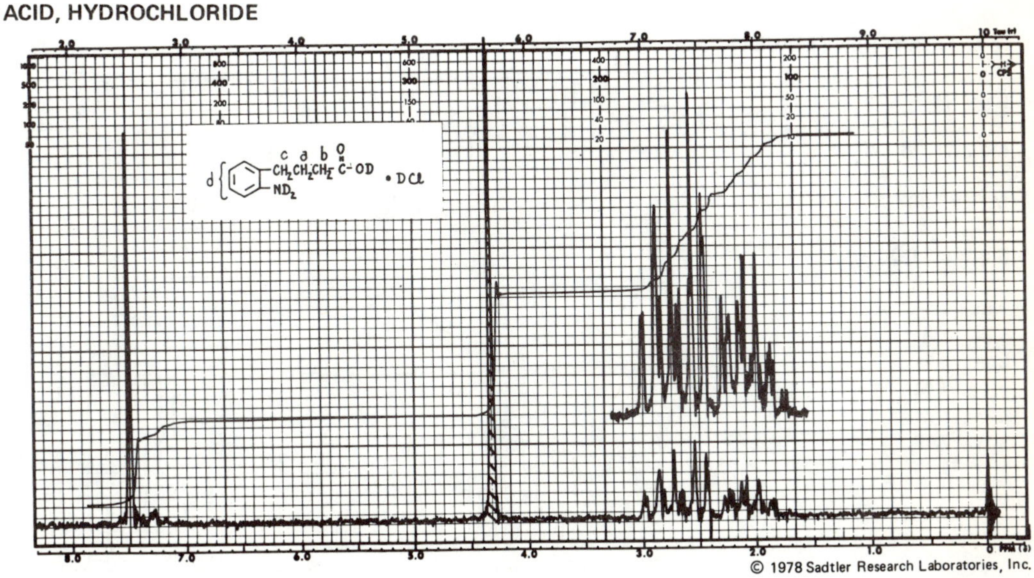

© 1978 Sadtler Research Laboratories, Inc.

(p-NITROPHENYL)ACETIC ACID

$C_8H_7NO_4$

Mol. Wt. 181.15

M.P. 154-156°C

Solvent: Trifluoroacetic acid

Assignments
a = 3.96
b = 7.60
c = 8.33
OH is unobserved

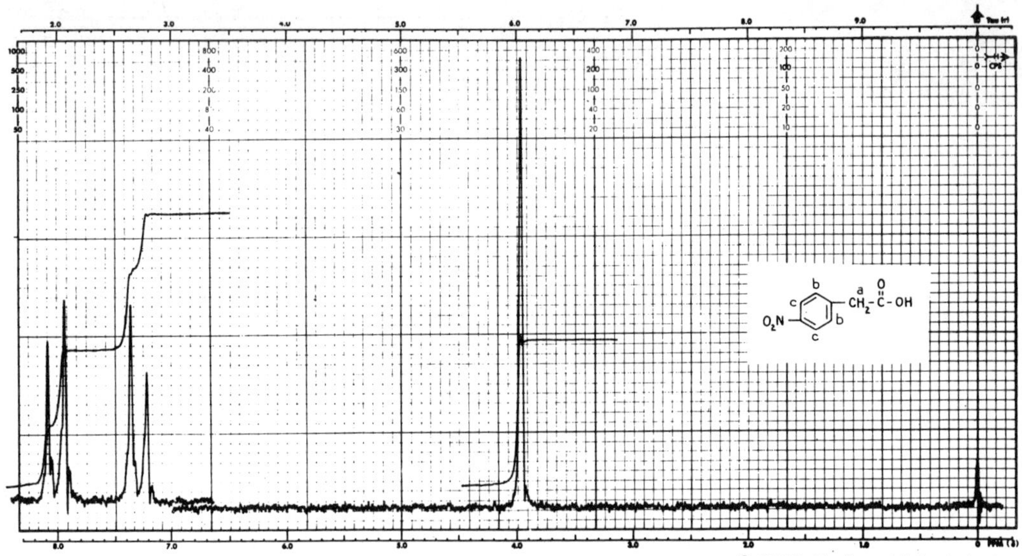

2561

© 1978 Sadtler Research Laboratories, Inc.

2-(p-CHLOROPHENOXY)-2-METHYLPROPIONIC ACID

$C_{10}H_{11}ClO_3$

Mol. Wt. 214.65

M.P. 120-121°C

Solvent: $CDCl_3$

Assignments
a = 1.59
b = 6.88
c = 7.24
d = 9.48

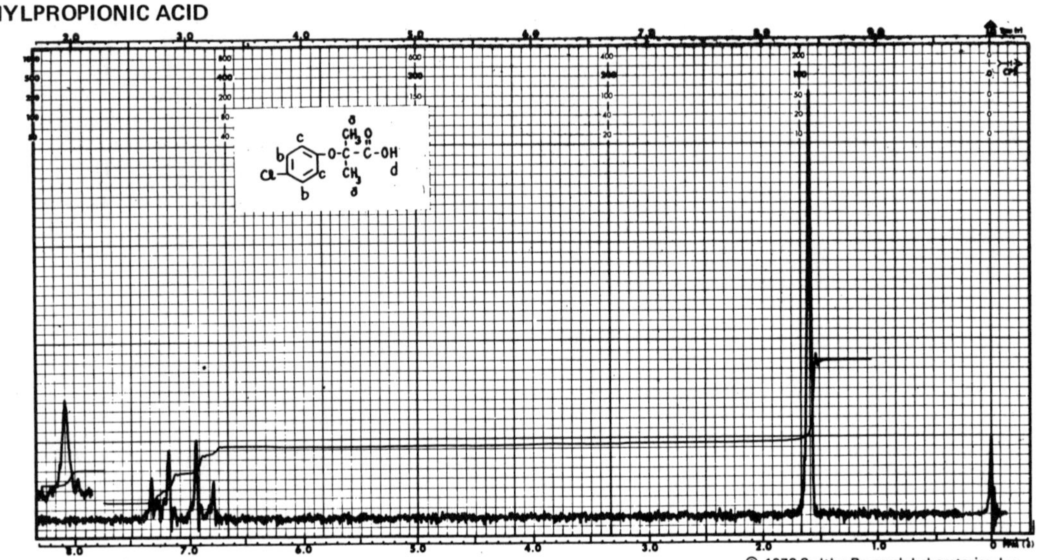

2562

© 1978 Sadtler Research Laboratories, Inc.

(p-ETHOXYPHENYL)ACETIC ACID

$C_{10}H_{12}O_3$

Mol. Wt. 180.20

M.P. 87-90°C

Solvent: $CDCl_3$

Assignments
a = 1.40
b = 3.59
c = 4.02
d = 6.84
e = 7.19
f = 10.70

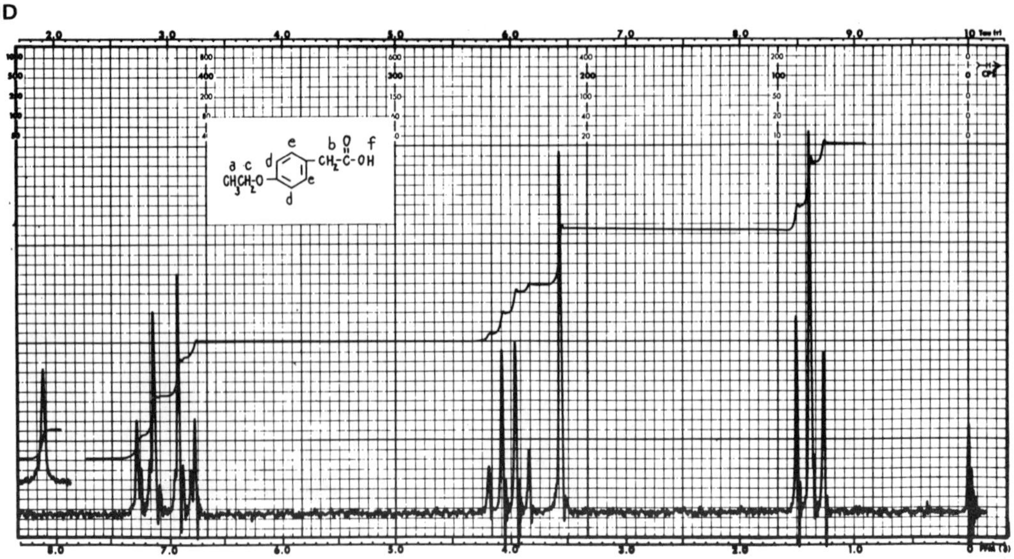

2563

© 1978 Sadtler Research Laboratories, Inc.

p-METHYLCINNAMIC ACID

$C_{10}H_{10}O_2$

Mol. Wt. 162.19

M.P. 197-199°C (lit.)

Solvent: Dimethyl sulfoxide-d_6

Assignments
a = 2.33
b = 6.49
c = 7.21
d = 7.59
e = 7.61
f = unobserved

2564

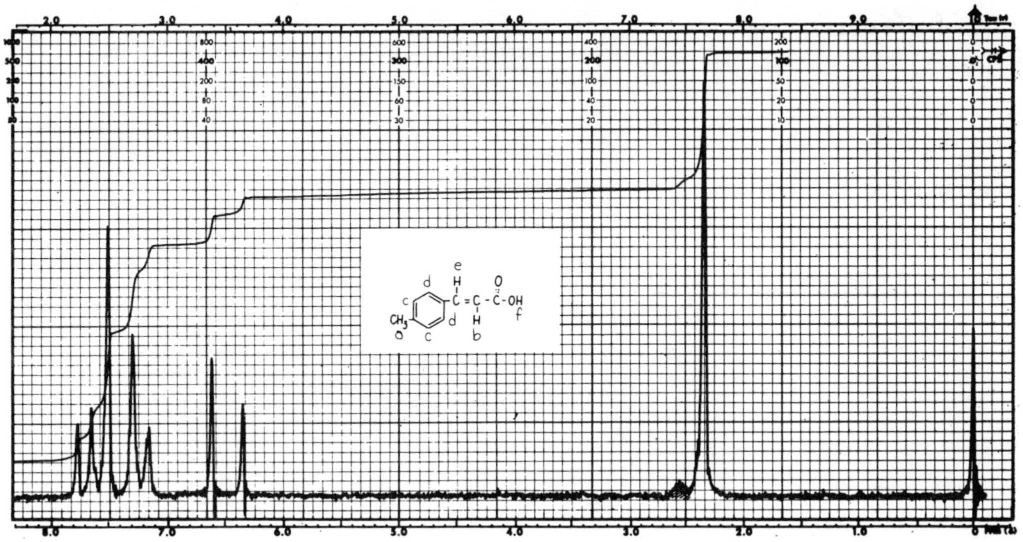

© 1978 Sadtler Research Laboratories, Inc.

(p-PHENYLENEDIOXY)DIACETIC ACID

$C_{10}H_{10}O_6$

Mol. Wt. 226.19

Solvent: Polysol-d

Assignments
a = 4.52
b = 6.83
c = 11.79

2565

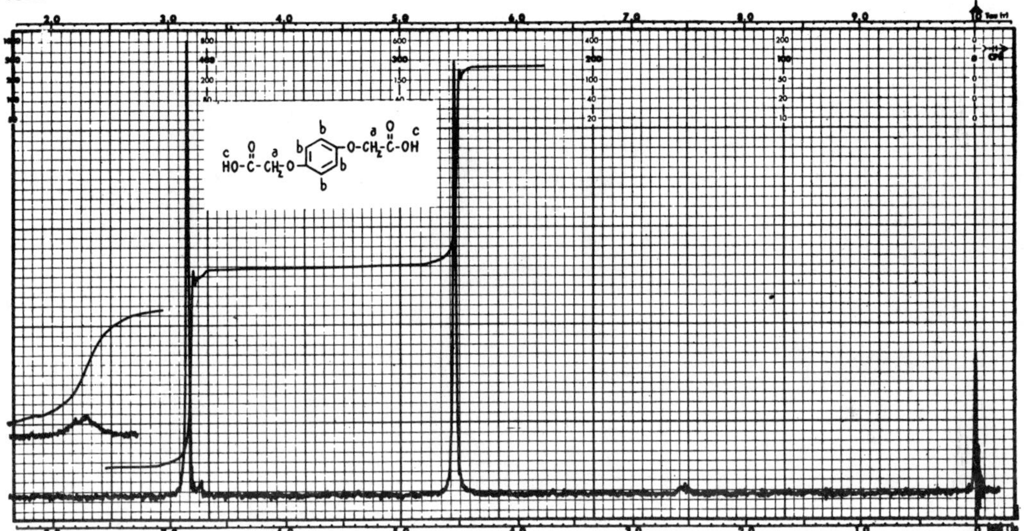

© 1978 Sadtler Research Laboratories, Inc.

BENZOIC ACID

$C_7H_6O_2$

Mol. Wt. 122.13

M.P. 121°C

Solvent: CCl_4

Assignments
a = 7.52
b = 8.14
c = 12.82

2566

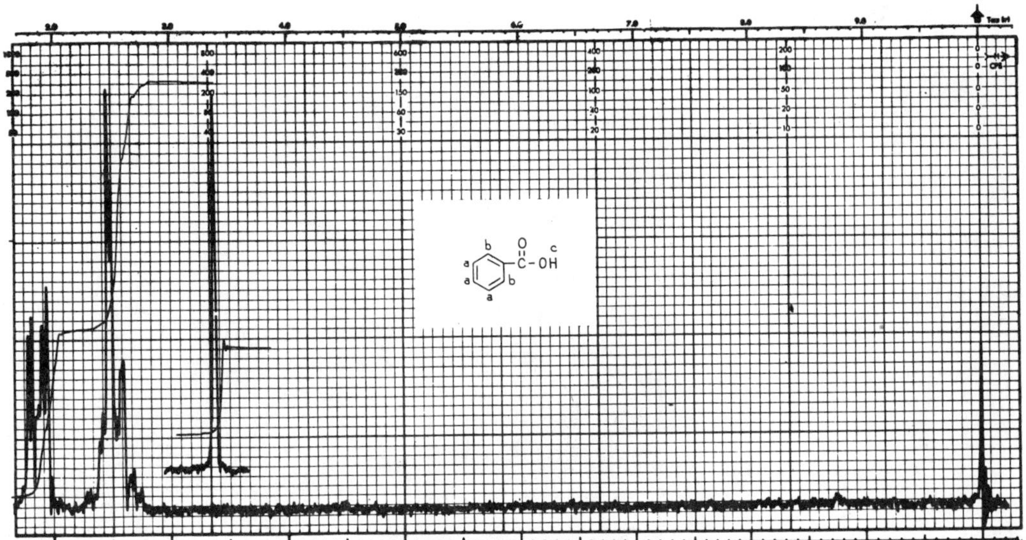

© 1978 Sadtler Research Laboratories, Inc.

p-TOLUIC ACID

$C_8H_8O_2$

Mol. Wt. 136.15

M.P. 178-180°C (lit.)

Solvent: Trifluoroacetic acid

Assignments
a = 2.47
b = 7.30
c = 8.02
OH under TFA

2567

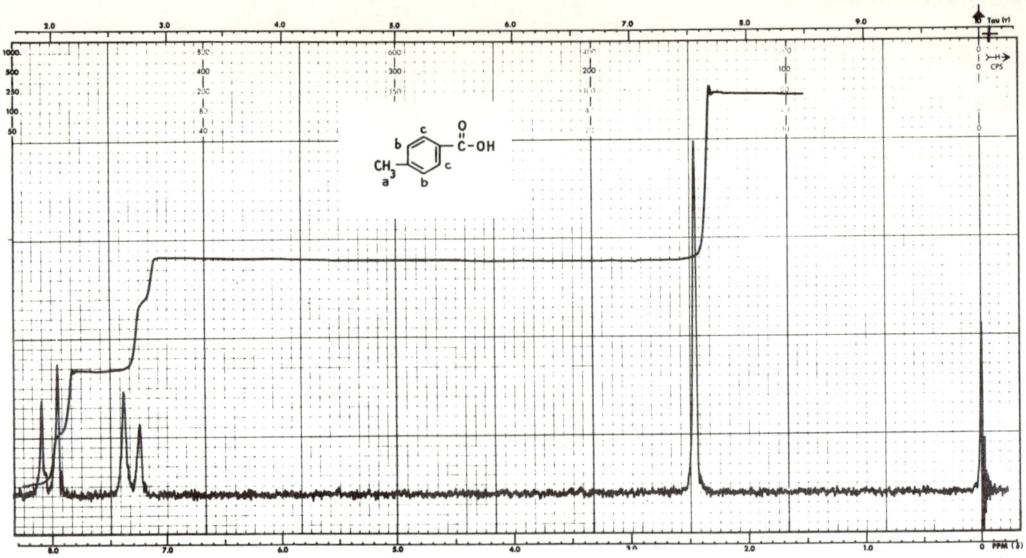

© 1978 Sadtler Research Laboratories, Inc.

---

a,a,a-TRIFLUORO-o-TOLUIC ACID

$C_8H_5F_3O_2$

Mol. Wt. 190.12

M.P. 109-110°C

Solvent: Polysol-d

Assignments
a = 7.30-8.00
b = 12.70

2568

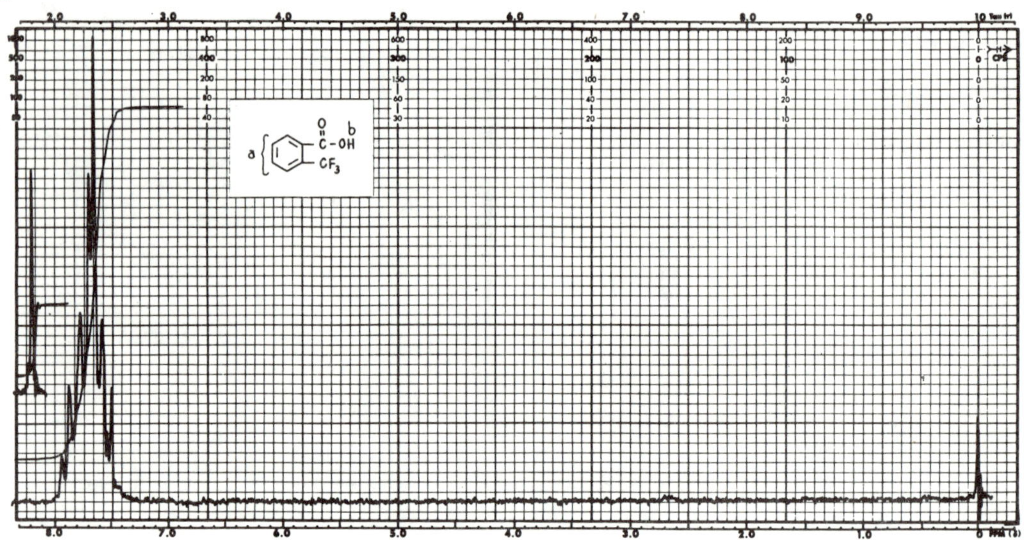

© 1978 Sadtler Research Laboratories, Inc.

---

o-FLUOROBENZOIC ACID

$C_7H_5FO_2$

Mol. Wt. 140.11

M.P. 124-126°C

Solvent: Polysol-d

Assignments
a = 6.90-7.82
b = 8.02
c = 12.57
d = 2.68 DMSO-d_5

2569

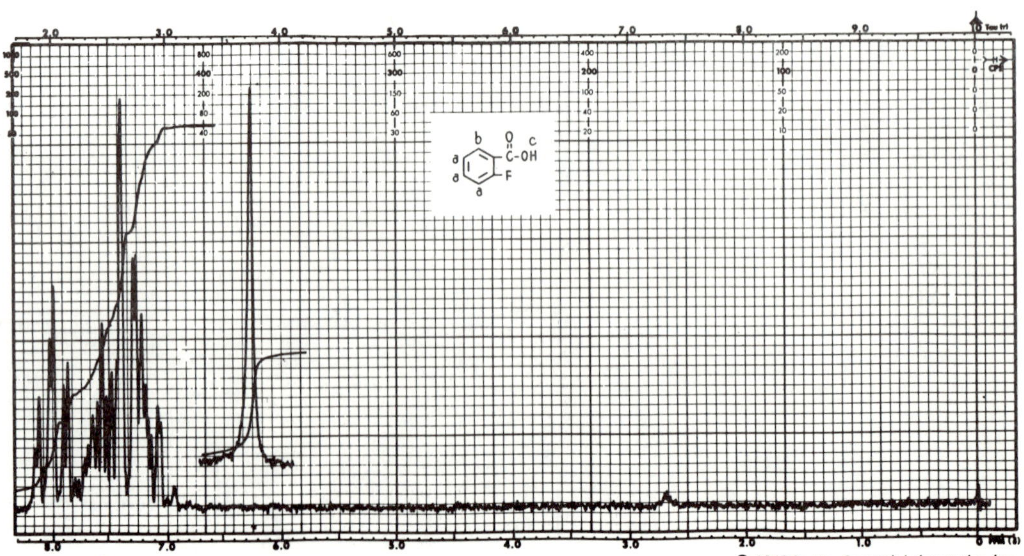

© 1978 Sadtler Research Laboratories, Inc.

p-FLUOROBENZOIC ACID

$C_7H_5FO_2$

Mol. Wt. 140.12

M.P. 180-182°C

Solvent: Acetone

Assignments
a = 7.27
b = 8.11
c = 9.72

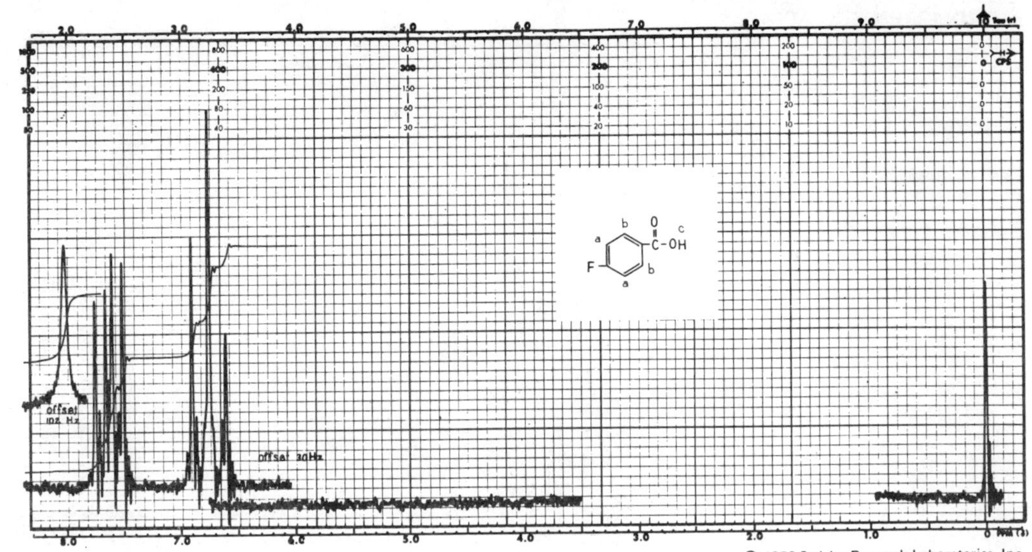

2 570

© 1978 Sadtler Research Laboratories, Inc.

---

p-CHLOROBENZOIC ACID

$C_7H_5ClO_2$

Mol. Wt. 156.57

M.P. 238-240°C

Solvent: Dimethyl sulfoxide

Assignments
a = 7.56
b = 8.00
c = 8.26 *
* H_2O present

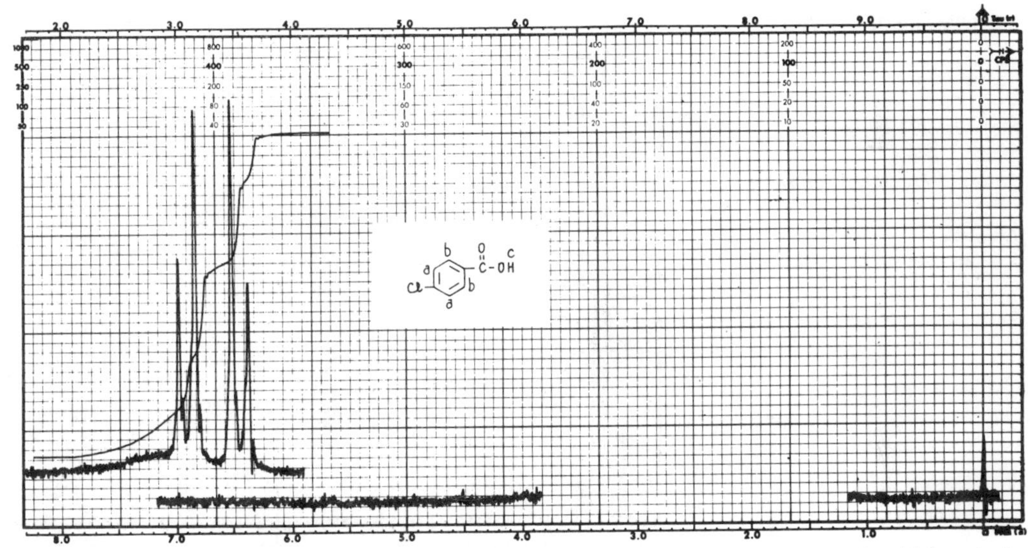

2571

© 1978 Sadtler Research Laboratories, Inc.

---

o-BROMOBENZOIC ACID

$C_7H_5BrO_2$

Mol. Wt. 201.02

Solvent: Dimethyl sulfoxide

Assignments
a = ca 7.42
b = ca 7.69
c = 9.11

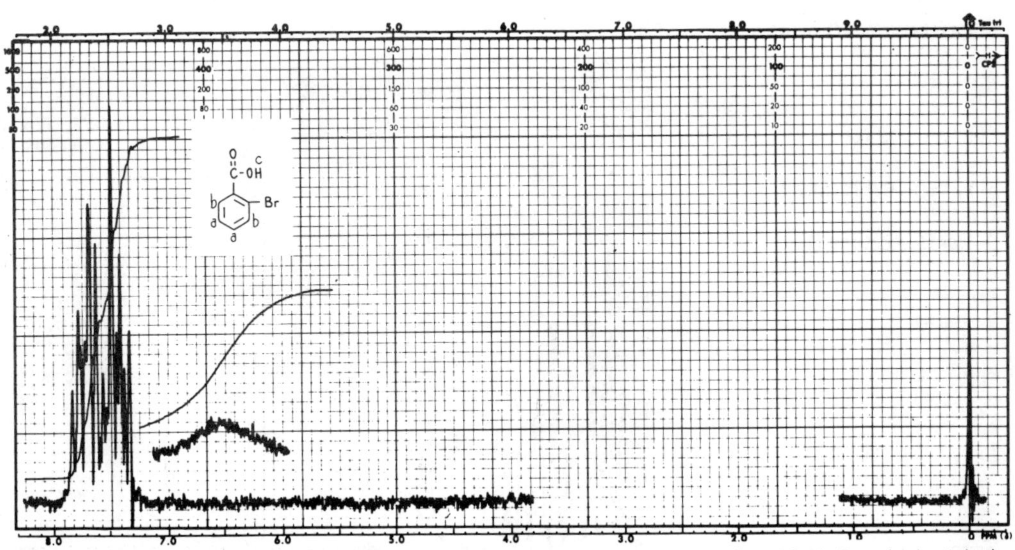

2572

© 1978 Sadtler Research Laboratories, Inc.

m-BROMOBENZOIC ACID

$C_7H_5BrO_2$

Mol. Wt. 201.03

M.P. 156-158°C

Solvent: Polysol-d

Assignments

a = 7.32
b = 7.64
c = 7.97
d = 8.15
e = 11.21

2573

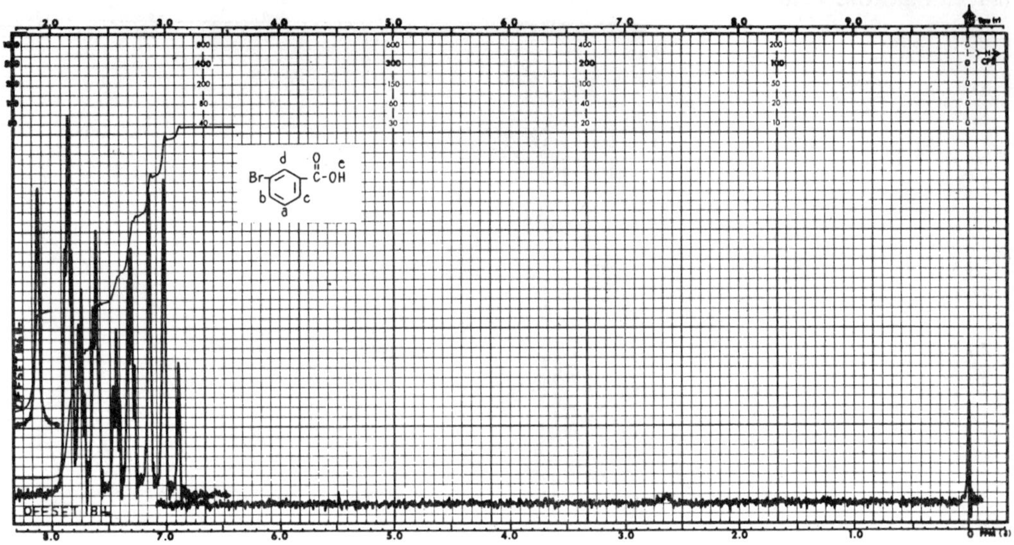

© 1978 Sadtler Research Laboratories, Inc.

p-BROMOBENZOIC ACID

$C_7H_5BrO_2$

Mol. Wt. 201.02

Solvent: Dimethyl sulfoxide-D_6

Assignments

a = ca 7.28
b = 7.71
c = 7.90
d = 2.56 DMSO-D_5

2574

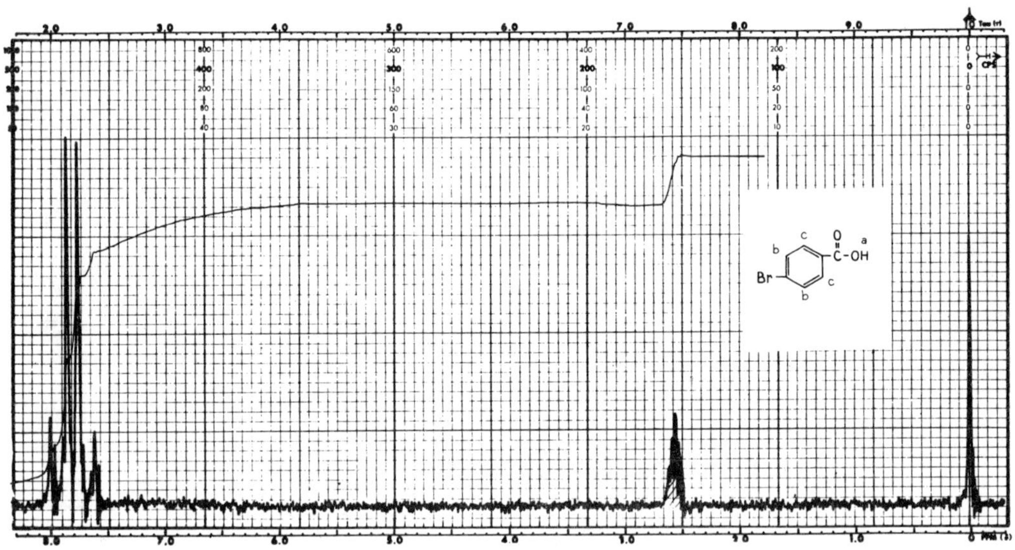

© 1978 Sadtler Research Laboratories, Inc.

o-IODOBENZOIC ACID

$C_7H_5IO_2$

Mol. Wt. 248.02

M.P. 160-162°C (lit.)

Solvent: Acetone-D_6

Assignments

a = 6.41
b = 7.23
c = 7.52
d = 7.90
e = 8.06
f = 2.08 acetone

2575

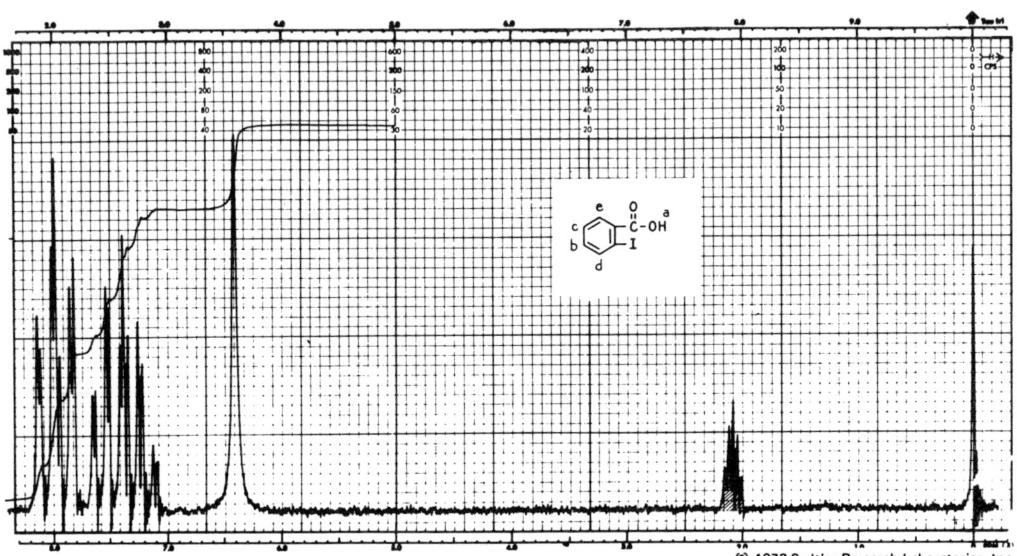

© 1978 Sadtler Research Laboratories, Inc.

m-IODOBENZOIC ACID

$C_7H_5IO_2$

Mol. Wt. 248.02

M.P. 185-187°C

Solvent: Polysol-d

Assignments
a = 7.20
b = 7.88
c = 8.00
d = 8.35
e = 11.86

2576

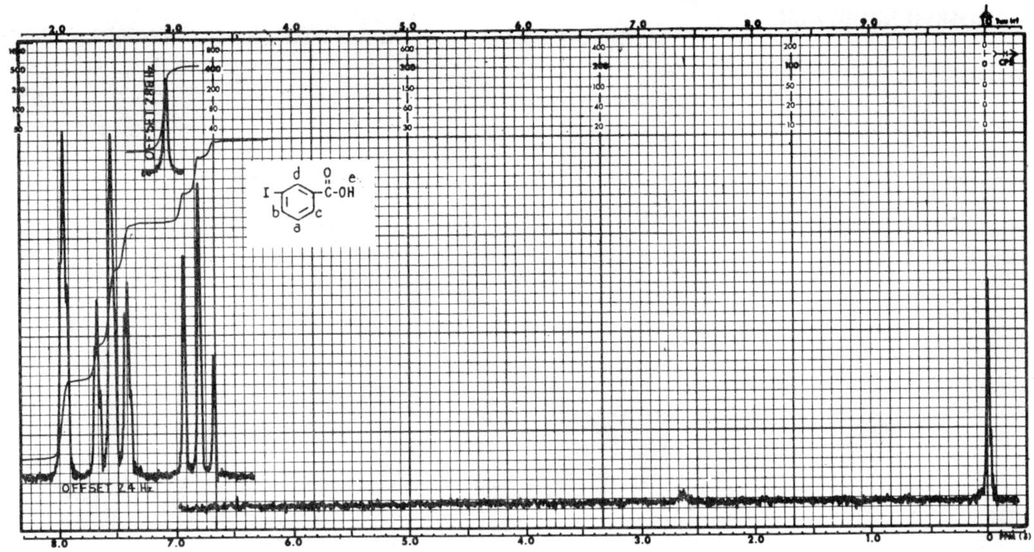

© 1978 Sadtler Research Laboratories, Inc.

p-IODOBENZOIC ACID

$C_7H_5IO_2$

Mol. Wt. 248.02

M.P. 265-270°C

Solvent: Dimethyl sulfoxide-D_6

Assignments
a = 7.71
b = 7.90
c = 2.58 solvent
OH is spread
out 450-650 cps

2577

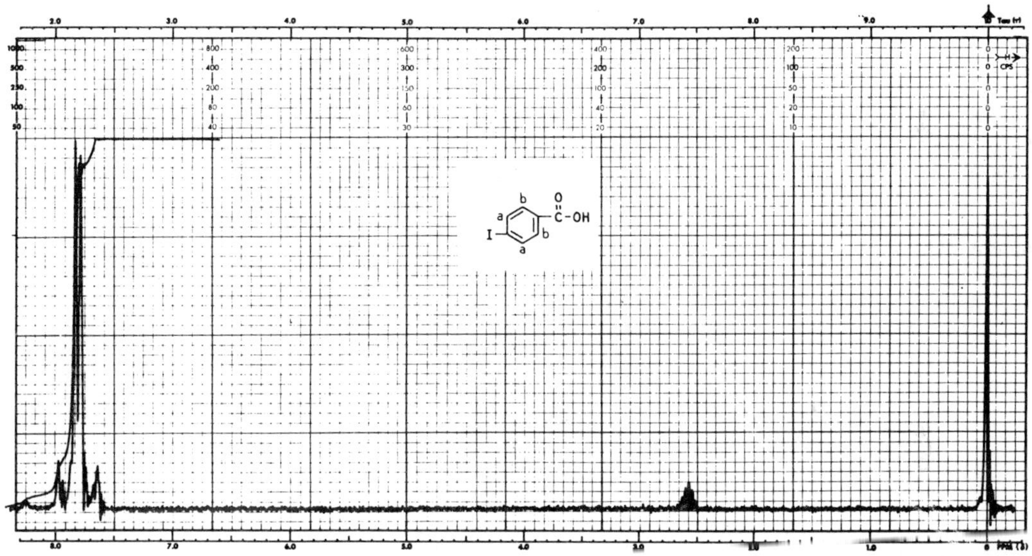

© 1978 Sadtler Research Laboratories, Inc.

p-AMINOBENZOIC ACID

$C_7H_7NO_2$

Mol. Wt. 137.14

M.P. 187-188°C

Solvent: Acetone

Assignments
a = 6.55
b = 6.76
c = 7.83

2578

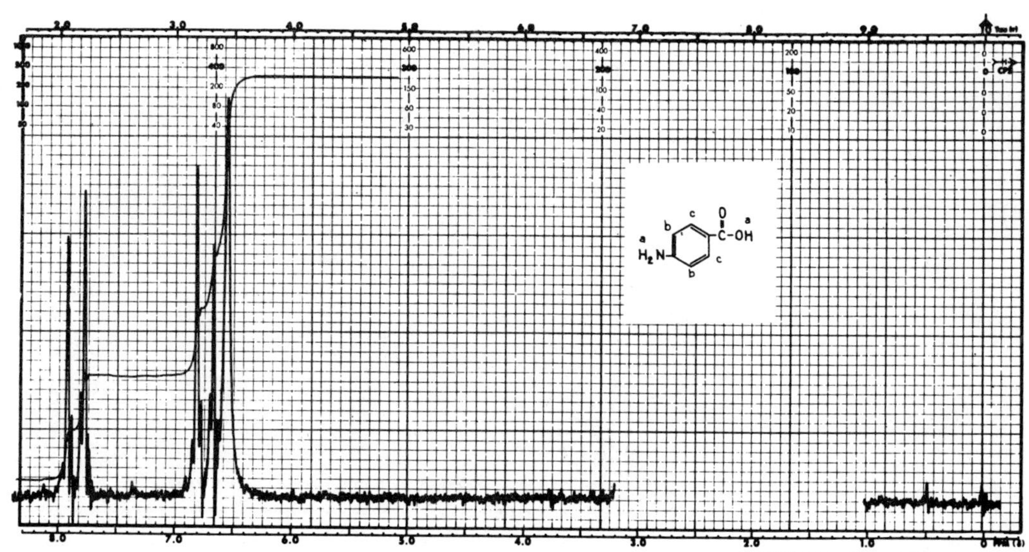

© 1978 Sadtler Research Laboratories, Inc.

N-PHENYLANTHRANILIC ACID

$C_{13}H_{11}NO_2$

Mol. Wt. 213.24

M.P. 185-187°C

Solvent: Dimethyl sulfoxide

Assignments
a = 6.81
b = 7.00-7.60
c = 8.00
d = 9.72

2579

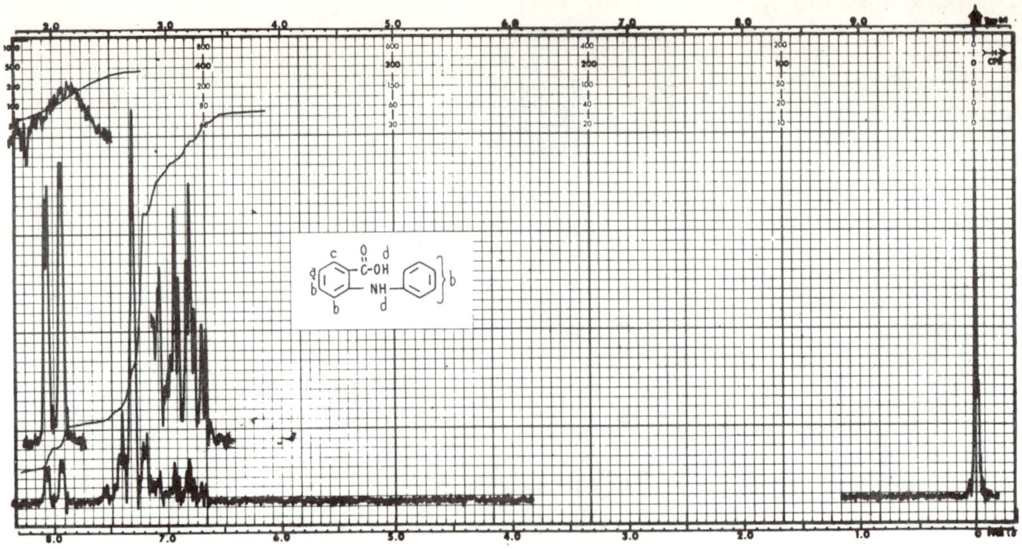

© 1978 Sadtler Research Laboratories, Inc.

m-CYANOBENZOIC ACID

$C_8H_5NO_2$

Mol. Wt. 147.13

M.P. 222-224°C

Solvent: Dimethyl sulfoxide

Assignments
a = 7.74
b = 8.10
c = 8.18-8.50
d = 10.32

2580

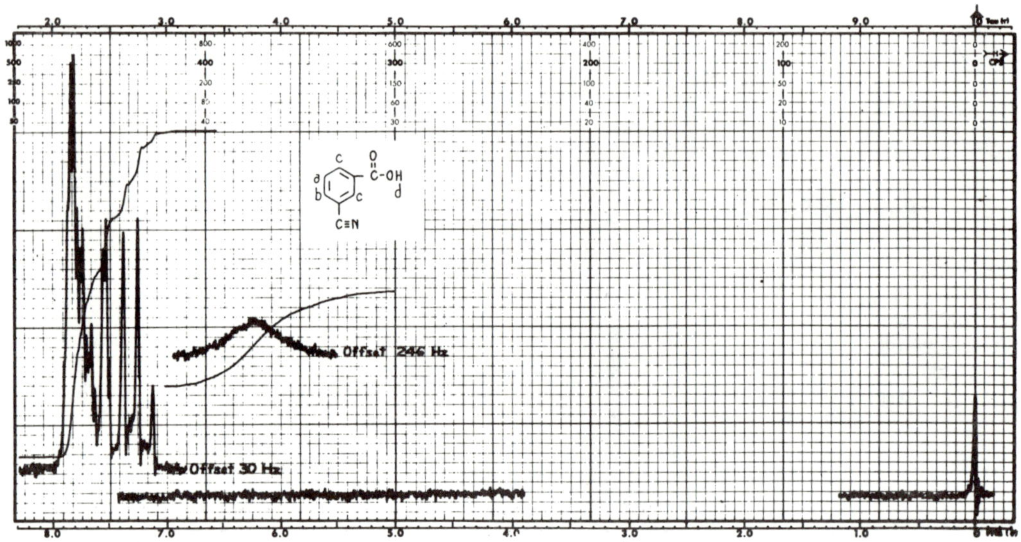

© 1978 Sadtler Research Laboratories, Inc.

o-NITROBENZOIC ACID

$C_7H_5NO_4$

Mol. Wt. 167.12

M.P. 147-148°C

Solvent: Acetone

Assignments
a = 7.75
b = 7.90
c = 10.67

2581

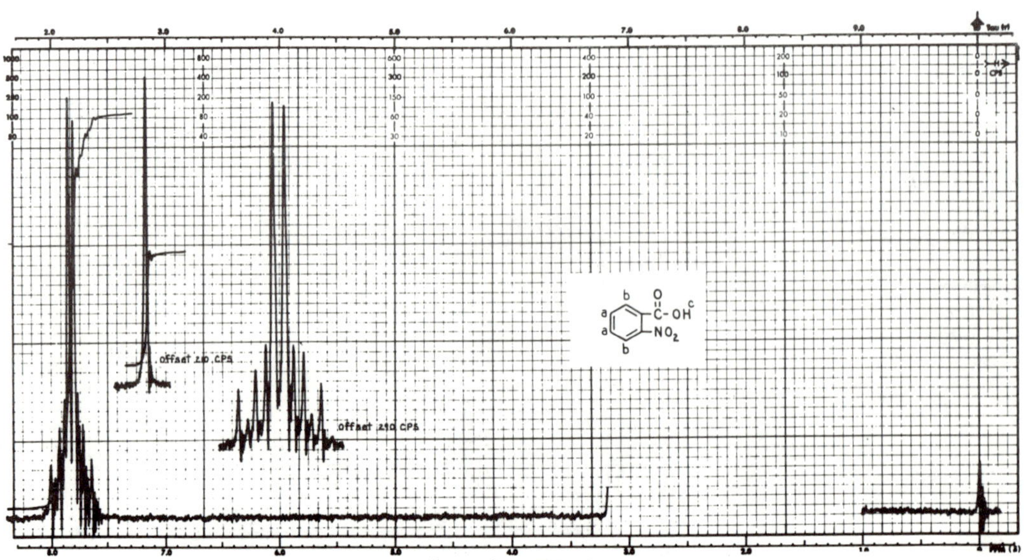

© 1978 Sadtler Research Laboratories, Inc.

p-NITROBENZOIC ACID

$C_7H_5NO_4$

Mol. Wt. 167.12

M.P. 241-243°C

Solvent: Dimethyl sulfoxide-d_6

Assignments
a = 8.21
b = 8.30
c = unrecorded *
d = 2.59 DMSO-d_5
* Occurs as a very broad band from 9-14 ppm

2582

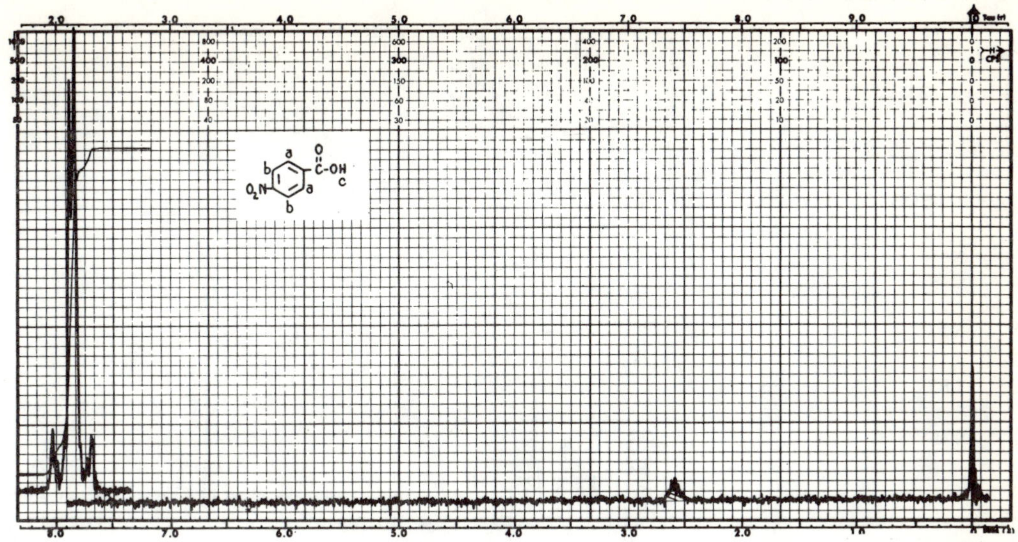

© 1978 Sadtler Research Laboratories, Inc.

---

m-ANISIC ACID

$C_8H_8O_3$

Mol. Wt. 152.15

M.P. 104-106°C

Solvent: Dimethyl sulfoxide-d_6

Assignments
a = 3.85
b = 7.20
c = 7.43
d = 7.50-7.80
e = 12.80

2583

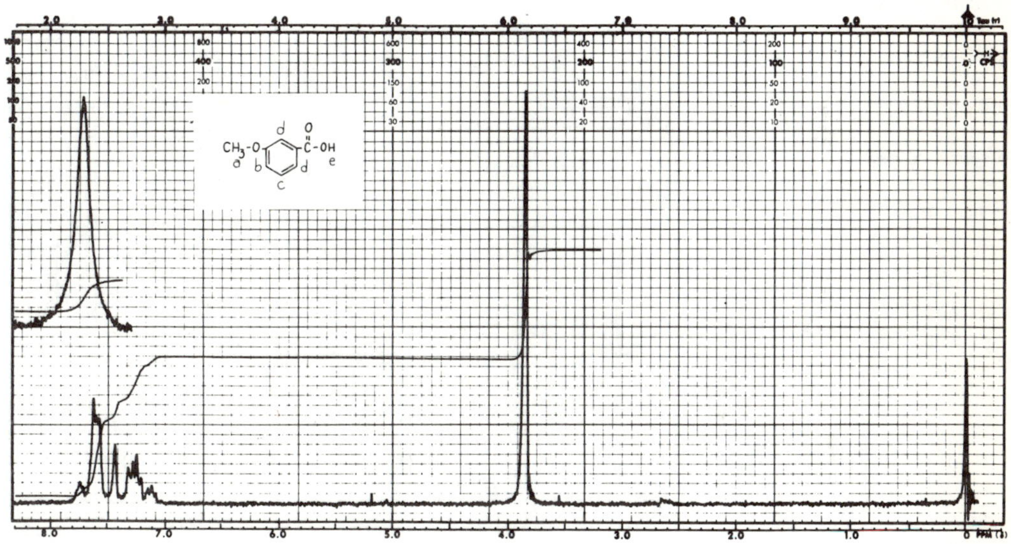

© 1978 Sadtler Research Laboratories, Inc.

---

p-ANISIC ACID

$C_8H_8O_3$

Mol. Wt. 152.15

M.P. 184-185°C

Solvent: Dimethyl sulfoxide-d_6

Assignments
a = 3.82
b = 7.00
c = 7.91
d = 2.56 DMSO-d_5
OH unobserved

2584

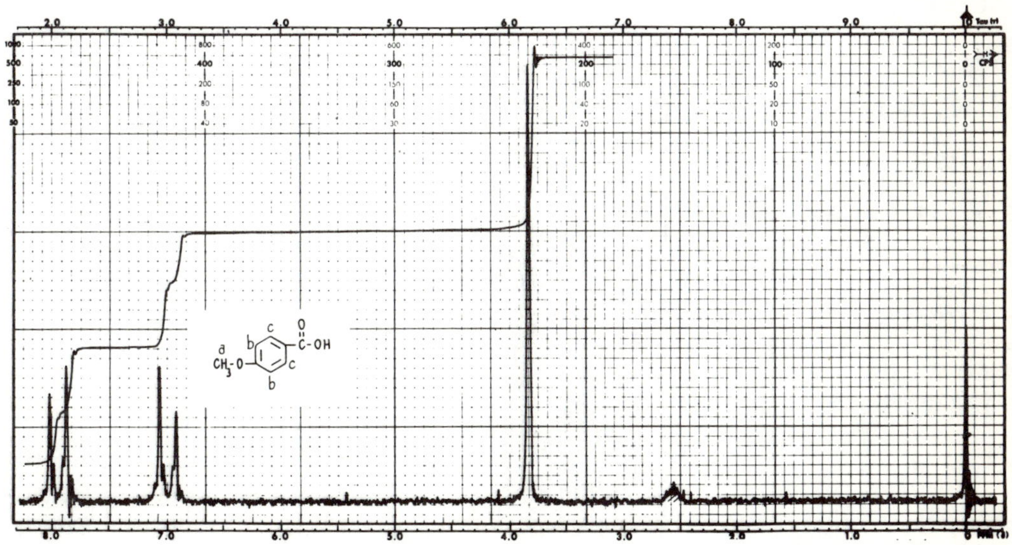

© 1978 Sadtler Research Laboratories, Inc.

o-ETHOXYBENZOIC ACID

$C_9H_{10}O_3$

Mol. Wt. 166.18

B.P. 175°C/15mm

Solvent: CCl_4

Assignments
a = 1.49
b = 4.28
c = 7.03
d = 7.11
e = 7.53
f = 8.08
g = 10.59

2585

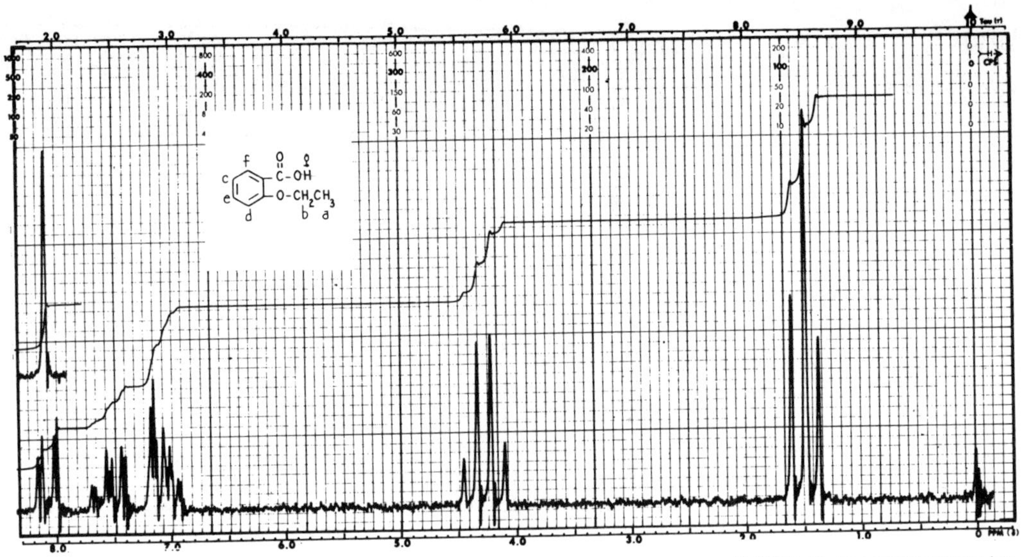

© 1978 Sadtler Research Laboratories, Inc.

p-PROPOXYBENZOIC ACID

$C_{10}H_{12}O_3$

Mol. Wt. 180 .21

Solvent: Polysol-d

Assignments
a = 1.02
b = 1.77
c = 3.91
d = 6.90
e = 7.95
f = 10.95

2586

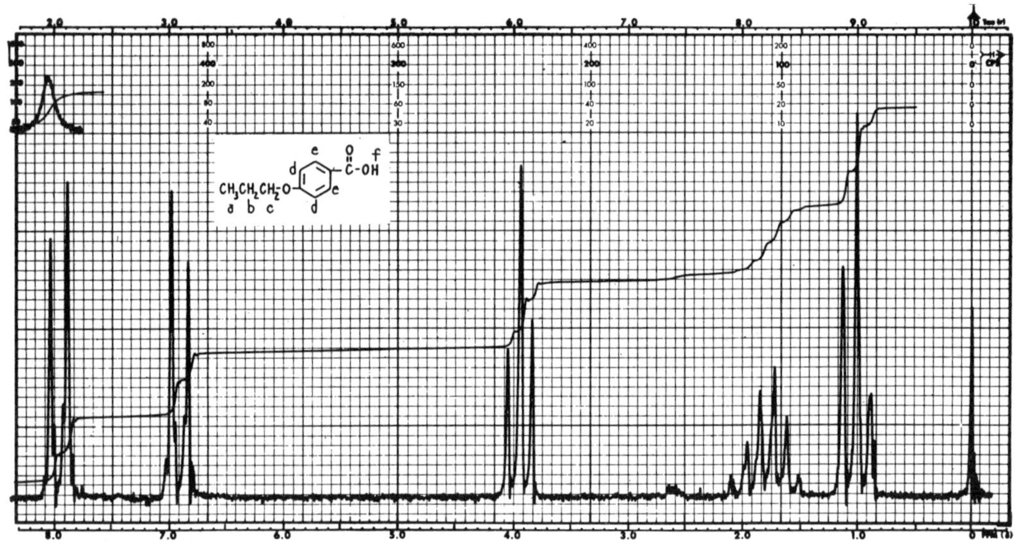

© 1978 Sadtler Research Laboratories, Inc.

SALICYLIC ACID

$C_7H_6O_3$

Mol. Wt. 138.12

M.P. 159°C

Solvent: Trifluoroacetic acid

Assignments
a = 7.02
b = 7.09
c = 7.59
d = 8.00
OH's are unobserved

2587

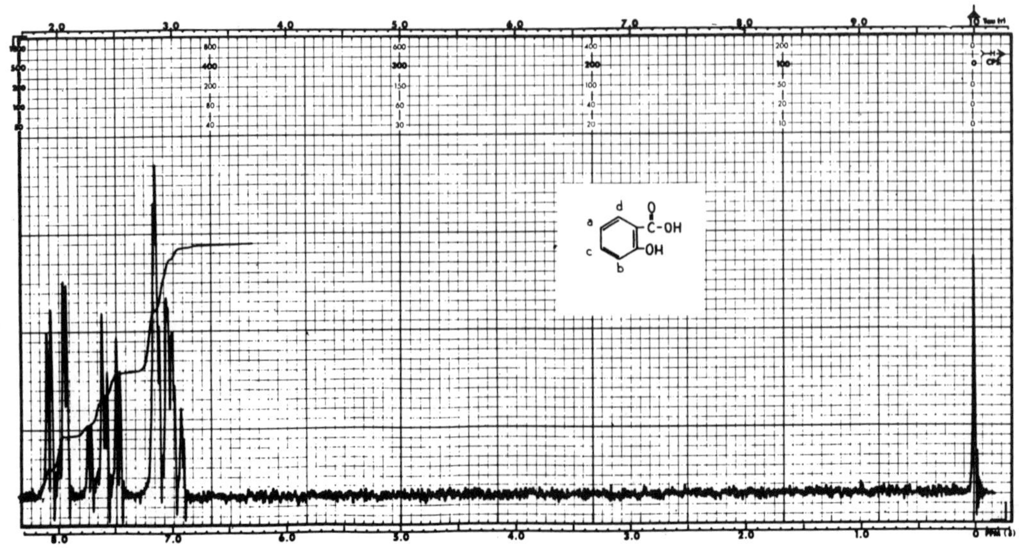

© 1978 Sadtler Research Laboratories, Inc.

p-HYDROXYBENZOIC ACID

$C_7H_6O_3$

Mol. Wt. 138.12

M.P. 215-216°C

Solvent: Acetone

Assignments

a = 7.07

b = 8.11

c = 9.76

2588

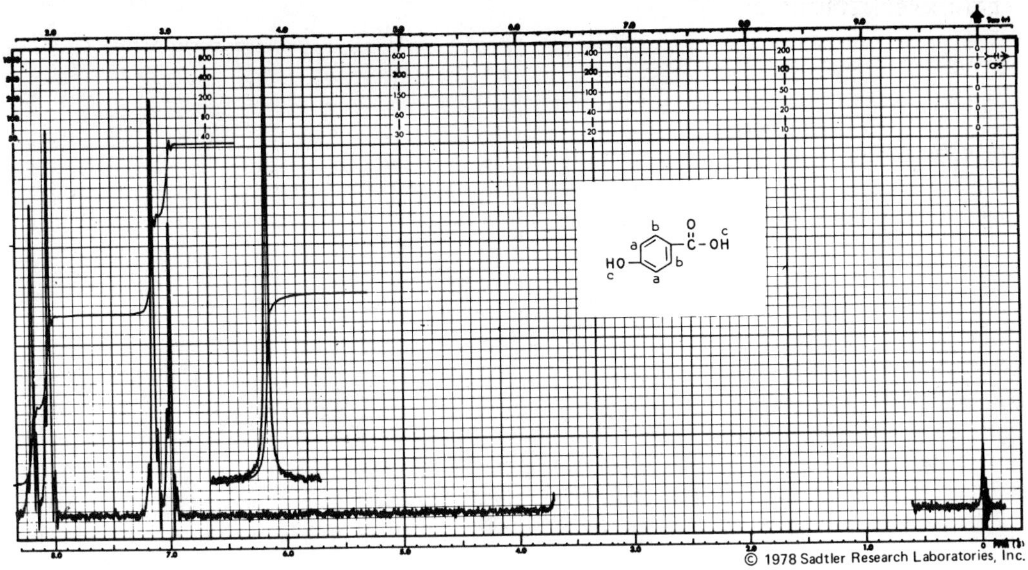

© 1978 Sadtler Research Laboratories, Inc.

4,4'-OXYDIBENZOIC ACID

$C_{14}H_{10}O_5$

Mol. Wt. 258.23

M.P. 300°C

Solvent: Dimethyl sulfoxide

Assignments

a = 7.20

b = 8.09

c = 12.02

2589

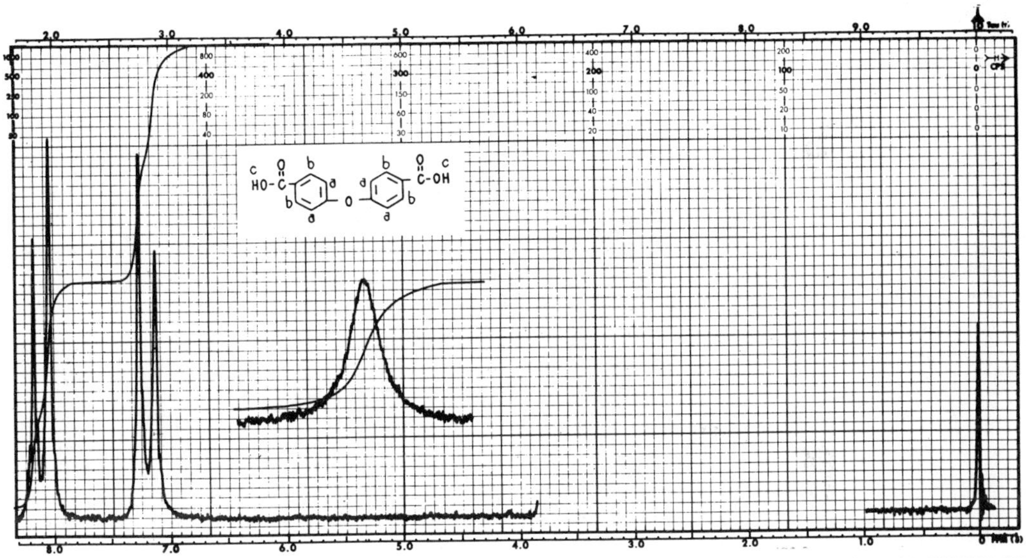

© 1978 Sadtler Research Laboratories, Inc.

PHTHALIC ACID

$C_8H_6O_4$

Mol. Wt. 166.13

M.P. 204-206°C (dec.)

Solvent: Polysol-d

Assignments

a = 7.51

b = 7.71

c = 11.90

2590

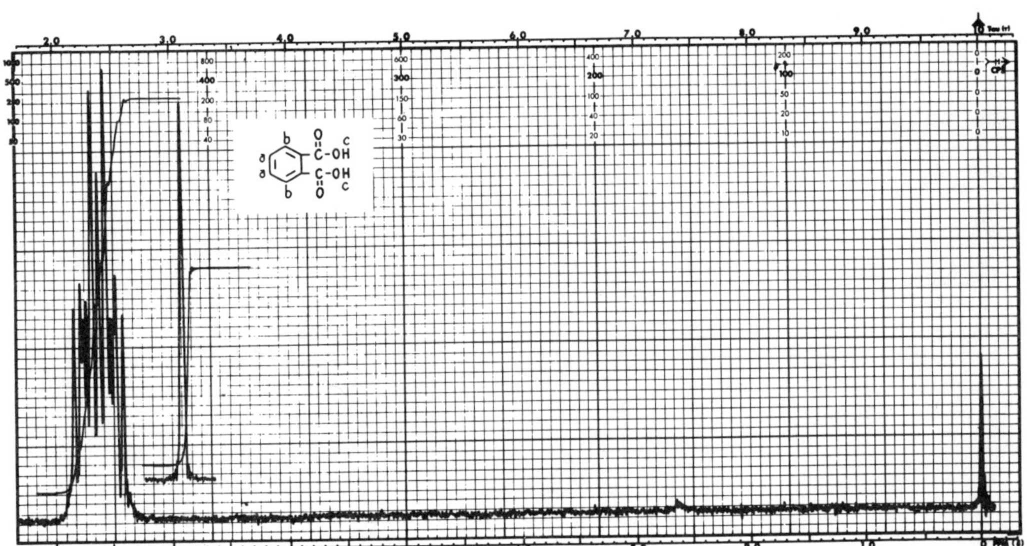

© 1978 Sadtler Research Laboratories, Inc.

ISOPHTHALIC ACID

$C_8H_6O_4$

Mol. Wt. 166.13

M.P. 344-347°C

Solvent: Polysol-d

Assignments
a = 7.50
b = 8.20
c = 8.64
d = 11.70
e = unobserved

2591

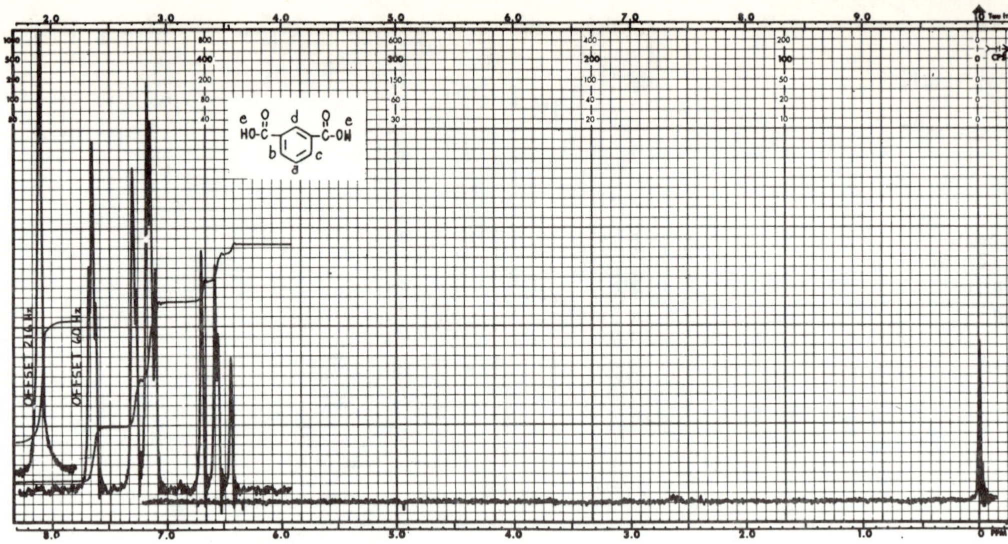

© 1978 Sadtler Research Laboratories, Inc.

TEREPHTHALIC ACID

$C_8H_6O_4$

Mol. Wt. 166.13

Solvent: Dimethy l sulfoxide-D_6

Assignments
a = 8.11
b = 9.41 *
c = 2.60 DMSO-D_5
* Includes H_2O

2592

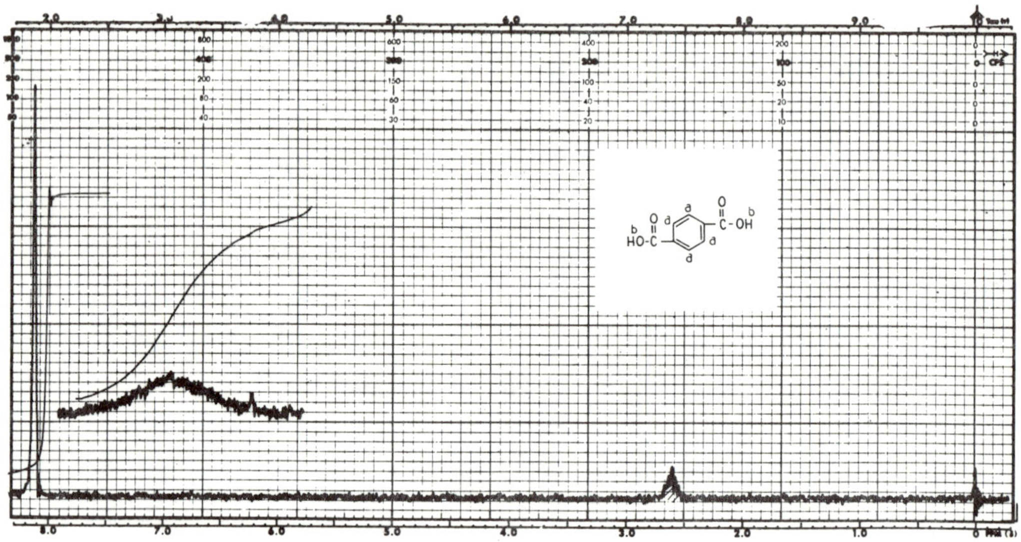

© 1978 Sadtler Research Laboratories, Inc.

2,6-DIMETHYLBENZOIC ACID

$C_9H_{10}O_2$

Mol. Wt. 150.18

M.P. 115-117°C

Solvent: $CDCl_3$

Assignments
a = 2.41
b = 6.90-7.35
c = 11.40

2593

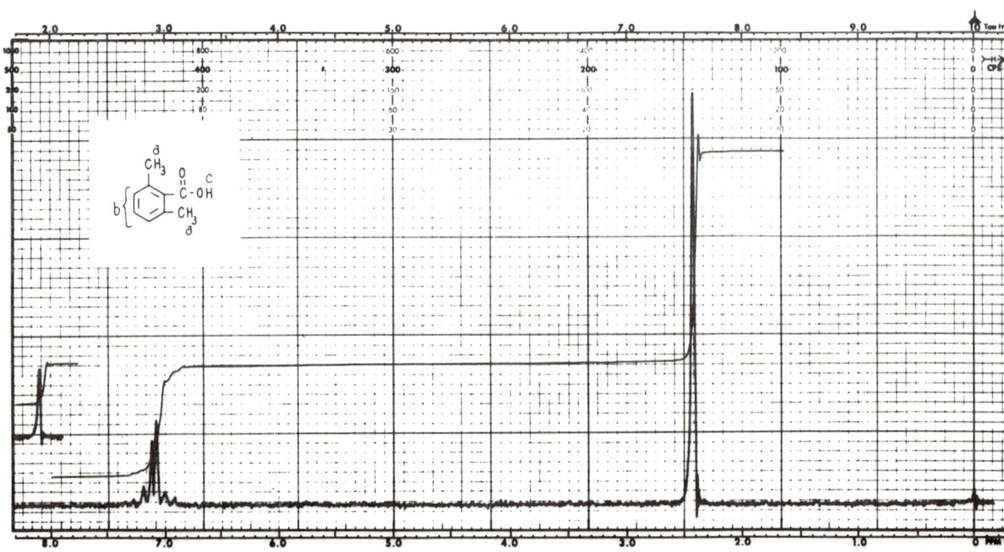

© 1978 Sadtler Research Laboratories, Inc.

2,4-DINITROBENZOIC ACID

$C_7H_4N_2O_6$

Mol. Wt. 212.12

M.P. 179-181°C

Solvent: Dimethyl sulfoxide

Assignments
a = 8.09
b = 8.59
c = 8.70
d = 9.61 *
* H_2O present

2594

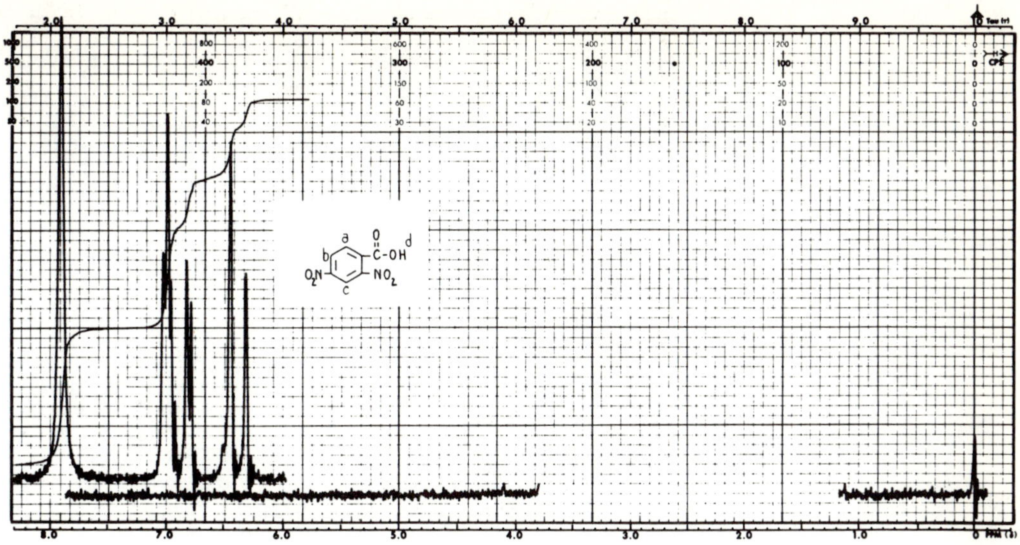

© 1978 Sadtler Research Laboratories, Inc.

6-AMINO-m-ANISIC ACID

$C_8H_9NO_3$

Mol. Wt. 167.17

Solvent: Polysol-d

Assignments
a = 3.70
b = 6.67
c = 6.90
d = 7.29
e = 7.87
f = 2.53 DMSO-d_5

2595

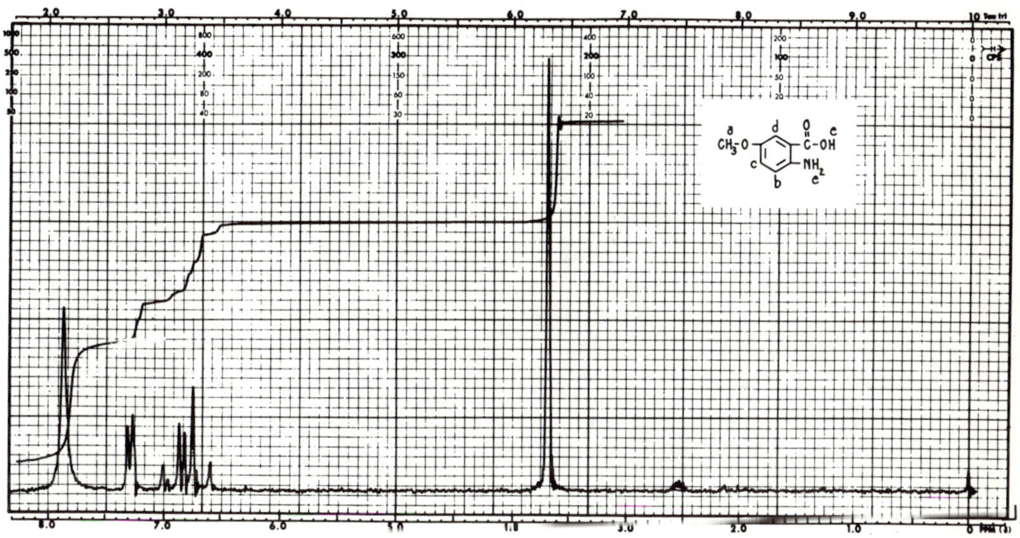

© 1978 Sadtler Research Laboratories, Inc.

2,4-CRESOTIC ACID

$C_8H_8O_3$

Mol. Wt. 152.15

Solvent: Polysol-d

Assignments
a = 2.31
b = 6.65
c = 6.71
d = 7.72
e = 10.29

2596

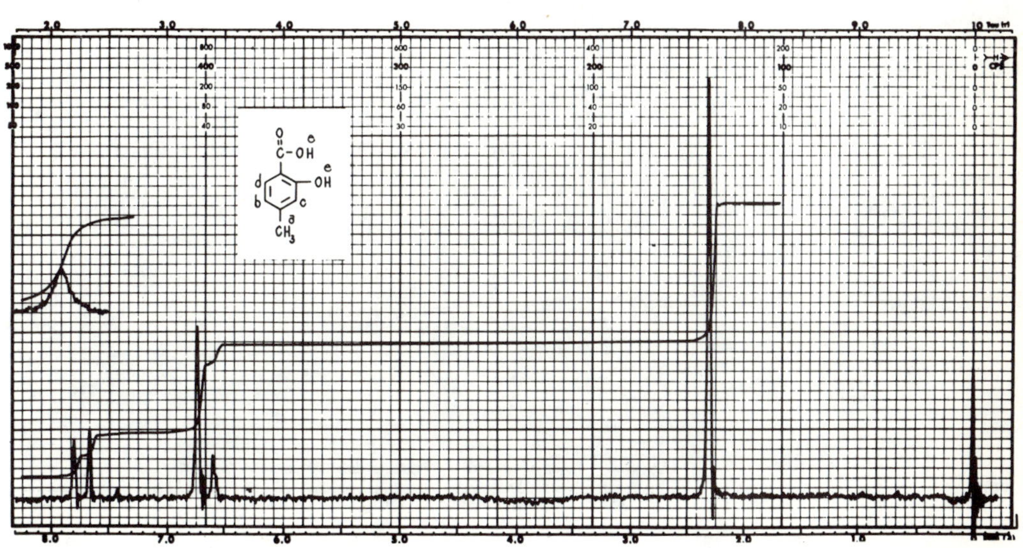

© 1978 Sadtler Research Laboratories, Inc.

3-NITROSALICYCLIC ACID

$C_7H_5NO_5$

Mol. Wt. 183.12

M.P. 147-149°C

Solvent: Polysol-d

Assignments
a = 7.00
b = 8.09
c = 12.49

2597

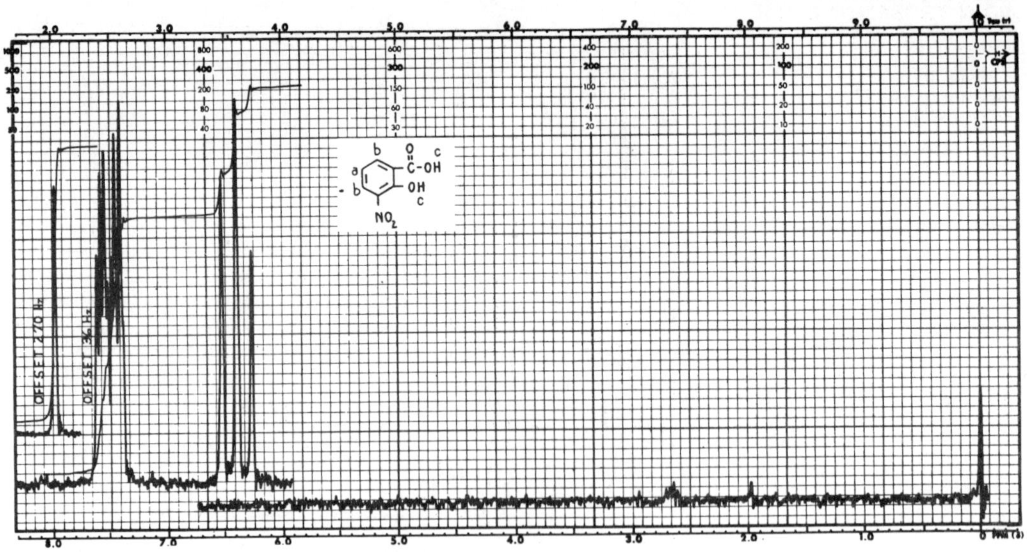

© 1978 Sadtler Research Laboratories, Inc.

VANILLIC ACID

$C_8H_8O_4$

Mol. Wt. 168.15

M.P. 210-212°C

Solvent: Dimethyl sulfoxide-D_6

Assignments
a = 3.92
b = 7.00
c = 7.59
d = 7.67
e = 10.63
f = 2.68 DMSO-D_5

2598

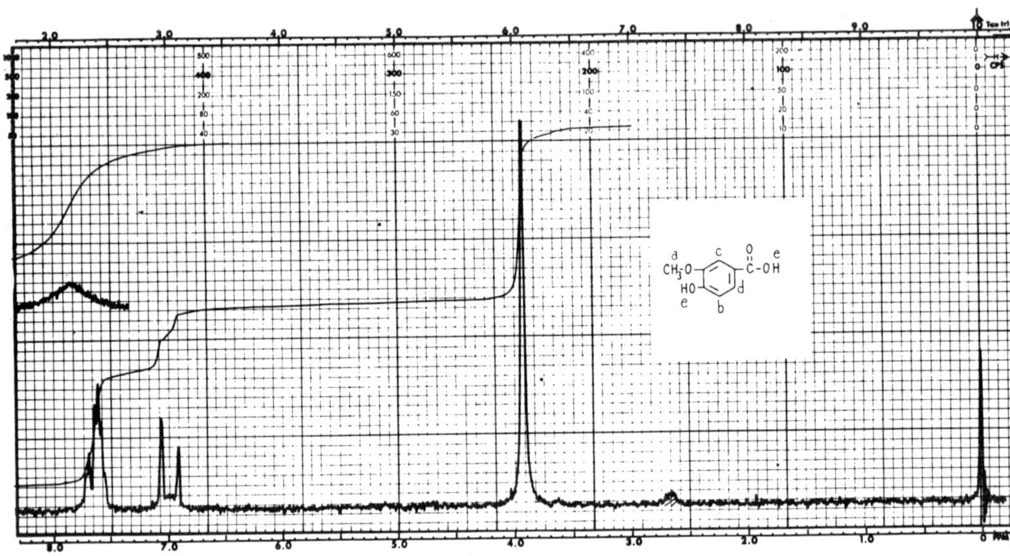

© 1978 Sadtler Research Laboratories, Inc.

β-RESORCYLIC ACID

$C_7H_6O_4$

Mol. Wt. 154.12

Solvent: Dimethyl sulfoxide

Assignments
a = 6.39
b = 6.45
c = 7.71
d = 10.20 *
* H_2O present

2599

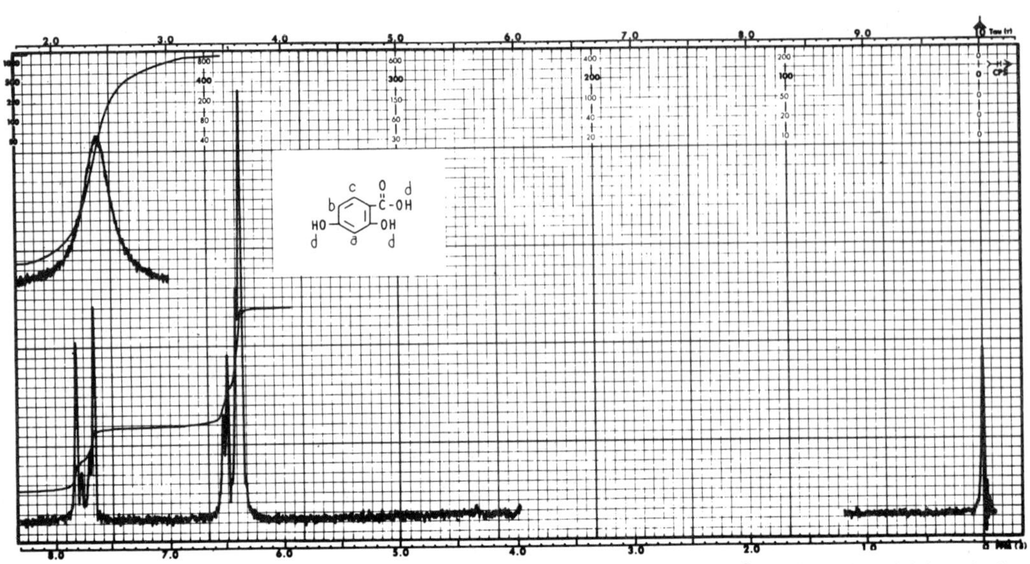

© 1978 Sadtler Research Laboratories, Inc.

PROTOCATECHUIC ACID

$C_7H_6O_4$

Mol. Wt. 154.12

M.P. 200-202°C

Solvent: DMSO-d_6

Assignments
a = 6.99
b = 7.52
c = 7.61
d = ca 9.27

2600

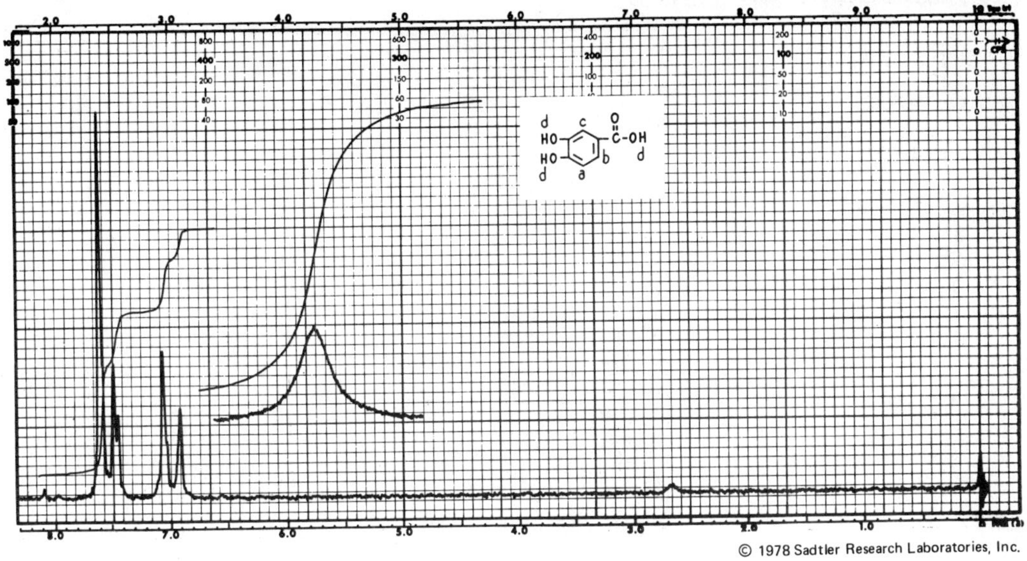

© 1978 Sadtler Research Laboratories, Inc.

4-NITROPHTHALIC ACID

$C_8H_5NO_6$

Mol. Wt. 211.13

M.P. 164-165°C

Solvent: D_2O

Assignments
a = 7.95
b = 8.46
c = 8.52
d = 4.99 HDO

2601

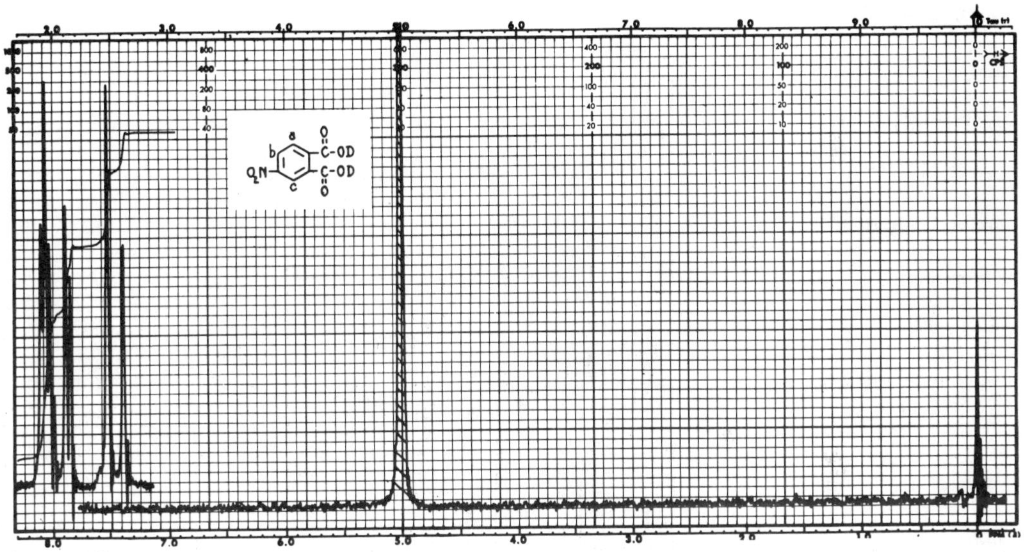

© 1978 Sadtler Research Laboratories, Inc.

5-NITROISOPHTHALIC ACID

$C_8H_5NO_6$

Mol. Wt. 211.13

M.P. 255°C (dec.) (lit.)

Solvent: DMSO

Assignments
a = 8.80
b = 12.01 *
* Plus H_2O

2602

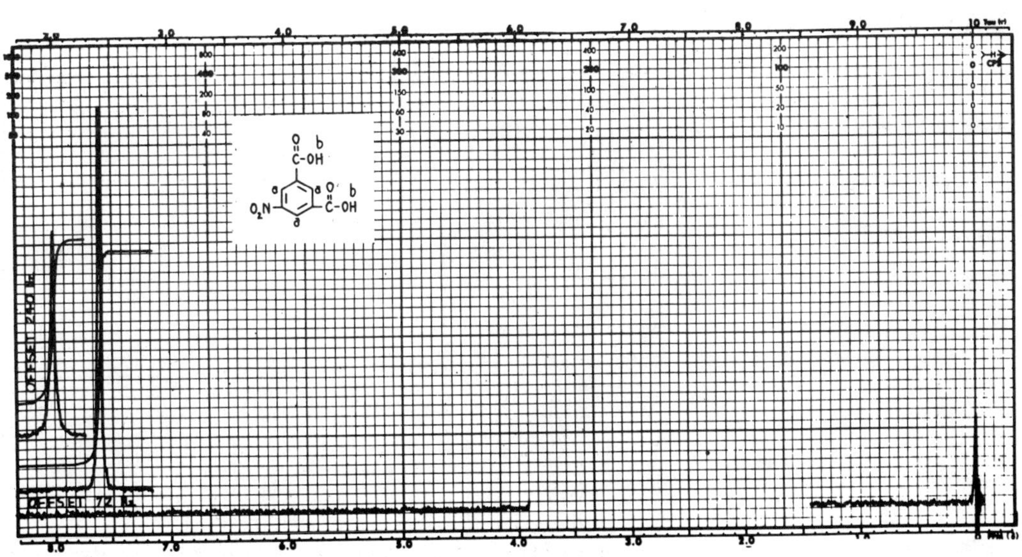

© 1978 Sadtler Research Laboratories, Inc.

1,3,5-BENZENETRICARBOXYLIC ACID

$C_9H_6O_6$

Mol. Wt. 210.14

Solvent: Dimethyl sulfoxide

Assignments
a = 8.72
b = 11.53*
*H_2O present

2603

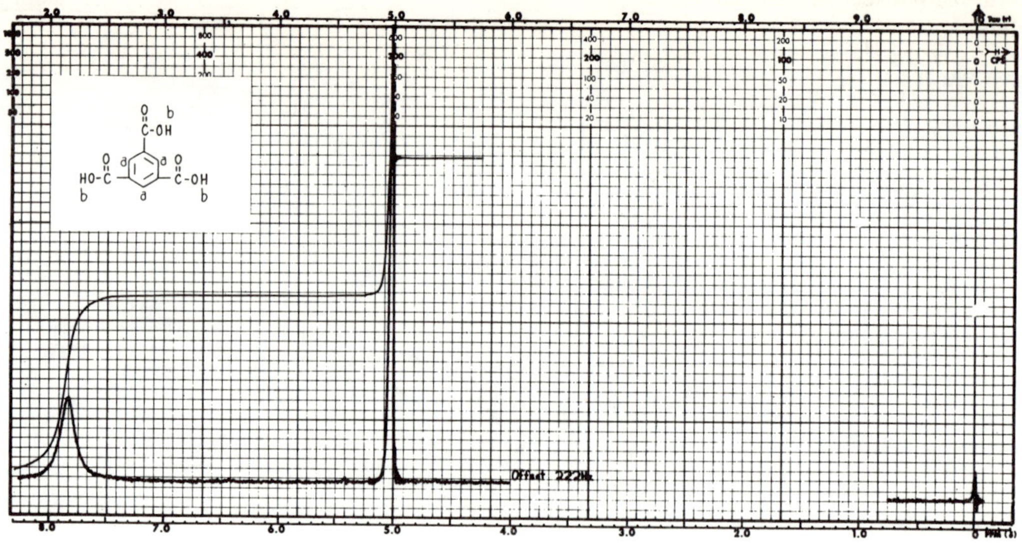

© 1978 Sadtler Research Laboratories, Inc.

3-FUROIC ACID

$C_5H_4O_3$

Mol. Wt. 112.09

M.P. 120-122°C

Solvent: Polysol-d

Assignments
a = 6.83
b = 7.60
c = 8.20
d = 9.19

2604

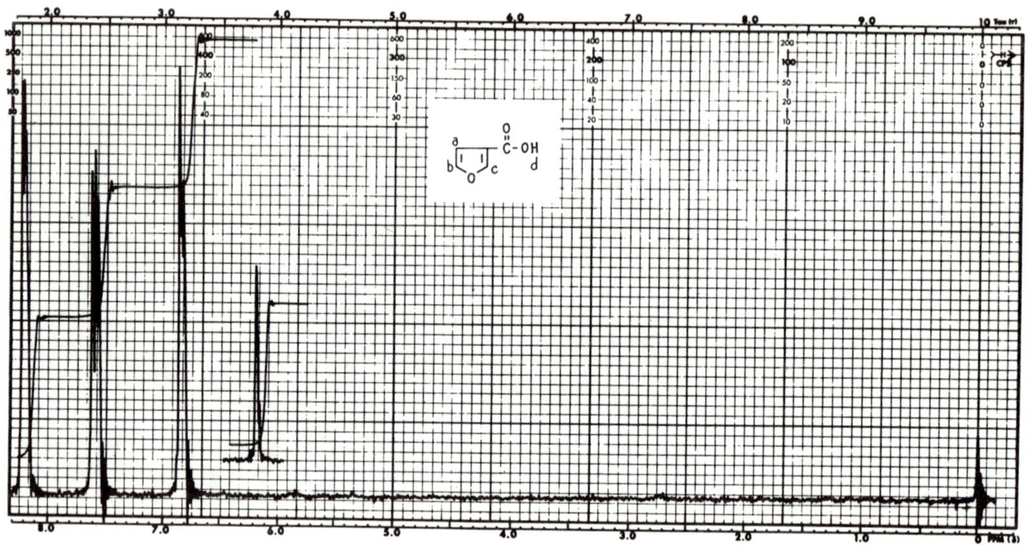

© 1978 Sadtler Research Laboratories, Inc.

2-FURANACRYLIC ACID

$C_7H_6O_3$

Mol. Wt. 138.12

M.P. 141-143°C

Solvent: $CDCl_3$

Assignments
a = 6.28
b = 6.48
c = 6.73
d = 7.49
e = 7.51
f = 10.32
g = 1.29 imp.

2605

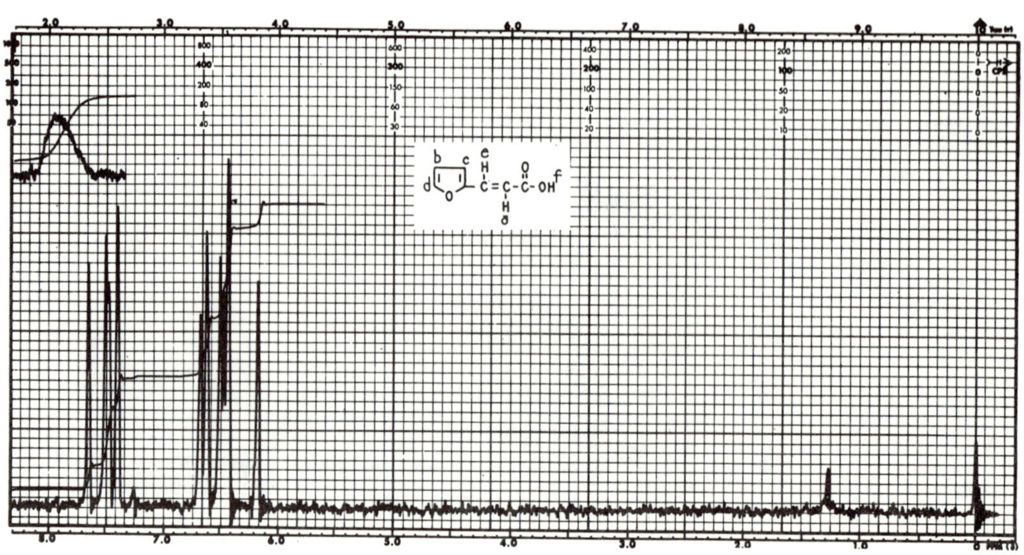

© 1978 Sadtler Research Laboratories, Inc.

NICOTINIC ACID, HYDROCHLORIDE

$C_6H_5NO_2 \cdot HCl$

Mol. Wt. 159.57

M.P. 275°C (dec.)

Solvent: D_2O

Assignments
a = 8.53
b = ca 9.28
c = 9.53
d = 4.90 HDO

2606

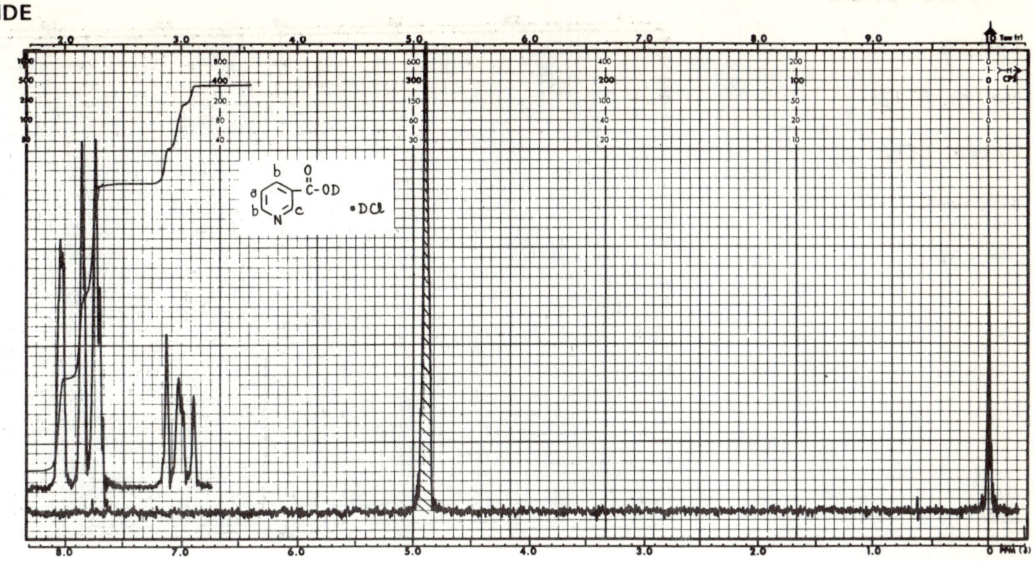

© 1978 Sadtler Research Laboratories, Inc.

3-INDOLEACETIC ACID

$C_{10}H_9NO_2$

Mol. Wt. 175.19

M.P. 166-168°C

Solvent: Dimethyl sulfoxide

Assignments
a = 3.64
b = 6.82-7.70
c = 10.66
d = 2.50 DMS
NH unobserved

2607

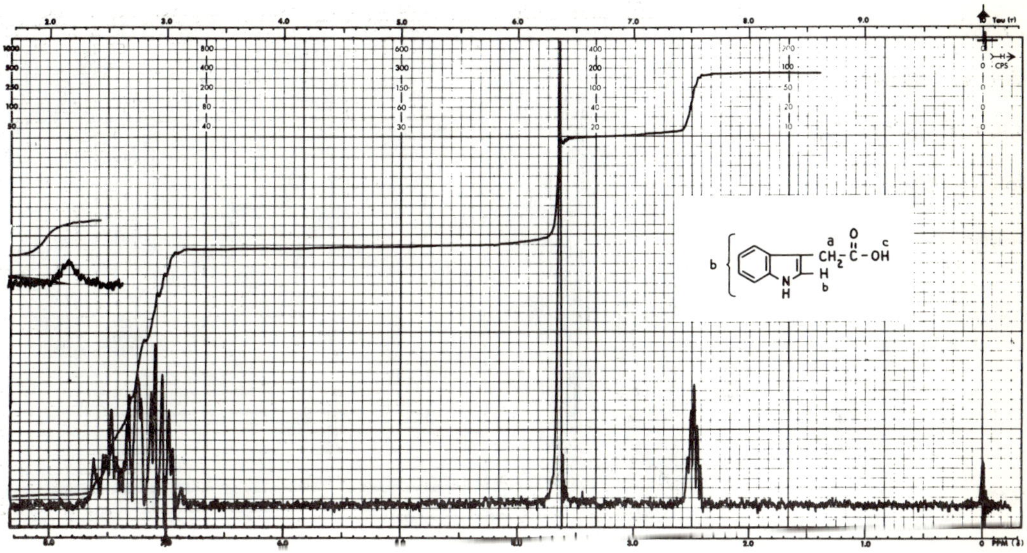

© 1978 Sadtler Research Laboratories, Inc.

PIPERONYLIC ACID

$C_8H_6O_4$

Mol. Wt. 166.13

M.P. 228°C (lit.)

Solvent: Polysol-d

Assignments
a = 6.08
b = 6.90
c = 7.38
d = 7.56
e = 10.80

2608

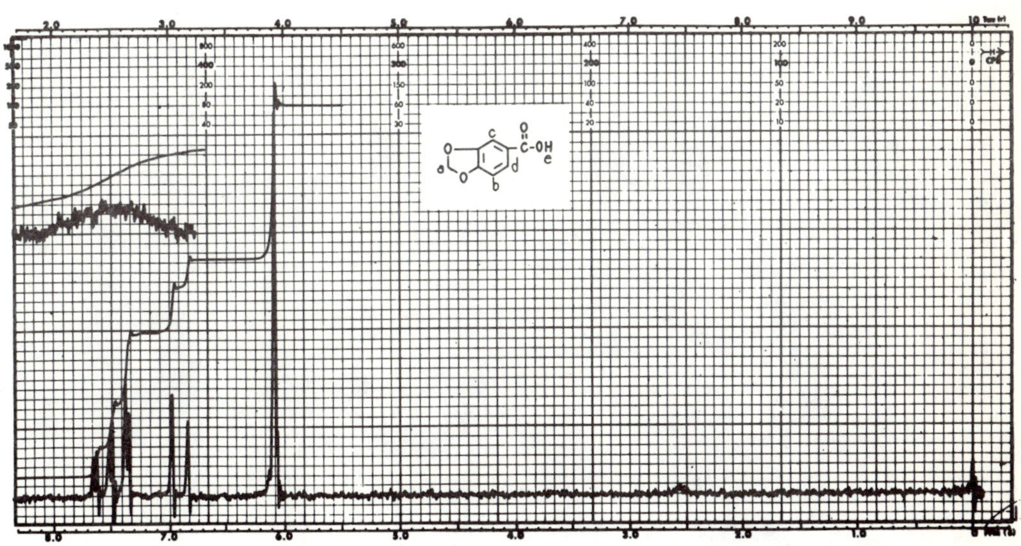

© 1978 Sadtler Research Laboratories, Inc.

THIOACETIC ACID

C_2H_4OS

Mol. Wt. 76.12

Solvent: CCl_4

Assignments
a = 2.37
b = 4.83

2609

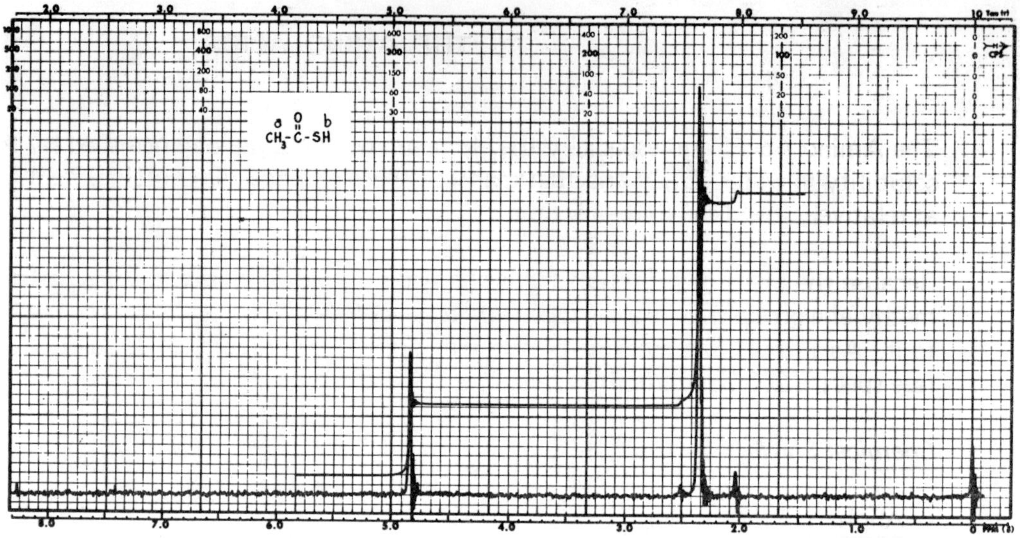

© 1978 Sadtler Research Laboratories, Inc.

THIOBENZOIC ACID

C_7H_6OS

Mol. Wt. 138.19

B.P. 103-105°C/12mm

Solvent: $CDCl_3$

Assignments
a = 5.35
b = 7.22-7.67
c = 7.90
d = 0.79 imp.

2610

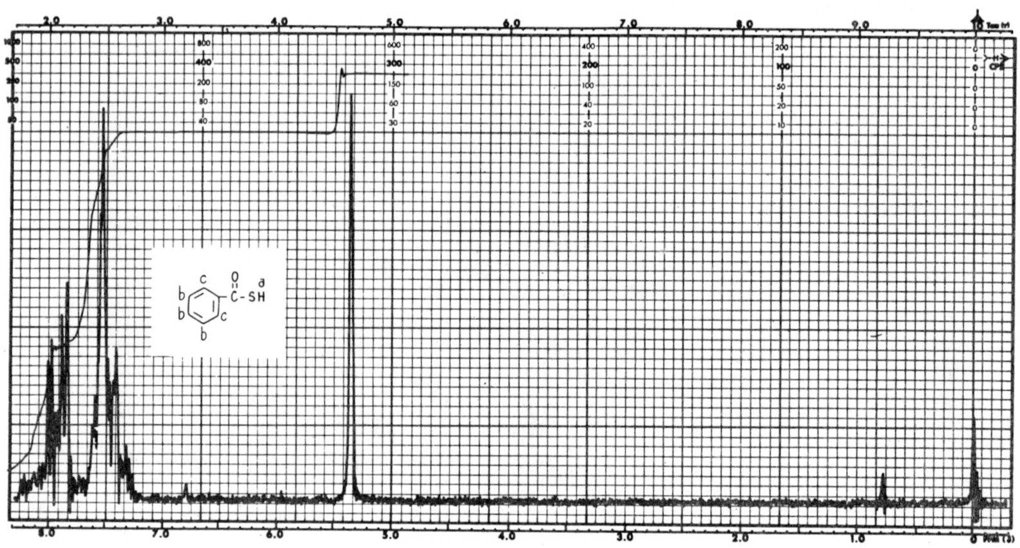

© 1978 Sadtler Research Laboratories, Inc.

THE AMINO ACIDS

The Amino Acids, and especially the alpha amino acids, are distinguished by their high degree of solubility in water (D_2O) and that many of these compounds contain a methine resonance band at relatively low field (3.3–4.5 ppm).

Because the methine proton of the alpha Amino Acids is an asymmetric carbon, an adjacent methylene group may display clear non-equivalence in chemical shift for the two hydrogens bonded to it. As a result, the methine proton may appear as a doublet of doublets rather than as a triplet (see spectra 2630 and 2631).

Chemical Shifts

Aliphatic Protons

Substituted Glycines

$HO{-}C(=O){-}CH_2$	$-N(X,Y)$	Solvent	Spectrum
3.58	$-NH_2$	D_2O	2611
3.62	$-NH{-}CH_3$	D_2O	2612
3.68	$-NH{-}(CH_2)_2{-}C{\equiv}N$	D_2O	2613
3.81	$-NH{-}\emptyset$	D_2O	2634
3.81	$-NH{-}C(=O){-}CH(NH_2){-}CH_3$	D_2O	2639
3.90	$-NH{-}C(=O){-}CH_2{-}NH_2$	D_2O	2638
3.92	$-NH{-}C(=O){-}CH_2{-}NH_2(HCl)$	D_2O	2646
3.98	$-N(CH_3)_2$	D_2O	2642

$HO{-}C(=O){-}CH(NH_2)$	$-X$		
3.30	$-(CH_2)_4{-}NH_2$	D_2O	2628
3.50	$-CH(OH){-}CH_3$	D_2O	2626
3.55	$-CH(CH_3)_2$	D_2O	2619
3.70	$-CH_3$	D_2O	2614
3.70	$-CH_2{-}CH_3$	D_2O	2615
3.71	$-R_3$	D_2O	2618
3.79	$-(CH_2)_2{-}C(=O){-}NH_2$	D_2O	2632
3.82	$-(CH_2)_3{-}NH_2(HCl)$	D_2O	2644
3.91	$-(CH_2)_2{-}NH_2$	D_2O	2643
4.00	$-CH_2{-}C(=O){-}NH_2$	D_2O	2630
4.47	$-CH_2{-}SH$	D_2O	2645

GLYCINE

$C_2H_5NO_2$

Mol. Wt. 75.07

M.P. 233°C (dec.) (lit.)

Solvent: D_2O

Assignments
a = 3.58
b = 4.90 HDO

2611

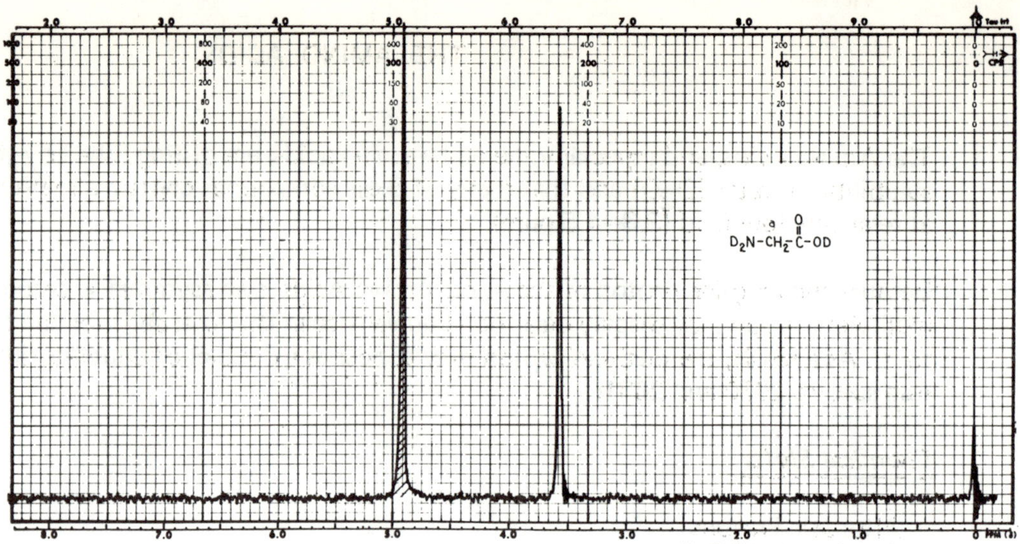

© 1978 Sadtler Research Laboratories, Inc.

SARCOSINE

$C_3H_7NO_2$

Mol. Wt. 89.09

M.P. ~ 212°C (dec.)

Solvent: D_2O

Assignments
a = 2.77
b = 3.62
c = 4.79 HDO

2612

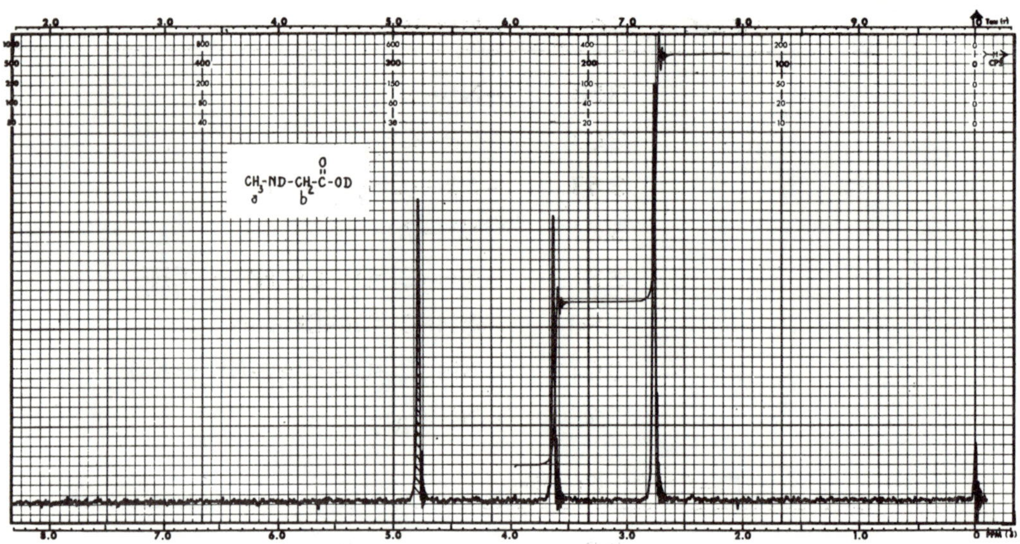

© 1978 Sadtler Research Laboratories, Inc.

N-(2-CYANOETHYL) GLYCINE

$C_5H_8N_2O_2$

Mol. Wt. 128.13

Solvent: D_2O

Assignments
a = 3.01
b = 3.43
c = 3.68
d = 4.70 HDO

2613

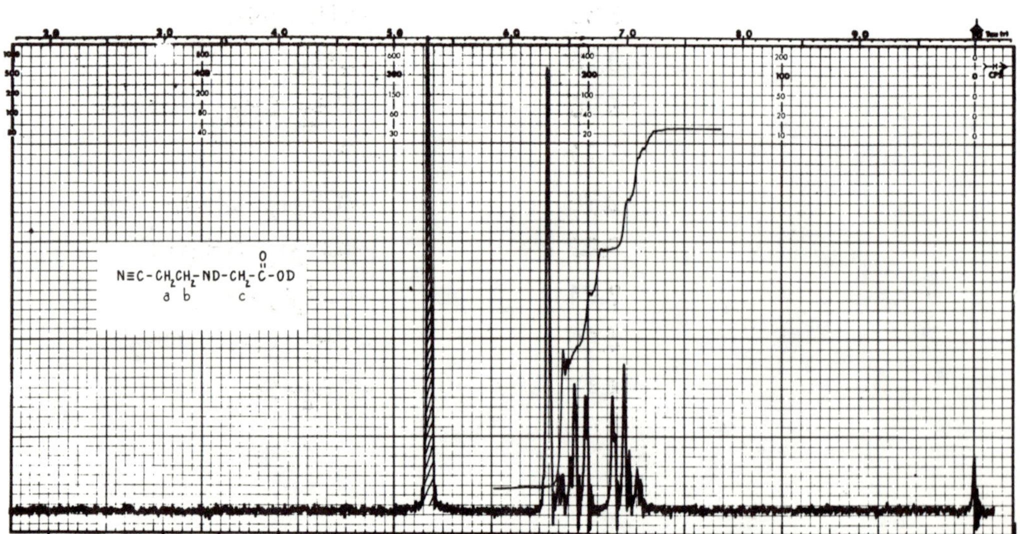

© 1978 Sadtler Research Laboratories, Inc.

L-ALANINE

$C_3H_7NO_2$

Mol. Wt. 89.09

Solvent: D_2O

Assignments
a = 1.48
b = 3.70
c = 4.65 HDO

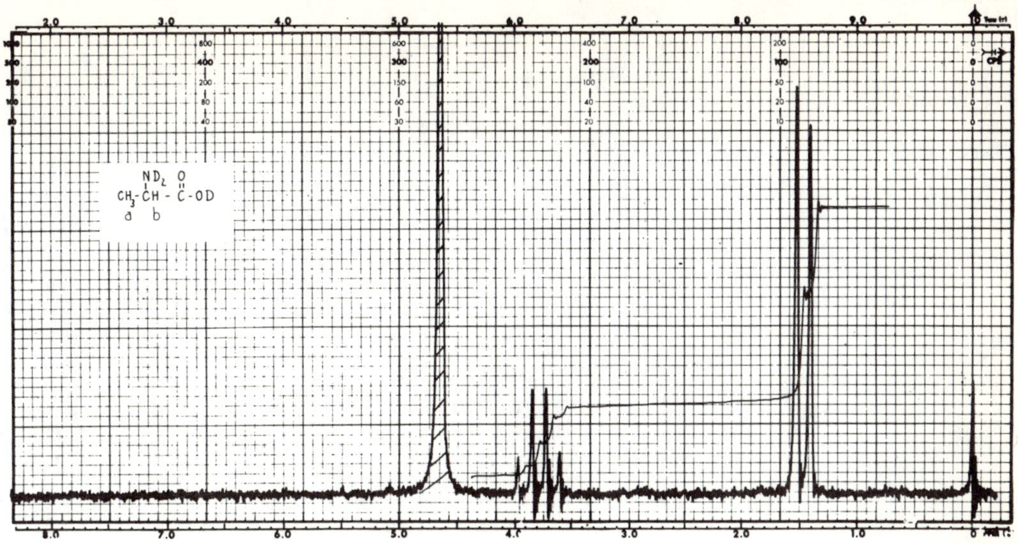

© 1978 Sadtler Research Laboratories, Inc.

2614

(±)-2-AMINOBUTYRIC ACID

$C_4H_9NO_2$

Mol. Wt. 103.12

M.P. 203°C (dec.) (lit.)

Solvent: D_2O

Assignments
a = 0.99
b = 1.88
c = 3.70
d = 4.69 HDO

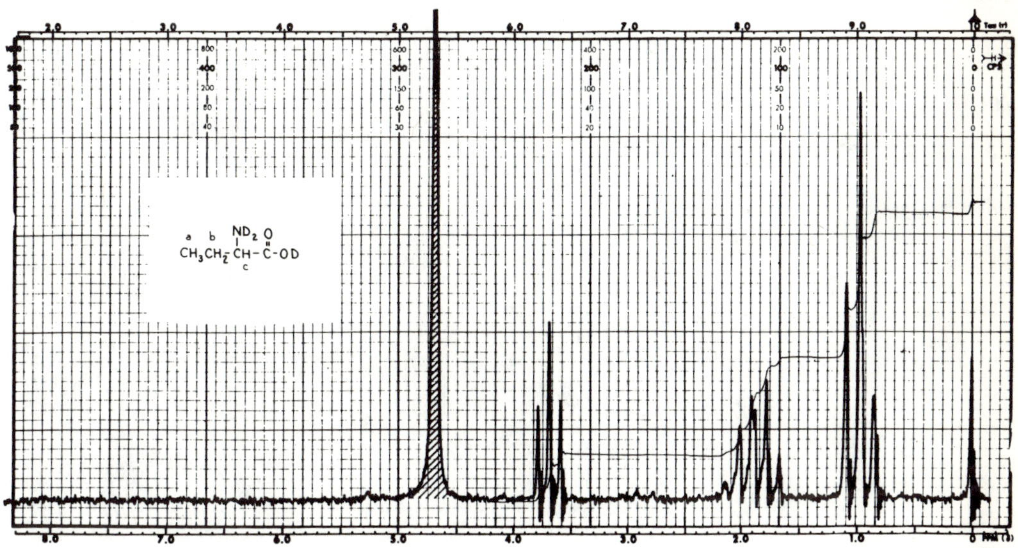

© 1978 Sadtler Research Laboratories, Inc.

2615

2-METHYLALANINE

$C_4H_9NO_2$

Mol. Wt. 103.12

Solvent: D_2O

Assignments
a = 1.50
b = 4.79 HDO

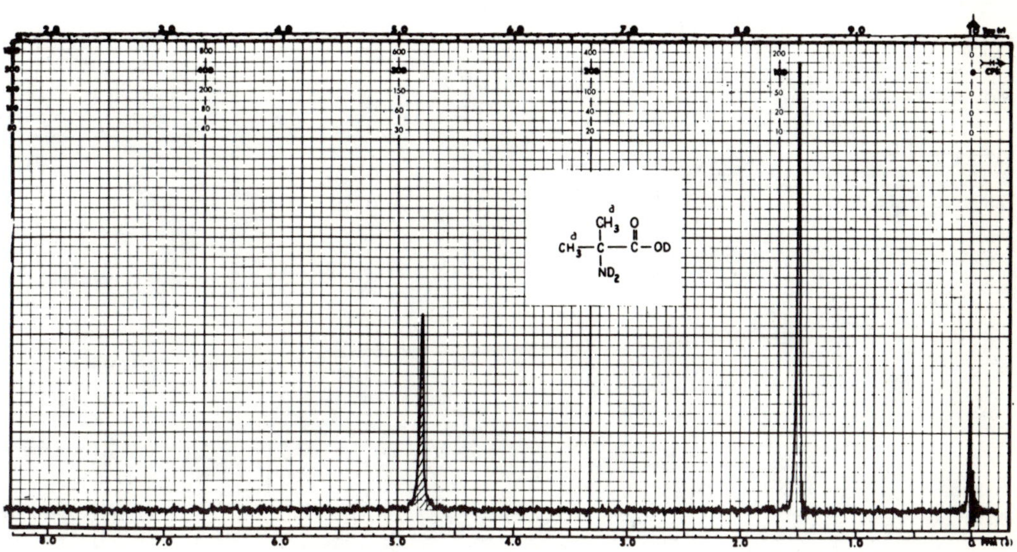

© 1978 Sadtler Research Laboratories, Inc.

2616

β-ALANINE

$C_3H_7NO_2$

Mol. Wt. 89.09

Solvent: D_2O

Assignments
a = 2.53
b = 3.18
c = 5.22 HDO

2617

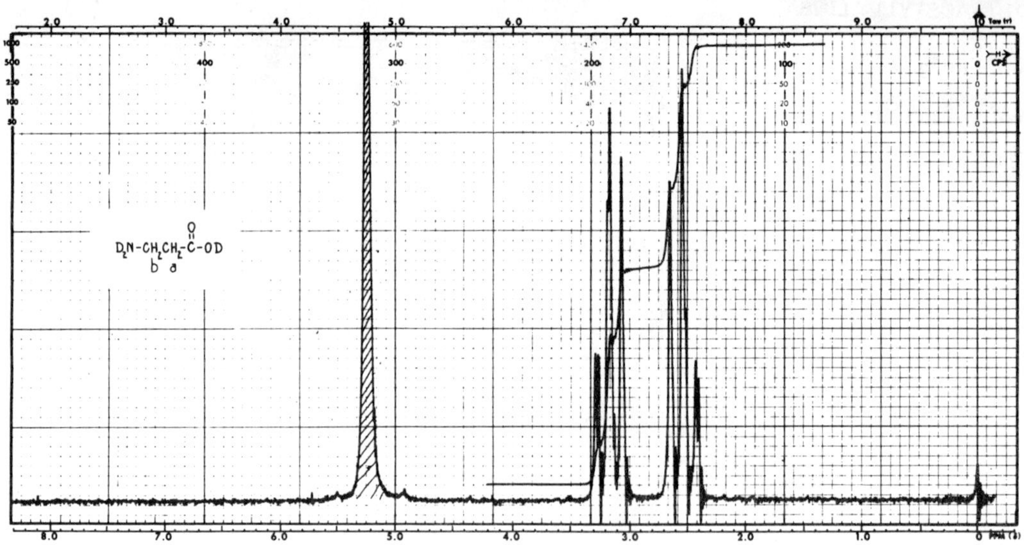

© 1978 Sadtler Research Laboratories, Inc.

NORVALINE

$C_5H_{11}NO_2$

Mol. Wt. 117.15

B.P. 148-150°C/10mm

Solvent: D_2O

Assignments
a = 0.95
b = 1.16-2.10
c = 3.71
d = 4.68 HDO

2618

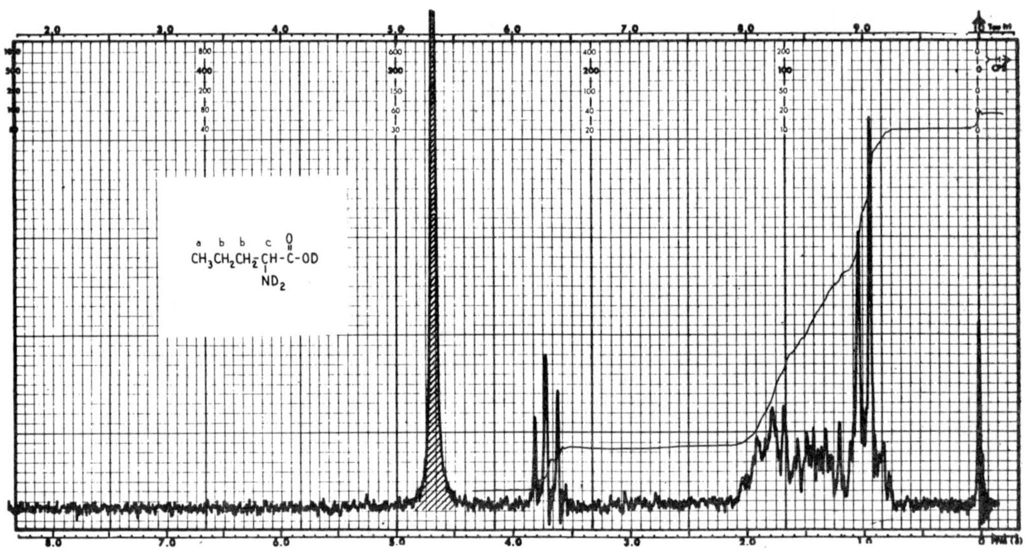

© 1978 Sadtler Research Laboratories, Inc.

DL-VALINE

$C_5H_{11}NO_2$

Mol. Wt. 117.15

M.P. 290.5°C

Solvent: D_2O

Assignments
a = 0.90
b = 1.05
c = 2.22
d = 3.55
e = 4.69 HDO

2619

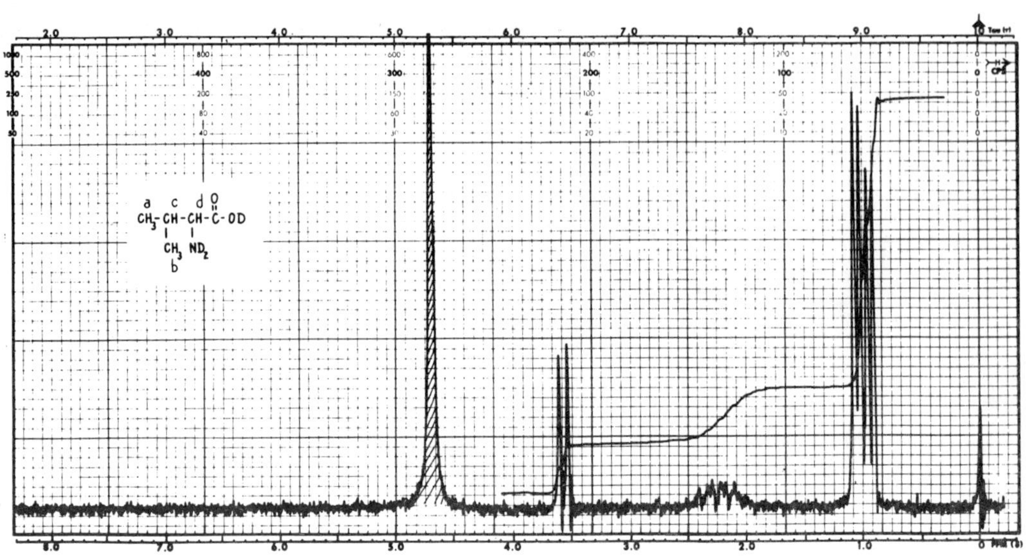

© 1978 Sadtler Research Laboratories, Inc.

DL-N-ACETYLVALINE

$C_7H_{13}NO_3$

Mol. Wt. 159.19

Solvent: Dimethyl sulfoxide-d_6

Assignments
a = 0.92
b = 1.92
c = 2.02
d = 4.19
e = 7.93
f = 11.85

2620

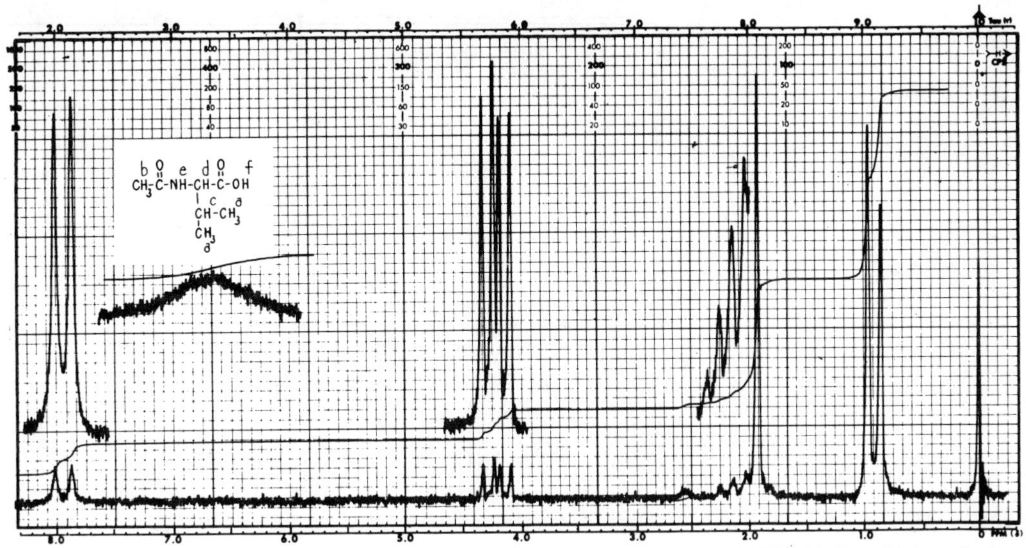

© 1978 Sadtler Research Laboratories, Inc.

N-ACETYL-L-LEUCINE

$C_8H_{15}NO_3$

Mol. Wt. 173.21

M.P. 185°C (lit.)

Solvent: Polysol-d

Assignments
a = 0.91
b = ca 1.59
c = 1.91
d = 4.30
e = 7.91
f = 11.01

2621

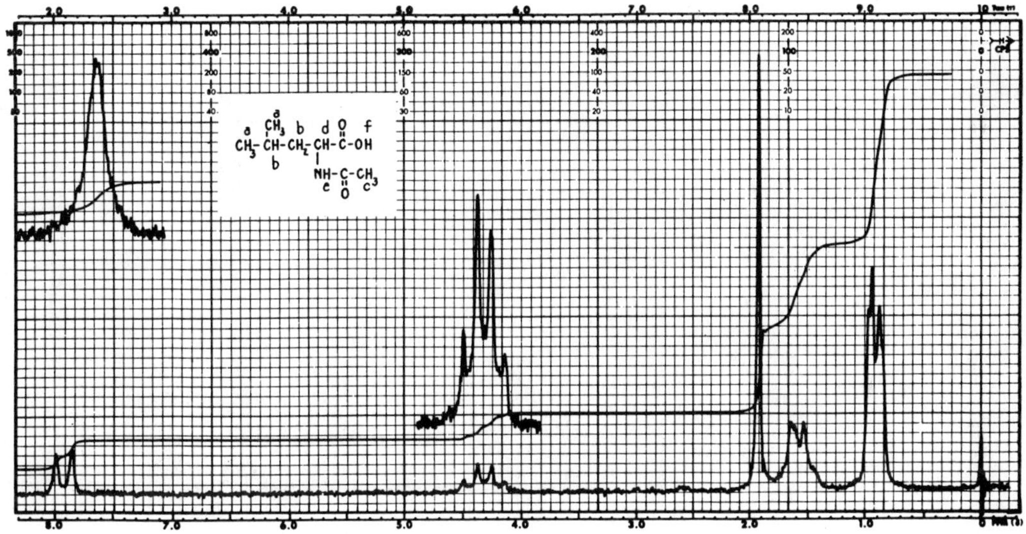

© 1978 Sadtler Research Laboratories, Inc.

7-AMINOHEPTANOIC ACID

$C_7H_{15}NO_2$

Mol. Wt. 145.20

Solvent: D_2O

Assignments
a = 1.10-1.90
b = 2.15
c = 2.98
d = 4.77 HDO

2622

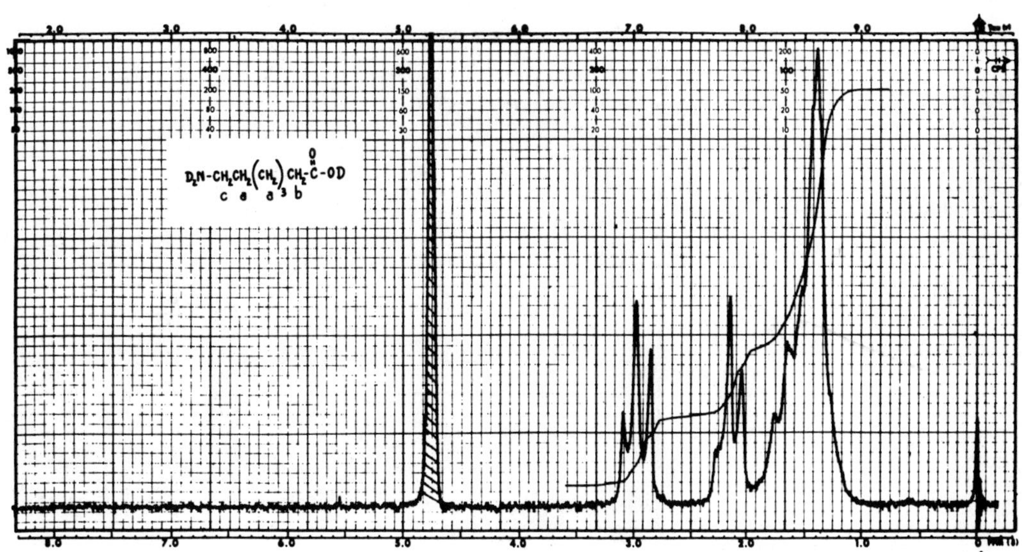

© 1978 Sadtler Research Laboratories, Inc.

8-AMINOOCTANOIC ACID

$C_8H_{17}NO_2$

Mol. Wt. 159.23

Solvent: D_2O

Assignments
a = 1.10-1.60
b = 1.64
c = 2.14
d = 2.98
e = 4.40 HDO

2623

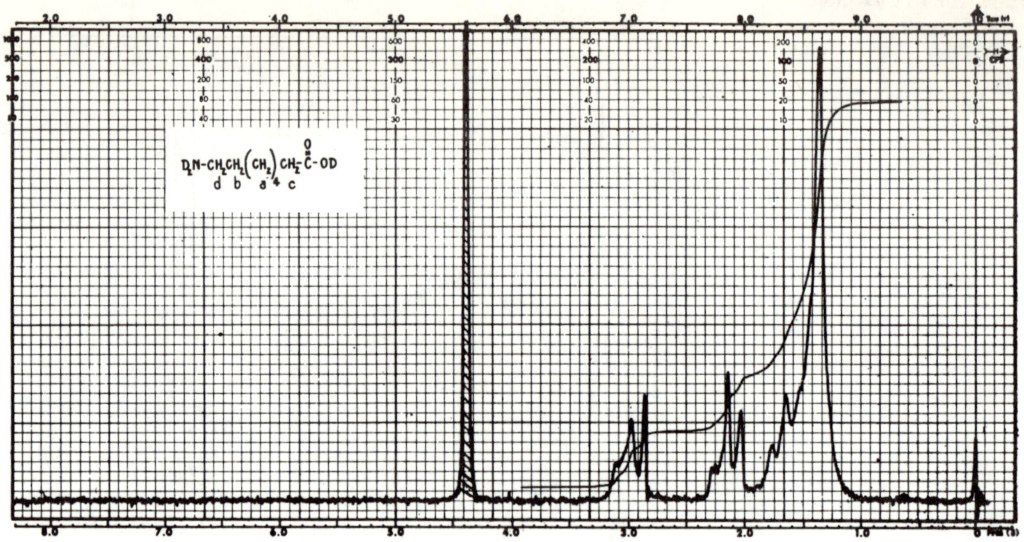

© 1978 Sadtler Research Laboratories, Inc.

L-(+)-CYSTEINE

$C_3H_7NO_2S$

Mol. Wt. 121.16

Solvent: D_2O

Assignments
a = 3.10
b = 4.03
c = 4.64 HDO

2624

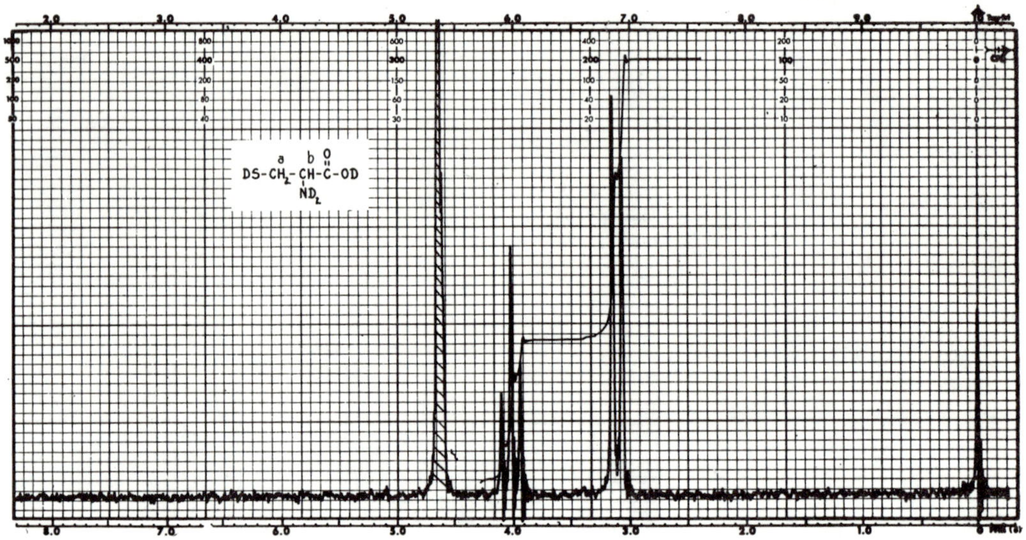

© 1978 Sadtler Research Laboratories, Inc.

2-METHYLSERINE

$C_4H_9NO_3$

Mol. Wt. 119.12

Solvent: D_2O

Assignments
a = 1.46
b = 3.63
c = 3.91

2625

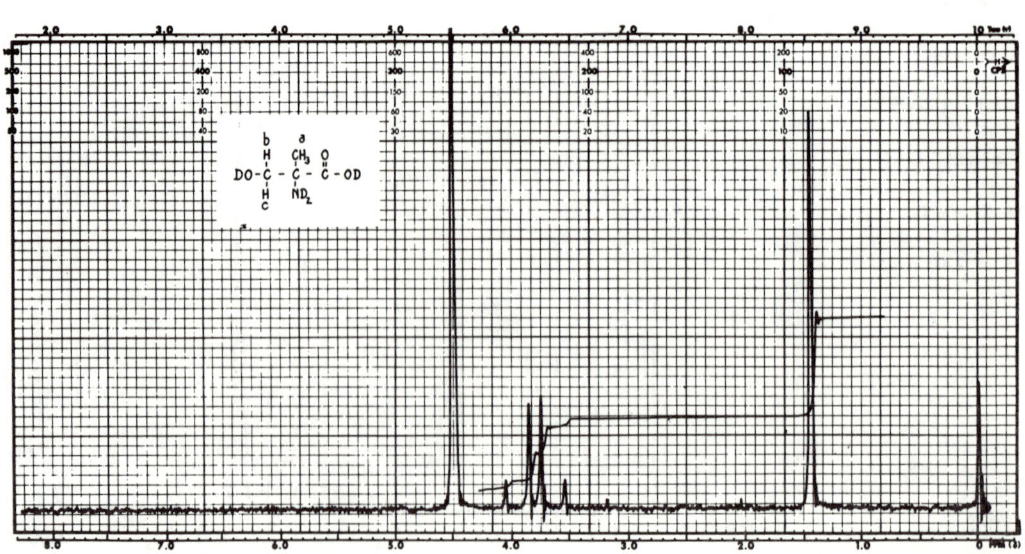

© 1978 Sadtler Research Laboratories, Inc.

DL-THREONINE

$C_4H_9NO_3$

Mol. Wt. 119.12

M.P. 229-230°C (lit.)

Solvent: D_2O

Assignments
a = 1.29
b = 3.50
c = 4.16
d = 4.63 HDO

2626

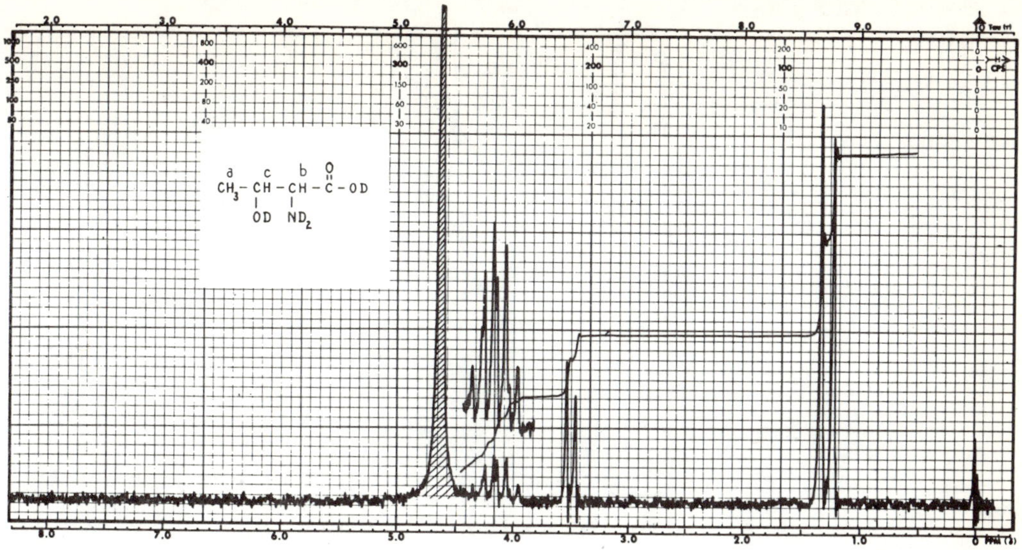

© 1978 Sadtler Research Laboratories, Inc.

N-ACETYL-DL-METHIONINE

$C_7H_{13}NO_3S$

Mol. Wt. 191.25

M.P. 114-116°C

Solvent: Polysol-d

Assignments
a = ca 1.94
b = 2.08
c = 2.54
d = 4.42
e = 8.06
f = 11.80

2627

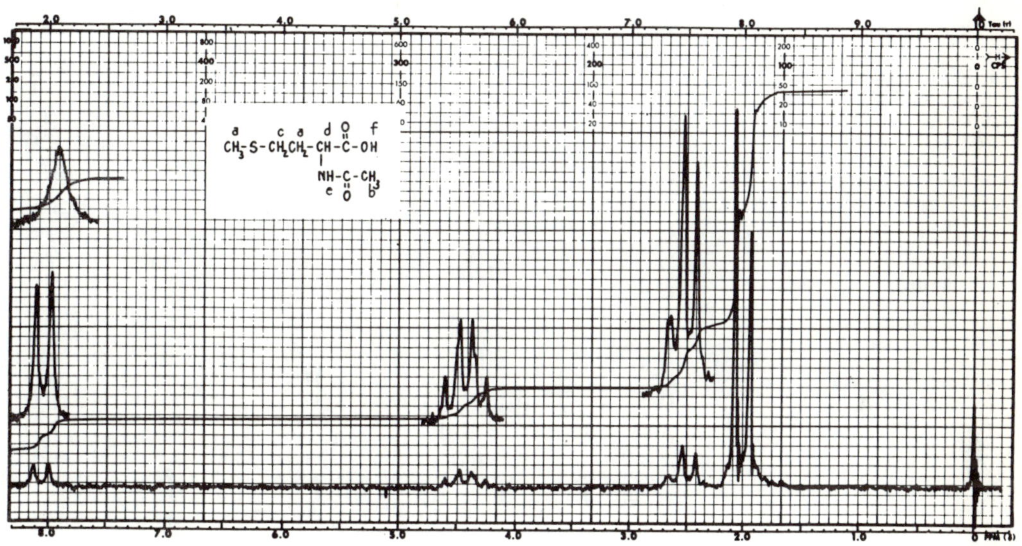

© 1978 Sadtler Research Laboratories, Inc.

DL-LYSINE

$C_6H_{14}N_2O_2$

Mol. Wt. 146.19

M.P. 224°C

Solvent: D_2O

Assignments
a = 1.11-2.00
b = 2.90
c = 3.30
d = 4.61 HDO
e = 3.81, 5.46 HDO side bands

2628

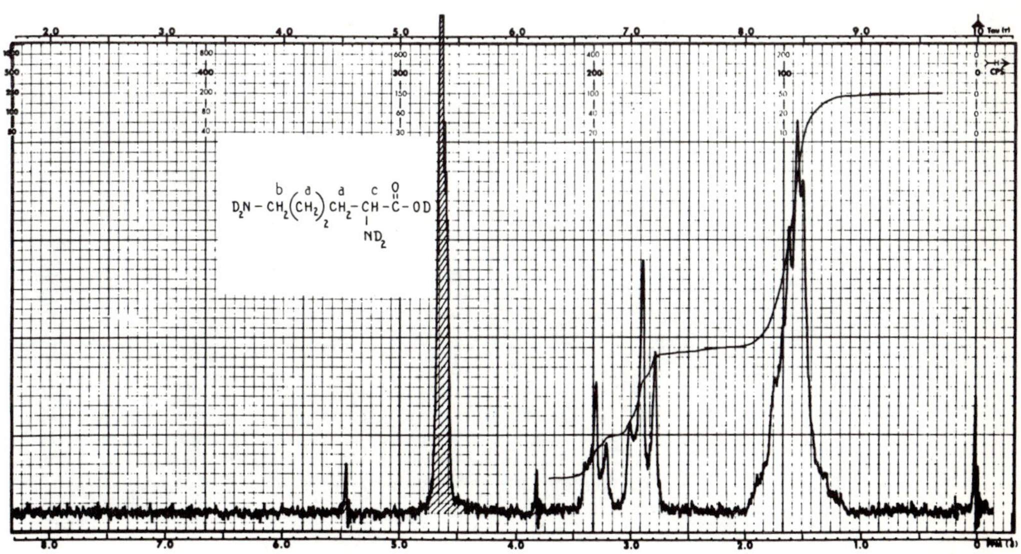

© 1978 Sadtler Research Laboratories, Inc.

ARGININE

$C_6H_{14}N_4O_2$

Mol. Wt. 174.20

M.P. 230°C

Solvent: D_2O

Assignments
a = 1.62
b = 3.22
c = 4.78 HDO

2629

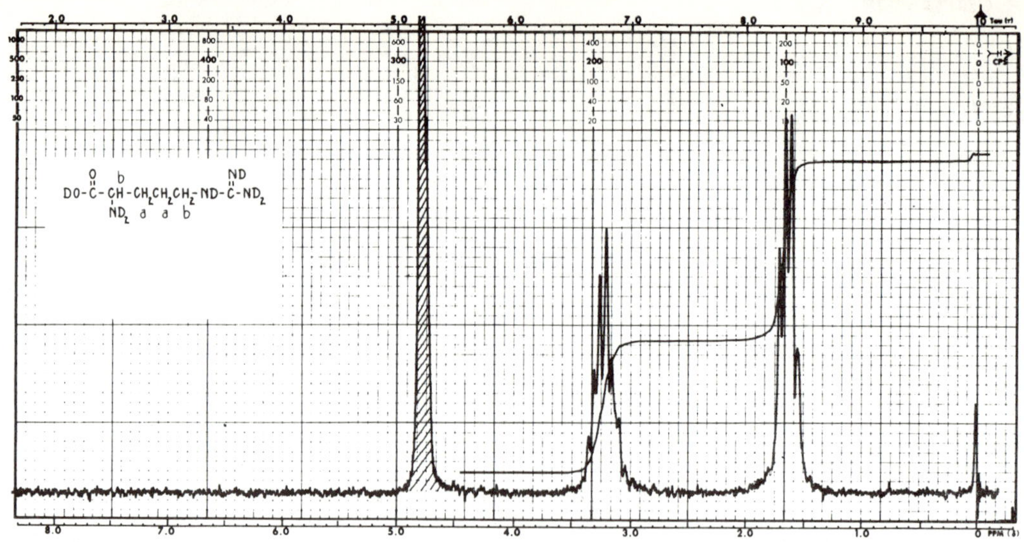

© 1978 Sadtler Research Laboratories, Inc.

DL-ASPARAGINE, HYDRATE

$C_4H_8N_2O_3 \cdot H_2O$

Mol. Wt. 150.14

Solvent: D_2O

Assignments
a = ca 2.84
b = ca 2.95
c = 4.00
d = 4.58 HDO

2630

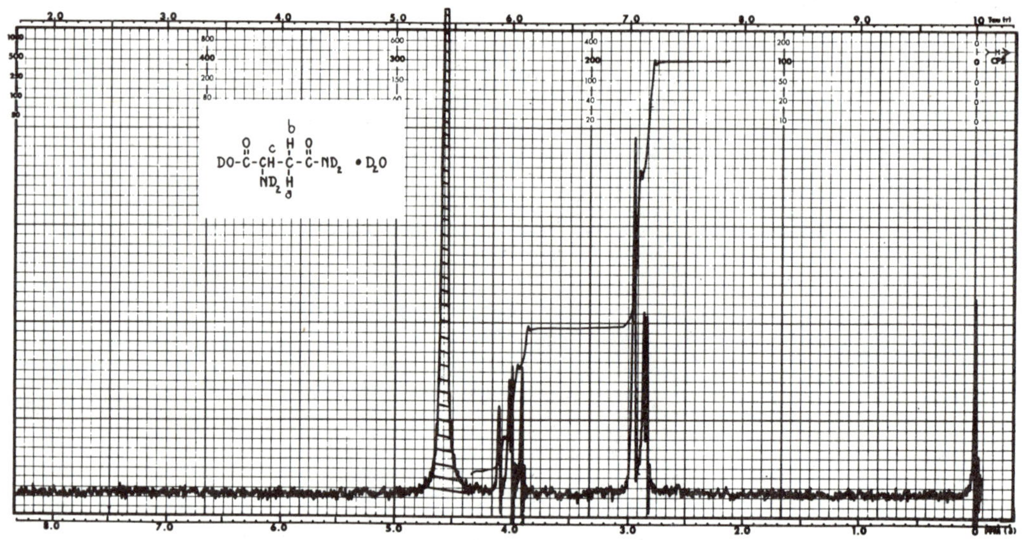

© 1978 Sadtler Research Laboratories, Inc.

L-ASPARAGINE, MONOHYDRATE

$C_4H_8N_2O_3 \cdot H_2O$

Mol. Wt. 150.14

Solvent: D_2O

Assignments
a = 2.83
b = 2.97
c = 4.02
d = 4.57 HDO

2631

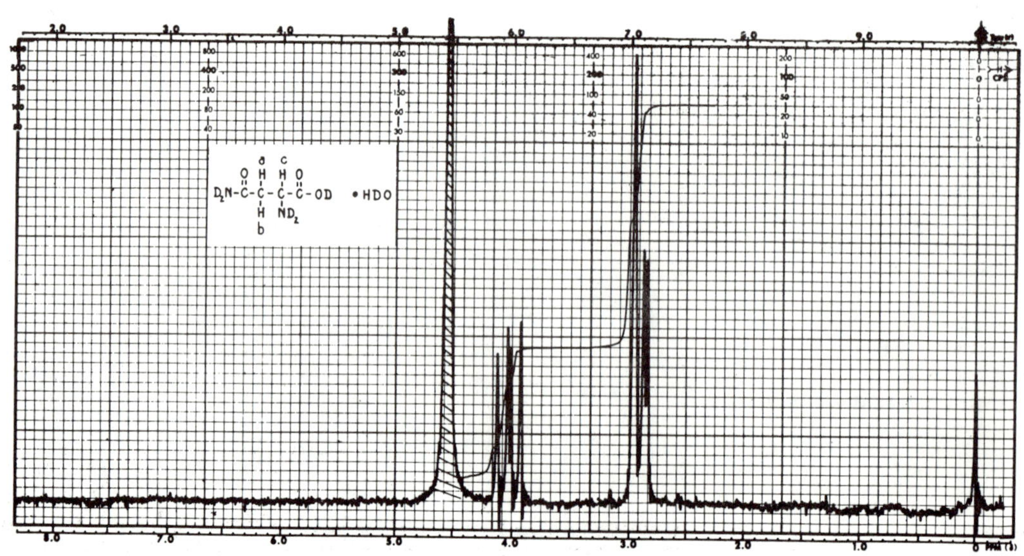

© 1978 Sadtler Research Laboratories, Inc.

DL-GLUTAMINE

$C_5H_{10}N_2O_3$

Mol. Wt. 146.15

Solvent: D_2O

Assignments
a = 2.20
b = 2.47
c = 3.79
d = 4.19 HDO

2632

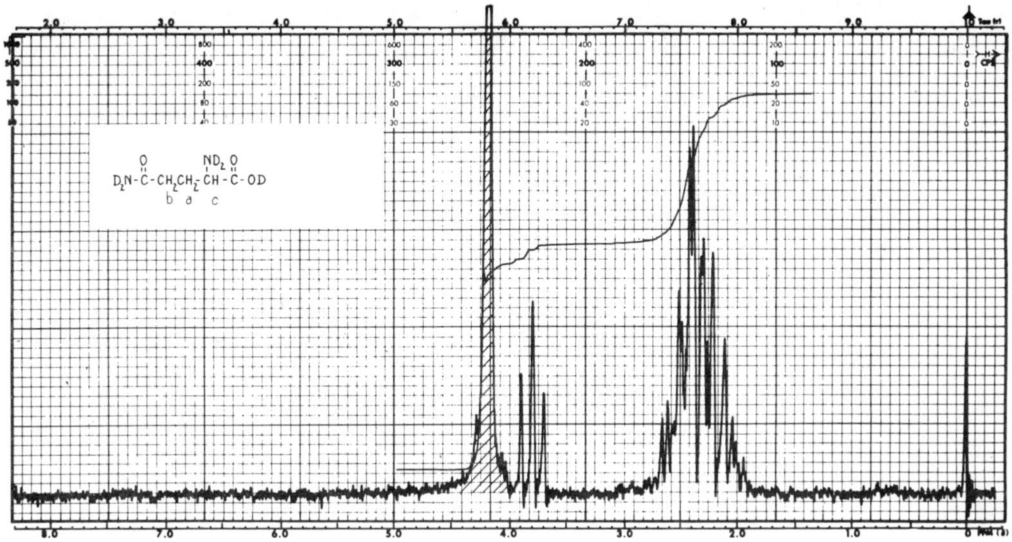

© 1978 Sadtler Research Laboratories, Inc.

L-GLUTAMIC ACID, 5-BENZYL ESTER

$C_{12}H_{15}NO_4$

Mol. Wt. 237.26

M.P. 181-182°C

Solvent: Trifluoroacetic acid

Assignments
a = 2.49
b = 2.89
c = 4.47
d = 5.22
e = 7.34
f = 7.70
g = unobserved *
* In exchange with solvent

2633

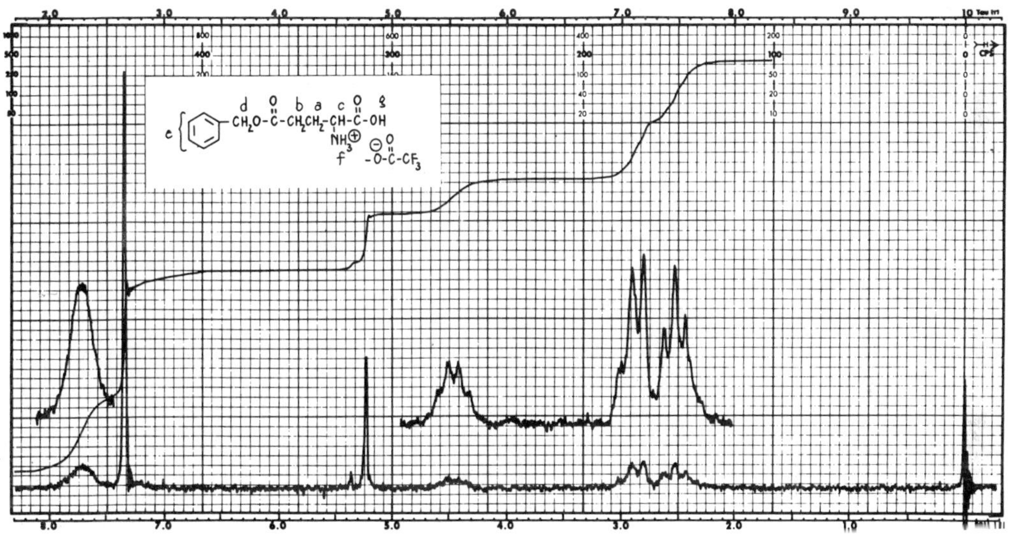

© 1978 Sadtler Research Laboratories, Inc.

N-PHENYLGLYCINE

$C_8H_9NO_2$

Mol. Wt. 151.17

M.P. 123-125°C

Solvent: Dimethyl sulfoxide-D_6

Assignments
a = 3.81
b = 6.60
c = ca 6.81
d = 7.11
e = 2.52 DMSO-D_5

2634

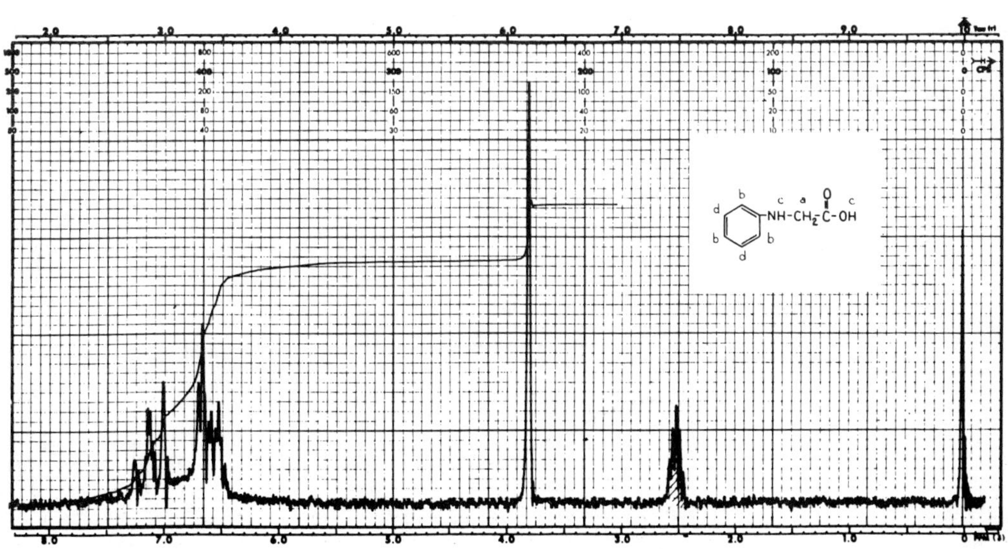

© 1978 Sadtler Research Laboratories, Inc.

2,2-DIPHENYLGLYCINE

$C_{14}H_{13}NO_2$

Mol. Wt. 227.27

Solvent: Trifluoroacetic acid

Assignments
a = ca 7.50
b = 8.17
c = unobserved *
* Under solvent band

2635

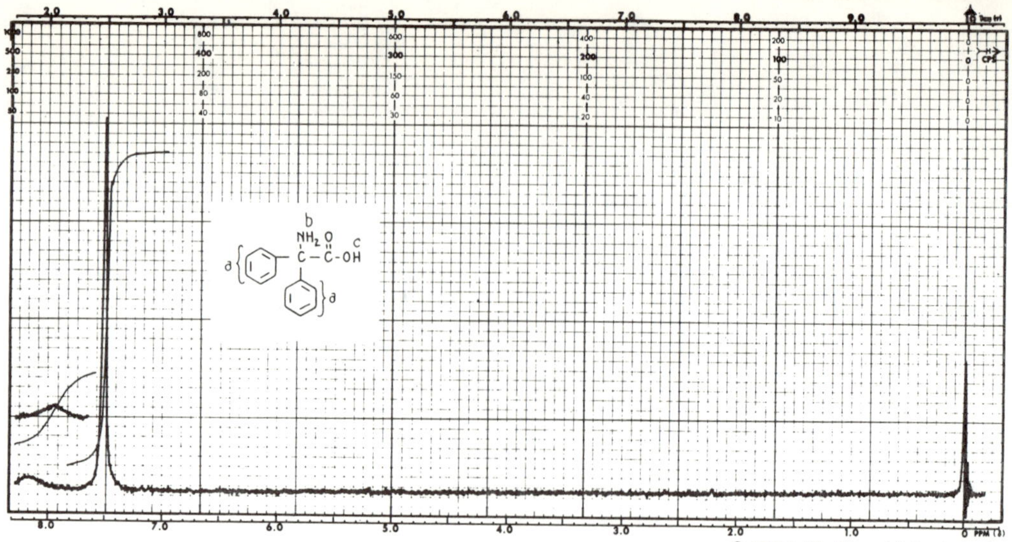

© 1978 Sadtler Research Laboratories, Inc.

L-PROLINE

$C_5H_9NO_2$

Mol. Wt. 115.13

M.P. 220-222°C

Solvent: D_2O

Assignments
a = 1.74-2.60
b = 3.39
c = 4.14
d = 4.69 HDO

2636

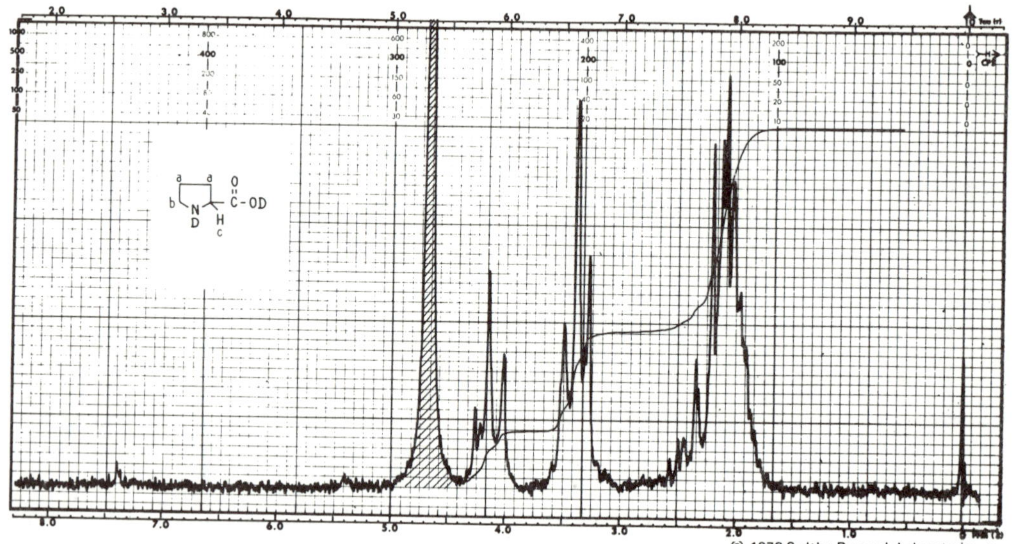

© 1978 Sadtler Research Laboratories, Inc.

L-(−)-4-HYDROXYPROLINE

$C_5H_9NO_3$

Mol. Wt. 131.13

Solvent: D_2O

Assignments
a = 2.00-2.80
b = ca 3.54
c = 4.50-4.90
d = 5.49 HDO
TFA added to shift
HDO downfield

2637

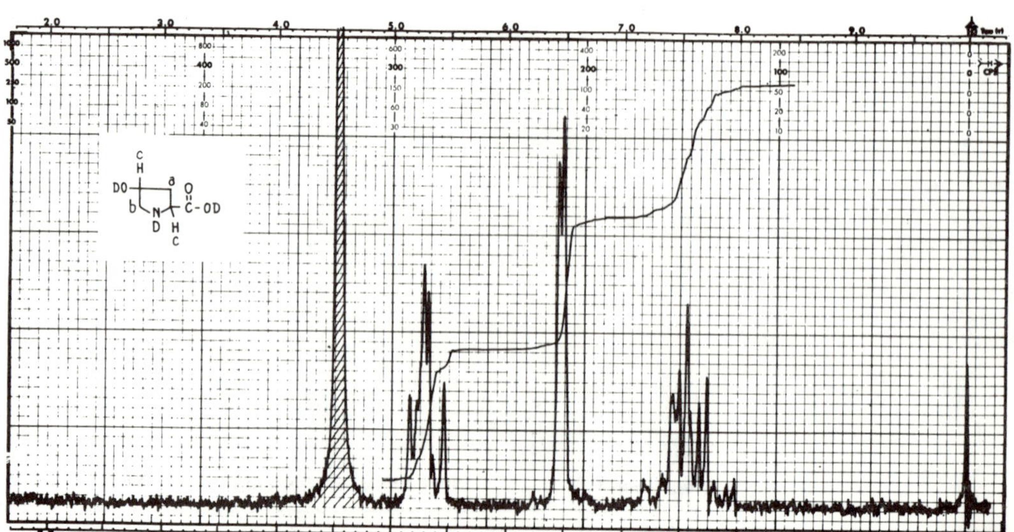

© 1978 Sadtler Research Laboratories, Inc.

N-GLYCYLGLYCINE

$C_4H_8N_2O_3$

Mol. Wt. 132.12

Solvent: D_2O

Assignments
a = 3.84 or b
b = 3.90 or a
c = 4.66 HDO

2638

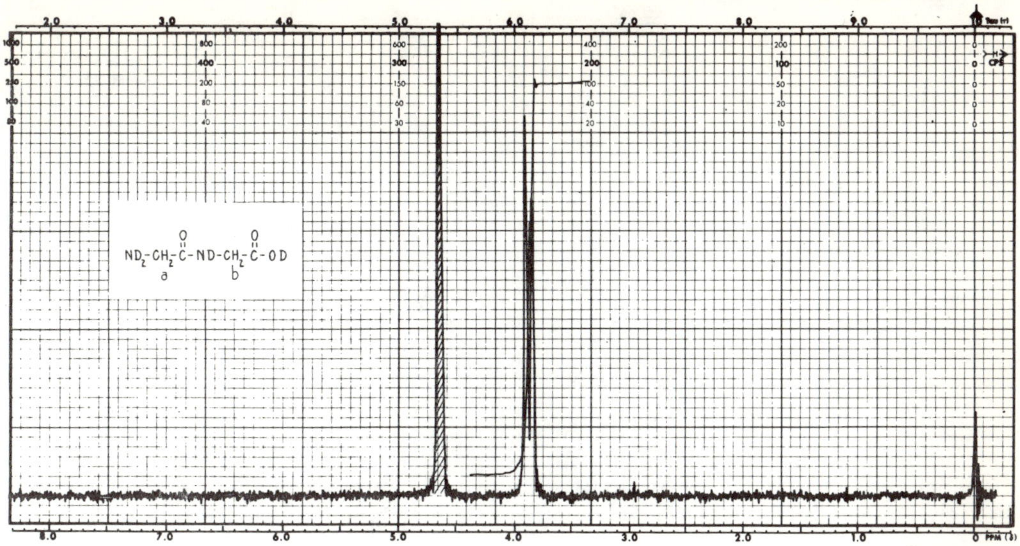

© 1978 Sadtler Research Laboratories, Inc.

DL-N-ALANYLGLYCINE

$C_5H_{10}N_2O_3$

Mol. Wt. 146.15

Solvent: D_2O

Assignments
a = 1.56
b = 3.81
c = 4.12
d = 4.56 HDO

2639

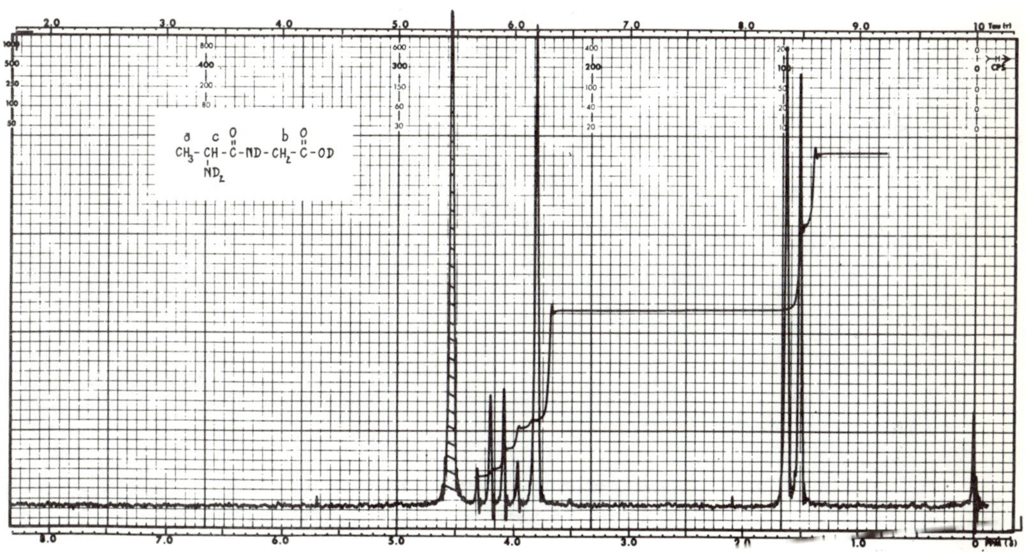

© 1978 Sadtler Research Laboratories, Inc.

DL-N-GLYCYLVALINE

$C_7H_{14}N_2O_3$

Mol. Wt. 174.20

M.P. 226°C (lit.)

Solvent: D_2O

Assignments
a = 0.88 or b
b = 0.90 or a
c = 2.06
d = 3.80
e = 4.01
f = 5.60 HDO

2640

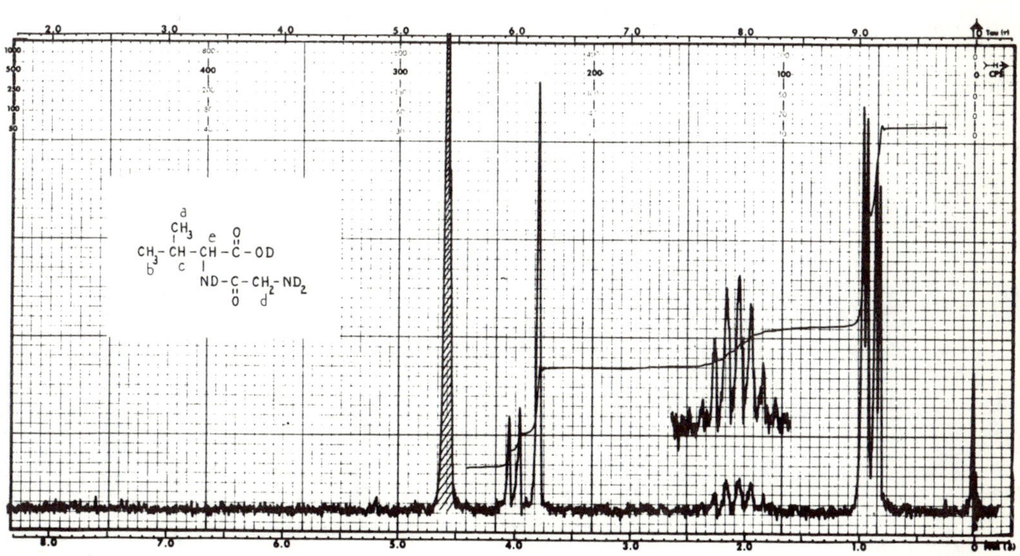

© 1978 Sadtler Research Laboratories, Inc.

DL-N-GLYCYLSERINE

$C_5H_{10}N_2O_4$

Mol. Wt. 162.15

M.P. 198°C (lit.)

Solvent: D_2O

Assignments
a = 3.98
b = 3.99
c = 4.63
d = 4.98 HDO

2641

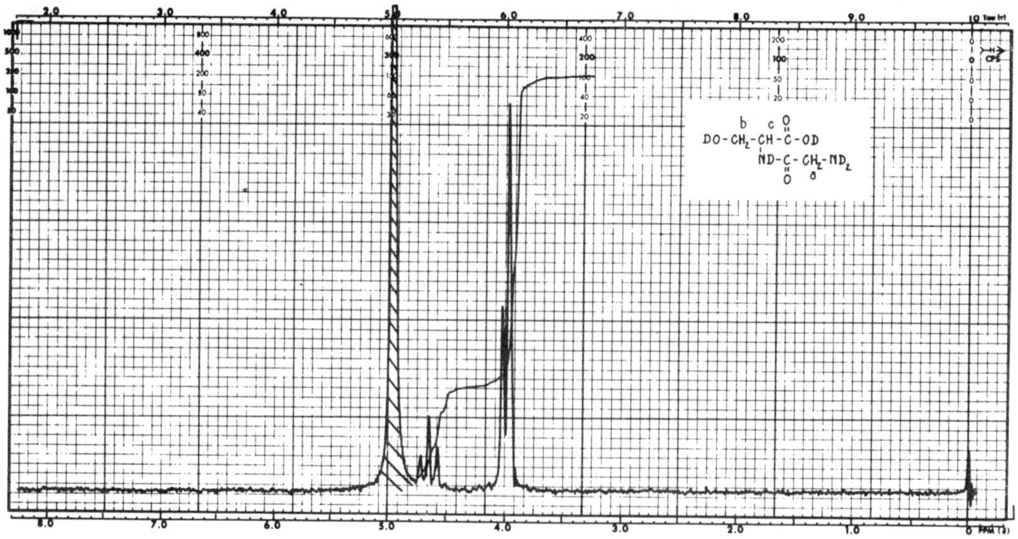

© 1978 Sadtler Research Laboratories, Inc.

N,N-DIMETHYLGLYCINE, HYDROCHLORIDE

$C_4H_9NO_2 \cdot HCl$

Mol. Wt. 139.58

Solvent: D_2O

Assignments
a = 2.97
b = 3.98
c = 4.65 HDO

2642

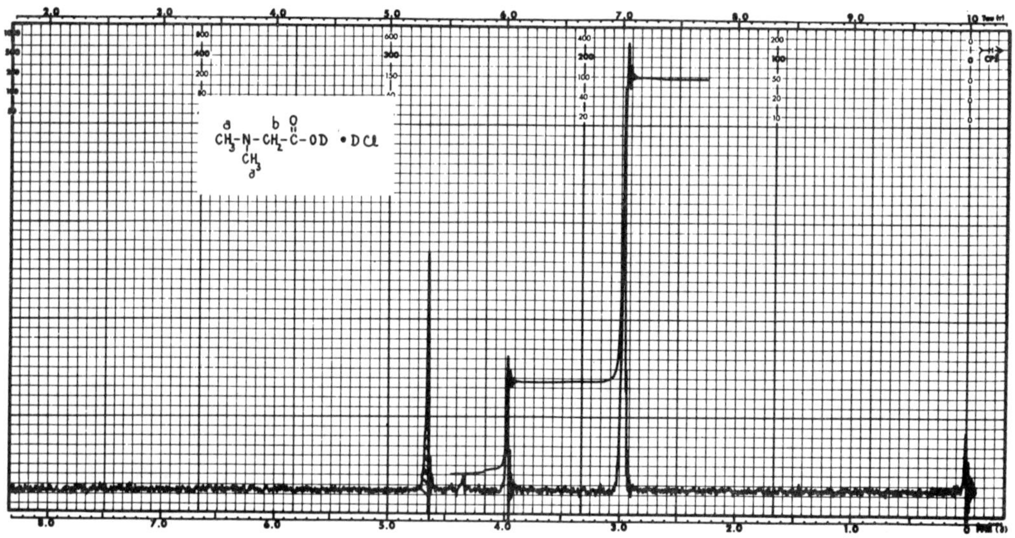

© 1978 Sadtler Research Laboratories, Inc.

L-2,4-DIAMINOBUTYRIC ACID, MONOHYDROCHLORIDE

$C_4H_{10}N_2O_2 \cdot HCl$

Mol. Wt. 154.60

Solvent: D_2O

Assignments
a = 2.30
b = 3.31
c = 3.91
d = 4.89 HDO

2643

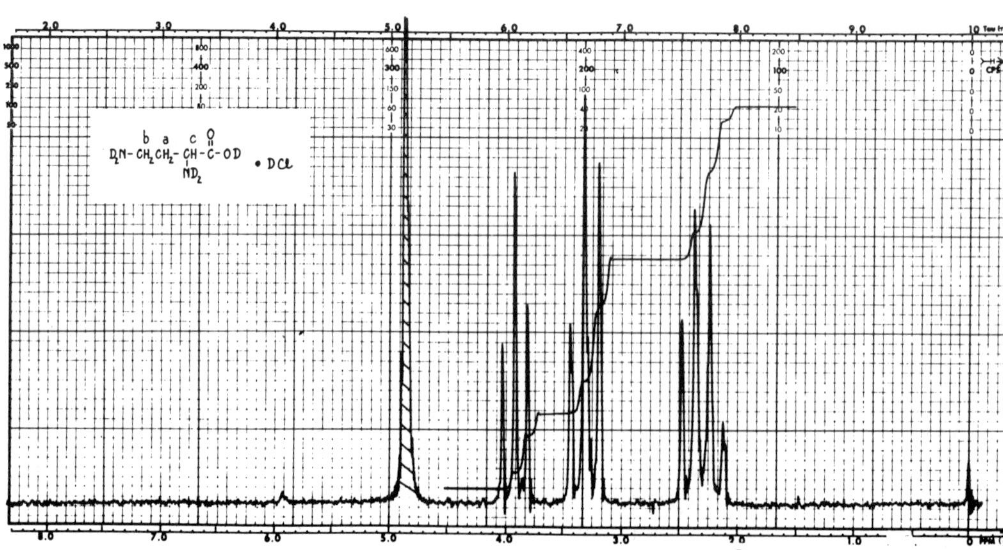

© 1978 Sadtler Research Laboratories, Inc.

L-(+)-ORNITHINE, MONOHYDROCHLORIDE

$C_5H_{12}N_2O_2 \cdot HCl$

Mol. Wt. 168.62

Solvent: D_2O

Assignments
a = 1.42-2.28
b = 3.10
c = 3.82
d = 4.98 HDO

2644

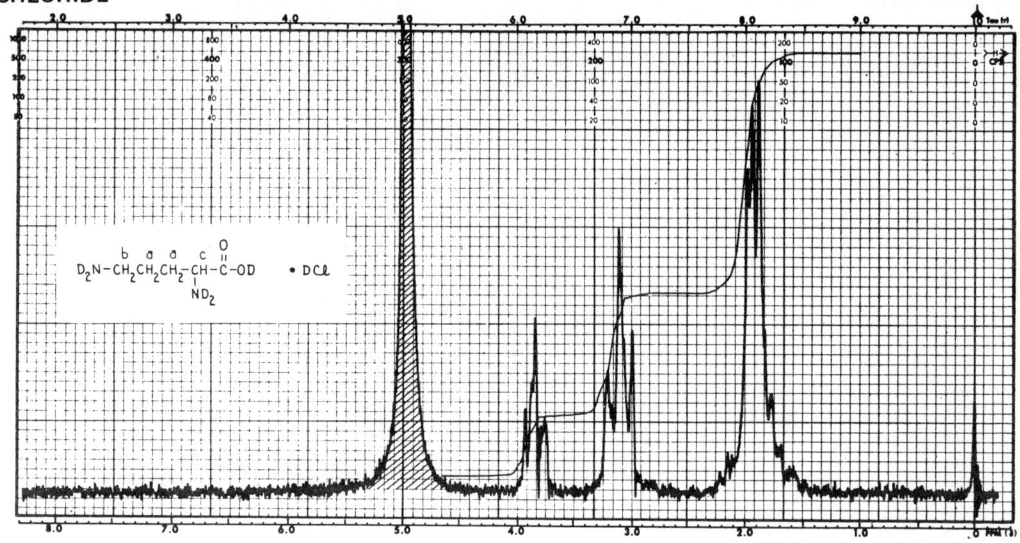

© 1978 Sadtler Research Laboratories, Inc.

L-(+)-CYSTEINE, HYDROCHLORIDE

$C_3H_7NO_2S \cdot HCl$

Mol. Wt. 157.62

M.P. ~ 178°C (dec.)

Solvent: D_2O

Assignments
a = 3.21
b = 4.47
c = 5.02 HDO

2645

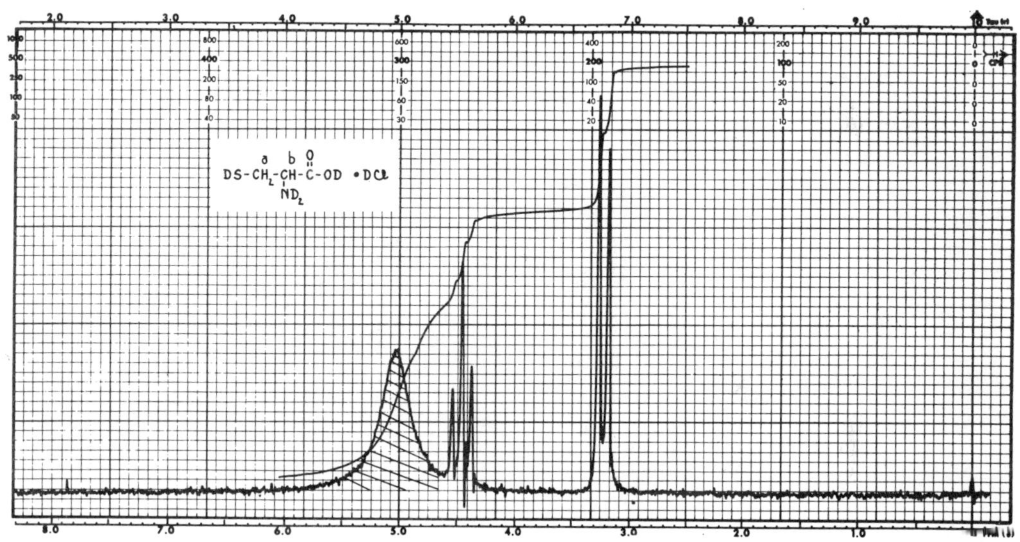

© 1978 Sadtler Research Laboratories, Inc.

N-GLYCYLGLYCINE, HYDROCHLORIDE

$C_4H_8N_2O_3 \cdot HCl$

Mol. Wt. 168.58

Solvent: D_2O

Assignments
a = 3.92
b = 4.10
c = 4.66 HDO

2646

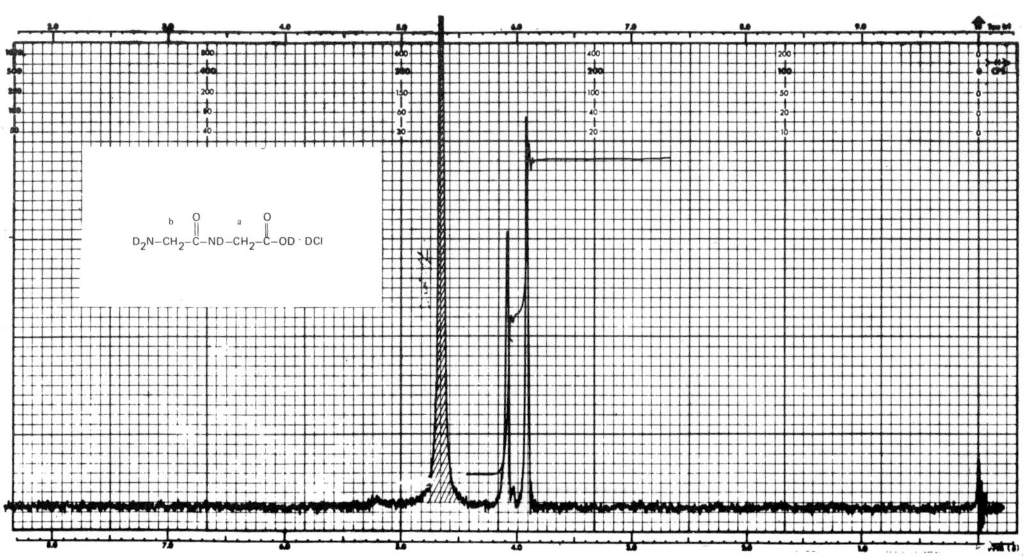

© 1978 Sadtler Research Laboratories, Inc.

NOTES

The Salts of the Carboxylic Acids are very similar in most respects to their corresponding Carboxylic acids with the exception of their increased solubility in water (D_2O). The chemical shifts listed below vary over a relatively wide range due to the different solvents employed. The solubility of the Carboxylic Acid Salts, in turn, is directly related to the metal which is present in the compound.

Chemical Shifts

Aliphatic Protons

HC–3	HC–2	HC–1		Solvent	Spectrum
		1.88	CH_3–C(=O)–O–K	D_2O	----
		1.90	CH_3–C(=O)–O–Na	DMSO-d_6	----
		1.90	CH_3–C(=O)–OH (H_2N–R)	$CDCl_3$	2685
		1.99	CH_3–C(=O)–OH (H_2N–OH)	D_2O	2687
	1.01		CH_3–CH_2–C(=O)–O–Na	D_2O	----
	(1.09	2.30)	(CH_3–CH_2–C(=O)–O$)_2$–Cd	D_2O	2649
0.90	1.56	2.17	CH_3–CH_2–CH_2–C(=O)–O–Na	D_2O	----

Substituted Acetic Acid Salts

	–CH_2	–X		
Na–O–C(=O)–	3.11	–C(=O)–O–Na	D_2O	2663
Na–O–C(=O)–	3.19	–N($CH_3$$)_2$	D_2O	2657
–Sn–O–C(=O)–	3.50	–Ø	CCl_4	----
Na–O–C(=O)–	3.51	–SH	D_2O	2658
K–O–C(=O)–	3.53	–Ø	D_2O	----
–Hg–O–C(=O)–	3.60	–Ø	Polysol	----
Na–O–C(=O)–	3.63	–I	D_2O	2656
Na–O–C(=O)–	4.25	–1–Naphthalene	D_2O	2674
Na–O–C(=O)–	4.73	–F	D_2O	2653

Olefinic Protons

The vinyl protons of Acrylic Acid salts appear as a higher-order series of peaks in the chemical shift range from 5.5–6.3 ppm. Characteristically, the two protons on the beta carbon are more strongly deshielded than the proton on the alpha carbon atom.

A series of 2-substituted Acrylic Acid Salts is presented below.

Olefinic Protons continued

M–O–C(=O)–	CH	=CH	–X		Solvent	Spectrum
Na–O–C(=O)–	6.02	6.02	–C(=O)–OH	(cis)	D_2O	----
Na–O–C(=O)–	6.10	6.10	–C(=O)–O–Na	(cis)	D_2O	2672
Na–O–C(=O)–	6.50	6.50	–C(=O)–O–Na	(trans)	D_2O	----
Na–O–C(=O)–	5.84	6.59	$-CH_3$		D_2O	----
$(R_4)_3$–Sn–O–C(=O)–	5.75	6.75	$-CH_3$	(trans)	CCl_4	----
$(R_4)_3$–Sn–O–C(=O)–	6.76	6.76	–C(=O)–O–Sn$(R_4)_3$	(trans)	$CDCl_3$	----
Na–O–C(=O)–	6.53	7.54	–Ø	(trans)	D_2O	----

Aromatic Protons

Salts of Benzoic Acid

para/meta	ortho		Solvent	Spectrum
7.3–7.7	7.91	Ø–C(=O)–O–Ba–	D_2O	----
7.3–7.6	7.98	Ø–C(=O)–O–Na	D_2O	2675
7.3–7.7	7.99	Ø–C(=O)–O–NH_4	D_2O	2683
7.2–7.7	8.05	Ø–C(=O)–O–Hg–	Polysol	----

Aldehydic Protons

Salts of Formic Acid

H–C(=O)–O	–X	Solvent	Spectrum
8.08	–Sn$(R_4)_3$	CCl_4	----
8.17	–Na	TFA	----
8.46	–Li	D_2O	----
8.48	–Ca–	D_2O	2647

FORMIC ACID, CALCIUM SALT

$C_2H_2CaO_4$

Mol. Wt. 130.12

Solvent: D_2O

Assignments
a = 8.48
b = 4.70 HDO

2647

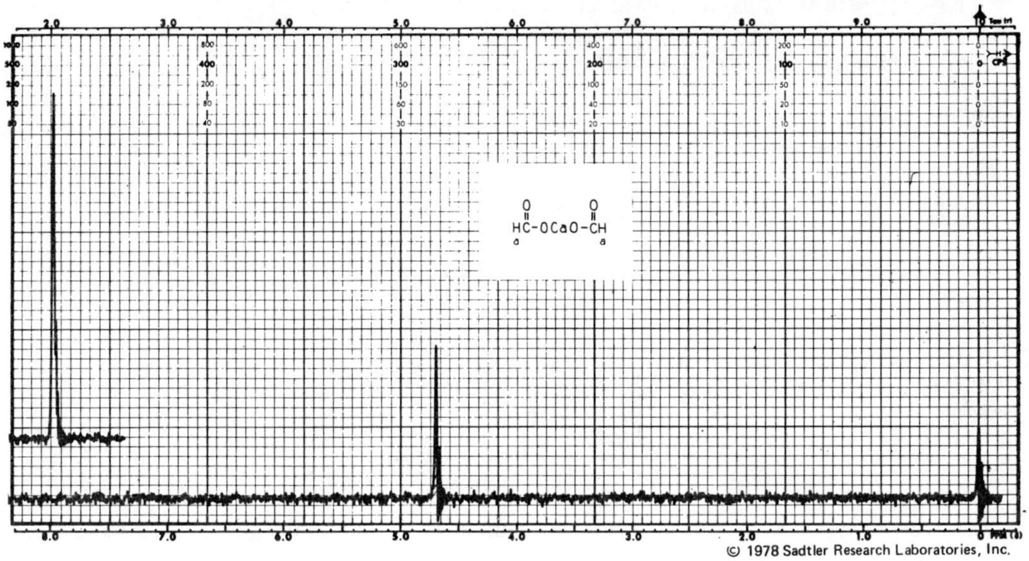

© 1978 Sadtler Research Laboratories, Inc.

LITHIUM FORMATE, HYDRATE

$CHLiO_2 \cdot H_2O$

Mol. Wt. 69.97

Solvent: D_2O

Assignments
a = 8.46
b = 4.72 HDO

2648

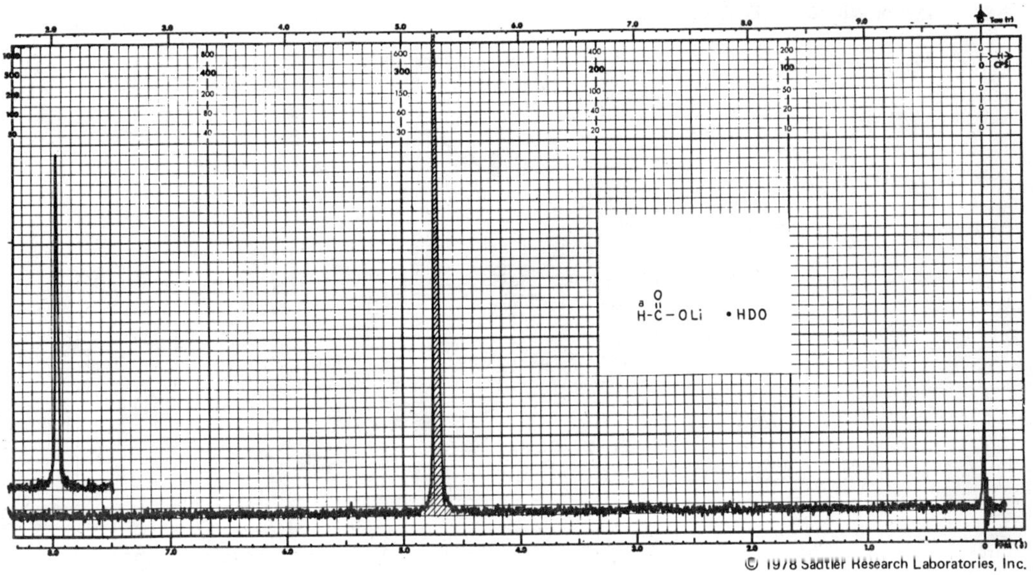

© 1978 Sadtler Research Laboratories, Inc.

PROPIONIC ACID, CADIUM SALT

$C_6H_{10}CdO_4$

Mol. Wt. 258.55

Solvent: D_2O

Assignments
a = 1.09
b = 2.30
c = 4.81 HDO
d = 2.00 imp.

2649

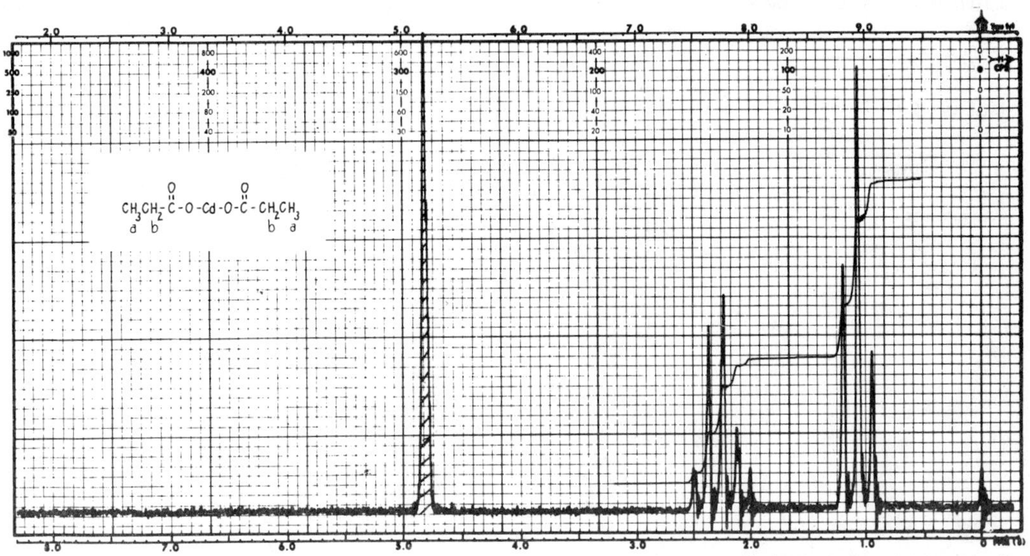

© 1978 Sadtler Research Laboratories, Inc.

2-ETHYLHEXANOIC ACID, SODIUM SALT

$C_8H_{15}NaO_2$

Mol. Wt. 166.20

Solvent: D_2O

Assignments
a = ca 0.89
b = 1.00-1.75
c = 2.07
d = 4.63 HDO

2650

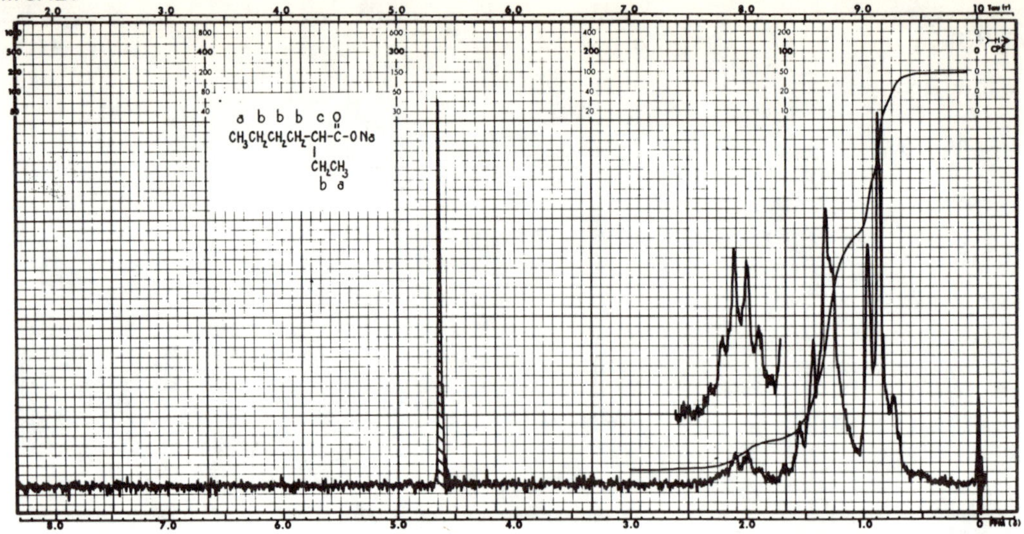

© 1978 Sadtler Research Laboratories, Inc.

2-ETHYLHEXANOIC ACID, CALCIUM SALT

$C_{16}H_{30}CaO_4$

Mol. Wt. 326.49

Solvent: Polysol-d

Assignments
a = 0.89
b = 1.00-1.75
c = 1.97
d = 2.56 DMSO-d_5
e = 3.71 H_2O

2651

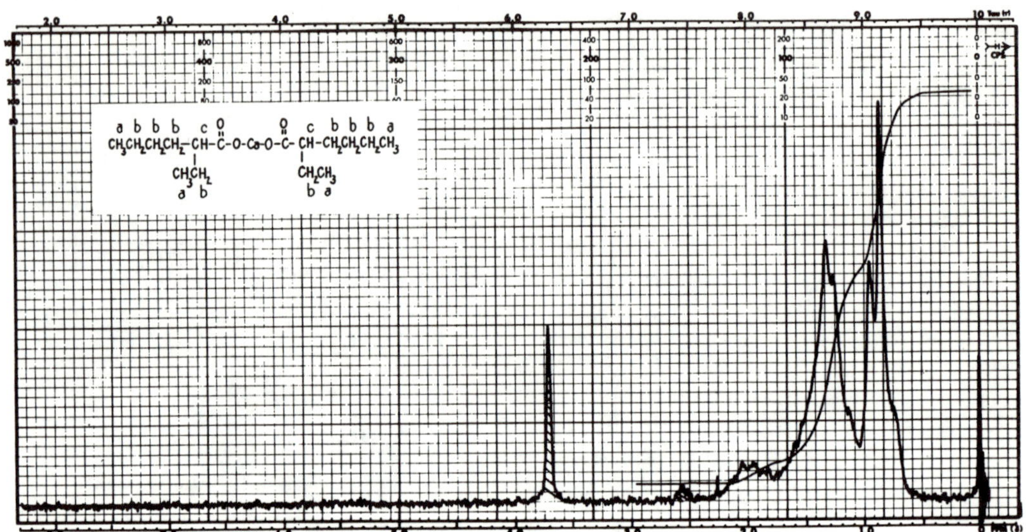

© 1978 Sadtler Research Laboratories, Inc.

LAURIC ACID, SODIUM SALT

$C_{12}H_{23}NaO_2$

Mol. Wt. 222.31

Solvent: D_2O

Assignments
a = 0.86
b = 1.29
c = 2.13

2652

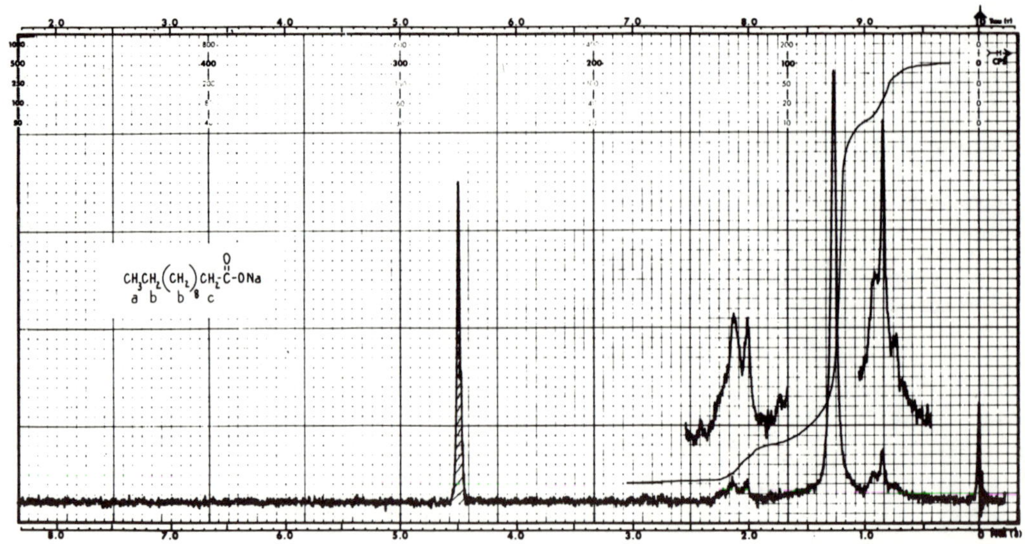

© 1978 Sadtler Research Laboratories, Inc.

FLUOROACETIC ACID, SODIUM SALT

$C_2H_2FNaO_2$

Mol. Wt. 100.03

Solvent: D_2O

Assignments
a = 4.73
b = 4.22 HDO

2653

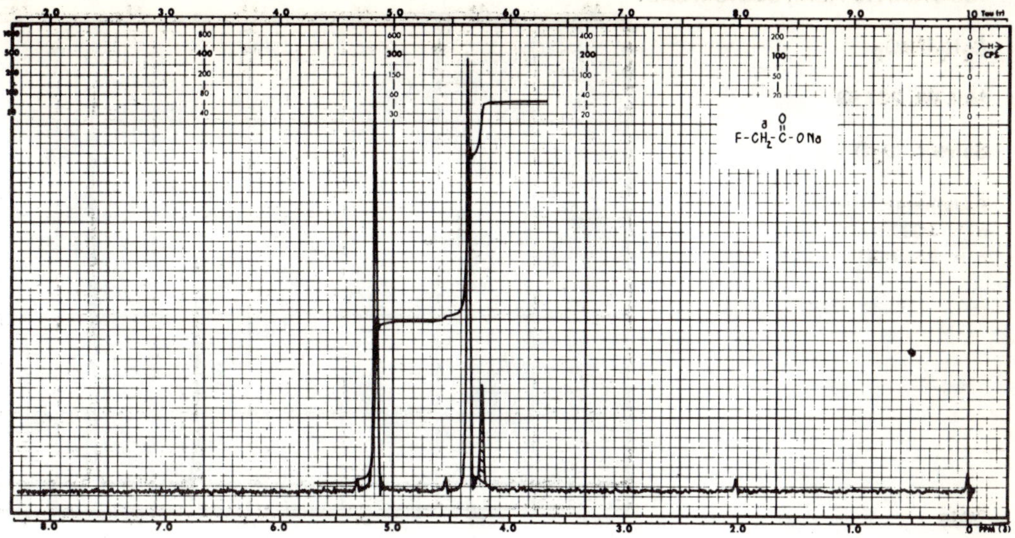

© 1978 Sadtler Research Laboratories, Inc.

DIFLUOROACETIC ACID, SODIUM SALT

$C_2HF_2NaO_2$

Mol. Wt. 118.02

Solvent: D_2O

Assignments
a = 5.91
b = 4.48 HDO

2654

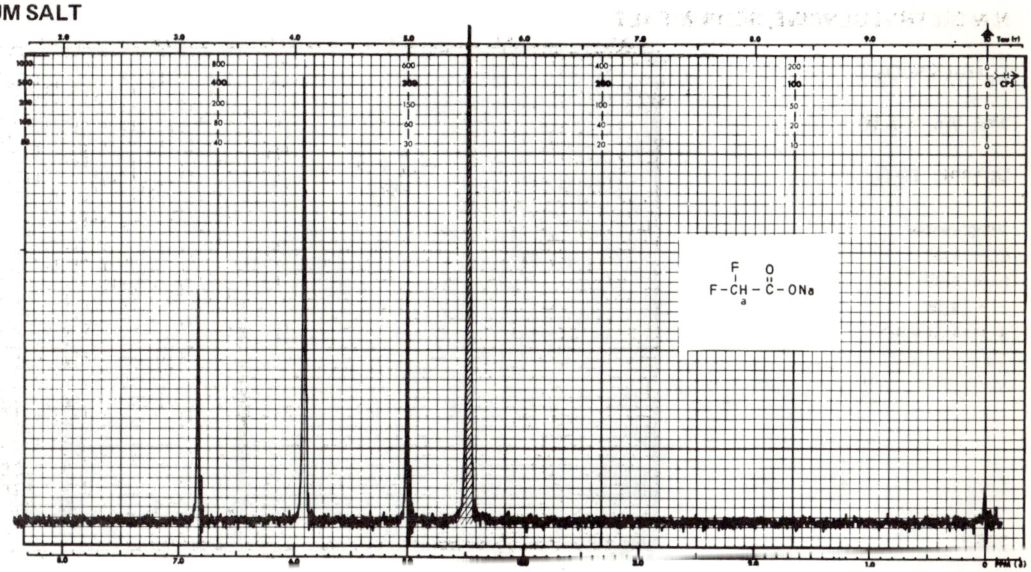

© 1978 Sadtler Research Laboratories, Inc.

2,3-DICHLORO-2-METHYLPROPIONIC ACID, SODIUM SALT

$C_4H_5Cl_2NaO_2$

Mol. Wt. 178.98

Solvent: D_2O

Assignments
a = 1.75
b = 3.93
c = 4.53 HDO
d = 4.07 imp.

2655

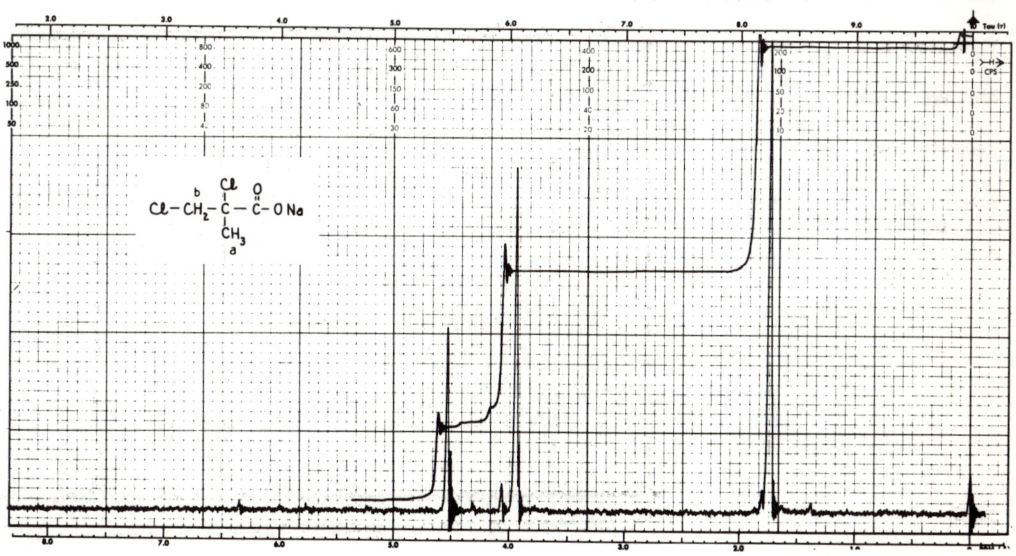

© 1978 Sadtler Research Laboratories, Inc.

IODOACETIC ACID, SODIUM SALT

$C_2H_2INaO_2$

Mol. Wt. 207.93

Solvent: D_2O

Assignments
a = 3.63
b = 4.54 HDO

2656

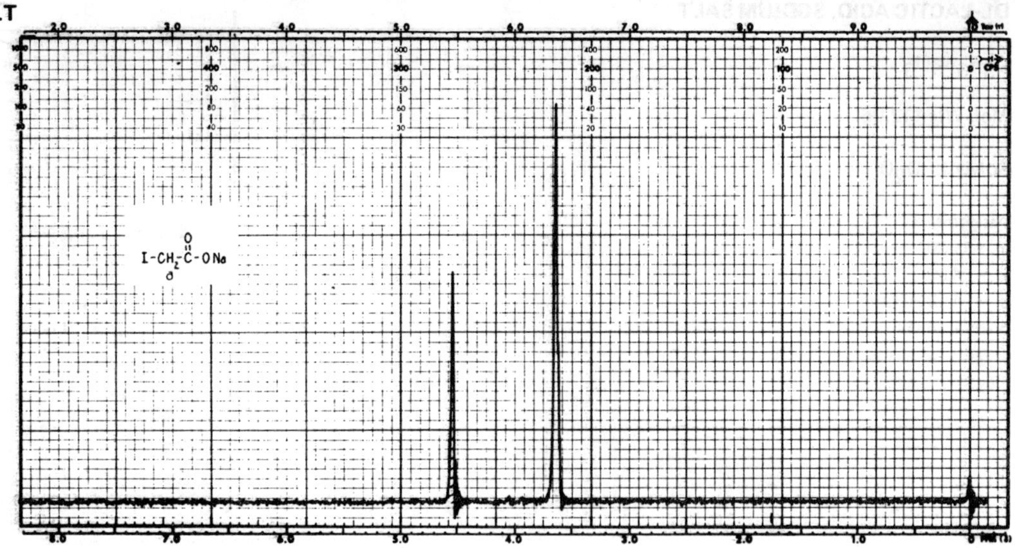

© 1978 Sadtler Research Laboratories, Inc.

N,N-DIETHYLGLYCINE, SODIUM SALT

$C_6H_{12}NNaO_2$

Mol. Wt. 153.16

Solvent: D_2O

Assignments
a = 1.06
b = 2.69
c = 3.19
d = 4.70 HDO

2657

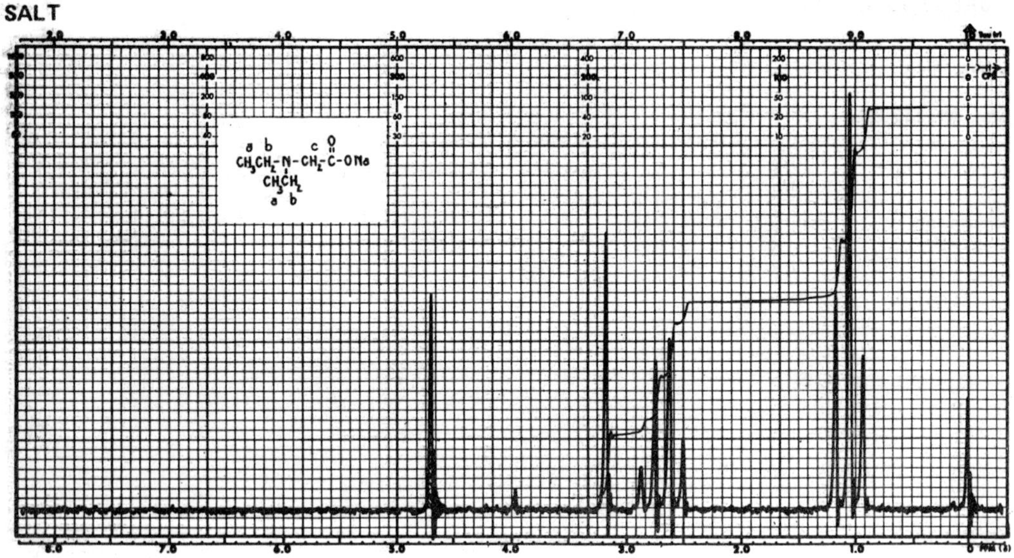

© 1978 Sadtler Research Laboratories, Inc.

MERCAPTOACETIC ACID, SODIUM SALT

$C_2H_3NaO_2S$

Mol. Wt. 114.10

Solvent: D_2O

Assignments
a = 3.51
b = 3.21 imp.
c = 4.61 HDO

2658

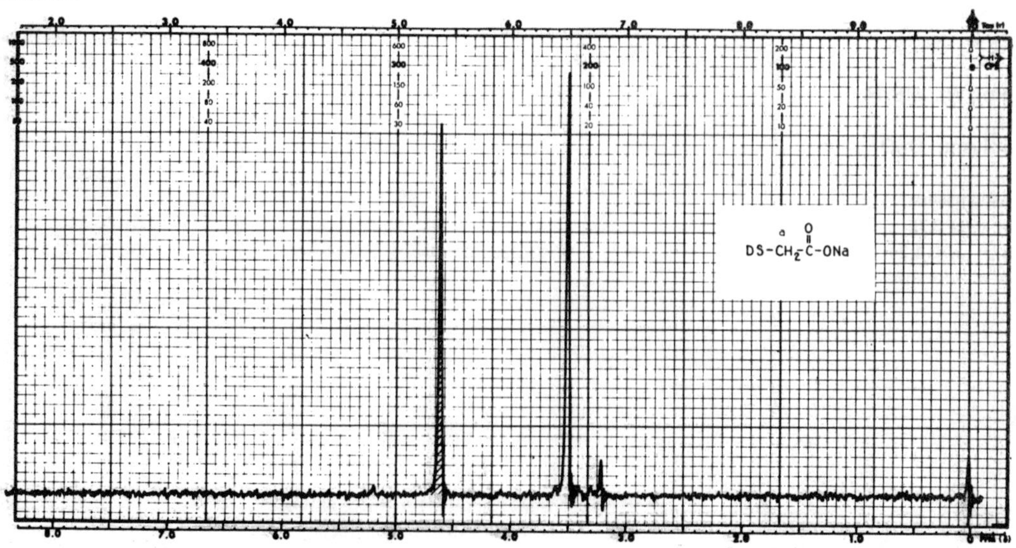

© 1978 Sadtler Research Laboratories, Inc.

DL-LACTIC ACID, SODIUM SALT

$C_3H_5NaO_3$

Mol. Wt. 112.06

Solvent: D_2O

Assignments

a = 1.30

b = 4.09

c = 4.62 HDO

d = 1.16 imp.

2659

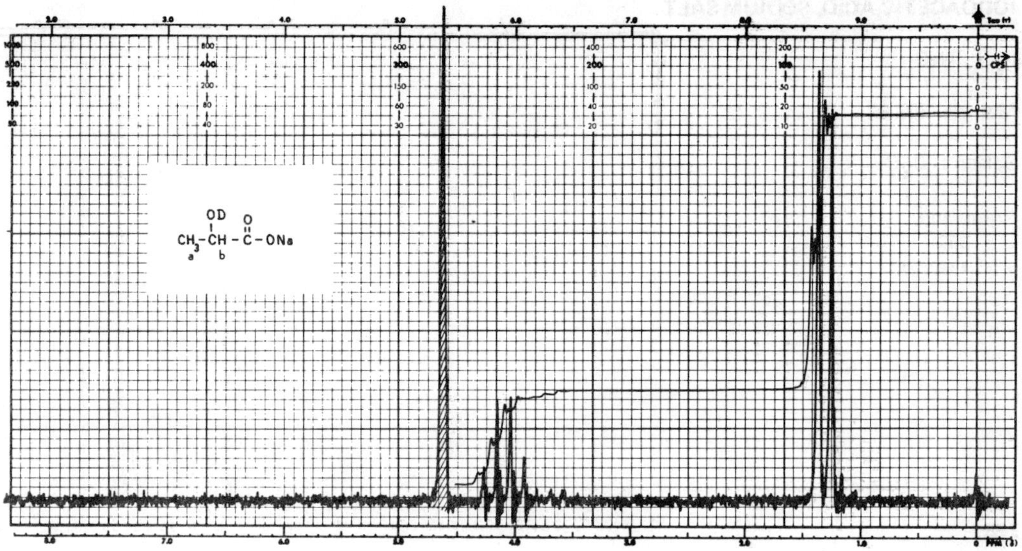

© 1978 Sadtler Research Laboratories, Inc.

HYDRACRYLIC ACID, SODIUM SALT

$C_3H_5NaO_3$

Mol. Wt. 112.06

Solvent: D_2O

Assignments

a = 2.41

b = 3.79

c = 4.69 HDO

2660

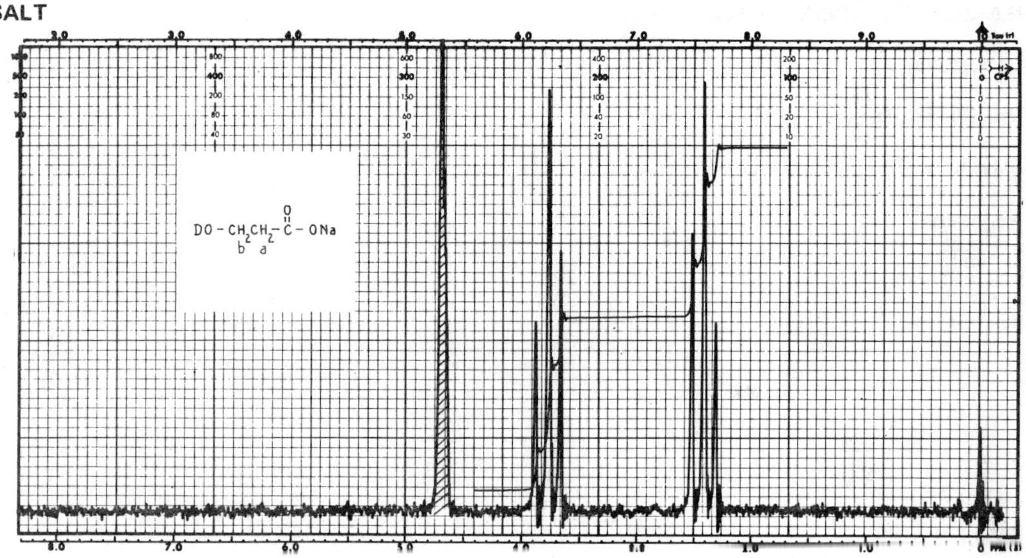

© 1978 Sadtler Research Laboratories, Inc.

GLUCONIC ACID, SODIUM SALT

$C_6H_{11}NaO_7$

Mol. Wt. 218.14

Solvent: D_2O

Assignments

a = ca 3.75

b = 3.70-4.20

c = 4.67 HDO

2661

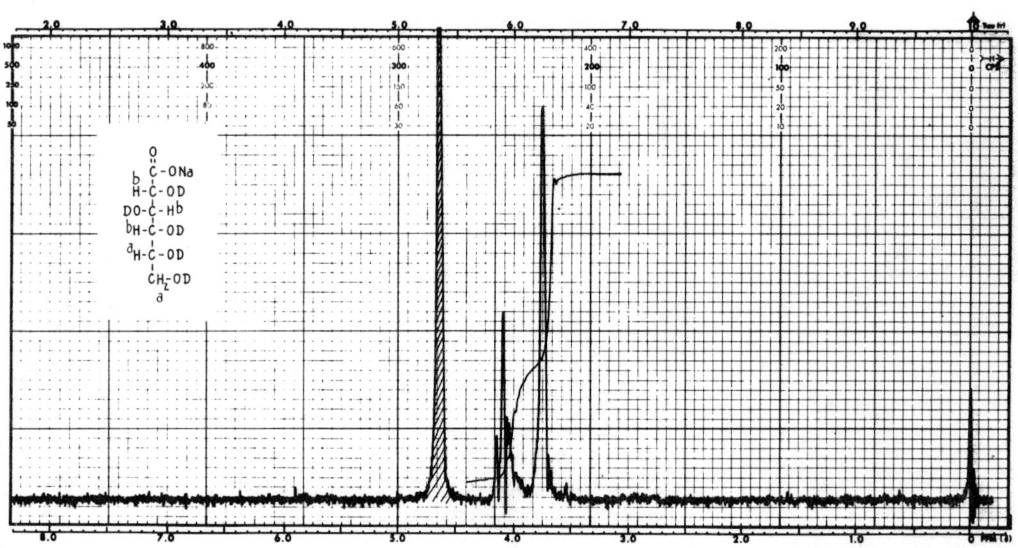

© 1978 Sadtler Research Laboratories, Inc.

UNIV. COLL. N.W. BANGOR LIBRARY

GLUCOHEPTONIC ACID, SODIUM SALT

$C_7H_{13}NaO_8$

Mol. Wt. 248.17

Solvent: D_2O

Assignments
a = 3.74
b = 3.60-4.10
c = 4.16
d = 4.69 HDO

2662

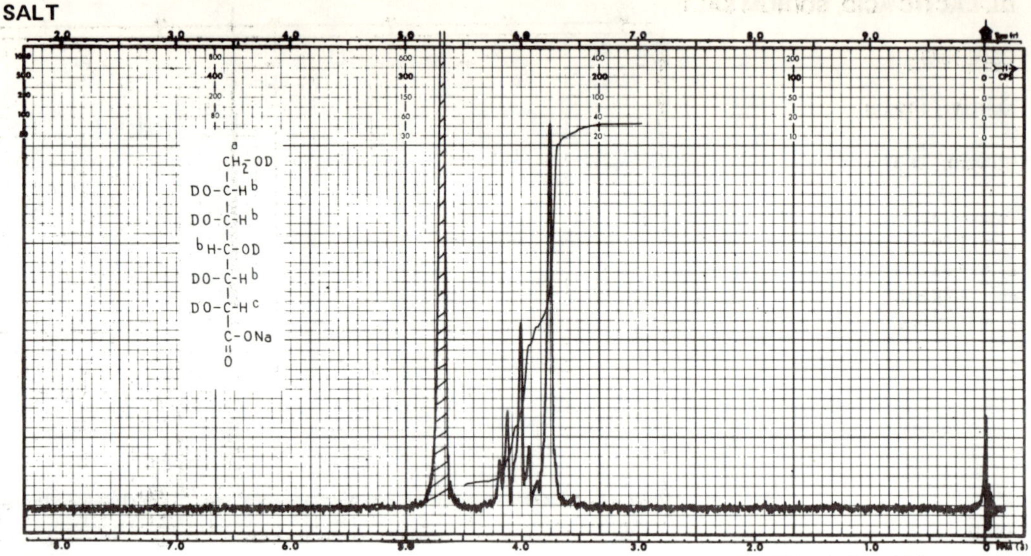

© 1978 Sadtler Research Laboratories, Inc.

MALONIC ACID, DISODIUM SALT, HYDRATE

$C_3H_2Na_2O_4 \cdot H_2O$

Mol. Wt. 166.04

Solvent: D_2O

Assignments
a = 3.11
b = 4.65 HDO

2663

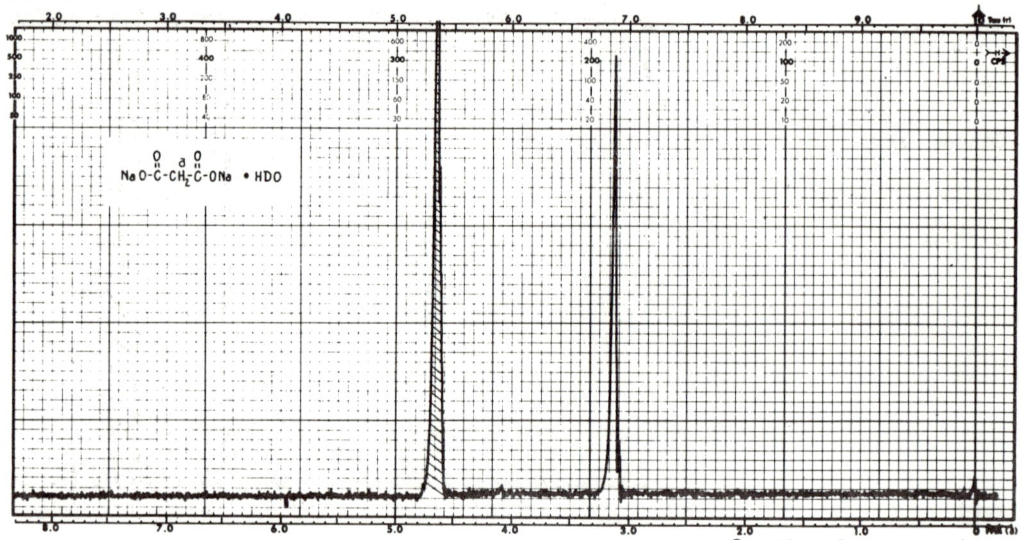

© 1978 Sadtler Research Laboratories, Inc.

SUCCINIC ACID, DISODIUM SALT

$C_4H_4Na_2O_4$

Mol. Wt. 162.05

Solvent: D_2O

Assignments
a = 2.38
b = 4.65 HDO

2664

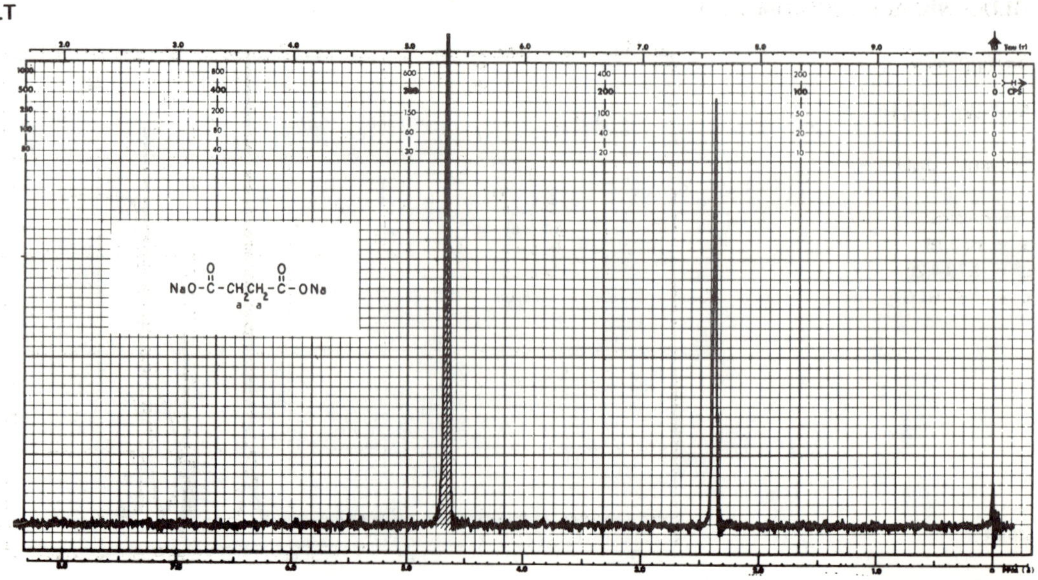

© 1978 Sadtler Research Laboratories, Inc.

NITRILOTRIACETIC ACID, DISODIUM SALT, HYDRATE

$C_6H_7NNa_2O_6 \cdot H_2O$

Mol. Wt. 253.12

M.P. 153°C (dec.)

Solvent: D_2O

Assignments
a = 3.74
b = 4.30 HDO

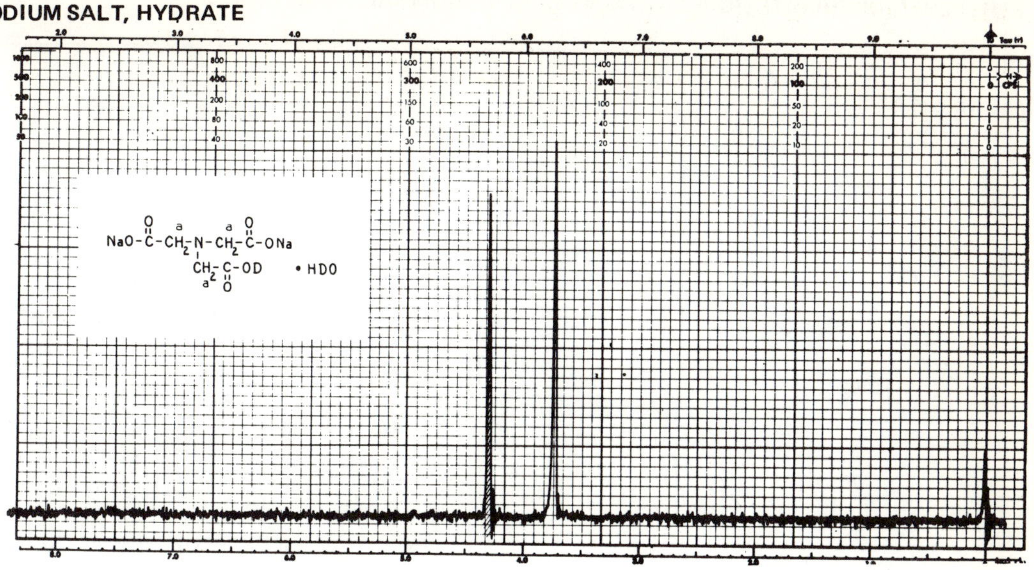

2665

© 1978 Sadtler Research Laboratories, Inc.

NITRILOTRIACETIC ACID, TRISODIUM SALT, HYDRATE

$C_6H_6NNa_3O_6 \cdot H_2O$

Mol. Wt. 275.10

M.P. > 320°C

Solvent: D_2O

Assignments
a = 3.09
b = 4.61 HDO

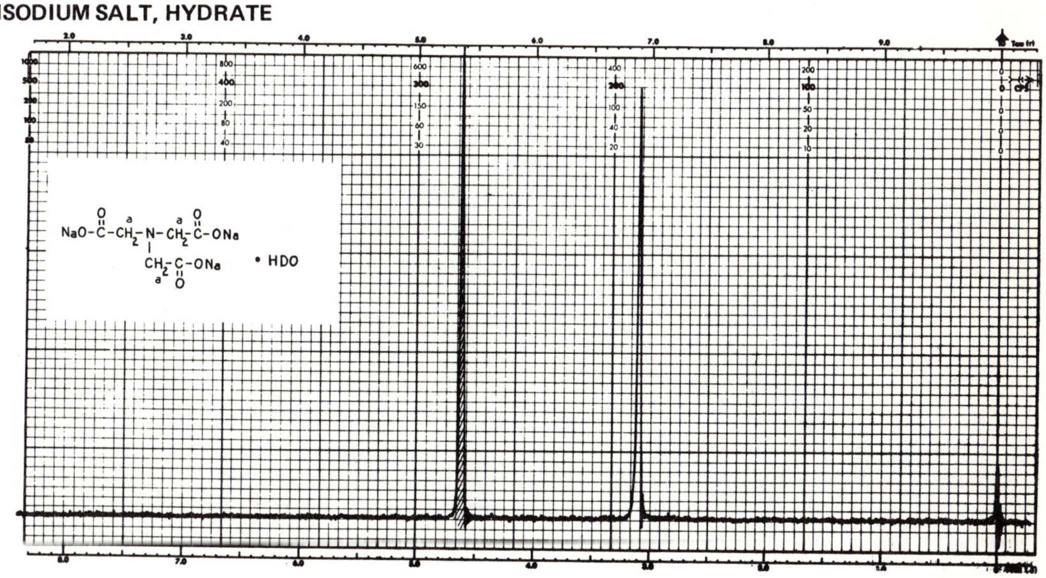

2666

© 1978 Sadtler Research Laboratories, Inc.

CITRIC ACID, POTASSIUM SALT, HYDRATE

$C_6H_5K_3O_7 \cdot H_2O$

Mol. Wt. 324.42

M.P. 230°C (dec.) (lit.)

Solvent: D_2O

Assignments
a = 2.44 or b
b = 2.70 or a
c = 4.63 HDO

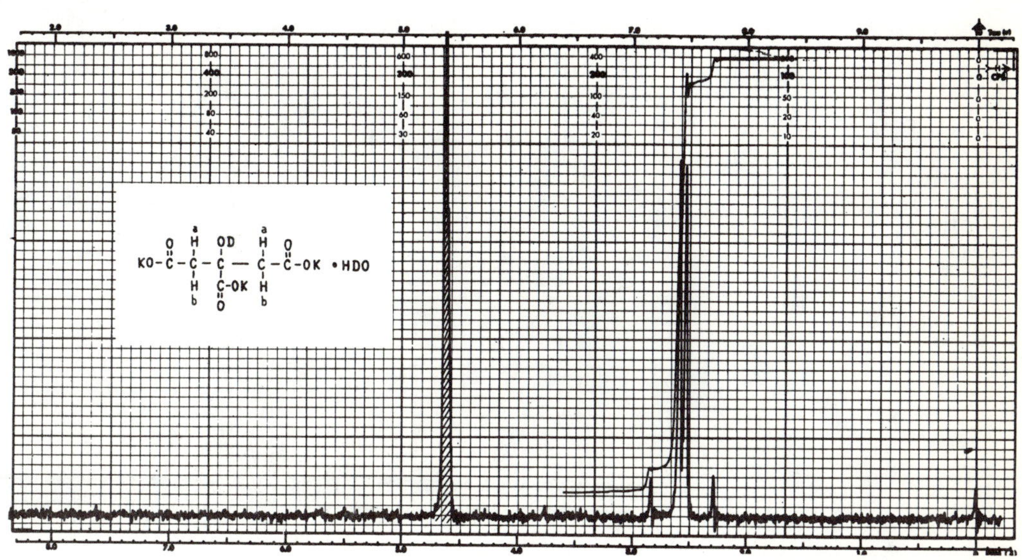

2667

© 1978 Sadtler Research Laboratories, Inc.

(ETHYLENEDINITRILO)TETRAACETIC ACID, DIPOTASSIUM SALT, DIHYDRATE

$C_{10}H_{16}N_2O_8K_2 \cdot 2H_2O$

Mol. Wt. 404.47

Solvent: D_2O

Assignments
a = 3.64
b = 3.90
c = 4.72 HDO

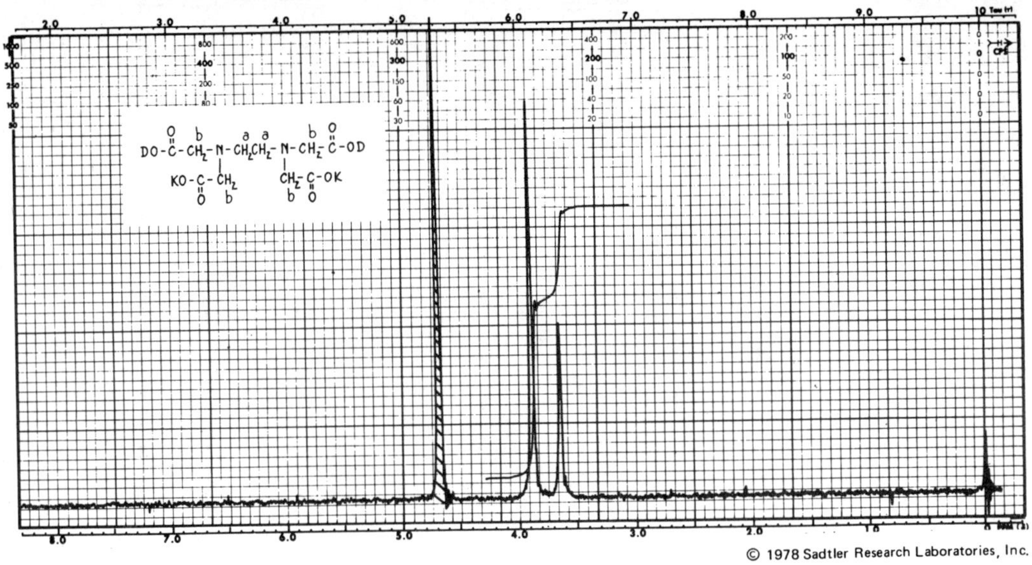

2668

© 1978 Sadtler Research Laboratories, Inc.

ACRYLIC ACID, CALCIUM SALT

$C_6H_6CaO_4$

Mol. Wt. 182.19

Solvent: D_2O

Assignments
a = 5.50-6.30
b = 4.80 HDO
c = 1.95 imp.

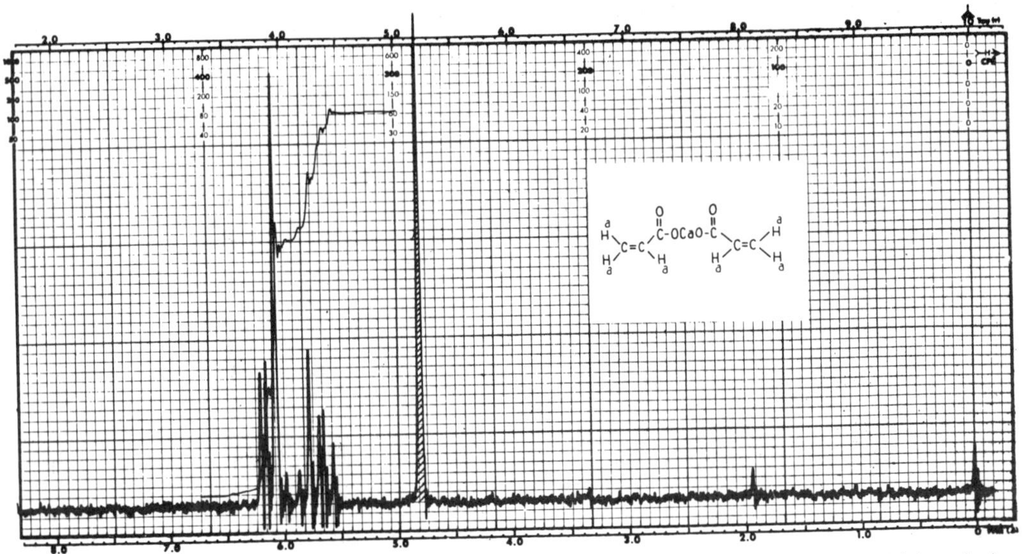

2669

© 1978 Sadtler Research Laboratories, Inc.

10-UNDECENOIC ACID, POTASSIUM SALT

$C_{11}H_{19}KO_2$

Mol. Wt. 222.37

Solvent: D_2O

Assignments
a = 1.31
b = 2.09
c = 4.91
d = 4.97
e = 5.78
f = 4.62 HDO

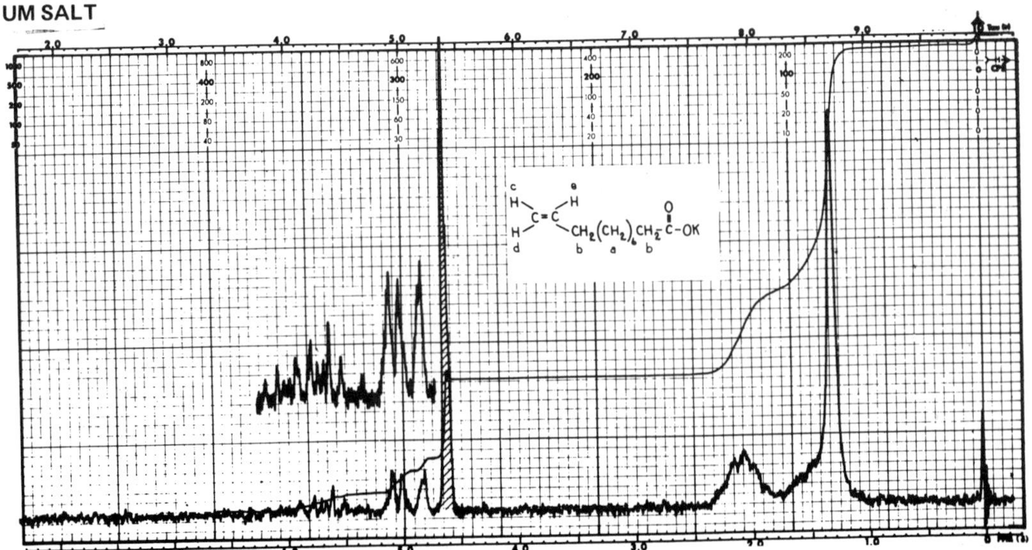

2670

© 1978 Sadtler Research Laboratories, Inc.

SORBIC ACID, POTASSIUM SALT

$C_6H_7KO_2$

Mol. Wt. 118.22

Solvent: D_2O

Assignments
a = 1.82
b = 5.83
c = 6.01-6.53
d = 7.00
e = 4.62 HDO

2671

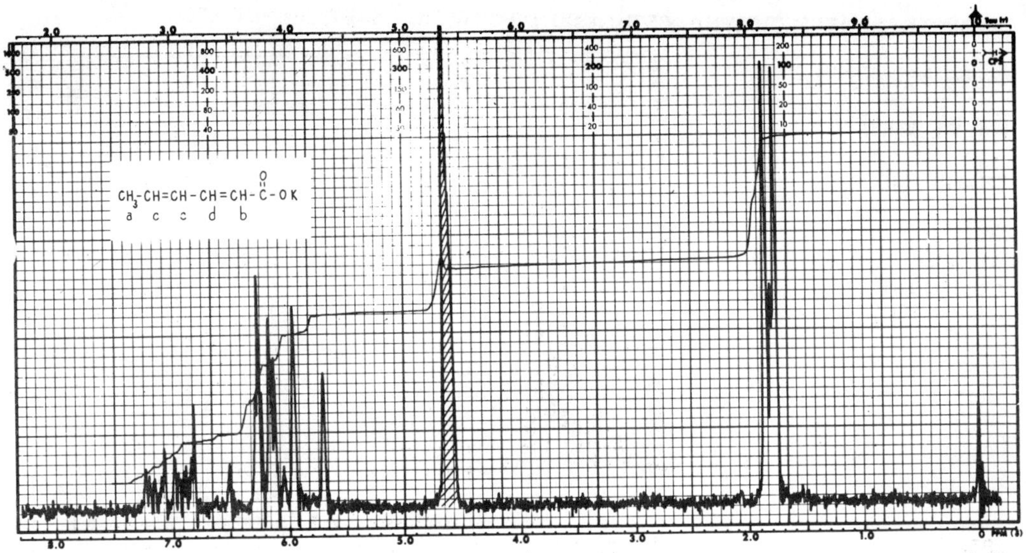

© 1978 Sadtler Research Laboratories, Inc.

MALEIC ACID, SODIUM SALT

$C_4H_2Na_2O_4$

Mol. Wt. 160.05

M.P. > 286°C

Solvent: D_2O

Assignments
a = 6.10
b = 4.64 HDO

2672

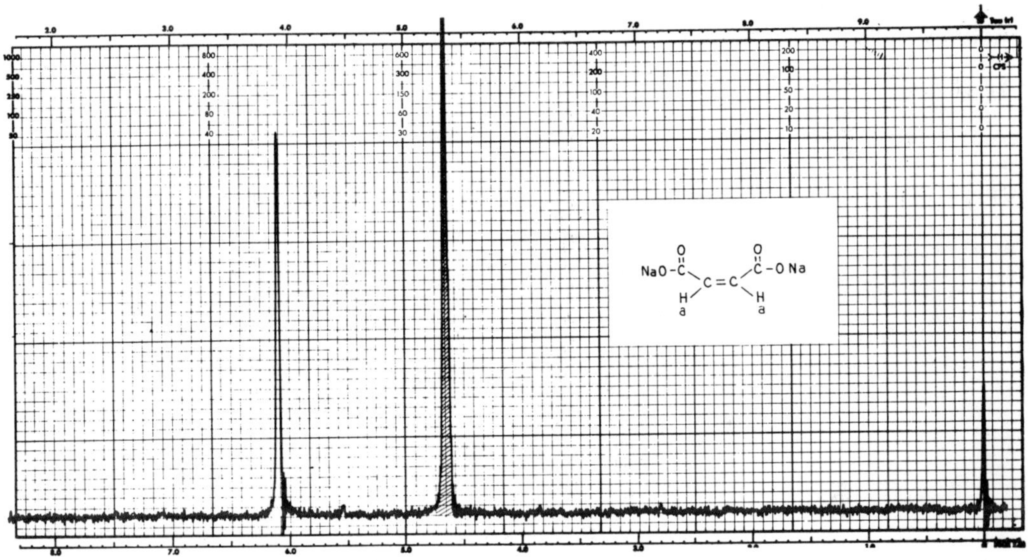

© 1978 Sadtler Research Laboratories, Inc.

PHENYLACETIC ACID, SODIUM SALT

$C_8H_7NaO_2$

Mol. Wt. 158.13

Solvent: D_2O

Assignments
a = 3.53
b = 7.32
c = 4.62 HDO

2673

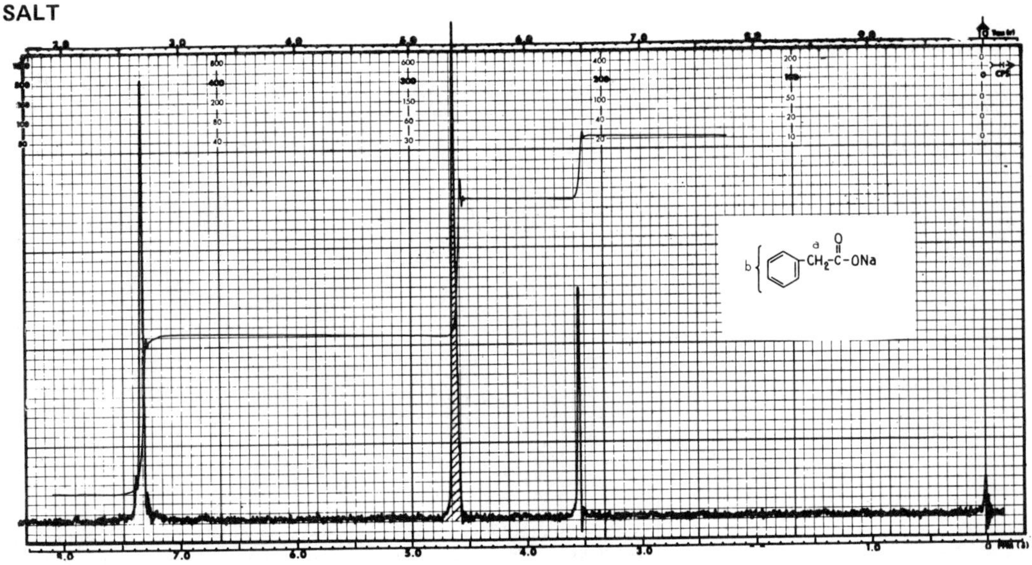

© 1978 Sadtler Research Laboratories, Inc.

1-NAPHTHALENEACETIC ACID, SODIUM SALT

$C_{12}H_9NaO_2$

Mol. Wt. 208.19

Solvent: D_2O

Assignments
a = 4.25
b = 7.60-8.50
c = 5.00 HDO

2674

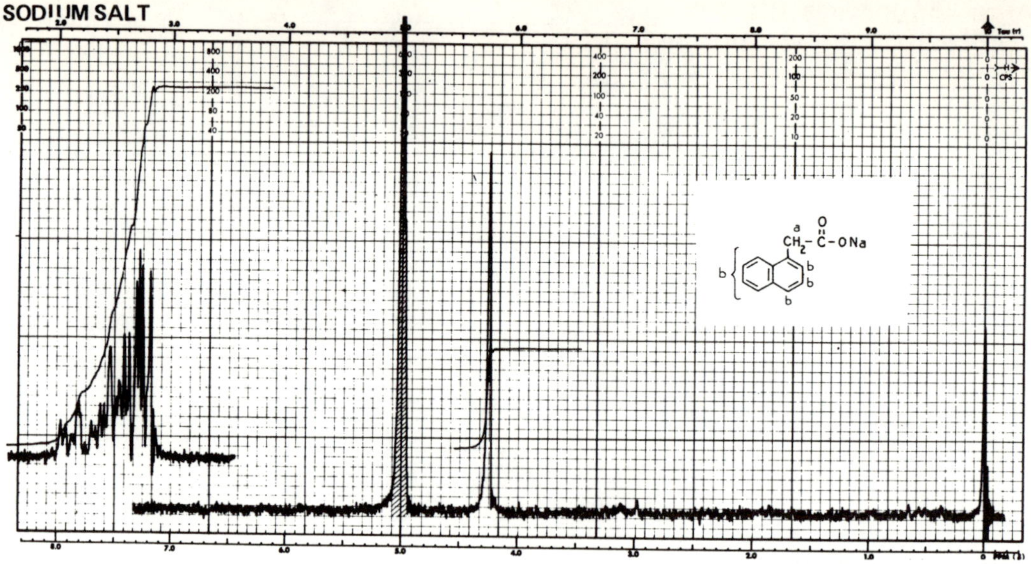

© 1978 Sadtler Research Laboratories, Inc.

BENZOIC ACID, SODIUM SALT

$C_7H_5NaO_2$

Mol. Wt. 144.11

Solvent: D_2O

Assignments
a = 7.31-7.64
b = 7.98
c = 4.75 HDO

2675

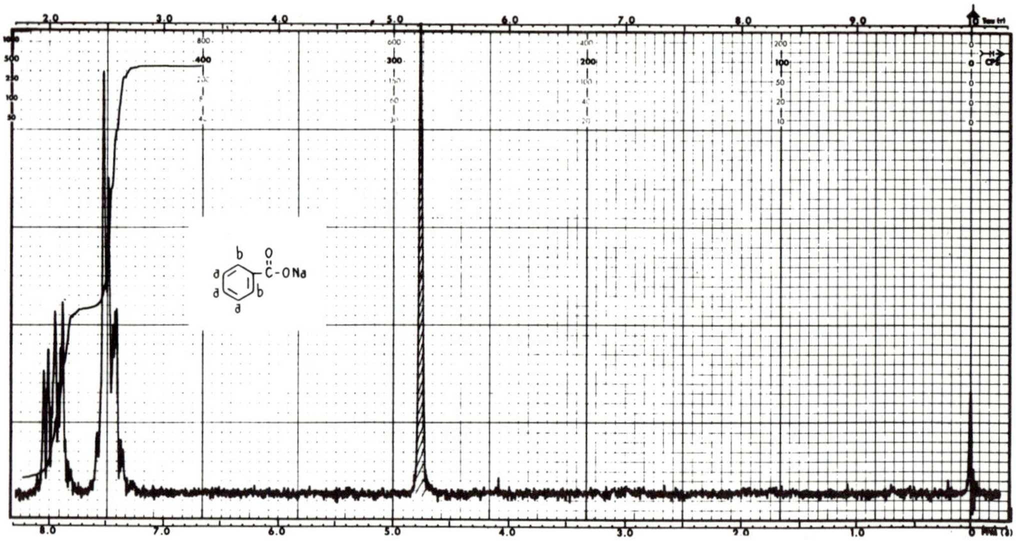

© 1978 Sadtler Research Laboratories, Inc.

m-AMINOBENZOIC ACID, SODIUM SALT

$C_7H_6NaNO_2$

Mol. Wt. 159.12

Solvent: D_2O

Assignments
a = 6.90
b = 7.27
c = 7.39
d = 7.45
e = 4.50 HDO

2676

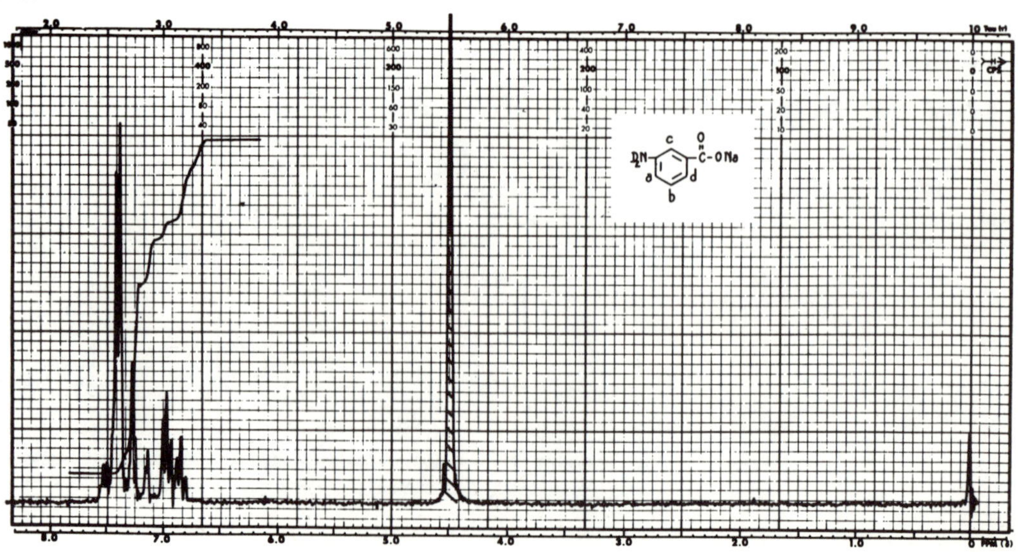

© 1978 Sadtler Research Laboratories, Inc.

p-AMINOBENZOIC ACID, SODIUM SALT

$C_7H_6NNaO_2$

Mol. Wt. 159.12

Solvent: D_2O

Assignments
a = 6.80
b = 7.82
c = 4.77 HDO

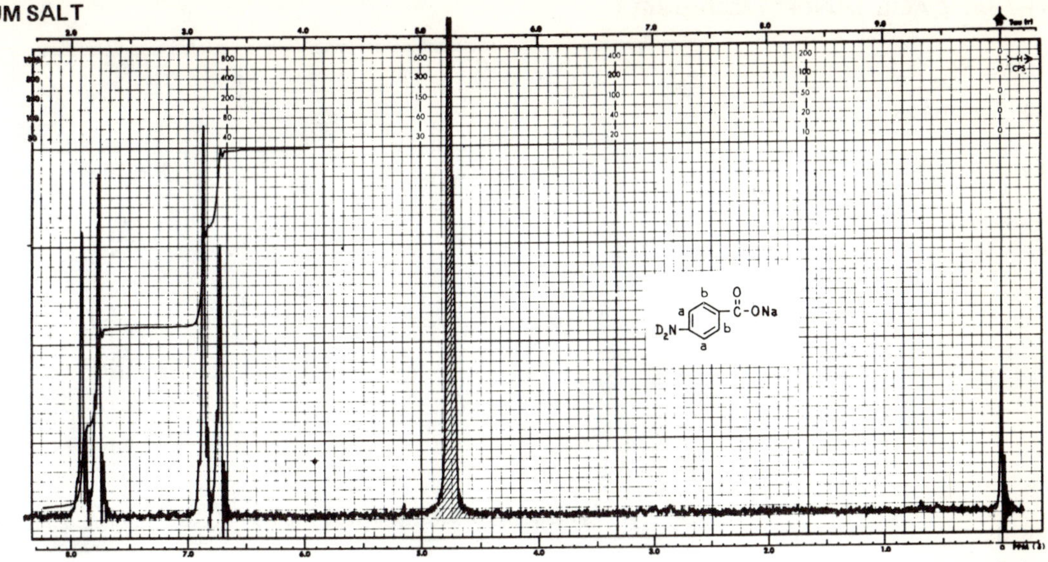

2677

© 1978 Sadtler Research Laboratories, Inc.

SODIUM SALICYLATE

$C_7H_5O_3Na$

Mol. Wt. 160.11

Solvent: D_2O

Assignments
a = ca 6.97
b = ca 7.00
c = 7.44
d = 7.92
e = 4.80 HDO

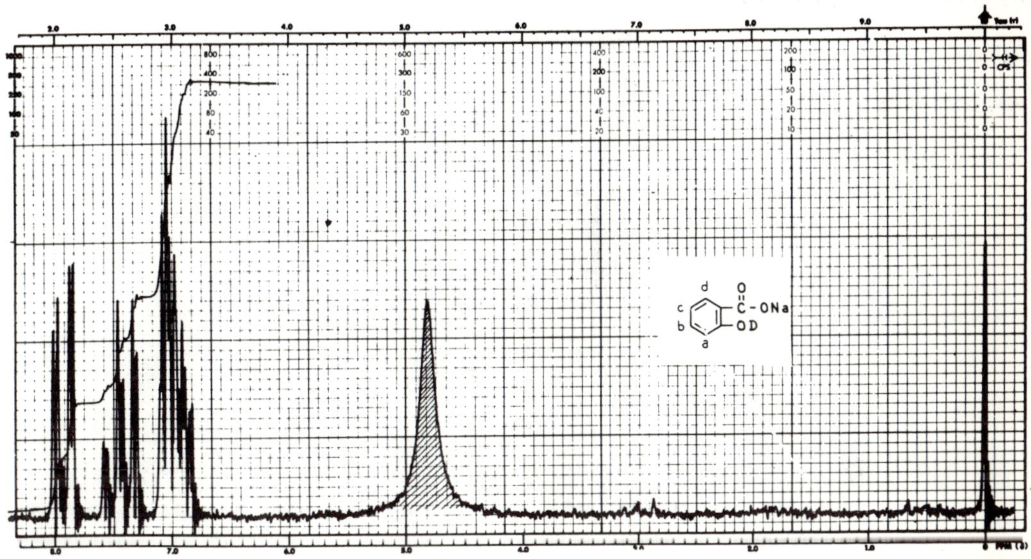

2678

© 1978 Sadtler Research Laboratories, Inc.

p-HYDROXYBENZOIC ACID, MONOSODIUM SALT

$C_7H_5NaO_3$

Mol. Wt. 160.11

Solvent: D_2O

Assignments
a = 7.06
b = 8.04
c = 5.13 HDO

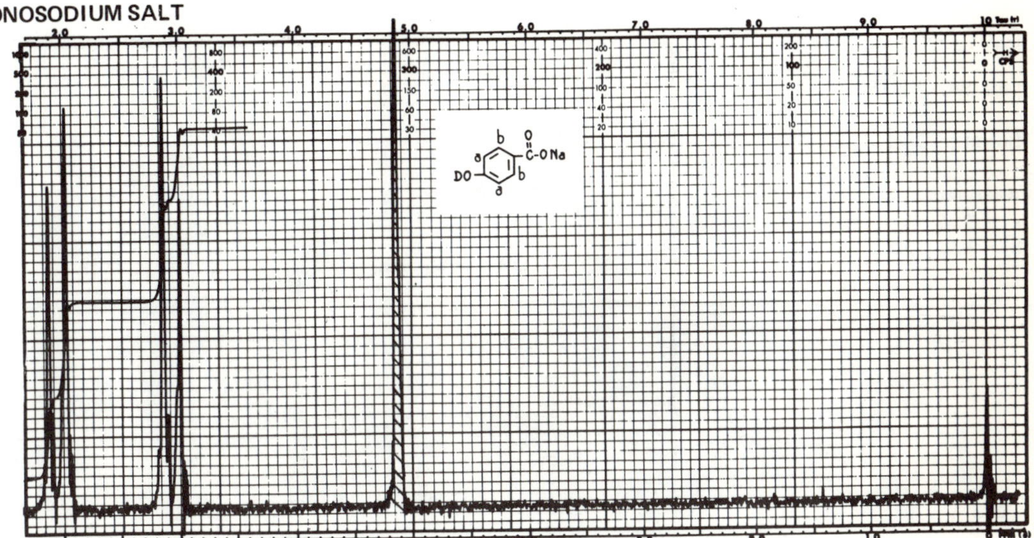

2679

© 1978 Sadtler Research Laboratories, Inc.

PHTHALIC ACID, MONOPOTASSIUM SALT

$C_8H_5KO_4$

Mol. Wt. 204.23

Solvent: D_2O

Assignments
a = 7.61
b = 8.00
c = 4.46 HDO

2680

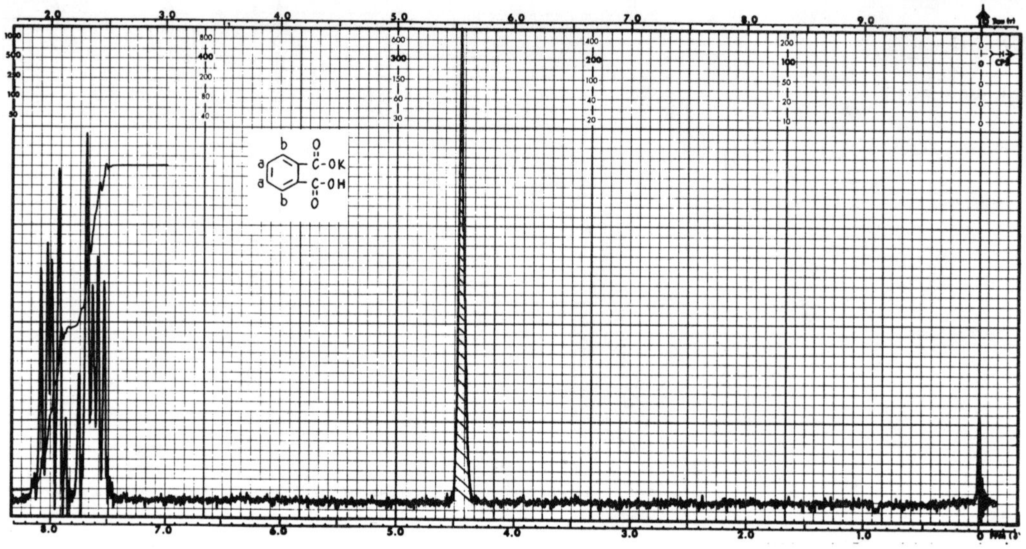

© 1978 Sadtler Research Laboratories, Inc.

2-FUROIC ACID, POTASSIUM SALT

$C_5H_3KO_3$

Mol. Wt. 150.18

Solvent: D_2O

Assignments
a = 6.59
b = 7.05
c = 7.64
d = 4.68 HDO

2681

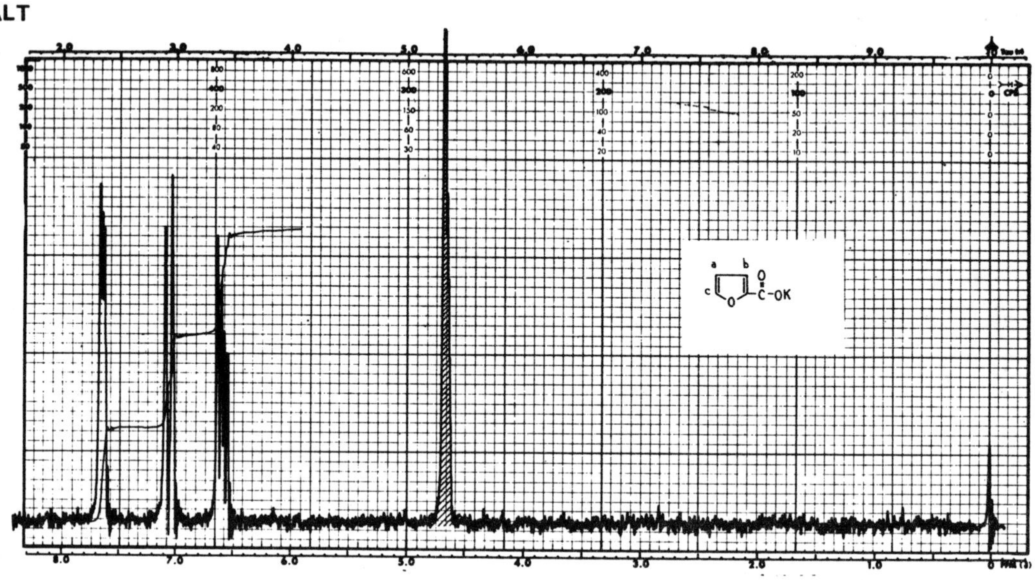

© 1978 Sadtler Research Laboratories, Inc.

SUCCINIC ACID, DIAMMONIUM SALT

$C_4H_{12}N_2O_4$

Mol. Wt. 152.15

Solvent: D_2O

Assignments
a = 2.50
b = 5.09 HDO

2682

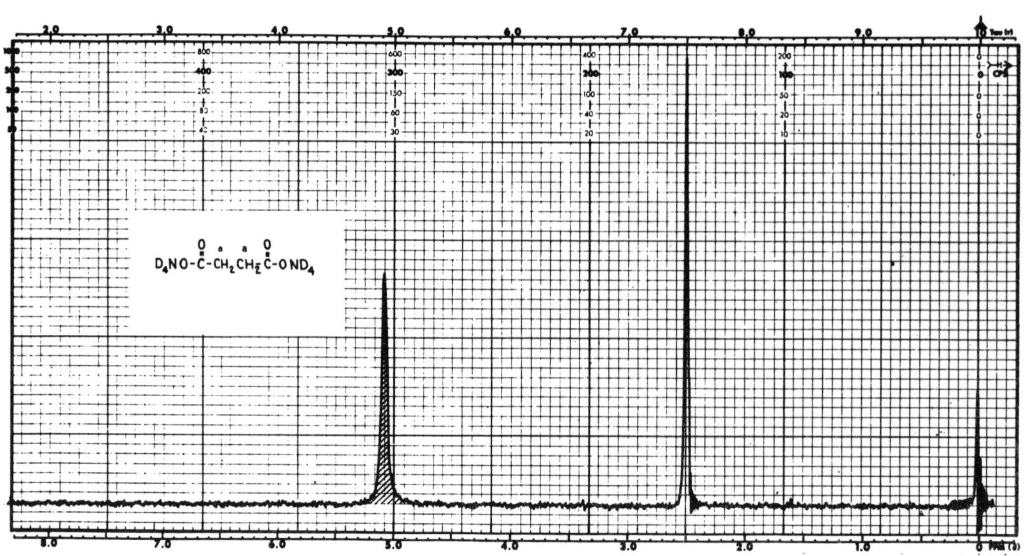

© 1978 Sadtler Research Laboratories, Inc.

BENZOIC ACID, AMMONIUM SALT

$C_7H_9NO_2$

Mol. Wt. 139.16

M.P. 194-195°C

Solvent: D_2O

Assignments
a = ca 7.52
b = 7.99
c = 4.94 HDO

2683

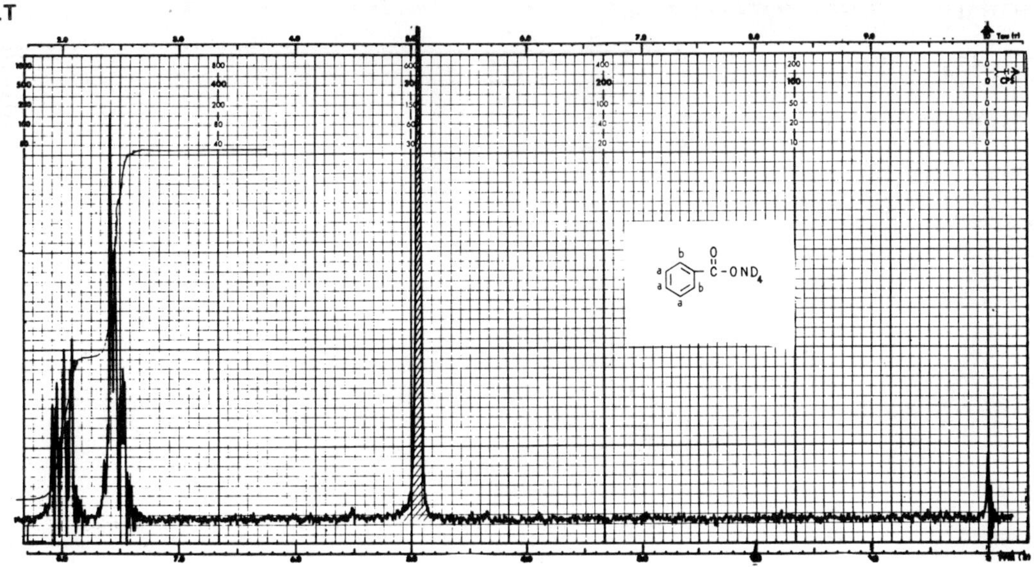

© 1978 Sadtler Research Laboratories, Inc.

SALICYLIC ACID, AMMONIUM SALT

$C_7H_9NO_3$

Mol. Wt. 155.15

Solvent: D_2O

Assignments
a = 6.97
b = 7.04
c = 7.46
d = 7.97
e = 5.30 HDO

2684

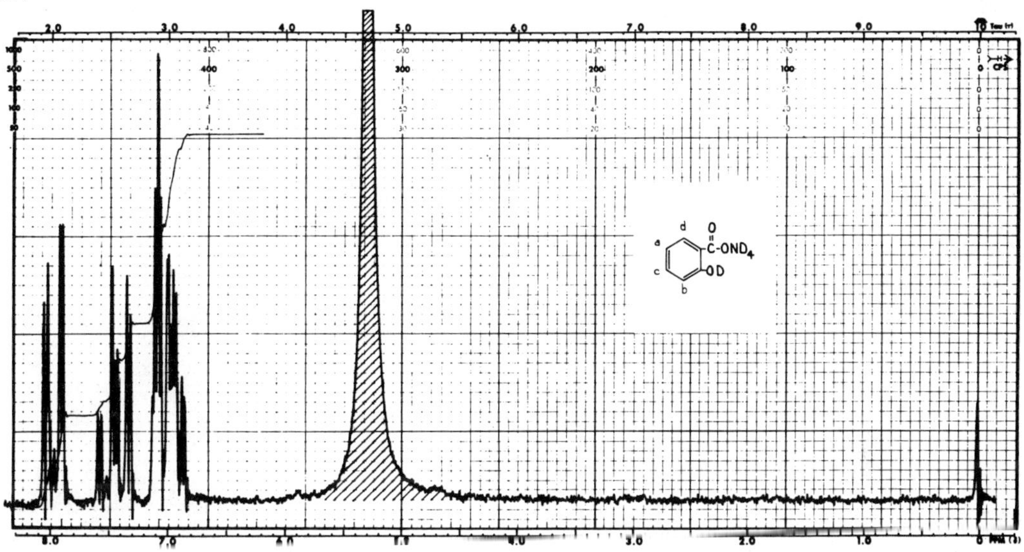

© 1978 Sadtler Research Laboratories, Inc.

DECYLAMINE, ACETATE(1:1) (SALT)

$C_{10}H_{23}N \cdot C_2H_4O_2$

Mol. Wt. 225.33

Solvent: $CDCl_3$

Assignments
a = 0.90
b = ca 1.27
c = 1.90
d = 2.80
e = 8.14

2685

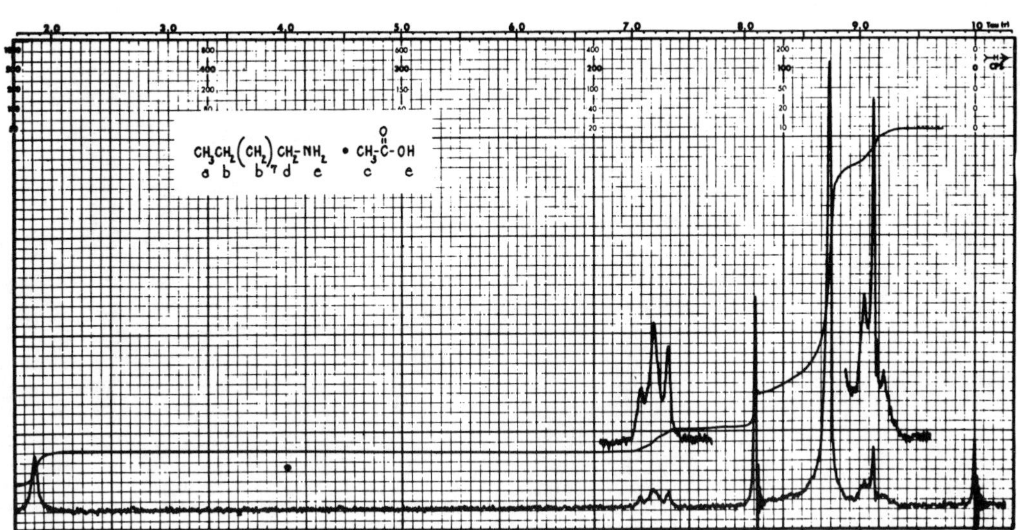

© 1978 Sadtler Research Laboratories, Inc.

DODECYLAMINE, ACETATE

$C_{12}H_{27}N \cdot C_2H_4O_2$

Mol. Wt. 245.41

Solvent: $CDCl_3$

Assignments
a = 0.90
b = 1.10-1.80
c = 1.91
d = 2.79
e = 7.98

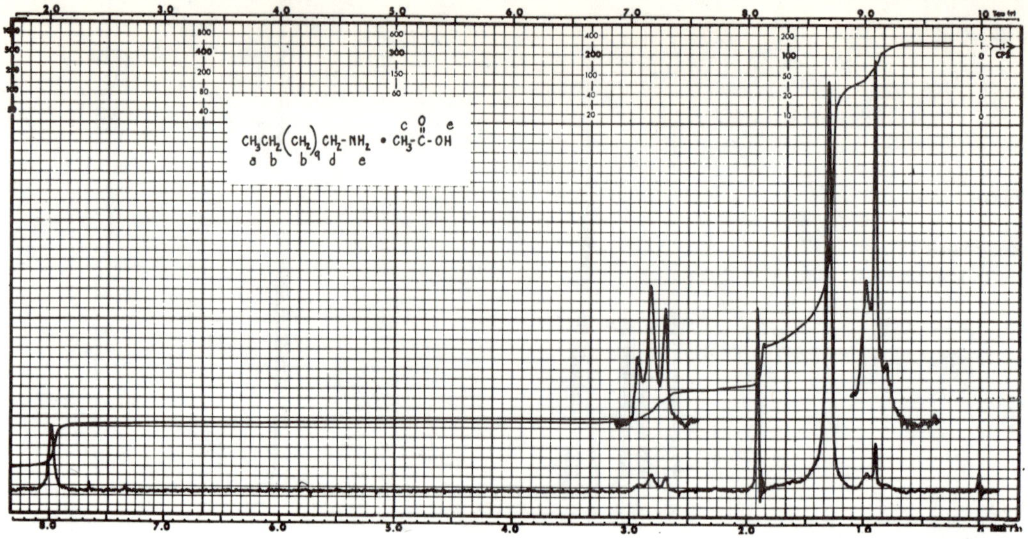

2686

© 1978 Sadtler Research Laboratories, Inc.

HYDROXYLAMINE, ACETATE (SALT)

$H_3NO \cdot C_2H_4O_2$

Mol. Wt. 93.08

Solvent: D_2O

Assignments
a = 1.99
b = 5.57 HDO

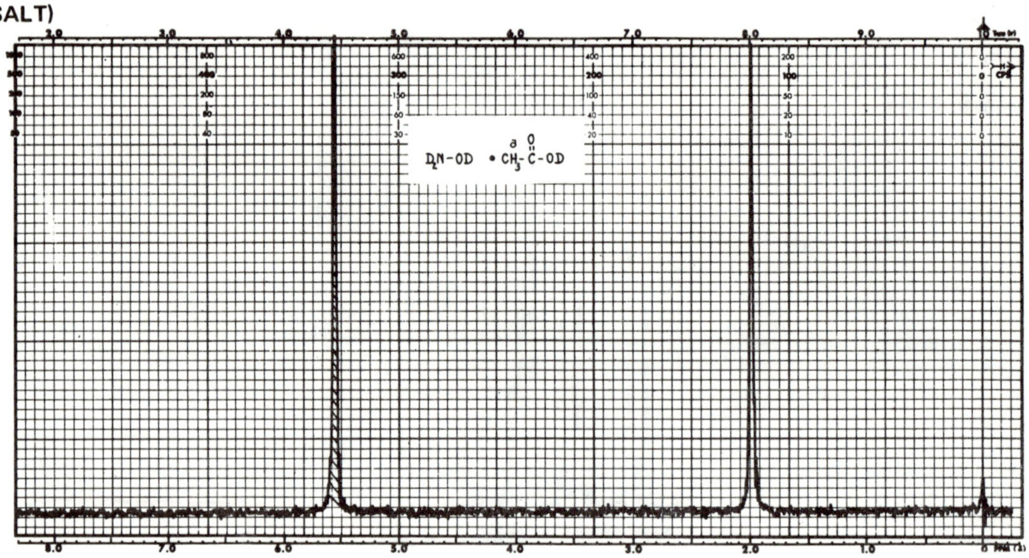

2687

© 1978 Sadtler Research Laboratories, Inc.

N,N-DIMETHYL-p-PHENYLENEDIAMINE, OXALATE (2:1)

$C_{16}H_{24}N_4 \cdot C_2H_2O_4$

Mol. Wt. 362.43

Solvent: Trifluoroacetic acid

Assignments
a = 3.52
b = 7.91
c = unobserved
* Under solvent band

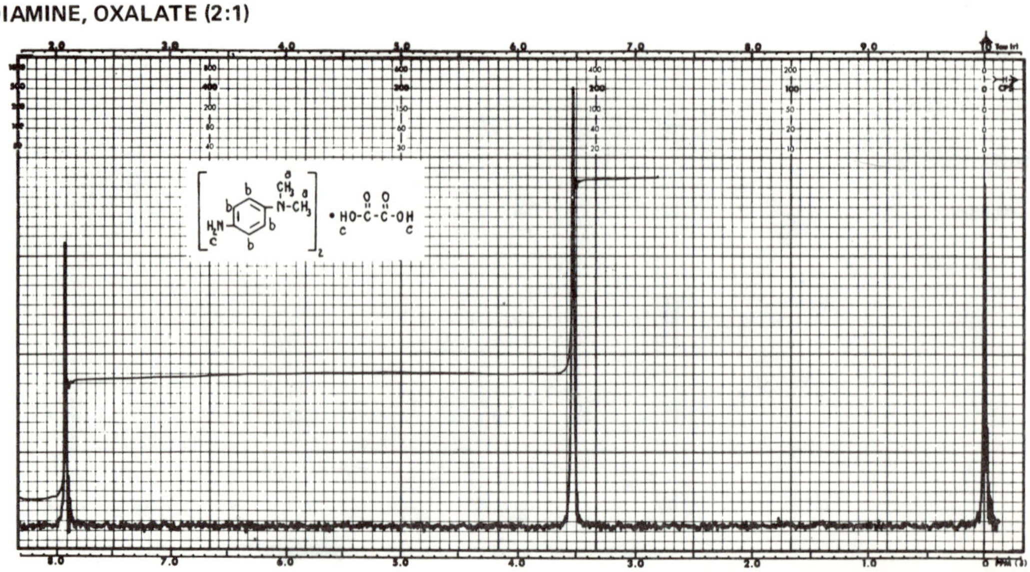

2688

© 1978 Sadtler Research Laboratories, Inc.

PIPERAZINE, ADIPATE (1:1)

$C_4H_{10}N_2 \cdot C_6H_{10}O_4$

Mol. Wt. 232.28

M.P. 256-257°C

Solvent: D_2O

Assignments

a = 1.60

b = 2.27

c = 3.30

d = 4.21 HDO

2689

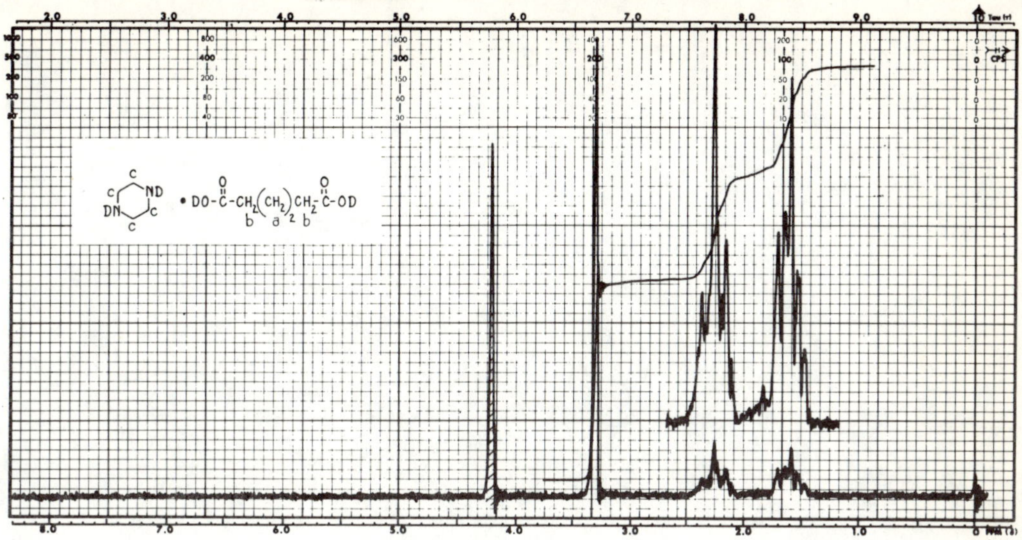

© 1978 Sadtler Research Laboratories, Inc.

ETHYLENEDIAMINE, MONO-(+)-TARTRATE

$C_2H_8N_2 \cdot C_4H_6O_6$

Mol. Wt. 210.19

M.P. 203°C (dec.)

Solvent: D_2O

Assignments

a = 3.41

b = 4.38

c = 4.79 HDO

2690

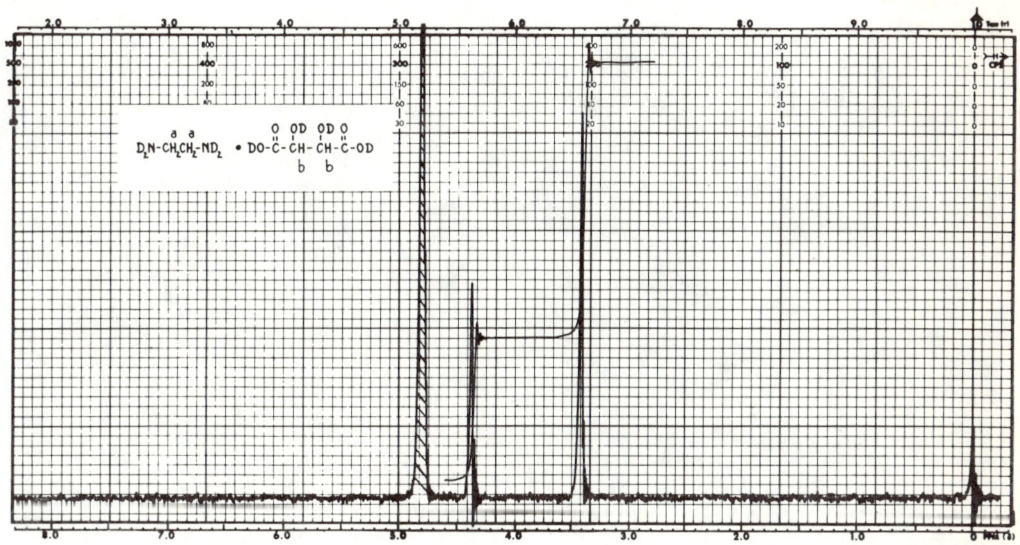

© 1978 Sadtler Research Laboratories, Inc.

NOTES

The Esters are one of the most important functional groups with a wide variety of combinations of aliphatic, olefinic and aromatic acids and alcohols. This section deals with the Aliphatic Esters of Aliphatic acids and Olefinic acids, and their substituted derivatives.

The carbonyl side of the ester functional group has a weakly deshielding effect on the protons of adjacent aliphatic groups, while the oxygen side of the linkage has a strongly deshielding effect.

The Esters are readily soluble in Carbon Tetrachloride and Deuterochloroform unless substituted by more polar functional groups.

Chemical Shifts

Aliphatic Protons

HC–3	HC–2	HC–1		Solvent	Spectrum
		1.90	$CH_3–C(=O)–O–C(CH_3)_3$	CCl_4	----
		1.95	$CH_3–C(=O)–O–R_{12}$	CCl_4	2697
		2.00	$CH_3–C(=O)O–CH_3$	CCl_4	2694
	1.10	2.27	$CH_3–CH_2–C(=O)–O–CH_3$	CCl_4	----
	1.10	2.27	$CH_3–CH_2–C(=O)–O–R_8$	CCl_4	----
	1.16	2.30	$CH_3–CH_2–C(=O)–O–CH(R)_2$	$CDCl_3$	2698
0.95	1.70	2.18	$CH_3–CH_2–CH_2–C(=O)–O–CH(CH_3)_2$	CCl_4	2700
0.91	1.61	2.21	$CH_3–CH_2–CH_2–C(=O)–O–CH_3$	CCl_4	----
	(1.13)	2.44	$(CH_3)_2–CH–C(=O)–O–R_4$	CCl_4	2702
	(1.14)	2.49	$(CH_3)_2–CH–C(=O)–O–CH_3$	CCl_4	----
	(1.17)	----	$(CH_3)_3–C–C(=O)–O–R_2$	CCl_4	2703
(0.96)	1.97	2.12	$(CH_3)_2–CH–CH_2–C(=O)–O–CH_3$	CCl_4	2699
(0.93)	1.99	2.18	$(CH_3)_2–CH–CH_2–C(=O)–O–R_5$	$CDCl_3$	----
		3.61	$CH_3–O–C(=O)–R_7$	CCl_4	2704
		3.65	$CH_3–O–C(=O)–CH_3$	CCl_4	2694
		3.70	$CH_3–O–C(=O)–CH=CH_2$	CCl_4	----
		4.00	$CH_3–O–C(=O)–CCl_3$	CCl_4	2722
	1.23	4.07	$CH_3–CH_2–O–C(=O)–C(CH_3)_3$	CCl_4	2703
	1.29	4.19	$CH_3–CH_2–O–C(=O)–R_4$	CCl_4	2701
	1.41	4.41	$CH_3–CH_2–O–C(=O)–CCl_3$	$CDCl_3$	2723
0.98	1.67	4.08	$CH_3–CH_2–CH_2–O–C(=O)–H$	CCl_4	----
0.98	1.78	4.30	$CH_3–CH_2–CH_2–O–C(=O)–CF_3$	$CDCl_3$	----
	(1.22)	4.92	$(CH_3)_2–CH–O–C(=O)–R_3$	CCl_4	2700
	(1.39)	5.20	$(CH_3)_2–CH–O–C(=O)–CF_3$	CCl_4	----
(0.96)	1.97	3.89	$(CH_3)_2–CH–CH_2–O–C(=O)–CH=CH_2$	CCl_4	2807

Olefinic Protons

The olefinic protons of Acrylic Acid Esters appear in the NMR spectrum as a higher-order ABC pattern in the chemical shift range from 5.6–6.1 ppm. The proton which is trans to the carbonyl group resonates at highest field, the geminal proton resides at slightly lower field, and the hydrogen which is cis to the carbonyl appears at lowest field.

Acrylate Esters

cis	trans	geminal		Solvent	Spectrum
6.34	5.72	6.09	CH_2=CH–C(=O)–O–R	CCl_4	2807

The Vinyl Esters of aliphatic carboxylic acids produce a much clearer pattern than the corresponding Acrylic Acid protons. The three olefinic protons of the Vinyl Esters produce resonance bands over the range from 4.3 to 7.5 ppm. The two protons bonded to the beta carbon resonate at highest field while the geminal proton which is bonded to the alpha carbon resonates at lower field.

Vinyl Esters of Aliphatic Acids

cis	trans	geminal			
4.82	4.51	7.23	CH_2=CH–O–C(=O)–R_{17}	$CDCl_3$	2802

2-Substituted Acrylate Esters

R–O–C(=O)–	CH	=CH	–X			
R_6–O–C(=O)–	6.79	6.79	–C(=O)–O–R_6	(trans)	CCl_4	2827
CH_3–O–C(=O)–	5.71	6.84	–R_7	(trans)	CCl_4	2815
CH_3–O–C(=O)–	6.86	6.86	–C(=O)–O–CH_3	(trans)	$CDCl_3$	2825
R_2–O–C(=O)–	5.79	6.90	–CH_3	(trans)	CCl_4	2814
CH_3–O–C(=O)–	6.59	7.03	–C(=O)–CH_3	(trans)	$CDCl_3$	2821
C_6–O–C(=O)–	6.40	7.67	–Ø	(trans)	CCl_4	2857

The Aldehydic Protons

The Aldehydic protons of the Formic Acid esters appears as a sharp singlet at low field in the range from 7.9 to 8.10 ppm.

H–C(=O)–O	–X		
7.90	–$(CH_2)_2$–Ø	CCl_4	2831
7.91	–R_7	CCl_4	2692
7.96	–R_8	CCl_4	2693
7.98	–CH_3	CCl_4	----
8.07	–CH_2–CH$(R_2)_2$	$CDCl_3$	2691
8.10	–$(CH_2)_2$–O–R	$CDCl_3$	----

2-ETHYL-1-BUTANOL, FORMATE

$C_7H_{14}O_2$

Mol. Wt. 130.19

Solvent: $CDCl_3$

Assignments
a = 0.90
b = 1.10-1.80
c = 4.10
d = 8.07
e = 2.10 imp.

2691

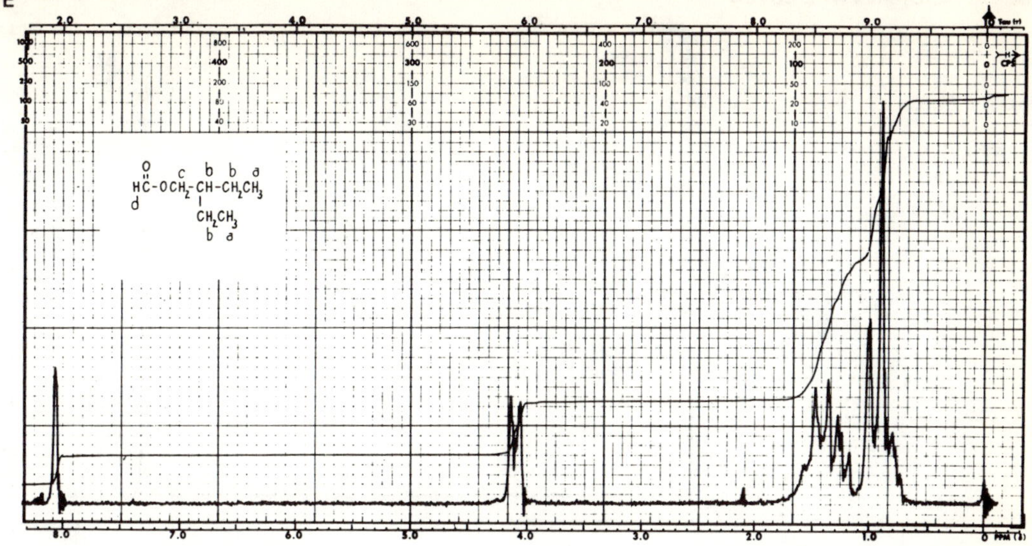

© 1978 Sadtler Research Laboratories, Inc.

FORMIC ACID, HEPTYL ESTER

$C_8H_{16}O_2$

Mol. Wt. 144.22

Solvent: CCl_4

Assignments
a = 0.90
b = 1.08-1.91
c = 4.10
d = 7.91

2692

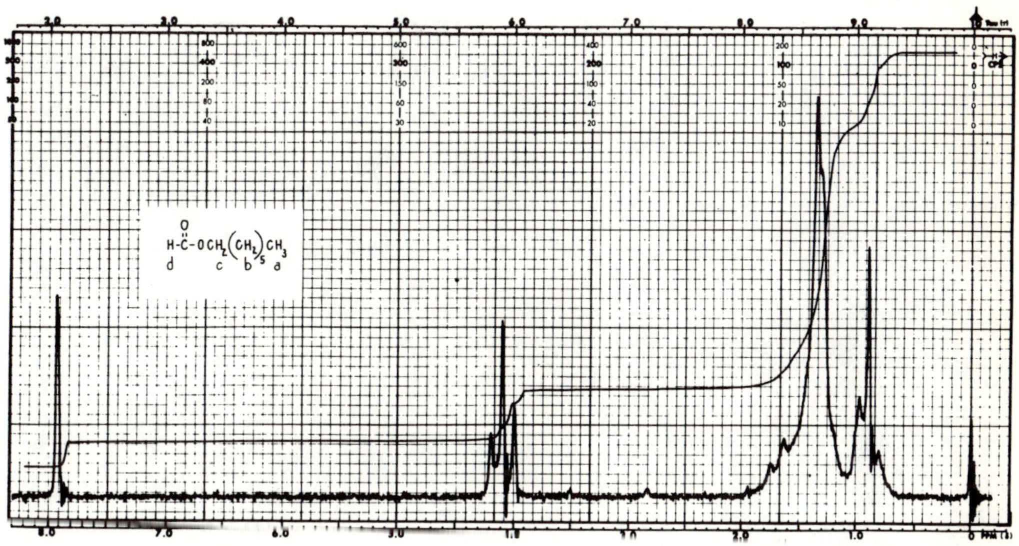

© 1978 Sadtler Research Laboratories, Inc.

FORMIC ACID, OCTYL ESTER

$C_9H_{18}O_2$

Mol. Wt. 158.23

B.P. 82°C/20mm

Solvent: CCl_4

Assignments
a = 0.90
b = 1.31
c = 4.10
d = 7.69

2693

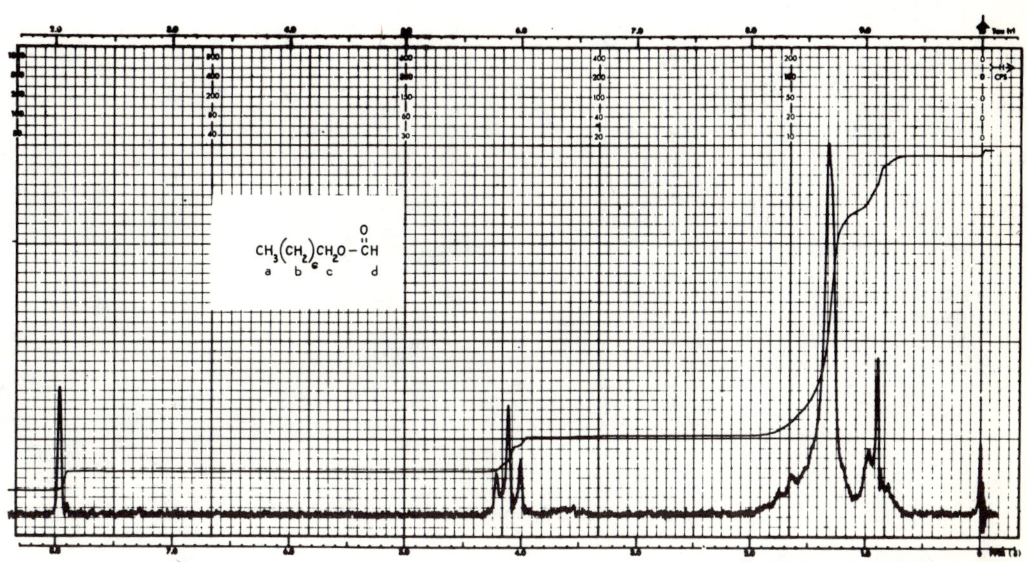

© 1978 Sadtler Research Laboratories, Inc.

ACETIC ACID, METHYL ESTER

$C_3H_6O_2$

Mol. Wt. 74.08

B.P. 55-57°C

Solvent: CCl_4

Assignments

a = 2.00

b = 3.65

2694

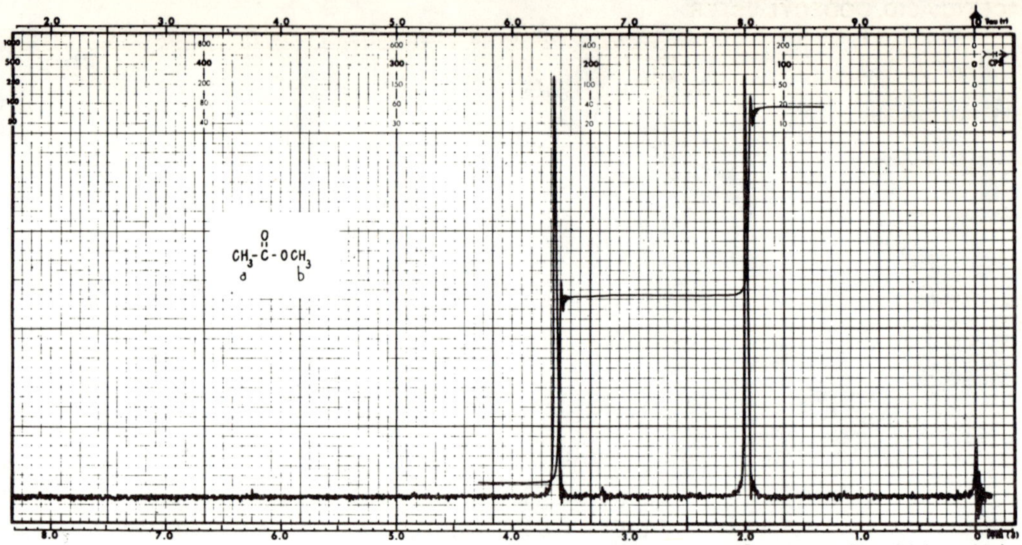

© 1978 Sadtler Research Laboratories, Inc.

ACETIC ACID, 2-ETHYLBUTYL ESTER

$C_8H_{16}O_2$

Mol. Wt. 144.22

B.P. 162.4°C

Solvent: CCl_4

Assignments

a = 0.90

b = 1.10-1.70

c = 1.96

d = 3.92

2695

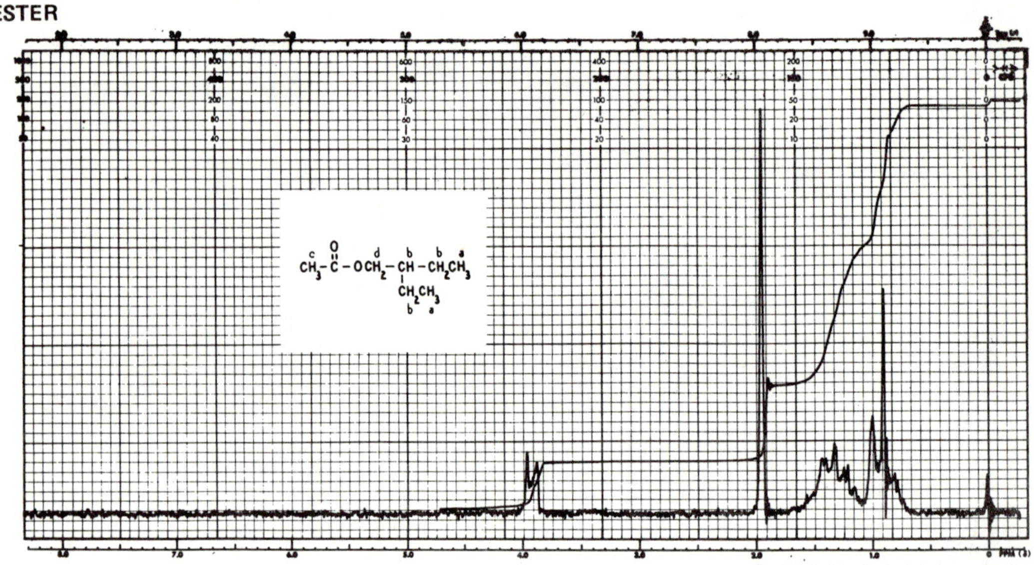

© 1978 Sadtler Research Laboratories, Inc.

ACETIC ACID, 2-ETHYLHEXYL ESTER

$C_{10}H_{20}O_2$

Mol. Wt. 172.27

B.P. 198.6°C

Solvent: CCl_4

Assignments

a = 0.90

b = 1.10-1.60

c = 1.94

d = 3.90

2696

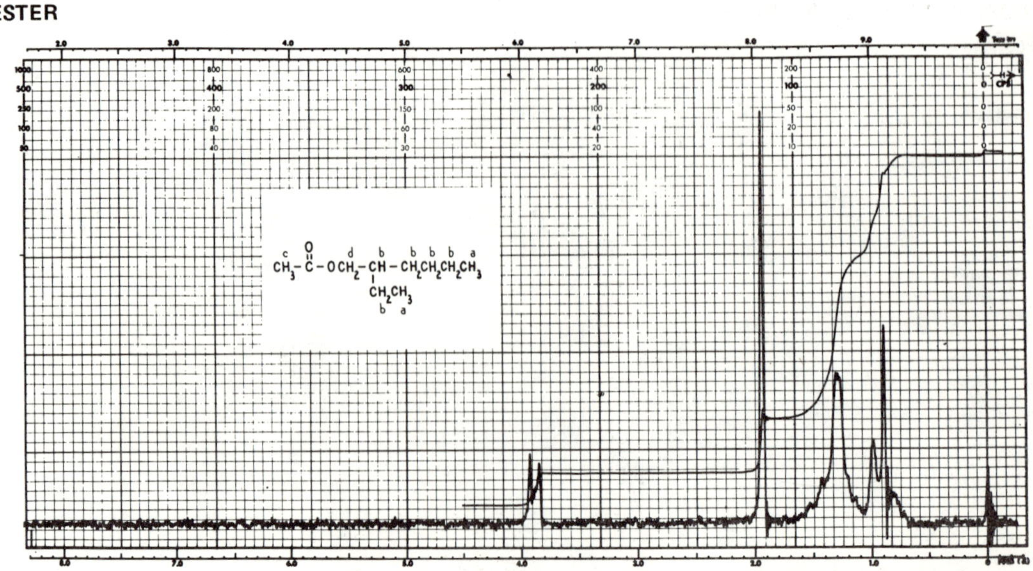

© 1978 Sadtler Research Laboratories, Inc.

ACETIC ACID, DODECYL ESTER

$C_{14}H_{28}O_2$

Mol. Wt. 228.38

B.P. 260-270°C

Solvent: CCl_4

Assignments
a = 0.89
b = 1.28
c = 1.95
d = 3.98

2697

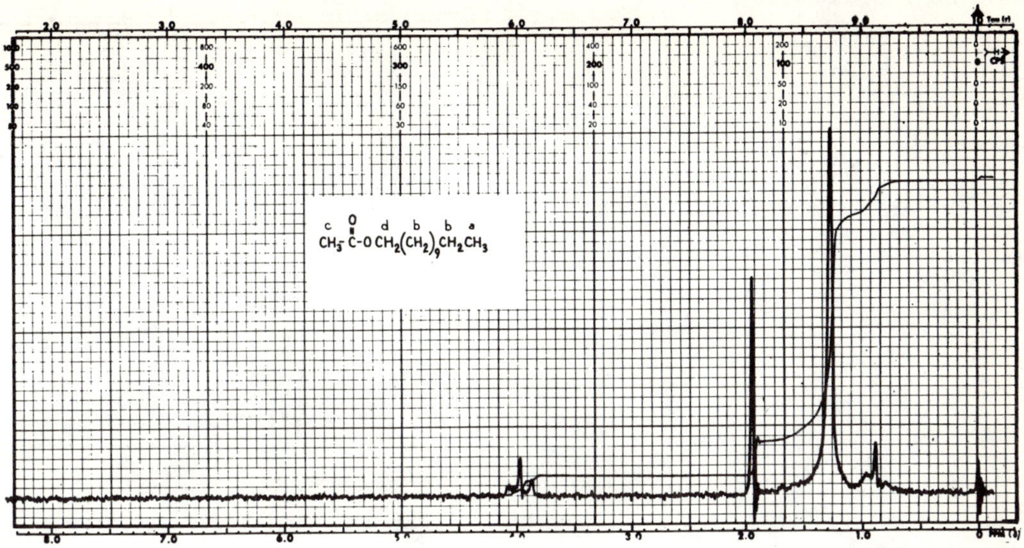

© 1978 Sadtler Research Laboratories, Inc.

PROPIONIC ACID, sec-BUTYL ESTER

$C_7H_{14}O_2$

Mol. Wt. 130.19

Solvent: $CDCl_3$

Assignments
a = 0.90
b = 1.16
c = 1.21
d = 1.55
e = 2.30
f = 4.85

2698

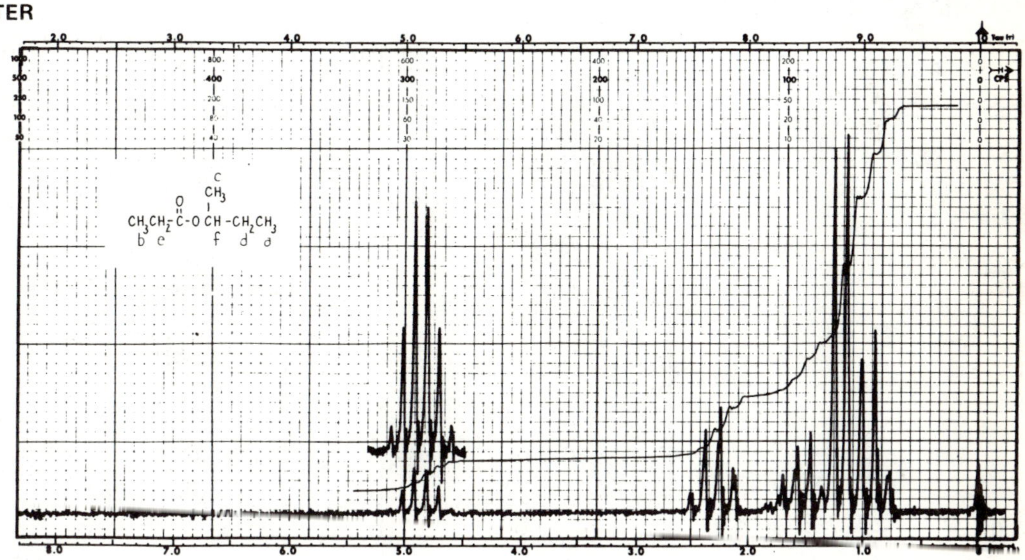

© 1978 Sadtler Research Laboratories, Inc.

ISOVALERIC ACID, METHYL ESTER

$C_6H_{12}O_2$

Mol. Wt. 116.16

B.P. 114-116°C

Solvent: CCl_4

Assignments
a = 0.96
b = 1.97
c = 2.12
d = 3.61

2699

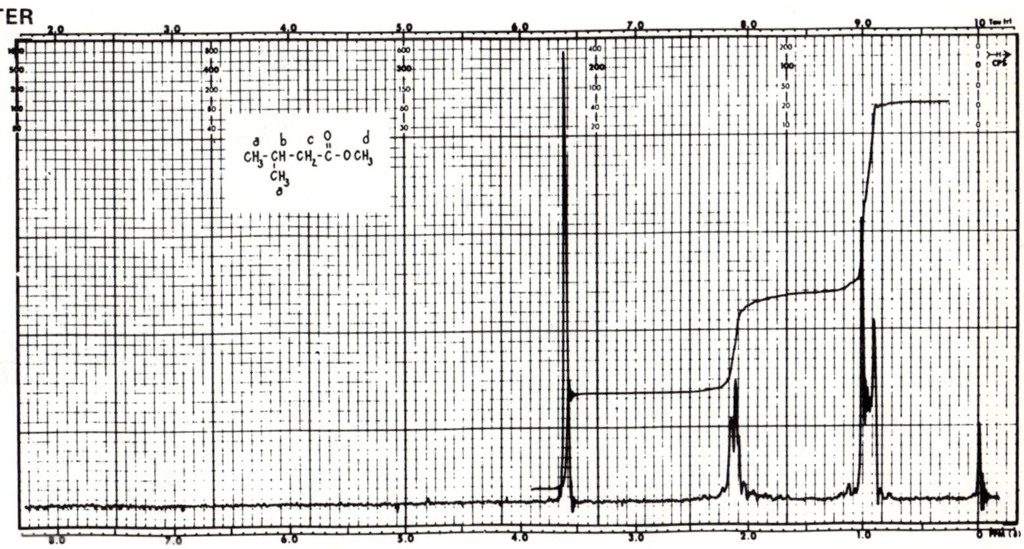

© 1978 Sadtler Research Laboratories, Inc.

BUTYRIC ACID, ISOPROPYL ESTER

$C_7H_{14}O_2$

Mol. Wt. 130.19

B.P. 133-135°C

Solvent: CCl_4

Assignments

a = 0.95
b = 1.22
c = 1.70
d = 2.18
e = 4.92

2700

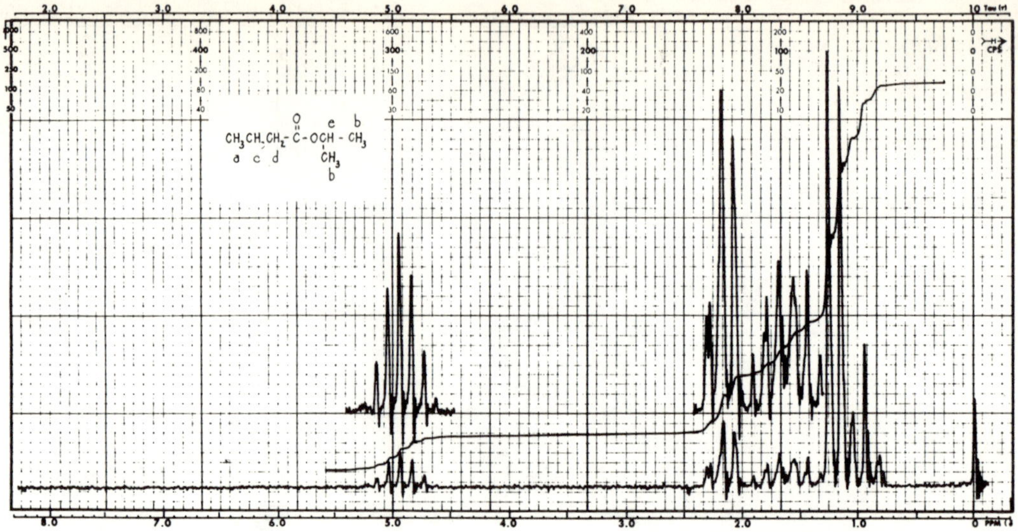

© 1978 Sadtler Research Laboratories, Inc.

VALERIC ACID, ETHYL ESTER

$C_7H_{14}O_2$

Mol. Wt. 130.19

B.P. 144-146°C

Solvent: CCl_4

Assignments

a = 0.98
b = 1.29
c = ca 1.54
d = 2.30
e = 4.19

2701

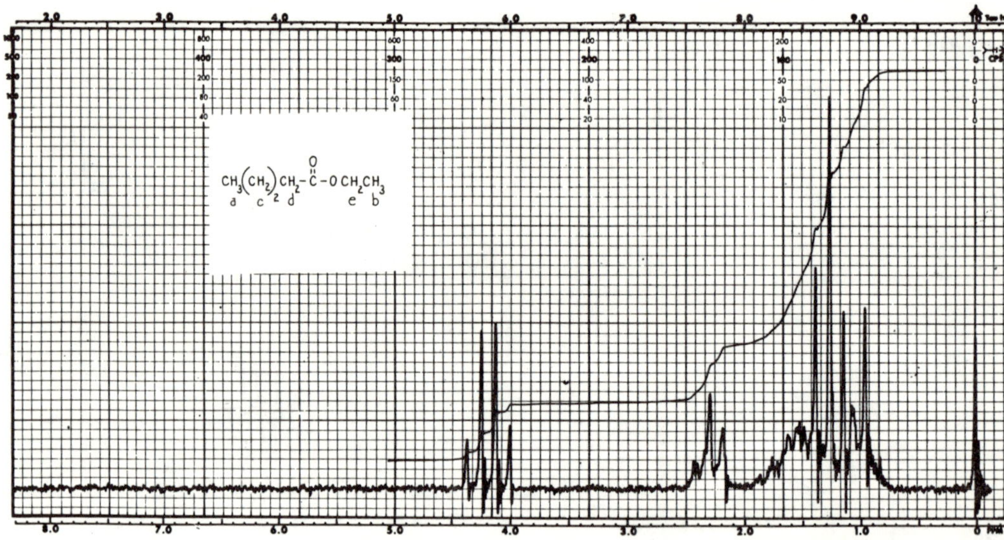

© 1978 Sadtler Research Laboratories, Inc.

ISOBUTYRIC ACID, BUTYL ESTER

$C_8H_{16}O_2$

Mol. Wt. 144.22

M.P. 149-151°C

Solvent: CCl_4

Assignments

a = 1.09
b = 1.13
c = 1.26-1.84
d = ca 2.44
e = 4.00

2702

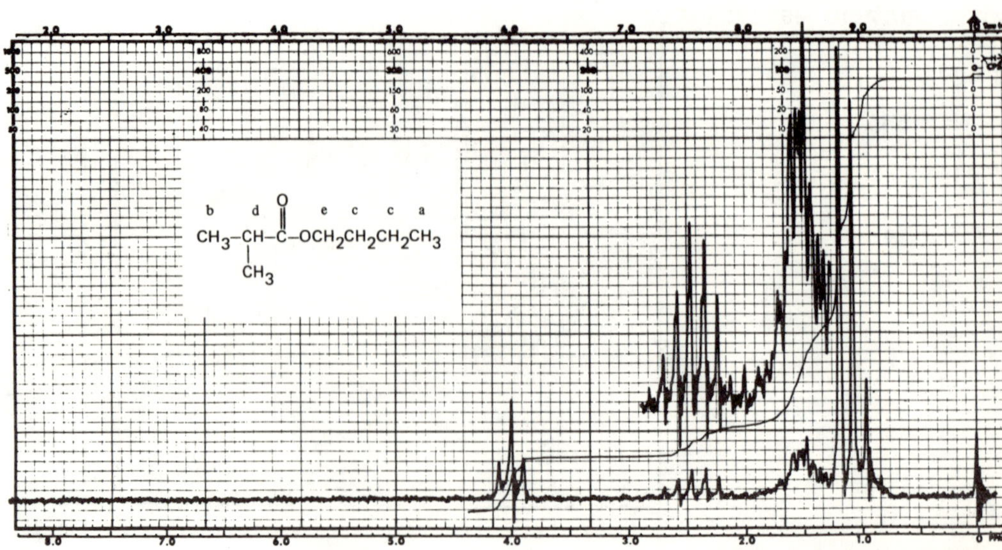

© 1978 Sadtler Research Laboratories, Inc.

PIVALIC ACID, ETHYL ESTER

$C_7H_{14}O_2$

Mol. Wt. 130.19

B.P. 116-117°C

Solvent: CCl_4

Assignments
a = 1.17
b = 1.23
c = 4.07

2703

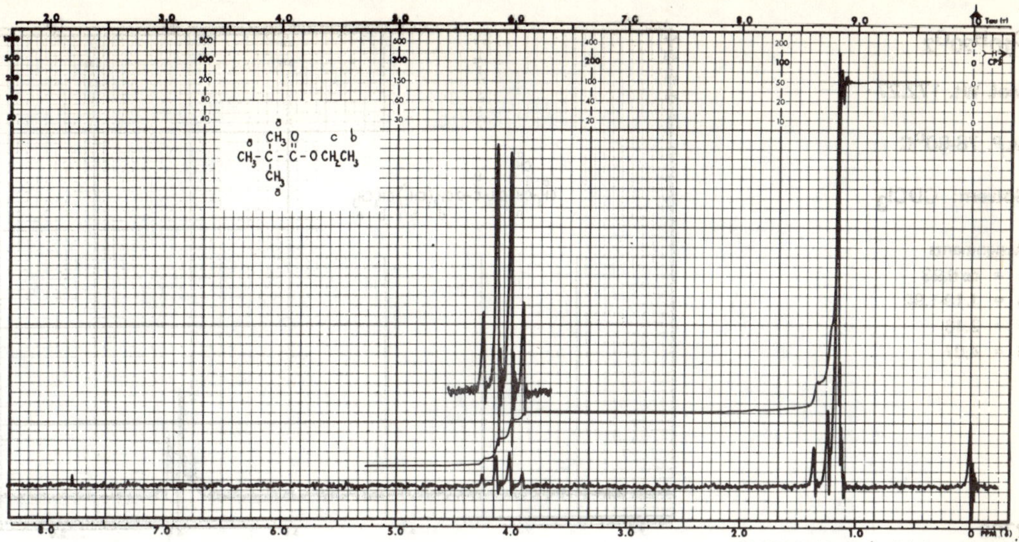

© 1978 Sadtler Research Laboratories, Inc.

OCTANOIC ACID, METHYL ESTER

$C_9H_{18}O_2$

Mol. Wt. 158.32

B.P. 61-63°C/5mm

Solvent: CCl_4

Assignments
a = 0.92
b = 1.33
c = 1.61
d = 2.25
e = 3.61

2704

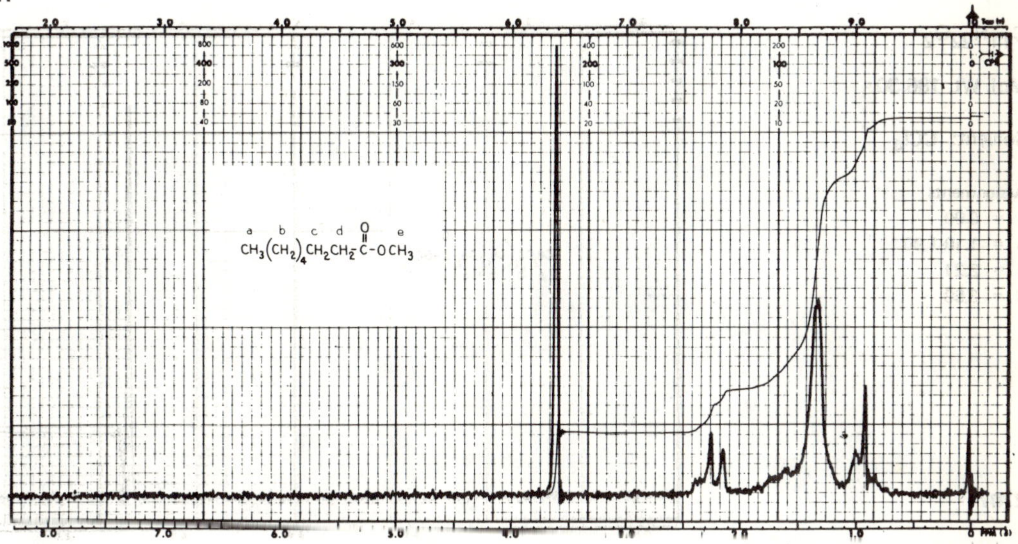

© 1978 Sadtler Research Laboratories, Inc.

VALERIC ACID, PENTYL ESTER

$C_{10}H_{20}O_2$

Mol. Wt. 172.27

B.P. 87-88°C/10mm

Solvent: CCl_4

Assignments
a = 0.92
b = 1.10-1.90
c = 2.22
d = 4.01

2705

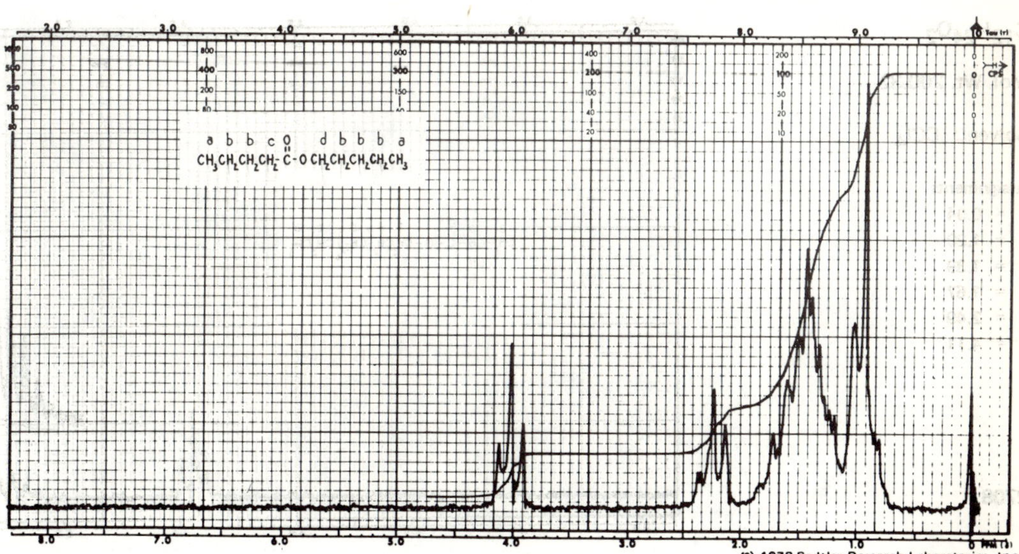

© 1978 Sadtler Research Laboratories, Inc.

ISOVALERIC ACID, PENTYL ESTER

$C_{10}H_{20}O_2$

Mol. Wt. 172.27

M.P. 78-80°C

Solvent: $CDCl_3$

Assignments
a = ca 0.93
b = 1.10-1.90
c = 2.18
d = 4.07

2706

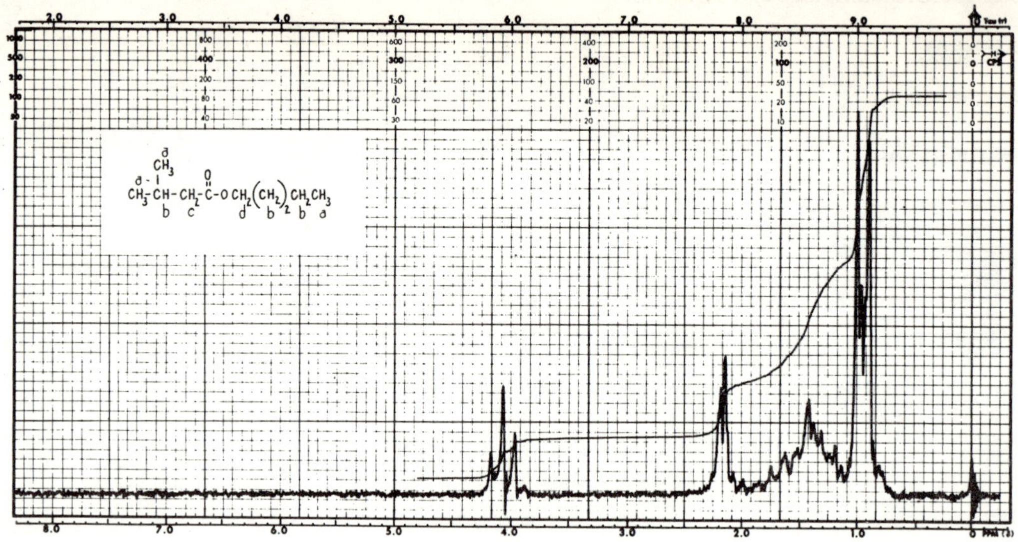

© 1978 Sadtler Research Laboratories, Inc.

DECANOIC ACID, METHYL ESTER

$C_{11}H_{22}O_2$

Mol. Wt. 186.30

Solvent: CCl_4

Assignments
a = 0.90
b = 1.10-1.90
c = 2.21
d = 3.59

2707

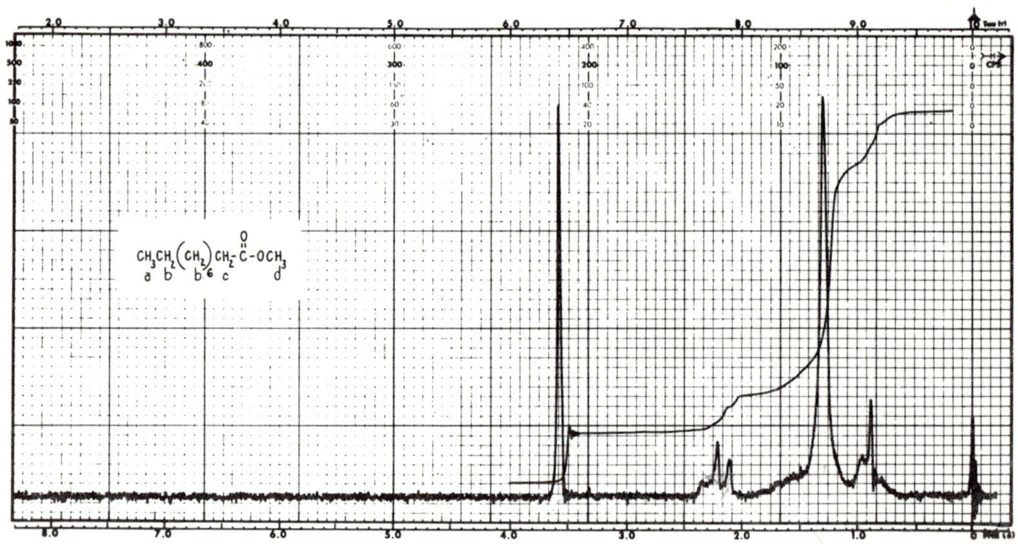

© 1978 Sadtler Research Laboratories, Inc.

2-METHYLPROPIONIC ACID, OCTYL ESTER

$C_{12}H_{24}O_2$

Mol. Wt. 200.32

Solvent: CCl_4

Assignments
a = 0.94
b = 1.20
c = 1.39
d = 1.67
e = 2.49
f = 4.11

2708

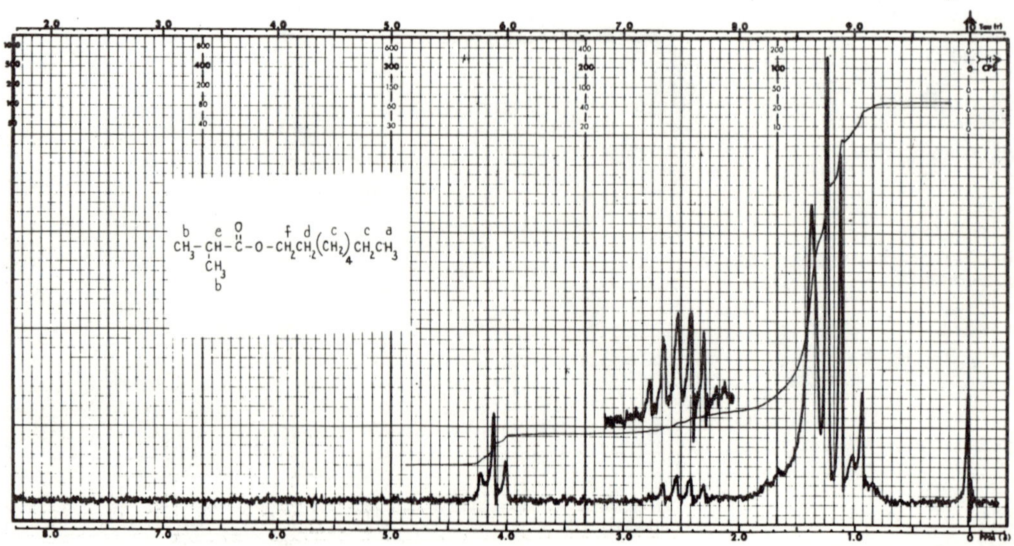

© 1978 Sadtler Research Laboratories, Inc.

HEPTANOIC ACID, ISOPENTYL ESTER

$C_{12}H_{24}O_2$

Mol. Wt. 200.32

Solvent: CCl_4

Assignments
a = 0.90
b = 0.96
c = 1.10-2.00
d = 2.20
e = 4.03

2709

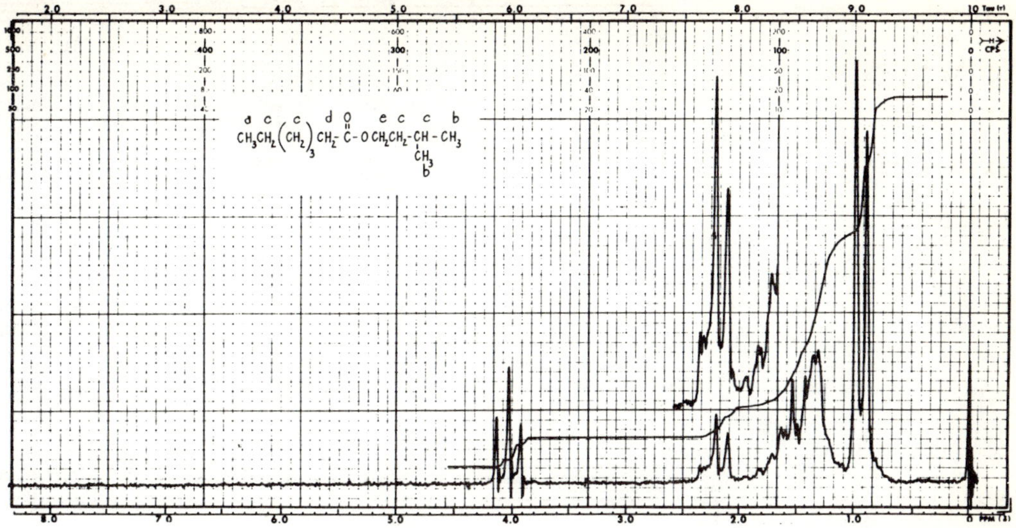

© 1978 Sadtler Research Laboratories, Inc.

HEPTADECANOIC ACID, METHYL ESTER

$C_{18}H_{36}O_2$

Mol. Wt. 248.49

M.P. 30°C

B.P. 184-187°C/9mm (lit.)

Solvent: CCl_4

Assignments
a = 0.89
b = 1.27
c = 2.23
d = 3.61
e = 2.09 imp.

2710

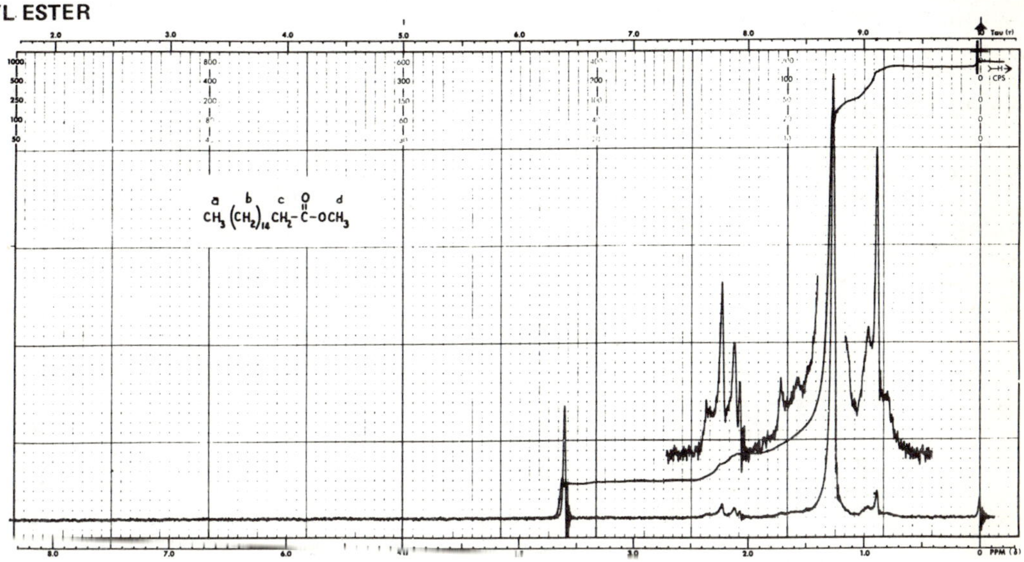

© 1978 Sadtler Research Laboratories, Inc.

LAURIC ACID, DODECYL ESTER

$C_{24}H_{48}O_2$

Mol. Wt. 368 .65

M.P. 27-29°C

Solve nt: $CDCl_3$

Assignments
a = 0.90
b = 1.10-2.00
c = 2.28
d = 4.05

2711

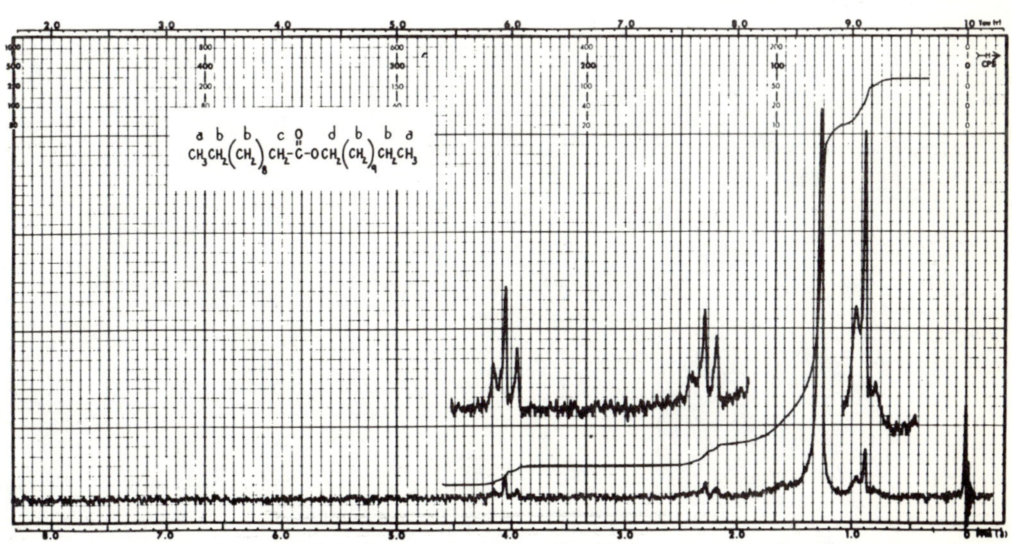

© 1978 Sadtler Research Laboratories, Inc.

TRICOSANOIC ACID, METHYL ESTER

$C_{24}H_{48}O_2$

Mol. Wt. 368.65

Solvent: $CDCl_3$

Assignments

a = 0.88
b = 1.27
c = ca 1.71
d = 2.30
e = 3.63

2712

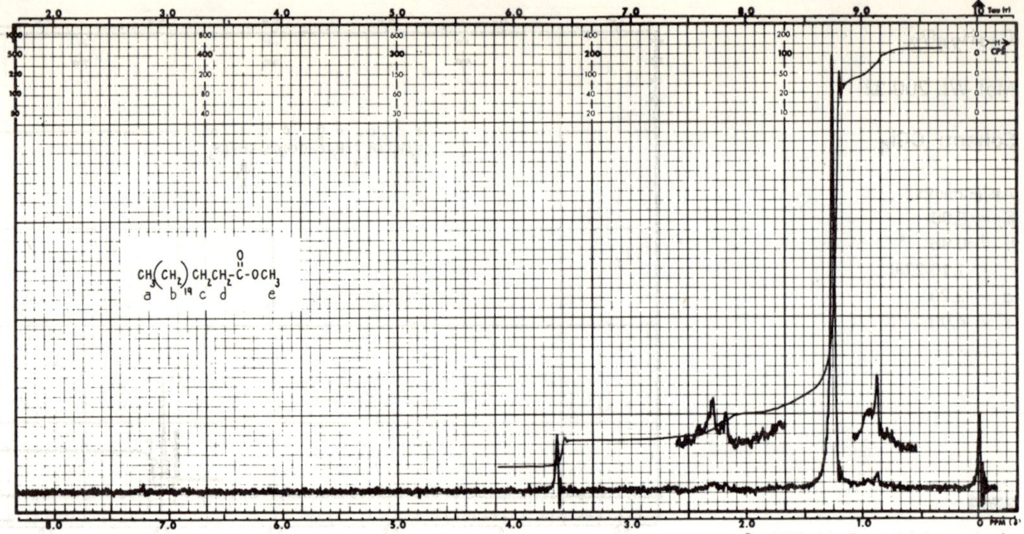

© 1978 Sadtler Research Laboratories, Inc.

ACETIC ACID, CYCLOOCTYL ESTER

$C_{10}H_{18}O_2$

Mol. Wt. 170.25

Solvent: CCl_4

Assignments

a = 1.30-2.05
b = 1.95
c = 4.83

2713

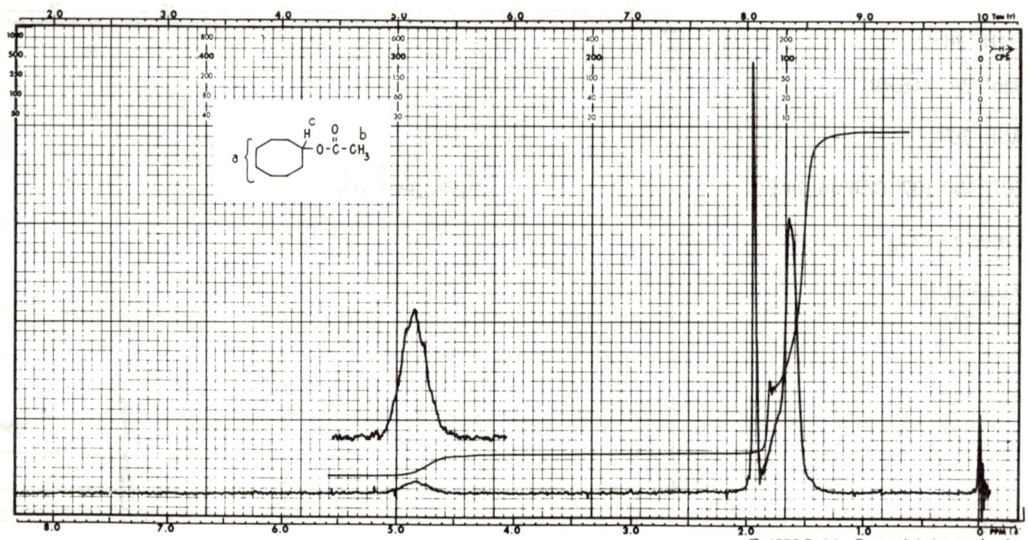

© 1978 Sadtler Research Laboratories, Inc.

BUTYRIC ACID, CYCLOHEXYL ESTER

$C_{10}H_{18}O_2$

Mol. Wt. 170.25

B.P. 86-89°C/10mm (lit.)

Solvent: CCl_4

Assignments

a = 0.93
b = 1.14-2.00
c = 2.20
d = 4.68

2714

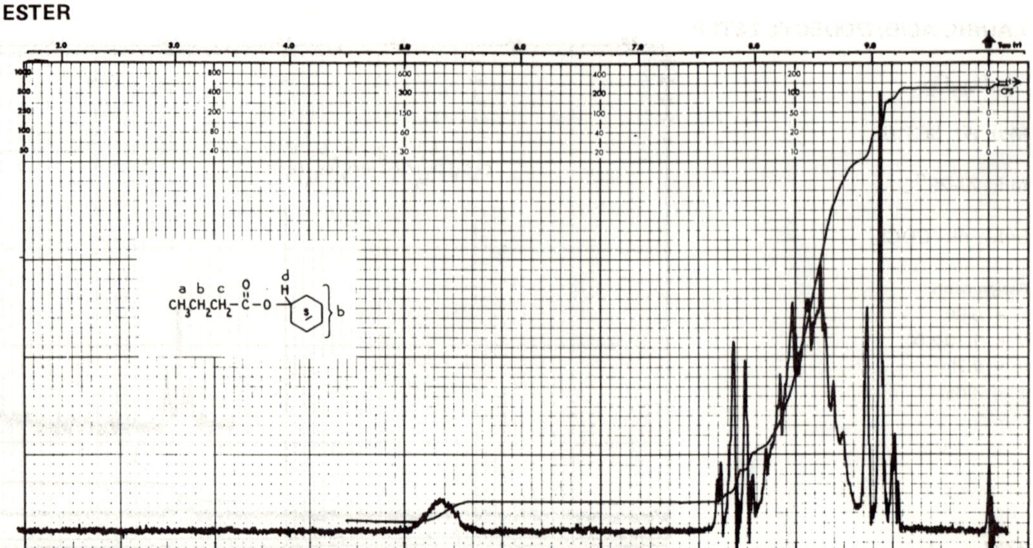

© 1978 Sadtler Research Laboratories, Inc.

ISOVALERIC ACID, CYCLOHEXYL ESTER

$C_{11}H_{20}O_2$

Mol. Wt. 184.28

Solvent: CCl_4

Assignments
a = 0.95
b = 1.10-2.50
c = ca 2.09
d = 4.70
e = 1.16 imp.

2715

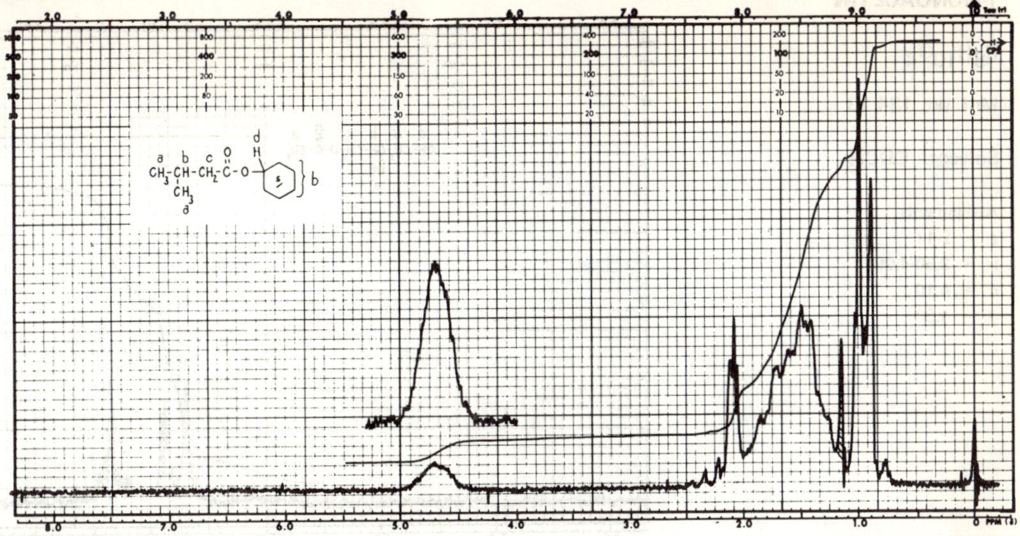

© 1978 Sadtler Research Laboratories, Inc.

ACETIC ACID, CYANOMETHYL ESTER

$C_4H_5NO_2$

Mol. Wt. 99.09

B.P. 71-72°C/10mm

Solvent: $CDCl_3$

Assignments
a = 2.14
b = 4.69

2716

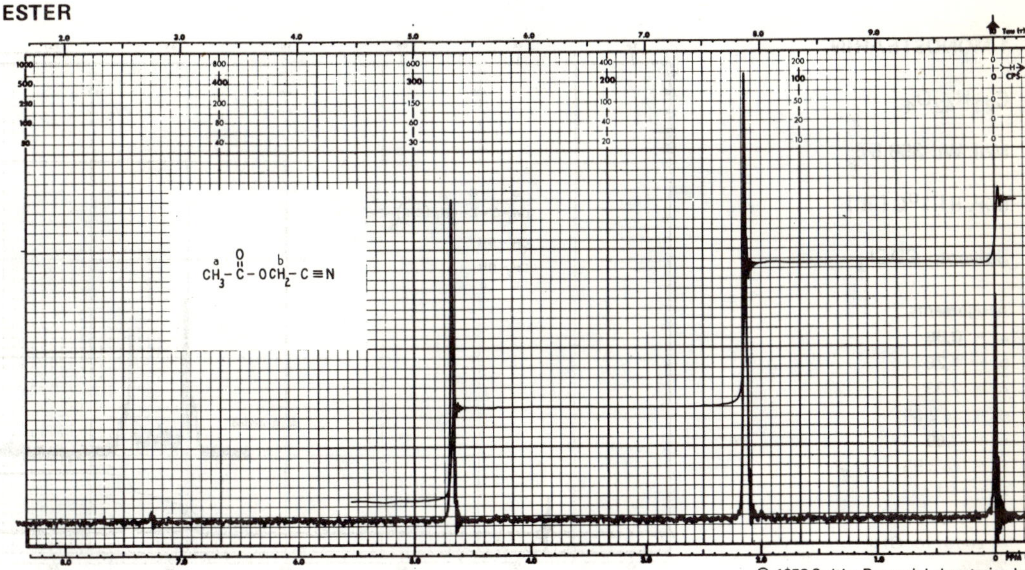

© 1978 Sadtler Research Laboratories, Inc.

2-(2-BUTOXYETHOXY)ETHANOL, ACETATE

$C_{10}H_{20}O_4$

Mol. Wt. 204.27

B.P. 235-250°C (lit.)

Solvent: CCl_4

Assignments
a = 0.92
b = 1.45
c = 1.98
d = 3.38
e = 3.50
f = 3.58
g = 4.12

2717

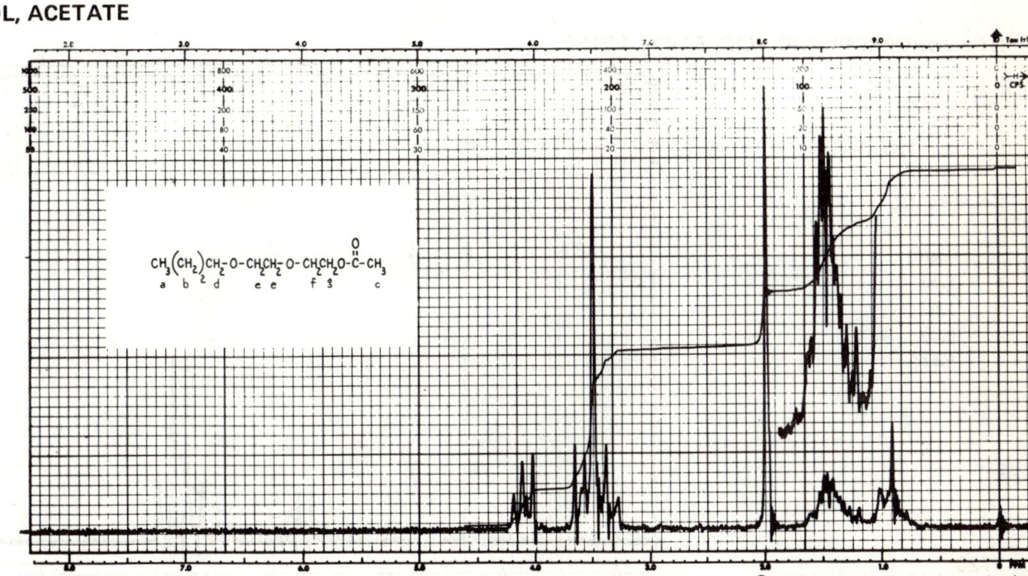

© 1978 Sadtler Research Laboratories, Inc.

1-MONOACETIN

$C_5H_{10}O_4$

Mol. Wt. 134.13

Solvent: $CDCl_3$

Assignments
a = 2.11
b = 3.50-4.00
c = 4.10
d = 4.30

2718

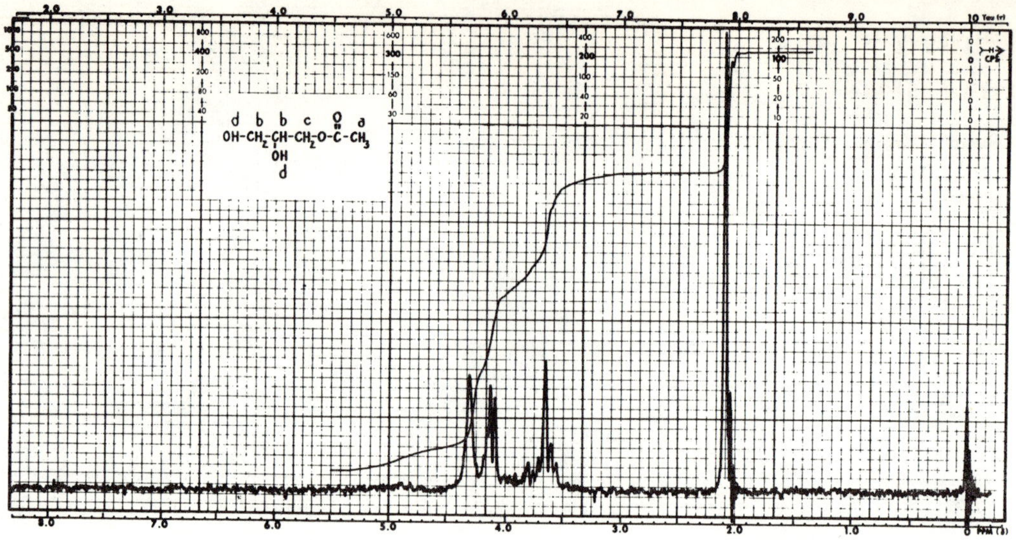

© 1978 Sadtler Research Laboratories, Inc.

1-MONOSTEARIN

$C_{21}H_{42}O_4$

Mol. Wt. 358.57

M.P. 81.1°C (lit.)

Solvent: $CDCl_3$

Assignments
a = 0.89
b = 1.26
c = 2.33
d = ca 2.38
e = 3.69
f = ca 3.93
g = 4.13
Trace of acid
added for OH

2719

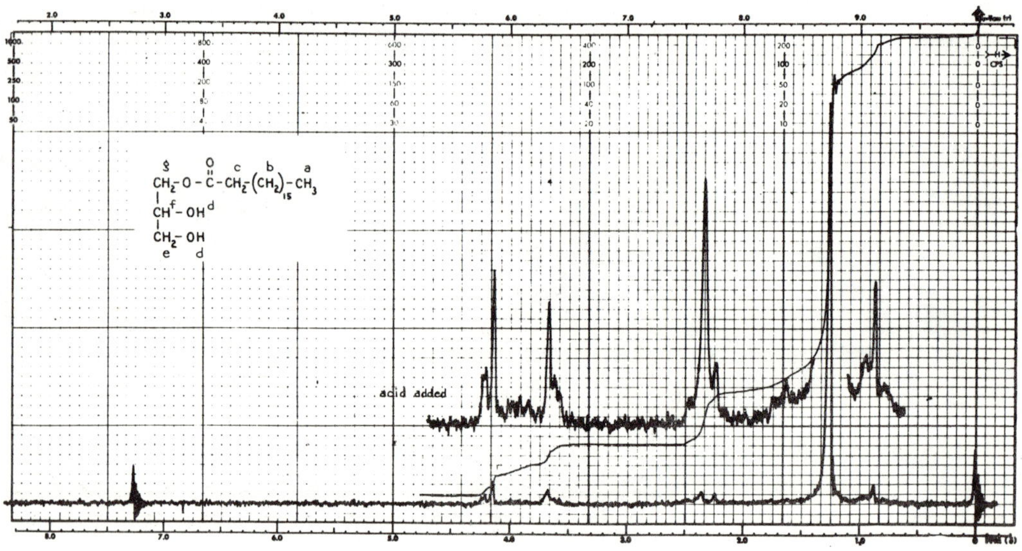

© 1978 Sadtler Research Laboratories, Inc.

2-FLUOROPROPIONIC ACID, ETHYL ESTER

$C_5H_9FO_2$

Mol. Wt. 120.12

Solvent: CCl_4

Assignments
a = 1.30
b = 1.50
c = 4.20
d = 4.87

2720

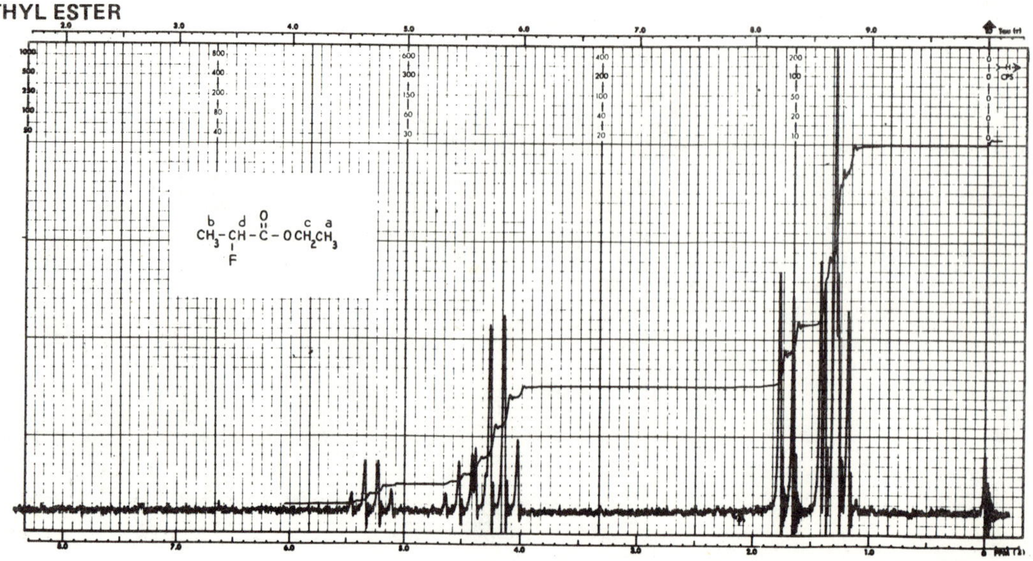

© 1978 Sadtler Research Laboratories, Inc.

TRIFLUOROACETIC ACID, PENTYL ESTER

$C_7H_{11}F_3O_2$

Mol. Wt. 184.16

Solvent: $CDCl_3$

Assignments
a = 0.90
b = ca 1.35
c = 1.75
d = 4.32

2721

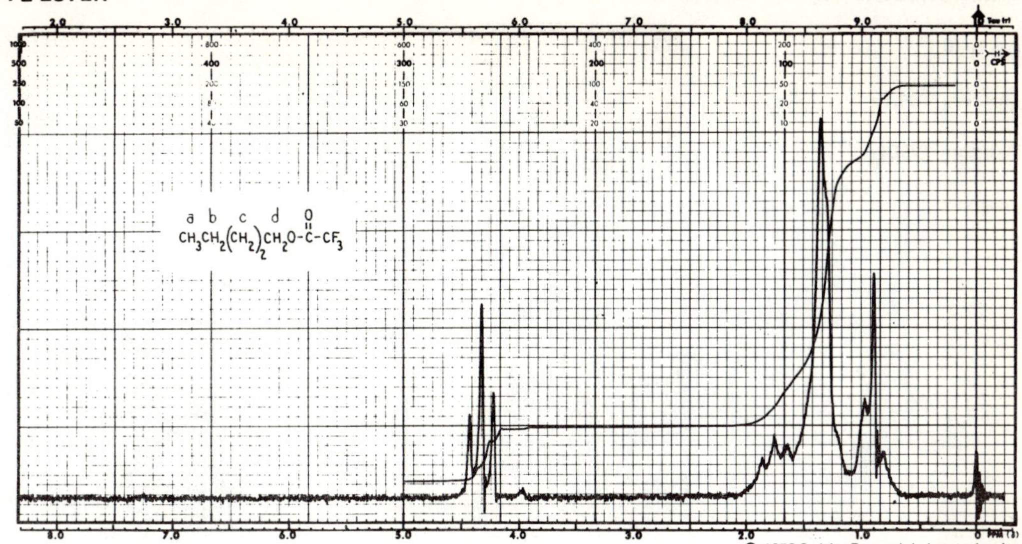

© 1978 Sadtler Research Laboratories, Inc.

TRICHLOROACETIC ACID, METHYL ESTER

$C_3H_3Cl_3O_2$

Mol. Wt. 177.42

Solvent: CCl_4

Assignments
a = 4.00

2722

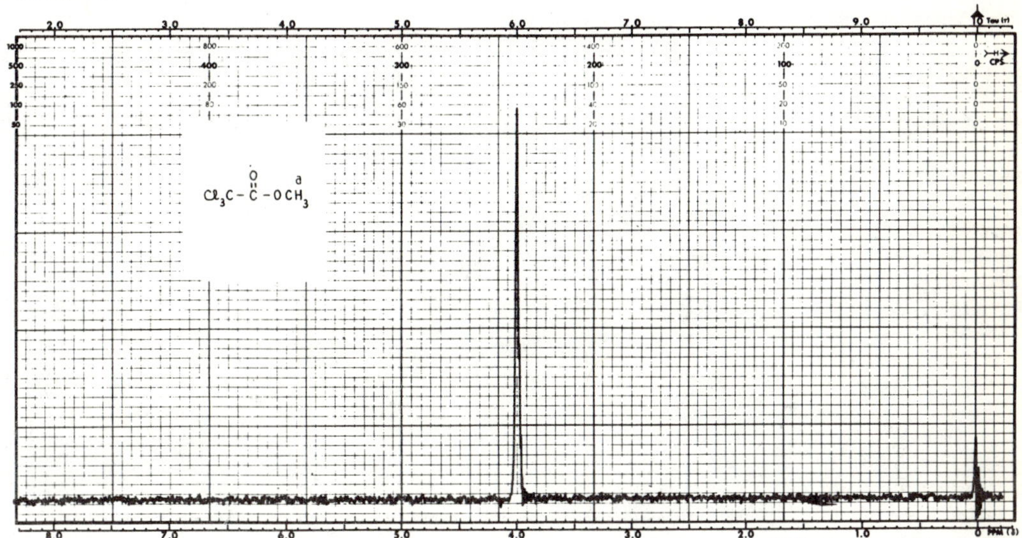

© 1978 Sadtler Research Laboratories, Inc.

TRICHLOROACETIC ACID, ETHYL ESTER

$C_4H_5Cl_3O_2$

Mol. Wt. 191.44

B.P. 53-55°C/10mm

Solvent: $CDCl_3$

Assignments
a = 1.41
b = 4.41

2723

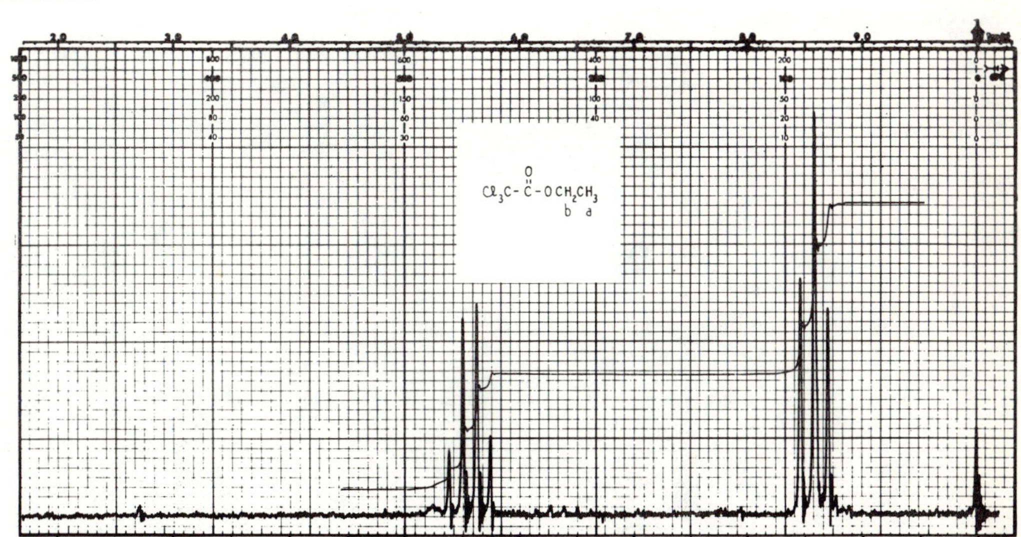

© 1978 Sadtler Research Laboratories, Inc.

3-CHLOROPROPIONIC ACID, ETHYL ESTER

$C_5H_9ClO_2$

Mol. Wt. 136.58

B.P. 83-85°C/55mm

Solvent: CCl_4

Assignments
a = 1.25
b = 2.71
c = 3.71
d = 4.14

2724

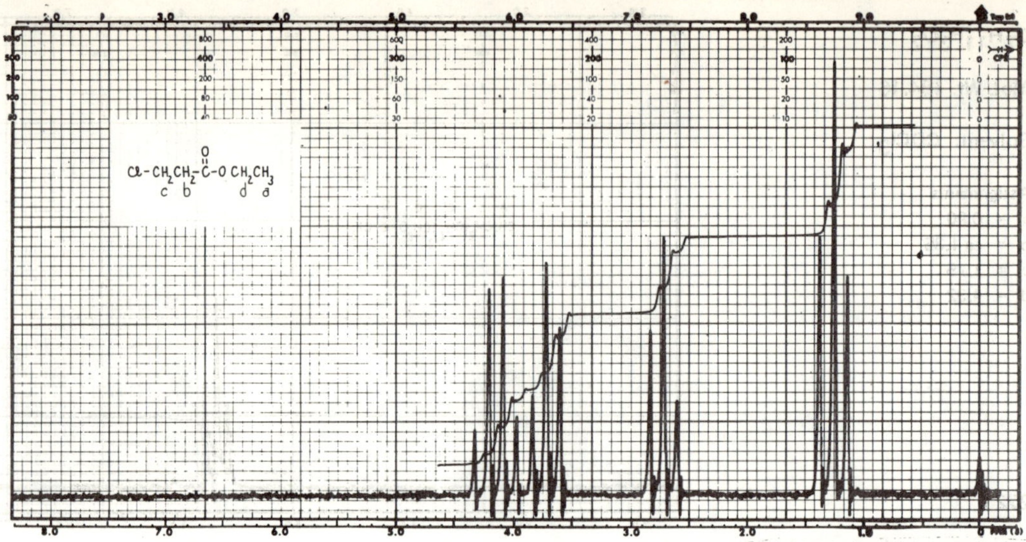

© 1978 Sadtler Research Laboratories, Inc.

3-CHLOROPROPIONIC ACID, ISOPROPYL ESTER

$C_6H_{11}ClO_2$

Mol. Wt. 150.16

Solvent: CCl_4

Assignments
a = 1.25
b = 2.70
c = 3.73
d = 5.05

2725

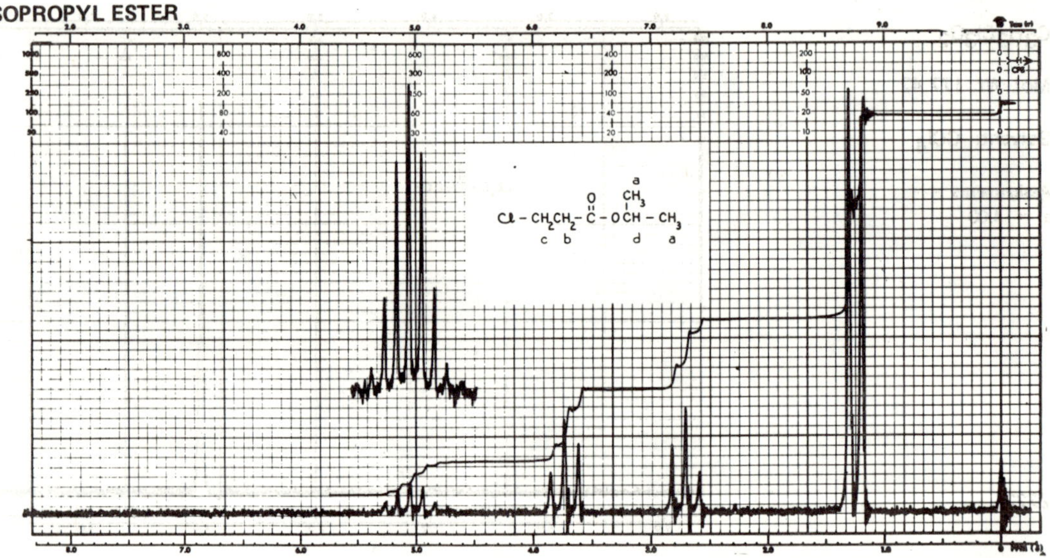

© 1978 Sadtler Research Laboratories, Inc.

2-(o-TOLUIDINO)ETHANOL

$C_4H_6Cl_2O_2$

Mol. Wt. 157.00

Solvent: CCl_4

Assignments
a = 3.50-4.15
b = 3.85
c = 4.38

2726

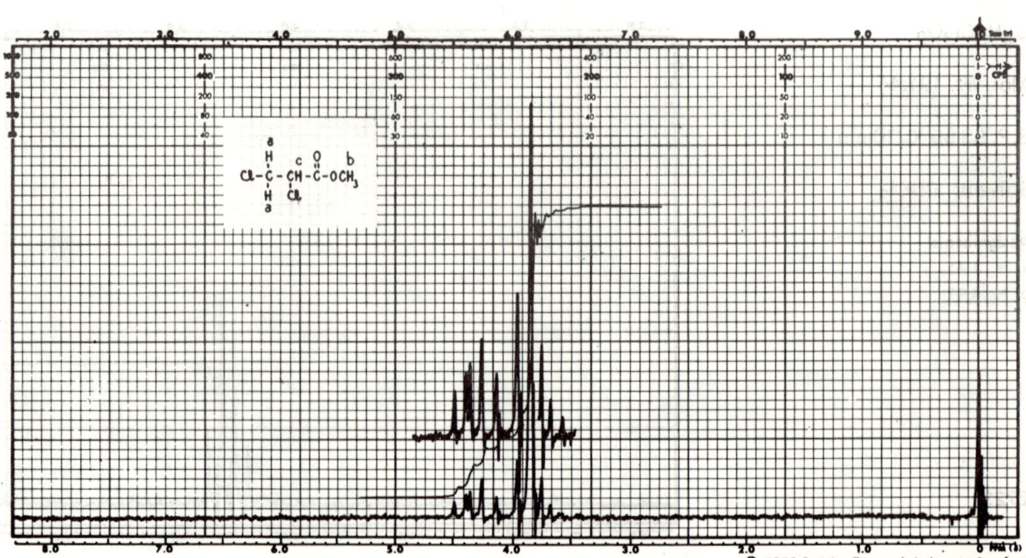

© 1978 Sadtler Research Laboratories, Inc.

6-CHLOROHEXANOIC ACID, ISOPROPYL ESTER

$C_9H_{17}ClO_2$

Mol. Wt. 192.69

B.P. 80-82°C/2.5mm

Solvent: CCl_4

Assignments
a = 1.21
b = ca 1.65
c = 2.23
d = 3.51
e = 4.92

2727

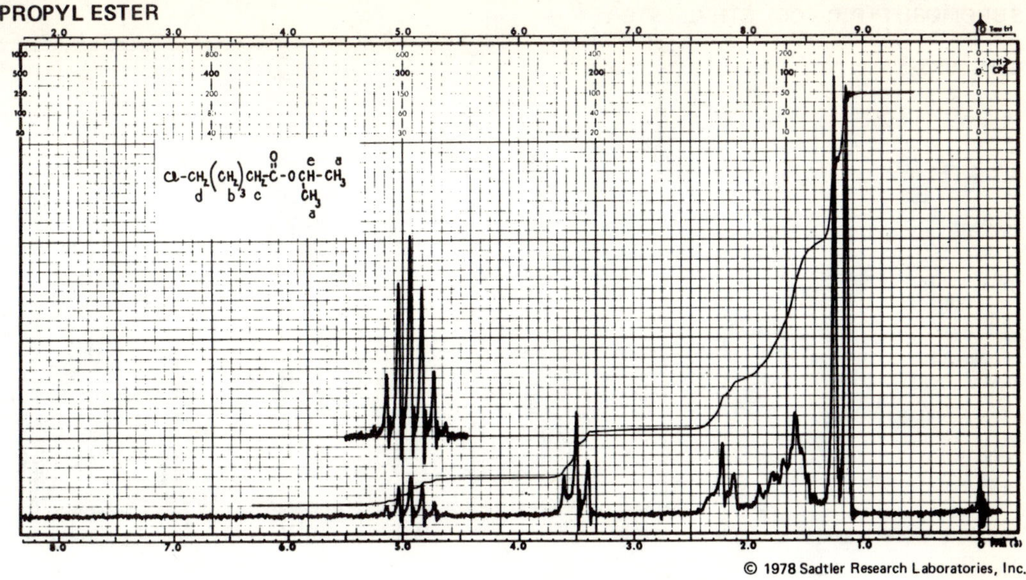

© 1978 Sadtler Research Laboratories, Inc.

BROMOACETIC ACID, METHYL ESTER

$C_3H_5BrO_2$

Mol. Wt. 152.98

B.P. 59-60°C/20mm (lit.)

Solvent: CCl_4

Assignments
a = 3.77

2728

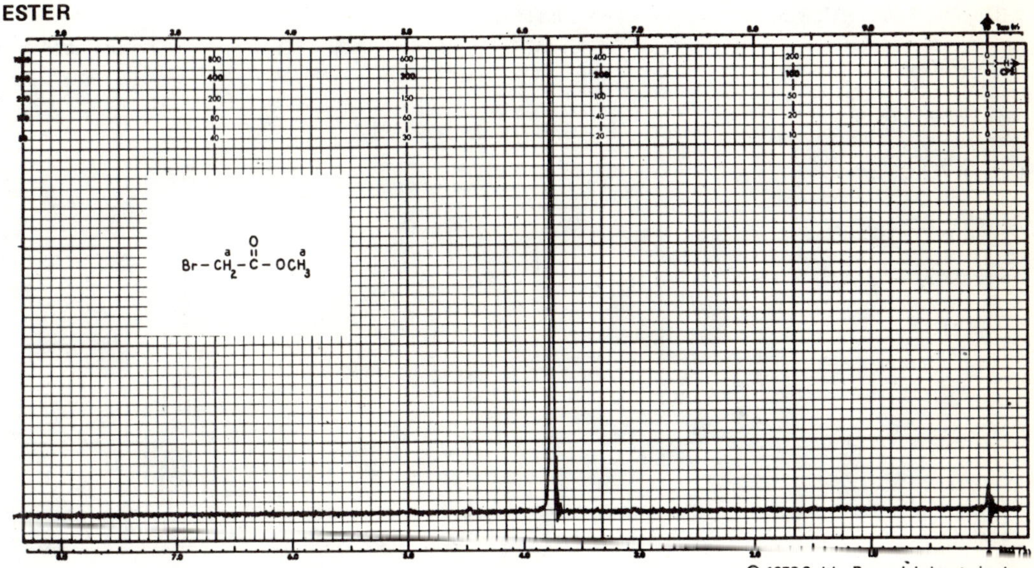

© 1978 Sadtler Research Laboratories, Inc.

2-BROMO-2-METHYLPROPIONIC ACID, ETHYL ESTER

$C_6H_{11}BrO_2$

Mol. Wt. 195.06

B.P. 65-67°C/11mm

Solvent: $CDCl_3$

Assignments
a = 1.31
b = 1.94
c = 4.21

2729

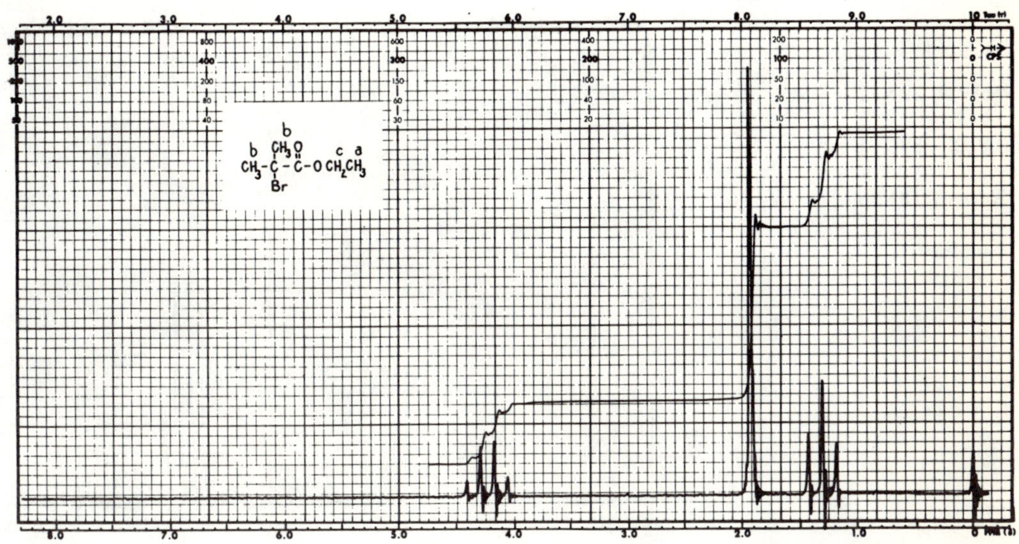

© 1978 Sadtler Research Laboratories, Inc.

3-BROMOBUTYRIC ACID, ETHYL ESTER

$C_6H_{11}BrO_2$

Mol. Wt. 195.06

B.P. 180-183°C

Solvent: CCl_4

Assignments
a = 1.29
b = 1.76
c = ca 2.82
d = 4.13
e = 4.40
f = 7.62 imp.

2730

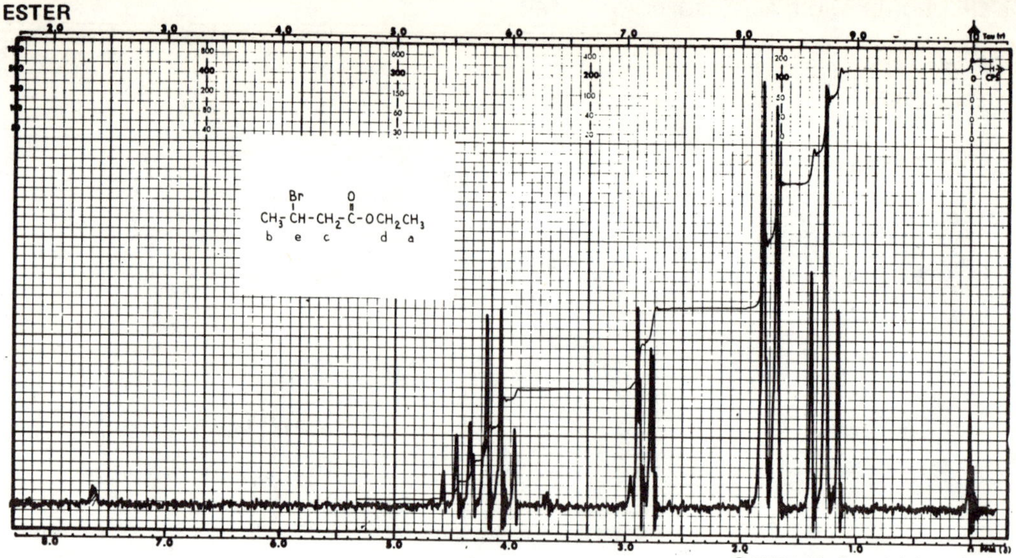

© 1978 Sadtler Research Laboratories, Inc.

2-BROMO-3-METHYLBUTYRIC ACID, ETHYL ESTER

$C_7H_{13}BrO_2$

Mol. Wt. 209.09

B.P. 75-78°C/12mm

Solvent: $CDCl_3$

Assignments
a = 1.02
b = 1.11
c = 1.30
d = 2.20
e = 4.05
f = 4.22

2731

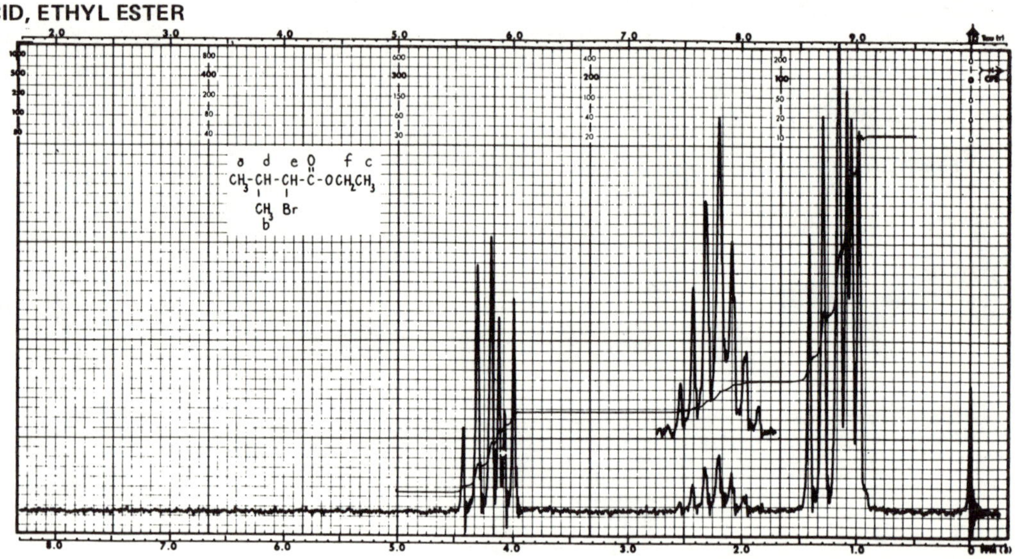

© 1978 Sadtler Research Laboratories, Inc.

2-BROMOHEXANOIC ACID, ETHYL ESTER

$C_8H_{15}BrO_2$

Mol. Wt. 223.12

B.P. 93-95°C/15mm

Solvent: $CDCl_3$

Assignments
a = 0.92
b = 1.30
c = 1.10-1.70
d = 2.04
e = 4.20
f = 4.22

2732

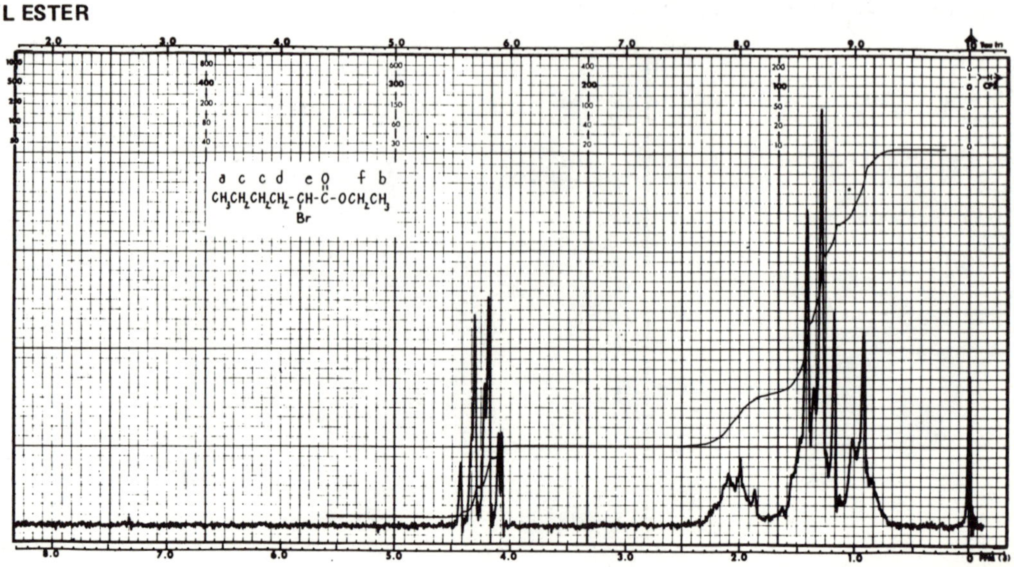

© 1978 Sadtler Research Laboratories, Inc.

2-BROMODECANOIC ACID, ETHYL ESTER

$C_{12}H_{23}BrO_2$

Mol. Wt. 279.22

Solvent: CCl_4

Assignments

a = 0.90
b = 1.29
c = ca 1.30
d = 1.93
e = 4.09
f = 4.19

2733

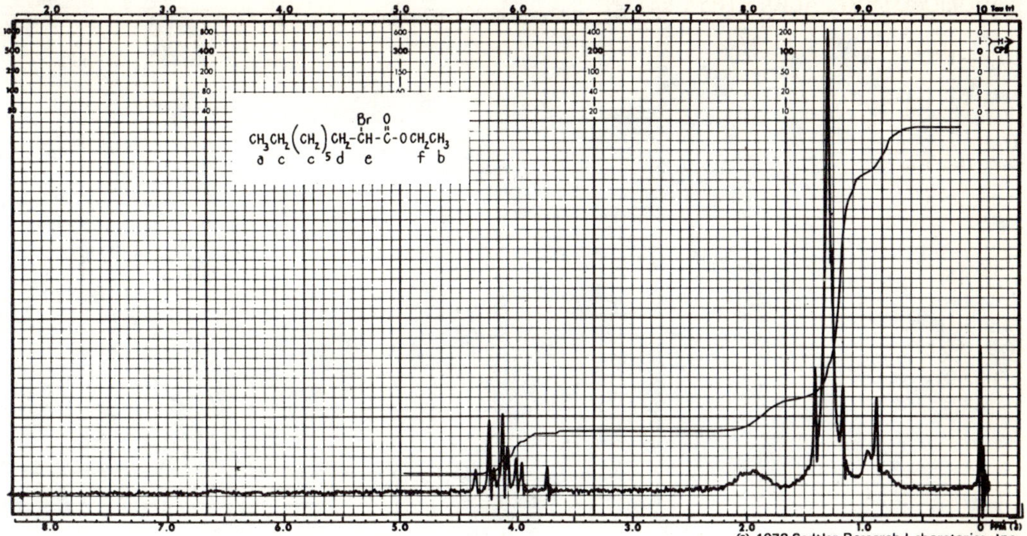

© 1978 Sadtler Research Laboratories, Inc.

IODOACETIC ACID, ETHYL ESTER

$C_4H_7IO_2$

Mol. Wt. 214.00

Solvent: CCl_4

Assignments

a = 1.29
b = 3.69
c = 4.14

2734

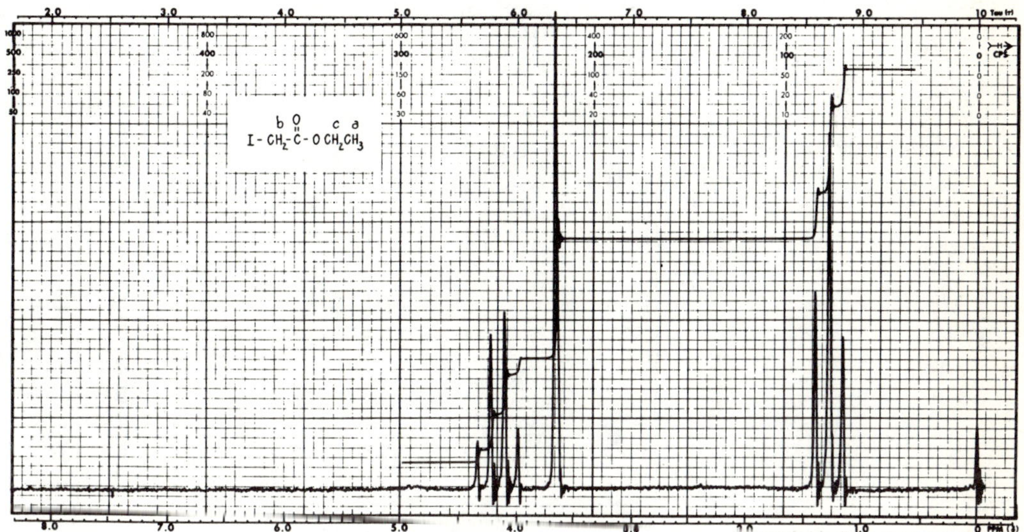

© 1978 Sadtler Research Laboratories, Inc.

DL-ALANINE, METHYL ESTER, HYDROCHLORIDE

$C_4H_9NO_2 \cdot HCl$

Mol. Wt. 139.58

Solvent: D_2O

Assignments

a = 0.57
b = 3.84
c = 4.20
d = 4.66 HDO

2735

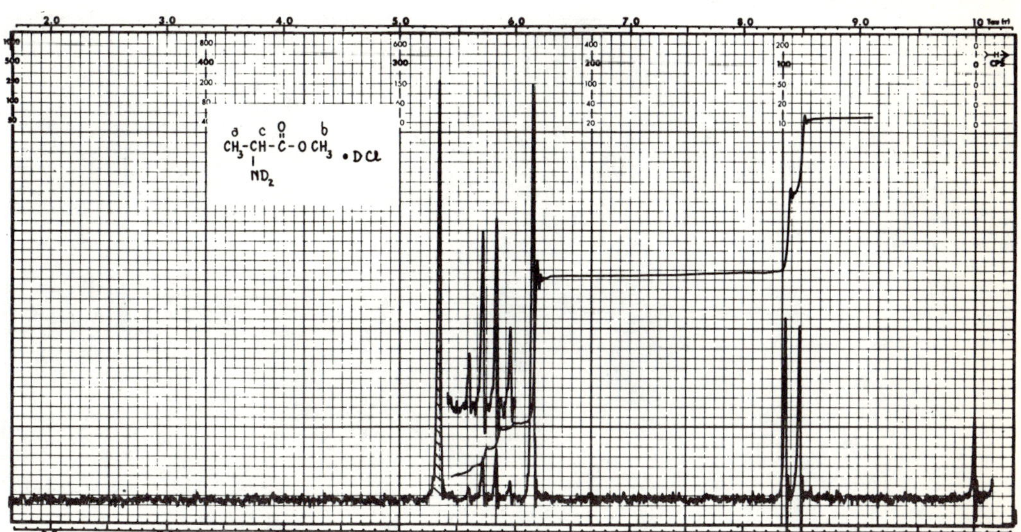

© 1978 Sadtler Research Laboratories, Inc.

DL-ALANINE, ETHYL ESTER, HYDROCHLORIDE

$C_5H_{11}NO_2 \cdot HCl$

Mol. Wt. 153.61

M.P. 83-85°C

Solvent: D_2O

Assignments
a = 1.30
b = 1.60
c = 4.21
d = 4.30
e = 4.49 HDO

2736

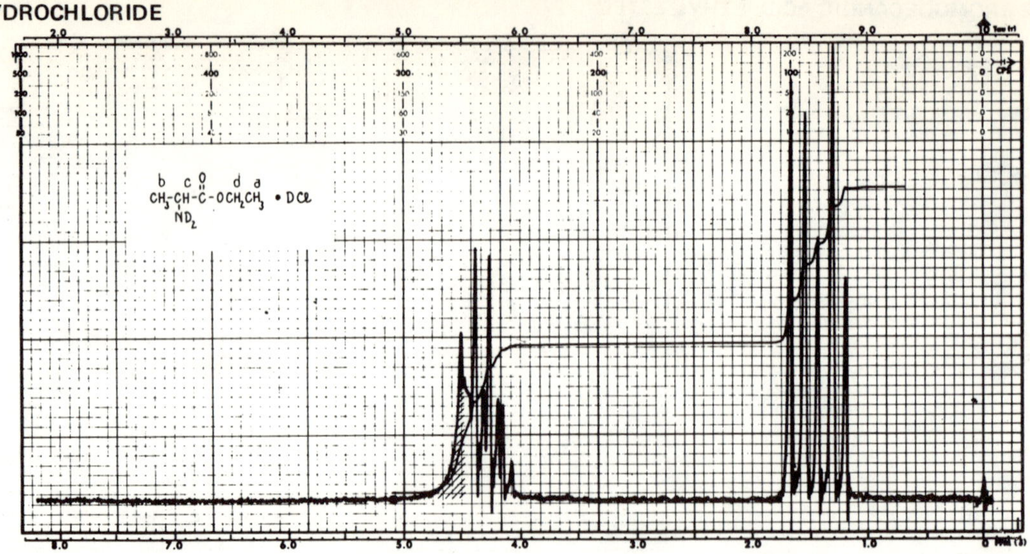

© 1978 Sadtler Research Laboratories, Inc.

3-CYANOPROPIONIC ACID, METHYL ESTER

$C_5H_7NO_2$

Mol. Wt. 113.12

B.P. 106-107°C/15mm

Solvent: CCl_4

Assignments
a = 2.62
b = 3.72

2737

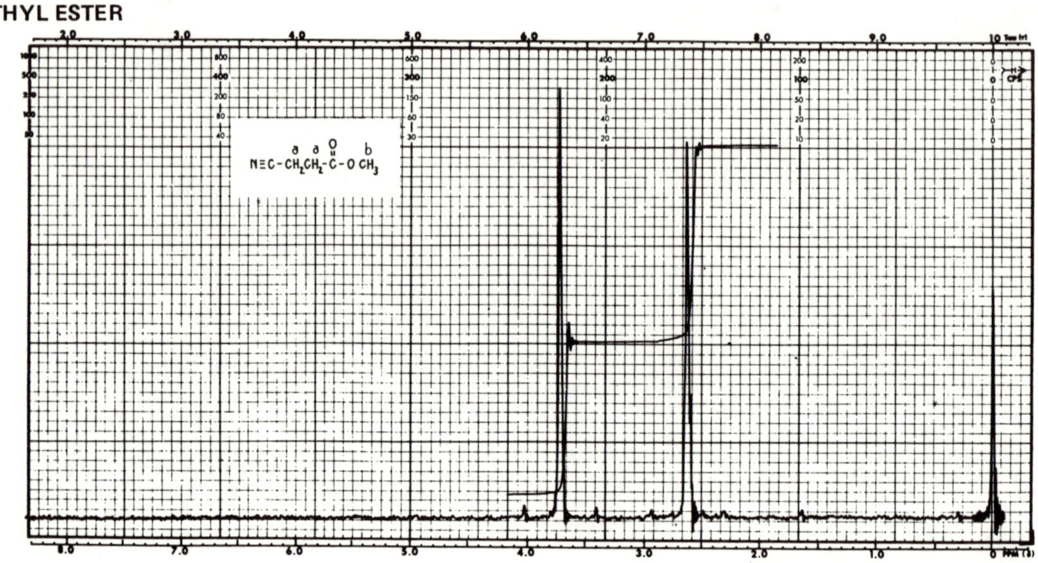

© 1978 Sadtler Research Laboratories, Inc.

3-MERCAPTOPROPIONIC ACID, METHYL ESTER

$C_4H_8O_2S$

Mol. Wt. 120.17

Solvent: $CDCl_3$

Assignments
a = 1.68
b = 2.40-3.00
c = 3.71

2738

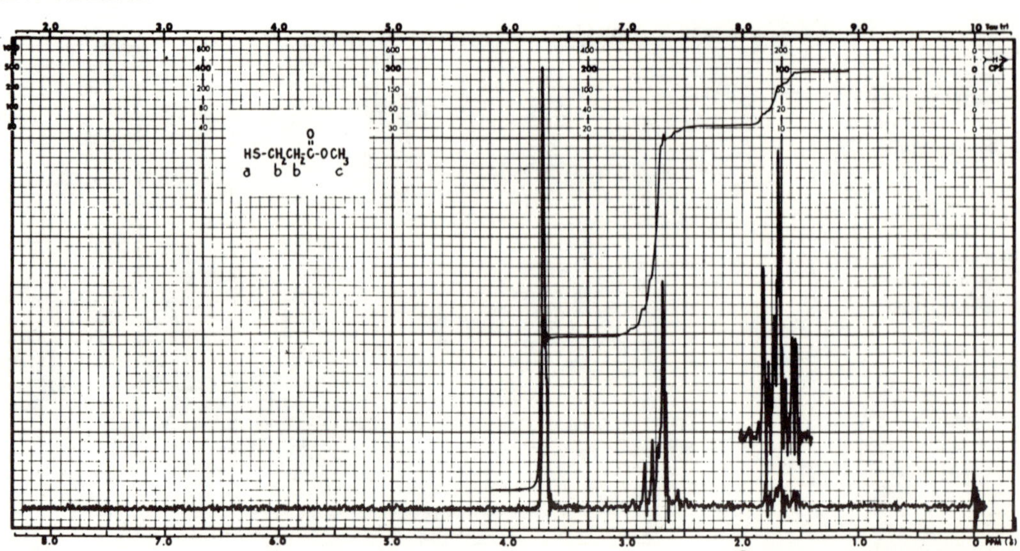

© 1978 Sadtler Research Laboratories, Inc.

METHOXYACETIC ACID, METHYL ESTER

$C_4H_8O_3$

Mol. Wt. 104.11

Solvent: CCl_4

Assignments
a = 3.38
b = 3.69
c = 3.90

2739

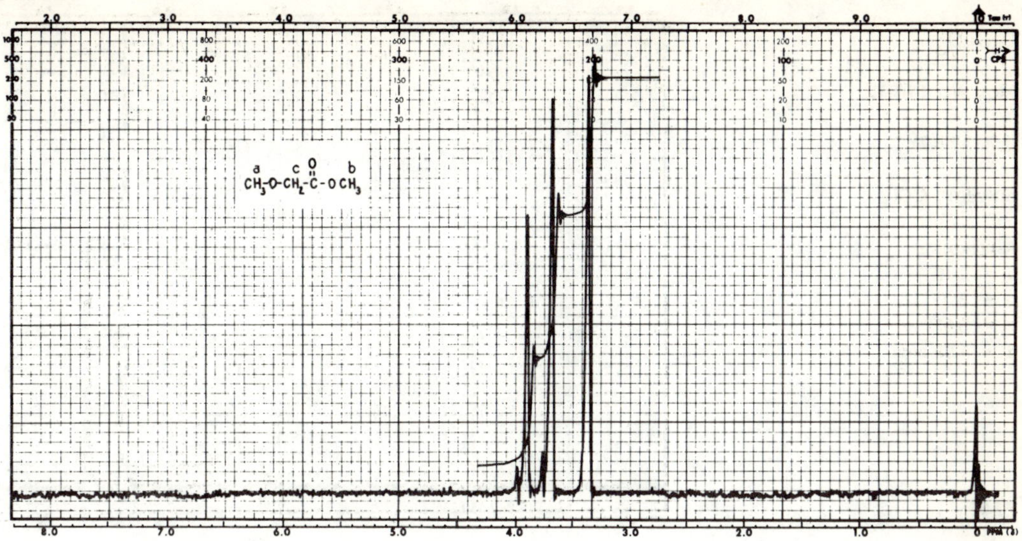

© 1978 Sadtler Research Laboratories, Inc.

DL-LACTIC ACID, METHYL ESTER

$C_4H_8O_3$

Mol. Wt. 104.11

B.P. 34-35°C/6mm

Solvent: CCl_4

Assignments
a = 1.36
b = 3.24
c = 3.74
d = 4.19

2740

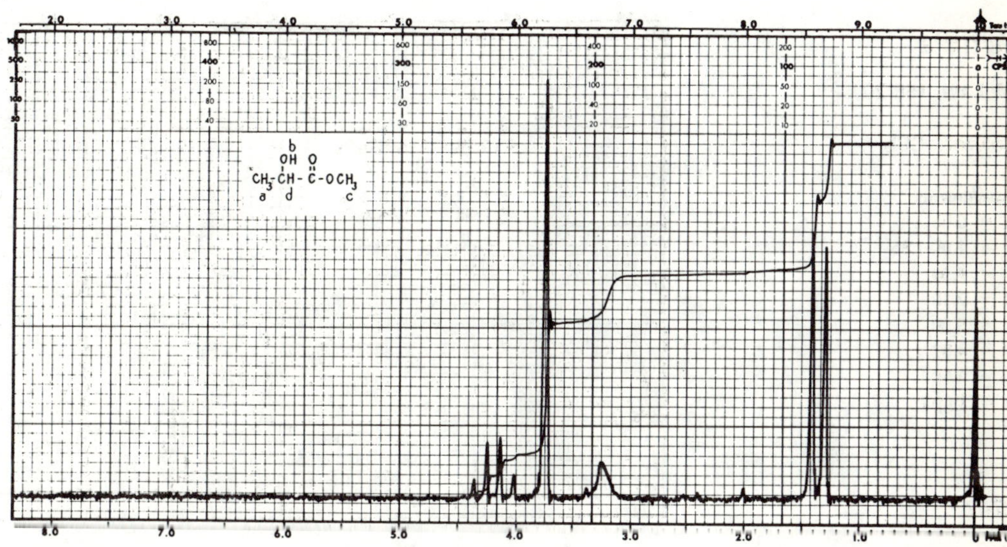

© 1978 Sadtler Research Laboratories, Inc.

2-HYDROXY-4-METHYLVALERIC ACID, ETHYL ESTER

$C_8H_{16}O_3$

Mol. Wt. 160.21

B.P. 80-82°C/10mm

Solvent: CCl_4

Assignments
a = 0.94
b = 1.30
c = ca 1.49
d = ca 1.80
e = 2.72
f = 4.04
g = 4.20
h = 2.09 imp.

2741

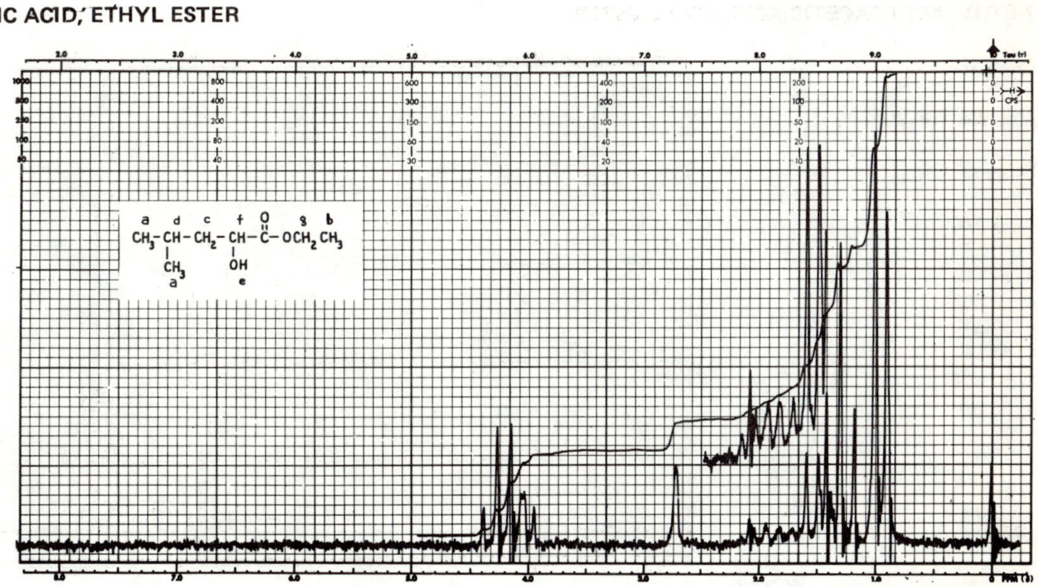

© 1978 Sadtler Research Laboratories, Inc.

ACETOACETIC ACID, METHYL ESTER

$C_5H_8O_3$

Mol. Wt. 116.12

B.P. 171.7°C

Solvent: CCl_4

Assignments
a = 2.19
b = 3.37
c = 3.67
d = 1.92 imp.

2742

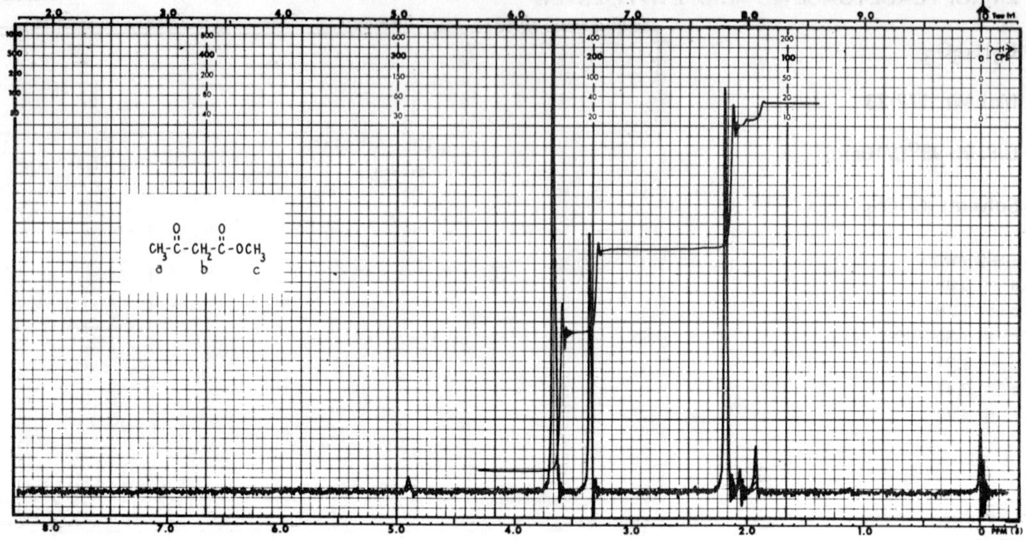

© 1978 Sadtler Research Laboratories, Inc.

ACETOACETIC ACID, BUTYL ESTER

$C_8H_{14}O_3$

Mol. Wt. 158.20

Solvent: CCl_4

Assignments
a = 0.95
b = 1.00-1.85
c = 1.90
d = 2.19
e = 3.30
f = 4.07
g = 4.88
h = 11.93

2743

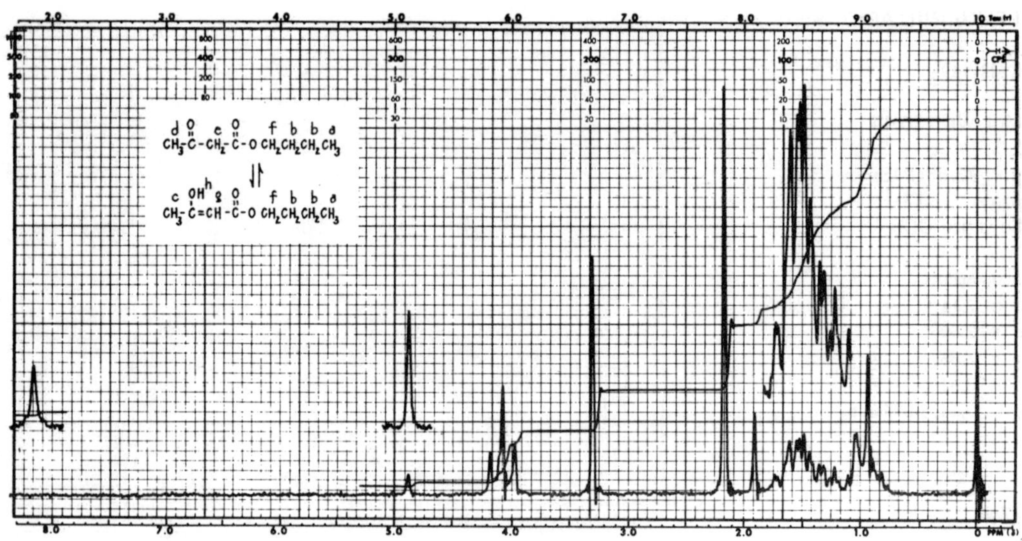

© 1978 Sadtler Research Laboratories, Inc.

2-ETHYLACETOACETIC ACID, ETHYL ESTER

$C_8H_{14}O_3$

Mol. Wt. 158.20

Solvent: CCl_4

Assignments
a = 0.93
b = 1.29
c = 1.82
d = 2.12
e = 3.20
f = 4.14
g = 3.70 imp.

2744

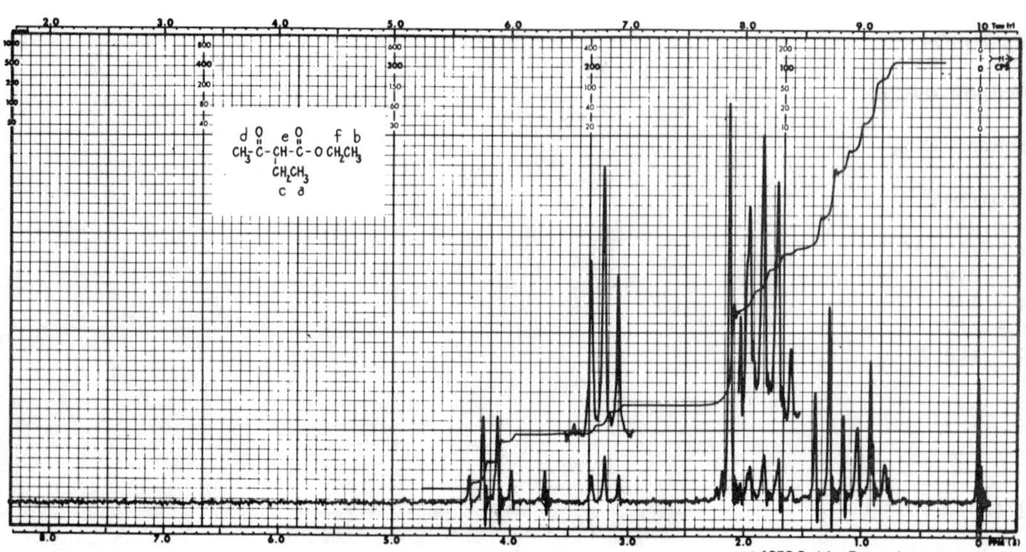

© 1978 Sadtler Research Laboratories, Inc.

2-PROPYLACETOACETIC ACID, ETHYL ESTER

$C_9H_{16}O_3$

Mol. Wt. 172.23

B.P. 95-98°C/10mm

Solvent: $CDCl_3$

Assignments
a = 0.95
b = 1.29
c = 1.30-2.10
d = 2.21
e = 3.47
f = 4.20

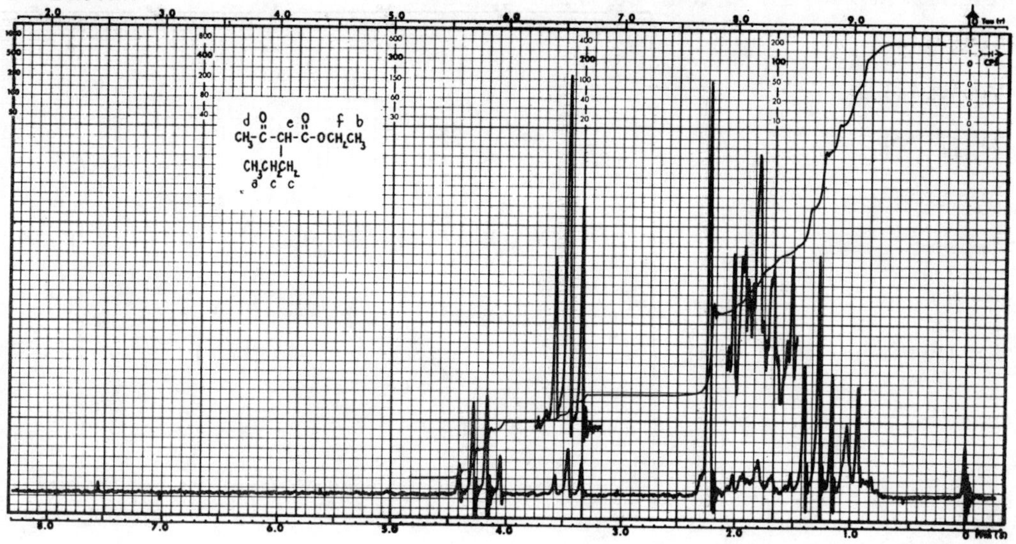

2745

© 1978 Sadtler Research Laboratories, Inc.

LEVULINIC ACID, METHYL ESTER

$C_6H_{10}O_3$

Mol. Wt. 130.14

B.P. 93.5°C/18mm

Solvent: CCl_4

Assignments
a = 2.11
b = 2.50
c = 2.66
d = 3.61
e = 3.34, 5.61 imps.

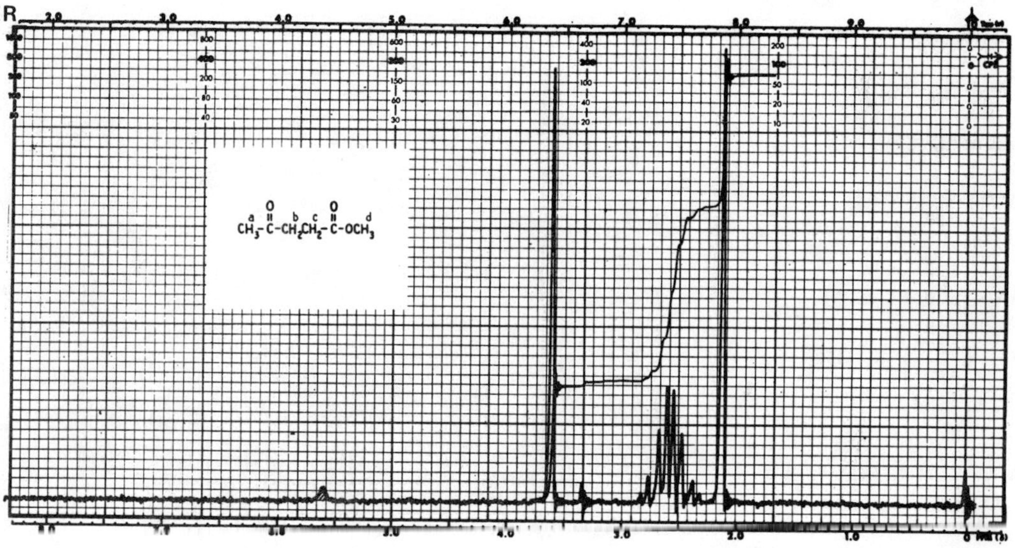

2746

© 1978 Sadtler Research Laboratories, Inc.

LEVULINIC ACID, ISOPENTYL ESTER

$C_{10}H_{18}O_3$

Mol. Wt. 186.25

Solvent: CCl_4

Assignments
a = 0.93
b = 1.25-1.95
c = 2.10
d = 2.45
e = 2.62
f = 4.03

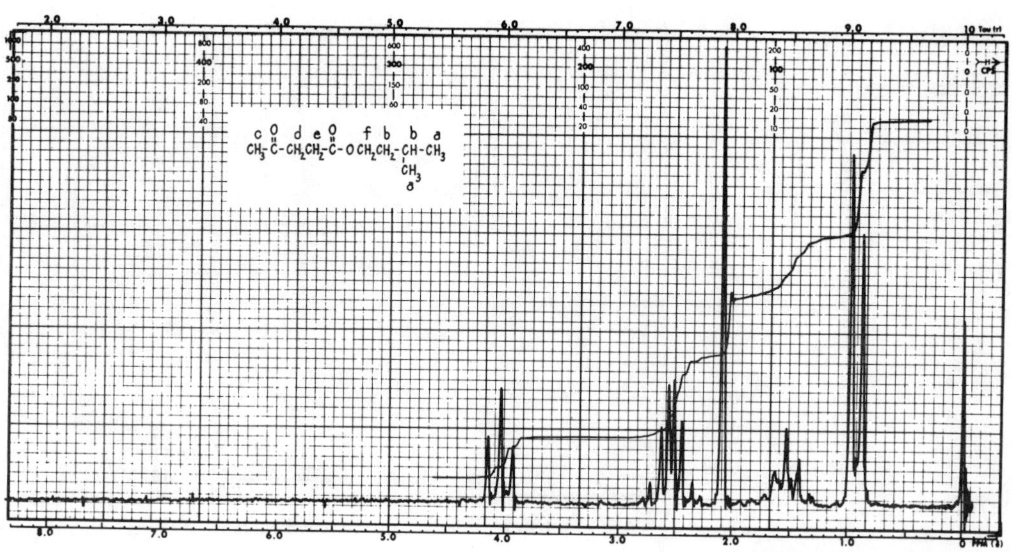

2747

© 1978 Sadtler Research Laboratories, Inc.

6-(CHLOROFORMYL)HEXANOIC ACID, METHYL ESTER

$C_8H_{13}ClO_3$

Mol. Wt. 192.64

Solvent: $CDCl_3$

Assignments
a = 1.10-2.10
b = 2.33
c = 2.94
d = 3.67

2748

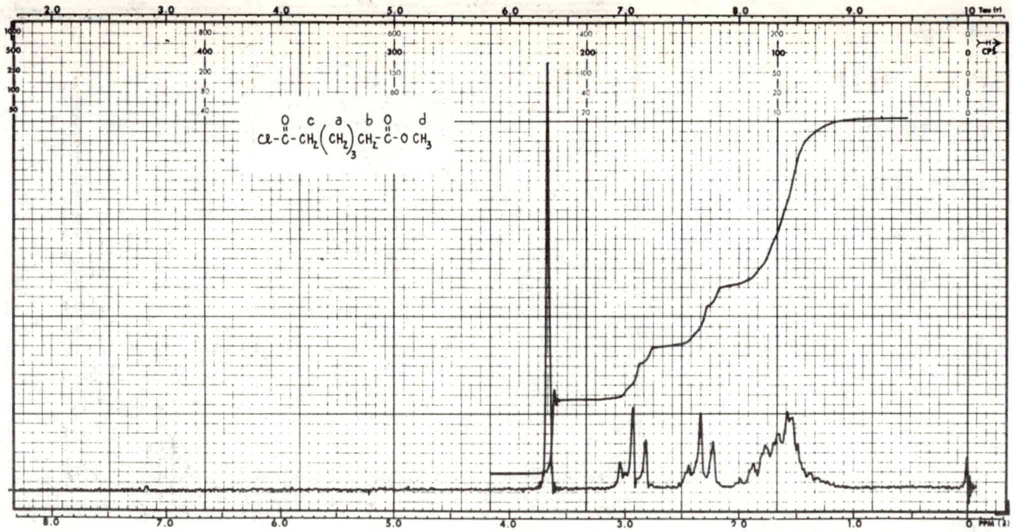

© 1978 Sadtler Research Laboratories, Inc.

SEBACIC ACID, MONOETHYL ESTER

$C_{12}H_{22}O_4$

Mol. Wt. 230.31

M.P. 34-36°C

Solvent: $CDCl_3$

Assignments
a = 1.27
b = 1.10-1.55
c = 1.63
d = ca 2.32
e = 4.13
f = 11.60

2749

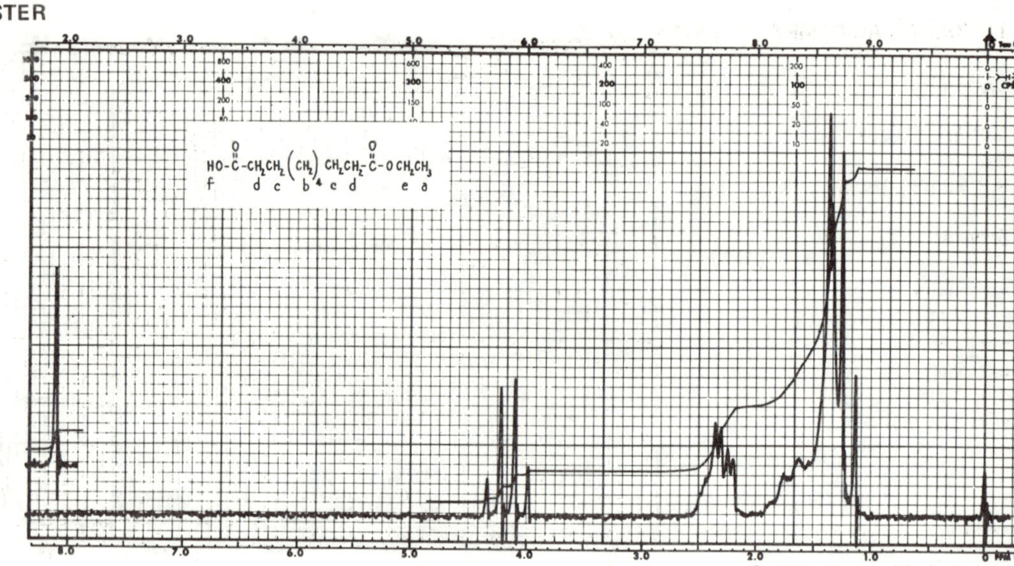

© 1978 Sadtler Research Laboratories, Inc.

MALONIC ACID, ETHYL ESTER, POTASSIUM SALT

$C_5H_7KO_4$

Mol. Wt. 170.21

Solvent: D_2O

Assignments
a = 1.26
b = 4.19
c = 4.55 HDO

2750

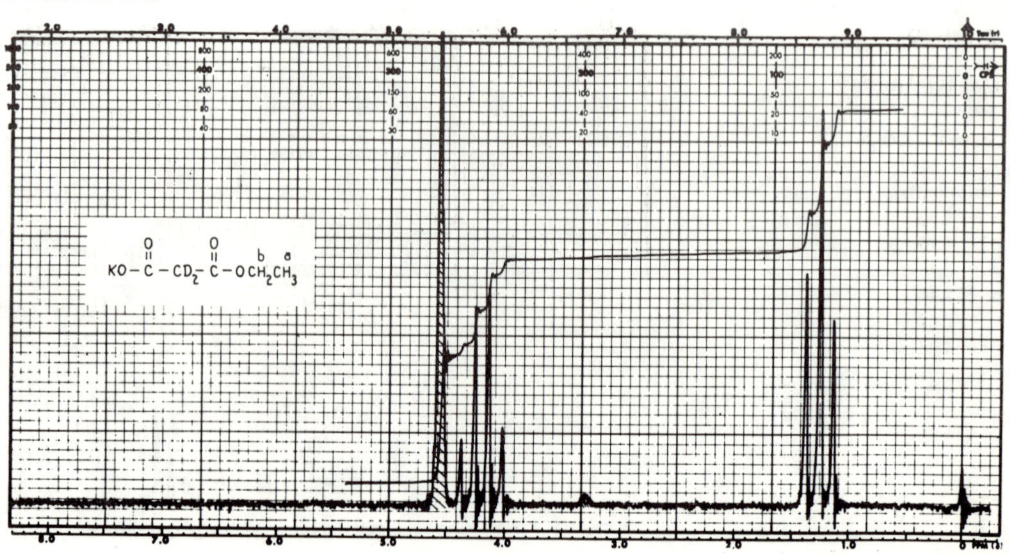

© 1978 Sadtler Research Laboratories, Inc.

HYDANTOIC ACID, ETHYL ESTER

$C_5H_{10}N_2O_3$

Mol. Wt. 146.15

M.P. 135-137°C

Solvent: Dimethyl sulfoxide-d_6

Assignments
a = 1.20
b = 3.79
c = 4.11
d = 5.61
e = 6.29
f = 2.52 DMSO-d_5
g = 3.31 H_2O

2751

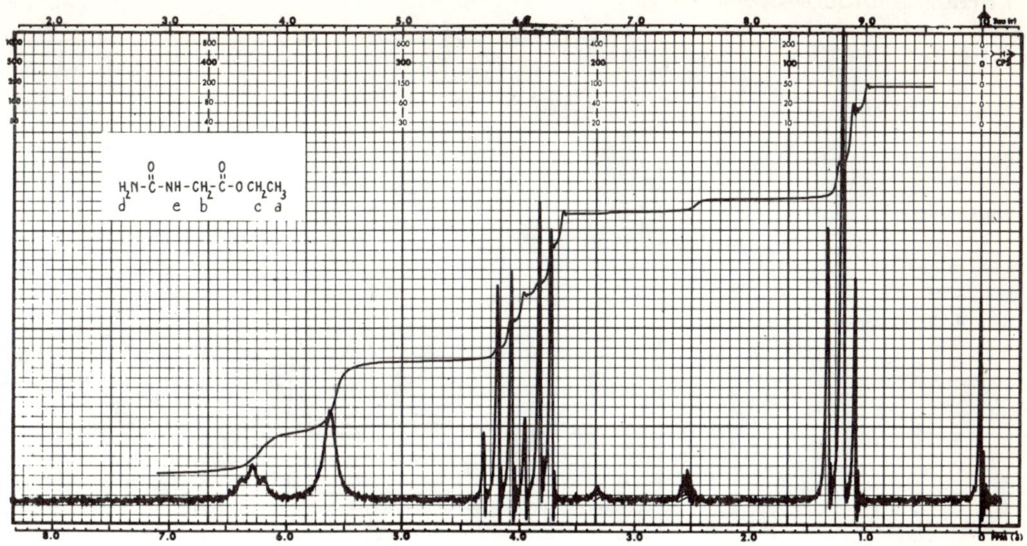

© 1978 Sadtler Research Laboratories, Inc.

ETHYLENE GLYCOL, DIACETATE

$C_6H_{10}O_4$

Mol. Wt. 146.14

B.P. 189-192°C

Solvent: CCl_4

Assignments
a = 2.01
b = 4.19

2752

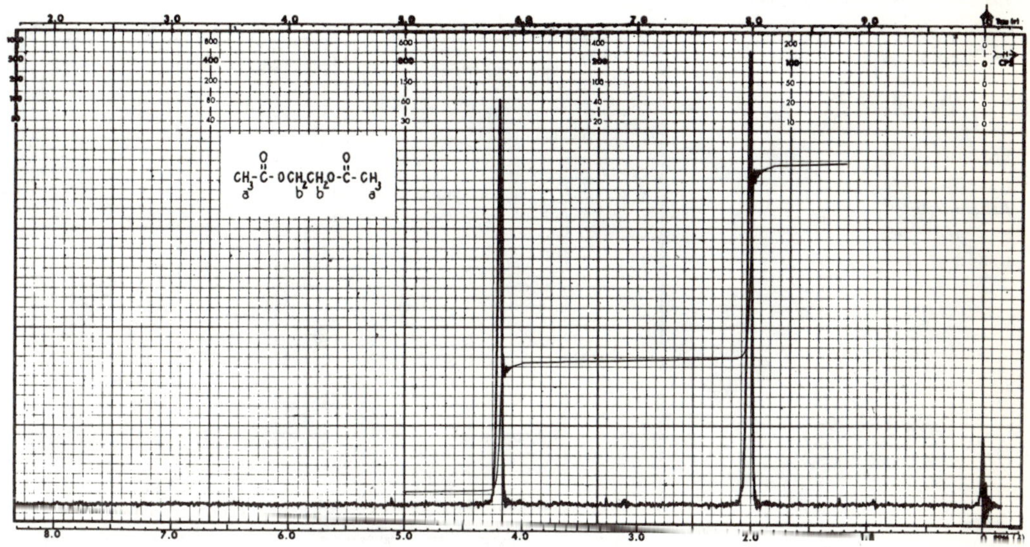

© 1978 Sadtler Research Laboratories, Inc.

ETHYLENE GLYCOL, DIPROPIONATE

$C_8H_{14}O_4$

Mol. Wt. 174.20

Solvent: $CDCl_3$

Assignments
a = 1.15
b = 2.37
c = 4.30

2753

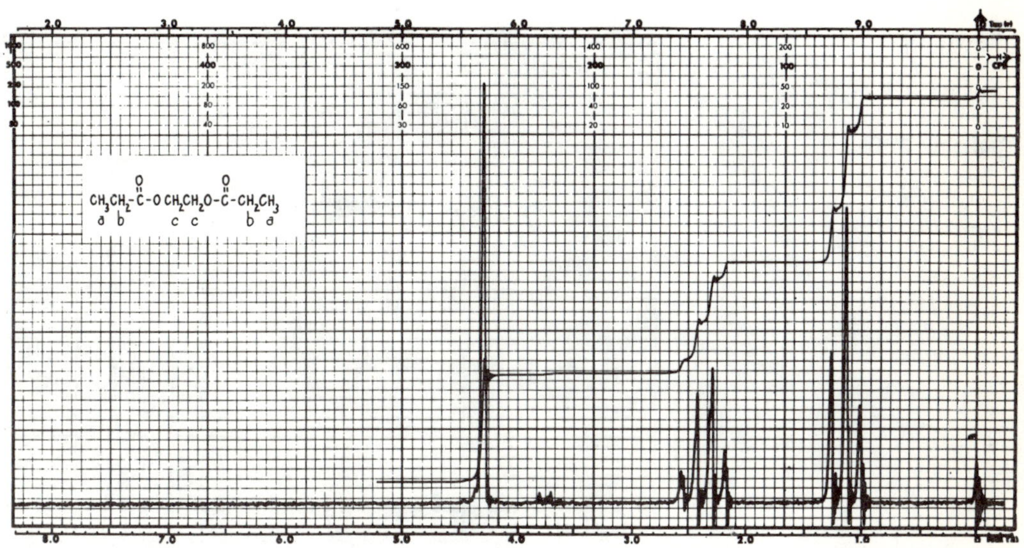

© 1978 Sadtler Research Laboratories, Inc.

1,2-PROPANEDIOL, DIACETATE

$C_7H_{12}O_4$

Mol. Wt. 160.17

Solvent: CCl_4

Assignments
a = 1.25
b = ca 2.07
c = 4.00
d = 4.13
e = 5.10

2754

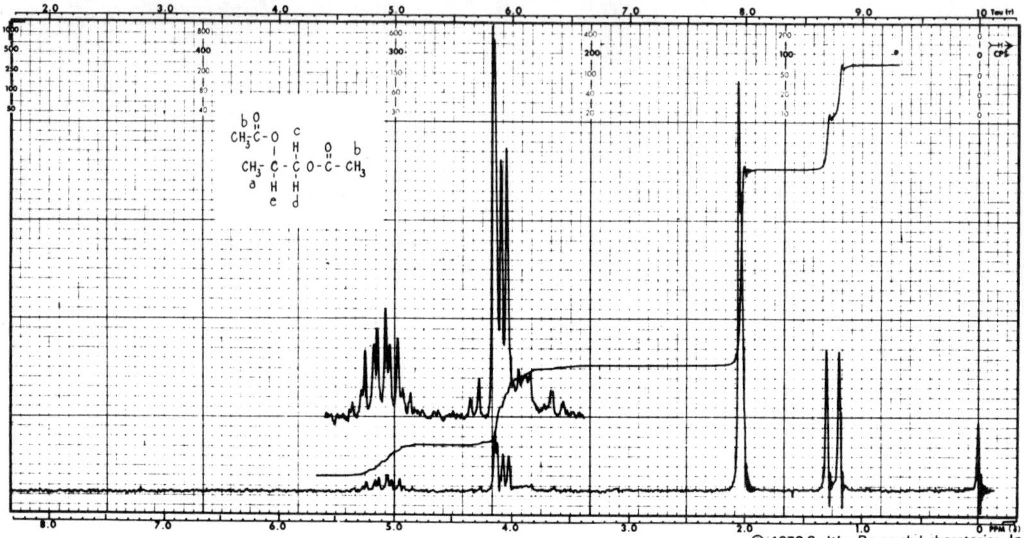

© 1978 Sadtler Research Laboratories, Inc.

1,5-PENTANEDIOL, DIACETATE

$C_9H_{16}O_4$

Mol. Wt. 188.23

Solvent: CCl_4

Assignments
a = 1.20-1.90
b = 1.97
c = 4.02

2755

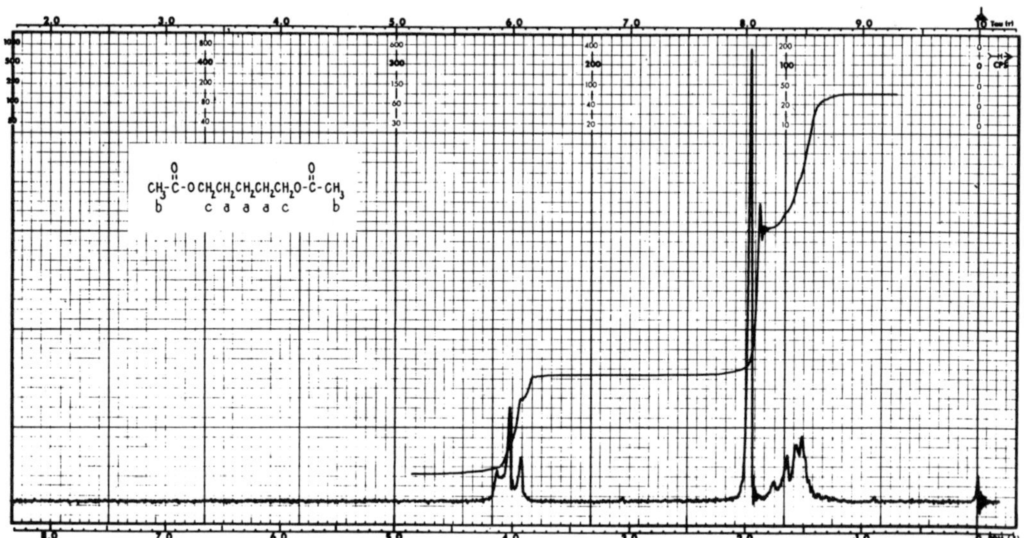

© 1978 Sadtler Research Laboratories, Inc.

TRIETHYLENE GLYCOL, DIACETATE

$C_{10}H_{18}O_6$

Mol. Wt. 234.25

Solvent: $CDCl_3$

Assignments
a = 2.08
b = 3.64
c = 3.68
d = 4.21

2756

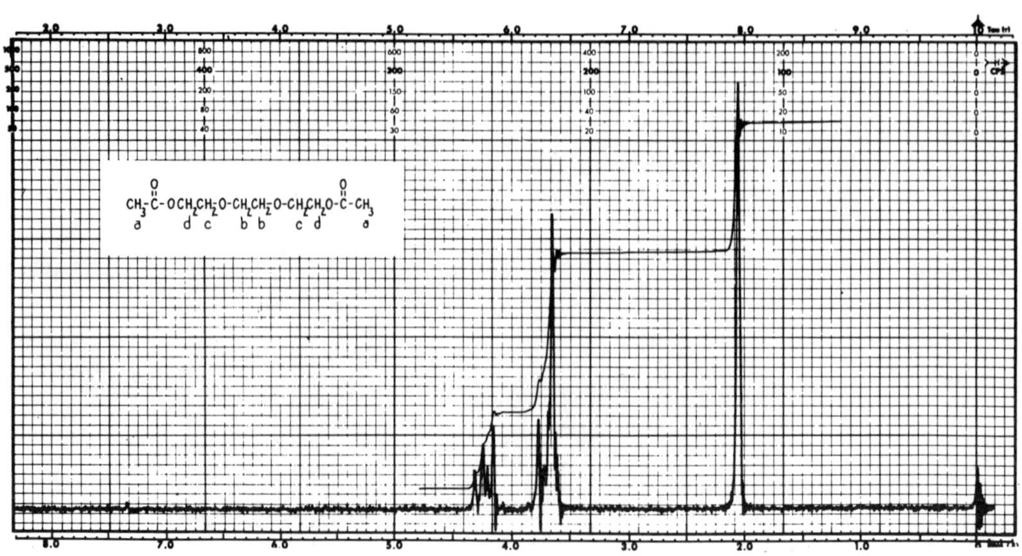

© 1978 Sadtler Research Laboratories, Inc.

TRIETHYLENE GLYCOL, BIS(2-ETHYLHEXANOATE)

$C_{22}H_{42}O_6$

Mol. Wt. 402.58

Solvent: $CDCl_3$

Assignments
a = 0.91
b = 1.10-1.91
c = 2.29
d = 3.66
e = 3.70
f = 4.25

2757

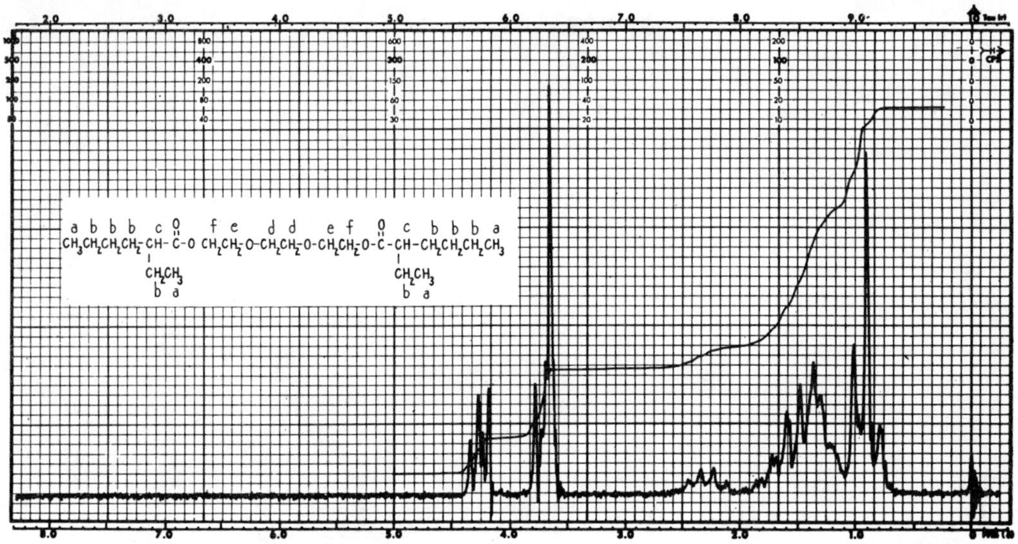

© 1978 Sadtler Research Laboratories, Inc.

1,3-DIPALMITIN

$C_{35}H_{68}O_5$

Mol. Wt. 568.93

M.P. 70°C (lit.)

Solvent: $CDCl_3$

Assignments
a = 0.89
b = 1.29
c = 1.62
d = 2.36
e = ca 2.50
f = 4.19

2758

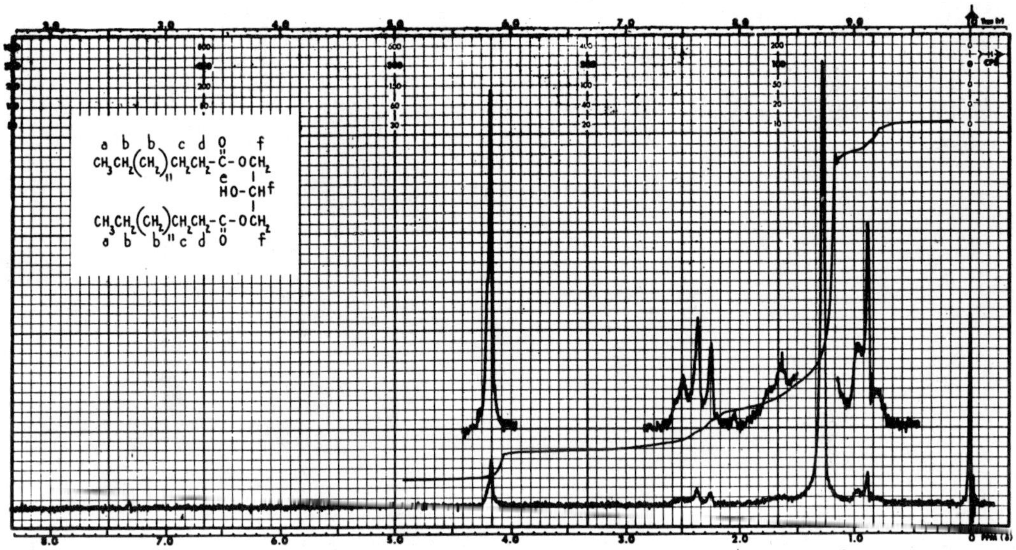

© 1978 Sadtler Research Laboratories, Inc.

1,3-DISTEARIN

$C_{39}H_{76}O_5$

Mol. Wt. 625.04

M.P. 78.8°C (lit.)

Solvent: $CDCl_3$

Assignments
a = 0.89
b = 1.26
c = ca 2.35
d = ca 4.13

2759

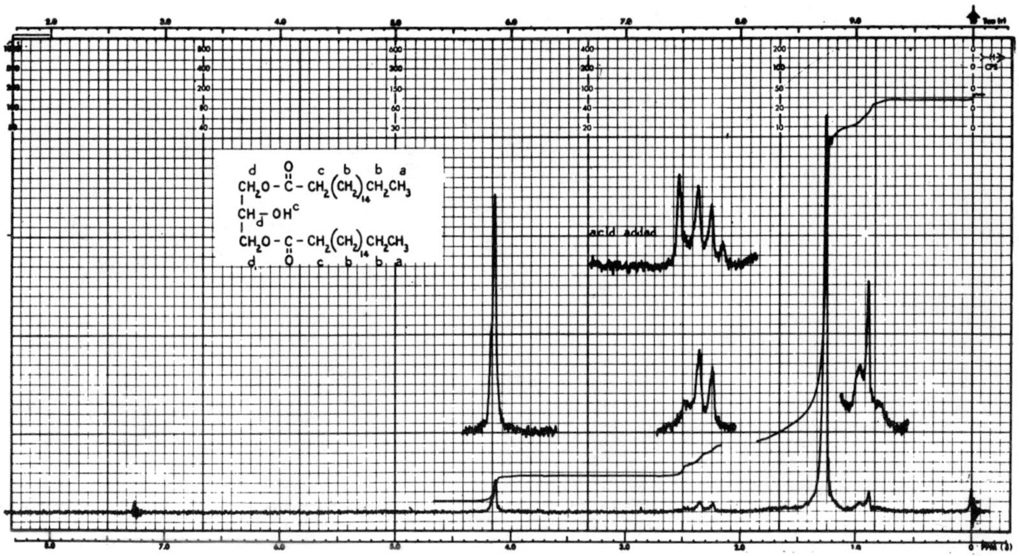

© 1978 Sadtler Research Laboratories, Inc.

TRIOCTANOIN

$C_{27}H_{50}O_6$

Mol. Wt. 470.70

M.P. 6-8°C

Solvent: CCl_4

Assignments
a = 0.90
b = 1.10-1.45
c = 1.59
d = 2.29
e = 4.10
f = 4.23
g = 5.19

2760

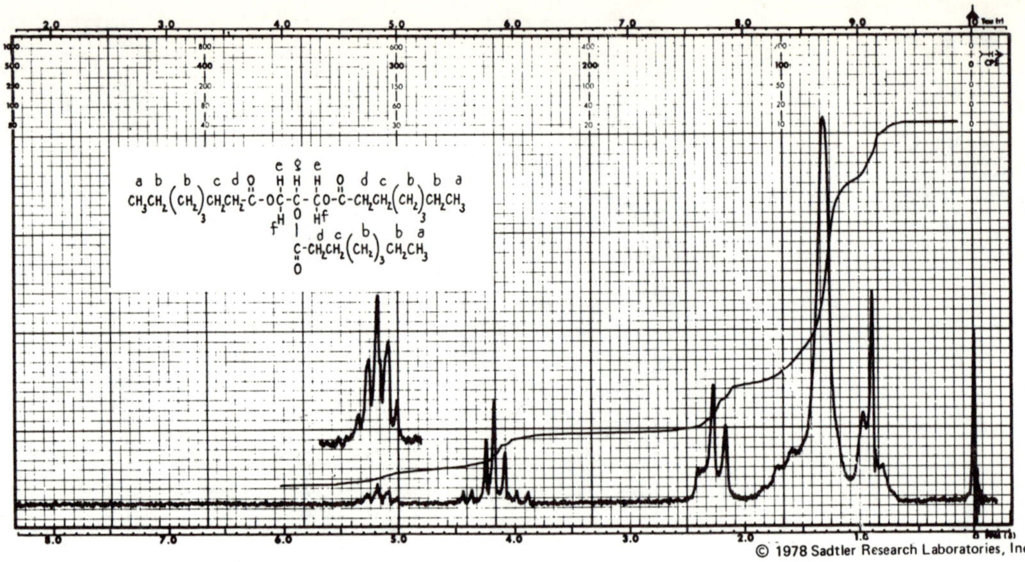

© 1978 Sadtler Research Laboratories, Inc.

TRISTEARIN

$C_{57}H_{110}O_6$

Mol. Wt. 891.51

Solvent: $CDCl_3$

Assignments
a = 0.89
b = ca 1.28
c = 1.61
d = 2.30
e = 4.15
f = 4.29
g = 5.16

2761

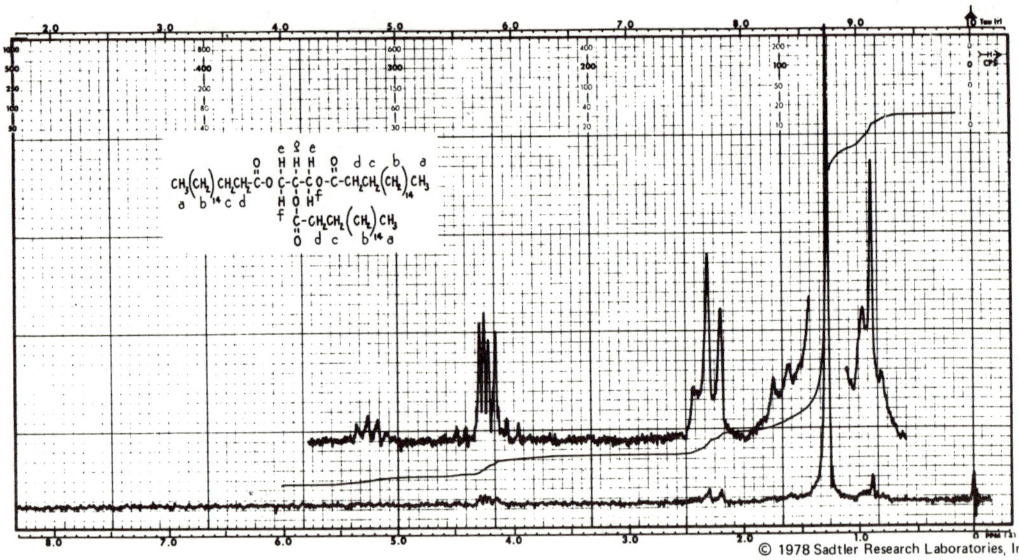

© 1978 Sadtler Research Laboratories, Inc.

CITRIC ACID, TRIS(2-ETHYLHEXYL) ESTER, ACETATE

$C_{32}H_{58}O_8$

Mol. Wt. 570.81

Solvent: CCl_4

Assignments
a = ca 0.93
b = 1.00-1.75
c = 1.99
d = 3.18
e = ca 3.98

2762

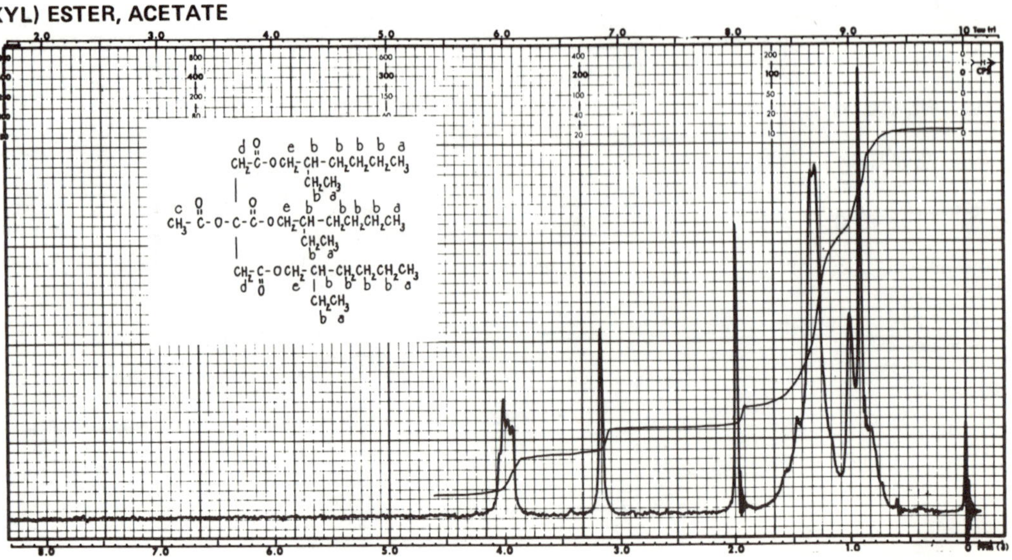

© 1978 Sadtler Research Laboratories, Inc.

PENTAERYTHRITOL, TETRALAURATE

$C_{53}H_{100}O_8$

Mol. Wt. 865.38

M.P. 52-53°C

Solvent: $CDCl_3$

Assignments
a = 0.89
b = 1.10-1.90
c = 2.30
d = 4.10

2763

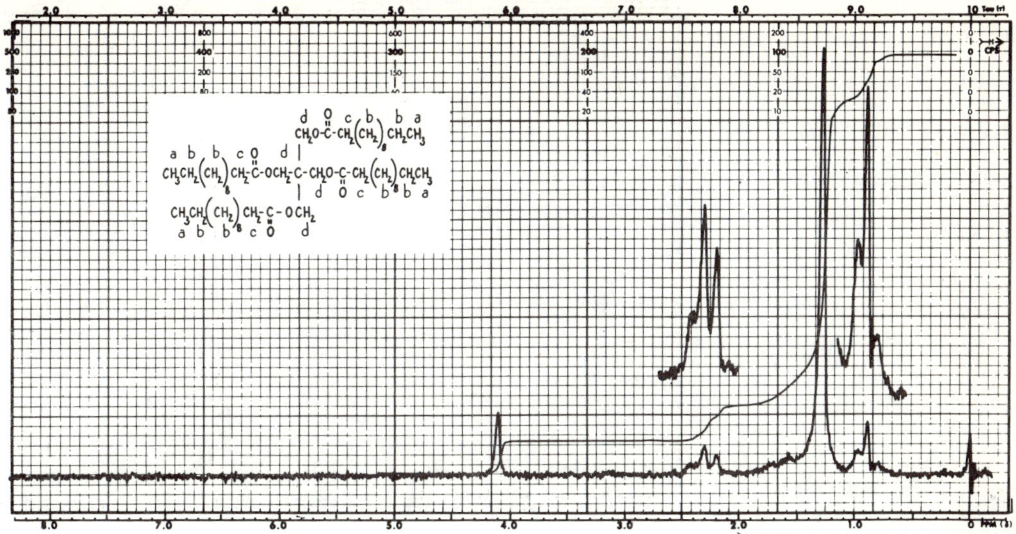

© 1978 Sadtler Research Laboratories, Inc.

SUCROSE, OCTAACETATE

$C_{28}H_{38}O_{19}$

Mol. Wt. 678.60

M.P. 89°C

Solvent: $CDCl_3$

Assignments
a = 2.00-2.20
b = 4.10-4.40
c = 4.75-5.80

2764

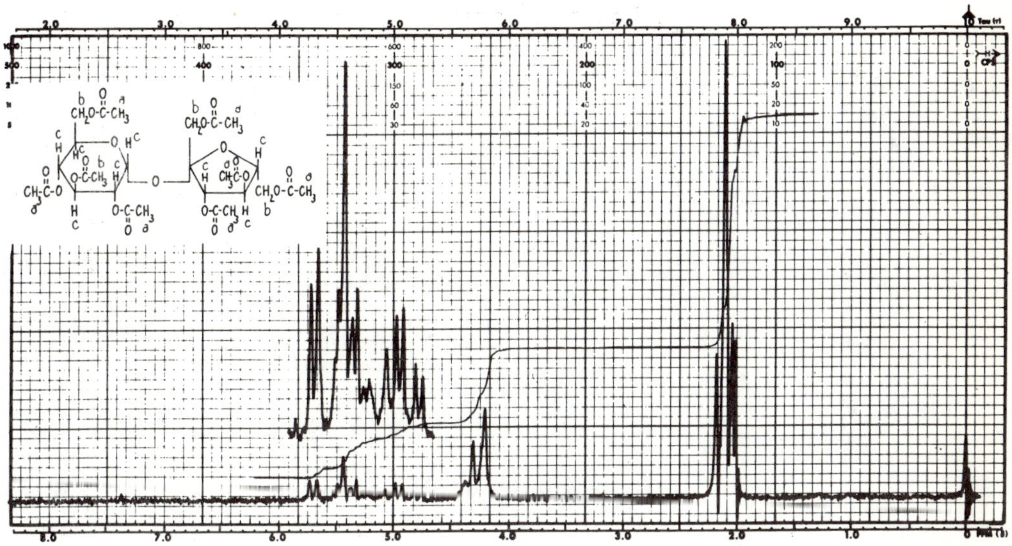

© 1978 Sadtler Research Laboratories, Inc.

CARBONIC ACID, DIBUTYL ESTER

$C_9H_{18}O_3$

Mol. Wt. 174.24

B.P. 90-92°C/15mm

Solvent: $CDCl_3$

Assignments
a = 0.93
b = 1.00-2.00
c = 4.11

2765

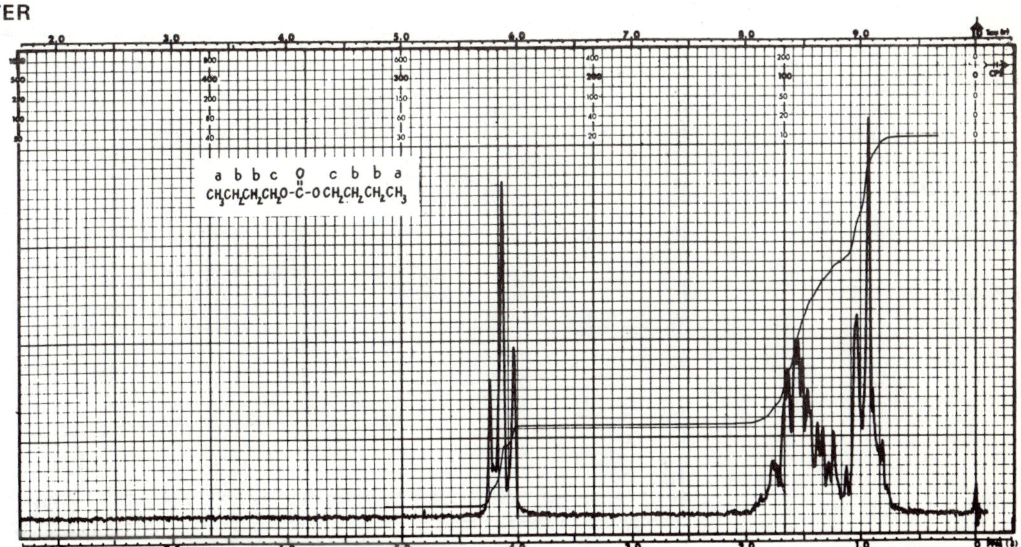

© 1978 Sadtler Research Laboratories, Inc.

OXALIC ACID, DIBUTYL ESTER

$C_{10}H_{18}O_4$

Mol. Wt. 202.25

B.P. 243.4°C (lit.)

Solvent: $CDCl_3$

Assignments
a = 0.96
b = 1.10-2.05
c = 4.27

2766

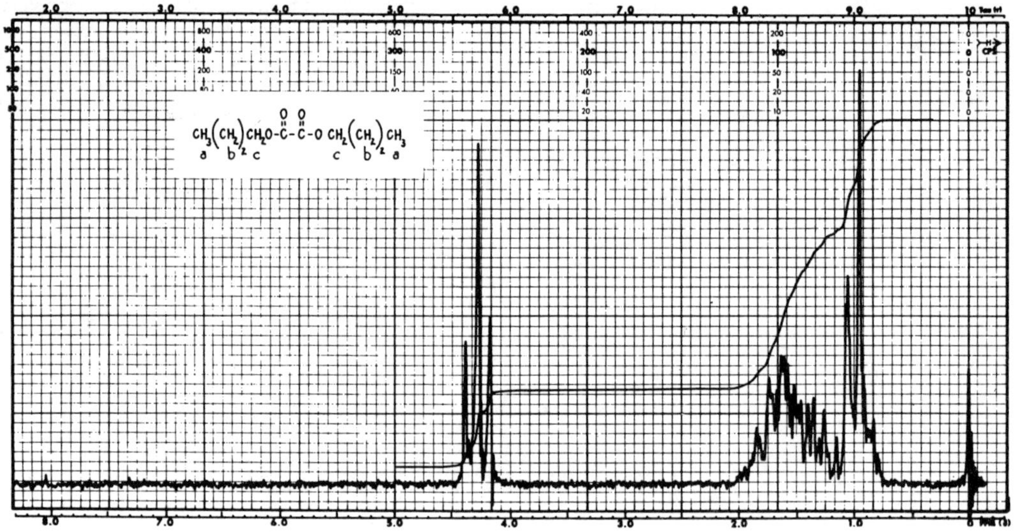

© 1978 Sadtler Research Laboratories, Inc.

MALONIC ACID, DIBUTYLESTER

$C_{11}H_{20}O_4$

Mol. Wt. 216.28

Solvent: CCl_4

Assignments
a = 0.95
b = 1.50
c = 3.22
d = 4.10

2767

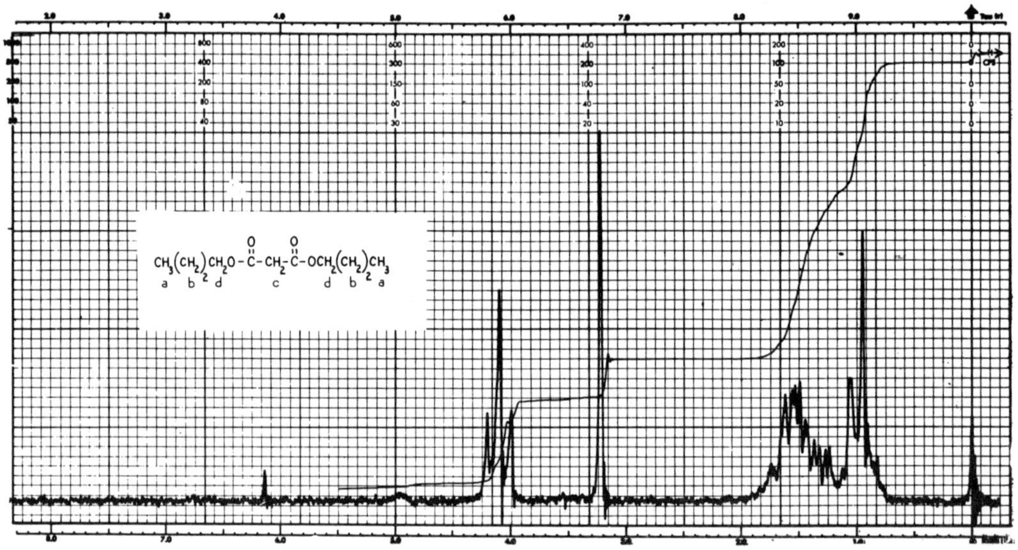

© 1978 Sadtler Research Laboratories, Inc.

MALONIC ACID, DI-tert-BUTYL ESTER

$C_{11}H_{20}O_4$

Mol. Wt. 216.28

B.P. 48-50°C/0.1mm

Solvent: CCl_4

Assignments
a = 1.44
b = 3.06

2768

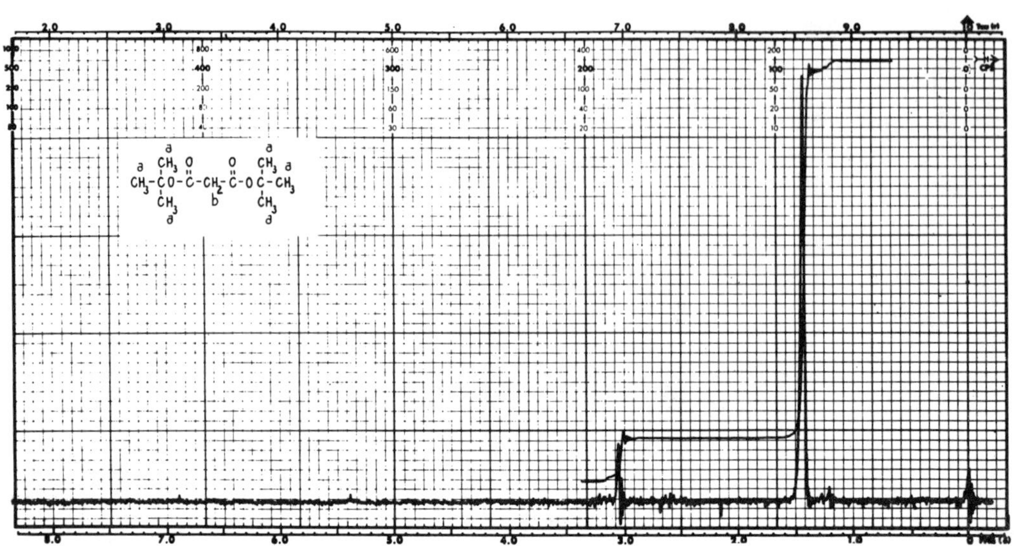

© 1978 Sadtler Research Laboratories, Inc.

MALONIC ACID, DIDODECYL ESTER

$C_{27}H_{52}O_4$

Mol. Wt. 440.71

M.P. 29-33°C

Solven t: $CDCl_3$

Assignments
a = 0.89
b = 1.10-1.90
c = 3.34
d = 4.13

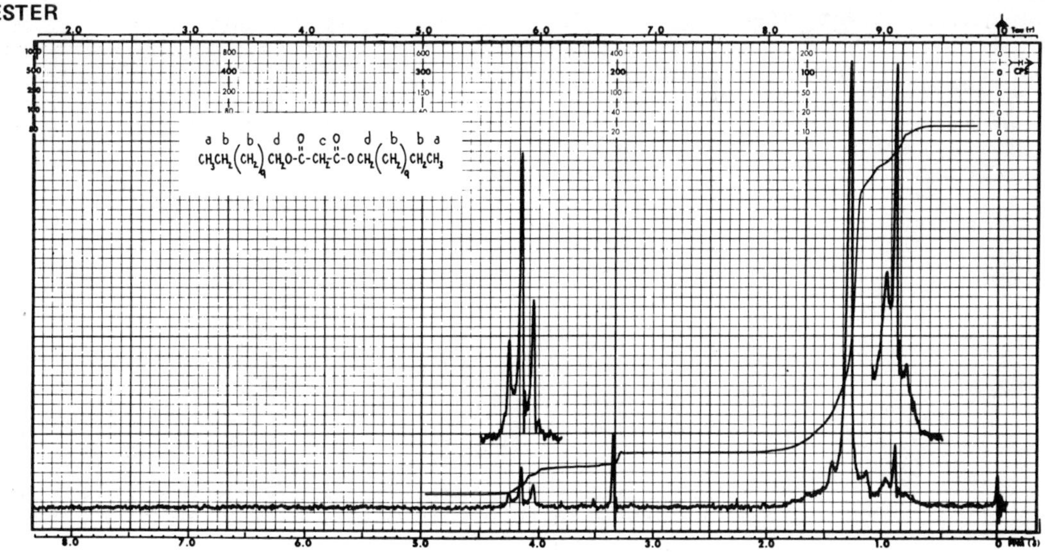

2769

© 1978 Sadtler Research Laboratories, Inc.

PENTYLMALONIC ACID, DIETHYL ESTER

$C_{12}H_{22}O_4$

Mol. Wt. 230.31

Solvent: CCl_4

Assignments
a = 0.90
b = 1.25
c = 0.65-1.55
d = 1.77
e = 3.14
f = 4.13

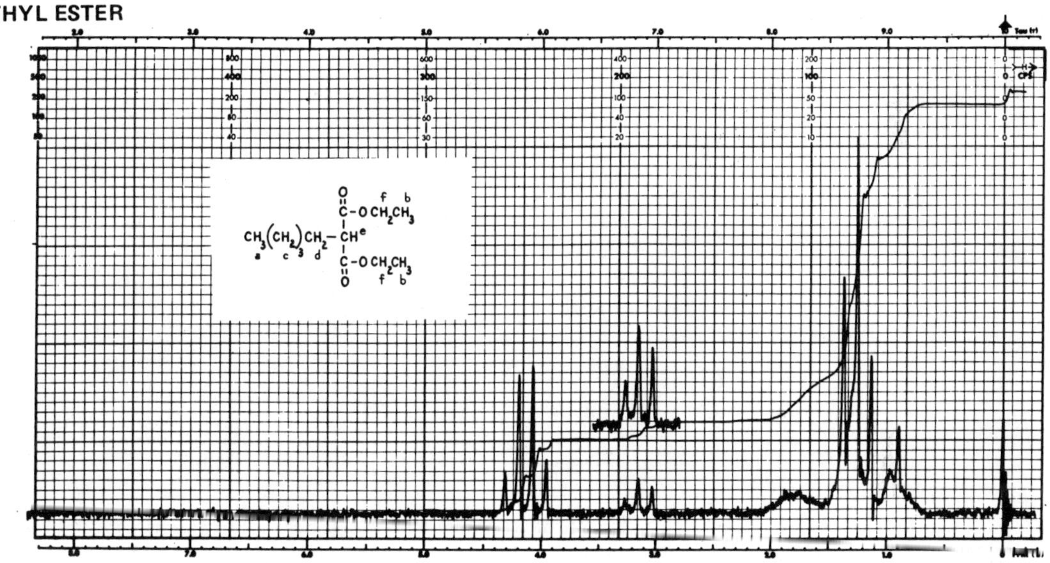

2770

© 1978 Sadtler Research Laboratories, Inc.

DIETHYLMALONIC ACID, DIMETHYL ESTER

$C_9H_{16}O_4$

Mol. Wt. 188.23

Solvent: CCl_4

Assignments
a = 0.80
b = 1.85
c = 3.67

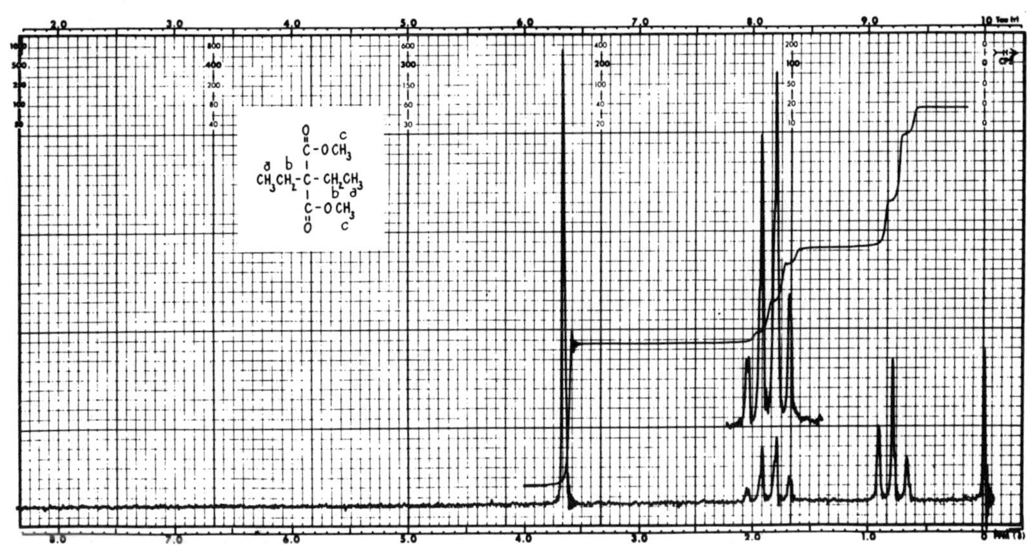

2771

© 1978 Sadtler Research Laboratories, Inc.

DIBROMOMALONIC ACID, DIETHYL ESTER

$C_7H_{10}Br_2O_4$

Mol. Wt. 317.97

B.P. 103-106°C/4mm

Solvent: CCl_4

Assignments
a = 1.37
b = 4.34

2772

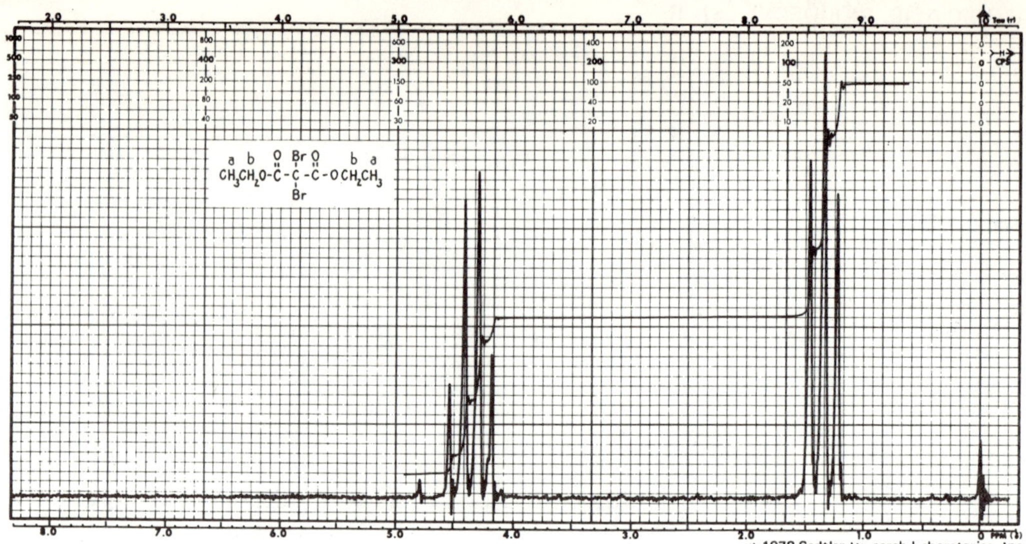

© 1978 Sadtler Research Laboratories, Inc.

SUCCINIC ACID, DIISOPROPYL ESTER

$C_{10}H_{18}O_4$

Mol. Wt. 202.25

Solvent: CCl_4

Assignments
a = 1.22
b = 2.49
c = 4.96

2773

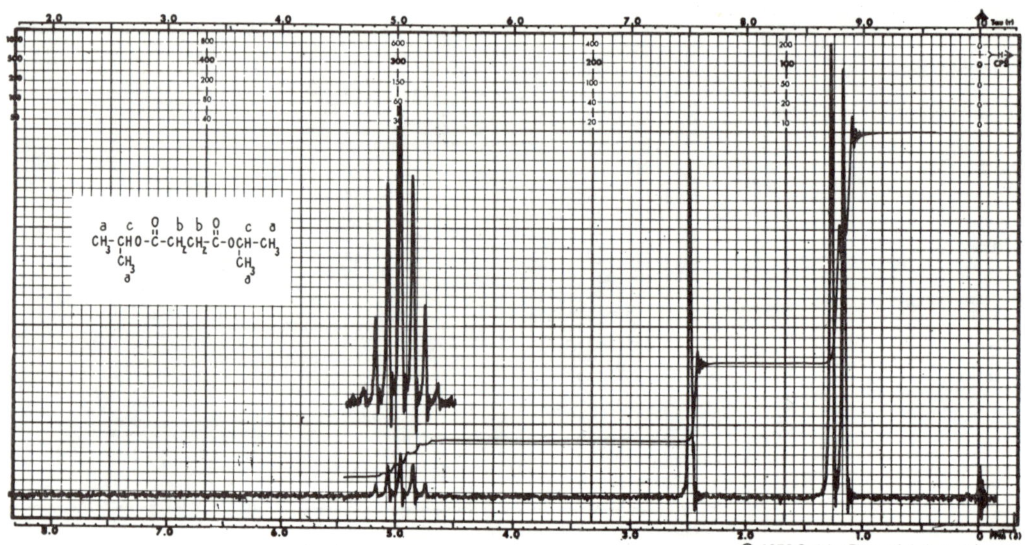

© 1978 Sadtler Research Laboratories, Inc.

SUCCINIC ACID, DIPENTYL ESTER

$C_{14}H_{26}O_4$

Mol. Wt. 258.36

B.P. 146-148°C/5mm

Solvent: CCl_4

Assignments
a = 0.93
b = 1.44
c = 2.51
d = 4.01

2774

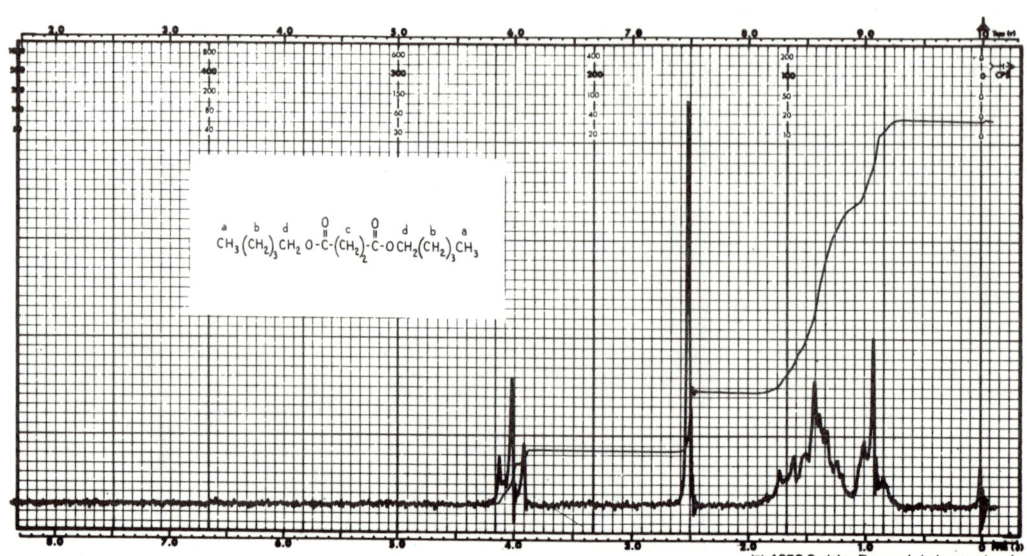

© 1978 Sadtler Research Laboratories, Inc.

GLUTARIC ACID, DIBUTYL ESTER

$C_{13}H_{24}O_4$

Mol. Wt. 244.33

B.P. 139-142°C/4mm

Solvent: CCl_4

Assignments
a = 0.95
b = 1.10-1.80
c = 1.85
d = 2.29
e = 4.00

2775

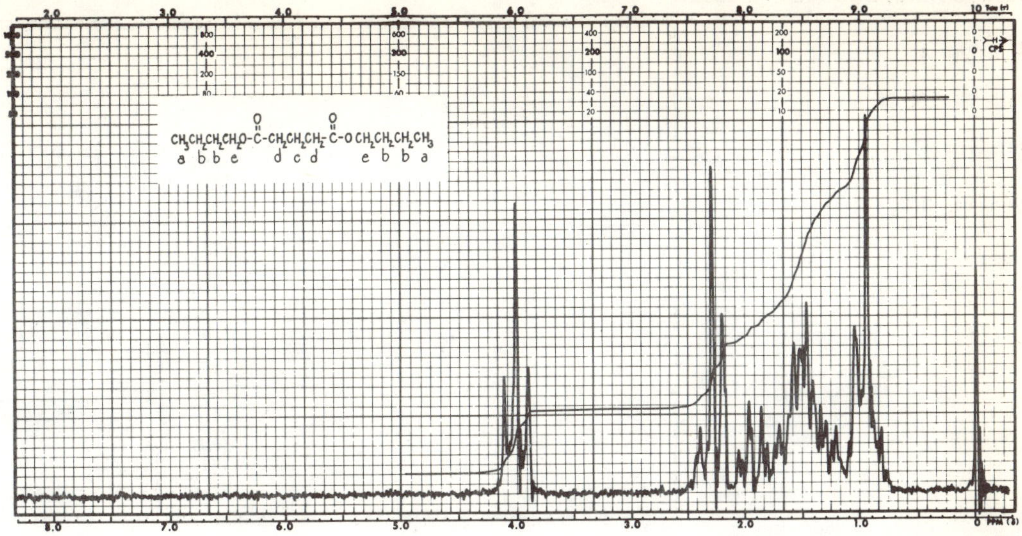

© 1978 Sadtler Research Laboratories, Inc.

3,3-DIMETHYLGLUTARIC ACID, DIMETHYL ESTER

$C_9H_{16}O_4$

Mol. Wt. 188.23

B.P. 104-105°C/16mm

Solvent: CCl_4

Assignments
a = 1.07
b = 2.35
c = 3.59

2776

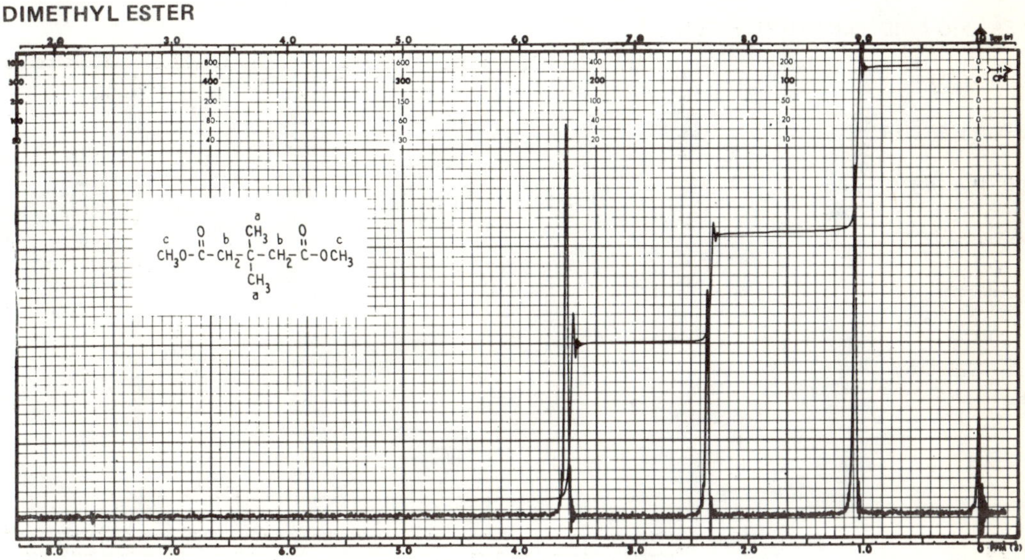

© 1978 Sadtler Research Laboratories, Inc.

ADIPIC ACID, DIETHYL ESTER

$C_{10}H_{18}O_4$

Mol. Wt. 202.25

B.P. 239-241°C/761mm (lit.)

Solvent: CCl_4

Assignments
a = 1.23
b = 1.61
c = 2.27
d = 4.06

2777

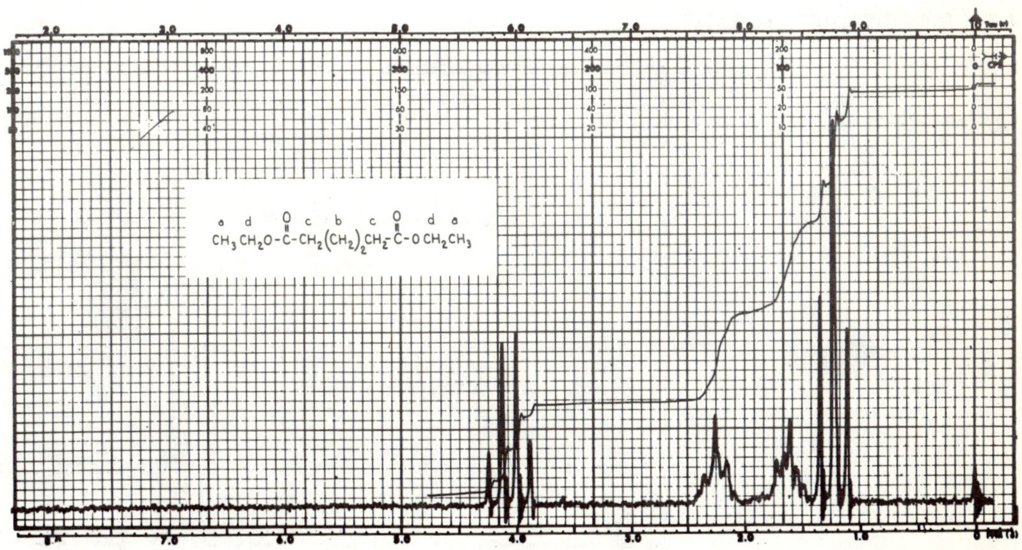

© 1978 Sadtler Research Laboratories, Inc.

ADIPIC ACID, DIISOBUTYL ESTER

$C_{14}H_{26}O_4$

Mol. Wt. 258.36

B.P. 278°C

Solvent: CCl_4

Assignments
- a = 0.96
- b = 1.50-2.10
- c = 2.29
- d = 3.80

2778

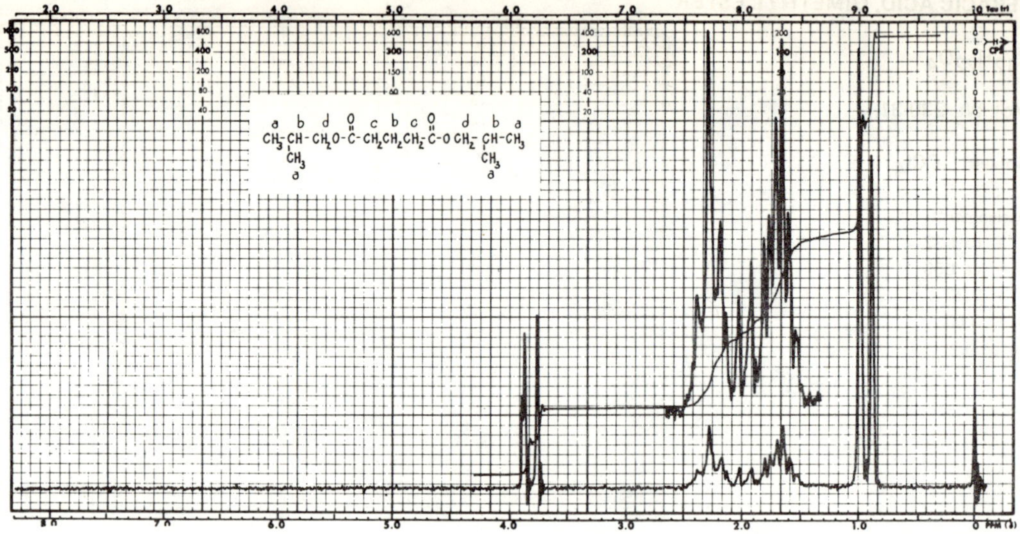

© 1978 Sadtler Research Laboratories, Inc.

AZELAIC ACID, DIMETHYL ESTER

$C_{11}H_{20}O_4$

Mol. Wt. 216.28

B.P. 154-156°C/10mm

Solvent: CCl_4

Assignments
- a = ca 1.40
- b = 1.61
- c = 2.28
- d = 3.61

2779

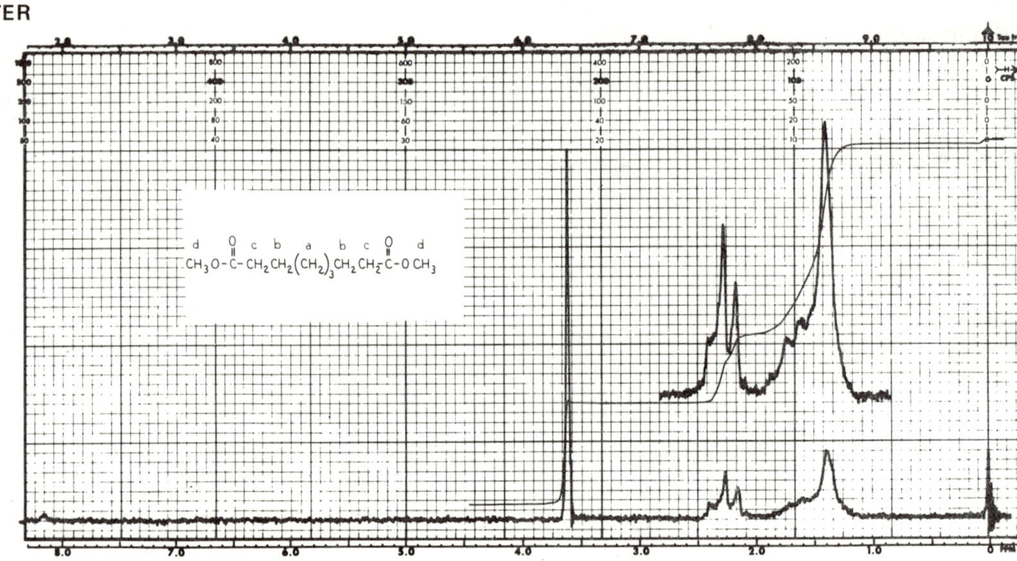

© 1978 Sadtler Research Laboratories, Inc.

AZALAIC ACID, DIETHYL ESTER

$C_{13}H_{24}O_4$

Mol. Wt. 244.33

Solvent: $CDCl_3$

Assignments
- a = 1.24
- b = ca 1.38
- c = 1.62
- d = 2.29
- e = 4.11
- f = 3.66 imp.

2780

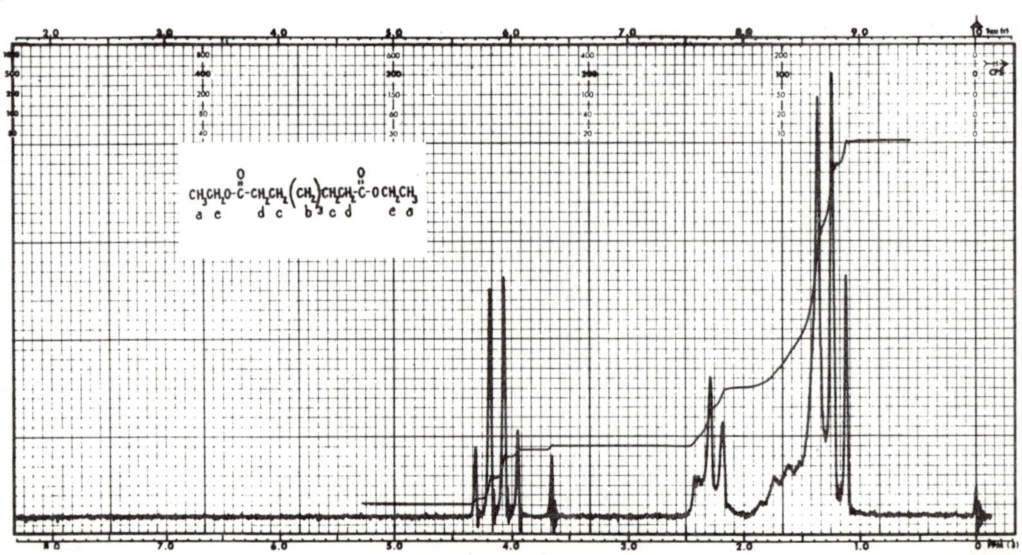

© 1978 Sadtler Research Laboratories, Inc.

SEBACIC ACID, DIMETHYL ESTER

$C_{12}H_{22}O_4$

Mol. Wt. 230.31

Solvent: CCl_4

Assignments
a = ca 1.32
b = ca 1.55
c = 2.23
d = 3.60

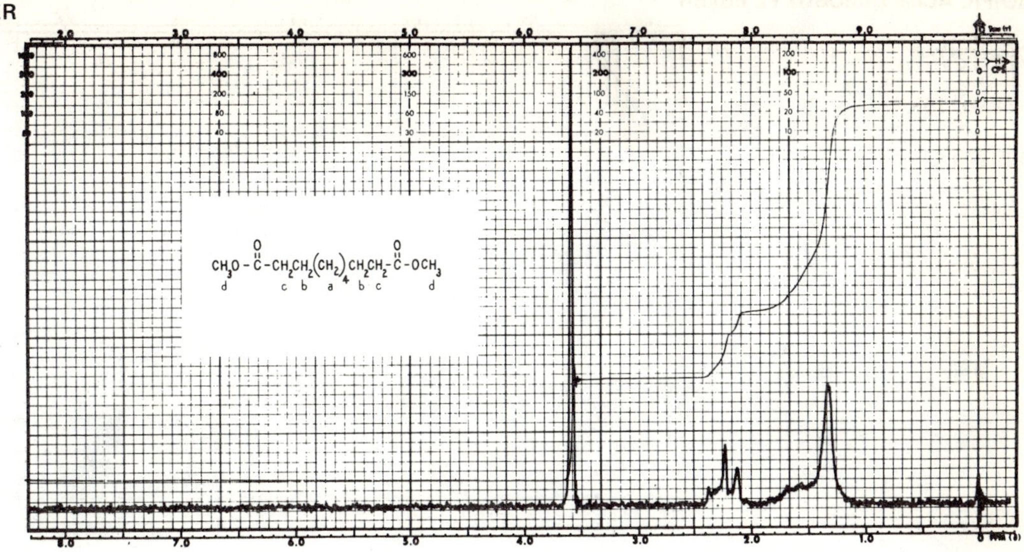

2781

© 1978 Sadtler Research Laboratories, Inc.

SEBACIC ACID, BIS(2-ETHYLHEXYL) ESTER

$C_{26}H_{50}O_4$

Mol. Wt. 426.69

B.P. 185-187°C/5mm

Solvent: $CDCl_3$

Assignments
a = 0.90
b = 1.10-2.00
c = 2.29
d = 4.02

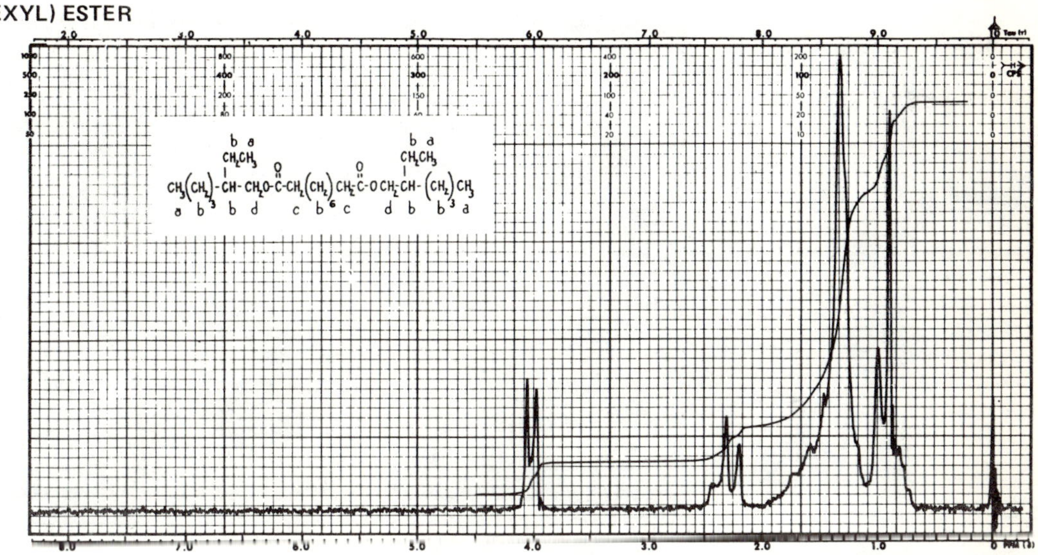

2782

© 1978 Sadtler Research Laboratories, Inc.

SEBACIC ACID, DIDODECYL ESTER

$C_{34}H_{66}O_4$

Mol. Wt. 538.90

M.P. 50-51°C

Solvent: $CDCl_3$

Assignments
a = 0.89
b = 1.10-1.90
c = 2.28
d = 4.05

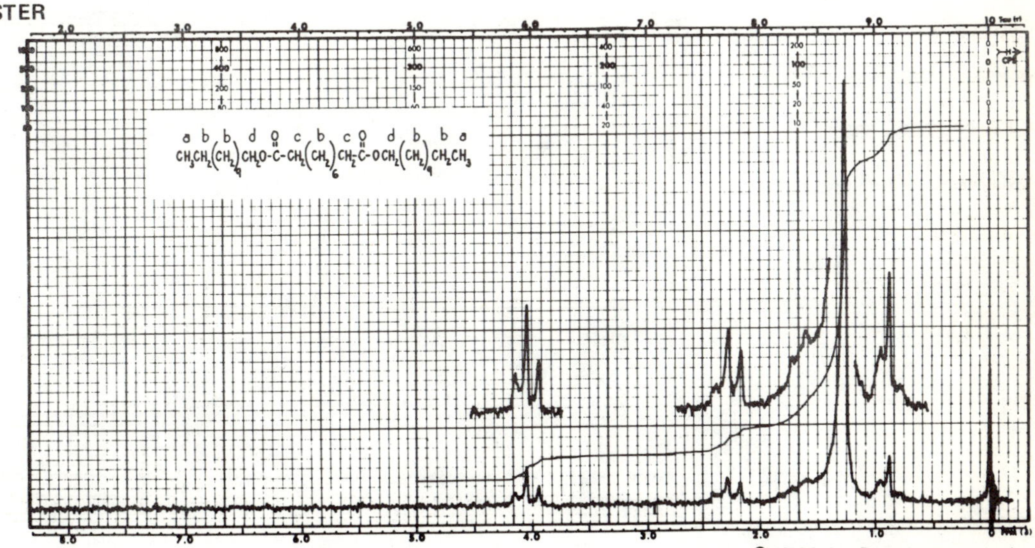

2783

© 1978 Sadtler Research Laboratories, Inc.

SEBACIC ACID, BIS(2-BUTOXYETHYL) ESTER

$C_{22}H_{46}O_6$

Mol. Wt. 406.61

Solvent: CCl_4

Assignments
a = 0.94
b = 1.15-1.91
c = 2.27
d = 3.29-3.68
e = 4.11

2784

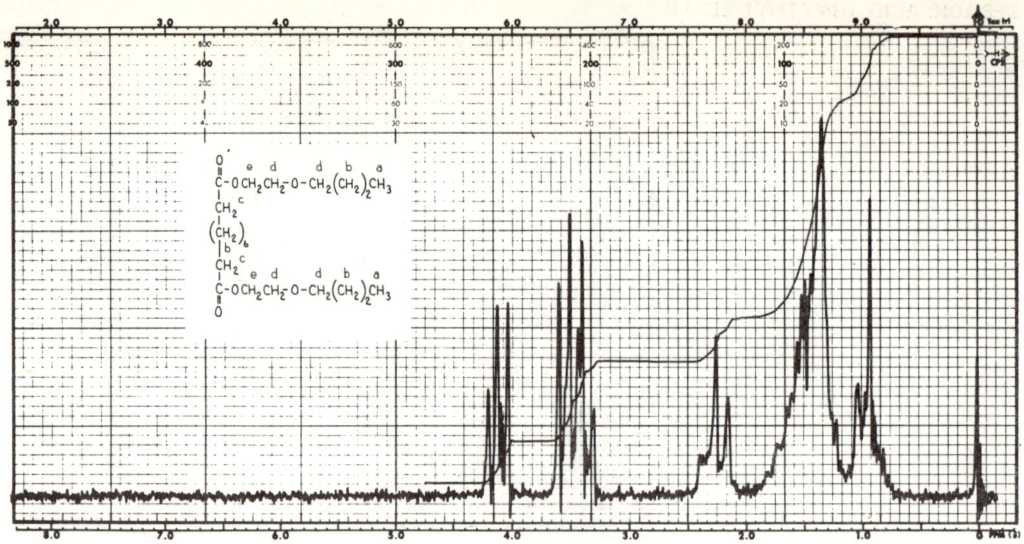

© 1978 Sadtler Research Laboratories, Inc.

DODECANEDIOIC ACID, DIMETHYL ESTER

$C_{14}H_{26}O_4$

Mol. Wt. 258.36

B.P. 168°C/9mm (lit.)

Solvent: $CDCl_3$

Assignments
a = ca 1.28
b = 1.59
c = 2.30
d = 3.65

2785

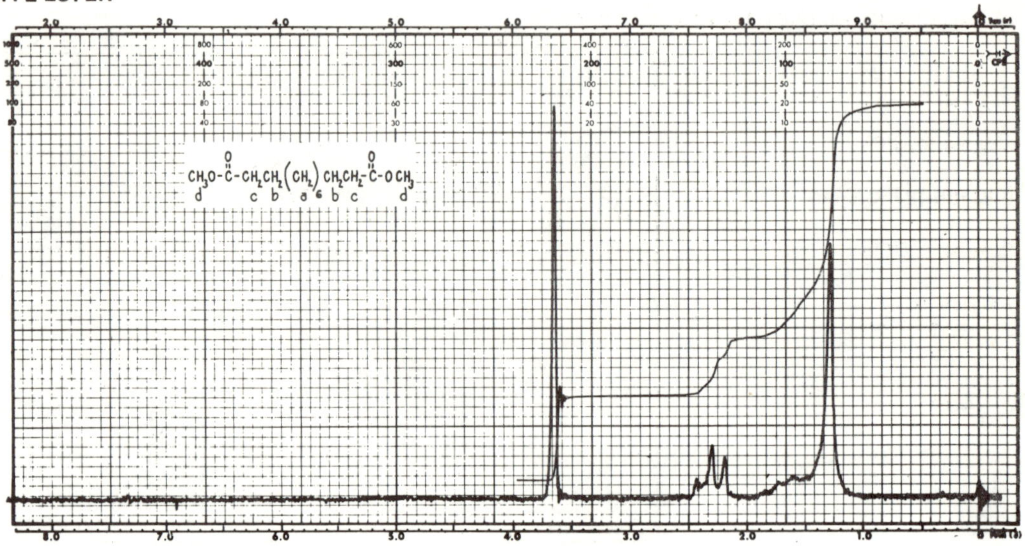

© 1978 Sadtler Research Laboratories, Inc.

1,2-CYCLOPROPANEDICARBOXYLIC ACID, DIETHYL ESTER

$C_9H_{14}O_4$

Mol. Wt. 186.21

B.P. 70-75°C/1mm

Solvent: CCl_4

Assignments
a = 1.22
b = 1.31
c = 2.01
d = 4.08

2786

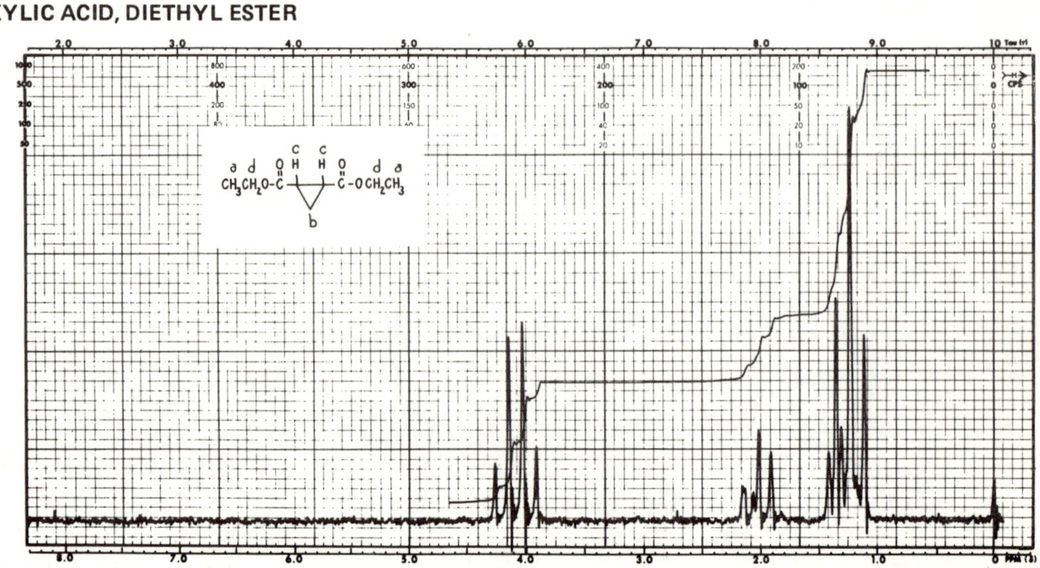

© 1978 Sadtler Research Laboratories, Inc.

1,3-CYCLOHEXANEDICARBOXYLIC ACID, DIMETHYL ESTER

$C_{10}H_{16}O_4$

Mol. Wt. 200.24

B.P. 130-132°C/15mm

Solvent: CCl_4

Assignments
a = 1.10-2.80
b = 3.60
c = 3.63

2787

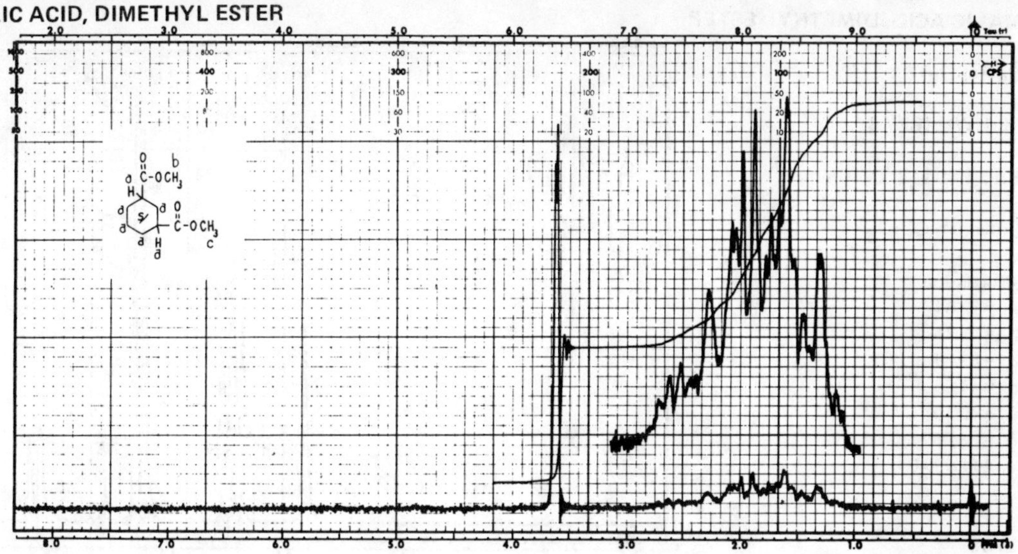

© 1978 Sadtler Research Laboratories, Inc.

1,2-CYCLOHEXANEDICARBOXYLIC ACID, BIS(2-ETHYLHEXYL) ESTER

$C_{24}H_{44}O_4$

Mol. Wt. 396.62

B.P. 216°C/5mm

Solvent: CCl_4

Assignments
a = 0.94
b = 1.10-2.30
c = 2.70
d = 3.94

2788

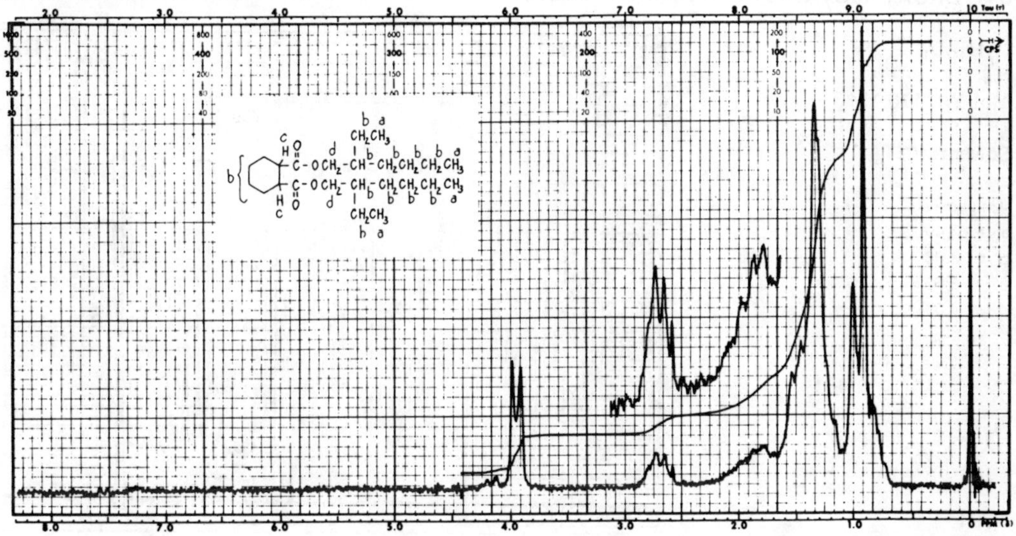

© 1978 Sadtler Research Laboratories, Inc.

IMINODIACETIC ACID, DIETHYL ESTER

$C_8H_{15}NO_4$

Mol. Wt. 189.21

B.P. 120-121°C/8mm

Solvent: CCl_4

Assignments
a = 1.29
b = 2.07
c = 3.38
d = 4.13

2789

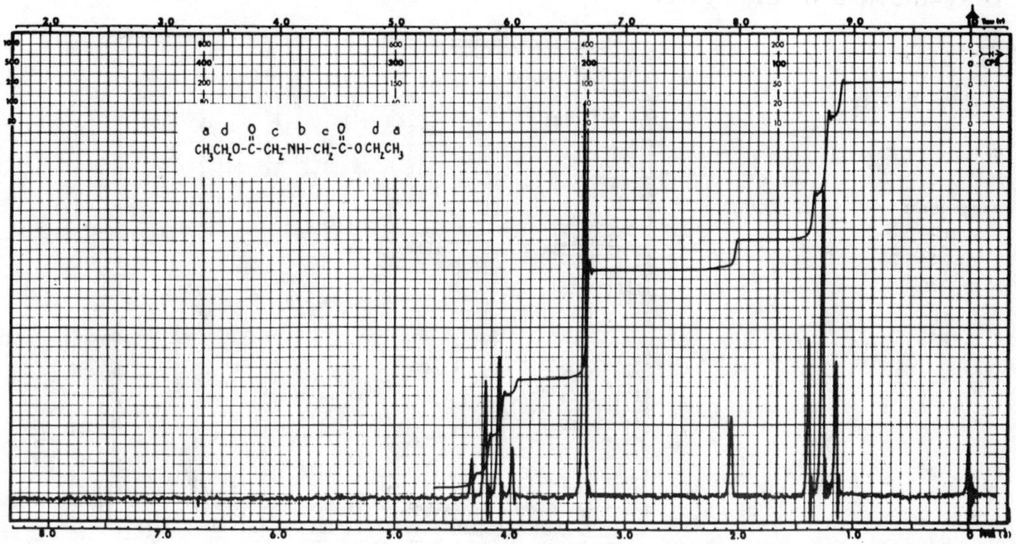

© 1978 Sadtler Research Laboratories, Inc.

MALIC ACID, DIMETHYL ESTER

$C_6H_{10}O_5$

Mol. Wt. 162.14

B.P. 111-112°C/10mm

Solvent: $CDCl_3$

Assignments
a = 2.84
b = 3.71
c = 3.79
d = 4.55
e = 5.19 *
f = 3.46 imp.
* Plus h_2O

2790

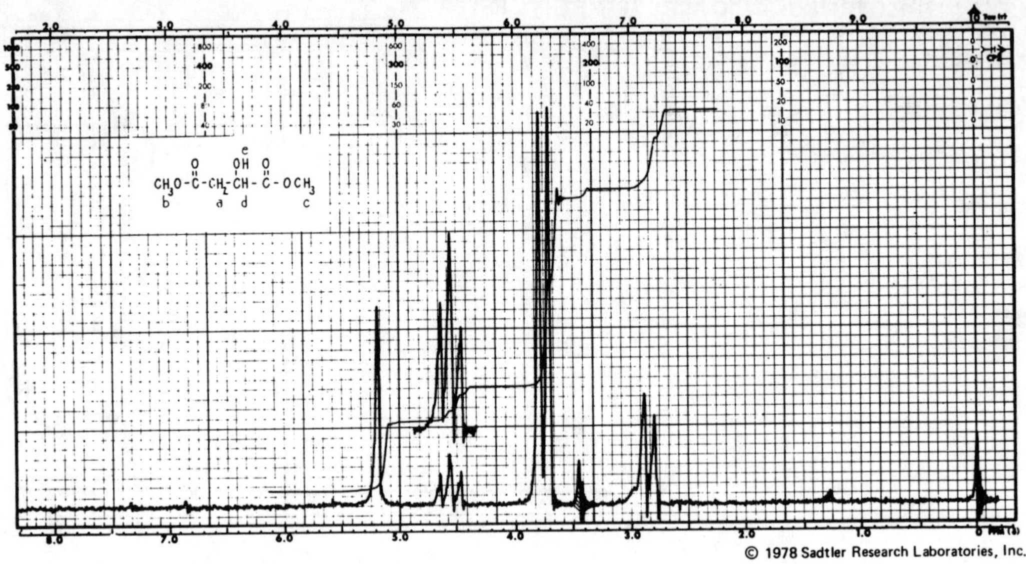

© 1978 Sadtler Research Laboratories, Inc.

3-HYDROXYGLUTARIC ACID, DIMETHYL ESTER

$C_7H_{12}O_5$

Mol. Wt. 176.17

Solvent: CCl_4

Assignments
a = 2.49
b = 2.90
c = 3.69
d = 4.31

2791

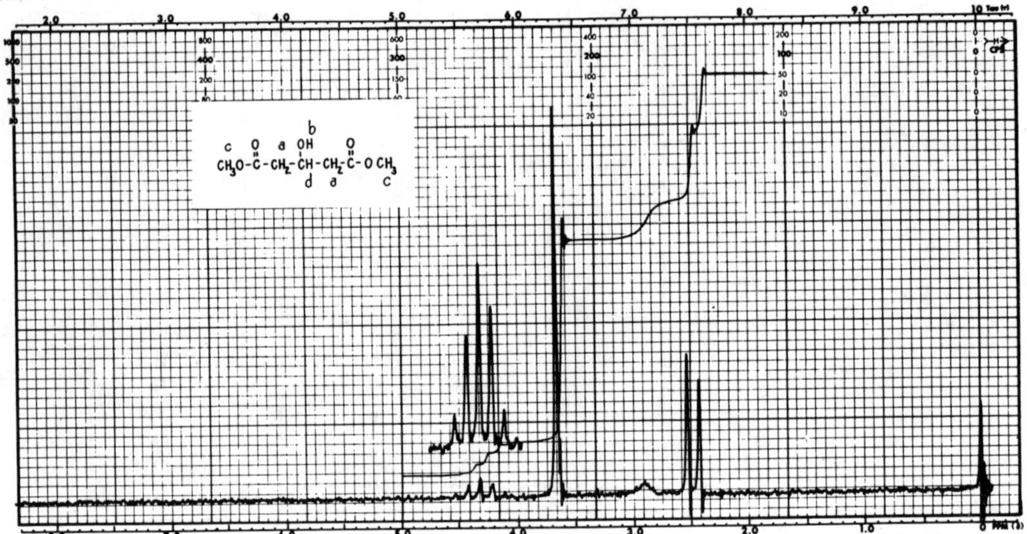

© 1978 Sadtler Research Laboratories, Inc.

TARTARIC ACID, DIBUTYL ESTER

$C_{12}H_{22}O_6$

Mol. Wt. 262.31

Solvent: CCl_4

Assignments
a = 0.99
b = 1.14-1.95
c = 3.70
d = 4.22
e = 4.42

2792

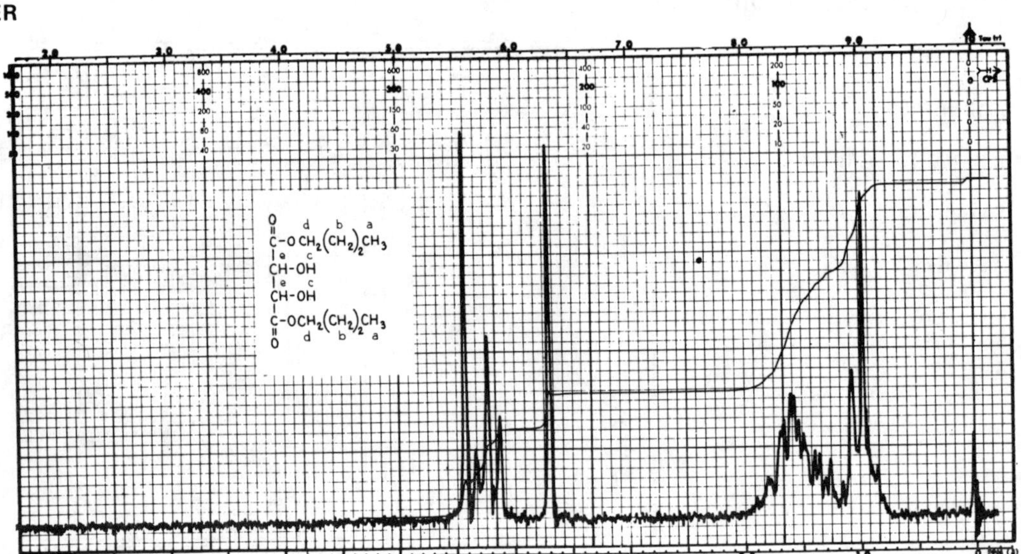

© 1978 Sadtler Research Laboratories, Inc.

3,3',3''-NITRILOTRIPROPIONIC ACID, TRIMETHYL ESTER

$C_{12}H_{21}NO_6$

Mol. Wt. 275.30

Solvent: $CDCl_3$

Assignments
a = 2.43
b = 2.79
c = 3.64

2793

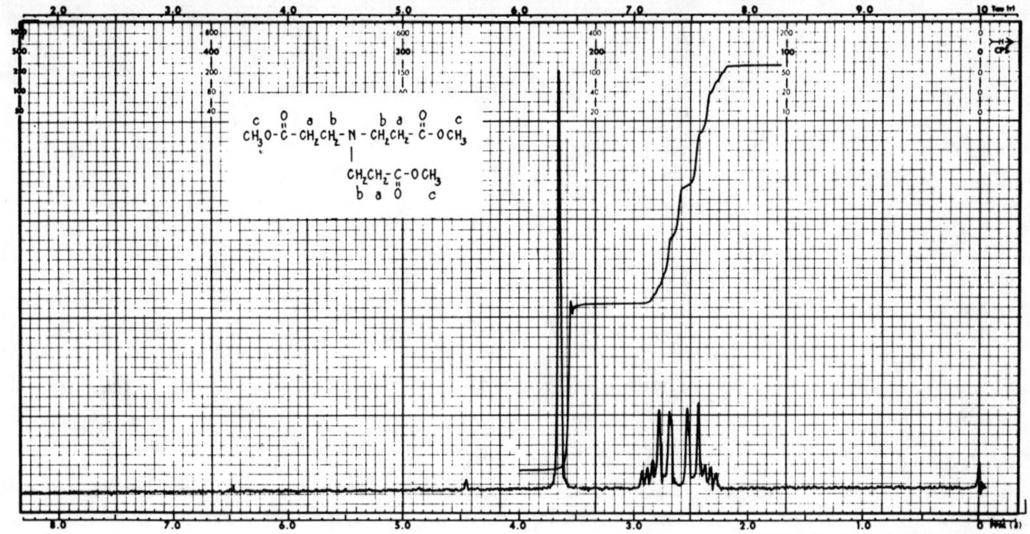

© 1978 Sadtler Research Laboratories, Inc.

NITRILOTRIACETIC ACID, TRIETHYL ESTER

$C_{12}H_{21}NO_6$

Mol. Wt. 275.30

B.P. 183-186°C/12mm

Solvent: $CDCl_3$

Assignments
a = 1.29
b = 3.67
c = 4.19

2794

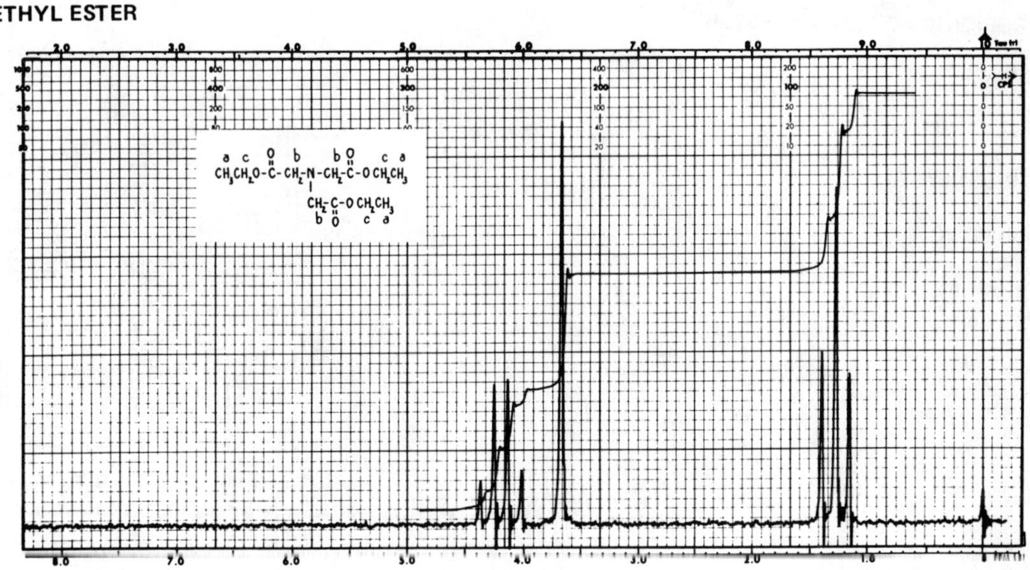

© 1978 Sadtler Research Laboratories, Inc.

1,1,2,2-ETHANETETRACARBOXYLIC ACID, TETRAETHYL ESTER

$C_{14}H_{22}O_8$

Mol. Wt. 318.33

M.P. 73-75°C

Solvent: $CDCl_3$

Assignments
a = 1.27
b = 4.09
c = 4.20

2795

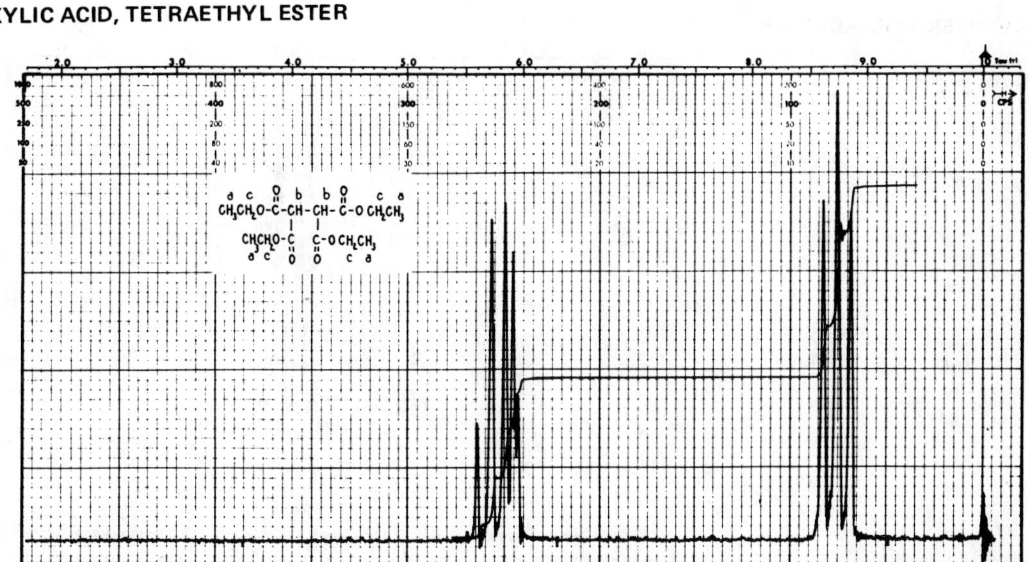

© 1978 Sadtler Research Laboratories, Inc.

1,1,4,4-BUTANETETRACARBOXYLIC ACID, TETRAMETHYL ESTER

$C_{12}H_{18}O_8$

Mol. Wt. 290.27

M.P. 76-78°C

Solvent: $CDCl_3$

Assignments
a = 1.96
b = 3.40
c = 3.75

2796

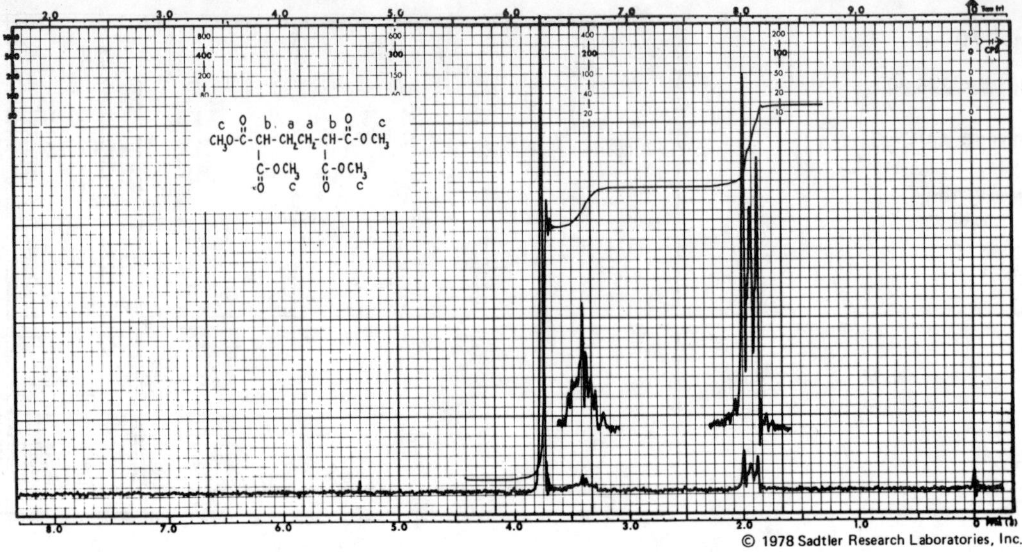

© 1978 Sadtler Research Laboratories, Inc.

PROPIONIC ACID, VINYL ESTER

$C_5H_8O_2$

Mol. Wt. 100.12

B.P. 93-95°C

Solvent: CCl_4

Assignments
a = 1.16
b = 2.36
c = 4.47
d = 4.77
e = 7.24

2797

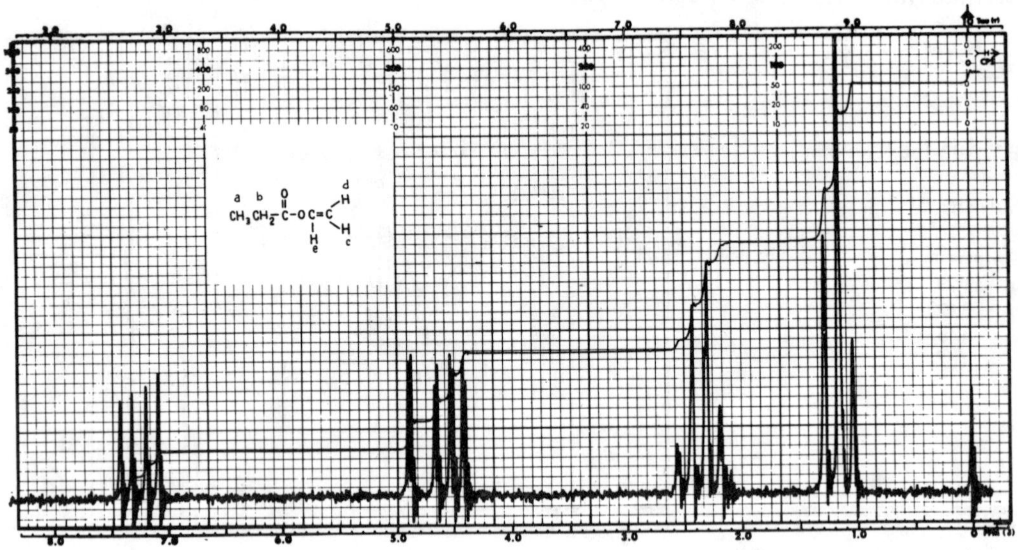

© 1978 Sadtler Research Laboratories, Inc.

4-PENTEN-1-OL, ACETATE

$C_7H_{12}O_2$

Mol. Wt. 128.17

Solvent: CCl_4

Assignments
a = 1.74
b = 1.96
c = 2.06
d = 4.00
e = 4.95
f = 5.00
g = 5.67

2798

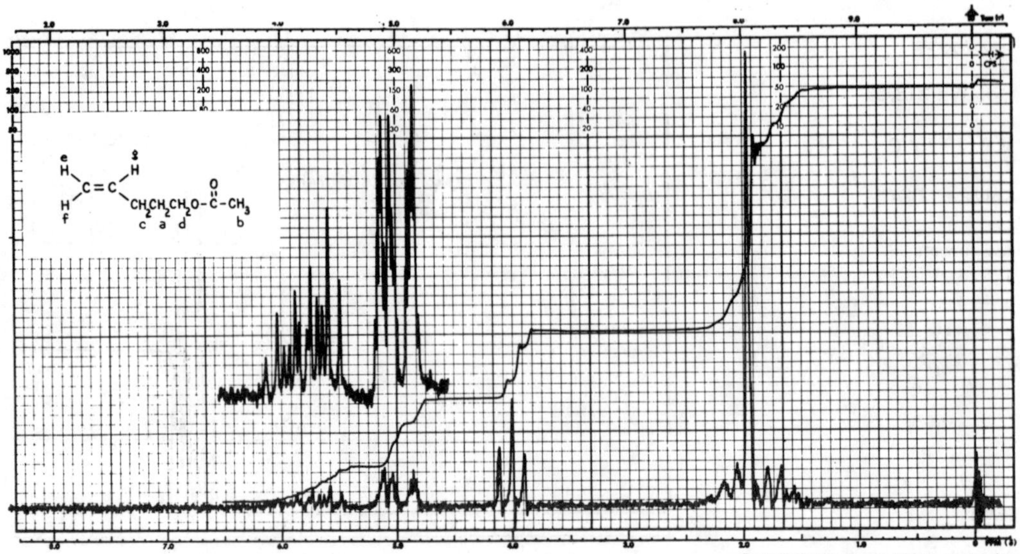

© 1978 Sadtler Research Laboratories, Inc.

2-HEXENOIC ACID, METHYL ESTER

$C_7H_{12}O_2$

Mol. Wt. 128.17

Solvent: CCl_4

Assignments
a = 0.94
b = 1.50
c = 2.16
d = 3.62
e = 5.71
f = 6.83

2799

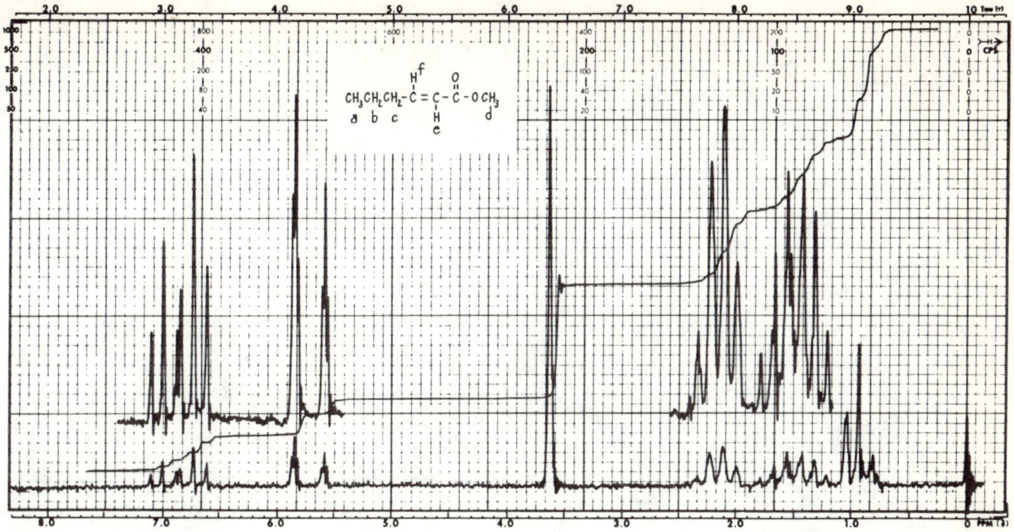

© 1978 Sadtler Research Laboratories, Inc.

trans-2-HEXEN-1-OL, ACETATE

$C_8H_{14}O_2$

Mol. Wt. 142.20

Solvent: CCl_4

Assignments
a = 0.91
b = 1.41
c = 1.97
d = ca 2.01
e = 4.41
f = ca 5.61

2800

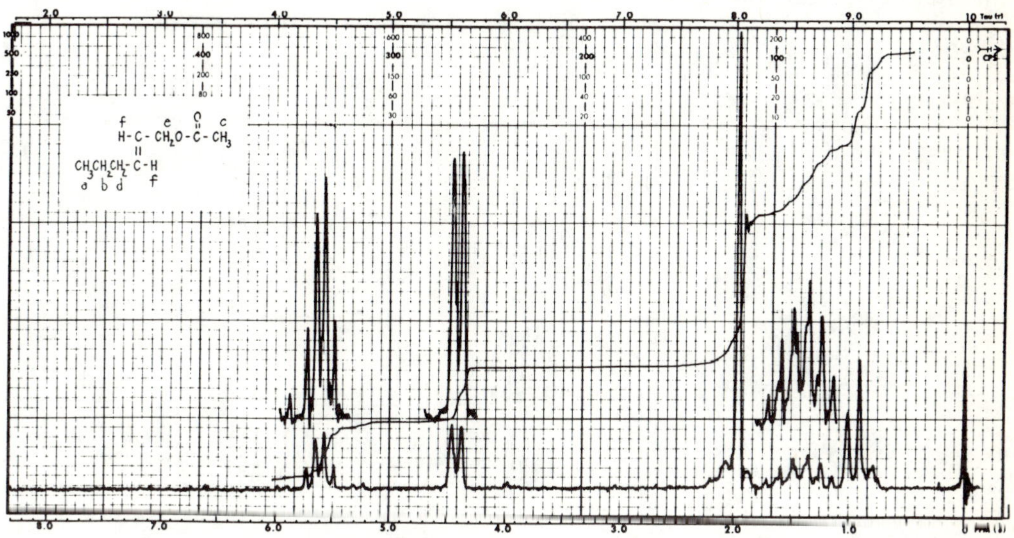

© 1978 Sadtler Research Laboratories, Inc.

NONANOIC ACID, ALLYL ESTER

$C_{12}H_{22}O_2$

Mol. Wt. 198.31

Solvent: CCl_4

Assignments
a = 0.90
b = 1.10-1.90
c = 2.24
d = 4.48
e = 5.13
f = 5.20
g = 5.82

2801

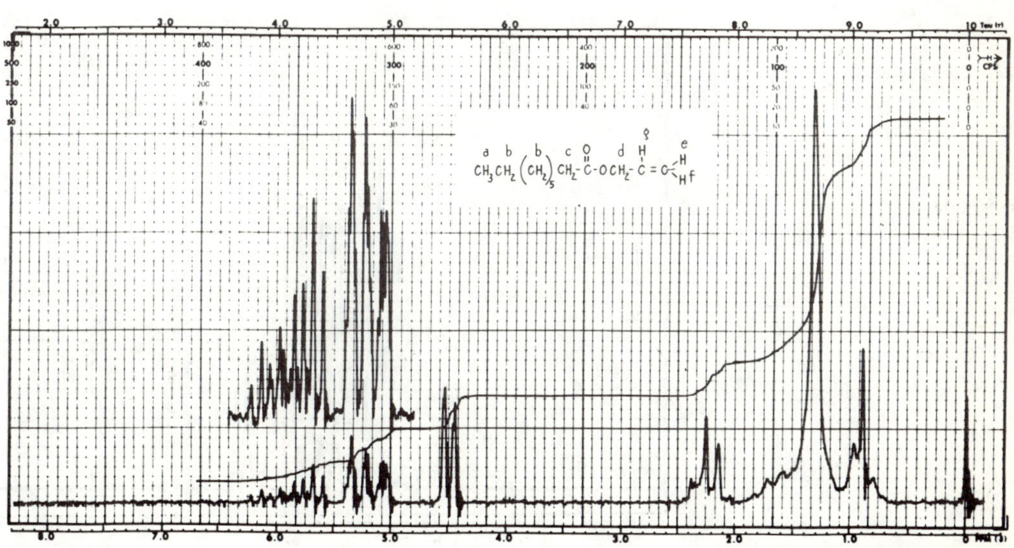

© 1978 Sadtler Research Laboratories, Inc.

STEARIC ACID, VINYL ESTER

$C_{20}H_{38}O_2$

Mol. Wt. 310.52

M.P. 28-30°C

Solvent: $CDCl_3$

Assignments
a = 0.89
b = 1.29
c = 2.38
d = 4.51
e = 4.82
f = 7.23

2802

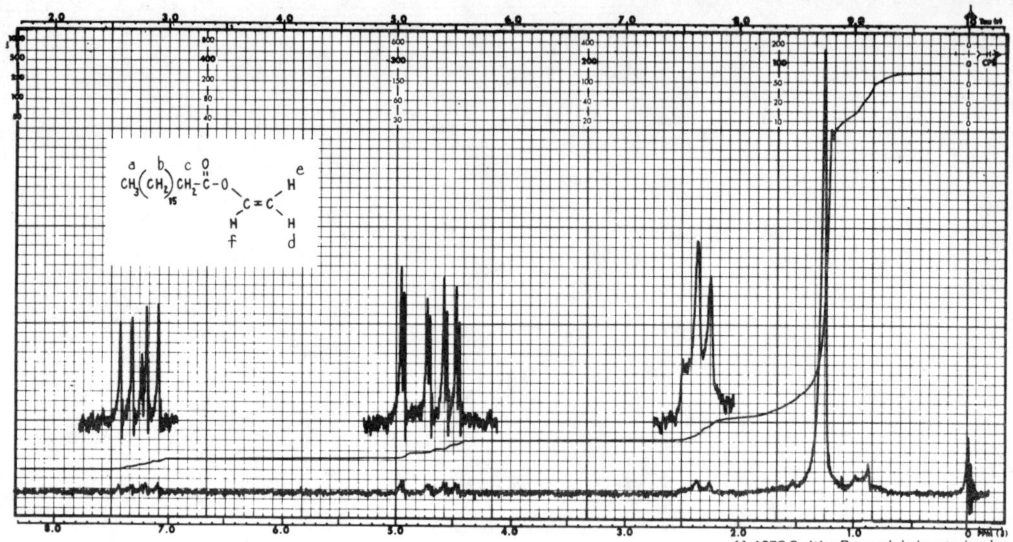

© 1978 Sadtler Research Laboratories, Inc.

CHLOROACETIC ACID, VINYL ESTER

$C_4H_5ClO_2$

Mol. Wt. 120.54

B.P. 37-38°C/16mm

Solvent: $CDCl_3$

Assignments
a = 4.11
b = 4.66
c = 4.94
d = 7.21

2803

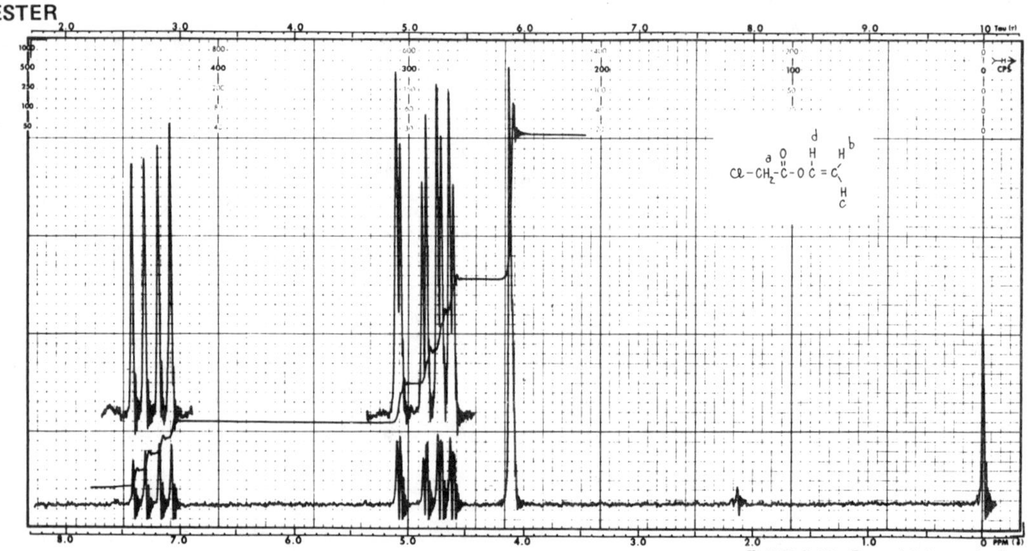

© 1978 Sadtler Research Laboratories, Inc.

CHLOROACETIC ACID, ALLYL ESTER

$C_5H_7ClO_2$

Mol. Wt. 134.56

Solvent: CCl_4

Assignments
a = 4.03
b = 4.62
c = 5.21
d = 5.30
e = 5.89

2804

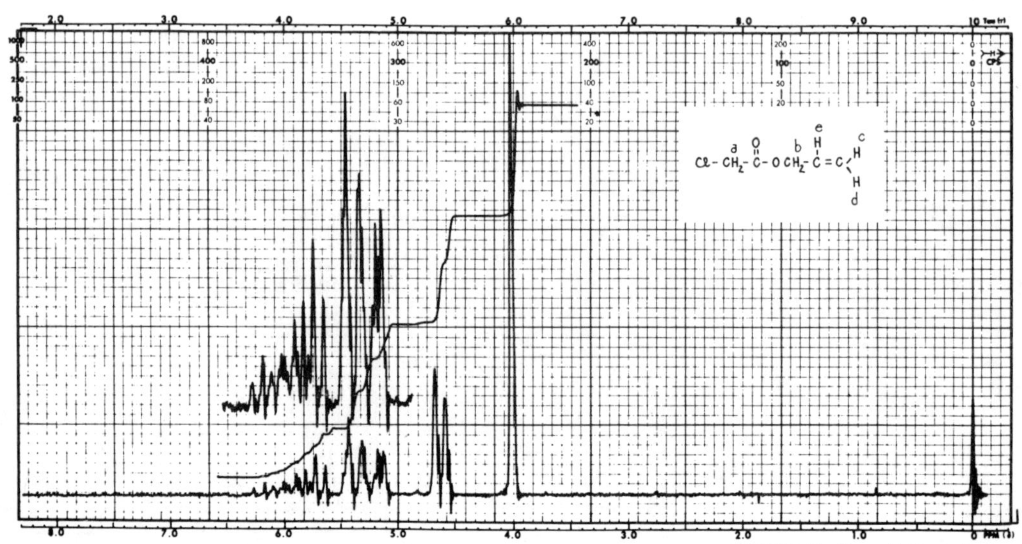

© 1978 Sadtler Research Laboratories, Inc.

OXALIC ACID, DIALLYL ESTER

$C_8H_{10}O_4$

Mol. Wt. 170.17

B.P. 217-219°C

Solvent: $CDCl_3$

Assignments
a = 4.71
b = 5.22
c = 5.33
d = 5.92

2805

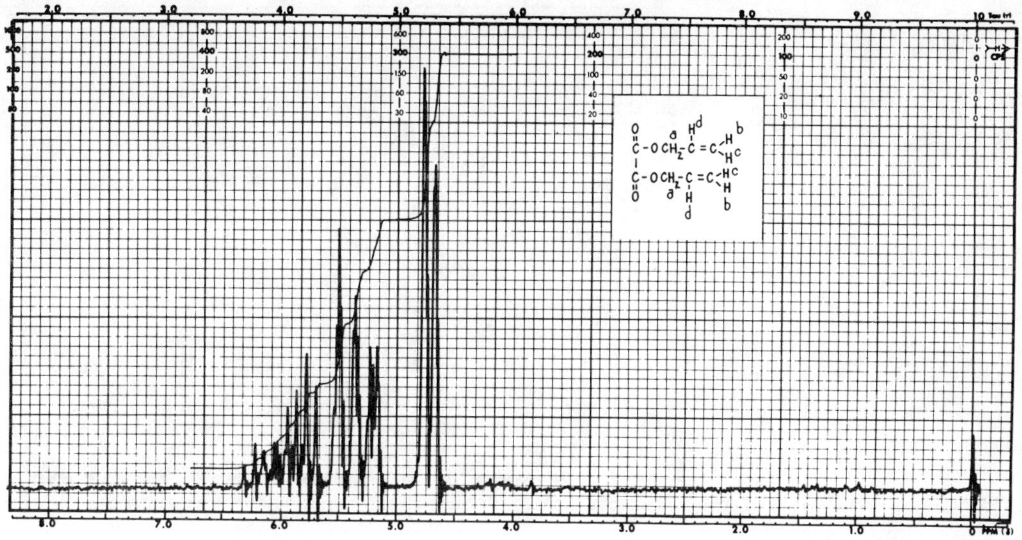

© 1978 Sadtler Research Laboratories, Inc.

SUCCINIC ACID, DIALLYL ESTER

$C_{10}H_{14}O_4$

Mol. Wt. 198.22

B.P. 105-108°C

Solvent: CCl_4

Assignments
a = 2.60
b = 4.53
c = 5.17
d = 5.23
e = 5.87

2806

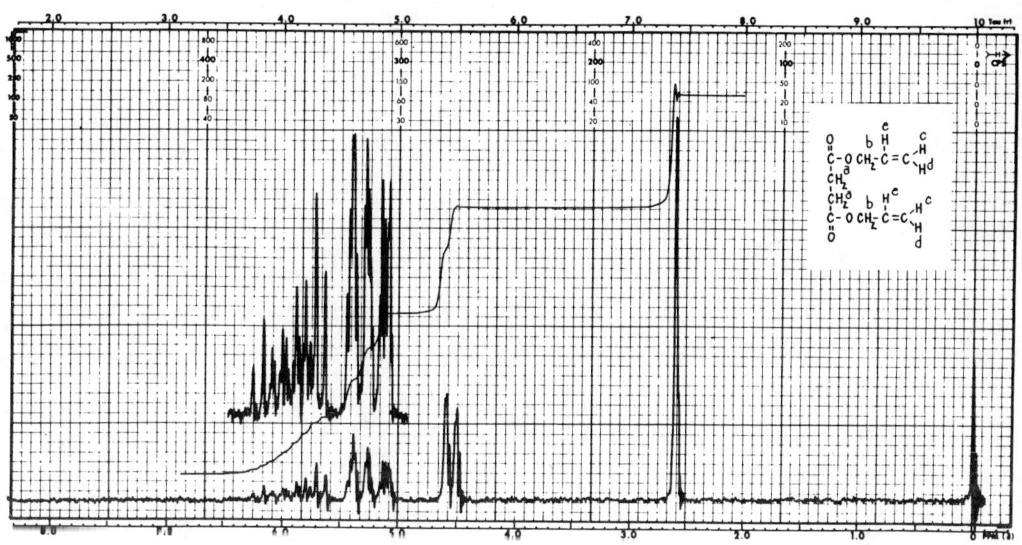

© 1978 Sadtler Research Laboratories, Inc.

ACRYLIC ACID, ISOBUTYL ESTER

$C_7H_{12}O_2$

Mol. Wt. 128.17

Solvent: CCl_4

Assignments
a = 0.96
b = 1.97
c = 3.89
d = 5.72
e = ca 6.09
f = 6.34

2807

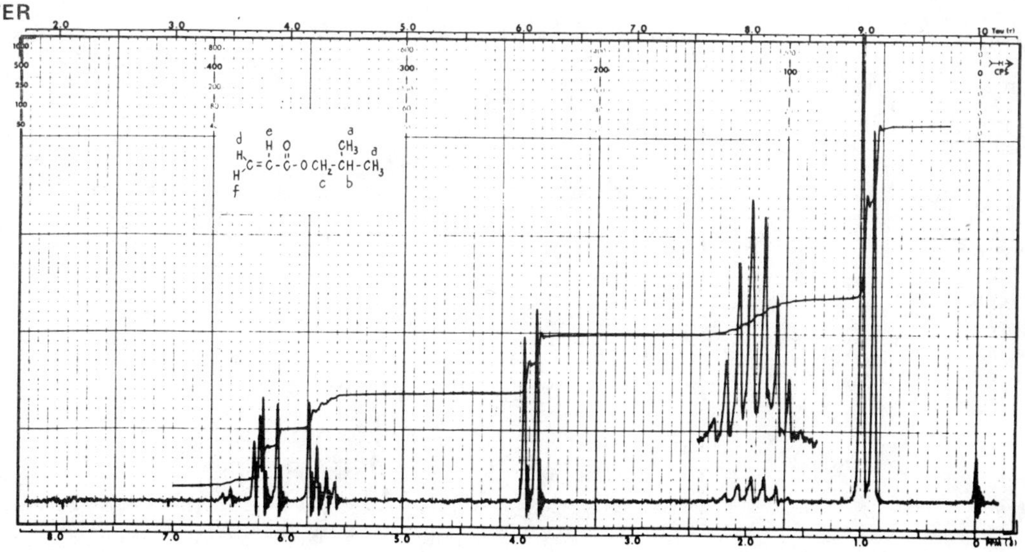

© 1978 Sadtler Research Laboratories, Inc.

ACRYLIC ACID, (2-ETHYLHEXYL) ESTER

$C_{11}H_{20}O_2$

Mol. Wt. 184.28

B.R. 123-127°C/60mm

Solvent: CCl_4

Assignments
a = ca 0.95
b = 1.10-1.87
c = 4.03
d = 5.73
e = ca 6.03
f = 6.37

2808

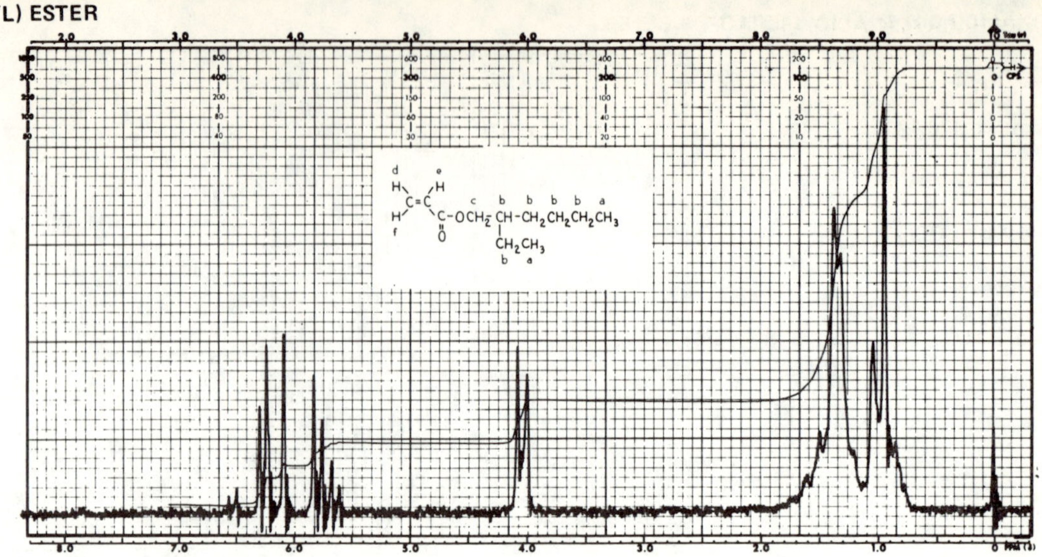

© 1978 Sadtler Research Laboratories, Inc.

ACRYLIC ACID, DODECYL ESTER

$C_{15}H_{28}O_2$

Mol. Wt. 240.39

Solvent: CCl_4

Assignments
a = 0.90
b = 1.30
c = 4.10
d = 5.71
e = 6.18
f = 6.39

2809

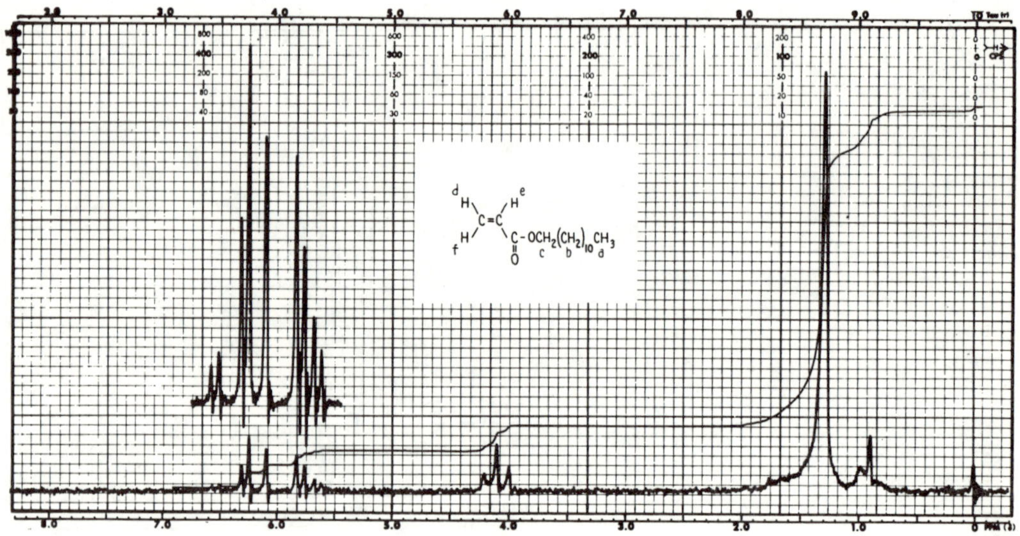

© 1978 Sadtler Research Laboratories, Inc.

METHACRYLIC ACID, ETHYL ESTER

$C_6H_{10}O_2$

Mol. Wt. 114.15

B.P. 118°C

Solvent: CCl_4

Assignments
a = 1.29
b = 1.91
c = 4.15
d = 5.49
e = 6.06

2810

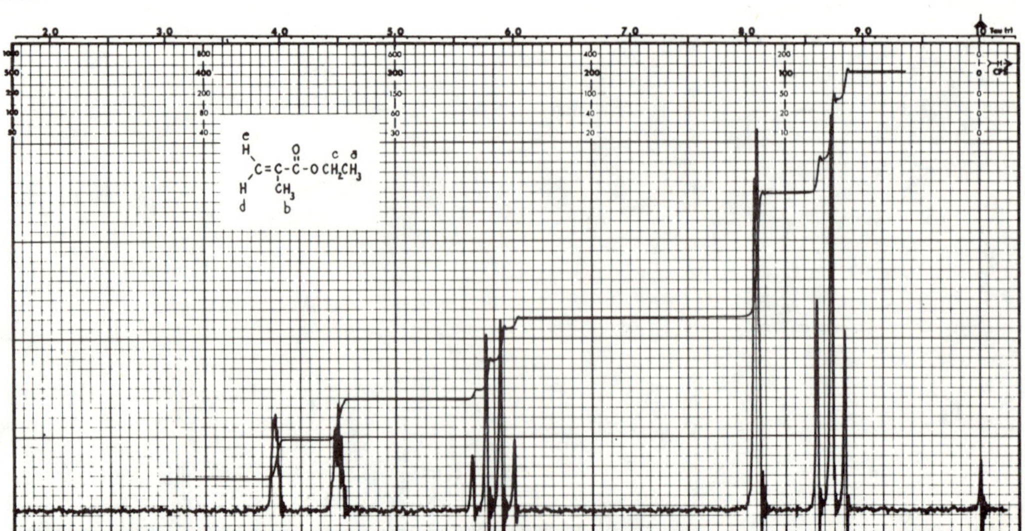

© 1978 Sadtler Research Laboratories, Inc.

METHACRYLIC ACID, ISOPROPYL ESTER

$C_7H_{12}O_2$

Mol. Wt. 128.17

B.P. 125-126°C

Solvent: CCl_4

Assignments
- a = 1.27
- b = 1.90
- c = 4.99
- d = 5.43
- e = 5.99

2811

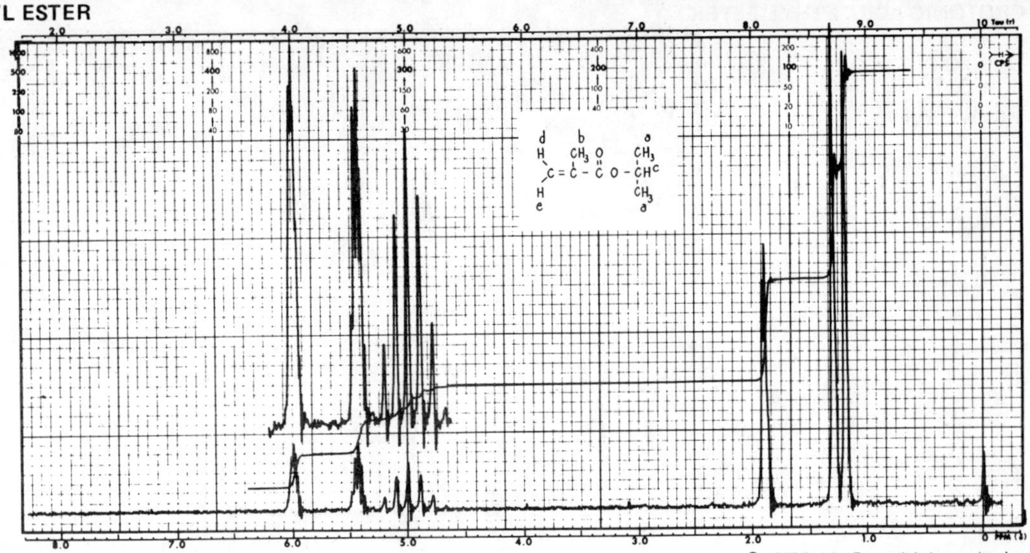

© 1978 Sadtler Research Laboratories, Inc.

METHACRYLIC ACID, ISOBUTYL ESTER

$C_8H_{14}O_2$

Mol. Wt. 142.20

Solvent: CCl_4

Assignment
- a = 0.98
- b = 1.89
- c = 1.92
- d = 3.89
- e = 5.47
- f = 6.04

2812

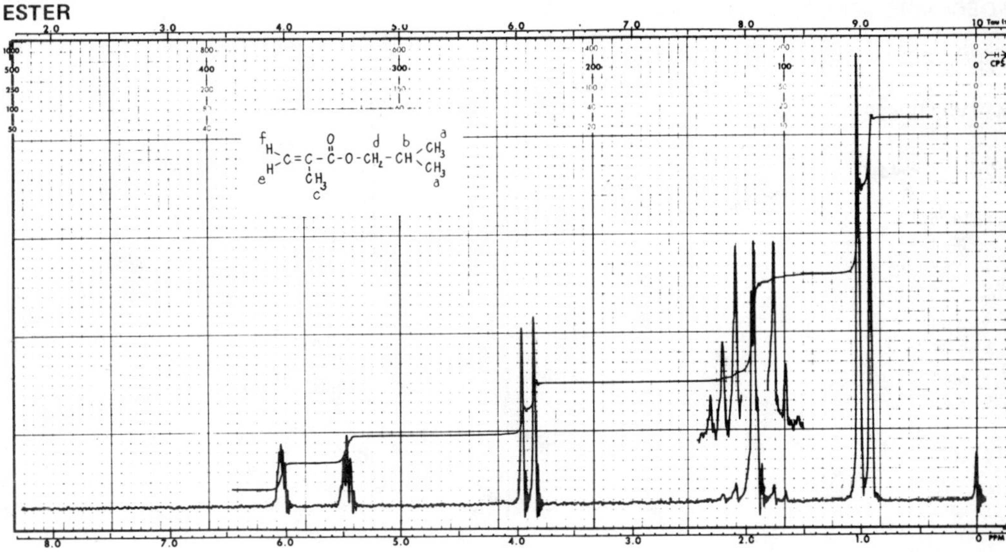

© 1978 Sadtler Research Laboratories, Inc.

METHACRYLIC ACID, DECYL ESTER

$C_{14}H_{26}O_2$

Mol. Wt. 226.36

B.P. 155-156°C/22mm

Solvent: CCl_4

Assignments
- a = 0.90
- b = 1.30
- c = 1.92
- d = 4.10
- e = 5.39
- f = 6.02

2813

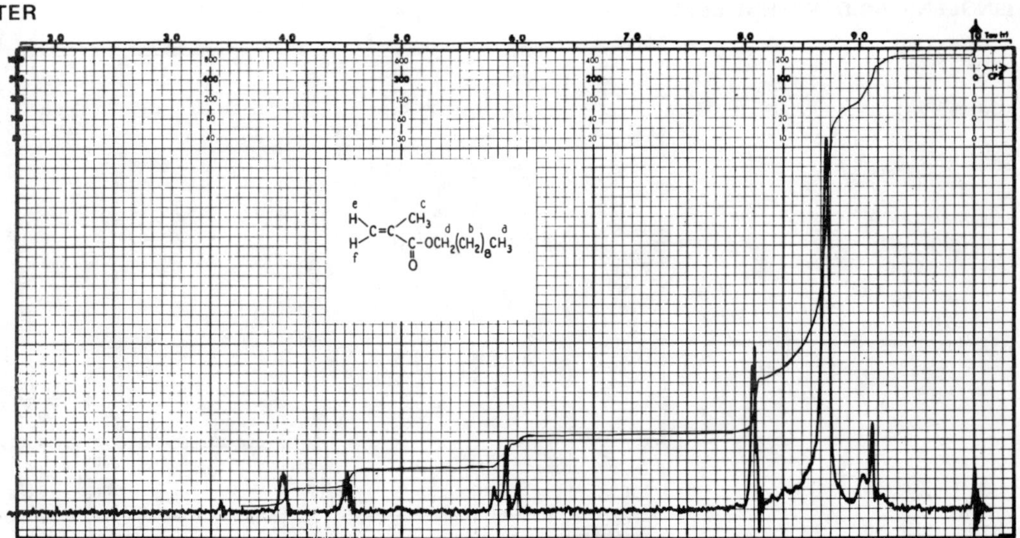

© 1978 Sadtler Research Laboratories, Inc.

CROTONIC ACID, ETHYL ESTER

$C_6H_{10}O_2$

Mol. Wt. 114.15

B.P. 138-139°C

Solvent: CCl_4

Assignments
a = 1.24
b = 1.89
c = 4.11
d = 5.79
e = 6.90

2814

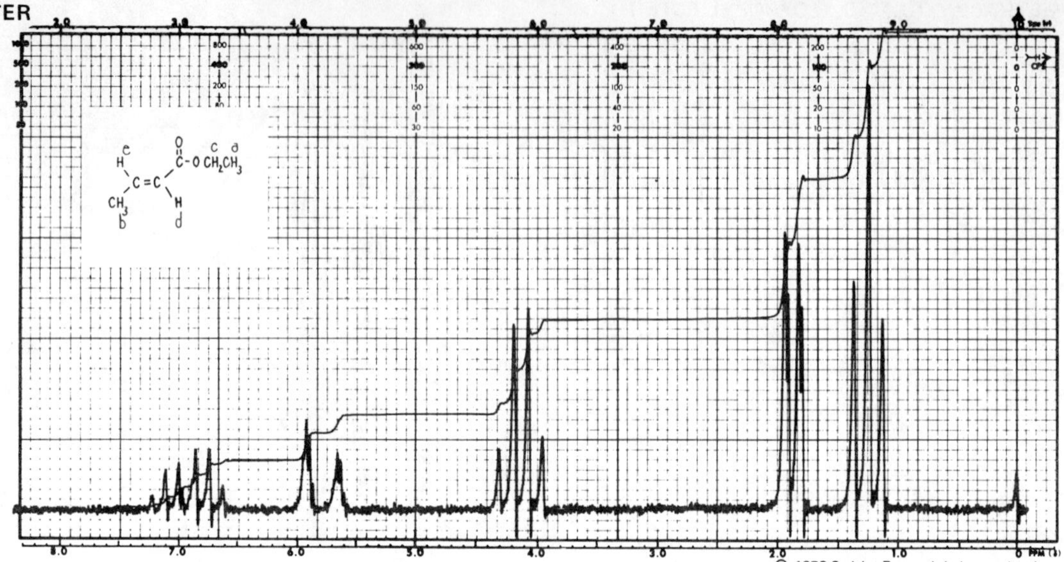

© 1978 Sadtler Research Laboratories, Inc.

2-DECENOIC ACID, METHYL ESTER

$C_{11}H_{20}O_2$

Mol. Wt. 184.28

Solvent: CCl_4

Assignments
a = 0.89
b = 1.10-1.75
c = 2.16
d = 3.63
e = 5.71
f = 6.84
g = 5.47 imp.

2815

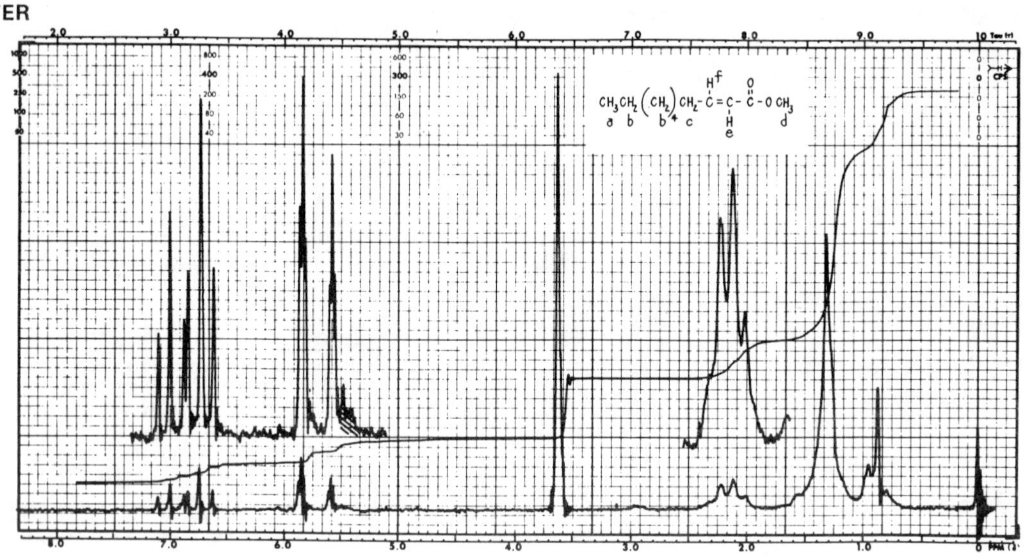

© 1978 Sadtler Research Laboratories, Inc.

LINOLENIC ACID, METHYL ESTER

$C_{19}H_{32}O_2$

Mol. Wt. 292 .47

Solvent: CCl_4

Assignments
a = 0.90
b = 1.10-1.80
c = 1.80-2.40
d = 2.71
e = 3.59
f = ca 5.29

2816

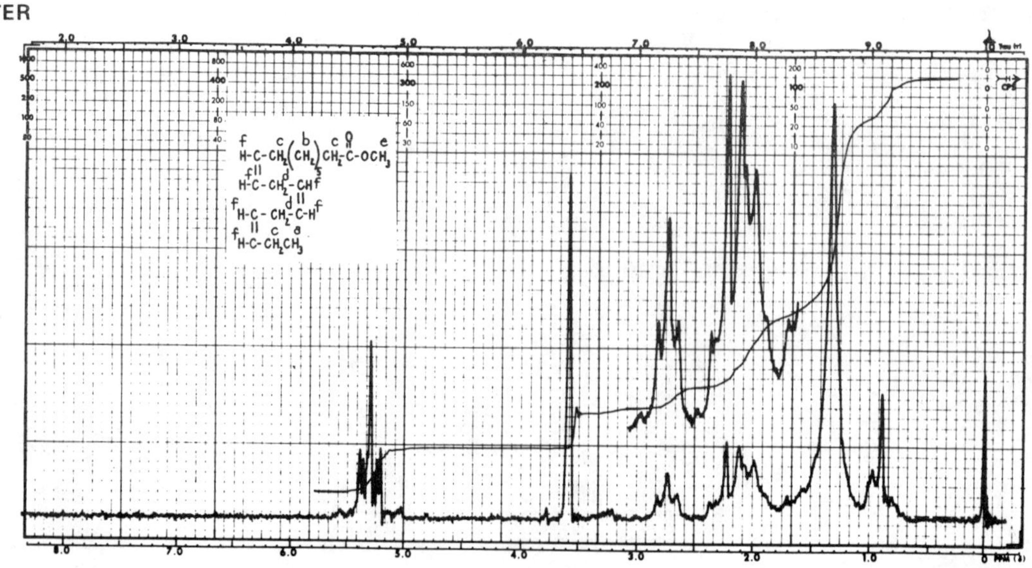

© 1978 Sadtler Research Laboratories, Inc.

OLEIC ACID, ETHYL ESTER

$C_{20}H_{38}O_2$

Mol. Wt. 310.52

B.P. 205-208°C/10mm

Solvent: CCl_4

Assignments
a = 0.88
b = 1.22
c = 1.29
d = 1.96
e = 2.20
f = 4.06
g = 5.28

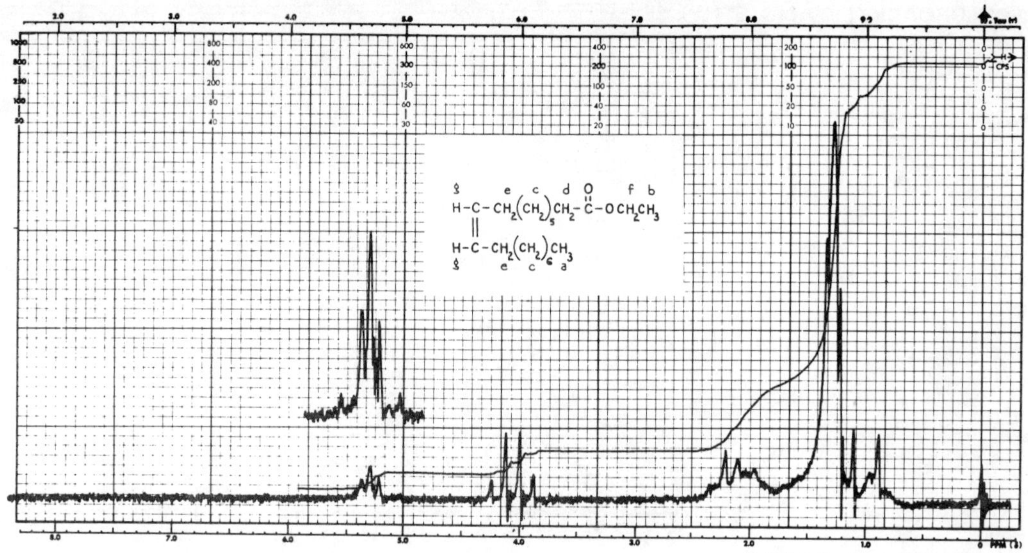

2817

© 1978 Sadtler Research Laboratories, Inc.

ELAIDIC ACID, ETHYL ESTER

$C_{20}H_{38}O_2$

Mol. Wt. 310.52

B.P. 185-187°C/0.1mm (lit.)

Solvent: CCl_4

Assignments
a = 0.88
b = 1.21
c = ca 1.28
d = 1.73-2.30
e = 4.04
f = 5.30

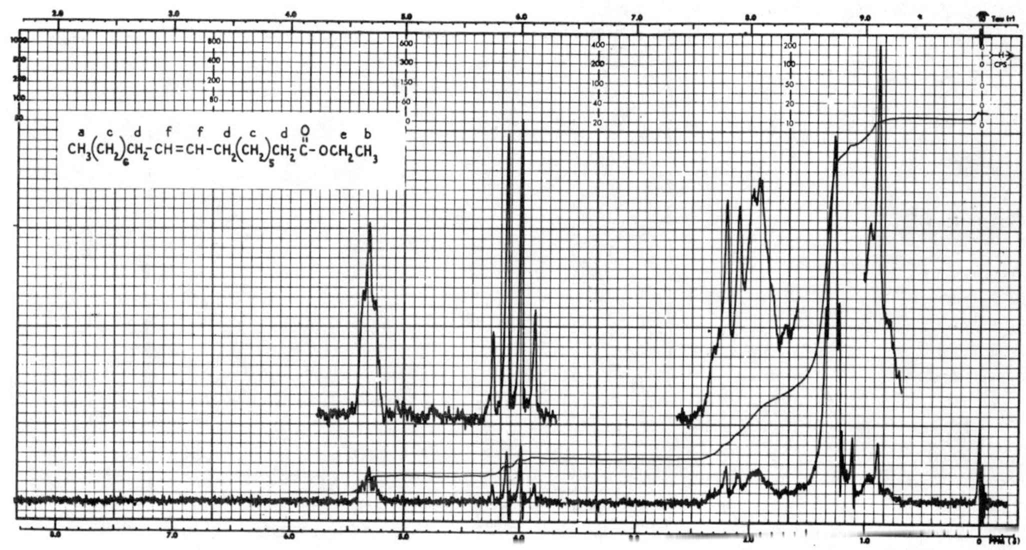

2818

© 1978 Sadtler Research Laboratories, Inc.

ELAIDIC ACID, BUTYL ESTER

$C_{22}H_{42}O_2$

Mol. Wt. 338.58

Solvent: $CDCl_3$

Assignments
a = 0.90
b = 1.10-1.80
c = ca 1.95
d = 2.29
e = 4.06
f = 5.38

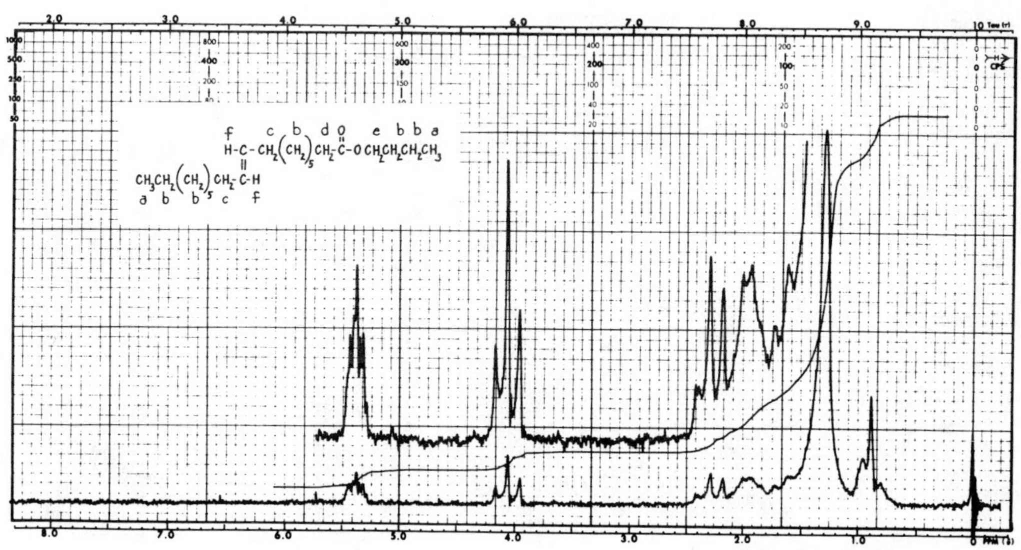

2819

© 1978 Sadtler Research Laboratories, Inc.

2-CHLOROACRYLIC ACID, ETHYL ESTER

$C_5H_7ClO_2$

Mol. Wt. 134.56

Solvent: CCl_4

Assignments
a = 1.33
b = 4.29
c = 5.99
d = 6.51
e = 3.83 imp.

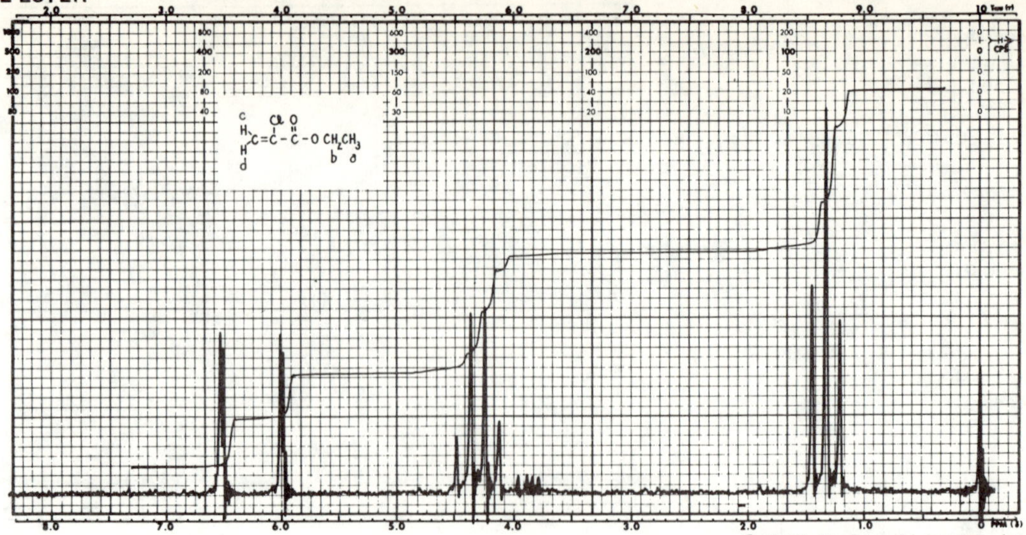

2820

© 1978 Sadtler Research Laboratories, Inc.

4-OXO-2-PENTENOIC ACID, METHYL ESTER

$C_6H_8O_3$

Mol. Wt. 128.13

M.P. 60°C

Solvent: $CDCl_3$

Assignments
a = 2.37
b = 3.82
c = 6.59
d = 7.03

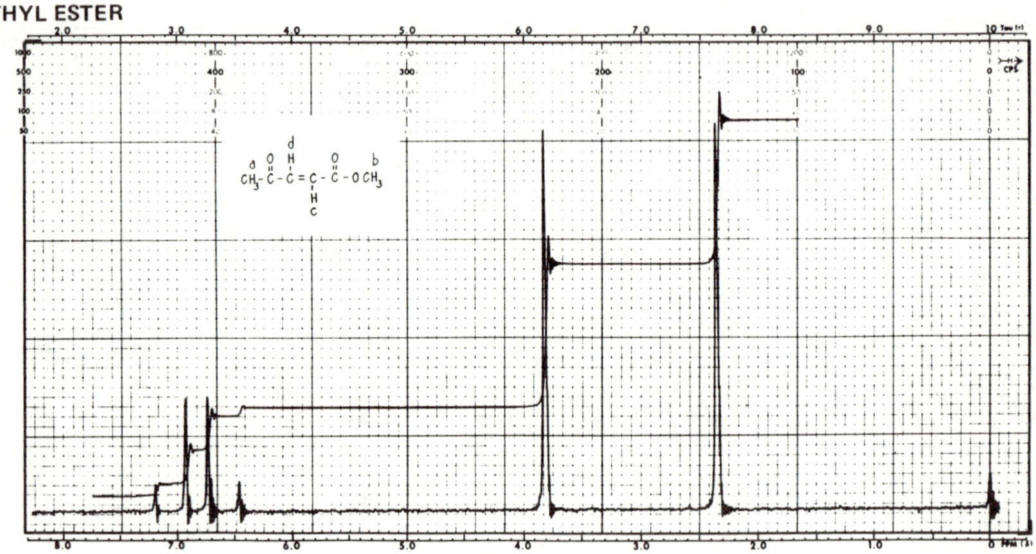

2821

© 1978 Sadtler Research Laboratories, Inc.

RICINOLEIC ACID, METHYL ESTER, ACETATE

$C_{21}H_{38}O_4$

Mol. Wt. 354.53

B.P. 181-184°C/0.14mm

Solvent: CCl_4

Assignments
a = 0.89
b = 1.06-1.78
c = 1.95
d = 1.95-2.42
e = 3.60
f = 4.77
g = ca 5.35

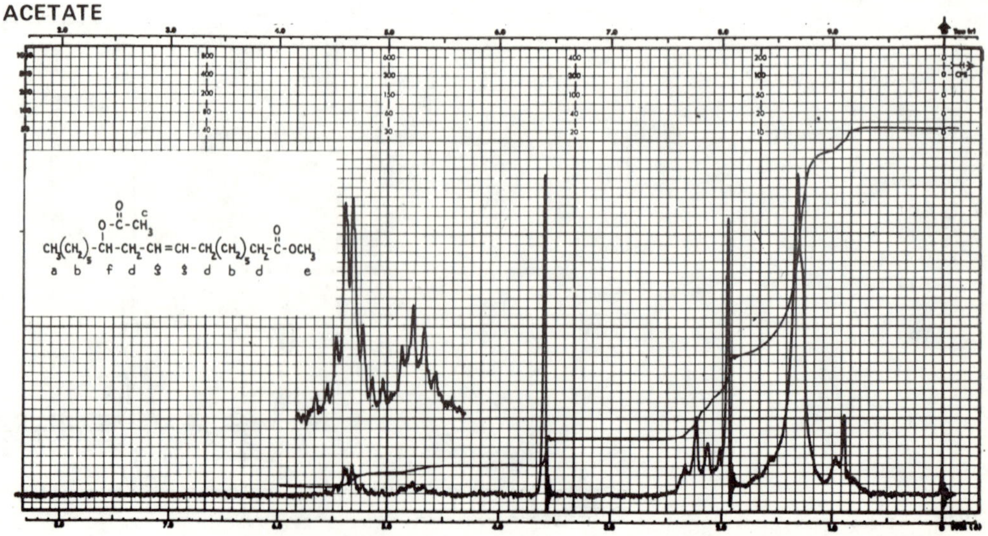

2822

© 1978 Sadtler Research Laboratories, Inc.

TRIOLEIN

$C_{57}H_{104}O_6$

Mol. Wt. 885.46

B.P. 235-240°C/18mm

Solvent: CCl_4

Assignments
a = 0.90
b = ca 1.30
c = ca 2.01
d = 2.28
e = 3.87-4.42
f = 5.00-5.60

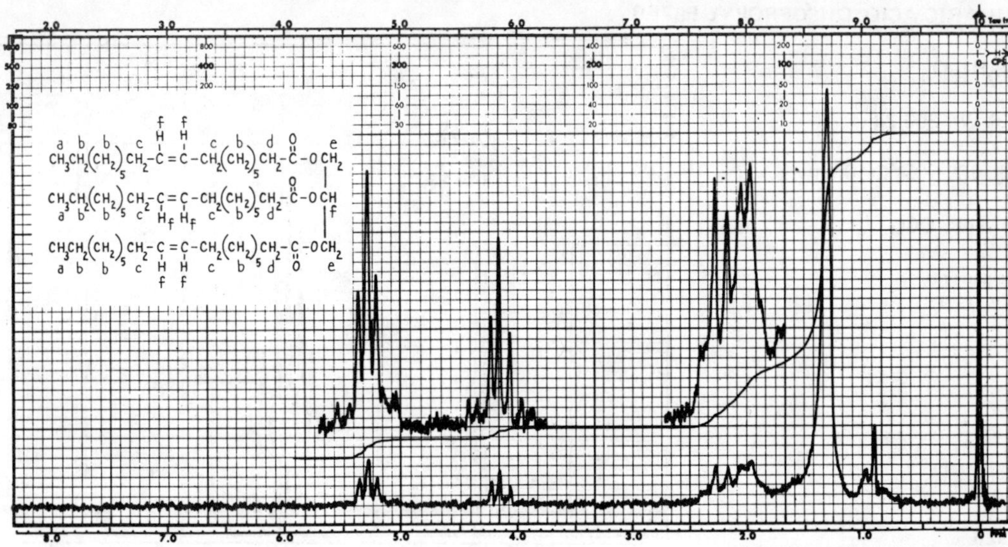

© 1978 Sadtler Research Laboratories, Inc.

2823

5-NORBORNENE-2-CARBOXYLIC ACID, ETHYL ESTER

$C_{10}H_{14}O_2$

Mol. Wt. 166.22

B.P. 86°C/10mm

Solvent: CCl_4

Assignments
a = 1.20 or b
b = 1.24 or a
c = 1.40-2.30
d = 2.69-3.28
e = 4.01 or f
f = 4.09 or e
g = 5.77-6.22

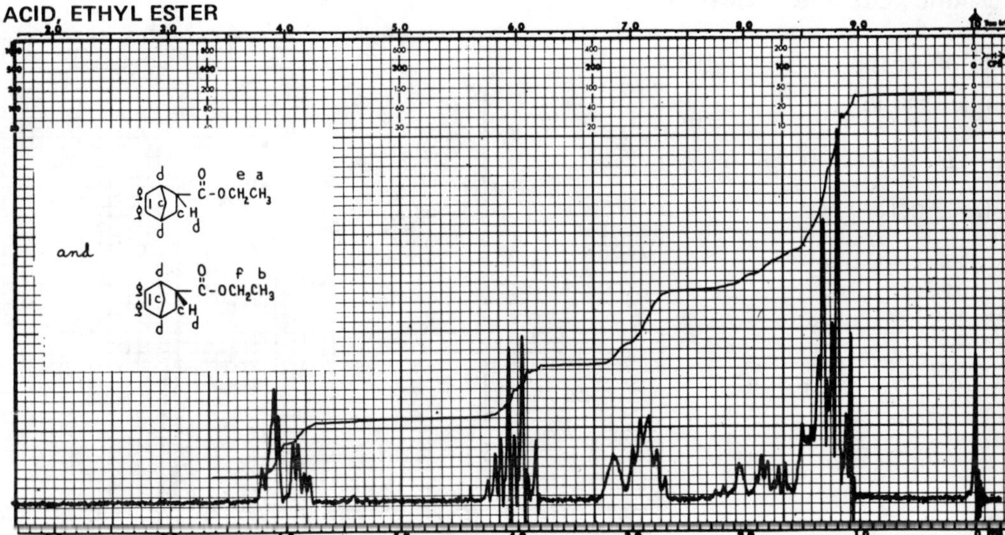

© 1978 Sadtler Research Laboratories, Inc.

2824

FUMARIC ACID, DIMETHYL ESTER

$C_6H_8O_4$

Mol. Wt. 144.13

B.P. 192°C

M.P. 102°C

Solvent: $CDCl_3$

Assignments
a = 3.80
b = 6.86

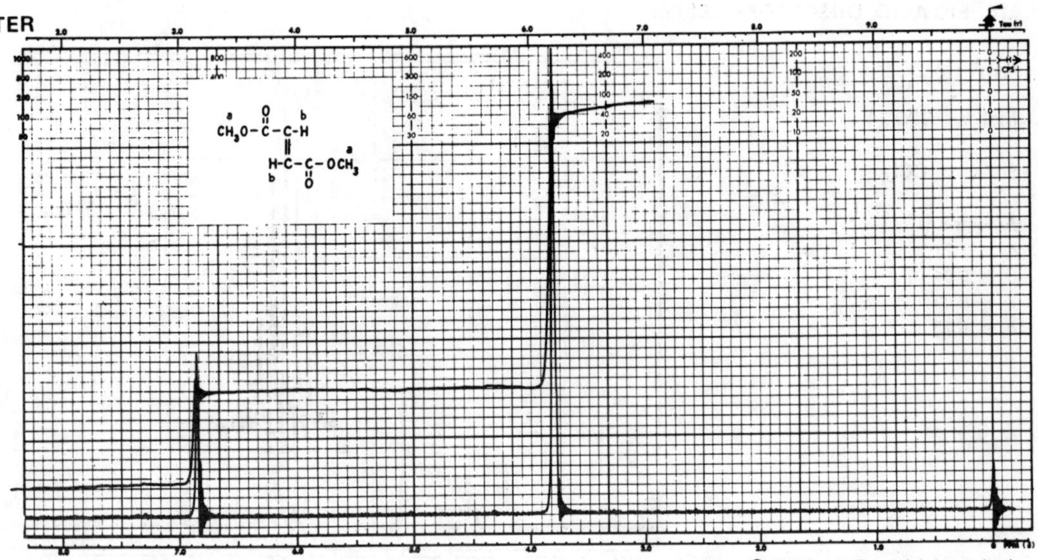

© 1978 Sadtler Research Laboratories, Inc.

2825

FUMARIC ACID, DIISOPROPYL ESTER

$C_{10}H_{16}O_4$

Mol. Wt. 200.24

B.P. 108-109°C/10mm

Solvent: CCl_4

Assignments
a = 1.29
b = 5.07
c = 6.71

2826

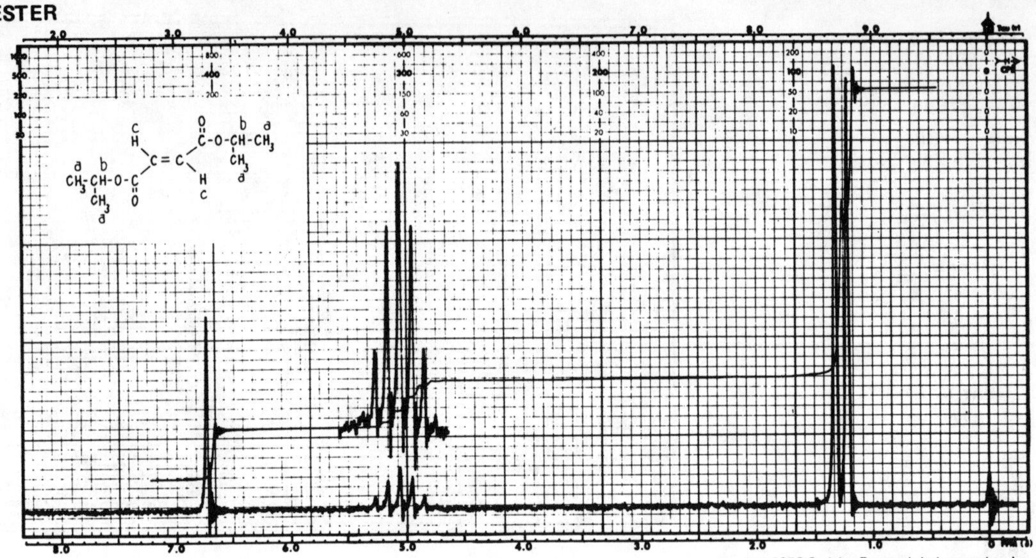

© 1978 Sadtler Research Laboratories, Inc.

FUMARIC ACID, DIHEXYL ESTER

$C_{16}H_{28}O_4$

Mol. Wt. 284.40

B.P. 134°C/0.4mm

Solvent: CCl_4

Assignments
a = 0.92
b = 1.10-1.99
c = 4.17
d = 6.79

2827

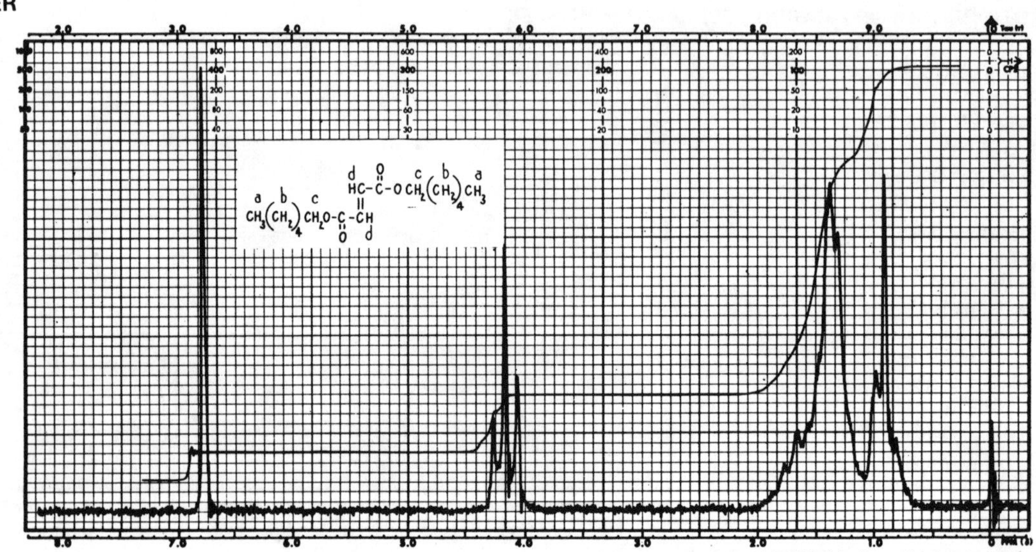

© 1978 Sadtler Research Laboratories, Inc.

MALEIC ACID, DIISOPROPYL ESTER

$C_{10}H_{16}O_4$

Mol. Wt. 200.24

Solvent: CCl_4

Assignments
a = 1.27
b = 5.04
c = 6.09
d = 6.73

2828

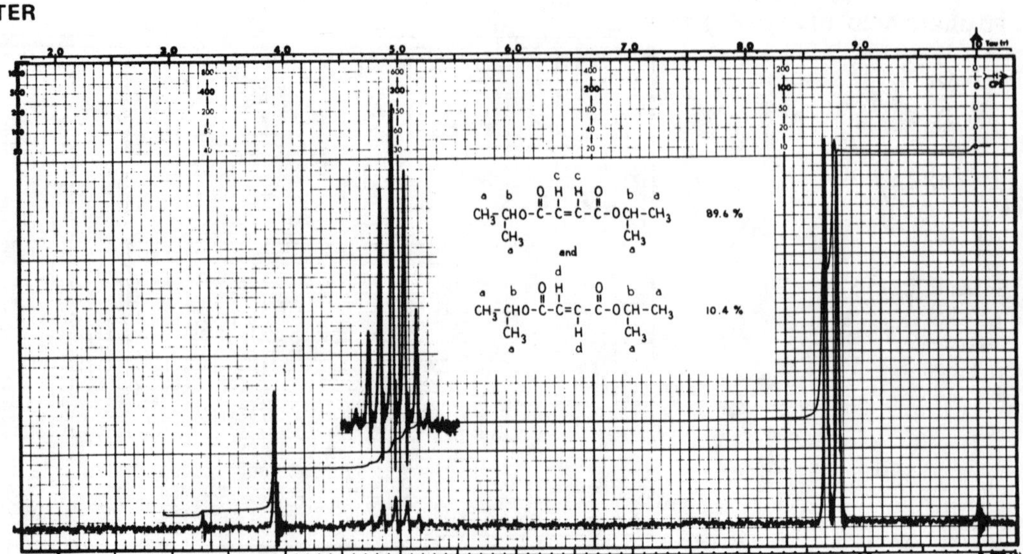

© 1978 Sadtler Research Laboratories, Inc.

DIALLYL MALEATE

$C_{10}H_{12}O_4$

Mol. Wt. 196.20

Solvent: CCl_4

Assignments
a = 4.58
b = 5.15
c = 5.23
d = 5.80
e = 6.15
f = 6.20, 6.39 imps.

2829

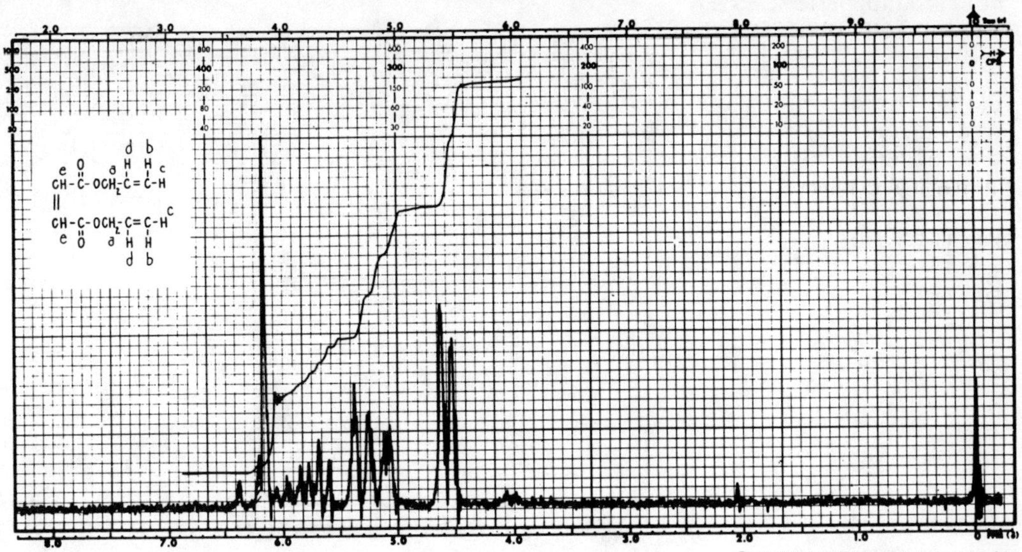

© 1978 Sadtler Research Laboratories, Inc.

ETHENETETRACARBOXYLIC ACID, TETRAETHYL ESTER

$C_{14}H_{20}O_8$

Mol. Wt. 316.31

M.P. 56-58°C

Solvent: $CDCl_3$

Assignments
a = 1.30
b = 4.30

2830

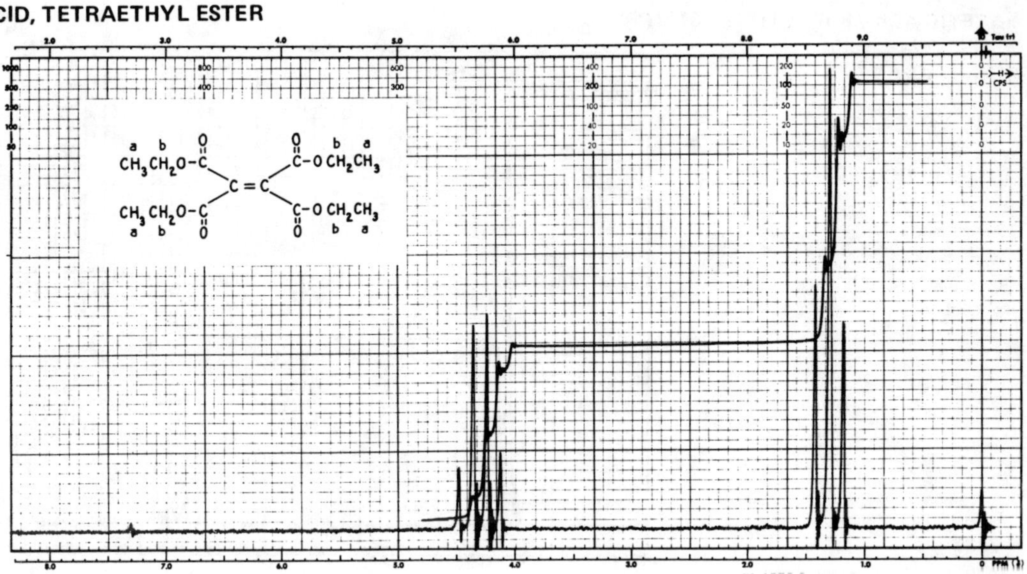

© 1978 Sadtler Research Laboratories, Inc.

FORMIC ACID, PHENETHYL ESTER

$C_9H_{10}O_2$

Mol. Wt. 150.18

Solvent: CCl_4

Assignments
a = 2.90
b = 4.30
c = 7.16
d = 7.90

2831

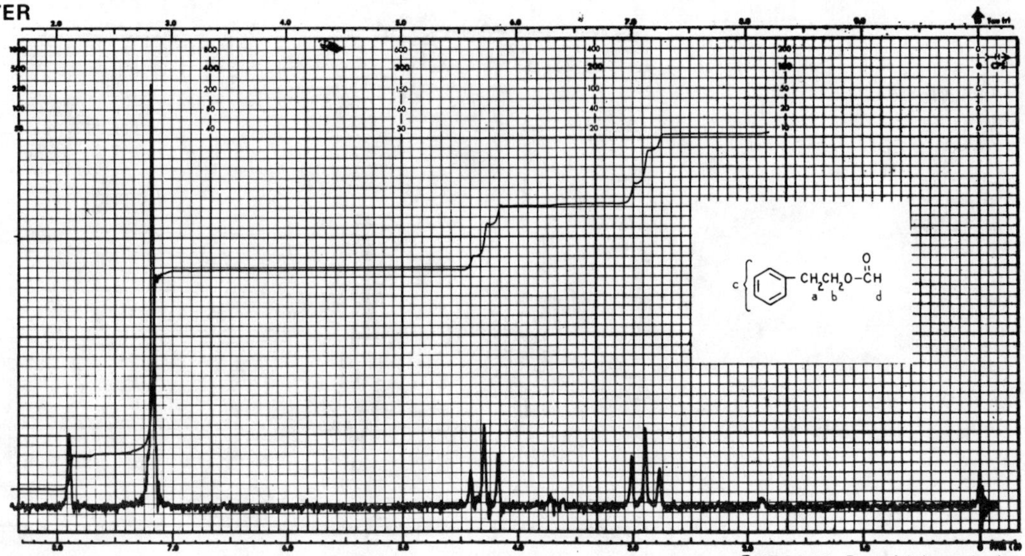

© 1978 Sadtler Research Laboratories, Inc.

ACETIC ACID, BENZYL ESTER

$C_9H_{10}O_2$

Mol. Wt. 150.18

Solvent: CCl_4

Assignments

a = 1.96

b = 4.99

c = 7.20

2832

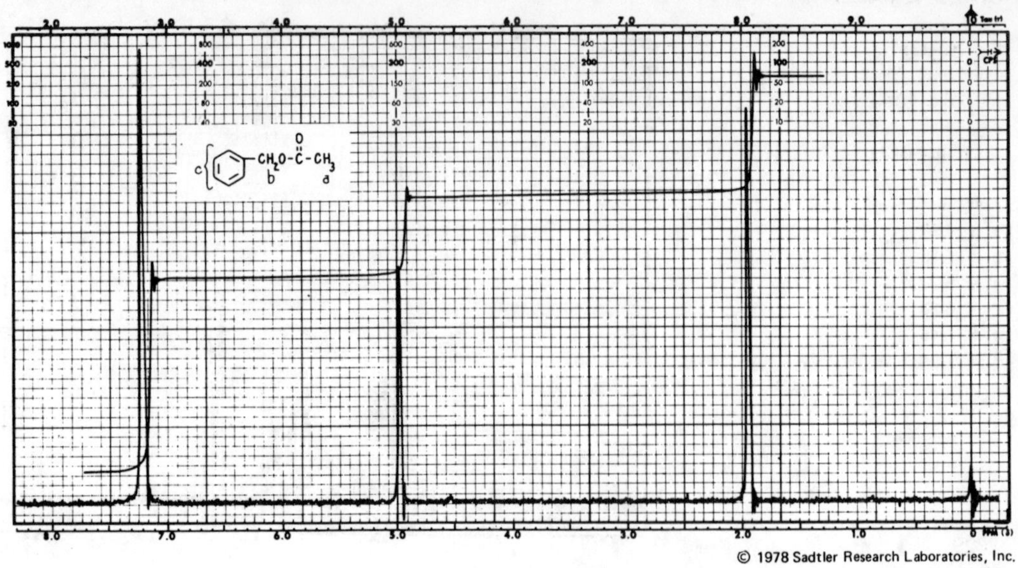

© 1978 Sadtler Research Laboratories, Inc.

ACETIC ACID, PHENETHYL ESTER

$C_{10}H_{12}O_2$

Mol. Wt. 164.21

Solvent: CCl_4

Assignments

a = 1.90

b = 2.82

c = 4.16

d = 7.11

2833

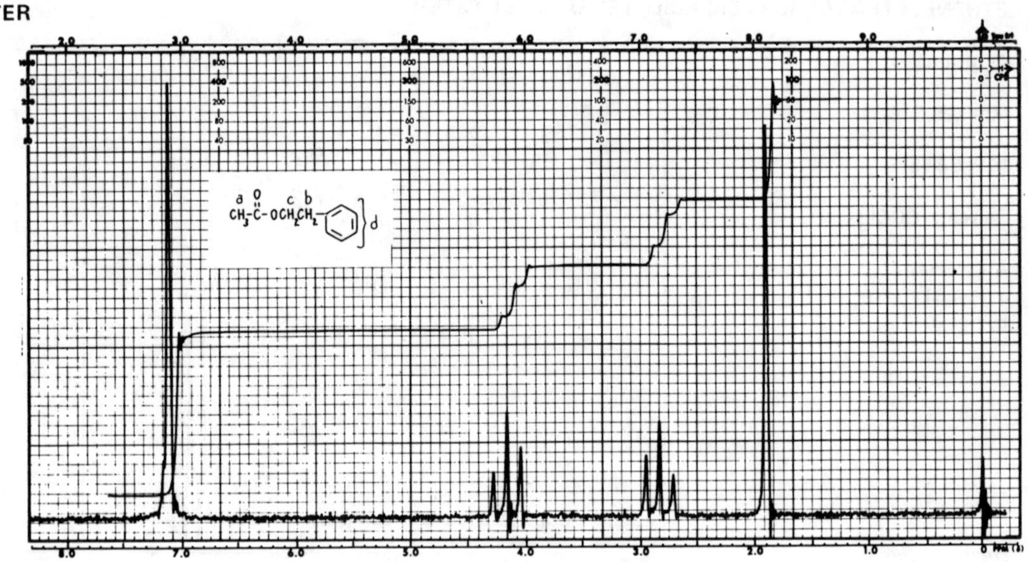

© 1978 Sadtler Research Laboratories, Inc.

α-METHYLBENZYL ALCOHOL, ACETATE

$C_{10}H_{12}O_2$

Mol. Wt. 164.21

Solvent: $CDCl_3$

Assignments

a = 1.52

b = 2.03

c = 5.90

d = 7.32

2834

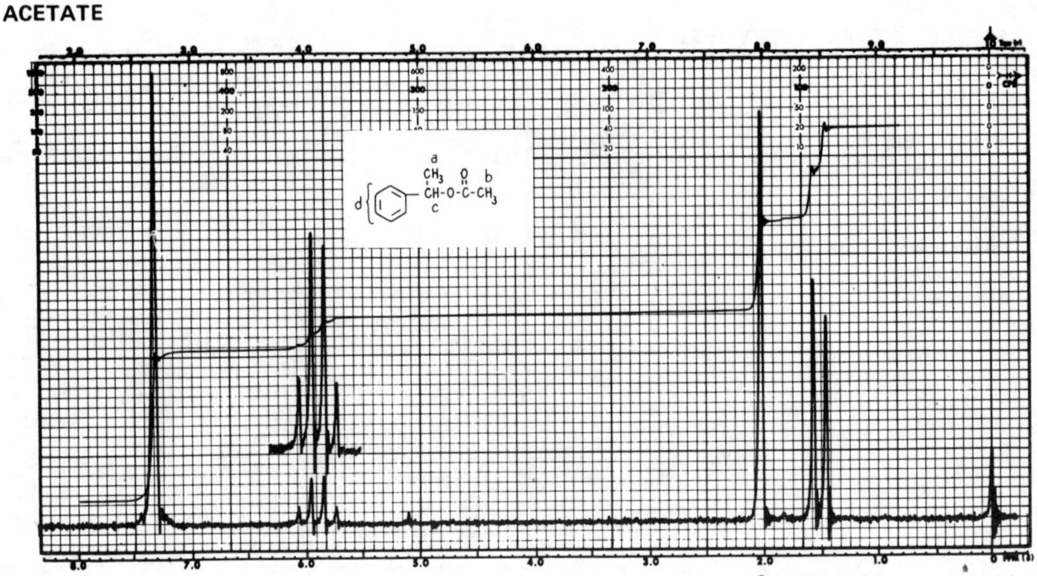

© 1978 Sadtler Research Laboratories, Inc.

PROPIONIC ACID, PHENETHYL ESTER

$C_{11}H_{14}O_2$

Mol. Wt. 178.23

Solvent: CCl_4

Assignments
a = 1.09
b = 2.22
c = 2.88
d = 4.22
e = 7.19

2835

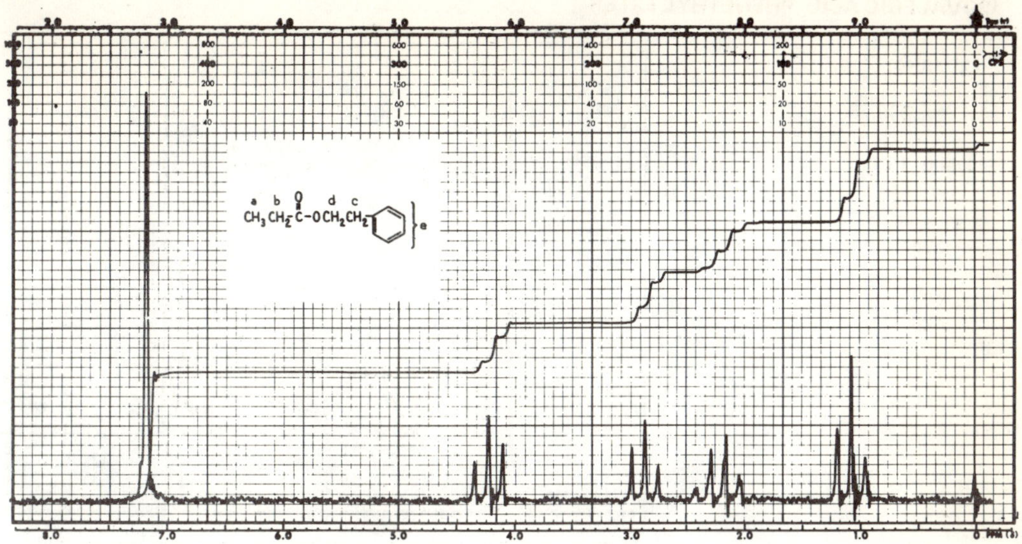

© 1978 Sadtler Research Laboratories, Inc.

ISOBUTYRIC ACID, PHENETHYL ESTER

$C_{12}H_{16}O_2$

Mol. Wt. 192.26

Solvent: CCl_4

Assignments
a = 1.12
b = 2.45
c = 2.89
d = 4.21
e = ca 7.18

2836

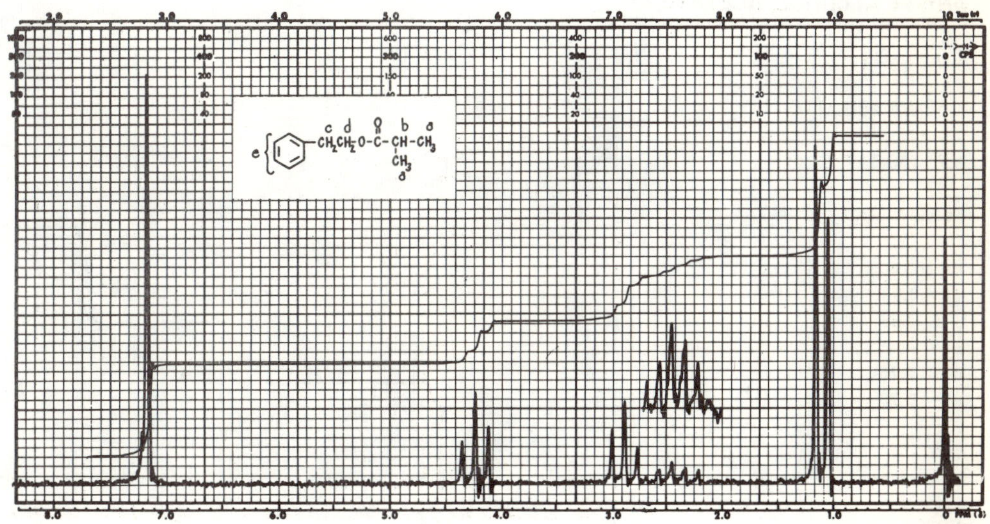

© 1978 Sadtler Research Laboratories, Inc.

ISOVALERIC ACID, BENZYL ESTER

$C_{12}H_{16}O_2$

Mol. Wt. 192.26

Solvent: CCl_4

Assignments
a = 0.90
b = 2.10
c = 4.90
d = 7.20

2837

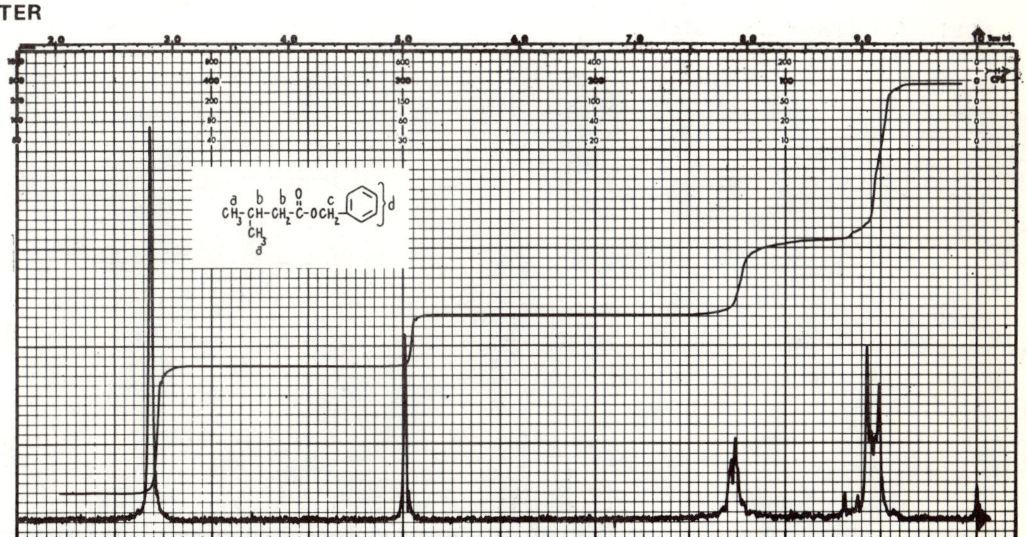

© 1978 Sadtler Research Laboratories, Inc.

ISOVALERIC ACID, PHENETHYL ESTER

$C_{13}H_{18}O_2$

Mol. Wt. 206.28

Solvent: CCl_4

Assignments
a = 0.90
b = 2.08
c = 2.08
d = 2.85
e = 4.20
f = 7.16
g = 1.00 imp.
h = 1.12 imp.

2838

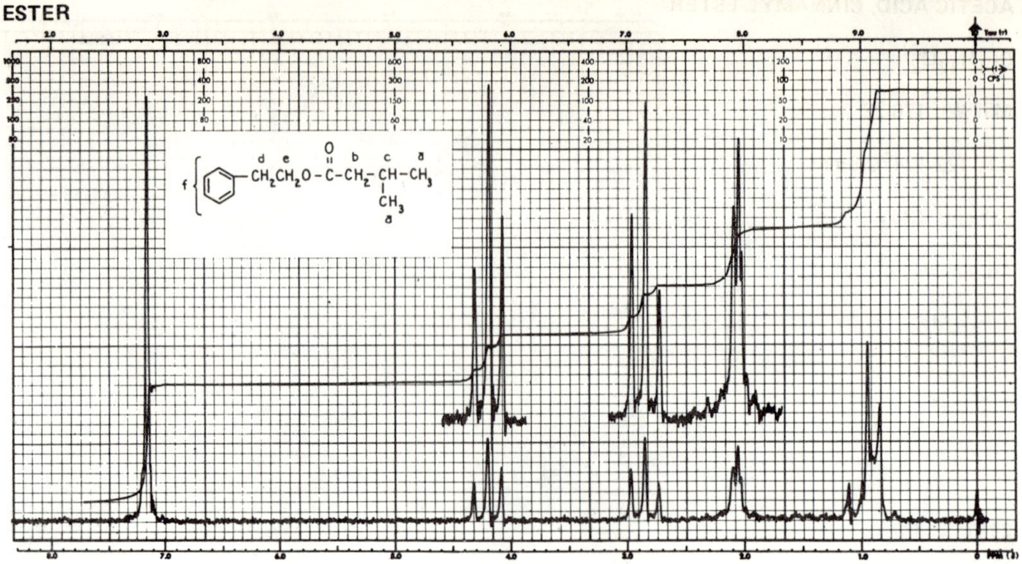

© 1978 Sadtler Research Laboratories, Inc.

ANISYL ALCOHOL, ACETATE

$C_{10}H_{12}O_3$

Mol. Wt. 180.21

B.P. 110-117°C

Solvent: CCl_4

Assignments
a = 1.93
b = 3.68
c = 4.91
d = 6.75
e = 7.20

2839

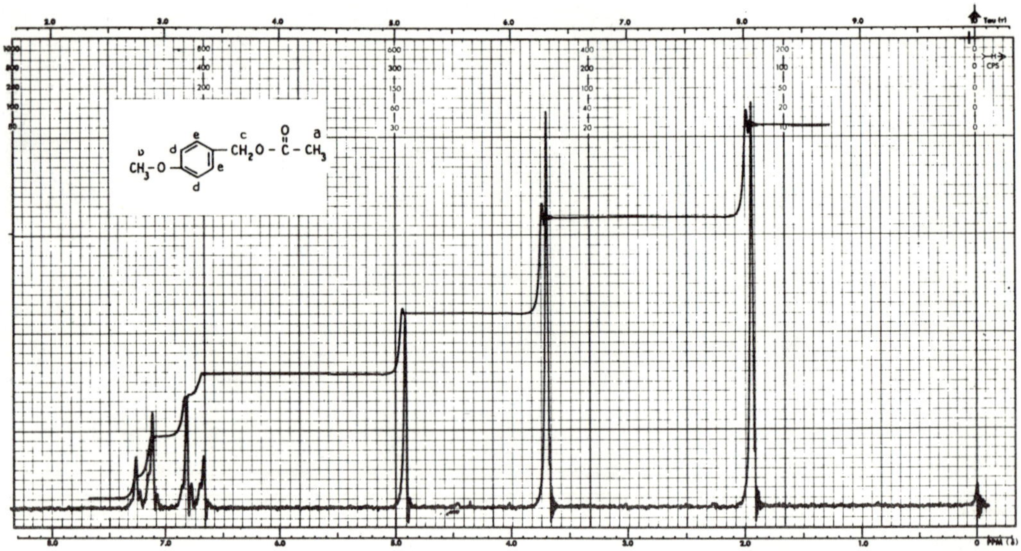

© 1978 Sadtler Research Laboratories, Inc.

ACETIC ACID, PHENACYL ESTER

$C_{10}H_{10}O_3$

Mol. Wt. 178.19

M.P. 47-48°C

Solvent: $CDCl_3$

Assignments
a = 2.13
b = 5.30
c = 7.20-7.70
d = 7.89

2840

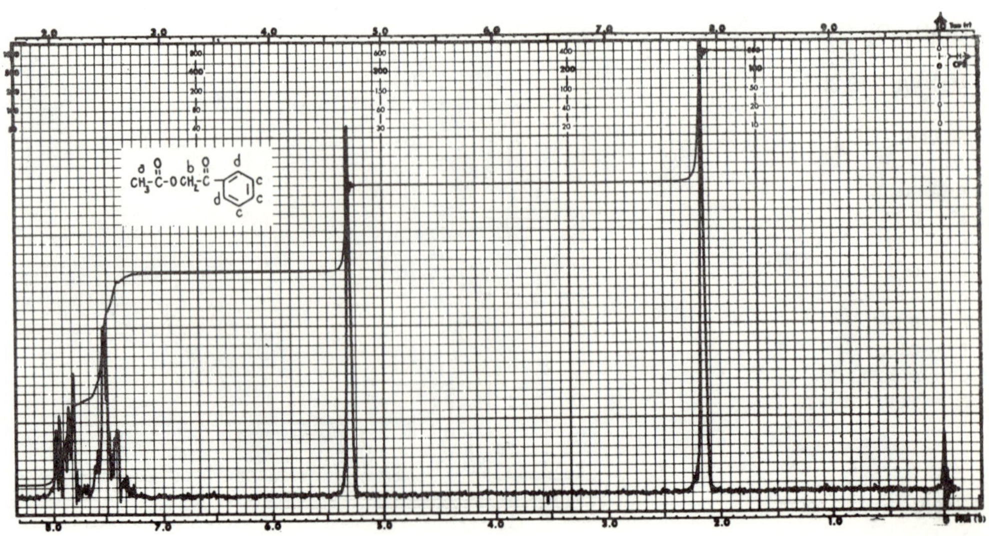

© 1978 Sadtler Research Laboratories, Inc.

ACETIC ACID, CINNAMYL ESTER

$C_{11}H_{12}O_2$

Mol. Wt. 176.22

B.P. 122-124°C/5mm

Solvent: CCl_4

Assignments
a = 1.99
b = 4.61
c = 6.16
d = 6.62
e = 7.26

2841

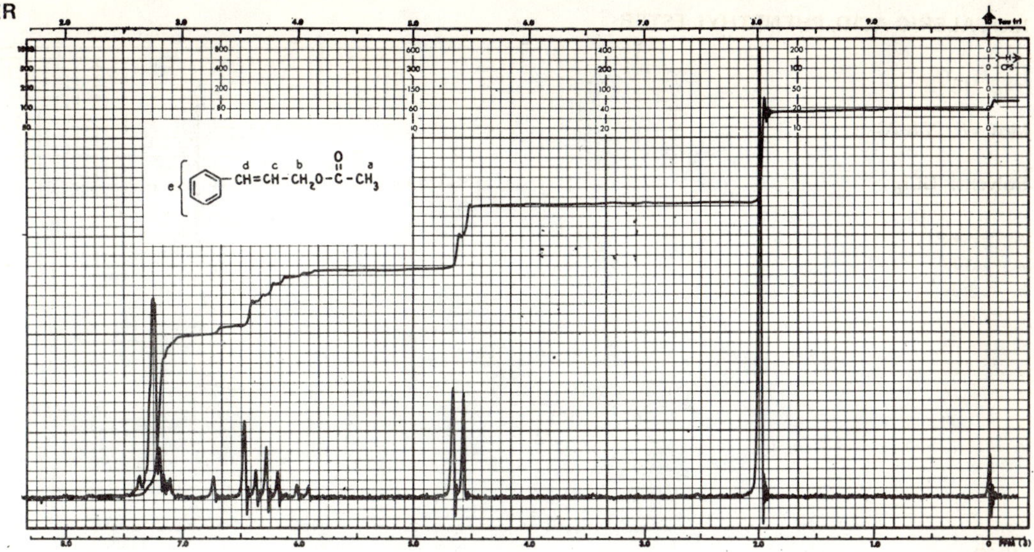

© 1978 Sadtler Research Laboratories, Inc.

BUTYRIC ACID, CINNAMYL ESTER

$C_{13}H_{16}O_2$

Mol. Wt. 204.27

Solvent: CCl_4

Assignments
a = 0.95
b = 1.67
c = 2.24
d = 4.66
e = 6.19
f = 6.61
g = 7.05-7.45

2842

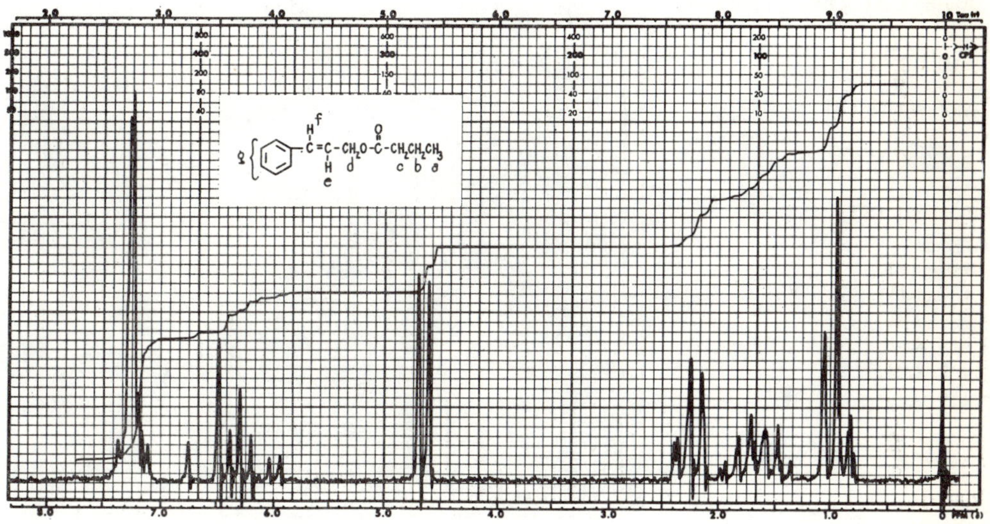

© 1978 Sadtler Research Laboratories, Inc.

ISOVALERIC ACID, CINNAMYL ESTER

$C_{14}H_{18}O_2$

Mol. Wt. 218.30

Solvent: CCl_4

Assignments
a = 0.97
b = 1.52
c = ca 2.12
d = 4.62
e = 6.18
f = 6.55
g = ca 7.20
h = 1.08, 1.20 imps.

2843

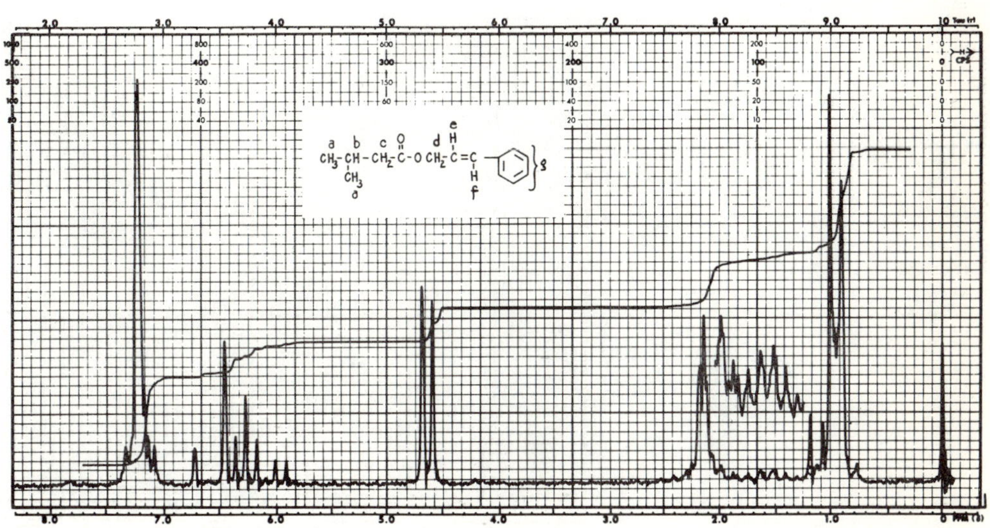

© 1978 Sadtler Research Laboratories, Inc.

BROMOACETIC ACID, BENZYL ESTER

$C_9H_9BrO_2$

Mol. Wt. 229.08

B.P. 85-90°C/1mm

Solvent: $CDCl_3$

Assignments
a = 3.83
b = 5.20
c = ca 7.38

2844

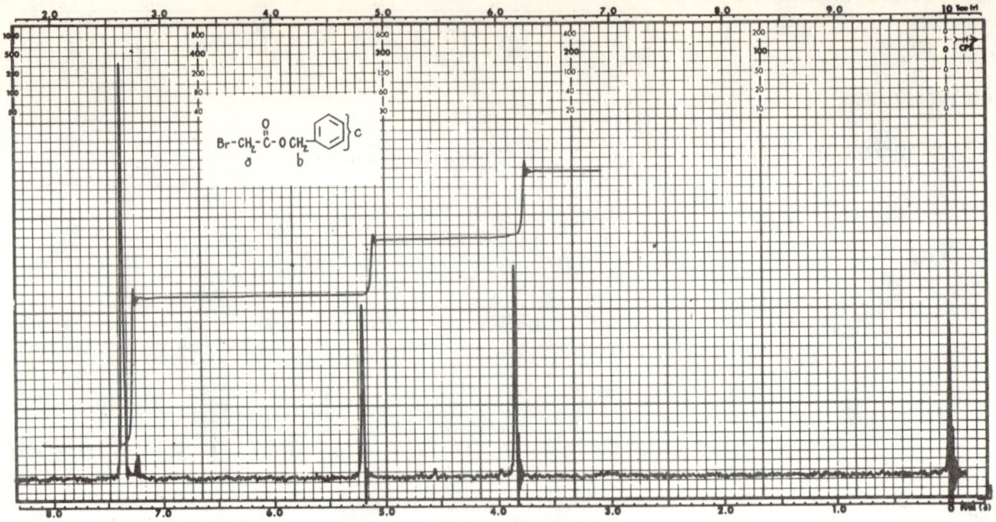

© 1978 Sadtler Research Laboratories, Inc.

ACRYLIC ACID, BENZYL ESTER

$C_{10}H_{10}O_2$

Mol. Wt. 162.19

B.P. 110-111°C/8mm

Solvent: CCl_4

Assignments
a = 5.11
b = 5.72
c = ca 6.19
d = 6.39
e = 7.29

2845

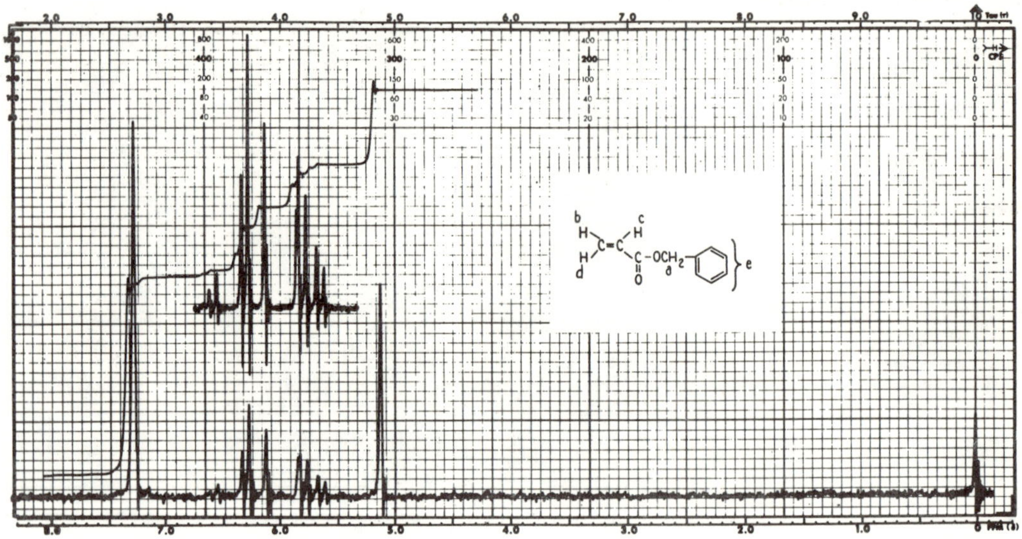

© 1978 Sadtler Research Laboratories, Inc.

METHACRYLIC ACID, BENZYL ESTER

$C_{11}H_{12}O_2$

Mol. Wt. 176.22

B.P. 231-233°C

Solvent: $CDCl_3$

Assignments
a = 1.94
b = 5.17
c = 5.52
d = 6.12
e = ca 7.31

2846

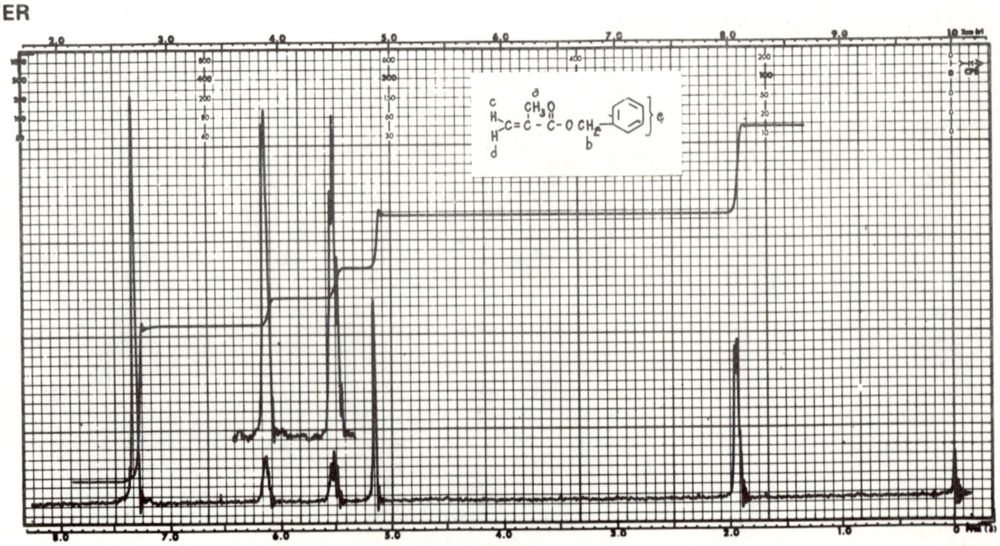

© 1978 Sadtler Research Laboratories, Inc.

PHENYLACETIC ACID, BUTYL ESTER

$C_{12}H_{16}O_2$

Mol. Wt. 192.26

B.P. 135-141°C/18mm

Solvent: CCl_4

Assignments
a = 0.91
b = 1.10-1.90
c = 3.51
d = 4.03
e = ca 7.21

2847

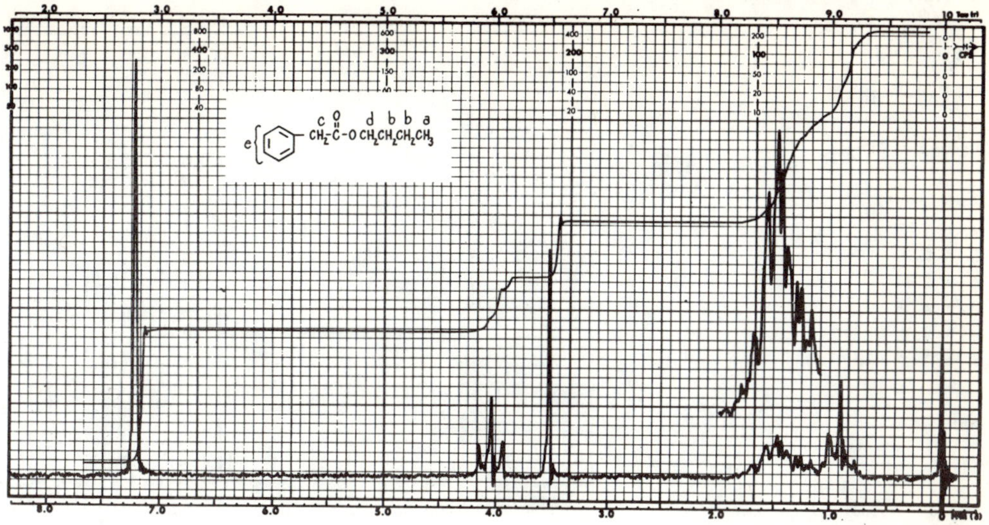

© 1978 Sadtler Research Laboratories, Inc.

PHENYLACETIC ACID, OCTYL ESTER

$C_{16}H_{24}O_2$

Mol. Wt. 248.37

Solvent: CCl_4

Assignments
a = 0.89
b = 1.10-1.90
c = 3.48
d = 3.99
e = ca 7.18

2848

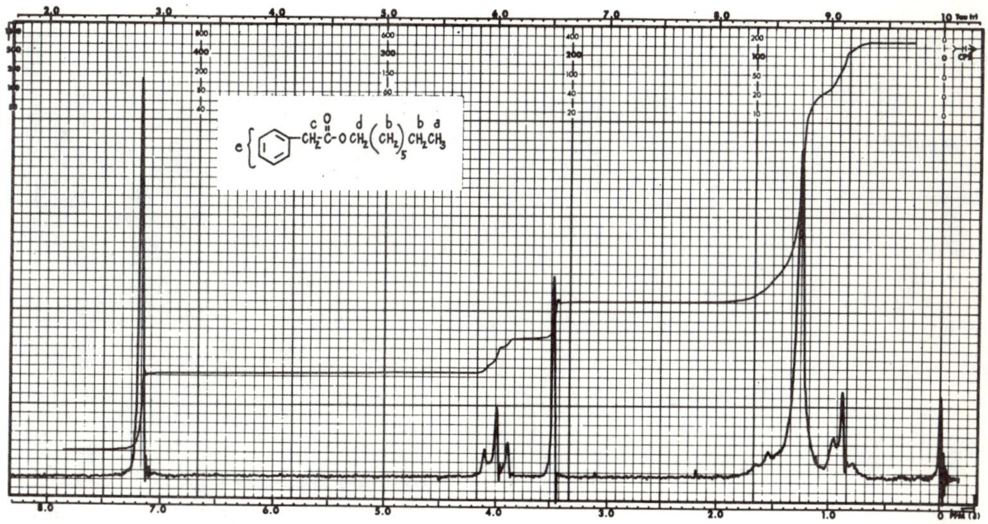

© 1978 Sadtler Research Laboratories, Inc.

4-PHENYLBUTYRIC ACID, METHYL ESTER

$C_{11}H_{14}O_2$

Mol. Wt. 178.23

Solvent: CCl_4

Assignments
a = 1.90
b = 2.21
c = 2.61
d = 3.60
e = ca 7.11

2849

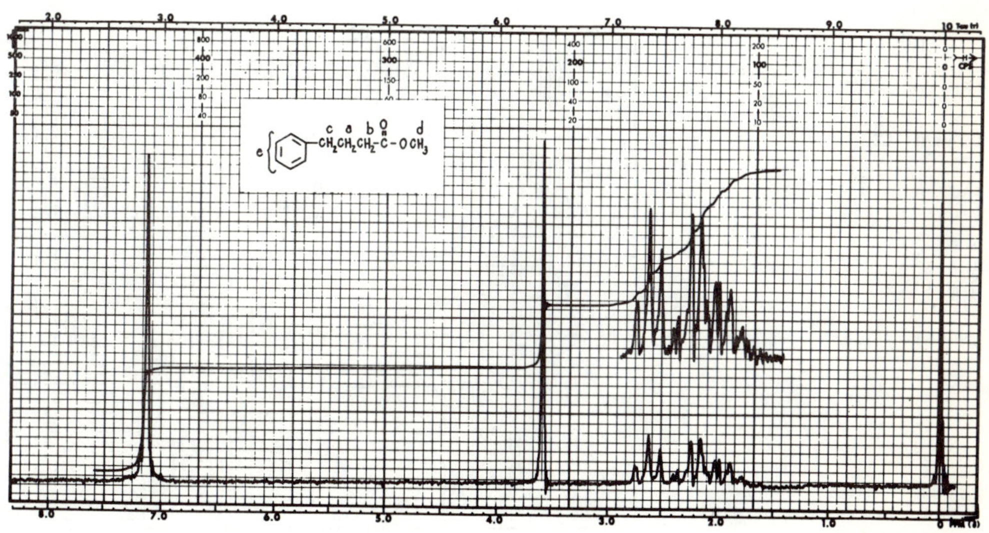

© 1978 Sadtler Research Laboratories, Inc.

3-PHENYLHYDRACRYLIC ACID, ETHYL ESTER

$C_{11}H_{14}O_3$

Mol. Wt. 194.23

Solvent: CCl_4

Assignments
a = 1.19
b = 2.59
c = 3.50
d = 4.06
e = 4.99
f = 7.05-7.45

2850

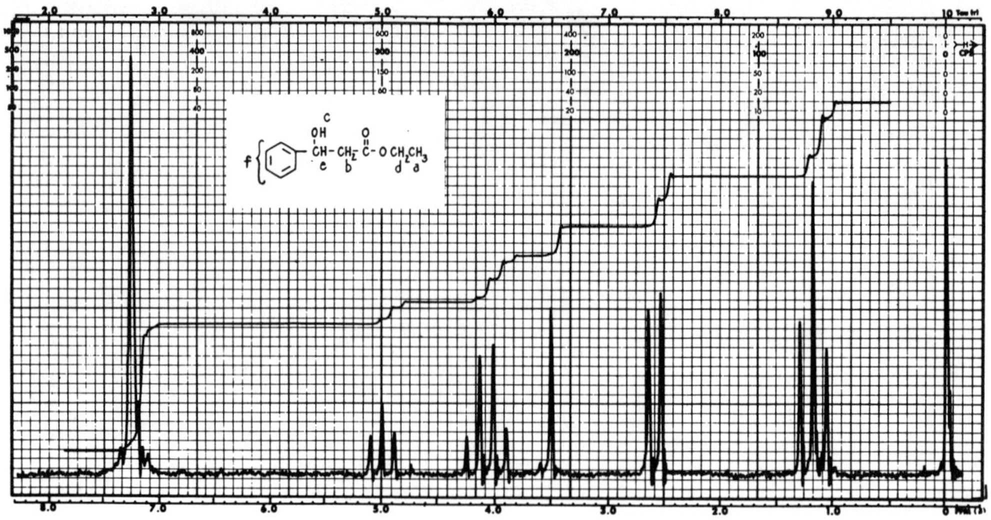

© 1978 Sadtler Research Laboratories, Inc.

HIPPURIC ACID, METHYL ESTER

$C_{10}H_{11}NO_3$

Mol. Wt. 193.20

M.P. 80°C

Solvent: $CDCl_3$

Assignments
a = 3.73
b = 4.20
c = 7.06
d = 7.23-7.58
e = 7.80

2851

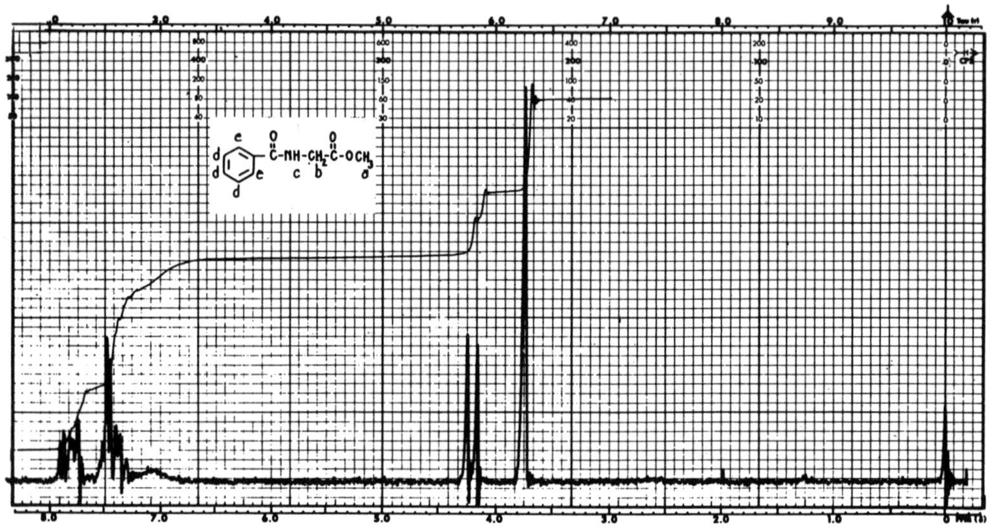

© 1978 Sadtler Research Laboratories, Inc.

METHYLPHENYLMALONIC ACID, DIETHYL ESTER

$C_{14}H_{18}O_4$

Mol. Wt. 250.30

Solvent: CCl_4

Assignments
a = 1.21
b = 1.77
c = 4.16
d = 7.27

2852

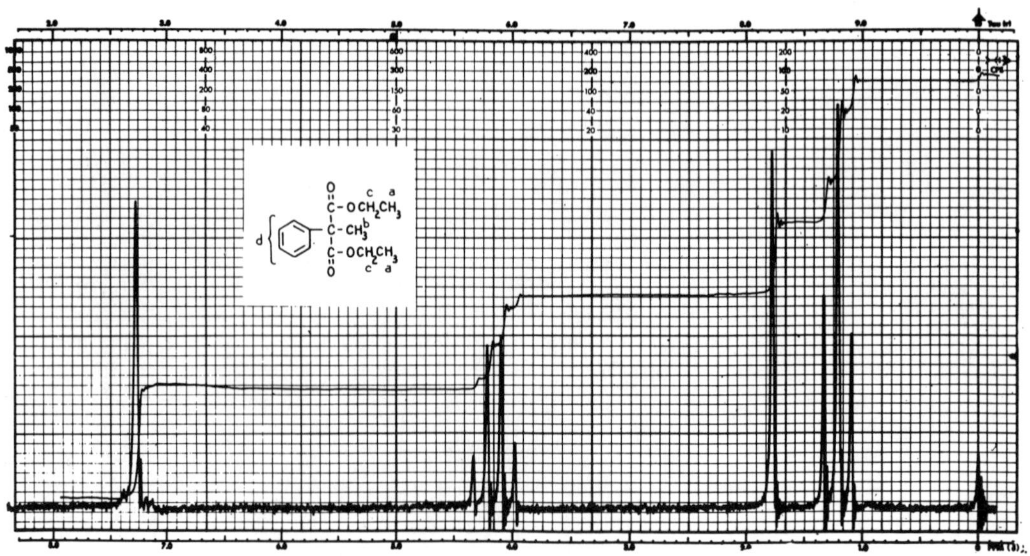

© 1978 Sadtler Research Laboratories, Inc.

3,3'-(BENZYLIMINO)DIPROPIONIC ACID, DIETHYL ESTER

$C_{17}H_{25}NO_4$

Mol. Wt. 307.39

Solvent: $CDCl_3$

Assignments
a = 1.20
b = 2.38
c = 2.73
d = 3.56
e = 4.05
f = 7.20

2853

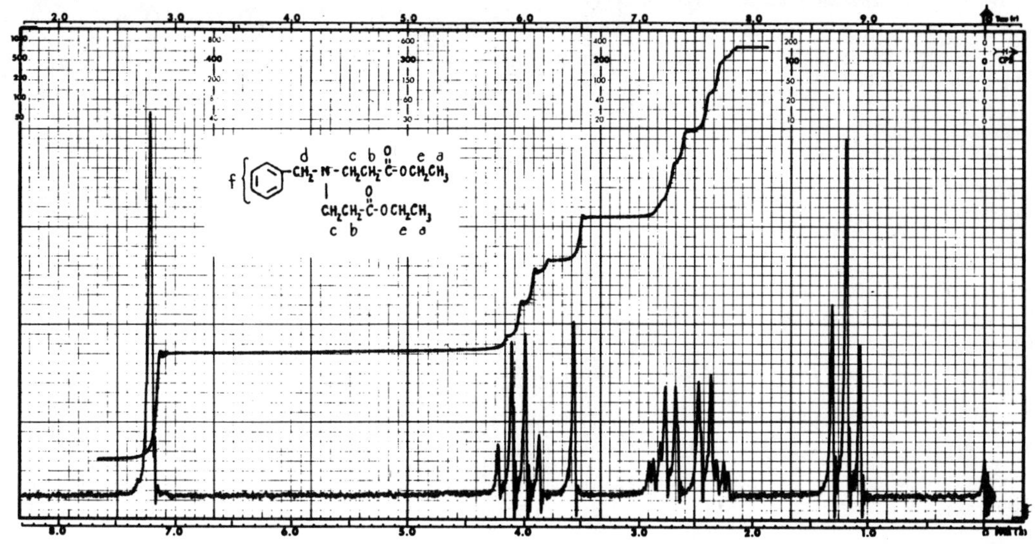

© 1978 Sadtler Research Laboratories, Inc.

o-TOLYLACETIC ACID, ETHYL ESTER

$C_{11}H_{14}O_2$

Mol. Wt. 178.23

Solvent: CCl_4

Assignments
a = 1.21
b = 2.30
c = 3.50
d = 4.08
e = 7.08

2854

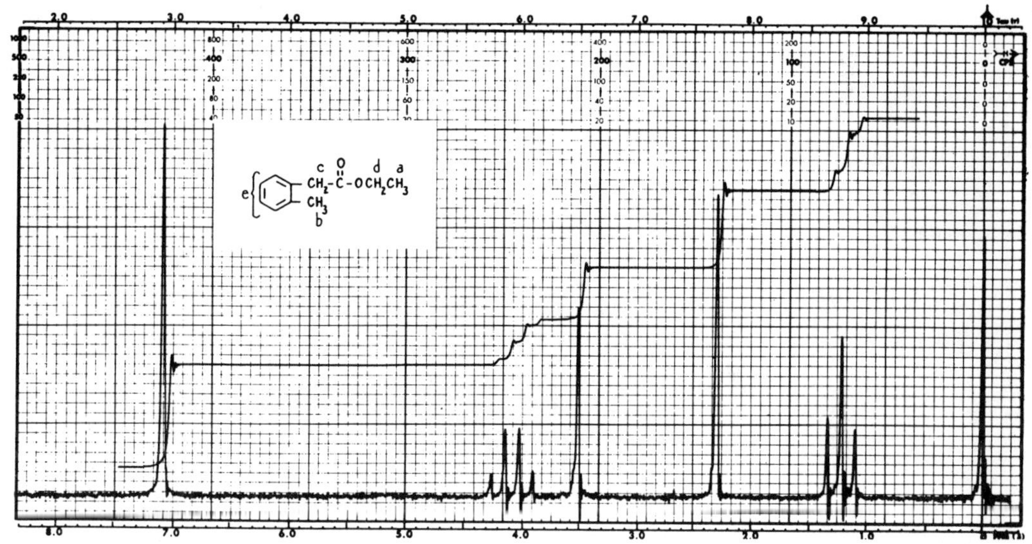

© 1978 Sadtler Research Laboratories, Inc.

FLUORENE-9-CARBOXYLIC ACID, METHYL ESTER

$C_{15}H_{12}O_2$

Mol. Wt. 224.26

M.P. 66-68°C

Solvent: $CDCl_3$

Assignments
a = 3.65
b = 4.80
c = ca 7.32
d = ca 7.64

2855

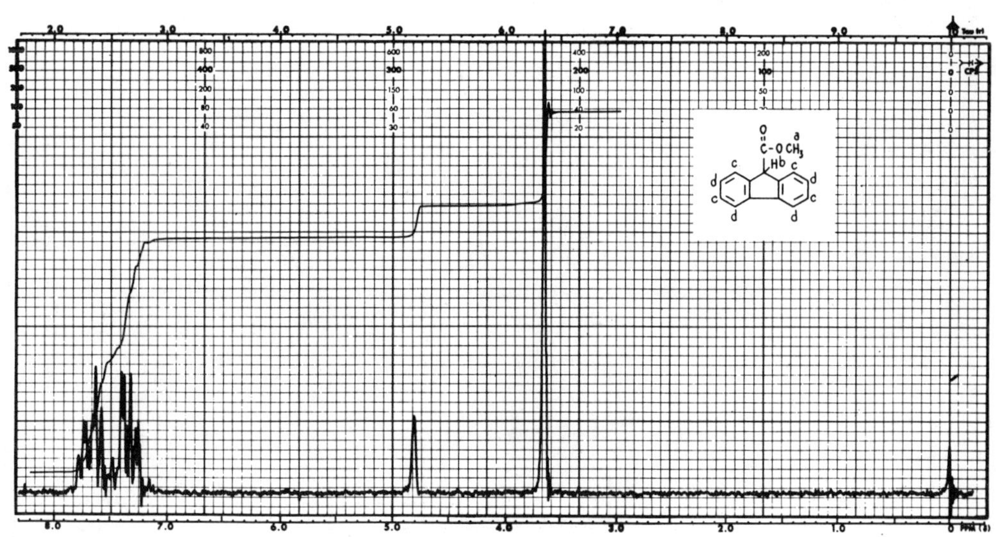

© 1978 Sadtler Research Laboratories, Inc.

β-METHYLCINNAMIC ACID, ETHYL ESTER

$C_{12}H_{14}O_2$

Mol. Wt. 190.24

B.P. 123-125°C/6mm

Solvent: CCl_4

Assignments
a = 1.23
b = 2.51
c = 4.11
d = 6.03
e = 7.05-7.55

2856

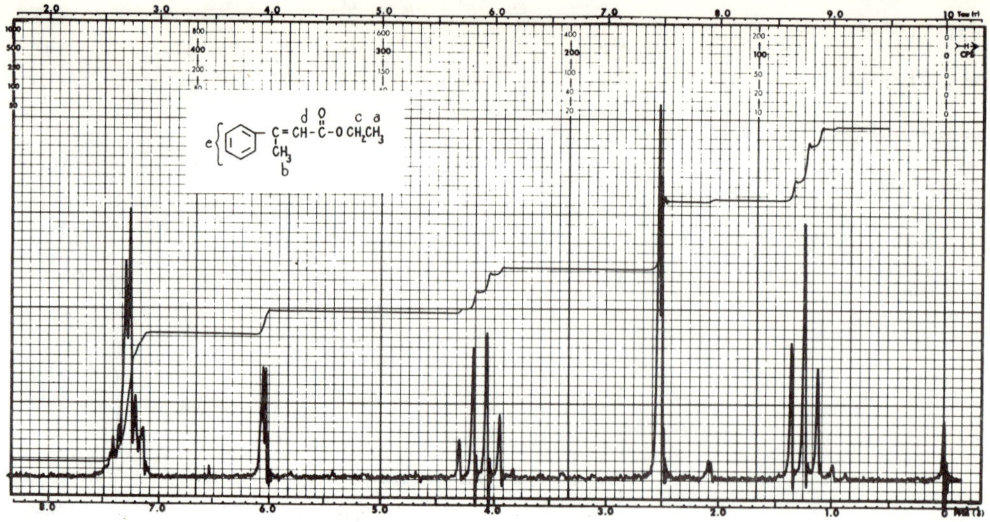

© 1978 Sadtler Research Laboratories, Inc.

CINNAMIC ACID, CYCLOHEXYL ESTER

$C_{15}H_{18}O_2$

Mol. Wt. 230.31

Solvent: CCl_4

Assignments
a = 1.08-2.18
b = 4.87
c = 6.40
d = 7.44
e = 7.67

2857

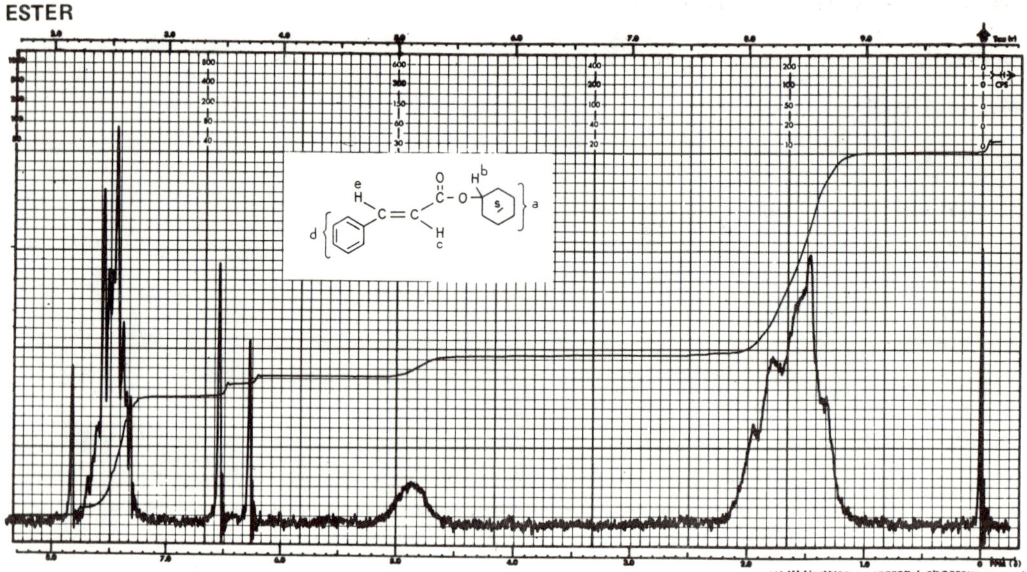

© 1978 Sadtler Research Laboratories, Inc.

α-CYANOCINNAMIC ACID, ETHYL ESTER

$C_{12}H_{11}NO_2$

Mol. Wt. 201.23

M.P. 48.5-49.5°C

Solvent: $CDCl_3$

Assignments
a = 1.36
b = 4.34
c = 7.20-7.60
d = 7.96
e = 8.20

2858

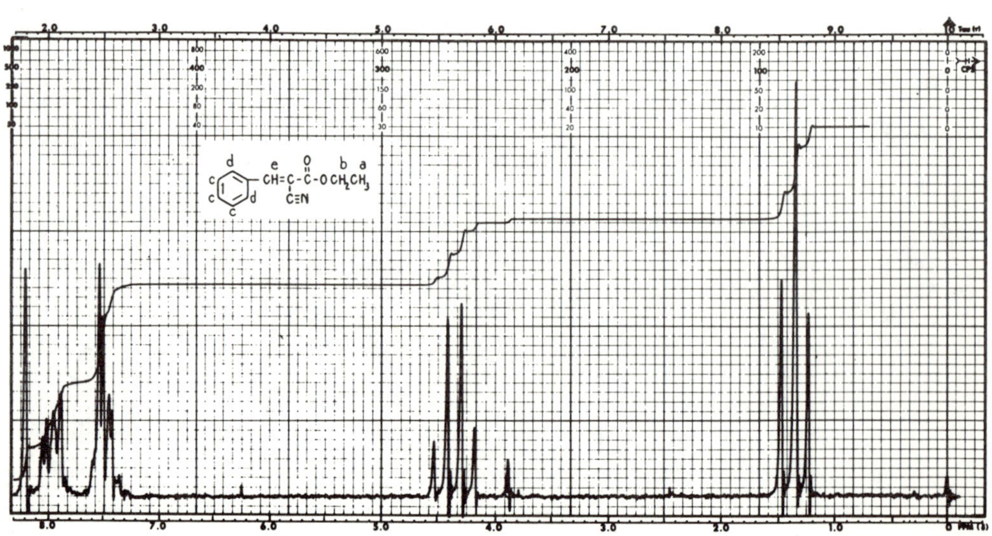

© 1978 Sadtler Research Laboratories, Inc.

(TRIPHENYLPHOSPHORANYLIDENE)ACETIC ACID, METHYL ESTER

$C_{21}H_{19}O_2P$

Mol. Wt. 334.36

M.P. 170-172°C

Solvent: Trifluoroacetic acid

Assignments
a = 3.77
b = 4.53
c = 7.50-8.10

2859

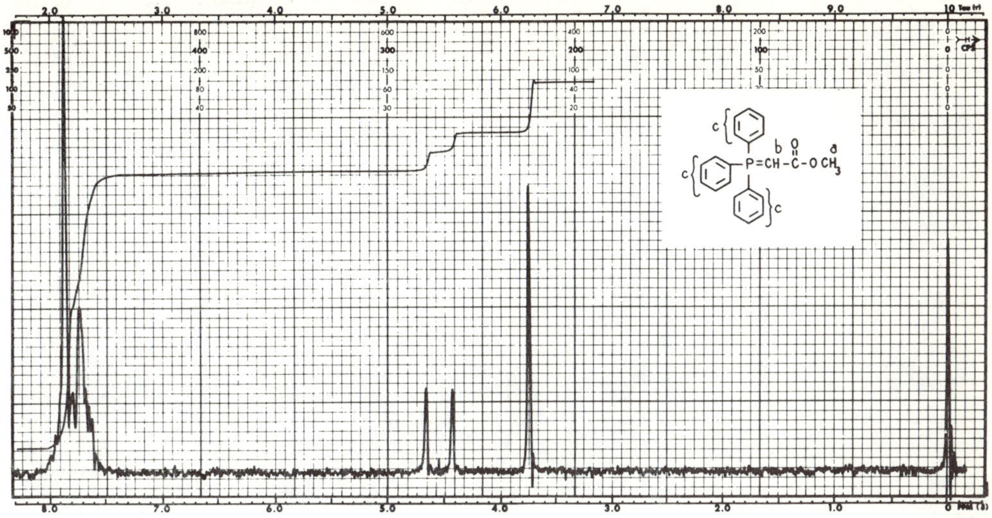

© 1978 Sadtler Research Laboratories, Inc.

PHENYLACETIC ACID, PHENETHYL ESTER

$C_{16}H_{16}O_2$

Mol. Wt. 240.30

Solvent: CCl_4

Assignments
a = 2.79
b = 3.44
c = 4.18
d = 7.15

2860

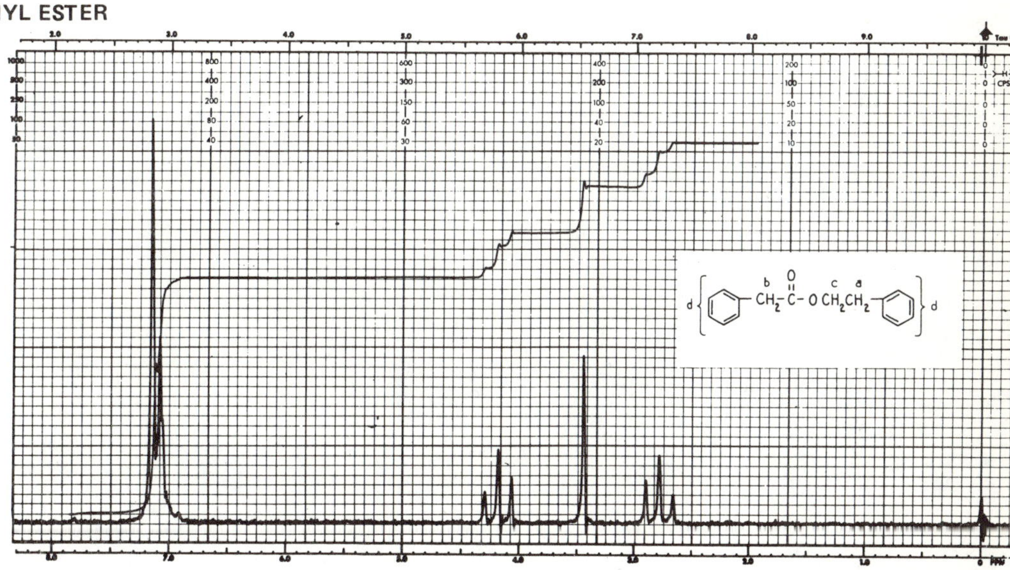

© 1978 Sadtler Research Laboratories, Inc.

CINNAMIC ACID, BENZYL ESTER

$C_{16}H_{14}O_2$

Mol. Wt. 238.27

M.P. 39°C

B.P. 244°C/25mm

Solvent: CCl_4

Assignments
a = 5.17
b = 6.38
c = 7.31
d = 7.63
e = 2.05 imp.

2861

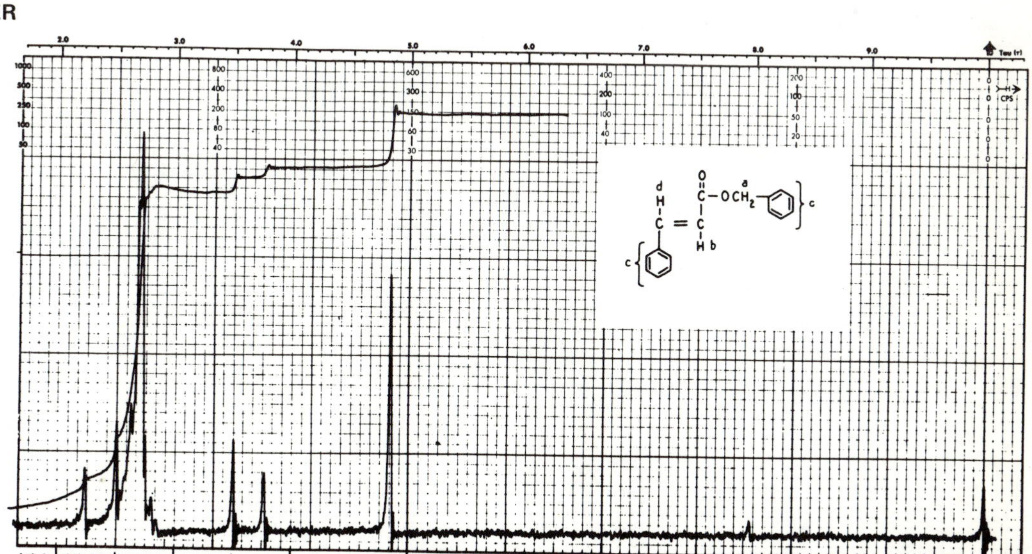

© 1978 Sadtler Research Laboratories, Inc.

CINNAMIC ACID, PHENETHYL ESTER

$C_{17}H_{16}O_2$

Mol. Wt. 252.32

M.P. 56-57°C

Solvent: CCl_4

Assignments
a = 2.94
b = 4.32
c = 6.30
d = 7.19
e = ca 7.31
f = 7.57

2862

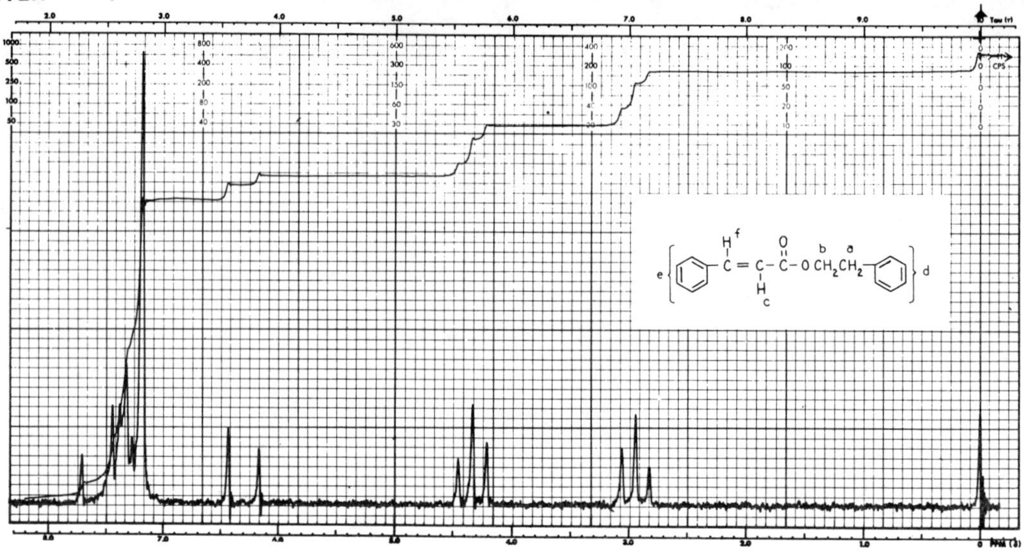

© 1978 Sadtler Research Laboratories, Inc.

THE AROMATIC ESTERS

The aliphatic esters of Benzoic Acid are more highly deshielded by the adjacent oxygen atom than those of the aliphatic and olefinic carboxylic acids.

The ortho aromatic protons are also strongly deshielded by the adjacent carbonyl group and resonate about 0.5 ppm to lower field than the meta and para protons.

The aromatic esters are readily soluble in the halogenated solvents normally used as solvents in NMR and do not exhibit any unusual solvent effects. These compounds do not display any special spin-spin couplings.

Chemical Shifts

Aliphatic Protons

HC–3	HC–2	HC–1		Solvent	Spectrum
		2.11	CH_3–C(=O)–O–Ø	$CDCl_3$	2930
		2.35	CH_3–C(=O)–O–Ø (4-NO_2)	$CDCl_3$	2926
	1.20	2.47	CH_3–CH_2–C(=O)–O–Ø	CCl_4	2921
	(1.27)	2.70	$(CH_3)_2$–CH–C(=O)–O–Ø	CCl_4	- - - -
	(1.34)	- - - -	$(CH_3)_3$–C–C(=O)–O–Ø	$CDCl_3$	2922
		3.82	CH_2–O–C(=O)–Ø	CCl_4	- - - -
	1.33	4.30	CH_3–CH_2–O–C(=O)–Ø	CCl_4	- - - -
1.00	1.73	4.22	CH_3–CH_2–CH_2–O–C(=O)–Ø	CCl_4	2863
	(1.33)	5.28	$(CH_3)_2$–CH–O–C(=O)–Ø	CCl_4	- - - -
(1.00)	2.08	4.10	$(CH_3)_2$–CH–CH_2–O–C(=O)–Ø	$CDCl_3$	2864

Aromatic Protons

The carbonyl side of the ester linkage strongly deshields the ortho aromatic protons producing a series of multiplets similar to those observed for the ketones and amides.

The oxygen side of the group has a much weaker shielding effect than the oxygen atom of the aliphatic ethers, resulting in a broad, complex higher-order series of multiplets in the chemical shift range from 6.9–7.5 ppm.

Benzoic Acid, Propyl Ester

para/meta	ortho			
7.1–7.6	8.01	Ø–C(=O)–O–R_3	CCl_4	2863

Aromatic Protons continued

Propionic Acid, Phenyl Ester

para/meta/ortho		Solvent	Spectrum
6.9–7.5	Ø–O–C(=O)–R_2	CCl_4	2921

Para Substituted Phenyl Esters

R–C(=O)–O–	ortho	meta	para	Solvent	Spectrum
	6.82	7.01	–CH_3	CCl_4	2924
	7.06	7.06	–F	$CDCl_3$	----
	7.09	7.09	–O–C(=O)–R	$CDCl_3$	----
	7.19	7.80	–C(=O)–H	CCl_4	----
	7.19	7.88	–N=N–Ø	$CDCl_3$	----
	7.32	8.29	–NO_2	$CDCl_3$	2926

Para Substituted Benzoic Acid Esters

R–O–C(=O)–	ortho	meta	para	Solvent	Spectrum
	7.83	6.63	–NH_2	$CDCl_3$	2881
	7.87	6.84	–OH	$CDCl_3$	2894
	7.99	6.88	–O–CH_3	$CDCl_3$	2888
	7.95	7.06	–N=C=O	$CDCl_3$	2884
	7.88	7.36	–$C(CH_3)_3$	CCl_4	2870
	7.95	7.38	–Cl	$CDCl_3$	2874
	7.83	7.49	–Br	CCl_4	2875
	7.82	7.70	–I	Polysol	----
	8.32	7.91	–N=O	$CDCl_3$	2883
	8.07	8.07	–C(=O)–O–R_2	$CDCl_3$	2907
	8.26	8.26	–NO_2	$CDCl_3$	2886

BENZOIC ACID, PROPYL ESTER

$C_{10}H_{12}O_2$

Mol. Wt. 164.21

B.P. 227-228°C

Solvent: CCl_4

Assignments
a = 1.00
b = 1.73
c = 4.22
d = 7.10-7.60
e = 8.01

2863

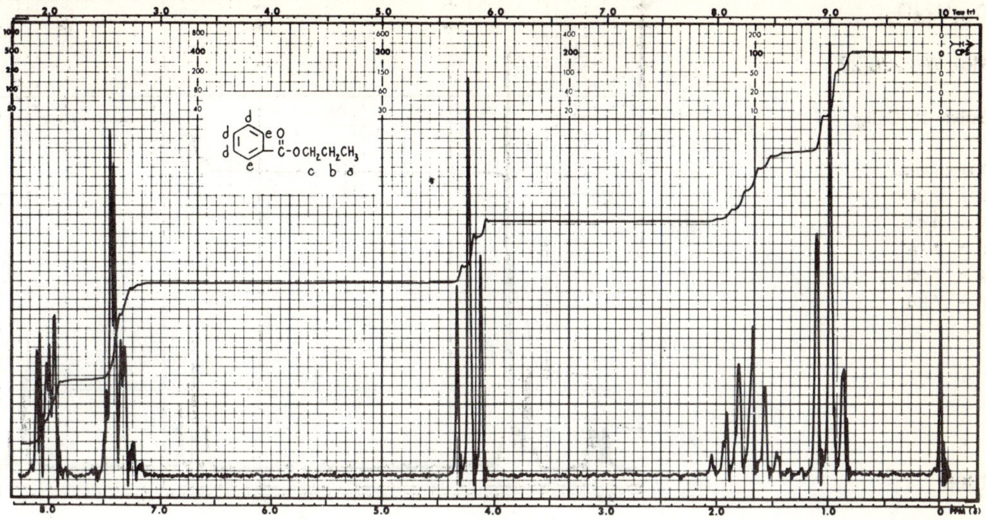

© 1978 Sadtler Research Laboratories, Inc.

BENZOIC ACID, ISOBUTYL ESTER

$C_{11}H_{14}O_2$

Mol. Wt. 178.22

B.P. 237°C

Solvent: $CDCl_3$

Assignments
a = 1.00
b = 2.08
c = 4.10
d = 7.48
e = 8.06

2004

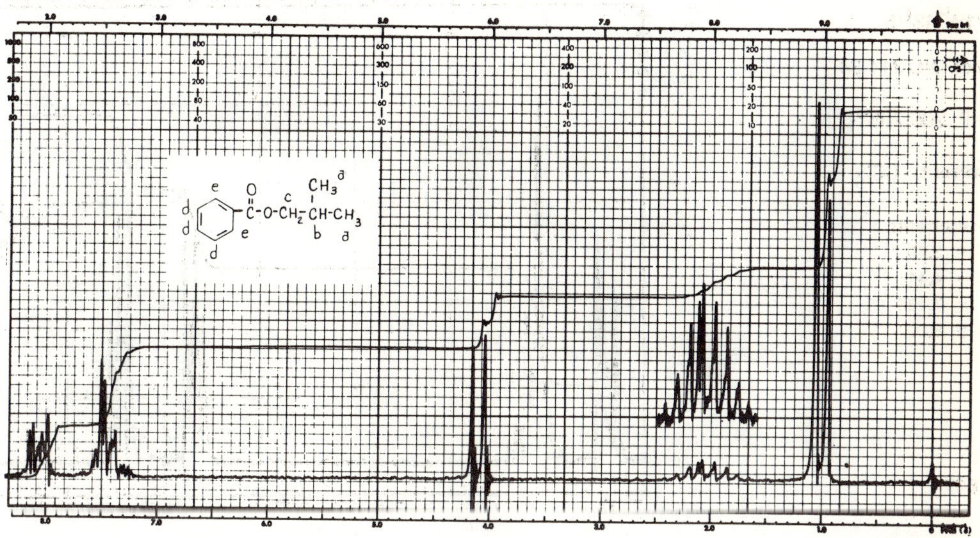

© 1978 Sadtler Research Laboratories, Inc.

BENZOIC ACID, 2-(DIMETHYLAMINO)ETHYL ESTER

$C_{11}H_{15}NO_2$

Mol. Wt. 193.25

B.P. 132-133°C/9mm

Solvent: $CDCl_3$

Assignments
a = 2.31
b = 2.70
c = 4.41
d = 7.20-7.60
e = 8.03

2865

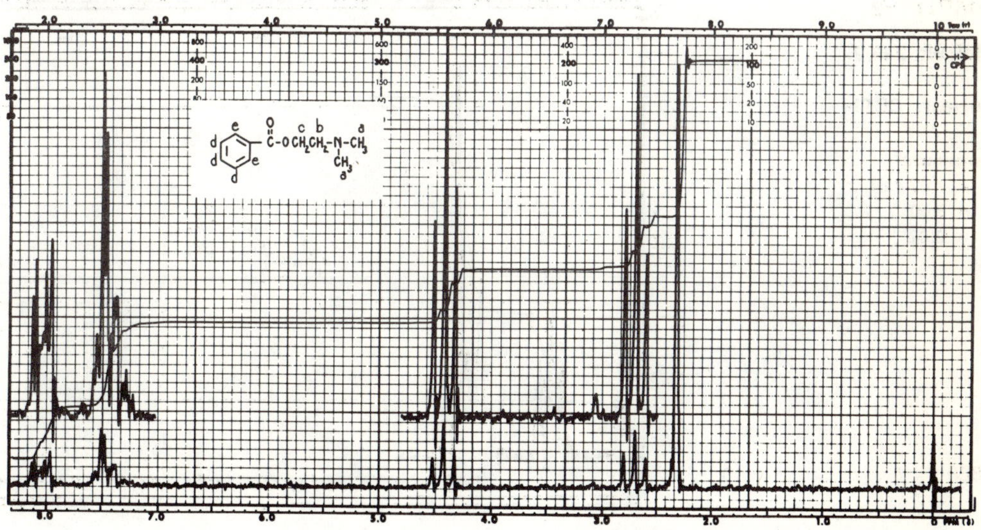

© 1978 Sadtler Research Laboratories, Inc.

BENZOIC ACID, PHENETHYL ESTER

$C_{15}H_{14}O_2$

Mol. Wt. 226.28

Solvent: CCl_4

Assignments
a = 2.96
b = 4.41
c = 7.15
d = 7.30
e = 7.94

2866

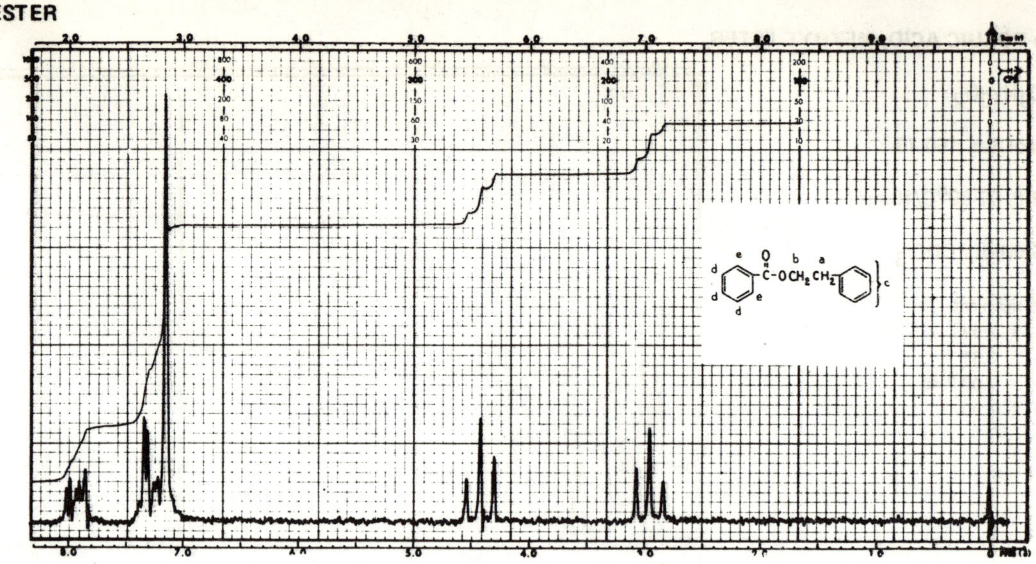

© 1978 Sadtler Research Laboratories, Inc.

DIETHYLENE GLYCOL, DIBENZOATE

$C_{18}H_{18}O_5$

Mol. Wt. 314.34

B.P. 200-205°C

Solvent: CCl_4

Assignments
a = 3.79
b = 4.41
c = 7.13-7.63
d = 8.00

2867

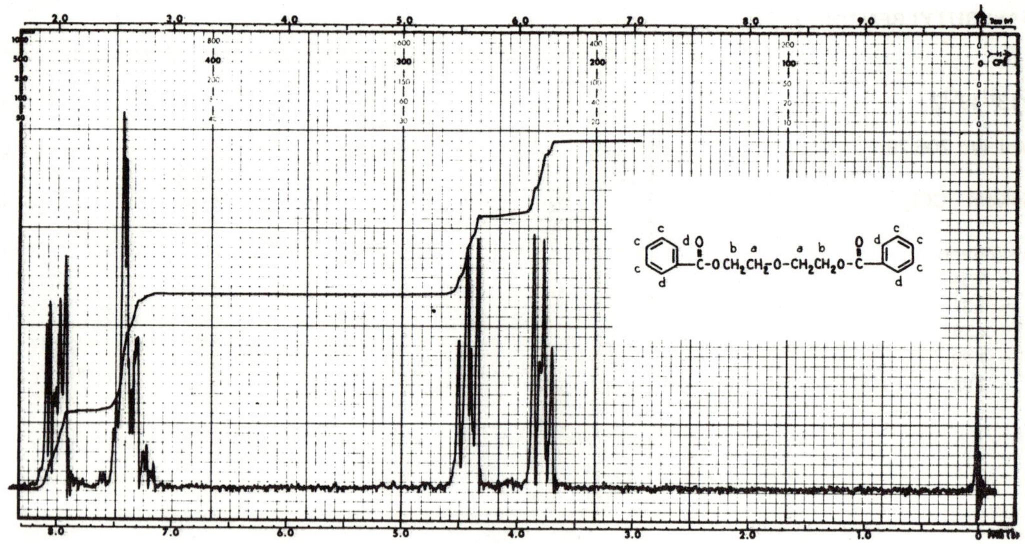

© 1978 Sadtler Research Laboratories, Inc.

BENZOIC ACID, 8-QUINOLINYL ESTER

$C_{16}H_{11}NO_2$

Mol. Wt. 249.27

Solvent: $CDCl_3$

Assignments
a = 7.02-7.54
b = 8.00
c = 8.62

2868

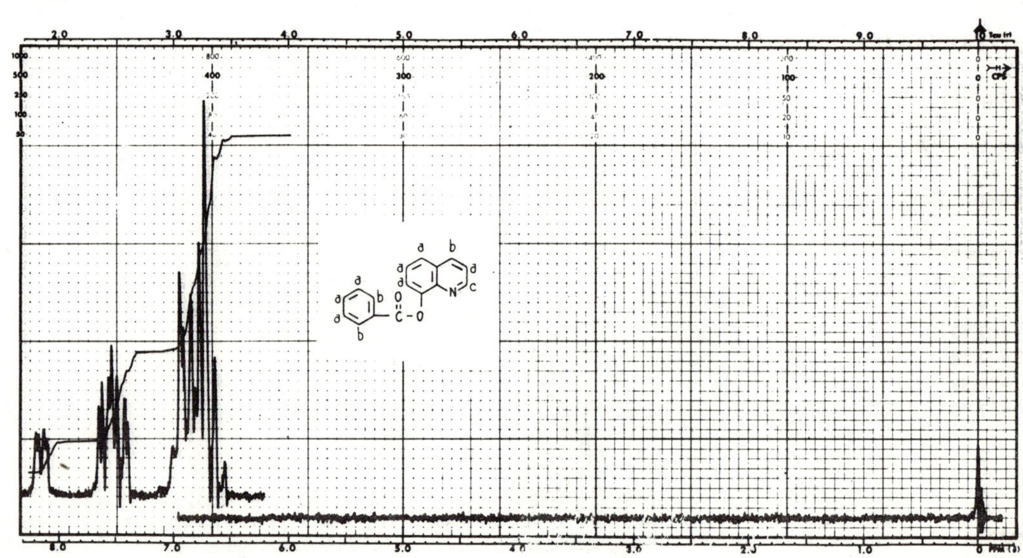

© 1978 Sadtler Research Laboratories, Inc.

o-TOLUIC ACID, METHYL ESTER

$C_9H_{10}O_2$

Mol. Wt. 150.18

B.P. 97-99°C/15mm

Solvent: $CDCl_3$

Assignments
a = 2.59
b = 3.81
c = 7.00-7.40
d = 7.90

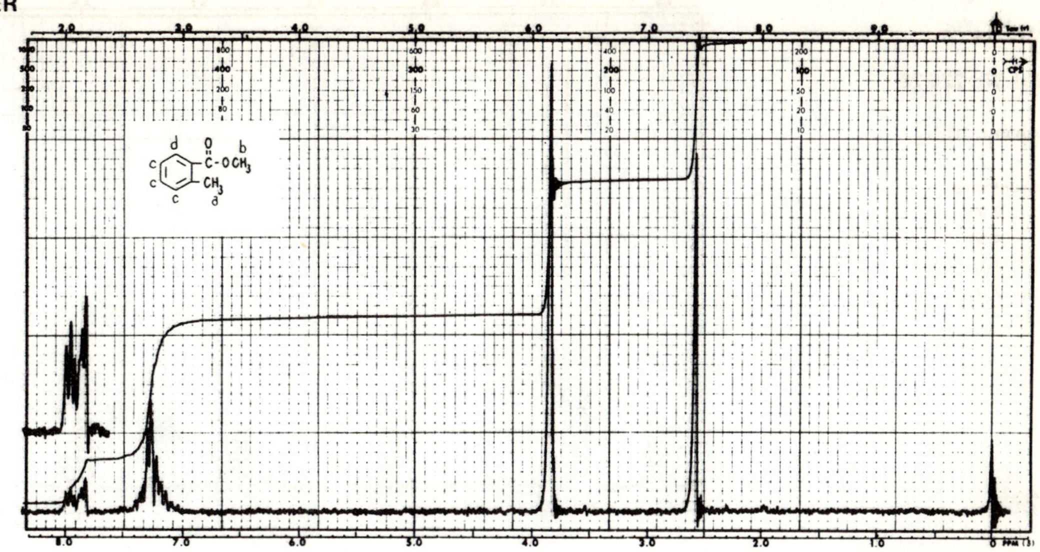

2869

© 1978 Sadtler Research Laboratories, Inc.

p-tert-BUTYLBENZOIC ACID, ETHYL ESTER

$C_{13}H_{18}O_2$

Mol. Wt. 206.29

Solvent: CCl_4

Assignments
a = 1.34
b = 1.37
c = 4.29
d = 7.36
e = 7.88

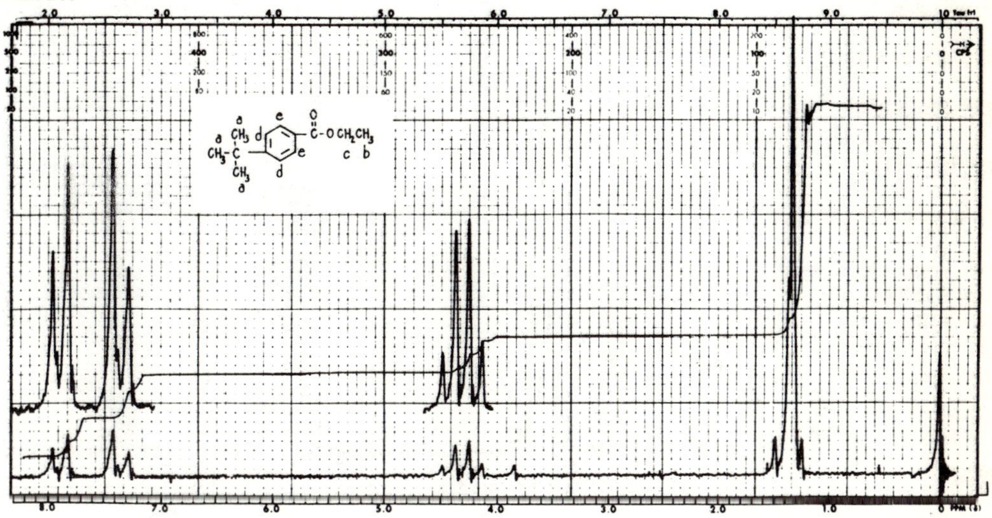

2870

© 1978 Sadtler Research Laboratories, Inc.

m-FLUOROBENZOIC ACID, ETHYL ESTER

$C_9H_9FO_2$

Mol. Wt. 168.17

Solvent: CCl_4

Assignments
a = 1.39
b = 4.31
c = 7.13
d = 7.28
e = 7.63
f = 7.76

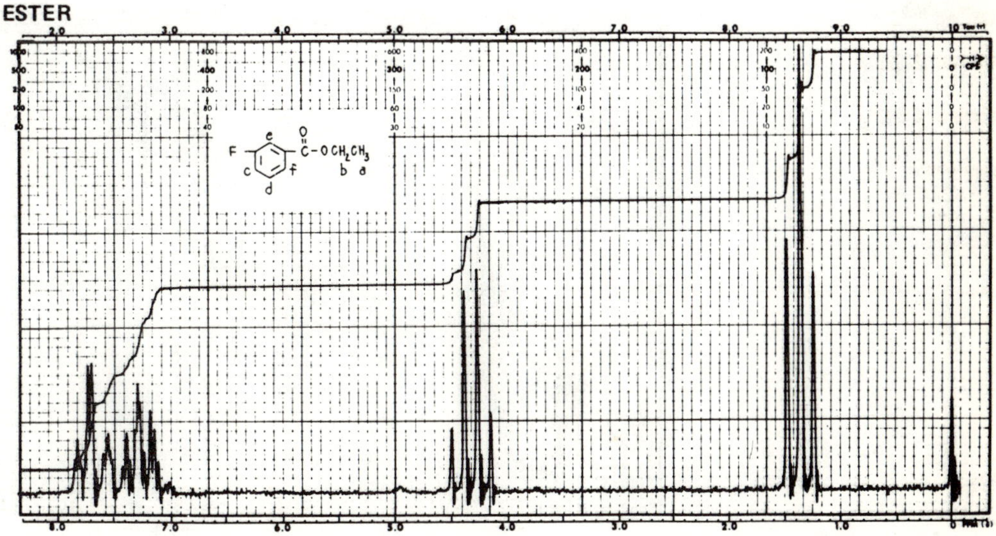

2871

© 1978 Sadtler Research Laboratories, Inc.

o-CHLOROBENZOIC ACID, METHYL ESTER

$C_8H_7ClO_2$

Mol. Wt. 170.60

B.P. 110-112°C/10mm

Solvent: CCl_4

Assignments
a = 3.85
b = 7.08-7.50
c = 7.73

2872

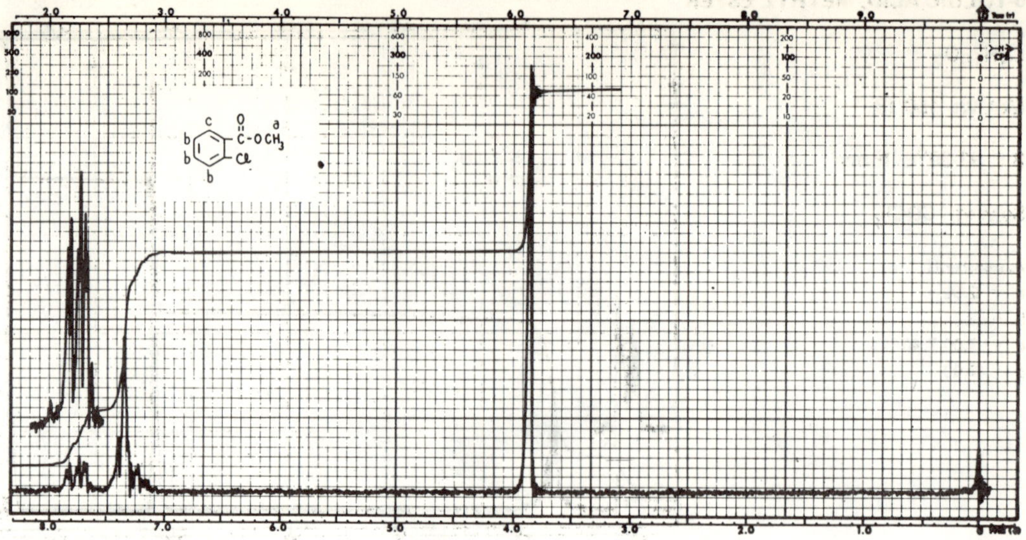

© 1978 Sadtler Research Laboratories, Inc.

m-CHLOROBENZOIC ACID, METHYL ESTER

$C_8H_7ClO_2$

Mol. Wt. 170.60

M.P. 18-19°C

Solvent: CCl_4

Assignments
a = 3.88
b = 7.30
c = 7.46
d = 7.89
e = 7.92

2873

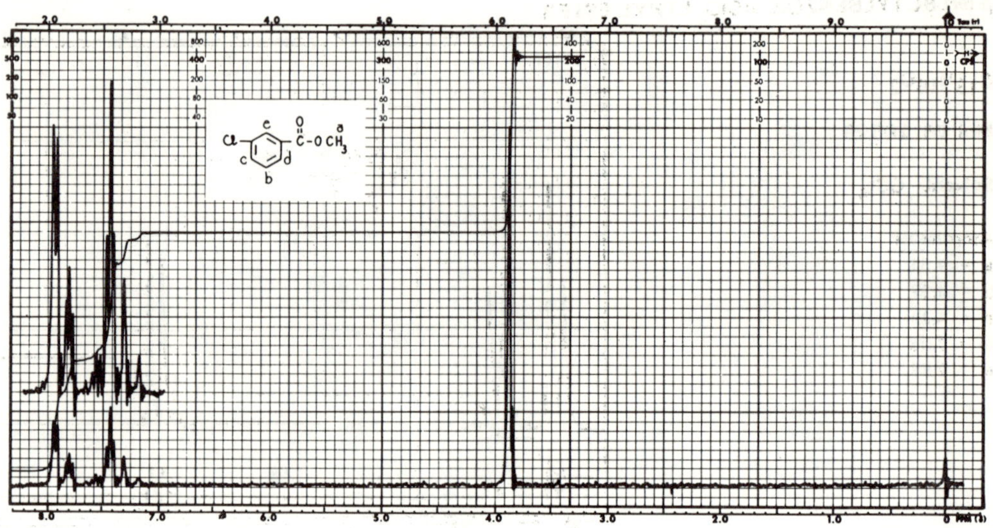

© 1978 Sadtler Research Laboratories, Inc.

p-CHLOROBENZOIC ACID, METHYL ESTER

$C_8H_7ClO_2$

Mol. Wt. 170.60

Solvent: $CDCl_3$

Assignments
a = 3.89
b = 7.38
c = 7.95

2874

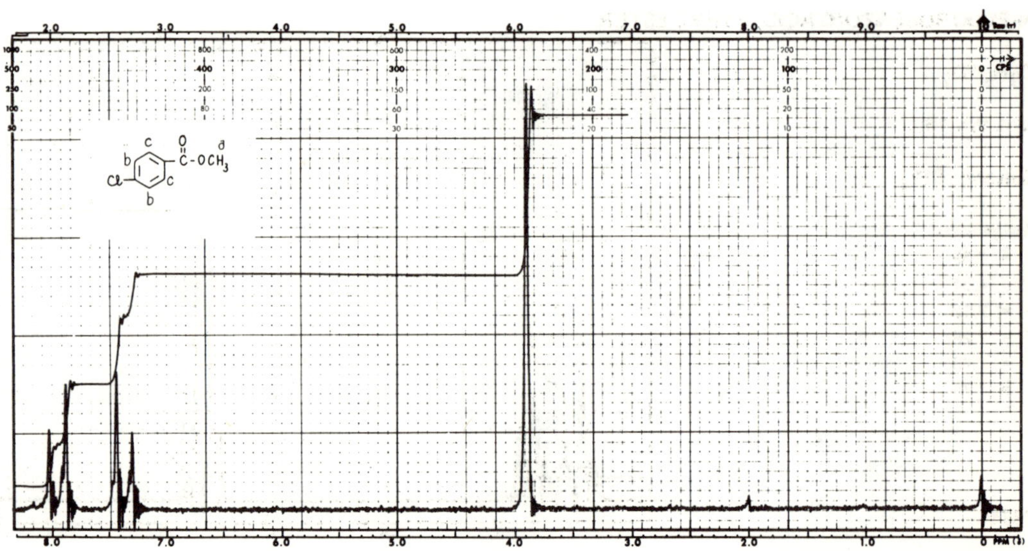

© 1978 Sadtler Research Laboratories, Inc.

p-BROMOBENZOIC ACID, ETHYL ESTER

$C_9H_9BrO_2$

Mol. Wt. 229.08

B.P. 115-116°C/5mm

Solvent: CCl_4

Assignments
a = 1.34
b = 4.31
c = 7.49
d = 7.83

2875

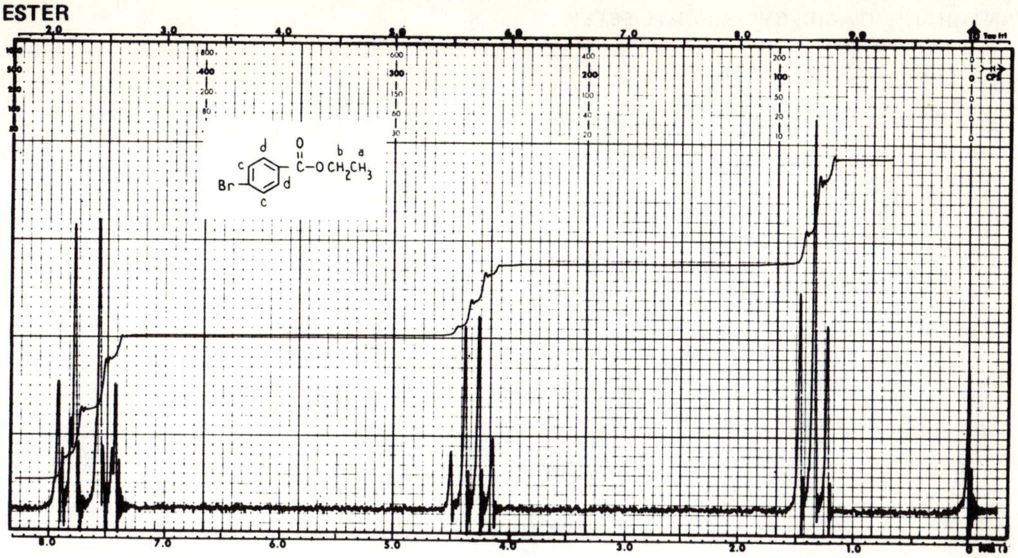

© 1978 Sadtler Research Laboratories, Inc.

---

m-IODOBENZOIC ACID, METHYL ESTER

$C_8H_7IO_2$

Mol. Wt. 262.05

M.P. 48-50°C

Solvent: $CDCl_3$

Assignments
a = 3.89
b = 7.00
c = 7.82
d = 7.94
e = 8.33

2876

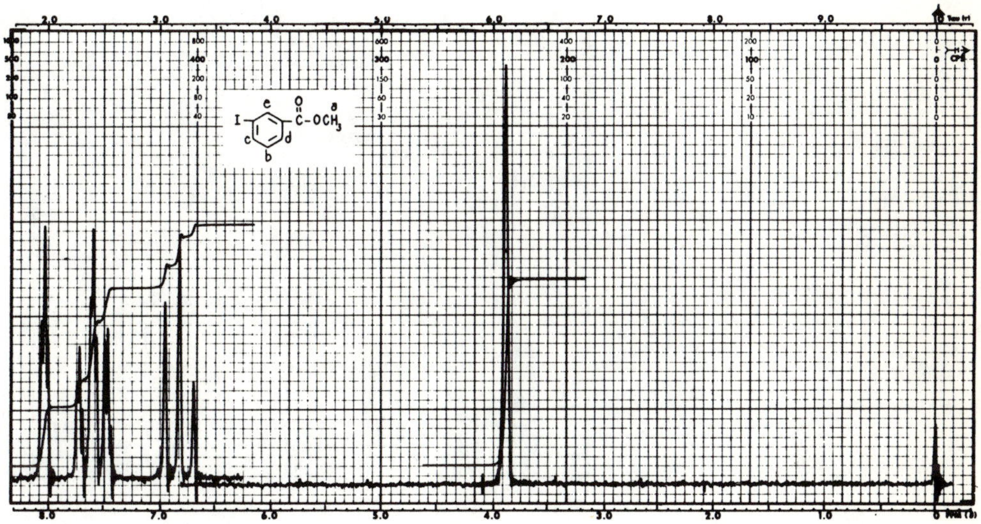

© 1978 Sadtler Research Laboratories, Inc.

ANTHRANILIC ACID, HEXYL ESTER

$C_{13}H_{19}NO_2$

Mol. Wt. 221.30

B.P. 194-195°C/16mm

Solvent: CCl_4

Assignments
a = 0.91
b = 1.11-1.96
c = 4.21
d = 5.73
e = 6.37-6.71
f = 7.13
g = 7.75
h = 3.71 imp.

2877

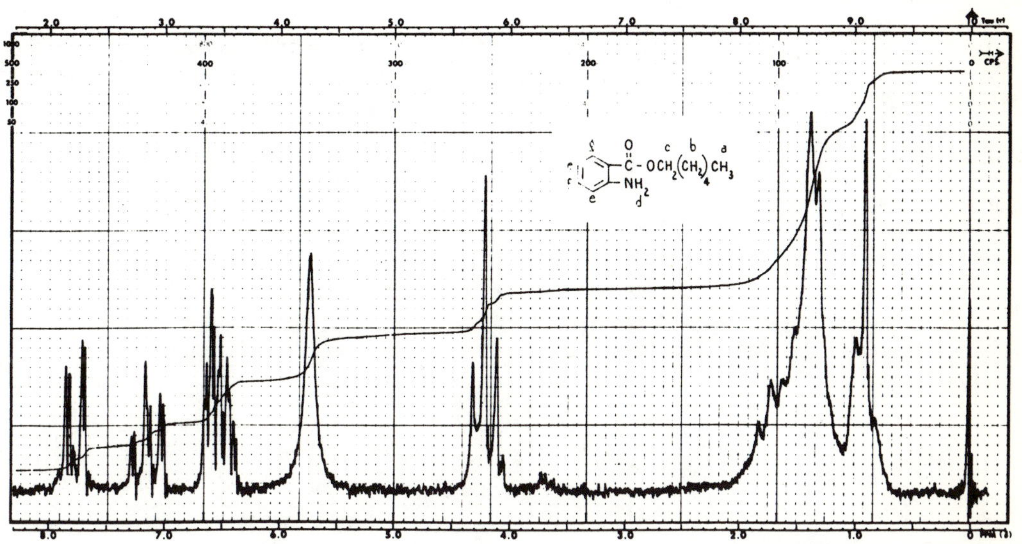

© 1978 Sadtler Research Laboratories, Inc.

ANTHRANILIC ACID, CYCLOHEXYL ESTER

$C_{13}H_{17}NO_2$

Mol. Wt. 219.29

Solvent: $CDCl_3$

Assignments
a = 0.90-2.20
b = 4.99
c = 5.71
d = 6.54
e = 6.59
f = 7.20
g = 7.89
h = 3.79 imp.

2878

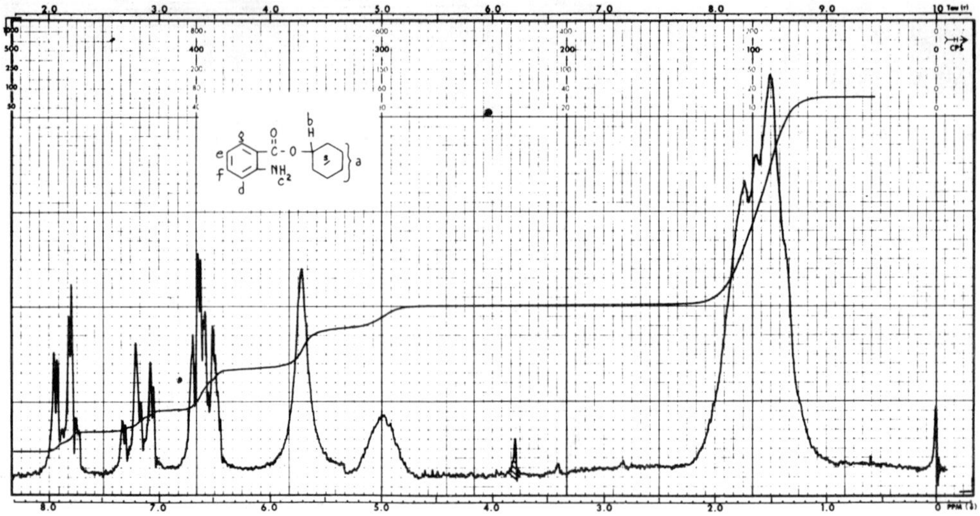

© 1978 Sadtler Research Laboratories, Inc.

ANTHRANILIC ACID, PENTYL ESTER

$C_{12}H_{17}NO_2$

Mol. Wt. 207.27

Solvent: DMSO-d_6

Assignments
a = 0.90
b = 1.10-2.00
c = 4.21
d = 6.32
e = 6.51
f = 6.76
g = 7.20
h = 7.75

2879

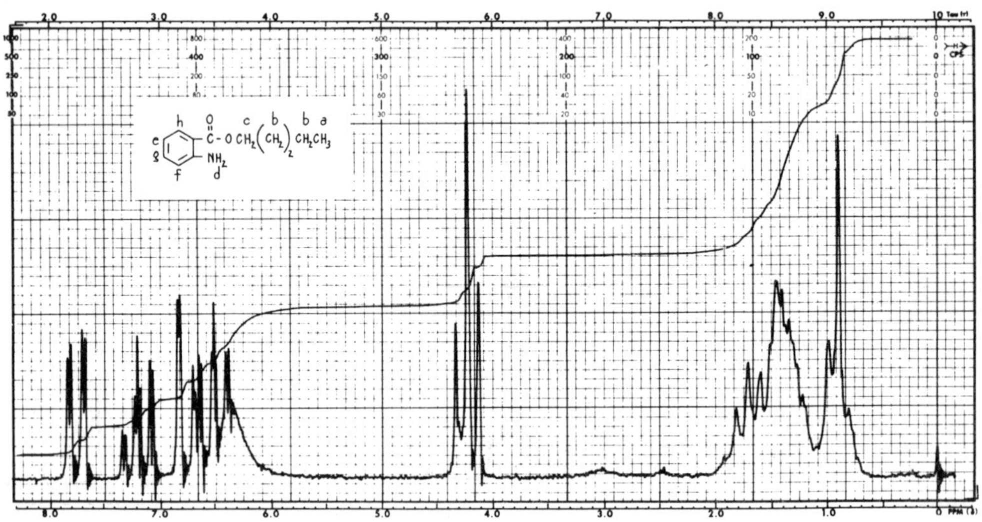

© 1978 Sadtler Research Laboratories, Inc.

m-AMINOBENZOIC ACID, METHYL ESTER

$C_8H_9NO_2$

Mol. Wt. 151.17

M.P. 52-54°C

Solvent: $CDCl_3$

Assignments
a = 3.80
b = 3.88
c = 6.77
d = 7.11
e = 7.30
f = 7.36

2880

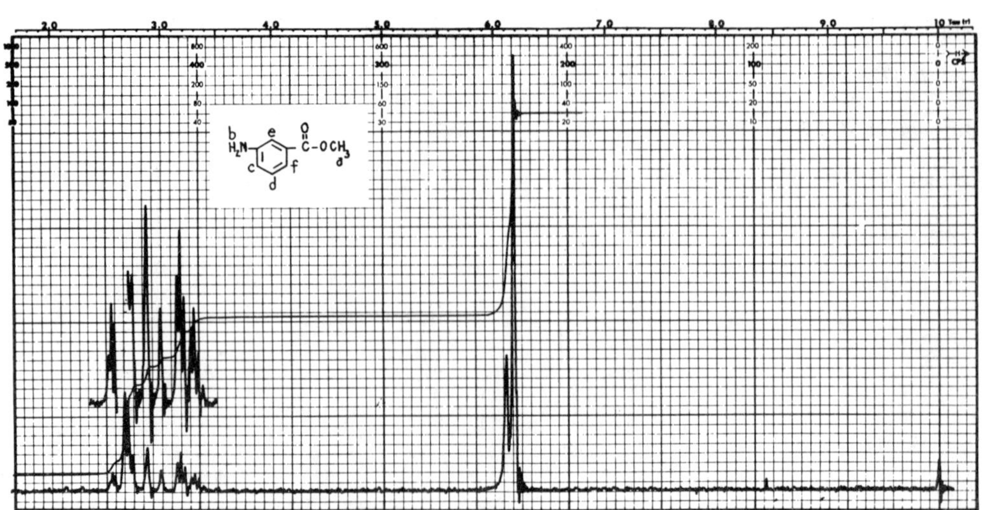

© 1978 Sadtler Research Laboratories, Inc.

p-AMINOBENZOIC ACID, METHYL ESTER

$C_8H_9NO_2$

Mol. Wt. 151.17

M.P. 111-113°C

Solvent: $CDCl_3$

Assignments
a = 3.82
b = 4.03
c = 6.63
d = 7.83

2881

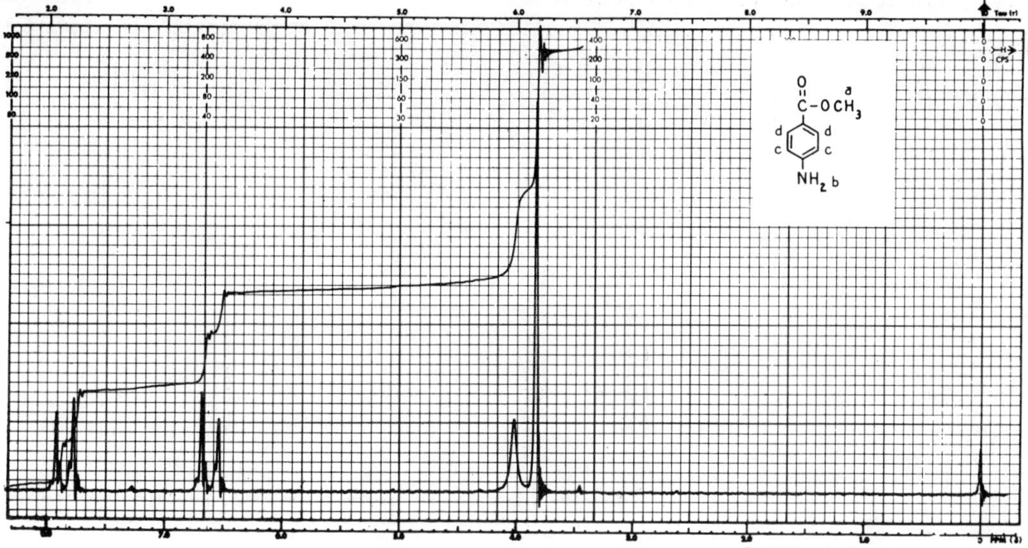

© 1978 Sadtler Research Laboratories, Inc.

p-AMINOBENZOIC ACID, ISOPROPYL ESTER

$C_{10}H_{13}NO_2$

Mol. Wt. 179.22

M.P. 85-86°C

Solvent: $CDCl_3$

Assignments
a = 1.32
b = 4.22
c = 5.21
d = 6.62
e = 7.86

2882

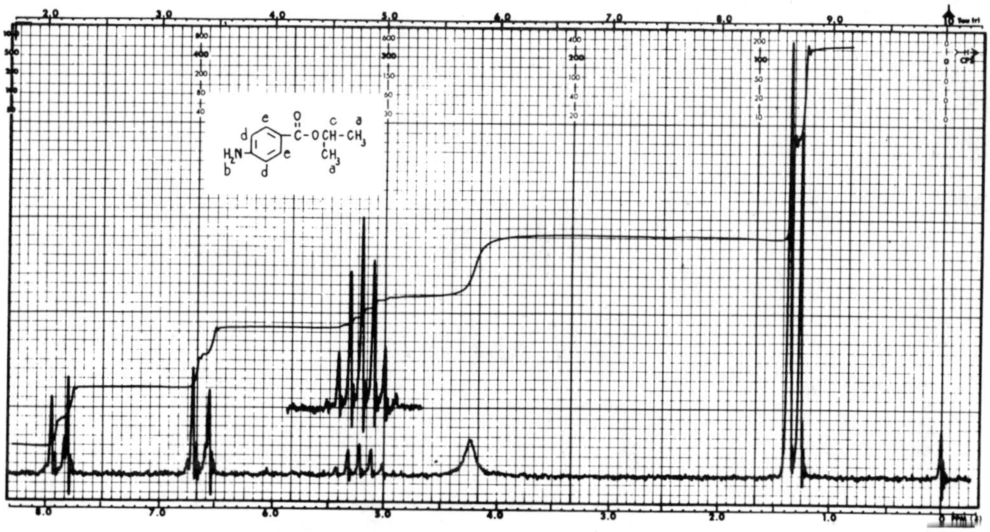

© 1978 Sadtler Research Laboratories, Inc.

p-NITROSOBENZOIC ACID, ETHYL ESTER

$C_9H_9NO_3$

Mol. Wt. 179.18

M.P. 79-81°C

Solvent: $CDCl_3$

Assignments
a = 1.45
b = 4.45
c = 7.91
d = 8.32

2883

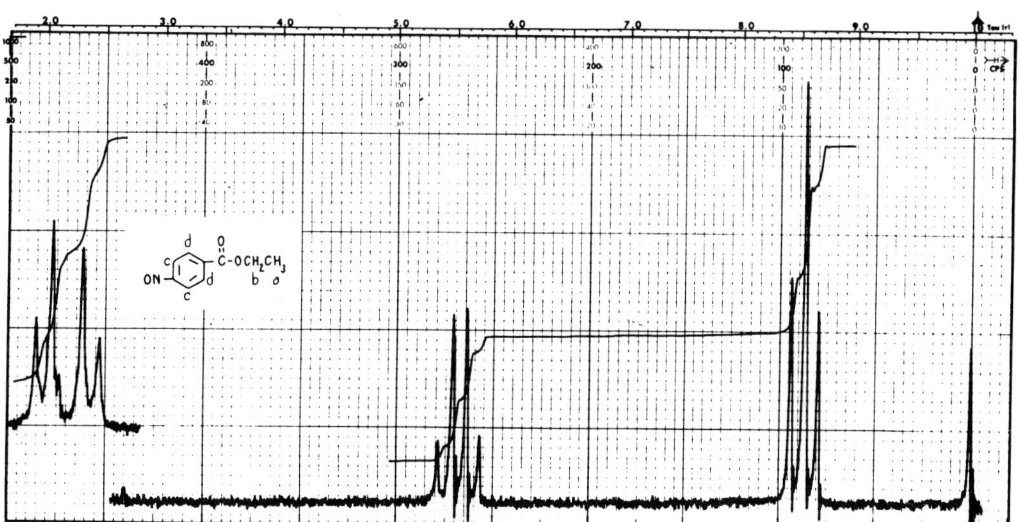

© 1978 Sadtler Research Laboratories, Inc.

p-ISOCYANATOBENZOIC ACID, ETHYL ESTER

$C_{10}H_9NO_3$

Mol. Wt. 191.19

M.P. 27-29°C

Solvent: $CDCl_3$

Assignments
a = 1.38
b = 4.34
c = 7.06
d = 7.95
e = 3.88 imp.

2884

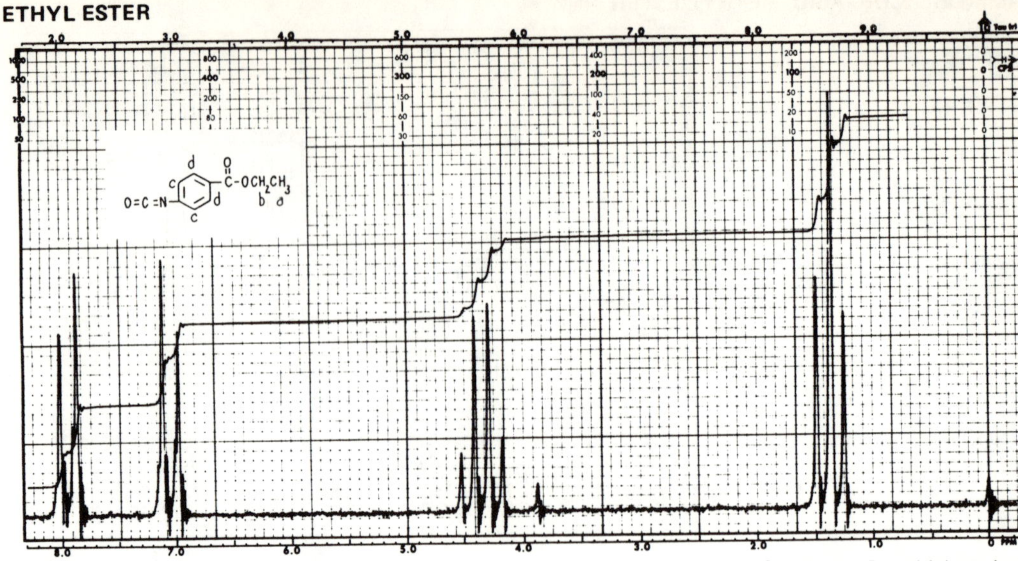

© 1978 Sadtler Research Laboratories, Inc.

m-NITROBENZOIC ACID, ETHYL ESTER

$C_9H_9NO_4$

Mol. Wt. 195.18

M.P. 41-43°C

Solvent: CCl_4

Assignments
a = 1.42
b = 4.40
c = 7.59
d = ca 10.00
e = 10.40

2885

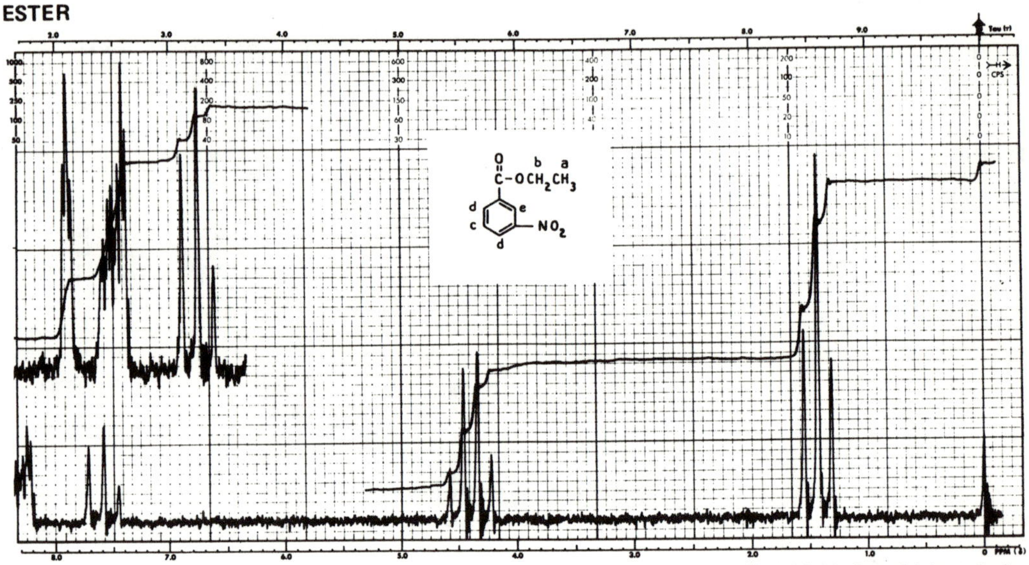

© 1978 Sadtler Research Laboratories, Inc.

p-NITROBENZOIC ACID, METHYL ESTER

$C_8H_7NO_4$

Mol. Wt. 181.15

M.P. 94-96°C

Solvent: $CDCl_3$

Assignments
a = 3.97
b = 8.26

2886

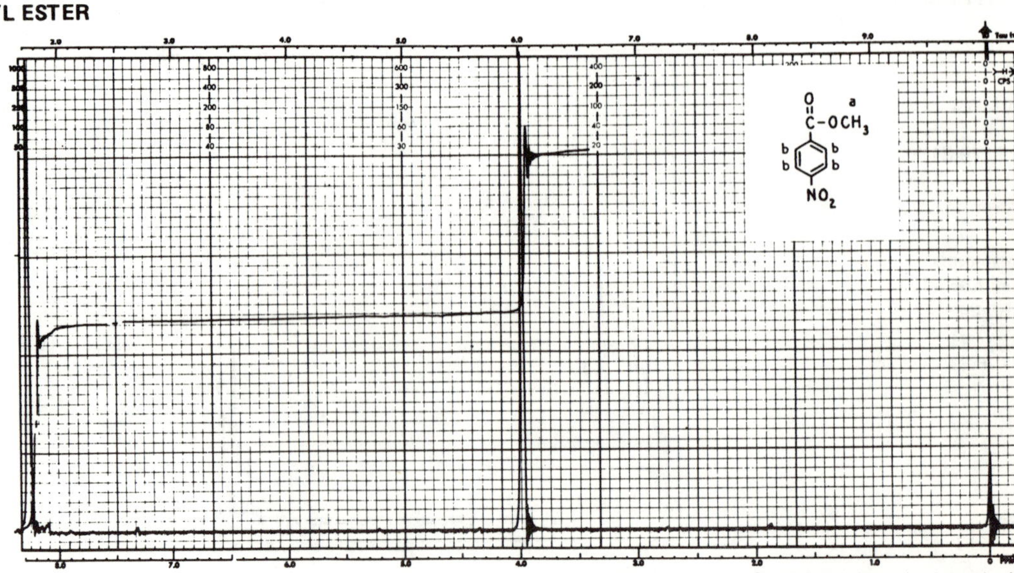

© 1978 Sadtler Research Laboratories, Inc.

o-ANISIC ACID, METHYL ESTER

$C_9H_{10}O_3$

Mol. Wt. 166.18

B.P. 117-118°C/8mm

Solvent: CCl_4

Assignments
a = 3.80
b = 6.75-7.07
c = 7.37
d = 7.69

2887

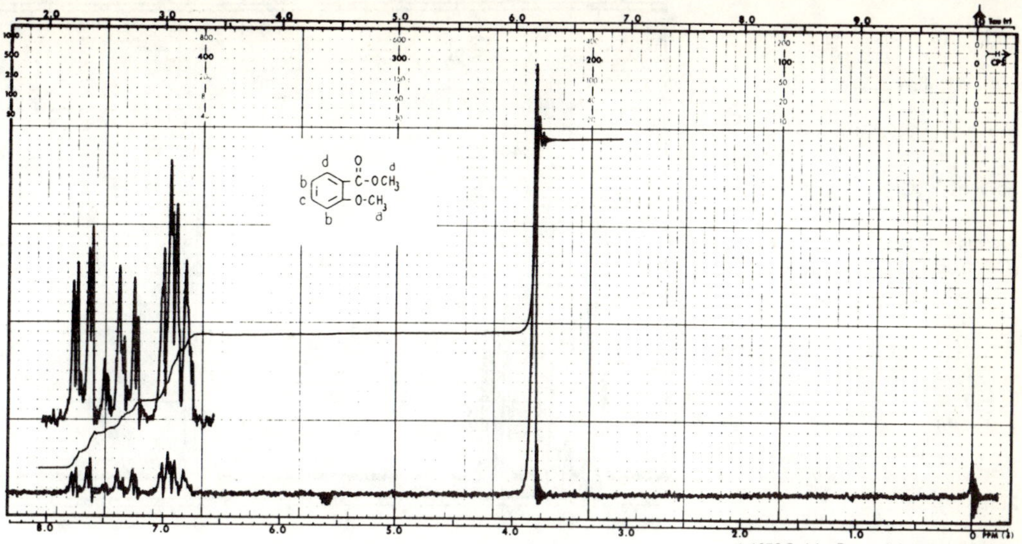

© 1978 Sadtler Research Laboratories, Inc.

p-ANISIC ACID, METHYL ESTER

$C_9H_{10}O_3$

Mol. Wt. 166.18

M.P. 49.5-50.5°C

Solvent: $CDCl_3$

Assignments
a = 3.82 or b
b = 3.87 or a
c = 6.88
d = 7.99

2888

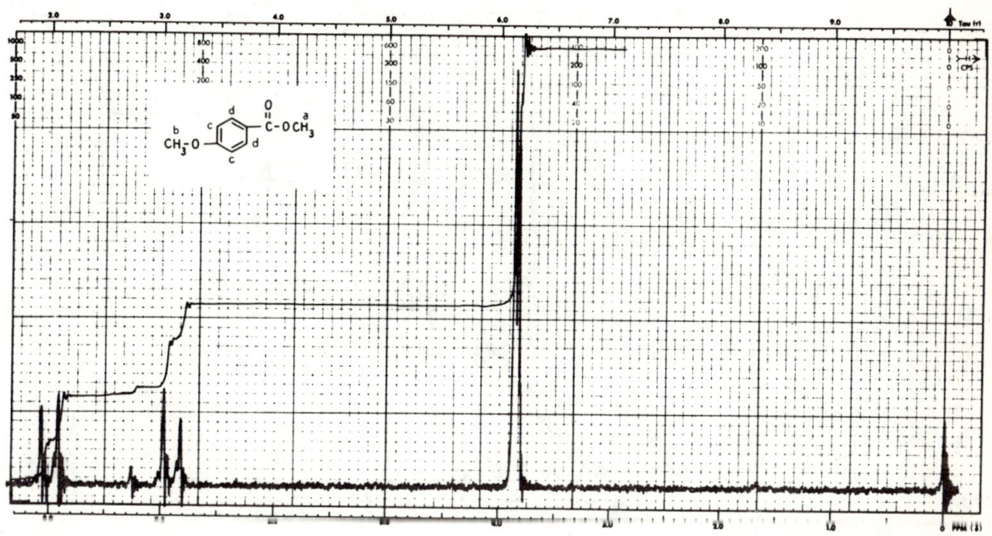

© 1978 Sadtler Research Laboratories, Inc.

N,N'-p-PHENYLENEBISACETAMIDE

$C_{10}H_{12}N_2O_2$

Mol. Wt. 192.22

M.P. 310-312°C

Solvent: Trifluoroacetic acid

Assignments
a = 2.53
b = 7.61
c = 9.45

2889

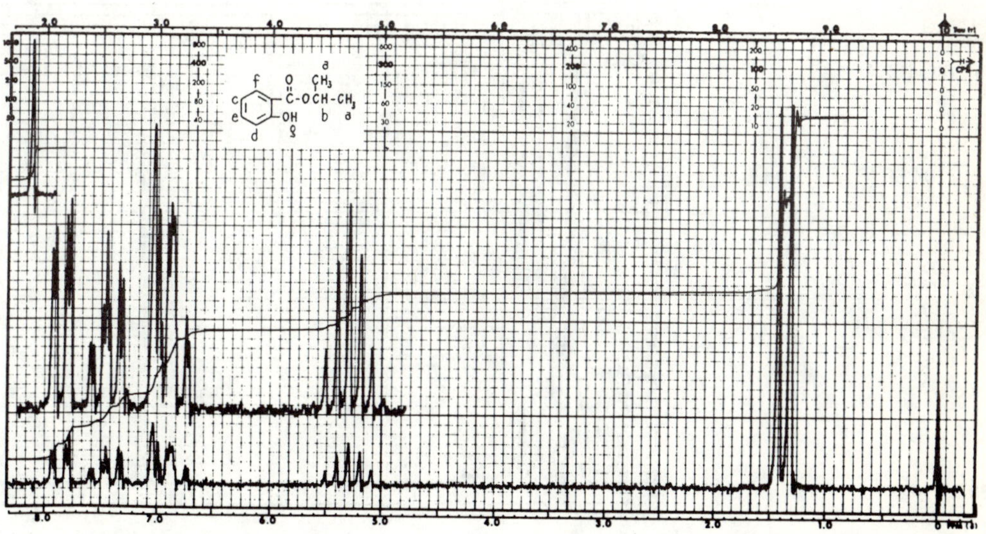

© 1978 Sadtler Research Laboratories, Inc.

SALICYLIC ACID, BUTYL ESTER

$C_{11}H_{14}O_3$

Mol. Wt. 194.23

B.P. 132-133°C/15mm

Solvent: CCl_4

Assignments
a = 0.98
b = ca 1.62
c = 4.29
d = 6.72
e = 6.85
f = 7.32
g = 7.70
OH is unobserved

2890

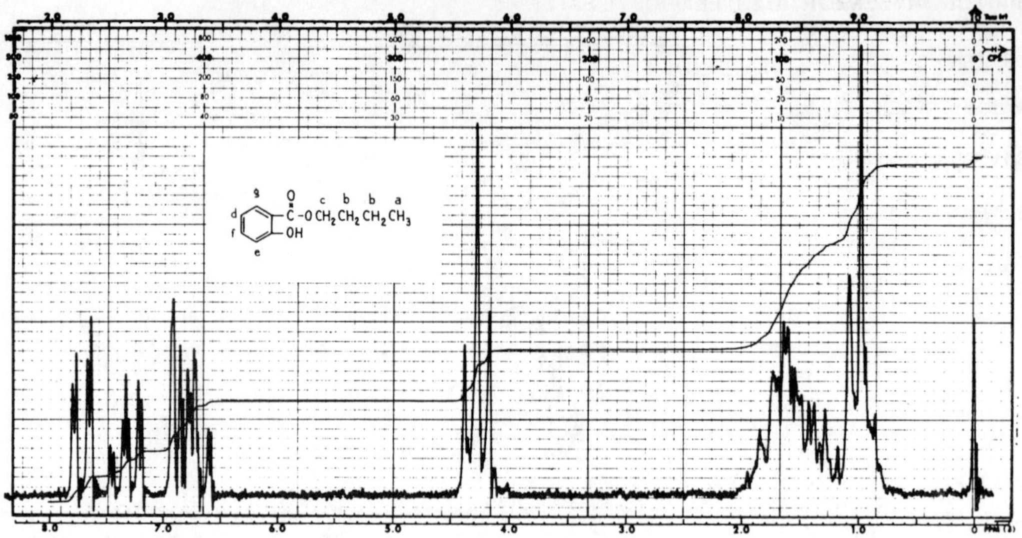

© 1978 Sadtler Research Laboratories, Inc.

SALICYLIC ACID, PHENETHYL ESTER

$C_{15}H_{14}O_3$

Mol. Wt. 242.26

Solvent: $CDCl_3$

Assignments
a = 3.07
b = 4.53
c = 6.68-7.05
d = 7.27
e = 7.40-7.90
f = 10.72
g = 1.52 imp.

2891

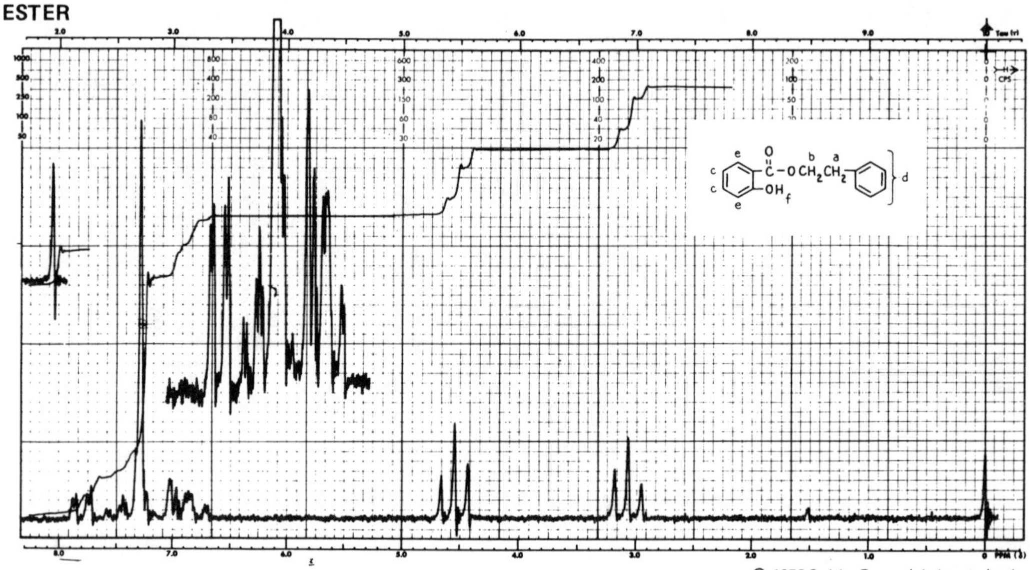

© 1978 Sadtler Research Laboratories, Inc.

m-HYDROXYBENZOIC ACID, METHYL ESTER

$C_8H_8O_3$

Mol. Wt. 152.15

M.P. 131°C

B.P. 270-280°C (dec.) (lit.)

Solvent: $CDCl_3$

Assignments
a = 3.91
b = ca 6.72
c = 6.95-7.48
d = 7.62

2892

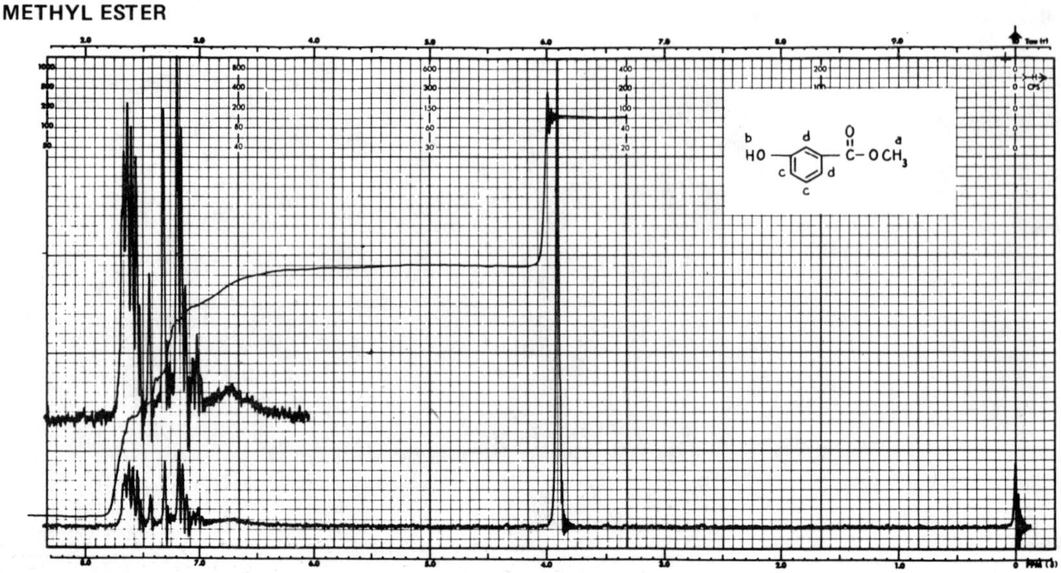

© 1978 Sadtler Research Laboratories, Inc.

p-HYDROXYBENZOIC ACID, ISOPROPYL ESTER

$C_{10}H_{12}O_3$

Mol. Wt. 180.21

M.P. 86°C (lit.)

Solvent: $CDCl_3$

Assignments

a = 1.36
b = 5.27
c = 6.93
d = 6.91
e = 7.96

2893

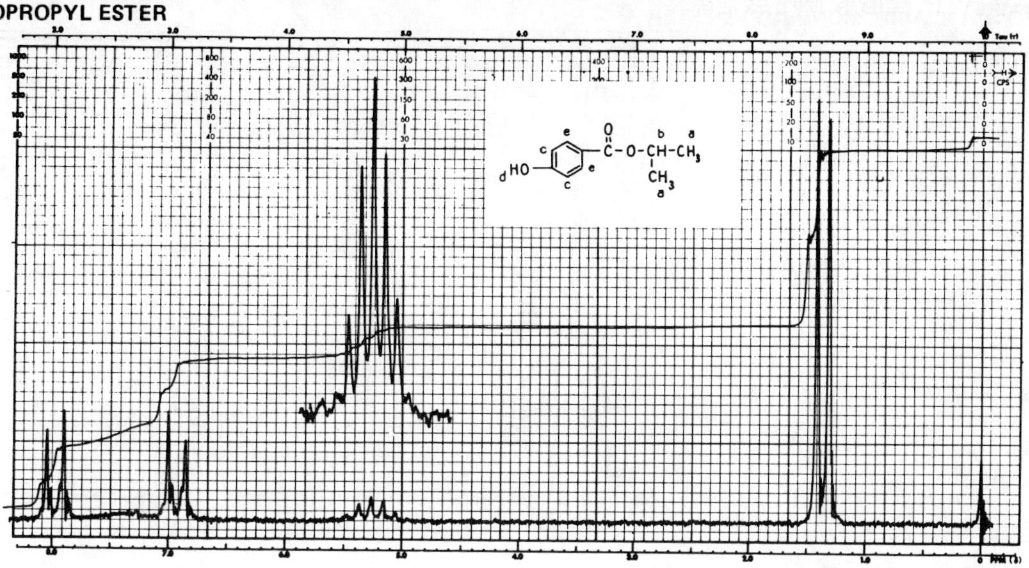

© 1978 Sadtler Research Laboratories, Inc.

p-HYDROXYBENZOIC ACID, BUTYL ESTER

$C_{11}H_{14}O_3$

Mol. Wt. 194.23

M.P. 67-69°C (lit.)

Solvent: CCl_4

Assignments

a = 0.97
b = 1.60
c = 4.27
d = 6.84
e = 7.90
f = 8.05

2894

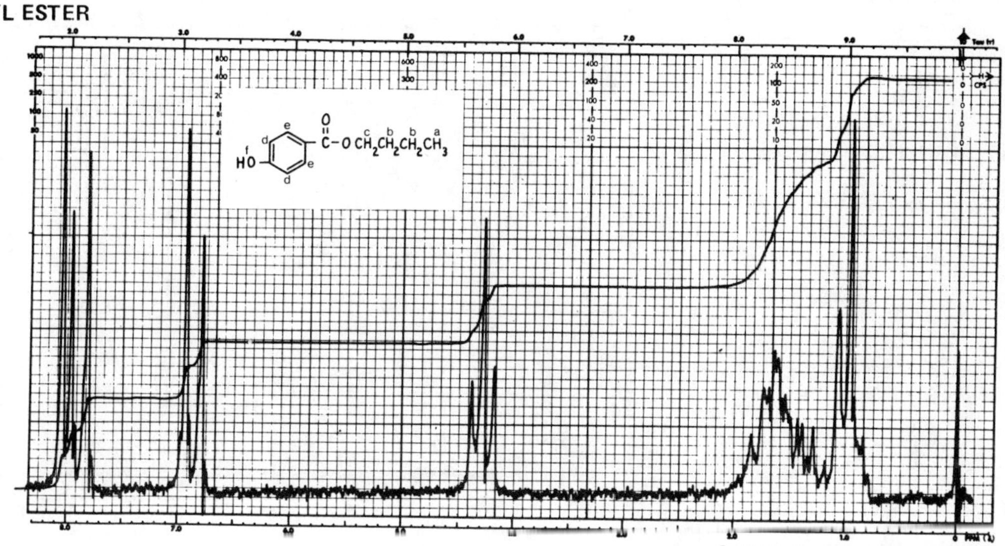

© 1978 Sadtler Research Laboratories, Inc.

p-HYDROXYBENZOIC ACID, DODECYL ESTER

$C_{19}H_{30}O_3$

Mol. Wt. 306.45

Solvent: CCl_4

Assignments

a = 0.89
b = 1.26
c = 4.30
d = 6.87
e = 7.70
f = 7.90
g = 2.16 imp.

2895

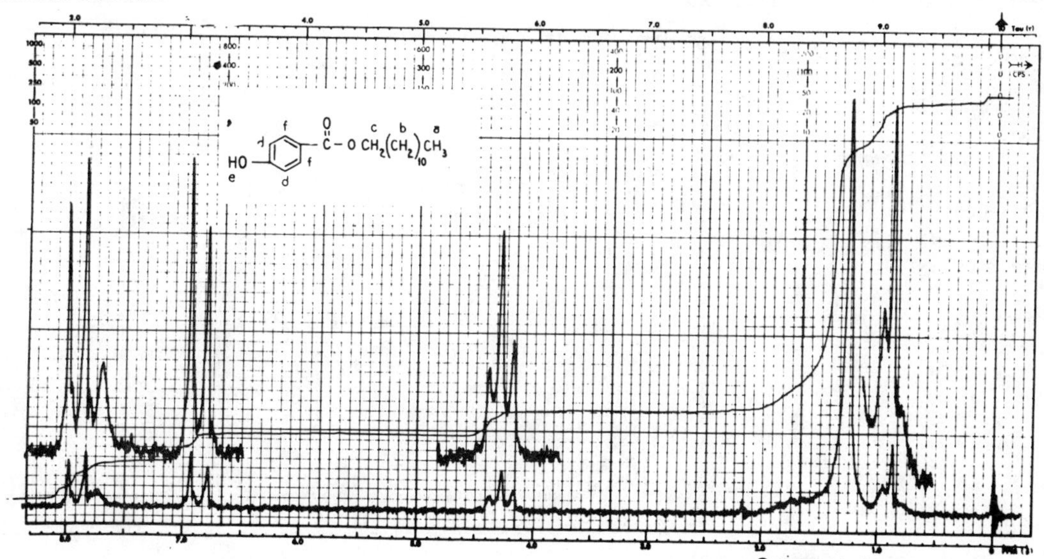

© 1978 Sadtler Research Laboratories, Inc.

PHTHALIC ACID, MONOBUTYL ESTER

$C_{12}H_{14}O_4$

Mol. Wt. 222.24

M.P. 73-78°C

Solvent: $CDCl_3$

Assignments
a = 0.95
b = 1.10-2.00
c = 4.37
d = 7.30-8.05
e = 12.48

2896

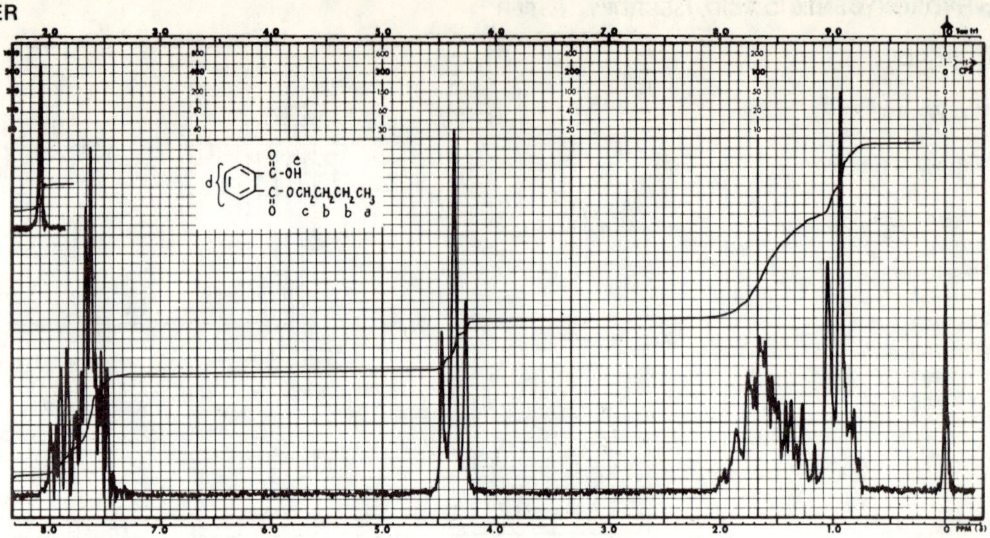

© 1978 Sadtler Research Laboratories, Inc.

PHTHALIC ACID, DIPENTYL ESTER

$C_{18}H_{26}O_4$

Mol. Wt. 306.41

B.P. 204-206°C/11mm

Solvent: CCl_4

Assignments
a = 0.92
b = 1.10-2.00
c = 4.21
d = 7.48
e = 7.62

2897

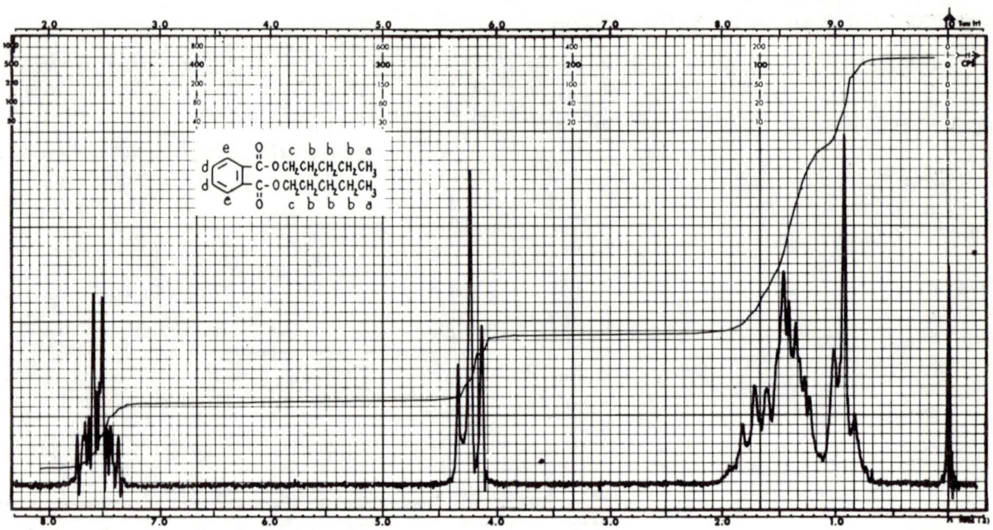

© 1978 Sadtler Research Laboratories, Inc.

PHTHALIC ACID, BIS(2-ETHYLHEXYL) ESTER

$C_{24}H_{38}O_4$

Mol. Wt. 390.57

B.P. 222-230°C/4mm

F.P. −55°C

Solvent: CCl_4

Assignments
a = 0.91
b = 1.10-1.90
c = 4.19
d = 7.41
e = 7.65

2898

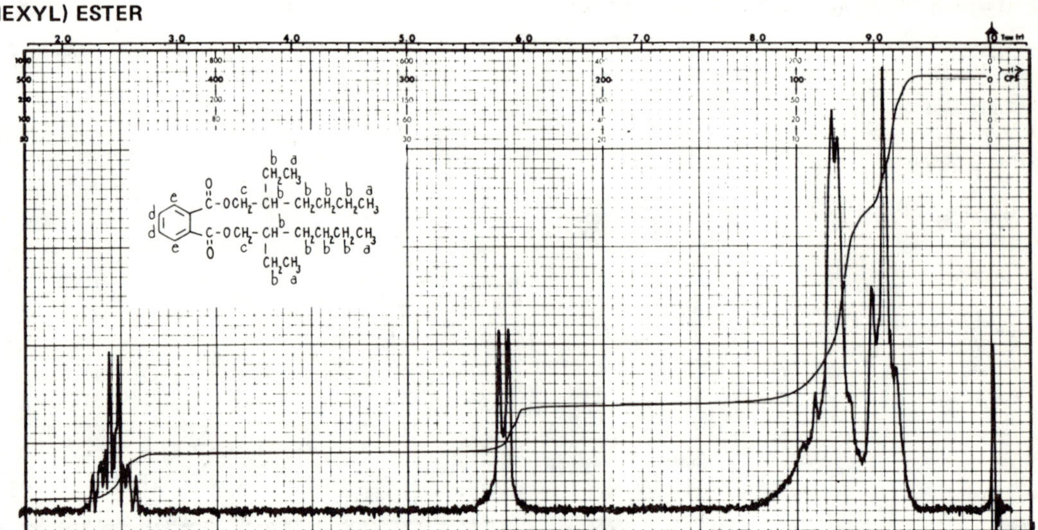

© 1978 Sadtler Research Laboratories, Inc.

PHTHALIC ACID, DIALLYL ESTER

$C_{14}H_{14}O_4$

Mol. Wt. 246.27

B.P. 160-163°C/4mm

Solvent: CCl_4

Assignments
a = 4.72
b = 5.21
c = 5.31
d = 5.90
e = 7.48
f = 7.67
g = 1.28 imp.

2899

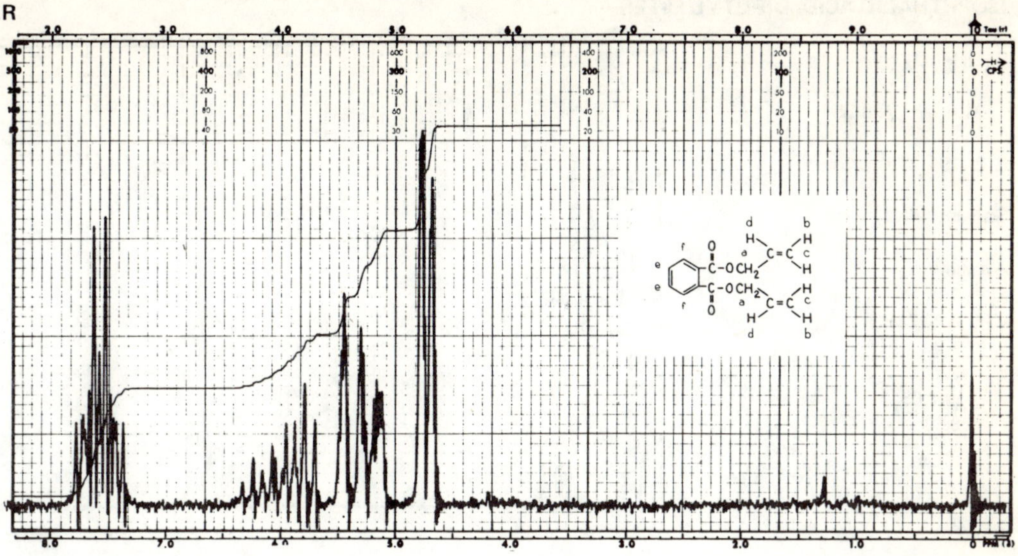

© 1978 Sadtler Research Laboratories, Inc.

TETRACHLOROPHTHALIC ACID, DIPROPYL ESTER

$C_{14}H_{14}Cl_4O_4$

Mol. Wt. 388.08

M.P. 24-27°C

Solvent: $CDCl_3$

Assignments
a = 1.01
b = 1.80
c = 4.30

2900

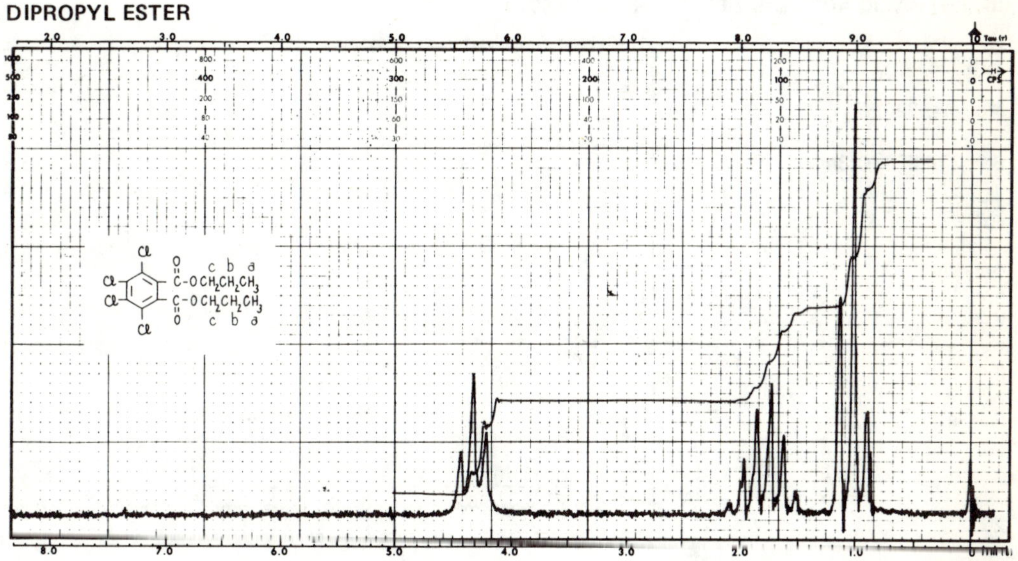

© 1978 Sadtler Research Laboratories, Inc.

ISOPHTHALIC ACID, DIMETHYL ESTER

$C_{10}H_{10}O_4$

Mol. Wt. 194.19

M.P. 64-65°C

Solvent: $CDCl_3$

Assignments
a = 3.94
b = 7.50
c = 8.19
d = 8.63

2901

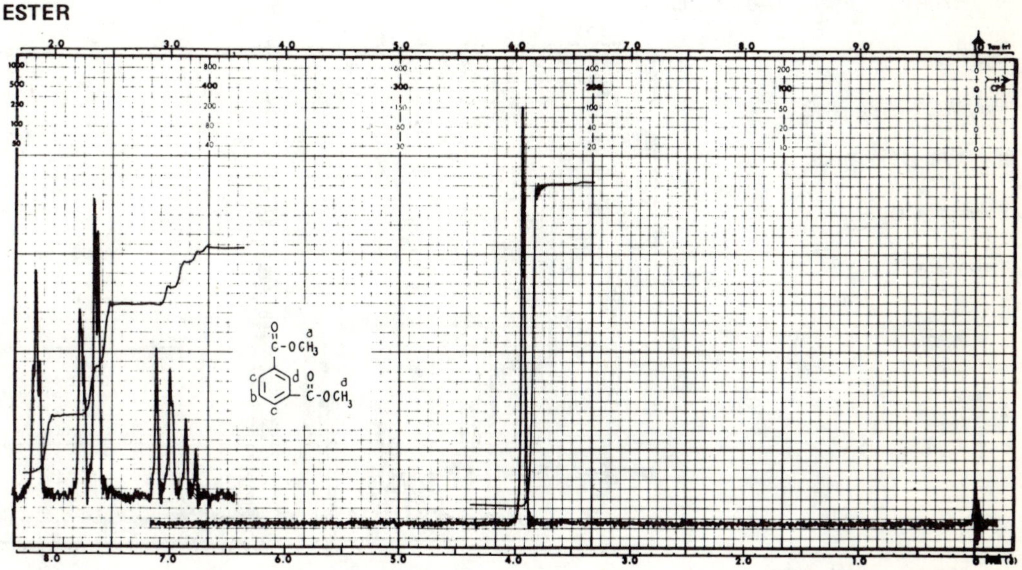

© 1978 Sadtler Research Laboratories, Inc.

ISOPHTHALIC ACID, DIBUTYL ESTER

$C_{16}H_{22}O_4$

Mol. Wt. 278.35

B.P. 191-193°C/3mm

Solvent: CCl_4

Assignments
a = 1.00
b = 1.20-2.08
c = 4.30
d = 7.42
e = ca 8.05
f = 8.53

2902

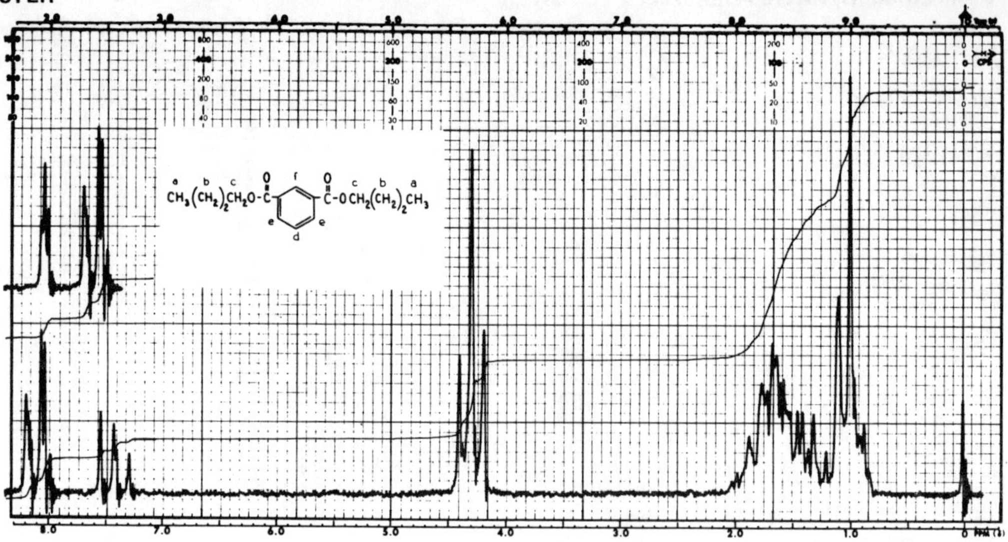

© 1978 Sadtler Research Laboratories, Inc.

ISOPHTHALIC ACID, BIS(2-ETHYLHEXYL) ESTER

$C_{24}H_{38}O_4$

Mol. Wt. 390.57

B.P. 224-226°C/3mm

Solvent: CCl_4

Assignments
a = 0.91
b = 1.00
c = 1.17-1.99
d = 4.22
e = 7.43
f = 8.12
g = 8.56

2903

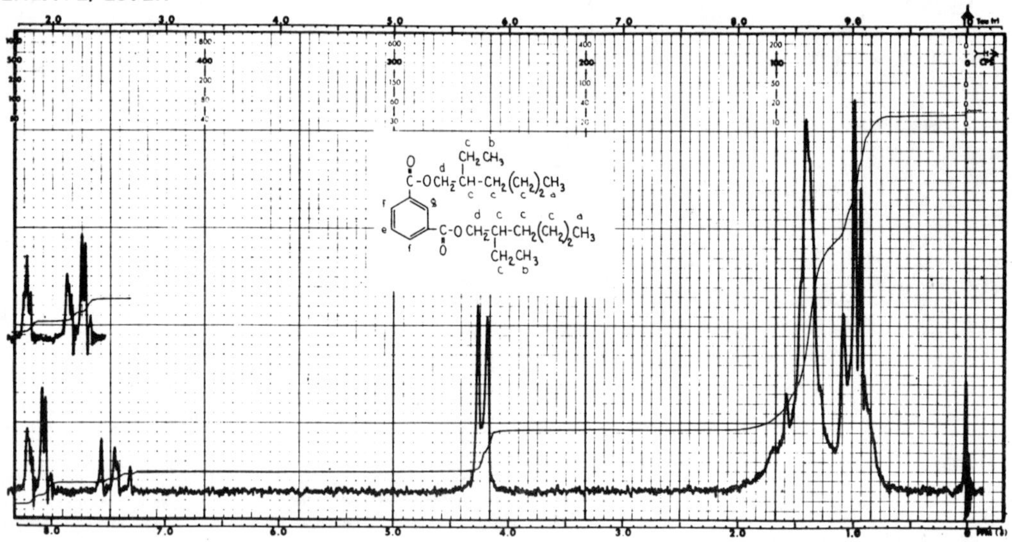

© 1978 Sadtler Research Laboratories, Inc.

ISOPHTHALIC ACID, DIALLYL ESTER

$C_{14}H_{14}O_4$

Mol. Wt. 246.27

B.P. 176-177°C/5mm

Solvent: $CDCl_3$

Assignments
a = 4.88
b = 5.31
c = 5.42
d = 6.01
e = 7.55
f = 8.25
g = 8.78

2904

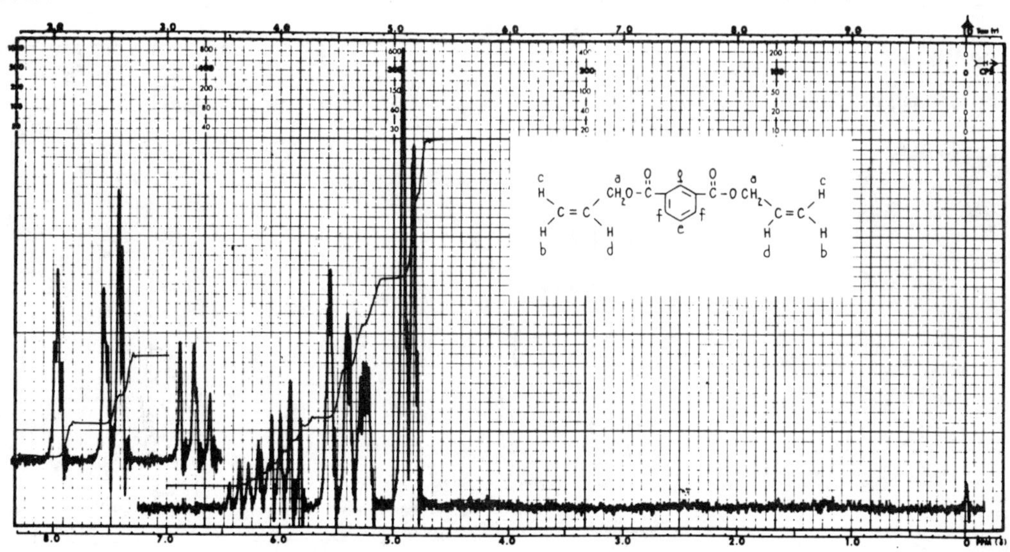

© 1978 Sadtler Research Laboratories, Inc.

5-AMINOISOPHTHALIC ACID, DIMETHYL ESTER

$C_{10}H_{11}NO_4$

Mol. Wt. 209.20

M.P. 180-183°C

Solvent: Polysol-d

Assignments
a = 3.89
b = 5.23
c = 7.49
d = 7.80

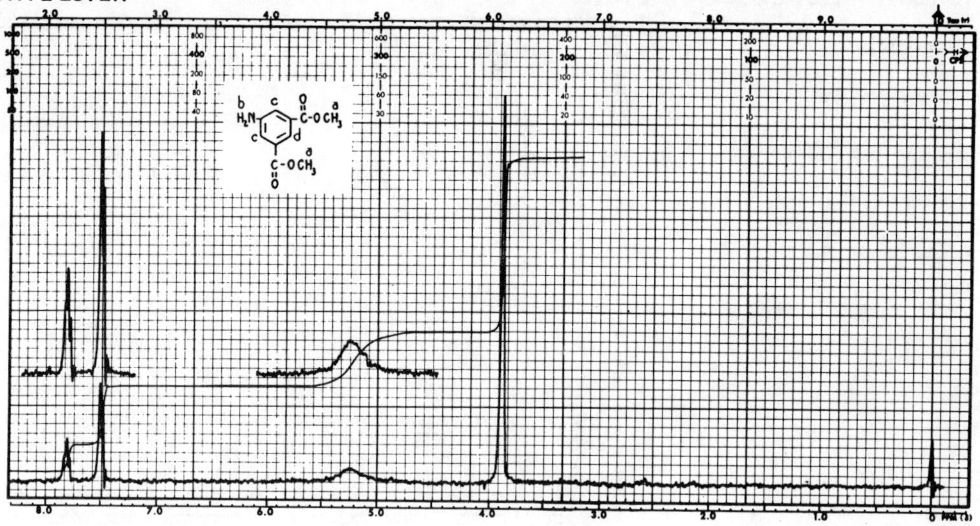

2905

© 1978 Sadtler Research Laboratories, Inc.

5-NITROISOPHTHALIC ACID, DIMETHYL ESTER

$C_{10}H_9NO_6$

Mol. Wt. 239.19

M.P. 122-124°C

Solvent: $CDCl_3$

Assignments
a = 4.05
b = 8.98

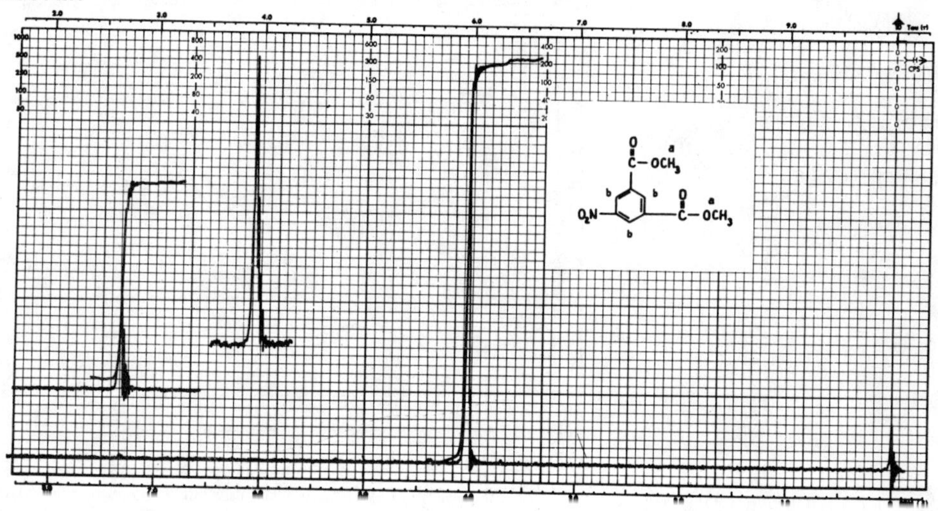

2906

© 1978 Sadtler Research Laboratories, Inc.

TEREPHTHALIC ACID, DIETHYL ESTER

$C_{12}H_{14}O_4$

Mol. Wt. 222.24

M.P. 43-44°C

Solvent: $CDCl_3$

Assignments
a = 1.40
b = 4.39
c = 8.07

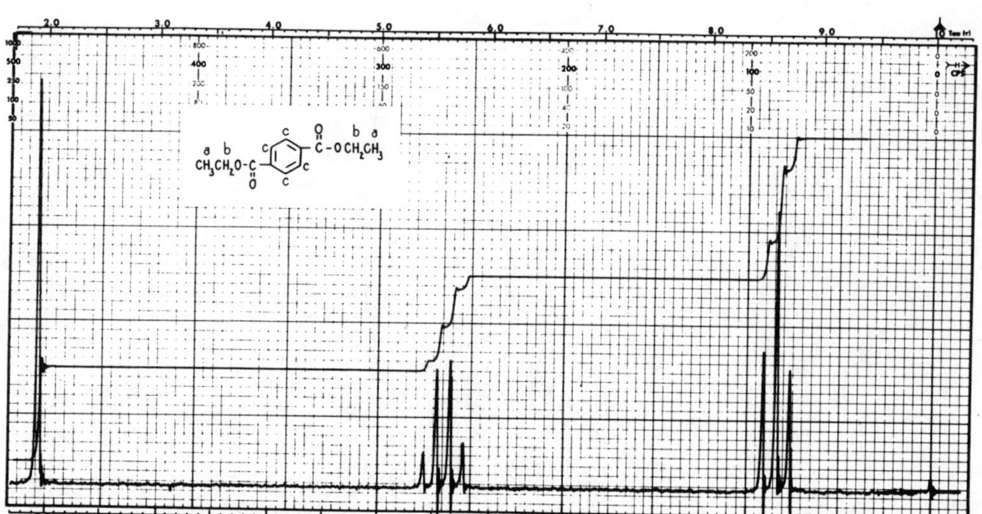

2907

© 1978 Sadtler Research Laboratories, Inc.

TEREPHTHALIC ACID, DIBUTYL ESTER

$C_{16}H_{22}O_4$

Mol. Wt. 278.35

B.P. 194-196°C/8mm

Solvent: CCl_4

Assignments
a = 1.00
b = 1.18-1.93
c = 4.28
d = 7.98

2908

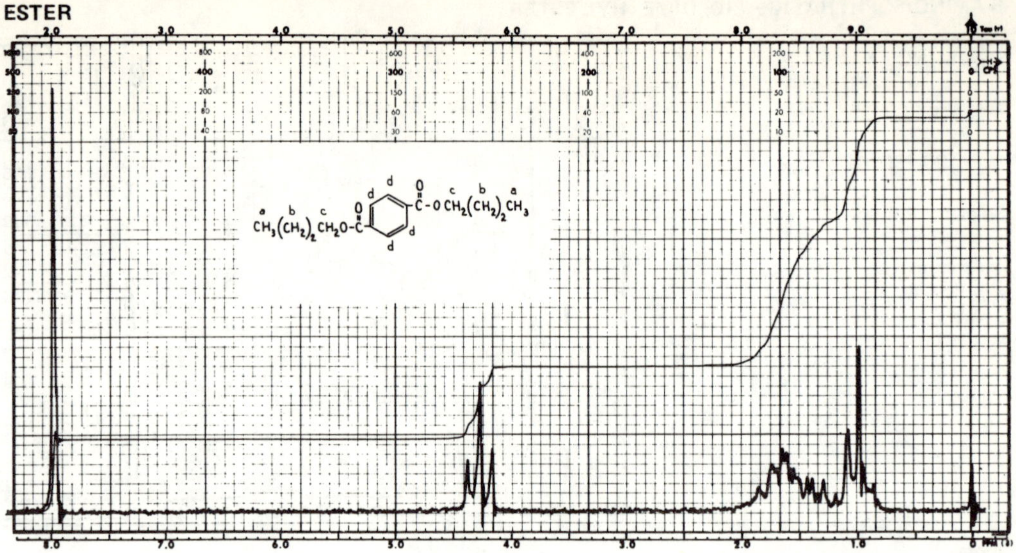

© 1978 Sadtler Research Laboratories, Inc.

TEREPHTHALIC ACID, DIOCTYL ESTER

$C_{24}H_{38}O_4$

Mol. Wt. 390.57

M.P. 42.5-43.5°C

Solvent: $CDCl_3$

Assignments
a = 0.90
b = 1.35
c = 1.80
d = 4.35
e = 8.11

2909

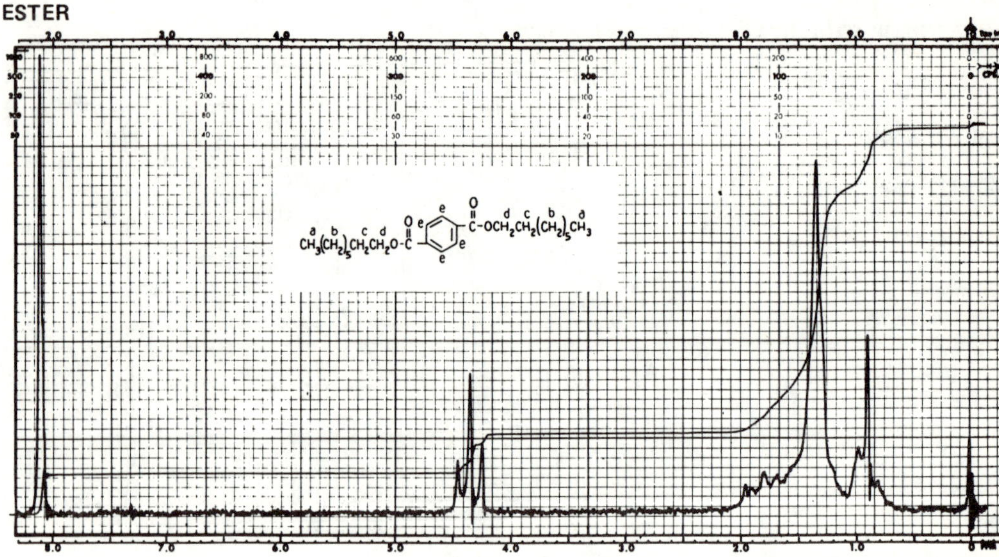

© 1978 Sadtler Research Laboratories, Inc.

2-AMINOTEREPHTHALIC ACID, DIMETHYL ESTER

$C_{10}H_{11}NO_4$

Mol. Wt. 209.20

M.P. 133-135°C

Solvent: Polysol-d

Assignments
a = 3.84
b = 6.11
c = 7.05
d = 7.43
e = 7.77

2910

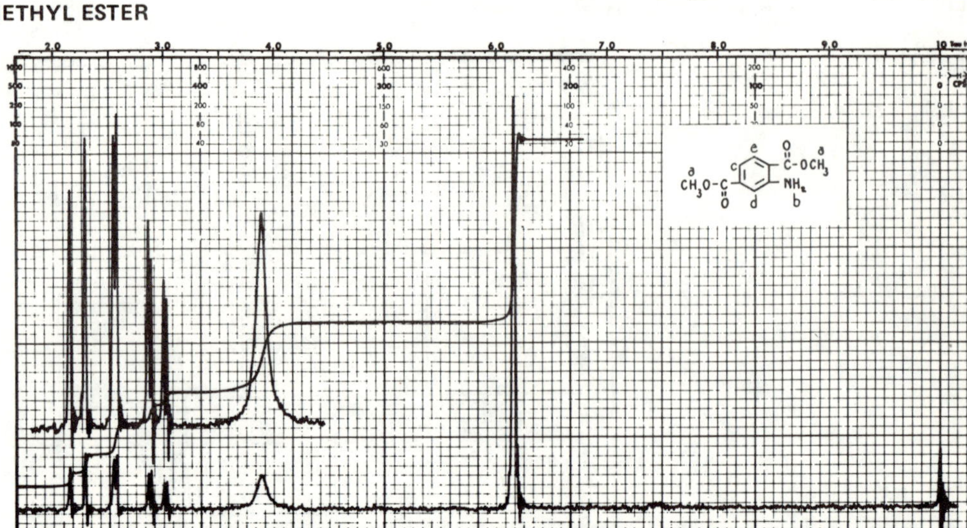

© 1978 Sadtler Research Laboratories, Inc.

3,5-DINITROBENZOIC ACID, ETHYL ESTER

$C_9H_8N_2O_6$

Mol. Wt. 240.17

M.P. 93°C

Solvent: $CDCl_3$

Assignments
a = 1.49
b = 4.51
c = ca 9.11
d = 4.07 imp.

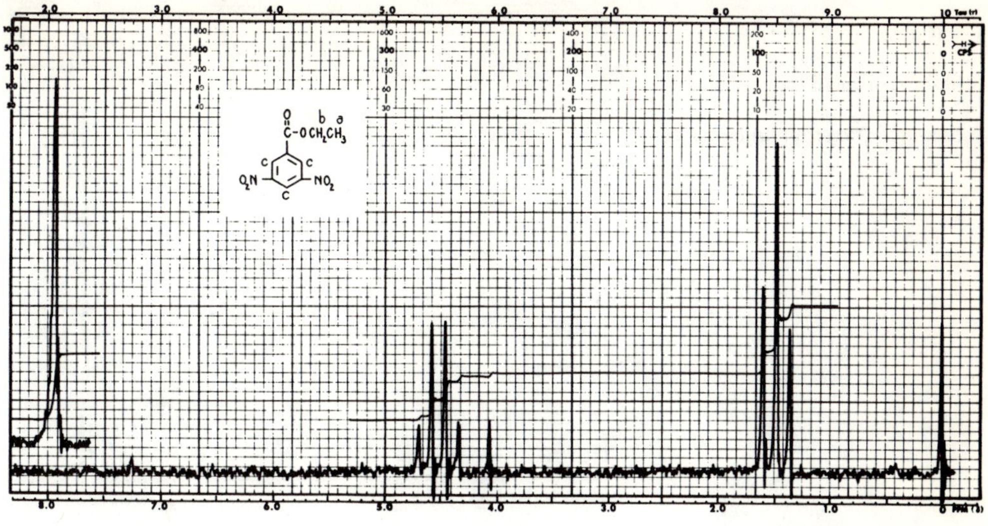

2911

© 1978 Sadtler Research Laboratories, Inc.

3,5-DINITROBENZOIC ACID, CYCLOHEXYL ESTER

$C_{13}H_{14}N_2O_6$

Mol. Wt. 294.27

M.P. 112°C

Solvent: $CDCl_3$

Assignments
a = 1.20-2.30
b = 5.16
c = ca 9.19

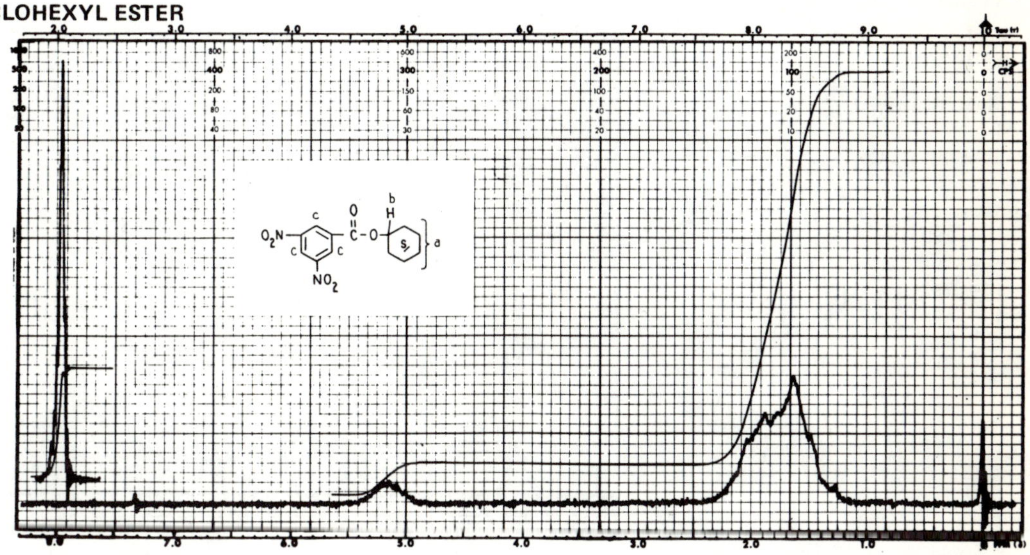

2912

© 1978 Sadtler Research Laboratories, Inc.

3,5-DINITROBENZOIC ACID, BENZYL ESTER

$C_{14}H_{10}N_2O_6$

Mol. Wt. 302.25

M.P. 112°C

Solvent: $CDCl_3$

Assignments
a = 5.44
b = 7.40
c = 9.13

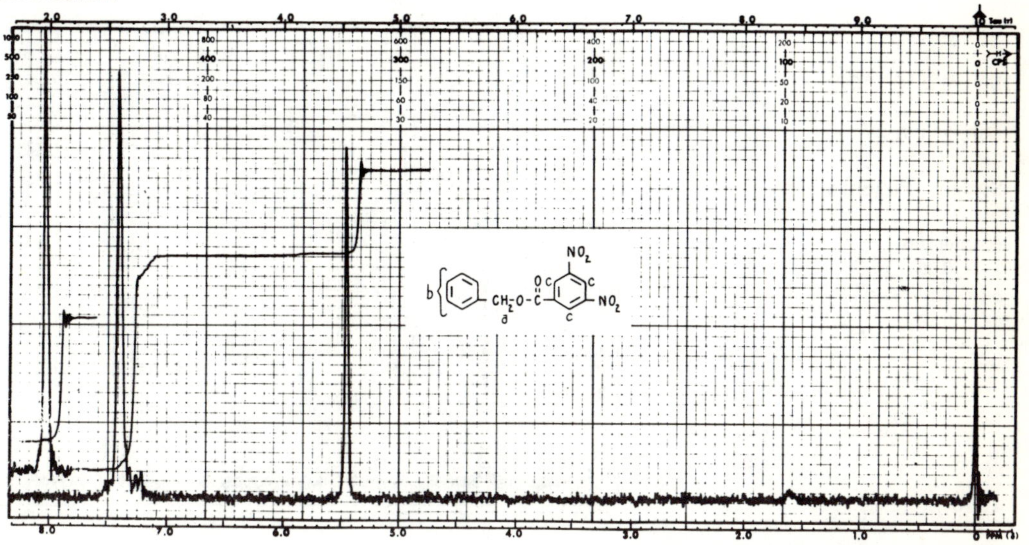

2913

© 1978 Sadtler Research Laboratories, Inc.

3,5-DIHYDROXYBENZOIC ACID, METHYL ESTER

$C_8H_8O_4$

Mol. Wt. 168.15

Solvent: Dimethyl sulfoxide-d_6

Assignments
a = 3.86
b = 6.58
c = 6.95
d = 9.59

2914

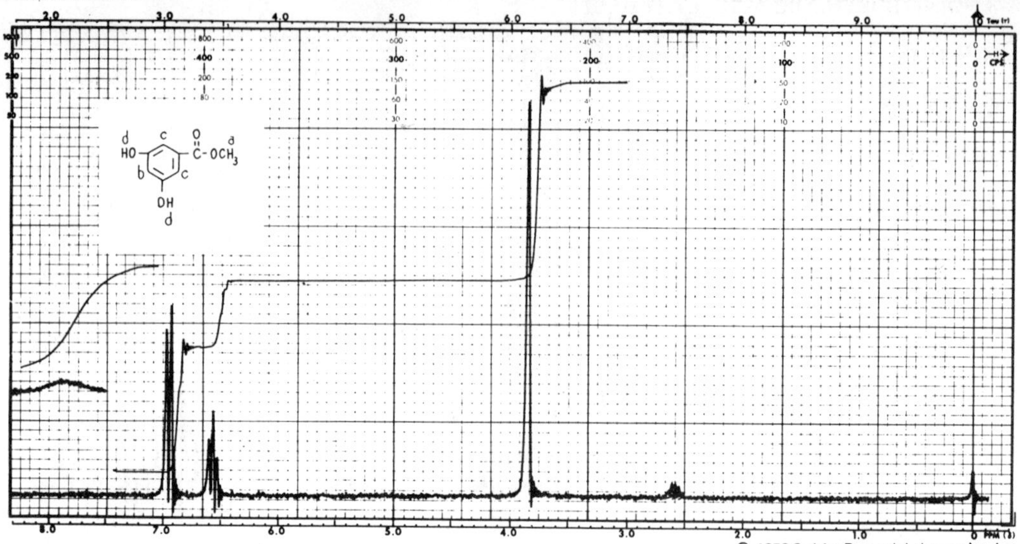

© 1978 Sadtler Research Laboratories, Inc.

SYRINGIC ACID, METHYL ESTER

$C_{10}H_{12}O_5$

Mol. Wt. 212.20

Solvent: $CDCl_3$

Assignments
a = 3.89
b = 3.92
c = 5.92
d = 7.33

2915

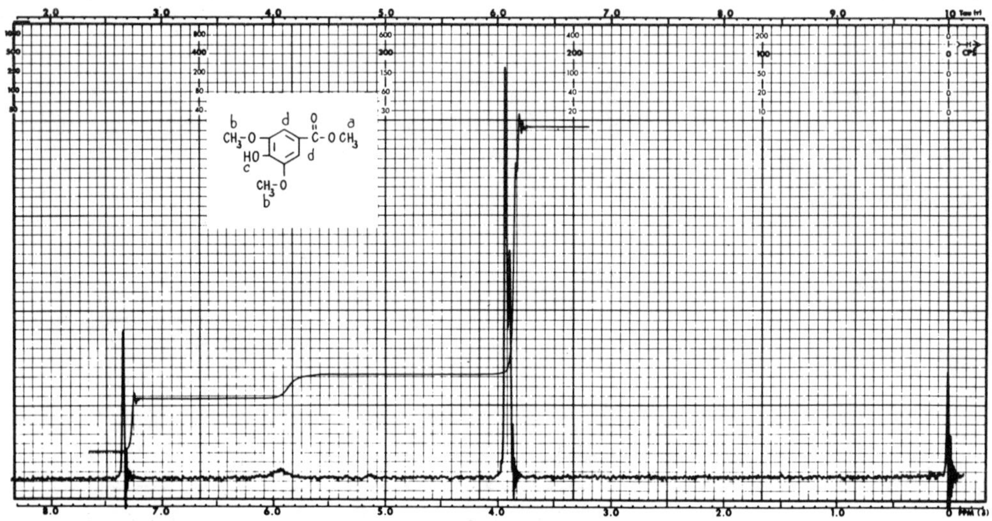

© 1978 Sadtler Research Laboratories, Inc.

GALLIC ACID, METHYL ESTER

$C_8H_8O_5$

Mol. Wt. 184.15

M.P. 201-203°C

Solvent: Polysol-d

Assignments
a = 3.80
b = 7.12
c = 8.19

2916

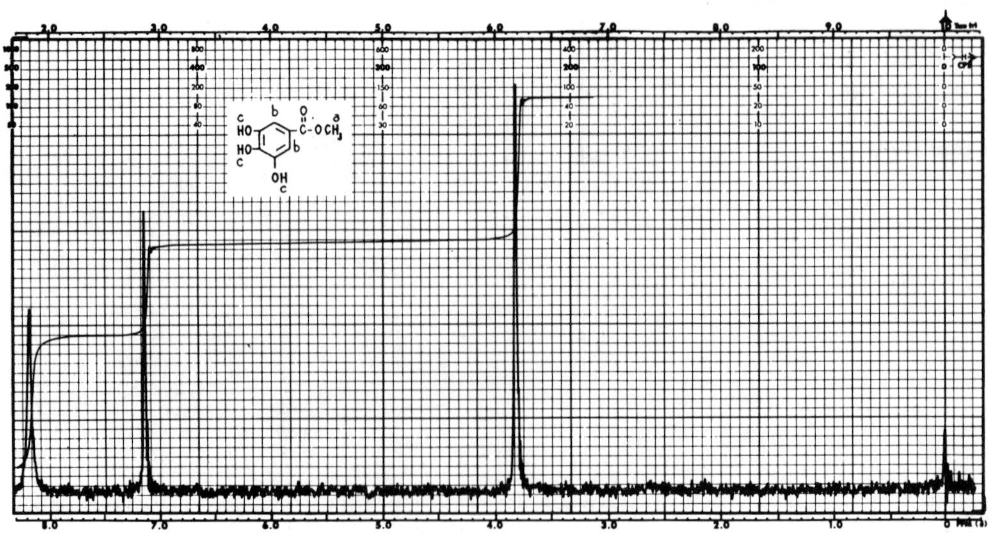

© 1978 Sadtler Research Laboratories, Inc.

GALLIC ACID, BUTYL ESTER

$C_{11}H_{14}O_5$

Mol. Wt. 226.23

M.P. 134°C

Solvent: Acetone-D_6

Assignments

a = 0.96
b = 1.17-1.89
c = 4.25
d = 7.20
e = 7.72
f = 2.10 Acetone-D_5

2917

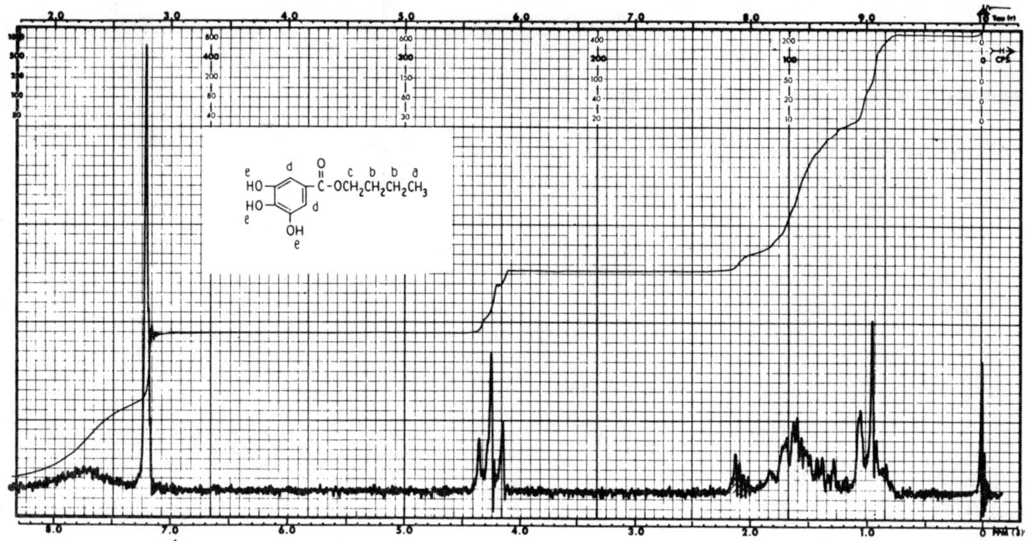

© 1978 Sadtler Research Laboratories, Inc.

GALLIC ACID, ISOBUTYL ESTER

$C_{11}H_{14}O_5$

Mol. Wt. 226.23

M.P. 129°C

Solvent: Acetone-D_6

Assignments

a = 1.00
b = 2.05
c = 4.04
d = ca 6.60
e = 7.23

2918

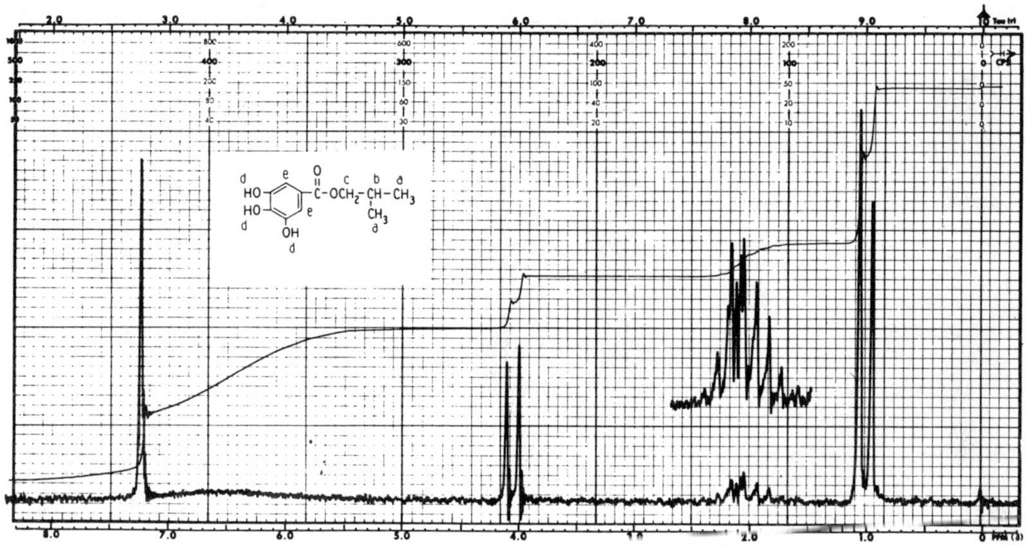

© 1978 Sadtler Research Laboratories, Inc.

GALLIC ACID, DODECYL ESTER

$C_{19}H_{30}O_5$

Mol. Wt. 338.45

M.P. 96-97°C

Solvent: Acetone-D_6

Assignments

a = 0.89
b = 1.29
c = 4.14
d = 6.47
e = 7.01
f = 2.01 Acetone-D_5

2919

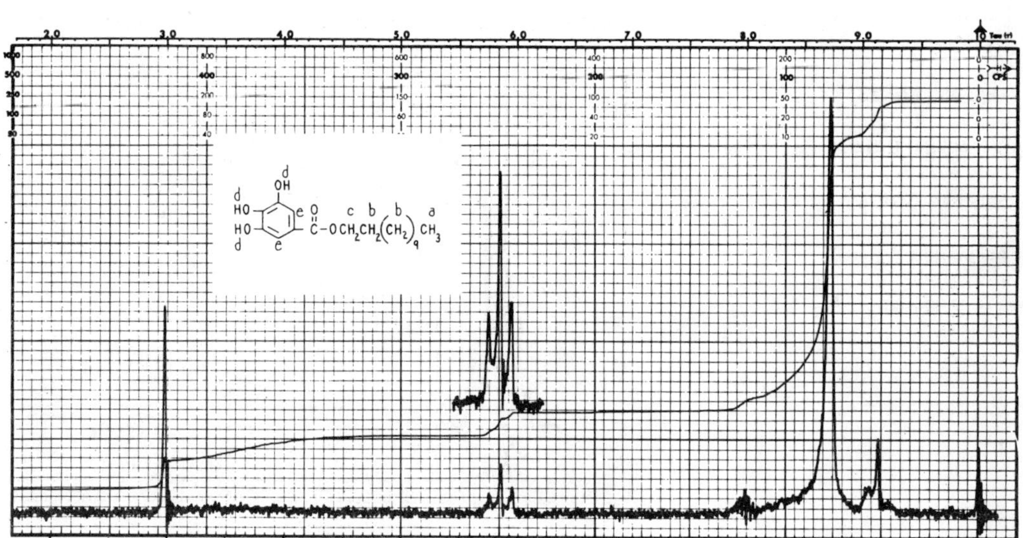

© 1978 Sadtler Research Laboratories, Inc.

PYROMELLITIC ACID, TETRAMETHYL ESTER

$C_{14}H_{14}O_8$

Mol. Wt. 310.26

M.P. 143-144°C

Solvent: $CDCl_3$

Assignments
a = 3.96
b = 8.04

2920

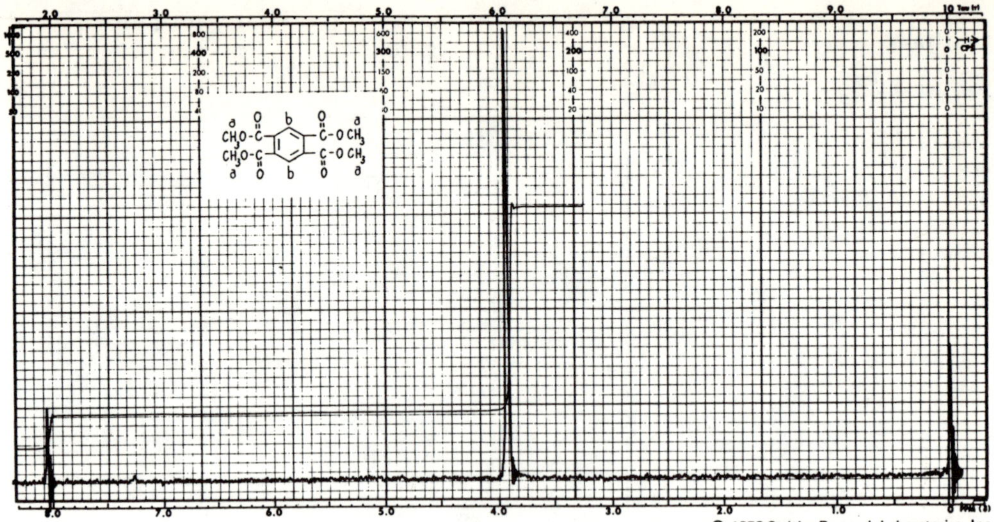

© 1978 Sadtler Research Laboratories, Inc.

PROPIONIC ACID, PHENYL ESTER

$C_9H_{10}O_2$

Mol. Wt. 150.18

B.P. 75-77°C/4mm

Solvent: CCl_4

Assignments
a = 1.20
b = 2.47
c = 6.90-7.50

2921

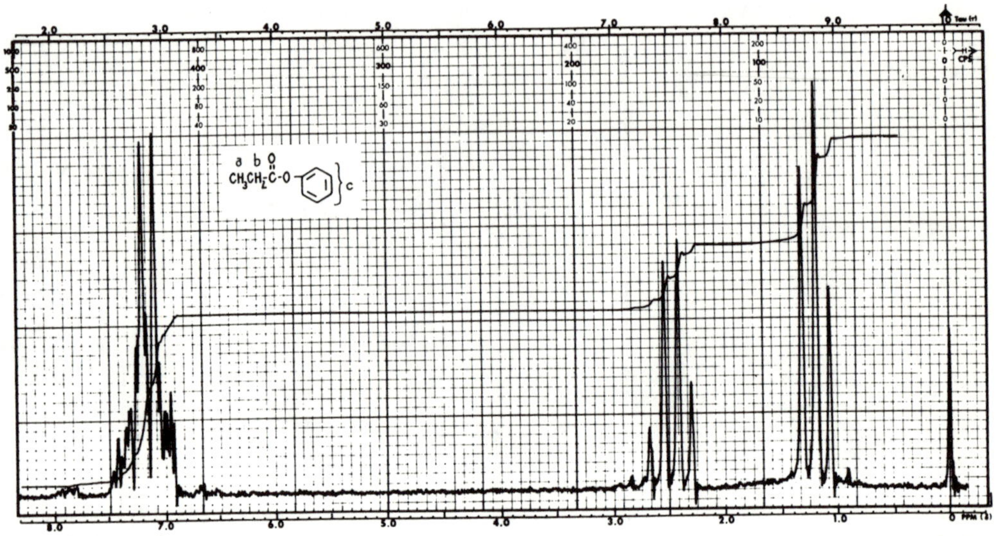

© 1978 Sadtler Research Laboratories, Inc.

PIVALIC ACID, PHENYL ESTER

$C_{11}H_{14}O_2$

Mol. Wt. 178.23

B.P. 85-87°C/5mm

Solvent: $CDCl_3$

Assignments
a = 1.34
b = 6.90-7.60

2922

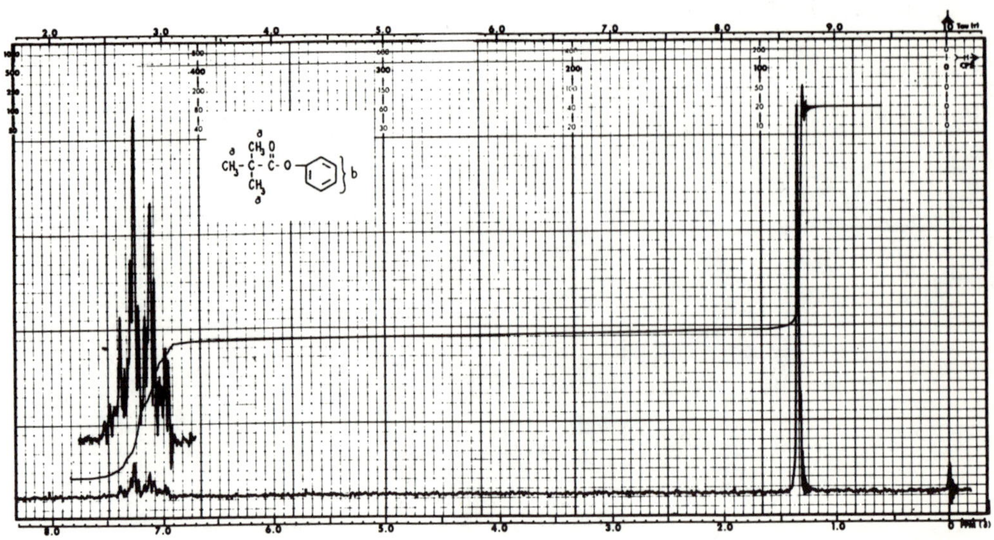

© 1978 Sadtler Research Laboratories, Inc.

TRIFLUOROACETIC ACID, PHENYL ESTER

$C_8H_5F_3O_2$

Mol. Wt. 190.12

B.P. 145-147°C

Solvent: $CDCl_3$

Assignments
a = 6.85-7.50

2923

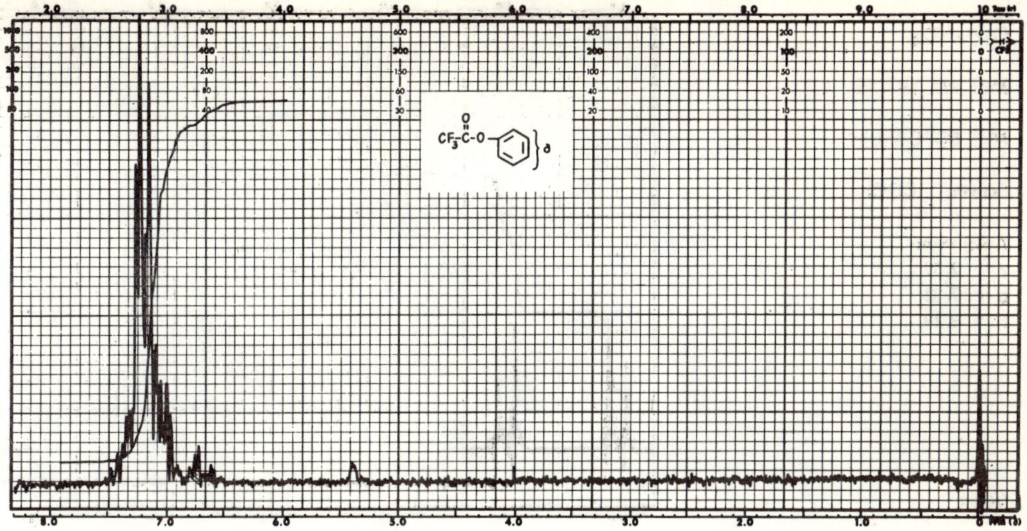

© 1978 Sadtler Research Laboratories, Inc.

ACETIC ACID, p-TOLYL ESTER

$C_9H_{10}O_2$

Mol. Wt. 150.18

Solvent: CCl_4

Assignments
a = 2.12
b = 2.28
c = 6.82
d = 7.01

2924

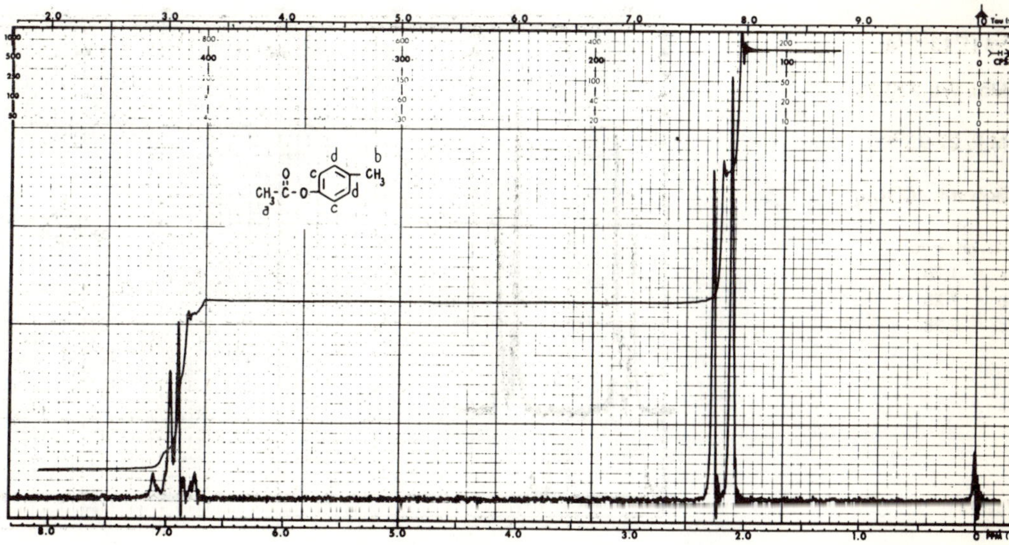

© 1978 Sadtler Research Laboratories, Inc.

OCTANOIC ACID, p-TOLYL ESTER

$C_{15}H_{22}O_2$

Mol. Wt. 234.34

Solvent: CCl_4

Assignments
a = 0.90
b = 1.36
c = 1.70
d = 2.32
e = 2.45
f = 6.88
g = 7.10

2925

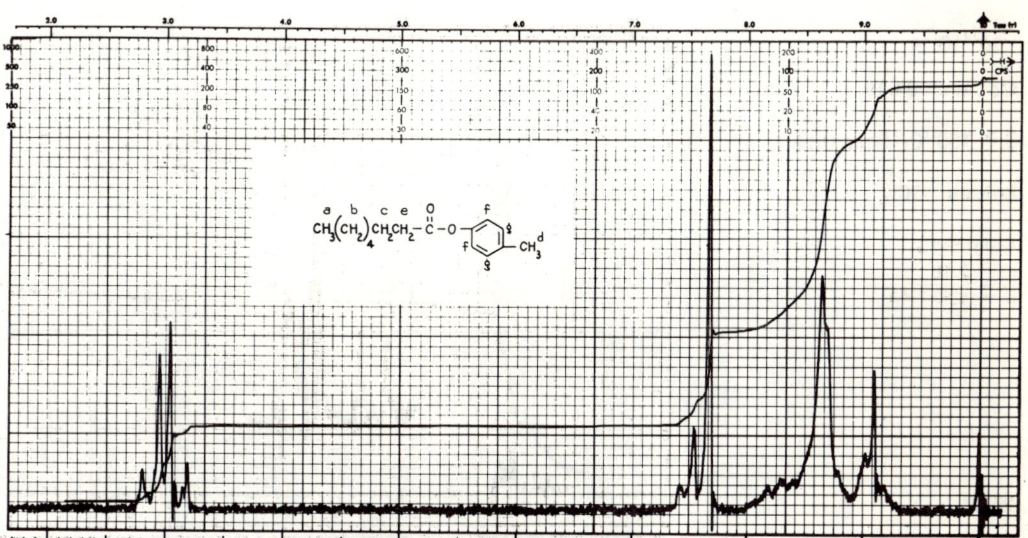

© 1978 Sadtler Research Laboratories, Inc.

ACETIC ACID, p-NITROPHENYL ESTER

$C_8H_7NO_4$

Mol. Wt. 181.15

M.P. 80°C (lit.)

Solvent: $CDCl_3$

Assignments
a = 2.35
b = 7.32
c = 8.29

2926

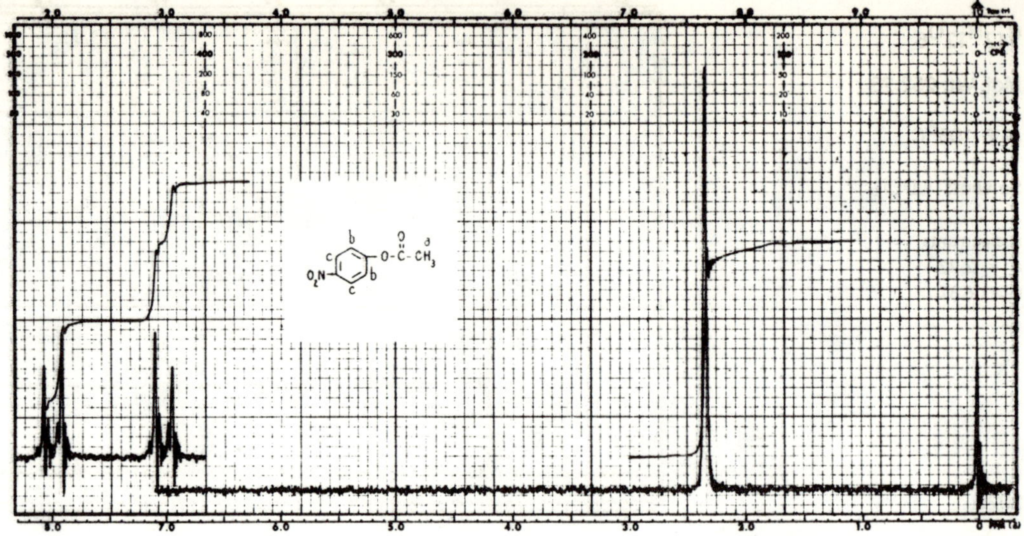

© 1978 Sadtler Research Laboratories, Inc.

PIVALIC ACID, p-NITROPHENYL ESTER

$C_{11}H_{13}NO_4$

Mol. Wt. 223.23

M.P. 92-95°C

Solvent: $CDCl_3$

Assignments
a = 1.38
b = 7.25
c = 8.28

2927

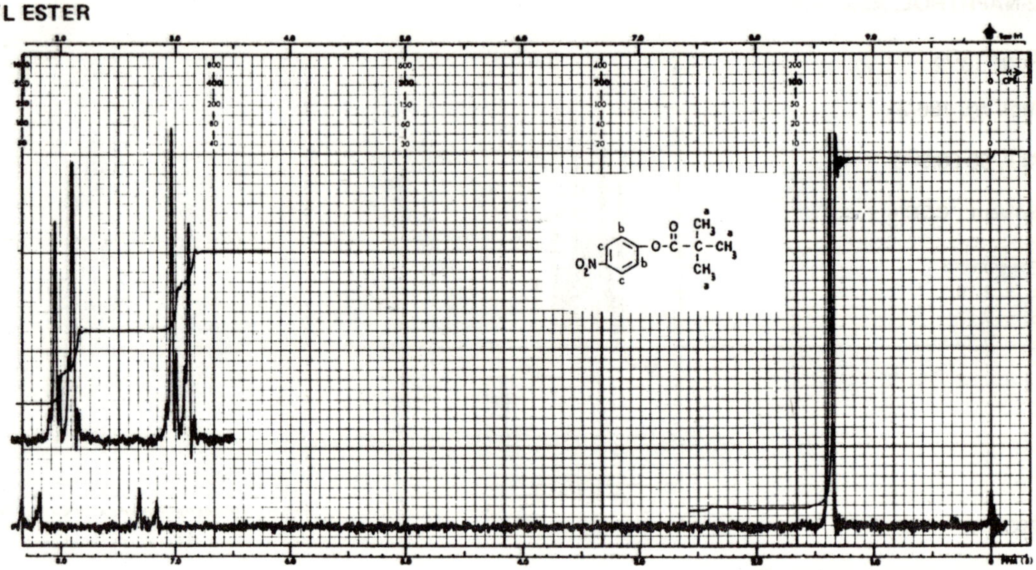

© 1978 Sadtler Research Laboratories, Inc.

RESORCINOL, DIACETATE

$C_{10}H_{10}O_4$

Mol. Wt. 194.19

B.P. 150°C/10mm (lit.)

Solvent: CCl_4

Assignments
a = 2.18
b = 6.80-7.43

2928

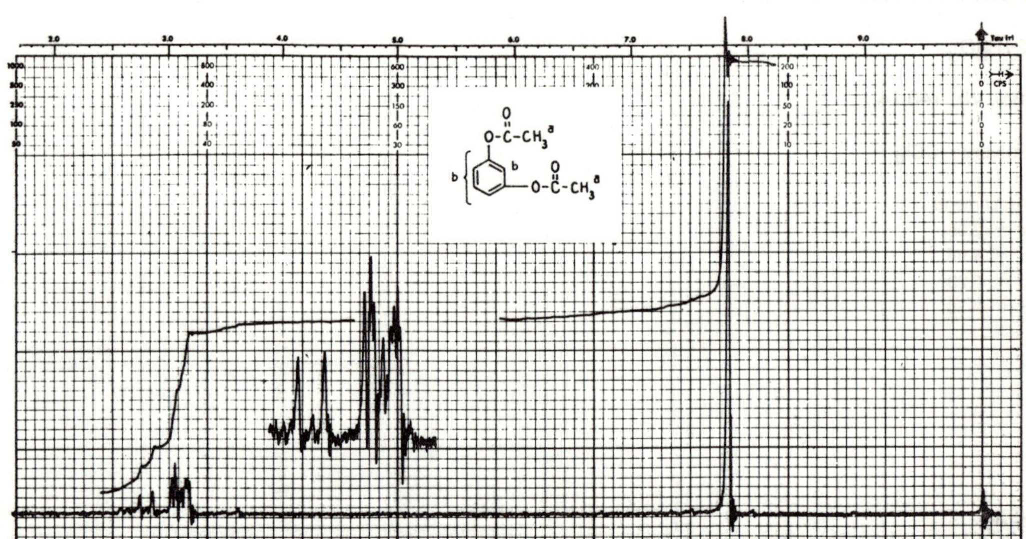

© 1978 Sadtler Research Laboratories, Inc.

PHLOROGLUCINOL, TRIACETATE

$C_{12}H_{12}O_6$

Mol. Wt. 252.23

M.P. 105°C

Solvent: $CDCl_3$

Assignments
a = 2.22
b = 6.87

2929

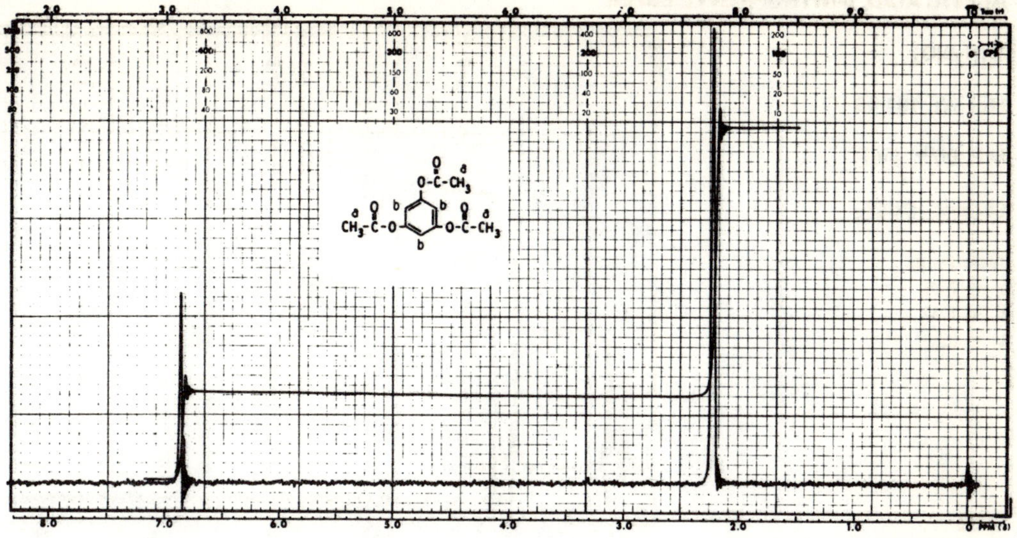

© 1978 Sadtler Research Laboratories, Inc.

2-NAPHTHOL, ACETATE

$C_{12}H_{10}O_2$

Mol. Wt. 186.21

M.P. 68-68.5°C

Solvent: $CDCl_3$

Assignments
a = 2.19
b = 7.17
c = 7.29-7.90

2930

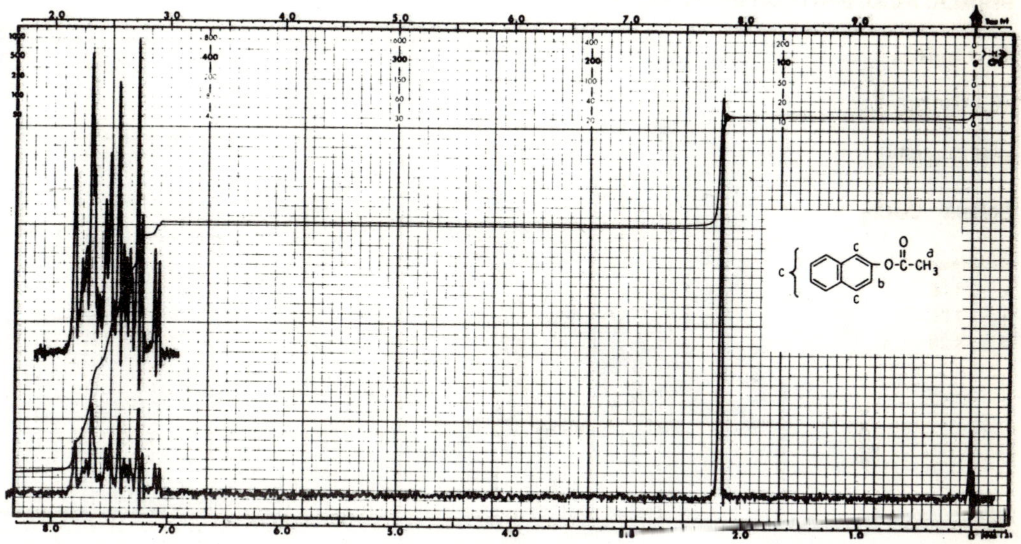

© 1978 Sadtler Research Laboratories, Inc.

FURFURYL ALCOHOL, ACETATE

$C_7H_8O_3$

Mol. Wt. 140.14

Solvent: CCl_4

Assignments
a = 2.00
b = 4.94
c = 6.29
d = 7.30

2931

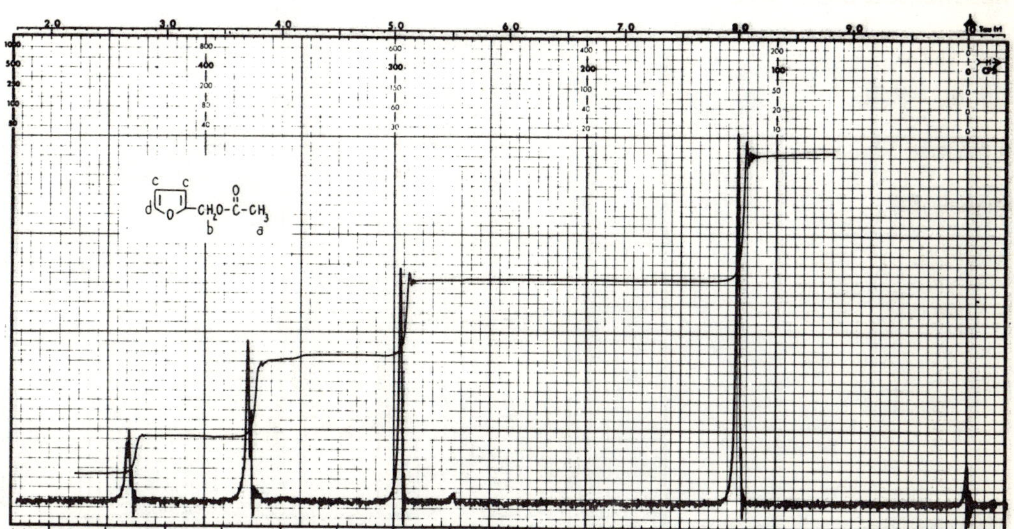

© 1978 Sadtler Research Laboratories, Inc.

PHENYLACETIC ACID, m-TOLYL ESTER

$C_{15}H_{14}O_2$

Mol. Wt. 226.28

M.P. 50°C

Solvent: CCl_4

Assignments
a = 2.21
b = 3.63
c = 6.60-7.38
d = 7.20

2932

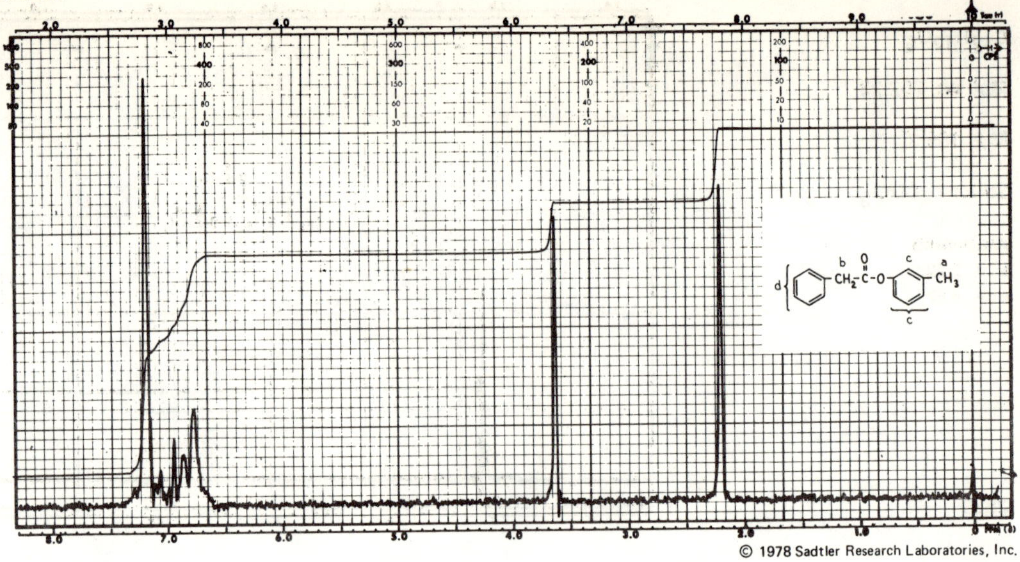

© 1978 Sadtler Research Laboratories, Inc.

PHENYLACETIC ACID, p-TOLYL ESTER

$C_{15}H_{14}O_2$

Mol. Wt. 226.28

Solvent: $CDCl_3$

Assignments
a = 2.29
b = 3.79
c = 6.90
d = 7.08
e = 7.30

2933

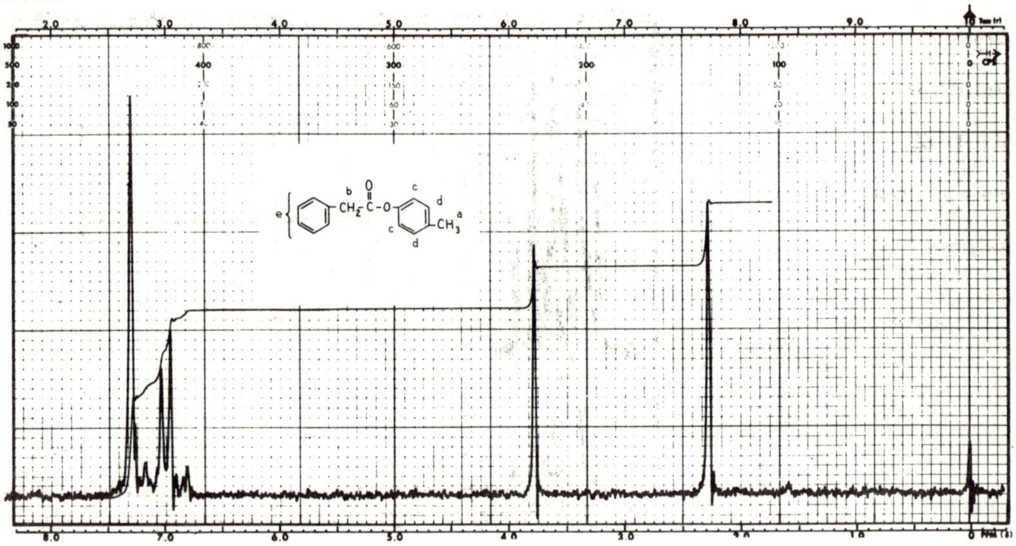

© 1978 Sadtler Research Laboratories, Inc.

2-FURANACRYLIC ACID, ETHYL ESTER

$C_9H_{10}O_3$

Mol. Wt. 166.18

Solvent: $CDCl_3$

Assignments
a = 1.29
b = 4.21
c = 6.28
d = 6.42
e = 6.60
f = 7.41
g = 7.49

2934

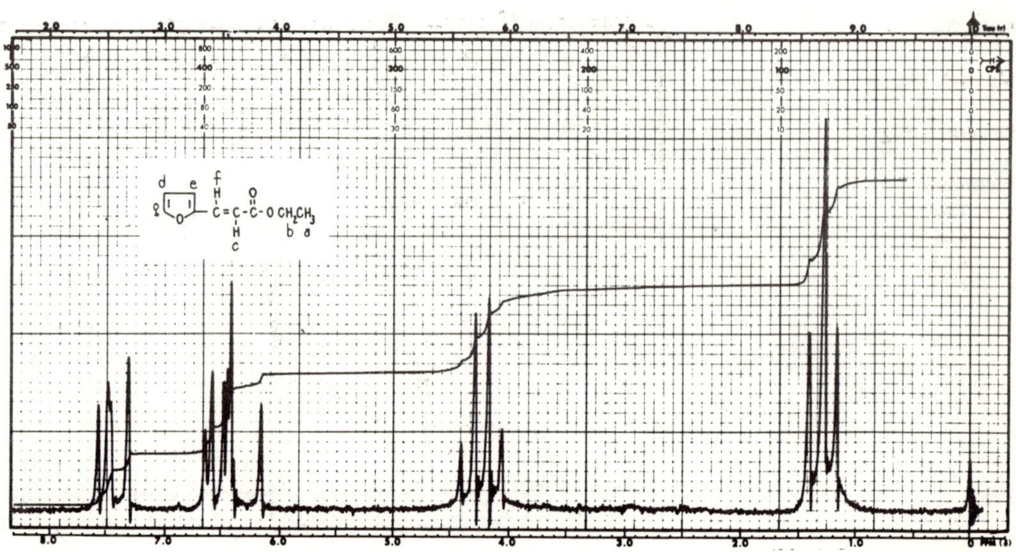

© 1978 Sadtler Research Laboratories, Inc.

2-FUROIC ACID, PROPYL ESTER

$C_8H_{10}O_3$

Mol. Wt. 154.17

B.P. 211°C (lit.)

Solvent: CCl_4

Assignments

a = 1.01
b = 1.72
c = 4.20
d = 6.47
e = 7.11
f = 7.58
g = 2.92, 3.52 imps.

2935

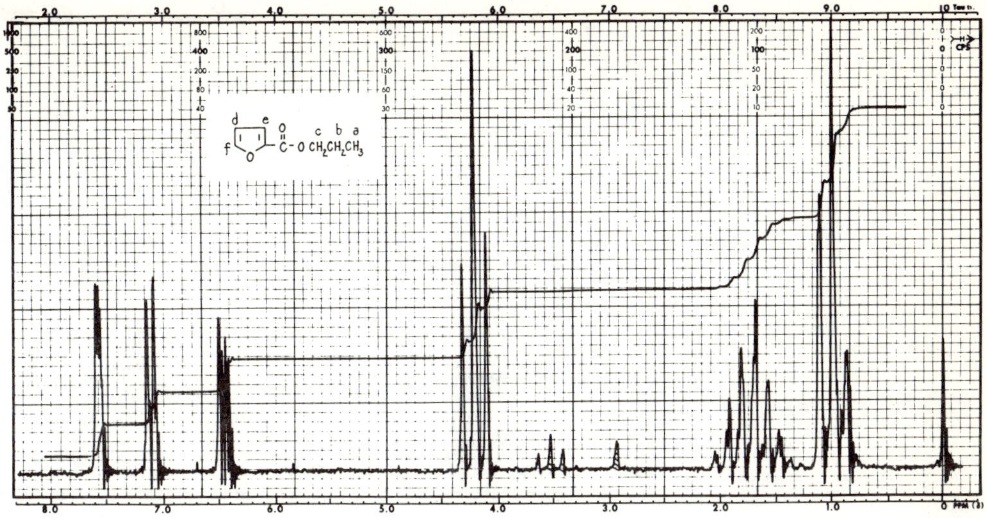

© 1978 Sadtler Research Laboratories, Inc.

2-FUROIC ACID, PENTYL ESTER

$C_9H_{14}O_3$

Mol. Wt. 170.21

Solvent: CCl_4

Assignments

a = 0.93
b = 1.10-2.00
c = 4.22
d = 6.44
e = 7.07
f = 7.53

2936

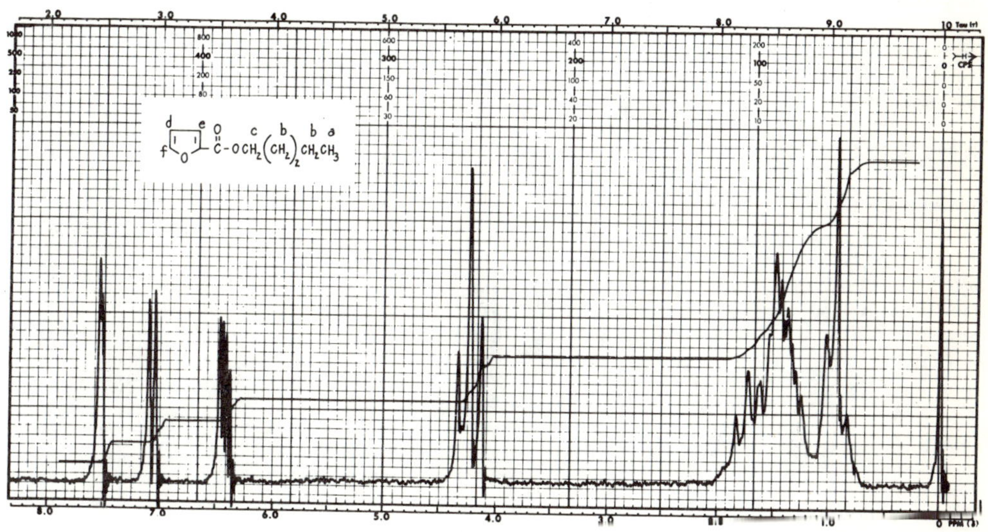

© 1978 Sadtler Research Laboratories, Inc.

5-NITRO-2-FUROIC ACID, ETHYL ESTER

$C_7H_7NO_5$

Mol. Wt. 185.13

Solvent: $CDCl_3$

Assignments

a = 1.40
b = 4.44
c = 7.32

2937

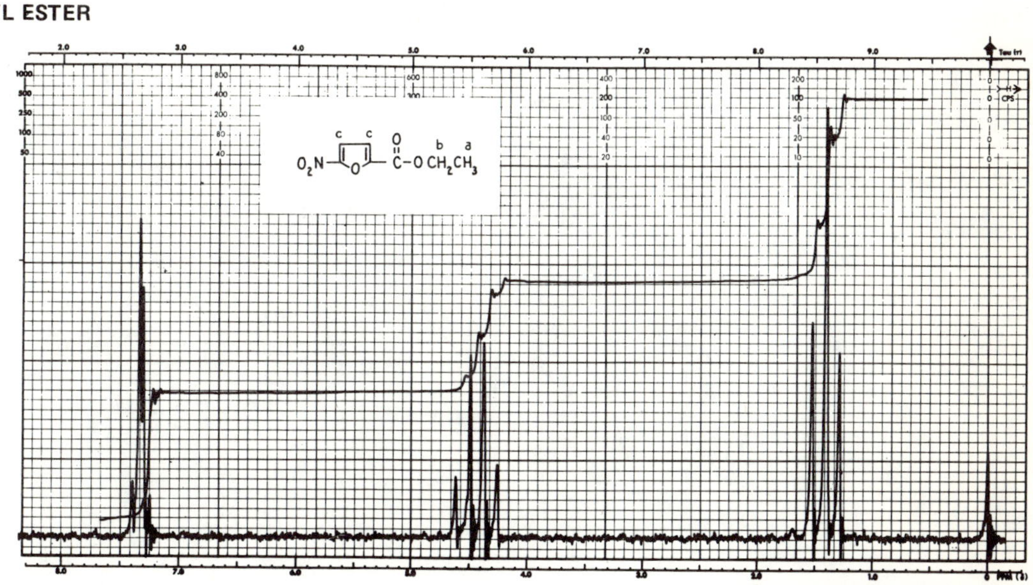

© 1978 Sadtler Research Laboratories, Inc.

2-PYRIDINEACETIC ACID, METHYL ESTER

$C_8H_9NO_2$

Mol. Wt. 151.17

B.P. 120°C/10mm (lit.)

Solvent: $CDCl_3$

Assignments
a = 3.67
b = 3.84
c = 7.15
d = 7.29
e = 7.61
f = 8.51

2938

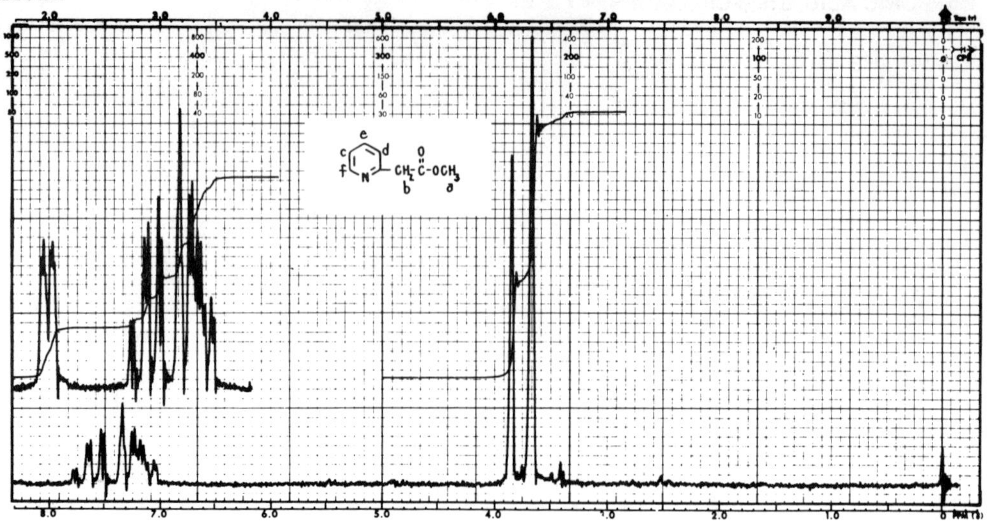

© 1978 Sadtler Research Laboratories, Inc.

NICOTINIC ACID, NONYL ESTER

$C_{15}H_{23}NO_2$

Mol. Wt. 249.36

B.P. 118-122°C/1mm

Solvent: $CDCl_3$

Assignments
a = 0.89
b = 1.10-1.60
c = 1.79
d = 4.35
e = 7.35
f = 8.29
g = 8.77
h = 9.24

2939

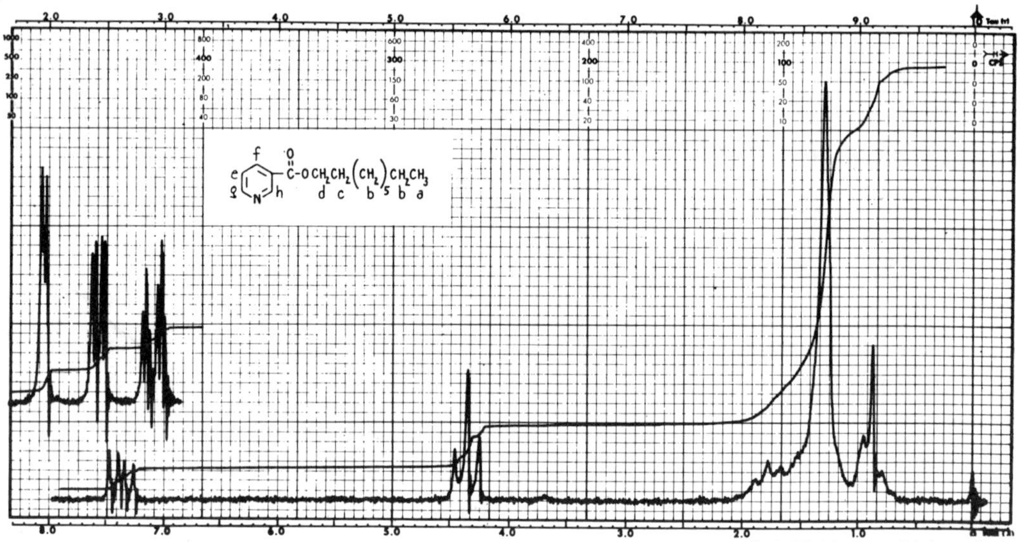

© 1978 Sadtler Research Laboratories, Inc.

CARBONIC ACID, DI-p-TOLYL ESTER

$C_{15}H_{14}O_3$

Mol. Wt. 242.28

M.P. 110-112°C

Solvent: CCl_4

Assignments
a = 2.31
b = 7.05

2940

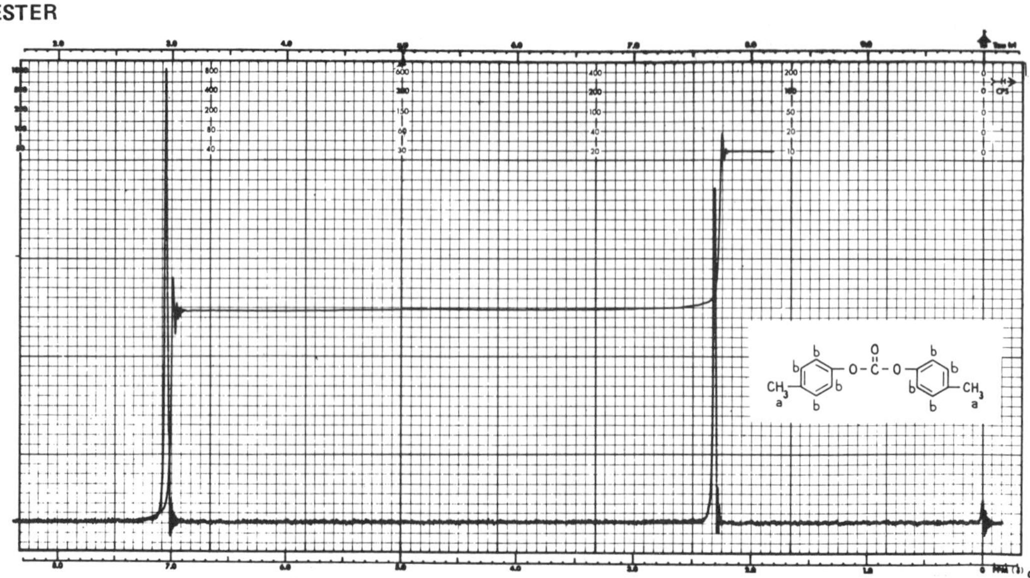

© 1978 Sadtler Research Laboratories, Inc.

CARBONIC ACID, BIS(p-CHLOROPHENYL) ESTER

$C_{13}H_8Cl_2O_3$

Mol. Wt. 283.11

M.P. 149-151°C

Solvent: $CDCl_3$

Assignments
a = 7.19
b = 7.35

2941

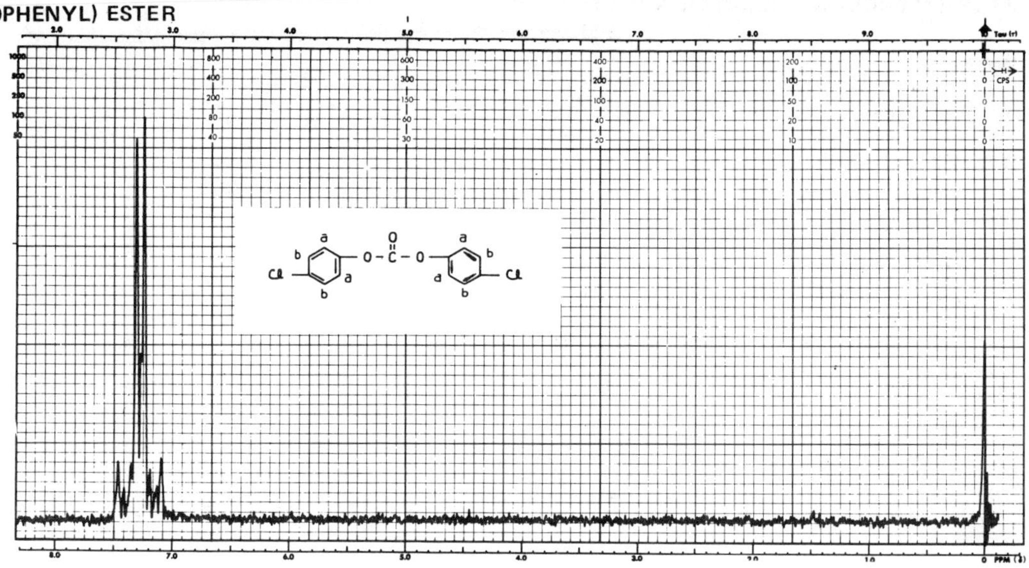

© 1978 Sadtler Research Laboratories, Inc.

BENZOIC ACID, BENZYL ESTER

$C_{14}H_{12}O_2$

Mol. Wt. 212.25

Solvent: CCl_4

Assignments
a = 5.21
b = 7.01-7.49
c = 7.98

2942

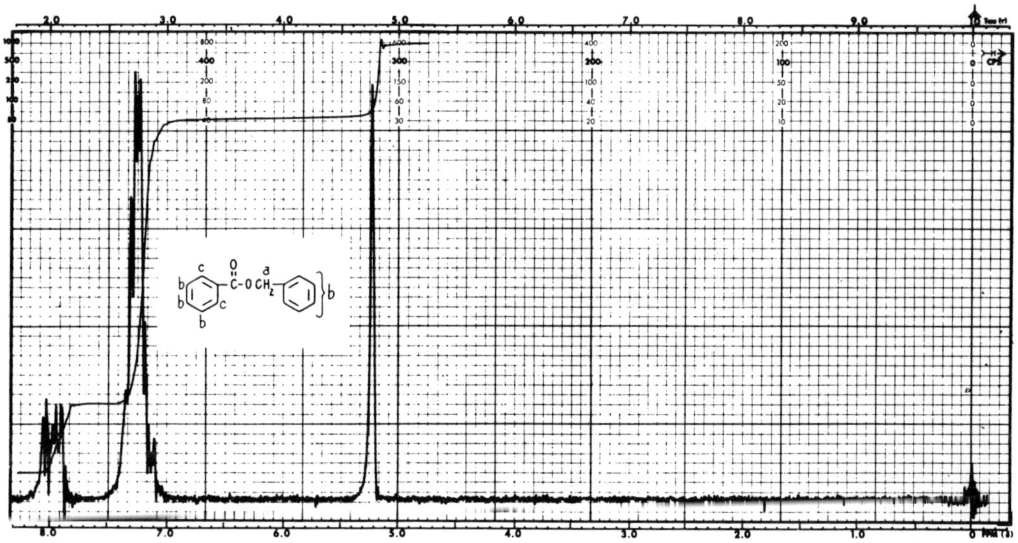

© 1978 Sadtler Research Laboratories, Inc.

BENZOIC ACID, p-TOLYL ESTER

$C_{14}H_{12}O_2$

Mol. Wt. 212.25

M.P. 71-72°C

Solvent: CCl_4

Assignments
a = 2.34
b = 7.07
c = 7.45
d = 8.12

2943

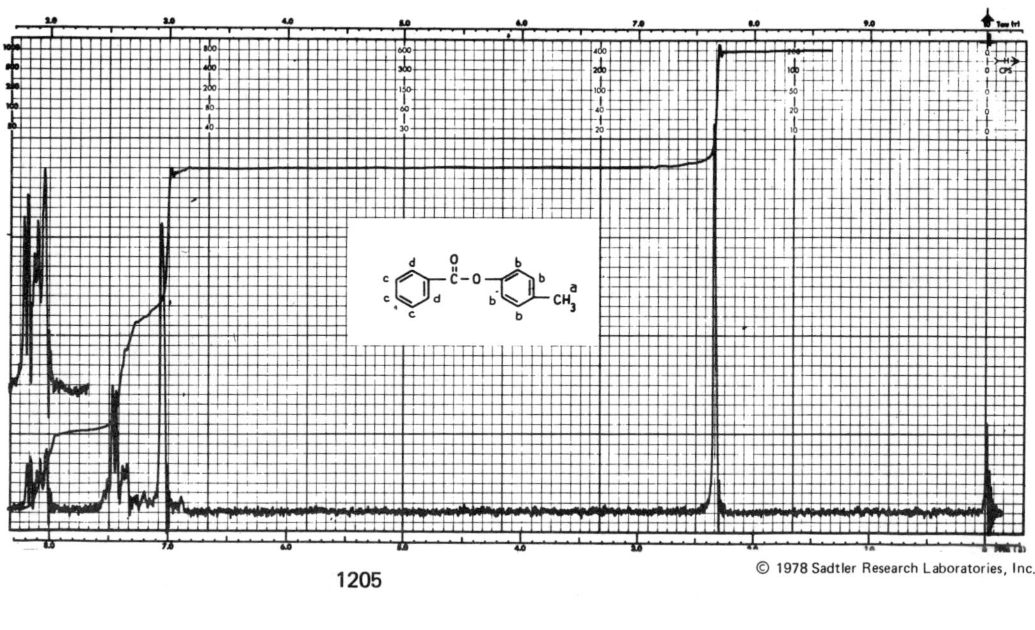

© 1978 Sadtler Research Laboratories, Inc.

ETHYLENE GLYCOL, DIBENZOATE

$C_{16}H_{14}O_4$

Mol. Wt. 270.29

Flash Point 186°C

Solvent: $CDCl_3$

Assignments
a = 4.63
b = 7.41
c = 8.04

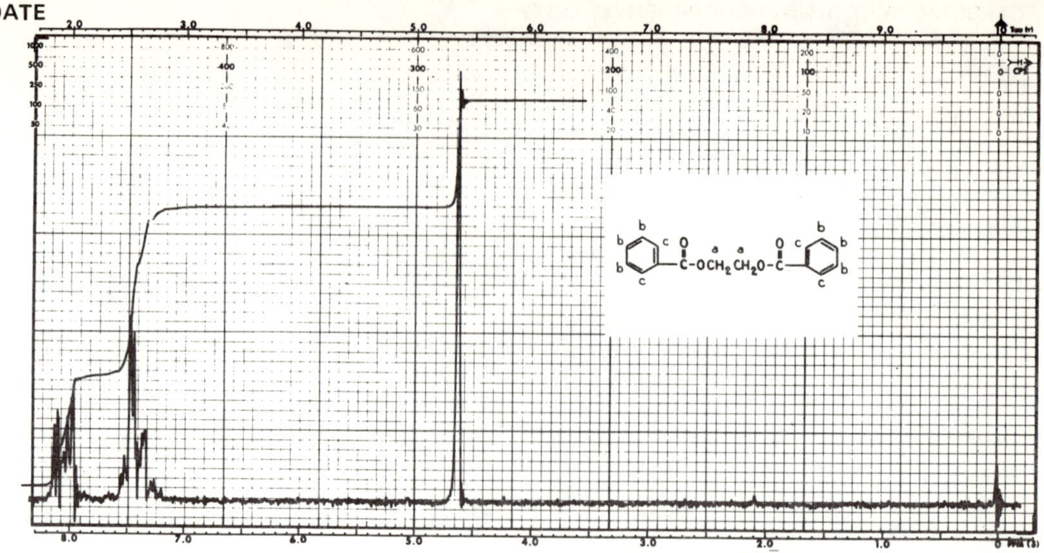

2944

© 1978 Sadtler Research Laboratories, Inc.

PYROCATECHOL, DIBENZOATE

$C_{20}H_{14}O_4$

Mol. Wt. 318.33

M.P. 82-83°C

Solvent: $CDCl_3$

Assignments
a = 7.08-7.63
b = 8.01

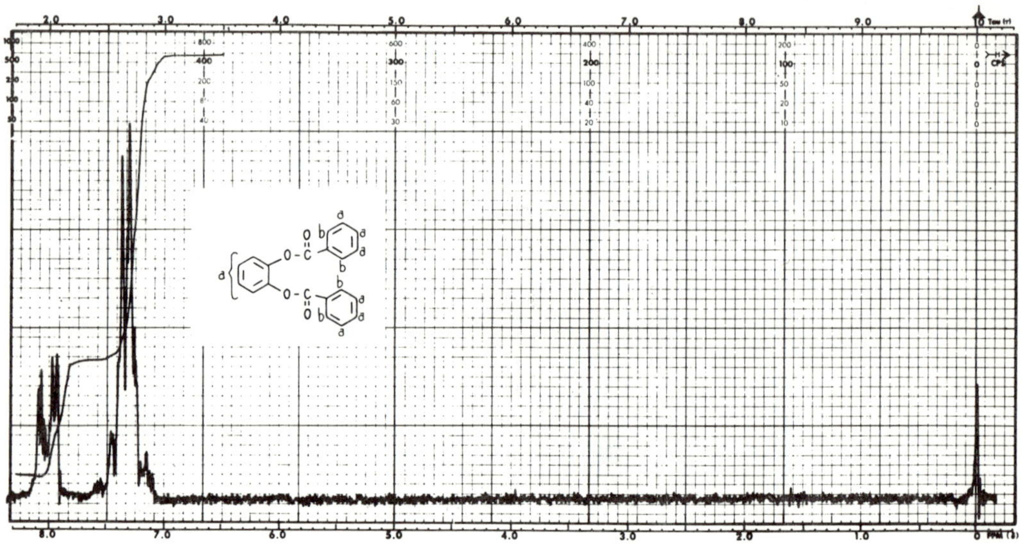

2945

© 1978 Sadtler Research Laboratories, Inc.

THE LACTONES

Due to their cyclic structure, the aliphatic cyclic esters often display non-equivalence in chemical shift for the two protons of the methylene group in the ring system.

This situation is clearly observed in spectrum 2951 in which the five ring hydrogens each display distinctly different chemical shifts. The aromatic protons of the aromatic lactones generally produce complex patterns in the chemical shift range from 6.5 to 8.0 ppm depending upon which group of the ester moiety is bonded to the aromatic ring. When the oxygen atom is bonded to the ring, the resonance bands tend to be in the high field portion of the range, 6.8–7.4 ppm for spectrum 2955. When the carbonyl group is bonded to the aromatic ring, then the resonance bands tend to occupy the lower half of the range.

The Lactones, like the other esters are readily soluble in the chlorinated NMR solvents, Carbon Tetrachloride and Deuterochloroform.

Chemical Shifts

Alicyclic Protons

–O–	HC–5	HC–4	HC–3	–C(=O)–	Substituent	Solvent	Spectrum
	4.40	1.5–2.1	2.1–2.7		5–Butyl	CCl_4	2947
	3.9–4.4	1.9, 2.4	2.60		3–CH_3	CCl_4	2946
	4.2, 4.6	3.69	2.6, 2.8		4–Ø	$CDCl_3$	2951
	4.3, 4.6	2.2–2.8	3.00		3–Br	CCl_4	2948
	4.29	1.9–3.0	3.71		3–C(=O)–CH_3	CCl_4	2950

4,5-DIHYDRO-3-METHYL-2(3H)-FURANONE

$C_5H_8O_2$

Mol. Wt. 100.12

Solvent: CCl_4

Assignments
a = 1.21
b = 1.68-2.82
c = 4.02-4.42
d = 3.30, 3.62 imps.

2946

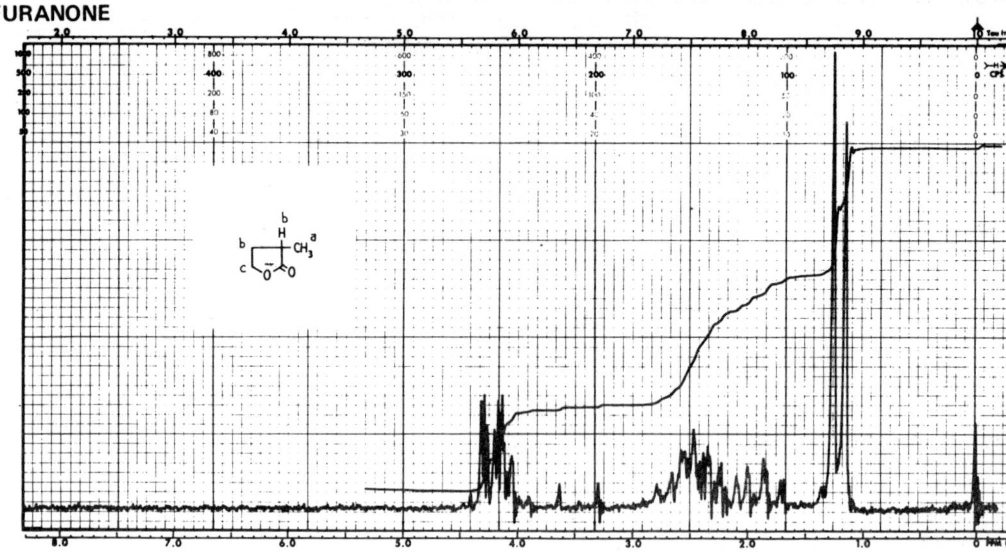

© 1978 Sadtler Research Laboratories, Inc.

4-HYDROXYOCTANOIC ACID, γ-LACTONE

$C_8H_{14}O_2$

Mol. Wt. 142.20

B.P. 84-85°C/5mm (lit.)

Solvent: CCl_4

Assignments
a = 0.92
b = 1.14-2.07
c = 2.07-2.66
d = ca 4.40

2947

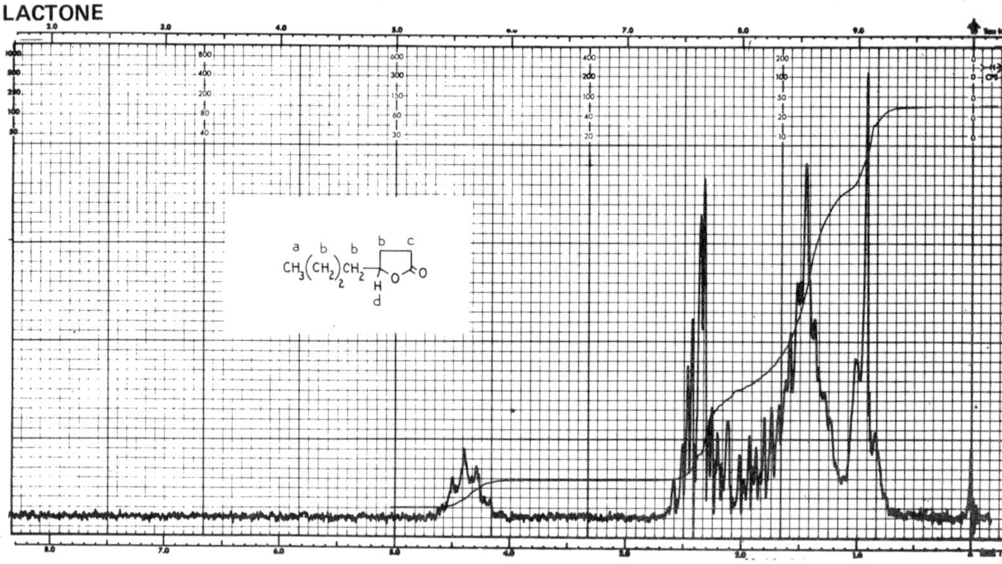

© 1978 Sadtler Research Laboratories, Inc.

3-BROMODIHYDRO-2(3H)-FURANONE

$C_4H_5BrO_2$

Mol. Wt. 164.99

Solvent: CCl_4

Assignments
a = 2.20-3.27
b = ca 4.43

2948

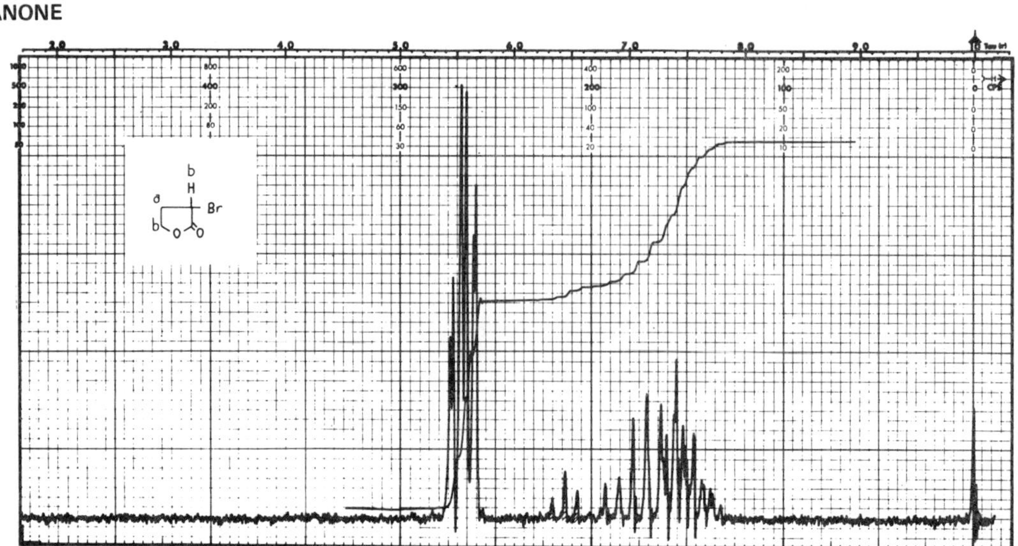

© 1978 Sadtler Research Laboratories, Inc.

2,4-DIHYDROXY-3,3-DIMETHYLBUTYRIC ACID, γ-LACTONE

$C_6H_{10}O_3$

Mol. Wt. 130.14

Solvent: $CDCl_3$

Assignments

a = 1.02 or b
b = 1.18 or a
c = 3.97
d = 4.22
e = 4.43
Trace of D_2O added

2949

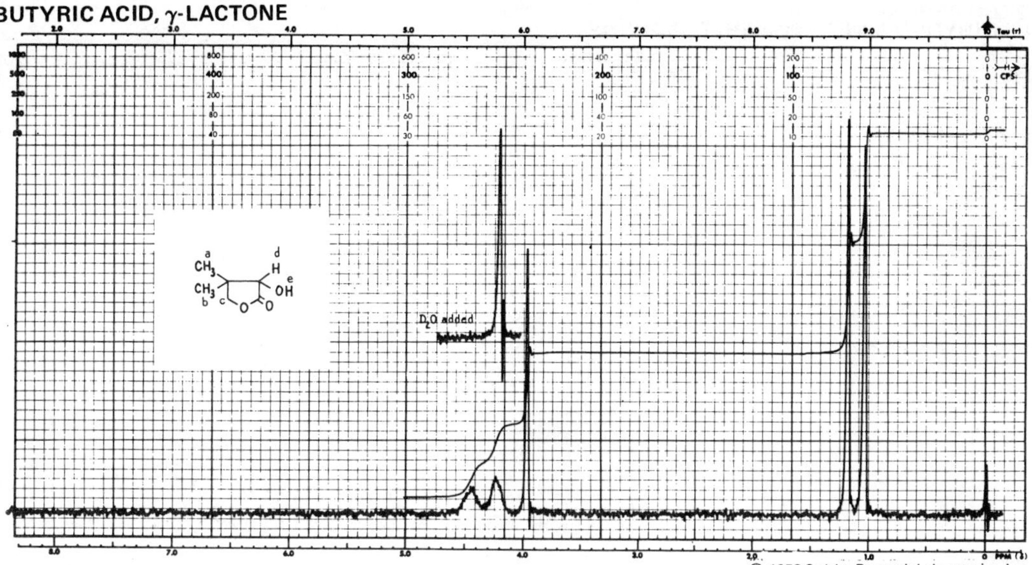

© 1978 Sadtler Research Laboratories, Inc.

3-ACETYLDIHYDRO-2(3H)-FURANONE

$C_6H_8O_3$

Mol. Wt. 128.13

B.P. 107-108°C/5mm

Solvent: CCl_4

Assignments

a = 1.92-3.02
b = 2.39
c = 3.71
d = 4.29

2950

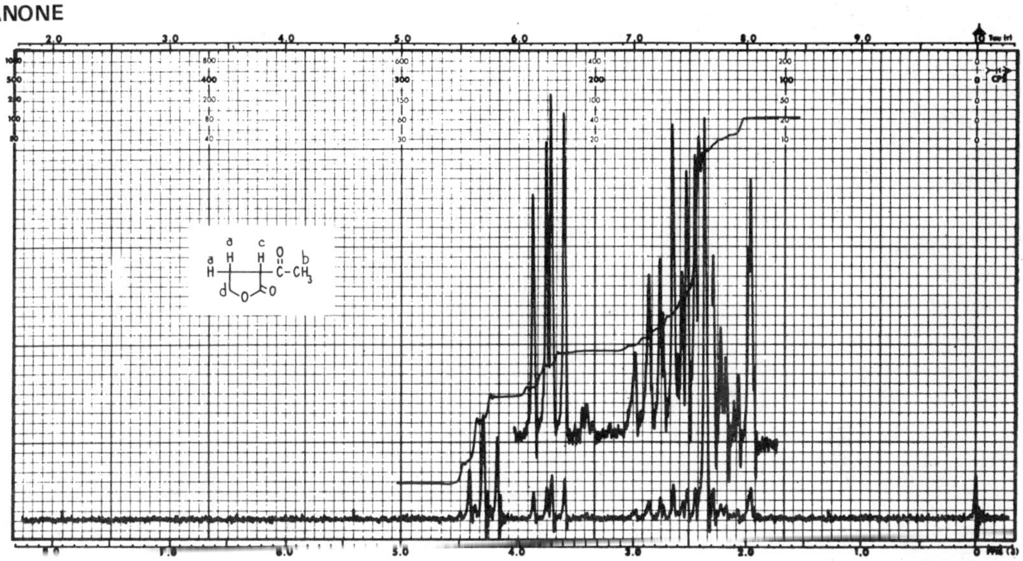

© 1978 Sadtler Research Laboratories, Inc.

DIHYDRO-4-PHENYL-2(3H)-FURANONE

$C_{10}H_{10}O_2$

Mol. Wt. 162.19

Solvent: $CDCl_3$

Assignments

a = 2.58 or b
b = 2.80 or a
c = 3.69
d = 4.16 or e
e = 4.58 or d
f = ca 7.25

2951

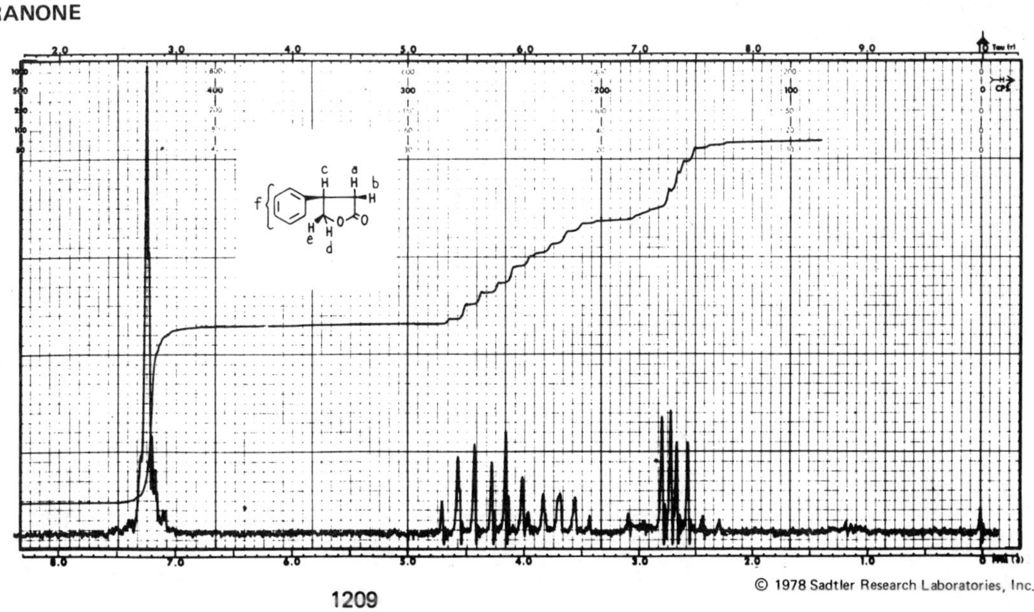

© 1978 Sadtler Research Laboratories, Inc.

2-OXEPANONE

$C_6H_{10}O_2$

Mol. Wt. 144.15

B.P. 118-119°C/15mm

Solvent: CCl_4

Assignments
a = 1.40-2.00
b = 2.59
c = 4.20

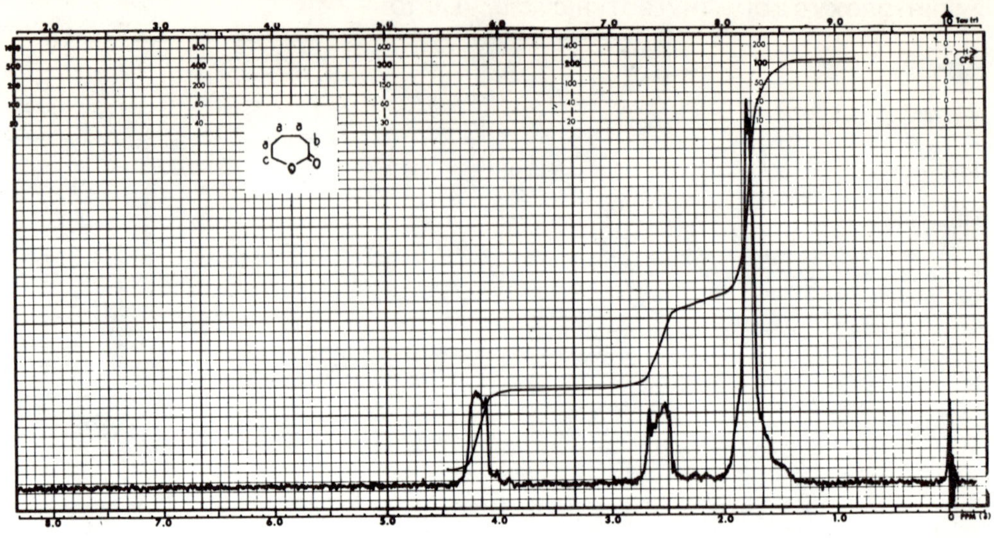

2952

© 1978 Sadtler Research Laboratories, Inc.

OXACYCLOP ENTADECAN-2-ONE

$C_{15}H_{28}O_2$

Mol. Wt. 240.39

Solvent: $CDCl_3$

Assignments
a = 1.20-2.00
b = 2.31
c = 4.11

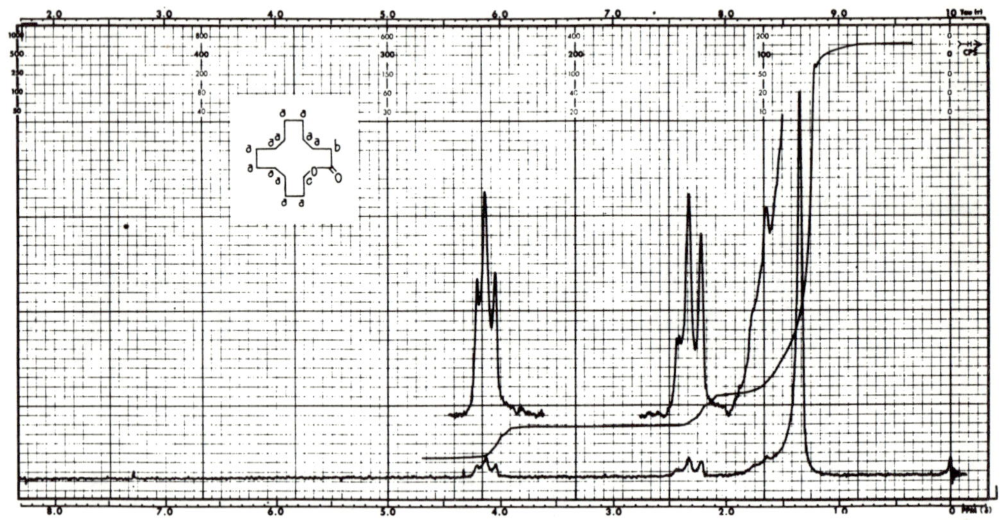

2953

© 1978 Sadtler Research Laboratories, Inc.

5-METHYL-2(3H)-FURANONE

$C_5H_6O_2$

Mol. Wt. 98.10

Solvent: CCl_4

Assignments
a = 1.99
b = 3.08
c = 5.10

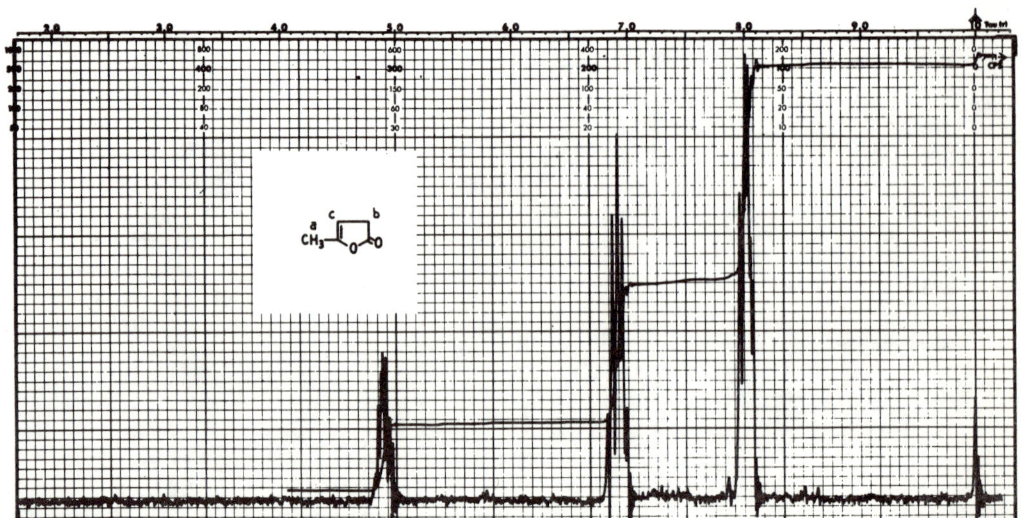

2954

© 1978 Sadtler Research Laboratories, Inc.

(o-HYDROXYPHENYL) ACETIC ACID, γ-LACTONE

$C_8H_6O_2$

Mol. Wt. 134.14

M.P. 49-51°C

Solvent: $CDCl_3$

Assignments
a = 3.59
b = 6.85-7.46

2955

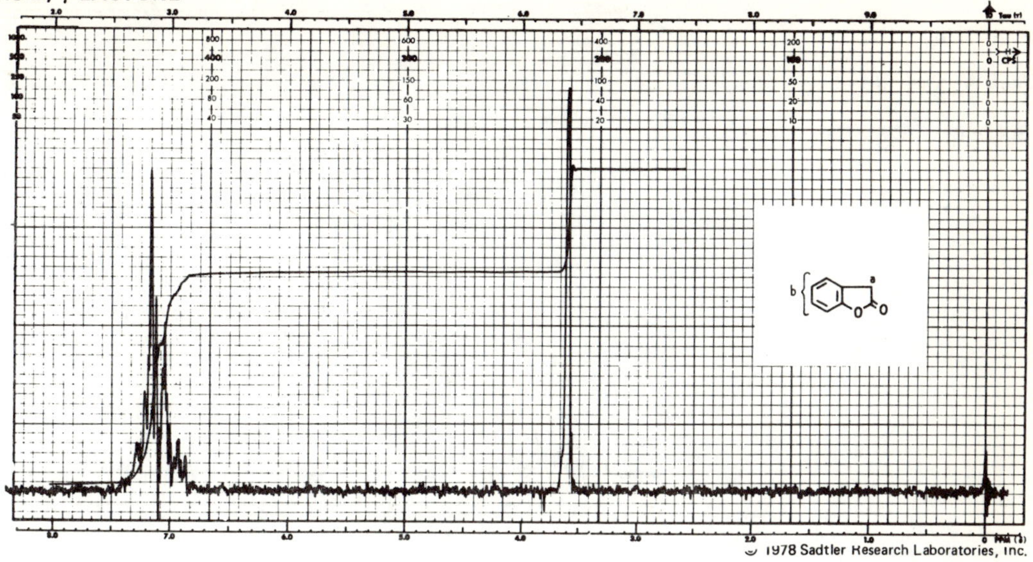

6-NITROPHTHALIDE

$C_8H_5NO_4$

Mol. Wt. 179.13

M.P. 139-141°C

Solvent: $CDCl_3$

Assignments
a = 5.48
b = 7.78
c = 8.61
d = 8.76

2956

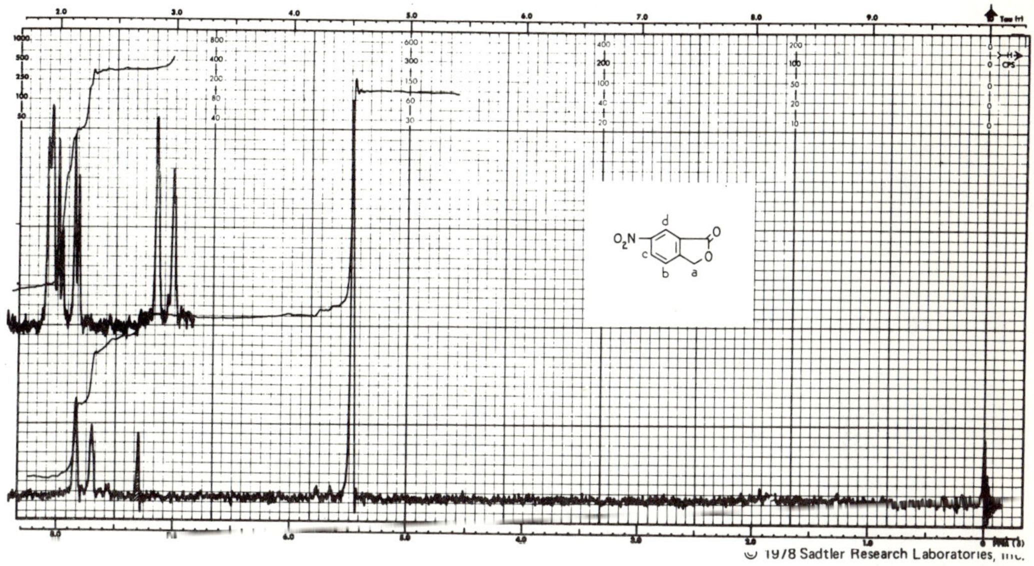

3-CHLOROCOUMARIN

$C_9H_5ClO_2$

Mol. Wt. 180.59

Solvent: $CDCl_3$

Assignments
a = 7.15-7.70
b = 7.88

2957

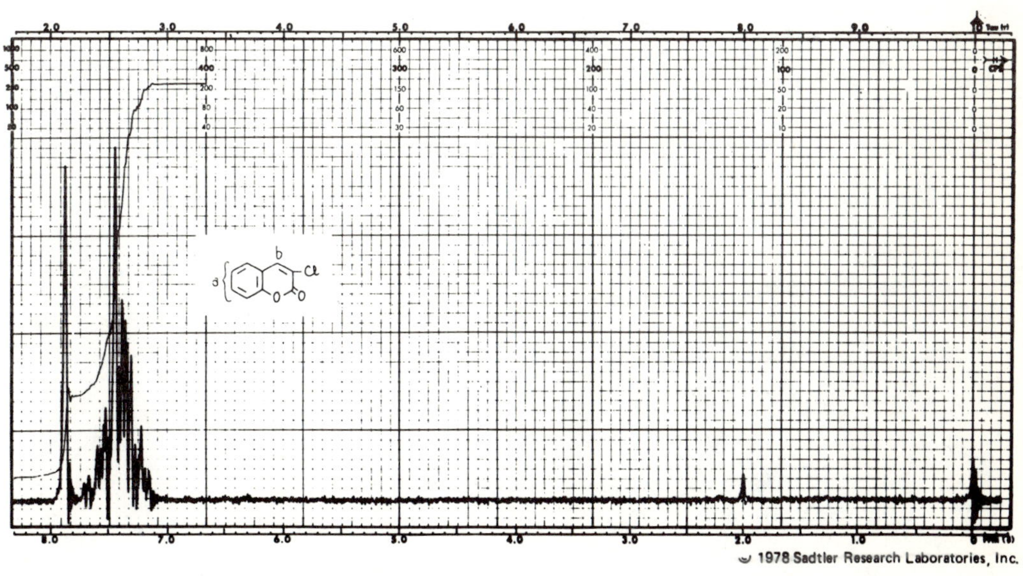

4-HYDROXYCOUMARIN, ACETATE

$C_{11}H_8O_4$

Mol. Wt. 204.18

Solvent: $CDCl_3$

Assignments
a = 2.43
b = 6.50
c = 7.22-7.79

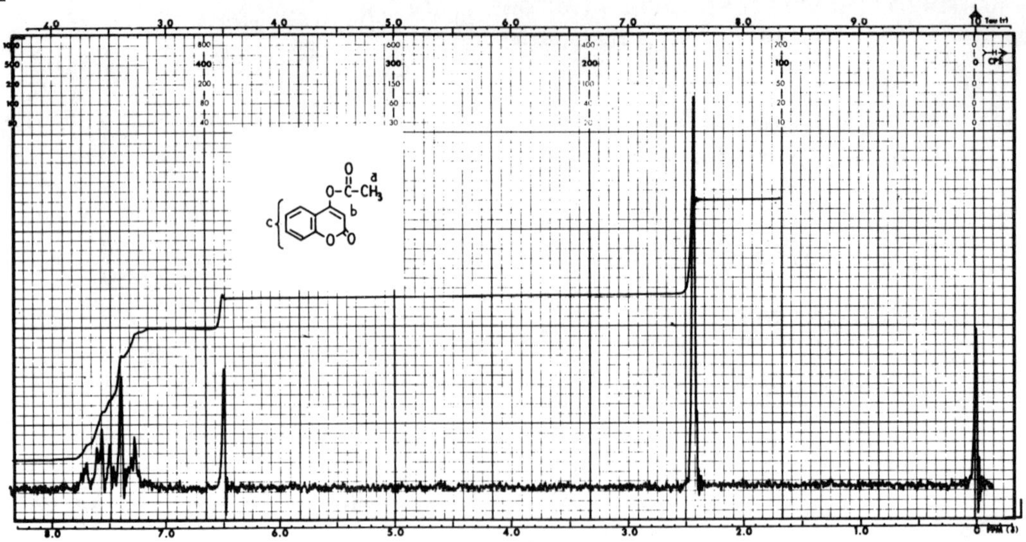

© 1978 Sadtler Research Laboratories, Inc.

2958

The aliphatic esters of Chloroformic Acid are notable in that their alpha hydrocarbon groups are more strongly deshielded than either the esters of simple aliphatic or aromatic carboxylic acids.

The Chloroformates are readily soluble in the chlorinated solvents. Their spectra display no unusual couplings nor coupling constants.

Chemical Shifts

Aliphatic Protons

HC–4	HC–3	HC–2	HC–1		Solvent	Spectrum
			3.93	CH_3–O–C(=O)–Cl	CCl_4	----
		1.40	4.39	CH_3–CH_2–O–C(=O)–Cl	$CDCl_3$	----
	1.00	1.76	4.22	CH_3–CH_2–CH_2–O–C(=O)–Cl	CCl_4	----
1.01	1.75	1.45	4.31	CH_3–CH_2–CH_2–CH_2–O–C(=O)–Cl	CCl_4	2959
	(0.99)	2.08	4.11	$(CH_3)_2$–CH–CH_2–O–C(=O)–Cl	$CDCl_3$	----

Aromatic Protons

Phenyl Chloroformate

para/meta/ortho		Solvent	Spectrum
7.0–7.6	Ø–O–C(=O)–Cl	CCl_4	2968

Para Substituted Phenyl Chloroformates

Cl–C(=O)–O– ortho	meta	para	Solvent	Spectrum
7.03	6.77	–O–CH_3	CCl_4	----
7.10	7.30	–Cl	CCl_4	----
7.07	7.33	–$C(CH_3)_3$	CCl_4	----
7.11	8.12	–NO_2	DMSO-d_6	2969

CHLOROFORMIC ACID, BUTYL ESTER

$C_5H_9ClO_2$

Mol. Wt. 136.58

B.P. 44-45°C/20mm

Solvent: CCl_4

Assignments
a = 1.01
b = 1.19-2.05
c = 4.31

2959

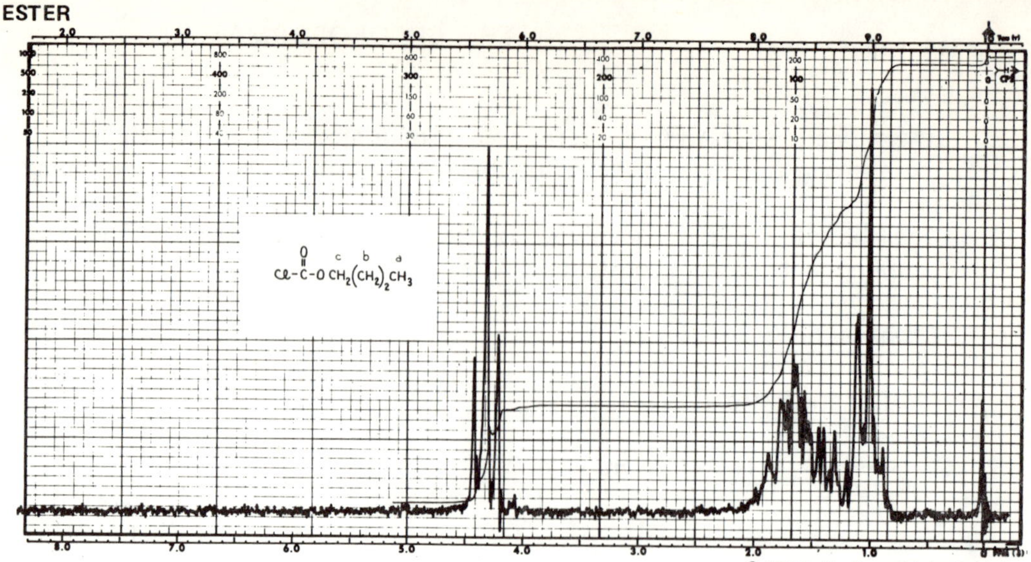

© 1978 Sadtler Research Laboratories, Inc.

CHLOROFORMIC ACID, 2-ETHYLHEXYL ESTER

$C_9H_{17}ClO_2$

Mol. Wt. 192.69

Solvent: CCl_4

Assignments
a = 0.93
b = ca 1.32
c = 4.20

2960

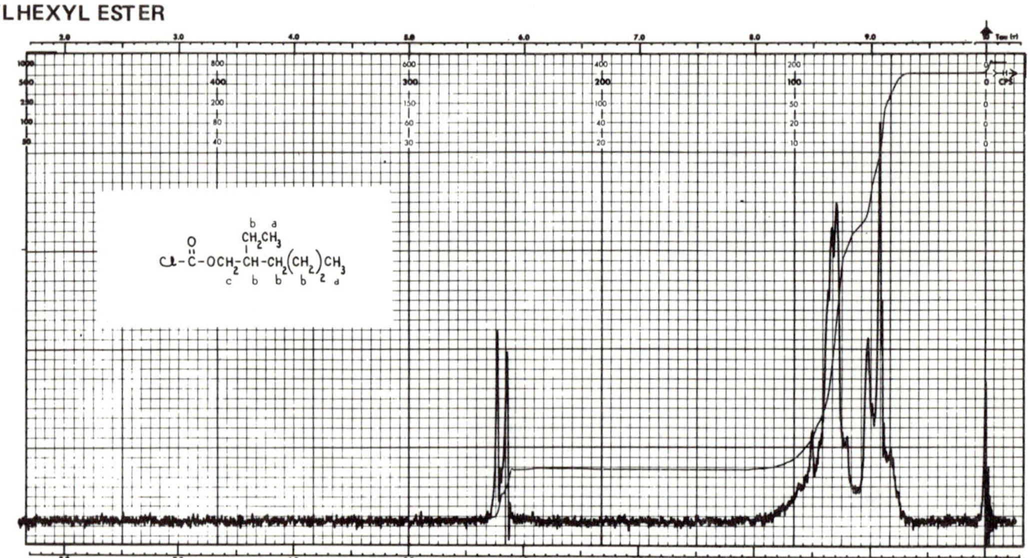

© 1978 Sadtler Research Laboratories, Inc.

CHLOROFORMIC ACID, DODECYL ESTER

$C_{13}H_{25}ClO_2$

Mol. Wt. 248.80

B.P. 130°C/5mm & 99°C/1mm

Solvent: CCl_4

Assignments
a = 0.89
b = 1.28
c = ca 1.71
d = 4.26

2961

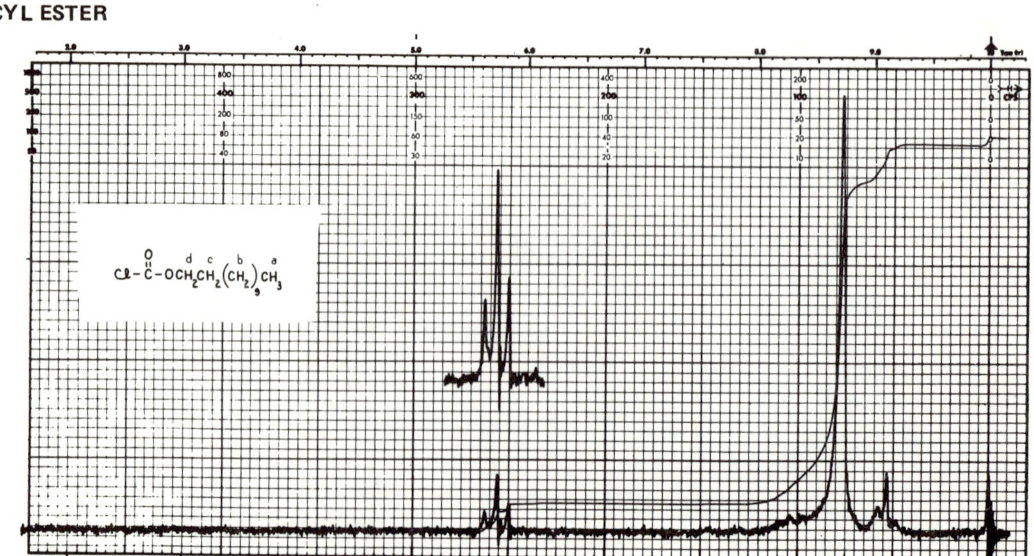

...978 Sadtler Research Laboratories, Inc.

CHLOROFORMIC ACID, 2-CHLOROETHYL ESTER

$C_3H_4Cl_2O_2$

Mol. Wt. 142.97

B.P. 52-54°C/12mm

Solvent: CCl_4

Assignments
a = 3.72
b = 4.54

2962

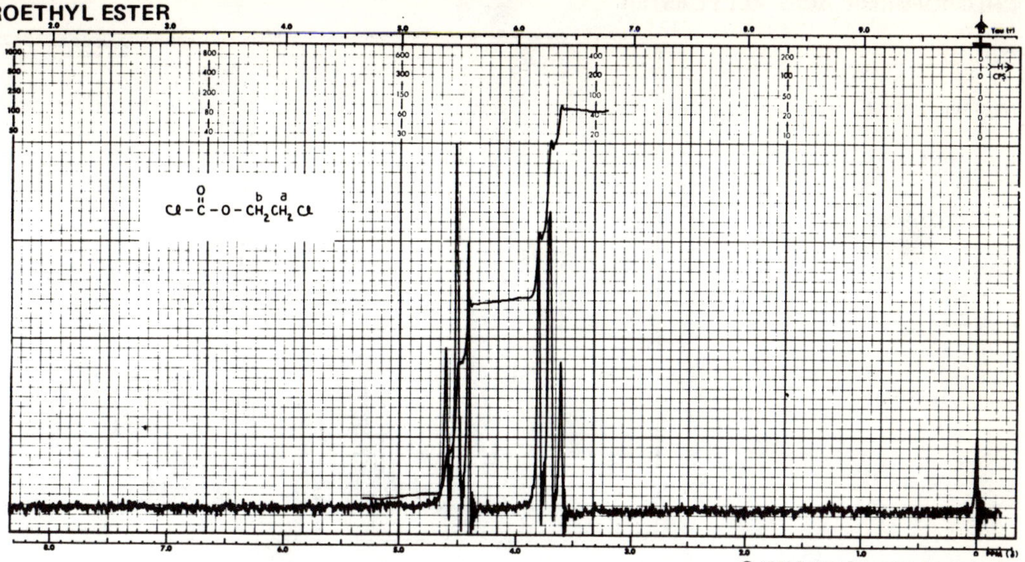

© 1978 Sadtler Research Laboratories, Inc.

CHLOROFORMIC ACID, 2,2,2-TRICHLOROETHYL ESTER

$C_3H_2Cl_4O_2$

Mol. Wt. 211.86

B.P. 171-172°C

Solvent: CCl_4

Assignments
a = 4.87

2963

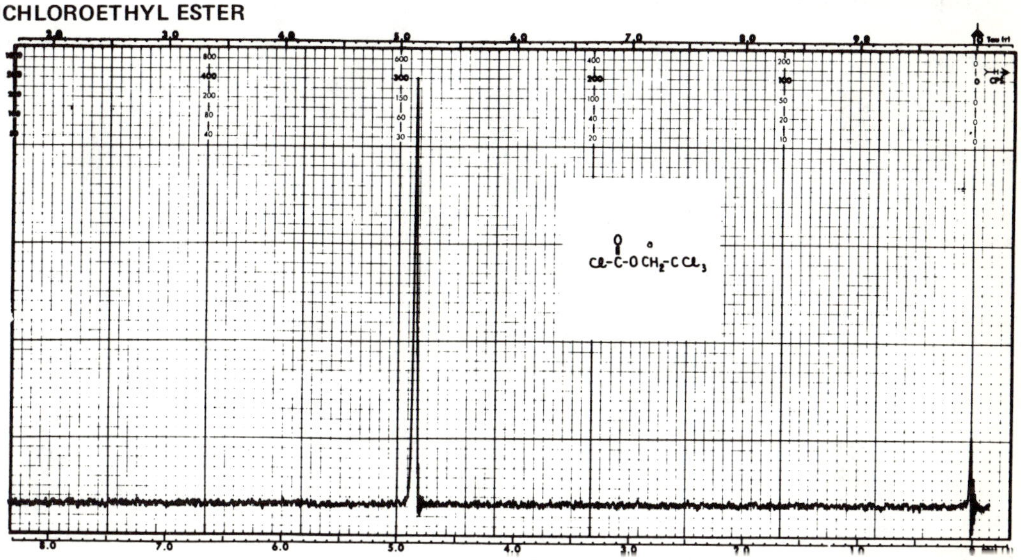

© 1978 Sadtler Research Laboratories, Inc.

2,2-DIMETHYL-1,3-PROPANEDIOL, BIS(CHLOROFORMATE)

$C_7H_{10}Cl_2O_4$

Mol. Wt. 229.06

B.P. 120-122°C/13mm

Solvent: CCl_4

Assignments
a = 1.08
b = 4.11

2964

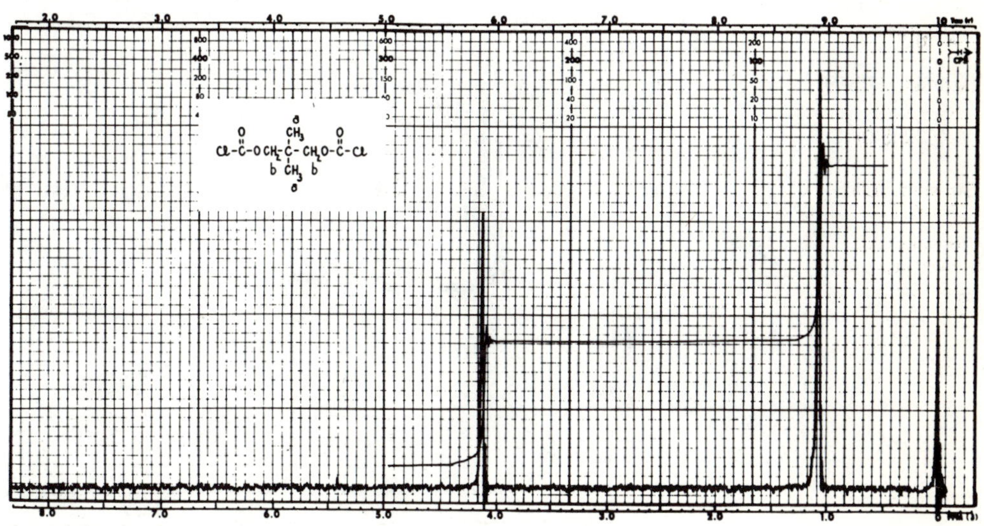

© 1978 Sadtler Research Laboratories, Inc.

CHLOROFORMIC ACID, ALLYL ESTER

$C_4H_5ClO_2$

Mol. Wt. 120.54

B.P. 112-114°C

Solvent: CCl_4

Assignments
a = 4.78
b = 5.39
c = 5.42
d = 5.90
e = 3.98, 4.07 imps.

2965

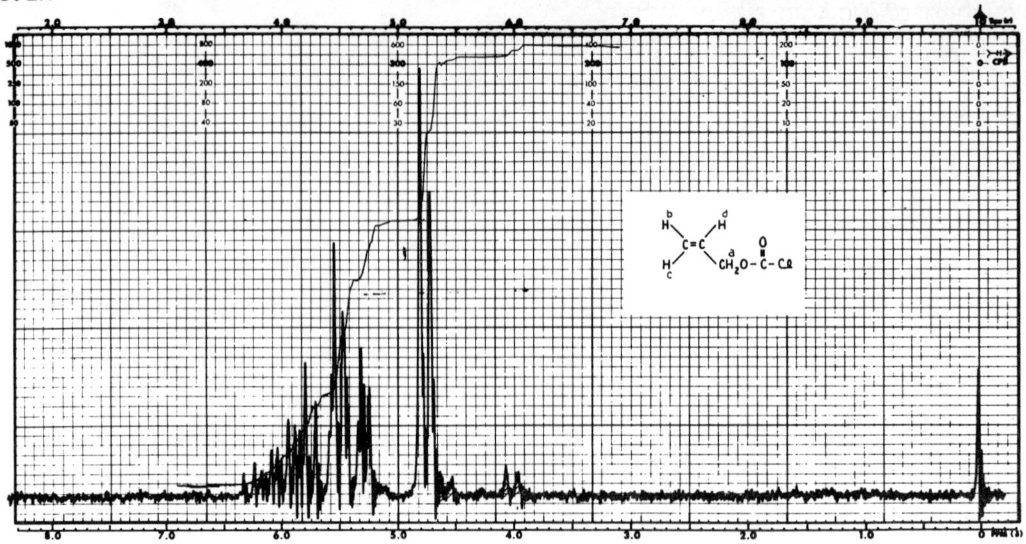

© 1978 Sadtler Research Laboratories, Inc.

CHLOROFORMIC ACID, 9-OCTADECENYL ESTER

$C_{19}H_{35}ClO_2$

Mol. Wt. 330.94

Solvent: CCl_4

Assignments
a = 0.89
b = 1.30
c = 1.97
d = 4.27
e = 5.30

2966

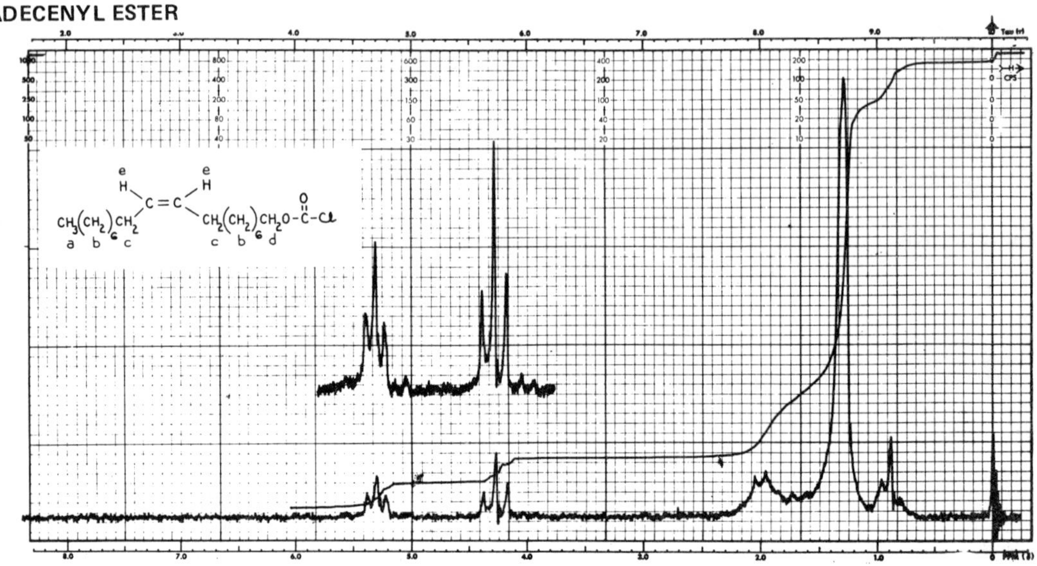

© 1978 Sadtler Research Laboratories, Inc.

CHLOROFORMIC ACID, p-NITROBENZYL ESTER

$C_8H_6ClNO_4$

Mol. Wt. 215.59

M.P. 32.5-34°C

Solvent: $CDCl_3$

Assignments
a = 5.41
b = 7.60
c = 8.27

2967

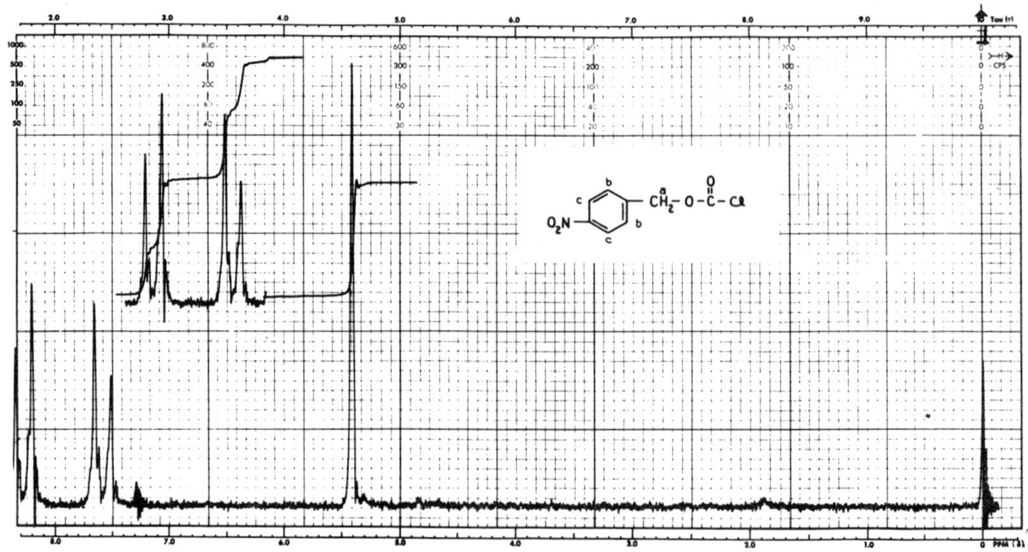

© 1978 Sadtler Research Laboratories, Inc.

CHLOROFORMIC ACID, PHENYL ESTER

$C_7H_5ClO_2$

Mol. Wt. 156.57

B.P. 192°C

Solvent: CCl_4

Assignments
a = ca 7.23

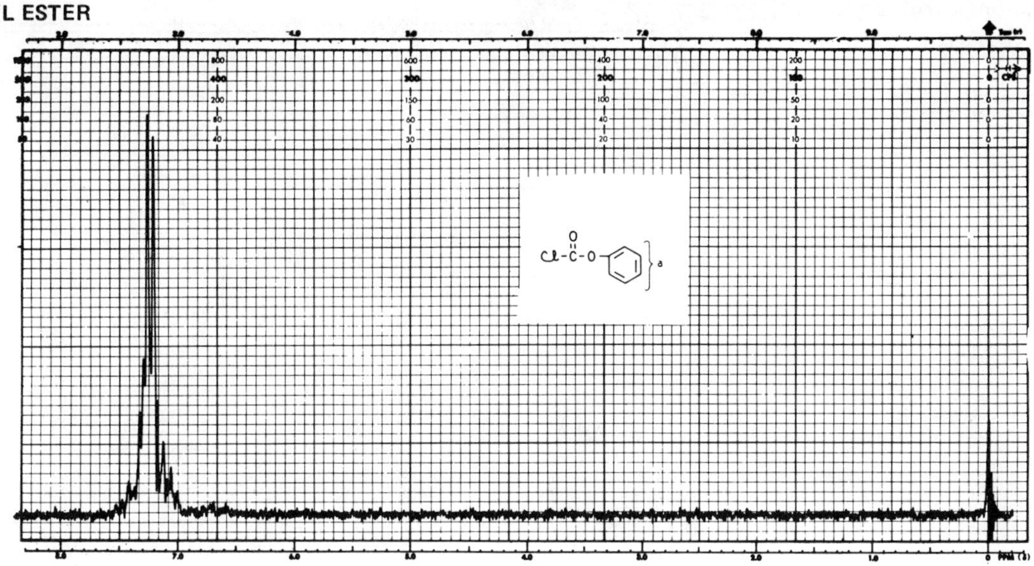

2968

© 1978 Sadtler Research Laboratories, Inc.

CHLOROFORMIC ACID, p-NITROPHENYL ESTER

$C_7H_4ClNO_4$

Mol. Wt. 201.57

M.P. 80-82°C

Solvent: Dimethyl sulfoxide-d_6

Assignments
a = 7.11
b = 8.12

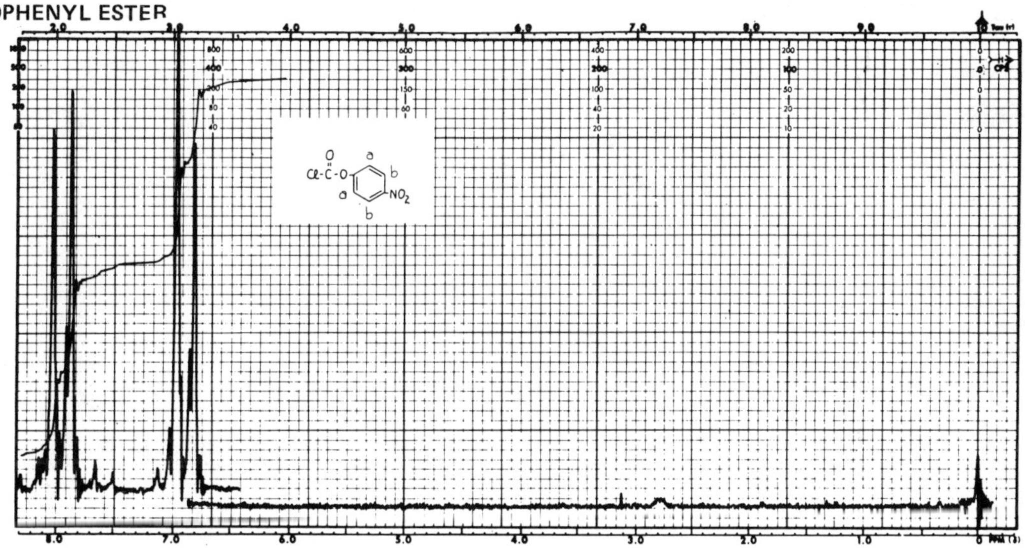

2969

© 1978 Sadtler Research Laboratories, Inc.

THIOACETIC ACID, S-ETHYL ESTER

C_4H_8OS

Mol. Wt. 104.17

B.P. 115-117°C

Solvent: CCl_4

Assignments
a = 1.23
b = 2.28
c = 2.83

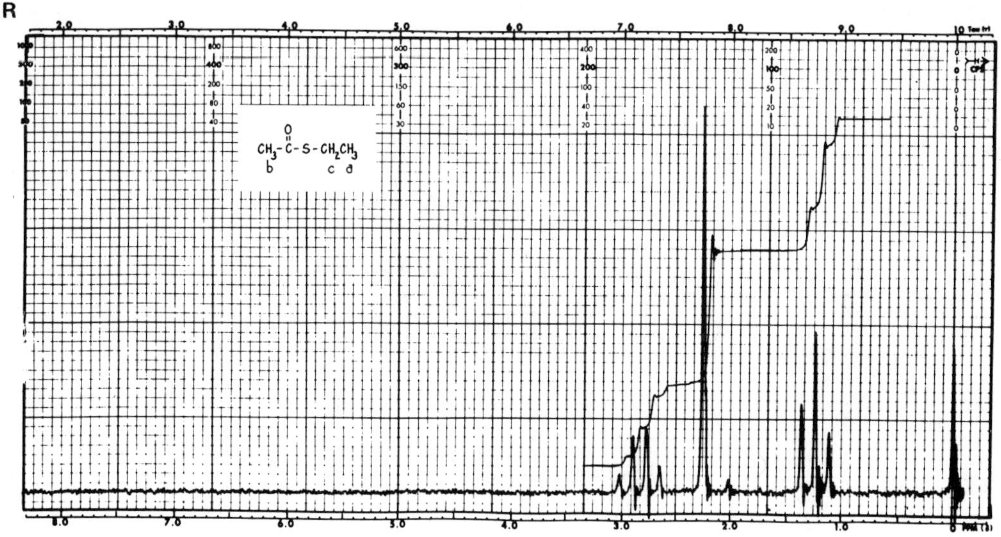

2970

© 1978 Sadtler Research Laboratories, Inc.

THIOBENZOIC ACID. S-BENZYL ESTER

$C_{14}H_{12}OS$

Mol. Wt. 228.32

M.P. 37-39°C

Solvent: $CDCl_3$

Assignments
a = 4.25
b = 7.00-7.50
c = 7.91

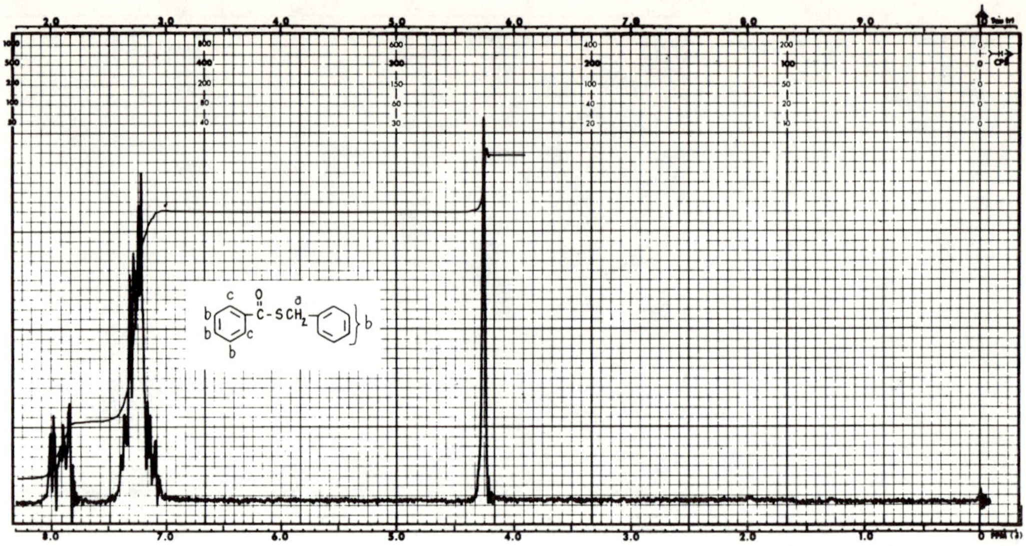

2971

© 1978 Sadtler Research Laboratories, Inc.

THIOCARBONIC ACID, O-ETHYL S-2-HYDROXYETHYL ESTER

$C_5H_{10}O_3S$

Mol. Wt. 150.20

B.P. 109-112°C/5mm

Solvent: CCl_4

Assignments
a = 1.31
b = 2.98
c = 3.61
d = 3.72
e = 4.27

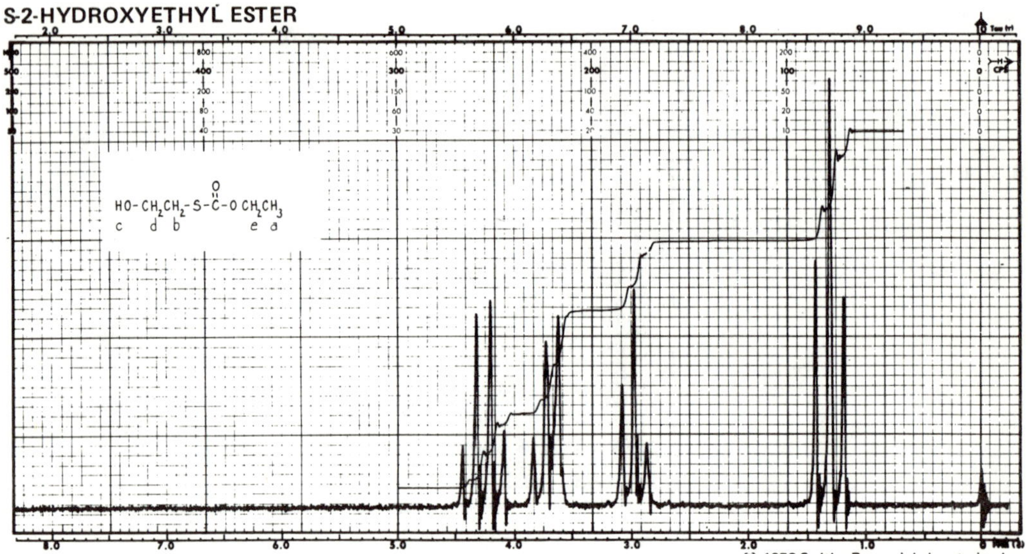

2972

© 1978 Sadtler Research Laboratories, Inc.

THE CARBAMATES

The NMR spectra of the Carbamates are often quite complex in appearance since up to three aliphatic or aromatic groups may be present in the molecule. In addition, the secondary Carbamates (R–NH–C(=O)–O–) often display coupling between the NH proton and the hydrogens bonded to the adjacent carbon atom. The coupling constant for this interaction is about 7.5 Hz (spectrum 2983).

Hydrocarbon groups adjacent to the nitrogen side of the linkage are deshielded to intermediate field (about 3 ppm) while those adjacent to the oxygen side are deshielded by an additional 1 ppm to about 4 ppm.

Chemical Shifts

Aliphatic Protons

HC–3	HC–2	HC–1	–NH–		Solvent	Spectrum
		(3.06)	----	$(CH_3)_2{-}N{-}C(=O){-}O{-}\emptyset$	$CDCl_3$	2991
		2.78	5.09	$CH_3{-}NH{-}C(=O){-}O{-}R_2$	$CDCl_3$	2980
	1.12	3.15	5.57	$CH_3{-}CH_2{-}NH{-}C(=O){-}O{-}R_2$	CCl_4	----
0.90	1.45	3.08	5.25	$CH_3{-}CH_2{-}CH_2{-}NH{-}C(=O){-}O{-}R_2$	CCl_4	2981
	(1.15)	3.71	4.66	$(CH_3)_2{-}CH{-}NH{-}C(=O){-}O{-}R_2$	$CDCl_3$	----

HC–3	HC–2	HC–1	NH/NH_2			
		3.60	5.30	$CH_3{-}O{-}C(=O){-}NH{-}R_4$	CCl_4	2982
		3.63	5.13	$CH_3{-}O{-}C(=O){-}NH_2$	$CDCl_3$	----
	1.20	4.03	5.25	$CH_3{-}CH_2{-}O{-}C(=O){-}NH{-}R_3$	CCl_4	2981
	1.21	4.07	----	$CH_3{-}CH_2{-}O{-}C(=O){-}N(R_4)_2$	CCl_4	2984
	1.22	4.08	5.20	$CH_3{-}CH_2{-}O{-}C(=O){-}NH_2$	$CDCl_3$	----
	1.24	4.11	5.09	$CH_3{-}CH_2{-}O{-}C(=O){-}NH{-}CH_3$	$CDCl_3$	2980
0.94	1.68	4.10	6.90	$CH_3{-}CH_2{-}CH_2{-}O{-}C(=O){-}NH{-}\emptyset$	$CDCl_3$	2987
	(1.22)	4.89	4.97	$(CH_3)_2{-}CH{-}O{-}C(=O){-}NH_2$	$CDCl_3$	2973
	(1.45)	----	4.88	$(CH_3)_3{-}C{-}O{-}C(=O){-}NH_2$	$CDCl_3$	2975

Aromatic Protons

Phenyl Carbamates

para/meta	ortho			
6.8–7.4	7.51	$\emptyset{-}O{-}C(=O){-}NH_2$	DMSO	2978

para/meta/ortho			
6.6–7.5	$\emptyset{-}NH{-}C(=O){-}O{-}R_3$	$CDCl_3$	2987

CARBAMIC ACID, ISOPROPYL ESTER

$C_4H_9NO_2$

Mol. Wt. 103.12

Solvent: $CDCl_3$

Assignments
a = 1.22
b = 4.89
c = 4.97

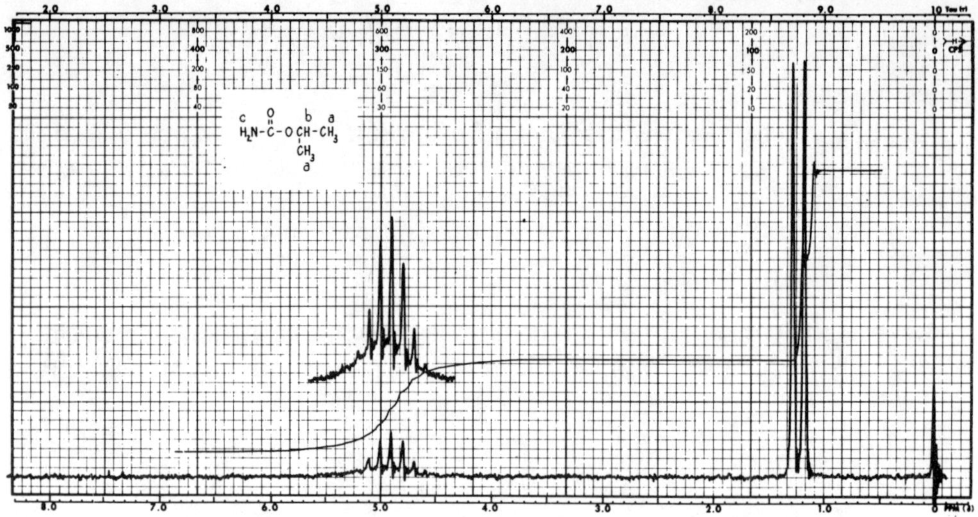

2973

© 1978 Sadtler Research Laboratories, Inc.

CARBAMIC ACID, BUTYL ESTER

$C_5H_{11}NO_2$

Mol. Wt. 117.15

M.P. 53-54°C

Solvent: $CDCl_3$

Assignments
a = 0.98
b = 1.12-1.91
c = 4.07
d = 5.37

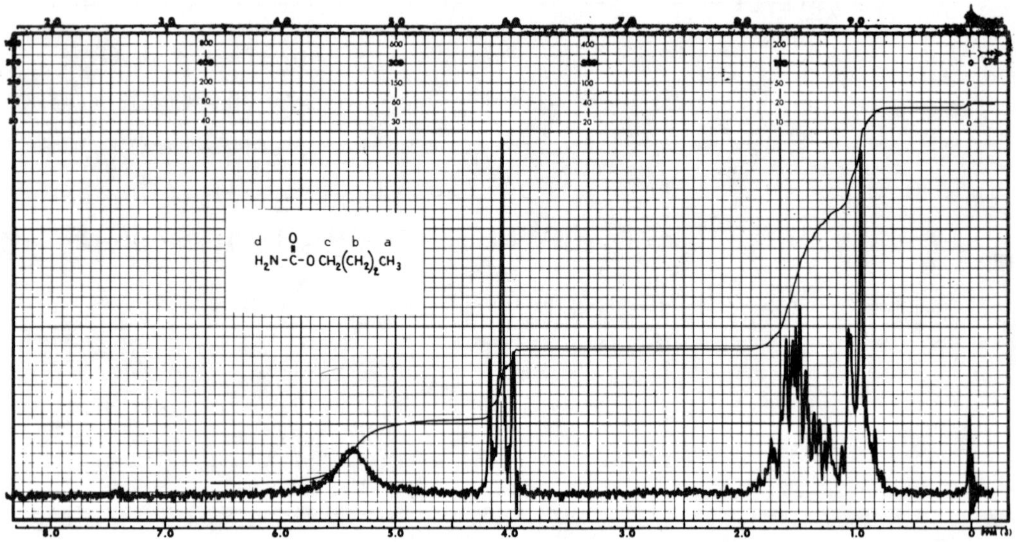

2974

© 1978 Sadtler Research Laboratories, Inc.

CARBAMIC ACID, tert-BUTYL ESTER

$C_5H_{11}NO_2$

Mol. Wt. 117.15

Solvent: $CDCl_3$

Assignments
a = 1.45
b = 4.88

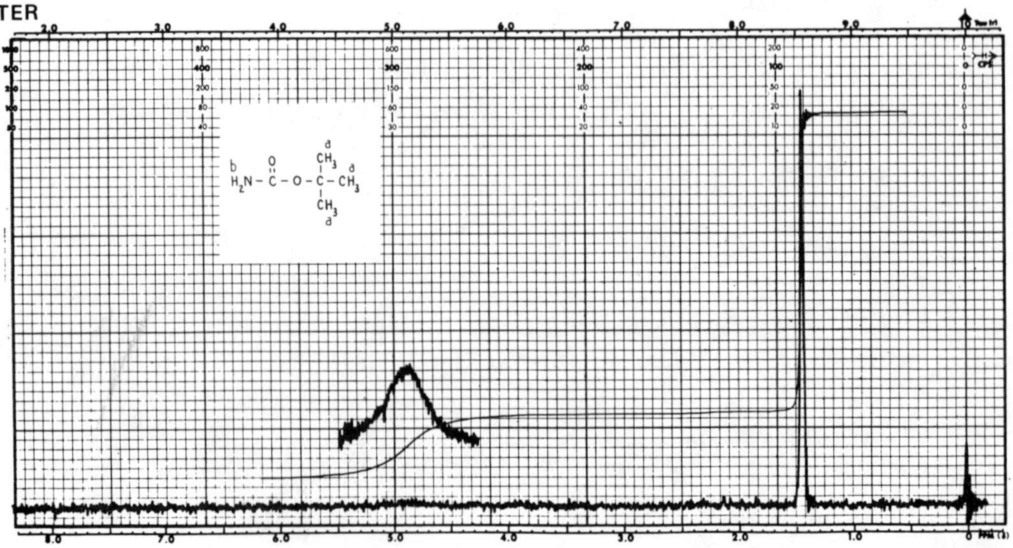

2975

© 1978 Sadtler Research Laboratories, Inc.

CARBAMIC ACID, DODECYL ESTER

$C_{13}H_{27}NO_2$

Mol. Wt. 229.37

M.P. 82.5-83.5°C

Solvent: $CDCl_3$

Assignments

a = 0.88
b = ca 1.28
c = 4.04
d = 4.92

2976

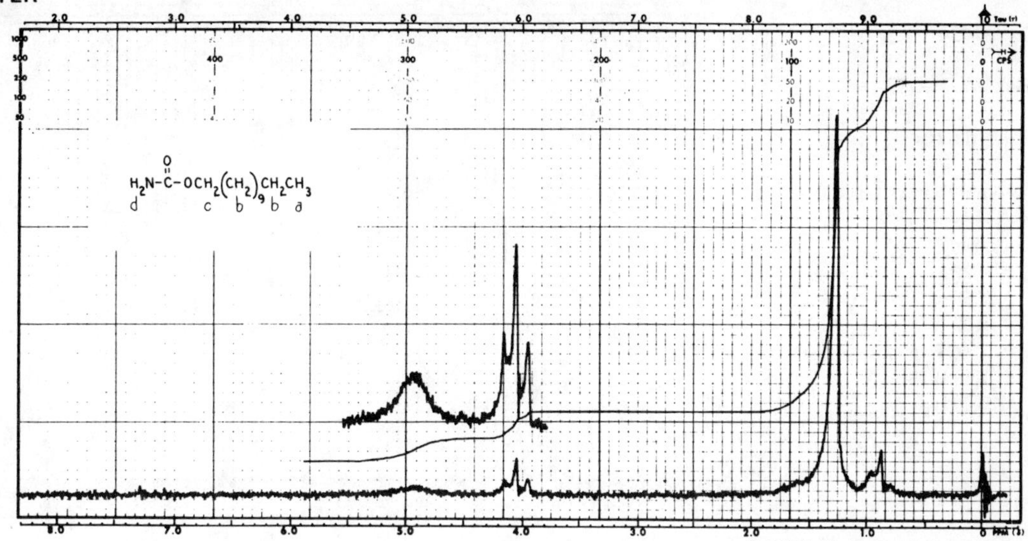

© 1978 Sadtler Research Laboratories, Inc.

CARBAMIC ACID, ALLYL ESTER

$C_4H_7NO_2$

Mol. Wt. 101.11

Solvent: CCl_4

Assignments

a = 4.48
b = 5.14
c = 5.21
d = ca 5.71
e = 5.79

2977

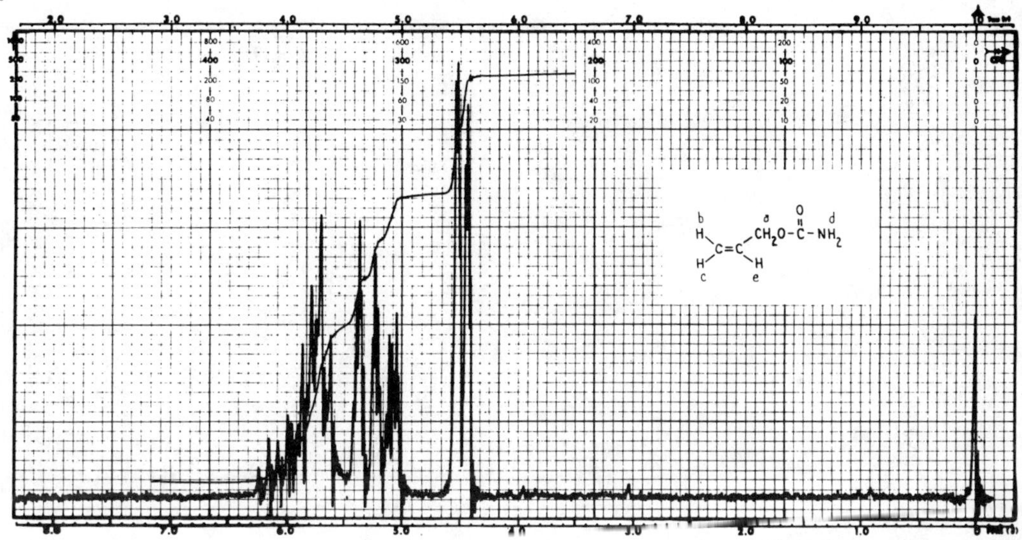

© 1978 Sadtler Research Laboratories, Inc.

CARBAMIC ACID, PHENYL ESTER

$C_7H_7NO_2$

Mol. Wt. 137.14

M.P. 241°C

Solvent: Dimethyl sulfoxide

Assignments

a = 6.82-7.40
b = 7.51
c = 8.64

2978

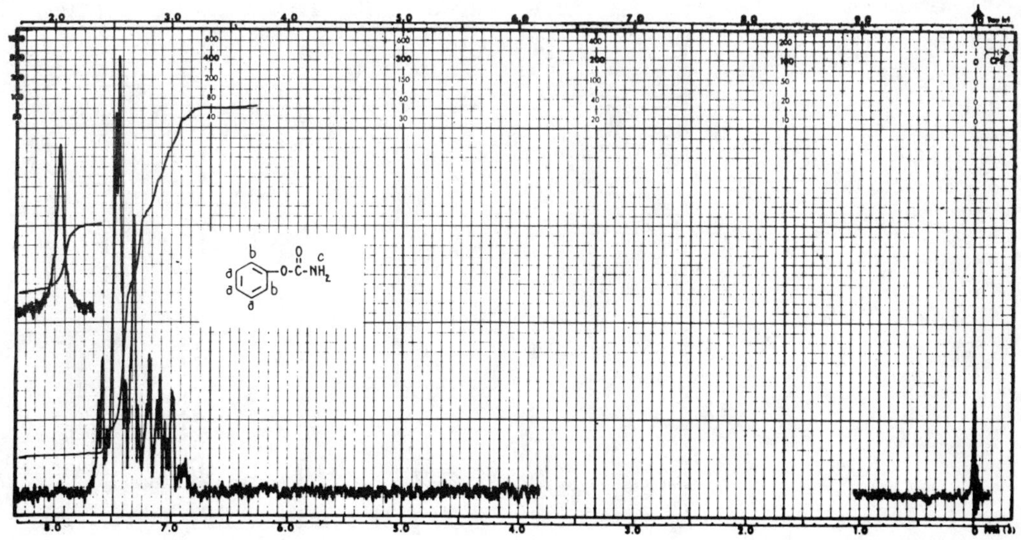

© 1978 Sadtler Research Laboratories, Inc.

CARBAMIC ACID, p-TOLYL ESTER

$C_8H_9NO_2$

Mol. Wt. 151.17

M.P. 163-164°C

Solvent: Dimethyl sulfoxide-d6

Assignments
a = 2.27
b = 6.80
c = 6.92
d = 7.14
e = 2.48 DMSO-D5
f = 3.20 H_2O

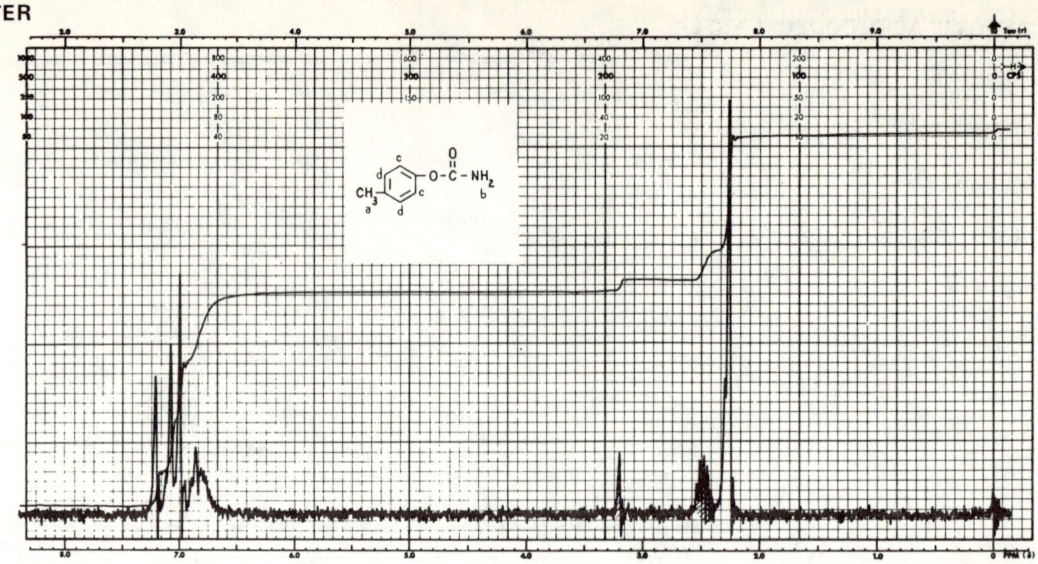

2979

© 1978 Sadtler Research Laboratories, Inc.

METHYLCARBAMIC ACID, ETHYL ESTER

$C_4H_9NO_2$

Mol. Wt. 103.12

B.P. 168-169°C

Solvent: $CDCl_3$

Assignments
a = 1.24
b = 2.78
c = 4.11
d = 5.09

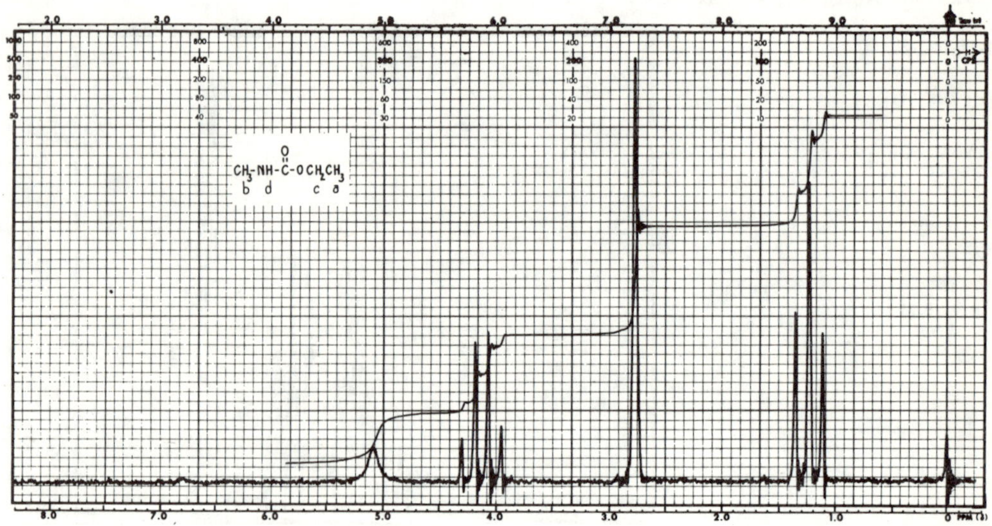

2980

© 1978 Sadtler Research Laboratories, Inc.

PROPYLCARBAMIC ACID, ETHYL ESTER

$C_6H_{13}NO_2$

Mol. Wt. 131.18

Solvent: CCl_4

Assignments
a = 0.90
b = 1.20
c = 1.45
d = 3.08
e = 4.03
f = 5.25

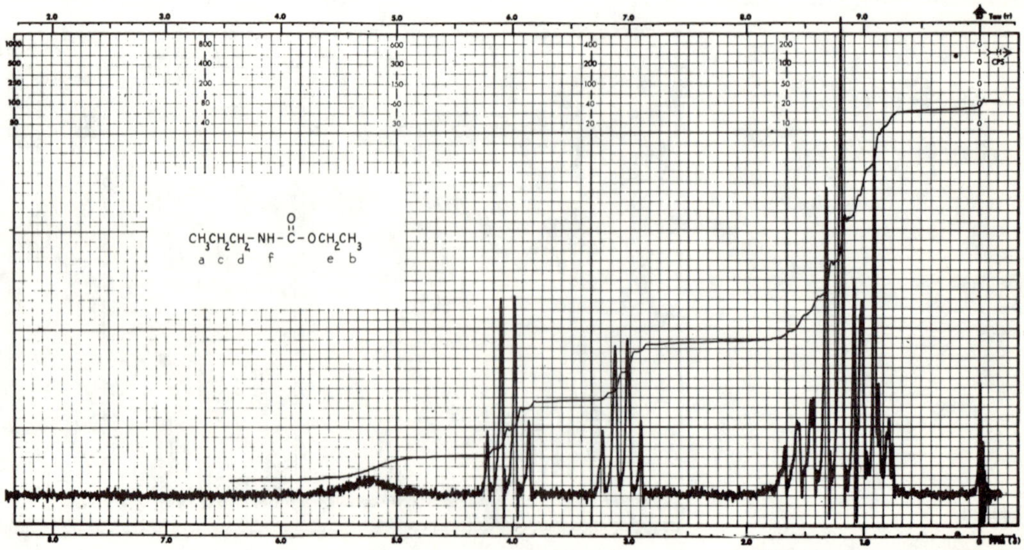

2981

© 1978 Sadtler Research Laboratories, Inc.

BUTYLCARBAMIC ACID, METHYL ESTER

$C_6H_{13}NO_2$

Mol. Wt. 131.18

Solvent: CCl_4

Assignments

a = 0.94
b = 1.12-1.64
c = 3.09
d = 3.60
e = 5.30

2982

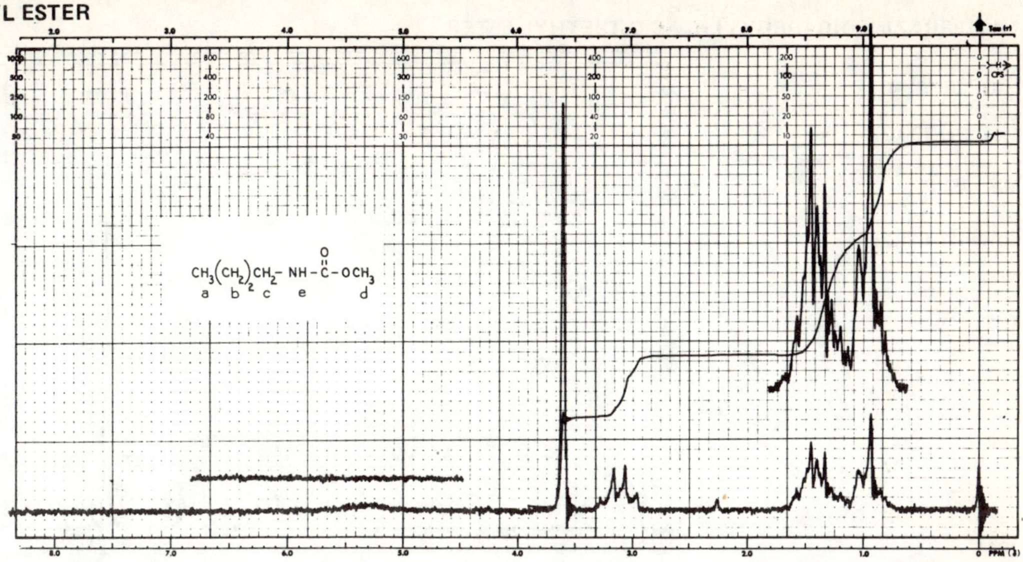

© 1978 Sadtler Research Laboratories, Inc.

N,N'-METHYLENEDICARBAMIC ACID, DIETHYL ESTER

$C_7H_{14}N_2O_4$

Mol. Wt. 190.20

Solvent: $CDCl_3$

Assignments

a = 1.23
b = 4.14
c = 4.51
d = 6.42

2983

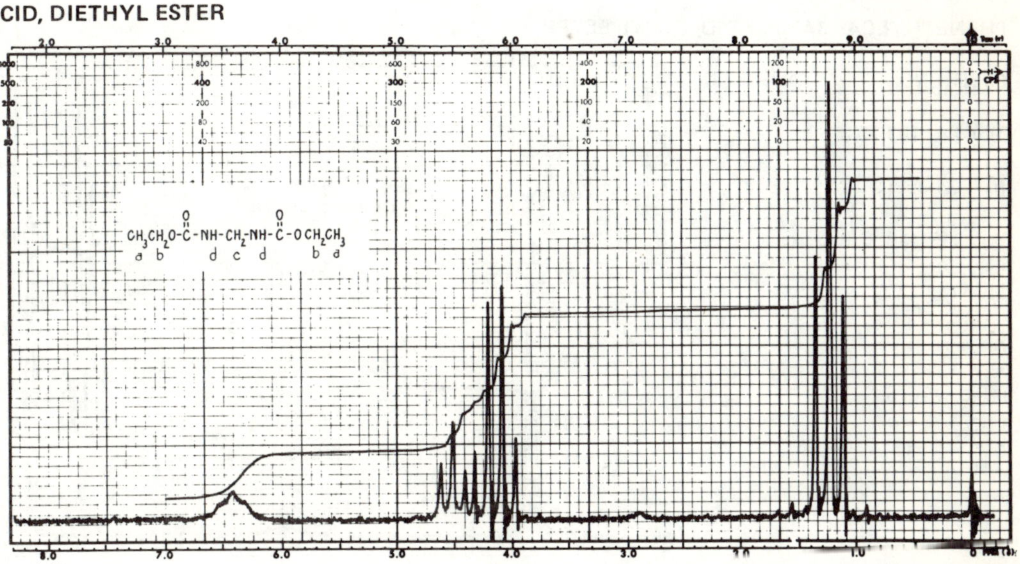

© 1978 Sadtler Research Laboratories, Inc.

DIBUTYLCARBAMIC ACID, ETHYL ESTER

$C_{11}H_{23}NO_2$

Mol. Wt. 201.31

B.P. 101-103°C/6mm

Solvent: CCl_4

Assignments

a = 0.92
b = 1.21
c = 1.20-1.75
d = 3.19
e = 4.07

2984

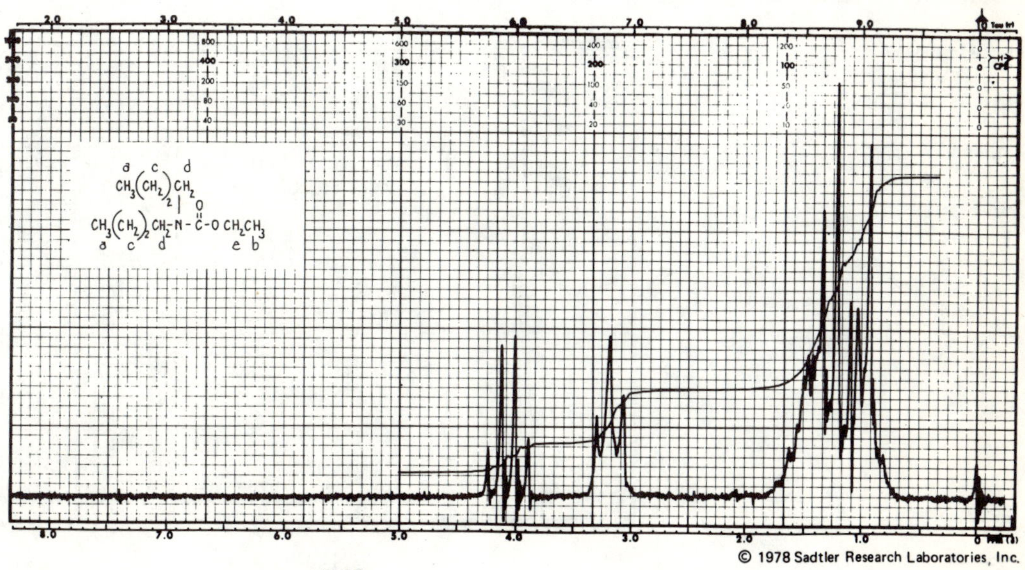

© 1978 Sadtler Research Laboratories, Inc.

1,4-PIPERAZINEDICARBOXYLIC ACID, DIETHYL ESTER

$C_{10}H_{18}N_2O_4$

Mol. Wt. 230.27

M.P. 45°C

Solvent: $CDCl_3$

Assignments
a = 1.26
b = 3.44
c = 4.14

2985

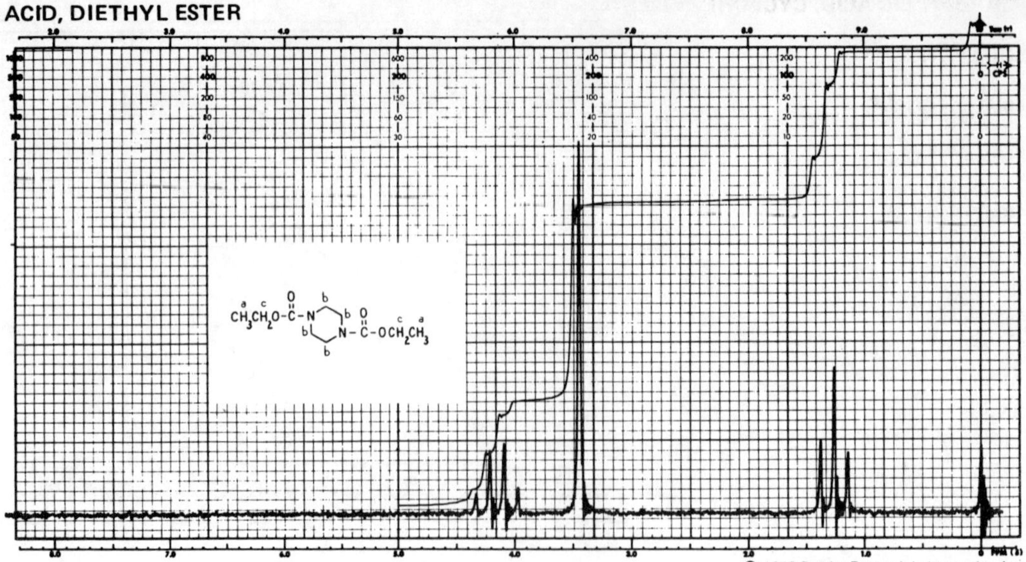

© 1978 Sadtler Research Laboratories, Inc.

PHENETHYLCARBAMIC ACID, ETHYL ESTER

$C_{11}H_{15}NO_2$

Mol. Wt. 193.25

M.P. 34-35°C

B.P. 115-118°C/0.2mm

Solvent: CCl_4

Assignments
a = 1.15
b = 2.73 or b
c = 3.32 or c
d = 4.00
e = ca 5.00
f = 7.13

2986

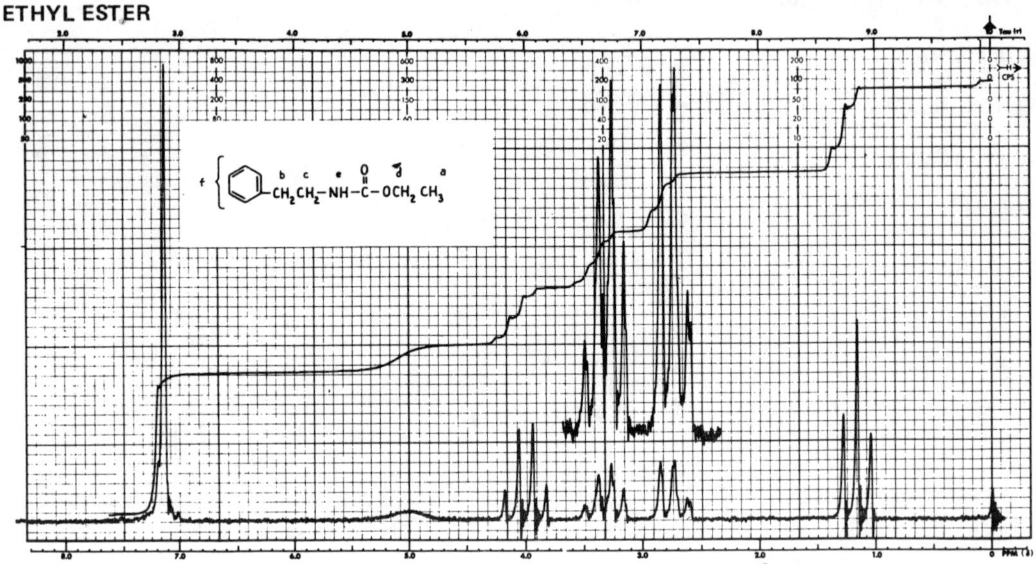

© 1978 Sadtler Research Laboratories, Inc.

CARBANILIC ACID, PROPYL ESTER

$C_{10}H_{13}NO_2$

Mol. Wt. 179.22

Solvent: $CDCl_3$

Assignments
a = 0.94
b = 1.68
c = 4.10
d = 6.60-7.50

2987

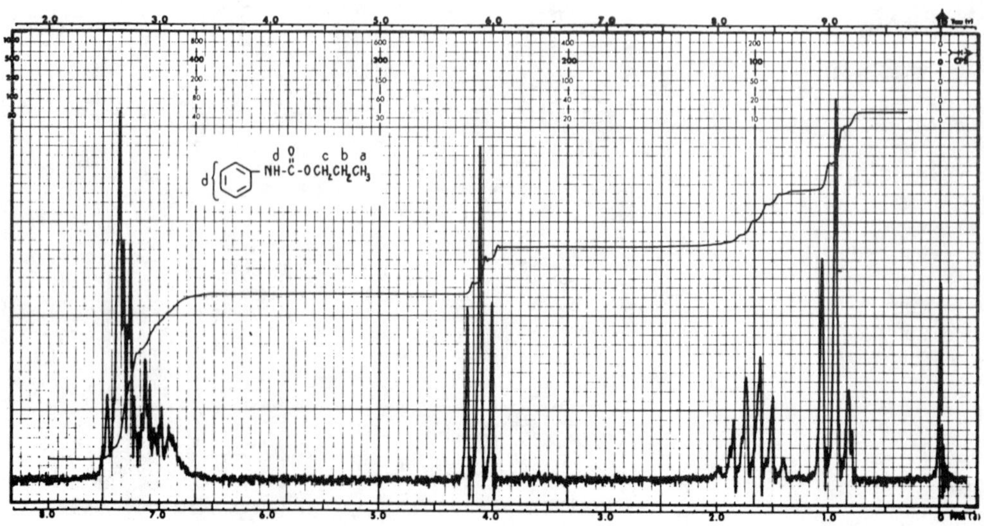

© 1978 Sadtler Research Laboratories, Inc.

CARBANILIC ACID, CYCLOHEXYL ESTER

$C_{13}H_{17}NO_2$

Mol. Wt. 219.29

Solvent: $CDCl_3$

Assignments
- a = 1.00-2.20
- b = 4.79
- c = 6.80-7.50
- d = unobserved

2988

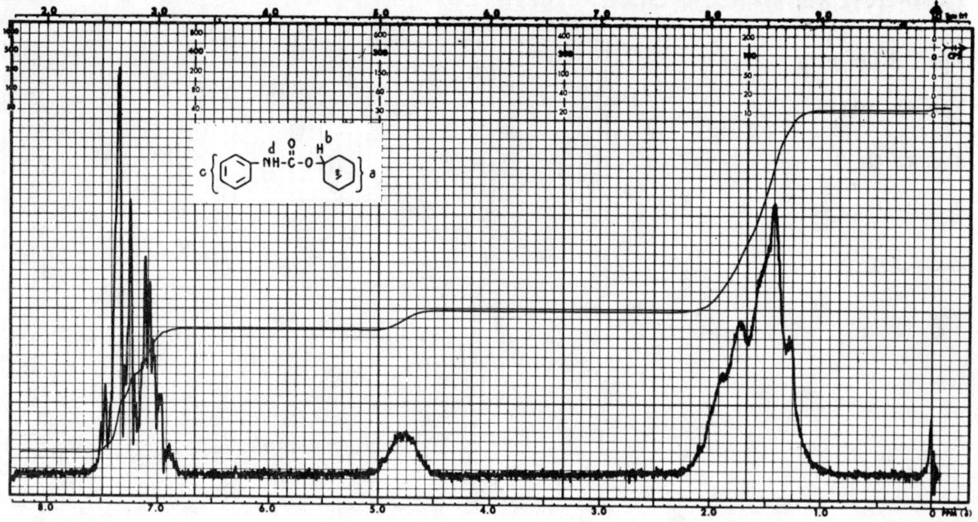

© 1978 Sadtler Research Laboratories, Inc.

CARBANILIC ACID, DIPHENYLMETHYL ESTER

$C_{20}H_{17}NO_2$

Mol. Wt. 303.36

M.P. 140°C

Solvent: $CDCl_3$

Assignments
- a = 6.89
- b = 6.75-7.50
- c = ca 7.30

2989

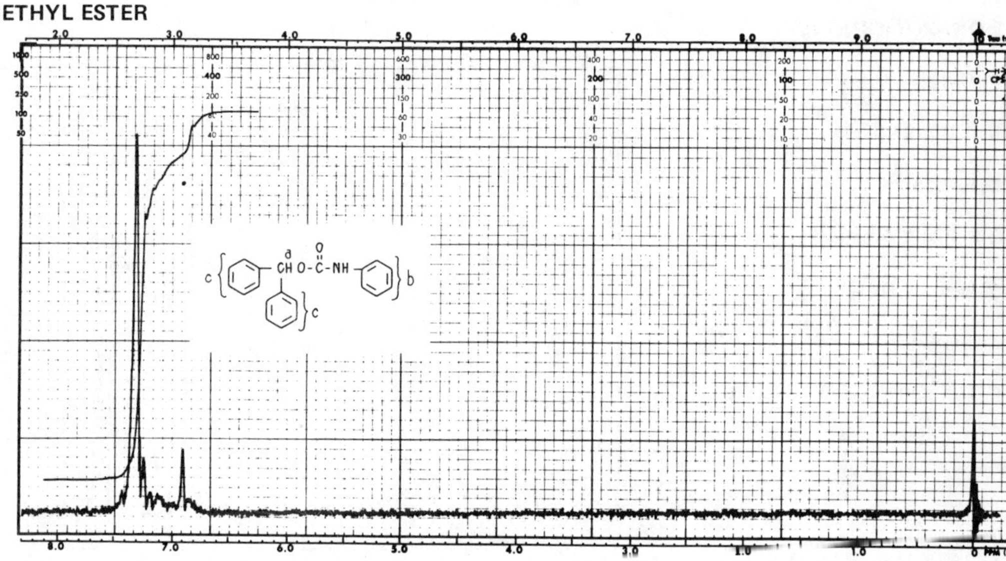

© 1978 Sadtler Research Laboratories, Inc.

N-BUTYLCARBANILIC ACID, ETHYL ESTER

$C_{13}H_{19}NO_2$

Mol. Wt. 221.30

Solvent: CCl_4

Assignments
- a = 0.89
- b = 1.11
- c = 1.10-1.70
- d = 3.61
- e = 4.03
- f = 6.90-7.40

2990

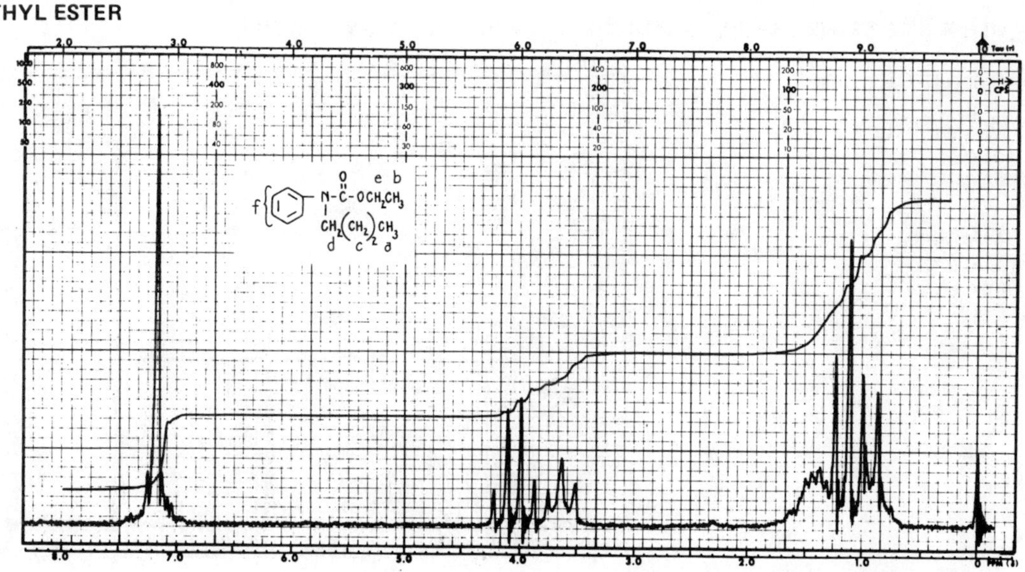

© 1978 Sadtler Research Laboratories, Inc.

DIMETHYLCARBAMIC ACID, PHENYL ESTER

$C_9H_{11}NO_2$

Mol. Wt. 165.19

M.P. 45-46°C

Solvent: $CDCl_3$

Assignments
a = 3.06
b = 7.28

2991

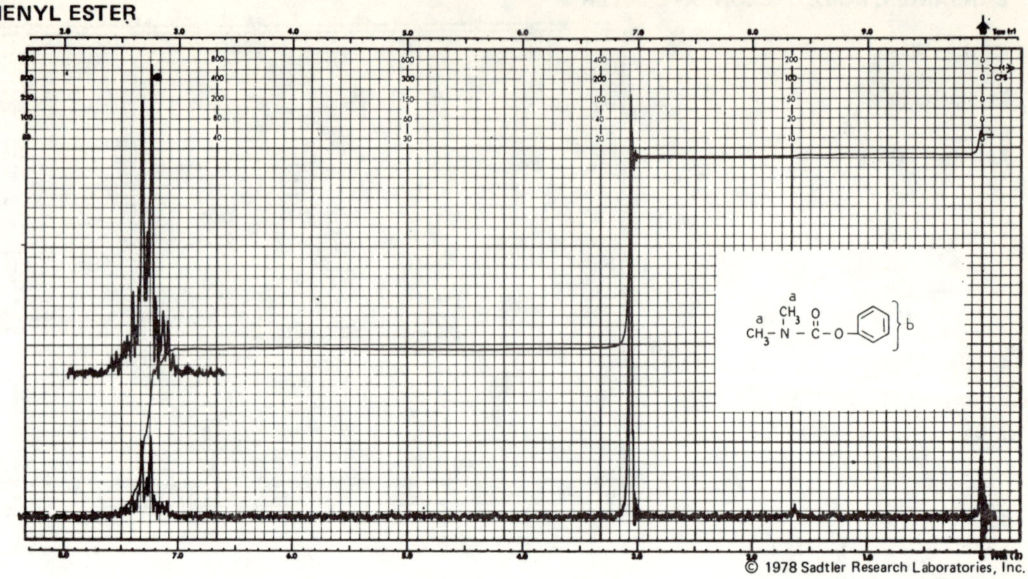

© 1978 Sadtler Research Laboratories, Inc.

2-OXAZOLIDINONE

$C_3H_5NO_2$

Mol. Wt. 87.08

M.P. 86-89°C

B.P. 220°C/48mm

Solvent: $CDCl_3$

Assignments
a = 3.62
b = 4.48
c = 6.42

2992

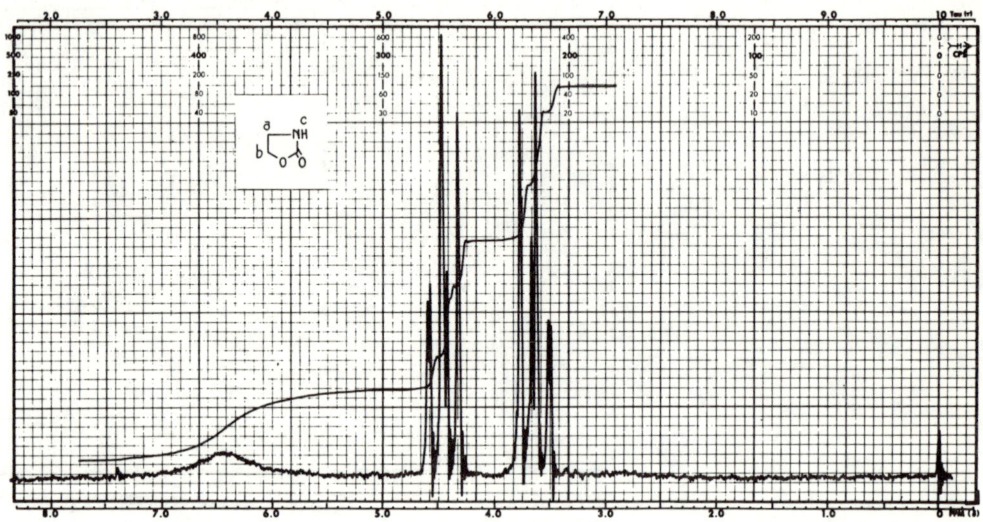

© 1978 Sadtler Research Laboratories, Inc.

s-TRIAZINE-1,3,5-(2H,4H,6H)-TRICARBOXYLIC ACID, TRIMETHYL ESTER

$C_9H_{15}N_3O_6$

Mol. Wt. 261.24

Solvent: $CDCl_3$

Assignments
a = 3.77
b = 5.11

2993

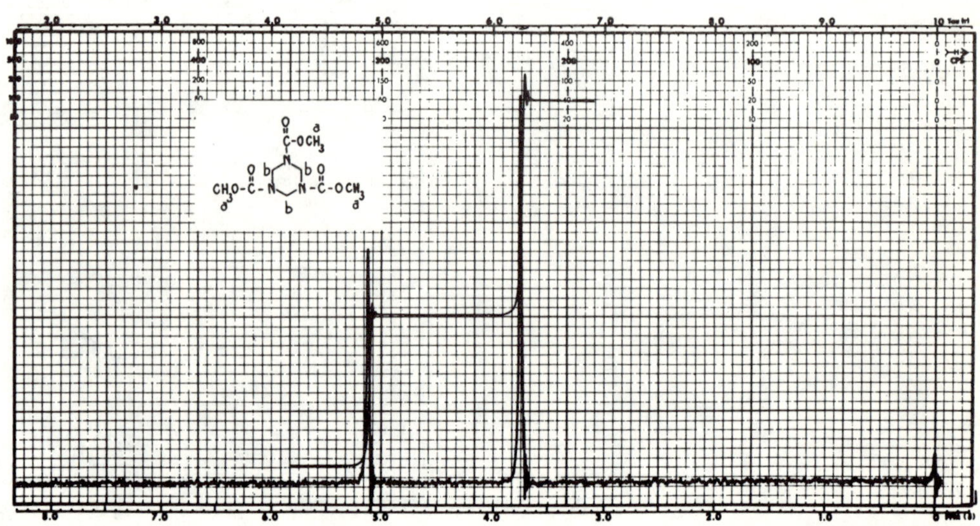

© 1978 Sadtler Research Laboratories, Inc.

3-PHENYL-2-OXAZOLIDINONE

$C_9H_9NO_2$

Mol. Wt. 163.18

Solvent: $CDCl_3$

Assignments
a = 3.90
b = 4.36
c = 6.90-7.60
d = 1.93 imp.

2994

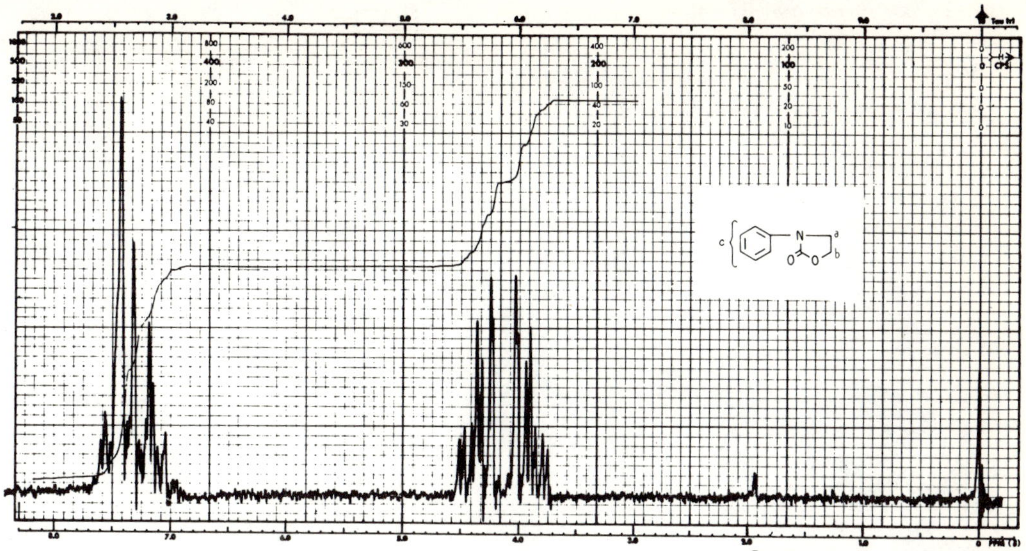

© 1978 Sadtler Research Laboratories, Inc.

BENZOYL PEROXIDE

$C_{14}H_{10}O_4$

Mol. Wt. 242.23

M.P. 103-105°C

Solvent: $CDCl_3$

Assignments
a = 7.20-7.80
b = 8.06

2995

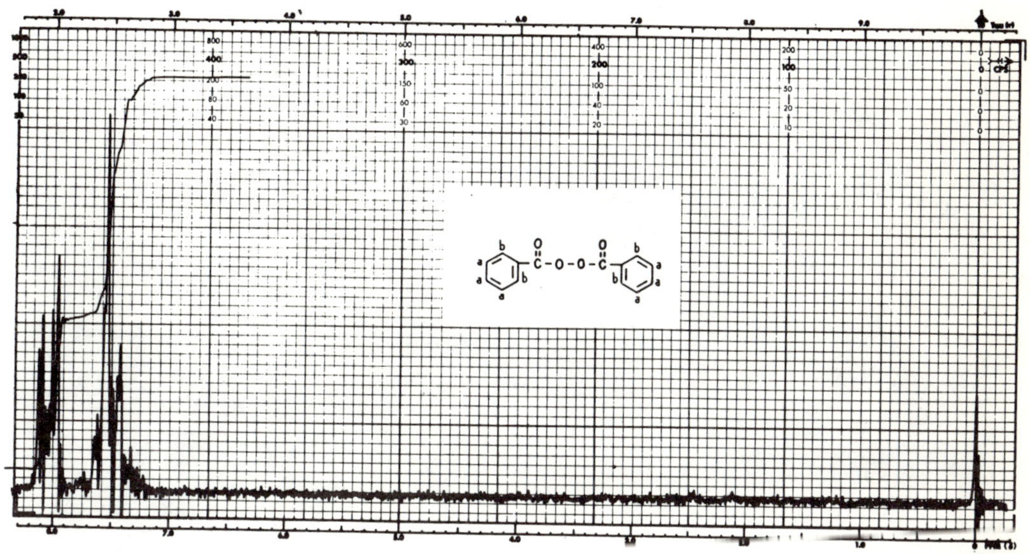

© 1978 Sadtler Research Laboratories, Inc.

NOTES

The chemical shifts of the Esters of Phosphonic and Phosphoric Acid are similar to those of the carboxylic acids, however, their spectra are distinguished by the spin-spin coupling interactions of the nearby hydrocarbon groups with the Phosphorus nucleus. As noted below, many of these couplings and their associated coupling constants are quite sensitive to structural and substituent differences. Both groups of compounds are quite soluble in Carbon Tetrachloride and Deuterochloroform and no unusual solvent effects have been noted for these two solvents.

THE PHOSPHONATES

Chemical Shifts

Aliphatic Protons

HC–3	HC–2	HC–1		Solvent	Spectrum
		(3.71)	$(CH_3–O)_2–P(=O)–CH_2–O–CH_3$	CCl_4	----
		(3.72)	$(CH_3–O)_2–P(=O)–H$	CCl_4	----
		(3.80)	$(CH_3–O)_2–P(=O)–CH_2–C(=O)–R$	$CDCl_3$	----
	(1.22	4.01)	$(CH_3–CH_2–O)_2–P(=O)–CH_2–Ø$	$CDCl_3$	----
	(1.29	4.04)	$(CH_3–CH_2–O)_2–P(=O)–CH_3$	CCl_4	----
	(1.36	4.08)	$(CH_3–CH_2–O)_2–P(=O)–H$	CCl_4	----
	(1.31	4.11)	$(CH_3–CH_2–O)_2–P(=O)–CH_2–CH_3$	$CDCl_3$	----
	(1.36	4.11)	$(CH_3–CH_2–O)_2–P(=O)–CH=CH_2$	$CDCl_3$	----
	(1.35	4.17)	$(CH_3–CH_2–O)_2–P(=O)–CH_2–C(=O)–R$	CCl_4	----
	(1.40	4.22)	$(CH_3–CH_2–O)_2–P(=O)–CH_2–C≡N$	$CDCl_3$	----
(0.99	1.65	3.51)	$(CH_3–CH_2–CH_2–O)_2–P(=O)–CH_2–Cl$	CCl_4	----
	(1.32	4.64)	$((CH_3)_2–CH–O)_2–P(=O)–H$	CCl_4	----
		1.33	$CH_3–P(=O)–(O–R_2)_2$	CCl_4	----
	1.19	1.91	$CH_3–CH_2–P(=O)–(O–R_2)_2$	$CDCl_3$	----

Substituted Methyl Phosphonates

$(R–O)_2–P(=O)–$	CH_2	$–X$	$J_{P(=O)–CH_2}$	
	1.91	$–CH_3$	0 Hz (?)	$CDCl_3$
	2.98	$–C≡N$	20.9 Hz	$CDCl_3$
	3.05	$–Ø$ $(4–NH_2)$	21.3 Hz	$CDCl_3$
	3.10	$–C(=O)–CH_3$	23.1 Hz	$CDCl_3$
	3.51	$–Cl$	11.5 Hz	CCl_4
	3.69	$–O–CH_3$	8.0 Hz	CCl_4

Chemical Shifts continued

Phosphonyl Protons

$H{-}P(=O)-(O{-}R)_2$		$J_{P(=O)-H}$	Solvent	Spectrum
4.40	$-(O{-}CH(CH_3)_2)_2$	411 Hz	CCl_4	
5.69	$-(O{-}R_2)_2$	342 Hz	CCl_4	
5.80	$-(O{-}CH_3)_2$	697 Hz	CCl_4	
6.11	$-(O{-}Ø)_2$	733 Hz	$CDCl_3$	
6.76	$-(O{-}R_{18})_2$	691 Hz	$CDCl_3$	

Aromatic Protons

The aromatic protons of the phenyl esters of Phosphonic Acid resonate as a broad, single peak or a complex series of bands centered at about 7.2 ppm.

Phenyl groups bonded to the Phosphorus nucleus display a strong deshielding of the ortho protons which resonate near 7.8 ppm and couple to the Phosphorus nucleus with a coupling constant of about 13 Hz.

Phenyl Phosphonates

para/meta	ortho			
7.3–7.6	7.80	$Ø{-}P(=O)-(OH)_2$	Acetone	- - - -

para/meta/ortho			
ca. 7.21	$(Ø{-}O)_2{-}P(=O)-R_{12}$	$CDCl_3$	- - - -
7.3–7.6	$(Ø{-}O)_2{-}P(=O)-H$	$CDCl_3$	- - - -

Coupling and Coupling Constants

$J_{P(=O)-C-O-CH_3}$ = 1.1 Hz

$J_{P(=O)-O-CH_2}$ = 7.3–8.1 Hz

$J_{P(=O)-O-CH_3}$ = 10.8 Hz

$J_{P(=O)-CH_2}$ = 11–24 Hz

$J_{P(=O)-CH_3}$ = 17.5 Hz

$J_{P(=O)-H}$ = 340–740 Hz

THE PHOSPHATES

Chemical Shifts

Aliphatic Protons

HC–3	HC–2	HC–1		Solvent	Spectrum
		(3.75)	$(CH_3–O)_2–P(=O)–O–Ø$	CCl_4	----
	(1.33	4.06)	$(CH_3–CH_2–O)_2–P(=O)–OH$	CCl_4	2997
	(1.31	4.11)	$(CH_3–CH_2–O)_2–P(=O)–O–Ø$	CCl_4	----
(0.99	1.70	3.99)	$(CH_3–CH_2–CH_2–O)_3–P(=O)$	$CDCl_3$	2999
(0.92	1.62	4.00)	$(CH_3–CH_2–CH_2–O)_2–P(=O)–O–Ø$	CCl_4	----

Aromatic Protons

para/meta/ortho			
ca. 7.20	$(Ø–O)_2–P(=O)–OH$	$CDCl_3$	----
ca. 7.21	$(Ø–O–P(=O)–(O–CH_3)_2$	CCl_4	----
6.8–7.4	$(Ø–O)_2–P(=O)–Ag$	Polysol	----

Exchangeable Protons

HO–			
8.83	$HO–P(=O)–(O–R_{16})_2$	$CDCl_3$	----
(10.19)	$(HO)_2–P(=O)–2–Naphthalene$	Polysol	----
11.06	$HO–P(=O)–(O–R_2)_2$	CCl_4	----
12.22	$HO–P(=O)–O–Ø)_2$	$CDCl_3$	----

Coupling and Coupling Constants

$J_{P(=O)–O–CH_3}$ = 10–12 Hz

$J_{P(=O)–O–CH_2}$ = 6–7 Hz

VINYLPHOSPHONIC ACID, BIS(2-CHLOROETHYL) ESTER

$C_6H_{11}Cl_2O_3P$

Mol. Wt. 233.03

B.P. 132°C/1mm

Solvent: CCl_4

Assignments
a = 3.71
b = ca 4.20
c = 5.65-6.88
d = 4.80 imp.

2996

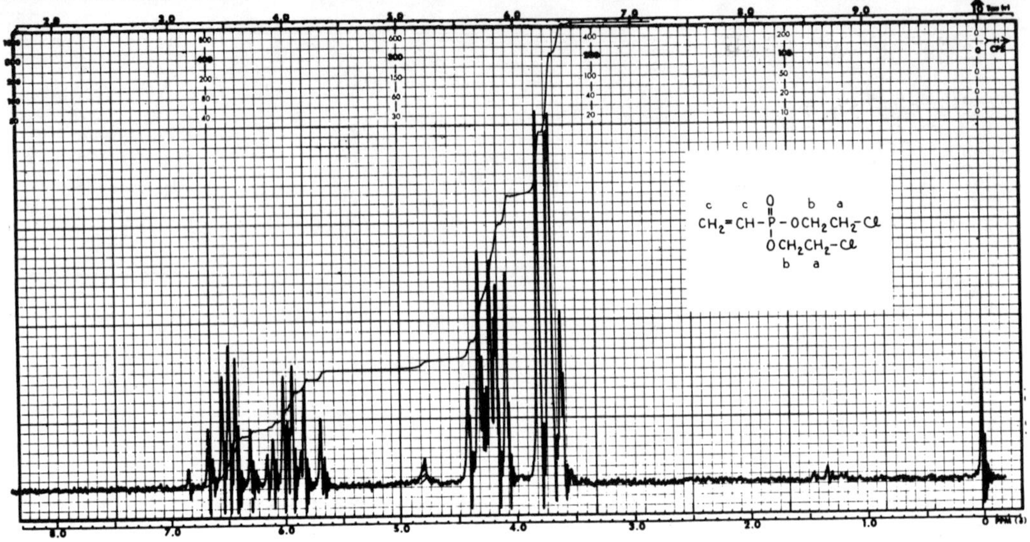

© 1978 Sadtler Research Laboratories, Inc.

PHOSPHORIC ACID, DIETHYL ESTER

$C_4H_{11}O_4P$

Mol. Wt. 154.10

M.P. 4-7°C

Solvent: CCl_4

Assignments
a = 1.33
b = 4.06
c = 11.06

2997

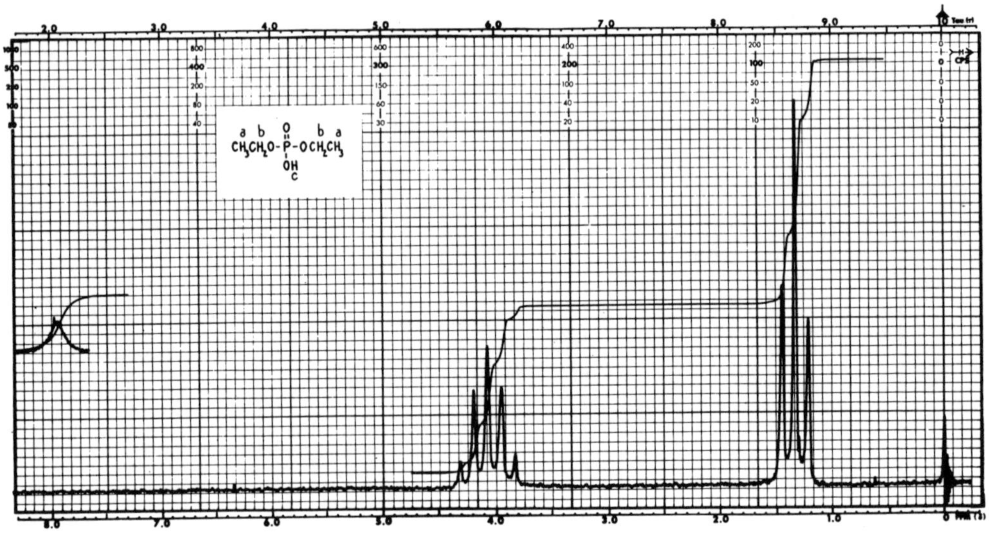

© 1978 Sadtler Research Laboratories, Inc.

N,N-DIETHYLPHOSPHORAMIDIC ACID, DIBUTYL ESTER

$C_{12}H_{28}NO_3P$

Mol. Wt. 265.34

Solvent: CCl_4

Assignments
a = 0.95
b = 1.06
c = 1.30-1.91
d = 2.91 or e
e = 3.10 or d
f = 3.86

2998

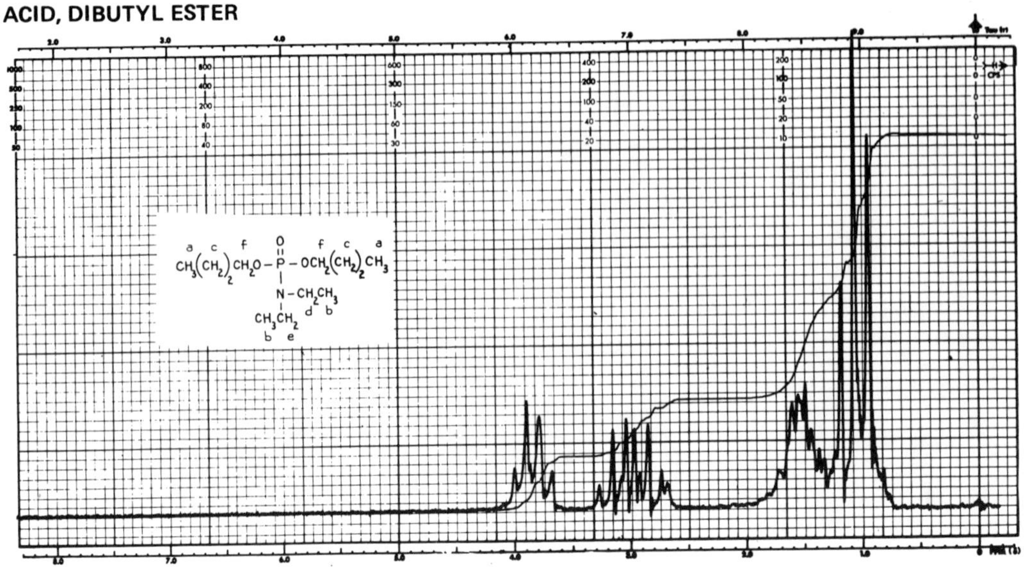

© 1978 Sadtler Research Laboratories, Inc.

PHOSPHORIC ACID, TRIPROPYL ESTER

$C_9H_{21}O_4P$

Mol. Wt. 224.24

Solvent: $CDCl_3$

Assignments
a = 0.99
b = 1.70
c = 3.99
d = 2.66 imp.

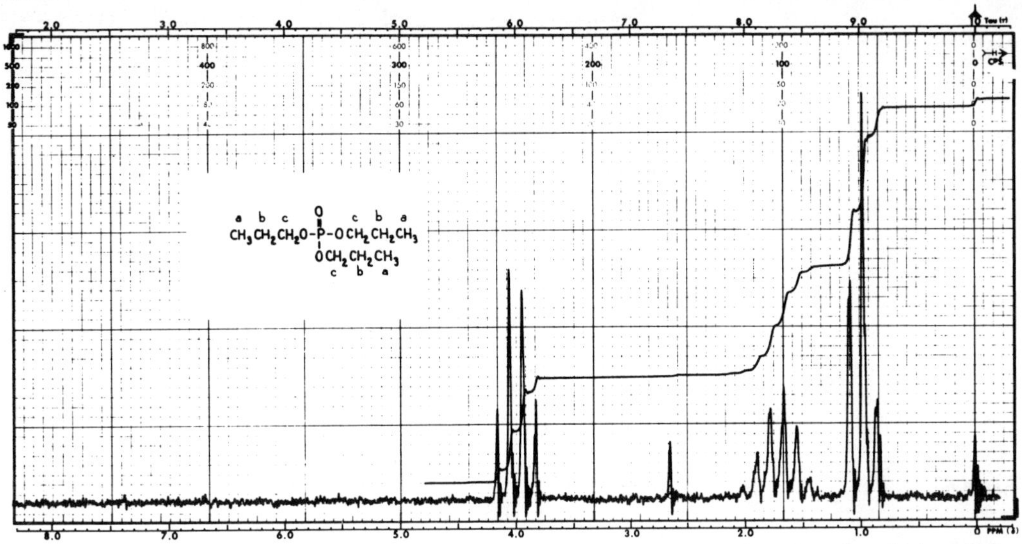

2999

© 1978 Sadtler Research Laboratories, Inc.

PHOSPHORIC ACID, TRIBENZYL ESTER

$C_{21}H_{21}O_4P$

Mol. Wt. 368.37

M.P. 64-65°C

Solvent: $CDCl_3$

Assignments
a = 4.98
b = 7.28

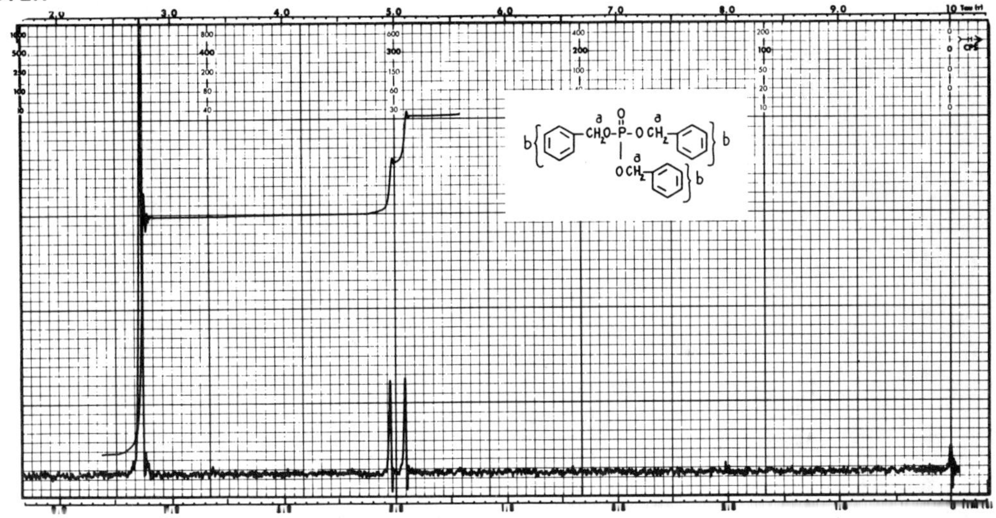

3000

© 1978 Sadtler Research Laboratories, Inc.

NOTES

ALPHABETICAL INDEX

ALPHABETICAL INDEX

ALPHABETICAL INDEX

ALPHABETICAL INDEX

ALPHABETICAL INDEX

ALPHABETICAL INDEX

ALPHABETICAL INDEX

ALPHABETICAL INDEX

ALPHABETICAL INDEX

ALPHABETICAL INDEX

ALPHABETICAL INDEX

ALPHABETICAL INDEX

ALPHABETICAL INDEX

ALPHABETICAL INDEX